Standard Atomic Weights, 1995

[Scaled to $A_r(^{12}C) = 12$]

The atomic weights of many elements are not invariant but depend on the origin and treatment of the material. The standard values of $A_r(E)$ and the uncertainties (in parentheses, following the last significant figure to which they are attributed) apply to elements of natural terrestrial origin. The footnotes to this Table elaborate the types of variation which may occur for individual elements and which may be larger than the listed uncertainties of values of $A_r(E)$. Names of elements with atomic number 104 to 111 are temporary.

Name	Symbol	Atomic Number	Atomic Weight	Footnotes	Name	Symbol	Atomic Number	Atomic Weight	Footnotes
Actinium*	Ac	89			Neon	Ne	10	20.1797(6)	g, m
Aluminium	Al	13	26.981538(2)		Neptunium*	Np	93		
Americium*	Am	95			Nickel	Ni	28	58.6934(2)	
Antimony	Sb	51	121.760(1)	g	Niobium	Nb	41	92.90638(2)	
Argon	Ar	18	39.948(1)	g, r	Nitrogen	N	7	14.00674(7)	g, r
Arsenic	As	33	74.92160(2)		Nobelium*	No	102		
Astatine*	At	85			Osmium	Os	76	190.23(3)	g
Barium	Ba	56	137.327(7)		Oxygen	O	8	15.9994(3)	g, r
Berkelium*	Bk	97			Palladium	Pd	46	106.42(1)	g
Beryllium	Be	4	9.012182(3)		Phosphorus	P	15	30.973762(4)	
Bismuth	Bi	83	208.98038(2)		Platinum	Pt	78	195.078(2)	
Boron	B	5	10.811(7)	g, m, r	Plutonium*	Pu	94		
Bromine	Br	35	79.904(1)		Polonium*	Po	84		
Cadmium	Cd	48	112.411(8)	g	Potassium	K	19	39.0983(1)	g
Calcium	Ca	20	40.078(4)	g	Praseodymium	Pr	59	140.90765(3)	
Californium*	Cf	98			Promethium*	Pm	61		
Carbon	C	6	12.0107(8)	g, r	Protactinium*	Pa	91	231.03588(2)	
Cerium	Ce	58	140.116(1)	g	Radium*	Ra	88		
Cesium	Cs	55	132.90545(2)		Radon*	Rn	86		
Chlorine	Cl	17	35.4527(9)	m	Rhenium	Re	75	186.207(1)	
Chromium	Cr	24	51.9961(6)		Rhodium	Rh	45	102.90550(2)	
Cobalt	Co	27	58.933200(9)		Rubidium	Rb	37	85.4678(3)	g
Copper	Cu	29	63.546(3)	r	Ruthenium	Ru	44	101.07(2)	g
Curium*	Cm	96			Samarium	Sm	62	150.36(3)	g
Dysprosium	Dy	66	162.50(3)	g	Scandium	Sc	21	44.955910(8)	
Einsteinium*	Es	99			Selenium	Se	34	78.96(3)	
Erbium	Er	68	167.26(3)	g	Silicon	Si	14	28.0855(3)	r
Europium	Eu	63	151.964(1)	g	Silver	Ag	47	107.8682(2)	g
Fermium*	Fm	100			Sodium	Na	11	22.989770(2)	
Fluorine	F	9	18.9984032(5)		Strontium	Sr	38	87.62(1)	g, r
Francium*	Fr	87			Sulfur	S	16	32.066(6)	g, r
Gadolinium	Gd	64	157.25(3)	g	Tantalum	Ta	73	180.9479(1)	
Gallium	Ga	31	69.723(1)		Technetium*	Tc	43		
Germanium	Ge	32	72.61(2)		Tellurium	Te	52	127.60(3)	g
Gold	Au	79	196.96655(2)		Terbium	Tb	65	158.92534(3)	
Hafnium	Hf	72	178.49(2)		Thallium	Tl	81	204.3833(2)	
Helium	He	2	4.002602(2)	g, r	Thorium*	Th	90	232.0381(1)	g
Holmium	Ho	67	164.93032(3)		Thulium	Tm	69	168.93421(2)	
Hydrogen	H	1	1.00794(7)	g, m, r	Tin	Sn	50	118.710(7)	g
Indium	In	49	114.818(3)		Titanium	Ti	22	47.867(1)	
Iodine	I	53	126.90447(3)		Tungsten	W	74	183.84(1)	
Iridium	Ir	77	192.217(3)		Unnilennium*	Une	109		
Iron	Fe	26	55.845(2)		Unnilhexium*	Unh	106		
Krypton	Kr	36	83.80(1)	g, m	Unniloctium*	Uno	108		
Lanthanum	La	57	138.9055(2)	g	Unnilpentium*	Unp	105		
Lawrencium*	Lr	103			Unnilquadium*	Unq	104		
Lead	Pb	82	207.2(1)	g, r	Unnilseptium*	Uns	107		
Lithium	Li	3	[6.941(2)]†	g, m, r	Ununnilium	Uun	110		
Lutetium	Lu	71	174.967(1)	g	Unununium	Uuu	111		
Magnesium	Mg	12	24.3050(6)		Uranium*	U	92	238.0289(1)	g, m
Manganese	Mn	25	54.938049(9)		Vanadium	V	23	50.9415(1)	
Mendelevium*	Md	101			Xenon	Xe	54	131.29(2)	g, m
Mercury	Hg	80	200.59(2)		Ytterbium	Yb	70	173.04(3)	g
Molybdenum	Mo	42	95.94(1)	g	Yttrium	Y	39	88.90585(2)	
Neodymium	Nd	60	144.24(3)	g	Zinc	Zn	30	65.39(2)	
					Zirconium	Zr	40	91.224(2)	g

*   Element has no stable nuclides.

†   Commercially available Li materials have atomic weights that range between 6.94 and 6.99; if a more accurate value is required, it must be determined for the specific material.

g   geological specimens are known in which the element has an isotopic composition outside the limits for normal material. The difference between the atomic weight of the element in such specimens and that given in the Table may exceed the stated uncertainty.

m   modified isotopic compositions may be found in commercially available material because it has been subjected to an undisclosed or inadvertent isotopic fractionation. Substantial deviations in atomic weight of the element from that given in the Table can occur.

r   range in isotopic composition of normal terrestrial material prevents a more precise $A_r(E)$ being given; the tabulated $A_r(E)$ value should be applicable to any normal material.

Source: International Union of Pure and Applied Chemistry. 1996. Atomic weights of the elements, 1996, *Pure Appl. Chem.* 68:2339.

Standard Methods

FOR THE

Examination of Water and Wastewater

20th Edition

Standard Methods

FOR THE

Examination of Water and Wastewater

20th Edition

1 9 9 8

PREPARED AND PUBLISHED JOINTLY BY

American Public Health Association
American Water Works Association
Water Environment Federation

JOINT EDITORIAL BOARD

Lenore S. Clesceri, WEF, Chair
Arnold E. Greenberg, APHA
Andrew D. Eaton, AWWA

MANAGING EDITOR
Mary Ann H. Franson

PUBLICATION OFFICE

American Public Health Association
1015 Fifteenth Street, NW
Washington, DC 20005-2605

30M12/98

ISBN 0-87553-235-7
ISSN 55-1979

Printed and bound in the United States of America.
    Composition: Maryland Composition Company, Glen Burnie, MD
        Set in: Times Roman
    Printing: United Book Press, Inc., Baltimore, Maryland
    Binding: United Book Press, Inc., Baltimore, Maryland
    Cover Design: Jane Perini, JEP Graphics, Potomac, Maryland

PREFACE TO THE TWENTIETH EDITION

The Nineteenth and Earlier Editions

The first edition of *Standard Methods* was published in 1905. Each subsequent edition presented significant improvements of methodology and enlarged its scope to include techniques suitable for examination of many types of samples encountered in the assessment and control of water quality and water pollution.

A brief history of *Standard Methods* is of interest because of its contemporary relevance. A movement for "securing the adoption of more uniform and efficient methods of water analysis" led in the 1880's to the organization of a special committee of the Chemical Section of American Association for the Advancement of Science. A report of this committee, published in 1889, was entitled: A Method, in Part, for the Sanitary Examination of Water, and for the Statement of Results, Offered for General Adoption.* Five topics were covered: (1) "free" and "albuminoid" ammonia; (2) oxygen-consuming capacity; (3) total nitrogen as nitrates and nitrites; (4) nitrogen as nitrites; and (5) statement of results.

In 1895, members of the American Public Health Association, recognizing the need for standard methods in the bacteriological examination of water, sponsored a convention of bacteriologists to discuss the problem. As a result, an APHA committee was appointed "to draw up procedures for the study of bacteria in a uniform manner and with special references to the differentiation of species." Submitted in 1897,† the procedures found wide acceptance.

In 1899, APHA appointed a Committee on Standard Methods of Water Analysis, charged with the extension of standard procedures to all methods involved in the analysis of water. The committee report, published in 1905, constituted the first edition of *Standard Methods* (then entitled *Standard Methods of Water Analysis*). Physical, chemical, microscopic, and bacteriological methods of water examination were included. In its letter of transmittal, the Committee stated:

> The methods of analysis presented in this report as "Standard Methods" are believed to represent the best current practice of American water analysts, and to be generally applicable in connection with the ordinary problems of water purification, sewage disposal and sanitary investigations. Analysts working on widely different problems manifestly cannot use methods which are identical, and special problems obviously require the methods best adapted to them; but, while recognizing these facts, it yet remains true that sound progress in analytical work will advance in proportion to the general adoption of methods which are reliable, uniform and adequate.
>
> It is said by some that standard methods within the field of applied science tend to stifle investigations and that they retard true progress. If such standards are used in the proper spirit, this ought not to be so. The Committee strongly desires that every effort shall be continued to improve the techniques of water analysis and especially to compare current methods with those herein recommended, where different, so that the results obtained may be still more accurate and reliable than they are at present.

Revised and enlarged editions were published by APHA under the title *Standard Methods of Water Analysis* in 1912 (Second Edition), 1917 (Third), 1920 (Fourth), and 1923 (Fifth). In 1925, the American Water Works Association joined APHA in publishing the Sixth Edition, which had the broader title, *Standard Methods of the Examination of Water and Sewage.* Joint publication was continued in the Seventh Edition, dated 1933.

In 1935, the Federation of Sewage Works Associations (now the Water Environment Federation) issued a committee report, "Standard Methods of Sewage Analysis."‡ With minor modifications, these methods were incorporated into the Eighth Edition (1936) of *Standard Methods*, which was thus the first to provide methods for the examination of "sewages, effluents, industrial wastes, grossly polluted waters, sludges, and muds." The Ninth Edition, appearing in 1946, likewise contained these methods, and in the following year the Federation became a full-fledged publishing partner. Since 1947, the work of the *Standard Methods* committees of the three associations—APHA, AWWA, and WEF—has been coordinated by a Joint Editorial Board, on which all three are represented.

The Tenth Edition (1955) included methods specific for examination of industrial wastewaters; this was reflected by a new title: *Standard Methods for the Examination of Water, Sewage and Industrial Wastes.* To describe more accurately and concisely the contents of the Eleventh Edition (1960), the title was shortened to *Standard Methods for the Examination of Water and Wastewater.* It remained unchanged in the Twelfth Edition (1965), the Thirteenth Edition (1971), the Fourteenth Edition (1976), and the Fifteenth Edition (1981).

In the Fourteenth Edition, the separation of test methods for water from those for wastewater was discontinued. All methods for a given component or characteristic appeared under a single heading. With minor differences, the organization of the Fourteenth Edition was retained for the Fifteenth and Sixteenth (1985) Editions. Two major policy decisions of the Joint Editorial Board were implemented for the Sixteenth Edition. First, the International System of Units (SI) was adopted except where prevailing field systems or practices require English units. Second, the use of trade names or proprietary materials was eliminated insofar as possible, to avoid potential claims regarding restraint of trade or commercial favoritism.

The organization of the Seventeenth Edition (1989) reflected a commitment to develop and retain a permanent numbering system. New numbers were assigned to all sections, and unused numbers were reserved for future use. All part numbers were expanded to multiples of 1000 instead of 100. The parts retained their identity from the previous edition, with the exception of Part 6000, which contained methods for the measurement of specific organic compounds. The more general procedures for organics were found in Part 5000.

---

* *J. Anal. Chem.* 3:398 (1889).
† *Proc. Amer. Pub. Health Assoc.* 23:56 (1897).

‡ *Sewage Works J.* 7:444 (1935).

The Seventeenth Edition also underwent a major revision in the introductory Part 1000. Sections dealing with statistical analysis, data quality, and methods development were greatly expanded. The section on reagent water was updated to include a classification scheme for various types of reagent water. At the beginning of each of the subsequent parts of the manual, sections were included that discussed quality assurance and other matters of general application within the specific subject area, to minimize repetition in the succeeding text.

The Eighteenth Edition (1992) underwent only minor revisions in the format from the 17th edition. A number of new methods were added in each section. The 18th Edition has many of its methods cited for compliance monitoring of both drinking water and wastewater.

In the Nineteenth Edition (1995), sections were added on laboratory safety and waste management in Part 1000. Substantial changes occurred throughout, adding new methodology and revisions to many of the sections.

The Twentieth Edition

The Twentieth Edition has maintained the trend of the Nineteenth Edition in continued renewal of Part 1000. Significant revision has occurred in the sections on data quality (1030), sampling (1060) and reagent water (1080).

In Part 2000 (physical and aggregate properties), odor (2150) has been revised to supply new tables for odor identification. The salinity (2520) formula has been made compatible with conductivity nomenclature and quality control procedures have been updated and strengthened.

Significant reworking of the introductory material has occurred in Part 3000 (metals); the introduction now includes a user guide to appropriate methods of metal analysis. A new section, inductively coupled plasma/mass spectrometry (ICP/MS), has been added. Anodic stripping voltammetry (3130) has been expanded to include zinc. The sections on ICP, sample preparation, and specific metal analyses have been revised.

Part 4000 (inorganic nonmetallic constituents) has been reviewed and includes new methods on flow injection analysis (4130), potassium permanganate (4500-KMnO$_4$), and capillary ion electrophoresis (4140). Ozone (4500-O$_3$) methods have been updated. Significant revisions also have been made in the nitrogen sections. Other sections have undergone minor revisions.

Part 5000 (aggregate organic constituents) has significantly revised sections on chemical oxygen demand (5220), total organic carbon (5310) (from the Nineteenth Edition supplement), and dissolved organic halogen (5320). Freon has been mostly replaced by hexane in the oil and grease section (5520).

In Part 6000 (individual organic compounds), a new section on volatile organic compounds has replaced a number of old sections and a major section on quality control has been added.

Various editorial changes were made in Part 7000 (radioactivity) and a revision in gamma-emitting radionuclides (7120) was made.

Part 8000 (toxicity testing) underwent major changes with new protocols for quality assurance (8020), P450 methodology (8070) from the Nineteenth Edition supplement, pore water test procedures (8080), protozoa (8310), rotifers (8420), *Daphnia* (8711), *Ceriodaphnia* (8712), mysids (8714), decapods (8740), echinoderm fertilization and development (8810), and fathead minnows (8911).

Other sections have been revised significantly and illustrations of many test organisms have been added.

Part 9000 (microbiological examination) has had major revisions to quality assurance and pathogenic bacteria (9260) and minor revisions in several other sections.

Part 10000 (biological examination) has undergone minor revisions. Some new figures and illustrations of organisms have been added.

Making Reagents

Following the instructions for making reagents may result in preparation of quantities larger than actually needed. In some cases these materials are toxic. To promote economy and minimize waste, the analyst should review needs and scale down solution volumes where appropriate. This conservative attitude also should extend to purchasing policies so that unused chemicals do not accumulate or need to be discarded as their shelf lives expire.

Selection and Approval of Methods

For each new edition both the technical criteria for selection of methods and the formal procedures for their approval and inclusion are reviewed critically. In regard to the approval procedures, it is considered particularly important to assure that the methods presented have been reviewed and are supported by the largest number of qualified people, so that they may represent a true consensus of expert opinion.

For the Fourteenth Edition a Joint Task Group was established for each test. This scheme has continued for each subsequent edition. Appointment of an individual to a Joint Task Group generally was based on the expressed interest or recognized expertise of the individual. The effort in every case was to assemble a group having maximum available expertise in the test methods of concern.

Each Joint Task Group was charged with reviewing the pertinent methods in the Nineteenth Edition along with other methods from the literature, recommending the methods to be included in the Twentieth Edition, and presenting those methods in the form of a proposed section manuscript. Subsequently, each section manuscript (except for Part 1000) was ratified by vote of those members of the Standard Methods Committee who asked to review sections in that part. Every negative vote and every comment submitted in the balloting was reviewed by the Joint Editorial Board. Relevant suggestions were referred appropriately for resolution. When negative votes on the first ballot could not be resolved by the Joint Task Group or the Joint Editorial Board, the section was reballoted among all who voted affirmatively or negatively on the original ballot. Only a few issues could not be resolved in this manner and the Joint Editorial Board made the final decision.

The general and quality assurance information presented in Part 1000 was treated somewhat differently. Again, Joint Task Groups were formed, given a charge, and allowed to produce a consensus draft. This draft was reviewed by the Joint Editorial Board Liaison and subsequently by the Joint Editorial Board. The draft sections were sent to the Standard Methods Committee and comments resulting from this review were used to develop the final draft.

The methods presented here, as in previous editions, are believed to be the best available and generally accepted procedures for the analysis of water, wastewaters, and related materials. They represent the recommendations of specialists, ratified by a large number of analysts and others of more general expertise, and as such are truly consensus standards, offering a valid and recognized basis for control and evaluation.

The technical criteria for selection of methods were applied by the Joint Task Groups and by the individuals reviewing their recommendations, with the Joint Editorial Board providing only general guidelines. In addition to the classical concepts of precision, bias, and minimum detectable concentration, selection of a method also must recognize such considerations as the time required to obtain a result, needs for specialized equipment and for special training of the analyst, and other factors related to the cost of the analysis and the feasibility of its widespread use.

Status of Methods

All methods in the Twentieth Edition are dated to assist users in identifying those methods that have been changed significantly between editions. The year the section was approved by the Standard Methods Committee is indicated in a footnote at the beginning of each section. Sections or methods that appeared in the Nineteenth Edition that are unchanged, or changed only editorially in the Twentieth Edition, show an approval date of 1993 or 1994. Sections or methods that were changed significantly, or that were reaffirmed by general balloting of the Standard Methods Committee, are dated 1996 or 1997. If an individual method within a section was revised, that method carries an approval date different from that of the rest of the section.

Methods in the Twentieth Edition are divided into fundamental classes: PROPOSED, SPECIALIZED, STANDARD, AND GENERAL. None of the methods in the Twentieth Edition have the specialized designation. Regardless of assigned class, all methods must be approved by the Standard Methods Committee. The four classes are described below:

1. PROPOSED—A PROPOSED method must undergo development and validation that meets the requirements set forth in Section 1040A of *Standard Methods*.
2. SPECIALIZED—A procedure qualifies as a SPECIALIZED method in one of two ways: a) The procedure must undergo development and validation and collaborative testing that meet the requirements set forth in Sections 1040B and C of *Standard Methods*, respectively; or b) The procedure is the "METHOD OF CHOICE" of the members of the Standard Methods Committee actively conducting the analysis and it has appeared in TWO PREVIOUS EDITIONS of *Standard Methods*.
3. STANDARD—A procedure qualifies as a STANDARD method in one of two ways: a) The procedure must undergo development and validation and collaborative testing that meet the requirements set forth in Sections 1040B and C of *Standard Methods*, respectively, and it is "WIDELY USED" by the members of the Standard Methods Committee; or b) The procedure is "WIDELY USED" by the members of the Standard Methods Committee and it has appeared in TWO PREVIOUS EDITIONS of *Standard Methods*.
4. GENERAL—A procedure qualifies as a GENERAL method if it has appeared in TWO PREVIOUS EDITIONS of *Standard Methods*.

Assignment of a classification to a method is done by the Joint Editorial Board. When making method classifications, the Joint Editorial Board evaluates the results of the survey on method use by the Standard Methods Committee that is conducted at the time of general balloting of the method. In addition, the Joint Editorial Board considers recommendations offered by Joint Task Groups and the Part Coordinator.

Methods categorized as "PROPOSED," "SPECIALIZED," and "GENERAL" are so designated in their titles; methods with no designation are "STANDARD."

Technical progress makes advisable the establishment of a program to keep *Standard Methods* abreast of advances in research and general practice. The Joint Editorial Board has developed the following procedure for effecting interim changes in methods between editions:

1. Any method given proposed status in the current edition may be elevated by action of the Joint Editorial Board, on the basis of adequate published data supporting such a change as submitted to the Board by the appropriate Joint Task Group. Notification of such a change in status shall be accomplished by publication in the official journals of the three associations sponsoring *Standard Methods*.
2. No method may be abandoned or reduced to a lower status during the interval between editions.
3. A new method may be adopted as proposed, specialized, or standard by the Joint Editorial Board between editions, such action being based on the usual consensus procedure. Such new methods may be published in supplements to editions of *Standard Methods*. It is intended that a supplement be published midway between editions.

Even more important to maintaining the current status of these standards is the intention of the sponsors and the Joint Editorial Board that subsequent editions will appear regularly at reasonably short intervals.

Reader comments and questions concerning this manual should be addressed to: Standard Methods Manager, American Water Works Association, 6666 West Quincy Avenue, Denver, CO 80235.

Acknowledgments

For the work in preparing the methods for the Twentieth Edition, the Joint Editorial Board gives full credit to the Standard Methods Committees of the American Water Works Association and of the Water Environment Federation and to the Committee on Laboratory Standards and Practices of the American Public Health Association. Full credit also is given to those individuals who were not members of the sponsoring societies. A list of all committee members follows these pages. Herbert J. Brass, U.S. Environmental Protection Agency, served as a liaison from EPA to the Joint Editorial Board; thanks are due for his interest and help.

The Joint Editorial Board expresses its appreciation to Fernando M. Treviño, former Executive Director, and Mohammad N. Akhter, M.D., current Executive Director, American Public Health Association, to John B. Mannion, former Executive Director, and Jack W. Hoffbuhr, current Executive Director, American Water Works Association, and to Quincalee Brown, Executive Director, Water Environment Federation, for their cooperation and advice in the development of this publication. Steven J. Posavec, Standard Methods Manager and Joint Editorial Board Secretary, provided a variety of important services that are vital to the preparation of a volume of this type. Ellen Meyer, Director of Publications, American Public Health Association, functioned as publisher. Judy Castagna, also with APHA, served as production manager. Special recognition for her valuable services is due to Mary Ann H. Franson, Managing Editor, who discharged most efficiently the extensive and detailed responsibilities on which this publication depends.

Joint Editorial Board

Lenore S. Clesceri, Water Environment Federation, Chair
Arnold E. Greenberg, American Public Health Association
Andrew D. Eaton, American Water Works Association

Standard Methods Committee and Joint Task Group Members

TABLE OF CONTENTS

PAGE

TABLES

PLATES

PART 1000

INTRODUCTION

1010 INTRODUCTION

1010 A. Scope and Application of Methods

The procedures described in these standards are intended for the examination of waters of a wide range of quality, including water suitable for domestic or industrial supplies, surface water, ground water, cooling or circulating water, boiler water, boiler feed water, treated and untreated municipal or industrial wastewater, and saline water. The unity of the fields of water supply, receiving water quality, and wastewater treatment and disposal is recognized by presenting methods of analysis for each constituent in a single section for all types of waters.

An effort has been made to present methods that apply generally. Where alternative methods are necessary for samples of different composition, the basis for selecting the most appropriate method is presented as clearly as possible. However, samples with extreme concentrations or otherwise unusual compositions or characteristics may present difficulties that preclude the direct use of these methods. Hence, some modification of a procedure may be necessary in specific instances. Whenever a procedure is modified, the analyst should state plainly the nature of modification in the report of results.

Certain procedures are intended for use with sludges and sediments. Here again, the effort has been to present methods of the widest possible application, but when chemical sludges or slurries or other samples of highly unusual composition are encountered, the methods of this manual may require modification or may be inappropriate.

Most of the methods included here have been endorsed by regulatory agencies. Procedural modification without formal approval may be unacceptable to a regulatory body.

The analysis of bulk chemicals received for water treatment is not included herein. A committee of the American Water Works Association prepares and issues standards for water treatment chemicals.

Part 1000 contains information that is common to, or useful in, laboratories desiring to produce analytical results of known quality, that is, of known accuracy and with known uncertainty in that accuracy. To accomplish this, apply the quality assurance methods described herein to the standard methods described elsewhere in this publication. Other sections of Part 1000 address laboratory equipment, laboratory safety, sampling procedures, and method development and validation, all of which provide necessary information.

1010 B. Statistics

1. Normal Distribution

If a measurement is repeated many times under essentially identical conditions, the results of each measurement, x, will be distributed randomly about a mean value (arithmetic average) because of uncontrollable or experimental error. If an infinite number of such measurements were to be accumulated, the individual values would be distributed in a curve similar to those shown in Figure 1010:1. The left curve illustrates the Gaussian or normal distribution, which is described precisely by the mean, μ, and the standard deviation, σ. The mean, or average, of the distribution is simply the sum of all values divided by the number of values so summed, i.e., $\mu = (\Sigma_i x_i)/n$. Because no measurements are repeated an infinite number of times, an *estimate* of the mean is made, using the same summation procedure but with n equal to a finite number of repeated measurements (10, or 20, or. . .). This estimate of μ is denoted by $\bar{x}$. The standard deviation of the normal distribution is defined as $\sigma = [\Sigma(x-\mu)^2/n]^{1/2}$. Again, the analyst can only estimate the standard deviation because the number of observations made is finite; the estimate of σ is denoted by s and is calculated as follows:

$$s = [\Sigma(x-\bar{x})^2/(n-1)]^{1/2}$$

The standard deviation fixes the width, or spread, of the normal distribution, and also includes a fixed fraction of the values making up the curve. For example, 68.27% of the measurements lie between $\mu \pm 1\sigma$, 95.45% between $\mu \pm 2\sigma$, and 99.70% between $\mu \pm 3\sigma$. It is sufficiently accurate to state that 95% of the values are within $\pm 2\sigma$ and 99% within $\pm 3\sigma$. When values are assigned to the $\pm \sigma$ multiples, they are confidence limits. For example, 10 $\pm$ 4 indicates that the confidence limits are 6 and 14, while values from 6 to 14 represent the confidence interval.

Another useful statistic is the standard error of the mean, σ_μ, which is the standard deviation divided by the square root of the number of values, or $\sigma/\sqrt{n}$. This is an estimate of the accuracy of

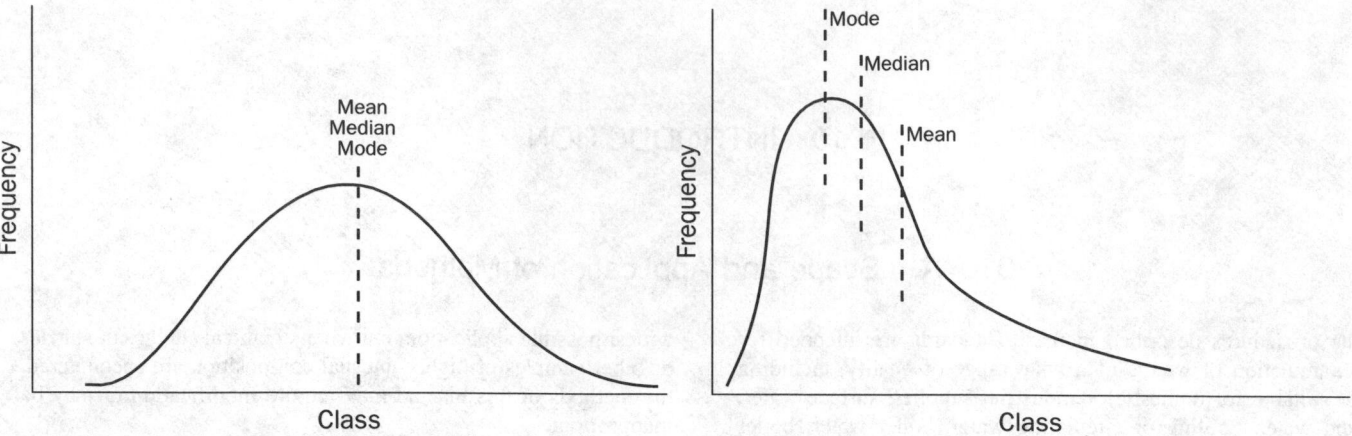

Figure 1010:1. Normal (left) and skewed (right) distributions.

the mean and implies that another sample from the same population would have a mean within some multiple of this. Multiples of this statistic include the same fraction of the values as stated above for σ. In practice, a relatively small number of average values is available, so the confidence intervals of the mean are expressed as $\bar{x} \pm ts/\sqrt{n}$ where t has the following values for 95% confidence intervals:

n	t	n	t
2	12.71	5	2.78
3	4.30	10	2.26
4	3.18	∞	1.96

The use of t compensates for the tendency of a small number of values to underestimate uncertainty. For $n > 15$, it is common to use $t = 2$ to estimate the 95% confidence interval.

Still another statistic is the relative standard deviation, σ/μ, with its estimate $s/\bar{x}$, also known as the coefficient of variation (CV), which commonly is expressed as a percentage. This statistic normalizes the standard deviation and sometimes facilitates making direct comparisons among analyses that include a wide range of concentrations. For example, if analyses at low concentrations yield a result of 10 ± 1.5 mg/L and at high concentrations 100 ± 8 mg/L, the standard deviations do not appear comparable. However, the percent relative standard deviations are $100 (1.5/10) = 15\%$ and $100 (8/100) = 8\%$, which indicate the smaller variability obtained by using this parameter.

2. Log-Normal Distribution

In many cases the results obtained from analysis of environmental samples will not be normally distributed, i.e., a graph of the data will be obviously skewed, as shown at right in Figure 1010:1, with the mode, median, and mean being distinctly different. To obtain a nearly normal distribution, convert the results to logarithms and then calculate $\bar{x}$ and s. The antilogarithms of these two values are estimates of the geometric mean and the geometric standard deviation, $\bar{x}_g$ and s_g.

3. Rejection of Data

Quite often in a series of measurements, one or more of the results will differ greatly from the other values. Theoretically, no result should be rejected, because it may indicate either a faulty technique that casts doubt on all results or the presence of a true variant in the distribution. In practice, reject the result of any analysis in which a known error has occurred. In environmental studies, extremely high and low concentrations of contaminants may indicate the existence of areas with problems or areas with no contamination, so they should not be rejected arbitrarily.

TABLE 1010:I. CRITICAL VALUES FOR 5% AND 1% TESTS OF DISCORDANCY FOR A SINGLE OUTLIER IN A NORMAL SAMPLE

Number of Measurements n	Critical Value	
	5%	1%
3	1.15	1.15
4	1.46	1.49
5	1.67	1.75
6	1.82	1.94
7	1.94	2.10
8	2.03	2.22
9	2.11	2.32
10	2.18	2.41
12	2.29	2.55
14	2.37	2.66
15	2.41	2.71
16	2.44	2.75
18	2.50	2.82
20	2.56	2.88
30	2.74	3.10
40	2.87	3.24
50	2.96	3.34
60	3.03	3.41
100	3.21	3.60
120	3.27	3.66

Source: BARNETT, V. & T. LEWIS. 1984. Outliers in Statistical Data. John Wiley & Sons, New York, N.Y.

An objective test for outliers has been described.[1] If a set of data is ordered from low to high: $x_L, x_2 \ldots x_H$, and the average and standard deviation are calculated, then suspected high or low outliers can be tested by the following procedure. First, calculate the statistic T:

$$T = (x_H - \bar{x})/s \text{ for a high value, or}$$
$$T = (\bar{x} - x_L)/s \text{ for a low value.}$$

Second, compare the value of T with the value from Table 1010:I for either a 5% or 1% level of significance. If the calculated T is larger than the table value for the number of measurements, n, then the x_H or x_L is an outlier at that level of significance.

Further information on statistical techniques is available elsewhere.[2,3]

4. References

1. BARNETT, V. & T. LEWIS. 1984. Outliers in Statistical Data. John Wiley & Sons, New York, N.Y.
2. NATRELLA, M.G. 1963. Experimental Statistics. National Bur. Standards Handbook 91, Washington, D.C.
3. SNEDECOR, G.W. & W.G. COCHRAN. 1980. Statistical Methods. Iowa State University Press, Ames.

1010 C. Glossary

1. Definition of Terms

The purpose of this glossary is to define concepts, not regulatory terms; it is not intended to be all-inclusive.

Accuracy—combination of bias and precision of an analytical procedure, which reflects the closeness of a measured value to a true value.

Bias—consistent deviation of measured values from the true value, caused by systematic errors in a procedure.

Calibration check standard—standard used to determine the state of calibration of an instrument between periodic recalibrations.

Confidence coefficient—the probability, %, that a measurement result will lie within the confidence interval or between the confidence limits.

Confidence interval—set of possible values within which the true value will lie with a specified level of probability.

Confidence limit—one of the boundary values defining the confidence interval.

Detection levels—Various levels in increasing order are:

Instrumental detection level (IDL)—the constituent concentration that produces a signal greater than five times the signal/noise ratio of the instrument. This is similar, in many respects, to "critical level" and "criterion of detection." The latter level is stated as 1.645 times the s of blank analyses.

Lower level of detection (LLD)—the constituent concentration in reagent water that produces a signal $2(1.645)s$ above the mean of blank analyses. This sets both Type I and Type II errors at 5%. Other names for this level are "detection level" and "level of detection" (LOD).

Method detection level (MDL)—the constituent concentration that, when processed through the complete method, produces a signal with a 99% probability that it is different from the blank. For seven replicates of the sample, the mean must be $3.14s$ above the blank where s is the standard deviation of the seven replicates. Compute MDL from replicate measurements one to five times the actual MDL. The MDL will be larger than the LLD because of the few replications and the sample processing steps and may vary with constituent and matrix.

Level of quantitation (LOQ)/minimum quantitation level (MQL)—the constituent concentration that produces a signal sufficiently greater than the blank that it can be detected within specified levels by good laboratories during routine operating conditions. Typically it is the concentration that produces a signal $10s$ above the reagent water blank signal.

Duplicate—usually the smallest number of replicates (two) but specifically herein refers to duplicate samples, i.e., two samples taken at the same time from one location.

Internal standard—a pure compound added to a sample extract just before instrumental analysis to permit correction for inefficiencies.

Laboratory control standard—a standard, usually certified by an outside agency, used to measure the bias in a procedure. For certain constituents and matrices, use National Institute of Standards and Technology (NIST) Standard Reference Materials when they are available.

Precision—measure of the degree of agreement among replicate analyses of a sample, usually expressed as the standard deviation.

Quality assessment—procedure for determining the quality of laboratory measurements by use of data from internal and external quality control measures.

Quality assurance—a definitive plan for laboratory operation that specifies the measures used to produce data of known precision and bias.

Quality control—set of measures within a sample analysis methodology to assure that the process is in control.

Random error—the deviation in any step in an analytical procedure that can be treated by standard statistical techniques.

Replicate—repeated operation occurring within an analytical procedure. Two or more analyses for the same constituent in an extract of a single sample constitute replicate extract analyses.

Surrogate standard—a pure compound added to a sample in the laboratory just before processing so that the overall efficiency of a method can be determined.

Type I error—also called alpha error, is the probability of deciding a constituent is present when it actually is absent.

Type II error—also called beta error, is the probability of not detecting a constituent when it actually is present.

1020 QUALITY ASSURANCE

1020 A. Introduction

This section applies primarily to chemical analyses. See Section 9020 for quality assurance and control for microbiological analyses.

Quality assurance (QA) is the definitive program for laboratory operation that specifies the measures required to produce defensible data of known precision and accuracy. This program will be defined in a documented laboratory quality system.

The laboratory quality system will consist of a QA manual, written procedures, work instructions, and records. The manual should include a quality policy that defines the statistical level of confidence used to express the precision and bias of data, as well as the method detection limits. Quality systems, which include QA policies and all quality control (QC) processes, must be in place to document and ensure the quality of analytical data produced by the laboratory and to demonstrate the competence of the laboratory. Quality systems are essential for any laboratory seeking accreditation under state or federal laboratory certification programs. Included in quality assurance are quality control (Section 1020B) and quality assessment (Section 1020C). See Section 1030 for evaluation of data quality.

1. Quality Assurance Planning

Establish a QA program and prepare a QA manual or plan. Include in the QA manual and associated documents the following items[1-4]: cover sheet with approval signatures; quality policy statement; organizational structure; staff responsibilities; analyst training and performance requirements; tests performed by the laboratory; procedures for handling and receiving samples; sample control and documentation procedures; procedures for achieving traceability of measurements; major equipment, instrumentation, and reference measurement standards used; standard operating procedures (SOPs) for each analytical method; procedures for generation, approval, and control of policies and procedures; procedures for procurement of reference materials and supplies; procedures for procurement of subcontractors' services; internal quality control activities; procedures for calibration, verification, and maintenance of instrumentation and equipment; data-verification practices including interlaboratory comparison and proficiency-testing programs; procedures to be followed for feedback and corrective action whenever testing discrepancies are detected; procedures for exceptions that permit departure from documented policies; procedures for system and performance audits and reviews; procedures for assessing data precision and accuracy and determining method detection limits; procedures for data reduction, validation, and reporting; procedures for records archiving; procedures and systems for control of the testing environment; and procedures for dealing with complaints from users of the data. Also define and include the responsibility for, and frequency of, management review and updates to the QA manual and associated documents.

On the title page, include approval signatures and a statement that the manual has been reviewed and determined to be appropriate for the scope, volume, and range of testing activities at the laboratory,[4] as well as an indication that management has made a commitment to assure that the quality systems defined in the QA manual are implemented and followed at all times.

In the QA manual, clearly specify and document the managerial responsibility, authority, quality goals, objectives, and commitment to quality. Write the manual so that it is clearly understood and ensures that all laboratory personnel understand their roles and responsibilities.

Implement and follow chain-of-custody procedures to ensure that chain of custody is maintained and documented for each sample. Institute procedures to permit tracing a sample and its derivatives through all steps from collection through analysis to reporting final results to the laboratory's client and disposal of the sample. Routinely practice adequate and complete documentation, which is critical to assure data defensibility and to meet laboratory accreditation/certification requirements, and ensure full traceability for all tests and samples.

Standard operating procedures (SOPs) describe the analytical methods to be used in the laboratory in sufficient detail that a competent analyst unfamiliar with the method can conduct a reliable review and/or obtain acceptable results. Include in SOPs, where applicable, the following items[2-5]: title of referenced, consensus test method; sample matrix or matrices; method detection level (MDL); scope and application; summary of SOP; definitions; interferences; safety considerations; waste management; apparatus, equipment, and supplies; reagents and standards; sample collection, preservation, shipment, and storage requirements; specific quality control practices, frequency, acceptance criteria, and required corrective action if acceptance criteria are not met; calibration and standardization; details on the actual test procedure, including sample preparation; calculations; qualifications and performance requirements for analysts (including number and type of analyses); data assessment/data management; references; and any tables, flowcharts, and validation or method performance data. At a minimum, validate a new SOP before use by first determining the MDL and performing an initial demonstration of capability using relevant regulatory guidelines.

Use and document preventive maintenance procedures for instrumentation and equipment. An effective preventive maintenance program will reduce instrument malfunctions, maintain more consistent calibration, be cost-effective, and reduce downtime. Include measurement traceability to National Institute of Standards and Technology (NIST) Standard Reference Materials (SRMs) or commercially available reference materials certified traceable to NIST SRMs in the QA manual or SOP to establish integrity of the laboratory calibration and measurement program. Formulate document-control procedures, which are essential to data defensibility, to cover the complete process of document generation, approval, distribution, storage, recall, archiving, and disposal. Maintain logbooks for each test or procedure performed with complete documentation on preparation and analysis of each sample, including sample identification, associated standards and QC samples, method reference, date/time of preparation/analysis, analyst, weights and volumes used, results obtained, and any problems encountered. Keep logbooks that document maintenance and calibration for each instrument or piece of equipment. Calibration procedures, corrective actions, internal quality control activities, performance audits, and data assessments for precision and accuracy (bias) are discussed in Sections 1020B and C.

Data reduction, validation, and reporting are the final steps in the data-generation process. The data obtained from an analytical instrument must first be subjected to the data reduction processes described in the applicable SOP before the final result can be obtained. Specify calculations and any correction factors, as well as the steps to be followed in generating the sample result, in the QA manual or SOP. Also specify all of the data validation steps to be followed before the final result is made available. Report results in standard units of mass, volume, or concentration as specified in the method or SOP. Report results below the MDL in accordance with the procedure prescribed in the SOP. Ideally, include a statement of uncertainty with each result. See references and bibliography for other useful information and guidance on establishing a QA program and developing an effective QA manual.

2. References

1. STANLEY, T.T. & S.S. VERNER. 1983. Interim Guidelines and Specifications for Preparing Quality Assurance Project Plans. EPA-600/4-83-004, U.S. Environmental Protection Agency, Washington, D.C.
2. QUALITY SYSTEMS COMMITTEE, NATIONAL ENVIRONMENTAL LABORATORY ACCREDITATION CONFERENCE. 1996. National Environmental Laboratory Accreditation Conference, 2nd Annual Meeting, Washington, D.C. [available online]. U.S. Environmental Protection Agency, Washington, D.C.
3. QUALITY SYSTEMS COMMITTEE, NATIONAL ENVIRONMENTAL LABORATORY ACCREDITATION CONFERENCE. 1997. National Environmental Laboratory Accreditation Conference, 2nd Interim Meeting, Bethesda, Md. [available online]. U.S. Environmental Protection Agency, Washington, D.C.
4. INTERNATIONAL ORGANIZATION FOR STANDARDIZATION. 1996. General Requirements for the Competence of Testing and Calibration Laboratories, ISO/IEC Guide 25-Draft Four. International Org. for Standardization, Geneva, Switzerland.
5. U.S. ENVIRONMENTAL PROTECTION AGENCY. 1995. Guidance for the Preparation of Standard Operating Procedures (SOPs) for Quality-Related Documents. EPA QA/G-6, Washington, D.C.

3. Bibliography

DELFINO, J.J. 1977. Quality assurance in water and wastewater analysis laboratories. *Water Sew. Works* 124:79.

INHORN, S.L., ed. 1978. Quality Assurance Practices for Health Laboratories. American Public Health Assoc., Washington, D.C.

STANLEY, T.W. & S.S. VERNER. 1983. Interim Guidelines and Specifications for Preparing Quality Assurance Project Plans. EPA-600/4-83-004, U.S. Environmental Protection Agency, Washington, D.C.

U.S. ENVIRONMENTAL PROTECTION AGENCY. 1997. Manual for the Certification of Laboratories Analyzing Drinking Water. EPA-815-B-97-001, U.S. Environmental Protection Agency, Washington, D.C.

INTERNATIONAL ORGANIZATION FOR STANDARDIZATION. 1990. General Requirements for the Competence of Testing and Calibration Laboratories, ISO/IEC Guide 25. International Org. for Standardization, Geneva, Switzerland.

U.S. ENVIRONMENTAL PROTECTION AGENCY. 1994. National Environmental Laboratory Accreditation Conference (NELAC) Notice of Conference and Availability of Standards. *Federal Register* 59, No. 231.

U.S. ENVIRONMENTAL PROTECTION AGENCY. 1995. Good Automated Laboratory Practices. U.S. Environmental Protection Agency, Research Triangle Park, N.C.

AMERICAN ASSOCIATION FOR LABORATORY ACCREDITATION. 1996. General Requirements for Accreditation. A2LA, American Assoc. Laboratory Accreditation, Gaithersburg, Md.

1020 B.   Quality Control

Include in each analytical method or SOP the minimum required QC for each analysis. A good quality control program consists of at least the following elements, as applicable: initial demonstration of capability, ongoing demonstration of capability, method detection limit determination, reagent blank (also referred to as method blank), laboratory-fortified blank (also referred to as blank spike), laboratory-fortified matrix (also referred to as matrix spike), laboratory-fortified matrix duplicate (also referred to as matrix spike duplicate) or duplicate sample, internal standard, surrogate standard (for organic analysis) or tracer (for radiochemistry), calibration, control charts, and corrective action, frequency of QC indicators, QC acceptance criteria, and definitions of a batch. Sections 1010 and 1030 describe calculations for evaluating data quality.

1. Initial Demonstration of Capability

The laboratory should conduct an initial demonstration of capability (IDC) at least once, by each analyst, before analysis of any sample, to demonstrate proficiency to perform the method and obtain acceptable results for each analyte. The IDC also is used to demonstrate that modifications to the method by the laboratory will produce results as precise and accurate as results produced by the reference method. As a minimum, include a reagent blank and at least four laboratory-fortified blanks (LFBs) at a concentration between 10 times the method detection level (MDL) and the midpoint of the calibration curve or other level as specified in the method. Run the IDC after analyzing all required calibration standards. Ensure that the reagent blank does not contain any analyte of interest at a concentration greater than half the MQL or other level as specified in the method. See Section 1010C, for definition of MQL. Ensure that precision and accuracy (percent recovery) calculated for the LFBs are within the acceptance criteria listed in the method of choice. If no acceptance criteria are provided, use 80 to 120% recovery and ≤20% relative standard deviation (RSD), as a starting point. If details of initial demonstration of capability are not provided in the method of choice, specify and reference the method or procedure used for demonstrating capability.

2. Ongoing Demonstration of Capability

The ongoing demonstration of capability, sometimes referred to as a "laboratory control sample or laboratory control standard," "quality control check sample," or "laboratory-fortified blank," is used to ensure that the laboratory remains in control during the period when samples are analyzed, and separates laboratory performance from method performance on the sample matrix. See ¶ 5 below for further details on the laboratory-fortified blank. Preferably obtain this sample from an external source (not the same stock as the calibration standards). Analyze QC check samples on a quarterly basis, at a minimum.

3. Method Detection Level Determination and Application

Determine the method detection level (MDL) for each analyte of interest and method to be used before data from any samples are reported, using the procedure described in Section 1030C. As a starting point for determining the concentration to use in MDL determination, use an estimate of five times the estimated detection limit. Perform MDL determinations as an iterative process. If calculated MDL is not within a factor of l0 of the value for the known addition, repeat determinations at a more suitable concentration. Conduct MDL determinations at least annually (or other specified frequency) for each analyte and method in use at the laboratory. Perform or verify MDL determination for each instrument. Perform MDL determinations over a period of at least 3 d for each part of the procedure. Calculate recoveries for MDL samples. Recoveries should be between 50 and 150% and %RSD values $\leq$ 20% or repeat the MDL determination. Maintain MDL and IDC data and have them available for inspection.

Apply the MDL to reporting sample results as follows:

- Report results below the MDL as "not detected."
- Report results between the MDL and MQL with qualification for quantitation.
- Report results above the MQL with a value and its associated error.

4. Reagent Blank

A reagent blank or method blank consists of reagent water (See Section 1080) and all reagents that normally are in contact with a sample during the entire analytical procedure. The reagent blank is used to determine the contribution of the reagents and the preparative analytical steps to error in the measurement. As a minimum, include one reagent blank with each sample set (batch) or on a 5% basis, whichever is more frequent. Analyze a blank after the daily calibration standard and after highly contaminated samples if carryover is suspected. Evaluate reagent blank results for the presence of contamination. If unacceptable contamination is present in the reagent blank, identify and eliminate source of contamination. Typically, sample results are suspect if analyte(s) in the reagent blank are greater than the MQL. Samples analyzed with an associated contaminated blank must be re-prepared and re-analyzed. Refer to the method of choice for specific acceptance criteria for the reagent blank. Guidelines for qualifying sample results with consideration to reagent blank results are as follows:

- If the reagent blank is less than the MDL and sample results are greater than the MQL, then no qualification is required.
- If the reagent blank is greater than the MDL but less than the MQL and sample results are greater than the MQL, then qualify the results to indicate that analyte was detected in the reagent blank.
- If the reagent blank is greater than the MQL, further corrective action and qualification is required.

5. Laboratory-Fortified Blank

A laboratory-fortified blank is a reagent water sample to which a known concentration of the analytes of interest has been added. A LFB is used to evaluate laboratory performance and analyte recovery in a blank matrix. As a minimum, include one LFB with each sample set (batch) or on a 5% basis, whichever is more frequent. The definition of a batch is typically method-specific. Process the LFB through all of the sample preparation and analysis steps. Use an added concentration of at least 10 times the MDL, the midpoint of the calibration curve, or other level as specified in the method. Prepare the addition solution from a different reference source than that used for calibration. Evaluate the LFB for percent recovery of the added analytes. If LFB results are out of control, take corrective action, including re-preparation and re-analysis of associated samples if required. Use the results obtained for the LFB to evaluate batch performance, calculate recovery limits, and plot control charts (see ¶ 12 below). Refer to the method of choice for specific acceptance criteria for the LFB.

6. Laboratory-Fortified Matrix

A laboratory-fortified matrix (LFM) is an additional portion of a sample to which known amounts of the analytes of interest are added before sample preparation. The LFM is used to evaluate analyte recovery in a sample matrix. As a minimum, include one LFM with each sample set (batch) or on a 5% basis, whichever is more frequent. Add a concentration of at least 10 times the MRL, the midpoint of the calibration curve, or other level as specified in the method to the selected sample(s). Preferably use the same concentration as for the LFB to allow the analyst to separate the effect of matrix from laboratory performance. Prepare the LFM from a reference source different from that used for calibration. Make the addition such that sample background levels do not adversely affect the recovery (preferably adjust LFM concentrations if the known sample is above five times the background level). For example, if the sample contains the analyte of interest, make the LFM sample at a concentration equivalent to the concentration found in the known sample. Evaluate the results obtained for LFMs for accuracy or percent recovery. If LFM results are out of control, take corrective action to rectify the effect or use another method or the method of standard addition. Refer to the method of choice for specific acceptance criteria for LFMs until the laboratory develops statistically valid, laboratory-specific performance criteria. Base sample batch acceptance on results of LFB analyses rather than LFMs alone, because the matrix of the LFM sample may interfere with the method performance.

7. Laboratory-Fortified Matrix Duplicate/Duplicate Sample

A LFM duplicate is a second portion of the sample described in ¶ 6 above to which a known amount of the analyte of interest is added before sample preparation. If sufficient sample volume is collected, this second portion of sample is added and processed in the same way as the LFM. If sufficient sample volume is not collected to analyze a LFM duplicate, use an additional portion of an alternate sample to obtain results for a duplicate sample to gather data on precision. As a minimum, include one LFM duplicate or one duplicate sample with each sample set (batch) or on a 5% basis, whichever is more frequent. Evaluate the results obtained for LFM duplicates for precision and accuracy (precision alone for duplicate samples). If LFM duplicate results are out of control, take corrective action to rectify the effect or use another method or the method of standard addition. If duplicate results are out of control, reprepare and reanalyze the sample and take additional corrective action as needed (such as reanalysis of sample batch). Refer to the method of choice for specific acceptance criteria for LFM duplicates or duplicate samples until the laboratory develops statistically valid, laboratory-specific performance criteria. If no limits are included in the method of choice, calculate preliminary limits from initial demonstration of capability. Base sample batch acceptance on results of LFB analyses rather than LFM duplicates alone, because the matrix of the LFM sample may interfere with the method performance.

8. Internal Standard

Internal standards (IS) are used for organic analyses by GC/MS, some GC analyses, and some metals analyses by ICP/MS. An internal standard is an analyte included in each standard and added to each sample or sample extract/digestate just before sample analysis. Internal standards should mimic the analytes of interest but not interfere with the analysis. Choose an internal standard having retention time or mass spectrum separate from the analytes of interest and eluting in a representative area of the chromatogram. Internal standards are used to monitor retention time, calculate relative response, and quantify the analytes of interest in each sample or sample extract/digestate. When quantifying by the internal standard method, measure all analyte responses relative to this internal standard, unless interference is suspected. If internal standard results are out of control, take corrective action, including reanalysis if required. Refer to the method of choice for specific internal standards and their acceptance criteria.

9. Surrogates and Tracers

Surrogates are used for organic analyses; tracers are used for radiochemistry analyses. Surrogates and tracers are used to evaluate method performance in each sample. A surrogate standard is a compound of a known amount added to each sample before extraction. Surrogates mimic the analytes of interest and are compound(s) unlikely to be found in environmental samples, such as fluorinated compounds or stable, isotopically labeled analogs of the analytes of interest. Tracers are a different isotope of the analyte or element of interest. Surrogates and tracers are introduced to samples before extraction to monitor extraction efficiency and percent recovery in each sample. If surrogate or tracer results are out of control, take corrective action, including repreparation and

reanalysis if required. Refer to the method of choice for specific surrogates or tracers and their acceptance criteria, until the laboratory develops statistically valid, laboratory-specific performance criteria.

10. Calibration

a. Instrument calibration: Perform instrument calibration, as well as maintenance, according to instrument manual instructions. Use instrument manufacturer's recommendations for calibration. Perform instrument performance checks, such as those for GC/MS analyses, according to method or SOP instructions.

b. Initial calibration: Perform initial calibration with a minimum of three concentrations of standards for linear curves, a minimum of five concentrations of standards for nonlinear curves, or as specified by the method of choice. Choose a lowest concentration at the reporting limit, and highest concentration at the upper end of the calibration range. Ensure that the calibration range encompasses the analytical concentration values expected in the samples or required dilutions. Choose calibration standard concentrations with no more than one order of magnitude between concentrations.

Use the following calibration functions as appropriate: response factor for internal standard calibration, calibration factor for external standard calibration, or calibration curve. Calibration curves may be linear through the origin, linear not through the origin, or nonlinear through or not through the origin. Some nonlinear functions can be linearized through mathematical transformations, e.g., log. The following acceptance criteria are recommended for the various calibration functions.

If response factors or calibration factors are used, the calculated %RSD for each analyte of interest must be less than the method-specified value. When using response factors (e.g., for GC/MS analysis), evaluate the performance or sensitivity of the instrument for the analyte of interest against minimum acceptance values for the response factors. Refer to the method of choice for the calibration procedure and acceptance criteria on the response factors or calibration factors for each analyte.

If linear regression is used, use the minimum correlation coefficient specified in the method. If the minimum correlation coefficient is not specified, then a minimum value of 0.995 is recommended. Compare each calibration point to the curve and recalculate. If any recalculated values are not within the method acceptance criteria, identify the source of outlier(s) and correct before sample quantitation. Alternately, a method's calibration can be judged against a reference method by measuring the method's "calibration linearity" or %RSD among the "response factors" at each calibration level or concentration.[2]

Use initial calibration, with any of the above functions (response factor, calibration factor, or calibration curve), for quantitation of the analytes of interest in samples. Use calibration verification, described in the next section, only for checks on the initial calibration and not for sample quantitation, unless otherwise specified by the method of choice. Perform initial calibration when the instrument is set up and whenever the calibration verification criteria are not met.

c. Calibration verification: Calibration verification is the periodic confirmation by analysis of a calibration standard that the instrument performance has not changed significantly from the initial calibration. Base this verification on time (e.g., every 12 h) or on the number of samples analyzed (e.g., after every 10 samples). Verify calibration by analyzing a single standard at a concentration near or at the midpoint of the calibration range. The evaluation of the calibration verification analysis is based either on allowable deviations from the values obtained in the initial calibration or from specific points on the calibration curve. If the calibration verification is out of control, take corrective action, including reanalysis of any affected samples. Refer to the method of choice for the frequency of calibration verification and the acceptance criteria for calibration verification.

11. QC Calculations

The following is a compilation of equations frequently used in QC calculations.

a. Initial calibrations:

Relative response factor (RRF):

$$RRF(x) = \frac{A_x}{A_{is}} \times \frac{C_{is}}{C_x}$$

where:

RRF = relative response factor,
A = peak area or height of characteristic ion measured,
C = concentration,
is = internal standard, and
x = analyte of interest.

Response factor (RF):

$$RF(x) = \frac{A_x}{C_x}$$

where:

RF = response factor,
A = peak area or height,
C = concentration, and
x = analyte of interest.

Calibration factor (CF):

$$CF = \frac{\text{peak area (or height) of standards}}{\text{mass injected}}$$

Relative standard deviation (%RSD):

$$\% \, RSD = \frac{s}{\bar{x}} \times 100\%$$

$$s = \sqrt{\sum_{i=1}^{n} \frac{(x_i - \bar{x})^2}{(n-1)}}$$

where:

s = standard deviation,
n = total number of values,
x_i = each individual value used to calculate mean, and
$\bar{x}$ = mean of n values.

b. Calibration verification:

% Difference (%D) for response factor:

$$\% \, D = \frac{\overline{RF_i} - RF_c}{\overline{RF_i}} \times 100\%$$

where:

$\overline{RF_i}$ = average RF or RRF from initial calibration, and
RF_c = relative RF or RRF from calibration verification standard.

% Difference (%D) for values:

$$\% D = \frac{\text{true value} - \text{found value}}{\text{true value}} \times 100\%$$

% Recovery:

$$\% \text{ Recovery} = \frac{\text{found value}}{\text{true value}} \times 100\%$$

c. Laboratory-fortified blank (laboratory control sample):

$$\% \text{ Recovery} = \frac{\text{found value}}{\text{true value}} \times 100\%$$

d. Surrogates:

$$\% \text{ Recovery} = \frac{\text{quantity measured}}{\text{quantity added}} \times 100\%$$

e. Laboratory-fortified matrix (LFM) sample (matrix spike sample):

$$\% \text{ Recovery} = \frac{(\text{LFM sample result} - \text{sample result})}{\text{known LFM added concentration}} \times 100\%$$

f. Duplicate sample:

Relative percent difference (RPD):

$$RPD = \frac{(\text{sample result} - \text{duplicate result})}{(\text{sample result} + \text{duplicate result})/2} \times 100\%$$

g. Method of standards addition:

$$\text{Sample concentrations} \times \text{mg/L} = \frac{S_2 \times V_1 \times C}{(S_1 - S_2) \times V_2}$$

where:

C = concentration of the standard solution, mg/L,
S_1 = signal for fortified portion,
S_2 = signal for unfortified portion,
V_1 = volume of standard addition, L, and
V_2 = volume of sample portion used for method of standard addition, L.

12. Control Charts

Two types of control charts commonly used in laboratories are as follows: accuracy or means charts for QC samples, including reagent blanks, laboratory control standards, calibration check standards, laboratory fortified blanks, laboratory fortified matrices, and surrogates; and precision or range charts, %RSD or relative percent difference (RPD), for replicate or duplicate analyses. These charts are essential tools for quality control. Computer-generated and maintained lists or databases with values, limits, and trending may be used as an alternate to control charts.

a. Accuracy (means) chart: The accuracy chart for QC samples is constructed from the average and standard deviation of a specified number of measurements of the analyte of interest. The ac-

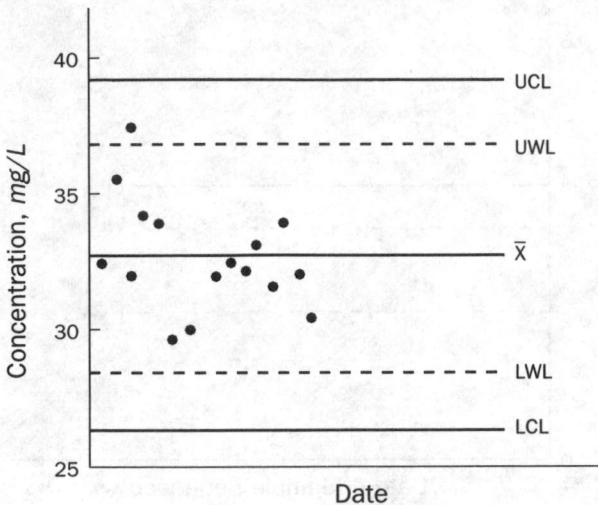
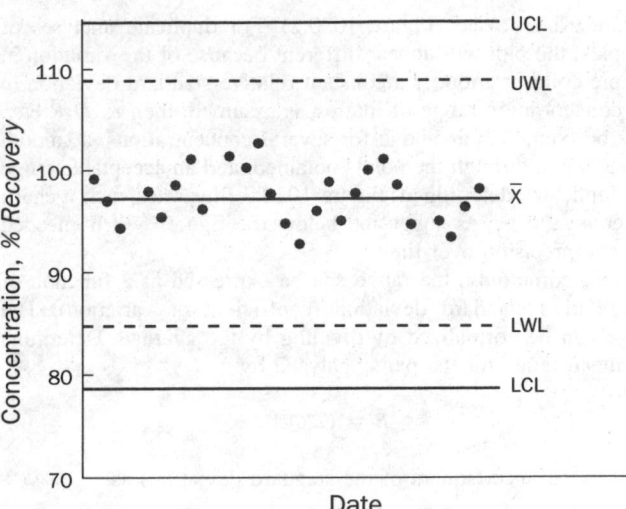

Figure 1020:1. Control charts for means.

curacy chart includes upper and lower warning levels (WL) and upper and lower control levels (CL). Common practice is to use $\pm 2s$ and $\pm 3s$ limits for the WL and CL, respectively, where s represents standard deviation. These values are derived from stated or measured values for reference materials. The number of measurements, n or n-1, used to determine the standard deviation, s, is specified relative to statistical confidence limits of 95% for WLs and 99% for CLs. Set up an accuracy chart by using either the calculated values for mean and standard deviation or the percent recovery. Percent recovery is necessary if the concentration varies. Construct a chart for each analytical method. Enter results on the chart each time the QC sample is analyzed. Examples of control charts for accuracy are given in Figure 1020:1.

b. Precision (range) chart: The precision chart also is constructed from the average and standard deviation of a specified number of measurements of the analyte of interest. If the standard deviation of the method is known, use the factors from Table 1020:I to construct the central line and warning and control limits as in Figure 1020:2. Perfect agreement between replicates or duplicates results in a difference of zero when the values are subtracted, so the baseline on the chart is zero. Therefore for precision charts, only upper warning limits and upper control limits are meaningful. The standard deviation is converted to the

range so that the analyst need only subtract the two results to plot the value on the precision chart. The mean range is computed as:

$$\overline{R} = D_2 s$$

the control limit as

$$CL = \overline{R} \pm 3s(R) = D_4 \overline{R}$$

and the warning limit as

$$WL = \overline{R} \pm 2s(R) = \overline{R} \pm 2/3(D_4 \overline{R} - \overline{R})$$

where:

D_2 = factor to convert s to the range (1.128 for duplicates, as given in Table 1020:I),
$s(R)$ = standard deviation of the range, and
D_4 = factor to convert mean range to $3s(R)$ (3.267 for duplicates, as given in Table 1020:I).

A precision chart is rather simple when duplicate analyses of

TABLE 1020:I. FACTORS FOR COMPUTING LINES ON RANGE
CONTROL CHARTS

Number of Observations n	Factor for Central Line (D_2)	Factor for Control Limits (D_4)
2	1.128	3.267
3	1.693	2.575
4	2.059	2.282
5	2.326	2.115
6	2.534	2.004

Source: ROSENSTEIN, M. & A. S. GOLDEN. 1964. Statistical Techniques for Quality Control of Environmental Radioassays. AQCS Rep. Stat-1. Public Health Serv., Winchester, Mass.

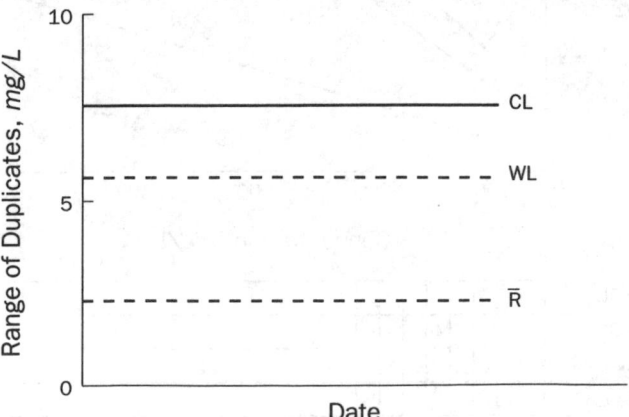

Figure 1020:2. Duplicate analyses of a standard.

a standard are used (Figure 1020:2). For duplicate analyses of samples, the plot will appear different because of the variation in sample concentration. If a constant relative standard deviation in the concentration range of interest is assumed, then $\overline{R}$, $D_4\overline{R}$ etc., may be computed as above for several concentrations, a smooth curve drawn through the points obtained, and an acceptable range for duplicates determined. Figure 1020:3 illustrates such a chart. A separate table, as suggested below the figure, will be needed to track precision over time.

More commonly, the range can be expressed as a function of the relative standard deviation (coefficient of variation). The range can be normalized by dividing by the average. Determine the mean range for the pairs analyzed by

$$\overline{R} = (\Sigma R_i)/n$$

and the variance (square of the standard deviation) as

$$s_R^2 = (\Sigma R_i^2 - n\overline{R}^2)/(n - 1)$$

Then draw lines on the chart at $\overline{R} + 2s_R$ and $\overline{R} + 3s_R$ and, for each duplicate analysis, calculate normalized range and enter the result on the chart. Figure 1020:4 is an example of such a chart.

c. Chart analyses: If the warning limits (WL) are at the 95% confidence level, 1 out of 20 points, on the average, would exceed that limit, whereas only 1 out of 100 would exceed the control

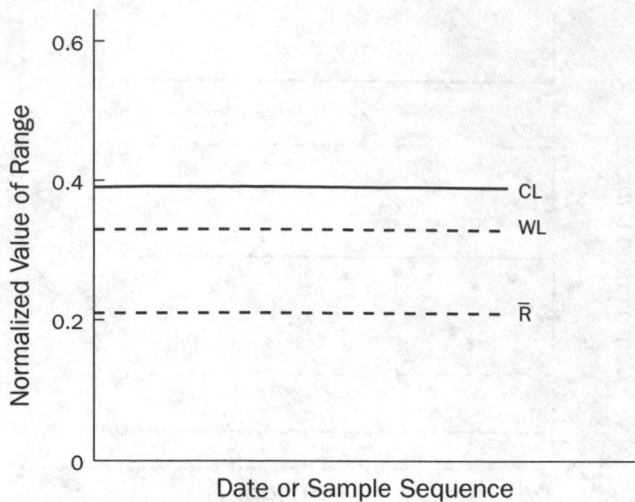

Figure 1020:4. Range chart for variable ranges.

limits (CL). Use the following guidelines, based on these statistical parameters, which are illustrated in Figure 1020:5:

Control limit—If one measurement exceeds a CL, repeat the analysis immediately. If the repeat measurement is within the CL, continue analyses; if it exceeds the CL, discontinue analyses and correct the problem.

Warning limit—If two out of three successive points exceed a WL, analyze another sample. If the next point is within the WL, continue analyses; if the next point exceeds the WL, evaluate potential bias and correct the problem.

Standard deviation—If four out of five successive points exceed 1s, or are in decreasing or increasing order, analyze another sample. If the next point is less than 1s, or changes the order, continue analyses; otherwise, discontinue analyses and correct the problem.

Trending—If seven successive samples are on the same side of the central line, discontinue analyses and correct the problem.

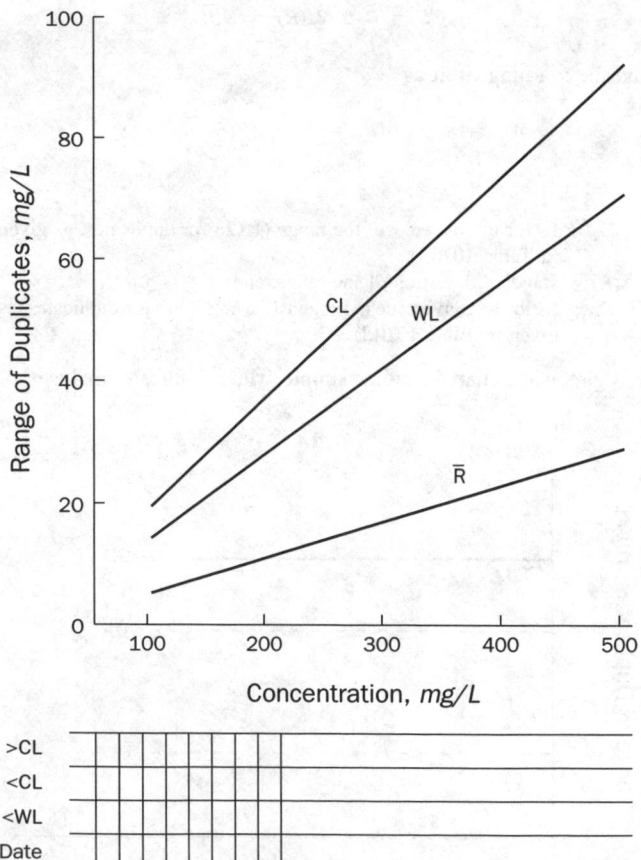

Figure 1020:3. Range chart for variable concentrations.

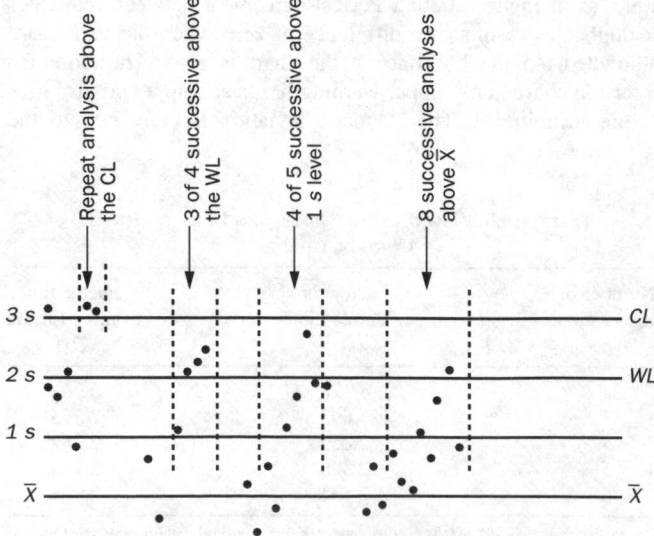

Figure 1020:5. Means control chart with out-of-control data (upper half).

The above considerations apply when the conditions are either above or below the central line, but not on both sides, e.g., four of five values must exceed either $+1s$ or $-1s$. After correcting the problem, reanalyze the samples analyzed between the last in-control measurement and the out-of-control one.

Another important function of the control chart is assessment of improvements in method precision. In the accuracy and precision charts, if measurements never or rarely exceed the WL, recalculate the WL and CL using the 10 to 20 most recent data points. Trends in precision can be detected sooner if running averages of 10 to 20 are kept. Trends indicate systematic error; random error is revealed when measurements randomly exceed warning or control limits.

13. QC Evaluation for Small Sample Sizes

Small sample sizes, such as for field blanks and duplicate samples, may not be suitable for QC evaluation with control charts. QC evaluation techniques for small sample sizes are discussed elsewhere.[3]

14. Corrective Action

Quality control data outside the acceptance limits or exhibiting a trend are evidence of unacceptable error in the analytical process. Take corrective action promptly to determine and eliminate the source of the error. Do not report data until the cause of the problem is identified and either corrected or qualified. Example data qualifiers are listed in Table 1020:II. Qualifying data does not eliminate the need to take corrective actions, but allows for the reporting of data of known quality when it is either not possible or practical to reanalyze the sample(s). Maintain records of all out-of-control events, determined causes, and corrective action taken. The goal of corrective action is not only to eliminate such events, but also to reduce repetition of the causes.

Corrective action begins with the analyst, who is responsible for knowing when the analytical process is out of control. The analyst should initiate corrective action when a QC check exceeds the acceptance limits or exhibits trending and should report an out-of-control event to the supervisor. Such events include QC outliers, hold-time failures, loss of sample, equipment malfunctions, and evidence of sample contamination. Recommended corrective action to be used when QC data are unacceptable are as follows:

- Check data for calculation or transcription error. Correct results if error occurred.
- Check to see if sample(s) was prepared and analyzed according to the approved method and SOP. If it was not, prepare and/or analyze again.
- Check calibration standards against an independent standard or reference material. If calibration standards fail, reprepare calibration standards and/or recalibrate instrument and reanalyze affected sample(s).

TABLE 1020:II. EXAMPLE DATA QUALIFIERS

Symbol	Explanation
B	Analyte found in reagent blank. Indicates possible reagent or background contamination.
E	Reported value exceeded calibration range.
J	Reported value is an estimate because concentration is less than reporting limit or because certain QC criteria were not met.
N	Organic constituents tentatively identified. Confirmation is needed.
PND	Precision not determined.
R	Sample results rejected because of gross deficiencies in QC or method performance. Re-sampling and/or re-analysis is necessary.
RND	Recovery not determined.
U	Compound was analyzed for, but not detected.

* Based on U.S. Environmental Protection Agency guidelines.[1]

- If a LFB fails, reanalyze another laboratory-fortified blank.
- If a second LFB fails, check an independent reference material. If the second source is acceptable, reprepare and reanalyze affected sample(s).
- If a LFM fails, check LFB. If the LFB is acceptable, qualify the data for the LFM sample or use another method or the method of standard addition.
- If a LFM and the associated LFB fail, reprepare and reanalyze affected samples.
- If reagent blank fails, analyze another reagent blank.
- If second reagent blank fails, reprepare and reanalyze affected sample(s).
- If the surrogate or internal standard known addition fails and there are no calculation or reporting errors, reprepare and reanalyze affected sample(s).

If data qualifiers are used to qualify samples not meeting QC requirements, the data may or may not be usable for the intended purposes. It is the responsibility of the laboratory to provide the client or end-user of the data with sufficient information to determine the usability of qualified data.

15. References

1. U.S. ENVIRONMENTAL PROTECTION AGENCY. 1990. Quality Assurance/ Quality Control Guidance for Removal Activities, Sampling QA/QC Plan and Data Validation Procedures. EPA-540/G-90/004, U.S. Environmental Protection Agency, Washington, D.C.
2. U.S. ENVIRONMENTAL PROTECTION AGENCY. 1997. 304h Streamlining Proposal Rule. *Federal Register*, March 28, 1997 (15034).
3. U.S. ENVIRONMENTAL PROTECTION AGENCY. 1994. National Functional Guidelines for Inorganic Data Review. EPA-540/R-94-013, U.S. Environmental Protection Agency, Contract Laboratory Program, Office of Emergency and Remedial Response, Washington, D.C.

1020 C. Quality Assessment

Quality assessment is the process used to ensure that quality control measures are being performed as required and to determine the quality of the data produced by the laboratory. It includes such items as proficiency samples, laboratory intercomparison samples, and performance audits. These are applied to test the precision, accuracy, and detection limits of the methods in use, and to assess adherence to standard operating procedure requirements.

1. Laboratory Check Samples (Internal Proficiency)

The laboratory should perform self-evaluation of its proficiency for each analyte and method in use by periodically analyzing laboratory check samples. Check samples with known amounts of the analytes of interest supplied by an outside organization or blind additions can be prepared independently within the laboratory to determine percent recovery of the analytes of interest by each method.

In general, method performance will have been established beforehand; acceptable percent recovery consists of values that fall within the established acceptance range. For example, if the acceptable range of recovery for a substance is 85 to 115%, then the analyst is expected to achieve a recovery within that range on all laboratory check samples and to take corrective action if results are outside of the acceptance range.

2. Laboratory Intercomparison Samples

A good quality assessment program requires participation in periodic laboratory intercomparison studies. Commercial and some governmental programs supply laboratory intercomparison samples containing one or multiple constituents in various matrices. The frequency of participation in intercomparison studies should be adjusted relative to the quality of results produced by the analysts. For routine procedures, semi-annual analyses are customary. If failures occur, take corrective action and analyze laboratory check samples more frequently until acceptable performance is achieved.

3. Compliance Audits

Compliance audits are conducted to evaluate whether the laboratory meets the applicable requirements of the SOP or consensus method claimed as followed by the laboratory. Compliance audits can be conducted by internal or external parties. A checklist can be used to document the manner in which a sample is treated from time of receipt to final reporting of the result. The goal of compliance audits is to detect any deviations from the SOP or consensus method so that corrective action can be taken on those deviations. An example format for a checklist is shown in Table 1020:III.

4. Laboratory Quality Systems Audits

A quality systems audit program is designed and conducted to address all program elements and provide a review of the quality

TABLE 1020:III. AUDIT OF A SOIL ANALYSIS PROCEDURE

Procedure	Comment	Remarks
1. Sample entered into logbook	yes	lab number assigned
2. Sample weighed	yes	dry weight
3. Drying procedure followed	no	maintenance of oven not done
4a. Balance calibrated	yes	once per year
b. Cleaned and zero adjusted	yes	weekly
5. Sample ground	yes	to pass 50 mesh
6. Ball mill cleaned	yes	should be after each sample

system. Quality systems audits should be conducted by a qualified auditor(s) who is knowledgeable about the section or analysis being audited. Audit all major elements of the quality system at least annually. Quality system audits may be conducted internally or externally; both types should occur on a regular scheduled basis and should be handled properly to protect confidentiality. Internal audits are used for self-evaluation and improvement. External audits are used for accreditation as well as education on client requirements and for approval of the end use of the data. Corrective action should be taken on all audit findings and its effectiveness reviewed at or before the next scheduled audit.

5. Management Review

Review and revision of the quality system, conducted by laboratory management, is vital to its maintenance and effectiveness. Management review should assess the effectiveness of the quality system and corrective action implementation, and should include internal and external audit results, performance evaluation sample results, input from end user complaints, and corrective actions.

6. Bibliography

JARVIS, A.M. & L. SIU. 1981. Environmental Radioactivity Laboratory Intercomparison Studies Program. EPA-600/4-81-004, U.S. Environmental Protection Agency, Las Vegas, Nev.

INTERNATIONAL ORGANIZATION FOR STANDARDIZATION. 1990. General Requirements for the Competence of Testing and Calibration Laboratories, ISO/IEC Guide 25. International Org. for Standardization, Geneva, Switzerland.

AMERICAN SOCIETY FOR TESTING AND MATERIALS. 1996. Standard Practice for Determination of Precision and Bias of Applicable Test Methods of Committee D-19 on Water. ASTM D2777-96, American Society Testing & Materials, West Conshohocken, Pa.

1030 DATA QUALITY

1030 A. Introduction

The role of the analytical laboratory is to produce measurement-based information that is technically valid, legally defensible, and of known quality. Quality assurance is aimed at optimizing the reliability of the measurement process. All measurements contain error, which may be systematic (with an unvarying magnitude) or random (with equal probability of being positive or negative and varying in magnitude). Determination of the systematic and random error components of an analytical method uniquely defines the analytical performance of that method.[1] Quality control (QC) procedures identify and control these sources of error.

1. Measures of Quality Control

Random error (precision) and systematic error (bias) are two routine indicators of measurement quality used by analysts to assess validity of the analytical process. Precision is the closeness of agreement between repeated measurements. A measurement has acceptable precision if the random errors are low. Accuracy is the closeness of a measurement to the true value. A measurement is acceptably accurate when both the systematic and random errors are low. QC results outside the acceptance limits, as set by the data quality objectives, are evidence of an analytical process that may be out of control due to determinant errors such as contaminated reagents or degraded standards.

2. Measurement Error and Data Use

Measurement error, whether random or systematic, reduces the usability of laboratory data. As a measured value decreases, its relative error (e.g., relative standard deviation) may increase and its usable information decrease. Reporting tools, such as detection or quantitation limits, frequently are used to establish a lower limit on usable information content.

Laboratory data may be used for such purposes as regulatory monitoring, environmental decision-making, and process control. The procedures used to extract information for these different purposes vary and may be diametrically opposed. For example, a measurement for regulatory monitoring may be appropriately qualified when below the detection level because the error bar is relatively large and may preclude a statistically sound decision. Data collected over a period of time, however, may be treated by statistical methods to provide a statistically sound decision even when many of the data are below detection levels.[2]

3. The Analyst's Responsibility

The analyst must understand the measures of quality control and how to apply them to the data quality objectives of process control, regulatory monitoring, and environmental field studies. It is important that the quality objectives for the data be clearly defined and detailed before sample analysis so that the data will be technically correct and legally defensible.

4. Reference

1. YOUDEN, W.J. 1975. Statistical Manual of the Association of Official Analytical Chemists. Assoc. Official Analytical Chemists, Arlington, Va.
2. OSBORN, K.E. 1995. You Can't Compute with Less Thans. *Water Environment Laboratory Solutions*, Water Environment Federation, Alexandria, Va.

1030 B. Measurement Uncertainty

1. Introduction

Even with the fullest possible extent of correction, every measurement has error that is ultimately unknown and unknowable. The description of this unknown error is "measurement uncertainty."

Reporting uncertainty along with a measurement result is good practice, and may spare the user from making unwarranted or risky decisions based only on the measurement.

Whereas measurement error (E) is the actual, unknown deviation of the measurement (M) from the unknown true value (T), measurement uncertainty (U) is the state of knowledge about this unknown deviation, and is often expressed as U, as in $M \pm U$. U may be defined as an uncertainty expression.[1,2] This section concerns the definition of U, how to compute it, a recommendation for reporting uncertainty, the interpretation and scope of uncertainty, and other ways of expressing measurement uncertainty.

2. Error

A measurement can be related to the unknown true value and unknown measurement error as follows:

$$M = T + E$$

This is a simple additive relationship. There are other plausible relationships between M and E, such as multiplicative or arbitrary functional relationships, which are not discussed here.

Because E is unknown, M must be regarded as an uncertain measurement. In some practical situations, a value may be treated as known. T^* may be, for example, a published reference value, a traceable value, or a consensus value. The purpose of the substitution may be for convenience or because the measurement process that produced T^* has less bias or variation than the one that produced M. For example, based on the average of many measurements, a vessel might be thought to contain $T^* = 50$ μg/L of salt in water. It then may be sampled and routinely measured, resulting in a reported concentration of $M = 51$ μg/L. The actual concentration may be $T = 49.9$ μg/L, resulting in $E = 51 - 49.9 = 1.1$ μg/L.

To generalize the nature of uncertainty, measurement error may be negligible or large in absolute terms (i.e., in the original units) or relative terms (i.e., unitless, $E \div T$ or T^*). The perceived acceptability of the magnitude of an absolute error depends on its intended use. For example, an absolute error of 1.1 μg/L may be inconsequential for an application where any concentration over 30 μg/L will be sufficient. However, if it is to be used instead as a standard for precision measurement (e.g., of pharmaceutical ingredients), 1.1 μg/L too much could be unacceptable.

3. Uncertainty

Reported measurement uncertainty will contain the actual measurement error with a stated level of confidence. For example, if $M \pm U$ is presented as a 95% confidence interval, approximately 95% of the time, the measurement error E will fall within the range of $\pm U$.

4. Bias

Bias is the systematic component of error. It is defined as the signed deviation between the limiting average measured value and the true value being measured as the number of measurements in the average tends to infinity and the uncertainty about the average tends to zero. For example, the reason the $T = 49.9$ μg/L salt solution is thought to be $T^* = 50$ μg/L could be a bias, $B = 0.1$ μg/L. The "leftover" error, $1.1 - 0.1 = 1.0$ μg/L, is the random component. This random component (also called stochastic error) changes with each measurement. The bias is fixed, and may be related to the laboratory method used to produce T^*. Usually, a recognized method will be used to produce or certify the traceable standard, a sample with a certificate stating the accepted true value T^*. The method may be the best method available or simply the most widely accepted method. It is chosen to have very low error, both bias and random. Such a traceable standard may be purchased from a standards organization such as NIST.

5. Bias and Random Variation

Measurement error, E, (and measurement uncertainty) can be split into two components, random and systematic:

$$E = Z + B$$

Random error, Z, is the component of the measurement error that changes from one measurement to the next, under certain conditions. Random measurement errors are assumed to be independent and have a distribution, often assumed to be Gaussian (i.e., they are normally distributed). The normal distribution of Z is characterized by the distribution mean, μ, and standard deviation, σ_E. In discussion of measurement error distribution, μ is

assumed to be zero because any non-zero component is part of bias, by definition. The population standard deviation, σ_E, can be used to characterize the random component of measurement error because the critical values of the normal distribution are well known and widely available. For example, about 95% of the normal distribution lies within the interval $\mu \pm 2\sigma_E$. Hence, if there is no measurement bias, and measurement errors are independent and normally distributed, $M \pm 2\sigma_E$ (95% confidence, assumed normal) is a suitable way to report a measurement and its uncertainty. More generally, normal probability tables and statistical software give the proportion of the normal distribution and thus the % confidence gained that is contained within $\pm k\sigma_E$ for any value of scalar k.

Usually, however, the population standard deviation, σ_E, is not known and must be estimated by the sample standard deviation, s_E. This estimate of the standard deviation is based on multiple observations and statistical estimation. In this case, the choice of the scalar k must be based not on the normal distribution function, but on the Student's t distribution, taking into account the number of degrees of freedom associated with s_E.

Systematic error (B) is all error that is not random, and typically is equated with bias. Systematic error also can contain outright mistakes (blunders) and lack of control (drifts, fluctuations, etc.).[3] In this manual, the terms "systematic error" and "bias" are used interchangeably.

Systematic uncertainty often is more difficult to estimate and make useful than is random uncertainty. Knowledge about bias is likely to be hard to obtain, and once obtained it is appropriately and likely to be exploited to make the measurement less biased. If measurement bias is known exactly (or nearly so), the user can subtract it from M to reduce total measurement error.

If measurement bias is entirely unknown, and could take on any value from a wide but unknown distribution of plausible values, users may adopt a worst-case approach and report an extreme bound, or they may simply ignore the bias altogether. For example, historical data may indicate that significant interlaboratory biases are present, or that every time a measurement system is cleaned, a shift is observed in QC measurements of standards. In the absence of traceable standards, it is hard for laboratory management or analysts to do anything other than ignore the potential problem.

The recommended practice is to conduct routine QA/QC measurements with a suite of internal standards. Plot measurements on control charts, and when an out-of-control condition is encountered, recalibrate the system with traceable standards. This permits the laboratory to publish a boundary on bias, assuming that the underlying behavior of the measurement system is somewhat predictable and acceptably small in scale in between QA/QC sampling (e.g., slow drifts and small shifts).

6. Repeatability, Reproducibility, and Sources of Bias and Variation

a. Sources and measurement: The sources of bias and variability in measurements are many; they include sampling error, sample preparation, interference by matrix or other measurement quantities/qualities, calibration error variation, software errors, counting statistics, deviations from method by analyst, instrument differences (e.g., chamber volume, voltage level), environmental changes (temperature, humidity, ambient light, etc.), contamination of sample or equipment (e.g., carryover and ambient contamination), variations in purity of solvent, reagent, catalyst, etc., stability and age of sample, analyte, or matrix, and warm-up or cool-down effects, or a tendency to drift over time.

The simplest strategy for estimating typical measurement bias is to measure a traceable (known) standard, then compute the difference between the measured value M and the known value T, assumed to be the true value being measured.

$$M - T = B + Z$$

The uncertainty in the measurement of the traceable standard is assumed to be small, although in practice there may be situations where this is not an appropriate assumption. If random measurement uncertainty is negligible (i.e., $Z \approx 0$), the difference, $M - T$, will provide an estimate of bias (B). If random uncertainty is not negligible, it can be observed and quantified by making a measurement repeatedly on the same test specimen (if the measurement process is not destructive). This may be part of a QA/QC procedure.

b. Repeatability: As quantified by the repeatability standard deviation (σ_{RPT}), repeatability is the minimal variability of a measurement system obtained by repeatedly measuring the same specimen while allowing *no* controllable sources of variability to affect the measurement. Repeatability also can be obtained by pooling sample standard deviations of measurements of J different specimens, as follows:

$$\sigma_{RPT} = \sqrt{\frac{1}{J} \cdot \sum_{i=1}^{J} \sigma_{RPT,i}^2}$$

Repeatability also is called "intrinsic measurement variability," and is considered an approximate lower boundary to the measurement standard deviation that will be experienced in practice. The repeatability standard deviation sometimes is used to compute uncertainty intervals, $\pm U$, that can be referred to as ultimate instrument variability, based on the Student's t distribution function ($\pm U = \pm k s_{RPT}$).

Common sense and application experience demonstrate that repeatability is an overly optimistic estimate to report as measurement uncertainty for routine measurement. In routine use, measurements will be subject to many sources of bias and variability that are intentionally eliminated or restrained during a repeatability study. In routine use, uncertainty in both bias (B) and variability (Z) are greater.

c. Reproducibility: As quantified by the reproducibility standard deviation (σ_{RPD}), reproducibility is the variability of a measurement system obtained by repeatedly measuring a sample while allowing (or requiring) selected sources of bias or variability to affect the measurement. With σ_{RPD}, provide list of known applicable sources of bias and variability, and whether or not they were varied.

Barring statistical variation (i.e., variation in estimates of variability, such as the noisiness in sample standard deviations), the reproducibility standard deviation always is greater than the repeatability standard deviation, because it has additional components. Typically, one or more of the following is varied in a reproducibility study: instrument, analyst, laboratory, or day. Preferably design a study tailored to the particular measurement system (see 1030B.7). If the sample is varied, compute reproducibility standard deviations separately for each sample, then pool

results if they are homogeneous. Treat factors varied in the study as random factors and assume them to be independent normal random variables with zero mean. However, this assumption often can be challenged, because the sample and possibly the target populations may be small (they may even be identical), and there may be a question of "representativeness." For example, six laboratories (or analysts, or instruments) may report usable measurements out of a total population of twenty capable of doing tandem mass spectrometry for a particular analyte and matrix. It is hard to know how representative the six are of the twenty, especially after a ranking and exclusion process that can follow a study, and whether the biases of the twenty are normally distributed (probably not discernible from six measurements, even if the six are representative).

It may be more appropriate to treat each factor with few, known factor values (i.e., choices such as laboratories) as fixed factors, to use the statistical term. Fixed factors have fixed effects. That is, each laboratory has a different bias, as might each analyst, each instrument, and each day, but these biases are not assumed to have a known (or knowable) distribution. Therefore, a small sample cannot be used to estimate distribution parameters, particularly a standard deviation. For example, assuming that variables are random, normal, and have zero mean may be inappropriate in an interlaboratory round-robin study. It must be assumed that every laboratory has some bias, but it is difficult to characterize the biases because of laboratory anonymity, the small number of laboratories contributing usable data, and other factors.

Because of these concerns about assumptions and the potential ambiguity of its definition, do not report reproducibility unless it is accompanied with study design and a list of known sources of bias and variability and whether or not they were varied.

7. Gage Repeatability and Reproducibility, and the Measurement Capability Study

Combining the concepts of repeatability and reproducibility, the Gage Repeatability and Reproducibility (Gage R&R) approach has been developed.[4] It treats all factors as random (including biases), and is based on the simplest nontrivial model:

$$Z = Z_{RPT} + Z_L$$

where:

Z_{RPT} = normally distributed random variable with mean equal to zero and variance equal to σ_{RPT}^2, and

Z_L = normally distributed random variable with mean equal to zero and with the variance of the factor (e.g., interlaboratory) biases, σ_L^2.

The overall measurement variation then is quantified by

$$\sigma_E = \sigma_{RPD} = \sqrt{\sigma_{RPT}^2 + \sigma_L^2}$$

Estimates for σ_{RPT} and σ_{RPD} usually are obtained by conducting a nested designed study and analyzing variance components of the results. This approach can be generalized to reflect good practice in conducting experiments. The following measurement capability study (MCS) procedure is recommended. The objective of such studies is not necessarily to quantify the contribution of every source of bias and variability, but to study those considered to be important, through systematic error budgeting.

To perform a measurement capability study to assess measurement uncertainty through systematic error budgeting, proceed as follows:

Identify sources of bias and variation that affect measurement error. This can be done with a cause-and-effect diagram, perhaps with source categories of: equipment, analyst, method (i.e., procedure and algorithm), material (i.e., aspects of the test specimens), and environment.

Select sources to study, either empirically or theoretically. Typically, study sources that are influential, that can be varied during the MCS, and that cannot be eliminated during routine measurement. Select models for the sources. Treat sources of bias as fixed factors, and sources of variation as random factors.

Design and conduct the study, allowing (or requiring) the selected sources to contribute to measurement error. Analyze the data graphically and statistically (e.g., by regression analysis, ANOVA, or variance components analysis). Identify and possibly eliminate outliers (observations with responses that are far out of line with the general pattern of the data), and leverage points (observations that exert high, perhaps undue, influence).

Refine the models, if necessary (e.g., based on residual analysis), and draw inferences for future measurements. For random effects, this probably would be a confidence interval; for fixed effects, a table of estimated biases.

8. Other Assessments of Measurement Uncertainty

In addition to the strictly empirical MCS approach to assessing measurement uncertainty, there are alternative procedures, discussed below in order of increasing empiricism.

a. Exact theoretical: Some measurement methods are tied closely to exact first-principles models of physics or chemistry. For example, measurement systems that count or track the position and velocity of atomic particles can have exact formulas for measurement uncertainty based on the known theoretical behavior of the particles.

b. Delta method (law of propagation of uncertainty): If the measurement result can be expressed as a function of input variables with known error distributions, the distribution of the measurement result sometimes can be computed exactly.

c. Linearized: The mathematics of the delta method may be difficult, so a linearized form of $M = T + E$ may be used instead, involving a first-order Taylor series expansion about key variables that influence E:

$$(M + \delta M) = T + \delta M/\delta G_1 + \delta M/\delta G_2 + \delta M/\delta G_3 + \ldots$$

for sources G_1, G_2, G_3, etc. of bias and variation that are continuous variables (or can be represented by continuous variables). The distribution of this expression may be simpler to determine, as it involves the linear combination of scalar multiples of the random variables.

d. Simulation: Another use of the delta method is to conduct computer simulation. Again assuming that the distributions of measurement errors in input variables are known or can be approximated, a computer (i.e., Monte Carlo) simulation can obtain empirically the distribution of measurement errors in the result. Typically, one to ten thousand sets of random deviates are generated (each set has one random deviate for each variable), and the value of M is computed and archived. The archived distribution is an empirical characterization of the uncertainty in M.

e. Sensitivity study (designed experiment): If the identities and distributions of sources of bias and variation are known and these sources are continuous factors, but the functional form of the relationship between them and M is not known, an empirical sensitivity study (i.e., MCS) can be conducted to estimate the low-order coefficients ($\delta M/\delta G$) for any factor G. This will produce a Taylor series approximation to the δM, which can be used to estimate the distribution of δM, as in ¶ c above.

f. Random effects study: This is the nested MCS and variance components analysis described in ¶ 7 above.

g. Passive empirical (QA/QC-type data): An even more empirical and passive approach is to rely solely on QA/QC or similar data. The estimated standard deviation of sample measurements taken on many different days, by different analysts, using different equipment, perhaps in different laboratories can provide a useful indication of uncertainty.

9. Statements of Uncertainty

Always report measurements with a statement of uncertainty and the basis for the statement.

Develop uncertainty statements as follows:[4-6]

Involve experts in the measurement principles and use of the measurement system, individuals familiar with sampling contexts, and potential measurement users to generate a cause-and-effect diagram for measurement error, with sources of bias and variation ("factors") identified and prioritized. Consult literature quantifying bias and variation. If needed, conduct one or more measurement capability studies incorporating those sources thought to be most important. In some cases, Gage R&R studies may be sufficient. These studies will provide "snapshot" estimates of bias and variation.

Institute a QA/QC program in which traceable or internal standards are measured routinely and the results are plotted on $\overline{X}$ and R control charts (or equivalent charts). React to out-of-control signals on the control charts. In particular, re-calibrate using traceable standards when the mean control chart shows a statistically significant change. Use the control charts, relevant literature, and the MCSs to develop uncertainty statements that involve both bias and variation.

10. References

1. YOUDEN, W.J. 1972. Enduring values. *Technometrics* 14:1.
2. HENRION, M. & B. FISCHHOFF. 1986. Assessing uncertainty in physical constant. *Amer. J. Phys.* 54:791.
3. CURRIE, L. 1995. Nomenclature in evaluation of analytical methods including detection and quantification capabilities. *Pure Appl. Chem.* 67:1699.
4. MANDEL, J. 1991. Evaluation and Control of Measurements. Marcel Dekker, New York, N.Y.
5. NATIONAL INSTITUTE OF STANDARDS AND TECHNOLOGY. 1994. Technical note TN 1297. National Inst. Standards & Technology.
6. INTERNATIONAL STANDARDS ORGANIZATION. 1993. Guide to the Expression of Uncertainty in Measurement. International Standards Org., Geneva, Switzerland.

1030 C. Method Detection Level

1. Introduction

Detection levels are controversial, principally because of inadequate definition and confusion of terms. Frequently, the instrumental detection level is used for the method detection level and *vice versa*. Whatever term is used, most analysts agree that the smallest amount that can be detected above the noise in a procedure and within a stated confidence level is the detection level. The confidence levels are set so that the probabilities of both Type I and Type II errors are acceptably small.

Current practice identifies several detection levels (see 1010C), each of which has a defined purpose. These are the instrument detection level (IDL), the lower level of detection (LLD), the method detection level (MDL), and the level of quantitation (LOQ). Occasionally the instrument detection level is used as a guide for determining the MDL. The relationship among these levels is approximately IDL:LLD:MDL:LOQ = 1:2:4:10.

2. Determining Detection Levels

An operating analytical instrument usually produces a signal (noise) even when no sample is present or when a blank is being analyzed. Because any QA program requires frequent analysis of blanks, the mean and standard deviation become well known; the blank signal becomes very precise, i.e., the Gaussian curve of the blank distribution becomes very narrow. The IDL is the constituent concentration that produces a signal greater than three standard deviations of the mean noise level or that can be determined by injecting a standard to produce a signal that is five times the signal-to-noise ratio. The IDL is useful for estimating the constituent concentration or amount in an extract needed to produce a signal to permit calculating an estimated method detection level.

The LLD is the amount of constituent that produces a signal sufficiently large that 99% of the trials with that amount will produce a detectable signal. Determine the LLD by multiple injections of a standard at near zero concentration (concentration no greater that five times the IDL). Determine the standard deviation by the usual method. To reduce the probability of a Type I error (false detection) to 5%, multiply s by 1.645 from a cumulative normal probability table. Also, to reduce the probability of a Type II error (false nondetection) to 5%, double this amount to 3.290. As an example, if 20 determinations of a low-level standard yielded a standard deviation of 6 µg/L, the LLD is 3.29 × 6 = 20 µg/L.[1]

The MDL differs from the LLD in that samples containing the constituent of interest are processed through the complete analytical method. The method detection level is greater than the LLD because of extraction efficiency and extract concentration factors. The MDL can be achieved by experienced analysts operating well-calibrated instruments on a nonroutine basis. For example, to determine the MDL, add a constituent to reagent water, or to the matrix of interest, to make a concentration near the estimated MDL.[2] Prepare and analyze seven portions of this solution over a period of at least 3 d to ensure that MDL determination is more representative than measurements performed sequentially. Include all sample processing steps in the determination. Calculate the standard deviation and compute the MDL. The replicate measurements should be in the range of one to five times the calculated MDL. From a table of the one-sided t distribution select the value of t for $7 - 1 = 6$ degrees of freedom and at the 99% level; this value is 3.14. The product 3.14 times s is the desired MDL.

Although the LOQ is useful within a laboratory, the practical quantitation limit (PQL) has been proposed as the lowest level achievable among laboratories within specified limits during routine laboratory operations.[3] The PQL is significant because different laboratories will produce different MDLs even though using the same analytical procedures, instruments, and sample matrices. The PQL is about five times the MDL and represents a practical and routinely achievable detection level with a relatively good certainty that any reported value is reliable.

3. Description of Levels

Figure 1030:1 illustrates the detection levels discussed above. For this figure it is assumed that the signals from an analytical instrument are distributed normally and can be represented by a normal (Gaussian) curve.[4] The curve labeled B is representative of the background or blank signal distribution. As shown, the distribution of the blank signals is nearly as broad as for the other distributions, that is $\sigma_B = \sigma_I = \sigma_L$. As blank analyses continue, this curve will become narrower because of increased degrees of freedom.

The curve labeled I represents the IDL. Its average value is located $k\sigma_B$ units distant from the blank curve, and k represents the value of t (from the one-sided t distribution) that corresponds to the confidence level chosen to describe instrument performance. For a 95% level and $n = 14$, $k = 1.782$ and for a 99% limit, $k = 2.68$. The overlap of the B and I curves indicates the probability of not detecting a constituent when it is present (Type II error).

The curve at the extreme right of Figure 1030:1 represents the LLD. Because only a finite number of determinations is used for calculating the IDL and LLD, the curves are broader than the blank but are similar, so it is reasonable to choose $\sigma_I = \sigma_L$. Therefore, the LLD is $k\sigma_I + k\sigma_L = 2k\sigma_L$ from the blank curve.

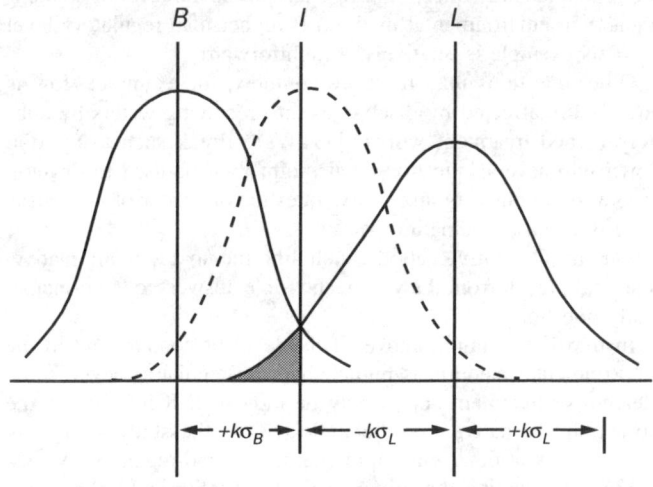

Figure 1030:1. Detection level relationship.

4. References

1. AMERICAN SOCIETY FOR TESTING AND MATERIALS. 1983. Standard Practice for Intralaboratory Quality Control Procedures and a Discussion on Reporting Low-Level Data. Designation D4210-83, American Soc. Testing & Materials, Philadelphia, Pa.

2. GLASER, J.A., D.L. FOERST, J.D. MCKEE, S.A. QUAVE & W.L. BUDDE. 1981. Trace analyses for wastewaters. *Environ. Sci. Technol.* 15:1426.

3. U.S. ENVIRONMENTAL PROTECTION AGENCY. 1985. National Primary Drinking Water Standards: Synthetic Organics, Inorganics, and Bacteriologicals. 40 CFR Part 141; *Federal Register* 50: No. 219, November 13, 1985.

4. OPPENHEIMER, J. & R. TRUSSELL. 1984. Detection limits in water quality analysis. *In* Proc. Water Quality Technology Conference (Denver, Colorado, December 2-5, 1984). American Water Works Assoc., Denver, Colo.

1030 D.   Data Quality Objectives

1. Introduction

Data quality objectives are systematic planning tools based on the scientific method. They are used to develop data collection designs and to establish specific criteria for the quality of data to be collected. The process helps planners identify decision-making points for data collection activities, to determine the decisions to be made based on the data collected, and to identify the criteria to be used for making each decision. This process documents the criteria for defensible decision-making before an environmental data collection activity begins.

2. Procedure

The data quality objective process comprises the stages explained in this section.

a. Stating the issue: Sometimes the reason for performing analyses is straightforward, e.g., to comply with a permit or other regulatory requirement. However, at times the reason is far more subjective; e.g., to gather data to support remedial decisions, or to track the changes in effluent quality resulting from process changes. A clear statement of the reason for the analyses is integral to establishing appropriate data quality objectives; this should include a statement of how the data are to be used, e.g., to determine permit compliance, to support decisions as to whether additional process changes will be necessary, etc.

b. Identifying possible decisions and actions: Initially, express the principal study question. For example: Is the level of contaminant A in environmental medium B higher than regulatory level C? This example is relatively straightforward.

Other questions may be more complex, for example: How is aquatic life affected by discharges into receiving waters by publicly owned treatment works (POTWs)? Break such a question down into several questions that might then be used to develop several decisions; organize these questions in order of consensus priority of all participating parties.

Identify alternative actions, including the no-action alternative, that could result from the various possible answers to the principal study questions.

In the first example above, if the level of contaminant in the environmental medium is higher than the regulatory level, some cleanup or treatment action may be indicated. If it is lower, the no-action alternative may be indicated, or the study team may wish to look at other environmental media and regulatory levels.

Finally, combine the principal study question with alternative actions into a decision statement. For the first example, the decision statement might be: Determine whether the mean level of contaminant A in environmental medium B exceeds the regulatory level C and requires remediation.

A multi-tiered decision statement might be: . . . if not, determine whether the maximum level of contaminant A in environmental medium D exceeds the regulatory level E and requires remediation.

c. Identifying inputs: Identify the information needed to make the necessary decision. Inputs may include measurements (including measurements of physical and chemical characteristics), data sources (historical), applicable action levels, or health effects concerns.

Identify and list the sources of information: previous data, historical records, regulatory guidance, professional judgment, scientific literature, and new data. Evaluate qualitatively whether any existing data are appropriate for the study. Existing data will be evaluated quantitatively later. Identify information needed to establish the action level. Define the basis for setting the action levels: they may be based on regulatory thresholds or standards or may be derived from issue-specific considerations, such as risk analysis. Determine only the criteria that will be used to set the numerical value. The actual numerical action level is determined later.

Confirm that the appropriate measurement methods exist to provide the necessary data. Assure that there are analytical methods for the parameters or contaminants of interest, and that they are appropriate for the matrix to be sampled. Consider the samples to be collected and the analytical methods to determine the potential for matrix interferences for each method. Assure that the limits of the method (e.g., detection limit, quantitation limit, reporting limit) are appropriate for the matrix (e.g., drinking water, wastewater, groundwater, leachate, soil, sediment, hazardous waste) and the parameter to be measured. Ensure that a laboratory is available to perform the analyses; determine its capacity, turnaround time, data product, and cost. Include this information as input to the decision-making process.

d. Identifying study limits: Identify both the geographical area and the time frame to which the decision will apply. Also define the scale of decision-making. Identify the smallest, most appropriate subsets of the total population for which decisions will be made. These subsets could be based on spatial or temporal boundaries. For example, while spatial boundaries of the issue may be a 300-acre site, samples may be collected from, and decisions made for, each square of a grid made up of 50-ft squares drawn on a site map. Also, while temporal boundaries of the issue may be identified (as the duration of storm events), samples may be collected at, and decisions made for, 2-h increments during a storm event. A decision resulting from this type of study might be to construct a stormwater bypass structure that would carry the first flow, which might contain the highest nutrient load, but would not necessarily carry the peak flow.

Identify any practical constraints on data collection. Identify any logistical issues that might interfere with the data-collection process, including seasonal conditions, daily variations, meteorological conditions, access conditions, availability of personnel, time, equipment, project budget, regulatory limits, appropriate analytical methods, matrix interferences, detection limits, reporting limits, site access limitations, and expertise.

e. Developing a decision rule: Define the parameter of interest, specify an action level, and integrate outputs from the previous data quality objective process steps into a single statement that describes a logical basis for choosing among alternative actions. A decision rule may be worded as follows, substituting case-specific information for the underlined words:

If *the factor of interest* within *the scale of decision making* is greater than *the action level*, then take *alternative action A*; otherwise take *alternative action B*.

The factor of interest is a descriptive measure (such as an instantaneous value, a mean, a median, or a proportion) that specifies the characteristic (such as calcium level in water, PCB level in soil, radon level in air) that the decision-maker would like to know about the statistical population affected by the potential decision (such as rivers or streams within a specific watershed, the specified depth of soil within a site boundary, or in basements or crawlspaces within a metropolitan area).

The scale of decision-making is the smallest, most appropriate subset for which separate decisions will be made (such as each stream segment/river mile or each square of a grid identified on a site map, or each section of township X, range Y of county Z).

The action level is a measurement threshold value of the parameter of interest that provides the criterion for choosing among alternative actions (such as a stream standard to protect aquatic life, a published regulatory standard, or a health-effects-related level).

Alternative action A is the alternative of choice if the action level is exceeded (such as initiate non-point-source controls, initiate cleanup of the soil to a specified depth, or distribute technical information to property owners). Noncompliance with the action level is the alternative hypothesis. (Either alternative action can be labeled A without making the decision rule any less valid.)

Alternative action B is the alternative of choice if the action level is not exceeded (such as continue routine monitoring, leave the soil in place, or provide a summary of the data collection activity to potential developers). Compliance with the action level is the null hypothesis that is generally the no-action alternative or baseline condition. Either alternative action can be labeled B without making the decision rule any less valid.

f. Specifying limits on decision errors: Establish limits on the decision error that the decision-maker will tolerate. Use these limits to establish performance goals for design of the data collection activity. Base limits on the consequences of making a wrong decision.

Decision-makers are interested in knowing the true state of some feature of the environment. Environmental data can be only an estimate of this true state; decisions therefore are based on environmental data that are in some degree of error. The goal is to develop a data-collection design that reduces the chances of

making a decision error to a level that is acceptable to the decision-maker. Sources of uncertainty include sample design error and measurement error; when combined, they represent the total study error.

Sample design error refers to the error inherent in using a portion of a population to represent the whole population. It is not practical, for example, to measure and record the concentration of an analyte at every point in a stream on a continuous basis; instead, measure analyte concentration at well-defined locations and time intervals to represent this analyte concentration continuum.

Measurement error refers to the error inherent in the measurement process. A measurement system does not measure, on a molecular level, the amount of an analyte in a sample; it measures an indicator of the amount of an analyte in a sample. This indicator might be the amount of a specific wavelength of light absorbed by a sample, the change in conductivity of a solution containing the analyte, or the amount of an analyte, in a gaseous or ionized form, that passes through a membrane.

Use data to choose between the one condition of the environment (the null hypothesis, H_0) and an alternative condition (the alternative hypothesis, H_a). A decision error occurs when the decision-maker rejects the null hypothesis when it is true (false-positive decision error) or fails to reject the null hypothesis when it is false (false-negative decision error).*

The null hypothesis usually is treated as the baseline condition that is presumed to be true in the absence of strong evidence to the contrary. Either condition may be selected as the null hypothesis, but if the null hypothesis is chosen carefully, it provides a way to guard against making the decision error that the decision-maker considers to have the more undesirable consequences.

While the possibility of a decision error never can be totally eliminated, it can be controlled by various means, including collecting a large number of samples (to control sampling design error), analyzing individual samples several times, or using more precise laboratory methods (to control measurement error). Better sampling designs also can be developed to collect data that more accurately represent the population of interest. Every study will use a different method of controlling decision errors, depending on the source of the largest components of total decision error in the data set and the ease of reducing those error components.

Reducing the probability of making decision errors generally increases study costs. In many cases, however, it is not necessary to control decision error within very small limits to meet the decision-maker's needs. If the consequences of decision errors are minor, a reasonable decision could be made on the basis of relatively crude data. If, on the other hand, consequences of decision errors are severe, the decision-maker will want to control sampling design and measurements within very small limits.

Factors used to judge data quality include precision, bias, representativeness, completeness, and comparability. Precision, bias, and completeness can be applied to the measurement (field and laboratory) system. Most analytical laboratories have systems to quantify these factors. Laboratory precision can be estimated through the analysis of laboratory replicates. Laboratory bias can be estimated by the analysis of standards, known additions, and

---

* Note that these definitions are not the same as false-positive or false-negative instrument readings, where similar terms commonly are used by laboratory or field personnel to describe a fault in a single result; false-positive and false-negative decision errors are defined in the context of hypothesis testing, where the terms are defined with respect to the null hypothesis.

performance evaluation (PE) samples. There is no common system in place to estimate field bias. A combination of field and laboratory completeness can be estimated through comparison of the number of analytical results provided by the laboratory with the number of analytical results specified in the sample design. Laboratory representativeness and comparability involve the analytical method used and the performance of the laboratory as compared to the performance of other laboratories (PE studies), which are not commonly quantified.

Precision, bias, representativeness, completeness, and comparability can be applied to the sample design: Precision would indicate how precisely this sample design reflects the total population. Bias would indicate how accurately this sample design reflects the total population. Representativeness would indicate to what extent the sample design is representative of the total population. Completeness would indicate how well the sample design reflects the complete population. Comparability would indicate the similarity of the sample design to other sample designs for similar situations. None of these usually is measured.

While data quality factors provide some insight into sample measurement errors, they do not provide any indication of sample design errors. These errors are additive, so that if precision were $\pm 90\%$, bias were $\pm 90\%$, and representativeness were $\pm 90\%$, combined uncertainty could be up to $\pm 27\%$:

$$(100\% \times 0.1) + (90\% \times 0.1) + (81\% \times 0.1) = 10\% + 9\% + 8\%$$
$$= 27\%$$

Because most errors are not quantifiable, a study usually is designed with a balance between acceptable decision errors and acceptable study cost.

g. Optimizing the design for collection: Identify the most resource-effective design for the study that will achieve the data quality objectives (DQOs). Use statistical techniques to develop alternative data collection designs and evaluate their efficiency in meeting the DQOs. To develop the optimal study design, it may be necessary to work through this step more than once after revisiting previous steps of the process.

Review the DQO outputs and existing environmental data, develop general data collection design alternatives, and formulate the mathematical expressions needed to solve the design issue for each data collection design alternative. Develop the following three mathematical expressions:

- A method for testing the statistical hypothesis and a sample size formula that corresponds to the method (e.g., Student's *t* test),
- A statistical model that describes the relationship of the measured value to the "true" value. Often the model will describe the components of error or bias believed to exist in the measured value, and
- A cost function that relates the number of samples to the total cost of sampling and analysis.

Select the optimal sample size that satisfies the DQOs for each data collection design alternative. Using the mathematical expressions specified above, calculate the optimal sample size that satisfies the DQOs. If no design will meet the limits on decision errors within the budget or other constraints, relax one or more constraints by, for example, increasing the budget for sampling and analysis, increasing the width of the region of uncertainty, increasing the tolerable decision error rates, relaxing other project constraints such as the schedule, or changing the boundaries; it may be possible to reduce sampling and analysis costs by changing or eliminating subgroups that will require separate decisions.

Select the most resource-effective data collection design that satisfies all of the DQOs and document the operational details and theoretical assumptions of the selected design in the sampling and analysis plan.

3. Bibliography

U.S. ENVIRONMENTAL PROTECTION AGENCY. 1994. Guidance for the Data Quality Objectives Process. EPA QA/G-4, Quality Assurance Management Staff, U.S. Environmental Protection Agency, Washington, D.C.

1030 E. Checking Correctness of Analyses

The following procedures for checking correctness of analyses are applicable specifically to water samples for which relatively complete analyses are made.[1] These include pH, conductivity, total dissolved solids (TDS), and major anionic and cationic constituents that are indications of general water quality.

The checks described do not require additional laboratory analyses. Three of the checks require calculation of the total dissolved solids and conductivity from measured constituents. Sum concentrations (in milligrams per liter) of constituents to calculate the total dissolved solids are as follows:

$$\text{Total dissolved solids} = 0.6 \text{ (alkalinity*)} + Na^+ + K^+$$
$$+ Ca^{2+} + Mg^{2+} + Cl^- + SO_4^{2-} + SiO_3^{2-} + NO_3^- + F^-$$

Calculate electrical conductivity from the equation:

$$G = \lambda C - (k_1\lambda + k_2)(C)^{3/2}$$

where:

G = conductivity of salt solution,
C = concentration of salt solution,
λ = equivalent conductance of salt solution at infinite dilution,
k_1, k_2 = constants for relaxation of ion cloud effect and electrophoretic effect relative to ion mobility.[1]

1. Anion-Cation Balance[2]

The anion and cation sums, when expressed as milliequivalents per liter, must balance because all potable waters are electrically neutral. The test is based on the percentage difference defined as follows:

$$\% \text{ difference} = 100 \frac{\Sigma \text{ cations} - \Sigma \text{ anions}}{\Sigma \text{ cations} + \Sigma \text{ anions}}$$

and the typical criteria for acceptance are as follows:

Anion Sum meq/L	Acceptable Difference
0–3.0	±0.2 meq/L
3.0–10.0	± 2%
10.0–800	5%

2. Measured TDS = Calculated TDS[2]

The measured total dissolved solids concentration should be higher than the calculated one because a significant contributor may not be included in the calculation. If the measured value is less than the calculated one, the higher ion sum and measured value are suspect; the sample should be reanalyzed. If the measured solids concentration is more than 20% higher than the calculated one, the low ion sum is suspect and selected constituents should be reanalyzed. The acceptable ratio is as follows:

$$1.0 < \frac{\text{measured TDS}}{\text{calculated TDS}} < 1.2$$

* As $CaCO_3$.

3. Measured EC = Calculated EC

If the calculated electrical conductivity (EC) is higher than the measured value, reanalyze the higher ion sum. If the calculated EC is less than the measured one, reanalyze the lower ion sum. The acceptable ratio is as follows:

$$0.9 < \frac{\text{calculated EC}}{\text{measured EC}} < 1.1$$

Some electrical conductivity values for ions commonly found in water are given in Table 1030:I.

4. Measured EC and Ion Sums

Both the anion and cation sums should be $^1/_{100}$ of the measured EC value. If either of the two sums does not meet this criterion, that sum is suspect; reanalyze the sample. The acceptable criteria are as follows:

$$100 \times \text{anion (or cation) sum, meq/L} = (0.9–1.1) \text{ EC}$$

5. Calculated TDS to EC Ratio

If the ratio of calculated TDS to conductivity falls below 0.55, the lower ion sum is suspect; reanalyze it. If the ratio is above 0.7, the higher ion sum is suspect; reanalyze it. If reanalysis causes no change in the lower ion sum, an unmeasured constituent, such as ammonia or nitrite, may be present at a significant concentration. If poorly dissociated calcium and sulfate ions are present, the TDS may be as high as 0.8 times the EC. The acceptable criterion is as follows:

$$\text{calculated TDS/conductivity} = 0.55–0.7$$

6. Measured TDS to EC Ratio

The acceptable criteria for this ratio are from 0.55 to 0.7. If the ratio of TDS to EC is outside these limits, measured TDS or measured conductivity is suspect; reanalyze.

TABLE 1030:I. CONDUCTIVITY FACTORS OF IONS COMMONLY FOUND IN WATER

Ion	Conductivity (25°C) μmhos/cm	
	Per me/L	Per mg/L
Bicarbonate	43.6	0.715
Calcium	52.0	2.60
Carbonate	84.6	2.82
Chloride	75.9	2.14
Magnesium	46.6	3.82
Nitrate	71.0	1.15
Potassium	72.0	1.84
Sodium	48.9	2.13
Sulfate	73.9	1.54

A more complete exposition[3] of the above quality-control checks has been published.

7. References

1. ROSSUM, J.R. 1975. Checking the accuracy of water analyses through the use of conductivity. *J. Amer. Water Works Assoc.* 67:204.

2. FRIEDMAN, L.C. & D.E. ERDMANN. 1982. Quality Assurance Practices for Analyses of Water and Fluvial Sediments. Tech. Water Resources Inc., Book 5, Chapter A6. U.S. Government Printing Off., Washington, D.C.

3. OPPENHEIMER, J. & A.D. EATON. 1986. Quality control and mineral analysis. *In* Proc. Water Quality Technology Conference (Houston, Texas, December 8-11, 1985). American Water Works Assoc., Denver, Colo.

1040   METHOD DEVELOPMENT AND EVALUATION

1040 A.   Introduction

Although standard methods are available from many nationally recognized sources, there may be occasions when they cannot be used or when no standard method exists for a particular constituent or characteristic. Therefore, method development may be required. Method development is the set of experimental procedures devised for measuring a known amount of a constituent in various matrices, in the case of chemical analyses; or a known characteristic (e.g., biological or toxicological) of various matrices.

1040 B.   Method Validation

Whether an entirely new method is developed by accepted research procedures or an existing method is modified to meet special requirements, validation by a three-step process is required: determination of single-operator precision and bias, analysis of independently prepared unknown samples, and determination of method ruggedness.

1. Single-Operator Characteristics

This part of the validation procedure requires determining the method detection level (MDL) as in Section 1030; the bias of the method, i.e., the systematic error of the method; and the precision obtainable by a single operator, i.e., the random error introduced in using the method. To make these determinations, analyze at least 7 but preferably 10 or more portions of a standard at each of several concentrations in each matrix that may be used. Use one concentration at, or slightly above, the MDL and one relatively high so that the range of concentrations for which the method is applicable can be specified.

The use of several concentrations to determine bias and precision will reveal the form of the relationship between these method characteristics and the concentration of the substance, the characteristic toxicity of the substance, or the biological factor of interest. This relationship may be constant, linear, or curvilinear and is a significant characteristic of the method that should be explained clearly. Table 1040:I shows calculation of precision and bias for a single concentration in a single matrix from eight replicate analyses of a standard with a known concentration of 1.30 mg/L.

TABLE 1040:I. PRECISION AND BIAS FOR A SINGLE CONCENTRATION IN A SINGLE MATRIX

Result mg/L	Difference (−1.30)	Squared Difference
1.23	−0.07	0.0049
1.21	−0.09	0.0081
1.30	0.0	0.0
1.59	0.29	0.0841
1.57	0.27	0.0729
1.21	−0.09	0.0081
1.53	0.23	0.0529
1.25	−0.05	0.0025
Sum	0.49	0.2335

The bias is $0.49/8 = 0.06$ mg/L and the precision is the square root of $0.2335/(8-1) = \sqrt{0.03336}$, or 0.18 mg/L (note that this is similar to the calculation for standard deviation).

2. Analysis of Unknown Samples

This step in the method validation procedure requires analysis of independently prepared standards where the value is unknown to the analyst. Analyze each unknown in replicate by following the standard operating procedure for the method. The mean amount recovered should be within three standard deviations (s) of the mean value of the standard but preferably within $2\ s$.

Obtain the unknowns from other personnel in the analyst's laboratory using either purchased analytical-grade reagents or standards available from National Institute of Standards and Technology (NIST). If available for the particular constituent, performance evaluation samples from EPA-Cincinnati are particularly useful.

3. Method Ruggedness

A test of the ruggedness, i.e., stability of the result produced when steps in the method are varied, is the final validation step. It is especially important to determine this characteristic of a method if it is to be proposed as a standard or reference method. A properly conducted ruggedness test will point out those procedural steps in which rigor is critical and those in which some leeway is permissible.

The Association of Official Analytical Chemists[1] has suggested a method for this test in which eight separate analyses can be used to determine the effect of varying seven different steps in an analytical procedure. To illustrate, suppose the effect of changing the factors in Table 1040:II is to be determined. To make the determination, denote the nominal factors by capital letters A through G and the variations by the corresponding lower-case letters. Then set up a table of the factors as in Table 1040:III.

If combination 1 is analyzed, the result will be s. If combination 2 is analyzed, the result will be t, and so on until all eight combinations have been analyzed. To determine the effect of varying a factor, find the four results where the factor was nominal (all caps) and the four where it was varied (all lower case) and compare the averages of the two groups. For example, to compare the effect of changing C to c, use results (s + u + w + y)/4 and (t + v + x + z)/4. Calculate all seven pairs to get seven differences, which can then be ranked to reveal those with a significant effect on the results. If there is no outstanding difference, calculate the average and standard deviation of the eight results s through z. The standard deviation is a realistic estimate of the precision of the method. This design tests main effects, not interactions.

4. Equivalency Testing

After a new method has been validated by the procedures listed above, it may be prudent to test the method for equivalency to

TABLE 1040:II. VARIATIONS IN FACTORS FOR METHOD RUGGEDNESS DETERMINATION

Factor	Nominal	Variation
Mixing time	10 min	12 min
Portion size	5 g	10 g
Acid concentration	$1M$	$1.1M$
Heat to	100°C	95°C
Hold heat for	5 min	10 min
Stirring	yes	no
pH adjust	6.0	6.5

TABLE 1040:III. FACTOR MATRIX FOR METHOD RUGGEDNESS DETERMINATION

Factor value	Combinations							
	1	2	3	4	5	6	7	8
A or a	A	A	A	A	a	a	a	a
B or b	B	B	b	b	B	B	b	b
C or c	C	c	C	c	C	c	C	c
D or d	D	D	d	d	d	d	D	D
E or e	E	e	E	e	e	E	e	E
F or f	F	f	f	F	F	f	f	F
G or g	G	g	g	G	g	G	G	g
Result	s	t	u	v	w	x	y	z

Source: YOUDEN, W. J. & E. H. STEINER. 1975. Statistical Manual of AOAC. Assoc. Official Analytical Chemists, Washington, D.C.

standard methods, unless none exist. This requires analysis of a minimum of three concentrations by the alternate and by the standard method. If the range of concentration is very broad, test more concentrations. Once an initial set of analyses (five or more) has been made at each chosen concentration, apply the following statistical steps:[2]

1. Test the distribution of data for normality and transform the data if necessary (Section 1010B).
2. Select an appropriate sample size based on an estimate of the standard deviation.[3]
3. Test the variances of the two methods using the F-ratio statistic.
4. Test the average values of the two methods using a Student-t statistic.

An explanation of each of these steps with additional techniques and examples has been published.[4] Because the number of analyses can be very large, the calculations become complex and familiarity with basic statistics is necessary. A listing of standard, reference, and equivalent methods for water analysis is available.[5]

5. References

1. YOUDEN, W.J. & E.H. STEINER. 1975. Statistical Manual of AOAC. Assoc. Official Analytical Chemists, Washington, D.C.
2. WILLIAMS, L.R. 1985. Harmonization of Biological Testing Methodology: A Performance Based Approach in Aquatic Toxicology and Hazard Assessment. 8th Symp. ASTM STP 891, R.C. Bahner & D.J. Hansen, eds. American Soc. Testing & Materials, Philadelphia, Pa.
3. NATRELLA, M.G. 1963. Experimental Statistics. National Bureau of Standards Handbook 91, Washington, D.C.
4. U.S. ENVIRONMENTAL PROTECTION AGENCY. 1983. Guidelines for Establishing Method Equivalency to Standard Methods. Rep. 600/X-83-037, Environmental Monitoring Systems Lab., Las Vegas, Nev.
5. U.S. ENVIRONMENTAL PROTECTION AGENCY. 1994. Guidelines establishing test procedures for the analysis of pollutants under the Clean Water Act. Final rule. 40 CFR Part 136; *Federal Register* 59:20:4504.

1040 C.   Collaborative Testing

Once a new or modified method has been developed and validated it is appropriate to determine whether the method should be made a standard method. The procedure to convert a method to standard status is the collaborative test.[1] In this test, different laboratories use the standard operating procedure to analyze a select number of samples to determine the method's bias and precision as would occur in normal practice.

In planning for a collaborative test, consider the following factors: a precisely written standard operating procedure, the number of variables to be tested, the number of levels to be tested, and the number of replicates required. Because method precision is estimated by the standard deviation, which itself is the result of many sources of variation, the variables that affect it must be tested. These may include the laboratory, operator, apparatus, and concentration range.

1. Variables

Test at least the following variables:

Laboratory—Involve at least three different laboratories, although more are desirable to provide a better estimate of the standard deviation;

Apparatus—Because model and manufacturer differences can be sources of error, analyze at least two replicates of each concentration per laboratory;

Operators—To determine overall precision, involve at least six analysts with not more than two from each laboratory;

Levels—If the method development has indicated that the relative standard deviation is constant, test three levels covering the range of the method. If it is not constant, use more levels spread uniformly over the operating range.

If matrix effects are suspected, conduct the test in each medium for which the method was developed. If this is not feasible, use appropriate grades of reagent water as long as this is stipulated in the resulting statement of method characteristics.

2. Number of Replicates

Calculate the number of replicates after the number of variables to be tested has been determined by using the formula:

$$r > 1 + (30/P)$$

where:
   r = number of replicates and
   P = the product of several variables.

The minimum number of replicates is two. As an example, if three levels of a substance are to be analyzed by single operators in six laboratories on a single apparatus, then P is calculated as follows:

$$P = 3 \times 1 \times 6 \times 1 = 18$$

and the number of replicates is

$$r > 1 + (30/18) > 2.7 \text{ or } r = 3.$$

3. Illustrative Collaborative Test

Send each of five laboratories four concentrations of a compound (4.3, 11.6, 23.4, and 32.7 mg/L) with instructions to analyze in triplicate using the procedure provided. Tabulate results as shown in Table 1040:IV below (the results for only one concentration are shown). Because there are no obviously aberrant values (use the method in Section 1010B to reject outliers), use all the data.

Calculate the average and standard deviation for each laboratory; use all 15 results to calculate a grand average and standard deviation. The difference between the average of each laboratory and the grand average reveals any significant bias, such as that shown for Laboratories 1 and 3. The difference between the grand average and the known value is the method bias, e.g., 33.0 − 32.7 = 0.3 mg/L or 0.9%. The relative standard deviation of the grand average (1.5 mg/L) is 4.5%, which is the method precision, and the s for each laboratory is the single-operator precision.

As noted in Table 1040:IV, the sum of the deviations from the known value for the laboratories was 1.3, so the average deviation (bias) was 1.3/5 = 0.26, rounded to 0.3, which is the same as the difference between the grand average and the known value.

For all four unknowns in this test, the percentage results indicated increasing bias and decreasing precision as the concentration decreased. Therefore, to describe the method in a formal statement, the precision would be given by a straight line with the formula $y = mx + b$; where y is the relative standard deviation, m is the slope of the line, x is the concentration, and b is

TABLE 1040:IV. SAMPLE COLLABORATIVE TEST RESULTS

Laboratory	Result mg/L	Experimental $x \pm s$	Deviation From Known	Deviation From Grand Average
1	32.7 35.2 36.3	34.7 ± 1.8	2.0	1.7
2	32.6 33.7 33.6	33.3 ± 0.6	0.6	0.3
3	30.6 30.6 32.4	31.2 ± 1.0	−1.5	−1.8
4	32.6 32.5 33.9	33.0 ± 0.8	0.3	0
5	32.4 33.4 32.9	32.6 ± 0.8	−0.1	−0.4
$(\Sigma x)/n$ = 33 s = 1.5			Σ = 1.3	Σ = −0.2

TABLE 1040:V. METHOD PRECISION AND BIAS

Known Amount mg/L	Amount Found mg/L	CV (% Standard Deviation)	Bias %
4.3	4.8	12.5	11.5
11.6	12.2	10.2	5.6
23.4	23.8	5.4	1.9
32.7	33	4.5	0.9

the relative standard deviation at concentration = 0. The values found from the collaborative test are shown in Table 1040:V.

These results indicate that the method is acceptable. However, concentrations of less than about 10 mg/L require greater care in analysis.

4. Reference

1. YOUDEN, W.J. & E.H. STEINER. 1975. Statistical Manual of the AOAC. Assoc. Official Analytical Chemists, Washington, D.C.

1050   EXPRESSION OF RESULTS

1050 A.   Units

This text uses the International System of Units (SI) and chemical and physical results are expressed in milligrams per liter (mg/L). See Section 7020D for expression of radioactivity results. Record only the significant figures. If concentrations generally are less than 1 mg/L, it may be more convenient to express results in micrograms per liter (μg/L). Use μg/L when concentrations are less than 0.1 mg/L.

Express concentrations greater than 10 000 mg/L in percent, 1% being equal to 10 000 mg/L when the specific gravity is 1.00. In solid samples and liquid wastes of high specific gravity, make a correction if the results are expressed as parts per million (ppm) or percent by weight:

$$\text{ppm by weight} = \frac{\text{mg/L}}{\text{sp gr}}$$

$$\% \text{ by weight} = \frac{\text{mg/L}}{10\,000 \times \text{sp gr}}$$

In such cases, if the result is given as milligrams per liter, state specific gravity.

The unit equivalents per million (epm), or the identical and less ambiguous term milligram-equivalents per liter, or milliequivalents per liter (me/L), can be valuable for making water treatment calculations and checking analyses by anion-cation balance.

Table 1050:I presents factors for converting concentrations of

TABLE 1050:I. CONVERSION FACTORS*
(Milligrams per Liter—Milliequivalents per Liter)

Ion (Cation)	me/L = mg/L ×	mg/L = me/L ×	Ion (Anion)	me/L = mg/L ×	mg/L = me/L ×
Al^{3+}	0.111 2	8.994	BO_2^-	0.023 36	42.81
B^{3+}	0.277 5	3.604	Br^-	0.012 52	79.90
Ba^{2+}	0.014 56	68.66	Cl^-	0.028 21	35.45
Ca^{2+}	0.049 90	20.04	CO_3^{2-}	0.033 33	30.00
Cr^{3+}	0.057 70	17.33	CrO_4^{2-}	0.017 24	58.00
			F^-	0.052 64	19.00
Cu^{2+}	0.031 47	31.77	HCO_3^-	0.016 39	61.02
Fe^{2+}	0.035 81	27.92	HPO_4^{2-}	0.020 84	47.99
Fe^{3+}	0.053 72	18.62	$H_2PO_4^-$	0.010 31	96.99
H^+	0.992 1	1.008	HS^-	0.030 24	33.07
K^+	0.025 58	39.10	HSO_3^-	0.012 33	81.07
			HSO_4^-	0.010 30	97.07
Li^+	0.144 1	6.941	I^-	0.007 880	126.9
Mg^{2+}	0.082 29	12.15	NO_2^-	0.021 74	46.01
Mn^{2+}	0.036 40	27.47	NO_3^-	0.016 13	62.00
Mn^{4+}	0.072 81	13.73	OH^-	0.058 80	17.01
Na^+	0.043 50	22.99	PO_4^{3-}	0.031 59	31.66
NH_4^+	0.055 44	18.04	S^{2-}	0.062 37	16.03
Pb^{2+}	0.009 653	103.6	SiO_3^{2-}	0.026 29	38.04
Sr^{2+}	0.022 83	43.81	SO_3^{2-}	0.024 98	40.03
Zn^{2+}	0.030 59	32.70	SO_4^{2-}	0.020 82	48.03

* Factors are based on ion charge and not on redox reactions that may be possible for certain of these ions. Cations and anions are listed separately in alphabetical order.

common ions from milligrams per liter to milliequivalents per liter, and vice versa. The term milliequivalent used in this table represents 0.001 of an equivalent weight. The equivalent weight, in turn, is defined as the weight of the ion (sum of the atomic weights of the atoms making up the ion) divided by the number of charges normally associated with the particular ion. The factors

for converting results from milligrams per liter to milliequivalents per liter were computed by dividing the ion charge by weight of the ion. Conversely, factors for converting results from milliequivalents per liter to milligrams per liter were calculated by dividing the weight of the ion by the ion charge.

1050 B.   Significant Figures

1. Reporting Requirements

To avoid ambiguity in reporting results or in presenting directions for a procedure, it is the custom to use "significant figures." All digits in a reported result are expected to be known definitely, except for the last digit, which may be in doubt. Such a number is said to contain only significant figures. If more than a single doubtful digit is carried, the extra digit or digits are not significant. If an analytical result is reported as "75.6 mg/L," the analyst should be quite certain of the "75," but may be uncertain as to whether the ".6" should be .5 or .7, or even .4 or .8, because of unavoidable uncertainty in the analytical procedure. If the standard deviation were known from previous work to be ± 2 mg/L, the analyst would have, or should have, rounded off the result to "76 mg/L" before reporting it. On the other hand, if the method were so good that a result of "75.61 mg/L" could have been conscientiously reported, then the analyst should not have rounded it off to 75.6.

Report only such figures as are justified by the accuracy of the work. Do not follow the all-too-common practice of requiring that quantities listed in a column have the same number of figures to the right of the decimal point.

2. Rounding Off

Round off by dropping digits that are not significant. If the digit 6, 7, 8, or 9 is dropped, increase preceding digit by one unit; if the digit 0, 1, 2, 3, or 4 is dropped, do not alter preceding digit. If the digit 5 is dropped, round off preceding digit to the nearest even number: thus 2.25 becomes 2.2 and 2.35 becomes 2.4.

3. Ambiguous Zeros

The digit 0 may record a measured value of zero or it may serve merely as a spacer to locate the decimal point. If the result of a sulfate determination is reported as 420 mg/L, the report recipient may be in doubt whether the zero is significant or not, because the zero cannot be deleted. If an analyst calculates a total residue of 1146 mg/L, but realizes that the 4 is somewhat doubtful and that therefore the 6 has no significance, the answer should be rounded off to 1150 mg/L and so reported but here, too, the report recipient will not know whether the zero is significant. Although the number could be expressed as a power of 10 (e.g., 11.5×10^2 or 1.15×10^3), this form is not used generally be-

cause it would not be consistent with the normal expression of results and might be confusing. In most other cases, there will be no doubt as to the sense in which the digit 0 is used. It is obvious that the zeros are significant in such numbers as 104 and 40.08. In a number written as 5.000, it is understood that all the zeros are significant, or else the number could have been rounded off to 5.00, 5.0, or 5, whichever was appropriate. Whenever the zero is ambiguous, it is advisable to accompany the result with an estimate of its uncertainty.

Sometimes, significant zeros are dropped without good cause. If a buret is read as "23.60 mL," it should be so recorded, and not as "23.6 mL." The first number indicates that the analyst took the trouble to estimate the second decimal place; "23.6 mL" would indicate a rather careless reading of the buret.

4. Standard Deviation

If, for example, a calculation yields a result of 1449 mg/L or 1451 mg/L with a standard deviation of ± 100 mg/L, report as 1449 ± 100 mg/L or 1451 ± 100 mg/L, respectively. Ensure that the number of significant figures in the standard deviation is not reduced if the value is 100 ± 1. This could cause incorrect rounding of data to 1400 or 1500 mg/L, respectively.

5. Calculations

As a practical operating rule, round off the result of a calculation in which several numbers are multiplied or divided to as few significant figures as are present in the factor with the fewest significant figures. Suppose that the following calculations must be made to obtain the result of an analysis:

$$\frac{56 \times 0.003\ 462 \times 43.22}{1.684}$$

A ten-place calculator yields an answer of "4.975 740 998." Round off this number to "5.0" because one of the measurements that entered into the calculation, 56, has only two significant figures. It was unnecessary to measure the other three factors to four significant figures because the "56" is the "weakest link in the chain" and limits accuracy of the answer. If the other factors were measured to only three, instead of four, significant figures, the answer would not suffer and the labor might be less.

When numbers are added or subtracted, the number that has the fewest decimal places, not necessarily the fewest significant figures, puts the limit on the number of places that justifiably may be carried in the sum or difference. Thus the sum

$$
\begin{array}{r}
0.0072 \\
12.02 \\
4.0078 \\
25.9 \\
\underline{4886} \\
4927.9350
\end{array}
$$

must be rounded off to "4928," no decimals, because one of the addends, 4886, has no decimal places. Notice that another addend, 25.9, has only three significant figures and yet it does not set a limit to the number of significant figures in the answer.

The preceding discussion is necessarily oversimplified. The reader is referred to mathematical texts for more detailed discussion.

1060 COLLECTION AND PRESERVATION OF SAMPLES

1060 A. Introduction

It is an old axiom that the result of any testing method can be no better than the sample on which it is performed. It is beyond the scope of this publication to specify detailed procedures for the collection of all samples because of varied purposes and analytical procedures. Detailed information is presented in specific methods. This section presents general considerations, applicable primarily to chemical analyses. See appropriate sections for samples to be used in toxicity testing and microbiological, biological, and radiological examinations.

The objective of sampling is to collect a portion of material small enough in volume to be transported conveniently and yet large enough for analytical purposes while still accurately representing the material being sampled. This objective implies that the relative proportions or concentrations of all pertinent components will be the same in the samples as in the material being sampled, and that the sample will be handled in such a way that no significant changes in composition occur before the tests are made.

Frequently the objective of sampling and testing is to demonstrate whether continuing compliance with specific regulatory requirements has been achieved. Samples are presented to the laboratory for specific determinations with the sampler being responsible for collecting a valid and representative sample. Because of the increasing importance placed on verifying the accuracy and representativeness of data, greater emphasis is placed on proper sample collection, tracking, and preservation techniques. Often, laboratory personnel help in planning a sampling program, in consultation with the user of the test results. Such consultation is essential to ensure selecting samples and analytical methods that provide a sound and valid basis for answering the questions that prompted the sampling and that will meet regulatory and/or project-specific requirements.

This section addresses the collection and preservation of water and wastewater samples; the general principles also apply to the sampling of solid or semisolid matrices.

1. General Requirements

Obtain a sample that meets the requirements of the sampling program and handle it so that it does not deteriorate or become contaminated or compromised before it is analyzed.

Ensure that all sampling equipment is clean and quality-assured before use. Use sample containers that are clean and free of contaminants. Bake at 450°C all bottles to be used for organic-analysis sampling.

Fill sample containers without prerinsing with sample; prerinsing results in loss of any pre-added preservative and sometimes can bias results high when certain components adhere to the sides of the container. Depending on determinations to be performed, fill the container full (most organic compound determinations) or leave space for aeration, mixing, etc. (microbiological and inorganic analyses). If a bottle already contains preservative, take care not to overfill the bottle, as preservative may be lost or diluted. Except when sampling for analysis of volatile organic compounds, leave an air space equivalent to approximately 1% of the container volume to allow for thermal expansion during shipment.

Special precautions (discussed below) are necessary for samples containing organic compounds and trace metals. Because many constituents may be present at low concentrations (micrograms or nanograms per liter), they may be totally or partially lost or easily contaminated when proper sampling and preservation procedures are not followed.

Composite samples can be obtained by collecting over a period of time, depth, or at many different sampling points. The details of collection vary with local conditions, so specific recommendations are not universally applicable. Sometimes it is more informative to analyze numerous separate samples instead of one composite so that variability, maxima, and minima can be determined.

Because of the inherent instability of certain properties and compounds, composite sampling for some analytes is not recommended where quantitative values are desired (examples include oil and grease, acidity, alkalinity, carbon dioxide, chlorine residual, iodine, hexavalent chromium, nitrate, volatile organic compounds, radon-222, dissolved oxygen, ozone, temperature, and pH). In certain cases, such as for BOD, composite samples are routinely required by regulatory agencies. Refrigerate composite samples for BOD and nitrite.

Sample carefully to ensure that analytical results represent the actual sample composition. Important factors affecting results are the presence of suspended matter or turbidity, the method chosen for removing a sample from its container, and the physical and chemical changes brought about by storage or aeration. Detailed procedures are essential when processing (blending, sieving, filtering) samples to be analyzed for trace constituents, especially metals and organic compounds. Some determinations can be invalidated by contamination during processing. Treat each sample individually with regard to the substances to be determined, the amount and nature of turbidity present, and other conditions that may influence the results.

Carefully consider the technique for collecting a representative sample and define it in the sampling plan. For metals it often is appropriate to collect both a filtered and an unfiltered sample to differentiate between total and dissolved metals present in the matrix. Be aware that some metals may partially sorb to filters. Beforehand, determine the acid requirements to bring the pH to <2 on a separate sample. Add the same relative amount of acid to all samples; use ultrapure acid preservative to prevent contamination. Be sure that the dilution caused by acidifying is negligible or sufficiently reproducible for a dilution correction factor. When filtered samples are to be collected, filter them, if possible, in the field, or at the point of collection before preservation with acid. Filter samples in a laboratory-controlled environment if field conditions could cause error or contamination; in this case filter as soon as possible. Often slight turbidity can be tolerated if experience shows that it will cause no interference in gravimetric or volumetric tests and that its influence can be corrected in colorimetric tests, where it has potentially the greatest interfering effect. Sample collector must state whether or not the sample has been filtered.

Make a record of every sample collected and identify every bottle with a unique sample number, preferably by attaching an appropriately inscribed tag or label. Document sufficient information to provide positive sample identification at a later date, including the unique sample identification number, the name of the sample collector, the date, hour, exact location, and, if possible, sample type (e.g., grab or composite), and any other data that may be needed for correlation, such as water temperature, weather conditions, water level, stream flow, post-collection conditions, etc. If space for all pertinent information for label or attached tag is insufficient, record information in a bound sample log book at the sampling site at the time of sample collection. Use waterproof ink to record all information (preferably with black, non-solvent-based ink). Fix sampling points by detailed description in the sampling plan, by maps, or with the aid of stakes, buoys, or landmarks in a manner that will permit their identification by other persons without reliance on memory or personal guidance. Global positioning systems (GPS) also are used and supply accurate sampling position data. Particularly when sample results are expected to be involved in litigation, use formal ''chain-of-custody'' procedures (see ¶ B.2 below), which trace sample history from collection to final reporting.

Before collecting samples from distribution systems, flush lines with three to five pipe volumes (or until water is being drawn from the main source) to ensure that the sample is representative of the supply, taking into account the volume of pipe to be flushed and the flow velocity. If the distribution system volume is unavailable, flush with tap fully open for at least 2 to 3 min before sampling. An exception to these guidelines, i.e., collecting a first draw sample, is when information on areas of reduced or restricted flow is desired or when samples for lead in drinking water are being collected.

Although well pumping protocols depend on the objectives of an investigation and other factors such as well characteristics and available equipment, a general rule is to collect samples from wells only after the well has been purged sufficiently (usually with three to ten well volumes) to ensure that the sample represents the groundwater. Purging stagnant water is critical. Sometimes it will be necessary to pump at a specified rate to achieve a characteristic drawdown, if this determines the zones from which the well is supplied; record purging rate and drawdown, if necessary. By using methods with minimal drawdown, purging volumes can be reduced significantly.

When samples are collected from a river or stream, observed results may vary with depth, stream flow, and distance from each shore. Selection of the number and distribution of sites at which samples should be collected depends on study objectives, stream characteristics, available equipment, and other factors. If equipment is available, take an integrated sample from top to bottom in the middle of the main channel of the stream or from side to side at mid-depth. If only grab or catch samples can be collected, preferably take them at various points of equal distance across the stream; if only one sample can be collected, take it in the middle of the main channel of the stream and at mid-depth. Integrated samples are described further in ¶ B.1c below.

Rivers, streams, lakes, and reservoirs are subject to considerable variations from normal causes such as seasonal stratification, diurnal variations, rainfall, runoff, and wind. Choose location, depth, and frequency of sampling depending on local conditions and the purpose of the investigation.

Use the following examples for general guidance. Avoid areas of excessive turbulence because of potential loss of volatile constituents and of potential presence of denser-than-air toxic vapors. Avoid sampling at weirs if possible because such locations tend to favor retrieval of lighter-than-water, immiscible compounds. Generally, collect samples beneath the surface in quiescent areas and open sampling container below surface with the mouth directed toward the current to avoid collecting surface scum unless oil and grease is a constituent of interest; then collect water at the surface. If composite samples are required, ensure that sample constituents are not lost during compositing because of improper handling of portions being composited. If samples will be analyzed for organic constituents, refrigerate composited portions. Do not composite samples for VOC analysis because some of the components will be lost through volatilization.

2. Safety Considerations

Because sample constituents may be toxic, take adequate precautions during sampling and sample handling. Toxic substances can enter through the skin and eyes and, in the case of vapors, also through the lungs. Ingestion can occur via direct contact of toxic materials with foods or by adsorption of vapors onto foods. Precautions may be limited to wearing gloves or may include coveralls, aprons, or other protective apparel. Often, the degree of protection provided by chemical protective clothing (CPC) is specific for different manufacturers and their product models[1]; ensure that the clothing chosen will offer adequate protection. Always wear eye protection (e.g., safety glasses with side shields or goggles). When toxic vapors may be present, sample only in well-ventilated areas, or use an appropriate respirator or self-contained breathing apparatus. In a laboratory, open sample containers in a fume hood. Never have food in the laboratory, near samples, or near sampling locations; always wash hands thoroughly before handling food.[2]

Always prohibit eating, drinking, or smoking near samples, sampling locations, and in the laboratory. Keep sparks, flames, and excessive heat sources away from samples and sampling locations. If flammable compounds are suspected or known to be present and samples are to be refrigerated, use only specially designed *explosion-proof* refrigerators.[2]

Collect samples safely, avoiding situations that may lead to accidents. When in doubt as to the level of safety precautions needed, consult a knowledgeable industrial hygienist or safety professional. Samples with radioactive contaminants may require other safety considerations; consult a health physicist.

Label adequately any sample known or suspected to be hazardous because of flammability, corrosivity, toxicity, oxidizing chemicals, or radioactivity, so that appropriate precautions can be taken during sample handling, storage, and disposal.

3. References

1. FORSBERG K. & L.H. KEITH. 1998. Instant Gloves and CPC Database. Instant Reference Sources, Inc. Austin, Tex.
2. WATER POLLUTION CONTROL FEDERATION. 1986. Removal of Hazardous Wastes in Wastewater Facilities—Halogenated Organics. Manual of Practice FD-11, Water Pollution Control Fed., Alexandria, Va.

1060 B. Collection of Samples

1. Types of Samples

a. Grab samples: Grab samples are single samples collected at a specific spot at a site over a short period of time (typically seconds or minutes). Thus, they represent a "snapshot" in both space and time of a sampling area. Discrete grab samples are taken at a selected location, depth, and time. Depth-integrated grab samples are collected over a predetermined part or the entire depth of a water column, at a selected location and time in a given body of water.

A sample can represent only the composition of its source at the time and place of collection. However, when a source is known to be relatively constant in composition over an extended time or over substantial distances in all directions, then the sample may represent a longer time period and/or a larger volume than the specific time and place at which it was collected. In such circumstances, a source may be represented adequately by single grab samples. Examples are protected groundwater supplies, water supplies receiving conventional treatment, some well-mixed surface waters, but rarely, wastewater streams, rivers, large lakes, shorelines, estuaries, and groundwater plumes.

When a source is known to vary with time, grab samples collected at suitable intervals and analyzed separately can document the extent, frequency, and duration of these variations. Choose sampling intervals on the basis of the expected frequency of changes, which may vary from as little as 5 min to as long as 1 h or more. Seasonal variations in natural systems may necessitate sampling over months. When the source composition varies in space (i.e., from location to location) rather than time, collect samples from appropriate locations that will meet the objectives of the study (for example, upstream and downstream from a point source, etc.).

The same principles apply to sampling wastewater sludges, sludge banks, and muds, although these matrices are not specifically addressed in this section. Take every possible precaution to obtain a representative sample or one conforming to a sampling program.

b. Composite samples: Composite samples should provide a more representative sampling of heterogeneous matrices in which the concentration of the analytes of interest may vary over short periods of time and/or space. Composite samples can be obtained by combining portions of multiple grab samples or by using specially designed automatic sampling devices. Sequential (time) composite samples are collected by using continuous, constant sample pumping or by mixing equal water volumes collected at regular time intervals. Flow-proportional composites are collected by continuous pumping at a rate proportional to the flow, by mixing equal volumes of water collected at time intervals that are inversely proportional to the volume of flow, or by mixing volumes of water proportional to the flow collected during or at regular time intervals.

Advantages of composite samples include reduced costs of analyzing a large number of samples, more representative samples of heterogeneous matrices, and larger sample sizes when amounts of test samples are limited. Disadvantages of composite samples include loss of analyte relationships in individual samples, potential dilution of analytes below detection levels, increased potential analytical interferences, and increased possibility of analyte interactions. In addition, use of composite samples may reduce the number of samples analyzed below the required statistical need for specified data quality objectives or project-specific objectives.

Do not use composite samples with components or characteristics subject to significant and unavoidable changes during storage. Analyze individual samples as soon as possible after collection and preferably at the sampling point. Examples are dissolved gases, residual chlorine, soluble sulfide, temperature, and pH. Changes in components such as dissolved oxygen or carbon dioxide, pH, or temperature may produce secondary changes in certain inorganic constituents such as iron, manganese, alkalinity, or hardness. Some organic analytes also may be changed by changes in the foregoing components. Use time-composite samples only for determining components that can be demonstrated to remain unchanged under the conditions of sample collection, preservation, and storage.

Collect individual portions in a wide-mouth bottle every hour (in some cases every half hour or even every 5 min) and mix at the end of the sampling period or combine in a single bottle as collected. If preservatives are used, add them to the sample bottle initially so that all portions of the composite are preserved as soon as collected.

Automatic sampling devices are available; however, do not use them unless the sample is preserved as described below. Composite samplers running for extended periods (weeks to months) should undergo routine cleaning of containers and sample lines to minimize sample growth and deposits.

c. Integrated (discharge-weighted) samples: For certain purposes, the information needed is best provided by analyzing mixtures of grab samples collected from different points simultaneously, or as nearly so as possible, using discharge-weighted methods such as equal-width increment (EWI) or equal discharge-increment (EDI) procedures and equipment. An example of the need for integrated sampling occurs in a river or stream that varies in composition across its width and depth. To evaluate average composition or total loading, use a mixture of samples representing various points in the cross-section, in proportion to their relative flows. The need for integrated samples also may exist if combined treatment is proposed for several separate wastewater streams, the interaction of which may have a significant effect on treatability or even on composition. Mathematical prediction of the interactions among chemical components may be inaccurate or impossible and testing a suitable integrated sample may provide more useful information.

Both lakes and reservoirs show spatial variations of composition (depth and horizontal location). However, there are conditions under which neither total nor average results are especially useful, but local variations are more important. In such cases, examine samples separately (i.e., do not integrate them).

Preparation of integrated samples usually requires equipment designed to collect a sample water uniformly across the depth profile. Knowledge of the volume, movement, and composition of the various parts of the water being sampled usually is required. Collecting integrated samples is a complicated and specialized process that must be described adequately in a sampling plan.

2. Chain-of-Custody Procedures

Properly designed and executed chain-of-custody forms will ensure sample integrity from collection to data reporting. This includes the ability to trace possession and handling of the sample from the time of collection through analysis and final disposition. This process is referred to as "chain-of-custody" and is required to demonstrate sample control when the data are to be used for regulation or litigation. Where litigation is not involved, chain-of-custody procedures are useful for routine control of samples.

A sample is considered to be under a person's custody if it is in the individual's physical possession, in the individual's sight, secured and tamper-proofed by that individual, or secured in an area restricted to authorized personnel. The following procedures summarize the major aspects of chain of custody. More detailed discussions are available.[1,2]

a. Sample labels (including bar-code labels): Use labels to prevent sample misidentification. Gummed paper labels or tags generally are adequate. Include at least the following information: a unique sample number, sample type, name of collector, date and time of collection, place of collection, and sample preservative. Also include date and time of preservation for comparison to date and time of collection. Affix tags or self-adhesive labels to sample containers before, or at the time of, sample collection.

b. Sample seals: Use sample seals to detect unauthorized tampering with samples up to the time of analysis. Use self-adhesive paper seals that include at least the following information: sample number (identical with number on sample label), collector's name, and date and time of sampling. Plastic shrink seals also may be used.

Attach seal in such a way that it is necessary to break it to open the sample container or the sample shipping container (e.g., a cooler). Affix seal to container before sample leaves custody of sampling personnel.

c. Field log book: Record all information pertinent to a field survey or sampling in a bound log book. As a minimum, include the following in the log book: purpose of sampling; location of sampling point; name and address of field contact; producer of material being sampled and address, if different from location; type of sample; and method, date, and time of preservation. If the sample is wastewater, identify process producing waste stream. Also provide suspected sample composition, including concentrations; number and volume of sample(s) taken; description of sampling point and sampling method; date and time of collection; collector's sample identification number(s); sample distribution and how transported; references such as maps or photographs of the sampling site; field observations and measurements; and signatures of personnel responsible for observations. Because sampling situations vary widely, it is essential to record sufficient information so that one could reconstruct the sampling event without reliance on the collector's memory. Protect log book and keep it in a safe place.

d. Chain-of-custody record: Fill out a chain-of-custody record to accompany each sample or group of samples. The record includes the following information: sample number; signature of collector; date, time, and address of collection; sample type; sample preservation requirements; signatures of persons involved in the chain of possession; and inclusive dates and times of possession.

e. Sample analysis request sheet: The sample analysis request sheet accompanies samples to the laboratory. The collector completes the field portion of such a form that includes most of the pertinent information noted in the log book. The laboratory portion of such a form is to be completed by laboratory personnel and includes: name of person receiving the sample, laboratory sample number, date of sample receipt, condition of each sample (i.e., if it is cold or warm, whether the container is full or not, color, if more than one phase is present, etc.), and determinations to be performed.

f. Sample delivery to the laboratory: Deliver sample(s) to laboratory as soon as practicable after collection, typically within 2 d. Where shorter sample holding times are required, make special arrangements to insure timely delivery to the laboratory. Where samples are shipped by a commercial carrier, include the waybill number in the sample custody documentation. Insure that samples are accompanied by a completed chain-of-custody record and a sample analysis request sheet. Deliver sample to sample custodian.

g. Receipt and logging of sample: In the laboratory, the sample custodian inspects the condition and seal of the sample and reconciles label information and seal against the chain-of-custody record before the sample is accepted for analysis. After acceptance, the custodian assigns a laboratory number, logs sample in the laboratory log book and/or computerized laboratory information management system, and stores it in a secured storage room or cabinet or refrigerator at the specified temperature until it is assigned to an analyst.

h. Assignment of sample for analysis: The laboratory supervisor usually assigns the sample for analysis. Once the sample is in the laboratory, the supervisor or analyst is responsible for its care and custody.

i. Disposal: Hold samples for the prescribed amount of time for the project or until the data have been reviewed and accepted. Document the disposition of samples. Ensure that disposal is in accordance with local, state, and U.S. EPA approved methods.

3. Sampling Methods

a. Manual sampling: Manual sampling involves minimal equipment but may be unduly costly and time-consuming for routine or large-scale sampling programs. It requires trained field technicians and is often necessary for regulatory and research investigations for which critical appraisal of field conditions and complex sample collection techniques are essential. Manually collect certain samples, such as waters containing oil and grease.

b. Automatic sampling: Automatic samplers can eliminate human errors in manual sampling, can reduce labor costs, may provide the means for more frequent sampling,[3] and are used increasingly. Be sure that the automatic sampler does not contaminate the sample. For example, plastic components may be incompatible with certain organic compounds that are soluble in the plastic parts or that can be contaminated (e.g., from phthalate esters) by contact with them. If sample constituents are generally known, contact the manufacturer of an automatic sampler regarding potential incompatibility of plastic components.

Program an automatic sampler in accordance with sampling needs. Carefully match pump speeds and tubing sizes to the type of sample to be taken.

c. Sorbent sampling: Use of solid sorbents, particularly membrane-type disks, is becoming more frequent. These methods offer advantages of rapid, inexpensive sampling if the analytes of interest can be adsorbed and desorbed efficiently and the water matrix is free of particulates that plug the sorbent.

4. Sample Containers

The type of sample container used is of utmost importance. Test sample containers and document that they are free of analytes of interest, especially when sampling and analyzing for very low analyte levels. Containers typically are made of plastic or glass, but one material may be preferred over the other. For example, silica, sodium, and boron may be leached from soft glass but not plastic, and trace levels of some pesticides and metals may sorb onto the walls of glass containers.[4] Thus, hard glass containers* are preferred. For samples containing organic compounds, do not use plastic containers except those made of fluorinated polymers such as polytetrafluoroethylene (PTFE).[3]

Some sample analytes may dissolve (be absorbed) into the walls of plastic containers; similarly, contaminants from plastic containers may leach into samples. Avoid plastics wherever possible because of potential contamination from phthalate esters. Container failure due to breakdown of the plastic is possible. Therefore, use glass containers for all organics analyses such as volatile organics, semivolatile organics, pesticides, PCBs, and oil and grease. Some analytes (e.g., bromine-containing compounds and some pesticides, polynuclear aromatic compounds, etc.) are light-sensitive; collect them in amber-colored glass containers to minimize photodegradation. Container caps, typically plastic, also

---

* Pyrex or equivalent.

can be a problem. Do not use caps with paper liners. Use foil or PTFE liners but be aware that metal liners can contaminate samples collected for metals analysis and they may also react with the sample if it is acidic or alkaline. Serum vials with PTFE-lined rubber or plastic septa are useful.

In rare situations it may be necessary to use sample containers not specifically prepared for use, or otherwise unsuitable for the particular situation; thoroughly document these deviations. Documentation should include type and source of container, and the preparation technique, e.g., acid washed with reagent water rinse. For QA purposes the inclusion of a bottle blank may be necessary.

5. Number of Samples

Because of variability from analytical and sampling procedures (i.e., population variability), a single sample is insufficient to reach any reasonable desired level of confidence. If an overall standard deviation (i.e., the standard deviation of combined sampling and analysis) is known, the required number of samples for a mobile matrix such as water may be estimated as follows:[4]

$$N \geq \left(\frac{ts}{U}\right)^2$$

where:

  N = number of samples,
  t = Student-t statistic for a given confidence level,
  s = overall standard deviation, and
  U = acceptable level of uncertainty.

To assist in calculations, use curves such as those in Figure 1060:1. As an example, if s is 0.5 mg/L, U is $\pm$ 0.2 mg/L, and a 95% confidence level is desired, approximately 25 to 30 samples must be taken.

The above equation assumes that total error (population variability) is known. Total variability consists of all sources of variability, including: the distribution of the analytes of interest within the sampling site, collection, preservation, preparation, and analysis of samples, and data handling and reporting. In simpler terms, error (variability) can be divided into sampling and analysis components. Sampling error due to population variability (including heterogeneous distribution of analytes in the environmental matrix) usually is much larger than analytical error components. Unfortunately, sampling error usually is not available and the analyst is left with only the published error of the measurement system (typically obtained by using a reagent water matrix under the best analytical conditions).

More accurate equations are available.[5] These are based on the Z distribution for determining the number of samples needed to estimate a mean concentration when variability is estimated in absolute terms using the standard deviation. The coefficient of variation (relative standard deviation) is used when variability is estimated in relative terms.

The number of random samples to be collected at a site can be influenced partly by the method that will be used. The values for standard deviation (SD) or relative standard deviation (RSD) may be obtained from each of the methods or in the literature.[6] However, calculations of estimated numbers of samples needed based only on this information will result in underestimated numbers of samples because only the analytical variances are considered, and the typically larger variances from the sampling operations are

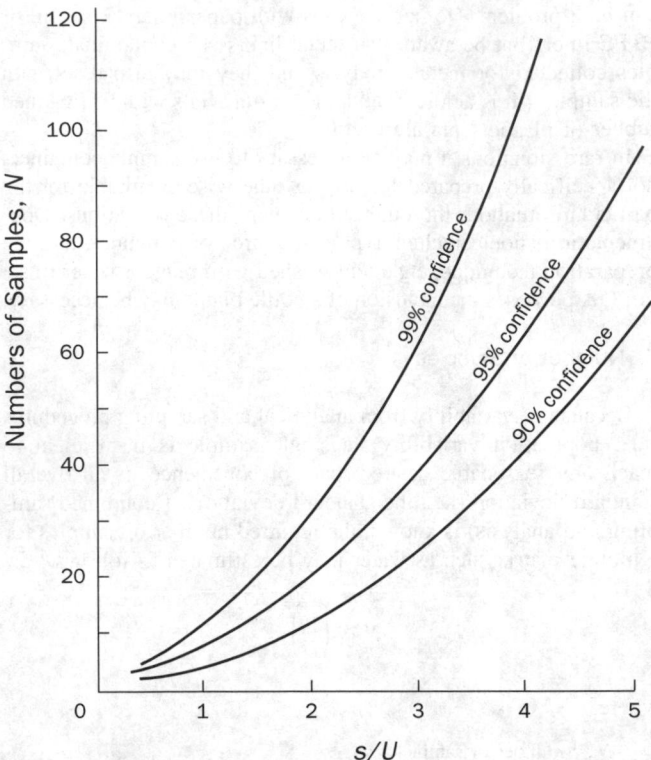

Figure 1060:1. Approximate number of samples required in estimating a mean concentration. Source: Methods for the Examination of Waters and Associated Materials: General Principles of Sampling and Accuracy of Results. 1980. Her Majesty's Stationery Off., London, England.

not included. Preferably, determine and use SDs or RSDs from overall sampling and analysis operations.

For estimates of numbers of samples needed for systematic sampling (e.g., drilling wells for sampling groundwater or for systematically sampling large water bodies such as lakes), equations are available[7] that relate number of samples to shape of grid, area covered, and space between nodes of grid. The grid spacing is a complex calculation that depends on the size and shape of any contaminated spot (such as a groundwater plume) to be identified, in addition to the geometric shape of the sampling grid.

See individual methods for types and numbers of quality assurance (QA) and quality control (QC) samples, e.g., for normal-level (procedural) or low-level (contamination) bias or for precision, involving sampling or laboratory analysis (either overall or individually). Estimates of numbers of QC samples needed to achieve specified confidence levels also can be calculated. Rates of false positives (Type I error) and false negatives (Type II error) are useful parameters for estimating required numbers of QC samples. A false positive is the incorrect conclusion that an analyte is present when it is absent. A false negative is the incorrect conclusion that an analyte is absent when it is present. If the frequency of false positives or false negatives desired to be detected is less than 10%, then

$$n = \frac{\ln \alpha}{\ln (1 - Y)}$$

where:

$\alpha = (1 - \text{desired confidence level})$, and
$Y = \text{frequency to detect } (<10\%)$.

If the frequency that is desirable to detect is more than 10%, iterative solution of a binomial equation is necessary.[5,8]

Equations are available as a computer program† for computing sample number by the Z distribution, for estimating samples needed in systematic sampling, and for estimating required number of QC samples.

6. Sample Volumes

Collect a 1-L sample for most physical and chemical analyses. For certain determinations, larger samples may be necessary. Table 1060:I lists volumes ordinarily required for analyses, but it is strongly recommended that the laboratory that will conduct the analyses also be consulted to verify the analytical needs of sampling procedures as they pertain to the goals and data quality objective of an investigation.

Do not use samples from the same container for multiple testing requirements (e.g., organic, inorganic, radiological, bacteriological, and microscopic examinations) because methods of collecting and handling are different for each type of test. Always collect enough sample volume in the appropriate container in order to comply with sample handling, storage, and preservation requirements.

7. References

1. U.S. ENVIRONMENTAL PROTECTION AGENCY. 1986. Test Methods for Evaluating Solid Waste: Physical/Chemical Methods, 3rd ed. Publ. No. SW-846, Off. Solid Waste and Emergency Response, Washington, D.C.
2. U.S. ENVIRONMENTAL PROTECTION AGENCY. 1982. NEIC Policies and Procedures. EPA-330/9/78/001/-R (rev. 1982).
3. WATER POLLUTION CONTROL FEDERATION. 1986. Removal of Hazardous Wastes in Wastewater Facilities—Halogenated Organics. Manual of Practice FD-11, Water Pollution Control Fed., Alexandria, Va.
4. Methods for the Examination of Waters and Associated Materials: General Principles of Sampling and Accuracy of Results. 1980. Her Majesty's Stationery Off., London, England.
5. KEITH, L.H., G.L. PATTON, D.L. LEWIS & P.G. EDWARDS. 1996. Determining numbers and kinds of analytical samples. Chapter 1 in Principles of Environmental Sampling, 2nd ed. ACS Professional Reference Book, American Chemical Soc., Washington, D.C.
6. KEITH, L.H. 1996. Compilation of EPA's Sampling and Analysis Methods, 2nd ed. Lewis Publ./CRC Press, Boca Raton, Fla.
7. GILBERT, R.O. 1987. Statistical Methods for Environmental Pollution Monitoring. Van Nostrand Reinhold, New York, N.Y.
8. GRANT, E.L. & R.S. LEAVENWORTH. 1988. Statistical Quality Control, 6th ed. McGraw-Hill, Inc., New York, N.Y.
9. U.S. ENVIRONMENTAL PROTECTION AGENCY. 1996. 40 CFR Part 136, Table II.
10. U.S. ENVIRONMENTAL PROTECTION AGENCY. 1992. Rules and Regulations. 40 CFR Parts 100-149.

† DQO-PRO, available (free) by downloading from American Chemical Society Division of Environmental Chemistry home page at http://acs.environmental.duq.edu/acsenv/envchem.htm, and also as part of the tutorial, Reliable Environmental Sampling and Analysis, Instant Reference Sources, Inc., http://instantref.com/inst.ref.htm.

TABLE 1060:I. SUMMARY OF SPECIAL SAMPLING AND HANDLING REQUIREMENTS*

Determination	Container†	Minimum Sample Size mL	Sample Type‡	Preservation§	Maximum Storage Recommended	Regulatory‖
Acidity	P, G(B)	100	g	Refrigerate	24 h	14 d
Alkalinity	P, G	200	g	Refrigerate	24 h	14 d
BOD	P, G	1000	g, c	Refrigerate	6 h	48 h
Boron	P (PTFE) or quartz	1000	g, c	HNO_3 to pH <2	28 d	6 months
Bromide	P, G	100	g, c	None required	28 d	28 d
Carbon, organic, total	G (B)	100	g, c	Analyze immediately; or refrigerate and add HCl, H_3PO_4, or H_2SO_4 to pH <2	7 d	28 d
Carbon dioxide	P, G	100	g	Analyze immediately	0.25 h	N.S.
COD	P, G	100	g, c	Analyze as soon as possible, or add H_2SO_4 to pH <2; refrigerate	7 d	28 d
Chloride	P, G	50	g, c	None required	N.S.	28 d
Chlorine, total, residual	P, G	500	g	Analyze immediately	0.25 h	0.25 h
Chlorine dioxide	P, G	500	g	Analyze immediately	0.25 h	N.S.
Chlorophyll	P, G	500	g	Unfiltered, dark, 4°C Filtered, dark, −20°C (Do not store in frost-free freezer)	24–48 h 28 d	
Color	P, G	500	g, c	Refrigerate	48 h	48 h
Specific conductance	P, G	500	g, c	Refrigerate	28 d	28 d
Cyanide						
Total	P, G	1000	g, c	Add NaOH to pH >12, refrigerate in dark#	24 h	14 d; 24 h if sulfide present
Amenable to chlorination	P, G	1000	g, c	Add 0.6 g ascorbic acid if chlorine is present and refrigerate	stat	14 d; 24 h if sulfide present
Fluoride	P	100	g, c	None required	28 d	28 d
Hardness	P, G	100	g, c	Add HNO_3 or H_2SO_4 to pH <2	6 months	6 months
Iodine	P, G	500	g	Analyze immediately	0.25 h	N.S.
Metals, general	P(A), G(A)	1000	g, c	For dissolved metals filter immediately, add HNO_3 to pH <2	6 months	6 months
Chromium VI	P(A), G(A)	1000	g	Refrigerate	24 h	24 h
Copper by colorimetry*			g, c			
Mercury	P(A), G(A)	1000	g, c	Add HNO_3 to pH <2, 4°C, refrigerate	28 d	28 d
Nitrogen						
Ammonia	P, G	500	g, c	Analyze as soon as possible or add H_2SO_4 to pH <2, refrigerate	7 d	28 d
Nitrate	P, G	100	g, c	Analyze as soon as possible; refrigerate	48 h	48 h (28 d for chlorinated samples)
Nitrate + nitrite	P, G	200	g, c	Add H_2SO_4 to pH <2, refrigerate	1–2 d	28 d
Nitrite	P, G	100	g, c	Analyze as soon as possible; refrigerate	none	48 h
Organic, Kjeldahl*	P, G	500	g, c	Refrigerate, add H_2SO_4 to pH <2	7 d	28 d
Odor	G	500	g	Analyze as soon as possible; refrigerate	6 h	N.S.
Oil and grease	G, wide-mouth calibrated	1000	g	Add HCl or H_2SO_4 to pH <2, refrigerate	28 d	28 d
Organic compounds						
MBAs	P, G	250	g, c	Refrigerate	48 h	N.S.
Pesticides*	G(S), PTFE-lined cap	1000	g, c	Refrigerate, add 1000 mg ascorbic acid/L if residual chlorine present	7 d	7 d until extraction; 40 d after extraction
Phenols	P, G, PTFE-lined cap	500	g, c	Refrigerate, add H_2SO_4 to pH <2	*	28 d until extraction
Purgeables* by purge and trap	G, PTFE-lined cap	2 × 40	g	Refrigerate; add HCl to pH <2; add 1000 mg ascorbic acid/L if residual chlorine present	7 d	14 d

TABLE 1060:I. CONT.

Determination	Container†	Minimum Sample Size mL	Sample Type‡	Preservation§	Maximum Storage Recommended	Regulatory‖
Base/neutrals & acids	G(S) amber	1000	g, c	Refrigerate	7 d	7 d until extraction; 40 d after extraction
Oxygen, dissolved	G, BOD bottle	300	g			
Electrode				Analyze immediately	0.25 h	0.25 h
Winkler				Titration may be delayed after acidification	8 h	8 h
Ozone	G	1000	g	Analyze immediately	0.25 h	N.S.
pH	P, G	50	g	Analyze immediately	0.25 h	0.25 h
Phosphate	G(A)	100	g	For dissolved phosphate filter immediately; refrigerate	48 h	N.S.
Phosphorus, total	P, G	100	g, c	Add H_2SO_4 to pH <2 and refrigerate	28 d	
Salinity	G, wax seal	240	g	Analyze immediately or use wax seal	6 months	N.S.
Silica	P (PTFE) or quartz	200	g, c	Refrigerate, do not freeze	28 d	28 d
Sludge digester gas	G, gas bottle	—	g	—	N.S.	
Solids[9]	P, G	200	g, c	Refrigerate	7 d	2–7 d; see cited reference
Sulfate	P, G	100	g, c	Refrigerate	28 d	28 d
Sulfide	P, G	100	g, c	Refrigerate; add 4 drops $2N$ zinc acetate/100 mL; add NaOH to pH >9	28 d	7 d
Temperature	P, G	—	g	Analyze immediately	0.25 h	0.25 h
Turbidity	P, G	100	g, c	Analyze same day; store in dark up to 24 h, refrigerate	24 h	48 h

* For determinations not listed, use glass or plastic containers; preferably refrigerate during storage and analyze as soon as possible.
† P = plastic (polyethylene or equivalent); G = glass; G(A) or P(A) = rinsed with 1 + 1 HNO_3; G(B) = glass, borosilicate; G(S) = glass, rinsed with organic solvents or baked.
‡ g = grab; c = composite.
§ Refrigerate = storage at 4°C ± 2°C; in the dark; analyze immediately = analyze usually within 15 min of sample collection.
‖ See citation[10] for possible differences regarding container and preservation requirements. N.S. = not stated in cited reference; stat = no storage allowed; analyze immediately.
\# If sample is chlorinated, see text for pretreatment.

1060 C. Sample Storage and Preservation

Complete and unequivocal preservation of samples, whether domestic wastewater, industrial wastes, or natural waters, is a practical impossibility because complete stability for every constituent never can be achieved. At best, preservation techniques only retard chemical and biological changes that inevitably continue after sample collection.

1. Sample Storage before Analysis

a. Nature of sample changes: Some determinations are more affected by sample storage than others. Certain cations are subject to loss by adsorption on, or ion exchange with, the walls of glass containers. These include aluminum, cadmium, chromium, copper, iron, lead, manganese, silver, and zinc, which are best collected in a separate clean bottle and acidified with nitric acid to a pH below 2.0 to minimize precipitation and adsorption on con-

tainer walls. Also, some organics may be subject to loss by adsorption to the walls of glass containers.

Temperature changes quickly; pH may change significantly in a matter of minutes; dissolved gases (oxygen, carbon dioxide) may be lost. Because changes in such basic water quality properties may occur so quickly, determine temperature, reduction-oxidation potential, and dissolved gases in situ and pH, specific conductance, turbidity, and alkalinity immediately after sample collection. Many organic compounds are sensitive to changes in pH and/or temperature resulting in reduced concentrations during storage.

Changes in the pH-alkalinity-carbon dioxide balance may cause calcium carbonate to precipitate, decreasing the values for calcium and total hardness.

Iron and manganese are readily soluble in their lower oxidation states but relatively insoluble in their higher oxidation states;

therefore, these cations may precipitate or they may dissolve from a sediment, depending on the redox potential of the sample. Microbiological activity may affect the nitrate-nitrite-ammonia content, phenol or BOD concentration, or the reduction of sulfate to sulfide. Residual chlorine is reduced to chloride. Sulfide, sulfite, ferrous iron, iodide, and cyanide may be lost through oxidation. Color, odor, and turbidity may increase, decrease, or change in quality. Sodium, silica, and boron may be leached from the glass container. Hexavalent chromium may be reduced to trivalent chromium.

Biological activity taking place in a sample may change the oxidation state of some constituents. Soluble constituents may be converted to organically bound materials in cell structures, or cell lysis may result in release of cellular material into solution. The well-known nitrogen and phosphorus cycles are examples of biological influences on sample composition.

Zero head-space is important in preservation of samples with volatile organic compounds and radon. Avoid loss of volatile materials by collecting sample in a completely filled container. Achieve this by carefully filling the bottle so that top of meniscus is above the top of the bottle rim. It is important to avoid spillage or air entrapment if preservatives such as HCl or ascorbic acid have already been added to the bottle. After capping or sealing bottle, check for air bubbles by inverting and gently tapping it; if one or more air bubbles are observed then, if practical, discard the sample and repeat refilling bottle with new sample until no air bubbles are observed (this cannot be done if bottle contained preservatives before it was filled).

Serum vials with septum caps are particularly useful in that a sample portion for analysis can be taken through the cap by using a syringe,[1] although the effect of pressure reduction in the headspace must be considered. Pulling a sample into a syringe under vacuum can result in low bias data for volatile compounds and the resulting headspace precludes taking further subsamples.

b. Time interval between collection and analysis: In general, the shorter the time that elapses between collection of a sample and its analysis, the more reliable will be the analytical results. For certain constituents and physical values, immediate analysis in the field is required. For composited samples it is common practice to use the time at the end of composite collection as the sample collection time.

Check with the analyzing laboratory to determine how much elapsed time may be allowed between sample collection and analysis; this depends on the character of the sample and the stability of the target analytes under the conditions of storage. Many regulatory methods limit the elapsed time between sample collection and analysis (see Table 1060:I). Changes caused by growth of microorganisms are greatly retarded by keeping the sample at a low temperature ($<4°C$ but above freezing). When the interval between sample collection and analysis is long enough to produce changes in either the concentration or the physical state of the constituent to be measured, follow the preservation practices given in Table 1060:I. Record time elapsed between sampling and analysis, and which preservative, if any, was added.

2. Preservation Techniques

To minimize the potential for volatilization or biodegradation between sampling and analysis, keep samples as cool as possible without freezing. Preferably pack samples in crushed or cubed ice or commercial ice substitutes before shipment. Avoid using dry ice because it will freeze samples and may cause glass containers to break. Dry ice also may effect a pH change in samples. Keep composite samples cool with ice or a refrigeration system set at 4°C during compositing. Analyze samples as quickly as possible on arrival at the laboratory. If immediate analysis is not possible, preferably store at 4°C.[1]

No single method of preservation is entirely satisfactory; choose the preservative with due regard to the determinations to be made. Use chemical preservatives only when they do not interfere with the analysis being made. When they are used, add them to the sample bottle initially so that all sample portions are preserved as soon as collected. Because a preservation method for one determination may interfere with another one, samples for multiple determinations may need to be split and preserved separately. All methods of preservation may be inadequate when applied to suspended matter. Do not use formaldehyde as a preservative for samples collected for chemical analysis because it affects many of the target analytes.

Methods of preservation are relatively limited and are intended generally to retard biological action, retard hydrolysis of chemical compounds and complexes, and reduce volatility of constituents.

Preservation methods are limited to pH control, chemical addition, the use of amber and opaque bottles, refrigeration, filtration, and freezing. Table 1060:I lists preservation methods by constituent. See Section 7010B for sample collection and preservation requirements for radionuclides.

The foregoing discussion is by no means exhaustive and comprehensive. Clearly it is impossible to prescribe absolute rules for preventing all possible changes. Additional advice will be found in the discussions under individual determinations, but to a large degree the dependability of an analytical determination rests on the experience and good judgment of the person collecting the sample. Numbers of samples required for confidence levels in data quality objectives, however, rely on statistical equations such as those discussed earlier.

3. Reference

1. WATER POLLUTION CONTROL FEDERATION. 1986. Removal of Hazardous Wastes in Wastewater Facilities—Halogenated Organics. Manual of Practice FD-11, Water Pollution Control Fed., Alexandria, Va.

4. Bibliography

KEITH, L.H., ed. 1996. Principles of Environmental Sampling, 2nd ed. ACS Professional Reference Book, American Chemical Soc., Washington, D.C.

1080   REAGENT WATER

1080 A.   Introduction

One of the most important aspects of analysis is the preparation of reagent water to be used for dilution of reagents and for blank analysis. Reagent water is water with no detectable concentration of the compound or element to be analyzed at the detection limit of the analytical method. Reagent water should be free of substances that interfere with analytical methods. The quality of water required is related directly to the analysis being made. Requirements for water quality may differ for organic, inorganic, and biological constituents depending on the use(s) for which the water is intended.

Any method of preparation of reagent water is acceptable pro-vided that the requisite quality can be met. Improperly maintained systems may add contaminants. Reverse osmosis, distillation, and deionization in various combinations all can produce reagent water when used in the proper arrangement. Ultrafiltration and/or ultraviolet treatment also may be used as part of the process. Section 1080 provides general guidelines for the preparation of reagent water. Table 1080:I lists commonly available processes for water purification and major classes of contaminants removed by purification.

For details on preparing water for microbiological tests, see Section 9020B.3c.

TABLE 1080:I. WATER PURIFICATION PROCESSES

Process	Major Classes of Contaminants*					
	Dissolved Ionized Solids	Dissolved Ionized Gases	Dissolved Organics	Particulates	Bacteria	Pyrogens/ Endotoxins
Distillation	G–E†	P	G	E	E	E
Deionization	E	E	P	P	P	P
Reverse osmosis	G‡	P	G	E	E	E
Carbon adsorption	P	P§	G–E ‖	P	P	P
Filtration	P	P	P	E	E	P
Ultrafiltration	P	P	G#	E	E	E
Ultraviolet oxidation	P	P	G–E**	P	G††	P

Permission to use this table from C3-A2, Vol. 11, No. 13, Aug. 1991, "Preparation and Testing of Reagent Water in the Clinical Laboratory - Second Edition" has been granted by the National Committee for Clinical Laboratory Standards. The complete current standard may be obtained from National Committee for Clinical Laboratory Standards, 771 E. Lancaster Ave., Villanova, PA 19085.

* E = Excellent (capable of complete or near total removal), G = Good (capable of removing large percentages), P = Poor (little or no removal).

† Resistivity of water purified by distillation is an order of magnitude less than water produced by deionization, due mainly to the presence of CO_2 and sometimes H_2S, NH_3, and other ionized gases if present in the feedwater.

‡ Resistivity of dissolved ionized solids in the product water depends on original feedwater resistivity.

§ Activated carbon removes chlorine by adsorption.

‖ When used in combination with other purification processes, special grades of activated carbon and other synthetic adsorbents exhibit excellent capabilities for removing organic contaminants. Their use, however, is targeted toward specific compounds and applications.

Ultrafilters have demonstrated usefulness in reducing specific feedwater organic contaminants based on the rated molecular weight cut-off of the membrane.

** 185 nm ultraviolet oxidation (batch process systems) is effective in removing trace organic contaminants when used as post-treatment. Feedwater makeup plays a critical role in the performance of these batch processors.

†† 254 nm UV sterilizers, while not physically removing bacteria, may have bactericidal or bacteriostatic capabilities limited by intensity, contact time, and flow rate.

1080 B.   Methods for Preparation of Reagent Water

1. Distillation

Prepare laboratory-grade distilled water by distilling water from a still of all-borosilicate glass, fused quartz, tin, or titanium. To remove ammonia distill from an acid solution. Remove CO_2 by boiling the water for 15 min and cooling rapidly to room tem-perature; exclude atmospheric CO_2 by using a tube containing soda lime or a commercially available CO_2-removing agent.*

Boiling the water may add other impurities by leaching impurities from the container. Freshly replaced filters, cartridges and resins initially can release impurities. Pretreat feedwater and provide peri-

_____

* Ascarite II, Fisher Scientific Co., or equivalent.

odic maintenance to minimize scale formation within the still. Pretreatment may be required where the feedwater contains significant concentrations of calcium, magnesium, and bicarbonate ions; it may involve demineralization via reverse osmosis or ion exchange.

2. Reverse Osmosis

Reverse osmosis is a process in which water is forced under pressure through a semipermeable membrane removing a portion of dissolved constituents and suspended impurities. Product water quality depends on feedwater quality.

Select the reverse osmosis membrane module appropriate to the characteristics of the feedwater. Obtain rejection data for contaminants in the feedwater at the operating pressure to be used in preparing reagent water. Set overall water production to make the most economical use of water without compromising the final quality of the permeate. Selection of spiral-wound or hollow fiber configurations depends on fouling potential of the feedwater. Regardless of configuration used, pretreatment may be required to minimize membrane fouling with colloids or particulates and to minimize introduction of chlorine, iron, and other oxidizing compounds that may degrade reverse osmosis membranes. Periodic flushing of the membrane modules is necessary.

3. Ion Exchange

Prepare deionized water by passing feedwater through a mixed-bed ion exchanger, consisting of strong anion and strong cation resins mixed together. When the system does not run continuously, recirculate product water through ion-exchange bed.

Use separate anion and cation resin beds in applications where resin regeneration is economically attractive. In such instances, position the anion exchanger downstream of the cation exchanger to remove leachates from the cation resin. Proper bed sizing is critical to the performance of the resins. In particular, set the length-to-diameter ratio of the bed in accordance with the maximum process flow rate to ensure that optimal face velocities are not exceeded and that sufficient residence time is provided.

In applications where the feedwater has significant quantities of organic matter, remove organics to minimize potential fouling of the resins. Possible pretreatments include prefiltration, distillation, reverse osmosis, or adsorption.

4. Adsorption

Adsorption is generally used to remove chlorine and organic impurities. It is accomplished typically with granular activated carbon. Efficiency of organics removal depends on the nature of the organic contaminants, the physical characteristics of the activated carbon, and the operating conditions. In general, organics adsorption efficiency is inversely proportional to solubility and may be inadequate for the removal of low-molecular-weight, polar compounds. Performance differences among activated carbons are attributable to the use of different raw materials and activation procedures. Select the appropriate activated carbon with regard to these differences. Even with optimum activated carbon, proper performance will not be attained unless the column is sized to give required face velocity and residence time at the maximum process flow rate.

Use of activated carbon may adversely affect resistivity. This effect may be controlled by use of reverse osmosis, mixed resins, or special adsorbents. To achieve the lowest level of organic contamination, use mixtures of polishing resins with special carbons in conjunction with additional treatment steps, such as reverse osmosis, natural carbons, ultraviolet oxidation, or ultrafiltration.

1080 C.   Reagent Water Quality

1. Quality Guidelines

Several guidelines for reagent water quality, based on contaminant levels, are available, but the final test is the appropriateness for the analysis. Table 1080:II lists some characteristics of various qualities of reagent water.

High-quality reagent water, having a minimum resistivity of 10 megohms-cm, 25°C (in line), typically is prepared by distillation, deionization, or reverse osmosis treatment of feedwater followed by polishing with a mixed-bed deionizer and passage through a 0.2 µm pore membrane filter. Alternatively treat by reverse osmosis followed by carbon adsorption and deionization. Determine quality at the time of production. Mixed-bed deionizers typically add small amounts of organic matter to water, especially if the beds are fresh. Resistivity should be >10 megohm-cm at 25°C, measured in-line. Resistivity measurements will not detect organics or nonionized contaminants, nor will they provide an accurate assessment of ionic contaminants at the microgram-per-liter level.

Medium-quality water typically is produced by distillation or deionization. Resistivity should be >1 megohm-cm at 25°C.

TABLE 1080:II. REAGENT WATER SPECIFICATIONS*

Quality Parameter	High	Medium	Low
Resistivity, megohm-cm at 25°C	>10	>1	0.1
Conductivity, µmho/cm at 25°C	<0.1	<1	10
SiO_2, mg/L	<0.05	<0.1	<1

Low-quality water should have a minimum resistivity of 0.1 megohm-cm, and may be used for glassware washing, preliminary rinsing of glassware, and as feedwater for production of higher-grade waters.

The pH of high- or medium-quality water cannot be measured accurately without contaminating the water. Measure other constituents as required for individual tests.

High-quality water cannot be stored without significant degradation; produce it continuously and use it immediately after processing. Medium-quality water may be stored, but keep storage to a minimum and provide quality consistent with the intended use. Store only in materials that protect the water from contamination, such as TFE and glass for organics analysis or plastics for metals. Store low-quality water in materials that protect the water from contamination.

2. Bibliography

AMERICAN SOCIETY FOR TESTING AND MATERIALS. 1991. Annual Book of ASTM Standards, Vol. 11-01, D 1193-91. American Soc. Testing & Materials, Philadelphia, Pa.

1090   LABORATORY OCCUPATIONAL HEALTH AND SAFETY

1090 A.   Introduction

1. General Discussion

Achievement of a safe and healthful workplace is the responsibility of the organization, the laboratory manager, the supervisory personnel and, finally, the laboratory personnel themselves. All laboratory employees must make every effort to protect themselves and their fellow workers by conscientiously adhering to the health and safety program that has been developed and documented specifically for their laboratory.

2. Organizing for Safety

a. Overall program: The responsibility for establishing and enforcing a laboratory health and safety (LH&S) program ultimately rests with the laboratory director. The LH&S program must, at the minimum, address how to protect oneself from the hazards of working with biological (1090H), chemical (1090J), and radiological (1090I) agents. Such a program is a necessary component of an overall laboratory quality system that provides for the health and safety of the entire laboratory staff. As a part of the quality system, all aspects of the LH&S program must be fully documented. Laboratory personnel must be trained. The LH&S program must be fully implemented and its application audited periodically. Appropriate records of all activities must be kept to document performance, meet appropriate regulatory requirements, and document the status of the LH&S program.

In the United States, the minimum standard of practice for health and safety activities is detailed in government documents.[1,2] Each laboratory should appoint as needed a chemical hygiene officer (CHO), a biological hygiene officer (BHO), a radiological hygiene officer (RHO), and, where appropriate or desired, a LH&S committee. The CHO, the committee, and laboratory management must develop, document, and implement a ''written'' laboratory hygiene plan (LHP), or chemical hygiene plan (CHP).

b. Specific responsibilities: Specific responsibilities applicable at various levels within the organization are as follows:

1) The chief executive officer (CEO) has ultimate responsibility for LH&S within the organization and must, with other managers and supervisors, provide continuous support for the LH&S program.

2) The supervisor has primary responsibility for the LH&S program in his or her work group.

3) The biological hygiene officer (BHO) has the responsibility to work with managers, supervisors, and other employees to develop and implement appropriate biological hygiene policies and practices; monitor procurement, use, and disposal of biological agents used in the laboratory; see that appropriate audits are conducted and that records are maintained; know the current legal requirements concerning working with biological agents; and seek ways to improve the biological hygiene program.

4) The chemical hygiene officer (CHO) has the same responsibilities as the biological hygiene officer, but with respect to chemicals, and also is responsible for helping supervisors (project directors) develop precautions and adequate facilities and for keeping material safety data sheets (MSDSs) available for review.

5) The radiological hygiene officer (RHO), referred to as radiation safety officer in most regulatory language, has the same responsibilities as the chemical hygiene officer, but with respect to radiological chemicals and exposure.

6) The laboratory supervisor has overall responsibility for chemical hygiene in the laboratory, including responsibility to ensure that workers know and follow the chemical hygiene rules, that protective equipment is available and in working order, and that appropriate training has been provided; performs regular, formal chemical hygiene and housekeeping inspections, including routine inspections of emergency equipment, and maintenance of appropriate records; knows the current legal requirements concerning regulated substances; specifies the required levels of protective apparel and equipment needed to perform the work; and ensures that facilities and training for use of any material being ordered are adequate.

7) The project director (or a director of a specific operation) has primary responsibility for biological, chemical, and/or radiological hygiene procedures as appropriate for all operations under his or her control.

8) The laboratory worker has the responsibility for planning and conducting each operation in accordance with the institutional chemical hygiene, biological hygiene, and radiological hygiene procedures, and for developing good personal chemical, biological, and radiological hygiene habits.

3. Records

Maintain records of all accidents including ''near-misses,'' medical care audits, inspections, and training for specified time periods that depend on the nature of the requirement. Keep records on standardized report forms containing sufficient information to enable an investigator to determine who was involved, what happened, when and where it happened, and what injuries or exposures, if any, resulted. Most importantly, these records should enable the formulation of appropriate corrective actions where warranted. The standard of practice for LH&S activities requires that a log (record) be kept of those accidents causing major disability. Record not only all accidents, but also "near-misses," to permit full evaluation of safety program effectiveness. Maintain a file detailing all of the recommendations for the LH&S program.

4. Information and Training[2]

The standard of practice for hazard communication or "right-to know" requires that employees be notified about hazards in the workplace.

Laboratory personnel must be under the direct supervision and regular observation of a technically qualified individual who must have knowledge of the hazards present, their health effects, and related emergency procedures. The supervisor must educate laboratory personnel in safe work practices at the time of initial assignment and when a new hazardous substance is introduced into the workplace. Personnel have a right to know what hazardous materials are present, the specific hazards created by those materials, and the required procedures to protect themselves against these hazards. The hazard communication standard [2] requires information and training on material safety data sheets (MSDSs), labeling, chemical inventory of all hazardous substances in the workplace, and informing contractors of hazardous substances.

Training dealing with health and safety techniques and work practices requires a concerted effort by management, and must be conducted on a routine basis by competent and qualified individuals to be effective. Records of training must be maintained.

5. References

1. OCCUPATIONAL SAFETY AND HEALTH ADMINISTRATION. Laboratory Standard. Occupational Exposure to Hazardous Chemicals in Laboratories. 29 CFR 1910.1450.
2. OCCUPATIONAL SAFETY AND HEALTH ADMINISTRATION. 1985. Hazard Communication. Final Rule. *Federal Register* 48:53280. 29 CFR 1910.1200.

6. Bibliography

DUX, J.P. & R.F. STALZER. 1988. Managing Safety in the Chemical Laboratory. Van Nostrand Reinhold Co., Inc., New York, N.Y.
FURR, A.K., ed. 1990. CRC Handbook of Laboratory Safety, 3rd ed. CRC Press, Inc., Boca Raton, Fla.

1090 B.   Safe Laboratory Practices

Use the information, rules, work practices, and/or procedures discussed below for essentially all laboratory work with chemicals.

1. General Rules

a. Accidents and spills:
1) Eye contact—Promptly flush eyes with water for a prolonged period (minimum of 15 min) and seek immediate medical attention.
2) Ingestion—Encourage victim to drink large amounts of water.
3) Skin contact—Promptly flush affected area with water for approximately 15 min and remove any contaminated clothing. If symptoms persist after washing, seek medical attention.
4) Clean-up—Promptly clean up spills, using appropriate protective apparel and equipment and proper disposal procedures.
5) Working alone—Avoid working alone in a building; do not work alone in a laboratory if the procedures to be conducted are hazardous.

b. Vigilance: Be alert to unsafe conditions and see that they are corrected when detected.

2. Work Practices/Rules

a. Work habits: Develop and encourage safe habits, avoid unnecessary exposure to chemicals by any route, and avoid working alone whenever possible.

b. Exhaust ventilation: Do not smell or taste chemicals. Vent any apparatus that may discharge toxic chemicals (vacuum pumps, distillation columns, etc.) into local exhaust devices.

c. Glove boxes: Inspect gloves and test glove boxes before use.

d. Cold and/or warm rooms: Do not allow release of toxic substances in cold rooms and/or warm rooms, because these rooms usually have no provisions for exhausting contaminants.

e. Use/choice of chemicals: Use only those chemicals for which the quality of the available ventilation system is appropriate.

f. Eating, smoking, and related activities: DO NOT eat, drink, smoke, chew gum, or apply cosmetics in areas where laboratory chemicals are present. Always wash hands before conducting these activities.

g. Food storage: DO NOT store, handle, or consume food or beverages in storage areas, refrigerators, or glassware and utensils that also are used for laboratory operations.

h. Equipment and glassware: Handle and store laboratory glassware with care to avoid damage. Do not use damaged glassware. Use extra care with Dewar flasks and other evacuated glass apparatus; shield or wrap them to contain chemicals and fragments should implosion occur. Use equipment for its designed purpose only.

i. Washing: Wash areas of exposed skin well before leaving the laboratory.

j. Horseplay: Avoid practical jokes or other behavior that might confuse, startle, or distract another worker.

k. Mouth suction: Do not use mouth suction for pipetting or starting a siphon.

l. Personal protective equipment: Do not wear personal protective clothing or equipment in nonlaboratory areas. Remove laboratory coats immediately on significant contamination with hazardous materials.

m. Personal apparel: Confine long hair and loose clothing. Wear shoes at all times in the laboratory but do not wear sandals or perforated shoes.

n. Personal housekeeping: Keep work area clean and uncluttered, with chemicals and equipment properly labeled and stored. Clean up work area on completion of an operation or at the end of each day.

o. Unattended operations: Leave lights on, place an appropriate sign on the door, and provide for containment of toxic substances in the event of failure of a utility service (such as cooling water) to an unattended operation.

3. Personal Protective Equipment

Carefully plan a program addressing the need for, use of, and the training with personal protective equipment. Such a program includes seeking information and advice about hazards, developing appropriate protective procedures, and proper positioning of equipment before beginning any new operations.

a. Eye protection: Wear appropriate eye protection (this applies to all persons, including visitors) where chemicals are stored or handled. Avoid use of contact lenses in the laboratory unless necessary; if contact lenses are used, inform supervisor so that special precautions can be taken.

b. Skin protection: Wear appropriate gloves when the potential for contact with toxic chemicals exists. Inspect gloves before each use, wash them before removal, and replace periodically.

c. Respiratory protection: Use appropriate respiratory equipment when engineering controls are unable to maintain air contaminant concentrations below the action levels, i.e., one half the permissible exposure limit (PEL)[1] or threshold limit value (TLV[7]), i.e., levels below which no irreversible health affects are expected. When work practices are used that are expected to cause routine exposures that exceed the PEL or TLV, respiratory protection is required to prevent overexposure to hazardous chemicals. If respirators are used or provided in the laboratory then the LH&S standard of practice requires that a complete respiratory protection plan (RPP) be in place. The minimum requirements for an RPP meeting the LH&S standard of practice are published.[1] Inspect respirators before use and check for proper fit.

d. Other protective equipment: Provide and use any other protective equipment and/or apparel as appropriate.

4. Engineering Controls

Fume hoods: Use the hood for operations that might result in the release of toxic chemical vapors or dust. As a rule of thumb, use a hood or other local ventilation device when working with any appreciably volatile substance with a TLV of less than 50 ppm. Confirm that hood performance is adequate before use. Open hood minimally during work. Keep hood door closed at all other times except when adjustments within the hood are being made. Keep stored materials in hoods to a minimum, and do not block vents or air flow. Provide at least an 8-cm space under and around all items used in hoods, and ensure that they are at least 15 cm from the front of the hood.

5. Waste Disposal

Ensure that the plan for each laboratory operation includes plans and training for waste disposal. Deposit chemical waste in appropriately labeled receptacles and follow all other waste disposal procedures of the Chemical Hygiene Plan (see 1090J). Do not discharge any of the following contaminants to the sewer: concentrated acids or bases; highly toxic, malodorous, or lachrymatory substances; substances that might interfere with the biological activity of wastewater treatment plants; and substances that may create fire or explosion hazards, cause structural damage, or obstruct flow. For further information on waste disposal, see Section 1100.

6. Work with Chemicals of Moderate Chronic or High Acute Toxicity

Examples of chemicals in this category include diisopropylfluorophosphate, hydrofluoric acid, and hydrogen cyanide. The following rules are intended to supplement the rules listed previously for routine laboratory operations. Their purpose is to minimize exposure to these toxic substances by any exposure route using all reasonable precautions. The precautions are appropriate for substances with moderate chronic or high acute toxicity used in significant quantities.

a. Location: Use and store these substances only in areas of restricted access with special warning signs. Always use a hood (previously evaluated to confirm adequate performance with a face velocity of at least 24 m/min) or other containment device for procedures that may result in the generation of aerosols or vapors containing the substance; trap released vapors to prevent their discharge with the hood exhaust.

b. Personal protection: Always avoid skin contact by use of gloves and long sleeves, and other protective apparel as appropriate. Always wash hands and arms immediately after working with these materials.

c. Records: Maintain records of the amounts of these materials on hand, amounts used, and the names of the workers involved.

d. Prevention of spills and accidents: Be prepared for accidents and spills.

Ensure that at least two people are present at all times if a compound in use is highly toxic or of unknown toxicity.

Store breakable containers of these substances in chemically resistant trays; also work and mount apparatus above such trays or cover work and storage surfaces with removable, absorbent, plastic-backed paper. If a major spill occurs outside the hood, evacuate the area; ensure that cleanup personnel wear suitable protective apparel and equipment.

e. Waste: Thoroughly decontaminate or incinerate contaminated clothing or shoes. If possible, chemically decontaminate by chemical conversion. Store contaminated waste in closed, suitably labeled, impervious containers (for liquids, in glass or plastic bottles half-filled with vermiculite).

7. Work with Chemicals of High Chronic Toxicity

Examples of chemicals in this category include (where they are used in quantities above a few milligrams, or a few grams, depending on the substance) dimethyl mercury, nickel carbonyl, benzo(a)pyrene, *N*-nitrosodiethylamine, and other substances with high carcinogenic potency. The following rules are intended to supplement the rules listed previously for routine laboratory operations.

a. Access: Conduct all transfers and work with these substances in a controlled area, i.e., a restricted-access hood, glove box, or portion of a laboratory, designated for use of highly toxic substances, for which all people with access are aware of the substances being used and necessary precautions.

b. Approvals: Prepare a plan for use and disposal of these materials and obtain the approval of the laboratory supervisor.

c. Non-contamination/decontamination: Protect vacuum pumps against contamination by scrubbers or HEPA filters and vent them into the hood. Decontaminate vacuum pumps or other contaminated equipment, including glassware, in the hood before removing them from the controlled area. Decontaminate the controlled area before routine work is resumed.

d. Exiting: On leaving a controlled area, remove any protective apparel (place it in an appropriately labeled container), and thoroughly wash hands, forearms, face, and neck.

e. Housekeeping: Use a wet mop or a vacuum cleaner equipped with a HEPA filter. Do not dry sweep if the toxic substance is a dry powder.

f. Medical surveillance: If using toxicologically significant quantities of such a substance on a regular basis (e.g., three times per week), consult a qualified physician about desirability of regular medical surveillance.

g. Records: Keep accurate records of the amounts of these substances stored and used, the dates of use, and names of users.

h. Signs and labels: Ensure that the controlled area is conspicuously marked with warning and restricted access signs and that all containers of these substances are appropriately labeled with identity and warning labels.

i. Spills: Ensure that contingency plans, equipment, and materials to minimize exposures of people and property are available in case of accident.

j. Storage: Store containers of these chemicals only in a ventilated, limited-access area.

k. Glove boxes: For a negative-pressure glove box, ensure that the ventilation rate is at least 2 volume changes/h and the pressure drop at least 1.3 cm of water. For a positive-pressure glove box, thoroughly check for leaks before each use. In either case, trap the exit gases or filter them through a HEPA filter and then release them into the hood.

l. Waste: Ensure that containers of contaminated waste (including washings from contaminated flasks) are transferred from the controlled area in a secondary container under the supervision of authorized personnel.

8. Physical Hazards

a. Electrical: Ensure that electrical wiring, connections, and apparatus conform to the requirement of the latest National Electrical Code. Fire, explosion, power outages, and electrical shocks are all serious hazards that may result from incorrect use of electrical devices. Ground all electrical equipment or use double-insulated equipment. Use ground fault interrupter circuit breakers to the maximum extent possible. Do not locate electrical receptacles inside fume hoods, and do not use equipment near volatile flammable solvents. Use approved safety refrigerators. Disconnect electrical equipment from the power supply before service or repair is attempted and never bypass safety interlocks. Attempting to repair equipment using employees not thoroughly acquainted

with electrical principles may present particularly dangerous situations.

b. Non-ionizing radiation: Non-ionizing radiation, also called electromagnetic radiation, is generally considered to be the radio frequency region of the radiation spectrum. For the purposes of dealing with personal exposures in laboratories, it also includes the microwave frequency region. Typical laboratory exposures to non-ionizing radiation usually include ultraviolet, visible, infrared, and microwave radiation.

For normal environmental conditions and for incident electromagnetic energy of frequencies from 10 MHz to 100 GHz, the radiation protection guide is 10 mW/cm^2. The radiation protection applies whether the radiation is continuous or intermittent. This means a power density of 10 mW/cm^2 for periods of 0.1 h or more, or an energy density of 1 mW-h/cm^2 during any 0.1-h period. These recommendations apply to both whole-body irradiation and partial body irradiation.

Ultraviolet radiation (UV) and lasers are used frequently. With properly constructed and operated instruments, it is not a significant hazard but can be harmful when used for controlling microorganisms in laboratory rooms or for sterilizing objects.

When using devices that generate or use non-ionizing radiation, observe the following precautions: Wear safety glasses or goggles with solid side pieces whenever there is a possibility of exposure to harmful (UV) radiation. Provide proper shielding (shiny metal surfaces reflect this energy). Shut off all these devices (UV lamps) when not in use. Post warning signs and install indicator lights to serve as a constant reminder when these types of devices are in use (UV lamps).

c. Mechanical: Shield or guard drive belts, pulleys, chain drivers, rotating shafts, and other types of mechanical power transmission apparatus. Laboratory equipment requiring this guarding includes vacuum pumps, mixers, blenders, and grinders.

Shield portable power tools. Guard equipment such as centrifuges, which have high-speed revolving parts, against "flyaways." Securely fasten equipment that has a tendency to vibrate (e.g., centrifuges and air compressors) to prevent the tendency to "walk" and locate them away from bottles and other items that may fall from shelves or benches because of the vibration.

d. Compressed gases: Gas cylinders may explode or "rocket" if improperly handled. Leaking cylinders may present an explosion hazard if the contents are flammable; they are an obvious health hazard if the contents are toxic; and they may lead to death by suffocation if the contents are inert gases. The Compressed Gas Association has published procedures governing use and storage of compressed gases. Transfer gas cylinders only with carts, hand trucks, or dollies. Secure gas cylinders properly during storage, transport, and use, and leave valve safety covers on cylinders during storage and transport. Avoid the use of adapters or couplers with compressed gas. Properly identify cylinder contents.

9. Chemical Hazards

a. General precautions: Chemical injuries may be external or internal. External injuries may result from skin exposure to caustic or corrosive substances such as acids, bases, or reactive salts. Take care to prevent accidents, such as splashes and container spills. Internal injuries may result from the toxic or corrosive effects of substances absorbed by the body. These internal injuries may result from inhalation, skin contact, or ingestion.

TABLE 1090:I. PERMISSIBLE EXPOSURE LIMITS, THRESHOLD LIMIT VALUES, SHORT-TERM EXPOSURE LIMITS, AND/OR CEILINGS FOR SOME INORGANIC CHEMICALS SPECIFIED IN *STANDARD METHODS*

Compound	Chemical Abstract No. CAS No.	PEL/TLV STEL(S)* or Ceiling(C) mg/m^3
Chromic acid and chromates†‡ (as CrO_3)	7440-47-3	0.1/0.05
Chromium, soluble chromic, chromous salts (as Cr)	7440-47-3	0.5/0.5
Chromium metal and insoluble salts	7440-47-3	1/0.5
Hydrogen chloride	7647-01-0	7.5(C)/7.5(C)
Hydrogen peroxide	7722-84-1	1.4/1.4
Lead‡	7439-92-1	−/0.15
Mercury†§	7439-97-6	0.1/0.05
Nitric acid	7697-37-2	5/5.2, 10(S)
Phosphoric acid	7664-38-2	1/1, 3(S)
Potassium hydroxide	1310-58-3	−/2(C)
Silver (metal and soluble compounds, as Ag)	7440-22-4	0.01/0.1 metal, 0.01 soluble as Ag
Sodium azide	26628-22-8	−/0.29(C)
Sodium hydroxide	1310-73-2	2(C)/2(C)
Sulfuric acid	7664-93-9	1/1, 3(S)

* Short-term exposure limit. See 29 CFR 1910.1028.
† (Suspect) carcinogen.
‡ Substance has a Biological Exposure Index (BEI).
§ Skin hazard.

TABLE 1090:II. PERMISSIBLE EXPOSURE LIMITS, THRESHOLD LIMIT VALUES, SHORT-TERM EXPOSURE LIMITS, AND/OR CEILINGS FOR ORGANIC SOLVENTS SPECIFIED IN *STANDARD METHODS*

Compound	Chemical Abstract No. CAS No.	PEL/TLV STEL(S)* or Ceiling(C) $ppm\ (v/v)$
Acetic acid	64-19-7	10/10, 15(S)
Acetone	67-64-1	1000/750, 1000(S)
Acetonitrile	75-05-8	40/40, 60(S)
Benzene†‡	71-43-2	10, 25(C), 50 peak 10 min/8 h/10
n-Butyl alcohol§	71-36-3	100/50(C)
tert-Butyl alcohol	75-65-0	100/100, 150(S)
Carbon disulfide‡	75-15-0	20, 30(C), 100 peak 30 min/8 h/10
Carbon tetrachloride†§	56-23-5	10, 25, 200 peak 5 min/4 h/5
Chloroform†	67-66-3	50(C)/10
Cyclohexanone§	108-94-1	50/50
Dioxane§ (diethylene dioxide)	123-91-1	100/25
Ethyl acetate	141-78-6	400/400
Ethyl alcohol	64-17-5	1000/1000
Ethyl ether (diethyl ether)	60-29-7	400/400, 500(S)
Ethylene glycol	107-21-1	−/50(C)
n-Hexane‡	110-54-3	100/50
Isoamyl alcohol (primary and secondary)	123-51-3	100/100, 125(S)
Isobutyl alcohol	78-83-1	100/50
Isopropyl alcohol	67-63-0	400/400, 500(S)
Isopropyl ether	108-20-3	500/250, 310(S)
Methyl alcohol§	67-56-1	200/200, 250(S)
2-Methoxyethanol§ (methyl cellosolve)	109-86-4	25/5
Methylene chloride†	75-09-2	500, 1000(C), 2000 peak 5 min/2 h/50
Pentane	109-66-0	1000/600, 750(S)
Perchloroethylene†‡ (tetrachloroethylene)	127-18-4	100, 200(C), 300 peak 5 min/3 h/50, 200(S)
n-Propyl alcohol§	71-23-8	200/200, 250(S)
Pyridine	110-86-1	5/5
Toluene‡§	108-88-3	200, 300(C), 500 peak 10 min/8 h/50
Xylenes‡ (o-, m-, p-isomers)	1330-20-7 (95-47-6, 108-38-3, 106-42-3)	100/100, 150(S)

* Short-term exposure limit. See 29 CFR 1910.1028.
† (Suspect) carcinogen.
‡ Substance has a Biological Exposure Index (BEI).
§ Skin hazard.

TABLE 1090:III. PERMISSIBLE EXPOSURE LIMITS, THRESHOLD LIMIT VALUES, SHORT-TERM EXPOSURE LIMITS, AND/OR CEILINGS FOR SOME OF THE REAGENTS SPECIFIED IN STANDARD METHODS

Compound	Chemical Abstract No. CAS No.	PEL/TLV STEL(S)* or Ceiling(C) *ppm (v/v)*
2-Aminoethanol (ethanolamine)	141-43-5	3/3, 6(S)
Benzidine†‡	92-87-5	Confirmed human carcinogen[1]
Benzyl chloride	100-44-7	1/1
Chlorobenzene	108-90-7	75/10
Diethanolamine	111-42-2	−/3
Naphthalene	91-20-3	10/10, 15(S)
Oxalic acid	144-62-7	1/1 mg/m³
Phenol‡	108-95-2	5/5
2-Chloro-6-(trichloromethyl) pyridine (nitrapyrin)	1929-82-4	
Total dust		15/10 mg/m³
Respirable fraction		5/− mg/m³

* Short-term exposure limit. See 29 CFR 1910.1028.
† (Suspect) carcinogen.
‡ Skin hazard.

Tables 1090:I, II, and III list PELs, TLVs, and/or short-term exposure limits and ceilings for some chemical materials specified in *Standard Methods*, as given in various published sources.[1-8] The PEL values reported in these tables are in some instances higher than the levels that some nations believe to be appropriate. Because the health and safety program should be driven by meeting best industrial hygiene practice, always use the lowest recommended exposure values when protecting human health.

In addition, pay careful attention to equipment corrosion that ultimately may lead to safety hazards from equipment failure.

b. Inorganic acids and bases: Many inorganic acids and bases have PELs and TLVs. Table 1090:I presents PELs (based on U.S. standards) and/or TLVs as well as short-term exposure limits and ceilings for some inorganic chemicals specified in *Standard Methods*. These PELs and TLVs indicate the maximum air concentration to which workers may be exposed. Fumes of these acids and bases are severe eye and respiratory system irritants. Liquid or solid acids and bases can quickly cause severe burns of the skin and eyes. When acids are heated to increase the rate of digestion of organic materials, they pose a significantly greater hazard because fumes are produced and the hot acid reacts very quickly with the skin.

Store acids and bases separately in well-ventilated areas and away from volatile organic and oxidizable materials. Use containers (rubber or plastic buckets) to transport acids and bases.

Work with strong acids and bases only in a properly functioning chemical fume hood. Slowly add acids and bases to water (with constant stirring) to avoid spattering. If skin contact is made, thoroughly flush the contaminated area with water and seek medical attention if irritation persists. Do not wear contaminated clothing until after it has been cleaned thoroughly. Leather items (e.g., belts and shoes) will retain acids even after rinsing with water and may cause severe burns if worn. If eye contact is made, immediately flush both eyes for at least 15 min with an eye wash and seek medical attention.

c. Perchloric acid and other highly reactive chemicals: Concentrated perchloric acid reacts violently or explosively on contact with organic material and may form explosive heavy metal perchlorates. Do not use laboratory fume hoods used with perchloric acid for organic reagents, particularly volatile solvents. In addition

to these hazards, perchloric acid produces severe burns when contact is made with the skin, eyes, or respiratory tract. Preferably provide a dedicated perchloric acid hood. Follow the manufacturer's instructions for proper cleaning, because exhaust ducts become coated and must be washed down regularly.

Use extreme caution when storing and handling highly reactive chemicals, such as strong oxidizers. Improper storage can promote heat evolution and explosion. Do not store strong oxidizers and reducers in close proximity.

d. Organic solvents and reagents: Most solvents specified in *Standard Methods* have PELs and/or TLVs as well as short-term exposure limits or ceilings for workplace exposures (see Table 1090:II).

Many organic reagents, unlike most organic solvents, do not have PELs/TLVs or short-term exposure limits and ceilings, but this does not mean that they are less hazardous. Table 1090:III contains PELs/TLVs or short-term exposure limits and ceilings for some reagents specified in *Standard Methods*.

Some compounds are suspect carcinogens and should be treated with extreme caution. These compounds include solvents and reagents such as benzene, carbon tetrachloride, chloroform, dioxane, perchloroethylene, and benzidine. Lists of chemicals with special hazardous characteristics are available from the Occupational Safety and Health Administration and the National Institute for Occupational Safety and Health. In the U.S., the lists of "Regulated Carcinogens" and of "Chemicals Having Substantial Evidence of Carcinogenicity" are especially important. Developing and following laboratory handling procedures for compounds on such authoritative lists should significantly reduce the potential for exposures.

Solvents used in the laboratory usually fall into several major categories: alcohols, chlorinated compounds, and hydrocarbons. Exposure to each of these classes of compounds can have a variety of health effects. Alcohols, in general, are intoxicants, capable of causing irritation of the mucous membranes and drowsiness. Chlorinated hydrocarbons cause narcosis and damage to the central nervous system and liver. Hydrocarbons, like the other two groups, are skin irritants and may cause dermatitis after prolonged skin exposure. Because of the volatility of these compounds, hazardous vapor concentrations can occur (fire or explosion hazard). Proper ventilation is essential.

The majority of organic reagents used in this manual fall into four major categories: acids, halogenated compounds, dyes and indicators, and pesticides. Most organic acids have irritant properties. They are predominantly solids from which aerosols may be produced. Dyes and indicators also present an aerosol problem. Handle pesticides with caution because they are poisons, and avoid contact with the skin. Wear gloves and protective clothing. The chlorinated compounds present much the same hazards as the chlorinated solvents (narcosis and damage to the central nervous system and liver). Proper labeling for the compound, including a date for disposal based on the manufacturer's recommendations, permits tracking chemical usage and disposal of outdated chemicals.

10. References

1. OCCUPATIONAL SAFETY AND HEALTH ADMINISTRATION. Respiratory Protection. 29 CFR 1910.134.
2. U.S. PUBLIC HEALTH SERVICE. 1980. Annual Report on Carcinogens. National Toxicology Program, Dep. Health & Human Services, U.S. Government Printing Off., Washington, D.C.
3. INTERNATIONAL AGENCY FOR RESEARCH ON CANCER. (Various dates). IARC monographs on the evaluation of the carcinogenic risk of chemicals to humans. World Health Org. Publications Center, Albany, N.Y.
4. NATIONAL INSTITUTE FOR OCCUPATIONAL SAFETY AND HEALTH. 1985. Occupational Health Guidelines for Chemical Hazards. NIOSH/OSHA-NIOSH Publ. No. 85-123, U.S. Government Printing Off., Washington, D.C.

5. U.S. PUBLIC HEALTH SERVICE, CENTERS FOR DISEASE CONTROL. 1993. Registry of Toxic Effects of Chemical Substances. U.S. Government Printing Off., Washington, D.C.
6. SAX, N.I. 1989. Dangerous Properties of Industrial Materials. Van Nostrand Reinhold, New York, N.Y.
7. AMERICAN CONFERENCE OF GOVERNMENTAL INDUSTRIAL HYGIENISTS. 1992. Threshold Limit Values for Chemical Substances and Physical Agents and Biological Exposure Indices. American Conf. Governmental Industrial Hygienists, Cincinnati, Ohio.
8. U.S. PUBLIC HEALTH SERVICE, CENTERS FOR DISEASE CONTROL. 1990. NIOSH Pocket Guide to Chemical Hazards. DHHS (NIOSH) Publ. No. 90-117, U.S. Government Printing Off., Washington, D.C.

11. Bibliography

SLINEY, D. 1980. Safety with Lasers and Other Optical Sources: A Comprehensive Handbook. Plenum Press, New York, N.Y.
COMMITTEE ON HAZARDOUS SUBSTANCES IN THE LABORATORY. 1981. Prudent Practices for Handling Hazardous Chemicals in Laboratories. National Academy Press, Washington, D.C.
COMMITTEE ON HAZARDOUS SUBSTANCES IN THE LABORATORY. 1983. Prudent Practices for Disposal of Chemicals from Laboratories. National Academy Press, Washington, D.C.
FURR, A.K., ed. 1990. CRC Handbook of Laboratory Safety, 3rd ed. CRC Press, Inc., Boca Raton, Fla.
COMPRESSED GAS ASSOCIATION, INC. 1990. Handbook of Compressed Gases. Van Nostrand Reinhold Co., New York, N.Y.
AMERICAN CONFERENCE OF GOVERNMENTAL INDUSTRIAL HYGIENISTS. 1992. Threshold Limit Values. American Conf. Governmental Industrial Hygienists, Cincinnati, Ohio.

1090 C.　Laboratory Facility/Fixed Equipment

1. Facility Design

The laboratory facility must have a general ventilation system with air intakes and exhausts located to avoid intake of contaminated air, well-ventilated stockrooms and/or storerooms, laboratory hoods and sinks, miscellaneous safety equipment including eyewash fountains and safety showers, and arrangements for the disposal of wastes and samples in accordance with applicable federal, state, and local regulations.

2. Facility and Fixed Equipment Maintenance

Maintain facilities and equipment with scheduled maintenance and continual surveillance to ensure proper operation. Give special attention to the adequacy of ventilation system.

a. Facility ventilation systems:

1) The general laboratory ventilation should provide a source of air for breathing and for input to local ventilation devices such as fume hoods. Do not rely on it for protection from exposure to toxic substances used during the working day. The system should direct air flow into the laboratory from nonlaboratory areas and then exhaust the air directly to the exterior of the building in a manner that will prevent its re-entry.

2) Laboratory fume hoods—As a minimum, provide at least 1 linear m of hood space per worker if workers spend most of their time working with chemicals or if they work with chemical sub-

stances with PELs or TLVs less than 100 ppm. Equip each hood with a continuous monitoring device to allow convenient confirmation of adequate hood performance before each use. If this is not possible, avoid work with substances with PELs or TLVs less than 100 ppm or with unknown toxicity, or provide other types of local ventilation devices.

3) Other local ventilation devices—Provide ventilated chemical/biological cabinets, canopy hoods and instrument/work station snorkels as needed. Many local ventilation devices require a separate exhaust duct, as do canopy hoods and snorkels.

4) Special ventilation areas/devices—It may be necessary to pass exhaust air from special ventilation areas or devices such as radiological hoods, glove boxes, and isolation rooms through HEPA filters, scrubbers, or other treatment before release into regular exhaust system. Ensure that cool rooms and warm rooms have provisions for rapid escape and for escape in the event of electrical failure.

5) Ventilation system modifications—Make alteration in the ventilation system only in consultation with an expert qualified in laboratory ventilation system design. Thoroughly test changes in the ventilation system to demonstrate adequate worker protection.

b. Facility ventilation system performance: A ventilation system rate of 4 to 12 room air changes per hour is considered adequate where local exhaust ventilation devices such as fume hoods are used as the primary method of control. General ventilation system air flow should not be turbulent and should be relatively uniform throughout the laboratory, with no high-velocity or static areas; air flow into and within laboratory fume hoods should not be excessively turbulent; fume hood face velocity should be adequate for the intended use (for general-purpose fume hoods this is typically 18 to 30 m/min). The effective protection provided by a fume hood depends on a number of factors including hood location and design, and cannot be determined solely on the basis of the face velocity.

c. Facility ventilation system evaluation: Evaluate performance characteristics (quality and quantity) of the ventilation system on installation, re-evaluate whenever a change in local ventilation devices is made, and monitor routinely. Schedule such monitoring with a frequency dictated by the type, age, condition, and any accessories associated with the device, but at least annually; monitor hoods at least quarterly. Document all ventilation system checks or actions such as flow checks, calibration, alterations, repairs, maintenance, or any other action that may determine or change flow efficiency or characteristics.

3. Bibliography

AMERICAN CONFERENCE OF GOVERNMENTAL INDUSTRIAL HYGIENISTS. 1982. Industrial Ventilation: A Manual of Recommended Practice, 17th ed. American Conf. Governmental Industrial Hygienists, Inc., Cincinnati, Ohio.

AMERICAN NATIONAL STANDARDS INSTITUTE & AMERICAN SOCIETY OF HEATING, REFRIGERATING, AND AIR CONDITIONING ENGINEERS. 1985. Method of Testing Performance of Laboratory Fume Hoods. ANSI/ASHRAE 110-1985, American National Standards Inst., Inc., New York, N.Y.

INSTITUTE OF ENVIRONMENTAL SCIENCES. 1986. Recommended Practice for Laminar Flow Clean Devices. RP-CC-002-86, Inst. Environmental Sciences, Mount Prospect, Ill.

NATIONAL SANITATION FOUNDATION. 1987. Class II Biohazard Cabinetry (Laminar Flow). Standard 49-1987, National Sanitation Found., Ann Arbor, Mich.

AMERICAN SOCIETY FOR TESTING AND MATERIALS. 1988. Standard Guide for Good Laboratory Practices in Laboratories Engaged in Sampling and Analysis of Water. ASTM D3856, American Soc. Testing & Materials, Philadelphia, Pa.

AMERICAN NATIONAL STANDARDS INSTITUTE & AMERICAN SOCIETY OF HEATING, REFRIGERATING, AND AIR CONDITIONING ENGINEERS. 1989. Ventilation for Acceptable Indoor Air Quality. ANSI/ASHRAE 62-1989, American National Standards Inst., Inc., New York, N.Y.

AMERICAN SOCIETY OF HEATING, REFRIGERATING, AND AIR CONDITIONING ENGINEERS. 1989. Handbook—Fundamentals. American Soc. Heating, Refrigerating & Air-Conditioning Engineers. Inc., Atlanta, Ga.

AMERICAN NATIONAL STANDARDS INSTITUTE. 1990. Emergency Eyewash and Shower Equipment. ANSI Z358.1-990, American National Standards Inst., New York, N.Y.

AMERICAN NATIONAL STANDARDS INSTITUTE. 1991. Fundamentals Governing the Design and Operation of Local Exhaust Systems. ANSI Z9.2-1979 (rev. 1991), American National Standards Inst., Inc., New York, N.Y.

AMERICAN NATIONAL STANDARDS INSTITUTE. 1991. Standard on Fire Protection for Laboratories Using Chemicals. NFPA 45, National Fire Protection Assoc., Quincy, Mass.

AMERICAN NATIONAL STANDARDS INSTITUTE & AMERICAN INDUSTRIAL HYGIENE ASSOCIATION. 1993. Standard for Laboratory Ventilation. ANSI/AIHA Z9.5, American Industrial Hygiene Assoc., Fairfax, Va.

AMERICAN NATIONAL STANDARDS INSTITUTE. 1993. Flammable and Combustible Liquids Code. NFPA 30, National Fire Protection Assoc., Quincy, Mass.

AMERICAN SOCIETY OF HEATING, REFRIGERATING, AND AIR CONDITIONING ENGINEERS. 1993. Handbook—Applications. American Soc. Heating, Refrigerating & Air-Conditioning Engineers, Inc., Atlanta, Ga.

1090 D.   Hazard Evaluation

1. Hazard Evaluation

Hazard evaluation refers to the assessment of whether an employee has been overexposed to a hazardous substance or if such an exposure episode is likely to occur and to what extent.

The evaluation does not require monitoring airborne concentrations of the hazardous substances involved. Such an assessment may be informal and simply involve considering, among other factors, the chemical and physical properties of the substance and the quantity of substance used. In addition, the exposure assessment may be sufficient to estimate the probability of an overexposure.

Specify, document, and use hazard assessment criteria. Base such criteria on the toxicity of the substances to be used, the exposure potential of the chemical procedures to be performed, and the capacity of the available engineering control systems.

In cases where continuous monitoring devices are used, include resulting exposure data in the exposure evaluation. Air monitoring *only* provides information for inhalation exposure. Other means are required to determine whether overexposure could have occurred as a result of ingestion, or dermal or eye contact.

2. Spills of Toxic or Hazardous Substances

Spills are usually the result of loss of containment due to equipment failure or breakage (uncontrolled releases). The loss of containment can result in overexposure episodes. Calculations using data from material safety data sheets, the chemical and physical properties of the substance, known laboratory air changes, and work-station air volume will allow assessment of the possibility of an overexposure episode.

3. Work Practice Assignment

Use the information calculated from these exposure assessments to develop the written work practices needed to protect the health of the employee while conducting the procedure.

4. Documentation of Hazard Assessments

Document, validate, and authenticate all hazard assessments, preferably using a standard form.

5. Bibliography

DOULL, J., C.D. KLAASSEN & M.O. AMDUR. 1980. Casarett and Doull's Toxicology: The Basic Science of Poisons, 2nd ed. Macmillan Publishing Co., Inc., New York, N.Y.

WILLIAMS, P.L. & J.L. BURSON. 1985. Industrial Toxicology: Safety and Health Applications in the Workplace. Van Nostrand Reinhold Co., New York, N.Y.

BATTELLE MEMORIAL INSTITUTE. 1985. Guidelines for Hazard Evaluation Procedures. American Inst. Chemical Engineers, New York, N.Y.

PLOG, B.A., ed. 1988. Fundamentals of Industrial Hygiene, 3rd ed. National Safety Council, Chicago, Ill.

OCCUPATIONAL SAFETY AND HEALTH ADMINISTRATION. 1988. Job Hazard Analysis. OSHA Publ. No. 3071, U.S. Dep. Labor, Occupational Safety & Health Admin., Washington, D.C.

OCCUPATIONAL SAFETY AND HEALTH ADMINISTRATION. 1988. Chemical Hazard Communication. OSHA Publ. No. 3084, U.S. Dep. Labor, Occupational Safety & Health Admin., Washington, D.C.

OCCUPATIONAL SAFETY AND HEALTH ADMINISTRATION. 1988. How to Prepare for Workplace Emergencies. OSHA Publ. No. 3088, U.S. Dep. Labor, Occupational Safety & Health Admin., Washington, D.C.

BRETHERICK, L., ed. 1990. Bretherick's Handbook of Reactive Chemical Hazards. Butterworths, London, U.K.

OCCUPATIONAL SAFETY AND HEALTH ADMINISTRATION. 1990. Hazard Communication Guidelines for Compliance. OSHA Publ. No. 3111, U.S. Dep. Labor, Occupational Safety & Health Admin., Washington, D.C.

HAWKINS, N.C., S.K. NORWOOD & J.C. ROCK, eds. 1991. A Strategy for Occupational Exposure Assessment. American Industrial Hygiene Assoc., Fairfax, Va.

CLAYTON, G.D. 1991. Patty's Industrial Hygiene and Toxicology. John Wiley & Sons, New York, N.Y.

Also see references 2 through 8, 1090B.10.

1090 E.   Personal Protective Equipment

1. Introduction

The employer must provide and maintain personal protective equipment (PPE) in condition that is sanitary and reliable against hazards in the workplace. All PPE also must be properly designed and constructed for the work to be performed. Several general references on PPE are available.[1-3].

It is essential to select PPE based on an assessment of the hazards[3] or potential hazards to which an employee is exposed, to insure that the correct PPE will be obtained. Use personal protective equipment only when it is not possible or feasible to provide engineering controls. Such personal protective equipment includes all clothing and other work accessories designed to create a barrier against workplace hazards.

The basic element of any personal protective equipment management program must be an in-depth evaluation of the equipment needed to protect against the hazards at the workplace. Management dedicated to the safety and health of employees must use that evaluation to set a standard operating procedure for personnel, then encourage those employees to use, maintain, and clean the equipment to protect themselves against those hazards.

Using personal protective equipment requires hazard awareness and training on the part of the user. Make employees aware that the equipment does not eliminate the hazard. If the equipment fails, exposure will occur. To reduce the possibility of failure, use equipment that is properly fitted and maintained in a clean and serviceable condition.

Selection of the proper piece of personal protective equipment for the job is important. Employers and employees must understand the equipment's purpose and its limitations. Do not alter or remove equipment even though an employee may find it uncomfortable (equipment may be uncomfortable simply because it does not fit properly).

2. Eye Protection

The LH&S standard of practice requires the use of eye and face protective equipment[3] where there is a reasonable probability of injury prevention through its use. Employers must provide a type of protector suitable for work to be performed, and employees must use the protectors. These requirements also apply to supervisors and management personnel, and to visitors while they are in hazardous areas.

Protectors must provide adequate protection against particular hazards for which they are designed, be reasonably comfortable when worn under the designated conditions, fit snugly without interfering with the movements or vision of the wearer, and be durable, easy to disinfect and clean, and kept in good repair.

In selecting the protector, consider the kind and degree of hazard. Where a choice of protectors is given, and the degree of protection required is not an important issue, worker comfort may be a deciding factor.

Persons using corrective glasses and those who are required to wear eye protection must wear glasses with protective lenses providing optical correction, goggles that can be worn over corrective glasses without disturbing the adjustment of the glasses, or goggles that incorporate corrective lenses mounted behind the protective lenses.

When limitations or precautions are indicated by the manufacturer, transmit them to the user and observe strictly. Safety glasses require special frames. Combinations of normal wire frames with safety lenses are not acceptable.

Design, construction, testing, and use of eye and face protection must be in accordance with national standards.[4]

3. Protective Work Gloves

Match glove material to the hazard: such materials as nitrile, neoprene, natural rubber, PVC, latex, and butyl rubber vary widely in chemical resistance. What may be safe with one chemical may prove harmful with another (see Tables 1090:IV and 1090:V). Glove thickness may be as important as glove material in some cases. Many organic reagents, unlike most organic solvents, do not have PELs/TLVs but this does not mean that they are less hazardous.

TABLE 1090:IV. GLOVE SELECTION FOR ORGANIC CHEMICAL HANDLING

Compound	Chemical Abstract No. CAS No.	Chemical Class	Butyl Rubber	Neoprene	Nitrile Rubber	PE	PVC	TFE	Viton
Ethers:		241							
Ethyl ether (diethyl ether)	60-29-7	241						X	
Isopropyl ether	108-20-3	241						X	
2-Methoxyethanol† (methyl cellosolve)	109-86-4	245	X						
Halogen compounds:		261							
Carbon tetrachloride*	56-23-5	261						X	X
Chloroform*	67-66-3	261						X	X
Methylene chloride	75-09-2	261						?	
Perchloroethylene* (tetrachloroethylene)	127-18-4	261							X
Hydrocarbons:		291							
n-Hexane	110-54-3	291			X			X	X
Pentane	109-66-0	291			X				X
Benzene*	71-43-2	292							X
Toluene	108-88-3	292						X	X
Xylenes (o-, m-, p-isomers)	1330-20-7	292						X	X
Hydroxyl compounds:		311							
n-Butyl alcohol	71-36-3	311	X	X	X	X		X	
Ethyl alcohol	64-17-5	311	X		X			X	
Methyl alcohol	67-56-1	311	X					X	
n-Propyl alcohol	71-23-8	311			X			X	
Isoamyl alcohol	123-51-3	311							
Isobutyl alcohol	78-83-1	311	X						
Isopropyl alcohol	67-63-0	312		X	X			X	
tert-Butyl alcohol (2,2-methylpropanol)	75-65-0	313	X						
Ethylene glycol	107-21-1	314	X	X	X	X	X	X	X
Ketones:		391							
Acetone	67-64-1	391	X						X
Cyclohexanone	108-94-1	391	X						
Heterocyclic compounds:		271							
Dioxane† (diethylene dioxide)	123-91-1	278	X					X	
Pyridine	110-86-1	271	X						
Miscellaneous organic compounds:									
Acetic acid	75-07-0	102	X	X			X	X	
Ethyl acetate	141-78-6	222	X					X	
Acetonitrile	75-05-8	431	X					X	
Carbon disulfide	75-15-0	502							X

* (Suspect) carcinogen.
† Skin hazard.

Evaluate physical properties of the glove material: In addition to chemical resistance, glove materials vary in physical toughness. Select the glove that provides the abrasion, tear, flame, and puncture resistance required for the job.

Maximize comfort and dexterity. Lined gloves absorb perspiration and help insulate the hand. Unlined gloves conform to the hand. Lighter-gauge gloves improve touch sensitivity and flexibility, heavier-gauge gloves add protection and strength.

Ensure a safe grip. Nonslip grips allow for easier and safer handling. Embossed, pebbled, etched, and dotted coatings improve grip in wet or dry working conditions.

Measure proper size and length. Loose-fitting gloves affect dexterity and can be hazardous. Tight-fitting gloves may cause hand fatigue and tend to wear out faster. Gloves should fit comfortably without restricting motion and they should be long enough to protect the wrist, forearm, elbow, or the entire arm, depending on the application.

4. Head Protection

Water and wastewater laboratories seldom require this kind of personal protection, but field work may require such protection.

Head injuries are caused by falling or flying objects or by bumping the head against a fixed object. Head protection, in the form of protective hats, must both resist penetration and absorb the shock of a blow. Make the shell of the hat of a material hard enough to resist the blow, and utilize a shock-absorbing lining composed of head band and crown straps to keep the shell away

TABLE 1090:V. GLOVE SELECTION FOR INORGANIC CHEMICAL HANDLING

Compound	Chemical Abstract No. (CAS No.)	Chemical Class	Suitable Glove Material					
			Butyl Rubber	Natural Rubber	Neoprene	Nitrile Rubber	PE	PVC
Inorganic acids:		370						
Chromic acid,* (Cr+6)	7440-47-3	370	X					X
Hydrochloric acid, 30–70% solutions	10035-10-6	370	X	X	X	X		X
Hydrochloric acid, <30% solutions	10035-10-6	370	X	X	X	X		X
Nitric acid, 30–70% solutions	7697-37-2	370	X		X			
Nitric acid, <30% solutions	7697-37-2	370	X	X	X	X		X
Phosphoric acid, >70% solutions	7664-38-2	370		X	X	X	X	X
Phosphoric acid, 30–70% solutions	7664-38-2	370		X	X	X		X
Sulfuric acid, >70% solutions	7664-93-9	370	X				X	
Sulfuric acid, 30–70% solutions	7664-93-9	370	X	X	X		X	X
Sulfuric acid, <30% solutions	7664-93-9	370	X	X	X		X	X
Inorganic bases:		380						
Ammonium hydroxide, 30–70% solutions	7664-41-7	380	X		X	X		
Ammonium hydroxide, <30% solutions	7664-41-7	380	X		X	X		X
Potassium hydroxide, 30–70% solutions	1310-58-3	380	X	X	X	X		X
Sodium hydroxide, >70% solutions	1310-73-2	380			X			X
Sodium hydroxide, 30–70% solutions	1310-73-2	380	X	X	X	X	X	X
Inorganic salt solutions:		340						
Dichromate solutions, <30%,* (Cr+6)	7440-47-3	340				X		
Inorganic miscellaneous:								
Hydrogen peroxide, 30–70% solutions	7722-39-3	300		X		X		X
Mercury†	7439-97-6	560			X			X

* (Suspect) carcinogen.
† Skin hazard.

from the wearer's skull. Protective materials used in helmets should be water-resistant and slow burning. Helmets consist essentially of a shell and suspension. Ventilation is provided by a space between the headband and the shell. Ensure that each helmet is accompanied by instructions explaining the proper method of adjusting and replacing the suspension and headband.

Visually inspect daily all components, shells, suspensions, headbands, sweatbands, and any accessories for signs of dents, cracks, penetration or any other damage that might reduce the degree of safety originally provided.

Do not store or carry helmets on the rear window deck of an automobile because sunlight and extreme heat may adversely affect the degree of protection.

Further information is available elsewhere.[3,5]

5. Hearing Protection

Exposure to high noise levels can cause hearing loss or impairment, and it can create physical and psychological stress. There is no cure for noise-induced hearing loss, so prevention of excessive noise exposure is the only way to avoid hearing damage. Specifically designed protection is required, depending on the type of noise encountered.

Use preformed or molded ear plugs fitted individually by a professional. Waxed cotton, foam, or fiberglass wool earplugs are self-forming. When properly inserted, they work as well as most molded earplugs. Plain cotton is ineffective as protection against hazardous noise.

Some earplugs are disposable, to be used one time and then thrown away. Clean nondisposable types after each use for proper protection.

Earmuffs need to make a perfect seal around the ear to be effective. Glasses, long sideburns, long hair, and facial movements, such as chewing, can reduce protection. Special equipment is available for use with glasses or beards.

More specific information on hearing conservation is available.[3]

6. Foot and Leg Protection

According to accident reviews most workers who suffered impact injuries to the feet were not wearing protective footwear. Furthermore, most of their employers did not require them to wear safety shoes. The typical foot injury was caused by objects falling less than 1.2 m and the median weight was about 30 kg. Most workers were injured while performing their normal job activities at their worksites.

Safety shoes should be sturdy and have an impact-resistant toe. In some shoes, metal insoles protect against puncture wounds. Additional protection, such as metatarsal guards, may be found in some types of footwear. Safety shoes come in a variety of styles and materials, such as leather and rubber boots and oxfords.

Safety footwear is classified according to its ability to meet minimum requirements for both compression and impact test. Those requirements and testing procedures and further information may be found elsewhere.[3,6]

7. References

1. AMERICAN CONFERENCE OF GOVERNMENTAL INDUSTRIAL HYGIENISTS. 1987. Guidelines for the Selection of Chemical Protective Clothing, 3rd ed. American Conf. Governmental Industrial Hygienists, Inc., Cincinnati, Ohio.

2. FORSBERG, K. & S.Z. MANSDORF. 1993. Quick Selection Guide To Chemical Protective Clothing, 2nd ed. Van Nostrand Reinhold, New York, N.Y.
3. OCCUPATIONAL SAFETY AND HEALTH ADMINISTRATION. General Requirements for Personal Protective Equipment. 29 CFR 1910.132.
4. AMERICAN NATIONAL STANDARDS INSTITUTE. 1968. Design, Construction, Testing, and Use of Eye and Face Protection. ANSI Z87 1-1968, American National Standards Inst., Inc., New York, N.Y.

5. AMERICAN NATIONAL STANDARDS INSTITUTE. 1986. Safety Requirements for Industrial Head Protection. ANSI Z89.1-1986. American National Standards Inst., Inc., New York, N.Y.
6. AMERICAN NATIONAL STANDARDS INSTITUTE. 1967 & 1983. Men's Safety-Toe Footware. ANSI Z41 1-1967 & Z41-1983, American National Standards Inst., Inc., New York, N.Y.

1090 F.   Worker Protection Medical Program

1. Preventive Medicine Program

The preventive medicine program should include inoculations to provide protection from tetanus and other diseases that are associated with the types of samples received and analyzed by the laboratory. The scope of this program depends on the diseases prevalent in the area where the samples originate. The program also must comply with the appropriate regulations.

2. Medical Surveillance

Routine surveillance may be indicated for anyone whose work involves routine handling of hazardous chemical or biological substances. Consult a qualified occupational health physician and/or toxicologist to determine whether a regular schedule of medical surveillance is indicated.

3. Environmental Monitoring

a. General: The initiation of environmental monitoring (exposure monitoring) associated with laboratory uses of hazardous chemical substances is triggered by exposures exceeding the action level (usually defined as one-half the PEL or TLV), PEL, or TLV. The employer is responsible for ensuring that employees' exposures to such substances do not exceed the PELs specified in the regulations dealing with air contaminants.[1]

b. Employee exposure determination: Determine a worker's exposure to any hazardous chemical substance if there is reason to believe that exposure levels for that substance routinely exceed the action level. Where there is no action level for a substance the worker exposure must not exceed the PEL or TLV. If the initial monitoring confirms that an employee exposure exceeds the action level, or in the absence of an action level, the PEL, the employer must immediately comply with the exposure monitoring provisions of the relevant national standard. Monitoring may be terminated in accordance with the relevant standard (if one exists) or when the exposures are found to be below the action level (one half the PEL or TLV) or in the absence of an action level, below the PEL or TLV. The workers are to be notified in accordance with national standard; if none exists, they should at least be notified within 15 working days after any monitoring results have become available to the employer, either by contacting the employee individually or by posting the results in an appropriate location accessible to employees.

4. Medical Consultation and Medical Examinations

All employees who work with hazardous chemicals should have an opportunity to receive medical attention (at no personal cost), including any follow up examinations that the examining physician determines to be necessary, under the following circumstances:

- Whenever an employee develops signs or symptoms associated with an exposure to a hazardous chemical that the employee may have been using.
- Where exposure monitoring reveals an exposure level routinely above the action level (or in the absence of an action level, the PEL or TLV). For a national regulated substance for which there are exposure monitoring and medical surveillance requirements, establish medical surveillance for the affected employee as prescribed by the particular standard.
- Whenever an uncontrolled event, such as a spill, leak, explosion, or other occurrence, takes place in the work area, resulting in the likelihood of a hazardous exposure. Provide the affected employee an opportunity for a medical consultation to determine the need for a medical examination.

All medical examinations and consultations should be performed by, or under the direct supervision of, a licensed physician, without cost to the employee or loss of pay, and at a reasonable time and place. Inform the physician of the identity of the hazardous chemical(s) to which the employee may have been exposed; the conditions under which the exposure occurred, including quantitative exposure data, if available; and the signs and symptoms of exposure that the employee is experiencing, if any. The employer must obtain from the examining physician a written opinion that includes any recommendation for further medical follow-up, the results of the medical examination and any associated tests, notice of any medical condition revealed during the examination that may place the employee at increased risk as a result of exposure to a hazardous chemical found in the workplace, and a statement that the employee has been informed by the physician of the results of the consultation or medical examination and any medical condition that may require further examination or treatment. The written opinion must not reveal specific findings or diagnoses unrelated to occupational exposure.

5. Reference

1. OCCUPATIONAL SAFETY AND HEALTH ADMINISTRATION. Air Contaminants. 29 CFR 1910.1000.

6. Bibliography

DOULL, J., C.D. KLAASSEN & M.O. AMDUR. 1980. Casarett and Doull's Toxicology: The Basic Science of Poisons, 2nd ed. Macmillan Publishing Co., Inc., New York, N.Y.

WILLIAMS, P.L. & J.L. BURSON. 1985. Industrial Toxicology: Safety and Health Applications in the Workplace. Van Nostrand Reinhold Co., New York, N.Y.

OCCUPATIONAL SAFETY AND HEALTH ADMINISTRATION. Access to Employee Exposure and Medical Records. 29 CFR 1910.20.

OCCUPATIONAL SAFETY AND HEALTH ADMINISTRATION. Occupational Exposure to Hazardous Chemicals in Laboratories. 29 CFR 1910.1450.

1090 G.   Provisions for Work with Particularly Hazardous Substances

The information outlined in the following paragraphs meets the LH&S standard of practice[1] and also represents good industrial hygiene practices.

1. Designated Area

Wherever appropriate, the employer must establish a "designated area," that is, an area that may be used for work with select carcinogens, reproductive toxins, or substances having a high degree of acute toxicity. A designated area may be the entire laboratory, an area of a laboratory, or a device such as a laboratory hood.

2. Select Carcinogen

In the U.S. a "select carcinogen" means any substance meeting at least one of the following criteria: the substance is regulated by OSHA as a carcinogen; it is listed under the category, "known to be carcinogenic," by the U.S. National Toxicology Program (NTP);[2] it is listed under Group 1 (carcinogenic to humans) by the International Agency for Research on Cancer (IARC);[3] or it is listed in either Group 2A or 2B by IARC[3] or under the category, "reasonably anticipated to be carcinogenic" by NTP,[2] and causes statistically significant tumor incidence in experimental animals after inhalation exposure of 6 to 7 h/d, 5 d/week, for significant portion of a lifetime to dosages of less than 10 mg/m^3, or after repeated skin application of less than 300 mg/kg of body weight/week, or after oral dosages of less than 50 mg/kg of body weight/d.

3. Use of Containment Devices

The work conducted and its scale must be appropriate to the physical facilities available and, especially, to the quality of ventilation.

The general laboratory ventilation system must be capable of providing air for breathing and for input to local ventilation devices. It should not be relied on for protection from toxic substances released into the laboratory, but should ensure that laboratory air is continually replaced, preventing increase of air concentrations of toxic substances during the working day, and that air flows into the laboratory from nonlaboratory areas and out to the exterior of the building.

4. References

1. OCCUPATIONAL SAFETY AND HEALTH ADMINISTRATION. Laboratory Standard. Occupational Exposure to Hazardous Chemicals in Laboratories. 29 CFR 1910.1450.
2. U.S. PUBLIC HEALTH SERVICE, NATIONAL TOXICOLOGY PROGRAM. 1980. Annual Report on Carcinogens. Dep. Health & Human Services, U.S. Government Printing Off., Washington, D.C.
3. INTERNATIONAL AGENCY FOR RESEARCH ON CANCER. (Various dates). IARC monographs on risk of chemicals to humans. World Health Org. Publications Center, Albany, N.Y.

5. Bibliography

NATIONAL INSTITUTES OF HEALTH. 1981. Guidelines for the Laboratory Use of Chemical Carcinogens. NIH Publ. No. 81-2385, U.S. Government Printing Off., Washington, D.C.
Also see 1090C.3.

1090 H.   Biological Safety

1. Introduction

The analysis of environmental samples involves worker contact with samples that may be contaminated with agents that present microbiological hazards. The majority of these agents involve exposures to pathogenic microorganisms or viruses that may produce human disease by accidental ingestion, inoculation, injection, or other means of cutaneous penetration. The primary means of exposure to these microbiological hazards involves hand-mouth contact while handling the samples, contaminated laboratory materials and/or aerosols created by incubating, pipetting, centrifuging, or blending of samples or cultures. Use the following rules, work practices and/or procedures to control or minimize exposure to these agents.

2. General Rules

Do not mix dilutions by blowing air through a pipet into a microbiological culture.

When working with grossly polluted samples, such as wastewater or high-density microbial cultures, use a pipetting device attached to a pipetting bulb to prevent accidental ingestion (never pipet by mouth).

Because untreated waters may contain waterborne pathogens, place all used pipets in a jar containing disinfectant solution for decontamination before glassware washing. Do not place used pipets on table tops, on laboratory carts, or in sinks without adequate decontamination.

3. Work Practices

Good personal hygiene practices are essential to control contact exposures. Frequently disinfect hands and working surfaces. Encourage immunization of laboratory staff against tetanus and possibly typhoid and other infectious agents to minimize risk of exposure.

Provide drinking water outside the laboratory, preferably from a foot-operated drinking fountain.

Eliminate flies and other insects to prevent contamination of sterile equipment, media, samples, and bacterial cultures and to prevent spread of infectious organisms to personnel.

Observe appropriate precautions in use of laboratory equipment. Use a leakproof blender tightly covered during operation to minimize contamination. Use a centrifuge tightly covered to minimize exposure if culture-containing tubes should shatter during centrifuging. The tube breakage produces a cleanup problem and microbiological aerosols. Conduct activities such as inserting a hot loop into a flask of broth culture in a manner that eliminates or minimizes the hazards due to aerosolized microorganisms. Sterilize contaminated materials (cultures, samples, used glassware, serological discards, etc.) by autoclaving before discarding them or processing for reuse. Preferably use specially marked biohazard bags for disposal. Dispose of contaminated broken glass in a specially marked container.

4. Procedures

Quaternary ammonium compounds that include a compatible detergent, or solutions of sodium hypochlorite are satisfactory disinfectants for pipet discard jars. Use the highest concentrations recommended for these commercial products provided that this concentration does not cause a loss of markings or fogging of pipets.

Sterilize biological waste materials to eliminate all infectious substances, and sterilize all contaminated equipment or apparatus before washing, storage, or disposal, preferably by autoclaving. When decontaminating materials in the autoclave, heat them to at least 121°C under a pressure of 103 kPa for a minimum of 30 min. The contact time is measured from time the contact chamber reaches 121°C. If the waste is contained in bags, add water to the contents to insure wet heat. Dry heat and chemical treatment also may be used for sterilizing nonplastic items. After sterilization, the wastes can be handled safely and disposed of by conventional disposal systems in accordance with local regulations.

5. Waste Disposal

Sterilize contaminated materials by autoclaving (see ¶ 3 above) before discarding them.

If combustible materials cannot be decontaminated, burn them with special precautions; permits for burning may be required. Use temporary storage for decay or permanent storage for treating radioactive wastes when alternatives are not available. Collect contaminated combustible wastes and animal carcasses in impermeable containers for disposal by incineration.

6. Bibliography

NATIONAL INSTITUTES OF HEALTH. 1981. Guidelines for the Laboratory Use of Chemical Carcinogens. NIH Publ. No. 81-2385, U.S. Government Printing Off., Washington, D.C.

INSTITUTE OF ENVIRONMENTAL SCIENCES. 1986. Recommended Practice for Laminar Flow Clean Devices. RP-CC-002-86. Inst. Environmental Sciences, Mount Prospect, Ill.

NATIONAL SANITATION FOUNDATION. 1987. Class II Biohazard Cabinetry (Laminar Flow). Standard 49-1987, National Sanitation Found., Ann Arbor, Mich.

AMERICAN SOCIETY FOR TESTING AND MATERIALS. 1988. Standard Guide for Good Laboratory Practices in Laboratories Engaged in Sampling and Analysis of Water. ASTM D3856, American Soc. Testing & Materials, Philadelphia, Pa.

FURR, A.K., ed. 1990. CRC Handbook of Laboratory Safety, 3rd ed. CRC Press, Inc., Boca Raton, Fla.

1090 I.   Radiological Safety

1. Introduction

This section discusses ionizing radiation safety related to gas chromatography detectors and specific analytical procedures (see Table 1090:VI). Ionizing radiation includes alpha particles, beta particles, gamma rays, and X-rays. Non-ionizing radiation safety is discussed elsewhere (1090B.8).

All persons are exposed to ionizing radiation. The average annual radiation dose to the whole body from cosmic, terrestrial, and internal sources, medical and dental X-rays, etc., is about 185 mrems/year (1.85 mSieverts/year). It is essential to prevent unnecessary continuous or intermittent occupational exposures, and to take steps to eliminate accidents that may result in dangerous radiation exposure.

Personnel who work in laboratories may be exposed to ionizing radiation sources using both procedures and instruments. Evaluate potential exposures and control the associated equipment and procedures using work practices developed to minimize and/or eliminate exposures.

Users of radioactive materials are responsible for compliance

TABLE 1090:VI. PROCEDURES INVOLVING POTENTIAL EXPOSURE TO IONIZING RADIATION

Standard Methods Section	Radionuclide	Type of Radiation	Comment
Part 6000 methods: GC with electron capture detectors	^{63}Ni or ^{3}H	Low-energy beta (^{63}Ni, 17 keV avg ^{3}H, 6 keV avg)	Internal hazard only. Requires survey techniques for low-energy beta radiation.
7110 Gross Alpha and Gross Beta Radioactivity	Alpha: Uranium, ^{230}Th, ^{239}Pu, ^{241}Am Beta: ^{137}Cs, ^{90}Sr	Alpha; beta and gamma; or beta only	Alpha radiation sources primarily an internal hazard. Beta and beta/gamma sources potential external hazards.
7500-Cs Radioactive Cesium	^{134}Cs, ^{137}Cs	Beta, gamma	
7500-I Radioactive Iodine	^{131}I	Beta, gamma	8-d half-life
7500-Ra Radium	^{226}Ra	Alpha, beta, gamma	^{226}Ra and ^{222}Rn daughters must be considered.
7500-Sr Total Radioactive Strontium and Strontium-90	^{89}Sr, ^{90}Sr	Beta-emitters	
7500-^{3}H Tritium	^{3}H	Low-energy beta (6 keV avg)	Internal hazard only
7500-U Uranium	U series	Alpha, beta, gamma	

with the requirements of their national nuclear regulatory body (in the United States, the Nuclear Regulatory Commission)[1] and/or related state regulations. In addition, administrative or local requirements may apply at specific facilities. The use of "exempt" quantities is regulated[1] even though the facility may be exempt from specific licensing requirements.

Radionuclides are used in laboratories to develop and evaluate analytical methods, to prepare counting standards, and to calibrate detectors and counting instruments (see Part 7000). Sealed sources, such as the nickel-63 detector cell used in electron capture gas chromatograph units, also are common.

2. Exposures

a. Exposure limits and control: The LH&S standard of practice[2] does not permit the use of personal protective equipment for allowing employee exposures above the limits specified by the NRC.[1] The NRC exposure limits are the maximum permissible exposures for 40 h in any workweek of 7 consecutive days.

The exposure limits may be adjusted proportionately (upward) for a period where the exposure is less than 40 h. However, the limit must be adjusted proportionately (downward) for periods where the exposure period is greater than 40 h.

Limiting exposure to ionizing radiation includes providing engineering (physical safeguards) and administrative (procedural) controls for using radioactive materials. Engineering controls include shields, barriers, and interlocks to limit external exposure, and exhaust ventilation systems and personal protective equipment to limit internal contamination. Administrative controls include conducting periodic surveys and reviews of activities, training in the use of radioactive materials, and documented procedures (see below).

Hazards associated with the use of devices, such as X-ray diffraction apparatus or an electron microscope, can be minimized or eliminated by following the manufacturer's operating instructions and the laboratory safety procedures.

b. Monitoring procedures and equipment: Radiochemical exposure monitoring may be done by collecting and analyzing wipe samples, using portable survey instruments, and/or by collecting and analyzing air samples. More than one technique usually is required.

Survey equipment may either integrate the response over time (e.g., exposure, absorbed dose), or results may be presented as a response rate (e.g., count rate or exposure rate). Typical choices include ion chambers, G-M counters, and scintillation detectors.

Thin-windowed GM-counters are suitable for wipe samples and for monitoring skin and clothing. An alpha scintillation monitor is needed to detect alpha-emitters. An excellent discussion of monitoring techniques for radioisotopes is available.[3]

c. Facility surveys: Conduct periodic surveys to assess the effectiveness of physical and procedural controls. Survey procedures generally use wipe tests for removable contamination and/or portable measurement devices for locating or measuring fixed and removable radioactivity.

1) Sealed sources—Check these sources for integrity by wipe tests at least every 6 months. Electron capture detectors using ^{63}Ni or ^{3}H require counting wipes by liquid scintillation or windowless gas-flow proportional counters to measure low-energy beta radiation effectively.

2) Work and storage areas—Survey these areas periodically to assess possible contamination or external radiation fields using portable survey instruments. The frequency of the surveys is dictated by the documented contamination record for the laboratory. Usually the radioactivities presented using the methods of Part 7000 are not measurable with routine survey instruments. Therefore use blanks in the analytical process in determining the presence of low-level contamination.

3) Documentation and records—Completely document each survey, identifying the personnel involved, the location, the type, model, and serial numbers of survey instruments used, the type and energy of radiations measured, the date and time of the survey, the instrument response to a check source, the instrument background count or exposure rate, and the results of each measurement.[4]

d. Personnel surveys and monitoring: Conduct and document surveys after routine use of unsealed radionuclide sources to confirm that personnel and the work area have not been contaminated by the process. Wear monitoring devices if there is a reasonable probability of exceeding 25% of the occupational exposure dose equivalent limit. Personal monitoring devices include film badges, thermoluminescent dosimeters, and solid state electronic dosimeters. The length of time the personal monitoring badges are worn before evaluation depends on the ability of the device to integrate the exposure over long periods, the probability and magnitude of the exposure, and the need to assure that the device is available and used.

Personnel performing procedures in *Standard Methods* would not be expected to receive exposures approaching 1.25 rem/quarter and may wish to consider 3-month wear periods if dosimetry is required.

Personal (external radiation) exposure is evaluated by using a personal dosimeter, preferably the film dosimeter (badge). The dosimeter badge measures the accumulated radiation over a period of time. Pocket ionization chambers, thermoluminescent dosimeters, and thimble chambers also may be used to supplement the film dosimeter.

Whole body or gamma spectrometry radiation detectors may be used to determine the presence of radioactive substances in the body, but these instruments are expensive and require the operator to be specially trained. Evaluate equipment and supplies that have been, or are suspected to have been, in contact with radioactive substances to determine if contamination is present. Because body waste may contain radioactive contaminants evaluate it also for the presence of contamination where personal exposures have been confirmed.

3. Work Practices

Each individual should be familiar with procedures for dealing with radiation emergencies from small spills to major accidents, depending on facility programs. Emergency procedures should include notifications required, containment methods, clean-up procedures, and survey techniques. Emergency supplies should be readily available for coping with major accidents.

Contamination is typically prevented through proper use of laboratory facilities and procedures. Procedures include the use of gloves, aprons, safety glasses, and other protective clothing to eliminate the possibility of skin contamination and transfer. Learn proper pipetting and weighing techniques before working with radioactive sources. Conduct work with unsealed radioactive sources in unobstructed work areas with adequate means of containing and absorbing potential spillage of liquids.

4. Procedures

Develop and implement a radiation safety plan and provide a copy to all persons working with radioactive materials or radiation-producing machines, and provide both lecture and practical training to all employees.

a. Safety plan elements: The recommended minimum plan should include procedures for obtaining authorization to use, order, handle, and store radionuclides; safe handling of unsealed radioactive material; safe response to radiation accidents; decontamination of personnel and facilities; personnel monitoring; laboratory monitoring; and disposal of radioactive materials.

b. Handling radioactive materials: Become knowledgeable about the hazards associated with the materials to be used. Plan work

activities to minimize the time spent handling radioactive materials or in using radioactive sources. Work as far from radioactive sources as possible, use shielding appropriate for the materials to be used, and use radioactive materials only in defined work areas. Wear protective clothing and dosimeters as appropriate. Monitor work areas to ensure maximum contamination control. Minimize the accumulation of waste materials in the work area. Use appropriate personal hygiene and self-monitor after using radioactive materials and after each decontamination procedure.

c. Training of users: Train personnel working with radioactive materials in radiation safety as part of the overall occupational health program. Address at least the following topics: characteristics of ionizing radiation and radioactive contamination; radiation dose limits; environmental radiation background; acute and chronic effects; internal and external modes of exposure; basic protective measures; responsibilities of employer and employees; radiation protection program responsibilities; posting, warning signs, and alarms; radiation monitoring programs; and emergency procedures.[5]

5. Waste Disposal

Generalized disposal criteria for radioactive wastes have been developed by the U.S. National Committee on Radiation Protection and Measurements.[4] Two general philosophies govern the disposal of radioactive wastes: dilution and dispersion to reduce the concentration of radionuclide by carrier dilution or dilution in a receiving medium, and concentration and confinement, usually involving reduction in waste volume with subsequent storage for decay purposes.

Airborne wastes can be treated by either method. Ventilation includes discharge from hooded operations to the atmosphere. Typical radioactive gases include iodine, krypton, and xenon. Iodine can be removed by scrubbing or by reaction with silver nitrate. Noble gases can be removed by absorption; standard techniques can be used for particulate. Dilution methods are suitable for liquids with low activity. Intermediate levels may be treated by various physical-chemical processes to separate the waste into a nonradioactive portion that can be disposed of by dilution and a high-activity portion to be stored. Solid wastes may consist of equipment, glassware, and other materials. When possible, decontaminate these materials and reuse. Decontamination usually results in a liquid waste.

Dispose of all waste in conformance with the requirements of the regulatory authority having jurisdiction.

Determine the laboratory's status and obtain approval before storing, treating and/or disposing of wastes.

6. References

1. NUCLEAR REGULATORY COMMISSION. Standards for Protection Against Radiation. 10 CFR Part 20.
2. OCCUPATIONAL SAFETY AND HEALTH ADMINISTRATION. Ionizing Radiation. 29 CFR 1910.96.
3. FURR, A.K., ed. 1990. CRC Handbook of Laboratory Safety, 3rd ed. CRC Press, Inc., Boca Raton, Fla.
4. NATIONAL COUNCIL ON RADIATION PROTECTION AND MEASUREMENTS. 1976. Environmental Radiation Measurements. Rep. No. 50, National Counc. Radiation Protection & Measurements, Washington, D.C.

7. Bibliography

NATIONAL COUNCIL ON RADIATION PROTECTION AND MEASUREMENTS. 1978. Instrumentation and Monitoring Methods for Radiation Protection. Rep. No. 57, National Counc. Radiation Protection & Measurements, Washington, D.C.

NATIONAL COUNCIL ON RADIATION PROTECTION AND MEASUREMENTS. 1978. A Handbook of Radioactivity Measurements Procedures. Rep. No. 58, National Counc. Radiation Protection & Measurements, Washington, D.C.

NATIONAL COUNCIL ON RADIATION PROTECTION AND MEASUREMENTS. 1978. Operational Radiation Safety Program. Rep. No. 59, National Counc. Radiation Protection & Measurements, Washington, D.C.

1090 J.   Chemical Hygiene Plan

1. Introduction

The information presented in this section describes the intent and lists the minimum requirements and critical elements of the OSHA laboratory standard.[1] This standard is performance-based and represents good industrial hygiene practice. Any organization that chooses not to follow or use the OSHA standards to meet that standard's requirements must demonstrate that the procedures it uses meet the minimum level of employee protection afforded by the OSHA standard.

Much of the information presented in the preceding subsections of 1090 was provided as guidance and should be used in developing and finalizing the laboratory chemical hygiene plan (CHP).

2. Requirements

Develop and implement a written CHP capable of protecting employees from health hazards associated with the chemicals used in the laboratory. This CHP must be capable of keeping exposures below the permissible exposure limits (PELs/TLVs) and also must be readily available to employees. The CHP must at minimum address the following elements, and must describe specific measures the employer will undertake to ensure laboratory employee protection.

a. *Standard operating procedures or work practices:* Include procedures and practices relevant to safety and health considerations. These are to be followed when laboratory work involves the use of hazardous chemicals. Include the information contained in MSDSs for hazardous chemicals when conducting a hazard assessment and developing work practices. Some of the guidance presented in an MSDS is intended for use in industrial settings where material is used in large quantities for a full work shift and not for the small-volume, short-duration exposures experienced in laboratories.

b. *Exposure hazard criteria and procedures:* These will be used to determine and implement control measures for reducing employee exposure to hazardous chemicals while conducting laboratory operations. They include engineering controls, the use of personal protective equipment, and hygiene practices. Pay particular attention to the selection of control measures for work activities that involve chemicals known to be extremely hazardous.

c. *Protective equipment performance procedures:* These include procedures for evaluating the performance of fume hoods and other protective equipment and specific measures to be taken to ensure proper and adequate performance of such equipment.

d. *Employee information and training:* This training must be timely, be refreshed periodically, evaluated for effectiveness, and documented.

e. *Approval procedures:* The plan must state the circumstances under which a specific laboratory operation, procedure, or activity requires prior approval before implementation.

f. *Employee medical consultation and medical examinations.*

g. *Safety personnel:* The plan must designate personnel responsible for implementation of the CHP. This should include the formal assignment of a Chemical Hygiene Officer and, where appropriate, the establishment of a Chemical Hygiene Committee.

h. *Additional employee protection:* The plan should include provisions for working with particularly hazardous substances. These substances include "select carcinogens" (see 1090G.2), reproductive toxins, and substances that have a high degree of acute toxicity. Give specific consideration to establishment of a designated area, use of containment devices such as fume hoods or glove boxes, procedures for safe removal of containment waste, decontamination procedures, emergency plans and procedure (test annually as a minimum), and employer review and evaluation of the effectiveness of the CHP at least annually and updating as necessary.

3. Reference

1. OCCUPATIONAL SAFETY AND HEALTH ADMINISTRATION. Occupational Exposure to Hazardous Chemicals in Laboratories. 29 CFR 1910.1450.

1100 WASTE MINIMIZATION AND DISPOSAL

1100 A. Introduction

Waste minimization and disposal are part of integrated hazardous materials management. It is important to become familiar with federal regulations regarding the use and disposal of hazardous materials prior to their purchase, storage, and use for water and wastewater analysis. Proper management of hazardous materials will reduce the amount of hazardous waste and associated disposal costs.

1100 B. Waste Minimization

1. General Considerations

Waste minimization or pollution prevention in the laboratory is the preferred approach in managing laboratory waste. Minimizing waste makes good economic sense: it reduces both costs and liabilities associated with waste disposal. For certain hazardous-waste generators it also is a regulatory requirement.

2. Waste Minimization Methods

Waste minimization methods include source reduction, recycling, and reclamation.[1] Waste treatment, which also may be considered a form of waste minimization, is addressed in 1100C.

Source reduction can be achieved through the purchase and use of smaller quantities of chemicals. While large-volume purchases may seem economical, the costs of disposing of expired-shelf-life materials also must be considered. Date chemical inventory and use oldest stock first, or if possible, use "just in time" material delivery. Commercial laboratories and chemical users in general can return samples or unopened chemicals to sender or supplier for recycling or disposal. Many suppliers will accept unopened containers of chemicals.

Substitute nonhazardous materials for hazardous chemicals where possible. Wherever possible use methods that do not require the use of hazardous chemicals or use micro-scale analytical methods.

Improving laboratory procedures, documentation, and training will increase awareness of waste minimization and proper disposal practices, and may allow different sections within a laboratory to share standards and stock chemicals. Evaluate hazardous materials storage and use areas for potential evaporation, spills, and leaks. Segregate waste streams where possible to keep nonhazardous waste from becoming hazardous waste through contact with hazardous waste. Segregation also facilitates treatment and disposal.

Transfer of unused stock chemicals to other areas of the laboratory where they may be used or to other institutions is a way to minimize waste.[2] Check with laboratory's legal counsel before transferring chemicals.

Recycling/reclamation has limited potential in water and wastewater laboratories. Volumes generated are generally too small for economical reclamation and purity requirements are often too great. However, organic solvents often can be distilled and recovered for reuse and mercury and silver can be recovered.[3]

3. References

1. ASHBROOK, P.C. & P.A. REINHARDT. 1985. Hazardous wastes in academia. *Environ. Sci. Technol.* 19:1150.
2. PINE, S.H. 1984. Chemical management: A method for waste reduction. *J. Chem. Educ.* 61:A45.
3. HENDRICKSON, K.J., M.M. BENJAMIN, J.F. FERGUSON & L. GOEBEL. 1984. Removal of silver and mercury from spent COD test solutions. *J. Water Pollut. Control. Fed.* 56(5):468.

4. Bibliography

AMERICAN CHEMICAL SOCIETY. 1985. Less is Better. Dep. Public Affairs, American Chemical Soc., Washington, D.C.

1100 C. Waste Treatment and Disposal

1. General Considerations

Stringent penalties exist for the improper disposal of hazardous wastes. Potential criminal and civil liability exists for both organizations and individuals. Specific requirements vary by state and local jurisdiction and are subject to change. Federal requirements for hazardous waste generators and transporters and for treatment, storage, and disposal facilities (TSDFs) are found in regulations pursuant to the Resource Conservation and Recovery Act of 1976 (RCRA) as amended by the Hazardous and Solid Waste Amendments of 1984 (HSWA). Many activities, in particular treatment, storage, and disposal of hazardous wastes, require a permit or license.[1,2]

Develop a plan for the safe and legal disposal of chemical and biological substances in conjunction with the laboratory supervisor and safety coordinator. Plan should address the proper transport, storage, treatment, and disposal of hazardous waste. Properly characterize composites and document wastes. Refer to Section 1090 on Safety with regard to protective equipment in the handling of hazardous materials.

2. Waste Treatment and Disposal Methods

Treatment can be used to reduce volume, mobility, and/or toxicity of hazardous waste where expertise and facilities are available. Treatment, even on a small scale, may require a permit. Consult with federal, state, and local regulatory officials.

Waste treatment methods include thermal, chemical, physical, and biological treatment, and combinations of these methods.[1]

a. Thermal treatment: Thermal treatment methods include incineration and sterilization. They involve using high temperatures to change the chemical, physical, or biological character or composition of the waste. Incineration is often used to destroy organic solvents and is preferred for infectious wastes, although sterilization through autoclaving and/or ultraviolet light also may be allowed. Check with local health department officials.

b. Chemical treatment: Methods include chemical reaction (oxidation/reduction, neutralization, ion exchange, chemical fixation, photolysis, coagulation, precipitation) of the waste material. Neutralization of acidic or alkaline wastes is the most common form of chemical treatment. Elementary neutralization of corrosive wastes is exempt from federal RCRA permitting requirements. Before discharge of wastes to a publicly owned treatment works (POTW), ensure that they contain no pollutants (other than corrosivity) exceeding the limits set by the POTW. The oxidation of cyanide to cyanate with a strong chemical oxidant is an example of a toxicity-reducing chemical treatment.

c. Physical treatment: Methods include solidification, compaction, photo-induced reaction, distillation, flocculation, sedimentation, flotation, aeration, filtration, centrifugation, reverse osmosis, ultrafiltration, gravity thickening, and carbon or resin adsorption. Physical treatment generally reduces volume or mobility of waste materials.

d. Biological treatment: Methods include using biosolids to destroy organic compounds, composting organic-rich wastes, and using bioreactors to promote decomposition. Biological treatment usually is economical on a scale larger than is possible in most water and wastewater laboratories.

e. Ultimate disposal: After waste minimization and treatment, remaining waste streams require disposal. Nonhazardous wastes that cannot be treated further can be discharged as wastewater, emitted to the atmosphere, or placed on or in the ground.

With extreme caution, it may be permissible to dispose of limited quantities (at certain concentrations) of laboratory wastes to the sanitary sewer system or to evaporate volatile wastes in chemical ventilation hoods. Obtain written permission of local, state, and federal authorities to dispose of waste in this manner. With increasing regulatory constraints imposed by RCRA, the Clean Air Act, and Clean Water Act, these disposal options are becoming increasingly limited. Wastes disposed of in this manner may contact other substances in the sewer or ventilation systems and produce hazardous reactions.

Most hazardous wastes generated in laboratories must be sent off site for further treatment and disposal. Exercise extreme care in selecting a reputable waste hauler and disposal firm. Many firms will assist laboratories in packaging and manifesting "lab packs," 19- to 208-L (5- to 55-gal) drums containing several smaller containers of wastes.[1] Liability does not disappear when the waste leaves the generator's facility. Ensure that the laboratory receives a copy of the completed manifest and certificate of treatment and/or disposal. If possible, visit the disposal facility in advance to observe how it will manage a waste.

Certain wastes require special handling. As mentioned previously, incinerate infectious waste or sterilize it before disposal. Before reuse, sterilize all nondisposable equipment that has come into contact with infectious waste.

Although most water and wastewater laboratories do not work with radiochemical wastes, some do. Handle radiochemical wastes with extreme care. Generalized disposal criteria for radioactive wastes have been developed by the National Council on Radiation Protection and Measurements.[3] Low-level radioactive waste must be in solid form for final disposal on land. Some firms will process liquid radioactive wastes into solids. Adding absorbent materials to liquid radioactive wastes is not permissible. Certain states allow low-level liquid radioactive waste to be discharged to a permitted publicly owned treatment works (POTW).

Other wastes that require special handling include polychlorinated biphenyls (PCBs), dioxin/furans and their precursors, petroleum products, and asbestos. Consult with federal and state officials before disposing of these wastes.

3. References

1. AMERICAN CHEMICAL SOCIETY. 1983. RCRA and Laboratories. Dep. Public Affairs, American Chemical Soc., Washington, D.C.
2. U.S. ENVIRONMENTAL PROTECTION AGENCY. 1990. Standards for Owners and Operators of Hazardous Waste Treatment, Storage, and Disposal Facilities. 40 CFR Part 264.
3. U.S. ENVIRONMENTAL PROTECTION AGENCY. Standards for Protection Against Radiation. 10 CFR Part 20.

4. Bibliography

NATIONAL ACADEMY OF SCIENCES, NATIONAL ACADEMY OF ENGINEERING & INSTITUTE OF MEDICINE. 1983. Prudent Practices for the Disposal of Chemicals from Laboratories. National Academy Press, Washington, D.C.
KROFTA, M. & L.K. WANG. 1985. Hazardous Waste Management in Institutions and Colleges. PB86-194180/AS, U.S. National Technical Information Serv., Springfield, Va.
SNIDER, E.H. 1992. Waste minimization. *In* L.K. Wang & M.H.S. Wang, eds., Handbook of Industrial Waste Treatment, p.1. Marcel Dekker, Inc., New York, N.Y.
DUFOUR, J.T. 1994. Hazardous Waste Management Guide for Laboratories. Dufour Group, Sacramento, Calif.

2010 INTRODUCTION

This part deals primarily with measurement of the physical properties of a sample, as distinguished from the concentrations of chemical or biological components. Many of the determinations included here, such as color, electrical conductivity, and turbidity, fit this category unequivocally. However, physical properties cannot be divorced entirely from chemical composition, and some of the techniques of this part measure aggregate properties resulting from the presence of a number of constituents. Others, for example, calcium carbonate saturation, are related to, or depend on, chemical tests. Also included here are tests for appearance, odor, and taste, which have been classified traditionally among physical properties, although the point could be argued. Finally, Section 2710, Tests on Sludges, includes certain biochemical tests. However, for convenience they are grouped with the other tests used for sludge.

With these minor exceptions, the contents of this part have been kept reasonably faithful to its name. Most of the methods included are either inherently or at least traditionally physical, as distinguished from the explicitly chemical, radiological, biological, or bacteriological methods of other parts.

2020 QUALITY ASSURANCE/QUALITY CONTROL

Part 2000 contains a variety of analytical methods, many of which are not amenable to standard quality-control techniques. General information on quality control is provided in Part 1000 and specific quality-control techniques are outlined in the individual methods. The following general guidelines may be applied to many of the methods in this part:

Evaluate analyst performance for each method. Determine competence by analyses of samples containing known concentrations.

Calibrate instruments and ensure that instrument measurements do not drift.

Assess the precision of analytical procedures by analyzing at least 10% of samples in duplicate. Analyze a minimum of one duplicate with each set of samples.

Determine bias of an analytical procedure in each sample batch by analysis of blanks, known additions with a frequency of at least 5% of samples, and, if possible, an externally provided standard.

2110 APPEARANCE*

To record the general physical appearance of a sample, use any terms that briefly describe its visible characteristics. These terms may state the presence of color, turbidity, suspended solids, organisms and their immature forms, sediment, floating material, and similar particulate matter detectable by the unaided eye. Use numerical values when they are available, as for color, turbidity, and suspended solids.

---

*Approved by Standard Methods Committee, 1993.

2120 COLOR*

2120 A. Introduction

Color in water may result from the presence of natural metallic ions (iron and manganese), humus and peat materials, plankton, weeds, and industrial wastes. Color is removed to make a water suitable for general and industrial applications. Colored industrial wastewaters may require color removal before discharge into watercourses.

1. Definitions

The term "color" is used here to mean true color, that is, the color of water from which turbidity has been removed. The term "apparent color" includes not only color due to substances in solution, but also that due to suspended matter. Apparent color is determined on the original sample without filtration or centrifugation. In some highly colored industrial wastewaters color is con-

---

*Approved by Standard Methods Committee, 1993.

tributed principally by colloidal or suspended material. In such cases both true color and apparent color should be determined.

2. Pretreatment for Turbidity Removal

To determine color by currently accepted methods, turbidity must be removed before analysis. Methods for removing turbidity without removing color vary. Filtration yields results that are reproducible from day to day and among laboratories. However, some filtration procedures also may remove some true color. Centrifugation avoids interaction of color with filter materials, but results vary with the sample nature and size and speed of the centrifuge. When sample dilution is necessary, whether it precedes or follows turbidity removal, it can alter the measured color.

Acceptable pretreatment procedures are included with each method. State the pretreatment method when reporting results.

3. Selection of Method

The visual comparison method is applicable to nearly all samples of potable water. Pollution by certain industrial wastes may produce unusual colors that cannot be matched. In this case use an instrumental method. A modification of the tristimulus and the spectrophotometric methods allows calculation of a single color value representing uniform chromaticity differences even when the sample exhibits color significantly different from that of platinum cobalt standards. For comparison of color values among laboratories, calibrate the visual method by the instrumental procedures.

4. Bibliography

OPTICAL SOCIETY OF AMERICA. 1943. Committee Report. The concept of color. *J. Opt. Soc. Amer.* 33:544.

JONES, H. et al. 1952. The Science of Color. Thomas Y. Crowell Co., New York, N.Y.

2120 B. Visual Comparison Method

1. General Discussion

a. Principle: Color is determined by visual comparison of the sample with known concentrations of colored solutions. Comparison also may be made with special, properly calibrated glass color disks. The platinum-cobalt method of measuring color is the standard method, the unit of color being that produced by 1 mg platinum/L in the form of the chloroplatinate ion. The ratio of cobalt to platinum may be varied to match the hue in special cases; the proportion given below is usually satisfactory to match the color of natural waters.

b. Application: The platinum-cobalt method is useful for measuring color of potable water and of water in which color is due to naturally occurring materials. It is not applicable to most highly colored industrial wastewaters.

c. Interference: Even a slight turbidity causes the apparent color to be noticeably higher than the true color; therefore remove turbidity before approximating true color by differential reading with different color filters[1] or by differential scattering measurements.[2] Neither technique, however, has reached the status of a standard method. Remove turbidity by centrifugation or by the filtration procedure described under Method C. Centrifuge for 1 h unless it has been demonstrated that centrifugation under other conditions accomplishes satisfactory turbidity removal.

The color value of water is extremely pH-dependent and invariably increases as the pH of the water is raised. When reporting a color value, specify the pH at which color is determined. For research purposes or when color values are to be compared among laboratories, determine the color response of a given water over a wide range of pH values.[3]

d. Field method: Because the platinum-cobalt standard method is not convenient for field use, compare water color with that of glass disks held at the end of metallic tubes containing glass comparator tubes filled with sample and colorless distilled water. Match sample color with the color of the tube of clear water plus the calibrated colored glass when viewed by looking toward a white surface. Calibrate each disk to correspond with the colors on the platinum-cobalt scale. The glass disks give results in substantial agreement with those obtained by the platinum-cobalt method and their use is recognized as a standard field procedure.

e. Nonstandard laboratory methods: Using glass disks or liquids other than water as standards for laboratory work is permissible only if these have been individually calibrated against platinum-cobalt standards. Waters of highly unusual color, such as those that may occur by mixture with certain industrial wastes, may have hues so far removed from those of the platinum-cobalt standards that comparison by the standard method is difficult or impossible. For such waters, use the methods in Sections 2120C and D. However, results so obtained are not directly comparable to those obtained with platinum-cobalt standards.

f. Sampling: Collect representative samples in clean glassware. Make the color determination within a reasonable period because biological or physical changes occurring in storage may affect color. With naturally colored waters these changes invariably lead to poor results.

2. Apparatus

a. Nessler tubes, matched, 50-mL, tall form.

b. pH meter, for determining sample pH (see Section 4500-H$^+$).

3. Preparation of Standards

a. If a reliable supply of potassium chloroplatinate cannot be purchased, use chloroplatinic acid prepared from metallic platinum. Do not use commercial chloroplatinic acid because it is very hygroscopic and may vary in platinum content. Potassium chloroplatinate is not hygroscopic.

b. Dissolve 1.246 g potassium chloroplatinate, K_2PtCl_6 (equivalent to 500 mg metallic Pt) and 1.00 g crystallized cobaltous chloride, $CoCl_2 \cdot 6H_2O$ (equivalent to about 250 mg metallic Co) in distilled water with 100 mL conc HCl and dilute to 1000 mL with distilled water. This stock standard has a color of 500 units.

c. If K$_2$PtCl$_6$ is not available, dissolve 500 mg pure metallic Pt in aqua regia with the aid of heat; remove HNO$_3$ by repeated evaporation with fresh portions of conc HCl. Dissolve this product, together with 1.00 g crystallized CoCl$_2 \cdot 6$H$_2$O, as directed above.

d. Prepare standards having colors of 5, 10, 15, 20, 25, 30, 35, 40, 45, 50, 60, and 70 by diluting 0.5, 1.0, 1.5, 2.0, 2.5, 3.0, 3.5, 4.0, 4.5, 5.0, 6.0, and 7.0 mL stock color standard with distilled water to 50 mL in nessler tubes. Protect these standards against evaporation and contamination when not in use.

4. Procedure

a. Estimation of intact sample: Observe sample color by filling a matched nessler tube to the 50-mL mark with sample and comparing it with standards. Look vertically downward through tubes toward a white or specular surface placed at such an angle that light is reflected upward through the columns of liquid. If turbidity is present and has not been removed, report as "apparent color." If the color exceeds 70 units, dilute sample with distilled water in known proportions until the color is within the range of the standards.

b. Measure pH of each sample.

5. Calculation

a. Calculate color units by the following equation:

$$\text{Color units} = \frac{A \times 50}{B}$$

where:

A = estimated color of a diluted sample and
B = mL sample taken for dilution.

b. Report color results in whole numbers and record as follows:

Color Units	Record to Nearest
1–50	1
51–100	5
101–250	10
251–500	20

c. Report sample pH.

6. References

1. KNIGHT, A.G. 1951. The photometric estimation of color in turbid waters. *J. Inst. Water Eng.* 5:623.
2. JULLANDER, I. & K. BRUNE. 1950. Light absorption measurements on turbid solutions. *Acta Chem. Scand.* 4:870.
3. BLACK, A.P. & R.F. CHRISTMAN. 1963. Characteristics of colored surface waters. *J. Amer. Water Works Assoc.* 55:753.

7. Bibliography

HAZEN, A. 1892. A new color standard for natural waters. *Amer. Chem. J.* 14:300.
HAZEN, A. 1896. The measurement of the colors of natural waters. *J. Amer. Chem. Soc.* 18:264.
Measurement of Color and Turbidity in Water. 1902. U.S. Geol. Surv., Div. Hydrog. Circ. 8, Washington, D.C.
RUDOLFS, W. & W.D. HANLON. 1951. Color in industrial wastes. *Sewage Ind. Wastes* 23:1125.
PALIN, A.T. 1955. Photometric determination of the colour and turbidity of water. *Water Water Eng.* 59:341.
CHRISTMAN, R.F. & M. GHASSEMI. 1966. Chemical nature of organic color in water. *J. Amer. Water Works Assoc.* 58:723.
GHASSEMI, M. & R.F. CHRISTMAN. 1968. Properties of the yellow organic acids of natural waters. *Limnol. Oceanogr.* 13:583.

2120 C. Spectrophotometric Method

1. General Discussion

a. Principle: The color of a filtered sample is expressed in terms that describe the sensation realized when viewing the sample. The hue (red, green, yellow, etc.) is designated by the term "dominant wavelength," the degree of brightness by "luminance," and the saturation (pale, pastel, etc.) by "purity." These values are best determined from the light transmission characteristics of the filtered sample by means of a spectrophotometer.

b. Application: This method is applicable to potable and surface waters and to wastewaters, both domestic and industrial.

c. Interference: Turbidity interferes. Remove by the filtration method described below.

2. Apparatus

a. Spectrophotometer, having 10-mm absorption cells, a narrow (10-nm or less) spectral band, and an effective operating range from 400 to 700 nm.

b. Filtration system, consisting of the following (see Figure 2120:1):

1) *Filtration flasks,* 250-mL, with side tubes.
2) *Walter crucible holder.*
3) *Glass Gooch filtering crucible with fritted disk,* pore size 40 to 60 μm.
4) *Calcined filter aid.**
5) *Vacuum system.*

3. Procedure

a. Preparation of sample: Bring two 50-mL samples to room temperature. Use one sample at the original pH; adjust pH of the other to 7.6 by using sulfuric acid (H$_2$SO$_4$) and sodium hydroxide (NaOH) of such concentrations that the resulting volume change does not exceed 3%. A standard pH is necessary because of the variation of color with pH. Remove excessive quantities of sus-

---

* Celite No. 505, Manville Corp., or equivalent.

pended materials by centrifuging. Treat each sample separately, as follows:

Thoroughly mix 0.1 g filter aid in a 10-mL portion of centrifuged sample and filter to form a precoat in the filter crucible. Direct filtrate to waste flask as indicated in Figure 2120:1. Mix 40 mg filter aid in a 35-mL portion of centrifuged sample. With vacuum still on, filter through the precoat and pass filtrate to waste flask until clear; then direct clear-filtrate flow to clean flask by means of the three-way stopcock and collect 25 mL for the transmittance determination.

b. Determination of light transmission characteristics: Thoroughly clean 1-cm absorption cells with detergent and rinse with distilled water. Rinse twice with filtered sample, clean external surfaces with lens paper, and fill cell with filtered sample.

Determine transmittance values (in percent) at each visible wavelength value presented in Table 2120:I, using the 10 ordinates marked with an asterisk for fairly accurate work and all 30 ordinates for increased accuracy. Set instrument to read 100% transmittance on the distilled water blank and make all determinations with a narrow spectral band.

4. Calculation

a. Tabulate transmittance values corresponding to wavelengths shown in Columns X, Y, and Z in Table 2120:I. Total each transmittance column and multiply totals by the appropriate factors (for 10 or 30 ordinates) shown at the bottom of the table, to obtain

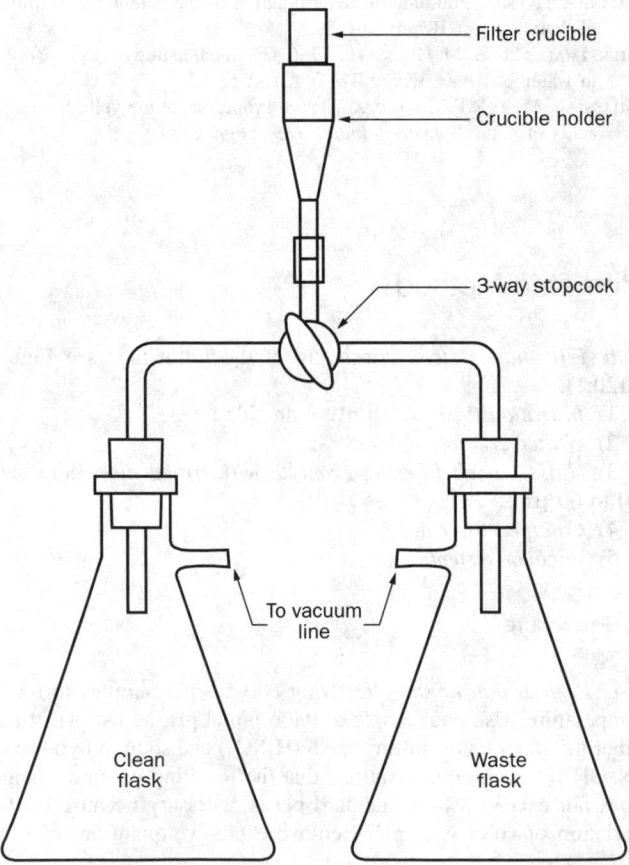

Figure 2120:1. Filtration system for color determinations.

(labels: Filter crucible; Crucible holder; 3-way stopcock; To vacuum line; Clean flask; Waste flask)

TABLE 2120:I. SELECTED ORDINATES FOR SPECTROPHOTOMETRIC COLOR DETERMINATIONS*

Ordinate No.	X	Y	Z
		Wavelength nm	
1	424.4	465.9	414.1
2*	435.5*	489.5*	422.2*
3	443.9	500.4	426.3
4	452.1	508.7	429.4
5*	461.2*	515.2*	432.0*
6	474.0	520.6	434.3
7	531.2	525.4	436.5
8*	544.3*	529.8*	438.6*
9	552.4	533.9	440.6
10	558.7	537.7	442.5
11*	564.1*	541.4*	444.4*
12	568.9	544.9	446.3
13	573.2	548.4	448.2
14*	577.4*	551.8*	450.1*
15	581.3	555.1	452.1
16	585.0	558.5	454.0
17*	588.7*	561.9*	455.9*
18	592.4	565.3	457.9
19	596.0	568.9	459.9
20*	599.6*	572.5*	462.0*
21	603.3	576.4	464.1
22	607.0	580.4	466.3
23*	610.9*	584.8*	468.7*
24	615.0	589.6	471.4
25	619.4	594.8	474.3
26*	624.2*	600.8*	477.7*
27	629.8	607.7	481.8
28	636.6	616.1	487.2
29*	645.9*	627.3*	495.2*
30	663.0	647.4	511.2

Factors When 30 Ordinates Used		
0.032 69	0.033 33	0.039 38

Factors When 10 Ordinates Used		
0.098 06	0.100 00	0.118 14

* Insert in each column the transmittance value (%) corresponding to the wavelength shown. Where limited accuracy is sufficient, use only the ordinates marked with an asterisk.

tristimulus values X, Y, and Z. The tristimulus value Y is *percent luminance.*

b. Calculate the trichromatic coefficients x and y from the tristimulus values X, Y, and Z by the following equations:

$$x = \frac{X}{X + Y + Z}$$

$$y = \frac{Y}{X + Y + Z}$$

Locate point (x, y) on one of the chromaticity diagrams in Figure 2120:2 and determine the dominant wavelength (in nanometers) and the purity (in percent) directly from the diagram.

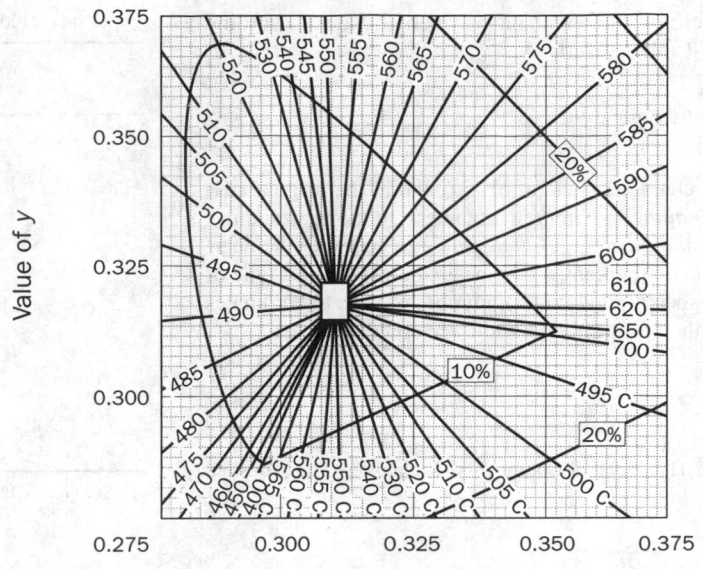

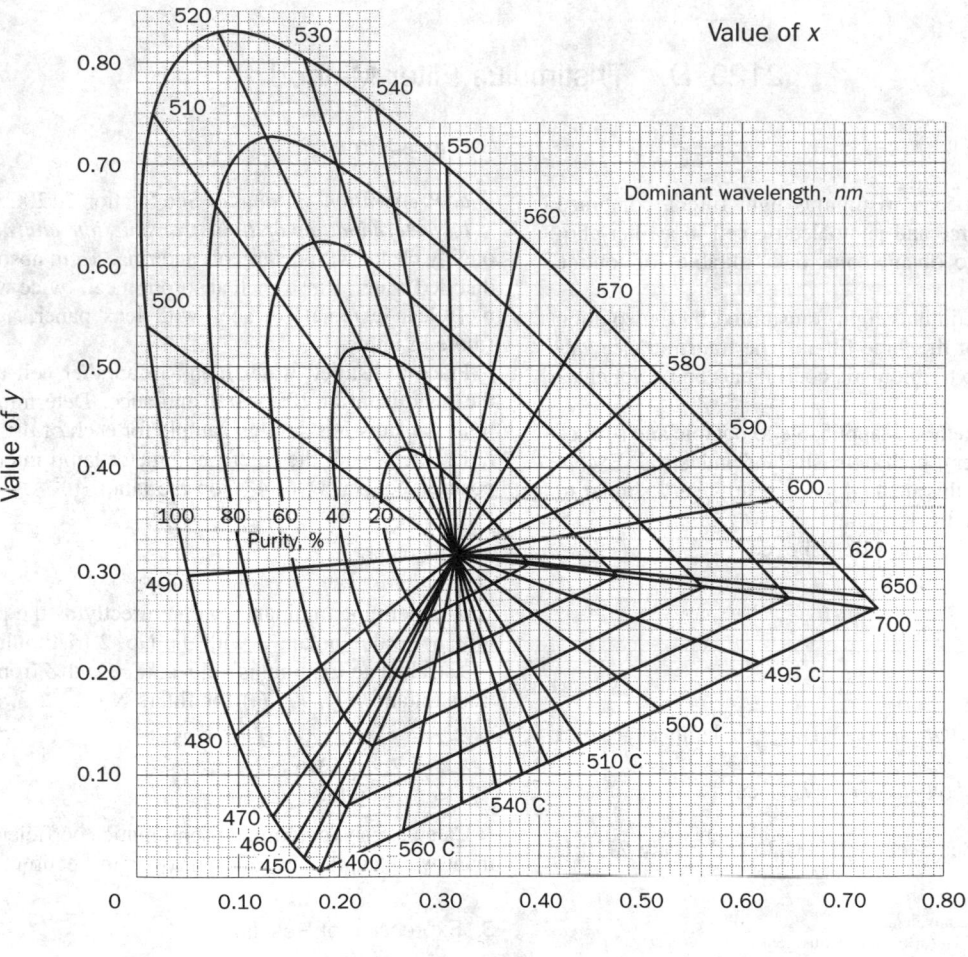

Figure 2120:2. Chromaticity diagrams.

Determine hue from the dominant-wavelength value, according to the ranges in Table 2120:II.

5. Expression of Results

Express color characteristics (at pH 7.6 and at the original pH) in terms of *dominant wavelength* (nanometers, to the nearest unit), *hue* (e.g., blue, blue-green, etc.), *luminance* (percent, to the nearest tenth), and *purity* (percent, to the nearest unit). Report type of instrument (i.e., spectrophotometer), number of selected ordinates (10 or 30), and the spectral band width (nanometers) used.

6. Bibliography

HARDY, A.C. 1936. Handbook of Colorimetry. Technology Press, Boston, Mass.

TABLE 2120:II. COLOR HUES FOR DOMINANT WAVELENGTH RANGES

Wavelength Range nm	Hue
400–465	Violet
465–482	Blue
482–497	Blue-green
497–530	Green
530–575	Greenish yellow
575–580	Yellow
580–587	Yellowish orange
587–598	Orange
598–620	Orange-red
620–700	Red
400–530c*	Blue-purple
530c–700*	Red-purple

* See Figure 2120:2 for significance of "c".

2120 D.   Tristimulus Filter Method

1. General Discussion

a. Principle: Three special tristimulus light filters, combined with a specific light source and photoelectric cell in a filter photometer, may be used to obtain color data suitable for routine control purposes.

The percentage of tristimulus light transmitted by the solution is determined for each of the three filters. The transmittance values then are converted to trichromatic coefficients and color characteristic values.

b. Application: This method is applicable to potable and surface waters and to wastewaters, both domestic and industrial. Except for most exacting work, this method gives results very similar to the more accurate Method C.

c. Interference: Turbidity must be removed.

2. Apparatus

a. Filter photometer.*

b. Filter photometer light source: Tungsten lamp at a color temperature of 3000°C.†

c. Filter photometer photoelectric cells, 1 cm.‡

d. Tristimulus filters.§

e. Filtration system: See Section 2120C.2b and Figure 2120:1.

---

* Fisher Electrophotometer or equivalent.
† General Electric lamp No. 1719 (at 6 V) or equivalent.
‡ General Electric photovoltaic cell, Type PV-1, or equivalent.
§ Corning CS-3-107 (No. 1), CS-4-98 (No. 2), and CS-5-70 (No. 3), or equivalent.

3. Procedure

a. Preparation of sample: See Section 2120C.3a.

b. Determination of light transmission characteristics: Thoroughly clean (with detergent) and rinse 1-cm absorption cells with distilled water. Rinse each absorption cell twice with filtered sample, clean external surfaces with lens paper, and fill cell with filtered sample.

Place a distilled water blank in another cell and use it to set the instrument at 100% transmittance. Determine percentage of light transmission through sample for each of the three tristimulus light filters, with the filter photometer lamp intensity switch in a position equivalent to 4 V on the lamp.

4. Calculation

a. Determine luminance value directly as the percentage transmittance value obtained with the No. 2 tristimulus filter.

b. Calculate tristimulus values *X*, *Y*, and *Z* from the percentage transmittance (T_1, T_2, T_3) for filters No. 1, 2, 3, as follows:

$$X = T_3 \times 0.06 + T_1 \times 0.25$$
$$Y = T_2 \times 0.316$$
$$Z = T_3 \times 0.374$$

Calculate and determine trichromatic coefficients *x* and *y*, dominant wavelength, hue, and purity as in Section 2120C.4b above.

5. Expression of Results

Express results as prescribed in Section 2120C.5.

2120 E.   ADMI Tristimulus Filter Method

1. General Discussion

a. Principle: This method is an extension of Tristimulus Method 2120D. By this method a measure of the sample color, independent of hue, may be obtained. It is based on use of the Adams-Nickerson chromatic value formula[1] for calculating single number color difference values, i.e., uniform color differences. For example, if two colors, A and B, are judged visually to differ from colorless to the same degree, their ADMI color values will be the same. The modification was developed by members of the American Dye Manufacturers Institute (ADMI).[2]

b. Application: This method is applicable to colored waters and wastewaters having color characteristics significantly different from platinum-cobalt standards, as well as to waters and wastewaters similar in hue to the standards.

c. Interference: Turbidity must be removed.

2. Apparatus

a. Filter photometer* equipped with CIE tristimulus filters (see 2120D.2*d*).

b. Filter photometer light source: Tungsten lamp at a color temperature of 3000°C (see 2120D.2*b*).

c. Absorption cells and appropriate cell holders: For color values less than 250 ADMI units, use cells with a 5.0-cm light path; for color values greater than 250, use cells with 1.0-cm light path.

d. Filtration system: See Section 2120C.2*b* and Figure 2120:1; or a centrifuge capable of achieving $1000 \times g$. (See Section 2120B.)

3. Procedure

a. Instrument calibration: Establish curves for each photometer; calibration data for one instrument cannot be applied to another one. Prepare a separate calibration curve for each absorption cell path length.

1) Prepare standards as described in 2120B.3. For a 5-cm cell length prepare standards having color values of 25, 50, 100, 200, and 250 by diluting 5.0, 10.0, 20.0, 40.0, and 50.0 mL stock color standard with distilled water to 100 mL in volumetric flasks. For the shorter pathlength, prepare appropriate standards with higher color values.

2) Determine light transmittance (see ¶ 3*c*, below) for each standard with each filter.

3) Using the calculations described in ¶ 3*d* below, calculate the tristimulus values (X_s, Y_s, Z_s) for each standard, determine the Munsell values, and calculate the intermediate value (DE).

4) Using the DE values for each standard, calculate a calibration factor F_n for each standard from the following equation:

$$F_n = \frac{(APHA)_n \, (b)}{(DE)_n}$$

where:

$(APHA)_n$ = APHA color value for standard n,
$(DE)_n$ = intermediate value calculated for standard n, and
b = cell light path, cm.

Placing $(DE)_n$ on the X axis and F_n on the Y axis, plot a curve for the standard solutions. Use calibration curve to derive the F value from DE values obtained with samples.

b. Sample preparation: Prepare two 100-mL sample portions (one at the original pH, one at pH 7.6) as described in Section 2120C.3*a*, or by centrifugation. (NOTE: Centrifugation is acceptable only if turbidity removal equivalent to filtration is achieved.)

c. Determination of light transmission characteristics: Thoroughly clean absorption cells with detergent and rinse with distilled water. Rinse each absorption cell twice with filtered sample. Clean external surfaces with lens paper and fill cell with sample. Determine sample light transmittance with the three filters to obtain the transmittance values: T_1 from Filter 1, T_2 from Filter 2, and T_3 from Filter 3. Standardize the instrument with each filter at 100% transmittance with distilled water.

d. Calculation of color values: Tristimulus values for samples are X_s, Y_s, and Z_s; for standards X_r, Y_r, and Z_r; and for distilled water X_c, Y_c, and Z_c. Munsell values for samples are V_{xs}, V_{ys}, and V_{zs}; for standards V_{xr}, V_{yr}, and V_{zr}; and for distilled water V_{xc}, Y_{yc}, and V_{zc}.

For each standard or sample calculate the tristimulus values from the following equations:

$$X = (T_3 \times 0.1899) + (T_1 \times 0.791)$$
$$Y = T_2$$
$$Z = T_3 \times 1.1835$$

Tristimulus values for the distilled water blank used to standardize the instrument are always:

$$X_c = 98.09$$
$$Y_c = 100.0$$
$$Z_c = 118.35$$

Convert the six tristimulus values (X_s, Y_s, Z_s, X_c, Y_c, Z_c) to the corresponding Munsell values using published tables 2, 3, 4† or by the equation given by Bridgeman.[3]

Calculate the intermediate value of DE from the equation:

$$DE = \{(0.23 \, \Delta V_y)^2 + [\Delta(V_x - V_y)]^2 + [0.4 \, \Delta(V_y - V_z)]^2\}^{1/2}$$

where:

$$V_y = V_{ys} - V_{yc}$$
$$\Delta(V_x - V_y) = (V_{xs} - V_{ys}) - (V_{xc} - V_{yc})$$
$$\Delta(V_y - V_z) = (V_{ys} - V_{zs}) - (V_{yc} - V_{zc})$$

when the sample is compared to distilled water.

With the standard calibration curve, use the DE value to determine the calibration factor F.

---

* Fisher Electrocolorimeter, Model 181, or equivalent.

† Instrumental Colour Systems, Ltd., 7 Bucklebury Place, Upper Woolhampton, Berkshire RG7 5UD, England.

Calculate the final ADMI color value as follows:

$$\text{ADMI value} = \frac{(F)\,(DE)}{b}$$

where:

b = absorption cell light path, cm.

Report ADMI color values at pH 7.6 and at the original pH.

4. Alternate Method

The ADMI color value also may be determined spectrophotometrically, using a spectrophotometer with a narrow (10-nm or less) spectral band and an effective operating range of 400 to 700 nm. This method is an extension of 2120C. Tristimulus values may be calculated from transmittance measurements, preferably by using the weighted ordinate method or by the selected ordinate method. The method has been described by Allen et al.,[2] who include work sheets and worked examples.

5. References

1. McLaren, K. 1970. The Adams-Nickerson colour-difference formula. *J. Soc. Dyers Colorists* 86:354.
2. Allen, W., W.B. Prescott, R.E. Derby, C.E. Garland, J.M. Peret & M. Saltzman. 1973. Determination of color of water and wastewater by means of ADMI color values. *Proc. 28th Ind. Waste Conf.*, Purdue Univ., Eng. Ext. Ser. No. 142:661.
3. Bridgeman, T. 1963. Inversion of the Munsell value equation. *J. Opt. Soc. Amer.* 53:499.

6. Bibliography

Judd, D.B. & G. Wyszecki. 1963. Color in Business, Science, and Industry, 2nd ed. John Wiley & Sons, New York, N.Y. (See Tables A, B, and C in Appendix.)
Wyszecki, G. & W.S. Stiles. 1967. Color Science. John Wiley & Sons, New York, N.Y. (See Tables 6.4, A, B, C, pp. 462-467.)

2130   TURBIDITY*

2130 A.   Introduction

1. Sources and Significance

Clarity of water is important in producing products destined for human consumption and in many manufacturing operations. Beverage producers, food processors, and potable water treatment plants drawing from a surface water source commonly rely on fluid-particle separation processes such as sedimentation and filtration to increase clarity and insure an acceptable product. The clarity of a natural body of water is an important determinant of its condition and productivity.

Turbidity in water is caused by suspended and colloidal matter such as clay, silt, finely divided organic and inorganic matter, and plankton and other microscopic organisms. Turbidity is an expression of the optical property that causes light to be scattered and absorbed rather than transmitted with no change in direction or flux level through the sample. Correlation of turbidity with the weight or particle number concentration of suspended matter is difficult because the size, shape, and refractive index of the particles affect the light-scattering properties of the suspension. When present in significant concentrations, particles consisting of light-absorbing materials such as activated carbon cause a negative interference. In low concentrations these particles tend to have a positive influence because they contribute to turbidity. The presence of dissolved, color-causing substances that absorb light may cause a negative interference. Some commercial instruments may have the capability of either correcting for a slight color interference or optically blanking out the color effect.

2. Selection of Method

Historically, the standard method for determination of turbidity has been based on the Jackson candle turbidimeter; however, the lowest turbidity value that can be measured directly on this device is 25 Jackson Turbidity Units (JTU). Because turbidities of water treated by conventional fluid-particle separation processes usually fall within the range of 0 to 1 unit, indirect secondary methods were developed to estimate turbidity. Electronic nephelometers are the preferred instruments for turbidity measurement.

Most commercial turbidimeters designed for measuring low turbidities give comparatively good indications of the intensity of light scattered in one particular direction, predominantly at right angles to the incident light. Turbidimeters with scattered-light detectors located at 90° to the incident beam are called nephelometers. Nephelometers are relatively unaffected by small differences in design parameters and therefore are specified as the standard instrument for measurement of low turbidities. Instruments of different make and model may vary in response.† However, interinstrument variation may be effectively negligible if good measurement techniques are used and the characteristics of the particles in the measured suspensions are similar. Poor measurement technique can have a greater effect on measurement error than small differences in instrument design. Turbidimeters of non-

---

* Approved by Standard Methods Committee, 1994.

† Nephelometers that instrument manufacturers claim meet the design specifications of this method may not give the same reading for a given suspension, even when each instrument has been calibrated using the manufacturer's manual. This differential performance is especially important when measurements are made for regulatory purposes. Consult regulatory authorities when selecting a nephelometer to be used for making measurements that will be reported for regulatory purposes.

standard design, such as forward-scattering devices, may be more sensitive than nephelometers to the presence of larger particles. While it may not be appropriate to compare their output with that of instruments of standard design, they still may be useful for process monitoring.

An additional cause of discrepancies in turbidity analysis is the use of suspensions of different types of particulate matter for instrument calibration. Like water samples, prepared suspensions have different optical properties depending on the particle size distributions, shapes, and refractive indices. A standard reference suspension having reproducible light-scattering properties is specified for nephelometer calibration.

Its precision, sensitivity, and applicability over a wide turbidity range make the nephelometric method preferable to visual methods. Report nephelometric measurement results as nephelometric turbidity units (NTU).

3. Storage of Sample

Determine turbidity as soon as possible after the sample is taken. Gently agitate all samples before examination to ensure a representative measurement. Sample preservation is not practical; begin analysis promptly. Refrigerate or cool to 4°C, to minimize microbiological decomposition of solids, if storage is required. For best results, measure turbidity immediately without altering the original sample conditions such as temperature or pH.

2130 B.   Nephelometric Method

1. General Discussion

a. Principle: This method is based on a comparison of the intensity of light scattered by the sample under defined conditions with the intensity of light scattered by a standard reference suspension under the same conditions. The higher the intensity of scattered light, the higher the turbidity. Formazin polymer is used as the primary standard reference suspension. The turbidity of a specified concentration of formazin suspension is defined as 4000 NTU.

b. Interference: Turbidity can be determined for any water sample that is free of debris and rapidly settling coarse sediment. Dirty glassware and the presence of air bubbles give false results. "True color," i.e., water color due to dissolved substances that absorb light, causes measured turbidities to be low. This effect usually is not significant in treated water.

2. Apparatus

a. Laboratory or process nephelometer consisting of a light source for illuminating the sample and one or more photoelectric detectors with a readout device to indicate intensity of light scattered at 90° to the path of incident light. Use an instrument designed to minimize stray light reaching the detector in the absence of turbidity and to be free from significant drift after a short warmup period. The sensitivity of the instrument should permit detecting turbidity differences of 0.02 NTU or less in the lowest range in waters having a turbidity of less than 1 NTU. Several ranges may be necessary to obtain both adequate coverage and sufficient sensitivity for low turbidities. Differences in instrument design will cause differences in measured values for turbidity even though the same suspension is used for calibration. To minimize such differences, observe the following design criteria:

1) Light source—Tungsten-filament lamp operated at a color temperature between 2200 and 3000°K.

2) Distance traversed by incident light and scattered light within the sample tube—Total not to exceed 10 cm.

3) Angle of light acceptance by detector—Centered at 90° to the incident light path and not to exceed ±30° from 90°. The detector and filter system, if used, shall have a spectral peak response between 400 and 600 nm.

b. Sample cells: Use sample cells or tubes of clear, colorless glass or plastic. Keep cells scrupulously clean, both inside and out, and discard if scratched or etched. Never handle them where the instrument's light beam will strike them. Use tubes with sufficient extra length, or with a protective case, so that they may be handled properly. Fill cells with samples and standards that have been agitated thoroughly and allow sufficient time for bubbles to escape.

Clean sample cells by thorough washing with laboratory soap inside and out followed by multiple rinses with distilled or deionized water; let cells air-dry. Handle sample cells only by the top to avoid dirt and fingerprints within the light path.

Cells may be coated on the outside with a thin layer of silicone oil to mask minor imperfections and scratches that may contribute to stray light. Use silicone oil with the same refractive index as glass. Avoid excess oil because it may attract dirt and contaminate the sample compartment of the instrument. Using a soft, lint-free cloth, spread the oil uniformly and wipe off excess. The cell should appear to be nearly dry with little or no visible oil.

Because small differences between sample cells significantly impact measurement, use either matched pairs of cells or the same cell for both standardization and sample measurement.

3. Reagents

a. Dilution water: High-purity water will cause some light scattering, which is detected by nephelometers as turbidity. To obtain low-turbidity water for dilutions, nominal value 0.02 NTU, pass laboratory reagent-grade water through a filter with pore size sufficiently small to remove essentially all particles larger than 0.1 μm;* the usual membrane filter used for bacteriological examinations is not satisfactory. Rinse collecting flask at least twice with filtered water and discard the next 200 mL.

Some commercial bottled demineralized waters have a low turbidity. These may be used when filtration is impractical or a good grade of water is not available to filter in the laboratory. Check turbidity of bottled water to make sure it is lower than the level that can be achieved in the laboratory.

---

* Nuclepore Corp., 7035 Commerce Circle, Pleasanton, CA, or equivalent.

b. Stock primary standard formazin suspension:

1) Solution I—Dissolve 1.000 g hydrazine sulfate, $(NH_2)_2 \cdot H_2SO_4$, in distilled water and dilute to 100 mL in a volumetric flask. CAUTION: *Hydrazine sulfate is a carcinogen; avoid inhalation, ingestion, and skin contact. Formazin suspensions can contain residual hydrazine sulfate.*

2) Solution II—Dissolve 10.00 g hexamethylenetetramine, $(CH_2)_6N_4$, in distilled water and dilute to 100 mL in a volumetric flask.

3) In a flask, mix 5.0 mL Solution I and 5.0 mL Solution II. Let stand for 24 h at 25 ± 3°C. This results in a 4000-NTU suspension. Transfer stock suspension to an amber glass or other UV-light-blocking bottle for storage. Make dilutions from this stock suspension. The stock suspension is stable for up to 1 year when properly stored.

c. Dilute turbidity suspensions: Dilute 4000 NTU primary standard suspension with high-quality dilution water. Prepare immediately before use and discard after use.

d. Secondary standards: Secondary standards are standards that the manufacturer (or an independent testing organization) has certified will give instrument calibration results equivalent (within certain limits) to the results obtained when the instrument is calibrated with the primary standard, i.e., user-prepared formazin. Various secondary standards are available including: commercial stock suspensions of 4000 NTU formazin, commercial suspensions of microspheres of styrene-divinylbenzene copolymer,† and items supplied by instrument manufacturers, such as sealed sample cells filled with latex suspension or with metal oxide particles in a polymer gel. The U.S. Environmental Protection Agency[1] designates user-prepared formazin, commercial stock formazin suspensions, and commercial styrene-divinylbenzene suspensions as "primary standards," and reserves the term "secondary standard" for the sealed standards mentioned above.

Secondary standards made with suspensions of microspheres of styrene-divinylbenzene copolymer typically are as stable as concentrated formazin and are much more stable than diluted formazin. These suspensions can be instrument-specific; therefore, use only suspensions formulated for the type of nephelometer being used. Secondary standards provided by the instrument manufacturer (sometimes called "permanent" standards) may be necessary to standardize some instruments before each reading and in other instruments only as a calibration check to determine when calibration with the primary standard is necessary.

All secondary standards, even so-called "permanent" standards, change with time. Replace them when their age exceeds the shelf life. Deterioration can be detected by measuring the turbidity of the standard after calibrating the instrument with a fresh formazin or microsphere suspension. If there is any doubt about the integrity or turbidity value of any secondary standard, check instrument calibration first with another secondary standard and then, if necessary, with user-prepared formazin. Most secondary standards have been carefully prepared by their manufacturer and should, if properly used, give good agreement with formazin. Prepare formazin primary standard only as a last resort. Proper application of secondary standards is specific for each make and model of nephelometer. Not all secondary standards have to be discarded when comparison with a primary standard shows that their turbidity value has changed. In some cases, the secondary

standard should be simply relabeled with the new turbidity value. Always follow the manufacturer's directions.

4. Procedure

a. General measurement techniques: Proper measurement techniques are important in minimizing the effects of instrument variables as well as stray light and air bubbles. Regardless of the instrument used, the measurement will be more accurate, precise, and repeatable if close attention is paid to proper measurement techniques.

Measure turbidity immediately to prevent temperature changes and particle flocculation and sedimentation from changing sample characteristics. If flocculation is apparent, break up aggregates by agitation. Avoid dilution whenever possible. Particles suspended in the original sample may dissolve or otherwise change characteristics when the temperature changes or when the sample is diluted.

Remove air or other entrained gases in the sample before measurement. Preferably degas even if no bubbles are visible. Degas by applying a partial vacuum, adding a nonfoaming-type surfactant, using an ultrasonic bath, or applying heat. In some cases, two or more of these techniques may be combined for more effective bubble removal. For example, it may be necessary to combine addition of a surfactant with use of an ultrasonic bath for some severe conditions. Any of these techniques, if misapplied, can alter sample turbidity; *use with care.* If degassing cannot be applied, bubble formation will be minimized if the samples are maintained at the temperature and pressure of the water before sampling.

Do not remove air bubbles by letting sample stand for a period of time because during standing, turbidity-causing particulates may settle and sample temperature may change. Both of these conditions alter sample turbidity, resulting in a nonrepresentative measurement.

Condensation may occur on the outside surface of a sample cell when a cold sample is being measured in a warm, humid environment. This interferes with turbidity measurement. Remove all moisture from the outside of the sample cell before placing the cell in the instrument. If fogging recurs, let sample warm slightly by letting it stand at room temperature or by partially immersing it in a warm water bath for a short time. Make sure samples are again well mixed.

b. Nephelometer calibration: Follow the manufacturer's operating instructions. Run at least one standard in each instrument range to be used. Make certain the nephelometer gives stable readings in all sensitivity ranges used. Follow techniques outlined in ¶s 2b and 4a for care and handling of sample cells, degassing, and dealing with condensation.

c. Measurement of turbidity: Gently agitate sample. Wait until air bubbles disappear and pour sample into cell. When possible, pour well-mixed sample into cell and immerse it in an ultrasonic bath for 1 to 2 s or apply vacuum degassing, causing complete bubble release. Read turbidity directly from instrument display.

d. Calibration of continuous turbidity monitors: Calibrate continuous turbidity monitors for low turbidities by determining turbidity of the water flowing out of them, using a laboratory-model nephelometer, or calibrate the instruments according to manufacturer's instructions with formazin primary standard or appropriate secondary standard.

---

† AMCO-AEPA-1 Standard, Advanced Polymer Systems, 3696 Haven Ave., Redwood City, CA, or equivalent.

5. Interpretation of Results

Report turbidity readings as follows:

Turbidity Range NTU	Report to the Nearest NTU
0–1.0	0.05
1–10	0.1
10–40	1
40–100	5
100–400	10
400–1000	50
>1000	100

When comparing water treatment efficiencies, do not estimate turbidity more closely than specified above. Uncertainties and discrepancies in turbidity measurements make it unlikely that results can be duplicated to greater precision than specified.

6. Reference

1. U.S. ENVIRONMENTAL PROTECTION AGENCY. 1993. Methods for Determination of Inorganic Substances in Environmental Samples.

EPA-600/R/93/100 - Draft. Environmental Monitoring Systems Lab., Cincinnati, Ohio.

7. Bibliography

HACH, C.C., R.D. VANOUS & J.M. HEER. 1985. Understanding Turbidity Measurement. Hach Co., Technical Information Ser., Booklet 11, Loveland, Colo.

KATZ, E.L. 1986. The stability of turbidity in raw water and its relationship to chlorine demand. *J. Amer. Water Works Assoc.* 78:72.

McCOY, W.F. & B.H. OLSON. 1986. Relationship among turbidity, particle counts and bacteriological quality within water distribution lines. *Water Res.* 20:1023.

BUCKLIN, K.E., G.A. McFETERS & A. AMIRTHARAJAH. 1991. Penetration of coliform through municipal drinking water filters. *Water Res.* 25:1013.

HERNANDEZ, E., R.A. BAKER & P.C. CRANDALL. 1991. Model for evaluating turbidity in cloudy beverages. *J. Food Sci.* 56:747.

HART, V.S., C.E. JOHNSON & R.D. LETTERMAN. 1992. An analysis of low-level turbidity measurements. *J. Amer. Water Works Assoc.,* 84(12):40.

LeCHEVALLIER, M.W. & W.D. NORTON. 1992. Examining relationship between particle counts and *Giardia, Cryptosporidium,* and turbidity. *J. Amer. Water Works Assoc.* 84(12):54.

2150 ODOR*

2150 A. Introduction

1. Discussion

Odor, like taste, depends on contact of a stimulating substance with the appropriate human receptor cell. The stimuli are chemical in nature and the term "chemical senses" often is applied to odor and taste. Water is a neutral medium, always present on or at the receptors that perceive sensory response. In its pure form, water is odor-free. Man and other animals can avoid many potentially toxic foods and waters because of adverse sensory response. These senses often provide the first warning of potential hazards in the environment.

Odor is recognized[1] as a quality factor affecting acceptability of drinking water (and foods prepared with it), tainting of fish and other aquatic organisms, and esthetics of recreational waters. Most organic and some inorganic chemicals contribute taste or odor. These chemicals may originate from municipal and industrial waste discharges, from natural sources such as decomposition of vegetable matter, or from associated microbial activity, and from disinfectants or their products.

The potential for impairment of the sensory quality of water

has increased as a result of expansion in the variety and quantity of waste materials, demand for water disposal of captured air pollutants, and increased reuse of available water supplies by a growing population. Domestic consumers and process industries such as food, beverage, and pharmaceutical manufacturers require water essentially free of tastes and odors.

Some substances, such as certain inorganic salts, produce taste without odor and are evaluated by taste testing (Section 2160). Many other sensations ascribed to the sense of taste actually are odors, even though the sensation is not noticed until the material is taken into the mouth. Because some odorous materials are detectable when present in only a few nanograms per liter, it is usually impractical and often impossible to isolate and identify the odor-producing chemical. The human nose is the practical odor-testing device used in this method. Odor tests are performed to provide qualitative descriptions and approximate quantitative measurements of odor intensity. The method for intensity measurement presented here is the *threshold odor* test, based on a method of limits.[2] This procedure, while not universally preferred,[3] has definite strengths.[4]

Sensory tests are useful as a check on the quality of raw and finished water and for control of odor through the treatment process.[2,3,5] They can assess the effectiveness of different treatments and provide a means of tracing the source of contamination.

---

* Approved by Standard Methods Committee, 1997.

Section 6040B provides an analytical procedure for quantifying several organic odor-producing compounds including geosmin and methylisoborneol.

2. References

1. U.S. ENVIRONMENTAL PROTECTION AGENCY. 1973. Proposed Criteria for Water Quality. Vol. 1, Washington, D.C.
2. AMERICAN SOCIETY FOR TESTING AND MATERIALS COMMITTEE E-18. 1996. Sensory Testing Methods: Second Edition. E. Chambers & M.B. Wolf, eds. American Soc. Testing & Materials, W. Conshohocken, Pa.
3. MALLEVIALLE, J. & I.H. SUFFET. 1987. Identification and Treatment of Tastes and Odors in Drinking Water. American Water Works Association Research Foundation and Lyonnaise des Eaux. AWWARF, Denver, Colo.
4. BRUVOLD, W.H. 1989. A Critical Review of Methods Used for the Sensory Evaluation of Water Quality. CRC Crit. Rev. Environ. Control. 19(4):291.
5. SUFFET, I., E. KAWCZYNSKI & J. MALLEVIALLE. 1995. Advances in Taste and Odor Treatment and Control. American Water Works Association Research Foundation and Lyonnaise des Eaux, AWWARF, Denver, Colo.

3. Bibliography

MONCRIEFF, R.W. 1946. The Chemical Senses. John Wiley & Sons, New York, N.Y.

BAKER, R.A. 1961. Problems of tastes and odors. J. Water Pollut. Control Fed. 33:1099.

BAKER, R.A. 1963. Odor effects of aqueous mixtures of organic chemicals. J. Water Pollut. Control Fed. 35:728.

ROSEN, A.A., R.T. SKEEL & M.B. ETTINGER. 1963. Relationship of river water odor to specific organic contaminants. J. Water Pollut. Control Fed. 35:777.

WRIGHT, R.H. 1964. The Science of Smell. Basic Books, New York, N.Y.

AMERINE, M.A., R.M. PANGBORN & E.B. ROESSLER. 1965. Principles of Sensory Evaluation of Food. Academic Press, New York, N.Y.

ROSEN, A.A. 1970. Report of research committee on tastes and odors. J. Amer. Water Works Assoc. 62:59.

GELDARD, F.A. 1972. The Human Senses. John Wiley & Sons, New York, N.Y.

MEILGAARD, M.C. 1988. Sensory evaluation techniques applied to water works samples. Water Quality Bull. 13(2/3):39. WHO Collaborating Centre on Surface and Groundwater Quality, Burlington, Ont., Canada

2150 B. Threshold Odor Test

1. General Discussion

a. Principle: Determine the threshold odor by diluting a sample with odor-free water until the least definitely perceptible odor is achieved. There is no absolute threshold odor concentration, because of inherent variation in individual olfactory capability. A given person varies in sensitivity over time. Day-to-day and within-day differences occur. Furthermore, responses vary as a result of the characteristic, as well as concentration, of odorant. The number of persons selected to measure threshold odor will depend on the objective of the tests, economics, and available personnel. Larger-sized panels are needed for sensory testing when the results must represent the population as a whole or when great precision is desired. Under such circumstances, panels of no fewer than five persons, and preferably ten or more, are recommended.[1] Measurement of threshold levels by one person is often a necessity at water treatment plants. Interpretation of the single tester result requires knowledge of the relative acuity of that person. Some investigators have used specific odorants, such as *m*-cresol or *n*-butanol, to calibrate a tester's response.[2] Others have used the University of Pennsylvania Smell Identification Test to assess an individual's ability to identify odors correctly.[3]

b. Application: This threshold method is applicable to samples ranging from nearly odorless natural waters to industrial wastes with threshold numbers in the thousands. There are no intrinsic difficulties with the highly odorous samples because they are reduced in concentration proportionately before being presented to the test observers.

c. Qualitative descriptions: A fully acceptable system for characterizing odor has not been developed despite efforts over more than a century. Nevertheless, Section 2170 (Flavor Profile Analysis) specifies a set of 23 odor reference standards that may be used if qualitative descriptions are important. These descriptors can be used with the Threshold Odor Test to standardize methods for sensory analysis.

d. Sampling and storage: Collect samples for odor testing in glass bottles with glass or TFE-lined closures. Complete tests as soon as possible after sample collection. If storage is necessary, collect at least 500 mL of sample in a bottle filled to the top; refrigerate, making sure that no extraneous odors can be drawn into the sample as it cools. Do not use plastic containers.

e. Dechlorination: Most tap waters and some wastewaters are chlorinated. Often it is desirable to determine the odor of the chlorinated sample as well as that of the same sample after dechlorination. Dechlorinate with thiosulfate in exact stoichiometric quantity as described under Nitrogen (Ammonia), Section 4500-NH$_3$.

f. Temperature: Threshold odor values vary with temperature. For most tap waters and raw water sources, a sample temperature of 60°C will permit detection of odors that otherwise might be missed; 60°C is the standard temperature for hot threshold odor tests. For some purposes—because the odor is too fleeting or there is excessive heat sensation—the hot odor test may not be applicable; where experience shows that a lower temperature is needed, use a standard test temperature of 40°C. For special purposes, other temperatures may be used. *Report temperature at which observations are made.*

2. Apparatus

To assure reliable threshold measurements, use odor-free glassware. Clean glassware shortly before use with nonodorous soap and acid cleaning solution and rinse with odor-free water. Reserve this glassware exclusively for threshold testing. Do not use rubber, cork, or plastic stoppers. Do not use narrow-mouth vessels.

a. Sample bottles, glass-stoppered or with TFE-lined closures.

b. Constant-temperature bath: A water bath or electric hot plate capable of temperature control of ± 1°C for odor tests at elevated temperatures. The bath must not contribute any odor to the odor flasks.

c. Odor flasks: Glass-stoppered, 500-mL (ST 32) erlenmeyer flasks, to hold sample dilutions during testing.

d. Pipets:

1) *Transfer and volumetric pipets or graduated cylinders:* 200-, 100-, 50-, and 25-mL.

2) *Measuring pipets:* 10-mL, graduated in tenths.

e. Thermometer: Zero to 110°C, chemical or metal-stem dial type.

3. Odor-Free Water

a. Sources: Prepare odor-free water by passing distilled, deionized, or tap water through activated carbon. If product water is not odor-free, rebuild or purify the system. In all cases verify quality of product water daily.

b. Odor-free-water generator (Figure 2150:1):* Make the PVC generator from a 2-ft length of 4-in. PVC pipe approved for use for drinking water purposes (e.g., Schedule 80, or National Water Council-approved in U.K.). Thread pipe end to accept threaded caps. Have a small threaded nipple in the cap center for water inlet or outlet. To retain the activated carbon, place coarse glass wool in top and bottom of generator. Regulate water flow to generator by a needle valve and a pressure regulator to provide the minimum pressure for the desired flow. Use activated carbon of approximately 12 to 40 mesh grain size.†

c. Generator operation: Pass tap or purified water through odor-free-water generator at rate of 100 mL/min. When generator is started, flush to remove carbon fines and discard product, or pre-rinse carbon.

Check quality of water obtained from the odor-free-water generator daily at 40 and 60°C before use. The life of the carbon will vary with the condition and amount of water filtered. Subtle odors of biological origin often are found if moist carbon filters stand idle between test periods. Detection of odor in the water coming through the carbon indicates that a change of carbon is needed.

4. Procedure

a. Precautions: Carefully select by preliminary tests the persons to make taste or odor tests. Although extreme sensitivity is not required, exclude insensitive persons and concentrate on observers who have a sincere interest in the test. Avoid extraneous odor stimuli such as those caused by smoking and eating before the test or those contributed by scented soaps, perfumes, and shaving lotions. Insure that the tester is free from colds or allergies that affect odor response. Limit frequency of tests to a number below the fatigue level by frequent rests in an odor-free atmosphere. Keep room in which tests are conducted free from distractions, drafts, and odor.[2] If necessary, set aside a special odor-free room ventilated by air that is filtered through activated carbon and maintained at a constant comfortable temperature and humidity.[4]

---

* For approximate metric dimensions in centimeters multiply dimensions in inches by 2.54.

† Nuchar WV-G, Westvaco, Covington, VA; Filtrasorb 200, Calgon Carbon Corporation, Pittsburgh, PA; or equivalent.

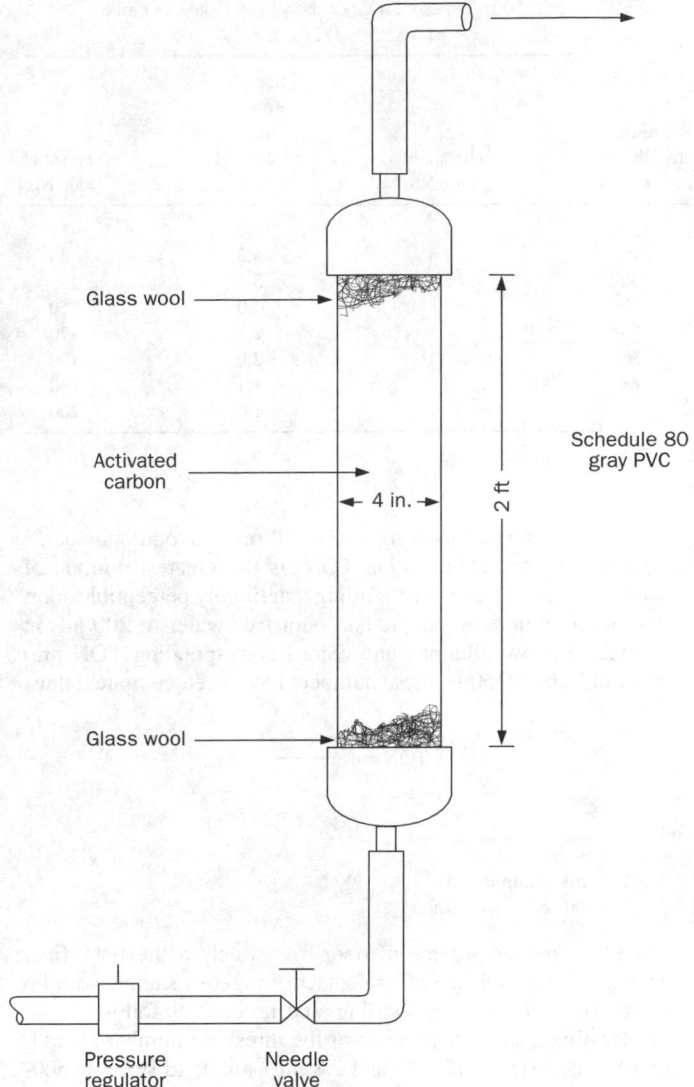

Figure 2150:1. Odor-free-water generator.

For precise work use a panel of five or more testers. Do not allow persons making odor measurements to prepare samples or to know dilution concentrations being evaluated. Familiarize testers with the procedure before they participate in a panel test. Present most dilute sample first to avoid tiring the senses with the concentrated sample. Keep temperature of samples during testing within 1°C of the specified temperature.

Because many raw and waste waters are colored or have decided turbidity that may bias results, use opaque or darkly colored odor flasks, such as red actinic erlenmeyer flasks.

b. Characterization: As part of the threshold test or as a separate test, direct each observer to describe the characteristic sample odor using odor reference standards (see Section 2170). Compile the consensus that may appear among testers and that affords a clue to the origin of the odorous component. The value of the characterization test increases as observers become more experienced with a particular category of odor, e.g., earthy, musty, chlorine.

TABLE 2150:I. THRESHOLD ODOR NUMBERS CORRESPONDING TO VARIOUS DILUTIONS

Sample Volume Diluted to 200 mL mL	Threshold Odor No.	Sample Volume Diluted to 200 mL mL	Threshold Odor No.
200	1	12.0	17
140	1.4	8.3	24
100	2	5.7	35
70	3	4.0	50
50	4	2.8	70
35	6	2.0	100
25	8	1.4	140
17	12	1.0	200

TABLE 2150:II. DILUTIONS FOR VARIOUS ODOR INTENSITIES

Sample Volume in Which Odor Is First Noted mL	Volumes to Be Diluted to 200 mL mL
200	200, 140, 100, 70, 50
50	50, 35, 25, 17, 12
12	12, 8.3, 5.7, 4.0, 2.8
2.8	Intermediate dilution

c. *Threshold measurement:*‡ The "threshold odor number," designated by the abbreviation TON, is the greatest dilution of sample with odor-free water yielding a definitely perceptible odor. Bring total volume of sample and odor-free water to 200 mL in each test. Follow dilutions and record corresponding TON presented in Table 2150:I. These numbers have been computed thus:

$$TON = \frac{A + B}{A}$$

where:

A = mL sample and
B = mL odor-free water.

1) Place proper volume of odor-free water in the flask first, add sample to water (avoiding contact of pipet or sample with lip or neck of flask), mix by swirling, and proceed as follows:

Determine approximate range of the threshold number by adding 200 mL, 50 mL, 12 mL, and 2.8 mL sample to separate 500-mL glass-stoppered erlenmeyer flasks containing odor-free water to make a total volume of 200 mL. Use a separate flask containing only odor-free water as reference for comparison. Heat dilutions and reference to desired test temperature.

2) Shake flask containing odor-free water, remove stopper, and sniff vapors. Test sample containing least amount of odor-bearing water in the same way. If odor can be detected in this dilution, prepare more dilute samples as described in ¶ 5) below. If odor cannot be detected in first dilution, repeat above procedure using sample containing next higher concentration of odor-bearing water, and continue this process until odor is detected clearly.

3) Based on results obtained in the preliminary test, prepare a set of dilutions using Table 2150:II as a guide. Prepare the five dilutions shown on the appropriate line and the three next most concentrated on the next line in Table 2150:II. For example, if odor was first noted in the flask containing 50 mL sample in the preliminary test, prepare flasks containing 50, 35, 25, 17, 12, 8.3, 5.7, and 4.0 mL sample, each diluted to 200 mL with odor-free

water. This array is necessary to challenge the range of sensitivities of the entire panel of testers.

Insert two or more blanks in the series near the expected threshold, but avoid any repeated pattern. Do not let tester know which dilutions are odorous and which are blanks. Instruct tester to smell each flask in sequence, beginning with the least concentrated sample, until odor is detected with certainty.

4) Record observations by indicating whether odor is noted in each test flask. For example:

mL Sample Diluted to 200 mL	12	0	17	25	0	35	50
Response	−	−	−	+	−	+	−

5) If the sample being tested requires more dilution than is provided by Table 2150:II, prepare an intermediate dilution consisting of 20 mL sample diluted to 200 L with odor-free water. Use this dilution for the threshold determination. Multiply TON obtained by 10 to correct for the intermediate dilution. In rare cases more than one tenfold intermediate dilution step may be required.

5. Calculation

The threshold odor number is the dilution ratio at which odor is just detectable. In the example above, (¶ 4c4), the first detectable odor occurred when 25 mL sample was diluted to 200 mL. Thus the threshold is 200 divided by 25, or 8. Table 2150:I lists the threshold numbers corresponding to common dilutions.

The smallest TON that can be observed is 1, as in the case where the odor flask contains 200 mL undiluted sample. If no odor is detected at this concentration, report "No odor observed" instead of a threshold number. (In special applications, fractional threshold numbers have been calculated.[6])

Anomalous responses sometimes occur; a low concentration may be called positive and a higher concentration in the series may be called negative. In such a case, designate the threshold as the point after which no further anomalies occur. For instance:

Increasing Concentration N

Response   −   −   +   −   +   +   +   +

#

Threshold

---

‡ There are numerous methods of arranging and presenting samples for odor determinations. The methods offered here are practical and economical of time and personnel. If extensive tests are planned and statistical analysis of data is required, become familiar with the more accurate methods that have been used extensively by flavor and allied industries.[5]

where:

- − signifies negative response and
- + signifies positive response.

Occasionally a flask contains residual odor or is contaminated inadvertently. For precise testing repeat entire threshold odor test to determine if the last flask marked " − " was actually a mislabelled blank of odor-free water or if the previous " + " was a contaminated sample.

Use appropriate statistical methods to calculate the most probable average threshold from large numbers of panel results. For most purposes, express the threshold of a group as the geometric mean of individual thresholds.

6. Interpretation of Results

A threshold number is not a precise value. In the case of the single observer it represents a judgment at the time of testing. Panel results are more meaningful because individual differences have less influence on the result. One or two observers can develop useful data if comparison with larger panels has been made to check their sensitivity. Do not make comparisons of data from time to time or place to place unless all test conditions have been standardized carefully and there is some basis for comparison of observed intensities.

7. References

1. AMERICAN SOCIETY FOR TESTING AND MATERIALS COMMITTEE E-18. 1996. Sensory Testing Methods: Second Edition. E. Chambers & M.B. Wolf, eds. American Soc. Testing & Materials, W. Conshohocken, Pa.
2. BAKER, R.A. 1962. Critical evaluation of olfactory measurement. *J. Water Pollut. Control Fed.* 34:582.
3. DOTY, R.L., P. SHAMAN, M.S. DANN & C.P. KIMMELMAN. 1984. University of Pennsylvania smell identification test: a rapid quantitative olfactory function test for the clinic. *Laryngoscope,* 94(2):176.
4. BAKER, R.A. 1963. Odor testing laboratory. *J. Water Pollut. Control Fed.* 35:1396.
5. INTERNATIONAL ORGANIZATION FOR STANDARDIZATION. 1985. ISO 6658 Sensory Analysis–Methodology—General Guidance. International Org. for Standardization, Geneva, Switzerland.
6. ROSEN, A.A., J.B. PETER & F.M. MIDDLETON. 1962. Odor thresholds of mixed organic chemicals. *J. Water Pollut. Control Fed.* 34:7.

8. Bibliography

HULBERT, R. & D. FEBEN. 1941. Studies on accuracy of threshold odor value. *J. Amer. Water Works Assoc.* 33:1945.
SPAULDING, C.H. 1942. Accuracy and application of threshold odor test. *J. Amer. Water Works Assoc.* 34:877.
THOMAS, H.A., JR. 1943. Calculation of threshold odor. *J. Amer. Water Works Assoc.* 35:751.
CARTWRIGHT, L.C., C.T. SNELL & P.H. KELLY. 1952. Organoleptic panel testing as a research tool. *Anal. Chem.* 24:503.
LAUGHLIN, H.F. 1954. Palatable level with the threshold odor test. *Taste Odor Control J.* 20:No. 8 (Aug.).
SHELLENBERGER, R.D. 1958. Procedures for determining threshold odor concentrations in aqueous solutions. *Taste Odor Control J.* 24:No. 5 (May).
LAUGHLIN, H.F. 1962. Influence of temperature in threshold odor evaluation. *Taste Odor Control J.* 28:No. 10 (Oct.).
The threshold odor test. 1963. *Taste Odor Control J.* 29:Nos. 6, 7, 8 (June, July, Aug.).
SUFFET, I.H. & S. SEGALL. 1971. Detecting taste and odor in drinking water. *J. Amer. Water Works Assoc.* 63:605.
STAHL, W.H., ed. 1973. Compilation of Odor and Taste Threshold Values Data. Amer. Soc. Testing & Materials Data Ser. DS 48, Philadelphia, Pa.
AMERICAN SOCIETY FOR TESTING AND MATERIALS. 1973. Annual Book of ASTM Standards. Part 23, D-1292-65, ASTM, Philadelphia, Pa.
AMERICAN SOCIETY FOR TESTING AND MATERIALS. 1986. Physical Requirement Guidelines for Sensory Evaluation Laboratories. STP 913, American Soc. Testing & Materials, Philadelphia, Pa.
AMERICAN SOCIETY FOR TESTING AND MATERIALS COMMITTEE E-679. 1990. Standard Practice for Determination of Odor and Taste Thresholds by a Forced-Choice Ascending Concentration Series Method of Limits. American Soc. Testing & Materials, Philadelphia, Pa.
AMERICAN SOCIETY FOR TESTING AND MATERIALS. 1991. E1432 Standard Practice for Defining and Calculating Individual and Group Sensory Thresholds from Forced Choice Data Sets of Intermediate Size. American Society for Testing and Materials, Philadelphia, Pa.
BRUVOLD, W.H. 1989. A critical review of methods used for the sensory evaluation of water quality. *Crit. Rev. Environ. Control* 19:291.
YOUNG, W.F., H. HORTH, R. CRANE, T. OGDEN & M. ARNOTT. 1996. Taste and odour threshold concentration of potential potable water contaminants. *Water Res.* 30:331.

2160   TASTE*

2160 A.   Introduction

1. General Discussion

Taste refers only to gustatory sensations called bitter, salty, sour, and sweet that result from chemical stimulation of sensory nerve endings located in the papillae of the tongue and soft palate.

---

* Approved by Standard Methods Committee, 1993.

Flavor refers to a complex of gustatory, olfactory, and trigeminal sensations resulting from chemical stimulation of sensory nerve endings located in the tongue, nasal cavity, and oral cavity.[1] Water samples taken into the mouth for sensory analysis always produce a flavor, although taste, odor, or mouth-feel may predominate, depending on the chemical substances present. Methods for sensory analysis presented herein require that the sample be taken into the mouth, that is, be tasted, but technically the sensory analysis requires evaluation of the complex sensation called flavor. As used here, taste refers to a method of sensory analysis in which samples are taken into the mouth but the resultant evaluations pertain to flavor.

Three methods have been developed for the sensory evaluation of water samples taken into the mouth: the flavor threshold test (FTT), the flavor rating assessment (FRA), and the flavor profile analysis (FPA) (Section 2170). The FTT is the oldest. It has been used extensively and is particularly useful for determining if the overall flavor of a sample of finished water is detectably different from a defined standard.[2] The FRA is especially valuable for determining if a sample of finished water is acceptable for daily consumption,[3] and the FPA is most useful for identifying and characterizing individual flavors in a water sample.[4]

Make flavor tests only on samples known to be safe for ingestion. Do not use samples that may be contaminated with bacteria, viruses, parasites, or hazardous chemicals, that contain dechlorinating agents such as sodium arsenite or that are derived from an unesthetic source. Do not make flavor tests on wastewaters or similar untreated effluents. Observe all sanitary and esthetic precautions with regard to apparatus and containers contacting the sample. Properly clean and sterilize containers before using them. Conduct analyses in a laboratory free from interfering background odors and if possible provide non-odorous carbon-filtered air at constant temperature and humidity. Use the procedure described in Section 2150 with respect to taste- and odor-free water to prepare dilution water and reference samples.

2. References

1. GELDARD, F.A. 1972. The Human Senses. John Wiley & Sons, New York, N.Y.
2. BAKER, R.A. 1961. Taste and Odor in Water: A Critical Review. Manufacturing Chemists' Assoc., Washington, D.C.
3. BRUVOLD, W.H. 1968. Scales for rating the taste of water. *J. Appl. Psychol.* 52:245.
4. MALLEVIALE, J. & I.H. SUFFET, eds. 1987. The Identification and Treatment of Tastes and Odors in Drinking Water. American Water Works Association Research Foundation, Denver, Colo.

2160 B.   Flavor Threshold Test (FTT)

1. General Discussion

Use the FTT to measure detectable flavor quantitatively. More precisely, use the method to compare the sample flavor objectively with that of specified reference water used as diluent.

The flavor threshold number (FTN) is the greatest dilution of sample with reference water yielding a definitely perceptible difference. The FTN is computed as follows:

$$FTN = \frac{A + B}{A}$$

where:

A = sample volume, mL, and
B = reference water (diluent) volume, mL.

Table 2160:I gives the FTNs corresponding to various dilutions.

2. Procedure

a. Panel selection: Carefully select by preliminary trials interested persons to make flavor tests. Exclude insensitive persons and insure that the testers are free from colds or allergies. Familiarize testers with the procedure before they participate in a panel test, but do not let them prepare samples or know dilution concentrations being evaluated. For precise work use a panel of five or more testers.

b. Taste characterization: Have each observer describe the characteristic sample flavor of the most concentrated sample. Compile the consensus that may appear among testers. The value of characterization increases as observers become more experienced with a particular flavor category such as chlorophenolic, grassy, or musty.

c. Preliminary test: To determine approximate range of the FTN, add 200-, 50-, 12-, and 4-mL sample portions to volumes of reference water (see Section 2150) designated in Table 2160:I in separate 300-mL glass beakers to make a total of 200 mL in each beaker, and mix gently with clean stirrer. Use separate beaker containing only reference water for comparison. Keep sample temperature during testing within 1°C of specified temperature. Present samples to each taster in a uniform manner, with the reference water presented first, followed by the most dilute sample. If a flavor can be detected in this dilution, prepare an intermediate sample by diluting 20 mL sample to 200 mL with reference water. Use this dilution for threshold determination and multiply FTN obtained by 10 to correct for intermediate dilution. In rare cases a higher intermediate dilution may be required.

If no flavor is detected in the most dilute sample, repeat using

TABLE 2160:I. FLAVOR THRESHOLD NUMBERS CORRESPONDING TO VARIOUS DILUTIONS

Sample Volume mL	Diluent Volume mL	Flavor Threshold No. FTN
200	0	1
100	100	2
70	130	3
50	150	4
35	165	6
25	175	8
17	183	12
12	188	17
8	192	25
6	194	33
4	196	50
3	197	67
2	198	100
1	199	200

the next concentration. Continue this process until flavor is detected clearly.

d. FTN determination: Based on results obtained in the preliminary test, prepare a set of dilutions using Table 2160:II as a guide. Prepare the seven dilutions shown on the appropriate line. This array is necessary to challenge the range of sensitivities of the entire panel of testers. If the sample being tested requires more dilution than is provided by Table 2160:II, make intermediate dilutions as directed in *c* above.

Use a clean 50-mL beaker filled to the 25-mL level or use an ordinary restaurant-style drinking glass for each dilution and reference sample. Do not use glassware used in sensory testing for other analyses. Between tests, sanitize containers in an automatic dishwasher supplied with water at not less than 60°C.

Maintain samples at $15 \pm 1°C$. However, if temperature of water in the distribution system is higher than 15°C, select an appropriate temperature. Specify temperature in reporting results.

Present series of samples to each tester in order of increasing concentration. Pair each sample with a known reference. Have tester taste sample by taking into the mouth whatever volume is comfortable, moving sample throughout the mouth, holding it for several seconds, and discharging it without swallowing. Have tester compare sample with reference and record whether a flavor or aftertaste is detectable. Insert two or more reference blanks in the series near the expected threshold, but avoid any repeated pattern. Do not let tester know which samples have flavor and which are blanks. Instruct tester to taste each sample in sequence, beginning with the least concentrated sample, until flavor is detected with certainty.

Record observations by indicating whether flavor is noted in each test beaker. For example:

mL Sample Diluted to 200 mL	6	8	12	0	17	25	35	0	50
Response	−	−	−	−	−	+	+	−	+

where:

- signifies negative response and
+ signifies positive response.

3. Calculation

The flavor threshold number is the dilution ratio at which flavor is just detectable. In the example above, the first detectable flavor occurred when 25 mL sample was diluted to 200 mL yielding a threshold number of 8 (Table 2160:I). Reference blanks do not influence calculation of the threshold.

TABLE 2160:II. DILUTIONS FOR DETERMINING THE FTN

Sample Volume in Which Taste Is First Noted *mL*	Volumes to be Diluted to 200 mL *mL*
200	200, 100, 70, 50, 35, 25, 17
50	50, 35, 25, 17, 12, 8, 6
12	12, 8, 6, 4, 3, 2, 1
4	Intermediate dilution

The smallest FTN that can be observed is 1, where the beaker contains 200 mL undiluted sample. If no flavor is detected at this concentration, report "No flavor observed" instead of a threshold number.

Anomalous responses sometimes occur; a low concentration may be called positive and a higher concentration in the series may be called negative. In such cases, designate the threshold as that point after which no further anomalies occur. The following illustrates an approach to an anomalous series (responses to reference blanks are excluded):

Increasing Concentration N

Response: −   +   −   +   +   +   +

Q

Threshold

Calculate mean and standard deviation of all FTNs if the distribution is reasonably symmetrical; otherwise, express the threshold of a group as the median or geometric mean of individual thresholds.

4. Interpretation of Results

An FTN is not a precise value. In the case of the single observer it represents a judgment at the time of testing. Panel results are more meaningful because individual differences have less influence on the test result. One or two observers can develop useful data if comparison with larger panels has been made to check their sensitivity. Do not make comparisons of data from time to time or place to place unless all test conditions have been standardized carefully and there is some basis for comparison of observed FTNs.

5. Bibliography

HULBERT, R. & D. FEBEN. 1941. Studies on accuracy of threshold odor value. *J. Amer. Water Works Assoc.* 33:1945.

SPAULDING, C.H. 1942. Accuracy and application of threshold odor test. *J. Amer. Water Works Assoc.* 34:877.

THOMAS, H.A. 1943. Calculation of threshold odor. *J. Amer. Water Works Assoc.* 35:751.

COX, G.J. & J. W. NATHANS. 1952. A study of the taste of fluoridated water. *J. Amer. Water Works Assoc.* 44:940.

LAUGHLIN, H.F. 1954. Palatable level with the threshold odor test. *Taste Odor Control J.* 20:1.

COX, G.J., J.W. NATHANS & N. VONAU. 1955. Subthreshold-to-taste thresholds of sodium, potassium, calcium and magnesium ions in water. *J. Appl. Physiol.* 8:283.

LOCKHART, E.E., C.L. TUCKER & M.C. MERRITT. 1955. The effect of water impurities on the flavor of brewed coffee. *Food Res.* 20:598.

CAMPBELL, C.L., R.K. DAWES, S. DEOLALKAR & M.C. MERRITT. 1958. Effects of certain chemicals in water on the flavor of brewed coffee. *Food Res.* 23:575.

MIDDLETON, F.M., A.A. ROSEN & R.H. BRUTTSCHELL. 1958. Taste and odor research tools for water utilities. *J. Amer. Water Works Assoc.* 50:231.

SHELLENBERGER, R.D. 1958. Procedures for determining threshold odor concentrations in aqueous solutions. *Taste Odor Control J.* 24:1.

COHEN, J.N., L.J. KAMPHAKE, E.K. HARRIS & R.L. WOODWARD. 1960. Taste threshold concentrations of metals in drinking water. *J. Amer. Water Works Assoc.* 52:660.

BAKER, R.A. 1961. Problems of tastes and odors. *J. Water Pollut. Control Fed.* 33:1099.

ROSEN, A.A., J.B. PETER & F.M. MIDDLETON. 1962. Odor thresholds of mixed organic chemicals. *J. Water Pollut. Control Fed.* 34:7.

BAKER, R.A. 1962. Critical evaluation of olfactory measurement. *J. Water Pollut. Control Fed.* 34:582.

BAKER, R.A. 1963. Threshold odors of organic chemicals. *J. Amer. Water Works Assoc.* 55:913.

BAKER, R.A. 1963. Odor testing laboratory. *J. Water Pollut. Control Fed.* 35:1396.

BAKER, R.A. 1963. Odor effects of aqueous mixtures of organic chemicals. *J. Water Pollut. Control Fed.* 35:728.

ROSEN, A.A., R.T. SKEEL & M.B. ETTINGER. 1963. Relationship of river water odor to specific organic contaminants. *J. Water Pollut. Control Fed.* 35:777.

BRYAN, P.E., L.N. KUZMINSKI, F.M. SAWYER & T.H. FENG. 1973. Taste thresholds of halogens in water. *J. Amer. Water Works Assoc.* 65:363.

2160 C. Flavor Rating Assessment (FRA)

1. General Discussion

When the purpose of the test is to estimate acceptability for daily consumption, use the flavor rating assessment described below. This procedure has been used with samples from public sources in laboratory research and consumer surveys to recommend standards governing mineral content in drinking water. Each tester is presented with a list of nine statements about the water ranging on a scale from very favorable to very unfavorable. The tester's task is to select the statement that best expresses his or her opinion. The individual rating is the scale number of the statement selected. The panel rating for a particular sample is an appropriate measure of central tendency of the scale numbers for all testers for that sample.

2. Samples

Sample finished water ready for human consumption or use experimentally treated water if the sanitary requirements given in Section 2160A.1 are met fully. Use taste- and odor-free water as described in Section 2150 and a solution of 2000 mg NaCl/L prepared with taste- and odor-free water as criterion samples.

3. Procedure

a. Panel selection and preparation: Give prospective testers thorough instructions and trial or orientation sessions followed by questions and discussion of procedures. In tasting samples, testers work alone. Select panel members on the basis of performance in these trial sessions. Do not let testers know the composition or source of specific samples.

b. Rating test: A single rating session may be used to evaluate up to 10 samples, including the criterion samples mentioned in ¶ 2 above. Allow at least 30 min rest between repeated rating sessions. For glassware requirements, see ¶ B.2*d*.

Present samples at a temperature that the testers will find pleasant for drinking water; maintain this temperature throughout testing. A temperature of 15°C is recommended, but in any case, do not let the test temperature exceed tap water temperatures customary at the time of the test. Specify test temperature in reporting results.

Independently randomize sample order for each tester. Instruct each to complete the following steps: 1) Taste about half the sample by taking water into the mouth, holding it for several seconds, and discharging it without swallowing; 2) Form an initial judgment on the rating scale; 3) Make a second tasting in a similar manner; 4) Make a final rating and record result on an appropriate data form; 5) Rinse mouth with reference water; 6) Rest 1 min before repeating Steps 1 through 5 on next sample.

c. Characterization: If characterization of flavor also is required, conduct a final rating session wherein each tester is asked to describe the flavor of each sample rated (see ¶ B.2*b*).

4. Calculation

Use the following scale for rating. Record ratings as integers ranging from one to nine, with one given the highest quality rating. Calculate mean and standard deviation of all ratings if the distribution is reasonably symmetrical, otherwise express the most typical rating of a group as the median or geometric mean of individual ratings.

Action tendency scale:

1) I would be very happy to accept this water as my everyday drinking water.

2) I would be happy to accept this water as my everyday drinking water.

3) I am sure that I could accept this water as my everyday drinking water.

4) I could accept this water as my everyday drinking water.

5) Maybe I could accept this water as my everyday drinking water.

6) I don't think I could accept this water as my everyday drinking water.

7) I could not accept this water as my everyday drinking water.

8) I could never drink this water.

9) I can't stand this water in my mouth and I could never drink it.

5. Interpretation of Results

Values representing the central tendency and dispersion of quality ratings for a laboratory panel are only estimates of these values for a defined consuming population.

6. Bibliography

BRUVOLD, W.H. & R.M. PANGBORN. 1966. Rated acceptability of mineral taste in water. *J. Appl. Psychol.* 50:22.

BRUVOLD, W.H., H.J. ONGERTH & R.C. DILLEHAY. 1967. Consumer attitudes toward mineral taste in domestic water. *J. Amer. Water Works Assoc.* 59:547.

DILLEHAY, R.C., W.H. BRUVOLD & J.P. SIEGEL. 1967. On the assessment of potability. *J. Appl. Psychol.* 51:89.

BRUVOLD, W.H. 1968. Mineral Taste in Domestic Water. Univ. California Water Resources Center, Los Angeles.

Bruvold, W.H. & H.J. Ongerth. 1969. Taste quality of mineralized water. *J. Amer. Water Works Assoc.* 61:170.

Bruvold, W.H. & W.R. Gaffey. 1969. Rated acceptability of mineral taste in water. II: Combinatorial effects of ions on quality and action tendency ratings. *J. Appl. Psychol.* 53:317.

Dillehay, R.C., W.H. Bruvold & J.P. Siegel. 1969. Attitude, object label, and stimulus factors in response to an attitude object. *J. Personal. Social Psychol.* 11:220.

Bruvold, W.H. 1970. Laboratory panel estimation of consumer assessments of taste and flavor. *J. Appl Psychol.* 54:326.

Pangborn, R.M., I.M. Trabue & R.C. Baldwin. 1970. Sensory examination of mineralized, chlorinated waters. *J. Amer. Water Works Assoc.* 62:572.

Pangborn, R.M., I.M. Trabue & A.C. Little. 1971. Analysis of coffee, tea and artifically flavored drinks prepared from mineralized waters. *J. Food Sci.* 36:355.

Bruvold, W.H. & P.C. Ward. 1971. Consumer assessment of water quality and the cost of improvements. *J. Amer. Water Works Assoc.* 63:3.

Pangborn, R.M. & L.L. Bertolero. 1972. Influence of temperature on taste intensity and degree of liking of drinking water. *J. Amer. Water Works Assoc.* 64:511.

Bruvold, W.H. 1975. Human perception and evaluation of water quality. *Crit. Rev. Environ. Control* 5:153.

Bruvold, W.H. 1976. Consumer Evaluation of the Cost and Quality of Domestic Water. Univ. California Water Resources Center, Davis.

2170   FLAVOR PROFILE ANALYSIS*

2170 A.   Introduction

1. Discussion

Flavor profile analysis (FPA) is a technique for identifying sample taste(s) and odor(s). For general information on taste see Section 2160; for information on odor see Section 2150.

FPA differs from threshold odor number because the sample is not diluted and each taste or odor attribute is individually characterized and assigned its own intensity rating. The single numerical rating obtained in measuring threshold odor is controlled by the most readily perceived odorant or mixture. Sample dilution

* Approved by Standard Methods Committee, 1997.

may change the odor attribute that is measured.[1,2] FPA determines the strength or intensity of each perceived taste or odor without dilution or treatment of the sample.

2. References

1. Mallevialle, J. & I.H. Suffet, eds. 1987. Identification and Treatment of Tastes and Odors in Drinking Water. American Water Works Assoc., Denver, Colo.
2. Bruchet, A., D. Khiari & I. Suffet. 1995. Monitoring and analysis. *In* I. Suffet, J. Mallevialle & E. Kawczynski, eds. Advances in Taste and Odor Treatment and Control. American Water Works Assoc. Research Foundation & Lyonnaise des Eaux, Denver, Colo.

2170 B.   Flavor Profile Analysis

1. General Discussion

a. Principle: Flavor profile analysis uses a group of four or five trained panelists to examine the sensory characteristics of samples. Flavor attributes are determined by tasting; odor attributes (aroma) are determined by sniffing the sample. The method allows more than one flavor, odor attribute, or feeling factor (e.g., drying, burning) to be determined per sample and each attribute's strength to be measured.

Panelists must be able to detect and recognize various odors present. Flavor profile analysis requires well-trained panelists and data interpreters. Reproducibility of results depends on the training and experience of the panelists.

Initially, panelists record their perceptions without discussion. Once each individual has made an independent assessment of a sample, the panel discusses its findings and reaches a consensus.[1]

b. Interference: Fatigue (adaptation) denotes the decrease in the analyst's sensory acuity with continued exposure to stimula-

tion.[1] Because odor mixtures are more complex than taste mixtures, olfactory adaptation is more serious.[2] Factors that can induce fatigue include odor intensity, type of odorants (some compounds such as geosmin and chlorine induce more fatigue than others), number of samples tested during a session, and rest interval between samples.[1] The rest interval also is important because it prevents carryover of odors between samples.

Background odors present during analysis affect results. Analyst illness, i.e., cold or allergy, can diminish or otherwise alter perception.

c. Application: Flavor profile analysis has been applied to drinking water sources, finished drinking water, sampling points within the drinking water treatment train, and bottled waters, and for investigating customer complaints.

d. Precautions: Some compounds are health hazards when inhaled or ingested. Do not use this method for industrial wastes or other samples suspected of containing high concentrations of hazardous compounds. Chemically analyze suspect samples to deter-

TABLE 2170:I. QUANTITATIVE ODOR REFERENCES*

Compound	Odor Characteristics	Stock Solution Concentration mg/L	Amount Placed Into 200 mL Pure Water at 25°C for Presentation
Geosmin	Earthy, red beets	0.2†	300 μL for 300 ng/L
2-Methylisoborneol	Earthy, peat-like, Brazil nut, soil	0.2†	200 μL for 200 ng/L
Free chlorine	Chlorinous	1.0	0.1 mL for 0.5 mg/L
		pH = 7	
Dichloramine	Swimming pool chlorine	pH = 7	
Monochloramine	Chlorinous	pH = 7	
trans,2-cis,6-Nonadienal	Cucumber, green vegetation	1000	1 μL for 5 μg/L
Styrene	Model airplane glue	1000	100 μL for 500 μg/L
Toluene	Glue, sweet solvency	1000	100 μL for 500 μg/L
cis-3-Hexenyl-1-acetate	Grassy	1000	100 μL for 500 μg/L
cis-3-Hexene-1-ol	Grassy, green apple	1000	100 μL for 500 μg/L
Cumene	Paste shoe polish, solvency	1000	20 μL for 100 μg/L
m-Xylene	Sweet solvency	1000	40 μL for 200 μg/L
Methylisobutyl ketone	Paint solvency	1000	200 μL for 1.0 mg/L
1,2,4-Trimethylbenzene	Shoe polish, coal tar	1000	50 μL for 250 μg/L
Indene	Glue, moth balls	1000	1 μL for 5 μg/L
Indan	Varnish, coal tar	1000	5 μL for 25 μg/L
Naphthalene	Sweet solvency	1000	1 μL for 5 μg/L
Benzofuran	Shoe polish, moth balls	1000	2 μL for 10 μg/L
2-Methyl-benzofuran	Moth balls, sweet solvency	1000	50 μL for 250 μg/L

NOTE: These compounds have actually been identified as causes of odors in raw and finished drinking water.

* Adapted from AMERICAN WATER WORKS ASSOCIATION. 1993. Flavor Profile Analysis: Screening and Training of Panelists. AWWA Manual, American Water Works Assoc., Denver, Colo.

† Compounds available only in solid form and must be dissolved in methanol.

mine whether hazardous chemicals are present before making flavor profile analysis. Do not taste untreated drinking water unless it is certain that no health threat (biological or chemical) exists.

Because flavor profile analysis includes discussion during which panelists reach consensus description, avoid including on the panel a person with a dominating personality.[3] The opinions of a senior member or panel leader also may have this undesired effect.[3]

2. Apparatus[4-8]

Reserve apparatus and glassware exclusively for flavor profile analysis. Prepare sample bottles by washing bottle and cap with detergent. Rinse 10 times with hot water. Optionally rinse with HCl (1:1). Rinse with odor-free water (see 2150B.3) at least three times. If there is residual odor, such as chalky, repeat cleaning.

Prepare 500-mL erlenmeyer flasks by either of the two methods described below provided that it imparts no odor to the sample. Make tests with freshly cleaned containers.

1) Without wearing rubber gloves wash flask with soapy water. Additionally, scrub outside of flask with scouring pad to remove body oils. Rinse 10 times with hot water and 3 times with odor-free water. To store flask add 100 to 200 mL odor-free water and stopper. Before use, rinse with 100 mL odor-free water. If there is residual odor, repeat cleaning.

2) Do not handle with rubber gloves. Before use, heat 200 mL odor-free water to boiling; lightly lay stopper over flask opening, permitting water vapor to escape. Discard boiled water and let flask cool to room temperature. If there is residual odor repeat or use cleaning alternative 1). After analyzing a sample, discard sample, rinse 10 times with hot tap water, add 200 mL odor-free water, stopper, and store.

For flasks with persistent odor use commercial cleaning mixture* or HCl (1:1).

a. *Sample bottles:* Glass, 1-L, with TFE-lined closure.

b. *Containers for odor analysis:*[4] Select container for odor analysis depending on panelists' preference, temperature of sample, and availability of container.

1) Panel preference[4-8]—Plastic cups, 7- or 8-ounce disposable, are convenient. Some panels have found that these plastic cups impart a plastic and/or floral aroma to samples. They prefer 500-mL erlenmeyer flasks with ST32 ground-glass stoppers. The panel should determine the acceptable, odor-free, test container.

2) Temperature[4]—If aroma samples are tested at 45°C, do not use plastic. If aroma samples are tested at 25°C, use the same plastic cup and sample for both aroma and flavor analysis.

3) Availability[4]—Containers must be consistently available because the panel responses may change when different containers are used.

c. *Containers for flavor analysis:*[4] Use either plastic disposable, e.g., polyethylene,† or glass containers. Do not use wax-lined or paper cups. If using glass containers verify that they contribute no taste to the sample. Wash glass containers as directed above.

d. *Watch glasses.*

e. *Constant-temperature water bath* capable of maintaining a temperature ± 1°C. The bath must not contribute any odor to the odor flasks or testing room.

f. *Thermometer,* graduated 0 to 110°C, chemical or metal-stem dial type.

g. *Syringes.*

h. *Ice chest.*

* Nochromix, Godax Laboratories, Inc., New York, NY, or equivalent.
† Plastic cup, 7- or 8-ounce, Solo or equivalent, any color except clear.

TABLE 2170:II. REPRESENTATIVE ODOR REFERENCE STANDARDS*

Compound	Odor Characteristics	Stock Solution Concentration mg/L	Amount Placed Into 200 mL Pure Water at 25°C for Presentation
2,3,6-Trichloroanisole	Leather, earthy	1000	4 μL for 20 μg/L
2,3-Diethylpyrazine	Mildew, damp basement	1000	2 μL for 10 μg/L
2-Isopropyl-3-methoxypyrazine	Potato bin, musty	1000	40 μL for 200 μg/L
Nonanal	Hay, sweet	1000	40 μL for 200 μg/L
Dimethyl sulfide	Decaying vegetation, canned corn	1000	1 μL for 5 μg/L
Dimethyl disulfide	Septic	1000	2 μL for 10 μg/L
Dimethyltrisulfide	Garlicky, oniony, septic	0.2	50 μL for 50 ng/L
Butyric acid	Putrid, sickening	1000	200 μL for 1 mg/L
trans,2-Nonenal	Cucumber with skin	1000	40 μL for 200 μg/L
Diphenyl ether	Geranium	1000†	20 μL for 100 μg/L
D-Limonene	Citrusy	1000	400 μL for 2 mg/L
Hexanal	Lettuce heart, pumpkin, green pistachio	1000	40 μL for 200 μg/L
Benzaldehyde	Sweet almond	1000	200 μL for 1 mg/L
Ethyl-2-methyl-butyrate	Fruity, pineapple	1000	1000 μL for 5 mg/L
2-Heptanone	Banana-like, sweet solventy	1000	100 μL for 500 μg/L
Hexachloro-1,3-butadiene	Sweet, minty, Vapo Rub	1000	800 μL for 4 mg/L
2-Isobutyl-3-methoxypyrazine	Green/bell pepper, musty	1000	40 μL for 200 μg/L
trans,2-trans,4-Decadienal	Rancid oily	1000	1000 μL for 5 mg/L
Butanol	Alcohol, solventy	1000	200 μL for 1 mg/L
Eucalyptol (cincole)	Topical ointment for chest colds	1000	40 μL for 200 μg/L
Pyridine	Sweet, alcohol, organic	1000	400 μL for 2 mg/L

NOTE: Possible causes of odors or chemical substitutes for causes of odors.
* Adapted from AMERICAN WATER WORKS ASSOCIATION. 1993. FLAVOR PROFILE ANALYSIS: SCREENING AND TRAINING OF PANELISTS. AWWA Manual, American Water Works Assoc., Denver, Colo.
* Compounds available only in solid form and must be dissolved in methanol.

i. *Odor-free testing room:*[3,5] Hold flavor profile analysis sessions in a clean, well-lit quiet, aroma-free, and temperature-controlled room. Seat panel members around a common table to facilitate exchange of responses during discussion. Place a blackboard or easel pad in the room so that all panelists are able to see it. Preferably use an easel pad with odor-free markers (such as a wax, china marker).

j. *Refrigerator* capable of maintaining a temperature of 4°C.

3. Reagents

a. *Odor-free water:* See Section 2150B.3.

b. *Crackers:* Use salt- and flavor-free crackers‡ to cleanse the palate during taste testing. Before tasting anything, use the crackers and taste-free water to cleanse the palate. Use crackers between samples to reduce carryover of perceptions.

c. *Odor and flavor standards:* Odor and flavor reference standards are being developed for use as a guide for qualitative descriptions (attributes). Aroma reference standards help panelists come to agreement on the description of specific aromas.[4-11] Tables 2170:I, II, and III list some of the aroma references in use. New references are being added continually. A taste and odor wheel has been developed from these lists.[2] For odor identification and training, place several drops of stock solutions (producing moderately strong odors) onto sterile cotton in 25-mL amber-colored vials with TFE-lined caps. Standards can be stored until odor changes.

d. *Taste standards:* Chemicals used for taste standards are sucrose (sweet), citric acid (sour), sodium chloride (salt), and caffeine (bitter). Table 2170:IV lists the chemicals used and their concentrations for basic taste standards at "slight," "moderate," and "strong" levels.[4,5] The taste standards provide reference points for both taste and odor intensity ratings. The panelists compare the intensity of what they are smelling or tasting to the intensity of the standards they have tasted. Make taste standards available during panel training. Because the tastes from all but the sweet standard tend to overpower any subsequent tasting, use only the sweet taste standard during actual sample analysis. If the panel meets as seldom as once a week, make sweet standards available so that panelists can "recalibrate" themselves. Panels that meet more than once a week may not need to "recalibrate." Make standards fresh each time they are used.[4]

4. Scale

The strength of a taste or odor is judged according to the following scale:

—	(odor-free)
T	(threshold)
2	(very weak)
4	(weak)
6	
8	(moderate)
10	
12	(strong)

5. Procedure

a. *Sample collection:* Collect sample in cleaned container. When sampling from a tap, remove all screens and aerators, min-

‡ Keebler Sea Toast or equivalent.

TABLE 2170:III. SUBSTITUTE ODOR REFERENCE STANDARDS*

Compound	Odor Characteristic	Preparation
Cloves	Spicy like cloves	Use supermarket brand of dried clove buds (spice). Add 3 clove buds to 200 mL pure water and swirl 1–2 min. Allow to stand overnight at room temperature, then discard the buds.
Dried grass	Hay	Place dried cut grass in erlenmeyer flask until half full.
Grass	Decaying vegetation	Weigh 2 g of fresh grass and mix into 200 mL pure water and let stand at room temperature. In 1–3 d, the odor will appear.
Grass	Septic	Allow the solution above for decaying vegetation to stand for an additional 1–2 weeks.
Rubber hose	Rubber hose	Boil a short section of rubber hose in 200 mL pure water for 5 min. Allow to cool and remove the hose.
Soap	Soapy	Place 5 g of chipped nonscented bar soap in 200 mL pure water.
Pencil shavings	Woody	Instruct panel member to sharpen a wood pencil and sniff the freshly exposed wood.

NOTE: Standards made from materials rather than chemicals.
* Adapted from AMERICAN WATER WORKS ASSOCIATION. 1993. Flavor Profile Analysis: Screening and Training of Panelists. AWWA Manual. American Water Works Assoc., Denver, Colo.

TABLE 2170:IV. BASIC TASTE STANDARDS

Chemical for Basic Tastes	Food or Beverage Corresponding to Intensity	Concentration %	Intensity Scale (1 to 12)*
Sweet: sugar	Canned fruit or vegetables	5.00	4 W
	Carbonated soda	10.00	8 M
	Syrup, jelly	15.00	12 S
Sour: citric acid	Fresh fruit jelly	0.05	4 W
	Some carbonated sodas	0.10	8 M
	Lemon juice	0.20	12 S
Salt: sodium chloride	Level in bread	0.40	4 W
	Dehydrated soup mix	0.70	8 M
	Soy sauce	1.00	12 S
Bitter: caffeine	Strong coffee	0.05	4 W
		0.10	8 M
		0.20	12 S
	or quinine hydrochloride dihydrate	0.001	4 W
		0.002	8 M
		0.004	12 S

* W = weak; M = moderate; S = strong.

imize turbulence. Flush tap at least 5 min. Reduce flow rate during sampling. Rinse bottle with sample, then fill it to the top, with no headspace. Chill or refrigerate sample immediately and analyze as soon after collection as possible, preferably within 24 h, but no longer than 48 h.

b. *Sample preparation:*[4] Pour samples into properly prepared glassware or acceptable disposable containers. Analyze samples at the same temperature. Adjust sample temperatures by placing samples in a water bath 15 min before analysis. Prepare a sample for each panelist. Examine odor samples before flavor samples.

1) Odor analysis—When using 500-mL erlenmeyer flasks for odor analysis, place 200 mL sample in the flask. Make transfer carefully to avoid loss of volatile components. When plastic cups are used, place 60 mL sample in cup and cover it with a watch glass.

2) Taste analysis—Bring sample to 25°C before pouring into containers for tasting. Cover with a watch glass if samples are not tasted immediately. If samples are tested at 25°C for odor, use the same sample for flavor analysis.

c. *Panelists:* Carefully screen and train panelists.[8]

d. *Pre-test considerations:* Notify panelists well in advance of panel session so a substitute can be found if necessary. Panelists who have colds or allergic attacks the day of the panel are unacceptable; they should ask the panel coordinator to find a substitute. Panelists must not smoke or eat for 15 to 30 min before the session. Wearing cologne or perfume or washing hands with scented soap before the session is not permitted.

e. *Sample analysis:* The panel consists of four or five members. If fewer than four panelists are available, store the sample until a full panel can meet.

1) Odor analysis—Heat samples to proper temperature. If erlenmeyer flasks are used, sample temperature is 45°C. If plastic cups are used, sample temperature is 25°C. Give each panelist his/her own sample. If a flask is used, hold it with one hand on the bottom and the other on the stopper. Do not touch flask neck. Gently swirl (do not shake) flask in a circular manner to ensure that volatile compounds are released into the headspace. Bring flask close to nose, remove stopper, and sniff at the flask opening. Record impressions of odor attributes in the order perceived and the assessment of each attribute's intensity. If a cup is used, gently swirl on the table top for a few seconds. Remove watchglass, and keeping hands away from the cup, sniff sample and record perceptions. CAUTION: *Sniff samples only if they are known to be toxicologically safe.* For both the flask and the plastic cup method, smell all samples before going on to the taste test. Do not discuss or interact with other panelists.

2) Taste analysis—When tasting, take sample into the mouth and roll it over the entire surface of the tongue. Slurping enhances the odor aspect of flavor. CAUTION: *Taste only samples known to be biologically and toxicologically safe.* Do not discuss or interact with other panelists.

3) Intensity—In both taste and odor analysis, each panelist determines intensity ratings by matching intensity of the flavor or aroma perceived with the defined intensities of the basic taste standards. Initially this may be difficult. Some panels have found it helpful to make basic taste standards available throughout the analysis so that any panelist who wishes to "recalibrate" may do so.

4) Re-examination of samples—First impressions are most important, particularly for intensity. The intensity rating may diminish upon re-examination because of fatigue and volatility of the odorant. However, if an aroma or flavor is difficult for a panelist to describe, the panelist may go back and re-evaluate the sample before recording the results.

5) Individual results—As soon as a sample is tasted or smelled, record individual results. Information recorded includes a descrip-

tion of aromas or flavors perceived, their intensity, and the order in which they were perceived.

6) Rest interval—Sniff odor-free water and rest at least 2 min between samples. Fatigue is a problem[1] that can be dealt with. When samples are expected to cause fatigue, the panel leader can increase the rest interval and limit the number of samples presented. The leader also may try to arrange samples such that samples known to be fatiguing are placed near the end of the sample row. Avoid juxtaposition of such samples. Also, use taste- and odor-free blanks between samples. Five to six samples per session is the maximum.

7) Discussion—When all panelists have had an opportunity to examine the sample, hold a discussion period. Each panelist's impression of the sample is stated and the panel leader records it to be seen by everyone. By examining the order of appearance, intensity, and description, the panel leader attempts to group responses together, soliciting comments from the panel members as to whether or not they agree. Sometimes panelists detect an aroma that they cannot describe; in the discussion they may see what another panelist called it and decide to agree with that description. With inexperienced panels, several different descriptors may be used for the same aroma. As the panel gains more experience, these differences tend to be reconciled. It is the responsibility of the panel leader to ensure that the panelists are provided with standards that duplicate the sample aromas. Descriptions that fewer than 50% of the panelists use are called "other notes" and are listed separately or are not included in the group results.

8) Recording panel results—Record the following: sample description; time of sampling; identity of panel member; flavor, aroma, and feeling factor descriptions; intensity rating; order of perception; range and average for each descriptor.

6. Calculation and Interpretation[1]

In early testing intensity was used to describe the strength of an odor in an ordinal scale from slight to strong.[12] If an odor was recognized, then it had an intensity. The rating scale has been modified to use numbers corresponding to the ordinal categories. Points (other than threshold) on the scale can be anchored by the use of taste standards. Thus a certain concentration of a taste standard is defined as a certain point on the rating scale and numerical ratings for nonthreshold intensities are anchored. Threshold ratings are not defined by any standards. Their meaning is different from that of the other ratings and they are difficult to manipulate mathematically.

Calculate averages if at least 50% of the panelists agree on a given description. If a panelist does not give that description, assign an intensity value of zero. For example:

Description	Intensity				
Panelist	I	II	III	IV	Average
Musty	2	4	0	2	2.0
Chlorinous	2	4	4	4	3.5

7. Quality Control[12]

Use odor-free samples and duplicates occasionally. The odor-free sample serves to detect guessing. Duplicates check reproduc-ibility. A standard concentration of an odor reference standard in odor-free water also can be used to check odor recognition and reproducibility of the intensity scale.

Use counterbalancing of samples so that each panelist receives the samples in a different order. This prevents error due to order effects such as fatigue, carryover, and expectation.

8. Precision and Bias

Initial studies of precision and bias have been completed.[12,13] An odor recognition and reproducibility study by untrained panelists has rated some of the odors of Tables 2170:I, II, and III as poor, fair, good, or excellent.[13]

9. References

1. SUFFET, I.H., B.M. BRADY, J.H.M. BARTELS, G.A. BURLINGAME, J. MALLEVIALLE & T. YOHE. 1988. Development of the flavor profile analysis method into a standard method for sensory analysis of water. *Water Sci. Technol.* 20:1.
2. AMERICAN SOCIETY FOR TESTING AND MATERIALS. 1968. Basic Principles of Sensory Evaluation. ASTM Spec. Tech. Publ. 433, American Soc. Testing & Materials, Philadelphia, Pa.
3. MEILGAARD, M.C. 1988. Sensory evaluation techniques applied to waterworks samples. *Water Qual. Bull.* 13(2-3):39. WHO Collaborating Centre on Surface and Groundwater Quality, Burlington, Ont., Canada.
4. BARTELS, J.H.M., B.M. BRADY & I.H. SUFFET. 1989. Appendix B *in* Taste and Odor in Drinking Water Supplies—Phases I and II. American Water Works Assoc. Research Foundation, Denver, Colo.
5. KRASNER, S.W., M.J. McGUIRE & V.B. FERGUSON. 1985. Tastes and odors: the flavor profile method. *J. Amer. Water Works Assoc.* 77:34.
6. BARTELS, J.H.M., B.M. BRADY & I.H. SUFFET. 1989. Appendix A *in* Taste and Odor in Drinking Water Supplies—Phases I and II. American Water Works Assoc. Research Foundation, Denver, Colo.
7. MALLEVIALLE, J. & I.H. SUFFET, eds. 1981. Identification and Treatment of Tastes and Odors in Drinking Water. Appendix A. American Water Works Assoc., Denver, Colo.
8. BARTELS, J.H.M., B.M. BRADY & I.H. SUFFET. 1987. Training panelists for the flavor profile analysis method. *J. Amer. Water Works Assoc.* 79:26.
9. SUFFET, I.H., J. HO, D. CHIOU, D. KHIARI & J. MALLEVIALLE. 1995. Taste-and-odor problems observed during drinking water treatment. *In* I.H. Suffet, J. Mallevialle & E. Kawczynski, eds. Advances in Taste and Odor Treatment and Control. American Water Works Assoc. Research Foundation & Lyonnaise des Eaux, Denver, Colo.
10. CAUL, J.F. 1957. The profile method of flavor analysis. *In* Advances in Food Research, Vol. 7. Academic Press, New York, N.Y.
11. BURLINGAME, G.A., D. KHIARI & I.H. SUFFET. 1991. Odor reference standards: The universal language. Proc. AWWA Water Quality Technology Conf., Miami, Fla., p. 7.
12. MENG, A. & I.H. SUFFET. 1992. Assessing the quality of flavor profile analysis data. *J. Amer. Water Works Assoc.* 84:89.
13. KRASNER, S.W. 1995. The use of reference materials in sensory analysis. *Water Sci. Technol.* 31(11):265.

10. Bibliography

AMERICAN WATER WORKS ASSOCIATION. 1993. Flavor Profile Analysis: Screening and Training of Panelists. AWWA Manual. American Water Works Assoc., Denver, Colo.

2310 ACIDITY*

2310 A. Introduction

Acidity of a water is its quantitative capacity to react with a strong base to a designated pH. The measured value may vary significantly with the end-point pH used in the determination. Acidity is a measure of an aggregate property of water and can be interpreted in terms of specific substances only when the chem-

ical composition of the sample is known. Strong mineral acids, weak acids such as carbonic and acetic, and hydrolyzing salts such as iron or aluminum sulfates may contribute to the measured acidity according to the method of determination.

Acids contribute to corrosiveness and influence chemical reaction rates, chemical speciation, and biological processes. The measurement also reflects a change in the quality of the source water.

*Approved by Standard Methods Committee, 1997.

2310 B. Titration Method

1. General Discussion

a. Principle: Hydrogen ions present in a sample as a result of dissociation or hydrolysis of solutes react with additions of standard alkali. Acidity thus depends on the end-point pH or indicator used. The construction of a titration curve by recording sample pH after successive small measured additions of titrant permits identification of inflection points and buffering capacity, if any, and allows the acidity to be determined with respect to any pH of interest.

In the titration of a single acidic species, as in the standardization of reagents, the most accurate end point is obtained from the inflection point of a titration curve. The inflection point is the pH at which curvature changes from convex to concave or vice versa.

Because accurate identification of inflection points may be difficult or impossible in buffered or complex mixtures, the titration in such cases is carried to an arbitrary end-point pH based on practical considerations. For routine control titrations or rapid preliminary estimates of acidity, the color change of an indicator may be used for the end point. Samples of industrial wastes, acid mine drainage, or other solutions that contain appreciable amounts of hydrolyzable metal ions such as iron, aluminum, or manganese are treated with hydrogen peroxide to ensure oxidation of any reduced forms of polyvalent cations, and boiled to hasten hydrolysis. Acidity results may be highly variable if this procedure is not followed exactly.

b. End points: Ideally the end point of the acidity titration should correspond to the stoichiometric equivalence point for neutralization of acids present. The pH at the equivalence point will depend on the sample, the choice among multiple inflection points, and the intended use of the data.

Dissolved carbon dioxide (CO_2) usually is the major acidic component of unpolluted surface waters; handle samples from such sources carefully to minimize the loss of dissolved gases. In a sample containing only carbon dioxide-bicarbonates-carbonates, titration to pH 8.3 at 25°C corresponds to stoichiometric neutralization of carbonic acid to bicarbonate. Because the color change of phenolphthalein indicator is close to pH 8.3, this value generally is accepted as a standard end point for titration of total acidity, including CO_2 and most weak acids. Metacresol purple also has an end point at pH 8.3 and gives a sharper color change.

For more complex mixtures or buffered solutions selection of an inflection point may be subjective. Consequently, use fixed end points of pH 3.7 and pH 8.3 for standard acidity determinations via a potentiometric titration in wastewaters and natural waters where the simple carbonate equilibria discussed above cannot be assumed. Bromphenol blue has a sharp color change at its end point of 3.7. The resulting titrations are identified, traditionally, as "methyl orange acidity" (pH 3.7) and "phenolphthalein" or total acidity (pH 8.3) regardless of the actual method of measurement.

c. Interferences: Dissolved gases contributing to acidity or alkalinity, such as CO_2, hydrogen sulfide, or ammonia, may be lost or gained during sampling, storage, or titration. Minimize such effects by titrating to the end point promptly after opening sample container, avoiding vigorous shaking or mixing, protecting sample from the atmosphere during titration, and letting sample become no warmer than it was at collection.

In the potentiometric titration, oily matter, suspended solids, precipitates, or other waste matter may coat the glass electrode and cause a sluggish response. Difficulty from this source is likely to be revealed in an erratic titration curve. Do *not* remove interferences from sample because they may contribute to its acidity. Briefly pause between titrant additions to let electrode come to equilibrium or clean the electrodes occasionally.

In samples containing oxidizable or hydrolyzable ions such as ferrous or ferric iron, aluminum, and manganese, the reaction rates at room temperature may be slow enough to cause drifting end points.

Do not use indicator titrations with colored or turbid samples that may obscure the color change at the end point. Residual free available chlorine in the sample may bleach the indicator. Eliminate this source of interference by adding 1 drop of $0.1M$ sodium thiosulfate ($Na_2S_2O_3$).

d. Selection of procedure: Determine sample acidity from the volume of standard alkali required to titrate a portion to a pH of 8.3 (phenolphthalein acidity) or pH 3.7 (methyl orange acidity of wastewaters and grossly polluted waters). Titrate at room temperature using a properly calibrated pH meter, electrically operated titrator, or color indicators.

Use the hot peroxide procedure (¶ 4a) to pretreat samples known or suspected to contain hydrolyzable metal ions or reduced forms of polyvalent cation, such as iron pickle liquors, acid mine drainage, and other industrial wastes.

Color indicators may be used for routine and control titrations in the absence of interfering color and turbidity and for preliminary titrations to select sample size and strength of titrant (¶ 4b).

e. Sample size: The range of acidities found in wastewaters is so large that a single sample size and normality of base used as titrant cannot be specified. Use a sufficiently large volume of titrant (20 mL or more from a 50-mL buret) to obtain relatively good volumetric precision while keeping sample volume sufficiently small to permit sharp end points. For samples having acidities less than about 1000 mg as calcium carbonate ($CaCO_3$)/L, select a volume with less than 50 mg $CaCO_3$ equivalent acidity and titrate with $0.02N$ sodium hydroxide (NaOH). For acidities greater than about 1000 mg as $CaCO_3$/L, use a portion containing acidity equivalent to less than 250 mg $CaCO_3$ and titrate with $0.1N$ NaOH. If necessary, make a preliminary titration to determine optimum sample size and/or normality of titrant.

f. Sampling and storage: Collect samples in polyethylene or borosilicate glass bottles and store at a low temperature. Fill bottles completely and cap tightly. Because waste samples may be subject to microbial action and to loss or gain of CO_2 or other gases when exposed to air, analyze samples without delay, preferably within 1 d. If biological activity is suspected analyze within 6 h. Avoid sample agitation and prolonged exposure to air.

2. Apparatus

a. Electrometric titrator: Use any commercial pH meter or electrically operated titrator that uses a glass electrode and can be read to 0.05 pH unit. Standardize and calibrate according to the manufacturer's instructions. Pay special attention to temperature compensation and electrode care. If automatic temperature compensation is not provided, titrate at 25 ± 5°C.

b. Titration vessel: The size and form will depend on the electrodes and the sample size. Keep the free space above the sample as small as practicable, but allow room for titrant and full immersion of the indicating portions of electrodes. For conventional-sized electrodes, use a 200-mL, tall-form Berzelius beaker without a spout. Fit beaker with a stopper having three holes, to accommodate the two electrodes and the buret. With a miniature combination glass-reference electrode use a 125-mL or 250-mL erlenmeyer flask with a two-hole stopper.

c. Magnetic stirrer.

d. Pipets, volumetric.

e. Flasks, volumetric, 1000-, 200-, 100-mL.

f. Burets, borosilicate glass, 50-, 25-, 10-mL.

g. Polyolefin bottle, 1-L.

3. Reagents

a. Carbon dioxide-free water: Prepare all stock and standard solutions and dilution water for the standardization procedure with distilled or deionized water that has been freshly boiled for 15 min and cooled to room temperature. The final pH of the water should be ≥ 6.0 and its conductivity should be <2 μmhos/cm.

b. Potassium hydrogen phthalate solution, approximately $0.05N$: Crush 15 to 20 g primary standard $KHC_8H_4O_4$ to about 100 mesh and dry at 120°C for 2 h. Cool in a desiccator. Weigh 10.0 ± 0.5 g (to the nearest mg), transfer to a 1-L volumetric flask, and dilute to 1000 mL.

c. Standard sodium hydroxide titrant, $0.1N$: Prepare solution approximately $0.1N$ as indicated under Preparation of Desk Reagents (see inside front cover). Standardize by titrating 40.00 mL $KHC_8H_4O_4$ solution (3b), using a 25-mL buret. Titrate to the inflection point (¶ 1a), which should be close to pH 8.7. Calculate normality of NaOH:

$$\text{Normality} = \frac{A \times B}{204.2 \times C}$$

where:

A = g $KHC_8H_4O_4$ weighed into 1-L flask,
B = mL $KHC_8H_4O_4$ solution taken for titration, and
C = mL NaOH solution used.

Use the measured normality in further calculations or adjust to $0.1000N$; 1 mL = 5.00 mg $CaCO_3$.

d. Standard sodium hydroxide titrant, $0.02N$: Dilute 200 mL $0.1N$ NaOH to 1000 mL and store in a polyolefin bottle protected from atmospheric CO_2 by a soda lime tube or tight cap. Standardize against $KHC_8H_4O_4$ as directed in ¶ 3c, using 15.00 mL $KHC_8H_4O_4$ solution and a 50-mL buret. Calculate normality as above (¶ 3c); 1 mL = 1.00 mg $CaCO_3$.

e. Hydrogen peroxide, H_2O_2, 30%.

f. Bromphenol blue indicator solution, pH 3.7 indicator: Dissolve 100 mg bromphenol blue, sodium salt, in 100 mL water.

g. Metacresol purple indicator solution, pH 8.3 indicator: Dissolve 100 mg metacresol purple in 100 mL water.

h. Phenolphthalein indicator solution, alcoholic, pH 8.3 indicator.

i. Sodium thiosulfate, $0.1M$: Dissolve 25 g $Na_2S_2O_3 \cdot 5H_2O$ and dilute to 1000 mL with distilled water.

4. Procedure

If sample is free from hydrolyzable metal ions and reduced forms of polyvalent cations, proceed with analysis according to b, c, or d. If sample is known or suspected to contain such substances, pretreat according to a.

a. Hot peroxide treatment: Pipet a suitable sample (see ¶ 1e) into titration flasks. Measure pH. If pH is above 4.0 add 5-mL increments of $0.02N$ sulfuric acid (H_2SO_4) (Section 2320B.3c) to reduce pH to 4 or less. Remove electrodes. Add 5 drops 30% H_2O_2 and boil for 2 to 5 min. Cool to room temperature and titrate with standard alkali to pH 8.3 according to the procedure of 4d.

b. Color change: Select sample size and normality of titrant according to criteria of ¶ 1e. Adjust sample to room temperature, if necessary, and with a pipet discharge sample into an erlenmeyer flask, while keeping pipet tip near flask bottom. If free residual chlorine is present add 0.05 mL (1 drop) $0.1M$ $Na_2S_2O_3$ solution, or destroy with ultraviolet radiation. Add 0.2 mL (5 drops) indicator solution and titrate over a white surface to a persistent color change characteristic of the equivalence point. Commercial indicator solutions or solids designated for the appropriate pH range (3.7 or 8.3) may be used. Check color at end point by adding the same concentration of indicator used with sample to a buffer solution at the designated pH.

c. Potentiometric titration curve:

1) Rinse electrodes and titration vessel with distilled water and drain. Select sample size and normality of titrant according to the criteria of ¶ 1e. Adjust sample to room temperature, if necessary, and with a pipet discharge sample while keeping pipet tip near the titration vessel bottom.

2) Measure sample pH. Add standard alkali in increments of 0.5 mL or less, such that a change of less than 0.2 pH units occurs per increment. After each addition, mix thoroughly but gently with a magnetic stirrer. Avoid splashing. Record pH when a constant reading is obtained. Continue adding titrant and measure pH until pH 9 is reached. Construct the titration curve by plotting observed pH values versus cumulative milliliters titrant added. A smooth curve showing one or more inflections should be obtained. A ragged or erratic curve may indicate that equilibrium was not reached between successive alkali additions. Determine acidity relative to a particular pH from the curve.

d. Potentiometric titration to pH 3.7 or 8.3: Prepare sample and titration assembly as specified in ¶ 4c1). Titrate to preselected end-point pH (¶ 1d) without recording intermediate pH values. As the end point is approached make smaller additions of alkali and be sure that pH equilibrium is reached before making the next addition.

5. Calculation

$$\text{Acidity, as mg } CaCO_3/L = \frac{[(A \times B) - (C \times D)] \times 50\,000}{\text{mL sample}}$$

where:

  A = mL NaOH titrant used,
  B = normality of NaOH,
  C = mL H_2SO_4 used (¶ 4a), and
  D = normality of H_2SO_4.

Report pH of the end point used, as follows: ''The acidity to pH ____ = ____ mg $CaCO_3$/L.'' If a negative value is obtained, report the value as negative. The absolute value of this negative value should be equivalent to the net alkalinity.

6. Precision and Bias

No general statement can be made about precision because of the great variation in sample characteristics. The precision of the titration is likely to be much greater than the uncertainties involved in sampling and sample handling before analysis.

Forty analysts in 17 laboratories analyzed synthetic water samples containing increments of bicarbonate equivalent to 20 mg $CaCO_3$/L. Titration according to the procedure of ¶ 4d gave a standard deviation of 1.8 mg $CaCO_3$/L, with negligible bias. Five laboratories analyzed two samples containing sulfuric, acetic, and formic acids and aluminum chloride by the procedures of ¶s 4b and 4d. The mean acidity of one sample (to pH 3.7) was 487 mg $CaCO_3$/L, with a standard deviation of 11 mg/L. The bromphenol blue titration of the same sample was 90 mg/L greater, with a standard deviation of 110 mg/L. The other sample had a potentiometric titration of 547 mg/L, with a standard deviation of 54 mg/L, while the corresponding indicator result was 85 mg/L greater, with a standard deviation of 56 mg/L. The major difference between the samples was the substitution of ferric ammonium citrate, in the second sample, for part of the aluminum chloride.

7. Bibliography

Winter, J.A. & M.R. Midgett. 1969. FWPCA Method Study 1. Mineral and Physical Analyses. Federal Water Pollution Control Admin., Washington, D.C.

Brown, E., M.W. Skougstad & M.J. Fishman. 1970. Methods for collection and analysis of water samples for dissolved minerals and gases. Chapter A1 *in* Book 5, Techniques of Water-Resources Investigations of United States Geological Survey. U.S. Geological Survey, Washington, D.C.

Snoeyink, V.L. & D. Jenkins. 1980. Water Chemistry. John Wiley & Sons, New York, N.Y.

2320   ALKALINITY*

2320 A.   Introduction

1. Discussion

Alkalinity of a water is its acid-neutralizing capacity. It is the sum of all the titratable bases. The measured value may vary significantly with the end-point pH used. Alkalinity is a measure of an aggregate property of water and can be interpreted in terms of specific substances only when the chemical composition of the sample is known.

Alkalinity is significant in many uses and treatments of natural waters and wastewaters. Because the alkalinity of many surface waters is primarily a function of carbonate, bicarbonate, and hydroxide content, it is taken as an indication of the concentration of these constitutents. The measured values also may include contributions from borates, phosphates, silicates, or other bases if these are present. Alkalinity in excess of alkaline earth metal concentrations is significant in determining the suitability of a water for irrigation. Alkalinity measurements are used in the interpre-

---

* Approved by Standard Methods Committee, 1997.

tation and control of water and wastewater treatment processes. Raw domestic wastewater has an alkalinity less than, or only slightly greater than, that of the water supply. Properly operating anaerobic digesters typically have supernatant alkalinities in the range of 2000 to 4000 mg calcium carbonate ($CaCO_3$)/L.[1]

2. Reference

1. POHLAND, F.G. & D.E. BLOODGOOD. 1963. Laboratory studies on mesophilic and thermophilic anaerobic sludge digestion. *J. Water Pollut. Control Fed.* 35:11.

2320 B. Titration Method

1. General Discussion

a. Principle: Hydroxyl ions present in a sample as a result of dissociation or hydrolysis of solutes react with additions of standard acid. Alkalinity thus depends on the end-point pH used. For methods of determining inflection points from titration curves and the rationale for titrating to fixed pH end points, see Section 2310B.1a.

For samples of low alkalinity (less than 20 mg $CaCO_3$/L) use an extrapolation technique based on the near proportionality of concentration of hydrogen ions to excess of titrant beyond the equivalence point. The amount of standard acid required to reduce pH exactly 0.30 pH unit is measured carefully. Because this change in pH corresponds to an exact doubling of the hydrogen ion concentration, a simple extrapolation can be made to the equivalence point.[1,2]

b. End points: When alkalinity is due entirely to carbonate or bicarbonate content, the pH at the equivalence point of the titration is determined by the concentration of carbon dioxide (CO_2) at that stage. CO_2 concentration depends, in turn, on the total carbonate species originally present and any losses that may have occurred during titration. The pH values in Table 2320:I are suggested as the equivalence points for the corresponding alkalinity concentrations as milligrams $CaCO_3$ per liter. "Phenolphthalein alkalinity" is the term traditionally used for the quantity measured by titration to pH 8.3 irrespective of the colored indicator, if any, used in the determination. Phenolphthalein or metacresol purple may be used for alkalinity titration to pH 8.3. Bromcresol green or a mixed bromcresol green-methyl red indicator may be used for pH 4.5.

TABLE 2320:I END-POINT pH VALUES

Test Condition	End Point pH	
	Total Alkalinity	Phenolphthalein Alkalinity
Alkalinity, mg $CaCO_3$/L:		
30	4.9	8.3
150	4.6	8.3
500	4.3	8.3
Silicates, phosphates known or suspected	4.5	8.3
Routine or automated analyses	4.5	8.3
Industrial waste or complex system	4.5	8.3

c. Interferences: Soaps, oily matter, suspended solids, or precipitates may coat the glass electrode and cause a sluggish response. Allow additional time between titrant additions to let electrode come to equilibrium or clean the electrodes occasionally. Do not filter, dilute, concentrate, or alter sample.

d. Selection of procedure: Determine sample alkalinity from volume of standard acid required to titrate a portion to a designated pH taken from ¶ 1b. Titrate at room temperature with a properly calibrated pH meter or electrically operated titrator, or use color indicators. If using color indicators, prepare and titrate an indicator blank.

Report alkalinity less than 20 mg $CaCO_3$/L only if it has been determined by the low-alkalinity method of ¶ 4d.

Construct a titration curve for standardization of reagents.

Color indicators may be used for routine and control titrations in the absence of interfering color and turbidity and for preliminary titrations to select sample size and strength of titrant (see below).

e. Sample size: See Section 2310B.1e for selection of size sample to be titrated and normality of titrant, substituting 0.02N or 0.1N sulfuric (H_2SO_4) or hydrochloric (HCl) acid for the standard alkali of that method. For the low-alkalinity method, titrate a 200-mL sample with 0.02N H_2SO_4 from a 10-mL buret.

f. Sampling and storage: See Section 2310B.1f.

2. Apparatus

See Section 2310B.2.

3. Reagents

a. Sodium carbonate solution, approximately 0.05N: Dry 3 to 5 g primary standard Na_2CO_3 at 250°C for 4 h and cool in a desiccator. Weigh 2.5 ± 0.2 g (to the nearest mg), transfer to a 1-L volumetric flask, fill flask to the mark with distilled water, and dissolve and mix reagent. Do not keep longer than 1 week.

b. Standard sulfuric acid or hydrochloric acid, 0.1N: Prepare acid solution of approximate normality as indicated under Preparation of Desk Reagents (see inside front cover). Standardize against 40.00 mL 0.05N Na_2CO_3 solution, with about 60 mL water, in a beaker by titrating potentiometrically to pH of about 5. Lift out electrodes, rinse into the same beaker, and boil gently for 3 to 5 min under a watch glass cover. Cool to room temperature, rinse cover glass into beaker, and finish titrating to the pH inflection point. Calculate normality:

$$\text{Normality, } N = \frac{A \times B}{53.00 \times C}$$

where:

A = g Na_2CO_3 weighed into 1-L flask,
B = mL Na_2CO_3 solution taken for titration, and
C = mL acid used.

Use measured normality in calculations or adjust to $0.1000N$; 1 mL $0.1000N$ solution = 5.00 mg $CaCO_3$.

c. Standard sulfuric acid or hydrochloric acid, 0.02N: Dilute 200.00 mL $0.1000N$ standard acid to 1000 mL with distilled or deionized water. Standardize by potentiometric titration of 15.00 mL $0.05N$ Na_2CO_3 according to the procedure of ¶ 3*b*; 1 mL = 1.00 mg $CaCO_3$.

d. Bromcresol green indicator solution, pH 4.5 indicator: Dissolve 100 mg bromcresol green, sodium salt, in 100 mL distilled water.

e. Mixed bromcresol green-methyl red indicator solution:[3] Use either the aqueous or the alcoholic solution:

1) Dissolve 100 mg bromcresol green sodium salt and 20 mg methyl red sodium salt in 100 mL distilled water.

2) Dissolve 100 mg bromcresol green and 20 mg methyl red in 100 mL 95% ethyl alcohol or isopropyl alcohol.

f. Metacresol purple indicator solution, pH 8.3 indicator: Dissolve 100 mg metacresol purple in 100 mL water.

g. Phenolphthalein solution, alcoholic, pH 8.3 indicator.

h. Sodium thiosulfate, 0.1N: See Section 2310B.3*i*.

4. Procedure

a. Color change: See Section 2310B.4*b*.

b. Potentiometric titration curve: Follow the procedure for determining acidity (Section 2310B.4*c*), substituting the appropriate normality of standard acid solution for standard NaOH, and continue titration to pH 4.5 or lower. Do not filter, dilute, concentrate, or alter the sample.

c. Potentiometric titration to preselected pH: Determine the appropriate end-point pH according to ¶ 1*b*. Prepare sample and titration assembly (Section 2310B.4*c*). Titrate to the end-point pH without recording intermediate pH values and without undue delay. As the end point is approached make smaller additions of acid and be sure that pH equilibrium is reached before adding more titrant.

d. Potentiometric titration of low alkalinity: For alkalinities less than 20 mg/L titrate 100 to 200 mL according to the procedure of ¶ 4*c*, above, using a 10-mL microburet and $0.02N$ standard acid solution. Stop the titration at a pH in the range 4.3 to 4.7 and record volume and exact pH. Carefully add additional titrant to reduce the pH exactly 0.30 pH unit and again record volume.

5. Calculations

a. Potentiometric titration to end-point pH:

$$\text{Alkalinity, mg } CaCO_3/L = \frac{A \times N \times 50\,000}{\text{mL sample}}$$

where:

A = mL standard acid used and
N = normality of standard acid

or

$$\text{Alkalinity, mg } CaCO_3/L = \frac{A \times t \times 1000}{\text{mL sample}}$$

where:

t = titer of standard acid, mg $CaCO_3$/mL.

Report pH of end point used as follows: "The alkalinity to pH _____ = _____ mg $CaCO_3$/L" and indicate clearly if this pH corresponds to an inflection point of the titration curve.

b. Potentiometric titration of low alkalinity:

Total alkalinity, mg $CaCO_3$/L

$$= \frac{(2\,B - C) \times N \times 50\,000}{\text{mL sample}}$$

where:

B = mL titrant to first recorded pH,
C = total mL titrant to reach pH 0.3 unit lower, and
N = normality of acid.

c. Calculation of alkalinity relationships: The results obtained from the phenolphthalein and total alkalinity determinations offer a means for stoichiometric classification of the three principal forms of alkalinity present in many waters. The classification ascribes the entire alkalinity to bicarbonate, carbonate, and hydroxide, and assumes the absence of other (weak) inorganic or organic acids, such as silicic, phosphoric, and boric acids. It further presupposes the incompatibility of hydroxide and bicarbonate alkalinities. Because the calculations are made on a stoichiometric basis, ion concentrations in the strictest sense are not represented in the results, which may differ significantly from actual concentrations especially at pH > 10. According to this scheme:

1) Carbonate (CO_3^{2-}) alkalinity is present when phenolphthalein alkalinity is not zero but is less than total alkalinity.

2) Hydroxide (OH^-) alkalinity is present if phenolphthalein alkalinity is more than half the total alkalinity.

3) Bicarbonate (HCO_3^-) alkalinity is present if phenolphthalein alkalinity is less than half the total alkalinity. These relationships may be calculated by the following scheme, where P is phenolphthalein alkalinity and T is total alkalinity (¶ 1*b*):

Select the smaller value of P or $(T-P)$. Then, carbonate alkalinity equals twice the smaller value. When the smaller value is P, the balance $(T-2P)$ is bicarbonate. When the smaller value is $(T-P)$, the balance $(2P-T)$ is hydroxide. All results are expressed as $CaCO_3$. The mathematical conversion of the results is shown in Table 2320:II. (A modification of Table 2320:II that is more accurate when $P \simeq {}^1/_2 T$ has been proposed.[4])

Alkalinity relationships also may be computed nomographically (see Carbon Dioxide, Section 4500-CO_2). Accurately measure pH, calculate OH^- concentration as milligrams $CaCO_3$ per liter, and calculate concentrations of CO_3^{2-} and HCO_3^- as milligrams $CaCO_3$ per liter from the OH^- concentration, and the phenolphthalein and total alkalinities by the following equations:

$$CO_3^{2-} = 2P - 2[OH^-]$$

$$HCO_3^- = T - 2P + [OH^-]$$

TABLE 2320:II. ALKALINITY RELATIONSHIPS*

Result of Titration	Hydroxide Alkalinity as CaCO$_3$	Carbonate Alkalinity as CaCO$_3$	Bicarbonate Concentration as CaCO$_3$
P = 0	0	0	T
P < ½T	0	2P	T − 2P
P = ½T	0	2P	0
P > ½T	2P − T	2(T − P)	0
P = T	T	0	0

*Key: P − phenolphthalein alkalinity; T − total alkalinity.

Similarly, if difficulty is experienced with the phenolphthalein end point, or if a check on the phenolphthalein titration is desired, calculate phenolphthalein alkalinity as CaCO$_3$ from the results of the nomographic determinations of carbonate and hydroxide ion concentrations:

$$P = 1/2 \, [CO_3^{2-}] + [OH^-]$$

6. Precision and Bias

No general statement can be made about precision because of the great variation in sample characteristics. The precision of the titration is likely to be much greater than the uncertainties involved in sampling and sample handling before the analysis.

In the range of 10 to 500 mg/L, when the alkalinity is due entirely to carbonates or bicarbonates, a standard deviation of 1 mg CaCO$_3$/L can be achieved. Forty analysts in 17 laboratories analyzed synthetic samples containing increments of bicarbonate equivalent to 120 mg CaCO$_3$/L. The titration procedure of ¶ 4b was used, with an end point pH of 4.5. The standard deviation was 5 mg/L and the average bias (lower than the true value) was 9 mg/L.[5]

Sodium carbonate solutions equivalent to 80 and 65 mg CaCO$_3$/L were analyzed by 12 laboratories according to the procedure of ¶ 4c.[6] The standard deviations were 8 and 5 mg/L, respectively, with negligible bias.[6] Four laboratories analyzed six samples having total alkalinities of about 1000 mg CaCO$_3$/L and containing various ratios of carbonate/bicarbonate by the procedures of both ¶ 4a and ¶ 4c. The pooled standard deviation was 40 mg/L, with negligible difference between the procedures.

7. References

1. LARSON, T.E. & L.M. HENLEY. 1955. Determination of low alkalinity or acidity in water. *Anal. Chem.* 27:851.
2. THOMAS, J.F.J. & J.J. LYNCH. 1960. Determination of carbonate alkalinity in natural waters. *J. Amer. Water Works Assoc.* 52:259.
3. COOPER, S.S. 1941. The mixed indicator bromocresol green-methyl red for carbonates in water. *Ind. Eng. Chem.*, Anal. Ed. 13:466.
4. JENKINS, S.R. & R.C. MOORE. 1977. A proposed modification to the classical method of calculating alkalinity in natural waters. *J. Amer. Water Works Assoc.* 69:56.
5. WINTER, J.A. & M.R. MIDGETT. 1969. FWPCA Method Study 1. Mineral and Physical Analyses. Federal Water Pollution Control Admin., Washington, D.C.
6. SMITH, R. 1980. Research Rep. No. 379, Council for Scientific and Industrial Research, South Africa.

8. Bibliography

AMERICAN SOCIETY FOR TESTING AND MATERIALS. 1982. Standard Methods for Acidity or Alkalinity of Water. Publ. D1067-70 (reapproved 1977), American Soc. Testing & Materials, Philadelphia, Pa.
SKOUGSTAD M.W., M.J. FISHMAN, L.C. FRIEDMAN, D.E. ERDMAN, & S.S. DUNCAN. 1979. Methods for determination of inorganic substances in water and fluvial sediments. *In* Techniques of Water-Resources Investigation of the United States Geological Survey. U.S. Geological Survey, Book 5, Chapter A1, Washington, D.C.

2330 CALCIUM CARBONATE SATURATION*

2330 A. Introduction

1. General Discussion

Calcium carbonate (CaCO$_3$) saturation indices commonly are used to evaluate the scale-forming and scale-dissolving tendencies of water. Assessing these tendencies is useful in corrosion control programs and in preventing CaCO$_3$ scaling in piping and equipment such as industrial heat exchangers or domestic water heaters.

Waters oversaturated with respect to CaCO$_3$ tend to precipitate CaCO$_3$. Waters undersaturated with respect to CaCO$_3$ tend to dissolve CaCO$_3$. Saturated waters, i.e., waters in equilibrium with

CaCO$_3$, have neither CaCO$_3$-precipitating nor CaCO$_3$-dissolving tendencies. Saturation represents the dividing line between "precipitation likely" and "precipitation not likely."

Several water quality characteristics must be measured to calculate the CaCO$_3$ saturation indices described here. Minimum requirements are total alkalinity (2320), total calcium (3500-Ca), pH (4500-H$^+$), and temperature (2550). The ionic strength also must be calculated or estimated from conductivity (2510) or total dissolved solids (2540C) measurements. Measure pH at the system's water temperature using a temperature-compensated pH meter. If pH is measured at a different temperature, for example in the laboratory, correct the measured pH.[1-3] In measuring pH and alkalinity, minimize CO$_2$ exchange between sample and atmosphere. Ideally, seal the sample from the atmosphere during

---

* Approved by Standard Methods Committee, 1993.

measurements[4]; at a minimum, avoid vigorous stirring of unsealed samples.

There are two general categories of $CaCO_3$ saturation indices: indices that determine whether a water has a *tendency* to precipitate $CaCO_3$ (i.e., is oversaturated) or to dissolve $CaCO_3$ (i.e., is undersaturated) and indices that estimate the *quantity* of $CaCO_3$ that can be precipitated from an oversaturated water and the amount that can be dissolved by an undersaturated water. Indices in the second category generally yield more information but are more difficult to determine.

2. Limitations

It is widely assumed that $CaCO_3$ will precipitate from oversaturated waters and that it cannot be deposited from undersaturated waters. Exceptions may occur. For example, $CaCO_3$ deposition from oversaturated waters is inhibited by the presence of phosphates (particularly polyphosphates), certain naturally occurring organics, and magnesium.[5-7] These materials can act as sequestering agents or as crystal poisons. Conversely, $CaCO_3$ deposits have been found in pipes conveying undersaturated water. This apparent contradiction is caused by high pH (relative to the bulk water pH) in the immediate vicinity of certain areas (cathodes) of corroding metal surfaces. A locally oversaturated condition may occur even if the bulk water is undersaturated. A small, but significant, amount of $CaCO_3$ can be deposited.

The calculations referred to here, even the most sophisticated computerized calculations, do not adequately describe these exceptions. For this reason, do not consider saturation indices as absolutes. Rather, view them as guides to the behavior of $CaCO_3$ in aqueous systems and supplement them, where possible, with experimentally derived information.

Similarly, the effects predicted by the indices do not always conform to expectations. The relationship between the indices and corrosion rates is a case in point. Conceptually, piping is protected when $CaCO_3$ is precipitated on its surfaces. $CaCO_3$ is believed to inhibit corrosion by clogging reactive areas and by providing a matrix to retain corrosion products, thus further sealing the surfaces. Waters with positive indices traditionally have been assumed to be protective while waters with negative indices have been assumed to be not protective, or corrosive. The expected relationship is observed sometimes,[8,9] but not always.[10,11] Unexpected results may be due in part to limited capability to predict $CaCO_3$ behavior. Also, water characteristics not directly involved in the calculation of the indices (e.g., dissolved oxygen, buffering intensity, chloride, sulfate, and water velocity) may influence corrosion rates appreciably.[9,12-16] Thus, do not estimate corrosion rates on the basis of $CaCO_3$ indices alone.

3. References

1. MERRILL, D.T. & R.L. SANKS. 1978, 1979. Corrosion control by deposition of $CaCO_3$ films: A practical approach for plant operators. *J. Amer. Water Works Assoc.* 70:592; 70:634 & 71:12.
2. LOEWENTHAL, R.E. & G. v. R. MARAIS. 1976. Carbonate Chemistry of Aquatic Systems: Theory and Applications. Ann Arbor Science Publishers, Ann Arbor, Mich.
3. MERRILL, D.T. 1976. Chemical conditioning for water softening and corrosion control. *In* R.L. Sanks, ed. Water Treatment Plant Design. Ann Arbor Science Publishers, Ann Arbor, Mich.
4. SCHOCK, M.R., W. MUELLER & R.W. BUELOW. 1980. Laboratory techniques for measurement of pH for corrosion control studies and water not in equilibrium with the atmosphere. *J. Amer. Water Works Assoc.* 72:304.
5. PYTKOWICZ, R.M. 1965. Rates of inorganic carbon nucleation. *J. Geol.* 73:196.
6. FERGUSON, J.F. & P.L. MCCARTY. 1969. Precipitation of Phosphate from Fresh Waters and Waste Waters. Tech. Rep. No. 120, Stanford Univ., Stanford, Calif.
7. MERRILL, D.T. & R.M. JORDEN. 1975. Lime-induced reactions in municipal wastewaters. *J. Water Pollut. Control Fed.* 47:2783.
8. DE MARTINI, F. 1938. Corrosion and the Langelier calcium carbonate saturation index. *J. Amer. Water Works Assoc.* 30:85.
9. LARSON, T.E. 1975. Corrosion by Domestic Waters. Bull. 59, Illinois State Water Survey.
10. JAMES M. MONTGOMERY, CONSULTING ENGINEERS, INC. 1985. Water Treatment Principles and Design. John Wiley & Sons, New York, N.Y.
11. STUMM, W. 1960. Investigations of the corrosive behavior of waters. *J. San. Eng. Div., Proc. Amer. Soc. Civil Eng.* 86:27.
12. PISIGAN, R.A., JR. & J.E. SINGLEY. 1985. Evaluation of water corrosivity using the Langelier index and relative corrosion rate models. *Materials Perform.* 24:26.
13. LANE, R.W. 1982. Control of corrosion in distribution and building water systems. *In* Proceedings AWWA Water Quality Technology Conf., Nashville, Tenn., Dec. 5–8, 1982.
14. SONTHEIMER, H., W. KOLLE & V.L. SNOEYINK. 1982. The siderite model of the formation of corrosion-resistant scales. *J. Amer. Water Works Assoc.* 73:572.
15. SCHOCK, M.R. & C.H. NEFF. 1982. Chemical aspects of internal corrosion; theory, prediction, and monitoring. *In* Proceedings AWWA Water Quality Technology Conf., Nashville, Tenn., Dec. 5–8, 1982.
16. AMERICAN WATER WORKS ASSOCIATION. 1986. Corrosion Control for Plant Operators. ISBW-O-89867-350-X. Denver, Colo.

2330 B.   Indices Indicating Tendency of a Water to Precipitate $CaCO_3$ or Dissolve $CaCO_3$

1. General Discussion

Indices that indicate $CaCO_3$ precipitation or dissolution tendencies define whether a water is oversaturated, saturated, or undersaturated with respect to $CaCO_3$. The most widely used indices are the Saturation Index (SI): the Relative Saturation (RS), also known as the Driving Force Index (DFI); and the Ryznar Index (RI). The SI is by far the most commonly used and will be described here. The RS and SI are related (see Equation 6, Section 2330D). The RI[1] has been used for many years, sometimes with good results. Because it is semi-empirical it may be less reliable than the SI.

2. Saturation Index by Calculation

SI is determined from Equation 1.

$$SI = pH - pH_s \qquad (1)$$

where:

  pH = measured pH and
  pH_s = pH of the water if it were in equilibrium with $CaCO_3$ at the existing calcium ion [Ca^{2+}] and bicarbonate ion [HCO_3^-] concentrations.

A positive SI connotes a water oversaturated with respect to $CaCO_3$. A negative SI signifies an undersaturated water. An SI of zero represents a water in equilibrium with $CaCO_3$.

  a. *Analytical solution for* pH_s: Determine pH_s as follows[2]:

$$pH_s = pK_2 - pK_s + p[Ca^{2+}] + p[HCO_3^-] + 5\ pf_m \qquad (2)$$

where:

  K_2 = second dissociation constant for carbonic acid, at the water temperature,
  K_s = solubility product constant for $CaCO_3$ at the water temperature,
  [Ca^{2+}] = calcium ion concentration, g-moles/L,
  [HCO_3^-] = bicarbonate ion concentration, g-moles/L, and
  f_m = activity coefficient for monovalent species at the specified temperature.

In Equation 2, p preceding a variable designates $-\log_{10}$ of that variable.

Calculate values of pK_2, pK_s, and pf_m required to solve Equation 2 from the equations in Table 2330:I. To save computation time, values for pK_2 and pK_s have been precalculated for selected temperatures (see Table 2330:II).

Table 2330:II gives several values for pK_s. Different isomorphs of $CaCO_3$ can form in aqueous systems, including calcite, aragonite, and vaterite. Each has somewhat different solubility properties. These differences can be accommodated when computing

pH_s simply by using the pK_s for the compound most likely to form. The form of $CaCO_3$ most commonly found in fresh water is calcite. Use the pK_s for calcite unless it is clear that a different form of $CaCO_3$ controls $CaCO_3$ solubility.

Estimate calcium ion concentration from total calcium measurements with Equation 3.

$$[Ca^{2+}] = Ca_t - Ca_{ip} \qquad (3)$$

where:

  Ca_t = total calcium, g-moles/L, and
  Ca_{ip} = calcium associated with ion pairs such as $CaHCO_3^+$, $CaSO_4^0$, and $CaOH^+$.

Calcium associated with ion pairs is not available to form $CaCO_3$.

Estimate [HCO_3^-], the bicarbonate ion concentration, from Equation 4.

$$[HCO_3^-] = \frac{Alk_t - Alk_o + 10^{(pf_m - pH)} - 10^{(pH + pf_m - pK_w)}}{1 + 0.5 \times 10^{(pH - pK_2)}} \qquad (4)$$

where:

  Alk_t = total alkalinity, as determined by acid titration to the carbonic acid end point, g-equivalents/L,
  K_w = dissociation constant for water, at the water temperature, and
  Alk_o = alkalinity contributed by NH_3^0, $H_3SiO_4^-$, HPO_4^{2-}, $B(OH)_4^-$, CH_3COO^- (acetate), HS^-, and ion pairs such as $CaHCO_3^+$ and $MgOH^+$. These contributions usually are small compared to the contributions of components normally considered (HCO_3^-, CO_3^{2-}, OH^-, and H^+).

Calculations can be simplified. For example, in Equation 4,

TABLE 2330:I. ESTIMATING EQUILIBRIUM CONSTANTS AND ACTIVITY COEFFICIENTS

Equation	Temperature Range	References
When complete mineral analysis is available:		
$I = {}^1\!/_2 \sum_{i=1}^{i} [X_i]\ Z_i^2$	—	3
When only conductivity is available:		
$I = 1.6 \times 10^{-5}C$	—	4
When only TDS is available:		
$I = TDS/40\ 000$	—	5
$pf_m = A\left[\dfrac{\sqrt{I}}{1 + \sqrt{I}} - 0.3I\right]$ (valid to $I < 0.5$)	—	3
$A = 1.82 \times 10^6\ (ET)^{-1.5}$	—	3
$E = \dfrac{60\ 954}{T + 116} - 68.937$	—	6
$pK_2 = 107.8871 + 0.032\ 528\ 49T - 5151.79/T - 38.925\ 61\ \log_{10}T + 563\ 713.9/T^2$	273–373	7
$pK_w = 4471/T + 0.017\ 06T - 6.0875$	280–338	8
$pK_{sc} = 171.9065 + 0.077\ 993T - 2839.319/T - 71.595\ \log_{10}T$	273–363	7
$pK_{sa} = 171.9773 + 0.077\ 993T - 2903.293/T - 71.595\ \log_{10}T$	273–363	7
$pK_{sv} = 172.1295 + 0.077\ 993T - 3074.688/T - 71.595\ \log_{10}T$	273–363	7

*I	= ionic strength	E	= dielectric constant
[X_i]	= concentration of component i, g-moles/L	T	= temperature, °K (°C + 273.2)
Z_i	= charge of species i	K_2	= second dissociation constant for carbonic acid
C	= conductivity, μmhos/cm	K_w	= dissociation constant for water
TDS	= total dissolved solids, mg/L	K_{sc}	= solubility product constant for calcite
pY	= $-\log_{10}$ of the value of any factor Y	K_{sa}	= solubility product constant for aragonite
f_m	= activity coefficient for monovalent species	K_{sv}	= solubility product constant for vaterite

TABLE 2330:II. PRECALCULATED VALUES FOR pK AND A AT SELECTED TEMPERATURES

Temperature °C	pK_2	pK_s Calcite	pK_s Aragonite	pK_s Vaterite	pK_w	A
5	10.55	8.39	8.24	7.77	14.73	0.494
10	10.49	8.41	8.26	7.80	14.53	0.498
15	10.43	8.43	8.28	7.84	14.34	0.502
20	10.38	8.45	8.31	7.87	14.16	0.506
25*	10.33	8.48	8.34	7.91	13.99	0.511
30	10.29	8.51	8.37	7.96	13.83	0.515
35	10.25	8.54	8.41	8.00	13.68	0.520
40	10.22	8.58	8.45	8.05	13.53	0.526
45	10.20	8.62	8.49	8.10	13.39	0.531
50	10.17	8.66	8.54	8.16	13.26	0.537
60	10.14	8.76	8.64	8.28	13.02	0.549
70	10.13	8.87	8.75	8.40	—	0.562
80	10.13	8.99	8.88	8.55	—	0.576
90	10.14	9.12	9.02	8.70	—	0.591

NOTE: All values determined from the equations of Table 2330:I.
A is used to calculate pf_m (see Table 2330:I).
* pf_m estimated from TDS values at 25°C care as follows:

TDS	pf_m
100	0.024
200	0.033
400	0.044
800	0.060
1000	0.066

terms containing exponents (e.g., $10^{(pH + pf_m - pk_w)}$) usually can be neglected for waters that are approximately neutral (pH 6.0 to 8.5) with alkalinity greater than about 50 mg/L as $CaCO_3$. The terms Ca_{ip} in Equation 3 and Alk_o in Equation 4 are difficult to calculate without computers. Therefore they usually are neglected for hand calculations. The simplified version of Equation 2 under such conditions is:

$$pH_s = pK_2 - pK_s + p[Ca_t] + p[Alk_t] + 5\ pf_m$$

1) Sample calculation—The calculation is best illustrated by working through an example. Assume that calcite controls $CaCO_3$ solubility and determine the SI for a water of the following composition:

Constituent	mg	÷ mg/mole	= g-moles/L
Calcium	152	40 000	3.80×10^{-3}
Magnesium	39	24 312	1.60×10^{-3}
Sodium	50	22 989	2.18×10^{-3}
Potassium	5	39 102	1.28×10^{-4}
Chloride	53	35 453	1.49×10^{-3}
Alkalinity (as $CaCO_3$)	130	50 000	2.60×10^{-3}*
Sulfate	430	96 060	4.48×10^{-3}
Silica (as SiO_2)	15	60 084	2.50×10^{-4}

*g-equivalents.
Water temperature = 20°C (293.2°K); pH = 9.00.

Before evaluating pf_m in Equation 2, determine the ionic

strength (I) and another constant (A). Estimate ionic strength from the first equation of Table 2330:I, assuming all the alkalinity is due to bicarbonate ion. Use the alkalinity concentration (2.60×10^{-3}) and the bicarbonate charge (-1) to calculate the contribution of alkalinity to ionic strength. Assume silica is mostly H_4SiO_4. Because H_4SiO_4 has zero charge, silica does not contribute to ionic strength.

$$
\begin{aligned}
I &= 0.5\ [4(3.80 \times 10^{-3}) + 4(1.60 \times 10^{-3}) + 2.18 \times 10^{-3} + \\
&\quad 1.28 \times 10^{-4} + 1.49 \times 10^{-3} + 2.60 \times 10^{-3} \\
&\quad + 4(4.48 \times 10^{-3})] \\
&= 2.29 \times 10^{-2}\ \text{g-moles/L}
\end{aligned}
$$

In the absence of a complete water analysis, estimate ionic strength from conductivity or total dissolved solids measurements (see alternative equations, Table 2330:I).

Estimate A from the equation in Table 2330:I, after first determining the dielectric constant E from the formula in the same table. Alternatively, use precalculated values of A in Table 2330:II. In Table 2330:II, $A = 0.506$ at 20°C.

Next estimate pf_m from the equation in Table 2330:I:

$$
\begin{aligned}
pf_m &= 0.506 \\
&\quad \times \left[\frac{\sqrt{2.29 \times 10^{-2}}}{1 + \sqrt{2.29 \times 10^{-2}}} - 0.3\ (2.29 \times 10^{-2}) \right] \\
&= 0.063
\end{aligned}
$$

Determine [HCO_3^-] from Equation 4. Neglect Alk_o, but because the pH exceeds 8.5, calculate the other terms. From Table 2330:II, $pK_2 = 10.38$ and $pK_w = 14.16$.

$$
\begin{aligned}
[HCO_3^-] &= \frac{2.60 \times 10^{-3} + 10^{(0.063 - 9.0)} - 10^{(9.0 + 0.063 - 14.16)}}{1 + 0.5 \times 10^{(9.0 - 10.38)}} \\
&= 2.54 \times 10^{-3}\ \text{g-moles/L}
\end{aligned}
$$

Therefore $p[HCO_3^-] = 2.60$.
Determine [Ca^{2+}] from Equation 3. Neglect Ca_{ip}.

$$[Ca^{2+}] = Ca_t = 3.80 \times 10^{-3}\ \text{g-moles/L}$$

Therefore $p[Ca^{2+}] = 2.42$.
From Table 2330:II, pK_s for calcite is 8.45.
Determine pH_s from Equation 2:

$$pH_s = 10.38 - 8.45 + 2.42 + 2.60 + 5\ (0.063) = 7.27$$

And finally, determine SI from Equation 1:

$$SI = 9.00 - 7.27 = 1.73$$

The positive SI indicates the water is oversaturated with respect to calcite.

2) Effect of neglecting Ca_{ip} and Alk_o—If Ca_{ip} is neglected, pH_s is underestimated and SI is overestimated by an amount equal to $p(1 - Y_{Ca_{ip}})$, where $Y_{Ca_{ip}}$ is the fraction of total calcium in ion pairs. For example, if $Y_{Ca_{ip}} = 0.30$ then the estimate for SI is 0.15 units too high. Similarly, if Alk_o is neglected, SI is overestimated by an amount equal to $p(1 - Y_{Alk_o})$, where Y_{Alk_o} is the fraction of total alkalinity contributed by species other than HCO_3^-, CO_3^{2-}, OH^-, and H^+. The effects of neglecting Ca_{ip} and Alk_o are additive.

Ca_{ip} and Alk_o may be neglected if the factors $Y_{Ca_{ip}}$ and Y_{Alk_o} are small and do not interfere with interpretation of the SI. The factors are small for waters of low and neutral pH, but they increase as pH values approach and exceed 9. At high pH values, however, the SI is typically much larger than its overestimate, so neglecting Ca_{ip} and Alk_o causes no problem. To return to the example above, when calculations were done with a computerized water chemistry code (SEQUIL) (see Table 2330:III) that considers Ca_{ip} and Alk_o, the SI was 1.48, i.e., 0.25 units lower than the result obtained by hand calculations. In this instance, neglecting Ca_{ip} and Alk_o did not interfere with interpreting the result. Both calculations showed the water to be strongly oversaturated.

The potential for misinterpretation is most acute in nearly-saturated waters of high sulfate concentration. Recirculating cooling water is an example. Calcium is sequestered by the robust $CaSO_4^0$ ion pair and the SI can be overestimated by as much as 0.3 to 0.5 units, even at neutral pH. Under these conditions, the SI may be thought to be zero (neither scale-forming or corrosive) when in fact it is negative.

Resolve this problem by determining pH_s using computerized water chemistry codes that consider ion pairs and the other forms of alkalinity. Section 2330D provides information about water chemistry codes. The most accurate calculations are obtained when a complete mineral analysis is provided.

An alternative but somewhat less rigorous procedure involves direct measurement of (Ca^{2+}), the calcium ion activity, with a calcium specific-ion electrode.[9] Use Equation 5 to determine $p[Ca^{2+}]$; then use $p[Ca^{2+}]$ in Equation 2.

$$p[Ca^{2+}] = p(Ca^{2+}) - 4pf_m \qquad (5)$$

This approach eliminates the need to determine Ca_{ip}. However, no equivalent procedure is available to bypass the determination of Alk_o.

b. Graphical solutions for pH_s: Caldwell-Lawrence diagrams can be used to determine pH_s.[10–12] The diagrams are particularly useful for estimating chemical dosages needed to achieve desired water conditions. Consult the references for descriptions of how to use the diagrams; see Section 2330D for additional information about the diagrams.

3. Saturation Index by Experimental Determination

a. Saturometry: Saturometers were developed to measure the relative saturation of seawater with respect to $CaCO_3$. A water of known calcium and pH is equilibrated with $CaCO_3$ in a sealed flask containing a pH electrode. The water temperature is controlled with a constant-temperature bath. During equilibration the pH decreases if $CaCO_3$ precipitates and increases if $CaCO_3$ dissolves. When the pH stops changing, equilibrium is said to have been achieved. The initial pH and calcium values and the final pH value are used to calculate the relative saturation (RS).[13] Equation 6, Section 2330D, may then be used to determine SI.

A major advantage of this method is that the approach to equilibrium can be tracked by measuring pH, thus minimizing uncertainty about the achievement of equilibrium. The method is most sensitive in the range of minimum buffering intensity (pH 7.5 to 8.5). The calculations do not consider ion pairs or noncarbonate alkalinity, except borate. The technique has been used for *in situ* oceanographic measurements[14] as well as in the laboratory.

The saturometry calculations discussed above use K_s of the $CaCO_3$ phase assumed to control solubility. Uncertainties occur if the identity of the controlling solid is unknown. Resolve these uncertainties by measuring K_s of the controlling solid. It is equal to the $CaCO_3$ activity product, $[Ca^{2+}] \times [CO_3^{2-}]$, at equilibrium. Calculate the latter from the equilibrium pH and initial calcium, alkalinity, and pH measurements.[15]

b. Alkalinity difference technique:[16] SI also can be determined by equilibrating water of known pH, calcium, and alkalinity with $CaCO_3$ in a sealed, constant-temperature system. The $CaCO_3$ activity product before equilibration is determined from initial calcium, pH, and alkalinity (or total carbonate) values. The $CaCO_3$ solubility product constant (K_s) equals the $CaCO_3$ activity product after equilibration, which is determined by using the alkalinity change that occurred during equilibration. RS is found by dividing the initial activity product by K_s. Calculate SI by using Equation 6 (see 2330D.1). The advantage of this method is that it makes no assumptions about the identity of the $CaCO_3$ phase. However it is more difficult to determine when equilibrium is achieved with this method than with the saturometry method.

Whatever the method used, use the temperatures that are the same as the temperature of the water source. Alternatively, correct test results to the temperature of the water source.[16]

4. References

1. RYZNAR, J.W. 1944. A new index for determining the amount of calcium carbonate formed by water. *J. Amer. Water Works Assoc.* 36:47.
2. SNOEYINK, V.L. & D. JENKINS. 1980. Water Chemistry. John Wiley & Sons, New York, N.Y.
3. STUMM, W. & J.J. MORGAN. 1981. Aquatic Chemistry, 2nd ed. John Wiley & Sons, New York, N.Y.
4. RUSSELL, L.L. 1976. Chemical Aspects of Groundwater Recharge with Wastewaters. Ph.D. thesis, Univ. California, Berkeley.
5. LANGELIER, W.F. 1936. The analytical control of anticorrosion water treatment. *J. Amer. Water Works Assoc.* 28:1500.
6. ROSSUM, J.R. & D.T. MERRILL. 1983. An evaluation of the calcium carbonate saturation indexes. *J. Amer. Water Works Assoc.* 75:95.
7. PLUMMER, L.N. & E. BUSENBERG. 1982. The solubilities of calcite, aragonite, and vaterite in CO_2-H_2O solutions between 0 and 90 degrees C, and an evaluation of the aqueous model for the system $CaCO_3$-CO_2-H_2O. *Geochim. Cosmochim. Acta* 46:1011.
8. ELECTRIC POWER RESEARCH INSTITUTE. 1982. Design and Operating Guidelines Manual for Cooling Water Treatment: Treatment of Recirculating Cooling Water. Section 4. Process Model Documentation and User's Manual. EPRI CS-2276, Electric Power Research Inst., Palo Alto, Calif.
9. GARRELS, R.M. & C.L. CHRIST. 1965. Solutions, Minerals, and Equilibria. Harper & Row, New York, N.Y.
10. MERRILL, D.T. & R.L. SANKS. 1978, 1979. Corrosion control by deposition of $CaCO_3$ films: A practical approach for plant operators. *J. Amer. Water Works Assoc.* 70:592; 70:634 & 71:12.
11. LOEWENTHAL, R.E. & G.v.R. MARAIS. 1976. Carbonate Chemistry of Aquatic Systems: Theory and Applications. Ann Arbor Science Publishers, Ann Arbor, Mich.
12. MERRILL, D.T. 1976. Chemical conditioning for water softening and corrosion control. *In* R.L. Sanks, ed., Water Treatment Plant Design. Ann Arbor Science Publishers, Ann Arbor, Mich.
13. BEN-YAAKOV, S. & I.R. KAPLAN. 1969. Determination of carbonate saturation of seawater with a carbonate saturometer. *Limnol. Oceanogr.* 14:874.
14. BEN-YAAKOV, S. & I.R. KAPLAN. 1971. Deep-sea in situ calcium carbonate saturometry. *J. Geophys. Res.* 76:72.

15. PLATH, D.C., K.S. JOHNSON & R.M. PYTKOWICZ. 1980. The solubility of calcite—probably containing magnesium—in seawater. *Mar. Chem.* 10:9.

16. BALZAR, W. 1980. Calcium carbonate saturometry by alkalinity difference. *Oceanol. Acta.* 3:237.

2330 C. Indices Predicting the Quantity of CaCO₃ That Can Be Precipitated or Dissolved

The Calcium Carbonate Precipitation Potential (CCPP) predicts both tendency to precipitate or to dissolve $CaCO_3$ and quantity that may be precipitated or dissolved. The CCPP also is known by other names, e.g., Calcium Carbonate Precipitation Capacity (CCPC).

The CCPP is defined as the quantity of $CaCO_3$ that theoretically can be precipitated from oversaturated waters or dissolved by undersaturated waters during equilibration.[1] The amount that actually precipitates or dissolves may be less, because equilibrium may not be achieved. The CCPP is negative for undersaturated waters, zero for saturated waters, and positive for oversaturated waters.

1. Calculating CCPP

The CCPP does not lend itself to hand calculations. Preferably calculate CCPP with computerized water chemistry models and Caldwell-Lawrence diagrams (see Section 2330D).

The most reliable calculations consider ion pairs and the contribution to alkalinity of other species besides HCO_3^-, CO_3^{2-}, OH^-, and H^+. Models that do not consider these factors overestimate the amount of $CaCO_3$ that can be precipitated and underestimate the amount of $CaCO_3$ that can be dissolved.

2. Experimental Determination of CCPP

Estimate CCPP by one of several experimental techniques.

a. Saturometry: See Section 2330B. The CCPP is determined as part of the RS calculation.

b. Alkalinity difference technique: See Section 2330B. The CCPP equals the difference between alkalinity (or calcium) values of the initial and equilibrated water, when they are expressed as $CaCO_3$.

c. Marble test: The marble test[1-5] is similar to the alkalinity difference technique. The CCPP equals the change in alkalinity (or calcium) values during equilibration, when they are expressed as $CaCO_3$.

d. Enslow test: The Enslow test[5] is a continuous version of the alkalinity difference or marble tests. Water is fed continuously to a leveling bulb or separatory funnel partly filled with $CaCO_3$. The effluent from this device is filtered through crushed marble so that the filtrate is assumed to be in equilibrium with $CaCO_3$. The CCPP equals the change in alkalinity (or calcium) values that occurs during passage through the apparatus.

e. Calcium carbonate deposition test:[6] The calcium carbonate deposition test (CCDT) is an electrochemical method that measures the electric current produced when dissolved oxygen is reduced on a rotating electrode. When an oversaturated water is placed in the apparatus, $CaCO_3$ deposits on the electrode. The deposits interfere with oxygen transfer and the current diminishes. The rate of $CaCO_3$ deposition is directly proportional to the rate at which the current declines. The CCDT and the CCPP are related, but they are not the same. The CCDT is a rate, and the CCPP is a quantity.

For realistic assessments of the CCPP (or CCDT) keep test temperature the same as the temperature of the water source. Alternatively, correct test results to the temperature of the water source.

3. References

1. MERRILL, D.T. & R.L. SANKS. 1978, 1979. Corrosion control by deposition of $CaCO_3$ films: A practical approach for plant operators. *J. Amer. Water Works Assoc.* 70:592; 70:634 & 71:12.
2. MERRILL, D.T. 1976. Chemical conditioning for water softening and corrosion control. *In* R.L. Sanks, ed. Water Treatment Plant Design. Ann Arbor Science Publishers, Ann Arbor, Mich.
3. DE MARTINI, F. 1938. Corrosion and the Langelier calcium carbonate saturation index. *J. Amer. Water Works Assoc.* 30:85.
4. HOOVER, C.P. 1938. Practical application of the Langelier method. *J. Amer. Water Works Assoc.* 30:1802.
5. DYE, J.F. & J.L. TUEPKER. 1971. Chemistry of the Lime-Soda Process. *In* American Water Works Association. Water Quality and Treatment. McGraw-Hill Book Co., New York, N.Y.
6. MCCLELLAND, N.I. & K.H. MANCY. 1979. CCDT bests Ryzner index as pipe $CaCO_3$ film predictor. *Water Sewage Works* 126:77.

2330 D. Diagrams and Computer Codes for CaCO₃ Indices

1. Description

Table 2330:III lists diagrams and computer codes that can be used to determine the SI and CCPP. It also provides a brief description of their characteristics.

Many computer codes do not calculate *SI* directly, but instead calculate the relative saturation (*RS*). When *RS* data are presented, calculate the *SI* from:[1]

$$SI = \log_{10}RS \qquad (6)$$

TABLE 2330:III. GRAPHS AND COMPUTER SOFTWARE THAT CAN BE USED TO CALCULATE CaCO$_3$ SATURATION INDICES*

Item†	CaCO$_3$ Indices		Approximate Temperature Range °C	Approximate Limit of Ionic Strength	Ion Pairs Considered?	Alk$_o$ Considered?‡	Minimum Equipment Required
	Basis for Calculation of SI	CCPP					
1. Caldwell-Lawrence diagrams[4]	pH$_{sa}$	P, D	2–25	0.030	No	No	Diagrams
2. ACAPP	RS	P, D	−10–110	6+	Yes	Yes	IBM-compatible PC, 512K bytes of RAM, MS DOS or PC DOS v.2.1 or higher
3. DRIVER	RS	P	7–65	2.5	Yes	Yes	Mainframe computer
4. INDEX C	pH$_{sa}$, pH$_{sb}$	P, D	0–50	0.5	No	No	Hewlett-Packard 41C calculator, with three memory modules
5. LEQUIL	RS	No	5–90	0.5	Yes	Yes	IBM-compatible PC, 256K RAM, Lotus 1-2-3 or work-alike, PC DOS or MS DOS v.2.0 or higher
6. MINTEQA1	RS	P, D	0–100	0.5	Yes	Yes	IBM-compatible PC, 512K bytes of RAM, PC DOS v.3.0 or higher, 10 megabyte hard disk drive, math coprocessor useful but not required
7. PHREEQE Standard	RS	P, D	0–100	0.5	Yes	Yes	IBM-compatible PC, known to work with 512K RAM, PC DOS or MS DOS v.2.11 or higher. Also available for mainframe computers.
For high-salinity waters	RS	P, D	0–80	7–8	Yes	Yes	IBM-compatible PC, 640K RAM recommended, with math coprocessor, MS DOS v.3.2 or higher
8. SEQUIL	RS	P, D	7–65	2.5	Yes	Yes	IBM-compatible PC, 512K bytes of RAM, MS DOS or PC DOS v.2.1 or higher.
9. SOLMINEQ.88	RS	P, D	0–350	6	Yes	Yes	IBM-compatible PC, 640K RAM, math coprocessor, PC DOS or MS DOS v.3.0 or higher. Also available for mainframe computer.
10. WTRCHEM	pH$_{sa}$	P, D	0–100	0.5	No	No	Any PC equipped with a BASIC interpreter, 5K RAM.
11. WATEQ4F	RS	No	0–100	0.5	Yes	Yes	IBM-compatible PC, known to work with 512K RAM, PC DOS or MS DOS v.2.11 or higher.

*SI = saturation index
CCPP = CaCO$_3$ precipitation potential
pH$_{sa}$ = alkalinity-based pH$_s$
pH$_{sb}$ = bicarbonate-based pH$_s$
P = calculates amount of CaCO$_3$ theoretically precipitated

D = calculates amount of CaCO$_3$ theoretically dissolved
RS = relative saturation
PC = personal computer
RAM = random access memory

† 1. Loewenthal and Marais[3] provide 10.2- by 11.4-cm diagrams, with documentation; Merrill[5] provides 10.2- by 16.5-cm diagrams, with documentation.
2. Radian Corp., 8501 MoPac Blvd., P.O. Box 201088, Austin, TX 78720-1088 Attn: J.G. Noblett (software and documentation).
3. Power Computing Co., 1930 Hi Line Dr., Dallas, TX, 74207 (software and documentation[6]).
4. Brown and Caldwell, P.O. Box 8045, Walnut Creek, CA 94596-1220 Attn: D.T. Merrill (software and documentation).
5. Illinois State Water Survey, Aquatic Chemistry Section, 2204 Griffith Dr., Champaign, IL 61820-7495 Attn: T.R. Holm (software and documentation).
6. Center for Exposure Assessment Modeling, Environmental Research Laboratory, Office of Research and Development. U.S. Environmental Protection Agency, Athens, GA 30613 (software and documentation[7]).
7. U.S. Geological Survey, National Center, MS 437, Reston, VA 22902, Chief of WATSTORE Program (provides software for mainframe version of standard code); U.S. Geological Survey, Water Resources Division, MS 420, 345 Middlefield Rd., Menlo Park, CA 94025 Attn: K. Nordstrom (provides software for personal computer version of standard code); National Water Research Institute, Canada Centre for Inland Waters, 867 Lakeshore Rd., Burlington, Ont., Canada L7R 4A6 Attn: A.S. Crowe (provides software and documentation [8,9] for personal computer versions of both standard and high-salinity codes); U.S. Geological Survey, Books and Open File Report Section, Box 25425, Federal Center, Denver, CO 80225 (provides documentation[8,10] for mainframe and personal computer versions of standard code).
8. Power Computing Company, 1930 Hi Line Dr., Dallas, TX 74207 (software and documentation[11]).
9. U.S. Geological Survey, Water Resources Division, MS 427, 345 Middlefield Rd., Menlo Park, CA 94025 Attn: Y.K. Kharaka (software and documentation[12]).
10. D.T. Merrill, Brown and Caldwell, P.O. Box 8045, Walnut Creek, CA 94596-1220 (code listing and documentation).
11. U.S. Geological Survey, Water Resources Division, MS 420, 345 Middlefield Rd., Menlo Park, CA 94025 Attn: K. Nordstrom (software), Books and Open File Report Section, Box 25425, Federal Center, Denver, CO 80225 (documentation[13]).
‡Codes differ in the species included in Alk$_o$.

where:

    RS = ratio of $CaCO_3$ activity product to $CaCO_3$ solubility product constant.

The diagrams and a few of the codes define pH_s as the pH the water would exhibit if it were in equilibrium with $CaCO_3$ at the existing calcium and total alkalinity concentrations.[2] This definition of pH_s differs from the definition following Equation 1 because alkalinity is used instead of bicarbonate. Within the pH range 6 to 9, alkalinity-based pH_s and bicarbonate-based pH_s are virtually equal, because total alkalinity is due almost entirely to bicarbonate ion. Above pH 9 they differ and Equation 6 no longer applies if SI is calculated with alkalinity-based pH_s. However, if SI is determined from bicarbonate-based pH_s, Equation 6 continues to apply.

Furthermore, calculating SI with alkalinity-based pH_s reverses the sign of the SI above pH values of approximately pK_2, i.e., a positive, not the usual negative, SI connotes an undersaturated water.[3] With bicarbonate-based pH_s or RS, sign reversal does not occur, thereby eliminating the confusing sign change. For these reasons, bicarbonate-based pH_s or RS is preferred. Table 2330:III lists the definition of pH_s used for each code.

Some models calculate only the amount of $CaCO_3$ that can be precipitated but not the amount of $CaCO_3$ that can be dissolved. Others calculate both.

The diagrams and codes can be used to determine many more parameters than the $CaCO_3$ saturation indices. A fee may be charged for computer software or graphs. The information in Table 2330:III describes parameters each code uses to calculate SI. Contact the sources listed below the table for current information.

2. References

1. SNOEYINK, V.L. & D. JENKINS. 1980. Water Chemistry. John Wiley & Sons, New York, N.Y.
2. LANGELIER, W.F. 1936. The analytical control of anticorrosion water treatment. *J. Amer. Water Works Assoc.* 28:1500.
3. LOEWENTHAL, R.E. & G.v.R. MARAIS. 1976. Carbonate Chemistry of Aquatic Systems: Theory and Applications. Ann Arbor Science Publishers, Ann Arbor, Mich.
4. MERRILL, D.T. & R.L. SANKS. 1978. Corrosion Control by Deposition of $CaCO_3$ Films: A Handbook of Practical Application and Instruction. American Water Works Assoc., Denver, Colo.
5. MERRILL, D.T. 1976. Chemical conditioning for water softening and corrosion control. *In* R.L. Sanks, ed., Water Treatment Plant Design. Ann Arbor Science Publishers, Ann Arbor, Mich.
6. ELECTRIC POWER RESEARCH INSTITUTE. 1982. Design and Operating Guidelines Manual for Cooling Water Treatment: Treatment of Recirculating Cooling Water. Section 4. Process Model Documentation and User's Manual. EPRI CS-2276, Electric Power Research Inst., Palo Alto, Calif.
7. BROWN, D.S. & J.D. ALLISON. 1987. MINTEQA1 Equilibrium Metal Speciation Model: A User's Manual. EPA-600/3-87-012.
8. PARKHURST, D.L., D.C. THORSTENSON & L.N. PLUMMER. 1980. PHREEQE—A Computer Program for Geochemical Calculations. USGS WRI 80-96 NTIS PB 81-167801.
9. CROWE, A.S. & F.J. LONGSTAFF. 1987. Extension of Geochemical Modeling Techniques to Brines: Coupling of the Pitzer Equations to PHREEQE. *In* Proceedings of Solving Groundwater Problems With Models, National Water Well Assoc., Denver, Colo. Feb. 10–12, 1987.
10. FLEMING, G.W. & L.N. PLUMMER. 1983. PHRQINPT—An Interactive Computer Program for Constructing Input Data Sets to the Geochemical Simulation Program PHREEQE. USGS WRI 83-4236.
11. ELECTRIC POWER RESEARCH INSTITUTE. SEQUIL—An Inorganic Aqueous Chemical Equilibrium Code for Personal Computers; Volume 1, User's Manual/Workbook, Version 1.0, GS-6234-CCML. Electric Power Research Inst., Palo Alto, Calif.
12. KHARAKA, Y.K., W.D. GUNTER, P. K. AGGARWAL, E.H. PERKINS & J.D. DEBRAAL. 1988. Solmineq.88—A Computer Program for Geochemical Modeling of Water–Rock Interactions. USGS WRIR 88-4227, Menlo Park, Calif.
13. BALL, J.W., D.K. NORDSTROM & D.W. ZACHMANN. 1987. WATEQ4F—A Personal Computer Fortran Translation of the Geochemical Model WATEQ2 With Revised Data Base. USGS OFR 87-50.

2340   HARDNESS*

2340 A.   Introduction

1. Definition

Originally, water hardness was understood to be a measure of the capacity of water to precipitate soap. Soap is precipitated chiefly by the calcium and magnesium ions present. Other polyvalent cations also may precipitate soap, but they often are in complex forms, frequently with organic constituents, and their role in water hardness may be minimal and difficult to define. In conformity with current practice, total hardness is defined as the sum of the calcium and magnesium concentrations, both expressed as calcium carbonate, in milligrams per liter.

When hardness numerically is greater than the sum of carbonate and bicarbonate alkalinity, that amount of hardness equivalent to the total alkalinity is called "carbonate hardness"; the amount of hardness in excess of this is called "noncarbonate hardness." When the hardness numerically is equal to or less than the sum of carbonate and bicarbonate alkalinity, all hardness is carbonate hardness and noncarbonate hardness is absent. The hardness may range from zero to hundreds of milligrams per liter, depending on the source and treatment to which the water has been subjected.

2. Selection of Method

Two methods are presented. Method B, hardness by calculation, is applicable to all waters and yields the higher accuracy. If

---

*Approved by Standard Methods Committee, 1997.

a mineral analysis is performed, hardness by calculation can be reported. Method C, the EDTA titration method, measures the calcium and magnesium ions and may be applied with appropriate modification to any kind of water. The procedure described affords a means of rapid analysis.

3. Reporting Results

When reporting hardness, state the method used, for example, "hardness (calc.)" or "hardness (EDTA)."

2340 B. Hardness by Calculation

1. Discussion

The preferred method for determining hardness is to compute it from the results of separate determinations of calcium and magnesium.

2. Calculation

Hardness, mg equivalent $CaCO_3/L$

$$= 2.497 \text{ [Ca, mg/L]} + 4118 \text{ [Mg, mg/L]}$$

2340 C. EDTA Titrimetric Method

1. General Discussion

a. Principle: Ethylenediaminetetraacetic acid and its sodium salts (abbreviated EDTA) form a chelated soluble complex when added to a solution of certain metal cations. If a small amount of a dye such as Eriochrome Black T or Calmagite is added to an aqueous solution containing calcium and magnesium ions at a pH of 10.0 ± 0.1, the solution becomes wine red. If EDTA is added as a titrant, the calcium and magnesium will be complexed, and when all of the magnesium and calcium has been complexed the solution turns from wine red to blue, marking the end point of the titration. Magnesium ion must be present to yield a satisfactory end point. To insure this, a small amount of complexometrically neutral magnesium salt of EDTA is added to the buffer; this automatically introduces sufficient magnesium and obviates the need for a blank correction.

The sharpness of the end point increases with increasing pH. However, the pH cannot be increased indefinitely because of the danger of precipitating calcium carbonate, $CaCO_3$, or magnesium hydroxide, $Mg(OH)_2$, and because the dye changes color at high pH values. The specified pH of 10.0 ± 0.1 is a satisfactory compromise. A limit of 5 min is set for the duration of the titration to minimize the tendency toward $CaCO_3$ precipitation.

b. Interference: Some metal ions interfere by causing fading or indistinct end points or by stoichiometric consumption of EDTA. Reduce this interference by adding certain inhibitors before titration. MgCDTA [see 2b3)], selectively complexes heavy metals, releases magnesium into the sample, and may be used as a substitute for toxic or malodorous inhibitors. It is useful only when the magnesium substituted for heavy metals does not contribute significantly to the total hardness. With heavy metal or polyphosphate concentrations below those indicated in Table 2340:I, use Inhibitor I or II. When higher concentrations of heavy metals are present, determine calcium and magnesium by a non-EDTA method (see Sections 3500-Ca and 3500-Mg) and obtain hardness by calculation. The figures in Table 2340:I are intended as a rough guide only and are based on using a 25-mL sample diluted to 50 mL.

TABLE 2340:I. MAXIMUM CONCENTRATIONS OF INTERFERENCES PERMISSIBLE WITH VARIOUS INHIBITORS*

Interfering Substance	Max. Interference Concentration mg/L	
	Inhibitor I	Inhibitor II
Aluminum	20	20
Barium	†	†
Cadmium	†	20
Cobalt	over 20	0.3
Copper	over 30	20
Iron	over 30	5
Lead	†	20
Manganese (Mn^{2+})	†	1
Nickel	over 20	0.3
Strontium	†	†
Zinc	†	200
Polyphosphate		10

* Based on 25-mL sample diluted to 50 mL.
† Titrates as hardness.

Suspended or colloidal organic matter also may interfere with the end point. Eliminate this interference by evaporating the sample to dryness on a steam bath and heating in a muffle furnace at 550°C until the organic matter is completely oxidized. Dissolve the residue in 20 mL $1N$ hydrochloric acid (HCl), neutralize to pH 7 with $1N$ sodium hydroxide (NaOH), and make up to 50 mL with distilled water; cool to room temperature and continue according to the general procedure.

c. Titration precautions: Conduct titrations at or near normal room temperature. The color change becomes impractically slow as the sample approaches freezing temperature. Indicator decomposition becomes a problem in hot water.

The specified pH may produce an environment conducive to $CaCO_3$ precipitation. Although the titrant slowly redissolves such precipitates, a drifting end point often yields low results. Completion of the titration within 5 min minimizes the tendency for $CaCO_3$ to precipitate. The following three methods also reduce precipitation loss:

1) Dilute sample with distilled water to reduce $CaCO_3$ concentration. This simple expedient has been incorporated in the procedure. If precipitation occurs at this dilution of 1 + 1 use modification 2) or 3). Using too small a sample contributes a systematic error due to the buret-reading error.

2) If the approximate hardness is known or is determined by a preliminary titration, add 90% or more of titrant to sample *before* adjusting pH with buffer.

3) Acidify sample and stir for 2 min to expel CO_2 *before* pH adjustment. Determine alkalinity to indicate amount of acid to be added.

2. Reagents

a. Buffer solution:

1) Dissolve 16.9 g ammonium chloride (NH_4Cl) in 143 mL conc ammonium hydroxide (NH_4OH). Add 1.25 g magnesium salt of EDTA (available commercially) and dilute to 250 mL with distilled water.

2) If the magnesium salt of EDTA is unavailable, dissolve 1.179 g disodium salt of ethylenediaminetetraacetic acid dihydrate (analytical reagent grade) and 780 mg magnesium sulfate ($MgSO_4 \cdot 7H_2O$) or 644 mg magnesium chloride ($MgCl_2 \cdot 6H_2O$) in 50 mL distilled water. Add this solution to 16.9 g NH_4Cl and 143 mL conc NH_4OH with mixing and dilute to 250 mL with distilled water. To attain the highest accuracy, adjust to exact equivalence through appropriate addition of a small amount of EDTA or $MgSO_4$ or $MgCl_2$.

Store Solution 1) or 2) in a plastic or borosilicate glass container for no longer than 1 month. Stopper tightly to prevent loss of ammonia (NH_3) or pickup of carbon dioxide (CO_2). Dispense buffer solution by means of a bulb-operated pipet. Discard buffer when 1 or 2 mL added to the sample fails to produce a pH of 10.0 ± 0.1 at the titration end point.

3) Satisfactory alternate "odorless buffers" also are available commercially. They contain the magnesium salt of EDTA and have the advantage of being relatively odorless and more stable than the NH_4Cl-NH_4OH buffer. They usually do not provide as good an end point as NH_4Cl-NH_4OH because of slower reactions and they may be unsuitable when this method is automated. Prepare one of these buffers by mixing 55 mL conc HCl with 400 mL distilled water and then, slowly and with stirring, adding 300 mL 2-aminoethanol (free of aluminum and heavier metals). Add 5.0 g magnesium salt of EDTA and dilute to 1 L with distilled water.

b. Complexing agents: For most waters no complexing agent is needed. Occasionally water containing interfering ions requires adding an appropriate complexing agent to give a clear, sharp change in color at the end point. The following are satisfactory:

1) *Inhibitor I:* Adjust acid samples to pH 6 or higher with buffer or $0.1N$ NaOH. Add 250 mg sodium cyanide (NaCN) in powder form. Add sufficient buffer to adjust to pH 10.0 ± 0.1. (CAUTION: *NaCN is extremely poisonous. Take extra precautions in its use.* Flush solutions containing this inhibitor down the drain with large quantities of water after insuring that no acid is present to liberate volatile poisonous hydrogen cyanide.)

2) *Inhibitor II:* Dissolve 5.0 g sodium sulfide nonahydrate ($Na_2S \cdot 9H_2O$) or 3.7 g $Na_2S \cdot 5H_2O$ in 100 mL distilled water. Exclude air with a tightly fitting rubber stopper. This inhibitor deteriorates through air oxidation. It produces a sulfide precipitate that obscures the end point when appreciable concentrations of heavy metals are present. Use 1 mL in ¶ 3b below.

3) *MgCDTA:* Magnesium salt of 1, 2-cyclohexanediaminetetraacetic acid. Add 250 mg per 100 mL sample and dissolve completely before adding buffer solution. Use this complexing agent to avoid using toxic or odorous inhibitors when interfering substances are present in concentrations that affect the end point but will not contribute significantly to the hardness value.

Commercial preparations incorporating a buffer and a complexing agent are available. Such mixtures must maintain pH 10.0 ± 0.1 during titration and give a clear, sharp end point when the sample is titrated.

c. Indicators: Many types of indicator solutions have been advocated and may be used if the analyst demonstrates that they yield accurate values. The prime difficulty with indicator solutions is deterioration with aging, giving indistinct end points. For example, alkaline solutions of Eriochrome Black T are sensitive to oxidants and aqueous or alcoholic solutions are unstable. In general, use the least amount of indicator providing a sharp end point. It is the analyst's responsibility to determine individually the optimal indicator concentration.

1) *Eriochrome Black T:* Sodium salt of 1-(1-hydroxy-2-naphthylazo)-5-nitro-2-naphthol-4-sulfonic acid; No. 203 in the Color Index. Dissolve 0.5 g dye in 100 g 2,2′,2″-nitrilotriethanol (also called triethanolamine) or 2-methoxymethanol (also called ethylene glycol monomethyl ether). Add 2 drops per 50 mL solution to be titrated. Adjust volume if necessary.

2) *Calmagite:* 1-(1-hydroxy-4-methyl-2-phenylazo)-2-naphthol-4-sulfonic acid. This is stable in aqueous solution and produces the same color change as Eriochrome Black T, with a sharper end point. Dissolve 0.10 g Calmagite in 100 mL distilled water. Use 1 mL per 50 mL solution to be titrated. Adjust volume if necessary.

3) Indicators 1 and 2 can be used in dry powder form if care is taken to avoid excess indicator. Prepared dry mixtures of these indicators and an inert salt are available commercially.

If the end point color change of these indicators is not clear and sharp, it usually means that an appropriate complexing agent is required. If NaCN inhibitor does not sharpen the end point, the indicator probably is at fault.

d. Standard EDTA titrant, 0.01M: Weigh 3.723 g analytical reagent-grade disodium ethylenediaminetetraacetate dihydrate, also called (ethylenedinitrilo)tetraacetic acid disodium salt (EDTA), dissolve in distilled water, and dilute to 1000 mL. Standardize against standard calcium solution (¶ 2e) as described in ¶ 3b below.

Because the titrant extracts hardness-producing cations from soft-glass containers, store in polyethylene (preferable) or borosilicate glass bottles. Compensate for gradual deterioration by periodic restandardization and by using a suitable correction factor.

e. Standard calcium solution: Weigh 1.000 g anhydrous $CaCO_3$ powder (primary standard or special reagent low in heavy metals, alkalis, and magnesium) into a 500-mL erlenmeyer flask. Place a funnel in the flask neck and add, a little at a time, 1 + 1 HCl until all $CaCO_3$ has dissolved. Add 200 mL distilled water and boil for a few minutes to expel CO_2. Cool, add a few drops of methyl red indicator, and adjust to the intermediate orange color by adding $3N$ NH_4OH or $1 + 1$ HCl, as required. Transfer quantitatively and dilute to 1000 mL with distilled water; 1 mL = 1.00 mg $CaCO_3$.

f. Sodium hydroxide, NaOH, 0.1N.

3. Procedure

a. Pretreatment of polluted water and wastewater samples: Use nitric acid-sulfuric acid or nitric acid-perchloric acid digestion (Section 3030).

b. Titration of sample: Select a sample volume that requires less than 15 mL EDTA titrant and complete titration within 5 min, measured from time of buffer addition.

Dilute 25.0 mL sample to about 50 mL with distilled water in a porcelain casserole or other suitable vessel. Add 1 to 2 mL buffer solution. Usually 1 mL will be sufficient to give a pH of 10.0 to 10.1. The absence of a sharp end-point color change in the titration usually means that an inhibitor must be added at this point (¶ 2*b* et seq.) or that the indicator has deteriorated.

Add 1 to 2 drops indicator solution or an appropriate amount of dry-powder indicator formulation [¶ 2*c*3)]. Add standard EDTA titrant slowly, with continuous stirring, until the last reddish tinge disappears. Add the last few drops at 3- to 5-s intervals. At the end point the solution normally is blue. Daylight or a daylight fluorescent lamp is recommended highly because ordinary incandescent lights tend to produce a reddish tinge in the blue at the end point.

If sufficient sample is available and interference is absent, improve accuracy by increasing sample size, as described in ¶ 3*c* below.

c. Low-hardness sample: For ion-exchanger effluent or other softened water and for natural waters of low hardness (less than 5 mg/L), take a larger sample, 100 to 1000 mL, for titration and add proportionately larger amounts of buffer, inhibitor, and indicator. Add standard EDTA titrant slowly from a microburet and run a blank, using redistilled, distilled, or deionized water of the same volume as the sample, to which identical amounts of buffer, inhibitor, and indicator have been added. Subtract volume of EDTA used for blank from volume of EDTA used for sample.

4. Calculation

$$\text{Hardness (EDTA) as mg } CaCO_3/L = \frac{A \times B \times 1000}{\text{mL sample}}$$

where:

A = mL titration for sample and
B = mg $CaCO_3$ equivalent to 1.00 mL EDTA titrant.

5. Precision and Bias

A synthetic sample containing 610 mg/L total hardness as $CaCO_3$ contributed by 108 mg Ca/L and 82 mg Mg/L, and the following supplementary substances: 3.1 mg K/L, 19.9 mg Na/L, 241 mg Cl^-/L, 0.25 mg NO_2^--N/L, 1.1 mg NO_3^--N/L, 259 mg SO_4^{2-}/L, and 42.5 mg total alkalinity/L (contributed by $NaHCO_3$) in distilled water was analyzed in 56 laboratories by the EDTA titrimetric method with a relative standard deviation of 2.9% and a relative error of 0.8%.

6. Bibliography

CONNORS, J.J. 1950. Advances in chemical and colorimetric methods. *J. Amer. Water Works Assoc.* 42:33.

DIEHL, H., C.A. GOETZ & C.C. HACH. 1950. The versenate titration for total hardness. *J. Amer. Water Works Assoc.* 42:40.

BETZ, J.D. & C.A. NOLL. 1950. Total hardness determination by direct colorimetric titration. *J. Amer. Water Works Assoc.* 42:49.

GOETZ, C.A., T.C. LOOMIS & H. DIEHL. 1950. Total hardness in water: The stability of standard disodium dihydrogen ethylenediaminetetraacetate solutions. *Anal. Chem.* 22:798.

DISKANT, E.M. 1952. Stable indicator solutions for complexometric determination of total hardness in water. *Anal. Chem.* 24:1856.

BARNARD, A.J., JR., W.C. BROAD & H. FLASCHKA. 1956 & 1957. The EDTA titration. *Chemist Analyst* 45:86 & 46:46.

GOETZ, C.A. & R.C. SMITH. 1959. Evaluation of various methods and reagents for total hardness and calcium hardness in water. *Iowa State J. Sci.* 34:81 (Aug. 15).

SCHWARZENBACH, G. & H. FLASCHKA. 1969. Complexometric Titrations, 2nd ed. Barnes & Noble, Inc., New York, N.Y.

2350   OXIDANT DEMAND/REQUIREMENT*

2350 A.   Introduction

1. Significance and Chemistry

Oxidants are added to water supplies and wastewater primarily for disinfection. Other beneficial uses include slime removal, oxidation of undesirable inorganic species (e.g., ferrous ion, reduced manganese, sulfide, and ammonia) and oxidation of organic constituents (e.g., taste- and odor-producing compounds). Oxidant demand refers to the difference between the added oxidant dose and the residual oxidant concentration measured after a prescribed contact time at a given pH and temperature. Oxidant requirement refers to the oxidant dose required to achieve a given oxidant residual at a prescribed contact time, pH, and temperature.

The fate of oxidants in water and wastewater is complex. For example, chlorine reacts with sample constituents by three general pathways: oxidation, addition, and substitution. First, chlorine can

---

* Approved by Standard Methods Committee, 1997.

oxidize reduced species, such as Fe^{2+}, Mn^{2+}, and sulfide. In these reactions, chlorine is reduced to inorganic chloride (Cl^-). Second, chlorine can add to olefins and other double-bond-containing organic compounds to produce chlorinated organic compounds. Third, chlorine can substitute onto chemical substrates. The addition and substitution reactions produce organochlorine species (e.g., chlorination of phenol to chlorophenols) or active chlorine species (e.g., chlorination of ammonia to produce monochloramine). Chlorine reacts with naturally occurring organic compounds by a combination of these mechanisms to generate such products as trihalomethanes. See Sections 4500-Cl (chlorine), 4500-ClO$_2$ (chlorine dioxide), and 4500-O$_3$ (ozone) for additional information.

Oxidant demand and oxidant requirement are significantly affected by the chemical and physical characteristics of the sample and the manner in which the oxidant consumption is measured. In particular, oxidant reactivity is influenced by temperature, pH, contact time, and oxidant dose. Oxidant demand and oxidant requirement are defined operationally by the analytical method used to determine the residual oxidant concentration. *Report sample temperature, pH, contact time, oxidant dose, and analytical method with oxidant demand or oxidant requirement.* Sample temperature strongly affects reaction kinetics and thus the demand exerted in a given contact time. Sample pH affects the form of the oxidant and the nature and extent of the demand. For example, ozone is unstable at high pH values and ozone demand is especially sensitive to sample pH. Oxidant demand increases with time and therefore the demand must be defined for a given contact time. Oxidant demand also is dependent on oxidant dose. Increasing oxidant dose usually will increase demand, but it is incorrect to assume that doubling the oxidant dose will double the oxidant demand. For these reasons, it is difficult to extrapolate oxidant demand data from one set of conditions to another. *Always study oxidant consumption under the range of conditions expected in the field.*

Oxidant consumption is used to evaluate oxidant demand and oxidant requirement. Report consumption values according to the objective of the study. For example, report chlorine *demand* as follows: "The sample dosed at 5.0 mg/L consumed 3.9 mg/L after 24 h at 20°C and pH 7.1, as measured by amperometric titration." By contrast, report ozone *requirement* as follows: "The sample required a dose of 2.1 mg/L to achieve an ozone residual of 0.5 mg/L after 20 min at 15°C and pH 6.5, as measured by the indigo method."

2. Selection of Method

Select a method to measure oxidant residuals used in the demand calculation that is specific and has adequate sensitivity. Some oxidant residual measurement techniques are subject to interferences from oxidation-produced oxidants. Interferences affect oxidant demand measurements because the concentrations of the interferents may change as the oxidant residual changes. Thus, calculate free chlorine demand in municipal wastewater as the difference between free chlorine dose and free chlorine residual measured after a desired contact time at a given temperature, pH, and chlorine dose for a specified analytical method. Chlorination of non-nitrified municipal wastewater probably produces chloramines. If the analytical method for free chlorine is subject to interferences from chloramines, then the free chlorine residual measurement will be too large (see Section 4500-Cl.A.3), and the resulting free chlorine demand value will be incorrectly low. It is sometimes difficult to predict the manner in which oxidant-produced oxidants will affect the demand measurement. The best approach is to use the analytical method most specific to the oxidant of interest but always indicate the method with the result.

The addition of reagents may cause loss of oxidant residual or other changes in oxidant demand. The loss of total chlorine upon addition of acid and KI is discussed in Section 4500-Cl.A.3a.

2350 B. Chlorine Demand/Requirement

1. General Discussion

a. Principle: The sample is divided into subsamples and each is dosed with the standardized oxidant (chlorine) solution to yield a series of increasing doses. After the appropriate contact time, oxidant residual, pH, and temperature are measured and the demand/requirement determined by difference between initial and final concentrations.

b. Selection of method: Chlorine consumption tests may be made to examine the demand or requirement for total chlorine, free chlorine, combined chlorine, monochloramine, or dichloramine. Specify the chlorine species consumed in the chlorine demand/requirement test. The analytical method should exhibit minimal interferences for the species examined. For demand/requirement studies with free chlorine, use only amperometric titration (Section 4500-Cl.D) or DPD methods (Sections 4500-Cl.F and 4500-Cl.G).

c. Interference: Refer to Section 4500-Cl.D.1b (amperometric titration), 4500-Cl.F.1d (DPD ferrous titrimetric method), or 4500-Cl.G.1b (DPD colorimetric method). Pay special attention

to interferences caused by oxidation products such as MnO_2, NH_2Cl, and $NHCl_2$. If the ammonia or organic nitrogen content of the water is significant, combined chlorine may form. See Section 4500-Cl for details. Under these conditions, expect interferences in the measurement of free chlorine by combined chlorine.

d. Minimum detectable concentration: Because it is calculated by difference, the minimum detectable chlorine demand/requirement is $\sqrt{2}$ times the minimum chlorine residual detectable by the analytical method. For minimum detectable chlorine residual, see Section 4500-Cl.F.1e (DPD ferrous titrimetric method) or 4500-Cl.G.1c (DPD colorimetric method). Minimum detectable demand also is influenced by amount of oxidant consumed relative to oxidant dose (see ¶ 6 below).

e. Sampling: Most reliable results are obtained on fresh samples that contain low amounts of suspended solids. If samples will be analyzed within 24 h of collection, refrigerate unacidified at 4°C immediately after collection. To preserve for up to 28 d, freeze unacidified samples at −20°C. Warm chilled samples to desired test condition before analysis.

2. Apparatus

See Section 4500-Cl.D (amperometric titration) or 4500-Cl.G (DPD colorimetric method).

3. Reagents

a. Chlorine-demand-free water: See Section 4500-Cl.C.3*m*. Alternatively, prepare dilutions, blanks, and dosing solutions from high-quality distilled water (preferably carbon-filtered redistilled water).

b. Acetic acid, conc (glacial).

c. Potassium iodide, KI, crystals.

d. Standard sodium thiosulfate titrant, 0.025N: See Section 4500-Cl.B.2*d*.

e. Starch indicator solution: See Section 4500-Cl.B.2*e*.

f. Reagents for determining residual chlorine: See Section 4500-Cl.D.3 (amperometric titration), 4500-Cl.F.2 (DPD ferrous titrimetric method), or 4500-Cl.G.3 (DPD colorimetric method).

g. Standard chlorine solution: Prepare by bubbling chlorine gas through distilled water or by diluting commercially available 5–7% (50 000–70 000 mg/L) sodium hypochlorite. Store in the dark or in a brown, glass-stoppered bottle. Standardize each day of use. A suitable strength of chlorine solution usually will be between 100 and 1000 mg/L, preferably about 100 times estimated chlorine demand. Use a solution of sufficient concentration so that adding the chlorine solution will not increase the volume of the treated portions by more than 5%.

Standardization—Place 2 mL acetic acid and 10 to 15 mL chlorine-demand-free water in a flask. Add about 1 g KI. Measure into the flask a suitable volume of chlorine solution. In choosing a convenient volume, note that 1 mL $0.025N$ $Na_2S_2O_3$ titrant is equivalent to about 0.9 mg chlorine as Cl_2. Select volumes that will require no more than 20 mL titrant.

Titrate with standardized $0.025N$ $Na_2S_2O_3$ titrant until the yellow iodine color almost disappears. Add 1 to 2 mL starch indicator solution and continue titrating until the blue color disappears.

Determine the blank by adding identical quantities of acid, KI, and starch indicator to a volume of chlorine-demand-free water corresponding to the sample used for titration. Perform whichever blank titration applies, according to Section 4500-Cl.B.3*d*. Calculate the chlorine stock concentration as described in Section 4500-Cl.B.4.

4. Procedure

Measure sample temperature and pH. Keep sample and sample portions at desired temperature and protect from light throughout the procedure. If pH adjustment is desired, prepare a blank in distilled water containing the same amount of buffer as in the sample. Carry the blank throughout the procedure.

Measure 5* equal sample portions of 200 mL† each into glass-stoppered bottles or flasks of ample capacity to permit mixing.

---

* The number of sample portions can be increased when working with samples of unknown demand and may be decreased when working with samples of familiar origin.

† Size of sample portions is not critical, but must be large enough to ensure reproducible results as well as provide volume sufficient to measure chlorine residual, pH, and temperature.

Add increasing amounts of standard chlorine solution (¶ 3*g*) to successive portions in the series. Try to bracket the estimated demand/requirement and satisfy criteria of ¶ 5*a*. Increase dosage between portions in increments of 0.1 mg/L for determining low demands/requirements and up to 1.0 mg/L or more for higher demands. Mix while adding. Dose sample portions according to a staggered schedule that will permit determining the residual after predetermined contact times.

Conduct test over desired contact period. Record contact time. At end of contact period, measure sample temperature, sample pH, and residual chlorine. Record residual measurement method used.

5. Calculation

a. Chlorine demand: Select sample portion with a residual at the end of the contact period that satisfies the following criteria:

1) $R_s < D_s - 1.4 R_{min}$,
2) $R_s > R_{min}$, and
3) Dose is most similar to the dosage range expected in the field

where:

R_s = residual after contact time, mg/L,
D_s = dose, mg/L, and
R_{min} = minimum residual measurable by the method, mg/L.

The first two criteria insure that the chlorine residual and demand are greater than their respective minimum detection limits. If no sample portion satisfies all criteria, repeat the test and adjust doses accordingly. Calculate chlorine demand as follows:

$$\text{Chlorine demand, mg/L} = (D_S - R_S) - (D_B - R_B)$$

where R_s and D_s are defined as above, and:

R_B = residual of blank after contact time, mg/L, and
D_B = blank dose, mg/L.

When reporting chlorine demand, include dose, contact time, sample temperature, sample pH, and analytical method.

b. Chlorine requirement: Report the chlorine dose that produced the target residual after the desired contact time. When reporting chlorine requirement, include the target residual, contact time, sample temperature, sample pH, and analytical method. Report chlorine demand of blank if it is greater than 10% of the difference between the requirement and the target residual.

6. Precision and Bias

For data on precision and bias of concentration measurements, see analytical method used. Because demand is calculated by difference, the uncertainty associated with the demand value will be greater than the uncertainty of the individual residual measurements. If the standard deviations of the dose measurement and residual measurements are the same, then the standard deviation and minimum detection limit of the oxidant demand will be $\sqrt{2}$ (approximately 1.4) times the standard deviation and minimum detection limit of the measurement technique, respectively.

The chlorine dose and amount consumed affect the precision and bias of demand calculation in two ways. First, the amount consumed must be sufficiently large, relative to the dose, to minimize errors associated with a value calculated from the difference of two numbers of approximately equal value. Second, the amount consumed must be small enough, relative to the dose, to prevent the residual concentration from being too small.

7. Bibliography

See Section 4500-Cl.D.7 or F.6, according to analytical method used.

2350 C.   Chlorine Dioxide Demand/Requirement

1. General Discussion

a. *Principle:* See Section 2350B.1. Chlorine dioxide consumption studies are made by dosing samples from a ClO_2 stock solution.

b. *Selection of method:* Use the amperometric method II (Section 4500-ClO_2.E) because of its high degree of accuracy and minimal interferences.

c. *Interference:* See Section 4500-ClO_2.E.1b.

d. *Minimum detectable concentration:* The minimum detectable chlorine dioxide demand/requirement is $\sqrt{2}$ times the minimum chlorine dioxide residual detectable by the analytical method (see Section 2350B.1d and B.6).

2. Apparatus

See Section 4500-ClO_2.E.2.

3. Reagents

See Section 4500-ClO_2.B.2 to prepare and standardize ClO_2.
See Section 4500-ClO_2.E.3 for reagents required to determine ClO_2 residual.

4. Procedure

Follow procedure of Section 2350B.4, using ClO_2 solution, rather than chlorine solution, for dosing sample portions.
Follow procedure of Section 4500-ClO_2.E.4 to measure ClO_2 residual.

5. Calculation

a. *Chlorine dioxide demand:* See Section 2350B.5a.
b. *Chlorine dioxide requirement:* See Section 2350B.5b.

6. Precision and Bias

See Section 2350B.6.

7. Bibliography

See Section 4500-ClO_2.E.6 and 7.

2350 D.   Ozone Demand/Requirement—Batch Method

1. General Discussion

a. *Principle:* See Section 2350B.1. Samples can be ozonated in batch and semi-batch modes. In batch ozone consumption studies, an ozone stock solution is used to add ozone to the samples. In semi-batch ozone consumption studies, a stream of ozone gas is added continuously to the sample.

Ozone decomposes at high pH. Thus, even pH-buffered distilled water has a non-zero ozone demand/requirement. Analyze blanks with all ozone consumption tests. Do not subtract the ozone demand of the blank from the ozone demand of the sample, but report it separately.

b. *Selection of method:* Ozone produces oxidants that interfere with iodometric methods. The indigo method (Section 4500-O_3.B) is recommended for measuring ozone residuals in ozone consumption studies. The indigo method measures only the demand for ozone; it does not measure the demand for ozone-produced oxidants such as the hydroxyl radical.

c. *Interference:* See Section 4500-O_3.B.1b.

d. *Minimum detectable concentration:* The minimum detectable ozone demand/requirement is $\sqrt{2}$ times the minimum ozone residual detectable by the analytical method (see Section 2350B.1d and B.6).

For minimum detectable ozone residuals, see Section 4500-O_3.B.1c.

2. Apparatus

a. *Ozone generator:* Use a laboratory-scale ozonator, capable of providing up to about 5% ozone in the gas phase at a gas flow of up to about 1 L/min.

b. *Apparatus for measuring residual ozone:* See Section 4500-O_3.B.2.

3. Reagents

a. *Ozone-demand-free water:* Ozonate reagent water (see Section 1080) for at least 1 h and purge with nitrogen gas for at least 1 h. CAUTION: *Conduct all laboratory ozonations under a vented hood.*

b. *Standard ozone solution:* Put about 800 mL of ozone-demand-free water in a 1-L flask. Bubble ozone (approximately 1 to 5% O_3 in the gas phase) through the water for about 30 min while stirring. At room temperature the ozone solution will contain about 10 to 20 mg O_3/L. If the flask is cooled in an ice bath throughout the procedure, the ozone concentration will be about 30 to 40 mg O_3/L. Standardize the ozone solution by the indigo method. Use a small sample volume (typically 1 mL) as directed in Section 4500-O_3.B.4a3).

4. Procedure

Follow procedure of Section 2350B.4, using standard ozone solution, rather than chlorine solution, for dosing sample portions. Carry a reagent blank through the procedure.

Follow procedure of Section 4500-O_3.B.4 to measure O_3 residual.

5. Calculation

a. Ozone demand: See Section 2350B.5a for selection of the proper sample portion. Calculate ozone demand in sample as follows:

$$\text{Ozone demand, mg/L} = D_s - R_s$$

where:

R_s = oxidant residual of sample after contact time, mg/L, and
D_s = sample oxidant dose, mg/L.

Calculate ozone demand in the blank separately.

$$\text{Ozone demand of blank, mg/L} = D_B - R_B$$

where:

R_B = oxidant residual of blank after contact time, mg/L, and
D_B = blank oxidant dose, mg/L.

Report the ozone demand and the ozone demand of the blank, ozone dose, contact time, sample temperature, sample pH, and analytical method.

b. Ozone requirement: See Section 2350B.5b.

6. Precision and Bias

See Section 2350B.6.

7. Bibliography

See Section 4500-O_3.B.7 and 8.

2350 E. Ozone Demand/Requirement—Semi-Batch Method

1. General Discussion

See Section 2350D.1.

The semi-batch method involves determination of ozone demand with the continuous addition of gaseous ozone to a batch reactor. The results obtained in this method depend on the mass-transfer characteristics of the reactor. In addition, some compounds that consume ozone may volatilize during the test.

2. Apparatus

All apparatus listed in Section 2350D.2 is required, plus:

a. Gas washing bottles, borosilicate glass, minimum volume 250 mL.

b. Tubing: Use only stainless steel or TFE tubing.

c. Glassware: Buret, 50 mL; beaker, 400 mL; graduated cylinder, 250 mL.

d. Wash bottle, 500 mL.

e. Magnetic stirrer (optional).

3. Reagents

a. Ozone-demand-free water: See Section 2350D.3a.

b. Sulfuric acid, H_2SO_4, 2N: Cautiously add 56 mL conc H_2SO_4 to 800 mL ozone-demand-free water in a 1-L volumetric flask. Mix thoroughly, cool, add up to mark with ozone-demand-free water.

c. Potassium iodide, KI: Dissolve 20 g KI in about 800 mL of ozone-demand-free water in a 1-L volumetric flask. Make up to mark with ozone-demand-free water.

d. Standard sodium thiosulfate titrant, $Na_2S_2O_3$, 0.1N: See Section 4500-Cl.B.2c.

e. Standard sodium thiosulfate titrant, $Na_2S_2O_3$, 0.005N: Dilute the proper volume (approximately 50 mL) of standardized 0.1N $Na_2S_2O_3$ to 1 L.

f. Starch indicator solution: See Section 4500-Cl.B.2e.

4. Procedure

Determine the output of the ozone generator by passing the ozone gas through two serial KI traps (Traps A and B) for about 10 min. For best results, keep gas flow below approximately 1 L/min. Each trap is a gas washing bottle containing a known volume (at least 200 mL) of 2% KI. Quantitatively transfer contents of each trap into a beaker, add 10 mL of 2N H_2SO_4, and titrate with standardized 0.005N $Na_2S_2O_3$ until the yellow iodine color almost disappears. Add 1 to 2 mL starch indicator solution and continue titrating to the disappearance of blue color.

Put a known volume (at least 200 mL) of sample in a separate gas washing bottle (label gas washing bottles to avoid contaminating the reaction vessel with iodide). Direct ozone gas through this reaction vessel. For ozone demand studies, direct gas stream leaving reaction vessel through a KI trap (Trap C) prepared as above. Ozonate sample for a given contact time. For ozone demand studies, turn ozonator off at end of contact time and pour contents of Trap C into a beaker. Add 10 mL 2N H_2SO_4 and titrate with 0.005N $Na_2S_2O_3$ as described above. For ozone requirement studies, remove a portion from the reaction vessel at the end of contact time and measure residual ozone concentration by the indigo method.

5. Calculation

a. Ozone dose:

$$\text{Ozone dose, mg/min} = \frac{(A + B) \times N \times 24}{T}$$

where:

A = mL titrant for Trap A,
B = mL titrant for Trap B,
N = normality of $Na_2S_2O_3$, and
T = ozonation time, min.

b. Ozone demand:

$$\text{Ozone demand, mg/min} = \text{ozone dose, mg/min} - \frac{C \times N \times 24}{T}$$

where:

C = mL titrant for Trap C.

Report sample ozone demand and blank ozone demand, ozone dose, ozonation time, sample temperature, sample pH, sample volume, and analytical method. Because the ozone transfer rate is highly dependent on experimental conditions, also report vessel volume, vessel type, gas flow rate, and sample volume.

c. Ozone requirement: The ozone requirement in the semibatch test is the ozone dose, mg/min, required to obtain the target ozone residual after the desired ozonation time. See Section 2350E.5a to calculate dose. When reporting ozone requirement, also include target oxidant residual as well as other experimental characteristics listed in ¶ *b* above.

6. Precision and Bias

See Section 2350B.6.

7. Bibliography

See Section 4500-O_3.B.7 and 8.

2510   CONDUCTIVITY*

2510 A.   Introduction

Conductivity, k, is a measure of the ability of an aqueous solution to carry an electric current. This ability depends on the presence of ions; on their total concentration, mobility, and valence; and on the temperature of measurement. Solutions of most inorganic compounds are relatively good conductors. Conversely, molecules of organic compounds that do not dissociate in aqueous solution conduct a current very poorly, if at all.

1. Definitions and Units of Expression

Conductance, G, is defined as the reciprocal of resistance, R:

$$G = \frac{1}{R}$$

where the unit of R is ohm and G is ohm^{-1} (sometimes written mho). Conductance of a solution is measured between two spatially fixed and chemically inert electrodes. To avoid polarization at the electrode surfaces the conductance measurement is made with an alternating current signal.[1] The conductance of a solution, G, is directly proportional to the electrode surface area, A, cm^2, and inversely proportional to the distance between the electrodes, L, cm. The constant of proportionality, k, such that:

$$G = k \left(\frac{A}{L}\right)$$

Approved by Standard Methods Committee, 1997.

is called "conductivity" (preferred to "specific conductance"). It is a characteristic property of the solution between the electrodes. The units of k are 1/ohm-cm or mho per centimeter. Conductivity is customarily reported in micromhos per centimeter (μmho/cm).

In the International System of Units (SI) the reciprocal of the ohm is the siemens (S) and conductivity is reported as millisiemens per meter (mS/m); 1 mS/m = 10 μmhos/cm and 1 μS/cm = 1 μmho/cm. To report results in SI units of mS/m divide μmhos/cm by 10.

To compare conductivities, values of k are reported relative to electrodes with A = 1 cm^2 and L = 1 cm. Absolute conductances, G_s, of standard potassium chloride solutions between electrodes of precise geometry have been measured; the corresponding standard conductivities, k_s, are shown in Table 2510:I.

The equivalent conductivity, Λ, of a solution is the conductivity per unit of concentration. As the concentration is decreased toward zero, Λ approaches a constant, designated as $\Lambda°$. With k in units of micromhos per centimeter it is necessary to convert concentration to units of equivalents per cubic centimeter; therefore:

$$\Lambda = 0.001k/\text{concentration}$$

where the units of Λ, k, and concentration are mho-cm^2/equivalent, μmho/cm, and equivalent/L, respectively. Equivalent conductivity, Λ, values for several concentrations of KCl are listed in Table 2510:I. In practice, solutions of KCl more dilute than 0.001M will not maintain stable conductivities because of absorp-

TABLE 2510:I. EQUIVALENT CONDUCTIVITY, Λ, AND CONDUCTIVITY, k, OF POTASSIUM CHLORIDE AT 25.0°C.*[2-4]

KCl Concentration M or equivalent/L	Equivalent Conductivity, Λ mho-cm^2/equivalent	Conductivity, k_s $\mu mho/cm$
0	149.9	
0.0001	148.9	14.9
0.0005	147.7	73.9
0.001	146.9	146.9
0.005	143.6	717.5
0.01	141.2	1 412
0.02	138.2	2 765
0.05	133.3	6 667
0.1	128.9	12 890
0.2	124.0	24 800
0.5	117.3	58 670
1	111.9	111 900

* Based on the absolute ohm, the 1968 temperature standard, and the dm^3 volume standard.[2] Values are accurate to ±0.1% or 0.1 $\mu mho/cm$, whichever is greater.

tion of atmospheric CO_2. Protect these dilute solutions from the atmosphere.

2. Measurement

a. Instrumental measurements: In the laboratory, conductance, G_s, (or resistance) of a standard KCl solution is measured and from the corresponding conductivity, k_s, (Table 2510:I) a cell constant, C, cm^{-1}, is calculated:

$$C = \frac{k_s}{G_s}$$

Most conductivity meters do not display the actual solution conductance, G, or resistance, R; rather, they generally have a dial that permits the user to adjust the internal cell constant to match the conductivity, k_s, of a standard. Once the cell constant has been determined, or set, the conductivity of an unknown solution,

$$k_u = CG_u$$

will be displayed by the meter.

Distilled water produced in a laboratory generally has a conductivity in the range 0.5 to 3 $\mu mhos/cm$. The conductivity increases shortly after exposure to both air and the water container.

The conductivity of potable waters in the United States ranges generally from 50 to 1500 $\mu mhos/cm$. The conductivity of domestic wastewaters may be near that of the local water supply, although some industrial wastes have conductivities above 10 000 $\mu mhos/cm$. Conductivity instruments are used in pipelines, channels, flowing streams, and lakes and can be incorporated in multiple-parameter monitoring stations using recorders.

Most problems in obtaining good data with conductivity monitoring equipment are related to electrode fouling and to inadequate sample circulation. Conductivities greater than 10 000 to 50 000 $\mu mho/cm$ or less than about 10 $\mu mho/cm$ may be difficult to measure with usual measurement electronics and cell capacitance. Consult the instrument manufacturer's manual or published references.[1,5,6]

Laboratory conductivity measurements are used to:

• Establish degree of mineralization to assess the effect of the total concentration of ions on chemical equilibria, physiological effect on plants or animals, corrosion rates, etc.

• Assess degree of mineralization of distilled and deionized water.

• Evaluate variations in dissolved mineral concentration of raw water or wastewater. Minor seasonal variations found in reservoir waters contrast sharply with the daily fluctuations in some polluted river waters. Wastewater containing significant trade wastes also may show a considerable daily variation.

• Estimate sample size to be used for common chemical determinations and to check results of a chemical analysis.

• Determine amount of ionic reagent needed in certain precipitation and neutralization reactions, the end point being denoted by a change in slope of the curve resulting from plotting conductivity against buret readings.

• Estimate total dissolved solids (mg/L) in a sample by multiplying conductivity (in micromhos per centimeter) by an empirical factor. This factor may vary from 0.55 to 0.9, depending on the soluble components of the water and on the temperature of measurement. Relatively high factors may be required for saline or boiler waters, whereas lower factors may apply where considerable hydroxide or free acid is present. Even though sample evaporation results in the change of bicarbonate to carbonate the empirical factor is derived for a comparatively constant water supply by dividing dissolved solids by conductivity.

• Approximate the milliequivalents per liter of either cations or anions in some waters by multiplying conductivity in units of micromhos per centimeter by 0.01.

b. Calculation of conductivity: For naturally occurring waters that contain mostly Ca^{2+}, Mg^{2+}, Na^+, K^+, HCO_3^-, SO_4^{2-}, and Cl^- and with TDS less than about 2500 mg/L, the following procedure can be used to calculate conductivity from measured ionic concentrations.[7] The abbreviated water analysis in Table 2510:II illustrates the calculation procedure.

At infinite dilution the contribution to conductivity by different kinds of ions is additive. In general, the relative contribution of each cation and anion is calculated by multiplying equivalent conductances, λ_+° and λ_-°, mho-cm^2/equivalent, by concentration in equivalents per liter and correcting units. Table 2510:III contains a short list of equivalent conductances for ions commonly found in natural waters.[8] Trace concentrations of ions generally make negligible contribution to the overall conductivity. A temperature coefficient of 0.02/deg is applicable to all ions, except H^+ (0.0139/deg) and OH^- (0.018/deg).

TABLE 2510:II. SAMPLE ANALYSIS ILLUSTRATING CALCULATION OF CONDUCTIVITY, k_{calc}, FOR NATURAL WATERS.[7]

Ions	mg/L	mM	$|z|\ \lambda_\pm^\circ$ mM	z^2mM
Ca	55	1.38	164.2	5.52
Mg	12	0.49	52.0	1.96
Na	28	1.22	61.1	1.22
K	3.2	0.08	5.9	0.08
HCO$_3$	170	2.79	124.2	2.79
SO$_4$	77	0.80	128.0	3.20
Cl	20	0.56	42.8	0.56
			578.2	15.33

TABLE 2510:III. EQUIVALENT CONDUCTANCES, λ°_+ AND λ°_-, (MHO-CM^2/EQUIVALENT) FOR IONS IN WATER AT 25.0 C.[8]

Cation	λ°_+	Anion	λ°_-
H$^+$	350	OH$^-$	198.6
1/2Ca^{2+}	59.5	HCO$_3^-$	44.5
1/2Mg^{2+}	53.1	1/2CO$_3^{2-}$	72
Na$^+$	50.1	1/2SO$_4^{2-}$	80.0
K$^+$	73.5	Cl$^-$	76.4
NH$_4^+$	73.5	Ac$^-$	40.9
1/2Fe^{2+}	54	F$^-$	54.4
1/3Fe^{3+}	68	NO$_3^-$	71.4
		H$_2$PO$_4^-$	33
		1/2HPO$_4^{2-}$	57

At finite concentrations, as opposed to infinite dilution, conductivity per equivalent decreases with increasing concentration (see Table 2510:I). For solutions composed of one anion type and one cation type, e.g., KCl as in Table 2510:I, the decrease in conductivity per equivalent with concentration can be calculated, ±0.1%, using an ionic-strength-based theory of Onsager.[9] When mixed salts are present, as is nearly always the case with natural and wastewaters, the theory is quite complicated.[10] The following semiempirical procedure can be used to calculate conductivity for naturally occurring waters:

First, calculate infinite dilution conductivity (Table 2510:II, Column 4):

$$k^\circ = \Sigma|z_i|(\lambda^\circ_{+i})(mM_i) + \Sigma|z_i|(\lambda^\circ_{-i})(mM_i)$$

where:

$|z_i|$ = absolute value of the charge of the i-th ion,
mM_i = millimolar concentration of the i-th ion, and
λ°_{+i}, λ°_{-i} = equivalent conductance of the i-th ion.

If mM is used to express concentration, the product, (λ°_+) (mM_i) or $(\lambda^\circ_-)(mM_i)$, corrects the units from liters to cm^3. In this case k° is 578.2 μmho/cm (Table 2510:II, Column 4).

Next, calculate ionic strength, IS in molar units:

$$IS = \Sigma z_i^2(mM_i)/2000$$

The ionic strength is 15.33/2000 = 0.00767 M (Table 2510:II, Column 5).

Calculate the monovalent ion activity coefficient, y, using the Davies equation for IS ≤ 0.5 M and for temperatures from 20 to 30°C.[9,11]

$$y = 10^{-0.5[IS^{1/2}/(1 + IS^{1/2}) - 0.3IS]}$$

In the present example $IS = 0.00767$ M and $y = 0.91$.

Finally, obtain the calculated value of conductivity, k_{calc}, from:

$$k_{calc} = k^\circ y^2$$

In the example being considered, $k_{calc} = 578.2 \times 0.91^2 = 478.8$ μmho/cm versus the reported value as measured by the USGS of 477 μmho/cm.

For 39 analyses of naturally occurring waters,[7,12] conductivities calculated in this manner agreed with the measured values to within 2%.

3. References

1. WILLARD, H.H., L.L. MERRITT & J.A. DEAN. 1974. Instrumental Methods of Analysis, 5th ed. D. Van Nostrand Company, New York, N.Y.
2. WU, Y.C., W.F. KOCH, W.J. HAMER & R.L. KAY. 1987. Review of electrolytic conductance standards. *J. Solution Chem.* 16:No.12.
3. JASPER, W.S. 1988. Secondary Standard Potassium Chloride Conductivity Solutions at 25°C. Corporate Metrology Laboratory, YSI Inc., Yellow Springs, Ohio.
4. ORGANISATION INTERNATIONALE DE MÉTROLOGIE LÉGALE. 1981. Standard Solutions Reproducing the Conductivity of Electrolytes, International Recommendation No. 56, 1st ed., June 1980. Bur. International de Métrologie Légale, Paris, France.
5. AMERICAN SOCIETY FOR TESTING AND MATERIALS. 1982. Standard test methods for electrical conductivity and resistivity of water. ASTM Designation D1125-82.
6. SCHOEMAKER, D.S., C.W. GARLAND & J.W. NIBLER. 1989. Experiments in Physical Chemistry, 5th ed. McGraw-Hill Book Co., New York, N.Y.
7. HAMILTON, C.E. 1978. Manual on Water. ASTM Spec. Tech. Publ. 442A, 4th ed. American Soc. Testing & Materials, Philadelphia, Pa.
8. DEAN, J.A. 1985. Lange's Handbook of Chemistry, 13th ed. McGraw-Hill Book Co., New York, N.Y.
9. ROBINSON, R.A. & R.H. STOKES. 1959. Electrolyte Solutions, 2nd ed. Academic Press, New York, N.Y.
10. HARNED, H.S. & B.B. OWEN. 1958. The Physical Chemistry of Electrolytic Solutions, 3rd ed. Reinhold Publishing Corp., New York, N.Y.
11. DAVIES, C.W. 1962. Ion Association. Elsevier Press, Amsterdam, The Netherlands.
12. TCHOBANOGLOUS, G. & E.D. SCHROEDER. 1985. Water Quality, Vol. 1. Addison-Wesley Publishing Company, Reading, Mass.

2510 B. Laboratory Method

1. General Discussion

See Section 2510A.

2. Apparatus

a. Self-contained conductivity instruments: Use an instrument capable of measuring conductivity with an error not exceeding 1% or 1 μmho/cm, whichever is greater.

b. Thermometer, capable of being read to the nearest 0.1°C and covering the range 23 to 27°C. Many conductivity meters are equipped to read an automatic temperature sensor.

c. Conductivity cell:

1) Platinum-electrode type—Conductivity cells containing plat-

inized electrodes are available in either pipet or immersion form. Cell choice depends on expected range of conductivity. Experimentally check instrument by comparing instrumental results with true conductivities of the KCl solutions listed in Table 2510:I. Clean new cells, not already coated and ready for use, with chromic-sulfuric acid cleaning mixture [see Section 2580B.3a2)] and platinize the electrodes before use. Subsequently, clean and replatinize them whenever the readings become erratic, when a sharp end point cannot be obtained, or when inspection shows that any platinum black has flaked off. To platinize, prepare a solution of 1 g chloroplatinic acid, $H_2PtCl_6 \cdot 6H_2O$, and 12 mg lead acetate in 100 mL distilled water. A more concentrated solution reduces the time required to platinize electrodes and may be used when time is a factor, e.g., when the cell constant is 1.0/cm or more. Immerse electrodes in this solution and connect both to the negative terminal of a 1.5-V dry cell battery. Connect positive side of battery to a piece of platinum wire and dip wire into the solution. Use a current such that only a small quantity of gas is evolved. Continue electrolysis until both cell electrodes are coated with platinum black. Save platinizing solution for subsequent use. Rinse electrodes thoroughly and when not in use keep immersed in distilled water.

2) *Nonplatinum-electrode type*—Use conductivity cells containing electrodes constructed from durable common metals (stainless steel among others) for continuous monitoring and field studies. Calibrate such cells by comparing sample conductivity with results obtained with a laboratory instrument. Use properly designed and mated cell and instrument to minimize errors in cell constant. Very long meter leads can affect performance of a conductivity meter. Under such circumstances, consult the manufacturer's manual for appropriate correction factors if necessary.

3. Reagents

a. Conductivity water: Any of several methods can be used to prepare reagent-grade water. The methods discussed in Section 1080 are recommended. The conductivity should be small compared to the value being measured.

b. Standard potassium chloride solution, KCl, 0.0100*M:* Dissolve 745.6 mg anhydrous KCl in conductivity water and dilute to 1000 mL in a class A volumetric flask at 25°C and store in a CO_2-free atmosphere. This is the standard reference solution, which at 25°C has a conductivity of 1412 μmhos/cm. It is satisfactory for most samples when the cell has a constant between 1 and 2 cm^{-1}. For other cell constants, use stronger or weaker KCl solutions listed in Table 2510:I. Care must be taken when using KCl solutions less than 0.001*M*, which can be unstable because of the influence of carbon dioxide on pure water. For low conductivity standards, Standard Reference Material 3190, with a certified conductivity of 25.0 μS/cm ± 0.3 μS/cm, may be obtained from NIST. Store in a glass-stoppered borosilicate glass bottle.

4. Procedure

a. Determination of cell constant: Rinse conductivity cell with at least three portions of 0.01*M* KCl solution. Adjust temperature of a fourth portion to 25.0 ± 0.1°C. If a conductivity meter displays resistance, *R*, ohms, measure resistance of this portion and note temperature. Compute cell constant, *C*:

$$C, cm^{-1} = (0.001412)(R_{KCl})[1 + 0.0191(t - 25)]$$

where:

R_{KCl} = measured resistance, ohms, and
  t = observed temperature, °C.

Conductivity meters often indicate conductivity directly. Commercial probes commonly contain a temperature sensor. With such instruments, rinse probe three times with 0.0100*M* KCl, as above. Adjust temperature compensation dial to 0.0191 C^{-1}. With probe in standard KCl solution, adjust meter to read 1412 μmho/cm. This procedure automatically adjusts cell constant internal to the meter.

b. Conductivity measurement: Thoroughly rinse cell with one or more portions of sample. Adjust temperature of a final portion to about 25°C. Measure sample resistance or conductivity and note temperature to ± 0.1°C.

5. Calculation

The temperature coefficient of most waters is only approximately the same as that of standard KCl solution; the more the temperature of measurement deviates from 25.0°C, the greater the uncertainty in applying the temperature correction. Report temperature-compensated conductivities as "μmho/cm k 25.0°C."

a. When sample resistance is measured, conductivity at 25°C is:

$$k = \frac{(1\,000\,000)(C)}{R_m[1 + 0.0191(t - 25)]}$$

where:

  k = conductivity, μmhos/cm,
  C = cell constant, cm^{-1},
R_m = measured resistance of sample, ohms, and
  t = temperature of measurement.

b. When sample conductivity is measured without internal temperature compensation conductivity at 25°C is:

$$k, \mu mho/cm = \frac{(k_m)}{1 + 0.0191(t - 25)}$$

where:

k_m = measured conductivity in units of μmho/cm at *t*°C, and other
  units are defined as above.

For instruments with automatic temperature compensation and readout directly in μmho/cm or similar units, the readout automatically is corrected to 25.0°C. Report displayed conductivity in designated units.

6. Precision and Bias

The precision of commercial conductivity meters is commonly between 0.1 and 1.0%. Reproducibility of 1 to 2% is expected after an instrument has been calibrated with such data as is shown in Table 2510:I.

2520   SALINITY*

2520 A.   Introduction

1. General Discussion

Salinity is an important unitless property of industrial and natural waters. It was originally conceived as a measure of the mass of dissolved salts in a given mass of solution. The experimental determination of the salt content by drying and weighing presents some difficulties due to the loss of some components. The only reliable way to determine the true or absolute salinity of a natural water is to make a complete chemical analysis. However, this method is time-consuming and cannot yield the precision necessary for accurate work. Thus, to determine salinity, one normally uses indirect methods involving the measurement of a physical property such as conductivity, density, sound speed, or refractive index. From an empirical relationship of salinity and the physical property detemined for a standard solution it is possible to calculate salinity. The resultant salinity is no more accurate than the empirical relationship. The precision of the measurement of a physical property will determine the precision in salinity. Following are the precisions of various physical measurements and the resultant salinity presently attainable with commercial instruments:

---

* Approved by Standard Methods Committee, 1993.

Property	Precision of Measurement	Precision of Salinity
Conductivity	± 0.0002	± 0.0002
Density	$\pm 3 \times 10^{-6} g/cm^3$	± 0.004
Sound speed	± 0.02 m/s	± 0.01

Although conductivity has the greatest precision, it responds only to ionic solutes. Density, although less precise, responds to all dissolved solutes.

2. Selection of Method

In the past, the salinity of seawater was determined by hydrometric and argentometric methods, both of which were included in previous editions of *Standard Methods* (see Sections 210B and C, 16th edition). In recent years the conductivity (2520B) and density (2520C) methods have been used because of their high sensitivity and precision. These two methods are recommended for precise field and laboratory work.

3. Quality Assurance

Calibrate salinometer or densimeter against standards of KCl or standard seawater. Expected precision is better than ± 0.01 salinity units with careful analysis and use of bracketing standards.

2520 B.   Electrical Conductivity Method

1. Determination

See Conductivity, Section 2510. Because of its high sensitivity and ease of measurement, the conductivity method is most commonly used to determine salinity.[1] For seawater measurements use the Practical Salinity Scale 1978.[2-5] This scale was developed relative to a KCl solution. A seawater with a conductivity, C, at 15°C equal to that of a KCl solution containing a mass of 32.4356 g in a mass of 1 kg of solution is defined as having a practical salinity of 35. This value was determined as an average of three independent laboratory studies. The salinity dependence of the conductivity ratio, R_t, as a function of temperature (t °C, International Practical Temperature Scale 1968) of a given sample to a standard $S = 35$ seawater is used to determine the salinity

$$S = a_0 + a_1 R_t^{1/2} + a_2 R_t + a_3 R_t^{3/2} + a_4 R_t^2 + a_5 R_t^{5/2} + \triangle S$$

where $\triangle S$ is given by

$$\triangle S = \left[\frac{t - 15}{1 + 0.0162(t - 15)} \right] (b_0 + b_1 R_t^{1/2}$$

$$+ b_2 R_t + b_3 R_t^{3/2} + b_4 R_t^2 + b_5 R_t^{5/2})$$

and:

$a_0 =$	0.0080	$b_0 =$	0.0005
$a_1 =$	-0.1692	$b_1 =$	-0.0056
$a_2 =$	25.3851	$b_2 =$	-0.0066
$a_3 =$	14.0941	$b_3 =$	-0.0375
$a_4 =$	-7.0261	$b_4 =$	0.0636
$a_5 =$	2.7081	$b_5 =$	-0.0144

valid from $S = 2$ to 42, where:

$$R_t = \frac{C \text{ (sample at } t)}{C \text{ (KCl solution at } t)}.$$

To measure the conductivity, use a conductivity bridge calibrated with standard seawater* with a known conductivity relative to KCl, following manufacturer's instructions and the procedures noted in Section 2510. If the measurements are to be made in estuarine waters, make secondary calibrations of weight-diluted seawater of known conductivity to ensure that the bridge is measuring true conductivities.

The Practical Salinity Scale recently has been extended to low

---

* Available from Standard Seawater Services, Institute of Oceanographic Services, Warmley, Godalming, Surrey, England.

salinities[6] to give an equation that is valid from 0 to 40 salinity. The equation is:

$$S = S_{PSS} - \frac{a_0}{1 + 1.5X + X^2} - \frac{b_0 f(t)}{1 + Y^{1/2} + Y^{3/2}}$$

where:

S_{PSS} = value determined from the Practical Salinity Scale given earlier,
$a_0 = 0.008$,
$b_0 = 0.0005$,
$X = 400R_t$,
$Y = 100R_t$, and
$f(t) = (t-15)/[1 + 0.0162 (t-15)]$

The practical salinity breaks with the old salinity-chlorinity relationship, $S = 1.806\ 55\ Cl$. Although the scale can be used for estuarine waters[7-10] and brines[11-13], there are limitations.[12,14-22]

2. References

1. LEWIS, E.L. 1978. Salinity: its definition and calculation. *J. Geophys. Res.* 83:466.
2. LEWIS, E.L. 1980. The practical salinity scale 1978 and its antecedents. *IEEEJ. Oceanic Eng.* OE-5:3.
3. BRADSHAW, A.L. & K.E. SCHLEICHER. 1980. Electrical conductivity of seawater. *IEEE J. Oceanic Eng.*OE-5:50.
4. CULKIN, F. & N.D. SMITH. 1980. Determination of the concentration of potassium chloride solution having the same electrical conductivity, at 15°C and infinite frequency, as standard seawater of salinity 35.000 °/oo (Chlorinity 19.37394 °/oo.*IEEE J. Oceanic Eng.* OE-5:22.
5. DAUPHINEE, T.M., J. ANCSIN, H.P. KLEIN & M.J.PHILLIPS. 1980. The effect of concentration and temperature on the conductivity ratio of potassium chloride solutions to standard seawater of salinity 35 °/oo (Cl.19.3740) at 15°C and 24°. *IEEE J. Oceanic Eng.*OE-5:17.
6. HILL, K.D., T.M. DAUPHINEE & D.J. WOODS. 1986. The Extension of the Practical Salinity Scale 1978 to low salinities. *IEEE J. Oceanic Eng.* OE-11:109.

7. MILLERO, F.J. 1975. The physical chemistry of estuarines. *In* T.M. Church, ed. Marine Chemistry in the Coastal Environment. American Chemical Soc. Symposium, Ser. 18.
8. MILLERO, F.J. 1978. The physical chemistry of Baltic Sea waters. *Thalassia Jugoslavica* 14:1.
9. MILLERO, F.J. 1984. The conductivity-salinity-chlorinity relationship for estuarine waters. *Limnol. Oceanogr.* 29:1318.
10. MILLERO, F.J. & K. KREMLING. 1976. The densities of Baltic Sea waters. *Deep-Sea Res.* 23:1129.
11. FERNANDEZ, F., F. VAZQUEZ & F.J. MILLERO. 1982. The density and composition of hypersaline waters of a Mexican lagoon. *Limnol. Oceanogr.* 27:315.
12. MILLERO, F.J. & P.V.CHETIRKIN. 1980. The density of Caspian Sea waters. *Deep-Sea Res.* 27:265.
13. MILLERO, F.J., A. MUCCI, J. ZULLIG & P. CHETIRKIN. 1982. The density of Red Sea brines. *Mar. Chem.* 11:477.
14. BREWER, P.G. & A. BRADSHAW. 1975. The effect of non-ideal composition of seawater on salinity and density. *J. Mar. Res.* 33:155.
15. CONNORS, D.N. & D.R. KESTER. 1974. Effect of major ion variations in the marine environment on the specific gravity-conductivity-chlorinity-salinity relationship. *Mar. Chem.* 2:301.
16. POISSON, A. 1980. The concentration of the KCl solution whose conductivity is that of standard seawater (35 °/oo) at 15°C. *IEEE J. Oceanic Eng.*OE-5:24.
17. POISSON, A. 1980. Conductivity/salinity/temperature relationship of diluted and concentrated standard seawater. *IEEE J.Oceanic Eng.* OE-5:17.
18. MILLERO, F.J., A.GONZALEZ & G.K. WARD. 1976. The density of seawater solutions at one atmosphere as a function of temperature and salinity. *J. Mar. Res.* 34:61.
19. MILLERO, F.J., A. GONZALEZ, P.G. BREWER & A.BRADSHAW. 1976. The density of North Atlantic and North Pacific deep waters. *Earth Planet Sci. Lett.* 32:468.
20. MILLERO, F.J., D. LAWSON & A.GONZALEZ. 1976. The density of artificial river and estuary waters. *J. Geophys. Res.* 81:1177.
21. MILLERO, F.J., P. CHETIRKIN & F. CULKIN. 1977. The relative conductivity and density of standard seawaters. *Deep-Sea Res.* 24:315.
22. MILLERO, F.J., D. FORSHT, D. MEANS, J. GRIESKES & K. KENYON. 1977. The density of North Pacific ocean waters. *J.Geophys. Res.* 83:2359.

2520 C.   Density Method

1. Determination

With a precise vibrating flow densimeter, it is possible to make rapid measurements of the density of natural waters. The measurements are made bypassing the sample through a vibrating tube encased in a constant-temperature jacket. The solution density (ρ) is proportional to the square of the period of the vibration (τ).

$$\rho = A + B\tau^2$$

where A and B are terms determined by calibration, B being determined by calibration with a densimeter with standard seawater. The difference between the density of the sample and that of pure water is given by:

$$\rho - \rho_0 = B(\tau^2 - \tau_0^2)$$

where τ and τ_0 are, respectively, the periods of the sample and water. The system is calibrated with two solutions of known density. Follow manufacturer's recommendations for calibration. These two solutions can be nitrogen gas and water or standard seawater and water. The salinity of the sample can be determined from the 1 atm international equation of state for seawater. This equation relates ($\rho - \rho_0$) to the practical salinity (S) as a function of temperature.[1]

$$\rho\ (kg/m^3) = \rho_0 + AS + BS^{3/2} + CS^2$$

where:

$A = 8.244\ 93 \times 10^{-1} - 4.0899 \times 10^{-3}t$
$\quad + 7.6438 \times 10^{-5}t^2 - 8.2467 \times 10^{-7}t^3 + 5.3875 \times 10^{-9}t^4$,
$B = -5.724\ 66 \times 10^{-3} + 1.0227 \times 10^{-4}t - 1.6546 \times 10^{-6}t^2$,
$C = 4.8314 \times 10^{-4}$,

and the density of water is given by:

$\rho_0 = 999.842\ 594 + 6.793\ 952 \times 10^{-2}t - 9.095\ 290 \times 10^{-3}\ t^2$
$\quad + 1.001\ 685 \times 10^{-4}t^3 - 1.120\ 083 \times 10^{-6}t^4 + 6.536\ 332$
$\quad \times 10^{-9}t^5$

Perform simple iteration by adjusting S until it gives the measured

$\rho - \rho_0$ at a given temperature. If the measurements are made at 25°C, the salinity can be determined from the following equation:

$$S = 1.3343\ (\rho - \rho_0) + 2.155\ 306 \times 10^{-4}$$
$$(\rho - \rho_0)^2 - 1.171\ 16 \times 10^{-5}\ (\rho - \rho_0)^3$$

which has a $\tau = 0.0012$ in S. Approximate salinities also can be determined from densities or specific gravities obtained with a hydrometer at a given temperature (Section 210B, 16th edition).

2. Reference

1. MILLERO, F.J. & A. POISSON. 1981. International one-atmosphere equation of state of seawater. *Deep Sea Res.* 28:625.

2520 D. Algorithm of Practical Salinity

Because all practical salinity measurements are carried out in reference to the conductivity of standard seawater (corrected to $S = 35$), it is the quantity R_t that will be available for salinity calculations. R_t normally is obtained directly by laboratory salinometers, but *in situ* measurements usually produce the quantity R, the ratio of the *in situ* conductivity to the standard conductivity at $S = 35$, $t = 15°C$, $p = 0$ (where p is the pressure above one standard atmosphere and the temperature is on the 1968 International Temperature Scale). R is factored into three parts, i.e.,

$$R = R_p r_t R_t$$

where:

R_p = ratio of *in situ* conductivity to conductivity of the same sample at the same temperature, but at $p = 0$ and r_t = ratio of conductivity of reference seawater, having a practical salinity of 35, at temperature t, to its conductivity at $t = 15°C$. From R_p and r_t calculate R_t using the *in situ* results, i.e.,

$$R_t = \frac{R}{R_p r_t}$$

R_p and r_t can be expressed as functions of the numerical values of the *in situ* parameters, R, t, and p, when t is expressed in °C and p in bars (10^5 Pa), as follows:

$$R_p = 1 + \frac{p(e_1 + e_2 p + e_3 p^2)}{1 + d_1 t + d_2 t^2 + (d_3 + d_4 t)R}$$

where:

$e_1 =$	2.070×10^{-4},	$d_1 =$	3.426×10^{-2},
$e_2 =$	-6.370×10^{-8},	$d_2 =$	4.464×10^{-4},
$e_3 =$	3.989×10^{-12},	$d_3 =$	4.215×10^{-1},
and		$d_4 =$	-3.107×10^{-3},

and

$$r_t = c_0 + c_1 t + c_2 t^2 + c_3 t^3 + c_4 t^4$$

where:

$c_0 = 0.676\ 609\ 7$,
$c_1 = 2.005\ 64 \times 10^{-2}$,
$c_2 = 1.104\ 259 \times 10^{-4}$,
$c_3 = -6.9698 \times 10^{-7}$, and
$c_4 = 1.0031 \times 10^{-9}$.

2530 FLOATABLES*

2530 A. Introduction

One important criterion for evaluating the possible effect of waste disposal into surface waters is the amount of floatable material in the waste. Two general types of floating matter are found: particulate matter that includes "grease balls," and liquid components capable of spreading as a thin, highly visible film over large areas. Floatable material in wastewaters is important because it accumulates on the surface, is often highly visible, is subject to wind-induced transport, may contain pathogenic bacteria and/or

viruses associated with individual particles, and can significantly concentrate metals and chlorinated hydrocarbons such as pesticides and PCBs. Colloidally dispersed oil and grease behave like other dispersed organic matter and are included in the material measured by the COD, BOD, and TOC tests. The floatable oil test indicates the readily separable fraction. The results are useful in designing oil and grease separators, in ascertaining the efficiency of operating separators, and in monitoring raw and treated wastewater streams. Many cities and districts have specified floatable oil and grease limits for wastewater discharged to sewers.

---

* Approved by Standard Methods Committee, 1993.

2530 B.   Particulate Floatables (GENERAL)

1. Discussion

a. Principle: This method depends on the gravity separation of particles having densities less than that of the surrounding water. Particles that collect on the surface and can be filtered out and dried at 103 to 105°C are defined by this test as floatable particles.

b. Application: This method is applicable to raw wastewater, treated primary and secondary effluent, and industrial wastewater. Because of the limited sensitivity, it is not applicable to tertiary effluents or receiving waters, whether freshwater or seawater.

c. Precautions: Even slight differences in sampling and handling during and after collection can give large differences in the measured amount of floatable material. Additionally, uniformity of the TFE* coating of the separation funnel is critical to obtaining reliable results. For a reproducible analysis treat all samples uniformly, preferably by mixing them in a standard manner, before flotation and use consistently prepared separation funnels as much as possible. Because the procedure relies on the difference in specific gravity between the liquid and the floating particles, temperature variations may affect the results. Conduct the test at a constant temperature the same as that of the receiving water body, and report temperature with results.

d. Minimum detectable concentration: The minimum reproducible detectable concentration is approximately 1 mg/L. Although the minimum levels that can be measured are below 1 mg/L, the results are not meaningful within the current established accuracy of the test.

2. Apparatus

a. Floatables sampler with mixer: Use a metal container of at least 5 L capacity equipped with a propeller mixer on a separate stand (Figure 2530:1), and with a 20-mm-ID bottom outlet cocked at an angle of 45° to the container wall in the direction of fluid movement. The 45° angle assures that even large particles will flow from the container into the flotation funnels where the sample is withdrawn. Fit exterior of bottom outlet with a short piece of tubing and a pinch clamp to allow unrestricted flow through the outlet. Coat inside of container with TFE as uniformly as possible, using a TFE spray to prevent oil and grease from sticking to the surface.

b. Flotation funnel: Use an Imhoff cone provided with a TFE stopcock at the bottom and extended at the top to a total volume of 3.5 L (Figure 2530:2). Coat inside of flotation funnel with TFE as uniformly as possible to prevent floatable grease particles sticking to the sides. Mount flotation funnels as shown in Figure 2530:3 with a light behind the bottom of the funnels to aid in reading levels.

c. Filter holder: Coat inside of top of a standard 500-mL membrane filter holder with TFE, again taking all possible precautions to obtain a uniform TFE coating.

d. Filters, glass fiber, fine porosity.†

e. Vacuum flask, 500 mL.

f. TFE coating: Follow instructions that accompany commercially available coating kits. Alternatively, have necessary glass-

---

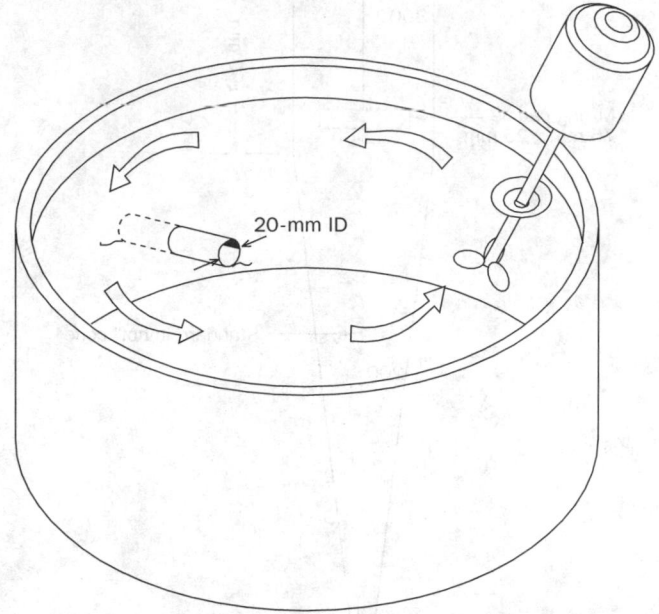

Figure 2530:1. Floatables sampler with mixer.

20-mm ID

ware coated commercially. Uniform coatings are key to the reliability of the test results, but in practice are difficult to obtain.

3. Procedure

a. Preparation of glass fiber filters: See Section 2540D.3a.

b. Sample collection and treatment: Collect sample in the floatables sampler at a point of complete mixing, transport to the laboratory, and place 3.0 L in the flotation funnel within 2 h after sample collection to minimize changes in the floatable material. While the flotation funnel is being filled, mix sampler contents with a small propeller mixer. Adjust mixing speed to provide uniform distribution of floating particles throughout the liquid but avoid extensive air entrapment through formation of a large vortex.

c. Correction for density and for concentration effects: When a receiving water has a density and ion concentration different from that of the waste, adjust sample density and ion concentration to that of the receiving water. For example, if the receiving water is ocean water, place 1.5 L sample in flotation funnel and add 1.5 L filtered seawater from the receiving area together with mixture of 39.8 g NaCl, 8.0 g $MgCl_2 \cdot 6H_2O$, and 2.3 g $CaCl_2 \cdot 2H_2O$. The final mixture contains the amount of floatables in a 1.5-L sample in a medium of approximately the same density and ion concentration as seawater.

d. Flotation: Mix flotation funnel contents at 40 rpm for 15 min using a paddle mixer (Figure 2530:3). Let settle for 5 min, mix at 100 rpm for 1 min, and let settle for 30 min. Discharge 2.8 L through bottom stopcock at a rate of 500 mL/min. Do not disturb the sample surface in the flotation funnel during discharge. With distilled water from a wash bottle, wash down any floatable material sticking to sides of stirring paddle and funnel. Let remaining 200 mL settle for 15 min and discharge settled solids and liquid down to the 40-mL mark on the Imhoff cone. Let settle

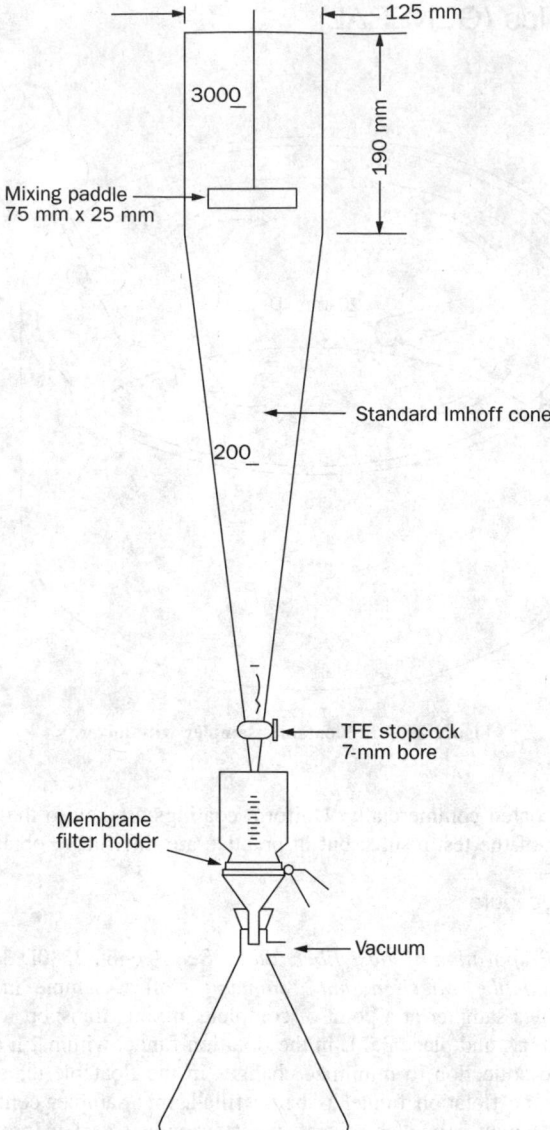

Figure 2530:2. Floatables flotation funnel and filter holder.

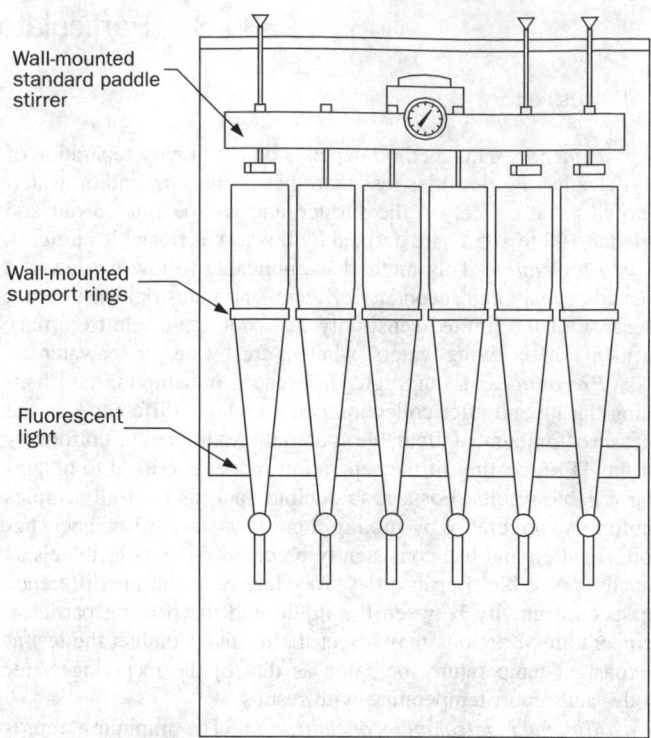

Figure 2530:3. Flotation funnels and mixing unit.

where:

> A = weight of filter + floatables, mg,
> B = weight of filter, mg, and
> C = sample volume, L. (Do not include volume used for density or concentration correction, if used.)

5. Precision and Bias

Precision varies with the concentration of suspended matter in the sample. There is no completely satisfactory procedure for determining the bias of the method for wastewater samples but approximate recovery can be determined by running a second test for floatables on all water discharged throughout the procedure, with the exception of the last 10 mL. Precision and bias are summarized in Table 2530:I. Experience with the method at one municipal treatment plant indicates that the practical lower limit of detection is approximately 1 mg/L.

TABLE 2530:I. COEFFICIENT OF VARIATION AND RECOVERY FOR PARTICULATE FLOATABLES TEST

Type of Wastewater	Average Floatables Concentration *mg/L*	No. of Samples	Coefficient of Variation %	Recovery %
Raw*	49	5	5.7	96
Raw	1.0	5	20	92
Primary effluent	2.7	5	15	91

* Additional floatable material added from skimmings of a primary sedimentation basin.

again for 10 min and discharge until only 10 mL liquid and the floating particles remain in funnel. Add 500 mL distilled water and stir by hand to separate entrapped settleable particles from the floatable particles. Let settle for 15 min, then discharge to the 40-mL mark. Let settle for 10 min, then discharge dropwise to the 10-mL mark. Filter remaining 10 mL and floating particles through a preweighed glass fiber filter. Wash sides of flotation funnel with distilled water to transfer all floatable material to filter.

e. Weighing: Dry and weigh glass fiber filter at 103 to 105° C for exactly 2 h (see Section 2540D.3*c*).

4. Calculation

$$\text{mg particulate floatables/L} = \frac{(A - B)}{C}$$

6. Bibliography

HEUKELEKIAN, H. & J. BALMAT. 1956. Chemical composition of the particulate fractions of domestic sewage. *Sewage Ind. Wastes* 31:413.

ENGINEERING-SCIENCE, INC. 1965. Determination and Removal of Floatable Material from Waste Water. Rep. for U.S. Public Health Serv. contracts WPD 12-01 (R1)-63 and WPD 12-02-64, Engineering-Science, Inc., Arcadia & Oakland, Calif.

HUNTER, J.V. & H. HEUKELEKIAN. 1965. Composition of domestic sewage fractions. *J. Water Pollut. Control Fed.* 37:1142.

NUSBAUM I. & L. BURTMAN. 1965. Determination of floatable matter in waste discharges. *J. Water Pollut. Control Fed.* 37:577.

SCHERFIG, J. & H. F. LUDWIG. 1967. Determination of floatables and hexane extractables in sewage. *In* Advances in Water Pollution Research, Vol. 3, p. 217, Water Pollution Control Federation, Washington, D.C.

SELLECK, R.E., L. W. BRACEWELL & R. CARTER. 1974. The Significance and Control of Wastewater Floatables in Coastal Waters. Rep. for U.S. Environmental Protection Agency contract R-800373, SERL Rep. No. 74-1, Sanitary Engineering Research Lab., Univ. California, Berkeley.

BRACEWELL, L.W. 1976. Contribution of Wastewater Discharges to Surface Films and Other Floatables on the Ocean Surface. Thesis, Univ. California, Berkeley.

BRACEWELL, L.W., R.E. SELLECK & R. CARTER. 1980. Contribution of wastewater discharges to ocean surface particulates. *J. Water Pollut. Control Fed.* 52:2230.

2530 C. Trichlorotrifluoroethane-Soluble Floatable Oil and Grease (GENERAL)

1. Discussion

The floatable oil and grease test does not measure a precise class of substances; rather, the results are determined by the conditions of the test. The fraction measured includes oil and grease, both floating and adhering to the sides of the test vessel. The adhering and the floating portions are of similar practical significance because it is assumed that most of the adhering portion would otherwise float under receiving water conditions. The results have been found to represent well the amount of oil removed in separators having overflow rates equivalent to test conditions.

2. Apparatus

a. Floatable oil tube (Figure 2530:4): Before use, carefully clean tube by brushing with a mild scouring power. Water must form a smooth film on inside of cleaned glass. Do not use lubricant on stopcock.

b. Conical flask, 300 mL.

3. Reagents

a. 1,1,2-trichloro-1,2,2-trifluoroethane*: See Section 5520B.3*b*.

b. Hydrochloric acid, HCl, 6*N*.

c. Filter paper.†

4. Procedure

a. Sampling: Collect samples at a place where there is a strong turbulence in the water and where floating material is not trapped at the surface. Fill floatable oil tube to mark by dipping into water. *Do not use samples taken to the laboratory in a bottle, because oil and grease cannot be redispersed to their original condition.*

b. Flotation: Support tube in a vertical position. Start flotation period at sampling site immediately after filling tube. The standard flotation time is 30 min. If a different time is used, state this

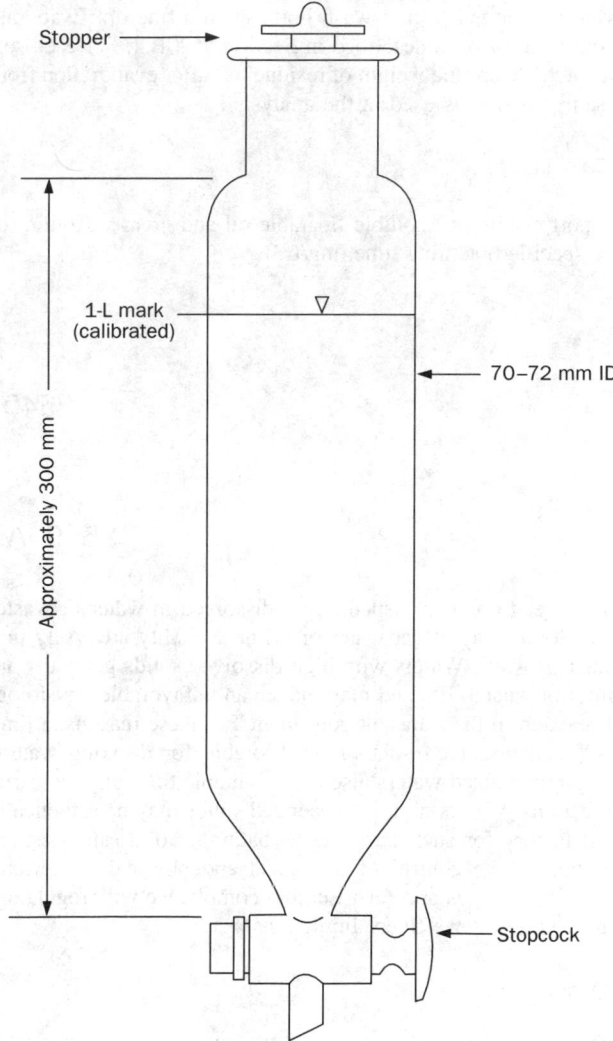

Figure 2530:4. Floatable oil tube, 1-L capacity.

Stopper

1-L mark (calibrated)

70–72 mm ID

Approximately 300 mm

Stopcock

---

* Freon or equivalent.
† Whatman No. 40 or equivalent.

variation in reporting results. At end of flotation period, discharge the first 900 mL of water carefully through bottom stopcock, stopping before any surface oil or other floating material escapes. Rotate tube slightly back and forth about its vertical axis to dislodge sludge from sides, and let settle for 5 min. Completely discharge sludge that has settled to the bottom or that comes down from the sides with the liquid. Scum on top of the liquid may mix with the water as it moves down the tube. If mixing occurs, stop drawing off water before any floatables have been lost. Let settle for 5 min before withdrawing remainder of water. After removing water, return tube to laboratory to complete test.

c. Extraction: Acidify to pH 2 or lower with a few drops of 6*N* HCl, add 50 to 100 mL trichlorotrifluoroethane, and shake vigorously. Let settle and draw off solvent into a clean dry beaker. Filter solvent through a dry filter paper into a tared 300-mL conical flask, taking care not to get any water on filter paper. Add a second 50-mL portion of trichlorotrifluoroethane and repeat extraction, settling, and filtration into the same 300-mL flask. A third extraction may be needed if the amount of floatables in sample exceeds 4 mg/L. Wash filter paper carefully with fresh solvent discharged from a wash bottle with a fine tip. Evaporate solvent from flask as described in Section 5520B.4. For each solvent batch, determine weight of residue left after evaporation from the same volume as used in the analysis.

5. Calculations

Report results as "soluble floatable oil and grease, 30 min (or other specified) settling time, mg/L."

Trichlorotrifluoroethane-soluble floatable oil and grease, 30 min settling time, mg/L

$$= \frac{(A - B) \times 1000}{\text{mL sample}}$$

where:

A = total gain in weight of tared flask, mg, and
B = calculated residue from solvent blank of the same volume as that used in the test, mg.

6. Precision and Bias

There is no standard against which bias of this test can be determined. Variability of replicates is influenced by sample heterogeneity. If large grease particles are present, the element of chance in sampling may be a major factor. One municipal wastewater discharge and two meat-packing plant discharges, both containing noticeable particles of grease, were analyzed in triplicate. Averages for the three wastewaters were 48, 57, and 25 mg/L; standard deviations averaged 11%. An oil refinery made duplicate determinations of its separator effluent on 15 consecutive days, obtaining results ranging from 5.1 to 11.2 mg/L. The average difference between pairs of samples was 0.37 mg/L.

7. Bibliography

POMEROY, R.D. 1953. Floatability of oil and grease in wastewaters. *Sewage Ind. Wastes* 25:1304.

2540　SOLIDS*

2540 A.　Introduction

Solids refer to matter suspended or dissolved in water or wastewater. Solids may affect water or effluent quality adversely in a number of ways. Waters with high dissolved solids generally are of inferior palatability and may induce an unfavorable physiological reaction in the transient consumer. For these reasons, a limit of 500 mg dissolved solids/L is desirable for drinking waters. Highly mineralized waters also are unsuitable for many industrial applications. Waters high in suspended solids may be esthetically unsatisfactory for such purposes as bathing. Solids analyses are important in the control of biological and physical wastewater treatment processes and for assessing compliance with regulatory agency wastewater effluent limitations.

1. Definitions

"Total solids" is the term applied to the material residue left in the vessel after evaporation of a sample and its subsequent drying in an oven at a defined temperature. Total solids includes

"total suspended solids," the portion of total solids retained by a filter, and "total dissolved solids," the portion that passes through the filter.

The type of filter holder, the pore size, porosity, area, and thickness of the filter and the physical nature, particle size, and amount of material deposited on the filter are the principal factors affecting separation of suspended from dissolved solids. "Dissolved solids" is the portion of solids that passes through a filter of 2.0 μm (or smaller) nominal pore size under specified conditions. "Suspended solids" is the portion retained on the filter.

"Fixed solids" is the term applied to the residue of total, suspended, or dissolved solids after heating to dryness for a specified time at a specified temperature. The weight loss on ignition is called "volatile solids." Determinations of fixed and volatile solids do not distinguish precisely between inorganic and organic matter because the loss on ignition is not confined to organic matter. It includes losses due to decomposition or volatilization of some mineral salts. Better characterization of organic matter can be made by such tests as total organic carbon (Section 5310), BOD (Section 5210), and COD (Section 5220).

---

* Approved by Standard Methods Committee, 1997.

"Settleable solids" is the term applied to the material settling out of suspension within a defined period. It may include floating material, depending on the technique (2540F.3*b*).

2. Sources of Error and Variability

Sampling, subsampling, and pipeting two-phase or three-phase samples may introduce serious errors. Make and keep such samples homogeneous during transfer. Use special handling to insure sample integrity when subsampling. Mix small samples with a magnetic stirrer. If suspended solids are present, pipet with wide-bore pipets. If part of a sample adheres to the sample container, consider this in evaluating and reporting results. Some samples dry with the formation of a crust that prevents water evaporation; special handling is required to deal with this. Avoid using a magnetic stirrer with samples containing magnetic particles.

The temperature at which the residue is dried has an important bearing on results, because weight losses due to volatilization of organic matter, mechanically occluded water, water of crystallization, and gases from heat-induced chemical decomposition, as well as weight gains due to oxidation, depend on temperature and time of heating. Each sample requires close attention to desiccation after drying. Minimize opening desiccator because moist air enters. Some samples may be stronger desiccants than those used in the desiccator and may take on water.

Residues dried at 103 to 105°C may retain not only water of crystallization but also some mechanically occluded water. Loss of CO_2 will result in conversion of bicarbonate to carbonate. Loss of organic matter by volatilization usually will be very slight. Because removal of occluded water is marginal at this temperature, attainment of constant weight may be very slow.

Residues dried at $180 \pm 2°C$ will lose almost all mechanically occluded water. Some water of crystallization may remain, especially if sulfates are present. Organic matter may be lost by volatilization, but not completely destroyed. Loss of CO_2 results from conversion of bicarbonates to carbonates and carbonates may be decomposed partially to oxides or basic salts. Some chloride and nitrate salts may be lost. In general, evaporating and drying water samples at 180°C yields values for dissolved solids closer to those obtained through summation of individually determined mineral species than the dissolved solids values secured through drying at the lower temperature.

To rinse filters and filtered solids and to clean labware use Type III water. Special samples may require a higher quality water; see Section 1080.

Results for residues high in oil or grease may be questionable because of the difficulty of drying to constant weight in a reasonable time.

To aid in quality assurance, analyze samples in duplicate. Dry samples to constant weight if possible. This entails multiple drying-cooling-weighing cycles for each determination.

Analyses performed for some special purposes may demand deviation from the stated procedures to include an unusual constituent with the measured solids. Whenever such variations of technique are introduced, record and present them with the results.

3. Sample Handling and Preservation

Use resistant-glass or plastic bottles, provided that the material in suspension does not adhere to container walls. Begin analysis as soon as possible because of the impracticality of preserving the sample. Refrigerate sample at 4°C up to the time of analysis to minimize microbiological decomposition of solids. Preferably do not hold samples more than 24 h. In no case hold sample more than 7 d. Bring samples to room temperature before analysis.

4. Selection of Method

Methods B through F are suitable for the determination of solids in potable, surface, and saline waters, as well as domestic and industrial wastewaters in the range up to 20 000 mg/L.

Method G is suitable for the determination of solids in sediments, as well as solid and semisolid materials produced during water and wastewater treatment.

5. Bibliography

THERIAULT, E.J. & H.H. WAGENHALS. 1923. Studies of representative sewage plants. *Pub. Health Bull.* No. 132.

U.S. ENVIRONMENTAL PROTECTION AGENCY. 1979. Methods for Chemical Analysis of Water and Wastes. Publ. 600/4-79-020, rev. Mar. 1983. Environmental Monitoring and Support Lab., U.S. Environmental Protection Agency, Cincinnati, Ohio.

2540 B.   Total Solids Dried at 103–105°C

1. General Discussion

a. Principle: A well-mixed sample is evaporated in a weighed dish and dried to constant weight in an oven at 103 to 105°C. The increase in weight over that of the empty dish represents the total solids. The results may not represent the weight of actual dissolved and suspended solids in wastewater samples (see above).

b. Interferences: Highly mineralized water with a significant concentration of calcium, magnesium, chloride, and/or sulfate may be hygroscopic and require prolonged drying, proper desiccation, and rapid weighing. Exclude large, floating particles or submerged agglomerates of nonhomogeneous materials from the sample if it is determined that their inclusion is not desired in the final result. Disperse visible floating oil and grease with a blender before withdrawing a sample portion for analysis. Because excessive residue in the dish may form a water-trapping crust, limit sample to no more than 200 mg residue (see 2540A.2).

2. Apparatus

a. Evaporating dishes: Dishes of 100-mL capacity made of one of the following materials:
   1) Porcelain, 90-mm diam.
   2) Platinum—Generally satisfactory for all purposes.

3) High-silica glass.*

b. *Muffle furnace* for operation at 550°C.

c. *Steam bath.*

d. *Desiccator,* provided with a desiccant containing a color indicator of moisture concentration or an instrumental indicator.

e. *Drying oven,* for operation at 103 to 105°C.

f. *Analytical balance,* capable of weighing to 0.1 mg.

g. *Magnetic stirrer* with TFE stirring bar.

h. *Wide-bore pipets.*†

i. *Graduated cylinder.*

j. *Low-form beaker.*‡

3. Procedure

a. *Preparation of evaporating dish:* If volatile solids are to be measured ignite clean evaporating dish at 550°C for 1 h in a muffle furnace. If only total solids are to be measured, heat clean dish to 103 to 105°C for 1 h. Store and cool dish in desiccator until needed. Weigh immediately before use.

b. *Sample analysis:* Choose a sample volume that will yield a residue between 2.5 and 200 mg. Pipet a measured volume of well-mixed sample, during mixing, to a preweighed dish. For homogeneous samples, pipet from the approximate midpoint of the container but not in the vortex. Choose a point both middepth and midway between wall and vortex. Evaporate to dryness on a steam bath or in a drying oven. Stir sample with a magnetic stirrer during transfer. If necessary, add successive sample portions to the same dish after evaporation. When evaporating in a drying

---

* Vycor, product of Corning Glass Works, Corning, NY, or equivalent.
† Kimble Nos. 37005 or 37034B, or equivalent.
‡ Class B or better.

oven, lower temperature to approximately 2°C below boiling to prevent splattering. Dry evaporated sample for at least 1 h in an oven at 103 to 105°C, cool dish in desiccator to balance temperature, and weigh. Repeat cycle of drying, cooling, desiccating, and weighing until a constant weight is obtained, or until weight change is less than 4% of previous weight or 0.5 mg, whichever is less. When weighing dried sample, be alert to change in weight due to air exposure and/or sample degradation. Analyze at least 10% of all samples in duplicate. Duplicate determinations should agree within 5% of their average weight.

4. Calculation

$$\text{mg total solids/L} = \frac{(A - B) \times 1000}{\text{sample volume, mL}}$$

where:

A = weight of dried residue + dish, mg, and
B = weight of dish, mg.

5. Precision

Single-laboratory duplicate analyses of 41 samples of water and wastewater were made with a standard deviation of differences of 6.0 mg/L.

6. Bibliography

SYMONS, G.E. & B. MOREY. 1941. The effect of drying time on the determination of solids in sewage and sewage sludges. *Sewage Works J.* 13:936.

2540 C. Total Dissolved Solids Dried at 180°C

1. General Discussion

a. *Principle:* A well-mixed sample is filtered through a standard glass fiber filter, and the filtrate is evaporated to dryness in a weighed dish and dried to constant weight at 180°C. The increase in dish weight represents the total dissolved solids. This procedure may be used for drying at other temperatures.

The results may not agree with the theoretical value for solids calculated from chemical analysis of sample (see above). Approximate methods for correlating chemical analysis with dissolved solids are available.[1] The filtrate from the total suspended solids determination (Section 2540D) may be used for determination of total dissolved solids.

b. *Interferences:* See 2540A.2 and 2540B.1. Highly mineralized waters with a considerable calcium, magnesium, chloride, and/or sulfate content may be hygroscopic and require prolonged drying, proper desiccation, and rapid weighing. Samples high in bicarbonate require careful and possibly prolonged drying at 180°C to insure complete conversion of bicarbonate to carbonate. Because excessive residue in the dish may form a water-trapping crust, limit sample to no more than 200 mg residue.

2. Apparatus

Apparatus listed in 2540B.2*a-h* is required, and in addition:

a. *Glass-fiber filter disks** without organic binder.

b. *Filtration apparatus:* One of the following, suitable for the filter disk selected:

1) *Membrane filter funnel.*

2) *Gooch crucible,* 25-mL to 40-mL capacity, with Gooch crucible adapter.

3) *Filtration apparatus* with reservoir and coarse (40- to 60-μm) fritted disk as filter support.†

c. *Suction flask,* of sufficient capacity for sample size selected.

d. *Drying oven,* for operation at 180 ± 2°C.

3. Procedure

a. *Preparation of glass-fiber filter disk:* If pre-prepared glass fiber filter disks are used, eliminate this step. Insert disk with

---

* Whatman grade 934AH; Gelman type A/E; Millipore type AP40; E-D Scientific Specialties grade 161; Environmental Express Pro Weigh; or other products that give demonstrably equivalent results. Practical filter diameters are 2.2 to 12.5 cm.
† Gelman No. 4201 or equivalent.

wrinkled side up into filtration apparatus. Apply vacuum and wash disk with three successive 20-mL volumes of reagent-grade water. Continue suction to remove all traces of water. Discard washings.

b. Preparation of evaporating dish: If volatile solids are to be measured, ignite cleaned evaporating dish at 550°C for 1 h in a muffle furnace. If only total dissolved solids are to be measured, heat clean dish to 180 ± 2°C for 1 h in an oven. Store in desiccator until needed. Weigh immediately before use.

c. Selection of filter and sample sizes: Choose sample volume to yield between 2.5 and 200 mg dried residue. If more than 10 min are required to complete filtration, increase filter size or decrease sample volume.

d. Sample analysis: Stir sample with a magnetic stirrer and pipet a measured volume onto a glass-fiber filter with applied vacuum. Wash with three successive 10-mL volumes of reagent-grade water, allowing complete drainage between washings, and continue suction for about 3 min after filtration is complete. Transfer total filtrate (with washings) to a weighed evaporating dish and evaporate to dryness on a steam bath or in a drying oven. If necessary, add successive portions to the same dish after evaporation. Dry evaporated sample for at least 1 h in an oven at 180 ± 2°C, cool in a desiccator to balance temperature, and weigh. Repeat drying cycle of drying, cooling, desiccating, and weighing until a constant weight is obtained or until weight change is less than 4% of previous weight or 0.5 mg, whichever is less. Analyze at least 10% of all samples in duplicate. Duplicate determinations should agree within 5% of their average weight. If volatile solids are to be determined, follow procedure in 2540E.

4. Calculation

$$\text{mg total dissolved solids/L} = \frac{(A - B) \times 1000}{\text{sample volume, mL}}$$

where:

A = weight of dried residue + dish, mg, and
B = weight of dish, mg.

5. Precision

Single-laboratory analyses of 77 samples of a known of 293 mg/L were made with a standard deviation of differences of 21.20 mg/L.

6. Reference

1. SOKOLOFF, V.P. 1933. Water of crystallization in total solids of water analysis. *Ind. Eng. Chem.,* Anal. Ed. 5:336.

7. Bibliography

HOWARD, C.S. 1933. Determination of total dissolved solids in water analysis. *Ind. Eng. Chem.,* Anal. Ed. 5:4.
U.S. GEOLOGICAL SURVEY. 1974. Methods for Collection and Analysis of Water Samples for Dissolved Minerals and Gases. Techniques of Water-Resources Investigations, Book 5, Chap. A1. U.S. Geological Surv., Washington, D.C.

2540 D.　Total Suspended Solids Dried at 103–105°C

1. General Discussion

a. Principle: A well-mixed sample is filtered through a weighed standard glass-fiber filter and the residue retained on the filter is dried to a constant weight at 103 to 105°C. The increase in weight of the filter represents the total suspended solids. If the suspended material clogs the filter and prolongs filtration, it may be necessary to increase the diameter of the filter or decrease the sample volume. To obtain an estimate of total suspended solids, calculate the difference between total dissolved solids and total solids.

b. Interferences: See 2540A.2 and 2540B.1. Exclude large floating particles or submerged agglomerates of nonhomogeneous materials from the sample if it is determined that their inclusion is not representative. Because excessive residue on the filter may form a water-entrapping crust, limit the sample size to that yielding no more than 200 mg residue. For samples high in dissolved solids thoroughly wash the filter to ensure removal of dissolved material. Prolonged filtration times resulting from filter clogging may produce high results owing to increased colloidal materials captured on the clogged filter.

2. Apparatus

Apparatus listed in Sections 2540B.2 and 2540C.2 is required, except for evaporating dishes, steam bath, and 180°C drying oven. In addition:

Aluminum weighing dishes.

3. Procedure

a. Preparation of glass-fiber filter disk: If pre-prepared glass fiber filter disks are used, eliminate this step. Insert disk with wrinkled side up in filtration apparatus. Apply vacuum and wash disk with three successive 20-mL portions of reagent-grade water. Continue suction to remove all traces of water, turn vacuum off, and discard washings. Remove filter from filtration apparatus and transfer to an inert aluminum weighing dish. If a Gooch crucible is used, remove crucible and filter combination. Dry in an oven at 103 to 105°C for 1 h. If volatile solids are to be measured, ignite at 550°C for 15 min in a muffle furnace. Cool in desiccator to balance temperature and weigh. Repeat cycle of drying or igniting, cooling, desiccating, and weighing until a constant weight is obtained or until weight change is less than 4% of the previous weighing or 0.5 mg, whichever is less. Store in desiccator until needed.

b. Selection of filter and sample sizes: Choose sample volume to yield between 2.5 and 200 mg dried residue. If volume filtered fails to meet minimum yield, increase sample volume up to 1 L. If complete filtration takes more than 10 min, increase filter diameter or decrease sample volume.

c. Sample analysis: Assemble filtering apparatus and filter and begin suction. Wet filter with a small volume of reagent-grade water to seat it. Stir sample with a magnetic stirrer at a speed to shear larger particles, if practical, to obtain a more uniform (preferably homogeneous) particle size. Centrifugal force may separate particles by size and density, resulting in poor precision when point of sample withdrawal is varied. While stirring, pipet a measured volume onto the seated glass-fiber filter. For homogeneous samples, pipet from the approximate midpoint of container but not in vortex. Choose a point both middepth and midway between wall and vortex. Wash filter with three successive 10-mL volumes of reagent-grade water, allowing complete drainage between washings, and continue suction for about 3 min after filtration is complete. Samples with high dissolved solids may require additional washings. Carefully remove filter from filtration apparatus and transfer to an aluminum weighing dish as a support. Alternatively, remove the crucible and filter combination from the crucible adapter if a Gooch crucible is used. Dry for at least 1 h at 103 to 105°C in an oven, cool in a desiccator to balance temperature, and weigh. Repeat the cycle of drying, cooling, desiccating, and weighing until a constant weight is obtained or until the weight change is less than 4% of the previous weight or 0.5 mg, whichever is less. Analyze at least 10% of all samples in duplicate. Duplicate determinations should agree within 5% of their average weight. If volatile solids are to be determined, treat the residue according to 2540E.

4. Calculation

$$\text{mg total suspended solids/L} = \frac{(A - B) \times 1000}{\text{sample volume, mL}}$$

where:

A = weight of filter + dried residue, mg, and
B = weight of filter, mg.

5. Precision

The standard deviation was 5.2 mg/L (coefficient of variation 33%) at 15 mg/L, 24 mg/L (10%) at 242 mg/L, and 13 mg/L (0.76%) at 1707 mg/L in studies by two analysts of four sets of 10 determinations each.

Single-laboratory duplicate analyses of 50 samples of water and wastewater were made with a standard deviation of differences of 2.8 mg/L.

6. Bibliography

DEGEN, J. & F.E. NUSSBERGER. 1956. Notes on the determination of suspended solids. *Sewage Ind. Wastes* 28:237.

CHANIN, G., E.H. CHOW, R.B. ALEXANDER & J. POWERS. 1958. Use of glass fiber filter medium in the suspended solids determination. *Sewage Ind. Wastes* 30:1062.

NUSBAUM, I. 1958. New method for determination of suspended solids. *Sewage Ind. Wastes* 30:1066.

SMITH, A.L. & A.E. GREENBERG. 1963. Evaluation of methods for determining suspended solids in wastewater. *J. Water Pollut. Control Fed.* 35:940.

WYCKOFF, B.M. 1964. Rapid solids determination using glass fiber filters. *Water Sewage Works* 111:277.

NATIONAL COUNCIL OF THE PAPER INDUSTRY FOR AIR AND STREAM IMPROVEMENT. 1975. A Preliminary Review of Analytical Methods for the Determination of Suspended Solids in Paper Industry Effluents for Compliance with EPA-NPDES Permit Terms. Spec. Rep. No. 75-01. National Council of the Paper Industry for Air & Stream Improvement, New York, N.Y.

NATIONAL COUNCIL OF THE PAPER INDUSTRY FOR AIR AND STREAM IMPROVEMENT. 1977. A Study of the Effect of Alternate Procedures on Effluent Suspended Solids Measurement. Stream Improvement Tech. Bull. No. 291, National Council of the Paper Industry for Air & Stream Improvement, New York, N.Y.

TREES, C.C. 1978. Analytical analysis of the effect of dissolved solids on suspended solids determination. *J. Water Pollut. Control Fed.* 50:2370.

2540 E.   Fixed and Volatile Solids Ignited at 550°C

1. General Discussion

a. Principle: The residue from Method B, C, or D is ignited to constant weight at 550°C. The remaining solids represent the fixed total, dissolved, or suspended solids while the weight lost on ignition is the volatile solids. The determination is useful in control of wastewater treatment plant operation because it offers a rough approximation of the amount of organic matter present in the solid fraction of wastewater, activated sludge, and industrial wastes.

b. Interferences: Negative errors in the volatile solids may be produced by loss of volatile matter during drying. Determination of low concentrations of volatile solids in the presence of high fixed solids concentrations may be subject to considerable error. In such cases, measure for suspect volatile components by another test, for example, total organic carbon (Section 5310). Highly alkaline residues may react with silica in sample or silica-containing crucibles.

2. Apparatus

See Sections 2540B.2, 2540C.2, and 2540D.2.

3. Procedure

Ignite residue produced by Method 2540B, C, or D to constant weight in a muffle furnace at a temperature of 550°C. Ignite a blank glass fiber filter along with samples. Have furnace up to temperature before inserting sample. Usually, 15 to 20 min ignition are required for 200 mg residue. However, more than one sample and/or heavier residues may overtax the furnace and necessitate longer ignition times. Let dish or filter disk cool partially in air until most of the heat has been dissipated. Transfer to a desiccator for final cooling in a dry atmosphere. Do not overload desiccator. Weigh dish or disk as soon as it has cooled to balance temperature. Repeat cycle of igniting, cooling, desiccating, and weighing until a constant weight is obtained or until weight change is less than 4% or 0.5 mg, whichever is less. Analyze at least 10% of all samples in duplicate. Duplicate determinations should agree within 5% of their average weight. Weight loss of the blank filter is an indication of unsuitability of a particular brand or type of filter for this analysis.

4. Calculation

$$\text{mg volatile solids/L} = \frac{(A - B) \times 1000}{\text{sample volume, mL}}$$

$$\text{mg fixed solids/L} = \frac{(B - C) \times 1000}{\text{sample volume, mL}}$$

where:

A = weight of residue + dish before ignition, mg,
B = weight of residue + dish or filter after ignition, mg, and
C = weight of dish or filter, mg.

5. Precision

The standard deviation was 11 mg/L at 170 mg/L volatile total solids in studies by three laboratories on four samples and 10 replicates. Bias data on actual samples cannot be obtained.

2540 F.   Settleable Solids

1. General Discussion

Settleable solids in surface and saline waters as well as domestic and industrial wastes may be determined and reported on either a volume (mL/L) or a weight (mg/L) basis.

2. Apparatus

The volumetric test requires only an Imhoff cone. The gravimetric test requires all the apparatus listed in Section 2540D.2 and a glass vessel with a minimum diameter of 9 cm.

3. Procedure

a. *Volumetric:* Fill an Imhoff cone to the 1-L mark with a well-mixed sample. Settle for 45 min, gently agitate sample near the sides of the cone with a rod or by spinning, settle 15 min longer, and record volume of settleable solids in the cone as milliliters per liter. If the settled matter contains pockets of liquid between large settled particles, estimate volume of these and subtract from volume of settled solids. The practical lower limit of measurement depends on sample composition and generally is in the range of 0.1 to 1.0 mL/L. Where a separation of settleable and floating materials occurs, do not estimate the floating material as settleable matter. Replicates usually are not required.

Where biological or chemical floc is present, the gravimetric method (3b) is preferred.

b. *Gravimetric:*
1) Determine total suspended solids as in Section 2540D.
2) Pour a well-mixed sample into a glass vessel of not less than 9 cm diam using not less than 1 L and sufficient sample to give a depth of 20 cm. Alternatively use a glass vessel of greater diameter and a larger volume of sample. Let stand quiescent for 1 h and, without disturbing the settled or floating material, siphon 250 mL from center of container at a point halfway between the surface of the settled material and the liquid surface. Determine total suspended solids (milligrams per liter) of this supernatant liquor (Section 2540D). These are the nonsettleable solids.

4. Calculation

mg settleable solids/L
 = mg total suspended solids/L − mg nonsettleable solids/L

5. Precision and Bias

Precision and bias data are not now available.

6. Bibliography

FISCHER, A.J. & G.E. SYMONS. 1944. The determination of settleable sewage solids by weight. *Water Sewage Works* 91:37.

2540 G.   Total, Fixed, and Volatile Solids in Solid and Semisolid Samples

1. General Discussion

a. *Applicability:* This method is applicable to the determination of total solids and its fixed and volatile fractions in such solid and semisolid samples as river and lake sediments, sludges separated from water and wastewater treatment processes, and sludge cakes from vacuum filtration, centrifugation, or other sludge dewatering processes.

b. *Interferences:* The determination of both total and volatile sol-

ids in these materials is subject to negative error due to loss of ammonium carbonate and volatile organic matter during drying. Although this is true also for wastewater, the effect tends to be more pronounced with sediments, and especially with sludges and sludge cakes. The mass of organic matter recovered from sludge and sediment requires a longer ignition time than that specified for wastewaters, effluents, or polluted waters. Carefully observe specified ignition time and temperature to control losses of volatile inorganic salts if these are a problem. Make all weighings quickly because wet samples tend to lose weight by evaporation. After drying or ignition, residues often are very hygroscopic and rapidly absorb moisture from the air. Highly alkaline residues may react with silica in the samples or silica-containing crucibles.

2. Apparatus

All the apparatus listed in Section 2540B.2 is required except that a magnetic stirrer and pipets are not used and a balance capable of weighing to 10 mg may be used.

3. Procedure

a. Total solids:

1) Preparation of evaporating dish—If volatile solids are to be measured, ignite a clean evaporating dish at 550°C for 1 h in a muffle furnace. If only total solids are to be measured, heat dish at 103 to 105°C for 1 h in an oven. Cool in desiccator, weigh, and store in desiccator until ready for use.

2) Sample analysis

a) Fluid samples—If the sample contains enough moisture to flow more or less readily, stir to homogenize, place 25 to 50 g in a prepared evaporating dish, and weigh. Evaporate to dryness on a water bath, dry at 103 to 105°C for 1 h, cool to balance temperature in an individual desiccator containing fresh desiccant, and weigh. Repeat heating, cooling, desiccating, and weighing procedure until the weight change is less than 4% or 50 mg, whichever is less. Analyze at least 10% of all samples in duplicate. Duplicate determinations should agree within 5% of their average weight.

b) Solid samples—If the sample consists of discrete pieces of solid material (dewatered sludge, for example), take cores from each piece with a No. 7 cork borer or pulverize the entire sample coarsely on a clean surface by hand, using rubber gloves. Place 25 to 50 g in a prepared evaporating dish and weigh. Place in an oven at 103 to 105°C overnight. Cool to balance temperature in a desiccator and weigh. Repeat drying (1 h), cooling, weighing, and desiccating steps until weight change is less than 4% or 50 mg, whichever is less. Analyze at least 10% of all samples in duplicate. Duplicate determinations should agree within 5% of their average weight.

b. Fixed and volatile solids: Transfer the dried residue from 2)a) above to a cool muffle furnace, heat furnace to 550°C, and ignite for 1 h. (If the residue contains large amounts of organic matter, first ignite it over a gas burner and under an exhaust hood in the presence of adequate air to lessen losses due to reducing conditions and to avoid odors in the laboratory.) Cool in desiccator to balance temperature and weigh. Repeat igniting (30 min), cooling, desiccating and weighing steps until the weight change is less than 4% or 50 mg, whichever is less. Analyze at least 10% of all samples in duplicate. Duplicate determinations should agree within 5% of their average weight.

4. Calculation

$$\% \text{ total solids} = \frac{(A - B) \times 100}{C - B}$$

$$\% \text{ volatile solids} = \frac{(A - D) \times 100}{A - B}$$

$$\% \text{ fixed solids} = \frac{(D - B) \times 100}{A - B}$$

where:

A = weight of dried residue + dish, mg,
B = weight of dish,
C = weight of wet sample + dish, mg, and
D = weight of residue + dish after ignition, mg.

5. Precision and Bias

Precision and bias data are not now available.

6. Bibliography

GOODMAN, B.L. 1964. Processing thickened sludge with chemical conditioners. Pages 78 et seq. *in* Sludge Concentration, Filtration and Incineration. Univ. Michigan Continued Education Ser. No. 113, Ann Arbor.

GRATTEAU, J.C. & R.I. DICK. 1968. Activated sludge suspended solids determinations. *Water Sewage Works* 115:468.

2550 TEMPERATURE*

2550 A. Introduction

Temperature readings are used in the calculation of various forms of alkalinity, in studies of saturation and stability with respect to calcium carbonate, in the calculation of salinity, and in general laboratory operations. In limnological studies, water temperatures as a function of depth often are required. Elevated temperatures resulting from discharges of heated water may have significant ecological impact. Identification of source of water supply, such as deep wells, often is possible by temperature measurements alone. Industrial plants often require data on water temperature for process use or heat-transmission calculations.

* Approved by Standard Methods Committee, 1993.

2550 B. Laboratory and Field Methods

1. Laboratory and Other Non-Depth Temperature Measurements

Normally, temperature measurements may be made with any good mercury-filled Celsius thermometer. As a minimum, the thermometer should have a scale marked for every 0.1°C, with markings etched on the capillary glass. The thermometer should have a minimal thermal capacity to permit rapid equilibration. Periodically check the thermometer against a precision thermometer certified by the National Institute of Standards and Technology (NIST, formerly National Bureau of Standards)* that is used with its certificate and correction chart. For field operations use a thermometer having a metal case to prevent breakage.

Thermometers are calibrated for total immersion or partial immersion. One calibrated for total immersion must be completely immersed to the depth of the etched circle around the stem just below the scale level.

2. Depth Temperature Measurements

Depth temperature required for limnological studies may be measured with a reversing thermometer, thermophone, or thermistor. The thermistor is most convenient and accurate; however, higher cost may preclude its use. Calibrate any temperature measurement devices with a NIST-certified thermometer before field use. Make readings with the thermometer or device immersed in water long enough to permit complete equilibration. Report results to the nearest 0.1 or 1.0°C, depending on need.

The thermometer commonly used for depth measurements is of the reversing type. It often is mounted on the sample collection apparatus so that a water sample may be obtained simultaneously.

---

* Some commercial thermometers may be as much as 3°C in error.

Correct readings of reversing thermometers for changes due to differences between temperature at reversal and temperature at time of reading. Calculate as follows:

$$\Delta T = \left[\frac{(T^1 - t)(T^1 + V_0)}{K} \right] \times \left[1 + \frac{(T^1 - t)(T^1 + V_0)}{K} \right] + L$$

where:

ΔT = correction to be added algebraically to uncorrected reading,
T^1 = uncorrected reading at reversal,
t = temperature at which thermometer is read,
V_0 = volume of small bulb end of capillary up to 0°C graduation,
K = constant depending on relative thermal expansion of mercury and glass (usual value of K = 6100), and
L = calibration correction of thermometer depending on T^1.

If series observations are made it is convenient to prepare graphs for a thermometer to obtain ΔT from any values of T^1 and t.

3. Bibliography

WARREN, H.F. & G.C. WHIPPLE. 1895. The thermophone—A new instrument for determining temperatures. *Mass. Inst. Technol. Quart.* 8: 125.

SVERDRUP, H.V., M.W. JOHNSON & R.H. FLEMING. 1942. The Oceans. Prentice-Hall, Inc., Englewood Cliffs, N.J.

AMERICAN SOCIETY FOR TESTING AND MATERIALS. 1949. Standard Specifications for ASTM Thermometers. No. E1-58, ASTM, Philadelphia, Pa.

REE, W.R. 1953. Thermistors for depth thermometry. *J. Amer. Water Works Assoc.* 45:259.

2560 PARTICLE COUNTING AND SIZE DISTRIBUTION (PROPOSED)*

2560 A. Introduction

1. General Discussion

Particles are ubiquitous in natural waters and in water and wastewater treatment streams. Particle counting and size distribution analysis can help to determine the makeup of natural waters, treatment plant influent, process water, and finished water. Similarly, it can aid in designing treatment processes, making decisions about changes in operations, and/or determining process efficiency. Methods for measuring particle size distribution included herein depend on electronic measurement devices because

---

* Approved by Standard Methods Committee, 1993.

manual methods are likely to be too slow for routine analysis. However, when particle size analysis is to include size distribution of large (>500-μm) aggregates, use direct microscopic counting and sizing. Principles of various types of instruments capable of producing both size and number concentration information on particulate dispersions are included. Unless explicitly stated otherwise, the term "size distribution" means an absolute size distribution, i.e., one that includes the number concentration or count.

In most particle-counting instruments, particles pass through a sensing zone where they are measured individually; the only exception included is the static type of light-scattering instrument. Instruments create an electronic pulse (voltage, current, or resistance) that is proportional to a characteristic size of the particle.

The instrument responses (pulse height, width, or area) are classified by magnitude and counted in each class to yield the particle size distribution.

2. Selection of Method

Three instrument types are included: electrical sensing zone instruments, light-blockage instruments, and light-scattering instruments.

Select instrument consistent with expected use of the particle size analysis. Instruments vary in the particle characteristic being sensed, lower and upper size limits of detection, degree of resolution of the size distribution, particle number concentration range that can be measured accurately, amount of shear to which a sample is subjected before measurement, amount of sample preparation, operator skill required, and the ease with which data can be obtained and manipulated into the desired forms. See Sections 2560B.1, C.1, and D.1, and manufacturers' literature for information on characteristics of each type of instrumentation.

Some instruments can be set up for either continuous-flow or batch sampling. Others can be used only for batch analysis. For instruments usable in both modes, check that no systematic differences in particle size distributions occur between continuous-flow measurements and batch samples taken at or near the intake point for continuous-flow samples.

3. Sample Collection and Handling

a. Batch samples: Use extreme care in obtaining, handling, and preparing batch samples to avoid changing total particle count and size distribution.

Choose representative times and locations for sampling. Ensure that particles are not subjected to greater physical forces during collection than in their natural setting. Collect samples from a body of water with submerged vessels to minimize turbulence and bubble entrainment. If sampling from particular depths, use standard samplers designed for that purpose. For flowing systems, make sure that the velocity into the opening of the sampling device is the same as that of the flowing stream (isokinetic sampling) and that the opening diameter is at least 50 times as large as the particles to be measured. For sampling from a tap, let water flow slowly and continuously down the side of the collection vessel.

Minimize particle contamination from the air, dilution water (or, for electrical sensing zone instruments, electrolyte solution) (see ¶ 4 below), and any vessel or glassware that comes in contact with the sample. Minimize exposure to air by keeping sample in a closed container and by minimizing time between sampling and analysis.

Preferably use glass bottles and other vessels with bottle cap liners of TFE.

Clean all glassware scrupulously by automatic dishwashing, vigorous hand brushing, and/or ultrasonication. Rinse glassware immediately before use with particle-free water. Between samples, rinse any part of the instrument that comes in contact with samples with either clean water or the upcoming sample. Alternatively, run multiple replicates and discard the first results.

To avoid breakup of aggregates of particles or flocs, sample and make dilutions very slowly using wide-bore pipets, needles, or other sampling devices; cut off pipet tips to avoid high velocities at the entrance. If sample dilution is required, add sample to dilution water, not vice versa, by submerging the pipet tip in the dilution water and releasing sample slowly. Use minimum intensity and duration of mixing adequate to dilute the suspension into the dilution water. Avoid mechanical stirrers inside the sample or ultrasonication. Simultaneously gently rotate and partially invert entire sample in a closed bottle. Use cylindrical dilution bottles to avoid sharp corners. Leave less than approximately 25% air space during mixing. To avoid sedimentation, make measurements immediately after mixing. Do not mix during measurement unless absolutely necessary to prevent sedimentation.

Most surface and ground waters contain relatively stable particles that aggregate slowly. Particle size distribution in biologically active waters or waters that have been treated with coagulants is more likely to change over short time periods. To minimize flocculation, minimize time between sampling and measurement. In highly flocculent systems, maximum holding time should be only a few minutes; for more stable samples, a few hours may be acceptable. Dilution slows flocculation kinetics and, in some cases, makes flocculation less likely. Make dilutions only immediately before measurement.

Samples measured at different temperature or pressure than when collected and those with biological activity may develop entrained air bubbles that interfere with measurement accuracy. If any gas bubbles are visible, let sample stand for a short time to degas naturally or use a mild vacuum to speed degassing. Use ultrasound to aid degassing, but only if floc breakup is not a problem.

Minimize time between sampling and analysis; if at all possible, make measurements immediately after sampling. If storage is unavoidable, refrigerate at 4°C but restore samples to room temperature, preferably in water bath, before measurement.

b. Continuous-flow: Using a particle counter as a continuous-flow monitor may be desirable. Many more samples can be processed by automated particle counting than by batch sample analysis. All the instruments mentioned in Methods B through D can be used in this mode, although instruments and samples not requiring dilution are easier to set up. For some instruments, this type of operation requires custom hardware; for others there is commercially available hardware.

Problems of batch samples are equally relevant to continuous analysis. Other critical considerations include selection of sampling point, velocity at entrance and throughout sample line, maintenance of stable sample flow rate, position of instrument's sensor in the sample line, and absence of flow-modifying devices upstream of the sensor.

Choose a sampling point that is representative and away from wall surfaces. Place entrance to sample line facing the flow, with the velocity at the opening nearly the same as the surrounding flow.

Sample preservation within the instrument is difficult because both deposition (or temporary holdup) of particles and floc breakup must be avoided. Deposition occurs by gravity settling of particles onto horizontal surfaces. Floc breakup occurs because of excessive shear forces. Minimize length of transmission lines, especially in horizontal components, and preferably have no horizontal lines. Avoid drastic changes in flow direction or velocity. Do not use flow-modifying devices such as pumps, fittings with irregular surfaces, sharp angle changes, flow controllers, etc., between the sampling point and the sensor. Locate pumps after the sensor. Preferably, provide curved sample lines (e.g., flexible tubing or bent glass tubing) rather than right-angle fittings.

For most instruments, flow rate is specified. Because calibration (see ¶ 5) can change with flow rate within the range, flow control within very narrow limits is essential. Calibrate and use the sensor at the same flow rate. For continuous particle size distribution measurements, flow control, not simply flow monitoring, is necessary. Do not make measurements at a flow rate different from that used for calibration. Provide a flow control system downstream of the sensor, maintain a constant rate, and do not introduce turbulence or pulsations before or through the sensor.

4. Dilution Water

Particle-free water is virtually impossible to obtain, but it is possible to produce water containing very few particles within the size range to be measured. Produce "particle-free" or "clean" water from distilled, deionized water or water taken from the same source as the samples. If all samples are from the same source or have similar chemical characteristics, preferably filter sample water to produce "particle-free" water with no change in chemical environment. Distilled, deionized water produced by ion exchange and cartridge filtration may produce acceptable "particle free" dilution water, but preferably use continuous closed-loop membrane filtration.

Dilution water preparation systems are available from particle-counter manufacturers. Alternatively, assemble a system similar to that shown in Figure 2560:1 (or any system that produces a water of sufficient quality). In the system shown, a pump draws the water from a bottle and puts it through the in-line filter. Use membrane filters with nominal pore sizes no more than 10% of the smallest particle size expected; alternatively use cartridge filters. Pass water through the filter several times. The system lets water be passed directly from the product-water bottle to the source-water bottle by opening of the clamps and three-way stopcock. Use glass tubing in the bottles, but use flexible tubing to allow draining product-water bottle into source-water bottle. Attach a glass wool or membrane filter to air inlets. Dilute samples by drawing water directly from the three-way stopcock into the bottle to be used in sample analysis.

A simpler system with one-pass filtration directly into the sample bottle may be adequate for many samples, depending on the

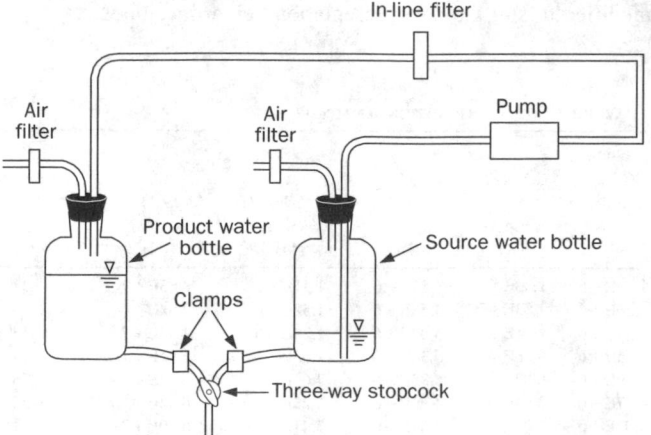

Figure 2560:1. Schematic of filtration apparatus for preparation of particle-free dilution water or electrolyte solution.

particle size range to be measured. Such a system would omit the product-water bottle, the connections between the two bottles and associated stopcock and clamps of Figure 2560:1. Dilution water is produced on demand and the effluent is put directly, without additional handling, into the sample container.

Guard against biological growth within filtration systems by frequent disassembly and adequate washing or replacement of components. For many samples, do not use chemical disinfectants because they might change the particle count and size distribution by their oxidizing potential.

5. Calibration

As a particle is detected in the sensing zone of any instrument, an electrical response is generated and sorted into a channel of the instrument based on its magnitude.

Calibrate by determining the channel number into which particles of known size are sorted by the instrument. Use spherical particles manufactured for this purpose. Calibration particles are available in suspension or as dry particles. Over time, suspensions are likely to undergo some aggregation; use ultrasound to break up flocs before calibration. Calibration particles are nearly monodisperse but do exhibit a small inherent (true) variance.

The precision (see Section 1030A.1) and resolution of the instrument influence its ability to sort particles into different channels. Resolution is a measure of the ability of an instrument to distinguish between particles of similar but different sizes. An instrument with high precision and good resolution will measure monodisperse particles in a very narrow size range; some instruments have sufficient precision and resolution so that calibration particles with extremely narrow size distributions (very small variance) are sorted into a few adjacent channels. In such cases, the true variance of the particle size and the measured variance will be nearly equal. An instrument with low precision but high resolution will yield a wide distribution on the same particles (i.e., measured variance > true variance). An instrument with high precision but low resolution (e.g., few channels) may yield a narrow measured distribution (e.g., all the particles in the same channel) even though the true distribution is broader (i.e., measured variance < true variance).

Use at least three sizes of calibration particles in reasonably similar number concentrations to calibrate a sensor. To analyze different-size particles in a mixture, ensure that the different sizes do not interfere with one another. Calibrate under conditions identical with those of sample measurements, e.g., settings on the instrument, flow rate, type of sample cell, dilution water or electrolyte solution, and mixing during measurement. Do not exceed the maximum concentration for the sensor during calibration.

The calibration curve depends on the characteristic of the particle measured by the instrument (diameter, area, or volume) and whether the pulses are sorted into channels on an arithmetic or logarithmic basis. For example, if an electrical sensing zone instrument (which responds to particle volume) is used in a logarithmic mode, the calibration curve will be the logarithm of particle volume vs. channel number.

Generally, increments between channels are equal on an arithmetic or logarithmic basis. In most light-blockage and light-scattering instruments, the lower and upper limits for each channel can be set by the user. Because most samples have broad particle size distributions, spanning at least one order of magnitude, and often two or three, preferably use larger increments for larger sizes. This is consistent with equal logarithmic spacing, although other less systematic intervals that preserve the characteristic of larger increments for larger sizes are permissible. Such intervals also are consistent with the resolution capabilities of most available instruments, which respond to a characteristic particle area or volume (proportional to square or cube of diameter, respectively), and therefore require greater resolution for larger particles than for smaller ones on a diameter basis.

At a given set of settings for a given sensor, each channel represents a certain average size and size increment. After measuring the channel number associated with several sizes of calibration particles, use a calibration equation to assign average sizes or, in some cases, the lower limit of size, to all other channels. Knowledge of the average size and both the arithmetic and logarithmic increment of each channel is necessary for reporting. In contrast to calibration particles, environmental particles are rarely spherical. Generally report measurements in terms of "equivalent spherical diameter," i.e., size of any particle taken as that of a sphere that would give the same response in the instrument. Because most sample particles are nonspherical and different instruments respond to different characteristics of particles, different measured particle size distributions result from different instruments.

Some light-scattering instruments calculate particle size from first principles and do not require calibration per se. For those instruments, periodically adjust the system optics. Make frequent checks that the instrument response is consistent by following the calibration procedures for other instruments.

6. Data Reporting

a. Particle concentrations: Number, surface, or volume concentration over a specified size range is particularly valuable as a summary factor for particle counting and size distribution. Its variation with some changes in an independent variable in a natural or engineered system may be of interest.

When reporting the number concentration (number per milliliter) of particles, report also the size range measured (both lower and upper size limits). The lower limit is particularly important, because most samples have large number concentrations near the lower limit of detection of currently available instruments. Never state or imply a lower size limit of zero.

In some studies a surface area concentration ($\mu m^2/mL$) or volume concentration ($\mu m^3/mL$) may be more relevant; convert

number concentration to these forms by multiplying by the area or volume, respectively, of a sphere with the mean diameter for each size class. (This is an approximation based on the assumption that particles are spherical.) In such cases, also report lower and upper limits of size class.

b. Tabulated size distributions: If the distribution itself (i.e., the variation of particle number, surface area, or volume concentration with particle size) is to be shown in tabular format, give for each instrument channel (or grouping of channels) the number concentration and the associated size range (lower and upper limits).

c. Graphical size distributions: For graphical reporting, preferably use the count information (on a number, surface area, or volume basis) as the ordinate and particle size as the abscissa.

For the count information, use absolute, rather than relative, scales because they indicate both concentration and size distribution. Also, preferably use differential, rather than cumulative, distributions because they show directly what size range contained the most particles. For ease of plotting, associate the number (or surface area or volume) concentration with a mean size for each channel, rather than with the size range (the lower and upper limit). To account for that change without losing information, normalize the data by dividing the concentration of particles in a given size class by the size interval for that class (on either an arithmetic or a logarithmic basis). Normalization prevents creation of artificial (or apparent) peaks or valleys in a distribution and ensures that the same distribution measured by different instruments with different size increments will be the same graphically. If data are not normalized, plot size distribution data (absolute and differential) as a discrete histogram (i.e., as a bar chart with the upper limit of one channel being identical to the lower limit of the next).

Preferably show particle sizes with a logarithmic scale. Most samples have broad distributions and most analyzers use larger increments of diameter with increasing size; these characteristics are consistent with a logarithmic scale, which also intrinsically avoids showing a zero size. Produce the logarithmic scale by showing the log of diameter on an arithmetic scale or the arithmetic values of diameter on a logarithmic scale.

d. Example calculations: A calculation layout (with example data) useful for preparing either tabular or graphical presentation is shown in Table 2560:I. This format may be abbreviated or modified to suit the data presentation being developed.

TABLE 2560:I. EXAMPLE CALCULATIONS FOR PARTICLE SIZE DISTRIBUTION ANALYSIS

Channel No. (A)	Lower Limit d_p μm (B)	Upper Limit d_p μm (C)	Mean Diam. d_{pi} μm (D)	Δd_{pi} μm (E)	$\log d_{pi}$ d_p in μm (F)	$\Delta \log d_{pi}$ (G)	Corrected Count (H)	Number Conc. ΔN_i $No./mL$ (I)	Volume Conc. ΔV_i $\mu m^3/mL$ (J)	$\Delta N_i/\Delta d_{pi}$ $No./$ $mL\text{-}\mu m$ (K)	$\Delta V_i/\Delta d_{pi}$ $\mu m^3/$ $mL\text{-}\mu m$ (L)	$\Delta N_i/\Delta \log d_{pi}$ $No./mL$ (M)	$\Delta V_i/\Delta \log d_{pi}$ $\mu m^3/mL$ (N)	$\log \Delta N_i/\Delta d_{pi}$ $\Delta N/\Delta d_p$ in $No./mL\text{-}\mu m$ (O)
6	2.95	3.39	3.16	0.44	0.50	0.06	5125	102 500	1.69E+6	2.33E+5	3.85E+6	1.71E+6	2.83E+7	5.37
7	3.39	3.89	3.63	0.50	0.56	0.06	4568	91 360	2.29E+6	1.83E+5	4.58E+6	1.52E+6	2.82E+7	5.26
8	3.89	4.47	4.17	0.58	0.62	0.06	3888	77 760	2.95E+6	1.34E+5	5.09E+6	1.30E+6	4.92E+7	5.13
9	4.47	5.13	4.79	0.66	0.68	0.06	3088	61 760	3.55E+6	9.36E+4	5.38E+6	1.03E+6	5.91E+7	4.97
10	5.13	5.89	5.50	0.76	0.74	0.06	2289	45 780	3.99E+6	6.02E+4	5.25E+6	7.63E+5	6.63E+7	4.78
11	5.89	6.76	6.31	0.87	0.80	0.06	1584	31 680	4.17E+6	3.64E+4	4.86E+6	5.28E+5	6.94E+7	4.56
12	6.76	7.76	7.24	1.00	0.86	0.06	1023	20 460	4.07E+6	2.05E+4	4.07E+6	3.41E+5	6.79E+7	4.31
13	7.76	8.91	8.32	1.15	0.92	0.06	631	12 620	3.81E+6	1.10E+4	3.31E+6	2.10E+5	6.33E+7	4.04
14	8.91	10.23	9.55	1.32	0.98	0.06	363	7 260	3.31E+6	5.50E+3	2.51E+6	1.21E+5	5.52E+7	3.74
15	10.23	11.75	10.96	1.52	1.04	0.06	199	3 980	2.74E+6	2.62E+3	1.80E+6	6.65E+4	4.59E+7	3.42

Columns A through G represent calibration information. Which of these values are set (or determined from calibration by the user) and which are calculated from this primary information depends on the type of instrument. In constructing such a table, preferably place primary information first. In most light-blockage and light-scattering instruments, the lower and upper size limits for each channel are known; calculate mean diameter as the arithmetic or logarithmic (geometric) mean of these limits, depending on whether instrument channels represent arithmetic or logarithmic increments. In electrical sensing zone instruments, the mean size (or log size) is determined directly from the calibration equation and increment width (arithmetic, Δd_{pi}, or logarithmic, $\Delta \log d_{pi}$) is pre-set; calculate lower and upper limits as mean size ± one half the arithmetic or logarithmic increment.

Columns H through O represent counting information for a sample. The corrected counts (Column H) are the direct counts from the particle size analyzer minus the background count for each channel. Adjust background count for dilution and differences in sample measured and dilution water measured for background count. Adjust this figure for sample dilution and volume analyzed to obtain the number concentration of particles (Column I). For example, for corrected count n, a 1:10 dilution, and 0.5 mL analyzed, number concentration is calculated as $(n \times 10)/0.5 = 20n$/mL. Volume concentration (Column J) may be approximated as the number of particles multiplied by the volume of a spherical particle with the mean diameter for the channel $(\pi d_p^3/6)$. Values in the remaining columns are calculated as shown in the table. The normalized, differential, absolute distribution functions obtained in Columns K (particle size distribution function) through O give the values needed for graphic presentations as follows:

K: particle numbers (arithmetic scale) vs. diameter (arithmetic scale)

L: particle volume (arithmetic scale) vs. diameter (arithmetic scale)

M: particle number (arithmetic scale) vs. diameter (logarithmic scale)

N: particle volume (arithmetic scale) vs. diameter (logarithmic scale)

O: log of particle number (arithmetic scale) vs. diameter (logarithmic scale).

(Alternatively, plot values from Column K directly on a logarithmic scale.)

If distribution K, L, M, or N is plotted, the area under the resulting graph between any two values of diameter represents the total number or volume concentration between those size limits. The logarithmic graph (O) shows the entire distribution better than other distributions, but can hide differences between similar distributions. It often is linear over specified size ranges.

7. Quality Control

See Sections 1020 and 1030. Observe precautions for sampling and handling discussed in ¶ 3 above. Particle counting and size analysis requires experienced analysts capable of judgment in recognizing unusual behavior of an instrument. Ability to recognize partial blockage of an aperture or sensor or electronic noise is essential.

Electronic noise can be detected directly in some instruments by making a particle count when no flow is being put through a sensor. Ensure that the noise is not environmental, i.e., due to other nearby instruments, poor grounding, or inconsistent electrical supply. Noise also can be created by excessive cable length between the sensor and instrument or worn cables and connectors. Perform noise checks periodically.

Analyze sample blanks, handled identically to real samples, daily. Discard data from all channels that yield counts in the blanks greater than 5% of the counts for real samples. This procedure accounts for both particle contamination and electronic noise. Develop a maximum acceptable total count for blanks for each sensor (or for each set of standard settings for one sensor); if blanks give more than this maximum, discontinue particle counting until the contamination or noise is eliminated. Set maximum so that the 5% rule is met for all channels of interest (i.e., down to some minimum size limit acceptable in the laboratory for the sensor in use and the samples being measured). Recognize that the lower size limit of measurement for every sensor and every instrument is more likely to be dictated by electronic noise and particle contamination than by electronic settings.

Test cleaning and rinsing for sample bottles by partially filling the container with clean water, swirling the water to contact all sides, and performing a particle count. When the container has been shown to contribute negligible counts, check cleanliness of the cap similarly. Develop and document the standard washing and rinsing procedure, verify its validity as indicated, and follow the procedure without exception.

Also develop standard procedure for mixing samples and dilution water. Test mixing by varying the mixing intensity and/or duration to determine the minimum that gives adequate reproducibility (a sign of uniform concentration) and to determine if greater mixing increases the total particle counts (a sign of floc breakup). Follow an established mixing procedure without exception.

To determine precision, make replicate measurements on at least 5% of the samples. Report standard deviation (or range) of the total particle number concentration for each sensor or the total of all sensors, if more than one is used. See Section 1030.

Particle counting and size distribution analyzers are calibrated for size, not concentration. Standards for calibrating (or checking) the ability of an instrument to measure particle concentration accurately are under development. Currently, only precision, not bias, can be measured well for particle concentration. Obtain an indication of bias by preparing standards of particles of known density at a known suspended solids concentration; multiply total measured particle volume concentration by the known density to estimate the suspended solids concentration from the particle measurements. Compare this estimate with the known suspended solids concentration as an indicator of the combined instrument and laboratory bias.

Periodically calibrate with particles traceable to the National Institute of Standards and Technology (NIST).

8. Bibliography

KAVANAUGH, M.C., C.H. TATE, A.R. TRUSSELL, R.R. TRUSSELL & G. TREWEEK. 1980. Use of particle size distribution measurements for selection and control of solid/liquid separation processes. *In* M.C. Kavanaugh & J. Leckie, eds. Particulates in Water. Advan. Chem. Ser. No. 189, American Chemical Soc., Washington, D.C.

LAWLER, D.F., C.R. O'MELIA & J.E. TOBIASON. 1980. Integral water treatment plant design: from particle size to plant performance. *In* M.C.

Kavanaugh & J. Leckie, eds. Particulates in Water. Advan. in Chem. Ser. No. 189, American Chemical Soc., Washington, D.C.

AMERICAN SOCIETY FOR TESTING AND MATERIALS. 1980. Standard method for determining the quality of calibration particles for automatic particle counters. F322-80, Annual Book of ASTM Standards. American Soc. Testing & Materials, Philadelphia, Pa.

STUMM, W. & J.J. MORGAN. 1981. Aquatic Chemistry, 2nd ed. John Wiley & Sons, New York, N.Y.

AMERICAN SOCIETY FOR TESTING AND MATERIALS. 1985. Standard practice for particle size analysis of particulated substances in the range of 0.2 to 75 micrometers by optical microscopy. E20-85, Annual Book of ASTM Standards. American Soc. Testing & Materials, Philadelphia, Pa.

AMERICAN SOCIETY FOR TESTING AND MATERIALS. 1987. Defining size calibration, resolution, and counting accuracy of a liquid-borne particle counter using near-monodisperse spherical particulate material. F658-87, Annual Book of ASTM Standards. American Soc. Testing & Materials, Philadelphia, Pa.

HARGESHEIMER, E.E., C.M. LEWIS & C.M. YENTSCH. 1992. Evaluation of Particle Counting as a Measure of Treatment Plant Performance. AWWA Research Foundation, Denver, Colo.

2560 B.   Electrical Sensing Zone Method

1. General Discussion

a. Principle: In electrical sensing zone instruments, particles are suspended in an electrolyte solution and pass through a small orifice. A constant current or potential is applied between electrodes on either side of the orifice; the change in resistance caused by the particle taking up volume in the orifice causes a change in potential or current (whichever is not being held constant electronically). The pulse is proportional to the particle volume.

The voltage or current pulses are amplified and sorted into size classes or channels based on their maximum height. Some instruments have fixed channels; others permit selection of number of size classes, size width of each channel, and/or lowest size to be measured.

b. Interferences: Interferences are caused by contamination with particles, the presence of gas bubbles, and electronic noise. See Sections 2560A.3a and b for sampling precautions, and 2560A.4 for particle-free water preparation. Minimize electronic noise as directed in 2560A.7. Increasing electrolyte strength is another method for minimizing electronic noise.

c. Detectable sizes and concentrations: The lower size limit of measurement depends on both electronic noise and aperture size. Manufacturers claim that the lower size limit (diameter) is approximately 2% of the aperture size, but not lower than approximately 0.4 μm, but values as low as approximately 0.7 μm have been reported. For the upper size limit, manufacturers state the maximum size is 40 to 60% of the aperture size, but a realistic upper limit for measuring flocs is 20% of the aperture size.

The minimum concentration limit depends on the ability to distinguish particles from background counts. For each channel, background count should be less than 5% of total count. In size ranges in which the count is very low (zero or nearly so), counts may be statistically unreliable. Determine if a larger sample volume provides more satisfactory (less scattered) data. Grouping data from several adjacent channels, thereby increasing the size increment, also can give a more accurate size distribution.

The maximum concentration limit depends on the rate at which the instrument can process different pulses and the need to avoid more than one particle in the sensing zone at the same time (coincidence). Follow manufacturers' instructions for upper limits on particle counts for each aperture size. Some instruments include a capability for correcting for overly concentrated samples, but do not use this feature because each particle is not measured individually and errors in absolute size distribution may result.

2. Apparatus

a. Particle counter and size-distribution analyzer.

b. Glassware: For glassware preparation, see Sections 2560A.3a and 2560A.7.

3. Reagents

a. Particle-free electrolyte solution: Prepare an electrolyte solution of NaCl, CaCl₂, NaNO₃, Ca(NO₃)₂, or some other simple inorganic salt at a concentration of 1 to 10% by weight. Pass solution through a continuous-flow membrane microfiltration system (see Section 2560A.4). Alternatively, use a commercially prepared electrolyte solution (filtration also may be required).

b. Calibration particles: See Section 2560A.5.

4. Procedure

a. Preparation: Let instrument warm up according to manufacturer's instructions. Select proper size aperture for particle sizes of interest; use more than one aperture if size distribution is wider than the range that can be measured with one aperture. Rinse aperture with acid solution, distilled water, and finally clean electrolyte solution. Install aperture. Choose settings on the instrument for each type of choice, including linear or logarithmic mode (spacing of sizes for each channel; logarithmic is preferable for environmental samples), desired size width of each channel, total number of channels and which channels are to be included, and method of starting and stopping counting. Counting can be started and stopped manually, by switches built into the manometer representing set sample volumes, by a set maximum count in any one channel, or stopped by time after a manual start. For reporting absolute measurements, choose a method that measures volume sampled.

b. Calibration: See Section 2560A.5. Calibrate each aperture with at least three particle sizes, determining the channel number at which the maximum count for each size is located. Handle calibration samples identically to those used for measurement. Because calibration can change, check it at each use of the instrument. Plot results as the volume of calibration particles vs. channel number (linear mode) or as the logarithm (base 10 or e) of the volume of calibration particles vs. channel number (logarithmic mode). The results should plot as a straight line. Use an equation for that line to assign an average size to each channel.

Convert results from particle volume to particle diameter for reporting.

c. Blank sample: Measure at least one blank sample of the electrolyte solution. Carry a sample bottle with only electrolyte solution through procedures identical to those used for samples. Subtract counts from the blank solution (i.e., background counts), channel by channel, from the counts of the samples. See Section 2560A.7.

d. Measurement of samples: Prepare diluted sample, ensuring that the volume of sample is no greater than the volume of the electrolyte solution (dilution water); if a more concentrated suspension is necessary, prepare dilution water with a higher electrolyte concentration. Mix gently (see Section 2560A.3*a*). Insert sample container into the instrument. Start vacuum to control the mercury column and start particle counting. During counting, con-tinuously watch the monitor provided on the instrument to check for blockage of the aperture by oversized particles. Check that the time for the specified volume (if the count is controlled by the stop and start switches on the manometer) is consistent with that found for particle-free electrolyte solution. When count is complete, check that the concentration did not exceed the manufacturer's recommendations. Repeat with a greater dilution if necessary. When a series of samples with various concentrations is to be measured and sample order is not critical, measure samples in order of expected increasing particle concentration, rinsing with next sample.

5. Calculation

See Section 2560A.6.

2560 C.   Light-Blockage Methods

1. General Discussion

a. Principle: In light-blockage instruments, a focused beam of light shines from one side of the measurement zone toward a photovoltaic cell on the other side. The illuminated volume of liquid constitutes the sensing zone. Particles pass through the sensing zone at a known velocity. The blockage of light by the particle creates a change in the voltage at the photovoltaic cell. In different instruments, different characteristics of the resulting signal are used to determine size: light obscuration instruments use pulse height (related to the cross-sectional area of the particle) while time-of-transition instruments use the pulse width (proportional to a characteristic length of the particle). In all cases, measurements usually are reported in terms of equivalent spherical diameter.

For light-obscuration instruments, consult manufacturer's literature to determine the relationship between height of the voltage pulse and particle size (and increment of particle size).

Time-of-transition devices typically utilize a laser beam as the light source. The beam may scan a stationary sample at a fixed velocity as it sweeps out an optical sensing zone. Alternatively, a suspension can be passed through a fixed optical sensing zone. In the latter case, the measured particle size is sensitive to flow rate through the sensor.

Devices using light blockage principles vary in the number of channels (size classes) into which particles are sorted. In some cases, these are pre-set by the manufacturer; in others, they can be set by the analyst. Different (interchangeable) sensors are available to measure different size ranges and different particle concentrations.

b. Interferences: Interferences are caused by contamination with particles, the presence of gas bubbles, and electronic noise. See Sections 2560A.3*a* and *b* for sampling precautions and 2560A.4 for particle-free dilution water preparation. Dilution water is necessary only for samples with particle number concentrations exceeding the capacity of the sensor to distinguish different particles. If dilution water is to be used, analyze it without sample and subtract particle counts in each channel from sample counts.

For devices in which sample flows through the sensor, some electronic noise can be detected by counting with clean water in the sensor and no flow (that is, with no true particle counts). If noise is present, it is likely to occur as a very large number of counts in the smallest size range (lowest channel). Eliminate this noise by re-setting the size range of the smallest channel (increasing the lowest size range to be measured) high enough to avoid significant counts in the lowest channel. For further measures to minimize electronic noise, see Section 2560A.7.

c. Detectable sizes and concentrations: The lower size limit of measurement depends on both electronic noise and sensor capability. Devices using the principle of light obscuration typically are limited to particles larger than approximately 1 μm. Devices using the time-of-transition principle can detect particles as small as 0.1 μm. The upper particle size limit usually is determined by the size of the orifice through which particles pass. Manufacturers state particle size limits (lower and upper) for each sensor. See Section 2560B.1*c*.

2. Apparatus

a. Particle counter and size distribution analyzer.
b. Glassware: See Sections 2560A.3*a* and 2560A.7.

3. Reagents

a. Particle-free dilution water: See Section 2560A.4.
b. Calibration particles: See Section 2560A.5.

4. Procedure

a. Preparation: Let instrument warm up according to manufacturer's instructions. Select sensor to measure size and concentration ranges expected; use more than one sensor if size distribution is wider than the range that can be measured with one sensor. Rinse sensor with a volume of particle-free dilution water equivalent to at least three times that used in measuring samples.

Choose settings on the instrument for each type of choice available: including total number of channels and which channels are to be included, absolute or relative counts, cumulative or differential counts, etc. Also choose method of starting and stopping counting based on total count, maximum count in one channel, duration of counting, or analysis of a specified volume of sample. Usually, the known-volume method is preferable because it lends itself to determination of absolute size distribution.

b. Calibration: See Section 2560A.5. Initial calibration of each sensor on an instrument requires several particle sizes to determine the relationship between particle size and channel number. Either this relationship may be controlled by varying the range of millivolt responses that are sorted into each channel or it is pre-set by the manufacturer.

After initial calibration, use at least two sizes of calibration particles for each sensor each time the instrument is used. For instruments in which the sample flows through a stationary sensing zone, measure flow rate of sample and adjust it to that used during initial calibration.

c. Blank sample: See Section 2560B.4c but substitute particle-free dilution water for electrolyte solution.

d. Measurement of samples: Dilute sample, if necessary to keep within manufacturer's guidelines. Mix gently (see Section 2560A.3c). For flow-type instruments with pressure-based samplers, insert sample container into the instrument; for vacuum-based samplers, insert sampler tubing into the sample container. Start counting only after sufficient sample has passed through the tubing connecting the sample and sensor to ensure that the sensor is receiving the sample. For instruments that scan a stationary sample, gently transfer sample into the measurement device and insert.

For all instruments, when the count is complete, check that the concentration did not exceed the manufacturer's recommendations. Repeat with a greater dilution if necessary.

Between samples, rinse with dilution water or the upcoming sample. When a series of samples with various concentrations is to be measured and sample order is not critical, measure samples in order of expected increasing particle concentration; preferably, rinse with the next sample.

5. Calculation

See Section 2560A.6.

2560 D. Light-Scattering Method

1. General Discussion

a. Principle: Light-scattering instruments may be either flow or static devices. In flow instruments, the direct path of the light beam through the flow cell is blocked by a particle as it flows through the measurement zone with the fluid, and light scattered over a fixed range of angles is collected and measured. Particle size is determined from the angle and intensity of scattering based on the principles of Fraunhofer diffraction and/or Mie scattering. In static instruments, the particles remain quiescent and a laser light beam scans part of the suspension. Scattered light is collected by a photovoltaic cell and the resulting response from all particles scanned is mathematically deconvoluted to generate the size distribution.

Available particle counters vary in the angle or range of angles at which scattered light is measured and in the number of channels (size classes) into which particles are sorted. For flow-type instruments, different (interchangeable) sensors are available to measure different size ranges and different particle concentrations. The particle sizes determined by these instruments are equivalent spherical diameters, i.e., they are determined from the amount of light that would have been scattered at the angle(s) built in to the instrument by a calibration sphere of that diameter.

b. Interferences: For both types of instruments, interferences are caused by contamination of the sample with particles, the presence of gas bubbles, and electronic noise. For static-type instruments, additional sources of interference can be surface contamination (particles, markings, or scratches) on the sample container and sample color. See Sections 2560A.3a and b for sampling precautions and 2560A.4 for particle-free dilution water preparation. Dilution water is necessary only for samples with particle number concentrations exceeding the capacity of the sensor to distinguish different particles. If dilution water is to be used, analyze it without sample and subtract particle counts in each channel from sample counts. To correct for color interference, consult the instrument manual.

In flow-type instruments, some electronic noise can be detected by counting with clean water in the sensor and no flow (that is, with no true particle counts). If noise is present, it is likely to occur as a very large number of counts in the smallest size range (lowest channel). Eliminate this noise by resetting the size range of the smallest channel (increasing the lowest size range to be measured) high enough to avoid significant counts in the lowest channel. For further measures to minimize electronic noise, see Section 2560A.7.

For static-type instruments, use a glass container that is optically clear and has no lettering or markings, at least in the light path. Minimize the effect of specific glassware by using the same sample container for all samples that will be compared and ensuring that the orientation of the sample container in the light beam is the same for every sample.

c. Detectable sizes and concentrations: The lower size limit of measurement depends on both electronic noise and sensor capability. For flow-type instruments, the upper particle size limit usually is determined by the size of the orifice through which particles pass. Manufacturers state particle size limits (lower and upper) for each sensor. See Section 2560B.1c.

Application of the Mie theory to the scattering data can extend the range of light-scattering instruments to as low as 0.1 μm. Alternatively, some instruments extend the lower limit of detectable diameters to 0.1 μm by augmenting Fraunhofer diffraction data with scattering intensity measurements at a fixed angle (often 90°) using two or three different wavelengths of incident light.

2. Apparatus

a. Particle counter and size distribution analyzer.
b. Glassware: See Sections 2560A.3a and 2560A.7.

3. Reagents

a. Particle-free dilution water: See Section 2560A.4.
b. Calibration particles: See Section 2560A.5.

4. Procedure

a. Preparation: Let instrument warm up according to manufacturer's instructions. For flow-type instruments, select sensor to measure size and concentration ranges expected; use more than one sensor if size distribution is wider than the range that can be measured with one sensor. Preferably use a sensor designed for the concentration range of the undiluted sample rather than diluting samples into the concentration range of the sensor. For both types of instruments, place in or pass through the sensor at least three times a volume of particle-free dilution water equivalent to that used in measuring samples, before the first sample.

Choose settings on the instrument for each type of choice available, including flow rate, total number of channels and which channels are to be included, absolute or relative counts, etc. Also choose method of starting and stopping counting: total count, maximum count in one channel, duration of counting, or analysis of a specified volume of sample. For flow-type instruments, the known-volume method is preferable because it lends itself to determination of absolute size distribution. For static-type analyzers, the analyst does not control sample size directly, but can control duration of measurement. Take care not to exceed maximum particle number concentration limits suggested by manufacturer.

Some flow-type instruments permit choice between vacuum or pressure systems for transporting the sample through the sensor. If floc preservation is essential, preferably use a vacuum system but ensure that bubbles do not form before the sensor; otherwise, use a pressure system with a pulseless gear pump.

b. Calibration: See Section 2560A.5. For flow-type instruments, initial calibration of each sensor on an instrument requires several particle sizes to determine the relationship between particle size and channel number. Either this relationship may be controlled by varying the range of millivolt responses that are sorted into each channel, or it is pre-set by the manufacturer. Static-type instruments are pre-set by the manufacturer because sizing is done by a software conversion of the light sensed. After initial calibration, use at least two sizes of calibration particles for each sensor each time the instrument is used. Although the calculation of particle sizes is based on first principles and not on this calibration, calibrate to ensure that the instrument is functioning properly. For flow-type instruments, measure flow rate through the sensor and adjust to that used during initial calibration.

c. Blank sample: See Section 2650B.4c, but substitute particle-free dilution water for electrolyte solution.

d. Measurement of samples: If dilution is necessary, dilute sample to keep within manufacturer's guidelines for the maximum counting rate. Mix gently (see Section 2560A.3c).

For flow-type instruments, initiate the flow from the sample through the sensor. Check that flow rate is the same as that used for sensor calibration. Start counting only after sufficient sample has passed through the tubing connecting the sample and sensor to ensure that the sensor is receiving the sample. When counting is complete, check that the concentration did not exceed the manufacturer's recommendations. Repeat with a greater dilution if necessary. For subsequent samples, rinse with dilution water or the next sample. When a series of samples with varying concentrations is to be measured and sample order is not critical, measure samples in order of expected increasing particle concentration; in this case, preferably rinse with next sample.

For static-type instruments, insert sample beaker or bottle and start particle counting. When counting is complete, check that the concentration did not exceed the manufacturer's recommendations by diluting and measuring again. If the second count is related to the first according to the ratio of the dilution factors of the two measurements, consider the first measurement acceptable. Repeat with a greater dilution if necessary. For subsequent samples, see directions in preceding paragraph.

5. Calculation

See Section 2560A.6.

2570 ASBESTOS*

2570 A. Introduction

1. Occurrence and Significance

The term "asbestos" describes a group of naturally occurring, inorganic, highly fibrous silicate minerals that are easily separated into long, thin, flexible fibers when crushed or processed. Included in the definition are the asbestiform (see ¶ 2 below) varieties of serpentine (chrysotile), riebeckite (crocidolite), grunerite (grunerite asbestos), anthophyllite (anthophyllite asbestos), tremolite (tremolite asbestos), and actinolite (actinolite asbestos).

Asbestos has been used widely as a thermal insulator and in filtration. The tiny, almost indestructible fibers penetrate lung tissue and linings of other body cavities, causing asbestosis and cancer in the lungs and mesothelioma in other cavity linings.

2. Definitions

Asbestiform—having a special type of fibrous habit (form) in which the fibers are separable into thinner fibers and ultimately into fibrils. This habit accounts for greater flexibility and higher tensile strength than other habits of the same mineral. More information on asbestiform mineralogy is available.[1,2]

Aspect ratio—ratio of the length of a fibrous particle to its average width.

Bundle—structure composed of three or more fibers in parallel arrangement with the fibers closer than one fiber diameter to each other.

Cluster—structure with fibers in a random arrangement such that all fibers are intermixed and no single fiber is isolated from the group; groupings of fibers must have more than two points touching.

---

* Approved by Standard Methods Committee, 1993.

Fiber (AHERA)—structure having a minimum length greater than or equal to 0.5 µm, an aspect ratio of 5:1 or greater, and substantially parallel sides.[3]

Fibril—single fiber that cannot be separated into smaller components without losing its fibrous properties or appearance.

Fibrous—composed of parallel, radiating, or interlaced aggregates of fibers, from which the fibers are sometimes separable. The crystalline aggregate of a mineral may be referred to as fibrous even if it is not composed of separable fibers, but has that distinct appearance. "Fibrous" is used in a general mineralogical way to describe aggregates of grains that crystallize in a needle-like habit and appear to be composed of fibers; it has a much more general meaning than "asbestos." While all asbestos minerals are fibrous, not all minerals having fibrous habits are asbestos.

Matrix—fiber or fibers with one end free and the other end embedded in, or hidden by, a particle. The exposed fiber must meet the fiber definition.

Structures—all the types of asbestos particles, including fibers, bundles, clusters, and matrices.

3. References

1. STEEL, E. & A. WYLIE. 1981. Mineralogical characteristics of asbestos. *In* P.H. Riordon, ed. Geology of Asbestos Deposits. Soc. Mining Engineers—American Inst. Mechanical Engineers, New York, N.Y.
2. ZUSSMAN, J. 1979. The mineralogy of asbestos. *In* Asbestos: Properties, Applications and Hazards. John Wiley & Sons, New York, N.Y.
3. U.S. ENVIRONMENTAL PROTECTION AGENCY. 1987. Asbestos-containing materials in schools: Final rule and notice. *Federal Register*, 40 CFR Part 763, Appendix A to Sub-part E, Oct. 30, 1987.

2570 B. Transmission Electron Microscopy Method

1. General Discussion

This method is used to determine the concentration of asbestos structures, expressed as the number of such structures per liter of water. Asbestos identification by transmission electron microscopy (TEM) is based on morphology, selected area electron diffraction (SAED), and energy dispersive X-ray analysis (EDXA). Information about structure size also is generated. Only asbestos structures containing fibers greater than or equal to 0.5 µm in length are counted. The concentrations of both fibrous asbestos structures greater than 10 µm in length and total asbestos structures per liter of water are determined. The fibrous asbestos structures greater than 10 µm in length are of specific interest for meeting the Federal Maximum Contaminant Level Goal (MCLG)

for drinking water, but in many cases the total asbestos concentration provides important additional information.

a. Principle: Sample portions are filtered through a membrane filter. A section of the filter is prepared and transferred to a TEM grid using the direct transfer method. The asbestiform structures are identified, sized, and counted by transmission electron microscopy (TEM), using selected area electron diffraction (SAED) and energy dispersive spectroscopy (EDS or EDXA) at a magnification of 15 000 to 20 000×.

b. Interferences: Certain minerals have properties (i.e., chemical or crystalline structure) that are very similar to those of asbestos minerals and may interfere with the analysis by causing false positives. Maintain references for the following materials in the laboratory for comparison with asbestos minerals, so that they are not misidentified as asbestos minerals: antigorite, attapulgite (palygorskite), halloysite, hornblende, pyroxenes, sepiolite, and vermiculite scrolls.

High concentrations of iron or other minerals in the water may coat asbestos fibers and prevent their full identification.

2. Sampling

a. Containers: Use new, pre-cleaned, capped bottles of glass or low-density (conventional) polyethylene, capable of holding at least 1 L. Do not use polypropylene bottles. Rinse bottles twice by filling approximately one-third full with fiber-free water and shaking vigorously for 30 s. Discard rinse water, fill bottles with fiber-free water, treat in ultrasonic bath (60 to 100 W) for 15 min, and rinse several times with fiber-free water.

Make blank determinations on the bottles before collecting a sample. Use one bottle in each batch or a minimum of one bottle in each 24 to test for background level when using polyethylene bottles. When sampling waters probably containing very low levels of asbestos, or for additional confidence in the bottle blanks, run additional blank determinations.

b. Collection: Follow general principles for water sampling (see Section 1060). Some specific considerations apply to asbestos fibers, which range in length from 0.1 μm to 20 μm or more. In large bodies of water, because of the range of sizes there may be a vertical distribution of particle sizes that may vary with depth. If a representative sample from a water supply tank or impoundment is required, take a carefully designated set of samples representing vertical as well as horizontal distribution and composite for analysis. When sampling from a distribution system, choose a commonly used faucet, remove all hoses or fittings, and let water run to waste for 1 to 3 min. (Often, the appropriate time to obtain a main's sample can be determined by waiting for a change in water temperature.) Because sediment may build up in valving works, do not adjust faucets or valves until all samples have been collected. Similarly, do not consider samples at hydrants and at dead ends of the distribution systems to be representative of the water in the system. As an additional precaution against contamination, rinse each bottle several times with the source water being sampled. For depth sampling, omit rinsing. Obtain a sample of approximately 800 mL from each sampling site, leaving some air space in each bottle. Using a waterproof marker, label each container with date, time, place, and sampler's initials.

c. Shipment: Ship water samples in a sealed container, but separate from any bulk or air samples intended for asbestos analysis. Preferably, ship in a cooler to retard bacterial or algal growth. Do not freeze the sample. In the laboratory, prepare sample within 48 h of collection.

3. Apparatus

a. High-efficiency particulate air (HEPA) filtered negative-flow hood.

b. Filter funnel assemblies, either 25 mm or 47 mm, of either of the following types:

1) *Disposable plastic units*, or

2) *Glass filtering unit.* With this type of unit, observe the following precautions: Never let unit dry after filtering. Immediately place it in detergent solution, scrub with a test-tube brush, and rinse several times in particle-free water. Periodically treat unit in detergent solution in an ultrasonic bath. Clean unit after each sample is filtered. Run a blank on particle-free water filtered through the glass filtering unit at frequent intervals to ensure absence of residual asbestos contamination.

c. Side-arm filter flask, 1000 mL.

d. Either:

1) *Mixed cellulose ester (MCE) membrane filters*, 25- or 47-mm diam, ≤ 0.45-μm and 5-μm pore size, or

2) *Polycarbonate (PC) filters*, 25- or 47-mm diam, ≤ 0.4-μm pore size.

e. Ultrasonic bath, tabletop model, 60 to 100 W.

f. Graduated pipet, disposable glass, 1, 5, and 10 mL.

g. Cabinet-type dessicator or *low-temperature drying oven.*

h. Cork borer, 7 mm.

i. Glass slides.

j. Petri dishes, glass, approximately 90-mm diam.

k. Mesh screen, stainless steel or aluminum, 30 to 40 mesh.

l. Ashless filter paper filters, 90-mm diam.

m. Exhaust or fume hood.

n. Scalpel blades.

o. Low-temperature plasma asher.

p. High-vacuum carbon evaporator with rotating stage, capable of less than 0.013 Pa pressure. Do not use units that evaporate carbon filaments in a vacuum generated only by oil rotary pump. Use carbon rods sharpened with a carbon rod sharpener to necks about 4 mm long and 1 mm in diam. Install rods in the evaporator so that the points are approximately 100 to 120 mm from surface of microscope slide held in the rotating device.

q. Lens tissue.

r. Copper TEM finder grids, 200 mesh. Use pre-calibrated grids, or determine grid opening area by either of the following methods:

1) Measure at least 20 grid openings in each of 20 random 200-mesh grids (total of 400 grid openings for every 1000 grids used) by placing the 20 grids on a glass slide and examining them under an optical microscope. Use a calibrated reticule to measure average length and width of 20 openings from each of the individual grids. From the accumulated data, calculate the average grid opening area.

2) Measure grid area at the TEM (¶ *s* below) at a calibrated screen magnification of between 15 000 and 20 000×. Measure one grid opening for each grid examined. Measure grid openings in both the x and y directions and calculate area.

s. Transmission electron microscope (TEM), 80 to 120 kV, equipped with energy dispersive X-ray system (EDXA), capable of performing electron diffraction with a fluorescent screen inscribed with calibrated gradations. The TEM must have a scanning transmission electron microscopy (STEM) attachment or be capable of producing a spot size of less than 250 nm diam at crossover. Calibrate routinely for magnification, camera constant, and EDXA settings according to procedures of *5d.*

4. Reagents and Materials

a. Acetone.

b. Dimethylformamide (DMF). CAUTION: *Toxic; use only in a fume hood.*

c. Glacial acetic acid. CAUTION: *Use in a fume hood.*

d. Chloroform.

e. 1-Methyl-2-pyrrolidinone.

f. Particle-free water: Use glass-distilled water or treat by reverse osmosis; filter through a filter with pore diam 0.45 μm or smaller.

g. *Non-asbestos standards* for minerals listed in ¶ 1b above.

h. *Asbestos standards* for minerals listed in Section 2570A.1.

5. Procedure

a. *Sample filtration:* Samples with high levels of organic contaminants may require pretreatment. A process using ultraviolet light and ozone bubbling is described elsewhere.[1] Drinking water samples prepared within 48 h of collection do not require pretreatment.

Under a HEPA hood, carefully wet-wipe exterior of sample bottle to remove any possible contamination before taking bottle into a clean preparation area separated from preparation areas for bulk or air-sample handling.

Prepare specimen in a clean HEPA filtered negative-pressure hood. Measure cleanliness of preparation area hoods by cumulative process blank concentrations (see ¶ b below).

If using a disposable plastic filter funnel unit, remove funnel assembly and discard top filter supplied with the apparatus, but be sure to retain the coarse polypropylene support pad in place. Assemble unit with the adapter and a properly sized neoprene stopper, and attach funnel to the 1000-mL side-arm vacuum flask. Moisten support pad with a few milliliters distilled water, place a 5.0-μm-pore-size MCE backing filter on support pad, and place an MCE or PC filter (≤0.45-μm-pore-size) on top of backing filter. After both filters are completely wet, apply vacuum, ensuring that filters are centered and pulled flat without air bubbles. If there are any irregularities on the filter surface, discard filters and repeat process. Replace funnel assembly. Return flask to atmospheric pressure. Alternatively, use glass filtering unit, following the same procedure to set up filters.

With flask at atmospheric pressure, add 20 mL particle-free water to funnel. Cover funnel with its plastic cover if the disposable filtering unit is used.

Briefly, by hand, shake capped bottle with sample suspension, then place it in tabletop ultrasonic bath and sonicate for 3.0 min. The water level in the bath should be approximately the same as that of the sample. After treatment, return sample bottle to work surface of HEPA hood. Carry out all preparation steps, until filters are ready for drying, in this hood.

Shake suspension vigorously by hand for 2 to 3 s. Estimate amount of liquid to be withdrawn to produce an adequate filter preparation. Experience has shown that a light staining of the filter surface will usually yield a suitable preparation. If the sample is relatively clean, use a volumetric cylinder to measure sample. If sample has a high particulate or asbestos content, withdraw a small volume (but at least 1 mL) with disposable glass pipet, inserting pipet halfway into sample.

NOTE: If, after examination in the TEM, the smallest volume measured (1.0 mL) yields an overloaded sample, make additional serial dilutions of the suspension. Shake suspension vigorously by hand for 2 to 3 s before taking serial dilution portion. Do not retreat in ultrasonic bath either original solution or any dilutions. Mix 10 mL sample with 90 mL particle-free water in a clean sample bottle to obtain a 1:10 serial dilution.

Uncover filter funnel and dispense mixture into the water in the funnel. Recover funnel and agitate but do not swirl the liquid. (One acceptable means of agitation is to inject 5 to 10 mL particle-free water from a squeeze bottle into the funnel.) Apply vacuum and filter. Discard pipet, if used.

Disassemble filtering unit and carefully remove filter with clean forceps. Place filter, particle side up, in a precleaned, labeled, disposable plastic petri dish or similar container.

To obtain an optimally loaded filter, make several filtrations with different sample portions. Use new disposable plastic funnel units or carefully cleaned glass units for each filtration. When additional filters are prepared, shake suspension without additional ultrasonic treatment before removing the sample portion. Each new filtration should represent at least a five-fold loading difference.

Dry MCE filters for at least 12 h (over desiccant) in airtight, cabinet-type desiccator. Alternatively, to shorten drying time for filters prepared using the acetone collapsing method, plug damp filter and attach to a glass slide as described in ¶ c1)a) below. Place the slide with filter plug(s) (up to eight plugs can be attached to one slide) on a bed of desiccant, cover, and place in desiccator for 1 to 2 h.

Place PC filters in a dessicator for at least 30 min before preparation; lengthy drying is not required.

b. *Sample blank preparation:* Prepare sample blanks that include both a process blank (50 mL particle-free water) for each set of samples analyzed and one unused filter from each new box of sample filters (MCE or PC). Blanks are considered contaminated if, after analysis, they are shown to contain more than 53 asbestos structures/mm^2. This corresponds to 3 or 4 asbestos structures found in 10 grid openings. Identify source of contamination before making any further analysis. Reject samples that were processed with the contaminated blanks and prepare new samples after source of contamination is found. Take special care with polycarbonate filters, because some have been shown to contain asbestos contamination.

c. *Specimen preparation:*

1) Mixed cellulose ester (MCE) filters

a) Filter fusing—Use either the acetone or the DMF-acetic acid method.

(1) Acetone fusing method—Remove a section from any quadrant of the sample and blank filters with a 7-mm cork borer. Place filter section (particle side up) on a clean microscope slide. Affix filter section to slide with a gummed page reinforcement or other suitable means. Label slide with a glass scribing tool or permanent marker.

Prepare a fusing dish as follows: Make a pad from five to six ashless paper filters and place in bottom of a glass petri dish. Place metal screen bridge on top of pad and saturate filter pads with acetone. Place slide on top of bridge and cover the petri dish. Wait approximately 5 min for sample filter to fuse and clear completely.

(2) DMF-acetic acid fusing method—Place drop of clearing solution (35% dimethylformamide [DMF], 15% glacial acetic acid, and 50% particle-free water by volume) on a clean microscope slide. CAUTION: *DMF is a toxic solvent; use only in a fume hood.* Use an amount of clearing solution that just saturates the filter. Using a clean scalpel blade, cut a wedge-shaped section of filter. A one-eighth filter section is sufficient. Carefully lay filter segment, sample surface upward, on top of solution. Bring filter and solution together at an angle of about 20 deg to help exclude air bubbles. Remove excess clearing solution with filter paper. Place slide in oven, on a slide-warmer, or on hot plate, in a fume hood, at 65 to 70°C for 5 to 10 min. The filter section should fuse and clear completely.

b) Plasma etching—Place microscope slide, with attached collapsed filter pieces, in a low-temperature plasma asher. Because plasma ashers vary greatly in their performance, both from unit to unit and between different positions in the asher barrel, it is difficult to specify operating conditions. Insufficient etching will result in a failure to expose embedded fibers; too much etching may result in the loss of particles from the filter surface. Calculate time for ashing on the basis of final sample observations in transmission electron microscope. Additional information on calibration is available.[2,3]

c) Carbon coating—Using high-vacuum carbon evaporator (¶ 3p), proceed as follows: Place glass slide holding filters on the rotation device and evacuate evaporator chamber to a pressure of less than 0.013 Pa. Perform evaporation in very short bursts, separated by 3 to 4 s to let electrodes cool. An experienced analyst can judge the thickness of the carbon film. Make initial tests on unused filters. If the carbon film is too thin, large particles will be lost from the TEM specimen, and there will be few complete and undamaged grid openings. A coating that is too thick will lead to a TEM image lacking in contrast and a compromised ability to obtain electron diffraction patterns. The carbon film should be as thin as possible and still remain intact on most of the grid openings of the TEM specimen.

d) Specimen washing—Prepare a Jaffe washer according to any published design.[1,4] One such washer consists of a simple stainless steel bridge contained in a glass petri dish. Place on the stainless steel bridge several pieces of lens tissue large enough to hang completely over the bridge and into the solvent. In a fume hood, fill petri dish with either acetone or DMF to one quarter of the level in the dish.

Place TEM grids, shiny side up, on a piece of lens tissue or filter paper so that individual grids can be easily picked up with forceps. Prepare from each sample three grids. Using a curved scalpel blade, excise three square (3-mm × 3-mm) pieces of carbon-coated MCE filter from random areas on the filter. Place each square filter piece, carbon-side up, on top of a TEM specimen grid (¶ 3r).

Place all three assemblies (filter/grid) for each sample on the same piece of saturated lens tissue in Jaffe washer. Place lid on the Jaffe washer and let system stand, preferably overnight.

Alternatively, place grids on a low-level (petri dish is filled only enough to wet paper on screen bridge) DMF Jaffe washer for 60 min. Then add enough solution of equal parts DMF/acetone to fill washer up to screen level. Remove grids after 30 min if they have cleared, i.e., all filter material has been removed from the carbon film, as determined by inspecting in the TEM. Let grids dry before placing in TEM.

2) Polycarbonate (PC) filters—Cover surface of a clean microscope slide with two strips of double-sided cellophane tape. Cut a strip of filter paper slightly narrower than width of slide. Position filter paper strip on center of length of slide. Using a clean, curved scalpel blade, cut a strip of the PC filter approximately 25 × 6 mm. Use a rocking motion of the scalpel blade to avoid tearing filter. Place PC strip, particle side up, on slide perpendicular to long axis of slide, making sure that the ends of the PC strip contact the double-sided cellophane tape. Each slide can hold several PC strips. Label filter paper next to each PC strip with sample number.

Carbon-coat filter strips as directed in ¶ 1)c) above. (Etching is not required.) Take special care to avoid overheating filter sections during carbon coating.

Prepare a Jaffe washer as described in ¶ 1)d) above, but fill washer with chloroform or 1-methyl-2-pyrrolidinone to the level of the screen. Using a clean curved scalpel blade, excise three 3-mm-square filter pieces from each PC strip. Place filter squares, carbon side up, on shiny side of a TEM grid (¶ 3r). Pick up grid and filter section together and place them on lens tissue in the Jaffe washer. Place lid on Jaffe washer and leave grids for at least 4 h. Best results are obtained with longer wicking times, up to 12 h. Carefully remove grids from the Jaffe washer and let dry in the grid box before placing them in a clean, marked grid box.

d. Instrument calibration: Calibrate instrumentation regularly, and keep a calibration record for each TEM in the laboratory, in accordance with the laboratory's quality assurance program. Record all calibrations in a log book along with dates of calibration and attached backup documentation.

Check TEM for both alignment and systems operation. Refer to manufacturer's operational manual for detailed instructions.

Calibrate camera length of TEM in electron diffraction (ED) operating mode before observing ED patterns of unknown samples. Measure camera length by using a carbon-coated grid on which a thin film of gold has been sputtered or evaporated. A thin film of gold may be evaporated directly on to a specimen grid containing asbestos fibers. This yields zone-axis ED patterns from the asbestos fibers superimposed on a ring pattern from the polycrystalline gold film. Optimize thickness of gold film so that only one or two sharp rings are obtained. Thicker gold films may mask weaker diffraction spots from the fibrous particles. Because unknown d-spacings of most interest in asbestos analysis are those lying closest to the transmitted beam, multiple gold rings from thicker films are unnecessary. Alternatively, use a gold standard specimen to obtain an average camera constant calculated for that particular instrument, which can then be used for ED patterns of unknowns taken during the corresponding period.

Calibrate magnification at the fluorescent screen at magnification used for structure counting. Use a grating replica (e.g., one containing at least 2160 lines/mm). Define a field of view on the fluorescent screen; the field must be measurable or previously inscribed with a scale or concentric circles (use metric scales). Place grating replica at the same distance from the objective lens as the specimen. For instruments that incorporate a eucentric tilting specimen stage, place all specimens and the grating replica at the eucentric position. Follow the instructions provided with the grating replica to calculate magnification. Frequency of calibration depends on service history of the microscope. Check calibration after any maintenance that involves adjustment of the power supply to the lens, the high-voltage system, or the mechanical disassembly of the electron optical column (apart from filament exchange).

Check smallest spot size of the TEM. At the crossover point, photograph spot size at a magnification of 25 000× (screen magnification 20 000×). An exposure time of 1 s usually is adequate. Measured spot size must be less than or equal to 250 nm.

Verify resolution and calibration of the EDXA as follows: Collect a standard EDXA Cu peak from the Cu grid and compare X-ray energy with channel number for Cu peak to make sure readings are within ± 10 eV. Collect a standard EDXA of crocidolite asbestos; elemental analysis of the crocidolite must resolve the Na peak. Collect a standard EDXA spectrum of chrysotile asbestos; elemental analysis of chrysotile must resolve both Si and Mg on a single chrysotile fiber.

e. Sample analysis: Carefully load TEM grid with grid bars oriented parallel/perpendicular to length of specimen holder. Use a hand lens or eye loupe if necessary. This procedure will line up the grid with the x and y translation directions of the microscope. Insert specimen holder into microscope.

Scan entire grid at low magnification ($250\times$ to $1000\times$) to determine its acceptability for high-magnification analysis. Grids are acceptable if the following conditions are met:

• The fraction of grid openings covered by the replica section is at least 50%.

• Relative to that section of the grid covered by the carbon replica, the fraction of intact grid openings is greater than 50%.

• The fractional area of undissolved filter is less than 10%.

• The fraction of grid squares with overlapping or folded replica film is less than 50%.

• At least 20 grid squares have no overlapping or folded replica, are less than 5% covered with holes, and have less than 5% opaque area due to incomplete filter dissolution.

If the grid meets these criteria, chose grid squares for analysis from various areas of the grid so that the entire grid is represented. To be suitable for analysis, an individual grid square must meet the following criteria:

• It must have less than 5% holes over its area.

• It must be less than 25% covered with particulate matter.

• It must be uniformly loaded.

Observe and record orientation of grid at 80 to $150\times$ on a grid map record sheet along with the location of the grid squares examined. If indexed grids are used, a grid map is not needed, but record identifying coordinates of the grid square.

At a screen magnification of 15 000 to $25\,000\times$, evaluate the grids for the most concentrated sample loading. Reject sample if it is estimated to contain more than about 25 asbestos structures per grid opening. Proceed to the next most concentrated sample until a set of grids is obtained that have less than 25 asbestos structures per grid opening.

Analyze a minimum of four grid squares for each sample.

Analyze approximately one-half of the predetermined sample area on one sample grid preparation and the remainder on a second sample grid preparation.

Use structure definitions given in Section 2570A.2 to enumerate asbestos structures. Record all data on count sheet. Record asbestos structures in two size categories: ≥ 0.5 µm to ≥ 10.0 µm and >10.0 µm. For fibers and bundles, record size category of greatest length of the structure. For matrices and clusters, record size category of the visible portion of the longest fiber or bundle involved with the structure, not the greatest overall dimension of the structure. No minimum or maximum width restrictions are applied to the fiber definition, as long as the minimum length and aspect ratio criteria are met. Record "NSD" (no structures detected) when no structures are found in the grid opening.

Record a typical electron diffraction pattern for each type of asbestos observed for each group of samples (or a minimum of every five samples) analyzed. Record micrograph number on count sheet. For chrysotile, record one X-ray spectrum for each tenth structure analyzed. For each type of amphibole, record one X-ray spectrum for each fifth structure analyzed. (More information on identification is available.[1,4]) Attach the printouts to the back of the count sheet. If the X-ray spectrum is stored, record file and disk number on count sheet.

Analytical sensitivity can be improved by increasing amount of liquid filtered, increasing number of grid openings analyzed, or decreasing size of filter used. Occasionally, because of high particle loadings or high asbestos concentration, the desired analytical sensitivity cannot be achieved in practice.

Unless a specific analytical sensitivity is desired, stop analysis on the 10th grid opening or the grid opening that contains the 100th asbestos structure, whichever comes first. If the analysis is stopped at the grid opening that contains the 100th asbestos structure, count entire grid square containing the 100th structure.

After analysis, remove grids from TEM, replace them in grid storage holder, and store for a minimum of one year from the date of the analysis for legal purposes. Sample filters also may be stored in the plastic petri dishes, if necessary. Prolonged storage of the remaining water sample is not recommended, because microbial growth may cause loss of asbestos structures to the sides of the storage container.

Report the following information for each water sample analyzed: asbestos concentration in structures per liter, for total structures and fibrous asbestos structures greater than 10 µm in length; types of asbestos present; number of asbestos structures counted; effective filtration area; average size of TEM grid openings counted; number of grid openings examined; size category for each structure counted. Include a copy of the TEM count sheet, either hand-written or computer-generated.

6. Calculations

Calculate amount of asbestos in a sample as follows:

$$\text{Asbestos concentration, structures/L} = \frac{N \times A_f \times D}{G \times A_G \times V_s}$$

where:

- N = number of asbestos structures counted,
- A_f = effective filter area of final sampling filter, mm^2,
- D = dilution factor (if applicable),
- G = number of grid openings counted,
- A_G = area of grid openings, mm^2, and
- V_s = volume of sample, L.

The same formula may be used to calculate asbestos fibers greater than 10 µm in length per liter based on the total number of fibers and bundles greater than 10 µm in length whether free or associated with matrices and clusters. Express final results as million structures per liter (MSL) and million fibers per liter (MFL).

7. Quality Control/Quality Assurance

Use the laboratory's quality-control checks to verify that system is performing according to accuracy and consistency specifications. Because of the difficulties in preparing known quantitative asbestos samples, routine quality-control testing focuses on reanalysis of samples (duplicate recounts). Reanalyze 1 out of every 10 samples, not including laboratory blanks.

In addition, set up quality assurance programs according to the criteria developed by Federal agencies.[2,3] These documents cover sample custody, sample preparation, blank checks for contamination, calibration, sample analysis, analyst qualifications, and technical facilities.

8. Precision and Bias

Precision measurements for intralaboratory comparisons have been found to have a relative standard deviation (RSD) of 13 to

22% for standard and environmental water samples, with an RSD of 8.4 to 29% for interlaboratory comparisons.[1] An earlier study found an interlaboratory reproducibility of 25 to 50% in standard samples.[5]

Accuracy measurements from inter- and intralaboratory studies have demonstrated an RSD of 17% for standard chrysotile suspensions, and an RSD of 16% for standard crocidolite suspensions.[1]

9. References

1. CHATFIELD, E.J. & M.J. DILLON. 1983. Analytical Method for the Determination of Asbestos in Water. EPA 600/4-83-043, U.S. Environmental Protection Agency, Athens, Ga.

2. NATIONAL INSTITUTE OF STANDARDS AND TECHNOLOGY & NATIONAL VOLUNTARY ACCREDITATION PROGRAM. 1989. Program Handbook for Airborne Asbestos Analysis. NISTIR 89-4137, U.S. Dep. Commerce, Gaithersburg, Md.

3. U.S. ENVIRONMENTAL PROTECTION AGENCY. 1987. Asbestos-containing materials in schools: Final rule and notice. *Federal Register*, 40 CFR Part 763, Appendix A to Sub-part E, Oct. 30, 1987.

4. YAMATE, G., S.C. AGARWALL & R.D. GIBBONS. 1984. Methodology for the Measurement of Airborne Asbestos by Electron Microscopy. USEPA draft report, Contract No. 68-02-3266, Research Triangle Park, N.C.

5. CHOPRA, K.S. 1977. Inter-laboratory measurements of amphibole and chrysotile fiber concentrations in water. *In* National Bureau of Standards Spec. Publ. 506, Proc. of the Workshop on Asbestos: Definitions and Measurement Methods. Gaithersburg, Md.

2580 OXIDATION-REDUCTION POTENTIAL (ORP)*

2580 A. Introduction

1. Significance

Oxidation and reduction (redox) reactions mediate the behavior of many chemical constituents in drinking, process, and wastewaters as well as most aquatic compartments of the environment.[1-5] The reactivities and mobilities of important elements in biological systems (e.g., Fe, S, N, and C), as well as those of a number of other metallic elements, depend strongly on redox conditions. Reactions involving both electrons and protons are pH- and Eh-dependent; therefore, chemical reactions in aqueous media often can be characterized by pH and Eh together with the activity of dissolved chemical species. Like pH, Eh represents an intensity factor. It does not characterize the capacity (i.e., poise) of the system for oxidation or reduction.

The potential difference measured in a solution between an inert indicator electrode and the standard hydrogen electrode should not be equated to Eh, a thermodynamic property, of the solution. The assumption of a reversible chemical equilibrium, fast electrode kinetics, and the lack of interfering reactions at the electrode surface are essential for such an interpretation. These conditions rarely, if ever, are met in natural water.

Thus, although measurement of Eh in water is relatively straightforward, many factors limit the interpretation of these values. These factors include irreversible reactions, electrode poisoning, the presence of multiple redox couples, very small exchange currents, and inert redox couples. Eh values measured in the field correlate poorly with Eh values calculated from the redox couples present. Nevertheless, measurement of redox potential, when properly performed and interpreted, is useful in developing a more complete understanding of water chemistry.

2. Sampling and Storage

Do not store samples; analyze on collection. Minimize both atmospheric contact and delay in analysis.

3. References

1. BAAS-BECKING, L.G.M., I.R. KAPLAN & D. MOORE. 1960. Limits of the natural environment in terms of pH and oxidation-reduction potentials. *J. Geol.* 68:243.

2. LANGMUIR, D. 1971. Eh-pH determination. *In* R.E. Carver, ed. Procedures in Sedimentary Petrology. Wiley-Interscience, New York, N.Y.

3. GARRELS, R.M. & C.L. CHRIST. 1965. Solutions, Minerals and Equilibria. Harper and Row, New York, N.Y.

4. BRECK, W.G. 1972. Redox potentials by equilibration. *J. Mar. Res.* 30:121.

5. BRICKER, O.P. 1982. Redox potential: Its measurement and importance in water systems. *In* R.A. Minear & L.H. Keith, eds. Water Analysis, Vol. 1, Inorganic Species. Academic Press, New York, N.Y.

4. Bibliography

GARRELS, R.M. 1960. Mineral Equilibria at Low Temperature and Pressure. Harper & Brothers, New York, N.Y.

POURBAIX, M.J.N. 1966. Atlas of Electrochemical Equilibriums in Aqueous Solutions (transl. J.A. Franklin, Centre Belge d'Etude de la Corrosion). Pergamon Press, Ltd., Oxford, England.

STUMM, W. 1967. Redox potential as an environmental parameter; conceptual significance and operational limitation. *In* Advances in Water Pollution Research. Proc. Int. Conf. Water Pollut. Res., 3rd, 1:283.

ADAMS, R.N. 1969. Electrochemistry at Solid Electrodes. Marcel Dekker, New York, N.Y.

BATES, R.G. 1973. Determination of pH, Theory and Practice, 2nd ed. John Wiley & Sons, New York, N.Y.

LINDBERG, R.D. & D.D. RUNNELLS. 1984. Groundwater redox reactions: an analysis of equilibrium state applied to Eh measurements and geochemical modeling. *Science* 225:925.

* Approved by Standard Methods Committee, 1997.

2580 B. Oxidation-Reduction Potential Measurement in Clean Water

1. General Discussion

a. Principle: Electrometric measurements are made by potentiometric determination of electron activity (or intensity) with an inert indicator electrode and a suitable reference electrode. Ideally, the indicator electrode will serve as either an electron donor or acceptor with respect to electroactive oxidized or reduced chemical species in solution. At redox equilibrium, the potential difference between the ideal indicator electrode and the reference electrode equals the redox potential of the system. However, inert indicator electrodes that behave ideally in all aqueous systems, particularly in natural waters, do not exist. Electrodes made of platinum are most commonly used for Eh measurements. They have limitations,[1] as do alternative materials such as gold and graphite.

The standard hydrogen reference electrode is fragile and impractical for routine laboratory and field use. Therefore, silver: silver-chloride or calomel reference electrodes are used commonly. The redox potential measurement is corrected for the difference between the potential of the reference electrode and that of the standard hydrogen electrode. See Section 4500-H$^+$, pH Value.

It is not possible to calibrate Eh electrodes over a range of redox potentials (as is done with pH electrodes). Instead, standard solutions that exhibit both chemical stability and known redox potentials for specific indicator electrodes are used to check electrode response at the temperature of measurement.

The potential of the platinum (Pt) Eh electrode versus the Ag/AgCl reference electrode with KCl electrolyte in ZoBell's solution ($3 \times 10^{-3}M$ potassium ferrocyanide and $3 \times 10^{-3}M$ potassium ferricyanide in 0.1M KCl)[2] has been measured as a function of temperature.[3] Good agreement was obtained between Eh values measured with this electrode pair in ZoBell's solution and those calculated from the stability constants at 8 to 85°C. The potential of the ZoBell's solution with this electrode configuration as a function of temperature can be calculated:[4]

$$Eh, \text{V} = 0.428 - 0.0022\,(T - 25)$$

where T = solution temperature, °C. Alternatively, select the value from Table 2580:I.

To determine the Eh of a sample relative to the standard hydrogen electrode, measure Eh of both sample and standard solution at the same temperature (within ±0.1°C). Then calculate Eh value of the sample:

$$Eh_{system} = E_{observed} + Eh_{ZoBell/reference} - Eh_{ZoBell\ observed}$$

where:

$E_{observed}$ = sample potential relative to reference electrode.
$Eh_{ZoBell/reference}$ = theoretical Eh of reference electrode and ZoBell's solution, relative to the standard hydrogen electrode (see Table 2580:I), and
$Eh_{ZoBell\,observed}$ = observed potential of ZoBell's solution, relative to the reference electrode.

The measurements described above can be applied analogously to other indicator electrode/reference electrode pairs and standard solutions.

TABLE 2580:I. POTENTIAL OF ZOBELL'S SOLUTION AS FUNCTION OF TEMPERATURE

T °C	E V	T °C	E V
1	0.481	16	0.448
2	0.479	17	0.446
3	0.476	18	0.443
4	0.474	19	0.441
5	0.472	20	0.439
6	0.470	21	0.437
7	0.468	22	0.435
8	0.465	23	0.432
9	0.463	24	0.430
10	0.461	25	0.428
11	0.459	26	0.426
12	0.457	27	0.424
13	0.454	28	0.421
14	0.452	29	0.419
15	0.450	30	0.417

b. Interferences: Specific interferences may be due to operation of either indicator or reference electrode, redox capacity or poise of the sample, sample preservation and handling, and temperature equilibration.

1) Sorption and poisoning effects on electrodes—Contamination of the electrode surface, salt bridge, or internal electrolyte in the case of reference electrodes, can lead to excessive drift, poor electrode response, and artifact potentials. Organic matter, sulfide, and bromide may cause these problems, particularly in long-term electrode use.[1,5-7] If excessive drift occurs or erratic performance of paired electrodes is observed in redox standard solutions after appropriate cleaning, refilling, or regeneration procedures, discard the faulty electrode and use a new one.

2) pH variations—Redox potential is sensitive to pH if hydrogen ion or hydroxide ion is involved in the redox half-cells. Cell potentials tend to increase as proton concentration increases (i.e., pH decreases) and Eh values drop as hydroxide concentrations increase (i.e., pH increase).

3) Sample handling and preservation—The sample poise will govern the resistance of the sample to change in redox potential; this phenomenon is analogous to the resistance to pH change afforded by buffer capacity. Except in concentrated process streams, sludges, leachates, and highly reducing or treated waters, the concentrations of oxidized or reduced species may be fairly low (i.e., $<10^{-4}M$). Under these conditions, handle reduced samples very carefully to avoid exposure to atmospheric oxygen. A closed cell sampling configuration may be used.[4,8] Samples cannot be stored or preserved; analyze at sampling.

4) Temperature equilibration—Obtain Eh standard solution reading for the electrode pair at a temperature as close as possible to that of the sample. Temperature determines the Eh reference potential for a particular solution and electrode pair. It also may affect the reversibility of the redox reaction, the magnitude of the exchange current, and the stability of the apparent redox potential reading, particularly in poorly poised solutions. Hold temperature constant for all measurements and report it with Eh results.

2. Apparatus

a. pH or millivolt meter: Use a pH meter or other type of high-impedance potentiometer capable of reading either pH or millivolts (mV). For most applications, a meter scale readable to ± 1400 mV is sufficient.

b. Reference electrode consisting of a half-cell providing a constant electrode potential. See Section 4500-H$^+$.B.2*b*.

c. Oxidation-reduction indicator electrode: The platinum electrode is used most commonly. A noble metal or graphite electrode may be useful for specific applications.

1) Noble metal electrode—Noble metal (i.e., gold or platinum) foil, wire, or billet types of electrode are inert and resistant to chemical reaction. Clean and polish electrode surface to insure reliable performance. Platinum electrodes may be cleaned by strong acid soaking,[9,10] hydrogen peroxide and detergent washing,[11] and anodic activation.[10] Abrasive polishing with crocus cloth, jeweler's rouge, or 400 to 600 grit wet/dry carborundum paper may be best.[5]

2) Graphite electrode—A wax-impregnated graphite (WIG) electrode may be used, especially in aqueous suspensions or soils.[12,13] The WIG electrode is more resistant to electrode poisoning than electrodes made of platinum wire.

d. Beakers: Preferably use polyethylene, TFE, or glass beakers.

e. Stirrer: Use a magnetic TFE-coated stirring-bar-type mixer.

f. Flow cell: Use for continuous flow measurements and for poorly buffered solutions.

3. Reagents

a. Standard redox solutions: Standardize the electrode system against redox solutions that provide stable known Eh values over a range of temperatures. Although standard solutions are available, they do not cover the anticipated range of Eh values. Commercially prepared solutions may be used, particularly in field testing. The composition and Eh values of standard solutions are shown in Table 2580:II. With reasonable care, these solutions are stable for several months.

b. Eh electrode cleaners: Use either:

1) Aqua regia—Mix 1 volume conc nitric acid with 3 volumes conc hydrochloric acid. Prepare fresh and dilute by at least 50% with water. Neutralize prior to discarding.

2) Chromic acid—Dissolve 5 g potassium dichromate, $K_2Cr_2O_7$, in 500 mL conc sulfuric acid. Conduct cleaning procedure in a fume hood.

4. Procedure

a. Instrument calibration: Follow manufacturer's instructions for using pH/millivolt meter and in preparing electrodes for use. Use a shorting lead to verify the zero point on the meter's millivolt scale. Equilibrate the standard solution to the temperature of the sample. Immerse electrodes in the gently stirred, standard solution in a beaker (or flow cell). Turn on meter, placing the function switch in the millivolt mode.

After several minutes for electrode equilibration, record reading to nearest millivolt. If the reading is more than ± 10 mV from the theoretical redox standard value at that temperature, replace reference electrode fluid and repeat the measurement. If that procedure fails to bring the reading to within ± 10 mV of the theoretical value, polish the sensing element of the indicator electrode with carborundum paper, crocus cloth, or jeweler's rouge. Rinse electrode thoroughly and recheck reading with a fresh portion of the standard solution. If the reading is within ± 10 mV of the theoretical value, record it and the temperature. If the reading is not within ± 10 mV, repeat the cleaning procedure or try another electrode. Then rinse the electrode with distilled water and proceed with the sample measurement. Recalibrate daily and more frequently if turbid, organic-rich, or high-dissolved-solids solutions are being measured.

b. Electrode cleaning procedure: Useful treatments for noble metal electrodes in restoring performance after long periods of use include immersion in warm (70°C) aqua regia for 1 to 2 min or 5 min in 6N HNO$_3$ after bringing to a boil. Alternatively treat with chromic acid solution followed by 6N HCl and rinse with water.

TABLE 2580:II. PREPARATION OF REDOX STANDARD SOLUTIONS

Standard Solution	Potentials of Pt Electrode vs. Selected Reference Electrodes at 25°C in Standard Solution					Weight of Chemicals Needed/1000 mL Aqueous Solution at 25°C
	Calomel	Silver:Silver Chloride Ag/AgCl			Standard Hydrogen	
	Hg/Hg$_2$Cl$_2$ saturated KCl	KCl 1.00M	KCl 4.00M	KCl saturated		
Light's solution[14]	+430	+439	+475	+476	+675	39.21 g ferrous ammonium sulfate, Fe(NH$_4$)$_2$(SO$_4$)$_2$ 6H$_2$O 48.22 g ferric ammonium sulfate, Fe(NH$_4$)(SO$_4$)$_2$·12H$_2$O 56.2 mL sulfuric acid, H$_2$SO$_4$, sp gr 1.84
ZoBell's solution*[2]	+183	+192	+228	+229	+428	1.4080 g potassium ferrocyanide, K$_4$Fe(CN)$_6$·3H$_2$O 1.0975 g potassium ferricyanide, K$_3$Fe(CN)$_6$ 7.4555 g potassium chloride, KCl

* Store in dark plastic bottle in a refrigerator.

c. Sample analysis: Check system for performance with the standard solution, rinse electrodes thoroughly with sample water, then immerse them in the gently stirred sample. Let equilibrate, record Eh value to the nearest millivolt, and temperature to $\pm 0.1°C$. Repeat with a second sample portion to confirm successive readings within ± 10 mV. Equilibration times vary and may take many minutes in poorly poised solutions. Successive readings that vary less than ± 10 mV over 10 min are adequate for most purposes. Make continuous flow or pumped sample measurements, particularly of poorly poised solutions, in a closed flow cell after external calibration of the electrode system.

See Table 2580:III for recommended combinations of electrodes, standards, and sample handling.

5. Trouble Shooting

a. Meter: Use a shorting lead to establish meter reading at zero millivolts whenever possible. If the meter cannot be zeroed, follow the manufacturer's instructions for service.

b. Electrodes: If the potentiometer is in good working order, the fault may be in the electrodes. Frequently, renewal of the filling solution for the salt bridge for the reference electrode is sufficient to restore electrode performance. Another useful check is to oppose the emf of a questionable reference electrode with that of the same type known to be in good order. Using an adapter, plug the good reference electrode into the indicator electrode jack of the potentiometer. Then plug the questionable electrode into the reference electrode jack. With the meter in the millivolt position, immerse electrodes in an electrolyte (e.g., KCl) solution and then into a redox standard solution. The two millivolt readings should be 0 ± 5 mV for both solutions. If different electrodes are used (e.g., silver:silver chloride versus calomel or vice versa), the reading should be 44 ± 5 mV for a good reference electrode.

Unless an indicator electrode has been poisoned, physically damaged, or shorted out, it usually is possible to restore function by proper cleaning.

6. Calculation

$$Eh_{system} = E_{observed} + E_{reference\ standard} - E_{reference\ observed}$$

Report temperature at which readings were made.

7. Precision and Bias

Standard solution measurements made at stable temperatures with a properly functioning electrode system should be accurate to within ± 10 mV. Calibration precision as reflected by the agreement of dual platinum electrodes versus an Ag:AgCl reference electrode for over a 2-year period has been estimated at ± 15 mV (i.e., one standard deviation) in ZoBell's solution (N = 78) at approximately 12°C. Precision on groundwater samples (N = 234) over the same period has been estimated at ± 22 mV (i.e., $\pm$ one standard deviation) in a closed flow cell.[15]

8. References

1. WHITFIELD, M. 1974. Thermodynamic limitations on the use of the platinum electrode in Eh measurements. *Limnol. Oceanogr.* 19:857.
2. ZOBELL, C.E. 1946. Studies on redox potential of marine sediments. *Amer. Assoc. Petroleum Geol.* 30:447.
3. NORDSTROM, D.K. 1977. Thermochemical equilibria of ZoBell's solution. *Geochim. Cosmochim. Acta* 41:1835.
4. WOOD, W.W. 1976. Guidelines for collection and field analysis of ground water samples for selected unstable constituents. Chapter D2 *in* Techniques of Water-Resources Investigations of the United States Geological Survey, Book 1. U.S. Geological Survey, Washington, D.C.
5. BRICKER, O.P. 1982. Redox potential: Its measurement and importance in water systems. *In* R.A. Minear & L.H. Keith, eds. Water Analysis, Vol. 1, Inorganic Species. Academic Press, New York, N.Y.
6. GERISCHER, H. 1950. Messungen der Austauschstromdichte beim Gleichgewichtspotential an einer Platinelektrode in Fe^{2+}/Fe^{3+} - Lösungen. *Z. Elektrochem.* 54:366.
7. BOULEGUE, J. & G. MICHARD. 1979. Sulfur speciations and redox processes in reducing environments. *In* E.A. Jenne, ed., ACS Symposium Ser., Vol. 93. American Chemical Soc., Washington, D.C.
8. GARSKE, E.E. & M.R. SCHOCK. 1986. An inexpensive flow-through cell and measurement system for monitoring selected chemical parameters in ground water. *Ground Water Monit. Rev.* 6:79.
9. STARKEY, R.L. & K.M. WIGHT. 1945. Anaerobic Corrosion in Soil. Tech. Rep. American Gas Assoc., New York, N.Y.
10. IVES, D.J.G. & G.J. JANZ. 1961. Reference Electrodes. Academic Press, New York, N.Y.
11. BOHN, H.L. 1971. Redox potentials. *Soil Sci.* 112:39.
12. SHAIKH, A.U., R.M. HAWK, R.A. SIMS & H.D. SCOTT. 1985. Graphite electrode for the measurement of redox potential and oxygen diffusion rate in soil. *Nucl. Chem. Waste Manage.* 5:237.
13. SHAIKH, A.U., R.M. HAWK, R.A. SIMS & H.D. SCOTT. 1985. Redox potential and oxygen diffusion rate as parameters for monitoring biodegradation of some organic wastes in soil. *Nucl. Chem. Waste Manage.* 5:337.
14. LIGHT, T.S. 1972. Standard solution for redox potential measurements. *Anal. Chem.* 44:1038.
15. BARCELONA, M.J., H.A. WEHRMANN, M.R. SCHOCK , M.E. SIEVERS & J.R. KARNY. 1989. Sampling Frequency for Ground-Water Quality Monitoring. EPA 600S4/89/032. Illinois State Water Survey-Univ. Illinois, rep. to USEPA-EMSL, Las Vegas, Nev.

TABLE 2580:III. RECOMMENDED COMBINATIONS FOR SELECTED SAMPLE TYPES

Sample Type	Indicator Electrode(s)	Reference Electrode	Type of Sample Cell
Process stream (low Br$^-$) (S^{2-})	Pt or Au	Calomel or silver: silver chloride	Closed continuous flow (dual indicator electrode)
(high Br$^-$)	Pt or Au	Calomel or silver: silver chloride with salt bridge (double junction reference electrode)	
Natural waters			
Surface waters	Pt or Au	Calomel or silver: silver chloride	Closed continuous flow (dual indicator electrode) or beaker
Groundwater	Pt or Au	Calomel or silver: silver chloride	Closed continuous flow (dual indicator electrode)
Soils, sludges	WIG, Pt wire	Calomel or silver: silver chloride	Beaker or soil core

2710 TESTS ON SLUDGES*

2710 A. Introduction

This section presents a series of tests uniquely applicable to sludges or slurries. The test data are useful in designing facilities for solids separation and concentration and for assessing operational behavior, especially of the activated sludge process.

___
* Approved by Standard Methods Committee, 1997.

2710 B. Oxygen-Consumption Rate

1. General Discussion

This test is used to determine the oxygen consumption rate of a sample of a biological suspension such as activated sludge. It is useful in laboratory and pilot-plant studies as well as in the operation of full-scale treatment plants. When used as a routine plant operation test, it often will indicate changes in operating conditions at an early stage. However, because test conditions are not necessarily identical to conditions at the sampling site, the observed measurement may not be identical with actual oxygen consumption rate.

2. Apparatus

a. Oxygen-consumption rate device: Either:

1) *Probe with an oxygen-sensitive electrode* (polarographic or galvanic), or

2) *Manometric or respirometric device* with appropriate readout and sample capacity of at least 300 mL. The device should have an oxygen supply capacity greater than the oxygen consumption rate of the biological suspension, or at least 150 mg/L·h.

b. Stopwatch or other suitable timing device.

c. Thermometer to read to $\pm$ 0.5°C.

3. Procedure

a. Calibration of oxygen-consumption rate device: Either:

1) Calibrate the oxygen probe and meter according to the method given in Section 4500-O.G, or

2) Calibrate the manometric or respirometric device according to manufacturer's instructions.

b. Volatile suspended solids determination: See Section 2540.

c. Preparation of sample: Adjust temperature of a suitable sample portion to that of the basin from which it was collected or to required evaluation temperature, and maintain constant during analysis. Record temperature. Increase DO concentration of sample by shaking it in a partially filled bottle or by bubbling air or oxygen through it.

d. Measurement of oxygen consumption rate:

1) Fill sample container to overflowing with an appropriate volume of a representative sample of the biological suspension to be tested.

2) If an oxygen-sensing probe is used, immediately insert it into a BOD bottle containing a magnetic stirring bar and the biological suspension. Displace enough suspension with probe to fill flared top of bottle and isolate its contents from the atmosphere. Activate probe stirring mechanism and magnetic stirrer. (NOTE: Adequate mixing is essential. For suspensions with high concentrations of suspended solids, i.e., >5000 mg/L, more vigorous mixing than that provided by the probe stirring mechanism and magnetic stirrer may be required.) If a manometric or respirometric device is used, follow manufacturer's instructions for startup.

3) After meter reading has stabilized, record initial DO and manometric or respirometric reading, and start timing device. Record appropriate DO, manometric, or respirometric data at time intervals of less than 1 min, depending on rate of consumption. Record data over a 15-min period or until DO becomes limiting, whichever occurs first. The oxygen probe may not be accurate below 1 mg DO/L. If a manometric or respirometric device is used, refer to manufacturer's instructions for lower limiting DO value. Low DO ($\leq$2 mg/L at the start of the test) may limit oxygen uptake by the biological suspension and will be indicated by a decreasing rate of oxygen consumption as the test progresses. Reject such data as being unrepresentative of suspension oxygen consumption rate and repeat test beginning with higher initial DO levels.

The results of this determination are quite sensitive to temperature variations and poor precision is obtained unless replicate determinations are made at the same temperature. When oxygen consumption is used as a plant control test, run periodic (at least monthly) replicate determinations to establish the precision of the technique. This determination also is sensitive to the time lag between sample collection and test initiation.

4. Calculations

If an oxygen probe is used, plot observed readings (DO, milligrams per liter) versus time (minutes) on arithmetic graph paper and determine the slope of the line of best fit. The slope is the oxygen consumption rate in milligrams per liter per minute.

If a manometric or respirometric device is used, refer to manufacturer's instructions for calculating the oxygen consumption rate.

Calculate specific oxygen consumption rate in milligrams per gram per hour as follows:

Specific oxygen consumption rate, (mg/g)/h

$$= \frac{\text{oxygen consumption rate, (mg/L)/min}}{\text{volatile suspended solids, g/L}} \times \frac{60 \text{ min}}{\text{h}}$$

5. Precision and Bias

Bias is not applicable. The precision for this test has not been determined.

6. Bibliography

UMBREIT, W.W., R.H. BURRIS & J.F. STAUFFER. 1964. Manometric Techniques. Burgess Publishing Co., Minneapolis, Minn.

2710 C. Settled Sludge Volume

1. General Discussion

The settled sludge volume of a biological suspension is useful in routine monitoring of biological processes. For activated sludge plant control, a 30-min settled sludge volume or the ratio of the 15-min to the 30-min settled sludge volume has been used to determine the returned-sludge flow rate and when to waste sludge. The 30-min settled sludge volume also is used to determine sludge volume index[1] (Section 2710D).

This method is inappropriate for dilute sludges because of the small volume of settled material. In such cases, use the volumetric test for settleable solids using an Imhoff cone (2540F). Results from 2540F are not comparable with those obtained with the procedure herein.

2. Apparatus

a. Settling column: Use 1-L graduated cylinder equipped with a stirring mechanism consisting of one or more thin rods extending the length of the column and positioned within two rod diameters of the cylinder wall. Provide a stirrer able to rotate the stirring rods at no greater than 4 rpm (peripheral tip speed of approximately 1.3 cm/s). See Figure 2710:1.

b. Stopwatch.

c. Thermometer.

3. Procedure

Place 1.0 L sample in settling column and distribute solids by covering the top and inverting cylinder three times. Insert stirring rods, activate stirring mechanism, start the stop watch, and let suspension settle. Continue stirring throughout test. Maintain suspension temperature during test at that in the basin from which the sample was taken.

Determine volume occupied by suspension at measured time intervals, e.g., 5, 10, 15, 20, 30, 45, and 60 min.

Report settled sludge volume of the suspension in milliliters for an indicated time interval.

Variations in suspension temperature, sampling and agitation methods, dimensions of settling column, and time between sampling and start of the determination significantly affect results.

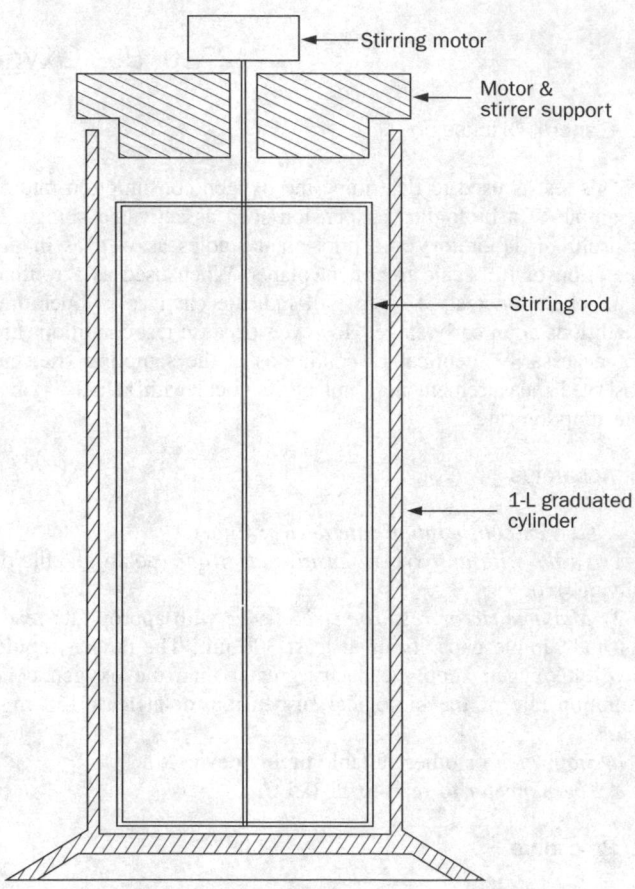

Figure 2710:1. Schematic diagram of settling vessel for settled sludge volume test.

4. Precision and Bias

Bias is not applicable. The precision for this test has not been determined.

5. Reference

1. DICK, R.I. & P.A. VESILIND. 1969. The SVI—What is It? *J. Water Pollut. Control Fed.* 41:1285.

2710 D.   Sludge Volume Index

1. General Discussion

The sludge volume index (SVI) is the volume in milliliters occupied by 1 g of a suspension after 30 min settling. SVI typically is used to monitor settling characteristics of activated sludge and other biological suspensions.[1] Although SVI is not supported theoretically,[2] experience has shown it to be useful in routine process control.

2. Procedure

Determine the suspended solids concentration of a well-mixed sample of the suspension (See Section 2540D).

Determine the 30 min settled sludge volume (See Section 2710C).

3. Calculations

$$\text{SVI} = \frac{\text{settled sludge volume (mL/L)} \times 1000}{\text{suspended solids (mg/L)}}$$

4. Precision and Bias

Precision is determined by the precision achieved in the suspended solids measurement, the settling characteristics of the suspension, and variables associated with the measurement of the settled sludge volume. Bias is not applicable.

5. References

1. DICK, R.I. & P.A. VESILIND. 1969. The SVI—What is it? *J. Water Pollut. Control Fed.* 41:1285.
2. FINCH, J. & H. IVES. 1950. Settleability indexes for activated sludge. *Sewage Ind. Wastes* 22:833.

6. Bibliography

DONALDSON, W. 1932. Some notes on the operation of sewage treatment works. *Sewage Works J.* 4:48.
MOHLMAN, F.W. 1934. The sludge index. *Sewage Works J.* 6:119.
RUDOLFS, W. & I.O. LACY. 1934. Settling and compacting of activated sludge. *Sewage Works J.* 6:647.

2710 E.   Zone Settling Rate

1. General Discussion

At high concentrations of suspended solids, suspensions settle in the zone-settling regime. This type of settling takes place under quiescent conditions and is characterized by a distinct interface between the supernatant liquor and the sludge zone. The height of this distinct sludge interface is measured with time. Zone settling data for suspensions that undergo zone settling, e.g., activated sludge and metal hydroxide suspensions, can be used in the design, operation, and evaluation of settling basins.[1-3]

2. Apparatus

a. Settling vessel: Use a transparent cylinder at least 1 m high and 10 cm in diameter. To reduce the discrepancy between laboratory and full-scale thickener results, use larger diameters and taller cylinders.[1,3] Attach a calibrated millimeter tape to outside of cylinder. Equip cylinder with a stirring mechanism, e.g., one or more thin rods positioned within two rod diameters of the internal wall of settling vessel. Stir suspension near vessel wall over the entire depth of suspension at a peripheral speed no greater than 1 cm/s. Greater speeds may interfere with the thickening process and yield inaccurate results.[4] Provide the settling vessel with a port in the bottom plate for filling and draining. See Figure 2710:2.

b. Stopwatch.

c. Thermometer.

3. Procedure

Maintain suspension in a reservoir in a uniformly mixed condition. Adjust temperature of suspension to that of the basin from which it was collected or to required evaluation temperature. Record temperature. Remove a well-mixed sample from reservoir and measure suspended solids concentration (Section 2540D).

Activate stirring mechanism. Fill settling vessel to a fixed height by pumping suspension from reservoir or by gravity flow. Fill at a rate sufficient to maintain a uniform suspended solids concentration throughout settling vessel at end of filling. The suspension should agglomerate, i.e., form a coarse structure with visible fluid channels, within a few minutes. If suspension does not agglomerate, test is invalid and should be repeated.

Record height of solids-liquid interface at intervals of about 1 min. Collect data for sufficient time to assure that suspension is exhibiting a constant zone-settling velocity and that any initial reflocculation period, characterized by an accelerating interfacial settling velocity, has been passed.

Zone settling rate is a function of suspended solids concentration, suspension characteristics, vessel dimensions, and laboratory artifacts. With the filling method described above and a sufficiently large cylinder, these artifacts should be minimized. However, even with careful testing suspensions often may behave erratically. Unpredictable behavior increases for sludges with high solids concentrations and poor settling characteristics, and in small cylinders.

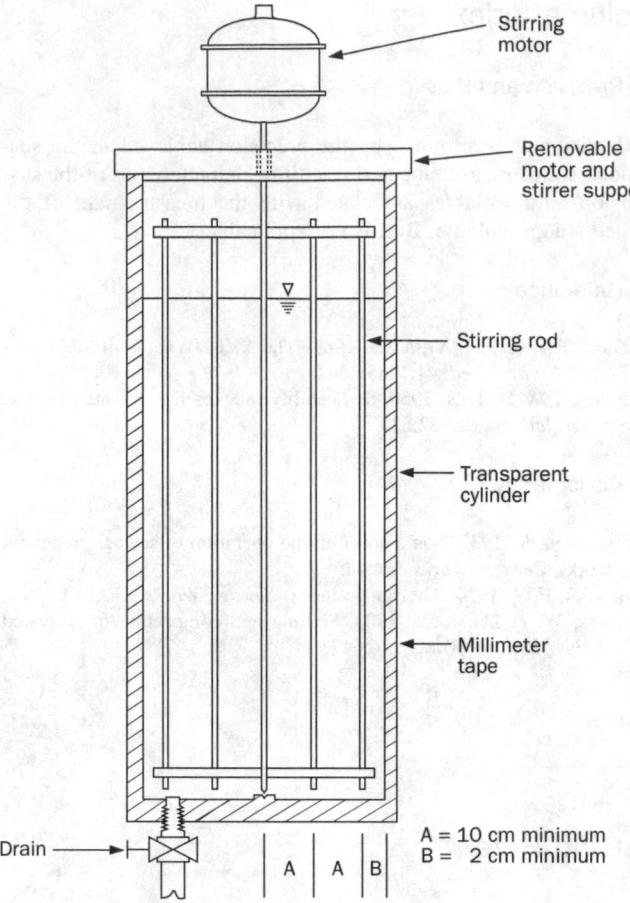

Stirring motor

Removable motor and stirrer support

Stirring rod

Transparent cylinder

Millimeter tape

Drain

A = 10 cm minimum
B = 2 cm minimum

A   A   B

Figure 2710:2. Schematic diagram of settling vessel for zone settling rate test.

4. Calculations

Plot interface height in centimeters vs. time in minutes.[1,3] Draw straight line through data points, ignoring initial shoulder or reflocculation period and compression shoulder. Calculate interfacial settling rate as slope of line in centimeters per minute.

5. Precision and Bias

Bias is not applicable. The precision for this test has not been determined.

6. References

1. Dick, R.I. 1972. Sludge treatment. *In* W.J. Weber, ed., Physicochemical Processes for Water Quality Control. Wiley-Interscience, New York, N.Y.
2. Dick, R.I. & K.W. Young. 1972. Analysis of thickening performance of final settling tanks. *Proc. 27th Ind. Waste Conf.*, Purdue Univ., Eng. Ext. Ser. No. 141, 33.
3. Vesilind, P.A. 1975. Treatment and Disposal of Wastewater Sludges. Ann Arbor Science Publishing Co., Ann Arbor, Mich.
4. Vesilind, P.A. 1968. Discussion of Evaluation of activated sludge thickening theories. *J. San. Eng. Div., Proc. Amer. Soc. Civil Eng.* 94: SA1, 185.

7. Bibliography

Dick, R.I. & R.B. Ewing. 1967. Evaluation of activated sludge thickening theories. *J. San. Eng. Div., Proc. Amer. Soc. Civil Eng.* 93:SA4, 9.
Dick, R.I. 1969. Fundamental aspects of sedimentation I & II. *Water Wastes Eng.* 3:47, 45, & 6:2.
Dick, R.I. 1970. Role of activated sludge final settling tanks. *J. San. Eng. Div., Proc. Amer. Soc. Civil Eng.* 96:SA2, 423.

2710 F.   Specific Gravity

1. General Discussion

The specific gravity of a sludge is the ratio of the masses of equal volumes of a sludge and distilled water. It is determined by comparing the mass of a known volume of a homogeneous sludge sample at a specific temperature to the mass of the same volume of distilled water at 4°C.

2. Apparatus

Container: A marked flask or bottle to hold a known sludge volume during weighing.

3. Procedure

Follow either *a* or *b*.

a. Record sample temperature, *T*. Weigh empty container and record weight, *W*. Fill empty container to mark with sample, weigh, and record weight, *S*. Fill empty container to mark with water, weigh, and record weight, *R*. Measure all masses to the nearest 10 mg.

b. If sample does not flow readily, add as much of it to container as possible without exerting pressure, record volume, weigh, and record mass, *P*. Fill container to mark with distilled water, taking care that air bubbles are not trapped in the sludge or container. Weigh and record mass, *Q*. Measure all masses to nearest 10 mg.

4. Calculation

Use *a* or *b*, matching choice of procedure above.

a. Calculate specific gravity, *SG*, from the formula

$$SG_{T/4°C} = \frac{\text{weight of sample}}{\text{weight of equal volume}} = \frac{S - W}{R - W} \times F$$
$$\text{of water at 4°C}$$

The values of the temperature correction factor *F* are given in Table 2710:I.

TABLE 2710:I. TEMPERATURE CORRECTION FACTOR

Temperature °C	Temperature Correction Factor
15	0.9991
20	0.9982
25	0.9975
30	0.9957
35	0.9941
40	0.9922
45	0.9903

b. Calculate specific gravity, *SG*, from the formula

$$SG_{T/4°C} = \frac{\text{weight of sample}}{\text{weight of equal volume of water at } 4°C} = \frac{(P - W)}{(R - W) - (Q - P)} \times F$$

For values of *F*, see Table 2710:I.

2710 G. Capillary Suction Time

1. General Discussion

The capillary suction time (CST) test determines rate of water release from sludge. It provides a quantitative measure, reported in seconds, of how readily a sludge releases its water. The results can be used to assist in sludge dewatering processes; to evaluate sludge conditioning aids and dosages; or, when used with a jar test and the settleable solids procedure, to evaluate coagulation effects on the rate of water release from sludges.

The test consists of placing a sludge sample in a small cylinder on a sheet of chromatography paper. The paper extracts liquid from the sludge by capillary action. The time required for the liquid to travel a specified distance is recorded automatically by monitoring the conductivity change occurring at two contact points appropriately spaced and in contact with the chromatography paper. The elapsed time is indicative of the water drainage rate. The CST test has been used as a relative indicator to characterize the performance of most sludge dewatering processes.

2. Apparatus

a. Test materials and apparatus may be fabricated (see Figure 2710:3) or are commercially available.* The unit includes a paper support block, stainless steel reservoir with 18-mm ID and 25-mm height, and a digital timer.

b. CST paper.†

c. Thermometer to read ± 0.5C.

d. Pipet, 10-mL, plastic with tip trimmed to allow free passage of sludge flocs.

3. Procedure

Turn on and reset CST meter. Dry CST test block and reservoir. Place a new CST paper on lower test block with rough side up and grain parallel to the 9-cm side. Add upper test block, insert sludge reservoir into test block, and seat it using light pressure and a quarter turn to prevent surface leaks. Measure and record temperature of sludge. Pipet 6.4 mL sludge into test cell reservoir; if pipetting is difficult because of sludge consistency, pour a rep-

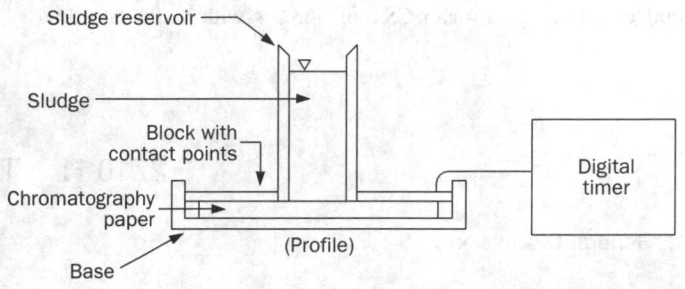

r₁ = 15.9 mm (0.625 in.)
r₂ = 22.2 mm (0.875 in.)

Figure 2710:3. Capillary suction time apparatus.

resentative sludge sample into the test cell until it is full. The CST device will begin time measurement as liquid being drawn into the paper reaches the inner pair of electrical contacts. Timing ends when the outer contact is reached. Record CST shown on digital display. Empty remaining sludge from reservoir and remove and discard used CST paper. Rinse and dry test block and reservoir. Repeat for a minimum of five determinations per sample to account for measurement variation and to allow identification of any faulty readings due to leaks or spills.

Variations in sludge temperature and sample volume can affect CST results. Ensure that all analyses are run under similar conditions. Sludge suspended solids concentration has a significant effect on test results. In evaluating sludge conditioners or moni-

---

* Venture Innovations, P.O. Box 53631, Lafayette, LA 70505; or Triton Electronics Ltd., Bigods Hall, Dunmow, Essex, England, CM63BE; or equivalent.
† Available from CST apparatus supplier or use Whatman No. 17 chromatography grade paper cut into 7- × 9-cm sections with grain parallel to long side.

toring operation of a dewatering process, avoid this effect by ensuring homogeneity among sludge samples. Comparison of CST data from different sludge samples from the same source (especially if taken on different days) cannot be made with confidence unless suspended solids concentrations are comparable. Make a rough correction for different solids contents by dividing the sludge's CST value by its corresponding solids concentration.

Characteristics of CST paper may vary between lots. If comparison of CST values for distilled water indicates such variations, subtract times for distilled water blanks from sample times to improve comparisons.

Record CST model used, paper type, sludge type, sludge temperature, and capillary suction time. Measure solids concentration and CST of distilled water using the same paper to provide useful information.

4. Precision and Bias

Ten tests conducted on an anaerobically digested pulp mill sludge resulted in a mean CST of 363.2 s with a standard deviation of 36.2 s. Twenty tests using an anaerobically digested municipal wastewater sludge gave a mean of 85.2 s with a standard deviation of 14.12 s. Triplicate analyses of 30 sample sets of conditioned and unconditioned alum sludge resulted in an average standard deviation of 1.0 s with means between 5 and 80 s. Method bias cannot be determined.

5. Bibliography

BASKERVILLE, R.C. & R.S. GALE. 1968. A simple automatic instrument for determining the filtrability of sewage sludges. *J. Inst. Water Pollut. Control* 67:233.

KAVANAGH, B.A. 1980. The dewatering of activated sludge: measurements of specific resistance to filtration and capillary suction time. *Water Pollut. Control* 79:388.

VESILIND, P.A. 1988. Capillary suction time as a fundamental measure of sludge dewaterability. *J. Water Pollut. Control Fed.* 60:215.

TILLER, F.M., Y.L. SHEN & A. ADIN. 1990. Capillary suction theory for rectangular cells. *Res. J. Water Pollut. Control Fed.* 62:130.

2710 H. Time-to-Filter

1. General Discussion

The time to filter (TTF) correlates with capillary suction time (CST) and is similar to the specific resistance to filtration if sludge solids content and filtrate viscosity do not vary among compared samples. The test requires approximately 200 mL sludge and can be used to assist in the daily operation of sludge dewatering processes or to evaluate sludge-conditioning polymers and dosages.

Testing with a smaller volume is possible in applications to evaluate water drainage rate subsequent to jar tests and settleable solids determination (see Section 2540F). In this case, drain collected sludge from one or more Imhoff cones after decanting as much supernatant as possible; use a small-volume TTF apparatus.

The test consists of placing a sludge sample in a Buchner funnel with a paper support filter, applying vacuum, and measuring the time required for 100 mL filtrate (or, for reduced sample volumes, 50% of original sample) to collect. While similar to the specific resistance to filtration test, the time-to-filter test is superior because of its ease of use and simplicity.

2. Apparatus

a. Time-to-filter large-volume or small-volume (Figure 2710:4) *assembly.*

*b. Filter paper**

c. Stopwatch.

3. Procedure

Place paper filter in funnel and make a firm seal by pre-wetting with a small volume of water with vacuum on. If using large-volume apparatus, take a 200-mL sample of sludge. With vacuum pump providing a constant vacuum of 51 kPa, pour sample into

---

* Whatman No. 1 or 2 or equivalent.

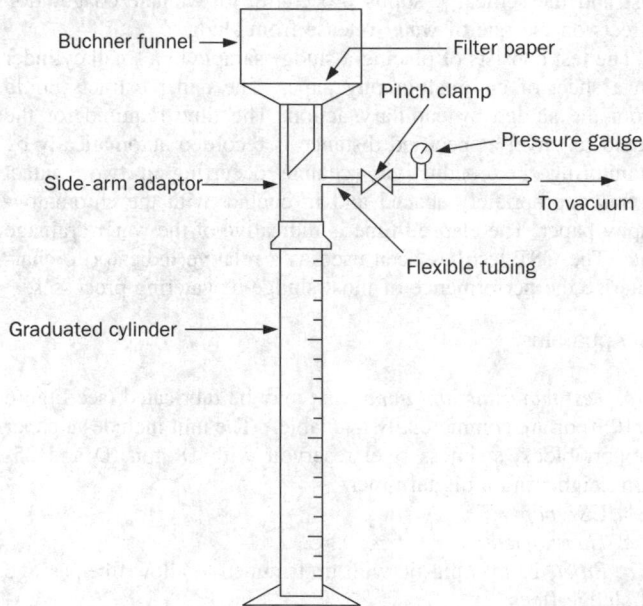

Figure 2710:4. TTF equipment. Large-volume equipment requires a 9-cm-diam Buchner funnel and a 250-mL graduated cylinder. Small-volume equipment requires a 2.5-cm-diam funnel and a 10-mL cylinder.

funnel. Start stopwatch or timer and determine time required for 100 mL of sample to collect in graduated cylinder. This is the time to filter. Make a minimum of three replicate determinations.

For the small-volume test, use 7 to 10 mL sludge. Record time required for 50% of sample to collect in graduated cylinder. Compare this time to filter only to other results using the same sample volume.

Sludge suspended solids concentration has a significant effect on test results. In evaluating sludge-conditioning products, compare results for which initial suspended solids concentrations are comparable. Make a rough correction for different solids contents by dividing the time-to-filter value by its corresponding solids concentration. However, variations in solids concentration occur in full-scale applications, and the time-to-filter results may be interpreted as indicating the overall rate of water release from sludges, including the effect of differing solids concentrations.

4. Precision and Bias

Variations in vacuum pressure, support filter type, sludge temperature, and sample volume can affect test results. Triplicate analyses of 18 sample sets of conditioned and unconditioned alum sludge resulted in an average method precision of 19 s (approx-

imately 4% of the average value) for the large-volume TTF test. Triplicate analyses of 9 sample sets of conditioned and unconditioned alum sludge resulted in a method precision of 9 s (approximately 6% of the average value) for the small-volume TTF test. Method bias, which refers to the agreement between the value determined by the test method and the real value, cannot be determined.

5. Bibliography

KNOCKE, W.R. & D.L. WAKELAND. 1983. Fundamental characteristics of water treatment plant sludges. *J. Amer. Water Works Assoc.* 113: 516.

DENTEL, S.K., T.A. BOBER, P.V. SHETTY & J.R. RESTA. 1986. Procedures Manual for Selection of Coagulant, Filtration, and Sludge Conditioning Aids in Water Treatment. Publ. 90515, American Water Works Assoc., Denver, Colo.

2720 ANAEROBIC SLUDGE DIGESTER GAS ANALYSIS*

2720 A. Introduction

Gas produced during the anaerobic decomposition of wastes contains methane (CH_4) and carbon dioxide (CO_2) as the major components with minor quantities of hydrogen (H_2), hydrogen sulfide (H_2S), nitrogen (N_2), and oxygen (O_2). It is saturated with water vapor. Common practice is to analyze the gases produced to estimate their fuel value and to check on the treatment process. The relative proportions of CO_2, CH_4, and N_2 are normally of most concern and the easiest to determine because of the relatively high percentages of these gases.

1. Selection of Method

Two procedures are described for gas analysis, the volumetric method (B), and the gas chromatographic method (C). The volumetric analysis is suitable for the determination of CO_2, H_2, CH_2, and O_2. Nitrogen is estimated indirectly by difference. Although the method is time-consuming, the equipment is relatively simple. Because no calibration is needed before use, the procedure is particularly appropriate when analyses are conducted infrequently.

The principal advantage of gas chromatography is speed. Commercial equipment is designed specifically for isothermal or temperature-programmed gas analysis and permits the routine separation and measurement of CO_2, N_2, O_2, and CH_4 in less than 15 to 20 min. The requirements for a recorder, pressure-regulated bottles of carrier gas, and certified standard gas mixtures for calibration raise costs to the point where infrequent analyses by this method may be uneconomical. The advantages of this system are freedom from the cumulative errors found in sequential volumetric measurements, adaptability to other gas component analyses, adaptability to intermittent on-line sampling and analysis, and the use of samples of 1 mL or less.[1]

2. Sample Collection

When the source of gas is some distance from the apparatus used for analysis, collect samples in sealed containers and bring to the instrument. Displacement collectors are the most suitable containers. Glass sampling bulbs (or tubes) with three-way glass or TFE stopcocks at each end, as indicated in Figure 2720:1, are particularly useful. These also are available with centrally located

---

* Approved by Standard Methods Committee, 1997.

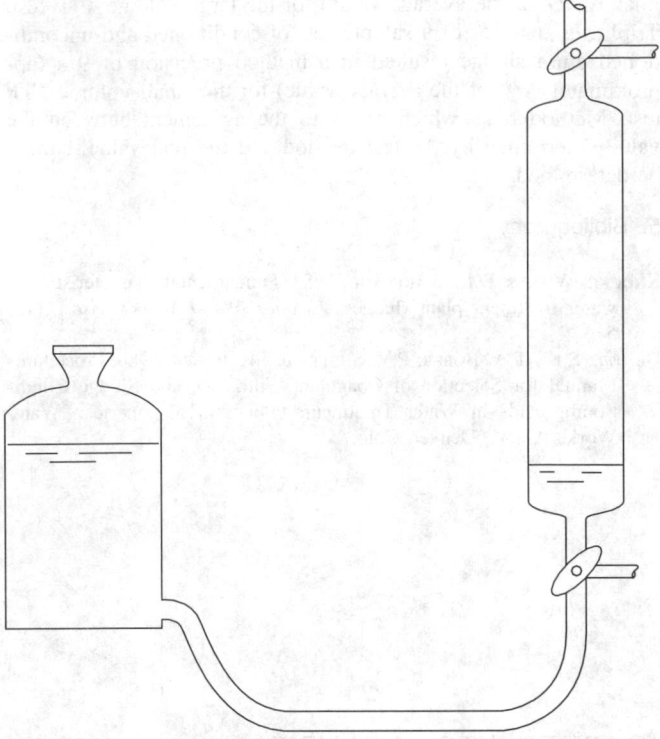

Figure 2720:1. Gas collection apparatus.

ports provided with septa for syringe transfer of samples. Replace septa periodically to prevent contamination by atmospheric gases. Connect one end of collector to gas source and vent three-way stopcock to the atmosphere. Clear line of air by passing 10 to 15 volumes of gas through vent and open stopcock to admit sample. If large quantities of gas are available, sweep air away by passing 10 to 15 volumes of gas through tube. If the gas supply is limited, fill the gas sampling bulb (tube) with an acidic salt solution.[2] Then completely displace the acidic salt solution in the gas sampling bulb (tube) with the sample gas. Because the acidic salt solution absorbs gases to some extent, fill the gas sampling bulb completely with the gas and seal off from any contact with displacement fluid during temporary storage. When transferring gas to the gas-analyzing apparatus, do not transfer any fluid.

3. References

1. GRUNE, W.N. & C.F. CHUEH. 1962–63. Sludge gas analysis using gas chromatograph. *Water Sewage Works* 109:468; 110:43, 77, 102, 127, 171, 220, and 254.
2. CHIN, K.K. & K.K. WONG. 1983. Thermophilic anaerobic digestion of palm oil mill effluent. *Water Res.* 17:993.

2720 B. Volumetric Method

1. General Discussion

a. Principle: This method may be used for the analysis either of digester gas or of methane in water (see Section 6211, Methane). A measured volume of gas is passed first through a solution of potassium hydroxide (KOH) to remove CO_2, next through a solution of alkaline pyrogallol to remove O_2, and then over heated cupric oxide, which removes H_2 by oxidation to water. After each of the above steps, the volume of gas remaining is measured; the decrease that results is a measure of the relative percentage of volume of each component in the mixture. Finally, CH_4 is determined by conversion to CO_2 and H_2O in a slow-combustion pipet or a catalytic oxidation assembly. The volume of CO_2 formed during combustion is measured to determine the fraction of methane originally present. Nitrogen is estimated by assuming that it represents the only gas remaining and equals the difference between 100% and the sum of the measured percentages of the other components.

When only CO_2 is measured, report only CO_2. No valid assumptions may be made about the remaining gases present without making a complete analysis.

Follow the equipment manufacturers' recommendations with respect to oxidation procedures.

CAUTION: *Do not attempt any slow-combustion procedure on digester gas because of the high probability of exceeding the explosive 5% by volume concentration of CH_4.*

2. Apparatus

Orsat-type gas-analysis apparatus, consisting of at least: (1) a water-jacketed gas buret with leveling bulb; (2) a CO_2-absorption pipet; (3) an O_2-absorption pipet; (4) a cupric oxide-hydrogen oxidation assembly; (5) a shielded catalytic CH_4-oxidation assembly or slow-combustion pipet assembly; and (6) a leveling bulb. With the slow-combustion pipet use a controlled source of current to heat the platinum filament electrically. Preferably use mercury as the displacement fluid; alternatively use aqueous Na_2SO_4-H_2SO_4 solution for sample collection. Use any commercially available gas analyzer having these units.

3. Reagents

a. Potassium hydroxide solution: Dissolve 500 g KOH in distilled water and dilute to 1 L.

b. Alkaline pyrogallol reagent: Dissolve 30 g pyrogallol (also called pyrogallic acid) in distilled water and make up to 100 mL. Add 500 mL KOH solution.

c. Oxygen gas: Use approximately 100 mL for each gas sample analyzed.

d. Displacement liquid (acidic salt solution): Dissolve 200 g Na_2SO_4 in 800 mL distilled water; add 30 mL conc H_2SO_4. Add a few drops of methyl orange indicator. When color fades, replace solution.

4. Procedure

a. Sample introduction: Transfer 5 to 10 mL gas sample into gas buret through a capillary-tube connection to the collector. Expel this sample to the atmosphere to purge the system. Transfer up to 100 mL gas sample to buret. Bring sample in buret to atmospheric or reference pressure by adjusting leveling bulb. Measure volume accurately and record as V_1.

b. Carbon dioxide absorption: Remove CO_2 from sample by passing it through the CO_2-absorption pipet charged with the KOH solution. Pass gas back and forth until sample volume remains constant. Before opening stopcocks between buret and any absorption pipet, make sure that the gas in the buret is under a slight positive pressure to prevent reagent in the pipet from contaminating stopcock or manifold. After absorption of CO_2, transfer sample to buret and measure volume. Record as V_2.

c. Oxygen absorption: Remove O_2 by passing sample through O_2-absorption pipet charged with alkaline pyrogallol reagent until sample volume remains constant. Measure volume and record as V_3. For digester gas samples, continue as directed in ¶ 4d. For CH_4 in water, store gas in CO_2 pipet and proceed ¶ 4e below.

d. Hydrogen oxidation: Remove H_2 by passing sample through CuO assembly maintained at a temperature in the range 290 to 300°C. When a constant volume has been obtained, transfer sample back to buret, cool, and measure volume. Record as V_4.

Waste to the atmosphere all but 20 to 25 mL of remaining gas. Measure volume and record as V_5. Store temporarily in CO_2-absorption pipet.

e. Methane oxidation: Purge inlet connections to buret with O_2 by drawing 5 to 10 mL into buret and expelling to the atmosphere. Oxidize CH_4 either by the catalytic oxidation process for digester gas and gas phase of water samples or by the slow-combustion process for gas phase of water samples.

1) Catalytic oxidation process—For catalytic oxidation of digester gas and gas phase of water samples, transfer 65 to 70 mL O_2 to buret and record this volume as V_6. Pass O_2 into CO_2-absorption pipet so that it will mix with sample stored there. Return this mixture to buret and measure volume. Record as V_7. This volume should closely approximate V_5 plus V_6. Pass O_2-sample mixture through catalytic oxidation assembly, which should be heated in accordance with the manufacturer's directions. Keep rate of gas passage less than 30 mL/min. After first pass, transfer mixture back and forth through the assembly between buret and reservoir at a rate not faster than 60 mL/min until a constant volume is obtained. Record as V_8.

2) Slow-combustion process—For slow combustion of the gas phase of water samples, transfer 35 to 40 mL O_2 to buret and record volume as V_6. Transfer O_2 to slow-combustion pipet and then transfer sample from CO_2-absorption pipet to buret. Heat platinum coil in combustion pipet to yellow heat while controlling temperature by adjusting current. Reduce pressure of O_2 in pipet to somewhat less than atmospheric pressure by means of the leveling bulb attached to the pipet. Pass sample into slow-combustion pipet at rate of approximately 10 mL/min. After the first pass, transfer sample and O_2 mixture back and forth between pipet and buret several times at a faster rate, allowing mercury in pipet to rise to a point just below heated coil. Collect sample in combustion pipet, turn off coil, and cool pipet and sample to room temperature with a jet of compressed air. Transfer sample to buret and measure volume. Record as V_8.

f. Measurement of carbon dioxide produced: Determine amount of CO_2 formed in the reaction by passing sample through CO_2-absorption pipet until volume remains constant. Record volume as V_9. Check accuracy of determination by absorbing residual O_2 from sample. After this absorption, record final volume as V_{10}.

5. Calculation

a. CH_4 and H_2 usually are the only combustible gases present in sludge digester gas. When this is the case, determine percentage by volume of each gas as follows:

$$\% \ CO_2 = \frac{(V_1 - V_2) \times 100}{V_1}$$

$$\% \ O_2 = \frac{(V_2 - V_3) \times 100}{V_1}$$

$$\%H_2 = \frac{(V_3 - V_4) \times 100}{V_1}$$

$$\% \ CH_4 = \frac{V_4 \times (V_8 - V_9) \times 100}{V_1 \times V_5}$$

$$\% \ N_2 = 100 - (\% \ CO_2 + \%O_2 + \%H_2 + \%CH_4)$$

b. Alternatively, calculate CH_4 by either of the two following equations:

$$\%CH_4 = \frac{V_4 \times (V_6 + V_{10} - V_9) \times 100}{2 \times V_1 \times V_5}$$

$$\%CH_4 = \frac{V_4 (V_7 - V_8) \times 100}{2 \times V_1 \times V_5}$$

Results from the calculations for CH_4 by the three equations should be in reasonable agreement. If not, repeat analysis after checking apparatus for sources of error, such as leaking stopcocks or connections. Other combustible gases, such as ethane, butane, or pentane, will cause a lack of agreement among the calculations; however, the possibility that digester gas contains a significant amount of any of these is remote.

6. Precision and Bias

A gas buret measures gas volume with a precision of 0.05 mL and a probable accuracy of 0.1 mL. With the large fractions of CO_2 and CH_4 normally present in digester gas, the overall error for their determination can be made less than 1%. The error in the determination of O_2 and H_2, however, can be considerable because of the small concentrations normally present. For a concentration as low as 1%, an error as large as 20% can be expected. When N_2 is present in a similar low-volume percentage, the error in its determination would be even greater, because errors in each of the other determinations would be reflected in the calculation for N_2.

7. Bibliography

YANT, W.F. & L.B. BERGER. 1936. Sampling of Mine Gases and the Use of the Bureau of Mines Portable Orsat Apparatus in Their Analysis. Miner's Circ. No. 34, U.S. Bur. Mines, Washington, D.C.

MULLEN, P.W. 1955. Modern Gas Analysis. Interscience Publishers, New York, N.Y.

2720 C.　Gas Chromatographic Method

1. General Discussion

See Section 6010C for discussions of gas chromatography.

2. Apparatus

a. Gas chromatograph: Use any instrument system equipped with a thermal conductivity detector (TCD), carrier-gas flow controllers, injector and column temperature setting dials, TCD current controller, attenuator, carrier-gas pressure gauge, injection port, signal output, and power switch. Some columns require temperature programming while others are isothermal. Preferably use a unit with a gas sampling loop and valve that allow automatic injection of a constant sample volume.

b. Sample introduction apparatus: An instrument equipped with gas-sampling valves is designed to permit automatic injection of a specific sample volume into the chromatograph. If such an instrument is not available, introduce samples with a 2-mL syringe fitted with a 27-gauge hypodermic needle. Reduce escape of gas by greasing plunger lightly with mineral oil or preferably by using a special gas-tight syringe. One-step separation of oxygen, nitrogen, methane, and carbon dioxide may be accomplished in concentric columns (column within a column) under isothermal conditions at room temperature instead of performing two column analyses. These concentric columns permit the simultaneous use of two different packings for the analysis of gas samples.

*c. Chromatographic column:** Select column on the basis of manufacturer's recommendations.† Report column and packing specifications and conditions of analyses with results.

A two-column system usually is required. A molecular sieve column separates H_2, O_2, N_2, CO, and CH_4 isothermally, but because CO_2 is adsorbed by the molecular sieve, a second column is needed to complete the analysis. A commonly used two-column system utilizes a Chromosorb 102 column and a Molecular Sieve 5A or 13X column to separate H_2, O_2, N_2, CH_4, and CO_2 isothermally.[1,2]

A single-column procedure can be used; however, it requires temperature programming. Columns packed with special porous spherical or granular packing materials effect separation with sharp, well-resolved peaks.[3] With temperature programming, columns packed with Chromosorb 102,[2] Carbosphere,[4] and Carbosieve[3] separate the gases listed.† Commercial equipment specifically designed for such operations is available.[2,4-6]

---

* Gas chromatographic methods are extremely sensitive to the materials used. Use of trade names in *Standard Methods* does not preclude the use of other existing or as-yet-undeveloped products that give *demonstrably* equivalent results.
† Commercially available columns and gases they will separate include:
Silica gel and activated alumina:
　Silica gel: H_2, air (O_2 + N_2), CO, CH_4, C_2H_6.
　Activated alumina: air (O_2 + N_2), CH_4, C_2H_6, C_3H_8.
Molecular sieves (zeolites):
　Molecular sieve 5A: H_2, O_2 − Ar, N_2, CH_4, CO.
　Molecular sieve 13X: O_2, N_2, CH_4, CO.
Porous polymers:
　Chromosorb 102: air (H_2, O_2, N_2, CO), CH_4, CO_2.
　Porapak Q: air - CO, CO_2.
　HayeSep Q: H_2, air (O_2 + N_2), CH_4, CO_2, C_2H_6, H_2S.
Carbon molecular sieves:
　Carbosphere: O_2, N_2, CO, CH_4, CO_2, C_2H_6 (these gases can be eluted isothermally at various temperatures or by temperature programming).
　Carbosieve S-II: H_2, air (O_2 + N_2), CO, CH_4, CO_2, C_2H_6 (temperature programming required).

d. Integrator/recorder: Use a 10–mV full-span strip chart recorder with the gas chromatograph. When minor components such as H_2 and H_2S are to be detected, a 1-mV full-span recorder is preferable.

Integrators that can easily detect very minute quantities of gases are available. Computerized data-processing systems, to record and manipulate the chromatographic signal, chromatographic base line, etc., also are available.

3. Reagents

a. Carrier gases: Preferably use helium for separating digester gases. It is impossible to detect less than 1% hydrogen when helium is used as the carrier gas.[1] Obtain linear TCD responses for molar concentrations of hydrogen between 0 and 60% by using an 8.5% hydrogen–91.5% helium carrier gas mixture.[7] To detect trace quantities of hydrogen use argon or nitrogen as carrier gas.[1]

b. Calibration gases: Use samples of CH_4, CO_2, and N_2 of known purity, or gas mixtures of certified composition, for calibration. Also use samples of O_2, H_2, and H_2S of known purity if these gases are to be measured. Preferably use custom-made gas mixtures to closely approximate digester gas composition.

c. Displacement liquids: See Section 2720B.3d.

4. Procedure

a. Preparation of gas chromatograph: Open main valve of carrier-gas cylinder and adjust carrier-gas flow rate to recommended values. To obtain accurate flow measurements, connect a soap-film flow meter to the TCD vent. Turn power on. Turn on oven heaters, if used, and detector current and adjust to desired values.

Set injection port and column temperatures as specified for the column being used. Set TCD current. Turn on recorder or data processor. Check that the injector/detector temperature has risen to the appropriate level and confirm that column temperature is stabilized. Set range and attenuation at appropriate positions.

The instrument is ready for use when the recorder yields a stable base line. Silica gel and molecular sieve columns gradually lose activity because of adsorbed moisture or materials permanently adsorbed at room temperature. If insufficient separations occur, reactivate by heating or repacking.

b. Calibration: For accurate results, prepare a calibration curve for each gas to be measured because different gas components do not give equivalent detector responses on either a weight or a molar basis. Calibrate with synthetic mixtures or with pure gases.

1) Synthetic mixtures—Use purchased gas mixtures of certified composition or prepare in the laboratory. Inject a standard volume of each mixture into the gas chromatograph and note response for each gas. Compute detector response, either as area under a peak or as height of peak, after correcting for attenuation. Read peak heights accurately and correlate with concentration of component in sample. Reproduce operating parameters exactly from one analysis to the next. If sufficient reproducibility cannot be obtained by this procedure, use peak areas for calibration. Preferably use peak areas when peaks are not symmetrical. Prepare calibration

curve by plotting either peak area or peak height against volume or mole percent for each component.

Modern integrators and data-processing systems are able to generate calibration tables for certified gas mixtures.

2) Pure gases—Introduce pure gases into chromatograph individually with a syringe. Inject sample volumes of 0.25, 0.5, 1.0 mL, etc., and plot detector response, corrected for attenuation, against gas volume.

When the analysis system yields a linear detector response with increasing gas component concentration from zero to the range of interest, run standard mixtures along with samples. If the same sample size is used, calculate gas concentration by direct proportions.

c. *Sample analysis:* If samples are to be injected with a syringe, equip sample collection container with a port closed by a rubber or silicone septum. To take a sample for analysis, expel air from barrel of syringe by depressing plunger and force needle through the septum. Withdraw plunger to take gas volume desired, pull needle from collection container, and inject sample rapidly into chromatograph.

When samples are to be injected through a gas-sampling valve, connect sample collection container to inlet tube. Let gas flow from collection tube through the valve to purge dead air space and fill sample tube. About 15 mL normally are sufficient to clear the lines and to provide a sample of 1 to 2 mL. Transfer sample from loop into carrier gas stream by following manufacturer's instructions. Bring samples to atmospheric pressure before injection.

When calibration curves have been prepared with a synthetic gas mixture of certified composition, use the same sample volume as that used during calibration. When calibration curves are prepared by the procedure using varying volumes of pure gases, inject any convenient gas sample volume up to about 2 mL.

Inject sample and standard gases in sequence to permit calculation of unknown gas concentration in volume (or mole) percent by direct comparison of sample and standard gas peak heights or areas. For more accurate analysis, make duplicate or triplicate injections of sample and standard gases.

5. Calculation

a. When calibration curves have been prepared with synthetic mixtures and the volume of the sample analyzed is the same as that used in calibration, read volume percent of each component directly from calibration curve after detector response for that component is computed.

b. When calibration curves are prepared with varying volumes of pure gases, calculate the percentage of each gas in the mixture as follows:

$$\text{Volume \%} = \frac{A}{B} \times 100$$

where:

A = partial volume of component (read from calibration curve) and
B = volume sample injected.

c. Where standard mixtures are run with samples and instrument response is linear from zero to the concentration range of interest:

$$\text{Volume \%} = \text{volume \% (std)} \frac{C}{D}$$

where:

C = recorder value of sample and
D = recorder value of standard.

d. Digester gases usually are saturated with water vapor that is not a digestion product. Therefore, apply corrections to calculate the dry volume percent of each digester gas component as follows:

$$\text{Dry volume \%} = \frac{\text{volume \% as above}}{1 - P_v}$$

where:

P_v = saturated water vapor pressure at room temperature and pressure, decimal %.

The saturation water vapor pressure can be found in common handbooks. The correction factor $\frac{1}{1 - P_v}$ is usually small and is often neglected by analysts.

6. Precision and Bias

Precision and bias depend on the instrument, the column, operating conditions, gas concentrations, and techniques of operation. The upper control limits for replicate analyses of a high-methane-content standard gas mixture (65.01% methane, 29.95% carbon dioxide, 0.99% oxygen, and 4.05% nitrogen) by a single operator were as follows: 65.21% for methane, 30.36% for carbon dioxide, 1.03% for oxygen, and 4.15% for nitrogen. The lower control limits for this same standard gas mixture were: 64.67% for methane, 29.88% for carbon dioxide, 0.85% for oxygen, and 3.85% for nitrogen. The observed relative standard deviations (RSD) were: 0.14% for methane, 0.26% for carbon dioxide, 3.2% for oxygen, and 1.25% for nitrogen.

In a similar analysis of another standard gas mixture (55% methane, 35% carbon dioxide, 10% diatomic gases—2% hydrogen, 6% nitrogen, 2% oxygen), typical upper control limits for precision of duplicate determinations were: 55.21% for methane, 35.39% for carbon dioxide, and 10.26% for the diatomic gases. The lower control limits for this standard gas mixture were: 54.78% for methane, 34.62% for carbon dioxide, and 9.73% for the diatomic gases. The observed RSDs for this analysis were: 0.13% for methane, 0.36% for carbon dioxide, and 0.88% for the diatomic gases.

A low-methane-content standard gas mixture (35.70% methane, 47.70% carbon dioxide, 3.07% oxygen, 8.16% nitrogen, and 5.37% hydrogen) was analyzed as above yielding upper control limits as follows: 34.4% for methane, 49.46% for carbon dioxide, 3.07% for oxygen, and 8.2% for nitrogen. The lower control limits were: 34.18% for methane, 49.11% for carbon dioxide, 2.83% for oxygen, and 7.98% for nitrogen. The observed RSDs were: 0.11% for methane, 0.16% for carbon dioxide, 1.37% for oxygen, and 0.46% for nitrogen.

With digester gas the sum of the percent CH_4, CO_2, and N_2 should approximate 100%. If it does not, suspect errors in collection, handling, storage, and injection of gas, or in instrumental operation or calibration.

7. References

1. SUPELCO, INC. 1983. Isothermal and Temperature Programmed Analyses of Permanent Gases and Light Hydrocarbons. GC Bull. 760F. Bellefonte, Pa.

2. MINDRUP, R. 1978. The analysis of gases and light hydrocarbons by gas chromatography. *J. Chromatogr. Sci.* 16:380.

3. SUPELCO, INC. 1983. Analyzing Mixtures of Permanent Gases and Light (C1 - C3) Hydrocarbons on a Single GC Column. GC Bull. 712F. Bellefonte, Pa.

4. ALLTECH ASSOCIATES, INC. 1989. Chromatography. Catalog No. 200. Deerfield, Ill.

5. SUPELCO, INC. 1983. Column Selection for Gas and Light Hydrocarbon Analysis. GC Bull. 786C. Bellefonte, Pa.

6. SUPELCO, INC. 1989. Supelco Chromatography Products. Catalog 27. Bellefonte, Pa.

7. PURCELL, J.E. & L.S. ETTRE. 1965. Analysis of hydrogen with thermal conductivity detectors. *J. Gas Chromatogr.* 3(2):69.

8. Bibliography

LEIBRAND, R.J. 1967. Atlas of gas analyses by gas chromatography. *J. Gas Chromatogr.* 5:518.

JEFFREY, P.G. & P.J. KIPPING. 1972. Gas Analysis by Gas Chromatography, 2nd ed. Pergamon Press Inc., Elmsford, N.Y.

BURGETT, C., L. GREEN & E. BONELLI. 1977. Chromatographic Methods in Gas Analysis. Hewlett-Packard Inc., Avondale, Pa.

THOMPSON, B. 1977. Fundamentals of Gas Analysis by Gas Chromatography. Varian Associates, Inc., Palo Alto, Calif.

2810 DISSOLVED GAS SUPERSATURATION*

2810 A. Introduction

Water can become supersaturated with atmospheric gases by various means, heating and air entrainment in spilled or pumped water being the most common. The primary sign of gas supersaturation is the formation of bubbles on submerged surfaces or within the vascular systems and tissues of aquatic organisms.

Gas supersaturation can limit aquatic life and interfere with water treatment processes. Levels of supersaturation lethal to aquatic organisms have been found in springs, rivers, wells, lakes, estuaries, and seawater. Gas supersaturation can be produced in pumped or processed water intended for drinking, fish hatchery supply, and laboratory bioassays. Seasonal and other temporal variations in supersaturation may occur in surface waters as a result of solar heating and photosynthesis. Because the rate of equilibration may be slow, supersaturation may persist in flowing water for days and excessive dissolved gas levels thus may persist far from the source of supersaturation.

Gas bubbles form only when the total dissolved gas pressure is greater than the sum of compensating pressures. Compensating pressures include water, barometric and, for organisms, tissue or blood pressure. The total dissolved gas pressure is equal to the sum of the partial pressures of all the dissolved gases, including water vapor. Typically, only nitrogen, oxygen, argon, carbon dioxide, and water vapor pressures need to be considered in most natural waters. Gas bubble disease, of fish or other aquatic organisms, is a result of excessive uncompensated gas pressure. A single supersaturated gas such as oxygen or nitrogen may not necessarily result in gas bubble disease because bubble formation depends largely on total dissolved gas pressure. The degree of gas saturation should be described in terms of pressures rather than concentration or volume units.

---

* Approved by Standard Methods Committee, 1996.

2810 B. Direct-Sensing Membrane-Diffusion Method

1. General Discussion

a. Principle: This method requires an instrument with a variable length of "gas permeable" tubing, connected to a pressure-measuring device. Dimethyl silicone rubber tubing often is used because it is highly permeable to dissolved gases, including water vapor. At steady state, the gauge pressure inside the tubing is equal to the difference in gas pressure (ΔP) between the total dissolved gas pressure and the ambient barometric pressure. When the water is in equilibrium with the atmosphere, ΔP equals zero. If ΔP is greater than zero, the water is supersaturated. Conversely, if ΔP is negative the water is undersaturated.

b. Working range: The working range of this method depends on the pressure-sensing device used, but typically will range from -150 to $+600$ mm Hg. Dissolved solids in wastewater will not interfere with this method. The practical depth range for these instruments is 1 to 10 m.

2. Apparatus

Several types of membrane-diffusion instruments are available commercially.* Alternatively, construct a unit from commercially available parts. Several units have been described, including a direct-reading instrument using pressure transducers and a digital readout,[1] an on-line unit that can activate an alarm system,[2] and an early model of the Weiss saturometer.[3] Each of these units has specific advantages and limitations; the instrument of choice will

---

* Common Sensing, Clark Fork, ID; Eco Enterprises, Seattle, WA; Novatech, Vancouver, BC, Canada; and Sweeney Aquametrics, Stony Creek, CT.

depend on the specific application. All these instruments are portable so that data collection is completed in the field.

Test the instrument for leaks according to the manufacturer's recommendation. Even a very small leak, difficult to detect and locate, will result in useless data. Calibrate the pressure-measuring device with a mercury manometer or certified pressure gauge. If a manometer is used, include fresh mercury that flows freely in the tubing. An alternative method for directly testing membrane-diffusion instruments in a small, closed chamber where induced ΔP levels can be compared against observed ΔP levels is available.[2]

Van Slyke-Neill[4] or gas chromatography methods[1] are inappropriate for calibration but they may be used to verify results. These methods measure individual gas concentrations and require further conversion to ΔP or partial pressure and suffer from sampling and sample handling problems.[5–7]

3. Procedure

At the start of each day, test the instrument for leaks and recalibrate. At a monitoring site, completely submerge the sensing element in the water, preferably below the hydrostatic compensation depth. This is the depth where the hydrostatic and total gas pressures are equal and as a result, bubbles will not form on the tubing. Bubble formation on the silicone rubber tubing seriously reduces accuracy. Compute hydrostatic compensation depth[5] as follows:

$$Z = \frac{\Delta P}{73.42}$$

where:

    Z = hydrostatic compensation depth, m, and
    ΔP = pressure difference between total dissolved gas pressure and the ambient barometric pressure, mm Hg.

The factor 73.42 is the hydrostatic pressure of fresh water at 20°C expressed in terms of mm Hg/m water depth. Because the variation of hydrostatic pressure with temperature and salinity is small, this equation can be used for all natural waters.

Dislodge formed bubbles on the tubing by gently striking the instrument or moving the instrument rapidly in the water. Movement of water across the silicone rubber tubing also facilitates establishing the equilibrium between gas pressure in the water and in the tubing.

Operate the instrument "bubble free" until a stable ΔP is observed. This may take from 5 to 30 min, depending on the ΔP, water temperature, water flow, and geometry of the system. The time response of the membrane-diffusion method is shown in Figure 2810:1 for "bubble-free" and "bubble" conditions.

If the instrument is used in heavily contaminated water containing oil or other organic compounds, clean the silicone tubing with a mild detergent according to the manufacturer's instructions. Silicone rubber tubing has been used in uncontaminated natural water for at least eight years without being adversely affected by attached algal growth.[2] The tubing can be damaged by abrasive grit, diatoms, biting aquatic organisms, certain organic compounds, and strong acids.[2]

Obtain the barometric pressure with each measurement by using a laboratory mercury barometer, a calibrated portable barometer, or pressure transducer. Barometric pressures reported by

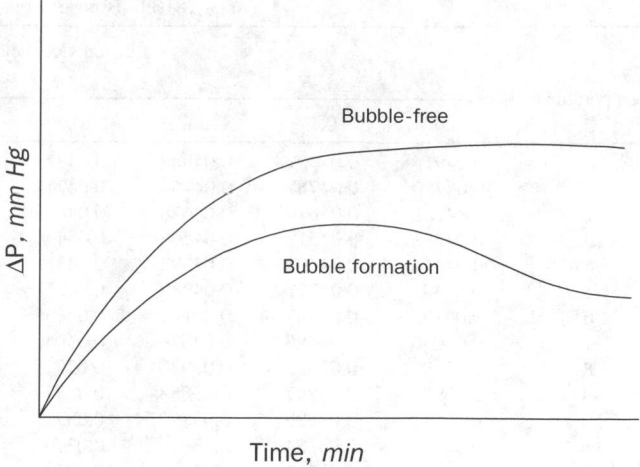

Figure 2810:1. Time response for the membrane-diffusion method.

weather agencies (or airports) are corrected to sea level and are unusable.

4. Calculation

a. Total gas pressure: Preferably report total gas pressure as ΔP.[2,6,8] Express pressure as millimeters of mercury.

Total gas pressure also has been reported as a percentage of local barometric pressure:

$$\text{TGP \%} = \left[\frac{P_b + \Delta P}{P_b} \right] \times 100$$

where:

    P_b = true local barometric pressure, mm Hg.

The reporting of total gas pressure as a percentage is not encouraged.[8]

b. Component gas pressures: When information on component gas supersaturation is needed, express data as partial pressures, differential pressures, or percent saturation.[5,8] This requires additional measurements of dissolved oxygen, temperature, and salinity† at the monitoring site. In a mixture of gases in a given volume, the partial pressure of a gas is the pressure that this gas would exert if it were the only gas present.

1) Oxygen partial pressure—Calculate partial pressure of oxygen as follows:

$$P_{O_2} = \frac{DO}{\beta_{O_2}} \times 0.5318$$

where:

    P_{O_2} = partial pressure of dissolved oxygen, mm Hg,
    β_{O_2} = Bunsen coefficient for oxygen (Table 2810:I), L/(L·atm), and
    DO = measured concentration of oxygen, mg/L.

Bunsen coefficients for marine waters are available.[5] The factor

---

† Methods for these variables may be found in Sections 4500-O, 2550, and 2520, respectively.

TABLE 2810:I. BUNSEN COEFFICIENT FOR OXYGEN IN FRESH WATER

Temperature °C	Bunsen Coefficient at Given Temperature (to nearest 0.1°C) L real gas at STP/(L · atmosphere)									
	0.0	0.1	0.2	0.3	0.4	0.5	0.6	0.7	0.8	0.9
0	0.04914	0.04901	0.04887	0.04873	0.04860	0.04847	0.04833	0.04820	0.04807	0.04793
1	0.04780	0.04767	0.04754	0.04741	0.04728	0.04716	0.04703	0.04680	0.04678	0.04665
2	0.04653	0.04640	0.04628	0.04615	0.04603	0.04591	0.04579	0.04567	0.04555	0.04543
3	0.04531	0.04519	0.04507	0.04495	0.04484	0.04472	0.04460	0.04449	0.04437	0.04426
4	0.04414	0.04403	0.04392	0.04381	0.04369	0.04358	0.04347	0.04336	0.04325	0.04314
5	0.04303	0.04292	0.04282	0.04271	0.04260	0.04250	0.04239	0.04229	0.04218	0.04206
6	0.04197	0.04187	0.04177	0.04166	0.04156	0.04146	0.04136	0.04126	0.04116	0.04106
7	0.04096	0.04086	0.04076	0.04066	0.04056	0.04047	0.04037	0.04027	0.04018	0.04008
8	0.03999	0.03989	0.03980	0.03971	0.03961	0.03952	0.03943	0.03933	0.03924	0.03915
9	0.03906	0.03897	0.03888	0.03879	0.03870	0.03861	0.03852	0.03843	0.03835	0.03826
10	0.03817	0.03809	0.03800	0.03791	0.03783	0.03774	0.03766	0.03757	0.03749	0.03741
11	0.03732	0.03724	0.03716	0.03707	0.03699	0.03691	0.03683	0.03675	0.03667	0.03659
12	0.03651	0.03643	0.03635	0.03627	0.03619	0.03611	0.03604	0.03596	0.03588	0.03581
13	0.03573	0.03565	0.03558	0.03550	0.03543	0.03535	0.03528	0.03520	0.03513	0.03505
14	0.03498	0.03491	0.03448	0.03476	0.03469	0.03462	0.03455	0.03448	0.03441	0.03433
15	0.03426	0.03419	0.03412	0.03406	0.03399	0.03392	0.03385	0.03378	0.03371	0.03364
16	0.03358	0.03351	0.03344	0.03338	0.03331	0.03324	0.03318	0.03311	0.03305	0.03298
17	0.03292	0.03285	0.03279	0.03272	0.03266	0.03260	0.03253	0.03247	0.03241	0.03235
18	0.03228	0.03222	0.03216	0.03210	0.03204	0.03198	0.03192	0.03186	0.03180	0.03174
19	0.03168	0.03162	0.03156	0.03150	0.03144	0.03138	0.03132	0.03126	0.03121	0.03115
20	0.03109	0.03103	0.03098	0.03092	0.03086	0.03081	0.03075	0.03070	0.03064	0.03059
21	0.03053	0.03048	0.03042	0.03037	0.03031	0.03026	0.03020	0.03015	0.03010	0.03004
22	0.02999	0.02994	0.02989	0.02983	0.02978	0.02973	0.02968	0.02963	0.02958	0.02952
23	0.02947	0.02942	0.02937	0.02932	0.02927	0.02922	0.02917	0.02912	0.02907	0.02902
24	0.02897	0.02893	0.02888	0.02883	0.02878	0.02873	0.02868	0.02864	0.02859	0.02854
25	0.02850	0.02845	0.02840	0.02835	0.02831	0.02826	0.02822	0.02817	0.02812	0.02808
26	0.02803	0.02799	0.02794	0.02790	0.02785	0.02781	0.02777	0.02772	0.02768	0.02763
27	0.02759	0.02755	0.02750	0.02746	0.02742	0.02737	0.02733	0.02729	0.02725	0.02720
28	0.02716	0.02712	0.02708	0.02704	0.02700	0.02695	0.02691	0.02687	0.02683	0.02679
29	0.02675	0.02671	0.02667	0.02663	0.02659	0.02655	0.02651	0.02647	0.02643	0.02639
30	0.02635	0.02632	0.02628	0.02624	0.02620	0.02616	0.02612	0.02609	0.02605	0.02601
31	0.02597	0.02594	0.02590	0.02586	0.02582	0.02579	0.02575	0.02571	0.02568	0.02564
32	0.02561	0.02557	0.02553	0.02550	0.02546	0.02543	0.02539	0.02536	0.02532	0.02529
33	0.02525	0.02522	0.02518	0.02515	0.02511	0.02508	0.02504	0.02501	0.02498	0.02494
34	0.02491	0.02488	0.02484	0.02481	0.02478	0.02474	0.02471	0.02468	0.02465	0.02461
35	0.02458	0.02455	0.02452	0.02448	0.02445	0.02442	0.02439	0.02436	0.02433	0.02429
36	0.02426	0.02423	0.02420	0.02417	0.02414	0.02411	0.02408	0.02405	0.02402	0.02399
37	0.02396	0.02393	0.02390	0.02387	0.02384	0.02381	0.02378	0.02375	0.02372	0.02369
38	0.02366	0.02363	0.02360	0.02358	0.02355	0.02352	0.02349	0.02346	0.02343	0.02341
39	0.02338	0.02335	0.02332	0.02329	0.02327	0.02324	0.02321	0.02318	0.02316	0.02313
40	0.02310	0.02308	0.02305	0.02302	0.02300	0.02297	0.02294	0.02292	0.02289	0.02286

Based on Benson and Krause.[9,10] $\beta = 9.9902 \times 10^{-4}\{\exp(9.7265 - 5.26895 \times 10^3/T + 1.00417 \times 10^6/T^2)\}$, where $T = 273.15 + °C$.

0.5318 equals $760/(1000\ K)$, where K is the ratio of molecular weight to molecular volume for oxygen gas.[5]

2) Nitrogen partial pressure—Estimate the partial pressure of nitrogen by subtracting the partial pressures of oxygen and water vapor from the total gas pressure.

$$P_{N_2} = P_b + \Delta P - P_{O_2} - P_{H_2O}$$

where:

P_{H_2O} = vapor pressure of water in mm Hg from Table 2810:II.

This term includes a small contribution from argon and any other gases present, including carbon dioxide and methane. The partial pressure of carbon dioxide is negligible in natural waters of pH > 7.0.

3) Nitrogen:oxygen partial pressure ratio—The ratio of the partial pressure of nitrogen to the partial pressure of oxygen (N_2:O_2) characterizes the relative contribution of the two gases to the total dissolved gas pressure. In water in equilibrium with air, this ratio is 3.77.

c. Differential pressures: The differential pressure of a gas is the difference between the partial pressures of that gas in water and air. The oxygen differential pressure may be calculated as

$$\Delta P_{O_2} = P_{O_2} - 0.209\ 46(P_b - P_{H_2O})$$

and the nitrogen differential pressure as

$$\Delta P_{N_2} = \Delta P - \Delta P_{O_2}$$

TABLE 2810:II. VAPOR PRESSURE OF FRESH WATER

Temperature °C	Vapor Pressure at Given Temperature (to nearest 0.1°C) mm Hg									
	0.0	0.1	0.2	0.3	0.4	0.5	0.6	0.7	0.8	0.9
0	4.58	4.61	4.64	4.68	4.71	4.75	4.78	4.82	4.85	4.89
1	4.92	4.96	4.99	5.03	5.07	5.10	5.14	5.18	5.21	5.25
2	5.29	5.33	5.36	5.40	5.44	5.48	5.52	5.56	5.60	5.64
3	5.68	5.72	5.76	5.80	5.84	5.88	5.92	5.97	6.01	6.05
4	6.09	6.14	6.18	6.22	6.27	6.31	6.36	6.40	6.44	6.49
5	6.54	6.58	6.63	6.67	6.72	6.77	6.81	6.86	6.91	6.96
6	7.01	7.05	7.10	7.15	7.20	7.25	7.30	7.35	7.40	7.45
7	7.51	7.56	7.61	7.66	7.71	7.77	7.82	7.87	7.93	7.98
8	8.04	8.09	8.15	8.20	8.26	8.31	8.37	8.43	8.48	8.54
9	8.60	8.66	8.72	8.87	8.84	8.89	8.95	9.02	9.08	9.14
10	9.20	9.26	9.32	9.39	9.45	9.51	9.58	9.64	9.70	9.77
11	9.83	9.90	9.97	10.03	10.10	10.17	10.23	10.30	10.37	10.44
12	10.51	10.58	10.65	10.72	10.76	10.86	10.93	11.00	11.07	11.15
13	11.22	11.29	11.37	11.44	11.52	11.59	11.67	11.74	11.82	11.90
14	11.98	12.05	12.13	12.21	12.29	12.37	12.45	12.53	12.61	12.69
15	12.78	12.86	12.94	13.05	13.11	13.19	13.28	13.36	13.45	13.54
16	13.62	13.71	13.80	13.89	13.97	14.06	14.15	14.24	14.33	14.43
17	14.52	14.61	14.70	14.80	14.89	14.98	15.08	15.17	15.27	15.37
18	15.46	15.56	15.66	15.76	15.86	15.96	16.06	16.16	16.26	16.36
19	16.46	16.57	16.67	16.77	16.88	16.98	17.09	17.20	17.30	17.41
20	17.52	17.63	17.74	17.85	17.96	18.07	18.18	18.29	18.41	18.52
21	18.64	18.75	18.87	18.98	19.10	19.22	19.33	19.45	19.57	19.69
22	19.81	19.93	20.05	20.48	20.60	20.42	20.55	20.67	20.80	20.93
23	21.05	21.18	21.31	21.44	21.57	21.70	21.83	21.96	22.09	22.23
24	22.36	22.50	22.63	22.77	22.90	23.04	23.18	23.32	23.46	23.60
25	23.74	23.88	24.03	24.17	24.31	24.46	24.60	24.75	24.90	25.04
26	25.19	25.34	25.49	25.64	25.80	25.95	26.10	26.26	26.41	26.57
27	26.72	26.88	27.04	27.20	27.36	27.52	27.68	27.84	28.00	28.17
28	28.33	28.50	28.66	28.83	29.00	29.17	29.34	29.51	29.68	29.85
29	30.03	30.80	30.37	30.55	30.73	30.91	31.08	31.26	31.44	31.62
30	31.81	31.99	32.17	32.36	32.54	32.73	32.92	33.11	33.30	33.49
31	33.68	33.87	34.06	34.26	34.45	34.65	34.85	35.05	35.24	35.44
32	35.65	35.85	36.05	36.25	36.46	36.67	36.87	37.08	37.29	37.50
33	37.71	37.92	38.14	38.35	38.57	38.78	39.00	39.22	39.44	39.66
34	39.88	40.10	40.33	40.55	40.78	41.01	41.23	41.46	41.69	41.92
35	42.16	42.39	42.63	42.86	43.10	43.34	43.58	43.82	44.06	44.30
36	44.55	44.79	45.04	45.28	45.53	45.78	46.03	46.29	46.54	46.79
37	47.05	47.31	47.56	47.82	48.08	48.35	48.61	48.87	49.14	49.41
38	49.67	49.94	50.21	50.49	50.76	51.03	51.31	51.59	51.87	52.14
39	52.43	52.71	52.99	53.28	53.56	53.85	54.14	54.43	54.72	55.01
40	55.31	55.60	55.90	56.20	56.50	56.80	57.10	57.41	57.71	58.02

Based on an equation presented by Green and Carritt.[11] This equation is cumbersome to use. The following equation[9] is adequate for most applications:

$$P_{H_2O} = 760\{\exp(11.8571 - 3,840.70/T - 216,961/T^2)\}, \text{ where } T = 273.15 + °C.[5]$$

d. *Percent of saturation:* In older literature, supersaturation values have been reported as percent saturation. This method of reporting component gases is discouraged but can be calculated as follows:

$$N_2(\%) = \left[\frac{P_{N_2}}{0.7902\,(P_b - P_{H_2O})}\right] \times 100$$

$$O_2(\%) = \left[\frac{P_{O_2}}{0.20946(P_b - P_{H_2O})}\right] \times 100$$

The following relationships are useful conversions:

$$TGP(\%) = 0.209\,46\,O_2(\%) + 0.7902\,N_2(\%)$$

$$\Delta P = 0.20946\left[\frac{O_2(\%)}{100} - 1\right][P_b - P_{H_2O}]$$
$$+ 0.7902\left[\frac{N_2(\%)}{100} - 1\right][P_b - P_{H_2O}]$$

$$\Delta P = \frac{DO}{\beta_{O_2}}(0.5318)(1 + N_2{:}O_2) - (P_b - P_{H_2O})$$

Use care with these relationships with older data because both TGP(%) and N_2(%) have been differently defined.[5]

5. Quality Control

The precision of the membrane-diffusion method depends primarily on the pressure-sensing instrument. For an experienced operator it is approximately ±1 to 2 mm Hg with an accuracy of ±3 to 5 mm Hg.[3,6] Air leaks, bubble formation, biofilm development, incomplete equilibration, or condensation produce negative errors while direct water leaks can result in positive errors in submersible units.

For accurate work, measure water temperature to the nearest ±0.1°C.

6. Reporting of Results

In reporting results, include the following data:
Sensor depth, m,
Barometric pressure, mm Hg,
Water temperature, °C,
Dissolved oxygen, mm Hg or mg/L,
Salinity, g/kg, and
ΔP, mm Hg.
If component gas information is needed add:
Partial pressure of oxygen, mm Hg,
Partial pressure of nitrogen, mm Hg, and
Nitrogen:oxygen partial pressure ratio
or
ΔP_{O_2}, mm Hg, and
ΔP_{N_2}, mm Hg.

7. Interpretation of Results

The biological effects of dissolved gas supersaturation depend on the species, age, depth in water column, length of exposure, temperature, and nitrogen:oxygen partial pressure ratio.[12] Safe limits generally are segregated into wild/natural circumstances, where behavior and hydrostatic pressure can modify the exposure by horizontal and vertical movements away from dangers, and captive environments such as aquaria, hatcheries, or laboratories, where conditions not only preclude escape but also include other significant stresses. Of these two realms, captive circumstances are more likely to cause illness or mortality from gas bubble disease and will do so sooner and at the lower ΔP levels.

In wild/natural circumstances, the limit of safe levels of gas supersaturation depends on the depth available to the species and/or species behavior, but this limit usually occurs at a ΔP between 50 and 150 mm Hg. Under captive conditions, the ΔP should be as close to zero as possible. For sensitive species and life stages, sublethal and lethal effects have been observed at ΔP of 10 to 50 mm Hg.[13]

8. References

1. D'AOUST, B.G., R. WHITE & H. SIEBOLD. 1975. Direct measurement of total dissolved gas partial pressure. *Undersea Biomed. Res.* 2:141.
2. BOUCK, G.R. 1982. Gasometer: an inexpensive device for continuous monitoring of dissolved gases and supersaturation. *Trans. Amer. Fish. Soc.* 111:505.
3. FICKEISEN, D.H., M.J. SCHNEIDER & J.C. MONTGOMERY. 1975. A comparative evaluation of the Weiss saturometer. *Trans. Amer. Fish. Soc.* 104:816.
4. BEININGEN, K.T. 1973. A Manual for Measuring Dissolved Oxygen and Nitrogen Gas Concentrations in Water with the Van Slyke-Neill Apparatus. Fish Commission of Oregon, Portland.
5. COLT, J. 1984. Computation of dissolved gas concentrations in water as functions of temperature, salinity, and pressure. Spec. Publ. 14, American Fisheries Soc., Bethesda, Md.
6. D'AOUST, B.G. & M.J.R. CLARK. 1980. Analysis of supersaturated air in natural waters and reservoirs. *Trans. Amer. Fish. Soc.* 109:708.
7. PIRIE, W.R. & W.A. HUBBERT. 1977. Assumptions in statistical analysis. *Trans. Amer. Fish. Soc.* 106:646.
8. COLT, J. 1983. The computation and reporting of dissolved gas levels. *Water Res.* 17:841.
9. BENSON, B.B. & D. KRAUSE. 1980. The concentration and isotopic fractionation in freshwater in equilibrium with the atmosphere. 1. Oxygen. *Limnol. Oceanogr.* 25:662.
10. BENSON, B.B. & D. KRAUSE. 1984. The concentration and isotopic fractionation of oxygen dissolved in freshwater and seawater in equilibrium with the atmosphere. *Limnol. Oceanogr.* 29:620.
11. GREEN, E.J. & D.E. CARRITT. 1967. New tables for oxygen saturation of seawater. *J. Mar. Res.* 25:140.
12. WEITKAMP, D.E. & M. KATZ. 1980. A review of dissolved gas supersaturation literature. *Trans. Amer. Fish. Soc.* 109:659.
13. COLT, J. 1986. The impact of gas supersaturation on the design and operation of aquatic culture systems. *Aquacult. Eng.* 5:49.

PART 3000

METALS

3010 INTRODUCTION

3010 A. General Discussion

1. Significance

The effects of metals in water and wastewater range from beneficial through troublesome to dangerously toxic. Some metals are essential to plant and animal growth while others may adversely affect water consumers, wastewater treatment systems, and receiving waters. The benefits versus toxicity of some metals depend on their concentrations in waters.

2. Types of Methods

Preliminary treatment is often required to present the metals to the analytical methodology in an appropriate form. Alternative methods for pretreatment of samples are presented in Section 3030.

Metals may be determined satisfactorily by a variety of methods, with the choice often depending on the precision and sensitivity required. Part 3000 describes colorimetric methods as well as instrumental methods, i.e., atomic absorption spectrometry, including flame, electrothermal (furnace), hydride, and cold vapor techniques; flame photometry; inductively coupled plasma emission spectrometry; inductively coupled plasma mass spectrometry, and anodic stripping voltammetry. Flame atomic absorption methods generally are applicable at moderate (0.1- to 10-mg/L) concentrations in clean and complex-matrix samples. Electrothermal methods generally can increase sensitivity if matrix problems do not interfere. Inductively coupled plasma emission techniques are applicable over a broad linear range and are especially sensitive for refractory elements. Inductively coupled plasma mass spectrometry offers significantly increased sensitivity for some elements (as low as 0.01 μg/L) in a variety of environmental matrices. Flame photometry gives good results at higher concentrations for several Group I and II elements. Anodic stripping offers high sensitivity for several elements in relatively clean matrices. Colorimetric methods are applicable to specific metal determinations where interferences are known not to compromise method accuracy; these methods may provide speciation information for some metals. Table 3010:I lists the methods available in Part 3000 for each metal.

3. Definition of Terms

a. *Dissolved metals:* Those metals in an unacidified sample that pass through a 0.45-μm membrane filter.

b. *Suspended metals:* Those metals in an unacidified sample that are retained by a 0.45-μm membrane filter.

c. *Total metals:* The concentration of metals determined in an unfiltered sample after vigorous digestion, or the sum of the concentrations of metals in the dissolved and suspended fractions. Note that total metals are defined operationally by the digestion procedure.

d. *Acid-extractable metals:* The concentration of metals in solution after treatment of an unfiltered sample with hot dilute mineral acid. To determine either dissolved or suspended metals, filter sample immediately after collection. Do not preserve with acid until after filtration.

TABLE 3010:I. APPLICABLE METHODS FOR ELEMENTAL ANALYSIS

Element	Flame Atomic Absorption (Direct)	Flame Atomic Absorption (Extracted)	Flame Photometry	Electrothermal Atomic Absorption	Hydride/Cold Vapor Atomic Absorption	Inductively Coupled Plasma (ICP)	ICP/Mass Spectrometry (ICP/MS)	Anodic Stripping Voltammetry	Alternative Methods†
Aluminum	3111D	3111E		3113B		3120A	3125		3500-Al.B
Antimony	3111B			3113B		3120A	3125		
Arsenic				3113B	3114B	3120A	3125		3500-As.B
Barium	3111D	3111E		3113B		3120A	3125		
Beryllium	3111D	3111E		3113B		3120A	3125		
Bismuth	3111B			3113B			3125*		
Boron						3120A	3125*		4500-B.B,C
Cadmium	3111B	3111C		3113B		3120A	3125	3130B	
Calcium	3111B,D	3111E				3120A	3125*		3500-Ca.B
Cesium	3111B						3125*		
Chromium	3111B	3111C		3113B		3120A	3125		3500-Cr.B,C
Cobalt	3111B	3111C		3113B		3120A	3125		
Copper	3111B	3111C		3113B		3120A	3125		3500-Cu.B,C
Gallium				3113B			3125*		
Germanium				3113B			3125*		
Gold	3111B			3113B			3125*		
Indium				3113B			3125*		
Iridium	3111B						3125*		

TABLE 3010:I. CONT.

Element	Flame Atomic Absorption (Direct)	Flame Atomic Absorption (Extracted)	Flame Photometry	Electrothermal Atomic Absorption	Hydride/Cold Vapor Atomic Absorption	Inductively Coupled Plasma (ICP)	ICP/Mass Spectrometry (ICP/MS)	Anodic Stripping Voltammetry	Alternative Methods†
Iron	3111B	3111C		3113B		3120A	3125*		3500-Fe.B
Lead	3111B	3111C		3113B		3120A	3125	3130B	3500-Pb.B
Lithium	3111B		3500-Li.B			3120A	3125*		
Magnesium	3111B					3120A	3125*		3500-Mg.B,C
Manganese	3111B	3111C		3113B		3120A	3125		3500-Mn.B
Mercury					3112B		3125*		
Molybdenum	3111D	3111E		3113B		3120A	3125		
Nickel	3111B	3111C		3113B		3120A	3125		
Osmium	3111D	3111E					3125*		
Palladium	3111B						3125*		
Platinum	3111B						3125*		
Potassium	3111B		3500-K.B			3120A	3125*		3500-K.C
Rhenium	3111D	3111E					3125*		
Rhodium	3111B						3125*		
Ruthenium	3111B						3125*		
Selenium				3113B	3114B,C	3120A	3125		3500-Se.C,D,E
Silicon	3111D	3111E				3120A	3125*		
Silver	3111B	3111C		3113B		3120A	3125		
Sodium	3111B		3500-Na.B			3120A	3125*		
Strontium	3111B		3500-Sr.B			3120A	3125		
Tellurium				3113B			3125*		
Thallium	3111B			3113B		3120A	3125		
Thorium	3111D	3111E					3125*		
Tin	3111B			3113B			3125*		
Titanium	3111D	3111E					3125*		
Uranium							3125		
Vanadium	3111D	3111E		3113B		3120A	3125		3500-V.B
Zinc	3111B	3111C		3113B		3120A	3125	3130B	3500-Zn.B

* Metal is not specifically mentioned in the method, but 3125 may be used successfully in most cases.
† Additional alternative methods for aluminum, beryllium, cadmium, mercury, selenium, silver, and zinc may be found in the 19th Edition of *Standard Methods*.

3010 B.   Sampling and Sample Preservation

Before collecting a sample, decide what fraction is to be analyzed (dissolved, suspended, total, or acid-extractable). This decision will determine in part whether the sample is acidified with or without filtration and the type of digestion required.

Serious errors may be introduced during sampling and storage because of contamination from sampling device, failure to remove residues of previous samples from sample container, and loss of metals by adsorption on and/or precipitation in sample container caused by failure to acidify the sample properly.

1. Sample Containers

The best sample containers are made of quartz or TFE. Because these containers are expensive, the preferred sample container is made of polypropylene or linear polyethylene with a polyethylene cap. Borosilicate glass containers also may be used, but avoid soft glass containers for samples containing metals in the microgram-per-liter range. Store samples for determination of silver in light-absorbing containers. Use only containers and filters that have been acid rinsed.

2. Preservation

Preserve samples immediately after sampling by acidifying with concentrated nitric acid (HNO_3) to pH <2. Filter samples for dissolved metals before preserving (see Section 3030). Usually 1.5 mL conc HNO_3/L sample (or 3 mL 1 + 1 HNO_3/L sample) is sufficient for short-term preservation. For samples with high buffer capacity, increase amount of acid (5 mL may be required for some alkaline or highly buffered samples). Use commercially available high-purity acid* or prepare high-purity acid by sub-boiling distillation of acid.

After acidifying sample, preferably store it in a refrigerator at approximately 4°C to prevent change in volume due to evaporation. Under these conditions, samples with metal concentrations of several milligrams per liter are stable for up to 6 months (except mercury, for which the limit is 5 weeks). For microgram-per-liter metal levels, analyze samples as soon as possible after sample collection.

_____
* Ultrex, J.T. Baker, or equivalent.

Alternatively, preserve samples for mercury analysis by adding 2 mL/L 20% (w/v) $K_2Cr_2O_7$ solution (prepared in 1 + 1 HNO_3). Store in a refrigerator not contaminated with mercury. (CAUTION: Mercury concentrations may increase in samples stored in plastic bottles in mercury-contaminated laboratories.)

3. Bibliography

STRUEMPLER, A.W. 1973. Adsorption characteristics of silver, lead, calcium, zinc and nickel on borosilicate glass, polyethylene and polypropylene container surfaces. *Anal. Chem.* 45:2251.

FELDMAN, C. 1974. Preservation of dilute mercury solutions. *Anal. Chem.* 46:99.

KING, W.G., J.M. RODRIGUEZ & C.M. WAI. 1974. Losses of trace concentrations of cadmium from aqueous solution during storage in glass containers. *Anal. Chem.* 46:771.

BATLEY, G.E. & D. GARDNER. 1977. Sampling and storage of natural waters for trace metal analysis. *Water Res.* 11:745.

SUBRAMANIAN, K.S., C.L. CHAKRABARTI, J.E. SUETIAS & I.S. MAINES. 1978. Preservation of some trace metals in samples of natural waters. *Anal. Chem.* 50:444.

BERMAN, S. & P. YEATS. 1985. Sampling of seawater for trace metals. *Crit. Rev. Anal. Chem.* 16:1.

WENDLANDT, E. 1986. Sample containers and analytical accessories made of modern plastics for trace analysis. *Gewaess. Wass. Abwass.* 86:79.

3010 C. General Precautions

1. Sources of Contamination

Avoid introducing contaminating metals from containers, distilled water, or membrane filters. Some plastic caps or cap liners may introduce metal contamination; for example, zinc has been found in black bakelite-type screw caps as well as in many rubber and plastic products, and cadmium has been found in plastic pipet tips. Lead is a ubiquitous contaminant in urban air and dust.

2. Contaminant Removal

Thoroughly clean sample containers with a metal-free nonionic detergent solution, rinse with tap water, soak in acid, and then rinse with metal-free water. For quartz, TFE, or glass materials, use 1 + 1 HNO_3, 1 + 1 HCl, or aqua regia (3 parts conc HCl + 1 part conc HNO_3) for soaking. For plastic material, use 1 + 1 HNO_3 or 1 + 1 HCl. Reliable soaking conditions are 24 h at 70°C. Chromic acid or chromium-free substitutes* may be used to remove organic deposits from containers, but rinse containers thoroughly with water to remove traces of chromium. Do not use chromic acid for plastic containers or if chromium is to be determined. Always use metal-free water in analysis and reagent preparation (see 3111B.3c). In these methods, the word "water" means metal-free water.

3. Airborne Contaminants

For analysis of microgram-per-liter concentrations of metals, airborne contaminants in the form of volatile compounds, dust, soot, and aerosols present in laboratory air may become significant. To avoid contamination use "clean laboratory" facilities such as commercially available laminar-flow clean-air benches or custom-designed work stations and analyze blanks that reflect the complete procedure.

4. Bibliography

MITCHELL, J.W. 1973. Ultrapurity in trace analysis. *Anal. Chem.* 45:492A.

GARDNER, M., D. HUNT & G. TOPPING. 1986. Analytical quality control (AQC) for monitoring trace metals in the coastal and marine environment. *Water Sci. Technol.* 18:35.

___
* Nochromix, Godax Laboratories, or equivalent.

3020 QUALITY ASSURANCE/QUALITY CONTROL

3020 A. Introduction

General information and recommendations for quality assurance (QA) and quality control (QC) are provided in Sections 1020 Quality Assurance, 1030 Data Quality, and 1040 Method Development and Evaluation. This section discusses QA/QC requirements that are common to the analytical methods presented in Part 3000. The requirements are recommended minimum QA/QC activities; refer to individual methods and regulatory program requirements for method-specific QA/QC requirements.

Always consider the overall purpose of analyses. QA/QC measures and substantiation for operational-control determinations may differ significantly from those for determinations of trace metals at water quality criteria levels. Levels of trace metals in environmental samples may be orders of magnitude lower than in potential sources of contamination.

Use replicates of measurable concentration to establish precision and known-additions recovery to determine bias. Use blanks, calibrations, control charts, known additions, standards, and other ancillary measurement tools as appropriate. Provide adequate documentation and record keeping to satisfy client requirements and performance criteria established by the laboratory.

3020 B. Quality Control Practices

1. Initial Quality Control

a. Initial demonstration of capability: Verify analyst capability before analyzing any samples and repeat periodically to demonstrate proficiency with the analytical method. Verify that the method being used provides sufficient sensitivity for the purpose of the measurement. Test analyst capability by analyzing at least four reagent water portions containing known additions of the analyte of interest. Confirm proficiency by generating analytical results that demonstrate precision and bias within acceptable limits representative of the analytical method.

b. Method detection limit (MDL): Before samples are analyzed, determine the MDL for each analyte by the procedures of Section 1030, or other applicable procedure.[1] Determine MDL at least annually for each method and major matrix category. Verify MDL for a new analyst or whenever instrument hardware or method operating conditions are modified. Analyze samples for MDL determinations over a 3- to 5-d period to generate a realistic value. Preferably use pooled data from several analysts rather than data from a single analyst.

c. Dynamic range (DR): Before using a new method, determine the dynamic range, i.e., the concentration range over which a method has an increasing response (linear or second-order), for each analyte by analyzing several standard solutions that bracket the range of interest. Each standard measurement should be within 10% of the true value for acceptance into the DR determination. Take measurements at both the low and high end of the calibration range to determine method suitability. Analytical instrumentation with curve-fitting features may allow utilization of nonlinear instrument response.

2. Calibration

a. Initial calibration: Calibrate initially with a minimum of a blank and three calibration standards of the analyte(s) of interest. Select calibration standards that bracket the expected concentration of the sample and that are within the method's dynamic range. The number of calibration points depends on the width of the dynamic range and the shape of the calibration curve. One calibration standard should be below the reporting limit for the method. As a general rule, differences between calibration standard concentrations should not be greater than one order of magnitude (i.e., 1, 10, 100, 1000). Apply linear or polynomial curve-fitting statistics, as appropriate, for analysis of the concentration-instrument response relationship. The appropriate linear or nonlinear correlation coefficient for standard concentration to instrument response should be ≥ 0.995. Use initial calibration for quantitation of analyte concentration in samples. Use calibration verification, ¶ *b* below, only for checks on the initial calibration and not for sample quantitation. Repeat initial calibration daily and whenever calibration verification acceptance criteria are not satisfied.

b. Calibration verification: Calibration verification is the periodic confirmation that instrument response has not changed significantly from the initial calibration. Verify calibration by analyzing a midpoint calibration standard (check standard) and calibration blank at the beginning and end of a sample run, periodically during a run (normally after each set of ten samples).

A check standard determination outside 90 to 110% of the expected concentration indicates a potential problem. If a check standard determination is outside 80 to 120% of the expected concentration, immediately cease sample analyses and initiate corrective action. Repeat initial calibration and sample determinations since the last acceptable calibration verification. Use calculated control limits (Section 1020B) to provide better indications of system performance and to provide tighter control limits.

c. Quality control sample: Analyze an externally generated quality control sample of known concentration at least quarterly and whenever new calibration stock solutions are prepared. Obtain this sample from a source external to the laboratory or prepare it from a source different from those used to prepare working standards. Use to validate the laboratory's working standards both qualitatively and quantitatively.

3. Batch Quality Control

a. Method blank (MB): A method blank (also known as reagent blank) is a portion of reagent water treated exactly as a sample, including exposure to all equipment, glassware, procedures, and reagents. The MB is used to assess whether analytes or interference are present within the analytical process or system. No analyte of interest should be present in the MB at a warning level based on the end user's requirements. Undertake immediate corrective action for MB measurements above the MDL. Include a minimum of one MB with each set of 20 or fewer samples.

b. Laboratory-fortified blank (LFB): The laboratory-fortified blank (also known as blank spike) is a method blank that has been fortified with a known concentration of analyte. It is used to evaluate ongoing laboratory performance and analyte recovery in a clean matrix. Prepare fortified concentrations approximating the midpoint of the calibration curve or lower with stock solutions prepared from a source different from those used to develop working standards. Calculate percent recovery, plot control charts, and determine control limits (Section 1020B) for these measurements. Ensure that the LFB meets performance criteria for the method. Establish corrective actions to be taken in the event that LFB does not satisfy acceptance criteria. Include a minimum of one LFB with each set of 20 or fewer samples.

c. Duplicates: Use duplicate samples of measurable concentration to measure precision of the analytical process. Randomly select routine samples to be analyzed twice. Process duplicate sample independently through entire sample preparation and analytical process. Include a minimum of one duplicate for each matrix type with each set of 20 or fewer samples.

d. Laboratory-fortified matrix (LFM)/laboratory-fortified matrix duplicate: Use LFM (also known as matrix spike) and LFM duplicate to evaluate the bias and precision, respectively, of the method as influenced by a specific matrix. Prepare by adding a known concentration of analytes to a randomly selected routing sample. Prepare addition concentrations to approximately double the concentration present in the original sample. If necessary, dilute sample to bring the measurement within the established calibration curve. Limit addition volume to 5% or less of sample volume. Calculate percent recovery and relative percent difference, plot control charts, and determine control limits (Section

1020B). Ensure that performance criteria for the method are satisfied. Process fortified samples independently through entire sample preparation and analytical process. Include a minimum of one LFM/LFM duplicate with each set of 20 or fewer samples.

e. Method of known additions: To analyze a new or unfamiliar matrix, use the method of known additions (Section 1020B) to demonstrate freedom from interference before reporting concentration data for the analyte. Verify absence of interferences by analyzing such samples undiluted and in a 1:10 dilution; results should be within 10% of each other. Limit known-addition volume to 10% or less of the sample volume.

4. Reference

1. U.S. ENVIRONMENTAL PROTECTION AGENCY. 1995. Definition and procedure for the determination of the method detection limit, revision 1.11. 40 CFR Part 136, Appendix B. *Federal Register* 5:23703.

3030   PRELIMINARY TREATMENT OF SAMPLES*

3030 A.   Introduction

Samples containing particulates or organic material generally require pretreatment before spectroscopic analysis. "Total metals" includes all metals, inorganically and organically bound, both dissolved and particulate. Colorless, transparent samples (primarily drinking water) having a turbidity of <1 NTU, no odor, and single phase may be analyzed directly by atomic absorption spectroscopy (flame or electrothermal vaporization) or inductively coupled plasma spectroscopy (atomic emission or mass spectrometry) for total metals without digestion. For further verification or if changes in existing matrices are encountered, compare digested and undigested samples to ensure comparable results. On collection, acidify such samples to pH <2 with conc nitric acid (1.5 mL HNO_3/L is usually adequate for drinking water) and analyze directly. Digest all other samples before determining total metals. To analyze for dissolved metals, filter sample, acidify filtrate, and store until analyses can be performed. To determine suspended metals, filter sample, digest filter and the material on it, and analyze. To determine acid-extractable metals, extract metals as indicated in Sections 3030E through K and analyze extract.

This section describes general pretreatment for samples in which metals are to be determined according to Sections 3110 through 3500-Zn with several exceptions. The special digestion techniques for mercury are given in Sections 3112B.4*b* and *c*, and those for arsenic and selenium in Sections 3114 and 3500-Se.

Take care not to introduce metals into samples during preliminary treatment. During pretreatment avoid contact with rubber, metal-based paints, cigarette smoke, paper tissues, and all metal products including those made of stainless steel, galvanized metal, and brass. Conventional fume hoods can contribute significantly to sample contamination, particularly during acid digestion in open containers. Keep vessels covered with watch glasses and turn spouts away from incoming air to reduce airborne contamination. Plastic pipet tips often are contaminated with copper, iron, zinc, and cadmium; before use soak in 2*N* HCl or HNO_3 for several days and rinse with deionized water. Avoid using colored plastics, which can contain metals. Use certified metal-free plastic containers and pipet tips when possible. Avoid using glass if analyzing for aluminum or silica.

Use metal-free water (see 3111B.3*c*) for all operations. Check reagent-grade acids used for preservation, extraction, and digestion for purity. If excessive metal concentrations are found, purify the acids by distillation or use ultra-pure acids. Inductively coupled plasma mass spectrometry (ICP-MS) may require use of ultra-pure acids and reagents to avoid measurable contamination. Process blanks through all digestion and filtration steps and evaluate blank results relative to corresponding sample results. Either apply corrections to sample results or take other corrective actions as necessary or appropriate.

---

* Approved by Standard Methods Committee, 1997.

3030 B.   Filtration for Dissolved and Suspended Metals

1. Filtration Procedures

If dissolved or suspended metals (see Section 3010A) are to be determined, filter sample at time of collection using a preconditioned plastic filtering device with either vacuum or pressure, containing a filter support of plastic or fluorocarbon, through a prewashed ungridded 0.4- to 0.45-μm-pore-diam membrane filter (polycarbonate or cellulose esters). Before use filter a blank consisting of metal-free (deionized) water to insure freedom from contamination. Precondition filter and filter device by rinsing with 50 mL deionized water. If the filter blank contains significant metals concentrations, soak membrane filters in approximately 0.5*N* HCl or 1*N* HNO_3 (recommended for electrothermal and ICP-MS analyses) and rinse with deionized water before use.

Before filtering, centrifuge highly turbid samples in acid-washed fluorocarbon or high-density plastic tubes to reduce loading on filters. Stirred, pressure filter units foul less readily than vacuum filters; filter at a pressure of 70 to 130 kPa. After filtration acidify filtrate to pH 2 with conc HNO_3 and store until analyses can be performed. If a precipitate forms on acidification, digest acidified filtrate before analysis as directed (see Section 3030E). Retain filter and digest it for direct determination of suspended metals.

If it is not possible to field-filter the sample without contaminating it, obtain sample in an "unpreserved" bottle as above and promptly cool to 4°C. Do not acid-preserve the sample. Then, without delay, filter sample under cleaner conditions in the laboratory.

Test pH of a portion of aqueous sample upon receipt in the laboratory to ensure that the sample has been properly filtered and acid-preserved.[1]

NOTE: Different filters display different sorption and filtration characteristics[2]; for trace analysis, test filter and filtration sytem to verify complete recovery of metals.

If suspended metals (see Section 3010A) are to be determined, filter sample as above for dissolved metals, but do not centrifuge before filtration. Retain filter and digest it for direct determination of suspended metals. Record sample volume filtered and include a filter in determination of the blank.

CAUTION: *Do not use perchloric acid to digest membrane filters.* (See 3030H for more information on handling $HClO_4$).

2. References

1. U.S. ENVIRONMENTAL PROTECTION AGENCY. 1994. Sample Preparation Procedure for Spectrochemical Determination of Total Recoverable Elements, Method 200.2. Environmental Monitoring Systems Lab., Cincinnati, Ohio.

2. HOROWITZ, A.J., K.R. LUM, J.R. GARBARINO, G.E.M. HALL, C. LEMIEUX & C.R. DEMAS. 1996. Problems with using filtration to define dissolved trace element concentrations in natural water samples. *Environ. Sci. Technol.* 30, 954.

3030 C.   Treatment for Acid-Extractable Metals

Extractable metals (see Section 3010A) are lightly adsorbed on particulate material. Because some sample digestion may be unavoidable use rigidly controlled conditions to obtain meaningful and reproducible results. Maintain constant sample volume, acid volume, and contact time. Express results as extractable metals and specify extraction conditions.

At collection, acidify entire sample with 5 mL conc HNO_3/L sample. To prepare sample, mix well, transfer 100 mL to a beaker or flask, and add 5 mL 1 + 1 high-purity HCl. Heat 15 min on a steam bath. Filter through a membrane filter (preconditioned as in Section 3030B) and carefully transfer filtrate to a tared volumetric flask. Adjust volume to 100 mL with metal-free water, mix, and analyze. If volume is greater than 100 mL, determine volume to nearest 0.1 mL by weight, analyze, and correct final concentration measurement by multiplying by the dilution factor (final volume ÷ 100).

3030 D.   Digestion for Metals

To reduce interference by organic matter and to convert metals associated with particulates to a form (usually the free metal) that can be determined by atomic absorption spectrometry or inductively-coupled plasma spectroscopy, use one of the digestion techniques presented below. Use the least rigorous digestion method required to provide acceptable and consistent recovery compatible with the analytical method and the metal being analyzed.[1–3]

1. Selection of Acid

Nitric acid will digest most samples adequately (Section 3030E). Nitrate is an acceptable matrix for both flame and electrothermal atomic absorption and the preferred matrix for ICP-MS.[4] Some samples may require addition of perchloric, hydrochloric, hydrofluoric, or sulfuric acid for complete digestion. These acids may interfere in the analysis of some metals and all provide a poorer matrix for both electrothermal and ICP-MS analysis. Confirm metal recovery for each digestion and analytical procedure used. Use Table 3030:I as a guide in determining which acids (in addition to HNO_3) to use for complete digestion. As a

TABLE 3030:I. ACIDS USED IN CONJUNCTION WITH HNO_3 FOR SAMPLE PREPARATION

Acid	Recommended for	May Be Helpful for	Not Recommended for
HCl	Ag	Sb, Ru, Sn	Th, Pb
H_2SO_4	Ti	—	Ag, Pb, Ba
$HClO_4$	—	Organic materials	—
HF	—	Siliceous materials	—

general rule, HNO_3 alone is adequate for clean samples or easily oxidized materials; HNO_3-H_2SO_4 or HNO_3-HCl digestion is adequate for readily oxidizable organic matter; HNO_3-$HClO_4$ or HNO_3-$HClO_4$-HF digestion is necessary for difficult-to-oxidize organic matter or minerals containing silicates. Although dry ashing is not generally recommended because of the loss of many volatile elements, it may be helpful if large amounts of organic matter are present.

2. Digestion Procedures

Dilute samples with Ag concentrations greater than 1 mg/L to contain less than 1 mg Ag/L for flame atomic absorption methods and 25 μg/L or less for electrothermal analysis.[2,5,6] To address problems with silver halide solubility in HNO_3, digest using method 3030F.3b.

Report digestion technique used.

Acid digestion techniques (Sections 3030E through I) generally yield comparable precision and bias for most sample types that are totally digested by the technique. Because acids used in digestion will add metals to the samples and blanks, minimize the volume of acids used.

Because the acid digestion techniques (3030E and F) normally are not total digestions, the microwave digestion procedure (3030K) may be used as an alternative. The microwave method is a closed-vessel procedure and thus is expected to provide improved precision when compared with hot-plate techniques. Microwave digestion is recommended for samples being analyzed by ICP-MS. The microwave digestion method is recommended for the analysis of Ag, Al, As, Ba, Be, Ca, Cd, Co, Cr, Cu, Fe, K, Mg, Mn, Mo, Na, Ni, Pb, Sb, Se, Tl, V, and Zn. Microwave digestion may be acceptable for additional analytes provided its performance for those elements is validated.

Suggested sample volumes are indicated below for flame atomic absorption spectrometry. Lesser volumes, to a minimum of 5 mL, are appropriate for graphite furnace, ICP, and ICP-MS. Do not subsample volumes less than 5 mL, especially when particulates are present. Instead dilute samples with elevated analyte concentrations after digestion. If the recommended volume exceeds digestion vessel capacity, add sample as evaporation proceeds. For samples containing particulates, wide-bore pipets may be useful for volume measurement and transfer.

When samples are concentrated during digestion (e.g., >100 mL sample used) determine metal recovery for each matrix digested, to verify method validity. Using larger samples will require additional acid, which also would increase the concentration of impurities.

Estimated Metal Concentration *mg/L*	Sample Volume* *mL*
<0.1	1000
0.1–10	100
10–100 +	10

*For flame atomic absorption spectrometry.

Report results as follows:

$$\text{Metal concentration, mg/L} = A \times \frac{B}{C}$$

where:

A = concentration of metal in digested solution, mg/L,
B = final volume of digested solution, mL, and
C = sample size, mL.

Prepare solid samples or liquid sludges with high solids contents on a weight basis. Mix sample and transfer a suitable amount (typically 1 g of a sludge with 15% total solids) directly into a preweighed digestion vessel. Reweigh and calculate weight of sample. Proceed with one of the digestion techniques presented below. However, as these digestion methods are predominantly for dissolved and extractable metals in aqueous samples, other approaches may be more appropriate for solid samples. For complete mineralization of solid samples, consult methods available elsewhere.[1,4,6,7] Report results on wet- or dry-weight basis as follows:

$$\text{Metal concentration, mg/kg (wet-weight basis)} = \frac{A \times B}{\text{g sample}}$$

$$\text{Metal concentration, mg/kg (dry-weight basis)} = \frac{A \times B}{\text{g sample}} \times \frac{100}{D}$$

where:

A = concentration of metal in digested solution, mg/L,
B = final volume of digested solution, mL, and
D = total solids, % (see Section 2540G).

Always prepare acid blanks for each type of digestion performed. Although it is always best to eliminate all relevant sources of contamination, a reagent blank prepared with the same acids and subjected to the same digestion procedure as the sample can correct for impurities present in acids and reagent water. However, blank correction is not recommended for any other sources of contamination such as impurities adsorbed on glassware.

3. References

1. BOUMANS, P.W.J.M., ed. 1987. Inductively Coupled Plasma Emission Spectroscopy, Part II: Applications and Fundamentals. John Wiley & Sons, New York, N.Y.

2. U.S. ENVIRONMENTAL PROTECTION AGENCY. 1992. Test Methods for Evaluating Solid Waste, Physical/Chemical Methods, SW-846, 3rd ed. Update 1, Methods 3005A, 3010A, 3020A & 3050A. Off. Solid Waste & Emergency Response, Washington, D.C.

3. HOENIG, M. & A.M. DE KERSABIEC. 1996. Sample preparation steps for analysis by atomic spectroscopy methods: Present status. *Spectrochim. Acta.* B51:1297.

4. JARVIS, K.E., A.L. GRAY & R.S. HOUK, eds. 1992. Sample preparation for ICP-MS. Chapter 7 *in* Handbook of Inductively Coupled Plasma Mass Spectrometry. Blackie, Glasgow & London, U.K.

5. U.S. ENVIRONMENTAL PROTECTION AGENCY. 1994. Sample Preparation Procedure for Spectrochemical Determination of Total Recoverable Elements, Method 200.2. Environmental Monitoring Systems Lab., Cincinnati, Ohio.

6. KINGSTON, H.M. & S. HASWELL, eds. 1997. Microwave Enhanced Chemistry: Fundamentals, Sample Preparation and Applications. American Chemical Soc., Washington, D.C.

7. BOCK, R. 1979. A Handbook of Decomposition Methods in Analytical Chemistry. Blackie, Glasgow, U.K.

3030 E. Nitric Acid Digestion

Because of the wide variation in concentration levels detected by various instrumental techniques and the need to deal adequately with sources of contamination at trace levels, this method presents one approach for high-level analytes (>0.1 mg/L) and another for trace levels (≤ 0.1 mg/L).

1. Digestion for Flame Atomic Absorption and High-Level Concentrations

a. Apparatus:
1) *Hot plate.*
2) *Conical (erlenmeyer) flasks,* 125-mL, or *Griffin beakers,* 150-mL, acid-washed and rinsed with water.
3) *Volumetric flasks,* 100-mL.
4) *Watch glasses,* ribbed and unribbed.
b. Reagent:
Nitric acid, HNO_3, conc, analytical or trace-metals grade.
c. Procedure: Transfer a measured volume (100 mL recommended) of well-mixed, acid-preserved sample appropriate for the expected metals concentrations to a flask or beaker (see 3030D for sample volume). In a hood, add 5 mL conc HNO_3. If a beaker is used, cover with a ribbed watch glass to minimize contamination. Boiling chips, glass beads, or Hengar granules may be added to aid boiling and minimize spatter when high concentration levels (>10 mg/L) are being determined. Bring to a slow boil and evaporate on a hot plate to the lowest volume possible (about 10 to 20 mL) before precipitation occurs. Continue heating and adding conc HNO_3 as necessary until digestion is complete as shown by a light-colored, clear solution. Do not let sample dry during digestion.

Wash down flask or beaker walls and watch glass cover (if used) with metal-free water and then filter if necessary (see Section 3030B). Transfer filtrate to a 100-mL volumetric flask with two 5-mL portions of water, adding these rinsings to the volumetric flask. Cool, dilute to mark, and mix thoroughly. Take portions of this solution for required metal determinations.

2. Digestion for Trace-Level (≤ 0.1 mg/L) Concentrations for ICP and ICP-MS[1]

a. Apparatus:
1) *Block heater,* dry, with temperature control.

2) *Polypropylene tubes**, graduated, round-bottom tubes with caps, 17×100 mm, acid-washed and rinsed with metal-free water. Preferably use tubes that simultaneously match the analysis instrument autosampler and the block digester. A fit with the centrifuge is secondary but also desirable.
3) *Pipetters,* assorted sizes or adjustable.
4) *Pipet tips.*
5) *Centrifuge.*
b. Reagent:
Nitric acid, HNO_3, conc, double distilled.†
c. Procedure: Soak new polypropylene tubes and caps overnight or for several days in $2N$ HNO_3. Triple rinse with metal-free water, and preferably dry in poly rackets or baskets in a low-temperature oven overnight. Store cleaned tubes in plastic bags before use. Pipet tips also may need to be cleaned; evaluate before use.

Pipet 10 mL well-mixed, acid-preserved sample into a precleaned, labeled tube with a macropipet. With a minimum volume change (<0.5 mL), add appropriate amount of analyte for matrix fortified samples. With a pipet, add 0.5 mL conc HNO_3 (or 1.0 mL 1 + 1 HNO_3) to all samples, blanks, standards, and quality control samples.

Place tubes in block heater in a hood and adjust temperature to 105°C. Drape caps over each tube to allow escape of acid vapors while preventing contamination. NOTE: Do not screw on caps at this time. Digest samples for a minimum of 2 h. Do not let samples boil. Add more conc nitric acid as necessary until digestion is complete by observation of a clear solution.

Remove tubes from heat and cool. Dilute back to original 10 mL volume with metal-free water. Adjust over-volume samples to next convenient gradation for calculations and note volume. (Apply concentration correction from Section 3030D.) If tubes contain particulates, centrifuge and decant clear portion into another precleaned tube. Tighten screw caps and store at 4°C until ready for analysis.

3. Reference

1. JARVIS, K.E., A.L. GRAY & R.S. HOUK, eds. 1992. Sample preparation for ICP-MS. Chapter 7 *in* Handbook of Inductively Coupled Plasma Mass Spectrometry. Blackie & Son, Ltd., Glasgow & London, U.K.

---
* Falcon tubes or equivalent.
† Ultrex, Optima grade or equivalent.

3030 F. Nitric Acid-Hydrochloric Acid Digestion

1. Apparatus

See 3030E.1*a*. The following also may be needed:
Steam bath.

2. Reagents

a. Nitric acid, HNO_3, conc, analytical grade or better (see Section 3030E).

b. Hydrochloric acid, HCl, 1 + 1.
c. Nitric acid, HNO_3, 1 + 1.

3. Procedure

a. Total HNO_3/HCl: Transfer a measured volume of well-mixed, acid-preserved sample appropriate for the expected metals concentrations to a flask or beaker (see 3030D for sample volume). In a hood add 3 mL conc HNO_3 and cover with a ribbed watch glass. Place flask or beaker on a hot plate and cautiously evaporate to less than 5 mL, making certain that sample does not boil and that no area of the bottom of the container is allowed to go dry. Cool. Rinse down walls of beaker and watch glass with a minimum of metal-free water and add 5 mL conc HNO_3. Cover container with a nonribbed watch glass and return to hot plate. Increase temperature of hot plate so that a gentle reflux action occurs. Continue heating, adding additional acid as necessary, until digestion is complete (generally indicated when the digestate is light in color or does not change in appearance with continued refluxing). Cool. Add 10 mL 1 + 1 HCl and 15 mL water per 100 mL anticipated final volume. Heat for an additional 15 min to dissolve any precipitate or residue. Cool, wash down beaker walls and watch glass with water, filter to remove insoluble material that could clog the nebulizer (see Section 3030B), and transfer filtrate to a 100-mL volumetric flask with rinsings. Alternatively centrifuge or let settle overnight. Adjust to volume and mix thoroughly.

b. Recoverable HNO_3/HCl: For this less rigorous digestion procedure, transfer a measured volume of well-mixed, acid-preserved sample to a flask or beaker. Add 2 mL 1 + 1 HNO_3 and 10 mL 1 + 1 HCl and cover with a ribbed watch glass. Heat on a steam bath or hot plate until volume has been reduced to near 25 mL, making certain sample does not boil. Cool and filter to remove insoluble material or alternatively centrifuge or let settle overnight. Quantitatively transfer sample to volumetric flask, adjust volume to 100 mL, and mix.

For trace-level digestion, use precautionary measures similar to those detailed in Section 3030E.

3030 G.   Nitric Acid-Sulfuric Acid Digestion

1. Apparatus

See 3030E.1*a*.

2. Reagents

a. Nitric acid, HNO_3, conc. (See 3030E for acid grades.)
b. Sulfuric acid, H_2SO_4, conc.

3. Procedure

Transfer a measured volume of well-mixed, acid-preserved sample appropriate for the expected metals concentrations to a flask or beaker (see 3030D for sample volume). Add 5 mL conc HNO_3 and cover with a ribbed watch glass. Bring to slow boil on hot plate and evaporate to 15 to 20 mL. Add 5 mL conc HNO_3 and 10 mL conc H_2SO_4, cooling flask or beaker between additions. Evaporate on a hot plate until dense white fumes of SO_3 just appear. If solution does not clear, add 10 mL conc HNO_3 and repeat evaporation to fumes of SO_3. Heat to remove all HNO_3 before continuing treatment. All HNO_3 will be removed when the solution is clear and no brownish fumes are evident. Do not let sample dry during digestion.

Cool and dilute to about 50 mL with water. Heat to almost boiling to dissolve slowly soluble salts. Filter if necessary, then complete procedure as directed in Section 3030E.1*c* beginning with, "Transfer filtrate . . ."

3030 H.   Nitric Acid-Perchloric Acid Digestion

1. Apparatus

See 3030E.1*a*. The following also are needed:
a. Safety shield.
b. Safety goggles.
c. Watch glasses.

2. Reagents

a. Nitric acid, HNO_3, conc.
b. Perchloric acid, $HClO_4$.
c. Ammonium acetate solution: Dissolve 500 g $NH_4C_2H_3O_2$ in 600 mL water.

3. Procedure

CAUTION: *Heated mixtures of $HClO_4$ and organic matter may explode violently. Avoid this hazard by taking the following precautions: (a) do not add $HClO_4$ to a hot solution containing organic matter; (b) always pretreat samples containing organic matter with HNO_3 before adding $HClO_4$; (c) avoid repeated fuming with $HClO_4$ in ordinary hoods (For routine operations, use a water pump attached to a glass fume eradicator. Stainless steel fume hoods with adequate water washdown facilities are available commercially and are acceptable for use with $HClO_4$); and (d) never let samples being digested with $HClO_4$ evaporate to dryness.*

Transfer a measured volume of well-mixed, acid-preserved sample appropriate for the expected metals concentrations to a flask or beaker (see 3030D for sample volume). In a hood add 5 mL conc HNO_3 and cover with a ribbed watch glass. Evaporate sample to 15 to 20 mL on a hot plate. Add 10 mL each of conc HNO_3 and $HClO_4$, cooling flask or beaker between additions. Evaporate gently on a hot plate until dense white fumes of $HClO_4$ just appear. If solution is not clear, keep solution just boiling until it clears. If necessary, add 10 mL conc HNO_3 to complete digestion. Cool, dilute to about 50 mL with water, and boil to expel any chlorine or oxides of nitrogen. Filter, then complete procedure as directed in 3030E.1c beginning with, ''Transfer filtrate . . .''

If lead is to be determined in the presence of high amounts of sulfate (e.g., determination of Pb in power plant fly ash samples), dissolve $PbSO_4$ precipitate as follows: Add 50 mL ammonium acetate solution to flask or beaker in which digestion was carried out and heat to incipient boiling. Rotate container occasionally to wet all interior surfaces and dissolve any deposited residue. Reconnect filter and slowly draw solution through it. Transfer filtrate to a 100-mL volumetric flask, cool, dilute to mark, mix thoroughly, and set aside for determination of lead.

3030 I.   Nitric Acid-Perchloric Acid-Hydrofluoric Acid Digestion

1. Apparatus

a. *Hot plate.*

b. *TFE beakers,* 250-mL, acid-washed and rinsed with water.

c. *Volumetric flasks,* 100-mL, polypropylene or other suitable plastic.

2. Reagents

a. *Nitric acid,* HNO_3, conc and 1 + 1.

b. *Perchloric acid,* $HClO_4$.

c. *Hydrofluoric acid,* HF, 48 to 51%.

3. Procedure

Caution: *See precautions for using $HClO_4$ in 3030H; handle HF with extreme care and provide adequate ventilation, especially for the heated solution. Avoid all contact with exposed skin. Provide medical attention for HF burns.*

Transfer a measured volume of well-mixed, acid-preserved sample appropriate for the expected metals concentrations into a 250-mL TFE beaker (see 3030D for sample volume). Evaporate on a hot plate to 15 to 20 mL. Add 12 mL conc HNO_3 and evaporate to near dryness. Repeat HNO_3 addition and evaporation. Let solution cool, add 20 mL $HClO_4$ and 1 mL HF, and boil until solution is clear and white fumes of $HClO_4$ have appeared. Cool, add about 50 mL water, filter, and proceed as directed in 3030E.1c beginning with, ''Transfer filtrate . . .''

3030 J.   Dry Ashing

The procedure appears in the Eighteenth Edition of *Standard Methods*. It has been deleted from subsequent editions.

3030 K.   Microwave-Assisted Digestion

1. Apparatus

a. *Microwave unit* with programmable power (minimum 545 W) to within ± 10 W of required power, having a corrosion-resistant, well-ventilated cavity and having all electronics protected against corrosion for safe operation. Use a unit having a rotating turntable with a minimum speed of 3 rpm to insure homogeneous distribution of microwave radiation. Use only laboratory-grade microwave equipment and closed digestion containers with pressure relief that are specifically designed for hot acid.[1]

b. *Vessels:* Construction requires an inner liner of perfluoroalkoxy (PFA) Teflon™,* other TFE, or composite fluorinated polymers,† capable of withstanding pressures of at least 760 ± 70 kPa (110 ± 10 psi), and capable of controlled pressure relief at the manufacturer's maximum pressure rating.

Acid wash all digestion vessels and rinse with water (¶ 2a). For new vessels or when changing between high- and low-concentration samples, clean by leaching with hot‡ hydrochloric acid (1:1) for a minimum of 2 h and then with hot nitric acid (1:1) for a minimum of 2 h; rinse with water and dry in a clean environment. Use this procedure whenever the previous use of digestion vessels is unknown or cross-contamination from vessels is suspected.

c. *Temperature feedback control system,* using shielded thermocouple, fiber-optic probe, or infrared detector.

d. *Bottles,* polyethylene, 125-mL, with caps.

---

* Or equivalent.

† Such as TFM™ or equivalent.

‡ At temperatures greater than 80°C, but not boiling.

e. Thermometer, accurate to ± 0.1°C.

f. Balance, large-capacity (1500 g), accurate to 0.1 g.

g. Filtration or centrifuge equipment (optional).

h. Plastic container with cover, 1-L, preferably made of PFA Teflon™§.

2. Reagents

a. Metal-free water: See Section 3111B.3c.

b. Nitric acid, HNO_3, conc, sub-boiling distilled. Non-sub-boiling acids can be used if they are shown not to contribute blanks.

3. Calibration of Microwave Unit

NOTE: For microwave units equipped with temperature feedback electronic controls, calibration of the microwave unit is not required provided performance specifications can be duplicated.

For cavity-type microwave equipment, evaluate absolute power (watts) by measuring the temperature rise in 1 kg water exposed to microwave radiation for a fixed time. With this measurement, the relationship between available power (W) and the partial power setting (%) of the unit can be estimated, and any absolute power in watts may be transferred from one unit to another. The calibration format required depends on type of electronic system used by manufacturer to provide partial microwave power. Few units have an accurate and precise linear relationship between percent power settings and absorbed power. Where linear circuits have been used, determine calibration curve by a three-point calibration method; otherwise, use the multiple-point calibration method.

a. Three-point calibration method: Measure power at 100% and 50% power using the procedure described in ¶ 3c and calculate power setting corresponding to required power in watts as specified in the procedure from the two-point line. Measure absorbed power at the calculated partial power setting. If the measured absorbed power does not correspond to the calculated power within ± 10 W, use the multiple-point calibration method, ¶ 3b. Use this point periodically to verify integrity of calibration.

b. Multiple-point calibration method: For each microwave unit, measure the following power settings: 100, 99, 98, 97, 95, 90, 80, 70, 60, 50, and 40% using the procedure described in ¶ 3c. These data are clustered about the customary working power ranges. Nonlinearity commonly is encountered at the upper end of the calibration curve. If the unit's electronics are known to have nonlinear deviations in any region of proportional power control, make a set of measurements that bracket the power to be used. The final calibration point should be at the partial power setting that will be used in the test. Check this setting periodically to evaluate the integrity of the calibration. If a significant change (± 10 W) is detected, re-evaluate entire calibration.

c. Equilibrate a large volume of water to room temperature (23 ± 2°C). Weigh 1 kg water (1000 g ± 1 g) or measure (1000 mL ± 1 mL) into a plastic, not glass, container, and measure the temperature to ± 0.1°C. Condition microwave unit by heating a glass beaker with 500 to 1000 mL tap water at full power for 5 min with the exhaust fan on. Loosely cover plastic container to reduce heat loss and place in normal sample path (at outer edge of rotating turntable); circulate continuously through the micro-

§ Or equivalent.

wave field for 120 s at desired power setting with exhaust fan on as it will be during normal operation. Remove plastic container and stir water vigorously. Use a magnetic stirring bar inserted immediately after microwave irradiation; record maximum temperature within the first 30 s to ± 0.1°C. Use a new sample for each additional measurement. If the water is reused, return both water and beaker to 23 ± 2°C. Make three measurements at each power setting. When any part of the high-voltage circuit, power source, or control components in the unit have been serviced or replaced, recheck calibration power. If power output has changed by more than ± 10 W, re-evaluate entire calibration.

Compute absorbed power by the following relationship:

$$P = \frac{(K)\ (Cp)\ (m)\ (\Delta T)}{t}$$

where:

P = apparent power absorbed by sample, W,

K = conversion factor for thermochemical calories sec^{-1} to watts (4.184),

Cp = heat capacity, thermal capacity, or specific heat (cal g^{-1} °C^{-1}) of water,

m = mass of water sample, g,

ΔT = final temperature minus initial temperature, °C, and

t = time, s.

For the experimental conditions of 120 s and 1 kg water (Cp at 25°C = 0.9997), the calibration equation simplifies to:

$$P = (\Delta T)\ (34.85)$$

Stable line voltage within the manufacturer's specification is necessary for accurate and reproducible calibration and operation. During measurement and operation it must not vary by more than ± 2 V. A constant power supply may be necessary if line voltage is unstable.

4. Procedure

CAUTION: *This method is designed for microwave digestion of waters only. It is not intended for the digestion of solids, for which high concentrations of organic compounds may result in high pressures and possibly unsafe conditions.*

CAUTION: *As a safety measure, never mix different manufacturers' vessels in the same procedure. Vessels constructed differently will retain heat at different rates; control of heating conditions assumes that all vessels have the same heat-transfer characteristics. Inspect casements for cracks and chemical corrosion. Failure to maintain the vessels' integrity may result in catastrophic failure.*

Both prescription controls and performance controls are provided for this procedure. Performance controls are the most general and most accurate. When equipment capability permits, use the performance criterion.

a. Performance criterion: The following procedure is based on heating acidified samples in two stages where the first stage is to reach 160 ± 4° in 10 min and the second stage is to permit a slow rise to 165 to 170°C during the second 10 min. This performance criterion is based on temperature feedback control system capability that is implemented in various ways by different manufacturers. Because the temperature of the acid controls the reaction, this is the essential condition that will reproduce results in this preparation method. Verification of temperature conditions inside the vessel at these specific times is sufficient to verify the critical procedural requirements.

b. Prescription criterion: For all PFA vessels without liners, a verified program that meets the performance-based temperature-time profile is 545 W for 10 min followed by 344 W for 10 min using five single-wall PFA Teflon™‖ digestion vessels.[2] Any verified program for a given microwave unit depends on unit power and operational power settings, heating times, number, type, and placement of digestion vessels within the unit, and sample and acid volumes. The change in power, time, and temperature profile is not directly proportional to the change in the number of sample vessels. Any deviations from the verified program conditions will require verification of the time-temperature profile to conform to the given two-stage profile. This may be done by laboratory personnel if suitable test equipment is available, or by the manufacturer of the microwave equipment.

c. General conditions: Weigh entire digestion vessel assembly to 0.1 g before use and record (*A*). Accurately transfer 45 mL of well-shaken sample into the digestion vessel. Pipet 5 mL conc HNO_3 into each vessel. Attach all safety equipment required for appropriate and safe vessel operation following manufacturer's specifications. Tighten cap to manufacturer's specifications. Weigh each capped vessel to the nearest 0.1 g (*B*).

Place appropriate number of vessels evenly distributed in the carousel. Treat sample blanks, known additions, and duplicates in the same manner as samples. For prescription control only, when fewer samples than the appropriate number are digested, fill remaining vessels with 45 mL water and 5 mL conc HNO_3 to obtain full complement of vessels for the particular program being used.

Place carousel in unit and seat it carefully on turntable. Program microwave unit to heat samples to 160 ± 4°C in 10 min and then, for the second stage, to permit a slow rise to 165 to 170°C for 10 min. Start microwave generator, making sure that turntable is turning and that exhaust fan is on.

At completion of the microwave program, let vessels cool for at least 5 min in the unit before removal. Cool samples further outside the unit by removing the carousel and letting them cool on a bench or in a water bath. When cooled to room temperature, weigh each vessel (to 0.1 g) and record weight (*C*).

If the net weight of sample plus acid decreased by more than 10%, discard sample.

_____
‖ Or equivalent.

Complete sample preparation by carefully uncapping and venting each vessel in a fume hood. Follow individual manufacturer's specifications for relieving pressure in individual vessel types. Transfer to acid-cleaned noncontaminating plastic bottles. If the digested sample contains particulates, filter, centrifuge, or settle overnight and decant.

5. Calculations

a. Dilution correction: Multiply results by 50/45 or 1.11 to account for the dilution caused by the addition of 5 mL acid to 45 mL sample.

b. Discarding of sample: To determine if the net weight of sample plus acid decreased by more than 10% during the digestion process, use the following calculation

$$\frac{[(B - A) - (C - A)]}{(B - A)} \times 100 > 10\% \ (1\% \text{ for multilayer vessels})$$

6. Quality Control

NOTE: When nitric acid digestion is used, recoveries of silver and antimony in some matrices may be unacceptably low. Verify recoveries using appropriate known additions.

Preferably include a quality-control sample in each loaded carousel. Prepare samples in batches including preparation blanks, sample duplicates, and pre-digestion known additions. Determine size of batch and frequency of quality-control samples by method of analysis and laboratory practice. The power of the microwave unit and batch size may prevent including one or more of the quality-control samples in each carousel. Do not group quality-control samples together but distribute them throughout the various carousels to give the best monitoring of digestion.

7. References

1. KINGSTON, H.M. & S. HASWELL, eds. 1997. Microwave Enhanced Chemistry: Fundamentals, Sample Preparation and Applications. American Chemical Soc., Washington, D.C.
2. U.S. ENVIRONMENTAL PROTECTION AGENCY. 1990. Microwave assisted acid digestion of aqueous samples and extracts. SW-846 Method 3015, Test Methods for Evaluating Solid Waste. U.S. Environmental Protection Agency, Washington, D.C.

3110 METALS BY ATOMIC ABSORPTION SPECTROMETRY

Because requirements for determining metals by atomic absorption spectrometry vary with metal and/or concentration to be determined, the method is presented as follows:

Section 3111, Metals by Flame Atomic Absorption Spectrometry, encompasses:

• Determination of antimony, bismuth, cadmium, calcium, cesium, chromium, cobalt, copper, gold, iridium, iron, lead, lithium, magnesium, manganese, nickel, palladium, platinum, potas-

sium, rhodium, ruthenium, silver, sodium, strontium, thallium, tin, and zinc by direct aspiration into an air-acetylene flame (3111B),

• Determination of low concentrations of cadmium, chromium, cobalt, copper, iron, lead, manganese, nickel, silver, and zinc by chelation with ammonium pyrrolidine dithiocarbamate (APDC), extraction into methyl isobutyl ketone (MIBK), and aspiration into an air-acetylene flame (3111C),

• Determination of aluminum, barium, beryllium, calcium,

molybdenum, osmium, rhenium, silicon, thorium, titanium, and vanadium by direct aspiration into a nitrous oxide-acetylene flame (3111D), and

• Determination of low concentrations of aluminum and beryllium by chelation with 8-hydroxyquinoline, extraction into MIBK, and aspiration into a nitrous oxide-acetylene flame (3111E).

Section 3112 covers determination of mercury by the cold vapor technique.

Section 3113 concerns determination of micro quantities of aluminum, antimony, arsenic, barium, beryllium, cadmium, chromium, cobalt, copper, iron, lead, manganese, molybdenum, nickel, selenium, silver, and tin by electrothermal atomic absorption spectrometry.

Section 3114 covers determination of arsenic and selenium by conversion to their hydrides and aspiration into an argon-hydrogen or nitrogen-hydrogen flame.

3111 METALS BY FLAME ATOMIC ABSORPTION SPECTROMETRY*

3111 A. Introduction

1. Principle

In flame atomic absorption spectrometry, a sample is aspirated into a flame and atomized. A light beam is directed through the flame, into a monochromator, and onto a detector that measures the amount of light absorbed by the atomized element in the flame. For some metals, atomic absorption exhibits superior sensitivity over flame emission. Because each metal has its own characteristic absorption wavelength, a source lamp composed of that element is used; this makes the method relatively free from spectral or radiation interferences. The amount of energy at the characteristic wavelength absorbed in the flame is proportional to the concentration of the element in the sample over a limited concentration range. Most atomic absorption instruments also are equipped for operation in an emission mode, which may provide better linearity for some elements.

2. Selection of Method

See Section 3110.

3. Interferences

a. Chemical interference: Many metals can be determined by direct aspiration of sample into an air-acetylene flame. The most troublesome type of interference is termed ''chemical'' and results from the lack of absorption by atoms bound in molecular combination in the flame. This can occur when the flame is not hot enough to dissociate the molecules or when the dissociated atom is oxidized immediately to a compound that will not dissociate further at the flame temperature. Such interferences may be reduced or eliminated by adding specific elements or compounds to the sample solution. For example, the interference of phosphate in the magnesium determination can be overcome by adding lanthanum. Similarly, introduction of calcium eliminates silica interference in the determination of manganese. However, silicon and metals such as aluminum, barium, beryllium, and vanadium require the higher-temperature, nitrous oxide-acetylene flame to dissociate their molecules. The nitrous oxide-acetylene

flame also can be useful in minimizing certain types of chemical interferences encountered in the air-acetylene flame. For example, the interference caused by high concentrations of phosphate in the determination of calcium in the air-acetylene flame is reduced in the nitrous oxide-acetylene flame.

MIBK extractions with APDC (see 3111C) are particularly useful where a salt matrix interferes, for example, in seawater. This procedure also concentrates the sample so that the detection limits are extended.

Brines and seawater can be analyzed by direct aspiration but sample dilution is recommended. Aspiration of solutions containing high concentrations of dissolved solids often results in solids buildup on the burner head. This requires frequent shutdown of the flame and cleaning of the burner head. Preferably use background correction when analyzing waters that contain in excess of 1% solids, especially when the primary resonance line of the element of interest is below 240 nm. Make more frequent recovery checks when analyzing brines and seawaters to insure accurate results in these concentrated and complex matrices.

Barium and other metals ionize in the flame, thereby reducing the ground state (potentially absorbing) population. The addition of an excess of a cation (sodium, potassium, or lithium) having a similar or lower ionization potential will overcome this problem. The wavelength of maximum absorption for arsenic is 193.7 nm and for selenium 196.0 nm—wavelengths at which the air-acetylene flame absorbs intensely. The sensitivity for arsenic and selenium can be improved by conversion to their gaseous hydrides and analyzing them in either a nitrogen-hydrogen or an argon-hydrogen flame with a quartz tube (see Section 3114).

b. Background correction: Molecular absorption and light scattering caused by solid particles in the flame can cause erroneously high absorption values resulting in positive errors. When such phenomena occur, use background correction to obtain accurate values. Use any one of three types of background correction: continuum-source, Zeeman, or Smith-Hieftje correction.

1) Continuum-source background correction—A continuum-source background corrector utilizes either a hydrogen-filled hollow cathode lamp with a metal cathode or a deuterium arc lamp. When both the line source hollow-cathode lamp and the continuum source are placed in the same optical path and are time-shared, the broadband background from the elemental signal is

---

* Approved by Standard Methods Committee, 1993

subtracted electronically, and the resultant signal will be background-compensated.

Both the hydrogen-filled hollow-cathode lamp and deuterium arc lamp have lower intensities than either the line source hollow-cathode lamp or electrodeless discharge lamps. To obtain a valid correction, match the intensities of the continuum source with the line source hollow-cathode or electrodeless discharge lamp. The matching may result in lowering the intensity of the line source or increasing the slit width; these measures have the disadvantage of raising the detection limit and possibly causing nonlinearity of the calibration curve. Background correction using a continuum source corrector is susceptible to interference from other absorbing lines in the spectral bandwidth. Miscorrection occurs from significant atomic absorption of the continuum source radiation by elements other than that being determined. When a line source hollow-cathode lamp is used without background correction, the presence of an absorbing line from another element in the spectral bandwidth will not cause an interference unless it overlaps the line of interest.

Continuum-source background correction will not remove direct absorption spectral overlap, where an element other than that being determined is capable of absorbing the line radiation of the element under study.

2) Zeeman background correction—This correction is based on the principle that a magnetic field splits the spectral line into two linearly polarized light beams parallel and perpendicular to the magnetic field. One is called the pi (π) component and the other the sigma (σ) component. These two light beams have exactly the same wavelength and differ only in the plane of polarization. The π line will be absorbed by both the atoms of the element of interest and by the background caused by broadband absorption and light scattering of the sample matrix. The σ line will be absorbed only by the background.

Zeeman background correction provides accurate background correction at much higher absorption levels than is possible with continuum source background correction systems. It also virtually eliminates the possibility of error from structured background. Because no additional light sources are required, the alignment and intensity limitations encountered using continuum sources are eliminated.

Disadvantages of the Zeeman method include reduced sensitivity for some elements, reduced linear range, and a "rollover" effect whereby the absorbance of some elements begins to decrease at high concentrations, resulting in a two-sided calibration curve.

3) Smith-Hieftje background correction—This correction is based on the principle that absorbance measured for a specific element is reduced as the current to the hollow cathode lamp is increased while absorption of nonspecific absorbing substances remains identical at all current levels. When this method is applied, the absorbance at a high-current mode is subtracted from the absorbance at a low-current mode. Under these conditions, any absorbance due to nonspecific background is subtracted out and corrected for.

Smith-Hieftje background correction provides a number of advantages over continuum-source correction. Accurate correction at higher absorbance levels is possible and error from structured background is virtually eliminated. In some cases, spectral interferences also can be eliminated. The usefulness of Smith-Hieftje background correction with electrodeless discharge lamps has not yet been established.

4. Sensitivity, Detection Limits, and Optimum Concentration Ranges

The sensitivity of flame atomic absorption spectrometry is defined as the metal concentration that produces an absorption of 1% (an absorbance of approximately 0.0044). The instrument detection limit is defined here as the concentration that produces absorption equivalent to twice the magnitude of the background fluctuation. Sensitivity and detection limits vary with the instrument, the element determined, the complexity of the matrix, and the technique selected. The optimum concentration range usually starts from the concentration of several times the detection limit and extends to the concentration at which the calibration curve starts to flatten. To achieve best results, use concentrations of samples and standards within the optimum concentration range of the spectrometer. See Table 3111:I for indication of concentration ranges measurable with conventional atomization. In many instances the concentration range shown in Table 3111:I may be extended downward either by scale expansion or by integrating the absorption signal over a long time. The range may be extended upward by dilution, using a less sensitive wavelength, rotating the burner head, or utilizing a microprocessor to linearize the calibration curve at high concentrations.

5. Preparation of Standards

Prepare standard solutions of known metal concentrations in water with a matrix similar to the sample. Use standards that bracket expected sample concentration and are within the method's working range. Very dilute standards should be prepared daily from stock solutions in concentrations greater than 500 mg/L. Stock standard solutions can be obtained from several commercial sources. They also can be prepared from National Institute of Standards and Technology (NIST) reference materials or by procedures outlined in the following sections.

For samples containing high and variable concentrations of matrix materials, make the major ions in the sample and the dilute standard similar. If the sample matrix is complex and components cannot be matched accurately with standards, use the method of standard additions, 3113B.4d2), to correct for matrix effects. If digestion is used, carry standards through the same digestion procedure used for samples.

6. Apparatus

a. *Atomic absorption spectrometer,* consisting of a light source emitting the line spectrum of an element (hollow-cathode lamp or electrodeless discharge lamp), a device for vaporizing the sample (usually a flame), a means of isolating an absorption line (monochromator or filter and adjustable slit), and a photoelectric detector with its associated electronic amplifying and measuring equipment.

b. *Burner:* The most common type of burner is a premix, which introduces the spray into a condensing chamber for removal of large droplets. The burner may be fitted with a conventional head containing a single slot; a three-slot Boling head, which may be preferred for direct aspiration with an air-acetylene flame; or a special head for use with nitrous oxide and acetylene.

c. *Readout*: Most instruments are equipped with either a digital or null meter readout mechanism. Most modern instruments are equipped with microprocessors or stand-alone control computers

TABLE 3111:I. ATOMIC ABSORPTION CONCENTRATION RANGES WITH DIRECT ASPIRATION ATOMIC ABSORPTION

Element	Wavelength nm	Flame Gases*	Instrument Detection Limit mg/L	Sensitivity mg/L	Optimum Concentration Range mg/L
Ag	328.1	A–Ac	0.01	0.06	0.1–4
Al	309.3	N–Ac	0.1	1	5–100
Au	242.8	A–Ac	0.01	0.25	0.5–20
Ba	553.6	N–Ac	0.03	0.4	1–20
Be	234.9	N–Ac	0.005	0.03	0.05–2
Bi	223.1	A–Ac	0.06	0.4	1–50
Ca	422.7	A–Ac	0.003	0.08	0.2–20
Cd	228.8	A–Ac	0.002	0.025	0.05–2
Co	240.7	A–Ac	0.03	0.2	0.5–10
Cr	357.9	A–Ac	0.02	0.1	0.2–10
Cs	852.1	A–Ac	0.02	0.3	0.5–15
Cu	324.7	A–Ac	0.01	0.1	0.2–10
Fe	248.3	A–Ac	0.02	0.12	0.3–10
Ir	264.0	A–Ac	0.6	8	—
K	766.5	A–Ac	0.005	0.04	0.1–2
Li	670.8	A–Ac	0.002	0.04	0.1–2
Mg	285.2	A–Ac	0.0005	0.007	0.02–2
Mn	279.5	A–Ac	0.01	0.05	0.1–10
Mo	313.3	N–Ac	0.1	0.5	1–20
Na	589.0	A–Ac	0.002	0.015	0.03–1
Ni	232.0	A–Ac	0.02	0.15	0.3–10
Os	290.9	N–Ac	0.08	1	—
Pb†	283.3	A–Ac	0.05	0.5	1–20
Pt	265.9	A–Ac	0.1	2	5–75
Rh	343.5	A–Ac	0.5	0.3	—
Ru	349.9	A–Ac	0.07	0.5	—
Sb	217.6	A–Ac	0.07	0.5	1–40
Si	251.6	N–Ac	0.3	2	5–150
Sn	224.6	A–Ac	0.8	4	10–200
Sr	460.7	A–Ac	0.03	0.15	0.3–5
Ti	365.3	N–Ac	0.3	2	5–100
V	318.4	N–Ac	0.2	1.5	2–100
Zn	213.9	A–Ac	0.005	0.02	0.05–2

* A–Ac = air-acetylene; N–Ac = nitrous oxide-acetylene..

† The more sensitive 217.0 nm wavelength is recommended for instruments with background correction capabilities.

Copyright ©ASTM. Reprinted with permission.

capable of integrating absorption signals over time and linearizing the calibration curve at high concentrations.

d. Lamps: Use either a hollow-cathode lamp or an electrodeless discharge lamp (EDL). Use one lamp for each element being measured. Multi-element hollow-cathode lamps generally provide lower sensitivity than single-element lamps. EDLs take a longer time to warm up and stabilize.

e. Pressure-reducing valves: Maintain supplies of fuel and oxidant at pressures somewhat higher than the controlled operating pressure of the instrument by using suitable reducing valves. Use a separate reducing valve for each gas.

f. Vent: Place a vent about 15 to 30 cm above the burner to remove fumes and vapors from the flame. This precaution protects laboratory personnel from toxic vapors, protects the instrument from corrosive vapors, and prevents flame stability from being affected by room drafts. A damper or variable-speed blower is desirable for modulating air flow and preventing flame disturbance. Select blower size to provide the air flow recommended by the instrument manufacturer. In laboratory locations with heavy particulate air pollution, use clean laboratory facilities (Section 3010C).

7. Quality Assurance/Quality Control

Some data typical of the precision and bias obtainable with the methods discussed are presented in Tables 3111:II and III.

Analyze a blank between sample or standard readings to verify baseline stability. Rezero when necessary.

To one sample out of every ten (or one sample from each group of samples if less than ten are being analyzed) add a known amount of the metal of interest and reanalyze to confirm recovery. The amount of metal added should be approximately equal to the amount found. If little metal is present add an amount close to the middle of the linear range of the test. Recovery of added metal should be between 85 and 115%.

TABLE 3111:II. INTERLABORATORY PRECISION AND BIAS DATA FOR ATOMIC ABSORPTION METHODS—DIRECT ASPIRATION AND EXTRACTED METALS

Metal	Conc.*	SD*	Relative SD %	Relative Error %	No. of Participants
Direct determination:					
Aluminum[1]	4.50	0.19	4.2	8.4	5
Barium[2]	1.00	0.089	8.9	2.7	11
Beryllium[1]	0.46	0.0213	4.6	23.0	11
Cadmium[3]	0.05	0.0108	21.6	8.2	26
Cadmium[1]	1.60	0.11	6.9	5.1	16
Calcium[1]	5.00	0.21	4.2	0.4	8
Chromium[1]	3.00	0.301	10.0	3.7	9
Cobalt[1]	4.00	0.243	6.1	0.5	14
Copper[3]	1.00	0.112	11.2	3.4	53
Copper[1]	4.00	0.331	8.3	2.8	15
Iron[1]	4.40	0.260	5.8	2.3	16
Iron[3]	0.30	0.0495	16.5	0.6	43
Lead[1]	6.00	0.28	4.7	0.2	14
Magnesium[3]	0.20	0.021	10.5	6.3	42
Magnesium[1]	1.10	0.116	10.5	10.0	8
Manganese[1]	4.05	0.317	7.8	1.3	16
Manganese[3]	0.05	0.0068	13.5	6.0	14
Nickel[1]	3.93	0.383	9.8	2.0	14
Silver[3]	0.05	0.0088	17.5	10.6	7
Silver[1]	2.00	0.07	3.5	1.0	10
Sodium[1]	2.70	0.122	4.5	4.1	12
Strontium[1]	1.00	0.05	5.0	0.2	12
Zinc[3]	0.50	0.041	8.2	0.4	48
Extracted determination:					
Aluminum[2]	300	32	10.7	0.7	15
Beryllium[2]	5	1.7	34.0	20.0	9
Cadmium[3]	50	21.9	43.8	13.3	12
Cobalt[1]	300	28.5	9.5	1.0	6
Copper[1]	100	71.7	71.7	12.0	8
Iron[1]	250	19.0	7.6	3.6	4
Manganese[1]	21.5	2.4	11.2	7.4	8
Molybdenum[1]	9.5	1.1	11.6	1.3	5
Nickel[1]	56.8	15.2	26.8	13.6	14
Lead[3]	50	11.8	23.5	19.0	8
Silver[1]	5.2	1.4	26.9	3.0	7

* For direct determinations, mg/L; for extracted determinations, μg/L.

Superscripts refer to reference numbers.

Source: AMERICAN SOCIETY FOR TESTING AND MATERIALS. 1986. Annual Book of ASTM Standards. Volume 11.01, Water and Environmental Technology. American Soc. Testing & Materials, Philadelphia, Pa. Copyright © ASTM. Reprinted with permission.

TABLE 3111:III. SINGLE-OPERATOR PRECISION AND RECOMMENDED CONTROL RANGES FOR ATOMIC ABSORPTION METHODS—DIRECT ASPIRATION AND EXTRACTED METALS

Metal	Conc.*	SD*	Relative SD %	No. of Participants	QC Std.*	Acceptable Range*
Direct determination:						
Aluminum[1]	4.50	0.23	5.1	15	5.00	4.3–5.7
Beryllium[1]	0.46	0.012	2.6	10	0.50	0.46–0.54
Calcium[1]	5.00	0.05	1.0	8	5.00	4.8–5.2
Chromium[1]	7.00	0.69	9.9	9	5.00	3.3–6.7
Cobalt[1]	4.00	0.21	5.3	14	4.00	3.4–4.6
Copper[1]	4.00	0.115	2.9	15	4.00	3.7–4.3
Iron[1]	5.00	0.19	3.8	16	5.00	4.4–5.6
Magnesium[1]	1.00	0.009	0.9	8	1.00	0.97–1.03
Nickel[4]	5.00	0.04	0.8	—	5.00	4.9–5.1
Silver[1]	2.00	0.25	12.5	10	2.00	1.2–2.8
Sodium[4]	8.2	0.1	1.2	—	5.00	4.8–5.2
Strontium[1]	1.00	0.04	4.0	12	1.00	0.87–1.13
Potassium[4]	1.6	0.2	12.5	—	1.6	1.0–2.2
Molybdenum[4]	7.5	0.07	0.9	—	10.0	9.7–10.3
Tin[4]	20.0	0.5	2.5	—	20.0	18.5–21.5
Titanium[4]	50.0	0.4	0.8	—	50.0	48.8–51.2
Vanadium	50.0	0.2	0.4	—	50.0	49.4–50.6
Extracted determination:						
Aluminum[1]	300	12	4.0	15	300	264–336
Cobalt[1]	300	20	6.7	6	300	220–380
Copper[1]	100	21	21	8	100	22–178
Iron[1]	250	12	4.8	4	250	180–320
Manganese[1]	21.5	202	10.2	8	25	17–23
Molybdenum[1]	9.5	1.0	10.5	5	10	5.5–14.5
Nickel[1]	56.8	9.2	16.2	14	50	22–78
Silver[1]	5.2	1.2	23.1	7	5.0	0.5–9.5

* For direct determinations, mg/L; for extracted determinations, μg/L.
Superscripts refer to reference numbers.
Source: AMERICAN SOCIETY FOR TESTING AND MATERIALS. 1986. Annual Book of ASTM Standards. Volume 11.01, Water and Environmental Technology. American Soc. Testing & Materials, Philadelphia, Pa. Copyright© ASTM. Reprinted with permission.

Analyze an additional standard solution after every ten samples or with each batch of samples, whichever is less, to confirm that the test is in control. Recommended concentrations of standards to be run, limits of acceptability, and reported single-operator precision data are listed in Table 3111:III.

See Section 3020 for additional recommended quality control procedures.

8. References

1. AMERICAN SOCIETY FOR TESTING AND MATERIALS. 1986. Annual Book of ASTM Standards, Volume 11.01, Water and Environmental Technology. American Soc. Testing & Materials, Philadelphia, Pa.
2. U.S. DEPARTMENT HEALTH, EDUCATION AND WELFARE. 1970. Water Metals No. 6, Study No. 37. U.S. Public Health Serv. Publ. No. 2029, Cincinnati, Ohio.
3. U.S. DEPARTMENT HEALTH, EDUCATION AND WELFARE. 1968. Water Metals No. 4, Study No. 30. U.S. Public Health Serv. Publ. No. 999-UTH-8, Cincinnati, Ohio.
4. U.S. ENVIRONMENTAL PROTECTION AGENCY. 1983. Methods for Chemical Analysis of Water and Wastes. Cincinnati, Ohio.

9. Bibliography

KAHN, H.L. 1968. Principles and Practice of Atomic Absorption. Advan. Chem. Ser. No. 73, Div. Water, Air & Waste Chemistry, American Chemical Soc., Washington, D.C.

RAMIRIZ-MUNOZ, J. 1968. Atomic Absorption Spectroscopy and Analysis by Atomic Absorption Flame Photometry. American Elsevier Publishing Co., New York, N.Y.

SLAVIN, W. 1968. Atomic Absorption Spectroscopy. John Wiley & Sons, New York, N.Y.

PAUS, P.E. 1971. The application of atomic absorption spectroscopy to the analysis of natural waters. Atomic Absorption Newsletter 10:69.

EDIGER, R.D. 1973. A review of water analysis by atomic absorption. Atomic Absorption Newsletter 12:151.

PAUS, P.E. 1973. Determination of some heavy metals in seawater by atomic absorption spectroscopy. Fresenius Zeitschr. Anal. Chem. 264:118.

BURRELL, D.C. 1975. Atomic Spectrometric Analysis of Heavy-Metal Pollutants in Water. Ann Arbor Science Publishers, Inc., Ann Arbor, Mich.

3111 B.   Direct Air-Acetylene Flame Method

1. General Discussion

This method is applicable to the determination of antimony, bismuth, cadmium, calcium, cesium, chromium, cobalt, copper, gold, iridium, iron, lead, lithium, magnesium, manganese, nickel, palladium, platinum, potassium, rhodium, ruthenium, silver, sodium, strontium, thallium, tin, and zinc.

2. Apparatus

Atomic absorption spectrometer and associated equipment: See Section 3111A.6. Use burner head recommended by the manufacturer.

3. Reagents

a. Air, cleaned and dried through a suitable filter to remove oil, water, and other foreign substances. The source may be a compressor or commercially bottled gas.

b. Acetylene, standard commercial grade. Acetone, which always is present in acetylene cylinders, can be prevented from entering and damaging the burner head by replacing a cylinder when its pressure has fallen to 689 kPa (100 psi) acetylene.

CAUTION: *Acetylene gas represents an explosive hazard in the laboratory. Follow instrument manufacturer's directions in plumbing and using this gas. Do not allow gas contact with copper, brass with >65% copper, silver, or liquid mercury; do not use copper or brass tubing, regulators, or fittings.*

c. Metal-free water: Use metal-free water for preparing all reagents and calibration standards and as dilution water. Prepare metal-free water by deionizing tap water and/or by using one of the following processes, depending on the metal concentration in the sample: single distillation, redistillation, or sub-boiling. Always check deionized or distilled water to determine whether the element of interest is present in trace amounts. (NOTE: *If the source water contains Hg or other volatile metals, single- or redistilled water may not be suitable for trace analysis because these metals distill over with the distilled water. In such cases, use sub-boiling to prepare metal-free water).*

d. Calcium solution: Dissolve 630 mg calcium carbonate, $CaCO_3$, in 50 mL of 1 + 5 HCl. If necessary, boil gently to obtain complete solution. Cool and dilute to 1000 mL with water.

e. Hydrochloric acid, HCl, 1%, 10%, 20% (all v/v), 1 + 5, 1 + 1, and conc.

f. Lanthanum solution: Dissolve 58.65 g lanthanum oxide, La_2O_3, in 250 mL conc HCl. Add acid slowly until the material is dissolved and dilute to 1000 mL with water.

g. Hydrogen peroxide, 30%.

h. Nitric acid, HNO_3, 2% (v/v), 1 + 1, and conc.

i. Aqua regia: Add 3 volumes conc HCl to 1 volume conc HNO_3.

j. Standard metal solutions: Prepare a series of standard metal solutions in the optimum concentration range by appropriate dilution of the following stock metal solutions with water containing 1.5 mL conc HNO_3/L. Stock standard solutions are available from a number of commercial suppliers. Alternatively, prepare as described below. Thoroughly dry reagents before use. In general,

use reagents of the highest purity. For hydrates, use fresh reagents.

1) *Antimony:* Dissolve 0.2669 g $K(SbO)C_4H_4O_6$ in water, add 10 mL 1 + 1 HCl and dilute to 1000 mL with water; 1.00 mL = 100 μg Sb.

2) *Bismuth:* Dissolve 0.100 g bismuth metal in a minimum volume of 1 + 1 HNO_3. Dilute to 1000 mL with 2% (v/v) HNO_3; 1.00 mL = 100 μg Bi.

3) *Cadmium:* Dissolve 0.100 g cadmium metal in 4 mL conc HNO_3. Add 8.0 mL conc HNO_3 and dilute to 1000 mL with water; 1.00 mL = 100 μg Cd.

4) *Calcium:* Suspend 0.2497 g $CaCO_3$ (dried at 180° for 1 h before weighing) in water and dissolve cautiously with a minimum amount of 1 + 1 HNO_3. Add 10.0 mL conc HNO_3 and dilute to 1000 mL with water; 1.00 mL = 100 μg Ca.

5) *Cesium:* Dissolve 0.1267 g cesium chloride, CsCl, in 1000 mL water; 1.00 mL = 100 μg Cs.

6) *Chromium:* Dissolve 0.1923 g CrO_3 in water. When solution is complete, acidify with 10 mL conc HNO_3 and dilute to 1000 mL with water; 1.00 mL = 100 μg Cr.

7) *Cobalt:* Dissolve 0.1000 g cobalt metal in a minimum amount of 1 + 1 HNO_3. Add 10.0 mL 1 + 1 HCl and dilute to 1000 mL with water; 1.00 mL = 100 μg Co.

8) *Copper:* Dissolve 0.100 g copper metal in 2 mL conc HNO_3, add 10.0 mL conc HNO_3 and dilute to 1000 mL with water; 1.00 mL = 100 μg Cu.

9) *Gold:* Dissolve 0.100 g gold metal in a minimum volume of aqua regia. Evaporate to dryness, dissolve residue in 5 mL conc HCl, cool, and dilute to 1000 mL with water; 1.00 mL = 100 μg Au.

10) *Iridium:* Dissolve 0.1147 g ammonium chloroiridate, $(NH_4)_2IrCl_6$, in a minimum volume of 1% (v/v) HCl and dilute to 100 mL with 1% (v/v) HCl; 1.00 mL = 500 μg Ir.

11) *Iron:* Dissolve 0.100 g iron wire in a mixture of 10 mL 1 + 1 HCl and 3 mL conc HNO_3. Add 5 mL conc HNO_3 and dilute to 1000 mL with water; 1.00 mL = 100 μg Fe.

12) *Lead:* Dissolve 0.1598 g lead nitrate, $Pb(NO_3)_2$, in a minimum amount of 1 + 1 HNO_3, add 10 mL conc HNO_3, and dilute to 1000 mL with water; 1.00 mL = 100 μg Pb.

13) *Lithium:* Dissolve 0.5323 g lithium carbonate, Li_2CO_3, in a minimum volume of 1 + 1 HNO_3. Add 10.0 mL conc HNO_3 and dilute to 1000 mL with water; 1.00 mL = 100 μg Li.

14) *Magnesium:* Dissolve 0.1658 g MgO in a minimum amount of 1 + 1 HNO_3. Add 10.0 mL conc HNO_3 and dilute to 1000 mL with water; 1.00 mL = 100 μg Mg.

15) *Manganese:* Dissolve 0.1000 g manganese metal in 10 mL conc HCl mixed with 1 mL conc HNO_3. Dilute to 1000 mL with water; 1.00 mL = 100 μg Mn.

16) *Nickel:* Dissolve 0.1000 g nickel metal in 10 mL hot conc HNO_3, cool, and dilute to 1000 mL with water; 1.00 mL = 100 μg Ni.

17) *Palladium:* Dissolve 0.100 g palladium wire in a minimum volume of aqua regia and evaporate just to dryness. Add 5 mL conc HCl and 25 mL water and warm until dissolution is complete. Dilute to 1000 mL with water; 1.00 mL = 100 μg Pd.

18) *Platinum:* Dissolve 0.100 g platinum metal in a minimum volume of aqua regia and evaporate just to dryness. Add 5 mL conc HCl and 0.1 g NaCl and again evaporate just to dryness.

Dissolve residue in 20 mL of 1 + 1 HCl and dilute to 1000 mL with water; 1.00 mL = 100 μg Pt.

19) *Potassium:* Dissolve 0.1907 g potassium chloride, KCl, (dried at 110°C) in water and make up to 1000 mL; 1.00 mL = 100 μg K.

20) *Rhodium:* Dissolve 0.386 g ammonium hexachlororhodate, $(NH_4)_3RhCl_6 \cdot 1.5H_2O$, in a minimum volume of 10% (v/v) HCl and dilute to 1000 mL with 10% (v/v) HCl; 1.00 mL = 100 μg Rh.

21) *Ruthenium:* Dissolve 0.205 g ruthenium chloride, $RuCl_3$, in a minimum volume of 20% (v/v) HCl and dilute to 1000 mL with 20% (v/v) HCl; 1.00 mL = 100 μg Ru.

22) *Silver:* Dissolve 0.1575 g silver nitrate, $AgNO_3$, in 100 mL water, add 10 mL conc HNO_3, and make up to 1000 mL; 1.00 mL = 100 μg Ag.

23) *Sodium:* Dissolve 0.2542 g sodium chloride, NaCl, dried at 140°C, in water, add 10 mL conc HNO_3 and make up to 1000 mL; 1.00 mL = 100 μg Na.

24) *Strontium:* Suspend 0.1685 g $SrCO_3$ in water and dissolve cautiously with a minimum amount of 1 + 1 HNO_3. Add 10.0 mL conc HNO_3 and dilute to 1000 mL with water: 1 mL = 100 μg Sr.

25) *Thallium:* Dissolve 0.1303 g thallium nitrate, $TlNO_3$, in water. Add 10 mL conc HNO_3 and dilute to 1000 mL with water; 1.00 mL = 100 μg Tl.

26) *Tin:* Dissolve 1.000 g tin metal in 100 mL conc HCl and dilute to 1000 mL with water; 1.00 mL = 1.00 mg Sn.

27) *Zinc:* Dissolve 0.100 g zinc metal in 20 mL 1 + 1 HCl and dilute to 1000 mL with water; 1.00 mL = 100 μg Zn.

4. Procedure

a. Sample preparation: Required sample preparation depends on the metal form being measured.

If dissolved metals are to be determined, see Section 3030B for sample preparation. If total or acid-extractable metals are to be determined, see Sections 3030C through K. For all samples, make certain that the concentrations of acid and matrix modifiers are the same in both samples and standards.

When determining Ca or Mg, dilute and mix 100 mL sample or standard with 10 mL lanthanum solution (¶ 3f) before aspirating. When determining Fe or Mn, mix 100 mL with 25 mL of Ca solution (¶ 3d) before aspirating. When determining Cr, mix 1 mL 30% H_2O_2 with each 100 mL before aspirating. Alternatively use proportionally smaller volumes.

b. Instrument operation: Because of differences between makes and models of atomic absorption spectrometers, it is not possible to formulate instructions applicable to every instrument. See manufacturer's operating manual. In general, proceed according to the following: Install a hollow-cathode lamp for the desired metal in the instrument and roughly set the wavelength dial according to Table 3111:I. Set slit width according to manufacturer's suggested setting for the element being measured. Turn on instrument, apply to the hollow-cathode lamp the current suggested by the manufacturer, and let instrument warm up until energy source stabilizes, generally about 10 to 20 min. Readjust current as necessary after warmup. Optimize wavelength by ad-justing wavelength dial until optimum energy gain is obtained. Align lamp in accordance with manufacturer's instructions.

Install suitable burner head and adjust burner head position. Turn on air and adjust flow rate to that specified by manufacturer to give maximum sensitivity for the metal being measured. Turn on acetylene, adjust flow rate to value specified, and ignite flame. Let flame stabilize for a few minutes. Aspirate a blank consisting of deionized water containing the same concentration of acid in standards and samples. Zero the instrument. Aspirate a standard solution and adjust aspiration rate of the nebulizer to obtain maximum sensitivity. Adjust burner both vertically and horizontally to obtain maximum response. Aspirate blank again and rezero the instrument. Aspirate a standard near the middle of the linear range. Record absorbance of this standard when freshly prepared and with a new hollow-cathode lamp. Refer to these data on subsequent determinations of the same element to check consistency of instrument setup and aging of hollow-cathode lamp and standard.

The instrument now is ready to operate. When analyses are finished, extinguish flame by turning off first acetylene and then air.

c. Standardization: Select at least three concentrations of each standard metal solution (prepared as in ¶ 3j above) to bracket the expected metal concentration of a sample. Aspirate blank and zero the instrument. Then aspirate each standard in turn into flame and record absorbance.

Prepare a calibration curve by plotting on linear graph paper absorbance of standards versus their concentrations. For instruments equipped with direct concentration readout, this step is unnecessary. With some instruments it may be necessary to convert percent absorption to absorbance by using a table generally provided by the manufacturer. Plot calibration curves for Ca and Mg based on original concentration of standards before dilution with lanthanum solution. Plot calibration curves for Fe and Mn based on original concentration of standards before dilution with Ca solution. Plot calibration curve for Cr based on original concentration of standard before addition of H_2O_2.

d. Analysis of samples: Rinse nebulizer by aspirating water containing 1.5 mL conc HNO_3/L. Aspirate blank and zero instrument. Aspirate sample and determine its absorbance.

5. Calculations

Calculate concentration of each metal ion, in micrograms per liter for trace elements, and in milligrams per liter for more common metals, by referring to the appropriate calibration curve prepared according to ¶ 4c. Alternatively, read concentration directly from the instrument readout if the instrument is so equipped. If the sample has been diluted, multiply by the appropriate dilution factor.

6. Bibliography

WILLIS, J.B. 1962. Determination of lead and other heavy metals in urine by atomic absorption spectrophotometry. *Anal. Chem.* 34:614.
Also see Section 3111A.8 and 9.

3111 C. Extraction/Air-Acetylene Flame Method

1. General Discussion

This method is suitable for the determination of low concentrations of cadmium, chromium, cobalt, copper, iron, lead, manganese, nickel, silver, and zinc. The method consists of chelation with ammonium pyrrolidine dithiocarbamate (APDC) and extraction into methyl isobutyl ketone (MIBK), followed by aspiration into an air-acetylene flame.

2. Apparatus

a. *Atomic absorption spectrometer and associated equipment:* See Section 3111A.6.

b. *Burner head,* conventional. Consult manufacturer's operating manual for suggested burner head.

3. Reagents

a. *Air:* See 3111B.3a.

b. *Acetylene:* See 3111B.3b.

c. *Metal-free water:* See 3111B.3c.

d. *Methyl isobutyl ketone (MIBK),* reagent grade. For trace analysis, purify MIBK by redistillation or by sub-boiling distillation.

e. *Ammonium pyrrolidine dithiocarbamate (APDC) solution:* Dissolve 4 g APDC in 100 mL water. If necessary, purify APDC with an equal volume of MIBK. Shake 30 s in a separatory funnel, let separate, and withdraw lower portion. Discard MIBK layer.

f. *Nitric acid,* HNO_3, conc, ultrapure.

g. *Standard metal solutions:* See 3111B.3j.

h. *Potassium permanganate solution,* $KMnO_4$, 5% (w/v) aqueous.

i. *Sodium sulfate,* Na_2SO_4, anhydrous.

j. *Water-saturated MIBK:* Mix one part purified MIBK with one part water in a separatory funnel. Shake 30 s and let separate. Discard aqueous layer. Save MIBK layer.

k. *Hydroxylamine hydrochloride solution,* 10% (w/v). This solution can be purchased commercially.

4. Procedure

a. *Instrument operation:* See Section 3111B.4b. After final adjusting of burner position, aspirate water-saturated MIBK into flame and gradually reduce fuel flow until flame is similar to that before aspiration of solvent.

b. *Standardization:* Select at least three concentrations of standard metal solutions (prepared as in 3111B.3j) to bracket expected sample metal concentration and to be, after extraction, in the optimum concentration range of the instrument. Adjust 100 mL of each standard and 100 mL of a metal-free water blank to pH 3 by adding $1N$ HNO_3 or $1N$ NaOH. For individual element extraction, use the following pH ranges to obtain optimum extraction efficiency:

Element	pH Range for Optimum Extraction
Ag	2–5 (complex unstable)
Cd	1–6
Co	2–10
Cr	3–9
Cu	0.1–8
Fe	2–5
Mn	2–4 (complex unstable)
Ni	2–4
Pb	0.1–6
Zn	2–6

NOTE: For Ag and Pb extraction the optimum pH value is 2.3 ± 0.2. The Mn complex deteriorates rapidly at room temperature, resulting in decreased instrument response. Chilling the extract to 0°C may preserve the complex for a few hours. If this is not possible and Mn cannot be analyzed immediately after extraction, use another analytical procedure.

Transfer each standard solution and blank to individual 200-mL volumetric flasks, add 1 mL APDC solution, and shake to mix. Add 10 mL MIBK and shake vigorously for 30 s. (The maximum volume ratio of sample to MIBK is 40.) Let contents of each flask separate into aqueous and organic layers, then carefully add water (adjusted to the same pH at which the extraction was carried out) down the side of each flask to bring the organic layer into the neck and accessible to the aspirating tube.

Aspirate organic extracts directly into the flame (zeroing instrument on a water-saturated MIBK blank) and record absorbance.

Prepare a calibration curve by plotting on linear graph paper absorbances of extracted standards against their concentrations before extraction.

c. *Analysis of samples:* Prepare samples in the same manner as the standards. Rinse atomizer by aspirating water-saturated MIBK. Aspirate organic extracts treated as above directly into the flame and record absorbances.

With the above extraction procedure only hexavalent chromium is measured. To determine total chromium, oxidize trivalent chromium to hexavalent chromium by bringing sample to a boil and adding sufficient $KMnO_4$ solution dropwise to give a persistent pink color while the solution is boiled for 10 min. Destroy excess $KMnO_4$ by adding 1 to 2 drops hydroxylamine hydrochloride solution to the boiling solution, allowing 2 min for the reaction to proceed. If pink color persists, add 1 to 2 more drops hydroxylamine hydrochloride solution and wait 2 min. Heat an additional 5 min. Cool, extract with MIBK, and aspirate.

During extraction, if an emulsion forms at the water-MIBK interface, add anhydrous Na_2SO_4 to obtain a homogeneous organic phase. In that case, also add Na_2SO_4 to all standards and blanks.

To avoid problems associated with instability of extracted metal complexes, determine metals immediately after extraction.

5. Calculations

Calculate the concentration of each metal ion in micrograms per liter by referring to the appropriate calibration curve.

6. Bibliography

ALLAN, J.E. 1961. The use of organic solvents in atomic absorption spectrophotometry. *Spectrochim. Acta* 17:467.

SACHDEV, S.L. & P.W. WEST. 1970. Concentration of trace metals by solvent extraction and their determination by atomic absorption spectrophotometry. *Environ. Sci. Technol.* 4:749.

3111 D.   Direct Nitrous Oxide-Acetylene Flame Method

1. General Discussion

This method is applicable to the determination of aluminum, barium, beryllium, calcium, molybdenum, osmium, rhenium, silicon, thorium, titanium, and vanadium.

2. Apparatus

a. Atomic absorption spectrometer and associated equipment: See Section 3111A.6.

b. Nitrous oxide burner head: Use special burner head as suggested in manufacturer's manual. At roughly 20-min intervals of operation it may be necessary to dislodge the carbon crust that forms along the slit surface with a carbon rod or appropriate alternative.

c. T-junction valve or other switching valve for rapidly changing from nitrous oxide to air, so that flame can be turned on or off with air as oxidant to prevent flashbacks.

3. Reagents

a. Air: See 3111B.3*a*.

b. Acetylene: See 3111B.3*b*.

c. Metal-free water: See 3111B.3*c*.

d. Hydrochloric acid, HCl, 1*N*, 1 + 1, and conc.

e. Nitric acid, HNO_3, conc.

f. Sulfuric acid, H_2SO_4, 1% (v/v).

g. Hydrofluoric acid, HF, 1*N*.

h. Nitrous oxide, commercially available cylinders. Fit nitrous oxide cylinder with a special nonfreezable regulator or wrap a heating coil around an ordinary regulator to prevent flashback at the burner caused by reduction in nitrous oxide flow through a frozen regulator. (Most modern atomic absorption instruments have automatic gas control systems that will shut down a nitrous oxide-acetylene flame safely in the event of a reduction in nitrous oxide flow rate.)

CAUTION: *Use nitrous oxide with strict adherence to manufacturer's directions. Improper sequencing of gas flows at startup and shutdown of instrument can produce explosions from flashback.*

i. Potassium chloride solution: Dissolve 250 g KCl in water and dilute to 1000 mL.

j. Aluminum nitrate solution: Dissolve 139 g $Al(NO_3)_3 \cdot 9H_2O$ in 150 mL water. Acidify slightly with conc HNO_3 to preclude possible hydrolysis and precipitation. Warm to dissolve completely. Cool and dilute to 200 mL.

k. Standard metal solutions: Prepare a series of standard metal solutions in the optimum concentration ranges by appropriate dilution of stock metal solutions with water containing 1.5 mL conc HNO_3/L. Stock standard solutions are available from a number of commercial suppliers. Alternatively, prepare as described below.

1) *Aluminum:* Dissolve 0.100 g aluminum metal in an acid mixture of 4 mL 1 + 1 HCl and 1 mL conc HNO_3 in a beaker. Warm gently to effect solution. Transfer to a 1-L flask, add 10 mL 1 + 1 HCl, and dilute to 1000 mL with water; 1.00 mL = 100 µg Al.

2) *Barium:* Dissolve 0.1516 g $BaCl_2$ (dried at 250° for 2 h), in about 10 mL water with 1 mL 1 + 1 HCl. Add 10.0 mL 1 + 1 HCl and dilute to 1000 mL with water; 1.00 mL = 100 µg Ba.

3) *Beryllium: Do not dry.* Dissolve 1.966 g $BeSO_4 \cdot 4H_2O$ in water, add 10.0 mL conc HNO_3, and dilute to 1000 mL with water; 1.00 mL = 100 µg Be.

4) *Calcium:* See 3111B.3*j*4).

5) *Molybdenum:* Dissolve 0.2043 g $(NH_4)_2 MoO_4$ in water and dilute to 1000 mL; 1.00 mL = 100 µg Mo.

6) *Osmium:* Obtain standard 0.1*M* osmium tetroxide solution* and store in glass bottle; 1.00 mL = 19.02 mg Os. Make dilutions daily as needed using 1% (v/v) H_2SO_4. CAUTION: *OsO_4 is extremely toxic and highly volatile.*

7) *Rhenium:* Dissolve 0.1554 g potassium perrhenate, $KReO_4$, in 200 mL water. Dilute to 1000 mL with 1% (v/v) H_2SO_4; 1.00 mL = 100 µg Re.

8) *Silica: Do not dry.* Dissolve 0.4730 g $Na_2SiO_3 \cdot 9H_2O$ in water. Add 10.0 mL conc HNO_3 and dilute to 1000 mL with water. 1.00 mL = 100 µg SiO_2. Store in polyethylene.

9) *Thorium:* Dissolve 0.238 g thorium nitrate, $Th(NO_3)_4 \cdot 4H_2O$ in 1000 mL water; 1.00 mL = 100 µg Th.

10) *Titanium:* Dissolve 0.3960 g pure (99.8 or 99.9%) titanium chloride, $TiCl_4$,† in a mixture of equal volumes of 1*N* HCl and 1*N* HF. Make up to 1000 mL with this acid mixture; 1.00 mL = 100 µg Ti.

11) *Vanadium:* Dissolve 0.2297 g ammonium metavanadate, NH_4VO_3, in a minimum amount of conc HNO_3. Heat to dissolve. Add 10 mL conc HNO_3, and dilute to 1000 mL with water; 1.00 mL = 100 µg V.

4. Procedure

a. Sample preparation: See Section 3111B.4*a*.

When determining Al, Ba, or Ti, mix 2 mL KCl solution into 100 mL sample or standard before aspiration. When determining

---

* GFS Chemicals, Inc., Columbus, OH, Cat. No. 64, or equivalent.
† Alpha Ventron, P.O. Box 299, 152 Andover St., Danvers, MA 01923, or equivalent.

Mo and V, mix 2 mL $Al(NO_3)_3 \cdot 9H_2O$ into 100 mL sample or standard before aspiration.

b. Instrument operation: See Section 3111B.4*b*. After adjusting wavelength, install a nitrous oxide burner head. Turn on acetylene (without igniting flame) and adjust flow rate to value specified by manufacturer for a nitrous oxide-acetylene flame. Turn off acetylene. With both air and nitrous oxide supplies turned on, set T-junction valve to nitrous oxide and adjust flow rate according to manufacturer's specifications. Turn switching valve to the air position and verify that flow rate is the same. Turn acetylene on and ignite to a bright yellow flame. With a rapid motion, turn switching valve to nitrous oxide. The flame should have a red cone above the burner. If it does not, adjust fuel flow to obtain red cone. After nitrous oxide flame has been ignited, let burner come to thermal equilibrium before beginning analysis.

Aspirate a blank consisting of deionized water containing 1.5 mL conc HNO_3/L and check aspiration rate. Adjust if necessary to a rate between 3 and 5 mL/ min. Zero the instrument. Aspirate a standard of the desired metal with a concentration near the midpoint of the optimum concentration range and adjust burner (both horizontally and vertically) in the light path to obtain maximum response. Aspirate blank again and re-zero the instrument. The instrument now is ready to run standards and samples.

To extinguish flame, turn switching valve from nitrous oxide to air and turn off acetylene. This procedure eliminates the danger of flashback that may occur on direct ignition or shutdown of nitrous oxide and acetylene. (See also discussion in 3111B.4*b*.)

c. Standardization: Select at least three concentrations of standard metal solutions (prepared as in ¶ 3*k*) to bracket the expected metal concentration of a sample. Aspirate each in turn into the flame and record absorbances.

Most modern instruments are equipped with microprocessors and digital readout which permit calibration in direct concentration terms. If instrument is not so equipped, prepare a calibration curve by plotting on linear graph paper absorbance of standards versus concentration. Plot calibration curves for Al, Ba, and Ti based on original concentration of standard before adding KCl solution. Plot calibration curves for Mo and V based on original concentration of standard before adding $Al(NO_3)_3$ solution.

d. Analysis of samples: Rinse atomizer by aspirating water containing 1.5 mL conc HNO_3/L and zero instrument. Aspirate a sample and determine its absorbance.

5. Calculations

Calculate concentration of each metal ion in micrograms per liter by referring to the appropriate calibration curve prepared according to ¶ 4*c*.

Alternatively, read the concentration directly from the instrument readout if the instrument is so equipped. If sample has been diluted, multiply by the appropriate dilution factor.

6. Bibliography

WILLIS, J.B. 1965. Nitrous oxide-acetylene flame in atomic absorption spectroscopy. *Nature* 207:715.
Also see Section 3111A.8 and 9.

3111 E. Extraction/Nitrous Oxide-Acetylene Flame Method

1. General Discussion

a. Application: This method is suitable for the determination of aluminum at concentrations less than 900 μg/L and beryllium at concentrations less than 30 μg/L. The method consists of chelation with 8-hydroxyquinoline, extraction with methyl isobutyl ketone (MIBK), and aspiration into a nitrous oxide-acetylene flame.

b. Interferences: Concentrations of Fe greater than 10 mg/L interfere by suppressing Al absorption. Iron interference can be masked by addition of hydroxylamine hydrochloride/1,10-phenanthroline. Mn concentrations up to 80 mg/L do not interfere if turbidity in the extract is allowed to settle. Mg forms an insoluble chelate with 8-hydroxyquinoline at pH 8.0 and tends to remove Al complex as a coprecipitate. However, the Mg complex forms slowly over 4 to 6 min; its interference can be avoided if the solution is extracted immediately after adding buffer.

2. Apparatus

Atomic absorption spectrometer and associated equipment: See Section 3111A.6.

3. Reagents

a. Air: See 3111B.3*a*.

b. Acetylene: See 3111B.3*b*.

c. Ammonium hydroxide, NH_4OH, conc.

d. Buffer: Dissolve 300 g ammonium acetate, $NH_4C_2H_3O_2$, in water, add 105 mL conc NH_4OH, and dilute to 1 L.

e. Metal-free water: See 3111B.2*c*.

f. Hydrochloric acid, HCl, conc.

g. 8-Hydroxyquinoline solution: Dissolve 20 g 8-hydroxyquinoline in about 200 mL water, add 60 mL glacial acetic acid, and dilute to 1 L with water.

h. Methyl isobutyl ketone: See 3111C.3*d*.

i. Nitric acid, HNO_3, conc.

j. Nitrous oxide: See 3111D.3*h*.

k. Standard metal solutions: Prepare a series of standard metal solutions containing 5 to 1000 μg/L by appropriate dilution of the stock metal solutions prepared according to 3111D.3*k*.

l. Iron masking solution: Dissolve 1.3 g hydroxylamine hydrochloride and 6.58 g 1,10-phenanthroline monohydrate in about 500 mL water and dilute to 1 L with water.

4. Procedure

a. Instrument operation: See Sections 3111B.4*b*, C.4*a*, and D.4*b*. After final adjusting of burner position, aspirate MIBK into flame and gradually reduce fuel flow until flame is similar to that before aspiration of solvent. Adjust wavelength setting according to Table 3111:I.

b. Standardization: Select at least three concentrations of standard metal solutions (prepared as in ¶ 3*k*) to bracket the expected metal concentration of a sample and transfer 100 mL of each (and 100 mL water blank) to four different 200-mL volumetric flasks. Add 2 mL 8-hydroxyquinoline solution, 2 mL masking solution (if required), and 10 mL buffer to one flask, immediately add 10 mL MIBK, and shake vigorously. The duration of shaking affects the forms of aluminum complexed. A fast, 10-s shaking time fa-vors monomeric Al, whereas 5 to 10 min of shaking also will complex polymeric species. Adjustment of the 8-hydroxyquinoline to sample ratio can improve recoveries of extremely high or low concentrations of aluminum. Treat each blank, standard, and sample in similar fashion. Continue as in Section 3111C.4*b*.

c. Analysis of samples: Rinse atomizer by aspirating water-saturated MIBK. Aspirate extracts of samples treated as above, and record absorbances.

5. Calculations

Calculate concentration of each metal in micrograms per liter by referring to the appropriate calibration curve prepared according to ¶ 4*b*.

3112 METALS BY COLD-VAPOR ATOMIC ABSORPTION SPECTROMETRY*

3112 A. Introduction

For general introductory material on atomic absorption spectrometric methods, see Section 3111A.

3112 B. Cold-Vapor Atomic Absorption Spectrometric Method

1. General Discussion

This method is applicable to the determination of mercury.

2. Apparatus

When possible, dedicate glassware for use in Hg analysis. Avoid using glassware previously exposed to high levels of Hg, such as those used in COD, TKN, or Cl⁻ analysis.

a. Atomic absorption spectrometer and associated equipment: See Section 3111A.6. Instruments and accessories specifically designed for measurement of mercury by the cold vapor technique are available commercially and may be substituted.

b. Absorption cell, a glass or plastic tube approximately 2.5 cm in diameter. An 11.4-cm-long tube has been found satisfactory but a 15-cm-long tube is preferred. Grind tube ends perpendicular to the longitudinal axis and cement quartz windows in place. Attach gas inlet and outlet ports (6.4 mm diam) 1.3 cm from each end.

c. Cell support: Strap cell to the flat nitrous-oxide burner head or other suitable support and align in light beam to give maximum transmittance.

d. Air pumps: Use any peristaltic pump with electronic speed control capable of delivering 2 L air/min. Any other regulated compressed air system or air cylinder also is satisfactory.

e. Flowmeter, capable of measuring an air flow of 2 L/min.

f. Aeration tubing, a straight glass frit having a coarse porosity for use in reaction flask.

g. Reaction flask, 250-mL erlenmeyer flask or a BOD bottle, fitted with a rubber stopper to hold aeration tube.

h. Drying tube, 150-mm × 18-mm-diam, containing 20 g Mg (ClO₄)₂. A 60-W light bulb with a suitable shade may be substituted to prevent condensation of moisture inside the absorption cell. Position bulb to maintain cell temperature at 10°C above ambient.

i. Connecting tubing, glass tubing to pass mercury vapor from reaction flask to absorption cell and to interconnect all other components. Clear vinyl plastic* tubing may be substituted for glass.

3. Reagents†

a. Metal-free water: See 3111B.3*c*.

b. Stock mercury solution: Dissolve 0.1354 g mercuric chloride, HgCl₂, in about 70 mL water, add 1 mL conc HNO₃, and dilute to 100 mL with water; 1.00 mL = 1.00 mg Hg.

* Approved by Standard Methods Commitee, 1993.

* Tygon or equivalent.
† Use specially prepared reagents low in mercury.

c. Standard mercury solutions: Prepare a series of standard mercury solutions containing 0 to 5 µg/L by appropriate dilution of stock mercury solution with water containing 10 mL conc HNO_3/L. Prepare standards daily.

d. Nitric acid, HNO_3, conc.

e. Potassium permanganate solution: Dissolve 50 g $KMnO_4$ in water and dilute to 1 L.

f. Potassium persulfate solution: Dissolve 50 g $K_2S_2O_8$ in water and dilute to 1 L.

g. Sodium chloride-hydroxylamine sulfate solution: Dissolve 120 g NaCl and 120 g $(NH_2OH)_2 \cdot H_2SO_4$ in water and dilute to 1 L. A 10% hydroxylamine hydrochloride solution may be substituted for the hydroxylamine sulfate.

h. Stannous ion (Sn^{2+}) solution: Use either stannous chloride, ¶ 1), or stannous sulfate, ¶ 2), to prepare this solution containing about 7.0 g Sn^{2+}/100 mL.

1) Dissolve 10 g $SnCl_2$ in water containing 20 mL conc HCl and dilute to 100 mL.

2) Dissolve 11 g $SnSO_4$ in water containing 7 mL conc H_2SO_4 and dilute to 100 mL.

Both solutions decompose with aging. If a suspension forms, stir reagent continuously during use. Reagent volume is sufficient to process about 20 samples; adjust volumes prepared to accommodate number of samples processed.

i. Sulfuric acid, H_2SO_4, conc.

4. Procedure

a. Instrument operation: See Section 3111B.4*b*. Set wavelength to 253.7 nm. Install absorption cell and align in light path to give

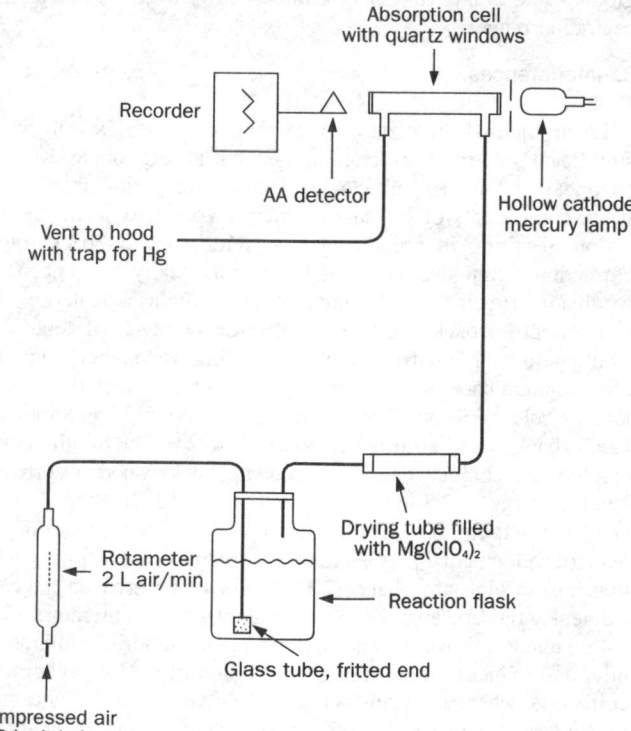

Figure 3112:1. Schematic arrangement of equipment for measurement of mercury by cold-vapor atomic absorption technique.

TABLE 3112:I. INTERLABORATORY PRECISION AND BIAS OF COLD-VAPOR ATOMIC ABSORPTION SPECTROMETRIC METHOD FOR MERCURY[1]

Form	Conc. µg/L	SD µg/L	Relative SD %	Relative Error %	No. of Participants
Inorganic	0.34	0.077	22.6	21.0	23
Inorganic	4.2	0.56	13.3	14.4	21
Organic	4.2	0.36	8.6	8.4	21

maximum transmission. Connect associated equipment to absorption cell with glass or vinyl plastic tubing as indicated in Figure 3112:1. Turn on air and adjust flow rate to 2 L/min. Allow air to flow continuously. Alternatively, follow manufacturer's directions for operation. NOTE: Fluorescent lighting may increase baseline noise.

b. Standardization: Transfer 100 mL of each of the 1.0, 2.0, and 5.0 µg/L Hg standard solutions and a blank of 100 mL water to 250-mL erlenmeyer reaction flasks. Add 5 mL conc H_2SO_4 and 2.5 mL conc HNO_3 to each flask. Add 15 mL $KMnO_4$ solution to each flask and let stand at least 15 min. Add 8 mL $K_2S_2O_8$ solution to each flask and heat for 2 h in a water bath at 95°C. Cool to room temperature.

Treating each flask individually, add enough NaCl-hydroxylamine solution to reduce excess $KMnO_4$, then add 5 mL $SnCl_2$ or $SnSO_4$ solution and immediately attach flask to aeration apparatus. As Hg is volatilized and carried into the absorption cell, absorbance will increase to a maximum within a few seconds. As soon as recorder returns approximately to the base line, remove stopper holding the frit from reaction flask, and replace with a flask containing water. Flush system for a few seconds and run the next standard in the same manner. Construct a standard curve by plotting peak height versus micrograms Hg.

c. Analysis of samples: Transfer 100 mL sample or portion diluted to 100 mL containing not more than 5.0 µg Hg/L to a reaction flask. Treat as in ¶ 4*b*. Seawaters, brines, and effluents high in chlorides require as much as an additional 25 mL $KMnO_4$ solution. During oxidation step, chlorides are converted to free chlorine, which absorbs at 253 nm. Remove all free chlorine before the Hg is reduced and swept into the cell by using an excess (25 mL) of hydroxylamine reagent.

Remove free chlorine by sparging sample gently with air or nitrogen after adding hydroxylamine reducing solution. Use a separate tube and frit to avoid carryover of residual stannous chloride, which could cause reduction and loss of mercury.

5. Calculation

Determine peak height of sample from recorder chart and read mercury value from standard curve prepared according to ¶ 4*b*.

6. Precision and Bias

Data on interlaboratory precision and bias for this method are given in Table 3112:I.

7. Reference

1. KOPP, J.F., M.C. LONGBOTTOM & L.B. LOBRING. 1972. "Cold vapor" method for determining mercury. *J. Amer. Water Works Assoc.* 64:20.

8. Bibliography

HATCH, W.R. & W.L. OTT. 1968. Determination of submicrogram quantities of mercury by atomic absorption spectrophotometry. *Anal. Chem.* 40:2085.

UTHE, J.F., F.A.J. ARMSTRONG & M.P. STAINTON. 1970. Mercury determination in fish samples by wet digestion and flameless atomic absorption spectrophotometry. *J. Fish. Res. Board Can.* 27:805.

FELDMAN, C. 1974. Preservation of dilute mercury solutions. *Anal. Chem.* 46:99.

BOTHNER, M.H. & D.E. ROBERTSON. 1975. Mercury contamination of sea water samples stored in polyethylene containers. *Anal. Chem.* 47: 592.

HAWLEY, J.E. & J.D. INGLE, JR. 1975. Improvements in cold vapor atomic absorption determination of mercury. *Anal. Chem.* 47:719.

LO, J.M. & C.M. WAL. 1975. Mercury loss from water during storage: Mechanisms and prevention. *Anal. Chem.* 47:1869.

EL-AWADY, A.A., R.B. MILLER & M.J. CARTER. 1976. Automated method for the determination of total and inorganic mercury in water and wastewater samples. *Anal. Chem.* 48:110.

ODA, C.E. & J.D. INGLE, JR. 1981. Speciation of mercury by cold vapor atomic absorption spectrometry with selective reduction. *Anal. Chem.* 53:2305.

SUDDENDORF, R.F. 1981. Interference by selenium or tellurium in the determination of mercury by cold vapor generation atomic absorption spectrometry. *Anal. Chem.* 53:2234.

HEIDEN, R.W. & D.A. AIKENS. 1983. Humic acid as a preservative for trace mercury (II) solutions stored in polyolefin containers. *Anal. Chem.* 55:2327.

CHOU, H.N. & C.A. NALEWAY. 1984. Determination of mercury by cold vapor atomic absorption spectrometry. *Anal. Chem.* 56:1737.

3113   METALS BY ELECTROTHERMAL ATOMIC ABSORPTION SPECTROMETRY*

3113 A.   Introduction

1. Applications

Electrothermal atomic absorption permits determination of most metallic elements with sensitivities and detection limits from 20 to 1000 times better than those of conventional flame techniques without extraction or sample concentration. This increase in sensitivity results from an increase in atom density within the furnace as compared to flame atomic absorption. Many elements can be determined at concentrations as low as 1.0 μg/L. An additional advantage of electrothermal atomic absorption is that only a very small volume of sample is required.

The electrothermal technique is used only at concentration levels below the optimum range of direct flame atomic absorption because it is subject to more interferences than the flame procedure and requires increased analysis time. The method of standard additions may be required to insure validity of data. Because of the high sensitivity of this technique, it is extremely susceptible to contamination; extra care in sample handling and analysis may be required.

2. Principle

Electrothermal atomic absorption spectroscopy is based on the same principle as direct flame atomization but an electrically heated atomizer or graphite furnace replaces the standard burner head. A discrete sample volume is dispensed into the graphite sample tube (or cup). Typically, determinations are made by heating the sample in three or more stages. First, a low current heats the tube to dry the sample. The second, or charring, stage destroys organic matter and volatilizes other matrix components at an intermediate temperature. Finally, a high current heats the tube to incandescence and, in an inert atmosphere, atomizes the element being determined. Additional stages frequently are added to aid in drying and charring, and to clean and cool the tube between samples. The resultant ground-state atomic vapor absorbs monochromatic radiation from the source. A photoelectric detector measures the intensity of transmitted radiation. The inverse of the transmittance is related logarithmically to the absorbance, which is directly proportional to the number density of vaporized ground-state atoms (the Beer-Lambert law) over a limited concentration range.

3. Interferences

Electrothermal atomization determinations may be subject to significant interferences from molecular absorption as well as chemical and matrix effects. Molecular absorption may occur when components of the sample matrix volatilize during atomization, resulting in broadband absorption. Several background correction techniques are available commercially to compensate for this interference. A continuum source such as a deuterium arc can correct for background up to absorbance levels of about 0.8. Continuum lamp intensity diminishes at long wavelengths and use of continuum background correction is limited to analytical wavelengths below 350 nm. Zeeman effect background correctors can handle background absorbance up to 1.5 to 2.0. The Smith-Hieftje correction technique can accommodate background absorbance levels as large as 2.5 to 3.0 (see Section 3111A.3). Both Zeeman and Smith-Hieftje background corrections are susceptible to rollover (development of a negative absorbance-concentration relationship) at high absorbances. The rollover absorbance for each element should be available in the manufacturer's literature. Curvature due to rollover should become apparent during calibration; dilution produces a more linear calibration plot. Use background correction when analyzing samples containing high concentrations of acid or dissolved solids and in determining elements for which an absorption line below 350 nm is used.

Matrix modification can be useful in minimizing interference and increasing analytical sensitivity. Determine need for a mod-

---

* Approved by Standard Methods Committee, 1993.

ifier by evaluating recovery of a sample with a known addition. Recovery near 100% indicates that sample matrix does not affect analysis. Chemical modifiers generally modify relative volatilities of matrix and metal. Some modifiers enhance matrix removal, isolating the metal, while other modifiers inhibit metal volatilization, allowing use of higher ashing/charring temperatures and increasing efficiency of matrix removal. Chemical modifiers are added at high concentration (percent level) and can lead to sample contamination from impurities in the modifier solution. Heavy use of chemical modifiers may reduce the useful life (normally 50 to 100 firings) of the graphite tube. Some specific chemical modifiers and approximate concentrations are listed in Table 3113:I.

Addition of a chemical modifier directly to the sample before analysis is restricted to inexpensive additives (e.g. phosphoric acid). Use of palladium salts for matrix modification normally requires methods of co-addition, in which sample and modifier are added consecutively to the furnace either manually or, preferably, with an automatic sampler. Palladium salts (nitrate is preferred, chloride is acceptable) are listed in Table 3113:I as a modifier for many metals. The palladium solution (50 to 2000 mg/L) generally includes citric or ascorbic acid, which aids reduction of palladium in the furnace. Citric acid levels of 1 to 2% are typical. Use of hydrogen (5%) in the coolant gas (available commercially as a mixture) also reduces palladium, eliminating need for organic reducing acids. CAUTION: *Do not mix hydrogen and other gases in the laboratory; hydrogen gas is very flammable—handle with caution.* Use low levels of palladium (50 to 250 mg/L) for normal samples and higher levels for complex samples. Addition of excess palladium modifier may widen atomization peaks; in such cases peak area measurements may provide higher quality results. The recommended mode of modifier use is through co-addition to the furnace of about 10 µL of the palladium (or other) modifier solution. Palladium may not be the best modifier in all cases and cannot be recommended unconditionally. Test samples requiring a modifier first with palladium; test other modifiers only if palladium is unsuccessful or to minimize modifier cost. See Section 3113B.3 for preparation of modifier solution.

Temperature ramping, i.e., gradual heating, can be used to decrease background interferences and permits analysis of samples with complex matrices. Ramping permits a controlled, continuous increase of furnace temperature in any of the various steps of the temperature sequence. Ramp drying is used for samples containing mixtures of solvents or for samples with a high salt content (to avoid spattering). If spattering is suspected, develop drying

ramp by visual inspection of the drying stage, using a mirror. Samples that contain a complex mixture of matrix components sometimes require ramp charring to effect controlled, complete thermal decomposition. Ramp atomization may minimize background absorption by permitting volatilization of the element being determined before the matrix. This is especially applicable in the determination of such volatile elements as cadmium and lead. Use of time-resolved absorbance profiles (available on most modern instruments) greatly aids method development. Changes in atomization, notably the element peak appearance time and magnitude of background and metal absorbances, can be monitored directly.

Improve analysis by using a graphite platform, inserted into the graphite tube, as the atomization site. The platform is not heated as directly by the current flowing through the graphite tube; thus the metal atomizes later and under more uniform conditions.

Use standard additions to compensate for matrix interferences. When making standard additions, determine whether the added metal and that in the sample behave similarly under the specified conditions. [See Section 3113B.4*d*2)]. In the extreme, test every sample for recovery (85 to 115% recovery desired) to determine if standard addition is needed. Test every sample type for recovery. Recovery of only 40 to 85% generally indicates that standard addition is required. Often, as long as the samples are from sources of consistent properties, a representative recovery can be used to characterize the analysis and determine the necessity of standard addition. Test samples of unknown origin or of complex composition (digestates, for example) individually for metal recovery. Ideally, chemical modifiers and graphite platforms render the sample fit to be analyzed using a standard analytical calibration curve. Always verify this assumption; however, a properly developed method with judicious use of chemical modifiers should eliminate the necessity for standard addition in all but the most extreme samples.

Chemical interaction of the graphite tube with various elements to form refractory carbides occurs at high charring and atomization temperatures. Elements that form carbides are barium, molybdenum, nickel, titanium, vanadium, and silicon. Carbide formation is characterized by broad, tailing atomization peaks and reduced sensitivity. Using pyrolytically coated tubes for these metals minimizes the problem.

4. Sensitivity, Detection Limits, and Optimum Concentration Range

Estimated detection limits and optimum concentration ranges are listed in Table 3113:II. These values may vary with the chemical form of the element being determined, sample composition, or instrumental conditions.

For a given sample, increased sensitivity may be achieved by using a larger sample volume or by reducing flow rate of the purge gas or by using gas interrupt during atomization. Note, however, that these techniques also will increase the effects of any interferences present. Sensitivity can be decreased by diluting the sample, reducing sample volume, increasing purge-gas flow, or using a less sensitive wavelength. Use of argon, rather than nitrogen, as the purge gas generally improves sensitivity and reproducibility. Hydrogen mixed with the inert gas may suppress chemical interference and increase sensitivity by acting as a reducing agent, thereby aiding in producing more ground-state atoms. Pyrolytically coated graphite tubes can increase sensitivity

TABLE 3113:I. POTENTIAL MATRIX MODIFIERS FOR ELECTROTHERMAL ATOMIC ABSORPTION SPECTROMETRY*

Modifier	Analyses for Which Modifier May Be Useful
1500 mg Pd/L + 1000 mg Mg(NO$_3$)$_2$/L[1]	Ag, As, Au, Bi, Cu, Ge, Mn, Hg, In, Sb, Se, Sn, Te, Tl
500–2000 mg Pd/L + reducing agent[2]†	Ag, As, Bi, Cd, Co, Cr, Cu, Fe, Hg, Mn, Ni, Pb, Sb
5000 mg Mg(NO$_3$)$_2$/L[1]	Be, Co, Cr, Fe, Mn, V
100–500 mg Pd/L[2]	As, Ga, Ge, Sn
50 mg Ni/L[2]	As, Se, Sb
2% PO$_4^{3-}$ + 1000 mg Mg (NO$_3$)$_2$/L[1]	Cd, Pb

*Assumes 10 µL modifier/10 µL sample.
†Citric acid (1–2%) preferred; ascorbic acid or H$_2$ acceptable.

TABLE 3113:II. DETECTION LEVELS AND CONCENTRATION RANGES FOR ELECTROTHERMAL ATOMIZATION ATOMIC ABSORPTION SPECTROMETRY

Element	Wavelength nm	Estimated Detection Limit μg/L	Optimum Concentration Range μg/L
Al	309.3	3	20–200
Sb	217.6	3	20–300
As	193.7	1	5–100
Ba	553.6	2	10–200
Be	234.9	0.2	1–30
Cd	228.8	0.1	0.5–10
Cr	357.9	2	5–100
Co	240.7	1	5–100
Cu	324.7	1	5–100
Fe	248.3	1	5–100
Pb*	283.3	1	5–100
Mn	279.5	0.2	1–30
Mo	313.3	1	3–60
Ni	232.0	1	5–100
Se	196.0	2	5–100
Ag	328.1	0.2	1–25
Sn	224.6	5	20–300

*The more sensitive 217.0-nm wavelength is recommended for instruments with background correction capabilities.

for the more refractory elements and are recommended. The optical pyrometer/maximum power accessory available on some instruments also offers increased sensitivity with lower atomization temperatures for many elements.

Using the Stabilized Temperature Platform Furnace (STPF) technique, which is a combination of individual techniques, also offers significant interference reduction with improved sensitivity. Sensitivity changes with sample tube age. Discard graphite tubes when significant variations in sensitivity or poor reproducibility are observed. The use of high acid concentrations, brine samples, and matrix modifiers often drastically reduces tube life. Preferably use the graphite platform in such situations.

5. References

1. PERKIN-ELMER CORP. 1991. Summary of Standard Conditions for Graphite Furnace. Perkin-Elmer Corp., Norwalk, Conn.
2. ROTHERY, E., ed. 1988. Analytical Methods for Graphite Tube Atomizers. Varian Techtron Pty, Ltd., Mulgrave, Victoria, Australia.

6. Bibliography

FERNANDEZ, F.J. & D.C. MANNING. 1971. Atomic absorption analyses of metal pollutants in water using a heated graphite atomizer. Atomic Absorption Newsletter 10:65.
SEGAR, D.A. & J.G. GONZALEZ. 1972. Evaluation of atomic absorption with a heated graphite atomizer for the direct determination of trace transition metals in sea water. Anal. Chim. Acta 58:7.
BARNARD, W.M. & M.J. FISHMAN. 1973. Evaluation of the use of heated graphite atomizer for the routine determination of trace metals in water. Atomic Absorption Newsletter 12:118.
KAHN, H.L. 1973. The detection of metallic elements in wastes and waters with the graphite furnace. Int. J. Environ. Anal. Chem. 3:121.
WALSH, P.R., J.L. FASCHING & R.A. DUCE. 1976. Matrix effects and their control during the flameless atomic absorption determination of arsenic. Anal. Chem. 48:1014.
HENN, E.L. 1977. Use of Molybdenum in Eliminating Matrix Interferences in Flameless Atomic Absorption. Spec. Tech. Publ. 618, American Soc. Testing & Materials, Philadelphia, Pa.
MARTIN, T.D. & J.F. KOPP. 1978. Methods for Metals in Drinking Water. U.S. Environmental Protection Agency, Environmental Monitoring and Support Lab., Cincinnati, Ohio.
HYDES, D.J. 1980. Reduction of matrix effects with a soluble organic acid in the carbon furnace atomic absorption spectrometric determination of cobalt, copper, and manganese in seawater. Anal. Chem. 52:289.
SOTERA, J.J. & H.L. KAHN. 1982. Background correction in AAS. Amer. Lab. 14:100.
SMITH, S.B. & G.M. HIEFTJE. 1983. A new background-correction method for atomic absorption spectrometry. Appl. Spectrosc. 37:419.
GROSSER, Z. 1985. Techniques in Graphite Furnace Atomic Absorption Spectrophotometry. Perkin-Elmer Corp., Ridgefield, Conn.
SLAVIN, W. & G.R. CARNICK. 1985. A survey of applications of the stabilized temperature platform furnace and Zeeman correction. Atomic Spectrosc. 6:157.
BRUEGGEMEYER, T. & F. FRICKE. 1986. Comparison of furnace & atomization behavior of aluminum from standard & thorium-treated L'vov platforms. Anal. Chem. 58:1143.

3113 B. Electrothermal Atomic Absorption Spectrometric Method

1. General Discussion

This method is suitable for determination of micro quantities of aluminum, antimony, arsenic, barium, beryllium, cadmium, chromium, cobalt, copper, iron, lead, manganese, molybdenum, nickel, selenium, silver, and tin. It is also applicable to analysis of bismuth, gallium, germanium, gold, indium, mercury, tellurium, thallium, and vanadium, but precision and accuracy data are not yet available.

2. Apparatus

a. Atomic absorption spectrometer: See Section 3111A.6a. The instrument must have background correction capability.

b. Source lamps: See Section 3111A.6d.

c. Graphite furnace: Use an electrically heated device with electronic control circuitry designed to carry a graphite tube or cup through a heating program that provides sufficient thermal energy to atomize the elements of interest. Furnace heat controllers with only three heating steps are adequate only for fresh waters with low dissolved solids content. For salt waters, brines, and other complex matrices, use a furnace controller with up to seven individually programmed heating steps. Fit the furnace into the sample compartment of the spectrometer in place of the conventional burner assembly. Use argon as a purge gas to minimize oxidation of the furnace tube and to prevent the formation of metallic oxides. Use graphite tubes with platforms to minimize interferences and to improve sensitivity.

d. Readout: See Section 3111A.6c.

e. Sample dispensers: Use microliter pipets (5 to 100 µL) or an automatic sampling device designed for the specific instrument.

f. Vent: See Section 3111A.6f.

g. Cooling water supply: Cool with tap water flowing at 1 to 4 L/min or use a recirculating cooling device.

h. Membrane filter apparatus: Use an all-glass filtering device and 0.45-µm or smaller-pore-diameter membrane filters. For trace analysis of aluminum, use polypropylene or TFE devices.

3. Reagents

a. Metal-free water: See Section 3111B.3c.

b. Hydrochloric acid, HCl, 1 + 1 and conc.

c. Nitric acid, HNO_3, 1 + 1 and conc.

d. Matrix modifier stock solutions:

1) *Magnesium nitrate,* 10 000 mg Mg/L: Dissolve 10.5 g $Mg(NO_3)_2 \cdot 6H_2O$ in water. Dilute to 100 mL.

2) *Nickel nitrate,* 10 000 mg Ni/L: Dissolve 4.96 g $Ni(NO_3)_2 \cdot 6H_2O$ in water. Dilute to 100 mL.

3) *Phosphoric acid,* 10% (v/v): Add 10 mL conc H_3PO_4 to water. Dilute to 100 mL.

4) *Palladium nitrate,* 4000 mg Pd/L: Dissolve 8.89 g $Pd(NO_3)_2 \cdot H_2O$ in water. Dilute to 1 L.

5) *Citric acid,* 4%: Dissolve 40 g citric acid in water. Dilute to 1 L.

NOTE: All of the modifier solutions recommended in Table 3113:I can be prepared with volumetric combination of the above solutions and water. For preparation of other matrix modifiers, see references or follow manufacturers' instructions.

e. Stock metal solutions: Refer to Sections 3111B and 3114.

f. Chelating resin: 100 to 200 mesh* purified by heating at 60°C in 10N NaOH for 24 h. Cool resin and rinse 10 times each with alternating portions of 1N HCl, metal-free water, 1N NaOH, and metal-free water.

g. Metal-free seawater (or brine): Fill a 1.4-cm-ID × 20-cm-long borosilicate glass column to within 2 cm of the top with purified chelating resin. Elute resin with successive 50-mL portions of 1N HCl, metal-free water, 1N NaOH, and metal-free water at the rate of 5 mL/min just before use. Pass salt water or brine through the column at a rate of 5 mL/min to extract trace metals present. Discard the first 10 bed volumes (300 mL) of eluate.

4. Procedures

a. Sample pretreatment: Before analysis, pretreat all samples as indicated below. Rinse all glassware with 1 + 1 HNO_3 and water. Carry out digestion procedures in a clean, dust-free labo-

ratory area to avoid sample contamination. For digestion of trace aluminum, use polypropylene or TFE utensils to avoid leachable aluminum from glassware.

1) Dissolved metals—See Section 3030B. For samples requiring arsenic and/or selenium analysis add 3 mL 30% hydrogen peroxide/100 mL sample and an appropriate volume of nickel nitrate solution (see Table 3113:I) before analysis. For all other metals no further pretreatment is required except for adding an optional matrix modifier.

2) Total recoverable metals (Al, Sb, Ba, Be, Cd, Cr, Co, Cu, Fe, Pb, Mn, Mo, Ni, Ag, and Sn)—NOTE: Sb and Sn are not recovered unless HCl is used in the digestion. See Section 3030D. Quantitatively transfer digested sample to a 100-mL volumetric flask, add an appropriate amount of matrix modifier (see Table 3113:I), and dilute to volume with water.

3) Total recoverable metals (As, Se)—Transfer 100 mL of shaken sample, 1 mL conc HNO_3, and 2 mL 30% H_2O_2 to a clean, acid-washed 250-mL beaker. Heat on a hot plate without allowing solution to boil until volume has been reduced to about 50 mL. Remove from hot plate and let cool to room temperature. Add an appropriate concentration of nickel (see Table 3113:I), and dilute to volume in a 100-mL volumetric flask with water. Substitution of palladium is uneconomical. Nickel may be deleted if palladium is co-added during analysis. Simultaneously prepare a digested blank by substituting water for sample and proceed with digestion as described above.

b. Instrument operation: Mount and align furnace device according to manufacturer's instructions. Turn on instrument and data collection system. Select appropriate light source and adjust to recommended electrical setting. Select proper wavelength and set all conditions according to manufacturer's instructions, including background correction. Background correction is important when elements are determined at short wavelengths or when sample has a high level of dissolved solids. Background correction normally is not necessary at wavelengths longer than 350 nm. If background correction above 350 nm is needed deuterium arc background correction is not useful and other types must be used.

Select proper inert- or sheath-gas flow. In some cases, it is desirable to interrupt the inert-gas flow during atomization. Such interruption results in increased sensitivity by increasing residence time of the atomic vapor in the optical path. Gas interruption also increases background absorption and intensifies interference effects, but modern background correction methods usually eliminate these problems. Consider advantages and disadvantages of this option for each matrix when optimizing analytical conditions.

To optimize graphite furnace conditions, carefully adjust furnace temperature settings to maximize sensitivity and precision and to minimize interferences. Follow manufacturer's instructions.

Use drying temperatures slightly above the solvent boiling point and provide enough time and temperature for complete evaporation without boiling or spattering.

Select atomization temperature by determining the lowest temperature providing maximum sensitivity without significantly eroding precision. Optimize by a series of successive determinations at various atomization temperatures using a standard solution giving an absorbance of 0.2 to 0.5.

The charring temperature must be high enough to maximize volatilization of interfering matrix components yet too low to volatilize the element of interest. With the drying and atomization

---

* Chelex 100, or equivalent, available from Bio-Rad Laboratories, Richmond, CA.

temperatures set to their optimum values, analyze a standard solution at a series of charring temperatures in increasing increments of 50 to 100°C. When the optimum charring temperature is exceeded, there will be a significant drop in sensitivity. Plot charring temperature versus sample absorbance: the optimum charring temperature is the highest temperature without reduced sensitivity. Verify optimization with major changes in sample matrix.

c. *Instrument calibration:* Prepare standard solutions for instrument calibration by diluting metal stock solutions. Prepare standard solutions fresh daily.

Prepare a blank and at least three calibration standards in the appropriate concentration range (see Table 3113:II) for correlating element concentration and instrument response. Match the matrix of the standard solutions to those of the samples as closely as possible. In most cases, this simply requires matching the acid

TABLE 3113:III. INTERLABORATORY SINGLE-ANALYST PRECISION DATA FOR ELECTROTHERMAL ATOMIZATION METHODS[1]

		Single-Analyst Precision % RSD					
Element	Concentration μg/L	Lab Pure Water	Drinking Water	Surface Water	Effluent 1	Effluent 2	Effluent 3
Al	28	66	108	70	—	—	66
	125	27	35	24	—	—	34
	11 000	11	—	—	22	—	—
	58 300	27	—	—	19	—	—
	460	9	—	—	—	30	—
	2 180	28	—	—	—	4	—
	10.5	20	13	13	13	56	18
	230	10	18	13	21	94	14
As	9.78	40	25	15	74	23	11
	227	10	6	8	11	15	6
Ba	56.5	36	21	29	59	23	27
	418	14	12	20	24	24	18
Be	0.45	18	27	15	30	2	11
	10.9	14	4	9	7	12	12
Cd	0.43	72	49	1	121	35	27
	12	11	17	22	14	11	15
Cr	9.87	24	33	10	23	15	10
	236	16	7	11	13	16	7
Co	29.7	10	17	10	19	24	12
	420	8	11	13	14	9	5
Cu	10.1	49	47	17	17	—	30
	234	8	15	6	21	—	11
	300	6	—	—	—	11	—
	1 670	11	—	—	—	6	—
Fe	26.1	144	52	153	—	—	124
	455	48	37	45	—	—	31
	1 030	17	—	—	30	—	—
	5 590	6	—	—	32	—	—
	370	14	—	—	—	19	—
	2 610	9	—	—	—	18	—
Pb	10.4	6	19	17	21	19	33
	243	17	7	17	18	12	16
Mn	0.44	187	180	—	—	—	275
	14.8	32	19	—	—	—	18
	91.0	15	—	—	48	—	—
	484.0	4	—	—	12	—	—
	111.0	12	—	—	—	21	—
	666.0	6	—	—	—	20	—
Ni	26.2	20	26	25	24	18	9
	461.0	15	11	9	8	11	4
Se	10.0	12	27	16	35	41	13
	235.0	6	6	15	6	13	14
Ag	8.48	10	—	—	15	27	16
	56.5	14	—	—	7	16	23
	0.45	27	166	48	—	—	—
	13.6	15	4	10	—	—	—

background of the samples. For seawaters or brines, however, use the metal-free matrix (¶ 3g) as the standard solution diluent. In addition, add the same concentration of matrix modifier (if required for sample analysis) to the standard solutions.

Inject a suitable portion of each standard solution, in order of increasing concentration. Analyze each standard solution in triplicate to verify method precision.

Construct an analytical curve by plotting the average peak absorbances or peak areas of the standard solution versus concentration on linear graph paper. Alternatively, use electronic instrument calibration if the instrument has this capability.

d. Sample analysis: Analyze all samples except those demonstrated to be free of matrix interferences (based on recoveries of 85% to 115% for known additions) using the method of standard additions. Analyze all samples at least in duplicate or until reproducible results are obtained. A variation of $\leq 10\%$ is considered acceptable reproducibility. Average replicate values.

TABLE 3113:IV. INTERLABORATORY OVERALL PRECISION DATA FOR ELECTROTHERMAL ATOMIZATION METHODS[1]

Element	Concentration μg/L	Overall Precision % RSD					
		Lab Pure Water	Drinking Water	Surface Water	Effluent 1	Effluent 2	Effluent 3
Al	28	99	114	124	—	—	131
	125	45	47	49	—	—	40
	11 000	19	—	—	43	—	—
	58 300	31	—	—	32	—	—
	460	20	—	—	—	47	—
	2 180	30	—	—	—	15	—
	10.5	37	19	22	50	103	39
	230	26	16	16	17	180	21
As	9.78	43	26	37	72	50	39
	227	18	12	13	20	15	14
Ba	56.5	68	38	43	116	43	65
	418	35	35	28	38	48	16
Be	0.45	28	31	15	67	50	35
	10.9	33	15	26	20	9	19
Cd	0.43	73	60	5	88	43	65
	12	19	25	41	26	20	27
Cr	9.87	30	53	24	60	41	23
	236	18	14	24	20	14	20
Co	29.7	13	26	17	18	21	17
	420	21	21	17	18	13	13
Cu	10.1	58	82	31	32	—	74
	234	12	33	19	21	—	26
	300	13	—	—	—	14	—
	1 670	12	—	—	—	13	—
Fe	26.1	115	93	306	—	—	204
	455	53	46	53	—	—	44
	1 030	32	—	—	25	—	—
	5 590	10	—	—	43	—	—
	370	28	—	—	—	22	—
	2 610	13	—	—	—	22	—
Pb	10.4	27	42	31	23	28	47
	243	18	19	17	19	19	25
Mn	0.44	299	272	—	—	—	248
	14.8	52	41	—	—	—	29
	91.0	16	—	—	45	—	—
	484.0	5	—	—	17	—	—
	111.0	15	—	—	—	17	—
	666.0	8	—	—	—	24	—
Ni	26.2	35	30	49	35	37	43
	461.0	23	22	15	12	21	17
Se	10.0	17	48	32	30	44	51
	235.0	16	18	18	17	22	34
Ag	8.48	23	—	—	16	35	34
	56.5	15	—	—	24	32	28
	0.45	57	90	368	—	—	—
	13.6	19	19	59	—	—	—

1) Direct determination—Inject a measured portion of pre-treated sample into the graphite furnace. Use the same volume as was used to prepare the calibration curve. Usually add modifier immediately after the sample, preferably using an automatic sampler or a micropipet. Some methods require modifier to be injected before the sample. Use the same volume and concentration of modifier for all standards and samples. Dry, char, and atomize according to the preset program. Repeat until reproducible results are obtained.

Compare the average absorbance value or peak area to the calibration curve to determine concentration of the element of interest. Alternatively, read results directly if the instrument is equipped with this capability. If absorbance (or concentration) or peak area of the sample is greater than absorbance (concentration) or peak area of the most concentrated standard solution, dilute sample and reanalyze. If very large dilutions are required, another technique (e.g., flame AA or ICP) may be more suitable for this sample. Large dilution factors magnify small errors on final cal-

TABLE 3113:V. INTERLABORATORY RELATIVE ERROR DATA FOR ELECTROTHERMAL ATOMIZATION METHODS[1]

Element	Concentration μg/L	Relative Error %					
		Lab Pure Water	Drinking Water	Surface Water	Effluent 1	Effluent 2	Effluent 3
Al	28.0	86	150	54	—	—	126
	125.0	4	41	39	—	—	30
	11 000.0	2	—	—	14	—	—
	58 300.0	12	—	—	7	—	—
	460.0	2	—	—	—	11	—
	2 180.0	11	—	—	—	9	—
Sb	10.5	30	32	28	24	28	36
	230.0	35	14	19	13	73	39
As	9.78	36	1	22	106	13	16
	227.0	3	7	10	19	6	13
Ba	56.5	132	54	44	116	59	40
	418.0	4	0	0	13	6	60
Be	0.45	40	16	11	16	10	15
	10.9	13	2	9	7	8	8
Cd	0.43	58	45	37	66	16	19
	12.0	4	6	5	22	18	3
Cr	9.87	10	9	4	2	5	15
	236.0	11	0	9	13	5	8
Co	29.7	7	7	1	6	3	13
	420.0	12	8	8	11	5	18
Cu	10.1	16	48	2	5	—	15
	234.0	8	7	0	4	—	19
	300.0	4	—	—	—	21	—
	1 670.0	6	—	—	—	2	—
Fe	26.1	85	60	379	—	—	158
	455.0	43	22	31	—	—	18
	1 030.0	8	—	—	8	—	—
	5 590.0	2	—	—	12	—	—
	370.0	4	—	—	—	11	—
	2 610.0	35	—	—	—	2	—
Pb	10.4	16	10	17	1	34	14
	243.0	5	15	8	18	15	29
Mn	0.44	332	304	—	—	—	556
	14.8	10	1	—	—	—	36
	91.0	31	—	—	10	—	—
	484.0	42	—	—	4	—	—
	111.0	1	—	—	—	29	—
	666.0	6	—	—	—	23	—
Ni	26.2	9	16	10	7	33	54
	461.0	15	19	18	31	16	18
Se	10.0	12	9	6	36	17	37
	235.0	7	7	0	13	10	17
Ag	8.48	12	—	—	1	51	20
	56.5	16	—	—	8	51	22
	0.45	34	162	534	—	—	—
	13.6	3	12	5	—	—	—

culation. Keep acid background and concentration of matrix modifier (if present in the solutions) constant. Dilute the sample in a blank solution of acid and matrix modifiers.

Proceed to ¶ 5a below.

2) Method of standard additions—Refer to ¶ 4c above. The method of standard additions is valid only when it falls in the linear portion of the calibration curve. Once instrument sensitivity has been optimized for the element of interest and the linear range for the element has been established, proceed with sample analyses.

Inject a measured volume of sample into furnace device. Dry, char or ash, and atomize samples according to preset program. Repeat until reproducible results are obtained. Record instrument response in absorbance or concentration as appropriate. Add a known concentration of the element of interest to a separate portion of sample so as not to change significantly the sample volume. Repeat the determination.

Add a known concentration (preferably twice that used in the first addition) to a separate sample portion. Mix well and repeat the determination.

Using linear graph paper, plot average absorbance or instrument response for the sample and the additions on the vertical axis against the concentrations of the added element on the horizontal axis, using zero as the concentration for the sample. Draw a straight line connecting the three points and extrapolate to zero absorbance. The intercept at the horizontal axis is the negative of the element concentration in the sample. The concentration axis to the left of the origin should be a mirror image of the axis to the right.

5. Calculations

a. Direct determination:

$$\mu g \text{ metal/L} = C \times F$$

where:

C = metal concentration as read directly from the instrument or from the calibration curve, μg/L, and
F = dilution factor.

b. Method of additions:

$$\mu g \text{ metal/L} = C \times F$$

where:

C = metal concentration as read from the method of additions plot, μg/L, and
F = dilution factor.

6. Precision and Bias

Data typical of the precision and bias obtainable are presented in Tables 3113:III, IV, and V.

7. Quality Control

See Section 3020 for specific quality control procedures to be followed during analysis. Although previous indications were that very low optimum concentration ranges were attainable for most metals (see Table 3113:II), data in Table 3113:III using variations of these protocols show that this may not be so. Exercise extreme care when applying this method to the lower concentration ranges. Verify analyst precision at the beginning of each analytical run by making triplicate analyses. Verify autosampler precision by checking volumes (by weight) delivered by the autosampler at routinely used injection volume settings.

8. Reference

1. COPELAND, T.R. & J.P. MANEY. 1986. EPA Method Study 31: Trace Metals by Atomic Absorption (Furnace Techniques). EPA-600/S4-85-070, U.S. Environmental Protection Agency, Environmental Monitoring and Support Lab., Cincinnati, Ohio.

9. Bibliography

RENSHAW, G.D. 1973. The determination of barium by flameless atomic absorption spectrophotometry using a modified graphite tube atomizer. *Atomic Absorption Newsletter* 12:158.

YANAGISAWA, M., T. TAKEUCHI & M. SUZUKI. 1973. Flameless atomic absorption spectrometry of antimony. *Anal. Chim. Acta* 64:381.

RATTONETTI, A. 1974. Determination of soluble cadmium, lead, silver and indium in rainwater and stream water with the use of flameless atomic absorption. *Anal. Chem.* 46:739.

HENN, E.L. 1975. Determination of selenium in water and industrial effluents by flameless atomic absorption. *Anal. Chem.* 47:428.

MARTIN, T.D. & J.F. KOPP. 1975. Determining selenium in water, wastewater, sediment and sludge by flameless atomic absorption spectrometry. *Atomic Absorption Newsletter* 14:109.

MARUTA, T., K. MINEGISHI & G. SUDOH. 1976. The flameless atomic absorption spectrometric determination of aluminum with a carbon atomization system. *Anal. Chim. Acta* 81:313.

CRANSTON, R.E. & J.W. MURRAY. 1978. The determination of chromium species in natural waters. *Anal. Chim. Acta* 99:275.

HOFFMEISTER, W. 1978. Determination of iron in ultrapure water by atomic absorption spectroscopy. *Z. Anal. Chem.* 50:289.

LAGAS, P. 1978. Determination of beryllium, barium, vanadium and some other elements in water by atomic absorption spectrometry with electrothermal atomization. *Anal. Chim. Acta* 98:261.

CARRONDO, M.J.T., J.N. LESTER & R. PERRY. 1979. Electrothermal atomic absorption determination of total aluminum in waters and waste waters. *Anal. Chim. Acta* 111:291.

NAKAHARA, T. & C.L. CHAKRABARTI. 1979. Direct determination of traces of molybdenum in synthetic sea water by atomic absorption spectrometry with electrothermal atomization and selective volatilization of the salt matrix. *Anal. Chim. Acta* 104:99.

TIMINAGA, M. & Y. UMEZAKI. 1979. Determination of submicrogram amounts of tin by atomic absorption spectrometry with electrothermal atomization. *Anal. Chim. Acta* 110:55.

3114 ARSENIC AND SELENIUM BY HYDRIDE GENERATION/ATOMIC ABSORPTION SPECTROMETRY*

3114 A. Introduction

For general introductory material on atomic absorption spectrometric methods, see Section 3111A.

Two methods are presented in this section: A manual method and a continuous-flow method especially recommended for sele-

_____
* Approved by Standard Methods Committee, 1997.

nium. Continuous-flow automated systems are preferable to manual hydride generators because the effect of sudden hydrogen generation on light-path transparency is removed and any blank response from contamination of the HCl reagent by the elements being determined is incorporated into the background base line.

3114 B. Manual Hydride Generation/Atomic Absorption Spectrometric Method

1. General Discussion

a. Principle: This method is applicable to the determination of arsenic and selenium by conversion to their hydrides by sodium borohydride reagent and transport into an atomic absorption atomizer.

Arsenous acid and selenous acid, the As(III) and Se(IV) oxidation states of arsenic and selenium, respectively, are instantaneously converted by sodium borohydride reagent in acid solution to their volatile hydrides. The hydrides are purged continuously by argon or nitrogen into a quartz cell heated electrically or by the flame of an atomic absorption spectrometer and converted to the gas-phase atoms. The sodium borohydride reducing agent, by rapid generation of the elemental hydrides in an appropriate reaction cell, minimizes dilution of the hydrides by the carrier gas and provides rapid, sensitive determinations of arsenic and selenium.

CAUTION: *Arsenic and selenium and their hydrides are toxic. Handle with care.*

At room temperature and solution pH values of 1 or less, arsenic acid, the As(V) oxidation state of arsenic, is reduced relatively slowly by sodium borohydride to As(III), which is then instantaneously converted to arsine. The arsine atomic absorption peaks commonly are decreased by one-fourth to one-third for As(V) when compared to As(III). Determination of total arsenic requires that all inorganic arsenic compounds be in the As(III) state. Organic and inorganic forms of arsenic are first oxidized to As(V) by acid digestion. The As(V) then is quantitatively reduced to As(III) with sodium or potassium iodide before reaction with sodium borohydride.

Selenic acid, the Se(VI) oxidation state of selenium, is not measurably reduced by sodium borohydride. To determine total selenium by atomic absorption and sodium borohydride, first reduce Se(VI) formed during the acid digestion procedure to Se(IV), being careful to prevent reoxidation by chlorine. Efficiency of reduction depends on temperature, reduction time, and HCl concentration. For 4N HCl, heat 1 h at 100°C. For 6N HCl, boiling for 10 min is sufficient.[1-3] Alternatively, autoclave samples in sealed containers at 121°C for 1 h. NOTE: Autoclaving in sealed containers may result in incomplete reduction, apparently

due to the buildup of chlorine gas. To obtain equal instrument responses for reduced Se(VI) and Se (IV) solutions of equal concentrations, manipulate HCl concentration and heating time. For further details, see Section 3500-Se.

b. Equipment selection: Certain atomic absorption atomizers and hydride reaction cells are available commercially for use with the sodium borohydride reagent. A functional manual system that can be constructed in the laboratory is presented in Figure 3114: 1. Irrespective of the hydride reaction cell-atomizer system selected, it must meet the following quality-control considerations: (*a*) it must provide a precise and reproducible standard curve between 0 and 20 μg As or Se/L and an instrumental detection limit between 0.1 and 0.5 μg As or Se/L; (*b*) when carried through the entire procedure, oxidation state couples [As (III) - As (V) or Se (IV) - Se (VI)] must cause equal instrument response; and (*c*) sample digestion must yield 80% or greater recovery of added cacodylic acid (dimethyl arsinic acid) and 90% or greater recovery of added As(III), As(V), Se(VI), or Se(IV).

Quartz atomization cells provide for the most sensitive arsenic and selenium hydride determinations. The quartz cell can be heated electrically or by an air-acetylene flame in an atomic absorption unit.

c. Digestion techniques: Waters and wastewaters may contain varying amounts of organic arsenic compounds and inorganic compounds of As(III), As(V), Se(IV), and Se(VI). To measure total arsenic and selenium in these samples requires sample digestion to solubilize particulate forms and oxidize reduced forms of arsenic and selenium and to convert any organic compounds to inorganic ones. Organic selenium compounds rarely have been demonstrated in water. It is left to the experienced analyst's judgment whether sample digestion is required.

Various alternative digestion procedures are provided in ¶s 4*c* and *d* below. Consider sulfuric-nitric-perchloric acid digestion (¶ 4*c*) or sulfuric-nitric acid digestion (3030F) as providing a measure of total recoverable arsenic rather than total arsenic because they do not completely convert certain organic arsenic compounds to As(V). The sulfuric-nitric-perchloric acid digestion effectively destroys organics and most particulates in untreated wastewaters or solid samples, but does not convert all organic arsenicals to As(V). The potassium persulfate digestion (¶ 4*d*) is

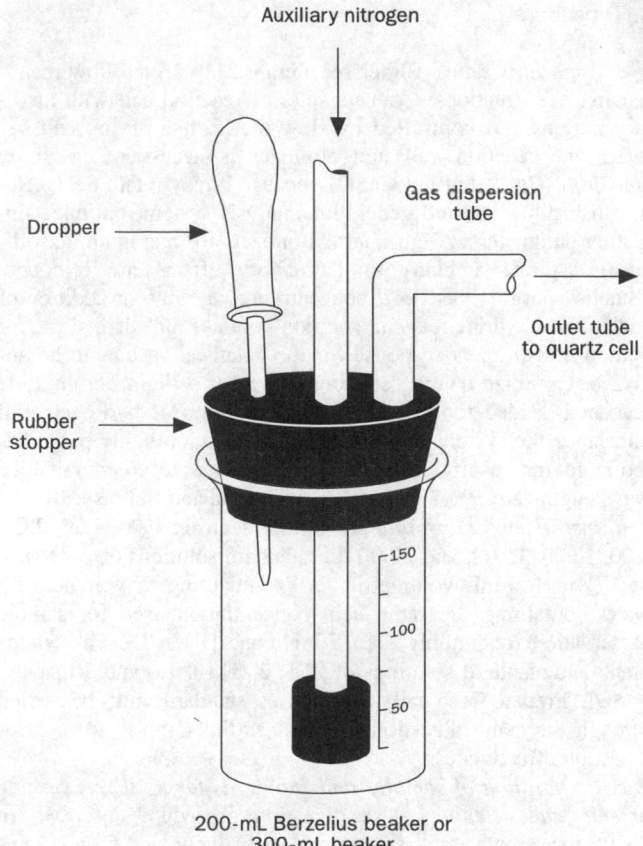

Auxiliary nitrogen

Dropper

Gas dispersion
tube

Outlet tube
to quartz cell

Rubber
stopper

150

100

50

200-mL Berzelius beaker or
300-mL beaker

Figure 3114:1. Manual reaction cell for producing As and Se hydrides.

effective for converting organic arsenic and selenium compounds to As(V) and Se(VI) in potable and surface waters and in most wastewaters.[4]

The HCl-autoclave reduction of Se(VI) described above and in ¶ 4f is an effective digestion procedure for total inorganic selenium; however, it has not been found effective for converting benzene substituted selenium compounds to inorganic selenium. In all cases, verify the effectiveness of digestion methods by carrying samples with known additions of organic As or Se(IV) through the entire procedure.

d. Interferences: Interferences are minimized because the As and Se hydrides are removed from the solution containing most potential interfering substances. Slight response variations occur when acid matrices are varied. Control these variations by treating standards and samples in the same manner. Low concentrations of noble metals (approximately 100 μg/L of Ag, Au, Pt, Pd, etc.), concentrations of copper, lead, and nickel at or greater than 1 mg/L, and concentrations between 0.1 and 1 mg/L of hydride-forming elements (Bi, Sb, Sn, and Te) may suppress the response of As and Se hydrides. Interference by transition metals depends strongly on HCl concentration. Interferences are less pronounced at 4 to 6N HCl than at lower concentrations.[5] The presence of As or Se in each other's matrices can cause similar suppression. Reduced nitrogen oxides resulting from HNO_3 digestion and nitrite also can suppress instrumental response for both elements. Large

concentrations of iodide interfere with the Se determination by reducing Se to its elemental form. Do not use any glassware for determining Se that has been used for iodide reduction of As(V).

To prevent chlorine gas produced in the reduction of Se(VI) to Se(IV) from reoxidizing the Se(IV), generate the hydride within a few hours of the reduction steps or purge the chlorine from the samples by sparging.[6]

Interferences depend on system design and defy quantitative description because of their synergistic effects. Certain waters and wastewaters can contain interferences in sufficient concentration to suppress absorption responses of As and Se. For representative samples in a given laboratory and for initial analyses of unknown wastewaters, add appropriate inorganic forms of As or Se to digested sample portions and measure recovery. If average recoveries are less than 90%, consider using alternative analytical procedures.

e. Detection limit and optimum concentration range: For both arsenic and selenium, analyzed by aspiration into a nitrogen-hydrogen flame after reduction, the method detection limit is 2 μg/L or lower and the optimum concentration range 2 to 20 μg/L.

2. Apparatus

a. Atomic absorption spectrometer equipped with air-acetylene flame and quartz cell with mounting bracket or an electrically heated quartz cell, As and Se electrodeless discharge lamps with power supply, background correction at measurement wavelengths, and appropriate strip-chart recorder. A good-quality 10-mV recorder with high sensitivity and a fast response time is needed.

b. Atomizer: Use one of the following:

1) *Cylindrical quartz cell,* 10 to 20 cm long, bracket-mountable above air-acetylene burner.

2) *Cylindrical quartz cell,* 10 to 20 cm long, electrically heated by external nichrome wire to 800 to 900°C.[7]

3) *Cylindrical quartz cell* with internal fuel rich hydrogen-oxygen (air) flame.[8]

The sensitivity of quartz cells deteriorates over several months of use. Sensitivity sometimes may be restored by treatment with 40% HF. CAUTION: *HF is extremely corrosive. Avoid all contact with exposed skin. Handle with care.*

c. Reaction cell for producing As or Se hydrides: See Figure 3114:1 for an example of a manual, laboratory-made system. A commercially available system is acceptable if it utilizes liquid sodium borohydride reagents; accepts samples digested in accordance with ¶s 4c, d, and e; accepts 4 to 6N HCl; and is efficiently and precisely stirred by the purging gas and/or a magnetic stirrer.

d. Eye dropper or syringe capable of delivering 0.5 to 3.0 mL sodium borohydride reagent. Exact and reproducible addition is required so that production of hydrogen gas does not vary significantly between determinations.

e. Vent: See Section 3111A.6f.

3. Reagents

a. Sodium borohydride reagent: Dissolve 8 g $NaBH_4$ in 200 mL 0.1N NaOH. Prepare fresh daily.

b. Sodium iodide prereductant solution: Dissolve 50 g NaI in 500 mL water. Prepare fresh daily. Alternatively use an equivalent KI solution.

c. Sulfuric acid, 18N.

d. Sulfuric acid, 2.5*N:* Cautiously add 35 mL conc H_2SO_4 to about 400 mL water, let cool, and adjust volume to 500 mL.

e. Potassium persulfate, 5% solution: Dissolve 25 g $K_2S_2O_8$ in water and dilute to 500 mL. Store in glass and refrigerate. Prepare weekly.

f. Nitric acid, HNO_3, conc.

g. Perchloric acid, $HClO_4$, conc.

h. Hydrochloric acid, HCl, conc.

i. Argon (or nitrogen), commercial grade.

j. Arsenic(III) solutions:

1) *Stock As(III) solution:* Dissolve 1.320 g arsenic trioxide, As_2O_3, in water containing 4 g NaOH. Dilute to 1 L; 1.00 mL = 1.00 mg As(III).

2) *Intermediate As(III) solution:* Dilute 10 mL stock As solution to 1000 mL with water containing 5 mL conc HCl; 1.00 mL = 10.0 μg As(III).

3) *Standard As(III) solution:* Dilute 10 mL intermediate As(III) solution to 1000 mL with water containing the same concentration of acid used for sample preservation (2 to 5 mL conc HNO_3); 1.00 mL = 0.100 μg As(III). Prepare diluted solutions daily.

k. Arsenic(V) solutions:

1) *Stock As(V) solution:* Dissolve 1.534 g arsenic pentoxide, As_2O_5, in distilled water containing 4 g NaOH. Dilute to 1 L; 1.00 mL = 1.00 mg As(V).

2) *Intermediate As(V) solution:* Prepare as for As(III) above; 1.00 mL = 10.0 μg As(V).

3) *Standard As(V) solution:* Prepare as for As(III) above; 1.00 mL = 0.100 μg As(V).

l. Organic arsenic solutions:

1) *Stock organic arsenic solution:* Dissolve 1.842 g dimethylarsinic acid (cacodylic acid), $(CH_3)_2AsOOH$, in water containing 4 g NaOH. Dilute to 1 L; 1.00 mL = 1.00 mg As. [NOTE: Check purity of cacodylic acid reagent against an intermediate arsenic standard (50 to 100 mg As/L) using flame atomic absorption.]

2) *Intermediate organic arsenic solution:* Prepare as for As(III) above; 1.00 mL = 10.0 μg As.

3) *Standard organic arsenic solution:* Prepare as for As(III) above; 1.00 mL = 0.100 μg As.

m. Selenium(IV) solutions:

1) *Stock Se(IV) solution:* Dissolve 2.190 g sodium selenite, Na_2SeO_3, in water containing 10 mL HCl and dilute to 1 L; 1.00 mL = 1.00 mg Se(IV).

2) *Intermediate Se(IV) solution:* Dilute 10 mL stock Se(IV) to 1000 mL with water containing 10 mL conc HCl; 1.00 mL = 10.0 μg Se(IV).

3) *Standard Se(IV) solution:* Dilute 10 mL intermediate Se(IV) solution to 1000 mL with water containing the same concentration of acid used for sample preservation (2 to 5 mL conc HNO_3). Prepare solution daily when checking the equivalency of instrument response for Se(IV) and Se(VI); 1.00 mL = 0.100 μg Se(IV).

n. Selenium(VI) solutions:

1) *Stock Se(VI) solution:* Dissolve 2.393 g sodium selenate, Na_2SeO_4, in water containing 10 mL conc HNO_3. Dilute to 1 L; 1.00 mL = 1.00 mg Se(VI).

2) *Intermediate Se(VI) solution:* Prepare as for Se(IV) above; 1.00 mL = 10.0 μg Se (VI).

3) *Standard Se(VI) solution:* Prepare as for Se(IV) above; 1.00 mL = 0.100 μg Se(VI).

4. Procedure

a. Apparatus setup: Either see Figure 3114:1 or follow manufacturer's instructions. Connect inlet of reaction cell with auxiliary purging gas controlled by flow meter. If a drying cell between the reaction cell and atomizer is necessary, use only anhydrous $CaCl_2$ but not $CaSO_4$ because it may retain SeH_2. Before using the hydride generation/analysis system, optimize operating parameters. Align quartz atomizers for maximum absorbance. Aspirate a blank until memory effects are removed. Establish purging gas flow, concentration and rate of addition of sodium borohydride reagent, solution volume, and stirring rate for optimum instrument response for the chemical species to be analyzed. Optimize quartz cell temperature. If sodium borohydride reagent is added too quickly, rapid evolution of hydrogen will unbalance the system. If the volume of solution being purged is too large, the absorption signal will be decreased. Recommended wavelengths are 193.7 and 196.0 nm for As and Se, respectively.

b. Instrument calibration standards: Transfer 0.00, 1.00, 2.00, 5.00, 10.00, 15.00, and 20.00 mL standard solutions of As(III) or Se(IV) to 100-mL volumetric flasks and bring to volume with water containing the same acid concentration used for sample preservation (commonly 2 to 5 mL conc HNO_3/L). This yields blank and standard solutions of 0, 1, 2, 5, 10, 15, and 20 μg As or Se/L. Prepare fresh daily. In all cases, standards must be carried through the same digestion protocol as the samples to monitor digestion effectiveness.

c. Preparation of samples and standards for total recoverable arsenic and selenium: Use digestion procedure described in 3030F for samples and standards. Alternatively, add 50 mL sample, As(III), or Se(IV) standard to 200-mL Berzelius beaker or 100-mL micro-kjeldahl flask. Add 7 mL 18*N* H_2SO_4 and 5 mL conc HNO_3. Add a small boiling chip or glass beads if necessary. Evaporate to SO_3 fumes. Maintain oxidizing conditions at all times by adding small amounts of HNO_3 to prevent solution from darkening. Maintain an excess of HNO_3 until all organic matter is destroyed. Complete digestion usually is indicated by a light-colored solution. Cool slightly, add 25 mL water and 1 mL conc $HClO_4$ and again evaporate to SO_3 fumes to expel oxides of nitrogen. CAUTION: *See Section 3030H for cautions on use of $HClO_4$.* Monitor effectiveness of either digestion procedure used by adding 5 mL of standard organic arsenic solution or 5 mL of a standard selenium solution to a 50-mL sample and measuring recovery, carrying standards and the sample with known addition through entire procedure. To report total recoverable arsenic as total arsenic, average recoveries of cacodylic acid must exceed 80%. After final evaporation of SO_3 fumes, dilute to 50 mL for arsenic measurements or to 30 mL for selenium measurements. For analysis of both elements in a single sample, increase sample volume to 100 mL and double the volumes of acids used in the digestion. Adjust final digestate volume to 100 mL. Use 50 mL for As and 30 mL for Se determinations, making appropriate volume corrections in calculating results.

d. Preparation of samples and standards for total arsenic and selenium: Add 50 mL undigested sample or standard to a 200-mL Berzelius beaker or 100-mL micro-kjeldahl flask. Add 1 mL 2.5*N* H_2SO_4 and 5 mL 5% $K_2S_2O_8$. Boil gently on a pre-heated hot plate for approximately 30 to 40 min or until a final volume of 10 mL is reached. Do not let sample go to dryness. Alternatively heat in an autoclave at 121°C for 1 h in capped containers. After manual digestion, dilute to 50 mL for subsequent arsenic measurements and to 30 mL for selenium measurements. Monitor effectiveness of digestion by measuring recovery of As or Se as above. If poor recovery of arsenic added as cacodylic acid is obtained, reanalyze using double the amount of $K_2S_2O_8$. For analysis of both elements in a single sample, increase sample volume to 100 mL and double the volumes of acids used in the digestion. Adjust final digestate volume to 100 mL. Use 50 mL for As and 30 mL for Se determinations, making appropriate volume corrections in calculating results.

e. Determination of arsenic with sodium borohydride: To 50 mL digested standard or sample in a 200-mL Berzelius beaker (see Figure 3114:1) add 5 mL conc HCl and mix. Add 5 mL NaI prereductant solution, mix, and wait at least 30 min. (NOTE: The NaI reagent has not been found necessary for certain hydride reaction cell designs if a 20 to 30% loss in instrument sensitivity is not important and variables of solution acid conditions, temperatures, and volumes for production of As(V) and arsine can be controlled strictly. Such control requires an automated delivery system; see Section 3114C.)

Attach one Berzelius beaker at a time to the rubber stopper containing the gas dispersion tube for the purging gas, the sodium borohydride reagent inlet, and the outlet to the quartz cell. Turn on strip-chart recorder and wait until the base line is established by the purging gas and all air is expelled from the reaction cell. Add 0.5 mL sodium borohydride reagent. After the instrument absorbance has reached a maximum and returned to the base line, remove beaker, rinse dispersion tube with water, and proceed to the next sample or standard. Periodically compare standard As(III) and As(V) curves for response consistency. Check for presence of chemical interferences that suppress instrument response for arsine by treating a digested sample with 10 µg/L As(III) or As(V) as appropriate. Average recoveries should be not less than 90%.

f. Determination of selenium with sodium borohydride: To 30 mL digested standard or sample in a 200-mL Berzelius beaker or 100-mL micro-kjeldahl flask, add 15 mL conc HCl and mix. Heat for a predetermined period at 90 to 100°C. Alternatively autoclave at 121°C in capped containers for 60 min, or heat for a predetermined time in open test tubes using a 90 to 100°C hot water bath or an aluminum block digester. Check effectiveness of the selected heating time by demonstrating equal instrument responses for calibration curves prepared either from standard Se(IV) or from Se(VI) solutions. Effective heat exposure for converting Se(VI) to Se(IV), with no loss of Se(IV), ranges between 5 and 60 min when open beakers or test tubes are used. Establish a heating time for effective conversion and apply this time to all samples and standards. Do not digest standard Se(IV) and Se(VI) solutions used for this check of equivalency. After prereduction of Se(VI) to Se(IV), attach Berzelius beakers, one at a time, to the purge apparatus. For each, turn on the strip-chart recorder and wait until the base line is established. Add 0.50 mL sodium borohydride reagent. After the instrument absorbance has reached a maximum and returned to the base line, remove beaker, rinse dispersion tube with water, and proceed to the next sample or standard. Check for presence of chemical interferences that suppress selenium hydride instrument response by treating a digested sample with 10 µg Se (IV)/L. Average recoveries should be not less than 90%.

5. Calculation

Construct a standard curve by plotting peak heights or areas of standards versus concentration of standards. Measure peak heights or areas of samples and read concentrations from curve. If sample was diluted (or concentrated) before sample digestion, apply an appropriate factor. On instruments so equipped, read concentrations directly after standard calibration.

6. Precision and Bias

Single-laboratory, single-operator data were collected for As(III) and organic arsenic by both manual and automated methods, and for the manual determination of selenium. Recovery values (%) from seven replicates are given below:

Method	As(III)	Org As	Se(IV)	Se(VI)
Manual with digestion	91.8	87.3	—	—
Manual without digestion	109.4	19.4	100.6	110.8
Automated with digestion	99.8	98.4	—	—
Automated without digestion	92.5	10.4	—	—

7. References

1. VIJAN, P.N. & D. LEUNG. 1980. Reduction of chemical interference and speciation studies in the hydride generation-atomic absorption method for selenium. *Anal. Chim. Acta* 120:141.
2. VOTH-BEACH, L.M. & D.E. SHRADER. 1985. Reduction of interferences in the determination of arsenic and selenium by hydride generation. *Spectroscopy* 1:60.
3. JULSHAMN, K., O. RINGDAL, K.-E. SLINNING & O.R. BRAEKKAN. 1982. Optimization of the determination of selenium in marine samples by atomic absorption spectrometry: Comparison of a flameless graphite furnace atomic absorption system with a hydride generation atomic absorption system. *Spectrochim. Acta.* 37B:473.
4. NYGAARD, D.D. & J.H. LOWRY. 1982. Sample digestion procedures for simultaneous determination of arsenic, antimony, and selenium by inductively coupled argon plasma emission spectrometry with hydride generation. *Anal. Chem.* 54:803.
5. WELZ, B. & M. MELCHER. 1984. Mechanisms of transition metal interferences in hydride generation atomic-absorption spectrometry. Part 1. Influence of cobalt, copper, iron and nickel on selenium determination. *Analyst* 109:569.
6. KRIVAN, V., K. PETRICK, B. WELZ & M. MELCHER. 1985. Radiotracer error-diagnostic investigation of selenium determination by hydride-generation atomic absorption spectrometry involving treatment with hydrogen peroxide and hydrochloric acid. *Anal. Chem.* 57:1703.
7. CHU, R.C., G.P. BARRON & P.A.W. BAUMGARNER. 1972. Arsenic determination at submicrogram levels by arsine evolution and flameless atomic absorption spectrophotometric technique. *Anal. Chem.* 44:1476.
8. SIEMER, D.D., P. KOTEEL & V. JARIWALA. 1976. Optimization of arsine generation in atomic absorption arsenic determinations. *Anal. Chem.* 48:836.

8. Bibliography

FERNANDEZ, F.J. & D.C. MANNING. 1971. The determination of arsenic at submicrogram levels by atomic absorption spectrophotometry. *Atomic Absorption Newsletter* 10:86.

JARREL-ASH CORPORATION. 1971. High Sensitivity Arsenic Determination by Atomic Absorption. Jarrel-Ash Atomic Absorption Applications Laboratory Bull. No. As-3.

MANNING, D.C. 1971. A high sensitivity arsenic-selenium sampling system for atomic absorption spectroscopy. *Atomic Absorption Newsletter* 10:123.

BRAMAN, R.S. & C.C. FOREBACK. 1973. Methylated forms of arsenic in the environment. *Science* 182:1247.

CALDWELL, J.S., R.J. LISHKA & E.F. MCFARREN. 1973. Evaluation of a low-cost arsenic and selenium determination of microgram-per-liter levels. *J. Amer. Water Works Assoc.*, 65:731.

AGGETT, J. & A.C. ASPELL. 1976. The determination of arsenic (III) and total arsenic by atomic-absorption spectroscopy. *Analyst* 101:341.

FIORINO, J.A., J.W. JONES & S.G. CAPAR. 1976. Sequential determination of arsenic, selenium, antimony, and tellurium in foods via rapid hydride evolution and atomic absorption spectrometry. *Anal. Chem.* 48: 120.

CUTTER, G.A. 1978. Species determination of selenium in natural waters. *Anal. Chem. Acta* 98:59.

GODDEN, R.G. & D.R. THOMERSON. 1980. Generation of covalent hydrides in atomic absorption spectroscopy. *Analyst* 105:1137.

BROWN, R.M., JR., R.C. FRY, J.L. MOYERS, S.J. NORTHWAY, M.B. DENTON & G.S. WILSON. 1981. Interference by volatile nitrogen oxides and transition-metal catalysis in the preconcentration of arsenic and selenium as hydrides. *Anal. Chem.* 53:1560.

SINEMUS, H.W., M. MELCHER & B. WELZ. 1981. Influence of valence state on the determination of antimony, bismuth, selenium, and tellurium in lake water using the hydride AA technique. *Atomic Spectrosc.* 2:81.

DEDINA, J. 1982. Interference of volatile hydride forming elements in selenium determination by atomic absorption spectrometry with hydride generation. *Anal. Chem.* 54:2097.

3114 C. Continuous Hydride Generation/Atomic Absorption Spectrometric Method

1. General Discussion

The continuous hydride generator offers the advantages of simplicity in operation, excellent reproducibility, low detection limits, and high sample volume throughput for selenium analysis following preparations as described in 3500-Se.B or 3114B.4c and d.

a. Principle: See Section 3114B.

b. Interferences: Free chlorine in hydrochloric acid is a common but difficult-to-diagnose interference. (The amount of chlorine varies with manufacturer and with each lot from the same manufacturer). Chlorine oxidizes the hydride and can contaminate the hydride generator to prevent recoveries under any conditions. When interference is encountered, or preferably before using each new bottle of HCl, eliminate chlorine from a 2.3-L bottle of conc HCl by bubbling with helium (commercial grade, 100 mL/min) for 3 h.

Excess oxidant (peroxide, persulfate, or permanganate) from the total selenium digestion can oxidize the hydride. Follow procedures in 3500-Se.B.2, 3, or 4 to ensure removal of all oxidizing agents before hydride generation.

Nitrite is a common trace constituent in natural and waste waters, and at levels as low as 10 μg/L nitrite can reduce the recovery of hydrogen selenide from Se(IV) by over 50%. Moreover, during the reduction of Se(VI) to Se(IV) by digestion with HCl (3500-Se.B.5), some nitrate is converted to nitrite, which subsequently interferes. When this interference is suspected, add sulfanilamide after sample acidification (or HCl digestion). The diazotization reaction between nitrite and sulfanilamide completely removes the interferent effect (i.e., the standard addition slope is normal).

2. Apparatus

a. Continuous hydride generator: The basic unit is composed of two parts: a precision peristaltic pump, which is used to meter and mix reagents and sample solutions, and the gas-liquid separator. At the gas-liquid separator a constant flow of argon strips out the hydrogen and metal hydride gases formed in the reaction and carries them to the heated quartz absorption cell (3114B.1b and 2b), which is supported by a metal bracket mounted on top of the regular air acetylene burner head. The spent liquid flows out of the separator via a constant level side drain to a waste bucket. Schematics and operating parameters are shown in Figure 3114:2.

Check flow rates frequently to ensure a steady flow; an uneven flow in any tubing will cause an erratic signal. Remove tubings from pump rollers when not in use. Typical flow rates are: sample, 7 mL/min; acid, 1 mL/min; borohydride reagent, 1 mL/min. Argon flow usually is pre-fixed, typically at 90 mL/min.

b. Atomic absorption spectrometric equipment: See Section 3111A.6.

3. Reagents

a. Hydrochloric acid, HCl, 5 + 1: Handle conc HCl under a fume hood. If necessary, remove free Cl_2 by stripping conc HCl with helium as described above.

b. Borohydride reagent: Dissolve 0.6 g $NaBH_4$ and 0.5 g NaOH in 100 mL water. CAUTION: *Sodium borohydride is toxic, flammable, and corrosive.*

c. Selenium reference standard solution, 1000 mg/L: Use commercially available standard; verify that selenium is Se(IV).

d. Intermediate standard solution, 1 mg/L: Dilute 1 mL reference standard solution to 1 L in a volumetric flask with distilled water.

e. Working standard solutions, 5, 10, 20, 30, and 40 μg/L: Dilute 0.5, 1.0, 2.0, 3.0, and 4.0 mL intermediate standard solution to 100 mL in a volumetric flask.

f. Sulfanilamide solution: Prepare a 2.5% (w/v) solution daily; add several drops conc HCl per 50 mL solution to facilitate dissolution.

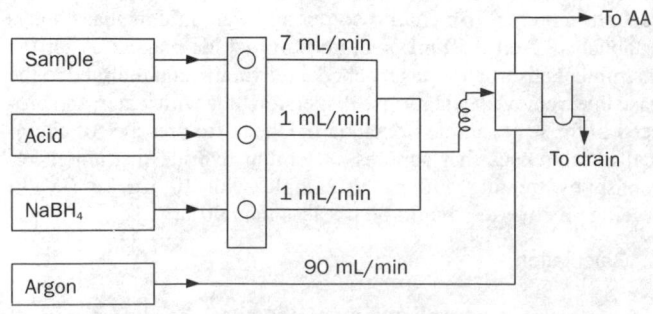

Figure 3114:2. Schematic of a continuous hydride generator.

4. Procedure

a. Sample preparation: See Section 3500-Se or 3114 B.4c and d for preparation steps for various Se fractions or total Se.

b. Preconditioning hydride generator: For newly installed tubing, turn on pump for at least 10 to 15 min before instrument calibration. Sample the highest standard for a few minutes to let volatile hydride react with the reactive sites in the transfer lines and on the quartz absorption cell surfaces.

c. Instrument calibration: Depending on total void volume in sample tubing, sampling time of 15 to 20 s generally is sufficient to obtain a steady signal. Between samples, submerge uptake tube in rinse water. Calibrate instrument daily after a 45-min lamp warmup time. Use either the hollow cathode or the electrodeless discharge lamp.

d. Antifoaming agents: Certain samples, particularly wastewater samples containing a high concentration of proteinaceous substances, can cause excessive foaming that could carry the liquid directly into the heated quartz absorption cell and cause splattering of salty deposits onto the windows of the spectrometer. Add a drop of antifoaming agent* to eliminate this problem.

e. Nitrite removal: After samples have been acidified, or after acid digestion, add 0.1 mL sulfanilamide solution per 10 mL sample and let react for 2 min.

f. Analysis: Follow manufacturer's instructions for operation of analytical equipment.

5. Calculation

Construct a calibration curve based on absorbance vs. standard concentration. Apply dilution factors on diluted samples.

6. Precision and Bias

Working standards were analyzed together with batches of water samples on a routine production basis. The standards were compounded using chemically pure sodium selenite and sodium selenate. The values of Se(IV) + Se(VI) were determined by

---

* Dow Corning or equivalent.

converting Se(VI) to Se(IV) by digestion with HCl. Results are tabulated below.

No. Analyses	Mean Se(IV) μg/L	Rel. Dev. %	Se(IV) + Se(VI) μg/L	Rel. Del. %
21	4.3	12	10.3	7
26	8.5	12	19.7	6
22	17.2	7	39.2	8
20	52.8	5	106.0	6

7. Bibliography

REAMER, D.C. & C. VEILOON. 1981. Preparation of biological materials for determination of selenium by hydride generation-AAS. *Anal. Chem.* 53:1192.

SINEMUS, H.W., M. MELCHER & B. WELZ. 1981. Influence of valence state on the determination of antimony, bismuth, selenium, and tellurium in lake water using the hydride AA technique. *Atomic Spectrosc.* 2:81.

RODEN, D.R. & D.E. TALLMAN. 1982. Determination of inorganic selenium species in groundwaters containing organic interferences by ion chromatography and hydride generation/atomic absorption spectrometry. *Anal. Chem.* 54:307.

CUTTER, G. 1983. Elimination of nitrite interference in the determination of selenium by hydride generation. *Anal. Chim. Acta* 149:391.

NARASAKI, H. & M. IKEDA. 1984. Automated determination of arsenic and selenium by atomic absorption spectrometry with hydride generation. *Anal. Chem.* 56:2059.

WELZ, B. & M. MELCHER. 1985. Decomposition of marine biological tissues for determination of arsenic, selenium, and mercury using hydride-generation and cold-vapor atomic absorption spectrometries. *Anal. Chem.* 57:427.

EBDON, L. & S.T. SPARKES. 1987. Determination of arsenic and selenium in environmental samples by hydride generation-direct current plasma-atomic emission spectrometry. *Microchem. J.* 36:198.

EBDON, L. & J.R. WILKINSON. 1987. The determination of arsenic and selenium in coal by continuous flow hydride-generation atomic absorption spectroscopy and atomic fluorescence spectrometry. *Anal. Chim. Acta.* 194:177.

VOTH-BEACH, L.M. & D.E. SHRADER. 1985. Reduction of interferences in the determination of arsenic and selenium by hydride generation. *Spectroscopy* 1:60.

3120 METALS BY PLASMA EMISSION SPECTROSCOPY*

3120 A. Introduction

1. General Discussion

Emission spectroscopy using inductively coupled plasma (ICP) was developed in the mid-1960's[1,2] as a rapid, sensitive, and convenient method for the determination of metals in water and wastewater samples.[3-6] Dissolved metals are determined in filtered and acidified samples. Total metals are determined after appropriate digestion. Care must be taken to ensure that potential

interferences are dealt with, especially when dissolved solids exceed 1500 mg/L.

2. References

1. GREENFIELD, S., I.L. JONES & C.T. BERRY. 1964. High-pressure plasma-spectroscopic emission sources. *Analyst* 89: 713.

2. WENDT, R.H. & V.A. FASSEL. 1965. Induction-coupled plasma spectrometric excitation source. *Anal. Chem.* 37:920.

---

* Approved by Standard Methods Committee, 1993.

3. U.S. ENVIRONMENTAL PROTECTION AGENCY. 1994. Method 200.7. Inductively coupled plasma-atomic emission spectrometric method for trace element analysis of water and wastes. Methods for the Determination of Metals in Environmental Samples–Supplement I. EPA 600/R-94-111, May 1994.

4. AMERICAN SOCIETY FOR TESTING AND MATERIALS. 1987. Annual Book of ASTM Standards, Vol. 11.01. American Soc. Testing & Materials, Philadelphia, Pa.

5. FISHMAN, M.J. & W.L. BRADFORD, eds. 1982. A Supplement to Methods for the Determination of Inorganic Substances in Water and Fluvial Sediments. Rep. No. 82-272, U.S. Geological Survey, Washington, D.C.

6. GARBARINO, J.R. & H.E. TAYLOR. 1985. Trace Analysis. Recent Developments and Applications of Inductively Coupled Plasma Emission Spectroscopy to Trace Elemental Analysis of Water. Volume 4. Academic Press, New York, N.Y.

3120 B.   Inductively Coupled Plasma (ICP) Method

1. General Discussion

a. Principle: An ICP source consists of a flowing stream of argon gas ionized by an applied radio frequency field typically oscillating at 27.1 MHz. This field is inductively coupled to the ionized gas by a water-cooled coil surrounding a quartz "torch" that supports and confines the plasma. A sample aerosol is generated in an appropriate nebulizer and spray chamber and is carried into the plasma through an injector tube located within the torch. The sample aerosol is injected directly into the ICP, subjecting the constituent atoms to temperatures of about 6000 to 8000°K.[1] Because this results in almost complete dissociation of molecules, significant reduction in chemical interferences is achieved. The high temperature of the plasma excites atomic emission efficiently. Ionization of a high percentage of atoms produces ionic emission spectra. The ICP provides an optically "thin" source that is not subject to self-absorption except at very high concentrations. Thus linear dynamic ranges of four to six orders of magnitude are observed for many elements.[2]

The efficient excitation provided by the ICP results in low detection limits for many elements. This, coupled with the extended dynamic range, permits effective multielement determination of metals.[3] The light emitted from the ICP is focused onto the entrance slit of either a monochromator or a polychromator that effects dispersion. A precisely aligned exit slit is used to isolate a portion of the emission spectrum for intensity measurement using a photomultiplier tube. The monochromator uses a single exit slit/photomultiplier and may use a computer-controlled scanning mechanism to examine emission wavelengths sequentially. The polychromator uses multiple fixed exit slits and corresponding photomultiplier tubes; it simultaneously monitors all configured wavelengths using a computer-controlled readout system. The sequential approach provides greater wavelength selection while the simultaneous approach can provide greater sample throughput.

b. Applicable metals and analytical limits: Table 3120:I lists elements for which this method applies, recommended analytical wavelengths, and typical estimated instrument detection limits using conventional pneumatic nebulization. Actual working detection limits are sample-dependent. Typical upper limits for linear calibration also are included in Table 3120:I.

c. Interferences: Interferences may be categorized as follows:

1) Spectral interferences—Light emission from spectral sources other than the element of interest may contribute to apparent net signal intensity. Sources of spectral interference include direct spectral line overlaps, broadened wings of intense spectral lines, ion-atom recombination continuum emission, molecular band emission, and stray (scattered) light from the emission of elements at high concentrations.[4] Avoid line overlaps by selecting alternate analytical wavelengths. Avoid or minimize other spectral interference by judicious choice of background correction positions. A wavelength scan of the element line region is useful for detecting potential spectral interferences and for selecting positions for background correction. Make corrections for residual spectral interference using empirically determined correction factors in conjunction with the computer software supplied by the spectrometer manufacturer or with the calculation detailed below. The empirical correction method cannot be used with scanning spectrometer systems if the analytical and interfering lines cannot be precisely and reproducibly located. In addition, if using a polychromator, verify absence of spectral interference from an element that could occur in a sample but for which there is no channel in the detector array. Do this by analyzing single-element solutions of 100 mg/L concentration and noting for each element channel the apparent concentration from the interfering substance that is greater than the element's instrument detection limit.

2) Nonspectral interferences

a) Physical interferences are effects associated with sample nebulization and transport processes. Changes in the physical properties of samples, such as viscosity and surface tension, can cause significant error. This usually occurs when samples containing more than 10% (by volume) acid or more than 1500 mg dissolved solids/L are analyzed using calibration standards containing ≤ 5% acid. Whenever a new or unusual sample matrix is encountered, use the test described in ¶ 4g. If physical interference is present, compensate for it by sample dilution, by using matrix-matched calibration standards, or by applying the method of standard addition (see ¶ 5d below).

High dissolved solids content also can contribute to instrumental drift by causing salt buildup at the tip of the nebulizer gas orifice. Using prehumidified argon for sample nebulization lessens this problem. Better control of the argon flow rate to the nebulizer using a mass flow controller improves instrument performance.

b) Chemical interferences are caused by molecular compound formation, ionization effects, and thermochemical effects associated with sample vaporization and atomization in the plasma. Normally these effects are not pronounced and can be minimized by careful selection of operating conditions (incident power, plasma observation position, etc.). Chemical interferences are highly dependent on sample matrix and element of interest. As with physical interferences, compensate for them by using matrix matched standards or by standard addition (¶ 5d). To determine the presence of chemical interference, follow instructions in ¶ 4g.

TABLE 3120:I. SUGGESTED WAVELENGTHS, ESTIMATED DETECTION LIMITS, ALTERNATE WAVELENGTHS, CALIBRATION CONCENTRATIONS, AND UPPER LIMITS

Element	Suggested Wavelength nm	Estimated Detection Limit μg/L	Alternate Wavelength* nm	Calibration Concentration mg/L	Upper Limit Concentration† mg/L
Aluminum	308.22	40	237.32	10.0	100
Antimony	206.83	30	217.58	10.0	100
Arsenic	193.70	50	189.04‡	10.0	100
Barium	455.40	2	493.41	1.0	50
Beryllium	313.04	0.3	234.86	1.0	10
Boron	249.77	5	249.68	1.0	50
Cadmium	226.50	4	214.44	2.0	50
Calcium	317.93	10	315.89	10.0	100
Chromium	267.72	7	206.15	5.0	50
Cobalt	228.62	7	230.79	2.0	50
Copper	324.75	6	219.96	1.0	50
Iron	259.94	7	238.20	10.0	100
Lead	220.35	40	217.00	10.0	100
Lithium	670.78	4§	—	5.0	100
Magnesium	279.08	30	279.55	10.0	100
Manganese	257.61	2	294.92	2.0	50
Molybdenum	202.03	8	203.84	10.0	100
Nickel	231.60	15	221.65	2.0	50
Potassium	766.49	100§	769.90	10.0	100
Selenium	196.03	75	203.99	5.0	100
Silica (SiO_2)	212.41	20	251.61	21.4	100
Silver	328.07	7	338.29	2.0	50
Sodium	589.00	30§	589.59	10.0	100
Strontium	407.77	0.5	421.55	1.0	50
Thallium	190.86‡	40	377.57	10.0	100
Vanadium	292.40	8	—	1.0	50
Zinc	213.86	2	206.20	5.0	100

* Other wavelengths may be substituted if they provide the needed sensitivity and are corrected for spectral interference.
† Defines the top end of the effective calibration range. Do not extrapolate to concentrations beyond highest standard.
‡ Available with vacuum or inert gas purged optical path.
§ Sensitive to operating conditions.

2. Apparatus

a. *ICP source:* The ICP source consists of a radio frequency (RF) generator capable of generating at least 1.1 KW of power, torch, tesla coil, load coil, impedance matching network, nebulizer, spray chamber, and drain. High-quality flow regulators are required for both the nebulizer argon and the plasma support gas flow. A peristaltic pump is recommended to regulate sample flow to the nebulizer. The type of nebulizer and spray chamber used may depend on the samples to be analyzed as well as on the equipment manufacturer. In general, pneumatic nebulizers of the concentric or cross-flow design are used. Viscous samples and samples containing particulates or high dissolved solids content (>5000 mg/L) may require nebulizers of the Babington type.[5]

b. *Spectrometer:* The spectrometer may be of the simultaneous (polychromator) or sequential (monochromator) type with air-path, inert gas purged, or vacuum optics. A spectral bandpass of 0.05 nm or less is required. The instrument should permit examination of the spectral background surrounding the emission lines used for metals determination. It is necessary to be able to measure and correct for spectral background at one or more positions on either side of the analytical lines.

3. Reagents and Standards

Use reagents that are of ultra-high-purity grade or equivalent. Redistilled acids are acceptable. Except as noted, dry all salts at 105°C for 1 h and store in a desiccator before weighing. Use deionized water prepared by passing water through at least two stages of deionization with mixed bed cation/anion exchange resins.[6] Use deionized water for preparing all calibration standards, reagents, and for dilution.

a. *Hydrochloric acid,* HCl, conc and 1 + 1.

b. *Nitric acid,* HNO_3, conc.

c. *Nitric acid,* HNO_3, 1 + 1: Add 500 mL conc HNO_3 to 400 mL water and dilute to 1 L.

d. *Standard stock solutions:* See 3111B, 3111D, and 3114B. CAUTION: *Many metal salts are extremely toxic and may be fatal if swallowed. Wash hands thoroughly after handling.*

1) *Aluminum:* See 3111D.3k1).

2) *Antimony:* See 3111B.3j1).

3) *Arsenic:* See 3114B.3k1).

4) *Barium:* See 3111D.3k2).

5) *Beryllium:* See 3111D.3k3).

6) *Boron: Do not dry* but keep bottle tightly stoppered and

store in a desiccator. Dissolve 0.5716 g anhydrous H_3BO_3 in water and dilute to 1000 mL; 1 mL = 100 μg B.

7) *Cadmium:* See 3111B.3*j*3).
8) *Calcium:* See 3111B.3*j*4).
9) *Chromium:* See 3111B.3*j*6).
10) *Cobalt:* See 3111B.3*j*7).
11) *Copper:* See 3111B.3*j*8).
12) *Iron:* See 3111B.3*j*11).
13) *Lead:* See 3111B.3*j*12).
14) *Lithium:* See 3111B.3*j*13).
15) *Magnesium:* See 3111B.3*j*14).
16) *Manganese:* See 3111B.3*j*15).
17) *Molybdenum:* See 3111D.3*k*4).
18) *Nickel:* See 3111B.3*j*16).
19) *Potassium:* See 3111B.3*j*19).
20) *Selenium:* See 3114B.3*n*1).
21) *Silica:* See 3111D.3*k*7).
22) *Silver:* See 3111B.3*j*22).
23) *Sodium:* See 3111B.3*j*23).
24) *Strontium:* See 3111B.3*j*24).
25) *Thallium:* See 3111B.3*j*25).
26) *Vanadium:* See 3111D.3*k*10).
27) *Zinc:* See 3111B.3*j*27).

e. Calibration standards: Prepare mixed calibration standards containing the concentrations shown in Table 3120:I by combining appropriate volumes of the stock solutions in 100-mL volumetric flasks. Add 2 mL 1 + 1 HNO_3 and 10 mL 1 + 1 HCl and dilute to 100 mL with water. Before preparing mixed standards, analyze each stock solution separately to determine possible spectral interference or the presence of impurities. When preparing mixed standards take care that the elements are compatible and stable. Store mixed standard solutions in an FEP fluorocarbon or unused polyethylene bottle. Verify calibration standards initially using the quality control standard; monitor weekly for stability. The following are recommended combinations using the suggested analytical lines in Table 3120:I. Alternative combinations are acceptable.

1) *Mixed standard solution I:* Manganese, beryllium, cadmium, lead, selenium, and zinc.

2) *Mixed standard solution II:* Barium, copper, iron, vanadium, and cobalt.

3) *Mixed standard solution III:* Molybdenum, silica, arsenic, strontium, and lithium.

4) *Mixed standard solution IV:* Calcium, sodium, potassium, aluminum, chromium, and nickel.

5) *Mixed standard solution V:* Antimony, boron, magnesium, silver, and thallium. If addition of silver results in an initial precipitation, add 15 mL water and warm flask until solution clears. Cool and dilute to 100 mL with water. For this acid combination limit the silver concentration to 2 mg/L. Silver under these conditions is stable in a tap water matrix for 30 d. Higher concentrations of silver require additional HCl.

f. Calibration blank: Dilute 2 mL 1 + 1 HNO_3 and 10 mL 1 + 1 HCl to 100 mL with water. Prepare a sufficient quantity to be used to flush the system between standards and samples.

g. Method blank: Carry a reagent blank through entire sample preparation procedure. Prepare method blank to contain the same acid types and concentrations as the sample solutions.

h. Instrument check standard: Prepare instrument check standards by combining compatible elements at a concentration of 2 mg/L.

i. Instrument quality control sample: Obtain a certified aqueous reference standard from an outside source and prepare according to instructions provided by the supplier. Use the same acid matrix as the calibration standards.

j. Method quality control sample: Carry the instrument quality control sample (¶ 3*i*) through the entire sample preparation procedure.

k. Argon: Use technical or welder's grade. If gas appears to be a source of problems, use prepurified grade.

4. Procedure

a. Sample preparation: See Section 3030F.

b. Operating conditions: Because of differences among makes and models of satisfactory instruments, no detailed operating instructions can be provided. Follow manufacturer's instructions. Establish instrumental detection limit, precision, optimum background correction positions, linear dynamic range, and interferences for each analytical line. Verify that the instrument configuration and operating conditions satisfy the analytical requirements and that they can be reproduced on a day-to-day basis. An atom-to-ion emission intensity ratio [Cu(I) 324.75 nm/ Mn(II) 257.61 nm] can be used to reproduce optimum conditions for multielement analysis precisely. The Cu/Mn intensity ratio may be incorporated into the calibration procedure, including specifications for sensitivity and for precision.[7] Keep daily or weekly records of the Cu and Mn intensities and/or the intensities of critical element lines. Also record settings for optical alignment of the polychromator, sample uptake rate, power readings (incident, reflected), photomultiplier tube attenuation, mass flow controller settings, and system maintenance.

c. Instrument calibration: Set up instrument as directed (¶ *b*). Warm up for 30 min. For polychromators, perform an optical alignment using the profile lamp or solution. Check alignment of plasma torch and spectrometer entrance slit, particularly if maintenance of the sample introduction system was performed. Make Cu/Mn or similar intensity ratio adjustment.

Calibrate instrument according to manufacturer's recommended procedure using calibration standards and blank. Aspirate each standard or blank for a minimum of 15 s after reaching the plasma before beginning signal integration. Rinse with calibration blank or similar solution for at least 60 s between each standard to eliminate any carryover from the previous standard. Use average intensity of multiple integrations of standards or samples to reduce random error.

Before analyzing samples, analyze instrument check standard. Concentration values obtained should not deviate from the actual values by more than ±5% (or the established control limits, whichever is lower).

d. Analysis of samples: Begin each sample run with an analysis of the calibration blank, then analyze the method blank. This permits a check of the sample preparation reagents and procedures for contamination. Analyze samples, alternating them with analyses of calibration blank. Rinse for at least 60 s with dilute acid between samples and blanks. After introducing each sample or blank let system equilibrate before starting signal integration. Examine each analysis of the calibration blank to verify that no carry-over memory effect has occurred. If carry-over is observed, repeat rinsing until proper blank values are obtained. Make appropriate dilutions and acidifications of the sample to determine concentrations beyond the linear calibration range.

TABLE 3120:II. ICP PRECISION AND BIAS DATA

Element	Concentration Range $\mu g/L$	Total Digestion* $\mu g/L$		Recoverable Digestion* $\mu g/L$	
Aluminum	69–4792	$X =$	$0.9273C + 3.6$	$X =$	$0.9380C + 22.1$
		$S =$	$0.0559X + 18.6$	$S =$	$0.0873X + 31.7$
		$SR =$	$0.0507X + 3.5$	$SR =$	$0.0481X + 18.8$
Antimony	77–1406	$X =$	$0.7940C - 17.0$	$X =$	$0.8908C + 0.9$
		$S =$	$0.1556X - 0.6$	$S =$	$0.0982X + 8.3$
		$SR =$	$0.1081X + 3.9$	$SR =$	$0.0682X + 2.5$
Arsenic	69–1887	$X =$	$1.0437C - 12.2$	$X =$	$1.0175C + 3.9$
		$S =$	$0.1239X + 2.4$	$S =$	$0.1288X + 6.1$
		$SR =$	$0.0874X + 6.4$	$SR =$	$0.0643X + 10.3$
Barium	9–377	$X =$	$0.7683C + 0.47$	$X =$	$0.8380C + 1.68$
		$S =$	$0.1819X + 2.78$	$S =$	$0.2540X + 0.30$
		$SR =$	$0.1285X + 2.55$	$SR =$	$0.0826X + 3.54$
Beryllium	3–1906	$X =$	$0.9629C + 0.05$	$X =$	$1.0177C - 0.55$
		$S =$	$0.0136X + 0.95$	$S =$	$0.0359X + 0.90$
		$SR =$	$0.0203X - 0.07$	$SR =$	$0.0445X - 0.10$
Boron	19–5189	$X =$	$0.8807C + 9.0$	$X =$	$0.9676C + 18.7$
		$S =$	$0.1150X + 14.1$	$S =$	$0.1320X + 16.0$
		$SR =$	$0.0742X + 23.2$	$SR =$	$0.0743X + 21.1$
Cadmium	9–1943	$X =$	$0.9874C - 0.18$	$X =$	$1.0137C - 0.65$
		$S =$	$0.0557X + 2.02$	$S =$	$0.0585X + 1.15$
		$SR =$	$0.0300X + 0.94$	$SR =$	$0.0332X + 0.90$
Calcium	17–47 170	$X =$	$0.9182C - 2.6$	$X =$	$0.9658C + 0.8$
		$S =$	$0.1228X + 10.1$	$S =$	$0.0917X + 6.9$
		$SR =$	$0.0189X + 3.7$	$SR =$	$0.0327X + 10.1$
Chromium	13–1406	$X =$	$0.9544C + 3.1$	$X =$	$1.0049C - 1.2$
		$S =$	$0.0499X + 4.4$	$S =$	$0.0698X + 2.8$
		$SR =$	$0.0009X + 7.9$	$SR =$	$0.0571X + 1.0$
Cobalt	17–2340	$X =$	$0.9209C - 4.5$	$X =$	$0.9278C - 1.5$
		$S =$	$0.0436X + 3.8$	$S =$	$0.0498X + 2.6$
		$SR =$	$0.0428X + 0.5$	$SR =$	$0.0407X + 0.4$
Copper	8–1887	$X =$	$0.9297C - 0.30$	$X =$	$0.9647C - 3.64$
		$S =$	$0.0442X + 2.85$	$S =$	$0.0497X + 2.28$
		$SR =$	$0.0128X + 2.53$	$SR =$	$0.0406X + 0.96$
Iron	13–9359	$X =$	$0.8829C + 7.0$	$X =$	$0.9830C + 5.7$
		$S =$	$0.0683X + 11.5$	$S =$	$0.1024X + 13.0$
		$SR =$	$-0.0046X + 10.0$	$SR =$	$0.0790X + 11.5$
Lead	42–4717	$X =$	$0.9699C - 2.2$	$X =$	$1.0056C + 4.1$
		$S =$	$0.0558X + 7.0$	$S =$	$0.0799X + 4.6$
		$SR =$	$0.0353X + 3.6$	$SR =$	$0.0448X + 3.5$
Magnesium	34–13 868	$X =$	$0.9881C - 1.1$	$X =$	$0.9879C + 2.2$
		$S =$	$0.0607X + 11.6$	$S =$	$0.0564X + 13.2$
		$SR =$	$0.0298X + 0.6$	$SR =$	$0.0268X + 8.1$
Manganese	4–1887	$X =$	$0.9417C + 0.13$	$X =$	$0.9725C + 0.07$
		$S =$	$0.0324X + 0.88$	$S =$	$0.0557X + 0.76$
		$SR =$	$0.0153X + 0.91$	$SR =$	$0.0400X + 0.82$
Molybdenum	17–1830	$X =$	$0.9682C + 0.1$	$X =$	$0.9707C - 2.3$
		$S =$	$0.0618X + 1.6$	$S =$	$0.0811X + 3.8$
		$SR =$	$0.0371X + 2.2$	$SR =$	$0.0529X + 2.1$
Nickel	17–47 170	$X =$	$0.9508C + 0.4$	$X =$	$0.9869C + 1.5$
		$S =$	$0.0604X + 4.4$	$S =$	$0.0526X + 5.5$
		$SR =$	$0.0425X + 3.6$	$SR =$	$0.0393X + 2.2$
Potassium	347–14 151	$X =$	$0.8669C - 36.4$	$X =$	$0.9355C - 183.1$
		$S =$	$0.0934X + 77.8$	$S =$	$0.0481X + 177.2$
		$SR =$	$-0.0099X + 144.2$	$SR =$	$0.0329X + 60.9$

TABLE 3120:II, CONT.

Element	Concentration Range μg/L	Total Digestion* μg/L	Recoverable Digestion* μg/L
Selenium	69–1415	$X = 0.9363C - 2.5$ $S = 0.0855X + 17.8$ $SR = 0.0284X + 9.3$	$X = 0.9737C - 1.0$ $S = 0.1523X + 7.8$ $SR = 0.0443X + 6.6$
Silicon	189–9434	$X = 0.5742C - 35.6$ $S = 0.4160X + 37.8$ $SR = 0.1987X + 8.4$	$X = 0.9737C - 60.8$ $S = 0.3288X + 46.0$ $SR = 0.2133X + 22.6$
Silver	8–189	$X = 0.4466C + 5.07$ $S = 0.5055X - 3.05$ $SR = 0.2086X - 1.74$	$X = 0.3987C + 8.25$ $S = 0.5478X - 3.93$ $SR = 0.1836X - 0.27$
Sodium	35–47 170	$X = 0.9581C + 39.6$ $S = 0.2097X + 33.0$ $SR = 0.0280X + 105.8$	$X = 1.0526C + 26.7$ $S = 0.1473X + 27.4$ $SR = 0.0884X + 50.5$
Thallium	79–1434	$X = 0.9020C - 7.3$ $S = 0.1004X + 18.3$ $SR = 0.0364X + 11.5$	$X = 0.9238C + 5.5$ $S = 0.2156X + 5.7$ $SR = -0.0106X + 48.0$
Vanadium	13–4698	$X = 0.9615C - 2.0$ $S = 0.0618X + 1.7$ $SR = 0.0220X + 0.7$	$X = 0.9551C + 0.4$ $S = 0.0927X + 1.5$ $SR = 0.0472X + 0.5$
Zinc	7–7076	$X = 0.9356C - 0.30$ $S = 0.0914X + 3.75$ $SR = -0.0130X + 10.07$	$X = 0.9500C + 1.22$ $S = 0.0597X + 6.50$ $SR = 0.0153X + 7.78$

*X = mean recovery, μg/L,
C = true value, μg/L,
S = multi-laboratory standard deviation, μg/L,
SR = single-analyst standard deviation, μg/L.

e. *Instrumental quality control:* Analyze instrument check standard once per 10 samples to determine if significant instrument drift has occurred. If agreement is not within ± 5% of the expected values (or within the established control limits, whichever is lower), terminate analysis of samples, correct problem, and recalibrate instrument. If the intensity ratio reference is used, resetting this ratio may restore calibration without the need for reanalyzing calibration standards. Analyze instrument check standard to confirm proper recalibration. Reanalyze one or more samples analyzed just before termination of the analytical run. Results should agree to within ± 5%, otherwise all samples analyzed after the last acceptable instrument check standard analysis must be reanalyzed.

Analyze instrument quality control sample within every run. Use this analysis to verify accuracy and stability of the calibration standards. If any result is not within ± 5% of the certified value, prepare a new calibration standard and recalibrate the instrument. If this does not correct the problem, prepare a new stock solution and a new calibration standard and repeat calibration.

f. *Method quality control:* Analyze the method quality control sample within every run. Results should agree to within ± 5% of the certified values. Greater discrepancies may reflect losses or contamination during sample preparation.

g. *Test for matrix interference:* When analyzing a new or unusual sample matrix verify that neither a positive nor negative nonlinear interference effect is operative. If the element is present at a concentration above 1 mg/L, use serial dilution with calibration blank. Results from the analyses of a dilution should be within ± 5% of the original result. Alternately, or if the concentration is either below 1 mg/L or not detected, use a post-digestion addition equal to 1 mg/L. Recovery of the addition should be either between 95% and 105% or within established control limits of ± 2 standard deviations around the mean. If a matrix effect causes test results to fall outside the critical limits, complete the analysis after either diluting the sample to eliminate the matrix effect while maintaining a detectable concentration of at least twice the detection limit or applying the method of standard additions.

5. Calculations and Corrections

a. *Blank correction:* Subtract result of an adjacent calibration blank from each sample result to make a baseline drift correction. (Concentrations printed out should include negative and positive values to compensate for positive and negative baseline drift. Make certain that the calibration blank used for blank correction has not been contaminated by carry-over.) Use the result of the method blank analysis to correct for reagent contamination. Alternatively, intersperse method blanks with appropriate samples. Reagent blank and baseline drift correction are accomplished in one subtraction.

b. *Dilution correction:* If the sample was diluted or concentrated in preparation, multiply results by a dilution factor (DF) calculated as follows:

$$DF = \frac{\text{Final weight or volume}}{\text{Initial weight or volume}}$$

c. *Correction for spectral interference:* Correct for spectral interference by using computer software supplied by the instrument manufacturer or by using the manual method based on interference correction factors. Determine interference correction factors by analyzing single-element stock solutions of appropriate concentrations under conditions matching as closely as possible those used for sample analysis. Unless analysis conditions can be reproduced accurately from day to day, or for longer periods, redetermine interference correction factors found to affect the results significantly each time samples are analyzed.[7, 8] Calculate interference correction factors (K_{ij}) from apparent concentrations observed in the analysis of the high-purity stock solutions:

$$K_{ij} = \frac{\text{Apparent concentration of element } i}{\text{Actual concentration of interfering element } j}$$

where the apparent concentration of element i is the difference between the observed concentration in the stock solution and the observed concentration in the blank. Correct sample concentrations observed for element i (already corrected for baseline drift), for spectral interferences from elements j, k, and l; for example:

Concentration of element i corrected for spectral interference

$$= \text{Observed concentration of } i - (K_{ij}) \begin{array}{c} \text{Observed} \\ \text{concentration} \\ \text{of interfering} \\ \text{element } j \end{array} - (K_{ik}) \begin{array}{c} \text{Observed} \\ \text{concentration} \\ \text{of interfering} \\ \text{element } k \end{array}$$

$$- (K_{il}) \begin{array}{c} \text{Observed} \\ \text{concentration} \\ \text{of interfering} \\ \text{element } l \end{array}$$

Interference correction factors may be negative if background correction is used for element i. A negative K_{ij} can result where an interfering line is encountered at the background correction wavelength rather than at the peak wavelength. Determine concentrations of interfering elements j, k, and l within their respective linear ranges. Mutual interferences (i interferes with j and j interferes with i) require iterative or matrix methods for calculation.

d. Correction for nonspectral interference: If nonspectral interference correction is necessary, use the method of standard additions. It is applicable when the chemical and physical form of the element in the standard addition is the same as in the sample, *or* the ICP converts the metal in both sample and addition to the same form; the interference effect is independent of metal concentration over the concentration range of standard additions; and the analytical calibration curve is linear over the concentration range of standard additions.

Use an addition not less than 50% nor more than 100% of the element concentration in the sample so that measurement preci-

sion will not be degraded and interferences that depend on element/interferent ratios will not cause erroneous results. Apply the method to all elements in the sample set using background correction at carefully chosen off-line positions. Multielement standard addition can be used if it has been determined that added elements are not interferents.

e. Reporting data: Report analytical data in concentration units of milligrams per liter using up to three significant figures. Report results below the determined detection limit as not detected less than the stated detection limit corrected for sample dilution.

6. Precision and Bias

As a guide to the generally expected precision and bias, see the linear regression equations in Table 3120:II.[9] Additional interlaboratory information is available.[10]

7. References

1. FAIRES, L.M., B.A. PALMER, R. ENGLEMAN, JR. & T.M. NIEMCZYK. 1984. Temperature determinations in the inductively coupled plasma using a Fourier transform spectrometer. *Spectrochim. Acta* 39B:819.
2. BARNES, R.M. 1978. Recent advances in emission spectroscopy: inductively coupled plasma discharges for spectrochemical analysis. *CRC Crit. Rev. Anal. Chem.* 7:203.
3. PARSONS, M.L., S. MAJOR & A.R. FORSTER. 1983. Trace element determination by atomic spectroscopic methods - State of the art. *Appl. Spectrosc.* 37:411.
4. LARSON, G.F., V.A. FASSEL, R. K. WINGE & R.N. KNISELEY. 1976. Ultratrace analysis by optical emission spectroscopy: The stray light problem. *Appl. Spectrosc.* 30:384.
5. GARBARINO, J.R. & H.E. TAYLOR. 1979. A Babington-type nebulizer for use in the analysis of natural water samples by inductively coupled plasma spectrometry. *Appl. Spectrosc.* 34:584.
6. AMERICAN SOCIETY FOR TESTING AND MATERIALS. 1988. Standard specification for reagent water, D1193-77 (reapproved 1983). Annual Book of ASTM Standards. American Soc. for Testing & Materials, Philadelphia, Pa.
7. BOTTO, R.I. 1984. Quality assurance in operating a multielement ICP emission spectrometer. *Spectrochim. Acta* 39B:95.
8. BOTTO, R.I. 1982. Long-term stability of spectral interference calibrations for inductively coupled plasma atomic emission spectrometry. *Anal. Chem.* 54:1654.
9. MAXFIELD, R. & B. MINDAK. 1985. EPA Method Study 27, Method 200. 7 (Trace Metals by ICP). EPA-600/S4-85/05. National Technical Information Serv., Springfield, Va.
10. GARBARINO, J.R., B.E. JONES, G. P. STEIN, W.T. BELSER & H.E. TAYLOR. 1985. Statistical evaluation of an inductively coupled plasma atomic emission spectrometric method for routine water quality testing. *Appl. Spectrosc.* 39:53.

3125 METALS BY INDUCTIVELY COUPLED PLASMA/MASS SPECTROMETRY*

3125 A. Introduction

1. General Discussion

This method is used for the determination of trace metals and metalloids in surface, ground, and drinking waters by inductively coupled plasma/mass spectrometry (ICP/MS). It may also be suitable for wastewater, soils, sediments, sludge, and biological samples after suitable digestion followed by dilution and/or cleanup.[1,2] Additional sources of information on quality assurance and other aspects of ICP/MS analysis of metals are available.[3-5]

The method is intended to be performance-based, allowing extension of the elemental analyte list, implementation of "clean" preparation techniques as they become available, and other appropriate modifications of the base method as technology evolves. Preferably validate modifications to the base method by use of the quality control standards specified in the method.

Instrument detection limits for many analytes are between 1 and 100 ng/L. The method is best suited for the determination of metals in ambient or pristine fresh-water matrices. More complex matrices may require some type of cleanup to reduce matrix effects to a manageable level. Various cleanup techniques are available to reduce matrix interferences and/or concentrate analytes of interest.[6-10]

This method is ideally used by analysts experienced in the use of ICP/MS, the interpretation of spectral and matrix interference, and procedures for their correction. Preferably demonstrate analyst proficiency through analysis of a performance evaluation sample before the generation of data.

2. References

1. MONTASER, A. & D.W. GOLIGHTLY, eds. 1992. Inductively Coupled Plasmas in Analytical Atomic Spectrometry, 2nd ed. VCH Publishers, Inc., New York, N.Y.

2. DATE, A.R. & A.L. GRAY. 1989. Applications of Inductively Coupled Plasma Mass Spectrometry. Blackie & Son, Ltd., Glasgow, U.K.

3. U.S. ENVIRONMENTAL PROTECTION AGENCY. 1994. Determination of trace elements in waters and wastes by inductively coupled plasma-mass spectrometry, Method 200.8. U.S. Environmental Protection Agency, Environmental Monitoring Systems Lab., Cincinnati, Ohio.

4. LONGBOTTOM, J.E., T.D. MARTIN, K.W. EDGELL, S.E. LONG, M.R. PLANTZ & B.E. WARDEN. 1994. Determination of trace elements in water by inductively coupled plasma-mass spectrometry: collaborative study. J. AOAC Internat. 77:1004.

5. U.S. ENVIRONMENTAL PROTECTION AGENCY. 1995. Method 1638: Determination of trace elements in ambient waters by inductively coupled plasma-mass spectrometry. U.S. Environmental Protection Agency, Off. Water, Washington, D.C.

6. MCLAREN, J.W., A.P. MYKYTIUK, S.N. WILLIE & S. S. BERMAN. 1985. Determination of trace metals in seawater by inductively coupled plasma mass spectrometry with preconcentration on silica-immobilized 8-hydroxyquinoline. Anal. Chem. 57:2907.

7. BURBA, P. & P.G. WILLMER. 1987. Multielement preconcentration for atomic spectroscopy by sorption of dithiocarbamate metal complexes (e.g., HMDC) on cellulose collectors. Fresenius Z. Anal. Chem. 329: 539.

8. WANG, X. & R.M. BARNES. 1989. Chelating resins for on-line flow injection preconcentration with inductively coupled plasma atomic emission spectroscopy. J. Anal. Atom. Spectrom. 4:509.

9. SIRIRAKS, A., H.M. KINGSTON & J.M. RIVIELLO. 1990. Chelation ion chromatography as a method for trace elemental analysis in complex environmental and biological samples. Anal. Chem. 62:1185.

10. PUGET SOUND WATER QUALITY AUTHORITY. 1996. Recommended Guidelines for Measuring Metals in Puget Sound Marine Water, Sediment and Tissue Samples. Appendix D: Alternate Methods for the Analysis of Marine Water Samples. Puget Sound Water Quality Authority, Olympia, Wash.

* Approved by Standard Methods Committee, 1997.

3125 B. Inductively Coupled Plasma/Mass Spectrometry (ICP/MS) Method

1. General Discussion

a. Principle: Sample material is introduced into an argon-based, high-temperature radio-frequency plasma, usually by pneumatic nebulization. Energy transfer from the plasma to the sample stream causes desolvation, atomization, and ionization of target elements. Ions generated by these energy-transfer processes are extracted from the plasma through a differential vacuum interface, and separated on the basis of their mass-to-charge ratio by a mass spectrometer. The mass spectrometer usually is of the quadrupole or magnetic sector type. The ions passing through the mass spectrometer are counted, usually by an electron multiplier detector, and the resulting information processed by a computer-based data-handling system.

b. Applicable elements and analytical limits: This method is suitable for aluminum, antimony, arsenic, barium, beryllium, cadmium, chromium, cobalt, copper, lead, manganese, molybdenum, nickel, selenium, silver, strontium, thallium, uranium, vanadium, and zinc. The method is also acceptable for other elemental analytes as long as the same quality assurance practices are followed. The basic element suite and recommended analytical masses are given in Table 3125:I.

Typical instrument detection limits (IDL)[1,2] for method analytes are presented in Table 3125:I. Determine the IDL and

TABLE 3125:I. RECOMMENDED ANALYTE MASSES, INSTRUMENTAL
DETECTION LIMITS (IDL), AND INTERNAL STANDARDS

Element	Analytical Mass	IDL μg/L	Recommended Internal Standard
Be	9	0.025	Li
Al	27	0.03	Sc
V	51	0.02	Sc
Cr	52	0.04	Sc
Cr	53	0.03	Sc
Mn	55	0.002	Sc
Co	59	0.002	Sc
Ni	60	0.004	Sc
Ni	62	0.025	Sc
Cu	63	0.003	Sc
Cu	65	0.004	Sc
Zn	66	0.017	Ge
Zn	68	0.020	Ge
As	75	0.025	Ge
Se	77	0.093	Ge
Se	82	0.064	Ge
Ag	107	0.003	In
Ag	109	0.002	In
Cd	111	0.006	In
Cd	114	0.003	In
Sb	121	0.07	In
Sb	123	0.07	In
Tl	203	0.03	Th
Tl	205	0.03	Th
Pb	208	0.005	Th
U	235	0.032	Th
U	238	0.001	Th
Mo	98	0.003†	In
Ba	135	0.008†	In
Sr	88	0.001‡	In

* IDLs were determined on a Perkin Elmer Elan 6000 ICP/MS using seven replicate analyses of a 1% nitric acid solution, at Manchester Environmental Laboratory, July 1996.
† From EPA Method 200.8 for the Analysis of Drinking Waters-Application Note, Order No. ENVA-300A, The Perkin Elmer Corporation, 1996.
‡ From Perkin Elmer Technical Summary TSMS-12.

method detection level (or limit) (MDL) for all analytes before method implementation. Section 1030 contains additional information and approaches for the evaluation of detection capabilities.

The MDL is defined in Section 1010C and elsewhere.[2] Determination of the MDL for each element is critical for complex matrices such as seawater, brines, and industrial effluents. The MDL will typically be higher than the IDL, because of background analyte in metals preparation and analysis laboratories and matrix-based interferences. Determine both IDL and MDL upon initial implementation of this method, and then yearly or whenever the instrument configuration changes or major maintenance occurs, whichever comes first.

Determine linear dynamic ranges (LDR) for all method analytes. LDR is defined as the maximum concentration of analyte above the highest calibration point where analyte response is within ± 10% of the theoretical response. When determining linear dynamic ranges, avoid using unduly high concentrations of analyte that might damage the detector. Determine LDR on multielement mixtures, to account for possible interelement effects. Determine LDR on initial implementation of this method, and then yearly.

c. Interferences: ICP/MS is subject to several types of interferences.

1) Isotopes of different elements that form ions of the same nominal mass-to-charge ratio are not resolved by the quadrupole mass spectrometer, and cause isobaric elemental interferences. Typically, ICP/MS instrument operating software will have all known isobaric interferences entered, and will perform necessary calculations automatically. Table 3125:II shows many of the commonly used corrections. Monitor the following additional masses: ^{83}Kr, ^{99}Ru, ^{118}Sn, and ^{125}Te. It is necessary to monitor these masses to correct for isobaric interference caused by ^{82}Kr on ^{82}Se, by ^{98}Ru on ^{98}Mo, by ^{114}Sn on ^{114}Cd, and by ^{123}Te on ^{123}Sb. Monitor ArCl at mass 77, to estimate chloride interferences. Verify that all elemental and molecular correction equations used in this method are correct and appropriate for the mass spectrometer used and sample matrix.

2) Abundance sensitivity is an analytical condition in which the tails of an abundant mass peak contribute to or obscure adjacent masses. Adjust spectrometer resolution to minimize these interferences.

3) Polyatomic (molecular) ion interferences are caused by ions consisting of more than one atom and having the same nominal

TABLE 3125:II. ELEMENTAL ABUNDANCE EQUATIONS AND COMMON
MOLECULAR ION CORRECTION EQUATIONS

Elemental and Molecular Equations*†

$$Li\ 6 = C\ 6$$
$$Be\ 9 = C\ 9$$
$$Al\ 27 = C\ 27$$
$$Sc\ 45 = C\ 45$$
$$V\ 51 = C\ 51 - (3.127)[(C\ 53) - (0.113 \times C\ 52)]$$
$$Cr\ 52 = C\ 52$$
$$Cr\ 53 = C\ 53$$
$$Mn\ 55 = C\ 55$$
$$Co\ 59 = C\ 59$$
$$Ni\ 60 = C\ 60$$
$$Ni\ 62 = C\ 62$$
$$Cu\ 63 = C\ 63$$
$$Cu\ 65 = C\ 65$$
$$Zn\ 66 = C\ 66$$
$$Zn\ 68 = C\ 68$$
$$As\ 75 = C\ 75 - (3.127)[(C\ 77) - (0.815 \times C\ 82)]$$
$$Se\ 77 = C\ 77$$
$$Se\ 82 = C\ 82 - (1.008696 \times C\ 83)$$
$$Sr\ 88 = C\ 88$$
$$Mo\ 98 = C\ 98 - (0.110588 \times C\ 101)$$
$$Rh\ 103 = C\ 103$$
$$Ag\ 107 = C\ 107$$
$$Ag\ 109 = C\ 109$$
$$Cd\ 111 = C\ 111 - (1.073)[(C\ 108) - (0.712 \times C\ 106)]$$
$$Cd\ 114 = C\ 114 - (0.02686 \times C\ 118)$$
$$Sb\ 121 = C\ 121$$
$$Sb\ 123 = C\ 123 - (0.127189 \times C\ 125)$$
$$Ba\ 135 = C\ 135$$
$$Ho\ 165 = C\ 165$$
$$Tl\ 203 = C\ 203$$
$$Tl\ 205 = C\ 205$$
$$Pb\ 208 = C\ 208 + (1 \times C\ 206) + (1 \times C\ 207)$$
$$Th\ 232 = C\ 232$$
$$U\ 238 = C\ 238$$

* C = calibration blank corrected counts at indicated mass.
† From EPA Method 200.8 for the Analysis of Drinking Waters − Application Note, Order No. ENVA-300A, The Perkin Elmer Corporation, 1996.

mass-to-charge ratio as the isotope of interest. Most of the common molecular ion interferences have been identified and are listed in Table 3125:III. Because of the severity of chloride ion interference on important analytes, particularly arsenic and selenium, hydrochloric acid is not recommended for use in preparation of any samples to be analyzed by ICP/MS. The mathematical corrections for chloride interferences only correct chloride to a concentration of 0.4%. Because chloride ion is present in most environmental samples, it is critical to use chloride correction equations for affected masses. A high-resolution ICP/MS may be used to resolve interferences caused by polyatomic ions. Polyatomic interferences are strongly influenced by instrument design and plasma operating conditions, and can be reduced in some cases by careful adjustment of nebulizer gas flow and other instrument operating parameters.

4) Physical interferences include differences in viscosity, surface tension, and dissolved solids between samples and calibration standards. To minimize these effects, dissolved solid levels in analytical samples should not exceed 0.5%. Dilute water and wastewater samples containing dissolved solids at or above 0.5% before analysis. Use internal standards for correction of physical interferences. Any internal standards used should demonstrate comparable analytical behavior to the elements being determined.

5) Memory interferences occur when analytes from a previous sample or standard are measured in the current sample. Use a sufficiently long rinse or flush between samples to minimize this type of interference. If memory interferences persist, they may be indications of problems in the sample introduction system. Severe memory interferences may require disassembly and cleaning of the entire sample introduction system, including the plasma torch, and the sampler and skimmer cones.

6) Ionization interferences result when moderate (0.1 to 1%) amounts of a matrix ion change the analyte signal. This effect, which usually reduces the analyte signal, also is known as "suppression." Correct for suppression by use of internal standardization techniques.

2. Apparatus

a. Inductively coupled plasma/mass spectrometer: Instrumentation, available from several manufacturers, includes a mass spectrometer detector, inductively coupled plasma source, mass flow controllers for regulation of ICP gas flows, peristaltic pump for sample introduction, and a computerized data acquisition and instrument control system. An x-y autosampler also may be used with appropriate control software.

b. Laboratory ware: Use precleaned plastic laboratory ware for standard and sample preparation. Teflon,* either tetrafluoroethylene hexafluoropropylene-copolymer (FEP), polytetrafluoroethylene (PTFE), or perfluoroalkoxy PTFE (PFA) is preferred for standard preparation and sample digestion, while high-density polyethylene (HDPE) and other dense, metal-free plastics may be acceptable for internal standards, known-addition solutions, etc. Check each new lot of autosampler tubes for suitability, and preclean autosampler tubes and pipettor tips (see Section 3010C.2).

c. Air displacement pipets, 10 to 100 μL, 100 to 1000 μL, and 1 to 10 mL size.

d. Analytical balance, accurate to 0.1 mg.

---

* Or equivalent.

TABLE 3125:III. COMMON MOLECULAR ION INTERFERENCES IN ICP/MS[1]

Molecular Ion	Mass	Element Measurement Affected by Interference
Background molecular ions:		
NH^+	15	—
OH^+	17	—
OH_2^+	18	—
C_2^+	24	Mg
CN^+	26	Mg
CO^+	28	Si
N_2^+	28	Si
N_2H^+	29	Si
NO^+	30	—
NOH^+	31	P
O_2^+	32	S
O_2H^+	33	—
$^{36}ArH^+$	37	Cl
$^{38}ArH^+$	39	K
$^{40}ArH^+$	41	—
CO_2^+	44	Ca
CO_2^+H	45	Sc
ArC^+, ArO^+	52	Cr
ArN^+	54	Cr
$ArNH^+$	55	Mn
ArO^+	56	Fe
ArH^+	57	Fe
$^{40}Ar^{36}Ar^+$	76	Se
$^{40}Ar^{38}Ar$	78	Se
$^{40}Ar_2^+$	80	Se
Matrix molecular ions:		
Bromide:		
$^{81}BrH^+$	82	Se
$^{79}BrO^+$	95	Mo
$^{81}BrO^+$	97	Mo
$^{81}BrOH^+$	09	Mo
$Ar^{81}Br^+$	121	Sb
Chloride:		
$^{35}ClO^+$	51	V
$^{35}ClOH^+$	52	Cr
$^{37}ClO^+$	53	Cr
$^{37}ClOH^+$	54	Cr
$Ar^{35}Cl^+$	75	As
$Ar^{37}Cl^+$	77	Se
Sulfate:		
$^{32}SO^+$	48	Ti
$^{32}SOH^+$	49	—
$^{34}SO^+$	50	V, Cr
$^{34}SOH^+$	51	V
SO_2^+, S_2^+	64	Zn
$Ar^{32}S^+$	72	Ge
$Ar^{34}S^+$	74	Ge
Phosphate:		
PO^+	47	Ti
POH^+	48	Ti
PO_2^+	63	Cu
ArP^+	71	Ga
Group I & II metals:		
$ArNa^+$	63	Cu
ArK^+	79	Br
$ArCa^+$	80	Se
Matrix oxides*		
TiO	62–66	Ni, Cu, Zn
ZrO	106–112	Ag, Cd
MoO	108–116	Cd
NbO	109	Ag

* Oxide interferences normally will be very small and will affect the method elements only when oxide-producing elements are present at relatively high concentrations, or when the instrument is improperly tuned or maintained. Preferably monitor Ti and Zr isotopes for soil, sediment, or solid waste samples, because these samples potentially contain high levels of these interfering elements.

e. Sample preparation apparatus, such as hot plates, microwave digestors, and heated sand baths. Any sample preparation device has the potential to introduce trace levels of target analytes to the sample.

f. Clean hood (optional), Class 100 (certified to contain less than 100 particles/m^3), for sample preparation and manipulation. Preferably perform all sample manipulations, digestions, dilutions, etc. in a certified Class 100 environment. Alternatively, handle samples in glove boxes, plastic fume hoods, or other environments where random contamination by trace metals can be minimized.

3. Reagents

a. Acids: Use ultra-high-purity grade (or equivalent) acids to prepare standards and to process sample. Redistilled acids are acceptable if each batch is demonstrated to be free from contamination by target analytes. Use extreme care in the handling of acids in the laboratory to avoid contamination of the acids with trace levels of metals.

1) *Nitric acid,* HNO$_3$, conc (specific gravity 1.41).

2) *Nitric acid,* 1 + 1: Add 500 mL conc HNO$_3$ to 500 mL reagent water.

3) *Nitric acid,* 2%: Add 20 mL conc HNO$_3$ to 100 mL reagent water; dilute to 1000 mL.

4) *Nitric acid,* 1%: Add 10 mL conc HNO$_3$ to 100 mL reagent water; dilute to 1000 mL.

b. Reagent water: Use water of the highest possible purity for blank, standard, and sample preparation (see Section 1080). Alternatively, use the procedure described below to produce water of acceptable quality. Other water preparation regimes may be used, provided that the water produced is metal-free. Reagent water containing trace amounts of analyte elements will cause erroneous results.

Produce reagent water using a softener/reverse osmosis unit with subsequent UV sterilization. After the general deionization system use a dual-column strong acid/strong base ion exchange system to polish laboratory reagent water before production of metal-free water. Use a multi-stage reagent water system, with two strong acid/strong base ion exchange columns and an activated carbon filter for organics removal for final polishing of laboratory reagent water. Use only high-purity water for preparation of samples and standards.

c. Stock, standard, and other required solutions: See 3120B.3*d* for preparation of standard stock solutions from elemental materials (pure metals, salts). Preferably, purchase high-purity commercially prepared stock solutions and dilute to required concentrations. Single- or multi-element stock solutions (1000 mg/L) of the following elements are required: aluminum, antimony, arsenic, barium, beryllium, cerium, cadmium, chromium, cobalt, copper, germanium, indium, lead, magnesium, manganese, molybdenum, nickel, rhodium, scandium, selenium, silver, strontium, terbium, thallium, thorium, uranium, vanadium, and zinc. Prepare internal standard stock separately from target element stock solution. The potential for incompatibility between target elements and/or internal standards exists, and could cause precipitation or other solution instability.

1) *Internal standard stock solution:* Lithium, scandium, germanium, indium, and thorium are suggested as internal standards. The following masses are monitored: ^{6}Li, ^{45}Sc, ^{72}Ge, ^{115}In, and ^{232}Th. Add to all samples, standards, and quality control (QC)

samples a level of internal standard that will give a suitable counts/second (cps) signal (for most internal standards, 200 000 to 500 000 cps; for lithium, 20 000 to 70 000 cps). Minimize error introduced by dilution during this addition by using an appropriately high concentration of internal standard mix solution. Maintain volume ratio for all internal standard additions.

Prepare internal standard mix as follows: Prepare a nominal 50-mg/L solution of ^{6}Li by dissolving 0.15 g ^{6}Li$_2$CO$_3$ (isotopically pure, i.e., 95% or greater purity†) in a minimal amount of 1:1 HNO$_3$. Pipet 5.0 mL 1000-mg/L scandium, germanium, indium, and thorium standards into the lithium solution, dilute resulting solution to 500.0 mL, and mix thoroughly. The resultant concentration of Sc, Ge, In, and Th will be 10 mg/L. Older instruments may require higher levels of internal standard to achieve acceptable levels of precision.

Other internal standards, such as rhodium, yttrium, terbium, holmium, and bismuth may also be used in this method. Ensure that internal standard mix used is stable and that there are no undesired interactions between elements.

Screen all samples for internal standard elements before analysis. The analysis of a few representative samples for internal standards should be sufficient. Analyze samples ''as received'' or ''as digested'' (before addition of internal standard), then add internal standard mix and reanalyze. Monitor counts at the internal standard masses. If the ''as received'' or ''as digested'' samples show appreciable detector counts (10% or higher of samples with added internal standard), dilute sample or use an alternate internal standard. If the internal standard response of the sample with the addition is not within 70 to 125% of the response for a calibration blank with the internal standard added, either dilute the sample before analysis, or use an alternate internal standard. During actual analysis, monitor internal standard masses and note all internal standard recoveries over 125% of internal standard response in calibration blank. Interpret results for these samples with caution.

The internal standard mix may be added to blanks, standards, and samples by pumping the solution so it is mixed with the sample stream in the sample introduction process.

2) *Instrument optimization/tuning solution,* containing the following elements: barium, beryllium, cadmium, cerium, cobalt, copper, germanium, indium, magnesium, rhodium, scandium, terbium, thallium, and lead. Prepare this solution in 2% HNO$_3$. This mix includes all common elements used in optimization and tuning of the various ICP/MS operational parameters. It may be possible to use fewer elements in this solution, depending on the instrument manufacturer's recommendations.

3) *Calibration standards,* 0, 5, 10, 20, 50, and 100 µg/L.‡ Other calibration regimes are acceptable, provided the full suite of quality assurance samples and standards is run to validate these method changes. Fewer standards may be used, and a two-point blank/mid range calibration technique commonly used in ICP optical methods should also produce acceptable results. Calibrate all analytes using the selected concentrations. Prepare all calibration standards and blanks in a matrix of 2% nitric acid. Add internal standard mix to all calibration standards to provide appropriate count rates for interference correction. NOTE: All standards and blanks used in this method have the internal standard mix added at the same ratio.

---

† Cambridge Isotope Laboratories or equivalent.
‡ Performance data for the method were obtained with these concentrations.

TABLE 3125:IV. SUGGESTED ANALYTICAL RUN SEQUENCE

Sample Type	Comments
Tuning/optimization standard	Check mass calibration and resolution
Tuning/optimization standard	Optimize instrument for maximum rhodium counts while keeping oxides, double charged ions, and background within instrument specifications
Rinse	—
Reagent blank	Check for contamination
Reagent blank	Calibration standard blank
5-μg/L standard	—
10-μg/L standard	—
20-μg/L standard	—
50-μg/L standard	—
100-μg/L standard	—
Rinse	—
Initial calibration verification, 50 μg/L	—
Initial calibration blank	—
0.30-μg/L standard	Low-level calibration verification
1.0-μg/L standard	Low-level calibration verification
External reference material	NIST 1643c or equivalent
Continuing calibration verification	—
Continuing blank calibration	—
Project sample method blank	—
Project sample laboratory-fortified blank	—
Project sample 1–4	—
Project sample 5	—
Project sample 5 with known addition	—
Project sample 5 duplicate with known addition	—
Continuing calibration verification	—
Continuing calibration blank	—

4) *Method blank*, consisting of reagent water (¶ 3*b*) taken through entire sample preparation process. For dissolved samples,

TABLE 3125:V. SUMMARY OF PERFORMANCE CRITERIA

Performance Characteristic	Criteria
Mass resolution	Manufacturer's specification
Mass calibration	Manufacturer's specification
Ba^{2+}/Ba^+	Manufacturer's specification
CeO/Ce	Manufacturer's specification
Background counts at mass 220	Manufacturer's specification
Correlation coefficient	≥ 0.995
Calibration blanks	< Reporting limit
Calibration verification standards	$\pm 10\%$ of true value
Laboratory fortified blank (control sample)	$\pm 30\%$ of true value
Precision	$\pm 20\%$ relative percent difference for lab duplicates
Known-addition recovery	75–125%
0.3 and 1.0 μg/L standards	Dependent on data quality objectives
Reference materials	Dependent on data quality objectives
Internal standard response	70–125% of response in calibration blank with known addition

take reagent water through same filtration and preservation processes used for samples. For samples requiring digestion, process reagent water with the same digestion techniques as samples. Add internal standard mix to method blank.

5) *Calibration verification standard:* Prepare a mid-range standard, from a source different from the source of the calibration standards, in 2% HNO_3, with equivalent addition of internal standard.

6) *Calibration verification blank:* Use 2% HNO_3.

7) *Laboratory fortified blank* (optional)*:* Prepare solution with 2% nitric acid and method analytes added at about 50 μg/L. This standard, sometimes called a laboratory control sample (LCS), is used to validate digestion techniques and known-addition levels.

8) *Reference materials:* Externally prepared reference material, preferably from National Institute of Standards and Technology (NIST) 1643 series or equivalent.

9) *Known-addition solution for samples:* Add stock standard to sample in such a way that volume change is less than 5%. In the absence of information on analyte levels in the sample, prepare known additions at around 50 μg/L. If analyte concentration levels are known, add at 50 to 200% of the sample levels. For samples undergoing digestion, make additions before digestion. For the determination of dissolved metals, make additions after filtration, preferably immediately before analysis.

10) *Low-level standards:* Use both a 0.3- and a 1.0-μg/L standard when expected analyte concentration is below 5 μg/L. Prepare both these standards in 2% nitric acid.

Prepare volumetrically a mixed standard containing the method analytes at desired concentration(s) (0.30 μg/L, 1.0 μg/L, or both). Prepare weekly in 100-mL quantities.

TABLE 3125:VI. QUALITY CONTROL ANALYSES FOR ICP/MS METHOD

Analysis	Frequency	Acceptance Criteria
Reference material [¶3*c*9)]	Greater of: once per sample batch, or 5%	Dependent on data quality objectives
Preparatory/method blank [¶ 3*c*4)]	Greater of: once per sample batch, or 5%	± Absolute value of instrument detection limit; ± absolute value of laboratory reporting limit or MDL is acceptable
Laboratory fortified blank [¶ 3*c*7)]	Greater of: once per sample batch, or 5%	± 30% of true value
Duplicate known-addition samples	Greater of: once per sample batch, or 5%	± 20% relative percent difference
Continuing calibration verification standards [¶ 3*c*5)]	10%	± 10% of known concentration
Continuing calibration verification blank [¶ 3*c*6)]	10%	± Absolute value of instrument detection limit; ± absolute value of laboratory reporting limit or MDL is acceptable

d. Argon: Use a prepurified grade of argon unless it can be demonstrated that other grades can be used successfully. The use of prepurified argon is usually necessary because of the presence of krypton as an impurity in technical argon. ^{82}Kr interferes with the determination of ^{82}Se. Monitor ^{83}Kr at all times.

4. Procedures

a. Sample preparation: See Sections 3010 and 3020 for general guidance regarding sampling and quality control. See Section 3030E for recommended sample digestion technique for all analytes except silver and antimony. If silver and antimony are target analytes, use method given in 3030F, paying special attention to interferences caused by chloride ion, and using all applicable elemental corrections. Alternative digestion techniques and additional guidance on sample preparation are available.[3,4]

Ideally use a ''clean'' environment for any sample handling, manipulation, or preparation. Preferably perform all sample manipulations in a Class 100 clean hood or room to minimize potential contamination artifacts in digested or filtered samples.

b. Instrument operating conditions: Follow manufacturer's standard operating procedures for initialization, mass calibration, gas flow optimization, and other instrument operating conditions. Maintain complete and detailed information on the operational status of the instrument whenever it is used.

c. Analytical run sequence: A suggested analytical run sequence, including instrument tuning/optimization, checking of re-agent blanks, instrument calibration and calibration verification, analysis of samples, and analysis of quality control samples and blanks, is given in Table 3125:IV.

d. Instrument tuning and optimization: Follow manufacturer's instructions for optimizing instrument performance. The most important optimization criteria include nebulizer gas flows, detector and lens voltages, radio-frequency forward power, and mass calibration. Periodically check mass calibration and instrument resolution. Ideally, optimize the instrument to minimize oxide formation and doubly-charged species formation. Measure the CeO/Ce ratio to monitor oxide formation, and measure doubly-charged species by determination of the Ba^{2+}/Ba^{+} ratio. Both these ratios should meet the manufacturer's criteria before instrument calibration. Monitor background counts at mass 220 after optimization and compare with manufacturer's criteria. A summary of performance criteria related to optimization and tuning, calibration, and analytical performance for this method is given in Table 3125:V.

e. Instrument calibration: After optimization and tuning, calibrate instrument using an appropriate range of calibration standards. Use appropriate regression techniques to determine calibration lines or curves for each analyte. For acceptable calibrations, correlation coefficients for regression curves are ideally 0.995 or greater.

Immediately after calibration, run initial calibration verification standard, ¶ 3*c*5); acceptance criteria are ±10% of known analyte concentration. Next run initial calibration verification blank, ¶

TABLE 3125:VII. METHOD PERFORMANCE WITH CALIBRATION VERIFICATION STANDARDS*

Element	Mass	Continuing Calibration Verification Standard (N = 44)				Initial Calibration Verification Standard (N = 12)			
		Mean Recovery %	Mean	Standard Deviation	Relative Standard Deviation %	Mean Recovery %	Mean	Standard Deviation	Relative Standard Deviation %
Be	9	98.71	49.35	3.43	6.94	100.06	50.03	1.90	3.80
Al	27	99.62	49.81	2.99	6.01	98.42	49.21	1.69	3.44
V	51	100.97	50.48	1.36	2.68	99.91	49.96	1.23	2.47
Cr	52	101.39	50.70	1.86	3.66	99.94	49.97	1.47	2.95
Cr	53	100.68	50.34	1.91	3.79	99.13	49.56	1.44	2.90
Mn	55	101.20	50.60	1.98	3.91	99.48	49.74	1.40	2.82
Co	59	101.67	50.83	2.44	4.79	99.44	49.72	1.61	3.24
Ni	60	99.97	49.99	2.14	4.28	97.98	48.99	1.70	3.47
Ni	62	99.79	49.89	2.09	4.18	97.57	48.79	1.32	2.71
Cu	63	100.51	50.25	2.19	4.36	97.87	48.93	1.63	3.33
Cu	65	100.39	50.19	2.26	4.51	98.34	49.17	1.58	3.20
Zn	66	101.07	50.53	1.93	3.82	98.75	49.38	0.87	1.76
Zn	68	100.42	50.21	1.89	3.77	97.75	48.87	0.50	1.02
As	75	100.76	50.38	1.15	2.28	98.83	49.41	0.89	1.80
Se	77	101.71	50.85	1.43	2.81	99.54	49.77	1.01	2.03
Se	82	101.97	50.98	1.50	2.95	99.76	49.88	0.94	1.89
Ag	107	101.50	50.75	1.68	3.30	99.27	49.63	1.17	2.36
Ag	109	101.65	50.83	1.68	3.31	99.66	49.83	1.54	3.08
Cd	111	100.92	50.46	1.94	3.84	98.61	49.30	1.36	2.77
Cd	114	100.90	50.45	2.07	4.10	99.20	49.60	1.41	2.84
Sb	121	100.14	50.07	2.39	4.77	99.38	49.69	1.38	2.78
Sb	123	99.98	49.99	2.48	4.97	99.09	49.54	1.34	2.71
Tl	203	101.36	50.68	1.64	3.23	100.05	50.02	1.01	2.01
Tl	205	102.40	51.20	1.93	3.78	101.23	50.62	1.45	2.87
Pb	208	101.21	50.61	1.65	3.25	99.33	49.67	0.84	1.69
U	238	101.54	50.77	1.93	3.80	99.80	49.90	1.36	2.72

* Single-laboratory, single-operator, single-instrument data, determined using a 50-µg/L standard prepared from sources independent of calibration standard source. Data acquired January-November 1996 during actual sample determinations. Performance of continuing calibration verification standards at different levels may vary. Perkin-Elmer Elan 6000 ICP/MS used for determination.

3c6); acceptance criteria are ideally ± the absolute value of the instrument detection limit for each analyte, but in practice, ± the absolute value of the laboratory reporting limit or the laboratory method detection limit for each analyte is acceptable. Verify low-level calibration by running 0.3- and/or 1.0-μg/L standards, if analyte concentrations are less than 5 μg/L.

f. Sample analysis: Ensure that all vessels and reagents are free from contamination. During analytical run (see Table 3125:IV), include quality control analyses according to schedule of Table 3125:VI, or follow project-specific QA/QC protocols.

Internal standard recoveries must be between 70% and 125% of internal standard response in the laboratory-fortified blank; otherwise, dilute sample, add internal standard mix, and reanalyze.

Make known-addition analyses for each separate matrix in a digestion or filtration batch.

5. Calculations and Corrections

Configure instrument software to report internal standard corrected results. For water samples, preferably report results in micrograms per liter. Report appropriate number of significant figures.

a. Correction for dilutions and solids: Correct all results for dilutions, and raise reporting limit for all analytes reported from the diluted sample by a corresponding amount. Similarly, if results for solid samples are to be determined, use Method 2540B to determine total solids. Report results for solid samples as mi-

crograms per kilogram, dry weight. Correct all results for solids content of solid samples. Use the following equation to correct solid or sediment sample results for dilution during digestion and moisture content:

$$R_{corr} = \frac{R_{uncorr} \times V}{W \times \% \ TS/100}$$

where:

R_{corr} = corrected result, μg/kg,
R_{uncorr} = uncorrected elemental result, μg/L,
V = volume of digestate (after digestion), L,
W = mass of the wet sample, kg, and
$\% \ TS$ = percent total solids determined in the solid sample.

b. Compensation for interferences: Use instrument software to correct for interferences listed previously for this method. See Table 3125:III for a listing of the most common molecular ion interferences.

c. Data reporting: Establish appropriate reporting limits for method analytes based on instrument detection limits and the laboratory blank. For regulatory programs, ensure that reporting limits for method analytes are a factor of three below relevant regulatory criteria.

If method blank contamination is typically random, sporadic, or otherwise not in statistical control, do not correct results for the method blank. Consider the correction of results for laboratory

TABLE 3125:VIII. METHOD PERFORMANCE FOR RECOVERY OF KNOWN ADDITION IN NATURAL WATERS*

| Element | Mass | Total Recoverable Metals† | | Dissolved Metals‡ | |
		Mean Recovery %	Relative Standard Deviation %	Mean Recovery %	Relative Standard Deviation %
Be	9	89.09	5.77	—	—
V	51	87.00	8.82	—	—
Cr	52	87.33	8.42	88.38	6.43
Cr	53	86.93	7.90	88.52	5.95
Mn	55	91.81	10.12	—	—
Co	59	87.67	8.92	—	—
Ni	60	85.07	8.42	89.31	5.70
Ni	62	84.67	8.21	89.00	5.82
Cu	63	84.13	8.46	88.55	8.33
Cu	65	84.37	8.05	88.26	7.80
Zn	66	86.14	23.01	95.59	13.81
Zn	68	81.95	20.31	91.94	13.27
As	75	90.43	4.46	97.30	8.84
Se	77	83.09	4.76	105.36	10.80
Se	82	83.42	4.73	105.36	10.75
Ag	107	—	—	91.98	5.06
Ag	109	—	—	92.25	4.96
Cd	111	91.37	5.47	96.91	6.03
Cd	114	91.47	6.04	97.03	5.42
Sb	121	94.40	5.24	—	—
Sb	123	94.56	5.36	—	—
Tl	203	97.24	5.42	—	—
Tl	205	98.14	6.21	—	—
Pb	208	96.09	7.08	100.69	7.28

* Single-laboratory, single-operator, single-instrument data. Samples were Washington State surface waters from various locations. Data acquired January-November 1996 during actual sample determinations. Performance of known additions at different levels may vary. Perkin-Elmer Elan 6000 ICP/MS used for determination.
† Known-addition level 20 μg/L. Additions made before preparation according to Section 3030E (modified by cleanhood digestion in TFE beakers). N = 20.
‡ Known-addition level for Cd and Pb 1 μg/L; for other analytes 10 μg/L. Additions made after filtration through 1:1 HNO₃ precleaned 0.45-μm filters. N = 28.

TABLE 3125:IX. METHOD PERFORMANCE WITH LOW-LEVEL CHECK STANDARDS*

Element	Mass	1.0-μg/L Standard				0.3-μg/L Standard			
		Mean Recovery %	Mean	Standard Deviation	Relative Standard Deviation %	Mean Recovery %	Mean	Standard Deviation	Relative Standard Deviation %
Be	9	97	0.97	0.06	6.24	95	0.284	0.03	12.11
Al	27	121	1.21	0.32	26.49	196	0.588	0.44	74.30
V	51	104	1.04	0.06	5.83	111	0.332	0.10	28.96
Cr	52	119	1.19	0.34	28.62	163	0.490	0.37	75.90
Cr	53	102	1.02	0.36	35.54	113	0.338	0.32	93.70
Mn	55	103	1.03	0.07	6.55	110	0.329	0.08	25.64
Co	59	103	1.03	0.07	6.42	102	0.307	0.04	12.53
Ni	60	101	1.01	0.05	5.24	107	0.321	0.05	14.14
Ni	62	102	1.02	0.06	5.42	109	0.326	0.05	15.94
Cu	63	107	1.07	0.09	8.78	118	0.355	0.06	18.29
Cu	65	107	1.07	0.10	9.05	117	0.352	0.06	17.69
Zn	66	117	1.17	0.51	43.52	182	0.547	0.68	124.13
Zn	68	116	1.16	0.50	42.90	179	0.537	0.66	122.12
As	75	97	0.97	0.05	5.23	101	0.302	0.06	18.29
Se	77	89	0.89	0.08	8.72	88	0.265	0.08	29.07
Se	82	92	0.92	0.14	15.50	106	0.317	0.14	43.91
Ag	107	101	1.01	0.05	4.53	94	0.282	0.04	15.74
Ag	109	103	1.03	0.07	6.57	92	0.277	0.04	13.68
Cd	111	98	0.98	0.04	3.80	96	0.288	0.03	8.74
Cd	114	100	1.00	0.03	3.39	98	0.293	0.03	8.70
Sb	121	94	0.94	0.05	5.28	93	0.280	0.06	21.89
Sb	123	94	0.94	0.05	5.36	93	0.278	0.06	22.39
Tl	203	101	1.01	0.04	3.57	98	0.294	0.03	11.89
Tl	205	104	1.04	0.05	5.15	100	0.300	0.03	10.43
Pb	208	104	1.04	0.04	3.65	104	0.312	0.03	11.13
U	238	106	1.06	0.05	4.64	102	0.307	0.03	9.92

* Single-laboratory, single-operator, single-instrument data. $N = 24$ for both standards.

method blanks only if it can be demonstrated that the concentration of analytes in the method blank is within statistical control over a period of months. Report all method blank data explicitly in a manner identical to sample reporting procedures.

d. Documentation: Maintain documentation for the following (where applicable): instrument tuning, mass calibration, calibration verification, analyses of blanks (method, field, calibration, and equipment blanks), IDL and MDL studies, analyses of samples and duplicates with known additions, laboratory and field duplicate information, serial dilutions, internal standard recoveries, and any relevant quality control charts.

Also maintain, and keep available for review, all raw data generated in support of the method.[5]

6. Method Performance

Table 3125:I presents instrument detection limit (IDL) data generated by this method; this represents optimal state-of-the-art instrument detection capabilities, not recommended method detection or reporting limits. Tables 3125:VII through IX contain single-laboratory, single-operator, single-instrument performance data generated by this method for calibration verification standards, low-level standards, and known-addition recoveries for fresh-water matrices. Performance data for this method for some analytes are not currently available. However, performance data for similar ICP/MS methods are available in the literature.[1,4]

7. References

1. U.S. ENVIRONMENTAL PROTECTION AGENCY. 1994. Determination of trace elements in waters and wastes by inductively coupled plasma-mass spectrometry, Method 200.8. U.S. Environmental Protection Agency, Environmental Monitoring Systems Lab., Cincinnati, Ohio.
2. U.S. ENVIRONMENTAL PROTECTION AGENCY. 1984. Definition and procedure for the determination of the method detection limit, revision 1.11. 40 CFR 136, Appendix B.
3. U.S. ENVIRONMENTAL PROTECTION AGENCY. 1991. Methods for the determination of metals in environmental samples. U.S. Environmental Protection Agency, Off. Research & Development, Washington D.C.
4. U.S. ENVIRONMENTAL PROTECTION AGENCY. 1995. Method 1638: Determination of trace elements in ambient waters by inductively coupled plasma mass spectrometry. U.S. Environmental Protection Agency, Off. Water, Washington, D.C.
5. U.S. ENVIRONMENTAL PROTECTION AGENCY. 1995. Guidance on the Documentation and Evaluation of Trace Metals Data Collected for Clean Water Act Compliance Monitoring. U.S. Environmental Protection Agency, Off. Water, Washington, D.C.

8. Bibliography

GRAY, A.L. 1974. A plasma source for mass analysis. *Proc. Soc. Anal. Chem.* 11:182.
HAYHURST, A.N. & N.R. TELFORD. 1977. Mass spectrometric sampling of ions from atmospheric pressure flames. I. Characteristics and calibration of the sampling system. *Combust. Flame.* 67.
HOUK, R.S., V.A. FASSEL, G.D. FLESCH, H.J. SVEC, A.L. GRAY & C.E. TAYLOR. 1980. Inductively coupled argon plasma as an ion source

for mass spectrometric determination of trace elements. *Anal. Chem.* 52:2283.

Douglas, D.J. & J.B. French. 1981. Elemental analysis with a microwave-induced plasma/quadrupole mass spectrometer system. *Anal. Chem.* 53:37.

Houk, R.S., V.A. Fassel & H.J. Svec. 1981. Inductively coupled plasma-mass spectrometry: Sample introduction, ionization, ion extraction and analytical results. *Dyn. Mass Spectrom.* 6:234.

Olivares, J.A. & R.S. Houk. 1985. Ion sampling for inductively coupled plasma mass spectrometry. *Anal. Chem.* 57:2674.

Houk, R.S. 1986. Mass spectrometry of inductively coupled plasmas. *Anal. Chem.* 58:97.

Thompson, J.J. & R.S. Houk. 1986. Inductively coupled plasma mass spectrometric detection for multielement flow injection analysis and elemental speciation by reversed-phase liquid chromatography. *Anal. Chem.* 58:2541.

Vaughan, M.A. & G. Horlick. 1986. Oxide, hydroxide, and doubly charged analyte species in inductively coupled plasma/mass spectrometry. *Appl. Spectrosc.* 40:434.

Garbarino, J.R. & H.E. Taylor. 1987. Stable isotope dilution analysis of hydrologic samples by inductively coupled plasma mass spectrometry. *Anal. Chem.* 59:1568.

Beauchemin, D., J.W. McLaren, A.P. Mykytiuk & S.S. Berman. 1987. Determination of trace metals in a river water reference material by inductively coupled plasma mass spectrometry. *Anal. Chem.* 59:778.

Thompson, J.J. & R.S. Houk. 1987. A study of internal standardization in inductively coupled plasma-mass spectrometry. *Appl. Spectrosc.* 41:801.

Jarvis, K.E., A.L. Gray & R.S. Houk. 1992. Inductively Coupled Plasma Mass Spectrometry. Blackie Academic & Professional, Chapman & Hall, New York, N.Y.

Taylor, D.B., H.M. Kingston, D.J. Nogay, D. Koller & R. Hutton. 1996. On-line solid-phase chelation for the determination of eight metals in environmental waters by ICP-MS. *JAAS* 11:187.

Kingston, H.M.S. & S. Haswell, eds. 1997. Microwave Enhanced Chemistry: Fundamentals, Sample Preparation, and Applications. ACS Professional Reference Book Ser., American Chemical Soc., Washington, D.C.

U.S. Environmental Protection Agency. 1998. Inductively coupled plasma-mass spectrometry, Method 6020. *In* Solid Waste Methods. SW846, Update 4, U.S. Environmental Protection Agency, Environmental Monitoring Systems Lab., Cincinnati, Ohio.

3130   METALS BY ANODIC STRIPPING VOLTAMMETRY*

3130 A.   Introduction

Anodic stripping voltammetry (ASV) is one of the most sensitive metal analysis techniques; it is as much as 10 to 100 times more sensitive than electrothermal atomic absorption spectroscopy for some metals. This corresponds to detection limits in the nanogram-per-liter range. The technique requires no sample extraction or preconcentration, it is nondestructive, and it allows simultaneous determination of four to six trace metals, utilizing inexpensive instrumentation. The disadvantages of ASV are that it is restricted to amalgam-forming metals, analysis time is longer than for spectroscopic methods, and interferences and high sensitivity can present severe limitations. The analysis should be performed only by analysts skilled in ASV methodology because of the interferences and potential for trace background contamination.

---

* Approved by Standard Methods Committee, 1997.

3130 B.   Determination of Lead, Cadmium, and Zinc

1. General Discussion

a. Principle: Anodic stripping voltammetry is a two-step electroanalytical technique. In the preconcentration step, metal ions in the sample solution are reduced at negative potential and concentrated into a mercury electrode. The concentration of the metal in the mercury is 100 to 1000 times greater than that of the metal ion in the sample solution. The preconcentration step is followed by a stripping step applying a positive potential scan. The amalgamated metal is oxidized rapidly and the accompanying current is proportional to metal concentration.

b. Detection limits and working range: The limit of detection for metal determination using ASV depends on the metal determined, deposition time, stirring rate, solution pH, sample matrix, working electrode (hanging mercury drop electrode, HMDE, or thin mercury film electrode, TMFE), and mode of the stripping potential scan (square wave or differential pulse). Cadmium, lead, and zinc are concentrated efficiently during pre-electrolysis because of their high solubility in mercury and thus have low detection limits (<1 µg/L). Long deposition times and high stirring rates increase the concentration of metal preconcentrated in the mercury phase and reduce detection limits. The effects of solution pH and matrix are more complicated. In general, add a high concentration of inert electrolyte to samples to maintain a high, constant ionic strength. Acidify sample to a low pH or add a pH buffer. If the pH buffer or other component of the sample matrix complexes the metal (3130B.1c), detection limits often are increased.

The choice of working electrode is determined largely by the working range of concentration required. The HMDE is best suited for analysis from approximately 1 µg/L to 10 mg/L, while the TMFE is superior for detection below 1 µg/L.

c. Interferences: Major interferences include intermetallic compound formation, overlapping stripping peaks, adsorption of organics, and complexation. Intermetallic compounds can form in the mercury phase when high concentrations of certain metals are present simultaneously. Zinc forms intermetallic compounds with cobalt and nickel, and both zinc and cadmium form intermetallic compounds with copper, silver, and gold. As a result, the stripping peak for the constituent metals may be severely depressed or shifted and additional peaks due to intermetallic compound stripping may be observed. Minimize or avoid intermetallic compound formation by use of a hanging mercury drop electrode instead of a thin film mercury electrode when metal concentrations are above 1 µg/L, application of a preconcentration potential sufficiently negative to reduce the desired but not the interfering metal, and use of a relatively short preconcentration period followed by a relatively large pulse modulation (50 mV) during the stripping stage. In general, suspect formation of intermetallic compounds if metals are present in concentrations above 1 mg/L. If metals are present at concentrations above 10 mg/L, do not use anodic stripping voltammetry. Concentrations above 10 mg/L usually can be quantitated by methods such as those given in Sections 3111 and 3120.

Separate overlapping stripping peaks by various methods, including appropriate choice of buffer and electrolyte.[1-3] If only one of the metal peaks is of interest, eliminate interfering peaks by selective complexation with a suitable ligand, such as EDTA. Judicious choice of preconcentration potential can result in the deposition of the selected metal but not the interfering metal in the mercury electrode. Selection of buffer/ligand also may help to distinguish metals during the preconcentration step. Alternatively, use "medium exchange," in which preconcentration is performed with the electrodes in the sample and stripping is performed in a different electrolyte solution. In this procedure, metals are deposited from the sample into the amalgam as usual, but they may be stripped into a medium that provides different stripping peak potentials for the overlapping metals.

Minimize interferences from adsorption of organic compounds and complexation by removal of the organic matter. Digest samples with high-purity acids as described in Section 3030. Make standard additions to determine if complexation or adsorption remains a problem. Analyze a metal-free solution with a matrix similar to that of the sample both before and after addition of known quantities of standard. Repeat procedure for sample. If the slope of the stripping current versus added metal is significantly different in the sample relative to the metal-free solution, digest sample further. The choice of stripping waveform also is important. While both square-wave and differential-pulse stripping attempt to minimize the contribution of adsorption currents to the total measured stripping current, square-wave stripping does this more effectively. Thus, use square-wave stripping instead of differential-pulse stripping when adsorption occurs.

2. Apparatus

a. Electrochemical analyzer: The basic electrochemical analyzer for ASV applications contains a three-electrode potentiostat, which very precisely controls potential applied to the working electrode relative to the reference electrode, and a sensitive current measuring device. It is capable of delivering potential pulses of various amplitudes and frequencies, and provides several scan rates and current ranges. More advanced ASV instruments offer automated timing, gas purge and stirring, and data processing routines including curve smoothing, baseline correction, and background subtraction.

Two variations of stripping waveforms are commonly used: differential pulse (DPASV) and square wave (SWASV) waveforms. The differential-pulse waveform consists of a series of pulses superimposed on a linear voltage ramp, while the square-wave waveform consists of a series of pulses superimposed on a staircase potential waveform. Square-wave stripping is significantly faster than differential-pulse stripping and is typically ten times more sensitive. Most commercially available ASV instruments perform both differential-pulse and square-wave stripping.

b. Electrodes and cell: Provide working, reference, and auxiliary electrodes. Working electrodes are either hanging mercury drop or thin mercury film electrodes. Hanging mercury drop electrodes must be capable of dispensing mercury in very precisely controlled drop sizes. Three types of electrodes meet this requirement: static mercury drop electrodes, controlled growth mercury drop electrodes, or Kemula-type electrodes. In any case, use a drop knocker to remove an old drop before dispensing a fresh mercury drop.

When the lowest detection limits are required, a thin mercury film electrode is preferred. This electrode consists of a rotating glassy carbon disk plated with mercury in situ during preconcentration of the analyte. A high-precision, constant-speed rotator controls the rotation rate of the electrode and provides reproducible mass transport.

Reference electrodes may be either saturated calomel or silver/silver chloride electrodes. Use a platinum wire for the auxiliary electrode.

Use cells constructed of glass, or preferably fused silica or TFE, because they are more resistant to solution adsorption or leaching. Cover cell with a lid that provides reproducible placement of the electrodes and gas purging tubes. Provide an additional hole in the cell lid for addition of standards. Most commercially available mercury drop electrodes include electrolytic cells, reference and auxiliary electrodes, and gas purging tubes.

Use a constant-speed stirring mechanism to provide reproducible mass transport in samples and standards.

Locate the cell in an area where temperature is relatively constant. Alternatively, use a constant-temperature water bath and cell jacket.

c. Oxygen-removal apparatus: Oxygen interferes in electrochemical analyses; remove it from solution before preconcentration by purging with nitrogen or argon. Provide two gas inlet tubes through the cell lid: one extends into the solution and the second purges the space above the solution. A gas outlet hole in the lid provides for removal of oxygen and excess purging gas.

d. Recording device: If the electrochemical analyzer is not equipped with a digital data acquisition system, use an XY plotter to record stripping voltammograms.

e. Timer: If preconcentration and equilibration periods are not controlled by the instrument, use an accurate timing device.

f. Polishing wheel: To obtain the high polish required for a glassy carbon disk electrode, use a motorized polishing wheel.

3. Reagents

CAUTION: *Follow proper practices for disposal of any solutions containing mercury.*

a. Metal-free water: Use deionized water to prepare buffers, electrolytes, standards, etc. Use water with at least 18 megohm-cm resistivity (see Section 1080).

b. Nitric acid, HNO_3, conc, high-purity.*

c. Nitric acid, HNO_3, 6N, 1.6N (10%), and 0.01N.

d. Purging gas (nitrogen or argon), high-purity. Remove traces of oxygen in nitrogen or argon gas before purging the solution. Pass gas through sequential scrubbing columns containing vanadous chloride in the first, deionized water in the second, and buffer (or electrolyte) solution in the third column.

e. Metal standards: Prepare stock solutions containing 1 mg metal/mL in polyethylene bottles. Purchase these solutions commercially or prepare as in Section 3111. Daily prepare dilutions of stock standards in a matrix similar to that of the samples to cover the concentration range desired.

f. Electrolyte/buffer: Use one of the following:

1) *Acetate buffer,* pH 4.5: Dissolve 16.4 g anhydrous sodium acetate, $NaC_2H_3O_2$, in 800 mL water. Adjust to pH 4.5 with high-purity glacial acetic acid.* Dilute to 1 L with water.

2) *Citrate buffer,* pH 3: Dissolve 42.5 g citric acid (monohydrate) in 700 mL water. Adjust to pH 3 with high-purity NH_4OH.* Dilute to 1 L with water.

3) *Phosphate buffer,* pH 6.8: Dissolve 24 g NaH_2PO_4 in 500 mL water. Adjust to pH 6.8 with 1N NaOH. Dilute to 1 L with water.

g. Mercury: Use commercially available triply distilled metallic mercury for hanging mercury drop electrodes. CAUTION: *Mercury vapors are highly toxic. Use only in well-ventilated area.*

h. Mercuric nitrate solution, $Hg(NO_3)_2$: For thin mercury film electrodes, dissolve 0.325 g $Hg(NO_3)_2$ in 100 mL 0.01N HNO_3.

i. Reference electrode filling solution: Available from electrode manufacturer.

j. Amalgamated zinc: Dissolve 2 g $Hg(NO_3)_2$ in 25 mL conc HNO_3; dilute to 250 mL with water. In a separate beaker, clean approximately 50 g mossy zinc by gently oxidizing the surface with 10% HNO_3 and rinse with water. Add $Hg(NO_3)_2$ solution to cleaned zinc and stir with a glass rod. If barely visible bubbles do not appear, add a small amount of 6N HNO_3. Zinc should rapidly acquire a shiny, metallic appearance. Decant solution and store for amalgamating future batches of zinc. Rinse amalgamated zinc copiously with water and transfer to a gas scrubbing column.

k. Hydrochloric acid, HCl, conc.

l. Vanadous chloride: Add 2 g ammonium metavanadate, NH_4VO_3, to 25 mL conc HCl and heat to boiling. Solution should turn blue-green. Dilute to 250 mL with water. Pour solution into gas scrubbing column packed with amalgamated zinc and bubble purging gas through it until the solution turns a clear violet color. When the violet color is replaced by a blue, green, or brown color, regenerate vanadous chloride by adding HCl.

m. Siliconizing solution: Preferably use commercially available solutions in sealed ampules for siliconizing capillaries used for hanging mercury drops. CAUTION: *Most commercial siliconizing reagents contain CCl_4, a toxic and cancer-suspect agent. Handle with gloves and avoid breathing vapors.*

n. Alumina suspensions, 1, 0.3, and 0.05 μm. Use commercially available alumina suspensions in water, or make a suspension by adding a small amount of water to the alumina.

o. Hydrofluoric acid, HF, 5%: Dilute 5 mL conc HF to 100 mL with water.

p. Methanol.

q. Sodium hydroxide, NaOH, 1N.

4. Procedure

a. Sample preparation and storage: Collect samples in precleaned, acid-soaked polyethylene or TFE bottles. Add 2 mL conc HNO_3/L sample and mix well. Cap tightly and store in refrigerator or freezer until ready for analysis.

b. Cell preparation: Soak clean cell in 6N HNO_3 overnight and rinse well with water before use.

c. Electrode preparation:

1) HMDE—Follow manufacturer's guidance for capillary cleaning. If not available, use the following procedure. Remove all mercury from the capillary. Aspirate the following through the capillary in the order listed: 6N HNO_3, water, 5% HF, water, methanol, and air. Dry capillary at 100°C for 1 h. Siliconize cooled capillary using a siliconizing solution. Between uses, fill capillary with clean mercury and immerse tip in clean mercury. If the capillary fails to suspend a drop of mercury, repeat cleaning.

2) TMFE—Polish glassy carbon disks used for thin mercury film electrodes to a high metallic sheen with alumina suspensions, progressively decreasing particle size from 1 μm to 0.05 μm. Use a motorized polishing wheel for best results. Completely rinse off all traces of alumina with water. Check disk frequently for etching or pitting; repolish as necessary to maintain reproducible mercury film deposition.

d. Instrumental conditions: Use the following conditions:

Variable	Value
Initial potential	−1.00 V (Pb, Cd); −1.20 V (Zn)
Final potential	0.00 V
Equilibration potential	−1.00 V (Pb, Cd); −1.20 V (Zn)
HMDE drop size	medium
TMFE rotation rate	2000 rpm
DPASV:	
Pulse amplitude	25 mV
Pulse period	0.5 s
Pulse width	50 ms
Sample width	17 ms
Scan rate	5 mV/s
SWASV:	
SW amplitude	25 mV
Step potential	4 mV
Frequency	100 Hz

e. Deoxygenation: Pipet 2 mL sample and 3 mL electrolyte/buffer into cell. If using a TMFE, add 10 μL $Hg(NO_3)_2$ solution. Place electrodes in cell and secure cell lid. Deoxygenate solution with purified purging gas for 10 min while stirring. When solution purge is completed, purge the space above the solution with purified purging gas. Continue head-space purge throughout analysis.

f. Preconcentration: If using a HMDE, dispense a new mercury drop. Start preconcentration, stirring and timing simultaneously. Precisely control and keep constant preconcentration times and stirring rates for solutions and standards. Generally use 120 s and a rotation rate of 2000 rpm for the TMFE.

After the metal is sufficiently concentrated in the amalgam, stop stirring or TMFE rotation for an equilibration period of precisely 30 s.

---

* Ultrex, Suprapur, Aristar, or equivalent.

g. Anodic stripping: After equilibration period, begin anodic stripping without stirring and make potential at the working electrode progressively more positive as a function of time. Monitor stripping current and plot as a function of applied potential in stripping voltammograms. Use peak current to quantify metal concentration and peak potential to identify the metal.

h. Add 5 to 50 μL standard solution and repeat analysis, beginning with deoxygenation of sample. Adjust volume of added standard solution to obtain 30 to 70% increase of the stripping peak. If 50 μL addition is not sufficient, use standard solution with a higher concentration of metal. Shorten deoxygenation step to 1 min after initial gas purge.

5. Calculations

Calculate the concentration of metal in the original sample using the following equation:

$$C_o = \frac{C_s \times V_s}{V_o} \times \frac{i_o}{(i_s - i_o)}$$

where:

C_o = concentration of metal in sample, mg/L,
C_s = concentration of metal in standard solution, mg/L,
i_o = stripping peak height in original sample,
i_s = stripping peak height in sample with standard addition,
V_o = volume of sample, mL, and
V_s = volume of standard solution added, mL.

6. Quality Control

Follow quality control guidelines outlined in Section 3020 with respect to use of additions, duplicates, and blanks for best results. Blanks are critical because of the high sensitivity of the method.

7. Precision and Bias

Table 3130:I gives precision data for analyses of samples with various matrices.

8. References

1. WANG, J. 1985. Stripping Analysis: Principles, Instrumentation, and Applications. VCH Publishers, Inc., Deerfield Beach, Fla.

2. BRAININA, K.H. & E. NEYMAN. 1993. Electroanalytical Stripping Methods. John Wiley & Sons, New York, N.Y.

3. KISSINGER, P.T. & W.R. HEINEMAN. 1996. Laboratory Techniques in Electroanalytical Chemistry, 2nd ed. Marcel Dekker, Inc., New York, N.Y.

4. MARTIN-GOLDBERG, M. 1989. Unpublished data. ResearchTriangle Institute, Metals Analysis Facility. ResearchTriangle Park, N.C.

5. BRULAND, K.W., K.H. COALE & L. MART. 1985. Analysis of seawater for dissolved Cd, Cu, and Pb: An intercomparison of voltammetric and atomic absorption methods. *Mar. Chem.*17:285.

6. OSTAPCZUK, P., P. VALENTA & H.W. NURNBERG. 1986. Square wave voltammetry—a rapid and reliable determination method of Zn, Cd, Pb, Ni, and Co in biological and environmental samples. *J. Electroanal. Chem.* 214:51.

7. CLARK, B.R., D.W. DePAOLI, D.R. McTAGGART & B.D. PATTON. 1988. An on-line voltammetric analyzer for trace metals in wastewater. *Anal. Chim. Acta* 215:13.

8. BRETT, C.M.A., A.M. OLIVEIRA-BRETT & L. TUGULEA. 1996. Anodic stripping voltammetry of trace metals by batch injection analysis. *Anal. Chim. Acta* 322:151.

9. Bibliography

E.G. & G. PRINCETON APPLIED RESEARCH CORP. 1980. Differential pulse anodic stripping voltammetry of water and wastewater. Application Note W-1. Princeton, N.J.

PETERSON, W.M. & R.V. WONG. 1981. Fundamentals of stripping voltammetry. *Amer. Lab.* 13(11):116.

E.G. & G. PRINCETON APPLIED RESEARCH CORP. 1982. Basics of voltammetry and polarography. Application Note P-2. Princeton, N.J.

NURNBERG, H.W. 1984. The voltammetric approach in trace metal chemistry of natural waters and atmospheric precipitation. *Anal. Chim. Acta* 164:1.

SHUMAN, M.S. & MARTIN-GOLDBERG, M. 1984. Electrochemical methods: Anodic stripping. *In* R. Minear & L. Keith, eds. Water Analysis: Inorganic Species. II:345. Academic Press, Orlando, Fla.

FLORENCE, T.M. 1986. Electrochemical approaches to trace element speciation in waters: A review. *Analyst* 111:489.

FLORENCE, T.M. 1992. Trace element speciation by anodic stripping voltammetry. *Analyst* 117:551.

TERCIER, M.-L. & J. BUFFLE. 1993. In situ voltammetric measurements in natural waters: Future prospects and challenges. *Electroanalysis* 5:187.

TABLE 3130:I. PRECISION OF CD, PB, AND ZN ANALYSIS BY ASV

Sample	Electrode	ASV Mode	Metal Concentration μg/L			RSD %		
			Cd	Pb	Zn	Cd	Pb	Zn
Tap water #1[4]	HMDE	SW	0.068	0.57	—	4.2	4.8	—
Tap water #2[4]	HMDE	SW	—	2.50	—	—	5.1	—
Seawater #1[5]	TFME	DP	0.0121	0.0086	—	10.7	8.1	—
Seawater #2[5]	TFME	DP	0.032	0.032	—	6.3	6.3	—
Soil extract #1[6]	HMDE	SW	189	11.8	—	2.5	5.6	—
Soil extract #1[6]	HMDE	DP	186	11.9	—	2.5	4.0	—
Deionized water[4]	HMDE	SW	0.13	0.79	—	5.5	2.2	—
Wastewater #1[7]	HMDE	DP	—	74	26	—	4.3	4.5
Wastewater #2[7]	HMDE	DP	—	47	86	—	5.2	6.3
Wastewater #3[7]	HMDE	DP	—	46	65	—	4.6	6.2
Wastewater #4[8]	TFME	SW	5.2	60	12	5.2	6.1	7.4

3500-Al ALUMINUM*

3500-Al A. Introduction

1. Occurrence and Significance

Aluminum (Al) is the second element in Group IIIA of the periodic table; it has an atomic number of 13, an atomic weight of 26.98, and a valence of 3. The average abundance in the earth's crust is 8.1%; in soils it is 0.9 to 6.5%; in streams it is 400 µg/L; in U.S. drinking waters it is 54 µg/L, and in groundwater it is <0.1 µg/L. Aluminum occurs in the earth's crust in combination with silicon and oxygen to form feldspars, micas, and clay minerals. The most important minerals are bauxite and corundum, which is used as an abrasive. Aluminum and its alloys are used for heat exchangers, aircraft parts, building materials, containers, etc. Aluminum potassium sulfate (alum) is used in water-treatment processes to flocculate suspended particles, but it may leave a residue of aluminum in the finished water.

Aluminum's occurrence in natural waters is controlled by pH and by very finely suspended mineral particles. The cation Al^{3+}

* Approved by Standard Methods Committee, 1993.

predominates at pH less than 4. Above neutral pH, the predominant dissolved form is $Al(OH)_4^-$. Aluminum is nonessential for plants and animals. Concentrations exceeding 1.5 mg/L constitute a toxicity hazard in the marine environment, and levels below 200 µg/L present a minimal risk. The United Nations Food and Agriculture Organization's recommended maximum level for irrigation waters is 5 mg/L. The possibility of a link between elevated aluminum levels in brain tissues and Alzheimer's disease has been raised. The proposed U.S. EPA secondary drinking water standard MCL is 0.05 mg/L.

2. Selection of Method

The atomic absorption spectrometric methods (3111D and E, and 3113B) and the inductively coupled plasma methods (3120 and 3125) are free from such common interferences as fluoride and phosphate, and are preferred. The Eriochrome cyanine R colorimetric method (B) provides a means for estimating aluminum with simpler instrumentation.

3500-Al B. Eriochrome Cyanine R Method

1. General Discussion

a. *Principle:* With Eriochrome cyanine R dye, dilute aluminum solutions buffered to a pH of 6.0 produce a red to pink complex that exhibits maximum absorption at 535 nm. The intensity of the developed color is influenced by the aluminum concentration, reaction time, temperature, pH, alkalinity, and concentration of other ions in the sample. To compensate for color and turbidity, the aluminum in one portion of sample is complexed with EDTA to provide a blank. The interference of iron and manganese, two elements commonly found in water when aluminum is present, is eliminated by adding ascorbic acid. The optimum aluminum range lies between 20 and 300 µg/L but can be extended upward by sample dilution.

b. *Interference:* Negative errors are caused by both fluoride and polyphosphates. When the fluoride concentration is constant, the percentage error decreases with increasing amounts of aluminum. Because the fluoride concentration often is known or can be determined readily, fairly accurate results can be obtained by adding the known amount of fluoride to a set of standards. A simpler correction can be determined from the family of curves in Figure 3500-Al:1. A procedure is given for the removal of complex phosphate interference. Orthophosphate in concentrations under 10 mg/L does not interfere. The interference caused by even small amounts of alkalinity is removed by acidifying the sample just beyond the neutralization point of methyl orange. Sulfate does not interfere up to a concentration of 2000 mg/L.

c. *Minimum detectable concentration:* The minimum aluminum concentration detectable by this method in the absence of fluorides and complex phosphates is approximately 6 µg/L.

d. *Sample handling:* Collect samples in clean, acid-rinsed bottles, preferably plastic, and examine them as soon as possible after collection. If only soluble aluminum is to be determined, filter a portion of sample through a 0.45-µm membrane filter; discard first 50 mL of filtrate and use succeeding filtrate for the determination. Do not use filter paper, absorbent cotton, or glass wool for filtering any solution that is to be tested for aluminum, because they will remove most of the soluble aluminum.

2. Apparatus

a. *Colorimetric equipment:* One of the following is required:

1) *Spectrophotometer,* for use at 535 nm, with a light path of 1 cm or longer.

2) *Filter photometer,* providing a light path of 1 cm or longer and equipped with a green filter with maximum transmittance between 525 and 535 nm.

3) *Nessler tubes,* 50-mL, tall form, matched.

b. *Glassware:* Treat all glassware with warm 1 + 1 HCl and rinse with aluminum-free distilled water to avoid errors due to materials absorbed on the glass. Rinse sufficiently to remove all acid.

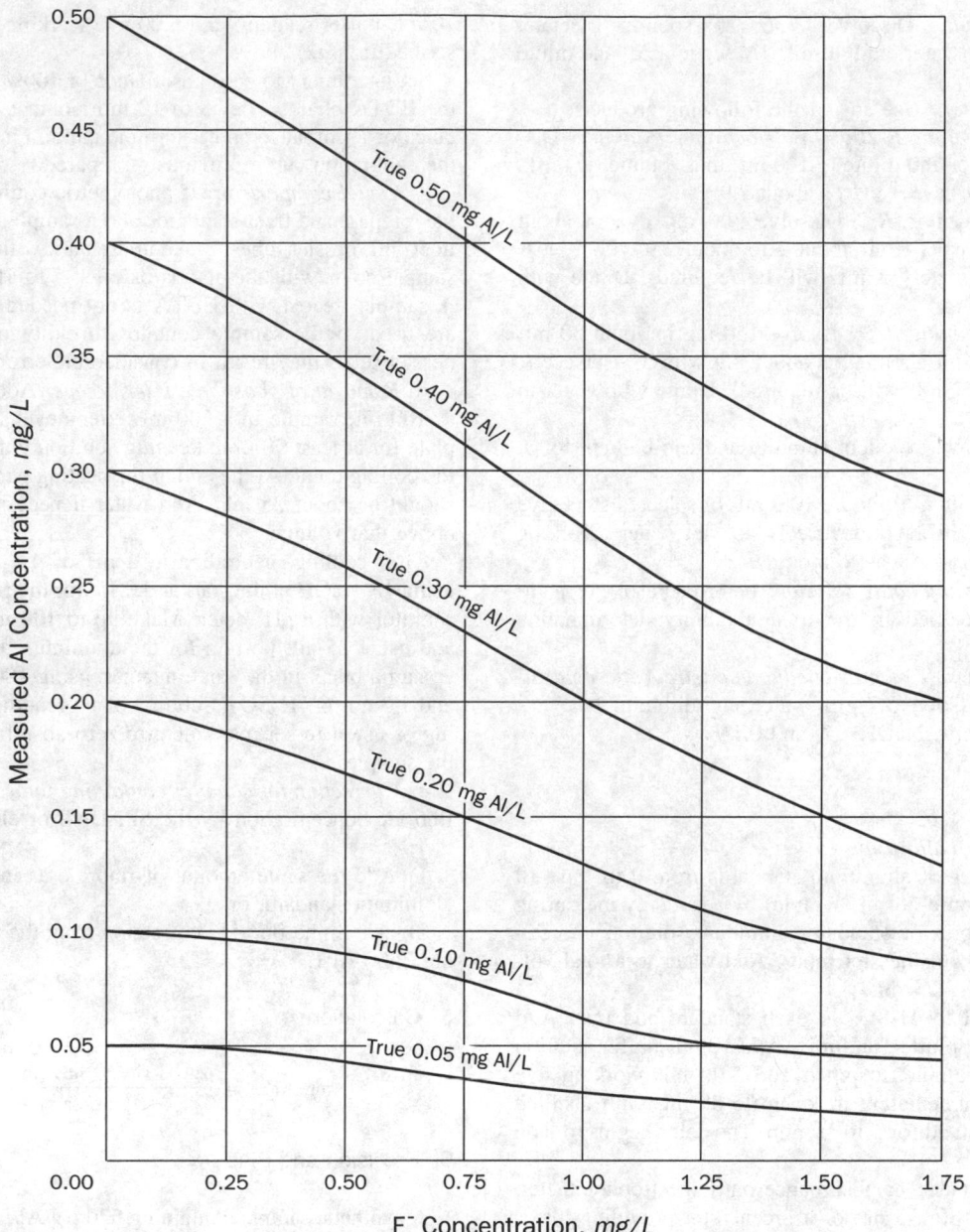

Figure 3500-Al:1. Correction curves for estimation of aluminum in the presence of fluoride. Above the mg F^-/L present, locate the point corresponding to the apparent mg Al/L measured. From this point interpolate between the curves shown, if the point does not fall directly on one of the curves, to read the true mg Al/L on the ordinate, which corresponds to 0.00 mg F^-/L. For example, an apparent 0.20 mg Al/L in a sample containing 1.00 mg F^-/L would actually be 0.30 mg Al/L if no fluoride were present to interfere.

3. Reagents

Use reagents low in aluminum, and aluminum-free distilled water.

a. Stock aluminum solution: Use either the metal (1) or the salt (2) for preparing stock solution; 1.00 mL = 500 μg Al:

1) Dissolve 500.0 mg aluminum metal in 10 mL conc HCl by heating gently. Dilute to 1000 mL with water, or

2) Dissolve 8.791 g aluminum potassium sulfate (also called potassium alum), $AlK(SO_4)_2 \cdot 12H_2O$, in water and dilute to 1000 mL. Correct this weight by dividing by the decimal fraction of assayed $AlK(SO_4)_2 \cdot 12H_2O$ in the reagent used.

b. Standard aluminum solution: Dilute 10.00 mL stock aluminum solution to 1000 mL with water; 1.00 mL = 5.00 μg Al. Prepare daily.

c. Sulfuric acid, H_2SO_4, 0.02N and 6N.

d. Ascorbic acid solution: Dissolve 0.1 g ascorbic acid in water and make up to 100 mL in a volumetric flask. Prepare fresh daily.

e. Buffer reagent: Dissolve 136 g sodium acetate, $NaC_2H_3O_2 \cdot 3H_2O$, in water, add 40 mL $1N$ acetic acid, and dilute to 1 L.

f. Stock dye solution: Use any of the following products:

1) *Solochrome cyanine R-200* or Eriochrome cyanine:†* Dissolve 100 mg in water and dilute to 100 mL in a volumetric flask. This solution should have a pH of about 2.9.

2) *Eriochrome cyanine R:‡* Dissolve 300 mg dye in about 50 mL water. Adjust pH from about 9 to about 2.9 with $1 + 1$ acetic acid (approximately 3 mL will be required). Dilute with water to 100 mL.

3) *Eriochrome cyanine R:§* Dissolve 150 mg in about 50 mL water. Adjust pH from about 9 to about 2.9 with $1 + 1$ acetic acid (approximately 2 mL will be required). Dilute with water to 100 mL.

Stock solutions have excellent stability and can be kept for at least a year.

g. Working dye solution: Dilute 10.0 mL of selected stock dye solution to 100 mL in a volumetric flask with water. Working solutions are stable for at least 6 months.

h. Methyl orange indicator solution, or bromcresol green indicator solution specified in the total alkalinity determination (Section 2320B.3*d*).

i. EDTA (sodium salt of ethylenediamine-tetraacetic acid dihydrate), $0.01M$: Dissolve 3.7 g in water, and dilute to 1 L.

j. Sodium hydroxide, NaOH, $1N$ and $0.1N$.

4. Procedure

a. Preparation of calibration curve:

1) Prepare a series of aluminum standards from 0 to 7 μg (0 to 280 μg/L based on a 25-mL sample) by accurately measuring the calculated volumes of standard aluminum solution into 50-mL volumetric flasks or nessler tubes. Add water to a total volume of approximately 25 mL.

2) Add 1 mL $0.02N$ H_2SO_4 to each standard and mix. Add 1 mL ascorbic acid solution and mix. Add 10 mL buffer solution and mix. With a volumetric pipet, add 5.00 mL working dye reagent and mix. Immediately make up to 50 mL with distilled water. Mix and let stand for 5 to 10 min. The color begins to fade after 15 min.

3) Read transmittance or absorbance on a spectrophotometer, using a wavelength of 535 nm or a green filter providing maximum transmittance between 525 and 535 nm. Adjust instrument to zero absorbance with the standard containing no aluminum.

Plot concentration of Al (micrograms Al in 50 mL final volume) against absorbance.

b. Sample treatment in absence of fluoride and complex phosphates: Place 25.0 mL sample, or a portion diluted to 25 mL, in a porcelain dish or flask, add a few drops of methyl orange indicator, and titrate with $0.02N$ H_2SO_4 to a faint pink color. Record reading and discard sample. To two similar samples at room temperature add the same amount of $0.02N$ H_2SO_4 used in the titration and 1 mL in excess.

To one sample add 1 mL EDTA solution. This will serve as a blank by complexing any aluminum present and compensating for color and turbidity. To both samples add 1 mL ascorbic acid,

* Arnold Hoffman & Co., Providence, RI.
† K & K Laboratories, K & K Lab. Div., Life Sciences Group, Plainview, NY.
‡ Pfaltz & Bauer, Inc., Stamford, CT.
§ EM Science, Gibbstown, NJ.

10 mL buffer reagent, and 5.00 mL working dye reagent as prescribed in ¶ *a*2) above.

Set instrument to zero absorbance or 100% transmittance using the EDTA blank. After 5 to 10 min contact time, read transmittance or absorbance and determine aluminum concentration from the calibration curve previously prepared.

c. Visual comparison: If photometric equipment is not available, prepare and treat standards and a sample, as described above, in 50-mL nessler tubes. Make up to mark with water and compare sample color with the standards after 5 to 10 min contact time. A sample treated with EDTA is not needed when nessler tubes are used. If the sample contains turbidity or color, the use of nessler tubes may result in considerable error.

d. Removal of phosphate interference: Add 1.7 mL $6N$ H_2SO_4 to 100 mL sample in a 200-mL erlenmeyer flask. Heat on a hot plate for at least 90 min, keeping solution temperature just below the boiling point. At the end of the heating period solution volume should be about 25 mL. Add water if necessary to keep it at or above that volume.

After cooling, neutralize to a pH of 4.3 to 4.5 with NaOH, using $1N$ NaOH at the start and $0.1N$ for the final fine adjustment. Monitor with a pH meter. Make up to 100 mL with water, mix, and use a 25-mL portion for the aluminum test.

Run a blank in the same manner, using 100 mL distilled water and 1.7 mL $6N$ H_2SO_4. Subtract blank reading from sample reading or use it to set instrument to zero absorbance before reading the sample.

e. Correction for samples containing fluoride: Measure sample fluoride concentration by the SPADNS or electrode method. Either:

1) Add the same amount of fluoride as in the sample to each aluminum standard, or

2) Determine fluoride correction from the set of curves in Figure 3500-Al:1.

5. Calculation

$$\text{mg Al/L} = \frac{\mu\text{g Al (in 50 mL final volume)}}{\text{mL sample}}$$

6. Precision and Bias

A synthetic sample containing 520 μg Al/L and no interference in distilled water was analyzed by the Eriochrome cyanine R method in 27 laboratories. Relative standard deviation was 34.4% and relative error 1.7%.

A second synthetic sample containing 50 μg Al/L, 500 μg Ba/L, and 5 μg Be/L in distilled water was analyzed in 35 laboratories. Relative standard deviation was 38.5% and relative error 22.0%.

A third synthetic sample containing 500 μg Al/L, 50 μg Cd/L, 110 μg Cr/L, 1000 μg Cu/L, 300 μg Fe/L, 70 μg Pb/L, 50 μg Mn/L, 150 μg Ag/L, and 650 μg Zn/L in distilled water was analyzed in 26 laboratories. Relative standard deviation was 28.8% and relative error 6.2%.

A fourth synthetic sample containing 540 μg Al/L and 2.5 mg polyphosphate/L in distilled water was analyzed in 16 laboratories that hydrolyzed the sample in the prescribed manner. Relative standard deviation was 44.3% and relative error 1.3%. In 12 laboratories that applied no corrective measures, the relative standard deviation was 49.2% and the relative error 8.9%.

A fifth synthetic sample containing 480 μg Al/L and 750 μg F/L in distilled water was analyzed in 16 laboratories that relied on the curve to correct for the fluoride content. Relative standard deviation was 25.5% and relative error 2.3%. The 17 laboratories that added fluoride to the aluminum standards showed a relative standard deviation of 22.5% and a relative error of 7.1%.

7. Bibliography

SHULL, K.E. & G.R. GUTHAN. 1967. Rapid modified Eriochrome cyanine R method for determination of aluminum in water. *J. Amer. Water Works Assoc.* 59:1456.

3500-Sb ANTIMONY

Antimony (Sb) is the fourth element in Group VA in the periodic table; it has an atomic number of 51, an atomic weight of 121.75, and valences of 3 and 5. The average abundance of Sb in the earth's crust is 0.2 ppm; in soils it is 1 ppm; in streams it is 1 μg/L, and in groundwaters it is <0.1 mg/L. Antimony is sometimes found native, but more commonly in stibnite (Sb_2S_3). It is used in alloys of lead and in batteries, bullets, solder, pyrotechnics, and semiconductors.

The common aqueous species are SbO_2^-, $HSbO_2$, and complexes with carbonate and sulfate. Soluble salts of antimony are toxic. The U.S. EPA primary drinking water standard MCL is 6 μg/L.

The electrothermal atomic absorption spectrometric method (3113B) or the inductively coupled plasma/mass spectrometric method (3125) are the methods of choice because of their sensitivity. Alternatively use the flame atomic absorption spectrometric method (3111B) or the inductively coupled plasma method (3120) when high sensitivity is not required.

3500-As ARSENIC*

3500-As A. Introduction

1. Occurrence and Significance

Arsenic (As) is the third element in Group VA of the periodic table; it has an atomic number of 33, an atomic weight of 74.92, and valences of 3 and 5. The average abundance of As in the earth's crust is 1.8 ppm; in soils it is 5.5 to 13 ppm; in streams it is less than 2 μg/L, and in groundwater it is generally less than 100 μg/L. It occurs naturally in sulfide minerals such as pyrite. Arsenic is used in alloys with lead, in storage batteries, and in ammunition. Arsenic compounds are widely used in pesticides and in wood preservatives.

Arsenic is nonessential for plants but is an essential trace element in several animal species. The predominant form between pH 3 and pH 7 is $H_2AsO_4^-$, between pH 7 and pH 11 it is $HAsO_4^{2-}$, and under reducing conditions it is $HAsO_2(aq)$ (or H_3AsO_3). Aqueous arsenic in the form of arsenite, arsenate, and organic arsenicals may result from mineral dissolution, industrial discharges, or the application of pesticides. The chemical form of arsenic depends on its source (inorganic arsenic from minerals, industrial discharges, and pesticides; organic arsenic from industrial discharges, pesticides, and biological action on inorganic arsenic).

Severe poisoning can arise from the ingestion of as little as 100 mg arsenic trioxide; chronic effects may result from the ac-cumulation of arsenic compounds in the body at low intake levels. Carcinogenic properties also have been imputed to arsenic compounds. The toxicity of arsenic depends on its chemical form. Arsenite is many times more toxic than arsenate. For the protection of aquatic life, the average concentration of As^{3+} in water should not exceed 72 μg/L and the maximum should not exceed 140 μg/L. The United Nations Food and Agriculture Organization's recommended maximum level for irrigation waters is 100 μg/L. The U.S. EPA primary drinking water standard MCL is 0.05 mg/L.

2. Selection of Method

Methods are available to identify and determine total arsenic, arsenite, and arsenate. Unpolluted fresh water normally does not contain organic arsenic compounds, but may contain inorganic arsenic compounds in the form of arsenate and arsenite. The electrothermal atomic absorption spectrometric method (3113B) is the method of choice in the absence of overwhelming interferences. The hydride generation-atomic absorption method (3114B) is preferred when interferences are present that cannot be overcome by standard electrothermal techniques (e.g., matrix modifiers, background correction). The silver diethyldithiocarbamate method (B), in which arsine is generated by reaction with sodium borohydride in acidic solution, is applicable to the determination of total inorganic arsenic when interferences are absent and when the sample contains no methylarsenic compounds. This method also provides the advantage of being able to identify and quantify arsenate

---

* Approved by Standard Methods Committee, 1997.

and arsenite separately by generating arsine at different pHs. The inductively coupled plasma (ICP) emission spectroscopy method (3120) is useful at higher concentrations (greater than 50 μg/L) while the ICP-mass spectrometric method (3125) is applicable at lower concentrations if chloride does not interfere. When measuring arsenic species, document that speciation does not change over time. No universal preservative for speciation measurements has been identified.

3500-As B. Silver Diethyldithiocarbamate Method

1. General Discussion

a. Principle: Arsenite, containing trivalent arsenic, is reduced selectively by aqueous sodium borohydride solution to arsine, AsH_3, in an aqueous medium of pH 6. Arsenate, methylarsonic acid, and dimethylarsenic acid are not reduced under these conditions. The generated arsine is swept by a stream of oxygen-free nitrogen from the reduction vessel through a scrubber containing glass wool or cotton impregnated with lead acetate solution into an absorber tube containing silver diethyldithiocarbamate and morpholine dissolved in chloroform. The intensity of the red color that develops is measured at 520 nm. To determine total inorganic arsenic in the absence of methylarsenic compounds, a sample portion is reduced at a pH of about 1. Alternatively, arsenate is measured in a sample from which arsenite has been removed by reduction to arsine gas at pH 6 as above. The sample is then acidified with hydrochloric acid and another portion of sodium borohydride solution is added. The arsine formed from arsenate is collected in fresh absorber solution.

b. Interferences: Although certain metals—chromium, cobalt, copper, mercury, molybdenum, nickel, platinum, silver, and selenium—influence the generation of arsine, their concentrations in water are seldom high enough to interfere, except in the instance of acid rock drainage. H_2S interferes, but the interference is removed with lead acetate. Antimony is reduced to stibine, which forms a colored complex with an absorption maximum at 510 nm and interferes with the arsenic determination. Methylarsenic compounds are reduced at pH 1 to methylarsines, which form colored complexes with the absorber solution. If methylarsenic compounds are present, measurements of total arsenic and arsenate are unreliable. The results for arsenite are not influenced by methylarsenic compounds.

c. Minimum detectable quantity: 1 μg arsenic.

2. Apparatus

a. Arsine generator, scrubber, and absorption tube: See Figure 3500-As:1. Use a 200-mL three-necked flask with a sidearm (19/22 or similar size female ground-glass joint) through which the inert gas delivery tube reaching almost to the bottom of the flask is inserted; a 24/40 female ground-glass joint to carry the scrubber; and a second side arm closed with a rubber septum, or preferably by a screw cap with a hole in its top for insertion of a TFE-faced silicone septum. Place a small magnetic stirring bar in the flask. Fit absorber tube (20 mL capacity) to the scrubber and fill with silver diethyldithiocarbamate solution. Do not use rubber or cork stoppers because they may absorb arsine. Clean glass equipment with concentrated nitric acid.

b. Fume hood: Use apparatus in a well-ventilated hood with flask secured on top of a magnetic stirrer.

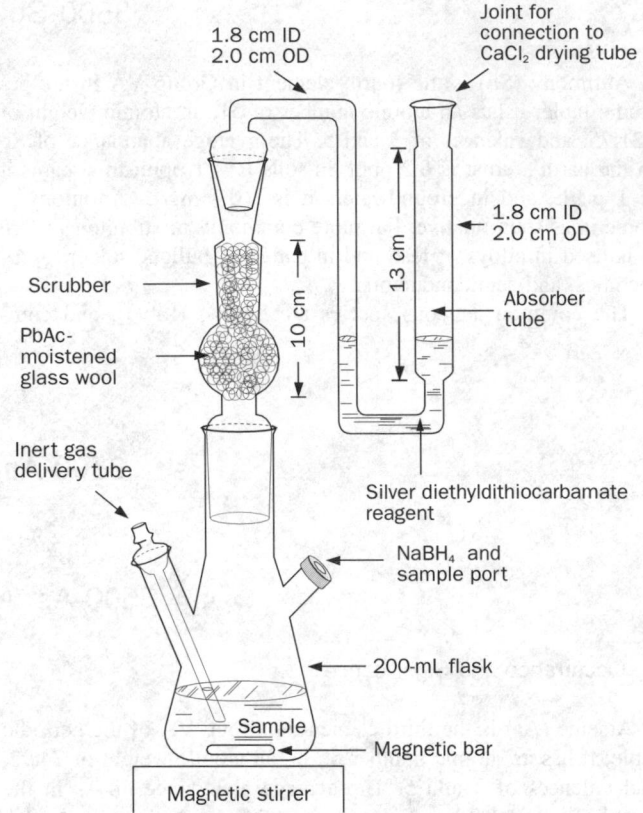

Figure 3500-As:1. Arsine generator and absorber assembly.

c. Photometric equipment:

1) *Spectrophotometer,* for use at 520 nm.

2) *Filter photometer,* with green filter having a maximum transmittance in the 500- to 540-nm range.

3) *Cells,* for spectrophotometer or filter photometer, 1-cm, clean, dry, and each equipped with a tightly fitting cover (TFE stopper) to prevent chloroform evaporation.

3. Reagents

a. Reagent water: See Section 1080A.

b. Acetate buffer, pH 5.5: Mix 428 mL 0.2*M* sodium acetate, $NaC_2H_3O_2$, and 72 mL 0.2*M* acetic acid, CH_3COOH.

c. Sodium acetate, 0.2*M*: Dissolve 16.46 g anhydrous sodium acetate or 27.36 g sodium acetate trihydrate, $NaC_2H_3O_2 \cdot 3H_2O$, in water. Dilute to 1000 mL with water.

d. Acetic acid, 0.2*M*: Dissolve 11.5 mL glacial acetic acid in water. Dilute to 1000 mL.

e. Sodium borohydride solution, 1%: Dissolve 0.4 g sodium hydroxide, NaOH (4 pellets), in 400 mL water. Add 4.0 g sodium borohydride, $NaBH_4$ (check for absence of arsenic). Shake to dissolve and to mix. Prepare fresh every few days.

f. Hydrochloric acid, HCl, 2*M*: Dilute 165 mL conc HCl to 1000 mL with water.

g. Lead acetate solution: Dissolve 10.0 g $Pb(CH_3COO)_2 \cdot 3H_2O$ in 100 mL water.

h. Silver diethyldithiocarbamate solution: Dissolve 1.0 mL morpholine (CAUTION: *Corrosive—avoid contact with skin*) in 70 mL chloroform, $CHCl_3$. Add 0.30 g silver diethyldithiocarbamate, $AgSCSN(C_2H_5)_2$; shake in a stoppered flask until most is dissolved. Dilute to 100 mL with chloroform. Filter and store in a tightly closed brown bottle in a refrigerator.

i. Standard arsenite solution: Dissolve 0.1734 g $NaAsO_2$ in water and dilute to 1000 mL with water. CAUTION: *Toxic—avoid contact with skin and do not ingest.* Dilute 10.0 mL to 100 mL with water; dilute 10.0 mL of this intermediate solution to 100 mL with water; 1.00 mL = 1.00 μg As.

j. Standard arsenate solution: Dissolve 0.416 g $Na_2HAsO_4 \cdot 7H_2O$ in water and dilute to 1000 mL. Dilute 10.0 mL to 100 mL with water; dilute 10 mL of this intermediate solution to 100 mL; 1.00 mL = 1.00 μg As.

4. Procedure

a. Arsenite:

1) Preparation of scrubber and absorber—Dip glass wool into lead acetate solution; remove excess by squeezing glass wool. Press glass wool between pieces of filter paper, then fluff it. Alternatively, if cotton is used treat it similarly but dry in a desiccator and fluff thoroughly when dry. Place a plug of loose glass wool or cotton in scrubber tube. Add 4.00 mL silver diethyldithiocarbamate solution to absorber tube (5.00 mL may be used to provide enough volume to rinse spectrophotometer cell).

2) Loading of arsine generator—Pipet not more than 70 mL sample containing not more than 20.0 μg As (arsenite) into the generator flask. Add 10 mL acetate buffer. If necessary, adjust total volume of liquid to 80 mL. Flush flask with nitrogen at the rate of 60 mL/min.

3) Arsine generation and measurement—While nitrogen is passing through the system, use a 30-mL syringe to inject through the septum 15 mL 1% sodium borohydride solution within 2 min. Stir vigorously with magnetic stirrer. Pass nitrogen through system for an additional 15 min to flush arsine into absorber solution. Pour absorber solution into a clean and dry spectrophotometric cell and measure absorbance at 520 nm against chloroform. Determine concentration from a calibration curve obtained with arsenite standards. If arsenate also is to be determined for this sample by using the same sample portion, save the liquid in the generator flask.

4) Preparation of standard curves—Treat standard arsenite solution containing 0.0, 1.0, 2.0, 5.0, 10.0, and 20.0 μg As as described

in ¶s 1) through 3) above. Plot absorbance versus micrograms arsenic in the standard.

b. Arsenate: After removal of arsenite as arsine, treat sample to convert arsenate to arsine:

If the lead acetate-impregnated glass wool has become ineffective in removing hydrogen sulfide (if it has become gray to black) replace glass wool [see ¶ *a*1)]. Pass nitrogen through system at the rate of 60 mL/min. Cautiously add 10 mL 2.0*N* HCl. Generate arsine as directed in ¶ 4*a*3) and prepare standard curves with standard solutions of arsenate according to procedure of ¶ 4*a*4).

c. Total inorganic arsenic: Prepare scrubber and absorber as directed in ¶ 4*a*1) and load arsine generator as directed in ¶ 4*a*2) using 10 mL 2.0*N* HCl instead of acetate buffer. Generate arsine and measure as directed in ¶ 4*a*3). Prepare standard curves according to ¶ 4*a*4). Curves obtained with standard arsenite solution are almost identical to those obtained with arsenate standard solutions. Therefore, use either arsenite or arsenate standards.

5. Calculation

Calculate arsenite, arsenate, and total inorganic arsenic from readings and calibration curves obtained in 4*a*, *b*, and *c*, respectively, as follows:

$$\text{mg As/L} = \frac{\mu\text{g As (from calibration curve)}}{\text{mL sample in generator flask}}$$

6. Precision and Bias

Interlaboratory comparisons are not available. The relative standard deviation of results obtained with arsenite/arsenate mixtures containing approximately 10 μg arsenic was less than 10%.

7. Bibliography

PEOPLES, S.A., J. LAKSO & T. LAIS. 1971. The simultaneous determination of methylarsonic acid and inorganic arsenic in urine. *Proc. West. Pharmacol. Soc.* 14:178.

AGGETT, J. & A.C. ASPELL. 1976. Determination of arsenic (III) and total arsenic by the silver diethyldithiocarbamate method. *Analyst* 101:912.

HOWARD, A.G. & M.H. ARBAB-ZAVAR. 1980. Sequential spectrophotometric determination of inorganic arsenic (III) and arsenic (V) species. *Analyst* 105:338.

PANDE, S.P. 1980. Morpholine as a substitute for pyridine in determination of arsenic in water. *J. Inst. Chem. (India)* 52:256.

IRGOLIC, K.J. 1986. Arsenic in the environment. *In* A. V. Xavier, ed. Frontiers in Bioinorganic Chemistry. VCH Publishers, Weinheim, Germany.

IRGOLIC, K.J. 1987. Analytical procedures for the determination of organic compounds of metals and metalloids in environmental samples. *Sci. Total Environ.* 64:61.

3500-Ba BARIUM

Barium (Ba) is the fifth element in Group IIA in the periodic table; it has an atomic number of 56, an atomic weight of 137.33, and a valence of 2. The average abundance of Ba in the earth's crust is 390 ppm; in soils it is 63 to 810 ppm; in streams it is 10 mg/L; in U.S. drinking waters it is 49 μg/L; and in groundwaters it is 0.05 to 1 mg/L. It is found chiefly in barite ($BaSO_4$) or in witherite ($BaCO_3$). Barium's main use is in mud slurries used in drilling oil and exploration wells, but it is also used in pigments, rat poisons, pyrotechnics, and in medicine.

The solubility of barium in natural waters is controlled by the solubility of $BaSO_4$, and somewhat by its adsorption on hydroxides. High concentrations of barium occur in some brines. Concentrations exceeding 1 mg/L constitute a toxicity hazard in the marine environment. The U.S. EPA primary drinking water standard MCL is 1 mg/L.

Perform analyses by the atomic absorption spectrometric methods (3111D or E), the electrothermal atomic absorption method (3113B), or the inductively coupled plasma methods (3120 or 3125).

3500-Be BERYLLIUM

Beryllium (Be) is the first element in Group IIA of the periodic table; it has an atomic number of 4, an atomic weight of 9.01, and a valence of 2. The average abundance of Be in the earth's crust is 2 ppm; in soils it is 0.8 to 1.3 ppm; in streams it is 0.2 μg/L, in U.S. drinking waters and in groundwaters it is typically <0.1 μg/L. Beryllium occurs in nature in deposits of beryls in granitic rocks. Beryllium is used in high-strength alloys of copper and nickel, windows in X-ray tubes, and as a moderator in nuclear reactors.

Beryllium solubility is controlled in natural waters by the solubility of beryllium hydroxides. The solubility at pH 6.0 is approximately 0.1 μg/L. It is nonessential for plants and animals.

Acute toxicity occurs at 130 μg/L, and chronic toxicity at 5 μg/L in freshwater species. The United Nations Food and Agriculture Organization recommended maximum level for irrigation waters is 100 μg/L. The U.S. EPA primary drinking water standard MCL for beryllium is 4 μg/L.

The atomic absorption spectrometric methods (3111D and E, and 3113B) and the inductively coupled plasma (ICP) methods (3120 and 3125) are the methods of choice. If atomic absorption or ICP instrumentation is not available, the aluminon colorimetric method detailed in the 19th Edition of *Standard Methods* may be used. This method has poorer precision and bias than the methods of choice.

3500-Bi BISMUTH

Bismuth (Bi) is the fifth element in Group VA in the periodic table; it has an atomic number of 83, an atomic weight of 208.98, and valences of 3 and 5. The average abundance of Bi in the earth's crust in 0.08 ppm; in streams it is <0.02 mg/L, and in groundwaters it is <0.1 mg/L. Bismuth occurs in association with lead and silver ores, and occasionally as the native element. ^{210}Bi, ^{212}Bi, and ^{214}Bi are naturally occuring radioisotopes produced in the decay of uranium and thorium. The metal is used in alloys of lead, tin, and cadmium, and in some pharmaceuticals.

In natural water, Bi^{3+} ion will occur, and complex ions with nitrate and chloride also might be expected. The iodide and telluride compounds are toxic by ingestion or inhalation.

Perform analyses by the atomic absorption spectrometric method (3111B) or by the electrothermal atomic absorption method (3113B). The inductively coupled plasma mass spectrometric method (3125) also may be applied successfully in most cases (with lower detection limits), even though bismuth is not specifically listed as an analyte in the method.

3500-B   BORON

See Section 4500-B.

3500-Cd   CADMIUM

Cadmium (Cd) is the second element in Group IIB of the periodic table; it has an atomic number of 48, an atomic weight of 112.41, and a valence of 2. The average abundance of Cd in the earth's crust is 0.16 ppm; in soils it is 0.1 to 0.5 ppm; in streams it is 1 μg/L, and in groundwaters it is from 1 to 10 μg/L. Cadmium occurs in sulfide minerals that also contain zinc, lead, or copper. The metal is used in electroplating, batteries, paint pigments, and in alloys with various other metals. Cadmium is usually associated with zinc at a ratio of about 1 part cadmium to 500 parts zinc in most rocks and soils.

The solubility of cadmium is controlled in natural waters by carbonate equilibria. Guidelines for maximum cadmium concentrations in natural water are linked to the hardness or alkalinity of the water (i.e., the softer the water, the lower the permitted level of cadmium). It is nonessential for plants and animals. Cadmium is extremely toxic and accumulates in the kidneys and liver, with prolonged intake at low levels sometimes leading to dysfunction of the kidneys. The United Nations Food and Agriculture Organization recommended maximum level for cadmium in irrigation waters is 10 μg/L. The U.S. EPA primary drinking water standard MCL is 10 μg/L.

The electrothermal atomic absorption spectrometric method (3113B) is preferred. The flame atomic absorption methods (3111B and C) and inductively coupled plasma methods (3120 and 3125) provide acceptable precision and bias, with higher detection limits. Anodic stripping voltammetry (3130B) can achieve superior detection limits, but is susceptible to interferences from copper, silver, gold, and organic compounds. When atomic absorption spectrometric or inductively coupled plasma apparatus is unavailable and the desired precision is not as great, the dithizone method detailed in the 19th Edition of *Standard Methods* is suitable.

3500-Ca   CALCIUM*

3500-Ca A.   Introduction

1. Occurrence and Significance

Calcium (Ca) is the third element in Group IIA of the periodic table; it has an atomic number of 20, an atomic weight of 40.08, and a valence of 2. The average abundance of Ca in the earth's crust is 4.9%; in soils it is 0.07 to 1.7%; in streams it is about 15 mg/L; and in groundwaters it is from 1 to >500 mg/L. The most common forms of calcium are calcium carbonate (calcite) and calcium-magnesium carbonate (dolomite). Calcium compounds are widely used in pharmaceuticals, photography, lime, de-icing salts, pigments, fertilizers, and plasters. Calcium carbonate solubility is controlled by pH and dissolved CO_2. The CO_2, HCO_3^-, and CO_3^{2-} equilibrium is the major buffering mechanism in fresh waters. Hardness is based on the concentration of calcium and magnesium salts, and often is used as a measure of potable water quality.

NOTE: Calcium is necessary in plant and animal nutrition and is an essential component of bones, shells, and plant structures. The presence of calcium in water supplies results from passage over deposits of limestone, dolomite, gypsum, and gypsiferous shale. Small concentrations of calcium carbonate combat corrosion of metal pipes by laying down a protective coating. Because precipitation of calcite in pipes and in heat-exchangers can cause damage, the amount of calcium in domestic and industrial waters is often controlled by water softening (e.g., ion exchange, reverse osmosis). Calcium carbonate saturation and water hardness are discussed in Sections 2330 and 2340, respectively.

Calcium contributes to the total hardness of water. Chemical softening treatment, reverse osmosis, electrodialysis, or ion exchange is used to reduce calcium and the associated hardness.

2. Selection of Method

The atomic absorption methods (3111B, D, and E) and inductively coupled plasma method (3120) are accurate means of determining calcium. The EDTA titration method gives good results

---

* Approved by Standard Methods Committee, 1997.

for control and routine applications, but for samples containing high P levels (>50 mg/L) only the atomic absorption or atomic emission methods are recommended because of interferences often encountered with EDTA indicators.

3500-Ca B. EDTA Titrimetric Method

1. General Discussion

a. Principle: When EDTA (ethylenediaminetetraacetic acid or its salts) is added to water containing both calcium and magnesium, it combines first with the calcium. Calcium can be determined directly, with EDTA, when the pH is made sufficiently high that the magnesium is largely precipitated as the hydroxide and an indicator is used that combines with calcium only. Several indicators give a color change when all of the calcium has been complexed by the EDTA at a pH of 12 to 13.

b. Interference: Under conditions of this test, the following concentrations of ions cause no interference with the calcium hardness determination: Cu^{2+}, 2 mg/L; Fe^{2+}, 20 mg/L; Fe^{3+}, 20 mg/L; Mn^{2+}, 10 mg/L; Zn^{2+}, 5 mg/L; Pb^{2+}, 5 mg/L; Al^{3+}, 5 mg/L; and Sn^{4+}, 5 mg/L. Orthophosphate precipitates calcium at the pH of the test. Strontium and barium give a positive interference and alkalinity in excess of 300 mg/L may cause an indistinct end point in hard waters.

2. Reagents

a. Sodium hydroxide, NaOH, 1*N.*

b. Indicators: Many indicators are available for the calcium titration. Some are described in the literature (see Bibliography); others are commercial preparations and also may be used. Murexide (ammonium purpurate) was the first indicator available for detecting the calcium end point; directions for its use are presented in this procedure. Individuals who have difficulty recognizing the murexide end point may find the indicator Eriochrome Blue Black R (color index number 202) or Solochrome Dark Blue an improvement because of the color change from red to pure blue. Eriochrome Blue Black R is sodium-1-(2-hydroxy-1-naphthylazo)-2-naphthol-4-sulfonic acid. Other indicators specifically designed for use as end-point detectors in EDTA titration of calcium may be used.

1) *Murexide (ammonium purpurate) indicator:* This indicator changes from pink to purple at the end point. Prepare by dissolving 150 mg dye in 100 g absolute ethylene glycol. Water solutions of the dye are not stable for longer than 1 d. A ground mixture of dye powder and sodium chloride (NaCl) provides a stable form of the indicator. Prepare by mixing 200 mg murexide with 100 g solid NaCl and grinding the mixture to 40 to 50 mesh. Titrate immediately after adding indicator because it is unstable under alkaline conditions. Facilitate end-point recognition by preparing a color comparison blank containing 2.0 mL NaOH solution, 0.2 g solid indicator mixture (or 1 to 2 drops if a solution is used), and sufficient standard EDTA titrant (0.05 to 0.10 mL) to produce an unchanging color.

2) *Eriochrome Blue Black R indicator:* Prepare a stable form of the indicator by grinding together in a mortar 200 mg powdered

3. Storage of Samples

The customary precautions are sufficient if care is taken to redissolve any calcium carbonate that may precipitate on standing.

dye and 100 g solid NaCl to 40 to 50 mesh. Store in a tightly stoppered bottle. Use 0.2 g of ground mixture for the titration in the same manner as murexide indicator. During titration the color changes from red through purple to bluish purple to a pure blue with no trace of reddish or purple tint. The pH of some (not all) waters must be raised to 14 (rather than 12 to 13) by the use of 8*N* NaOH to get a good color change.

c. Standard EDTA titrant, 0.01M: Prepare standard EDTA titrant and standardize against standard calcium solution as described in Section 2340C to obtain EDTA/$CaCO_3$ equivalence. Standard EDTA titrant, 0.0100*M*, is equivalent to 1000 mg $CaCO_3$/1.00 mL; use titrated equivalent for *B* in the calculations in 4.

3. Procedure

a. Pretreatment of water and wastewater samples: Follow the procedure described in Section 3030E or I if samples require preliminary digestion.

b. Sample preparation: Because of the high pH used in this procedure, titrate immediately after adding alkali and indicator. Use 50.0 mL sample, or a smaller portion diluted to 50 mL so that the calcium content is about 5 to 10 mg. Analyze hard waters with alkalinity higher than 300 mg $CaCO_3$/L by taking a smaller portion and diluting to 50 mL. Alternatively, adjust sample pH into the acid range (pH <6), boil for 1 min to dispel CO_2, and cool before beginning titration.

c. Titration: Add 2.0 mL NaOH solution or a volume sufficient to produce a pH of 12 to 13. Stir. Add 0.1 to 0.2 g indicator mixture selected (or 1 to 2 drops if a solution is used). Add EDTA titrant slowly, with continuous stirring to the proper end point. When using murexide, check end point by adding 1 to 2 drops of titrant in excess to make certain that no further color change occurs.

4. Calculation

$$\text{mg Ca/L} = \frac{A \times B \times 400.8}{\text{mL sample}}$$

$$\text{Calcium hardness as mg } CaCO_3/L = \frac{A \times B \times 1000}{\text{mL sample}}$$

where:

A = mL titrant for sample and
B = mg $CaCO_3$ equivalent to 1.00 mL EDTA titrant at the calcium indicator end point.

5. Precision and Bias

A synthetic sample containing 108 mg Ca/L, 82 mg Mg/L, 3.1 mg K/L, 19.9 mg Na/L, 241 mg Cl^-/L, 1.1 mg NO_3^--N/L,

0.25 mg NO_2^--N/L, 259 mg SO_4^{2-}/L, and 42.5 mg total alkalinity/L (contributed by $NaHCO_3$) in distilled water was analyzed in 44 laboratories by the EDTA titrimetric method, with a relative standard deviation of 9.2% and a relative error of 1.9%.

6. Bibliography

DIEHL, H. & J.L. ELLINGBOE. 1956. Indicator for titration of calcium in the presence of magnesium using disodium dihydrogen ethylenediamine tetraacetate. *Anal. Chem.* 28:882.

HILDEBRAND, G.P. & C.N. REILLEY. 1957. New indicator for complexometric titration of calcium in the presence of magnesium. *Anal. Chem.* 29:258.

PATTON, J. & W. REEDER. 1956. New indicator for titration of calcium with (ethylenedinitrilo) tetraacetate. *Anal. Chem.* 28:1026.

SCHWARZENBACH, G. 1957. Complexometric Titrations. Interscience Publishers, New York, N.Y.

FURMAN, N.H. 1962. Standard Methods of Chemical Analysis, 6th ed. D. Van Nostrand Co., Inc., Princeton, N.J.

KATZ, H. & R. NAVONE. 1964. Method for simultaneous determination of calcium and magnesium. *J. Amer. Water Works Assoc.* 56:121.

3500-Cs   CESIUM

Cesium (Cs) is the sixth element in Group IA of the periodic table; it has an atomic number of 55, an atomic weight of 132.90, and a valence of 1. The average abundance of Cs in the earth's crust is 2.6 ppm; in soils it is 1 to 5 ppm; in streams it is 0.02 mg/L; and in groundwaters it is generally <0.1 mg/L. Cesium is found in lepidolite and in the water of certain mineral springs. ^{137}Cs, with a 33-year half-life, is widely dispersed on the earth's surface as a result of the radioactive fallout from the atmospheric testing of nuclear weapons. Cesium compounds are used in photoelectric cells, as a catalyst, and in brewing. Some cesium compounds are fire hazards.

Perform analyses by the flame atomic absorption spectrometric method (3111B). The inductively coupled plasma/mass spectrometric method (3125) also may be applied successfully in most cases (with lower detection limits), even though cesium is not specifically listed as an analyte in the method.

3500-Cr   CHROMIUM*

3500-Cr A.   Introduction

1. Occurrence and Significance

Chromium (Cr) is the first element in Group VIB in the periodic table; it has an atomic number of 24, an atomic weight of 51.99, and valences of 1 through 6. The average abundance of Cr in the earth's crust is 122 ppm; in soils Cr ranges from 11 to 22 ppm; in streams it averages about 1 μg/L, and in groundwaters it is generally 100 μg/L. Chromium is found chiefly in chrome-iron ore ($FeO \cdot Cr_2O_3$). Chromium is used in alloys, in electroplating, and in pigments. Chromate compounds frequently are added to cooling water for corrosion control.

In natural waters trivalent chromium exists as Cr^{3+}, $Cr(OH)^{2+}$, $Cr(OH)_2^+$, and $Cr(OH)_4^-$; in the hexavalent form chromium exists as CrO_4^{2-} and as $Cr_2O_7^{2-}$. Cr^{3+} would be expected to form strong complexes with amines, and would be adsorbed by clay minerals. Chromium may exist in water supplies in both the hexavalent and the trivalent state although the trivalent form rarely occurs in potable water.

Chromium is considered nonessential for plants, but an essential trace element for animals. Hexavalent compounds have been shown to be carcinogenic by inhalation and are corrosive to tissue. The chromium guidelines for natural water are linked to the hardness or alkalinity of the water (i.e., the softer the water, the lower the permitted level for chromium). The United Nations Food and Agriculture Organization recommended maximum level for irrigation waters is 100 μg/L. The U.S. EPA primary drinking water standard MCL is 0.1 mg/L for total chromium.

2. Selection of Method

The colorimetric method (B) is useful for the determination of hexavalent chromium in a natural or treated water in the range from 100 to 1000 μg/L. This range can be extended by appropriate sample dilution or concentration and/or use of longer cell paths. The ion chromatographic method with photometric detection (C) is suitable for determining dissolved hexavalent chromium in drinking water, groundwater, and industrial wastewater effluents. The electrothermal atomic absorption spectrometric method (3113B) is suitable for determining low levels of total chromium (< 50 μg/L) in water and wastewater, and the flame atomic absorption spectrometric methods (3111B and C) and the inductively coupled plasma methods (3120 and 3125) are appropriate for measuring concentrations up to milligram-per-liter levels.

3. Sample Handling

If only the dissolved metal content is desired, filter sample through a 0.45-μm membrane filter at time of collection, and after

---

* Approved by Standard Methods Committee, 1997.

filtration acidify filtrate with conc nitric acid (HNO_3) to pH <2. If only dissolved hexavalent chromium is desired, adjust pH of filtrate to 8 or above with $1N$ sodium hydroxide solution and refrigerate. If the total chromium content is desired, acidify un-

filtered sample at time of collection with conc HNO_3 to pH <2. If total hexavalent chromium is desired, adjust the pH of unfiltered sample to 8 or above with $1N$ sodium hydroxide and refrigerate.

3500-Cr B.   Colorimetric Method

1. General Discussion

a. Principle: This procedure measures only hexavalent chromium (Cr^{6+}). Therefore, to determine total chromium convert all the chromium to the hexavalent state by oxidation with potassium permanganate. NOTE: The oxidation process may not provide total conversion of all chromium species to Cr^{6+}.[1-3] For total chromium determination, acid-digest the sample (see Section 3030) and follow with a suitable instrumental analysis technique. The hexavalent chromium is determined colorimetrically by reaction with diphenylcarbazide in acid solution. A red-violet colored complex of unknown composition is produced. The reaction is very sensitive, the molar absorptivity based on chromium being about 40 000 L g^{-1} cm^{-1} at 540 nm. To determine total chromium, digest the sample with a sulfuric-nitric acid mixture and then oxidize with potassium permanganate before reacting with the diphenylcarbazide.

b. Interferences: The reaction with diphenylcarbazide is nearly specific for chromium. Hexavalent molybdenum and mercury salts will react to form color with the reagent but the intensities are much lower than that for chromium at the specified pH. Concentrations as high as 200 mg Mo or Hg/L can be tolerated. Vanadium interferes strongly but concentrations up to 10 times that of chromium will not cause trouble. Potential interference from permanganate is eliminated by prior reduction with azide. Iron in concentrations greater than 1 mg/L may produce a yellow color but the ferric ion (Fe^{3+}) color is not strong and no difficulty is encountered normally if the absorbance is measured photometrically at the appropriate wavelength. Interfering amounts of molybdenum, vanadium, iron, and copper can be removed by extraction of the cupferrates of these metals into chloroform ($CHCl_3$). A procedure for this extraction is provided but do not use it unless necessary, because residual cupferron and $CHCl_3$ in the aqueous solution complicate the later oxidation. Therefore, follow the extraction by additional treatment with acid fuming to decompose these compounds.

2. Apparatus

a. Colorimetric equipment: One of the following is required:

1) *Spectrophotometer,* for use at 540 nm, with a light path of 1 cm or longer.

2) *Filter photometer,* providing a light path of 1 cm or longer and equipped with a greenish yellow filter having maximum transmittance near 540 nm.

b. Separatory funnels, 125-mL, Squibb form, with glass or TFE stopcock and stopper.

c. Acid-washed glassware: New and unscratched glassware will minimize chromium adsorption on glass surfaces during the

oxidation procedure. Do not use glassware previously treated with chromic acid. Thoroughly clean other used glassware and new glassware with nitric or hydrochloric acid to remove chromium traces.

3. Reagents

Use reagent water (see Section 1080) for reagent preparation and analytical procedure.

a. Stock chromium solution: Dissolve 141.4 mg $K_2Cr_2O_7$ in water and dilute to 100 mL; 1.00 mL = 500 µg Cr.

b. Standard chromium solution: Dilute 1.00 mL stock chromium solution to 100 mL; 1.00 mL = 5.00 µg Cr.

c. Nitric acid, HNO_3, conc.

d. Sulfuric acid, H_2SO_4, conc, $18N$, and $6N$.

e. Sulfuric acid, H_2SO_4, $0.2N$: Dilute 17 mL $6N$ H_2SO_4 to 500 mL with water.

f. Phosphoric acid, H_3PO_4, conc.

g. Methyl orange indicator solution.

h. Hydrogen peroxide, H_2O_2, 30%.

i. Ammonium hydroxide, NH_4OH, conc.

j. Potassium permanganate solution: Dissolve 4 g $KMnO_4$ in 100 mL water.

k. Sodium azide solution: Dissolve 0.5 g NaN_3 in 100 mL water.

l. Diphenylcarbazide solution: Dissolve 250 mg 1,5-diphenylcarbazide (1,5-diphenylcarbohydrazide) in 50 mL acetone. Store in a brown bottle. Discard when solution becomes discolored.

m. Chloroform, $CHCl_3$: Avoid or redistill material that comes in containers with metal or metal-lined caps.

n. Cupferron solution: Dissolve 5 g cupferron, $C_6H_5N(NO)ONH_4$, in 95 mL water.

o. Sodium hydroxide, $1N$: Dissolve 40 g NaOH in 1 L water. Store in plastic bottle.

4. Procedure

a. Preparation of calibration curve: To compensate for possible slight losses of chromium during digestion or other analytical operations, treat standards by the same procedure as the sample. Accordingly, pipet measured volumes of standard chromium solution (5 µg/mL) ranging from 2.00 to 20.0 mL, to give standards for 10 to 100 µg Cr, into 250-mL beakers or conical flasks. Depending on pretreatment used in ¶ *b* below, proceed with subsequent treatment of standards as if they were samples, also carrying out cupferron treatment of standards if this is required for samples.

Develop color as for samples, transfer a suitable portion of each colored solution to a 1-cm absorption cell, and measure absorbance at 540 nm, using reagent water as reference. Correct absorbance readings of standards by subtracting absorbance of a reagent blank carried through the method.

Construct a calibration curve by plotting corrected absorbance values against micrograms chromium in 102 mL final volume.

b. Treatment of sample: If sample has been filtered and/or only hexavalent chromium is desired, start analysis within 24 h of collection and proceed to ¶ 4*e*. NOTE: Recent evidence[4] suggests that preserved samples can be held for 30 d without substantial changes to Cr^{6+} concentrations. If total dissolved chromium is desired and there are interfering amounts of molybdenum, vanadium, copper, or iron present, proceed to ¶ 4*c*. If interferences are not present, proceed to ¶ 4*d*.

If sample is unfiltered and total chromium is desired, digest with HNO_3 and H_2SO_4 as in Section 3030G. If interferences are present, proceed to ¶s 4*c*, 4*d*, and 4*e*. If there are no interferences, proceed to ¶s 4*d* and 4*e*.

c. Removal of molybdenum, vanadium, iron, and copper with cupferron: Pipet a portion of sample containing 10 to 100 μg Cr into a 125-mL separatory funnel. Dilute to about 40 mL with water and chill in an ice bath. Add 5 mL ice-cold cupferron solution, shake well, and let stand in ice bath for 1 min. Extract in separatory funnel with three successive 5-mL portions of $CHCl_3$; shake each portion thoroughly with aqueous solution, let layers separate, and withdraw and discard $CHCl_3$ extract. Transfer extracted aqueous solution to a 125-mL conical flask. Wash separatory funnel with a small amount of water and add wash water to flask. Boil for about 5 min to volatilize $CHCl_3$ and cool. Add 5 mL HNO_3 and 3 mL H_2SO_4. Boil samples to the appearance of SO_3 fumes. Cool slightly, carefully add 5 mL HNO_3, and again boil to fumes to complete decomposition of organic matter. Cool, wash sides of flask, and boil once more to SO_3 fumes, assuming elimination of all HNO_3. Cool and add 25 mL water.

d. Oxidation of trivalent chromium: Pipet a portion of digested sample with or without interferences removed, and containing 10 to 100 μg Cr, into a 125-mL conical flask. Add several drops of methyl orange indicator, then add conc NH_4OH until solution just begins to turn yellow. Add 1 + 1 H_2SO_4 dropwise until it is acidic, plus 1 mL (20 drops) in excess. Adjust volume to about 40 mL, add two or more acid-washed glass beads, and heat to boiling. Add 2 drops $KMnO_4$ solution to give a dark red color. If fading occurs, add $KMnO_4$ dropwise to maintain an excess of about 2 drops. Boil for 2 min longer. Add 1 mL NaN_3 solution and continue boiling gently. If red color does not fade completely after boiling for approximately 30 s, add another 1 mL NaN_3 solution. Continue boiling for 1 min after color has faded completely, then cool.

e. Color development and measurement: Add 0.25 mL (5 drops) H_3PO_4. Use 0.2N H_2SO_4 and a pH meter to adjust solution to pH 1.0 ± 0.3. NOTE: Recent work[5] identifies the optimum pH range for color development to be 1.6 to 2.2; the matter of optimum pH range is currently being considered by *Standard Methods*. Transfer solution to a 100-mL volumetric flask, dilute to 100 mL, and mix. Add 2.0 mL diphenylcarbazide solution, mix, and let stand 5 to 10 min for full color development. Transfer an appropriate portion to a 1-cm absorption cell and measure its absorbance at 540 nm, using reagent water as reference. Correct absorbance reading of sample by subtracting absorbance of a blank carried through the method (see also note below). From the corrected absorbance, determine micrograms chromium present by reference to the calibration curve.

NOTE: If the solution is turbid after dilution to 100 mL in ¶ *e* above, take an absorbance reading before adding carbazide reagent and correct absorbance reading of final colored solution by subtracting the absorbance measured previously.

5. Calculation

For digested samples:

$$mg\ Cr/L = \frac{\mu g\ Cr\ (in\ 102\ mL\ final\ volume)}{A \times B} \times 100$$

where:

A = mL original sample, and
B = mL portion from 100 mL digested sample.

For undigested samples:

$$mg\ Cr/L = \frac{\mu g\ Cr\ (in\ 102\ mL\ final\ volume)}{A}$$

6. Precision and Bias

Collaborative test data from 16 laboratories were obtained on reagent water, tap water, 10% NaCl solution, treated water from synthetic organic industrial waste, EPA extraction leachate, process water, lake water, and effluent from a steel pickle liquor treatment plant.[6] The test data yielded the following relationships:

Reagent water:

$$S_t = 0.037x + 0.006$$
$$S_o = 0.022x + 0.004$$

Drinking or wastewater:

$$S_t = 0.067x + 0.004$$
$$S_o = 0.037x + 0.002$$

Leachate:

$$S_t = 0.032x + 0.007$$
$$S_o = 0.017x + 0.004$$

where:

S_t = overall precision,
S_o = single-operator precision, and
x = chromium concentration, mg/L.

7. References

1. BARTLETT, R. & B. JAMES. 1988. Mobility and bioavailability of chromium in soils. *In* J.O. Nriagu & E. Noeboer, eds. Chromium in the Natural and Human Environments, p. 267. Wiley-Interscience, New York, N.Y.
2. VITALE, R.J., G.R. MUSSOLINE, J.C. PETURA & B.R. JAMES. 1994. Hexavalent chromium extraction from soils: Evaluation of an alkaline digestion method. *J. Environ. Qual.* 23:1249.

3. VITALE, R.J., G.R. MUSSOLINE, K.A. RINEHIMER, J.C. PETURA & B.R. JAMES. 1997. Extraction of sparingly soluble chromate from soils: Evaluation of methods and Eh-pH effects. *Environ. Sci. Technol.* 31: 390.

4. U.S. ENVIRONMENTAL PROTECTION AGENCY. 1996. Determination of hexavalent chromium by ion chromatography. Method 1636. EPA 821-R-96-003, U.S. Environmental Protection Agency, Washington, D.C.

5. VITALE, R.J., G.R. MUSSOLINE & K.A. RINEHIMER. 1997. Environmental monitoring of chromium in air, soil, and water. *Regul. Toxicol. Pharmac.* 26:S80.

6. AMERICAN SOCIETY FOR TESTING AND MATERIALS. 1986. Chromium, total. ASTM D1687-86, Annual Book of ASTM Standards, Vol. 11.01. American Soc. Testing & Materials, Philadelphia, Pa.

8. Bibliography

ROWLAND, G.P., JR. 1939. Photoelectric colorimetry—Optical study of permanganate ion and of chromium-diphenylcarbazide system. *Anal. Chem.* 11:442.

SALTZMAN, B.E. 1952. Microdetermination of chromium with diphenyl-carbazide by permanganate oxidation. *Anal. Chem.* 24:1016.

URONE, P.F. 1955. Stability of colorimetric reagent for chromium, 5-diphenylcarbazide, in various solvents. *Anal. Chem.* 27:1354.

ALLEN, T.L. 1958. Microdetermination of chromium with 1,5- diphenyl-carbohydrazide. *Anal. Chem.* 30:447.

SANDELL, E.B. 1959. Colorimetric Determination of Traces of Metals, 3rd ed. Interscience Publishers, New York, N.Y.

3500-Cr C.   Ion Chromatographic Method

1. General Discussion

a. Principle: This method is applicable to determination of dissolved hexavalent chromium in drinking water, groundwater, and industrial wastewater effluents. An aqueous sample is filtered and its pH adjusted to 9 to 9.5 with a concentrated buffer. This pH adjustment reduces the solubility of trivalent chromium and preserves the hexavalent chromium oxidation state. The sample is introduced into the instrument's eluent stream of ammonium sulfate and ammonium hydroxide. Trivalent chromium in solution is separated from the hexavalent chromium by the column. After separation, hexavalent chromium reacts with an azide dye to produce a chromogen that is measured at 530 nm. Hexavalent chromium is identified on the basis of retention time.

Although this method was developed using specific commercial equipment, use of another manufacturer's equipment should be acceptable if appropriate adjustments are made.

b. Interferences: Interferences may come from several sources. Use a good grade of salts for the buffer because trace amounts of chromium may be included.

Several soluble species of trivalent chromium in the sample may be oxidized to the hexavalent form in an alkaline medium in the presence of such oxidants as hydrogen peroxide, ozone, and manganese dioxide. The hexavalent form can be reduced to the trivalent in the presence of reducing species in an acid medium.

High ionic concentration may cause column overload. Samples high in chloride and/or sulfate might show this phenomenon, which is characterized by a change in peak geometry.

Interfering organic compounds are removed by the guard column.

c. Minimum detectable concentrations: The method detection limits obtained in a single laboratory with a 250-μL loop were as follows:

Reagent water	0.4 μg/L
Drinking water	0.3 μg/L
Groundwater	0.3 μg/L
Primary wastewater effluent	0.3 μg/L
Electroplating waste	0.3 μg/L

d. Sample preservation and holding time: Filter sample through a 0.45-μm filter. Use a portion of sample to rinse syringe filter unit and filter, then collect the required volume of filtrate. Adjust pH to 9 to 9.5 by adding buffer solution dropwise while checking pH with a pH meter.

Ship and store sample at 4°C. Bring to room temperature before analysis. Analyze samples within 24 h of collection.

2. Apparatus

a. Ion chromatograph equipped with a pump capable of precisely delivering a flow of 1 to 5 mL/min. The metallic parts of the pump must not contact sample, eluent, or reagent. Sample loops should be available or the instrument should be capable of delivering from 50 to 250-μL injections of sample. The visible absorption cell should not contain metallic parts that contact the eluent-sample flow. The cell must be usable at 530 nm. Use plastic pressurized containers to deliver eluent and post-column reagent. Use high-purity helium (99.995%) to pressurize the eluent and post-column reagent vessel.

b. Guard column, to be placed before the separator column, containing an adsorbent capable of adsorbing organic compounds and particulates that would damage or interfere with the analysis or equipment.*

c. Separator column, packed with a high-capacity anion-exchange resin capable of resolving chromate from other sample constituents.†

d. Recorder, integrator, or computer for receiving signals from the detector as a function of time.

e. Labware: Soak all reusable labware (glass, plastic, etc.) including sample containers, overnight in laboratory-grade detergent, rinse, and soak for 4 h in a mixture of nitric acid (1 part), hydrochloric acid (2 parts), and reagent water (9 parts). Rinse with tap water and reagent water. NOTE: Never use chromic acid cleaning solution.

f. Syringe, equipped with male luer-type fitting and a capacity of at least 3 mL.

---

* IonPac NGl, Dionex, 4700 Lakeside Drive, Sunnyvale, CA 94086, or equivalent.
† IonPac AS7, Dionex, or equivalent.

3. Reagents

a. Reagent water: Deionized or distilled water free from interferences at the minimum detection limit of each constituent, filtered through a 0.2-μm membrane filter and having a conductance of less than 0.1 μS/cm. Use for preparing all reagents.

b. Cr(VI) stock solution, 100 mg Cr^{6+}/L: Prepare from primary standard grade potassium dichromate. Dissolve 0.1414 g $K_2Cr_2O_7$ in water and dilute to 500 mL in a volumetric flask. pH adjustment is not required. Store in plastic. CAUTION: *Hexavalent chromium is toxic and a suspected carcinogen; handle with care.*

c. Eluent: Dissolve 33 g ammonium sulfate $(NH_4)_2SO_4$, in 500 mL water and add 6.5 mL conc ammonium hydroxide, NH_4OH. Dilute to 1 L with water.

d. Post-column reagent: Dissolve 0.5 g 1,5-diphenylcarbazide in 100 mL HPLC-grade methanol. Add with stirring to 500 mL water containing 28 mL conc H_2SO_4. Dilute to 1 L with water. Reagent is stable for 4 or 5 d; prepare only as needed.

e. Buffer solution: Dissolve 33 g ammonium sulfate, $(NH_4)_2SO_4$, in 75 mL water and add 6.5 mL conc ammonium hydroxide, NH_4OH. Dilute to 100 mL with water.

4. Procedure

a. Instrument setup: Establish ion chromatograph operating conditions as indicated in Table 3500-Cr:I. Set flow rate of the eluent pump at 1.5 mL/min and adjust pressure of reagent delivery module so that the system flow rate, measured after the detector, is 2.0 mL/min. Measure system flow rate using a graduated cylinder and stopwatch. Allow approximately 30 min after adjustment before measuring flow.

Use an injection loop size based on required sensitivity. A 50-μL loop is sufficient, although a 250-μL loop was used to determine the method detection limit.

b. Calibration: Before sample analysis, construct a calibration curve using a minimum of a blank and three standards that bracket the expected sample concentration range. Prepare calibration standards from the stock standard (3*b*) by appropriate dilution with water in volumetric flasks. Adjust to pH 9 to 9.5 with buffer solution (3*e*) before final dilution. Injection volumes of standards should be about 10 times the injection loop volume to insure complete loop flushing.

c. Sample analysis: Bring chilled, pH-adjusted sample to ambient temperature. Fill a clean syringe with sample, attach a 0.45-μm syringe filter, and inject 10 times the sample loop volume

into the instrument. Dilute any sample that has a concentration greater than the highest calibration standard.

5. Calculation

Determine area or height of the Cr(VI) peak in the calibration standard chromatograms. Calculate a calibration line by regressing peak area (or height) against standard concentration in mg/L. A correlation coefficient less than 0.995 may indicate a problem with the analysis.

For samples, measure area (or height) of Cr(VI) peak in sample chromatogram, as determined by retention time. Calculate Cr(VI) concentration by interpolating from the calibration line. Correct data for any dilutions made.

Currently available instrumentation automates the entire measurement process (peak measurement, calibration, and sample measurements and calculations). Ensure that enough quality control samples are analyzed to monitor the instrumental processes.

6. Quality Control

a. Initial demonstration of performance: Before sample analysis, set up instrument and analyze enough known samples to determine estimates for the method detection limit and linear calibration range. Use the initial demonstration of performance to characterize instrument performance, i.e., method detection levels (MDLs) and linear calibration range.

b. Initial and continuing calibration performance: Initially, after every 10 samples, and after the final sample, analyze an independent check sample and a calibration blank. The concentration of the calibration check sample should be near the mid-calibration range; prepare from a source independent of the calibration standards. Use acceptance criteria for check standard recovery and calibration standard concentration based on project goals for precision and accuracy. Typical values for recovery of the check standard range from 90 to 110%. The acceptance criteria for the calibration blank are typically set at ± the nominal MDL.

c. Reagent blank analysis: Analyze one laboratory reagent blank with each batch of samples. Significant Cr(VI) detected in the reagent blank is a sign of contamination. Identify source and eliminate contamination.

d. Laboratory-fortified matrix (also known as matrix spike) analysis: To a portion of a sample, add a known quantity of Cr(VI). After analysis, calculate percent recovery of the known addition. If the recovery falls outside the control limits (typically 75 to 125%), the matrix may be interfering with the results. Perform additional testing to determine whether analysis by the method of standard additions will overcome the interference. Analyze fortified matrix samples as frequently as dictated by project goals and anticipated similarity of matrices in the sample set.

e. Laboratory control sample: Analyze a laboratory control sample (LCS) from an external source with every sample batch. Process LCS and samples identically, including filtering and pH adjustment. Base acceptance criteria for LCS recovery on project goals for precision and bias. Typical values for acceptable recovery range from 90 to 110%.

7. Precision and Bias

The instrument operating conditions and data from a single-laboratory test of the method are shown in Tables 3500-Cr:I and 3500-Cr:II, respectively.

TABLE 3500-Cr:I. ION CHROMATOGRAPHIC CONDITIONS

Variable	Value
Guard column	Dionex IonPac NG1
Separator column	Dionex IonPac AS7
Eluent	250 mM $(NH_4)_2SO_4$
	100 mM NH_4OH
Eluent flow rate	1.5 mL/min
Post-column reagent	2 mM diphenylcarbohydrazide
	10% v/v CH_3OH
	1N H_2SO_4
Post-column reagent flow rate	0.5 mL/min
Detector	Visible @ 530 nm
Retention time	3.8 min

TABLE 3500-Cr:II. SINGLE-LABORATORY PRECISION AND BIAS

Simple Type	Conc.* $\mu g/L$	Mean Recovery %	RPD†
Reagent water	100	100	0.8
	1000	100	0.0
Drinking water	100	105	6.7
	1000	98	1.5
Groundwater	100	98	0.0
	1000	96	0.8
Primary wastewater effluent	100	100	0.7
	1000	104	2.7
Electroplating effluent	100	99	0.4
	1000	101	0.4

*Sample fortified at this concentration level.
†RPD - relative percent difference between fortified duplicates.

Multilaboratory test data are shown in Table 3500-Cr:III.‡ Fifteen laboratories analyzed samples ranging from 1.2 to 960 μg Cr/L.

For reagent water matrix:

$$S_o = 0.033x + 0.106$$
$$S_t = 0.050x + 0.559$$
$$\text{Mean recovery} = 1.04x + 0.183$$

where:

S_o = single-operator precision,
S_t = overall precision, and
x = added amount.

For wastewater matrix:

$$S_o = 0.041x + 0.039$$
$$S_t = 0.059x + 1.05$$
$$\text{Mean recovery} = 0.989x - 0.41$$

‡ The multilaboratory precision and bias data cited in this method were the result of a collaborative study carried out jointly between U.S. EPA Environmental Monitoring Systems Laboratory (Cincinnati) and Committee D-19 of ASTM.

TABLE 3500-Cr:III. MULTILABORATORY DETERMINATION OF BIAS FOR HEXAVALENT CHROMIUM*

Water	Amount Added $\mu g/L$	Amount Found $\mu g/L$	S_t	S_o	Bias %
Reagent	6.00	6.68	1.03	0.53	+11.3
	8.00	8.64	1.10		+8.0
	16.0	17.4	2.25	0.77	+8.8
	20.0	21.4	2.31		+7.0
	100	101	1.91	3.76	+1.0
	140	143	5.52		+2.1
	800	819	24.3	12.7	+2.4
	960	966	18.5		+7.3
Waste	6.0	5.63	1.17	0.55	−6.2
	8.0	7.31	1.91		−8.6
	16.0	15.1	2.70	1.85	−5.6
	20.0	19.8	1.01		−1.0
	100	98.9	4.36	3.31	−1.1
	140	138	8.39		−1.4
	800	796	60.6	27.1	−0.5
	960	944	72.1		−1.7

*Each Youden pair was used to calculate one laboratory data point (S_o).

The eleven water samples consisted of a reagent water blank and five Youden pairs. The nine wastewater samples consisted of a wastewater blank and four Youden pairs.

8. Bibliography

U.S. ENVIRONMENTAL PROTECTION AGENCY. 1991. Methods for the Determination of Metals in Environmental Samples. Method 218.6, EPA-600/4-91-010, Environmental Monitoring Systems Lab., Cincinnati, Ohio.

DIONEX. 1990. Technical Note No. 26. Dionex, Sunnyvale, Calif.

EDGEL, K.W., J.A. LONGBOTTOM & R.A. JOYCE. 1994. Determination of dissolved hexavalent chromium in drinking water, ground water, and industrial wastewater effluents by ion chromatography: Collaborative study. J. Assoc. Offic. Anal. Chem. 77:994.

3500-Co   COBALT

Cobalt (Co) is the second element in Group VIII in the periodic table; it has an atomic number of 27, an atomic weight of 58.93, and valences of 1, 2, and 3. The average abundance of Co in the earth's crust is 29 ppm; in soils it is 1.0 to 14 ppm; in streams it is 0.2 $\mu g/L$; and in groundwaters it is 1 to 10 $\mu g/L$. Cobalt occurs only sparingly in ores, usually as the sulfide or the arsenide. It is widely used in alloys of various steels, in electroplating, in fertilizers, and in porcelain and glass.

The solubility of cobalt is controlled by coprecipitation or adsorption by oxides or manganese and iron, by carbonate precipi-

tation, and by the formation of complex ions. Cobalt dust is flammable and is toxic by inhalation. Cobalt is considered essential for algae and some bacteria, nonessential for higher plants, and an essential trace element for animals. The United Nations Food and Agriculture Organization recommended maximum level for irrigation waters is 100 $\mu g/L$.

Perform analyses by the flame atomic absorption spectrometric methods (3113B and C), by the electrothermal atomic absorption method (3113B), or by the inductively coupled plasma methods (3120 and 3125).

3500-Cu COPPER*

3500-Cu A. Introduction

1. Occurrence and Significance

Copper (Cu) is the first element in Group IB in the periodic table; it has an atomic number of 29, an atomic weight of 63.54, and valences of 1 and 2. The average abundance of Cu in the earth's crust is 68 ppm; in soils it is 9 to 33 ppm; in streams it is 4 to 12 $\mu g/L$; and in groundwater it is <0.1 mg/L. Copper occurs in its native state, but is also found in many minerals, the most important of which are those containing sulfide compounds (e.g., chalcopyrite), but also those with oxides and carbonates. Copper is widely used in electrical wiring, roofing, various alloys, pigments, cooking utensils, piping, and in the chemical industry. Copper salts are used in water supply systems to control biological growths in reservoirs and distribution pipes and to catalyze the oxidation of manganese. Copper forms a number of complexes in natural waters with inorganic and organic ligands. Among the common aqueous species are Cu^{2+}, $Cu(OH)_2$, and $CuHCO_3^+$. Corrosion of copper-containing alloys in pipe fittings may introduce measurable amounts of copper into the water in a pipe system.

Copper is considered an essential trace element for plants and animals. Some compounds are toxic by ingestion or inhalation.

* Approved by Standard Methods Committee, 1993.

The United Nations Food and Agriculture Organization recommended maximum level for irrigation waters is 200 $\mu g/L$. Under the lead-copper rule, the U.S. EPA drinking water 90th percentile action level is 1.3 mg/L.

2. Selection of Method

The atomic absorption spectrometric methods (3111B and C), the inductively coupled plasma methods (3120 and 3125), and the neocuproine method (B) are recommended because of their freedom from interferences. The electrothermal atomic absorption method (3113B) also may be used with success with an appropriate matrix modifier. The bathocuproine method (C) may be used for potable waters.

3. Sampling and Storage

Copper ion tends to be adsorbed on the surface of sample containers. Therefore, analyze samples as soon as possible after collection. If storage is necessary, use 0.5 mL 1 + 1 HCl/100 mL sample, or acidify to pH <2 with HNO_3, to prevent this adsorption.

3500-Cu B. Neocuproine Method

1. General Discussion

a. Principle: Cuprous ion (Cu^+) in neutral or slightly acidic solution reacts with 2,9-dimethyl-1,10-phenanthroline (neocuproine) to form a complex in which 2 moles of neocuproine are bound by 1 mole of Cu^+ ion. The complex can be extracted by a number of organic solvents, including a chloroform-methanol ($CHCl_3$-CH_3OH) mixture, to give a yellow solution with a molar absorptivity of about 8000 at 4571 nm. The reaction is virtually specific for copper; the color follows Beer's law up to a concentration of 0.2 mg Cu/25 mL solvent; full color development is obtained when the pH of the aqueous solution is between 3 and 9; the color is stable in $CHCl_3$-CH_3OH for several days.

The sample is treated with hydroxylamine-hydrochloride to reduce cupric ions to cuprous ions. Sodium citrate is used to complex metallic ions that might precipitate when the pH is raised. The pH is adjusted to 4 to 6 with NH_4OH, a solution of neocuproine in methanol is added, and the resultant complex is extracted into $CHCl_3$. After dilution of the $CHCl_3$ to an exact volume with CH_3OH, the absorbance of the solution is measured at 4571 nm.

b. Interference: Large amounts of chromium and tin may interfere. Avoid interference from chromium by adding sulfurous acid to reduce chromate and complex chromic ion. In the presence of much tin or excessive amounts of other oxidizing ions, use up to 20 mL additional hydroxylamine-hydrochloride solution.

Cyanide, sulfide, and organic matter interfere but can be removed by a digestion procedure.

c. Minimum detectable concentration: The minimum detectable concentration, corresponding to 0.01 absorbance or 98% transmittance, is 3 μg Cu when a 1-cm cell is used and 0.6 μg Cu when a 5-cm cell is used.

2. Apparatus

a. Colorimetric equipment: One of the following is required:

1) Spectrophotometer, for use at 4571 nm, providing a light path of 1 cm or longer.

2) Filter photometer, providing a light path of 1 cm or longer and equipped with a narrow-band violet filter having maximum transmittance in the range 450 to 4601 nm.

b. Separatory funnels, 125-mL, Squibb form, with glass or TFE stopcock and stopper.

3. Reagents

a. Redistilled water, copper-free: Because most ordinary distilled water contains detectable amounts of copper, use redistilled water, prepared by distilling singly distilled water in a resistant-glass still, or distilled water passed through an ion-exchange unit, to prepare all reagents and dilutions.

b. Stock copper solution: To 200.0 mg polished electrolytic copper wire or foil in a 250-mL conical flask, add 10 mL water and 5 mL conc HNO_3. After the reaction has slowed, warm gently to complete dissolution of the copper and boil to expel oxides of nitrogen, using precautions to avoid loss of copper. Cool, add about 50 mL water, transfer quantitatively to a 1-L volumetric flask, and dilute to the mark with water; 1 mL = 200 μg Cu.

c. Standard copper solution: Dilute 50.00 mL stock copper solution to 500 mL with water; 1.00 mL = 20.0 μg Cu.

d. Sulfuric acid, H_2SO_4, conc.

e. Hydroxylamine-hydrochloride solution: Dissolve 50 g $NH_2OH \cdot HCl$ in 450 mL water.

f. Sodium citrate solution: Dissolve 150 g $Na_3C_6H_5O_7 \cdot 2H_2O$ in 400 mL water. Add 5 mL $NH_2OH \cdot HCl$ solution and 10 mL neocuproine reagent. Extract with 50 mL $CHCl_3$ to remove copper impurities and discard $CHCl_3$ layer.

g. Ammonium hydroxide, NH_4OH, 5*N*: Dilute 330 mL conc NH_4OH (28-29%) to 1000 mL with water. Store in a polyethylene bottle.

h. Congo red paper, or other pH test paper showing a color change in the pH range of 4 to 6.

i. Neocuproine reagent: Dissolve 100 mg 2,9-dimethyl-1,10-phenanthroline hemihydrate* in 100 mL methanol. This solution is stable under ordinary storage conditions for a month or more.

j. Chloroform, $CHCl_3$: Avoid or redistill material that comes in containers with metal-lined caps.

k. Methanol, CH_3OH, reagent grade.

l. Nitric acid, HNO_3, conc.

m. Hydrochloric acid, HCl, conc.

4. Procedure

a. Preparation of calibration curve: Pipet 50 mL water into a 125-mL separatory funnel for use as a reagent blank. Prepare standards by pipetting 1.00 to 10.00 mL (20.0 to 200 μg Cu) standard copper solution into a series of 125-mL separatory funnels, and dilute to 50 mL with water. Add 1 mL conc H_2SO_4 and use the extraction procedure given in ¶ 4*b* below.

Construct a calibration curve by plotting absorbance versus micrograms of copper.

To prepare a calibration curve for smaller amounts of copper, dilute 10.0 mL standard copper solution to 100 mL. Carry 1.00-

* GFS Chemicals, Inc., Columbus, OH, or equivalent.

to 10.00-mL volumes of this diluted standard through the previously described procedure, but use 5-cm cells to measure absorbance.

b. Treatment of sample: Transfer 100 mL sample to a 250-mL beaker, add 1 mL conc H_2SO_4 and 5 mL conc HNO_3. Add a few boiling chips and cautiously evaporate to dense white SO_3 fumes on a hot plate. If solution remains colored, cool, add another 5 mL conc HNO_3, and again evaporate to dense white fumes. Repeat, if necessary, until solution becomes colorless.

Cool, add about 80 mL water, and bring to a boil. Cool and filter into a 100-mL volumetric flask. Make up to 100 mL with water using mostly beaker and filter washings.

Pipet 50.0 mL or other suitable portion containing 4 to 200 μg Cu, from the solution obtained from preliminary treatment, into a 125-mL separatory funnel. Dilute, if necessary, to 50 mL with water. Add 5 mL $NH_2OH \cdot HCl$ solution and 10 mL sodium citrate solution, and mix thoroughly. Adjust pH to approximately 4 by adding 1-mL increments of NH_4OH until Congo red paper is just definitely red (or other suitable pH test paper indicates a value between 4 and 6).

Add 10 mL neocuproine reagent and 10 mL $CHCl_3$. Stopper and shake vigorously for 30 s or more to extract the copper-neocuproine complex into the $CHCl_3$. Let mixture separate into two layers and withdraw lower $CHCl_3$ layer into a 25-mL volumetric flask, taking care not to transfer any of the aqueous layer. Repeat extraction of the water layer with an additional 10 mL $CHCl_3$ and combine extracts. Dilute combined extracts to 25 mL with CH_3OH, stopper, and mix thoroughly.

Transfer an appropriate portion of extract to a suitable absorption cell (1 cm for 40 to 200 μg Cu; 5 cm for lesser amounts) and measure absorbance at 4571 nm or with a 450- to 460-nm filter. Use a sample blank prepared by carrying 50 mL water through the complete digestion and analytical procedure.

Determine micrograms copper in final solution by reference to the appropriate calibration curve.

5. Calculation

$$\text{mg Cu/L} = \frac{\text{μg Cu (in 25 mL final volume)}}{\text{mL portion taken for extraction}}$$

6. Bibliography

Smith, G.F. & W.H. McCurdy. 1952. 2,9-Dimethyl-1,10-phenanthroline: New specific in spectrophotometric determination of copper. *Anal. Chem.* 24:371.

Luke, C.L. & M.E. Campbell. 1953. Determination of impurities in germanium and silicon. *Anal. Chem.* 25:1586.

Gahler, A.R. 1954. Colorimetric determination of copper with neocuproine. *Anal. Chem.* 26:577.

Fulton, J.W. & J. Hastings. 1956. Photometric determinations of copper in aluminum and lead-tin solder with neocuproine. *Anal. Chem.* 28:174.

Frank, A.J., A.B. Goulston & A.A. Deacutis. 1957. Spectrophotometric determination of copper in titanium. *Anal. Chem.* 29:750.

3500-Cu C. Bathocuproine Method

1. General Discussion

a. Principle: Cuprous ion forms a water-soluble orange-colored chelate with bathocuproinc disulfonate (2,9-dimethyl-4,7-diphenyl-1,10-phenanthrolinedisulfonic acid, disodium salt). While the color forms over the pH range 3.5 to 11.0, the recommended pH range is between 4 and 5. The sample is buffered at a pH of about 4.3 and reduced with hydroxylamine hydrochloride. The absorbance is measured at 4841 nm. The method can be applied to copper concentrations up to at least 5 mg/L with a sensitivity of 20 μg/L.

b. Interference: The following substances can be tolerated with an error of less than ± 2%:

Substance	Concentration mg/L
Cations	
Aluminum	100
Beryllium	10
Cadmium	100
Calcium	1000
Chromium (III)	10
Cobalt (II)	5
Iron (II)	100
Iron (III)	100
Lithium	500
Magnesium	100
Manganese (II)	500
Nickel (II)	500
Sodium	1000
Strontium	200
Thorium (IV)	100
Zinc	200
Anions	
Chlorate	1000
Chloride	1000
Fluoride	500
Nitrate	200
Nitrite	200
Orthophosphate	1000
Perchlorate	1000
Sulfate	1000
Compounds	
Residual chlorine	1
Linear alkylate sulfonate (LAS)	40

Cyanide, thiocyanate, persulfate, and EDTA also can interfere.

c. Minimum detectable concentration: 20 μg/L with a 5-cm cell.

2. Apparatus

a. Colorimetric equipment: One of the following, with a light path of 1 to 5 cm (unless nessler tubes are used):

1) *Spectrophotometer,* for use at 4841 nm.

2) *Filter photometer,* equipped with a blue-green filter exhibiting maximum light transmission near 4841 nm.

3) *Nessler tubes,* matched, 100-mL, tall form.

b. Acid-washed glassware: Rinse all glassware with conc HCl and then with copper-free water.

3. Reagents

a. Copper-free water: See Method B, ¶ 3*a.*

b. Stock copper solution: Prepare as directed in Method B, ¶ 3*b,* but use 20.00 mg copper wire or foil; 1.00 mL = 20.00 μg Cu.

c. Standard copper solution: Dilute 250 mL stock copper solution to 1000 mL with water; 1.00 mL = 5.00 μg Cu. Prepare daily.

d. Hydrochloric acid, HCl, 1 + 1.

e. Hydroxylamine hydrochloride solution: See Method B, ¶ 3*e.*

f. Sodium citrate solution: Dissolve 300 g $Na_3C_6H_5O_7 \cdot 2H_2O$ in water and make up to 1000 mL.

g. Disodium bathocuproine disulfonate solution: Dissolve 1.000 g $C_{12}H_4N_2(CH_3)_2(C_6H_4)_2 (SO_3Na)_2$ in water and make up to 1000 mL.

4. Procedure

Pipet 50.0 mL sample, or a suitable portion diluted to 50.0 mL, into a 250-mL erlenmeyer flask. In separate 250-mL erlenmeyer flasks, prepare a 50.0-mL water blank and a series of 50.0-mL copper standards containing 5.0, 10.0, 15.0, 20.0, and 25.0 μg Cu. To sample, blank, and standards add, mixing after each addition, 1.00 mL 1 + 1 HCl, 5.00 mL $NH_2OH \cdot HCl$ solution, 5.00 mL sodium citrate solution, and 5.00 mL disodium bathocuproine disulfonate solution. Transfer to cells and read sample absorbance against the blank at 4841 nm. Plot absorbance against micrograms Cu in standards for the calibration curve. Estimate concentration from the calibration curve.

5. Calculation

$$\text{mg Cu/L} = \frac{\mu g \text{ Cu (in 66 mL final volume)}}{\text{mL sample}}$$

6. Precision and Bias

A synthetic sample containing 1000 μg Cu/L, 500 μg Al/L, 50 μg Cd/L, 110 μg Cr/L, 300 μg Fe/L, 70 μg Pb/L, 50 μg Mn/L, 150 μg Ag/L, and 650 μg Zn/L was analyzed in 33 laboratories by the bathocuproine method, with a relative standard deviation of 4.1% and a relative error of 0.3%.

7. Bibliography

SMITH, G.F. & D.H. WILKINS. 1953. New colorimetric reagent specific for copper. *Anal. Chem.* 25:510.

BORCHARDT, L.G. & J.P. BUTLER. 1957. Determination of trace amounts of copper. *Anal. Chem.* 29:414.

ZAK, B. 1958. Simple procedure for the single sample determination of serum copper and iron. *Clinica Chim. Acta* 3:328.

BLAIR, D. & H. DIEHL. 1961. Bathophenanthrolinedisulfonic acid and bathocuproinedisulfonic acid, water soluble reagents for iron and copper. *Talanta* 7:163.

3500-Ga GALLIUM

Gallium (Ga) is the third element in Group IIIA in the periodic table; it has an atomic number of 31, an atomic weight of 67.72, and valences of 1, 2, and 3. The average abundance of Ga in the earth's crust is 19 ppm; in soils it is 1.9 to 29 ppm; in streams it is 0.09 µg/L; and in groundwaters it is <0.1 mg/L. Gallium occurs in many zinc ores, and nearly always in bauxite. Gallium compounds are used in semiconducting devices.

The element exists as Ga^{3+} in natural water, and its solubility is controlled by formation of the hydroxide. It is considered nonessential for plants and animals.

Perform analyses by the electrothermal atomic absorption method (3113B). The inductively coupled plasma/mass spectrometric method (3125) also may be applied successfully in most cases (with lower detection limits), even though gallium is not specifically listed as an analyte in the method.

3500-Ge GERMANIUM

Germanium (Ge) is the third element in Group IVA in the periodic table; it has an atomic number of 32, an atomic weight bnof 72.59, and valences of 2 and 4. The average abundance of Ge in the earth's crust is 1.5 ppm; in streams it is 0.03 to 0.1 µg/L; and in groundwaters it is <0.1 mg/L. Germanium is found in germanite, in certain zinc ores, and in elevated levels in certain hot spring waters. Germanium alloys are used in transistors, gold alloys, phosphors, and semiconducting devices.

Germanium is present in natural waters in the tetravalent state, and its distribution in natural waters probably is controlled by adsorption on clay mineral surfaces. It is nonessential for plants and animals.

Perform analyses by the electrothermal atomic absorption method (3113B). The inductively coupled plasma/mass spectrometric method (3125) also may be applied successfully in most cases (with lower detection limits), even though germanium is not specifically listed as an analyte in the method.

3500-Au GOLD

Gold (Au) is the third element in Group IB in the periodic table; it has an atomic number of 79, an atomic weight of 196.97, and valences of 1 and 3. The average abundance of Au in the earth's crust is 0.004 ppm; in streams it is 2 µg/L; and in groundwater it is <0.1 mg/L. Gold occurs in the native form, and is associated with quartz or pyrite. The main uses of gold are in jewelry, dentistry, electronics, and the aerospace industry.

Gold solubility is restricted to acidic waters in the presence of oxidizing agents and chloride, or in alkaline solutions in the presence of hydrogen sulfide. Its solubility may be influenced by natural organic acids. Compounds of gold containing thiosulfate and cyanide have some human toxicity.

Perform analyses by the atomic absorption spectrometric method (3111B) or by the electrothermal atomic absorption method (3113B). The inductively coupled plasma mass spectrometric method (3125) also may be applied successfully in most cases (with lower detection limits), even though gold is not specifically listed as an analyte in the method.

3500-In INDIUM

Indium (In) is the fourth element in Group IIIA in the periodic table; it has an atomic number of 49, an atomic weight of 114.82, and valences of 1, 2, and 3. The average abundance of indium in the earth's crust is 0.19 ppm; in streams it is <0.01 µg/L; and in groundwaters it is <0.1 mg/L. Indium often occurs in combination with zinc ores, and sometimes with pyrites and siderite. Indium is used in alloys for bearings, brazing, solder, and in electrical devices.

Indium exists as In^{3+} and as a number of complex ions. Its solubility is controlled by the formation of the insoluble hydroxide. The metal and its compounds are toxic by inhalation.

Perform analyses by the electrothermal atomic absorption method (3113B). The inductively coupled plasma mass spectrometric method (3125) also may be applied successfully in most cases (with lower detection limits), even though indium is not specifically listed as an analyte in the method.

3500-Ir IRIDIUM

Iridium (Ir) is the eighth element in Group VIII of the periodic table; it has an atomic number of 77, an atomic weight of 192.2, and valences of 1, 3, and 4. The average abundance of Ir in the earth's crust is probably <0.001 ppm, and in groundwaters it is <0.1 mg/L. Iridium occurs uncombined with platinum and other metals. It is used in alloys with platinum in catalysts, thermocouples, electrodes, and wires.

The aqueous chemistry is controlled by complex compounds, although the solubility in natural waters is relatively unknown.

Perform analyses by the flame atomic absorption spectrometric method (3111B). The inductively coupled plasma/mass spectrometric method (3125) also may be applied successfully in most cases (with lower detection limits), even though iridium is not specifically listed as an analyte in the method.

3500-Fe IRON*

3500-Fe A. Introduction

1. Occurrence and Significance

Iron (Fe) is the first element in Group VIII of the periodic table; it has an atomic number of 26, an atomic weight of 55.85, and common valences of 2 and 3 (and occasionally valences of 1, 4, and 6). The average abundance of Fe in the earth's crust is 6.22%; in soils Fe ranges from 0.5 to 4.3%; in streams it averages about 0.7 mg/L; and in groundwater it is 0.1 to 10 mg/L. Iron occurs in the minerals hematite, magnetite, taconite, and pyrite. It is widely used in steel and in other alloys.

The solubility of ferrous ion (Fe^{2+}) is controlled by the carbonate concentration. Because groundwater is often anoxic, any soluble iron in groundwater is usually in the ferrous state. On exposure to air or addition of oxidants, ferrous iron is oxidized to the ferric state (Fe^{3+}) and may hydrolyze to form red, insoluble hydrated ferric oxide. In the absence of complex-forming ions, ferric iron is not significantly soluble unless the pH is very low.

Elevated iron levels in water can cause stains in plumbing, laundry, and cooking utensils, and can impart objectionable tastes and colors to foods. The United Nations Food and Agriculture Organization recommended level for irrigation waters is 5 mg/L. The U.S. EPA secondary drinking water standard MCL is 0.3 mg/L.

2. Selection of Method

Sensitivity and detection limits for the atomic absorption spectrometric methods (3111B and C), the inductively coupled plasma method (3120), and the phenanthroline colorimetric procedure described here (B) are similar and generally adequate for analysis of natural or treated waters. Lower detection limits can be achieved with electrothermal atomic absorption spectrometry (3113B) when an appropriate matrix modifier is used. The complexing reagents used in the colorimetric procedures are specific for ferrous iron but the atomic absorption procedures are not. However, because of the instability of ferrous iron, which is

changed easily to the ferric form in solutions in contact with air, determination of ferrous iron requires special precautions and may need to be done in the field at the time of sample collection.

The procedure for determining ferrous iron using 1,10-phenanthroline (3500-Fe.B.4c) has a somewhat limited applicability; avoid long storage time or exposure of samples to light. A rigorous quantitative distinction between ferrous and ferric iron can be obtained with a special procedure using bathophenanthroline. Spectrophotometric methods using bathophenanthroline[1–6] and other organic complexing reagents such as ferrozine[7] or TPTZ[8] are capable of determining iron concentrations as low as 1 μg/L. A chemiluminescence procedure[9] is stated to have a detection limit of 5 ng/L. Additional procedures are described elsewhere.[10–13]

3. Sampling and Storage

Plan in advance the methods of collecting, storing, and pretreating samples. Clean sample container with acid and rinse with reagent water. Equipment for membrane filtration of samples in the field may be required to determine iron in solution (dissolved iron). Dissolved iron, considered to be that passing through a 0.45-μm membrane filter, may include colloidal iron. The value of the determination depends greatly on the care taken to obtain a representative sample. Iron in well or tap water samples may vary in concentration and form with duration and degree of flushing before and during sampling. When taking a sample portion for determining iron in suspension, shake the sample bottle often and vigorously to obtain a uniform suspension of precipitated iron. Use particular care when colloidal iron adheres to the sample bottle. This problem can be acute with plastic bottles.

For a precise determination of total iron, use a separate container for sample collection. Treat with acid at the time of collection to place the iron in solution and prevent adsorption or deposition on the walls of the sample container. Take account of the added acid in measuring portions for analysis. The addition of acid to the sample may eliminate the need for adding acid before digestion (3500-Fe.B.4a).

---

* Approved by Standard Methods Committee, 1997.

4. References

1. LEE, G.F. & W. STUMM. 1960. Determination of ferrous iron in the presence of ferric iron using bathophenanthroline. *J. Amer. Water Works Assoc.* 52:1567.
2. GHOSH, M.M., J.T. O'CONNOR & R.S. ENGELBRECHT. 1967. Bathophenanthroline method for the determination of ferrous iron. *J. Amer. Water Works Assoc.* 59:897.
3. BLAIR, D. & H. DIEHL. 1961. Bathophenanthroline-disulfonic acid and bathocuproine-disulfonic acid, water soluble reagents for iron and copper. *Talanta* 7:163.
4. SHAPIRO, J. 1966. On the measurement of ferrous iron in natural waters. *Limnol. Oceanogr.* 11:293.
5. McMAHON, J.W. 1967. The influence of light and acid on the measurement of ferrous iron in lake water. *Limnol. Oceanogr.* 12:437.
6. McMAHON, J.W. 1969. An acid-free bathophenanthroline method for measuring dissolved ferrous iron in lake water. *Water Res.* 3:743.
7. GIBBS, C. 1976. Characterization and application of ferrozine iron reagent as a ferrous iron indicator. *Anal. Chem.* 48:1197.
8. DOUGAN, W.K. & A.L. WILSON. 1973. Absorbtiometric determination of iron with TPTZ. *Water Treat. Exam.* 22:110.
9. SEITZ, W.R. & D.M. HERCULES. 1972. Determination of trace amounts of iron (II) using chemiluminescence analysis. *Anal. Chem.* 44:2143.
10. MOSS, M.L. & M.G. MELLON. 1942. Colorimetric determination of iron with 2,2'-bipyridine and with 2,2',2"-tripyridine. *Ind. Eng. Chem., Anal. Ed.* 14:862.
11. WELCHER, F.J. 1947. Organic Analytical Reagents. D. Van Nostrand Co., Princeton, N.J., Vol. 3, pp. 100–104.
12. MORRIS, R.L. 1952. Determination of iron in water in the presence of heavy metals. *Anal. Chem.* 24:1376.
13. DOIG, M.T., III & D.F. MARTIN. 1971. Effect of humic acids on iron analyses in natural water. *Water Res.* 5:689.

3500-Fe B. Phenanthroline Method

1. General Discussion

a. Principle: Iron is brought into solution, reduced to the ferrous state by boiling with acid and hydroxylamine, and treated with 1,10-phenanthroline at pH 3.2 to 3.3. Three molecules of phenanthroline chelate each atom of ferrous iron to form an orange-red complex. The colored solution obeys Beer's law; its intensity is independent of pH from 3 to 9. A pH between 2.9 and 3.5 insures rapid color development in the presence of an excess of phenanthroline. Color standards are stable for at least 6 months.

b. Interference: Among the interfering substances are strong oxidizing agents, cyanide, nitrite, and phosphates (polyphosphates more so than orthophosphate), chromium, zinc in concentrations exceeding 10 times that of iron, cobalt and copper in excess of 5 mg/L, and nickel in excess of 2 mg/L. Bismuth, cadmium, mercury, molybdate, and silver precipitate phenanthroline. The initial boiling with acid converts polyphosphates to orthophosphate and removes cyanide and nitrite that otherwise would interfere. Adding excess hydroxylamine eliminates errors caused by excessive concentrations of strong oxidizing reagents. In the presence of interfering metal ions, use a larger excess of phenanthroline to replace that complexed by the interfering metals. Where excessive concentrations of interfering metal ions are present, the extraction method may be used.

If noticeable amounts of color or organic matter are present, it may be necessary to evaporate the sample, gently ash the residue, and redissolve in acid. The ashing may be carried out in silica, porcelain, or platinum crucibles that have been boiled for several hours in 6N HCl. The presence of excessive amounts of organic matter may necessitate digestion before use of the extraction procedure.

c. Minimum detectable concentration: Dissolved or total concentrations of iron as low as 10 μg/L can be determined with a spectrophotometer using cells with a 5 cm or longer light path. Carry a blank through the entire procedure to allow for correction.

2. Apparatus

a. Colorimetric equipment: One of the following is required:

1) *Spectrophotometer,* for use at 510 nm, providing a light path of 1 cm or longer.
2) *Filter photometer,* providing a light path of 1 cm or longer and equipped with a green filter having maximum transmittance near 510 nm.
3) *Nessler tubes,* matched, 100-mL, tall form.

b. Acid-washed glassware: Wash all glassware with conc hydrochloric acid (HCl) and rinse with reagent water before use to remove deposits of iron oxide.

c. Separatory funnels: 125-mL, Squibb form, with groundglass or TFE stopcocks and stoppers.

3. Reagents

Use reagents low in iron. Use reagent water (see 1080 and 3111B.3c) in preparing standards and reagent solutions and in procedure. Store reagents in glass-stoppered bottles. The HCl and ammonium acetate solutions are stable indefinitely if tightly stoppered. The hydroxylamine, phenanthroline, and stock iron solutions are stable for several months. The standard iron solutions are not stable; prepare daily as needed by diluting the stock solution. Visual standards in nessler tubes are stable for several months if sealed and protected from light.

a. Hydrochloric acid, HCl, conc, containing less than 0.5 ppm iron.

b. Hydroxylamine solution: Dissolve 10 g NH$_2$OH·HCl in 100 mL water.

c. Ammonium acetate buffer solution: Dissolve 250 g NH$_4$C$_2$H$_3$O$_2$ in 150 mL water. Add 700 mL conc (glacial) acetic acid. Because even a good grade of NH$_4$C$_2$H$_3$O$_2$ contains a significant amount of iron, prepare new reference standards with each buffer preparation.

d. Sodium acetate solution: Dissolve 200 g NaC$_2$H$_3$O$_2$·3H$_2$O in 800 mL water.

e. Phenanthroline solution: Dissolve 100 mg 1,10-phenanthroline monohydrate, C$_{12}$H$_8$N$_2$·H$_2$O, in 100 mL water by stirring and heating to 80°C. Do not boil. Discard the solution if it darkens.

Heating is unnecessary if 2 drops conc HCl are added to the water. (NOTE: One milliliter of this reagent is sufficient for no more than 100 μg Fe.)

f. Potassium permanganate, 0.1M: Dissolve 0.316 KMnO$_4$ in reagent water and dilute to 100 mL.

g. Stock iron solution: Use metal (1) or salt (2) for preparing the stock solution.

1) Use electrolytic iron wire, or "iron wire for standardizing," to prepare the solution. If necessary, clean wire with fine sandpaper to remove any oxide coating and to produce a bright surface. Weigh 200.0 mg wire and place in a 1000-mL volumetric flask. Dissolve in 20 mL 6N sulfuric acid (H$_2$SO$_4$) and dilute to mark with water; 1.00 mL = 200 μg Fe.

2) If ferrous ammonium sulfate is preferred, slowly add 20 mL conc H$_2$SO$_4$ to 50 mL water and dissolve 1.404 g Fe(NH$_4$)$_2$(SO$_4$)$_2$·6H$_2$O. Add 0.1M potassium permanganate (KMnO$_4$) dropwise until a faint pink color persists. Dilute to 1000 mL with water and mix; 1.00 mL = 200 μg Fe.

h. Standard iron solutions: Prepare daily for use.

1) Pipet 50.00 mL stock solution into a 1000-mL volumetric flask and dilute to mark with water; 1.00 mL = 10.0 μg Fe.

2) Pipet 5.00 mL stock solution into a 1000-mL volumetric flask and dilute to mark with water; 1.00 mL = 1.00 μg Fe.

i. Diisopropyl or isopropyl ether. CAUTION: *Ethers may form explosive peroxides; test before using.*

4. Procedure

a. Total iron: Mix sample thoroughly and measure 50.0 mL into a 125-mL erlenmeyer flask. If this sample volume contains more than 200 μg iron use a smaller accurately measured portion and dilute to 50.0 mL. Add 2 mL conc HCl and 1 mL NH$_2$OH·HCl solution. Add a few glass beads and heat to boiling. To insure dissolution of all the iron, continue boiling until volume is reduced to 15 to 20 mL. (If the sample is ashed, take up residue in 2 mL conc HCl and 5 mL water.) Cool to room temperature and transfer to a 50- or 100-mL volumetric flask or nessler tube. Add 10 mL NH$_4$C$_2$H$_3$O$_2$ buffer solution and 4 mL phenanthroline solution, and dilute to mark with water. Mix thoroughly and allow a minimum of 10 min for maximum color development.

b. Dissolved iron: Immediately after collection filter sample through a 0.45-μm membrane filter into a vacuum flask containing 1 mL conc HCl/100 mL sample. Analyze filtrate for total dissolved iron (¶ 4*a*) and/or dissolved ferrous iron (¶ 4*c*). (This procedure also can be used in the laboratory if it is understood that normal sample exposure to air during shipment may result in precipitation of iron.)

Calculate suspended iron by subtracting dissolved from total iron.

c. Ferrous iron: Determine ferrous iron at sampling site because of the possibility of change in the ferrous-ferric ratio with time in acid solutions. To determine ferrous iron only, acidify a separate sample with 2 mL conc HCl/100 mL sample at time of collection. Fill bottle directly from sampling source and stopper. Immediately withdraw a 50-mL portion of acidified sample and add 20 mL phenanthroline solution and 10 mL NH$_4$C$_2$H$_3$O$_2$ solution with vigorous stirring. Dilute to 100 mL and measure color intensity within 5 to 10 min. Do not expose to sunlight. (Color development is rapid in the presence of excess phenanthroline. The phenanthroline volume given is suitable for less than 50 μg total iron; if larger amounts are present, use a correspondingly larger volume of phenanthroline or a more concentrated reagent.)

Calculate ferric iron by subtracting ferrous from total iron.

d. Color measurement: Prepare a series of standards by accurately pipetting calculated volumes of standard iron solutions [use solution described in ¶ *h*.2) to measure 1- to 10-μg portions] into 125-mL erlenmeyer flasks, diluting to 50 mL by adding measured volumes of water, and carrying out the steps in ¶ 4*a* beginning with transfer to a 100-mL volumetric flask or nessler tube.

For visual comparison, prepare a set of at least 10 standards, ranging from 1 to 100 μg Fe in the final 100-mL volume. Compare colors in 100-mL tall-form nessler tubes.

For photometric measurement, use Table 3500-Fe:I as a rough guide for selecting proper light path at 510 nm. Read standards against water set at zero absorbance and plot a calibration curve, including a blank (see ¶ 3*c* and General Introduction).

If samples are colored or turbid, carry a second set of samples through all steps of the procedure without adding phenanthroline. Instead of water, use the prepared blanks to set photometer to zero absorbance and read each sample developed with phenanthroline against the corresponding blank without phenanthroline. Translate observed photometer readings into iron values by means of the calibration curve. This procedure does *not* compensate for interfering ions.

e. Samples containing organic interferences: Digest samples containing substantial amounts of organic substances according to the directions given in Sections 3030G or H.

1) If a digested sample has been prepared according to the directions given in Section 3030G or H, pipet 10.0 mL or other suitable portion containing 20 to 500 μg Fe into a 125-mL separatory funnel. If the volume taken is less than 10 mL, add water to make up to 10 mL. To the separatory funnel add 15 mL conc HCl for a 10-mL aqueous volume; or, if the portion taken was greater than 10.0 mL, add 1.5 mL conc HCl/mL sample. Mix, cool, and proceed with 4*e*3) below.

2) To prepare a sample solely for determining iron, measure a suitable volume containing 20 to 500 μg Fe and carry it through the digestion procedure described in either Section 3030G or H. However, use only 5 mL H$_2$SO$_4$ or HClO$_4$ and omit H$_2$O$_2$. When digestion is complete, cool, dilute with 10 mL water, heat almost to boiling to dissolve slowly soluble salts, and, if the sample is still cloudy, filter through a glass-fiber, sintered-glass, or porcelain filter, washing with 2 to 3 mL water. Quantitatively transfer filtrate or clear solution to a 25-mL volumetric flask and make up to 25 mL with water. Empty flask into a 125-mL separatory funnel, rinse flask with 5 mL conc HCl and add to the funnel. Add 25 mL conc HCl measured with the same flask. Mix and cool to room temperature.

TABLE 3500-Fe:I. SELECTION OF LIGHT PATH LENGTH FOR VARIOUS IRON CONCENTRATIONS

Fe μg		Light Path *cm*
50-mL Final Volume	100-mL Final Volume	
50–200	100–400	1
25–100	50–200	2
10–40	20–80	5
5–20	10–40	10

3) Extract the iron from the HCl solution in the separatory funnel by shaking for 30 s with 25 mL isopropyl ether (CAUTION). Draw off lower acid layer into a second separatory funnel. Extract acid solution again with 25 mL isopropyl ether, drain acid layer into a suitable clean vessel, and add ether layer to the ether in the first funnel. Pour acid layer back into second separatory funnel and re-extract with 25 mL isopropyl ether. Withdraw and discard acid layer and add ether layer to first funnel. Persistence of a yellow color in the HCl solution after three extractions does not signify incomplete separation of iron because copper, which is not extracted, gives a similar yellow color.

Shake combined ether extracts with 25 mL water to return iron to aqueous phase and transfer lower aqueous layer to a 100-mL volumetric flask. Repeat extraction with a second 25-mL portion of water, adding this to the first aqueous extract. Discard ether layer.

4) Add 1 mL $NH_2OH \cdot HCl$ solution, 10 mL phenanthroline solution, and 10 mL $NaC_2H_3O_2$ solution. Dilute to 100 mL with water, mix thoroughly, and let stand for a minimum of 10 min. Measure absorbance at 510 nm using a 5-cm absorption cell for amounts of iron less than 100 μg or 1-cm cell for quantities from 100 to 500 μg. As reference, use either water or a sample blank prepared by carrying the specified quantities of acids through the entire analytical procedure. If water is used as reference, correct sample absorbance by subtracting absorbance of a sample blank.

Determine micrograms of iron in the sample from the absorbance (corrected, if necessary) by reference to the calibration curve prepared by using a suitable range of iron standards containing the same amounts of phenanthroline, hydroxylamine, and sodium acetate as the sample.

5. Calculation

When the sample has been treated according to 4a, b, c, or 4e2):

$$\text{mg Fe/L} = \frac{\mu g \text{ Fe (in 100 mL final volume)}}{\text{mL sample}}$$

When the sample has been treated according to 4e1):

$$\text{mg Fe/L} = \frac{\mu g \text{ Fe (in 100 mL final volume)}}{\text{mL sample}} \times \frac{100}{\text{mL portion}}$$

Report details of sample collection, storage, and pretreatment if they are pertinent to interpretation of results.

6. Precision and Bias

Precision and bias depend on the method of sample collection and storage, the method of color measurement, the iron concentration, and the presence of interfering color, turbidity, and foreign ions. In general, optimum reliability of visual comparison in nessler tubes is not better than 5% and often only 10%, whereas, under optimum conditions, photometric measurement may be reliable to 3% or 3 μg, whichever is greater. The sensitivity limit for visual observation in nessler tubes is approximately 1 μg Fe. Sample variability and instability may affect precision and bias of this determination more than will the errors of analysis. Serious divergences have been found in reports of different laboratories because of variations in methods of collecting and treating samples.

A synthetic sample containing 300 μg Fe/L, 500 μg Al/L, 50 μg Cd/L, 110 μg Cr/L, 470 μg Cu/L, 70 μg Pb/L, 120 μg Mn/L, 150 μg Ag/L, and 650 μg Zn/L in distilled water was analyzed in 44 laboratories by the phenanthroline method, with a relative standard deviation of 25.5% and a relative error of 13.3%.

7. Bibliography

CHRONHEIM, G. & W. WINK. 1942. Determination of divalent iron (by o-nitrosophenol). Ind. Eng. Chem., Anal. Ed. 14:447.

MEHLIG, R.P. & R.H. HULETT. 1942. Spectrophotometric determination of iron with o-phenanthroline and with nitro-o-phenanthroline. Ind. Eng. Chem., Anal. Ed. 14:869.

CALDWELL, D.H. & R.B. ADAMS. 1946. Colorimetric determination of iron in water with o-phenanthroline. J. Amer. Water Works Assoc. 38: 727.

WELCHER, F.J. 1947. Organic Analytical Reagents. D. Van Nostrand Co., Princeton, N.J., Vol. 3, pp. 85–93.

KOLTHOFF, I.M., T.S. LEE & D.L. LEUSSING. 1948. Equilibrium and kinetic studies on the formation and dissociation of ferroin and ferrin. Anal. Chem. 20:985.

RYAN, J.A. & G.H. BOTHAM. 1949. Iron in aluminum alloys: Colorimetric determination using 1,10-phenanthroline. Anal. Chem. 21:1521.

REITZ, L.K., A.S. O'BRIEN & T.L. DAVIS. 1950. Evaluation of three iron methods using a factorial experiment. Anal. Chem. 22:1470.

SANDELL, E.B. 1959. Chapter 22 in Colorimetric Determination of Traces of Metals, 3rd ed. Interscience Publishers, New York, N.Y.

SKOUGSTAD, M.W., M.J. FISHMAN, L.C. FRIEDMAN, D.E. ERDMANN & S.S. DUNCAN. 1979. Methods for Determination of Inorganic Substances in Water and Fluvial Sediment. Chapter A1 in Book 5, Techniques of Water Resources Investigations of the United States Geological Survey. U.S. Geological Surv., Washington, D.C.

3500-Pb LEAD*

3500-Pb A. Introduction

1. Occurrence and Significance

Lead (Pb) is the fifth element in Group IVA in the periodic table; it has an atomic number of 82, an atomic weight of 207.19, and valences of 2 and 4. The average abundance of Pb in the earth's crust is 13 ppm; in soils it ranges from 2.6 to 25 ppm; in streams it is 3 $\mu g/L$, and in groundwaters it is generally <0.1 mg/L. Lead is obtained chiefly from galena (PbS). It is used in batteries, ammunition, solder, piping, pigments, insecticides, and alloys. Lead also was used in gasoline for many years as an anti-knock agent in the form of tetraethyl lead.

The common aqueous species are Pb^{2+} and hydroxide and carbonate complexes. Lead in a water supply may come from industrial, mine, and smelter discharges or from the dissolution of plumbing and plumbing fixtures. Tap waters that are inherently noncorrosive or not suitably treated may contain lead resulting from an attack on lead service pipes, lead interior plumbing, brass fixtures and fittings, or solder pipe joints.

_____

*Approved by Standard Methods Committee, 1997.

Lead is nonessential for plants and animals. It is toxic by ingestion and is a cumulative poison. The Food and Drug Administration regulates lead content in food and in house paints. Under the lead-copper rule, the U.S. EPA drinking water 90th percentile action level is 15 $\mu g/L$.

2. Selection of Method

The atomic absorption spectrometric method (3111B) has a relatively high detection limit in the flame mode and requires an extraction procedure (3111C) for the low concentrations common in potable water. The electrothermal atomic absorption (AA) method (3113B) is more sensitive for low concentrations and does not require extraction. The inductively coupled plasma/mass spectrometric method (3125) is even more sensitive than the electrothermal AA method. The inductively coupled plasma method (3120) has a sensitivity similar to that of the flame atomic absorption method. Anodic stripping voltammetry (3130B) can achieve superior detection limits, but is susceptible to interferences from copper, silver, gold, and organic compounds. The dithizone method (B) is sensitive and specific as a colorimetric procedure.

3500-Pb B. Dithizone Method

1. General Discussion

a. Principle: An acidified sample containing microgram quantities of lead is mixed with ammoniacal citrate-cyanide reducing solution and extracted with dithizone in chloroform ($CHCl_3$) to form a cherry-red lead dithizonate. The color of the mixed color solution is measured photometrically.[1,2] Sample volume taken for analysis may be 2 L when digestion is used.

b. Interference: In a weakly ammoniacal cyanide solution (pH 8.5 to 9.5) dithizone forms colored complexes with bismuth, stannous tin, and monovalent thallium. In strongly ammoniacal citrate-cyanide solution (pH 10 to 11.5) the dithizonates of these ions are unstable and are extracted only partially.[3] This method uses a high pH, mixed color, single dithizone extraction. Interference from stannous tin and monovalent thallium is reduced further when these ions are oxidized during preliminary digestion. A modification of the method allows detection and elimination of bismuth interference. Excessive quantities of bismuth, thallium, and tin may be removed.[4]

Dithizone in $CHCl_3$ absorbs at 510 nm; control its interference by using nearly equal concentrations of excess dithizone in samples, standards, and blank.

The method is without interference for the determination of 0.0 to 30.0 μg Pb in the presence of 20 μg Tl^+, 100 μg Sn^{2+}, 200 μg In^{3+}, and 1000 μg each of Ba^{2+}, Cd^{2+}, Co^{2+}, Cu^{2+},

Mg^{2+}, Mn^{2+}, Hg^{2+}, Sr^{2+}, Zn^{2+}, Al^{3+}, Sb^{3+}, As^{3+}, Cr^{3+}, Fe^{3+}, V^{3+}, PO_4^{3-}, and SO_4^{2-}. Gram quantities of alkali metals do not interfere. A modification is provided to avoid interference from excessive quantities of bismuth or tin.

c. Preliminary sample treatment: At time of collection acidify with conc HNO_3 to pH <2 but avoid excess HNO_3. Add 5 mL 0.1N iodine solution to avoid losses of volatile organo-lead compounds during handling and digesting of samples. Prepare a blank of lead-free water and carry through the procedure.

d. Digestion of samples: Unless digestion is shown to be unnecessary, digest all samples for dissolved or total lead as described in 3030H or K.

e. Minimum detectable concentration: 1.0 μg Pb/10 mL dithizone solution.

2. Apparatus

a. Spectrophotometer for use at 510 nm, providing a light path of 1 cm or longer.

b. pH meter.

c. Separatory funnels: 250-mL Squibb type. Clean all glassware, including sample bottles, with 1 + 1 HNO_3. Rinse thoroughly with reagent water.

d. Automatic dispensing burets: Use for all reagents to minimize indeterminate contamination errors.

3. Reagents

Prepare all reagents in lead-free water.

a. Stock lead solution: Dissolve 0.1599 g lead nitrate, $Pb(NO_3)_2$ (minimum purity 99.5%), in approximately 200 mL water. Add 10 mL conc HNO_3 and dilute to 1000 mL with water. Alternatively, dissolve 0.1000 g pure Pb metal in 20 mL 1 + 1 HNO_3 and dilute to 1000 mL with water; 1.00 mL = 100 μg Pb.

b. Working lead solution: Dilute 2.0 mL stock solution to 100 mL with water; 1 mL = 2.00 μg Pb.

c. Nitric acid, HNO_3, 1 + 4: Dilute 200 mL conc HNO_3 to 1 L with water.

d. Ammonium hydroxide, NH_4OH, 1 + 9: Dilute 10 mL conc NH_4OH to 100 mL with water.

e. Citrate-cyanide reducing solution: Dissolve 400 g dibasic ammonium citrate, $(NH_4)_2HC_6H_5O_7$, 20 g anhydrous sodium sulfite, Na_2SO_3, 10 g hydroxylamine hydrochloride, $NH_2OH \cdot HCl$, and 40 g potassium cyanide, KCN (CAUTION: *Poison*) in water and dilute to 1 L. Mix this solution with 2 L conc NH_4OH. *Do not pipet by mouth. Prepare solution in a fume hood.*

f. Stock dithizone solution: The dithizone concentration in the stock dithizone solutions is based on having a 100% pure dithizone reagent. Some commercial grades of dithizone are contaminated with the oxidation product diphenylthiocarbodiazone or with metals. Purify dithizone as directed below. For dithizone solutions not stronger than 0.001% (w/v), calculate the exact concentration by dividing the absorbance of the solution in a 1.00-cm cell at 606 nm by 40.6×10^3, the molar absorptivity.

In a fume hood, dissolve 100 mg dithizone in 50 mL $CHCl_3$ in a 150-mL beaker and filter through a 7-cm-diam paper.* Receive filtrate in a 500-mL separatory funnel or in a 125-mL erlenmeyer flask under slight vacuum; use a filtering device designed to handle the $CHCl_3$ vapor. Wash beaker with two 5-mL portions $CHCl_3$, and filter. Wash the paper with three 5-mL portions $CHCl_3$, adding final portion dropwise to edge of paper. If filtrate is in flask, transfer with $CHCl_3$ to a 500-mL separatory funnel.

Add 100 mL 1 + 99 NH_4OH to separatory funnel and shake moderately for 1 min; excessive agitation produces slowly breaking emulsions. Let layers separate, swirling funnel gently to submerge $CHCl_3$ droplets held on surface of aqueous layer. Transfer $CHCl_3$ layer to 250-mL separatory funnel, retaining the orange-red aqueous layer in the 500-mL funnel. Repeat extraction, receiving $CHCl_3$ layer in another 250-mL separatory funnel and transferring aqueous layer, using 1 + 99 NH_4OH, to the 500-mL funnel holding the first extract. Repeat extraction, transferring the aqueous layer to 500-mL funnel. Discard $CHCl_3$ layer.

To combined extracts in the 500-mL separatory funnel add 1 + 1 HCl in 2-mL portions, mixing after each addition, until dithizone precipitates and solution is no longer orange-red. Extract precipitated dithizone with three 25-mL portions $CHCl_3$. Dilute combined extracts to 1000 mL with $CHCl_3$; 1.00 mL = 100 μg dithizone.

g. Dithizone working solution: Dilute 100 mL stock dithizone solution to 250 mL with $CHCl_3$; 1 mL = 40 μg dithizone.

h. Special dithizone solution: Dissolve 250 mg dithizone in 250 mL $CHCl_3$. This solution may be prepared without purification because all extracts using it are discarded.

i. Sodium sulfite solution: Dissolve 5 g anhydrous Na_2SO_3 in 100 mL water.

j. Iodine solution: Dissolve 40 g KI in 25 mL water, add 12.7 g resublimed iodine, and dilute to 1000 mL.

4. Procedure

a. With sample digestion: CAUTION: *Perform the following procedure (excluding use of spectrophotometer) in a fume hood.* To a digested sample containing not more than 1 mL conc acid add 20 mL 1 + 4 HNO_3 and filter through lead-free filter paper† and filter funnel directly into a 250-mL separatory funnel. Rinse digestion beaker with 50 mL water and add to filter. Add 50 mL ammoniacal citrate-cyanide solution, mix, and cool to room temperature. Add 10 mL dithizone working solution, shake stoppered funnel vigorously for 30 s, and let layers separate. Insert lead-free cotton in stem of separatory funnel and draw off lower layer. Discard 1 to 2 mL $CHCl_3$ layer, then fill absorption cell. Measure absorbance of extract at 510 nm, using dithizone working solution, ¶ 3g, to zero spectrophotometer.

b. Without sample digestion: To 100 mL acidified sample (pH 2) in a 250-mL separatory funnel add 20 mL 1 + 4 HNO_3 and 50 mL citrate-cyanide reducing solution; mix. Add 10 mL dithizone working solution and proceed as in ¶ 4a.

c. Calibration curve: Plot concentration of at least five standards and a blank against absorbance. Determine concentration of lead in extract from curve. All concentrations are μg Pb/10 mL final extract.

d. Removal of excess interferences: The dithizonates of bismuth, tin, and thallium differ from lead dithizonate in maximum absorbance. Detect their presence by measuring sample absorbance at 510 nm and at 465 nm. Calculate corrected absorbance of sample at each wavelength by subtracting absorbance of blank at same wavelength. Calculate ratio of corrected absorbance at 510 nm to corrected absorbance at 465 nm. The ratio of corrected absorbances for lead dithizonate is 2.08 and for bismuth dithizonate is 1.07. If the ratio for the sample indicates interference, i.e., is markedly less than 2.08, proceed as follows with a new 100-mL sample: If the sample has not been digested, add 5 mL Na_2SO_3 solution to reduce iodine preservative. Adjust sample to pH 2.5 using a pH meter and 1 + 4 HNO_3 or 1 + 9 NH_4OH as required. Transfer sample to 250-mL separatory funnel, extract with a minimum of three 10-mL portions special dithizone solution, or until the $CHCl_3$ layer is distinctly green. Extract with 20-mL portions $CHCl_3$ to remove dithizone (absence of green). Add 20 mL 1 + 4 HNO_3, 50 mL citrate-cyanide reducing solution, and 10 mL dithizone working solution. Extract as in ¶ 4a and measure absorbance.

5. Calculation

$$\text{mg Pb/L} = \frac{\mu g \text{ Pb (in 10 mL, from calibration curve)}}{\text{mL sample}}$$

6. Precision and Bias

Single-operator precision in recovering 0.0104 mg Pb/L from Mississippi River water was 6.8% relative standard deviation and

---

* Whatman No. 42 or equivalent.

† Whatman No. 541 or equivalent.

− 1.4% relative error. At the level of 0.026 mg Pb/L, recovery was made with 4.8% relative standard deviation and 15% relative error.

7. References

1. SNYDER, L.J. 1947. Improved dithizone method for determination of lead—mixed color method at high pH. *Anal. Chem.* 19:684.

2. SANDELL, E.B. 1959. Colorimetric Determination of Traces of Metals, 3rd ed. Interscience, New York, N.Y.
3. WICHMANN, H.J. 1939. Isolation and determination of trace metals—the dithizone system. *Ind. Eng. Chem.*, Anal. Ed. 11:66.
4. AMERICAN SOCIETY FOR TESTING AND MATERIALS. 1977. Annual Book of ASTM Standards. Part 26, Method D3112-77, American Soc. Testing & Materials, Philadelphia, Pa.

3500-Li LITHIUM*

3500-Li A. Introduction

1. Occurrence and Significance

Lithium (Li) is the second element in Group IA of the periodic table; it has an atomic number of 3, an atomic weight of 6.94, and a valence of 1. The average abundance of Li in the earth's crust is 18 ppm; in soils it is 14 to 32 ppm; in streams it is 3 μg/L, and in groundwaters it is <0.1 mg/L. The more important minerals containing lithium are lepidolite, spodumene, petalite, and amblygonite. Lithium compounds are used in pharmaceuticals, soaps, batteries, welding flux, ceramics, reducing agents (e.g., lithium aluminum hydride), and cosmetics.

Many lithium salts are only slightly soluble, and the metal's concentration in water is controlled by incorporation in clay minerals of soils. Lithium is considered nonessential for plants and

animals, but it is essential for some microorganisms. Some lithium salts are toxic by ingestion. The United Nations Food and Agriculture Organization recommended maximum level for lithium in irrigation waters is 2.5 mg/L.

2. Selection of Method

The atomic absorption spectrometric method (3111B) and the inductively coupled plasma method (3120) are preferred. The flame emission photometric method (B) also is available for laboratories not equipped to use preferred methods. The inductively coupled plasma/mass spectrometric method (3125) may be applied successfully in most cases (with lower detection limits), even though lithium is not specifically listed as an analyte in the method.

* Approved by Standard Methods Committee, 1997.

3500-Li B. Flame Emission Photometric Method

1. General Discussion

a. Principle: Lithium can be determined in trace amounts by flame photometric methods at a wavelength of 670.8 nm.

b. Interference: A molecular band of strontium hydroxide with an absorption maximum at 671.0 nm interferes in the flame photometric determination of lithium. Ionization of lithium can be significant in both the air-acetylene and nitrous oxide-acetylene flames and can be suppressed by adding potassium. See Section 3500-Na.B.1*b* for additional information on minimizing interferences in flame photometry.

c. Minimum detectable concentration: The minimum lithium concentration detectable is approximately 0.1 μg/L for reagent water analyzed on an atomic absorption spectrophotometer in the emission mode with an air-acetylene flame, or 0.03 μg/L with a nitrous oxide-acetylene flame.

d. Sampling and storage: Preferably collect sample in a poly-

ethylene bottle, although borosilicate glass containers also may be used. At time of collection adjust sample to pH <2 with nitric acid (HNO_3).

2. Apparatus

Flame photometer: A flame photometer or an atomic absorption spectrometer operating in the emission mode using a lean air-acetylene flame is recommended.

3. Reagents

Use reagent water (see 3111B.3*c*) in reagent preparation and analysis.

a. Potassium ionization suppressant: Dissolve 95.35 g KCl dried at 110°C and dilute to 1000 mL with water; 1.00 mL = 50 mg K.

b. Stock lithium solution: Dissolve 152.7 mg anhydrous lithium chloride, LiCl, in water and dilute to 250 mL; 1.00 mL = 100 μg Li. Dry salt overnight in an oven at 105°C. Cool in a desiccator and weigh immediately after removal from desiccator. Alternatively, purchase prepared stock from a reputable supplier.

c. Standard lithium solution: Dilute 10.00 mL stock LiCl solution to 500 mL with water; 1.00 mL = 2.0 μg Li.

4. Procedure

a. Pretreatment of polluted water and wastewater samples: Choose digestion method appropriate to matrix (see Section 3030).

b. Suppressing ionization: If necessary, filter sample through medium-porosity paper, add 1.0 mL potassium ionization suppressant to 50 mL volumetric flask, and dilute with sample for flame photometric determination. Sample solution will be in a 0.1% K matrix.

c. Treatment of standard solutions: Prepare dilutions of the Li standard solution to bracket sample concentration or to establish at least three points on a calibration curve of emission intensity against Li concentration. Prepare standards by adding appropriate volumes of standard lithium solution to 25 mL water + 1.0 mL potassium ionization suppressant reagent in a 50-mL volumetric flask. Dilute to 50.0 mL and mix. Both samples and standards will be in a 0.1% K matrix to suppress ionization of lithium.

d. Flame photometric measurement: Determine lithium concentration by direct intensity measurements at a wavelength of 670.8 nm. The bracketing method (Section 3500-Na.B.4*f*) can be used with some photometric instruments, while the construction of a calibration curve is necessary with others. Run sample, water, and lithium standard as nearly simultaneously as possible. For best results, average several readings on each solution.

Follow the manufacturer's instructions for instrument operation.

5. Calculation

$$\mu g\ Li/L = (\mu g\ Li/L\ in\ portion\ analyzed) \times D$$

where:

D = dilution ratio

$$= \frac{mL\ sample\ +\ mL\ water}{mL\ sample}$$

6. Quality Control

Process a QC standard through entire analytical protocol as a way of determining systematic bias. The control limits for precision of duplicate determinations at concentrations (in water) of 4.0 μg/L and 10.0 μg/L were 4.09 ± 0.056 μg/L and 9.96 ± 0.094 μg/L, respectively. The single-operator RSD was 1.38% for a lithium solution containing 10 μg/L.

7. Bibliography

FISHMAN, M.J. 1962. Flame photometric determination of lithium in water. *J. Amer. Water Works Assoc.* 54:228.

PICKETT, E.E. & S.R. KOIRTYOHANN. 1968. The nitrous oxide-acetylene flame in emission analysis-I. General characteristics. *Spectrochem. Acta.* 23B:235.

KOIRTYOHANN, S.R. & E.E. PICKETT. 1968. The nitrous oxide-acetylene flame in emission analysis-II. Lithium and the alkaline earths. *Spectrochem. Acta*, 23B:673.

URE, A.M. & R.L. MITCHELL. 1975. Lithium, sodium, potassium, rubidium, and cesium. *In* J.A. Dean & T.C. Rains, eds. Flame Emission and Atomic Absorption Spectrometry. Dekker, New York, N.Y.

THOMPSON, K.C. & R.J. REYNOLDS. 1978. Atomic Absorption Fluorescence, and Flame Spectroscopy—A Practical Approach, 2nd ed. John Wiley & Sons, New York, N.Y.

WILLARD, H.H., L.L. MERRIT, J.A. DEAN & F.A. SETTLE., JR. 1981. Instrumental Methods of Analysis, 6th ed. Wadsworth Publishing Co., Belmont, Calif.

3500-Mg MAGNESIUM*

3500-Mg A. Introduction

1. Occurrence and Significance

Magnesium (Mg) is the second element in Group IIA of the periodic table; it has an atomic number of 12, an atomic weight of 24.30, and a valence of 2. The average abundance of Mg in the earth's crust is 2.1%; in soils it is 0.03 to 0.84%; in streams it is 4 mg/L, and in groundwaters it is >5 mg/L. Magnesium occurs commonly in the minerals magnesite and dolomite. Magnesium is used in alloys, pyrotechnics, flash photography, drying agents, refractories, fertilizers, pharmaceuticals, and foods.

The common aqueous species is Mg^{2+}. The carbonate equilibrium reactions for magnesium are more complicated than for calcium, and conditions for direct precipitation of dolomite in natural waters are not common. Important contributors to the hardness of a water, magnesium salts break down when heated, forming scale in boilers. Chemical softening, reverse osmosis, or ion exchange reduces magnesium and associated hardness to acceptable levels.

Magnesium is an essential element in chlorophyll and in red blood cells. Some salts of magnesium are toxic by ingestion or

---

* Approved by Standard Methods Committee, 1997.

inhalation. Concentrations greater than 125 mg/L also can have a cathartic and diuretic effect.

2. Selection of Method

The methods presented are applicable to waters and wastewaters. Direct determinations can be made with the atomic absorption spectrometric method (3111B) and inductively coupled plasma method (3120). The inductively coupled plasma mass spectrometric method (3125) also may be applied successfully in most cases (with lower detection limits), even though magnesium is not specifically listed as an analyte in the method. These methods can be applied to most concentrations encountered, although sample dilution may be required. Choice of method is largely a matter of personal preference and analyst experience. A calculation method (B) also is available.

3500-Mg B.   Calculation Method

Magnesium may be estimated as the difference between hardness and calcium as $CaCO_3$ if interfering metals are present in noninterfering concentrations in the calcium titration (Section 3500-Ca.B) and suitable inhibitors are used in the hardness titration (Section 2340C).

$$mg\ Mg/L = [total\ hardness\ (as\ mg\ CaCO_3/L) \\ - calcium\ hardness\ (as\ mg\ CaCO_3/L)] \times 0.243$$

3500-Mn   MANGANESE*

3500-Mn A.   Introduction

1. Occurrence and Significance

Manganese (Mn) is the first element in Group VIIB in the periodic table; it has an atomic number of 25, an atomic weight of 54.94, and common valences of 2, 4, and 7 (and more rarely, valences of 1, 3, 5, and 6). The average abundance of Mn in the earth's crust is 1060 ppm; in soils it is 61 to 1010 ppm; in streams it is 7 μg/L, and in groundwaters it is <0.1 mg/L. Manganese is associated with iron minerals, and occurs in nodules in ocean, fresh waters, and soils. The common ores are pyrolusite (MnO_2) and psilomelane. Manganese is used in steel alloys, batteries, and food additives.

The common aqueous species are the reduced Mn^{2+} and the oxidized Mn^{4+}. The aqueous chemistry of manganese is similar to that of iron. Since groundwater is often anoxic, any soluble manganese in groundwater is usually in the reduced state (Mn^{2+}). Upon exposure to air or other oxidants, groundwater containing manganese usually will precipitate black MnO_2. Elevated manganese levels therefore can cause stains in plumbing/laundry, and cooking utensils. It is considered an essential trace element for plants and animals. The United Nations Food and Agriculture Organization recommended maximum level for manganese in irrigation waters is 0.2 mg/L. The U.S. EPA secondary drinking water standard MCL is 50 μg/L.

2. Selection of Method

The atomic absorption spectrometric methods (3111B and C), the electrothermal atomic absorption method (3113B), and the inductively coupled plasma methods (3120 and 3125) permit direct determination with acceptable sensitivity and are the methods of choice. Of the various colorimetric methods, the persulfate method (B) is preferred because the use of mercuric ion can control interference from a limited chloride ion concentration.

3. Sampling and Storage

Manganese may exist in a soluble form in a neutral water when first collected, but it oxidizes to a higher oxidation state and precipitates or becomes adsorbed on the container walls. Determine manganese very soon after sample collection. When delay is unavoidable, total manganese can be determined if the sample is acidified at the time of collection with HNO_3 to pH <2. See Section 3010B.

---

* Approved by Standard Methods Committee, 1993.

3500-Mn B. Persulfate Method

1. General Discussion

a. Principle: Persulfate oxidation of soluble manganous compounds to form permanganate is carried out in the presence of silver nitrate. The resulting color is stable for at least 24 h if excess persulfate is present and organic matter is absent.

b. Interference: As much as 0.1 g chloride (Cl^-) in a 50-mL sample can be prevented from interfering by adding 1 g mercuric sulfate ($HgSO_4$) to form slightly dissociated complexes. Bromide and iodide still will interfere and only trace amounts may be present. The persulfate procedure can be used for potable water with trace to small amounts of organic matter if the period of heating is increased after more persulfate has been added.

For wastewaters containing organic matter, use preliminary digestion with nitric and sulfuric acids (HNO_3 and H_2SO_4) (see Section 3030G). If large amounts of Cl^- also are present, boiling with HNO_3 helps remove it. Interfering traces of Cl^- are eliminated by $HgSO_4$ in the special reagent.

Colored solutions from other inorganic ions are compensated for in the final colorimetric step.

Samples that have been exposed to air may give low results due to precipitation of manganese dioxide (MnO_2). Add 1 drop 30% hydrogen peroxide (H_2O_2) to the sample, after adding the special reagent, to redissolve precipitated manganese.

c. Minimum detectable concentration: The molar absorptivity of permanganate ion is about 2300 L g^{-1} cm^{-1}. This corresponds to a minimum detectable concentration (98% transmittance) of 210 μg Mn/L when a 1-cm cell is used or 42 μg Mn/L when a 5-cm cell is used.

2. Apparatus

Colorimetric equipment: One of the following is required:

a. Spectrophotometer, for use at 525 nm, providing a light path of 1 cm or longer.

b. Filter photometer, providing a light path of 1 cm or longer and equipped with a green filter having maximum transmittance near 525 nm.

c. Nessler tubes, matched, 100-mL, tall form.

3. Reagents

a. Special reagent: Dissolve 75 g $HgSO_4$ in 400 mL conc HNO_3 and 200 mL distilled water. Add 200 mL 85% phosphoric acid (H_3PO_4), and 35 mg silver nitrate ($AgNO_3$). Dilute the cooled solution to 1 L.

b. Ammonium persulfate, $(NH_4)_2S_2O_8$, solid.

c. Standard manganese solution: Prepare a 0.1N potassium permanganate ($KMnO_4$) solution by dissolving 3.2 g $KMnO_4$ in distilled water and making up to 1 L. Age for several weeks in sunlight or heat for several hours near the boiling point, then filter through a fine fritted-glass filter crucible and standardize against sodium oxalate as follows:

Weigh several 100- to 200-mg samples of $Na_2C_2O_4$ to 0.1 mg and transfer to 400-mL beakers. To each beaker, add 100 mL distilled water and stir to dissolve. Add 10 mL 1 + 1 H_2SO_4 and heat rapidly to 90 to 95°C. Titrate rapidly with the $KMnO_4$ solution to be standardized, while stirring, to a slight pink end-point color that persists for at least 1 min. Do not let temperature fall below 85°C. If necessary, warm beaker contents during titration; 100 mg $Na_2C_2O_4$ will consume about 15 mL permanganate solution. Run a blank on distilled water and H_2SO_4.

$$\text{Normality of } KMnO_4 = \frac{\text{g } Na_2C_2O_4}{(A - B) \times 0.067\ 01}$$

where:

A = mL titrant for sample and
B = mL titrant for blank.

Average results of several titrations. Calculate volume of this solution necessary to prepare 1 L of solution so that 1.00 mL = 50.0 μg Mn, as follows:

$$\text{mL } KMnO_4 = \frac{4.55}{\text{normality } KMnO_4}$$

To this volume add 2 to 3 mL conc H_2SO_4 and $NaHSO_3$ solution dropwise, with stirring, until the permanganate color disappears. Boil to remove excess SO_2, cool, and dilute to 1000 mL with distilled water. Dilute this solution further to measure small amounts of manganese.

d. Standard manganese solution (alternate): Dissolve 1.000 g manganese metal (99.8% min.) in 10 mL redistilled HNO_3. Dilute to 1000 mL with 1% (v/v) HCl; 1 mL = 1.000 mg Mn. Dilute 10 mL to 200 mL with distilled water; 1 mL = 0.05 mg Mn. Prepare dilute solution daily.

e. Hydrogen peroxide, H_2O_2, 30%.

f. Nitric acid, HNO_3, conc.

g. Sulfuric acid, H_2SO_4, conc.

h. Sodium nitrite solution: Dissolve 5.0 g $NaNO_2$ in 95 mL distilled water.

i. Sodium oxalate, $Na_2C_2O_4$, primary standard.

j. Sodium bisulfite: Dissolve 10 g $NaHSO_3$ in 100 mL distilled water.

4. Procedure

a. Treatment of sample: If a digested sample has been prepared according to directions for reducing organic matter and/or excessive chlorides in Section 3030G, pipet a portion containing 0.05 to 2.0 mg Mn into a 250-mL conical flask. Add distilled water, if necessary, to 90 mL and proceed as in ¶ *b*.

b. To a suitable sample portion add 5 mL special reagent and 1 drop H_2O_2. Concentrate to 90 mL by boiling or dilute to 90 mL. Add 1 g $(NH_4)_2S_2O_8$, bring to a boil, and boil for 1 min. Do not heat on a water bath. Remove from heat source, let stand 1 min, then cool under the tap. (Boiling too long results in decomposition of excess persulfate and subsequent loss of permanganate color; cooling too slowly has the same effect.) Dilute to 100 mL with distilled water free from reducing substances and mix. Prepare standards containing 0, 5.00, . . . 1500 μg Mn by treating various amounts of standard Mn solution in the same way.

c. Nessler tube comparison: Use standards prepared as in ¶ 4*b* and containing 5 to 100 μg Mn/100 mL final volume. Compare samples and standards visually.

d. Photometric determination: Use a series of standards from 0 to 1500 µg Mn/100 mL final volume. Make photometric measurements against a distilled water blank. The following table shows light path length appropriate for various amounts of manganese in 100 mL final volume:

Mn Range µg	Light Path cm
5–200	15
20–400	5
50–1000	2
100–1500	1

Prepare a calibration curve of manganese concentration vs. absorbance from the standards and determine Mn in the samples from the curve. If turbidity or interfering color is present, make corrections as in ¶ 4e.

e. Correction for turbidity or interfering color: Avoid filtration because of possible retention of some permanganate on the filter paper. If visual comparison is used, the effect of turbidity only can be estimated and no correction can be made for interfering colored ions. When photometric measurements are made, use the following "bleaching" method, which also corrects for interfering color: As soon as the photometer reading has been made, add 0.05 mL H_2O_2 solution directly to the sample in the optical cell. Mix and, as soon as the permanganate color has faded completely and no bubbles remain, read again. Deduct absorbance of bleached solution from initial absorbance to obtain absorbance due to Mn.

5. Calculation

a. When all of the original sample is taken for analysis:

$$\text{mg Mn/L} = \frac{\mu g \text{ Mn/100 mL}}{\text{mL sample}} \times \frac{100}{\text{mL portion}}$$

b. When a portion of the digested sample (100 mL final volume) is taken for analysis:

$$\text{mg Mn/L} = \frac{\mu g \text{ Mn (in 100 mL final volume)}}{\text{mL sample}}$$

6. Precision and Bias

A synthetic sample containing 120 µg Mn/L, 500 µg Al/L, 50 µg Cd/L, 110 µg Cr/L, 470 µg Cu/L, 300 µg Fe/L, 70 µg Pb/L, 150 µg Ag/L, and 650 µg Zn/L in distilled water was analyzed in 33 laboratories by the persulfate method, with a relative standard deviation of 26.3% and a relative error of 0%.

A second synthetic sample, similar in all respects except for 50 µg Mn/L and 1000 µg Cu/L, was analyzed in 17 laboratories by the persulfate method, with a relative standard deviation of 50.3% and a relative error of 7.2%.

7. Bibliography

RICHARDS, M.D. 1930. Colorimetric determination of manganese in biological material. *Analyst* 55:554.

NYDAHL, F. 1949. Determination of manganese by the persulfate method. *Anal. Chem. Acta.* 3:144.

MILLS, S.M. 1950. Elusive manganese. *Water Sewage Works* 97:92.

SANDELL, E.B. 1959. Colorimetric Determination of Traces of Metals, 3rd ed. Interscience Publishers, New York, N.Y., Chapter 26.

3500-Hg MERCURY

Mercury (Hg) is the third element in Group IIB in the periodic table; it has an atomic number of 80, an atomic weight of 200.59, and valences of 1 and 2. The average abundance of Hg in the earth's crust is 0.09 ppm; in soils it is 30 to 160 ppb; in streams it is 0.07 µg/L, and in groundwaters it is 0.5 to 1 µg/L. Mercury occurs free in nature, but the chief source is cinnibar (HgS). Mercury is used in amalgams, mirror coatings, vapor lamps, paints, measuring devices (thermometers, barometers, manometers), pharmaceuticals, pesticides, and fungicides. It is often used in paper mills as a mold retardant for paper.

The common aqueous species are Hg^{2+}, $Hg(OH)_2^0$, Hg^0, and stable complexes with organic ligands. Inorganic mercury can be methylated in sediments when sulfides are present to form dimethyl mercury, $(CH_3)_2Hg$, which is very toxic and can concentrate in the aquatic food chain. Mercury poisoning occurred in Japan in the 1950s as the result of consumption of shellfish that had accumulated mercury. In times past, mercury was used in the haberdashery industry to block hats (the cause of the "mad hatter" syndrome).

Mercury is considered nonessential for plants and animals. The U.S. EPA primary drinking water standard MCL is 2 µg/L.

The cold-vapor atomic absorption method (3112B) is the method of choice for all samples. The inductively coupled plasma mass spectrometric method (3125) also may be applied successfully in some cases, even though mercury is not specifically listed as an analyte in the method. The dithizone method detailed in the 19th edition of *Standard Methods* can be used for determining high levels of mercury (>2 µg/L) in potable waters.

Because mercury can be lost readily from samples, preserve them by treating with HNO_3 to reduce the pH to <2 (see Section 1060). Glass storage containers are preferred to plastic, because they can extend the holding time to 30 d, rather than only the 14 d allowed in plastic containers.

3500-Mo MOLYBDENUM

Molybdenum (Mo) is the second element in Group VIB in the periodic table; it has an atomic number of 42, and atomic weight of 95.95, and valences of 2, 3, 4, 5, and 6. The average abundance of Mo in the earth's crust is 1.2 ppm; in soils it is 2.5 ppm; in streams it is 1 μg/L, and in groundwaters it is <0.1 mg/L. Molybdenum occurs naturally as molybdenite (MoS_2) and wulfenite ($PbMoO_4$). It is used in alloys, ink pigments, catalysts, and lubricants.

The common aqueous species are $HMoO_4^-$, MoO_4^{2-}, and organic complexes. It is considered an essential trace element for plants and animals. The United Nations Food and Agriculture Organization recommended maximum level for irrigation waters is 0.01 mg/L.

Use one of the flame atomic absorption spectrometric methods (3111D or E), the electrothermal atomic absorption spectrometric method (3113B), or one of the inductively coupled plasma methods (3120 or 3125).

3500-Ni NICKEL

Nickel (Ni) is the third element in Group VIII in the periodic table; it has an atomic number of 28, an atomic weight of 58.69, and a common valence of 2 and less commonly 1, 3, or 4. The average abundance of Ni in the earth's crust is 1.2 ppm; in soils it is 2.5 ppm; in streams it is 1 μg/L, and in groundwaters it is <0.1 mg/L. Nickel is obtained chiefly from pyrrhotite and garnierite. Nickel is used in alloys, magnets, protective coatings, catalysts, and batteries.

The common aqueous species is Ni^{2+}. In reducing conditions insoluble sulfides can form, while in aerobic conditions nickel complexes with hydroxide, carbonates, and organic ligands can form. It is suspected to be an essential trace element for some plants and animals. The United Nations Food and Agriculture Organization recommended maximum level for irrigation waters is 200 μg/L. The U.S. EPA primary drinking water standard MCL is 0.1 mg/L.

The atomic absorption spectrometric methods (3111B and C), the inductively coupled plasma methods (3120 and 3125), and the electrothermal atomic absorption spectrometric method (3113B) are the methods of choice for all samples.

3500-Os OSMIUM

Osmium (Os) is the seventh element in Group VIII in the periodic table; it has an atomic number of 76, an atomic weight of 190.2, and valences of 3, 4, and 6, and less commonly 1, 2, 5, 7, and 8. The average abundance of Os in the earth's crust is probably <0.005 ppm, and in groundwaters it is <0.01 mg/L. Osmium occurs in iridosime and in platinum-bearing river sands.

Osmium is used as a hardener with iridium and as a catalyst with platinum.

The aqueous chemistry is controlled by complex compounds, although the solubility in natural waters is relatively unknown.

Analyze by flame atomic absorption methods (3111D and E).

3500-Pd PALLADIUM

Palladium (Pd) is the sixth element in Group VIII of the periodic table; it has an atomic number of 46, an atomic weight of 106.42, and valences of 2 and 4. Palladium occurs with platinum in nature. It is used in alloys to make electrical relays, catalysts, in the making of "white gold," and in protective coatings.

Palladium has no known toxic effects. The United Nations Food and Agriculture Organization recommended maximum level for irrigation waters is 5 mg/L.

Preferably analyze by flame atomic absorption method (3111B). The inductively coupled plasma-mass spectrometric method (3125) also may be applied successfully in most cases (with lower detection limits), even though palladium is not specifically listed as an analyte in the method.

3500-Pt PLATINUM

Platinum (Pt) is the ninth element in Group VIII of the periodic table; it has an atomic number of 78, an atomic weight of 195.1, and valences of 2 and 4. The average abundance of Pt in the earth's crust is probably <0.01 ppm, and in groundwaters it is <0.1 mg/L. Platinum is usually found in its native state, but also may be found as sperrylite ($PtAs_2$). Platinum is used as a catalyst and in laboratoryware, jewelry, and surgical wire.

The aqueous chemistry in natural waters is relatively unknown,

although its solubility is probably controlled by complex compounds. In powder form, platinum can be flammable, and its soluble salts are toxic by inhalation.

Preferably analyze by flame atomic absorption method (3111B). The inductively coupled plasma mass spectrometric method (3125) also may be applied successfully in most cases (with lower detection limits), even though platinum is not specifically listed as an analyte in the method.

3500-K POTASSIUM*

3500-K A. Introduction

1. Occurrence and Significance

Potassium (K) is the fourth element in Group IA of the periodic table; it has an atomic number of 19, an atomic weight of 39.10, and a valence of 1. The average abundance of K in the earth's crust is 1.84%; in soils it has a range of 0.1 to 2.6%; in streams it is 2.3 mg/L, and in groundwaters it has a range of 0.5 to 10 mg/L. Potassium is commonly associated with aluminosilicate minerals such as feldspars. ^{40}K is a naturally occurring radioactive isotope with a half-life of 1.3×10^9 years. Potassium compounds are used in glass, fertilizers, baking powder, soft drinks, explosives, electroplating, and pigments. Potassium is an essential element in both plant and human nutrition, and occurs in groundwaters as a result of mineral dissolution, from decomposing plant material, and from agricultural runoff.

The common aqueous species is K^+. Unlike sodium, it does not remain in solution, but is assimilated by plants and is incorporated into a number of clay-mineral structures.

* Approved by Standard Methods Committee, 1997.

2. Selection of Method

Methods for the determination of potassium include flame atomic absorption (3111B), inductively coupled plasma (3120), flame photometry (B), and selective ion electrode (C). The inductively coupled plasma/mass spectrometric method (3125) usually may be applied successfully (with lower detection limits), even though potassium is not specifically listed as an analyte in the method. The preferred methods are rapid, sensitive, and accurate; selection depends on instrument availability and analyst choice.

3. Storage of Samples

Do not store samples in soft-glass bottles because of the possibility of contamination from leaching of the glass. Use acid-washed polyethylene or borosilicate glass bottles. Adjust sample to pH < 2 with nitric acid. This will dissolve potassium salts and reduce adsorption on vessel walls.

3500-K B. Flame Photometric Method

1. General Discussion

a. Principle: Trace amounts of potassium can be determined in either a direct-reading or internal-standard type of flame photometer at a wavelength of 766.5 nm. Because much of the information pertaining to sodium applies equally to the potassium determination, carefully study the entire discussion dealing with

the flame photometric determination of sodium (Section 3500-Na.B) before making a potassium determination.

b. Interference: Interference in the internal-standard method may occur at sodium-to-potassium ratios of 5:1 or greater. Calcium may interfere if the calcium-to-potassium ratio is 10:1 or more. Magnesium begins to interfere when the magnesium-to-potassium ratio exceeds 100:1.

c. Minimum detectable concentration: Potassium levels of approximately 0.1 mg/L can be determined.

2. Apparatus

See Section 3500-Na.B.2.

3. Reagents

To minimize potassium pickup, store all solutions in plastic bottles. Shake each container thoroughly to dissolve accumulated salts from walls before pouring.

a. Reagent water: See Section 1080. Use this water for preparing all reagents and calibration standards, and as dilution water.

b. Stock potassium solution: Dissolve 1.907 g KCl dried at 110°C and dilute to 1000 mL with water; 1 mL = 1.00 mg K.

c. Intermediate potassium solution: Dilute 10.0 mL stock potassium solution with water to 100 mL; 1.00 mL = 0.100 mg K. Use this solution to prepare calibration curve in potassium range of 1 to 10 mg/L.

d. Standard potassium solution: Dilute 10.0 mL intermediate potassium solution with water to 100 mL; 1.00 mL = 0.010 mg K. Use this solution to prepare calibration curve in potassium range of 0.1 to 1.0 mg/L.

e. Standard lithium solution: See Section 3500-Na.B.3*e*.

4. Procedure

Make determination as described in Section 3500-Na.B.4, but measure emission intensity at 766.5 nm.

5. Calculation

See Section 3500-Na.B.5.

6. Precision and Bias

A synthetic sample containing 3.1 mg K^+/L, 108 mg Ca^{2+}/L, 82 mg Mg^{2+}/L, 19.9 mg Na^+/L, 241 mg Cl^-/L, 0.25 mg NO_2^--N/L, 1.1 mg NO_3^--N/L, 259 mg SO_4^{2-}/L, and 42.5 mg total alkalinity/L (contributed by $NaHCO_3$) was analyzed in 33 laboratories by the flame photometric method, with a relative standard deviation of 15.5% and a relative error of 2.3%.

7. Bibliography

MEHLICH, A. & R.J. MONROE. 1952. Report on potassium analyses by means of flame photometer methods. *J. Assoc. Offic. Agr. Chem.* 35:588.
Also see 3500-Na.B.7.

3500-K C. Potassium-Selective Electrode Method

1. General Discussion

a. Principle: Potassium ion is measured potentiometrically by using a potassium ion-selective electrode and a double-junction, sleeve-type reference electrode. The analysis is performed with either a pH meter having an expanded millivolt scale capable of being read to the nearest 0.1 mV or a specific ion meter having a direct concentration scale for potassium.

Before measurement, an ionic strength adjustor reagent is added to both standards and samples to maintain a constant ionic strength. The electrode response is measured in standard solutions with potassium concentrations spanning the range of interest using a calibration line derived either by the instrument meter or manually. The electrode response in sample solutions is measured following the same procedure and potassium concentration determined from the calibration line or instrument direct readout.

b. Interferences: Although most sensitive to potassium, the potassium electrode will respond to other cations at high concentrations; this can result in a positive bias. Table 3500-K:I lists the concentration of common cations causing a 10% error at various concentrations of potassium chloride with a background ionic strength of 0.12*N* sodium chloride. Of the cations listed, ammonium ion is most often present in samples at concentrations high enough to result in a significant bias. It can be converted to gaseous ammonia by adjusting to pH > 10.

An electrode exposed to interfering cations tends to drift and respond sluggishly. To restore normal performance soak electrode

TABLE 3500-K:I. CONCENTRATION OF CATIONS INTERFERING AT VARIOUS CONCENTRATIONS OF POTASSIUM

	Concentration Causing 10% Error mg/L		
Cation	K conc = 1 mg/L	K conc = 10 mg/L	K conc = 100 mg/L
Cs^+	1.0	10	100
NH_4^+	2.7	27	270
Tl^+	31.4	314	3 140
Ag^+	2 765	27 650	276 500
$Tris^+$	3 105	31 050	310 500
Li^+	356	3 560	35 600
Na^+	1 179	11 790	117 900
H^+	3.6*	2.6*	1.6*

* pH.

for 1 h in distilled water and then for several hours in a standard potassium solution.

c. Detection limits: Samples containing from 0.1 to 1000 mg K^+/L may be analyzed. To measure higher concentrations dilute the sample.

2. Apparatus

a. Expanded-scale or digital pH meter or ion-selective meter.
b. Potassium ion-selective electrode.

c. Sleeve-type double-junction reference electrode: Fill outer sleeve with reference electrode filling solution (see ¶ 3*b*). Fill inner sleeve with inner filling solution provided with the electrode.

d. pH electrode.

e. Mixer, magnetic, with a TFE-coated stirring bar.

3. Reagents

a. Ionic strength adjustor (ISA): Dissolve 29.22 g NaCl in reagent water and dilute to 100 mL.

b. Reference electrode outer sleeve filling solution: Dilute 2 mL ISA solution to 100 mL with reagent water.

c. Stock potassium solution: See 3500-K.B.3*b*.

d. Sodium hydroxide, NaOH, 6*N*.

e. Reagent water: See Section 1080.

4. Procedure

a. Preparation of standards: Prepare a series of standards containing 100.0, 10.0, 1.0, and 0.1 mg K^+/L by making serial dilutions of the stock potassium solution as in Section 3500-K.B.3*c* and 3*d*.

b. Instrument calibration: Fill reference electrode according to the manufacturer's instructions using reference electrode filling solution. Transfer 100 mL 0.1 mg K^+/L standard into a 150-mL beaker and add 2 mL ISA. Raise pH to about 11. Stir gently with magnetic mixer. Immerse electrodes, wait approximately 2 min for potential stabilization and record meter reading. Thoroughly rinse electrodes and blot dry. Repeat for each standard solution in order of increasing concentration. Prepare calibration curve on semilogarithmic graph paper by plotting observed potential in millivolts (linear scale) against concentration (log scale). Alternatively, calculate calibration line by regression analysis.

c. Analysis of samples: Transfer 100 mL sample into a 150-mL beaker and follow procedure applied to standards in ¶ 4*b* above. From the measured response, calculate K^+ concentration from calibration curve.

5. Precision

Reproducibility of potential measured, over the method's range, can be expected to be ± 0.4 mV, corresponding to about ± 2.5% in concentration.

6. Quality Assurance

The slope of the calibration line should be −56 mV/10-fold concentration change. If the slope is outside the range of −56 ±3 mV, the electrode may require maintenance (replace filling solutions). If the proper electrode response cannot be obtained, replace electrode.

Analyze an independent check standard with a mid-range potassium concentration throughout analysis of a series, initially, every ten samples, and after final sample. If the value has changed by more than 5%, recalibrate electrode. Analyze a reagent blank at the same frequency. Readings must represent a lower concentration than the lowest concentration standard (0.1 mg/L).

7. Bibliography

PIODA, L., V. STANKOVA & W. SIMON. 1969. Highly selective potassium ion responsive liquid membrane electrode. *Anal. Lett.* 2 (12): 665.

MIDGLEY, D. & K. TORRANCE. 1978. Potentiometric Water Analysis. John Wiley & Sons, New York, N.Y.

BAILEY, P.L. 1980. Analysis with Ion-Selective Electrodes. Heyden & Son Ltd., Philadelphia, Pa.

3500-Re   RHENIUM

Rhenium (Re) is the third element in Group VIIB in the periodic table; it has an atomic number of 75, an atomic weight of 186.21, and valences of 1 through 7, with 7 being the most stable. The average abundance of Re in the earth's crust is 7 ppm, and in groundwaters it is <0.1 mg/L. Rhenium is found in columbite, tantalite, and wolframite, as well as in molybdenum ore concentrates. It is used in tungsten-molybdenum-based alloys, thermo-couples, filaments, and flash bulbs. Rhenium in the powder form can be flammable.

For analysis methods, see flame atomic absorption methods (3111D and E). The inductively coupled plasma mass spectrometric method (3125) also may be applied successfully in most cases (with lower detection limits), even though rhenium is not specifically listed as an analyte in the method.

3500-Rh   RHODIUM

Rhodium (Rh) is the fifth element in Group VIII in the periodic table; it has an atomic number of 45, an atomic weight of 102.91, and valences of 1 through 6, the most common being 1 and 3. Rhodium is found in its native state in platinum-bearing sands. It is used in platinum alloys for thermocouples, electrical contacts, and jewelry.

The aqueous chemistry in natural waters is relatively unknown.

The metal is flammable in the powder form, and its salts are toxic by inhalation.

For analysis see flame atomic absorption method (3111B). The inductively coupled plasma/mass spectrometric method (3125) also may be applied successfully in most cases (with lower detection limits), even though rhodium is not specifically listed as an analyte in the method.

3500-Ru RUTHENIUM

Ruthenium (Ru) is the fourth element in Group VIII in the periodic table; it has an atomic number of 44, an atomic weight of 101.07, and valences of 1 through 7, the most common being 2, 3, and 4. The average abundance of Ru in the earth's crust is probably <0.01 ppm, and in groundwaters it is <0.1 mg/L. It occurs in its native state in platinum-bearing river sands. It is used in jewelry with platinum, in electrical contacts, and as a catalyst.

The aqueous chemistry in natural waters is relatively unknown. Ruthenium has no known toxic effects.

For analysis see flame atomic absorption method (3111B). The inductively coupled plasma mass spectrometric method (3125) also may be applied successfully in most cases (with lower detection limits), even though ruthenium is not specifically listed as an analyte in the method.

3500-Se SELENIUM*

3500-Se A. Introduction

1. Occurrence and Significance

Selenium (Se) is the third element in Group VIA in the periodic table; it has an atomic number of 34, an atomic weight of 78.96, and valences of 2, 4, or 6. The average abundance of Se in the earth's crust is 0.2 ppm; in soils it is 0.27 to 0.74 ppm; in streams it is 0.2 μg/L, and in groundwaters it is <0.1 mg/L. Selenium is used in electronics, ceramics, and shampoos.

The inorganic fraction of dissolved selenium consists predominantly of selenium as the selenate ion (SeO_4^{2-}), designated here as Se(VI), and selenium as the selenite ion (SeO_3^{2-}), Se(IV). Other common aqueous species include Se^{2-}, HSe^-, and Se^0. Selenium is considered a nonessential trace element for most plants, but is an essential trace nutrient for most animals, and selenium deficiency diseases are well known in veterinary medicine. Above trace levels, ingested selenium is toxic to animals and may be toxic to humans. While the selenium concentration of most natural waters is low, the pore water in seleniferous soils in semiarid areas may contain up to hundreds or thousands of micrograms dissolved selenium per liter. Certain plants that grow in such areas accumulate large concentrations of selenium and may poison livestock that graze on them. Water drained from such soil may cause severe environmental pollution and wildlife toxicity. Selenopolysulfide ions (SSe^{2-}) may occur in the presence of hydrogen sulfide in waterlogged, anoxic soils. Selenium derived from microbial degradation of seleniferous organic matter includes selenite, selenate, and the volatile organic compounds dimethylselenide and dimethyldiselenide. Nonvolatile organic selenium compounds may be released to water by microbial processes. Soluble selenium may be leached from coal ash and fly ash at electric power plants that burn seleniferous coal.

The United Nations Food and Agriculture Organization recommended maximum level for selenium in irrigation waters is 20 μg/L. The U.S. EPA primary drinking water standard MCL is 50 μg/L.

2. Selection of Method

The selenium methods using hydride generation atomic absorption (3114B and C), electrothermal atomic absorption (3113B), and derivatization colorimetry (C) are the most sensitive currently available. For determination of selenium at higher concentrations, the inductively coupled plasma methods (3120 and 3125) may be used.

By using suitable preparatory steps to convert other chemical species to Se(IV), it is possible to distinguish the chemical species in the sample. In drinking water and most surface and ground waters, Se(IV), Se(VI), and particulate selenium frequently are the only significant species. However, when speciation is important, for example, when a new matrix is being analyzed, the general analytical scheme shown in Figure 3500-Se:1 may be carried out as follows: Determine volatile selenium by stripping sample with nitrogen or air and collecting selenium in alkaline hydrogen peroxide (see Method D). To obtain an estimate of selenium in suspended particles, determine total selenium, filter sample, and make a second determination of total selenium. In any case, filter sample. Occasionally, a filtered sample may have the odor of hydrogen sulfide and a yellow color; such a sample may contain selenopolysulfides, which may be estimated by comparing results of total selenium analyses before and after acidification, stripping with nitrogen, settling for 10 min, and refiltration. Determine selenite, Se(IV), by analyzing filtered water sample directly by Methods 3114B or C, or by 3500-Se.C or E. In principle, sample digestion with HCl will convert Se(VI) to Se(IV), and the value determined will equal the sum of the two species. In practice, samples frequently contain an unknown masking agent that produces an unduly low result. Test for this effect by analyzing samples with known additions of both species. If recovery is good, the HCl digestion followed by analyses will yield reliable results. If recovery is poor and organic selenium is to be determined subsequently, attempt to remove the interference by sample pretreatment with resin (B.1). Interference also can be eliminated by digestion with an oxidizing agent (B.2, 3, and 4), but these procedures prevent distinguishing of Se(VI) and organic selenium and also oxidize many organic selenium compounds. To measure nonvolatile organic selenium compounds, use Method E.

---

* Approved by Standard Methods Committee, 1993.

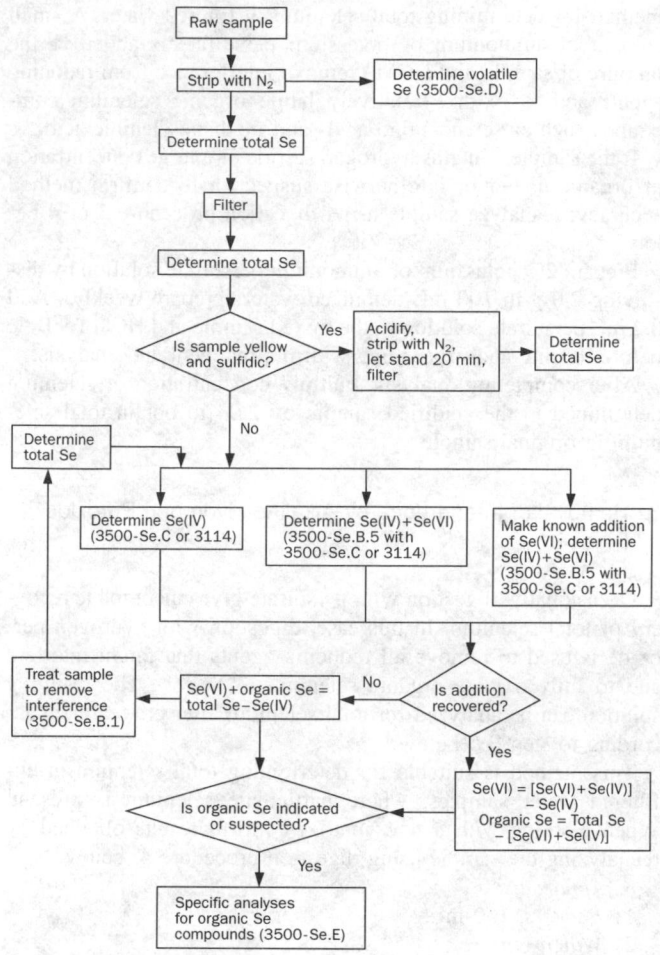

Figure 3500-Se:1. General scheme for speciation of selenium in water.

The choice of digestion method for oxidizing interferences and organic selenium depends on sample matrix. The methods described in B.2, 3, and 4, in order of increasing complexity and digestive ability, use ammonium (or potassium) persulfate, hydrogen peroxide, and potassium permanganate. Ammonium persulfate digestion is adequate for most filtered ground, drinking, and surface water. Hydrogen peroxide digestion may be required if organic selenium compounds are present, and potassium permanganate digestion may be needed with unfiltered samples or those containing refractory organic selenium compounds. Confirm results obtained with one digestion method by using a more rigorous method when characterizing a new matrix.

3. Interferences

Interferences are found in certain reagents, as well as in samples. Recognition of the presence of an interferant is critical, especially when unknown sample matrices are being analyzed. Routinely add Se(IV) and Se(VI) to test for interference. If present, characterize the interference and correct by the method of standard additions. A slope less than one indicates interference. In cases of mild interference (recoveries reduced by 25% or less), the standard additions method will largely correct determined values.

Because the hydride atomic absorption method is extremely sensitive, samples frequently need to be diluted to bring them within the linear range of the instrument. Diluting a filtered water sample frequently will eliminate sample-related interferences. Include full reagent blanks in each run to ensure absence of contamination from reagents. Hydride generator atomic absorption is susceptible to common interference problems related to nitrite in the sample or free chlorine in the reagent HCl (see Methods 3114B and C).

4. Bibliography

U.S. NATIONAL ACADEMY OF SCIENCES. 1976. Selenium: Medical and Biological Effects of Environmental Pollutants. National Academy of Sciences, Washington, D.C.

CUTTER, G.A. 1978. Species determination of selenium in natural waters. *Anal. Chim. Acta* 98:59.

ROBBERECHT, H. & R. VAN GRIEKEN. 1982. Selenium in environmental waters: Determination, speciation and concentration levels. *Talanta* 29:823.

KUBOTA, J. & E.E. CARY. 1982. Cobalt, molybdenum and selenium. *In* A.L. Page et al., eds. Methods of soil analysis, Part 2, 2nd ed. *Agronomy* 9:485.

RAPTIS, S. et al. 1983. A survey of selenium in the environment and a critical review of its determination at trace levels. *Fresenius Z. Anal. Chem.* 316:105.

CUTTER, G. 1983. Elimination of nitrite interference in the determination of selenium by hydride generation. *Anal. Chim. Acta* 149:391.

CAMPBELL, A. 1984. Critical evaluation of analytical methods for the determination of trace elements in various matrices. Part 1. Determination of selenium in biological materials and water. *Anal. Chem.* 56:645.

LEMLY, A.D. 1985. Ecological basis for regulating aquatic emissions from the power industry: The case with selenium. *Regul. Toxic. Pharm.* 5:465.

OHLENDORF, H.M., D.J. HOFFMAN, M.K. SAIKI & T.W. ALDRICH. 1986. Embryonic mortality and abnormalities of aquatic birds: Apparent impacts of selenium from irrigation drainwater. *Sci. Total Environ.* 52:49.

3500-Se B.   Sample Preparation

1. Removal of Organic and Iron Interference by Resin Pretreatment

Interferences are common in selenium analysis, particularly when chemical speciation is attempted. Routine pretreatment of

water samples as described here is not necessary. The methods in this section should be tried when poor recovery of standard additions indicates a problem.

Many waters contain iron and/or dissolved organic matter (humic acid) in quantities sufficient to interfere. Reduction of Se(VI) to Se(IV) usually is nonquantitative. When Se(VI) standard additions show poor recovery, treat the sample before analysis. To remove dissolved organic compounds, pass an acidified sample through a resin. Because dissolved organic selenium compounds also may be removed by this treatment, also determine total selenium in the untreated water sample (see 2, 3, or 4 below). To remove iron use a strong base ion exchange resin.* Iron is removed as the anionic chloro complex. In this treatment the acidity and ion exchanger do not alter speciation; complete speciation of selenium is possible.

a. Apparatus:

1) *Chromatography column* for organics removal, glass, about 0.8 cm ID × 30 cm long, with fluorocarbon metering valve.

2) *Chromatography column* for ion exchange, disposable polyethylene.†

3) *pH meter.*

b. Reagents:

1) *Organics-removal resin:* Thoroughly rinse 16 to 50 mesh resin‡ with deionized water and remove resin fines by decanting. Rinse three times with pH 12 solution. Store resin in pH 12 solution and refrigerate to prevent bacterial growth.

2) *Anion exchange resin:* Add 100 to 200 mesh anion exchange resin* to a beaker and thoroughly rinse with deionized water. Cover resin with 4N HCl, stir, and let settle. Decant and repeat acid rinse twice more. Store resin in 4N HCl.

3) *Hydrochloric acid,* conc: Before use, bubble helium through the acid for 3 h at rate of 100 mL/min. (CAUTION: *Use a fume hood.*)

4) *pH 1.6 solution:* Adjust pH of deionized water to 1.6 with HCl.

5) *pH 12 solution:* Adjust pH of deionized water to 12 with KOH.

c. Procedure:

1) Organic removal—Place 5 cm washed resin in a 0.8-cm-ID column. Precondition column, at 1 mL/min, with 30 mL pH 12 solution and 20 mL pH 1.6 solution. Using HCl and a pH meter adjust sample to pH 1.6 to 1.8. Pass sample through preconditioned column at rate of 1 mL/min. Discard first 10 mL and use next 11 to 50 mL collected for Se(IV) determinations by Methods 3114B or C, or 3500-Se.C preceded, if Se(VI) also is to be determined, by preparatory step B.5. If more than 50 mL sample are needed, use another column or use a column with twice as much resin.

2) Iron removal—Place 4 cm prepared resin in a small chromatographic column (add resin to column filled with 4N HCl to avoid air bubbles). Rinse column with 10 mL 4N HCl at flow rate < 6 mL/min. Let solution drain to top of resin, but do not let the column run dry. Adjust sample to 4N HCl and pour into column. Discard first 10 mL and collect the next 11 to 100 mL for Se(IV) analysis by Methods 3114B or C, or 3500-Se.C preceded, if necessary, by preparatory step B.2 and if Se(VI) also is to be determined, by preparatory step B.5 below.

2. Removal of Interference by Persulfate Digestion

The combination of this procedure with step B.5 below and Methods 3114B or C, or 3500-Se.C is, in most cases, the preferred method for determining total selenium in filtered water. A small amount of ammonium or potassium persulfate is added to the mixture of sample and HCl to remove interference from reducing agents and to oxidize relatively labile organic selenium compounds such as selenoamino acids and methaneseleninic acid.

If the sample contains hydrogen sulfide or a large concentration of organic matter or is otherwise suspect or to confirm method accuracy, reanalyze sample using digestion procedure 3 or 4 below.

Prepare 2% potassium or ammonium persulfate solution by dissolving 2.0 g in 100 mL deionized water (prepare weekly). Add 0.2 mL persulfate solution to the mixed sample and HCl of ¶ B.5c before heating and proceeding with pretreatment and analysis.

After completing analysis multiply concentration of selenium determined in the acidified sample by 2.04 to obtain total selenium in original sample.

3. Removal of Interference by Alkaline Hydrogen Peroxide Digestion

Occasionally, digestion with persulfate gives incomplete recovery of total selenium. In this case, digestion with hydrogen peroxide is used to remove all reducing agents that might interfere and to fully oxidize organic selenium to Se(VI). The resulting solution can be analyzed for total selenium after pretreatment according to step B.5 below.

This method is suitable for determining total selenium in unfiltered water samples, where particulate selenium is present. When working with a new matrix, confirm results obtained by reanalyzing the sample using digestion procedure 4, below.

a. Apparatus:

1) *Beakers,* 150-mL.

2) *Watch glasses.*

3) *Hot plate.*

4) *Pipetter,* 1-mL, and tips.

5) *Graduated cylinder,* 25-mL.

b. Reagents:

1) *Hydrogen peroxide,* H_2O_2, 30%. Keep refrigerated.

2) *Sodium hydroxide,* NaOH, 1N.

3) *Hydrochloric acid,* HCl, 1.5N: Dilute 125 mL conc HCl to 1 L with deionized water.

c. Procedure: Add 2 mL 30% H_2O_2 and 1 mL 1 N NaOH to 25 mL sample in a beaker. Cover beaker to control spattering and simmer on hot plate until fine bubbles characteristic of H_2O_2 decomposition subside and are replaced by ordinary boiling. Add 1 mL 1.5N HCl to redissolve any precipitate that may have formed, let cool, and pour into graduated cylinder. Rinse beaker with deionized water into graduated cylinder and make volume up to 25 mL. Proceed to B.5 and chosen analytical method.

4. Removal of Interference by Permanganate Digestion

This digestion method utilizes potassium permanganate to oxidize selenium and remove interfering organic compounds. Excess $KMnO_4$ and MnO_2 are removed by reaction with hydroxylamine. HCl digestion is included here, because it is conveniently performed in the same reaction vial. Selenite may then be determined directly by Methods 3114B or C, or 3500-Se.C.

Verify recovery for the given matrix. This method gives good recovery even with heavily contaminated water samples that con-

---

* Bio-Rad AG1-X8 or equivalent.
† Bio-Rad Econo-Columns or equivalent.
‡ Amberlite XAD-8, Supelco, or equivalent.

tain organic selenium compounds, dissolved organic matter, and visible suspended material.

Permanganate may oxidize chloride ion to free chlorine. Part of the chlorine (which interferes with hydride analysis) is removed by reaction with hydroxylamine, but the best way to eliminate free chlorine is by prolonged heating in an open vial during the digestion step. Excess hydroxylamine may reduce recovery by reducing selenium to Se(0).

a. Apparatus:

1) *Oven with thermostat,* for continuous operation at $110 \pm 5°C$.

2) *Digestion vials,* 40-mL, with fluorocarbon-lined screw caps.

3) *Metal support rack* to hold 40 digestion vials.

b. Reagents:

1) *Hydrochloric acid,* HCl, conc. [See B.1*b*3)].

2) *Hydrochloric acid,* HCl, 10*N:* Dilute 1000 mL conc HCl to 1200 mL with deionized water.

3) *Sulfuric acid,* H_2SO_4, conc and 25%. NOTE: Many brands of H_2SO_4 are contaminated with selenium. Run reagent blanks when starting a new bottle. Make 25% v/v solution by adding 250 mL conc H_2SO_4 to 500 mL deionized water, and diluting to 1 L.

4) *Potassium permanganate solution,* $KMnO_4$, 5% (w/v): Dissolve 50 g $KMnO_4$ in 1000 mL deionized water.

5) *Hydroxylamine hydrochloride solution:* Dissolve 100 g $NH_2OH \cdot HCl$ in 1000 mL deionized water.

c. Procedure: Pipe 5 mL sample into digestion vial, add 5 mL 25% H_2SO_4 and 1 mL $KMnO_4$ solution. Screw cap on and place in preheated oven at 110°C for 1 h. Remove tray with vials from the oven and cool to room temperature. Open vial, carefully add a few drops hydroxylamine hydrochloride solution, mix, and wait until sample is decolorized and residual manganese dioxide is dissolved. Avoid excess hydroxylamine solution, which can cause a low reading. Add 10 mL conc HCl to the sample and heat vial 60 min at 95°C without cap. Let cool to room temperature. Transfer sample to a 25-mL volumetric flask or graduated cylinder, rinse vial into flask, dilute to mark, and mix well. Proceed to analyze by Methods 3114B or C, or 3500-Se.C. If Method 3114 B or C is used, multiply spectrometer readings by the dilution factor as follows:

$$\text{Concentration, } \mu g/L = \frac{\text{final volume}}{\text{volume of sample}} \times \text{reading}$$

5. Reduction of Se(VI) to Se(IV) by Hydrochloric Acid Digestion

Se(VI) is reduced to Se(IV) by digestion with HCl. Determine Se(IV) + Se(VI) by either hydride generation atomic absorption spectrometer (Method 3114B or C) or as an organic derivative (Method C).

Test any given sample matrix to ensure recovery of added Se(VI). If recovery is poor, try to remove interference by the procedure of ¶ B.1, above, or if at least 75% recovery is achieved, use the method of standard additions. The method described here is of limited utility for direct analysis of water samples, but is useful as a step in determining total selenium in a sample where selenium has been oxidized to Se(VI).

a. Apparatus:

1) *Dispenser,* bottle type, 5-mL, suitable for dispensing concentrated HCl.

2) *Pipetters,* 0.2- and 5-mL.

3) *Screw-cap culture tubes,* borosilicate glass, 25- × 150-mm.

4) *Boiling water bath,* suitable for heating culture tubes; a 1-L beaker on a hot plate is suitable.

b. Reagents:

1) *Sodium selenate additions solution:* Dilute 1000 mg/L stock selenate solution with deionized water to prepare a solution of 1 to 10 mg/L, such that the concentration of the additions solution will be approximately 50 times greater than anticipated total selenium in the sample to be analyzed.

2) *Hydrochloric acid,* HCl, conc: See B.1*b*3).

c. Procedure: Calibrate acid dispenser using water. Preheat water bath. Pipet 5 mL filtered sample into a culture tube. Add 5 mL conc HCl. Loosely cap tube (do not tighten) and place in boiling water bath for 20 min. Let tube cool and tighten cap. Determine total Se(IV) by Methods 3114B or C, or 3500-Se.C.

Add 0.200 mL additions solution with a microliter pipet to sample and proceed as above. Analyze a deionized water blank and a blank with the addition to ensure absence of contamination and to determine the true value of the addition.

Multiply the concentration of selenium determined in the acidified sample by 2.00 to obtain total concentration of Se(IV) + Se(VI). Multiply reading obtained for sample with addition by 2.04.

6. Bibliography

JANGHORBANI, M., B. TING, A. NAHAPETLAN & R. YOUNG. 1982. Conversion of urinary selenium to selenium IV by wet oxidation. *Anal. Chem.* 54:1188.

ADELOJU, S.B. & A.M. BOND. 1984. Critical evaluation of some wet digestion methods for the stripping voltammetric determination of selenium in biological materials. *Anal. Chem.* 56:2397.

BRIMMER, S.P., W.R. FAWCETT & K.A. KULHAVY. 1987. Quantitative reduction of selenate ion to selenite in aqueous samples. *Anal. Chem.* 59:1470.

3500-Se C. Colorimetric Method

1. General Discussion

a. Principle: This method is specific to determining selenite ion in aqueous solution. Selenite ion reacts with 2,3-diaminonaphthalene to produce a brightly colored and strongly fluorescent piazselenol compound, which is extracted in cyclohexane and measured colorimetrically.

The optimum pH for formation of the piazselenol complex is approximately 1.5 but should not be above 2.5 because above pH 2, the rate of formation of the colored compound is critically dependent on pH. When indicators are used to adjust pH, results frequently are erratic; results can be improved when pH is monitored electrochemically.

b. Interference: No inorganic compounds are known to give a positive interference. Colored organic compounds extractable by cyclohexane may be encountered, but usually they are absent or can be removed by oxidizing the sample (see B.2, 3, or 4) or by treating it to remove dissolved organics (see B.1). Negative interference results from compounds that reduce the concentration of diaminonaphthalene by oxidizing it. Addition of EDTA eliminates negative interference from at least 2.5 mg Fe^{2+}.

c. Minimum detectable quantity: 10 µg Se/L.

2. Apparatus

a. Colorimetric equipment: A spectrophotometer, for use at 480 nm, providing a light path of 1 cm or longer.

b. Separatory funnel, 250-mL, preferably with a fluorocarbon stopcock.

c. Thermostatically controlled water bath (50°C) with cover.

d. pH meter.

e. Centrifuge, with rotor for 50-mL tubes (optional).

f. Centrifuge bottles, 60-mL, screw-capped, fluorocarbon.

g. Shaker, suitable for separatory funnel (optional).

3. Reagents

Use reagent water (see Section 1080) in preparing reagents.

a. Selenium standard reference solution: Dissolve 2.190 g sodium selenite, Na_2SeO_3, in water containing 10 mL HCl and dilute to 1 L. 1.00 mL = 1.00 mg Se(IV).

b. Working standard selenium solutions: Dilute selenium reference standard solution with water or suitable background solution to produce a series of working standards spanning the concentration range of interest.

c. Hydrochloric acid: HCl, conc and 0.1N.

d. Ammonium hydroxide, NH_4OH, 50% v/v.

e. Cyclohexane, C_6H_{12}.

f. 2,3-Diaminonaphthalene (DAN) solution: Dissolve 200 mg DAN in 200 mL 0.1N HCl. Shake 5 min. Extract three times with 25-mL portions of cyclohexane, retaining aqueous phase and discarding organic portions. Filter into opaque container* and store in cool, dark place for no longer than 8 h. CAUTION: *Toxic, handle with extreme care.*

g. Hydroxylamine-EDTA solution (HA-EDTA): Dissolve 4.5 g Na_2EDTA in approximately 450 mL water. Add 12.5 g hydroxylamine hydrochloride ($NH_2OH \cdot HCl$) and adjust volume to 500 mL.

4. Procedure

a. Formation of piazselenol: Add 2 mL HA-EDTA solution to 10 mL sample in 60-mL centrifuge bottle (filtered if Se(IV) is to be determined; oxidized using Method B.2, 3, or 4, then reduced using Method B.5 for total Se). Adjust to pH 1.5 ± 0.3 with 0.1N HCl and 50% NH_4OH, using a pH meter. Add 5 mL DAN solution and heat in a covered water bath at 50°C for 30 min.

b. Extraction of piazselenol: Cool and add 2.0 mL cyclohexane. Cap container securely and shake vigorously for 5 min. Let solution stand for 5 min or until cyclohexane layer becomes well separated. If separation is slow, centrifuge for 5 min at 2000 rpm. Place bottle in a clamp on a ringstand at a 45° angle to the vertical. Remove aqueous phase using a disposable pipet attached to a vacuum line. Transfer organic phase to a small capped container using a clean disposable pipet, or to the spectrophotometer cuvette if absorbance is to be read immediately.

c. Determination of absorbance: Read absorbance at 480 nm using a zero standard. The piazselenol color is very stable but evaporation of the cyclohexane concentrates the color unless the container is capped. CAUTION: *Avoid inhaling cyclohexane vapors.* Beer's Law is obeyed up to 2 mg/L.

5. Calculation

Construct calibration curve using at least a three-point standard curve to bracket the expected sample concentration. Plot absorbance vs. concentration. Correct for digestion blank and any reagent blank.

6. Precision and Bias

Three standard reference materials (wheat flour, water, and a commercial standard) were used to evaluate Se recovery.[1] The wheat flour sample was digested using HNO_3 and $HClO_4$ to convert total selenium to Se(VI), digested with HCl to convert Se(VI) to Se(IV) and finally, the colorimetric method was used. Results were as follows:

Standard	Selenium Concentration µg Se/L	
	Expected	Recovered*
NBS, SRM 1567, wheat flour†	1097 ± 197	1113 ± 8
NBS, SRM 1543ib, water	9.7 ± 0.5	8.7 ± 0
Fisher Certified AAS Standard	1002 ± 8	1002 ± 0

* Analyses in triplicate.
† Dry weight basis.

7. Reference

1. HOLTZCLAW, K.M., R.H. NEAL, G. SPOSITO & S.J. TRAINA. 1987. A sensitive colorimetric method for the quantitation of selenite in soil solutions and natural waters. *Soil Sci. Soc. Amer. J.* 51:75.

---

* Use Whatman No. 42 filter paper, or equivalent.

8. Bibliography

HOSTE, J. & J. GILLIS. 1955. Spectrophotometric determination of traces of selenium with 3,3'-diaminobenzidine. *Anal. Chim. Acta.* 12:158.

CHENG, K. 1956. Determination of traces of selenium. *Anal. Chem.* 28:1738.

MAGIN, G.B. et al. 1960. Suggested modified method for colorimetric determination of selenium in natural waters. *J. Amer. Water Works Assoc.* 52:1199.

ROSSUM, J.R. & P.A. VILLARRUZ. 1962. Suggested methods for determining selenium in water. *J. Amer. Water Works Assoc.* 54:746.

OLSEN, O.E. 1973. Simplified spectrophotometric analysis of plants for selenium. *J. Assoc. Offic. Anal. Chem.* 56:1073.

3500-Se D. Determination of Volatile Selenium

1. General Discussion

Dimethylselenide and dimethyldiselenide are low boiling, extremely malodorous organic compounds sparingly soluble in water. They are produced by microbial processes in seleniferous soil and decaying seleniferous organic matter, and occasionally are present in natural waters. They are readily air stripped from a water sample and can be collected with high efficiency in an alkaline solution of hydrogen peroxide, which oxidizes them quantitatively to Se(VI). Total selenium is determined by digestion with HCl and analysis by the hydride atomic absorption (3114B or C) or colorimetric (3500-Se.C) methods.

Either nitrogen or air may be used to strip the sample. Preferably use nitrogen if air-sensitive compounds (e.g., selenopolysulfides) are suspected.

Because volatile selenium can be lost in the course of sample collection and handling, preferably air-strip the sample in the field immediately after it is collected. After boiling to decompose H_2O_2, return the alkaline peroxide solution to the laboratory for analysis.

2. Apparatus

All apparatus required for selenate reduction (B.5) and Methods 3114B or C, or 3500-Se.C, plus:

a. Gas washing bottles, borosilicate glass, 250-mL, with coarse porous glass gas dispersion frit. Mark 100-mL level on side of bottle.

b. Rotameter, to measure 3 L/min air flow.

c. Gas flow regulator.

d. Hot plate.

e. Graduated cylinder, 100-mL.

f. Beakers, 250-mL.

g. Rubber tubing, to interconnect gas washing bottles and other gas equipment.

h. Rubber gloves.

3. Reagents

All reagents required for selenate reduction (B.5) and Methods 3114B or C, or 3500-Se.C, plus:

a. Hydrogen peroxide, 30%. Refrigerate.

b. Sodium hydroxide solution, NaOH, 1*N*.

c. Compressed air or nitrogen.

4. Procedure

Set up air-flow train in this order: Regulated air supply→rotameter→gas washing bottle 1→gas washing bottle 2.

Prepare alkaline peroxide solution immediately before use by pouring 20 mL 30% H_2O_2 into a 100-mL graduated cylinder, adding 50 to 60 mL deionized water and 5 mL 1*N* NaOH, and making up to 100 mL. CAUTION: *Alkaline H_2O_2 is unstable. Do not keep in glass bottle; hold at about 0 °C in oversized plastic bottle. Solution is corrosive; protect eyes and skin.* Pour into gas washing bottle 2. Pour approximately 100 mL freshly collected sample into gas washing bottle 1. Do not attempt to measure sample volume accurately before volatile selenium determination, as unnecessary handling may cause volatile selenium to be lost.

Connect and check all air lines, turn on air and adjust flow to 3 L/min. Strip for 30 min or more. After 30 min, turn off air, disconnect gas washing bottle 2, and place it on the hot plate. Adjust heat to produce a gentle simmering of oxygen bubbles from decomposition of H_2O_2. Continue heating until the characteristic effervescence of oxygen subsides and is replaced by ordinary boiling. Remove from hot plate and let cool. Pour solution into beaker. (Volume will be very near to 100 mL, and correction will usually be unnecessary.) Analyze for total selenium using HCl digestion (B.5) and Methods 3114B or C, or 3500-Se.C. Once boiled, this solution may be safely stored and transported in plastic bottles. Measure volume of sample in gas washing bottle 1.

5. Calculation

The concentration of volatile selenium compounds in the original water sample can be calculated as:

$$C = \frac{100}{\text{volume of original sample}} \times \text{conc of Se in solution}$$

6. Precision and Bias

Approximately 90% of dimethylselenide in samples will be recovered with 30 min air stripping. The recovery of dimethyldiselenide is not known. Loss of gases to the atmosphere during sampling and handling that precede analysis may cause a significant negative error.

3500-Se E.   Determination of Nonvolatile Organic Selenium Compounds

1. General Discussion

In principle, the total amount of dissolved organic selenium plus polysulfidic selenium may be estimated by comparing "total Se," determined by oxidation and HCl digestion (B.2 or 3 and B.5, or B.4), followed by Methods 3114B or C, or 3500-Se.C, with Se(IV) + Se(VI) determined by HCl digestion (B.5) and Method 3114C, or 3500-Se.C. In practice, this will give a meaningful estimate only if a known addition of Se(VI) is fully recovered. Even if recovery is good, this estimate may be unreliable, because it is the difference of two (frequently larger) numbers determined by slightly different methods. Comparing total Se before and after treatment with resin [B.1c1)] gives a similarly unreliable estimate of nonvolatile organic Se.

It is preferable to separate and directly determine nonvolatile organic selenium. One method involves adsorption of dissolved organic matter onto a C-18 reverse phase HPLC resin, elution with an organic solvent, and determination of selenium in this fraction. While this technique is relatively simple, it is affected by pH and small organic molecules (e.g., individual selenium containing amino acids) are not retained by the resins. Adjust sample pH to 1.5 to 2.0 before using the column, but because the latter problem cannot be solved easily, the use of organic adsorbents provides only an estimate of organic selenium concentration.

Alternatively, isolate specific compounds and determine their selenium content. In some natural waters selenium may be associated with dissolved polypeptides or small proteins, and even small amounts of free selenoamino acids may be present. Because selenoamino acids are the most toxic form of the element, a direct determination is sometimes desirable.

To determine selenium in dissolved peptides, hydrolyze with acid and isolate the free amino acids via ligand exchange chromatography. Elute the selenoamino acids from the column and determine selenium. Selenoamino acids are unstable during acid hydrolysis and even using nonoxidizing methyl sulfonic acid and nitrogen-purged glass ampules, selenoamino acid recoveries are only 50 to 80%. This method is good only for estimating protein-bound selenium. A somewhat more reliable estimate of free selenoamino acids and selenium associated with small oligopeptides is obtained by performing a similar procedure without the hydrolysis step.

While these methods are too intricate for routine use, semi-quantitative, and sensitive only to certain classes of organic compounds, at present they are the only ones available with any practical experience.

Imperfect separation of organic selenium compounds from inorganic forms of selenium may cause interference. In parallel with the actual determination, always perform the procedure using a solution compounded to resemble the actual matrix and containing a similar amount of selenium, but in the form of Se(IV) and Se(VI) to determine degree of interference.

2. Apparatus

a. *Rotary evaporator,* with temperature-control bath and 30-mL pear-shaped flasks.

b. *Glass ampule sealing apparatus,* or oxygen-gas torch.

c. *Heating block,* 100°C, or pressure cooker.

d. *Glass chromatography columns,* 15 cm long, 0.7 cm ID.*

e. *Glass syringe,* 50-mL.

f. *Glass ampules,* 10- or 20- mL. Clean by heating in a muffle furnace at 400°C for 24 h.

g. *pH meter.*

3. Reagents

a. *Hydrochloric acid,* HCl, 1N.

b. *Methyl sulfonic acid,* conc.

c. *Ammonium hydroxide,* NH_4OH, 1.5N: Dilute 100 mL conc NH_4OH to 1 L with deionized water.

d. *Sodium hydroxide,* NaOH pellets and 1N solution.

e. *pH 1.6 solution:* Adjust pH of deionized water to 1.6 using HCl.

f. *pH 9.0 solution:* Adjust pH of deionized water to 9.0 using NaOH.

g. *Copper sulfate solution,* $CuSO_4$, 1M: Dissolve 25 g $CuSO_4 \cdot 5H_2O$ in deionized water and dilute to 100 mL.

h. *Methanol,* purified.

i. *C-18 cartridges*† Using a glass syringe, pass the following sequence of reagents through the cartridge to clean the resin (6 mL/min or less): 10 mL deionized water; 20 mL 1N HCl; 10 mL deionized water; 20 mL methanol; 10 mL deionized water; and 20 mL pH 1.6 solution. Refrigerate but do not freeze cleaned cartridges.

j. *Ligand-exchange chromatographic resin,* 100/200 mesh: Rinse resin‡ with deionized water to remove fines. Then rinse resin three times in the following sequence: 1N HCl; 1.5N NH_4OH; and deionized water. Store resin wet.

k. *Copper-treated ligand-exchange chromatographic resin,* 100/200 mesh: Rinse resin‡ with deionized water to remove fines. Then rinse three times with 1N HCl, followed by deionized water. Using NaOH adjust pH of supernatant above the resin to about 7. Add $CuSO_4$ solution to the resin and stir. After settling, decant supernatant and add more $CuSO_4$ solution. Decant $CuSO_4$ solution and rinse with deionized water until no copper is noticeable in the supernatant. Rinse resin three times with 1.5N NH_4OH and three times with deionized water. Store resin wet.

4. Procedures

a. *Extractable organic selenium:* Adjust sample (5 to 50 mL) to pH 1.5 to 2.0 using HCl and place in a clean glass syringe with an attached cleaned C-18 cartridge. Push sample through the cartridge at a rate of 6 mL/min. After removing the cartridge, draw 2 mL pH 1.6 solution into syringe as a rinse, reattach cartridge, and push the rinse through the cartridge. Repeat two additional times. The cartridge can be refrigerated for storage. To elute organic selenium, push 10 mL methanol through the cartridge at rate of 2 mL/min and collect eluate in a 30 mL pear-shaped flask. Remove methanol by rotary evaporation, with the water bath temperature less than 40°C. Use deionized water to solubilize and transfer the residue into the vessel used for total selenium digestions. Determine total dissolved selenium by di-

---

* Bio-Rad glass Econo-Columns or equivalent.
† Sep-Pak, Waters Associates, or equivalent.
‡ Chelex 100 (ammonium form) Bio Rad, or equivalent.

gestion with persulfate (B.2) or peroxide (B.3), reduction of Se(VI) (B.5), and analysis by Methods 3114B or C, or 3500-Se.C.

b. Hydrolysis of protein-bound selenium: Place filtered sample in a 10- or 20-mL glass ampule (depending on desired volume), and add conc methyl sulfonic acid to adjust concentration to 4*M*. Purge acidified sample with nitrogen for 10 min and seal top with a torch. Heat sealed vial at 100°C for 24 h in a heating block or pressure cooker. Transfer cooled hydrolysis solution with deionized water rinses to a 50-mL beaker and place in an ice bath. Using NaOH pellets and 1*N* NaOH, adjust to pH 9.0, taking care not to allow solution to heat to boiling.

c. Determination of selenoamino acids: If sample is not hydrolyzed as in ¶ 4*b*, filter, and adjust pH to 9.0 using 1*N* NaOH.

Fill an empty chromatographic column with deionized water and add ammonium-form resin to a depth of 2 cm. Add copper-treated resin to form a 12-cm length of resin. (The ammonium-form resin removes any copper that bleeds from the copper-treated resin above it). Rinse with deionized water until the pH of the effluent is 9.0; maintain flow through the column by gravity. Pass sample through column, rinse sample beaker with 5 mL

pH 9 solution, and place the rinse on column (after the last of the sample reaches the top of the resin). Rinse beaker twice more. Discard flow through column.

Place clean beaker under column and add 20 mL 1.5*N* NH$_4$OH to the column. Neutralize NH$_4$OH eluate with 2.5 mL conc HCl. Determine total dissolved selenium by digestion with persulfate (B.2) or peroxide (B.3), reduction of Se(VI) (B.5) and analysis by Methods 3114B or C, or 3500-Se.C.

5. Precision and Bias

These procedures are only semiquantitative. Typical relative standard deviation is 12% for the C-18 isolation of dissolved organic selenium, and 15% for protein-bound selenium.

6. Bibliography

CUTTER, G. 1982. Selenium in reducing waters. *Science* 217:829.
COOKE, T.D. & K.W. BRULAND. 1987. Aquatic chemistry of selenium: evidence of biomethylation. *Environ. Sci. Technol.* 21:1214.

3500-Ag   SILVER

Silver (Ag) is the second element in Group IB of the periodic table; it has an atomic number of 47, an atomic weight of 107.87, and valences of 1 and 2. The average abundance of Ag in the earth's crust is 0.08 ppm; in soils it is <0.01 to 0.5 ppm; in streams it is 0.3 μg/L; in U.S. drinking waters it is 0.23 μg/L, and in groundwater it is <0.1 μg/L. Silver occurs in its native state and in combination with many nonmetallic elements such as argentite (Ag$_2$S) and horn silver (AgCl). Lead and copper ores also may yield considerable silver. Silver is widely used in photography, silverware, jewelry, mirrors, and batteries. Silver iodide has been used in the seeding of clouds, and silver oxide to a limited extent is used as a disinfectant for water.

In acidic water Ag$^+$ would predominate, and in high-chloride water a series of complexes would be expected. Silver is nonessential for plants and animals. Silver can cause argyria, a permanent, blue-gray discoloration of the skin and eyes that imparts a ghostly appearance. Concentrations in the range of 0.4 to 1 mg/L have caused pathological changes in the kidneys, liver, and spleen of rats. Toxic effects on fish in fresh water have been observed at concentrations as low as 0.17 μg/L. For freshwater aquatic life, total recoverable silver should not exceed 1.2 mg/L.

The atomic absorption spectrometric methods (3111B and C) and the inductively coupled plasma methods (3120 and 3125) are preferred. The electrothermal atomization method (3113B) is the most sensitive for determining silver in natural waters. The dithizone method detailed in the 19th edition of *Standard Methods* can be used when an atomic absorption spectrometer is unavailable. A method suitable for analysis of silver in industrial or other wastewaters at levels above 1 mg/L is available.[1]

If total silver is to be determined, acidify sample with conc nitric acid (HNO$_3$) to pH <2 at time of collection. If sample contains particulate matter and only the "dissolved" metal content is to be determined, filter through a 0.45-μm membrane filter at time of collection. After filtration, acidify filtrate with HNO$_3$ to pH <2. Complete analysis as soon after collection as possible. Some samples may require special storage and digestion; see Section 3030D.

Reference

1. U.S. ENVIRONMENTAL PROTECTION AGENCY. 1994. Approved Inorganic Test Procedures. *Federal Register* 59(20):4504.

3500-Na   SODIUM*

3500-Na A.   Introduction

1. Occurrence and Significance

Sodium (Na) is the third element in Group IA of the periodic table; it has an atomic number of 11, an atomic weight of 22.99, and a valence of 1. The average abundance of Na in the earth's crust is 2.5%; in soils it is 0.02 to 0.62%; in streams it is 6.3 mg/L, and in groundwaters it is generally >5 mg/L. Sodium occurs with silicates and with salt deposits. Sodium compounds are used in many applications, including caustic soda, salt, fertilizers, and water treatment chemicals.

Sodium is very soluble, and its monovalent ion Na^+ can reach concentrations as high as 15 000 mg/L in equilibrium with sodium bicarbonate. The ratio of sodium to total cations is important in agriculture and human physiology. Soil permeability can be harmed by a high sodium ratio. In large concentrations it may affect persons with cardiac difficulties. A limiting concentration of 2 to 3 mg/L is recommended in feedwaters destined for high-pressure boilers. When necessary, sodium can be removed by the hydrogen-exchange process or by distillation. The U.S. EPA advisory limit for sodium in drinking water is 20 mg/L.

_____
* Approved by Standard Methods Committee, 1997.

2. Selection of Method

Method 3111B uses an atomic absorption spectrometer in the flame absorption mode. Method 3120C uses inductively coupled plasma; this method is not as sensitive as the other methods, but usually this is not important. Method 3500-Na.B uses either a flame photometer or an atomic absorption spectrometer in the flame emission mode. The inductively coupled plasma/mass spectrometric method (3125) also may be applied successfully in most cases (with lower detection limits), even though sodium is not specifically listed as an analyte in the method. When all of these instruments are available, the choice will depend on factors including relative quality of the instruments, precision and sensitivity required, number of samples and analytes per sample, matrix effects, and relative ease of instrument operation. If an atomic absorption spectrometer is used, operation in the emission mode is preferred.

3. Storage of Sample

Store alkaline samples or samples containing low sodium concentrations in polyethylene bottles to eliminate the possibility of sample contamination due to leaching of the glass container.

3500-Na B.   Flame Emission Photometric Method

1. General Discussion

a. Principle: Trace amounts of sodium can be determined by flame emission photometry at 589 nm. Sample is nebulized into a gas flame under carefully controlled, reproducible excitation conditions. The sodium resonant spectral line at 589 nm is isolated by interference filters or by light-dispersing devices such as prisms or gratings. Emission light intensity is measured by a phototube, photomultiplier, or photodiode. The light intensity at 589 nm is approximately proportional to the sodium concentration. Alignment of the wavelength dispersing device and wavelength readout may not be precise. The appropriate wavelength setting, which may be slightly more or less than 589 nm, can be determined from the maximum emission intensity when aspirating a sodium standard solution, and then used for emission measurements. The calibration curve may be linear but has a tendency to level off or even reverse at higher concentrations. Work in the linear to near-linear range.

Minimize interference by incorporation of one or more of the following:

1) Operate at the lowest practical concentration range.

2) Add releasing agents, such as strontium or lanthanum at 1000 mg/L, to suppress ionization and anion interference. Among common anions capable of causing interference are Cl^-, SO_4^{2-} and HCO_3^- in relatively large amounts.

3) Matrix-match standards and samples by adding identical amounts of interfering substances present in the sample to calibration standards.

4) Apply an experimentally determined correction in those instances where the sample contains a single important interference.

5) Remove interfering ions.

6) Remove burner-clogging particulate matter from the sample by filtration through a filter paper of medium retentiveness.

7) Use the standard addition technique as described in the flame photometric method for strontium (3500-Sr.B). The method involves preparing a calibration curve using the sample matrix as a diluent, and determining the sample concentration either mathematically or graphically.

8) Use the internal standard technique. Potassium and calcium interfere with sodium determination by the internal-standard method if the potassium-to-sodium ratio is ≥5:1 and the calcium-to-sodium ratio is ≥10:1. When these ratios are exceeded, determine calcium and potassium concentrations and matrix-match sodium calibration standards by addition of approximately equivalent concentrations of interfering ions. Interference from

magnesium is not significant until the magnesium-to-sodium ratio exceeds 100, a rare occurrence.

c. Minimum detectable concentration: Better flame photometers or atomic absorption spectrometers operating in the emission mode can be used to determine sodium levels approximating 5 μg/L.

2. Apparatus

a. Flame photometer (either direct-reading or internal-standard type) or atomic absorption spectrometer operating in the flame emission mode.

b. Glassware: Rinse all glassware with $1 + 15$ HNO_3 followed by several portions of reagent water (¶ 3a).

3. Reagents

To minimize sodium contamination, store all solutions in plastic bottles. Use small containers to reduce the amount of dry element that may be picked up from the bottle walls when the solution is poured. Shake each container vigorously to wash accumulated salts from walls before pouring solution.

a. Reagent water: See Section 1080. Use reagent water to prepare all reagents and calibration standards, and as dilution water.

b. Stock sodium solution: Dissolve 2.542 g NaCl dried at 140°C to constant weight and dilute to 1000 mL with water; 1.00 mL = 1.00 mg Na.

c. Intermediate sodium solution: Dilute 10.00 mL stock sodium solution with water to 100.0 mL; 1.00 mL = 0.10 mg Na (1.00 mL = 100 μg Na). Use this intermediate solution to prepare calibration curve in sodium range of 1 to 10 mg/L.

d. Standard sodium solution: Dilute 10.00 mL intermediate sodium solution with water to 100 mL; 1.00 mL = 10.0 μg Na. Use this solution to prepare calibration curve in sodium range of 0.1 to 1.0 mg/L.

4. Procedure

a. Pretreatment of polluted water and wastewater samples: Follow the procedures described in Section 3030.

b. Instrument operation: Because of differences between makes and models of instruments, it is impossible to formulate detailed operating instructions. Follow manufacturer's recommendation for selecting proper photocell and wavelength, adjusting slit width and sensitivity, appropriate fuel and oxidant gas pressures, and the steps for warm-up, correcting for interferences and flame background, rinsing of burner, igniting flame, and measuring emission intensity.

c. Direct-intensity measurement: Prepare a blank and sodium calibration standards in stepped amounts in any of the following applicable ranges: 0 to 1.0, 0 to 10, or 0 to 100 mg/L. Determine emission intensity at 589 nm. Aspirate calibration standards and samples enough times to secure a reliable average reading for each. Construct a calibration curve from the sodium standards. Determine sodium concentration of sample from the calibration curve. Where a large number of samples must be run routinely, the calibration curve provides sufficient accuracy. If greater precision and less bias are desired and time is available, use the bracketing approach described in ¶ 4d below.

d Bracketing approach: From the calibration curve, select and prepare sodium standards that immediately bracket the emission intensity of the sample. Determine emission intensities of the bracketing standards (one sodium standard slightly less and the other slightly greater than the sample) and the sample as nearly simultaneously as possible. Repeat the determination on bracketing standards and sample. Calculate the sodium concentration by the equation in ¶ 5b and average the findings.

5. Calculation

a. For direct reference to the calibration curve:

$$\text{mg Na/L} = (\text{mg Na/L in portion}) \times D$$

b. For the bracketing approach:

$$\text{mg Na/L} = \left[\frac{(B - A)(s - a)}{(b - a)} + A \right] D$$

where:

B = mg Na/L in upper bracketing standard,
A = mg Na/L in lower bracketing standard,
b = emission intensity of upper bracketing standard,
a = emission intensity of lower bracketing standard,
s = emission intensity of sample, and
D = dilution ratio,

$$= \frac{\text{mL sample} + \text{mL water}}{\text{mL sample}}$$

6. Precision and Bias

A synthetic sample containing 19.9 mg Na^+/L, 108 mg Ca^{2+}/L, 82 mg Mg^{2+}/L, 3.1 mg K^+/L, 241 mg Cl^-/L, 0.25 mg NO_2^--N/L, 1.1 mg NO_3^--N/L, 259 mg SO_4^{2-}/L, and 42.5 mg total alkalinity/L (as $CaCO_3$) was analyzed in 35 laboratories by the flame photometric method, with a relative standard deviation of 17.3% and a relative error of 4.0%.

7. Bibliography

WEST, P.W., P. FOLSE & D. MONTGOMERY. 1950. Application of flame spectrophotometry to water analysis. *Anal. Chem.* 22:667.

COLLINS, C.G. & H. POLKINHORNE. 1952. An investigation of anionic interference in the determination of small quantities of potassium and sodium with a new flame photometer. *Analyst* 77:430.

MELOCHE, V.W. 1956. Flame photometry. *Anal. Chem.* 28:1844.

BURRIEL-MARTI, F. & J. RAMIREZ-MUNOZ. 1957. Flame Photometry: A Manual of Methods and Applications. D. Van Nostrand Co., Princeton, N.J.

DEAN, J.A. 1960. Flame Photometry. McGraw-Hill Publishing Co., New York, N.Y.

URE, A.M. & R.L. MITCHELL. 1975. Lithium, sodium, potassium, rubidium, and cesium. *In* J.A. Dean & T.C. Rains, eds. Flame Emission and Atomic Absorption Spectrometry. Dekker, New York, N.Y.

THOMPSON, K.C. & R.J. REYNOLDS. 1978. Atomic Absorption, Fluorescence, and Flame Spectroscopy—A Practical Approach, 2nd ed. John Wiley & Sons, New York, N.Y.

WILLARD, H.H., L.L. MERRIT, JR., J.A. DEAN & F.A. SETTLE, JR. 1981. Instrumental Methods of Analysis, 6th ed. Wadsworth Publishing Co., Belmont, Calif.

AMERICAN SOCIETY FOR TESTING AND MATERIALS. 1988. Method D 1428-82: Standard test methods for sodium and potassium in water and water-formed deposits by flame photometry. Annual Book of ASTM Standards, Vol. 11.01. American Soc. Testing & Materials, Philadelphia, Pa.

3500-Sr STRONTIUM*

3500-Sr A.   Introduction

1. Occurrence and Significance

Strontium (Sr) is the fourth element in Group IIA of the periodic table; it has an atomic number of 38, an atomic weight of 87.62, and a valence of 2. The average abundance of Sr in the earth's crust is 384 ppm; in soils Sr ranges from 3.6 to 160 ppm; in streams it averages 50 μg/L, and in groundwaters it ranges from 0.01 to 10 mg/L. Strontium is found chiefly in celestite ($SrSO_4$) and in strontianite ($SrCO_3$). Strontium compounds are used in pigments, pyrotechnics, ceramics, and flares. ^{90}Sr is a fission product of nuclear reactor fuels, and was widely distributed on the earth's surface as a result of fallout from nuclear weapons testing.

The common aqueous species is Sr^{2+}. The solubility of strontium is controlled by carbonate and sulfate. Some compounds are toxic by ingestion and inhalation. Although there is no U.S. EPA drinking water standard MCL for concentration

_____
* Approved by Standard Methods Committee, 1997.

of strontium, strontium-90 measurements are required when the gross beta activity of a water sample is greater than 50 pCi/L. The U.S. EPA primary drinking water standard MCL for ^{90}Sr is 8 pCi/L.

A method for determination of ^{90}Sr is found in Section 7500-Sr.

2. Selection of Method

The atomic absorption spectrometric method (3111B) and inductively coupled plasma methods (3120 and 3125) are preferred. The flame emission photometric method (B) also is available for those laboratories that do not have the equipment needed for one of the preferred methods.

3. Sampling and Storage

Polyethylene bottles are preferable for sample storage, although borosilicate glass containers also may be used. At time of collection adjust sample to pH <2 with nitric acid (HNO_3).

3500-Sr B.   Flame Emission Photometric Method

1. General Discussion

a. Principle: The flame photometric method can be used for the determination of strontium in the concentration range prevalent in natural waters. The strontium emission is measured at a wavelength of 460.7 nm, while the background intensity is measured at a wavelength of 466 nm. The difference in readings obtained at these two wavelengths measures the light intensity emitted by strontium.

b. Interference: Emission intensity is a linear function of strontium concentration and concentration of other constituents. The standard addition technique distributes the same ions throughout the standards and the sample, thereby equalizing the radiation effect of possible interfering substances. A very low pH (<1) could produce an interference, but sample dilution should eliminate this interference.

c. Minimum detectable concentration: Strontium levels of about 0.2 mg/L can be detected by the flame photometric method without prior sample concentration.

2. Apparatus

Spectrophotometer, equipped with photomultiplier tube and flame accessories; or an atomic absorption spectrophotometer capable of operation in flame emission mode.

3. Reagents

a. Stock strontium solution: Dissolve 2.415 g strontium nitrate, $Sr(NO_3)_2$, dried to constant weight at 140°C, in 1000 mL 1% (v/v) HNO_3; 1.00 mL = 1.00 mg Sr.

b. Standard strontium solution: Dilute 25.00 mL stock strontium solution to 1000 mL with water; 1.00 mL = 25.0 μg Sr. Use this solution for preparing Sr standards in the 0.2- to 25-mg/L range.

c. Nitric acid, HNO_3, conc.

4. Procedure

a. Pretreatment of polluted water and wastewater samples: Select an appropriate procedure from Section 3030.

b. Preparation of strontium standards: Dilute samples, if necessary, to contain less than 400 mg Ca or Ba/L and less than 40 mg Sr/L. Add 25.0 mL sample (or a lesser but consistent volume to keep all standards in the linear range of the instrument) to 25.0 mL of each of a series of four or more strontium standards containing from 0 mg/L to a concentration exceeding that of the sample. For most natural waters 0, 2.0, 5.0, and 10.0 mg Sr/L standards are sufficient. A broader range curve might be preferable for brines. Dilute the brine sufficiently to eliminate burner splatter and clogging.

c. Concentration of low-level strontium samples: Concentrate samples containing less than 2 mg Sr/L. Polluted water or wastewater samples can be concentrated during digestion by starting with a larger volume (see Section 3030D). For other samples, add 3 to 5 drops conc HNO_3 to 250 mL sample and evaporate to about 25 mL. Cool and make up to 50.0 mL with distilled water. Proceed as in ¶ *b*. The HNO_3 concentration in the sample prepared for atomization can approach 0.4 mL/50 mL without producing interference.

d. Flame photometric measurement: Measure emission intensity of prepared samples (standards plus sample) at wavelengths of 460.7 and 466 nm. Follow manufacturer's instructions for correct instrument operation. Use a fuel-rich nitrous oxide-acetylene flame, if possible.

5. Calculation

a. Using a calculator or computer with linear regression capability, enter the net intensity (reading at 460.7 nm minus reading at 466 nm) versus concentration added to the sample and solve the equation for zero emissions. The negative of this number multiplied by any dilution factor is the sample concentration.

b. Plot net intensity (reading at 460.7 nm minus reading at 466 nm) against strontium concentration added to the sample. Because the plot forms a straight calibration line that intersects the ordinate, strontium concentration can be calculated from the equation:

$$mg \ Sr/L = \frac{A - B}{C} \times \frac{D}{E}$$

where:

A = sample emission-intensity reading of sample plus 0 mg/L at 460.7 nm,
B = background radiation reading at 466 nm, and
C = slope of calibration line.

Use the ratio D/E only when E mL of sample are concentrated to a final volume D mL (typically 50 mL).

c. Graphical method: Strontium concentration also can be evaluated by the graphical method illustrated in Figure 3500-Sr: 1. Plot net intensity against strontium concentration added to sample. If the line intersects the ordinate at Y emissions, the strontium concentration is where the abscissa value of the point on the calibration line has an ordinate value of 2Y emissions due to the two-fold dilution with standards (if sample and standards are mixed in equal volumes). The calibration line in the example intersects the ordinate at 12. Thus, Y = 12 and 2Y = 24. The strontium concentration of the sample is the abscissa value of the

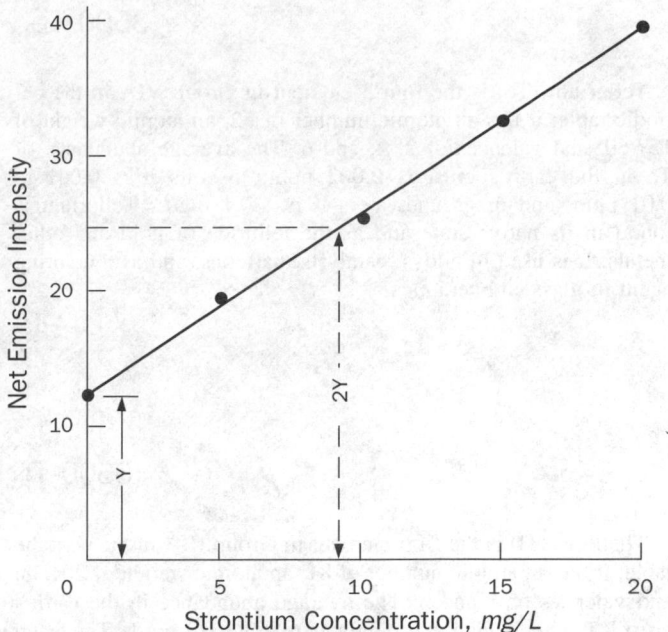

Figure 3500-Sr:1. Graphical method of computing strontium concentration.

point on the calibration line having an ordinate value of 24. In the example, the strontium concentration is 9.0 mg/L.

d. Report strontium concentrations below 10 mg/L to the nearest 0.1 mg/L and above 10 mg/L to the nearest whole number.

6. Quality Control

See Part 1000 and Section 3020 for specific quality control procedures to be followed during sample preparation and analysis.

7. Precision and Bias

Strontium concentrations in the range 12.0 to 16.0 mg/L can be determined with an accuracy within ± 1 to 2 mg/L.

8. Bibliography

Chow, T.J. & T.G. Thompson. 1955. Flame photometric determination of strontium in sea water. *Anal. Chem.* 27:18.

Nichols, M.S. & D.R. McNall. 1957. Strontium content of Wisconsin municipal waters. *J. Amer. Water Works Assoc.* 49:1493.

Horr, C.A. 1959. A survey of analytical methods for the determination of strontium in natural water. U.S. Geol. Surv. Water Supply Pap. No. 1496A.

3500-Te TELLURIUM

Tellurium (Te) is the fourth element in Group VIA in the periodic table; it has an atomic number of 52, an atomic weight of 127.60, and valences of 2, 4, and 6. The average abundance of Te in the earth's crust is 0.002 ppm; in soils it is 0.001 to 0.01 ppm; and in groundwaters it is <0.1 mg/L. Tellurium is found in its native state and as the telluride of gold and other metals. It is used in alloys, catalysts, batteries, and as a coloring agent in glass and ceramics.

The common aqueous species is TeO_3^{2-}. The metal and its compounds are toxic by inhalation.

Perform analyses by the electrothermal atomic absorption method (3113B). The inductively coupled plasma mass spectrometric method (3125) also may be applied successfully in most cases (with lower detection limits), even though tellurium is not specifically listed as an analyte in the method.

3500-Tl THALLIUM

Thallium (Tl) is the fifth element in Group IIIA in the periodic table; it has an atomic number of 81, an atomic weight of 204.38, and valences of 1 and 3. The average abundance in the earth's crust is 0.07 ppm, and in groundwaters it is <0.1 mg/L. The metal occurs chiefly in pyrites. Thallium is used in the production of glasses and rodenticides, in photoelectric applications, and in electrodes for dissolved oxygen meters.

The common aqueous species is Tl^+. It is nonessential for plants and animals. Compounds of thallium are toxic on contact with moisture, and by inhalation. The U.S. EPA primary drinking water standard MCL is 2 μg/L.

For analysis, use one of the atomic absorption spectrometric methods (3111B or 3113B), or one of the inductively coupled plasma methods (3120 or 3125), depending upon sensitivity requirements.

3500-Th THORIUM

Thorium (Th) is the first element in the actinium series of the periodic table; it has an atomic number of 90, an atomic weight of 232.04, and a valence of 4. The average abundance in the earth's crust is 8.1 ppm; in soils it is 13 ppm; in streams it is 0.1 μg/L, and in groundwaters it is <0.1 mg/L. Thorium is a radioactive element, with ^{232}Th having a half-life of 1.4×10^{10} years. It is widely distributed in the earth, with the principal mineral being monazite. Thorium is used in sun lamps, photoelectric cells, incandescent lighting, and gas mantles.

The aqueous chemistry of thorium is controlled by the Th^{4+} ion, which forms a set of complex species with hydroxides. Thorium's radioactive decay isotopes are dangerous when inhaled or ingested as thorium dust particles.

Either of the flame atomic absorption spectrometric methods (3111D or E) may be used for analysis. The inductively coupled plasma mass spectrometric method (3125) also may be applied successfully in most cases (with lower detection limits), even though thorium is not specifically listed as an analyte in the method.

3500-Sn TIN

Tin (Sn) is the fourth element in Group IVA in the periodic table; it has an atomic number of 50, an atomic weight of 118.69, and valences of 2 and 4. The average abundance in the earth's crust is 2.1 ppm; in soils it is 10 ppm; in streams it is 0.1 μg/L, and in groundwaters it is <0.1 mg/L. Tin is found mostly in the mineral cassiterite (SnO_2), in association with granitic rocks. Tin is used in reducing agents, solder, bronze, pewter, and coatings for various metals.

The common aqueous species are Sn^{4+}, $Sn(OH)_4$, $SnO(OH)_2$, and $SnO(OH)_3^-$. Tin is adsorbed to suspended solids, sulfides, and hydroxides. Tin can be methylated in sediments. Tributyl tin undergoes biodegradation quickly. Organo-tin compounds are toxic. Tin is considered nonessential for plants and animals.

Either the flame atomic absorption method (3111B) or the electrothermal atomic absorption method (3113B) may be used successfully for analyses, depending upon the sensitivity desired. The inductively coupled plasma/mass spectrometric method (3125) also may be applied successfully in most cases (with lower detection limits), even though tin is not specifically listed as an analyte in the method.

3500-Ti TITANIUM

Titanium (Ti) is the first element in Group IVB in the periodic table; it has an atomic number of 22, an atomic weight of 47.88, and valences of 2, 3, and 4. The average abundance of Ti in the earth's crust is 0.6%; in soils it is 1700 to 6600 ppm; in streams it is 3 $\mu g/L$, and in groundwaters it is <0.1 mg/L. The element is commonly associated with iron minerals. Titanium is used in alloys for aircraft, marine, and food-handling equipment. Compounds of the metal are used in pigments and as a reducing agent.

Titanium species are usually insoluble in natural waters, with the Ti^{4+} species being the most common ion when found. Some compounds are toxic by ingestion and the pure metal is flammable.

Either of the flame atomic absorption spectrometric methods (3111D or E) may be used. The inductively coupled plasma mass spectrometric method (3125) also may be applied successfully in most cases (with lower detection limits), even though titanium is not specifically listed as an analyte in the method.

3500-U URANIUM

Uranium (U) is the third element in the actinide series of the periodic table; it has an atomic number of 92, an atomic weight of 238.04, and valences of 3, 4, and 6. The average abundance of U in the earth's crust is 2.3 ppm, and in soils it is 1.8 ppm. Concentrations of uranium in drinking waters usually are expressed in terms of picocuries per liter, but that is now being replaced by Becquerel per liter (Bq/L). The approximate conversion factor, assuming equilibrium between ^{234}U and ^{238}U, is 1 μg uranium equals 0.67 pCi. The mean concentration of uranium in drinking water is 1.8 pCi/L. The chief ore is uraninite, or pitchblende, uranous uranate $[U(UO_4)_2]$. Uranium is known mainly for its use in the nuclear industry, but has also been used in glass, ceramics, and photography.

Uranium compounds are radioactive and are thereby toxic by inhalation and ingestion. There are three natural radioisotopes of uranium. Uranium-238 has a half-life of 4.5×10^9 years, represents 99% of uranium's natural abundance, and is not fissionable, but can be used to form plutonium-239, which is fissionable. Uranium-235 has a half-life of 7.1×10^8 years, represents 0.75% of uranium's natural abundance, is readily fissionable, and was the energy source in the original atomic bombs. Uranium-234 has a half-life of 2.5×10^5 years and represents only 0.006% of uranium's natural abundance.

The common forms in natural water are U^{4+} and UO_2^{2+}. In natural waters below pH 5, UO_2^{2+} would dominate; in the pH range of 5 to 10, soluble carbonate complexes predominate. Although there is no U.S. EPA drinking water standard MCL for uranium, an analysis for uranium is required if the gross alpha activity of a water sample is greater than 15 pCi/L.

Perform analyses by the inductively coupled plasma/mass spectrometry method (3125) or by one of the methods in 7500-U (for regulatory compliance purposes).

3500-V VANADIUM*

3500-V A. Introduction

1. Occurrence and Significance

Vanadium (V) is the first element in Group VB in the periodic table; it has an atomic number of 23, an atomic weight of 50.94, and valences of 2, 3, 4, and 5. The average abundance of V in the earth's crust is 136 ppm; in soils it ranges from 15 to 110 ppm; in streams it averages about 0.9 $\mu g/L$, and in groundwaters it is generally <0.1 mg/L. Though relatively rare, vanadium is found in a variety of minerals; most important among these are vanadinite $[Pb_5(VO_4)_3Cl]$, and patronite (possibly VS_4), occurring chiefly in Peru. Vanadium complexes have been noted in coal and petroleum deposits. Vanadium is used in steel alloys and as a catalyst in the production of sulfuric acid and synthetic rubber.

The dominant form in natural waters is V^{5+}. It is associated with organic complexes and is insoluble in reducing environments. It is considered nonessential for most higher plants and animals, although it may be an essential trace element for some algae and microorganisms. Laboratory and epidemiological evidence suggests that vanadium may play a beneficial role in the prevention of heart disease. In water supplies in New Mexico, which has a low incidence of heart disease, vanadium has been found in concentrations of 20 to 150 $\mu g/L$. In a state where incidence of heart disease is high, vanadium was not found in water supplies. However, vanadium pentoxide dust causes gastrointestinal and respiratory disturbances. The United Nations Food and

---

* Approved by Standard Methods Committee, 1997.

Agriculture Organization recommended maximum level for irrigation waters is 0.1 mg/L.

2. Selection of Method

The atomic absorption spectrometric methods (3111D and E), the electrothermal atomic absorption method (3113B), the induc-

tively coupled plasma methods (3120 and 3125), and gallic acid method (3500-V.B) are suitable for potable water samples. The atomic absorption spectrometric and inductively coupled plasma methods are preferred for polluted samples. The electrothermal atomic absorption method also may be used successfully with an appropriate matrix modifier.

3500-V B.   Gallic Acid Method

1. General Discussion

a. Principle: The concentration of trace amounts of vanadium in water is determined by measuring the catalytic effect it exerts on the rate of oxidation of gallic acid by persulfate in acid solution. Under the given conditions of concentrations of reactants, temperature, and reaction time, the extent of oxidation of gallic acid is proportional to the concentration of vanadium. Vanadium is determined by measuring the absorbance of the sample at 415 nm and comparing it with that of standard solutions treated identically.

b. Interference: The substances listed in Table 3500-V:I will interfere in the determination of vanadium if the specified concentrations are exceeded. This is not a serious problem for Cr^{6+}, Co^{2+}, Mo^{6+}, Ni^{2+}, Ag^+, and U^{6+} because the tolerable concentration is greater than that commonly encountered in fresh water. However, in some samples the tolerable concentration of Cu^{2+}, Fe^{2+}, and Fe^{3+} may be exceeded. Because of the high sensitivity of the method, interfering substances in concentrations only slightly above tolerance limits can be rendered harmless by dilution.

Traces of Br^- and I^- interfere seriously and dilution alone will not always reduce the concentration below tolerance limits. Mercuric ion may be added to complex these halides and minimize their interference; however, mercuric ion itself interferes if in excess. Adding 350 μg mercuric nitrate, $Hg(NO_3)_2$, per sample permits determination of vanadium in the presence of up to 100 mg Cl^-/L, 250 μg Br^-/L, and 250 μg I^-/L. Dilute samples containing high concentrations of these ions to concentrations below the values given above and add $Hg(NO_3)_2$.

TABLE 3500-V:I. CONCENTRATION AT WHICH VARIOUS IONS INTERFERE IN THE DETERMINATION OF VANADIUM

Ion	Concentration mg/L
Cr^{6+}	1.0
Co^{2+}	1.0
Cu^{2+}	0.05
Fe^{2+}	0.3
Fe^{3+}	0.5
Mo^{6+}	0.1
Ni^{2+}	3.0
Ag^+	2.0
U^{6+}	3.0
Br^-	0.1
Cl^-	100.0
I^-	0.001

c. Minimum detectable concentration: 0.025 μg V in approximately 13 mL final volume or approximately 2 μg V/L.

2. Apparatus

a. Water bath, capable of being operated at 25 ± 0.5°C.
b. Colorimetric equipment: One of the following is required:
1) *Spectrophotometer,* for measurements at 415 nm, with a light path of 1 to 5 cm.
2) *Filter photometer,* providing a light path of 1 to 5 cm and equipped with a violet filter with maximum transmittance near 415 nm.

3. Reagents

Use reagent water (see Section 1080) in preparation of reagents, for dilutions, and as blanks.
a. Stock vanadium solution: Dissolve 229.6 mg ammonium metavanadate, NH_4VO_3, in a volumetric flask containing approximately 800 mL water and 15 mL 1 + 1 nitric acid (HNO_3). Dilute to 1000 mL; 1.00 mL = 100 μg V.
b. Intermediate vanadium solution: Dilute 1.00 mL stock vanadium solution with water to 100 mL; 1.00 mL = 1.00 μg V.
c. Standard vanadium solution: Dilute 1.00 mL intermediate vanadium solution with water to 100 mL; 1.00 mL = 0.010 μg V.
d. Mercuric nitrate solution: Dissolve 350 mg $Hg(NO_3)_2 \cdot H_2O$ in 1000 mL water.
e. Ammonium persulfate-phosphoric acid reagent: Dissolve 2.5 g $(NH_4)_2S_2O_8$ in 25 mL water. Bring just to a boil, remove from heat, and add 25 mL conc H_3PO_4. Let stand approximately 24 h before use. Discard after 48 h.
f. Gallic acid solution: Dissolve 2 g $H_6C_7O_5$ in 100 mL warm water, heat to a temperature just below boiling, and filter through filter paper.* Prepare a fresh solution for each set of samples.

4. Procedure

a. Preparation of standards and sample: Prepare both blank and sufficient standards by diluting 0- to 8.0-mL portions (0 to 0.08 μg V) of standard vanadium solution to 10 mL with water. Pipet sample (10.00 mL maximum) containing less than 0.08 μg V into a suitable container and adjust volume to 10.0 mL with water. Filter colored or turbid samples. Add 1.0 mL $Hg(NO_3)_2$ solution to each blank, standard, and sample. Place containers in

---

* Whatman No. 42 or equivalent.

a water bath regulated to $25 \pm 0.5°C$ and allow 30 to 45 min for samples to come to the bath temperature.

b. Color development and measurement: Add 1.0 mL ammonium persulfate-phosphoric acid reagent (temperature equilibrated), swirl to mix thoroughly, and return to water bath. Add 1.0 mL gallic acid solution (temperature equilibrated), swirl to mix thoroughly, and return to water bath. Add gallic acid to successive samples at intervals of 30 s or longer to permit accurate control of reaction time. Exactly 60 min after adding gallic acid, remove sample from water bath and measure its absorbance at 415 nm, using water as a reference. Subtract absorbance of blank from absorbance of each standard and sample. Construct a calibration curve by plotting absorbance values of standards versus micrograms vanadium. Determine amount of vanadium in a sample by referring to the corresponding absorbance on the calibration curve. Prepare a calibration curve with each set of samples.

5. Calculation

$$\text{mg V/L} = \frac{\mu g \text{ V (in 13 mL final volume)}}{\text{original sample volume, mL}}$$

6. Precision and Bias

In a synthetic sample containing 6 μg V/L, 40 μg As/L, 250 μg Be/L, 240 μg B/L, and 20 μg Se/L in distilled water, vanadium was measured in 22 laboratories with a relative standard deviation of 20% and no relative error.

7. Bibliography

FISHMAN, M.J. & M.V. SKOUGSTAD. 1964. Catalytic determination of vanadium in water. *Anal. Chem.* 36:1643.

3500-Zn   ZINC*

3500-Zn A.   Introduction

1. Occurrence and Significance

Zinc (Zn) is the first element in Group IIB in the periodic table; it has an atomic number of 30, an atomic weight of 65.38, and a valence of 2. The average abundance of Zn in the earth's crust is 76 ppm; in soils it is 25 to 68 ppm; in streams it is 20 μg/L, and in groundwaters it is <0.1 mg/L. The solubility of zinc is controlled in natural waters by adsorption on mineral surfaces, carbonate equilibrium, and organic complexes. Zinc is used in a number of alloys such as brass and bronze, and in batteries, fungicides, and pigments. Zinc is an essential growth element for plants and animals but at elevated levels it is toxic to some species of aquatic life. The United Nations Food and Agriculture Organization recommended level for zinc in irrigation waters is 2 mg/L. The U.S. EPA secondary drinking water standard MCL is 5 mg/L. Concentrations above 5 mg/L can cause a bitter astringent taste and an opalescence in alkaline waters. Zinc most commonly enters the domestic water supply from deterioration of galvanized iron and dezincification of brass. In such cases lead and cadmium also may be present because they are impurities of the zinc used in galvanizing. Zinc in water also may result from industrial waste pollution.

2. Selection of Method

The atomic absorption spectrometric methods (3111B and C) and inductively coupled plasma methods (3120 and 3125) are preferred. The zincon method (B), suitable for analysis of both potable and polluted waters, may be used if instrumentation for the preferred methods is not available..

3. Sampling and Storage

See Section 3010B.2 for sample handling and storage.

---

* Approved by Standard Methods Committee, 1997.

3500-Zn B.   Zincon Method

1. General Discussion

a. Principle: Zinc forms a blue complex with 2-carboxy-2'-hydroxy-5'-sulfoformazyl benzene (zincon) in a solution buffered to pH 9.0. Other heavy metals likewise form colored complexes with zincon. Cyanide is added to complex zinc and heavy metals. Cyclohexanone is added to free zinc selectively from its cyanide complex so that it can be complexed with zincon to form a blue color. Sodium ascorbate reduces manganese interference. The developed color is stable except in the presence of copper (see table below).

b. Interferences: The following ions interfere at concentrations exceeding those listed:

Ion	mg/L	Ion	mg/L
Cd^{2+}	1	Cr^{3+}	10
Al^{3+}	5	Ni^{2+}	20
Mn^{2+}	5	Cu^{2+}	30
Fe^{3+}	7	Co^{2+}	30
Fe^{2+}	9	CrO_4^{2-}	50

c. Minimum detectable concentration: 0.02 mg Zn/L.

2. Apparatus

a. Colorimetric equipment: One of the following is required:

1) *Spectrophotometer,* for measurements at 620 nm, providing a light path of 1 cm or longer.

2) *Filter photometer,* providing a light path of 1 cm or longer and equipped with a red filter having maximum transmittance near 620 nm. Deviation from Beer's Law occurs when the filter band pass exceeds 20 nm.

b. Graduated cylinders, 50-mL, with ground-glass stoppers, Class B or better.

c. Erlenmeyer flasks, 50-mL.

d. Filtration apparatus: 0.45-μm filters and filter holders.

3. Reagents

a. Metal-free water: See Section 3111B.3c. Use water for rinsing apparatus and preparing solutions and dilutions.

b. Stock zinc solution: Dissolve 1000 mg (1.000 g) zinc metal in 10 mL 1 + 1 HNO_3. Dilute and boil to expel oxides of nitrogen. Dilute to 1000 mL; 1.00 mL = 1.00 mg Zn.

c. Standard zinc solution: Dilute 10.00 mL stock zinc solution to 1000 mL; 1.00 mL = 10.00 μg Zn.

d. Sodium ascorbate, fine granular powder, USP.

e. Potassium cyanide solution: Dissolve 1.00 g KCN in approximately 50 mL water and dilute to 100 mL. CAUTION: *Potassium cyanide is a deadly poison. Avoid skin contact or inhalation of vapors. Do not pipet by mouth or bring in contact with acids.*

f. Buffer solution, pH 9.0: Dissolve 8.4 g NaOH pellets in about 500 mL water. Add 31.0 g H_3BO_3 and swirl or stir to dissolve. Dilute to 1000 mL with water and mix thoroughly.

g. Zincon reagent: Dissolve 100 mg zincon (2-carboxy-2'-hydroxy-5'-sulfoformazyl benzene) in 100 mL methanol. Because zincon dissolves slowly, stir and/or let stand overnight.

h. Cyclohexanone, purified.

i. Hydrochloric acid, HCl, conc and 1N.

j. Sodium hydroxide, NaOH, 6N and 1N.

4. Procedure

a. Preparation of colorimetric standards: Accurately deliver 0, 0.5, 1.0, 3.0, 5.0, 10.0, and 14.0 mL standard zinc solution to a series of 50-mL graduated mixing cylinders. Dilute each to 20.0 mL to yield solutions containing 0, 0.25, 0.5, 1.5, 2.5, 5.0, and 7.0 mg Zn/L, respectively. (Lower-range standards may be prepared to extend the quantitation range. Longer optical path cells can be used. Verify linearity of response in this lower concentration range.) Add the following to each solution in sequence, mixing thoroughly after each addition: 0.5 g sodium ascorbate, 5.0 mL buffer solution, 2.0 mL KCN solution, and 3.0 mL zincon solution. Pipet 20.0 mL of the solution into a clean 50-mL erlenmeyer flask. Reserve remaining solution to zero the instrument. Add 1.0 mL cyclohexanone to the erlenmeyer flask. Swirl for 10 s and note time. Transfer portions of both solutions to clean sample cells. Use solution without cyclohexanone to zero colorimeter. Read and record absorbance for solution with cyclohexanone after 1 min. The calibration curve does not pass through zero because of the color enhancement effect of cyclohexanone on zincon.

b. Treatment of samples: To determine readily acid-extractable total zinc, add 1 mL conc HCl to 50 mL sample and mix thoroughly. Filter and adjust to pH 7. To determine dissolved zinc, filter sample through a 0.45-μm membrane filter. Adjust to pH 7 with 1N NaOH or 1N HCl if necessary after filtering.

c. Sample analysis: Cool samples to less than 30°C if necessary. Analyze 20.0 mL of prepared sample as described in ¶ 4a above, beginning with "Add the following to each solution . . ." If the zinc concentration exceeds 7 mg Zn/L prepare a sample dilution and analyze a 20.0-mL portion.

5. Calculation

Read zinc concentration (in milligrams per liter) directly from the calibration curve.

6. Precision and Bias

A synthetic sample containing 650 μg Zn/L, 500 μg Al/L, 50 μg Cd/L, 110 μg Cr/L, 470 μg Cu/L, 300 μg Fe/L, 70 μg Pb/L, 120 μg Mn/L, and 150 μg Ag/L in doubly demineralized water was analyzed in a single laboratory. A series of 10 replicates gave a relative standard deviation of 0.96% and a relative error of 0.15%. A wastewater sample from an industry in Standard Industrial Classification (SIC) No. 3333, primary smelting and refining of zinc, was analyzed by 10 different persons. The mean zinc concentration was 3.36 mg Zn/L and the relative standard deviation was 1.7%. The relative error compared to results from an atomic absorption analysis of the same sample was −1.0%.

7. Bibliography

PLATTE, J.A. & V.M. MARCY. 1959. Photometric determination of zinc with zincon. *Anal. Chem.* 31:1226.

RUSH, R.M. & J.H. YOE. 1954. Colorimetric determination of zinc and copper with 2-carboxy-2'-hydroxy-5'-sulfoformazyl-benzene. *Anal. Chem.* 26:1345.

MILLER, D.G. 1979. Colorimetric determination of zinc with zincon and cyclohexanone. *J. Water Pollut. Control Fed.* 51:2402.

PANDE, S.P. 1980. Study on the determination of zinc in drinking water. *J. IWWA XII* (3):275.

4010   INTRODUCTION

The analytical methods included in this part make use of classical wet chemical techniques and their automated variations and such modern instrumental techniques as ion chromatography. Methods that measure various forms of chlorine, nitrogen, and phosphorus are presented. The procedures are intended for use in the assessment and control of receiving water quality, the treatment and supply of potable water, and the measurement of operation and process efficiency in wastewater treatment. The methods also are appropriate and applicable in evaluation of environmental water-quality concerns. The introduction to each procedure contains reference to special field sampling conditions, appropriate sample containers, proper procedures for sampling and storage, and the applicability of the method.

4020   QUALITY ASSURANCE/QUALITY CONTROL

4020 A.   Introduction

Without quality control results there is no confidence in analytical results reported from tests. As described in Part 1000 and Section 3020, essential quality control measurements include: method calibration, standardization of reagents, assessment of individual capability to perform the analysis, performance of blind check samples, determination of the sensitivity of the test procedure (method detection level), and daily evaluation of bias, precision, and the presence of laboratory contamination or other analytical interference. Details of these procedures, expected ranges of results, and frequency of performance should be formalized in a written Quality Assurance Manual and Standard Operating Procedures.

For a number of the procedures contained in Part 4000, the traditional determination of bias using a known addition to either a sample or a blank, is not possible. Examples of these procedures include pH, dissolved oxygen, residual chlorine, and carbon dioxide. The inability to perform a reliable known addition does not relieve the analyst of the responsibility for evaluating test bias. Analysts are encouraged to purchase certified ready-made solutions of known levels of these constituents as a means of measuring bias. In any situation, evaluate precision through analysis of sample duplicates.

Participate in a regular program (at a minimum, annually, and preferably semi-annually) of proficiency testing (PT)/performance evaluation (PE) studies. The information and analytical confidence gained in the routine performance of the studies more than offset any costs associated with these studies. An unacceptable result on a PT study sample is often the first indication that a test protocol is not being followed successfully. Investigate circumstances fully to find the cause. Within many jurisdictions, participation in PT studies is a required part of laboratory certification.

Many of the methods contained in Part 4000 include specific quality-control procedures. These are considered to be the minimum quality controls necessary to successful performance of the method. Additional quality control procedures can and should be used. Section 4020B describes a number of QC procedures that are applicable to many of the methods.

4020 B.   Quality Control Practices

1. Initial Quality Control

See Section 3020B.1.

2. Calibration

See Section 3020B.2. Most methods for inorganic nonmetals do not have wide dynamic ranges. Standards for initial calibration therefore should be spaced more closely than one order of magnitude under these circumstances. Verify calibration by analyzing a midpoint or lower calibration standard and blank as directed. Alternatively, verify calibration with two standards, one near the low end and one near the high end, if the blank is used to zero the instrument.

3. Batch Quality Control

See Sections 3020B.3a through d.

4110 DETERMINATION OF ANIONS BY ION CHROMATOGRAPHY*

4110 A.   Introduction

Because of rapid changes in technology, this section is currently undergoing substantial revision.

Determination of the common anions such as bromide, chloride, fluoride, nitrate, nitrite, phosphate, and sulfate often is desirable to characterize a water and/or to assess the need for specific treatment. Although conventional colorimetric, electrometric, or titrimetric methods are available for determining individual an-

ions, only ion chromatography provides a single instrumental technique that may be used for their rapid, sequential measurement. Ion chromatography eliminates the need to use hazardous reagents and it effectively distinguishes among the halides (Br^-, Cl^-, and F^-) and the oxy-ions (SO_3^{2-}, SO_4^{2-} or NO_2^-, NO_3^-).

This method is applicable, after filtration to remove particles larger than 0.2 μm, to surface, ground, and wastewaters as well as drinking water. Some industrial process waters, such as boiler water and cooling water, also may be analyzed by this method.

---

* Approved by Standard Methods Committee, 1997.

4110 B.   Ion Chromatography with Chemical Suppression of Eluent Conductivity

1. General Discussion

a. Principle: A water sample is injected into a stream of carbonate-bicarbonate eluent and passed through a series of ion exchangers. The anions of interest are separated on the basis of their relative affinities for a low capacity, strongly basic anion exchanger (guard and separator columns). The separated anions are directed through a hollow fiber cation exchanger membrane (fiber suppressor) or micromembrane suppressor bathed in continuously flowing strongly acid solution (regenerant solution). In the suppressor the separated anions are converted to their highly conductive acid forms and the carbonate-bicarbonate eluent is converted to weakly conductive carbonic acid. The separated anions in their acid forms are measured by conductivity. They are identified on the basis of retention time as compared to standards. Quantitation is by measurement of peak area or peak height.

b. Interferences: Any substance that has a retention time coinciding with that of any anion to be determined and produces a detector response will interfere. For example, relatively high concentrations of low-molecular-weight organic acids interfere with the determination of chloride and fluoride by isocratic analyses. A high concentration of any one ion also interferes with the resolution, and sometimes retention, of others. Sample dilution or gradient elution overcomes many interferences. To resolve uncertainties of identification or quantitation use the method of known additions. Spurious peaks may result from contaminants in reagent water, glassware, or sample processing apparatus. Because small sample volumes are used, scrupulously avoid contamination. Modifications such as preconcentration of samples, gradient elution, or reinjection of portions of the eluted sample may alleviate some interferences but require individual validation for precision and bias.

c. Minimum detectable concentration: The minimum detectable concentration of an anion is a function of sample size and conductivity scale used. Generally, minimum detectable concentrations are near 0.1 mg/L for Br^-, Cl^-, NO_3^-, NO_2^-, PO_4^{3-}, and SO_4^{2-} with a 100-μL sample loop and a 10-μS/cm full-scale setting on the conductivity detector. Lower values may be

achieved by using a higher scale setting, an electronic integrator, or a larger sample size.

d. Limitations: This method is not recommended for the determination of F^- in unknown matrices. Equivalency studies have indicated positive or negative bias and poor precision in some samples. Recent interlaboratory studies show acceptable results. Two effects are common: first, F^- is difficult to quantitate at low concentrations because of the major negative contribution of the ''water dip'' (corresponding to the elution of water); second, the simple organic acids (formic, carbonic, etc.) elute close to fluoride and will interfere. Determine precision and bias before analyzing samples. F^- can be determined accurately by ion chromatography using special techniques such as dilute eluent or gradient elution using an NaOH eluent or alternative columns.

2. Apparatus

a. Ion chromatograph, including an injection valve, a sample loop, guard column, separator column, and fiber or membrane suppressors, a temperature-compensated small-volume conductivity cell and detector (6 μL or less), and a strip-chart recorder capable of full-scale response of 2 s or less. An electronic peak integrator is optional. Use an ion chromatograph capable of delivering 2 to 5 mL eluent/min at a pressure of 1400 to 6900 kPa.

b. Anion separator column, with styrene divinylbenzene-based low-capacity pellicular anion-exchange resin capable of resolving Br^-, Cl^-, NO_3^-, NO_2^-, PO_4^{3-}, and SO_4^{2-}.*

c. Guard column, identical to separator column† to protect separator column from fouling by particulates or organics.

d. Fiber suppressor or membrane suppressor:‡ Cation-exchange membrane capable of continuously converting eluent and separated anions to their acid forms. Alternatively, use continuously regenerated suppression systems.

---

* Dionex P/N 37041 or equivalent.
† Dionex P/N 37042 or equivalent.
‡ Dionex P/N 037072 (micro membrane—high capacity/low volume—suppressor) or equivalent.

3. Reagents

a. Deionized or distilled water free from interferences at the minimum detection limit of each constituent, filtered through a 0.2-μm membrane filter to avoid plugging columns, and having a conductance of < 0.1 μS/cm.

b. Eluent solution, sodium bicarbonate-sodium carbonate, 0.0017M NaHCO$_3$-0.0018M Na$_2$CO$_3$: Dissolve 0.5712 g NaHCO$_3$ and 0.7632 g Na$_2$CO$_3$ in water and dilute to 4 L.

c. Regenerant solution, H$_2$SO$_4$, 0.025N: Dilute 2.8 mL conc H$_2$SO$_4$ to 4 L.

d. Standard anion solutions, 1000 mg/L: Prepare a series of standard anion solutions by weighing the indicated amount of salt, dried to a constant weight at 105°C, to 1000 mL. Store in plastic bottles in a refrigerator; these solutions are stable for at least 1 month. Verify stability.

Anion§	Salt	Amount g/L
Cl$^-$	NaCl	1.6485
Br$^-$	NaBr	1.2876
NO$_3^-$	NaNO$_3$	1.3707 (226 mg NO$_3^-$-N/L)
NO$_2^-$	NaNO$_2$	1.4998‖ (304 mg NO$_2^-$-N/L)
PO$_4^{3-}$	KH$_2$PO$_4$	1.4330 (326 mg PO$_4^{3-}$-P/L)
SO$_4^{2-}$	K$_2$SO$_4$	1.8141

§ Expressed as compound.

‖ Do not oven-dry, but dry to constant weight in a desiccator.

e. Combined working standard solution, high range: Combine 12 mL of standard anion solutions, 1000 mg/L (¶ *d*) of NO$_2^-$, NO$_3^-$, HPO$_4^{2-}$, and Br$^-$, 20 mL of Cl$^-$, and 80 mL of SO$_4^{2-}$. Dilute to 1000 mL and store in a plastic bottle protected from light. Solution contains 12 mg/L each of NO$_2^-$, NO$_3^-$, HPO$_4^{2-}$, and Br$^-$, 20 mg/L of Cl$^-$, and 80 mg/L of SO$_4^{2-}$. Prepare fresh daily.

f. Combined working standard solution, low range: Dilute 25 mL of the high-range mixture (¶ *e*) to 100 mL and store in a plastic bottle protected from light. Solution contains 3 mg/L each of NO$_2^-$, NO$_3^-$, HPO$_4^{2-}$, and Br$^-$, 5 mg/L Cl$^-$, and 20 mg/L of SO$_4^{2-}$. Prepare fresh daily.

g. Alternative combined working standard solutions: Prepare appropriate combinations according to anion concentration to be determined. If NO$_2^-$ and PO$_4^{3-}$ are not included, the combined working standard is stable for 1 month. Dilute solutions containing NO$_2^-$ and PO$_4^{3-}$ must be made daily.

4. Procedure

a. System equilibration: Turn on ion chromatograph and adjust eluent flow rate to approximate the separation achieved in Figure 4110:1 (about 2 mL/min). Adjust detector to desired setting (usually 10 to 30 μS) and let system come to equilibrium (15 to 20 min). A stable base line indicates equilibrium conditions. Adjust detector offset to zero out eluent conductivity; with the fiber or membrane suppressor adjust the regeneration flow rate to maintain stability, usually 2.5 to 3 mL/min.

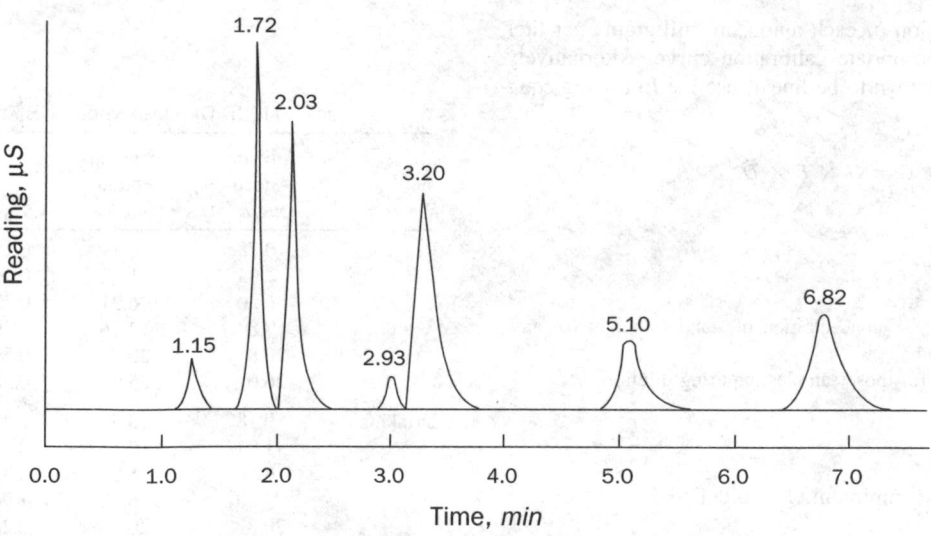

Figure 4110:1. Typical inorganic anion separation. Eluent: 1.7 mM NaHCO$_3$, 1.8 mM Na$_2$CO$_3$; sample loop: 50 μL; flow: 2.0 mL/min; column: Dionex AG4A plus AS4A.

Anion	Time	Conc mg/L	Anion	Time	Conc mg/L
Fluoride	1.15	2	Nitrate	3.20	10
Chloride	1.72	20	Orthophosphate	5.10	10
Nitrite	2.03	10	Sulfate	6.82	20
Bromide	2.93	5			

b. Calibration: Inject standards containing a single anion or a mixture and determine approximate retention times. Observed times vary with conditions but if standard eluent and anion separator column are used, retention always is in the order F^-, Cl^-, NO_2^-, Br^-, NO_3^-, HPO_4^{2-}, and SO_4^{2-}. Inject at least three different concentrations (one near the minimum reporting limit) for each anion to be measured and construct a calibration curve by plotting peak height or area against concentration on linear graph paper. Recalibrate whenever the detector setting, eluent, or regenerant is changed. To minimize the effect of the "water dip"# on F^- analysis, analyze standards that bracket the expected result or eliminate the water dip by diluting the sample with eluent or by adding concentrated eluent to the sample to give the same HCO_3^-/CO_3^{2-} concentration as in the eluent. If sample adjustments are made, adjust standards and blanks identically.

If linearity is established for a given detector setting, single standard calibration is acceptable. Record peak height or area and retention time for calculation of the calibration factor, F. However, a calibration curve will result in better precision and bias. HPO_4^{2-} is nonlinear below 1.0 mg/L.

c. Sample analysis: Remove sample particulates, if necessary, by filtering through a prewashed 0.2-μm-pore-diam membrane filter. Using a prewashed syringe of 1 to 10 mL capacity equipped with a male luer fitting inject sample or standard. Inject enough sample to flush sample loop several times: for 0.1 mL sample loop inject at least 1 mL. Switch ion chromatograph from load to inject mode and record peak heights and retention times on strip chart recorder. After the last peak (SO_4^{2-}) has appeared and the conductivity signal has returned to base line, another sample can be injected.

5. Calculations

Calculate concentration of each anion, in milligrams per liter, by referring to the appropriate calibration curve. Alternatively, when the response is shown to be linear, use the following equation:

$$C = H \times F \times D$$

where:

C = mg anion/L,
H = peak height or area,
F = response factor = concentration of standard/height (or area) of standard, and
D = dilution factor for those samples requiring dilution.

6. Quality Control

See Section 4020 for minimum QC guidelines.

7. Precision and Bias

The data in Tables 4110:I through VII were produced in a joint validation study with EPA and ASTM participation. Nineteen laboratories participated and used known additions of six prepared concentrates in three waters (reagent, waste, and drinking) of their choice.

Water dip occurs because water conductivity in sample is less than eluent conductivity (eluent is diluted by water).

TABLE 4110:I. DETERMINATION OF BIAS FOR FLUORIDE

Water	Amount Added mg/L	Amount Found mg/L	S_t	S_0	Bias %
Reagent	0.26	0.25	0.08	0.11	−3.8
	0.34	0.29	0.11		−14.7
	2.12	2.12	0.07	0.12	0.0
	2.55	2.48	0.14		−2.7
	6.79	6.76	0.20	0.19	−0.4
	8.49	8.46	0.30		−0.4
Drinking	0.26	0.24	0.08	0.05	−7.7
	0.34	0.34	0.11		0.0
	2.12	2.09	0.18	0.06	−1.4
	2.55	2.55	0.16		0.0
	6.79	6.84	0.54	0.25	+0.7
	8.49	8.37	0.75		−1.4
Waste	0.26	0.25	0.15	0.06	−3.8
	0.34	0.32	0.08		−5.9
	2.12	2.13	0.22	0.15	+0.5
	2.55	2.48	0.16		−2.7
	6.79	6.65	0.41	0.20	−2.1
	8.49	8.27	0.36		−2.6

Source: American Society for Testing and Materials. 1992. Method D4327. Annual Book of ASTM Standards, Vol. 11.01 Water. American Soc. Testing & Materials, Philadelphia, Pa.

TABLE 4110:II. DETERMINATION OF BIAS FOR CHLORIDE

Water	Amount Added mg/L	Amount Found mg/L	S_t	S_0	Bias %
Reagent	0.78	0.79	0.17	0.29	+1.3
	1.04	1.12	0.46		+7.7
	6.50	6.31	0.27	0.14	−2.9
	7.80	7.76	0.39		−0.5
	20.8	20.7	0.54	0.62	−0.5
	26.0	25.9	0.58		−0.4
Drinking	0.78	0.54	0.35	0.20	−30.8
	1.04	0.51	0.38		−51.0
	6.50	5.24	1.35	1.48	−19.4
	7.80	6.02	1.90		−22.8
	20.8	20.0	2.26	1.14	−3.8
	26.0	24.0	2.65		−7.7
Waste	0.78	0.43	0.32	0.39	−44.9
	1.04	0.65	0.48		−37.5
	6.50	4.59	1.82	0.83	−29.4
	7.80	5.45	2.02		−30.1
	20.8	18.3	2.41	1.57	−11.8
	26.0	23.0	2.50		−11.5

Source: American Society for Testing and Materials. 1992. Method D4327. Annual Book of ASTM Standards, Vol. 11.01 Water. American Soc. Testing & Materials, Philadelphia, Pa.

TABLE 4110:III. DETERMINATION OF BIAS FOR NITRITE NITROGEN

Water	Amount Added mg/L	Amount Found mg/L	S_t	S_0	Bias %
Reagent	0.36	0.37	0.04	0.04	+2.8
	0.48	0.48	0.06		0.0
	3.00	3.18	0.12	0.06	+6.0
	3.60	3.83	0.12		+6.4
	9.60	9.84	0.36	0.26	+2.5
	12.0	12.1	0.27		+0.6
Drinking	0.36	0.30	0.13	0.03	−16.7
	0.48	0.40	0.14		−16.7
	3.00	3.02	0.23	0.12	+0.7
	3.60	3.62	0.22		+0.6
	9.60	9.59	0.44	0.28	−0.1
	12.0	11.6	0.59		−3.1
Waste	0.36	0.34	0.06	0.04	−5.6
	0.48	0.46	0.07		−4.2
	3.00	3.18	0.13	0.10	+6.0
	3.60	3.76	0.18		+4.4
	9.60	9.74	0.49	0.26	+1.5
	12.0	12.0	0.56		+0.3

Source: American Society for Testing and Materials. 1992. Method D4327. Annual Book of ASTM Standards, Vol. 11.01 Water. American Soc. Testing & Materials, Philadelphia, Pa.

TABLE 4110:V. DETERMINATION OF BIAS FOR NITRATE NITROGEN

Water	Amount Added mg/L	Amount Found mg/L	S_t	S_0	Bias %
Reagent	0.42	0.42	0.04	0.02	0.0
	0.56	0.56	0.06		0.0
	3.51	3.34	0.15	0.08	−4.8
	4.21	4.05	0.28		−3.8
	11.2	11.1	0.47	0.34	−1.1
	14.0	14.4	0.61		+2.6
Drinking	0.42	0.46	0.08	0.03	+9.5
	0.56	0.58	0.09		+3.6
	3.51	3.45	0.27	0.10	−1.7
	4.21	4.21	0.38		0.0
	11.2	11.5	0.50	0.48	+2.3
	14.0	14.2	0.70		+1.6
Waste	0.42	0.36	0.07	0.06	−14.6
	0.56	0.40	0.16		−28.6
	3.51	3.19	0.31	0.07	−9.1
	4.21	3.84	0.28		−8.8
	11.2	10.9	0.35	0.51	−3.0
	14.0	14.1	0.74		+0.4

Source: American Society for Testing and Materials. 1992. Method D4327. Annual Book of ASTM Standards, Vol. 11.01 Water. American Soc. Testing & Materials, Philadelphia, Pa.

TABLE 4110:IV. DETERMINATION OF BIAS FOR BROMIDE

Water	Amount Added mg/L	Amount Found mg/L	S_t	S_0	Bias %
Water	0.63	0.69	0.11	0.05	+9.5
	0.84	0.85	0.12		+1.2
	5.24	5.21	0.22	0.21	−0.6
	6.29	6.17	0.35		−1.9
	16.8	17.1	0.70	0.36	+1.6
	21.0	21.3	0.93		+1.5
Drinking	0.63	0.63	0.13	0.04	0.0
	0.84	0.81	0.13		−3.6
	5.24	5.11	0.23	0.13	−2.5
	6.29	6.18	0.30		−1.7
	16.8	17.0	0.55	0.57	+0.9
	21.0	20.9	0.65		−0.4
Waste	0.63	0.63	0.15	0.09	0.0
	0.84	0.85	0.15		+1.2
	5.24	5.23	0.36	0.11	−0.2
	6.29	6.27	0.46		−0.3
	16.8	16.6	0.69	0.43	−1.0
	21.0	21.1	0.63		+0.3

Source: American Society for Testing and Materials. 1992. Method D4327. Annual Book of ASTM Standards, Vol. 11.01 Water. American Soc. Testing & Materials, Philadelphia, Pa.

TABLE 4110:VI. DETERMINATION OF BIAS FOR ORTHOPHOSPHATE

Water	Amount Added mg/L	Amount Found mg/L	S_t	S_0	Bias %
Reagent	0.69	0.69	0.06	0.06	0.0
	0.92	0.98	0.15		+6.5
	5.77	5.72	0.36	0.18	−0.9
	6.92	6.78	0.42		−2.0
	18.4	18.8	1.04	0.63	+2.1
	23.1	23.2	0.35		+0.4
Drinking	0.69	0.70	0.17	0.17	+1.4
	0.92	0.96	0.20		+4.3
	5.77	5.43	0.52	0.40	−5.9
	6.92	6.29	0.72		−9.1
	18.4	18.0	0.68	0.59	−2.2
	23.1	22.6	1.07		−2.0
Waste	0.68	0.64	0.26	0.09	−7.2
	0.92	0.82	0.28		−10.9
	5.77	5.18	0.66	0.34	−10.2
	6.92	6.24	0.74		−9.8
	18.4	17.6	2.08	1.27	−4.1
	23.1	22.4	0.87		−3.0

Source: American Society for Testing and Materials. 1992. Method D4327. Annual Book of ASTM Standards, Vol. 11.01 Water. American Soc. Testing & Materials, Philadelphia, Pa.

TABLE 4110:VII. DETERMINATION OF BIAS FOR SULFATE

Water	Amount Added mg/L	Amount Found mg/L	S_t	S_0	Bias %
Reagent	2.85	2.83	0.32	0.52	−0.7
	3.80	3.83	0.92		+0.8
	23.8	24.0	1.67	0.68	+0.8
	28.5	28.5	1.56		−0.1
	76.0	76.8	3.42	2.33	+1.1
	95.0	95.7	3.59		+0.7
Drinking	2.85	1.12	0.37	0.41	−60.7
	3.80	2.26	0.97		−40.3
	23.8	21.8	1.26	0.51	−8.4
	28.5	25.9	2.48		−9.1
	76.0	74.5	4.63	2.70	−2.0
	95.0	92.3	5.19		−2.8
Waste	2.85	1.89	0.37	0.24	−33.7
	3.80	2.10	1.25		−44.7
	23.8	20.3	3.19	0.58	−14.7
	28.5	24.5	3.24		−14.0
	76.0	71.4	5.65	3.39	−6.1
	95.0	90.3	6.80		−5.0

Source: American Society for Testing and Materials. 1992. Method D4327. Annual Book of ASTM Standards, Vol. 11.01 Water. American Soc. Testing & Materials, Philadelphia, Pa.

8. Bibliography

SMALL, H., T. STEVENS & W. BAUMAN. 1975. Novel ion exchange chromatographic method using conductimetric detection. *Anal. Chem.* 47:1801.

JENKE, D. 1981. Anion peak migration in ion chromatography. *Anal. Chem.* 53:1536.

BYNUM, M.I., S. TYREE & W. WEISER. 1981. Effect of major ions on the determination of trace ions by ion chromatography. *Anal. Chem.* 53: 1935.

WEISS, J. 1986. Handbook of Ion Chromatography. E.L. Johnson, ed. Dionex Corp., Sunnyvale, Calif.

PFAFF, J.D., C.A. BROCKHOFF & J.W. O'DELL. 1994. The Determination of Inorganic Anions in Water by Ion Chromatography. Method 300.0A, U.S. Environmental Protection Agency, Environmental Monitoring Systems Lab., Cincinnati, Ohio.

4110 C.   Single-Column Ion Chromatography with Electronic Suppression of Eluent Conductivity and Conductimetric Detection

1. General Discussion

a. Principle: A small portion of a filtered, homogeneous, aqueous sample or a sample containing no particles larger than 0.45 μm is injected into an ion chromatograph. The sample merges with the eluent stream and is pumped through the ion chromatographic system. Anions are separated on the basis of their affinity for the active sites of the column packing material. Conductivity detector readings (either peak area or peak height) are used to compute concentrations.

b. Interferences: Any two species that have similar retention times can be considered to interfere with each other. This method has potential coelution interference between short-chain acids and fluoride and chloride. Solid-phase extraction cartridges can be used to retain organic acids and pass inorganic anions. The interference-free solution then can be introduced into the ion chromatograph for separation.

This method is usable but not recommended for fluoride. Acetate, formate, and carbonate interfere in determining fluoride under the conditions listed in Table 4110:VIII. Filtering devices may be used to remove organic materials for fluoride measurements; simultaneously, use a lower eluent flow rate.

Chlorate and bromide coelute under the specified conditions. Determine whether other anions in the sample coelute with the anions of interest.

Additional interference occurs when anions of high concentrations overlap neighboring anionic species. Minimize this by sample dilution with reagent water.

TABLE 4110:VIII. DETECTION LIMITS FOR ANIONS IN REAGENT WATER*

Anion	Retention Time min	MDL†[1] mg/L
Cl^-	2.3	0.035
NO_2^-	3.1	0.022
Br^-	4.2	0.110
NO_3^-	5.3	0.035
PO_4^{3-}	5.7	0.110
SO_4^{2-}	8.3	0.350

* Standard conditions as defined in text.

† MDL calculated from the peak height in mm taken from chart recorder.

Best separation is achieved with sample pH between 5 and 9. When samples are injected the eluent pH will seldom change unless the sample pH is very low. Raise sample pH by adding a small amount of a hydroxide salt to enable the eluent to control pH.

Because method sensitivity is high, avoid contamination by reagent water and equipment. Determine any background or interference due to the matrix when adding the QC sample into any matrix other than reagent water.

c. Minimum detectable concentration: The minimum detectable concentration of an anion is a function of sample volume and the signal-to-noise ratio of the detector-recorder combination. Generally, minimum detectable concentrations are about 0.1 mg/L for the anions with an injection volume of 100 μL. Preconcentrators or using larger injection volumes can reduce detection limits to nanogram-per-liter levels for the common anions. However, co-elution is a possible problem with large injection volumes. Determine method detection limit for each anion of interest.

d. Prefiltration: If particularly contaminated samples are run, prefilter before or during injection. If the guard column becomes contaminated, follow manufacturer's suggestions for cleanup.

2. Apparatus

a. Ion chromatograph, complete with all required accessories including syringes, analytical columns, gases, detector, and a data system. Required accessories are listed below.

b. Filter device, 0.45 μm, placed before separator column to protect it from fouling by particulates or organic constituents.*

c. Anion separator column, packed with low-capacity anion-exchange resin capable of resolving fluoride, chloride, nitrite, bromide, nitrate, orthophosphate, and sulfate.†

d. Conductivity detector, flow-through, with integral heat-exchange unit allowing automatic temperature control and with separate working and reference electrodes.

e. Pump, constant flow rate controlled, high-pressure liquid chromatographic type, to deliver 1.5 mL/min.

f. Data system, including one or more computer, integrator, or strip chart recorder compatible with detector output voltage.

g. Sample injector: Either an automatic sample processor or a manual injector. If manual injector is used, provide several glass syringes of > 200 μL capacity. The automatic device must be compatible and able to inject a minimum sample volume of 100 μL.

3. Reagents

a. Reagent water: Distilled or deionized water of 18 megohm-cm resistivity containing no particles larger than 0.20 μm.

b. Borate/gluconate concentrate: Combine 16.00 g sodium gluconate, 18.00 g boric acid, 25.00 g sodium tetraborate decahydrate, and 125 mL glycerin in 600 mL reagent water. Mix and dilute to 1 L with reagent water.

c. Eluent solution, 0.0110M borate, 0.0015M gluconate, 12% (v/v) acetonitrile: Combine 20 mL borate/gluconate concentrate, 120 mL HPLC-grade acetonitrile, and 20 mL HPLC-grade *n*-butanol, and dilute to 1 L with reagent water. Use an in-line filter before the separator column to assure freedom from particulates. If the base line drifts, degas eluent with an inert gas such as helium or argon.

d. Stock standard solutions: See 4110B.3*e.*

e. Combined working standard solutions, high-range: See 4110B.3*e.*

f. Combined working standard solutions, low-range: See 4110B.3*f.*

4. Procedure

a. System equilibration: Set up ion chromatograph in accordance with the manufacturer's directions. Install guard and separator columns and begin pumping eluent until a stable base line is achieved. The background conductivity of the eluent solution is 278 μS ± 10%.

---

* Waters P/N 84560 or equivalent.
† Waters P/N 07355 or equivalent.

b. Calibration: Determine retention time for each anion by injecting a standard solution containing only the anion of interest and noting the time required for a peak to appear. Retention times vary with operating conditions and with anion concentration. Late eluters show the greatest variation. The shift in retention time is

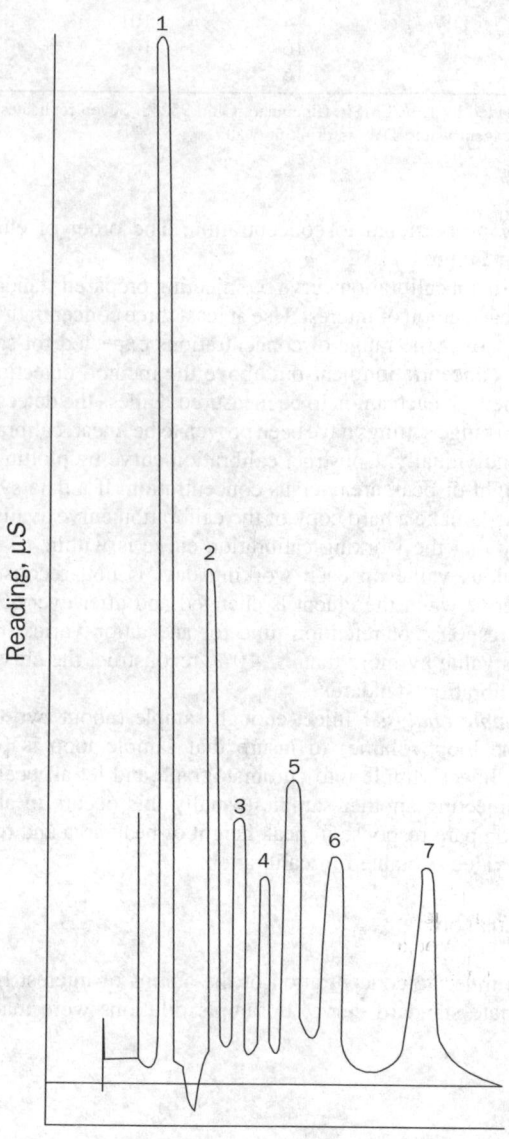

Figure 4110:2. Typical inorganic anion separation. Eluent: borate/gluconate; flow rate: 1.5 mL/min; injection volume: 100 μL.

Anion	Time min	Conc mg/L
1. System—peak	—	—
2. Cl^-	2.3	40
3. NO_2^-	3.1	10
4. Br^-	4.2	40
5. NO_3^-	5.3	20
6. PO_4^{3-}	5.7	40
7. SO_4^{2-}	8.3	80

TABLE 4110:IX. SINGLE-COLUMN CHROMATOGRAPHY SINGLE-OPERATOR PRECISION AND BIAS*

Anion	Sample Type†	Amount Added mg/L	Mean Recovery %	SD mg/L	Anion	Sample Type†	Amount Added mg/L	Mean Recovery %	SD mg/L
Cl^-	RW	16	105	1.6	NO_3^-	RW	8	103	0.75
	DW	16	98	1.9		DW	8	87	1.9
NO_2^-	RW	4	101	0.10	PO_4^{3-}	RW	16	113	0.92
	DW	4	101	0.43		DW	16	110	1.6
Br^-	RW	16	104	0.75	SO_4^{2-}	RW	32	101	0.42
	DW	16	98	2.3		DW	32	94	4.8

*Data provided by EPA/EMSL, Cincinnati, Ohio 95268. Seven replicates were analyzed for each anion and sample type.
†RW = reagent water; DW = drinking water.

inversely proportional to concentration. The order of elution is shown in Figure 4110:2.

Construct a calibration curve by injecting prepared standards including each anion of interest. Use at least three concentrations plus a blank. Cover the range of concentrations expected for samples. Use one concentration near but above the method detection limit established for each anion to be measured. Unless the detector's attenuation range settings have been proven to be linear, calibrate each setting individually. Construct calibration curve by plotting either peak height or peak area versus concentration. If a data system is being used, make a hard copy of the calibration curve available.

Verify that the working calibration curve is within ± 10% of the previous value on each working day; if not, reconstruct it. Also, verify when the eluent is changed and after every 20 samples. If response or retention time for any anion varies from the previous value by more than ± 10%, reconstruct the curve using fresh calibration standards.

c. Sample analysis: Inject enough sample (about two to three times the loop volume) to insure that sample loop is properly flushed. Inject sample into chromatograph and let all peaks elute before injecting another sample (usually this occurs in about 20 min). Compare response in peak height or peak area and retention time to values obtained in calibration.

5. Calculation

Determine the concentration of the anions of interest from the appropriate standard curve. If sample dilutions were made, calculate concentration:

$$C = A \times F$$

where:

C = anion concentration, mg/L,
A = mg/L from calibration curve, and
F = dilution factor.

6. Quality Control

a. If columns other than those listed in 4110C.2*c* are used, demonstrate that the resolution of all peaks is similar to that shown in Figure 4110:2.

b. Generate accuracy and precision data with this method by using a reference standard of known concentration prepared independently of the laboratory making the analysis. Compare with data in Precision and Bias, below.

c. Analyze a quality control sample at least every 10 samples. Follow general guidelines from Section 4020.

7. Precision and Bias

Precision and bias data are given in Table 4110:IX.

8. Reference

1. GLASER, J., D. FOERST, G. McKEE, S. QUAVE & W. BUDDE. 1981. Trace analyses for wastewater. *Environ. Sci. Technol.* 15:1426.

4120 SEGMENTED CONTINUOUS FLOW ANALYSIS*

4120 A. Introduction

1. Background and Applications

Air-segmented flow analysis (SFA) is a method that automates a large number of wet chemical analyses. An SFA analyzer can be thought of as a "conveyor belt" system for wet chemical

analysis, in which reagents are added in a "production-line" manner. Applications have been developed to duplicate manual procedures precisely. SFA was first applied to analysis of sodium and potassium in human serum, with a flame photometer as the detection device, by removing protein interferences with a selectively porous membrane (dialyzer).

The advantages of segmented flow, compared to the manual

* Approved by Standard Methods Committee, 1997.

method, include reduced sample and reagent consumption, improved repeatability, and minimal operator contact with hazardous materials. A typical SFA system can analyze 30 to 120 samples/h. Reproducibility is enhanced by the precise timing and repeatability of the system. Because of this, the chemical reactions do not need to go to 100% completion. Decreasing the number of manual sample/solution manipulations reduces labor costs, improves workplace safety, and improves analytical precision. Complex chemistries using dangerous chemicals can be carried out in sealed systems. Unstable reagents can be made up in situ. An SFA analyzer uses smaller volumes of reagents and samples than manual methods, producing less chemical waste needing disposal.

SFA is not limited to single-phase colorimetric determinations. Segmented-flow techniques often include analytical procedures such as mixing, dilution, distillation, digestion, dialysis, solvent extractions, and/or catalytic conversion. In-line distillation methods are used for the determinations of ammonia, fluoride, cyanide, phenols, and other volatile compounds. In-line digestion can be used for the determination of total phosphorous, total cyanide, and total nitrogen (kjeldahl $+ NO_2 + NO_3$). Dialysis membranes are used to eliminate interferences such as proteins and color, and other types of membranes are available for various analytical needs. SFA also is well-suited for automated liquid/liquid extractions, such as in the determination of MBAS. Packed-bed ion exchange columns can be used to remove interferences and enhance sensitivity and selectivity of the detection.

Specific automated SFA methods are described in the sections for the analytes of interest.

2. Bibliography

BEGG, R.D. 1971. Dynamics of continuous segmented flow analysis. *Anal. Chem.* 43:854.

THIERS, R.E., A.H. REED & K. DELANDER. 1971. Origin of the lag phase of continuous flow curves. *Clin. Chem.* 17:42.

FURMAN, W.B. 1976. Continous Flow Analysis. Theory and Practice. Marcel Dekker, Inc., New York, N.Y.

COAKLEY, W.A. 1978. Handbook of Automated Analysis. Marcel Dekker, Inc., New York, N.Y.

SNYDER, L.R. 1980. Continuous flow analysis: present and future. *Anal. Chem. Acta* 114:3.

4120 B.   Segmented Flow Analysis Method

1. General Discussion

a. Principle: A rudimentary system (Figure 4120:1) contains four basic components: a sampling device, a liquid transport device such as a peristaltic pump, the analytical cartridge where the chemistry takes place, and the detector to quantify the analyte.

In a generalized system, samples are loaded onto an automatic sampler. The sampler arm moves the sample pickup needle between the sample cup and a wash reservoir containing a solution closely matching the sample matrix and free of the analyte. The wash solution is pumped continuously through the reservoir to eliminate cross-contamination. The sample is pumped to the analytical cartridge as a discrete portion separated from the wash by an air-bubble created during the sampler arm's travel from wash reservoir to sample cup and back.

In the analytical cartridge, the system adds the sample to the reagent(s) and introduces proportionately identical air-bubbles to reagent or sample stream. Alternatively, another gas or immiscible fluid can be substituted for air. The analyzer then proportions the analyte sample into a number of analytical segments depending on sample time, wash time, and segmentation frequency. Relative flow and initial reagent concentration determine the amount and concentration of each reagent added. The micro-circulation pattern enhances mixing, as do mixing coils, which swirl the analytical system to utilize gravitational forces. Chemical reactions, solvent separation, catalytic reaction, dilution, distillation, heating, and/or special applications take place in their appropriate sections of the analytical cartridge as the segmented stream flows toward the detector.

A typical SFA detector is a spectrophotometer that measures the color development at a specific wavelength. Other detectors, such as flame photometers and ion-selective electrodes, can be used. SFA detectors utilize flow-through cells, and typically send their output to a computerized data-collection system and/or a chart-recorder. The baseline is the reading when only the reagents and wash water are flowing through the system. Because gas bubbles are compressible, highly reflective, and electrically nonconductive, they severely distort the signal in the detector; therefore, many systems remove the bubbles before the optical light path. However, if the system removes the bubbles at any point within the system, the segregated liquids will be able to interact and pool. This interaction can cause cross-contamination or loss of wash, and decreases the rate at which samples can be processed. Real-time analog or digital data reconstruction techniques known as curve regeneration can remove the effect of pooling at the flow-cell debubbler and/or any other unsegmented zones of the system. "Bubble-gating" is a technique that does not remove the bubbles, but instead uses analog or digital processing to remove the distortion caused by the bubbles. Bubble-gating requires a sufficiently fast detector response time and requires that the volume of the measurement cell be smaller than the volume of the individual liquid segment.

b. Sample dispersion and interferences: Theoretically, the output of the detector is square-wave. Several carryover processes can deform the output exponentially. The first process, longitudinal dispersion, occurs as a result of laminar flow. Segmentation of the flow with air bubbles minimizes the dispersion and mixing

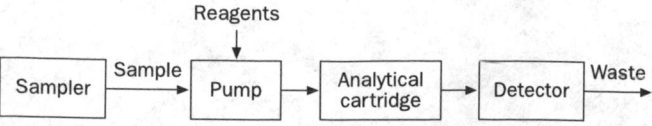

Figure 4120:1 Schematic of a segmented flow analyzer.

between segments. The second process is axial or lag-phase dispersion. It arises from stagnant liquid film that wets the inner surfaces of the transmission tubing. Segmented streams depend on wet surfaces for hydraulic stability. The back-pressure within non-wet tubing increases in direct proportion to the number of bubbles it contains and causes surging and bubble breakup. Corrective measures include adding specific wetting agents (surfactants) to reagents and minimizing the length of transmission tubing.

Loose or leaking connections are another cause of carryover and can cause poor reproducibility. Wrap TFE tape around leaking screw fittings. When necessary, slightly flange the ends of types of tubing that require it for a tight connection. For other connections, sleeve one size of tubing over another size. Use a noninterfering lubricant for other tubing connections. Blockages in the tubing can cause back-pressure and leaks. Clean out or replace any blocked tubing or connection. A good indicator for problems is the bubble pattern; visually inspect the system for any abnormal bubble pattern that may indicate problems with flow.

For each analysis, check individual method for compounds that can interfere with color development and/or color reading. Other possible interferences include turbidity, color, and salinity. Turbid and/or colored samples may require filtration. In another interference-elimination technique, known as matrix correction, the solution is measured at two separate wavelengths, and the result at the interference wavelength is subtracted from that at the analytical wavelength.

2. Apparatus

a. Tubing and connections: Use mini- or micro-bore tubing on analytical cartridges. Replace flexible tubing that becomes discolored, develops a ''sticky'' texture, or loses ability to spring back into shape immediately after compression. Also see manufacturer's manual and specific methods.

b. Electrical equipment and connections: Make electrical connections with screw terminals or plug-and-socket connections. Use shielded electrical cables. Use conditioned power or a universal power supply if electrical current is subject to fluctuations. See manufacturer's manual for additional information.

c. Automated analytical equipment: Dedicate a chemistry manifold and tubing to each specific chemistry. See specific methods and manufacturer's manual for additional information.

d. Water baths: When necessary, use a thermostatically controlled heating/cooling bath to decrease analysis time and/or improve sensitivity. Several types of baths are available; the most common are coils heated or cooled by water or oil. Temperature-

controlled laboratories reduce drift in temperature-sensitive chemistries if water baths are not used.

3. Reagents

Prepare reagents according to specific methods and manufacturer's instructions. If required, filter or degas a reagent. Use reagent water (see Section 1080) if available; if not, use a grade of water that is free of the analyte and interfering substances. Run blanks to demonstrate purity of the water used to prepare reagents and wash SFA system. Minimize exposure of reagents to air, and refrigerate if necessary. If reagents are made in large quantities, preferably decant a volume sufficient for one analytical run into a smaller container. If using a wetting agent, add it to the reagent just before the start of the run. Reagents and wetting agents have a limited shelf-life. Old reagents or wetting agents can produce poor reproducibility and distorted peaks. Do not change reagent solutions or add reagent to any reagent reservoirs during analysis. Always start with a sufficient quantity to last through the analytical run.

4. Procedure

For specific operating instructions, consult manufacturer's directions and methods for analytes of interest. At startup of a system, pump reagents and wash water through system until system has reached equilibrium (bubble pattern smooth and consistent) and base line is stable. Meanwhile, load samples and standards into sample cups or tubes and type corresponding tags into computer table. When ready, command computer to begin run. Most systems will run the highest standard to trigger the beginning of the run, followed by a blank to check return to base line, and then a set of standards covering the analytical range (sampling from lowest to highest concentration). Construct a curve plotting concentration against absorbance or detector reading and extrapolate results (many systems will do this automatically). Run a new curve daily immediately before use. Calculation and interpretation of results depend on individual chemistry and are analogous to the manual method. Insert blanks and standards periodically to check and correct for any drift of base line and/or sensitivity. Some systems will run a specific standard periodically as a ''drift,'' and automatically will adjust sample results. At end of a run, let system flush according to manufacturer's recommendations.

5. Quality Control

See Section 4020 and individual methods for quality control methods and precision and bias data.

4130 INORGANIC NONMETALS BY FLOW INJECTION ANALYSIS*

4130 A. Introduction

1. Principle

Flow injection analysis (FIA) is an automated method of introducing a precisely measured portion of liquid sample into a continuously flowing carrier stream. The sample portion usually is injected into the carrier stream by either an injection valve with a fixed-volume sample loop or an injection valve in which a fixed time period determines injected sample volume. As the sample portion leaves the injection valve, it disperses into the carrier stream and forms an asymmetric Gaussian gradient in analyte concentration. This concentration gradient is detected continuously by either a color reaction or another analyte-specific detector through which the carrier and gradient flow.

When a color reaction is used as the detector, the color reaction reagents also flow continuously into the carrier stream. Each color reagent merges with the carrier stream and is added to the analyte gradient in the carrier in a proportion equal to the relative flow rates of the carrier stream and merging color reagent. The color reagent becomes part of the carrier after it is injected and has the effect of modifying or derivatizing the analyte in the gradient. Each subsequent color reagent has a similar effect, finally resulting in a color gradient proportional to the analyte gradient. When the color gradient passes through a flow cell placed in a flow-through absorbance detector, an absorbance peak is formed. The area of this peak is proportional to the analyte concentration in the injected sample. A series of calibration standards is injected to generate detector response data used to produce a calibration curve. It is important that the FIA flow rates, injected sample portion volume, temperature, and time the sample is flowing through the system ("residence time") be the same for calibration standards and unknowns. Careful selection of flow rate, injected sample volume, frequency of sample injection, reagent flow rates, and residence time determines the precise dilution of the sample's original analyte concentration into the useful concentration range of the color reaction. All of these parameters ultimately determine the sample throughput, dynamic range of the method, reaction time of the color reaction discrimination against slow interference reactions, signal-to-noise ratio, and method detection level (MDL).

2. Applications

FIA enjoys the advantages of all continuous-flow methods: There is a constantly measured reagent blank, the "base line" against which all samples are measured; high sample throughput encourages frequent use of quality control samples; large numbers of samples can be analyzed in batches; sample volume measurement, reagent addition, reaction time, and detection occur reproducibly without the need for discrete measurement and transfer vessels such as cuvettes, pipets, and volumetric flasks; and all samples share a single reaction manifold or vessel consisting of inert flow tubing.

Specific FIA methods are presented as Sections $4500\text{-}Br^-$.D, $4500\text{-}Cl^-$.G, $4500\text{-}CN^-$.N and O, $4500\text{-}F^-$.G, $4500\text{-}NH_3$.H, $4500\text{-}NO_3^-$.I, 4500-N.B, $4500\text{-}N_{org}$.D, 4500-P.G, H, and I, $4500\text{-}SiO_2$.F, $4500\text{-}SO_4^{2-}$.G, and $4500\text{-}S^{2-}$.I.

---

*Approved by Standard Methods Committee, 1997.

4130 B. Quality Control

When FIA methods are used, follow a formal laboratory quality control program. The minimum requirements consist of an initial demonstration of laboratory capability and periodic analysis of laboratory reagent blanks, fortified blanks, and other laboratory solutions as a continuing check on performance. Maintain performance records that define the quality of the data generated.

See Section 1020, Quality Assurance, and Section 4020 for the elements of such a quality control program.

4140   INORGANIC ANIONS BY CAPILLARY ION ELECTROPHORESIS (PROPOSED)*

4140 A.   Introduction

Determination of common inorganic anions such as fluoride, chloride, bromide, nitrite, nitrate, orthophosphate, and sulfate is a significant component of water quality analysis. Instrumental techniques that can determine multiple analytes in a single analysis, i.e., ion chromatography (Section 4110) and capillary ion electrophoresis, offer significant time and operating cost savings over traditional single-analyte wet chemical analysis.

Capillary ion electrophoresis is rapid (complete analysis in less than 5 min) and provides additional anion information, i.e., organic acids, not available with isocratic ion chromatography (IC). Operating costs are significantly less than those of ion chroma-

_____
* Approved by Standard Methods Committee, 1997.

tography. Capillary ion electrophoresis can detect all anions present in the sample matrix, providing an anionic "fingerprint."

Anion selectivity of capillary ion electrophoresis is different from that of IC and eliminates many of the difficulties present in the early portion of an IC chromatogram. For example, sample matrix neutral organics, water, and cations do not interfere with anion analysis, and fluoride is well resolved from monovalent organic acids. Sample preparation typically is dilution with reagent water and removal of suspended solids by filtration. If necessary, hydrophobic sample components such as oil and grease can be removed with the use of HPLC solid-phase extraction cartridges without biasing anion concentrations.

4140 B.   Capillary Ion Electrophoresis with Indirect UV Detection

1. General Discussion

a. Principle: A buffered aqueous electrolyte solution containing a UV-absorbing anion salt (sodium chromate) and an electroosmotic flow modifier (OFM) is used to fill a 75-μm-ID silica capillary. An electric field is generated by applying 15 kV of applied voltage using a negative power supply; this defines the detector end of the capillary as the anode. Sample is introduced at the cathodic end of the capillary and anions are separated on the basis of their differences in mobility in the electric field as they migrate through the capillary. Cations migrate in the opposite direction and are not detected. Water and neutral organics are not attracted towards the anode; they migrate after the anions and thus do not interfere with anion analysis. Anions are detected as they displace charge-for-charge the UV-absorbing electrolyte anion (chromate), causing a net decrease in UV absorbance in the analyte anion zone compared to the background electrolyte. Detector polarity is reversed to provide positive mv response to the data system (Figure 4140:1). As in chromatography, the analytes are identified by their migration time and quantitated by using time-corrected peak area relative to standards. After the analytes of interest are detected, the capillary is purged with fresh electrolyte, eliminating the remainder of the sample matrix before the next analysis.

b. Interferences: Any anion that has a migration time similar to the analytes of interest can be considered an interference. This method has been designed to minimize potential interference typically found in environmental waters, groundwater, drinking water, and wastewater.

Formate is a common potential interference with fluoride; it is a common impurity in reagent water, has a migration time similar to that of fluoride, and is an indicator of loss of water purification

system performance and TOC greater than 0.1 mg/L. The addition of 5 mg formate/L in the mixed working anion standard, and to sample where identification of fluoride is in question, aids in the correct identification of fluoride.

Generally, a high concentration of any one ion may interfere with resolution of analyte anions in close proximity. Dilution in reagent water usually is helpful. Modifications in the electrolyte formulation can overcome resolution problems but require individual validation for precision and bias. This method is capable of interference-free resolution of a 1:100 differential of Br^- to Cl^-, and NO_2^- and NO_3^- to SO_4^{2-}, and 1:1000 differential of Cl^- and SO_4^{2-}.

Dissolved ferric iron in the mg/L range gives a low bias for PO_4. However, transition metals do not precipitate with chromate because of the alkaline electrolyte pH.

c. Minimum detectable concentrations: The minimum detectable concentration for an anion is a function of sample size. Generally, for a 30-s sampling time, the minimum detectable concentrations are 0.1 mg/L (Figure 4140:2). According to the method for calculating MDL given in Section 1030, the calculated detection limits are below 0.1 mg/L. These detection limits can be compromised by analyte impurities in the electrolyte.

d. Limitations: Samples with high ionic strength may show a decrease in analyte migration time. This variable is addressed by using normalized migration time with respect to a reference peak, chloride, for identification, and using time-corrected area for quantitation. With electrophoresis, published data indicate that analyte peak area is a function of migration time. At high analyte anion concentrations, peak shape becomes asymmetrical; this phenomenon is typical and is different from that observed in ion chromatography.

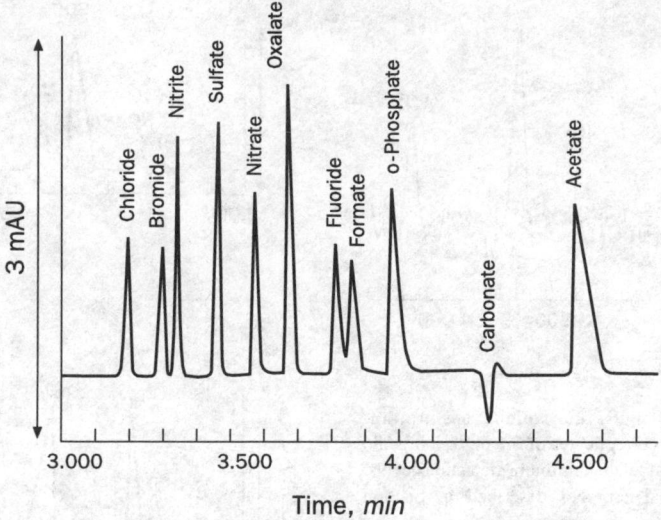

Figure 4140:1. Electropherogram of the inorganic anions and typically found organic acids using capillary ion electrophoresis and chromate electrolyte.
Electrolyte: 4.7 mN Na_2CrO_4/4.0 mM TTAOH/10 mM CHES/0.1 mM calcium gluconate; capillary: 75-μm-ID $\times$ 375-μm-OD $\times$ 60-cm length, uncoated silica; voltage: 15 kV using a negative power supply; current: 14 $\pm$ 1 μA; sampling: hydrostatic at 10 cm for 30 s; detection: indirect UV with Hg lamp and 254-nm filter.

Anion	Conc mg/L	Migration Time min	Migration Time Ratio to Cl	Peak Area	Time-Corrected Peak Area
Chloride	2.0	3.200	1.000	1204	376.04
Bromide	4.0	3.296	1.030	1147	348.05
Nitrite	4.0	3.343	1.045	2012	601.72
Sulfate	4.0	3.465	1.083	1948	562.05
Nitrate	4.0	3.583	1.120	1805	503.69
Oxalate	5.0	3.684	1.151	3102	842.14
Fluoride	1.0	2.823	1.195	1708	446.65
Formate	5.0	3.873	1.210	1420	366.61
o-Phosphate	4.0	4.004	1.251	2924	730.25
Carbonate & bicarbonate		4.281	1.338		
Acetate	5.0	4.560	1.425	3958	868.01

2. Apparatus

*a. Capillary ion electrophoresis (CIE) system:** Various commercial instruments are available that integrate a negative high-voltage power supply, electrolyte reservoirs, covered sample carousel, hydrostatic sampling mechanism, capillary purge mechanism, self-aligning capillary holder, and UV detector capable of 254-nm detection in a single temperature-controlled compartment at 25°C. Optimal detection limits are attained with a fixed-wavelength UV detector with Hg lamp and 254-nm filter.

b. Capillary: 75-μm-ID $\times$ 375-μm-OD $\times$ 60-cm-long fused silica capillary with a portion of its outer coating removed to act as the UV detector window. Capillaries can be purchased premade* or on a spool and prepared as needed.

*c. Data system:** HPLC-based integrator or computer. Optimum performance is attained with a computer data system and electrophoresis-specific data processing including data acquisition at 20 points/s, migration times determined at midpoint of peak width, identification based on normalized migration times with respect to a reference peak, and time-corrected peak area.

3. Reagents

a. Reagent water: See Section 1080. Ensure that water is analyte-free. The concentration of dissolved organic material will influence overall performance; preferably use reagent water with <50 μg TOC/L.

b. Chromate electrolyte solution: Prepare as directed from individual reagents, or purchase electrolyte preformulated.

1) *Sodium chromate concentrate, 100 mM:* In a 1-L volumetric flask dissolve 23.41 g sodium chromate tetrahydrate, $Na_2CrO_4 \cdot 4H_2O$, in 500 mL water and dilute to 1 L with water. Store in a capped glass or plastic container at ambient temperature; this reagent is stable for 1 year.

2) *Electroosmotic flow modifier concentrate, 100 mM:* In a 100-mL volumetric flask dissolve 3.365 g tetradecyltrimethyl

___
* Waters Corp. or equivalent.

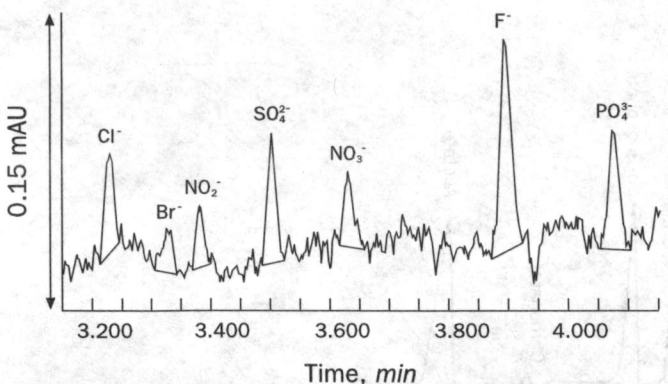

Figure 4140:2. Electropherogram of 0.1 mg/L inorganic anions at minimum detection level. Seven replicates of 0.1 mg/L inorganic anion standard used to calculate minimum detection levels, as mg/L, using analytical protocol described in Section 1030E.

Chloride	= 0.046	Nitrite	= 0.072
Nitrate	= 0.084	Phosphate	= 0.041
Bromide	= 0.090	Sulfate	= 0.032
Fluoride	= 0.020		

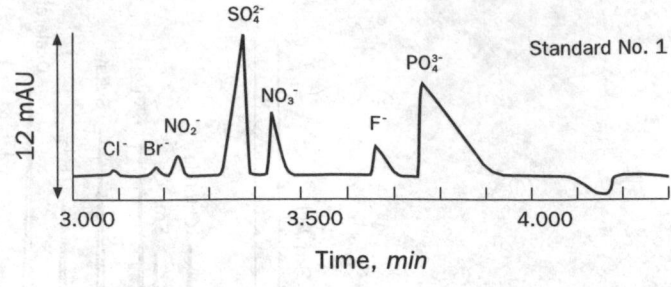

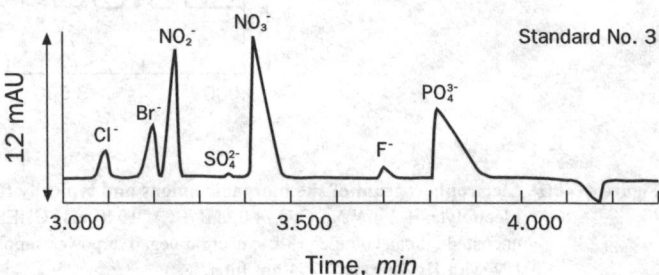

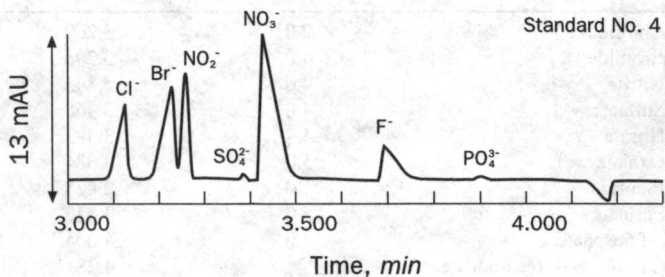

ammonium bromide (TTAB), mol wt 336.4, in 50 mL water and dilute to 100 mL. Store in a capped glass or plastic container at ambient temperature; this reagent is stable for 1 year.

3) *Buffer concentrate,* 100 m*M:* In a 1-L volumetric flask dissolve 20.73 g 2-[*N*-cyclohexylamino]-ethane sulfonate (CHES), mol wt 207.29, in 500 mL water and dilute to 1 L. Store in a capped glass or plastic container at ambient temperature; this reagent is stable for 1 year.

4) *Calcium gluconate concentrate,* 1 m*M:* In a 1-L volumetric flask dissolve 0.43 g calcium gluconate, mol wt 430.38, in 500 mL water and dilute to 1 L. Store in a capped glass or plastic container at ambient temperature; this reagent is stable for 1 year.

5) *Sodium hydroxide solution,* NaOH, 100 m*M:* In a 1-L plastic volumetric flask dissolve 4 g sodium hydroxide, NaOH, in 500 mL water and dilute to 1 L. Store in a capped plastic container at ambient temperature; this reagent is stable for 1 month.

6) *Chromate electrolyte solution:* Prerinse an anion exchange cartridge in the hydroxide form with 10 mL 100-m*M* NaOH fol-

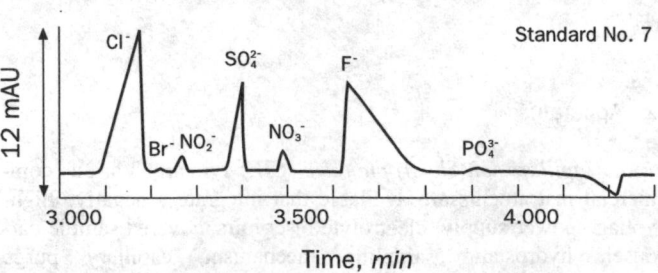

Figure 4140:3. Representative electropherograms of Youden anion standards. For composition of standards, see Table 4140:I.

TABLE 4140:I. COLLABORATIVE DESIGN AS FOUR YOUDEN PAIR SETS

	Anion Concentration in Individual Youden Pair Standards *mg/L*							
Anion	1	2	3	4	5	6	7	8
Cl^-	0.7	2.0	3.0	15.0	40.0	20.0	50.0	0.5
Br^-	2.0	3.0	15.0	40.0	20.0	50.0	0.7	0.5
NO_2^-	3.0	40.0	20.0	15.0	50.0	0.5	2.0	0.7
SO_4^{2-}	40.0	50.0	0.5	0.7	2.0	3.0	15.0	20.0
NO_3^-	15.0	20.0	40.0	50.0	0.5	0.7	2.0	3.0
F^-	2.0	0.7	0.5	3.0	10.0	7.0	20.0	25.0
PO_4^{3-}	50.0	40.0	20.0	0.5	3.0	2.0	0.7	15.0

* The collaborative design is intended to demonstrate performance between 0.1 and 50 mg anion/L, except for fluoride between 0.1 and 25 mg/L. The concentrations among anions are varied so as not to have any one standard at all low or all high anion concentrations.

lowed by 10 mL water; discard the washings. Slowly pass 4 mL 100-mM TTAB concentrate through the cartridge into a 100-mL volumetric flask. Rinse cartridge with 10 mL water and add to flask. (NOTE: This step is needed to convert the TTAB from the bromide form into the hydroxide form TTAOH. The step can be eliminated if commercially available 100 mM TTAOH is used.)

To the 100-mL volumetric flask containing the TTAOH add 4.7 mL sodium chromate concentrate, 10 mL CHES buffer concentrate, and 10 mL calcium gluconate concentrate. Mix and dilute to 100 mL with water. The pH should be 9 $\pm$ 0.1; final solution is 4.7 mM sodium chromate, 4 mM TTAOH, 10 mM CHES, and 0.1 mM calcium gluconate. Filter and degas through a 0.45-μm aqueous membrane, using a vacuum apparatus. Store any remaining electrolyte in a capped plastic container at ambient temperature for up to 1 month.

c. Standard anion solution, 1000 mg/L: Prepare a series of individual standard anion solutions by adding the indicated amount of salt, dried to constant weight at 105°C, to 100 mL with water. Store in plastic bottles; these solutions are stable for 3 months. (Alternatively, purchase individual certified 1000-mg/L anion standards and store following manufacturer's directions.)

Anion	Salt	Amount g/100mL
Chloride	NaCl	0.1649
Bromide	NaBr	0.1288
Formate	NaCO$_2$H	0.1510
Fluoride	NaF	0.2210
Nitrite	NaNO$_2$	0.1499* (1000 mg NO$_2^-$/L = 304.3 mg NO$_2^-$ − N/L)
Nitrate	NaNO$_3$	0.1371 (1000 mg NO$_3^-$/L = 225.8 mg NO$_3^-$ − N/L)
Phosphate	Na$_2$HPO$_4$†	0.1500 (1000 mg PO$_4^{3-}$/L = 326.1 mg PO$_4^{3-}$ − P/L)
Sulfate	Na$_2$SO$_4$†	0.1480 (1000 mg SO$_4^{2-}$/L = 676.3 mg SO$_4^{2-}$ − S/L)

* Do not oven-dry, but dry to constant weight in a desiccator over phosphorous pentoxide.
† Potassium salts can be used, but with corresponding modification of salt amounts.

d. Mixed working anion standard solutions: Prepare at least three different working anion standard solutions that bracket the expected sample range, from 0.1 to 50 mg/L. Add 5 mg formate/L to all standards. Use 0.1 mL standard anion solution/100 mL working anion solution (equal to 1 mg anion/L). (Above 50 mg/L each anion, chloride, bromide, nitrite, sulfate, and nitrate are no longer baseline-resolved. Analytes that are not baseline-resolved may give a low bias. If the analytes are baseline-resolved, quantitation is linear to 100 mg/L.) Store in plastic containers in the refrigerator; prepare fresh standards weekly. Figure 4140:3 shows representative electropherograms of anion standards and Table 4140:I gives the composition of the standards.

e. Calibration verification sample: Use a certified performance evaluation standard, or equivalent, within the range of the mixed working anion standard solutions analyzed as an unknown. Refer to Section 4020.

f. Analyte known-addition sample: To each sample matrix add a known amount of analyte, and use to evaluate analyte recovery.

4. Procedure

a. Capillary conditioning: Set up CIE system according to manufacturer's instructions. Rinse capillary with 100 mM NaOH

TABLE 4140:II. ANION MIGRATION TIME REPRODUCIBILITY FROM YOUDEN PAIR STANDARDS

Youden Standard	Anion Midpoint Migration Time, Average of Triplicate Samplings *min*						
	Cl$^-$	Br$^-$	NO$_2^-$	SO$_4^{2-}$	NO$_3^-$	F$^-$	PO$_4^{3-}$
1	3.132	3.226	3.275	3.405	3.502	3.761	3.906
2	3.147	3.239	3.298	3.431	3.517	3.779	3.931
3	3.138	3.231	3.283	3.411	3.497	3.771	3.925
4	3.158	3.244	3.307	3.434	3.510	3.781	3.963
5	3.184	3.271	3.331	3.435	3.551	3.787	3.981
6	3.171	3.260	3.312	3.418	3.537	3.776	3.964
7	3.191	3.272	3.315	3.437	3.544	3.773	3.978
8	3.152	3.248	3.294	3.418	3.526	3.739	3.954
SD	0.021	0.015	0.018	0.012	0.20	0.015	0.027
%RSD	0.67%	0.46%	0.55%	0.36%	0.56%	0.40%	0.68%

* Average SD = 0.018 min = 1.1 s; average %RSD = 0.53%.

for 5 min. Place fresh degassed electrolyte into both reservoirs and purge capillary with electrolyte for 3 min to remove all previous solutions and air bubbles. Apply voltage of 15 kV and note the current; if the expected 14 $\pm$ 1 μA is observed, the CIE system is ready for use. Zero UV detector to 0.000 absorbance.

b. Analysis conditions: Program CE system to apply constant current of 14 μA for the run time. Use 30 s hydrostatic sampling time for all standard and sample introduction. Analysis time is 5 min.

c. Analyte migration time calibration: Determine migration time of each analyte daily using the midrange mixed working anion standard. Perform duplicate analysis to insure migration time stability. Use the midpoint of peak width, defined as midpoint between the start and stop integration marks, as the migration time for each analyte; this accounts for the observed nonsymmetrical peak shapes. (Use of peak apex may result in analyte misidentification.) The migration order is always Cl$^-$, Br$^-$, NO$_2^-$, SO$_4^{2-}$, NO$_3^-$, F$^-$, and PO$_4^{3-}$. Dissolved HCO$_3^-$ is the last peak in the standard (see Figure 4140:1). Set analyte migration time window as 2% of the migration time determined above, except for Cl$^-$, which is set at 10%. Chloride is always the first peak and is used as the reference peak for analyte qualitative identification; identify anions on the basis of normalized migration times with respect to the reference peak, or migration time ratio. (See Figure 4140:1 and Table 4140:II.)

d. Analyte response calibration: Analyze all three mixed working anion standards in duplicate. Plot time-corrected peak area for each analyte versus concentration using a linear regression through zero. (In capillary electrophoresis peak area is a function of analyte migration time, which may change during analyses. Time-corrected peak area is a well-documented CE normalization routine, i.e., peak area divided by migration time. (NOTE: *Do not use analyte peak height.*) Calibration is accepted as linear if regression coefficient of variation, R^2, is greater than 0.995. Linearity calibration curves for anions are shown in Figures 4140:4 through 6.

e. Sample analysis: After initial calibration run samples in the following order: calibration verification sample, reagent blank, 10 unknown samples, calibration verification sample, reagent blank, etc. Filter samples containing high concentrations of suspended solids. If peaks are not baseline-resolved, dilute sample 1:5 with water and repeat analysis for unresolved analyte quantitation. Re-

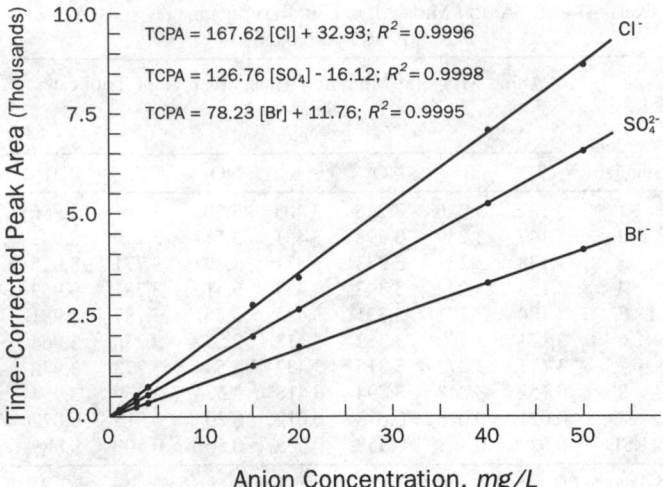

Figure 4140:4. **Linearity calibration curve for chloride, bromide, and sulfate.** Three data points were used per concentration; based on Youden pair design.

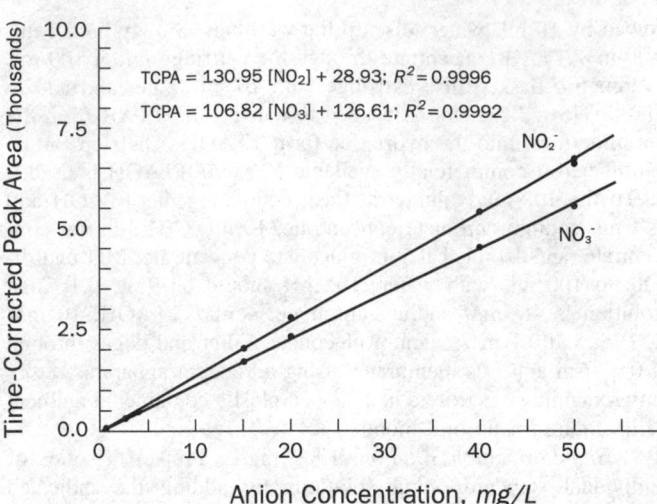

Figure 4140:6. **Linearity calibration curve for nitrite and nitrate.** Three data points were used per concentration; based on Youden pair design.

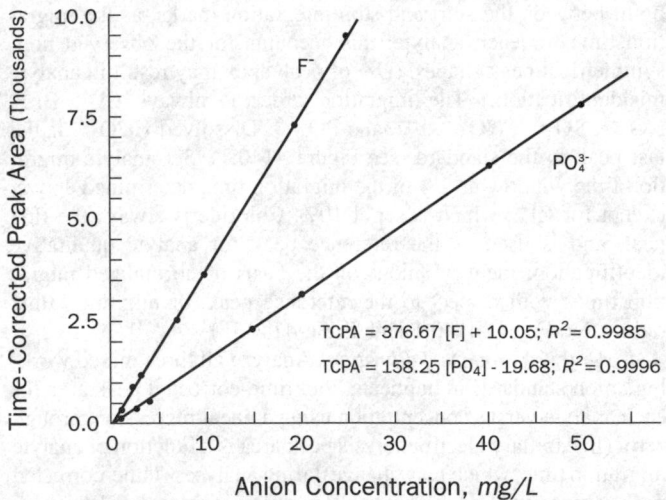

Figure 4140:5. **Linearity calibration curve for fluoride and *o*-phosphate.** Three data points were used per concentration; based on Youden pair design.

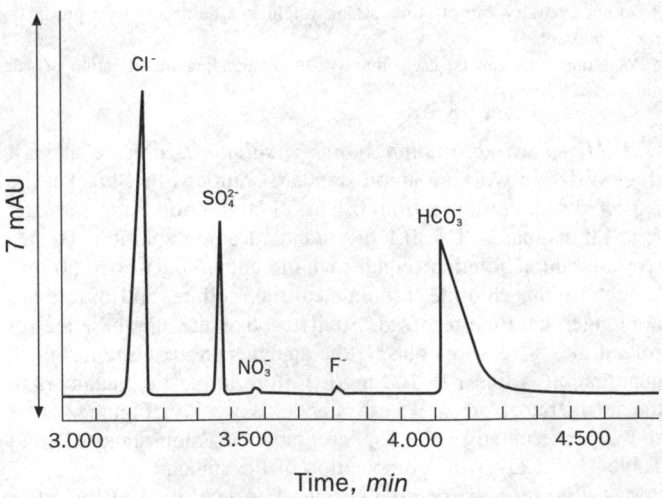

Figure 4140:7. **Electropherogram of typical drinking water.**

Chloride	= 24.72 mg/L	Fluoride	< 0.10 mg/L
Sulfate	= 7.99 mg/L	Carbonate &	
Nitrate	= 0.36 mg/L	bicarbonate	= natural

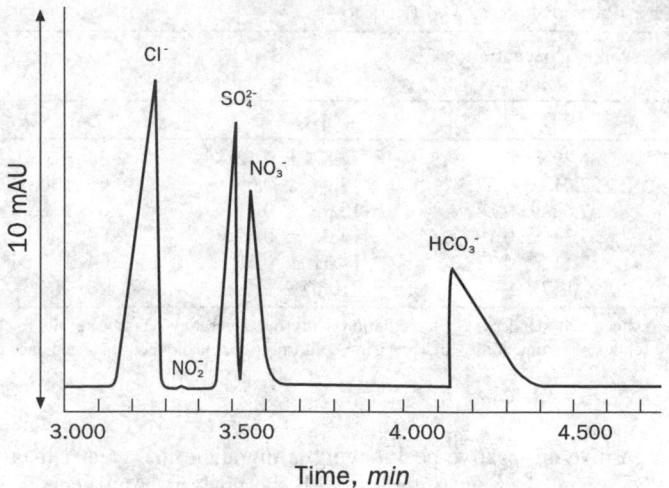

Figure 4140:8. Electropherogram of typical municipal wastewater discharge, undiluted.

Chloride	= 93.3 mg/L	Nitrate	< 40.8 mg/L
Nitrate	= 0.46 mg/L	Carbonate &	
Sulfate	= 60.3 mg/L	bicarbonate = natural	

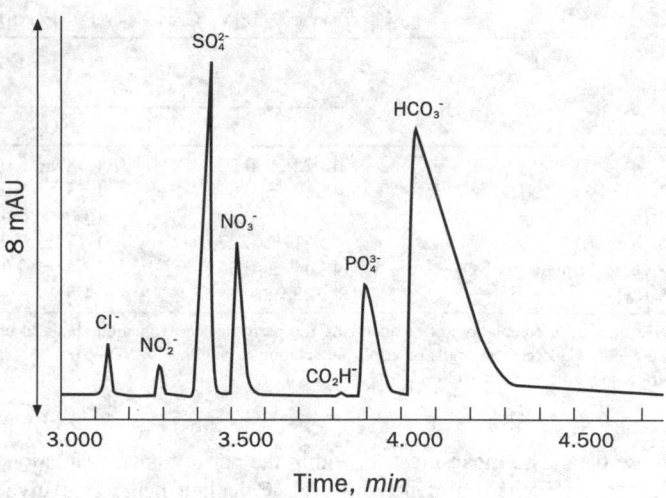

Figure 4140:9. Electropherogram of typical industrial wastewater discharge, undiluted.

Chloride	= 2.0 mg/L	Formate	< 0.05 mg/L
Nitrite	= 1.6 mg/L	Phosphate	= 12.3 mg/L
Sulfate	= 34.7 mg/L	Carbonate &	
Nitrate	= 16.5 mg/L	bicarbonate = natural	

solved analytes in the undiluted sample are considered correct quantitation. Electropherograms of typical samples are shown in Figures 4140:7 through 9.

5. Calculation

Relate the time-corrected peak area for each sample analyte with the calibration curve to determine concentration of analyte. If the sample was diluted, multiply anion concentration by the dilution factor to obtain original sample concentration, as follows:

$$C = A \times F$$

where:

C = analyte concentration in original sample, mg/L,
A = analyte concentration from calibration curve, mg/L, and
F = scale factor or dilution factor. (For a 1:5 sample dilution, $F = 5$.)

6. Quality Control

a. Analytical performance check: Unless analyst has already demonstrated ability to generate data with acceptable precision and bias by this method, proceed as follows: Analyze seven replicates of a certified performance evaluation standard containing the analytes of interest. Calculate mean and standard deviation of

TABLE 4140:III. COMPARISON OF CAPILLARY ION ELECTROPHORESIS AND OTHER METHODS

Source	Statistic	Value for Given Anion mg/L					
		Cl^-	NO_2^-	SO_4^{2-}	NO_3^-	F^-	PO_4^{3-}
Performance evaluation standard	True value	43.00	1.77	37.20	15.37	2.69	6.29
Wet chemical and ion chromatography methods	Measured mean	43.30	1.77	37.00	15.42	2.75	6.38
	Measured SD	3.09	0.07	2.24	1.15	0.26	0.21
CIE using chromate electrolyte‡	Average (n = 18)	43.34	1.64	37.11	14.41	2.64	6.34
	CIE/mean	1.003	0.927	1.003	0.935	0.959	0.993
	CIE/true value	1.008	0.927	0.996	0.938	0.981	1.008

* Purchased from APG Laboratories, June 1996; diluted 1:100 with deionized water.
† Measured result is the average from numerous laboratories using approved *Standard Methods* and EPA wet chemistry and ion chromatography methods.
‡ CIE results determined in July 1996 with proposed EPA and ASTM method, operationally identical to 4140; they are the average from four laboratories using the Youden pair standards for quantitation. These data can be considered known addition of the performance evaluation standard in reagent water; they conform to quality control acceptance limits given in Section 1020.

TABLE 4140:IV. CAPILLARY ION ELECTROPHORESIS REPRODUCIBILITY AND PRECISION*

Laboratory No.	Anion Concentration mg/L				
	Cl^-	NO_2^-	SO_4^{2-}	NO_3^-	F^-
1 ($n = 5$)	43.22 ± 0.22	1.58 ± 0.09	36.39 ± 0.33	14.57 ± 0.12	2.54 ± 0.10
2 ($n = 5$)	43.68 ± 0.61	1.58 ± 0.08	37.01 ± 0.37	13.94 ± 0.09	2.69 ± 0.02
3 ($n = 5$)	43.93 ± 0.39	1.60 ± 0.06	37.68 ± 0.24	15.05 ± 0.11	2.69 ± 0.03
4 ($n = 3$)	42.51 ± 0.22	1.78 ± 0.06	37.34 ± 0.19	14.06 ± 0.07	2.69 ± 0.02
Average (mean ± SD)	43.34 ± 0.36	1.64 ± 0.07	37.11 ± 0.28	14.41 ± 0.10	2.64 ± 0.04
%RSD	0.83%	4.5%	0.77%	0.67%	1.61%

* Results from four laboratories analyzing the performance evaluation standard using the Youden pair standards for quantitation. Only one laboratory reported results for PO_4^{3-} as 6.34 ± 0.02 mg/L on triplicate samplings yielding %RSD of 0.07%. Calculated replicate reproducibility and precision conform to the quality control acceptance limits given in Section 1020.

these data. The mean must be within the performance evaluation standard's 95% confidence interval. Calculate percent relative standard deviation (RSD) for these data as (SD × 100) / mean; % RSD should conform to acceptance limit given in Section 1020B.

b. *Calibration verification:* Analyze an independent, certified performance evaluation standard at the beginning and end of the analyses, or if many samples are analyzed, after every 10 samples. The determined analyte concentration should be within ± 10% of the true value, and the migration time of the Cl^- reference peak should be within 5% of the calibrated migration time. If the Cl^- reference peak differs by more than 5% of the calibrated migration time, repeat capillary conditioning and recalibrate before proceeding.

c. *Water blank analysis:* At the beginning of every set of analyses run a water blank to demonstrate that the water is free of analyte anions. Dissolved bicarbonate will always be observed as a positive or negative peak having a migration time greater than PO_4^{3-} and does not interfere with the analysis. Any negative peak indicates the presence of an anion impurity in the electrolyte; a positive peak indicates the presence of an impurity in the reagent water. If this is noted, discard electrolyte and prepare electrolyte and sample dilutions again with water from a different source.

d. *Analyte recovery verification:* For each sample matrix analyzed, e.g., drinking water, surface water, groundwater, or wastewater, analyze duplicate known-addition samples (¶ 3f). Analyte recoveries should conform to acceptance limits given in Section 1020B.

e. *Blind check sample:* Analyze an unknown certified performance evaluation check sample at least once every 6 months to verify method accuracy.

f. *Sample duplicates:* Analyze one or more sample duplicates every 10 samples.

TABLE 4140:V. CAPILLARY ION ELECTROPHORESIS KNOWN-ADDITION RECOVERY AND PRECISION OF PERFORMANCE EVALUATION STANDARD WITH DRINKING WATER

Variable	Value for Given Anion					
	Cl^-	NO_2^-	SO_4^{2-}	NO_3^-	F^-	PO_4^{3-}
Milford drinking water (MDW) $n = 3$, concentration as mg/L	24.72 ± 0.18	Not detected	7.99 ± 0.07	0.36 ± 0.05	Not detected	Not detected
%RSD	0.73	—	0.91	13.3	—	—
Performance evaluation standard (PES), concentration as mg/L	43.00	1.77	37.20	15.37	2.69	6.29
MDW + PES,* $n = 3$, concentration as mg/L	66.57 ± 0.34	1.74 ± 0.03	45.19 ± 0.17	15.42 ± 0.12	2.62 ± 0.07	5.55 ± 0.31
%RSD	0.51%	1.85%	0.38%	0.79%	2.69%	5.52%
% Recovery	97.9%	98.3%	100.2%	98.1%	97.4%	88.2%

* Performance evaluation standard diluted 1:100 with Milford drinking water. Calculated analyte recovery and precision conform to the quality control acceptance limits given in Section 1020.

7. Precision and Bias

Table 4140:III compares results of capillary ion electrophoresis with those of other approved methods. Precision and bias data are given in Tables 4140:IV and 4140:V. Comparison of other methods and capillary ion electrophoresis for wastewater effluent, drinking water, and landfill leachates are given in Table 4140:VI.

8. Bibliography

ROMANO, J. & J. KROL. 1993. Capillary electrophoresis, an environmental method for the determination of anions in water. *J. Chromatogr.* 640:403.

JANDIK, P. & G. BONN. 1993. Capillary Electrophoresis of Small Molecules and Ions. VCH Publishers, New York, N.Y.

TABLE 4140:VI. COMPARISON OF CAPILLARY ION ELECTROPHORESIS WITH CHROMATE ELECTROLYTE WITH OTHER METHODS FOR THE DETERMINATION OF ANIONS

| Anion | Matrix | Sample No. | Concentration by Given Method* mg/L | | |
			Wet Chemical or Other Method†	IC‡	CIE
Chloride	Effluent	1	—	149	147
		2	—	162	161
		3	—	152	151
		4	—	139	139
		5	—	111	110
		6	—	109	107
		7	—	3.6	3.5
	Drinking water	1	5.5	5.1	5.0
		2	5.5	5.0	5.0
		3	5.3	5.2	5.2
		4	5.5	5.1	5.1
		5	5.3	5.0	5.1
		6	5.3	4.9	4.9
		7	5.5	4.9	4.9
	Landfill leachate	1	0.1	<0.1	ND
		2	230	245	240
Fluoride	Effluent	1	1.7	1.2	1.5
		2	0.9	0.6	0.6
		3	0.8	0.5	0.6
		4	0.8	0.4	0.7
		5	0.9	0.5	0.8
		6	0.9	0.5	0.7
		7	<0.1	ND	<0.1
	Drinking water	1	1.2	0.9	0.9
		2	1.3	0.9	0.9
		3	1.3	0.9	0.9
		4	1.3	0.9	0.9
		5	1.3	0.9	0.9
		6	0.9	0.6	0.6
		7	1.3	0.9	0.9
	Landfill leachate	1	<0.2	ND	ND
		2	16	10.6	10.9

TABLE 4140:VI. CONT.

Anion	Matrix	Sample No.	Concentration by Given Method* mg/L		
			Wet Chemical or Other Method†	IC‡	CIE
Sulfate	Effluent	1	98	87.5	86.4
		2	110	95.3	95.9
		3	130	118	115
		4	130	139	136
		5	110	113	110
		6	100	107	106
		7	6	5.6	5.8
	Drinking water	1	6	5.8	6.0
		2	6	5.8	6.0
		3	6	5.9	6.1
		4	6	5.9	6.1
		5	5	5.8	6.2
		6	4	3.0	3.4
		7	5	5.8	6.1
	Landfill leachate	1	<1	ND	ND
		2	190	211	201
Nitrite & nitrate§ (as N)	Effluent	1	0.3	ND	ND
		2	—	ND	ND
		3	—	ND	ND
		4	—	ND	0.5
		5	—	2.1	2.4
		6	2.4	1.9	2.2
		7	0.7	0.3	0.4
	Drinking water	1	0.6	0.3	0.4
		2	0.6	0.3	0.4
		3	0.4	0.3	0.4
		4	0.6	0.3	0.3
		5	0.6	0.3	0.4
		6	0.3	0.1	0.1
		7	0.5	0.3	0.4
	Landfill leachate	1	—	ND	ND
		2	—	ND	ND
Orthophosphate (as P)	Effluent	1	3.4	ND	2.8
		2	4.9	ND	4.4
		3	4.7	ND	4.5
		4	5.3	ND	4.2
		5	3.0	ND	3.0
		6	2.9	ND	2.3
		7	<0.1	ND	0.04
	Drinking water	1	<0.1	ND	ND
		2	<0.1	ND	ND
		3	—	ND	ND
		4	<0.1	ND	ND
		5	<0.1	ND	ND
		6	—	ND	ND
		7	—	ND	ND
	Landfill leachate	1	<0.1	ND	0.1
		2	2.2	1.6	1.4

* — = test not performed; ND = not detected.
† Methods used were: chloride—iodometric method (4500-Cl.C); fluoride—ion selective electrode method (4500-F.C); sulfate—turbidimetric method (4500-SO$_4$.E); nitrite + nitrate (total)—cadmium reduction method (4500-NO$_3$.E); orthophosphate—ascorbic acid method (4500-P.E).
‡ Single-column ion chromatography with direct conductivity detection (4110C).
§ Each technique gave separate nitrite and nitrate values; because of their interconvertability, results were added for comparison purposes.

4500-B BORON*

4500-B A. Introduction

1. Occurrence and Significance

Boron (B) is the first element in Group IIIA of the periodic table; it has an atomic number of 5, an atomic weight of 10.81, and a valence of 3. The average abundance of B in the earth's crust is 9 ppm; in soils it is 18 to 63 ppm; in streams it is 10 μg/L; and in groundwaters it is 0.01 to 10 mg/L. The most important mineral is borax, which is used in the preparation of heat-resistant glasses, detergents, porcelain enamels, fertilizers, and fiberglass.

The most common form of boron in natural waters is H_3BO_3. Although boron is an element essential for plant growth, in excess of 2.0 mg/L in irrigation water, it is deleterious to certain plants and some plants may be affected adversely by concentrations as low as 1.0 mg/L (or even less in commercial greenhouses). Drinking waters rarely contain more than 1 mg B/L and generally less than 0.1 mg/L, concentrations considered innocuous for human consumption. Seawater contains approximately 5 mg B/L and this element is found in saline estuaries in association with other seawater salts.

_____
* Approved by Standard Methods Committee, 1993.

The ingestion of large amounts of boron can affect the central nervous system. Protracted ingestion may result in a clinical syndrome known as borism.

2. Selection of Method

Preferably, perform analyses by the inductively coupled plasma method (3120). The inductively coupled plasma/mass spectrometric method (3125) also may be applied successfully in most cases (with lower detection limits), even though boron is not specifically listed as an analyte in the method.

The curcumin method (B) is applicable in the 0.10- to 1.0-mg/L range, while the carmine method (C) is suitable for the determination of boron concentration in the 1- to 10-mg/L range. The range of these methods can be extended by dilution or concentration of the sample.

3. Sampling and Storage

Store samples in polyethylene bottles or alkali-resistant, boron-free glassware.

4500-B B. Curcumin Method

1. General Discussion

a. Principle: When a sample of water containing boron is acidified and evaporated in the presence of curcumin, a red-colored product called rosocyanine is formed. The rosocyanine is taken up in a suitable solvent and the red color is compared with standards visually or photometrically.

b. Interference: NO_3^--N concentrations above 20 mg/L interfere. Significantly high results are possible when the total of calcium and magnesium hardness exceeds 100 mg/L as calcium carbonate ($CaCO_3$). Moderate hardness levels also can cause a considerable percentage error in the low boron range. This interference springs from the insolubility of the hardness salts in 95% ethanol and consequent turbidity in the final solution. Filter the final solution or pass the original sample through a column of strongly acidic cation-exchange resin in the hydrogen form to remove interfering cations. The latter procedure permits application of the method to samples of high hardness or solids content. Phosphate does not interfere.

c. Minimum detectable quantity: 0.2 μg B.

2. Apparatus

a. Colorimetric equipment: One of the following is required:

1) *Spectrophotometer,* for use at 540 nm, with a minimum light path of 1 cm.

2) *Filter photometer,* equipped with a green filter having a maximum transmittance near 540 nm, with a minimum light path of 1 cm.

b. Evaporating dishes, 100- to 150-mL capacity, of high-silica glass,* platinum, or other suitable material.

c. Water bath, set at 55 $\pm$ 2°C.

d. Glass-stoppered volumetric flasks, 25- and 50-mL capacity.

e. Ion-exchange column, 50 cm long by 1.3 cm in diameter.

3. Reagents

Store all reagents in polyethylene or boron-free containers.

a. Stock boron solution: Dissolve 571.6 mg anhydrous boric acid, H_3BO_3, in distilled water and dilute to 1000 mL; 1.00 mL = 100 μg B. Because H_3BO_3 loses weight on drying at 105°C, use a reagent meeting ACS specifications and keep the bottle tightly stoppered to prevent entrance of atmospheric moisture.

b. Standard boron solution: Dilute 10.00 mL stock boron solution to 1000 mL with distilled water; 1.00 mL = 1.00 μg B.

c. Curcumin reagent: Dissolve 40 mg finely ground curcumin† and 5.0 g oxalic acid in 80 mL 95% ethyl alcohol. Add 4.2 mL conc HCl, make up to 100 mL with ethyl alcohol in a 100-mL volumetric flask, and filter if reagent is turbid (isopropyl alcohol,

_____
* Vycor, manufactured by Corning Glass Works, or equivalent.
† Eastman No. 1179 or equivalent.

95%, may be used in place of ethyl alcohol). This reagent is stable for several days if stored in a refrigerator.

d. Ethyl or isopropyl alcohol, 95%.

e. Reagents for removal of high hardness and cation interference:

1) *Strongly acidic cation-exchange resin.*

2) *Hydrochloric acid, HCl, 1 + 5.*

4. Procedure

a. Precautions: Closely control such variables as volumes and concentrations of reagents, as well as time and temperature of drying. Use evaporating dishes identical in shape, size, and composition to insure equal evaporation time because increasing the time increases intensity of the resulting color.

b. Preparation of calibration curve: Pipet 0 (blank), 0.25, 0.50, 0.75, and 1.00 µg boron into evaporating dishes of the same type, shape, and size. Add distilled water to each standard to bring total volume to 1.0 mL. Add 4.0 mL curcumin reagent to each and swirl gently to mix contents thoroughly. Float dishes on a water bath set at 55 ± 2°C and let them remain for 80 min, which is usually sufficient for complete drying and removal of HCl. Keep drying time constant for standards and samples. After dishes cool to room temperature, add 10 mL 95% ethyl alcohol to each dish and stir gently with a polyethylene rod to insure complete dissolution of the red-colored product.

Wash contents of dish into a 25-mL volumetric flask, using 95% ethyl alcohol. Make up to mark with 95% ethyl alcohol and mix thoroughly by inverting. Read absorbance of standards and samples at a wavelength of 540 nm after setting reagent blank at zero absorbance. The calibration curve is linear from 0 to 1.00 µg boron. Make photometric readings within 1 h of drying samples.

c. Sample treatment: For waters containing 0.10 to 1.00 mg B/L, use 1.00 mL sample. For waters containing more than 1.00 mg B/L, make an appropriate dilution with boron-free distilled water, so that a 1.00-mL portion contains approximately 0.50 µg boron.

Pipet 1.00 mL sample or dilution into an evaporating dish. Unless the calibration curve is being determined at the same time, prepare a blank and a standard containing 0.50 µg boron and run in conjunction with the sample. Proceed as in ¶ 4b, beginning with "Add 4.0 mL curcumin reagent...." If the final solution is turbid, filter through filter paper‡ before reading absorbance. Calculate boron content from calibration curve.

d. Visual comparison: The photometric method may be adapted to visual estimation of low boron concentrations, from 50 to 200 µg/L, as follows: Dilute the standard boron solution 1 + 3 with distilled water; 1.00 mL = 0.20 µg B. Pipet 0, 0.05, 0.10, 0.15, and 0.20 µg boron into evaporating dishes as indicated in ¶ 4b. At the same time add an appropriate volume of sample (1.00 mL or portion diluted to 1.00 mL) to an identical evaporating dish.

‡ Whatman No. 30 or equivalent.

The total boron should be between 0.05 and 0.20 µg. Proceed as in ¶ 4b, beginning with "Add 4.0 mL curcumin reagent...." Compare color of samples with standards within 1 h of drying samples.

e. Removal of high hardness and cation interference: Prepare an ion-exchange column of approximately 20 cm × 1.3 cm diam. Charge column with a strongly acidic cation-exchange resin. Backwash column with distilled water to remove entrained air bubbles. Keep the resin covered with liquid at all times. Pass 50 mL 1 + 5 HCl through column at a rate of 0.2 mL acid/mL resin in column/min and wash column free of acid with distilled water.

Pipet 25 mL sample, or a smaller sample of known high boron content diluted to 25 mL, onto the resin column. Adjust rate of flow to about 2 drops/s and collect effluent in a 50-mL volumetric flask. Wash column with small portions of distilled water until flask is filled to mark. Mix and transfer 2.00 mL into evaporating dish. Add 4.0 mL curcumin reagent and complete the analysis as described in ¶ 4b preceding.

5. Calculation

Use the following equation to calculate boron concentration from absorbance readings:

$$\text{mg B/L} = \frac{A_2 \times C}{A_1 \times S}$$

where:

A_1 = absorbance of standard,
A_2 = absorbance of sample,
C = µg B in standard taken, and
S = mL sample.

6. Precision and Bias

A synthetic sample containing 240 µg B/L, 40 µg As/L, 250 µg Be/L, 20 µg Se/L, and 6 µg V/L in distilled water was analyzed in 30 laboratories by the curcumin method with a relative standard deviation of 22.8% and a relative error of 0%.

7. Bibliography

SILVERMAN, L. & K. TREGO. 1953. Colorimetric microdetermination of boron by the curcumin-acetone solution method. *Anal. Chem.* 25: 1264.

DIRLE, W.T., E. TRUOG & K.C. BERGER. 1954. Boron determination in soils and plants—Simplified curcumin procedure. *Anal. Chem.* 26: 418.

LUKE, C.L. 1955. Determination of traces of boron in silicon, germanium, and germanium dioxide. *Anal. Chem.* 27:1150.

LISHKA, R.J. 1961. Comparison of analytical procedures for boron. *J. Amer. Water Works Assoc.* 53:1517.

BUNTON, N.G. & B.H. TAIT. 1969. Determination of boron in waters and effluents using curcumin. *J. Amer. Water Works Assoc.* 61:357.

4500-B C. Carmine Method

1. General Discussion

a. Principle: In the presence of boron, a solution of carmine or carminic acid in concentrated sulfuric acid changes from a bright red to a bluish red or blue, depending on the concentration of boron present.

b. Interference: The ions commonly found in water and wastewater do not interfere.

c. Minimum detectable quantity: 2 μg B.

2. Apparatus

Colorimetric equipment: One of the following is required:

a. Spectrophotometer, for use at 585 nm, with a minimum light path of 1 cm.

b. Filter photometer, equipped with an orange filter having a maximum transmittance near 585 nm, with a minimum light path of 1 cm.

3. Reagents

Store all reagents in polyethylene or boron-free containers.

a. Standard boron solution: Prepare as directed in Method B, ¶ 3b.

b. Hydrochloric acid, HCl, conc and 1 + 11.

c. Sulfuric acid, H_2SO_4, conc.

d. Carmine reagent: Dissolve 920 mg carmine N.F. 40, or carminic acid, in 1 L conc H_2SO_4. (If unable to zero spectrophotometer, dilute carmine 1 + 1 with conc H_2SO_4 to replace above reagent.)

4. Procedure

a. Preliminary sample treatment: If sample contains less than 1 mg B/L, pipet a portion containing 2 to 20 μg B into a platinum dish, make alkaline with 1*N* NaOH plus a slight excess, and evaporate to dryness on a steam or hot water bath. If necessary, destroy any organic material by ignition at 500 to 550°C. Acidify cooled residue (ignited or not) with 2.5 mL 1 + 11 HCl and triturate with a rubber policeman to dissolve. Centrifuge if necessary to obtain a clear solution. Pipet 2.00 mL clear concentrate into a small flask or 30-mL test tube. Treat reagent blank identically.

b. Color development: Prepare a series of boron standard solutions (100, 250, 500, 750, and 1000 μg) in 100 mL with distilled water. Pipet 2.00 mL of each standard solution into a small flask or 30-mL test tube.

Treat blank and calibration standards exactly as the sample. Add 2 drops (0.1 mL) conc HCl, carefully introduce 10.0 mL conc H_2SO_4, mix, and let cool to room temperature. Add 10.0 mL carmine reagent, mix well, and after 45 to 60 min measure absorbance at 585 nm in a cell of 1-cm or longer light path, using the blank as reference.

To avoid error, make sure that no bubbles are present in the optical cell while photometric readings are being made. Bubbles may appear as a result of incomplete mixing of reagents. Because carmine reagent deteriorates, check calibration curve daily.

5. Calculation

$$\text{mg B/L} = \frac{\mu\text{g B (in approx. 22 mL final volume)}}{\text{mL sample}} \times 1.25$$

6. Precision and Bias

A synthetic sample containing 180 μg B/L, 50 μg As/L, 400 μg Be/L, and 50 μg Se/L in distilled water was analyzed in nine laboratories by the carmine method with a relative standard deviation of 35.5% and a relative error of 0.6%.

7. Bibliography

HATCHER, J.T. & L.V. WILCOX. 1950. Colorimetric determination of boron using carmine. *Anal. Chem.* 22:567.

4500-Br⁻ BROMIDE*

4500-Br⁻ A. Introduction

1. Occurrence

Bromide occurs in varying amounts in ground and surface waters in coastal areas as a result of seawater intrusion and sea-spray-affected precipitation. The bromide content of ground waters and stream baseflows also can be affected by connate water. Industrial and oil-field brine discharges can contribute to the bromide in water sources. Under normal circumstances, the bromide content of most drinking waters is small, seldom exceeding 1 mg/L. Even levels of <100 μg/L can lead to formation of bromate or brominated by-products in disinfected waters.

2. Selection of Method

Described here are a colorimetric procedure suitable for the determination of bromide in most drinking waters and a flow injection analysis method. Bromide preferably is determined by the ion chromatography method (4110) or by capillary ion electrophoresis (4140).

* Approved by Standard Methods Committee, 1997.

4500-Br⁻ B. Phenol Red Colorimetric Method

1. General Discussion

a. Principle: When a sample containing bromide ions (Br⁻) is treated with a dilute solution of chloramine-T in the presence of phenol red, the oxidation of bromide and subsequent bromination of the phenol red occur readily. If the reaction is buffered to pH 4.5 to 4.7, the color of the brominated compound will range from reddish to violet, depending on the bromide concentration. Thus, a sharp differentiation can be made among various concentrations of bromide. The concentration of chloramine-T and timing of the reaction before dechlorination are critical.

b. Interference: Most materials present in ordinary tap water do not interfere, but oxidizing and reducing agents and higher concentrations of chloride and bicarbonate can interfere. Free chlorine in samples should be destroyed as directed in Section 5210B.4*e*2); analyze bromide in a portion of dechlorinated sample. Addition of substantial chloride to the pH buffer solution (see ¶ 3*a* below) can eliminate chloride interference for waters with very low bromide/chloride ratios, such as those affected by dissolved road salt. Small amounts of dissolved iodide do not interfere, but small concentrations of ammonium ion interfere substantially. Sample dilution may reduce interferences to acceptable levels for some saline and waste waters. However, if two dilutions differing by a factor of at least five do not give comparable values, the method is inapplicable. Bromide concentration in diluted samples must be within the range of the method (0.1 to 1 mg/L).

c. Minimum detectable concentration: 0.1 mg Br⁻/L.

2. Apparatus

a. Colorimetric equipment: One of the following is required:

1) *Spectrophotometer,* for use at 590 nm, providing a light path of at least 2 cm.

2) *Filter photometer,* providing a light path of at least 2 cm and equipped with an orange filter having a maximum transmittance near 590 nm.

3) *Nessler tubes,* matched, 100-mL, tall form.

b. Acid-washed glassware: Wash all glassware with 1 + 6 HNO_3 and rinse with distilled water to remove all trace of adsorbed bromide.

3. Reagents

a. Acetate buffer solution: Dissolve 90 g NaCl and 68 g sodium acetate trihydrate, $NaC_2H_3O_2 \cdot 3H_2O$, in distilled water. Add 30 mL conc (glacial) acetic acid and make up to 1 L. The pH should be 4.6 to 4.7.

b. Phenol red indicator solution: Dissolve 21 mg phenolsulfonephthalein sodium salt and dilute to 100 mL with distilled water.

c. Chloramine-T solution: Dissolve 500 mg chloramine-T, sodium p-toluenesulfonchloramide, and dilute to 100 mL with distilled water. Store in a dark bottle and refrigerate.

d. Sodium thiosulfate, 2M: Dissolve 49.6 g $Na_2S_2O_3 \cdot 5H_2O$ or 31.6 g $Na_2S_2O_3$ and dilute to 100 mL with distilled water.

e. Stock bromide solution: Dissolve 744.6 mg anhydrous KBr in distilled water and make up to 1000 mL; 1.00 mL = 500 µg Br⁻.

f. Standard bromide solution: Dilute 10.00 mL stock bromide solution to 1000 mL with distilled water; 1.00 mL = 5.00 µg Br⁻.

4. Procedure

a. Preparation of bromide standards: Prepare at least six standards, 0, 0.20, 0.40, 0.60, 0.80 and 1.00 mg Br⁻/L, by diluting 0.0, 2.00, 4.00, 6.00, 8.00, and 10.00 mL standard bromide solution to 50.00 mL with distilled water. Treat standards the same as samples in ¶ 4*b*.

b. Treatment of sample: Add 2 mL buffer solution, 2 mL phenol red solution, and 0.5 mL chloramine-T solution to 50.0 mL sample or two separate sample dilutions (see 1*b* above) such that the final bromide concentration is in the range of 0.1 to 1.0 mg Br⁻/L. Mix thoroughly immediately after each addition. Exactly 20 min after adding chloramine-T, dechlorinate by adding, with mixing, 0.5 mL $Na_2S_2O_3$ solution. Compare visually in nessler tubes against bromide standards prepared simultaneously, or preferably read in a photometer at 590 nm against a reagent blank. Determine the bromide values from a calibration curve of mg Br⁻/L (in 55 mL final volume) against absorbance. A 2.54-cm light path yields an absorbance value of approximately 0.36 for 1 mg Br⁻/L.

5. Calculation

$$mg\ Br^-/L = mg\ Br^-/L \text{ (from calibration curve)} \times \text{dilution factor (if any)}.$$
Results are based on 55 mL final volume for samples and standards.

6. Bibliography

STENGER, V.A. & I.M. KOLTHOFF. 1935. Detection and colorimetric estimation of microquantities of bromide. *J. Amer. Chem. Soc.* 57:831.

HOUGHTON, G.U. 1946. The bromide content of underground waters. *J. Soc. Chem. Ind.* (London) 65:227.

GOLDMAN, E. & D. BYLES. 1959. Suggested revision of phenol red method for bromide. *J. Amer. Water Works Assoc.* 51:1051.

SOLLO, F.W., T.E. LARSON & F.F. McGURK. 1971. Colorimetric methods for bromine. *Environ. Sci. Technol.* 5:240.

WRIGHT, E.R., R.A. SMITH & F.G. MESSICK. 1978. *In* D.F. Boltz & J.A. Howell, eds. Colorimetric Determination of Nonmetals, 2nd ed. Wiley-Interscience, New York, N.Y.

BASEL, C.L., J.D. DEFREESE & D.O. WHITTEMORE. 1982. Interferences in automated phenol red method for determination of bromide in water. *Anal. Chem.* 54:2090.

4500-Br⁻ C.   (Reserved)

4500-Br⁻ D.   Flow Injection Analysis (PROPOSED)

1. General Discussion

a. Principle: Bromide is oxidized to bromine by chloramine-T, followed by substitution of bromine on phenol red to produce bromphenol blue. The absorbance measured at 590 nm is proportional to the concentration of bromide in the sample. Sodium thiosulfate is added to reduce interference from chloride.

This method is suitable for the determination of bromide in waters containing up to 20 000 mg Cl⁻/L, including drinking, ground, and surface waters, and domestic and industrial wastes. The method determines total bromide, or, if the sample is filtered through a 0.45-μm-pore-size filter, the result is called "dissolved bromide." The difference between total bromide and dissolved bromide is called "insoluble bromide."

Also see Section 4500-Br⁻.A and 4130, Flow Injection Analysis (FIA).

b. Interferences: Remove large or fibrous particulates by filtering sample through glass wool. Guard against contamination from reagents, water, glassware, and the sample preservation process.

Chloride interference is reduced by the addition of sodium thiosulfate. Chloramine-T dissociates in aqueous solution to form hypochlorous acid, which can then react with chloride, causing substitution of chloride at positions ortho to the hydroxy groups on phenol red, just as in bromination. Sodium thiosulfate reacts with chlorine to reduce this interferent to a selectivity (ratio of analyte to interferent concentration) of >28 000.

2. Apparatus

Flow injection analysis equipment consisting of:
a. FIA injection valve with sample loop or equivalent.
b. Multichannel proportioning pump.
c. FIA manifold with flow cell (Figure 4500-Br⁻:1). Relative flow rates only are shown. Tubing volumes are given as an example only; they may be scaled down proportionally. Use manifold tubing of an inert material such as TFE.*
d. Absorbance detector, 590 nm, 10-nm bandpass.
e. Valve control and data acquisition system.

3. Reagents

Use reagent water (>10 megohm) to prepare carrier and all solutions. As an alternative to preparing reagents by weight/weight, use weight/volume.
a. Chloramine-T: To a tared 1-L container add 0.40 g chloramine-T hydrate (mol wt 227.65) and 999 g water. Cap and invert container to dissolve. Discard after 1 week.
b. Phenol red: To a tared 1-L container add 929 g water and 30.0 g glacial acetic acid. Swirl contents of container. Add 41.0 g

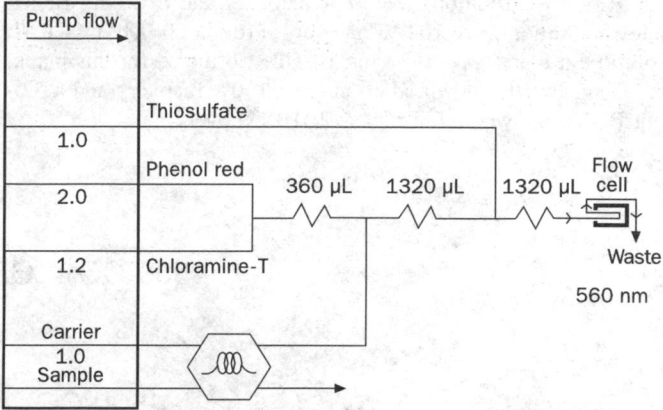

Figure 4500-Br⁻:1. FIA bromide manifold.

sodium acetate and swirl container until it is dissolved. Add 0.040 g phenol red. Mix with a magnetic stirrer. Discard after 1 week.
c. Thiosulfate: To a tared 1-L container, add 724 g water and 500 g sodium thiosulfate pentahydrate, $Na_2S_2O_3 \cdot 5H_2O$. Dissolve by adding the solid slowly while stirring. The solid should be completely dissolved within 30 min. Gentle heating may be required. Discard after 1 week.
d. Stock bromide standard, 100.0 mg Br⁻/L: To a 1-L volumetric flask add 0.129 g sodium bromide, NaBr. Dissolve in sufficient water, dilute to mark, and invert to mix.
e. Stock bromide standard, 10.0 mg Br⁻/L: To a 500-mL volumetric flask add 50 mL stock standard (¶ 3*d*). Dilute to mark and invert to mix. Prepare fresh monthly.
f. Standard bromide solutions: Prepare bromide standards for the calibration curve in the desired concentration range, using the stock standard (¶ *e*), and diluting with water.

4. Procedure

Set up a manifold equivalent to that in Figure 4500-Br⁻:1 and follow method supplied by manufacturer, or laboratory standard operating procedure for this method. Follow quality control guidelines outlined in Section 4020.

5. Calculations

Prepare standard curves by plotting absorbance of standards processed through the manifold vs. bromide concentration. The calibration curve gives a good fit to a second-order polynomial.

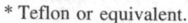 * Teflon or equivalent.

6. Precision and Bias

a. Precision: With a 300-μL sample loop, ten replicates of a 5.0-mg Br^-/L standard gave a mean of 5.10 mg Br^-/L and a relative standard deviation of 0.73%.

b. Bias: With a 300-μ/L sample loop, solutions of sodium chloride were fortified in triplicate with bromide and mean blanks and recoveries were measured. From a 10 000-mg Cl^-/L solution, a blank gave 0.13 mg Br^-/L. Corrected for this blank, a 1.0-mg Br^-/L known addition gave 98% recovery and a 5.0-mg Br^-/L known addition gave 102 % recovery. From a 20 000 mg Cl^-/L solution, a blank gave 0.27 mg Br^-/L. Corrected for this blank, a 1.0-mg Br^-/L known addition gave 100% recovery and a 5.0-mg Br^-/L known addition gave 101% recovery.

c. MDL: Using a published MDL method[1] and a 300-μL sample loop, analysts ran 21 replicates of a 0.5-mg Br^-/L standard. These gave a mean of 0.468 mg Br^-/L, a standard deviation of 0.030 mg Br^-/L, and an MDL of 0.07 mg Br^-/L. A lower MDL may be obtained by increasing the sample loop volume and increasing the ratio of carrier flow rate to reagents flow rate.

7. Reference

1. U.S. Environmental Protection Agency. 1989. Definition and procedure for the determination of method detection limits. Appendix B to CFR 136 rev. 1.11 amended June 30, 1986. 49 CFR 43430.
2. Anagnostopoulou, P. & M. Koupparis. 1986. Automated FIA phenol red method for determination of bromide. *Anal. Chem.* 58:322.

4500-CO₂ CARBON DIOXIDE*

4500-CO₂ A. Introduction

1. Occurrence and Significance

Surface waters normally contain less than 10 mg free carbon dioxide (CO_2) per liter while some groundwaters may easily exceed that concentration. The CO_2 content of a water may contribute significantly to corrosion. Recarbonation of a supply during the last stages of water softening is a recognized treatment process. The subject of saturation with respect to calcium carbonate is discussed in Section 2330.

2. Selection of Method

A nomographic and a titrimetric method are described for the estimation of free CO_2 in drinking water. The titration may be performed potentiometrically or with phenolphthalein indicator. Properly conducted, the more rapid, simple indicator method is satisfactory for field tests and for control and routine applications if

it is understood that the method gives, at best, only an approximation.

The nomographic method (B) usually gives a closer estimation of the total free CO_2 when the pH and alkalinity determinations are made immediately and correctly at the time of sampling. The pH measurement preferably should be made with an electrometric pH meter, properly calibrated with standard buffer solutions in the pH range of 7 to 8. The error resulting from inaccurate pH measurements grows with an increase in total alkalinity. For example, an inaccuracy of 0.1 in the pH determination causes a CO_2 error of 2 to 4 mg/L in the pH range of 7.0 to 7.3 and a total alkalinity of 100 mg $CaCO_3$/L. In the same pH range, the error approaches 10 to 15 mg/L when the total alkalinity is 400 mg as $CaCO_3$/L.

Under favorable conditions, agreement between the titrimetric and nomographic methods is reasonably good. When agreement is not precise and the CO_2 determination is of particular importance, state the method used.

The calculation of the total CO_2, free and combined, is given in Method D.

* Approved by Standard Methods Committee, 1997.

4500-CO₂ B. Nomographic Determination of Free Carbon Dioxide and the Three Forms of Alkalinity*

1. General Discussion

Diagrams and nomographs enable the rapid calculation of the CO_2, bicarbonate, carbonate, and hydroxide content of natural and treated waters. These graphical presentations are based on equations relating the ionization equilibria of the carbonates and water.

If pH, total alkalinity, temperature, and total mineral content are known, any or all of the alkalinity forms and CO_2 can be determined nomographically.

A set of charts, Figures 4500-CO₂:1–4,† is presented for use

* See also Alkalinity, Section 2320.

† Copies of the nomographs in Figures 4500-CO₂:1–4, enlarged to several times the size shown here, may be obtained as a set from Standard Methods Manager, The American Water Works Association, 6666 West Quincy Ave., Denver, CO 80235.

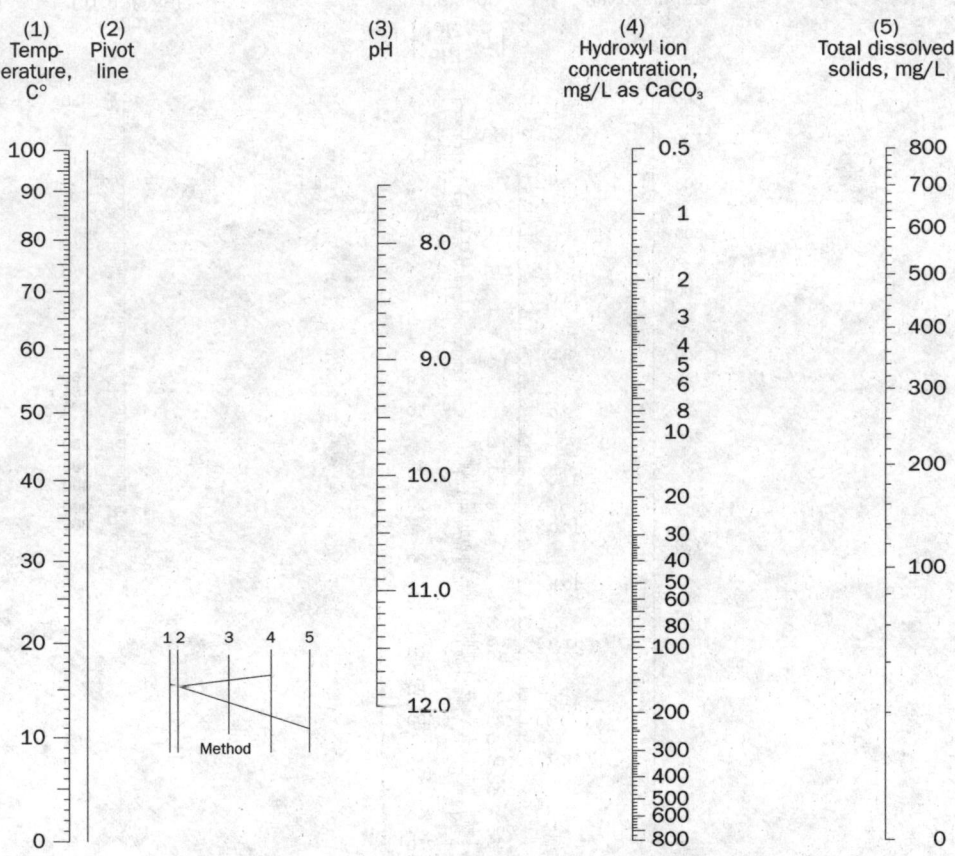

Figure 4500-CO₂:1. Nomograph for evaluation of hydroxide ion concentration. To use: Align temperature (Scale 1) and total dissolved solids (Scale 5); pivot on Line 2 to proper pH (Scale 3); read hydroxide ion concentration, as mg CaCO₃/L, on Scale 4. (Example: For 13°C temperature, 240 mg total dissolved solids/L, pH 9.8, the hydroxide ion concentration is found to be 1.4 mg as CaCO₃/L.)

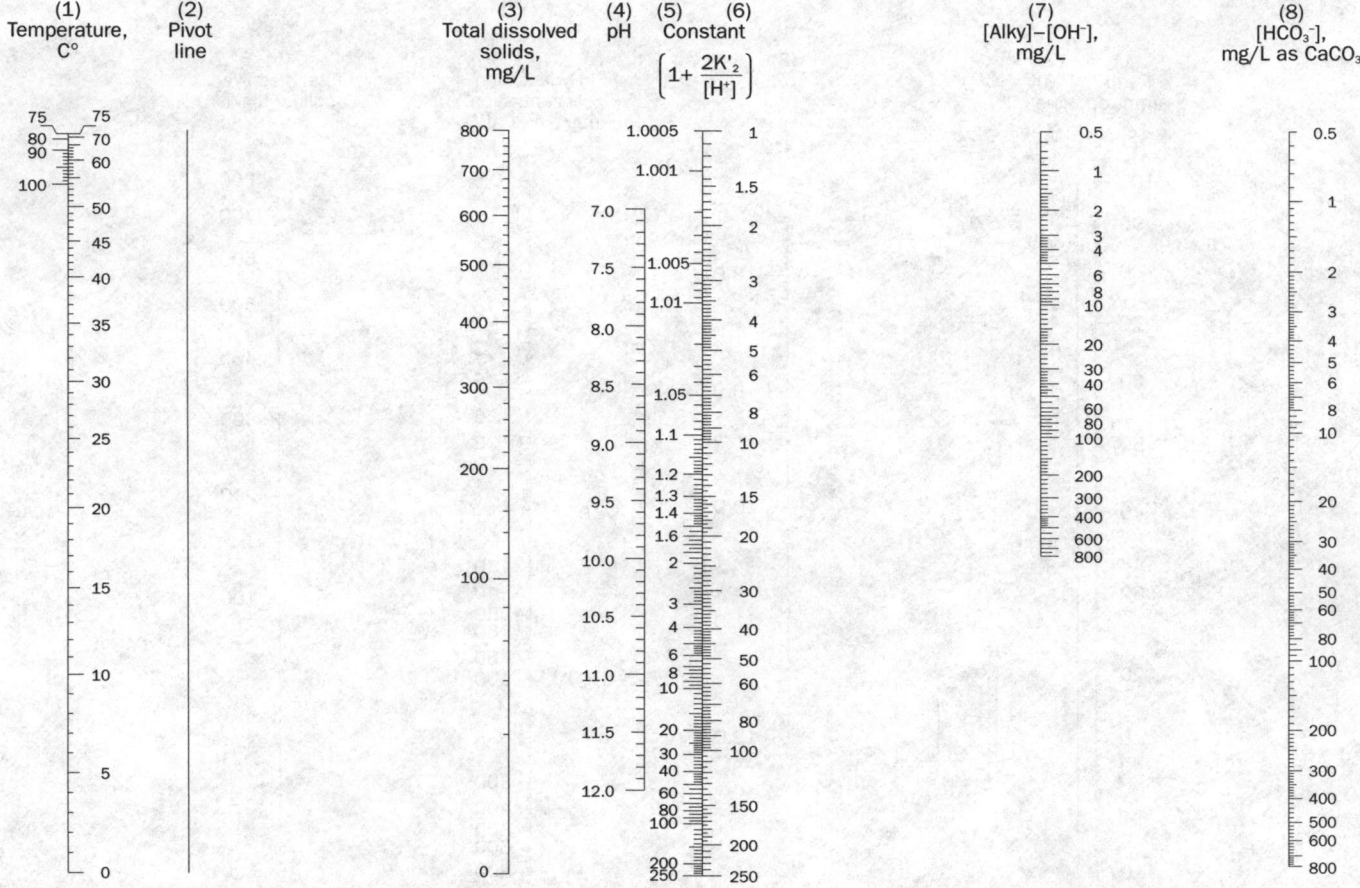

Figure 4500-CO₂:2. Nomograph for evaluation of bicarbonate alkalinity. To use: Align temperature (Scale 1) and total dissolved solids (Scale 3); pivot on Line 2 to proper pH (Scale 4) and read constant on Scale 5; locate constant on Scale 6 and align with nonhydroxide alkalinity (found with aid of Figure 4500-CO₂:1) on Scale 7; read bicarbonate alkalinity on Scale 8. (Example: For 13°C temperature, 240 mg total dissolved solids/L, pH 9.8, and 140 mg alkalinity/L, the bicarbonate content is found to be 90 mg as CaCO₃/L.)

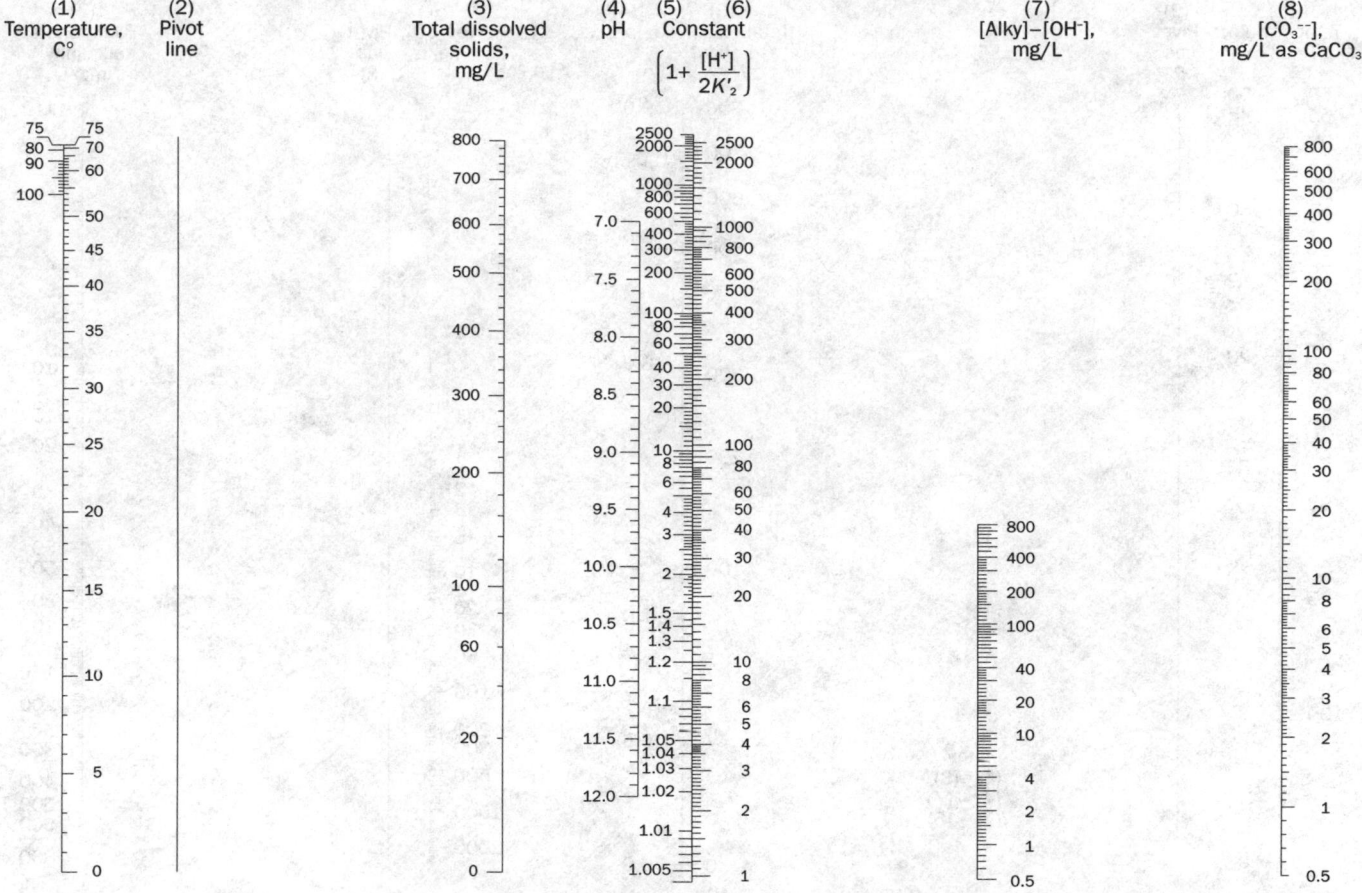

Figure 4500-CO₂:3. Nomograph for evaluation of carbonate alkalinity. To use: Align temperature (Scale 1) and total dissolved solids (Scale 3); pivot on Line 2 to proper pH (Scale 4) and read constant on Scale 5; locate constant on Scale 6 and align with nonhydroxide alkalinity (found with aid of Figure 4500-CO₂:1) on Scale 7; read carbonate alkalinity on Scale 8. (Example: For 13°C temperature, 240 mg total dissolved solids/L, pH 9.8, and 140 mg alkalinity/L, the carbonate content is found to be 50 mg as CaCO₃/L.)

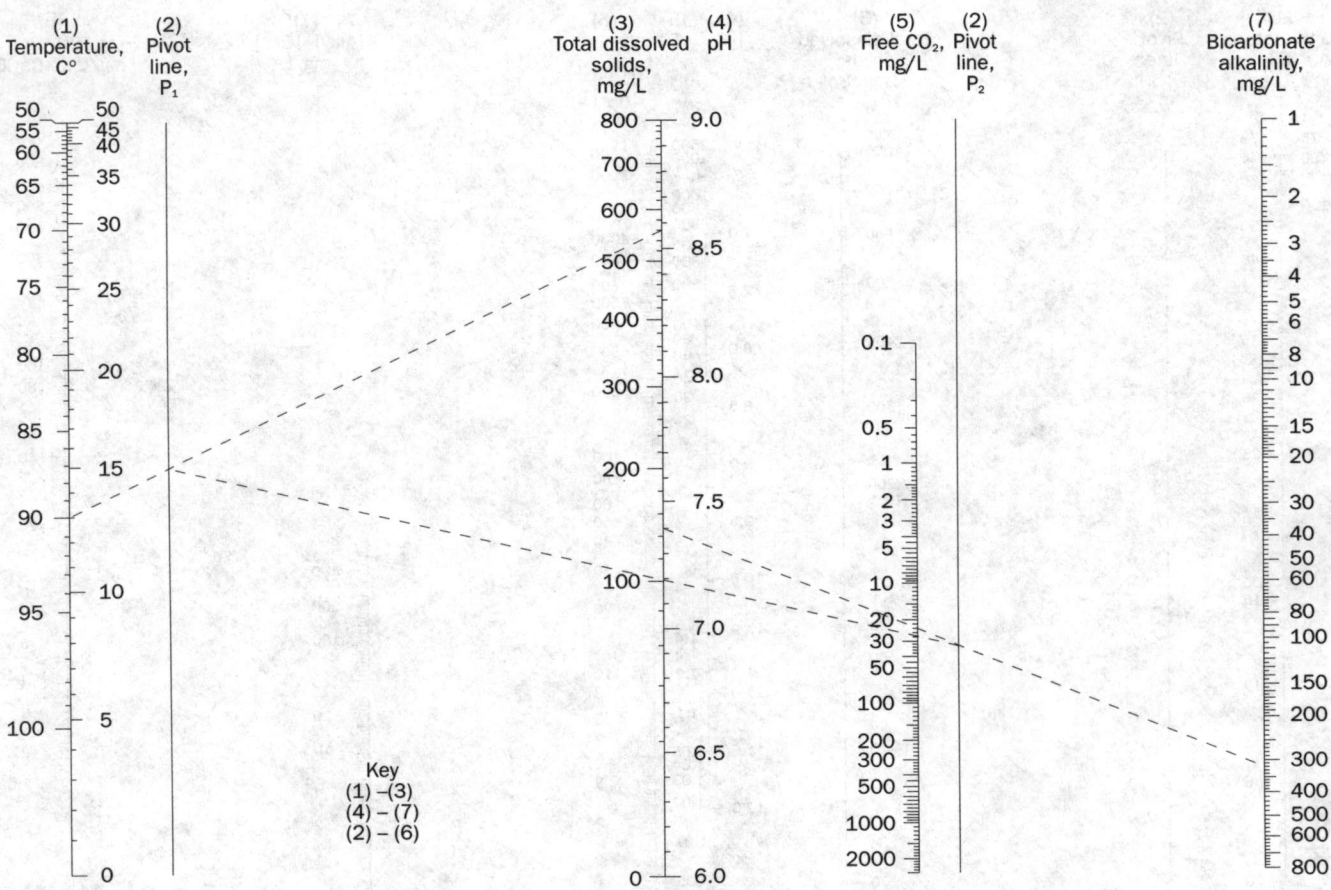

Figure 4500-CO₂:4. Nomograph for evaluation of free carbon dioxide content. To use: Align temperature (Scale 1) and total dissolved solids (Scale 3), which determines Point P₁ on Line 2; align pH (Scale 4) and bicarbonate alkalinity (Scale 7), which determines Point P₂ on Line 6; align P₁ and P₂ and read free carbon dioxide on Scale 5. (Example: For 13°C temperature, 560 mg total dissolved solids/L, pH 7.4, and 320 mg alkalinity/L, the free carbon dioxide content is found to be 28 mg/L.)

where their accuracy for the individual water supply is confirmed. The nomographs and the equations on which they are based are valid only when the salts of weak acids other than carbonic acid are absent or present in extremely small amounts.

Some treatment processes, such as superchlorination and coagulation, can affect significantly pH and total-alkalinity values of a poorly buffered water of low alkalinity and low total-dissolved-mineral content. In such instances the nomographs may not be applicable.

2. Precision and Bias

The precision possible with the nomographs depends on the size and range of the scales. With practice, the recommended nomographs can be read with a precision of 1%. However, the overall bias of the results depends on the bias of the analytical data applied to the nomographs and the validity of the theoretical equations and the numerical constants on which the nomographs are based. An approximate check of the bias of the calculations can be made by summing the three forms of alkalinity. Their sum should equal the total alkalinity.

3. Bibliography

MOORE, E.W. 1939. Graphic determination of carbon dioxide and the three forms of alkalinity. *J. Amer. Water Works Assoc.* 31:51.

4500-CO$_2$ C.   Titrimetric Method for Free Carbon Dioxide

1. General Discussion

a. Principle: Free CO$_2$ reacts with sodium carbonate or sodium hydroxide to form sodium bicarbonate. Completion of the reaction is indicated potentiometrically or by the development of the pink color characteristic of phenolphthalein indicator at the equivalence pH of 8.3. A 0.01N sodium bicarbonate (NaHCO$_3$) solution containing the recommended volume of phenolphthalein indicator is a suitable color standard until familiarity is obtained with the color at the end point.

b. Interference: Cations and anions that quantitatively disturb the normal CO$_2$-carbonate equilibrium interfere with the determination. Metal ions that precipitate in alkaline solution, such as aluminum, chromium, copper, and iron, contribute to high results. Ferrous ion should not exceed 1.0 mg/L. Positive errors also are caused by weak bases, such as ammonia or amines, and by salts of weak acids and strong bases, such as borate, nitrite, phosphate, silicate, and sulfide. Such substances should not exceed 5% of the CO$_2$ concentration. The titrimetric method for CO$_2$ is inapplicable to samples containing acid mine wastes and effluent from acid-regenerated cation exchangers. Negative errors may be introduced by high total dissolved solids, such as those encountered in seawater, or by addition of excess indicator.

c. Sampling and storage: Even with a careful collection technique, some loss in free CO$_2$ can be expected in storage and transit. This occurs more frequently when the gas is present in large amounts. Occasionally a sample may show an increase in free CO$_2$ content on standing. Consequently, determine free CO$_2$ immediately at the point of sampling. Where a field determination is impractical, fill completely a bottle for laboratory examination. Keep the sample, until tested, at a temperature lower than that at which the water was collected. Make the laboratory examination as soon as possible to minimize the effect of CO$_2$ changes.

2. Apparatus

See Section 2310B.2.

3. Reagents

See Section 2310B.3.

4. Procedure

Follow the procedure given in Section 2310B.4*b*, phenolphthalein, or 2310B.4*d*, using end-point pH 8.3.

5. Calculation

$$\text{mg CO}_2/\text{L} = \frac{A \times N \times 44\,000}{\text{mL sample}}$$

where:

A = mL titrant and
N = normality of NaOH.

6. Precision and Bias

Precision and bias of the titrimetric method are on the order of $\pm 10\%$ of the known CO$_2$ concentration.

4500-CO₂ D.   Carbon Dioxide and Forms of Alkalinity by Calculation

1. General Discussion

When the total alkalinity of a water (Section 2320) is due almost entirely to hydroxides, carbonates, or bicarbonates, and the total dissolved solids (Section 2540) is not greater than 500 mg/L, the alkalinity forms and free CO_2 can be calculated from the sample pH and total alkalinity. The calculation is subject to the same limitations as the nomographic procedure given above and the additional restriction of using a single temperature, 25°C. The calculations are based on the ionization constants:

$$K_1 = \frac{[H^+][HCO_3^-]}{[H_2CO_3^*]} \quad (K_1 = 10^{-6.36})$$

and

$$K_2 = \frac{[H^+][CO_3^{2-}]}{[HCO_3^-]} \quad (K_2 = 10^{-10.33})$$

where:

$$[H_2CO_3^*] = [H_2CO_3] + [CO_2(aq)]$$

Activity coefficients are assumed equal to unity.

2. Calculation

Compute the forms of alkalinity and sample pH and total alkalinity using the following equations:

a. Bicarbonate alkalinity:

$$HCO_3^- \text{ as mg } CaCO_3/L = \frac{T - 5.0 \times 10^{(pH-10)}}{1 + 0.94 \times 10^{(pH-10)}}$$

where:

T = total alkalinity, mg $CaCO_3/L$

b. Carbonate alkalinity:

$$CO_3^{2-} \text{ as mg } CaCO_3/L = 0.94 \times B \times 10^{(pH-10)}$$

where:

B = bicarbonate alkalinity, from *a*.

c. Hydroxide alkalinity:

$$OH^- \text{ as mg } CaCO_3/L = 5.0 \times 10^{(pH-10)}$$

d. Free carbon dioxide:

$$\text{mg } CO_2/L = 2.0 \times B \times 10^{(6-pH)}$$

where:

B = bicarbonate alkalinity, from *a*.

e. Total carbon dioxide:

$$\text{mg total } CO_2/L = A + 0.44 (2B + C)$$

where:

A = mg free CO_2/L,
B = bicarbonate alkalinity from *a*, and
C = carbonate alkalinity from *b*.

3. Bibliography

DYE, J.F. 1958. Correlation of the two principal methods of calculating the three kinds of alkalinity. *J. Amer. Water Works Assoc.* 50:812.

4500-CN⁻   CYANIDE*

4500-CN⁻ A.   Introduction

1. General Discussion

"Cyanide" refers to all of the CN groups in cyanide compounds that can be determined as the cyanide ion, CN^-, by the methods used. The cyanide compounds in which cyanide can be obtained as CN^- are classed as simple and complex cyanides.

Simple cyanides are represented by the formula $A(CN)_x$, where A is an alkali (sodium, potassium, ammonium) or a metal, and x, the valence of A, is the number of CN groups. In aqueous solutions of simple alkali cyanides, the CN group is present as CN^- and molecular HCN, the ratio depending on pH and the dissociation constant for molecular HCN ($pK_a \sim 9.2$). In most natural waters HCN greatly predominates.[1] In solutions of simple metal cyanides, the CN group may occur also in the form of complex metal-cyanide anions of varying stability. Many simple metal cyanides are sparingly soluble or almost insoluble [CuCN, AgCN, $Zn(CN)_2$], but they form a variety of highly soluble, complex metal cyanides in the presence of alkali cyanides.

Complex cyanides have a variety of formulae, but the alkali-metallic cyanides normally can be represented by $A_yM(CN)_x$. In

* Approved by Standard Methods Committee, 1997.

this formula, A represents the alkali present y times, M the heavy metal (ferrous and ferric iron, cadmium, copper, nickel, silver, zinc, or others), and x the number of CN groups; x is equal to the valence of A taken y times plus that of the heavy metal. Initial dissociation of each of these soluble, alkali-metallic, complex cyanides yields an anion that is the radical $M(CN)_x^{y-}$. This may dissociate further, depending on several factors, with the liberation of CN⁻ and consequent formation of HCN.

The great toxicity to aquatic life of molecular HCN is well known;[2-5] it is formed in solutions of cyanide by hydrolytic reaction of CN⁻ with water. The toxicity of CN⁻ is less than that of HCN; it usually is unimportant because most of the free cyanide (CN group present as CN⁻ or as HCN) exists as HCN,[2-5] as the pH of most natural waters is substantially lower than the pK_a for molecular HCN. The toxicity to fish of most tested solutions of complex cyanides is attributable mainly to the HCN resulting from dissociation of the complexes.[2,4,5] Analytical distinction between HCN and other cyanide species in solutions of complex cyanides is possible.[2,5-9,10]

The degree of dissociation of the various metallocyanide complexes at equilibrium, which may not be attained for a long time, increases with decreased concentration and decreased pH, and is inversely related to the highly variable stability of the complexes.[2,4,5] The zinc- and cadmium-cyanide complexes are dissociated almost totally in very dilute solutions; thus these complexes can result in acute toxicity to fish at any ordinary pH. In equally dilute solutions there is much less dissociation for the nickel-cyanide complex and the more stable cyanide complexes formed with copper (I) and silver. Acute toxicity to fish from dilute solutions containing copper-cyanide or silver-cyanide complex anions can be due to the toxicity of the undissociated ions, although the complex ions are much less toxic than HCN.[2,5]

The iron-cyanide complex ions are very stable and not materially toxic; in the dark, acutely toxic levels of HCN are attained only in solutions that are not very dilute and have been aged for a long time. However, these complexes are subject to extensive and rapid photolysis, yielding toxic HCN, on exposure of dilute solutions to direct sunlight.[2,11] The photodecomposition depends on exposure to ultraviolet radiation, and therefore is slow in deep, turbid, or shaded receiving waters. Loss of HCN to the atmosphere and its bacterial and chemical destruction concurrent with its production tend to prevent increases of HCN concentrations to harmful levels. Regulatory distinction between cyanide complexed with iron and that bound in less stable complexes, as well as between the complexed cyanide and free cyanide or HCN, can, therefore, be justified.

Historically, the generally accepted physicochemical technique for industrial waste treatment of cyanide compounds is alkaline chlorination:

$$NaCN + Cl_2 \rightleftharpoons CNCl + NaCl \qquad (1)$$

The first reaction product on chlorination is cyanogen chloride (CNCl), a highly toxic gas of limited solubility. The toxicity of CNCl may exceed that of equal concentrations of cyanide.[2,3,12] At an alkaline pH, CNCl hydrolyzes to the cyanate ion (CNO⁻), which has only limited toxicity.

There is no known natural reduction reaction that may convert CNO⁻ to CN⁻.[13] On the other hand, breakdown of toxic CNCl is pH- and time-dependent. At pH 9, with no excess chlorine present, CNCl may persist for 24 h.[14,15]

$$CNCl + 2NaOH \rightleftharpoons NaCNO + NaCl + H_2O \qquad (2)$$

CNO⁻ can be oxidized further with chlorine at a nearly neutral pH to CO_2 and N_2:

$$2NaCNO + 4NaOH + 3Cl_2 \rightleftharpoons 6NaCl$$
$$+ 2CO_2 + N_2 + 2H_2O \qquad (3)$$

CNO⁻ also will be converted on acidification to NH_4^+:

$$2NaCNO + H_2SO_4 + 4H_2O \rightleftharpoons (NH_4)_2SO_4 + 2NaHCO_3 \qquad (4)$$

The alkaline chlorination of cyanide compounds is relatively fast, but depends equally on the dissociation constant, which also governs toxicity. Metal cyanide complexes, such as nickel, cobalt, silver, and gold, do not dissociate readily. The chlorination reaction therefore requires more time and a significant chlorine excess.[16] Iron cyanides, because they do not dissociate to any degree, are not oxidized by chlorination. There is correlation between the refractory properties of the noted complexes, in their resistance to chlorination and lack of toxicity.

Thus, it is advantageous to differentiate between *total cyanide* and *cyanides amenable to chlorination*. When total cyanide is determined, the almost nondissociable cyanides, as well as cyanide bound in complexes that are readily dissociable and complexes of intermediate stability, are measured. Cyanide compounds that are amenable to chlorination include free cyanide as well as those complex cyanides that are potentially dissociable, almost wholly or in large degree, and therefore, potentially toxic at low concentrations, even in the dark. The chlorination test procedure is carried out under rigorous conditions appropriate for measurement of the more dissociable forms of cyanide.

The free and potentially dissociable cyanides also may be estimated when using the *weak acid dissociable* procedure. These methods depend on a rigorous distillation, but the solution is only slightly acidified, and elimination of iron cyanides is insured by the earlier addition of precipitation chemicals to the distillation flask and by the avoidance of ultraviolet irradiation.

The *cyanogen chloride* procedure is common with the colorimetric test for cyanides amenable to chlorination. This test is based on the addition of chloramine-T and subsequent color complex formation with pyridine-barbituric acid solution. Without the addition of chloramine-T, only existing CNCl is measured. CNCl is a gas that hydrolyzes to CNO⁻; sample preservation is not possible. Because of this, spot testing of CNCl levels may be best. This procedure can be adapted and used when the sample is collected.

There may be analytical requirements for the determination of CNO⁻, even though the reported toxicity level is low. On acidification, CNO⁻ decomposes to ammonia (NH₃).[3] Molecular ammonia and metal-ammonia complexes are toxic to aquatic life.[17]

Thiocyanate (SCN⁻) is not very toxic to aquatic life.[2,18] However, upon chlorination, toxic CNCl is formed, as discussed above.[2,3,12] At least where subsequent chlorination is anticipated, the determination of SCN⁻ is desirable. Thiocyanate is biodegradable; ammonium is released in this reaction. Although the typical detoxifying agents used in cyanide poisoning induce thiocyanate formation, biochemical cyclic reactions with cyanide are possible, resulting in detectable levels of cyanide from exposure to thiocyanate.[18] Thiocyanate may be analyzed in samples properly preserved for determination of cyanide; however, thiocyanate also can be preserved in samples by acidification with H_2SO_4 to pH ≤2.

2. Cyanide in Solid Waste

a. Soluble cyanide: Determination of soluble cyanide requires sample leaching with distilled water until solubility equilibrium is established. One hour of stirring in distilled water should be satisfactory. Cyanide analysis is then performed on the leachate. Low cyanide concentration in the leachate may indicate presence of sparingly soluble metal cyanides. The cyanide content of the leachate is indicative of residual solubility of insoluble metal cyanides in the waste.

High levels of cyanide in the leachate indicate soluble cyanide in the solid waste. When 500 mL distilled water are stirred into a 500-mg solid waste sample, the cyanide concentration (mg/L) of the leachate multiplied by 1000 will give the solubility level of the cyanide in the solid waste in milligrams per kilogram. The leachate may be analyzed for total cyanide and/or cyanide amenable to chlorination.

b. Insoluble cyanide: The insoluble cyanide of the solid waste can be determined with the total cyanide method by placing a 500-mg sample with 500 mL distilled water in the distillation flask and in general following the distillation procedure (Section 4500-CN⁻.C). In calculating, multiply by 1000 to give the cyanide content of the solid sample in milligrams per kilogram. Insoluble iron cyanides in the solid can be leached out earlier by stirring a weighed sample for 12 to 16 h in a 10% NaOH solution. The leached and wash waters of the solid waste will give the iron cyanide content with the distillation procedure. Prechlorination will have eliminated all cyanide amenable to chlorination. Do not expose sample to sunlight.

3. Selection of Method

a. Total cyanide after distillation: After removal of interfering substances, the metal cyanide is converted to HCN gas, which is distilled and absorbed in sodium hydroxide (NaOH) solution.[19] Because of the catalytic decomposition of cyanide in the presence of cobalt at high temperature in a strong acid solution,[20,21] cobalticyanide is not recovered completely. Indications are that cyanide complexes of the noble metals, i.e., gold, platinum, and palladium, are not recovered fully by this procedure either. Distillation also separates cyanide from other color-producing and possibly interfering organic or inorganic contaminants. Subsequent analysis is for the simple salt, sodium cyanide (NaCN). Some organic cyanide compounds, such as cyanohydrins, are decomposed by the distillation. Aldehydes convert cyanide to cyanohydrins.

The absorption liquid is analyzed by a titrimetric, colorimetric, or cyanide-ion-selective electrode procedure:

1) The titration method (D) is suitable for cyanide concentrations above 1 mg/L.

2) The colorimetric methods (E, N, and O) are suitable for cyanide concentrations as low as 1 to 5 μg/L under ideal conditions. Method N uses flow injection analysis of the distillate. Method O uses flow injection analysis following transfer through a semipermeable membrane for separating gaseous cyanide, and colorimetric analysis. Method E uses conventional colorimetric analysis of the distillate from Method C.

3) The ion-selective electrode method (F) using the cyanide ion electrode is applicable in the concentration range of 0.05 to 10 mg/L.

b. Cyanide amenable to chlorination:

1) Distillation of two samples is required, one that has been chlorinated to destroy all amenable cyanide present and the other unchlorinated. Analyze absorption liquids from both tests for total cyanide. The observed difference equals cyanides amenable to chlorination.

2) The colorimetric methods, by conversion of amenable cyanide and SCN⁻ to CNCl and developing the color complex with pyridine-barbituric acid solution, are used for the determination of the total of these cyanides (H, N, and O). Repeating the test with the cyanide masked by the addition of formaldehyde provides a measure of the SCN⁻ content. When subtracted from the earlier results this provides an estimate of the amenable CN⁻ content. This method is useful for natural and ground waters, clean metal finishing, and heat treating effluents. Sanitary wastes may exhibit interference.

3) The *weak acid dissociable cyanides* procedure also measures the cyanide amenable to chlorination by freeing HCN from the dissociable cyanide. After being collected in a NaOH absorption solution, CN⁻ may be determined by one of the finishing procedures given for the total cyanide determination. An automated procedure (O) also is presented.

It should be noted that although cyanide amenable to chlorination and weak acid dissociable cyanide appear to be identical, certain industrial effluents (e.g., pulp and paper, petroleum refining industry effluents) contain some poorly understood substances that may produce interference. Application of the procedure for cyanide amenable to chlorination yields negative values. For natural waters and metal-finishing effluents, the direct colorimetric determination appears to be the simplest and most economical.

c. Cyanogen chloride: The colorimetric method for measuring cyanide amenable to chlorination may be used, but omit the chloramine-T addition. The spot test also may be used.

d. Spot test for sample screening: This procedure allows a quick sample screening to establish whether more than 50 μg/L cyanide amenable to chlorination is present. The test also may be used to estimate the CNCl content at the time of sampling.

e. Cyanate: CNO⁻ is converted to ammonium carbonate, $(NH_4)_2CO_3$, by acid hydrolysis at elevated temperature. Ammonia (NH_3) is determined before the conversion of the CNO⁻ and again afterwards. The CNO⁻ is estimated from the difference in NH_3 found in the two tests. [22–24] Measure NH_3 by either:

1) The selective electrode method, using the NH_3 gas electrode (4500-NH₃.D); or

2) The colorimetric method, using the phenate method for NH_3 (Section 4500-NH₃.F or G).

f. Thiocyanate: Use the colorimetric determination with ferric nitrate as a color-producing compound.

4. References

1. MILNE, D. 1950. Equilibria in dilute cyanide waste solutions. *Sewage Ind. Wastes* 23:904.

2. DOUDOROFF, P. 1976. Toxicity to fish of cyanides and related compounds. A review. EPA 600/3-76-038, U.S. Environmental Protection Agency, Duluth, Minn.

3. DOUDOROFF, P. & M. KATZ. 1950. Critical review of literature on the toxicity of industrial wastes and their components to fish. *Sewage Ind. Wastes* 22:1432.

4. DOUDOROFF, P. 1956. Some experiments on the toxicity of complex cyanides to fish. *Sewage Ind. Wastes* 28:1020.

5. DOUDOROFF, P., G. LEDUC & C.R. SCHNEIDER. 1966. Acute toxicity to fish of solutions containing complex metal cyanides, in relation to

concentrations of molecular hydrocyanic acid. *Trans. Amer. Fish. Soc.* 95:6.

6. SCHNEIDER, C.R. & H. FREUND. 1962. Determination of low level hydrocyanic acid. *Anal. Chem.* 34:69.

7. CLAEYS R. & H. FREUND. 1968. Gas chromatographic separation of HCN. *Environ. Sci. Technol.* 2:458.

8. MONTGOMERY, H.A.C., D.K. GARDINER & J.G. GREGORY. 1969. Determination of free hydrogen cyanide in river water by a solvent-extraction method. *Analyst* 94:284.

9. NELSON, K.H. & L. LYSYJ. 1971. Analysis of water for molecular hydrogen cyanide. *J. Water Pollut. Control Fed.* 43:799.

10. BRODERIUS, S.J. 1981. Determination of hydrocyanic acid and free cyanide in aqueous solution. *Anal. Chem.* 53:1472.

11. BURDICK, G.E. & M. LIPSCHUETZ. 1948. Toxicity of ferro and ferri-cyanide solutions to fish. *Trans. Amer. Fish. Soc.* 78:192.

12. ZILLICH, J.A. 1972. Toxicity of combined chlorine residuals to fresh-water fish. *J. Water Pollut. Control Fed.* 44:212.

13. RESNICK, J.D., W. MOORE & M.E. ETTINGER. 1958. The behavior of cyanates in polluted waters. *Ind. Eng. Chem.* 50:71.

14. PETTET, A.E.J. & G.C. WARE. 1955. Disposal of cyanide wastes. *Chem. Ind.* 1955:1232.

15. BAILEY, P.L. & E. BISHOP. 1972. Hydrolysis of cyanogen chloride. *Analyst* 97:691.

16. LANCY, L. & W. ZABBAN. 1962. Analytical methods and instrumentation for determining cyanogen compounds. Spec. Tech. Publ. 337, American Soc. Testing & Materials, Philadelphia, Pa.

17. CALAMARI, D. & R. MARCHETTI. 1975. Predicted and observed acute toxicity of copper and ammonia to rainbow trout. *Progr. Water Technol.* 7(3-4):569.

18. WOOD, J.L. 1975. Biochemistry. Chapter 4 *in* A.A. Newman, ed. Chemistry and Biochemistry of Thiocyanic Acid and its Derivatives. Academic Press, New York, N.Y.

19. SERFASS, E.J. & R.B. FREEMAN. 1952. Analytical method for the determination of cyanides in plating wastes and in effluents from treatment processes. *Plating* 39:267.

20. LESCHBER, R. & H. SCHLICHTING. 1969. Über die Zersetzlichkeit Komplexer Metallcyanide bei der Cyanidbestimmung in Abwasser. *Z. Anal. Chem.* 245:300.

21. BASSETT, H., JR. & A.S. CORBET. 1924. The hydrolysis of potassium ferricyanide and potassium cobalticyanide by sulfuric acid. *J. Chem. Soc.* 125:1358.

22. DODGE, B.F. & W. ZABBAN. 1952. Analytical methods for the determination of cyanates in plating wastes. *Plating* 39:381.

23. GARDNER, D.C. 1956. The colorimetric determination of cyanates in effluents. *Plating* 43:743.

24. Procedures for Analyzing Metal Finishing Wastes. 1954. Ohio River Valley Sanitation Commission, Cincinnati, Ohio.

4500-CN⁻ B.   Preliminary Treatment of Samples

CAUTION—*Use care in manipulating cyanide-containing samples because of toxicity. Process in a hood or other well-ventilated area. Avoid contact, inhalation, or ingestion.*

1. General Discussion

The nature of the preliminary treatment will vary according to the interfering substance present. Sulfides, fatty acids, and oxidizing agents are removed by special procedures. Most other interfering substances are removed by distillation. The importance of the distillation procedure cannot be overemphasized.

2. Preservation of Samples

Oxidizing agents, such as chlorine, decompose most cyanides. Test by placing a drop of sample on a strip of potassium iodide (KI)-starch paper previously moistened with acetate buffer solution, pH 4 (Section 4500-Cl.C.3e). If a bluish discoloration is noted, add 0.1 g sodium arsenite (NaAsO₂)/L sample and retest. Repeat addition if necessary. Sodium thiosulfate or ascorbic acid also may be used, but avoid an excess greater than 0.1 g Na₂S₂O₃/L. Manganese dioxide, nitrosyl chloride, etc., if present, also may cause discoloration of the test paper. If possible, carry out this procedure before preserving sample as described below. If the following test indicates presence of sulfide, oxidizing compounds would not be expected.

Oxidized products of sulfide convert CN⁻ to SCN⁻ rapidly, especially at high pH.[1] Test for S²⁻ by placing a drop of sample on lead acetate test paper previously moistened with acetic acid buffer solution, pH 4 (Section 4500-Cl.C.3e). Darkening of the paper indicates presence of S²⁻. Add lead acetate, or if the S²⁻ concentration is too high, add powdered lead carbonate [Pb(CO₃)₂] to avoid significantly reducing pH. Repeat test until a drop of treated sample no longer darkens the acidified lead acetate test paper. Filter sample before raising pH for stabilization. When particulate, metal cyanide complexes are suspected, filter solution before removing S²⁻. Reconstitute sample by returning filtered particulates to the sample bottle after S²⁻ removal. Homogenize particulates before analyses.

Aldehydes convert cyanide to cyanohydrin. Longer contact times between cyanide and the aldehyde and the higher ratios of aldehyde to cyanide both result in increasing losses of cyanide that are not reversible during analysis. If the presence of aldehydes is suspected, stabilize with NaOH at time of collection and add 2 mL 3.5% ethylenediamine solution per 100 mL of sample.

Because most cyanides are very reactive and unstable, analyze samples as soon as possible. If sample cannot be analyzed immediately, add NaOH pellets or a strong NaOH solution to raise sample pH to 12 to 12.5, add dechlorinating agent if sample is disinfected, and store in a closed, dark bottle in a cool place.

To analyze for CNCl collect a separate sample and omit NaOH addition because CNCl is converted rapidly to CNO⁻ at high pH. Make colorimetric estimation immediately after sampling.

3. Interferences

a. Oxidizing agents may destroy most of the cyanide during storage and manipulation. Add NaAsO₂ or Na₂S₂O₃ as directed above; avoid excess Na₂S₂O₃.

b. Sulfide will distill over with cyanide and, therefore, adversely affect colorimetric, titrimetric, and electrode procedures. Test for and remove S^{2-} as directed above. Treat 25 mL more than required for the distillation to provide sufficient filtrate volume.

c. *Fatty acids* that distill and form soaps under alkaline titration conditions make the end point almost impossible to detect. Remove fatty acids by extraction.[2] Acidify sample with acetic acid (1 + 9) to pH 6.0 to 7.0. (CAUTION—*Perform this operation in a hood as quickly as possible.*) Immediately extract with iso-octane, hexane, or $CHCl_3$ (preference in order named). Use a solvent volume equal to 20% of sample volume. One extraction usually is adequate to reduce fatty acid concentration below the interference level. Avoid multiple extractions or a long contact time at low pH to minimize loss of HCN. When extraction is completed, immediately raise pH to >12 with NaOH solution.

d. *Carbonate* in high concentration may affect the distillation procedure by causing the violent release of carbon dioxide with excessive foaming when acid is added before distillation and by reducing pH of the absorption solution. Use calcium hydroxide to preserve such samples.[3] Add calcium hydroxide slowly, with stirring, to pH 12 to 12.5. After precipitate settles, decant supernatant liquid for determining cyanide.

Insoluble complex cyanide compounds will not be determined. If such compounds are present, filter a measured amount of well-mixed treated sample through a glass fiber or membrane filter (47-mm diam or less). Rinse filter with dilute (1 to 9) acetic acid until effervescence ceases. Treat entire filter with insoluble material as insoluble cyanide (4500-CN^-.A.2b) or add to filtrate before distillation.

e. *Other possible interferences* include substances that might contribute color or turbidity. In most cases, distillation will remove these.

Note, however, that the strong acid distillation procedure requires using sulfuric acid with various reagents. With certain wastes, these conditions may result in reactions that otherwise would not occur in the aqueous sample. As a quality control measure, periodically conduct addition and recovery tests with industrial waste samples.

f. *Aldehydes* convert cyanide to cyanohydrin, which forms nitrile under the distillation conditions. Only direct titration without distillation can be used, which reveals only non-complex cyanides. Formaldehyde interference is noticeable in concentrations exceeding 0.5 mg/L. Use the following spot test to establish absence or presence of aldehydes (detection limit 0.05 mg/L):[4-6]

1) Reagents

a) *MBTH indicator solution:* Dissolve 0.05 g 3-methyl, 2-benzothiazolone hydrazone hydrochloride in 100 mL water. Filter if turbid.

b) *Ferric chloride oxidizing solution:* Dissolve 1.6 g sulfamic acid and 1 g $FeCl_3·6H_2O$ in 100 mL water.

c) *Ethylenediamine solution,* 3.5%: Dilute 3.5 mL pharmaceutical-grade anhydrous $NH_2CH_2CH_2NH_2$ to 100 mL with water.

2) Procedure—If the sample is alkaline, add 1 + 1 H_2SO_4 to 10 mL sample to adjust pH to less than 8. Place 1 drop of sample and 1 drop distilled water for a blank in separate cavities of a white spot plate. Add 1 drop MBTH solution and then 1 drop $FeCl_3$ oxidizing solution to each spot. Allow 10 min for color development. The color change will be from a faint green-yellow to a deeper green with blue-green to blue at higher concentrations of aldehyde. The blank should remain yellow.

To minimize aldehyde interference, add 2 mL of 3.5% ethylenediamine solution/100 mL sample. This quantity overcomes the interference caused by up to 50 mg/L formaldehyde.

When using a known addition in testing, 100% recovery of the CN^- is not necessarily to be expected. Recovery depends on the aldehyde excess, time of contact, and sample temperature.

g. *Glucose and other sugars,* especially at the pH of preservation, lead to cyanohydrin formation by reaction of cyanide with aldose.[7] Reduce cyanohydrin to cyanide with ethylenediamine (see f above). MBTH is not applicable.

h. *Nitrite* may form HCN during distillation in Methods C, G, and L, by reacting with organic compounds.[8,9] Also, NO_3^- may reduce to NO_2^-, which interferes. To avoid NO_2^- interference, add 2 g sulfamic acid to the sample before distillation. *Nitrate* also may interfere by reacting with SCN^-.[10]

i. *Some sulfur compounds* may decompose during distillation, releasing S, H_2S, or SO_2. Sulfur compounds may convert cyanide to thiocyanate and also may interfere with the analytical procedures for CN^-. To avoid this potential interference, add 50 mg $PbCO_3$ to the absorption solution before distillation. Filter sample before proceeding with the colorimetric or titrimetric determination.

Absorbed SO_2 forms Na_2SO_3 which consumes chloramine-T added in the colorimetric determination. The volume of chloramine-T added is sufficient to overcome 100 to 200 mg SO_3^{2-}/L. Test for presence of chloramine-T after adding it by placing a drop of sample on KI-starch test paper; add more chloramine-T if the test paper remains blank, or use Method F.

Some wastewaters, such as those from coal gasification or chemical extraction mining, contain high concentrations of sulfites. Pretreat sample to avoid overloading the absorption solution with SO_3^{2-}. Titrate a suitable sample iodometrically (Section 4500-O) with dropwise addition of 30% H_2O_2 solution to determine volume of H_2O_2 needed for the 500 mL distillation sample. Subsequently, add H_2O_2 dropwise while stirring, but in only such volume that not more than 300 to 400 mg SO_3^{2-}/L will remain. Adding a lesser quantity than calculated is required to avoid oxidizing any CN^- that may be present.

j. *Alternate procedure:* The strong acid distillation procedure uses concentrated acid with magnesium chloride to dissociate metal-cyanide complexes. In some instances, particularly with industrial wastes, it may be susceptible to interferences such as those from conversion of thiocyanate to cyanide in the presence of an oxidant, e.g., nitrate. If such interferences are present use a ligand displacement procedure with a mildly acidic medium with EDTA to dissociate metal-cyanide complexes.[10] Under such conditions thiocyanate is relatively stable and many oxidants, including nitrate, are weaker.

If any cyanide procedure is revised to meet specific requirements, obtain recovery data by the addition of known amounts of cyanide.

4. References

1. LUTHY, R.G. & S. G. BRUCE, JR. 1979. Kinetics of reaction of cyanide and reduced sulfur species in aqueous solution. *Environ. Sci. Technol.* 13:1481.

2. KRUSE, J.M. & M.G. MELLON. 1951. Colorimetric determination of cyanides. *Sewage Ind. Wastes* 23:1402.

3. LUTHY, R.G., S.G. BRUCE, R.W. WALTERS & D.V. NAKLES. 1979. Cyanide and thiocyanate in coal gasification wastewater. *J. Water Pollut. Control Fed.* 51:2267.

4. SAWICKI, E., T.W. STANLEY, T.R. HAUSER & W. ELBERT. 1961. The 3-methyl-2-benzothiazolone hydrazone test. Sensitive new methods

for the detection, rapid estimation, and determination of aliphatic aldehydes. *Anal. Chem.* 33:93.

5. HAUSER, T.R. & R.L. CUMMINS. 1964. Increasing sensitivity of 3-methyl-2-benzothiazone hydrazone test for analysis of aliphatic aldehydes in air. *Anal. Chem.* 36:679.

6. Methods of Air Sampling and Analysis, 1st ed. 1972. Inter Society Committee, Air Pollution Control Assoc., pp. 199-204.

7. RAAF, S.F., W.G. CHARACKLIS, M.A. KESSICK & C.H. WARD. 1977. Fate of cyanide and related compounds in aerobic microbial systems. *Water Res.* 11:477.

8. RAPEAN, J.C., T. HANSON & R. A. JOHNSON. 1980. Biodegradation of cyanide-nitrate interference in the standard test for total cyanide. Proc. 35th Ind. Waste Conf., Purdue Univ., Lafayette, Ind., p. 430.

9. CASEY, J.P. 1980. Nitrosation and cyanohydrin decomposition artifacts in distillation test for cyanide. Extended Abs. American Chemical Soc., Div. Environmental Chemistry, Aug. 24-29, 1980, Las Vegas, Nev.

10. CSIKAI, N.J. & A.J. BARNARD, JR. 1983. Determination of total cyanide in thiocyanate-containing waste water. *Anal. Chem.* 55:1677.

4500-CN⁻ C. Total Cyanide after Distillation

1. General Discussion

Hydrogen cyanide (HCN) is liberated from an acidified sample by distillation and purging with air. The HCN gas is collected by passing it through an NaOH scrubbing solution. Cyanide concentration in the scrubbing solution is determined by titrimetric, colorimetric, or potentiometric procedures.

2. Apparatus

The apparatus is shown in Figure 4500-CN⁻:1. It includes:

a. Boiling flask, 1 L, with inlet tube and provision for water-cooled condenser.

b. Gas absorber, with gas dispersion tube equipped with medium-porosity fritted outlet.

c. Heating element, adjustable.

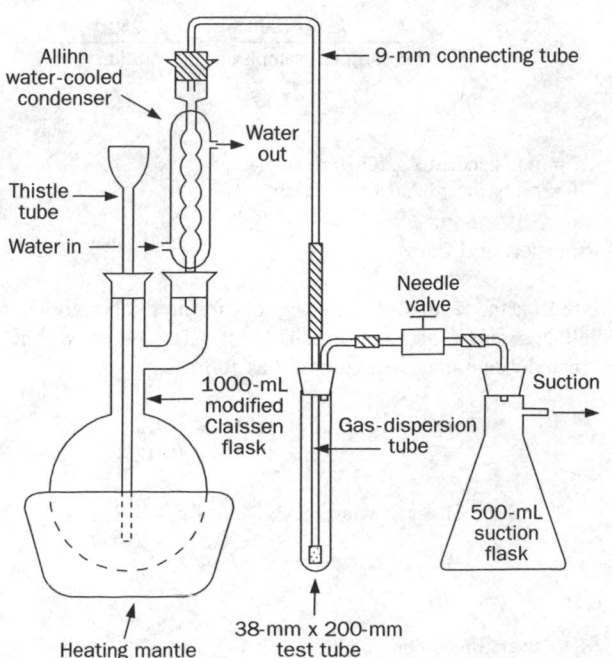

Allihn water-cooled condenser

9-mm connecting tube

Water out

Thistle tube

Water in

Needle valve

1000-mL modified Claissen flask

Gas-dispersion tube

Suction

500-mL suction flask

Heating mantle

38-mm x 200-mm test tube

Figure 4500-CN⁻:1. Cyanide distillation apparatus.

d. Ground glass ST joints, TFE-sleeved or with an appropriate lubricant for the boiling flask and condenser. Neoprene stopper and plastic threaded joints also may be used.

3. Reagents

a. Sodium hydroxide solution: Dissolve 40 g NaOH in water and dilute to 1 L.

b. Magnesium chloride reagent: Dissolve 510 g $MgCl_2 \cdot 6H_2O$ in water and dilute to 1 L.

c. Sulfuric acid, H_2SO_4, 1 + 1.

d. Lead carbonate, $PbCO_3$, powdered.

e. Sulfamic acid, NH_2SO_3H.

4. Procedure

a. Add 500 mL sample, containing not more than 10 mg CN⁻/L (diluted if necessary with distilled water) to the boiling flask. If a higher CN⁻ content is anticipated, use the spot test (4500-CN⁻.K) to approximate the required dilution. Add 10 mL NaOH solution to the gas scrubber and dilute, if necessary, with distilled water to obtain an adequate liquid depth in the absorber. Do not use more than 225 mL total volume of absorber solution. When S^{2-} generation from the distilling flask is anticipated add 50 or more mg powdered $PbCO_3$ to the absorber solution to precipitate S^{2-}. Connect the train, consisting of boiling flask air inlet, flask, condenser, gas washer, suction flask trap, and aspirator. Adjust suction so that approximately 1 air bubble/s enters the boiling flask. This air rate will carry HCN gas from flask to absorber and usually will prevent a reverse flow of HCN through the air inlet. If this air rate does not prevent sample backup in the delivery tube, increase air-flow rate to 2 air bubbles/s. Observe air purge rate in the absorber where the liquid level should be raised not more than 6.5 to 10 mm. Maintain air flow throughout the reaction.

b. Add 2 g sulfamic acid through the air inlet tube and wash down with distilled water.

c. Add 50 mL 1 1 H_2SO_4 through the air inlet tube. Rinse tube with distilled water and let air mix flask contents for 3 min. Add 20 mL $MgCl_2$ reagent through air inlet and wash down with stream of water. A precipitate that may form redissolves on heating.

d. Heat with rapid boiling, but do not flood condenser inlet or permit vapors to rise more than halfway into condenser. Adequate refluxing is indicated by a reflux rate of 40 to 50 drops/min from the condenser lip. Reflux for at least 1 h. Discontinue heating but continue air flow for 15 min. Cool and quantitatively transfer absorption solution to a 250-mL volumetric flask. Rinse absorber and its connecting tubing sparingly with distilled water and add to flask. Dilute to volume with distilled water and mix thoroughly.

e. Determine cyanide concentration in the absorption solution by procedure of 4500-CN⁻. D, E, or F.

f. Distillation gives quantitative recovery of even refractory cyanides such as iron complexes. To obtain complete recovery of cobalticyanide use ultraviolet radiation pretreatment.[1,2] If incomplete recovery is suspected, distill again by refilling the gas washer with a fresh charge of NaOH solution and refluxing 1 h more. The cyanide from the second reflux, if any, will indicate completeness of recovery.

g. As a quality control measure, periodically test apparatus, reagents, and other potential variables in the concentration range of interest. As an example at least $100 \pm 4\%$ recovery from 1 mg CN⁻/L standard should be obtained.

5. References

1. Casapieri, P., R. Scott & E.A. Simpson. 1970. The determination of cyanide ions in waters and effluents by an Auto Analyzer procedure. *Anal. Chim. Acta* 49:188.

2. Goulden, P.D., K.A. Badar & P. Brooksbank. 1972. Determination of nanogram quantities of simple and complex cyanides in water. *Anal. Chem.* 44:1845.

4500-CN⁻ D.   Titrimetric Method

1. General Discussion

a. Principle: CN⁻ in the alkaline distillate from the preliminary treatment procedure is titrated with standard silver nitrate (AgNO₃) to form the soluble cyanide complex, $Ag(CN)_2^-$. As soon as all CN⁻ has been complexed and a small excess of Ag⁺ has been added, the excess Ag⁺ is detected by the silver-sensitive indicator, *p*-dimethylaminobenzalrhodanine, which immediately turns from a yellow to a salmon color.[1] The distillation has provided a 2:1 concentration. The indicator is sensitive to about 0.1 mg Ag/L. If titration shows that CN⁻ is below 1 mg/L, examine another portion colorimetrically or potentiometrically.

2. Apparatus

Koch microburet, 10-mL capacity.

3. Reagents

a. Indicator solution: Dissolve 20 mg *p*-dimethylaminobenzalrhodanine in 100 mL acetone.

b. Standard silver nitrate titrant: Dissolve 3.27 g AgNO₃ in 1 L distilled water. Standardize against standard NaCl solution, using the argentometric method with K_2CrO_4 indicator, as directed in Chloride, Section 4500-Cl⁻.B.

Dilute 500 mL AgNO₃ solution according to the titer found so that 1.00 mL is equivalent to 1.00 mg CN⁻.

c. Sodium hydroxide dilution solution: Dissolve 1.6 g NaOH in 1 L distilled water.

4. Procedure

a. From the absorption solution take a measured volume of sample so that the titration will require approximately 1 to 10 mL AgNO₃ titrant. Dilute to 100 mL using the NaOH dilution solution or to some other convenient volume to be used for all titrations. For samples with low cyanide concentration (≤ 5 mg/L) do not dilute. Add 0.5 mL indicator solution.

b. Titrate with standard AgNO₃ titrant to the first change in color from a canary yellow to a salmon hue. Titrate a blank containing the same amount of alkali and water, i.e., 100 mL NaOH dilution solution (or volume used for sample). As the analyst becomes accustomed to the end point, blank titrations decrease from the high values usually experienced in the first few trials to 1 drop or less, with a corresponding improvement in precision.

5. Calculation

$$\text{mg CN}^-/\text{L} = \frac{(A - B) \times 1000}{\text{mL original sample}} \times \frac{250}{\text{mL portion used}}$$

where:

A = mL standard AgNO₃ for sample and
B = mL standard AgNO₃ for blank.

6. Precision and Bias[2]

Based on the results of six operators in three laboratories, the overall and single-operator precision of this method within its designated range may be expressed as follows:

$$\text{Reagent water: } S_T = 0.04x + 0.038$$
$$S_o = 0.01x + 0.018$$

$$\text{Selected water matrices: } S_T = 0.06x + 0.711$$
$$S_o = 0.04x + 0.027$$

where:

S_T = overall precision, mg/L,
S_o = single-operator precision, mg/L, and
x = cyanide concentration, mg/L.

Recoveries of known amounts of cyanide from reagent water and selected water matrices are:

Medium	Added mg/L	Recovered mg/L	n	S_T	Bias	% Bias
Reagent	2.00	2.10	18	0.1267	0.10	5
water	5.00	4.65	18	0.2199	−0.35	−7
	5.00	5.18	18	0.2612	0.18	4
Selected	2.00	2.80	18	0.8695	0.80	40
water	5.00	5.29	18	1.1160	0.29	6
matrices	5.00	5.75	18	0.9970	0.75	15

7. References

1. RYAN, J.A. & G.W. CULSHAW. 1944. The use of *p*-dimethylamino-benzylidene rhodanine as an indicator for the volumetric determination of cyanides. *Analyst* 69:370.
2. AMERICAN SOCIETY FOR TESTING & MATERIALS. 1987. Research Rep. D2036:19-1131. American Soc. Testing & Materials, Philadelphia, Pa.

4500-CN⁻ E. Colorimetric Method

1. General Discussion

a. Principle: CN⁻ in the alkaline distillate from preliminary treatment is converted to CNCl by reaction with chloramine-T at pH <8 without hydrolyzing to CNO⁻.[1] (CAUTION—*CNCl is a toxic gas; avoid inhalation.*) After the reaction is complete, CNCl forms a red-blue color on addition of a pyridine-barbituric acid reagent. Maximum color absorbance in aqueous solution is between 575 and 582 nm. To obtain colors of comparable intensity, have the same salt content in sample and standards.

b. Interference: All known interferences are eliminated or reduced to a minimum by distillation.

2. Apparatus

Colorimetric equipment: One of the following is required:

a. Spectrophotometer, for use at 578 nm, providing a light path of 10 mm or longer.

b. Filter photometer, providing a light path of at least 10 mm and equipped with a red filter having maximum transmittance at 570 to 580 nm.

3. Reagents

a. Chloramine-T solution: Dissolve 1.0 g white, water-soluble powder in 100 mL water. Prepare weekly and store in refrigerator.

b. Stock cyanide solution: Dissolve approximately 1.6 g NaOH and 2.51 g KCN in 1 L distilled water. (CAUTION—*KCN is highly toxic; avoid contact or inhalation.*) Standardize against standard silver nitrate (AgNO₃) titrant as described in Section 4500-CN⁻D.4, using 25 mL KCN solution. Check titer weekly because the solution gradually loses strength; 1 mL = 1 mg CN⁻.

c. Standard cyanide solution: Based on the concentration determined for the KCN stock solution (¶ 3*b*) calculate volume required (approximately 10 mL) to prepare 1 L of a 10 μg CN⁻/mL solution. Dilute with the NaOH dilution solution. Dilute 10 mL of the 10 μg CN⁻/mL solution to 100 mL with the NaOH dilution solution; 1.0 mL = 1.0 μg CN⁻. Prepare fresh daily and keep in a glass-stoppered bottle. (CAUTION—*Toxic; take care to avoid ingestion.*)

d. Pyridine-barbituric acid reagent: Place 15 g barbituric acid in a 250-mL volumetric flask and add just enough water to wash sides of flask and wet barbituric acid. Add 75 mL pyridine and mix. Add 15 mL conc hydrochloric acid (HCl), mix, and cool to room temperature. Dilute to volume and mix until barbituric acid is dissolved. The solution is stable for approximately 6 months if stored in an amber bottle under refrigeration; discard if precipitate develops.

e. Acetate buffer: Dissolve 410 g sodium acetate trihydrate, NaC₂H₃O₂·3H₂O, in 500 mL of water. Add glacial acetic acid to adjust to pH 4.5, approximately 500 mL.

f. Sodium hydroxide dilution solution: Dissolve 1.6 g NaOH in 1 L distilled water.

4. Procedure

a. Preparation of standard curve: Pipet a series of standards containing 1 to 10 μg CN⁻ into 50-mL volumetric flasks (0.02 to 0.2 μg CN⁻/mL). Dilute to 40 mL with NaOH dilution solution. Use 40 mL of NaOH dilution solution as blank. Develop and measure absorbance in 10-mm cells as described in ¶ *b* for both standards and blank. For concentrations lower than 0.02 μg CN⁻/mL use 100-mm cells.

Recheck calibration curve periodically and each time a new reagent is prepared.

b. Color development: Pipet a portion of absorption solution into a 50-mL volumetric flask and dilute to 40 mL with NaOH dilution solution. Add 1 mL acetate buffer and 2 mL chloramine-T solution, stopper, and mix by inversion twice. Let stand exactly 2 min.

Add 5 mL pyridine-barbituric acid reagent, dilute to volume with distilled water, mix thoroughly, and let stand exactly 8 min. Measure absorbance against distilled water at 578 nm.

Measure absorbance of blank (0.0 mg CN⁻/L) using 40 mL NaOH dilution solution and procedures for color development.

5. Calculation

Use the linear regression feature available on most scientific calculators, or compute slope and intercept of standard curve as follows:

$$m = \frac{n \sum ca - \sum c \sum a}{n \sum a^2 - (\sum a)^2}$$

$$b = \frac{\sum a^2 \sum c - \sum a \sum uc}{n \sum a^2 - (\sum a)^2}$$

where:

a = absorbance of standard solution,
c = concentration of CN^- in standard, mg/L,
n = number of standard solutions,
m = slope of standard curve, and
b = intercept on c axis.

Include the blank concentration, 0.0 mg CN^-/L and blank absorbance in the calculations above.

$$CN^-, mg/L = (ma_1 + b) \times \frac{50}{X} \times \frac{250}{Y}$$

where:

X = absorption solution, mL,
Y = original sample, mL, and
a_1 = absorbance of sample solution.

6. Precision and Bias[2]

Based on the results of nine operators in nine laboratories, the overall and single-operator precision of this method within its designated ranges may be expressed as follows:

Reagent water: $S_T = 0.06x + 0.003$
$S_o = 0.11x + 0.010$

Selected water matrices: $S_T = 0.04x + 0.018$
$S_o = 0.04x + 0.008$

where:

S_T = overall precision, mg/L,
S_o = single-operator precision, mg/L, and
x = cyanide concentration, mg/L.

Recoveries of known amounts of cyanide from reagent water and selected water matrices (coke plant and refinery wastes, sewage, and surface water) are:

Medium	Added mg/L	Recovered mg/L	n	S_T	Bias	% Bias
Reagent	0.060	0.060	26	0.0101	0.000	0
water	0.500	0.480	23	0.0258	−0.020	−4
	0.900	0.996	27	0.0669	0.096	11
Selected	0.060	0.060	25	0.0145	0.000	0
water	0.500	0.489	26	0.0501	−0.011	−3
matrices	0.900	0.959	24	0.0509	0.059	7

7. References

1. AMUS, E. & H. GARSCHAGEN. 1953. Über die Verwendung der Barbitsäure für die photometrische Bestimmund von Cyanid und Rhodanid. *Z. Anal. Chem.* 138:414.
2. AMERICAN SOCIETY FOR TESTING & MATERIALS. 1987. Research Rep. D2036:19-1131. American Soc. Testing & Materials, Philadelphia, Pa.

4500-CN⁻ F.   Cyanide-Selective Electrode Method

1. General Discussion

CN^- in the alkaline distillate from the preliminary treatment procedures can be determined potentiometrically by using a CN^--selective electrode in combination with a double-junction reference electrode and a pH meter having an expanded millivolt scale, or a specific ion meter. This method can be used to determine CN^- concentration in place of either the colorimetric or titrimetric procedures in the concentration range of 0.05 to 10 mg CN^-/L.[1-3] If the CN^--selective electrode method is used, the previously described titration screening step can be omitted.

2. Apparatus

a. *Expanded-scale pH meter or specific-ion meter.*
b. *Cyanide-ion-selective electrode.**
c. *Reference electrode,* double-junction.
d. *Magnetic mixer* with TFE-coated stirring bar.
e. *Koch microburet,* 10-mL capacity.

3. Reagents

a. *Stock standard cyanide solution:* See Section 4500-CN^-.E.3*b*.

_____
* Orion Model 94-06A or equivalent.

b. *Sodium hydroxide dilution solution:* Dissolve 1.6 g NaOH in water and dilute to 1 L.
c. *Standard cyanide solution:* Dilute a calculated volume (approximately 25 mL) of stock KCN solution, based on the determined concentration, to 1000 mL with NaOH diluent. Mix thoroughly; 1 mL 25 µg CN^-.
d. *Dilute standard cyanide solution:* Dilute 100.0 mL standard CN^- solution to 1000 mL with NaOH diluent; 1.00 mL = 2.5 µg CN^-. Prepare daily and keep in a dark, glass-stoppered bottle.
e. *Potassium nitrate solution:* Dissolve 100 g KNO₃ in water and dilute to 1 L. Adjust to pH 12 with KOH. This is the outer filling solution for the double-junction reference electrode.

4. Procedure

a. *Calibration:* Using Koch microburet and standard CN^- solution, prepare four (or more) additional solutions containing 2.5, 0.25, 0.125, and 0.025 µg CN^-/mL in NaOH dilution solution. Transfer approximately 100 mL of each of these standard solutions into a 250-mL beaker prerinsed with a small portion of standard being tested. Immerse CN^- and double-junction reference electrodes. Mix well on a magnetic stirrer at 25°C, maintaining as closely as possible the same stirring rate for all solutions.

Always progress from the lowest to the highest concentration of standard because otherwise equilibrium is reached only slowly.

The electrode membrane dissolves in solutions of high CN^- concentration; do not use with a concentration above 25 µg CN^-/mL. After making measurements remove electrode and soak in water.

After equilibrium is reached (at least 5 min and not more than 10 min), record potential (millivolt) readings. Plot CN^- concentration on logarithmic axis of semilogarithmic paper versus potential developed in solution on linear axis. A straight line with a slope of approximately 59 mV per decade indicates that the instrument and electrodes are operating properly. Record slope of line obtained (millivolts/decade of concentration). The slope may vary somewhat from the theoretical value of 59.2 mV per decade because of manufacturing variation and reference electrode (liquid-junction) potentials. The slope should be a straight line and is the basis for calculating sample concentration. Follow manufacturer's instructions for direct-reading ion meters.

b. *Measurement of sample:* Place 100 mL of absorption liquid obtained in Section 4500-CN⁻.C.4d (or an accurately measured portion diluted to 100 mL with NaOH dilution solution) into a 250-mL beaker. When measuring low CN^- concentrations, first rinse beaker and electrodes with a small volume of sample. Immerse CN^- and double-junction reference electrodes and mix on a magnetic stirrer at the same stirring rate used for calibration. After equilibrium is reached (at least 5 min and not more than 10 min), record values indicated on ion meter or found from graph prepared as above. Calculate concentration as directed below.

5. Calculations

$$\text{mg } CN^-/L = \mu g \ CN^-/mL \text{ from graph or meter} \times \frac{100}{x} \times \frac{250}{y}$$

where:

  x = volume of absorption solution, mL, and
  y = volume of original sample, mL.

6. Precision and Bias[4]

The precision of the CN^--ion-selective electrode method using the absorption solution from total cyanide distillation has been found in collaborative testing to be linear within its designated range.

Based on the results of six operators in five laboratories, the overall and single-operator precision of this method within its designated range may be expressed as follows:

Reagent water: $S_T = 0.06x + 0.003$
$S_o = 0.03x + 0.008$

Selected water matrices: $S_T = 0.05x + 0.008$
$S_o = 0.03x + 0.012$

where:

  S_T = overall precision, mg/L,
  S_o = single-operator precision, mg/L, and
  x = cyanide concentration, mg/L.

Recoveries of known amounts of cyanide from reagent water and selected water matrices are:

Medium	Added mg/L	Recovered mg/L	n	S_T	Bias	% Bias
Reagent	0.060	0.059	18	0.0086	−0.001	2
water	0.500	0.459	18	0.0281	−0.041	−8
	0.900	0.911	18	0.0552	0.011	1
	5.00	5.07	18	0.297	0.07	1
Selected	0.060	0.058	14	0.0071	−0.002	−3
water	0.500	0.468	21	0.0414	−0.032	−6
matrices	0.900	0.922	19	0.0532	0.022	2
	5.00	5.13	20	0.2839	0.13	3

7. References

1. ORION RESEARCH, INC. 1975. Cyanide Ion Electrode Instruction Manual. Cambridge, Mass.
2. FRANT, M.S., J.W. ROSS & J.H. RISEMAN. 1972. An electrode indicator technique for measuring low levels of cyanide. *Anal. Chem.* 44:2227.
3. SEKERKA, J. & J.F. LECHNER. 1976. Potentiometric determination of low levels of simple and total cyanides. *Water Res.* 10:479.
4. AMERICAN SOCIETY FOR TESTING & MATERIALS. 1987. Research Rep. D2036:19-1131. American Soc. Testing & Materials, Philadelphia, Pa.

4500-CN⁻ G.   Cyanides Amenable to Chlorination after Distillation

1. General Discussion

This method is applicable to the determination of cyanides amenable to chlorination.

After part of the sample is chlorinated to decompose the cyanides, both the chlorinated and the untreated sample are subjected to distillation as described in Section 4500-CN⁻.C. The difference between the CN^- concentrations found in the two samples is expressed as cyanides amenable to chlorination.

Some unidentified organic chemicals may oxidize or form breakdown products during chlorination, giving higher results for cyanide after chlorination than before chlorination. This may lead

to a negative value for cyanides amenable to chlorination after distillation for wastes from, for example, the steel industry, petroleum refining, and pulp and paper processing. Where such interferences are encountered use Method 4500-CN⁻.I for determining dissociable cyanide.

Protect sample from exposure to ultraviolet radiation, and perform manipulations under incandescent light, to prevent photodecomposition of some metal-cyanide complexes by ultraviolet light.

2. Apparatus

a. *Distillation apparatus:* See Section 4500-CN⁻.C.2
b. *Apparatus for determining cyanide* by either the titrimetric

method, Section 4500-CN⁻.D.2, the colorimetric method, Section 4500-CN⁻.E.2, or the electrode method, Section 4500-CN⁻.F.2.

3. Reagents

a. All reagents listed in Section 4500-CN⁻.C.3.

b. All reagents listed in Section 4500-CN⁻.D.3, 4500-CN⁻.E.3, or 4500-CN⁻.F.3, depending on method of estimation.

c. Calcium hypochlorite solution: Dissolve 5 g Ca(OCl)$_2$ in 100 mL distilled water. Store in an amber-colored glass bottle in the dark. Prepare monthly.

d. Potassium iodide(KI)-starch test paper.

4. Procedure

a. Divide sample into two equal portions of 500 mL (or equal portions diluted to 500 mL) and chlorinate one as in ¶ *b* below. Analyze both portions for CN⁻. The difference in determined concentrations is the cyanide amenable to chlorination.

b. Place one portion in a 1-L beaker covered with aluminum foil or black paper. Keep beaker covered with a wrapped watch glass during chlorination. Add Ca(OCl)$_2$ solution dropwise to sample while agitating and maintaining pH between 11 and 12 by adding NaOH solution. Test for chlorine by placing a drop of treated sample on a strip of KI-starch paper. A distinct blue color indicates sufficient chlorine (approximately 50 to 100 mg Cl$_2$/L). Maintain excess residual chlorine for 1 h while agitating. If necessary, add more Ca(OCl)$_2$ and/or NaOH.

c. After 1 h remove any residual chlorine by dropwise addition of NaAsO$_2$ solution (2 g/100 mL) or by addition of 8 drops H$_2$O$_2$ (3%) followed by 4 drops Na$_2$S$_2$O$_3$ solution (500 g/L). Test with KI-starch paper until there is no color change.

d. Distill both chlorinated and unchlorinated samples as in Section 4500-CN⁻.C. Test according to Methods D, E, or F.

5. Calculation

$$\text{mg CN}^- \text{ amenable to chlorination/L} = G - H$$

where:

G = mg CN⁻/L found in unchlorinated portion of sample and
H = mg CN⁻/L found in chlorinated portion of sample.

For samples containing significant quantities of iron cyanides, it is possible that the second distillation will give a higher value for CN⁻ than the test for total cyanide, leading to a negative result. When the difference is within the precision limits of the method, report, "no determinable quantities of cyanide amenable to chlorination." If the difference is greater than the precision limit, ascertain the cause such as presence of interferences, manipulation of the procedure, etc., or use Method I.

6. Precision and Bias[1]

The precision and bias information given in this section may not apply to waters of untested matrices.

a. Precision:

1) *Colorimetric*—Based on the results of eight operators in seven laboratories, the overall and single-operator precision of this test method within its designated range may be expressed as follows:

$$\text{Reagent water: } S_T = 0.18x + 0.005$$
$$S_o = 0.06x + 0.003$$

$$\text{Selected water matrices: } S_T = 0.20x + 0.009$$
$$S_o = 0.05x + 0.005$$

2) *Titrimetric*—Based on the results of six operators in three laboratories, the overall and single-operator precision of this test method within its designated range may be expressed as follows:

$$\text{Reagent water: } S_T = 0.01x + 0.439$$
$$S_o = 0.241 - 0.03x$$

$$\text{Selected water matrices: } S_T = 0.12x + 0.378$$
$$S_o = 0.209 - 0.01x$$

where:

S_T = overall precision, mg/L,
S_o = single-operator precision, mg/L, and
x = cyanide concentration, mg CN⁻/L

b. Bias: Recoveries of known amount of cyanide amenable to chlorination from reagent water and selected water matrices are shown below:

Medium	Technique	Added mg/L	Recovered mg/L	n	S_T	Bias	% Bias
Reagent water	Colorimetric	0.008	0.009	21	0.0033	0.001	13
		0.019	0.023	20	0.0070	0.004	21
		0.080	0.103	20	0.0304	0.018	23
		0.191	0.228	21	0.0428	0.037	19
	Titrimetric	1.00	0.73	18	0.350	−0.27	−27
		1.00	0.81	18	0.551	−0.19	−19
		4.00	3.29	18	0.477	−0.71	−18
Selected water matrices	Colorimetric	0.008	0.013	17	0.0077	0.005	63
		0.019	0.025	18	0.0121	0.006	32
		0.080	0.100	18	0.0372	0.020	25
		0.191	0.229	18	0.0503	0.038	20
	Titrimetric	1.00	1.20	18	0.703	0.20	20
		1.00	1.10	18	0.328	0.10	10
		4.00	3.83	18	0.818	−0.17	−4

7. Reference

1. AMERICAN SOCIETY FOR TESTING & MATERIALS. 1987. Research Rep. D2036:19-1131. American Soc. Testing & Materials, Philadelphia, Pa.

4500-CN⁻ H. Cyanides Amenable to Chlorination without Distillation (Short-Cut Method)

1. General Discussion

This method covers the determination of HCN and of CN complexes that are amenable to chlorination and also thiocyanates (SCN⁻). The procedure does not measure cyanates (CNO⁻) or iron cyanide complexes, but does determine cyanogen chloride (CNCl). It may be modified for use in presence of thiocyanates. The method requires neither lengthy distillation nor the chlorination of one sample before distillation. The recovery of CN⁻ from metal cyanide complexes will be comparable to that in Methods G and I.

The cyanides are converted to CNCl by chloramine-T after the sample has been heated. In the absence of nickel, copper, silver, and gold cyanide complexes or SCN⁻, the CNCl may be developed at room temperature. The pyridine-barbituric acid reagent produces a red-blue color in the sample. The color can be estimated visually against standards or photometrically at 578 nm. The dissolved salt content in the standards used for the development of the calibration curve should be near the salt content of the sample, including the added NaOH and phosphate buffer.

The method's usefulness is limited by thiocyanate interference. Although the procedure allows the specific determination of CN⁻ amenable to chlorination (see 4500-CN⁻.H.2 and 5) by masking the CN⁻ content and thereby establishing a correction for the thiocyanide content, the ratio of SCN⁻ to CN⁻ should not exceed 3 to be applicable. In working with unknown samples, screen the sample for SCN⁻ by the spot test (4500-CN⁻.K).

2. Interferences

a. Remove interfering agents as described in Section 4500-CN⁻.B with the exception of NO₂⁻ and NO₃⁻ (4500-CN⁻.B.3h).

b. The SCN⁻ ion reacts with chloramine-T to give a positive error equivalent to its concentration. The procedure allows the separate determination of SCN⁻ and subtraction of this value from the results for the total. Use the spot test (4500-CN⁻.K) for SCN⁻ when its presence is suspected. If the SCN⁻ content is more than three times the CN⁻ content, use Method G or I.

c. Reducing chemical compounds, such as SO₃²⁻, may interfere by consuming chlorine in the chloramine-T addition. A significant excess of chlorine is provided, but the procedure prescribes a test (4500-CN⁻.H.5d) to avoid this interference.

d. Color and turbidity may interfere with the colorimetric determination. Overcome this interference by extraction with chloroform (4500-CN⁻.B.3c) but omit reduction of the pH. Otherwise, use Method G or I.

Compensation for color and turbidity may be made by determining absorbance of a second sample solution to which all reagents except chloramine-T have been added.

e. Color intensity and absorption are affected by wide variations in total dissolved solids content of the sample.

For samples containing high concentrations of dissolved solids (3000 to 10 000 mg/L), add 6 g NaCl/L NaOH solution (1.6 g/L) used to prepare standards. For samples containing dissolved solids concentrations greater than 10 000 mg/L, add sufficient NaCl to the NaOH solution to approximate the dissolved solids content.

3. Apparatus

a. *Apparatus listed in 4500-CN⁻.E.2.*
b. *Hot water bath.*

4. Reagents

a. *Reagents listed in Sections 4500-CN⁻.B and E.3.*
b. *Sodium chloride,* NaCl, crystals.
c. *Sodium carbonate,* Na₂CO₃, crystals.
d. *Sulfuric acid solution,* H₂SO₄, 1N.
e. *EDTA solution,* 0.05M: Dissolve 18.5 g disodium salt of ethylenediamine tetraacetic acid in water and dilute to 1 L.
f. *Formaldehyde solution,* 10%: Dilute 27 mL formaldehyde (37% pharmaceutical grade) to 100 mL.
g. *Phosphate buffer:* Dissolve 138 g sodium dihydrogen phosphate monohydrate, NaH₂PO₄·H₂O, in water and dilute to 1 L. Refrigerate.

5. Procedure

a. Calibrate as directed in Section 4500-CN⁻.E.1a and 4a. For samples with more than 3000 mg total dissolved solids/L, prepare a calibration curve from standards and blank NaOH solutions containing 6 g NaCl/L. Samples containing total dissolved solids exceeding 10 000 mg/L require appropriate standards and a new calibration curve.

b. Adjust pH of 50 mL sample to between 11.4 and 11.8. If acid is needed, add a small amount (0.2 to 0.4 g) of sodium carbonate and stir to dissolve. Then add HCl solution (1 + 9) dropwise while stirring. For raising the pH, use NaOH solution (40 g/L).

c. Pipet 20.0 mL of adjusted sample into a 50-mL volumetric flask. If the cyanide concentration is greater than 0.3 mg/L, use a smaller portion and dilute to 20 mL with NaOH solution. Do not exceed the concentration limit of 0.3 mg/L.

d. To insure uniform color development, both in calibration and testing, maintain a uniform temperature. Immerse flasks in a water bath held at 27 ± 1°C for 10 min before adding reagents and keep samples in water bath until all reagents have been added.

Add 4 mL phosphate buffer and swirl to mix. Add one drop of EDTA solution, and mix.

e. Add 2 mL chloramine-T solution and swirl to mix. Place 1 drop of sample on potassium iodide-starch test paper that has been moistened previously with acetate buffer solution. Repeat the chloramine-T addition if required. After exactly 3 min, add 5 mL pyridine-barbituric acid reagent and swirl to mix.

f. Remove samples from water bath, dilute to volume, and mix. Allow 8 min from the addition of the pyridine-barbituric acid reagent for color development.

Determine absorbance at 578 nm in a 1.0-cm cell versus distilled water.

Calculate concentration of cyanide, mg/L in the original sample following instructions given in 4500-CN⁻ E.

g. If the presence of thiocyanate is suspected, pipet a second 20-mL portion of pH-adjusted sample into a 50-mL volumetric flask. Add 3 drops 10% formaldehyde solution. Mix and let stand

10 min. Place in a water bath at $27 \pm 1°C$ for an additional 10 min before the addition of the reagent chemicals and hold in the bath until all reagents have been added.

Continue with *b* above.

Calculate the concentration of cyanide, as milligrams per liter, in the original sample following instructions given in 4500-CN⁻. E.

h. In the presence of thiocyanate, cyanide amenable to chlorination is equal to the difference between the concentrations of cyanide obtained in *f* and *g*.

6. Calculation

See 4500-CN⁻.E.5.

Deduct SCN⁻ value from the results found when the CN⁻ has not been masked by formaldehyde addition (total) for cyanide content.

7. Precision and Bias[1]

This precision and bias information may not apply to waters of untested matrices.

a. Precision: Based on the results of 14 operators in nine laboratories, the overall and single-operator precision of this test method within its designated range may be expressed as follows:

$$\text{Reagent water: } S_T = 0.10x + 0.006$$
$$S_o = 0.07x + 0.005$$

$$\text{Selected water matrices: } S_T = 0.11x + 0.007$$
$$S_o = 0.02x + 0.005$$

where:

S_T = overall precision, mg/L,
S_o = single-operator precision, mg/L, and
x = cyanide concentration, mg/L.

b. Bias: Recoveries of known amounts of cyanide from reagent water and selected water matrices including creek waters, diluted sewage (1 to 20), and industrial wastewater are shown below.

Medium	Added mg/L CN⁻	SCN⁻	Recovered mg/L	n	S_T	Bias	% Bias
Reagent	0.005		0.007	42	0.0049	0.002	40
water	0.027		0.036	41	0.0109	0.009	25
	0.090		0.100	42	0.0167	0.010	11
	0.090	0.080	0.080	39	0.0121	−0.010	11
	0.270		0.276	42	0.0320	0.006	2
Selected	0.005		0.003	40	0.0042	−0.002	40
water	0.027		0.026	42	0.0093	−0.001	4
matrices	0.090		0.087	42	0.0202	−0.003	3
	0.090	0.080	0.068	37	0.0146	−0.022	24
	0.270		0.245	41	0.0319	−0.025	9

8. Reference

1. AMERICAN SOCIETY FOR TESTING & MATERIALS. 1987. Research Rep. D2036:19-1074. American Soc. Testing & Materials, Philadelphia, Pa.

4500-CN⁻ I.   Weak Acid Dissociable Cyanide

1. General Discussion

Hydrogen cyanide (HCN) is liberated from a slightly acidified (pH 4.5 to 6.0) sample under the prescribed distillation conditions. The method does not recover CN⁻ from tight complexes that would not be amenable to oxidation by chlorine. The acetate buffer used contains zinc salts to precipitate iron cyanide as a further assurance of the selectivity of the method. In other respects the method is similar to 4500-CN⁻.C.

2. Interferences

See 4500-CN⁻.B.3.

Protect sample and apparatus from ultraviolet light to prevent photodecomposition of some metal-cyanide complexes and an increase in concentration of weak acid dissociable cyanide.

If procedure is used to determine low concentrations of cyanide in samples of ferri- and ferrocyanide, add more, e.g., fivefold excess, zinc acetate solution before adding acid and distilling.

3. Apparatus

See Section 4500-CN⁻.C.2 and Figure 4500-CN⁻:1, and also Section 4500-CN⁻.D.2, 4500-CN⁻.E.2, or 4500-CN⁻.F.2, depending on method of estimation.

4. Reagents

a. Reagents listed in Section 4500-CN⁻.C.3.

b. Reagents listed in Section 4500-CN⁻.D.3, 4500-CN⁻.E.3, or 4500-CN⁻.F.3, depending on method of estimation.

c. Acetic acid, 1 + 9: Mix 1 volume of glacial acetic acid with 9 volumes of water.

d. Acetate buffer: Dissolve 410 g sodium acetate trihydrate ($NaC_2H_3O_2 \cdot 3H_2O$) in 500 mL water. Add glacial acetic acid to yield a solution pH of 4.5 (approximately 500 mL).

e. Zinc acetate solution, 100 g/L: Dissolve 120 g $Zn(C_2H_3O_2)_2 \cdot 2H_2O$ in 500 mL water. Dilute to 1 L.

f. Methyl red indicator.

5. Procedure

Follow procedure described in 4500-CN⁻.C.4, but with the following modifications:

a. Do not add sulfamic acid, because NO_2^- and NO_3^- do not interfere.

b. Instead of H_2SO_4 and $MgCl_2$ reagents, add 20 mL each of the acetate buffer and zinc acetate solutions through air inlet tube. Also add 2 to 3 drops methyl red indicator. Rinse air inlet tube with water and let air mix contents. If the solution is not pink, add acetic acid $(1 + 9)$ dropwise through air inlet tube until a pink color persists.

c. Follow instructions beginning with 4500-CN⁻.C.4*d*.

d. For determining CN⁻ in the absorption solution, use the preferred finish method (4500-CN⁻.D, E, or F).

6. Precision and Bias[1]

The precision and bias information given in this section may not apply to waters of untested matrices.

a. Precision:

1) Colorimetric—Based on the results of nine operators in nine laboratories, the overall and single-operator precision of this test method within its designated range may be expressed as follows:

$$\text{Reagent water: } S_T = 0.09x + 0.010$$
$$S_o = 0.08x + 0.005$$

$$\text{Selected water matrices: } S_T = 0.08x + 0.012$$
$$S_o = 0.05x + 0.008$$

2) Electrode—Based on the results of six operators in five laboratories, the overall and single-operator precision of this test method within its designated range may be expressed as follows:

$$\text{Reagent water: } S_T = 0.09x + 0.004$$
$$S_o = 0.02x - 0.009$$

$$\text{Selected water matrices: } S_T = 0.08x + 0.005$$
$$S_o = 0.02x + 0.004$$

3) Titrimetric—Based on the results of six operators in three laboratories, the overall and single-operator precision of this test method within its designated range may be expressed as follows:

$$\text{Reagent water: } S_T = 0.532 - 0.10x$$
$$S_o = 0.151 - 0.01x$$

$$\text{Selected water matrices: } S_T = 0.604 - 0.06x$$
$$S_o = 0.092 + 0.02x$$

where:

S_T = overall precision,
S_o = single-operator precision, and
x = cyanide concentration, mg/L.

b. Bias: Recoveries of known amounts of cyanide from reagent water and selected water matrices are shown below.

Medium	Technique	Added mg/L	Recovered mg/L	n	S_T	Bias	% Bias
Reagent water	Colorimetric	0.030	0.030	25	0.0089	0.000	0
		0.100	0.117	27	0.0251	0.017	17
		0.400	0.361	27	0.0400	−0.039	−10
	Electrode	0.030	0.030	21	0.0059	0.000	0
		0.100	0.095	21	0.0163	−0.005	−5
		0.400	0.365	21	0.0316	−0.035	−9
		1.000	0.940	21	0.0903	−0.060	−6
	Titrimetric	1.00	1.35	18	0.4348	0.35	35
		1.00	1.38	18	0.3688	0.38	38
		4.00	3.67	18	0.1830	−0.33	−8
Selected water matrices	Colorimetric	0.030	0.029	15	0.0062	0.001	3
		0.100	0.118	24	0.0312	0.018	18
		0.400	0.381	23	0.0389	−0.019	−5
	Electrode	0.030	0.029	20	0.0048	−0.001	−3
		0.100	0.104	21	0.0125	0.004	4
		0.400	0.357	21	0.0372	−0.043	−11
		1.000	0.935	21	0.0739	−0.065	−7
	Titrimetric	1.00	1.55	18	0.5466	0.55	55
		1.00	1.53	18	0.4625	0.53	53
		4.00	3.90	18	0.3513	−0.10	−3

7. Reference

1. AMERICAN SOCIETY FOR TESTING & MATERIALS. 1987. Research Rep. D2036:19-1131. American Soc. Testing & Materials, Philadelphia, Pa.

4500-CN⁻ J. Cyanogen Chloride

1. General Discussion

Cyanogen chloride (CNCl) is the first reaction product when cyanide compounds are chlorinated. It is a volatile gas, only slightly soluble in water, but highly toxic even in low concentrations. (CAUTION: *Avoid inhalation or contact.*) A mixed pyridine-barbituric acid reagent produces a red-blue color with CNCl.

Because CNCl hydrolyzes to cyanate (CNO⁻) at a pH of 12 or more, collect a separate sample for CNCl analysis (See Section 4500-CN⁻.B.2) in a closed container without sodium hydroxide (NaOH). A quick test with a spot plate or comparator as soon as the sample is collected may be the only procedure for avoiding hydrolysis of CNCl due to time lapse between sampling and analysis.

If starch-iodide (KI) test paper indicates presence of chlorine or other oxidizing agents, add sodium thiosulfate ($Na_2S_2O_3$) immediately as directed in Section 4500-CN⁻.B.2.

2. Apparatus

See Section 4500-CN⁻.E.2.

3. Reagents

a. Reagents listed in Sections 4500-CN⁻.E.3 and 4500-CN⁻.H.4.

b. Phosphate buffer: Dissolve 138 g sodium dihydrogen phosphate monohydrate, $NaH_2PO_4 \cdot H_2O$, in water and dilute to 1 L. Refrigerate.

4. Procedure

a. Preparation of standard curve: Pipet a series of standards containing 1 to 10 μg CN^- into 50-mL volumetric flasks (0.02 to 0.2 μg CN^-/mL). Dilute to 20 mL with NaOH dilution solution. Use 20 mL of NaOH dilution solution for the blank. Add 2 mL chloramine-T solution and 4 mL phosphate buffer; stopper and mix by inversion two or three times. Add 5 mL pyridine-barbituric acid reagent, dilute to volume with water, mix thoroughly, and let stand exactly 8 min for color development. Measure absorbance at 578 nm in a 10-mm cell using distilled water as a reference. Calculate slope and intercept of the curve.

b. If sample pH is above 8, reduce it to 8.0 to 8.5 by careful addition of phosphate buffer. Measure 20 mL sample portion into 50-mL volumetric flask. If more than 0.20 mg $CNCl\text{-}CN^-$/L is present use a smaller portion diluted to 20 mL with water. Add 1 mL phosphate buffer, stopper and mix by inversion *one* time. Let stand 2 min. Add 5 mL pyridine-barbituric acid reagent, stopper and mix by inversion *one* time. Let color develop 3 min, dilute to volume with water, mix thoroughly, and let stand an additional 5 min. Measure absorbance at 578 nm in 10-mm cell using distilled water as a reference.

5. Calculation

Compute slope (m) and intercept (b) of standard curve as directed in 4500-CN^-.E.5.

$$\text{Cyanogen chloride as } CN^-, \text{ mg/L} = (ma_1 + b) \times \frac{50}{\text{mL sample}}$$

where:

a_1 = absorbance of sample solution.

6. Precision[1]

Cyanogen chloride is unstable and round-robin testing is not possible. Single-operator precision is as follows:

Six operators made 70 duplicate analyses on samples of different concentrations within the applicable range of the method. The overall single-operator precision within its designated range may be expressed as follows:

$$\log S_o = (0.5308 \log c) - 1.9842$$
$$\log R = (0.5292 \log c) - 1.8436$$

where:

c = mg $CNCl\text{-}CN^-$/L,
S_o = single-operator precision in the range of the method (precision is dependent on concentration), and
R = range between duplicate determinations.

The collaborative test data were obtained on reagent-grade water. For other matrices, these data may not apply.

7. Reference

1. AMERICAN SOCIETY FOR TESTING & MATERIALS. 1989. Research Rep. D4165:19-1100. American Soc. Testing & Materials, Philadelphia, Pa.

4500-CN^- K.   Spot Test for Sample Screening

1. General Discussion

The spot test procedure permits quick screening to establish whether more than 50 μg/L of cyanide amenable to chlorination is present. The test also establishes the presence or absence of cyanogen chloride (CNCl). With practice and dilution, the test reveals the approximate concentration range of these compounds by the color development compared with that of similarly treated standards.

When chloramine-T is added to cyanides amenable to chlorination, CNCl is formed. CNCl forms a red-blue color with the mixed reagent pyridine-barbituric acid. When testing for CNCl omit the chloramine-T addition. (CAUTION: *CNCl is a toxic gas; avoid inhalation.*)

The presence of formaldehyde in excess of 0.5 mg/L interferes with the test. A spot test for the presence of aldehydes and a method for removal of this interference are given in Section 4500-CN^-.B.3.

Thiocyanate (SCN^-) reacts with chloramine-T, thereby creating a positive interference. The CN^- can be masked with formaldehyde and the sample retested. This makes the spot test specific for SCN^-. In this manner it is possible to determine whether the spot discoloration is due to the presence of CN^-, SCN^-, or both.

2. Apparatus

a. Porcelain spot plate with 6 to 12 cavities.
b. Dropping pipets.
c. Glass stirring rods.

3. Reagents

a. Chloramine-T solution: See Section 4500-CN^-.E.3*a*.
b. Stock cyanide solution: See Section 4500-CN^-.E.3*b*.
c. Pyridine-barbituric acid reagent: See Section 4500-CN^-.E.3*d*.
d. Hydrochloric acid, HCl, 1 + 9.
e. Phenolphthalein indicator aqueous solution.
f. Sodium carbonate, Na_2CO_3, anhydrous.
g. Formaldehyde, 37%, pharmaceutical grade.

4. Procedure

If the solution to be tested has a pH greater than 10, neutralize a

20- to 25-mL portion. Add about 250 mg Na$_2$CO$_3$ and swirl to dissolve. Add 1 drop phenolphthalein indicator. Add 1 + 9 HCl dropwise with constant swirling until the solution becomes colorless. Place 3 drops sample and 3 drops distilled water (for blanks) in separate cavities of the spot plate. To each cavity, add 1 drop chloramine-T solution and mix with a clean stirring rod. Add 1 drop pyridine-barbituric acid solution to each cavity and again mix. After 1 min, the sample spot will turn pink to red if 50 µg/L or more of CN⁻ are present. The blank spot will be faint yellow because of the color of the reagents. Until familiarity with the spot test is gained, use, in place of the water blank, a standard solution containing 50 µg CN⁻/L for color comparison. This standard can be made by diluting the stock cyanide solution (¶ 3b).

If SCN⁻ is suspected, test a second sample pretreated as follows: Heat a 20- to 25-mL sample in a water bath at 50°C; add 0.1 mL formaldehyde and hold for 10 min. This treatment will mask up to 5 mg CN⁻/L, if present. Repeat spot testing procedure. Color development indicates presence of SCN⁻. Comparing color intensity in the two spot tests is useful in judging relative concentration of CN⁻ and SCN⁻. If deep coloration is produced, serial dilution of sample and additional testing may allow closer approximation of the concentrations.

4500-CN⁻ L. Cyanates

1. General Discussion

Cyanate (CNO⁻) may be of interest in analysis of industrial waste samples because the alkaline chlorination process used for the oxidation of cyanide yields cyanate in the second reaction.

Cyanate is unstable at neutral or low pH; therefore, stabilize the sample as soon as collected by adding sodium hydroxide (NaOH) to pH >12. Remove residual chlorine by adding sodium thiosulfate (Na$_2$S$_2$O$_3$) (see Section 4500-CN⁻.B.2).

a. *Principle:* Cyanate hydrolyzes to ammonia when heated at low pH.

$$2NaCNO + H_2SO_4 + 4H_2O \;N\; (NH_4)_2SO_4 + 2NaHCO_3$$

The ammonia concentration must be determined on one sample portion before acidification. The ammonia content before and after hydrolysis of cyanate may be measured by phenate (4500-NH$_3$.F), or ammonia-selective electrode (4500-NH$_3$.D) method.[1] The test is applicable to cyanate compounds in natural waters and industrial waste.

b. *Interferences:*

1) Organic nitrogenous compounds may hydrolyze to ammonia (NH$_3$) upon acidification. To minimize this interference, control acidification and heating closely.

2) Reduce oxidants that oxidize cyanate to carbon dioxide and nitrogen with Na$_2$S$_2$O$_3$ (see Section 4500-CN⁻.G).

3) Industrial waste containing organic material may contain unknown interferences.

c. *Detection limit:* 1 to 2 mg CNO⁻/L.

2. Apparatus

a. *Expanded-scale pH meter or selective-ion meter.*
b. *Ammonia-selective electrode.**
c. *Magnetic mixer,* with TFE-coated stirring bar.

* Orion Model 95-10, EIL Model 8002-2, Beckman Model 39565, or equivalent.

d. *Heat barrier:* Use a 3-mm-thick insulator under beaker to insulate against heat produced by stirrer motor.

3. Reagents

a. *Stock ammonium chloride solution:* Dissolve 3.819 g anhydrous NH$_4$Cl, dried at 100°C, in water, and dilute to 1 L; 1.00 mL = 1.00 mg N = 1.22 mg NH$_3$.

b. *Standard ammonium chloride solution:* From the stock NH$_4$Cl solution prepare standard solutions containing 1.0, 10.0, and 100.0 mg NH$_3$-N/L by diluting with ammonia-free water.

c. *Sodium hydroxide, 10N:* Dissolve 400 g NaOH in water and dilute to 1 L.

d. *Sulfuric acid solution, H$_2$SO$_4$, 1 + 1.*

e. *Ammonium chloride solution:* Dissolve 5.4 g NH$_4$Cl in distilled water and dilute to 1 L. (Use only for soaking electrodes.)

4. Procedure

a. *Calibration:* Daily, calibrate the ammonia electrode as in 4500-NH$_3$.F.4b and c using standard NH$_4$Cl solutions.

b. *Treatment of sample:* Dilute sample, if necessary, so that the CNO⁻ concentration is 1 to 200 mg/L or NH$_3$-N is 0.5 to 100 mg/L. Take or prepare at least 200 mL. From this 200 mL, take a 100-mL portion and, following the calibration procedure, establish the potential (millivolts) developed from the sample. Check electrode reading with prepared standards and adjust instrument calibration setting daily. Record NH$_3$-N content of untreated sample (B).

Acidify 100 mL of prepared sample by adding 0.5 mL 1 + 1 H$_2$SO$_4$ to a pH of 2.0 to 2.5. Heat sample to 90 to 95°C and maintain temperature for 30 min. Cool to room temperature and restore to original volume by adding ammonia-free water. Pour into a 150-mL beaker, immerse electrode, start magnetic stirrer, then add 1 mL 10N NaOH solution. With pH paper check that pH is greater than 11. If necessary, add more NaOH until pH 11 is reached.

After equilibrium has been reached (30 s) record the potential reading. Estimate NH$_3$-N content from calibration curve.

5. Calculations

$$\text{mg NH}_3\text{-N derived from CNO}^-/\text{L} = A - B$$

where:

A = mg NH_3-N/L found in the acidified and heated sample portion
and
B = mg NH_3-N/L found in untreated portion.

$$\text{mg CNO}^-/\text{L} = 3.0 \times (A - B)$$

6. Precision

No data on precision of this method are available. See 4500-NH_3.A.4 for precision of ammonia-selective electrode method.

7. Reference

1. THOMAS, R.F. & R.L. BOOTH. 1973. Selective electrode determination of ammonia in water and wastes. *Environ. Sci. Technol.* 7:523.

4500-CN⁻ M. Thiocyanate

1. General Discussion

When wastewater containing thiocyanate (SCN^-) is chlorinated, highly toxic cyanogen chloride (CNCl) is formed. At an acidic pH, ferric ion (Fe^{3+}) and SCN^- form an intense red color suitable for colorimetric determination.

a. Interference:

1) Hexavalent chromium (Cr^{6+}) interferes and is removed by adding ferrous sulfate ($FeSO_4$) after adjusting to pH 1 to 2 with nitric acid (HNO_3). Raising the pH to 9 with $1N$ sodium hydroxide (NaOH) precipitates Fe^{3+} and Cr^{3+}, which are then filtered out.

2) Reducing agents that reduce Fe^{3+} to Fe^{2+}, thus preventing formation of ferric thiocyanate complex, are destroyed by adding a few drops of hydrogen peroxide (H_2O_2). Avoid excess H_2O_2 to prevent reaction with SCN^-.

3) Industrial wastes may be highly colored or contain various interfering organic compounds. To eliminate these interferences,[1] use the pretreatment procedure given in ¶ 4c below. It is the analyst's responsibility to validate the method's applicability without pretreatment (¶ 4b). If in doubt, pretreat sample before proceeding with analysis (¶ 4c).

4) If sample contains cyanide amenable to chlorination and would be preserved for the cyanide determination at a high pH, sulfide could interfere by converting cyanide to SCN^-. To preserve SCN^- and CN^-, precipitate the sulfide by adding lead salts according to 4500-CN^-.B.2 before adding alkali; filter to remove precipitate.

5) Thiocyanate is biodegradable. Preserve samples at pH <2 by adding mineral acid and refrigerate.

6) If interferences from industrial wastes are not removed as directed in ¶ 4c below, consider adopting a solvent extraction technique with colorimetric or atomic absorption analysis of the extract.[2,3]

b. Application: 0.1 to 2.0 mg SCN^-/L in natural or wastewaters. For higher concentrations, use a portion of diluted sample.

2. Apparatus

a. Spectrophotometer or filter photometer, for use at 460 nm, providing a light path of 5 cm.

b. Glass adsorption column: Use a 50-mL buret with a glass-wool plug, and pack with macroreticular resin (¶ 3f) approximately 40 cm high. For convenience, apply a powder funnel of the same diameter as the buret to the top with a short piece of plastic tubing.

3. Reagents

a. Ferric nitrate solution: Dissolve 404 g $Fe(NO_3)_3 \cdot 9H_2O$ in about 800 mL distilled water. Add 80 mL conc HNO_3 and dilute to 1 L.

b. Nitric acid solution, 0.1N: Mix 6.4 mL conc HNO_3 in about 800 mL distilled water and dilute to 1 L.

c. Stock thiocyanate solution: Dissolve 1.673 g potassium thiocyanate (KSCN) in distilled water and dilute to 1000 mL; 1.00 mL 1.00 mg SCN^-.

d. Standard thiocyanate solution: Dilute 10 mL stock solution to 1 L with distilled water; 1.00 mL = 0.01 mg SCN^-.

e. Sodium hydroxide solution, 4 g/L: Dissolve 4 g NaOH in about 800 mL distilled water and dilute to 1 L.

f. Macroreticular resin, 18 to 50 mesh:* The available resin may not be purified. Some samples have shown contamination with waxes and oil, giving poor permeability and adsorption. Purify as follows:

Place sufficient resin to fill the column or columns in a beaker and add 5 times the resin volume of acetone. Stir gently for 1 h. Pour off fines and acetone from settled resin and add 5 times the resin volume of hexane. Stir for 1 h. Pour off fines and hexane and add 5 times the resin volume of methanol. Stir for 15 min. Pour off methanol and add 3 times the resin volume of 0.1N NaOH. Stir for 15 min. Pour off NaOH solution and add 3 times the resin volume of 0.1N HNO_3. Stir for 15 min. Pour off HNO_3 solution and add 3 times the resin volume of distilled water. Stir for 15 min. Drain excess water and use purified resin to fill the column. Store excess purified resin after covering it with distilled water. Keep in a closed jar.

g. Methyl alcohol.

---

* Amberlite® XAD-7, or equivalent.

4. Procedure

a. Preparation of calibration curve: Prepare a series of standards containing from 0.02 mg to 0.40 mg SCN⁻ by pipetting measured volumes of standard KSCN solution into 200-mL volumetric flasks and diluting with water. Mix well. Develop color according to ¶ *b* below. Plot absorbance against SCN⁻ concentration expressed as mg/50 mL sample. The absorbance plot should be linear.

b. Color development: Use a filtered sample or portion from a diluted solution so that the concentration of SCN⁻ is between 0.1 and 2 mg/L. Adjust pH to 2 with conc HNO_3 added dropwise. Pipet 50-mL portion to a beaker, add 2.5 mL ferric nitrate, and mix.

Fill a 5-cm absorption cell and measure absorbance against a reagent blank at 460 nm or close to the maximum absorbance found with the instrument being used. Measure absorbance of the developed color against a reagent blank within 5 min from adding the reagent. (The color develops within 30 s and fades on standing in light.)

c. Sample pretreatment:

1) Color and various organic compounds interfere with absorbance measurement. At pH 2, macroreticular resin removes these interfering materials by adsorption without affecting thiocyanate.

2) To prepare the adsorption column, fill it with resin, rinse with 100 mL methanol, and follow by rinses with 100 mL $0.1N$ NaOH, 100 mL $0.1N$ HNO_3, and finally with 100 mL distilled water. If previously purified resin is used, omit these preparatory steps.

3) When washing, regenerating, or passing a sample through the column, as solution level approaches resin bed, add and drain five separate 5-mL volumes of solution or water (depending on which is used in next step) to approximate bed height. After last 5-mL volume, fill column with remaining liquid. This procedure prevents undue mixing of solutions and helps void the column of the previous solution.

4) Acidify 150 mL sample (or a dilution) to pH 2 by adding conc HNO_3 dropwise while stirring. Pass it through the column at a flow rate not to exceed 20 mL/min. If the resin becomes packed and the flow rate falls to 4 to 5 mL/min, use gentle pressure through a manually operated hand pump or squeeze bulb on the column. In this case, use a separator funnel for the liquid reservoir instead of the powder funnel. Alternatively use a vacuum bottle as a receiver and apply gentle vacuum. Do not let liquid level drop below the adsorbent in the column.

5) When passing a sample through the column, measure 90 mL of sample in a graduated cylinder, and from this use the five 5-mL additions as directed in ¶ 3), then pour the remainder of the 90 mL into the column. Add rest of sample and collect 60 mL eluate to be tested after the first 60 mL has passed through the column.

6) Prepare a new calibration curve using standards prepared according to ¶ 4*a*, but acidify standards according to ¶ 4*b*, and pass them through the adsorption column. Develop color and measure absorbance according to ¶ 4*b* against a reagent blank prepared by passing acidified, distilled water through the adsorption column.

7) Pipet 50 mL from the collected eluate to a beaker, add 2.5 mL ferric nitrate solution, and mix. Measure absorbance according to ¶ 4*b* against a reagent blank [see ¶ 6) above].

8) From the measured absorbance value, determine thiocyanate content of the sample or dilution using the absorbance plot.

9) Each day the column is in use, test a mid-range standard to check absorption curve.

10) Regenerate column between samples by rinsing with 100 mL $0.1N$ NaOH; 50 mL $0.1N$ HNO_3; and 100 mL water. Insure that the water has rinsed empty glass section of the buret. Occasionally rinse with 100 mL methanol for complete regeneration. Adsorbed weak organic acids and thiocyanate residuals from earlier tests are eluted by the NaOH rinse. Leave the column covered with the last rinse water for storage.

5. Calculation

Compute slope (*m*) and intercept (*b*) of standard curve as directed in 4500-CN⁻.E.5.

Calculate thiocyanate concentration as follows:

$$\text{mg SCN}^-/\text{L} = (ma_1 + b) \times \text{dilution factor}$$

where:

a_1 = absorbance of sample solution.

6. Precision and Bias[4]

a. Precision: Based on the results of twelve operators in nine laboratories, at four levels of concentration, the precision of the test method within its designated range is linear with concentration and may be expressed as follows:

$$\text{Reagent water: } S_T = 0.093x + 0.0426$$
$$S_o = 0.045x + 0.010$$

$$\text{Water matrix: } S_T = 0.055x + 0.0679$$
$$S_o = 0.024x + 0.182$$

where:

S_T = overall precision, mg/L,
S_o = pooled single-operator precision, mg/L, and
x = thiocyanate concentration, mg/L.

b. Bias: Recoveries of known amounts of thiocyanate from reagent water and selected water matrices including natural waters, laboratory effluent, steel mill effluent, and dechlorinated and treated sanitary effluents were as follows:

Medium	Added mg/L	Recovered mg/L	n	S_T	Bias	% Bias
Reagent	1.42	1.411	30	0.181	−0.009	−0.6
water	0.71	0.683	27	0.091	−0.027	−4
	0.35	0.329	30	0.084	−0.021	−6
	0.07	0.068	30	0.052	−0.002	−3
Selected	1.42	1.408	26	0.151	−0.012	−0.8
water	0.71	0.668	29	0.096	−0.042	−6
matrices	0.35	0.320	29	0.085	−0.030	−9
	0.07	0.050	29	0.079	−0.020	−29

For other matrices these data may not apply.

7. References

1. SPENCER, R.R., J. LEENHEER & V.C. MARTI. 1980. Automated colorimetric determination of thiocyanate, thiosulfate and tetrathionate in water. 94th Annu. Meeting. Assoc. Official Agricultural Chemists, Washington, D.C. 1981.
2. DANCHICK, R.S. & D.F. BOLTZ. 1968. Indirect spectrophotometric and atomic absorption spectrometric methods in determination of thiocyanate. *Anal. Chem.* 43:2215.
3. LUTHY, R.G. 1978. Manual of Methods: Preservation and Analysis of Coal Gasification Wastewaters. FE-2496-16, U.S. Dep. Energy, National Technical Information Serv., Springfield, Va.
4. AMERICAN SOCIETY FOR TESTING & MATERIALS. 1989. Research Rep. D4193:19-1099. American Soc. Testing & Materials, Philadelphia, Pa.

4500-CN⁻ N.   Total Cyanide after Distillation, by Flow Injection Analysis (PROPOSED)

1. General Discussion

a. Principle: Total cyanides are digested and steam-distilled from the sample as in 4500-CN⁻.C, cyanides amenable to chlorination are digested and steam-distilled from the sample as in 4500-CN⁻.G, or weak acid dissociable cyanides are digested and steam-distilled from the sample as in 4500-CN⁻.I, by using the apparatus described in 4500-CN⁻.C or an equivalent distillation apparatus. In any case, the distillate should consist of cyanide in $0.25M$ NaOH. The cyanide in this distillate is converted to cyanogen chloride, CNCl, by reaction with chloramine-T at pH less than 8. The CNCl then forms a red-blue dye by reacting with pyridine-barbituric acid reagent. The absorbance of this red dye is measured at 570 nm and is proportional to the total or weak acid dissociable cyanide in the sample.

Also see Sections 4500-CN⁻.A and E, and Section 4130, Flow Injection Analysis (FIA).

b. Interferences: Remove large or fibrous particulates by filtering sample through glass wool. Guard against contamination from reagents, water, glassware, and the sample preservation process.

Nonvolatile interferences are eliminated or minimized by the distillation procedure. Some of the known interferences are aldehydes, nitrate-nitrite, and oxidizing agents such as chlorine, thiocyanate, thiosulfate, and sulfide. Multiple interferences may require the analysis of a series of laboratory fortified sample matrices (LFM) to verify the suitability of the chosen treatment. See Section 4500-CN⁻.B for a discussion of preliminary treatment of samples to be distilled.

2. Apparatus

Flow injection analysis equipment consisting of:

a. FIA injection valve with sample loop or equivalent.

b. Multichannel proportioning pump.

c. FIA manifold (Figure 4500-CN⁻:2) with tubing heater and flow cell. Relative flow rates only are shown. Tubing volumes are given as an example only; they may be scaled down proportionally. Use manifold tubing of an inert material such as TFE.*

---

* Teflon or equivalent.

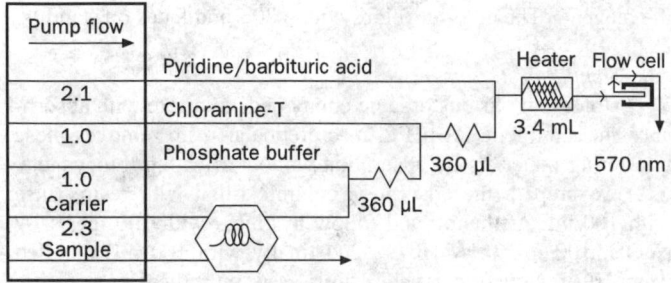

Figure 4500-CN⁻:2. FIA cyanide manifold.

d. Absorbance detector, 570 nm, 10-nm bandpass.

e. Injection valve control and data acquisition system.

3. Reagents

Use reagent water (>10 megohm) for all solutions. To prevent bubble formation, degas carrier and all reagents with helium. Pass He at 140 kPa (20 psi) through a helium degassing tube. Bubble He through 1 L solution for 1 min. As an alternative to preparing reagents by weight/weight, use weight/volume.

a. Carrier solution, $0.25M$: In a 1-L plastic container dissolve 10.0 g NaOH in 1.00 L water.

b. Phosphate buffer, $0.71M$: To a 1-L tared container add 97.0 g potassium phosphate, monobasic, anhydrous, KH_2PO_4, and 975 g water. Stir or shake until dissolved. Prepare fresh monthly.

c. Chloramine-T: Dissolve 2.0 g chloramine-T hydrate (mol wt 227.65) in 500 mL water. Prepare fresh daily.

d. Pyridine/barbituric acid: In fume hood, place 15.0 g barbituric acid in a tared 1-L container and add 100 g water, rinsing down sides of beaker to wet the barbituric acid. Add 73 g pyridine (C_5H_5N) with stirring and mix until barbituric acid dissolves. Add 18 g conc HCl, then an additional 825 g water, and mix. Prepare fresh weekly.

e. Stock cyanide standard, 100 mg CN⁻/L: In a 1-L container, dissolve 2.0 g potassium hydroxide (KOH) in approximately 800 mL water. Add 0.250 g potassium cyanide (KCN). CAUTION:

TABLE 4500-CN⁻:I. RESULTS OF SINGLE-LABORATORY STUDIES WITH SELECTED MATRICES

Matrix	Sample/Blank Designation	Known Addition $mg\ CN^-/L$	Recovery %	Relative Standard Deviation %
Wastewater treatment plant influent	Reference sample*	—	94	—
	Blank†	0.050	96	—
		0.10	96	—
	Site A‡	0	—	<0.5
		0.010	104	—
		0.020	104	—
	Site B‡	0	—	<0.5
		0.010	101	—
		0.020	106	—
	Site C‡	0	—	<0.5
		0.010	103	—
		0.020	108	—
Wastewater treatment plant effluent	Reference sample*	—	95	—
	Blank†	0.050	88	—
		0.10	95	—
	Site A‡	0	—	<0.5
		0.010	112	—
		0.020	106	—
	Site B‡	0	—	<0.5
		0.010	110	—
		0.020	105	—
	Site C‡	0	—	<0.5
		0.010	101	—
		0.020	106	—
Landfill leachate	Reference sample*	—	98	—
	Blank†	0.050	96	—
		0.10	98	—
	Site A‡	0	—	<0.5
		0.050	114	—
		0.10	106	—
	Site B‡	0	—	<0.5
		0.050	104	—
		0.10	104	—
	Site C‡	0	—	<0.5
		0.050	103	—
		0.10	107	—

* U.S. EPA QC sample, 0.498 mg CN⁻/L, diluted five-fold.
† Determined in duplicate.
‡ Samples diluted five-fold. Samples without known additions determined four times; sample with known additions determined in duplicate; typical relative differences between duplicates <0.5%.

KCN is highly toxic. Avoid inhalation of dust or contact with the solid or solutions. Make to final weight of 1000 g with water and mix. Prepare fresh weekly or standardize weekly using procedure in Section 4500-CN⁻.D.4.

f. Standard cyanide solution: Prepare cyanide standards in the desired concentration range, using the stock cyanide standard (¶ 3e) and diluting with the 0.25M NaOH carrier (¶ 3a).

4. Procedure

Set up a manifold equivalent to that in Figure 4500-CN⁻:2 and follow method supplied by manufacturer or laboratory standard operating procedure for this method. Follow quality control guidelines outlined in Section 4020.

5. Calculation

Prepare standard curves by plotting absorbance of standards processed through manifold versus cyanide concentration. The calibration curve is linear.

6. Precision and Bias

a. Recovery and relative standard deviation: The results of single-laboratory studies with various matrices are given in Table 4500-CN⁻:I.

b. MDL without distillation: Using a published MDL method,[1] analysts ran 21 replicates of an undistilled 0.010-mg CN⁻/L standard with a 780-μL sample loop. These gave a mean of 0.010 mg CN⁻/L, a standard deviation of 0.00012 mg CN⁻/L, and an MDL of 0.0003 mg CN⁻/L. A lower MDL may be obtained by increasing the sample loop volume and increasing the ratio of carrier flow rate to reagent flow rate.

c. MDL with distillation: Using a published MDL method,[1] analysts ran 21 replicates of a 0.0050-mg CN⁻/L standard distilled using the distillation device† equivalent to the apparatus specified in 4500-CN⁻.C. When the 0.25M NaOH distillates were determined with a 780-μL sample loop, they gave a mean of 0.0045 mg CN⁻/L, a standard deviation of 0.0002 mg CN⁻/L, and an MDL of 0.0006 mg CN⁻/L.

d. Precision study: Ten injections of an undistilled 0.050-mg CN⁻/L standard gave a relative standard deviation of 0.21%.

7. Reference

1. U.S. ENVIRONMENTAL PROTECTION AGENCY. 1989. Definition and procedure for the determination of method detection limits. Appendix B to 40 CFR 136 rev. 12.11 amended June 30, 1986. 49 CFR 43430.

---

† MICRO DIST, Lachat Instruments, Milwaukee, WI.

4500-CN⁻ O. Total Cyanide and Weak Acid Dissociable Cyanide by Flow Injection Analysis (PROPOSED)

1. General Discussion

a. Principle: Total cyanide consists of various metal-cyanide complexes. To break down or digest these complexes to yield HCN, the sample is mixed with heated phosphoric acid and then irradiated with ultraviolet radiation. The resulting "donor stream" contains the product HCN (aq). This donor stream is passed over a silicone rubber gas permeation membrane. The HCN from the donor stream is extracted by the membrane as HCN (g) and is trapped in a parallel "acceptor stream" that consists of dilute sodium hydroxide, the equivalent of the distillate resulting from the digesting distillations in the sample preparation methods 4500-CN⁻.C, G, and I.

As in 4500-CN⁻.N, the cyanide in this acceptor stream or distillate is converted to cyanogen chloride, CNCl, by reaction with chloramine-T at pH less than 8. The CNCl then forms a red-blue dye by reacting with pyridine-barbituric acid reagent. The absorbance of this red dye is measured at 570 nm and is proportional to the total or weak acid dissociable cyanide in the sample.

The weak acid dissociable (WAD) cyanide method is similar except that ultraviolet radiation and phosphoric acid are not used in the donor stream. Instead, a solution of dihydrogen phosphate is used as the donor stream.

Also see Sections 4500-CN⁻.A, E, and N and Section 4130, Flow Injection Analysis (FIA).

b. Interferences: Remove large or fibrous particulates by filtering the sample through glass wool. Guard against contamination from reagents, water, glassware, and the sample preservation process.

Nonvolatile interferences are eliminated or minimized by the gas-permeable membrane.

Multiple interferences may require the analysis of a series of sample matrices with known additions to verify the suitability of the chosen treatment. See Section 4500-CN⁻.B for a discussion of preliminary treatment of samples that will be distilled.

1) Total cyanide interferences—Sulfide up to a concentration of 10 mg/L and thiocyanate up to a concentration of 20 mg/L do not interfere in the determination of 100 µg CN⁻/L. When a sample containing nitrate at 100 mg NO₃⁻-N/L and 20 mg/L thiocyanate was treated with sulfamic acid, the determined value was 138.2 µg CN⁻/L for a known concentration of 100 µg CN⁻/L. When pretreated with ethylenediamine, a sample containing 50 mg formaldehyde/L did not interfere in the determination of cyanide.

2) WAD interferences—Sulfide up to 10 mg/L and thiocyanate up to 50 mg/L do not interfere in the determination of 0.1 mg/L cyanide.

2. Apparatus

Flow injection analysis equipment consisting of:
a. FIA injection valve with sample loop or equivalent.
b. Multichannel proportioning pump.
c. FIA manifold (Figure 4500-CN⁻:3) with tubing heater, in-line ultraviolet digestion fluidics, a gas-permeable silicone rubber membrane and its holder, and flow cell. In Figure 4500-CN⁻:3, relative flow rates only are shown. The tubing volumes are given as an example only; they may be scaled down proportionally. Use

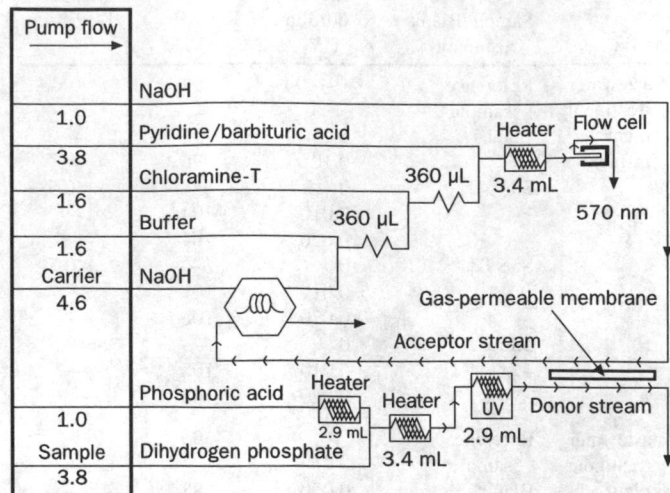

Figure 4500-CN⁻:3. FIA in-line total and WAD cyanide manifold.

manifold tubing of an inert material such as TFE. The ultraviolet unit should consist of TFE tubing irradiated by a mercury discharge ultraviolet lamp emitting radiation at 254 nm.

d. Absorbance detector, 570 nm, 10-nm bandpass.
e. Injection valve control and data acquisition system.

3. Reagents

Use reagent water (>10 megohm) for all solutions. To prevent bubble formation, degas carrier and all reagents with helium. Pass He at 140 kPa (20 psi) through a helium degassing tube. Bubble He through 1 L of solution for 1 min. As an alternative to preparing reagents by weight/weight, use weight/volume.

a. Phosphoric acid donor stream (total cyanide): To a 1-L volumetric flask, add approximately 700 mL water, then add 30 mL conc phosphoric acid, H₃PO₄. Mix and let solution cool. Dilute to mark. Prepare fresh monthly.

b. Dihydrogen phosphate donor stream (WAD cyanide): To a tared 1-L container add 97 g anhydrous potassium dihydrogen phosphate, KH₂PO₄, and 975 g water. Stir for 2 h or until the potassium phosphate has gone into solution. Degas with helium. Prepare fresh monthly.

c. NaOH acceptor stream, carrier, and diluent (total and WAD cyanide), 0.025M NaOH: To a 1-L container add 1.0 g sodium hydroxide (NaOH) and 999 g water. Mix with a magnetic stirrer for about 5 min. Cover with a laboratory film. Degas with helium. Prepare fresh daily.

d. Buffer (total and WAD cyanide), 0.71M phosphate: To a 1-L tared container add 97.0 g potassium phosphate, monobasic, anhydrous, KH₂PO₄, and 975 g water. Stir or shake until dissolved. Prepare fresh monthly.

e. Chloramine-T solution (total and WAD cyanide): Dissolve 3 g chloramine-T hydrate in 500 mL water. Degas with helium. Prepare fresh daily. NOTE: Chloramine-T is an air-sensitive solid. Preferably discard this chemical 6 months after opening.

f. Pyridine/barbituric acid solution (total and WAD cyanide): In the fume hood, place 15.0 g barbituric acid in a tared 1-L container and add 100 g water, rinsing down the sides of the beaker to wet the barbituric acid. Add 73 g pyridine (C_5H_5N) with stirring and mix until the barbituric acid dissolves. Add 18 g conc HCl, then add an additional 825 mL water and mix. Prepare fresh weekly.

g. Stock cyanide standard, 100 mg CN^-/L: In a 1-L container dissolve 2.0 g potassium hydroxide, KOH, in approximately 800 mL water. Add 0.250 g potassium cyanide, KCN. CAUTION: *KCN is highly toxic. Avoid inhalation of dust or contact with the solid or solutions.* Make to final weight of 1000 g with water and invert three times to mix. Prepare fresh weekly or standardize weekly using the procedure in Section 4500-CN^--D.4.

h. Standard cyanide solutions: Prepare cyanide standards in the desired concentration range, using the stock cyanide standard (¶ 3g) and diluting with the NaOH standards diluent (¶ 3c).

4. Procedure

Set up a manifold equivalent to that in Figure 4500-CN^-:3 and follow the method supplied by the manufacturer or laboratory standard operating procedure for this method. Follow quality control guidelines outlined in Section 4020.

5. Calculations

Prepare standard curves by plotting absorbance of standards processed through the manifold vs. cyanide concentration. The calibration curve is linear.

6. Precision and Bias

a. MDL, total cyanide: A 420-μL sample loop was used in the total cyanide method. Using a published MDL method[1], analysts ran 21 replicates of a 10.0-μg CN^-/L standard. These gave a mean of 9.69 μg CN^-/L, a standard deviation of 0.86 μg CN^-/L, and an MDL of 2.7 μg CN^-/L.

b. MDL, WAD cyanide: A 420-μL sample loop was used in the WAD cyanide method. Using a published MDL method[1], analysts ran 21 replicates of a 10.0-μg CN^-/L standard. These gave a mean of 11.5 μg CN^-/L, a standard deviation of 0.73 μg CN^-/L, and an MDL of 2.3 μg CN^-/L.

c. Precision study, total cyanide: Seven injections of a 100.0-μg CN^-/L standard gave a relative standard deviation (RSD) of 1.0%.

d. Precision study, WAD cyanide: Ten injections of a 200.0-μg CN^-/L standard gave an RSD of 1.3%.

e. Recovery of total cyanide: Two injections each were made of solutions of potassium ferricyanide and potassium ferrocyanide, both at a concentration equivalent to 100 μg CN^-/L. Both gave an average recovery of 98%.

7. Reference

1. U.S. ENVIRONMENTAL PROTECTION AGENCY. 1989. Definition and procedure for the determination of method detection limits. Appendix B to 40 CFR 136 rev. 12.11 amended June 30, 1986. 49 CFR 43430.

4500-Cl CHLORINE (RESIDUAL)*

4500-Cl A. Introduction

1. Effects of Chlorination

The chlorination of water supplies and polluted waters serves primarily to destroy or deactivate disease-producing microorganisms. A secondary benefit, particularly in treating drinking water, is the overall improvement in water quality resulting from the reaction of chlorine with ammonia, iron, manganese, sulfide, and some organic substances.

Chlorination may produce adverse effects. Taste and odor characteristics of phenols and other organic compounds present in a water supply may be intensified. Potentially carcinogenic chloroorganic compounds such as chloroform may be formed. Combined chlorine formed on chlorination of ammonia- or amine-bearing waters adversely affects some aquatic life. To fulfill the primary purpose of chlorination and to minimize any adverse effects, it is essential that proper testing procedures be used with a foreknowledge of the limitations of the analytical determination.

2. Chlorine Forms and Reactions

Chlorine applied to water in its molecular or hypochlorite form initially undergoes hydrolysis to form free chlorine consisting of aqueous molecular chlorine, hypochlorous acid, and hypochlorite ion. The relative proportion of these free chlorine forms is pH- and temperature-dependent. At the pH of most waters, hypochlorous acid and hypochlorite ion will predominate.

Free chlorine reacts readily with ammonia and certain nitrogenous compounds to form combined chlorine. With ammonia, chlorine reacts to form the chloramines: monochloramine, dichloramine, and nitrogen trichloride. The presence and concentrations of these combined forms depend chiefly on pH, temperature, initial chlorine-to-nitrogen ratio, absolute chlorine demand, and reaction time. Both free and combined chlorine may be present simultaneously. Combined chlorine in water supplies may be formed in the treatment of raw waters containing ammonia or by the addition of ammonia or ammonium salts. Chlorinated wastewater effluents, as well as certain chlorinated industrial effluents,

---

* Approved by Standard Methods Committee, 1993.

normally contain only combined chlorine. Historically, the principal analytical problem has been to distinguish between free and combined forms of chlorine.

3. Selection of Method

In two separate but related studies, samples were prepared and distributed to participating laboratories to evaluate chlorine methods. Because of poor accuracy and precision and a high overall (average) total error in these studies, all orthotolidine procedures except one were dropped in the 14th edition of this work. The useful stabilized neutral orthotolidine method was deleted from the 15th edition because of the toxic nature of orthotolidine. The leuco crystal violet (LCV) procedure was dropped from the 17th edition because of its relative difficulty and the lack of comparative advantages.

a. Natural and treated waters: The iodometric methods (B and C) are suitable for measuring total chlorine concentrations greater than 1 mg/L, but the amperometric end point of Methods C and D gives greater sensitivity. All acidic iodometric methods suffer from interferences, generally in proportion to the quantity of potassium iodide (KI) and H^+ added.

The amperometric titration method (D) is a standard of comparison for the determination of free or combined chlorine. It is affected little by common oxidizing agents, temperature variations, turbidity, and color. The method is not as simple as the colorimetric methods and requires greater operator skill to obtain the best reliability. Loss of chlorine can occur because of rapid stirring in some commercial equipment. Clean and conditioned electrodes are necessary for sharp end points.

A low-level amperometric titration procedure (E) has been added to determine total chlorine at levels below 0.2 mg/L. This method is recommended only when quantification of such low residuals is necessary. The interferences are similar to those found with the standard amperometric procedure (D). The DPD methods (Methods F and G) are operationally simpler for determining free chlorine than the amperometric titration. Procedures are given for estimating the separate mono- and dichloramine and combined fractions. High concentrations of monochloramine interfere with the free chlorine determination unless the reaction is stopped with arsenite or thioacetamide. In addition, the DPD methods are subject to interference by oxidized forms of manganese unless compensated for by a blank.

The amperometric and DPD methods are unaffected by dichloramine concentrations in the range of 0 to 9 mg Cl as Cl_2/L in the determination of free chlorine. Nitrogen trichloride, if present, may react partially as free chlorine in the amperometric, DPD, and FACTS methods. The extent of this interference in the DPD methods does not appear to be significant.

The free chlorine test, syringaldazine (FACTS, Method H) was developed specifically for free chlorine. It is unaffected by significant concentrations of monochloramine, dichloramine, nitrate, nitrite, and oxidized forms of manganese.[1]

Sample color and turbidity may interfere in all colorimetric procedures.

Organic contaminants may produce a false free chlorine reading in most colorimetric methods (see ¶ 3b below). Many strong oxidizing agents interfere in the measurement of free chlorine in all methods. Such interferences include bromine, chlorine dioxide, iodine, permanganate, hydrogen peroxide, and ozone. However, the reduced forms of these compounds—bromide, chloride, io-dide, manganous ion, and oxygen, in the absence of other oxidants, do not interfere. Reducing agents such as ferrous compounds, hydrogen sulfide, and oxidizable organic matter generally do not interfere.

b. Wastewaters: The determination of total chlorine in samples containing organic matter presents special problems. Because of the presence of ammonia, amines, and organic compounds, particularly organic nitrogen, residual chlorine exists in a combined state. A considerable residual may exist in this form, but at the same time there may be appreciable unsatisfied chlorine demand. Addition of reagents in the determination may change these relationships so that residual chlorine is lost during the analysis. Only the DPD method for total chlorine is performed under neutral pH conditions. In wastewater, the differentiation between free chlorine and combined chlorine ordinarily is not made because wastewater chlorination seldom is carried far enough to produce free chlorine.

The determination of residual chlorine in industrial wastes is similar to that in domestic wastewater when the waste contains organic matter, but may be similar to the determination in water when the waste is low in organic matter.

None of these methods is applicable to estuarine or marine waters because the bromide is converted to bromine and bromamines, which are detected as free or total chlorine. A procedure for estimating this interference is available for the DPD method.

Although the methods given below are useful for the determination of residual chlorine in wastewaters and treated effluents, select the method in accordance with sample composition. Some industrial wastes, or mixtures of wastes with domestic wastewater, may require special precautions and modifications to obtain satisfactory results.

Determine free chlorine in wastewater by any of the methods provided that known interfering substances are absent or appropriate correction techniques are used. The amperometric method is the method of choice because it is not subject to interference from color, turbidity, iron, manganese, or nitrite nitrogen. The DPD method is subject to interference from high concentrations of monochloramine, which is avoided by adding thioacetamide immediately after reagent addition. Oxidized forms of manganese at all levels encountered in water will interfere in all methods except in the free chlorine measurement of amperometric titrations and FACTS, but a blank correction for manganese can be made in Methods F and G.

The FACTS method is unaffected by concentrations of monochloramine, dichloramine, nitrite, iron, manganese, and other interfering compounds normally found in domestic wastewaters.

For total chlorine in samples containing significant amounts of organic matter, use either the DPD methods (F and G), amperometric, or iodometric back titration method (C) to prevent contact between the full concentration of liberated iodine and the sample. With Method C, do not use the starch-iodide end point if the concentration is less than 1 mg/L. In the absence of interference, the amperometric and starch-iodide end points give concordant results. The amperometric end point is inherently more sensitive and is free of interference from color and turbidity, which can cause difficulty with the starch-iodide end point. On the other hand, certain metals, surface-active agents, and complex anions in some industrial wastes interfere in the amperometric titration and indicate the need for another method for such wastewaters. Silver in the form of soluble silver cyanide complex, in concentrations of 1.0 mg Ag/L, poisons the cell at pH 4.0 but not at 7.0.

The silver ion, in the absence of the cyanide complex, gives extensive response in the current at pH 4.0 and gradually poisons the cell at all pH levels. Cuprous copper in the soluble copper cyanide ion, in concentrations of 5 mg Cu/L or less, poisons the cell at pH 4.0 and 7.0. Although iron and nitrite may interfere with this method, minimize the interference by buffering to pH 4.0 before adding KI. Oxidized forms of manganese interfere in all methods for total chlorine including amperometric titration. An unusually high content of organic matter may cause uncertainty in the end point.

Regardless of end-point detection, either phenylarsine oxide or thiosulfate may be used as the standard reducing reagent at pH 4. The former is more stable and is preferred.

The DPD titrimetric and colorimetric methods (F and G, respectively) are applicable to determining total chlorine in polluted waters. In addition, both DPD procedures and the amperometric free titration method allow for estimating monochloramine and dichloramine fractions. Because all methods for total chlorine depend on the stoichiometric production of iodine, waters containing iodine-reducing substances may not be analyzed accurately by these methods, especially where iodine remains in the solution for a significant time. This problem occurs in Methods B and D. The back titration procedure (C) and Methods F and G cause immediate reaction of the iodine generated so that it has little chance to react with other iodine-reducing substances.

In all colorimetric procedures, compensate for color and turbidity by using color and turbidity blanks.

A method (I) for total residual chlorine using a potentiometric iodide electrode is proposed. This method is suitable for analysis of chlorine residuals in natural and treated waters and wastewater effluents. No differentiation of free and combined chlorine is possible. This procedure is an adaptation of other iodometric techniques and is subject to the same inferences.

4. Sampling and Storage

Chlorine in aqueous solution is not stable, and the chlorine content of samples or solutions, particularly weak solutions, will decrease rapidly. Exposure to sunlight or other strong light or agitation will accelerate the reduction of chlorine. Therefore, start chlorine determinations immediately after sampling, avoiding excessive light and agitation. Do not store samples to be analyzed for chlorine.

5. Reference

1. COOPER, W.J., N.M. ROSCHER & R.A. SLIFER. 1982. Determining free available chlorine by DPD-colorimetric, DPD-steadifac (colorimetric) and FACTS procedures. *J. Amer. Water Works Assoc.* 74:362.

6. Bibliography

MARKS, H.C., D.B. WILLIAMS & G.U. GLASGOW. 1951. Determination of residual chlorine compounds. *J. Amer. Water Works Assoc.* 43:201.

NICOLSON, N.J. 1965. An evaluation of the methods for determining residual chlorine in water, Part 1. Free chlorine. *Analyst* 90:187.

WHITTLE, G.P. & A. LAPTEFF, JR. 1973. New analytical techniques for the study of water disinfection. *In* Chemistry of Water Supply, Treatment, and Distribution, p. 63. Ann Arbor Science Publishers, Ann Arbor, Mich.

GUTER, W.J., W.J. COOPER & C.A. SORBER. 1974. Evaluation of existing field test kits for determining free chlorine residuals in aqueous solutions. *J. Amer. Water Works Assoc.* 66:38.

4500-Cl B. Iodometric Method I

1. General Discussion

a. Principle: Chlorine will liberate free iodine from potassium iodide (KI) solutions at pH 8 or less. The liberated iodine is titrated with a standard solution of sodium thiosulfate ($Na_2S_2O_3$) with starch as the indicator. Titrate at pH 3 to 4 because the reaction is not stoichiometric at neutral pH due to partial oxidation of thiosulfate to sulfate.

b. Interference: Oxidized forms of manganese and other oxidizing agents interfere. Reducing agents such as organic sulfides also interfere. Although the neutral titration minimizes the interfering effect of ferric and nitrite ions, the acid titration is preferred because some forms of combined chlorine do not react at pH 7. Use only acetic acid for the acid titration; sulfuric acid (H_2SO_4) will increase interferences; *never use hydrochloric acid (HCl).* See Section A.3 for discussion of other interferences.

c. Minimum detectable concentration: The minimum detectable concentration approximates 40 μg Cl as Cl_2/L if 0.01N $Na_2S_2O_3$ is used with a 1000-mL sample. Concentrations below 1 mg/L cannot be determined accurately by the starch-iodide end point used in this method. Lower concentrations can be measured with the amperometric end point in Methods C and D.

2. Reagents

a. Acetic acid, conc (glacial).

b. Potassium iodide, KI, crystals.

c. Standard sodium thiosulfate, 0.1N: Dissolve 25 g $Na_2S_2O_3 \cdot 5H_2O$ in 1 L freshly boiled distilled water and standardize against potassium bi-iodate or potassium dichromate after at least 2 weeks storage. This initial storage is necessary to allow oxidation of any bisulfite ion present. Use boiled distilled water and add a few milliliters chloroform ($CHCl_3$) to minimize bacterial decomposition.

Standardize 0.1N $Na_2S_2O_3$ by one of the following:

1) Iodate method—Dissolve 3.249 g anhydrous potassium bi-iodate, $KH(IO_3)_2$, primary standard quality, or 3.567 g KIO_3 dried at 103 ± 2°C for 1 h, in distilled water and dilute to 1000 mL to yield a 0.1000N solution. Store in a glass-stoppered bottle.

To 80 mL distilled water, add, with constant stirring, 1 mL conc H_2SO_4, 10.00 mL 0.1000N $KH(IO_3)_2$, and 1 g KI. Titrate immediately with 0.1N $Na_2S_2O_3$ titrant until the yellow color of the liberated iodine almost is discharged. Add 1 mL starch indicator solution and continue titrating until the blue color disappears.

2) *Dichromate method*—Dissolve 4.904 g anhydrous potassium dichromate, $K_2Cr_2O_7$, of primary standard quality, in distilled water and dilute to 1000 mL to yield a 0.1000N solution. Store in a glass-stoppered bottle.

Proceed as in the iodate method, with the following exceptions: Substitute 10.00 mL 0.1000N $K_2Cr_2O_7$ for iodate and let reaction mixture stand 6 min in the dark before titrating with 0.1N $Na_2S_2O_3$ titrant.

$$\text{Normality Na}_2\text{S}_2\text{O}_3 = \frac{1}{\text{mL Na}_2\text{S}_2\text{O}_3 \text{ consumed}}$$

d. Standard sodium thiosulfate titrant, 0.01N or 0.025N: Improve the stability of 0.01N or 0.025N $Na_2S_2O_3$ by diluting an aged 0.1N solution, made as directed above, with freshly boiled distilled water. Add 4 g sodium borate and 10 mg mercuric iodide/L solution. For accurate work, standardize this solution daily in accordance with the directions given above, using 0.01N or 0.025N iodate or $K_2Cr_2O_7$. Use sufficient volumes of these standard solutions so that their final dilution is not greater than 1 + 4. To speed up operations where many samples must be titrated use an automatic buret of a type in which rubber does not come in contact with the solution. Standard titrants, 0.0100N and 0.0250N, are equivalent, respectively, to 354.5 μg and 886.3 μg Cl as Cl_2/1.00 mL.

e. Starch indicator solution: To 5 g starch (potato, arrowroot, or soluble), add a little cold water and grind in a mortar to a thin paste. Pour into 1 L of boiling distilled water, stir, and let settle overnight. Use clear supernate. Preserve with 1.25 g salicylic acid, 4 g zinc chloride, or a combination of 4 g sodium propionate and 2 g sodium azide/L starch solution. Some commercial starch substitutes are satisfactory.

f. Standard iodine, 0.1N: See C.3*g*.

g. Dilute standard iodine, 0.0282N: See C.3*h*.

3. Procedure

a. Volume of sample: Select a sample volume that will require no more than 20 mL 0.01N $Na_2S_2O_3$ and no less than 0.2 mL for the starch-iodide end point. For a chlorine range of 1 to 10 mg/L, use a 500-mL sample; above 10 mg/L, use proportionately less sample. Use smaller samples and volumes of titrant with the amperometric end point.

b. Preparation for titration: Place 5 mL acetic acid, or enough to reduce the pH to between 3.0 and 4.0, in a flask or white porcelain casserole. Add about 1 g KI estimated on a spatula. Pour sample in and mix with a stirring rod.

c. Titration: Titrate away from direct sunlight. Add 0.025N or 0.01N $Na_2S_2O_3$ from a buret until the yellow color of the liberated iodine almost is discharged. Add 1 mL starch solution and titrate until blue color is discharged.

If the titration is made with 0.025N $Na_2S_2O_3$ instead of 0.01N, then, with a 1-L sample, 1 drop is equivalent to about 50 μg/L. It is not possible to discern the end point with greater accuracy.

d. Blank titration: Correct result of sample titration by determining blank contributed by oxidizing or reducing reagent impurities. The blank also compensates for the concentration of iodine bound to starch at the end point.

Take a volume of distilled water corresponding to the sample used for titration in ¶s 3*a*–*c*, add 5 mL acetic acid, 1 g KI, and

1 mL starch solution. Perform blank titration as in 1) or 2) below, whichever applies.

1) If a blue color develops, titrate with 0.01N or 0.025N $Na_2S_2O_3$ to disappearance of blue color and record result. B (see ¶ 4, below) is negative.

2) If no blue color occurs, titrate with 0.0282N iodine solution until a blue color appears. Back-titrate with 0.01N or 0.025N $Na_2S_2O_3$ and record the difference. B is positive.

Before calculating the chlorine concentration, subtract the blank titration of ¶ 1) from the sample titration; or, if necessary, add the net equivalent value of the blank titration of ¶ 2).

4. Calculation

For standardizing chlorine solution for temporary standards:

$$\text{mg Cl as Cl}_2/\text{mL} = \frac{(A \pm B) \times N \times 35.45}{\text{mL sample}}$$

For determining total available residual chlorine in a water sample:

$$\text{mg Cl as Cl}_2/\text{L} = \frac{(A \pm B) \times N \times 35\,450}{\text{mL sample}}$$

where:

A = mL titration for sample,
B = mL titration for blank (positive or negative), and
N = normality of $Na_2S_2O_3$.

5. Precision and Bias

Published studies[1,2] give the results of nine methods used to analyze synthetic water samples without interferences; variations of some of the methods appear in this edition. More current data are not now available.

6. References

1. Water Chlorine (Residual) No. 1. 1969. *Analytical Reference Service Rep. No. 35*, U.S. Environmental Protection Agency, Cincinnati, Ohio.
2. Water Chlorine (Residual) No. 2. 1971. *Analytical Reference Service Rep. No. 40*, U.S. Environmental Protection Agency, Cincinnati, Ohio.

7. Bibliography

LEA, C. 1933. Chemical control of sewage chlorination. The use and value of orthotolidine test. *J. Soc. Chem. Ind.* (London) 52:245T.

AMERICAN WATER WORKS ASSOCIATION. 1943. Committee report. Control of chlorination. *J. Amer. Water Works Assoc.* 35:1315.

MARKS, H.C., R. JOINER & F.B. STRANDSKOV. 1948. Amperometric titration of residual chlorine in sewage. *Water Sewage Works* 95:175.

STRANDSKOV, F.B. , H.C. MARKS & D.H. HORCHIER. 1949. Application of a new residual chlorine method to effluent chlorination. *Sewage Works J.* 21:23.

NUSBAUM, I. & L.A. MEYERSON. 1951. Determination of chlorine demands and chlorine residuals in sewage. *Sewage Ind. Wastes* 23:968.

MARKS, H.C., & N.S. CHAMBERLIN. 1953. Determination of residual chlorine in metal finishing wastes. *Anal. Chem.* 24:1885.

4500-Cl C. Iodometric Method II

1. General Discussion

a. Principle: In this method, used for wastewater analysis, the end-point signal is reversed because the unreacted standard reducing agent (phenylarsine oxide or thiosulfate) remaining in the sample is titrated with standard iodine or standard iodate, rather than the iodine released being titrated directly. This indirect procedure is necessary regardless of the method of end-point detection, to avoid contact between the full concentration of liberated iodine and the wastewater.

When iodate is used as a back titrant, use only phosphoric acid. Do not use acetate buffer.

b. Interference: Oxidized forms of manganese and other oxidizing agents give positive interferences. Reducing agents such as organic sulfides do not interfere as much as in Method B. Minimize iron and nitrite interference by buffering to pH 4.0 before adding potassium iodide (KI). An unusually high content of organic matter may cause some uncertainty in the end point. Whenever manganese, iron, and other interferences definitely are absent, reduce this uncertainty and improve precision by acidifying to pH 1.0. Control interference from more than 0.2 mg nitrite/L with phosphoric acid-sulfamic acid reagent. A larger fraction of organic chloramines will react at lower pH along with interfering substances. See Section A.3 for a discussion of other interferences.

2. Apparatus

For a description of the amperometric end-point detection apparatus and a discussion of its use, see D.2*a*.

3. Reagents

a. Standard phenylarsine oxide solution, 0.005 64N: Dissolve approximately 0.8 g phenylarsine oxide powder in 150 mL 0.3N NaOH solution. After settling, decant 110 mL into 800 mL distilled water and mix thoroughly. Bring to pH 6 to 7 with 6N HCl and dilute to 950 mL with distilled water. CAUTION: *Severe poison, cancer suspect agent.*

Standardization—Accurately measure 5 to 10 mL freshly standardized 0.0282N iodine solution into a flask and add 1 mL KI solution. Titrate with phenylarsine oxide solution, using the amperometric end point (Method D) or starch solution (see B.2*e*) as an indicator. Adjust to 0.005 64N and recheck against the standard iodine solution; 1.00 mL = 200 µg available chlorine. (CAUTION: *Toxic—take care to avoid ingestion.*)

b. Standard sodium thiosulfate solution, 0.1N: See B.2*c*.

c. Standard sodium thiosulfate solution, 0.005 64N: Prepare by diluting 0.1N Na$_2$S$_2$O$_3$. For maximum stability of the dilute solution, prepare by diluting an aged 0.1N solution with freshly boiled distilled water (to minimize bacterial action) and add 4 g Na$_4$B$_4$O$_7$/L. To inhibit mold formation optionally add either 10 mg HgI$_2$ or 2 drops toluene per liter of solution. Standardize daily as directed in B.2*c* using 0.005 64N K$_2$Cr$_2$O$_7$ or iodate solution. Use sufficient volume of sample so that the final dilution does not exceed 1 + 2. Use an automatic buret of a type in which rubber does not come in contact with the solution. 1.00 mL = 200 µg available chlorine.

d. Potassium iodide, KI, crystals.

e. Acetate buffer solution, pH 4.0: Dissolve 146 g anhydrous NaC$_2$H$_3$O$_2$, or 243 g NaC$_2$H$_3$O$_2$·3H$_2$O, in 400 mL distilled water, add 480 g conc acetic acid, and dilute to 1 L with chlorine-demand-free water.

f. Standard arsenite solution, 0.1N: Accurately weigh a stoppered weighing bottle containing approximately 4.95 g arsenic trioxide, As$_2$O$_3$. Transfer without loss to a 1-L volumetric flask and again weigh bottle. Do not attempt to brush out adhering oxide. Moisten As$_2$O$_3$ with water and add 15 g NaOH and 100 mL distilled water. Swirl flask contents gently to dissolve. Dilute to 250 mL with distilled water and saturate with CO$_2$, thus converting all NaOH to NaHCO$_3$. Dilute to mark, stopper, and mix thoroughly. This solution will preserve its titer almost indefinitely. (CAUTION: *Severe poison. Cancer suspect agent.*)

$$\text{Normality} = \frac{\text{g As}_2\text{O}_3}{49.455}$$

g. Standard iodine solution, 0.1N: Dissolve 40 g KI in 25 mL chlorine-demand-free water, add 13 g resublimed iodine, and stir until dissolved. Transfer to a 1-L volumetric flask and dilute to mark.

Standardization—Accurately measure 40 to 50 mL 0.1N arsenite solution into a flask and titrate with 0.1N iodine solution, using starch solution as indicator. To obtain accurate results, insure that the solution is saturated with CO$_2$ at end of titration by passing current of CO$_2$ through solution for a few minutes just before end point is reached, or add a few drops of HCl to liberate sufficient CO$_2$ to saturate solution. Alternatively standardize against Na$_2$S$_2$O$_3$; see B.2*c*1).

Optionally, prepare 0.1000N iodine solution directly as a standard solution by weighing 12.69 g primary standard resublimed iodine. Because I$_2$ may be volatilized and lose from both solid and solution, transfer the solid immediately to KI as specified above. Never let solution stand in open containers for extended periods.

h. Standard iodine titrant, 0.0282N: Dissolve 25 g KI in a little distilled water in a 1-L volumetric flask, add correct amount of 0.1N iodine solution exactly standardized to yield a 0.0282N solution, and dilute to 1 L with chlorine-demand-free water. For accurate work, standardize daily according to directions in ¶ 3*g* above, using 5 to 10 mL of arsenite or Na$_2$S$_2$O$_3$ solution. Store in amber bottles or in the dark; protect solution from direct sunlight at all times and keep from all contact with rubber.

i. Starch indicator: See B.2*e*.

j. Standard iodate titrant, 0.005 64N: Dissolve 201.2 mg primary standard grade KIO$_3$, dried for 1 h at 103°C, or 183.3 mg primary standard anhydrous potassium bi-iodate in distilled water and dilute to 1 L.

k. Phosphoric acid solution, H$_3$PO$_4$, 1 + 9.

l. Phosphoric acid-sulfamic acid solution: Dissolve 20 g NH$_2$SO$_3$H in 1 L 1 + 9 phosphoric acid.

m. Chlorine-demand-free water: Prepare chlorine-demand-free water from good-quality distilled or deionized water by adding sufficient chlorine to give 5 mg/L free chlorine. After standing 2 d this solution should contain at least 2 mg/L free chlorine; if not, discard and obtain better-quality water. Remove remaining free chlorine by placing container in sunlight or irradiating with an ultraviolet lamp. After several hours take sample, add KI, and measure total chlorine with a colorimetric method using a nessler tube to increase sensitivity. Do not use before last trace of free and combined chlorine has been removed.

Distilled water commonly contains ammonia and also may contain reducing agents. Collect good-quality distilled or deionized water in a sealed container from which water can be drawn by gravity. To the air inlet of the container add an H_2SO_4 trap consisting of a large test tube half filled with $1 + 1$ H_2SO_4 connected in series with a similar but empty test tube. Fit both test tubes with stoppers and inlet tubes terminating near the bottom of the tubes and outlet tubes terminating near the top of the tubes. Connect outlet tube of trap containing H_2SO_4 to the distilled water container, connect inlet tube to outlet of empty test tube. The empty test tube will prevent discharge to the atmosphere of H_2SO_4 due to temperature-induced pressure changes. Stored in such a container, chlorine-demand-free water is stable for several weeks unless bacterial growth occurs.

4. Procedure

a. Preparation for titration:

1) Volume of sample—For chlorine concentration of 10 mg/L or less, titrate 200 mL. For greater chlorine concentrations, use proportionately less sample and dilute to 200 mL with chlorine-demand-free water. Use a sample of such size that not more than 10 mL phenylarsine oxide solution is required.

2) Preparation for titration—Measure 5 mL 0.005 64N phenylarsine oxide or thiosulfate for chlorine concentrations from 2 to 5 mg/L, and 10 mL for concentrations of 5 to 10 mg/L, into a flask or casserole for titration with standard iodine or iodate. Start stirring. For titration by amperometry or standard iodine, also add excess KI (approximately 1 g) and 4 mL acetate buffer solution or enough to reduce the pH to between 3.5 and 4.2.

b. Titration: Use one of the following:

1) Amperometric titration—Add 0.0282N iodine titrant in small increments from a 1-mL buret or pipet. Observe meter needle response as iodine is added: the pointer remains practically stationary until the end point is approached, whereupon each iodine increment causes a temporary deflection of the microammeter, with the pointer dropping back to its original position. Stop titration at end point when a small increment of iodine titrant gives a definite pointer deflection upscale and the pointer does not re-turn promptly to its original position. Record volume of iodine titrant used to reach end point.

2) Colorimetric (iodine) titration—Add 1 mL starch solution and titrate with 0.0282N iodine to the first appearance of blue color that persists after complete mixing.

3) Colorimetric (iodate) titration—To suitable flask or casserole add 200 mL chlorine-demand-free water and add, with agitation, the required volume of reductant, an excess of KI (approximately 0.5 g), 2 mL 10% H_3PO_4 solution, and 1 mL starch solution in the order given, and titrate immediately* with 0.005 64N iodate solution to the first appearance of a blue color that persists after complete mixing. Designate volume of iodate solution used as A. Repeat procedure, substituting 200 mL sample for the 200 mL chlorine-demand-free water. If sample is colored or turbid, titrate to the first change in color, using for comparison another portion of sample with H_3PO_4 added. Designate this volume of iodate solution as B.

5. Calculation

a. Titration with standard iodine:

$$\text{mg Cl as } Cl_2/L = \frac{(A - 5B) \times 200}{C}$$

where:

A = mL 0.005 64N reductant,
B = mL 0.0282 N I_2, and
C = mL sample.

b. Titration with standard iodate:

$$\text{mg Cl as } Cl_2/L = \frac{(A - B) \times 200}{C}$$

where:

A = mL $Na_2S_2O_3$,
B = mL iodate required to titrate $Na_2S_2O_3$, and
C = mL sample.

6. Bibliography

See B.7.

---

* Titration may be delayed up to 10 min without appreciable error if H_3PO_4 is not added until immediately before titration.

4500-Cl D.   Amperometric Titration Method

1. General Discussion

Amperometric titration requires a higher degree of skill and care than the colorimetric methods. Chlorine residuals over 2 mg/L are measured best by means of smaller samples or by dilution with water that has neither residual chlorine nor a chlorine demand. The method can be used to determine total chlorine and can differentiate between free and combined chlorine. A further differentiation into monochloramine and dichloramine fractions is possible by control of KI concentration and pH.

a. Principle: The amperometric method is a special adaptation of the polarographic principle. Free chlorine is titrated at a pH between 6.5 and 7.5, a range in which the combined chlorine reacts slowly. The combined chlorine, in turn, is titrated in the presence of the proper amount of KI in the pH range 3.5 to 4.5. When free chlorine is determined, the pH must not be greater than 7.5 because the reaction becomes sluggish at higher pH values, nor less than 6.5 because at lower pH values some combined chlorine may react even in the absence of iodide. When combined chlorine is determined, the pH must not be less than 3.5 because of increased interferences at lower pH values, nor greater than 4.5 because the iodide reaction is not quantitative at higher pH values. The tendency of monochloramine to react more readily with iodide than does dichloramine provides a means for further differentiation. The addition of a small amount of KI in the neutral pH range enables estimation of monochloramine content. Lowering the pH into the acid range and increasing the KI concentration allows the separation determination of dichloramine.

Organic chloramines can be measured as free chlorine, monochloramine, or dichloramine, depending on the activity of chlorine in the organic compound.

Phenylarsine oxide is stable even in dilute solution and each mole reacts with two equivalents of halogen. A special amperometric cell is used to detect the end point of the residual chlorine-phenylarsine oxide titration. The cell consists of a nonpolarizable reference electrode that is immersed in a salt solution and a readily polarizable noble-metal electrode that is in contact with both the salt solution and the sample being titrated. In some applications, end-point selectivity is improved by adding $+200$ mV to the platinum electrode versus silver, silver chloride. Another approach to end-point detection uses dual platinum electrodes, a mercury cell with voltage divider to impress a potential across the electrodes, and a microammeter. If there is no chlorine residual in the sample, the microammeter reading will be comparatively low because of cell polarization. The greater the residual, the greater the microammeter reading. The meter acts merely as a null-point indicator—that is, the actual meter reading is not important, but rather the relative readings as the titration proceeds. The gradual addition of phenylarsine oxide causes the cell to become more and more polarized because of the decrease in chlorine. The end point is recognized when no further decrease in meter reading can be obtained by adding more phenylarsine oxide.

b. Interference: Accurate determinations of free chlorine cannot be made in the presence of nitrogen trichloride, NCl_3, or chlorine dioxide, which titrate partly as free chlorine. When present, NCl_3 can titrate partly as free chlorine and partly as dichloramine, contributing a positive error in both fractions at a rate of approximately 0.1%/min. Some organic chloramines also can be titrated in each step. Monochloramine can intrude into the free chlorine fraction and dichloramine can interfere in the monochloramine fraction, especially at high temperatures and prolonged titration times. Free halogens other than chlorine also will titrate as free chlorine. Combined chlorine reacts with iodide ions to produce iodine. When titration for free chlorine follows a combined chlorine titration, which requires addition of KI, erroneous results may occur unless the measuring cell is rinsed thoroughly with distilled water between titrations.

Interference from copper has been noted in samples taken from copper pipe or after heavy copper sulfate treatment of reservoirs, with metallic copper plating out on the electrode. Silver ions also poison the electrode. Interference occurs in some highly colored waters and in waters containing surface-active agents. Very low temperatures slow response of measuring cell and longer time is required for the titration, but precision is not affected. A reduction in reaction rate is caused by pH values above 7.5; overcome this by buffering all samples to pH 7.0 or less. On the other hand, some substances, such as manganese, nitrite, and iron, do not interfere. The violent stirring of some commercial titrators can lower chlorine values by volatilization. When dilution is used for samples containing high chlorine content, take care that the dilution water is free of chlorine and ammonia and possesses no chlorine demand.

See A.3 for a discussion of other interferences.

2. Apparatus

a. End-point detection apparatus, consisting of a cell unit connected to a microammeter, with necessary electrical accessories. The cell unit includes a noble-metal electrode of sufficient surface area, a salt bridge to provide an electrical connection without diffusion of electrolyte, and a reference electrode of silver-silver chloride in a saturated sodium chloride solution connected into the circuit by means of the salt bridge. Numerous commercial systems are available.

Keep platinum electrode free of deposits and foreign matter. Vigorous chemical cleaning generally is unnecessary. Occasional mechanical cleaning with a suitable abrasive usually is sufficient. Keep salt bridge in good operating condition; do not allow it to become plugged nor permit appreciable flow of electrolyte through it. Keep solution surrounding reference electrode free of contamination and maintain it at constant composition by insuring an adequate supply of undissolved salt at all times. A cell with two metal electrodes polarized by a small DC potential also may be used. (See Bibliography.)

b. Agitator, designed to give adequate agitation at the noble-metal electrode surface to insure proper sensitivity. Thoroughly clean agitator and exposed electrode system to remove all chlorine-consuming contaminants by immersing them in water containing 1 to 2 mg/L free chlorine for a few minutes. Add KI to the same water and let agitator and electrodes remain immersed for 5 min. After thorough rinsing with chlorine-demand-free water or the sample to be tested, sensitized electrodes and agitator are ready for use. Remove iodide reagent completely from cell.

c. Buret: Commercial titrators usually are equipped with suitable burets (1 mL). Manual burets are available.*

d. Glassware, exposed to water containing at least 10 mg/L chlorine for 3 h or more before use and rinsed with chlorine-demand-free water.

3. Reagents

a. Standard phenylarsine oxide titrant: See C.3a.

b. Phosphate buffer solution, pH 7: Dissolve 25.4 g anhydrous KH_2PO_4 and 34.1 g anhydrous Na_2HPO_4 in 800 mL distilled water. Add 2 mL sodium hypochlorite solution containing 1% chlorine and mix thoroughly. Protect from sunlight for 2 d. Determine that free chlorine still remains in the solution. Then expose to sunlight until no chlorine remains. If necessary, carry out the final dechlorination with an ultraviolet lamp. Determine that no total chlorine remains by adding KI and measuring with one of the colorimetric tests. Dilute to 1 L with distilled water and filter if any precipitate is present.

c. Potassium iodide solution: Dissolve 50 g KI and dilute to 1 L with freshly boiled and cooled distilled water. Store in the dark in a brown glass-stoppered bottle, preferably in the refrigerator. Discard when solution becomes yellow.

d. Acetate buffer solution, pH 4: See C.3e.

---

* Kimax 17110-F, 5 mL, Kimble Products, Box 1035, Toledo, OH, or equivalent.

4. Procedure

a. Sample volume: Select a sample volume requiring no more than 2 mL phenylarsine oxide titrant. Thus, for chlorine concentrations of 2 mg/L or less, take a 200-mL sample; for chlorine levels in excess of 2 mg/L, use 100 mL or proportionately less.

b. Free chlorine: Unless sample pH is known to be between 6.5 and 7.5, add 1 mL pH 7 phosphate buffer solution to produce a pH of 6.5 to 7.5. Titrate with standard phenylarsine oxide titrant, observing current changes on microammeter. Add titrant in progressively smaller increments until all needle movement ceases. Make successive buret readings when needle action becomes sluggish, signaling approach of end point. Subtract last very small increment that causes no needle response because of overtitration. Alternatively, use a system involving continuous current measurements and determine end point mathematically.

Continue titrating for combined chlorine as described in ¶ 4c below or for the separate monochloramine and dichloramine fractions as detailed in ¶s 4e and 4f.

c. Combined chlorine: To sample remaining from free-chlorine titration add 1.00 mL KI solution and 1 mL acetate buffer solution, in that order. Titrate with phenylarsine oxide titrant to the end point, as above. Do not refill buret but simply continue titration after recording figure for free chlorine. Again subtract last increment to give amount of titrant actually used in reaction with chlorine. (If titration was continued without refilling buret, this figure represents total chlorine. Subtracting free chlorine from total gives combined chlorine.) Wash apparatus and sample cell thoroughly to remove iodide ion to avoid inaccuracies when the titrator is used subsequently for a free chlorine determination.

d. Separate samples: If desired, determine total chlorine and free chlorine on separate samples. If sample pH is between 3.5 and 9.5 and total chlorine alone is required, treat sample immediately with 1 mL KI solution followed by 1 mL acetate buffer solution, and titrate with phenylarsine oxide titrant as described in ¶ 4c preceding.

e. Monochloramine: After titrating for free chlorine, add 0.2 mL KI solution to same sample and, without refilling buret, continue titration with phenylarsine oxide titrant to end point. Subtract last increment to obtain net volume of titrant consumed by monochloramine.

f. Dichloramine: Add 1 mL acetate buffer solution and 1 mL KI solution to same sample and titrate final dichloramine fraction as described above.

5. Calculation

Convert individual titrations for free chlorine, combined chlorine, total chlorine, monochloramine, and dichloramine by the following equation:

$$\text{mg Cl as Cl}_2/\text{L} = \frac{A \times 200}{\text{mL sample}}$$

where:

A = mL phenylarsine oxide titration.

6. Precision and Bias

See B.5.

7. Bibliography

FOULK, C.W. & A.T. BAWDEN. 1926. A new type of endpoint in electrometric titration and its application to iodimetry. *J. Amer. Chem. Soc.* 48:2045.

MARKS, H.C. & J.R. GLASS. 1942. A new method of determining residual chlorine. *J. Amer. Water Works Assoc.* 34:1227.

HALLER, J.F. & S.S. LISTEK. 1948. Determination of chloride dioxide and other active chlorine compounds in water. *Anal. Chem.* 20:639.

MAHAN, W.A. 1949. Simplified amperometric titration apparatus for determining residual chlorine in water. *Water Sewage Works* 96:171.

KOLTHOFF, I.M. & J.J. LINGANE. 1952. Polarography, 2nd ed. Interscience Publishers, New York, N.Y.

MORROW, J.J. 1966. Residual chlorine determination with dual polarizable electrodes. *J. Amer. Water Works Assoc.* 58:363.

4500-Cl E.   Low-Level Amperometric Titration Method

1. General Discussion

Detection and quantification of chlorine residuals below 0.2 mg/L require special modifications to the amperometric titration procedure. With these modifications chlorine concentrations at the 10-μg/L level can be measured. It is not possible to differentiate between free and combined chlorine forms. Oxidizing agents that interfere with the amperometric titration method (D) will interfere.

a. Principle: This method modifies D by using a more dilute titrant and a graphical procedure to determine the end point.

b. Interference: See D.1b.

2. Apparatus

See D.2.

3. Reagents

a. Potassium bi-iodate, 0.002 256N: Dissolve 0.7332 g anhydrous potassium bi-iodate, $KH(IO_3)_2$, in 500 mL chlorine-free distilled water and dilute to 1000 mL. Dilute 10.00 mL to 100.0 mL with chlorine-free distilled water. Use only freshly prepared solution for the standardization of phenylarsine oxide.

b. *Potassium iodide,* KI crystals.

c. *Low-strength phenylarsine oxide titrant,* 0.000 564N: Dilute 10.00 mL of 0.005 64N phenylarsine oxide (see C.3a) to 100.0 mL with chlorine-demand-free water (see C.3m).

Standardization—Dilute 5.00 mL 0.002 256N potassium bi-iodate to 200 mL with chlorine-free water. Add approximately 1.5 g KI and stir to dissolve. Add 1 mL acetate buffer and let stand in the dark for 6 min. Titrate using the amperometric titrator and determine the equivalence point as indicated below.

$$\text{Normality} = 0.002256 \times 5/A$$

where:

A = mL phenylarsine oxide titrant required to reach the equivalence point of standard bi-iodate.

d. *Acetate buffer solution,* pH 4: See C.3e.

4. Procedure

Select a sample volume requiring no more than 2 mL phenylarsine oxide titrant. A 200-mL sample will be adequate for samples containing less than 0.2 mg total chlorine/L.

Before beginning titration, rinse buret with titrant several times. Rinse sample container with distilled water and then with sample. Add 200 mL sample to sample container and approximately 1.5 g KI. Dissolve, using a stirrer or mixer. Add 1 mL acetate buffer and place container in end-point detection apparatus. When the current signal stabilizes, record the reading. Initially adjust meter to a near full-scale deflection. Titrate by adding small, known,

volumes of titrant. After each addition, record cumulative volume added and current reading when the signal stabilizes. If meter reading falls to near or below 10% of full-scale deflection, record low reading, readjust meter to near full-scale deflection, and record difference between low amount and readjusted high deflection. Add this value to all deflection readings for subsequent titrant additions. Continue adding titrant until no further meter deflection occurs. If fewer than three titrant additions were made before meter deflection ceased, discard sample and repeat analysis using smaller titrant increments.

Determine equivalence point by plotting total meter deflection against titrant volume added. Draw straight line through the first several points in the plot and a second, horizontal straight line corresponding to the final total deflection in the meter. Read equivalence point as the volume of titrant added at the intersection of these two lines.

5. Calculation

$$\text{mg Cl as Cl}_2/\text{L} = \frac{A \times 200 \times N}{B \times 0.00564}$$

where:

A = mL titrant at equivalence point,
B = sample volume, mL, and
N = phenylarsine oxide normality.

6. Bibliography

BROOKS, A. S. & G. L. SEEGERT. 1979. Low-level chlorine analysis by amperometric titration. *J. Water Pollut. Control Fed.* 51:2636.

4500-Cl F.   DPD Ferrous Titrimetric Method

1. General Discussion

a. *Principle:* N,N-diethyl-p-phenylenediamine (DPD) is used as an indicator in the titrimetric procedure with ferrous ammonium sulfate (FAS). Where complete differentiation of chlorine species is not required, the procedure may be simplified to give only free and combined chlorine or total chlorine.

In the absence of iodide ion, free chlorine reacts instantly with DPD indicator to produce a red color. Subsequent addition of a small amount of iodide ion acts catalytically to cause monochloramine to produce color. Addition of iodide ion to excess evokes a rapid response from dichloramine. In the presence of iodide ion, part of the nitrogen trichloride (NCl₃) is included with dichloramine and part with free chlorine. A supplementary procedure based on adding iodide ion before DPD permits estimating proportion of NCl₃ appearing with free chlorine.

Chlorine dioxide (ClO₂) appears, to the extent of one-fifth of its total chlorine content, with free chlorine. A full response from ClO₂, corresponding to its total chlorine content, may be obtained if the sample first is acidified in the presence of iodide ion and subsequently is brought back to an approximately neutral pH by adding bicarbonate ion. Bromine, bromamine, and iodine react with DPD indicator and appear with free chlorine.

Addition of glycine before determination of free chlorine converts free chlorine to unreactive forms, with only bromine and iodine residuals remaining. Subtractions of these residuals from the residual measured without glycine permits differentiation of free chlorine from bromine and iodine.

b. *pH control:* For accurate results careful pH control is essential. At the proper pH of 6.2 to 6.5, the red colors produced may be titrated to sharp colorless end points. *Titrate as soon as the red color is formed in each step.* Too low a pH in the first step tends to make the monochloramine show in the free-chlorine step and the dichloramine in the monochloramine step. Too high a pH causes dissolved oxygen to give a color.

c. *Temperature control:* In all methods for differentiating free chlorine from chloramines, higher temperatures increase the tendency for chloramines to react and lead to increased apparent free-chlorine results. Higher temperatures also increase color fading. Complete measurements rapidly, especially at higher temperature.

d. *Interference:* The most significant interfering substance likely to be encountered in water is oxidized manganese. To correct for this, place 5 mL buffer solution and 0.5 mL sodium arsenite solution in the titration flask. Add 100 mL sample and mix. Add 5 mL DPD indicator solution, mix, and titrate with standard FAS titrant until red color is discharged. Subtract reading from

Reading *A* obtained by the normal procedure as described in ¶ 3*a*1) of this method or from the total chlorine reading obtained in the simplified procedure given in ¶ 3*a*4). If the combined reagent in powder form (see below) is used, first add KI and arsenite to the sample and mix, then add combined buffer-indicator reagent.

As an alternative to sodium arsenite use a 0.25% solution of thioacetamide, adding 0.5 mL to 100 mL sample.

Interference by copper up to approximately 10 mg Cu/L is overcome by the EDTA incorporated in the reagents. EDTA enhances stability of DPD indicator solution by retarding deterioration due to oxidation, and in the test itself, provides suppression of dissolved oxygen errors by preventing trace metal catalysis.

Chromate in excess of 2 mg/L interferes with end-point determination. Add barium chloride to mask this interference by precipitation.

High concentrations of combined chlorine can break through into the free chlorine fraction. *If free chlorine is to be measured in the presence of more than 0.5 mg/L combined chlorine, use the thioacetamide modification.* If this modification is not used, a color-development time in excess of 1 min leads to progressively greater interference from monochloramine. Adding thioacetamide (0.5 mL 0.25% solution to 100 mL) immediately after mixing DPD reagent with sample completely stops further reaction with combined chlorine in the free chlorine measurement. Continue immediately with FAS titration to obtain free chlorine. Obtain total chlorine from the normal procedure, i.e., without thioacetamide.

Because high concentrations of iodide are used to measure combined chlorine and only traces of iodide greatly increase chloramine interference in free chlorine measurements, take care to avoid iodide contamination by rinsing between samples or using separate glassware.

See A.3 for a discussion of other interferences.

e. Minimum detectable concentration: Approximately 18 μg Cl as Cl_2/L. This detection limit is achievable under ideal conditions; normal working detection limits typically are higher.

2. Reagents

a. Phosphate buffer solution: Dissolve 24 g anhydrous Na_2HPO_4 and 46 g anhydrous KH_2PO_4 in distilled water. Combine with 100 mL distilled water in which 800 mg disodium ethylenediamine tetraacetate dihydrate (EDTA) have been dissolved. Dilute to 1 L with distilled water and optionally add either 20 mg $HgCl_2$ or 2 drops toluene to prevent mold growth. Interference from trace amounts of iodide in the reagents can be negated by optional addition of 20 mg $HgCl_2$ to the solution. (CAUTION: *$HgCl_2$ is toxic—take care to avoid ingestion.*)

b. N,N-Diethyl-p-phenylenediamine (DPD) indicator solution: Dissolve 1 g DPD oxalate,* or 1.5 g DPD sulfate pentahydrate,† or 1.1 g anhydrous DPD sulfate in chlorine-free distilled water containing 8 mL 1 + 3 H_2SO_4 and 200 mg disodium EDTA. Make up to 1 L, store in a brown glass-stoppered bottle in the dark, and discard when discolored. Periodically check solution blank for absorbance and discard when absorbance at 515 nm exceeds 0.002/cm. (The buffer and indicator sulfate are available

---

* Eastman chemical No. 7102 or equivalent.
† Available from Gallard-Schlesinger Chemical Mfg. Corp., 584 Mineola Avenue, Carle Place, NY 11514, or equivalent.

commercially as a combined reagent in stable powder form.) CAUTION: *The oxalate is toxic—take care to avoid ingestion.*

c. Standard ferrous ammonium sulfate (FAS) titrant: Dissolve 1.106 g $Fe(NH_4)_2(SO_4)_2 \cdot 6H_2O$ in distilled water containing 1 mL 1 + 3 H_2SO_4 and make up to 1 L with freshly boiled and cooled distilled water. This standard may be used for 1 month, and the titer checked by potassium dichromate. For this purpose add 10 mL 1 + 5 H_2SO_4, 5 mL conc H_3PO_4, and 2 mL 0.1% barium diphenylamine sulfonate indicator to a 100-mL sample of FAS and titrate with potassium dichromate to a violet end point that persists for 30 s. FAS titrant equivalent to 100 μg Cl as Cl_2/1.00 mL requires 20.00 mL dichromate for titration.

d. Potassium iodide, KI, crystals.

e. Potassium iodide solution: Dissolve 500 mg KI and dilute to 100 mL, using freshly boiled and cooled distilled water. Store in a brown glass-stoppered bottle, preferably in a refrigerator. Discard when solution becomes yellow.

f. Potassium dichromate solution, 0.691 g to 1000 mL.

g. Barium diphenylaminesulfonate, 0.1%: Dissolve 0.1 g $(C_6H_5NHC_6H_4\text{-}4\text{-}SO_3)_2Ba$ in 100 mL distilled water.

h. Sodium arsenite solution: Dissolve 5.0 g $NaAsO_2$ in distilled water and dilute to 1 L. (CAUTION: *Toxic—take care to avoid ingestion.*)

i. Thioacetamide solution: Dissolve 250 mg CH_3CSNH_2 in 100 mL distilled water. (CAUTION: *Cancer suspect agent. Take care to avoid skin contact or ingestion.*)

j. Chlorine-demand-free water: See C.3*m*.

k. Glycine solution: Dissolve 20 g glycine (aminoacetic acid) in sufficient chlorine-demand-free water to bring to 100 mL total volume. Store under refrigerated conditions and discard if cloudiness develops.

l. Barium chloride crystals, $BaCl_2 \cdot 2H_2O$.

3. Procedure

The quantities given below are suitable for concentrations of total chlorine up to 5 mg/L. If total chlorine exceeds 5 mg/L, use a smaller sample and dilute to a total volume of 100 mL. Mix usual volumes of buffer reagent and DPD indicator solution, or usual amount of DPD powder, with distilled water *before* adding sufficient sample to bring total volume to 100 mL. (If sample is added before buffer, test does not work.)

If chromate is present (>2 mg/L) add and mix 0.2 g $BaCl_2 \cdot 2H_2O$/100 mL sample before adding other reagents. If, in addition, sulfate is >500 mg/L, use 0.4 g $BaCl_2 \cdot 2H_2O$/100 mL sample.

a. Free chlorine or chloramine: Place 5 mL each of buffer reagent and DPD indicator solution in titration flask and mix (or use about 500 mg DPD powder). Add 100 mL sample, or diluted sample, and mix.

1) Free chlorine—Titrate rapidly with standard FAS titrant until red color is discharged (Reading *A*).

2) Monochloramine—Add one very small crystal of KI (about 0.5 mg) or 0.1 mL (2 drops) KI solution and mix. Continue titrating until red color is discharged again (Reading *B*).

3) Dichloramine—Add several crystals KI (about 1 g) and mix to dissolve. Let stand for 2 min and continue titrating until red color is discharged (Reading *C*). For dichloramine concentrations greater than 1 mg/L, let stand 2 min more if color driftback indicates slightly incomplete reaction. When dichloramine concen-

trations are not expected to be high, use half the specified amount of KI.

Total

4) Simplified procedure for free and combined chlorine or total chlorine—Omit 2) above to obtain monochloramine and dichloramine together as combined chlorine. To obtain total chlorine in one reading, add full amount of KI at the start, with the specified amounts of buffer reagent and DPD indicator, and titrate after 2 min standing.

b. Nitrogen trichloride: Place one very small crystal of KI (about 0.5 mg) or 0.1 mL KI solution in a titration flask. Add 100 mL sample and mix. Add contents to a second flask containing 5 mL each of buffer reagent and DPD indicator solution (or add about 500 mg DPD powder direct to the first flask). Titrate rapidly with standard FAS titrant until red color is discharged (Reading *N*).

Free + bromine

c. Free chlorine in presence of bromine or iodine: Determine free chlorine as in ¶ 3*a*1). To a second 100-mL sample, add 1 mL glycine solution before adding DPD and buffer. Titrate according to ¶ 3*a*1). Subtract the second reading from the first to obtain Reading A.

4. Calculation

For a 100-mL sample, 1.00 mL standard FAS titrant = 1.00 mg Cl as Cl_2/L.

Reading	NCl_3 Absent	NCl_3 Present
A	Free Cl	Free Cl
B − A	NH_2Cl	NH_2Cl
C − B	$NHCl_2$	$NHCl_2 + \frac{1}{2}NCl_3$
N	—	Free Cl + $\frac{1}{2}NCl_3$
2(N − A)	—	NCl_3
C − N	—	$NHCl_2$

In the event that monochloramine is present with NCl_3, it will be included in *N*, in which case obtain NCl_3 from $2(N-B)$.

Chlorine dioxide, if present, is included in *A* to the extent of one-fifth of its total chlorine content.

In the simplified procedure for free and combined chlorine, only *A* (free Cl) and *C* (total Cl) are required. Obtain combined chlorine from $C-A$.

The result obtained in the simplified total chlorine procedure corresponds to *C*.

5. Precision and Bias

See B.5.

6. Bibliography

PALIN, A.T. 1957. The determination of free and combined chlorine in water by the use of diethyl-p-phenylene diamine. *J. Amer. Water Works Assoc.* 49:873.

PALIN, A.T. 1960. Colorimetric determination of chlorine dioxide in water. *Water Sewage Works* 107:457.

PALIN, A.T. 1961. The determination of free residual bromine in water. *Water Sewage Works* 108:461.

NICOLSON, N.J. 1963, 1965, 1966. Determination of chlorine in water, Parts 1, 2, and 3. Water Res. Assoc. Tech. Pap. Nos. 29, 47, and 53.

PALIN, A.T. 1967. Methods for determination, in water, of free and combined available chlorine, chlorine dioxide and chlorite, bromine, iodine, and ozone using diethyl-p-phenylenediamine (DPD). *J. Inst. Water Eng.* 21:537.

PALIN, A.T. 1968. Determination of nitrogen trichloride in water. *J. Amer. Water Works Assoc.* 60:847.

PALIN, A.T. 1975. Current DPD methods for residual halogen compounds and ozone in water. *J. Amer. Water Works Assoc.* 67:32.

Methods for the Examination of Waters and Associated Materials. Chemical Disinfecting Agents in Water and Effluents, and Chlorine Demand. 1980. Her Majesty's Stationery Off., London, England.

4500-Cl G.   DPD Colorimetric Method

1. General Discussion

a. Principle: This is a colorimetric version of the DPD method and is based on the same principles. Instead of titration with standard ferrous ammonium sulfate (FAS) solution as in the titrimetric method, a colorimetric procedure is used.

b. Interference: See A.3 and F.1*d*. Compensate for color and turbidity by using sample to zero photometer. Minimize chromate interference by using the thioacetamide blank correction.

c. Minimum detectable concentration: Approximately 10 μg Cl as Cl_2/L. This detection limit is achievable under ideal conditions; normal working detection limits typically are higher.

2. Apparatus

a. Photometric equipment: One of the following is required:

1) *Spectrophotometer,* for use at a wavelength of 515 nm and providing a light path of 1 cm or longer.

2) *Filter photometer,* equipped with a filter having maximum transmission in the wavelength range of 490 to 530 nm and providing a light path of 1 cm or longer.

b. Glassware: Use separate glassware, including separate spectrophotometer cells, for free and combined (dichloramine) measurements, to avoid iodide contamination in free chlorine measurement.

3. Reagents

See F.2*a, b, c, d, e, h, i,* and *j.*

4. Procedure

a. Calibration of photometric equipment: Calibrate instrument with chlorine or potassium permanganate solutions.

1) Chlorine solutions—Prepare chlorine standards in the range of 0.05 to 4 mg/L from about 100 mg/L chlorine water standardized as follows: Place 2 mL acetic acid and 10 to 25 mL chlorine-

demand-free water in a flask. Add about 1 g KI. Measure into the flask a suitable volume of chlorine solution. In choosing a convenient volume, note that 1 mL 0.025N $Na_2S_2O_3$ titrant (see B.2d) is equivalent to about 0.9 mg chlorine. Titrate with standardized 0.025N $Na_2S_2O_3$ titrant until the yellow iodine color almost disappears. Add 1 to 2 mL starch indicator solution and continue titrating to disappearance of blue color.

Determine the blank by adding identical quantities of acid, KI, and starch indicator to a volume of chlorine-demand-free water corresponding to the sample used for titration. Perform blank titration A or B, whichever applies, according to B.3d.

$$\text{mg Cl as } Cl_2/\text{mL} = \frac{(A + B) \times N \times 35.45}{\text{mL sample}}$$

where:

N = normality of $Na_2S_2O_3$,
A = mL titrant for sample,
B = mL titrant for blank (to be added or subtracted according to required blank titration. See B.3d).

Use chlorine-demand-free water and glassware to prepare these standards. Develop color by first placing 5 mL phosphate buffer solution and 5 mL DPD indicator reagent in flask and then adding 100 mL chlorine standard with thorough mixing as described in b and c below. Fill photometer or colorimeter cell from flask and read color at 515 nm. Return cell contents to flask and titrate with standard FAS titrant as a check on chlorine concentration.

2) *Potassium permanganate solutions*—Prepare a stock solution containing 891 mg $KMnO_4$/1000 mL. Dilute 10.00 mL stock solution to 100 mL with distilled water in a volumetric flask. When 1 mL of this solution is diluted to 100 mL with distilled water, a chlorine equivalent of 1.00 mg/L will be produced in the DPD reaction. Prepare a series of $KMnO_4$ standards covering the chlorine equivalent range of 0.05 to 4 mg/L. Develop color by first placing 5 mL phosphate buffer and 5 mL DPD indicator reagent in flask and adding 100 mL standard with thorough mixing as described in b and c below. Fill photometer or colorimeter cell from flask and read color at 515 nm. Return cell contents to flask and titrate with FAS titrant as a check on any absorption of permanganate by distilled water.

Obtain all readings by comparison to color standards or the standard curve before use in calculation.

b. Volume of sample: Use a sample volume appropriate to the photometer or colorimeter. The following procedure is based on using 10-mL volumes; adjust reagent quantities proportionately for other sample volumes. Dilute sample with chlorine-demand-free water when total chlorine exceeds 4 mg/L.

c. Free chlorine: Place 0.5 mL each of buffer reagent and DPD indicator reagent in a test tube or photometer cell. Add 10 mL sample and mix. Read color immediately (Reading A).

d. Monochloramine: Continue by adding one very small crystal of KI (about 0.1 mg) and mix. If dichloramine concentration is expected to be high, instead of small crystal add 0.1 mL (2 drops) freshly prepared KI solution (0.1 g/100 mL). Read color immediately (Reading B).

e. Dichloramine: Continue by adding several crystals of KI (about 0.1 g) and mix to dissolve. Let stand about 2 min and read color (Reading C).

f. Nitrogen trichloride: Place a very small crystal of KI (about 0.1 mg) in a clean test tube or photometer cell. Add 10 mL sample and mix. To a second tube or cell add 0.5 mL each of buffer and indicator reagents; mix. Add contents to first tube or cell and mix. Read color immediately (Reading N).

g. Chromate correction using thioacetamide: Add 0.5 mL thioacetamide solution (F.2i) to 100 mL sample. After mixing, add buffer and DPD reagent. Read color immediately. Add several crystals of KI (about 0.1 g) and mix to dissolve. Let stand about 2 min and read color. Subtract the first reading from Reading A and the second reading from Reading C and use in calculations.

h. Simplified procedure for total chlorine: Omit Step d above to obtain monochloramine and dichloramine together as combined chlorine. To obtain total chlorine in one reading, add the full amount of KI at the start, with the specified amounts of buffer reagent and DPD indicator. Read color after 2 min.

5. Calculation

Reading	NCl_3 Absent	NCl_3 Present
A	Free Cl	Free Cl
$B - A$	NH_2Cl	NH_2Cl
$C - B$	$NHCl_2$	$NHCl_2 + \frac{1}{2}NCl_3$
N	—	Free Cl + $\frac{1}{2}NCl_3$
$2(N - A)$	—	NCl_3
$C - N$	—	$NHCl_2$

In the event that monochloramine is present with NCl_3, it will be included in Reading N, in which case obtain NCl_3 from $2(N - B)$.

6. Bibliography

See F.6.

4500-Cl H.   Syringaldazine (FACTS) Method

1. General Discussion

a. Principle: The free (available) chlorine test, syringaldazine (FACTS) measures free chlorine over the range of 0.1 to 10 mg/L. A saturated solution of syringaldazine (3,5-dimethoxy-4-hydroxybenzaldazine) in 2-propanol is used. Syringaldazine is oxidized by free chlorine on a 1:1 molar basis to produce a colored product with an absorption maximum of 530 nm. The color product is only slightly soluble in water; therefore, at chlorine concentrations greater than 1 mg/L, the final reaction mixture must contain 2-propanol to prevent product precipitation and color fading.

The optimum color and solubility (minimum fading) are obtained in a solution having a pH between 6.5 and 6.8. At a pH less than 6, color development is slow and reproducibility is poor. At a pH greater than 7, the color develops rapidly but fades quickly. A buffer is required to maintain the reaction mixture pH at approximately 6.7. Take care with waters of high acidity or alkalinity to assure that the added buffer maintains the proper pH.

Temperature has a minimal effect on the color reaction. The maximum error observed at temperature extremes of 5 and 35°C is ± 10%.

b. Interferences: Interferences common to other methods for determining free chlorine do not affect the FACTS procedure. Monochloramine concentrations up to 18 mg/L, dichloramine concentrations up to 10 mg/L, and manganese concentrations (oxidized forms) up to 1 mg/L do not interfere. Trichloramine at levels above 0.6 mg/L produces an apparent free chlorine reaction. Very high concentrations of monochloramine (≥35 mg/L) and oxidized manganese (≥2.6 mg/L) produce a color with syringaldazine slowly. Ferric iron can react with syringaldazine; however, concentrations up to 10 mg/L do not interfere. Nitrite (≤250 mg/L), nitrate (≤100 mg/L), sulfate (≤1000 mg/L), and chloride (≤1000 mg/L) do not interfere. Waters with high hardness (≥500 mg/L) will produce a cloudy solution that can be compensated for by using a blank. Oxygen does not interfere.

Other strong oxidizing agents, such as iodine, bromine, and ozone, will produce a color.

c. Minimum detectable concentration: The FACTS procedure is sensitive to free chlorine concentrations of 0.1 mg/L or less.

2. Apparatus

Colorimetric equipment: One of the following is required:

a. Filter photometer, providing a light path of 1 cm for chlorine concentrations ≤1 mg/L or a light path from 1 to 10 mm for chlorine concentrations above 1 mg/L; also equipped with a filter having a band pass of 500 to 560 nm.

b. Spectrophotometer, for use at 530 nm, providing the light paths noted above.

3. Reagents

a. Chlorine-demand-free water: See C.3*m*. Use to prepare reagent solutions and sample dilutions.

b. Syringaldazine indicator: Dissolve 115 mg 3,5-dimethoxy-4-hydroxybenzaldazine* in 1 L 2-propanol.

c. 2-Propanol: To aid in dissolution use ultrasonic agitation or gentle heating and stirring. Redistill reagent-grade 2-propanol to remove chlorine demand. Use a 30.5-cm Vigreux column and take the middle 75% fraction. Alternatively, chlorinate good-quality 2-propanol to maintain a free residual overnight; then expose to UV light or sunlight to dechlorinate. CAUTION: *2-Propanol is extremely flammable.*

d. Buffer: Dissolve 17.01 g KH_2PO_4 in 250 mL water; pH should be 4.4. Dissolve 17.75 g Na_2HPO_4 in 250 mL water; the pH should be 9.9. Mix equal volumes of these solutions to obtain FACTS buffer, pH 6.6. Verify pH with pH meter. For waters containing considerable hardness or high alkalinity other pH 6.6 buffers can be used, for example, 23.21 g maleic acid and 16.5 mL 50% NaOH per liter of water.

e. Hypochlorite solution: Dilute household hypochlorite solution, which contains about 30 000 to 50 000 mg Cl equivalent/L, to a strength between 100 and 1000 mg/L. Standardize as directed in G.4*a*1).

4. Procedure

a. Calibration of photometer: Prepare a calibration curve by making dilutions of a standardized hypochlorite solution (¶ 3*e*). Develop and measure colors as described in ¶ 4*b*, below. Check calibration regularly, especially as reagent ages.

b. Free chlorine analysis: Add 3 mL sample and 0.1 mL buffer to a 5-mL-capacity test tube. Add 1 mL syringaldazine indicator, cap tube, and invert twice to mix. Transfer to a photometer tube or spectrophotometer cell and measure absorbance. Compare absorbance value obtained with calibration curve and report corresponding value as milligrams free chlorine per liter.

5. Bibliography

BAUER, R. & C. RUPE. 1971. Use of syringaldazine in a photometric method for estimating "free" chlorine in water. *Anal. Chem.* 43: 421.

COOPER, W.J., C.A. SORBER & E.P. MEIER. 1975. A rapid, free, available chlorine test with syringaldazine (FACTS). *J. Amer. Water Works Assoc.* 67:34.

COOPER, W.J., P.H. GIBBS, E.M. OTT & P. PATEL. 1983. Equivalency testing of procedures for measuring free available chlorine: amperometric titration, DPD, and FACTS. *J. Amer. Water Works Assoc.* 75:625.

---

* Aldrich No. 17, 753-9, Aldrich Chemical Company, Inc., 1001 West St. Paul Ave., Milwaukee, WI 53233, or equivalent.

4500-Cl I. Iodometric Electrode Technique

1. General Discussion

a. Principle: This method involves the direct potentiometric measurement of iodine released on the addition of potassium io-

dide to an acidified sample. A platinum-iodide electrode pair is used in combination with an expanded-scale pH meter.

b. Interference: All oxidizing agents that interfere with other iodometric procedures interfere. These include oxidized manganese and iodate, bromine, and cupric ions. Silver and mercuric ions above 10 and 20 mg/L interfere.

2. Apparatus

a. Electrodes: Use either a combination electrode consisting of a platinum electrode and an iodide ion-selective electrode or two individual electrodes. Both systems are available commercially.

b. pH/millivolt meter: Use an expanded-scale pH/millivolt meter with 0.1 mV readability or a direct-reading selective ion meter.

3. Reagents

a. pH 4 buffer solution: See C.3*e.*

b. Chlorine-demand-free water: See C.3*m.*

c. Potassium iodide solution: Dissolve 42 g KI and 0.2 g Na_2CO_3 in 500 mL chlorine-demand-free, distilled water. Store in a dark bottle.

d. Standard potassium iodate 0.002 81N: Dissolve 0.1002 g KIO_3 in chlorine-demand-free, distilled water and dilute to 1000 mL. Each 1.0 mL, when diluted to 100 mL, produces a solution equivalent to 1 mg/L as Cl_2.

4. Procedure

a. Standardization: Pipet into three 100-mL stoppered volumetric flasks 0.20, 1.00, and 5.00 mL standard iodate solution. Add to each flask, and a fourth flask to be used as a reagent blank, 1 mL each of acetate buffer solution and KI solution. Stopper, swirl to mix, and let stand 2 min before dilution. Dilute each standard to 100 mL with chlorine-demand-free distilled water. Stopper, invert flask several times to mix, and pour into separate 150-mL beakers. Stir gently without turbulence, using a magnetic stirrer, and immerse electrode(s) in the 0.2-mg/L (0.2-mL) standard. Wait for the potential to stabilize and record potential in mV. Rinse electrodes with chlorine-demand-free water and repeat for each standard and for the reagent blank. Prepare a calibration curve by plotting, on semilogarithmic paper, potential (linear axis) against concentration. Determine apparent chlorine concentration in the reagent blank from this graph (Reading B).

b. Analysis: Select a volume of sample containing no more than 0.5 mg chlorine. Pipet 1 mL acetate buffer solution and 1 mL KI into a 100-mL glass-stoppered volumetric flask. Stopper, swirl and let stand for at least 2 min. Adjust sample pH to 4 to 5, if necessary (mid-range pH paper is adequate for pH measurement), by adding acetic acid. Add pH-adjusted sample to volumetric flask and dilute to mark. Stopper and mix by inversion several times. Let stand for 2 min. Pour into a 150-mL beaker, immerse the electrode(s), wait for the potential to stabilize, and record. If the mV reading is greater than that recorded for the 5-mg/L standard, repeat analysis with a smaller volume of sample.

5. Calculation

Determine chlorine concentration (mg/L) corresponding to the recorded mV reading from the standard curve. This is Reading A. Determine total residual chlorine from the following:

$$\text{Total residual chlorine} = A \times 100/V$$

where V = sample volume, mL. If total residual chlorine is below 0.2 mg/L, subtract apparent chlorine in reagent blank (Reading B) to obtain the true total residual chlorine value.

6. Bibliography

DIMMOCK, N.A. & D. MIDGLEY. 1981. Determination of Total Residual Chlorine in Cooling Water with the Orion 97-70 Ion Selective Electrode. Central Electricity Generating Board (U.K.) Report RD/L/2159N81.

JENKINS, R.L. & R.B. BAIRD. 1979. Determination of total chlorine residual in treated wastewaters by electrode. *Anal. Letters* 12:125.

SYNNOTT, J.C. & A.M. SMITH. 1985. Total Residual Chlorine by Ion-Selective Electrode—from Bench Top to Continuous Monitor. Paper presented at 5th International Conf. on Chemistry for Protection of the Environment, Leuven, Belgium.

4500-Cl⁻   CHLORIDE*

4500-Cl⁻   A.   Introduction

1. Occurrence

Chloride, in the form of chloride (Cl^-) ion, is one of the major inorganic anions in water and wastewater. The salty taste produced by chloride concentrations is variable and dependent on the chemical composition of water. Some waters containing 250 mg Cl^-/L may have a detectable salty taste if the cation is sodium.

On the other hand, the typical salty taste may be absent in waters containing as much as 1000 mg/L when the predominant cations are calcium and magnesium.

The chloride concentration is higher in wastewater than in raw water because sodium chloride (NaCl) is a common article of diet and passes unchanged through the digestive system. Along the sea coast, chloride may be present in high concentrations because of leakage of salt water into the sewerage system. It also may be increased by industrial processes.

* Approved by Standard Methods Committee, 1997.

A high chloride content may harm metallic pipes and structures, as well as growing plants.

2. Selection of Method

Six methods are presented for the determination of chloride. Because the first two are similar in most respects, selection is largely a matter of personal preference. The argentometric method (B) is suitable for use in relatively clear waters when 0.15 to 10 mg Cl⁻ are present in the portion titrated. The end point of the mercuric nitrate method (C) is easier to detect. The potentiometric method (D) is suitable for colored or turbid samples in which color-indicated end points might be difficult to observe. The potentiometric method can be used without a pretreatment step for samples containing ferric ions (if not present in an amount greater than the chloride concentration), chromic, phosphate, and ferrous and other heavy-metal ions. The ferricyanide method (E) is an automated technique. Flow injection analysis (G), an automated colorimetric technique, is useful for analyzing large numbers of samples. Preferably determine chloride by ion chromatography (Section 4110). Chloride also can be determined by the capillary ion electrophoresis method (Section 4140). Methods (C and G) in which mercury, a highly toxic reagent, is used require special disposal practices to avoid improper sewage discharges. Follow appropriate regulatory procedures (see Section 1090).

3. Sampling and Storage

Collect representative samples in clean, chemically resistant glass or plastic bottles. The maximum sample portion required is 100 mL. No special preservative is necessary if the sample is to be stored.

4500-Cl⁻ B.   Argentometric Method

1. General Discussion

a. Principle: In a neutral or slightly alkaline solution, potassium chromate can indicate the end point of the silver nitrate titration of chloride. Silver chloride is precipitated quantitatively before red silver chromate is formed.

b. Interference: Substances in amounts normally found in potable waters will not interfere. Bromide, iodide, and cyanide register as equivalent chloride concentrations. Sulfide, thiosulfate, and sulfite ions interfere but can be removed by treatment with hydrogen peroxide. Orthophosphate in excess of 25 mg/L interferes by precipitating as silver phosphate. Iron in excess of 10 mg/L interferes by masking the end point.

2. Apparatus

a. Erlenmeyer flask, 250-mL.
b. Buret, 50-mL.

3. Reagents

a. Potassium chromate indicator solution: Dissolve 50 g K_2CrO_4 in a little distilled water. Add $AgNO_3$ solution until a definite red precipitate is formed. Let stand 12 h, filter, and dilute to 1 L with distilled water.

b. Standard silver nitrate titrant, 0.0141*M* (0.0141*N*): Dissolve 2.395 g $AgNO_3$ in distilled water and dilute to 1000 mL. Standardize against NaCl by the procedure described in ¶ *4b* below; 1.00 mL = 500 μg Cl⁻. Store in a brown bottle.

c. Standard sodium chloride, 0.0141*M* (0.0141*N*): Dissolve 824.0 mg NaCl (dried at 140°C) in distilled water and dilute to 1000 mL; 1.00 mL = 500 μg Cl⁻.

d. Special reagents for removal of interference:

1) *Aluminum hydroxide suspension:* Dissolve 125 g aluminum potassium sulfate or aluminum ammonium sulfate, $AlK(SO_4)_2·12H_2O$ or $AlNH_4(SO_4)_2·12H_2O$, in 1 L distilled water. Warm to 60°C and add 55 mL conc ammonium hydroxide (NH_4OH) slowly with stirring. Let stand about 1 h, transfer to a large bottle, and wash precipitate by successive additions, with thorough mixing and decanting with distilled water, until free from chloride. When freshly prepared, the suspension occupies a volume of approximately 1 L.

2) *Phenolphthalein indicator solution.*
3) *Sodium hydroxide,* NaOH, 1*N*.
4) *Sulfuric acid,* H_2SO_4, 1*N*.
5) *Hydrogen peroxide,* H_2O_2, 30%.

4. Procedure

a. Sample preparation: Use a 100-mL sample or a suitable portion diluted to 100 mL. If the sample is highly colored, add 3 mL $Al(OH)_3$ suspension, mix, let settle, and filter.

If sulfide, sulfite, or thiosulfate is present, add 1 mL H_2O_2 and stir for 1 min.

b. Titration: Directly titrate samples in the pH range 7 to 10. Adjust sample pH to 7 to 10 with H_2SO_4 or NaOH if it is not in this range. For adjustment, preferably use a pH meter with a non-chloride-type reference electrode. (If only a chloride-type electrode is available, determine amount of acid or alkali needed for adjustment and discard this sample portion. Treat a separate portion with required acid or alkali and continue analysis.) Add 1.0 mL K_2CrO_4 indicator solution. Titrate with standard $AgNO_3$ titrant to a pinkish yellow end point. Be consistent in end-point recognition.

Standardize $AgNO_3$ titrant and establish reagent blank value by the titration method outlined above. A blank of 0.2 to 0.3 mL is usual.

5. Calculation

$$mg\ Cl^-/L = \frac{(A - B) \times N \times 35\ 450}{mL\ sample}$$

where:

A = mL titration for sample,
B = mL titration for blank, and
N = normality of AgNO₃.

$$mg\ NaCl/L = (mg\ Cl^-/L) \times 1.65$$

6. Precision and Bias

A synthetic sample containing 241 mg Cl^-/L, 108 mg Ca/L, 82 mg Mg/L; 3.1 mg K/L, 19.9 mg Na/L, 1.1 mg NO_3^--N/L, 0.25 mg NO_2^-- N/L, 259 mg SO_4^{2-}/L, and 42.5 mg total alkalinity/L (contributed by $NaHCO_3$) in distilled water was analyzed in 41 laboratories by the argentometric method, with a relative standard deviation of 4.2% and a relative error of 1.7%.

7. Bibliography

HAZEN, A. 1889. On the determination of chlorine in water. *Amer. Chem. J.* 11:409.

KOLTHOFF, I.M. & V.A. STENGER. 1947. Volumetric Analysis, 2nd ed. Vol. 2. Interscience Publishers, New York, N.Y., pp. 242–245, 256–258.

PAUSTIAN, P. 1987. A novel method to calculate the Mohr chloride titration. *In* Advances in Water Analysis and Treatment, Proc. 14th Annu. AWWA Water Quality Technology Conf., November 16-20, 1986, Portland, Ore., p. 673. American Water Works Assoc., Denver, Colo.

4500-Cl⁻ C. Mercuric Nitrate Method

1. General Discussion

a. Principle: Chloride can be titrated with mercuric nitrate, $Hg(NO_3)_2$, because of the formation of soluble, slightly dissociated mercuric chloride. In the pH range 2.3 to 2.8, diphenylcarbazone indicates the titration end point by formation of a purple complex with the excess mercuric ions. Xylene cyanol FF serves as a pH indicator and end-point enhancer. Increasing the strength of the titrant and modifying the indicator mixtures extend the range of measurable chloride concentrations.

b. Interference: Bromide and iodide are titrated with $Hg(NO_3)_2$ in the same manner as chloride. Chromate, ferric, and sulfite ions interfere when present in excess of 10 mg/L.

2. Apparatus

a. Erlenmeyer flask, 250-mL.
b. Microburet, 5-mL with 0.01-mL graduation intervals.

3. Reagents

a. Standard sodium chloride, 0.0141*M* (0.0141*N*): See Method B, ¶ 3c above.

b. Nitric acid, HNO₃, 0.1*N.*

c. Sodium hydroxide, NaOH, 0.1*N.*

d. Reagents for chloride concentrations below 100 mg/L:

1) *Indicator-acidifier reagent:* The HNO_3 concentration of this reagent is an important factor in the success of the determination and can be varied as indicated in a) or b) to suit the alkalinity range of the sample. Reagent a) contains sufficient HNO_3 to neutralize a total alkalinity of 150 mg as $CaCO_3/L$ to the proper pH in a 100-mL sample. Adjust amount of HNO_3 to accommodate samples of alkalinity different from 150 mg/L.

a) Dissolve, in the order named, 250 mg s-diphenylcarbazone, 4.0 mL conc HNO_3, and 30 mg xylene cyanol FF in 100 mL 95% ethyl alcohol or isopropyl alcohol. Store in a dark bottle in a refrigerator. This reagent is not stable indefinitely. Deterioration causes a slow end point and high results.

b) Because pH control is critical, adjust pH of highly alkaline or acid samples to 2.5 ± 0.1 with 0.1*N* HNO_3 or NaOH, not with sodium carbonate (Na_2CO_3). Use a pH meter with a nonchloride type of reference electrode for pH adjustment. If only the usual chloride-type reference electrode is available for pH adjustment, determine amount of acid or alkali required to obtain a pH of 2.5 ± 0.1 and discard this sample portion. Treat a separate sample portion with the determined amount of acid or alkali and continue analysis. Under these circumstances, omit HNO_3 from indicator reagent.

2) *Standard mercuric nitrate titrant,* 0.007 05*M* (0.0141*N*): Dissolve 2.3 g $Hg(NO_3)_2$ or 2.5 g $Hg(NO_3)_2 \cdot H_2O$ in 100 mL distilled water containing 0.25 mL conc HNO_3. Dilute to just under 1 L. Make a preliminary standardization by following the procedure described in ¶ 4a. Use replicates containing 5.00 mL standard NaCl solution and 10 mg sodium bicarbonate ($NaHCO_3$) diluted to 100 mL with distilled water. Adjust titrant to 0.0141*N* and make a final standardization; 1.00 mL = 500 μg Cl^-. Store away from light in a dark bottle.

e. Reagent for chloride concentrations greater than 100 mg/L:

1) *Mixed indicator reagent:* Dissolve 0.50 g diphenylcarbazone powder and 0.05 g bromphenol blue powder in 75 mL 95% ethyl or isopropyl alcohol and dilute to 100 mL with the same alcohol.

2) *Strong standard mercuric nitrate titrant,* 0.0705*M* (0.141*N*) Dissolve 25 g $Hg(NO_3)_2 \cdot H_2O$ in 900 mL distilled water containing 5.0 mL conc HNO_3. Dilute to just under 1 L and standardize by following the procedure described in ¶ 4b. Use replicates containing 25.00 mL standard NaCl solution and 25 mL distilled water. Adjust titrant to 0.141*N* and make a final standardization; 1.00 mL = 5.00 mg Cl^-.

4. Procedure

a. Titration of chloride concentrations less than 100 mg/L: Use a 100-mL sample or smaller portion so that the chloride content is less than 10 mg.

Add 1.0 mL indicator-acidifier reagent. (The color of the solution should be green-blue at this point. A light green indicates pH less than 2.0; a pure blue indicates pH more than 3.8.) For most potable waters, the pH after this addition will be 2.5 ± 0.1. For highly alkaline or acid waters, adjust pH to about 8 before adding indicator-acidifier reagent.

Titrate with $0.0141N$ $Hg(NO_3)_2$ titrant to a definite purple end point. The solution turns from green-blue to blue a few drops before the end point.

Determine blank by titrating 100 mL distilled water containing 10 mg $NaHCO_3$.

b. Titration of chloride concentrations greater than 100 mg/L: Use a sample portion (5 to 50 mL) requiring less than 5 mL titrant to reach the end point. Measure into a 150-mL beaker. Add approximately 0.5 mL mixed indicator reagent and mix well. The color should be purple. Add $0.1N$ HNO_3 dropwise until the color just turns yellow. Titrate with strong $Hg(NO_3)_2$ titrant to first permanent dark purple. Titrate a distilled water blank using the same procedure.

5. Calculation

$$\text{mg Cl}^-/\text{L} = \frac{(A - B) \times N \times 35\,450}{\text{mL sample}}$$

where:

A = mL titration for sample,
B = mL titration for blank, and
N = normality of $Hg(NO_3)_2$.

$$\text{mg NaCl/L} = (\text{mg Cl}^-/\text{L}) \times 1.65$$

6. Precision and Bias

A synthetic sample containing 241 mg Cl^-/L, 108 mg Ca/L, 82 mg Mg/L, 3.1 mg K/L, 19.9 mg Na/L, 1.1 mg NO_3^--N/L, 0.25 mg NO_2^--N/L, 259 mg SO_4^{2-}/L, and 42.5 mg total alkalinity/L (contributed by $NaHCO_3$) in distilled water was analyzed in 10 laboratories by the mercurimetric method, with a relative standard deviation of 3.3% and a relative error of 2.9%.

7. Bibliography

KOLTHOFF, I.M. & V.A. STENGER. 1947. Volumetric Analysis, 2nd ed. Vol. 2. Interscience Publishers, New York, N.Y., pp. 334-335.
DOMASK, W.C. & K.A. KOBE. 1952. Mercurimetric determination of chlorides and water-soluble chlorohydrins. *Anal. Chem.* 24:989.
GOLDMAN, E. 1959. New indicator for the mercurimetric chloride determination in potable water. *Anal. Chem.* 31:1127.

4500-Cl⁻ D. Potentiometric Method

1. General Discussion

a. Principle: Chloride is determined by potentiometric titration with silver nitrate solution with a glass and silver-silver chloride electrode system. During titration an electronic voltmeter is used to detect the change in potential between the two electrodes. The end point of the titration is that instrument reading at which the greatest change in voltage has occurred for a small and constant increment of silver nitrate added.

b. Interference: Iodide and bromide also are titrated as chloride. Ferricyanide causes high results and must be removed. Chromate and dichromate interfere and should be reduced to the chromic state or removed. Ferric iron interferes if present in an amount substantially higher than the amount of chloride. Chromic ion, ferrous ion, and phosphate do not interfere.

Grossly contaminated samples usually require pretreatment. Where contamination is minor, some contaminants can be destroyed simply by adding nitric acid.

2. Apparatus

a. Glass and silver-silver chloride electrodes: Prepare in the laboratory or purchase a silver electrode coated with AgCl for use with specified instruments. Instructions on use and care of electrodes are supplied by the manufacturer.

b. Electronic voltmeter, to measure potential difference between electrodes: A pH meter may be converted to this use by substituting the appropriate electrode.

c. Mechanical stirrer, with plastic-coated or glass impeller.

3. Reagents

a. Standard sodium chloride solution, $0.0141M$ $(0.0141N)$: See ¶ 4500-Cl⁻.B.3c.

b. Nitric acid, HNO_3, conc.

c. Standard silver nitrate titrant, $0.0141M$ $(0.0141N)$: See ¶ 4500-Cl⁻.B.3b.

d. Pretreatment reagents:

1) *Sulfuric acid,* H_2SO_4, 1 + 1.
2) *Hydrogen peroxide,* H_2O_2, 30%.
3) *Sodium hydroxide,* NaOH, $1N$.

4. Procedure

a. Standardization: The various instruments that can be used in this determination differ in operating details; follow the manufacturer's instructions. Make necessary mechanical adjustments. Then, after allowing sufficient time for warmup (10 min), balance internal electrical components to give an instrument setting of 0 mV or, if a pH meter is used, a pH reading of 7.0.

1) Place 10.0 mL standard NaCl solution in a 250-mL beaker, dilute to about 100 mL, and add 2.0 mL conc HNO_3. Immerse stirrer and electrodes.

2) Set instrument to desired range of millivolts or pH units. Start stirrer.

3) Add standard $AgNO_3$ titrant, recording scale reading after each addition. At the start, large increments of $AgNO_3$ may be added; then, as the end point is approached, add smaller and equal increments (0.1 or 0.2 mL) at longer intervals, so that the exact

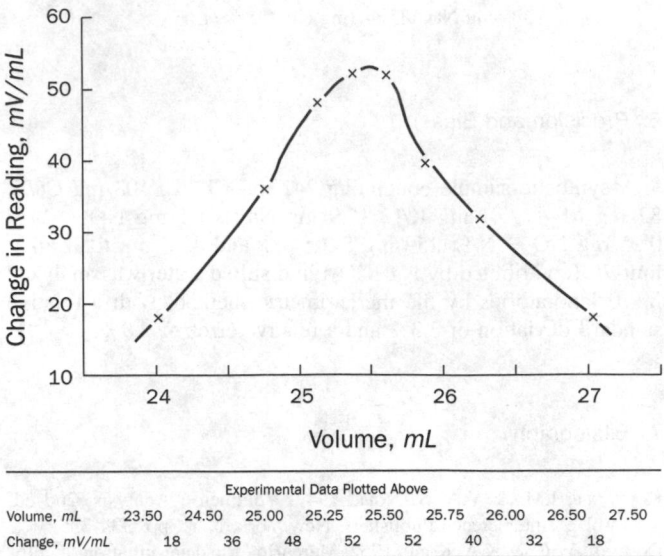

			Experimental Data Plotted Above					
Volume, mL	23.50	24.50	25.00	25.25	25.50	25.75	26.00	26.50	27.50
Change, mV/mL	18	36	48	52	52	40	32	18

Figure 4500-Cl⁻:1. Example of differential titration curve (end point is 25.5 mL).

end point can be determined. Determine volume of $AgNO_3$ used at the point at which there is the greatest change in instrument reading per unit addition of $AgNO_3$.

4) Plot a differential titration curve if the exact end point cannot be determined by inspecting the data. Plot change in instrument reading for equal increments of $AgNO_3$ against volume of $AgNO_3$ added, using average of buret readings before and after each addition. The procedure is illustrated in Figure 4500-Cl⁻:1.

b. Sample analysis:

1) Pipet 100.0 mL sample, or a portion containing not more than 10 mg Cl⁻, into a 250-mL beaker. In the absence of interfering substances, proceed with ¶ 3) below.

2) In the presence of organic compounds, sulfite, or other interferences (such as large amounts of ferric iron, cyanide, or sulfide) acidify sample with H_2SO_4, using litmus paper. Boil for 5 min to remove volatile compounds. Add more H_2SO_4, if necessary, to keep solution acidic. Add 3 mL H_2O_2 and boil for 15 min, adding chloride-free distilled water to keep the volume above 50 mL. Dilute to 100 mL, add NaOH solution dropwise until alkaline to litmus, then 10 drops in excess. Boil for 5 min, filter into a 250-mL beaker, and wash precipitate and paper several times with hot distilled water.

3) Add conc HNO_3 dropwise until acidic to litmus paper, then 2.0 mL in excess. Cool and dilute to 100 mL if necessary. Immerse stirrer and electrodes and start stirrer. Make any necessary adjustments according to the manufacturer's instructions and set selector switch to appropriate setting for measuring the difference of potential between electrodes.

4) Complete determination by titrating according to ¶ 4a4). If an end-point reading has been established from previous determinations for similar samples and conditions, use this predetermined end point. For the most accurate work, make a blank titration by carrying chloride-free distilled water through the procedure.

5. Calculation

$$\text{mg Cl}^-/\text{L} = \frac{(A - B) \times N \times 35\,450}{\text{mL sample}}$$

where:

A = mL $AgNO_3$,
B = mL blank, and
N = normality of titrant.

6. Precision and Bias

In the absence of interfering substances, the precision and bias are estimated to be about 0.12 mg for 5 mg Cl⁻, or 2.5% of the amount present. When pretreatment is required to remove interfering substances, the precision and bias are reduced to about 0.25 mg for 5 mg Cl⁻, or 5% of amount present.

7. Bibliography

KOLTHOFF, I.M. & N.H. FURMAN. 1931. Potentiometric Titrations, 2nd ed. John Wiley & Sons, New York, N.Y.

REFFENBURG, H.B. 1935. Colorimetric determination of small quantities of chlorides in water. *Ind. Eng. Chem.*, Anal. Ed. 7:14.

CALDWELL, J.R. & H.V. MEYER. 1935. Chloride determination. *Ind. Eng. Chem.*, Anal. Ed. 7:38.

SERFASS, E.J. & R.F. MURACA. 1954. Procedures for Analyzing Metal-Finishing Wastes. Ohio River Valley Water Sanitation Commission, Cincinnati, Ohio, p. 80.

FURMAN, N.H., ed. 1962. Standard Methods of Chemical Analysis, 6th ed. D. Van Nostrand Co., Princeton, N.J., Vol. I.

WALTON, H.F. 1964. Principles and Methods of Chemical Analysis. Prentice-Hall, Inc., Englewood Cliffs, N.J.

WILLARD, H.H., L.L. MERRITT & J.A. DEAN. 1965. Instrumental Methods of Analysis, 4th ed. D. Van Nostrand Co., Princeton, N.J.

4500-Cl⁻ E.   Automated Ferricyanide Method

1. General Discussion

a. Principle: Thiocyanate ion is liberated from mercuric thiocyanate by the formation of soluble mercuric chloride. In the presence of ferric ion, free thiocyanate ion forms a highly colored ferric thiocyanate, of which the intensity is proportional to the chloride concentration.

b. Interferences: Remove particulate matter by filtration or centrifugation before analysis. Guard against contamination from reagents, water, glassware, and sample preservation process. No chemical interferences are significant.

c. Application: The method is applicable to potable, surface, and saline waters, and domestic and industrial wastewaters. The concentration range is 1 to 200 mg Cl⁻/L; it can be extended by dilution.

2. Apparatus

a. Automated analytical equipment: An example of the continuous-flow analytical instrument consists of the interchangeable components shown in Figure 4500-Cl⁻:2.

b. Filters, 480-nm.

3. Reagents

a. Stock mercuric thiocyanate solution: Dissolve 4.17 g $Hg(SCN)_2$ in about 500 mL methanol, dilute to 1000 mL with methanol, mix, and filter through filter paper.

b. Stock ferric nitrate solution: Dissolve 202 g $Fe(NO_3)_3 \cdot 9H_2O$ in about 500 mL distilled water, then carefully add 21 mL conc HNO_3. Dilute to 1000 mL with distilled water and mix. Filter through paper and store in an amber bottle.

c. Color reagent: Add 150 mL stock $Hg(SCN)_2$ solution to 150 mL stock $Fe(NO_3)_3$ solution. Mix and dilute to 1000 mL with distilled water. Add 0.5 mL polyoxyethylene 23 lauryl ether.*

d. Stock chloride solution: Dissolve 1.6482 g NaCl, dried at 140°C, in distilled water and dilute to 1000 mL; 1.00 mL = 1.00 mg Cl⁻.

e. Standard chloride solutions: Prepare chloride standards in the desired concentration range, such as 1 to 200 mg/L, using stock chloride solution.

4. Procedure

Set up manifold as shown in Figure 4500-Cl⁻:2 and follow general procedure described by the manufacturer.

5. Calculation

Prepare standard curves by plotting response of standards processed through the manifold against chloride concentrations in

_____
* Brij 35, available from ICI Americas, Wilmington, DE, or equivalent.

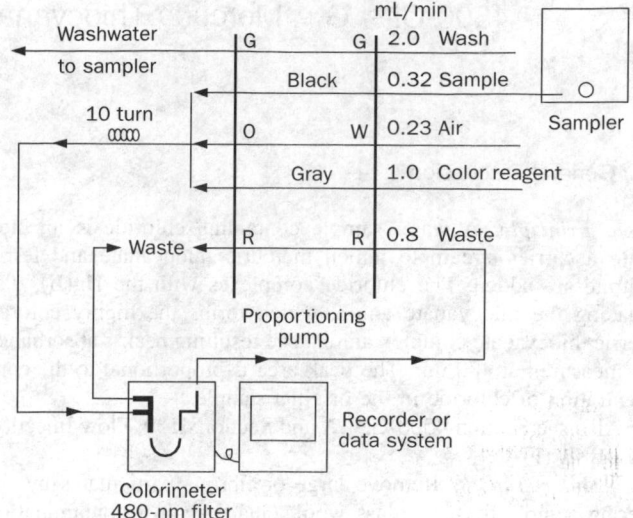

Figure 4500-Cl⁻:2. Flow scheme for automated chloride analysis.

standards. Compute sample chloride concentration by comparing sample response with standard curve.

6. Precision and Bias

With an automated system in a single laboratory six samples were analyzed in septuplicate. At a concentration ranging from about 1 to 50 mg Cl⁻/L the average standard deviation was 0.39 mg/L. The coefficient of variation was 2.2%. In two samples with added chloride, recoveries were 104% and 97%.

7. Bibliography

ZALL, D.M., D. FISHER & M.D. GARNER. 1956. Photometric determination of chlorides in water. *Anal. Chem.* 28:1665.

O'BRIEN, J.E. 1962. Automatic analysis of chlorides in sewage. *Wastes Eng.* 33:670.

4500-Cl⁻ F. (Reserved)

4500-Cl⁻ G. Mercuric Thiocyanate Flow Injection Analysis (PROPOSED)

1. General Discussion

a. Principle: A water sample containing chloride is injected into a carrier stream to which mercuric thiocyanate and ferric nitrate are added. The chloride complexes with the Hg(II), displacing the thiocyanate anion, which forms the highly colored ferric thiocyanate complex anion. The resulting peak's absorbance is measured at 480 nm. The peak area is proportional to the concentration of chloride in the original sample.

Also see Section 4500-Cl⁻.A and Section 4130, Flow Injection Analysis (FIA).

b. Interferences: Remove large or fibrous particulates by filtering sample through glass wool. Guard against contamination from reagents, water, glassware, and the sample preservation process.

Substances such as sulfite and thiosulfate, which reduce iron(III) to iron(II) and mercury(II) to mercury(I), can interfere. Halides, which also form strong complexes with mercuric ion (e.g., Br⁻, I⁻), give a positive interference.

2. Apparatus

Flow injection analysis equipment consisting of:
a. FIA injection valve with sample loop.
b. Multichannel proportioning pump.
c. FIA manifold with flow cell (Figure 4500-Cl⁻:3). Relative flow rates only are shown. Tubing volumes are given as an example only; they may be scaled down proportionally. Use manifold tubing of an inert material such as TFE.*
d. Absorbance detector, 480 nm, 10-nm bandpass.
e. Valve control and data acquisition system.

3. Reagents

Use reagent water (>10 megohm) to prepare carrier and all solutions.

a. Stock mercuric thiocyanate solution: In a 1-L volumetric flask, dissolve 4.17 g mercuric thiocyanate, Hg(SCN)₂, in about 500 mL methanol. Dilute to mark with methanol and mix. CAUTION: *Mercuric thiocyanate is toxic. Wear gloves!*

* Teflon or equivalent.

TABLE 4500-Cl⁻ I: RESULTS OF SINGLE-LABORATORY STUDIES WITH SELECTED MATRICES

Matrix	Sample/Blank Designation	Known Addition mg Cl⁻/L	Recovery %	Relative Standard Deviation %
Wastewater treatment plant influent	Reference sample*	—	101	—
	Blank†	10	104	—
		20	102	—
	Site A‡	0	—	0.4
		10	92	—
		20	101	—
	Site B‡	0	—	0.2
		10	97	—
		20	106	—
	Site C‡	0	—	0.4
		10	102	—
		20	102	—
Wastewater treatment plant effluent	Reference sample*	—	101	—
	Blank†	10	104	—
		20	102	—
	Site A‡	0	—	0.3
		10	98	—
		20	101	—
	Site B‡	0	—	0.2
		10	99	—
		20	103	—
	Site C‡	0	—	0.4
		10	91	—
		20	97	—
Landfill leachate‖	Reference sample*	—	100	—
	Blank†	10	101	—
		20	100	—
	Site A§	0	—	0.3
		10	97	—
		20	103	—
	Site B§	0	—	0.2
		10	89	—
		20	103	—
	Site C§	0	—	0.5
		10	89	—
		20	103	—

* U.S. EPA nutrient QC sample, 51.7 mg Cl⁻/L.
† Determined in duplicate.
‡ Samples diluted 5-fold. Samples without known additions determined four times; samples with known additions determined in duplicate. Typical relative difference between duplicates 0.2%.
§ Sample from Site A diluted 50-fold, those from B and C 100-fold. Samples without known additions determined four times; samples with known additions determined in duplicate; typical relative difference between duplicates 0.5%.

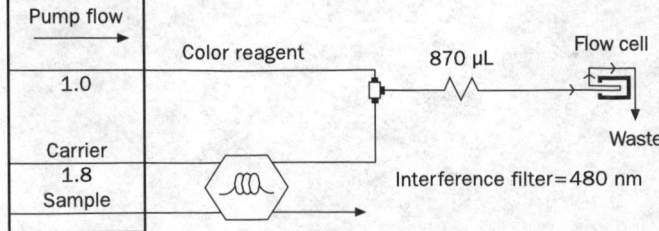

Figure 4500-Cl⁻:3. FIA chloride manifold.

b. Stock ferric nitrate reagent, 0.5M: In a 1-L volumetric flask, dissolve 202 g ferric nitrate, $Fe(NO_3)_3 \cdot 9H_2O$, in approximately 800 mL water. Add 25 mL conc HNO_3 and dilute to mark. Invert to mix.

c. Color reagent: In a 500-mL volumetric flask, mix 75 mL stock mercuric thiocyanate solution with 75 mL stock ferric nitrate reagent and dilute to mark with water. Invert to mix. Vacuum filter through a 0.45-μm membrane filter. The color reagent also is available as a commercially prepared solution that is stable for several months.

d. Stock chloride standard, 1000 mg Cl⁻/L: In a 105°C oven, dry 3 g primary standard grade sodium chloride, NaCl, overnight. In a 1-L volumetric flask, dissolve 1.648 g primary standard grade sodium chloride in about 500 mL water. Dilute to mark and invert to mix.

e. Standard chloride solutions: Prepare chloride standards for the calibration curve in the desired concentration range, using the stock standard (¶ 3d), and diluting with water.

4. Procedure

Set up a manifold equivalent to that in Figure 4500-Cl⁻:3 and follow method supplied by manufacturer, or laboratory standard operating procedure for this method. Follow quality control procedures described in Section 4020.

5. Calculations

Prepare standard curves by plotting absorbance of standards processed through the manifold versus chloride concentration.

The calibration curve gives a good fit to a second-order polynomial.

6. Precision and Bias

a. Recovery and relative standard deviation: The results of single-laboratory studies with various matrices are given in Table 4500-Cl⁻:I.

b. MDL: A 100-μL sample loop was used in the method described above. Using a published MDL method[1] analysts ran 21 replicates of a 1.0-mg Cl⁻/L standard. These gave a mean of 1.19 mg Cl⁻/L, a standard deviation of 0.027 mg Cl⁻/L, and an MDL of 0.07 mg Cl⁻/L. This is only an estimate because the ratio of standard to the MDL is above guidelines (see Section 1030). A lower MDL may be obtained by increasing the sample loop volume and increasing the ratio of carrier flow rate to reagents flow rate. A higher MDL may be obtained by decreasing the sample loop volume and decreasing this ratio.

7. Reference

1. U.S. ENVIRONMENTAL PROTECTION AGENCY. 1989. Definition and Procedure for the Determination of Method Detection Limits. Appendix B to 40 CFR 136 rev. 1.11 amended June 30, 1986. 49 CFR 43430.

4500-ClO₂   CHLORINE DIOXIDE*

4500-ClO₂ A.   Introduction

Because the physical and chemical properties of chlorine dioxide resemble those of chlorine in many respects, read the entire discussion of Residual Chlorine (Section 4500-Cl) before attempting a chlorine dioxide determination.

1. Occurrence and Significance

Chlorine dioxide, ClO_2, has been used widely as a bleaching agent in the paper and pulp industry. It has been applied to water supplies to combat tastes and odors due to phenolic-type wastes, actinomycetes, and algae, as well as to oxidize soluble iron and manganese to a more easily removable form. It is a disinfectant, and some results suggest that it may be stronger than free chlorine or hypochlorite.

Chlorine dioxide is a deep yellow, volatile, unpleasant-smelling gas that is toxic and under certain conditions may react explo-

sively. It should be handled with care in a vented area. The use of odor to warn of exposure to concentrations of health significance may not be adequate.

There are several methods of generating ClO_2; for laboratory purposes the acidification of a solution of sodium chlorite followed by suitable scrubbing and capture of the released gaseous ClO_2 is the most practical. CAUTION: *Sodium chlorite is a powerful oxidizer; keep out of direct contact with oxidizable material to avoid possibility of explosion.*

2. Selection of Method

The iodometric method (B) gives a very precise measure of total available strength of a solution in terms of its ability to liberate iodine from iodide. However, ClO_2, chlorine, chlorite, and hypochlorite are not distinguished easily by this technique. It is designed primarily, and best used, for standardizing ClO_2 so-

---
* Approved by Standard Methods Committee, 1993.

lutions needed for preparation of temporary standards. It often is inapplicable to industrial wastes.

The amperometric methods (C and E) are useful when a knowledge of the various chlorine fractions in a water sample is desired. They distinguish various chlorine compounds of interest with good accuracy and precision, but require specialized equipment and considerable analytical skill.

The *N,N*-diethyl-*p*-phenylenediamine (DPD) method (D) has the advantages of a relatively easy-to-perform colorimetric test with the ability to distinguish between ClO_2 and some forms of chlorine. This technique is not as accurate as the amperometric method, but should yield results adequate for many common applications. NOTE: Reports in the literature indicate that the DPD method is subject to interference from monochloramine and chloraminoacetic acid, and the chlorite anion.[1]

3. Sampling and Storage

Determine ClO_2 promptly after collecting the sample. Do not expose sample to sunlight or strong artificial light and do not aerate to mix. Most of these methods can be performed on site, with prior calibration in the laboratory. Minimum ClO_2 losses occur when the determination is completed immediately at the site of sample collection.

4. Reference

1. CHISWELL, B. & K.R. O'HALLORAN. 1991. Use of Lissamine Green B as a spectrophotometric reagent for the determination of low residuals of chlorine dioxide. *Analyst* 116:657.

5. Bibliography

INGOLS, R.S. & G.M. RIDENOUR. 1948. Chemical properties of chlorine dioxide in water treatment. *J. Amer. Water Works Assoc.* 40:1207.
PALIN, A.T. 1948. Chlorine dioxide in water treatment. *J. Inst. Water Eng.* 11:61.
HODGDEN, H.W. & R.S. INGOLS. 1954. Direct colorimetric method for determination of chlorine dioxide in water. *Anal. Chem.* 26:1224.
FEUSS, J.V. 1964. Problems in determination of chlorine dioxide residuals. *J. Amer. Water Works Assoc.* 56:607.
MASSCHELEIN, W. 1966. Spectrophotometric determination of chlorine dioxide with acid chrome violet K. *Anal. Chem.* 38:1839.
MASSCHELEIN, W. 1969. Les Oxydes de Chlore et le Chlorite de Sodium. Dunod, Paris, Chapter XI.

4500-ClO₂ B.   Iodometric Method

1. General Discussion

a. Principle: A pure solution of ClO_2 is prepared from gaseous ClO_2 by slowly adding dilute H_2SO_4 to a sodium chlorite (NaClO₂) solution. Contaminants such as chlorine are removed from the gas stream by a NaClO₂ scrubber; the gas is passed into distilled water in a steady stream of air. See CAUTION, ¶ A.1.

ClO_2 releases free iodine from a KI solution acidified with acetic acid or H_2SO_4. The liberated iodine is titrated with a standard solution of sodium thiosulfate (Na₂S₂O₃), with starch as the indicator.

b. Interference: There is little interference in this method, but temperature and strong light affect solution stability. Minimize ClO_2 losses by storing stock ClO_2 solution in a dark refrigerator and by preparing and titrating dilute ClO_2 solutions for standardization purposes at the lowest practicable temperature and in subdued light.

c. Minimum detectable concentration: One drop (0.05 mL) of 0.01N (0.01M) Na₂S₂O₃ is equivalent to 20 µg ClO_2/L (or 40 µg/L in terms of available chlorine) when a 500-mL sample is titrated.

2. Reagents

All reagents listed for the determination of residual chlorine in Section 4500-Cl.B.2*a-g* are required. Also needed are the following:

a. Stock chlorine dioxide solution: Prepare a gas generating and absorbing system as illustrated in Figure 4500-ClO₂:1. Connect aspirator flask, 500-mL capacity, with rubber tubing to a source of purified compressed air. Let air bubble through a layer

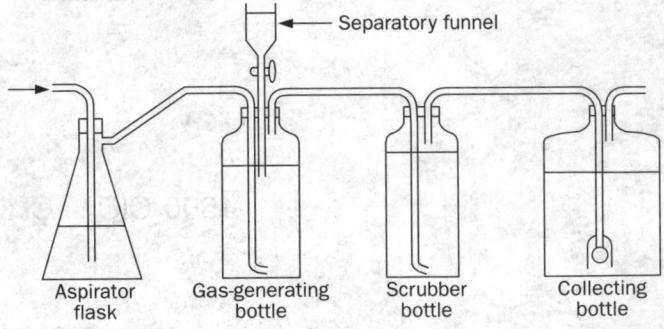

Figure 4500-ClO₂:1. Chlorine dioxide generation and absorption system.

of 300 mL distilled water in flask and then pass through a glass tube ending within 5 mm of the bottom of the 1-L gas-generating bottle. Conduct evolved gas via glass tubing through a scrubber bottle containing saturated NaClO₂ solution or a tower packed with flaked NaClO₂, and finally, via glass tubing, into a 2-L borosilicate glass collecting bottle where the gas is absorbed in 1500 mL distilled water. Provide an air outlet tube on collecting bottle for escape of air. Select for gas generation a bottle constructed of strong borosilicate glass and having a mouth wide enough to permit insertion of three separate glass tubes: the first leading almost to the bottom for admitting air, the second reaching below the liquid surface for gradual introduction of H_2SO_4, and the third near the top for exit of evolved gas and air. Fit to second tube a graduated cylindrical separatory funnel to contain H_2SO_4. Locate this system in a fume hood with an adequate shield.

Dissolve 10 g NaClO$_2$ in 750 mL distilled water and place in generating bottle. Carefully add 2 mL conc H$_2$SO$_4$ to 18 mL distilled water and mix. Transfer to funnel. Connect flask to generating bottle, generating bottle to scrubber, and the latter to collecting bottle. Pass a smooth current of air through the system, as evidenced by the bubbling rate in all bottles.

Introduce 5-mL increments of H$_2$SO$_4$ from funnel into generating bottle at 5-min intervals. Continue air flow for 30 min after last portion of acid has been added.

Store yellow stock solution in glass-stoppered dark-colored bottle in a dark refrigerator. The concentration of ClO$_2$ thus prepared varies between 250 and 600 mg/L, corresponding to approximately 500 to 1200 mg free chlorine/L.

b. Standard chlorine dioxide solution: Use this solution for preparing temporary ClO$_2$ standards. Dilute required volume of stock ClO$_2$ solution to desired strength with chlorine-demand-free water (see Section 4500-Cl.C.3*m*). Standardize solution by titrating with standard 0.01*N* (0.01*M*) or 0.025*N* (0.025*M*) Na$_2$S$_2$O$_3$ titrant in the presence of KI, acid, and starch indicator by following the procedure given in ¶ 3 below. A full or nearly full bottle of chlorine or ClO$_2$ solution retains its titer longer than a partially full one. When repeated withdrawals reduce volume to a critical level, standardize diluted solution at the beginning, midway in the series of withdrawals, and at the end of the series. Shake contents thoroughly before drawing off needed solution from middle of the glass-stoppered dark-colored bottle. Prepare this solution frequently.

3. Procedure

Select volume of sample, prepare for titration, and titrate sample and blank as described in Section 4500-Cl.B.3. The only exception is the following: *Let ClO$_2$ react in the dark with acid and KI for 5 min before starting titration.*

4. Calculations

Express ClO$_2$ concentrations in terms of ClO$_2$ or as free chlorine content. Free chlorine is defined as the total oxidizing power of ClO$_2$ measured by titrating iodine released by ClO$_2$ from an acidic solution of KI. Calculate result in terms of chlorine itself.

For standardizing ClO$_2$ solution:

$$\text{mg ClO}_2/\text{mL} = \frac{(A \pm B) \times N \times 13.49}{\text{mL sample titrated}}$$

For determining ClO$_2$ temporary standards:

$$\text{mg ClO}_2 \text{ as Cl}_2/\text{mL} = \frac{(A \pm B) \times N \times 35.45}{\text{mL sample titrated}}$$

where:

A = mL titration for sample,
B = mL titration for blank (positive or negative, see 4500-Cl.B.3*d*), and
N = normality of Na$_2$S$_2$O$_3$ = molarity of Na$_2$S$_2$O$_3$.

5. Bibliography

Post, M.A. & W.A. Moore. 1959. Determination of chlorine dioxide in treated surface waters. *Anal. Chem.* 31:1872.

4500-ClO$_2$ C.　Amperometric Method I

1. General Discussion

a. Principle: The amperometric titration of ClO$_2$ is an extension of the amperometric method for chlorine. By performing four titrations with phenylarsine oxide, free chlorine (including hypochlorite and hypochlorous acid), chloramines, chlorite, and ClO$_2$ may be determined separately. The first titration step consists of conversion of ClO$_2$ to chlorite and chlorate through addition of sufficient NaOH to produce a pH of 12, followed by neutralization to a pH of 7 and titration of free chlorine. In the second titration KI is added to a sample that has been treated similarly with alkali and had the pH readjusted to 7; titration yields free chlorine and monochloramine. The third titration involves addition of KI and pH adjustment to 7, followed by titration of free chlorine, monochloramine, and one-fifth of the available ClO$_2$. In the fourth titration, addition of sufficient H$_2$SO$_4$ to lower the pH to 2 enables all available ClO$_2$ and chlorite, as well as the total free chlorine, to liberate an equivalent amount of iodine from the added KI and thus be titrated.

b. Interference: The interferences described in Section 4500-Cl.D.1*b* apply also to determination of ClO$_2$.

2. Apparatus

The apparatus required is given in Sections 4500-Cl.D.2*a* through *d*.

3. Reagents

All reagents listed for the determination of chlorine in Section 4500-Cl.D.3 are required. Also needed are the following:
a. Sodium hydroxide, NaOH, 6*N* (6*M*).
b. Sulfuric acid, H$_2$SO$_4$, 6*N* (3*M*), 1 + 5.

4. Procedure

Minimize effects of pH, time, and temperature of reaction by standardizing all conditions.

a. Titration of free available chlorine (hypochlorite and hypochlorous acid): Add sufficient 6*N* (6*M*) NaOH to raise sample pH to 12. After 10 min, add 6*N* (3*M*) H$_2$SO$_4$ to lower pH to 7. Titrate with standard phenylarsine oxide titrant to the amperometric end point as given in Section 4500-Cl.D. Record result as *A*.

b. Titration of free available chlorine and chloramine: Add 6*N* (6*M*) NaOH to raise sample pH to 12. After 10 min, add 6*N* (3*M*) H$_2$SO$_4$ to reduce pH to 7. Add 1 mL KI solution. Titrate with standard phenylarsine oxide titrant to the amperometric end point. Record result as *B*.

c. Titration of free available chlorine, chloramine, and one-fifth of available ClO$_2$: Adjust sample pH to 7 with pH 7 phosphate buffer solution. Add 1 mL KI solution. Titrate with standard phenylarsine oxide titrant to the amperometric end point. Record result as C.

d. Titration of free available chlorine, chloramines, ClO₂, and chlorite: Add 1 mL KI solution to sample. Add sufficient $6N$ ($3M$) H_2SO_4 to lower pH to 2. After 10 min, add sufficient $6N$ ($6M$) NaOH to raise pH to 7. Titrate with standard phenylarsine oxide titrant to the amperometric end point. Record result as D.

5. Calculation

Convert individual titrations (A, B, C, and D) into chlorine concentration by the following equation:

$$\text{mg Cl as } Cl_2/L = \frac{E \times 200}{\text{mL sample}}$$

where:

E = mL phenylarsine oxide titration for each individual sample A, B, C, or D.

Calculate ClO₂ and individual chlorine fractions as follows:

$$\text{mg } ClO_2 \text{ as } ClO_2/L = 1.9 \, (C - B)$$
$$\text{mg } ClO_2 \text{ as } Cl_2/L = 5 \, (C - B)$$
$$\text{mg free available chlorine/L} = A$$
$$\text{mg chloramine/L as chlorine} = B - A$$
$$\text{mg chlorite/L as chlorine} = 4B - 5C + D$$

6. Bibliography

HALLER, J.F. & S.S. LISTEK. 1948. Determination of chlorine dioxide and other active chlorine compounds in water. *Anal. Chem.* 20:639.

4500-ClO₂ D. DPD Method

1. General Discussion

a. Principle: This method is an extension of the N,N-diethyl-p-phenylenediamine (DPD) method for determining free chlorine and chloramines in water. ClO₂ appears in the first step of this procedure but only to the extent of one-fifth of its total available chlorine content corresponding to reduction of ClO₂ to chlorite ion. If the sample is then acidified in the presence of iodide the chlorite also reacts. When neutralized by subsequent addition of bicarbonate, the color thus produced corresponds to the total available chlorine content of the ClO₂. If chlorite is present in the sample, this will be included in the step involving acidification and neutralization. Chlorite that did not result from ClO₂ reduction by the procedure will cause a positive error equal to twice this chlorite concentration. In evaluating mixtures of these various chloro-compounds, it is necessary to suppress free chlorine by adding glycine before reacting the sample with DPD reagent. Differentiation is based on the fact that glycine converts free chlorine instantaneously into chloroaminoacetic acid but has no effect on ClO₂.

b. Interference: The interference by oxidized manganese described in Section 4500-Cl.F.1d applies also to ClO₂ determination. Manganese interference appears as an increase in the first titrations after addition of DPD, with or without KI, and irrespective of whether there has been prior addition of glycine. Titration readings must be corrected suitably. Interference by chromate in wastewaters may be corrected similarly.

Iron contributed to the sample by adding ferrous ammonium sulfate (FAS) titrant may activate chlorite so as to interfere with the first end point of the titration. Suppress this effect with additional EDTA, disodium salt.

Exercise caution in the selection of this method, because of interferences from monochloramine and chloraminoacetic acid and the chlorite anion.

2. Reagents

Reagents required in addition to those for the DPD free-combined chlorine method as listed in Section 4500-Cl.F.2 are as follows:

a. Glycine solution: Dissolve 10 g NH_2CH_2COOH in 100 mL distilled water.

b. Sulfuric acid solution: Dilute 5 mL conc H_2SO_4 to 100 mL with distilled water.

c. Sodium bicarbonate solution: Dissolve 27.5 g $NaHCO_3$ in 500 mL distilled water.

d. EDTA: Disodium salt of ethylenediamine tetraacetic acid, solid.

3. Procedure

For samples containing more than 5 mg/L total available chlorine follow the dilution procedure given in Section 4500-Cl.F.3.

a. Chlorine dioxide: Add 2 mL glycine solution to 100 mL sample and mix. Place 5 mL each of buffer reagent and DPD indicator solution in a separate titration flask and mix (or use about 500 mg DPD powder). Add about 200 mg EDTA, disodium salt. Then add glycine-treated sample and mix. Titrate rapidly with standard FAS titrant until red color is discharged (Reading G).

b. Free available chlorine and chloramine: Using a second 100-mL sample follow the procedures of Section 4500-Cl.F.3a adding about 200 mg EDTA, disodium salt, initially with the DPD reagents (Readings A, B, and C).

c. Total available chlorine including chlorite: After obtaining Reading C add 1 mL H_2SO_4 solution to the same sample in titration flask, mix, and let stand about 2 min. Add 5 mL $NaHCO_3$ solution, mix, and titrate (Reading D).

d. Colorimetric procedure: Instead of titration with standard FAS solution, colorimetric procedures may be used to obtain the readings at each stage. Calibrate colorimeters with standard permanganate solution as directed in Section 4500-Cl.G.4a. Use of additional EDTA, disodium salt, with the DPD reagents is not required in colorimetric procedures.

4. Calculations

For 100 mL sample, 1 mL FAS solution = 1 mg available chlorine/L.

In the absence of chlorite:

$$\text{Chlorine dioxide} = 5G \text{ (or } 1.9G \text{ expressed as ClO}_2)$$
$$\text{Free available chlorine} = A - G$$
$$\text{Monochloramine} = B - A$$
$$\text{Dichloramine} = C - B$$
$$\text{Total available chlorine} = C + 4G$$

If the step leading to Reading *B* is omitted, monochloramine and dichloramine are obtained together when:

$$\text{Combined available chlorine} = C - A$$

If it is desired to check for presence of chlorite in sample, obtain Reading *D*. Chlorite is indicated if *D* is greater than *C* + 4*G*.

In the presence of chlorite:

$$\text{Chlorine dioxide} = 5G \text{ (or } 1.9G \text{ expressed as ClO}_2)$$
$$\text{Chlorite} = D - (C + 4G)$$
$$\text{Free available chlorine} = A - G$$
$$\text{Monochloramine} = B - A$$
$$\text{Dichloramine} = C - B$$
$$\text{Total available chlorine} = D$$

If *B* is omitted,

$$\text{Combined available chlorine} = C - A$$

5. Bibliography

PALIN, A.T. 1960. Colorimetric determination of chlorine dioxide in water. *Water Sewage Works* 107:457.

PALIN, A.T. 1967. Methods for the determination, in water, of free and combined available chlorine, chlorine dioxide and chlorite, bromine, iodine, and ozone using diethyl-p-phenylenediamine (DPD). *J. Inst. Water Eng.* 21:537.

PALIN, A.T. 1974. Analytical control of water disinfection with special reference to differential DPD methods for chlorine, chlorine dioxide, bromine, iodine and ozone. *J. Inst. Water Eng.* 28:139.

PALIN, A.T. 1975. Current DPD methods for residual halogen compounds and ozone in water. *J. Amer. Water Works Assoc.* 67:32.

4500-ClO$_2$ E.   Amperometric Method II

1. General Discussion

a. Principle: Like Amperometric Method I (Section 4500-ClO$_2$.C), this procedure entails successive titrations of combinations of chlorine species. Subsequent calculations determine the concentration of each species. The equilibrium for reduction of the chlorine species of interest by iodide is pH-dependent.

The analysis of a sample for chlorine, chlorine dioxide, chlorite, and chlorate requires the following steps: determination of all of the chlorine (free plus combined) and one-fifth of the chlorine dioxide at pH 7; lowering sample pH to 2 and determination of the remaining four-fifths of the ClO$_2$ and all of the chlorite (the chlorite measured in this step comes from the chlorite originally present in the sample and that formed in the first titration); preparation of a second sample by purging with nitrogen to remove ClO$_2$ and by reacting with iodide at pH 7 to remove any chlorine remaining; lowering latter sample pH to 2 and determination of all chlorite present (this chlorite only comes from the chlorite originally present in the sample); and, in a third sample, determination of all of the relevant, oxidized chlorine species— chlorine, chlorine dioxide, chlorite, and chlorate—after reduction in hydrochloric acid.[1]

This procedure can be applied to concentrated solutions (10 to 100 mg/L) or dilute solutions (0.1 to 10 mg/L) by appropriate selection of titrant concentration and sample size.

b. Interferences: At pH values above 4, significant iodate formation is possible if iodine is formed in the absence of iodide;[2] this results in a negative bias in titrating the first and second samples. Acidification of these samples causes reduction of iodate to iodine and a positive bias. To prevent formation of iodate add 1 g KI granules to stirred sample.

A positive bias results from oxidation of iodide to iodine by dissolved oxygen in strongly acidic solutions.[1] To minimize this bias, use bromide as the reducing agent in titrating the third sample (bromide is not oxidized by oxygen under these conditions). After reaction is completed, add iodide, which will be oxidized to iodine by the bromine formed from the reduction of the original chlorine species. Add iodide carefully so that bromine gas is not lost. Rapid dilution of the sample with sodium phosphate decreases sample acidity and minimizes oxidation of iodide by oxygen. The pH of the solution to be titrated should be between 1.0 and 2.0. Carry a blank through the procedure as a check on iodide oxidation.

The potential for interferences from manganese, copper, and nitrate is minimized by buffering the sample to pH $\geq$4.[3,4] For the method presented here, the low pH required for the chlorite and chlorate analyses provides conditions favorable to manganese, copper, and nitrite interferences.

2. Apparatus

a. Titrators: See Section 4500-Cl.D.2a through *d*. Amperometric titrators with a platinum-platinum electrode system are more stable and require less maintenance. (NOTE: Chlorine dioxide may attack adhesives used to connect the platinum plate to the electrode, resulting in poor readings.)

If a potentiometric titrator is used, provide a platinum sensing electrode and a silver chloride reference electrode for end-point detection.

b. Glassware: Store glassware used in this method separately from other laboratory glassware and do not use for other purposes because ClO$_2$ reacts with glass to form a hydrophobic surface coating. To satisfy any ClO$_2$ demand, before first use immerse all glassware in a strong ClO$_2$ solution (200 to 500 mg/L) for 24 h and rinse only with water between uses.

c. Sampling: ClO$_2$ is volatile and will vaporize easily from aqueous solution. When sampling a liquid stream, minimize contact with air by placing a flexible sample line to reach the bottom of the sample container, letting several container volumes overflow, slowly removing sample line, and capping container with minimum headspace. Protect from sunlight. Remove sample portions with a volumetric pipet with pipet tip placed at bottom of container. Drain pipet by placing its tip below the surface of reagent or dilution water.

3. Reagents

a. Standard sodium thiosulfate, 0.100N (0.100M): See Section 4500-Cl.B.2c.

b. Standard phenylarsine oxide, 0.005 64N (0.005 64M): See Section 4500-Cl.C.3a. (Weigh out 1.25 g phenylarsine oxide and standardize to 0.005 64M.)

c. Phosphate buffer solution, pH 7: See Section 4500-Cl.D.3b.

d. Potassium iodide, KI, granules.

e. Saturated sodium phosphate solution: Prepare a saturated solution of Na$_2$HPO$_4$·12H$_2$O with cold deionized-distilled water.

f. Potassium bromide solution, 5%: Dissolve 5 g KBr and dilute to 100 mL. Store in a brown glass-stoppered bottle. Make fresh weekly.

g. Hydrochloric acid, HCl, conc.

h. Hydrochloric acid, HCl, 2.5N (2.5M): Cautiously add 200 mL conc HCl, with mixing, to distilled water, diluting to 1000 mL.

i. Purge gas: Use nitrogen gas for purging ClO$_2$ from samples. Assure that gas is free of contaminants and pass it through a 5% KI scrub solution. Discard solution at first sign of color.

4. Procedure

Use either sodium thiosulfate or phenylarsine oxide as titrant. Select concentration on basis of concentration range expected. The total mass of oxidant species should be no greater than about 15 mg. Make appropriate sample dilutions if necessary. A convenient volume for titration is 200 to 300 mL. Preferably analyze all samples and blanks in triplicate.

Minimize effects of pH, time, and temperature of reaction by standardizing all conditions.

a. Titration of residual chlorine and one-fifth of available ClO$_2$: Place 1 mL pH 7 phosphate buffer in beaker and add distilled-deionized dilution water if needed. Introduce sample with minimum aeration and add 1 g KI granules while stirring. Titrate to end point (see Section 4500-Cl.D). Record reading A = mL titrant/mL sample.

b. Titration of four-fifths of available ClO$_2$ and chlorite: Continuing with same sample, add 2 mL 2.5N (2.5M) HCl. Let stand in the dark for 5 min. Titrate to end point. Record reading B = mL titrant/mL sample.

c. Titration of nonvolatilized chlorine: Place 1 mL pH 7 phosphate buffer in purge vessel and add distilled-deionized dilution water if needed. Add sample and purge with nitrogen gas for 15 min. Use a gas-dispersion tube to give good gas-liquid contact.

Add 1 g KI granules while stirring and titrate to end point. Record reading C = mL titrant/mL sample.

d. Titration of chlorite: Continuing with same sample, add 2 mL 2.5N (2.5M) HCl. Let stand in the dark for 5 min. Titrate to end point, and record reading D = mL titrant/mL sample.

e. Titration of chlorine, ClO$_2$, chlorate, and chlorite: Add 1 mL KBr and 10 mL conc HCl to 50-mL reaction flask and mix. Carefully add 15 mL sample, with minimum aeration. Mix and stopper immediately. Let stand in the dark for 20 min. Rapidly add 1 g KI granules and shake vigorously for 5 s. Rapidly transfer to titration flask containing 25 mL saturated Na$_2$HPO$_4$ solution. Rinse reaction flask thoroughly and add rinse water to titration flask. Final titration volume should be about 200 to 300 mL. Titrate to end point.

Repeat procedure of preceding paragraph using distilled-deionized water in place of sample to determine blank value.

Record reading E = (mL titrant sample − mL titrant blank)/mL sample.

NOTE: The 15-mL sample volume can be adjusted to provide an appropriate dilution, but maintain the ratio of sample to HCl.

5. Calculations

Because the combining power of the titrants is pH-dependent, all calculations are based on the equivalents of reducing titrant required to react with equivalents of oxidant present. Use Table 4500-ClO$_2$:I to obtain the equivalent weights to be used in the calculations.

In the following equations, N is the normality of the titrant used in equivalents per liter and A through E are as defined previously.

$$\text{Chlorite, mg ClO}_2^-/\text{L} = D \times N \times 16\,863$$
$$\text{Chlorate, mg ClO}_2^{2-}/\text{L} = [E - (A + B)] \times N \times 13\,909$$
$$\text{Chlorine dioxide, mg ClO}_2/\text{L} = (5/4) \times (B - D) \times N \times 13\,490$$
$$\text{Chlorine, mg Cl}_2/\text{L} = \{A - [(B - D)/4]\} \times N \times 35\,453$$

6. References

1. AIETA, E.M., P.V. ROBERTS & M. HERNANDEZ. 1984. Determination of chlorine dioxide, chlorine, chlorite, and chlorate in water. *J. Amer. Water Works Assoc.* 76:64.
2. WONG, G. 1982. Factors affecting the amperometric determination of trace quantities of total residual chlorine in seawater. *Environ. Sci. Technol.* 16:11.
3. WHITE, G. 1972. Handbook of Chlorination. Van Nostrand Reinhold Co., New York, N.Y.
4. JOLLEY, R. & J. CARPENTER. 1982. Aqueous Chemistry of Chlorine: Chemistry, Analysis, and Environmental Fate of Reactive Oxidant Species. ORNL/TM-788, Oak Ridge National Lab., Oak Ridge, Tenn.

TABLE 4500-ClO$_2$:I. EQUIVALENT WEIGHTS FOR CALCULATING CONCENTRATIONS ON THE BASIS OF MASS

pH	Species	Molecular Weight *mg/mol*	Electrons Transferred	Equivalent Weight *mg/eq*
7	Chlorine dioxide	67 452	1	67 452
2, 0.1	Chlorine dioxide	67 452	5	13 490
7, 2, 0.1	Chlorine	70 906	2	35 453
2, 0.1	Chlorite	67 452	4	16 863
0.1	Chlorate	83 451	6	13 909

7. Bibliography

AIETA, E.M. 1985. Amperometric analysis of chlorine dioxide, chlorine and chlorite in aqueous solution. Presented at American Water Works Assoc. Water Quality Technology Conf. 13, Houston, Texas.

GORDON, G. 1982. Improved methods of analysis for chlorate, chlorite, and hypochlorite ions. Presented at American Water Works Assoc. Water Quality Technology Conf., Nashville, Tenn.

TANG, T.F. & G. GORDON. 1980. Quantitative determination of chloride, chlorite, and chlorate ions in a mixture by successive potentiometric titrations. *Anal. Chem.* 52:1430.

4500-F⁻ FLUORIDE*

4500-F⁻ A. Introduction

A fluoride concentration of approximately 1.0 mg/L in drinking water effectively reduces dental caries without harmful effects on health. Fluoride may occur naturally in water or it may be added in controlled amounts. Some fluorosis may occur when the fluoride level exceeds the recommended limits. In rare instances the naturally occurring fluoride concentration may approach 10 mg/L; such waters should be defluoridated.

Accurate determination of fluoride has increased in importance with the growth of the practice of fluoridation of water supplies as a public health measure. Maintenance of an optimal fluoride concentration is essential in maintaining effectiveness and safety of the fluoridation procedure.

1. Preliminary Treatment

Among the methods suggested for determining fluoride ion (F⁻) in water, the electrode and colorimetric methods are the most satisfactory. Because both methods are subject to errors due to interfering ions (Table 4500:F⁻:I), it may be necessary to distill the sample as directed in Section 4500-F⁻.B before making the determination. When interfering ions are not present in excess of the tolerance of the method, the fluoride determination may be made directly without distillation.

2. Selection of Method

The electrode methods (C and G) are suitable for fluoride concentrations from 0.1 to more than 10 mg/L. Adding the prescribed buffer frees the electrode method from most interferences that adversely effect the SPADNS colorimetric method and necessitate preliminary distillation. Some substances in industrial wastes, such as fluoborates, may be sufficiently concentrated to present problems in electrode measurements and will not be measured without a preliminary distillation. Fluoride measurements can be made with an ion-selective electrode and either an expanded-scale pH meter or a specific ion meter, usually without distillation, in the time necessary for electrode equilibration.

The SPADNs method (D) has a linear analytical range of 0 to 1.40 mg F⁻/L. Use of a nonlinear calibration can extend the range to 3.5 mg F⁻/L. Color development is virtually instantaneous. Color

---

* Approved by Standard Methods Committee, 1997.

TABLE 4500-F⁻:I. CONCENTRATION OF SOME SUBSTANCES CAUSING 0.1-MG/L ERROR AT 1.0 MG F/L IN FLUORIDE METHODS

Substance	Method C (Electrode)		Method D (SPADNS)	
	Conc *mg/L*	Type of Error*	Conc *mg/L*	Type of Error*
Alkalinity			5 000	
(CaCO₃)	7 000	+		−
Aluminum (Al³⁺)	3.0	−	0.1†	−
Chloride (Cl⁻)	20 000		7 000	+
Chlorine	5 000			Remove completely with arsenite
Color & turbidity				Remove or compensate for
Iron	200	−	10	−
Hexametaphosphate				
([NaPO₃]₆)	50 000		1.0	+
Phosphate				
(PO₄³⁻)	50 000		16	+
Sulfate				
(SO₄²⁻)	50 000	−	200	−

* + denotes positive error
 − denotes negative error
 Blank denotes no measurable error.
† On immediate reading. Tolerance increases with time: after 2 h, 3.0; after 4 h, 30.

determinations are made photometrically, using either a filter photometer or a spectrophotometer. A curve developed from standards is used for determining the fluoride concentration of a sample.

Fluoride also may be determined by the automated complexone method, Method E.

Ion chromatography (Section 4110) is an acceptable method if weaker eluents are used to separate fluoride from interfering peaks or fluoride can be determined by capillary ion electrophoresis (Section 4140).

The flow injection method (G) is a convenient automated technique for analyzing large numbers of samples.

3. Sampling and Storage

Preferably use polyethylene bottles for collecting and storing samples for fluoride analysis. Glass bottles are satisfactory if previously they have not contained high-fluoride solutions. Always rinse bottle with a portion of sample.

For the SPADNs method, never use an excess of dechlorinating agent. Dechlorinate with sodium arsenite rather than sodium thiosulfate when using the SPADNS method because the latter may produce turbidity that causes erroneous readings.

4500-F⁻ B.   Preliminary Distillation Step

1. Discussion

Fluoride can be separated from other nonvolatile constituents in water by conversion to hydrofluoric or fluosilicic acid and subsequent distillation. The conversion is accomplished by using a strong, high-boiling acid. To protect against glassware etching, hydrofluoric acid is converted to fluosilicic acid by using soft glass beads. Quantitative fluoride recovery is approached by using a relatively large sample. Acid and sulfate carryover are minimized by distilling over a controlled temperature range.

Distillation will separate fluoride from most water samples. Some tightly bound fluoride, such as that in biological materials, may require digestion before distillation, but water samples seldom require such drastic treatment. Distillation produces a distillate volume equal to that of the original water sample so that usually it is not necessary to incorporate a dilution factor when

expressing analytical results. The distillate will be essentially free of substances that might interfere with the fluoride determination if the apparatus used is adequate and distillation has been carried out properly. The only common volatile constituent likely to cause interference with colorimetric analysis of the distillate is chloride. When the concentration of chloride is high enough to interfere, add silver sulfate to the sulfuric acid distilling mixture to minimize the volatilization of hydrogen chloride.

Heating an acid-water mixture can be hazardous if precautions are not taken: *Mix acid and water thoroughly before heating.* Use of a quartz heating mantle and a magnetic stirrer in the distillation apparatus simplifies the mixing step.

2. Apparatus

a. Distillation apparatus consisting of a 1-L round-bottom long-neck borosilicate glass boiling flask, a connecting tube, an efficient condenser, a thermometer adapter, and a thermometer that can be read to 200°C. Use standard taper joints for all connections in the direct vapor path. Position the thermometer so that the bulb always is immersed in boiling mixture. The apparatus should be disassembled easily to permit adding sample. Substituting a thermoregulator and necessary circuitry for the thermometer is acceptable and provides some automation.

Alternative types of distillation apparatus may be used. Carefully evaluate any apparatus for fluoride recovery and sulfate carryover. The critical points are obstructions in the vapor path and trapping of liquid in the adapter and condenser. (The condenser should have a vapor path with minimum obstruction. A double-jacketed condenser, with cooling water in the outer jacket and the inner spiral tube, is ideal, but other condensers are acceptable if they have minimum obstructions. Avoid using Graham-type condensers.) Avoid using an open flame as a heat source if possible, because heat applied to the boiling flask above the liquid level causes superheating of vapor and subsequent sulfate carryover.

CAUTION: *Regardless of apparatus used, provide for thorough mixing of sample and acid; heating a non-homogenous acid-water mixture will result in bumping or possibly a violent explosion.*

The preferred apparatus is illustrated in Figure 4500-F⁻:1.

b. Quartz hemispherical heating mantle, for full-voltage operation.

c. Magnetic stirrer, with TFE-coated stirring bar.

d. Soft glass beads.

3. Reagents

a. Sulfuric acid, H_2SO_4, conc, reagent grade.

b. Silver sulfate, Ag_2SO_4, crystals, reagent grade.

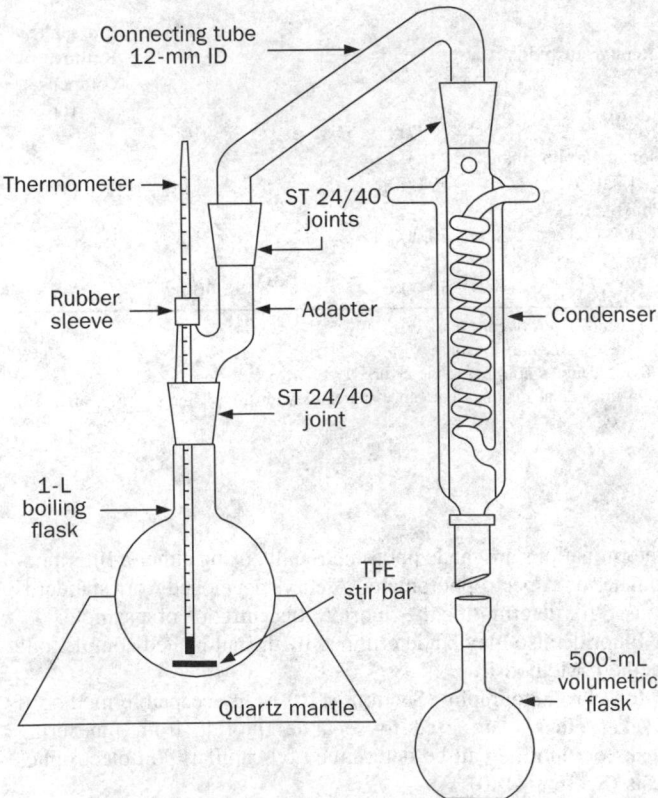

Figure 4500-F⁻:1. Direct distillation apparatus for fluoride.

Connecting tube 12-mm ID

Thermometer

ST 24/40 joints

Rubber sleeve

Adapter

Condenser

ST 24/40 joint

1-L boiling flask

TFE stir bar

500-mL volumetric flask

Quartz mantle

4. Procedure

a. Place 400 mL distilled water in the distilling flask and, with the magnetic stirrer operating, carefully add 200 mL conc H_2SO_4. Keep stirrer in operation throughout distillation. Add a few glass beads and connect the apparatus as shown in Figure 4500-F⁻:1, making sure all joints are tight. Begin heating and continue until flask contents reach 180°C (because of heat retention by the mantle, it is necessary to discontinue heating when the temperature reaches 178°C to prevent overheating). Discard distillate. This process removes fluoride contamination and adjusts the acid-water ratio for subsequent distillations.

b. After the acid mixture remaining in the steps outlined in ¶ 4*a*, or previous distillations, has cooled to 80°C or below, add 300 mL sample, with stirrer operating, and distill until the temperature reaches 180°C. To prevent sulfate carryover, turn off heat before 178°C. Retain the distillate for analysis.

c. Add Ag_2SO_4 to the distilling flask at the rate of 5 mg/mg Cl⁻ when the chloride concentration is high enough to interfere (see Table 4500-F⁻:I).

d. Use H_2SO_4 solution in the flask repeatedly until contaminants from samples accumulate to such an extent that recovery is affected or interferences appear in the distillate. Check acid suitability periodically by distilling standard fluoride samples and analyzing for both fluoride and sulfate. After distilling samples containing more than 3 mg F⁻/L, flush still by adding 300 mL distilled water, redistill, and combine the two fluoride distillates. If necessary, repeat flushing until the fluoride content of the last distillate is at a minimum. Include additional fluoride recovered with that of the first distillation. After periods of inactivity, similarly flush still and discard distillate.

5. Interpretation of Results

The recovery of fluoride is quantitative within the accuracy of the methods used for its measurement.

6. Bibliography

BELLACK, E. 1958. Simplified fluoride distillation method. *J. Amer. Water Works Assoc.* 50:530.
BELLACK, E. 1961. Automatic fluoride distillation. *J. Amer. Water Works Assoc.* 53:98.
ZEHNPFENNIG, R.G. 1976. Letter to the editor. *Environ. Sci. Technol.* 10: 1049.

4500-F⁻ C.   Ion-Selective Electrode Method

1. General Discussion

a. Principle: The fluoride electrode is an ion-selective sensor. The key element in the fluoride electrode is the laser-type doped lanthanum fluoride crystal across which a potential is established by fluoride solutions of different concentrations. The crystal contacts the sample solution at one face and an internal reference solution at the other. The cell may be represented by:

$$Ag|AgCl, Cl^- (0.3M), F^- (0.001M) |LaF_3| \text{ test}$$
solution|reference electrode

The fluoride electrode can be used with a standard calomel reference electrode and almost any modern pH meter having an expanded millivolt scale. Calomel electrodes contain both metallic and dissolved mercury; therefore, dispose of them only in approved sites or recycle. For this reason, the Ag/AgCl reference electrode is preferred.

The fluoride electrode measures the ion activity of fluoride in solution rather than concentration. Fluoride ion activity depends on the solution total ionic strength and pH, and on fluoride complexing species. Adding an appropriate buffer provides a nearly uniform ionic strength background, adjusts pH, and breaks up complexes so that, in effect, the electrode measures concentration.

b. Interference: Table 4500-F⁻:I lists common interferences. Fluoride forms complexes with several polyvalent cations, notably aluminum and iron. The extent to which complexation takes place depends on solution pH, relative levels of fluoride, and complexing species. However, CDTA (cyclohexylenediaminetetraacetic acid), a component of the buffer, preferentially will complex interfering cations and release free fluoride ions. Concentrations of aluminum, the most common interference, up to 3.0 mg/L can be complexed preferentially. In acid solution, F⁻ forms a poorly ionized HF·HF complex but the buffer maintains a pH above 5 to minimize hydrogen fluoride complex formation. In alkaline solution hydroxide ion also can interfere with electrode response to fluoride ion whenever the hydroxide ion concentration is greater than one-tenth the concentration of fluoride ion. At the pH maintained by the buffer, no hydroxide interference occurs.

Fluoborates are widely used in industrial processes. Dilute solutions of fluoborate or fluoboric acid hydrolyze to liberate fluoride ion but in concentrated solutions, as in electroplating wastes, hydrolysis does not occur completely. Distill such samples or measure fluoborate with a fluoborate-selective electrode. Also distill the sample if the dissolved solids concentration exceeds 10 000 mg/L.

2. Apparatus

a. Expanded-scale or digital pH meter or ion-selective meter.
b. Sleeve-type reference electrode: Do not use fiber-tip reference electrodes because they exhibit erratic behavior in very dilute solutions.
c. Fluoride electrode.
d. Magnetic stirrer, with TFE-coated stirring bar.
e. Timer.

3. Reagents

a. Stock fluoride solution: Dissolve 221.0 mg anhydrous sodium fluoride, NaF, in distilled water and dilute to 1000 mL; 1.00 mL = 100 μg F⁻.

b. Standard fluoride solution: Dilute 100 mL stock fluoride solution to 1000 mL with distilled water; 1.00 mL = 10.0 μg F⁻.

c. Fluoride buffer: Place approximately 500 mL distilled water in a 1-L beaker and add 57 mL glacial acetic acid, 58 g NaCl, and 4.0 g 1,2 cyclohexylenediaminetetraacetic acid (CDTA).* Stir to dissolve. Place beaker in a cool water bath and add slowly 6N NaOH (about 125 mL) with stirring, until pH is between 5.3 and 5.5. Transfer to a 1-L volumetric flask and add distilled water to the mark. This buffer, as well as a more concentrated version, is available commercially. In using the concentrated buffer follow the manufacturer's directions.

4. Procedure

a. Instrument calibration: No major adjustment of any instrument normally is required to use electrodes in the range of 0.2 to 2.0 mg F⁻/L. For those instruments with zero at center scale adjust calibration control so that the 1.0 mg F⁻/L standard reads at the center zero (100 mV) when the meter is in the expanded-scale position. This cannot be done on some meters that do not have a millivolt calibration control. To use a selective-ion meter follow the manufacturer's instructions.

b. Preparation of fluoride standards: Prepare a series of standards by diluting with distilled water 5.0, 10.0, and 20.0 mL of standard fluoride solution to 100 mL with distilled water. These standards are equivalent to 0.5, 1.0, and 2.0 mg F⁻/L.

c. Treatment of standards and sample: In 100-mL beakers or other convenient containers add by volumetric pipet from 10 to 25 mL standard or sample. Bring standards and sample to same temperature, preferably room temperature. Add an equal volume of buffer. The total volume should be sufficient to immerse the electrodes and permit operation of the stirring bar.

d. Measurement with electrode: Immerse electrodes in each of the fluoride standard solutions and measure developed potential while stirring on a magnetic stirrer. Avoid stirring before immersing electrodes because entrapped air around the crystal can produce erroneous readings or needle fluctuations. Let electrodes remain in the solution 3 min (or until reading is constant) before taking a final millivolt reading. A layer of insulating material between stirrer and beaker minimizes solution heating. Withdraw electrodes, rinse with distilled water, and *blot dry* between readings. (CAUTION: Blotting may poison electrode if not done gently.) Repeat measurements with samples.

When using an expanded-scale pH meter or selective-ion meter, frequently recalibrate the electrode by checking potential reading

---

* Also known as 1,2 cyclohexylenedinitrilotetraacetic acid.

of the 1.00-mg F⁻/L standard and adjusting the calibration control, if necessary, until meter reads as before.

If a direct-reading instrument is not used, plot potential measurement of fluoride standards against concentration on two-cycle semilogarithmic graph paper. Plot milligrams F⁻ per liter on the logarithmic axis (ordinate), with the lowest concentration at the bottom of the graph. Plot millivolts on the abscissa. From the potential measurement for each sample, read the corresponding fluoride concentration from the standard curve.

The known-additions method may be substituted for the calibration method described. Follow the directions of the instrument manufacturer.

Selective-ion meters may necessitate using a slightly altered procedure, such as preparing 1.00 and 10.0 mg F⁻/L standards or some other concentration. Follow the manufacturer's directions. Commercial standards, often already diluted with buffer, frequently are supplied with the meter. Verify the stated fluoride concentration of these standards by comparing them with standards prepared by the analyst.

5. Calculation

$$\text{mg F}^-/\text{L} = \frac{\mu\text{g F}^-}{\text{mL sample}}$$

6. Precision and Bias

A synthetic sample containing 0.850 mg F⁻/L in distilled water was analyzed in 111 laboratories by the electrode method, with a relative standard deviation of 3.6% and a relative error of 0.7%.

A second synthetic sample containing 0.750 mg F⁻/L, 2.5 mg $(NaPO_3)_6$/L, and 300 mg alkalinity/L added as $NaHCO_3$, was analyzed in 111 laboratories by the electrode method, with a relative standard deviation of 4.8% and a relative error of 0.2%.

A third synthetic sample containing 0.900 mg F⁻/L, 0.500 mg Al/L, and 200 mg SO_4^{2-}/L was analyzed in 13 laboratories by the electrode method, with a relative standard deviation of 2.9% and a relative error of 4.9%.

7. Bibliography

FRANT, M.S. & J.W. ROSS, JR. 1968. Use of total ionic strength adjustment buffer for electrode determination of fluoride in water supplies. *Anal. Chem.* 40:1169.

HARWOOD, J.E. 1969. The use of an ion-selective electrode for routine analysis of water samples. *Water Res.* 3:273.

4500-F⁻ D.   SPADNS Method

1. General Discussion

a. Principle: The SPADNS colorimetric method is based on the reaction between fluoride and a zirconium-dye lake. Fluoride reacts with the dye lake, dissociating a portion of it into a colorless complex anion (ZrF_6^{2-}); and the dye. As the amount of fluoride increases, the color produced becomes progressively lighter.

The reaction rate between fluoride and zirconium ions is influenced greatly by the acidity of the reaction mixture. If the proportion of acid in the reagent is increased, the reaction can be made almost instantaneous. Under such conditions, however, the effect of various ions differs from that in the conventional alizarin methods. The selection of dye for this rapid fluoride method is governed largely by the resulting tolerance to these ions.

b. Interference: Table 4500-F⁻:I lists common interferences. Because these are neither linear in effect nor algebraically additive, mathematical compensation is impossible. Whenever any one substance is present in sufficient quantity to produce an error of 0.1 mg/L or whenever the total interfering effect is in doubt, distill the sample. Also distill colored or turbid samples. In some instances, sample dilution or adding appropriate amounts of interfering substances to the standards may be used to compensate for the interference effect. If alkalinity is the only significant interference, neutralize it with either hydrochloric or nitric acid. Chlorine interferes and provision for its removal is made.

Volumetric measurement of sample and reagent is extremely important to analytical accuracy. Use samples and standards at the same temperature or at least within 2°C. Maintain constant temperature throughout the color development period. Prepare different calibration curves for different temperature ranges.

2. Apparatus

Colorimetric equipment: One of the following is required:

a. Spectrophotometer, for use at 570 nm, providing a light path of at least 1 cm.

b. Filter photometer, providing a light path of at least 1 cm and equipped with a greenish yellow filter having maximum transmittance at 550 to 580 nm.

3. Reagents

a. Standard fluoride solution: Prepare as directed in the electrode method, Section 4500-F⁻.C.3*b.*

b. SPADNS solution: Dissolve 958 mg SPADNS, sodium 2-(parasulfophenylazo)-1,8-dihydroxy-3,6-naphthalene disulfonate, also called 4,5-dihydroxy-3-(parasulfophenylazo)-2,7-naphthalenedisulfonic acid trisodium salt, in distilled water and dilute to 500 mL. This solution is stable for at least 1 year if protected from direct sunlight.

c. Zirconyl-acid reagent: Dissolve 133 mg zirconyl chloride octahydrate, $ZrOCl_2 \cdot 8H_2O$, in about 25 mL distilled water. Add 350 mL conc HCl and dilute to 500 mL with distilled water.

d. Acid zirconyl-SPADNS reagent: Mix equal volumes of SPADNS solution and zirconyl-acid reagent. The combined reagent is stable for at least 2 years.

e. Reference solution: Add 10 mL SPADNS solution to 100 mL distilled water. Dilute 7 mL conc HCl to 10 mL and add to the diluted SPADNS solution. The resulting solution, used for setting the instrument reference point (zero), is stable for at least 1 year. Alternatively, use a prepared standard of 0 mg F⁻/L as a reference.

f. Sodium arsenite solution: Dissolve 5.0 g $NaAsO_2$ and dilute to 1 L with distilled water. (CAUTION: *Toxic—avoid ingestion.*)

4. Procedure

a. Preparation of standard curve: Prepare fluoride standards in the range of 0 to 1.40 mg F⁻/L by diluting appropriate quantities of standard fluoride solution to 50 mL with distilled water. Pipet 5.00 mL each of SPADNS solution and zirconyl-acid reagent, or 10.00 mL mixed acid-zirconyl-SPADNS reagent, to each standard and mix well. Avoid contamination. Set photometer to zero absorbance with the reference solution and obtain absorbance readings of standards. Plot a curve of the milligrams fluoride-absorbance relationship. Prepare a new standard curve whenever a fresh reagent is made

or a different standard temperature is desired. As an alternative to using a reference, set photometer at some convenient point (0.300 or 0.500 absorbance) with the prepared 0 mg F⁻/L standard.

b. Sample pretreatment: If the sample contains residual chlorine, remove it by adding 1 drop (0.05 mL) $NaAsO_2$ solution/0.1 mg residual chlorine and mix. (Sodium arsenite concentrations of 1300 mg/L produce an error of 0.1 mg/L at 1.0 mg F⁻/L.)

c. Color development: Use a 50.0-mL sample or a portion diluted to 50 mL with distilled water. Adjust sample temperature to that used for the standard curve. Add 5.00 mL each of SPADNS solution and zirconyl-acid reagent, or 10.00 mL acid-zirconyl-SPADNS reagent; mix well and read absorbance, first setting the reference point of the photometer as above. If the absorbance falls beyond the range of the standard curve, repeat using a diluted sample.

5. Calculation

$$mg\ F^-/L = \frac{A}{mL\ sample} \times \frac{B}{C}$$

where:

A = μg F⁻ determined from plotted curve,
B = final volume of diluted sample, mL, and
C = volume of diluted sample used for color development, mL.

When the prepared 0 mg F⁻/L standard is used to set the photometer, alternatively calculate fluoride concentration as follows:

$$mg\ F^-/L = \frac{A_0 - A_x}{A_0 - A_1}$$

where:

A_0 = absorbance of the prepared 0 mg F⁻/L standard,
A_1 = absorbance of a prepared 1.0 mg F⁻/L standard, and
A_x = absorbance of the prepared sample.

6. Precision and Bias

A synthetic sample containing 0.830 mg F⁻/L and no interference in distilled water was analyzed in 53 laboratories by the SPADNS method, with a relative standard deviation of 8.0% and a relative error of 1.2%. After direct distillation of the sample, the relative standard deviation was 11.0% and the relative error 2.4%.

A synthetic sample containing 0.570 mg F⁻/L, 10 mg Al/L, 200 mg SO_4^{2-}/L, and 300 mg total alkalinity/L was analyzed in 53 laboratories by the SPADNS method without distillation, with a relative standard deviation of 16.2% and a relative error of 7.0%. After direct distillation of the sample, the relative standard deviation was 17.2% and the relative error 5.3%.

A synthetic sample containing 0.680 mg F⁻/L, 2 mg Al/L, 2.5 mg $(NaPO_3)_6$/L, 200 mg SO_4^{2-}/L, and 300 mg total alkalinity/L was analyzed in 53 laboratories by direct distillation and SPADNS methods with a relative standard deviation of 2.8% and a relative error of 5.9%.

7. Bibliography

BELLACK, E. & P.J. SCHOUBOE. 1968. Rapid photometric determination of fluoride with SPADNS-zirconium lake. *Anal. Chem.* 30:2032.

4500-F⁻ E. Complexone Method

1. General Discussion

a. Principle: The sample is distilled in the automated system, and the distillate is reacted with alizarin fluorine blue-lanthanum reagent to form a blue complex that is measured colorimetrically at 620 nm.

b. Interferences: Interferences normally associated with the determination of fluoride are removed by distillation.

c. Application: This method is applicable to potable, surface, and saline waters as well as domestic and industrial wastewaters. The range of the method, which can be modified by using the adjustable colorimeter, is 0.1 to 2.0 mg F⁻/L.

2. Apparatus

An example of the required continuous-flow analytical instrument consists of the interchangeable components in the number and manner indicated in Figure 4500-F⁻:2.

3. Reagents

a. Standard fluoride solution: Prepare in appropriate concentrations from 0.10 to 2.0 mg F⁻/L using the stock fluoride solution (see Section 4500-F⁻.C.3a).

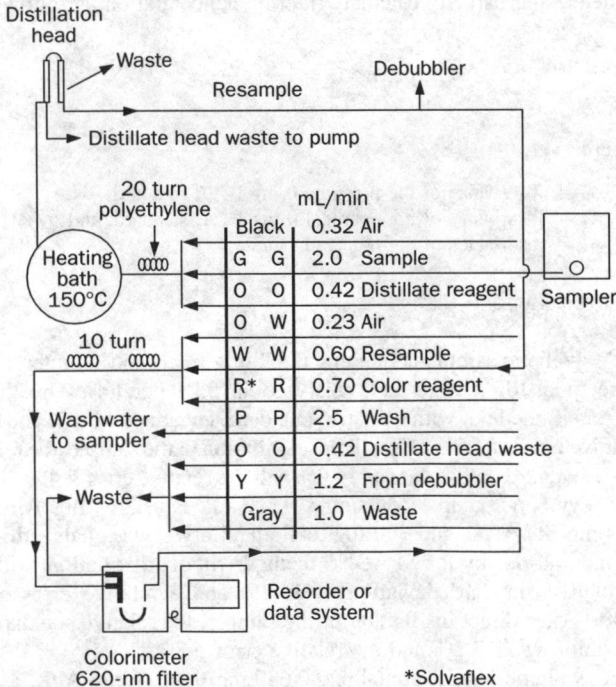

Distillation head

Waste

Resample

Debubbler

Distillate head waste to pump

20 turn polyethylene

Heating bath 150°C

10 turn

Washwater to sampler

Waste

	mL/min	
Black	0.32	Air
G G	2.0	Sample
O O	0.42	Distillate reagent
O W	0.23	Air
W W	0.60	Resample
R* R	0.70	Color reagent
P P	2.5	Wash
O O	0.42	Distillate head waste
Y Y	1.2	From debubbler
Gray	1.0	Waste

Sampler

Recorder or data system

Colorimeter 620-nm filter

*Solvaflex

Figure 4500-F⁻:2. Fluoride manifold.

b. Distillation reagent: Add 50 mL conc H_2SO_4 to about 600 mL distilled water. Add 10.00 mL stock fluoride solution (see Section 4500-F⁻.C.3a; 1.00 mL = 100 μg F⁻) and dilute to 1000 mL.

c. Acetate buffer solution: Dissolve 60 g anhydrous sodium acetate, $NaC_2H_3O_2$, in about 600 mL distilled water. Add 100 mL conc (glacial) acetic acid and dilute to 1 L.

d. Alizarin fluorine blue stock solution: Add 960 mg alizarin fluorine,* $C_{14}H_7O_4 \cdot CH_2N(CH_2 \cdot COOH)_2$, to 100 mL distilled water. Add 2 mL conc NH_4OH and mix until dye is dissolved. Add 2 mL conc (glacial) acetic acid, dilute to 250 mL and store in an amber bottle in the refrigerator.

e. Lanthanum nitrate stock solution: Dissolve 1.08 g $La(NO_3)_3$ in about 100 mL distilled water, dilute to 250 mL, and store in refrigerator.

f. Working color reagent: Mix in the following order: 300 mL acetate buffer solution, 150 mL acetone, 50 mL tertiary butanol, 36 mL alizarin fluorine blue stock solution, 40 mL lanthanum nitrate stock solution, and 2 mL polyoxyethylene 23 lauryl ether.† Dilute to 1 L with distilled water. This reagent is stable for 2 to 4 d.

4. Procedure

No special handling or preparation of sample is required.

Set up manifold as shown in Figure 4500-F⁻:2 and follow the manufacturer's instructions.

5. Calculation

Prepare standard curves by plotting response of standards processed through the manifold against constituent concentrations in standards. Compute sample concentrations by comparing sample response with standard curve.

6. Precision and Bias

In a single laboratory four samples of natural water containing from 0.40 to 0.82 mg F⁻/L were analyzed in septuplicate. Average precision was ± 0.03 mg F⁻/L. To two of the samples, additions of 0.20 and 0.80 mg F⁻/L were made. Average recovery of the additions was 98%.

7. Bibliography

WEINSTEIN, L.H., R.H. MANDL, D.C. McCUNE, J.S. JACOBSON & A.E. HITCHCOCK. 1963. A semi-automated method for the determination of fluorine in air and plant tissues. *Boyce Thompson Inst.* 22:207.

---

* J.T. Baker Catalog number J-112 or equivalent.
† Brij-35, available from ICI Americas, Wilmington, DE, or equivalent.

4500-F$^-$ F. (Reserved)

Section 4500-F$^-$ G. Ion-Selective Electrode Flow Injection Analysis (PROPOSED)

1. General Discussion

a. Principle: Fluoride is determined potentiometrically by using a combination fluoride-selective electrode in a flow cell. The fluoride electrode consists of a lanthanum fluoride crystal across which a potential is developed by fluoride ions. The reference cell is a Ag/AgCl/Cl$^-$ cell. The reference junction is of the annular liquid-junction type and encloses the fluoride-sensitive crystal.

Also see Section 4500-F$^-$.C and Section 4130, Flow Injection Analysis (FIA).

b. Interferences: Remove large or fibrous particulates by filtering sample through glass wool. Guard against contamination from reagents, water, glassware, and the sample preservation process.

The polyvalent cations Si^{4+}, Al^{3+}, and Fe^{3+} interfere by forming complexes with fluoride. As part of the buffer reagent, 1,2-cyclohexyldiaminetetraacetic acid (CDTA) is added to preferentially complex these cations and eliminate this interference when these concentrations do not exceed 3.0 mg Al^{3+}/L and 20 mg Fe^{3+}/L.

Some interferents are removed by distillation; see Section 4500-F$^-$.B. Drinking water samples generally do not require sample distillation.

2. Apparatus

Flow injection analysis equipment consisting of:
a. FIA injection valve with sample loop or equivalent.
b. Multichannel proportioning pump.
c. FIA manifold (Figure 4500-F$^-$:3) with tubing heater and ion-selective electrode flow cell. In Figure 4500-F$^-$:3, relative flow rates only are shown. Tubing volumes are given as an example only; they may be scaled down proportionally. Use manifold tubing of an inert material such as TFE.

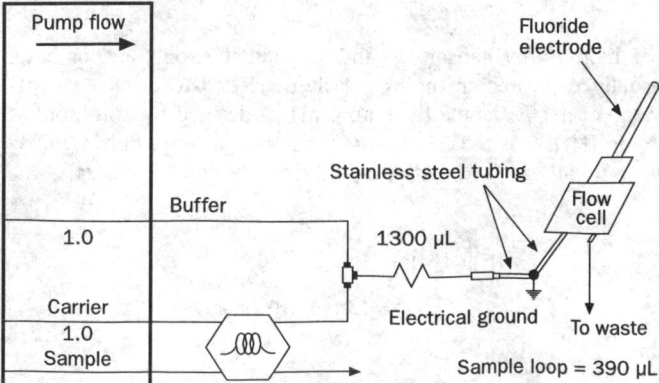

Figure 4500-F$^-$:3. FIA fluoride manifold.

TABLE 4500-F$^-$:II. RESULTS OF SINGLE-LABORATORY STUDIES WITH SELECTED MATRICES

Matrix	Sample/Blank Designation	Known Addition mg F$^-$/L	Recovery %	Relative Standard Deviation %
Wastewater treatment plant influent	Reference sample*	—	101	—
	Blank†	1.0	91	—
		2.0	97	—
	Site A‡	0	—	4.8
		1.0	93	—
		2.0	82	—
	Site B‡	0	—	6.4
		1.0	96	—
		2.0	86	—
	Site C‡§	0	—	15
		1.0	99	—
		2.0	86	—
Wastewater treatment plant effluent	Reference sample*	—	103	—
	Blank†	1.0	97	—
		2.0	97	—
	Site A‡	0	—	ND
		1.0	ND	—
		2.0	ND	—
	Site B‡	0	—	<0.1
		1.0	80	—
		2.0	78	—
	Site C‡	0	—	<0.1
		1.0	93	—
		2.0	91	—
Landfill leachate‖	Reference sample*	—	99	—
	Blank†	1.0	87	—
		2.0	88	—
	Site A‡	0	—	13
		1.0	74	—
		2.0	68	—
	Site B‡	0	—	10
		1.0	68	—
		2.0	73	—
	Site C‡	0	—	32
		1.0	66	—
		2.0	79	—

ND = not detectable.
* U.S. EPA QC sample, 1.81 mg F$^-$/L.
† Determined in duplicate.
‡ Samples without known additions determined four times; samples with known additions determined in duplicate. Typical difference between duplicates for influent 5%, for effluent 6%.
§ Mean concentration 0.18 mg F$^-$/L.
‖ All sites had mean concentration of <0.2 mg F$^-$/L.

d. Combination ion-selective electrode.

e. Injection valve control and data acquisition system.

3. Reagents

Use reagent water (>10 megohm) for all solutions. To prevent bubble formation, degas carrier and buffer with helium. Pass He at 140 kPa (20 psi) through a helium degassing tube. Bubble He through 1 L solution for 1 min.

a. Carrier, 1.0 mg F^-/L: Add 10 mL or 10 g stock fluoride standard (¶ 3*d*) to 990 mL water and mix well.

b. Buffer: To a tared 1-L polyethylene container add 929.5 g water, 59.8 g glacial acetic acid, 30.0 g sodium hydroxide, NaOH, 58.0 g sodium chloride, NaCl, 0.5 g stock fluoride standard (¶ 3*d*), and 4.0 g 1,2-cyclohexyldiaminetetraacetic acid (CDTA) (also called trans-1,2-diaminocyclohexane). Stir on a magnetic stir plate until all material has dissolved.

c. Electrode conditioning solution: To a tared 1-L container, add 534 g buffer (¶ 3*b*) and 500 g carrier (¶ 3*a*). Shake or stir to mix thoroughly. Store fluoride electrode in this solution when it is not in use.

d. Stock fluoride standard, 100.0 mg F^-/L: In a 1-L volumetric flask, dissolve 0.2210 g sodium fluoride, NaF, in approximately 950 mL water. Dilute to mark with water and mix well. Store in a polyethylene bottle.

e. Standard fluoride solutions: Prepare fluoride standards in the desired concentration range, using the stock standard (¶ 3*d*), and diluting with water. A blank or zero concentration standard cannot be prepared for this method because it will give an undefined response from the fluoride electrode.

4. Procedure

Set up a manifold equivalent to that in Figure 4500-F^-:3 and follow method supplied by manufacturer or laboratory standard operating procedure for this method. Follow quality control procedures outlined in Section 4020.

5. Calculations

Prepare standard curves by plotting the electrode response to standards processed through the manifold vs. fluoride concentration. Standards greater than 1.0 mg F^-/L will give positive peaks, standards less than 1.0 mg F^-/L will give negative peaks, and the 1.0 mg F^-/L standard having the same concentration as the carrier will give no peak. The calibration curve gives a good fit to a second-order polynomial.

It is not necessary to plot the response versus log[F^-]; if this is done the calibration curve will still be a second-order polynomial because there is a concentration-dependent kinetic effect in the flowing stream electrode system.

6. Precision and Bias

The samples used in the studies described below were not distilled.

a. Recovery and relative standard deviation: The results of single-laboratory studies with various matrices are given in Table 4500-F^-:II.

b. MDL: A 390-μL sample loop was used in the method described above. Ten replicates of a 1.0-mg F^-/L standard were run to obtain an MDL of 0.02 mg F^-/L.

c. Precision: Ten replicate standards of 2.0 mg F^-/L gave a % RSD of 0.5%.

4500-H⁺   pH VALUE*

4500-H⁺ A.   Introduction

1. Principles

Measurement of pH is one of the most important and frequently used tests in water chemistry. Practically every phase of water supply and wastewater treatment, e.g., acid-base neutralization, water softening, precipitation, coagulation, disinfection, and corrosion control, is pH-dependent. pH is used in alkalinity and carbon dioxide measurements and many other acid-base equilibria. At a given temperature the *intensity* of the acidic or basic character of a solution is indicated by pH or hydrogen ion activity. Alkalinity and acidity are the acid- and base-neutralizing capacities of a water and usually are expressed as milligrams $CaCO_3$

---

* Approved by Standard Methods Committee, 1996.

per liter. Buffer capacity is the amount of strong acid or base, usually expressed in moles per liter, needed to change the pH value of a 1-L sample by 1 unit. pH as defined by Sorenson[1] is $-\log[H^+]$; it is the "intensity" factor of acidity. Pure water is very slightly ionized and at equilibrium the ion product is

$$[H^+][OH^-] = K_w$$
$$= 1.01 \times 10^{-14} \text{ at } 25°C \qquad (1)$$

and

$$[H^+] = [OH^-]$$
$$= 1.005 \times 10^{-7}$$

where:

$[H^+]$ = activity of hydrogen ions, moles/L,
$[OH^-]$ = activity of hydroxyl ions, moles/L, and
K_w = ion product of water.

Because of ionic interactions in all but very dilute solutions, it is necessary to use the "activity" of an ion and not its molar concentration. Use of the term pH assumes that the activity of the hydrogen ion, a_H^+, is being considered. The *approximate* equivalence to molarity, $[H^+]$ can be presumed only in very dilute solutions (ionic strength <0.1).

A logarithmic scale is convenient for expressing a wide range of ionic activities. Equation 1 in logarithmic form and corrected to reflect activity is:

$$(-\log_{10} a_{H^+}) + (-\log_{10} a_{OH^-}) = 14 \qquad (2)$$

or

$$pH + pOH = pK_w$$

where:

$pH† = \log_{10} a_{H^+}$ and
$pOH = \log_{10} a_{OH^-}$.

---

† p designates $-\log_{10}$ of a number.

Equation 2 states that as pH increases pOH decreases correspondingly and vice versa because pK_w is constant for a given temperature. At 25°C, pH 7.0 is neutral, the activities of the hydrogen and hydroxyl ions are equal, and each corresponds to an approximate activity of 10^{-7} moles/L. The neutral point is temperature-dependent and is pH 7.5 at 0°C and pH 6.5 at 60°C.

The pH value of a highly dilute solution is approximately the same as the negative common logarithm of the hydrogen ion concentration. Natural waters usually have pH values in the range of 4 to 9, and most are slightly basic because of the presence of bicarbonates and carbonates of the alkali and alkaline earth metals.

2. Reference

1. SORENSON, S. 1909. Über die Messung und die Bedeutung der Wasserstoff ionen Konzentration bei Enzymatischen Prozessen. *Biochem. Z.* 21:131.

4500-H⁺ B.   Electrometric Method

1. General Discussion

a. Principle: The basic principle of electrometric pH measurement is determination of the activity of the hydrogen ions by potentiometric measurement using a standard hydrogen electrode and a reference electrode. The hydrogen electrode consists of a platinum electrode across which hydrogen gas is bubbled at a pressure of 101 kPa. Because of difficulty in its use and the potential for poisoning the hydrogen electrode, the glass electrode commonly is used. The electromotive force (emf) produced in the glass electrode system varies linearly with pH. This linear relationship is described by plotting the measured emf against the pH of different buffers. Sample pH is determined by extrapolation.

Because single ion activities such as a_H^+ cannot be measured, pH is defined operationally on a potentiometric scale. The pH measuring instrument is calibrated potentiometrically with an indicating (glass) electrode and a reference electrode using National Institute of Standards and Technology (NIST) buffers having assigned values so that:

$$pH_B = -\log_{10} a_{H^+}$$

where:

pH_B = assigned pH of NIST buffer.

The operational pH scale is used to measure sample pH and is defined as:

$$pH_x = pH_B \pm \frac{F(E_x - E_s)}{2.303\ RT}$$

where:

pH_x = potentiometrically measured sample pH,
F = Faraday: 9.649×10^4 coulomb/mole,
E_x = sample emf, V,
E_s = buffer emf, V,
R = gas constant; 8.314 joule/(mole °K), and
T = absolute temperature, °K.

NOTE: Although the equation for pH_x appears in the literature with a plus sign, the sign of emf readings in millivolts for most pH meters manufactured in the U.S. is negative. The choice of negative sign is consistent with the IUPAC Stockholm convention concerning the sign of electrode potential.[1,2]

The activity scale gives values that are higher than those on Sorenson's scale by 0.04 units:

$$pH\ (activity) = pH\ (Sorenson) + 0.04$$

The equation for pH_x assumes that the emf of the cells containing the sample and buffer is due solely to hydrogen ion activity unaffected by sample composition. In practice, samples will have varying ionic species and ionic strengths, both affecting H^+ activity. This imposes an experimental limitation on pH measurement; thus, to obtain meaningful results, the differences between E_x and

E_s should be minimal. Samples must be dilute aqueous solutions of simple solutes ($<0.2M$). (Choose buffers to bracket the sample.) Determination of pH cannot be made accurately in nonaqueous media, suspensions, colloids, or high-ionic-strength solutions.

b. Interferences: The glass electrode is relatively free from interference from color, turbidity, colloidal matter, oxidants, reductants, or high salinity, except for a sodium error at pH > 10. Reduce this error by using special "low sodium error" electrodes.

pH measurements are affected by temperature in two ways: mechanical effects that are caused by changes in the properties of the electrodes and chemical effects caused by equilibrium changes. In the first instance, the Nernstian slope increases with increasing temperature and electrodes take time to achieve thermal equilibrium. This can cause long-term drift in pH. Because chemical equilibrium affects pH, standard pH buffers have a specified pH at indicated temperatures.

Always report temperature at which pH is measured.

2. Apparatus

a. pH meter consisting of potentiometer, a glass electrode, a reference electrode, and a temperature-compensating device. A circuit is completed through the potentiometer when the electrodes are immersed in the test solution. Many pH meters are capable of reading pH or millivolts and some have scale expansion that permits reading to 0.001 pH unit, but most instruments are not that precise.

For routine work use a pH meter accurate and reproducible to 0.1 pH unit with a range of 0 to 14 and equipped with a temperature-compensation adjustment.

Although manufacturers provide operating instructions, the use of different descriptive terms may be confusing. For most instruments, there are two controls: intercept (set buffer, asymmetry, standardize) and slope (temperature, offset); their functions are shown diagramatically in Figures 4500-H$^+$:1 and 2. The intercept control shifts the response curve laterally to pass through the isopotential point with no change in slope. This permits bringing the instrument on scale (0 mV) with a pH 7 buffer that has no change in potential with temperature.

The slope control rotates the emf/pH slope about the isopotential point (0 mV/pH 7). To adjust slope for temperature without disturbing the intercept, select a buffer that brackets the sample with pH 7 buffer and adjust slope control to pH of this buffer. The instrument will indicate correct millivolt change per unit pH at the test temperature.

b. Reference electrode consisting of a half cell that provides a constant electrode potential. Commonly used are calomel and silver: silver-chloride electrodes. Either is available with several types of liquid junctions.

The liquid junction of the reference electrode is critical because at this point the electrode forms a salt bridge with the sample or buffer and a liquid junction potential is generated that in turn affects the potential produced by the reference electrode. Reference electrode junctions may be annular ceramic, quartz, or asbestos fiber, or the sleeve type. The quartz type is most widely used. The asbestos fiber type is not recommended for strongly basic solutions. Follow the manufacturer's recommendation on use and care of the reference electrode.

Refill nonsealed electrodes with the correct electrolyte to proper level and make sure junction is properly wetted.

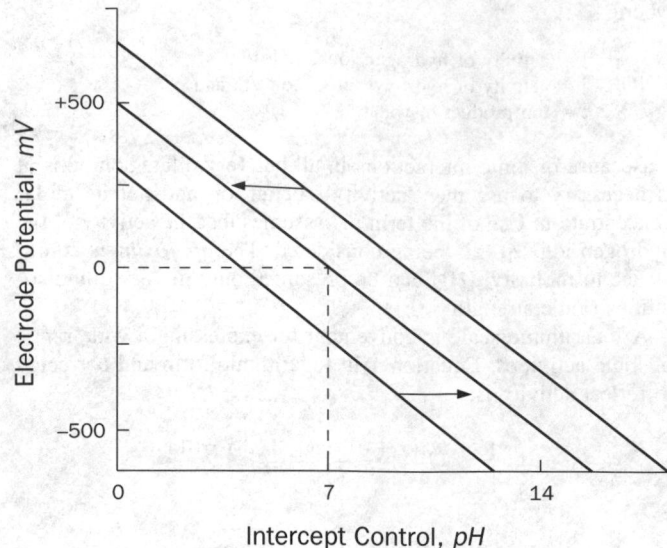

Figure 4500-H$^+$:1. Electrode potential vs. pH. Intercept control shifts response curve laterally.

c. Glass electrode: The sensor electrode is a bulb of special glass containing a fixed concentration of HCl or a buffered chloride solution in contact with an internal reference electrode. Upon immersion of a new electrode in a solution the outer bulb surface becomes hydrated and exchanges sodium ions for hydrogen ions to build up a surface layer of hydrogen ions. This, together with the repulsion of anions by fixed, negatively charged silicate sites, produces at the glass-solution interface a potential that is a function of hydrogen ion activity in solution.

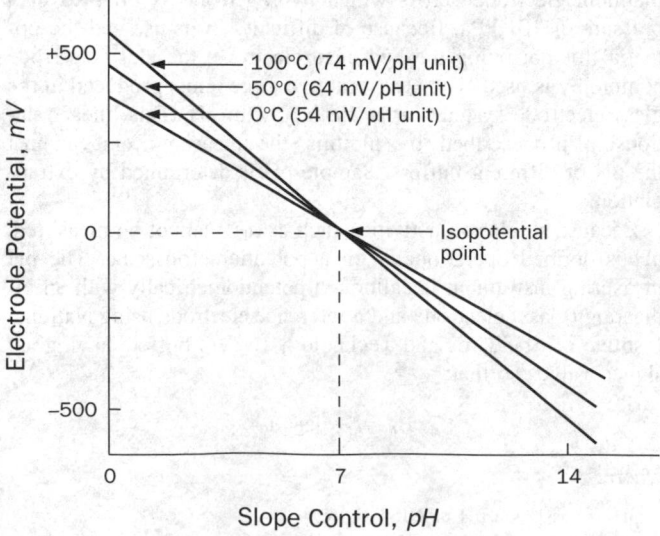

Figure 4500-H$^+$:2. Typical pH electrode response as a function of temperature.

Several types of glass electrodes are available. Combination electrodes incorporate the glass and reference electrodes into a single probe. Use a "low sodium error" electrode that can operate at high temperatures for measuring pH over 10 because standard glass electrodes yield erroneously low values. For measuring pH below 1 standard glass electrodes yield erroneously high values; use liquid membrane electrodes instead.

d. Beakers: Preferably use polyethylene or TFE* beakers.

e. Stirrer: Use either a magnetic, TFE-coated stirring bar or a mechanical stirrer with inert plastic-coated impeller.

f. Flow chamber: Use for continuous flow measurements or for poorly buffered solutions.

3. Reagents

a. General preparation: Calibrate the electrode system against standard buffer solutions of known pH. Because buffer solutions may deteriorate as a result of mold growth or contamination, prepare fresh as needed for accurate work by weighing the amounts of chemicals specified in Table 4500-H⁺:I, dissolving in distilled water at 25°C, and diluting to 1000 mL. This is particularly important for borate and carbonate buffers.

Boil and cool distilled water having a conductivity of less than 2 μmhos/cm. To 50 mL add 1 drop of saturated KCl solution suitable for reference electrode use. If the pH of this test solution is between 6.0 and 7.0, use it to prepare all standard solutions.

Dry KH_2PO_4 at 110 to 130°C for 2 h before weighing but do not heat unstable hydrated potassium tetroxalate above 60°C nor dry the other specified buffer salts.

Although ACS-grade chemicals generally are satisfactory for preparing buffer solutions, use certified materials available from the National Institute of Standards and Technology when the greatest accuracy is required. For routine analysis, use commercially available buffer tablets, powders, or solutions of tested quality. In preparing buffer solutions from solid salts, insure complete solution.

_____
* Teflon or equivalent.

As a rule, select and prepare buffer solutions classed as primary standards in Table 4500-H⁺:I; reserve secondary standards for extreme situations encountered in wastewater measurements. Consult Table 4500- H⁺:II for accepted pH of standard buffer solutions at temperatures other than 25°C. In routine use, store buffer solutions and samples in polyethylene bottles. Replace buffer solutions every 4 weeks.

b. Saturated potassium hydrogen tartrate solution: Shake vigorously an excess (5 to 10 g) of finely crystalline $KHC_4H_4O_6$ with 100 to 300 mL distilled water at 25°C in a glass-stoppered bottle. Separate clear solution from undissolved material by decantation or filtration. Preserve for 2 months or more by adding one thymol crystal (8 mm diam) per 200 mL solution.

c. Saturated calcium hydroxide solution: Calcine a well-washed, low-alkali grade $CaCO_3$ in a platinum dish by igniting for 1 h at 1000°C. Cool, hydrate by slowly adding distilled water with stirring, and heat to boiling. Cool, filter, and collect solid $Ca(OH)_2$ on a fritted glass filter of medium porosity. Dry at 110°C, cool, and pulverize to uniformly fine granules. Vigorously shake an excess of fine granules with distilled water in a stoppered polyethylene bottle. Let temperature come to 25°C after mixing. Filter supernatant under suction through a sintered glass filter of medium porosity and use filtrate as the buffer solution. Discard buffer solution when atmospheric CO_2 causes turbidity to appear.

d. Auxiliary solutions: 0.1N NaOH, 0.1N HCl, 5N HCl (dilute five volumes 6N HCl with one volume distilled water), and acid potassium fluoride solution (dissolve 2 g KF in 2 mL conc H_2SO_4 and dilute to 100 mL with distilled water).

4. Procedure

a. Instrument calibration: In each case follow manufacturer's instructions for pH meter and for storage and preparation of electrodes for use. Recommended solutions for short-term storage of electrodes vary with type of electrode and manufacturer, but generally have a conductivity greater than 4000 μmhos/cm. Tap water is a better substitute than distilled water, but pH 4 buffer

TABLE 4500-H⁺:I. PREPARATION OF pH STANDARD SOLUTIONS[3]

Standard Solution (molality)	pH at 25°C	Weight of Chemicals Needed/1000 mL Aqueous Solution at 25°C
Primary standards:		
Potassium hydrogen tartrate (saturated at 25°C)	3.557	> 7 g $KHC_4H_4O_6$*
0.05 potassium dihydrogen citrate	3.776	11.41 g $KH_2C_6H_5O_7$
0.05 potassium hydrogen phthalate	4.004	10.12 g $KHC_8H_4O_4$
0.025 potassium dihydrogen phosphate + 0.025 disodium hydrogen phosphate	6.863	3.387 g KH_2PO_4 + 3.533 g Na_2HPO_4†
0.008 695 potassium dihydrogen phosphate + 0.030 43 disodium hydrogen phosphate	7.415	1.179 g KH_2PO_4 + 4.303 g Na_2HPO_4†
0.01 sodium borate decahydrate (borax)	9.183	3.80 g $Na_2B_4O_7 \cdot 10H_2O$†
0.025 sodium bicarbonate + 0.025 sodium carbonate	10.014	2.092 g $NaHCO_3$ + 2.640 g Na_2CO_3
Secondary standards:		
0.05 potassium tetroxalate dihydrate	1.679	12.61 g $KH_3C_4O_8 \cdot 2H_2O$
Calcium hydroxide (saturated at 25°C)	12.454	> 2 g $Ca(OH)_2$*

* Approximate solubility.
† Prepare with freshly boiled and cooled distilled water (carbon-dioxide-free).

TABLE 4500-H⁺:II. STANDARD pH VALUES[3]

Temperature °C	Primary Standards							Secondary Standards	
	Tartrate (Saturated)	Citrate (0.05M)	Phthalate (0.05M)	Phosphate (1:1)	Phosphate (1:3.5)	Borax (0.01M)	Bicarbonate-Carbonate (0.025M)	Tetroxalate (0.05M)	Calcium Hydroxide (Saturated)
0			4.003	6.982	7.534	9.460	10.321	1.666	
5			3.998	6.949	7.501	9.392	10.248	1.668	
10			3.996	6.921	7.472	9.331	10.181	1.670	
15			3.996	6.898	7.449	9.276	10.120	1.672	
20			3.999	6.878	7.430	9.227	10.064	1.675	
25	3.557	3.776	4.004	6.863	7.415	9.183	10.014	1.679	12.454
30	3.552		4.011	6.851	7.403	9.143	9.968	1.683	
35	3.549		4.020	6.842	7.394	9.107	9.928	1.688	
37			4.024	6.839	7.392	9.093			
40	3.547		4.030	6.836	7.388	9.074	9.891	1.694	
45	3.547		4.042	6.832	7.385	9.044	9.859	1.700	
50	3.549		4.055	6.831	7.384	9.017	9.831	1.707	
55	3.554		4.070					1.715	
60	3.560		4.085					1.723	
70	3.580		4.12					1.743	
80	3.609		4.16					1.766	
90	3.650		4.19					1.792	
85	3.674		4.21					1.806	

is best for the single glass electrode and saturated KCl is preferred for a calomel and Ag/AgCl reference electrode. Saturated KCl is the preferred solution for a combination electrode. Keep electrodes wet by returning them to storage solution whenever pH meter is not in use.

Before use, remove electrodes from storage solution, rinse, blot dry with a soft tissue, place in initial buffer solution, and set the isopotential point (¶ 2a above). Select a second buffer within 2 pH units of sample pH and bring sample and buffer to same temperature, which may be the room temperature, a fixed temperature such as 25°C, or the temperature of a fresh sample. Remove electrodes from first buffer, rinse thoroughly with distilled water, blot dry, and immerse in second buffer. Record temperature of measurement and adjust temperature dial on meter so that meter indicates pH value of buffer at test temperature (this is a slope adjustment).

Use the pH value listed in the tables for the buffer used at the test temperature. Remove electrodes from second buffer, rinse thoroughly with distilled water and dry electrodes as indicated above. Immerse in a third buffer below pH 10, approximately 3 pH units different from the second; the reading should be within 0.1 unit for the pH of the third buffer. If the meter response shows a difference greater than 0.1 pH unit from expected value, look for trouble with the electrodes or potentiometer (see ¶s 5a and b below).

The purpose of standardization is to adjust the response of the glass electrode to the instrument. When only occasional pH measurements are made standardize instrument before each measurement. When frequent measurements are made and the instrument is stable, standardize less frequently. If sample pH values vary widely, standardize for each sample with a buffer having a pH within 1 to 2 pH units of the sample.

b. Sample analysis: Establish equilibrium between electrodes and sample by stirring sample to insure homogeneity; stir gently to minimize carbon dioxide entrainment. For buffered samples or those of high ionic strength, condition electrodes after cleaning by dipping them into sample for 1 min. Blot dry, immerse in a fresh portion of the same sample, and read pH.

With dilute, poorly buffered solutions, equilibrate electrodes by immersing in three or four successive portions of sample. Take a fresh sample to measure pH.

5. Trouble Shooting

a. Potentiometer: To locate trouble source disconnect electrodes and, using a short-circuit strap, connect reference electrode terminal to glass electrode terminal. Observe change in pH when instrument calibration knob is adjusted. If potentiometer is operating properly, it will respond rapidly and evenly to changes in calibration over a wide scale range. A faulty potentiometer will fail to respond, will react erratically, or will show a drift upon adjustment. Switch to the millivolt scale on which the meter should read zero. If inexperienced, do not attempt potentiometer repair other than maintenance as described in instrument manual.

b. Electrodes: If potentiometer is functioning properly, look for the instrument fault in the electrode pair. Substitute one electrode at a time and cross-check with two buffers that are about 4 pH units apart. A deviation greater than 0.1 pH unit indicates a faulty electrode. Glass electrodes fail because of scratches, deterioration, or accumulation of debris on the glass surface. Rejuvenate electrode by alternately immersing it three times each in 0.1N HCl and 0.1N NaOH. If this fails, immerse tip in KF solution for 30 s. After rejuvenation, soak in pH 7.0 buffer overnight. Rinse and store in pH 7.0 buffer. Rinse again with distilled water before use. Protein coatings can be removed by soaking glass electrodes in a 10% pepsin solution adjusted to pH 1 to 2.

To check reference electrode, oppose the emf of a questionable reference electrode against another one of the same type that is known to be good. Using an adapter, plug good reference electrode into glass electrode jack of potentiometer; then plug questioned electrode into reference electrode jack. Set meter to read millivolts and take readings with both electrodes immersed in the same electrolyte (KCl) solution and then in the same buffer solution. The millivolt readings should be 0 ± 5 mV for both solutions. If different electrodes are used, i.e., silver: silver-chloride against calomel or vice versa, the reading will be 44 ± 5 mV for a good reference electrode.

Reference electrode troubles generally are traceable to a clogged junction. Interruption of the continuous trickle of electrolyte through the junction causes increase in response time and drift in reading. Clear a clogged junction by applying suction to the tip or by boiling tip in distilled water until the electrolyte flows freely when suction is applied to tip or pressure is applied to the fill hole. Replaceable junctions are available commercially.

6. Precision and Bias

By careful use of a laboratory pH meter with good electrodes, a precision of ± 0.02 pH unit and an accuracy of ± 0.05 pH unit can be achieved. However, ± 0.1 pH unit represents the limit of accuracy under normal conditions, especially for measurement of water and poorly buffered solutions. For this reason, report pH values to the nearest 0.1 pH unit. A synthetic sample of a Clark and Lubs buffer solution of pH 7.3 was analyzed electrometrically by 30 laboratories with a standard deviation of ± 0.13 pH unit.

7. References

1. BATES, R.G. 1978. Concept and determination of pH. *In* I.M. Kolthoff & P.J. Elving, eds. Treatise on Analytical Chemistry. Part 1, Vol. 1, p. 821. Wiley-Interscience, New York, N. Y.
2. LICHT, T.S. & A.J. DE BETHUNE. 1957. Recent developments concerning the signs of electrode potentials. *J. Chem. Educ.* 34:433.
3. DURST, R.A. 1975. Standard Reference Materials: Standardization of pH Measurements. NBS Spec. Publ. 260-53, National Bur. Standards, Washington, D.C.

8. Bibliography

CLARK, W.M. 1928. The Determination of Hydrogen Ions, 3rd ed. Williams & Wilkins Co., Baltimore, Md.
DOLE, M. 1941. The Glass Electrode. John Wiley & Sons, New York, N.Y.
BATES, R.G. & S.F. ACREE. 1945. pH of aqueous mixtures of potassium dihydrogen phosphate and disodium hydrogen phosphate at 0 to 60°C. *J. Res. Nat. Bur. Standards* 34:373.
LANGELIER, W.F. 1946. Effect of temperature on the pH of natural water. *J. Amer. Water Works Assoc.* 38:179.
FELDMAN, I. 1956. Use and abuse of pH measurements. *Anal. Chem.* 28:1859.
BRITTON, H.T.S. 1956. Hydrogen Ions, 4th ed. D. Van Nostrand Co., Princeton, N.J.
KOLTHOFF, I.M. & H.A. LAITINEN. 1958. pH and Electrotitrations. John Wiley & Sons, New York, N.Y.
KOLTHOFF, I.M. & P.J. ELVING. 1959. Treatise on Analytical Chemistry. Part I, Vol. 1, Chapter 10. Wiley-Interscience, New York, N.Y.
BATES, R.G. 1962. Revised standard values for pH measurements from 0 to 95°C. *J. Res. Nat. Bur. Standards* 66A:179.
AMERICAN WATER WORKS ASSOCIATION. 1964. Simplified Procedures for Water Examination. Manual M12, American Water Works Assoc., New York, N.Y.
WINSTEAD, M. 1967. Reagent Grade Water: How, When and Why? American Soc. Medical Technologists, The Steck Company, Austin, Tex.
STAPLES, B.R. & R.G. BATES. 1969. Two new standards for the pH scale. *J. Res. Nat. Bur. Standards* 73A:37.
BATES, R.G. 1973. Determination of pH, Theory and Practice, 2nd ed. John Wiley & Sons, New York, N.Y.

4500-I IODINE*

4500-I A. Introduction

1. Uses and Forms

Elemental iodine is not a natural constituent of natural waters. Iodine may be added to potable and swimming pool waters as a disinfectant. For wastewaters, iodine has had limited application. Use of iodine generally is restricted to personal or remote water supplies where ease of application, storage stability, and an inertness toward organic matter are important considerations. Some swimming pool waters are treated with iodine to lessen eye burn among swimmers and to provide a stable disinfectant residual less affected by adverse environmental conditions.

Iodine is applied in the elemental form or produced in situ by the simultaneous addition of an iodide salt and a suitable oxidant. In the latter case, an excess of iodide may be maintained to serve

* Approved by Standard Methods Committee, 1997.

as a reservoir for iodine production; the determination of iodide is desirable for disinfectant control (see Iodide, Section 4500-I⁻).

Elemental I_2 can undergo hydrolysis to form hypoiodous acid (HOI), which can dissociate to form hypoiodite (OI⁻) under strongly basic conditions. Hypoiodous acid/hypoiodite ion may further disproportionate to form iodate. In the presence of excess iodide, iodine may react with iodide to form tri-iodide ion (I_3^-). The rate and the extent to which these reactions may occur depend on pH and the concentration of iodide in the solution. Basic conditions favor formation of hypoiodite and iodate. Acidic conditions and the presence of iodide favor formation of iodine and tri-iodide ion. Thus, the relative concentrations of these iodine species in the resulting solution can be quite variable. Hypoiodous acid/hypoiodite also can act as an iodinating agent, reacting with organic compounds to form iodinated organic compounds. Elemental I_2, hypoiodous acid, hypoiodite ion, and tri-iodide ion are considered active iodine. There is no generally accepted method for the determination of each of these species individually. Most analytical methods use the oxidizing power of all forms of active iodine for its determination and the results usually are expressed as an equivalent concentration of elemental iodine. The effects of iodate or dissolved organic iodine on these methods have not been thoroughly investigated.

2. Selection of Method

For potable and swimming pool waters treated with elemental iodine, both the amperometric titration and leuco crystal violet colorimetric methods give acceptable results. However, oxidized forms of manganese interfere with the leuco crystal violet method.

Where the iodide and chloride ion concentrations are above 50 mg/L and 200 mg/L, respectively, interference in color production may occur in the leuco crystal violet method and the amperometric method is preferred. However, because of the extreme sensitivity of the leuco crystal violet method, this interference may be eliminated by sample dilution to obtain halogen ion concentrations less than 50 mg/L.

For wastewaters or highly polluted waters, organic constituents normally do not interfere with either the amperometric or leuco crystal violet procedures. Determine which of the methods yields the more acceptable results, because specific substances present in these waters may interfere in one method but not in the other. Certain metallic cations such as copper and silver interfere in the amperometric titration procedure. Iodate, which is a naturally occurring species of iodine in marine waters, will also interfere in the amperometric titration by reacting with excess iodide under acidic conditions to form I_2 and/or I_3^-. The rate of the reaction is most pH-dependent between pH 3 and 5. Thus, the magnitude of this interference may depend on the concentration of iodate present and the analytical conditions. The leuco crystal violet method is relatively free of interference from these and other cations and anions with the exceptions noted previously.

For waters containing iodine coexisting with free chlorine, combined chlorine, or other excess oxidants, of the methods described only the leuco crystal violet method can determine iodine specifically. This condition occurs in the in-situ production of iodine by the reaction of iodide and excess oxidant. Under these conditions, the amperometric method would continue to titrate the iodine produced in a cyclic reaction until exhaustion of the oxidant.

4500-I B. Leuco Crystal Violet Method

1. General Discussion

The leuco crystal violet method determines aqueous iodine present as elemental iodine and hypoiodous acid. Excess common oxidants do not interfere. While the method utilizes the sum of the oxidative power of all forms of active iodine residuals, the results are expressed as the equivalent concentration of iodine. The method also is capable of determining the sum of iodine and free iodide concentrations; the free iodide concentration can be determined by difference (see Iodide, Section 4500-I⁻).

a. Principle: Mercuric chloride added to aqueous elemental iodine solutions causes essentially complete hydrolysis of iodine and the stoichiometric production of hypoiodous acid. The compound 4,4′,4″-methylidynetris (N,N-dimethylaniline), also known by the common name of leuco crystal violet, reacts instantaneously with the hypoiodous acid to form crystal violet dye. The absorbance of this dye is highly pH-dependent. The maximum absorbance is produced in the pH range of 3.5 to 4.0 and is measured at a wavelength of 592 nm. Below a pH of 3.5, the absorbance drops precipitously. Above a pH of about 4.7, the excess leuco crystal violet in the sample precipitates and masks the absorbance of the crystal violet dye. Accurate pH control is essential to maximize precision. The absorbance follows Beer's law over a wide range of iodine concentrations and the developed color is stable for several hours.

In the presence of certain excess oxidants such as free chlorine or chloramines, the iodine residual will exist exclusively in the form of hypoiodous acid. The leuco crystal violet is relatively insensitive to the combined forms of chlorine while any free chlorine is converted to chloramine by reaction with an ammonium salt incorporated in the test reagents. All the hypoiodous acid is determined. As hypoiodous acid, the weight concentration value found, expressed as an equivalent elemental I_2 concentration, is equal to twice that of an elemental I_2 solution of the same weight concentration.

b. Interference: Oxidized forms of manganese interfere by oxidizing the indicator to crystal violet dye and yield apparent high iodine concentrations.

Iodide and chloride ion concentrations above 50 mg/L and 200 mg/L, respectively, interfere by inhibiting full color production. Dilute the sample to eliminate this interference.

Combined chlorine residuals normally do not interfere provided that the test is completed within 5 min after adding the indicator solution. Eliminate interference from free chlorine by adding an ammonium salt buffer to form combined chlorine.

c. Minimum detectable concentration: 10 μg I as I_2/L.

2. Apparatus

a. Colorimetric equipment: One of the following is required:
1) *Filter photometer,* with a light path of 1 cm or longer,

equipped with an orange filter having maximum transmittance near 592 nm.

2) *Spectrophotometer,* for use at 592 nm, with a light path of 1 cm or longer.

b. Volumetric flasks, 100-mL, with plastic caps or ground-glass stoppers.

c. Glassware: Completely remove reducing substances from glassware or plastic containers, including containers for storage of reagent solutions (see Section 4500-Cl.D.2*d*).

3. Reagents

a. Iodine-demand-free water: See Section 4500-I⁻.B.3*a*. Prepare all stock iodine and reagent solutions with iodine-demand-free water.

b. Stock iodine solution: Prepare a saturated iodine solution by dissolving 20 g elemental iodine in 300 mL water. Let stand several hours. Decant iodine solution and dilute 170 mL to 2000 mL. Standardize solution by titrating with standard sodium thiosulfate ($Na_2S_2O_3$) titrant as described in Section 4500-Cl.B.3*b* and *c* or amperometrically as in Section 4500-I.C.

Calculate iodine concentration:

$$\text{mg I as } I_2/\text{mL} = \text{normality of iodine solution} \times 126.9$$

Prepare a working solution of 10 µg I as I_2/mL by appropriate dilution of the standardized stock solution.

c. Citric buffer solution, pH 3.8: See Section 4500-I⁻.B.3*c*.

d. Leuco crystal violet indicator: See Section 4500-I⁻.B.3*d*.

e. Sodium thiosulfate solution: See Section 4500-I⁻.B.3*f*.

4. Procedure

a. Preparation of temporary iodine standards: For greater accuracy, standardize working solution immediately before use by the amperometric titration method (Method C). Prepare standards in the range of 0.1 to 6.0 mg I as I_2/L by adding 1 to 60 mL working solution to 100 mL glass-stoppered volumetric flasks, in increments of 1 mL or larger. Adjust these volumes if the measured iodine concentration of working solution varies by 5% or more from 10 µg I as I_2/mL.

Measure 50.0 mL of each diluted iodine working solution into a 100-mL glass-stoppered volumetric flask. Add 1.0 mL citric buffer solution, gently swirl to mix, and let stand for at least 30 s. Add 1.0 mL leuco crystal violet indicator and swirl to develop color. Dilute to 100 mL and mix.

b. Photometric calibration: Transfer colored temporary standards of known iodine concentrations to cells of 1-cm light path and read absorbance in a photometer or spectrophotometer at a wavelength of 592 nm against a distilled water reference. Plot absorbance values against iodine concentrations to construct a curve that follows Beer's law.

c. Color development of iodine sample: Measure 50.0 mL sample into a 100-mL volumetric flask and treat as described for preparation of temporary iodine standards, ¶ 4*a*. Match test sample visually with temporary standards or read absorbance photometrically and refer to standard calibration curve for the iodine equivalent.

d. Samples containing >6.0 mg I as I_2/L: Place approximately 25 mL water in a 100-mL volumetric flask. Add 1.0 mL citric buffer solution and a measured volume of 25 mL or less of sample. Mix and let stand for at least 30 s. Add 1.0 mL leuco crystal violet indicator, mix, and dilute to mark. Match visually with standards or read absorbance photometrically and compare with calibration curve from which the initial iodine is obtained by applying the dilution factor. Select one of the following sample volumes to remain within optimum iodine range:

Iodine mg/L	Sample Volume Required mL
6.0–12.0	25.0
12.0–30	10.0
30–60	5.0

e. Samples containing both chlorine and iodine: For samples containing free or combined chlorine and iodine, follow procedure given in ¶ 4*c* or *d* above but read absorbance within 5 min after adding leuco crystal violet indicator.

f. Compensation for turbidity and color: Compensate for natural color or turbidity by adding 5 mL $Na_2S_2O_3$ solution to a 50-mL sample. Add reagents to sample as described previously and use as blank to set zero absorbance on the photometer. Measure all samples in relation to this blank and, from calibration curve, determine concentrations of iodine.

5. Bibliography

BLACK, A.P. & G.P. WHITTLE. 1967. New methods for the colorimetric determination of halogen residuals. Part I. Iodine, iodide, and iodate. *J. Amer. Water Works Assoc.* 59:471.

4500-I C. Amperometric Titration Method

1. General Discussion

The amperometric titration method for iodine is a modification of the amperometric method for residual chlorine (see Section 4500-Cl.D). Iodine residuals over 7 mg/L are best measured with smaller samples or by dilution. In most cases the titration results represent free iodine because combined iodine rarely is encountered.

a. Principle: The principle of the amperometric method as described for the determination of total residual chlorine is applicable to the determination of residual iodine. Iodine is determined using buffer solution, pH 4.0, and potassium iodide (KI) solution. Maintain pH at 4.0 because at pH values less than 3.5 substances such as oxidized forms of manganese interfere, while at pH values greater than 4.5, the reaction is not quantitative. Adding KI improves the sharpness of the end point.

b. Interference: Free chlorine and the interferences described in Section 4500-Cl.D.1*b* also interfere in the iodine determination.

2. Apparatus

See Section 4500-Cl.D.2*a* through *d*.

3. Reagents

With the exception of phosphate buffer solution, pH 7.0, all reagents listed for the determination of residual chlorine in Section 4500-Cl.D.3 are required. Standardized phenylarsine oxide solution (1 mL = 1 mg chlorine/L for a 200-mL sample) is equivalent to 3.58 mg I as I_2/mL for a 200-mL sample.

4. Procedure

a. Sample volume: Select a sample volume that will require no more than 2 mL phenylarsine oxide titrant. For iodine concentrations of 7 mg/L or less, take a 200-mL volume; for iodine levels above 7 mg/L, use 100 mL or proportionately less diluted to 200 mL with water.

b. Free iodine: To the sample add 1 mL KI solution and 1 mL acetate buffer, pH 4.0 solution. Titrate with phenylarsine oxide titrant to the end point described in Section 4500-Cl.D.4.

5. Calculation

Calculate the iodine concentration by the following equation:

$$\text{mg I as } I_2/L = \frac{A \times 3.58 \times 200}{\text{mL sample}}$$

where:

A = mL phenylarsine oxide titration to the end point.

6. Bibliography

MARKS, H.C. & J.R. GLASS. 1942. A new method of determining residual chlorine. *J. Amer. Water Works Assoc.* 34:1227.

4500-I⁻   IODIDE*

4500-I⁻   A.   Introduction

1. Occurrence

Iodide is found in natural waters at concentrations ranging from 40 µg I⁻/L in coastal surface seawater to <1 µg I⁻/L in deep ocean water and fresh water. Higher concentrations may be found in brines, certain industrial wastes, and waters treated with iodine. Iodide is thermodynamically unstable relative to iodate in oxygenated waters.

2. Selection of Method

The leuco crystal violet method (B) is applicable to iodide concentrations of 50 to 6000 µg/L. The catalytic reduction method (C) is applicable to iodide concentrations of 80 µg I⁻/L or less. The voltammetric method (D) is the most sensitive method. It can be used for samples with iodide concentrations of 0.13 to 10.2 µg I⁻/L. It is also species-specific. It is insensitive to iodate, iodine, and most organic iodine compounds. It requires minimal sample manipulation, aside from an occasional dilution for samples with high concentrations of iodide. Thus, the concentrations of iodide in many types of water samples may be determined directly with the voltammetric method.

The choice of method depends on the sample and concentration to be determined. The high chloride concentrations of brines, seawater, and many estuarine waters will interfere with color development in the leuco crystal violet method. In the presence of iodine, the leuco crystal violet method gives the sum of iodine and iodide. Iodide may be determined by the difference after concentration of iodine has been estimated independently (see Section 4500-I). In the catalytic reduction method, As(III), under acidic conditions, is a strong reducing agent and will reduce the oxidized forms of iodine to iodide. Thus, this method measures not only iodide, but also the sum of all the inorganic iodine species including iodide, iodate, hypoiodous acid, hypoiodite ion, and elemental iodine. Because iodate is the thermodynamically stable form of dissolved iodine in oxygenated natural waters and is frequently the dominant species of dissolved iodine, the catalytic reduction method is likely to overestimate the concentration of iodide. This method works well only under exactly reproducible conditions.

* Approved by Standard Methods Committee, 1997.

4500-I$^-$ B. Leuco Crystal Violet Method

1. General Discussion

a. Principle: Iodide is selectively oxidized to iodine by the addition of potassium peroxymonosulfate, $KHSO_5$. The iodine produced reacts instantaneously with the colorless indicator reagent containing 4,4′,4″-methylidynetris (*N,N*-dimethylaniline), also known as leuco crystal violet, to produce the highly colored crystal violet dye. The developed color is sufficiently stable for the determination of an absorbance value and adheres to Beer's law over a wide range of iodine concentrations. Absorbance is highly pH-dependent, and must be measured within the pH range of 3.5 to 4.0 at a wavelength of 592 nm. Accurate control of pH is essential for maximum precision. (See Section 4500-I.B.1*a*.) Follow the general principles for quality control (Section 4020).

b. Interference: Chloride concentrations greater than 200 mg/L may interfere with color development. Reduce these interferences by diluting sample to contain less than 200 mg Cl$^-$/L.

2. Apparatus

a. Colorimetric equipment: One of the following is required:

1) *Filter photometer*, providing a light path of 1 cm or longer, equipped with an orange filter having maximum transmittance near 592 nm.

2) *Spectrophotometer*, for use at 592 nm, providing a light path of 1 cm or longer.

b. Volumetric flasks: 100-mL with plastic caps or ground-glass stoppers.

c. Glassware: Completely remove any reducing substances from all glassware or plastic containers, including containers for storing reagent solutions (see Section 4500-Cl.D.2*d*).

3. Reagents

a. Iodine-demand-free water: Prepare a 1-m ion-exchange column of 2.5 to 5 cm diam, containing strongly acid cation and strong basic anion exchange resins. If a commercial analytical-grade mixed-bed resin is used, verify that compounds that react with iodine are removed. Pass distilled water at a slow rate through the resin bed and collect in clean container that will protect the treated water from undue exposure to the atmosphere.

Prepare all stock iodide and reagent solutions with iodine-demand-free water.

b. Stock iodide solution: Dissolve 1.3081 g KI in water and dilute to 1000 mL; 1 mL = 1 mg I$^-$.

c. Citric buffer solution, pH 3.8:

1) *Citric acid:* Dissolve 192.2 g $C_6H_8O_7$ or 210.2 g $C_6H_8O_7 \cdot H_2O$ and dilute to 1 L with water.

2) *Ammonium hydroxide*, 2*N*: Add 131 mL conc NH_4OH to about 700 mL water and dilute to 1 L. Store in a polyethylene bottle.

3) *Final buffer solution*: Slowly add, with mixing, 350 mL 2*N* NH_4OH solution to 670 mL citric acid. Add 80 g ammonium dihydrogen phosphate ($NH_4H_2PO_4$) and stir to dissolve.

d. Leuco crystal violet indicator: Measure 200 mL water and 3.2 mL conc sulfuric acid (H_2SO_4) into a brown glass container of at least 1-L capacity. Introduce a magnetic stirring bar and mix at moderate speed. Add 1.5 g 4,4′,4″-methylidynetris (*N,N*-di-methylaniline)* and with a small amount of water wash down any reagent adhering to neck or sides of container. Mix until dissolved.

To 800 mL water, add 2.5 g mercuric chloride ($HgCl_2$) and stir to dissolve. With mixing, add $HgCl_2$ solution to leuco crystal violet solution. For maximum stability, adjust pH of final solution to 1.5 or less, adding, if necessary, conc H_2SO_4 dropwise. Store in a brown glass bottle away from direct sunlight. Discard after 6 months. Do not use a rubber stopper.

e. Potassium peroxymonosulfate solution: Obtain $KHSO_5$ as a commercial product,† which is a stable powdered mixture containing 42.8% $KHSO_5$ by weight and a mixture of $KHSO_4$ and K_2SO_4. Dissolve 1.5 g powder in water and dilute to 1 L.

f. Sodium thiosulfate solution: Dissolve 5.0 g $Na_2S_2O_3 \cdot H_2O$ in water and dilute to 1 L.

4. Procedure

a. Preparation of temporary iodine standards: Add suitable portions of stock iodide solution, or of dilutions of stock iodide solution, to water to prepare a series of 0.1 to 6.0 mg I$^-$/L in increments of 0.1 mg/L or larger.

Measure 50.0 mL dilute KI standard solution into a 100-mL glass-stoppered volumetric flask. Add 1.0 mL citric buffer and 0.5 mL $KHSO_5$ solution. Swirl to mix and let stand approximately 1 min. Add 1.0 mL leuco crystal violet indicator, mix, and dilute to 100 mL. For best results, read absorbance as described below within 5 min after adding indicator solution.

b. Photometric calibration: Transfer colored temporary standards of known iodide concentrations to cells of 1-cm light path and read absorbance in a photometer or spectrophotometer at a wavelength of 592 nm against a water reference. Plot absorbance values against iodide concentrations to construct a curve that follows Beer's law.

c. Color development of sample: Measure a 50.0-mL sample into a 100-mL volumetric flask and treat as described for preparation of temporary iodine standards, ¶ 4*a*. Read absorbance photometrically and refer to standard calibration curve for iodide equivalent.

d. Samples containing >6.0 mg I$^-$/L: Place approximately 25 mL water in a 100-mL volumetric flask. Add 1.0 mL citric buffer and a measured volume of 25 mL or less of sample. Add 0.5 mL $KHSO_5$ solution. Swirl to mix and let stand approximately 1 min. Add 1.0 mL leuco crystal violet indicator, mix, and dilute to 100 mL.

Read absorbance photometrically and compare with calibration curve from which the initial iodide concentration is obtained by applying the dilution factor. Select one of the following sample volumes to remain within the optimum iodide range.

Iodide mg/L	Sample Volume Required mL
6.0–12.0	25.0
12.0–30	10.0
30–60	5.0

---

* Eastman chemical No. 3651 or equivalent.
† Oxone, E.I. duPont de Nemours and Co., Inc., Wilmington, DE, or equivalent.

e. Determination of iodide in the presence of iodine: On separate samples determine (1) total iodide and iodine, and (2) iodine. The iodide concentration is the difference between the iodine determined and the total iodine-iodide obtained. Determine iodine by not adding KHSO$_5$ solution in the iodide method and by comparing the absorbance value to the calibration curve developed for iodide.

f. Compensation for turbidity and color: Compensate for natural color or turbidity by adding 5 mL Na$_2$S$_2$O$_3$ solution to a 50-mL sample. Add reagents to sample as described previously and use as the blank to set zero absorbance on photometer. Measure all samples in relation to this blank and, from the calibration curve, determine concentrations of iodide or total iodine-iodide.

5. Bibliography

BLACK, A.P. & G.P. WHITTLE. 1967. New methods for the colorimetric determination of halogen residuals. Part I. Iodine, iodide, and iodate. *J. Amer. Water Works Assoc.* 59:471.

4500-I⁻ C. Catalytic Reduction Method

1. General Discussion

a. Principle: Iodide can be determined by using its ability to catalyze the reduction of ceric ions by arsenious acid. The effect is nonlinearly proportional to the amount of iodide present. The reaction is stopped after a specific time interval by the addition of ferrous ammonium sulfate. The resulting ferric ions are directly proportional to the remaining ceric ions and develop a relatively stable color complex with potassium thiocyanate.

Pretreatment by digestion with chromic acid and distillation is necessary to estimate the nonsusceptible bound forms of iodine.

b. Interferences: The formation of noncatalytic forms of iodine and the inhibitory effects of silver and mercury are reduced by adding an excess of sodium chloride (NaCl) that sensitizes the reaction. Iodate, hypoiodous acid/hypoiodite ion, and elemental iodine interfere. Under acidic conditions, As(III) may reduce these forms of inorganic iodine to iodide and include them as iodide in the subsequent detection of iodide.

2. Apparatus

a. Water bath, capable of temperature control to 30 ± 0.5°C.

b. Colorimetric equipment: One of the following is required:

1) *Spectrophotometer*, for use at wavelengths of 510 or 525 nm and providing a light path of 1 cm.

2) *Filter photometer*, providing a light path of 1 cm and equipped with a green filter having maximum transmittance near 525 nm.

c. Test tubes, 2 × 15 cm.

d. Stopwatch.

3. Reagents

Store all stock solutions in tightly stoppered containers in the dark. Prepare all reagent solutions in distilled water.

a. Distilled water, containing less than 0.3 μg total I/L.

b. Sodium chloride solution: Dissolve 200.0 g NaCl in water and dilute to 1 L. Recrystallize the NaCl if an interfering amount of iodine is present, using a water-ethanol mixture.

c. Arsenious acid: Dissolve 4.946 g As$_2$O$_3$ in water, add 0.20 mL conc H$_2$SO$_4$, and dilute to 1000 mL.

d. Sulfuric acid, H$_2$SO$_4$, conc.

e. Ceric ammonium sulfate: Dissolve 13.38 g Ce(NH$_4$)$_4$(SO$_4$)$_4$·4H$_2$O in water, add 44 mL conc H$_2$SO$_4$, and make up to 1 L.

f. Ferrous ammonium sulfate reagent: Dissolve 1.50 g Fe(NH$_4$)$_2$(SO$_4$)$_2$·6H$_2$O in 100 mL distilled water containing 0.6 mL conc H$_2$SO$_4$. Prepare daily.

g. Potassium thiocyanate solution: Dissolve 4.00 g KSCN in 100 mL water.

h. Stock iodide solution: Dissolve 261.6 mg anhydrous KI in water and dilute to 1000 mL; 1.00 mL = 200 μg I⁻.

i. Intermediate iodide solution: Dilute 20.00 mL stock iodide solution to 1000 mL with water; 1.00 mL = 4.00 μg I⁻.

j. Standard iodide solution: Dilute 25.00 mL intermediate iodide solution to 1000 mL with water; 1.00 mL = 0.100 μg I⁻.

4. Procedure

a. Sample size: Add 10.00 mL sample, or a portion made up to 10.00 mL with water, to a 2- × 15-cm test tube. If possible, keep iodide content in the range 0.2 to 0.6 μg. Use thoroughly clean glassware and apparatus.

b. Color measurement: Add reagents in the following order: 1.00 mL NaCl solution, 0.50 mL As$_2$O$_3$ solution, and 0.50 mL conc H$_2$SO$_4$.

Place reaction mixture and ceric ammonium sulfate solution in 30°C water bath and let come to temperature equilibrium. Add 1.0 mL ceric ammonium sulfate solution, mix by inversion, and start stopwatch to time reaction. Use an inert clean test tube stopper when mixing. After 15 ± 0.1 min remove sample from water bath and add immediately 1.00 mL ferrous ammonium sulfate reagent with mixing, whereupon the yellow ceric ion color should disappear. Then add, with mixing, 1.00 mL KSCN solution. Replace sample in water bath. Within 1 h after adding thiocyanate read absorbance in a photometric instrument. Maintain temperature of solution and cell compartment at 30 ± 0.5°C until absorbance is determined. If several samples are run, start reactions at 1-min intervals to allow time for additions of ferrous ammonium sulfate and thiocyanate. (If temperature control of cell compartment is not possible, let final solution come to room temperature and measure absorbance with cell compartment at room temperature.)

c. Calibration standards: Treat standards containing 0, 0.2, 0.4, 0.6, and 0.8 μg I⁻/10.00 mL of solution as in ¶ 4b above. Run with each set of samples to establish a calibration curve.

5. Calculation

$$mg\ I^-/L = \frac{\mu g\ I\ (in\ 15\ mL\ final\ volume)}{mL\ sample}$$

6. Precision and Bias

Results obtained by this method are reproducible on samples of Los Angeles source waters, and have been reported to be ac-curate to ±0.3 μg I⁻/L on samples of Yugoslavian water con-taining from 0 to 14.0 μg I⁻/L. Follow general principles for quality control (Section 4020).

7. Bibliography

ROGINA, B. & M. DUBRAVCIC. 1953. Microdetermination of iodides by arresting the catalytic reduction of ceric ions. *Analyst* 78:594.

DUBRAVCIC, M. 1955. Determination of iodine in natural waters (sodium chloride as a reagent in the catalytic reduction of ceric ions). *Analyst* 80:295.

4500-I⁻ D. Voltammetric Method

1. General Discussion

a. Principle: Iodide is deposited onto the surface of a static mercury drop electrode (SMDE) as mercurous iodide under an applied potential for a specified period of time. The deposited mercurous iodide is reduced by a cathodic potential scan. This reaction gives rise to a current peak at about −0.33 V relative to the saturated calomel electrode. The height of the current peak is directly proportional to the concentration of iodide in solution, which is quantified by the method of internal standard additions.

b. Interferences: Sulfide can interfere. Remove it as hydrogen sulfide by acidifying the sample and then purging it with air. Adjust pH of sample back to about pH 8 before analysis.

2. Apparatus

a. Voltammetric analyzer system, consisting of a potentiostat, static mercury drop electrode (SMDE), stirrer, and plotter, that can be operated in the cathodic stripping square wave voltam-metry-SMDE mode with adjustable deposition potential, deposi-tion time, equilibration time, scan rate, scan range, scan incre-ment, pulse height, frequency, and drop size.

A saturated calomel electrode is used as the reference electrode through a salt bridge.

b. Glassware: Wash glassware and other surfaces contacting the sample or reagents with 10% (v/v) HCl (low in iodide); thor-oughly rinse with reagent water (see Section 1080) before use.

3. Reagents

Use chemicals low in iodide whenever available.

a. Oxygen-free water: Remove oxygen in reagent water (see Section 1080) by bubbling it with argon gas while boiling it for 20 min in an erlenmeyer flask. Let water cool while argon bub-bling continues. Tightly stopper flask and store water under ni-trogen. Prepare water immediately before use.

b. Alkaline pyrogallol solution: Dissolve 30 g pyrogallol in 200 mL oxygen-free water. Dissolve 120 g potassium hydroxide (KOH) in 400 mL oxygen-free water. Mix 300 mL KOH solution with 100 mL pyrogallol solution.

c. Sodium sulfite solution, 1M: Dissolve 1.26 g sodium sulfite, Na₂SO₃, in oxygen-free water and dilute to 10 mL.

d. Sodium sulfite solution, 0.1M: Dilute 5.0 mL 1M sodium sulfite solution to 50 mL. Prepare fresh daily.

e. Oxygen-free argon gas: Bubble argon gas (at least 99.99% pure) through a series of three traps containing, respectively, alkaline pyrogallol solution, 0.1M sodium sulfite solution, and oxygen-free water.

f. Standard iodide solution: Dry several grams potassium io-dide, KI, in an oven at 80°C overnight. Dissolve 1.660 g dried KI in reagent water (see Section 1080) and dilute to 500 mL. Dilute 5 mL solution to 500 mL, and dilute 5 mL of the latter solution to 500 mL.

g. Polyethylene glycol p-isooctylphenyl ether (PEG-IOPE) so-lution, 0.2%: Dilute 0.2 mL commercially available reagent* to 100 mL in reagent water (see Section 1080).

4. Procedure

a. Sample measurement: Transfer 10 mL sample, 0.05 mL PEG-IOPE solution, and 0.2 mL 1M Na₂SO₃ solution (which also acts as the supporting electrolyte in fresh-water samples) to po-larographic cell containing a magnetic stirrer. Purge solution with oxygen-free argon gas for 1 min. Set electrode at SMDE mode. Record a voltammogram in the cathodic stripping square wave voltammetry mode under the following conditions: deposition po-tential, −0.15 V; deposition time, 60 s; equilibration time, 5 s; scan rate, 200 mV/s; scan range, 0.15 to −0.6 V; scan increment, 2 mV; pulse height, 20 mV; frequency, 100 Hz; and the largest drop size. Measure magnitude of current peak above baseline at center of peak at an applied potential of about −0.33 V relative to saturated calomel electrode in the voltammogram.

b. Internal standard additions: Add 0.1 mL 2 μM standard KI solution to the cell. Purge solution with oxygen-free argon gas for 0.5 min. Record a voltammogram under conditions described in ¶ 4a and again determine magnitude of current peak. Repeat procedure twice, for a total of three additions.

c. Blank determination: Determine method reagent blank by treating reagent water as a sample.

5. Calculation

For the *j*th addition of the standard KI (*j* = 0, 1, 2, 3), compute

---

* Triton X-100, Catalog No. T9284, Sigma-Aldrich Corp., P.O. Box 14508, St. Louis, MO 63178.

the following variables:

$$Y_j = I_j (V_x + jV_s + V_c)$$
$$X_j = jV_sC_s$$

where:

I_j = height of jth peak, nA,
V_x = sample volume, mL,
V_s = volume of standard KI added during each internal addition, mL,
V_c = total volume of PEG-IOPE solution and sodium sulfite added during analysis, mL, and
C_s = concentration of iodide in standard KI solution, nM.

Determine slope, B, and intercept, A, of line relating Y_j to X_j by linear least squares method. Calculate concentration of iodide in sample as:

$$C_x = \frac{A}{B \times V_x}$$

where:

C_x = concentration of iodide, nM, and other terms are as defined above.

If there is a reagent blank, subtract the reagent blank from C_x to get true concentration in sample.

Multiply C_x (or blank-corrected C_x) by 0.1269 to obtain concentration in μg/L.

6. Precision

In one laboratory, using seawater samples with a concentration of iodide of about 6 μg I/L, the precision was about ±5%. Follow general principles of quality control as in Section 4020.

7. Bibliography

LUTHER, G.W., III, C.B. SWARTZ & W.J. ULLMAN. 1988. Direct determination of iodide in seawater by cathodic stripping square wave voltammetry. *Anal. Chem.* 60:1721.
WONG, G.T.F. & L.S. ZHANG. 1992. Chemical removal of oxygen with sulfite for the polarographic or voltammetric determination of iodate or iodide in seawater. *Mar. Chem.* 38:109.
WONG, G.T.F. & L.S. ZHANG. 1992. Determination of total inorganic iodine in seawater by cathodic stripping square wave voltammetry. *Talanta* 39:355.

4500-IO$_3^-$ IODATE*

4500-IO$_3^-$ A.   Introduction

1. Occurrence

Iodate is found in natural waters at concentrations ranging from 60 μg I/L in deep ocean water to undetectable (3 μg I/L) in estuarine water and fresh water. Iodate is the thermodynamically stable form of dissolved inorganic iodine in waters containing dissolved oxygen; it is absent in anoxic waters.

2. Selection of Method

The differential pulse polarographic method is species-specific and highly sensitive. It is applicable to iodate concentrations of 3 to at least 130 μg I/L and can determine iodate in the presence

* Approved by Standard Methods Committee, 1997.

of other iodine species such as iodide and organic iodine. It can be used for the direct determination of iodate in many types of water samples.

3. Sampling and Storage

Collect representative samples in clean glass or plastic bottles. Clean sample bottles with 10% (v/v) hydrochloric acid (low in iodate) and thoroughly rinse them with reagent water (see Section 1080) before use. Most samples can be analyzed directly without further treatment. Highly turbid samples may be filtered through glass fiber filters before analysis. For storage of up to 2 d, refrigerate sample at 4°C. For longer storage, freeze sample and store at or below −5°C. Frozen samples can be stored for at least 1 month.

4500-IO$_3^-$ B.   Polarographic Method

1. General Discussion

a. Principle: Under mildly basic conditions, iodate is reduced to iodide at a dropping mercury electrode by a cathodic potential scan. This reaction gives rise to a current peak centered around

−1.1 V relative to the saturated calomel electrode. The height of the current peak is directly proportional to the concentration of iodate, which is quantified by the method of standard additions.

b. Interferences: Dissolved oxygen and zinc interfere. Remove dissolved oxygen by bubbling oxygen-free argon gas through sample and by reacting oxygen with added sodium sulfite. Remove interference from zinc by complexing with EDTA (ethylene diaminetetraacetate).

2. Apparatus

a. Polarographic analyzer system: A polarographic analyzer system, consisting of a potentiostat, a static mercury drop electrode (SMDE), a stirrer, and a plotter, that can be operated in the difference pulse polarography-SMDE mode with adjustable drop time, scan increment, pulse height, scan range, and drop size.

Use a saturated calomel electrode as the reference electrode.

b. Glassware: Acid-wash glassware and other surfaces contacting sample or reagents with 10% (v/v) HCl (low in iodate); thoroughly rinse with reagent water (see Section 1080) before use.

3. Reagents

Use chemicals low in iodate whenever available.

a. Oxygen-free water: See Section 4500-I⁻.D.3a.

b. Alkaline pyrogallol solution: See Section 4500-I⁻.D.3b.

c. Oxygen-free argon gas: See Section 4500-I⁻.D.3e.

d. Sodium sulfite solution: See Section 4500-I⁻.D.3c.

e. Standard iodate solution, 25 μM: Dry several grams potassium iodate, KIO_3, in an oven at 80°C overnight. Dissolve 1.070 g dried KIO_3 in reagent water (see Section 1080) and dilute to 1000 mL. Dilute 5 mL of this solution to 1000 mL.

f. Na_2EDTA solution, 0.1M: Dissolve 3.722 g $Na_2EDTA \cdot 2H_2O$ (disodium ethylenediaminetetraacetate) in reagent water (see Section 1080) and dilute to 100 mL.

g. Supporting electrolyte: Dissolve 54.8 g sodium chloride, 0.30 g potassium bromide, and 1.05 g sodium bicarbonate in reagent water (see Section 1080) to form a final volume of 250 mL.

4. Procedure

a. Sample measurement: Transfer 5 mL sample and 0.5 mL supporting electrolyte to polarographic cell containing a magnetic stirrer. Check pH of solution to make sure it is about 8. (For marine waters with salinities above 15, the supporting electrolyte is not needed.) Remove dissolved oxygen by bubbling sample

rigorously with oxygen-free argon gas for 0.5 min with stirring. Add 0.1 mL 1M sodium sulfite solution to sample and purge, with stirring, with oxygen-free argon gas for one additional minute. Add 0.01 mL 0.1M disodium EDTA and purge, with stirring, with oxygen-free argon for another 0.5 min. Set electrode in the SMDE mode. Record a polarogram in the differential pulse polarography mode under the following conditions: drop time, 1 s; scan increment, 6 mV; pulse height, 0.06 V; and scan range, −0.65 to −1.35 V. Use a medium drop size that allows mercury droplets to be formed and dislodged from the dropping mercury electrode at a steady and consistent rate. Measure height of current peak above base line at an applied potential of about −1.1 V relative to the saturated calomel electrode.

b. Internal standard additions: Add 0.05 mL 25 μM standard iodate solution to cell. Purge solution, with stirring, with oxygen-free argon gas for 0.5 min.

Record a polarogram under conditions described in ¶ 4a and determine height of current peak again. Repeat this procedure two additional times.

c. Blank determination: Determine method reagent blank by treating reagent water as a sample.

5. Calculation

Follow calculations given in Section 4500-I⁻.D.5, with substitution of appropriate compounds in the definitions of terms.

6. Precision

In one laboratory, analyzing seawater samples with a concentration of iodate of 60 μg I/L, the precision was about ±3%. Follow general principles for quality control (see Section 4020).

7. Bibliography

HERRING, J.R. & P.S. LISS. 1974. A new method for the determination of iodine species in seawater. *Deep-Sea Res.* 21:777.

TAKAYANAGI, K. & G.T.F. WONG. 1986. The oxidation of iodide to iodate for the polarographic determination of total iodine in natural waters. *Talanta* 33:451.

WONG, G.T.F. & L.S. ZHANG. 1992. Chemical removal of oxygen with sulfite for the polarographic or voltammetric determination of iodate or iodide in seawater. *Mar. Chem.* 38:109.

4500-N   NITROGEN*

4500-N A.   Introduction

In waters and wastewaters the forms of nitrogen of greatest interest are, in order of decreasing oxidation state, nitrate, nitrite, ammonia, and organic nitrogen. All these forms of nitrogen, as well as nitrogen gas (N_2), are biochemically interconvertible and are components of the nitrogen cycle. They are of interest for many reasons.

Organic nitrogen is defined functionally as organically bound nitrogen in the trinegative oxidation state. It does not include all organic nitrogen compounds. Analytically, organic nitrogen and ammonia can be determined together and have been referred to as "kjeldahl nitrogen," a term that reflects the technique used in their determination. Organic nitrogen includes such natural materials as proteins and peptides, nucleic acids and urea, and numerous synthetic organic materials. Typical organic nitrogen con-

---

* Approved by Standard Methods Committee, 1997.

centrations vary from a few hundred micrograms per liter in some lakes to more than 20 mg/L in raw sewage.

Total oxidized nitrogen is the sum of nitrate and nitrite nitrogen. Nitrate generally occurs in trace quantities in surface water but may attain high levels in some groundwater. In excessive amounts, it contributes to the illness known as methemoglobinemia in infants. A limit of 10 mg nitrate as nitrogen/L has been imposed on drinking water to prevent this disorder. Nitrate is found only in small amounts in fresh domestic wastewater but in the effluent of nitrifying biological treatment plants nitrate may be found in concentrations of up to 30 mg nitrate as nitrogen/ L. It is an essential nutrient for many photosynthetic autotrophs and in some cases has been identified as the growth-limiting nutrient.

Nitrite is an intermediate oxidation state of nitrogen, both in the oxidation of ammonia to nitrate and in the reduction of nitrate. Such oxidation and reduction may occur in wastewater treatment plants, water distribution systems, and natural waters. Nitrite can enter a water supply system through its use as a corrosion inhibitor in industrial process water. Nitrite is the actual etiologic agent of methemoglobinemia. Nitrous acid, which is formed from nitrite in acidic solution, can react with secondary amines (RR′NH) to form nitrosamines (RR′N-NO), many of which are known to be carcinogens. The toxicologic significance of nitrosation reactions in vivo and in the natural environment is the subject of much current concern and research.

Ammonia is present naturally in surface and wastewaters. Its concentration generally is low in groundwaters because it adsorbs to soil particles and clays and is not leached readily from soils. It is produced largely by deamination of organic nitrogen-containing compounds and by hydrolysis of urea. At some water treatment plants ammonia is added to react with chlorine to form a combined chlorine residual. Ammonia concentrations encountered in water vary from less than 10 μg ammonia nitrogen/L in some natural surface and groundwaters to more than 30 mg/L in some wastewaters.

In this manual, organic nitrogen is referred to and reported as organic N, nitrate nitrogen as NO_3^--N, nitrite nitrogen as NO_2^--N, and ammonia nitrogen as NH_3-N.

Total nitrogen can be determined through oxidative digestion of all digestible nitrogen forms to nitrate, followed by quantitation of the nitrate. Two procedures, one using a persulfate/UV digestion (4500-N.B), and the other using persulfate digestion (4500-N.C) are presented. The procedures give good results for total nitrogen, composed of organic nitrogen (including some aromatic nitrogen-containing compounds), ammonia, nitrite, and nitrate. Molecular nitrogen is not determined and recovery of some industrial nitrogen-containing compounds is low.

Chloride ions do not interfere with persulfate oxidation, but the rate of reduction of nitrate to nitrite (during subsequent nitrate analysis by cadmium reduction) is significantly decreased by chlorides. Ammonium and nitrate ions adsorbed on suspended pure clay or silt particles should give a quantitative yield from persulfate digestion. If suspended matter remains after digestion, remove it before the reduction step.

If suspended organic matter is dissolved by the persulfate digestion reagent, yields comparable to those from true solutions are obtained; if it is not dissolved, the results are unreliable and probably reflect a negative interference. The persulfate method is not effective in wastes with high organic loadings. Dilute such samples and re-analyze until results from two dilutions agree.

4500-N B.   In-Line UV/Persulfate Digestion and Oxidation with Flow Injection Analysis (PROPOSED)

1. General Discussion

a. Principle: Nitrogen compounds are digested and oxidized in-line to nitrate by use of heated alkaline persulfate and ultraviolet radiation. The digested sample is injected onto the manifold where its nitrate is reduced to nitrite by a cadmium granule column. The nitrite then is determined by diazotization with sulfanilamide under acidic conditions to form a diazonium ion. The diazonium ion is coupled with *N*-(1-naphthyl)ethylenediamine dihydrochloride. The resulting pink dye absorbs at 540 nm and is proportional to total nitrogen.

This method recovers nearly all forms of organic and inorganic nitrogen, reduced and oxidized, including ammonia, nitrate, and nitrite. It differs from the total kjeldahl nitrogen method described in Section 4500-N$_{org}$.D, which does not recover the oxidized forms of nitrogen. This method recovers nitrogen components of biological origin such as amino acids, proteins, and peptides as ammonia, but may not recover the nitrogenous compounds of some industrial wastes such as amines, nitro-compounds, hydrazones, oximes, semicarbazones, and some refractory tertiary amines.

See Section 4500-N.A for a discussion of the various forms of nitrogen found in waters and wastewaters, Sections 4500-N$_{org}$.A and B for a discussion of total nitrogen methods, and Section 4130, Flow Injection Analysis (FIA). Also see Section 4500-N.C for a similar, batch total nitrogen method that uses only persulfate.

b. Interferences: Remove large or fibrous particulates by filtering sample though glass wool. Guard against contamination from reagents, water, glassware, and the sample preservation process.

2. Apparatus

Flow injection analysis equipment consisting of:
a. FIA injection valve with sample loop or equivalent.
b. Multichannel proportioning pump.
c. FIA manifold (Figure 4500-N:1) with tubing heater, in-line ultraviolet digestion fluidics including a debubbler consisting of a gas-permeable TFE membrane and its holder, and flow cell. In Figure 4500-N:1, relative flow rates only are shown. Tubing volumes are given as an example only; they may be scaled down

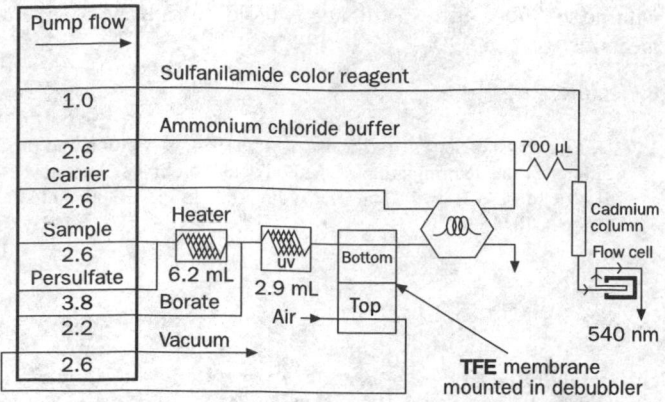

Figure 4500-N:1. FIA in-line total nitrogen manifold.

proportionally. Use manifold tubing of an inert material such as TFE. The block marked "UV" should consist of TFE tubing irradiated by a mercury discharge ultraviolet lamp emitting radiation at 254 nm.

d. Absorbance detector, 540 nm, 10-nm bandpass.

e. Injection valve control and data acquisition system.

3. Reagents

Use reagent water (>10 megohm) to prepare carrier and for all solutions. To prevent bubble formation, degas carrier and all reagents with helium. Pass He at 140 kPa (20 psi) through a helium degassing tube. Bubble He through 1 L solution for 1 min. As an alternative to preparing reagents by weight/weight, use weight/volume.

a. Borate solution, $Na_2B_4O_7·10H_2O$: In a 1-L volumetric flask dissolve 38.0 g $Na_2B_4O_7·10H_2O$ and 3.0 g sodium hydroxide, NaOH, in approximately 900 mL water, using a magnetic stirring bar. Gentle heating will speed dissolution. Adjust to pH 9.0 with NaOH or conc hydrochloric acid (HCl). Dilute to mark and invert to mix.

b. Persulfate solution, $K_2S_2O_8$: Potassium persulfate solid reagent usually contains nitrogen contamination. Higher contamination levels result in larger blank peaks.

To a tared 1-L container, add 975 g water and 49 g $K_2S_2O_8$. Add a magnetic stirring bar, dissolve persulfate, and dilute to mark. Invert to mix.

c. Ammonium chloride buffer: CAUTION: *Fumes. Use a hood.* To a 1-L volumetric flask add 500 mL water, 105 mL conc HCl, and 95 mL conc ammonium hydroxide, NH₄OH. Dissolve, dilute to mark, and invert to mix. Adjust to pH 8.5 with 1*N* HCl or 1*N* NaOH solution.

d. Sulfanilamide color reagent: To a tared, dark, 1-L container add 876 g water, 170 g 85% phosphoric acid, H_3PO_4, 40.0 g sulfanilamide, and 1.0 g *N*-(1-naphthyl)ethylenediamine dihydrochloride (NED). Shake to wet solids and stir for 30 min to dissolve. Store in a dark bottle and discard when solution turns dark pink.

e. Cadmium column: See Section 4500-NO₃⁻.I.3*c, d,* and *e.*

f. Stock nitrate standard, 1000 mg N/L: In a 1-L volumetric flask dissolve 7.221 g potassium nitrate, KNO₃ (dried at 60°C for

1 h), or 4.93 g sodium nitrite, NaNO₂, in about 800 mL water. Dilute to mark and invert to mix. When refrigerated the standard may be stored for up to 3 months.

g. Standard solutions: Prepare nitrate standards in the desired concentration range, using stock nitrate standards (¶ 3*f*), and diluting with water.

4. Procedure

Set up a manifold equivalent to that in Figure 4500-N:1 and follow method supplied by manufacturer, or laboratory standard operating procedure for this method.

Carry both standards and samples through this procedure. If samples have been preserved with sulfuric acid, preserve standards similarly. Samples may be homogenized. Turbid samples may be filtered, since digestion effectiveness on nitrogen-containing particles is unknown; however, organic nitrogen may be lost in the filtration.

5. Calculation

Prepare standard curves by plotting absorbance of standards processed through the manifold versus nitrogen concentration. The calibration curve is linear.

Verify digestion efficiency by determining urea, glutamic acid, or nicotinic acid standards (4500-N.C.3*d*) at regular intervals. In the concentration range of the method, the recovery of these compounds should be >95%.

6. Quality Control

See Section 4130B.

7. Precision and Bias

a. MDL: Using a 70-μL sample loop and a published MDL method,[1] analysts ran 21 replicates of a 0.20-mg N/L standard. These gave a mean of 0.18 mg N/L, a standard deviation of 0.008 mg N/L, and MDL of 0.020 mg N/L.

b. Precision study: Ten injections each of a 4.00-mg N/L standard and of a 10.0-mg N/L standard both gave a relative standard deviation of 0.6%.

TABLE 4500-N:I. RECOVERIES OF TOTAL NITROGEN

Compound	Mean Recovery %	
	10 mg N/L	4 mg N/L
Ammonium chloride	98.1	99.7
Sodium nitrite	100.5	101.8
Glycine	101.0	100.8
Glutamic acid	99.7	99.2
Ammonium *p*-toluenesulfonate	99.6	97.4
Glycine *p*-toluenesulfonate	101.4	102.3
Nicotinic acid	98.6	102.0
Urea	94.9	98.0
EDTA	89.4	89.4

c. Recovery of total nitrogen: Table 4500-N:I shows recoveries for various nitrogen compounds determined at 10 mg N/L and 4.0 mg N/L. All compounds were determined in triplicate.

d. Ammonia recoveries from wastewater treatment plant effluent with known additions: To a sample of wastewater treatment plant effluent, ammonium chloride was added at two concentrations, 2.50 and 5.00 mg N/L, and analyses were made in triplicate to give mean recoveries of 96% and 95%, respectively. A sample with no additions also was diluted twofold in triplicate to give a mean recovery of 99%.

8. Reference

1. U.S. ENVIRONMENTAL PROTECTION AGENCY. 1984. Definition and procedure for the determination of method detection limits. Appendix B to 40 CFR 136 Rev. 1.11 amended June 30, 1986. 49 CFR 43430, October 26, 1984.

4500-N C. Persulfate Method

1. General Discussion

The persulfate method determines total nitrogen by oxidation of all nitrogenous compounds to nitrate. Should ammonia, nitrate, and nitrite be determined individually, "organic nitrogen" can be obtained by difference.

a. Principle: Alkaline oxidation at 100 to 110°C converts organic and inorganic nitrogen to nitrate. Total nitrogen is determined by analyzing the nitrate in the digestate.

b. Selection of nitrate measurement method: Automated or manual cadmium reduction may be used to determine total nitrogen levels below 2.9 mg N/L. Results summarized in Table 4500-N:II were obtained using automated cadmium reduction.

2. Apparatus

a. Autoclave, or hotplate and pressure cooker capable of developing 100 to 110°C for 30 min.

b. Glass culture tubes: * 30-mL screw-capped (polypropylene linerless caps), 20 mm OD × 150 mm long. Clean before initial use by autoclaving with digestion reagent.

c. Apparatus for nitrate determination: See Section 4500-NO$_3^-$.E or F.

d. Automated analytical equipment: An example of the continuous-flow analytical instrument consists of components shown in Figure 4500-NO$_3^-$:2.

3. Reagents

a. Ammonia-free and nitrate-free water: Prepare by ion exchange or distillation methods as directed in 4500-NH$_3$.B.3*a* and 4500-NO$_3^-$.B.3*a*.

b. Stock nitrate solution: Prepare as directed in 4500-NO$_3^-$.B.3*b*.

c. Intermediate nitrate solution: Prepare as directed in 4500-NO$_3^-$.B.3*c*.

d. Stock glutamic acid solution: Dry glutamic acid, C$_3$H$_5$NH$_2$(COOH)$_2$, in an oven at 105°C for 24 h. Dissolve 1.051 g in water and dilute to 1000 mL; 1.00 mL = 100 μg N. Preserve with 2 mL CHCl$_3$/L.

e. Intermediate glutamic acid solution: Dilute 100 mL stock glutamic acid solution to 1000 mL with water; 1.00 mL = 10.0 μg N. Preserve with 2 mL CHCl$_3$/L.

* 18-415, Comar, Inc., Vineland, NJ, or equivalent.

f. Digestion reagent: Dissolve 20.1 g low nitrogen (<0.001% N) potassium persulfate, K$_2$S$_2$O$_8$, and 3.0 g NaOH in water and dilute to 1000 mL just before use.

g. Borate buffer solution: Dissolve 61.8 g boric acid, H$_3$BO$_3$, and 8.0 g NaOH in water and dilute to 1000 mL.

h. Copper sulfate solution: Dissolve 2.0 g CuSO$_4$·5H$_2$O in 90 mL water and dilute to 100 mL.

i. Ammonium chloride solution: Dissolve 10.0 g NH$_4$Cl in 1 L water. Adjust to pH 8.5 by adding three or four NaOH pellets as necessary or NaOH solution before bringing to volume. This reagent is stable for 2 weeks when refrigerated.

j. Color reagent: Combine 1500 mL water, 200.0 mL conc phosphoric acid, H$_3$PO$_4$, 20.0 g sulfanilamide, and 1.0 g N-(1-naphthyl)-ethylenediamine dihydrochloride. Dilute to 2000 mL. Add 2.0 mL polyoxyethylene 23 lauryl ether.† Store at 4°C in the

† Brij-35, available from ICI Americas, Inc., Wilmington, DE, or equivalent.

TABLE 4500-N:II. PRECISION DATA FOR TOTAL NITROGEN, PERSULFATE METHOD, BASED ON TRIPLICATE ANALYSES OF NICOTINIC ACID

Lab/ Analyst	Nicotinic Acid mg N/L	Recovery of N %	Standard Deviation mg/L	Relative Standard Deviation %
1/1	0.5	104	0.019	3.82
1/2	0.5	99.7	0.012	2.44
2/1	0.5	97.7	0.035	7.02
2/2	0.5	87.8	0.024	4.89
3/1	0.5	105	0.072	13.7
3/2	0.5	95.3	0.015	3.20
1/1	1.0	98.5	0.023	2.32
1/2	1.0	99.3	0.022	2.21
2/1	1.0	113	0.053	5.31
2/2	1.0	97.1	0.031	3.10
3/1	1.0	96.2	0.019	1.97
3/2	1.0	102	0.025	2.46
1/1	2.0	100	0.030	1.50
1/2	2.0	97.8	0.014	0.7
2/1	2.0	104	0.069	3.45
2/2	2.0	100	0.080	3.98
3/1	2.0	95.3	0.078	4.12
3/2	2.0	98.3	0.015	0.75

dark. Prepare fresh reagent every 6 weeks. Alternatively, prepare proportionally smaller volumes to minimize waste.

4. Procedure

a. Calibration curve: Prepare NO_3^- calibration standards in the range 0 to 2.9 mg NO_3^--N/L by diluting to 100 mL the following volumes of intermediate nitrate solution: 0, 1.00, 2.00, 4.00 . . . 29.0 mL. Treat standards in the same manner as samples.

b. Digestion check standard: Prepare glutamic acid digestion check standard of 2.9 mg N/L by diluting, to 100 mL, a 29.0-mL volume of intermediate glutamic acid solution. Treat digestion check standard in the same manner as samples.

c. Digestion: Samples preserved with acid cannot be analyzed by this method. To a culture tube, add 10.0 mL sample or standard or a portion diluted to 10.0 mL. Add 5.0 mL digestion reagent. Cap tightly. Mix by inverting twice. Heat for 30 min in an autoclave or pressure cooker at 100 to 110°C. Slowly cool to room temperature. Add 1.0 mL borate buffer solution. Mix by inverting at least twice.

d. Blank: Carry a reagent blank through all steps of the procedure and apply necessary corrections to the results.

e. Nitrate measurement: Determine nitrate by cadmium reduction. Set up manifold as shown in Figure 4500-NO_3^-:2, but use reagents specified in 4500-N_{org}.C.3.

5. Calculation

Prepare the standard curve by plotting the absorbances or peak heights of the nitrate calibration standards carried through the digestion procedure against their nitrogen concentrations. Compute organic N sample concentration by comparing sample absorbance or peak height with the standard curve.

6. Precision and Bias

See Table 4500-N:II.

7. Bibliography

D'ELIA, C.F., P.A. STEUDLER & N. CORWIN. 1977. Determination of total nitrogen in aqueous samples using persulfate digestion. *Limnol. Oceanogr.* 22:760.

NYDAHL, F. 1978. On the peroxodisulphate oxidation of total nitrogen in waters to nitrate. *Water Res.* 12:1123.

VALDERRAMA, J.C. 1981. The simultaneous analysis of total nitrogen and total phosphorus in natural waters. *Mar. Chem.* 10:109.

EBINA, J., T. TSUTSUI & T. SHIRAI. 1983. Simultaneous determination of total nitrogen and total phosphorus in water using peroxodisulfate oxidation. *Water Res.* 17:1721.

AMEEL, J.J., R.P. AXLER & C.J. OWEN. 1993. Persulfate digestion for determination of total nitrogen and phosphorus in low-nutrient waters. *Amer. Environ. Lab.* 10/93:1.

4500-NH₃   NITROGEN (AMMONIA)*

4500-NH₃ A.   Introduction

1. Selection of Method

The two major factors that influence selection of the method to determine ammonia are concentration and presence of interferences. In general, direct manual determination of low concentrations of ammonia is confined to drinking waters, clean surface or groundwater, and good-quality nitrified wastewater effluent. In other instances, and where interferences are present or greater precision is necessary, a preliminary distillation step (B) is required.

A titrimetric method (C), an ammonia-selective electrode method (D), an ammonia-selective electrode method using known addition (E), a phenate method (F), and two automated versions of the phenate method (G and H) are presented. Methods D, E, F, G, and H may be used either with or without sample distillation. The data presented in Tables 4500-NH₃:I and III should be helpful in selecting the appropriate method of analysis.

Nesslerization has been dropped as a standard method, although it has been considered a classic water quality measurement for more than a century. The use of mercury in this test warrants its deletion because of the disposal problems.

The distillation and titration procedure is used especially for NH₃-N concentrations greater than 5 mg/L. Use boric acid as the absorbent following distillation if the distillate is to be titrated.

The ammonia-selective electrode method is applicable over the range from 0.03 to 1400 mg NH₃-N/L.

The manual phenate method is applicable to both fresh water and seawater and is linear to 0.6 mg NH₃-N/L. Distill into sulfuric acid (H_2SO_4) absorbent for the phentate method when interferences are present.

The automated phenate method is applicable over the range of 0.02 to 2.0 mg NH₃-N/L.

2. Interferences

Glycine, urea, glutamic acid, cyanates, and acetamide hydrolyze very slowly in solution on standing but, of these, only urea and cyanates will hydrolyze on distillation at pH of 9.5. Hydrolysis amounts to about 7% at this pH for urea and about 5% for cyanates. Volatile alkaline compounds such as hydrazine and amines will influence titrimetric results. Residual chlorine reacts with ammonia; remove by sample pretreatment. If a sample is likely to contain residual chlorine, immediately upon collection, treat with dechlorinating agent as in 4500-NH₃.B.3*d.*

---

* Approved by Standard Methods Committee, 1997.

3. Storage of Samples

Most reliable results are obtained on fresh samples. If samples are to be analyzed within 24 h of collection, refrigerate unacidified at 4°C. For preservation for up to 28 d, freeze at -20°C unacidified, or preserve samples by acidifying to pH <2 and storing at 4°C. If acid preservation is used, neutralize samples with NaOH or KOH immediately before making the determination. CAUTION: Although acidification is suitable for certain types of samples, it produces interferences when exchangeable ammonium is present in unfiltered solids.

4. Bibliography

THAYER, G.W. 1970. Comparison of two storage methods for the analysis of nitrogen and phosphorus fractions in estuarine water. *Chesapeake Sci.* 11:155.

SALLEY, B.A., J.G. BRADSHAW & B.J. NEILSON. 1986. Results of Comparative Studies of Presevation Techniques for Nutrient Analysis on Water Samples. Virginia Institute of Marine Science, Gloucester Point.

4500-NH$_3$ B. Preliminary Distillation Step

1. General Discussion

The sample is buffered at pH 9.5 with a borate buffer to decrease hydrolysis of cyanates and organic nitrogen compounds. It is distilled into a solution of boric acid when titration is to be used or into H_2SO_4 when the phenate method is used. The ammonia in the distillate can be determined either colorimetrically by the phenate method or titrimetrically with standard H_2SO_4 and a mixed indicator or a pH meter. The choice between the colorimetric and the acidimetric methods depends on the concentration of ammonia. Ammonia in the distillate also can be determined by the ammonia-selective electrode method, using $0.04N$ H_2SO_4 to trap the ammonia.

2. Apparatus

a. Distillation apparatus: Arrange a borosilicate glass flask of 800- to 2000-mL capacity attached to a vertical condenser so that the outlet tip may be submerged below the surface of the receiving acid solution. Use an all-borosilicate-glass apparatus or one with condensing units constructed of block tin or aluminum tubes.

b. pH meter.

3. Reagents

a. Ammonia-free water: Prepare by ion-exchange or distillation methods:

1) Ion exchange—Prepare ammonia-free water by passing distilled water through an ion-exchange column containing a strongly acidic cation-exchange resin mixed with a strongly basic anion-exchange resin. Select resins that will remove organic compounds that interfere with the ammonia determination. Some anion-exchange resins tend to release ammonia. If this occurs, prepare ammonia-free water with a strongly acidic cation-exchange resin. Regenerate the column according to the manufacturer's instructions. Check ammonia-free water for the possibility of a high blank value.

2) Distillation—Eliminate traces of ammonia in distilled water by adding 0.1 mL conc H_2SO_4 to 1 L distilled water and redistilling. Alternatively, treat distilled water with sufficient bromine or chlorine water to produce a free halogen residual of 2 to 5 mg/L and redistill after standing at least 1 h. Discard the first 100 mL distillate. Check redistilled water for the possibility of a high blank.

It is very difficult to store ammonia-free water in the laboratory without contamination from gaseous ammonia. However, if storage is necessary, store in a tightly stoppered glass container to which is added about 10 g ion-exchange resin (preferably a strongly acidic cation-exchange resin)/L ammonia-free water. For use, let resin settle and decant ammonia-free water. If a high blank value is produced, replace the resin or prepare fresh ammonia-free water.

Use ammonia-free distilled water for preparing all reagents, rinsing, and sample dilution.

b. Borate buffer solution: Add 88 mL $0.1N$ NaOH solution to 500 mL approximately $0.025M$ sodium tetraborate ($Na_2B_4O_7$) solution (9.5 g $Na_2B_4O_7 \cdot 10$ H_2O/L) and dilute to 1 L.

c. Sodium hydroxide, $6N$.

d. Dechlorinating reagent: Dissolve 3.5 g sodium thiosulfate ($Na_2S_2O_3 \cdot 5H_2O$) in water and dilute to 1 L. Prepare fresh weekly. Use 1 mL reagent to remove 1 mg/L residual chlorine in 500-mL sample.

e. Neutralization agent.

1) *Sodium hydroxide,* NaOH, $1N$.

2) *Sulfuric acid,* H_2SO_4, $1N$.

f. Absorbent solution, plain boric acid: Dissolve 20 g H_3BO_3 in water and dilute to 1 L.

g. Indicating boric acid solution: See Section 4500-NH$_3$.C.3*a* and *b.*

h. Sulfuric acid, $0.04N$: Dilute 1.0 mL conc H_2SO_4 to 1 L.

4. Procedure

a. Preparation of equipment: Add 500 mL water and 20 mL borate buffer, adjust pH to 9.5 with $6N$ NaOH solution, and add to a distillation flask. Add a few glass beads or boiling chips and use this mixture to steam out the distillation apparatus until distillate shows no traces of ammonia.

b. Sample preparation: Use 500 mL dechlorinated sample or a known portion diluted to 500 mL with water. When NH$_3$-N concentration is less than 100 µg/L, use a sample volume of 1000 mL. Remove residual chlorine by adding, at the time of collection, dechlorinating agent equivalent to the chlorine residual. If necessary, neutralize to approximately pH 7 with dilute acid or base, using a pH meter.

Add 25 mL borate buffer solution and adjust to pH 9.5 with 6N NaOH using a pH meter.

c. Distillation: To minimize contamination, leave distillation apparatus assembled after steaming out and until just before starting sample distillation. Disconnect steaming-out flask and immediately transfer sample flask to distillation apparatus. Distill at a rate of 6 to 10 mL/min with the tip of the delivery tube below the surface of acid receiving solution. Collect distillate in a 500-mL erlenmeyer flask containing 50 mL indicating boric acid solution for titrimetric method. Distill ammonia into 50 mL 0.04N H$_2$SO$_4$ for the ammonia-selective electrode method and for the phenate method. Collect at least 200 mL distillate. Lower distillation receiver so that the end of the delivery tube is free of contact with the liquid and continue distillation during the last minute or two to cleanse condenser and delivery tube. Dilute to 500 mL with water.

When the phenate method is used for determining NH$_3$-N, neutralize distillate with 1N NaOH solution.

d. Ammonia determination: Determine ammonia by the titrimetric method (C), the ammonia-selective electrode methods (D and E), or the phenate methods (F and G).

5. Bibliography

NICHOLS, M.S. & M.E. FOOTE. 1931. Distillation of free ammonia from buffered solutions. *Ind. Eng. Chem.*, Anal. Ed. 3:311.

GRIFFIN, A.E. & N.S. CHAMBERLIN. 1941. Relation of ammonia nitrogen to breakpoint chlorination. *Amer. J. Pub. Health* 31:803.

PALIN, A.T. 1950. Symposium on the sterilization of water. Chemical aspects of chlorination. *J. Inst. Water Eng.* 4:565.

TARAS, M.J. 1953. Effect of free residual chlorination of nitrogen compounds in water. *J. Amer. Water Works Assoc.* 45:47.

4500-NH$_3$ C. Titrimetric Method

1. General Discussion

The titrimetric method is used only on samples that have been carried through preliminary distillation (see Section 4500-NH$_3$.B). The following table is useful in selecting sample volume for the distillation and titration method.

Ammonia Nitrogen in Sample *mg/L*	Sample Volume *mL*
5–10	250
10–20	100
20–50	50.0
50–100	25.0

2. Apparatus

Distillation apparatus: See Section 4500-NH$_3$.B.2a and b.

3. Reagents

Use ammonia-free water in making all reagents and dilutions.

a. Mixed indicator solution: Dissolve 200 mg methyl red indicator in 100 mL 95% ethyl or isopropyl alcohol. Dissolve 100 mg methylene blue in 50 mL 95% ethyl or isopropyl alcohol. Combine solutions. Prepare monthly.

b. Indicating boric acid solution: Dissolve 20 g H$_3$BO$_3$ in water, add 10 mL mixed indicator solution, and dilute to 1 L. Prepare monthly.

c. Standard sulfuric acid titrant, 0.02N: Prepare and standardize as directed in Alkalinity, Section 2320B.3c. For greatest accuracy, standardize titrant against an amount of Na$_2$CO$_3$ that has been incorporated in the indicating boric acid solution to reproduce the actual conditions of sample titration; 1.00 mL = 14 × normality × 1000 μg N. (For 0.02N, 1.00 mL = 280 μg N.)

4. Procedure

a. Proceed as described in Section 4500-NH$_3$.B using indicating boric acid solution as absorbent for the distillate.

b. Sludge or sediment samples: Rapidly weigh to within ±1% an amount of wet sample, equivalent to approximately 1 g dry weight, in a weighing bottle or crucible. Wash sample into a 500-mL kjeldahl flask with water and dilute to 250 mL. Proceed as in ¶ 4a but add a piece of paraffin wax to distillation flask and collect only 100 mL distillate.

c. Titrate ammonia in distillate with standard 0.02N H$_2$SO$_4$ titrant until indicator turns a pale lavender.

d. Blank: Carry a blank through all steps of the procedure and apply the necessary correction to the results.

5. Calculation

a. Liquid samples:

$$\text{mg NH}_3\text{-N/L} = \frac{(A - B) \times 280}{\text{mL sample}}$$

b. Sludge or sediment samples:

$$\text{mg NH}_3\text{-N/kg} = \frac{(A - B) \times 280}{\text{g dry wt sample}}$$

where:

A = volume of H$_2$SO$_4$ titrated for sample, mL, and
B = volume of H$_2$SO$_4$ titrated for blank, mL.

6. Precision and Bias

Three synthetic samples containing ammonia and other constituents dissolved in distilled water were distilled and analyzed by titration.

Sample 1 contained 200 μg NH_3-N/L, 10 mg Cl^-/L, 1.0 mg NO_3^--N/L, 1.5 mg organic N/L, 10.0 mg PO_4^{3-}/L, and 5.0 mg silica/L. The relative standard deviation and relative error for the 21 participating laboratories were 69.8% and 20%, respectively.

Sample 2 contained 800 μg NH_3-N/L, 200 mg Cl^-/L, 1.0 mg NO_3^--N/L, 0.8 mg organic N/L, 5.0 mg PO_4^{3-}/L, and 15.0 mg silica/L. The relative standard deviation and relative error for the 20 participating laboratories were 28.6% and 5%, respectively.

Sample 3 contained 1500 μg NH_3-N/L, 400 mg Cl^-/L, 1.0 mg NO_3^--N/L, 0.2 mg organic N/L, 0.5 mg PO_4^{3-}/L, and 30.0 mg silica/L. The relative standard deviation and relative error for the 21 participating laboratories were 21.6%, and 2.6%, respectively.

7. Bibliography

MEEKER, E.W. & E.C. WAGNER. 1933. Titration of ammonia in the presence of boric acid. *Ind. Eng. Chem.*, Anal. Ed. 5:396.

WAGNER, E.C. 1940. Titration of ammonia in the presence of boric acid. *Ind. Eng. Chem.*, Anal. Ed. 12:711.

4500-NH_3 D. Ammonia-Selective Electrode Method

1. General Discussion

a. Principle: The ammonia-selective electrode uses a hydrophobic gas-permeable membrane to separate the sample solution from an electrode internal solution of ammonium chloride. Dissolved ammonia ($NH_{3(aq)}$ and NH_4^+) is converted to $NH_{3(aq)}$ by raising pH to above 11 with a strong base. $NH_{3(aq)}$ diffuses through the membrane and changes the internal solution pH that is sensed by a pH electrode. The fixed level of chloride in the internal solution is sensed by a chloride ion-selective electrode that serves as the reference electrode. Potentiometric measurements are made with a pH meter having an expanded millivolt scale or with a specific ion meter.

b. Scope and application: This method is applicable to the measurement of 0.03 to 1400 mg NH_3-N/L in potable and surface waters and domestic and industrial wastes. High concentrations of dissolved ions affect the measurement, but color and turbidity do not. Sample distillation is unnecessary. Use standard solutions and samples that have the same temperature and contain about the same total level of dissolved species. The ammonia-selective electrode responds slowly below 1 mg NH_3-N/L; hence, use longer times of electrode immersion (2 to 3 min) to obtain stable readings.

c. Interference: Amines are a positive interference. This may be enhanced by acidification. Mercury and silver interfere by complexing with ammonia, unless the NaOH/EDTA solution (3c) is used.

d. Sample preservation: Refrigerate at 4°C for samples to be analyzed within 24 h. Preserve samples high in organic and nitrogenous matter, and any other samples for longer storage, by lowering pH to 2 or less with conc H_2SO_4.

2. Apparatus

a. Electrometer: A pH meter with expanded millivolt scale capable of 0.1 mV resolution between −700 mV and +700 mV or a specific ion meter.

b. Ammonia-selective electrode.

c. Magnetic stirrer, thermally insulated, with TFE-coated stirring bar.

* Orion Model 95-12, EIL Model 8002-2, Beckman Model 39565, or equivalent.

3. Reagents

a. Ammonia-free water: See Section 4500-NH_3.B.3a. Use for making all reagents.

b. Sodium hydroxide, 10N.

c. NaOH/EDTA solution, 10N: Dissolve 400 g NaOH in 800 mL water. Add 45.2 g ethylenediaminetetraacetic acid, tetrasodium salt, tetrahydrate ($Na_4EDTA \cdot 4 H_2O$) and stir to dissolve. Cool and dilute to 1000 mL.

d. Stock ammonium chloride solution: Dissolve 3.819 g anhydrous NH_4Cl (dried at 100°C) in water, and dilute to 1000 mL; 1.00 mL = 1.00 mg N = 1.22 mg NH_3.

e. Standard ammonium chloride solutions: See ¶ 4a below.

4. Procedure

a. Preparation of standards: Prepare a series of standard solutions covering the concentrations of 1000, 100, 10, 1, and 0.1 mg NH_3-N/L by making decimal dilutions of stock NH_4Cl solution with water.

b. Electrometer calibration: Place 100 mL of each standard solution in a 150-mL beaker. Immerse electrode in standard of lowest concentration and mix with a magnetic stirrer. Limit stirring speed to minimize possible loss of ammonia from the solution. Maintain the same stirring rate and a temperature of about 25°C throughout calibration and testing procedures. Add a sufficient volume of 10N NaOH solution (1 mL usually is sufficient) to raise pH above 11. If the presence of silver or mercury is possible, use NaOH/EDTA solution in place of NaOH solution. If it is necessary to add more than 1 mL of either NaOH or NaOH/EDTA solution, note volume used, because it is required for subsequent calculations. Keep electrode in solution until a stable millivolt reading is obtained. Do not add NaOH solution before immersing electrode, because ammonia may be lost from a basic solution. Repeat procedure with remaining standards, proceeding from lowest to highest concentration. Wait until the reading has stablized (at least 2 to 3 min) before recording millivolts for standards and samples containing ≤ 1 mg NH_3-N/L.

c. Preparation of standard curve: Using semilogarithmic graph paper, plot ammonia concentration in milligrams NH_3-N per liter on the log axis vs. potential in millivolts on the linear axis starting with the lowest concentration at the bottom of the scale. If the electrode is functioning properly a tenfold change of

NH₃-N concentration produces a potential change of about 59 mV.

d. Calibration of specific ion meter: Refer to manufacturer's instructions and proceed as in ¶s 4a and b.

e. Measurement of samples: Dilute if necessary to bring NH₃-N concentration to within calibration curve range. Place 100 mL sample in 150-mL beaker and follow procedure in ¶ 4b above. Record volume of 10N NaOH added. Read NH₃-N concentration from standard curve.

5. Calculation

$$\text{mg NH}_3\text{-N/L} = A \times B \times \left[\frac{100 + D}{100 + C}\right]$$

TABLE 4500-NH₃:I. PRECISION AND BIAS OF AMMONIA-SELECTIVE ELECTRODE

Level mg/L	Matrix	Mean Recovery %	Precision Overall % RSD	Precision Single Operator % RSD
0.04	Distilled water	200	125	25
	Effluent water	100	75	0
0.10	Distilled water	180	50	10
	Effluent water	470	610	10
0.80	Distilled water	105	14	5
	Effluent water	105	38	7.5
20	Distilled water	95	10	5
	Effluent water	95	15	10
100	Distilled water	98	5	2
	Effluent water	97	—	—
750	Distilled water	97	10.4	1.6
	Effluent water	99	14.1	1.3

Source: AMERICAN SOCIETY FOR TESTING AND MATERIALS. Method 1426-79. American Soc. Testing & Materials, Philadelphia, Pa.

where:

A = dilution factor,
B = concentration of NH₃-N/L, mg/L, from calibration curve,
C = volume of 10N NaOH added to calibration standards, mL, and
D = volume of 10N NaOH added to sample, mL.

6. Precision and Bias

For the ammonia-selective electrode in a single laboratory using surface water samples at concentrations of 1.00, 0.77, 0.19, and 0.13 mg NH₃-N/L, standard deviations were ±0.038, ±0.017, ±0.007, and ±0.003, respectively. In a single laboratory using surface water samples at concentrations of 0.10 and 0.13 mg NH₃-N/L, recoveries were 96% and 91%, respectively. The results of an interlaboratory study involving 12 laboratories using the ammonia-selective electrode on distilled water and effluents are summarized in Table 4500-NH₃:I.

7. Bibliography

BANWART, W.L., J.M. BREMNER & M.A. TABATABAI. 1972. Determination of ammonium in soil extracts and water samples by an ammonia electrode. *Comm. Soil Sci. Plant Anal.* 3:449.

MIDGLEY, C. & K. TERRANCE. 1972. The determination of ammonia in condensed steam and boiler feed-water with a potentiometric ammonia probe. *Analyst* 97:626.

BOOTH, R.L. & R.F. THOMAS. 1973. Selective electrode determination of ammonia in water and wastes. *Environ. Sci. Technol.* 7:523.

U.S. ENVIRONMENTAL PROTECTION AGENCY. 1979. Methods for Chemical Analysis of Water and Wastes. EPA-600/4-79-020, National Environmental Research Center, Cincinnati, Ohio.

AMERICAN SOCIETY FOR TESTING AND MATERIALS. 1979. Method 1426–79. American Soc. Testing & Materials, Philadelphia, Pa.

4500-NH₃ E. Ammonia-Selective Electrode Method Using Known Addition

1. General Discussion

a. Principle: When a linear relationship exists between concentration and response, known addition is convenient for measuring occasional samples because no calibration is needed. Because an accurate measurement requires that the concentration at least double as a result of the addition, sample concentration must be known within a factor of three. Total concentration of ammonia can be measured in the absence of complexing agents down to 0.8 mg NH₃-N/L or in the presence of a large excess (50 to 100 times) of complexing agent. Known addition is a convenient check on the results of direct measurement.

b. See Section 4500-NH₃.D.1 for further discussion.

2. Apparatus

Use apparatus specified in Section 4500-NH₃.D.2.

3. Reagents

Use reagents specified in Section 4500-NH₃.D.3.

Add standard ammonium chloride solution approximately 10 times as concentrated as samples being measured.

4. Procedure

a. Dilute 1000 mg/L stock solution to make a standard solution about 10 times as concentrated as the sample concentrate.

b. Add 1 mL 10N NaOH to each 100 mL sample and immediately immerse electrode. When checking a direct measurement, leave electrode in 100 mL of sample solution. Use magnetic stirring throughout. Measure mV reading and record as E_1.

c. Pipet 10 mL of standard solution into sample. Thoroughly stir and immediately record new mV reading as E_2.

TABLE 4500-NH₃:II.* VALUES OF Q VS. ΔE (59 MV SLOPE) FOR 10% VOLUME CHANGE

ΔE	Q	ΔE	Q	ΔE	Q	ΔE	Q	ΔE	Q
5.0	0.297	9.0	0.178	16.0	0.0952	24.0	0.0556	32.0	0.0354
5.1	0.293	9.1	0.176	16.2	0.0938	24.2	0.0549	32.2	0.0351
5.2	0.288	9.2	0.174	16.4	0.0924	24.4	0.0543	32.4	0.0347
5.3	0.284	9.3	0.173	16.6	0.0910	24.6	0.0536	32.6	0.0343
5.4	0.280	9.4	0.171	16.8	0.0897	24.8	0.0530	32.8	0.0340
5.5	0.276	9.5	0.169	17.0	0.0884	25.0	0.0523	33.0	0.0335
5.6	0.272	9.6	0.167	17.2	0.0871	25.2	0.0517	33.2	0.0333
5.7	0.268	9.7	0.165	17.4	0.0858	25.4	0.0511	33.4	0.0329
5.8	0.264	9.8	0.164	17.6	0.0846	25.6	0.0505	33.6	0.0326
5.9	0.260	9.9	0.162	17.8	0.0834	25.8	0.0499	33.8	0.0323
6.0	0.257	10.0	0.160	18.0	0.0822	26.0	0.0494	34.0	0.0319
6.1	0.253	10.2	0.157	18.2	0.0811	26.2	0.0488	34.2	0.0316
6.2	0.250	10.4	0.154	18.4	0.0799	26.4	0.0482	34.4	0.0313
6.3	0.247	10.6	0.151	18.6	0.0788	26.6	0.0477	34.6	0.0310
6.4	0.243	10.8	0.148	18.8	0.0777	26.8	0.0471	34.8	0.0307
6.5	0.240	11.0	0.145	19.0	0.0767	27.0	0.0466	35.0	0.0304
6.6	0.237	11.2	0.143	19.2	0.0756	27.2	0.0461	36.0	0.0289
6.7	0.234	11.4	0.140	19.4	0.0746	27.4	0.0456	37.0	0.0275
6.8	0.231	11.6	0.137	19.6	0.0736	27.6	0.0450	38.0	0.0261
6.9	0.228	11.8	0.135	19.8	0.0726	27.8	0.0445	39.0	0.0249
7.0	0.225	12.0	0.133	20.0	0.0716	28.0	0.0440	40.0	0.0237
7.1	0.222	12.2	0.130	20.2	0.0707	28.2	0.0435	41.0	0.0226
7.2	0.219	12.4	0.128	20.4	0.0698	28.4	0.0431	42.0	0.0216
7.3	0.217	12.6	0.126	20.6	0.0689	28.6	0.0426	43.0	0.0206
7.4	0.214	12.8	0.123	20.8	0.0680	28.8	0.0421	44.0	0.0196
7.5	0.212	13.0	0.121	21.0	0.0671	29.0	0.0417	45.0	0.0187
7.6	0.209	13.2	0.119	21.2	0.0662	29.2	0.0412	46.0	0.0179
7.7	0.207	13.4	0.117	21.4	0.0654	29.4	0.0408	47.0	0.0171
7.8	0.204	13.6	0.115	21.6	0.0645	29.6	0.0403	48.0	0.0163
7.9	0.202	13.8	0.113	21.8	0.0637	29.8	0.0399	49.0	0.0156
8.0	0.199	14.0	0.112	22.0	0.0629	30.0	0.0394	50.0	0.0149
8.1	0.197	14.2	0.110	22.2	0.0621	30.2	0.0390	51.0	0.0143
8.2	0.195	14.4	0.108	22.4	0.0613	30.4	0.0386	52.0	0.0137
8.3	0.193	14.6	0.106	22.6	0.0606	30.6	0.0382	53.0	0.0131
8.4	0.190	14.8	0.105	22.8	0.0598	30.8	0.0378	54.0	0.0125
8.5	0.188	15.0	0.103	23.0	0.0591	31.0	0.0374	55.0	0.0120
8.6	0.186	15.2	0.1013	23.2	0.0584	31.2	0.0370	56.0	0.0115
8.7	0.184	15.4	0.0997	23.4	0.0576	31.4	0.0366	57.0	0.0110
8.8	0.182	15.6	0.0982	23.6	0.0569	31.6	0.0362	58.0	0.0105
8.9	0.180	15.8	0.0967	23.8	0.0563	31.8	0.0358	59.0	0.0101

* Orion Research Inc. Instruction Manual, Ammonia Electrode, Model 95-12, Boston, MA. 02129.

5. Calculation

a. $\Delta E = E_1 - E_2$.

b. From Table 4500-NH₃:II find the concentration ratio, Q, corresponding to change in potential, ΔE. To determine original total sample concentration, multiply Q by the concentration of the added standard:

$$C_o = Q\,C_s$$

where:

C_o = total sample concentration, mg/L,
Q = reading from known-addition table, and
C_s = concentration of added standard, mg/L.

c. To check a direct measurement, compare results of the two methods. If they agree within ±4%, the measurements probably are good. If the known-addition result is much larger than the direct measurement, the sample may contain complexing agents.

6. Precision and Bias

In 38 water samples analyzed by both the phenate and the known-addition ammonia-selective electrode method, the electrode method yielded a mean recovery of 102% of the values obtained by the phenate method when the NH₃-N concentrations varied between 0.30 and 0.78 mg/L. In 57 wastewater samples similarly compared, the electrode method yielded a mean recovery of 108% of the values obtained by the phenate method using distillation when the NH₃-N concentrations varied between 10.2 and 34.7 mg N/L. In 20 instances in which two to four replicates of these samples were analyzed, the mean standard deviation was 1.32 mg N/L. In three measurements at a sewer outfall, distillation did not change statistically the value obtained by the electrode method. In 12 studies using standards in the 2.5- to 30-mg N/L range, average recovery by the phenate method was 97% and by the electrode method 101%.

4500-NH₃ F. Phenate Method

1. General Discussion

a. Principle: An intensely blue compound, indophenol, is formed by the reaction of ammonia, hypochlorite, and phenol catalyzed by sodium nitroprusside.

b. Interferences: Complexing magnesium and calcium with citrate eliminates interference produced by precipitation of these ions at high pH. There is no interference from other trivalent forms of nitrogen. Remove interfering turbidity by distillation or filtration. If hydrogen sulfide is present, remove by acidifying samples to pH 3 with dilute HCl and aerating vigorously until sulfide odor no longer can be detected.

2. Apparatus

Spectrophotometer for use at 640 nm with a light path of 1 cm or greater.

3. Reagents

a. Phenol solution: Mix 11.1 mL liquified phenol ($\geq$89%) with 95% v/v ethyl alcohol to a final volume of 100 mL. Prepare weekly. CAUTION: *Wear gloves and eye protection when handling phenol; use good ventilation to minimize all personnel exposure to this toxic volatile substance.*

b. Sodium nitroprusside, 0.5% w/v: Dissolve 0.5 g sodium nitroprusside in 100 mL deionized water. Store in amber bottle for up to 1 month.

c. Alkaline citrate: Dissolve 200 g trisodium citrate and 10 g sodium hydroxide in deionized water. Dilute to 1000 mL.

d. Sodium hypochlorite, commercial solution, about 5%. This solution slowly decomposes once the seal on the bottle cap is broken. Replace about every 2 months.

e. Oxidizing solution: Mix 100 mL alkaline citrate solution with 25 mL sodium hypochlorite. Prepare fresh daily.

f. Stock ammonium solution: See Section 4500-NH₃.D.3*d.*

g. Standard ammonium solution: Use stock ammonium solution and water to prepare a calibration curve in a range appropriate for the concentrations of the samples.

4. Procedure

To a 25-mL sample in a 50-mL erlenmeyer flask, add, with thorough mixing after each addition, 1 mL phenol solution, 1 mL sodium nitroprusside solution, and 2.5 mL oxidizing solution. Cover samples with plastic wrap or paraffin wrapper film. Let color develop at room temperature (22 to 27°C) in subdued light for at least 1 h. Color is stable for 24 h. Measure absorbance at 640 nm. Prepare a blank and at least two other standards by diluting stock ammonia solution into the sample concentration range. Treat standards the same as samples.

5. Calculations

Prepare a standard curve by plotting absorbance readings of standards against ammonia concentrations of standards. Compute

sample concentration by comparing sample absorbance with the standard curve.

6. Precision and Bias

For the manual phenate method, reagent water solutions of ammonium sulfate were prepared and analyzed by two analysts in each of three laboratories. Results are summarized in Table 4500-NH₃:III.

7. Bibliography

SOLORZANO, L. 1969. Determination of ammonia in natural waters by the phenolhypochlorite method. *Limnol. Oceanogr.* 14:799.

PARSONS, T.R., Y. MAITA & C.M. LALLI. 1984. A Manual of Chemical and Biological Methods for Seawater Analysis. Pergamon Press, Elmsford, N.Y.

TABLE 4500-NH₃:III. PRECISION DATA FOR MANUAL PHENATE METHOD BASED ON TRIPLICATE ANALYSES OF AMMONIUM SULFATE

Lab/ Analyst	NH₃-N Concentration mg/L	Optical Density	Relative Standard Deviation %
1/1	0.1	0.129	1.55
1/2	0.1	0.114	9.66
2/1	0.1	0.100	10.2
2/2	0.1	0.122	2.36
3/1	0.1	0.112	3.61
3/2	0.1	0.107	1.94
1/1	0.3	0.393	0.39
1/2	0.3	0.364	0.32
2/1	0.3	0.372	2.64
2/2	0.3	0.339	0.90
3/1	0.3	0.370	0.31
3/2	0.3	0.373	0.46
1/1	0.5	0.637	0.77
1/2	0.5	0.630	0.56
2/1	0.5	0.624	1.65
2/2	0.5	0.618	0.86
3/1	0.5	0.561	0.27
3/2	0.5	0.569	0.91

4500-NH₃ G. Automated Phenate Method

1. General Discussion

a. Principle: Alkaline phenol and hypochlorite react with ammonia to form indophenol blue that is proportional to the ammonia concentration. The blue color formed is intensified with sodium nitroprusside.

b. Interferences: Seawater contains calcium and magnesium ions in sufficient concentrations to cause precipitation during analysis. Adding EDTA and sodium potassium tartrate reduces the problem. Eliminate any marked variation in acidity or alka-

linity among samples because intensity of measured color is pH-dependent. Likewise, insure that pH of wash water and standard ammonia solutions approximates that of sample. For example, if sample has been preserved with 0.8 mL conc H_2SO_4/L, include 0.8 mL conc H_2SO_4/L in wash water and standards. Remove interfering turbidity by filtration. Color in the samples that absorbs in the photometric range used for analysis interferes.

c. Application: Ammonia nitrogen can be determined in potable, surface, and saline waters as well as domestic and industrial wastewaters over a range of 0.02 to 2.0 mg/L when photometric measurement is made at 630 to 660 nm in a 10- to 50-mm tubular flow cell at rates of up to 60 samples/h. Determine higher concentrations by diluting the sample.

2. Apparatus

Automated analytical equipment. An example of the required continuous-flow analytical instrument consists of the interchangeable components shown in Figure 4500-NH$_3$:1.

3. Reagents

a. Ammonia-free distilled water: See Section 4500-NH$_3$.B.3*a*. Use for preparing all reagents and dilutions.

b. Sulfuric acid, H_2SO_4, 5N, air scrubber solution: Carefully add 139 mL conc H_2SO_4 to approximately 500 mL water, cool to room temperature, and dilute to 1 L.

c. Sodium phenate solution: In a 1-L erlenmeyer flask, dissolve 93 mL liquid (≥89%) phenol in 500 mL water. In small increments and with agitation, cautiously add 32 g NaOH. Cool flask under running water and dilute to 1 L. CAUTION: *Minimize exposure of personnel to this compound by wearing gloves and eye protection, and using proper ventilation.*

d. Sodium hypochlorite solution: Dilute 250 mL bleach solution containing 5.25% NaOCl to 500 mL with water.

e. EDTA reagent: Dissolve 50 g disodium ethylenediamine tetraacetate and approximately six pellets NaOH in 1 L water. For salt-water samples where EDTA reagent does not prevent precipitation of cations, use sodium potassium tartrate solution prepared as follows:

Sodium potassium tartrate solution: To 900 mL water add 100 g NaKC$_4$H$_4$O$_6$·4H$_2$O, two pellets NaOH, and a few boiling chips, and boil gently for 45 min. Cover, cool, and dilute to 1 L. Adjust pH to 5.2 ± 0.05 with H$_2$SO$_4$. Let settle overnight in a cool place and filter to remove precipitate. Add 0.5 mL polyoxyethylene 23 lauryl ether* solution and store in stoppered bottle.

f. Sodium nitroprusside solution: Dissolve 0.5 g Na$_2$(NO)Fe(CN)$_5$·2H$_2$O in 1 L water.

---

* Brij-35, available from ICI Americas, Wilmington, DE.

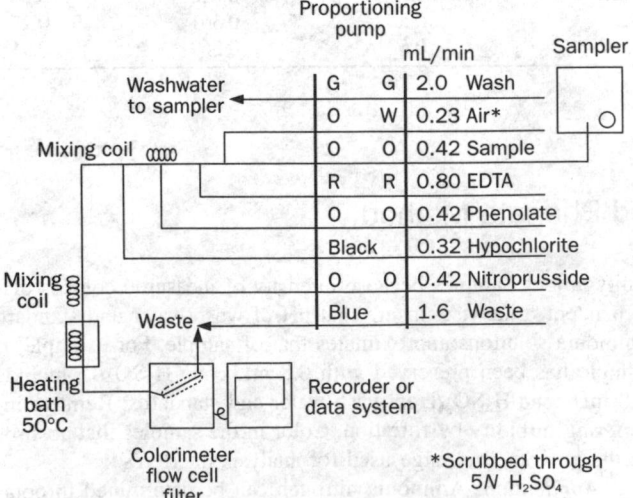

Figure 4500-NH$_3$:1. Ammonia manifold.

g. Ammonia standard solutions: See Sections 4500-NH$_3$.D.3*c* and *d*. Use standard ammonia solution and water to prepare the calibration curve in the appropriate ammonia concentration range. To analyze saline waters use substitute ocean water of the following composition to prepare calibration standards:

Constituent	Concentration g/L
NaCl	24.53
MgCl$_2$	5.20
CaCl$_2$	1.16
KCl	0.70
SrCl$_2$	0.03
Na$_2$SO$_4$	4.09
NaHCO$_3$	0.20
KBr	0.10
H$_3$BO$_3$	0.03
NaF	0.003

Subtract blank background response of substitute seawater from standards before preparing standard curve.

4. Procedure

a. Eliminate marked variation in acidity or alkalinity among samples. Adjust pH of wash water and standard ammonia solutions to approximately that of sample.

b. Set up manifold and complete system as shown in Figure 4500-NH$_3$:1.

c. Obtain a stable base line with all reagents, feeding wash water through sample line.

d. Typically, use a 60/h, 6:1 cam with a common wash.

5. Calculation

Prepare standard curves by plotting response of standards processed through the manifold against NH$_3$-N concentrations in standards. Compute sample NH$_3$-N concentration by comparing sample response with standard curve.

6. Precision and Bias

For an automated phenate system in a single laboratory using surface water samples at concentrations of 1.41, 0.77, 0.59, and 0.43 mg NH$_3$-N/L, the standard deviation was ±0.005 mg/L, and at concentrations of 0.16 and 1.44 mg NH$_3$-N/L, recoveries were 107 and 99%, respectively.

7. Bibliography

HILLER, A. & D. VAN SLYKE. 1933. Determination of ammonia in blood. *J. Biol. Chem.* 102:499.

FIORE, J. & J.E. O'BRIEN. 1962. Ammonia determination by automatic analysis. *Wastes Eng.* 33:352.

AMERICAN SOCIETY FOR TESTING AND MATERIALS. 1966. Manual on Industrial Water and Industrial Waste Water, 2nd ed. American Soc. Testing & Materials, Philadelphia, Pa.

O'CONNOR, B., R. DOBBS, B. VILLIERS & R. DEAN. 1967. Laboratory distillation of municipal waste effluents. *J. Water Pollut. Control Fed.* 39:25.

4500-NH₃ H. Flow Injection Analysis (PROPOSED)

1. General Discussion

a. Principle: A water sample containing ammonia or ammonium cation is injected into an FIA carrier stream to which a complexing buffer, alkaline phenol, and hypochlorite are added. This reaction, the Berthelot reaction, produces the blue indophenol dye. The blue color is intensified by the addition of nitroferricyanide. The resulting peak's absorbance is measured at 630 nm. The peak area is proportional to the concentration of ammonia in the original sample.

Also see Section 4500-NH₃.F and Section 4130, Flow Injection Analysis (FIA).

b. Interferences: Remove large or fibrous particulates by filtering sample through glass wool. Guard against contamination from reagents, water, glassware, and the sample preservation process.

Also see Section 4500-NH₃.A. Some interferents are removed by distillation; see Section 4500-NH₃.B.

2. Apparatus

Flow injection analysis equipment consisting of:

a. FIA injection valve with sample loop or equivalent.

b. Multichannel proportioning pump.

c. FIA manifold (Figure 4500-NH₃:2) with tubing heater and flow cell. In Figure 4500-NH₃:2, relative flow rates only are shown. Tubing volumes are given as an example only; they may be scaled down proportionally. Use manifold tubing of an inert material such as TFE.*

d. Absorbance detector, 630 nm, 10-nm bandpass.

e. Injection valve control and data acquisition system.

3. Reagents

Use reagent water (>10 megohm) to prepare carrier and all solutions. To prevent bubble formation, degas carrier and buffer with helium. Use He at 140 kPa (20 psi) through a helium degassing tube. Bubble He through 1 L solution for 1 min. As an alternative to preparing reagents by weight/weight, use weight/volume.

---

* Teflon or equivalent.

TABLE 4500-NH₃:IV. RESULTS OF SINGLE-LABORATORY STUDIES WITH SELECTED MATRICES

Matrix	Sample/Blank Designation	Known Addition mg NH₃-N/L	Recovery %	Relative Standard Deviation %
Wastewater treatment plant influent	Reference sample*	—	104	
	Blank†	0.4	100	—
		0.8	102	—
	Site A‡§	0	—	0.5
		0.4	108	—
		0.8	105	—
	Site B‡§	0	—	0.5
		0.4	105	—
		0.8	105	—
	Site C‡§	0	—	1.1
		0.4	109	—
		0.8	107	—
Wastewater treatment plant effluent	Reference sample*	—	106	—
	Blank†	0.4	105	—
		0.8	105	—
	Site A‡‖	0	—	ND
		0.4	90	—
		0.8	88	—
	Site B‡‖	0	—	<0.1
		0.4	93	—
		0.8	94	—
	Site C‡‖	0	—	<0.1
		0.4	89	—
		0.8	91	—
Landfill leachate‖	Reference sample*	—	106	—
	Blank†	0.4	105	—
		0.8	106	—
	Site A‡#	0	—	1.9
		0.4	125	—
		0.8	114	—
	Site B‡#	0	—	0.4
		0.4	96	—
		0.8	106	—
	Site C‡#	0	—	0.2
		0.4	102	—
		0.8	107	—

ND = not detectable.

* U.S. EPA QC sample, 1.98 mg N/L.

† Determined in duplicate.

‡ Samples without known additions determined four times; samples with known additions determined in duplicate.

§ Site A and C samples diluted 20-fold, Site B sample diluted 100-fold. Typical relative difference between duplicates 0.2%.

‖ Samples not diluted. Typical relative difference between duplicates <1%.

Site A and C samples diluted 50-fold; Site B sample diluted 150-fold. Typical relative difference between duplicates 0.3%.

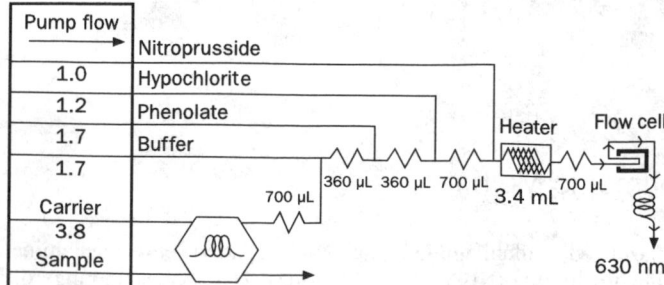

Figure 4500-NH₃:2. FIA ammonia manifold.

a. Buffer: To a 1-L tared container add 50.0 g disodium ethylenediamine tetraacetate and 5.5 g sodium hydroxide, NaOH. Add 968 mL water. Mix with a magnetic stirrer until dissolved.

b. Phenolate: CAUTION: *Wear gloves. Phenol causes severe burns and is rapidly absorbed into the body through the skin.* To a tared 1-L container, add 888 g water. Add 94.2 g 88% liquefied phenol or 83 g crystalline phenol, C_6H_5OH. While stirring, slowly add 32 g NaOH. Cool and invert to mix thoroughly. Do not degas.

c. Hypochlorite: To a tared 500-mL container add 250 g 5.25% sodium hypochlorite, NaOCl bleach solution† and 250 g water. Stir or shake to mix.

d. Nitroprusside: To a tared 1-L container add 3.50 g sodium nitroprusside (sodium nitroferricyanide), $Na_2Fe(CN)_5NO \cdot 2H_2O$, and 1000 g water. Invert to mix.

e. Stock ammonia standard, 1000 mg N/L: In a 1-L volumetric flask dissolve 3.819 g ammonium chloride, NH_4Cl, that has been dried for 2 h at 110°C, in about 800 mL water. Dilute to mark and invert to mix.

f. Standard ammonia solutions: Prepare ammonia standards in desired concentration range, using the stock standard (¶ 3e), and diluting with water.

4. Procedure

Set up a manifold equivalent to that in Figure 4500-NH₃:2 and follow method supplied by manufacturer or laboratory standard

---

† Clorox, The Clorox Company, Pleasanton, CA, or equivalent.

operating procedure for this method. Follow quality control procedures described in Section 4020.

5. Calculations

Prepare standard curves by plotting the absorbance of standards processed through the manifold versus ammonia concentration. The calibration curve is linear.

6. Precision and Bias

a. Recovery and relative standard deviation: The results of single-laboratory studies with various matrices are given in Table 4500-NH₃:IV.

b. MDL: A 650-µL sample loop was used in the method described above. Using a published MDL method,[1] analysts ran 21 replicates of a 0.020-mg N/L standard. These gave a mean of 0.0204 mg N/L, a standard deviation of 0.0007 mg N/L, and an MDL of 0.002 mg N/L.

7. Reference

1. U.S. ENVIRONMENTAL PROTECTION AGENCY. 1989. Definition and Procedure for the Determination of Method Detection Limits. Appendix B to 40 CFR 136 rev. 1.11 amended June 30, 1986. 49 CFR 43430.

4500-NO₂⁻   NITROGEN (NITRITE)*

4500-NO₂⁻ A.   Introduction

1. Occurrence and Significance

For a discussion of the chemical characteristics, sources, and effects of nitrite nitrogen, see Section 4500-N.

---

* Approved by Standard Methods Committee, 1993.

2. Selection of Method

The colorimetric method (B) is suitable for concentrations of 5 to 1000 µg NO_2^--N/L (See ¶ B.1a). Nitrite values can be obtained by the automated method given in Section 4500-NO₃⁻.E with the Cu-Cd reduction step omitted. Additionally, nitrite nitrogen can be determined by ion chromatography (Section 4110), and by flow injection analysis (see Sections 4130 and 4500-NO₃⁻.I).

4500-NO₂⁻ B.   Colorimetric Method

1. General Discussion

a. Principle: Nitrite (NO_2^-) is determined through formation of a reddish purple azo dye produced at pH 2.0 to 2.5 by coupling

diazotized sulfanilamide with *N*-(1-naphthyl)-ethylenediamine dihydrochloride (NED dihydrochloride). The applicable range of the method for spectrophotometric measurements is 10 to 1000 µg NO_2^--N/L. Photometric measurements can be made in the range 5 to 50 μg N/L if a 5-cm light path and a green color filter are used. The color system obeys Beer's law up to 180 μg N/ L with a 1-cm light path at 543 nm. Higher NO$_2^-$ concentrations can be determined by diluting a sample.

b. Interferences: Chemical incompatibility makes it unlikely that NO$_2^-$, free chlorine, and nitrogen trichloride (NCl$_3$) will co-exist. NCl$_3$ imparts a false red color when color reagent is added. The following ions interfere because of precipitation under test conditions and should be absent: Sb^{3+}, Au^{3+}, Bi^{3+}, Fe^{3+}, Pb^{2+}, Hg^{2+}, Ag$^+$, chloroplatinate (PtCl$_6^{2-}$), and metavanadate (VO$_3^{2-}$). Cupric ion may cause low results by catalyzing decomposition of the diazonium salt. Colored ions that alter the color system also should be absent. Remove suspended solids by filtration.

c. Storage of sample: Never use acid preservation for samples to be analyzed for NO$_2^-$. Make the determination promptly on fresh samples to prevent bacterial conversion of NO$_2^-$ to NO$_3^-$ or NH$_3$. For short-term preservation for 1 to 2 d, freeze at −20°C or store at 4°C.

2. Apparatus

Colorimetric equipment: One of the following is required:

a. Spectrophotometer, for use at 543 nm, providing a light path of 1 cm or longer.

b. Filter photometer, providing a light path of 1 cm or longer and equipped with a green filter having maximum transmittance near 540 nm.

3. Reagents

a. Nitrite-free water: If it is not known that the distilled or demineralized water is free from NO$_2^-$, use either of the following procedures to prepare nitrite-free water:

1) Add to 1 L distilled water one small crystal each of KMnO$_4$ and either Ba(OH)$_2$ or Ca(OH)$_2$. Redistill in an all-borosilicate-glass apparatus and discard the initial 50 mL of distillate. Collect the distillate fraction that is free of permanganate; a red color with DPD reagent (Section 4500-Cl.F.2*b*) indicates the presence of permanganate.

2) Add 1 mL conc H$_2$SO$_4$ and 0.2 mL MnSO$_4$ solution (36.4 g MnSO$_4$·H$_2$O/100 mL distilled water) to each 1 L distilled water, and make pink with 1 to 3 mL KMnO4 solution (400 mg KMnO$_4$/ L distilled water). Redistill as described in the preceding paragraph.

Use nitrite-free water in making all reagents and dilutions.

b. Color reagent: To 800 mL water add 100 mL 85% phosphoric acid and 10 g sulfanilamide. After dissolving sulfanilamide completely, add 1 g *N*-(1-naphthyl)-ethylenediamine dihydrochloride. Mix to dissolve, then dilute to 1 L with water. Solution is stable for about a month when stored in a dark bottle in refrigerator.

c. Sodium oxalate, 0.025*M* (0.05*N*): Dissolve 3.350 g Na$_2$C$_2$O$_4$, primary standard grade, in water and dilute to 1000 mL.

d. Ferrous ammonium sulfate, 0.05*M* (0.05*N*): Dissolve 19.607 g Fe(NH$_4$)$_2$ (SO$_4$)$_2$·6H$_2$O plus 20 mL conc H$_2$SO$_4$ in water and dilute to 1000 mL. Standardize as in Section 5220B.3*d*.

e. Stock nitrite solution: Commercial reagent-grade NaNO$_2$ assays at less than 99%. Because NO$_2^-$ is oxidized readily in the presence of moisture, use a fresh bottle of reagent for preparing the stock solution and keep bottles tightly stoppered against the free access of air when not in use. To determine NaNO$_2$ content, add a known excess of standard 0.01*M* (0.05*N*) KMnO$_4$ solution (see ¶ *h* below), discharge permanganate color with a known quantity of standard reductant such as 0.025*M* Na$_2$C$_2$O$_4$ or 0.05*M* Fe(NH$_4$)$_2$(SO$_4$)$_2$, and back-titrate with standard permanganate solution.

1) Preparation of stock solution—Dissolve 1.232 g NaNO$_2$ in water and dilute to 1000 mL; 1.00 mL = 250 μg N. Preserve with 1 mL CHCl$_3$.

2) Standardization of stock nitrite solution—Pipet, in order, 50.00 mL standard 0.01*M* (0.05*N*) KMnO$_4$, 5 mL conc H$_2$SO$_4$, and 50.00 mL stock NO$_2^-$ solution into a glass-stoppered flask or bottle. Submerge pipet tip well below surface of permanganate-acid solution while adding stock NO$_2^-$ solution. Shake gently and warm to 70 to 80°C on a hot plate. Discharge permanganate color by adding sufficient 10-mL portions of standard 0.025*M* Na$_2$C$_2$O$_4$. Titrate excess Na$_2$C$_2$O$_4$ with 0.01*M* (0.05*N*) KMnO$_4$ to the faint pink end point. Carry a water blank through the entire procedure and make the necessary corrections in the final calculation as shown in the equation below.

If standard 0.05*M* ferrous ammonium sulfate solution is substituted for Na$_2$C$_2$O$_4$, omit heating and extend reaction period between KMnO$_4$ and Fe^{2+} to 5 min before making final KMnO$_4$ titration.

Calculate NO$_2^-$-N content of stock solution by the following equation:

$$ A = \frac{[(B \times C) - (D \times E)] \times 7}{F} $$

where:

A = mg NO$_2^-$-N/mL in stock NaNO$_2$ solution,
B = total mL standard KMnO$_4$ used,
C = normality of standard KMnO$_4$,
D = total mL standard reductant added,
E = normality of standard reductant, and
F = mL stock NaNO$_2$ solution taken for titration.

Each 1.00 mL 0.01*M* (0.05*N*) KMnO$_4$ consumed by the NaNO$_2$ solution corresponds to 1750 μg NO$_2^-$-N.

f. Intermediate nitrite solution: Calculate the volume, G, of stock NO$_2^-$ solution required for the intermediate NO$_2^-$ solution from $G = 12.5/A$. Dilute the volume G (approximately 50 mL) to 250 mL with water; 1.00 mL = 50.0 μg N. Prepare daily.

g. Standard nitrite solution: Dilute 10.00 mL intermediate NO$_2^-$ solution to 1000 mL with water; 1.00 mL = 0.500 μg N. Prepare daily.

h. Standard potassium permanganate titrant, 0.01*M* (0.05*N*): Dissolve 1.6 g KMnO$_4$ in 1 L distilled water. Keep in a brown glass-stoppered bottle and age for at least 1 week. Carefully decant or pipet supernate without stirring up any sediment. Standardize this solution frequently by the following procedure:

Weigh to the nearest 0.1 mg several 100- to 200-mg samples of anhydrous Na$_2$C$_2$O$_4$ into 400-mL beakers. To each beaker, in turn, add 100 mL distilled water and stir to dissolve. Add 10 mL 1 + 1 H$_2$SO$_4$ and heat rapidly to 90 to 95°C. Titrate rapidly with permanganate solution to be standardized, while stirring, to a slight pink end-point color that persists for at least 1 min. Do not let temperature fall below 85°C. If necessary, warm beaker contents during titration; 100 mg will consume about 6 mL solution. Run a blank on distilled water and H_2SO_4.

$$\text{Normality of KMnO}_4 = \frac{\text{g Na}_2\text{C}_2\text{O}_4}{(A - B) \times 0.335\ 05}$$

where:

A = mL titrant for sample and
B = mL titrant for blank.

Average the results of several titrations.

4. Procedure

a. *Removal of suspended solids:* If sample contains suspended solids, filter through a 0.45-μm-pore-diam membrane filter.

b. *Color development:* If sample pH is not between 5 and 9, adjust to that range with $1N$ HCl or NH_4OH as required. To 50.0 mL sample, or to a portion diluted to 50.0 mL, add 2 mL color reagent and mix.

c. *Photometric measurement:* Between 10 min and 2 h after adding color reagent to samples and standards, measure absorbance at 543 nm. As a guide use the following light paths for the indicated NO_2^--N concentrations:

Light Path Length cm	NO_2^--N μg/L
1	2–25
5	2–6
10	<2

5. Calculation

Prepare a standard curve by plotting absorbance of standards against NO_2^--N concentration. Compute sample concentration directly from curve.

6. Precision and Bias

In a single laboratory using wastewater samples at concentrations of 0.04, 0.24, 0.55, and 1.04 mg NO^-_3 + NO_2^--N/L, the standard deviations were ±0.005, ± 0.004, ± 0.005, and ± 0.01, respectively. In a single laboratory using wastewater samples at concentrations of 0.24, 0.55, and 1.05 mg NO_3^- + NO_2^--N/L, the recoveries were 100%, 102%, and 100%, respectively.[1]

7. Reference

1. U.S. Environmental Protection Agency. 1979. Methods for Chemical Analysis of Water and Wastes. Method 353. 3. U.S. Environmental Protection Agency, Washington, D.C.

8. Bibliography

Boltz, D.F., ed. 1958. Colorimetric Determination of Nonmetals. Interscience Publishers, New York, N.Y.

Nydahl, F. 1976. On the optimum conditions for the reduction of nitrate by cadmium. *Talanta* 23:349.

4500-NO_2^- C. (Reserved)

4500-NO_3^- NITROGEN (NITRATE)*

4500-NO_3^- A. Introduction

1. Selection of Method

Determination of nitrate (NO_3^-) is difficult because of the relatively complex procedures required, the high probability that interfering constituents will be present, and the limited concentration ranges of the various techniques.

An ultraviolet (UV) technique (Method B) that measures the absorbance of NO_3^- at 220 nm is suitable for screening uncontaminated water (low in organic matter).

Screen a sample; if necessary, then select a method suitable for its concentration range and probable interferences. Nitrate may be determined by ion chromatography (Section 4110) or capillary ion electrophoresis (Section 4140). Applicable ranges for other methods are: nitrate electrode method (D), 0.14 to 1400 mg NO_3^--N/L; cadmium reduction method (E), 0.01 to 1.0 mg NO_3^--N/L; automated cadmium reduction methods (F and I), 0.001 to 10 mg NO_3^--N/L. For higher NO_3^--N concentrations, dilute into the range of the selected method.

Colorimetric methods (B, E) require an optically clear sample. Filter turbid sample through 0.45-μm-pore-diam membrane filter. Test filters for nitrate contamination.

* Approved by Standard Methods Committee, 1997.

2. Storage of Samples

Start NO$_3^-$ determinations promptly after sampling. If storage is necessary, store for up to 2 d at 4°C; disinfected samples are stable much longer without acid preservation. For longer storage of unchlorinated samples, preserve with 2 mL conc H$_2$SO$_4$/L and store at 4°C. NOTE: When sample is preserved with acid, NO$_3^-$ and NO$_2^-$ cannot be determined as individual species.

4500-NO$_3^-$ B.   Ultraviolet Spectrophotometric Screening Method

1. General Discussion

a. Principle: Use this technique only for screening samples that have low organic matter contents, i.e., uncontaminated natural waters and potable water supplies. The NO$_3^-$ calibration curve follows Beer's law up to 11 mg N/L.

Measurement of UV absorption at 220 nm enables rapid determination of NO$_3^-$. Because dissolved organic matter also may absorb at 220 nm and NO$_3^-$ does not absorb at 275 nm, a second measurement made at 275 nm may be used to correct the NO$_3^-$ value. The extent of this empirical correction is related to the nature and concentration of organic matter and may vary from one water to another. Consequently, this method is not recommended if a significant correction for organic matter absorbance is required, although it may be useful in monitoring NO$_3^-$ levels within a water body with a constant type of organic matter. Correction factors for organic matter absorbance can be established by the method of additions in combination with analysis of the original NO$_3^-$ content by another method. Sample filtration is intended to remove possible interference from suspended particles. Acidification with 1N HCl is designed to prevent interference from hydroxide or carbonate concentrations up to 1000 mg CaCO$_3$/L. Chloride has no effect on the determination.

b. Interference: Dissolved organic matter, surfactants, NO$_2^-$, and Cr^{6+} interfere. Various inorganic ions not normally found in natural water, such as chlorite and chlorate, may interfere. Inorganic substances can be compensated for by independent analysis of their concentrations and preparation of individual correction curves. For turbid samples, see ¶ A.1.

2. Apparatus

Spectrophotometer, for use at 220 nm and 275 nm with matched silica cells of 1-cm or longer light path.

3. Reagents

a. Nitrate-free water: Use redistilled or distilled, deionized water of highest purity to prepare all solutions and dilutions.

b. Stock nitrate solution: Dry potassium nitrate (KNO$_3$) in an oven at 105°C for 24 h. Dissolve 0.7218 g in water and dilute to 1000 mL; 1.00 mL = 100 µg NO$_3^-$-N. Preserve with 2 mL CHCl$_3$/L. This solution is stable for at least 6 months.

c. Intermediate nitrate solution: Dilute 100 mL stock nitrate solution to 1000 mL with water; 1.00 mL = 10.0 µg NO$_3^-$-N. Preserve with 2 mL CHCl$_3$/L. This solution is stable for 6 months.

d. Hydrochloric acid solution, HCl, 1N.

4. Procedure

a. Treatment of sample: To 50 mL clear sample, filtered if necessary, add 1 mL HCl solution and mix thoroughly.

b. Preparation of standard curve: Prepare NO$_3^-$ calibration standards in the range 0 to 7 mg NO$_3^-$-N/L by diluting to 50 mL the following volumes of intermediate nitrate solution: 0, 1.00, 2.00, 4.00, 7.00 ... 35.0 mL. Treat NO$_3^-$ standards in same manner as samples.

c. Spectrophotometric measurement: Read absorbance or transmittance against redistilled water set at zero absorbance or 100% transmittance. Use a wavelength of 220 nm to obtain NO$_3^-$ reading and a wavelength of 275 nm to determine interference due to dissolved organic matter.

5. Calculation

For samples and standards, subtract two times the absorbance reading at 275 nm from the reading at 220 nm to obtain absorbance due to NO$_3^-$. Construct a standard curve by plotting absorbance due to NO$_3^-$ against NO$_3^-$-N concentration of standard. Using corrected sample absorbances, obtain sample concentrations directly from standard curve. NOTE: If correction value is more than 10% of the reading at 220 nm, do not use this method.

6. Bibliography

HOATHER, R.C. & R.F. RACKMAN. 1959. Oxidized nitrogen and sewage effluents observed by ultraviolet spectrophotometry. *Analyst* 84:549.

GOLDMAN, E. & R. JACOBS. 1961. Determination of nitrates by ultraviolet absorption. *J. Amer. Water Works Assoc.* 53:187.

ARMSTRONG, F.A.J. 1963. Determination of nitrate in water by ultraviolet spectrophotometry. *Anal. Chem.* 35:1292.

NAVONE, R. 1964. Proposed method for nitrate in potable waters. *J. Amer. Water Works Assoc.* 56:781.

4500-NO₃⁻ C. (Reserved)

4500-NO₃⁻ D. Nitrate Electrode Method

1. General Discussion

a. Principle: The NO_3^- ion electrode is a selective sensor that develops a potential across a thin, porous, inert membrane that holds in place a water-immiscible liquid ion exchanger. The electrode responds to NO_3^- ion activity between about 10^{-5} and 10^{-1} M (0.14 to 1400 mg NO_3^--N/L). The lower limit of detection is determined by the small but finite solubility of the liquid ion exchanger.

b. Interferences: Chloride and bicarbonate ions interfere when their weight ratios to NO_3^--N are >10 or >5, respectively. Ions that are potential interferences but do not normally occur at significant levels in potable waters are NO_2^-, CN^-, S^{2-}, Br^-, I^-, ClO_3^-, and ClO_4^-. Although the electrodes function satisfactorily in buffers over the range pH 3 to 9, erratic responses have been noted where pH is not held constant. Because the electrode responds to NO_3^- activity rather than concentration, ionic strength must be constant in all samples and standards. Minimize these problems by using a buffer solution containing Ag_2SO_4 to remove Cl^-, Br^-, I^-, S^{2-}, and CN^-, sulfamic acid to remove NO_2^-, a buffer at pH 3 to eliminate HCO_3^- and to maintain a constant pH and ionic strength, and $Al_2(SO_4)_3$ to complex organic acids.

2. Apparatus

a. pH meter, expanded-scale or digital, capable of 0.1 mV resolution.

*b. Double-junction reference electrode.** Fill outer chamber with $(NH_4)_2SO_4$ solution.

c. Nitrate ion electrode:† Carefully follow manufacturer's instructions regarding care and storage.

d. Magnetic stirrer: TFE-coated stirring bar.

3. Reagents

a. Nitrate-free water: Prepare as described in ¶ B.3*a*. Use for all solutions and dilutions.

b. Stock nitrate solution: Prepare as described in ¶ B.3*b*.

c. Standard nitrate solutions: Dilute 1.0, 10, and 50 mL stock nitrate solution to 100 mL with water to obtain standard solutions of 1.0, 10, and 50 mg NO_3^--N/L, respectively.

d. Buffer solution: Dissolve 17.32 g $Al_2(SO_4)_3\cdot18H_2O$, 3.43 g Ag_2SO_4, 1.28 g H_3BO_3, and 2.52 g sulfamic acid (H_2NSO_3H), in

about 800 mL water. Adjust to pH 3.0 by slowly adding 0.10*N* NaOH. Dilute to 1000 mL and store in a dark glass bottle.

e. Sodium hydroxide, NaOH, 0.1*N*.

f. Reference electrode filling solution: Dissolve 0.53 g $(NH_4)_2SO_4$ in water and dilute to 100 mL.

4. Procedure

a. Preparation of calibration curve: Transfer 10 mL of 1 mg NO_3^--N/L standard to a 50-mL beaker, add 10 mL buffer, and stir with a magnetic stirrer. Immerse tips of electrodes and record millivolt reading when stable (after about 1 min). Remove electrodes, rinse, and blot dry. Repeat for 10-mg NO_3^--N/L and 50-mg NO_3^--N/L standards. Plot potential measurements against NO_3^--N concentration on semilogarithmic graph paper, with NO_3^--N concentration on the logarithmic axis (abscissa) and potential (in millivolts) on the linear axis (ordinate). A straight line with a slope of $+57 \pm 3$ mV/decade at 25°C should result. Recalibrate electrodes several times daily by checking potential reading of the 10 mg NO_3^--N standard and adjusting the calibration control until the reading plotted on the calibration curve is displayed again.

b. Measurement of sample: Transfer 10 mL sample to a 50-mL beaker, add 10 mL buffer solution, and stir (for about 1 min) with a magnetic stirrer. Measure standards and samples at about the same temperature. Immerse electrode tips in sample and record potential reading when stable (after about 1 min). Read concentration from calibration curve.

5. Precision

Over the range of the method, precision of ±0.4 mV, corresponding to 2.5% in concentration, is expected.

6. Bibliography

LANGMUIR, D. & R.I. JACOBSON. 1970. Specific ion electrode determination of nitrate in some fresh waters and sewage effluents. *Environ. Sci. Technol.* 4:835.

KEENEY, D.R., B.H. BYRNES & J.J. GENSON. 1970. Determination of nitrate in waters with nitrate-selective ion electrode. *Analyst* 95:383.

MILHAM, P.J., A.S. AWAD, R.E. PAULL & J.H. BULL. 1970. Analysis of plants, soils and waters for nitrate by using an ion-selective electrode. *Analyst* 95:751.

SYNNOTT, J.C., S.J. WEST & J.W. ROSS. 1984. Comparison of ion-selective electrode and gas-sensing electrode technique for measurement of nitrate in environmental samples. *In* Pawlowski et al., eds., Studies in Environmental Science, No. 23, Chemistry for Protection of the Environment. Elsevier Press, New York, N.Y.

---

* Orion Model 90-02, or equivalent.
† Orion Model 93-07, Corning Model 476134, or equivalent.

4500-NO$_3^-$ E. Cadmium Reduction Method

1. General Discussion

a. Principle: NO$_3^-$ is reduced almost quantitatively to nitrite (NO$_2^-$) in the presence of cadmium (Cd). This method uses commercially available Cd granules treated with copper sulfate (CuSO$_4$) and packed in a glass column.

The NO$_2^-$ produced thus is determined by diazotizing with sulfanilamide and coupling with N-(1-naphthyl)-ethylenediamine dihydrochloride to form a highly colored azo dye that is measured colorimetrically. A correction may be made for any NO$_2^-$ present in the sample by analyzing without the reduction step. The applicable range of this method is 0.01 to 1.0 mg NO$_3^-$-N/L. The method is recommended especially for NO$_3^-$ levels below 0.1 mg N/L where other methods lack adequate sensitivity.

b. Interferences: Suspended matter in the column will restrict sample flow. For turbid samples, see ¶ A.1. Concentrations of iron, copper, or other metals above several milligrams per liter lower reduction efficiency. Add EDTA to samples to eliminate this interference. Oil and grease will coat the Cd surface. Remove by pre-extraction with an organic solvent (see Section 5520). Residual chlorine can interfere by oxidizing the Cd column, reducing its efficiency. Check samples for residual chlorine (see DPD methods in Section 4500-Cl). Remove residual chlorine by adding sodium thiosulfate (Na$_2$S$_2$O$_3$) solution (Section 4500-NH$_3$.B.3*d*). Sample color that absorbs at about 540 nm interferes.

2. Apparatus

a. Reduction column: Purchase or construct the column* (Figure 4500-NO$_3^-$:1) from a 100-mL volumetric pipet by removing the top portion. The column also can be constructed from two pieces of tubing joined end to end: join a 10-cm length of 3-cm-ID tubing to a 25-cm length of 3.5-mm-ID tubing. Add a TFE stopcock with metering valve[1] to control flow rate.

b. Colorimetric equipment: One of the following is required:

1) *Spectrophotometer,* for use at 543 nm, providing a light path of 1 cm or longer.

2) *Filter photometer,* with light path of 1 cm or longer and equipped with a filter having maximum transmittance near 540 nm.

3. Reagents

a. Nitrate-free water: See ¶ B.3*a*. The absorbance of a reagent blank prepared with this water should not exceed 0.01. Use for all solutions and dilutions.

b. Copper-cadmium granules: Wash 25 g new or used 20- to 100-mesh Cd granules† with 6*N* HCl and rinse with water. Swirl Cd with 100 mL 2% CuSO$_4$ solution for 5 min or until blue color partially fades. Decant and repeat with fresh CuSO$_4$ until a brown colloidal precipitate begins to develop. Gently flush with water to remove all precipitated Cu.

c. Color reagent: Prepare as directed in Section 4500-NO$_2^-$.B.3*b*.

---

* Tudor Scientific Glass Co., 555 Edgefield Road, Belvedere, SC 29841, Cat. TP-1730, or equivalent.
† EM Laboratories, Inc., 500 Exec. Blvd., Elmsford, NY, Cat. 2001, or equivalent.

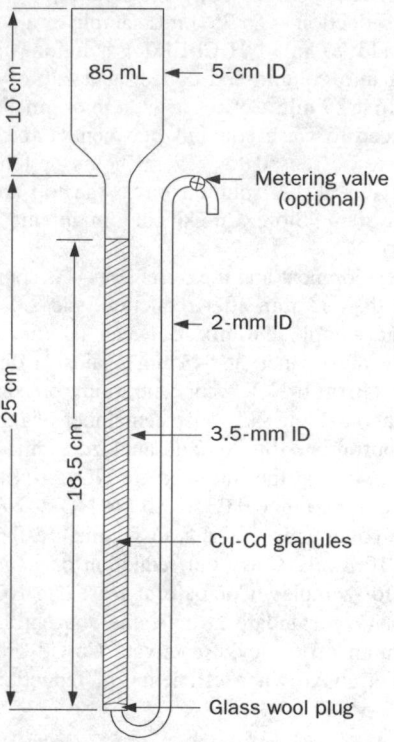

Figure 4500-NO$_3^-$:1. Reduction column.

d. Ammonium chloride-EDTA solution: Dissolve 13 g NH$_4$Cl and 1.7 g disodium ethylenediamine tetraacetate in 900 mL water. Adjust to pH 8.5 with conc NH$_4$OH and dilute to 1 L.

e. Dilute ammonium chloride-EDTA solution: Dilute 300 mL NH$_4$Cl-EDTA solution to 500 mL with water.

f. Hydrochloric acid, HCl, 6*N*.

g. Copper sulfate solution, 2%: Dissolve 20 g CuSO$_4$.5H$_2$O in 500 mL water and dilute to 1 L.

h. Stock nitrate solution: Prepare as directed in ¶ B.3*b*.

i. Intermediate nitrate solution: Prepare as directed in ¶ B.3*c*.

j. Stock nitrite solution: See Section 4500-NO$_2^-$.B.3*e*.

k. Intermediate nitrite solution: See Section 4500-NO$_2^-$.B.3*f*.

l. Working nitrite solution: Dilute 50.0 mL intermediate nitrite solution to 500 mL with nitrite-free water; 1.00 mL = 5 µg NO$_2^-$-N.

4. Procedure

a. Preparation of reduction column: Insert a glass wool plug into bottom of reduction column and fill with water. Add sufficient Cu-Cd granules to produce a column 18.5 cm long. Maintain water level above Cu-Cd granules to prevent entrapment of air. Wash column with 200 mL dilute NH$_4$Cl-EDTA solution. Activate column by passing through it, at 7 to 10 mL/min, at least 100 mL of a solution composed of 25% 1.0 mg NO$_3^-$-N/L standard and 75% NH$_4$Cl-EDTA solution.

b. Treatment of sample:

1) Turbidity removal—For turbid samples, see ¶ A.1.

2) *pH adjustment*—Adjust pH to between 7 and 9, as necessary, using a pH meter and dilute HCl or NaOH. This insures a pH of 8.5 after adding NH₄Cl- EDTA solution.

3) *Sample reduction*—To 25.0 mL sample or a portion diluted to 25.0 mL, add 75 mL NH₄Cl-EDTA solution and mix. Pour mixed sample into column and collect at a rate of 7 to 10 mL/min. Discard first 25 mL. Collect the rest in original sample flask. There is no need to wash columns between samples, but if columns are not to be reused for several hours or longer, pour 50 mL dilute NH₄Cl-EDTA solution on to the top and let it pass through the system. Store Cu-Cd column in this solution and never let it dry.

4) *Color development and measurement*—As soon as possible, and not more than 15 min after reduction, add 2.0 mL color reagent to 50 mL sample and mix. Between 10 min and 2 h afterward, measure absorbance at 543 nm against a distilled water-reagent blank. NOTE: If NO_3^- concentration exceeds the standard curve range (about 1 mg N/L), use remainder of reduced sample to make an appropriate dilution and analyze again.

c. Standards: Using the intermediate NO_3^-- N solution, prepare standards in the range 0.05 to 1.0 mg NO_3^--N/L by diluting the following volumes to 100 mL in volumetric flasks: 0.5, 1.0, 2.0, 5.0, and 10.0 mL. Carry out reduction of standards exactly as described for samples. Compare at least one NO_2^- standard to a reduced NO_3^- standard at the same concentration to verify reduction column efficiency. Reactivate Cu-Cd granules as described in ¶ 3*b* above when efficiency of reduction falls below about 75%.

5. Calculation

Obtain a standard curve by plotting absorbance of standards against NO_3^--N concentration. Compute sample concentrations directly from standard curve. Report as milligrams oxidized N per liter (the sum of NO_3^--N plus NO_2^--N) unless the concentration of NO_2^--N is separately determined and subtracted.

6. Precision and Bias

In a single laboratory using wastewater samples at concentrations of 0.04, 0.24, 0.55, and 1.04 mg NO_3^- + NO_2^--N/L, the standard deviations were ±0.005, ±0.004, ±0.005, and ±0.01, respectively. In a single laboratory using wastewater with additions of 0.24, 0.55, and 1.05 mg NO_3^- + NO_2^--N/L, the recoveries were 100%, 102%, and 100%, respectively.[2]

7. References

1. WOOD, E.D., F.A.J. ARMSTRONG & F.A. RICHARDS. 1967. Determination of nitrate in sea water by cadmium-copper reduction to nitrite. *J. Mar. Biol. Assoc. U.K.* 47:23.
2. U.S. ENVIRONMENTAL PROTECTION AGENCY. 1979. Methods for Chemical Analysis of Water and Wastes, Method 353.3. U.S. Environmental Protection Agency, Washington, D.C.

8. Bibliography

STRICKLAND, J.D.H. & T.R. PARSONS. 1972. A Practical Handbook of Sea Water Analysis, 2nd ed. Bull. No. 167, Fisheries Research Board Canada, Ottawa, Ont.
NYDAHL, F. 1976. On the optimum conditions for the reduction of nitrate by cadmium. *Talanta* 23:349.
AMERICAN SOCIETY FOR TESTING AND MATERIALS. 1987. Annual Book of ASTM Standards, Vol. 11.01. American Soc. Testing & Materials, Philadelphia, Pa.

4500-NO_3^- F. Automated Cadmium Reduction Method

1. General Discussion

a. Principle: See ¶ E.1*a*.

b. Interferences: Sample turbidity may interfere. Remove by filtration before analysis. Sample color that absorbs in the photometric range used for analysis also will interfere.

c. Application: Nitrate and nitrite, singly or together in potable, surface, and saline waters and domestic and industrial wastewaters, can be determined over a range of 0.5 to 10 mg N/L.

2. Apparatus

Automated analytical equipment: An example of the continuous-flow analytical instrument consists of the components shown in Figure 4500-NO_3^-:2.

3. Reagents

a. Deionized distilled water: See ¶ B.3*a*.

b. Copper sulfate solution: Dissolve 20 g CuSO₄·5H₂O in 500 mL water and dilute to 1 L.

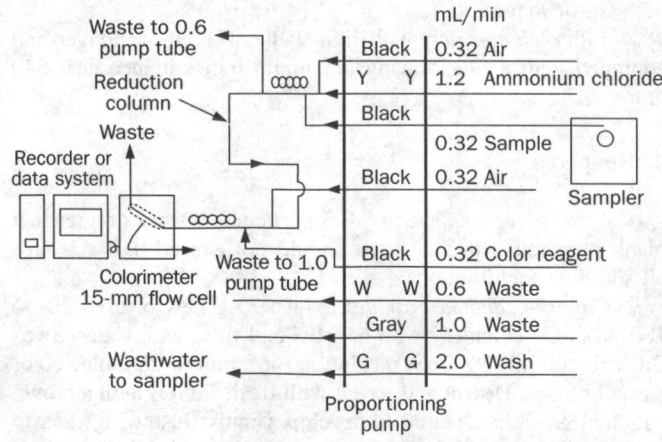

Figure 4500-NO_3^-:2. Nitrate-nitrite manifold.

c. Wash solution: Use water for unpreserved samples. For samples preserved with H_2SO_4, add 2 mL conc H_2SO_4/L wash water.

d. Copper-cadmium granules: See ¶ E.3*b*.

e. Hydrochloric acid, HCl, conc.

f. Ammonium hydroxide, NH_4OH, conc.

g. Color reagent: To approximately 800 mL water, add, while stirring, 100 mL conc H_3PO_4, 40 g sulfanilamide, and 2 g *N*-(1-naphthyl)-ethylenediamine dihydrochloride. Stir until dissolved and dilute to 1 L. Store in brown bottle and keep in the dark when not in use. This solution is stable for several months.

h. Ammonium chloride solution: Dissolve 85 g NH_4Cl in water and dilute to 1 L. Add 0.5 mL polyoxyethylene 23 lauryl ether.*

i. Stock nitrate solution: See ¶ B.3*b*.

j. Intermediate nitrate solution: See ¶ B.3*c*.

k. Standard nitrate solutions: Using intermediate NO$_3$$^-$-N solution and water, prepare standards for calibration curve in appropriate nitrate range. Compare at least one NO$_2$$^-$ standard to a NO$_3$$^-$ standard at the same concentration to verify column reduction efficiency. To examine saline waters prepare standard solutions with the substitute ocean water described in Section 4500-NH$_3$.G.3*g*.

l. Standard nitrite solution: See 4500-NO$_2$$^-$.B.3*g*.

4. Procedure

Set up manifold as shown in Figure 4500-NO$_3$$^-$:2 and follow general procedure described by the manufacturer.

If sample pH is below 5 or above 9, adjust to between 5 and 9 with either conc HCl or conc NH_4OH.

5. Calculation

Prepare standard curves by plotting response of standards processed through the manifold against NO$_3$$^-$-N concentration in standards. Compute sample NO$_3$$^-$-N concentration by comparing sample response with standard curve.

---

* Brij-35, available from ICI Americas, Inc., Wilmington, DE, or equivalent.

6. Precision and Bias

Data obtained in three laboratories with an automated system based on identical chemical principles but having slightly different configurations are given in the table below. Analyses were conducted on four natural water samples containing exact increments of inorganic nitrate:

Increment as NO$_3$$^-$-N μg/L	Standard Deviation μg N/L	Bias %	Bias μg N/L
290	12	+ 5.75	+ 17
350	92	+ 18.10	+ 63
2310	318	+ 4.47	+ 103
2480	176	- 2.69	- 67

In a single laboratory using surface water samples at concentrations of 100, 200, 800, and 2100 μg N/L, the standard deviations were 0, ±40, ±50, and ±50 μg/L, respectively, and at concentrations of 200 and 2200 μg N/L, recoveries were 100 and 96%, respectively.

Precision and bias for the system described herein are believed to be comparable.

7. Bibliography

FIORE, J. & J.E. O'BRIEN. 1962. Automation in sanitary chemistry—Parts 1 and 2. Determination of nitrates and nitrites. *Wastes Eng.* 33:128 & 238.

ARMSTRONG, F.A., C.R. STEARNS & J.D. STRICKLAND. 1967. The measurement of upwelling and subsequent biological processes by means of the Technicon AutoAnalyzer and associated equipment. *Deep Sea Res.* 14:381.

U.S. ENVIRONMENTAL PROTECTION AGENCY. 1979. Methods for Chemical Analysis of Water and Wastes. U.S. Environmental Protection Agency, Washington, D.C.

AMERICAN SOCIETY FOR TESTING AND MATERIALS. 1987. Annual Book of ASTM Standards, Vol. 11.01. American Soc. Testing & Materials, Philadelphia, Pa.

4500-NO$_3$$^-$ G. (Reserved)

4500-NO$_3$$^-$ H. Automated Hydrazine Reduction Method

1. General Discussion

a. Principle: NO$_3$$^-$ is reduced to NO$_2$$^-$ with hydrazine sulfate. The NO$_2$$^-$ (originally present) plus reduced NO$_3$$^-$ is determined by diazotization with sulfanilamide and coupling with *N*-(1-naphthyl)-ethylenediamine dihydrochloride to form a highly colored azo dye that is measured colorimetrically.

b. Interferences: Sample color that absorbs in the photometric range used will interfere. Concentrations of sulfide ion of less than 10 mg/L cause variations of NO$_3$$^-$ and NO$_2$$^-$ concentrations of ±10%.

c. Application: NO$_3$$^-$ + NO$_2$$^-$ in potable and surface water and in domestic and industrial wastes can be determined over a range of 0.01 to 10 mg N/L.

2. Apparatus

Automated analytical equipment: An example of the continuous-flow analytical instrument consists of the components shown in Figure 4500-NO_3^-:3.

3. Reagents

a. Color developing reagent: To approximately 500 mL water add 200 mL conc phosphoric acid and 10 g sulfanilamide. After sulfanilamide is dissolved completely, add 0.8 g *N*-(1-naphthyl)-ethylenediamine dihydrochloride. Dilute to 1 L with water, store in a dark bottle, and refrigerate. Solution is stable for approximately 1 month.

b. Copper sulfate stock solution: Dissolve 2.5 g $CuSO_4 \cdot 5H_2O$ in water and dilute to 1 L.

c. Copper sulfate dilute solution: Dilute 20 mL stock solution to 2 L.

d. Sodium hydroxide stock solution, 10N: Dissolve 400 g NaOH in 750 mL water, cool, and dilute to 1 L.

e. Sodium hydroxide, 1.0N: Dilute 100 mL stock NaOH solution to 1 L.

f. Hydrazine sulfate stock solution: Dissolve 27.5 g $N_2H_4 \cdot H_2SO_4$ in 900 mL water and dilute to 1 L. This solution is stable for approximately 6 months. CAUTION: *Toxic if ingested. Mark container with appropriate warning.*

g. Hydrazine sulfate dilute solution: Dilute 22 mL stock solution to 1 L.

h. Stock nitrate solution: See ¶ B.3*b*.

i. Intermediate nitrate solution: See ¶ B.3*c*.

j. Standard nitrate solutions: Prepare NO_3^- calibration standards in the range 0 to 10 mg/L by diluting to 100 mL the following volumes of stock nitrate solution: 0, 0.5, 1.0, 2.0 . . . 10.0 mL. For standards in the range of 0.01 mg/L use intermediate nitrate solution. Compare at least one nitrite standard to a nitrate standard at the same concentration to verify the efficiency of the reduction.

k. Standard nitrite solution: See ¶s 4500-NO_2^-.B.3*e, f,* and *g.*

4. Procedure

Set up manifold as shown in Figure 4500-NO_3^-:3 and follow general procedure described by manufacturer. Run a 2.0-mg NO_3^--N/L and a 2.0-mg NO_2^--N/L standard through the system to check for 100% reduction of nitrate to nitrite. The two peaks should be of equal height; if not, adjust concentration of the hydrazine sulfate solution: If the NO_3^- peak is lower than the NO_2^- peak, increase concentration of hydrazine sulfate until they are equal; if the NO_3^- peak is higher than the NO_2^- reduce concentration of hydrazine sulfate. When correct concentration has been determined, no further adjustment should be necessary.

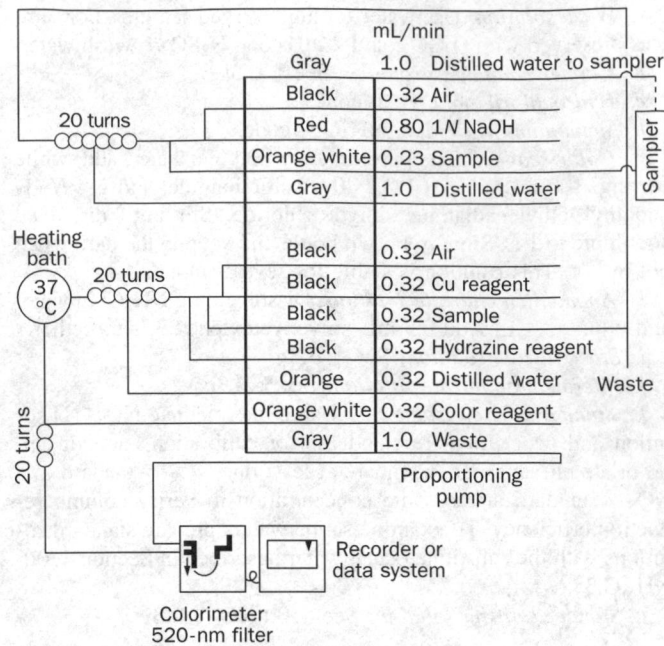

Figure 4500-NO_3^-:3. Nitrate-nitrite manifold.

5. Calculation

Prepare a standard curve by plotting response of processed standards against known concentrations. Compute concentrations of samples by comparing response with standard curve.

6. Precision and Bias

In a single laboratory using drinking water, surface water, and industrial waste at concentrations of 0.39, 1.15, 1.76, and 4.75 mg NO_3^--N/L, the standard deviations were ± 0.02, ± 0.01, ± 0.02, and ± 0.03, respectively. In a single laboratory using drinking water at concentrations of 0.75 and 2.97 mg NO_3^--N/L, the recoveries were 99% and 101%.[1]

7. Reference

1. U.S. ENVIRONMENTAL PROTECTION AGENCY. 1979. Methods for Chemical Analysis of Water and Wastes. U.S. Environmental Protection Agency, Washington, D.C.

8. Bibliography

KAMPHAKE, L., S. HANNAH & J. COHEN. 1967. Automated analysis for nitrate by hydrazine reduction. *Water Res.* 1:205.

4500-NO$_3^-$ I. Cadmium Reduction Flow Injection Method (PROPOSED)

1. General Discussion

a. Principle: The nitrate in the sample is reduced quantitatively to nitrite by passage of the sample through a copperized cadmium column. The resulting nitrite plus any nitrite originally in the sample is determined as a sum by diazotizing the nitrite with sulfanilamide followed by coupling with *N*-(1-naphthyl)ethylenediamine dihydrochloride. The resulting water-soluble dye has a magenta color; absorbance of the color at 540 nm is proportional to the nitrate + nitrite in the sample. This sum also is known as total oxidized nitrogen (TON).

Nitrite alone can be determined by removing the cadmium column, recalibrating the method, and repeating the sample analyses. A TON and a nitrite FIA method also can be run in parallel for a set of samples. In this arrangement, the concentrations determined in the nitrite method can be subtracted from the corresponding concentrations determined in the TON method to give the resulting nitrate concentrations of the samples.

Also see Section 4500-NO$_2^-$ and Section 4130, Flow Injection Analysis (FIA).

b. Interferences: Remove large or fibrous particulates by filtering sample through glass wool. Guard against nitrate and nitrite contamination from reagents, water, glassware, and the sample preservation process.

Residual chlorine can interfere by oxidizing the cadmium reduction column. Samples that contain large concentrations of oil and grease will coat the surface of the cadmium. Eliminate this interference by pre-extracting sample with an organic solvent.

Low results would be obtained for samples that contain high concentrations of iron, copper, or other metals. In this method, EDTA is added to the buffer to reduce this interference.

Also see Section 4500-NO$_2^-$.B.1*b* and *c* and Section 4500-NO$_3^-$.A.2 and B.1*b*.

2. Apparatus

Flow injection analysis equipment consisting of:
a. FIA injection valve with sample loop or equivalent.
b. Multichannel proportioning pump.
c. FIA manifold (Figure 4500-NO$_3^-$:4) with flow cell. Relative flow rates only are shown in Figure 4500-NO$_3^-$:4. Tubing volumes are given as an example only; they may be scaled down proportionally. Use manifold tubing of an inert material such as TFE.
d. Absorbance detector, 540 nm, 10-nm bandpass.
e. Injection valve control and data acquisition system.

3. Reagents

Use reagent water (>10 megohm) to prepare carrier and all solutions. To prevent bubble formation, degas carrier and buffer with helium. Pass He at 140 kPa (20 psi) through a helium degassing tube. Bubble He through 1 L solution for 1 min. As an alternative to preparing reagents by weight/weight, use weight/volume.

a. Ammonium chloride buffer: CAUTION: *Fumes. Use a hood.* To a tared 1-L container add 800.0 g water, 126 g conc hydrochloric acid, HCl, 55.6 g ammonium hydroxide, NH$_4$OH, and

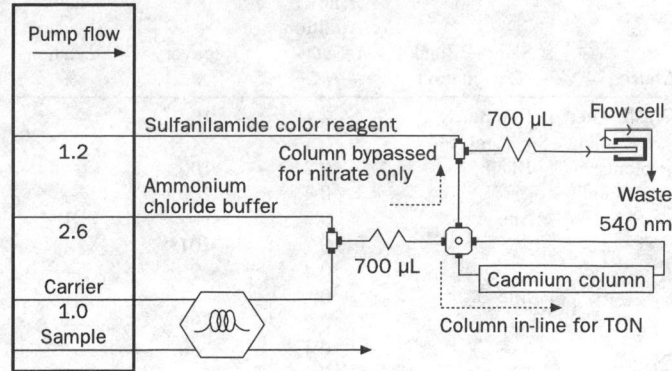

Figure 4500-NO$_3^-$:4. FIA nitrate + nitrite manifold.

1.0 g disodium EDTA. Shake until dissolved. The pH of this buffer should be 8.5.

b. Sulfanilamide color reagent: To a tared, dark 1-L container add 876 g water, 170 g 85% phosphoric acid, H$_3$PO$_4$, 40.0 g sulfanilamide, and 1.0 g *N*-(1-naphthyl)ethylenediamine dihydrochloride (NED). Shake until wetted and stir with stir bar for 30 min until dissolved. This solution is stable for 1 month.

c. Hydrochloric acid, HCl, 1*M*: To a 100-mL container, add 92 g water, then add 9.6 g conc HCl. Stir or shake to mix.

d. Copper sulfate solution, 2%: To a 1-L container, add 20 g copper sulfate pentahydrate, CuSO$_4$·5H$_2$O, to 991 g water. Stir or shake to dissolve.

e. Copperized cadmium granules: Place 10 to 20 g coarse cadmium granules (0.3- to 1.5-mm-diam) in a 250-mL beaker. Wash with 50 mL acetone, then water, then two 50-mL portions 1*M* HCl (¶ 3*c*). Rinse several times with water. CAUTION: *Cadmium is toxic and carcinogenic. Collect and store all waste cadmium. When handling cadmium, wear gloves and follow the precautions described on the cadmium's Material Safety Data Sheet.*

Add 100 mL 2% copper sulfate solution (¶ 3*d*) to cadmium prepared above. Swirl for about 5 min, then decant the liquid and repeat with a fresh 100-mL portion of the 2% copper sulfate solution. Continue this process until the blue aqueous copper color persists. Decant and wash with at least five portions of ammonium chloride buffer (¶ 3*a*) to remove colloidal copper. The cadmium should be black or dark gray. The copperized cadmium granules may be stored in a bottle under ammonium chloride buffer.

f. Stock nitrate standard, 200 mg N/L: In a 1-L volumetric flask dissolve 1.444 g potassium nitrate, KNO$_3$, in about 600 mL water. Add 2 mL chloroform. Dilute to mark and invert to mix. This solution is stable for 6 months.

g. Stock nitrite standard, 200.0 mg N/L: In a 1-L volumetric flask dissolve 0.986 g sodium nitrite, NaNO$_2$, or 1.214 g potassium nitrite, KNO$_2$, in approximately 800 mL water. Add 2 mL chloroform. Dilute to mark and invert to mix. Refrigerate.

h. Standard solution: Prepare nitrate or nitrite standards in the desired concentration range, using the stock standards (¶ 3*f* or *g*), and diluting with water.

TABLE 4500-NO$_3^-$:I. RESULTS OF SINGLE-LABORATORY STUDIES WITH SELECTED MATRICES

Matrix	Sample/Blank Designation	Known Addition mg NO$_3^-$-N/L	Recovery %	Relative Standard Deviation %
Wastewater treatment plant influent	Reference sample*	—	102	—
	Blank†	0.2	100	—
		0.4	100	—
	Site A‡	0	—	<1
		0.2	101	—
		0.4	96	—
	Site B‡	0	—	<1
		0.2	90	—
		0.4	88	—
	Site C‡	0	—	<1
		0.2	95	—
		0.4	95	—
Wastewater treatment plant effluent	Reference sample*	—	102	—
	Blank†	0.2	100	—
		0.4	95	—
	Site A‡	0	—	0.9
		0.2	95	—
		0.4	102	—
	Site B‡	0	—	0.7
		0.2	91	—
		0.4	101	—
	Site C‡	0	—	0.5
		0.2	91	—
		0.4	96	—
Landfill leachate	Reference sample*	—	98	—
	Blank†	0.2	100	—
		0.4	98	—
	Site A‡	0	—	<1
		0.2	104	—
		0.4	96	—
	Site B‡	0	—	<1
		0.2	95	—
		0.4	94	—
	Site C‡	0	—	<1
		0.2	91	—
		0.4	93	—

* U.S. EPA QC sample, 1.98 mg N/L.
† Determined in duplicate.
‡ Samples without known additions determined four times; samples with known additions determined in duplicate. Typical difference between duplicates: influent, 2%; effluent, 1%; leachate, <1%.

4. Procedure

Set up a manifold equivalent to that in Figure 4500-NO$_3^-$:4 and pack column with copperized cadmium granules. Follow methods supplied by column and instrument manufacturer or laboratory's standard operating procedure for this method. Follow quality control procedures outlined in Section 4020.

5. Calculations

Prepare standard curves by plotting absorbance of standards processed through the manifold versus TON or nitrite concentration. The calibration curve is linear.

If TON includes measurable nitrite concentrations, it is important that the cadmium column be 100% efficient. If the efficiency is less, the nitrite in the sample will give a positive percent error equal to the difference from 100%, causing an error in TON and nitrate determinations. To measure efficiency of the cadmium column, prepare two calibration curves, one using nitrate standards and one using equimolar nitrite standards. The column efficiency is:

Column efficiency = 100%
$\times$ (slope of nitrate curve/slope of nitrite curve)

Determine column efficiency at least weekly.

6. Precision and Bias

In the studies described below, nitrate was measured. There was no significant concentration of nitrite in the samples.

a. Recovery and relative standard deviation: Table 4500-NO$_3^-$:I gives results of single-laboratory studies.

b. MDL: A 800-μL sample loop was used in the method described above. Using a published MDL method,[1] analysts ran 21 replicates of a 2.00-μg N/L standard. These gave a mean of 1.82 μg N/L, a standard deviation of 0.098 μg N/L, and MDL of 0.25 μg N/L. A lower MDL may be obtained by increasing the sample loop volume and increasing the ratio of carrier flow rate to reagent flow rate.

7. Reference

1. U.S. ENVIRONMENTAL PROTECTION AGENCY. 1984. Definition and procedure for the determination of method detection limits. Appendix B to 40 CFR 136 rev. 1.11 amended June 30, 1986. 49 CFR 43430.

4500-N$_{org}$ NITROGEN (ORGANIC)*

4500-N$_{org}$ A.   Introduction

1. Selection of Method

The kjeldahl methods (B and C) determine nitrogen in the tri-negative state. They fail to account for nitrogen in the form of azide, azine, azo, hydrazone, nitrate, nitrite, nitrile, nitro, nitroso, oxime, and semi-carbazone. "Kjeldahl nitrogen" is the sum of organic nitrogen and ammonia nitrogen.

The major factor that influences the selection of a macro- or semi-micro-kjeldahl method to determine organic nitrogen is its concentration. The macro-kjeldahl method is applicable for samples containing either low or high concentrations of organic nitrogen but requires a relatively large sample volume for low concentrations. In the semi-micro-kjeldahl method, which is applicable to samples containing high concentrations of organic nitrogen, the sample volume should be chosen to contain organic plus ammonia nitrogen in the range of 0.2 to 2 mg.

The block digestion method (D) is a micro method with an automated analysis step capable of measuring organic nitrogen as low as 0.1 mg/L when blanks are carefully controlled.

2. Storage of Samples

The most reliable results are obtained on fresh samples. If an immediate analysis is not possible, preserve samples for kjeldahl digestion by acidifying to pH 1.5 to 2.0 with concentrated H_2SO_4 and storing at 4°C. Do not use $HgCl_2$ because it will interfere with ammonia removal.

3. Interferences

a. Nitrate: During kjeldahl digestion, nitrate in excess of 10 mg/L can oxidize a portion of the ammonia released from the digested organic nitrogen, producing N_2O and resulting in a negative interference. When sufficient organic matter in a low state of oxidation is present, nitrate can be reduced to ammonia, resulting in a positive interference. The conditions under which significant interferences occur are not well defined and there is no proven way to eliminate the interference in conjunction with the kjeldahl methods described herein.

b. Inorganic salts and solids: The acid and salt content of the kjeldahl digestion reagent is intended to produce a digestion tem-

* Approved by Standard Methods Committee, 1997.

perature of about 380°C. If the sample contains a very large quantity of salt or inorganic solids that dissolve during digestion, the temperature may rise above 400°C, at which point pyrolytic loss of nitrogen begins to occur. To prevent an excessive digestion temperature, add more H_2SO_4 to maintain the acid-salt balance. Not all salts cause precisely the same temperature rise, but adding 1 mL H_2SO_4/g salt in the sample gives reasonable results. Add the extra acid and the digestion reagent to both sample and reagent blank. Too much acid will lower the digestion temperature below 380°C and result in incomplete digestion and recovery. If necessary, add sodium hydroxide-sodium thiosulfate before the final distillation step to neutralize the excess acid.

Large amounts of salt or solids also may cause bumping during distillation. If this occurs, add more dilution water after digestion.

c. Organic matter: During kjeldahl digestion, H_2SO_4 oxidizes organic matter to CO_2 and H_2O. If a large amount of organic matter is present, a large amount of acid will be consumed, the ratio of salt to acid will increase, and the digestion temperature will increase. If enough organic matter is present, the temperature will rise above 400°C, resulting in pyrolytic loss of nitrogen. To prevent this, add to the digestion flask 10 mL conc H_2SO_4/3 g COD. Alternately, add 50 mL more digestion reagent/g COD. Additional sodium hydroxide-sodium thiosulfate reagent may be necessary to keep the distillation pH high. Because reagents may contain traces of ammonia, treat the reagent blank identically with the samples.

4. Use of a Catalyst

Mercury has been the catalyst of choice for kjeldahl digestion. Because of its toxicity and problems associated with legal disposal of mercury residues, a less toxic catalyst is recommended. Digestion of some samples may be complete or nearly complete without the use of a catalyst. Effective digestion results from the use of a reagent having a salt/acid ratio of 1 g/mL with copper as catalyst (B.3a), and specified temperature (B.2a) and time (B.4c). If a change is made in the reagent formula, report the change and indicate percentage recovery relative to the results for similar samples analyzed using the previous formula.

Before results are considered acceptable, determine nitrogen recovery from samples with known additions of nicotinic acid, to test completeness of digestion; and with ammonium chloride to test for loss of nitrogen.

4500-N$_{org}$ B. Macro-Kjeldahl Method

1. General Discussion

a. Principle: In the presence of H_2SO_4, potassium sulfate (K_2SO_4), and cupric sulfate ($CuSO_4$) catalyst, amino nitrogen of many organic materials is converted to ammonium. Free ammonia also is converted to ammonium. After addition of base, the ammonia is distilled from an alkaline medium and absorbed in boric or sulfuric acid. The ammonia may be determined colorimetrically, by ammonia-selective electrode, or by titration with a standard mineral acid.

b. Selection of ammonia measurement method: The sensitivity of colorimetric methods makes them particularly useful for determining organic nitrogen levels below 5 mg/L. The titrimetric and selective electrode methods of measuring ammonia in the distillate are suitable for determining a wide range of organic nitrogen concentrations. Selective electrode methods and automated colorimetric methods may be used for measurement of ammonia in digestate without distillation. Follow equipment manufacturer's instructions.

2. Apparatus

a. Digestion apparatus: Kjeldahl flasks with a total capacity of 800 mL yield the best results. Digest over a heating device adjusted so that 250 mL water at an initial temperature of 25°C can be heated to a rolling boil in approximately 5 min. For testing, preheat heaters for 10 min if gas-operated or 30 min if electric. A heating device meeting this specification should provide the temperature range of 375 to 385°C for effective digestion.

b. Distillation apparatus: See Section 4500-NH$_3$.B.2a.

c. Apparatus for ammonia determination: See Section 4500-NH$_3$.C.2, D.2, F.2, or G.2.

3. Reagents

Prepare all reagents and dilutions in ammonia-free water.

All of the reagents listed for the determination of Nitrogen (Ammonia), Section 4500-NH$_3$.C.3, D.3, F.3, or G.3, are required, plus the following:

a. Digestion reagent: Dissolve 134 g K_2SO_4 and 7.3 g $CuSO_4$ in about 800 mL water. Carefully add 134 mL conc H_2SO_4. When it has cooled to room temperature, dilute the solution to 1 L with water. Mix well. Keep at a temperature close to 20°C to prevent crystallization.

b. Sodium hydroxide-sodium thiosulfate reagent: Dissolve 500 g NaOH and 25 g $Na_2S_2O_3 \cdot 5H_2O$ in water and dilute to 1 L.

c. Borate buffer solution: See Section 4500-NH$_3$.B.3b.

d. Sodium hydroxide, NaOH, 6N.

4. Procedure

a. Selection of sample volume and sample preparation: Place a measured volume of sample in an 800-mL kjeldahl flask. Select sample size from the following tabulation:

Organic Nitrogen in Sample mg/L	Sample Size mL
0–1	500
1–10	250
10–20	100
20–50	50.0
50–100	25.0

If necessary, dilute sample to 300 mL, neutralize to pH 7, and dechlorinate as described in Section 4500-NH$_3$.B.4b.

b. Ammonia removal: Add 25 mL borate buffer and then 6N NaOH until pH 9.5 is reached. Add a few glass beads or boiling chips such as Hengar Granules #12 and boil off 300 mL. If desired, distill this fraction and determine ammonia nitrogen. Alternately, if ammonia has been determined by the distillation method, use residue in distilling flask for organic nitrogen determination.

For sludge and sediment samples, weigh wet sample in a crucible or weighing bottle, transfer contents to a kjeldahl flask, and determine kjeldahl nitrogen. Follow a similar procedure for ammonia nitrogen and organic nitrogen determined by difference. Determinations of organic and kjeldahl nitrogen on dried sludge and sediment samples are not accurate because drying results in loss of ammonium salts. Measure dry weight of sample on a separate portion.

c. Digestion: Cool and add carefully 50 mL digestion reagent (or substitute 6.7 mL conc H_2SO_4, 6.7 g K_2SO_4, and 0.365 g $CuSO_4$) to distillation flask. Add a few glass beads and, after mixing, heat under a hood or with suitable ejection equipment to remove acid fumes. Boil briskly until the volume is greatly reduced (to about 25 to 50 mL) and copious white fumes are observed (fumes may be dark for samples high in organic matter). Then continue to digest for an additional 30 min. As digestion continues, colored or turbid samples will become transparent and pale green. After digestion, let cool, dilute to 300 mL with water, and mix. Tilt flask away from personnel and carefully add 50 mL sodium hydroxide-thiosulfate reagent to form an alkaline layer at flask bottom. Connect flask to a steamed-out distillation apparatus and swirl flask to insure complete mixing. The pH of the solution should exceed 11.0.

d. Distillation: Distill and collect 200 mL distillate. Use 50 mL indicating boric acid as absorbent solution when ammonia is to be determined by titration. Use 50 mL 0.04N H_2SO_4 solution as absorbent for manual phenate or electrode methods. Extend tip of condenser well below level of absorbent solution and do not let temperature in condenser rise above 29°C. Lower collected distillate free of contact with condenser tip and continue distillation during last 1 or 2 min to cleanse condenser.

e. Final ammonia measurement: Use the titration, ammonia-selective electrode, manual phenate, or automated phenate method, Sections 4500-NH$_3$.C, D, F, and G, respectively.

f. Standards: Carry a reagent blank and standards through all steps of the procedure.

5. Calculation

See Section 4500-NH$_3$.C.5, D.5, F.5, or G.5.

TABLE 4500-N$_{org}$:I. PRECISION DATA FOR KJELDAHL NITROGEN METHOD BASED ON MEAN OF TRIPLICATE ANALYSES OF NICOTINIC ACID

Lab/ Analyst	Nicotinic Acid mg N/L	Recovery of N %	Standard Deviation mg/L	Relative Standard Deviation %
1/1	5	93.3	0.16	3.46
1/2	5	101	0.16	3.17
1/1	10	87.7	0.16	1.84
1/2	10	91.5	0.28	3.06
1/1	20	95.7	0.16	0.84
1/2	20	95.7	0.58	3.03
2/1	0.5	97.4	0.005	1.04
2/2	0.5	95.3	0.027	5.46
3/1	0.5	87.3	0.130	29.9
4/1	0.5	113	0.235	41.7
2/1	1.0	103	0.012	1.15
2/2	1.0	101	0.046	4.63
3/1	1.0	84.3	0.081	9.66
4/1	1.0	99.3	0.396	39.9
2/1	2.0	104	0	0
2/2	2.0	99.2	0.029	1.44
3/1	2.0	89.2	0.071	3.98
4/1	2.0	112	0.139	6.18

6. Precision and Bias

Two analysts in one laboratory prepared reagent water solutions of nicotinic acid and digested them by the macro-kjeldahl method. Ammonia in the distillate was determined by titration. Results are summarized in Table 4500-N$_{org}$:I.

7. Bibliography

KJELDAHL, J. 1883. A new method for the determination of nitrogen in organic matter. *Z. Anal. Chem.* 22:366.

PHELPS, E.B. 1905. The determination of organic nitrogen in sewage by the Kjeldahl process. *J. Infect. Dis.* (Suppl.) 1:225.

MCKENZIE, H.A. & H.S. WALLACE. 1954. The Kjeldahl determination of nitrogen: A critical study of digestion conditions. *Aust. J. Chem.* 7: 55.

MORGAN, G.B., J.B. LACKEY & F.W. GILCREAS. 1957. Quantitative determination of organic nitrogen in water, sewage, and industrial wastes. *Anal. Chem.* 29:833.

BOLTZ, D.F., ed. 1978. Colorimetric Determination of Nonmetals. Interscience Publishers, New York, N.Y.

JONES, M. & D. BRADSHAW. 1989. Copper: An alternative to mercury; more effective than zirconium in kjeldahl digestion of ecological materials. *Commun. Soil Sci. Plant Anal.* 20:1513.

4500-N$_{org}$ C. Semi-Micro-Kjeldahl Method

1. General Discussion

See Section 4500-N$_{org}$.B.1.

2. Apparatus

a. Digestion apparatus: Use kjeldahl flasks with a capacity of 100 mL in a semi-micro-kjeldahl digestion apparatus* equipped with heating elements to accommodate kjeldahl flasks and a suction outlet to vent fumes. The heating elements should provide the temperature range of 375 to 385°C for effective digestion.

b. Distillation apparatus: Use an all-glass unit equipped with a steam-generating vessel containing an immersion heater† (Figure 4500-N$_{org}$:1).

c. pH meter.

d. Apparatus for ammonia determination: See Section 4500-NH$_3$.C.2, D.2, F.2, or G.2.

3. Reagents

All of the reagents listed for the determination of Nitrogen (Ammonia) (Section 4500-NH$_3$.B.3) and Nitrogen (Organic) macro-kjeldahl (Section 4500-N$_{org}$.B.3) are required. Prepare all reagents and dilutions with ammonia-free water.

4. Procedure

a. Selection of sample volume: Determine the sample size from the following tabulation:

Organic Nitrogen in Sample mg/L	Sample Size mL
4–40	50
8–80	25
20–200	10
40–400	5

For sludge and sediment samples weigh a portion of wet sample containing between 0.2 and 2 mg organic nitrogen in a crucible or weighing bottle. Transfer sample quantitatively to a 100-mL beaker by diluting it and rinsing the weighing dish several times with small quantities of water. Make the transfer using as small a quantity of water as possible and do not exceed a total volume of 50 mL. Measure dry weight of sample on a separate portion.

b. Ammonia removal: Pipet 50 mL sample or an appropriate volume diluted to 50 mL with water into a 100-mL beaker. Add 3 mL borate buffer and adjust to pH 9.5 with 6*N* NaOH, using a pH meter. Quantitatively transfer sample to a 100-mL kjeldahl flask and boil off 30 mL. Alternatively, if ammonia removal is not required, digest samples directly as described in ¶ *c* below. Distillation following this direct digestion yields kjeldahl nitrogen concentration rather than organic nitrogen.

c. Digestion: Carefully add 10 mL digestion reagent to kjeldahl flask containing sample. Add five or six glass beads (3- to 4-mm size) to prevent bumping during digestion. Set each heating unit

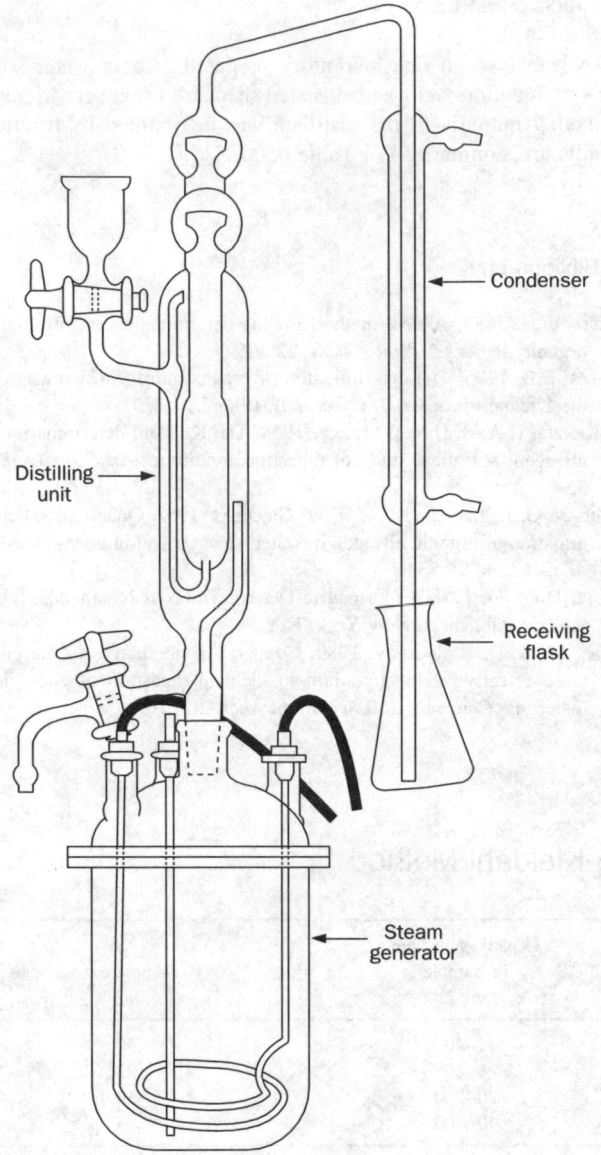

Figure 4500-N$_{org}$:1. Micro-kjeldahl distillation apparatus.

on the micro-kjeldahl digestion apparatus to its medium setting and heat flasks under a hood or with suitable ejection equipment to remove fumes of SO_3. Continue to boil briskly until solution becomes transparent and pale green and copious fumes are observed. Then turn each heating unit up to its maximum setting and digest for an additional 30 min. Cool. Quantitatively transfer digested sample by diluting and rinsing several times into micro-kjeldahl distillation apparatus so that total volume in distillation apparatus does not exceed 30 mL. Add 10 mL sodium hydroxide-thiosulfate reagent and turn on steam.

d. Distillation: Control rate of steam generation to boil contents in distillation unit so that neither escape of steam from tip of condenser nor bubbling of contents in receiving flask occurs. Distill and collect 30 to 40 mL distillate below surface of 10 mL absorbent solution contained in a 125-mL erlenmeyer flask. Use indicating boric acid for a titrimetric finish. Use 10 mL 0.04N H_2SO_4 solution for collecting distillate for the phenate or electrode methods. Extend tip of condenser well below level of absorbent solution and do not let temperature in condenser rise above 29°C. Lower collected distillate free of contact with condenser tip and continue distillation during last 1 or 2 min to cleanse condenser.

e. Standards: Carry a reagent blank and standards through all steps of procedure and apply necessary correction to results.

f. Final ammonia measurement: Use the titration, ammonia-selective electrode, manual phenate, or automated phenate method, Sections 4500-NH$_3$.C, D, F, and G, respectively.

5. Calculation

See Section 4500-NH$_3$.C.5, D.5, F.5, or G.5.

6. Precision and Bias

No data on the precision and bias of the semi-micro-kjeldahl method are available.

7. Bibliography

See Section 4500-N$_{org}$.B.7.

4500-N$_{org}$ D.   Block Digestion and Flow Injection Analysis (PROPOSED)

1. General Discussion

a. Principle: Samples of drinking, ground, and surface waters and of domestic and industrial wastes are digested in a block digestor with sulfuric acid and copper sulfate as a catalyst. The digestion recovers nitrogen components of biological origin, such as amino acids, proteins, and peptides, as ammonia, but may not recover the nitrogenous compounds of some industrial wastes such as amines, nitro compounds, hydrazones, oximes, semicar-

bazones, and some refractory tertiary amines. Nitrate is not recovered. See Section 4500-N for a discussion of the various forms of nitrogen found in waters and wastewaters, Sections 4500-N$_{org}$.A and B for a discussion of kjeldahl nitrogen methods, and Section 4130, Flow Injection Analysis (FIA).

The digested sample is injected onto the FIA manifold where its pH is controlled by raising it to a known, basic pH by neutralization with a concentrated buffer. This in-line neutralization converts the ammonium cation to ammonia, and also prevents undue influence of the sulfuric acid matrix on the pH-sensitive color reaction that follows. The ammonia thus produced is heated with salicylate and hypochlorite to produce a blue color that is proportional to the ammonia concentration. The color is intensified by adding sodium nitroprusside. The presence of EDTA in the buffer prevents precipitation of calcium and magnesium. The resulting peak's absorbance is measured at 660 nm. The peak area is proportional to the concentration of total kjeldahl nitrogen in the original sample.

b. Interferences: Remove large or fibrous particulates by filtering the sample through glass wool.

The main source of interference is ammonia. Ammonia is an airborne contaminant that is removed rapidly from ambient air by the digestion solution. Guard against ammonia contamination in reagents, water, glassware, and the sample preservation process. It is particularly important to prevent ammonia contamination in the sulfuric acid used for the digestion. Open sulfuric acid bottles away from laboratories in which ammonia or ammonium chloride have been used as reagents and store sulfuric acid away from such reagents. Ensure that the open ends of the block digestor's tubes can be covered to prevent ammonia from being scrubbed from the fume hood make-up air during the digestion.

If a sample consumes more than 10% of the sulfuric acid during digestion, the pH-dependent color reaction will show a matrix effect. The color reaction buffer will accommodate a range of 5.4% $\pm$ 0.4% H_2SO_4 (v/v) in the diluted digested sample. Sample matrices with a high concentration of carbohydrates or other organic material may consume more than 10% of the acid during digestion. If this effect is suspected, titrate digested sample with standardized sodium hydroxide to determine whether more than 10% of the sulfuric acid has been consumed during digestion. The block digestor also should have a means to prevent loss of sulfuric acid from the digestion tubes during the digestion period.

Also see Sections 4500-N$_{org}$.A and B.

2. Apparatus

Digestion and flow injection analysis equipment consisting of:

a. Block digestor capable of maintaining a temperature of 380°C for 2 h.

b. Digestion tubes capable of being heated to 380°C for 2 h and having a cover to prevent ammonia contamination and loss of sulfuric acid.

c. FIA injection valve with sample loop or equivalent.

d. Multichannel proportioning pump.

e. FIA manifold (Figure 4500-N$_{org}$:2) with tubing heater and flow cell. Relative flow rates only are shown in Figure 4500-N$_{org}$:

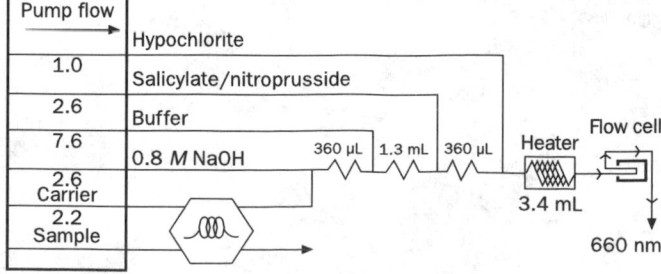

Figure 4500-N$_{org}$:2. FIA total kjeldahl nitrogen manifold.

2. Tubing volumes are given as an example only; they may be scaled down proportionally. Use manifold tubing of an inert material such as TFE.

f. Absorbance detector, 660 nm, 10-nm bandpass.

g. Injection valve control and data acquisition system.

3. Reagents

Use reagent water (>10 megohm) for all solutions. To prevent bubble formation, degas carrier and buffer with helium. Pass He at 140 kPa (20 psi) through a helium degassing tube. Bubble He through 1 L solution for 1 min. As an alternative to preparing reagents by weight/weight, use weight/volume.

a. Digestion solution: In a 1-L volumetric flask, dissolve 134.0 g potassium sulfate, K_2SO_4, and 7.3 g copper sulfate, $CuSO_4$, in 800 mL water. Then add slowly while swirling 134 mL conc sulfuric acid, H_2SO_4. Let cool, dilute to mark, and invert to mix.

b. Carrier and diluent: To a tared 1-L container, add 496 g digestion solution (¶ 3a) and 600 g water. Shake until dissolved.

c. Sodium hydroxide, NaOH, 0.8M: To a tared 1-L plastic container, add 32.0 g NaOH and 985.0 g water. Stir or shake until dissolved.

d. Buffer: To a tared 1-L container add 941 g water. Add and completely dissolve 35.0 g sodium phosphate dibasic heptahydrate, $Na_2HPO_4 \cdot 7H_2O$. Add 20.0 g disodium EDTA (ethylenediaminetetracetic acid disodium salt). The EDTA will not dissolve but will form a turbid solution. Finally, add 50 g NaOH. Stir or shake until dissolved.

e. Salicylate/nitroprusside: To a tared 1-L dark container, add 150.0 g sodium salicylate (salicylic acid sodium salt), $C_6H_4(OH)(COO)Na$, 1.00 g sodium nitroprusside (sodium nitroferricyanide dihydrate), $Na_2Fe(CN)_5NO \cdot 2H_2O$, and 908 g water. Stir or shake until dissolved. Prepare fresh monthly.

f. Hypochlorite: To a tared 250-mL container, add 16 g commercial 5.25% sodium hypochlorite bleach solution* and 234 g deionized water. Shake to mix.

g. Stock standard, 250 mg N/L: In a 1-L volumetric flask dissolve 0.9540 g ammonium chloride, NH$_4$Cl (dried for 2 h at 110°C), in about 800 mL water. Dilute to mark and invert to mix.

h. Standard ammonia solutions: Prepare ammonia standards in desired concentration range, using the stock standard (¶ 3g) and diluting with water.

i. Simulated digested standards: To prepare calibration standards without having to digest the standards prepared in ¶ 3h, proceed as follows:

Stock standard, 5.00 mg N/L: To a tared 250-mL container add about 5.0 g stock standard (250 mg N/L). Divide actual weight of solution added by 0.02 and make up to this resulting total weight with diluent (¶ 3b). Shake to mix. Prepare working standards from this stock standard, diluting with the diluent (¶ 3b), not water.

4. Procedure

a. Digestion procedure: Carry both standards and samples through this procedure.

To a 75-mL block digestor tube add 25.0 mL sample or standard and then add 10 mL digestion solution (¶ 3a) and mix. Add

---

* Regular Clorox, The Clorox Company, Pleasanton, CA, or equivalent.

four alundum granules to each tube for smooth boiling. Place tubes in preheated block digestor for 1 h at 200°C. After 1 h, increase block temperature to 380°C and continue to digest for 1 h at 380°C. Remove tubes from block and let cool for about 10 min. Dilute each to 25.0 mL with water and mix with vortex mixer. Cover tubes to prevent ammonia contamination.

b. FIA analysis: Set up a manifold equivalent to that in Figure 4500-N$_{org}$:2 and analyze digested standards and samples by method supplied by manufacturer or laboratory standard operating procedure. Follow quality control protocols described in Section 4020.

5. Calculations

Prepare standard curves by plotting absorbance of standards processed through the manifold versus ammonia concentration. The calibration curve is linear.

6. Precision and Bias

a. Recovery and relative standard deviation: Table 4500-N$_{org}$: II gives results of single-laboratory studies.

b. MDL: A 130-μL sample loop was used in the method described above. Using a published MDL method,[1] analysts ran 21 replicates of a 0.1-mg N/L standard. These gave a mean of 0.103 mg N/L, a standard deviation of 0.014 mg N/L, and MDL of 0.034 mg N/L. A lower MDL may be obtained by increasing the sample loop volume and increasing the ratio of carrier flow rate to reagents flow rate.

7. Reference

1. U.S. Environmental Protection Agency. 1984. Definition and procedure for the determination of method detection limits. Appendix B to 40 CFR 136 rev. 1.11 amended June 30, 1986. 49 CFR 43430.

TABLE 4500-N$_{ORG}$:II. RESULTS OF SINGLE-LABORATORY STUDIES WITH SELECTED MATRICES

Matrix	Sample/Blank Designation	Known Addition mg N/L	Recovery %	Relative Standard Deviation %
Wastewater treatment plant influent	Reference sample*	—	97	—
	Blank†	3.0	97	—
		6.0	99	—
	Site A‡§	0	—	3.3
		3.0	91	—
		6.0	95	—
	Site B‡§	0	—	3.6
		3.0	115	—
		6.0	93	—
	Site C‡§	0	—	5.1
		3.0	97	—
		6.0	107	—
Wastewater treatment plant effluent	Reference sample*	—	92	—
	Blank†	3.0	97	—
		6.0	100	—
	Site A‡‖	0	—	5.4
		3.0	94	—
		6.0	100	—
	Site B‡‖	0	—	4.1
		3.0	119	—
		6.0	81	—
	Site C‡‖	0	—	7.3
		3.0	93	—
		6.0	105	—
Landfill leachate	Reference sample*	—	96	—
	Blank†	3.0	101	—
		6.0	99	—
	Site A‡#	0	—	3.3
		3.0	95	—
		6.0	98	—
	Site B‡#	0	—	4.4
		3.0	134	—
		6.0	85	—
	Site C‡#	0	—	3.8
		3.0	98	—
		6.0	105	—

* U.S. EPA nutrient QC sample, 1.52 mg N/L.
† Determined in duplicate.
‡ Samples without known additions determined four times; samples with known additions determined in duplicate.
§ Sample dilutions: A - 5-fold; B - 10-fold; C - 5-fold. Typical relative difference between duplicates 3%.
‖ Sample dilutions: A - none; B - 2-fold; C - none. Typical relative difference between duplicates 1%.
Sample dilutions: A, B, and C - 25-fold. Typical relative difference between duplicates 4%.

4500-O   OXYGEN (DISSOLVED)*

4500-O A.   Introduction

1. Significance

Dissolved oxygen (DO) levels in natural and wastewaters depend on the physical, chemical, and biochemical activities in the water body. The analysis for DO is a key test in water pollution and waste treatment process control.

2. Selection of Method

Two methods for DO analysis are described: the Winkler or iodometric method and its modifications and the electrometric

___
* Approved by Standard Methods Committee, 1993.

method using membrane electrodes. The iodometric method[1] is a titrimetric procedure based on the oxidizing property of DO while the membrane electrode procedure is based on the rate of diffusion of molecular oxygen across a membrane.[2] The choice of procedure depends on the interferences present, the accuracy desired, and, in some cases, convenience or expedience.

3. References

1. WINKLER, L.W. 1888. The determination of dissolved oxygen in water. *Berlin. Deut. Chem. Ges.* 21:2843.
2. MANCY, K.H. & T. JAFFE. 1966. Analysis of Dissolved Oxygen in Natural and Waste Waters. Publ. No. 999-WP-37, U.S. Public Health Serv., Washington, D.C.

4500-O B.   Iodometric Methods

1. Principle

The iodometric test is the most precise and reliable titrimetric procedure for DO analysis. It is based on the addition of divalent manganese solution, followed by strong alkali, to the sample in a glass-stoppered bottle. DO rapidly oxidizes an equivalent amount of the dispersed divalent manganous hydroxide precipitate to hydroxides of higher valency states. In the presence of iodide ions in an acidic solution, the oxidized manganese reverts to the divalent state, with the liberation of iodine equivalent to the original DO content. The iodine is then titrated with a standard solution of thiosulfate.

The titration end point can be detected visually, with a starch indicator, or electrometrically, with potentiometric or dead-stop techniques.[1] Experienced analysts can maintain a precision of ± 50 µg/L with visual end-point detection and a precision of ± 5 µg/L with electrometric end-point detection.[1,2]

The liberated iodine also can be determined directly by simple absorption spectrophotometers.[3] This method can be used on a routine basis to provide very accurate estimates for DO in the microgram-per-liter range provided that interfering particulate matter, color, and chemical interferences are absent.

2. Selection of Method

Before selecting a method consider the effect of interferences, particularly oxidizing or reducing materials that may be present in the sample. Certain oxidizing agents liberate iodine from iodides (positive interference) and some reducing agents reduce iodine to iodide (negative interference). Most organic matter is oxidized partially when the oxidized manganese precipitate is acidified, thus causing negative errors.

Several modifications of the iodometric method are given to minimize the effect of interfering materials.[2] Among the more commonly used procedures are the azide modification,[4] the permanganate modification,[5] the alum flocculation modification,[6] and the copper sulfate-sulfamic acid flocculation modification.[7,8] The azide modification (C) effectively removes interference caused by nitrite, which is the most common interference in biologically treated effluents and incubated BOD samples. Use the permanganate modification (D) in the presence of ferrous iron. When the sample contains 5 or more mg ferric iron salts/L, add potassium fluoride (KF) as the first reagent in the azide modification or after the permanganate treatment for ferrous iron. Alternately, eliminate Fe(III) interference by using 85 to 87% phosphoric acid (H_3PO_4) instead of sulfuric acid (H_2SO_4) for acidification. This procedure has not been tested for Fe(III) concentrations above 20 mg/L.

Use the alum flocculation modification (E) in the presence of suspended solids that cause interference and the copper sulfate-sulfamic acid flocculation modification (F) on activated-sludge mixed liquor.

3. Collection of Samples

Collect samples very carefully. Methods of sampling are highly dependent on source to be sampled and, to a certain extent, on method of analysis. Do not let sample remain in contact with air or be agitated, because either condition causes a change in its gaseous content. Samples from any depth in streams, lakes, or reservoirs, and samples of boiler water, need special precautions to eliminate changes in pressure and temperature. Procedures and equipment have been developed for sampling waters under pressure and unconfined waters (e.g., streams, rivers, and reservoirs). Sampling procedures and equipment needed are described in American Society for Testing and Materials Special Technical Publication No. 148-1 and in U.S. Geological Survey Water Supply Paper No. 1454.

Collect surface water samples in narrow-mouth glass-stoppered BOD bottles of 300-mL capacity with tapered and pointed ground-glass stoppers and flared mouths. Avoid entraining or dissolving atmospheric oxygen. In sampling from a line under pressure, attach a glass or rubber tube to the tap and extend to bottom of bottle. Let bottle overflow two or three times its volume and replace stopper so that no air bubbles are entrained.

Suitable samplers for streams, ponds, or tanks of moderate depth are of the APHA type shown in Figure 4500-O:1. Use a Kemmerer-type sampler for samples collected from depths greater than 2 m. Bleed sample from bottom of sampler through a tube extending to bottom of a 250- to 300-mL BOD bottle. Fill bottle to overflowing (overflow for approximately 10 s), and prevent turbulence and formation of bubbles while filling. Record sample temperature to nearest degree Celsius or more precisely.

4. Preservation of Samples

Determine DO immediately on all samples containing an appreciable oxygen or iodine demand. Samples with no iodine demand may be stored for a few hours without change after adding manganous sulfate ($MnSO_4$) solution, alkali-iodide solution, and H_2SO_4, followed by shaking in the usual way. Protect stored samples from strong sunlight and titrate as soon as possible.

For samples with an iodine demand, preserve for 4 to 8 h by adding 0.7 mL conc H_2SO_4 and 1 mL sodium azide solution (2 g NaN_3/100 mL distilled water) to the BOD bottle. This will arrest biological activity and maintain DO if the bottle is stored at the temperature of collection or water-sealed and kept at 10 to 20°C. As soon as possible, complete the procedure, using 2 mL $MnSO_4$ solution, 3 mL alkali-iodide solution, and 2 mL conc H_2SO_4.

5. References

1. POTTER, E.C. & G.E. EVERITT. 1957. Advances in dissolved oxygen microanalysis. *J. Appl. Chem.* 9:642.
2. MANCY, K.H. & T. JAFFE. 1966. Analysis of Dissolved Oxygen in Natural and Waste Waters. Publ. No. 99-WP-37, U.S. Public Health Serv., Washington, D.C.
3. OULMAN, C.S. & E.R. BAUMANN. 1956. A colorimetric method for determining dissolved oxygen. *Sewage Ind. Wastes* 28:1461.
4. ALSTERBERG, G. 1925. Methods for the determination of elementary oxygen dissolved in water in the presence of nitrite. *Biochem. Z.* 159: 36.

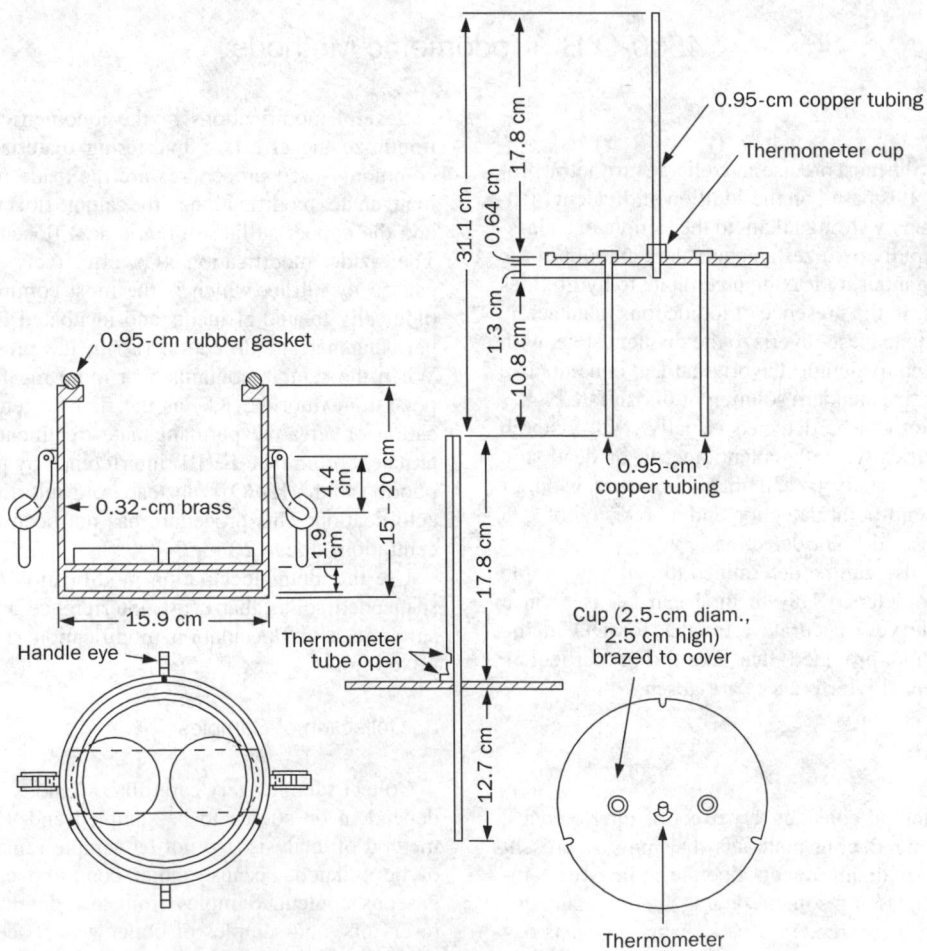

Figure 4500-O:1. DO and BOD sampler assembly.

5. RIDEAL, S. & G.G. STEWART. 1901. The determination of dissolved oxygen in waters in the presence of nitrites and of organic matter. *Analyst* 26:141.

6. RUCHHOFT, C.C. & W.A. MOORE. 1940. The determination of biochemical oxygen demand and dissolved oxygen of river mud suspensions. *Ind. Eng. Chem.*, Anal. Ed. 12:711.

7. PLACAK, O.R. & C.C. RUCHHOFT. 1941. Comparative study of the azide and Rideal-Stewart modifications of the Winkler method in the determination of biochemical oxygen demand. *Ind. Eng. Chem.*, Anal. Ed. 13:12.

8. RUCHHOFT, C.C. & O.R. PLACAK. 1942. Determination of dissolved oxygen in activated-sludge sewage mixtures. *Sewage Works J.* 14:638.

4500-O C.   Azide Modification

1. General Discussion

Use the azide modification for most wastewater, effluent, and stream samples, especially if samples contain more than 50 μg NO_2^--N/L and not more than 1 mg ferrous iron/L. Other reducing or oxidizing materials should be absent. If 1 mL KF solution is added before the sample is acidified and there is no delay in titration, the method is applicable in the presence of 100 to 200 mg ferric iron/L.

2. Reagents

a. Manganous sulfate solution: Dissolve 480 g $MnSO_4 \cdot 4H_2O$, 400 g $MnSO_4 \cdot 2H_2O$, or 364 g $MnSO_4 \cdot H_2O$ in distilled water, filter, and dilute to 1 L. The $MnSO_4$ solution should not give a color with starch when added to an acidified potassium iodide (KI) solution.

b. Alkali-iodide-azide reagent:

1) For saturated or less-than-saturated samples—Dissolve 500 g NaOH (or 700 g KOH) and 135 g NaI (or 150 g KI) in distilled water and dilute to 1 L. Add 10 g NaN_3 dissolved in 40 mL distilled water. Potassium and sodium salts may be used interchangeably. This reagent should not give a color with starch solution when diluted and acidified.

2) For supersaturated samples—Dissolve 10 g NaN_3 in 500 mL distilled water. Add 480 g sodium hydroxide (NaOH) and 750 g sodium iodide (NaI), and stir until dissolved. There will be a white turbidity due to sodium carbonate (Na_2CO_3), but this will do no harm. CAUTION—*Do not acidify this solution because toxic hydrazoic acid fumes may be produced.*

c. Sulfuric acid, H_2SO_4, conc: One milliliter is equivalent to about 3 mL alkali-iodide-azide reagent.

d. Starch: Use either an aqueous solution or soluble starch powder mixtures.

To prepare an aqueous solution, dissolve 2 g laboratory-grade soluble starch and 0.2 g salicylic acid, as a preservative, in 100 mL hot distilled water.

e. Standard sodium thiosulfate titrant: Dissolve 6.205 g $Na_2S_2O_3 \cdot 5H_2O$ in distilled water. Add 1.5 mL 6N NaOH or 0.4 g solid NaOH and dilute to 1000 mL. Standardize with bi-iodate solution.

f. Standard potassium bi-iodate solution, 0.0021M: Dissolve 812.4 mg $KH(IO_3)_2$ in distilled water and dilute to 1000 mL.

Standardization—Dissolve approximately 2 g KI, free from iodate, in an erlenmeyer flask with 100 to 150 mL distilled water. Add 1 mL 6N H_2SO_4 or a few drops of conc H_2SO_4 and 20.00 mL standard bi-iodate solution. Dilute to 200 mL and titrate liberated iodine with thiosulfate titrant, adding starch toward end of titration, when a pale straw color is reached. When the solutions are of equal strength, 20.00 mL 0.025M $Na_2S_2O_3$ should be required. If not, adjust the $Na_2S_2O_3$ solution to 0.025M.

3. Procedure

a. To the sample collected in a 250- to 300-mL bottle, add 1 mL $MnSO_4$ solution, followed by 1 mL alkali-iodide-azide reagent. If pipets are dipped into sample, rinse them before returning them to reagent bottles. Alternatively, hold pipet tips just above liquid surface when adding reagents. Stopper carefully to exclude air bubbles and mix by inverting bottle a few times. When precipitate has settled sufficiently (to approximately half the bottle volume) to leave clear supernate above the manganese hydroxide floc, add 1.0 mL conc H_2SO_4. Restopper and mix by inverting several times until dissolution is complete. Titrate a volume corresponding to 200 mL original sample after correction for sample loss by displacement with reagents. Thus, for a total of 2 mL (1 mL each) of $MnSO_4$ and alkali-iodide-azide reagents in a 300-mL bottle, titrate 200 × 300/(300 − 2) = 201 mL.

b. Titrate with 0.025M $Na_2S_2O_3$ solution to a pale straw color. Add a few drops of starch solution and continue titration to first disappearance of blue color. If end point is overrun, back-titrate with 0.0021M bi-iodate solution added dropwise, or by adding a measured volume of treated sample. Correct for amount of bi-iodate solution or sample. Disregard subsequent recolorations due to the catalytic effect of nitrite or to traces of ferric salts that have not been complexed with fluoride.

4. Calculation

a. For titration of 200 mL sample, 1 mL 0.025M $Na_2S_2O_3$ = 1 mg DO/L.

b. To express results as percent saturation at 101.3 kPa, use the solubility data in Table 4500-O:I. Equations for correcting solubilities to barometric pressures other than mean sea level and for various chlorinities are given below the table.

5. Precision and Bias

DO can be determined with a precision, expressed as a standard deviation, of about 20 μg/L in distilled water and about 60 μg/

TABLE 4500-O:I SOLUBILITY OF OXYGEN IN WATER EXPOSED TO WATER-SATURATED AIR AT ATMOSPHERIC PRESSURE (101.3 kPa)[1]

Temperature °C	Oxygen Solubility mg/L						Temperature °C	Oxygen Solubility mg/L					
	Chlorinity: 0	5.0	10.0	15.0	20.0	25.0		Chlorinity: 0	5.0	10.0	15.0	20.0	25.0
0.0	14.621	13.728	12.888	12.097	11.355	10.657	26.0	8.113	7.711	7.327	6.962	6.615	6.285
1.0	14.216	13.356	12.545	11.783	11.066	10.392	27.0	7.968	7.575	7.201	6.845	6.506	6.184
2.0	13.829	13.000	12.218	11.483	10.790	10.139	28.0	7.827	7.444	7.079	6.731	6.400	6.085
3.0	13.460	12.660	11.906	11.195	10.526	9.897	29.0	7.691	7.317	6.961	6.621	6.297	5.990
4.0	13.107	12.335	11.607	10.920	10.273	9.664	30.0	7.559	7.194	6.845	6.513	6.197	5.896
5.0	12.770	12.024	11.320	10.656	10.031	9.441	31.0	7.430	7.073	6.733	6.409	6.100	5.806
6.0	12.447	11.727	11.046	10.404	9.799	9.228	32.0	7.305	6.957	6.624	6.307	6.005	5.717
7.0	12.139	11.442	10.783	10.162	9.576	9.023	33.0	7.183	6.843	6.518	6.208	5.912	5.631
8.0	11.843	11.169	10.531	9.930	9.362	8.826	34.0	7.065	6.732	6.415	6.111	5.822	5.546
9.0	11.559	10.907	10.290	9.707	9.156	8.636	35.0	6.950	6.624	6.314	6.017	5.734	5.464
10.0	11.288	10.656	10.058	9.493	8.959	8.454	36.0	6.837	6.519	6.215	5.925	5.648	5.384
11.0	11.027	10.415	9.835	9.287	8.769	8.279	37.0	6.727	6.416	6.119	5.835	5.564	5.305
12.0	10.777	10.183	9.621	9.089	8.586	8.111	38.0	6.620	6.316	6.025	5.747	5.481	5.228
13.0	10.537	9.961	9.416	8.899	8.411	7.949	39.0	6.515	6.217	5.932	5.660	5.400	5.152
14.0	10.306	9.747	9.218	8.716	8.242	7.792	40.0	6.412	6.121	5.842	5.576	5.321	5.078
15.0	10.084	9.541	9.027	8.540	8.079	7.642	41.0	6.312	6.026	5.753	5.493	5.243	5.005
16.0	9.870	9.344	8.844	8.370	7.922	7.496	42.0	6.213	5.934	5.667	5.411	5.167	4.933
17.0	9.665	9.153	8.667	8.207	7.770	7.356	43.0	6.116	5.843	5.581	5.331	5.091	4.862
18.0	9.467	8.969	8.497	8.049	7.624	7.221	44.0	6.021	5.753	5.497	5.252	5.017	4.793
19.0	9.276	8.792	8.333	7.896	7.483	7.090	45.0	5.927	5.665	5.414	5.174	4.944	4.724
20.0	9.092	8.621	8.174	7.749	7.346	6.964	46.0	5.835	5.578	5.333	5.097	4.872	4.656
21.0	8.915	8.456	8.021	7.607	7.214	6.842	47.0	5.744	5.493	5.252	5.021	4.801	4.589
22.0	8.743	8.297	7.873	7.470	7.087	6.723	48.0	5.654	5.408	5.172	4.947	4.730	4.523
23.0	8.578	8.143	7.730	7.337	6.963	6.609	49.0	5.565	5.324	5.094	4.872	4.660	4.457
24.0	8.418	7.994	7.591	7.208	6.844	6.498	50.0	5.477	5.242	5.016	4.799	4.591	4.392
25.0	8.263	7.850	7.457	7.083	6.728	6.390							

NOTE:

1. The table provides three decimal places to aid interpolation. When computing saturation values to be used with measured values, such as in computing DO deficit in a receiving water, precision of measured values will control choice of decimal places to be used.

2. Equations are available to compute DO concentration in fresh water[1-3] and in seawater[1] at equilibrium with water-saturated air. Figures and tables also are available.[3]

Calculate the equilibrium oxygen concentration, C^*, from equation:

$$\ln C^* = -139.344\,11 + (1.575\,701 \times 10^5/T) - (6.642\,308 \times 10^7/T^2)$$
$$+ (1.243\,800 \times 10^{10}/T^3) - (8.621\,949 \times 10^{11}/T^4)$$
$$- \text{Chl}\,[(3.1929) \times 10^{-2}) - (1.9428 \times 10^1/T)$$
$$+ (3.8673 \times 10^3/T^2)]$$

where:

C^* = equilibrium oxygen concentration at 101.325 kPa, mg/L,
T = temperature (°K) = °C + 273.150, (°C is between 0.0 and 40.0 in the equation; the table is accurate up to 50.0), and
Chl = Chlorinity (see definition in Note 4, below).

Example 1: At 20°C and 0.000 Chl, ln C^* = 2.207 442 and C^* = 9.092 mg/L;

Example 2: At 20°C and 15.000 ChL,
ln C^* = (2.207 442) − 15.000 (0.010 657)
= 2.0476 and C^* = 7.749 mg/L.

When salinity is used, replace the chlorinity term ($-$Chl[...]) by:
$- S(1.7674 \times 10^{-2}) - (1.0754 \times 10^1/T) + (2.1407 \times 10^3/T^2)$

where:

S = salinity (see definition in Note 4, below).

3. For nonstandard conditions of pressure:

$$C_p = C^*P \left[\frac{(1 - P_{wv}/P)(1 - \theta P)}{(1 - P_{wv})(1 - \theta)} \right]$$

where:

C_p = equilibrium oxygen concentration at nonstandard pressure, mg/L,
C^* = equilibrium oxygen concentration at standard pressure of 1 atm, mg/L,

P = nonstandard pressure, atm,
P_{wv} = partial pressure of water vapor, atm, computed from: $\ln P_{wv} = 11.8571 - (3840.70/T) - (216\,961/T^2)$,
T = temperature, °K,
$\theta = 0.000\,975 - (1.426 \times 10^{-5}t) + (6.436 \times 10^{-8}t^2)$, and
t = temperature, °C.

N.B.: Although not explicit in the above, the quantity in brackets in the equation for C_p has dimensions of atm^{-1} per Reference 4, so that P multiplied by this quantity is dimensionless.

Also, the equation for ln P_{wv} is strictly valid for fresh water only, but for practical purposes no error is made by neglecting the effect of salinity. An equation for P_{wv} that includes the salinity factor may be found in Reference 1.

Example 3: At 20°C, 0.000 Chl, and 0.700 atm,
$C_p = C^* P (0.990\,092) = 6.30$ mg/L.

4. Definitions:

Salinity: Although salinity has been defined traditionally as the total solids in water after all carbonates have been converted to oxides, all bromide and iodide have been replaced by chloride, and all organic matter has been oxidized (see Section 2520), the new scale used to define salinity is based on the electrical conductivity of seawater relative to a specified solution of KCl in water.[5] The scale is dimensionless and the traditional dimension of parts per thousand (i.e., g/kg of solution) no longer applies.

Chlorinity: Chlorinity is defined in relation to salinity as follows:

Salinity = 1.806 55 × chlorinity

Although chlorinity is not equivalent to chloride concentration, the factor for converting a chloride concentration in seawater to include bromide, for example, is only 1.0045 (based on the relative molecular weights and amounts of the two ions). Therefore, for practical purposes, chloride concentration (in g/kg of solution) is nearly equal to chlorinity in seawater. For wastewater, it is necessary to know the ions responsible for the solution's electrical conductivity to correct for their effect on oxygen solubility and use of the tabular value. If this is not done, the equation is inappropriate unless the relative composition of the wastewater is similar to that of seawater.

L in wastewater and secondary effluents. In the presence of appreciable interference, even with proper modifications, the standard deviation may be as high as 100 μg/L. Still greater errors may occur in testing waters having organic suspended solids or heavy pollution. Avoid errors due to carelessness in collecting samples, prolonging the completion of test, or selecting an unsuitable modification.

6. References

1. BENSON, B.B. & D. KRAUSE, JR. 1984. The concentration and isotopic fractionation of oxygen dissolved in freshwater and seawater in equilibrium with the atmosphere. *Limnol. Oceanogr.* 29:620.

2. BENSON, B.B. & D. KRAUSE, JR. 1980. The concentration and isotopic fractionation of gases dissolved in fresh water in equilibrium with the atmosphere: I. Oxygen. *Limnol. Oceanogr.* 25:662.

3. MORTIMER, C.H. 1981. The oxygen content of air-saturated fresh waters over ranges of temperature and atmospheric pressure of limnological interest. *Int. Assoc. Theoret. Appl. Limnol.*, Communication No. 22, Stuttgart, West Germany.

4. SULZER, F. & W.M. WESTGARTH. 1962. Continuous D. O. recording in activated sludge. *Water Sewage Works* 109: 376.

5. UNITED NATIONS EDUCATIONAL, SCIENTIFIC & CULTURAL ORGANIZATION. 1981. Background Papers and Supporting Data on the Practical Salinity Scale 1978. Tech. Paper Mar. Sci. No. 37.

4500-O D.   Permanganate Modification

1. General Discussion

Use the permanganate modification only on samples containing ferrous iron. Interference from high concentrations of ferric iron (up to several hundred milligrams per liter), as in acid mine water, may be overcome by the addition of 1 mL potassium fluoride (KF) and azide, provided that the final titration is made immediately after acidification.

This procedure is ineffective for oxidation of sulfite, thiosulfate, polythionate, or the organic matter in wastewater. The error with samples containing 0.25% by volume of digester waste from the manufacture of sulfite pulp may amount to 7 to 8 mg DO/L. With such samples, use the alkali-hypochlorite modification.[1] At best, however, the latter procedure gives low results, the deviation amounting to 1 mg/L for samples containing 0.25% digester wastes.

2. Reagents

All the reagents required for Method C, and in addition:

a. Potassium permanganate solution: Dissolve 6.3 g $KMnO_4$ in distilled water and dilute to 1 L.

b. Potassium oxalate solution: Dissolve 2 g $K_2C_2O_4 \cdot H_2O$ in 100 mL distilled water; 1 mL will reduce about 1.1 mL permanganate solution.

c. Potassium fluoride solution: Dissolve 40 g $KF \cdot 2H_2O$ in distilled water and dilute to 100 mL.

3. Procedure

a. To a sample collected in a 250- to 300-mL bottle add, below the surface, 0.70 mL conc H_2SO_4, 1 mL $KMnO_4$ solution, and 1 mL KF solution. Stopper and mix by inverting. Never add more than 0.7 mL conc H_2SO_4 as the first step of pretreatment. Add acid with a 1-mL pipet graduated to 0.1 mL. Add sufficient $KMnO_4$ solution to obtain a violet tinge that persists for 5 min. If the permanganate color is destroyed in a shorter time, add additional $KMnO_4$ solution, but avoid large excesses.

b. Remove permanganate color completely by adding 0.5 to 1.0 mL $K_2C_2O_4$ solution. Mix well and let stand in the dark to facilitate the reaction. Excess oxalate causes low results; add only enough $K_2C_2O_4$ to decolorize the $KMnO_4$ completely without an excess of more than 0.5 mL. Complete decolorization in 2 to 10 min. If it is impossible to decolorize the sample without adding a large excess of oxalate, the DO result will be inaccurate.

c. From this point the procedure closely parallels that in Section 4500-O.C.3. Add 1 mL $MnSO_4$ solution and 3 mL alkali-iodide-azide reagent. Stopper, mix, and let precipitate settle a short time; acidify with 2 mL conc H_2SO_4. When 0.7 mL acid, 1 mL KF solution, 1 mL $KMnO_4$ solution, 1 mL $K_2C_2O_4$ solution, 1 mL $MnSO_4$ solution, and 3 mL alkali-iodide-azide (or a total of 7.7 mL reagents) are used in a 300-mL bottle, take $200 \times 300/(300 - 7.7) = 205$ mL for titration.

This correction is slightly in error because the $KMnO_4$ solution is nearly saturated with DO and 1 mL would add about 0.008 mg oxygen to the DO bottle. However, because precision of the method (standard deviation, 0.06 mL thiosulfate titration, or 0.012 mg DO) is 50% greater than this error, a correction is unnecessary. When substantially more $KMnO_4$ solution is used routinely, use a solution several times more concentrated so that 1 mL will satisfy the permanganate demand.

4. Reference

1. THERIAULT, E.J. & P.D. MCNAMEE. 1932. Dissolved oxygen in the presence of organic matter, hypochlorites, and sulfite wastes. *Ind. Eng. Chem.*, Anal. Ed. 4:59.

4500-O E.   Alum Flocculation Modification

1. General Discussion

Samples high in suspended solids may consume appreciable quantities of iodine in acid solution. The interference due to solids may be removed by alum flocculation.

2. Reagents

All the reagents required for the azide modification (Section 4500-O.C.2) and in addition:

a. Alum solution: Dissolve 10 g aluminum potassium sulfate, $AlK(SO_4)_2 \cdot 12H_2O$, in distilled water and dilute to 100 mL.

b. Ammonium hydroxide, NH_4OH, conc.

3. Procedure

Collect sample in a glass-stoppered bottle of 500 to 1000 mL capacity, using the same precautions as for regular DO samples. Add 10 mL alum solution and 1 to 2 mL conc NH_4OH. Stopper and invert gently for about 1 min. Let sample settle for about 10 min and siphon clear supernate into a 250- to 300-mL DO bottle until it overflows. Avoid sample aeration and keep siphon submerged at all times. Continue sample treatment as in Section 4500-O.C.3 or an appropriate modification.

4500-O F.   Copper Sulfate-Sulfamic Acid Flocculation Modification

1. General Discussion

This modification is used for biological flocs such as activated sludge mixtures, which have high oxygen utilization rates.

2. Reagents

All the reagents required for the azide modification (Section 4500-O.C.2) and, in addition:

Copper sulfate-sulfamic acid inhibitor solution: Dissolve 32 g technical-grade NH_2SO_2OH without heat in 475 mL distilled water. Dissolve 50 g $CuSO_4 \cdot 5H_2O$ in 500 mL distilled water. Mix the two solutions and add 25 mL conc acetic acid.

3. Procedure

Add 10 mL $CuSO_4$-NH_2SO_2OH inhibitor to a 1-L glass-stoppered bottle. Insert bottle in a special sampler designed so that bottle fills from a tube near bottom and overflows only 25 to 50% of bottle capacity. Collect sample, stopper, and mix by inverting. Let suspended solids settle and siphon relatively clear supernatant liquor into a 250- to 300-mL DO bottle. Continue sample treatment as rapidly as possible by the azide (Section 4500-O.C.3) or other appropriate modification.

4500-O G.   Membrane Electrode Method

1. General Discussion

Various modifications of the iodometric method have been developed to eliminate or minimize effects of interferences; nevertheless, the method still is inapplicable to a variety of industrial and domestic wastewaters.[1] Moreover, the iodometric method is not suited for field testing and cannot be adapted easily for continuous monitoring or for DO determinations in situ.

Polarographic methods using the dropping mercury electrode or the rotating platinum electrode have not been reliable always for the DO analysis in domestic and industrial wastewaters because impurities in the test solution can cause electrode poisoning or other interferences.[2,3] With membrane-covered electrode systems these problems are minimized, because the sensing element is protected by an oxygen-permeable plastic membrane that serves as a diffusion barrier against impurities.[4-6] Under steady-state conditions the current is directly proportional to the DO concentration.*

Membrane electrodes of the polarographic[4] as well as the galvanic[5] type have been used for DO measurements in lakes and reservoirs,[8] for stream survey and control of industrial effluents,[9,10] for continuous monitoring of DO in activated sludge units,[11] and for estuarine and oceanographic studies.[12] Being completely submersible, membrane electrodes are suited for analysis in situ. Their portability and ease of operation and mainte-

---

* Fundamentally, the current is directly proportional to the activity of molecular oxygen.[7]

nance make them particularly convenient for field applications. In laboratory investigations, membrane electrodes have been used for continuous DO analysis in bacterial cultures, including the BOD test.[5,13]

Membrane electrodes provide an excellent method for DO analysis in polluted waters, highly colored waters, and strong waste effluents. They are recommended for use especially under conditions that are unfavorable for use of the iodometric method, or when that test and its modifications are subject to serious errors caused by interferences.

a. Principle: Oxygen-sensitive membrane electrodes of the polarographic or galvanic type are composed of two solid metal electrodes in contact with supporting electrolyte separated from the test solution by a selective membrane. The basic difference between the galvanic and the polarographic systems is that in the former the electrode reaction is spontaneous (similar to that in a fuel cell), while in the latter an external source of applied voltage is needed to polarize the indicator electrode. Polyethylene and fluorocarbon membranes are used commonly because they are permeable to molecular oxygen and are relatively rugged.

Membrane electrodes are commercially available in some variety. In all these instruments the "diffusion current" is linearly proportional to the concentration of molecular oxygen. The current can be converted easily to concentration units (e.g., milligrams per liter) by a number of calibration procedures.

Membrane electrodes exhibit a relatively high temperature coefficient largely due to changes in the membrane permeability.[6] The effect of temperature on the electrode sensitivity, ϕ (microamperes per milligram per liter), can be expressed by the following simplified relationship:[6]

$$\log \phi = 0.43 \, mt + b$$

where:

t = temperature, °C,
m = constant that depends on the membrane material, and
b = constant that largely depends on membrane thickness.

If values of ϕ and m are determined for one temperature (ϕ_0 and t_0), it is possible to calculate the sensitivity at any desired temperature (ϕ and t) as follows:

$$\log \phi = \log \phi_0 + 0.43 \, m \, (t - t_0)$$

Nomographic charts for temperature correction can be constructed easily[7] and are available from some manufacturers. An example is shown in Figure 4500-O:2, in which, for simplicity, sensitivity is plotted versus temperature on semilogarithmic coordinates. Check one or two points frequently to confirm original calibration. If calibration changes, the new calibration should be parallel to the original, provided that the same membrane material is used.

Temperature compensation also can be made automatically by using thermistors in the electrode circuit.[4] However, thermistors may not compensate fully over a wide temperature range. For certain applications where high accuracy is required, use calibrated nomographic charts to correct for temperature effect.

To use the DO membrane electrode in estuarine waters or in wastewaters with varying ionic strength, correct for effect of salting-out on electrode sensitivity.[6,7] This effect is particularly significant for large changes in salt content. Electrode sensitivity

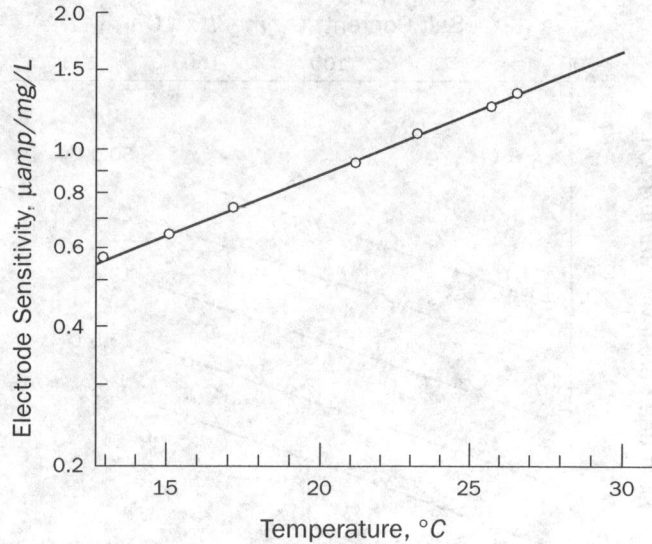

Figure 4500-O:2. Effect of temperature on electrode sensitivity.

varies with salt concentration according to the following relationship:

$$\log \phi_s = 0.43 \, m_s C_s + \log \phi_0$$

where:

ϕ_S, ϕ_0 = sensitivities in salt solution and distilled water, respectively,
C_S = salt concentration (preferably ionic strength), and
m_S = constant (salting-out coefficient).

If ϕ_0 and m_S are determined, it is possible to calculate sensitivity for any value of C_S. Conductivity measurements can be used to approximate salt concentration (C_S). This is particularly applicable to estuarine waters. Figure 4500-O:3 shows calibration curves for sensitivity of varying salt solutions at different temperatures.

b. Interference: Plastic films used with membrane electrode systems are permeable to a variety of gases besides oxygen, although none is depolarized easily at the indicator electrode. Prolonged use of membrane electrodes in waters containing such gases as hydrogen sulfide (H_2S) tends to lower cell sensitivity. Eliminate this interference by frequently changing and calibrating the membrane electrode.

c. Sampling: Because membrane electrodes offer the advantage of analysis in situ they eliminate errors caused by sample handling and storage. If sampling is required, use the same precautions suggested for the iodometric method.

2. Apparatus

Oxygen-sensitive membrane electrode, polarographic or galvanic, with appropriate meter.

3. Procedure

a. Calibration: Follow manufacturer's calibration procedure exactly to obtain guaranteed precision and accuracy. Generally, calibrate membrane electrodes by reading against air or a sample

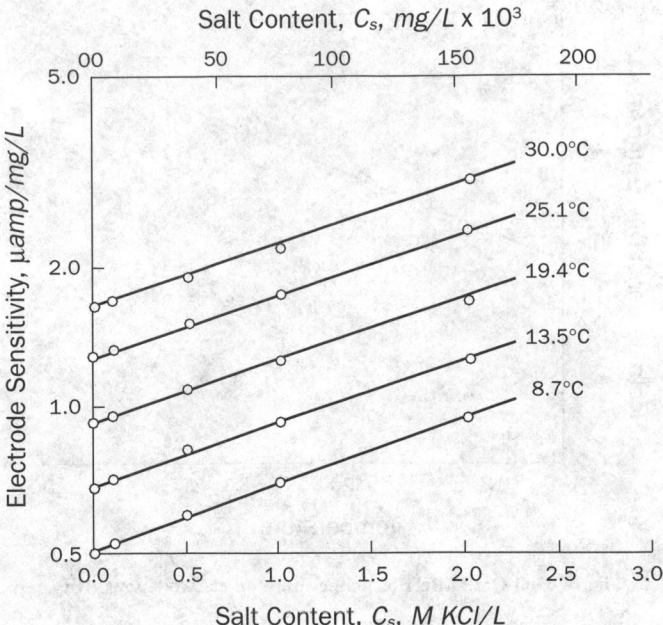

Figure 4500-O:3. The salting-out effect at different temperatures.

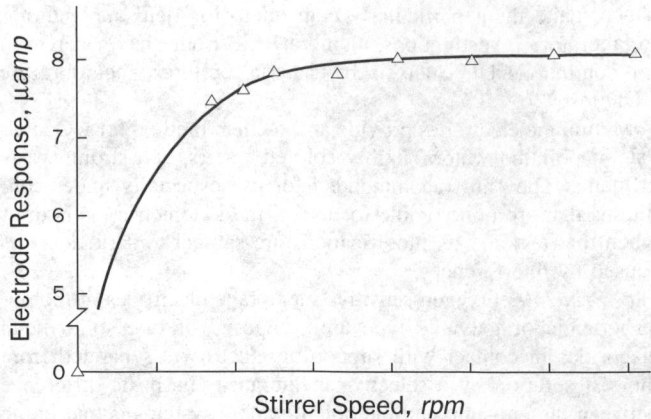

Figure 4500-O:4. Typical trend of effect of stirring on electrode response.

of known DO concentration (determined by iodometric method) as well as in a sample with zero DO. (Add excess sodium sulfite, Na_2SO_3, and a trace of cobalt chloride, $CoCl_2$, to bring DO to zero.) Preferably calibrate with samples of water under test. Avoid an iodometric calibration where interfering substances are suspected. The following illustrate the recommended procedures:

1) Fresh water—For unpolluted samples where interfering substances are absent, calibrate in the test solution or distilled water, whichever is more convenient.

2) Salt water—Calibrate directly with samples of seawater or waters having a constant salt concentration in excess of 1000 mg/L.

3) Fresh water containing pollutants or interfering substances—Calibrate with distilled water because erroneous results occur with the sample.

4) Salt water containing pollutants or interfering substances—Calibrate with a sample of clean water containing the same salt content as the sample. Add a concentrated potassium chloride (KCl) solution (see Conductivity, Section 2510 and Table 2510: I) to distilled water to produce the same specific conductance as that in the sample. For polluted ocean waters, calibrate with a sample of unpolluted seawater.

5) Estuary water containing varying quantities of salt—Calibrate with a sample of uncontaminated seawater or distilled or tap water. Determine sample chloride or salt concentration and revise calibration to account for change of oxygen solubility in the estuary water.[7]

b. *Sample measurement:* Follow all precautions recommended by manufacturer to insure acceptable results. Take care in changing membrane to avoid contamination of sensing element and also trapping of minute air bubbles under the membrane, which can lead to lowered response and high residual current. Provide sufficient sample flow across membrane surface to overcome erratic response (see Figure 4500-O:4 for a typical example of the effect of stirring).

c. *Validation of temperature effect:* Check frequently one or two points to verify temperature correction data.

4. Precision and Bias

With most commercially available membrane electrode systems an accuracy of ±0.1 mg DO/L and a precision of ±0.05 mg DO/L can be obtained.

5. References

1. McKEOWN, J.J., L.C. BROWN & G.W. GOVE. 1967. Comparative studies of dissolved oxygen analysis methods. *J. Water Pollut. Control Fed.* 39:1323.
2. LYNN, W.R. & D.A. OKUN. 1955. Experience with solid platinum electrodes in the determination of dissolved oxygen. *Sewage Ind. Wastes* 27:4.
3. MANCY, K.H. & D.A. OKUN. 1960. Automatic recording of dissolved oxygen in aqueous systems containing surface active agents. *Anal. Chem.* 32:108.
4. CARRITT, D.E. & J.W. KANWISHER. 1959. An electrode system for measuring dissolved oxygen. *Anal. Chem.* 31:5.
5. MANCY, K.H. & W.C. WESTGARTH. 1962. A galvanic cell oxygen analyzer. *J. Water Pollut. Control Fed.* 34:1037.
6. MANCY, K.H., D.A. OKUN & C.N. REILLEY. 1962. A galvanic cell oxygen analyzer. *J. Electroanal. Chem.* 4:65.
7. MANCY, K.H. & T. JAFFE. 1966. Analysis of Dissolved Oxygen in Natural and Waste Waters. Publ. No. 999-WP-37, U.S. Public Health Serv., Washington, D.C.
8. WEISS, C.M. & R.T. OGLESBY. 1963. Instrumentation for monitoring water quality in reservoirs. American Water Works Assoc. 83rd Annual Conf., New York, N.Y.
9. CLEARY, E.J. 1962. Introducing the ORSANCO robot monitor. *Proc. Water Quality Meas. Instrum.* Publ. No. 108, U.S. Public Health Serv., Washington, D.C.
10. MACKERETH, F.J.H. 1964. An improved galvanic cell for determination of oxygen concentrations in fluids. *J. Sci. Instrum.* 41:38.
11. SULZER, F. & W.M. WESTGARTH. 1962. Continuous D.O. recording in activated sludge. *Water Sewage Works* 109:376.
12. DUXBURY, A.C. 1963. Calibration and use of a galvanic type oxygen electrode in field work. *Limnol. Oceanogr.* 8:483.
13. LIPNER, H.J., L.R. WITHERSPOON & V.C. CHAMPEAUS. 1964. Adaptation of a galvanic cell for microanalysis of oxygen. *Anal. Chem.* 36:204.

4500-O₃   OZONE (RESIDUAL)*

4500-O₃ A.   Introduction

1. Sources

Ozone, a potent germicide, is used also as an oxidizing agent for the oxidation of organic compounds that produce taste and odor in drinking water, for the destruction of organic coloring matter, and for the oxidation of reduced iron or manganese salts to insoluble oxides.

2. Selection of Method

Ozone residual in water is determined by the indigo method. Residual ozone decays rapidly. Depending on water quality, the

---

* Approved by Standard Methods Committee, 1997.

ozone residual half-life may be several seconds to a few minutes. Methods also are available for determining ozone in process gases.[1,2]

3. References

1. RAKNESS, K.L., G. GORDON, B. LANGLAIS, W. MASSCHELEIN, N. MATSUMOTO, Y. RICHARD, C.M. ROBSON & I. SOMIYA. 1996. Guideline for measurement of ozone concentration in the process gas from an ozone generator. *Ozone: Sci. Eng.* 18:209.
2. RAKNESS, K.L., L.D. DEMERS, B.D. BLANK & D.J. HENRY. 1996. Gas phase ozone concentration comparisons from a commercial UV meter and KI wet-chemistry tests. *Ozone: Sci. Eng.* 18:231.

4500-O₃ B.   Indigo Colorimetric Method

1. General Discussion

The indigo colorimetric method is quantitative, selective, and simple; it replaces methods based on the measurement of total oxidant. The method is applicable to lake water, river infiltrate, manganese-containing groundwaters, extremely hard groundwaters, and even biologically treated domestic wastewaters.

a. Principle: In acidic solution, ozone rapidly decolorizes indigo. The decrease in absorbance is linear with increasing concentration. The proportionality constant at 600 nm is 0.42 ± 0.01/cm/mg/L ($\Delta E = 20\,000/M \cdot cm$) compared to the ultraviolet absorption of pure ozone of $E = 2950/M \cdot cm$ at 258 nm).[1]

b. Interferences: Hydrogen peroxide (H_2O_2) and organic peroxides decolorize the indigo reagent very slowly. H_2O_2 does not interfere if ozone is measured in less than 6 h after adding reagents. Organic peroxides may react more rapidly. Fe(III) does not interfere. Mn(II) does not interfere but it is oxidized by ozone to forms that decolorize the reagent. Correct for this decolorization by making the measurement relative to a blank in which the ozone has been destroyed selectively. Without the corrective procedure, 0.1 mg/L ozonated manganese gives a response of about 0.08 mg/L apparent ozone. Chlorine also interferes. Low concentrations of chlorine (<0.1 mg/L) can be masked by malonic acid. Bromine, which can be formed by oxidation of Br⁻, interferes (1 mole HOBr corresponds to 0.4 mole ozone). In the presence of HOBr or chlorine in excess of 0.1 mg/L, an accurate measurement of ozone cannot be made with this method.

c. Minimum detectable concentration: For the spectrophotometric procedure using thermostated cells and a high-quality photometer, the low-range procedure will measure down to 2 μg O₃/L. The practical lower limit for residual measurement is 10 to 20 μg/L.

d. Sampling: React sample with indigo as quickly as possible, because the residual may decay rapidly. Avoid loss of ozone residual due to off-gassing during sample collection. Do not run sample down side of flask. Add sample so that completely decolorized zones are eliminated quickly by swirling or stirring.

2. Apparatus

Photometer: Spectrophotometer or filter colorimeter for use at 600 ± 10 nm.

3. Reagents

a. Indigo stock solution: Add about 500 mL distilled water and 1 mL conc phosphoric acid to a 1-L volumetric flask. With stirring, add 770 mg potassium indigo trisulfonate, $C_{16}H_7N_2O_{11}S_3K_3$ (use only high-grade reagent, commercially available at about 80 to 85% purity). Fill to mark with distilled water. A 1:100 dilution exhibits an absorbance of 0.20 ± 0.010 cm at 600 nm. The stock solution is stable for about 4 months when stored in the dark. Discard when absorbance of a 1:100 dilution falls below 0.16/cm. Do not change concentration of dye for higher ranges of ozone residual. Volume of dye used may be adjusted.

b. Indigo reagent I: To a 1-L volumetric flask add 20 mL indigo stock solution, 10 g sodium dihydrogen phosphate (NaH_2PO_4), and 7 mL conc phosphoric acid. Dilute to mark. Prepare solution fresh when its absorbance decreases to less than 80% of its initial value, typically within a week.

c. Indigo reagent II: Proceed as with indigo reagent I, but add 100 mL indigo stock solution instead of 20 mL.

d. Malonic acid reagent: Dissolve 5 g malonic acid in water and dilute to 100 mL.

e. Glycine reagent: Dissolve 7 g glycine in water and dilute to 100 mL.

4. Procedure

a. Spectrophotometric, volumetric procedure:

1) Concentration range 0.01 to 0.1 mg O_3/L—Add 10.0 mL indigo reagent I to each of two 100-mL volumetric flasks. Fill one flask (blank) to mark with distilled water. Fill other flask to mark with sample. Measure absorbance of both solutions at ± 10 nm as soon as possible but at least within 4 h. Preferably use 10-cm cells. Calculate the ozone concentration from the difference between the absorbances found in sample and blank (¶ 5g below). (NOTE: A maximum delay of 4 h before spectrophotometric reading can be tolerated only for drinking water samples. For other sample types that cannot be read immediately, determine the relationship between time and absorbance.)

2) Range 0.05 to 0.5 mg O_3/L—Proceed as above using 10.0 mL indigo reagent II instead of reagent I. Preferably measure absorbance in 4- or 5-cm cells.

3) Concentrations greater than 0.3 mg O_3/L—Proceed using indigo reagent II, but for these higher ozone concentrations use a correspondingly smaller sample volume. Dilute resulting mixture to 100 mL with distilled water.

4) Control of interferences—In presence of low chlorine concentration (<0.1 mg/L), place 1 mL malonic acid reagent in both flasks before adding sample and/or filling to mark. Measure absorbance as soon as possible, within 60 min (Br^-, Br_2, and HOBr are only partially masked by malonic acid).

In presence of manganese prepare a blank solution using sample, in which ozone is selectively destroyed by addition of glycine. Place 0.1 mL glycine reagent in 100-mL volumetric flask (blank) and 10.0 mL indigo reagent II in second flask (sample). Pipet exactly the same volume of sample into each flask. Adjust dose so that decolorization in second flask is easily visible but complete bleaching does not result (maximum 80 mL).

Insure that pH of glycine/sample mixture in blank flask (before adding indigo) is not below 6 because reaction between ozone and glycine becomes very slow at low pH. Stopper flasks and mix by carefully inverting. Add 10.0 mL indigo reagent II to blank flask only 30 to 60 s after sample addition. Fill both flasks to the mark with ozone-free water and mix thoroughly. Measure absorbance of both solutions at comparable contact times of approximately 30 to 60 min (after this time, residual manganese oxides further discolor indigo only slowly and the drift of absorbance in blank and sample become comparable). Reduced absorbance in blank flask results from manganese oxides while that in sample flask is due to ozone plus manganese oxide.

5) Calibration—Because ozone is unstable, base measurements on known and constant loss of absorbance of the indigo reagent ($f = 0.42 + 0.01$/cm/mg O_3/L). For maximum accuracy analyze the lot of potassium indigo trisulfonate (no commercial lot has been found to deviate from $f = 0.42$) using the iodometric procedure.

When using a filter photometer, readjust the conversion factor, f, by comparing photometer sensitivity with absorbance at 600 nm by an accurate spectrophotometer.

b. Spectrophotometric, gravimetric procedure:

1) Add 10.0 mL indigo reagent II to 100-mL volumetric flask and fill flask (blank) to mark with distilled water. Obtain tare weight of a second flask (volumetric or erlenmeyer). Add 10.0 mL indigo reagent II to second flask. Fill directly with sample (do not run water down side), and swirl second flask until blue solution has turned to a light blue color. Weigh flask containing indigo and sample.

2) Preferably using 10-cm cells, measure absorbance of both solutions at 600 ± 10 nm as soon as possible, but at least within 4 h. NOTE: A maximum delay of 4 h before spectrophotometric reading is suitable only for drinking water samples. For other sample types, test the time drift.

5. Calculations

a. Spectrophotometric, volumetric method:

$$mg\ O_3/L = \frac{100 \times \Delta A}{f \times b \times V}$$

where:

ΔA = difference in absorbance between sample and blank,
b = path length of cell, cm,
V = volume of sample, mL (normally 90 mL), and
f = 0.42.

The factor f is based on a sensitivity factor of 20 000/cm for the change of absorbance (600 nm) per mole of added ozone per liter. It was calibrated by iodometric titration. The UV absorbance of ozone in pure water may serve as a secondary standard: the factor $f = 0.42$ corresponds to an absorption coefficient for aqueous ozone, E = 2950/$M\cdot$cm at 258 nm.

b. Spectrophotometric, gravimetric method:

$$mg\ O_3/L = \frac{(A_B \times 100) - (A_S \times V_T)}{f \times V_S \times b}$$

where:

A_B, A_S = absorbance of blank and sample, respectively,
V_S = volume of sample, mL = [(final weight − tare weight) g × 1.0 mL/g] − 10 mL,
V_T = total volume of sample plus indigo, mL = (final weight − tare weight) g × 1.0 mL/g,
b = path length of cell, cm, and
f = 0.42 (see ¶ a above).

6. Precision and Bias

For the spectrophotometric volumetric procedure in the absence of interferences, the relative error is less than 5% without special sampling setups. In laboratory testing this may be reduced to 1%. No data are available for the spectrophotometric gravimetric procedure.

Because this method is based on the differences in absorbance between the sample and blank (ΔA) the method is not applicable in the presence of chlorine. If the manganese content exceeds the ozone, precision is reduced. If the ratio of manganese to ozone is less than 10:1, ozone concentrations above 0.02 mg/L may be determined with a relative error of less than 20%.

7. Reference

1. HOIGNÉ, J. & H. BADER. 1980. Bestimmung von Ozon und Chlordioxid im Wasser mit der Indigo-Methode. *Vom Wasser* 55:261.

8. Bibliography

BADER, H. & J. HOIGNÉ. 1981. Determination of ozone in water by the indigo method. *Water Res.* 15:449.

BADER, H. & J. HOIGNÉ. 1982. Determination of ozone in water by the indigo method. A submitted standard method. *Ozone: Sci. Eng.* 4:169.

GILBERT, E. & J. HOIGNÉ. 1983. Messung von Ozon in Wasserwerken; Vergleich der DPD-und Indigo-Methode. *GWF-Wasser/Abwasser* 124:527.

HAAG, W.R. & J. HOIGNÉ. 1983. Ozonation of bromide-containing waters: kinetics of formation of hypobromous acid and bromate. *Environ. Sci. Technol.* 17:261.

STRAKA, M.R., G. GORDON & G.E. PACEY. 1985. Residual aqueous ozone determination by gas diffusion flow injection analysis. *Anal. Chem.* 57:1799.

GORDON, G. & G.E. PACEY. 1986. An introduction to the chemical reactions of ozone pertinent to its analysis. *In* R.G. Rice, L.J. Bollyky & W.J. Lacy, eds. Analytical Aspects of Ozone Treatment of Water and Wastewater—A Monograph. Lewis Publishers, Inc., Chelsea, Mich.

COFFEY, B.C., K.G. GRAFF, A.A. MOFIDI & J.T. GRAMITH. 1995. On-line monitoring of ozone disinfection effectiveness within an over-under baffled contactor. Proc. Annu. Conf., American Water Works Assoc., Anaheim, Calif. American Water Works Assoc., Denver, Colo.

METROPOLITAN WATER DISTRICT OF SOUTHERN CALIFORNIA. 1996. Demonstration-Scale Evaluation of Ozone/PEROXONE.'' American Water Works Assoc. Research Foundation.

4500-P   PHOSPHORUS*

4500-P A.   Introduction

1. Occurrence

Phosphorus occurs in natural waters and in wastewaters almost solely as phosphates. These are classified as orthophosphates, condensed phosphates (pyro-, meta-, and other polyphosphates), and organically bound phosphates. They occur in solution, in particles or detritus, or in the bodies of aquatic organisms.

These forms of phosphate arise from a variety of sources. Small amounts of orthophosphate or certain condensed phosphates are added to some water supplies during treatment. Larger quantities of the same compounds may be added when the water is used for laundering or other cleaning, because these materials are major constituents of many commercial cleaning preparations. Phosphates are used extensively in the treatment of boiler waters. Orthophosphates applied to agricultural or residential cultivated land as fertilizers are carried into surface waters with storm runoff and to a lesser extent with melting snow. Organic phosphates are formed primarily by biological processes. They are contributed to sewage by body wastes and food residues, and also may be formed from orthophosphates in biological treatment processes or by receiving water biota.

Phosphorus is essential to the growth of organisms and can be the nutrient that limits the primary productivity of a body of water. In instances where phosphate is a growth-limiting nutrient, the discharge of raw or treated wastewater, agricultural drainage, or certain industrial wastes to that water may stimulate the growth of photosynthetic aquatic micro- and macroorganisms in nuisance quantities.

Phosphates also occur in bottom sediments and in biological sludges, both as precipitated inorganic forms and incorporated into organic compounds.

2. Definition of Terms

Phosphorus analyses embody two general procedural steps: (*a*) conversion of the phosphorus form of interest to dissolved orthophosphate, and (*b*) colorimetric determination of dissolved orthophosphate. The separation of phosphorus into its various forms is defined analytically but the analytical differentiations have been selected so that they may be used for interpretive purposes.

Filtration through a 0.45-μm-pore-diam membrane filter separates dissolved from suspended forms of phosphorus. No claim is made that filtration through 0.45-μm filters is a true separation of suspended and dissolved forms of phosphorus; it is merely a convenient and replicable analytical technique designed to make a gross separation.

Membrane filtration is selected over depth filtration because of the greater likelihood of obtaining a consistent separation of particle sizes. Prefiltration through a glass fiber filter may be used to increase the filtration rate.

Phosphates that respond to colorimetric tests without preliminary hydrolysis or oxidative digestion of the sample are termed ''reactive phosphorus.'' While reactive phosphorus is largely a measure of orthophosphate, a small fraction of any condensed phosphate present usually is hydrolyzed unavoidably in the procedure. Reactive phosphorus occurs in both dissolved and suspended forms.

Acid hydrolysis at boiling-water temperature converts dissolved and particulate condensed phosphates to dissolved orthophosphate. The hydrolysis unavoidably releases some phosphate from organic compounds, but this may be reduced to a minimum by judicious selection of acid strength and hydrolysis time and temperature. The term ''acid-hydrolyzable phosphorus'' is preferred over ''condensed phosphate'' for this fraction.

The phosphate fractions that are converted to orthophosphate only by oxidation destruction of the organic matter present are

---

* Approved by Standard Methods Committee, 1997.

considered ''organic'' or ''organically bound'' phosphorus. The severity of the oxidation required for this conversion depends on the form—and to some extent on the amount—of the organic phosphorus present. Like reactive phosphorus and acid-hydrolyzable phosphorus, organic phosphorus occurs both in the dissolved and suspended fractions.

The total phosphorus as well as the dissolved and suspended phosphorus fractions each may be divided analytically into the three chemical types that have been described: reactive, acid-hydrolyzable, and organic phosphorus. Figure 4500-P:1 shows the steps for analysis of individual phosphorus fractions. As indicated, determinations usually are conducted only on the unfiltered and filtered samples. Suspended fractions generally are determined by difference; however, they may be determined directly by digestion of the material retained on a glass-fiber filter.

3. Selection of Method

a. Digestion methods: Because phosphorus may occur in combination with organic matter, a digestion method to determine

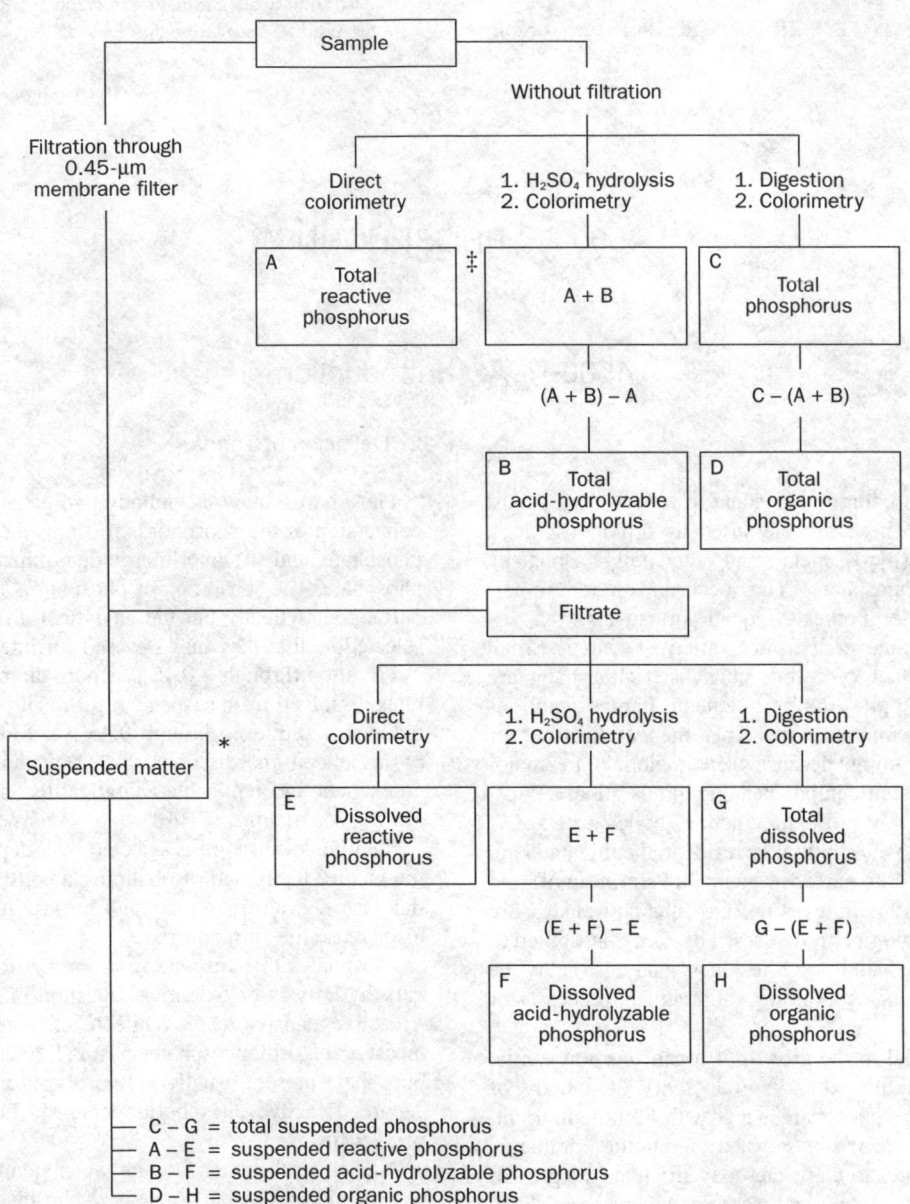

Figure 4500-P:1. Steps for analysis of phosphate fractions.

* Direct determination of phosphorus on the membrane filter containing suspended matter will be required where greater precision than that obtained by difference is desired. Digest filter with HNO_3 and follow by perchloric acid. Then perform colorimetry.

† Total phosphorus measurements on highly saline samples may be difficult because of precipitation of large quantities of salt as a result of digestion techniques that drastically reduce sample volume. For total phosphorus analyses on such samples, directly determine total dissolved phosphorus and total suspended phosphorus and add the results.

‡ In determination of total dissolved or total suspended reactive phosphorus, anomalous results may be obtained on samples containing large amounts of suspended sediments. Very often results depend largely on the degree of agitation and mixing to which samples are subjected during analysis because of a time-dependent desorption of orthophosphate from the suspended particles.

total phosphorus must be able to oxidize organic matter effectively to release phosphorus as orthophosphate. Three digestion methods are given in Section 4500-P.B.3, 4, and 5. The perchloric acid method, the most drastic and time-consuming method, is recommended only for particularly difficult samples such as sediments. The nitric acid-sulfuric acid method is recommended for most samples. By far the simplest method is the persulfate oxidation technique. Persulfate oxidation is coupled with ultraviolet light for a more efficient digestion in an automated in-line digestion/determination by flow injection analysis (4500-P.I). It is recommended that persulfate oxidation methods be checked against one or more of the more drastic digestion techniques and be adopted if identical recoveries are obtained.

After digestion, determine liberated orthophosphate by Method C, D, E, F, G, or H. The colorimetric method used, rather than the digestion procedure, governs in matters of interference and minimum detectable concentration.

b. Colorimetric method: Three methods of orthophosphate determination are described. Selection depends largely on the concentration range of orthophosphate. The vanadomolybdophosphoric acid method (C) is most useful for routine analysis in the range of 1 to 20 mg P/L. The stannous chloride method (D) or the ascorbic acid method (E) is more suited for the range of 0.01 to 6 mg P/L. An extraction step is recommended for the lower levels of this range and when interferences must be overcome. Automated versions of the ascorbic acid method (F, G, and H)

TABLE 4500-P:I. PRECISION AND BIAS DATA FOR MANUAL PHOSPHORUS METHODS

Method	Phosphorus Concentration			No. of Laboratories	Relative Standard Deviation %	Relative Error %
	Ortho-phosphate $\mu g/L$	Poly-phosphate $\mu g/L$	Total $\mu g/L$			
Vanadomolybdophosphoric acid	100			45	75.2	21.6
	600			43	19.6	10.8
	7000			44	8.6	5.4
Stannous chloride	100			45	25.5	28.7
	600			44	14.2	8.0
	7000			45	7.6	4.3
Ascorbic acid	100			3	9.1	10.0
	600			3	4.0	4.4
	7000			3	5.2	4.9
Acid hydrolysis + vanadomolybdophosphoric acid		80		37	106.8	7.4
		300		38	66.5	14.0
		3000		37	36.1	23.5
Acid hydrolysis + stannous chloride		80		39	60.1	12.5
		300		36	47.6	21.7
		3000		38	37.4	22.8
Persulfate + vanadomolybdophosphoric acid			210	32	55.8	1.6
			990	32	23.9	2.3
			10 230	31	6.5	0.3
Sulfuric-nitric acids + vanadomolybdophosphoric acid			210	23	65.6	20.9
			990	22	47.3	0.6
			10 230	20	7.0	0.4
Perchloric acid + vanadomolybdophosphoric acid			210	4	33.5	45.2
			990	5	20.3	2.6
			10 230	6	11.7	2.2
Persulfate + stannous chloride			210	29	28.1	9.2
			990	30	14.9	12.3
			10 230	29	11.5	4.3
Sulfuric-nitric acids + stannous chloride			210	20	20.8	1.2
			990	17	8.8	3.2
			10 230	19	7.5	0.4

also are presented. Careful attention to procedure may allow application of these methods to very low levels of phosphorus, such as those found in unimpaired fresh water.

Ion chromatography (4110) and capillary ion electrophoresis (4140) are useful for determination of orthophosphate in undigested samples.

4. Precision and Bias

To aid in method selection, Table 4500-P:I presents the results of various combinations of digestions, hydrolysis, and colorimetric techniques for three synthetic samples of the following compositions:

Sample 1: 100 µg orthophosphate phosphorus (PO_4^{3-}-P/L), 80 µg condensed phosphate phosphorus/L (sodium hexametaphosphate), 30 µg organic phosphorus/L (adenylic acid), 1.5 mg NH_3-N/L, 0.5 mg NO_3-N/L, and 400 mg Cl^-/L.

Sample 2: 600 µg PO_4^{3-}-P/L, 300 µg condensed phosphate phosphorus/L (sodium hexametaphosphate), 90 µg organic phosphorus/L (adenylic acid), 0.8 mg NH_3-N/L, 5.0 mg NO_3^--N/L, and 400 mg Cl^-/L.

Sample 3: 7.00 mg PO_4^{3-}-P/L, 3.00 mg condensed phosphate phosphorus/L (sodium hexametaphosphate), 0.230 mg organic phosphorus/L (adenylic acid), 0.20 mg NH_3-N/L, 0.05 mg NO_3^--N/L, and 400 mg Cl^-/L.

5. Sampling and Storage

If dissolved phosphorus forms are to be differentiated, filter sample immediately after collection. Preserve by freezing at or below $-10°C$. In some cases 40 mg $HgCl_2$/L may be added to the samples, especially when they are to be stored for long periods before analysis. CAUTION: *HgCl_2 is a hazardous substance; take appropriate precautions in disposal; use of HgCl_2 is not encouraged.* Do not add either acid or $CHCl_3$ as a preservative when phosphorus forms are to be determined. If total phosphorus alone is to be determined, add H_2SO_4 or HCl to pH<2 and cool to 4°C, or freeze without any additions.

Do not store samples containing low concentrations of phosphorus in plastic bottles unless kept in a frozen state because phosphates may be adsorbed onto the walls of plastic bottles.

Rinse all glass containers with hot dilute HCl, then rinse several times in reagent water. Never use commercial detergents containing phosphate for cleaning glassware used in phosphate analysis.

6. Bibliography

BLACK, C.A., D.D. EVANS, J.L. WHITE, L.E. ENSMINGER & F.E. CLARK, eds. 1965. Methods of Soil Analysis, Part 2, Chemical and Microbiological Properties. American Soc. Agronomy, Madison, Wisc.

JENKINS, D. 1965. A study of methods suitable for the analysis and preservation of phosphorus forms in an estuarine environment. SERL Rep. No. 65-18, Sanitary Engineering Research Lab., Univ. California, Berkeley.

LEE, G.F. 1967. Analytical chemistry of plant nutrients. *In* Proc. Int. Conf. Eutrophication, Madison, Wisc.

FITZGERALD, G.P. & S.L. FAUST. 1967. Effect of water sample preservation methods on the release of phosphorus from algae. *Limnol. Oceanogr.* 12:332.

4500-P B.   Sample Preparation

For information on selection of digestion method (¶s 3 through 5 below), see 4500-P.A.3*a*.

1. Preliminary Filtration

Filter samples for determination of dissolved reactive phosphorus, dissolved acid-hydrolyzable phosphorus, and total dissolved phosphorus through 0.45-µm membrane filters. A glass fiber filter may be used to prefilter hard-to-filter samples.

Wash membrane filters by soaking in distilled water before use because they may contribute significant amounts of phosphorus to samples containing low concentrations of phosphate. Use one of two washing techniques: (*a*) soak 50 filters in 2 L distilled water for 24 h; (*b*) soak 50 filters in 2 L distilled water for 1 h, change distilled water, and soak filters an additional 3 h. Membrane filters also may be washed by running several 100-mL portions of distilled water through them. This procedure requires more frequent determination of blank values to ensure consistency in washing and to evaluate different lots of filters.

2. Preliminary Acid Hydrolysis

The acid-hydrolyzable phosphorus content of the sample is defined operationally as the difference between reactive phosphorus as measured in the untreated sample and phosphate found after mild acid hydrolysis. Generally, it includes condensed phosphates such as pyro-, tripoly-, and higher-molecular-weight species such as hexametaphosphate. In addition, some natural waters contain organic phosphate compounds that are hydrolyzed to orthophosphate under the test conditions. Polyphosphates generally do not respond to reactive phosphorus tests but can be hydrolyzed to orthophosphate by boiling with acid.

After hydrolysis, determine reactive phosphorus by a colorimetric method (C, D, or E). Interferences, precision, bias, and sensitivity will depend on the colorimetric method used.

a. Apparatus:

Autoclave or pressure cooker, capable of operating at 98 to 137 kPa.

b. Reagents:

1) *Phenolphthalein indicator aqueous solution.*

2) *Strong acid solution:* Slowly add 300 mL conc H_2SO_4 to about 600 mL distilled water. When cool, add 4.0 mL conc HNO_3 and dilute to 1 L.

3) *Sodium hydroxide,* NaOH, 6*N*.

c. Procedure: To 100-mL sample or a portion diluted to 100 mL, add 0.05 mL (1 drop) phenolphthalein indicator solution. If a red color develops, add strong acid solution dropwise, to just discharge the color. Then add 1 mL more.

Boil gently for at least 90 min, adding distilled water to keep the volume between 25 and 50 mL. Alternatively, heat for 30 min in an autoclave or pressure cooker at 98 to 137 kPa. Cool, neutralize to a faint pink color with NaOH solution, and restore to the original 100-mL volume with distilled water.

Prepare a calibration curve by carrying a series of standards containing orthophosphate (see colorimetric method C, D, or E) through the hydrolysis step. Do not use orthophosphate standards without hydrolysis, because the salts added in hydrolysis cause an increase in the color intensity in some methods.

Determine reactive phosphorus content of treated portions, using Method C, D, or E. This gives the sum of polyphosphate and orthophosphate in the sample. To calculate its content of acid-hydrolyzable phosphorus, determine reactive phosphorus in a sample portion that has not been hydrolyzed, using the same colorimetric method as for treated sample, and subtract.

3. Perchloric Acid Digestion

a. Apparatus:

1) *Hot plate:* A 30- × 50-cm heating surface is adequate.

2) *Safety shield.*

3) *Safety goggles.*

4) *Erlenmeyer flasks,* 125-mL, acid-washed and rinsed with distilled water.

b. Reagents:

1) *Nitric acid,* HNO_3, conc.

2) *Perchloric acid,* $HClO_4 \cdot 2H_2O$, purchased as 70 to 72% $HClO_4$, reagent-grade.

3) *Sodium hydroxide,* NaOH, 6N.

4) *Methyl orange indicator solution.*

5) *Phenolphthalein indicator aqueous solution.*

c. Procedure: CAUTION—*Heated mixtures of $HClO_4$ and organic matter may explode violently. Avoid this hazard by taking the following precautions: (a) Do not add $HClO_4$ to a hot solution that may contain organic matter. (b) Always initiate digestion of samples containing organic matter with HNO_3. Complete digestion using the mixture of HNO_3 and $HClO_4$. (c) Do not fume with $HClO_4$ in ordinary hoods. Use hoods especially constructed for $HClO_4$ fuming or a glass fume eradicator* connected to a water pump. (d) Never let samples being digested with $HClO_4$ evaporate to dryness.*

Measure sample containing the desired amount of phosphorus (this will be determined by whether Method C, D, or E is to be used) into a 125-mL erlenmeyer flask. Acidify to methyl orange with conc HNO_3, add another 5 mL conc HNO_3, and evaporate on a steam bath or hot plate to 15 to 20 mL.

Add 10 mL each of conc HNO_3 and $HClO_4$ to the 125-mL conical flask, cooling the flask between additions. Add a few boiling chips, heat on a hot plate, and evaporate gently until dense white fumes of $HClO_4$ just appear. If solution is not clear, cover neck of flask with a watch glass and keep solution barely boiling until it clears. If necessary, add 10 mL more HNO_3 to aid oxidation.

Cool digested solution and add 1 drop aqueous phenolphthalein solution. Add 6N NaOH solution until the solution just turns pink. If necessary, filter neutralized solution and wash filter liberally with distilled water. Make up to 100 mL with distilled water.

Determine the PO_4^{3-}-P content of the treated sample by Method C, D, or E.

---

* GFS Chemical Co., Columbus, OH, or equivalent.

Prepare a calibration curve by carrying a series of standards containing orthophosphate (see Method C, D, or E) through digestion step. Do not use orthophosphate standards without treatment.

4. Sulfuric Acid-Nitric Acid Digestion

a. Apparatus:

1) *Digestion rack:* An electrically or gas-heated digestion rack with provision for withdrawal of fumes is recommended. Digestion racks typical of those used for micro-kjeldahl digestions are suitable.

2) *Micro-kjeldahl flasks.*

b. Reagents:

1) *Sulfuric acid,* H_2SO_4, conc.

2) *Nitric acid,* HNO_3, conc.

3) *Phenolphthalein indicator aqueous solution.*

4) *Sodium hydroxide,* NaOH, 1N.

c. Procedure: Into a micro-kjeldahl flask, measure a sample containing the desired amount of phosphorus (this is determined by the colorimetric method used). Add 1 mL conc H_2SO_4 and 5 mL conc HNO_3.

Digest to a volume of 1 mL and then continue until solution becomes colorless to remove HNO_3.

Cool and add approximately 20 mL distilled water, 0.05 mL (1 drop) phenolphthalein indicator, and as much 1N NaOH solution as required to produce a faint pink tinge. Transfer neutralized solution, filtering if necessary to remove particulate material or turbidity, into a 100-mL volumetric flask. Add filter washings to flask and adjust sample volume to 100 mL with distilled water.

Determine phosphorus by Method C, D, or E, for which a separate calibration curve has been constructed by carrying standards through the acid digestion procedure.

5. Persulfate Digestion Method

a. Apparatus:

1) *Hot plate:* A 30- × 50-cm heating surface is adequate.

2) *Autoclave:* An autoclave or pressure cooker capable of developing 98 to 137 kPa may be used in place of a hot plate.

3) *Glass scoop,* to hold required amounts of persulfate crystals.

b. Reagents:

1) *Phenolphthalein indicator aqueous solution.*

2) *Sulfuric acid solution:* Carefully add 300 mL conc H_2SO_4 to approximately 600 mL distilled water and dilute to 1 L with distilled water.

3) *Ammonium persulfate,* $(NH_4)_2S_2O_8$, solid, or potassium persulfate, $K_2S_2O_8$, solid.

4) *Sodium hydroxide,* NaOH, 1N.

c. Procedure: Use 50 mL or a suitable portion of thoroughly mixed sample. Add 0.05 mL (1 drop) phenolphthalein indicator solution. If a red color develops, add H_2SO_4 solution dropwise to just discharge the color. Then add 1 mL H_2SO_4 solution and either 0.4 g solid $(NH_4)_2S_2O_8$ or 0.5 g solid $K_2S_2O_8$.

Boil gently on a preheated hot plate for 30 to 40 min or until a final volume of 10 mL is reached. Organophosphorus compounds such as AMP may require as much as 1.5 to 2 h for complete digestion. Cool, dilute to 30 mL with distilled water, add 0.05 mL (1 drop) phenolphthalein indicator solution, and neutralize to a faint pink color with NaOH. Alternatively, heat for 30 min in an autoclave or pressure cooker at 98 to 137 kPa. Cool, add 0.05 mL (1 drop) phenolphthalein indicator solution, and neutralize to a faint pink color with NaOH. Make up to 100 mL with distilled water. In some samples a precipitate may form at this stage, but do not filter. For any subsequent subdividing of the sample, shake well. The precipitate (which is possibly a calcium phosphate) redissolves under the acid conditions of the colorimetric reactive phosphorus test. Determine phosphorus by Method C, D, or E, for which a separate calibration curve has been constructed by carrying standards through the persulfate digestion procedure.

6. Bibliography

Lee, G.F., N.L. Clesceri & G.P. Fitzgerald. 1965. Studies on the analysis of phosphates in algal cultures. *J. Air Water Pollut.* 9:715.

Shannon, J.E. & G.F. Lee. 1966. Hydrolysis of condensed phosphates in natural waters. *J. Air Water Pollut.* 10:735.

Gales, M.E., Jr., E.C. Julian & R.C. Kroner. 1966. Method for quantitative determination of total phosphorus in water. *J. Amer. Water Works Assoc.* 58:1363.

4500-P C. Vanadomolybdophosphoric Acid Colorimetric Method

1. General Discussion

a. Principle: In a dilute orthophosphate solution, ammonium molybdate reacts under acid conditions to form a heteropoly acid, molybdophosphoric acid. In the presence of vanadium, yellow vanadomolybdophosphoric acid is formed. The intensity of the yellow color is proportional to phosphate concentration.

b. Interference: Positive interference is caused by silica and arsenate only if the sample is heated. Negative interferences are caused by arsenate, fluoride, thorium, bismuth, sulfide, thiosulfate, thiocyanate, or excess molybdate. Blue color is caused by ferrous iron but this does not affect results if ferrous iron concentration is less than 100 mg/L. Sulfide interference may be removed by oxidation with bromine water. Ions that do not interfere in concentrations up to 1000 mg/L are Al^{3+}, Fe^{3+}, Mg^{2+}, Ca^{2+}, Ba^{2+}, Sr^{2+}, Li^+, Na^+, K^+, NH_4^+, Cd^{2+}, Mn^{2+}, Pb^{2+}, Hg^+, Hg^{2+}, Sn^{2+}, Cu^{2+}, Ni^{2+}, Ag^+, U^{4+}, Zr^{4+}, AsO_3^-, Br^-, CO_3^{2-}, ClO_4^-, CN^-, IO_3^-, SiO_4^{4-}, NO_3^-, NO_2^-, SO_4^{2-}, SO_3^{2-}, pyrophosphate, molybdate, tetraborate, selenate, benzoate, citrate, oxalate, lactate, tartrate, formate, and salicylate. If HNO_3 is used in the test, Cl^- interferes at 75 mg/L.

c. Minimum detectable concentration: The minimum detectable concentration is 200 µg P/L in 1-cm spectrophotometer cells.

2. Apparatus

a. Colorimetric equipment: One of the following is required:

1) *Spectrophotometer,* for use at 400 to 490 nm.

2) *Filter photometer,* provided with a blue or violet filter exhibiting maximum transmittance between 400 and 470 nm.

The wavelength at which color intensity is measured depends on sensitivity desired, because sensitivity varies tenfold with wavelengths 400 to 490 nm. Ferric iron causes interference at low wavelengths, particularly at 400 nm. A wavelength of 470 nm usually is used. Concentration ranges for different wavelengths are:

P Range mg/L	Wavelength nm
1.0– 5.0	400
2.0–10	420
4.0–18	470

b. Acid-washed glassware: Use acid-washed glassware for determining low concentrations of phosphorus. Phosphate contamination is common because of its absorption on glass surfaces. Avoid using commercial detergents containing phosphate. Clean all glassware with hot dilute HCl and rinse well with distilled water. Preferably, reserve the glassware only for phosphate determination, and after use, wash and keep filled with water until needed. If this is done, acid treatment is required only occasionally.

*c. Filtration apparatus and filter paper.**

3. Reagents

a. Phenolphthalein indicator aqueous solution.

b. Hydrochloric acid, HCl, 1 + 1. H_2SO_4, $HClO_4$, or HNO_3 may be substituted for HCl. The acid concentration in the determination is not critical but a final sample concentration of 0.5N is recommended.

c. Activated carbon.† Remove fine particles by rinsing with distilled water.

d. Vanadate-molybdate reagent:

1) *Solution A:* Dissolve 25 g ammonium molybdate, $(NH_4)_6Mo_7O_{24}\cdot4H_2O$, in 300 mL distilled water.

2) *Solution B:* Dissolve 1.25 g ammonium metavanadate, NH_4VO_3, by heating to boiling in 300 mL distilled water. Cool and add 330 mL conc HCl. Cool Solution B to room temperature, pour Solution A into Solution B, mix, and dilute to 1 L.

e. Standard phosphate solution: Dissolve in distilled water 219.5 mg anhydrous KH_2PO_4 and dilute to 1000 mL; 1.00 mL = 50.0 µg PO_4^{3-}-P.

4. Procedure

a. Sample pH adjustment: If sample pH is greater than 10, add 0.05 mL (1 drop) phenolphthalein indicator to 50.0 mL sample and discharge the red color with 1 + 1 HCl before diluting to 100 mL.

b. Color removal from sample: Remove excessive color in sample by shaking about 50 mL with 200 mg activated carbon in an erlenmeyer flask for 5 min and filter to remove carbon. Check each batch of carbon for phosphate because some batches produce high reagent blanks.

---

* Whatman No. 42 or equivalent.

† Darco G60 or equivalent.

c. Color development in sample: Place 35 mL or less of sample, containing 0.05 to 1.0 mg P, in a 50-mL volumetric flask. Add 10 mL vanadate-molybdate reagent and dilute to the mark with distilled water. Prepare a blank in which 35 mL distilled water is substituted for the sample. After 10 min or more, measure absorbance of sample versus a blank at a wavelength of 400 to 490 nm, depending on sensitivity desired (see ¶ *2a* above). The color is stable for days and its intensity is unaffected by variation in room temperature.

d. Preparation of calibration curve: Prepare a calibration curve by using suitable volumes of standard phosphate solution and proceeding as in ¶ *4c*. When ferric ion is low enough not to interfere, plot a family of calibration curves of one series of standard solutions for various wavelengths. This permits a wide latitude of concentrations in one series of determinations. Analyze at least one standard with each set of samples.

5. Calculation

$$\text{mg P/L} = \frac{\text{mg P(in 50 mL final volume)} \times 1000}{\text{mL sample}}$$

6. Precision and Bias

See Table 4500-P:I.

7. Bibliography

KITSON, R.E. & M.G. MELLON. 1944. Colorimetric determination of phosphorus as molybdovanadophosphoric acid. *Ind. Eng. Chem.*, Anal. Ed. 16:379.

BOLTZ, D.F. & M.G. MELLON. 1947. Determination of phosphorus, germanium, silicon, and arsenic by the heteropoly blue method. *Ind. Eng. Chem.*, Anal. Ed. 19:873.

GREENBERG, A.E., L.W. WEINBERGER & C.N. SAWYER. 1950. Control of nitrite interference in colorimetric determination of phosphorus. *Anal. Chem.* 22:499.

YOUNG, R.S. & A. GOLLEDGE. 1950. Determination of hexametaphosphate in water after threshold treatment. *Ind. Chem.* 26:13.

GRISWOLD, B.L., F.L. HUMOLLER & A.R. MCINTYRE. 1951. Inorganic phosphates and phosphate esters in tissue extracts. *Anal. Chem.* 23: 192.

BOLTZ, D.F., ed. 1958. Colorimetric Determination of Nonmetals. Interscience Publishers, New York, N.Y.

AMERICAN WATER WORKS ASSOCIATION. 1958. Committee report. Determination of orthophosphate, hydrolyzable phosphate, and total phosphate in surface waters. *J. Amer. Water Works Assoc.* 50:1563.

JACKSON, M.L. 1958. Soil Chemical Analysis. Prentice-Hall, Englewood Cliffs, N.J.

ABBOT, D.C., G.E. EMSDEN & J.R. HARRIS. 1963. A method for determining orthophosphate in water. *Analyst* 88:814.

PROFT, G. 1964. Determination of total phosphorus in water and wastewater as molybdovanadophosphoric acid. *Limnologica* 2:407.

4500-P D. Stannous Chloride Method

1. General Discussion

a. Principle: Molybdophosphoric acid is formed and reduced by stannous chloride to intensely colored molybdenum blue. This method is more sensitive than Method C and makes feasible measurements down to 7 µg P/L by use of increased light path length. Below 100 µg P/L an extraction step may increase reliability and lessen interference.

b. Interference: See Section 4500-P.C.1*b*.

c. Minimum detectable concentration: The minimum detectable concentration is about 3 µg P/L. The sensitivity at 0.3010 absorbance is about 10 µg P/L for an absorbance change of 0.009.

2. Apparatus

The same apparatus is required as for Method C, except that a pipetting bulb is required for the extraction step. Set spectrophotometer at 625 nm in the measurement of benzene-isobutanol extracts and at 690 nm for aqueous solutions. If the instrument is not equipped to read at 690 nm, use a wavelength of 650 nm for aqueous solutions, with somewhat reduced sensitivity and precision.

3. Reagents

a. Phenolphthalein indicator aqueous solution.

b. Strong-acid solution: Prepare as directed in Section 4500-P.B.2*b*2).

c. Ammonium molybdate reagent I: Dissolve 25 g $(NH_4)_6Mo_7O_{24} \cdot 4H_2O$ in 175 mL distilled water. Cautiously add 280 mL conc H_2SO_4 to 400 mL distilled water. Cool, add molybdate solution, and dilute to 1 L.

d. Stannous chloride reagent I: Dissolve 2.5 g fresh $SnCl_2 \cdot 2H_2O$ in 100 mL glycerol. Heat in a water bath and stir with a glass rod to hasten dissolution. This reagent is stable and requires neither preservatives nor special storage.

e. Standard phosphate solution: Prepare as directed in Section 4500-P.C.3*e*.

f. Reagents for extraction:

1) *Benzene-isobutanol solvent:* Mix equal volumes of benzene and isobutyl alcohol. (CAUTION—*This solvent is highly flammable.*)

2) *Ammonium molybdate reagent II:* Dissolve 40.1 g $(NH_4)_6Mo_7O_{24} \cdot 4H_2O$ in approximately 500 mL distilled water. Slowly add 396 mL ammonium molybdate reagent I. Cool and dilute to 1 L.

3) *Alcoholic sulfuric acid solution:* Cautiously add 20 mL conc H_2SO_4 to 980 mL methyl alcohol with continuous mixing.

4) *Dilute stannous chloride reagent II:* Mix 8 mL stannous chloride reagent I with 50 mL glycerol. This reagent is stable for at least 6 months.

4. Procedure

a. Preliminary sample treatment: To 100 mL sample containing not more than 200 µg P and free from color and turbidity, add 0.05 mL (1 drop) phenolphthalein indicator. If sample turns pink, add strong acid solution dropwise to discharge the color. If more than 0.25 mL (5 drops) is required, take a smaller sample and dilute to 100 mL with distilled water after first discharging the pink color with acid.

b. Color development: Add, with thorough mixing after each addition, 4.0 mL molybdate reagent I and 0.5 mL (10 drops) stannous chloride reagent I. Rate of color development and intensity of color depend on temperature of the final solution, each 1°C increase producing about 1% increase in color. Hence, hold samples, standards, and reagents within 2°C of one another and in the temperature range between 20 and 30°C.

c. Color measurement: After 10 min, but before 12 min, using the same specific interval for all determinations, measure color photometrically at 690 nm and compare with a calibration curve, using a distilled water blank. Light path lengths suitable for various concentration ranges are as follows:

Approximate P Range mg/L	Light Path cm
0.3–2	0.5
0.1–1	2
0.007–0.2	10

Always run a blank on reagents and distilled water. Because the color at first develops progressively and later fades, maintain equal timing conditions for samples and standards. Prepare at least one standard with each set of samples or once each day that tests are made. The calibration curve may deviate from a straight line at the upper concentrations of the 0.3 to 2.0-mg/L range.

d. Extraction: When increased sensitivity is desired or interferences must be overcome, extract phosphate as follows: Pipet a 40-mL sample, or one diluted to that volume, into a 125-mL se-paratory funnel. Add 50.0 mL benzene-isobutanol solvent and 15.0 mL molybdate reagent II. Close funnel at once and shake vigorously for exactly 15 s. If condensed phosphate is present, any delay will increase its conversion to orthophosphate. Remove stopper and withdraw 25.0 mL of separated organic layer, using a pipet with safety bulb. Transfer to a 50-mL volumetric flask, add 15 to 16 mL alcoholic H_2SO_4 solution, swirl, add 0.50 mL (10 drops) dilute stannous chloride reagent II, swirl, and dilute to the mark with alcoholic H_2SO_4. Mix thoroughly. After 10 min, but before 30 min, read against the blank at 625 nm. Prepare blank by carrying 40 mL distilled water through the same procedure used for the sample. Read phosphate concentration from a calibration curve prepared by taking known phosphate standards through the same procedure used for samples.

5. Calculation

Calculate as follows:

a. Direct procedure:

$$mg\ P/L = \frac{mg\ P\ (\text{in approximately 104.5 mL final volume}) \times 1000}{mL\ sample}$$

b. Extraction procedure:

$$mg\ P/L = \frac{mg\ P\ (\text{in 50 mL final volume}) \times 1000}{mL\ sample}$$

6. Precision and Bias

See Table 4500-P:I.

4500-P E.   Ascorbic Acid Method

1. General Discussion

a. Principle: Ammonium molybdate and potassium antimonyl tartrate react in acid medium with orthophosphate to form a heteropoly acid—phosphomolybdic acid—that is reduced to intensely colored molybdenum blue by ascorbic acid.

b. Interference: Arsenates react with the molybdate reagent to produce a blue color similar to that formed with phosphate. Concentrations as low as 0.1 mg As/L interfere with the phosphate determination. Hexavalent chromium and NO_2^- interfere to give results about 3% low at concentrations of 1 mg/L and 10 to 15% low at 10 mg/L. Sulfide (Na_2S) and silicate do not interfere at concentrations of 1.0 and 10 mg/L.

c. Minimum detectable concentration: Approximately 10 μg P/L. P ranges are as follows:

Approximate P Range mg/L	Light Path cm
0.30–2.0	0.5
0.15–1.30	1.0
0.01–0.25	5.0

2. Apparatus

a. Colorimetric equipment: One of the following is required:

1) *Spectrophotometer,* with infrared phototube for use at 880 nm, providing a light path of 2.5 cm or longer.

2) *Filter photometer,* equipped with a red color filter and a light path of 0.5 cm or longer.

b. Acid-washed glassware: See Section 4500-P.C.2b.

TABLE 4500-P:II. COMPARISON OF PRECISION AND BIAS OF ASCORBIC ACID METHODS

Ascorbic Acid Method	Phosphorus Concentration, Dissolved Orthophosphate $\mu g/L$	No. of Labora-tories	Relative Standard Deviation %		Relative Error %	
			Distilled Water	River Water	Distilled Water	River Water
13th Edition[1]	228	8	3.87	2.17	4.01	2.08
Current method[2]	228	8	3.03	1.75	2.38	1.39

3. Reagents

a. Sulfuric acid, H_2SO_4, *5N:* Dilute 70 mL conc H_2SO_4 to 500 mL with distilled water.

b. Potassium antimonyl tartrate solution: Dissolve 1.3715 g $K(SbO)C_4H_4O_6 \cdot \frac{1}{2}H_2O$ in 400 mL distilled water in a 500-mL volumetric flask and dilute to volume. Store in a glass-stoppered bottle.

c. Ammonium molybdate solution: Dissolve 20 g $(NH_4)_6Mo_7O_{24} \cdot 4H_2O$ in 500 mL distilled water. Store in a glass-stoppered bottle.

d. Ascorbic acid, 0.1*M:* Dissolve 1.76 g ascorbic acid in 100 mL distilled water. The solution is stable for about 1 week at 4°C.

e. Combined reagent: Mix the above reagents in the following proportions for 100 mL of the combined reagent: 50 mL 5*N* H_2SO_4, 5 mL potassium antimonyl tartrate solution, 15 mL ammonium molybdate solution, and 30 mL ascorbic acid solution. *Mix after addition of each reagent.* Let all reagents reach room temperature before they are mixed and mix in the order given. If turbidity forms in the combined reagent, shake and let stand for a few minutes until turbidity disappears before proceeding. The reagent is stable for 4 h.

f. Stock phosphate solution: See Section 4500-P.C.3*e.*

g. Standard phosphate solution: Dilute 50.0 mL stock phosphate solution to 1000 mL with distilled water; 1.00 mL = 2.50 μg P.

4. Procedure

a. Treatment of sample: Pipet 50.0 mL sample into a clean, dry test tube or 125-mL erlenmeyer flask. Add 0.05 mL (1 drop) phenolphthalein indicator. If a red color develops add 5*N* H_2SO_4 solution dropwise to just discharge the color. Add 8.0 mL combined reagent and mix thoroughly. After at least 10 min but no more than 30 min, measure absorbance of each sample at 880 nm, using reagent blank as the reference solution.

b. Correction for turbidity or interfering color: Natural color of water generally does not interfere at the high wavelength used. For highly colored or turbid waters, prepare a blank by adding all reagents except ascorbic acid and potassium antimonyl tartrate

to the sample. Subtract blank absorbance from absorbance of each sample.

c. Preparation of calibration curve: Prepare individual calibration curves from a series of six standards within the phosphate ranges indicated in ¶ 1*c* above. Use a distilled water blank with the combined reagent to make photometric readings for the calibration curve. Plot absorbance vs. phosphate concentration to give a straight line passing through the origin. Test at least one phosphate standard with each set of samples.

5. Calculation

$$\text{mg P/L} = \frac{\text{mg P (in approximately 58 mL final volume)} \times 1000}{\text{mL sample}}$$

6. Precision and Bias

The precision and bias values given in Table 4500-P:I are for a single-solution procedure given in the 13th edition. The present procedure differs in reagent-to-sample ratios, no addition of solvent, and acidity conditions. It is superior in precision and bias to the previous technique in the analysis of both distilled water and river water at the 228-μg P/L level (Table 4500-P:II).

7. References

1. EDWARDS, G.P., A.H. MOLOF & R.W. SCHNEEMAN. 1965. Determination of orthophosphate in fresh and saline waters. *J. Amer. Water Works Assoc.* 57:917.
2. MURPHY, J. & J. RILEY. 1962. A modified single solution method for the determination of phosphate in natural waters. *Anal. Chim. Acta* 27:31.

8. Bibliography

SLETTEN, O. & C.M. BACH. 1961. Modified stannous chloride reagent for orthophosphate determination. *J. Amer. Water Works Assoc.* 53: 1031.
STRICKLAND, J.D.H. & T.R. PARSONS. 1965. A Manual of Sea Water Analysis, 2nd ed. Fisheries Research Board of Canada, Ottawa.

4500-P F. Automated Ascorbic Acid Reduction Method

1. General Discussion

a. Principle: Ammonium molybdate and potassium antimonyl tartrate react with orthophosphate in an acid medium to form an antimony-phosphomolybdate complex, which, on reduction with ascorbic acid, yields an intense blue color suitable for photometric measurement.

b. Interferences: As much as 50 mg Fe^{3+}/L, 10 mg Cu/L, and 10 mg SiO_2/L can be tolerated. High silica concentrations cause positive interference.

In terms of phosphorus, the results are high by 0.005, 0.015, and 0.025 mg/L for silica concentrations of 20, 50, and 100 mg/L, respectively. Salt concentrations up to 20% (w/v) cause an error of less than 1%. Arsenate (AsO_4^{3-}) is a positive interference.

Eliminate interference from NO_2^- and S^{2-} by adding an excess of bromine water or a saturated potassium permanganate ($KMnO_4$) solution. Remove interfering turbidity by filtration before analysis. Filter samples for total or total hydrolyzable phosphorus only after digestion. Sample color that absorbs in the photometric range used for analysis also will interfere. See also Section 4500-P.E.1*b*.

c. Application: Orthophosphate can be determined in potable, surface, and saline waters as well as domestic and industrial wastewaters over a range of 0.001 to 10.0 mg P/L when photometric measurements are made at 650 to 660 or 880 nm in a 15-mm or 50-mm tubular flow cell. Determine higher concentrations by diluting sample. Although the automated test is designed for orthophosphate only, other phosphorus compounds can be converted to this reactive form by various sample pretreatments described in Section 4500-P.B.1, 2, and 5.

2. Apparatus

a. Automated analytical equipment: An example of the continuous-flow analytical instrument consists of the interchangeable components shown in Figure 4500-P:2. A flow cell of 15 or 50 mm and a filter of 650 to 660 or 880 nm may be used.

b. Hot plate or autoclave.

c. Acid-washed glassware: See Section 4500-P.C.2*b*.

3. Reagents

a. Potassium antimonyl tartrate solution: Dissolve 0.3 g $K(SbO)C_4H_4O_6 \cdot \frac{1}{2}H_2O$ in approximately 50 mL distilled water and dilute to 100 mL. Store at 4°C in a dark, glass-stoppered bottle.

b. Ammonium molybdate solution: Dissolve 4 g $(NH_4)_6Mo_7O_{24} \cdot 4H_2O$ in 100 mL distilled water. Store in a plastic bottle at 4°C.

c. Ascorbic acid solution: See Section 4500-P.E.3*d*.

d. Combined reagent: See Section 4500-P.E.3*e*.

e. Dilute sulfuric acid solution: Slowly add 140 mL conc H_2SO_4 to 600 mL distilled water. When cool, dilute to 1 L.

f. Ammonium persulfate, $(NH_4)_2S_2O_8$, crystalline.

g. Phenolphthalein indicator aqueous solution.

h. Stock phosphate solution: Dissolve 439.3 mg anhydrous

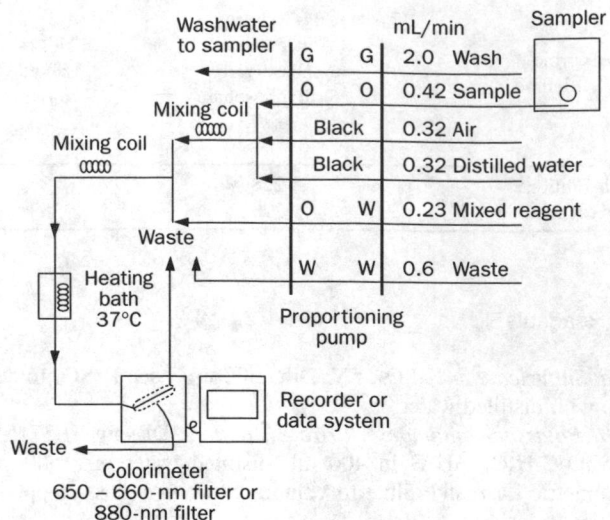

Figure 4500-P:2. Phosphate manifold for automated analytical system.

KH_2PO_4, dried for 1 h at 105°C, in distilled water and dilute to 1000 mL; 1.00 mL = 100 µg P.

i. Intermediate phosphate solution: Dilute 100.0 mL stock phosphate solution to 1000 mL with distilled water; 1.00 mL = 10.0 µg P.

j. Standard phosphate solutions: Prepare a suitable series of standards by diluting appropriate volumes of intermediate phosphate solution.

4. Procedure

Set up manifold as shown in Figure 4500-P:2 and follow the general procedure described by the manufacturer.

Add 0.05 mL (1 drop) phenolphthalein indicator solution to approximately 50 mL sample. If a red color develops, add H_2SO_4 (¶ 3*e*) dropwise to just discharge the color.

5. Calculation

Prepare standard curves by plotting response of standards processed through the manifold against P concentration in standards. Compute sample P concentration by comparing sample response with standard curve.

6. Precision and Bias

Six samples were analyzed in a single laboratory in septuplicate. At an average PO_4^{3-} concentration of 0.340 mg/L, the average deviation was 0.015 mg/L. The coefficient of variation was 6.2%. In two samples with added PO_4^{3-}, recoveries were 89 and 96%.

7. Bibliography

HENRIKSEN, A. 1966. An automatic method for determining orthophosphate in sewage and highly polluted waters. *Analyst* 91:652.

LOBRING, L.B. & R.L. BOOTH. 1973. Evaluation of the AutoAnalyzer II; A progress report. *In* Advances in Automated Analysis: 1972 Technicon International Congress. Vol. 8, p. 7. Mediad, Inc., Tarrytown, N.Y.

U.S. ENVIRONMENTAL PROTECTION AGENCY. 1979. Methods for Chemical Analysis of Water and Wastes. EPA-600/4-79-020, National Environmental Research Center, Cincinnati, Ohio.

U.S. ENVIRONMENTAL PROTECTION AGENCY. MDQARL Method Study 4, Automated Methods. National Environmental Research Center, Cincinnati, Ohio (in preparation).

4500-P G.   Flow Injection Analysis for Orthophosphate
(PROPOSED)

1. General Discussion

a. Principle: The orthophosphate ion (PO_4^{3-}) reacts with ammonium molybdate and antimony potassium tartrate under acidic conditions to form a complex. This complex is reduced with ascorbic acid to form a blue complex that absorbs light at 880 nm. The absorbance is proportional to the concentration of orthophosphate in the sample.

Also see Sections 4500-P.A, B, and F, and Section 4130, Flow Injection Analysis (FIA).

b. Interferences: Remove large or fibrous particulates by filtering sample through glass wool. Guard against contamination from reagents, water, glassware, and the sample preservation process.

Silica forms a pale blue complex that also absorbs at 880 nm. This interference is generally insignificant because a silica concentration of approximately 30 mg/L would be required to produce a 0.005 mg P/L positive error in orthophosphate.

Concentrations of ferric iron greater than 50 mg/L cause a negative error due to competition with the complex for the reducing agent ascorbic acid. Treat samples high in iron with sodium bisulfite to eliminate this interference, as well as the interference due to arsenates.

Glassware contamination is a problem in low-level phosphorus determinations. Wash glassware with hot dilute HCl and rinse with reagent water. Commercial detergents are rarely needed but, if they are used, use special phosphate-free preparations.

Also see Section 4500-P.F.

2. Apparatus

Flow injection analysis equipment consisting of:

a. FIA injection valve with sample loop or equivalent.

b. Multichannel proportioning pump.

c. FIA manifold (Figure 4500-P:3) with tubing heater and flow cell. Relative flow rates only are shown in Figure 4500-P:3. Tubing volumes are given as an example only; they may be scaled down proportionally. Use manifold tubing of an inert material such as TFE.

d. Absorbance detector, 880 nm, 10-nm bandpass.

e. Injection valve control and data acquisition system.

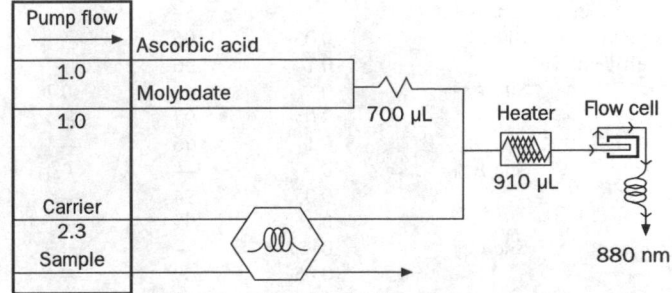

Figure 4500-P:3. FIA orthophosphate manifold.

3. Reagents

Use reagent water (>10 megohm) to prepare carrier and all solutions. To prevent bubble formation, degas carrier and buffer with helium. Pass He at 140 kPa (20 psi) through a helium degassing tube. Bubble He through 1 L solution for 1 min. As an alternative to preparing reagents by weight/weight, use weight/volume.

a. Stock ammonium molybdate solution: To a tared 1-L container add 40.0 g ammonium molybdate tetrahydrate $[(NH_4)_6Mo_7O_{24}\cdot4H_2O]$ and 983 g water. Mix with a magnetic stirrer for at least 4 h. Store in plastic and refrigerate.

b. Stock antimony potassium tartrate solution: To a 1-L dark, tared container add 3.0 g antimony potassium tartrate (potassium antimonyl tartrate hemihydrate), $K(SbO)C_4H_4O_6\cdot\frac{1}{2}H_2O$, and 995 g water. Mix with a magnetic stirrer until dissolved. Store in a dark bottle and refrigerate.

c. Working molybdate color reagent: To a tared 1-L container add 680 g water, then add 64.4 g conc sulfuric acid. (CAUTION: *This solution becomes very hot!*) Swirl to mix. When mixture can be handled comfortably, add 213 g stock ammonium molybdate solution (¶ 3a) and 72.0 g stock antimony potassium tartrate solution (¶ 3b). Shake and degas with helium.

d. Ascorbic acid solution: To a tared 1-L container, add 60.0 g granular ascorbic acid and 975 g water. Stir or shake until dissolved. Degas this reagent with helium, then add 1.0 g dodecyl

TABLE 4500-P:III. RESULTS OF SINGLE-LABORATORY STUDIES WITH
SELECTED MATRICES

Matrix	Sample/Blank Designation	Known Addition mg P/L	Recovery %	Relative Standard Deviation %
Wastewater treatment plant influent	Reference sample*	—	101	—
	Blank†	0.05	96	—
		0.1	95	—
	Site A‡§	0	—	0.7
		0.05	98	—
		0.1	101	—
	Site B‡§	0	—	5
		0.05	75	—
		0.1	91	—
	Site C‡§	0	—	0.6
		0.05	88	—
		0.1	97	—
Wastewater treatment plant effluent	Reference sample*	—	100	—
	Blank†	0.05	96	—
		0.1	96	—
	Site A‡‖	0	—	0.7
		0.05	94	—
		0.1	96	—
	Site B‡‖	0	—	0.3
		0.05	94	—
		0.1	99	—
	Site C‡‖	0	—	0.5
		0.05	109	—
		0.1	107	—
Landfill leachate	Reference sample*	—	98	—
	Blank†	0.05	94	—
		0.1	95	—
	Site A‡#	0	—	0.9
		0.05	105	—
		0.1	106	—
	Site B‡#	0	—	6.7
		0.05	89	—
		0.1	94	—
	Site C‡#	0	—	0.9
		0.05	110	—
		0.1	109	—

* U.S. EPA QC sample, 0.109 mg P/L.

† Determined in duplicate.

‡ Samples without known additions determined four times; samples with known additions determined in duplicate.

§ Sample dilutions: A - 5-fold; B - 100-fold; C - 10-fold. Typical relative difference between duplicates 0.5%.

‖ Sample dilutions: A - 5-fold; B - 20-fold; C - 10-fold. Typical relative difference between duplicates 0.3%.

Sample dilutions: A - 20-fold; B - 10-fold; C - 20-fold. Typical relative difference between duplicates 1%.

sulfate, $CH_3(CH_2)_{11}OSO_3Na$, stirring gently to mix. Prepare fresh weekly.

e. Stock orthophosphate standard, 25.00 mg P/L: In a 1-L volumetric flask dissolve 0.1099 g primary standard grade anhydrous potassium phosphate monobasic (KH_2PO_4) that has been dried for 1 h at 105°C in about 800 mL water. Dilute to mark with water and invert to mix.

f. Standard orthophosphate solutions: Prepare orthophosphate standards in desired concentration range, using stock standard (¶ 3e) and diluting with water.

4. Procedure

Set up a manifold equivalent to that in Figure 4500-P:3 and follow method supplied by manufacturer or laboratory standard operating procedure. Use quality control protocols outlined in Section 4020.

5. Calculations

Prepare standard curves by plotting absorbance of standards processed through the manifold versus orthophosphate concentration. The calibration curve is linear.

6. Precision and Bias

a. Recovery and relative standard deviation: Table 4500-P:III gives results of single-laboratory studies.

b. MDL: A 700-µL sample loop was used in the method described above. Using a published MDL method,[1] analysts ran 21 replicates of a 5.0-µg P/L standard. These gave a mean of 5.26 µg P/L, a standard deviation of 0.264 µg P/L, and MDL of 0.67 µg P/L.

7. Reference

1. U.S. ENVIRONMENTAL PROTECTION AGENCY. 1984. Definition and procedure for the determination of method detection limits. Appendix B to 40 CFR 136 Rev. 1.11 amended June 30, 1986. 49 CFR 43430.

4500-P H. Manual Digestion and Flow Injection Analysis for Total Phosphorus (PROPOSED)

1. General Discussion

a. Principle: Polyphosphates are converted to the orthophosphate form by a sulfuric acid digestion and organic phosphorus is converted to orthophosphate by a persulfate digestion. When the resulting solution is injected onto the manifold, the orthophosphate ion (PO_4^{3-}) reacts with ammonium molybdate and antimony potassium tartrate under acidic conditions to form a complex. This complex is reduced with ascorbic acid to form a blue complex that absorbs light at 880 nm. The absorbance is proportional to the concentration of total phosphorus in the sample.

See Section 4500-P.A for a discussion of the various forms of phosphorus found in waters and wastewaters, Section 4500-P.B for a discussion of sample preparation and digestion, and Section 4130, Flow Injection Analysis (FIA).

b. Interferences: See 4500-P.G.1*b*.

2. Apparatus

Digestion and flow injection analysis equipment consisting of:
a. Hotplate or autoclave.
b. FIA injection valve with sample loop or equivalent.
c. Multichannel proportioning pump.
d. FIA manifold (Figure 4500-P:4) with tubing heater and flow cell. Relative flow rates only are shown in Figure 4500-P:4. Tubing volumes are given as an example only; they may be scaled down proportionally. Use manifold tubing of an inert material such as TFE.

3. Reagents

Use reagent water (>10 megohm) for all solutions. To prevent bubble formation, degas carrier and buffer with helium. Pass He

at 140 kPa (20 psi) through a helium degassing tube. Bubble He through 1 L solution for 1 min. As an alternative to preparing reagents by weight/weight, use weight/volume.

Prepare reagents listed in 4500-P.G.3*a, b, d, e,* and *f,* and in addition:

a. Sulfuric acid carrier, H_2SO_4, 0.13*M*: To a tared 1-L container add 993 g water, then add 13.3 g conc H_2SO_4. Shake carefully to mix. Degas daily. Prepare fresh weekly.

b. Molybdate color reagent: To a tared 1-L container add 694 g water, then add 38.4 g conc H_2SO_4. (CAUTION: *The solution becomes very hot!*) Swirl to mix. When mixture can be handled comfortably, add 72.0 g stock antimony potassium tartrate (¶ G.3*b*) and 213 g stock ammonium molybdate (¶ G.3*a*). Shake to mix, and degas.

4. Procedure

See Section 4500-P.B.4 or 5 for digestion procedures. Carry both standards and samples through the digestion. The resulting solutions should be about 0.13*M* in sulfuric acid to match the concentration of the carrier. If the solutions differ more than 10% from this concentration, adjust concentration of carrier's sulfuric acid to match that of digested samples.

Set up a manifold equivalent to that in Figure 4500-P:4 and analyze digested samples and standards by following method supplied by manufacturer or laboratory's standard operating procedure. Use quality control protocols outlined in Section 4020.

5. Calculations

Prepare standard curves by plotting absorbance of standards processed through the manifold versus phosphorus concentration. The calibration curve is linear.

6. Precision and Bias

a. MDL: A 780-μL sample loop was used in the method described above. Using a published MDL method,[1] analysts ran 21 replicates of a 3.5-μg P/L standard. These gave a mean of 3.53 μg P/L, a standard deviation of 0.82 μg P/L, and MDL of 2.0 μg P/L. The MDL is limited mainly by the precision of the digestion.

b. Precision study: Ten injections of a 100.0-μg P/L standard gave a percent relative standard deviation of 0.3%.

7. Reference

1. U.S. ENVIRONMENTAL PROTECTION AGENCY. 1984. Definition and procedure for the determination of method detection limits. Appendix B to 40 CFR 136 Rev. 1.11 amended June 30, 1986. 49 CFR 43430.

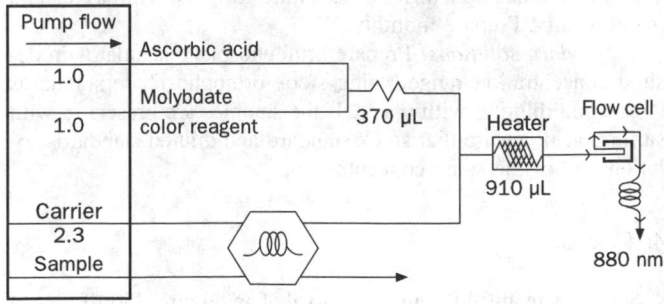

Figure 4500-P:4. FIA total phosphorus manifold.

4500-P I. In-line UV/Persulfate Digestion and Flow Injection Analysis for Total Phosphorus (PROPOSED)

1. General Discussion

a. Principle: Organic phosphorus is converted in-line to orthophosphate by heat, ultraviolet radiation, and persulfate digestion. At the same time, inorganic polyphosphates are converted to orthophosphate by in-line sulfuric acid digestion. The digestion processes occur before sample injection. A portion of the digested sample is then injected and its orthophosphate concentration determined by the flow injection method described in Section 4500-P.H.1.

See Section 4500-P.A for a discussion of the various forms of phosphorus found in waters and wastewaters, Section 4500-P.B for a discussion of sample preparation and digestion, and Section 4130, Flow Injection Analysis (FIA).

b. Interferences: See 4500-P.G.1*b*.

2. Apparatus

Flow injection analysis equipment consisting of:

a. FIA injection valve with sample loop or equivalent.

b. Multichannel proportioning pump.

c. FIA manifold (Figure 4500-P:5) with tubing heater, in-line ultraviolet digestion fluidics including a debubbler consisting of a gas-permeable TFE membrane and its holder, and flow cell. Relative flow rates only are shown in Figure 4500-P:5. Tubing volumes are given as an example only; they may be scaled down proportionally. Use manifold tubing of an inert material such as TFE. The block marked "UV" should consist of TFE tubing irradiated by a mercury discharge ultraviolet lamp emitting radiation at 254 nm.

d. Absorbance detector, 880 nm, 10-nm bandpass.

e. Injection valve control and data acquisition system.

3. Reagents

Use reagent water (>10 megohm) for all solutions. To prevent bubble formation, degas carrier and all reagents with helium. Pass He at 140 kPa (20 psi) through a helium degassing tube. Bubble He through 1 L solution for 1 min. As an alternative to preparing reagents by weight/weight, use weight/volume.

a. Digestion reagent 1: To a tared 1-L container, add 893.5 g water, then slowly add 196.0 g sulfuric acid, H_2SO_4. CAUTION: *This solution becomes very hot!* Prepare weekly. Degas before using.

b. Digestion reagent 2: To a tared 1-L container, add 1000 g water, then add 26 g potassium persulfate, $K_2S_2O_8$. Mix with a magnetic stirrer until dissolved. Prepare weekly. Degas before using.

c. Sulfuric acid carrier, 0.71*M*: To a tared 1-L container, slowly add 70 g H_2SO_4 to 962 g water. Add 5 g sodium chloride, NaCl. Let cool, then degas with helium. Add 1.0 g sodium dodecyl sulfate. Invert to mix. Prepare weekly.

d. Stock ammonium molybdate: To a tared 1-L container add 40.0 g ammonium molybdate tetrahydrate, $(NH_4)_6Mo_7O_{24}·4H_2O$, and 983 g water. Mix with a magnetic stirrer for at least 4 h. The solution can be stored in plastic for up to 2 months if refrigerated.

e. Stock antimony potassium tartrate: To a 1-L dark, plastic, tared container add 3.0 g antimony potassium tartrate (potassium antimonyl tartrate trihydrate), $C_8H_4K_2O_{12}Sb_2·3H_2O$, and 995 g water. Mix with a magnetic stirrer until dissolved. The solution can be stored in a dark plastic container for up to 2 months if refrigerated.

f. Molybdate color reagent: To a tared 1-L container add 715 g water, then 213 g stock ammonium molybdate (¶ 3*e*) and 72.0 g stock antimony potassium tartrate (¶ 3*f*). Add and dissolve 22.8 g sodium hydroxide, NaOH. Shake and degas with helium. Prepare weekly.

g. Ascorbic acid: To a tared 1-L container add 70.0 g ascorbic acid and 975 g water. Mix with a magnetic stirrer until dissolved. Degas with helium. Add 1.0 g sodium dodecyl sulfate. Mix with a magnetic stirrer. Prepare fresh every 2 d.

h. Stock orthophosphate standard, 1000 mg P/L: In a 1-L volumetric flask dissolve 4.396 g primary standard grade anhydrous potassium phosphate monobasic, KH_2PO_4 (dried for 1 h at 105°C), in about 800 mL water. Dilute to mark with water and invert to mix. Prepare monthly.

i. Standard solutions: Prepare orthophosphate standards in desired concentration range, using stock orthophosphate standards (¶ 3*i*), and diluting with water. If the samples are preserved with sulfuric acid, ensure that stock standard and diluted standards solutions are of the same concentration.

4. Procedure

Set up a manifold equivalent to that in Figure 4500-P:5 and follow method supplied by manufacturer or laboratory's standard operating procedure. Use quality control procedures described in Section 4020.

5. Calculations

Prepare standard curves by plotting absorbance of standards processed through manifold versus phosphorus concentration. The calibration curve is linear.

Verify digestion efficiency by determining tripolyphosphate and trimethylphosphate standards at regular intervals. In the con-

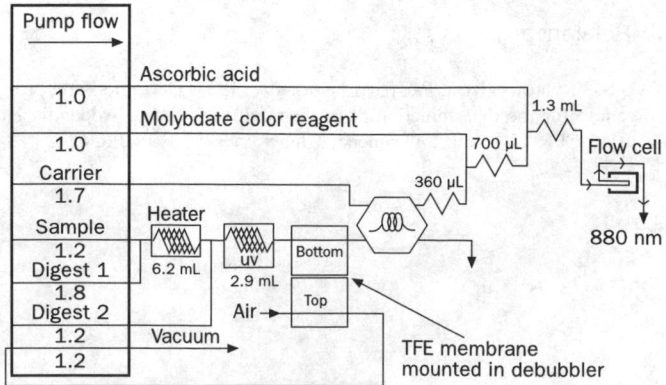

Figure 4500-P:5. FIA in-line total phosphorus manifold.

TABLE 4500-P:IV. RECOVERIES OF TOTAL PHOSPHORUS

Compound	Known Concentration mg P/L	Mean Concentration Recovered mg P/L	Recovery %	Relative Standard Deviation %
Sodium	10	8.99	90.3	0.5
pyrophosphate	2	1.81	90.2	0.6
	0.2	0.19	93.4	1.0
Phenylphosphate	10	10.10	101.5	0.2
	2	2.12	105.0	5.6
	0.2	0.20	101.3	1.0
Trimethylphosphate	10	8.99	90.3	0.2
	2	1.86	92.7	0.3
	0.2	0.18	95.3	1.1
Sodium	10	10.61	106.7	1.0
tripolyphosphate	2	2.14	106.6	0.2
	0.2	0.22	108.9	0.9

TABLE 4500-P:V. COMPARISON OF MANUAL AND IN-LINE TOTAL PHOSPHORUS METHODS

Samples	Concentration by Manual Persulfate Digestion Method mg P/L	Concentration by In-Line Digestion Method mg P/L	Relative Difference %
Influent (I2)	5.93	5.52	−6.9
Influent (I3)	5.03	4.50	−10.5
Influent (I5)	2.14	2.11	−1.4
Influent (I6)	1.88	1.71	−9.0
Effluent (E1)	3.42	2.87	−16.1
Effluent (E2)	3.62	3.55	−1.9
Effluent (E3)	3.26	3.34	+2.4
Effluent (E4)	8.36	8.16	−2.4
Effluent (E5)	0.65	0.71	+9.2
Effluent (E6)	0.74	0.81	+9.5
Phenylphosphate	1.95	1.91	−2.1
Trimethylphosphate	1.87	1.80	−3.7
Sodium pyrophosphate	1.90	1.73	−8.9
Sodium tripolyphosphate	1.84	1.73	−6.0

centration range of the method, the recovery of either of these compounds should be >95%.

6. Precision and Bias

a. MDL: A 390-μL sample loop was used in the method described above. Using a published MDL method,[1] analysts ran 21 replicates of a 0.10-mg P/L orthophosphate standard. These gave a mean of 0.10 mg P/L, a standard deviation of 0.003 mg P/L, and MDL of 0.007 mg P/L.

b. Precision of recovery study: Ten injections of a 10.0-mg P/L trimethylphosphate standard gave a mean percent recovery of 98% and a percent relative standard deviation of 0.8%.

c. Recovery of total phosphorus: Two organic and two inorganic complex phosphorus compounds were determined in triplicate at three concentrations. The results are shown in Table 4500-P:IV.

d. Comparison of in-line digestion with manual digestion method: Samples from a wastewater treatment plant influent and effluent and total phosphorus samples at 2.0 mg P/L were determined in duplicate with both manual persulfate digestion followed by the method in Section 4500-P.H and in-line digestion method. Table 4500-P:V gives the results of this comparison, and Figure 4500-P:6 shows the correlation between manual and in-line total phosphorus methods.

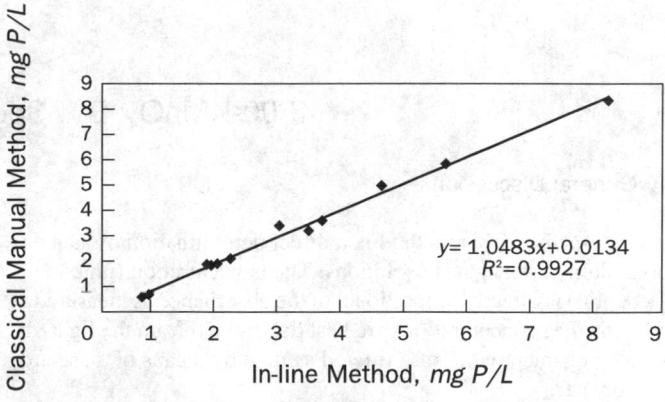

$$y = 1.0483x + 0.0134$$
$$R^2 = 0.9927$$

Figure 4500-P:6. Correlation between manual and in-line total phosphorus methods.

4500-KMnO₄   POTASSIUM PERMANGANATE*

4500-KMnO₄   A.   Introduction

1. Occurrence and Significance

Potassium permanganate, $KMnO_4$, has been widely used in both potable and nonpotable water sources. It has been applied to water supplies to remove taste, odor, color, iron, manganese, and sulfides and to control trihalomethanes (THMs) and zebra mussels. Municipal and industrial waste treatment facilities use potassium permanganate for odor control, toxic pollutant destruction, bio-augmentation, and grease removal.

Potassium permanganate is produced as a dark black-purple crystalline material. It has a solubility in water of 6 g/L at 20°C. The color of potassium permanganate solutions ranges from faint pink (dilute) to deep purple (concentrated). Under normal conditions the solid material is stable. However, as with all oxidizing agents, avoid contact with acids, peroxides, and all combustible organic or readily oxidizable materials.

2. Sampling and Storage

If kept dry, solid potassium permanganate may be stored indefinitely. Potassium permanganate solutions, made in oxidant-demand-free water, are stable for long periods of time if kept in an amber bottle out of direct sunlight.[1] For samples obtained from other water sources (those having an oxidant demand), analyze potassium permanganate on site, as soon as possible after sample collection.

3. Reference

1. DAY, R.A. & A.L. UNDERWOOD. 1986. Quantitative Analysis, 5th ed. Prentice-Hall, Englewood Cliffs, N.J.

---

* Approved by Standard Methods Committee, 1997.

4500-KMnO₄   B.   Spectrophotometric Method

1. General Discussion

a. Principle: This method is a direct determination of aqueous potassium permanganate solutions. The concentration (pink to violet color) is directly proportional to the absorbance as measured at 525 nm. The concentrations are best determined from the light-absorbing characteristics of a filtered sample by means of a spectrophotometer.

b. Application: This method is applicable to ground and surface waters.

c. Interference: Turbidity and manganese dioxide interfere. Remove by the filtration methods described below. Other color-producing compounds also interfere. Compensate for color by using an untreated sample to zero the spectrophotometer.

d. Minimum detectable concentration: As a guide, use the following light paths for the indicated $KMnO_4$ concentrations measured at 525 nm. Sample dilution may be required, depending on the initial concentration.

Cell Path Length cm	Range mg KMnO₄/L	Expected Absorbance for 1-mg/L KMnO₄ Solution
1	0.5–100	0.016
2.5	0.2–25	0.039
5	0.1–20	0.078

2. Apparatus

a. Photometric equipment: Use one of the following:

1) *Spectrophotometer,* for use at a wavelength of 525 nm and providing a light path of 1 cm or longer.

2) *Filter photometer,* equipped with a filter having a maximum transmittance at or near 525 nm and providing a light path of 1 cm or longer.

*b. Filtration apparatus.**

c. Filters: Use 0.22-μm filters that do not react with $KMnO_4$ (or smallest glass fiber filters available).

3. Reagents

Use potassium-permanganate-demand-free water (¶ *e* below) for all reagent preparation and dilutions.

a. Calcium chloride solution, $CaCl_2$, 1M: Dissolve 111 g $CaCl_2$ in water and dilute to 1 L.

b. Sulfuric acid, H_2SO_4, 20%: Add 20 g conc H_2SO_4 slowly, with stirring, to 80 mL water. After cooling, adjust final volume to 100 mL.

c. Sodium oxalate, $Na_2C_2O_4$, primary standard.

d. Sodium thiosulfate solution, $Na_2S_2O_3 \cdot 5H_2O$, 0.019M: Dissolve 0.471 g $Na_2S_2O_3 \cdot 5H_2O$ in water and dilute to 100 mL.

e. Potassium-permanganate-demand-free water: Add one small crystal $KMnO_4$ to 1 L distilled or deionized water; let stand.

---

* Millipore or equivalent.

After 1 to 2 d, a residual pink color should be present; if not, discard and obtain better-quality water or increase permanganate added. Redistill in an all-borosilicate-glass apparatus and discard initial 50 mL distillate. Collect distillate fraction that is free of permanganate: a red color with DPD reagent (see Section 4500-Cl.F.2b) indicates presence of permanganate.

f. Potassium permanganate standard, $KMnO_4$, 0.006*M:* Dissolve 1.000 g $KMnO_4$ in water and dilute to 1000 mL. Standardize as follows: Accurately weigh about 0.1 g primary standard grade sodium oxalate and dissolve it in 150 mL water in a 250-mL erlenmeyer flask. Add 20 mL 20% H_2SO_4 and heat to 70 to 80°C. Titrate the warm oxalate solution with the potassium permanganate standard until a pink coloration persists (60 s). Calculate the potassium permanganate concentration:

$$\text{mg } KMnO_4/L = \frac{W \times 1000}{2.1197 \times V}$$

where:

  W = weight of sodium oxalate, mg, and
  V = mL $KMnO_4$ titrant.

4. Procedure

a. Calibration of photometric equipment: Prepare calibration curve by diluting standardized potassium permanganate solution. Make dilutions appropriate for the cell path length and range desired. Plot absorbance (*y* axis) versus $KMnO_4$ concentration (*x* axis). Calculate a best-fit line through the points. Preferably perform a calibration check with a known $KMnO_4$ standard before any analysis to ensure that equipment is in proper working order.

b. Potassium permanganate analysis: Check zero on spectrophotometer at 525 nm with deionized water. If the water is soft (i.e.,<40 mg/L hardness as $CaCO_3$), add 1 mL $CaCl_2$ solution/L sample (111 mg/L as $CaCl_2$) to aid in removal of any colloidal manganese dioxide and suspended solids. Pass 50 mL sample through a 0.22-μm filter. Rinse spectrophotometer cell with two or three portions of filtrate. Fill cell and check that no air bubbles are present in the solution or on the sides of the cell. Measure absorbance at 525 nm (Reading A). For best results, minimize time between filtration and reading absorbance. To 100 mL sample, add 0.1 mL $CaCl_2$ solution. Add 0.1 mL sodium thiosulfate solution per 1 mg/L $KMnO_4$ (based on Reading A). Pass through a 0.22-μm filter and measure absorbance (Reading B).

5. Calculation

$$\text{Correct absorbance} = A - B$$

where:

  A = absorbance of sample, and
  B = absorbance of blank.

Compare corrected absorbance value obtained with the calibration curve and report the corresponding value as milligrams potassium permanganate per liter.

6. Precision and Bias

Based on the results obtained by eight analysts in a single laboratory, the overall precision (pooled standard deviation) was determined to be 0.035 mg/L for a 1-cm cell. The mean recovery for 12 measurements at 4 initial concentrations was 98%. The method detection level (MDL), as determined using Method 1030, was 0.083 mg $KMnO_4$/L.

4500-SiO₂   SILICA*

4500-SiO₂ A.   Introduction

1. Occurrence and Significance

Silicon does not occur free in nature, but rather as free silica (SiO_2) in coarsely crystalline (quartz, rock crystal, amethyst, etc.) and microcrystalline (flint, chert, jasper, etc.) varieties of quartz, the major component of sand and sandstone. Silicon is found in combination with other elements in silicates, represented by feldspar, hornblende, mica, asbestos, and other clay minerals. Silicates also occur in rocks such as granite, basalt, and shale. Silicon therefore is usually reported as silica (SiO_2) when rocks, sediments, soils, and water are analyzed. The average abundance of silica in different rock types is 7 to 80%, in typical soils 50 to 80%, and in surface and groundwater 14 mg/L.

The common aqueous forms of silica are H_4SiO_4 and $H_3SiO_4^-$. In the presence of magnesium, it can form scale deposits in boilers and in steam turbines. It is considered a nonessential trace element for most plants, but essential for most animals. Chronic exposure to silica dust can be toxic. There is no U.S. EPA drinking water standard MCL for silica.

A more complete discussion of the occurrence and chemistry of silica in natural waters is available.[1]

2. Selection of Method

Perform analyses by the electrothermal atomic absorption method (3113B) or one of the colorimetric methods (C, D, E, or F), depending on the fraction to be measured. The inductively coupled plasma mass spectrometric method (3125) or the inductively coupled plasma method (3120) also may be applied successfully in most cases (with lower detection limits), even though silica is not specifically listed as an analyte in the method.

_____
* Approved by Standard Methods Committee, 1997.

Methods 3120 and 3125 determine total silica. Methods C, D, E, and F determine molybdate-reactive silica. As noted in Section 4500-SiO$_2$.C.4, it is possible to convert other forms of silica to the molybdate-reactive form for determination by these methods. Method 3111D determines more than one form of silica. It will determine all dissolved silica and some colloidally dispersed silica. The determination of silica present in micrometer and sub-micrometer particles will depend on the size distribution, composition, and structure of the particles; thus Method 3111D cannot be said to determine total silica.

Method C is recommended for relatively pure waters containing from 0.4 to 25 mg SiO$_2$/L. As with most colorimetric methods, the range can be extended, if necessary, by diluting, by concentrating, or by varying the light path. Interferences due to tannin, color, and turbidity are more severe with this method than with Method D. Moreover, the yellow color produced by Method C has a limited stability and attention to timing is necessary. When applicable, however, it offers greater speed and simplicity than Method D because one reagent fewer is used; one timing step is eliminated; and many natural waters can be analyzed without dilution, which is not often the case with Method D. Method D is recommended for the low range, from 0.04 to 2 mg SiO$_2$/L. This range also can be extended if necessary. Such extension may be desirable if interference is expected from tannin, color, or turbidity. A combination of factors renders Methods D, E, and F less susceptible than Method C to those interferences; also, the blue color in Methods D, E, and F is more stable than the yellow color in Method C. However, many samples will require dilution because of the high sensitivity of the method. Permanent artificial color standards are not available for the blue color developed in Method D.

The yellow color produced by Method C and the blue color produced by Methods D, E, and F are affected by high concentrations of salts. With seawater the yellow color intensity is decreased by 20 to 35% and the blue color intensity is increased by 10 to 15%. When waters of high ionic strength are analyzed by these methods, use silica standards of approximately the same ionic strengths.[2]

Method E or F may be used where large numbers of samples are analyzed regularly. Method 3111D is recommended for broad-range use. Although Method 3111D is usable from 1 to 300 mg SiO$_2$/L, optimal results are obtained from about 20 to 300 mg/L. The range can be extended upward by dilution if necessary. This method is rapid and does not require any timing step.

The inductively coupled plasma method (3120) also may be used in analyses for silica.

3. Sampling and Storage

Collect samples in bottles of polyethylene, other plastic, or hard rubber, especially if there will be a delay between collection and analysis. Borosilicate glass is less desirable, particularly with waters of pH above 8 or with seawater, in which cases a significant amount of silica in the glass can dissolve. Freezing to preserve samples for analysis of other constituents can lower soluble silica values by as much as 20 to 40% in waters that have a pH below 6. Do not acidify samples for preservation because silica precipitates in acidic solutions.

4. References

1. HEM, J.D. 1985. Study and interpretation of the Chemical Characteristics of Natural Water, 3rd ed. U.S. Geol. Surv. Water Supply Pap. No. 2254.
2. FANNING, K.A. & M.E.Q. PILSON. 1973. On the spectrophotometric determination of dissolved silica in natural waters. *Anal. Chem.* 45:136.

5. Bibliography

ROY, C.J. 1945. Silica in natural waters. *Amer. J. Sci.* 243:393.
VAIL, J.G. 1952. The Soluble Silicates, Their Properties and Uses. Reinhold Publishing Corp., New York, N.Y. Vol. 1, pp. 95–97, 100–161.

4500-SiO$_2$ B. (Reserved)

4500-SiO$_2$ C. Molybdosilicate Method

1. General Discussion

a. Principle: Ammonium molybdate at pH approximately 1.2 reacts with silica and any phosphate present to produce heteropoly acids. Oxalic acid is added to destroy the molybdophosphoric acid but not the molybdosilicic acid. Even if phosphate is known to be absent, the addition of oxalic acid is highly desirable and is a mandatory step in both this method and Method D. The intensity of the yellow color is proportional to the concentration of "molybdate-reactive" silica. In at least one of its forms, silica does not react with molybdate even though it is capable of passing through filter paper and is not noticeably turbid. It is not known to what extent such "unreactive" silica occurs in waters. Terms such as "colloidal," "crystalloidal," and "ionic" have been used to distinguish among various forms of silica but such terminology cannot be substantiated. "Molybdate-unreactive" silica can be converted to the "molybdate-reactive" form by heating or fusing with alkali. Molybdate-reactive or unreactive does not imply reactivity, or lack of it, toward *other* reagents or processes.

b. Interference: Because both apparatus and reagents may contribute silica, avoid using glassware as much as possible and use reagents low in silica. Also, make a blank determination to correct for silica so introduced. In both this method and Method D, tannin, large amounts of iron, color, turbidity, sulfide, and phosphate interfere. Treatment with oxalic acid eliminates interference from phosphate and decreases interference from tannin. If necessary, use photometric compensation to cancel interference from color or turbidity.

c. Minimum detectable concentration: Approximately 1 mg SiO_2/L can be detected in 50-mL nessler tubes.

2. Apparatus

a. Platinum dishes, 100-mL.

b. Colorimetric equipment: One of the following is required:

1) *Spectrophotometer,* for use at 410 nm, providing a light path of 1 cm or longer. See Table 4500-SiO₂:I for light path selection.

2) *Filter photometer,* providing a light path of 1 cm or longer and equipped with a violet filter having maximum transmittance near 410 nm.

3) *Nessler tubes,* matched, 50-mL, tall form.

3. Reagents

For best results, set aside and use batches of chemicals low in silica. Use distilled reagent water in making reagents and dilutions. Store all reagents in plastic containers to guard against high blanks.

a. Sodium bicarbonate, $NaHCO_3$, powder.

b. Sulfuric acid, H_2SO_4, 1N.

c. Hydrochloric acid, HCl, 1 + 1.

d. Ammonium molybdate reagent: Dissolve 10 g $(NH_4)_6Mo_7O_{24}·4H_2O$ in water, with stirring and gentle warming, and dilute to 100 mL. Filter if necessary. Adjust to pH 7 to 8 with silica-free NH_4OH or NaOH and store in a polyethylene bottle to stabilize. (If the pH is not adjusted, a precipitate gradually forms. If the solution is stored in glass, silica may leach out and cause high blanks.) If necessary, prepare silica-free NH_4OH by passing gaseous NH_3 into distilled water contained in a plastic bottle.

e. Oxalic acid solution: Dissolve 7.5 g $H_2C_2O_4·H_2O$ in water and dilute.

f. Stock silica solution: Dissolve 4.73 g sodium metasilicate nonahydrate, $Na_2SiO_3·9H_2O$, in water and dilute to 1000 mL. For work of highest accuracy, analyze 100.0-mL portions by the gravimetric method.[1] Store in a tightly stoppered plastic bottle.

TABLE 4500-SiO₂.I. SELECTION OF LIGHT PATH LENGTH FOR VARIOUS SILICA CONCENTRATIONS

Light Path cm	Method C Silica in 55 mL Final Volume μg	Method D Silica in 55 mL Final Volume μg	
		650 nm Wavelength	815 nm Wavelength
1	200–1300	40–300	20–100
2	100–700	20–150	10–50
5	4–250	7–50	4–20
10	20–130	4–30	2–10

g. Standard silica solution: Dilute 10.00 mL stock solution to 1000 mL with water; 1.00 mL = 10.0 μg SiO_2. Calculate exact concentration from concentration of stock silica solution. Store in a tightly stoppered plastic bottle.

h. Permanent color solutions:

1) *Potassium chromate solution:* Dissolve 630 mg K_2CrO_4 in water and dilute to 1 L.

2) *Borax solution:* Dissolve 10 g sodium borate decahydrate, $Na_2B_4O_7·10H_2O$, in water and dilute to 1 L.

4. Procedure

a. Color development: To 50.0 mL sample add in rapid succession 1.0 mL 1 + 1 HCl and 2.0 mL ammonium molybdate reagent. Mix by inverting at least six times and let stand for 5 to 10 min. Add 2.0 mL oxalic acid solution and mix thoroughly. Read color after 2 min but before 15 min, measuring time from addition of oxalic acid. Because the yellow color obeys Beer's law, measure photometrically or visually.

b. To detect the presence of molybdate-unreactive silica, digest sample with $NaHCO_3$ before color development. This digestion is not necessarily sufficient to convert all molybdate-unreactive silica to the molybdate-reactive form. Complex silicates and higher silica polymers may require extended fusion with alkali at high temperatures or digestion under pressure for complete conversion. Omit digestion if all the silica is known to react with molybdate.

Prepare a clear sample by filtration if necessary. Place 50.0 mL, or a smaller portion diluted to 50 mL, in a 100-mL platinum dish. Add 200 mg silica-free $NaHCO_3$ and digest on a steam bath for 1 h. Cool and add slowly, with stirring, 2.4 mL 1N H_2SO_4. Do not interrupt analysis but proceed *at once* with remaining steps. Transfer quantitatively to a 50-mL nessler tube and make up to mark with water. (Tall-form 50-mL nessler tubes are convenient for mixing even if the solution subsequently is transferred to an absorption cell for photometric measurement.)

c. Preparation of standards: If $NaHCO_3$ pretreatment is used, add to the standards (approximately 45 mL total volume) 200 mg $NaHCO_3$ and 2.4 mL 1N H_2SO_4, to compensate both for the slight amount of silica introduced by the reagents and for the effect of the salt on color intensity. Dilute to 50.0 mL.

d. Correction for color or turbidity: Prepare a special blank for every sample that needs such correction. Carry two identical portions of each such sample through the procedure, including $NaHCO_3$ treatment if this is used. To one portion add all reagents as directed in ¶ 4a preceding. To the other portion add HCl and oxalic acid but no molybdate. Adjust photometer to zero absorbance with the blank containing no molybdate before reading absorbance of molybdate-treated sample.

e. Photometric measurement: Prepare a calibration curve from a series of approximately six standards to cover the optimum ranges cited in Table 4500-SiO₂:I. Follow direction of ¶ 4a above on suitable portions of standard silica solution diluted to 50.0 mL in nessler tubes. Set photometer at zero absorbance with water and read all standards, including a reagent blank, against water. Plot micrograms silica in the final (55 mL) developed solution against photometer readings. Run a reagent blank and at least one standard with each group of samples to confirm that the calibration curve previously established has not shifted.

f. Visual comparison: Make a set of permanent artificial color standards, using K_2CrO_4 and borax solutions. Mix liquid volumes specified in Table 4500-SiO_2:II and place them in well-stoppered, appropriately labeled 50-mL nessler tubes. Verify correctness of these permanent artificial standards by comparing them visually against standards prepared by analyzing portions of the standard silica solution. Use permanent artificial color standards only for visual comparison.

5. Calculation

$$\text{mg } SiO_2/L = \frac{\mu g \ SiO_2 \ (\text{in 55 mL final volume})}{\text{mL sample}}$$

Report whether $NaHCO_3$ digestion was used.

6. Precision and Bias

A synthetic sample containing 5.0 mg SiO_2/L, 10 mg Cl^-/L, 0.20 mg NH_3-N/L, 1.0 mg NO_3^--N/L, 1.5 mg organic N/L, and 10.0 mg PO_4^{3-}/L in distilled water was analyzed in 19 laboratories by the molybdosilicate method with a relative standard deviation of 14.3% and a relative error of 7.8%.

Another synthetic sample containing 15.0 mg SiO_2/L, 200 mg Cl^-/L, 0.800 mg NH_3-N/L, 1.0 mg NO_3^--N/L, 0.800 mg organic N/L, and 5.0 mg PO_4^{3-}/L in distilled water was analyzed in 19

TABLE 4500-SiO_2:II. PREPARATION OF PERMANENT COLOR STANDARDS FOR VISUAL DETERMINATION OF SILICA

Values in Silica μg	Potassium Chromate Solution mL	Borax Solution mL	Water mL
0	0.0	25	30
100	1.0	25	29
200	2.0	25	28
400	4.0	25	26
500	5.0	25	25
750	7.5	25	22
1000	10.0	25	20

laboratories by the molybdosilicate method, with a relative standard deviation of 8.4% and a relative error of 4.2%.

A third synthetic sample containing 30.0 mg SiO_2/L, 400 mg Cl^-/L, 1.50 mg NH_3-N/L, 1.0 mg NO_3^--N/L, 0.200 mg organic N/L, and 0.500 mg PO_4^{3-}/L, in distilled water was analyzed in 20 laboratories by the molybdosilicate method, with a relative standard deviation of 7.7% and a relative error of 9.8%.

All results were obtained after sample digestion with $NaHCO_3$.

7. Reference

1. EATON, A.D., L.S. CLESCERI & A.E. GREENBERG, eds. 1995. Standard Methods for the Examination of Water and Wastewater., 19th ed. American Public Health Assoc., American Water Works Assoc., & Water Environment Fed., Washington, D.C.

8. Bibliography

DIENERT, F. & F. WANDENBULCKE. 1923. On the determination of silica in waters. *Bull. Soc. Chim. France* 33:1131, *Compt. Rend.* 176:1478.

DIENERT, F. & F. WANDENBULCKE. 1924. A study of colloidal silica. *Compt. Rend.* 178:564.

SWANK, H.W. & M.G. MELLON. 1934. Colorimetric standards for silica. *Ind. Eng. Chem.,* Anal. Ed. 6:348.

TOURKY, A.R. & D.H. BANGHAM. 1936. Colloidal silica in natural waters and the "silicomolybdate" colour test. *Nature* 138:587.

BIRNBAUM, N. & G.H. WALDEN. 1938. Co-precipitation of ammonium silicomolybdate and ammonium phosphomolybdate. *J. Amer. Chem. Soc.* 60:66.

KAHLER, H.L. 1941. Determination of soluble silica in water: A photometric method. *Ind. Eng. Chem.,* Anal. Ed. 13:536.

NOLL, C.A. & J.J. MAGUIRE. 1942. Effect of container on soluble silica content of water samples. *Ind. Eng. Chem.,* Anal. Ed. 14:569.

SCHWARTZ, M.C. 1942. Photometric determination of silica in the presence of phosphates. *Ind. Eng. Chem.,* Anal. Ed. 14:893.

GUTTER, H. 1945. Influence of pH on the composition and physical aspects of the ammonium molybdates. *Compt. Rend.* 220:146.

MILTON, R.F. 1951. Formation of silicomolybdate. *Analyst* 76:431.

KILLEFFER, D.H. & A. LINZ. 1952. Molybdenum Compounds, Their Chemistry and Technology. Interscience Publishers, New York, N.Y. pp. 1–2, 42–45, 67–82, 87–92.

STRICKLAND, J.D.H. 1952. The preparation and properties of silico molybdic acid. *J. Amer. Chem. Soc.* 74:862, 868, 872.

CHOW, D.T.W. & R.J. ROBINSON. 1953. The forms of silicate available for colorimetric determination. *Anal. Chem.* 25:646.

4500-SiO_2 D.   Heteropoly Blue Method

1. General Discussion

a. Principle: The principles outlined under Method C, ¶ 1*a*, also apply to this method. The yellow molybdosilicic acid is reduced by means of aminonaphtholsulfonic acid to heteropoly blue. The blue color is more intense than the yellow color of Method C and provides increased sensitivity.

b. Interference: See Section 4500-SiO_2.C.1*b*.

c. Minimum detectable concentration: Approximately 20 μg

SiO_2/L can be detected in 50-mL nessler tubes and 50 μg SiO_2/L spectrophotometrically with a 1-cm light path at 815 nm.

2. Apparatus

a. Platinum dishes, 100-mL.

b. Colorimetric equipment: One of the following is required:

1) *Spectrophotometer,* for use at approximately 815 nm. The color system also obeys Beer's law at 650 nm, with appreciably

reduced sensitivity. Use light path of 1 cm or longer. See Table 4500-SiO$_2$:I for light path selection.

2) *Filter photometer,* provided with a red filter exhibiting maximum transmittance in the wavelength range of 600 to 815 nm. Sensitivity improves with increasing wavelength. Use light path of 1 cm or longer.

3) *Nessler tubes,* matched, 50-mL, tall form.

3. Reagents

For best results, set aside and use batches of chemicals low in silica. Store all reagents in plastic containers to guard against high blanks. Use distilled water that does not contain detectable silica after storage in glass.

All of the reagents listed in Section 4500-SiO$_2$.C.3 are required, and in addition:

Reducing agent: Dissolve 500 mg 1-amino-2-naphthol-4-sulfonic acid and 1 g Na$_2$SO$_3$ in 50 mL distilled water, with gentle warming if necessary; add this to a solution of 30 g NaHSO$_3$ in 150 mL distilled water. Filter into a plastic bottle. Discard when solution becomes dark. Prolong reagent life by storing in a refrigerator and away from light. Do not use aminonaphtholsulfonic acid that is incompletely soluble or that produces reagents that are dark even when freshly prepared.*

4. Procedure

a. Color development: Proceed as in 4500-SiO$_2$.C.4a up to and including the words, "Add 2.0 mL oxalic acid solution and mix thoroughly." Measuring time from the moment of adding oxalic acid, wait at least 2 min but not more than 15 min, add 2.0 mL reducing agent, and mix thoroughly. After 5 min, measure blue color photometrically or visually. If NaHCO$_3$ pretreatment is used, follow 4500-SiO$_2$.C.4b.

b. Photometric measurement: Prepare a calibration curve from a series of approximately six standards to cover the optimum range indicated in Table 4500-SiO$_2$:I. Carry out the steps described above on suitable portions of standard silica solution diluted to 50.0 mL in nessler tubes; pretreat standards if NaHCO$_3$ digestion is used (see 4500-SiO$_2$.C.4b). Adjust photometer to zero absorbance with distilled water and read all standards, including a reagent blank, against distilled water. If necessary to correct for color or turbidity in a sample, see 4500-SiO$_2$.C.4d. To the special blank add HCl and oxalic acid, but no molybdate or reducing

---

* Eastman No. 360 has been found satisfactory.

agent. Plot micrograms silica in the final 55 mL developed solution against absorbance. Run a reagent blank and at least one standard with each group of samples to check the calibration curve.

c. Visual comparison: Prepare a series of not less than 12 standards, covering the range 0 to 120 µg SiO$_2$, by placing the calculated volumes of standard silica solution in 50-mL nessler tubes, diluting to mark with distilled water, and developing color as described in ¶ *a* preceding.

5. Calculation

$$\text{mg SiO}_2/\text{L} = \frac{\mu g \text{ SiO}_2 \text{ (in 55 mL final volume)}}{\text{mL sample}}$$

Report whether NaHCO$_3$ digestion was used.

6. Precision and Bias

A synthetic sample containing 5.0 mg SiO$_2$/L, 10 mg Cl$^-$/L, 0.200 mg NH$_3$-N/L, 1.0 mg NO$_3^-$-N/L, 1.5 mg organic N/L, and 10.0 mg PO$_4^{3-}$/L in distilled water was analyzed in 11 laboratories by the heteropoly blue method, with a relative standard deviation of 27.2% and a relative error of 3.0%.

A second synthetic sample containing 15 mg SiO$_2$/L, 200 mg Cl$^-$/L, 0.800 mg NH$_3$-N/L, 1.0 mg NO$_3^-$-N/L, 0.800 mg organic N/L, and 5.0 mg PO$_4^{3-}$/L in distilled water was analyzed in 11 laboratories by the heteropoly blue method, with a relative standard deviation of 18.0% and a relative error of 2.9%.

A third synthetic sample containing 30.0 mg SiO$_2$/L, 400 mg Cl$^-$/L, 1.50 mg NH$_3$-N/L, 1.0 mg NO$_3^-$-N/L, 0.200 mg organic N/L, and 0.500 mg PO$_4^{3-}$/L in distilled water was analyzed in 10 laboratories by the heteropoly blue method with a relative standard deviation of 4.9% and a relative error of 5.1%.

All results were obtained after sample digestion with NaHCO$_3$.

7. Bibliography

BUNTING, W.E. 1944. Determination of soluble silica in very low concentrations. *Ind. Eng. Chem.,* Anal. Ed. 16:612.

STRAUB, F.G. & H. GRABOWSKI. 1944. Photometric determination of silica in condensed steam in the presence of phosphates. *Ind. Eng. Chem.,* Anal. Ed. 16:574.

BOLTZ, D.F. & M.G. MELLON. 1947. Determination of phosphorus, germanium, silicon, and arsenic by the heteropoly blue method. *Ind. Eng. Chem.,* Anal. Ed. 19:873.

MILTON, R.F. 1951. Estimation of silica in water. *J. Appl. Chem.* (London) 1:(Supplement No. 2) 126.

CARLSON, A.B. & C.V. BANKS. 1952. Spectrophotometric determination of silicon. *Anal. Chem.* 24:472.

4500-SiO₂ E. Automated Method for Molybdate-Reactive Silica

1. General Discussion

a. Principle: This method is an adaptation of the heteropoly blue method (Method D) utilizing a continuous-flow analytical instrument.

b. Interferences: See Section 4500-SiO₂.C.1*b.* If particulate matter is present, filter sample or use a continuous filter as an integral part of the system.

c. Application: This method is applicable to potable, surface, domestic, and other waters containing 0 to 20 mg SiO₂/L. The range of concentration can be broadened to 0 to 80 mg/L by substituting a 15-mm flow cell for the 50-mm flow cell shown in Figure 4500-SiO₂:1.

2. Apparatus

Automated analytical equipment: An example of the continuous-flow analytical instrument consists of the interchangeable components shown in Figure 4500-SiO₂:1.

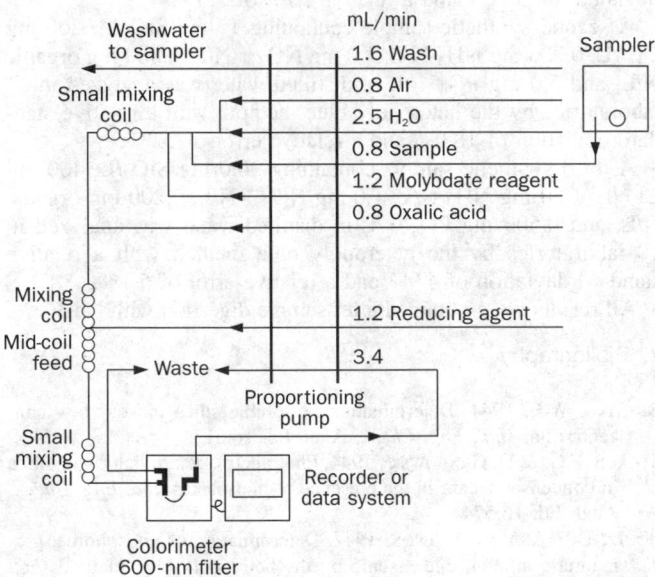

Figure 4500-SiO₂:1. Silica manifold.

3. Reagents

a. Sulfuric acid, H_2SO_4, $0.05M$ ($0.1N$).

b. Ammonium molybdate reagent: Dissolve 10 g $(NH_4)_6Mo_7O_{24} \cdot 4H_2O$ in 1 L $0.05M$ H_2SO_4. Filter and store in an amber plastic bottle.

c. Oxalic acid solution: Dissolve 50 g oxalic acid in 900 mL distilled water and dilute to 1 L.

d. Reducing agent: Dissolve 120 g $NaHSO_3$ and 4 g Na_2SO_3 in 800 mL warm distilled water. Add 2 g 1-amino-2-naphthol-4-sulfonic acid, mix well, and dilute to 1 L. Filter into amber plastic bottle for storage.

To prepare working reagent, dilute 100 mL to 1 L with distilled water. Make working reagent daily.

e. Standard silica solution: See 4500-SiO₂.C.3*g.*

4. Procedure

Set up manifold as shown in Figure 4500-SiO₂:1 and follow the general procedure described by the manufacturer. Determine absorbance at 660 nm. Use quality control procedures given in Section 4020.

5. Calculation

Prepare standard curves by plotting response of standards processed through the manifold against SiO₂ concentration in standards. Compute sample SiO₂ concentration by comparing sample response with standard curve.

6. Precision and Bias

For 0 to 20 mg SiO₂/L, when a 50-mm flow cell was used at 40 samples/h, the detection limit was 0.1 mg/L, sensitivity (concentration giving 0.398 absorbance) was 7.1 mg/L, and the coefficient of variation (95% confidence level at 7.1 mg/L) was 1.6%. For 0 to 80 mg SiO₂/L, when a 15-mm flow cell was used at 50 samples/h, detection limit was 0.5 mg/L, sensitivity was 31 mg/ L, and coefficient of variation at 31 mg/L was 1.5%.

4500-SiO₂ F. Flow Injection Analysis for Molybdate-Reactive Silicate (PROPOSED)

1. General Discussion

Silicate reacts with molybdate under acidic conditions to form yellow beta-molybdosilicic acid. This acid is subsequently reduced with stannous chloride to form a heteropoly blue complex that has an absorbance maximum at 820 nm. Oxalic acid is added to reduce the interference from phosphate.

Collect samples in polyethylene or other plastic bottles and refrigerate at 4°C. Chemical preservation for silica is not recommended. Adding acid may cause polymerization of reactive silicate species. Freezing decreases silicate concentrations, especially at concentrations greater than 100 μg SiO₂/L. If filtration is required, preferably use a 0.45-μm TFE filter. Samples may be held for 28 d.

Also see Sections 4500-SiO$_2$.A, D, and E, and Section 4130, Flow Injection Analysis (FIA).

b. Interferences: Remove large or fibrous particulates by filtering sample through inert filter.

The interference due to phosphates is reduced by the addition of oxalic acid as a reagent on the flow injection manifold. By the following method, a solution of 1000 µg P/L was determined as 20 µg SiO$_2$/L. Verify extent of phosphate interference by determining a solution of phosphate at the highest concentration that is expected to occur.

Tannin and large amounts of iron or sulfides are interferences. Remove sulfides by boiling an acidified sample. Add disodium EDTA to eliminate interference due to iron. Treat with oxalic acid to decrease interference from tannin.

Sample color and turbidity can interfere. Determine presence of these interferences by analyzing samples without the presence of the molybdate.

Avoid silica contamination by storing samples, standards, and reagents in plastic. Do not use glass-distilled water for reagents or standards.

2. Apparatus

Flow injection analysis equipment consisting of:

a. FIA injection valve with sample loop or equivalent.

b. Multichannel proportioning pump.

c. FIA manifold (Figure 4500-SiO$_2$:2) with tubing heater and flow cell. Relative flow rates only are shown in Figure 4500-SiO$_2$:2. Tubing volumes are given as an example only; they may be scaled down proportionally. Use manifold tubing of an inert material such as TFE.*

d. Absorbance detector, 820 nm, 10-nm bandpass.

e. Injection valve control and data acquisition system.

3. Reagents

Use reagent water (>10 megohm) to prepare carrier and all solutions. To prevent bubble formation, degas carrier and buffer with helium. Pass He at 140 kPa (20 psi) through a helium degassing tube. Bubble He through 1 L solution for 1 min. All reagents can also be prepared on a weight/volume basis if desired.

a. Molybdate: To a tared 500-mL container add 20.0 g ammonium molybdate tetrahydrate [(NH$_4$)$_6$Mo$_7$O$_{24}$·4H$_2$O]. Add 486 g warm water and 14.8 g conc sulfuric acid (H$_2$SO$_4$). Stir or shake until dissolved. Store in plastic and refrigerate. Prepare fresh monthly and discard if precipitate or blue color is observed.

b. Oxalic acid: To a tared 500-mL container add 50.0 g oxalic acid (HO$_2$CCO$_2$H·2H$_2$O) and 490 g water. Stir or shake until dissolved. Store in plastic.

c. Stannous chloride: To a tared 1-L container, add 978 g water. Add 40.0 g conc H$_2$SO$_4$. Dissolve 2.0 g hydroxylamine

---

* Teflon or equivalent.

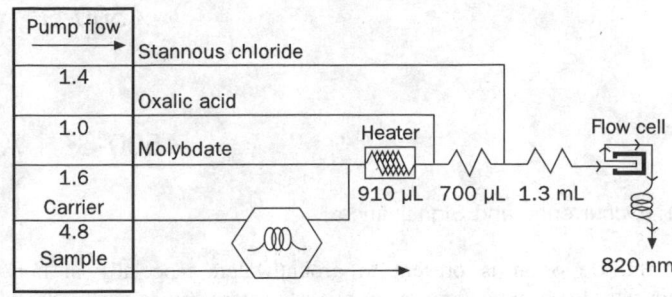

Figure 4500-SiO$_2$:2. FIA manifold.

hydrochloride in this solution. Then dissolve 0.30 g stannous chloride. Prepare fresh weekly.

d. Stock silicate standard, 100 mg SiO$_2$/L: In a 1-L volumetric flask dissolve 0.473 g sodium metasilicate nonahydrate (Na$_2$SiO$_3$·9H$_2$O) in approximately 800 mL water. Dilute to mark and invert three times. Alternatively, use a commercially prepared standard solution, especially if nonstoichiometry of the solid metasilicate is suspected; the original degree of polymerization of the sodium metasilicate, which depends on storage time, can affect free silica concentration of the resulting solution.

e. Standard silicate solutions: Prepare silicate standards in the desired concentration range, using the stock standard (¶ 3*d*), and diluting with water.

4. Procedure

Set up a manifold equivalent to that in Figure 4500-SiO$_2$:2 and follow method supplied by manufacturer or laboratory standard operating procedure.

5. Calculations

Prepare standard curves by plotting absorbance of standards processed through the manifold versus silicate concentration. The calibration curve is linear.

6. Precision and Bias

Twenty-one replicates of a 5.0-µg SiO$_2$/L standard were analyzed with a 780-µL sample loop by a published MDL method.[1] These gave a mean of 4.86 µg SiO$_2$/L, a standard deviation of 0.31 µg SiO$_2$/L, and an MDL of 0.78 µg SiO$_2$/L.

7. Quality Control

Follow procedures outlined in Section 4020.

8. Reference

1. U.S. Environmental Protection Agency. 1989. Definition and Procedure for the Determination of Method Detection Limits. Appendix B to 40 CFR 136 rev. 1.11 amended June 30, 1986. 49 CFR 43430.

4500-S²⁻ SULFIDE*

4500-S²⁻ A. Introduction

1. Occurrence and Significance

Sulfide often is present in groundwater, especially in hot springs. Its common presence in wastewaters comes partly from the decomposition of organic matter, sometimes from industrial wastes, but mostly from the bacterial reduction of sulfate. Hydrogen sulfide escaping into the air from sulfide-containing wastewater causes odor nuisances. The threshold odor concentration of H_2S in clean water is between 0.025 and 0.25 µg/L. Gaseous H_2S is very toxic and has claimed the lives of numerous workers in sewers. At levels toxic to humans it interferes with the olfactory system, giving a false sense of the safe absence of H_2S. It attacks metals directly and indirectly has caused serious corrosion of concrete sewers because it is oxidized biologically to H_2SO_4 on the pipe wall. Dissolved H_2S is toxic to fish and other aquatic organisms.

2. Categories of Sulfides

From an analytical standpoint, three categories of sulfide in water and wastewater are distinguished.

---

* Approved by Standard Methods Committee, 1997.

a. Total sulfide includes dissolved H_2S and HS^-, as well as acid-soluble metallic sulfides present in suspended matter. The S^{2-} is negligible, amounting to less than 0.5% of the dissolved sulfide at pH 12, less than 0.05% at pH 11, etc. Copper and silver sulfides are so insoluble that they do not respond in ordinary sulfide determinations; they can be ignored for practical purposes.

b. Dissolved sulfide is that remaining after suspended solids have been removed by flocculation and settling.

c. Un-ionized hydrogen sulfide may be calculated from the concentration of dissolved sulfide, the sample pH, and the practical ionization constant of H_2S.

Figure 4500-S²⁻:1 shows analytical flow paths for sulfide determinations under various conditions and options.

3. Sampling and Storage

Take samples with minimum aeration. Either analyze samples immediately after collection or preserve for later analysis with zinc acetate solution. To preserve a sample for a total sulfide determination put zinc acetate and sodium hydroxide solutions into bottle before filling it with sample. Use 4 drops of 2N zinc acetate soution per 100 mL sample. Increase volume of zinc acetate solution if the sulfide concentration is expected to be greater

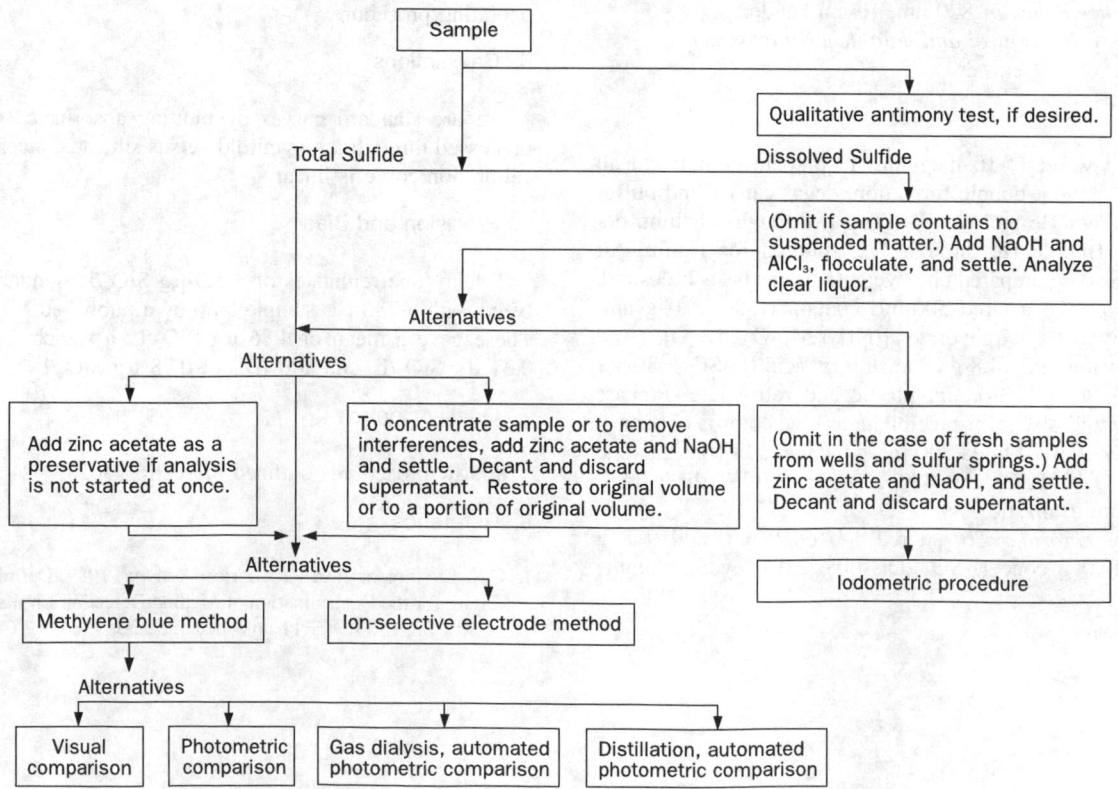

Figure 4500-S²⁻:1. Analytical flow paths for sulfide determinations.

than 64 mg/L. The final pH should be at least 9. Add more NaOH if necessary. Fill bottle completely and stopper.

4. Qualitative Tests

A qualitative test for sulfide often is useful. It is advisable in the examination of industrial wastes containing interfering substances that may give a false negative result in the methylene blue method (D).

a. Antimony test: To about 200 mL sample, add 0.5 mL saturated solution of potassium antimony tartrate and 0.5 mL 6*N* HCl in excess of phenolphthalein alkalinity.

Yellow antimony sulfide (Sb_2S_3) is discernible at a sulfide concentration of 0.5 mg/L. Comparisons with samples of known sulfide concentration make the technique roughly quantitative. The only known interferences are metallic ions such as lead, which hold the sulfide so firmly that it does not produce Sb_2S_3, and dithionite, which decomposes in acid solution to produce sulfide.

b. Silver-silver sulfide electrode test: Dilute sample 1:1 with alkaline antioxidant reagent (see G.3*a* below). Measure electrode potential relative to a double-junction reference electrode and estimate the sulfide concentration from an old calibration curve or the example calibration curve in the electrode manual. This gives a reasonable estimate of sulfide concentration if the electrode is in good condition.

c. Lead acetate paper and silver foil tests: Confirm odors attributed to H$_2$S with lead acetate paper. On exposure to the vapor of a slightly acidified sample, the paper becomes blackened by formation of PbS. A strip of silver foil is more sensitive than lead acetate paper. Clean the silver by dipping in NaCN solution and rinse. CAUTION: *NaCN is toxic, handle with care.* Silver is suitable particularly for long-time exposure in the vicinity of possible H$_2$S sources because black Ag$_2$S is permanent whereas PbS slowly oxidizes.

5. Selection of Quantitative Methods

Iodine oxidizes sulfide in acid solution. A titration based on this reaction is an accurate method for determining sulfide at concentrations above 1 mg/L if interferences are absent and if loss of H$_2$S is avoided. The iodometric method (F) is useful for standardizing the methylene blue colorimetric methods (D, E, and I) and is suitable for analyzing samples freshly taken from wells or springs. The method can be used for wastewater and partly oxidized water from sulfur springs if interfering substances are removed first. The automated methylene blue method with distillation (I) is useful for a variety of samples containing more than 1 mg S$^-$/L.

The methylene blue method (D) is based on the reaction of sulfide, ferric chloride, and dimethyl-*p*-phenylenediamine to produce methylene blue. Ammonium phosphate is added after color development to remove ferric chloride color. The procedure is applicable at sulfide concentrations between 0.1 and 20.0 mg/L. The automated methlyene blue method (E) is similar to Method D. A gas dialysis technique separates the sulfide from the sample matrix. Gas dialysis eliminates most interferences, including turbidity and color. The addition of the antioxidant ascorbic acid improves sulfide recoveries. The method is applicable at sulfide concentrations between 0.002 and 0.100 mg/L.

Potentiometric methods utilizing a silver electrode (G) may be suitable. From the potential of the electrode relative to a reference electrode an estimate can be made of the sulfide concentration, but careful attention to details of procedures and frequent standardizations are needed to secure good results. The electrode is useful particularly as an end-point indicator for titration of dissolved sulfide with silver nitrate. The ion-selective electrode method is unaffected by sample color or turbidity and is applicable for concentrations greater than 0.03 mg/L.

6. Preparation of Sulfide Standards

Take care in preparing reliable stock solutions of sulfide for calibration and quality control. Prepare sulfide standards from sodium sulfide nonahydrate (Na$_2$S·9H$_2$O) crystals. These crystals usually have excess water present on the surface, in addition to a layer of contamination from oxidation products (polysulfides, polythionates, and sulfate) of sulfide reacting with atmospheric oxygen. Further, solutions of sulfide are prone to ready oxidation by dissolved and atmospheric oxygen. Use reagent water to prepare sulfide standards and sample dilutions. Boil and degas with either argon or nitrogen while cooling. Purchase the smallest amount of solid standards possible and keep no longer than 1 year. Preferably handle and store solid sulfide standards and stock solutions in an inert atmosphere glove bag or glove box to reduce contamination due to oxidation.

Preferably remove single crystals of Na$_2$S·9H$_2$O from reagent bottle with nonmetallic tweezers; quickly rinse in degassed reagent water to remove surface contamination. Blot crystal dry with a tissue, then rapidly transfer to a tared, stoppered weighing bottle containing 5 to 10 mL degassed reagent water. Repeat procedure until desired amount of sodium sulfide is in weighing bottle. Determine amount of Na$_2$S·9H$_2$O in weighing bottle by difference, then multiply the weight by 0.133 to determine the amount of S^{2-}. Avoid excess agitation and mixing of the solution with atmospheric oxygen. Quantitatively transfer and dilute entire contents of weighing bottle to an appropriate size volumetric flask with degassed reagent water to prepare a known concentration sulfide stock solution (3.750 g Na$_2$S·9H$_2$O diluted to a final volume of 500 mL will give a stock solution of which 1.00 mL = 1.00 mg S^{2-}). Alternatively, purchase precertified stock solutions of sulfide. Verify concentration of stock solution daily using the iodometric method (F). Store stock solution with minimum headspace for no more than 1 week.

7. Bibliography

CRUSE, H. & R.D. POMEROY. 1969. Hydrogen sulfide odor threshold. *J. Amer. Water Works Assoc.* 61:677.

KARCHMER, J.H., ed. 1970. The Analytical Chemistry of Sulfur and Its Compounds. Wiley-Interscience, New York, N.Y.

NICKLESS, G., ed. 1970. Inorganic Sulphur Chemistry. Elsevier Publ., Amsterdam, The Netherlands.

U.S. ENVIRONMENTAL PROTECTION AGENCY. 1974. Process Design Manual for Sulfide Control in Sanitary Sewerage Systems. Publ. 625/1-74-005.

BAGARINAO, T. 1992. Sulfide as an environmental factor and toxicant: Tolerance and adaptations in aquatic organisms. *Aquat. Toxicol.* 24:21.

4500-S²⁻ B. Separation of Soluble and Insoluble Sulfides

Unless the sample is entirely free from suspended solids (dissolved sulfide equals total sulfide), to measure dissolved sulfide first remove insoluble matter. This can be done by producing an aluminum hydroxide floc that is settled, leaving a clear supernatant for analysis.

1. Apparatus

Glass bottles with stoppers: Use 100 mL if sulfide will be determined by the methylene blue method and 500 to 1000 mL if by the iodometric method.

2. Reagents

a. Sodium hydroxide solution, NaOH, 6*N*.
b. Aluminum chloride solution: Because of the hygroscopic and caking tendencies of this chemical, purchase 100-g bottles of $AlCl_3 \cdot 6H_2O$. Dissolve contents of a previously unopened 100-g bottle in 144 mL distilled water.

3. Procedure

a. To a 100-mL glass bottle add 0.2 mL (nominally 4 drops) 6*N* NaOH. Fill bottle with sample and immediately add 0.2 mL (4 drops) $AlCl_3$ solution. Stopper bottle with no air under stopper. Rotate back and forth about a transverse axis vigorously for 1 min or longer to flocculate contents. Vary volumes of these added chemicals to get good clarification without using excessively large amounts and to produce a pH of 6 to 9. If a 500- or 1000-mL bottle is used, add proportionally larger amounts of reagents.
b. Let settle until reasonably clear supernatant can be drawn off. With proper flocculation, this may take 5 to 15 min. Do not wait longer than necessary.
c. Either analyze the supernatant immediately or preserve with 2*N* zinc acetate (see Section 4500-S²⁻.C).

4500-S²⁻ C. Sample Pretreatment to Remove Interfering Substances or to Concentrate the Sulfide

The iodometric method suffers interference from reducing substances that react with iodine, including thiosulfate, sulfite, and various organic compounds, both solid and dissolved.

Strong reducing agents also interfere in the methylene blue method (D) by preventing formation of the blue color. Thiosulfate at concentrations about 10 mg/L may retard color formation or completely prevent it. Ferrocyanide produces a blue color. Sulfide itself prevents the reaction if its concentration is very high, in the range of several hundred milligrams per liter. To avoid the possibility of false negative results, use the antimony method to obtain a qualitative result in industrial wastes likely to contain sulfide but showing no color by the methylene blue method. Iodide, which is likely to be present in oil-field wastewaters, may diminish color formation if its concentration exceeds 2 mg/L. Many metals (e.g., Hg, Cd, Cu) form insoluble sulfides and give low recoveries.

Eliminate interferences due to sulfite, thiosulfate, iodide, and many other soluble substances, but not ferrocyanide, by first precipitating ZnS, removing the supernatant, and replacing it with distilled water. Use the same procedure, even when not needed for removal of interferences, to concentrate sulfide. The automated methylene blue method (E) is relatively free from interferences because gas dialysis separates the sulfide from the sample matrix.

1. Apparatus

Glass bottles with stoppers: See Section 4500-S²⁻.B.1.

2. Reagents

a. Zinc acetate solution: Dissolve 220 g $Zn(C_2H_3O_2)_2 \cdot 2H_2O$ in 870 mL water; this makes 1 L solution.

b. Sodium hydroxide solution, NaOH, 6*N*.

3. Procedure

a. Put 0.20 mL (4 drops) zinc acetate solution and 0.10 mL (2 drops) 6*N* NaOH into a 100-mL glass bottle, fill with sample, and add 0.10 mL (2 drops) 6*N* NaOH solution. Stopper with no air bubbles under stopper and mix by rotating back and forth vigorously about a transverse axis. For the iodometric procedure, use a 500-mL bottle or other convenient size, with proportionally larger volumes of reagents. Vary volume of reagents added according to sample so that the resulting precipitate is not excessively bulky and settles readily. Add enough NaOH to raise the pH above 9. Let precipitate settle for 30 min. The treated sample is relatively stable and can be held for several hours. However, if much iron is present, oxidation may be fairly rapid.
b. If the iodometric method is to be used, collect precipitate on a glass fiber filter and continue at once with titration according to the procedure of Method F. If the methylene blue method (D) is used, let precipitate settle for 30 min and decant as much supernatant as possible without loss of precipitate. Refill bottle with distilled water, shake to resuspend precipitate, and quickly withdraw a sample. If interfering substances are present in high concentration, settle, decant, and refill a second time. If sulfide concentration is known to be low, add only enough water to bring volume to one-half or one-fifth of original volume. Use this technique for analyzing samples of very low sulfide concentrations. After determining the sulfide concentration colorimetrically, multiply the result by the ratio of final to initial volume. No concentration or pretreatment steps to remove interferences are necessary for Method E.

4500-S^{2-} D. Methylene Blue Method

1. Apparatus

a. Matched test tubes, approximately 125 mm long and 15 mm OD.

b. Droppers, delivering 20 drops/mL methylene blue solution. To obtain uniform drops hold dropper in a vertical position and let drops form slowly.

c. If photometric rather than visual color determination will be used, either:

1) *Spectrophotometer,* for use at a wavelength of 664 nm with cells providing light paths of 1 cm and 1 mm, or other path lengths, or

2) *Filter photometer,* with a filter providing maximum transmittance near 660 nm.

2. Reagents

a. Amine-sulfuric acid stock solution: Dissolve 27 g *N,N*-dimethyl-*p*-phenylenediamine oxalate* in an iced mixture of 50 mL conc H_2SO_4 and 20 mL distilled water. Cool and dilute to 100 mL with distilled water. Use fresh oxalate because an old supply may be oxidized and discolored to a degree that results in interfering colors in the test. Store in a dark glass bottle. When this stock solution is diluted and used in the procedure with a sulfide-free sample, it first will be pink but then should become colorless within 3 min.

b. Amine-sulfuric acid reagent: Dilute 25 mL amine-sulfuric acid stock solution with 975 mL 1 + 1 H_2SO_4. Store in a dark glass bottle.

c. Ferric chloride solution: Dissolve 100 g $FeCl_3 \cdot 6H_2O$ in 40 mL water.

d. Sulfuric acid solution, H_2SO_4, 1 + 1.

e. Diammonium hydrogen phosphate solution: Dissolve 400 g $(NH_4)_2HPO_4$ in 800 mL distilled water.

f. Methylene blue solution I: Use USP grade dye or one certified by the Biological Stain Commission. The dye content should be reported on the label and should be 84% or more. Dissolve 1.0 g in distilled water and make up to 1 L. This solution will be approximately the correct strength, but because of variation between different lots of dye, standardize against sulfide solutions of known strength and adjust its concentration so that 0.05 mL (1 drop) = 1.0 mg sulfide/L.

Standardization—Prepare five known-concentration sulfide standards ranging from 1 to 8 mg/L as described in 4500-S^{2-}.A.6, or proceed as follows: Put several grams of clean, washed crystals of $Na_2S \cdot 9H_2O$ into a small beaker. Add somewhat less than enough water to cover crystals. Stir occasionally for a few minutes, then pour solution into another vessel. This solution reacts slowly with oxygen but the change is insignificant if anal-ysis is performed within a few hours. Prepare solution daily. To 1 L distilled water add 1 drop of Na_2S solution and mix. Immediately determine sulfide concentration by the methylene blue procedure and by the iodometric procedure. Repeat, using more than 1 drop Na_2S solution or smaller volumes of water, until at least five tests have been made, with a range of sulfide concentrations between 1 and 8 mg/L. Calculate average percent error of the methylene blue result as compared to the iodometric result. If the average error is negative, that is, methylene blue results are lower than iodometric results, dilute methylene blue solution by the same percentage, so that a greater volume will be used in matching colors. If methylene blue results are high, increase solution strength by adding more dye.

g. Methylene blue solution II: Dilute 10.00 mL of adjusted methylene blue solution I to 100 mL with reagent water.

3. Procedure

a. Color development: Transfer 7.5 mL sample to each of two matched test tubes, using a special wide-tip pipet or filling to marks on test tubes. If sample has been preserved with zinc acetate, shake vigorously before taking subsample. Add to Tube A 0.5 mL amine-sulfuric acid reagent and 0.15 mL (3 drops) $FeCl_3$ solution. Mix immediately by inverting slowly, only once. (Excessive mixing causes low results by loss of H_2S as a gas before it has had time to react). To Tube B add 0.5 mL 1 + 1 H_2SO_4 and 0.15 mL (3 drops) $FeCl_3$ solution and mix. The presence of S^{2-} will be indicated by the appearance of blue color in Tube A. Color development usually is complete in about 1 min, but a longer time often is required for fading out of the initial pink color. Wait 3 to 5 min and add 1.6 mL $(NH_4)_2HPO_4$ solution to each tube. Wait 3 to 15 min and make color comparisons. If zinc acetate was used, wait at least 10 min before making a visual color comparison.

b. Color determination:

1) Visual color estimation—Add methylene blue solution I or II, depending on sulfide concentration and desired accuracy, dropwise, to the second tube, until color matches that developed in first tube. If the concentration exceeds 20 mg/L, repeat test with a portion of sample diluted tenfold.

With methylene blue solution I, adjusted so that 0.05 mL (1 drop) = 1.0 mg S^{2-}/L when 7.5 mL of sample are used:

$$\text{mg } S^{2-}/L = \text{no. drops solution I} + 0.1 \text{ (no. drops solution II)}$$

2) Photometric color measurement—A cell with a light path of 1 cm is suitable for measuring sulfide concentrations from 0.1 to 2.0 mg/L. Use shorter or longer light paths for higher or lower concentrations. This method is suitable for sample concentrations up to 20 mg/L. Zero instrument with a portion of treated sample from Tube B. Prepare calibration curves on basis of colorimetric tests made on Na_2S solutions simultaneously analyzed by the iodometric method, plotting concentration vs. absorbance. A linear relationship between concentration and absorbance can be assumed from 0 to 1.0 mg/L.

Read sulfide concentration from calibration curve.

---

* Eastman catalog No. 5672 has been found satisfactory for this purpose.

4. Precision and Bias

In a study by two chemists working in the same laboratory, the standard deviation estimated from 34 sets of duplicate sulfide measurements was 0.04 mg/L for concentrations between 0.2 and 1.5 mg/L. The average recoveries of known additions were 92% for 40 samples containing 0.5 to 1.5 mg/L and 89% for samples containing less than 0.1 mg/L.

5. Bibliography

POMEROY, R.D. 1936. The determination of sulfides in sewage. *Sewage Works J.* 8:572.

NUSBAUM, I. 1965. Determining sulfides in water and waste water. *Water Sewage Works* 112:113.

4500-S²⁻ E. Gas Dialysis, Automated Methylene Blue Method

1. Apparatus

a. Automated analytical equipment: An example of the continuous-flow analytical instrument consists of the interchangeable components shown in Figure 4500-S²⁻:2.

The sampler is equipped with a mixer to stir samples before analysis and the gas dialysis membrane, which is maintained at room temperature, separates H₂S from the sample matrix.

2. Reagents

a. N,N-dimethyl-p-phenylenediamine stock solution: Dissolve 1 g N,N-dimethyl-p-phenylenediamine dihydrochloride in 500 mL 6N HCl. Prepare fresh monthly. Store in an amber bottle.

b. N,N-dimethyl-p-phenylenediamine working solution: Dilute 190 mL N,N-dimethyl-p-phenylenediamine stock solution to 1 L. Store in an amber bottle. Prepare weekly.

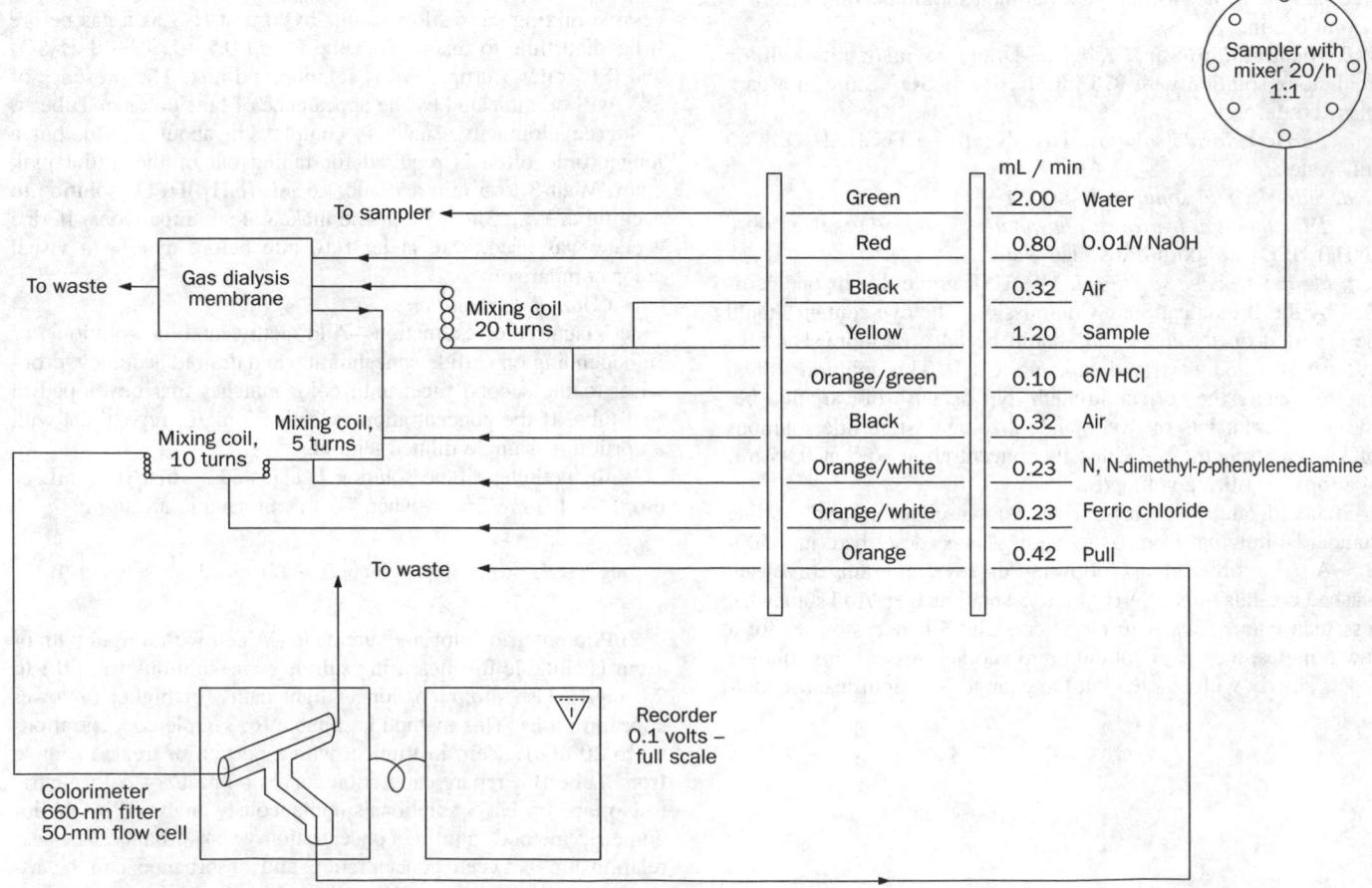

Figure 4500-S²⁻:2. Sulfide manifold.

c. Ferric chloride stock solution: Dissolve 13.5 g FeCl$_3$·6H$_2$O in 500 mL 5N HCl. Store in an amber bottle. Prepare fresh monthly.

d. Working ferric chloride solution: Dilute 190 mL ferric chloride stock solution to 1 L. Store in an amber bottle. Prepare fresh weekly.

e. Hydrochloric acid, HCl, 6N:

f. Sodium hydroxide stock solution, NaOH, 1N.

g. Sodium hydroxide, NaOH, 0.01N: Dilute 10 mL NaOH stock solution to 1 L.

h. Sulfide stock solution, 1.00 mg S^{2-}/1.00 mL: See 4500-S^{2-}.A.6.

i. Sulfide intermediate standard solution: Dilute 10 mL sulfide stock solution to 1 L with water. Prepare fresh daily. Standardize by iodometric titration method, 4500-S^{2-}.F. 1 mL ≈ 0.01 mg S^{2-}.

j. Sulfide tertiary standard solution: Dilute 50 mL sulfide intermediate solution to 500 mL with 0.01N NaOH. Prepare fresh daily. Use standardization value from ¶ 2i to determine exact concentration. 1.00 mL ≈ 0.001 mg S^{2-}.

k. Working sulfide standard solutions: Prepare a suitable series of standards by diluting appropriate volumes of sulfide tertiary standing solutions with 0.01N NaOH. Prepare fresh daily.

l. Zinc acetate preservative solution: Dissolve 220 g Zn(C$_2$H$_3$O$_2$)$_2$·2H$_2$O in 870 mL water (this makes 1 L solution).

3. Procedure

For unpreserved, freshly collected samples and sulfide working standards, add, in order, 4 drops 2N zinc acetate, 0.5 mL 6N NaOH, and 400 mg ascorbic acid/100 mL. For preserved samples, add 0.5 mL 6N NaOH and 400 mg ascorbic acid/100 mL. Shake well.

Let precipitate settle for at least 30 min. Pour a portion of well-mixed sample or working standard into a sample cup. Set up manifold as shown in Figure 4500-S^{2-}:2 and follow the general procedure described by the manufacturer. Determine absorbance at 660 nm.

4. Calculation

Prepare standard curves by plotting peak heights of standards processed through the manifold against S^{2-} concentration in the standards. Compute S^{2-} sample concentration by comparing sample response with standard curve.

5. Precision and Bias

In a single laboratory, samples with S^{2-} concentrations of 0.012, 0.015, 0.034, and 0.085 mg/L had standard deviations of 0.001, 0.001, 0.001, and 0.001 mg/L, respectively, with coefficients of variation of 8.3%, 6.3%, 2.9%, and 1.2%, respectively. In two environmental samples with added S^{2-}, recoveries were 104.2% and 97.6%.

6. Bibliography

FRANCOM, D., L.R. GOODWIN & F.P. DIEKEN. 1989. Determination of low level sulfides in environmental waters by automated gas dialysis/methylene blue colorimetry. *Anal. Lett.* 22:2587.

4500-S^{2-} F. Iodometric Method

1. Reagents

a. Hydrochloric acid, HCl, 6N.

b. Standard iodine solution, 0.0250N: Dissolve 20 to 25 g KI in a little water and add 3.2 g iodine. After iodine has dissolved, dilute to 1000 mL and standardize against 0.0250N Na$_2$S$_2$O$_3$, using starch solution as indicator.

c. Standard sodium thiosulfate solution, 0.0250N: See Section 4500-O.C.2e.

d. Starch solution: See Section 4500-O.C.2d.

2. Procedure

a. Measure from a buret into a 500-mL flask an amount of iodine solution estimated to be an excess over the amount of sulfide present. Add distilled water, if necessary, to bring volume to about 20 mL. Add 2 mL 6N HCl. Pipet 200 mL sample into flask, discharging sample under solution surface. If iodine color disappears, add more iodine until color remains. Back-titrate with Na$_2$S$_2$O$_3$ solution, adding a few drops of starch solution as end point is approached, and continuing until blue color disappears.

b. If sulfide was precipitated with zinc and ZnS filtered out, return filter with precipitate to original bottle and add about 100 mL water. Add iodine solution and HCl and titrate as in ¶ 2a above.

3. Calculation

One milliliter 0.0250N iodine solution reacts with 0.4 mg S^{2-}:

$$\text{mg S}^{2-}/\text{L} = \frac{[(A \times B) - (C \times D)] \times 16\,000}{\text{mL sample}}$$

where:

   A = mL iodine solution,
   B = normality of iodine solution,
   C = mL Na$_2$S$_2$O$_3$ solution, and
   D = normality of Na$_2$S$_2$O$_3$ solution.

4. Precision

The precision of the end point varies with the sample. In clean waters it should be determinable within 1 drop, which is equivalent to 0.1 mg/L in a 200-mL sample.

4500-S²⁻ G. Ion-Selective Electrode Method

1. General Discussion

a. Principle: The potential of a silver/sulfide ion-selective electrode (ISE) is related to the sulfide ion activity. An alkaline antioxidant reagent (AAR) is added to samples and standards to inhibit oxidation of sulfide by oxygen and to provide a constant ionic strength and pH. Use of the AAR allows calibration in terms of total dissolved sulfide concentration. All samples and standards must be at the same temperature. Sulfide concentrations between 0.032 mg/L ($1 \times 10^{-6}M$) and 100 mg/L can be measured without preconcentration. For lower concentrations, preconcentration is necessary.

b. Interferences: Humic substances may interfere with Ag/S-ISE measurements. For highly colored water (high concentration of humic substances), use the method of standard additions to check results. Sulfide is oxidized by dissolved oxygen. Sulfide oxidation may cause potential readings to drift in the direction of decreasing concentration, i.e., to more positive values. Flush surface of samples and standards with nitrogen to minimize contact with atmospheric oxygen for low-level measurements. Temperature changes may cause potentials to drift either upward or downward. Therefore, let standards and samples come to the same temperature. If samples cannot be analyzed immediately, preserve dissolved sulfide by precipitating with zinc acetate (4500-S²⁻.C).

2. Apparatus

a. Silver/sulfide electrode:*

b. Double-junction reference electrode.

c. Electrode polishing strips.†

d. pH meter with millivolt scale, capable of 0.1-mV resolution. Meters that can be calibrated in concentration and that perform standard-additions calculations are available.

e. Electrochemical cell: Make suitable cell from a 150-mL beaker and a sheet of rigid plastic (PVC or acrylic) with holes drilled to allow insertion of the electrodes and a tube for flushing the headspace with nitrogen. Alternatively, purchase a polarographic cell with gas transfer tube.‡

f. Gas dispersion tube: Use to deaerate water for preparing reagents and standards.

g. Magnetic stirrer and stirring bar: Use a piece of styrofoam or cardboard to insulate the cell from the magnetic stirrer.

3. Reagents

a. Alkaline antioxidant reagent (AAR): To approximately 600 mL deaerated reagent water (DRW) in a 1-L volumetric flask, add 80 g NaOH, 35 g ascorbic acid, and 67 g Na₂H₂EDTA. Swirl to dissolve and dilute to 1 L. The color of freshly prepared AAR will range from colorless to yellow. Store in a tightly capped brown glass bottle. Discard when solution becomes brown.

b. Lead perchlorate, 0.1M: Dissolve 4.60 g Pb(ClO₄)₂·3H₂O in 100 mL reagent water. Standardize by titrating with Na₂H₂EDTA. Alternatively, use commercially available 0.1M Pb(ClO₄)₂ solutions.

c. Sulfide stock solution, 130 mg/L: See 4500-S²⁻.A.6, and dilute 13.0 mL of 1.00 mg S²⁻/mL stock to 100.0 mL with AAR. Alternatively, add 500 mL AAR and 10 g Na₂S·9H₂O to a 1-L volumetric flask; dissolve. Dilute to 1 L with DRW. Use deaerated artificial seawater (DASW), Table 8010:III, or 0.7M NaCl if sulfide concentrations are to be determined in seawater. Standardize stock solution by titrating with 0.1M Pb(ClO₄)₂. Pipet 50 mL sulfide stock solution into the electrochemical cell. (Use 10 mL with a small-volume polarographic cell.) Insert Ag/S electrode and reference electrode and read initial potential. Titrate with 0.1M Pb(ClO₄)₂. Let electrode potential stabilize and record potential after each addition. Locate equivalence point as in Section 4500-Cl⁻.D.4a. Alternatively, linearize the titration curve.[1] Calculate the function F_1 for points before the equivalence point.

$$F_1 = (V_o + V)10^{\frac{E}{m}}$$

where:

V_o = volume of stock solution, mL,
V = titrant volume, mL,
E = potential, mV, and
m = slope of calibration curve, mV/log unit.

Plot F_1 as a function of titrant volume. Extrapolate to find the intersection with the x-axis; that is, the equivalence point. Calculate sulfide concentration in the stock solution from:

$$C = \frac{V_{eq}[Pb]}{V_o}$$

where:

C = sulfide concentration, mg/L,
V_{eq} = equivalence volume, mL,
$[Pb]$ = concentration of Pb in titrant, mg/L, and
V_o = volume of stock solution, mL.

Store stock solution in a tightly capped bottle for 1 week or less. The stock solution also can be standardized iodometrically (see Section 4500-S²⁻.E). CAUTION: *Store in a fume hood.*

d. Sulfide standards: Prepare sulfide standards daily by serial dilution of stock. Add AAR and Zn(C₂H₃O₂)₂ solutions to 100-mL volumetric flasks. Add sulfide solutions and dilute to volume with DRW (or DASW). Refer to Table 4500-S²⁻:I for volumes. Prepare at least one standard with a concentration less than the lowest sample concentration.

* Orion 941600 or equivalent.
† Orion 948201 or equivalent.
‡ EG&G Princeton Applied Research K0066, K0060, G0028, or equivalent.

TABLE 4500-S²⁻:I. DILUTION OF SULFIDE STOCK SOLUTION FOR PREPARATION OF STANDARDS (100 mL TOTAL VOLUME)

Dilution	Alkaline Antioxidant Reagent mL	Sulfide Solution	Sulfide Solution mL	1M Zinc Acetate mL
1:10	45	Stock	10	0.15
1:100	50	Stock	1	0.15
1:1 000	45	1:100	10	0.14
1:10 000	50	1:100	1	0.15

4. Procedure

Check electrode performance and calibrate daily. Check electrode potential in a sulfide standard every 2 h. The procedure depends on the sulfide concentration and the time between sample collection and sulfide determination. If the total sulfide concentration is greater than 0.03 mg/L ($1 \times 10^{-6}M$) and the time delay is only a few minutes, sulfide can be determined directly. Otherwise, precipitate ZnS and filter as described in Section 4500-S²⁻.C.

a. Check electrode performance: Pipet 50 mL AAR, 50 mL DWR, and 1 mL sulfide stock solution into the measurement cell. Place Ag/S and reference electrodes in the solution and read potential. Add 10 mL stock solution and read potential. The change in potential should be -28 ± 2 mV. If it is not, follow the troubleshooting procedure in the electrode manual.

b. Calibration: Place electrodes in the most dilute standard but use calibration standards that bracket the sulfide concentrations in the samples. Record potential when the rate of change is less than 0.3 mV/min. (This may take up to 30 min for very low sulfide concentrations, i.e., less than 0.03 mg/L.) Rinse electrodes, blot dry with a tissue, and read potential of the next highest standard. For a meter that can be calibrated directly in concentration, follow manufacturer's directions. For other meters, plot potential as a function of the logarithm (base 10) of the sulfide concentration. For potentials in the linear range, calculate the slope and intercept of the linear portion of the calibration plot.

c. Sulfide determination by comparison with calibration curve, no ZnS precipitation: Add 40 mL AAR, 0.15 mL (3 drops) zinc acetate, and 50 mL sample to a 100-mL volumetric flask. Dilute to 100 mL with AAR. Pour into the electrochemical cell and insert the electrodes. Record potential when the rate of change is less than 0.3 mV/min. Read sulfide concentration from the calibration curve. Alternatively, for potentials in the linear range, calculate the sulfide concentration from:

$$S_{Tot} = 10^{\frac{E-b}{m}}$$

where:

E = electrode potential and
b and m are the intercept and slope of the calibration curve. For a meter that can be calibrated directly in concentration, follow the manufacturer's directions.

d. Sulfide determination by comparison with calibration curve, with ZnS precipitation: Place filter with ZnS precipitate in a 150-mL beaker containing a stir bar. Wash sample bottle with 50 mL AAR and 20 mL DRW and pour the washings into the beaker. Stir to dissolve precipitate. Remove filter with forceps while rinsing it into the beaker with a minimum amount of DRW. Quantitatively transfer to a 100-mL volumetric flask and dilute to mark

with DRW. Pour into the electrochemical cell and place the electrodes in the solution. Measure potential as in ¶ 4c above. Calculate sulfide concentration (¶ 4c).

e. Sulfide determination by standard addition with or without ZnS precipitation: Measure the Ag/S-ISE electrode potential as in ¶ c or d above. Add sulfide stock solution and measure potential again. Calculate sulfide concentration as follows:

$$C_o = \frac{fC_s}{(1+f)10^{\frac{E_s - E_o}{m}} - 1}$$

where:

C_o and C_s = sulfide concentrations in sample and known addition,
E_o and E_s = potentials measured for sample and known addition,
m = slope of calibration curve (approximately 28 mV/log S²⁻, and
f = ratio of known-addition volume to sample volume.

f. Sulfide determination by titration: Use the same procedure as for standardizing the sulfide stock solution (¶ 3c). The minimum sulfide concentration for determination by titration is 0.3 mg/L ($10^{-5}M$).

5. Precision

For sulfide determination by comparison with the calibration curve, the relative standard deviation varies with the sulfide concentration. RSD values of 23% for 0.0091 mg/L and 5% for 0.182 mg/L have been reported.[2] (0.0091 μg/L was below the range for which the potential varied linearly with the logarithm of the sulfide concentration, i.e., the Nernstian range.) For sulfide determination by standard addition, the precision is greatest if the amount of sulfide added is as large as possible while staying within the linear range.[3]

6. References

1. GRAN, G. 1952. Determination of the equivalence point in potentiometric titrations. Part II. *Analyst* 77:661.
2. BAUMANN, E. 1974. Determination of parts per billion sulfide in water with the sulfide-selective electrode. *Anal. Chem.* 46:1345.
3. RATZLAFF, K.L. 1979. Optimizing precision in standard addition measurement. *Anal. Chem.* 51:232.

7. Bibliography

ORION RESEARCH, INC. 1980. Instruction Manual for Silver-Sulfide Electrode.
VIVIT, D.V., J.W. BALL & E.A. JENNE. 1984. Specific-ion electrode determinations of sulfide preconcentrated from San Francisco Bay waters. *Environ. Geol. Water Sci.* 6:79.

4500-S²⁻ H.   Calculation of Un-ionized Hydrogen Sulfide

Hydrogen sulfide (H_2S) and bisulfide ion (HS^-), which together constitute dissolved sulfide, are in equilibrium with hydrogen ions:

$$H_2S \rightleftharpoons H^+ + HS^-$$

The conditional ionization constant, which is valid for the temperature and ionic strength of the water of interest, relates the concentrations of H_2S and HS^-:

$$K_1' = \frac{[H^+][HS^-]}{[H_2S]}$$

The conditional constant is used to calculate the distribution of dissolved sulfide between the two species. The conditional ionization constant of H_2S is approximately 7.0. It differs from 7.0 by less than 0.2 log units for the ionic strengths and temperatures likely to be encountered in water-quality monitoring. The fraction of sulfide present as H_2S can be estimated with an error of less than 40% from Figure 4500-S²⁻:3. If more accuracy is needed, use the methods given below.

1. Calculation for Fresh Water and Brackish Water ($I < 0.1M$)

Calculate the dissociation constant for zero ionic strength (pK_1) and the temperature of interest.[1] If the temperature is 25°C, then pK_1 is 6.98. Otherwise:

$$pK_1(T) = 32.55 + 1519.44/T - 15.672 \log_{10}T + 0.02722T$$

where T is temperature (°K, i.e., $T°C + 273.15$). Next, calculate the ionic strength I as in Table 2330:I, the Debye-Hückel A parameter, and the negative logarithm of the monovalent ion activity coefficient (pf_m):

$$A = 0.7083 - 2.277 \times 10^{-3}T + 5.399 \times 10^{-6}T^2$$

$$pf_m = A\left(\frac{\sqrt{I}}{1 + \sqrt{I}} - 0.3I\right)$$

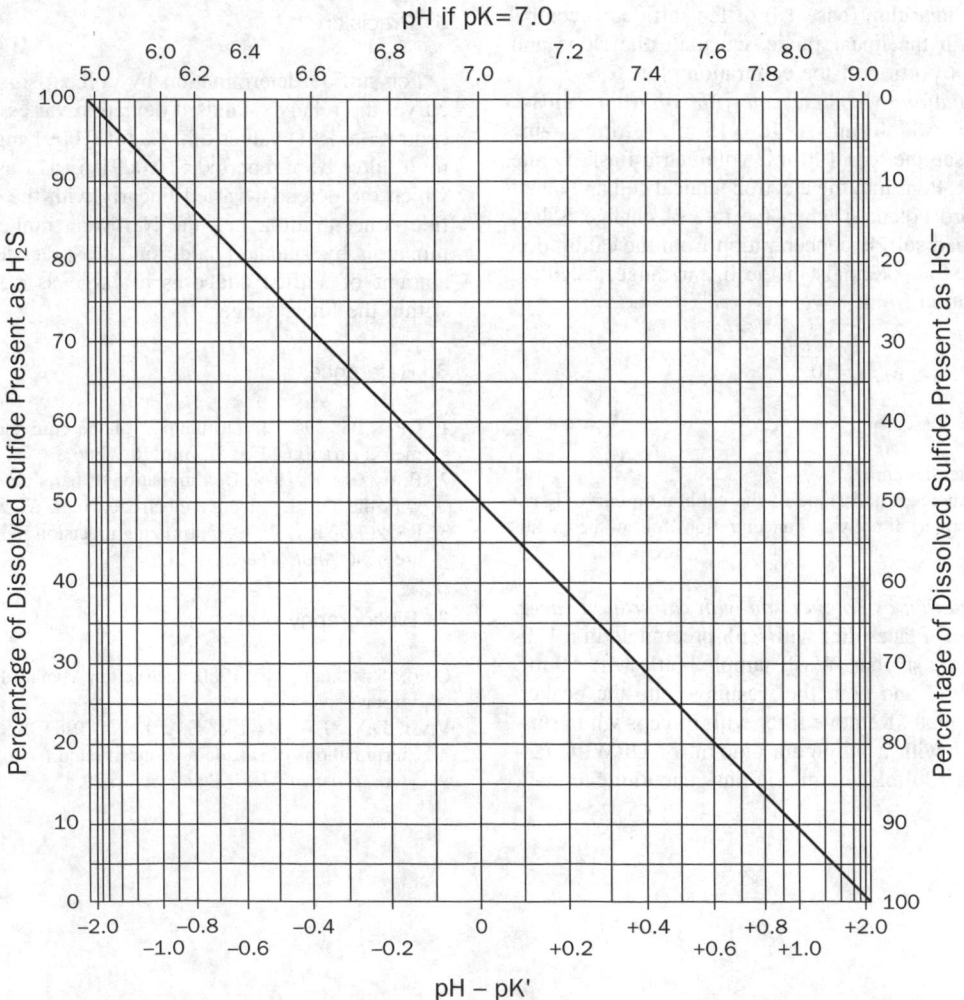

Figure 4500-S²⁻:3. Proportions of H₂S and HS⁻ in dissolved sulfide.

Calculate the conditional ionization constant, K_1', and the hydrogen ion concentration, $[H^+]$:

$$K_1' = 10^{-pK_1 + 2pf_m}$$

$$[H^+] = 10^{-pH + pf_m}$$

Finally, calculate the un-ionized hydrogen sulfide concentration, $[H_2S]$, from the total sulfide concentration, S_T:

$$[H_2S] = \frac{S_T}{1 + \dfrac{K_1'}{[H^+]}}$$

Sample calculation: Total sulfide concentration 0.32 mg/L ($1.0 \times 10^{-5}M$), pH 6.75, ionic strength 0.02M, temperature 15.5°C.

$$pK_1 = 32.55 \times \frac{1519.44}{288.65} - 15.672 \times \log_{10}(288.65)$$
$$+ 0.02722 \times 288.65 = 7.11$$

$$A = 0.7083 - 2.277 \times 10^{-3} \times 288.65 + 5.399 \times 10^{-6} \times (288.65)^2$$
$$= 0.501$$

$$pf_m = 0.501 \times \left(\frac{\sqrt{0.02}}{1 + \sqrt{0.02}} - 0.3 \times 0.02 \right)$$
$$= 0.059$$

$$K_1' = 10^{-7.11 + 2 \times 0.059}$$
$$= 1.014 \times 10^{-7}$$

$$[H^+] = 10^{-6.75 + 0.059}$$
$$= 2.037 \times 10^{-7}$$

$$[H_2S] = \frac{1 \times 10^{-5}}{1 + \dfrac{1.014 \times 10^{-7}}{2.037 \times 10^{-7}}}$$
$$= 6.68 \times 10^{-6}M$$
$$= 0.21 \text{ mg/L (as S)}$$

2. Calculation for Seawater and Estuarine Water

This procedure differs only in calculating the conditional ionization constant, which can be calculated accurately.[1] The (potentially) largest source of error in calculating un-ionized hydrogen sulfide in seawater is the hydrogen ion concentration. Calibrate the pH electrode in artificial seawater at the temperature of the water of interest.[2] Alternatively, if the pH electrode is calibrated using NIST buffers (as in Section 4500-H^+), measure pH of dilute acid (10^{-4}–$10^{-3}N$ HNO$_3$, HCl, or HClO$_4$) in artificial seawater diluted to the salinity of the water of interest and at the temperature of interest and calculate a correction factor.[3] (Prepare artificial seawater as in Table 8010:III, substituting NaCl for NaF, NaHCO$_3$, and Na$_2$SiO$_3$·9H$_2$O on an equimolar basis.)

Calculate pK_1' as outlined in Section 4500-S^{2-}.H.1. Calculate the coefficients A and B[1] (A and B are not Debye-Hückel parameters):

$$A = -0.2391 + \frac{35.685}{T}$$

$$B = 0.0109 - \frac{0.3776}{T}$$

Calculate pK_1':

$$pK_1' = pK_1 + A\sqrt{S} + BS$$

where:

S = salinity, g/kg.

Calculate K_1':

$$K_1' = 10^{-pK_1'}$$

Sample calculation: Total sulfide concentration 0.32 mg/L ($1 \times 10^{-5}M$), pH 6.75, salinity 35 g/kg ($I = 0.7M$), temperature 15.5°C.

$$A = -0.2391 + \frac{35.685}{288.65}$$
$$= -0.115$$

$$B = 0.0109 - \frac{0.3776}{288.65}$$
$$= 0.009\,59$$

From 4500-S^{2-}.H.1, $pK_1 = 7.11$.

$$pK_1' = 7.11 - 0.115\sqrt{35} + 0.00959 \times 35$$
$$= 6.77$$

$$K_1' = 10^{-6.77}$$
$$= 1.70 \times 10^{-7}$$

$$pf_m = 0.501 \times \left(\frac{\sqrt{0.7}}{1 + \sqrt{0.7}} - 0.3 \times 0.7 \right)$$
$$= 0.12$$

$$[H^+] = 10^{-6.75 + 0.12}$$
$$= 2.34 \times 10^{-7}$$

$$[H_2S] = \frac{1 \times 10^{-5}}{1 + \dfrac{1.7 \times 10^{-7}}{2.3 \times 10^{-7}}}$$
$$= 5.8 \times 10^{-6}M$$
$$= 0.19 \text{ mg/L (as S)}$$

3. References

1. MILLERO, F.J. 1986. The thermodynamics and kinetics of the hydrogen sulfide system in natural waters. *Mar. Chem.* 18:121.
2. MILLERO, F.J. 1986. The pH of estuarine waters. *Limnol. Oceanogr.* 31:839.
3. SIGEL, H., A.D. ZUBERBUHLER & O. YAMAUCHI. 1991. Comments on potentiometric pH titrations and the relationship between pH-meter reading and hydrogen ion concentration. *Anal. Chim. Acta.* 255:63.

4. Bibliography

ARCHER, D.G. & P. WANG. 1990. The dielectric constant of water and Debye-Hückel Limiting Law Slopes. *J. Phys. Chem. Ref. Data* 12: 817.

4500-S²⁻ I. Distillation, Methylene Blue Flow Injection Analysis (PROPOSED)

1. General Discussion

a. Principle: Water and wastewater samples are distilled into a sodium hydroxide trapping solution and the distillate is analyzed. Hydrogen sulfide (H_2S) reacts in acid media and in the presence of ferric chloride with two molecules of N, N-dimethyl-p-phenylenediamine to form methylene blue. The resulting color is read at 660 nm.

b. Sample preservation: Because H_2S oxidizes rapidly, analyze samples and standards without delay. To preserve samples, add 4 drops $2M$ zinc acetate to 100 mL sample and adjust pH to >9 with $6M$ NaOH, then cool to 4°C. Samples are distilled into a trapping solution resulting in $0.25M$ NaOH matrix.

Also see Sections 4500-S²⁻.A, B, and E, and Section 4130, Flow Injection Analysis (FIA).

c. Interferences: This method measures total sulfide, which is defined as the acid-soluble sulfide fraction of a sample. Total sulfide includes both acid-soluble sulfides such as H_2S, and acid-soluble metal sulfides present in suspended matter. This method does not measure acid-insoluble sulfides such as CuS.

Most nonvolatile interferences are eliminated by distillation. Strong reducing agents inhibit color formation at concentrations of several hundred milligrams per liter. Iodide interferes at concentrations greater than 2 mg I/L.

Also see Section 4500-S²⁻.A and B.

2. Apparatus

a. Distillation apparatus consisting of a glass or polypropylene micro-distillation device* capable of distilling 6 mL or more of sample into a $0.25M$ NaOH final concentration trapping solution.

b. Flow injection analysis equipment consisting of:
1) *FIA injection valve* with sample loop or equivalent.
2) *Multichannel proportioning pump.*
3) *FIA manifold* (Figure 4500-S²⁻:4) with cation exchange column and flow cell. Relative flow rates only are shown in Figure 4500-S²⁻:4. Tubing volumes are given as an example only; they may be scaled down proportionally. Use manifold tubing of an inert material such as TFE.
4) *Absorbance detector,* 660 nm, 10-nm bandpass.
5) *Injection valve control and data acquisition system.*

3. Reagents

Use reagent water (>10 megohm) for all solutions. To prevent bubble formation, degas carrier and buffer with helium. Pass He at 140 kPa (20 psi) through a helium degassing tube. Bubble He through 1 L solution for 1 min.

a. Sodium hydroxide carrier and diluent, NaOH, $0.25M$: In a 2-L volumetric flask, dissolve 20 g NaOH in approximately 1800 mL water. Dilute to mark and mix with a magnetic stirrer until dissolved. Store in a plastic container.

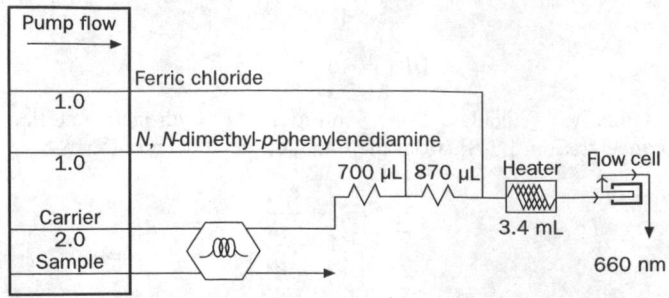

Figure 4500-S²⁻:4. FIA sulfide manifold.

b. Hydrochloric acid, HCl, $3M$: To a tared 1-L container, add 752 g water and then slowly add 295 g conc HCl. Invert to mix.

c. Hydrochloric acid, HCl, $0.20M$: To a tared 1-L container, add 983.5 g water. Then add 19.7 g conc HCl. Invert to mix.

d. N,N-dimethyl-p-phenylenediamine: In a 1-L volumetric flask dissolve 1.0 g N,N-dimethyl-p-phenylenediamine dihydrochloride, $(CH_3)_2NC_6H_4NH_2 \cdot 2HCl$, in about 800 mL $3M$ HCl (¶ 3b). Dilute to mark and invert to mix. If solution appears dark, it is likely that the N,N-dimethyl-p-phenylenediamine dihydrochloride is decomposed; discard, and use fresh reagent.

e. Ferric chloride: In a 500-mL volumetric flask dissolve 6.65 g ferric chloride hexahydrate, $FeCl_3 \cdot 6H_2O$, in about 450 mL $0.20M$ HCl (¶ 3c). Dilute to mark with water and invert to mix.

f. Stock sulfide standard, 100 mg S²⁻/L: In a 1-L volumetric flask dissolve 0.7491 g sodium sulfide nonahydrate, $Na_2S \cdot 9H_2O$, in approximately 900 mL NaOH diluent (¶ 3a). Dilute to mark and invert to mix.

g. Standard solutions: Prepare sulfide standards in desired concentration range, using stock standard (¶ 3f), and diluting with NaOH diluent (¶ 3a).

h. Sulfuric acid distillation releasing solution, H_2SO_4, $9M$: To a tared 500-mL container, add 150.0 g water, then add slowly while swirling, in increments of 40 g, 276 g conc H_2SO_4. CAUTION: *Solution will become very hot. Allow to cool before using.*

4. Procedure

a. Distillation: This procedure is designed for the determination of sulfides in aqueous solutions, solid waste materials, or effluents. To preserve and remove sulfide from interfering substances, distill samples immediately after collection.

Follow manufacturer's instructions for use of distillation apparatus. Add sufficient $9M$ H_2SO_4 (¶ 3h) to sample to dissolve ZnS (s), digest total sulfides, and release the sulfide as hydrogen sulfide gas. Immediately place sample on-line with the receiving vessel or collector tube and distill hydrogen sulfide and water in the sample into a $0.25M$ trapping solution.

b. Flow injection analysis: Set up a manifold equivalent to that in Figure 4500-S²⁻:4 and follow method supplied by the manufacturer or laboratory standard operating procedure. The carrier concentration should be identical to the final concentration of NaOH in the trapping solution from the distillation procedure (¶ 4a). Follow quality control protocols outlined in Section 4020.

* Lachat Instruments MICRO DIST or equivalent.

5. Calculations

Prepare standard curves by plotting absorbance of standards processed through the manifold versus sulfide concentration.

6. Precision and Bias

a. MDL: A 200-μL sample loop was used in the method described above. Using a published method,[1] analysts ran 21 replicates of a 10.0-mg S^{2-}/L standard. These gave a mean of 9.0 mg S^{2-}/L, a standard deviation of 0.23 mg S^{2-}/L, and MDL of 0.58 mg S^{2-}/L. A higher MDL may be obtained by decreasing sample loop volume.

b. Precision: Ten injections of a distilled 50-mg S^{2-}/L standard gave a mean of 49.4 mg S^{2-}/L, a standard deviation of 0.27 mg S^{2-}/L, and percent relative standard deviation of 0.54.

7. Reference

1. U.S. ENVIRONMENTAL PROTECTION AGENCY. 1984. Definition and procedure for the determination of method detection limits. Appendix B to 40 CFR 136 Rev. 1.11 amended June 30, 1986. 49 CFR 43430.

4500-SO$_3^{2-}$ SULFITE*

4500-SO$_3^{2-}$ A. Introduction

1. Occurrence

Sulfite ions (SO$_3^{2-}$) may occur in boilers and boiler feedwaters treated with sulfite for dissolved oxygen control, in natural waters or wastewaters as a result of industrial pollution, and in treatment plant effluents dechlorinated with sulfur dioxide (SO$_2$). Excess sulfite ion in boiler waters is deleterious because it lowers the pH and promotes corrosion. Control of sulfite ion in wastewater treatment and discharge may be important environmentally, principally because of its toxicity to fish and other aquatic life and its rapid oxygen demand.

2. Selection of Method

The iodometric titration method is suitable for relatively clean waters with concentrations above 2 mg SO$_3^{2-}$/L. The phenanthroline colorimetric determination, following evolution of sulfite from the sample matrix as SO$_2$, is preferred for low levels of sulfite.

* Approved by Standard Methods Committee, 1996.

4500-SO$_3^{2-}$ B. Iodometric Method

1. General Discussion

a. Principle: An acidified sample containing sulfite (SO$_3^{2-}$) is titrated with a standardized potassium iodide-iodate titrant. Free iodine, liberated by the iodide-iodate reagent, reacts with SO$_3^{2-}$. The titration endpoint is signalled by the blue color resulting from the first excess of iodine reacting with a starch indicator.

b. Interferences: The presence of other oxidizable materials, such as sulfide, thiosulfate, and Fe^{2+} ions, can cause apparently high results for sulfite. Some metal ions, such as Cu^{2+}, may catalyze the oxidation of SO$_3^{2-}$ to SO$_4^{2-}$ when the sample is exposed to air, thus leading to low results. NO$_2^-$ will react with SO$_3^{2-}$ in the acidic reaction medium and lead to low sulfite results unless sulfamic acid is added to destroy nitrite. Addition of EDTA as a complexing agent at the time of sample collection inhibits Cu^{2+} catalysis and promotes oxidation of ferrous to ferric iron before analysis. Sulfide and thiosulfate ions normally would be expected only in samples containing certain industrial discharges, but must be accounted for if present. Sulfide may be removed by adding about 0.5 g zinc acetate and analyzing the supernatant of the settled sample. However, thiosulfate may have to be determined by an independent method (e.g., the formaldehyde/iodometric method[1]), and then the sulfite determined by difference.

c. Minimum detectable concentration: 2 mg SO$_3^{2-}$/L.

2. Reagents

a. Sulfuric acid: H$_2$SO$_4$, 1 + 1.

b. Standard potassium iodide-iodate titrant, 0.002083M: Dissolve 0.4458 g primary-grade anhydrous KIO$_3$ (dried for 4 h at 120°C), 4.35 g KI, and 310 mg sodium bicarbonate (NaHCO$_3$) in distilled water and dilute to 1000 mL; 1.00 mL = 500 μg SO$_3^{2-}$.

c. Sulfamic acid, NH$_2$SO$_3$H, crystalline.

d. EDTA reagent: Dissolve 2.5 g disodium EDTA in 100 mL distilled water.

e. Starch indicator: To 5 g starch (potato, arrowroot, or soluble) in a mortar, add a little cold distilled water and grind to a paste. Add mixture to 1 L boiling distilled water, stir, and let settle overnight. Use clear supernatant. Preserve by adding either 1.3 g salicylic acid, 4 g $ZnCl_2$, or a combination of 4 g sodium propionate and 2 g sodium azide to 1 L starch solution.

3. Procedure

a. Sample collection: Collect a fresh sample, taking care to minimize contact with air. Fix cooled samples ($<50°C$) immediately by adding 1 mL EDTA solution/100 mL sample. Cool hot samples to 50°C or below. Do not filter.

b. Titration: Add 1 mL H_2SO_4 and 0.1 g NH_2SO_3H crystals to a 250-mL erlenmeyer flask or other suitable titration vessel. Accurately measure 50 to 100 mL EDTA-stabilized sample into flask, keeping pipet tip below liquid surface. Add 1 mL starch indicator solution. Titrate immediately with standard $KI-KIO_3$ titrant, while swirling flask, until a faint permanent blue color develops. Analyze a reagent blank using distilled water instead of sample.

4. Calculation

$$mg\ SO_3^{2-}/L = \frac{(A - B) \times M \times 6 \times 40\,000}{mL\ sample}$$

where:

A = mL titrant for sample,
B = mL titrant for blank, and
M = molarity of $KI-KIO_3$ titrant.

5. Precision and Bias

Three laboratories analyzed five replicate portions of a standard sulfite solution and of secondary treated wastewater effluent to which sulfite was added. The data are summarized below. Individual analyst's precision ranged from 0.7 to 3.6% standard deviation ($N = 45$).

Sample	$\overline{X}$ mg/L	Standard Deviation, σ mg/L	Relative Standard Deviation %
Standard, 6.3 mg SO_3^{2-}/L	4.5	0.25	5.5
Secondary effluent with 2.0 mg SO_3^{2-}/L	2.1	0.28	13.4
Secondary effluent with 4.0 mg SO_3^{2-}/L	3.6	0.17	4.8

6. Reference

1. KURTENACKER, A. 1924. The aldehyde-bisulfite reaction in mass analysis. *Z. Anal. Chem.* 64:56.

4500-SO_3^{2-} C. Phenanthroline Method

1. General Discussion

a. Principle: An acidified sample is purged with nitrogen gas and the liberated SO_2 is trapped in an absorbing solution containing ferric ion and 1,10-phenanthroline. Ferric iron is reduced to the ferrous state by SO_2, producing the orange tris(1,10-phenanthroline) iron(II) complex. After excess ferric iron is removed with ammonium bifluoride, the phenanthroline complex is measured colorimetrically at 510 nm.[1]

b. Interferences: See Section 4500-SO_3^{2-}.B.1*b*.

c. Minimum detectable concentration: 0.01 mg SO_3^{2-}/L.

2. Apparatus

a. Apparatus for evolution of SO_2: Figure 4500-SO_3^{2-}:1 shows the following components:

1) *Gas flow meter,* with a capacity to measure 2 L/min of pure nitrogen gas.

2) *Gas washing bottle,* 250-mL, with coarse-porosity, 12- mm-diam fritted cylinder gas dispersion tube.

3) *Tubing connectors,* quick-disconnect, polypropylene.

4) *Tubing,* flexible PVC, for use in all connections.

5) *Nessler tube,* 100-mL.

b. Colorimetric equipment: One of the following is required:

1) *Spectrophotometer,* for use at 510 nm, providing a light path of 1 cm or longer.

2) *Filter photometer,* providing a light path of 1 cm or longer

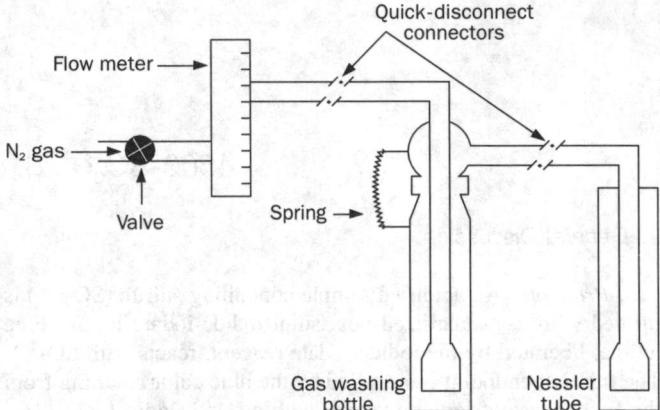

Figure 4500-SO_3^{2-}:1. Apparatus for evolution of SO_2 from samples for colorimetric analysis.

and equipped with a green filter having maximum transmittance near 510 nm.

3. Reagents

a. 1,10-phenanthroline solution, 0.03M: Dissolve 5.95 g 1,10-phenanthroline in 100 mL 95% ethanol. Dilute to 1 L with distilled water. Discard if solution becomes colored.

b. *Ferric ammonium sulfate solution, 0.01M:* Dissolve 4.82 g $NH_4Fe(SO_4)_2 \cdot 12H_2O$ in 1 L distilled water to which has been added 1 mL conc H_2SO_4 to suppress ferric hydrolysis. Filter through a glass fiber filter if insoluble matter is visible. If necessary, adjust volume of acid so that a mixture of 10 parts of phenanthroline solution and one part of ferric ammonium sulfate solution will have a pH between 5 and 6.

c. *Ammonium bifluoride, 5%:* Dissolve 25 g NH_4HF_2 in 500 mL distilled water. Store in a polyethylene bottle and dispense with a plastic pipet.

d. *Potassium tetrachloromercurate, (TCM), K_2HgCl_4, 0.04M:* Dissolve 10.86 g $HgCl_2$, 5.96 g KCl, and 0.066 g disodium EDTA in distilled water and dilute to 1 L. Adjust pH to 5.2. This reagent normally is stable for 6 months, but discard if a precipitate forms.[2]

e. *Dilute TCM-stabilized sulfite standard:* Dissolve 0.5 g Na_2SO_3 in 500 mL distilled water. Standardize on the day of preparation, but wait at least 30 min to allow the rate of oxidation to slow. Determine molarity by titrating with standard 0.0125M potassium iodide-iodate titrant using starch indicator (see Section 4500-SO_3^{2-}.B). Calculate molarity of working standard as follows:

$$\text{Molarity of } SO_3^{2-} \text{ standard} = \frac{(A - B) \times M}{\text{mL sample}}$$

where:

- A = titrant for sample, mL,
- B = titrant for blank, mL,
- M = molarity of potassium iodide-iodate titrant.

Because stock Na_2SO_3 solution is unstable, immediately after standardization, pipet 10 mL into a 500-mL volumetric flask partially filled with TCM and dilute to mark with TCM. Calculate the concentration of this dilute sulfite solution by multiplying the stock solution concentration by 0.02. This TCM-stabilized standard is stable for 30 d if stored at 5°C. Discard as soon as any precipitate is noticed at the bottom.

f. *Standard potassium iodide-iodate titrant, 0.0125M:* See Section 4500-SO_3^{2-}.B.2b.

g. *Hydrochloric acid,* 1 + 1.

h. *Octyl alcohol,* reagent-grade.

i. *Sulfamic acid, 10%:* Dissolve 10 g NH_2SO_3H in 100 mL distilled water. This reagent can be kept for a few days if protected from air.

j. *EDTA reagent:* See Section 4500-SO_3^{2-}.B.2d.

4. Procedure

a. *Sample collection:* Collect a fresh sample taking care to minimize contact with air. Fix cooled samples (<50°C) immediately by adding 1 mL EDTA solution for each 100 mL sample.

b. *SO_2 evolution:* Prepare the absorbing solution by adding 5 mL 1,10-phenanthroline solution, 0.5 mL ferric ammonium sulfate solution, 25 mL distilled water, and 5 drops octyl alcohol (to act as defoamer) to a 100-mL nessler tube; insert a gas dispersion tube. Add 1 mL sulfamic acid solution to the gas washing bottle and 100 mL of sample or a portion containing less than 100 μg SO_3^{2-} diluted to 100 mL. Add 10 mL 1 + 1 HCl and immediately connect the gas washing bottle to the gas train as shown in

Figure 4500-SO_3^{2-}:1. Place a spring or rubber band on the gas washing bottle to keep the top securely closed during gas flow. Adjust nitrogen flow to 2.0 L/min and purge for 60 min.

c. *Colorimetric measurement:* After exactly 60 min, turn off nitrogen flow, disconnect nessler tube, and immediately add 1 mL ammonium bifluoride solution. Remove gas dispersion tube after rinsing it with distilled water into the tube and forcing the rinse water into the nessler tube with a rubber bulb. Dilute to 50 mL in the nessler tube and mix by rapidly moving the tube in a circular motion. Do not let rubber stoppers or PVC tubing come in contact with the absorbing solution. After at least 5 min from the time of adding ammonium bifluoride, read the absorbance versus distilled water at 510 nm using either a 5-cm cell for a range of 0 to 30 μg SO_3^{2-} per portion or a 1-cm cell for a range of 0 to 100 μg SO_3^{2-}. Avoid transferring octyl alcohol into the cell by letting it rise to the surface of the absorbing solution and transferring the clear lower solution to the cell with a pipet. Make a calibration curve by analyzing a procedure blank and at least three standards. Run at least one standard with each set of samples. For maximum accuracy hold samples and standards at the same temperature and keep the time interval from start of purging to the addition of ammonium bifluoride constant. This is easier to do if several gas trains are used simultaneously in parallel. If ambient temperatures are subject to frequent fluctuation, a water bath may be used to control color development at a fixed temperature.

5. Calculation

$$\text{mg } SO_3^{2-}/L = \frac{\mu g \ SO_3^{2-} \text{ from calibration curve}}{\text{mL sample}}$$

6. Precision and Bias

Three laboratories analyzed five replicate portions of a standard sulfite solution and of secondary treated wastewater effluent to which sulfite was added. The data are summarized below. Individual analyst's precision ranged from 4.1 to 10.5% standard deviation ($N = 45$).

Sample	$\overline{X}$ mg/L	Standard Deviation, σ mg/L	Relative Standard Deviation %
Standard, 4.7 mg SO_3^{2-}/L	3.7	0.78	21
Secondary effluent with 0.12 mg SO_3^{2-}/L	0.12	0.03	25
Secondary effluent with 4.0 mg SO_3^{2-}/L	3.7	0.30	8.0

7. References

1. STEPHENS, B.G. & F. LINDSTROM. 1964. Spectrophotometric determination of sulfur dioxide suitable for atmospheric analysis. *Anal. Chem.* 36:1308.

2. WEST, P. W. & G.C. GAEKE. 1956. Fixation of sulfur dioxide as sulfitomercurate and subsequent colorimetric determination. *Anal. Chem.* 28:1816.

4500-SO$_4^{2-}$ SULFATE*

4500-SO$_4^{2-}$ A.   Introduction

1. Occurrence

Sulfate (SO$_4^{2-}$) is widely distributed in nature and may be present in natural waters in concentrations ranging from a few to several thousand milligrams per liter. Mine drainage wastes may contribute large amounts of SO$_4^{2-}$ through pyrite oxidation. Sodium and magnesium sulfate exert a cathartic action.

2. Selection of Method

The ion chromatographic method (4110) and capillary ion electrophoresis (CIE—see Section 4140) are suitable for sulfate con-

---

* Approved by Standard Methods Committee, 1997.

centrations above 0.1 mg/L. The gravimetric methods (C and D) are suitable for SO$_4^{2-}$ concentrations above 10 mg/L. The turbidimetric method (E) is applicable in the range of 1 to 40 mg SO$_4^{2-}$ /L. The automated methylthymol blue methods (F and G) are the procedures for analyzing large numbers of samples for sulfate alone when the equipment is available; over 30 samples can be analyzed per hour. Methods C, D, F, G, 4110, or CIE (4140) are preferred for accurate results.

3. Sampling and Storage

In the presence of organic matter certain bacteria may reduce SO$_4^{2-}$ to S^{2-}. To avoid this, store samples at 4°C.

4500-SO$_4^{2-}$ B.   (Reserved)

4500-SO$_4^{2-}$ C.   Gravimetric Method with Ignition of Residue

1. General Discussion

a. Principle: Sulfate is precipitated in a hydrochloric acid (HCl) solution as barium sulfate (BaSO$_4$) by the addition of barium chloride (BaCl$_2$).

The precipitation is carried out near the boiling temperature, and after a period of digestion the precipitate is filtered, washed with water until free of Cl$^-$, ignited or dried, and weighed as BaSO$_4$.

b. Interference: The gravimetric determination of SO$_4^{2-}$ is subject to many errors, both positive and negative. In potable waters where the mineral concentration is low, these may be of minor importance.

1) Interferences leading to high results—Suspended matter, silica, BaCl$_2$ precipitant, NO$_3^-$, SO$_3^{2-}$ and occluded mother liquor in the precipitate are the principal factors in positive errors. Suspended matter may be present in both the sample and the precipitating solution; soluble silicate may be rendered insoluble and SO$_3^{2-}$ may be oxidized to SO$_4^{2-}$ during analysis. Barium nitrate [Ba(NO$_3$)$_2$], BaCl$_2$, and water are occluded to some extent with the BaSO$_4$ although water is driven off if the temperature of ignition is sufficiently high.

2) Interferences leading to low results—Alkali metal sulfates frequently yield low results. This is true especially of alkali hydrogen sulfates. Occlusion of alkali sulfate with BaSO$_4$ causes substitution of an element of lower atomic weight than barium in

the precipitate. Hydrogen sulfates of alkali metals act similarly and, in addition, decompose on being heated. Heavy metals, such as chromium and iron, cause low results by interfering with the complete precipitation of SO$_4^{2-}$ and by formation of heavy metal sulfates. BaSO$_4$ has small but significant solubility, which is increased in the presence of acid. Although an acid medium is necessary to prevent precipitation of barium carbonate and phosphate, it is important to limit its concentration to minimize the solution effect.

2. Apparatus

a. Steam bath.

b. Drying oven, equipped with thermostatic control.

c. Muffle furnace, with temperature indicator.

d. Desiccator.

e. Analytical balance, capable of weighing to 0.1 mg.

f. Filter: Use one of the following:

1) *Filter paper,* acid-washed, ashless hard-finish, sufficiently retentive for fine precipitates.

2) *Membrane filter,* with a pore size of about 0.45 μm.

g. Filtering apparatus, appropriate to the type of filter selected. (Coat membrane filter holder with silicone fluid to prevent precipitate from adhering.)

3. Reagents

a. Methyl red indicator solution: Dissolve 100 mg methyl red sodium salt in distilled water and dilute to 100 mL.

b. Hydrochloric acid, HCl, 1 + 1.

c. Barium chloride solution: Dissolve 100 g BaCl$_2$·2H$_2$O in 1 L distilled water. Filter through a membrane filter or hard-finish filter paper before use; 1 mL is capable of precipitating approximately 40 mg SO$_4^{2-}$.

d. Silver nitrate-nitric acid reagent: Dissolve 8.5 g AgNO$_3$ and 0.5 mL conc HNO$_3$ in 500 mL distilled water.

*e. Silicone fluid.**

4. Procedure

a. Removal of silica: If the silica concentration exceeds 25 mg/L, evaporate sample nearly to dryness in a platinum dish on a steam bath. Add 1 mL HCl, tilt, and rotate dish until the acid comes in complete contact with the residue. Continue evaporation to dryness. Complete drying in an oven at 180°C and if organic matter is present, char over flame of a burner. Moisten residue with 2 mL distilled water and 1 mL HCl, and evaporate to dryness on a steam bath. Add 2 mL HCl, take up soluble residue in hot water, and filter. Wash insoluble silica with several small portions of hot distilled water. Combine filtrate and washings. Discard residue.

b. Precipitation of barium sulfate: Adjust volume of clarified sample to contain approximately 50 mg SO$_4^{2-}$ in a 250-mL volume. Lower concentrations of SO$_4^{2-}$ may be tolerated if it is impracticable to concentrate sample to the optimum level, but in such cases limit total volume to 150 mL. Adjust pH with HCl to pH 4.5 to 5.0, using a pH meter or the orange color of methyl red indicator. Add 1 to 2 mL HCl. Heat to boiling and, while

_____

* "Desicote" (Beckman), or equivalent.

stirring gently, slowly add warm BaCl$_2$ solution until precipitation appears to be complete; then add about 2 mL in excess. If amount of precipitate is small, add a total of 5 mL BaCl$_2$ solution. Digest precipitate at 80 to 90°C, preferably overnight but for not less than 2 h.

c. Filtration and weighing: Mix a small amount of ashless filter paper pulp with the BaSO$_4$, quantitatively transfer to a filter, and filter at room temperature. The pulp aids filtration and reduces the tendency of the precipitate to creep. Wash precipitate with small portions of warm distilled water until washings are free of Cl$^-$ as indicated by testing with AgNO$_3$-HNO$_3$ reagent. Place filter and precipitate in a weighed platinum crucible and ignite at 800°C for 1 h. Do not let filter paper flame. Cool in desiccator and weigh.

5. Calculation

$$\text{mg SO}_4^{2-}/\text{L} = \frac{\text{mg BaSO}_4 \times 411.6}{\text{mL sample}}$$

6. Precision and Bias

A synthetic sample containing 259 mg SO$_4^{2-}$/L, 108 mg Ca^{2+}/L, 82 mg Mg^{2+}/L, 3.1 mg K$^+$/L, 19.9 mg Na$^+$/L, 241 mg Cl$^-$/L, 0.250 mg NO$_2^-$-N/L, 1.1 mg NO$_3^-$-N/L, and 42.5 mg total alkalinity/L (contributed by NaHCO$_3$) was analyzed in 32 laboratories by the gravimetric method, with a relative standard deviation of 4.7% and a relative error of 1.9%.

7. Bibliography

HILLEBRAND, W.F. et al. 1953. Applied Inorganic Analysis, 2nd ed. John Wiley & Sons, New York, N.Y.

KOLTHOFF, I.M., E.J. MEEHAN, E.B. SANDELL & S. BRUCKENSTEIN. 1969. Quantitative Chemical Analysis, 4th ed. Macmillan Co., New York, N.Y.

4500-SO$_4^{2-}$ D.   Gravimetric Method with Drying of Residue

1. General Discussion

See Method C, preceding.

2. Apparatus

With the exception of the filter paper, all of the apparatus cited in Section 4500-SO$_4^{2-}$.C.2 is required, plus the following:

a. Filters: Use one of the following:

1) *Fritted-glass filter,* fine ("F") porosity, with a maximum pore size of 5 μm.

2) *Membrane filter,* with a pore size of about 0.45 μm.

b. Vacuum oven.

3. Reagents

All the reagents listed in Section 4500-SO$_4^{2-}$.C.3 are required.

4. Procedure

a. Removal of interference: See Section 4500-SO$_4^{2-}$.C.4a.

b. Precipitation of barium sulfate: See Section 4500-SO$_4^{2-}$.C.4b.

c. Preparation of filters:

1) Fritted glass filter—Dry to constant weight in an oven maintained at 105°C or higher, cool in desiccator, and weigh.

2) Membrane filter—Place filter on a piece of filter paper or a watch glass and dry to constant weight* in a vacuum oven at 80°C, while maintaining a vacuum of at least 85 kPa or in a conventional oven at a temperature of 103 to 105°C. Cool in desiccator and weigh membrane only.

_____

* Constant weight is defined as a change of not more than 0.5 mg in two successive operations consisting of heating, cooling in desiccator, and weighing.

d. Filtration and weighing: Filter $BaSO_4$ at room temperature. Wash precipitate with several small portions of warm distilled water until washings are free of Cl^-, as indicated by testing with $AgNO_3$-HNO_3 reagent. If a membrane filter is used add a few drops of silicone fluid to the suspension before filtering, to prevent adherence of precipitate to holder. Dry filter and precipitate by the same procedure used in preparing filter. Cool in a desiccator and weigh.

5. Calculation

$$\text{mg } SO_4{}^{2-}/L = \frac{\text{mg } BaSO_4 \times 411.6}{\text{mL sample}}$$

6. Bibliography

See Section 4500-$SO_4{}^{2-}$.C.7.

4500-$SO_4{}^{2-}$ E.   Turbidimetric Method

1. General Discussion

a. Principle: Sulfate ion ($SO_4{}^{2-}$) is precipitated in an acetic acid medium with barium chloride ($BaCl_2$) so as to form barium sulfate ($BaSO_4$) crystals of uniform size. Light absorbance of the $BaSO_4$ suspension is measured by a photometer and the $SO_4{}^{2-}$ concentration is determined by comparison of the reading with a standard curve.

b. Interference: Color or suspended matter in large amounts will interfere. Some suspended matter may be removed by filtration. If both are small in comparison with the $SO_4{}^{2-}$ concentration, correct for interference as indicated in ¶ *4d* below. Silica in excess of 500 mg/L will interfere, and in waters containing large quantities of organic material it may not be possible to precipitate $BaSO_4$ satisfactorily.

In potable waters there are no ions other than $SO_4{}^{2-}$ that will form insoluble compounds with barium under strongly acid conditions. Make determination at room temperature; variation over a range of 10°C will not cause appreciable error.

c. Minimum detectable concentration: Approximately 1 mg $SO_4{}^{2-}$/L.

2. Apparatus

a. Magnetic stirrer: Use a constant stirring speed. It is convenient to incorporate a fixed resistance in series with the motor operating the magnetic stirrer to regulate stirring speed. Use magnets of identical shape and size. The exact speed of stirring is not critical, but keep it constant for each run of samples and standards and adjust it to prevent splashing.

b. Photometer: One of the following is required, with preference in the order given:

1) *Nephelometer.*

2) *Spectrophotometer,* for use at 420 nm, providing a light path of 2.5 to 10 cm.

3) *Filter photometer,* equipped with a violet filter having maximum transmittance near 420 nm and providing a light path of 2.5 to 10 cm.

c. Stopwatch or electric timer.

d. Measuring spoon, capacity 0.2 to 0.3 mL.

3. Reagents

a. Buffer solution A: Dissolve 30 g magnesium chloride, $MgCl_2 \cdot 6H_2O$, 5 g sodium acetate, $CH_3COONa \cdot 3H_2O$, 1.0 g po-

tassium nitrate, KNO_3, and 20 mL acetic acid, CH_3COOH (99%), in 500 mL distilled water and make up to 1000 mL.

b. Buffer solution B (required when the sample $SO_4{}^{2-}$ concentration is less than 10 mg/L): Dissolve 30 g $MgCl_2 \cdot 6H_2O$, 5 g $CH_3COONa \cdot 3H_2O$, 1.0 g KNO_3, 0.111 g sodium sulfate, Na_2SO_4, and 20 mL acetic acid (99%) in 500 mL distilled water and make up to 1000 mL.

c. Barium chloride, $BaCl_2$, crystals, 20 to 30 mesh. In standardization, uniform turbidity is produced with this mesh range and the appropriate buffer.

d. Standard sulfate solution: Prepare a standard sulfate solution as described in 1) or 2) below; 1.00 mL = 100 μg $SO_4{}^{2-}$.

1) Dilute 10.4 mL standard 0.0200N H_2SO_4 titrant specified in Alkalinity, Section 2320B.3c, to 100 mL with distilled water.

2) Dissolve 0.1479 g anhydrous Na_2SO_4 in distilled water and dilute to 1000 mL.

4. Procedure

a. Formation of barium sulfate turbidity: Measure 100 mL sample, or a suitable portion made up to 100 mL, into a 250-mL erlenmeyer flask. Add 20 mL buffer solution and mix in stirring apparatus. While stirring, add a spoonful of $BaCl_2$ crystals and begin timing immediately. Stir for 60 $\pm$ 2 s at constant speed.

b. Measurement of barium sulfate turbidity: After stirring period has ended, pour solution into absorption cell of photometer and measure turbidity at 5 $\pm$ 0.5 min.

c. Preparation of calibration curve: Estimate $SO_4{}^{2-}$ concentration in sample by comparing turbidity reading with a calibration curve prepared by carrying $SO_4{}^{2-}$ standards through the entire procedure. Space standards at 5-mg/L increments in the 0- to 40-mg/L $SO_4{}^{2-}$ range. Above 40 mg/L accuracy decreases and $BaSO_4$ suspensions lose stability. Check reliability of calibration curve by running a standard with every three or four samples.

d. Correction for sample color and turbidity: Correct for sample color and turbidity by running blanks to which $BaCl_2$ is not added.

5. Calculation

$$\text{mg } SO_4{}^{2-}/L = \frac{\text{mg } SO_4{}^{2-} \times 1000}{\text{mL sample}}$$

If buffer solution A was used, determine $SO_4{}^{2-}$ concentration directly from the calibration curve after subtracting sample absorbance before adding $BaCl_2$. If buffer solution B was used subtract SO$_4^{2-}$ concentration of blank from apparent SO$_4^{2-}$ concentration as determined above; because the calibration curve is not a straight line, this is not equivalent to subtracting blank absorbance from sample absorbance.

6. Precision and Bias

With a turbidimeter,* in a single laboratory with a sample having a mean of 7.45 mg SO$_4^{2-}$/L, a standard deviation of 0.13

---

* Hach 2100 A.

mg/L and a coefficient of variation of 1.7% were obtained. Two samples dosed with sulfate gave recoveries of 85 and 91%.

7. Bibliography

SHEEN, R.T., H.L. KAHLER & E.M. ROSS. 1935. Turbidimetric determination of sulfate in water. *Ind. Eng. Chem.,* Anal. Ed. 7:262.

THOMAS, J.F. & J.E. COTTON. 1954. A turbidimetric sulfate determination. *Water Sewage Works* 101:462.

ROSSUM, J.R. & P. VILLARRUZ. 1961. Suggested methods for turbidimetric determination of sulfate in water. *J. Amer. Water Works Assoc.* 53: 873.

4500-SO$_4^{2-}$ F. Automated Methylthymol Blue Method

1. General Discussion

a. Principle: Barium sulfate is formed by the reaction of the SO$_4^{2-}$ with barium chloride (BaCl$_2$) at a low pH. At high pH excess barium reacts with methylthymol blue to produce a blue chelate. The uncomplexed methylthymol blue is gray. The amount of gray uncomplexed methylthymol blue indicates the concentration of SO$_4^{2-}$.

b. Interferences: Because many cations interfere, use an ion-exchange column to remove interferences.

Molybdenum, often used to treat cooling waters, has been shown to cause a strong positive bias with this method, even with as little as 1 mg Mo/L.

c. Application: This method is applicable to potable, ground, surface, and saline waters as well as domestic and industrial wastewaters over a range from about 10 to 300 mg SO$_4^{2-}$/L.

2. Apparatus

a. Automated analytical equipment: An example of the required continuous-flow analytical instrument consists of the interchangeable components shown in Figure 4500-SO$_4^{2-}$:1.

b. Ion-exchange column: Fill a piece of 2-mm-ID glass tubing about 20 cm long with the ion-exchange resin.* To simplify filling column put resin in distilled water and aspirate it into the tubing, which contains a glass-wool plug. After filling, plug other end of tube with glass wool. Avoid trapped air in the column.

3. Reagents

a. Barium chloride solution: Dissolve 1.526 g BaCl$_2$·2H$_2$O in 500 mL distilled water and dilute to 1 L. Store in a polyethylene bottle.

b. Methylthymol blue reagent: Dissolve 118.2 mg methylthymol blue† in 25 mL BaCl$_2$ solution. Add 4 mL 1*N* HCl and 71 mL distilled water and dilute to 500 mL with 95% ethanol. Store in a brown glass bottle. Prepare fresh daily.

c. Buffer solution, pH 10.1: Dissolve 6.75 g NH$_4$Cl in 500 mL distilled water. Add 57 mL conc NH$_4$OH and dilute to 1 L with

---

* Ion-exchange resin Bio-Rex 70, 20-50 mesh, sodium form, available from Bio-Rad Laboratories, Richmond, CA 94804, or equivalent.
† Eastman Organic Chemicals, Rochester, NY 14615. No. 8068 3′,3″ bis [N,N-bis(carboxymethyl)-aminolmethyl] thymolsulfonphthalein pentasodium salt.

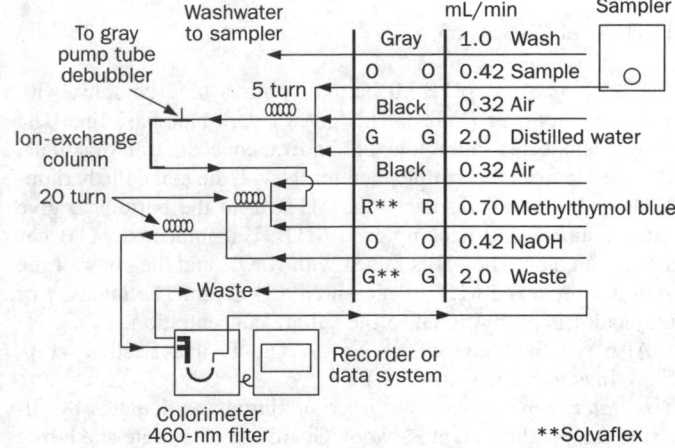

Figure 4500-SO$_4^{2-}$:1. Sulfate manifold.

distilled water. Adjust pH to 10.1 and store in a polyethylene bottle. Prepare fresh monthly.

d. EDTA reagent: Dissolve 40 g tetrasodium ethylenediaminetetraacetate in 500 mL pH 10.1 buffer solution. Dilute to 1 L with pH 10.1 buffer solution and store in a polyethylene bottle.

e. Sodium hydroxide solution, 0.36*N:* Dissolve 7.2 g NaOH in 250 mL distilled water. Cool and make up to 500 mL with distilled water.

f. Stock sulfate solution: Dissolve 1.479 g anhydrous Na$_2$SO$_4$ in 500 mL distilled water and dilute to 1000 mL; 1.00 mL = 1.00 mg SO$_4^{2-}$.

g. Standard sulfate solutions: Prepare in appropriate concentrations from 10 to 300 mg SO$_4^{2-}$/L, using the stock sulfate solution.

4. Procedure

Set up the manifold as shown in Figure 4500-SO$_4^{2-}$:1 and follow the general procedure described by the manufacturer.

After use, rinse methylthymol blue and NaOH reagent lines in water for a few minutes, rinse them in the EDTA solution for 10 min, and then rinse in distilled water.

5. Calculation

Prepare standard curves by plotting peak heights of standards processed through the manifold against SO_4^{2-} concentrations in standards. Compute sample SO_4^{2-} concentration by comparing sample peak height with standard curve.

6. Precision and Bias

In a single laboratory a sample with an average concentration of about 28 mg SO_4^{2-}/L had a standard deviation of 0.68 mg/L

and a coefficient of variation of 2.4%. In two samples with added SO_4^{2-}, recoveries were 91% and 100%.

7. Bibliography

LAZRUS, A.L., K.C. HILL & J.P. LODGE. 1965. A new colorimetric micro-determination of sulfate ion. *In* Automation in Analytical Chemistry. Technicon Symposium.

COLOROS, E., M. R. PANESAR & F.P. PERRY. 1976. Linearizing the calibration curve in determination of sulfate by the methylthymol blue method. *Anal. Chem.* 48:1693.

4500-SO_4^{2-} G.   Methylthymol Blue Flow Injection Analysis (PROPOSED)

1. General Discussion

a. Principle: At pH 13.0 barium forms a blue complex with methylthymol blue (MTB). This gives a dark blue base line. The sample is injected into a low, but known, concentration of sulfate. The sulfate from the sample then reacts with the ethanolic barium-MTB solution and displaces the MTB from the barium to give barium sulfate and uncomplexed MTB. Uncomplexed MTB has a grayish color. The pH is raised with NaOH and the color of the gray uncomplexed MTB is measured at 460 nm. The intensity of gray color is proportional to the sulfate concentration.

Also see Sections 4500-SO_4^{2-}.A and F, and Section 4130, Flow Injection Analysis (FIA).

b. Interferences: Remove large or fibrous particulates by filtering sample through glass wool. Guard against nitrate and nitrite contamination from reagents, water, glassware, and the sample preservation process.

A cation-exchange column removes multivalent cations such as Ca^{2+} and Mg^{2+}. A midrange sulfate standard containing a typical level of hardness as $CaCO_3$ can be run periodically to check the performance of the column. Any decrease in peak height from that of a sulfate standard without added $CaCO_3$ indicates the need to regenerate or replace the resin.

Neutralize samples that have pH less than 2. High acid concentrations can displace multivalent cations from the column.

Orthophosphate forms a precipitate with barium at high pH. If samples are known to be high in orthophosphate, make a recovery study using added amounts of sulfate, or run a sample blank containing only the orthophosphate matrix.

Also see Section 4500-SO_4^{2-}.F.1*b*.

2. Apparatus

Flow injection analysis equipment consisting of:

a. FIA injection valve with sample loop or equivalent.

b. Multichannel proportioning pump.

c. FIA manifold (Figure 4500-SO_4^{2-}:2) with cation-exchange column and flow cell. Relative flow rates only are shown in Figure 4500-SO_4^{2-}:2. Tubing volumes are given as an example only; they may be scaled down proportionally. Use manifold tubing of an inert material such as TFE.

d. Absorbance detector, 460 nm, 10-nm bandpass.

e. Injection valve control and data acquisition system.

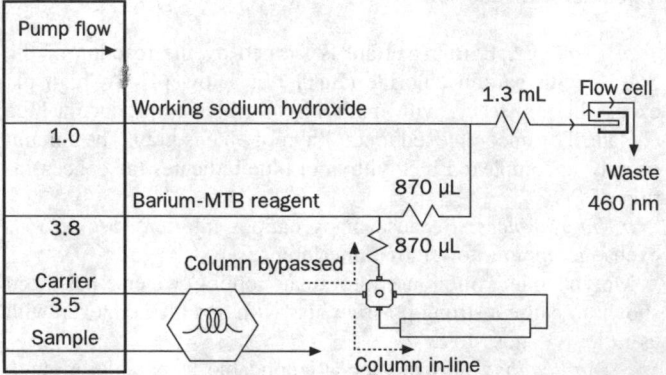

Figure 4500-SO_4^{2-}:2. FIA manifold.

3. Reagents

Use reagent water (>10 megohm) for all solutions. To prevent bubble formation, degas carrier and buffer with helium. Pass He at 140 kPa (20 psi) through a helium degassing tube. Bubble He through 1 L solution for 1 min. As an alternative to preparing reagents by weight/weight, use weight/volume.

a. Carrier solution, 0.30 mg SO_4^{2-}/L: To a tared 1-L container, add 0.30 g 1000 mg/L stock sulfate standard (¶ 3*h*) and 999.7 g water. Shake or stir to mix. Degas with helium.

b. Barium chloride solution, 6.24m*M*: To a tared 1-L container, add 1.526 g barium chloride dihydrate, $BaCl_2·2H_2O$, and 995 g water. Shake or stir until dissolved. Degas with helium.

c. Hydrochloric acid, HCl, 1.0*M*: To a tared 1-L container, add 913 g water and 99.6 g conc HCl (specific gravity 1.20, 37%). CAUTION: *Fumes.* Shake or stir to mix well. Degas with helium.

d. Barium—MTB color reagent: NOTE: The purity of the methylthymol blue and the denaturants in the alcohol are critical. Use the sources specified below, or test the material from alternative sources for suitability before using.

To a tared 500-mL dry brown plastic bottle, place 0.236 g methylthymol blue, 3,3'-bis[*N,N*-di(carboxymethyl)amino-methyl]-thymol-sulfonephthalein, pentasodium salt. Add 50 g barium chloride solution (¶ 3*b*), which may be used to aid in transfer of the dye. Swirl to dissolve. Add 4.0 g of 1.0*M* HCl (¶ 3*c*) and mix. The solution should turn orange. Add 71 g water and 321 g ethanol (ethyl alcohol, specially denatured anhydrous alcohol*). Stir or shake to mix well. The pH should be 2.5. Prepare solution the day before use and store, refrigerated, in a brown plastic bottle. Let warm to room temperature before using, then degas with helium.

e. Stock sodium hydroxide solution, NaOH, 50% (w/v) solution: To a glass 1-L container add 500 g water and 500.0 g NaOH. Dilute to 1 L. CAUTION: *The solution becomes very hot.* Shake or stir until dissolved. Cool to ambient. Store in a plastic bottle.

f. Working sodium hydroxide solution, NaOH, 0.18*M:* To a tared plastic 1-L container add 982 g water and 19.8 g stock NaOH solution (¶ 3*e*). Shake or stir to mix. Degas with helium.

g. Cation exchange column preparation: Prepare approximately 0.5 g ion exchange resin,† 50 to 100 mesh, by mixing with sufficient water to make a slurry. Remove one end fitting from the threaded glass column. Fill column with water and aspirate slurry into column or let it settle into column by gravity. Take care to avoid trapping air bubbles in column or fittings at this point and during all subsequent operations. When resin has settled, replace end fitting. To ensure a good seal, remove any resin particles from the threads of glass, column end, and end fitting. To store column, join ends of the TFE tubing.

To test column effectiveness, make up two midrange standards, one of only sodium sulfate and the other with an identical concentration of sodium sulfate but with hardness typical of the samples. If the column is depleted, the standard with hardness will give a lower response because the divalent Mg^{2+} and Ca^{2+} cations are complexing with the free MTB. If depletion has occurred, repack column with fresh resin.

h. Stock sulfate standard, 1000 mg SO$_4{}^{2-}$/L: Dry approximately 2 g sodium sulfate, Na$_2$SO$_4$, at 105° overnight. Cool in a desiccator. In a 1-L volumetric flask, add 1.479 g of dried sodium sulfate to about 800 mL water. Dissolve by swirling, dilute to mark, and mix by inversion.

i. Standard solutions: Prepare sulfate standards in desired concentration range, using the stock standard (¶ 3*h*), and diluting with water.

4. Procedure

Set up a manifold equivalent to that in Figure 4500-SO$_4{}^{2-}$:2 and follow method supplied by manufacturer or laboratory standard operating procedure. Follow quality control protocols outlined in Section 4020.

5. Calculations

Prepare standard curves by plotting adsorbance of standards processed through the manifold versus sulfate concentration.

6. Precision and Bias

a. Recovery and relative standard deviation: Table 4500-SO$_4{}^{2-}$:I gives results of single-laboratory studies.

b. MDL: A 180-μL sample loop was used in the method described above. Using a published MDL method,[1] analysts ran 21

* Aldrich 24, 511-9, or equivalent.
† BioRex 70 or equivalent.

TABLE 4500-SO$_4{}^{2-}$:I. RESULTS OF SINGLE-LABORATORY STUDIES WITH SELECTED MATRICES

Matrix	Sample/Blank Designation	Known Addition mg SO$_4{}^{2-}$/L	Recovery %	Relative Standard Deviation %
Wastewater treatment plant influent	Reference sample*	—	99	—
	Blank†	10.0	99	—
		20.0	99	—
	Site A‡	0	—	0.7
		10.0	109	—
		20.0	110	—
	Site B‡	0	—	0.7
		10.0	106	—
		20.0	112	—
	Site C‡	0	—	1.9
		10.0	104	—
		20.0	107	—
Wastewater treatment plant effluent	Reference sample*	—	99	—
	Blank†	10.0	95	—
		20.0	99	—
	Site A‡	0	—	0.9
		10.0	108	—
		20.0	108	—
	Site B‡	0	—	2.4
		10.0	107	—
		20.0	107	—
	Site C‡	0	—	0.6
		10.0	97	—
		20.0	104	—
Landfill leachate	Reference sample*	—	100	—
	Blank†	10.0	100	—
		20.0	99	—
	Site A‡	0	—	0.7
		10.0	106	—
		20.0	110	—
	Site B‡	0	—	0.5
		10.0	106	—
		20.0	107	—
	Site C‡	0	—	0.9
		10.0	101	—
		20.0	103	—

* U.S. EPA QC sample, 20.0 mg SO$_4{}^{2-}$/L.
† Determined in duplicate.
‡ Samples without known additions determined four times; samples with known additions determined in duplicate. Typical relative difference between duplicates 1%. Sample dilutions: Influent and effluent, all sites - 5-fold; leachate A - 100-fold; B - 50-fold; C - 10-fold.

replicates of a 5.00-mg SO$_4{}^{2-}$/L standard. These gave a mean of 4.80 mg SO$_4{}^{2-}$/L, a standard deviation of 0.69 mg SO$_4{}^{2-}$/L, and MDL of 1.8 mg SO$_4{}^{2-}$/L.

7. Reference

1. U.S. ENVIRONMENTAL PROTECTION AGENCY. 1984. Definition and procedure for the determination of method detection limits. Appendix B to 40 CFR 136 rev. 1.11 amended June 30, 1986. 49 CFR 43430.

PART 5000

AGGREGATE ORGANIC CONSTITUENTS

5010 INTRODUCTION

5010 A. General Discussion

Analyses for organic matter in water and wastewater can be classified into two general types of measurements: those that quantify an aggregate amount of organic matter comprising organic constituents with a common characteristic and those that quantify individual organic compounds. The latter can be found in Part 6000. The former, described here in Part 5000, have been grouped into four categories: oxygen-demanding substances, organically bound elements, classes of compounds, and formation potentials.

Methods for total organic carbon and chemical oxygen demand are used to assess the total amount of organics present. Gross fractions of the organic matter can be identified analytically, as in the measurement of BOD, which is an index of the biodegradable organics present, oil and grease, which represents material extractable from a sample by a nonpolar solvent, or dissolved organic halide (DOX), which measures organically bound halogens. Trihalomethane formation potential is an aggregate measure of the total concentration of trihalomethanes formed upon chlorination of a water sample.

Analyses of organics are made to assess the concentration and general composition of organic matter in raw water supplies, wastewaters, treated effluents, and receiving waters; and to determine the efficiency of treatment processes.

5010 B. Sample Collection and Preservation

The sampling, field treatment, preservation, and storage of samples taken for organic matter analysis are covered in detail in the individual introductions to the methods. If possible, analyze samples immediately because preservatives often interfere with the tests. Otherwise, store at a low temperature (4°C) immediately after collection to preserve most samples. Use chemical preservatives only when they are shown not to interfere with the examinations to be made (see Section 1060). Never use preservatives for samples to be analyzed for BOD. When preservatives are used, add them to the sample bottle initially so that all portions are preserved as soon as collected. No single method of preservation is entirely satisfactory; choose the preservative with due regard to the determinations that are to be made. All methods of preservation may be inadequate when applied to samples containing significant amounts of suspended matter.

5020 QUALITY ASSURANCE/QUALITY CONTROL

Part 1000 contains important information relevant to analyses included in Part 5000. Give particular attention to Sections 1020B (Quality Control), 1060 (Collection and Preservation of Samples), 1080 (Reagent Water), and 1090 (Laboratory Occupational Health and Safety), all of which are critical for many of the Part 5000 methods.

Take special precautions when analyses are performed by independent laboratories. Reliable use of independent laboratories deserves the same quality assurance procedures observed for in-house analyses: replicate samples, samples with known additions, and blanks.

Preparation of samples with known additions may not be feasible for certain analyses. In such cases, consider using a mixture, in varying ratios, of several samples. Use the reported concentrations in the samples and the proportions in which they were mixed to calculate the expected concentration in the mixture. Examine laboratory performance using externally prepared standards and check samples (see Section 1020B).

Reagent water (Section 1080) should give satisfactory results for most of the analyses in Part 5000, but additional purification steps may be needed for certain methods, such as dissolved organic halogen (DOX) and disinfection by-product formation potential (DBPFP).

5210   BIOCHEMICAL OXYGEN DEMAND (BOD)*

5210 A.   Introduction

1. General Discussion

The biochemical oxygen demand (BOD) determination is an empirical test in which standardized laboratory procedures are used to determine the relative oxygen requirements of wastewaters, effluents, and polluted waters. The test has its widest application in measuring waste loadings to treatment plants and in evaluating the BOD-removal efficiency of such treatment systems. The test measures the molecular oxygen utilized during a specified incubation period for the biochemical degradation of organic material (carbonaceous demand) and the oxygen used to oxidize inorganic material such as sulfides and ferrous iron. It also may measure the amount of oxygen used to oxidize reduced forms of nitrogen (nitrogenous demand) unless their oxidation is prevented by an inhibitor. The seeding and dilution procedures provide an estimate of the BOD at pH 6.5 to 7.5.

Measurements of oxygen consumed in a 5-d test period (5-d BOD or BOD_5, 5210B), oxygen consumed after 60 to 90 d of incubation (ultimate BOD or UBOD, 5210C), and continuous oxygen uptake (respirometric method, 5210D) are described here. Many other variations of oxygen demand measurements exist, including using shorter and longer incubation periods and tests to determine rates of oxygen uptake. Alternative seeding, dilution, and incubation conditions can be chosen to mimic receiving-water conditions, thereby providing an estimate of the environmental effects of wastewaters and effluents.

The UBOD measures the oxygen required for the total degradation of organic material (ultimate carbonaceous demand) and/or the oxygen to oxidize reduced nitrogen compounds (ultimate nitrogenous demand). UBOD values and appropriate kinetic descriptions are needed in water quality modeling studies such as UBOD: BOD_5 ratios for relating stream assimilative capacity to regulatory requirements; definition of river, estuary, or lake deoxygenation kinetics; and instream ultimate carbonaceous BOD (UCBOD) values for model calibration.

2. Carbonaceous Versus Nitrogenous BOD

A number of factors, for example, soluble versus particulate organics, settleable and floatable solids, oxidation of reduced iron and sulfur compounds, or lack of mixing may affect the accuracy and precision of BOD measurements. Presently, there is no way to include adjustments or corrections to account for the effect of these factors.

Oxidation of reduced forms of nitrogen, such as ammonia and organic nitrogen, can be mediated by microorganisms and exert nitrogenous demand. Nitrogenous demand historically has been considered an interference in the determination of BOD, as clearly evidenced by the inclusion of ammonia in the dilution water. The interference from nitrogenous demand can now be prevented by an inhibitory chemical.[1] If an inhibiting chemical is not used, the oxygen demand measured is the sum of carbonaceous and nitrogenous demands.

Measurements that include nitrogenous demand generally are not useful for assessing the oxygen demand associated with organic material. Nitrogenous demand can be estimated directly from ammonia nitrogen (Section 4500-NH_3); and carbonaceous demand can be estimated by subtracting the theoretical equivalent of the reduced nitrogen oxidation from uninhibited test results. However, this method is cumbersome and is subject to considerable error. Chemical inhibition of nitrogenous demand provides a more direct and more reliable measure of carbonaceous demand.

The extent of oxidation of nitrogenous compounds during the 5-d incubation period depends on the concentration and type of microorganisms capable of carrying out this oxidation. Such organisms usually are not present in raw or settled primary sewage in sufficient numbers to oxidize sufficient quantities of reduced nitrogen forms in the 5-d BOD test. Many biological treatment plant effluents contain sufficient numbers of nitrifying organisms to cause nitrification in BOD tests. Because oxidation of nitrogenous compounds can occur in such samples, inhibition of nitrification as directed in 5210B.4e6) is recommended for samples of secondary effluent, for samples seeded with secondary effluent, and for samples of polluted waters.

Report results as carbonaceous biochemical oxygen demand ($CBOD_5$) when inhibiting the nitrogenous oxygen demand. When nitrification is not inhibited, report results as BOD_5.

3. Dilution Requirements

The BOD concentration in most wastewaters exceeds the concentration of dissolved oxygen (DO) available in an air-saturated sample. Therefore, it is necessary to dilute the sample before incubation to bring the oxygen demand and supply into appropriate balance. Because bacterial growth requires nutrients such as nitrogen, phosphorus, and trace metals, these are added to the dilution water, which is buffered to ensure that the pH of the incubated sample remains in a range suitable for bacterial growth. Complete stabilization of a sample may require a period of incubation too long for practical purposes; therefore, 5 d has been accepted as the standard incubation period.

If the dilution water is of poor quality, the BOD of the dilution water will appear as sample BOD. This effect will be amplified by the dilution factor. A positive bias will result. The methods included below (5210B and 5210C) contain both a dilution-water check and a dilution-water blank. Seeded dilution waters are checked further for acceptable quality by measuring their consumption of oxygen from a known organic mixture, usually glucose and glutamic acid.

The source of dilution water is not restricted and may be distilled, tap, or receiving-stream water free of biodegradable organics and bioinhibitory substances such as chlorine or heavy metals. Distilled water may contain ammonia or volatile organics; deionized waters often are contaminated with soluble organics leached from the resin bed. Use of copper-lined stills or copper fittings

---

* Approved by Standard Methods Committee, 1997.

attached to distilled water lines may produce water containing excessive amounts of copper (see Section 3500-Cu).

4. Reference

1. YOUNG, J.C. 1973. Chemical methods for nitrification control. *J. Water Pollut. Control Fed.* 45:637.

5. Bibliography

THERIAULT, E.J., P.D. McNAMEE & C.T. BUTTERFIELD. 1931. Selection of dilution water for use in oxygen demand tests. *Pub. Health Rep.* 46:1084.

LEA, W.L. & M.S. NICHOLS. 1937. Influence of phosphorus and nitrogen on biochemical oxygen demand. *Sewage Works J.* 9:34.

RUCHHOFT, C.C. 1941. Report on the cooperative study of dilution waters made for the Standard Methods Committee of the Federation of Sewage Works Associations. *Sewage Works J.* 13:669.

MOHLMAN, F.W., E. HURWITZ, G.R. BARNETT & H.K. RAMER. 1950. Experience with modified methods for BOD. *Sewage Ind. Wastes* 22:31.

5210 B.   5-Day BOD Test

1. General Discussion

a. Principle: The method consists of filling with sample, to overflowing, an airtight bottle of the specified size and incubating it at the specified temperature for 5 d. Dissolved oxygen is measured initially and after incubation, and the BOD is computed from the difference between initial and final DO. Because the initial DO is determined shortly after the dilution is made, all oxygen uptake occurring after this measurement is included in the BOD measurement.

b. Sampling and storage: Samples for BOD analysis may degrade significantly during storage between collection and analysis, resulting in low BOD values. Minimize reduction of BOD by analyzing sample promptly or by cooling it to near-freezing temperature during storage. However, even at low temperature, keep holding time to a minimum. Warm chilled samples to $20 \pm 3°C$ before analysis.

1) Grab samples—If analysis is begun within 2 h of collection, cold storage is unnecessary. If analysis is not started within 2 h of sample collection, keep sample at or below 4°C from the time of collection. Begin analysis within 6 h of collection; when this is not possible because the sampling site is distant from the laboratory, store at or below 4°C and report length and temperature of storage with the results. In no case start analysis more than 24 h after grab sample collection. When samples are to be used for regulatory purposes make every effort to deliver samples for analysis within 6 h of collection.

2) Composite samples—Keep samples at or below 4°C during compositing. Limit compositing period to 24 h. Use the same criteria as for storage of grab samples, starting the measurement of holding time from end of compositing period. State storage time and conditions as part of the results.

2. Apparatus

a. Incubation bottles: Use glass bottles having 60 mL or greater capacity (300-mL bottles having a ground-glass stopper and a flared mouth are preferred). Clean bottles with a detergent, rinse thoroughly, and drain before use. As a precaution against drawing air into the dilution bottle during incubation, use a water seal. Obtain satisfactory water seals by inverting bottles in a water bath or by adding water to the flared mouth of special BOD bottles. Place a paper or plastic cup or foil cap over flared mouth of bottle to reduce evaporation of the water seal during incubation.

b. Air incubator or water bath, thermostatically controlled at $20 \pm 1°C$. Exclude all light to prevent possibility of photosynthetic production of DO.

3. Reagents

Prepare reagents in advance but discard if there is any sign of precipitation or biological growth in the stock bottles. Commercial equivalents of these reagents are acceptable and different stock concentrations may be used if doses are adjusted proportionally.

a. Phosphate buffer solution: Dissolve 8.5 g KH_2PO_4, 21.75 g K_2HPO_4, 33.4 g $Na_2HPO_4·7H_2O$, and 1.7 g NH_4Cl in about 500 mL distilled water and dilute to 1 L. The pH should be 7.2 without further adjustment. Alternatively, dissolve 42.5 g KH_2PO_4 or 54.3 g K_2HPO_4 in about 700 mL distilled water. Adjust pH to 7.2 with 30% NaOH and dilute to 1 L.

b. Magnesium sulfate solution: Dissolve 22.5 g $MgSO_4·7H_2O$ in distilled water and dilute to 1 L.

c. Calcium chloride solution: Dissolve 27.5 g $CaCl_2$ in distilled water and dilute to 1 L.

d. Ferric chloride solution: Dissolve 0.25 g $FeCl_3·6H_2O$ in distilled water and dilute to 1 L.

e. Acid and alkali solutions, 1N, for neutralization of caustic or acidic waste samples.

1) Acid—Slowly and while stirring, add 28 mL conc sulfuric acid to distilled water. Dilute to 1 L.

2) Alkali—Dissolve 40 g sodium hydroxide in distilled water. Dilute to 1 L.

f. Sodium sulfite solution: Dissolve 1.575 g Na_2SO_3 in 1000 mL distilled water. This solution is not stable; prepare daily.

g. Nitrification inhibitor, 2-chloro-6-(trichloromethyl) pyridine.*

h. Glucose-glutamic acid solution: Dry reagent-grade glucose and reagent-grade glutamic acid at 103°C for 1 h. Add 150 mg glucose and 150 mg glutamic acid to distilled water and dilute to 1 L. Prepare fresh immediately before use.

---

* Nitrification Inhibitor, Formula 2533, Hach Co., Loveland, CO, or equivalent.

i. Ammonium chloride solution: Dissolve 1.15 g NH₄Cl in about 500 mL distilled water, adjust pH to 7.2 with NaOH solution, and dilute to 1 L. Solution contains 0.3 mg N/mL.

j. Dilution water: Use demineralized, distilled, tap, or natural water for making sample dilutions.

4. Procedure

a. Preparation of dilution water: Place desired volume of water (¶ 3*j*) in a suitable bottle and add 1 mL each of phosphate buffer, MgSO₄, CaCl₂, and FeCl₃ solutions/L of water. Seed dilution water, if desired, as described in ¶ 4*d*. Test dilution water as described in ¶ 4*h* so that water of assured quality always is on hand.

Before use bring dilution water temperature to 20 ± 3°C. Saturate with DO by shaking in a partially filled bottle or by aerating with organic-free filtered air. Alternatively, store in cotton-plugged bottles long enough for water to become saturated with DO. Protect water quality by using clean glassware, tubing, and bottles.

b. Dilution water storage: Source water (¶ 3*j*) may be stored before use as long as the prepared dilution water meets quality control criteria in the dilution water blank (¶ 4*h*). Such storage may improve the quality of some source waters but may allow biological growth to cause deterioration in others. Preferably do not store prepared dilution water for more than 24 h after adding nutrients, minerals, and buffer unless dilution water blanks consistently meet quality control limits. Discard stored source water if dilution water blank shows more than 0.2 mg/L DO depletion in 5 d.

c. Glucose-glutamic acid check: Because the BOD test is a bioassay its results can be influenced greatly by the presence of toxicants or by use of a poor seeding material. Distilled waters frequently are contaminated with copper; some sewage seeds are relatively inactive. Low results always are obtained with such seeds and waters. Periodically check dilution water quality, seed effectiveness, and analytical technique by making BOD measurements on a mixture of 150 mg glucose/L and 150 mg glutamic acid/L as a ''standard'' check solution. Glucose has an exceptionally high and variable oxidation rate but when it is used with glutamic acid, the oxidation rate is stabilized and is similar to that obtained with many municipal wastes. Alternatively, if a particular wastewater contains an identifiable major constituent that contributes to the BOD, use this compound in place of the glucose-glutamic acid.

Determine the 5-d 20°C BOD of a 2% dilution of the glucose-glutamic acid standard check solution using the techniques outlined in ¶s 4*d-j*. Adjust concentrations of commercial mixtures to give 3 mg/L glucose and 3 mg/L glutamic acid in each GGA test bottle. Evaluate data as described in ¶ 6, Precision and Bias.

d. Seeding:

1) *Seed source*—It is necessary to have present a population of microorganisms capable of oxidizing the biodegradable organic matter in the sample. Domestic wastewater, unchlorinated or otherwise-undisinfected effluents from biological waste treatment plants, and surface waters receiving wastewater discharges contain satisfactory microbial populations. Some samples do not contain a sufficient microbial population (for example, some untreated industrial wastes, disinfected wastes, high-temperature wastes, or wastes with extreme pH values). For such wastes seed the dilution water or sample by adding a population of microor-

ganisms. The preferred seed is effluent or mixed liquor from a biological treatment system processing the waste. Where such seed is not available, use supernatant from domestic wastewater after settling at room temperature for at least 1 h but no longer than 36 h. When effluent or mixed liquor from a biological treatment process is used, inhibition of nitrification is recommended.

Some samples may contain materials not degraded at normal rates by the microorganisms in settled domestic wastewater. Seed such samples with an adapted microbial population obtained from the undisinfected effluent or mixed liquor of a biological process treating the waste. In the absence of such a facility, obtain seed from the receiving water below (preferably 3 to 8 km) the point of discharge. When such seed sources also are not available, develop an adapted seed in the laboratory by continuously aerating a sample of settled domestic wastewater and adding small daily increments of waste. Optionally use a soil suspension or activated sludge, or a commercial seed preparation to obtain the initial microbial population. Determine the existence of a satisfactory population by testing the performance of the seed in BOD tests on the sample. BOD values that increase with time of adaptation to a steady high value indicate successful seed adaptation.

2) *Seed control*—Determine BOD of the seeding material as for any other sample. This is the *seed control.* From the value of the seed control and a knowledge of the seeding material dilution (in the dilution water) determine seed DO uptake. Ideally, make dilutions of seed such that the largest quantity results in at least 50% DO depletion. A plot of DO depletion, in milligrams per liter, versus milliters of seed for all bottles having a 2-mg/L depletion and a 1.0-mg/L minimum residual DO should present a straight line for which the slope indicates DO depletion per milliliter of seed. The DO-axis intercept is oxygen depletion caused by the dilution water and should be less than 0.1 mg/L (¶ 4*h*). Alternatively, divide DO depletion by volume of seed in milliliters for each seed control bottle having a 2-mg/L depletion and a 1.0-mg/L residual DO. Average the results for all bottles meeting minimum depletion and residual DO criteria. The DO uptake attributable to the seed added to each bottle should be between 0.6 and 1.0 mg/L, but the amount of seed added should be adjusted from this range to that required to provide glucose-glutamic acid check results in the range of 198 ± 30.5 mg/L. To determine DO uptake for a test bottle, subtract DO uptake attributable to the seed from total DO uptake (see ¶ 5).

Techniques for adding seeding material to dilution water are described for two sample dilution methods (¶ 4*f*).

e. Sample pretreatment: Check pH of all samples before testing unless previous experience indicates that pH is within the acceptable range.

1) *Samples containing caustic alkalinity (pH >8.5) or acidity (pH <6.0)*—Neutralize samples to pH 6.5 to 7.5 with a solution of sulfuric acid (H₂SO₄) or sodium hydroxide (NaOH) of such strength that the quantity of reagent does not dilute the sample by more than 0.5%. The pH of dilution water should not be affected by the lowest sample dilution. Always seed samples that have been pH-adjusted.

2) *Samples containing residual chlorine compounds*—If possible, avoid samples containing residual chlorine by sampling ahead of chlorination processes. If the sample has been chlorinated but no detectable chlorine residual is present, seed the dilution water. If residual chlorine is present, dechlorinate sample and seed the dilution water (¶ 4*f*). Do not test chlorinated/dechlorinated samples without seeding the dilution water. In some

samples chlorine will dissipate within 1 to 2 h of standing in the light. This often occurs during sample transport and handling. For samples in which chlorine residual does not dissipate in a reasonably short time, destroy chlorine residual by adding Na_2SO_3 solution. Determine required volume of Na_2SO_3 solution on a 100- to 1000-mL portion of neutralized sample by adding 10 mL of 1 + 1 acetic acid or 1 + 50 H_2SO_4, 10 mL potassium iodide (KI) solution (10 g/100 mL) per 1000 mL portion, and titrating with Na_2SO_3 solution to the starch-iodine end point for residual. Add to neutralized sample the relative volume of Na_2SO_3 solution determined by the above test, mix, and after 10 to 20 min check sample for residual chlorine. (NOTE: Excess Na_2SO_3 exerts an oxygen demand and reacts slowly with certain organic chloramine compounds that may be present in chlorinated samples.)

3) *Samples containing other toxic substances*—Certain industrial wastes, for example, plating wastes, contain toxic metals. Such samples often require special study and treatment.

4) *Samples supersaturated with DO*—Samples containing more than 9 mg DO/ L at 20°C may be encountered in cold waters or in water where photosynthesis occurs. To prevent loss of oxygen during incubation of such samples, reduce DO to saturation at 20°C by bringing sample to about 20°C in partially filled bottle while agitating by vigorous shaking or by aerating with clean, filtered compressed air.

5) *Sample temperature adjustment*—Bring samples to 20 ± 1°C before making dilutions.

6) *Nitrification inhibition*—If nitrification inhibition is desired add 3 mg 2-chloro-6-(trichloro methyl) pyridine (TCMP) to each 300-mL bottle before capping or add sufficient amounts to the dilution water to make a final concentration of 10 mg/L. (NOTE: Pure TCMP may dissolve slowly and can float on top of the sample. Some commercial formulations dissolve more readily but are not 100% TCMP; adjust dosage accordingly.) Samples that may require nitrification inhibition include, but are not limited to, biologically treated effluents, samples seeded with biologically treated effluents, and river waters. Note the use of nitrogen inhibition in reporting results.

f. Dilution technique: Make several dilutions of sample that will result in a residual DO of at least 1 mg/L and a DO uptake of at least 2 mg/L after a 5-d incubation. Five dilutions are recommended unless experience with a particular sample shows that use of a smaller number of dilutions produces at least two bottles giving acceptable minimum DO depletion and residual limits. A more rapid analysis, such as COD, may be correlated approximately with BOD and serve as a guide in selecting dilutions. In the absence of prior knowledge, use the following dilutions: 0.0 to 1.0% for strong industrial wastes, 1 to 5% for raw and settled wastewater, 5 to 25% for biologically treated effluent, and 25 to 100% for polluted river waters.

Prepare dilutions either in graduated cylinders or volumetric glassware, and then transfer to BOD bottles or prepare directly in BOD bottles. Either dilution method can be combined with any DO measurement technique. The number of bottles to be prepared for each dilution depends on the DO technique and the number of replicates desired.

When using graduated cylinders or volumetric flasks to prepare dilutions, and when seeding is necessary, add seed either directly to dilution water or to individual cylinders or flasks before dilution. Seeding of individual cylinders or flasks avoids a declining ratio of seed to sample as increasing dilutions are made. When dilutions are prepared directly in BOD bottles and when seeding

is necessary, add seed directly to dilution water or directly to the BOD bottles. When a bottle contains more than 67% of the sample after dilution, nutrients may be limited in the diluted sample and subsequently reduce biological activity. In such samples, add the nutrient, mineral, and buffer solutions (¶ *3a* through *e*) directly to individual BOD bottles at a rate of 1 mL/L (0.33 mL/300-mL bottle) or use commercially prepared solutions designed to dose the appropriate bottle size.

1) *Dilutions prepared in graduated cylinders or volumetric flasks*—If the azide modification of the titrimetric iodometric method (Section 4500-O.C) is used, carefully siphon dilution water, seeded if necessary, into a 1- to 2-L-capacity flask or cylinder. Fill half full without entraining air. Add desired quantity of carefully mixed sample and dilute to appropriate level with dilution water. Mix well with a plunger-type mixing rod; avoid entraining air. Siphon mixed dilution into two BOD bottles. Determine initial DO on one of these bottles. Stopper the second bottle tightly, water-seal, and incubate for 5 d at 20°C. If the membrane electrode method is used for DO measurement, siphon dilution mixture into one BOD bottle. Determine initial DO on this bottle and replace any displaced contents with sample dilution to fill the bottle. Stopper tightly, water-seal, and incubate for 5 d at 20°C.

2) *Dilutions prepared directly in BOD bottles*—Using a wide-tip volumetric pipet, add the desired sample volume to individual BOD bottles of known capacity. Add appropriate amounts of seed material either to the individual BOD bottles or to the dilution water. Fill bottles with enough dilution water, seeded if necessary, so that insertion of stopper will displace all air, leaving no bubbles. For dilutions greater than 1:100 make a primary dilution in a graduated cylinder before making final dilution in the bottle. When using titrimetric iodometric methods for DO measurement, prepare two bottles at each dilution. Determine initial DO on one bottle. Stopper second bottle tightly, water-seal, and incubate for 5 d at 20°C. If the membrane electrode method is used for DO measurement, prepare only one BOD bottle for each dilution. Determine initial DO on this bottle and replace any displaced contents with dilution water to fill the bottle. Stopper tightly, water-seal, and incubate for 5 d at 20°C. Rinse DO electrode between determinations to prevent cross-contamination of samples.

Use the azide modification of the iodometric method (Section 4500-O.C) or the membrane electrode method (Section 4500-O.G) to determine initial DO on all sample dilutions, dilution water blanks, and where appropriate, seed controls.

If the membrane electrode method is used, the azide modification of the iodometric method (Method 4500-O.C) is recommended for calibrating the DO probe.

g. Determination of initial DO: If the sample contains materials that react rapidly with DO, determine initial DO immediately after filling BOD bottle with diluted sample. If rapid initial DO uptake is insignificant, the time period between preparing dilution and measuring initial DO is not critical but should not exceed 30 min.

h. Dilution water blank: Use a dilution water blank as a rough check on quality of unseeded dilution water and cleanliness of incubation bottles. Together with each batch of samples incubate a bottle of unseeded dilution water. Determine initial and final DO as in ¶s *4g* and *j*. The DO uptake should not be more than 0.2 mg/L and preferably not more than 0.1 mg/L Discard all dilution water having a DO uptake greater than 0.2 mg/L and either eliminate source of contamination or select an alternate dilution water source..

i. Incubation: Incubate at 20°C ± 1°C BOD bottles containing desired dilutions, seed controls, dilution water blanks, and glucose-glutamic acid checks. Water-seal bottles as described in ¶ 4*f*.

j. Determination of final DO: After 5 d incubation determine DO in sample dilutions, blanks, and checks as in ¶ 4*g*.

5. Calculation

For each test bottle meeting the 2.0-mg/L minimum DO depletion and the 1.0-mg/L residual DO, calculate BOD_5 as follows:

When dilution water is not seeded:

$$BOD_5, mg/L = \frac{D_1 - D_2}{P}$$

When dilution water is seeded:

$$BOD_5, mg/L = \frac{(D_1 - D_2) - (B_1 - B_2)f}{P}$$

where:

   D_1 = DO of diluted sample immediately after preparation, mg/L,
   D_2 = DO of diluted sample after 5 d incubation at 20°C, mg/L,
   P = decimal volumetric fraction of sample used,
   B_1 = DO of seed control before incubation, mg/L (¶ 4*d*),
   B_2 = DO of seed control after incubation mg/L (¶ 4*d*), and
   f = ratio of seed in diluted sample to seed in seed control = (% seed in diluted sample)/(% seed in seed control).

If seed material is added directly to sample or to seed control bottles:

f = (volume of seed in diluted sample)/(volume of seed in seed control)

Report results as $CBOD_5$ if nitrification is inhibited.

If more than one sample dilution meets the criteria of a residual DO of at least 1 mg/L and a DO depletion of at least 2 mg/L and there is no evidence of toxicity at higher sample concentrations or the existence of an obvious anomaly, average results in the acceptable range.

In these calculations, do not make corrections for DO uptake by the dilution water blank during incubation. This correction is unnecessary if dilution water meets the blank criteria stipulated above. If the dilution water does not meet these criteria, proper corrections are difficult ; do not record results or, as a minimum, mark them as not meeting quality control criteria.

6. Precision and Bias

There is no measurement for establishing bias of the BOD procedure. The glucose-glutamic acid check prescribed in ¶ 4*c* is intended to be a reference point for evaluation of dilution water quality, seed effectiveness, and analytical technique. Single-laboratory tests using a 300-mg/L mixed glucose-glutamic acid solution provided the following results:

Number of months: 14
Number of triplicates: 421
Average monthly recovery: 204 mg/L
Average monthly standard deviation: 10.4 mg/L

In a series of interlaboratory studies,[1] each involving 2 to 112 laboratories (and as many analysts and seed sources), 5-d BOD measurements were made on synthetic water samples containing a 1:1 mixture of glucose and glutamic acid in the total concentration range of 3.3 to 231 mg/L. The regression equations for mean value, $\overline{X}$, and standard deviation, S, from these studies were:

$$\overline{X} = 0.658 \text{ (added level, mg/L)} + 0.280 \text{ mg/L}$$
$$S = 0.100 \text{ (added level, mg/L)} + 0.547 \text{ mg/L}$$

For the 300-mg/L mixed primary standard, the average 5-d BOD would be 198 mg/L with a standard deviation of 30.5 mg/L. When nitrification inhibitors are used, GGA test results falling outside the 198 ± 30.5 control limit quite often indicate use of incorrect amounts of seed. Adjust amount of seed added to the GGA test to achieve results falling within this range.

a. Control limits: Because of many factors affecting BOD tests in multilaboratory studies and the resulting extreme variability in test results, one standard deviation, as determined by interlaboratory tests, is recommended as a control limit for individual laboratories. Alternatively, for each laboratory, establish its control limits by performing a minimum of 25 glucose-glutamic acid checks (¶ 4*c*) over a period of several weeks or months and calculating the mean and standard deviation. Use the mean ± 3 standard deviations as the control limit for future glucose-glutamic acid checks. Compare calculated control limits to the single-laboratory tests presented above and to interlaboratory results. If control limits are outside the range of 198 ± 30.5, re-evaluate the control limits and investigate source of the problem. If measured BOD for a glucose-glutamic acid check is outside the accepted control limit range, reject tests made with that seed and dilution water.

b. Working range and detection limit: The working range is equal to the difference between the maximum initial DO (7 to 9 mg/L) and minimum DO residual of 1 mg/L multiplied by the dilution factor. A lower detection limit of 2 mg/L is established by the requirement for a minimum DO depletion of 2 mg/L.

7. Reference

1. U.S. ENVIRONMENTAL PROTECTION AGENCY , OFFICE OF RESEARCH AND DEVELOPMENT. 1986. Method-by-Method Statistics from Water Pollution (WP) Laboratory Performance Evaluation Studies. Quality Assurance Branch, Environmental Monitoring and Support Lab., Cincinnati, Ohio.

8. Bibliography

YOUNG, J.C., G.N. MCDERMOTT & D. JENKINS . 1981. Alterations in the BOD procedure for the 15th edition of Standard Methods for the Examination of Water and Wastewater. *J. Water Pollut. Control Fed.* 53:1253.

5210 C. Ultimate BOD Test (PROPOSED)

1. General Discussion

The ultimate BOD test is an extension of the 5-d dilution BOD test as described in 5210B but with a number of specific test requirements and differences in application. The user should be familar with the 5210B procedure before conducting tests for UBOD.

a. Principle: The method consists of placing a single sample dilution in full, airtight bottles and incubating under specified conditions for an extended period depending on wastewater, effluent, river, or estuary quality.[1] Dissolved oxygen (DO) is measured (with probes) initially and intermittently during the test. From the DO versus time series, UBOD is calculated by an appropriate statistical technique. For improved accuracy, run tests in triplicate.

Bottle size and incubation time are flexible to accommodate individual sample characteristics and laboratory limitations. Incubation temperature, however, is 20°C. Most effluents and some naturally occurring surface waters contain materials with oxygen demands exceeding the DO available in air-saturated water. Therefore, it is necessary either to dilute the sample or to monitor DO frequently to ensure that low DO or anaerobic conditions do not occur. When DO concentrations approach 2 mg/L, the sample should be reaerated.

Because bacterial growth requires nutrients such as nitrogen, phosphorus, and trace metals, the necessary amounts may be added to the dilution water together with buffer to ensure that pH remains in a range suitable for bacterial growth and seed to provide an adequate bacterial population. However, if the result is being used to estimate the rate of oxidation of naturally occurring surface waters, addition of nutrients and seed probably accelerates the decay rate and produces misleading results. If only UBOD is desired, it may be advantageous to add supplemental nutrients that accelerate decay and reduce the test duration. When nutrients are used, they also should be used in the dilution water blank. Because of the wide range of water and wastewater characteristics and varied applications of UBOD data, no specific nutrient or buffer formulations are included.

The extent of oxidation of nitrogenous compounds during the prescribed incubation period depends on the presence of microorganisms capable of carrying out this oxidation. Such organisms may not be present in wastewaters in sufficient numbers to oxidize significant quantities of reduced nitrogen. This situation may be reversed in naturally occurring surface waters. Erratic results may be obtained when a nitrification inhibitor is used;[2] therefore, the specified method precludes use of a nitrogen inhibitor unless prior experimental evidence on the particular sample suggests that it is acceptable.* Monitor NO_2^--N and NO_3^--N to compute the oxygen equivalency of the nitrification reaction. When these values are subtracted from the DO vs. time series, the carbonaceous BOD time series can be constructed.[3]

b. Sampling and storage: See Section 5210B.1*b.*

2. Apparatus

a. Incubation bottles: Glass bottles with ground-glass stoppers,† 2-L (or larger) capacity. Glass serum bottles of 4- to 10-L capacity are available. Alternatively use nonground-glass bottles with nonbiodegradable plastic caps as a plug insert. Do not reuse the plugs because discoloration occurs with continued use. Replace plugs every 7 to 14 d. Do not use rubber stoppers that may exert an oxygen demand. Clean bottles with a detergent and wash with dilute HCl (3*N*) to remove surface films and precipitated inorganic salts; rinse thoroughly with DI water before use. Cover top of bottles with paper after rinsing to prevent dust from collecting. To prevent drawing air into the sample bottle during incubation, use a water seal. If the bottle does not have a flared mouth, construct a water seal by making a watertight dam around the stopper (or plug) and fill with water from the reservoir as necessary. Cover dam with clean aluminum foil to retard evaporation. If a 2-L BOD bottle is used, fill reservoir with sample and cover with a polyethylene cap before incubation.

Place a clean magnetic stirring bar in each bottle to mix contents before making DO measurement or taking a subsample. Do not remove the magnets until the test is complete.

Alternatively use a series of 300-mL BOD bottles as described in 5210B, if larger bottles are not available or incubation space is limited.

b. Reservoir bottle: 4-L or larger glass bottle. Close with screw plastic cap or non-rubber plug.

c. Incubator or water bath, thermostatically controlled at 20 ± 1°C. Exclude all light to prevent the possibility of photosynthetic production of DO.

d. Oxygen-sensitive membrane electrode: See Section 4500-O.G.2.

3. Procedure

a. River water samples: Preferably fill large BOD bottle (>2 L, or alternatively 6 or more 300-mL BOD bottles) with sample at 20°C. Add no nutrients, seed, or nitrification inhibitor if in-bottle decay rates will be used to estimate in-stream rates. Do not dilute sample unless it is known by pretesting or by experience to have a high ultimate BOD (>20 mg/L).

Measure DO in each bottle, stopper, and make an airtight seal. Incubate at 20°C in the dark.

Measure DO in each bottle at intervals of at least 2 to 5 d over a period of 30 to 60 d (minimum of 6 to 8 readings) or longer under special circumstances. To avoid oxygen depletion in samples containing NH_3-N, measure DO more frequently until nitrification has taken place. If DO falls to about 2 mg/L, reaerate as directed below. Replace sample lost by the cap and DO probe displacement by adding 1 to 2 mL sample from the reservoir bottle.

When DO approaches 2 mg/L, reaerate. Pour a small amount of sample into a clean vessel and reaerate the remainder directly in the bottle by vigorous shaking or bubbling with purified air (medical grade). Refill bottle from the storage reservoir and meas-

---

* Some analysts have reported satisfactory results with 2-chloro-6-(trichloromethyl) pyridine (Nitrification Inhibitor, Formula 2533, Hach Co., Loveland, CO, or equivalent).

† Wheaton 2-L BOD bottle No. 227580, 1000 North Tenth St., Millville, NJ, or equivalent.

TABLE 5210:I. UBOD RESULTS FOR WASTEWATER SAMPLE

Day	(1) Average DO* mg/L	(2) Average Blank DO† mg/L	(3) Accumulated DO Consumed by Sample‡ mg/L	(4) Average NO₃-N mg/L	(5) NBOD mg/L§	(6) CBOD mg/L‖
0	8.1	—	0	0.0	0	0
3	5.6	—	2.5	—	0	2.5
5	3.5/8.0	—	4.6	0.0	0	4.6
7	6.2	—	6.4	—	0.23	6.2
10	3.2/8.2	—	9.4	0.10	0.46	8.9
15	4.3	—	13.3	—	0.58	12.7
18	2.7/8.1	—	14.9	0.15	0.69	14.2
20	6.6	—	16.4	—	0.80	15.6
25	5.4	—	17.6	0.20	0.92	16.7
30	2.6/8.2	—	20.4	—	0.92	19.5
40	5.3	—	23.3	0.20	0.92	22.4
50	3.1/8.0	—	25.5	—	0.92	24.6
60	4.5	—	29.0	—	0.92	28.1
70	3.3/8.1	—	30.2	—	0.92	29.3
90	5.4	—	32.9	0.20	0.92	32.0

*Two readings indicate concentrations before and after reaeration.
†None was used.
‡Column (1)–blank correction (none needed in the example).
§Column (4) × 4.57 (linear interpolation between values).
‖[Column (3)–Column (5)] × dilution factor

Ultimate CBOD = 34.5 mg/L; CBOD decay rate = 0.03/d (calculated with first-order equation from 5210C.4).

ure DO. This concentration becomes the initial DO for the next measurement. If using 300-mL BOD bottles, pour all of the sample from the several bottles used into a clean vessel, reaerate, and refill the small bottles.

Analyze for nitrate plus nitrite nitrogen (NO_3^--N + NO_2^--N) (see Sections 4500-NO_2^- and 4500-NO_3^-) on Days 0, 5, 10, 15, 20, and 30. Alternatively, determine NO_2^--N and NO_3^--N each time DO is determined, thereby producing corresponding BOD and nitrogen determinations. If the ultimate demand occurs at a time greater than 30 d, make additional analyses at 30-d intervals. Remove 10 to 20 mL from the bottle for these analyses. Refill bottle as necessary from the reservoir bottle. Preserve NO_2^--N + NO_3^--N subsample with H_2SO_4 to pH <2 and refrigerate. If the purpose of the UBOD test is to assess the UBOD and not to provide data for rate calculations, measure nitrate nitrogen concentration only at Day 0 and on the last day of the test (kinetic rate estimates are not useful when the nitrification reaction is not followed).

Calculate oxygen consumption during each time interval and make appropriate corrections for nitrogenous oxygen demand. Correct by using 3.43 × the NH_3-N to NO_2^--N conversion plus 1.14 × the NO_2^--N to NO_3^--N conversion to reflect the stoichiometry of the oxidation of NH_4^+ to NO_2^- or NO_3^-.

When using a dilution water blank, subtract DO uptake of the blank from the total DO consumed. High-quality reagent water without nutrients typically will consume a maximum of 1 mg DO/L in a 30- to 90-d period. If DO uptake of the dilution water is greater than 0.5 mg/L for a 20-d period, or 1 mg/L for a 90-d period, report the magnitude of the correction and try to obtain higher-quality dilution water for use with subsequent UBOD tests.

When the weekly DO consumption drops below 1 to 2% of the total accumulative consumption, calculate the ultimate BOD using a nonlinear regression method.

b. Wastewater treatment plant samples: Use high-quality reagent water (see Section 1080) for dilution water. Add no nitrification inhibitors if decay rates are desired. If seed and nutrients are necessary, add the same amounts of each to the dilution water blank. Use minimal sample dilution. As a rule of thumb, the ultimate BOD of the diluted sample should be in the range of 20 to 30 mg/L. Dilution to this level probably will require two or three sample reaerations during the incubation period to avoid having dissolved oxygen concentrations fall below 2 mg/L.

Use 2-L or larger BOD bottles (alternatively, multiple 300-mL BOD bottles) for each dilution. Add desired volume of sample to each bottle and fill with dilution water.

Fill a BOD bottle with dilution water to serve as a dilution water blank. Treat blank the same as all samples. Follow procedure given in ¶ 3a) and incubate for at least as long as UBOD test.

4. Calculations

An example of results obtained for a wastewater sample, undiluted, without seed and nutrients, is given in Table 5210:I.

UBOD can be estimated by using a first-order model described as follows:

$$BOD_t = UBOD (1 - e^{-kt})$$

where:

BOD_t = oxygen uptake measured at time t, mg/L, and
k = first-order oxygen uptake rate.

The data in Table 5210:I were analyzed with a nonlinear regression technique applied to the above first-order model.[4] However, a first-order kinetic model may not always be the best choice. Significantly better statistical fits usually are obtained with alternative kinetic models including sum of two first-order and logistic function models.[1,3–8]

5. Precision and Bias

The precision of the ultimate BOD test was assessed with a series of replicate tests in a single laboratory. Interlaboratory studies have not been conducted.

Reference	Replicate No.	UBOD mg/L	Precision Summary*
2	1	154	μ = 151 mg/L
	2	154	
	3	145	CV = 3.5%
5	1	10.3	
	2	11.1	
	3	9.6	μ = 10.0 mg/L
	4	9.9	
	5	9.8	CV = 5.8%
	6	9.6	
6	1	12.8	μ = 12.4 mg/L
	2	12.6	
	3	12.6	CV = 4.4%
	4	11.6	

*μ = mean,
CV = coefficient of variation.

Bias was assessed by determining the BOD of a known concentration of glucose (150 mg/L) and glutamic acid (150 mg/L). This solution has a UBOD of 321 mg/L to 308 mg/L, depending on extent of nitrification. The results of the study conducted in triplicate were:[1]

Estimated* UBOD mg/L	Theoretical BOD mg/L	Percent Difference
276	308/321	−10/−14
310	308/321	+1/−3
303	308/321	−2/−6

*By statistical model.

6. References

1. MARTONE, C.H. 1976. Studies Related to the Determination of Biodegradability and Long Term BOD. M.S. thesis, Dep. Civil Engineering, Tufts Univ., Medford, Mass.
2. NATIONAL COUNCIL OF THE PAPER INDUSTRY FOR AIR AND STREAM IMPROVEMENT, INC. 1986. A Review of the Separation of Carbonaceous and Nitrogenous BOD in Long-Term BOD Measurements. Tech. Bull. No. 461, New York, N.Y.
3. NATIONAL COUNCIL OF THE PAPER INDUSTRY FOR AIR AND STREAM IMPROVEMENT, INC. 1982. A Review of Ultimate BOD and Its Kinetic Formulation for Pulp and Paper Mill Effluents. Tech. Bull. No. 382, New York, N.Y.
4. BARNWELL, T. 1980. Least Squares Estimates of BOD Parameters. *J. Environ. Eng. Div., Proc. Amer. Soc. Civil Eng.* 107(EE6):1197.
5. NATIONAL COUNCIL OF THE PAPER INDUSTRY FOR AIR AND STREAM IMPROVEMENT, INC. 1982. A Proposal to Examine the Effect of Mixing on Long Term BOD Test. NE82-01, New York, N.Y.
6. NATIONAL COUNCIL OF THE PAPER INDUSTRY FOR AIR AND STREAM IMPROVEMENT, INC. 1982. A Study of the Selection, Calibration, and Verification of Mathematical Water Quality Models. Tech. Bull. No. 367, New York, N.Y.
7. NATIONAL COUNCIL OF THE PAPER INDUSTRY FOR AIR AND STREAM IMPROVEMENT, INC. 1987. User's Manual for Parameter Estimation for First Order Ultimate BOD Decay, BODFO. Tech. Bull. No. 529, New York, N.Y.
8. CHU, W.S. & E.W. STRECKER. 1972. Parameter Identification In Water System Model. Dep. Civil Engineering, Univ. Washington, Seattle.

5210 D.   Respirometric Method (PROPOSED)

1. General Discussion

a. Principle: Respirometric methods provide direct measurement of the oxygen consumed by microorganisms from an air or oxygen-enriched environment in a closed vessel under conditions of constant temperature and agitation.

b. Uses: Respirometry measures oxygen uptake more or less continuously over time. Respirometric methods are useful for assessing: biodegradation of specific chemicals; treatability of organic industrial wastes; the effect of known amounts of toxic compounds on the oxygen-uptake reaction of a test wastewater or organic chemical; the concentration at which a pollutant or a wastewater measurably inhibits biological degradation; the effect of various treatments such as disinfection, nutrient addition, and pH adjustment on oxidation rates; the oxygen requirement for essentially complete oxidation of biologically oxidizable matter; the need for using adapted seed in other biochemical oxygen-uptake measurements, such as the dilution BOD test; and stability of sludges.

Respirometric data typically will be used comparatively, that is, in a direct comparison between oxygen uptakes from two test samples or from a test sample and a control. Because of inherent differences among uses, among seed cultures, among applications of results, and among instruments, a single procedure for respirometric tests applicable to all cases cannot be defined. Therefore, only basic recommendations and guidelines for overall test setup and procedure are given. Follow manufacturer's instructions for operating details for specific commercial instruments.

c. Types of respirometers: Four principal types of commercial respirometers are available. Manometric respirometers relate oxygen uptake to the change in pressure caused by oxygen consumption while maintaining a constant volume. Volumetric respirometers measure oxygen uptake in incremental changes in gas volume while maintaining a constant pressure at the time of reading. Electrolytic respirometers monitor the amount of oxygen produced by electrolysis of water to maintain a constant oxygen pressure within the reaction vessel. Direct-input respirometers deliver oxygen to the sample from a pure oxygen supply through metering on demand as detected by minute pressure differences. Most respirometers have been instrumented to permit data collection and processing by computer. Reaction-vessel contents are mixed by using a magnetic or mechanical stirring device or by bubbling the gaseous phase within the reaction vessel through the liquid phase. All respirometers remove carbon dioxide produced during biological growth by suspending a concentrated adsorbent (granular or solution) within the closed reaction chamber or by recirculating the gas phase through an external scrubber.

d. Interferences: Evolution of gases other than CO_2 may introduce errors in pressure or volume measurements; this is uncommon in the presence of dissolved oxygen. Incomplete CO_2 absorption will introduce errors if appropriate amounts and concentrations of alkaline absorbent are not used. Temperature fluctuations or inadequate mixing will introduce error. Fluctuations in barometric pressure can cause errors with some respirometers. Become familiar with the limits of the instrument to be used.

e. Minimum detectable concentration: Most commercial respirometers can detect oxygen demand in increments as small as 0.1 mg but test precision depends on the total amount of oxygen consumed at the time of reading, the precision of pressure or volume measurement, and the effect of temperature and barometric pressure changes. Upper limits of oxygen uptake rate are determined by the ability to transfer oxygen into the solution from the gas phase, which typically is related to mixing intensity. Transfer limits typically range from less than 10 mg O_2/L/h for low-intensity mixing to above 100 mg O_2/L/h for high-intensity mixing.

f. Relationship to dilution BOD: Variations in waste composition, substrate concentration, mixing, and oxygen concentrations from one wastewater source to another generally preclude use of a general relationship between oxygen uptake by respirometers and the 5-d, 20°C, BOD (see 5210B, above). Reasonably accurate correlations may be possible for a specific wastewater. The incubation period for respirometric measurements need not be 5 d because equally valid correlations can be made between the 5-d dilution BOD and respirometric oxygen uptake at any time after 2 d.[1,2] The point of common dilution and respirometric BOD seems to occur at about 2 to 3 d incubation for municipal wastewaters. Correlations between respirometric measurements and 5-d BOD for industrial wastes and specific chemicals are less certain. Respirometric measurements also can provide an indication of the ultimate biochemical oxygen demand (UBOD) (see Section 5210C). In many cases, it is reasonable to consider that the 28- to 30-d oxygen uptake is essentially equal to the UBOD.[3]

More commonly, respirometers are used as a diagnostic tool. The continuous readout of oxygen consumption in respirometric measurements indicates lag, toxicity, or any abnormalities in the biodegradation reaction. The change in the normal shape of an oxygen-uptake curve in the first few hours may help to identify the effect of toxic or unusual wastes entering a treatment plant in time to make operating corrections.

g. Relationship to other test methods and protocols: This method supports most of the protocols and guidelines established by the European Organization for Economic Co-operation and Development[3] (OECD) that require measurement of oxygen uptake.

h. Sampling and storage:

1) Grab samples—If analysis is begun within 2 h of sample collection, cold storage is unnecessary. Otherwise, keep sample at or below 4°C from the time of collection. Begin analysis within 6 h of collection; when this is not possible, store at or below 4°C and report length and temperature of storage. Never start analysis more than 24 h after grab sample collection.

2) Composite samples—Keep samples at or below 4°C during compositing. Limit compositing period to 24 h. Use the same criteria as for storage of grab samples, starting the measurement of holding time from the end of the compositing period. State storage time and conditions with results.

2. Apparatus

a. Respirometer system: Use commercial apparatus and check manufacturer's instructions for specific system requirements, reaction vessel type and volume, and instrument operating characteristics.

b. Incubator or water bath: Use a constant-temperature room, incubator chamber, or water bath to control temperature to ± 1°C. Exclude all light to prevent oxygen formation by algae in the sample. Use red, actinic-coated bottles for analysis outside of a darkened incubator.

3. Reagents

Formulations of reagent solutions are given for 1-L volumes, but smaller or larger volumes may be prepared according to need. Discard any reagent showing signs of biological growth or chemical precipitation. Stock solutions can be sterilized by autoclaving to provide longer shelf life.

a. Distilled water: Use only high-quality water distilled from a block tin or all-glass still (see Section 1080). Deionized water may be used but often contains high bacterial counts. The water must contain less than 0.01 mg heavy metals/L and be free of chlorine, chloramines, caustic alkalinity, organic material, or acids. Make all reagents with this water. When other waters are required for special-purpose testing, state clearly their source and quality characteristics.

b. Phosphate buffer solution, 1.5N: Dissolve 207 g sodium dihydrogen phosphate, $NaH_2PO_4 \cdot H_2O$, in water. Neutralize to pH 7.2 with 6N KOH (¶ 3g below) and dilute to 1 L.

c. Ammonium chloride solution, 0.71N: Dissolve 38.2 g ammonium chloride, NH_4Cl, in water. Neutralize to pH 7.0 with KOH. Dilute to 1.0 L; 1 mL = 10 mg N.

d. Calcium chloride solution, 0.25N: Dissolve 27.7 g $CaCl_2$ in water and dilute to 1 L; 1 mL = 10 mg Ca.

e. Magnesium sulfate solution, 0.41N: Dissolve 101 g $MgSO_4 \cdot 7H_2O$ in water and dilute to 1 L; 1 mL = 10 mg Mg.

f. Ferric chloride solution, 0.018N: Dissolve 4.84 g $FeCl_3 \cdot 6H_2O$ in water and dilute to 1 L; 1 mL = 1.0 mg Fe.

g. Potassium hydroxide solution, 6N: Dissolve 336 g KOH in about 700 mL water and dilute to 1 L. CAUTION: *Add KOH to water slowly and use constant mixing to prevent excessive heat buildup.* Alternately, use commercial solutions containing 30 to 50% KOH by weight.

h. Acid solutions, 1N: Add 28 mL conc H_2SO_4 or 83 mL conc HCl to about 700 mL water. Dilute to 1 L.

i. Alkali solution, 1N: Add 40 g NaOH to 700 mL water. Dilute to 1 L.

j. Nitrification inhibitor: Reagent-grade 2-chloro-6-(trichloromethyl) pyridine (TCMP) or equivalent.[3]*

---

* Formula 2533, Hach Chemical Co., Loveland, CO, or equivalent. NOTE: Some commercial formulations are not pure TCMP. Check with supplier to verify compound purity and adjust dosages accordingly.

k. Glucose-glutamic acid solution: Dry reagent-grade glucose and reagent-grade glutamic acid at 103°C for 1 h. Add 15.0 g glucose and 15.0 g glutamic acid to distilled water and dilute to 1 L. Neutralize to pH 7.0 using 6N potassium hydroxide (¶ 3g). This solution may be stored for up to 1 week at 4°C.

l. Electrolyte solution (for electrolytic respirometers): Use manufacturer's recommended solution.

m. Sodium sulfite solution, 0.025N: Dissolve 1.575 g Na_2SO_3 in about 800 mL water. Dilute to 1 L. This solution is not stable; prepare daily or as needed.

n. Trace element solution: Dissolve 40 mg $MnSO_4·4H_2O$, 57 mg H_3BO_3, 43 mg $ZnSO_4·7H_2O$, 35 mg $(NH_4)_6 Mo_7O_{24}$, and 100 mg Fe-chelate ($FeCl_3$-EDTA) in about 800 mL water. Dilute to 1 L. Sterilize at 120°C and 200 kPa (2 atm) pressure for 20 min.

o. Yeast extract solution:[3] Add 15 mg laboratory- or pharmaceutical-grade brewer's yeast extract to 100 mL water. Make this solution fresh immediately before each test in which it is used.

p. Nutrient solution:[3] Add 2.5 mL phosphate buffer solution (3b), 0.65 mL ammonium chloride solution (3c), 1.0 mL calcium chloride solution (3d), 0.22 mL magnesium sulfate solution (3a), 0.1 mL ferric chloride solution (3f), 1 mL trace element solution (3n), and 1 mL yeast extract solution (3o) to about 900 mL water. Dilute to 1 L. This nutrient solution and those of ¶s n and o above are specifically formulated for use with the OECD method.[3] (NOTE: A 10:1 concentrated nutrient solution can be made and diluted accordingly.)

4. Procedure

a. Instrument operation: Follow respirometer manufacturer's instructions for assembly, testing, calibration, and operation of the instrument. NOTE: The manufacturer's stated maximum and minimum limits of measurement are not always the same as the instrument output limits. Make sure that test conditions are within the limits of measurement.

b. Sample volume: Sample volume or concentration of organic chemicals to be added to test vessels is a function of expected oxygen uptake characteristics and oxygen transfer capability of the instrument. Small volumes or low concentrations may be required for high-strength wastes. Large volumes may be required for low-strength wastes to improve accuracy.

c. Data recording interval: Set instrument to give data readings at suitable intervals. Intervals of 15 min to 6 h typically are used.

d. Sample preparation:

1) Homogenization—If sample contains large settleable or floatable solids, homogenize it with a blender and transfer representative test portions while all solids are in suspension. If there is a concern for changing sample characteristics, skip this step.

2) pH adjustment—Neutralize samples to pH 7.0 with H_2SO_4 or NaOH of such strength (¶s 3h and i) that reagent quantity does not dilute the sample more than 0.5%.

3) Dechlorination—Avoid analyzing samples containing residual chlorine by collecting the samples ahead of chlorination processes. If residual chlorine is present, aerate as described in ¶ 5) below or let stand in light for 1 to 2 h. If a chlorine residual persists, add Na_2SO_3 solution. Determine required volume of Na_2SO_3 solution by adding 10 mL 1 + 1 acetic acid or 1 + 50 H_2SO_4 and 10 mL potassium iodide solution (10 g/100 mL) to a portion of the sample. Titrate with 0.025N Na_2SO_3 solution to the starch-iodine end point (see Section 4500-Cl.B). Add to the neutralized sample a proportional volume of Na_2SO_3 solution determined above, mix, and after 10 to 20 min check for residual chlorine. Re-seed the sample (see ¶ 4h below).

4) Samples containing toxic substances—Certain industrial wastes contain toxic metals or organic compounds. These often require special study and treatment.[3]

5) Initial oxygen concentration—If samples contain dissolved oxygen concentrations above or below the desired concentration, agitate or aerate with clean and filtered compressed air for about 1 h immediately before testing. Minimum and maximum actual DO concentrations will vary with test objectives. In some cases, pure oxygen may be added to respirometer vessels to increase oxygen levels above ambient.

6) Temperature adjustment—Bring samples and dilution water to desired test temperature (±1°C) before making dilutions or transferring to test vessels.

e. Sample dilution: Use distilled water or water from other appropriate sources free of organic matter. In some cases, receiving stream water may be used for dilution. Add desired sample volume to test vessels using a wide-tip volumetric pipet or other suitable volumetric glassware. Add dilution water to bring sample to about 80% of desired final volume. Add appropriate amounts of nutrients, minerals, buffer, nitrification inhibitor if desired, and seed culture as described in ¶s 4f and h below. Dilute sample to desired final volume. The number of test vessels to prepare for each dilution depends on test objectives and number of replicates desired.

f. Nutrients, minerals, and buffer: Add sufficient ammonia nitrogen to provide a COD:N:P ratio of 100:5:1 or a TOC:N:P ratio of 30:5:1. Add 2 mL each of calcium, magnesium, ferric chloride, and trace mineral solutions to each liter of diluted sample unless sufficient amounts of these minerals are present in the original sample. Phosphorus requirements will be met by the phosphate buffer if it is used (1 mL/50 mg/L COD or ultimate BOD of diluted sample usually is sufficient to maintain pH between 6.8 and 7.2). Be cautious in adding phosphate buffer to samples containing metal salts because metal phosphates may precipitate and show less toxic or beneficial effect than when phosphate is not present. For OECD-compatible tests, substitute the nutrient, mineral, and buffer amounts listed in ¶ 3p for the above nutrient/mineral/buffer quantities.

g. Nitrification inhibition: If nitrification inhibition is desired, add 10 mg 2-chloro-6-(trichloromethyl) pyridine (TCMP)/L sample in the test vessel. Samples that may nitrify readily include biologically treated effluents, samples seeded with biologically treated effluents, and river waters.[4]

h. Seeding: See 5210B.4d1) for seed preparation. Use sufficient amounts of seed culture to prevent major lags in the oxygen uptake reaction but not so much that the oxygen uptake of the seed exceeds about 10% of the oxygen uptake of the seeded sample.

Determine the oxygen uptake of the seeding material as for any other sample. This is the seed control. Typically, the seed volume in the seed control should be 10 times the volume used in seeded samples.

i. Incubation: Incubate samples at 20°C or other suitable temperature ±1.0°C. Take care that the stirring device does not increase the temperature of the sample.

5. Calculations

To convert instrument readings to oxygen uptake, refer to manufacturer's procedures.

Correct oxygen uptake for seed and dilution by the following equation:

$$C = [A - B(S_A/S_B)](1000/N_A)$$

where:

C = corrected oxygen uptake of sample, mg/L,
A = measured oxygen uptake in seeded sample, mg,
B = measured oxygen uptake in seed control, mg,
S_A = volume of seed in Sample A, mL,
S_B = volume of seed in Sample B, mL, and
N_A = volume of undiluted sample in Sample A, mL.

6. Quality Control

Periodically use the following procedure to check distilled water quality, instrument quality, instrument function, and analytical technique by making oxygen uptake measurements using a mixture of glucose and glutamic acid as a standard check solution.

Adjust water for sample formulation to test temperature and saturate with DO by aerating with clean, organic-free filtered air. Protect water quality by using clean glassware, tubing, and bottles.

Prepare a *test solution* by adding 10 mL glucose-glutamic acid solution (3k); 6 mL phosphate buffer (3b); 2 mL each of ammonium chloride (3c), magnesium sulfate (3e), calcium chloride (3d), ferric chloride (3f), and trace element solution (3n) to approximately 800 mL water. Add 10 mg nitrification inhibitor (TCMP)/L. Add sufficient seed from a suitable source as described in ¶ 4h to give a lag time less than 6 h (usually 25 mL supernatant from settled primary effluent/L test solution is sufficient). Dilute to 1 L. Adjust temperature to 20 ± 1°C.

Prepare a *seed blank* by diluting 500 mL or more of the seed solution to 800 mL with distilled water. Add the same amount of buffer, nutrients, and TCMP as in the test solution, and dilute to 1 L. Adjust temperature to 20 ± 1°C.

Place test solution and seed blank solution in separate reaction vessels of respirometer and incubate for 5 d at 20°C. Run at least three replicates of each. The seed-corrected oxygen uptake after 5 d incubation should be 260 ± 30 mg/L. If the value of the check is outside this range, repeat the test using a fresh seed culture and seek the cause of the problem.

7. Precision and Bias

a. Precision: No standard is available to check the accuracy of respirometric oxygen uptake measurements. To obtain laboratory precision data, use a glucose-glutamic acid mixture (¶ 6 above) having a known theoretical maximum oxygen uptake value. Tests with this and similar organic compound mixtures have shown that the standard deviation, expressed as the coefficient of variation, C_v, is approximately 5% for samples having total oxygen uptakes of 50 to 100 mg/L and 3% for more concentrated samples.[1,2] Individual instruments have different readability limits that can affect precision. The minimum response or sensitivity of most commercial respirometers ranges from 0.05 to 1 mg oxygen. Check manufacturer's specifications for sensitivity of the instrument at hand.

b. Control limits: To establish laboratory control limits, perform a minimum of 25 glucose-glutamic acid checks over a period of several weeks or months and calculate mean and standard deviation. If measured oxygen uptake in 5 d at 20°C is outside the 260 ± 30 mg/L range, re-evaluate procedure to identify source of error. For other samples, use the mean ± 3 standard deviations as the control limit.

c. Working range and detection limits: The working range and detection limits are established by the limits of each commercial instrument. Refer to manufacturer's specifications.

8. References

1. YOUNG J.C. & E.R. BAUMANN. 1976. The electrolytic respirometer—Factors affecting oxygen uptake measurements. *Water Res.* 10:1031.
2. YOUNG, J.C. & E.R. BAUMANN. 1976. The electrolytic respirometer—Use in water pollution control plant laboratories. *Water Res.* 10:1141.
3. ORGANIZATION FOR ECONOMIC CO-OPERATION AND DEVELOPMENT. 1981. Method 5.2, Annex V, Part C *in* OECD Guidelines for Testing of Chemicals. OECD, Paris, France.
4. YOUNG, J.C. 1973. Chemical methods for nitrification control. *J. Water Poll. Control Fed.* 45:637.

9. Bibliography

HEUKELEKIAN, H. 1947. Use of direct method of oxygen utilization in waste treatment studies. *Sew. Works J.* 19:375.
CALDWELL, D.H. & W.F. LANGELIER. 1948. Manometric measurement of the biochemical oxygen demand of sewage. *Sew. Works J.* 20:202.
GELLMAN, I. & H. HEUKELEKIAN. 1951. Studies of biochemical oxidation by direct methods. *Sew. Ind. Wastes.* 23:1267.
JENKINS, D. 1960. The use of manometric methods in the study of sewage and trade wastes. *In* Waste Treatment, p. 99. Pergamon Press, Oxford, U.K.
MONTGOMERY, H.A.C. 1967. The Determination of Biochemical Oxygen Demand by Respirometric Methods. *Water Res.* 1:631.
CEDENA, F., A. DROHOBYCZAR, M.I. BEACH & D. BARNES. 1988. A novel approach to simplified respirometric oxygen demand determinations. *Proc. 43rd Ind. Waste Conf.*, Purdue Univ., Lafayette, Ind.

5220 CHEMICAL OXYGEN DEMAND (COD)*

5220 A. Introduction

Chemical oxygen demand (COD) is defined as the amount of a specified oxidant that reacts with the sample under controlled conditions. The quantity of oxidant consumed is expressed in terms of its oxygen equivalence. Because of its unique chemical properties, the dichromate ion ($Cr_2O_7^{2-}$) is the specified oxidant in Methods 5220B, C, and D; it is reduced to the chromic ion (Cr^{3+}) in these tests. Both organic and inorganic components of a sample are subject to oxidation, but in most cases the organic component predominates and is of the greater interest. If it is desired to measure either organic or inorganic COD alone, additional steps not described here must be taken to distinguish one from the other. COD is a defined test; the extent of sample oxidation can be affected by digestion time, reagent strength, and sample COD concentration.

COD often is used as a measurement of pollutants in wastewater and natural waters. Other related analytical values are biochemical oxygen demand (BOD), total organic carbon (TOC), and total oxygen demand (TOD). In many cases it is possible to correlate two or more of these values for a given sample. BOD is a measure of oxygen consumed by microorganisms under specific conditions; TOC is a measure of organic carbon in a sample; TOD is a measure of the amount of oxygen consumed by all elements in a sample when complete (total) oxidation is achieved.

In a COD analysis, hazardous wastes of mercury, hexavalent chromium, sulfuric acid, silver, and acids are generated. Methods 5220C and D reduce these waste problems but may be less accurate and less representative. (See ¶ 2 below.)

1. Selection of Method

The open reflux method (B) is suitable for a wide range of wastes where a large sample size is preferred. The closed reflux methods (C and D) are more economical in the use of metallic salt reagents and generate smaller quantities of hazardous waste, but require homogenization of samples containing suspended solids to obtain reproducible results. Ampules and culture tubes with premeasured reagents are available commercially. Measurements of sample volumes as well as reagent volumes and concentrations are critical. Consequently, obtain specifications as to limits of error for premixed reagents from manufacturer before use.

Determine COD values of >50 mg O_2/L by using procedures 5220B.4a, C.4, or D.4. Use procedure 5220B.4b to determine, with lesser accuracy, COD values from 5 to 50 mg O_2/L.

2. Interferences and Limitations

Oxidation of most organic compounds is 95 to 100% of the theoretical value. Pyridine and related compounds resist oxidation and volatile organic compounds will react in proportion to their contact with the oxidant. Straight-chain aliphatic compounds are oxidized more effectively in the presence of a silver sulfate catalyst.

---

* Approved by Standard Methods Committee, 1997.

The most common interferent is the chloride ion. Chloride reacts with silver ion to precipitate silver chloride, and thus inhibits the catalytic activity of silver. Bromide, iodide, and any other reagent that inactivates the silver ion can interfere similarly. Such interferences are negative in that they tend to restrict the oxidizing action of the dichromate ion itself. However, under the rigorous digestion procedures for COD analyses, chloride, bromide, or iodide can react with dichromate to produce the elemental form of the halogen and the chromic ion. Results then are in error on the high side. The difficulties caused by the presence of the chloride can be overcome largely, though not completely, by complexing with mercuric sulfate ($HgSO_4$) before the refluxing procedure. Although 1 g $HgSO_4$ is specified for 50 mL sample, a lesser amount may be used where sample chloride concentration is known to be less than 2000 mg/L, as long as a 10:1 weight ratio of $HgSO_4$:Cl^- is maintained. Do not use the test for samples containing more than 2000 mg Cl^-/L. Techniques designed to measure COD in saline waters are available.[1,2]

Halide interferences may be removed by precipitation with silver ion and filtration before digestion. This approach may introduce substantial errors due to the occlusion and carrydown of COD matter from heterogenous samples.

Ammonia and its derivatives, in the waste or generated from nitrogen-containing organic matter, are not oxidized. However, elemental chlorine reacts with these compounds. Hence, corrections for chloride interferences are difficult.

Nitrite (NO_2^-) exerts a COD of 1.1 mg O_2/mg NO_2^--N. Because concentrations of NO_2^- in waters rarely exceed 1 or 2 mg NO_2^--N/L, the interference is considered insignificant and usually is ignored. To eliminate a significant interference due to NO_2^-, add 10 mg sulfamic acid for each mg NO_2^--N present in the sample volume used; add the same amount of sulfamic acid to the reflux vessel containing the distilled water blank.

Reduced inorganic species such as ferrous iron, sulfide, manganous manganese, etc., are oxidized quantitatively under the test conditions. For samples containing significant levels of these species, stoichiometric oxidation can be assumed from known initial concentration of the interfering species and corrections can be made to the COD value obtained.

The silver, hexavalent chromium, and mercury salts used in the COD determinations create hazardous wastes. The greatest problem is in the use of mercury. If the chloride contribution to COD is negligible, $HgSO_4$ can be omitted. Smaller sample sizes (see 5220C and D) reduce the waste. Recovery of the waste material may be feasible if allowed by regulatory authority.[3]

3. Sampling and Storage

Preferably collect samples in glass bottles. Test unstable samples without delay. If delay before analysis is unavoidable, preserve sample by acidification to pH ≤ 2 using conc H_2SO_4. Blend (homogenize) all samples containing suspended solids before analysis. If COD is to be related to BOD, TOC, etc., ensure that all tests receive identical pretreatment. Make preliminary dilutions for wastes containing a high COD to reduce the error inherent in measuring small sample volumes.

4. References

1. BURNS, E.R. & C. MARSHALL. 1965. Correction for chloride interference in the chemical oxygen demand test. *J. Water Pollut. Control Fed.* 37:1716.

2. BAUMANN, F.I. 1974. Dichromate reflux chemical oxygen demand: A proposed method for chloride correction in highly saline waters. *Anal. Chem.* 46:1336.

3. HOLM, T.R. 1996. Treatment of Spent Chemical Oxygen Demand Solutions for Safe Disposal. Illinois State Water Survey, Champaign.

4. AMERICAN SOCIETY FOR TESTING AND MATERIALS. 1995. Standard Test Methods for Chemical Oxygen Demand. (Dichromate Oxygen Demand) of Water. D1252-95, American Soc. Testing & Materials, Philadelphia, Pa.

5220 B. Open Reflux Method

1. General Discussion

a. Principle: Most types of organic matter are oxidized by a boiling mixture of chromic and sulfuric acids. A sample is refluxed in strongly acid solution with a known excess of potassium dichromate ($K_2Cr_2O_7$). After digestion, the remaining unreduced $K_2Cr_2O_7$ is titrated with ferrous ammonium sulfate to determine the amount of $K_2Cr_2O_7$ consumed and the oxidizable matter is calculated in terms of oxygen equivalent. Keep ratios of reagent weights, volumes, and strengths constant when sample volumes other than 50 mL are used. The standard 2-h reflux time may be reduced if it has been shown that a shorter period yields the same results. Some samples with very low COD or with highly heterogeneous solids content may need to be analyzed in replicate to yield the most reliable data. Results are further enhanced by reacting a maximum quantity of dichromate, provided that some residual dichromate remains.

2. Apparatus

a. Reflux apparatus, consisting of 500- or 250-mL erlenmeyer flasks with ground-glass 24/40 neck and 300-mm jacket Liebig, West, or equivalent condenser with 24/40 ground-glass joint, and a hot plate having sufficient power to produce at least 1.4 W/cm^2 of heating surface, or equivalent.

b. Blender.

c. Pipets, Class A and wide-bore.

3. Reagents

a. Standard potassium dichromate solution, 0.04167M: Dissolve 12.259 g $K_2Cr_2O_7$, primary standard grade, previously dried at 150°C for 2 h, in distilled water and dilute to 1000 mL. This reagent undergoes a six-electron reduction reaction; the equivalent concentration is 6 × 0.04167M or 0.2500N.

b. Sulfuric acid reagent: Add Ag_2SO_4, reagent or technical grade, crystals or powder, to conc H_2SO_4 at the rate of 5.5 g Ag_2SO_4/kg H_2SO_4. Let stand 1 to 2 d to dissolve. Mix.

c. Ferroin indicator solution: Dissolve 1.485 g 1,10-phenanthroline monohydrate and 695 mg $FeSO_4 \cdot 7H_2O$ in distilled water and dilute to 100 mL. This indicator solution may be purchased already prepared.*

d. Standard ferrous ammonium sulfate (FAS) titrant, approximately 0.25M: Dissolve 98 g $Fe(NH_4)_2(SO_4)_2 \cdot 6H_2O$ in distilled

---

* GFS Chemicals, Inc., Columbus, OH, or equivalent.

water. Add 20 mL conc H_2SO_4, cool, and dilute to 1000 mL. Standardize this solution daily against standard $K_2Cr_2O_7$ solution as follows:

Dilute 25.00 mL standard $K_2Cr_2O_7$ to about 100 mL. Add 30 mL conc H_2SO_4 and cool. Titrate with FAS titrant using 0.10 to 0.15 mL (2 to 3 drops) ferroin indicator.

Molarity of FAS solution

$$= \frac{\text{Volume } 0.04167M \text{ } K_2Cr_2O_7 \text{ solution titrated, mL}}{\text{Volume FAS used in titration, mL}} \times 0.2500$$

e. Mercuric sulfate, $HgSO_4$, crystals or powder.

f. Sulfamic acid: Required only if the interference of nitrites is to be eliminated (see 5220A.2 above).

g. Potassium hydrogen phthalate (KHP) standard, $HOOCC_6H_4COOK$: Lightly crush and then dry KHP to constant weight at 110°C. Dissolve 425 mg in distilled water and dilute to 1000 mL. KHP has a theoretical COD[1] of 1.176 mg O_2/mg and this solution has a theoretical COD of 500 μg O_2/ mL. This solution is stable when refrigerated, but not indefinitely. Be alert to development of visible biological growth. If practical, prepare and transfer solution under sterile conditions. Weekly preparation usually is satisfactory.

4. Procedure

a. Treatment of samples with COD of >50 mg O_2/L: Blend sample if necessary and pipet 50.00 mL into a 500-mL refluxing flask. For samples with a COD of >900 mg O_2/L, use a smaller portion diluted to 50.00 mL. Add 1 g $HgSO_4$, several glass beads, and very slowly add 5.0 mL sulfuric acid reagent, with mixing to dissolve $HgSO_4$. Cool while mixing to avoid possible loss of volatile materials. Add 25.00 mL 0.04167M $K_2Cr_2O_7$ solution and mix. Attach flask to condenser and turn on cooling water. Add remaining sulfuric acid reagent (70 mL) through open end of condenser. Continue swirling and mixing while adding sulfuric acid reagent. CAUTION: *Mix reflux mixture thoroughly before applying heat to prevent local heating of flask bottom and a possible blowout of flask contents.*

Cover open end of condenser with a small beaker to prevent foreign material from entering refluxing mixture and reflux for 2 h. Cool and wash down condenser with distilled water. Disconnect reflux condenser and dilute mixture to about twice its volume with distilled water. Cool to room temperature and titrate excess $K_2Cr_2O_7$ with FAS, using 0.10 to 0.15 mL (2 to

3 drops) ferroin indicator. Although the quantity of ferroin indicator is not critical, use the same volume for all titrations. Take as the end point of the titration the first sharp color change from blue-green to reddish brown that persists for 1 min or longer. Duplicate determinations should agree within 5% of their average. Samples with suspended solids or components that are slow to oxidize may require additional determinations. The blue-green may reappear. In the same manner, reflux and titrate a blank containing the reagents and a volume of distilled water equal to that of sample.

b. Alternate procedure for low-COD samples: Follow procedure of ¶ 4a, with two exceptions: (*i*) use standard $0.004167M$ $K_2Cr_2O_7$, and (*ii*) titrate with standardized $0.025M$ FAS. Exercise extreme care with this procedure because even a trace of organic matter on the glassware or from the atmosphere may cause gross errors. If a further increase in sensitivity is required, concentrate a larger volume of sample before digesting under reflux as follows: Add all reagents to a sample larger than 50 mL and reduce total volume to 150 mL by boiling in the refluxing flask open to the atmosphere without the condenser attached. Compute amount of $HgSO_4$ to be added (before concentration) on the basis of a weight ratio of 10:1, $HgSO_4:Cl^-$, using the amount of Cl^- present in the original volume of sample. Carry a blank reagent through the same procedure. This technique has the advantage of concentrating the sample without significant losses of easily digested volatile materials. Hard-to-digest volatile materials such as volatile acids are lost, but an improvement is gained over ordinary evaporative concentration methods. Duplicate determinations are not expected to be as precise as in 5220B.4a.

c. Determination of standard solution: Evaluate the technique and quality of reagents by conducting the test on a standard potassium hydrogen phthalate solution.

5. Calculation

$$COD \text{ as mg } O_2/L = \frac{(A - B) \times M \times 8000}{mL \text{ sample}}$$

where:

A = mL FAS used for blank,
B = mL FAS used for sample,
M = molarity of FAS, and
8000 = milliequivalent weight of oxygen × 1000 mL/L.

6. Precision and Bias

A set of synthetic samples containing potassium hydrogen phthalate and NaCl was tested by 74 laboratories. At a COD of 200 mg O_2/L in the absence of chloride, the standard deviation was ± 13 mg/L (coefficient of variation, 6.5%). At COD of 160 mg O_2/L and 100 mg Cl^-/L, the standard deviation was ± 14 mg/L (coefficient of variation, 10.8%).

7. Reference

1. PITWELL, L.R. 1983. Standard COD. *Chem. Brit.* 19:907.

8. Bibliography

MOORE, W.A., R.C. KRONER & C.C. RUCHHOFT. 1949. Dichromate reflux method for determination of oxygen consumed. *Anal. Chem.* 21:953.
MEDALIA, A.I. 1951. Test for traces of organic matter in water. *Anal. Chem.* 23:1318.
MOORE, W.A., F.J. LUDZACK & C.C. RUCHHOFT. 1951. Determination of oxygen-consumed values of organic wastes. *Anal. Chem.* 23:1297.
DOBBS, R.A. & R.T. WILLIAMS. 1963. Elimination of chloride interference in the chemical oxygen demand test. *Anal. Chem.* 35:1064.

5220 C.   Closed Reflux, Titrimetric Method

1. General Discussion

a. Principle: See 5220B.1a.

b. Interferences and limitations: See 5220A.2. Volatile organic compounds are more completely oxidized in the closed system because of longer contact with the oxidant. Before each use inspect culture-tube caps for breaks in the TFE liner. Select culture-tube size according to block heater capacity and degree of sensitivity desired. Use the 25- × 150-mm tube for samples with low COD content because a larger volume sample can be treated.

This procedure is applicable to COD values between 40 and 400 mg/L. Obtain higher values by dilution. Alternatively, use higher concentrations of dichromate digestion solution to determine greater COD values. COD values of 100 mg/L or less can be obtained by using a more dilute dichromate digestion solution or a more dilute FAS titrant. Overall accuracy can be improved by using an FAS titrant which is less than the 0.10M solution specified below. Higher dichromate concentrations or reduced FAS concentrations probably require titrations to be done in a

separate vessel, rather than in the digestion vessel, because of the volumes of titrant required.

2. Apparatus

a. Digestion vessels: Preferably use borosilicate culture tubes, 16- × 100-mm, 20- × 150-mm, or 25- × 150-mm, with TFE-lined screw caps. Alternatively, use borosilicate ampules, 10-mL capacity, 19- to 20-mm diam.

Digestion vessels with premixed reagents and other accessories are available from commercial suppliers. Contact supplier for specifications.*

b. Block heater or similar device to operate at 150 ± 2°C, with holes to accommodate digestion vessels. Use of culture tubes probably requires the caps to be outside the vessel to protect caps from heat. CAUTION: *Do not use an oven because of the possibility of leaking samples generating a corrosive and possibly explosive*

---

* Hach Co., Bioscience, Inc., or equivalent.

atmosphere. Also, culture tube caps may not withstand the 150°C temperature in an oven.

   c. Microburet.

   d. Ampule sealer: Use only a mechanical sealer to insure strong, consistent seals.

3. Reagents

   a. Standard potassium dichromate digestion solution, 0.01667M: Add to about 500 mL distilled water 4.903 g $K_2Cr_2O_7$, primary standard grade, previously dried at 150°C for 2 h, 167 mL conc H_2SO_4, and 33.3 g $HgSO_4$. Dissolve, cool to room temperature, and dilute to 1000 mL.

   b. Sulfuric acid reagent: See Section 5220B.3*b*.

   c. Ferroin indicator solution: See Section 5220B.3*c*. Dilute this reagent by a factor of 5 (1 + 4).

   d. Standard ferrous ammonium sulfate titrant (FAS), approximately 0.10M: Dissolve 39.2 g $Fe(NH_4)_2(SO_4)_2 \cdot 6H_2O$ in distilled water. Add 20 mL conc H_2SO_4, cool, and dilute to 1000 mL. Standardize solution daily against standard $K_2Cr_2O_7$ digestion solution as follows:

   Pipet 5.00 mL digestion solution into a small beaker. Add 10 mL reagent water to substitute for sample. Cool to room temperature. Add 1 to 2 drops diluted ferroin indicator and titrate with FAS titrant.

$$\text{Molarity of FAS solution} = \frac{\text{Volume } 0.01667M\ K_2Cr_2O_7 \text{ solution titrated, mL}}{\text{Volume FAS used in titration, mL}} \times 0.1000$$

   e. Sulfamic acid: See Section 5220B.3*f*.

   f. Potassium hydrogen phthalate standard: See Section 5220B.3*g*.

4. Procedure

   Wash culture tubes and caps with 20% H_2SO_4 before first use to prevent contamination. Refer to Table 5220:I for proper sample and reagent volumes. Make volumetric measurements as accurate as practical; use Class A volumetric ware. The most critical volumes are of the sample and digestion solution. Use a microburet for titrations. Measure H_2SO_4 to ±0.1 mL. The use of hand-held pipettors with non-wetting (polyethylene) pipet tips is practical and adequate. Place sample in culture tube or ampule and add digestion solution. Carefully run sulfuric acid reagent down inside of vessel so an acid layer is formed under the sample-digestion solution layer. Tightly cap tubes or seal ampules, and invert each several times to mix completely. CAUTION: *Wear face shield and protect hands from heat produced when contents of vessels are mixed. Mix thoroughly before applying heat to prevent local heating of vessel bottom and possible explosive reaction.*

   Place tubes or ampules in block digester preheated to 150°C and reflux for 2 h behind a protective shield. CAUTION: *These sealed vessels may be under pressure from gases generated dur-*

TABLE 5220:I. SAMPLE AND REAGENT QUANTITIES FOR VARIOUS DIGESTION VESSELS

Digestion Vessel	Sample mL	Digestion Solution mL	Sulfuric Acid Reagent mL	Total Final Volume mL
Culture tubes:				
16 × 100 mm	2.50	1.50	3.5	7.5
20 × 150 mm	5.00	3.00	7.0	15.0
25 × 150 mm	10.00	6.00	14.0	30.0
Standard 10-mL ampules	2.50	1.50	3.5	7.5

ing digestion. Wear face and hand protection when handling. If sulfuric acid is omitted or reduced in concentration, very high and dangerous pressures will be generated at 150°C. Cool to room temperature and place vessels in test tube rack. Some mercuric sulfate may precipitate out but this will not affect the analysis. Remove culture tube caps and add small TFE-covered magnetic stirring bar. If ampules are used, transfer contents to a larger container for titrating. Add 0.05 to 0.10 mL (1 to 2 drops) ferroin indicator and stir rapidly on magnetic stirrer while titrating with standardized 0.10M FAS. The end point is a sharp color change from blue-green to reddish brown, although the blue-green may reappear within minutes. In the same manner reflux and titrate a blank containing the reagents and a volume of distilled water equal to that of the sample.

5. Calculation

$$\text{COD as mg } O_2/L = \frac{(A - B) \times M \times 8000}{\text{mL sample}}$$

where:

    A = mL FAS used for blank,
    B = mL FAS used for sample,
    M = molarity of FAS, and
 8000 = milliequivalent weight of oxygen × 1000 mL/L.

   Preferably analyze samples in duplicate because of small sample size. Samples that are inhomogeneous may require multiple determinations for accurate analysis. Results should agree within ±5% of their average unless the condition of the sample dictates otherwise.

6. Precision and Bias

   Sixty synthetic samples containing potassium hydrogen phthalate and NaCl were tested by six laboratories. At an average COD of 195 mg O_2/L in the absence of chloride, the standard deviation was ±11 mg O_2/L (coefficient of variation, 5.6%). At an average COD of 208 mg O_2/L and 100 mg Cl^-/L, the standard deviation was ±10 mg O_2/L (coefficient of variation, 4.8%).

5220 D. Closed Reflux, Colorimetric Method

1. General Discussion

a. Principle: See Section 5220B.1a. When a sample is digested, the dichromate ion oxidizes COD material in the sample. This results in the change of chromium from the hexavalent (VI) state to the trivalent (III) state. Both of these chromium species are colored and absorb in the visible region of the spectrum. The dichromate ion ($Cr_2O_7^{2-}$) absorbs strongly in the 400-nm region, where the chromic ion (Cr^{3+}) absorption is much less. The chromic ion absorbs strongly in the 600-nm region, where the dichromate has nearly zero absorption. In $9M$ sulfuric acid solution, the approximate molar extinction coefficients for these chromium species are as follows: Cr^{3+} — 50 L/mole cm at 604 nm; $Cr_2O_7^{2-}$ — 380 L/mole cm at 444 nm; Cr^{3+} — 25 L/mole cm at 426 nm. The Cr^{3+} ion has a minimum in the region of 400 nm. Thus a working absorption maximum is at 420 nm.

For COD values between 100 and 900 mg/L, increase in Cr^{3+} in the 600-nm region is determined. Higher values can be obtained by sample dilution. COD values of 90 mg/L or less can be determined by following the decrease in $Cr_2O_7^{2-}$ at 420 nm. The corresponding generation of Cr^{3+} gives a small absorption increase at 420 nm, but this is compensated for in the calibration procedure.

b. Interferences and limitations: See Section 5220C.1b.

For this procedure to be applicable, all visible light-absorbing interferents must be absent or be compensated for. This includes insoluble suspended matter as well as colored components. If either type of interference occurs, the test is not necessarily lost because COD can be determined titrimetrically as in 5220C.

2. Apparatus

a. See Section 5220C.2. Ensure that reaction vessels are of optical quality. Other types of absorption cells with varying path lengths may be used. Use the extinction coefficients of the ions of interest for this approach.

b. Spectrophotometer, for use at 600 nm and/or 420 nm with access opening adapter for ampule or 16-, 20-, or 25-mm tubes. Verify that the instrument operates in the region of 420 nm and 600 nm. Values slightly different from these may be found, depending on the spectral bandpass of the instrument.

3. Reagents

a. Digestion solution, high range: Add to about 500 mL distilled water 10.216 g $K_2Cr_2O_7$, primary standard grade, previously dried at 150°C for 2 h, 167 mL conc H_2SO_4, and 33.3 g $HgSO_4$. Dissolve, cool to room temperature, and dilute to 1000 mL.

b. Digestion solution, low range: Prepare as in 3a, but use only 1.022 g potassium dichromate.

c. Sulfuric acid reagent: See Section 5220B.3b.

d. Sulfamic acid: See Section 5220B.3f.

e. Potassium hydrogen phthalate standard: See Section 5220B.3g.

4. Procedure

a. Treatment of samples: Measure suitable volume of sample and reagents into tube or ampule as indicated in Table 5220:I.

Prepare, digest, and cool samples, blank, and one or more standards as directed in Section 5220C.4. *Note the safety precautions*. It is critical that the volume of each component be known and that the total volume be the same for each reaction vessel. If volumetric control is difficult, transfer digested sample, dilute to a known volume, and read. Premixed reagents in digestion tubes are available commercially.

b. Measurement of dichromate reduction: Cool sample to room temperature slowly to avoid precipitate formation. Once samples are cooled, vent, if necessary, to relieve any pressure generated during digestion. Mix contents of reaction vessels to combine condensed water and dislodge insoluble matter. Let suspended matter settle and ensure that optical path is clear. Measure absorption of each sample blank and standard at selected wavelength (420 nm or 600 nm). At 600 nm, use an undigested blank as reference solution. Analyze a digested blank to confirm good analytical reagents and to determine the blank COD; subtract blank COD from sample COD. Alternately, use digested blank as the reference solution once it is established that the blank has a low COD.

At 420 nm, use reagent water as a reference solution. Measure all samples, blanks, and standards against this solution. The absorption measurement of an undigested blank containing dichromate, with reagent water replacing sample, will give initial dichromate absorption. Any digested sample, blank, or standard that has a COD value will give lower absorbance because of the decrease in dichromate ion. Analyze a digested blank with reagent water replacing sample to ensure reagent quality and to determine the reagents' contribution to the decrease in absorbance during a given digestion. The difference between absorbances of a given digested sample and the digested blank is a measure of the sample COD. When standards are run, plot differences of digested blank absorbance and digested standard absorbance versus COD values for each standard.

c. Preparation of calibration curve: Prepare at least five standards from potassium hydrogen phthalate solution with COD equivalents to cover each concentration range. Make up to volume with reagent water; use same reagent volumes, tube, or ampule size, and digestion procedure as for samples. Prepare calibration curve for each new lot of tubes or ampules or when standards prepared in ¶ 4a differ by ≥5% from calibration curve. Curves should be linear. However, some nonlinearity may occur, depending on instrument used and overall accuracy needed.

5. Calculation

If samples, standards, and blanks are run under same conditions of volume and optical path length, calculate COD as follows:

$$\text{COD as mg } O_2/\text{L} = \frac{\text{mg } O_2 \text{ in final volume} \times 1000}{\text{mL sample}}$$

Preferably analyze samples in duplicate because of small sample size. Samples that are inhomogeneous may require multiple determinations for accurate analysis. These should not differ from their average by more than ±5% for the high-level COD test unless the condition of the sample dictates otherwise. In the low-level procedure, results below 25 mg/L may tend to be qualitative rather than quantitative.

6. Precision and Bias

Forty-eight synthetic samples containing potassium hydrogen phthalate and NaCl were tested by five laboratories. At an average COD of 193 mg O_2/L in the absence of chloride, the standard deviation was ± 17 mg O_2/L (coefficient of variation 8.7%). At an average COD of 212 mg O_2/L and 100 mg Cl^-/L, the standard deviation was ± 20 mg O_2/L (coefficient of variation, 9.6%). Additional QA/QC data for both high- and low-level procedures may be found elsewhere.[1]

7. Reference

1. AMERICAN SOCIETY FOR TESTING AND MATERIALS. 1995. Standard test methods for chemical oxygen demand (dichromate oxygen demand) of water. D1252-95, ASTM Annual Book of Standards. American Soc. Testing & Materials, Philadelphia, Pa.

8. Bibliography

JIRKA, A.M. & M.J. CARTER. 1975. Micro semi-automated analysis of surface and wastewaters for chemical oxygen demand. *Anal. Chem.* 47:1397.
HIMEBAUGH, R.R. & M.J. SMITH. 1979. Semi-micro tube method for chemical oxygen demand. *Anal. Chem.* 51:1085.

5310   TOTAL ORGANIC CARBON (TOC)*

5310 A.   Introduction

1. General Discussion

The organic carbon in water and wastewater is composed of a variety of organic compounds in various oxidation states. Some of these carbon compounds can be oxidized further by biological or chemical processes, and the biochemical oxygen demand (BOD), assimilable organic carbon (AOC), and chemical oxygen demand (COD) methods may be used to characterize these fractions. Total organic carbon (TOC) is a more convenient and direct expression of total organic content than either BOD, AOC, or COD, but does not provide the same kind of information. If a repeatable empirical relationship is established between TOC and BOD, AOC, or COD for a specific source water then TOC can be used to estimate the accompanying BOD, AOC, or COD. This relationship must be established independently for each set of matrix conditions, such as various points in a treatment process. Unlike BOD or COD, TOC is independent of the oxidation state of the organic matter and does not measure other organically bound elements, such as nitrogen and hydrogen, and inorganics that can contribute to the oxygen demand measured by BOD and COD. TOC measurement does not replace BOD, AOC, and COD testing.

Measurement of TOC is of vital importance to the operation of water treatment and waste treatment plants. Drinking water TOCs range from less than 100 μg/L to more than 25,000 μg/L. Wastewater may contain very high levels of organic compounds (TOC > 100 mg/L). Some of these applications may include waters with substantial ionic impurities as well as organic matter.

In many applications, the presence of organic contaminants may degrade ion-exchange capacity, serve as a nutrient source for undesired biological growth, or be otherwise detrimental to the process for which the water is to be utilized. For drinking waters in particular, organic compounds may react with disinfectants to produce potentially toxic and carcinogenic compounds.

To determine the quantity of organically bound carbon, the organic molecules must be broken down and converted to a single molecular form that can be measured quantitatively. TOC methods utilize high temperature, catalysts, and oxygen, or lower temperatures ($<100°C$) with ultraviolet irradiation, chemical oxidants, or combinations of these oxidants to convert organic carbon to carbon dioxide (CO_2). The CO_2 may be purged from the sample, dried, and transferred with a carrier gas to a nondispersive infrared analyzer or coulometric titrator. Alternatively, it may be separated from the sample liquid phase by a membrane selective to CO_2 into a high-purity water in which corresponding increase in conductivity is related to the CO_2 passing the membrane.

2. Fractions of Total Carbon

The methods and instruments used in measuring TOC analyze fractions of total carbon (TC) and measure TOC by two or more determinations. These fractions of total carbon are defined as: inorganic carbon—the carbonate, bicarbonate, and dissolved CO_2; total organic carbon (TOC)—all carbon atoms covalently bonded in organic molecules; dissolved organic carbon (DOC)—the fraction of TOC that passes through a 0.45-μm-pore-diam filter; suspended organic carbon—also referred to as particulate organic carbon, the fraction of TOC retained by a 0.45-μm filter; purgeable organic carbon—also referred to as volatile organic carbon, the fraction of TOC removed from an aqueous solution by gas stripping under specified conditions; and nonpurgeable organic carbon—the fraction of TOC not removed by gas stripping.

In most water samples, the inorganic carbon fraction is many times greater than the TOC fraction. Eliminating or compensating for inorganic carbon interferences requires determinations of both TC and inorganic carbon to measure TOC. Inorganic carbon interference can be eliminated by acidifying samples to pH 2 or less to convert inorganic carbon species to CO_2. Subsequent purging of the sample with a purified gas or vacuum degassing removes the CO_2 by volatilization. Sample purging also removes purgeable organic carbon so that the organic carbon measurement

*Approved by Standard Methods Committee, 1996.

made after eliminating inorganic carbon interferences is actually a nonpurgeable organic carbon determination: determine purgeable organic carbon to measure TOC. In many surface and ground waters the purgeable organic carbon contribution to TOC is negligible. Therefore, in practice, the nonpurgeable organic carbon determination is substituted for TOC.

Alternatively, inorganic carbon interference may be compensated for by separately measuring total carbon (TC) and inorganic carbon. The difference between TC and inorganic carbon is TOC.

The purgeable fraction of TOC is a function of the specific conditions and equipment employed. Sample temperature and salinity, gas-flow rate, type of gas diffuser, purging-vessel dimensions, volume purged, and purging time affect the division of TOC into purgeable and nonpurgeable fractions. When separately measuring purgeable organic carbon and nonpurgeable organic carbon on the same sample, use identical conditions for purging during the purgeable organic carbon measurement as in purging to prepare the nonpurgeable organic carbon portion for analysis. Consider the conditions of purging when comparing purgeable organic carbon or nonpurgeable organic carbon data from different laboratories or different instruments.

3. Selection of Method

The high-temperature combustion method (B) is suitable for samples with higher levels of TOC that would require dilution for the various persulfate methods (Method C or Method D). Generally, it also will determine organic carbon from compounds that are chemically refractory and not determined by Method C or Method D. High-temperature combustion may be desirable for samples containing high levels of suspended organic carbon, which may not be efficiently oxidized by persulfate and/or UV methods. Interlaboratory studies have shown biases on the order of 1 mg/L using older high-temperature instruments. With newer instruments, detection limits as low as 10 µg/L have been reported. Some high-temperature combustion instruments are not designed for levels below 1 mg/L. The high-temperature methods accumulate nonvolatile residues in the analyzer, whereas, in Method C, residuals are drained from the analyzer. Method C generally provides better sensitivity for lower-level (<1 mg/L) samples. Persulfate and/or UV oxidation are useful for TOC as low as 10 µg/L. Because the range of sensitivity of the methods overlaps, other factors may dictate method choice in the range of 1 mg/L to 50 mg/L. A method may be chosen on the basis of desired precision, ease of use, cost, etc. Method D generally is equivalent to Method C, but the equipment for Method D is no longer manufactured.

To qualify a particular instrument for use, demonstrate that the single-user precision and bias given in each method can be reproduced. Also, preferably demonstrate the overall precision by conducting in-house studies with more than one operator.

Evaluate the selected method to ensure that data quality objectives are attained. Evaluate method detection limit in a matrix as similar as possible to the unknowns as described in Section 1030. Be aware that instrument blanks are handled in a variety of ways in TOC analyzers and that the true magnitude of the blank may not be readily apparent to the analyst. Some instruments "zero out" much of the blank by adjusting the zero on the detector. Others enter blank values in units such as mv responses rather than absolute concentrations, whereas other instruments accumulate the total blank in the system during a blank run. Carefully observe the variability of low-level measurements and check it any time reagents or instrument operations are changed. The following methods note that when a water blank is run there is a contribution to the observed blank value from the level of carbon in the blank water.

The methods show expected single-operator and multiple-laboratory precision. These equations are based on referenced interlaboratory studies that in some cases were performed on older equipment. The range of testing is important to observe because the error and bias generally will be some significant fraction of the low standard. Consult references to determine type of equipment and conditions of the interlaboratory study. Determine the performance of the instrument being used by analyzing waters with matrices similar to those of unknowns, using the procedures outlined in Section 1040B.

4. Bibliography

FORD, D.L. 1968. Total organic carbon as a wastewater parameter. *Pub. Works* 99:89.

FREDERICKS, A.D. 1968. Concentration of organic carbon in the Gulf of Mexico: Off. Naval Res. Rep. 68–27T.

WILLIAMS, P.M. 1969. The determination of dissolved organic carbon in seawater: A comparison of two methods. *Limnol. Oceanogr.* 14:297.

CROLL, B.T. 1972. Determination of organic carbon in water. *Chem. Ind.* (London) 110:386.

SHARP, J. 1973. Total organic carbon in seawater—comparison of measurements using persulfate oxidation and high temperature combustion. *Mar. Chem.* 1:211.

BORTLIZ, J. 1976. Instrumental TOC analysis. *Vom Wasser* 46:35.

VAN STEENDEREN, R.A. 1976. Parameters which influence the organic carbon determination in water. *Water SA* 2:156.

TAKAHASHI, Y. 1979. Analysis Techniques for Organic Carbon and Organic Halogen. Proc. EPA/NATO-CCMS Conf. on Adsorption Techniques. Reston, Va.

STANDING COMMITTEE OF ANALYSIS, DEPARTMENT OF THE ENVIRONMENT, NATIONAL WATER COUNCIL 1980. The instrumental determination of total organic carbon, total oxygen demand and related determinants. 1979. Her Majesty's Stationery Off., London.

WERSHAW, R.L., M.J. FISHMAN, R.R. GRABBE & L.E. LOWE, eds. 1983. Methods for the Determination of Organic Substances in Water and Fluvial Sediments. U.S. Geological Survey Techniques of Water Resources Investigations, Book 5. Laboratory Analysis. Chapter A3.

KAPLAN, L.A. 1992. Comparison of high-temperature and persulfate oxidation methods for determination of dissolved organic carbon in freshwaters. *Limnol. Oceanogr.* 37:1119.

BENNER, R. & J.I. HEDGES. 1993. A test of the accuracy of freshwater DOC measurements by high-temperature catalytic oxidation and UV-promoted persulfate oxidation. *Mar. Chem.* 41:75.

WANGERSKY, P.J. 1993. Dissolved organic carbon methods: A critical review. *Mar. Chem.* 41:61.

HUTTE, R., R. GODEC & K. O'NEILL. 1994. Transferring NASA-sponsored advances in total organic carbon measurement to industrial applications. *Microcontamination* 12(4):21.

KAPLAN, L.A. 1994. A field and laboratory procedure to collect, process, and preserve freshwater samples for dissolved organic carbon analysis. *Limnol. Oceanogr.* 39:1470.

5310 B. High-Temperature Combustion Method

1. General Discussion

The high-temperature combustion method has been used for a wide variety of samples, but its utility is dependent on particle size reduction because it uses small-orifice syringes.

a. Principle: The sample is homogenized and diluted as necessary and a microportion is injected into a heated reaction chamber packed with an oxidative catalyst such as cobalt oxide, platinum group metals, or barium chromate. The water is vaporized and the organic carbon is oxidized to CO_2 and H_2O. The CO_2 from oxidation of organic and inorganic carbon is transported in the carrier-gas streams and is measured by means of a nondispersive infrared analyzer, or titrated coulometrically.

Because total carbon is measured, inorganic carbon must be removed by acidification and sparging or measured separately and TOC obtained by difference.

Measure inorganic carbon by injecting the sample into a reaction chamber where it is acidified. Under acidic conditions, all inorganic carbon is converted to CO_2, which is transferred to the detector and measured. Under these conditions organic carbon is not oxidized and only inorganic carbon is measured.

Alternatively, convert inorganic carbonates to CO_2 with acid and remove the CO_2 by purging before sample injection. The sample contains only the nonpurgeable organic carbon fraction of total carbon: a purgeable organic carbon determination also is necessary to measure TOC.

b. Interference: Removal of carbonate and bicarbonate by acidification and purging with purified gas results in the loss of volatile organic substances. The volatiles also can be lost during sample blending, particularly if the temperature is allowed to rise. Another important loss can occur if large carbon-containing particles fail to enter the needle used for injection. Filtration, although necessary to eliminate particulate organic matter when only DOC is to be determined, can result in loss or gain of DOC, depending on the physical properties of the carbon-containing compounds and the adsorption or desorption of carbonaceous material on the filter. Check filters for their contribution to DOC by analyzing a filtered blank. Note that any contact with organic material may contaminate a sample. Avoid contaminated glassware, plastic containers, and rubber tubing. Analyze sample treatment, system, and reagent blanks.

Combustion temperatures above 950°C are required to decompose some carbonates. Systems that use lower temperatures must destroy carbonates by acidification. Elemental carbon may not be oxidized at lower temperatures but generally it is not present in water samples nor is it formed during combustion of dilute samples. The advantage of using lower temperatures (680°C) is that fusion of dissolved salts is minimized, resulting in lower blank values. Gases evolved from combustion, such as water, halide compounds, and nitrogen oxides, may interfere with the detection system. Consult manufacturers' recommendations regarding proper selection of scrubber materials and check for any matrix interferences.

The major limitation to high-temperature techniques is the magnitude and variability of the blank. Instrument manufacturers have developed new catalysts and procedures that yield lower blanks, resulting in lower detection levels.

c. Minimum detectable concentration: 1 mg C/L or less, depending on the instrument used. This can be achieved with most high-temperature combustion analyzers although instrument performance varies. The minimum detectable concentration may be reduced by concentrating the sample, or by increasing the portion taken for analysis.

d. Sampling and storage: If possible, rinse bottles with sample before filling and carry field blanks through sampling procedure to check for any contamination that may occur. Collect and store samples in glass bottles protected from sunlight and seal with TFE-backed septa. Before use, wash bottles with acid, seal with aluminum foil, and bake at 400°C for at least 1 h. Wash uncleaned TFE septa with detergent, rinse repeatedly with organic-free water, wrap in aluminum foil, and bake at 100°C for 1 h. Check performance of new or cleaned septa by running appropriate blanks. Preferably use thick silicone rubber-backed TFE septa with open ring caps to produce a positive seal. Less rigorous cleaning may be acceptable if the concentration range is relatively high. Check bottle blanks with each set of sample bottles to determine effectiveness or necessity of cleaning. Preserve samples that cannot be examined immediately by holding at 4°C with minimal exposure to light and atmosphere. Acidification with phosphoric or sulfuric acid to a pH ≤ 2 at the time of collection is especially desirable for unstable samples, and may be used on all samples: acid preservation, however, invalidates any inorganic carbon determination on the samples.

2. Apparatus

a. Total organic carbon analyzer, using combustion techniques.

b. Sampling, injection, and sample preparation accessories, as prescribed by instrument manufacturer.

c. Sample blender or homogenizer.

d. Magnetic stirrer and TFE-coated stirring bars.

e. Filtering apparatus and 0.45-μm-pore-diam filters. Preferably use HPLC syringe filters with no detectable TOC blank. Glass fiber or silver membrane filters also can be used. Rinse filters before use and monitor filter blanks.

3. Reagents

a. Reagent water: Prepare reagents, blanks, and standard solutions from reagent water with a TOC value less than 2 $\times$ the MDL (see Sections 1030 and 1080).

b. Acid: Phosphoric acid, H_3PO_4. Alternatively use sulfuric acid, H_2SO_4.

c. Organic carbon stock solution: Dissolve 2.1254 g anhydrous primary-standard-grade potassium biphthalate, $C_8H_5KO_4$, in carbon-free water and dilute to 1000 mL; 1.00 mL = 1.00 mg carbon. Prepare laboratory control standards using any other appropriate organic-carbon-containing compound of adequate purity, stability, and water solubility. Preserve by acidifying with H_3PO_4 or H_2SO_4 to pH ≤ 2, and store at 4°C.

d. Inorganic carbon stock solution: Dissolve 4.4122 g anhydrous sodium carbonate, Na_2CO_3, in water, add 3.497 g anhydrous sodium bicarbonate, $NaHCO_3$, and dilute to 1000 mL; 1.00 mL = 1.00 mg carbon. Alternatively, use any other inorganic carbonate compound of adequate purity, stability, and water solubility. Keep tightly stoppered. Do not acidify.

e. Carrier gas: Purified oxygen or air, CO_2-free and containing less than 1 ppm hydrocarbon (as methane).

f. Purging gas: Any gas free of CO_2 and hydrocarbons.

4. Procedure

a. Instrument operation: Follow manufacturer's instructions for analyzer assembly, testing, calibration, and operation. Adjust to optimum combustion temperature before using instrument; monitor temperature to insure stability.

b. Sample treatment: If a sample contains gross solids or insoluble matter, homogenize until satisfactory replication is obtained. Analyze a homogenizing blank consisting of reagent water carried through the homogenizing treatment.

If inorganic carbon must be removed before analysis, transfer a representative portion (10 to 15 mL) to a 30-mL beaker, add acid to reduce pH to 2 or less, and purge with gas for 10 min. Inorganic carbon also may be removed by stirring the acidified sample in a beaker while directing a stream of purified gas into the beaker. Because volatile organic carbon will be lost during purging of the acidified solution, report organic carbon as total nonpurgeable organic carbon. Check efficiency of inorganic carbon removal for each sample matrix by splitting a sample into two portions and adding to one portion an inorganic carbon level similar to that of the sample. The TOC values should agree; if they do not, adjust sample container, sample volume, pH, purge gas flow rate, and purge time to obtain complete removal of inorganic carbon.

If the available instrument provides for a separate determination of inorganic carbon (carbonate, bicarbonate, free CO_2) and total carbon, omit decarbonation and determine TOC by difference between TC and inorganic carbon.

If dissolved organic carbon is to be determined, filter sample through 0.45-μm-pore-diam filter; analyze a filtering blank.

c. Sample injection: Withdraw a portion of prepared sample using a syringe fitted with a blunt-tipped needle. Select sample volume according to manufacturer's direction. Stir samples containing particulates with a magnetic stirrer. Select needle size consistent with sample particulate size. Other sample injection techniques, such as sample loops, may be used. Inject samples and standards into analyzer according to manufacturer's directions and record response. Repeat injection until consecutive measurements are obtained that are reproducible to within ± 10%.

d. Preparation of standard curve: Prepare standard organic and inorganic carbon series by diluting stock solutions to cover the expected range in samples within the linear range of the instrument. Dilute samples higher than the linear range of the instrument in reagent water. Inject and record peak height or area of these standards and a dilution water blank. Plot carbon concentration in milligrams per liter against corrected peak height or area on rectangular coordinate paper. This is unnecessary for instruments provided with a digital readout of concentration.

With most TOC analyzers, it is not possible to determine separate blanks for reagent water, reagents, and the entire system. In addition, some TOC analyzers produce a variable and erratic blank that cannot be corrected reliably. In many laboratories, reagent water is the major contributor to the blank value. Correcting only the instrument response of standards (which contain reagent water + reagents + system blank) creates a positive error, while also correcting samples (which contain only reagents and system blank contributions) for

the reagent water blank creates a negative error. Minimize errors by using reagent water and reagents low in carbon.

Inject samples and procedural blanks (consisting of reagent water taken through any pre-analysis steps—values are typically higher than those for reagent water) and determine sample organic carbon concentrations directly from the readout or measurements by comparing corrected instrument response to the calibration curve. Instruments with coulometric detectors do not require calibration curves. Regularly analyze laboratory control samples to confirm performance of the instrument (see Quality Control, below). These detectors accumulate the system blank; therefore, monitor system blank regularly.

5. Calculations

Calculate corrected instrument response of standards and samples by subtracting the reagent-water blank instrument response from that of the standard and sample. Prepare a standard curve of corrected instrument response vs. TOC concentration. Subtract procedural blank from each sample instrument response and compare to standard curve to determine carbon content. Apply appropriate dilution factor when necessary. Subtract inorganic carbon from total carbon when TOC is determined by difference.

NOTE: The reagent water blank may include an instrument contribution not dependent on reagent-water carbon, and a true response due to reagent-water carbon. When reagent-water carbon is a significant fraction of reagent-water blank, a negative error no larger than reagent-water blank is introduced in the sample values. If TOC analyzer design permits isolation of each of the contributions to the total blank, apply appropriate blank corrections to instrument response of standards (reagent blank, water blank, system blank) and sample (reagent blank and system blank).

6. Quality Control

Determine instrument detection limit according to Section 1030.

After every tenth analysis, analyze a blank and a laboratory control sample prepared from a source of material other than the calibration standards, at a level similar to the analytical samples. Preferably prepare the laboratory control sample in a matrix similar to that of the samples. Alternatively, periodically make known additions to samples to ensure recovery from unknown matrices.

7. Precision

The difficulty of sampling particulate matter on unfiltered samples limits the precision of the method to approximately 5 to 10%.

Interlaboratory studies of high-temperature combustion methods have been conducted in the range above 2 mg/L.[1] The resulting equation for single-operator precision on matrix water is:

$$S_o = 0.027x + 0.29$$

Overall precision is:

$$S_t = 0.044x + 1.49$$

where:

S_o = single-operator precision,
S_t = overall precision, and
x = TOC concentration, mg/L.

8. Reference

1. AMERICAN SOCIETY FOR TESTING AND MATERIALS. 1994. Standard Test Method for Total and Organic Carbon in Water by High Temperature Oxidation and by Coulometric Detection. D4129–88. Annual Book of ASTM Standards. American Soc. Testing & Materials, Philadelphia, Pa.

9. Bibliography

KATZ, J., S. ABRAHAM & N. BAKER. 1954. Analytical procedure using a combined combustion-diffusion vessel: Improved method for combustion of organic compounds in aqueous solution. *Anal. Chem.* 26: 1503.

VAN HALL, C.E., J. SAFRANKO & V. A. STENGER. 1963. Rapid combustion method for the determination of organic substances in aqueous solutions. *Anal. Chem.* 35:315.

SCHAFFER, R.B. et al. 1965. Application of a carbon analyzer in waste treatment. *J. Water Pollut. Control Fed.* 37:1545.

VAN HALL, C.E., D. BARTH & V. A. STENGER. 1965. Elimination of carbonates from aqueous solutions prior to organic carbon determinations. *Anal. Chem.* 37:769.

BUSCH, A.W. 1966. Energy, total carbon and oxygen demand. *Water Resour. Res.* 2:59.

BLACKMORE, R.H. & D. VOSHEL. 1967. Rapid determination of total organic carbon (TOC) in sewage. *Water Sewage Works* 114:398.

WILLIAMS, R.T. 1967. Water-pollution instrumentation—Analyzer looks for organic carbon. *Instrum. Technol.* 14:63.

AMERICAN SOCIETY FOR TESTING AND MATERIALS. 1994. Standard Test Method for Total Organic Carbon in Water. D2579-93. Annual Book of ASTM Standards. American Soc. Testing & Materials, Philadelphia, Pa.

5310 C.   Persulfate-Ultraviolet or Heated-Persulfate Oxidation Method

1. General Discussion

Many instruments utilizing persulfate oxidation of organic carbon are available. They depend either on heat or ultraviolet irradiation activation of the reagents. These oxidation methods provide rapid and precise measurement of trace levels of organic carbon in water.

a. Principle: Organic carbon is oxidized to carbon dioxide, CO_2, by persulfate in the presence of heat or ultraviolet light. The CO_2 produced may be purged from the sample, dried, and transferred with a carrier gas to a nondispersive infrared (NDIR) analyzer, or be coulometrically titrated, or be separated from the liquid stream by a membrane that allows the specific passage of CO_2 to high-purity water where a change in conductivity is measured and related to the CO_2 passing the membrane.

Some instruments utilize an ultraviolet lamp submerged in a continuously gas-purged reactor that is filled with a constant-feed persulfate solution. The samples are introduced serially into the reactor by an autosampler or they are injected manually. The CO_2 produced is sparged continuously from the solution and is carried in the gas stream to an infrared analyzer that is specifically tuned to the absorptive wavelength of CO_2. The instrument's microprocessor calculates the area of the peaks produced by the analyzer, compares them to the peak area of the calibration standard stored in its memory, and prints out a calculated organic carbon value in milligrams per liter.

Other UV-persulfate instruments use continuous-flow injection of the sample into the instrument. Removal of inorganic carbon by vacuum degassing is provided optionally. The sample is acidified and persulfate added. Sample flow is split; one channel passes to a delay coil while the other passes through the UV reactor. The CO_2 from each stream is separated from the sample stream by membranes selectively permeable to CO_2 that allow the CO_2 to pass into high-purity water where change in conductivity is measured. CO_2 from the non-UV-irradiated stream represents inorganic carbon. CO_2 from the irradiated stream represents TC. The instrument automatically converts the detector signals to unit of concentration (mg/L or μg/L). The TOC is calculated as the difference between the TC and inorganic carbon channels.

Heated-persulfate instruments utilize a digestion vessel heated to 95 to 100°C. Samples are added by direct injection, loop injection, line injection, or autosampler. After inorganic carbon is removed by acidification and sparging, a measured amount of persulfate solution is added to the sample. After an oxidation period, the resulting CO_2 is sparged from the solution and carried to an infrared analyzer specifically tuned to the absorptive wavelength of CO_2. The instrument's microprocessor converts the detector signal to organic carbon concentrations in mg/L based on stored calibration data.

b. Interferences: See Section 5310B.1. Insufficient acidification will result in incomplete release of CO_2.

The intensity of the ultraviolet light reaching the sample matrix may be reduced by highly turbid samples or with aging of the ultraviolet source, resulting in sluggish or incomplete oxidation. Large organic particles or very large or complex organic molecules such as tannins, lignins, and humic acid may be oxidized slowly because persulfate oxidation is rate-limited. However, oxidation of many large biological molecules such as proteins and monoclonal antibodies proceeds rapidly. Because the efficiency of conversion of organic carbon to CO_2 may be affected by many factors, check efficiency of oxidation with selected model compounds representative of the compounds of interest in a matrix representative of the sample.

Some instruments give low results for certain difficult-to-oxidize compounds under certain conditions. The following compounds are difficult to oxidize, are sufficiently soluble in water, and can be mixed and measured accurately at trace levels: urea, nicotinic acid, pyridine, *n*-butanol, acetic acid, leucine, acetonitrile, octoxynol-9, tartaric acid, 1,10-phenanthroline, 1- glutonic acid, 2-propanol, and sodium dodecylbenzenesulfonate. Use these compounds as matrix additions to evaluate oxidation efficiency.

Persulfate oxidation of organic molecules is slowed in samples containing significant concentrations of chloride by the preferential oxidation of chloride; at concentrations above 0.05% chloride, oxidation of organic matter may be inhibited. To remove this interference add mercuric nitrate* to the persulfate solution in UV-persulfate instruments, or extend reaction time and/or increase amount of persulfate solution in heated-persulfate instruments.

With any organic carbon measurement, contamination during sample handling and treatment is a likely source of interference. This is especially true of trace analysis. Take extreme care in sampling, handling, and analysis of samples below 1 mg TOC/L.

c. *Minimum detectable concentration:* Concentration of 0.01 mg TOC/L can be measured by some instruments if scrupulous attention is given to minimizing sample contamination and method background. See Section 1030 for procedures to evaluate the MDL for a specific instrument. Use the high-temperature combustion method (B) for high concentrations of TOC or dilute the sample, ensuring that the dilution process does not contaminate the sample.

d. *Sampling and storage:* See Section 5310B.1d.

2. Apparatus

a. *Total organic carbon analyzer* utilizing persulfate oxidation principle.

b. *Sampling and injection accessories,* as specified by the instrument manufacturer.

3. Reagents

a. *Reagents listed in Section 5310B.3.*

b. *Persulfate solution:* Different instrument manufacturers recommend different forms and concentrations of peroxydisulfate. Typical preparations are as follows:

1) *Sodium peroxydisulfate, 10%:* Dissolve 100 g reagent in water; bring volume to 1 L.

2) *Ammonium peroxydisulfate, 15%:* Dissolve 150 g reagent in water; bring volume to 1 L.

3) *Potassium peroxydisulfate, 2%:* Dissolve 20 g reagent in water; bring volume to 1 L.

Check blank values from reagents and, if values are high, purify reagent or use a higher-purity source.

4. Procedure

a. *Instrument operation:* Follow manufacturer's instructions for assembly, testing, calibration, and operation.

b. *Sample preparation:* If a sample contains gross particulates or insoluble matter, homogenize until a representative portion can be withdrawn through the syringe needle, autosampler tubing, or sample inlet system of continuous on-line monitor.

If dissolved organic carbon is to be determined, filter sample and a reagent water blank through 0.45-μm filter. HPLC syringe filters have been found to pass water without contamination. Glass fiber or silver membrane filters also can be used. Check filter blanks regularly.

To determine nonpurgeable organic carbon, transfer 15 to 30 mL sample to a flask or test tube and acidify to a pH of 2. Purge according to manufacturer's recommendations. In some instruments this is performed internally. Check efficiency of inorganic carbon removal for each sample matrix by splitting a sample into two portions; to one of the portions, add inorganic carbon to a level similar to that of the sample. The TOC values should agree.

If the values do not agree, adjust conditions such as sample container, sample volume, pH, purge-gas flow rate, and purge time to obtain complete removal of inorganic carbon.

c. *Sample injection:* See Section 5310B.4c.

d. *Standard curve preparation:* Prepare an organic carbon standard series over the range of organic carbon concentrations in the samples. Run standards and blanks and record analyzer's response. Determine instrument response for each standard and blank. Unless carbon dioxide is trapped and desorbed, producing consistent peak heights, determinations based on peak height may be inadequate because of differences in the rate of oxidation of standards and samples. Correct instrument response of standards by subtracting reagent water blank and plot organic carbon concentration in milligrams per liter against corrected instrument response. For instruments providing a digital computation of concentration, this is not necessary. Be sure that the instrument's algorithm includes blank correction and linearity of response. Analyze standards having concentrations above and below those determined in the samples, preferably prepared in a similar matrix, to confirm proper instrument operation.

5. Calculation

See Section 5310B.5, or use instrument manufacturer's procedure.

6. Quality Control

See Section 5310B.6.

7. Precision and Bias

Interlaboratory studies of persulfate and/or UV with NDIR detection methods have been conducted in the range of 0.1 mg/L to 4 000 mg/L of carbon.[1] The resulting equation for organic carbon, single-operator precision is:

$$S_o = 0.04x + 0.1$$

Overall precision is expressed as:

$$S_t = 0.08x + 0.1$$

where:

S_o = single-operator precision,
S_t = overall precision, and
x = TOC concentration, mg/L.

An interlaboratory study was conducted for the membrane conductivity method,† covering samples with 1 to 25 mg/L organic carbon concentrations. The resulting equation for single-operator precision is:

$$S_o = 0.012x - 0.022$$

Overall precision is expressed as:

$$S_t = 0.027x + 0.09$$

where terms are defined as above.

---

*NOTE: If mercuric nitrate is used to complex the chloride, use an appropriate disposal method for the treated waste to prevent mercury contamination.

†Data may be obtained from *Standard Methods* manager, American Water Works Association.

8. Reference

1. AMERICAN SOCIETY FOR TESTING AND MATERIALS. 1994. Standard Test Method for Total Carbon in Water by Ultraviolet, or Persulfate Oxidation, or Both, and Infrared Detection. D4839-88. Annual Book of ASTM Standards. American Soc. Testing & Materials, Philadelphia, Pa.

9. Bibliography

BEATTIE, J., C. BRICKER & D. GARVIN. 1961. Photolytic determination of trace amounts of organic material in water. Anal. Chem. 33:1890.

ARMSTRONG, F.A.J., P.M. WILLIAMS & J.D.H. STRICKLAND. 1966. Photooxidation of organic matter in sea water by ultraviolet radiation, analytical and other applications. Nature 211:481.

BRAMER, H.C., M.J. WALSH & S.C. CARUSO. 1966. Instrument for monitoring trace organic compounds in water. Water Sewage Works 113:275.

DOBBS, R.A., R.H. WISE & R.B. DEAN. 1967. Measurement of organic carbon in water using the hydrogen flame ionization detector. Anal. Chem. 39P:1255.

MONTGOMERY, H.A.C. & N.S. THOM. 1967. The determination of low concentrations of organic carbon in water. Analyst 87:689.

ARMSTRONG, F.A.J. & S. TIBBITS. 1968. Photochemical combustion of organic matter in seawater for nitrogen, phosphorus and carbon determination. J. Mar. Biol. Assoc. U.K. 48:143.

JONES, R.H. & A.F. DAGEFORD. 1968. Application of a high sensitivity total organic carbon analyzer. Proc. Instr. Soc. Amer. 6:43.

TAKAHASHI, Y., R.T. MOORE & R.J. JOYCE. 1972. Direct determination of organic carbon in water by reductive pyrolysis. Amer. Lab. 4:31.

COLLINS, K.J. & P.J. LeB. WILLIAMS. 1977. An automated photochemical method for the determination of dissolved organic carbon in sea and estuarine waters. Mar. Chem. 5:123.

GRAVELET-BLONDIN, L.R., H.R. VAN VLIET & P.A. MYNHARDT. 1980. An automated method for the determination of dissolved organic carbon in fresh water. Water SA 6:138.

OAKE, R.J. 1981. A Review of Photo-Oxidation for the Determination of Total Organic Carbon in Water. Water Research Center, Technical Rep. (TR 160). Medmenham, England.

VAN STEENDEREN, R.A. & J.S. LIN. 1981. Determination of dissolved organic carbon in water. Anal. Chem. 53:2157.

AIKEN, G.R. 1992. Chloride interference in the analysis of dissolved organic carbon by the wet oxidation method. Environ. Sci. Technol. 26:2435.

GODEC, R., K. O'NEIL, R. HUTTE. 1992. New Technology for TOC Analysis in Water, Ultrapure Water 9(9):17.

5310 D.   Wet-Oxidation Method

1. General Discussion

The wet-oxidation method is suitable for the analyses of water, water-suspended sediment mixtures, seawaters, brines, and wastewaters containing at least 0.1 mg nonpurgeable organic carbon/L. The method is not suitable for the determination of volatile organic constituents.

a. Principle: The sample is acidified, purged to remove inorganic carbon, and oxidized with persulfate in an autoclave at temperatures from 116 to 130°C. The resultant carbon dioxide (CO_2) is measured by nondispersive infrared spectrometry.

b. Interferences: See Section 5310B.1 and C.1.

c. Minimum detectable concentrations: High concentrations of reducing agents may interfere. Concentration of 0.10 mg TOC/L can be measured if scrupulous attention is given to minimizing sample contamination and method background. Use the high-temperature combustion method (B) for high concentrations of TOC.

d. Sampling and storage: See Section 5310B.1d.

2. Apparatus

a. Ampules, precombusted, 10-mL, glass.
b. Ampule purging and sealing unit.
c. Autoclave.
d. Carbon analyzer.
e. Homogenizer.

3. Reagents

In addition to the reagents specified in Section 5310B.3a, c, e, and f, the following reagents are required:

a. Phosphoric acid solution, H_3PO_4, 1.2N: Add 83 mL H_3PO_4 (85%) to water and dilute to 1 L with water. Store in a tightly stoppered glass bottle.

b. Potassium persulfate, reagent-grade, granular. Avoid using finely divided forms.

4. Procedure

Follow manufacturer's instructions for instrument assembly, testing calibration, and operation. Add 0.5 mL 1.2N H_3PO_4, solution to precombusted ampules.

To analyze for dissolved organic carbon, follow the filtration procedure in Method B. Homogenize sample to produce a uniform suspension. Rinse homogenizer with reagent water after each use. Pipet water sample (10.0 mL maximum) into an ampule. Adjust smaller volumes to 10 mL with reagent water. Prepare one reagent blank (10 mL reagent water plus acid and oxidant) for every 15 to 20 water samples. Prepare standards covering the range of 0.1 to 40 mg C/L by diluting the carbon standard solution. Immediately place filled ampules on purging and seating unit and purge them at rate of 60 mL/min for 6 min with purified oxygen. Add 0.2 g potassium persulfate using a dipper calibrated to deliver 0.2 g to the ampule. Seal samples according to the manufacturer's instructions. Place sealed samples, blanks, and a set of standards in ampule racks in an autoclave and digest 4 h at temperature between 116 and 130°C.

Set sensitivity range of carbon analyzer by adjusting the zero and span controls in accordance with the manufacturer's instructions. Break combusted ampules in the cutter assembly of the carbon analyzer, sweep CO_2 into the infrared cell with nitrogen gas, and record area of each CO_2 peak. CAUTION: *Because combusted ampules are under positive pressure, handle with care to prevent explosion.*

5. Calculations

Prepare an analytical standard curve by plotting peak area of each standard versus concentration (mg/L) of organic carbon standards. The relationship between peak area and carbon concentration is curvilinear. Define operating curves each day samples are analyzed.

Report nonpurgeable organic carbon concentration as follows: 0.1 mg/L to 0.9 mg/L, one significant figure; 1.0 mg/L and above, two significant figures.

6. Quality Control

See Section 5310B.6.

7. Precision and Bias

Multiple determinations of four different concentrations of aqueous potassium acid phthalate samples at 2.00, 5.00, 10.0, and 40.0 mg C/L resulted in mean values of 2.2, 5.3, 9.9, and 38 mg/L and standard deviations of 0.13, 0.15, 0.11, and 1.4, respectively.

Precision also may be expressed in terms of percent relative standard deviation as follows:

Number of Replicates	Mean mg/L	Relative Standard Deviation %
9	2.2	5.9
10	5.3	2.8
10	9.9	1.1
10	38.0	3.7

8. Bibliography

WILLIAMS, P.M. 1969. The determination of dissolved organic carbon in seawater: A comparison of two methods. *Limnol. Oceanogr.* 14:297.

OCEANOGRAPHY INTERNATIONAL CORP. 1970. The total carbon system operating manual. College Station, Tex.

MACKINNON, M.D. 1981. The measurement of organic carbon in seawater. *In* E.K. Duursma & R. Dawson, eds., Marine Organic Chemistry: Evolution, Composition, Interactions and Chemistry of Organic Matter in Seawater. Elsevier Scientific Publishing Co., New York, N.Y.

5320   DISSOLVED ORGANIC HALOGEN*

5320 A.   Introduction

Dissolved organic halogen (DOX) is a measurement used to estimate the total quantity of dissolved halogenated organic material in a water sample. This is similar to literature references to "total organic halogen" (TOX), "adsorbable organic halogen" (AOX), and carbon-adsorbable organic halogen (CAOX). The presence of halogenated organic molecules is indicative of disinfection by-products and other synthetic chemical contamination. Halogenated compounds that contribute to a DOX result include, but are not limited to: the trihalomethanes (THMs); organic solvents such as trichloroethene, tetrachloroethene, and other halogenated alkanes and alkenes; chlorinated and brominated pesticides and herbicides; polychlorinated biphenyls (PCBs); chlorinated aromatics such as hexachlorobenzene and 2,4-dichlorophenol; and high-molecular-weight, partially chlorinated aquatic humic substances. Compound-specific methods such as gas chromatography typically are more sensitive than DOX measurements.

The adsorption-pyrolysis-titrimetric method for DOX measures only the total molar amount of dissolved organically bound halogen retained on the activated carbon adsorbent; it yields no information about the structure or nature of the organic compounds to which the halogens are bound or about the individual halogens present. It is sensitive to organic chloride, bromide, and iodide, but does not detect fluorinated organics.

DOX measurement is an inexpensive and useful method for screening large numbers of samples before specific (and often more complex) analyses; for extensive field surveying for pollution by certain classes of synthetic organic compounds in natural waters; for mapping the extent of organohalide contamination in groundwater; for monitoring the breakthrough of some synthetic organic compounds in water treatment processes; and for estimating the level of formation of chlorinated organic by-products after disinfection. When used as a screening tool, a large positive (i.e., above background measurements) DOX test result indicates the need for identifying and quantifying specific substances. In saline or brackish waters the high inorganic halogen concentrations interfere. The possibility of overestimating DOX concentration because of inorganic halide interference always should be considered when interpreting results.

* Approved by Standard Methods Committee, 1997.

5320 B. Adsorption-Pyrolysis-Titrimetric Method

1. General Discussion

a. Principle: The method consists of four processes. First, dissolved organic material is separated from inorganic halides and concentrated from aqueous solution by adsorption onto activated carbon. Second, inorganic halides present on the activated carbon are removed by competitive displacement by nitrate ions. Third, the activated carbon with adsorbed organic material is introduced into a furnace that pyrolyzes organic carbon to carbon dioxide (CO_2) and the bound halogens to hydrogen halide (HX). Fourth, the HX is transported in a carrier gas stream to a microcoulometric titration cell where the amount of halide is quantified by measuring the current produced by silver-ion precipitation of the halides. The microcoulometric detector operates by maintaining a constant silver-ion concentration in a titration cell. An electric potential is applied to a solid silver electrode to produce silver ions in the cell solution. As hydrogen halide from the pyrolysis furnace enters the cell in the carrier gas, it is partitioned into the acetic acid solution where it precipitates as silver halide. The current that is produced is integrated over the period of the pyrolysis. The integrated area under the curve is proportional to the number of moles of halogen recovered. The mass concentration of organic halides is reported as an equivalent concentration of organically bound chloride in micrograms per liter. Because this DOX procedure relies on activated carbon to adsorb organic halides, it also has been referred to as carbon-adsorbable organic halogen (CAOX). Because of the poor adsorption efficiency of some organic compounds containing halogen and the desorption of some halogen-containing compounds during the removal of adsorbed inorganic halogen, this method does not measure total organic halogen.

When a sample is purged with inert gas before activated carbon adsorption, analysis of that sample determines the nonpurgeable dissolved organic halogen (NPDOX) fraction of DOX. The purgeable organic halogen concentration (POX) may be estimated by subtracting the NPDOX value from the DOX value. Alternatively, the POX fraction may be determined directly by purging the sample with carrier gas and introducing that gas stream and the volatilized organics directly into the pyrolysis furnace. Thus, depending on approach, the analysis of POX, DOX, and NPDOX may be determined directly or by difference. Finally, because the POX often is dominated by the THMs, they may be determined as directed in Section 6200 and used to estimate POX. However, this approach is not included here as a standardized procedure.

b. Interferences: The method is applicable only to aqueous samples free of visible particulate matter. Different instruments vary in tolerance of small amounts of suspended matter. Inorganic substances such as chloride, chlorite, chlorate, bromate, bromide, and iodide will adsorb on activated carbon to an extent dependent on their original concentration in the aqueous solution and the volume of sample adsorbed.[1] Positive interference will result if inorganic halides are not removed. Treating the activated carbon with a concentrated aqueous solution of nitrate ion causes competitive desorption from the activated carbon of inorganic halide species and washes inorganic halides from other surfaces. However, if the inorganic halide concentration is greater than 10 000 times[2] the concentration of organic halides, the DOX results may be affected significantly. In general, this procedure may not be applicable to samples with inorganic halide concentrations above 500 mg Cl^-/L, based on activated carbon quality testing results. Therefore, consider both the results of mineral analysis for inorganic halides and the results of the activated carbon quality test (see ¶ 5, below) when interpreting results.

Halogenated organic compounds that are weakly adsorbed on activated carbon are recovered only partially. These include certain alcohols and acids (e.g., chloroethanol), and such compounds as chloroacetic acid, that can be removed from activated carbon by the nitrate ion wash. However, for most halogenated organic molecules, recovery is very good; the activated carbon adsorbable organic halide (CAOX) therefore is a good estimate of true DOX.

Failure to acidify samples with nitric acid or sulfuric acid may result in reduced adsorption efficiency for some halogenated organic compounds and may intensify the inorganic halide interference. However, acidification may result in precipitation loss of humic acids and any DOX associated with that fraction. Further, if the water contains residual chlorine, reduce it before adsorption to eliminate positive interference resulting from continued chlorination reactions with organic compounds adsorbed on the activated carbon surface or with the activated carbon surface itself. The sulfite dechlorinating agent may cause decomposition of a small fraction of the DOX if nitric acid is used; this decomposition is avoided if sulfuric acid is used. Do not add acid in great excess.

Highly volatile components of the POX fraction may be lost during sampling, shipment, sample storage, sample handling, and sample preparation, or during sample adsorption. A laboratory quality-control program to ensure sample integrity from time of sampling until analysis is vital. During sample filtration for the analysis of samples containing undissolved solids, major losses of POX can be expected. Syringe-type filtration systems can minimize losses. Analyze for POX before sample filtration and analyze for NPDOX after filtration; the sum of POX and NPDOX is the total DOX. In preparing samples for DOX analysis, process a blank and a standard solution to determine effect of this procedure on DOX measurement. If an insignificant loss of POX occurs during the removal of particulate matter by filtration, DOX may be measured directly.

Granular activated carbon used to concentrate organic material from the sample can be a major source of variability in the analysis and has a dramatic effect on the minimum detectable concentration. Ideally, activated carbon should have a low halide content, readily release adsorbed inorganic halides on nitrate washing, be homogeneous, and readily adsorb *all* organic halide compounds even in the presence of large excesses of other organic material. An essential element of quality control for DOX requires testing and monitoring of activated carbon (see ¶ 5 below). Non-homogeneous activated carbon or activated carbon with a high background value affects the method reliability at low concentrations of DOX. A high and/or variable blank value raises the minimum detectable concentration. Random positive bias, in part because of the ease of activated carbon contamination during use, may necessitate analyzing duplicates of each sample. Because activated carbon from different sources may vary widely in the ease of releasing inorganic halides, test for this quality before using activated carbon. Proper quantification also may be affected by the adsorptive capacity of the activated carbon. If excessive organic loading occurs, some DOX may break through and not be recovered. For this reason, make serial adsorptions of each sample portion and individual analyses.

c. Sampling and storage: Collect and store samples in amber glass bottles with TFE-lined caps. If amber bottles are not available, store samples in the dark. To prepare sample bottles, acid wash, rinse with deionized water, seal with aluminum foil, and bake at 400°C for at least 1 h. If bottle blanks without baking show no detectable DOX, baking may be omitted. Wash septa with detergent, rinse repeatedly in organic-free, deionized water, wrap in aluminum foil, and bake for 1 h at 100°C. Preferably use thick silicone rubber-backed TFE septa and open ring caps to produce a positive seal that prevents loss of POX and contamination. Store sealed sample bottles in a clean environment until use. Completely fill sample bottles but take care not to volatilize any organic halogen compounds. Preserve samples that cannot be analyzed promptly by acidifying with concentrated nitric acid or sulfuric acid to pH 2. Refrigerate samples at 4°C with minimal exposure to light. Reduce any residual chlorine by adding sodium sulfite crystals (minimum: 5 mg/L). Add 4 drops conc H_2SO_4 plus sodium sulfite crystals to bottles shipped to the field. NOTE: Some organic chloramines are not completely dechlorinated by sodium sulfite, particularly at pH > 7. This may affect reported concentrations.[1] Analyze all samples within 14 d.

d. Minimum detectable concentration: For nonsaline waters free of particulate matter, 5 to 10 µg organic Cl^-/L is considered a typical range for detection limits. The minimum detectable concentration may be influenced by the analytical repeatability, equipment used, activated carbon quality, and the analyst. Determine the detection limit for each procedure, instrument, and analyst.

2. Apparatus

a. Adsorption assembly, including gas-tight sample reservoir, activated carbon-packed adsorption columns, column housings, and nitrate solution reservoir. In particular, note the following:

1) *Noncombustible insulating material (microcolumn method only):* Form into plugs to hold activated carbon in columns. NOTE: *Do not touch with fingers.*

2) *Activated carbon columns (microcolumn method only):* Pack 40 ± 5 mg activated carbon (¶ 3k) into dry glass tubing approximately 2 to 3 mm ID × 6 mm OD × 40 to 50 mm long. NOTE: *Protect these columns from all sources of halogenated organic vapors.* Clean glass tubes before use with a small-diameter pipe cleaner to remove residual carbon, then soak in chromate cleaning solution for 15 min and dry at 400°C. Rinse between steps with deionized water. NOTE: Use prepacked columns with caution, because of occasional reported contamination.

b. Analyzer assembly, including carrier gas source, boat sampler, and pyrolysis furnace, that can oxidatively pyrolyze halogenated organics at a temperature of 800 to 900°C to produce hydrogen halides and deliver them to the titration cell with a minimum overall efficiency of 90% for 2,4,6-trichlorophenol; including a microcoulometric titration system with integrator, digital display, and data system or chart recorder connection; including (optional) purging apparatus.

c. Chart recorder or microprocessor, controlled data system.

d. Batch adsorption equipment: Use instrument manufacturer's purge vessel or similar purging flask, erlenmeyer flasks (100 to 250 mL), and high-speed stirrers.

e. Filtering apparatus and filters: Use 0.45-µm-pore diam filters, preferably HPLC syringe filters or similar, with no detectable DOX blank. Rinsed glass-fiber filters are satisfactory for sample filtration. Preferably use membrane filters for separating activated carbon from aqueous phase.

3. Reagents and Materials

Use chemicals of ACS reagent grade or other grades if it can be demonstrated that the reagent is of sufficiently high purity to permit its use without lessening accuracy of the determination.

a. Carbon dioxide, argon, or nitrogen, as recommended by the equipment manufacturer, purity 99.99%.

b. Oxygen, purity 99.99%.

c. Aqueous acetic acid, 70 to 85%, as recommended by the equipment manufacturer.

d. Sodium chloride standard, NaCl: Dissolve 0.1648 g NaCl and dilute to 100 mL with reagent water; 1 µL = 1 µg Cl^-.

e. Ammonium chloride standard, NH_4Cl: Dissolve 0.1509 g NH_4Cl and dilute to 100 mL with reagent water; 1 µL = 1 µg Cl^-.

f. Trichlorophenol stock solution: Dissolve 1.856 g trichlorophenol and dilute to 100 mL with methanol; 1 µL = 10 µg Cl^-.

g. Trichlorophenol standard solution: Make a 1:20 dilution of the trichlorophenol stock solution with methanol; 1 µL = 0.5 µg Cl^-.

h. Trichloroacetic acid stock solution: Dilute 199.44 mg trichloroacetic acid in 1000 mL reagent water; 1 mL = 130 µg Cl^-.

i. Trichloroacetic acid standard solution: Dilute 2.0 mL trichloroacetic acid stock solution into 1000 mL with reagent water; 1 mL = 0.260 µg Cl^-.

j. Chloroform standard solution, $CHCl_3$: Dilute 100 mg $CHCl_3$ to 100 mL with methanol; 1 µL = 1 µg $CHCl_3$.

k. Blank standard: Use reagent water. Reagent water preferably is carbon-filtered, deionized water that has been heated and purged.

l. Nitrate wash solution, 0.08M: Dilute 8.2 g KNO_3 to 1000 mL with reagent water. Adjust to pH 2 with HNO_3. 1 L = 5000 mg NO_3^-.

m. Activated carbon, 100 to 200 mesh: Ideally use activated carbon having a very low apparent halide background that readily releases adsorbed inorganic halides on nitrate washing, and reliably adsorbs organic halides in the presence of a large excess of other organic compounds.* See ¶ 5 below for preparation and evaluation of activated carbon. CAUTION: *Protect activated carbon from contact with halogenated organic vapors.*

n. Sodium sulfite, Na_2SO_3, crystals.

o. Nitric acid, HNO_3, conc, or *sulfuric acid,* H_2SO_4, conc.

4. Procedure

Use either the microcolumn (4a) or batch adsorption (4b) method to determine DOX (as CAOX). If present, determine POX separately (¶4c). The microcolumn method utilizes small glass columns packed with activated carbon through which the sample is passed under positive pressure to adsorb the organic halogen compounds. The batch adsorption method uses a small quantity of activated carbon that is added to the sample. After stirring,

---

* Westvaco or Calgon Filtrasorb 400 or equivalent.

activated carbon is removed by filtration, washed with nitrate, and analyzed. The batch adsorption procedure typically is run on samples that have had POX analyzed directly [¶4c1)], yielding NPDOX directly as well.

a. Microcolumn procedure:

1) *Apparatus setup*—Adjust equipment in accordance with the manufacturer's instructions. Make several injections of NaCl solution directly into the titration cell [¶ 5c1)] as a microcoulometer/titration cell check at the start of each day.

2) *Sample pretreatment for DOX analysis*—If the sample has not been acidified during collection, adjust pH to 2 with HNO_3 or H_2SO_4. If the samples contain undissolved solids, filter through a glass-fiber filter (other means of removing particulate matter may be used, if it can be demonstrated that they do not cause significant interferences). Also filter a blank and standard. Analyze these to determine the contribution of filtration to the organic halogen measurement. Vacuum filtration will cause some loss of volatile organic halogen. Analyze for POX (¶4c) before filtration and NPDOX after filtration, unless it is shown that POX losses during filtration are insignificant for a specific water type.

3) *Sample adsorption*—Transfer a representative portion of sample to the cleaned sample reservoir with two activated carbon adsorption columns in series attached by the column housings to the reservoir outlet. Seal the reservoir. Adjust to produce a flow rate of about 3 mL/min. When the desired volume has been processed, stop the flow, detach the activated carbon housings and columns, and rinse the sample reservoir twice with reagent-grade water. Vary volume processed to produce optimum quantities of adsorbed DOX on the columns. Suggested volumes are as follows:

Volume Processed mL	Instrument Optimum Range $\mu g\ Cl^-$	Conc of DOX in Waters $\mu g/L$
100	0.5–50	5–50
50	12.5–50	250–1000
25	12.5–50	500–2000

If possible, avoid using volumes greater than 100 mL because the maximum adsorptive capacity of the activated carbon may be exceeded, leading to adsorbate breakthrough and loss of DOX. Larger sample volumes processed lead to an increased quantity of inorganic halide accumulated on the activated carbon and may result in a positive interference. Do not use a sample less than 25 mL to minimize volumetric errors. For samples exceeding 2000 µg DOX/L dilute before adsorption. Protect columns from the atmosphere until DOX is determined.

4) *Inorganic halide removal*—Attach columns through which sample has been processed in series to the nitrate wash reservoir and pass 2 to 5 mL NO_3^- solution through the columns at a rate of approximately 1 mL/min.

5) *DOX determination*—After concentrating sample on activated carbon and removing inorganic halogens by nitrate washing, pyrolyze contents of each microcolumn and determine organic halogen content. Remove top glass microcolumn from the column housing, taking care not to contaminate the sample with inorganic halides. Using a clean ejector rod, eject the activated carbon and noncombustible insulating material plugs into the sample boat. Prepare sample boat during the preceding 4 h by heating at 400

to 800°C for at least 4 min in an oxygen-rich atmosphere (i.e., in the pyrolysis furnace). Remove residual ash. Place ejector rod on the plug of the effluent end of the carbon microcolumn and place the influent end of the carbon microcolumn in the quartz boat first. Seal sample inlet tube and let instrument stabilize. After NO_3^- wash avoid contact with inorganic halides. Wear latex gloves while carrying out this procedure. Preferably clean work area frequently with deionized water.

Pyrolyze the activated carbon and determine halide content. Repeat for each microcolumn. Check for excess breakthrough (¶ 5b) and repeat analysis as necessary.

6) *Replicates*—When DOX determination is used strictly as a screening tool, total replication is not necessary. Single-operator precision (% CV) is expected to be less than 15% for tap water and wastewater (Table 5320:I). If system performance is consistently worse as demonstrated by routine QA duplicates, or if quality objectives dictate, run replicates of each sample by repeating steps 3, 4, and 5.

7) *Blanks*—Analyze one method blank [¶ 5e2)] with each set of ten samples. Preferably analyze the method blank before starting the sample set and run a blank after the last set of the day.

8) *Preparation and analysis of calibration standard*—Run daily calibration standards in accordance with ¶ 5c3) for POX analysis or ¶ 5c5) for microcolumn-adsorption DOX analysis. Accompany by a suitable blank [¶ 5e3) or ¶ 5e4)]. Be certain that analytical conditions and procedures (e.g., purging temperature) are the same for the analysis of calibration standards as for the analysis of samples.

b. Batch adsorption procedure:

1) *Apparatus setup*—Adjust equipment in accordance with the manufacturer's instructions.

2) *Sample pretreatment*—Adjust sample pH to 2 with conc HNO_3 or H_2SO_4 [see ¶ 4a2)].

3) *Sample adsorption*—Prepare carbon suspension by adding high-quality activated carbon to high-purity, deionized, granular activated carbon (GAC)-treated water to produce a uniform suspension of 10 mg carbon/mL. To an erlenmeyer flask, transfer prepurged sample of optimum size from a purging flask standardized in the same manner as the instrument's purging vessel. Add 20 mg activated carbon (2 mL carbon suspension). Using a high-speed mixer (20 000 rpm), stir for 45 min in an organohalide vapor-free environment. Filter through a membrane filter under vacuum or pressure, and collect filtrate. Remove flask containing filtrate. Wash carbon cake and filter with 10 mL NO_3^- wash solution. Add portions of wash solution serially to keep activated carbon and NO_3^- solution in contact for 15 min. Using clean instruments, transfer carbon cake and membrane filter to pyrolysis unit sample boat. Let instrument stabilize, pyrolyze, and determine the halide content of the first serial filter.

Add 20 mg more activated carbon to filtrate in erlenmeyer flask. Repeat carbon mixing, filtering, and washing procedures. Pyrolyze and determine halide content of second serial filter. If the second value is greater than 10% of the total value (first plus second), perform the NPDOX determination on an additional sample portion.

c. POX procedure (optional) (direct purge): Adjust apparatus [¶ 4a1)]. Select sample volume by comparing expected POX value (if known) with optimum instrument range. Using a gastight syringe, inject sample through septum into purge vessel, and purge as recommended by equipment manufacturer. Carefully control gas flow rate, sample temperature, and purging time. The maximum POX that can be determined is:

$$POX_{max}, \mu m/L = \frac{0.5 \times 1000}{mL\ sample \times 35.5}$$

If replicates are analyzed, sampling from replicate sample bottles may minimize variability due to volatilization losses.

5. Quality Control

a. Activated carbon quality: Purchase activated carbon ready for use or prepare activated carbon by milling and sieving high-quality activated carbon. Use only 100- to 200-mesh carbon in the microcolumn method. During preparation, take care not to expose the activated carbon to organic vapors. Use of a clean room is helpful. Prepare only small quantities (a month's supply or less) at one time. Discard the activated carbon if its DOX background concentration has increased significantly from the time of preparation or if the background is greater than 1 μg apparent organic Cl$^-$/40 mg activated carbon. Uniformity of activated carbon is important; therefore, after sieving small portions, combine and mix thoroughly. Transfer representative portions to clean glass bottles with ground-glass stoppers or with rubber-backed TFE septa and open ring caps. Store bottles in a gas-purged, evacuated, sealed desiccator.

Test each newly prepared batch of activated carbon to ensure adequate quality before use. Use only activated carbon meeting the guidelines outlined below.

1) Check activated carbon particle size by applying deionized water to two 40-mg activated carbon microcolumns. If flow rate is significantly less than 3 mL/min, resieve activated carbon to remove excess fines.

2) Analyze a pair of method blanks, ¶ 5*e*2). Reject carbon if the apparent organic halogen exceeds 1.2 μg/40 mg activated carbon.

If the activated carbon originated from a previously untested batch from a commercial supplier, test it for adsorption efficiency and inorganic halide rejection.

3) Adsorb replicate 100-mL portions of solutions containing 100, 500, and 1000 mg inorganic Cl$^-$/L deionized water. Wash with nitrate solution and analyze. The apparent organic halogen yield should not increase by more than 0.50 μg over the value determined in 2) above. A greater increase indicates significant interference at that concentration.

b. Serial adsorption: Each aqueous standard and sample is serially adsorbed on activated carbon in both procedures given above. Of the net organic halide, 90% or more should be adsorbed on the first activated carbon portion and the remaining 10% or less on the second. If, upon separate analysis of the two serial activated carbon portions, the second shows more than 10% of the net (after subtracting the method blank), reanalyze sample. Inorganic halogen interference or organic breakthrough are the most common reasons for a high second activated carbon value. Sample dilution before adsorption may improve recovery on the first activated carbon in series, but the minimum detectable concentration will be affected.

c. Standards: The standards used in routine analysis, quality control testing, and isolating specific causes during corrective maintenance include:

1) Sodium chloride standard (¶ 3*d*)—Use to check functioning of the titration cell and microcoulometer by injecting directly into the acetic acid solution of the titration cell. By examining the height and shape of the peak produced on the chart recorder and from the integrated value, problems associated with the cell and coulometer may be isolated. Use this standard at startup each day and after cell cleaning throughout the day. At daily startup consecutive duplicates should be within 3% of the historical mean. Depending on sample loading and number of analyses performed, it may be necessary to clean the titration cell several times per day. After cleaning, cell performance may be very unstable; therefore, inject a single NaCl standard before analyzing an instrument calibration standard [see ¶ 4) below]. Do *not* introduce NaCl standards into the pyrolysis furnace by application to the sample boat.

2) Ammonium chloride standard (¶ 3*e*)—Apply this standard to the sample boat to check for loss of halide in the pyrolysis furnace and entrance of the titration cell. Typically, this may be necessary when injection of a NaCl standard indicates proper titration cell and microcoulometer function but the recovery of the calibration standard is poor: suspect either poor conversion of organic chloride to hydrogen chloride or loss of hydrogen halide after conversion but before partitioning into the cell solution. To isolate the possible loss of hydrogen halides inject NH$_4$Cl standard directly onto the quartz sample boat. Recovery should be better than 95%, with a single peak of uniform shape produced. Use only a new quartz sample boat free of any residue; an encrusted boat dramatically reduces recovery. Use this standard for corrective maintenance problem isolation but not for routine analyses.

3) Purgeable organic halide calibration standards—For the POX analysis use aqueous chloroform solutions for instrument calibration. Also for POX analysis an aqueous bromoform standard can be used initially to insure acceptable purging conditions. Develop a standard curve over the dynamic range of the microcoulometer and check daily as in ¶ 5*c*5). Recovery of chloroform and bromoform should exceed 90% and 80%, respectively.

4) Instrument calibration standard—Direct injection of trichlorophenol working standard onto the nitrate-washed method blank in concentrations over the working range of the instrument determines linearity and calibration of the analyzer module. After checking for proper microcoulometer function by injecting NaCl standard, pyrolyze duplicate instrument calibration standards and then duplicate method blanks. The net response to the calibration standards should be within 3% of the calibration curve value. If not, check for loss of halide in the pyrolysis furnace using the ammonium chloride standard [¶ 5*c*2)].

5) Nonvolatile organic halide calibration standards—Develop an initial standard curve by analyzing aqueous solutions of 2,4,6-trichlorophenol, trichloroacetic acid (commonly formed during chlorination), or another appropriate halogenated organic compound over the dynamic range of the microcoulometer. This dynamic range typically is from 0.5 to 50 μg chloride, but will vary between microcoulometers and titration cells. Construct an initial calibration curve using five calibration standards in range of 0.5 to 50 μg organic chloride; recheck calibration curve after changes in an instrument's configuration, such as replacement of a titration cell or major instrument maintenance. Daily, analyze a calibration standard to check proper function of the instrumentation and procedures. Select check standard in the concentration range of samples to be analyzed that day. When sample filtration is used to remove particulate matter, also use this pretreatment with the calibration standard. If DOX recovery is less than 90%, analyze a set of instrument calibration standards [¶ 5*c*4)].

d. Standard addition recovery: During routine analyses, ideally make standard additions to every tenth sample. Where the compounds constituting the DOX are known, use standards of these compounds. Where the compounds constituting the DOX are wholly or partially unknown, use standards reflecting the relative abundance of the halogens, the molecular size, and the volatility of the halogenated compounds presumed to be present. Recovery of 90% or more of the added amount indicates that the analyses are in control. Do not base acceptance of data on standard addition recoveries.

e. Blanks: High precision and accuracy of the background or blank value is important to the accurate measurement of DOX. Make blank measurements daily. Blanks that may be required are:

1) Reagent water blank—Analyze each batch of organic-free reagent water. The blank should have less than the minimum detectable concentration. Use this blank to insure that the standards, equipment, and procedures are not contributing to the DOX. Once reagent water blank is demonstrated, it can be used to determine method blank and POX blank as described below.

2) Method blank—Analyze activated carbon that has been nitrate-washed. Analyze method blanks daily before sample analysis and after at least each 10 to 14 sample pyrolyses.

3) Purgeable organic halogen blank—Analyze organic-free, pre-purged, reagent water to determine the POX blank.

6. Calculation

Calculate the net organic halide content as chloride (C_4) of each replicate of each sample and standard:

$$C_4 = \frac{C_1 - C_3 + C_2 - C_3}{V}$$

where:

C_1 = organic halide as Cl^- on the first activated carbon column or activated carbon cake, μg,
C_2 = organic halide as Cl^- on the second activated carbon column or activated carbon cake, μg,
C_3 = mean of method blanks on the same day and same instrument, μg X as Cl^-,
C_4 = uncorrected net organic halide as Cl^- of absorbed sample, μg organic halide as Cl^-/L, and
V = volume of sample absorbed, L.

If $C_2 \le C_3$, then use:

$$C_4 = \frac{C_1 - C_3}{V}$$

If applicable, calculate net purgeable organic halide as Cl^- content (P_3):

$$P_3 = \frac{P_1 - P_2}{V}$$

where:

P_1 = sample purgeable organic halide as Cl^-, μg,
P_2 = blank purgeable organic halide as Cl^-, μg,
P_3 = uncorrected net purgeable organic halide as Cl^-, μg X as Cl^-/L, and
V = volume of sample or standard purged, L.

Report sample results and percent recovery of the corresponding calibration standards [¶ 5c3) or ¶ 5c5)]. Also report the calibration standard curve if it is significantly nonlinear.

7. Precision and Bias

Precision and bias depend on specific procedures, equipment, and analyst. Develop and routinely update precision and bias data

TABLE 5320:I. INTRALABORATORY, SINGLE-OPERATOR, DISSOLVED ORGANIC HALOGEN (MICROCOLUMN PROCEDURE)—PRECISION AND BIAS DATA

Characteristic of Analysis	Tap Water	Tap Water + 43.5 μg Organic Chloride	Ground Water (50:1)	Wastewater	Wastewater + 1000 μg Organic Chloride
Concentration determined, μg Cl^-/L:					
Replicate 1	38.5	89.0	123.6	186.0	1178.0
Replicate 2	36.7	90.9	124.8	195.0	1183.0
Replicate 3	43.1	88.4	125.2	195.0	1185.5
Replicate 4	35.9	90.1	123.3	204.0	1196.5
Replicate 5	41.1	91.7	125.3	185.0	1183.0
Replicate 6	48.5	93.0	127.0	236.5	1204.0
Replicate 7	52.8	97.0	123.5	204.0	1138.0
Mean, μg Cl^-/L	42.37	91.5	124.7	200.8	1181.1
Standard deviation:					
μg Cl^-/L	±6.29	±3.0	±1.3	±17.47	±21.04
%	15	3	1	9	2
Value of blank + standard addition, μg Cl^-/L	—	85.87	—	—	1200.8
Recovery, %	—	107	—	—	98
Error, %	—	7	—	—	2

for each procedure, each instrument configuration, and each analyst. Table 5320:I shows sample calculations of precision expressed as the standard deviation among replicates and bias in the recovery of 2,4,6-trichlorophenol.

8. Reference

1. STANBORO, W.D. & M.J. LENKEVICH. 1982. Slowly dechlorinated organic chloramines. *Science* 215:967.
2. SYMONS, J.M. & R. XIA. 1995. Interference of Br^-, BrO_3^-, and ClO_3^- in the DOX determination. *J. Amer. Water Works Assoc.* 87:81.

9. Bibliography

KUHN, W. 1974. Thesis, Univ. Karlsruhe, West Germany.
KUHN, W., F. FUCHS & H. SONTHEIMER. 1977. Untersuchungen zur Bestimmung des organisch gebundenen Chlors mit Hilfe eines neuartigen Anreicherungsverfahrens. *Z. Wasser-Abwasser Forsch.* 10(6): 162.

DRESSMAN, R.C., B.A. NAJAR & R. REDZIKOWSKI. 1979. The analysis of organohalides (OX) in water as a group parameter. Proc. 7th Annual Water Quality Technology Conf., Philadelphia, Pa. American Water Works Assoc., Denver, Colo.
TAKAHASHI, Y., et al. 1980. Measurement of total organic halides (TOX) and purgeable organic halides (POX) in water using carbon adsorption and microcoulometric detection. Proc. Symp. on Chemistry and Chemical Analysis of Water and Waste Water Intended for Reuse, Houston, Tex. American Chemical Soc., Washington, D.C.
DRESSMAN, R.C. 1980. Total Organic Halide, Method 450.1—Interim. Drinking Water Research Div., Municipal Environmental Research Lab., U.S. Environmental Protection Agency, Cincinnati, Ohio.
JEKEL, M.R. & P.V. ROBERTS. 1980. Total organic halogen as a parameter for the characterization of reclaimed waters: measurement, occurrence, formation, and removal. *Environ. Sci. Technol.* 14:970.
DRESSMAN, R.C. & A. STEVENS. 1983. Analysis of organohalides in water—An evaluation update. *J. Amer. Water Works Assoc.* 75:431.
RECKHOW, D.A., C. HULL, E. LEHAN, J.M. SYMONS, H-S KIM, Y-M HANG, L. SIMMS & R.C. DRESSMAN. 1990. The determination of total organic halide in water: A comparative study of two instruments. *Internat. J. Environ. Chem.* 38:1.

5510 AQUATIC HUMIC SUBSTANCES*

5510 A. Introduction

1. General Discussion

Aquatic humic substances (AHS) are heterogeneous, yellow to black, organic materials that include most of the naturally occurring dissolved organic matter in water. Aquatic humic substances have been shown to produce trihalomethanes (THMs) on chlorination and to affect the transport and fate of other organic and inorganic species through partition/adsorption, catalytic, and photolytic reactions.

Humic substances, the major fraction of soil organic matter, are mixtures; their chemical composition is poorly understood. They have been classified into three fractions based on water "solubility"†: humin is the fraction not soluble in water at any pH value; humic acid is not soluble under acidic conditions (pH < 2) but becomes soluble at higher pH; and fulvic acid is soluble at all pH conditions.

AHS have the solubility characteristics of fulvic acids but they should not be referred to as such unless they have been fractionated by precipitation at pH < 2. Avoid using the terms "humic acid" and "tannic acid" to describe AHS because they represent other classifications of natural organic materials.

The heterogeneity of AHS requires an operational definition. Isolation by the methods included herein most likely will be incomplete and compounds that are not AHS may be isolated incidentally. Users of these methods are cautioned in the interpretation of results; the bibliography suggests several sources for more information.

Measurement of AHS begins by separation of the sample into dissolved (containing AHS) and particulate organic carbon fractions. Although there is no distinct size that separates these two groups, 0.45 μm is used as the compromise between acceptable flow rate and rejection of small colloidal materials. Low-pressure liquid chromatography serves to concentrate these materials and to isolate them from interfering substances. AHS are quantified by measuring dissolved organic carbon (DOC), Method 5310.

2. Selection of Method

Concentration/isolation of AHS may be achieved by sorption on the nonpolar resin XAD-8 (Method 5510C) or by anion-exchange on diethylaminoethyl (DEAE) cellulose (Method 5510B). In a collaborative study with seven laboratories using deionized water fortified with about 10 mg AHS/L (previously isolated with XAD), the DEAE method gave better recoveries. Nevertheless, the XAD method has been used extensively; refer to the discussions of interferences and minimum detectable concentrations to assist in method selection. Both methods require further quality control development.

3. Bibliography

CHRISTMAN, R. F. & E. T. GJESSING, eds. 1983. Aquatic and Terrestrial Humic Substances. Ann Arbor Science, Ann Arbor, Mich.
MILES, C. J., J. R. TUSCHALL, JR. & P. L. BREZONIK. 1983. Isolation of aquatic humus with diethylaminoethyl cellulose. *Anal. Chem.* 55: 410.
THURMAN, E. M. 1984. Determination of aquatic humic substances in natural waters. *In* E. L. Meyer, ed. Selected Papers in the Hydrologic Sciences. U.S. Geological Survey Water Supply Paper 2262, p. 47.

---

* Approved by Standard Methods Committee, 1993.
† "Solubility" is here used as a general description of whether or not the material can be uniformly dispersed in an aqueous phase rather than as an expression of equilibrium between a pure solute and its aqueous solution.

TUSCHALL, J. R., JR. & G. GEORGE. 1984. Selective Isolation of Dissolved Organic Matter from Aquatic Systems. UILUWRC-84-190, Water Resources Center, Univ. Illinois, Urbana.

AIKEN, G. R., D. M. McKNIGHT, R. L. WERSHAW & P. L. MacCARTHY, eds. 1985. Humic Substances in Soil, Sediment, and Water. John Wiley & Sons, New York, N.Y.

5510 B. Diethylaminoethyl (DEAE) Method

1. General Discussion

a. Principle: AHS are concentrated by column chromatography on diethylaminoethyl (DEAE) cellulose and measured as dissolved organic carbon (DOC). AHS are weak organic acids that bind to anion-exchange materials, such as DEAE cellulose, at neutral pH values. The method is based on the assumption that AHS are the major dissolved organic acids present.

b. Interferences: Any carbonaceous nonhumic materials that are concentrated and isolated by the chromatographic method will interfere (false positive response). Substances that have been shown to interfere include fatty acids, phenols, surfactants, proteinaceous materials, and DOC leached from cellulose.

c. Minimum detectable concentration: Estimated limit of detection is 1.1 mg/L using a 50-mL water sample. The detection limit can be decreased by increasing sample volume. The major limitation is blank contamination.

d. Standard substance: Eliminate documentation of false negatives by analyses of a sample of known humic concentration at regular intervals (at least once per batch of samples).

2. Apparatus

a. Membrane filtration apparatus: Use an all-glass filtering device and 0.45-μm silver membrane filters. Consult manufacturer's specifications for filter details. Do not use filters that sorb AHS or are contaminated with detergents and other organic material.

b. Glass column, approximately 1 × 20 cm with silanized glass wool.

c. Dye-impregnated paper or strips for approximate pH measurements.

d. Organic carbon analyzer capable of measuring concentrations as low as 0.1 mg/L (see Section 5310).

e. Buchner funnel and filter paper. *

3. Reagents

a. Water, DOC-free: Preferably use activated-carbon-filtered, redistilled water.

b. DEAE cellulose, exchange capacity 0.22–1.0 meq/g.† Do not use high-exchange-capacity cellulose, which may decrease recovery of AHS. Take care not to overload low-exchange-capacity cellulose.

c. Hydrochloric acid, HCl, 0.1N: Add 8.3 mL conc HCl to 1000 mL water.

d. Hydrochloric acid, HCl, 0.5N: Add 41.5 mL conc HCl to 1000 mL water.

e. Sodium hydroxide, NaOH, 0.1N: Dissolve 4.0 g NaOH in 1000 mL water.

f. Sodium hydroxide, NaOH, 0.5N: Dissolve 20 g NaOH in 1000 mL water.

g. DOC standards: See Section 5310.

h. Potassium chloride, KCl, 0.01N: Add 0.75 g KCl to 1000 mL water.

i. Phosphoric acid, H_3PO_4, conc.

4. Procedure

a. Sample concentration and preservation: AHS are sensitive to biodegradation and photodegradation. Collect and store samples in organic-free glass containers. Filter at least duplicate portions through a 0.45-μm silver membrane filter as soon after collection as possible. Store samples in the dark at 4°C.

Use care to avoid overloading chromatographic columns and losing AHS. A rough guideline for sample volume selection is as follows:

Sample DOC mg/L	Sample Volume mL
0–2	250
2–10	50
10–50	25

b. Preparation of DEAE cellulose: Add 70 g DEAE cellulose to 1000 mL 0.5N HCl and stir gently for 1 h. Rinse cellulose with water in a Buchner funnel until funnel effluent pH is about 4. Resuspend DEAE in 1000 mL 0.5N NaOH and stir for 1 h. Rinse in a Buchner funnel with water until pH is about 6. Remove fines by suspending the treated DEAE in a 1000-mL graduated cylinder filled with water. Let mixture stand undisturbed for 1 h, then decant and discard the supernatant. Repeat removal of fines. Filter remaining DEAE using a Buchner funnel and store in a refrigerated glass container. Avoid prolonged storage, which may lead to microbial contamination.

c. Chromatography: Add 10 mL water to about 1 g DEAE to make a slurry. Carefully pipet enough into a 1- × 20-cm column fitted with a small (0.5-cm) glass-wool plug to make a 1-cm-deep column bed. Avoid getting DEAE on the sides of the column. Carefully place another 0.5-cm glass-wool plug on top of the bed. Rinse column with 50 mL 0.01N KCl (adjusted to pH 6 with 0.1N HCl or NaOH) just before sample concentration.

Adjust sample to pH 6 and pass it through the column at a flow rate of about 2 mL/min. Rinse with 5 mL water (pH 6). Elute AHS by adding about 3 mL 0.1N NaOH to the top of the column. Start collecting column effluent when it appears colored. (This will occur after about 1 mL has passed out of the column). Collect eluate in a graduated, conical test tube until it becomes colorless (about 2 mL). Acidify with conc H_3PO_4 to a pH of 2 or less

---

* Whatman No. 1 or equivalent.
† Whatman pre-swollen microgranular DE 52 or DE 51, or equivalent.

(about 2 to 3 drops) and remove dissolved carbon dioxide (inorganic carbon) by purging with nitrogen for 10 min. Avoid exposure of alkaline samples to air (i.e. acidify immediately) to minimize contamination with CO_2. Determine volume and DOC of acidified eluate.

Process two portions of water and a second portion of sample by the same procedure. Pack a fresh column of DEAE for each sample and each control (DEAE cannot be reused).

5. Calculation

Calculate the concentration of AHS as:

$$AHS, \text{ mg DOC/L} = [(A - B) \times C]/D$$

where:

A = average DOC concentration of the two sample NaOH eluates, mg C/L,

B = average DOC concentration of the two control NaOH eluates, mg C/L,

C = volume of eluate, L, and

D = volume of sample, L.

Multiplication of AHS, mg DOC/L, by 2 converts concentration to AHS, mg/L, if it is assumed that AHS contain 50% carbon. This will be the minimum concentration of AHS because recoveries are less than 100%.

6. Precision and Bias

For seven single-operator analyses, the relative standard deviation of triplicate samples (about 10 mg/L as AHS) ranged from 2.5 to 14.4% with an average of 4.9% ($n = 7$).

For seven single-operator analyses, recoveries ranged from 59.3 to 97.3% with an average of 77.4% and a relative standard deviation of 18.1%.

5510 C.   XAD Method

1. General Discussion

a. Principle: AHS are concentrated by column chromatography on XAD resin and measured as dissolved organic carbon (DOC). Acidification of AHS decreases polarity, allowing partition into the nonpolar XAD matrix. The method is based on the assumption that AHS are the major dissolved organic acids present.

b. Interferences: Any carbonaceous nonhumic materials that are concentrated and isolated by the chromatographic method will interfere. This includes fatty acids, phenols, surfactants, proteinaceous materials, and DOC leached from the resin, chromatography pump, or tubing.

c. Minimum detectable concentration: Estimated limit of detection is 1.4 mg/L using a 50-mL water sample. The detection limit can be decreased by increasing sample volume. The major limitation is blank contaminations.

2. Apparatus

See Section 5510B.2*a*, *c*, and *d*. In addition, the following are required:

a. *Glass column*, 0.2 × 25 cm with silanized glass wool.

b. *Pump*, with inert internal parts and tubing, capable of flow rates of 0.2 to 1.0 mL/min.*

c. TFE tubing, 0.2 cm ID.

d. Extraction apparatus, Soxhlet.

3. Reagents

In addition to reagents *a*, *c*, *e*, *g*, and *i* of Section 5510B:

a. XAD resin,† approximately 250-μm size.

---

* Pump parts may be of stainless steel or TFE.
† XAD-7 or equivalent.

b. Hexane.

c. Methanol.

d. Acetonitrile.

4. Procedure

a. Sample collection and preservation: See Section 5510B.4*a*.

b. Preparation of XAD resin: Clean resin by successive washing with 0.1*N* NaOH for 5 d. Extract resin sequentially in a Soxhlet extractor with hexane, methanol, acetonitrile, and methanol, for 24 h each. Pack clean resin into a 0.2- × 25-cm glass column that has a 2-mm length of glass wool in one end. After filling, cap column with another 2-mm length of glass wool.

Wet dry column with methanol. When the air has been displaced, pump distilled water through the column until the effluent concentration of DOC decreases to 0.5 mg/L (approximately 20 bed volumes).

c. Chromatography: Preclean column with three cycles of 0.1*N* NaOH and 0.1*N* HCl just before pumping sample into column. Leave column saturated with 0.1*N* HCl. Acidify sample to pH 2.0 with concentrated HCl, and pump it onto the column at rate of 1.0 mL/min. Save column effluent for DOC analysis. Significant concentrations of DOC in the effluent can indicate that the column was overloaded and that a smaller sample volume should be used. Colored organic acids adsorb to the top of the column. Back-elute (reverse flow) the column with 0.1*N* NaOH at 0.2 mL/min and collect eluate in a graduated, conical test tube until it becomes colorless (about 2 mL). Acidify with conc H_3PO_4 to a pH of 2 or less (about 2 to 3 drops) and remove dissolved carbon dioxide (inorganic carbon) by purging with nitrogen for 10 min. Avoid exposure of alkaline samples to air (i.e. acidify immediately) to minimize contamination with CO_2. Determine volume and DOC of acidified column effluent.

After eluting and collecting AHS from the column with back-elution using 0.1N NaOH, continue rinsing with about 20 bed-volumes of the basic solution. Rinse with water for about 20 bed volumes. Repeat the triplicate acid/base column precleaning procedure described above, then reuse the column to analyze a replicate sample. Process two portions of water by the same procedure to serve as controls.

The XAD column may be reused to analyze subsequent samples and controls if the triplicate acid/base precleaning procedure is repeated immediately before analysis of each replicate. Replace the column if recovery is poor or the resin becomes discolored.

5. Calculation

Calculate the concentration of AHS as given in 5510B.5.

6. Precision and Bias

For seven single-operator analyses, the relative standard deviation of triplicate samples (about 10 mg/L as AHS) ranged from 0.9 to 20.7% with an average of 5.4% ($n = 7$).

For seven single-operator analyses, recoveries ranged from 15.1 to 71.0% with an average of 51.6% and a relative standard deviation of 35.1%.

5520   OIL AND GREASE*

5520 A.   Introduction

In the determination of oil and grease, an absolute quantity of a specific substance is not measured. Rather, groups of substances with similar physical characteristics are determined quantitatively on the basis of their common solubility in an organic extracting solvent. "Oil and grease" is defined as any material recovered as a substance soluble in the solvent. It includes other material extracted by the solvent from an acidified sample (such as sulfur compounds, certain organic dyes, and chlorophyll) and not volatilized during the test. The 12th edition of *Standard Methods* prescribed the use of petroleum ether as the solvent for natural and treated waters and *n*-hexane for polluted waters. The 13th edition added trichlorotrifluoroethane as an optional solvent for all sample types. In the 14th through the 17th editions, only trichlorotrifluoroethane was specified. However, because of environmental problems associated with chlorofluorocarbons, an alternative solvent (80% *n*-hexane and 20% methyl-*tert*-butyl ether) was included for gravimetric methods in the 19th edition. In the 20th edition, trichlorotrifluoroethane has been dropped from all gravimetric procedures (retained for 5520C, an infrared method), and replaced by *n*-hexane. Solvent-recovery techniques are included and solvent recycling is strongly recommended.

It is important to understand that, unlike some constituents that represent distinct chemical elements, ions, compounds, or groups of compounds, oils and greases are defined by the method used for their determination. In a detailed study involving many complex organic matrices, it was shown that either *n*-hexane or 80/20 *n*-hexane/methyl-*tert*-butyl ether gave results that were not statistically different from results obtained with trichlorotrifluoroethane.[1] Although 5520B allows either solvent system for extraction of wastewaters, note that for certain regulatory purposes U.S. EPA currently recommends only *n*-hexane.[2]

The methods presented here are suitable for biological lipids and mineral hydrocarbons. They also may be suitable for most industrial wastewaters or treated effluents containing these materials, although sample complexity may result in either low or high results because of lack of analytical specificity. The method is not applicable to measurement of low-boiling fractions that volatilize at temperatures below 85°C.

1. Significance

Certain constituents measured by the oil and grease analysis may influence wastewater treatment systems. If present in excessive amounts, they may interfere with aerobic and anaerobic biological processes and lead to decreased wastewater treatment efficiency. When discharged in wastewater or treated effluents, they may cause surface films and shoreline deposits leading to environmental degradation.

A knowledge of the quantity of oil and grease present is helpful in proper design and operation of wastewater treatment systems and also may call attention to certain treatment difficulties.

In the absence of specially modified industrial products, oil and grease is composed primarily of fatty matter from animal and vegetable sources and from hydrocarbons of petroleum origin. The portion of oil and grease from each of these two major sources can be determined with Method 5520F. A knowledge of the relative composition of a sample minimizes the difficulty in determining the major source of the material and simplifies the correction of oil and grease problems in wastewater treatment plant operation and stream pollution abatement.

2. Selection of Method

For liquid samples, three methods are presented: the partition-gravimetric method (B), the partition-infrared method (C), and the Soxhlet method (D). Method C is designed for samples that might contain volatile hydrocarbons that otherwise would be lost in the solvent-removal operations of the gravimetric procedure. Method D is the method of choice when relatively polar, heavy petroleum fractions are present, or when the levels of nonvolatile greases may challenge the solubility limit of the solvent. For low levels of oil and grease (<10 mg/L), Method C is the method of

---

* Approved by Standard Methods Committee, 1997.

choice because gravimetric methods do not provide the needed precision.

Method E is a modification of the Soxhlet method and is suitable for sludges and similar materials. Method F can be used in conjunction with Methods B, C, D, or E to obtain a hydrocarbon measurement in addition to, or instead of, the oil and grease measurement. This method makes use of silica gel to separate hydrocarbons from the total oil and grease on the basis of polarity.

3. Sample Collection, Preservation, and Storage

Collect a representative grab sample in a wide-mouth glass bottle that has been washed with soap, rinsed with water, and finally rinsed with solvent to remove any residues that might interfere with the analysis. As an alternative to solvent rinsing, cap bottle with aluminum foil and bake at 200 to 250°C for at least 1 h. Use PTFE-lined caps for sample bottles; clean liners as above, but limit temperature to 110 to 200°C. Collect a separate sample for an oil and grease determination. Do not overfill the sample container and do not subdivide the sample in the laboratory. Collect replicate samples for replicate analyses or known-addition QA checks. Collect replicates either in rapid succession, in parallel, or in one large container with mechanical stirring (in the latter case, siphon individual portions). Typically, collect wastewater samples of approximately 1 L. If sample concentration is expected to be greater than 1000 mg extractable material/L, collect proportionately smaller volumes. If analysis is to be delayed for more than 2 h, acidify to pH 2 or lower with either 1:1 HCl or 1:1 H_2SO_4 and refrigerate. When information is required about average grease concentration over an extended period, examine individual portions collected at prescribed time intervals to eliminate losses of grease on sampling equipment during collection of a composite sample.

In sampling sludges, take every possible precaution to obtain a representative sample. When analysis cannot be made within 2 h, preserve samples with 1 mL conc HCl/80 g sample and refrigerate. Never preserve samples with $CHCl_3$ or sodium benzoate.

4. Interferences

a. Organic solvents have the ability to dissolve not only oil and grease but also other organic substances. Any filterable solvent-soluble substances (e.g., elemental sulfur, complex aromatic compounds, hydrocarbon derivatives of chlorine, sulfur, and nitrogen, and certain organic dyes) that are extracted and recovered are defined as oil and grease. No known solvent will dissolve selectively only oil and grease. Heavier residuals of petroleum may contain a significant portion of materials that are not solvent-extractable. The method is entirely empirical; duplicate results

with a high degree of precision can be obtained only by strict adherence to all details.

b. For Methods 5520B, D, E, and F, solvent removal results in the loss of short-chain hydrocarbons and simple aromatics by volatilization. Significant portions of petroleum distillates from gasoline through No. 2 fuel oil are lost in this process. Adhere strictly to sample drying time, to standardize gradual loss of weight due to volatilization. For Methods 5520B, D, E, and F, during the cooling of the distillation flask and extracted material, a gradual increase in weight may be observed, presumably due to the absorption of water if a desiccator is not used. For Method 5520C use of an infrared detector offers a degree of selectivity to overcome some coextracted interferences (¶ 4a). For Methods 5520D and E, use exactly the specified rate and time of extraction in the Soxhlet apparatus because of varying solubilities of different greases. For Method 5520F, the more polar hydrocarbons, such as complex aromatic compounds and hydrocarbon derivatives of chlorine, sulfur, and nitrogen, may be adsorbed by the silica gel. Extracted compounds other than hydrocarbons and fatty matter also interfere.

c. Alternative techniques may be needed for some samples if intractable emulsions form that cannot be broken by centrifugation. Such samples may include effluents from pulp/paper processing and zeolite manufacturing. Determine such modifications on a case-by-case basis.

d. Some sample matrices can increase the amount of water partitioned into the organic extraction fluid. When the extraction solvent from this type of sample is dried with sodium sulfate, the drying capacity of the sodium sulfate can be exceeded, thus allowing sodium sulfate to dissolve and pass into the tared flask. After drying, sodium sulfate crystals will be visible in the flask. The sodium sulfate that passes into the flask becomes a positive interference in gravimetric methods. If crystals are observed in the tared flask after drying, redissolve any oil and grease with 30 mL of extraction solvent and drain the solvent through a funnel containing a solvent-rinsed filter paper into a clean, tared flask. Rinse the first flask twice more, combining all solvent in the new flask, and treat as an extracted sample.

e. Silica gel fines may give positive interferences in 5520F if they pass through the filter. Use filters with smaller pores if this occurs with a particular batch of silica gel.

5. References

1. U.S. ENVIRONMENTAL PROTECTION AGENCY. 1995. Report of the Method 1664 Validation Studies. EPA-821-R-95-036, U.S. Environmental Protection Agency, Washington, D.C.
2. U.S. ENVIRONMENTAL PROTECTION AGENCY. 1995. Method 1664. EPA-821-B-94-004B, U.S. Environmental Protection Agency, Washington, D.C.

5520 B. Partition-Gravimetric Method

1. General Discussion

Dissolved or emulsified oil and grease is extracted from water by intimate contact with an extracting solvent. Some extractables, especially unsaturated fats and fatty acids, oxidize readily; hence, special precautions regarding temperature and solvent vapor displacement are included to minimize this effect. Organic solvents shaken with some samples may form an emulsion that is very difficult to break. This method includes a means for handling such emulsions. Recovery of solvents is discussed. Solvent recovery can reduce both vapor emissions to the atmosphere and costs.

2. Apparatus

    a. Separatory funnel, 2-L, with TFE* stopcock.
    b. Distilling flask, 125-mL.
    c. Liquid funnel, glass.
    d. Filter paper, 11-cm diam.†
    e. Centrifuge, capable of spinning at least four 100-mL glass centrifuge tubes at 2400 rpm or more.
    f. Centrifuge tubes, 100-mL, glass.
    g. Water bath, capable of maintaining 85°C.
    h. Vacuum pump or other source of vacuum.
    i. Distilling adapter with drip tip. Setup of distillate recovery apparatus is shown in Figure 5520:1. Alternatively, use commercially available solvent recovery equipment.
    j. Ice bath.
    k. Waste receptacle, for used solvent.
    l. Desiccator.

3. Reagents

    a. Hydrochloric or sulfuric acid, 1:1: Mix equal volumes of either acid and reagent water.
    b. n-*Hexane,* boiling point 69°C. The solvent should leave no measurable residue on evaporation; distill if necessary. Do not use any plastic tubing to transfer solvent between containers.
    c. Methyl-tert-butyl ether (MTBE), boiling point 55°C to 56°C. The solvent should leave no measurable residue on evaporation; distill if necessary. Do not use any plastic tubing to transfer solvent between containers.
    d. Sodium sulfate, Na$_2$SO$_4$, anhydrous crystal.
    e. Solvent mixture, 80% n-hexane/20% MTBE, v/v.

4. Procedure

When a sample is brought into the laboratory, either mark sample bottle at the water meniscus or weigh the bottle, for later determination of sample volume. If sample has not been acidified previously (see Section 5520A.3), acidify with either 1:1 HCl or 1:1 H$_2$SO$_4$ to pH 2 or lower (generally, 5 mL is sufficient for 1 L sample). Using liquid funnel, transfer sample to a separatory funnel. Carefully rinse sample bottle with 30 mL extracting solvent (either 100% n-hexane, ¶ *3b*, or solvent mixture, ¶ *3e*) and add solvent washings to separatory funnel. Shake vigorously for 2 min. Let layers separate. Drain aqueous layer and small amount of organic layer into original sample container. Drain solvent layer through a funnel containing a filter paper and 10 g Na$_2$SO$_4$, both of which have been solvent-rinsed, into a clean, tared distilling flask. If a clear solvent layer cannot be obtained and an emulsion of more than about 5 mL exists, drain emulsion and solvent layers into a glass centrifuge tube and centrifuge for 5 min at approximately 2400 rpm. Transfer centrifuged material to an appropriate separatory funnel and drain solvent layer through a funnel with a filter paper and 10 g Na$_2$SO$_4$, both of which have been prerinsed, into a clean, tared distilling flask. Recombine

---

* Teflon or equivalent.
† Whatman No. 40 or equivalent.

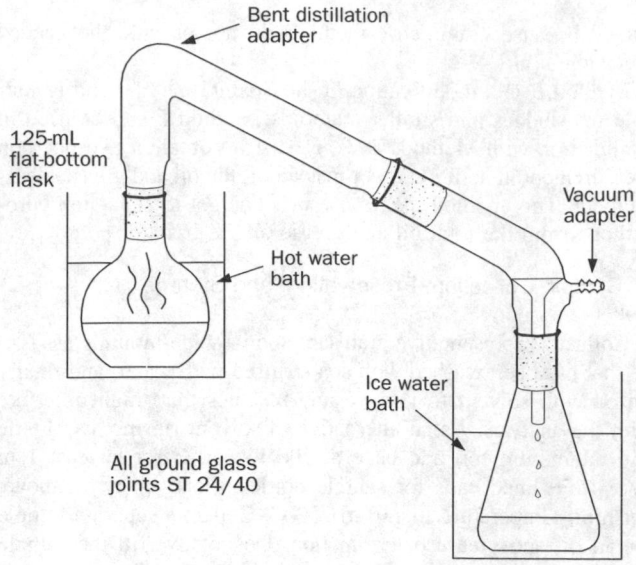

Figure 5520:1. Distillate recovery apparatus.

aqueous layers and any remaining emulsion or solids in separatory funnel. For samples with <5 mL of emulsion, drain only the clear solvent through a funnel with pre-moistened filter paper and 10 g Na$_2$SO$_4$. Recombine aqueous layers and any remaining emulsion or solids in separatory funnel. Extract twice more with 30 mL solvent each time, but first rinse sample container with each solvent portion. Repeat centrifugation step if emulsion persists in subsequent extraction steps. Combine extracts in tared distilling flask, and include in flask a final rinsing of filter and Na$_2$SO$_4$ with an additional 10 to 20 mL solvent. Distill solvent from flask in a water bath at 85°C for either solvent system. To maximize solvent recovery, fit distillation flask with a distillation adapter equipped with a drip tip and collect solvent in an ice-bath-cooled receiver (Figure 5520:1). When visible solvent condensation stops, remove flask from water bath. Cover water bath and dry flasks on top of cover, with water bath still at 85°C, for 15 min. Draw air through flask with an applied vacuum for the final 1 min. Cool in desiccator for at least 30 min and weigh. To determine initial sample volume, either fill sample bottle to mark with water and then pour water into a 1-L graduated cylinder, or weigh empty container and cap and calculate the sample volume by difference from the initial weight (assuming a sample density of 1.00).

5. Calculation

If the organic solvent is free of residue, the gain in weight of the tared distilling flask is due to oil and grease. Total gain in weight, *A*, of tared flask, less calculated residue from solvent blank, *B*, is the amount of oil and grease in the sample:

$$\text{mg oil and grease/L} = \frac{(A - B) \times 1000}{\text{mL sample}}$$

6. Precision and Bias

Method B with 80:20 hexane/MTBE mixture was tested by a single laboratory on a raw wastewater sample. The oil and grease concentration was 22.4 mg/L. When samples were dosed with 30 mg Fisher Heavy Mineral Oil, recovery of added oil was 84.2% with a standard deviation of 1.2 mg/L. Method B was tested with *n*-hexane as solvent. The method detection limit was determined to be 1.4 mg/L.[1] When reagent water was fortified with hexadecane and stearic acid each at approximately 20 mg/L, initial precision and recovery limit standards were 10% and 83 to 101%, respectively. Acceptable recovery limits for laboratory-fortified matrix/laboratory-fortified matrix duplicate and ongoing laboratory control standards are 79 to 114%, with a relative percent difference limit of 18%.

7. References

1. U.S. ENVIRONMENTAL PROTECTION AGENCY. 1995. Report of the Method 1664 Validation Studies. EPA-821-R-95-036, U.S. Environmental Protection Agency, Washington, D.C.
2. U.S. ENVIRONMENTAL PROTECTION AGENCY. 1995. Method 1664. EPA-821-B-94-004B, U.S. Environmental Protection Agency, Washington, D.C.

8. Bibliography

KIRSCHMAN, H.D. & R. POMEROY. 1949. Determination of oil in oil field waste waters. *Anal. Chem.* 21:793.

5520 C. Partition-Infrared Method

1. General Discussion

a. Principle: The use of trichlorotrifluoroethane as extraction solvent allows absorbance of the carbon-hydrogen bond in the infrared to be used to measure oil and grease. Elimination of the evaporation step permits infrared detection of many relatively volatile hydrocarbons. Thus, the lighter petroleum distillates, with the exception of gasoline, may be measured accurately. With adequate instrumentation, as little as 0.2 mg oil and grease/L can be measured.

b. Definitions: A "known oil" is defined as a sample of oil and/or grease that represents the only material of that type used or manufactured in the processes represented by a wastewater. An "unknown oil" is defined as one for which a representative sample of the oil or grease is not available for preparation of a standard.

2. Apparatus

a. Separatory funnel, 2-L, with TFE* stopcock.

b. Volumetric flask, 100-mL.

c. Liquid funnel, glass.

d. Filter paper, 11-cm diam.†

e. Centrifuge, capable of spinning at least four 100-mL glass centrifuge tubes at 2400 rpm or more.

f. Centrifuge tubes, 100-mL, glass.

g. Infrared spectrophotometer, double-beam, recording.

h. Cells, near-infrared silica.

3. Reagents

a. Hydrochloric acid, HCl, 1 + 1.

b. Trichlorotrifluoroethane (1,1,2-trichloro-1,2,2-trifluoroethane), boiling point 47°C. The solvent should leave no measurable residue on evaporation; distill if necessary. Do not use any plastic tubing to transfer solvent between containers.

c. Sodium sulfate, Na_2SO_4, anhydrous, crystal.

d. Reference oil: Prepare a mixture, by volume, of 37.5% isooctane, 37.5% hexadecane, and 25.0% benzene. Store in sealed container to prevent evaporation.

4. Procedure

Refer to Section 5520B.4 for sample handling and for method of dealing with sample emulsions. After carefully transferring sample to a separatory funnel, rinse sample bottle with 30 mL trichlorotrifluoroethane and add solvent washings to funnel. Shake vigorously for 2 min. Let layers separate. Drain all but a very small portion of the lower trichlorotrifluoroethane layer through a funnel containing a filter paper and 10 g Na_2SO_4, both of which have been solvent-rinsed, into a clean, 100-mL volumetric flask. If a clear solvent layer cannot be obtained and an emulsion of more than about 5 mL exists, see Section 5520B.4. Extract twice more with 30 mL solvent each time, but first rinse sample container with each solvent portion. Repeat centrifugation step if emulsion persists in subsequent extraction steps. Combine extracts in volumetric flask, and include in flask a final rinsing of filter and Na_2SO_4 with an additional 10 to 20 mL solvent. Adjust final volume to 100 mL with solvent.

Prepare a stock solution of known oil by rapidly transferring about 1 mL (0.5 to 1.0 g) of the oil or grease to a tared 100-mL volumetric flask. Stopper flask and weigh to nearest milligram. Add solvent to dissolve and dilute to mark. If the oil identity is unknown (5520C.1*b*) use the reference oil (5520C.3*d*) as the standard. Using volumetric techniques, prepare a series of standards over the range of interest. Select a pair of matched near-infrared silica cells. A 1-cm-path-length cell is appropriate for a working range of about 4 to 40 mg. Scan standards and samples from 3200 cm^{-1} to 2700 cm^{-1} with solvent in the reference beam and record results on absorbance paper. Measure absorbances of samples and standards by constructing a straight base line over the scan range and measuring absorbance of the peak maximum at 2930 cm^{-1} and subtracting baseline absorbance at that point. If the absorbance exceeds 0.8 for a sample, select a shorter path length or dilute as required. Use scans of standards to prepare a calibration curve.

5. Calculation

$$\text{mg oil and grease/L} = \frac{A \times 1000}{\text{mL sample}}$$

---

* Teflon or equivalent.
† Whatman No. 40 or equivalent.

where:

A = mg of oil or grease in extract as determined from calibration curve.

6. Precision and Bias

Method C was used by a single laboratory to test a wastewater sample. By this method the oil and grease concentration was 17.5 mg/L. When 1-L sample portions were dosed with 14.0 mg of a mixture of No. 2 fuel oil and Wesson oil, the recovery of added oils was 99% with a standard deviation of 1.4 mg.

7. Bibliography

GRUENFELD, M. 1973. Extraction of dispersed oils from water for quantitative analysis by infrared spectrophotometry. *Environ. Sci. Technol.* 7:636.

5520 D. Soxhlet Extraction Method

1. General Discussion

Soluble metallic soaps are hydrolyzed by acidification. Any oils and solid or viscous grease present are separated from the liquid samples by filtration. After extraction in a Soxhlet apparatus with solvent, the residue remaining after solvent evaporation is weighed to determine the oil and grease content. Compounds volatilized at or below 103°C will be lost when the filter is dried.

2. Apparatus

a. *Extraction apparatus,* Soxhlet, with 125-mL extraction flask.

b. *Extraction thimble,* paper, solvent-extracted.

c. *Electric heating mantle.*

d. *Vacuum pump* or other source of vacuum.

e. *Vacuum filtration apparatus.*

f. *Buchner funnel,* 12-cm.

g. *Filter paper,* 11-cm diam.*

h. *Muslin cloth disks,* 11-cm diam, solvent-extracted.

i. *Glass beads or glass wool,* solvent-extracted.

j. *Water bath,* capable of maintaining 85°C.

k. *Distilling adapter* with drip tip. See 5520B.2*i* and Figure 5520:1.

l. *Ice bath.*

m. *Waste receptacle,* for used solvent.

n. *Desiccator.*

3. Reagents

a. *Hydrochloric acid,* HCl, 1 + 1.

b. n-*Hexane:* See Section 5520B.3*b.*

c. *Methyl-tert-butyl ether* (MTBE): See Section 5520B.3*c.*

d. *Diatomaceous-silica filter aid suspension,*† 10 g/L distilled water.

e. *Solvent mixture,* 80% n-hexane/20% MTBE, v/v.

4. Procedure

When sample is brought into the laboratory, either mark sample bottle at the meniscus or weigh bottle for later determination of volume. If sample has not been acidified previously (see Section 5520A.3), acidify with 1:1 HCl or 1:1 H_2SO_4 to pH 2 or lower (generally, 5 mL is sufficient). Prepare filter consisting of a muslin cloth disk overlaid with filter paper. Wet paper and muslin and press down edges of paper. Using vacuum, pass 100 mL filter aid suspension through prepared filter and wash with 1 L distilled water. Apply vacuum until no more water passes filter. Filter acidified sample. Apply vacuum until no more water passes through filter. Using forceps, transfer entire filter to a watch glass. Add material adhering to edges of muslin cloth disk. Wipe sides and bottom of collecting vessel and Buchner funnel with pieces of filter paper soaked in extraction solvent, taking care to remove all films caused by grease and to collect all solid material. Add pieces of filter paper to material on watch glass. Roll all filter material containing sample and fit into an extraction thimble. Add any pieces of material remaining on watch glass. Wipe watch glass with a filter paper soaked in extraction solvent and place in extraction thimble. Dry filled thimble in a hot-air oven at 103°C for 30 min. Fill thimble with glass wool or small glass beads. Weigh extraction flask and add 100 mL extraction solvent (n-hexane, ¶ 3*b,* or solvent mixture, ¶ 3*e*). Extract oil and grease in a Soxhlet apparatus, at a rate of 20 cycles/h for 4 h. Time from first cycle. For stripping and recovery of solvent, cooling extraction flask before weighing, and determining initial sample volume, see Section 5520B.4.

5. Calculation

See Section 5520B.5.

6. Precision and Bias

In analyses of synthetic samples containing various amounts of Crisco and Shell S.A.E. No. 20 oil, an average recovery of 98.7% was obtained, with a standard deviation of 1.86%. Ten replicates each of two wastewater samples yielded standard deviations of 0.76 mg and 0.48 mg.

7. Bibliography

HATFIELD, W.D. & G.E. SYMONS. 1945. The determination of grease in sewage. *Sewage Works J.* 17:16.

GILCREAS, F.W., W.W. SANDERSON & R.P. ELMER. 1953. Two new methods for the determination of grease in sewage. *Sewage Ind. Wastes* 25:1379.

ULLMANN, W.W. & W.W. SANDERSON. 1959. A further study of methods for the determination of grease in sewage. *Sewage Ind. Wastes* 31:8.

---

* Whatman No. 40 or equivalent.

† Hyflo Super-Cel, Manville Corp., or equivalent.

5520 E. Extraction Method for Sludge Samples

1. General Discussion

Drying acidified sludge by heating leads to low results. Magnesium sulfate monohydrate is capable of combining with 75% of its own weight in water in forming $MgSO_4 \cdot 7H_2O$ and is used to dry sludge. After drying, the oil and grease can be extracted with an organic solvent.

2. Apparatus

a. *Beaker,* 150-mL, glass.

b. *Mortar and pestle,* porcelain.

c. *Extraction apparatus,* Soxhlet.

d. *Extraction thimble,* paper, solvent-extracted.

e. *Glass beads or glass wool,* solvent-extracted.

f. *Electric heating mantle.*

g. *Vacuum pump* or other source of vacuum.

h. *Liquid funnel,* glass.

i. *Grease-free cotton:* Extract nonabsorbent cotton with solvent.

j. *Water bath,* capable of maintaining 85°C.

k. *Distilling adapter* with drip tip. See 5520.2*i* and Figure 5520:1.

l. *Ice bath.*

m. *Waste receptacle,* for used solvent.

n. *Desiccator.*

3. Reagents

a. *Hydrochloric acid,* HCl, conc.

b. n-*Hexane:* See Section 5520B.3*b.*

c. *Methyl-*tert-*butyl ether (MTBE):* See Section 5520B.3*c.*

d. *Magnesium sulfate monohydrate:* Prepare $MgSO_4 \cdot H_2O$ by drying a thin layer overnight in an oven at 150°C.

e. *Solvent mixture,* 80% n-hexane/20% MTBE, v/v.

4. Procedure

When sample is brought into the laboratory, if it has not been acidified previously (Section 5520A.3), add 1 mL conc HCl/80 g sample. In a 150-mL beaker weigh out a sample of wet sludge, 20 ± 0.5 g, for which the dry-solids content is known. Acidify to pH 2.0 or lower (generally, 0.3 mL conc HCl is sufficient). Add 25 g $MgSO_4 \cdot H_2O$. Stir to a smooth paste and spread on sides of beaker to facilitate subsequent sample removal. Let stand until solidified, 15 to 30 min. Remove solids and grind in a porcelain mortar. Add powder to a paper extraction thimble. Wipe beaker and mortar with small pieces of filter paper moistened with solvent and add to thimble. Fill thimble with glass wool or small glass beads. Tare extraction flask, and add 100 mL extraction solvent (¶ 3*b* or *e*). Extract in a Soxhlet apparatus at a rate of 20 cycles/h for 4 h. If any turbidity or suspended matter is present in the extraction flask, remove by filtering through grease-free cotton into another weighed flask. Rinse flask and cotton with solvent. For solvent stripping and recovery, and cooling the extraction flask before weighing, see Section 5520B.4.

5. Calculation

Oil and grease as % of dry solids

$$= \frac{\text{gain in weight of flask, g} \times 100}{\text{weight of wet solids, g} \times \text{dry solids fraction}}$$

6. Precision

The examination of six replicate samples of sludge yielded a standard deviation of 4.6%.

5520 F. Hydrocarbons

1. General Discussion

Silica gel has the ability to adsorb polar materials. If a solution of hydrocarbons and fatty materials in a nonpolar solvent is mixed with silica gel, the fatty acids are removed selectively from solution. The materials not eliminated by silica gel adsorption are designated hydrocarbons by this test.

2. Apparatus

a. *Magnetic stirrer.*

b. *Magnetic stirring bars,* TFE-coated.

c. *Liquid funnel,* glass.

d. *Filter paper,* 11-cm diam.*

e. *Desiccator.*

3. Reagents

a. n-*Hexane:* See Section 5520B.3*b.*

b. *Trichlorotrifluoroethane:* See Section 5520C.3*b.*

c. *Silica gel,* 100 to 200 mesh.† Dry at 110°C for 24 h and store in a tightly sealed container.

4. Procedure

Use the oil and grease extracted by Method B, C, D, or E for this test. When only hydrocarbons are of interest, introduce this procedure in any of the previous methods before final measurement. When hydrocarbons are to be determined after total oil

---

* Whatman No. 40 or equivalent.

† Davidson Grade 923 or equivalent.

and grease has been measured, redissolve the extracted oil and grease in trichlorofluoroethane (Method C) or 100 mL n-hexane. To 100 mL solvent add 3.0 g silica gel/100 mg total oil and grease, up to a total of 30.0 g silica gel (1000 mg total oil and grease). For samples with more than 1000 mg total oil and grease use a measured volume of the 100 mL solvent dissolved sample, add appropriate amount of silica gel for amount of total oil and grease in the sample portion, and bring volume to 100 mL. Stopper container and stir on a magnetic stirrer for 5 min. For infrared measurement of hydrocarbons no further treatment is required before measurement as described in Method C. For gravimetric determinations, filter solution through filter paper pre-moistened with solvent, wash silica gel and filter paper with 10 mL solvent, and combine with filtrate. For solvent stripping and recovery, and for cooling extraction flask before weighing, see Section 5520B.4.

5. Calculation

Calculate hydrocarbon concentration, in milligrams per liter, as in oil and grease (Method B, C, D, or E).

6. Precision and Bias

The following data, obtained on synthetic samples, are indicative for natural animal, vegetable, and mineral products, but cannot be applied to the specialized industrial products previously discussed.

For hydrocarbon determinations on 10 synthetic solvent extracts containing known amounts of a wide variety of petroleum products, average recovery was 97.2%. Similar synthetic extracts of Wesson oil, olive oil, Crisco, and butter gave 0.0% recovery as hydrocarbons measured by infrared analysis.

Using reagent water fortified with approximately 20 mg/L each of hexadecane and stearic acid, initial hydrocarbon recovery limits based on hexadecane of 83 to 116% were developed, with a precision limit of 13%. Laboratory-fortified matrix/laboratory-fortified matrix duplicate gave recovery limits of 66 to 114% with a relative percent difference of 24%.

7. Bibliography

U.S. ENVIRONMENTAL PROTECTION AGENCY. 1995. Report of the Method 1664 Validation Studies. EPA-821-R-95-036, U.S. Environmental Protection Agency, Washington, D.C.

5530 PHENOLS*

5530 A. Introduction

Phenols, defined as hydroxy derivatives of benzene and its condensed nuclei, may occur in domestic and industrial wastewaters, natural waters, and potable water supplies. Chlorination of such waters may produce odorous and objectionable-tasting chlorophenols. Phenol removal processes in water treatment include superchlorination, chlorine dioxide or chloramine treatment, ozonation, and activated carbon adsorption.

1. Selection of Method

The analytical procedures offered here use the 4-aminoantipyrine colorimetric method that determines phenol, ortho- and meta-substituted phenols, and, under proper pH conditions, those para-substituted phenols in which the substitution is a carboxyl, halogen, methoxyl, or sulfonic acid group. The 4-aminoantipyrine method does not determine those para-substituted phenols where the substitution is an alkyl, aryl, nitro, benzoyl, nitroso, or aldehyde group. A typical example of these latter groups is paracresol, which may be present in certain industrial wastewaters and in polluted surface waters.

The 4-aminoantipyrine method is given in two forms: Method C, for extreme sensitivity, is adaptable for use in water samples containing less than 1 mg phenol/L. It concentrates the color in a nonaqueous solution. Method D retains the color in the aqueous solution. Because the relative amounts of various phenolic compounds in a given sample are unpredictable, it is not possible to

provide a universal standard containing a mixture of phenols. For this reason, phenol (C_6H_5OH) itself has been selected as a standard for colorimetric procedures and any color produced by the reaction of other phenolic compounds is reported as phenol. Because substitution generally reduces response, this value represents the minimum concentration of phenolic compounds. A gas-liquid chromatographic procedure is included in Section 6420B and may be applied to samples or concentrates to quantify individual phenolic compounds.

2. Interferences

Interferences such as phenol-decomposing bacteria, oxidizing and reducing substances, and alkaline pH values are dealt with by acidification. Some highly contaminated wastewaters may require specialized techniques for eliminating interferences and for quantitative recovery of phenolic compounds.

Eliminate major interferences as follows (see Section 5530B for reagents):

Oxidizing agents, such as chlorine and those detected by the liberation of iodine on acidification in the presence of potassium iodide (KI)—Remove immediately after sampling by adding excess ferrous sulfate ($FeSO_4$). If oxidizing agents are not removed, the phenolic compounds will be oxidized partially.

Sulfur compounds—Remove by acidifying to pH 4.0 with H_3PO_4 and aerating briefly by stirring. This eliminates the interference of hydrogen sulfide (H_2S) and sulfur dioxide (SO_2).

---

* Approved by Standard Methods Committee, 1993.

Oils and tars—Make an alkaline extraction by adjusting to pH 12 to 12.5 with NaOH pellets. Extract oil and tar from aqueous solution with 50 mL chloroform (CHCl₃). Discard oil- or tar-containing layer. Remove excess CHCl₃ in aqueous layer by warming on a water bath before proceeding with the distillation step.

3. Sampling

Sample in accordance with the instructions of Section 1060.

4. Preservation and Storage of Samples

Phenols in concentrations usually encountered in wastewaters are subject to biological and chemical oxidation. Preserve and store samples at 4°C or lower unless analyzed within 4 h after collection.

Acidify with 2 mL conc H_2SO_4/L.

Analyze preserved and stored samples within 28 d after collection.

5. Bibliography

ETTINGER, M.B., S. SCHOTT & C.C. RUCHHOFT. 1943. Preservation of phenol content in polluted river water samples previous to analysis. *J. Amer. Water Works Assoc.* 35:299.

CARTER, M.J. & M.T. HUSTON. 1978. Preservation of phenolic compounds in wastewaters. *Environ. Sci. Technol.* 12:309.

NEUFELD, R.D. & S.B. POLADINO. 1985. Comparison of 4-aminoantipyrine and gas-liquid chromatography techniques for analysis of phenolic compounds. *J. Water Pollut. Control Fed.* 57:1040.

5530 B.   Cleanup Procedure

1. Principle

Phenols are distilled from nonvolatile impurities. Because the volatilization of phenols is gradual, the distillate volume must ultimately equal that of the original sample.

2. Apparatus

a. Distillation apparatus, all-glass, consisting of a 1-L borosilicate glass distilling apparatus with Graham condenser.*

b. pH meter.

3. Reagents

Prepare all reagents with distilled water free of phenols and chlorine.

a. Phosphoric acid solution, H_3PO_4, 1 + 9: Dilute 10 mL 85% H_3PO_4 to 100 mL with water.

b. Methyl orange indicator solution.

c. Special reagents for turbid distillates:

1) *Sulfuric acid,* H_2SO_4, 1N.

2) *Sodium chloride,* NaCl.

3) *Chloroform,* CHCl₃, or *methylene chloride,* CH_2Cl_2.

4) *Sodium hydroxide,* NaOH, 2.5N: Dilute 41.7 mL 6N NaOH to 100 mL or dissolve 10 g NaOH pellets in 100 mL water.

4. Procedure

a. Measure 500 mL sample into a beaker, adjust pH to approximately 4.0 with H_3PO_4 solution using methyl orange indicator or a pH meter, and transfer to distillation apparatus. Use a 500-mL graduated cylinder as a receiver. Omit adding H_3PO_4 and adjust pH to 4.0 with 2.5N NaOH if sample was preserved as described in 5530A.4.

b. Distill 450 mL, stop distillation and, when boiling ceases, add 50 mL warm water to distilling flask. Continue distillation until a total of 500 mL has been collected.

c. One distillation should purify the sample adequately. Occasionally, however, the distillate is turbid. If so, acidify with H_3PO_4 solution and distill as described in ¶ 4b. If second distillate is still turbid, use extraction process described in ¶ 4d before distilling sample.

d. Treatment when second distillate is turbid: Extract a 500-mL portion of original sample as follows: Add 4 drops methyl orange indicator and make acidic to methyl orange with 1N H_2SO_4. Transfer to a separatory funnel and add 150 g NaCl. Shake with five successive portions of CHCl₃, using 40 mL in the first portion and 25 mL in each successive portion. Transfer CHCl₃ layer to a second separatory funnel and shake with three successive portions of 2.5N NaOH solution, using 4.0 mL in the first portion and 3.0 mL in each of the next two portions. Combine alkaline extracts, heat on a water bath until CHCl₃ has been removed, cool, and dilute to 500 mL with distilled water. Proceed with distillation as described in ¶s 4a and b.

NOTE: CH_2Cl_2 may be used instead of CHCl₃, especially if an emulsion forms when the CHCl₃ solution is extracted with NaOH.

---

* Corning No. 3360 or equivalent.

5530 C. Chloroform Extraction Method

1. General Discussion

a. *Principle:* Steam-distillable phenols react with 4-aminoantipyrine at pH 7.9 ± 0.1 in the presence of potassium ferricyanide to form a colored antipyrine dye. This dye is extracted from aqueous solution with $CHCl_3$ and the absorbance is measured at 460 nm. This method covers the phenol concentration range from 1.0 μg/L to over 250 μg/L with a sensitivity of 1 μg/L.

b. *Interference:* All interferences are eliminated or reduced to a minimum if the sample is preserved, stored, and distilled in accordance with the foregoing instructions.

c. *Minimum detectable quantity:* The minimum detectable quantity for clean samples containing no interferences is 0.5 μg phenol when a 25-mL $CHCl_3$ extraction with a 5-cm cell or a 50-mL $CHCl_3$ extraction with a 10-cm cell is used in the photometric measurement. This quantity is equivalent to 1 μg phenol/L in 500 mL distillate.

2. Apparatus

a. *Photometric equipment:* A spectrophotometer for use at 460 nm equipped with absorption cells providing light paths of 1 to 10 cm, depending on the absorbances of the colored solutions and the individual characteristics of the photometer.

b. *Filter funnels:* Buchner type with fritted disk.*

c. *Filter paper:* Alternatively use an appropriate 11-cm filter paper for filtering $CHCl_3$ extracts instead of the Buchner-type funnels and anhydrous Na_2SO_4.

d. *pH meter.*

e. *Separatory funnels,* 1000-mL, Squibb form, with ground-glass stoppers and TFE stopcocks. At least eight are required.

3. Reagents

Prepare all reagents with distilled water free of phenols and chlorine.

a. *Stock phenol solution:* Dissolve 100 mg phenol in freshly boiled and cooled distilled water and dilute to 100 mL. CAUTION—*Toxic; handle with extreme care.* Ordinarily this direct weighing yields a standard solution; if extreme accuracy is required, standardize as follows:

1) To 100 mL water in a 500-mL glass-stoppered conical flask, add 50.0 mL stock phenol solution and 10.0 mL bromate-bromide solution. Immediately add 5 mL conc HCl and swirl gently. If brown color of free bromine does not persist, add 10.0-mL portions of bromate-bromide solution until it does. Keep flask stoppered and let stand for 10 min; then add approximately 1 g KI. Usually four 10-mL portions of bromate-bromide solution are required if the stock phenol solution contains 1000 mg phenol/L.

2) Prepare a blank in exactly the same manner, using distilled water and 10.0 mL bromate-bromide solution. Titrate blank and sample with 0.025M sodium thiosulfate, using starch solution indicator.

3) Calculate the concentration of phenol solution as follows:

$$\text{mg phenol/L} = 7.842 \; [(A \times B) - C]$$

where:

    A = mL thiosulfate for blank,
    B = mL bromate-bromide solution used for sample divided by 10, and
    C = mL thiosulfate used for sample.

b. *Intermediate phenol solution:* Dilute 1.00 mL stock phenol solution in freshly boiled and cooled distilled water to 100 mL; 1 mL = 10.0 μg phenol. Prepare daily.

c. *Standard phenol solution:* Dilute 50.0 mL intermediate phenol solution to 500 mL with freshly boiled and cooled distilled water; 1 mL = 1.0 μg phenol. Prepare within 2 h of use.

d. *Bromate-bromide solution:* Dissolve 2.784 g anhydrous $KBrO_3$ in water, add 10 g KBr crystals, dissolve, and dilute to 1000 mL.

e. *Hydrochloric acid,* HCl, conc.

f. *Standard sodium thiosulfate titrant,* 0.025M: See Section 4500-O.C.2e.

g. *Starch solution:* See Section 4500-O.C.2d.

h. *Ammonium hydroxide,* NH_4OH, 0.5N: Dilute 35 mL fresh, conc NH_4OH to 1 L with water.

i. *Phosphate buffer solution:* Dissolve 104.5 g K_2HPO_4 and 72.3 g KH_2PO_4 in water and dilute to 1 L. The pH should be 6.8.

j. *4-Aminoantipyrine solution:* Dissolve 2.0 g 4-aminoantipyrine in water and dilute to 100 mL. Prepare daily.

k. *Potassium ferricyanide solution:* Dissolve 8.0 g $K_3Fe(CN)_6$ in water and dilute to 100 mL. Filter if necessary. Store in a brown glass bottle. Prepare fresh weekly.

l. *Chloroform,* $CHCl_3$.

m. *Sodium sulfate,* anhydrous Na_2SO_4, granular.

n. *Potassium iodide,* KI, crystals.

4. Procedure

Ordinarily, use Procedure a; however, Procedure b may be used for infrequent analyses.

a. Place 500 mL distillate, or a suitable portion containing not more than 50 μg phenol, diluted to 500 mL, in a 1-L beaker. Prepare a 500-mL distilled water blank and a series of 500-mL phenol standards containing 5, 10, 20, 30, 40, and 50 μg phenol.

Treat sample, blank, and standards as follows: Add 12.0 mL 0.5N NH_4OH and *immediately* adjust pH to 7.9 ± 0.1 with phosphate buffer. Under some circumstances, a higher pH may be required.† About 10 mL phosphate buffer are required. Transfer to a 1-L separatory funnel, add 3.0 mL aminoantipyrine solution, mix well, add 3.0 mL $K_3Fe(CN)_6$ solution, mix well, and let color develop for 15 min. The solution should be clear and light yellow.

Extract immediately with $CHCl_3$, using 25 mL for 1- to 5-cm cells and 50 mL for a 10-cm cell. Shake separatory funnel at least 10 times, let $CHCl_3$ settle, shake again 10 times, and let $CHCl_3$ settle again. Filter each $CHCl_3$ extract through filter paper or fritted glass funnels containing a 5-g layer of anhydrous Na_2SO_4. Collect dried extracts in clean cells for absorbance measurements; do not add more $CHCl_3$ or wash filter papers or funnels with $CHCl_3$.

Read absorbance of sample and standards against the blank at 460 nm. Plot absorbance against micrograms phenol concentra

---

* 15-mL Corning No. 36060 or equivalent.

† For NPDES permit analyses, pH 10 ± 0.1 is required.

tion. Construct a separate calibration curve for each photometer and check each curve periodically to insure reproducibility.

b. For infrequent analyses prepare only one standard phenol solution. Prepare 500 mL standard phenol solution of a strength approximately equal to the phenolic content of that portion of original sample used for final analysis. Also prepare a 500-mL distilled water blank.

Continue as described in ¶ a, above, but measure absorbances of sample and standard phenol solution against the blank at 460 nm.

5. Calculation

a. For Procedure a:

$$\mu g \ phenol/L = \frac{A}{B} \times 1000$$

where:

A = µg phenol in sample, from calibration curve, and
B = mL original sample.

b. For Procedure b, calculate the phenol content of the original sample:

$$\mu g \ phenol/L = \frac{C \times D \times 1000}{E \times B}$$

where:

C = µg standard phenol solution,
D = absorbance reading of sample,
E = absorbance of standard phenol solution, and
B = mL original sample.

6. Precision and Bias

Because the "phenol" value is based on C_6H_5OH, this method yields only an approximation and represents the minimum amount of phenols present. This is true because the phenolic reactivity to 4-aminoantipyrine varies with the types of phenols present.

In a study of 40 refinery wastewaters analyzed in duplicate at concentrations from 0.02 to 6.4 mg/L the average relative standard deviation was ± 12%. Data are not available for precision at lower concentrations.

7. Bibliography

SCOTT, R.D. 1931. Application of a bromine method in the determination of phenols and cresols. *Ind. Eng. Chem.*, Anal. Ed. 3:67.

EMERSON, E., H.H. BEACHAM & L.C. BEEGLE. 1943. The condensation of aminoantipyrine. II. A new color test for phenolic compounds. *J. Org. Chem.* 8:417.

ETTINGER, M.B. & R.C. KRONER. 1949. The determination of phenolic materials in industrial wastes. *Proc. 5th Ind. Waste Conf.*, Purdue Univ., p. 345.

ETTINGER, M.B., C.C. RUCHHOFT & R.J. LISHKA. 1951. Sensitive 4-aminoantipyrine method for phenolic compounds. *Anal. Chem.* 23:1783.

DANNIS, M. 1951. Determination of phenols by the aminoantipyrine method. *Sewage Ind. Wastes* 23:1516.

MOHLER, E.F., JR. & L.N. JACOB. 1957. Determination of phenolic-type compounds in water and industrial waste waters: Comparison of analytical methods. *Anal. Chem.* 29:1369.

BURTSCHELL, R.H., A.A. ROSEN, F.M. MIDDLETON & M.B. ETTINGER. 1959. Chlorine derivatives of phenol causing taste and odor. *J. Amer. Water Works Assoc.* 51:205.

GORDON, G.E. 1960. Colorimetric determination of phenolic materials in refinery waste waters. *Anal. Chem.* 32:1325.

OCHYNSKI, F.W. 1960. The absorptiometric determination of phenol. *Analyst* 85:278.

FAUST, S.D. & O.M. ALY. 1962. The determination of 2,4-dichlorophenol in water. *J. Amer. Water Works Assoc.* 54:235.

FAUST, S.D. & E.W. MIKULEWICZ. 1967. Factors influencing the condensation of 4-aminoantipyrine with derivatives of hydroxybenzene. II. Influence of hydronium ion concentration on absorptivity. *Water Res.* 1:509.

FAUST, S.D. & P.W. ANDERSON. 1968. Factors influencing the condensation of 4-aminoantipyrine with derivations of hydroxy benzene. III. A study of phenol content in surface waters. *Water Res.* 2:515.

SMITH, L.S. 1976. Evaluation of Instrument for the (Ultraviolet) Determination of Phenol in Water. EPA-600/4-76-048, U.S. Environmental Protection Agency, Cincinnati, Ohio.

5530 D. Direct Photometric Method

1. General Discussion

a. *Principle:* Steam-distillable phenolic compounds react with 4-aminoantipyrine at pH 7.9 ± 0.1 in the presence of potassium ferricyanide to form a colored antipyrine dye. This dye is kept in aqueous solution and the absorbance is measured at 500 nm.

b. *Interference:* Interferences are eliminated or reduced to a minimum by using the distillate from the preliminary distillation procedure.

c. *Minimum detectable quantity:* This method has less sensitivity than Method C. The minimum detectable quantity is 10 µg phenol when a 5-cm cell and 100 mL distillate are used.

2. Apparatus

a. *Photometric equipment:* Spectrophotometer equipped with absorption cells providing light paths of 1 to 5 cm for use at 500 nm.

b. *pH meter.*

3. Reagents

See Section 5530C.3.

4. Procedure

Place 100 mL distillate, or a portion containing not more than 0.5 mg phenol diluted to 100 mL, in a 250-mL beaker.

Prepare a 100-mL distilled water blank and a series of 100-mL phenol standards containing 0.1, 0.2, 0.3, 0.4, and 0.5 mg phenol. Treat sample, blank, and standards as follows: Add 2.5 mL 0.5N NH$_4$OH solution and immediately adjust to pH 7.9 ± 0.1 with phosphate buffer. Add 1.0 mL 4-aminoantipyrine solution, mix well, add 1.0 mL K$_3$Fe(CN)$_6$ solution, and mix well.

After 15 min, transfer to cells and read absorbance of sample and standards against the blank at 500 nm.

5. Calculation

a. Use of calibration curve: Estimate sample phenol content from photometric readings by using a calibration curve constructed as directed in Section 5530C.4*a*.

$$\text{mg phenol/L} = \frac{A}{B} \times 1000$$

where:

A = mg phenol in sample, from calibration curve, and
B = mL original sample.

b. Use of single phenol standard:

$$\text{mg phenol/L} = \frac{C \times D \times 1000}{E \times B}$$

where:

C = mg standard phenol solution,
D = absorbance of sample, and
E = absorbance of standard phenol solution.

6. Precision and Bias

Precision and bias data are not available.

5540 SURFACTANTS*

5540 A. Introduction

1. Occurrence and Significance

Surfactants enter waters and wastewaters mainly by discharge of aqueous wastes from household and industrial laundering and other cleansing operations. A surfactant combines in a single molecule a strongly hydrophobic group with a strongly hydrophilic one. Such molecules tend to congregate at the interfaces between the aqueous medium and the other phases of the system such as air, oily liquids, and particles, thus imparting properties such as foaming, emulsification, and particle suspension.

The surfactant hydrophobic group generally is a hydrocarbon radical (R) containing about 10 to 20 carbon atoms. The hydrophilic groups are of two types, those that ionize in water and those that do not. Ionic surfactants are subdivided into two categories, differentiated by the charge. An anionic surfactant ion is negatively charged, e.g., $(RSO_3)^-Na^+$, and a cationic one is positively charged, e.g., $(RMe_3N)^+Cl^-$. Nonionizing (nonionic) surfactants commonly contain a polyoxyethylene hydrophilic group ($ROCH_2CH_2OCH_2CH_2......OCH_2CH_2OH$, often abbreviated RE_n, where n is the average number of -OCH_2CH_2- units in the hydrophilic group). Hybrids of these types exist also.

In the United States, ionic surfactants amount to about two thirds of the total surfactants used and nonionics to about one third. Cationic surfactants amount to less than one tenth of the ionics and are used generally for disinfecting, fabric softening, and various cosmetic purposes rather than for their detersive properties. At current detergent and water usage levels the surfactant content of raw domestic wastewater is in the range of about 1 to 20 mg/L. Most domestic wastewater surfactants are dissolved in

equilibrium with proportional amounts adsorbed on particulates. Primary sludge concentrations range from 1 to 20 mg adsorbed anionic surfactant per gram dry weight.[1] In environmental waters the surfactant concentration generally is below 0.1 mg/L except in the vicinity of an outfall or other point source of entry.[2]

2. Analytical Precautions

Because of inherent properties of surfactants, special analytical precautions are necessary. Avoid foam formation because the surfactant concentration is higher in the foam phase than in the associated bulk aqueous phase and the latter may be significantly depleted. If foam is formed, let it subside by standing, or collapse it by other appropriate means, and remix the liquid phase before sampling. Adsorption of surfactant from aqueous solutions onto the walls of containers, when concentrations below about 1 mg/L are present, may seriously deplete the bulk aqueous phase. Minimize adsorption errors, if necessary, by rinsing container with sample, and for anionic surfactants by adding alkali phosphate (e.g., 0.03N KH$_2$PO$_4$).[3]

3. References

1. SWISHER, R.D. 1987. Surfactant Biodegradation, 2nd ed. Marcel Dekker, New York, N.Y.
2. ARTHUR D. LITTLE, INC. 1977. Human Safety and Environmental Aspects of Major Surfactants. Rep. No. PB-301193, National Technical Information Serv., Springfield, Va.
3. WEBER, W.J., JR., J.C. MORRIS & W. STUMM. 1962. Determination of alkylbenzenesulfonate by ultraviolet spectrophotometry. *Anal. Chem.* 34:1844.

*Approved by Standard Methods Committee, 1993.

5540 B.   Surfactant Separation by Sublation

1. General Discussion

a. Principle: The sublation process isolates the surfactant, regardless of type, from dilute aqueous solution, and yields a dried residue relatively free of nonsurfactant substances. It is accomplished by bubbling a stream of nitrogen up through a column containing the sample and an overlying layer of ethyl acetate. The surfactant is adsorbed at the water-gas interfaces of the bubbles and is carried into the ethyl acetate layer. The bubbles escape into the atmosphere leaving behind the surfactant dissolved in ethyl acetate. The solvent is separated, dehydrated, and evaporated, leaving the surfactant as a residue suitable for analysis. This procedure is the same as that used by the Organization for Economic Co-operation and Development (OECD),[1] following the development by Wickbold.[2,3]

b. Interferences: The sublation method is specific for surfactants, because any substance preferentially adsorbed at the water-gas interface is by definition a surfactant. Although nonsurfactant substances largely are rejected in this separation process, some amounts will be carried over mechanically into the ethyl acetate.

c. Limitations: The sublation process separates only dissolved surfactants. If particulate matter is present it holds back an equilibrium amount of adsorbed surfactant. As sublation removes the initially dissolved surfactant, the particulates tend to re-equilibrate and their adsorbed surfactants redissolve. Thus, continued sublation eventually should remove substantially all adsorbed surfactant. However, if the particulates adsorb the surfactant tightly, as sewage particulates usually do, complete removal may take a very long time. The procedure given herein calls for preliminary filtration and measures only dissolved surfactant. Determine adsorbed surfactant content by analyzing particulates removed by filtration; no standard method is available now.

d. Operating conditions: Make successive 5-min sublations from 1 L of sample containing 5 g NaHCO₃ and 100 g NaCl. Under the conditions specified, extensive transfer of surfactant occurs in the first sublation and is substantially complete in the second.[2-4]

e. Quantitation: Quantitate the surfactant residue by the procedures in 5540C or D. Direct weighing of the residue is not useful because the weight of surfactant isolated generally is too low, less than a milligram, and varied amounts of mechanically entrained nonsurfactants may be present. The procedure is applicable to water and wastewater samples.

2. Apparatus

a. Sublator: A glass column with dimensions as shown in Figure 5540:1. For the sintered glass disk use a coarse-porosity frit (designation "c"–nominal maximum pore diam 40 to 60 μm as measured by ASTM E-128) of the same diameter as the column internal diameter. Volume between disk and upper stopcock should be approximately 1 L.

b. Gas washing bottle, as indicated in Figure 5540:1, working volume 100 mL or more.

c. Separatory funnel, working volume 250 mL, preferably with inert TFE stopcock.

d. Filtration equipment, suitable for 1-L samples, using medium-porosity qualitative-grade filter paper.

e. Gas flowmeter, for measuring flows up to 1 L/min.

3. Reagents

a. Nitrogen, standard commercial grade.

b. Ethyl acetate: CAUTION: *Ethyl acetate is flammable and its vapors can form explosive mixtures with air.*

c. Sodium bicarbonate, NaHCO₃.

d. Sodium chloride, NaCl.

e. Water, surfactant-free.

4. Procedure

a. Sample size: Select a sample to contain not more than 1 to 2 mg surfactant.[4] For most waters the sample volume will be about 1 L; for wastewater use a smaller volume.

b. Filtration: Filter sample through medium-porosity qualitative filter paper. Wash filter paper by discarding the first few hundred milliliters of filtrate.

c. Assembly: Refer to Figure 5540:1.

Connect nitrogen cylinder through flowmeter to inlet of gas washing bottle. Connect gas outlet at top of sublator to a gas scrubber or other means for disposing of ethyl acetate vapor (e.g., vent to a hood or directly outdoors). In the absence of a flowmeter, ensure proper gas flow rate by measuring volume of gas leaving the sublator, with a water-displacement system.

d. Charging: Fill gas washing bottle about two-thirds full with ethyl acetate. Rinse sublation column with ethyl acetate and discard rinse. Place measured filtered sample in sublator and add 5 g NaHCO₃, 100 g NaCl, and sufficient water to bring the level up to or slightly above the upper stopcock (about 1 L total volume). If sample volume permits, add salts as a solution in 400 mL water or dissolve them in the sample and quantitatively transfer to the sublator. Add 100 mL ethyl acetate by running it carefully down the wall of the sublator to form a layer on top of the sample.

e. Sublation: Start the nitrogen flow, increasing the rate carefully to 1 L/min initially but do not exceed a rate at which the liquid phases begin vigorous intermixing at their interface. Avoid overly vigorous intermixing, which will lead to back-extraction of the surfactant into the aqueous phase and to dissolution of ethyl acetate. Continue sublation for 5 min at 1 L/min. If a lower flow rate is necessary to avoid phase intermixing, prolong sublation time proportionally. If the volume of the upper phase has decreased by more than about 20%, repeat the operation on a new sample but avoid excessive intermixing at the interface. Draw off entire ethyl acetate layer through upper stopcock into the separatory funnel; return any transferred water layer to the sublator. Filter ethyl acetate layer into a 250-mL beaker through a dry, medium-porosity, qualitative filter paper (prewashed with ethyl acetate to remove any adventitious surfactant) to remove any remaining aqueous phase.

Repeat process of preceding paragraph with a second 100-mL layer of ethyl acetate, using the same separatory funnel and filter, and finally rinse sublator wall with another 20 mL, all into the original beaker.

Evaporate ethyl acetate from the beaker on a steam bath in a hood, blowing a gentle stream of nitrogen or air over the liquid

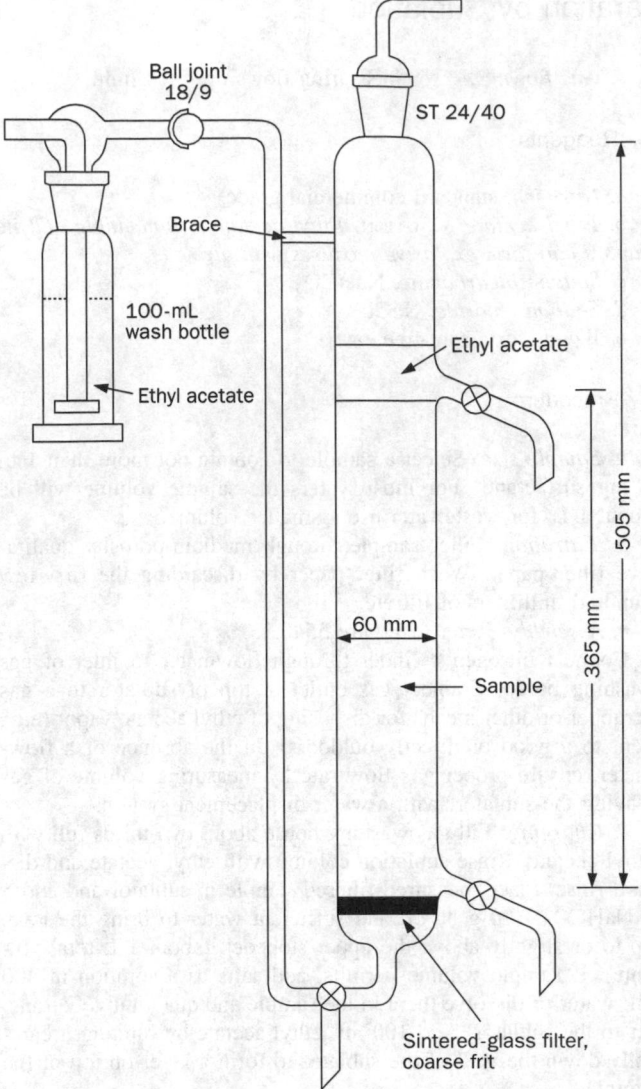

Figure 5540:1. Sublation apparatus.[1] See Section 5540B.2a and b and 4c.
Bottom stopcock: TFE plug, 4-mm bore; side stopcocks: TFE plug, 2-mm bore.

surface to speed evaporation and to minimize active boiling. Evaporate the first 100 mL during the second sublation to avoid overfilling the beaker. To avoid possible solute volatilization, discontinue heating after removing the ethyl acetate. The sublated surfactant remains in the beaker as a film of residue.

Draw off aqueous layer in the sublator and discard, using the stopcock just above the sintered disk to minimize disk fouling.

5. Precision and Bias

Estimates of the efficiency of surfactant transfer and recovery in the sublation process include the uncertainties of the analytical methods used in quantitating the surfactant. At present the analytical methods are semiquantitative for surfactant at levels below 1 mg/L in environmental samples.

With various known surfactants at 0.2 to 2 mg/L and appropriate analytical methods, over 90% of added surfactant was re-

TABLE 5540:I. SURFACTANT RECOVERY BY SUBLATION

Variable	MBAS	CTAS
Sample volume, mL	200–300	500
Concentration without sublation, mg/L	2.2–4.7	—
Concentration found in sublate,* mg/L	1.8–4.4	0.3–0.6
Recovery in sublate, %	87 ± 16†	—
Amount in second sublate,‡ mg	0.02 ± 0.02†	0.08 ± 0.01†
Amount added, mg	0.05–0.10§	0.50–0.67‖
Recovery in sublation,# %	94 ± 17†	92 ± 6†

* Two 5-min sublations.
† Average ± SD (n = 8).
‡ Two more 5-min sublations.
§ Reference LAS.
‖ Linear alcohol ethoxylate $C_{12-18}E_{11}$.
\# Fifth and sixth 5-min sublations.

covered in one 5-min sublation from 10% NaCl. Without NaCl, recovery of nonionics was over 90% but recovery of anionics and cationics was only 2 to 25%.[4]

Five laboratories studied the recovery of five anionic surfactant types from concentrations of 0.05, 0.2, 1.0, and 5.0 mg/L in aqueous solutions.[5] The amount in each solution was determined directly by methylene blue analysis and compared with the amount recovered in the sublation process, also analyzed by methylene blue. The overall average recovery was 95.9% with a standard deviation of ± 7.4 (n = 100). The extreme individual values for recovery were 65% and 115% and the other 98 values ranged from 75% to 109%. Recovery did not depend on surfactant concentration (average recoveries ranging from 94.7% at 5.0 mg/L to 96.8% at 1.0 mg/L) nor on the surfactant type (average recoveries ranging from 94.7% to 96.6%). Average recoveries at the five laboratories ranged from 90.0% to 98.0%.

Application of the sublation method in three laboratories to eight different samples of raw wastewater in duplicate gave the results shown in Table 5540:I. Methylene blue active substances (MBAS) recovery in double sublation averaged 87 ± 16% of that determined directly on the filtered wastewater; these results would have been influenced by any nonsurfactant MBAS that might have been present. Repeating double sublation on the spent aqueous phase yielded another 0.02 mg MBAS and another 0.08 mg cobalt thiocyanate active substances (CTAS). Adding 0.05 to 0.10 mg of known linear alkylbenzene sulfonate (LAS) or 0.50 to 0.67 mg of known linear alcohol-based $C_{12-18}E_{11}$ to the same sublator contents and again running double sublation resulted in over 90% recovery of the amount added.

6. References

1. ORGANIZATION FOR ECONOMIC CO-OPERATION AND DEVELOPMENT ENVIRONMENTAL DIRECTORATE. 1976. Proposed Method for the Determination of the Biodegradability of Surfactants Used in Synthetic Detergents. Org. for Economic Co-Operation & Development, Paris.
2. WICKBOLD, R. 1971. Enrichment and separation of surfactants from surface waters through transport in the gas/water interface. Tenside 8: 61.
3. WICKBOLD, R. 1972. Determination of nonionic surfactants in river- and wastewaters. Tenside 9:173.
4. KUNKEL, E., G. PEITSCHER & K. ESPETER. 1977. New developments in trace- and microanalysis of surfactants. Tenside 14:199.
5. DIVO, C., S. GAFA, T. LA NOCE, A. PARIS, C. RUFFO & M. SANNA. 1980. Use of nitrogen blowing technique in microdetermination of anionic surfactants. Riv. Ital. Sost. Grasse 57:329.

5540 C. Anionic Surfactants as MBAS

1. General Discussion

a. Definition and principle: Methylene blue active substances (MBAS) bring about the transfer of methylene blue, a cationic dye, from an aqueous solution into an immiscible organic liquid upon equilibration. This occurs through ion pair formation by the MBAS anion and the methylene blue cation. The intensity of the resulting blue color in the organic phase is a measure of MBAS. Anionic surfactants are among the most prominent of many substances, natural and synthetic, showing methylene blue activity. The MBAS method is useful for estimating the anionic surfactant content of waters and wastewaters, but the possible presence of other types of MBAS always must be kept in mind.

This method is relatively simple and precise. It comprises three successive extractions from acid aqueous medium containing excess methylene blue into chloroform ($CHCl_3$), followed by an aqueous backwash and measurement of the blue color in the $CHCl_3$ by spectrophotometry at 652 nm. The method is applicable at MBAS concentrations down to about 0.025 mg/L.

b. Anionic surfactant responses: Soaps do not respond in the MBAS method. Those used in or as detergents are alkali salts of C_{10-20} fatty acids $[RCO_2]^-Na^+$, and though anionic in nature they are so weakly ionized that an extractable ion pair is not formed under the conditions of the test. Nonsoap anionic surfactants commonly used in detergent formulations are strongly responsive. These include principally surfactants of the sulfonate type $[RSO_3]^-Na^+$, the sulfate ester type $[ROSO_3]^-Na^+$, and sulfated nonionics $[RE_nOSO_3]^-Na^+$. They are recovered almost completely by a single $CHCl_3$ extraction.

Linear alkylbenzene sulfonate (LAS) is the most widely used anionic surfactant and is used to standardize the MBAS method. LAS is not a single compound, but may comprise any or all of 26 isomers and homologs with structure $[R'C_6H_4SO_3]^-Na^+$, where R′ is a linear secondary alkyl group ranging from 10 to 14 carbon atoms in length. The manufacturing process defines the mixture, which may be modified further by the wastewater treatment process.

Sulfonate- and sulfate-type surfactants respond together in MBAS analysis, but they can be differentiated by other means. The sulfate type decomposes upon acid hydrolysis; the resulting decrease in MBAS corresponds to the original sulfate surfactant content while the MBAS remaining corresponds to the sulfonate surfactants. Alkylbenzene sulfonate can be identified and quantified by infrared spectrometry after purification.[1] LAS can be distinguished from other alkylbenzene sulfonate surfactants by infrared methods.[2] LAS can be identified unequivocally and its detailed isomer-homolog composition determined by desulfonation-gas chromatography.[3]

c. Interferences: Positive interferences result from all other MBAS species present; if a direct determination of any individual MBAS species, such as LAS, is sought, all others interfere. Substances such as organic sulfonates, sulfates, carboxylates and phenols, and inorganic thiocyanates, cyanates, nitrates, and chlorides also may transfer more or less methylene blue into the chloroform phase. The poorer the extractability of their ion pairs, the more effective is the aqueous backwash step in removing these positive interferences; interference from chloride is eliminated almost entirely and from nitrate largely so by the backwash. Because of the varied extractability of nonsurfactant MBAS, deviations in $CHCl_3$ ratio and backwashing procedure may lead to significant differences in the total MBAS observed, although the recovery of sulfonate- and sulfate-type surfactants will be substantially complete in all cases.

Negative interferences can result from the presence of cationic surfactants and other cationic materials, such as amines, because they compete with the methylene blue in the formation of ion pairs. Particulate matter may give negative interference through adsorption of MBAS. Although some of the adsorbed MBAS may be desorbed and paired during the $CHCl_3$ extractions, recovery may be incomplete and variable.

Minimize interferences by nonsurfactant materials by sublation if necessary (Section 5540B). Other countermeasures are nonstandard. Remove interfering cationic surfactants and other cationic materials by using a cation-exchange resin under suitable conditions.[3] Handle adsorption of MBAS by particulates preferably by filtering and analyzing the insolubles. With or without filtration, adsorbed MBAS can be desorbed by acid hydrolysis; however, MBAS originating in any sulfate ester-type surfactant present is destroyed simultaneously.[1] Sulfides, often present in raw or primary treated wastewater, may react with methylene blue to form a colorless reduction product, making the analysis impossible. Eliminate this interference by prior oxidation with hydrogen peroxide.

d. Molecular weight: Test results will appear to differ if expressed in terms of weight rather than in molar quantities. Equimolar amounts of two anionic surfactants with different molecular weights should give substantially equal colors in the $CHCl_3$ layer, although the amounts by weight may differ significantly. If results are to be expressed by weight, as generally is desirable, the average molecular weight of the surfactant measured must be known or a calibration curve made with that particular compound must be used. Because such detailed information generally is lacking, report results in terms of a suitable standard calibration curve, for example "0.65 mg MBAS/L (calculated as LAS, mol wt 318)."

e. Minimum detectable quantity: About 10 μg MBAS (calculated as LAS).

f. Application: The MBAS method has been applied successfully to drinking water samples. In wastewater, industrial wastes, and sludge, numerous materials normally present can interfere seriously if direct determination of MBAS is attempted. Most nonsurfactant aqueous-phase interferences can be removed by sublation. The method is linear over an approximate range of 10 to 200 μg of MBAS standard. This may vary somewhat, depending on source of standard material.

2. Apparatus

a. Colorimetric equipment: One of the following is required:

1) *Spectrophotometer,* for use at 652 nm, providing a light path of 1 cm or longer.

2) *Filter photometer,* providing a light path of 1 cm or longer and equipped with a red color filter exhibiting maximum transmittance near 652 nm.

b. Separatory funnels: 500-mL, preferably with inert TFE stopcocks and stoppers.

3. Reagents

a. Stock LAS solution: Weigh an amount of the reference material* equal to 1.00 g LAS on a 100% active basis. Dissolve in water and dilute to 1000 mL; 1.00 mL = 1.00 mg LAS. Store in a refrigerator to minimize biodegradation. If necessary, prepare weekly.

b. Standard LAS solution: Dilute 10.00 mL stock LAS solution to 1000 mL with water; 1.00 mL = 10.0 μg LAS. Prepare daily.

c. Phenolphthalein indicator solution, alcoholic.

d. Sodium hydroxide, NaOH, 1N.

e. Sulfuric acid, H_2SO_4, 1N and 6N.

f. Chloroform, $CHCl_3$: CAUTION: *Chloroform is toxic and a suspected carcinogen. Take appropriate precautions against inhalation and skin exposure.*

g. Methylene blue reagent: Dissolve 100 mg methylene blue† in 100 mL water. Transfer 30 mL to a 1000-mL flask. Add 500 mL water, 41 mL 6N H_2SO_4, and 50 g sodium phosphate, monobasic, monohydrate, $NaH_2PO_4 \cdot H_2O$. Shake until dissolved. Dilute to 1000 mL.

h. Wash solution: Add 41 mL 6N H_2SO_4 to 500 mL water in a 1000-mL flask. Add 50 g $NaH_2PO_4 \cdot H_2O$ and shake until dissolved. Dilute to 1000 mL.

i. Methanol, CH_3OH. CAUTION: *Methanol vapors are flammable and toxic; take appropriate precautions.*

j. Hydrogen peroxide, H_2O_2, 30%.

k. Glass wool: Pre-extract with $CHCl_3$ to remove interferences.

l. Water, reagent-grade, MBAS-free. Use for making all reagents and dilutions.

4. Procedure

a. Preparation of calibration curve: Prepare an initial calibration curve consisting of at least five standards covering the referenced (¶ 1f) or desired concentration range. Provided that linearity is demonstrated over the range of interest ($r = 0.995$ or better) run daily check standards at the reporting limit and a concentration above the expected samples' concentration. Check standard results should be within 25% of original value at the reporting limit and 10% of original value for all others. Otherwise, prepare a new calibration curve.

Prepare a series of separatory funnels for a reagent blank and selected standards. Pipet portions of standard LAS solution (¶ 3b) into funnels. Add sufficient water to make the total volume 100 mL in each separatory funnel. Treat each standard as described in ¶s 4d and e following, and plot a calibration curve of absorbance vs. micrograms LAS taken, specifying the molecular weight of the LAS used.

b. Sample size: For direct analysis of waters and wastewaters, select sample volume on the basis of expected MBAS concentration:

Expected MBAS Concentration mg/L	Sample Taken mL
0.025–0.080	400
0.08 –0.40	250
0.4 –2.0	100

* For sources of suitable reference material, contact *Standard Methods* manager.
† Eastman No. P573 or equivalent.

If expected MBAS concentration is above 2 mg/L, dilute sample containing 40 to 200 μg MBAS to 100 mL with water.

For analysis of samples purified by sublation, dissolve sublate residue (Section 5540B.4e) in 10 to 20 mL methanol, quantitatively transfer the entire amount (or a suitable portion if more than 200 μg MBAS is expected) to 25 to 50 mL water, evaporate without boiling until methanol is gone, adding water as necessary to avoid going to dryness, and dilute to about 100 mL with water.

c. Peroxide treatment: If necessary to avoid decolorization of methylene blue by sulfides, add a few drops of 30% H_2O_2.

d. Ion pairing and extraction:

1) Add sample to a separatory funnel. Make alkaline by dropwise addition of 1N NaOH, using phenolphthalein indicator. Discharge pink color by dropwise addition of 1N H_2SO_4.

2) Add 10 mL $CHCl_3$ and 25 mL methylene blue reagent. Rock funnel vigorously for 30 s and let phases separate. Alternatively, place a magnetic stirring bar in the separatory funnel; lay funnel on its side on a magnetic mixer and adjust speed of stirring to produce a rocking motion. Excessive agitation may cause emulsion formation. To break persistent emulsions add a small volume of isopropyl alcohol (<10 mL); add same volume of isopropyl alcohol to all standards. Some samples require a longer period of phase separation than others. Before draining $CHCl_3$ layer, swirl gently, then let settle.

3) Draw off $CHCl_3$ layer into a second separatory funnel. Rinse delivery tube of first separatory funnel with a small amount of $CHCl_3$. Repeat extraction two additional times, using 10 mL $CHCl_3$ each time. If blue color in water phase becomes faint or disappears, discard and repeat, using a smaller sample.

4) Combine all $CHCl_3$ extracts in the second separatory funnel. Add 50 mL wash solution and shake vigorously for 30 s. Emulsions do not form at this stage. Let settle, swirl, and draw off $CHCl_3$ layer through a funnel containing a plug of glass wool into a 100-mL volumetric flask; filtrate must be clear. Extract wash solution twice with 10 mL $CHCl_3$ each and add to flask through the glass wool. Rinse glass wool and funnel with $CHCl_3$. Collect washings in volumetric flask, dilute to mark with $CHCl_3$, and mix well.

e. Measurement: Determine absorbance at 652 nm against a blank of $CHCl_3$.

5. Calculation

From the calibration curve (¶ 4a) read micrograms of apparent LAS (mol wt __) corresponding to the measured absorbance.

$$\text{mg MBAS/L} = \frac{\mu g \text{ apparent LAS}}{\text{mL original sample}}$$

Report as "MBAS, calculated as LAS, mol wt __."

6. Precision and Bias

A synthetic sample containing 270 μg LAS/L in distilled water was analyzed in 110 laboratories with a relative standard deviation of 14.8% and a relative error of 10.6%.

A tap water sample to which was added 480 μg LAS/L was analyzed in 110 laboratories with a relative standard deviation of 9.9% and a relative error of 1.3%.

A river water sample with 2.94 mg LAS/L added was analyzed in 110 laboratories with a relative standard deviation of 9.1% and a relative error of 1.4%.[4]

7. References

1. AMERICAN PUBLIC HEALTH ASSOCIATION, AMERICAN WATER WORKS ASSOCIATION & WATER POLLUTION CONTROL FEDERATION. 1981. Carbon adsorption-infrared method. *In* Standard Methods for the Examination of Water and Wastewater, 15th ed. American Public Health Assoc., Washington, D.C.
2. OGDEN, C.P., H.L. WEBSTER & J. HALLIDAY. 1961. Determination of biologically soft and hard alkylbenzenesulfonates in detergents and sewage. *Analyst* 86:22.
3. OSBURN, Q.W. 1986. Analytical methodology for LAS in waters and wastes. *J. Amer. Oil Chem. Soc.* 63:257.
4. LISHKA, R.J. & J.H. PARKER. 1968. Water Surfactant No. 3, Study No. 32. U.S. Public Health Serv. Publ. No. 999-UIH-11, Cincinnati, Ohio.

8. Bibliography

BARR, T., J. OLIVER & W.V. STUBBINGS. 1948. The determination of surface-active agents in solution. *J. Soc. Chem. Ind.* (London) 67:45.
EPTON, S.R. 1948. New method for the rapid titrimetric analysis of sodium alkyl sulfates and related compounds. *Trans. Faraday Soc.* 44: 226.
EVANS, H.C. 1950. Determination of anionic synthetic detergents in sewage. *J. Soc. Chem. Ind.* (London) 69:Suppl. 2, S76.
DEGENS, P.N., JR., H.C. EVANS, J.D. KOMMER & P.A. WINSOR. 1953. Determination of sulfate and sulfonate anion-active detergents in sewage. *J. Appl. Chem.* (London) 3:54.
AMERICAN WATER WORKS ASSOCIATION. 1954. Task group report. Characteristics and effects of synthetic detergents. *J. Amer. Water Works Assoc.* 46:751.
EDWARDS, G.P. & M.E. GINN. 1954. Determination of synthetic detergents in sewage. *Sewage Ind. Wastes* 26:945.
LONGWELL, J. & W.D. MANIECE. 1955. Determination of anionic detergents in sewage, sewage effluents, and river water. *Analyst* 80:167.
MOORE, W.A. & R.A. KOLBESON. 1956. Determination of anionic detergents in surface waters and sewage with methyl green. *Anal. Chem.* 28:161.
ROSEN, A.A., F.M. MIDDLETON & N. TAYLOR. 1956. Identification of synthetic detergents in foams and surface waters. *J. Amer. Water Works Assoc.* 48:1321.
SALLEE, E.M., et al. 1956. Determination of trace amounts of alkyl benzenesulfonates in water. *Anal. Chem.* 28:1822.
AMERICAN WATER WORKS ASSOCIATION. 1958. Task group report. Determination of synthetic detergent content of raw water supplies. *J. Amer. Water Works Assoc.* 50:1343.
MCGUIRE, O.E., F. KENT, LaR. L. MILLER & G.J. PAPENMEIER. 1962. Field test for analysis of anionic detergents in well waters. *J. Amer. Water Works Assoc.* 54:665.
ABBOTT, D.C. 1962. The determination of traces of anionic surface-active materials in water. *Analyst* 87:286.
ABBOTT, D.C. 1963. A rapid test for anionic detergents in drinking water. *Analyst* 88:240.
REID, V.W., G.F. LONGMAN & E. HEINERTH. 1967. Determination of anionic-active detergents by two-phase titration. *Tenside* 4:292.
WANG, L.K., P.J. PANZARDI, W.W. SHUSTER & D. AULENBACH. 1975. Direct two-phase titration method for analyzing anionic nonsoap surfactants in fresh and saline waters. *J. Environ. Health* 38:159.

5540 D.   Nonionic Surfactants as CTAS

1. General Discussion

a. Definition and principle: Cobalt thiocyanate active substances (CTAS) are those that react with aqueous cobalt thiocyanate solution to give a cobalt-containing product extractable into an organic liquid in which it can be measured. Nonionic surfactants exhibit such activity, as may other natural and synthetic materials; thus, estimation of nonionic surfactants as CTAS is possible only if substantial freedom from interfering CTAS species can be assured.

The method requires sublation to remove nonsurfactant interferences and ion exchange to remove cationic and anionic surfactants, partition of CTAS into methylene chloride from excess aqueous cobalt thiocyanate by a single extraction, and measurement of CTAS in the methylene chloride by spectrophotometry at 620 nm. Lower limit of detectability is around 0.1 mg CTAS, calculated as $C_{12-18}E_{11}$. Beyond the sublation step the procedure is substantially identical to that of the Soap and Detergent Association (SDA).[1]

b. Nonionic surfactant responses: For pure individual molecular species the CTAS response is negligible up to about RE_5, where it increases sharply and continues to increase more gradually for longer polyether chains.[2,3] Fewer than about six oxygens in the molecule do not supply enough cumulative coordinate bond strength to hold the complex together. Commercial nonionic surfactants generally range from about RE_7 to RE_{15}; however, each such product, because of synthesis process constraints, is actually a mixture of many individual species ranging from perhaps RE_0 to RE_{2n} in a Poisson distribution averaging RE_n.

The hydrophobes used for nonionic surfactants in the U.S. household detergent industry are mainly linear primary and linear secondary alcohols with chain lengths ranging from about 12 to about 18 carbon atoms. Nonionics used in industrial operations include some based on branched octyl- and nonylphenols. These products give strong CTAS responses that may differ from each other, on a weight basis, by as much as a factor of 2. Specifically, eight such products showed responses from 0.20 to 0.36 absorbance units/mg by the SDA procedure.[1]

As with anionic surfactants measured as MBAS, the nonionic surfactants found in water and wastewater might have CTAS responses at least as varied as their commercial precursors because the proportions of the individual molecular species will have been changed by biochemical and physicochemical removal at varied rates, and further because their original molecular structures may have been changed by biodegradation processes.

c. Reference nonionic surfactant: Until it is practical to determine the nature and molecular composition of an unknown mixed CTAS, and to calculate or determine the CTAS responses of its component species, exact quantitation of uncharacterized CTAS in a sample in terms of weight is not possible. Instead, express the analytical result in terms of some arbitrarily chosen reference nonionic surfactant, i.e., as the weight of the reference that gives the same amount of CTAS response. The reference is the nonionic surfactant $C_{12-18}E_{11}$, derived from a mixture of linear primary alcohols ranging from 12 to 18 carbon atoms in chain length by reaction with ethylene oxide in a molar ratio of 1:11. $C_{12-18}E_{11}$ is reasonably representative of nonionic surfactants in commercial use; its CTAS response is about 0.21 absorbance units/mg.

If the identity of the nonionic surfactant in the sample is known, use that same material in preparing the calibration curve.

d. Interferences: Both anionic and cationic surfactants may show positive CTAS response[1,4] but both are removed in the ion-exchange step. Sublation removes nonsurfactant interferences. Physical interferences occur if some of the CTAS is adsorbed on particulate matter. Avoid such interference by filtering out the particulates for the sublation step; this will measure only dissolved CTAS.

e. Minimum detectable quantity: About 0.1 mg CTAS, calculated as $C_{12-18}E_{11}$, which corresponds to 0.1 mg/L in a 1-L sample.

f. Application: The method is suitable for determining dissolved nonionic surfactants of the ethoxylate type in most aqueous systems.

2. Apparatus

a. Sublation apparatus: See Section 5540B.2.

b. Ion-exchange column, glass, about 1- × 30-cm. Slurry anion-exchange resin in methanol and pour into column to give a bed about 10 cm deep. Insert plug of glass wool and then add a 10-cm bed of cation-exchange resin on top in the same manner. One column may be used for treating up to six sublated samples before repacking.

c. Spectrophotometer and 2.0-cm stoppered cells, suitable for measuring absorbance at 620 nm.

d. Separatory funnels, 125-mL, preferably with TFE stopcock and stopper.

e. Extraction flasks, Soxhlet type, 150-mL.

3. Reagents

a. Sublation reagents: See Section 5540B.3.

b. Anion-exchange resin, polystyrene-quaternary ammonium-type,* 50- to 100-mesh, hydroxide form. To convert chloride form to hydroxide, elute with 20 bed volumes of 1N NaOH and wash with methanol until free alkali is displaced.

c. Cation-exchange resin, polystyrene-sulfonate type,† 50- to 100-mesh, hydrogen form.

d. Cobaltothiocyanate reagent: Dissolve 30 g $Co(NO_3)_2 \cdot 6H_2O$ and 200 g NH_4SCN in water and dilute to 1 L. This reagent is stable for at least 1 month at room temperature.

e. Reference nonionic surfactant, $C_{12-18}E_{11}$: Reaction product of C_{12-18} linear primary alcohol with ethylene oxide in 1:11 molar ratio.‡

f. Reference nonionic surfactant stock solution, methanolic, approximately 2 mg nonionic/mL methanol: Quantitatively transfer entire contents (approximately 1 g nonionic) from preweighed ampule into 500-mL volumetric flask, thoroughly rinse ampule with methanol, make up to volume with methanol, and reweigh dried ampule. Calculate concentration in milligrams per milliliter as in ¶ 5a. Because of possible phase separation, use all material in the ampule.

---

* Bio-Rad, AGl-X2, or equivalent.
† Bio-Rad AG 50W-X8, or equivalent.
‡ For sources of suitable reference material, contact *Standard Methods* manager.

g. Reference nonionic surfactant standard solution, methanolic, approximately 0.1 mg nonionic/mL methanol: Dilute 10.00 mL stock solution to 200 mL with methanol. Exact concentration is 1/20 that of the stock solution.

h. Sodium hydroxide, NaOH: 1N.

i. Glass wool: Pre-extract with chloroform or methylene chloride.

j. Methanol, CH_3OH: CAUTION: *Methanol vapors are flammable and toxic; take appropriate precautions.*

k. Methylene chloride, CH_2Cl_2: CAUTION: *Methylene chloride vapors are toxic; take adequate precautions.*

l. Water: Use distilled or deionized, CTAS-free water for making reagents and dilutions.

4. Procedure

a. Purification by sublation: Proceed according to Section 5540B, using sample containing no more than 2 mg CTAS. (NOTE: For samples of known character containing no interfering materials, omit this step.)

b. Ion-exchange removal of anionic and cationic surfactants: Dissolve sublation residue in 5 to 10 mL methanol and transfer quantitatively to ion-exchange column. Elute with methanol at 1 drop/s into a clean, dry 150-mL extraction flask until about 125 mL is collected. Evaporate methanol on a steam bath aided by a gentle stream of clean, dry nitrogen or air, taking care to avoid loss by entrainment; remove from heat as soon as the methanol is completely evaporated. (NOTE: With samples of known character containing no anionic or cationic materials, omit step *b*.)

c. CTAS calibration curve: Into a series of 150-mL extraction flasks containing 10 to 20 mL methanol place 0.00, 5.00, 10.00, 20.00, and 30.00 mL reference nonionic surfactant standard solution and evaporate just to dryness. Continue as in ¶s 4d and e, below, and plot a calibration curve of absorbance against milligrams of reference nonionic taken, specifying its identity (e.g., $C_{12-18}E_{11}$ and lot number).

d. Cobalt complexing and extraction: Charge a 125-mL separatory funnel with 5 mL cobaltothiocyanate reagent. With precautions against excessive and variable evaporation of the methylene chloride, dissolve residue from ion-exchange operation, ¶ 4b, by adding 10.00 mL methylene chloride and swirling for a few seconds. Immediately transfer by pouring into the separatory funnel. *Do not rinse flask.* (NOTE: Because of the volatility of methylene chloride, rigidly standardize these operations with respect to handling and elapsed time; alternatively, evaporate the methanol in 200-mL erlenmeyer flasks to be stoppered with glass or TFE stoppers during dissolution. Transfer as directed here is incomplete, but in this case it will not introduce error because the loss of nonionics is exactly compensated for by the diminished volume of the organic layer in the extraction.) Shake separatory funnel vigorously for 60 s and let layers separate. Run lower layer into a 2.0-cm cell through a funnel containing a plug of pre-extracted glass wool and stopper. Be sure filtrate is absolutely clear. (NOTE: If desired, clarify by running the lower layer into a 12-mL centrifuge tube, stopper, spin at or above 1000 × g for 3 min, and transfer to the cell by a Pasteur pipet; use same procedure for both calibration and samples.)

e. Measurement: Determine absorbance at 620 nm against a blank of methylene chloride. (NOTE: If haze develops in the cell, warm slightly with a hot air gun or heat lamp to clarify.)

5. Calculations

a. Nonionic surfactant in reference nonionic stock solution ¶ *3f:*

$$\text{mg nonionic/mL methanol} = \text{mg reference sample/500 mL}$$

b. Nonionic surfactant in sample: From the calibration curve read milligrams of reference nonionic corresponding to the measured absorbance:

$$\text{mg CTAS/L} = \text{mg apparent nonionic/L sample}$$

Report as "CTAS, calculated as nonionic surfactant $C_{12-18}E_{11}$."

6. Precision and Bias

Twenty-four samples of 6.22% w/v solution of reference nonionic surfactant $C_{12-18}E_{11}$ were analyzed in three laboratories by CTAS alone, without sublation or ion exchange. The overall relative standard deviation was about 3%. Results of the three laboratories individually were:

Laboratory	% w/w ± SD
A	6.08 ± 0.14 ($n = 36$)
B	6.56 ± 0.17 ($n = 6$)
C	6.25 ± 0.14 ($n = 36$)
Overall	6.20 ± 0.19 ($n = 78$)

Samples of raw wastewater were freed of surfactants by four successive sublations, then 0.50 or 0.67 mg reference nonionic surfactant $C_{12-18}E_{11}$ was added and carried through the entire sequence of sublation, ion exchange, and CTAS extraction. Recoveries averaged 92% with overall standard deviation around 6%:

Laboratory	% Recovery ± SD
A	87 ± 4 ($n = 4$)
B	97 ± 1 ($n = 4$)
Overall	92 ± 6 ($n = 8$)

The above data relate to the bias and precision of the method when applied to a known nonionic surfactant. When the nature of the nonionic surfactant is unknown, there is greater uncertainty. The response of the reference $C_{12-18}E_{11}$ is about 0.21 absorbance units/mg, while that of the eight nonionic types mentioned under ¶ *1b* ranged from 0.20 to 0.36, and environmental nonionics might differ still more. If the nonionic surfactant in the sample has a response of 0.42, the result calculated in terms of milligrams $C_{12-18}E_{11}$ would be double the actual milligrams of the unknown nonionic.

7. References

1. SOAP AND DETERGENT ASSOCIATION ANALYTICAL SUBCOMMITTEE. 1977. Analytical methods for nonionic surfactants in laboratory biodegradation and environmental studies. *Environ. Sci. Technol.* 11: 1167.
2. CRABB, N.T. & H.E. PERSINGER. 1964. The determination of polyoxyethylene nonionic surfactants in water at the parts per million level. *J. Amer. Oil Chem. Soc.* 41:752.
3. CRABB, N.T. & H.E. PERSINGER. 1968. A determination of the apparent molar absorption coefficients of the cobalt thiocyanate complexes of nonylphenol ethylene oxide adducts. *J. Amer. Oil Chem. Soc.* 45:611.
4. GREFF, R.A., E.A. SETZKORN & W.D. LESLIE. 1965. A colorimetric method for the determination of parts per million of nonionic surfactants. *J. Amer Oil Chem. Soc.* 42:180.

8. Bibliography

TABAK, H.H. & R.L. BUNCH. 1981. Measurement of non-ionic surfactants in aqueous environments. Proc. 36th Ind. Waste Conf., p. 888. Purdue Univ., Lafayette, Ind.

5550 TANNIN AND LIGNIN*

5550 A. Introduction

Lignin is a plant constituent that often is discharged as a waste during the manufacture of paper pulp. Another plant constituent, tannin, may enter the water supply through the process of vegetable matter degradation or through the wastes of the tanning industry. Tannin also is applied in the so-called internal treatment of boiler waters, where it reduces scale formation by causing the production of a more easily handled sludge.

---

* Approved by Standard Methods Committee, 1993.

5550 B.   Colorimetric Method

1. General Discussion

a. Principle: Both lignin and tannin contain aromatic hydroxyl groups that react with Folin phenol reagent (tungstophosphoric and molybdophosphoric acids) to form a blue color suitable for estimation of concentrations up to at least 9 mg/L. However, the reaction is not specific for lignin or tannin, nor for compounds containing aromatic hydroxyl groups, inasmuch as many other reducing materials, both organic and inorganic, respond similarly.

b. Applicability: This method is generally suitable for the analysis of any organic chemical that will react with Folin phenol reagent to form measurable blue color at the concentration of interest. However, many compounds are reactive (see ¶ 1*c*) and each yields a different molar extinction coefficient (color intensity). Hence, the analyst must demonstrate conclusively the absence of interfering substances.

c. Interferences: Any substance able to reduce Folin phenol reagent will produce a false positive response. Organic chemicals known to interfere include hydroxylated aromatics, proteins, humic substances, nucleic acid bases, fructose, and amines. Inorganic substances known to interfere include iron (II), manganese (II), nitrite, cyanide, bisulfite, sulfite, sulfide, hydrazine, and hydroxylamine hydrochloride. Both 2 mg ferrous iron/L and 125 mg sodium sulfite/L individually produce a color equivalent to 1 mg tannic acid/L.

d. Minimum detectable concentrations: Approximately 0.025 mg/L for phenol and tannic acid and 0.1 mg/L for lignin with a 1-cm-path-length spectrophotometer.

2. Apparatus

Colorimetric equipment: One of the following is required:

a. Spectrophotometer, for use at 700 nm. A light path of 1 cm or longer yields satisfactory results.

b. Filter photometer, provided with a red filter exhibiting maximum transmittance in the wavelength range of 600 to 700 nm. Sensitivity improves with increasing wavelength. A light path of 1 cm or longer yields satisfactory results.

c. Nessler tubes, matched, 100-mL, tall form, marked at 50-mL volume.

3. Reagents

a. Folin phenol reagent: Transfer 100 g sodium tungstate, $Na_2WO_4 \cdot 2H_2O$, and 25 g sodium molybdate, $Na_2MoO_4 \cdot 2H_2O$, together with 700 mL distilled water, to a 2000-mL flat-bottom boiling flask. Add 50 mL 85% H_3PO_4 and 100 mL conc HCl. Connect to a reflux condenser and boil gently for 10 h. Add 150 g Li_2SO_4, 50 mL distilled water, and a few drops of liquid bromine. Boil without condenser for 15 min to remove excess bromine. Cool to 25°C, dilute to 1 L, and filter. Store finished reagent, which should have no greenish tint, in a tightly stoppered bottle to protect against reduction by air-borne dust and organic materials.

Alternatively, purchase commercially prepared Folin phenol reagent and use before the recommended expiration date.

b. Carbonate-tartrate reagent: Dissolve 200 g Na_2CO_3 and 12 g sodium tartrate, $Na_2C_4H_4O_6 \cdot 2H_2O$, in 750 mL hot distilled water, cool to 20°C, and dilute to 1 L.

c. Stock solution: The nature of the substance present in the sample dictates the choice of chemical used to prepare the standard, because each substance produces a different color intensity. Weigh 1.000 g tannic acid, tannin, lignin, or other compound being used for boiler water treatment or known to be a contaminant of the water sample. Dissolve in distilled water and dilute to 1000 mL. If the identity of the compound in the water sample is not known, use phenol and report results as "substances reducing Folin phenol reagent" in mg phenol/L. Interpret such results with caution.

Note that tannin and lignin are not individual chemical species of known molecular weight and structure; rather, they are substances containing a spectrum of chemicals of different molecular weights. Their chemical properties depend on source and method of isolation. If a particular substance is being added to the water, use it to prepare the stock solution.

d. Standard solution: Dilute 10.00 mL or 50.00 mL stock solution to 1000 mL with distilled water; 1.00 mL = 10.0 or 50.0 μg active ingredient.

4. Procedure

Bring 50-mL portions of clear sample and standards to a temperature above 20°C and maintain within a ± 2°C range. Add in rapid succession 1 mL Folin phenol reagent and 10 mL carbonate-tartrate reagent. Allow 30 min for color development. Compare visually against simultaneously prepared standards in matched Nessler tubes or make photometric readings against a reagent blank prepared at the same time. Use the following guide for instrumental measurement at a wavelength of 700 nm:

Tannic Acid in 61-mL Final Volume	Lignin in 61-mL Final Volume	Light Path
μg	μg	cm
50–600	100–1500	1
10–150	30–400	5

Report results in mg/L of the compound known to be present or as "substances reducing Folin phenol reagent" in mg phenol/L.

5. Precision and Bias

In a single laboratory analyzing seven replicates for phenol at 0.1 mg/L the precision was ± 7% and recovery was 107%.

6. Bibliography

FOLIN, O. & V. CIOCALTEU. 1927. On tyrosine and tryptophane determinates in proteins. *J. Biol. Chem.* 73:627.

BERK, A.A. & W.C. SCHROEDER. 1942. Determination of tannin substances in boiler water. *Ind. Eng. Chem.,* Anal. Ed. 14:456.

KLOSTER, M.B. 1974. Determination of tannin and lignin. *J. Amer. Water Works Assoc.* 66:44.

BOX, J.D. 1983. Investigation of the Folin-Ciocalteu phenol reagent for the determination of polyphenolic substances in natural waters. *Water Res.* 17:511; discussion by S.J. Randtke & R.A. Larson. 1984. *Water Res.* 18:1597.

5560   ORGANIC AND VOLATILE ACIDS*

5560 A.   Introduction

The measurement of organic acids, either by adsorption and elution from a chromatographic column or by distillation, can be used as a control test for anaerobic digestion. The chromatographic separation method is presented for organic acids (B), while a method using distillation (C) is presented for volatile acids. Alternative methods using GC or IC are available in the literature and may provide better speciation information for specific situations, but have not yet been recommended as standard methods.

Volatile fatty acids are classified as water-soluble fatty acids that can be distilled at atmospheric pressure. These volatile acids can be removed from aqueous solution by distillation, despite their high boiling points, because of co-distillation with water. This group includes water-soluble fatty acids with up to six carbon atoms.

The distillation method is empirical and gives incomplete and somewhat variable recovery. Factors such as heating rate and proportion of sample recovered as distillate affect the result, requiring the determination of a recovery factor for each apparatus and set of operating conditions. However, it is suitable for routine control purposes. Removing sludge solids from the sample reduces the possibility of hydrolysis of complex materials to volatile acids.

_____
* Approved by Standard Methods Committee, 1996.

5560 B.   Chromatographic Separation Method for Organic Acids

1. General Discussion

a. Principle: An acidified aqueous sample containing organic acids is adsorbed on a column of silicic acid and the acids are eluted with *n*-butanol in chloroform ($CHCl_3$). The eluate is collected and titrated with standard base. All short-chain (C_1 to C_6) organic acids are eluted by this solvent system and are reported collectively as total organic acids.

b. Interference: The $CHCl_3$-butanol solvent system is capable of eluting organic acids other than the volatile acids and also some synthetic detergents. Besides the so-called volatile acids, crotonic, adipic, pyruvic, phthalic, fumaric, lactic, succinic, malonic, gallic, aconitic, and oxalic acids; alkyl sulfates; and alkyl-aryl sulfonates are adsorbed by silicic acid and eluted.

c. Precautions: Basic alcohol solutions decrease in strength with time, particularly when exposed repeatedly to the atmosphere. These decreases usually are accompanied by the appearance of a white precipitate. The magnitude of such changes normally is not significant in process control if tests are made within a few days of standardization. To minimize this effect, store standard sodium hydroxide (NaOH) titrant in a tightly stoppered borosilicate glass bottle and protect from atmospheric carbon dioxide (CO_2) by attaching a tube of CO_2-absorbing material, as described in the inside front cover. For more precise analyses, standardize titrant or prepare before each analysis.

Although the procedure is adequate for routine analysis of most sludge samples, volatile-acids concentrations above 5000 mg/L may require an increased amount of organic solvent for quantitative recovery. Elute with a second portion of solvent and titrate to reveal possible incomplete recoveries.

2. Apparatus

a. Centrifuge or filtering assembly.

b. Crucibles, Gooch or medium-porosity fritted-glass, with filtering flask and vacuum source. Use crucibles of sufficient size (30 to 35 mL) to hold 12 g silicic acid.

c. Separatory funnel, 1000-mL.

3. Reagents

a. Silicic acid, specially prepared for chromatography, 50 to 200 mesh: Remove fines by slurrying in distilled water and decanting supernatant after settling for 15 min. Repeat several times. Dry washed acid in an oven at 103°C until *absolutely dry,* then store in a desiccator.

b. Chloroform-butanol reagent: Mix 300 mL reagent-grade $CHCl_3$, 100 mL *n*-butanol, and 80 mL 0.5N H_2SO_4 in a separatory funnel. Let water and organic layers separate. Drain off lower organic layer through a fluted filter paper into a dry bottle. CAUTION: *Chloroform has been classified as a cancer suspect agent. Use hood for preparation of reagent and conduct of test.*

c. Thymol blue indicator solution: Dissolve 80 mg thymol blue in 100 mL absolute methanol.

d. Phenolphthalein indicator solution: Dissolve 80 mg phenolphthalein in 100 mL absolute methanol.

e. Sulfuric acid, H_2SO_4, conc.

f. Standard sodium hydroxide, NaOH, 0.02N: Dilute 20 mL 1.0N NaOH stock solution to 1 L with absolute methanol. Prepare stock in water and standardize in accordance with the methods outlined in Section 2310B.3*d.*

4. Procedure

a. Pretreatment of sample: Centrifuge or vacuum-filter enough sludge to obtain 10 to 15 mL clear sample in a small test tube or beaker. Add a few drops of thymol blue indicator solution, then conc H_2SO_4 dropwise, until definitely red to thymol blue (pH = 1.0 to 1.2).

b. Column chromatography: Place 12 g silicic acid in a Gooch or fritted-glass crucible and apply suction to pack column. Tamp column while applying suction to reduce channeling when the sample is applied. With a pipet, distribute 5.0 mL acidified sample as uniformly as possible over column surface. Apply suction momentarily to draw sample into silicic acid. Release vacuum as soon as last portion of sample has entered column. Quickly add 65 mL $CHCl_3$-butanol reagent and apply suction. Discontinue suction just before the last of reagent enters column. Do not reuse columns.

c. Titration: Remove filter flask and purge eluted sample with N_2 gas or CO_2-free air immediately before titrating. (Obtain CO_2-free air by passing air through a CO_2 absorbant.*)

Titrate sample with standard $0.02N$ NaOH to phenolphthalein end point, using a fine-tip buret and taking care to avoid aeration. The fine-tip buret aids in improving accuracy and precision of the titration. Use N_2 gas or CO_2-free air delivered through a small glass tube to purge and mix sample and to prevent contact with atmospheric CO_2 during titration.

d. Blank: Carry a distilled water blank through steps ¶s 4a through 4c.

---

* Ascarite or equivalent.

5. Calculation

$$\text{Total organic acids (mg as acetic acid/L)} = \frac{(a - b) \times N \times 60\,000}{\text{mL sample}}$$

where:

a = mL NaOH used for sample,
b = mL NaOH used for blank, and
N = normality of NaOH.

6. Precision

Average recoveries of about 95% are obtained for organic acid concentrations above 200 mg as acetic acid/L. Individual tests generally vary from the average by approximately 3%. A greater variation results when lower concentrations of organic acids are present. Titration precision expressed as the standard deviation is about ± 0.1 mL (approximately ± 24 mg as acetic acid/L).

7. Bibliography

MUELLER, H.F., A.M. BUSWELL & T.E. LARSON. 1956. Chromatographic determination of volatile acids. *Sewage Ind. Wastes* 28:255.

MUELLER, H.F., T.E. LARSON & M. FERRETTI. 1960. Chromatographic separation and identification of organic acids. *Anal. Chem.* 32:687.

WESTERHOLD, A.F. 1963. Organic acids in digester liquor by chromatography. *J. Water Pollut. Control Fed.* 35:1431.

HATTINGH, W.H.J. & F.V. HAYWARD. 1964. An improved chromatographic method for the determination of total volatile fatty acid content in anaerobic digester liquors. *Int. J. Air Water Pollut.* 8:411.

POHLAND, F.G. & B.H. DICKSON, JR. 1964. Organic acids by column chromatography. *Water Works Wastes Eng.* 1:54.

5560 C.   Distillation Method

1. General Discussion

a. Principle: This technique recovers acids containing up to six carbon atoms. Fractional recovery of each acid increases with increasing molecular weight. Calculations and reporting are on the basis of acetic acid. The method often is applicable for control purposes. Because it is empirical, carry it out exactly as described. Because the still-heating rate, presence of sludge solids, and final distillate volume affect recovery, determine a recovery factor.

b. Interference: Hydrogen sulfide (H_2S) and CO_2 are liberated during distillation and will be titrated to give a positive error. Eliminate this error by discarding the first 15 mL of distillate and account for this in the recovery factor. Resides on glassware from some synthetic detergents have been reported to interfere; use water and dilute acid rinse cycles to prevent this problem.

2. Apparatus

a. Centrifuge, with head to carry four 50-mL tubes or 250-mL bottles.

b. Distillation flask, 500-mL capacity.

c. Condenser, about 76 cm long.

d. Adapter tube.

e. pH meter or recording titrator: See Section 2310B.2a.

f. Distillation assembly: Use a conventional distilling apparatus. To minimize fluctuations in distillation rate, supply heat with a variable-wattage electrical heater.

3. Reagents

a. Sulfuric acid, H_2SO_4, 1 + 1.

b. Standard sodium hydroxide titrant, $0.1N$: See Section 2310B.3c.

c. Phenolphthalein indicator solution.

d. Acetic acid stock solution, 2000 mg/L: Dilute 1.9 mL conc CH_3COOH to 1000 mL with deionized water. Standardize against $0.1N$ NaOH.

4. Procedure

a. Recovery factor: To determine the recovery factor, f, for a given apparatus, dilute an appropriate volume of acetic acid stock solution to 250 mL in a volumetric flask to approximate the expected sample concentration and distill as for a sample. Calculate the recovery factor

$$f = \frac{a}{b}$$

where:

a = volatile acid concentration recovered in distillate, mg/L, and
b = volatile acid concentration in standard solution used, mg/L.

b. Sample analysis: Centrifuge 200 mL sample for 5 min. Pour off and combine supernatant liquors. Place 100 mL supernatant liquor, or smaller portion diluted to 100 mL, in a 500-mL distillation flask. Add 100 mL distilled water, four to five clay chips or similar material to prevent bumping, and 5 mL H_2SO_4. Mix so that acid does not remain on bottom of flask. Connect flask to a condenser and adapter tube and distill at the rate of about 5 mL/min. Discard the first 15 mL and collect exactly 150 mL distillate in a 250-mL graduated cylinder. Titrate with $0.1N$ NaOH, using phenolphthalein indicator, a pH meter, or an automatic titrator. The end points of these three methods are, respectively, the first pink coloration that persists on standing a short time, pH 8.3, and the inflection point of the titration curve (see Section 2310). Titration at 95°C produces a stable end point.

Distill and analyze a blank and reference standard with each sample batch to insure system performance.

5. Calculation

$$\text{mg volatile acids as acetic acid/L} = \frac{\text{mL NaOH} \times N \times 60\,000}{\text{mL sample} \times f}$$

where:

N = normality of NaOH, and
f = recovery factor.

6. Bibliography

OLMSTEAD, W.H., W.M. WHITAKER & C.W. DUDEN. 1929–1930. Steam distillation of the lower volatile fatty acids from a saturated salt solution. *J. Biol. Chem.* 85:109.

OLMSTEAD, W.H., C.W. DUDEN, W.M. WHITAKER & R.F. PARKER. 1929–1930. A method for the rapid distillation of the lower volatile fatty acids from stools. *J. Biol. Chem.* 85:115.

BUSWELL, A.M. & S.L. NEAVE. 1930. Laboratory studies of sludge digestion. *Ill. State Water Surv. Bull.* 30:76.

HEUKELEKIAN, H. & A.J. KAPLOVSKY. 1949. Improved method of volatile-acid recovery from sewage sludges. *Sewage Works J.* 21:974.

KAPLOVSKY, A.J. 1951. Volatile-acid production during the digestion of seeded, unseeded, and limed fresh solids. *Sewage Ind. Wastes* 23:713.

5710 FORMATION OF TRIHALOMETHANES AND OTHER DISINFECTION BY-PRODUCTS*

5710 A. Introduction

Trihalomethanes (THMs) are produced during chlorination of water. Only four THM compounds normally are found: chloroform ($CHCl_3$), bromodichloromethane ($CHBrCl_2$), dibromochloromethane ($CHBr_2Cl$), and bromoform ($CHBr_3$). Additional chlorination by-products can be formed (including haloacetic acids and halonitriles; for example, see 5710D) during the relatively slow organic reactions that occur between free chlorine and naturally occurring organic precursors such as humic and fulvic acids. The formation potentials of these additional by-products also can be determined, but different quenching agents and different analytical procedures may be needed. Predictive models for estimating/calculating THM formation exist, but because eventual THM concentrations cannot be calculated precisely from conventional analyses, methods to determine the potential for forming THMs are useful in evaluating water treatment processes or water sources or for predicting THM concentrations in a distribution system.

To obtain reproducible and meaningful results, control such variables as temperature, reaction time, chlorine dose and residual, and pH. THM formation is enhanced by elevated temperatures and alkaline pH and by increasing concentrations of free chlorine residuals, although THM formation tends to level off at free chlorine residuals of 3 mg/L and above; a longer reaction time generally increases THM formation.[1,2]

Low concentrations of bromide exist in most natural waters and are responsible for the formation of brominated organic compounds. Figure 5710:1 shows that an oxidant ratio of about 40 times more chlorine than bromine (on a molar basis, = 40 on the

---

* Approved by Standard Methods Committee, 1994.

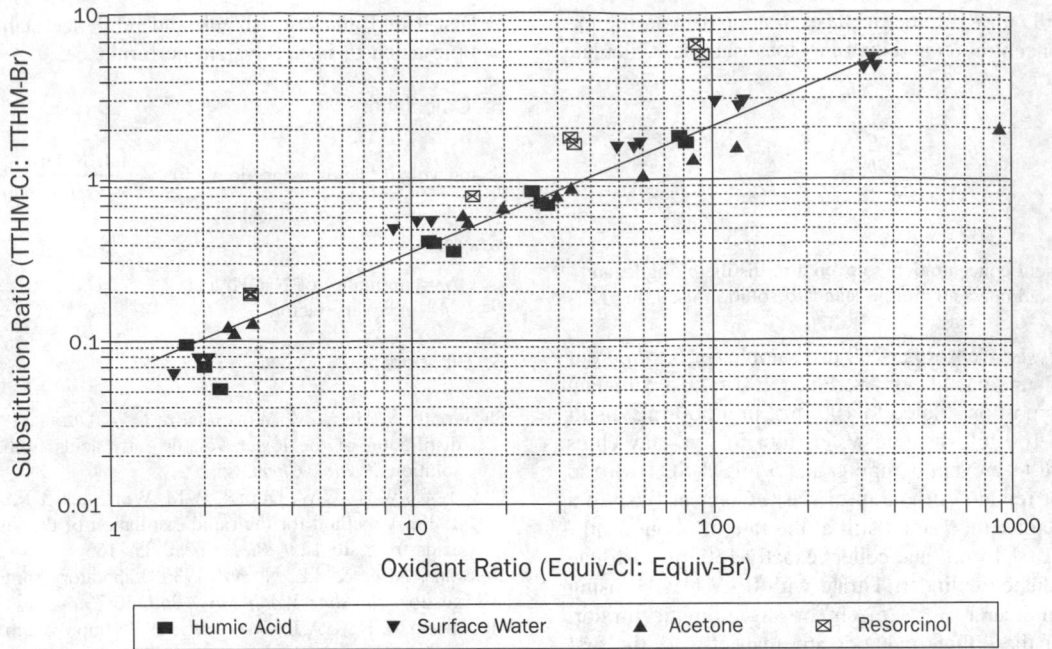

Figure 5710:1. Effect of changing molar oxidant ratios of free chlorine: free bromine on molar ratios of substituted organic chloride: organic bromide, using four different precursor substrates. Reaction times varied between 1 and 7 d. Standard conditions were used at pH = 7.0, except that the free chlorine residual after 7 d storage for the surface water was 17 mg/L instead of the 3 to 5 mg/L range for the other three substrates.

x axis) is required to form equimolar amounts of substituted organic chloride and bromide (= 1 on the y axis); small amounts of bromide also can increase the molar yield of THMs.[3]

The possible addition of organic precursors contained in reagent solutions cannot be accounted for accurately without a great deal of extra work; therefore, sample dilutions resulting from reagent additions (approximately 2%) are ignored in the final calculations. However, sample dilution may need to be taken into account if other volumes are used. Sample dilution also changes the concentrations of bromide and organic matter, potentially leading to speciation changes.

1. Definition of Terms

See Figures 5710:2a and b for the relationship among the following definitions.

Total trihalomethane (TTHM_T) is the sum of all four THM compound concentrations (see Section 5710B.5) produced at any time T (usually days). $TTHM_0$ is the total THM concentration at the time of sampling. $TTHM_0$ concentrations can range from nondetectable, which usually means the sample has not been chlorinated, to several hundred micrograms per liter if the sample has been chlorinated. $TTHM_7$ is the sum of all four THM compound concentrations produced during reactions of sample precursors with excess free chlorine over a 7-d reaction time.

Standard reaction conditions (see Section 5710B) are as follows: free chlorine residual at least 3 mg/L and not more than 5 mg/L at the end of a 7-d reaction (incubation) period, with sample incubation temperature of 25 ± 2°C, and pH controlled at 7.0 ±

0.2 with phosphate buffer. Standard conditions are not intended to simulate water treatment processes but are most useful for estimating the concentration of THM precursors, as well as for measuring the effectiveness of water-treatment options for reducing levels of THM precursors in the raw water.

Special applications permit different test conditions, but they must be stated explicitly when reporting results.

Trihalomethane formation potential (THMFP or Δ THMFP) is the difference between the final $TTHM_T$ concentration and the initial $TTHM_0$ concentration. If sample does not contain chlorine at the time of collection, $TTHM_0$ will be close to zero and the term THMFP may be used. If sample does contain chlorine at the time of collection, because of formation of THMs, use the term Δ THMFP (the increase of THM concentration during storage) when reporting the difference between TTHM concentrations.

The term "THMFP" often has been equated to the final TTHM concentration, even if the sample had contained chlorine when collected. To use this definition, explicitly define the term when reporting data.

Simulated distribution system trihalomethane (SDS-THM), Section 5710C, is the concentration of TTHMs in a sample that has been disinfected comparably to finished drinking water and under the same conditions and time as in a water distribution system. It includes pre-existing THMs plus those produced during storage. This method can be used in conjunction with laboratory, pilot, or full-scale studies of treatment processes to estimate expected concentration of THMs in a distribution system. Do not use SDS-THM to estimate the precursor removal efficiency of a treatment process, because THM yields are highly variable at low chlorine

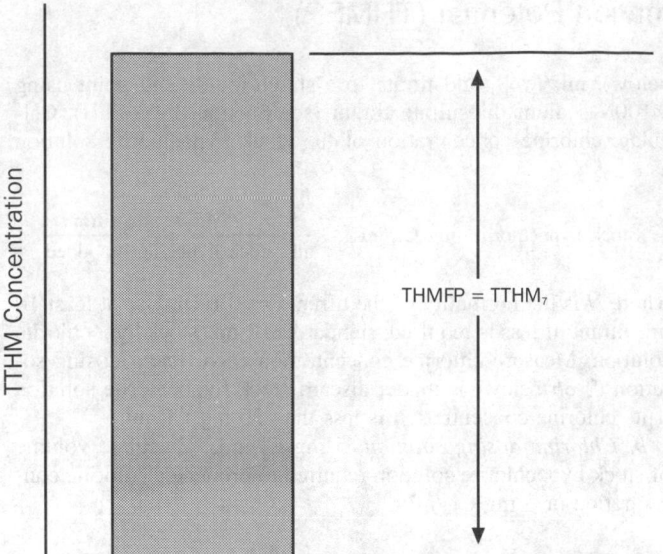

Figure 5710:2a. Relationships between definitions used in the formation potential test, for a sample that did not contain free chlorine at the time of sampling. Total THM concentration at the time of sampling ($TTHM_0$) was very close to or equal to zero; therefore, the THM formation potential for the 7-d reaction time (THMFP, with a free chlorine residual of at least 3 mg/L) was essentially equal to the total THM concentration in the sample at the end of the reaction storage time ($TTHM_7$).

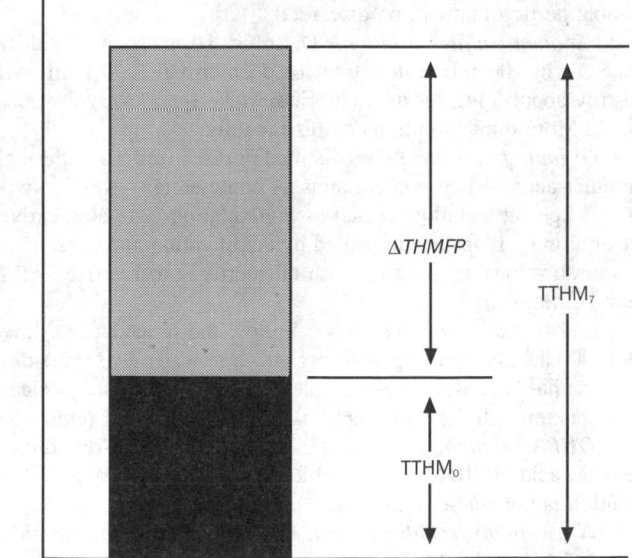

Figure 5710:2b. Relationships between definitions used in the formation potential test, for a sample that already contained free chlorine at the time of sampling. Total THM concentration at the time of sampling ($TTHM_0$) was a significant fraction of the final value obtained after 7-d storage ($TTHM_7$) with an excess of at least 3 mg/L free chlorine. Δ THMFP is the difference between these two values.

residual concentrations. For SDS-type testing, low chlorine residuals (< 1 mg Cl_2/L) are often encountered, thereby resulting in lower THM formation than would be obtained for higher chlorine residuals. THM yields at higher chlorine concentrations (> 3 mg/L) tend to level off and become relatively independent of variations in free chlorine residuals.

2. Sampling and Storage

Collect samples in 1-L glass bottles sealed with TFE-lined screw caps. If multiple tests will be performed for each sample, or if many different analyses will be performed for each sample (see Section 5710D), collect 4 L instead. Further, if multiple reaction time periods will be used to study reaction rates, process a separate sample bottle (taken from one large sample) for each time period. One liter is enough sample to determine chlorine demand and duplicate THM analyses. Use only freshly collected samples and process immediately. If this is not possible, store samples at 4°C and analyze as soon as possible. Significant sample degradation can occur in unpreserved samples within 24 h.

If the sample has been chlorinated previously, collect the sample with minimum turbulence and fill the sample bottle completely to avoid loss of THMs already present. Determine the zero-time THM concentration ($TTHM_0$), if desired, on another sample collected at the same time and dechlorinated immediately with fresh sodium sulfite solution, crystals, or sodium thiosulfate.

3. References

1. STEVENS, A.A. & J.M. SYMONS. 1977. Measurement of trihalomethanes and precursor concentration changes. *J. Amer. Water Works Assoc.* 69:546.
2. SYMONS, J.M., A.A. STEVENS, R.M. CLARK, E.E. GELDREICH, O.T. LOVE, JR. & J. DEMARCO. 1981. Treatment Techniques for Controlling Trihalomethanes in Drinking Water. EPA-600/2-81-56, U.S. Environmental Protection Agency, Cincinnati, Ohio.
3. SYMONS, J.M., S.W. KRASNER, L. SIMMS & M. SCLIMENTI. 1993. Measurement of trihalomethane and precursor concentrations revisited: The impact of bromide ion. *J. Amer. Water Works Assoc.* 85(1):51.

5710 B.   Trihalomethane Formation Potential (THMFP)

1. General Discussion

a. Principle: Under standard conditions, samples are buffered at pH 7.0 ± 0.2, chlorinated with an excess of free chlorine, and stored at 25 ± 2°C for 7 d to allow the reaction to approach completion. As a minimum, pH is buffered at a defined value and a free chlorine residual of 3 to 5 mg Cl_2/L exists at the end of the reaction time. THM concentration is determined by using liquid-liquid extraction (see Section 6232B) or purge and trap (see Section 6200).

b. Interference: If the water was exposed to free chlorine before sample collection (e.g., in a water treatment plant), a fraction of precursor material may have been converted to THM. Take special precautions to avoid loss of volatile THMs by minimizing turbulence and filling sample bottle completely.

Interference will be caused by any organic THM-precursor materials present in the reagents or adsorbed on glassware. Heat nonvolumetric glassware to 400°C for 1 h, unless routine analysis of blanks demonstrates that this precaution is unnecessary. Reagent impurity is difficult to control. It usually is traceable to reagent water containing bromide ion or organic impurities. Use high-grade reagent water as free of organic contamination and chlorine demand as possible. If anion exchange is used to remove bromide or organic ions, follow such treatment by activated carbon adsorption (see Section 1080).

Other interferences include volatile organic compounds (VOCs), including THMs and chlorine-demanding substances. VOCs may co-elute with THMs during analysis. THMs or other interfering substances that are present as the result of a chemical spill, etc., will bias the results.

Nitrogenous species and other constituents may interfere in the determination of free residual chlorine. Add enough free chlorine to oxidize chorine-demanding substances and leave a free chlorine residual of at least 3 mg/L, but not more than 5 mg/L, at the end of the incubation period. A free chlorine residual of at least 3 mg/L decreases the likelihood that a combined residual will be mistaken for a free residual and assures that THM formation occurs under conditions that are reasonably independent of variations in chlorine residual concentrations.

c. Minimum detectable quantity: The sensitivity of the method is determined by the analytical procedure used for THM.

2. Apparatus

a. Incubator, to maintain temperature of 25 ± 2°C.

b. Bottles, glass, with TFE-lined screw caps to contain 245 to 255 mL, 1-L, 4-L.

c. Vials, glass, 25- or 40-mL with TFE-lined screw caps.

d. pH meter, accurate to within ± 0.1 unit.

3. Reagents

Prepare aqueous reagents in organic-free water (¶ 3e below) unless chlorine-demand-free water (¶ 3f below) is specified.

a. Standardized stock hypochlorite solution: Dilute 1 mL, using a 1-mL volumetric pipet, 5% aqueous sodium hypochlorite (NaOCl, to be referred to as stock hypochlorite) solution to approximately 25 mL with chlorine-demand-free water (see ¶ 3f below), mix well, and titrate to a starch-iodide end point using 0.100N sodium thiosulfate titrant (see Section 4500-Cl.B). Calculate chlorine concentration of the stock hypochlorite solution as:

$$\text{Stock hypochlorite, mg } Cl_2/mL = \frac{N \times 35.45 \times mL \text{ titrant}}{mL \text{ stock hypochlorite added}}$$

where N is the normality of the titrant (= 0.100). Use at least 10 mL titrant; if less is required, standardize 2 mL stock hypochlorite solution. Measure chlorine concentration each time a dosing solution (¶ 3b below) is made; discard stock hypochlorite solution if its chlorine concentration is less than 20 mg Cl_2/mL.

b. Chlorine dosing solution, 5 mg Cl_2/mL: Calculate volume of stock hypochlorite solution required to produce a chlorine concentration of 5 mg Cl_2/mL:

$$mL \text{ required} = \frac{1250}{\text{stock hypochlorite conc. mg } Cl_2/mL}$$

Dilute this volume of stock hypochlorite solution in a 250-mL volumetric flask to the mark with chlorine-demand-free water. Mix and transfer to an amber bottle, seal with a TFE-lined screw cap, and refrigerate. Keep away from sunlight. Discard if the chlorine concentration drops below 4.7 mg Cl_2/mL; this will occur if the "initial chlorine concentration" (as determined in ¶ 4a below) drops below 94 mg Cl_2/L.

c. Phosphate buffer: Dissolve 68.1 g potassium dihydrogen phosphate (anhydrous), KH_2PO_4, and 11.7 g sodium hydroxide, NaOH, in 1 L water. Refrigerate when not in use. If a precipitate develops, filter through a glass fiber filter. After buffer is added to a sample, a pH of 7.0 should result. Check before use with a sample portion that can be discarded.

d. Sodium sulfite solution: Dissolve 10 g sodium sulfite, Na_2SO_3, in 100 mL water. Use for dechlorination: 0.1 mL will destroy about 5 mg residual chlorine. Make fresh every 2 weeks. NOTE: More dilute solutions oxidize readily.

e. Organic-free water: Pass distilled or deionized water through granular-activated-carbon columns. A commercial system may be used.* Special techniques such as preoxidation, activated carbon adsorption (perhaps accompanied by acidification and subsequent reneutralization), or purging with an inert gas to remove THMs may be necessary.

f. Chlorine-demand-free water: Follow the procedure outlined in 4500-Cl.C.3m, starting with organic-free water. After residual chlorine has been destroyed completely, purge by passing a clean, inert gas through the water until all THMs have been removed.

g. DHBA solution: Dissolve 0.078 g anhydrous 3,5-dihydroxybenzoic acid (DHBA) in 2 L chlorine-demand-free water. This solution is not stable; make fresh before each use.

h. Nitric or hydrochloric acid, HNO_3 or HCl, concentrations of 1:1, 1.0N, and 0.1N.

i. Sodium hydroxide, NaOH, 1.0N and 0.1N.

j. Borate buffer (optional): Dissolve 30.9 g anhydrous boric acid, H_3BO_3, and 10.8 g sodium hydroxide, NaOH, in 1 L water. Filter any precipitate that may form with a glass fiber filter. This solution will keep sample pH at 9.2; check before using. NOTE:

---

* Milli-Q, Millipore Corp., or equivalent.

Waters containing significant amounts of calcium may precipitate calcium phosphate (or carbonate) at higher pH values.

k. Mixed buffer (optional): Mix equal amounts of phosphate and borate buffer solutions, then adjust pH as desired before using with samples. Determine the amount of acid or base needed on a separate sample that can be discarded. This mixed buffer is reasonably effective in the pH range of 6 to 11.

4. Procedure

a. Chlorine demand determination: Determine or accurately estimate the 7-d sample chlorine demand. A high chlorine dose is specified below to drive the reaction close to completion quickly. The following procedure yields only a rough estimate of chlorine demand; other techniques may be used.

Pipet 5 mL chlorine dosing solution into a 250-mL bottle, fill completely with chlorine-demand-free water, and cap with a TFE-lined screw cap. Shake well. Titrate 100 mL with 0.025N sodium thiosulfate to determine the initial chlorine concentration (C_I). This should be about 100 mg Cl_2/L. Pipet 5 mL phosphate buffer and 5 mL chlorine dosing solution into a second 250-mL bottle, fill completely with sample, and seal with a TFE-lined screw cap. Store in the dark for at least 4 h at 25°C. After storage, determine chlorine residual (C_R). Calculate chlorine demand (D_{Cl}) as follows:

$$D_{Cl} = C_I - C_R$$

where:

D_{Cl} = chlorine demand, mg Cl_2/L,
$\quad C_R$ = chlorine residual of sample after at least 4 h storage, mg Cl_2/L, and
$\quad C_I$ = initial (dosed) chlorine concentration, mg Cl_2/L.

b. Sample chlorination: If sample contains more than 200 mg/L alkalinity or acidity, adjust pH to 7.0 ± 0.2 using 0.1 or 1.0N HNO$_3$, HCl, or NaOH and a pH meter. With a graduated pipet, transfer appropriate volume of the 5 mg Cl_2/mL chlorine dosing solution, V_D, into sample bottle:

$$V_D = \frac{D_{Cl} + 3}{5} \times \frac{V_S}{1000}$$

where:

$\quad V_S$ = volume of sample bottle, mL, and
$\quad V_D$ = volume of dosing solution required, mL.

Add 5 mL phosphate buffer solution if using a 250-mL sample bottle (or 1 mL buffer/50 mL sample) and fill completely with sample. Immediately seal with a TFE-lined screw cap, shake well, and store in the dark at 25 ± 2°C for 7 d. Analyze a reagent blank (¶ 4d) with each batch of samples. To increase the likelihood of achieving the desired chlorine residual concentration (3 to 5 mg/L) at the end of the 7-d reaction period, dose several sample portions to provide a range of chlorine concentrations, with each chlorine dose differing in increments of 2 mg Cl_2/L.

c. Sample analysis: After the 7-d reaction period, place 0.1 mL sulfite reducing solution in a 25-mL vial and gently and completely fill vial with sample. If free chlorine residual has not been determined previously, measure it using a method accurate to 0.1

mg/L and able to distinguish free and combined chlorine (see Section 4500-Cl). Adjust pH to the value required by the method chosen for chlorine analysis. [NOTE: If other by-products are to be measured, a different quenching agent may be needed (see 5710D.4). Also, if sample portions have been dosed with different chlorine concentrations, first determine the free chlorine residual and select only that portion having the required chlorine residual concentration of 3 to 5 mg/L for further processing.] If THMs will not be analyzed immediately, lower the pH to <2 by adding 1 or 2 drops of 1:1 HCl to the reduced sample in the vial. Seal vial with TFE-lined screw cap. Store samples at 4°C until ready for THM analysis (preferably no longer than 7 d). Let sample reach room temperature before beginning analysis.

d. Reagent blank: Add 1 mL chlorine dosing solution to 50 mL phosphate buffer, mix, and completely fill a 25-mL vial, seal with a TFE-lined screw cap, and store with samples. (NOTE: This reagent blank is for quality control of reagent solutions only and is not a true blank, because the reagent concentrations in this blank are considerably higher than those in samples. THM concentrations in the reagent blank will be biased high and cannot be subtracted from sample values. Make no further dilutions before the reaction because the reagent water itself might contribute to THM formation.) After reaction for 7 d, pipet 1 mL sulfite reducing solution into a 250-mL bottle and add, without stirring, 5.0 mL reacted reagent mixture. Immediately fill bottle with organic-free water that has been purged free of THMs and seal with a TFE-lined screw cap. Mix. Analyze a portion of this reagent blank for THMs using the same method used for samples. The sum of all THM compounds in the reagent blank should be less than 5 μg TTHM as CHCl$_3$/L.

The reagent blank is a rough measure of THMs contributed by reagents added to the samples, but it cannot be used as a correction factor. If the reagent blank is greater than 5% of the sample value or greater than 5 μg TTHM/L, whichever is larger, additional treatment for reagent water is necessary. See Section 1080. It also may be necessary to obtain reagents of higher purity. Analyze a reagent blank each time samples are analyzed and each time fresh reagents are prepared.

5. Calculation

Report concentration of each of the four common THM compounds separately because it is desirable to know their relative concentrations. Larger amounts of bromine-substituted compounds, relative to chlorine-substituted compounds, indicate a higher concentration of dissolved bromide in the water (see Figure 5710:1). Also report free chlorine concentration at end of reaction time along with the incubation time, temperature, and pH.

THM concentrations may be reported as a single value as micrograms CHCl$_3$ per liter (μg CHCl$_3$/L), or micromoles per liter (μM). Do not use the simple sum of mass units micrograms per liter except when required for regulatory reporting. Compute TTHM concentration using one of the following equations:

To report TTHM in units of μg CHCl$_3$/L:

$$TTHM = A + 0.728B + 0.574C + 0.472D$$

where:

$\quad A$ = μg CHCl$_3$/L,
$\quad B$ = μg CHBrCl$_2$/L,

C = μg CHBr$_2$Cl/L, and
D = μg CHBr$_3$/L.

To report TTHM in units of μM as CHCl$_3$:

$$TTHM = \frac{TTHM, \text{μg CHCl}_3\text{/L}}{119}$$

To report TTHM on a weight basis as μg/L (not used except for regulatory purposes):

$$TTHM = A + B + C + D$$

To report a change of TTHM concentration over 7 d:

$$\Delta\ THMFP = TTHM_7 - TTHM_0$$

Finally, if TTHM$_0$ = 0, then:

$$THMFP = TTHM_7 = \Delta\ THMFP$$

Do not make blank correction or a correction for sample dilution resulting from addition of reagents. If conditions differ from pH of 7, 25°C, 7-d reaction time, and 3 to 5 mg/L chlorine residual, report these nonstandard test conditions with the results. Nonstandard test conditions may mimic water quality conditions in a specific distribution system or may be relevant to other investigations (see Section 5710C).

6. Quality Control

a. Use dihydroxy-benzoic acid solution (DHBA) as a quality-control check, especially for the presence of interfering bromides in reagents or reagent water.

Dilute 1.0 mL chlorine dosing solution to 1000 mL with chlorine-demand-free water (diluted chlorine dosing solution). Pipet 5 mL phosphate buffer solution (pH = 7.0) into each of two 250-mL bottles; add 1.00 mL DHBA solution to one bottle and fill both bottles completely with diluted chlorine dosing solution; seal with TFE-lined screw caps. Store in the dark for 7 d at 25 ± 2°C, and analyze as directed in ¶ 4c.

b. The THM concentration of the solution containing the added DHBA minus the THM concentration of the blank (i.e., the bottle that does not contain added DHBA, which is a true blank for this application only and differs from the reagent blank discussed in ¶ 4b above) should be about 119 μg/L THM as CHCl$_3$, with essentially no contribution from bromide-containing THMs. If there is a significant contribution from brominated THMs, 10% or more of the total THM, it may be necessary to remove bromide from the reagent water or to obtain higher-purity reagents containing less bromide. Determine source of bromide and correct the problem. If the THM concentration of the water blank exceeds 20 μg/L, treat reagent water to reduce contamination.

7. Precision and Bias

The precision of this method is determined by the analytical precision and bias of the method used for measuring THM as well as the control of variables such as pH, chlorine residual, temperature, sample homogeneity, etc. Method bias can be determined only for synthetic solutions (e.g., the DHBA solution), because THM formation potential is not an intrinsic property of the sample but rather a quantity defined by this method.

Table 5710:I presents single-operator precision and bias data for samples processed under standard conditions. The values were obtained by analyzing DHBA solutions and blanks. The expected

TABLE 5710:I. SINGLE-OPERATOR PRECISION AND BIAS DATA FOR THMFP*

Sample	THM μg/L CHCl$_3$	CHCl$_2$Br	CHClBr$_2$	CHBr$_3$	THMFP μg/L as CHCl$_3$	Recovery %
Blank 1	0.8	—	—	—	0.8	—
Blank 2	1.9	—	—	—	1.9	—
Blank 3	0.1	0.1	—	—	0.2	—
Blank 4	0.7	—	—	—	0.7	—
Blank 5	0.5	—	—	—	0.5	—
Blank 6	0.7	—	—	—	0.7	—
Average					0.8	—
Standard deviation					± 0.6	—
DHBA 1	114.1	0.1	—	—	114.2	97.8
DHBA 2	113.2	—	—	—	113.2	96.9
DHBA 3	107.8	—	—	—	107.8	92.2
DHBA 4	108.3	—	—	—	108.3	92.7
DHBA 5	109.6	0.1	—	—	109.7	93.9
DHBA 6	111.8	0.1	—	—	111.9	95.8
DHBA 7	112.6	—	—	—	112.6	96.4
Average					111.1†	95.1
Standard deviation					± 2.5	2.2

* Source: MOORE, L., Unpublished data. U.S. Environmental Protection Agency, Cincinnati, Ohio.
† Expected value = 116.

TABLE 5710:II. SINGLE-OPERATOR PRECISION AND BIAS DATA FOR TTHM (pH = 9.2)*

Sample	THM $\mu g/L$				TTHM $\mu g\ CHCl_3/L$	Recovery %
	CHCl₃	CHCl₂Br	CHClBr₂	CHBr₃		
Blank 1	3.0	0.3	—	—	3.2	—
Blank 2	1.7	0.1	—	—	1.8	—
Blank 3	1.3	0.1	—	—	1.4	—
Blank 4	1.6	0.1	—	—	1.7	—
Blank 5	2.3	0.2	—	—	2.4	—
Blank 6	2.6	0.1	—	—	2.7	—
Blank 7	2.5	0.2	—	—	2.6	—
Average					2.3	—
Standard deviation (relative standard deviation)					± 0.6 (± 26.1%)	
DHBA 1	45.4	3.3	0.1	—	47.9	98.3
DHBA 2	51.0	3.9	0.1	—	53.9	111.2
DHBA 3	39.2	3.0	0.1	—	41.4	84.3
DHBA 4	48.3	3.6	0.1	—	51.0	105.0
DHBA 5	47.6	3.7	0.1	—	50.4	103.7
DHBA 6	43.4	3.2	0.1	—	45.8	93.8
DHBA 7	46.0	3.6	0.1	—	48.7	100.0
Average					48.4	99.5
Standard deviation (relative standard deviation)					± 3.7 (± 7.6%)	
RWS 1†	33.1	17.2	11.3	0.5	52.3	—
RWS 2	31.7	16.1	10.6	0.5	49.7	—
RWS 3	38.7	18.4	11.7	0.6	59.1	—
RWS 4	35.1	18.0	11.7	0.8	55.3	—
RWS 5	36.0	17.9	11.7	0.6	56.0	—
RWS 6	38.7	18.7	11.7	0.6	59.3	—
RWS 7	37.7	18.1	11.2	0.6	57.6	—
Average					55.6	
Standard deviation (relative standard deviation)					± 3.3 (± 5.9%)	

* Source: MOORE, L., Unpublished data. U.S. Environmental Protection Agency, Cincinnati, Ohio.
† RWS = raw water sample, filtered and diluted 1 part filtrate, 2 parts organic-free water.

value for the samples listed is 116 µg/L TTHM (as CHCl₃), rather than 119 µg/L, because the DHBA reagent used was only 97% pure. Percent recovery was calculated by the formula:

$$\% \ recovery = \frac{DHBA\ sample - average\ blank}{116} \times 100$$

Table 5710:II presents the same data set, except that the pH of samples and blanks was adjusted to 9.2 with borate buffer. Also included are results for single-operator precision with filtered river-water samples that had been diluted with 2 parts organic-free water to 1 part filtrate, again using borate-buffered samples at a pH of 9.2.

5710 C.   Simulated Distribution System Trihalomethanes (SDS-THM)

1. General Discussion

a. Principle: The SDS-THM testing method uses bench-scale techniques to provide an estimate of the THMs formed in a distribution system after disinfection.[1] It may be used to estimate the THM concentration at any point in a distribution system or to evaluate the formation of other disinfectant by-products (5710D).

However, to measure efficiency of any unit treatment process for precursor removal, see Section 5710B.

The term "disinfection," rather than "chlorination," is used because free chlorine residuals are not necessarily provided in all distribution systems. For example, monochloramine residuals may be used.

SDS-THM concentrations measured by this procedure generally will be lower than THM concentrations measured by procedures in 5710B because disinfectant concentrations used in SDS-type samples are intended to mimic conditions in a distribution system and are almost always lower than disinfectant concentrations used with standardized formation potential procedures.

Two types of SDS procedures may be used: (1) a simple storage method that requires only the addition, at the end of the desired storage time, of a quenching agent, sodium sulfite, to a sample collected from the entry to a distribution system; and (2) a comprehensive method that involves one or more steps described in Section 5710B, with appropriate modifications.

SDS procedures are not standard procedures in the traditional sense. Test variables are modified to mimic local distribution-system conditions with bench-top procedures. These conditions include temperature, pH, disinfectant dose and residual, bromide ion concentration, and reaction time (corresponding to the residence time of water within the distribution system). However, the method used to simulate a distribution system can be standardized according to specific needs.

2. Apparatus

See 5710B.2.

Temperature control: Appropriate equipment, such as a water bath or incubator, to control sample storage temperature, capable of a range of temperature adjustments required.

3. Reagents

See 5710B.3, and also 5710D.3, if applicable.

4. Procedure

a. Simple storage procedure: Measure and report both initial and final values for all variables, especially pH, temperature, $TTHM_0$ (if desired), and residual disinfectant concentration using a method accurate to 0.1 mg/L and able to distinguish among the various disinfectant forms—see Section 4500-Cl for chlorine analyses.

Collect treated, disinfected water from the clearwell of a treatment plant or other suitable location in either a 1-L or a 250-mL completely filled bottle. Seal with a TFE-lined screw cap and store at selected temperature for selected length of time. The storage time frequently chosen is the maximum residence time within the distribution system, and the temperature is representative of the distribution system.

A second sample for immediate quenching provides an initial THM concentration if desired ($TTHM_0$). Quench by adding 2 drops (0.1 mL) sodium sulfite solution to a 25-mL glass vial, and gently and completely fill vial with sample. [NOTE: If other by-products will be analyzed, a different quenching agent may be needed (see 5710D.4).]

At end of storage period, quench a portion of stored sample with sodium sulfite solution. Report all values (pH, temperature, and residual disinfectant concentration) together with THM results. Ideally, residual disinfectant concentration after storage equals residual disinfectant concentration found in the distribution system.

b. Bench-top procedure: Use any or all steps given in Section 5710B, except that all variables, such as disinfectant dose, residual concentrations, temperature, pH, and storage time are adjusted to simulate distribution system conditions. For better sample pH control, use a buffer such as the mixed-buffer solution (5710B.3k). If a buffer is used, adjust buffer pH to the appropriate value before adding to the sample. Add buffer to a sample portion to be discarded after titrating to the desired pH with either NaOH or HCl solution; add this determined amount of acid or base to the buffer solution before adding to actual samples. When measuring disinfectant residuals, adjust sample pH to that required by the analytical method for the residual disinfectant, because the buffer capacity of the sample may be greater than the amount of buffer required by the method. For example, DPD Methods 4500-Cl.F and 4500-Cl.G require control of the pH within the range from 6.2 to 6.5 for stable color development, but the sample pH might be buffered to a value of 8.3, requiring adjustment of the sample pH to the appropriate range by addition of mineral acid before color development with DPD reagents.

Also add disinfectant to sample if it does not already contain enough to provide the desired disinfectant residual for the chosen storage time and temperature. Use the chlorine demand procedure (5710B.4a) as a guide, or develop correlations between disinfectant use and TOC or other variables.

Process sample generally following procedures in 5710B. Measure and report both initial and final values of all test variables, especially temperature, pH, and residual disinfectant concentration, as well as all THM results.

5. Calculation

See 5710B.5. Report $TTHM_T$ values for SDS-type samples in any appropriate units, except that the concept of "formation potential" is not applicable. Use prefix "SDS-" to distinguish between "SDS-TTHM" and "THMFP" results. Also report each compound concentration (e.g., "$\mu g/L$ SDS_T-$CHCl_3$").

6. Quality Control

See 5710B.6 for check on reagent purity or as a check on analytical precision and control by using a reaction with a pure, organic compound under more stable, standardized conditions. (This applies only to the formation of THM, using free chlorine.)

7. Precision and Bias

No data are available. More variability of results is expected for SDS-type testing than for samples tested with the standard conditions of 5710B. A larger number of replicates will be required for SDS procedures, as opposed to standard conditions, to obtain reliable estimates of effects of treatment changes and techniques.

8. Reference

1. KOCH, B., S.W. KRASNER, M.J. SCLIMENTI & W.K. SCHIMPFF. 1991. Predicting the formation of DBPs by the simulated distribution system. *J. Amer. Water Works Assoc.* 83 (10):62.

5710 D.   Formation of Other Disinfection By-Products (DBPs)

1. General Discussion

a. Principle: The techniques and principles discussed in 5710A through C may be applied to other disinfection by-products (DBPs) and total/dissolved organic halogen (TOX/DOX) as well as for trihalomethanes (THMs). Although all the DBPs listed in this method may result from chlorination reactions, some may be formed by disinfectants other than free chlorine.

This method may be extended to cover formation potentials (5710B) and distribution system simulations (5710C) for additional by-products formed by reactions between other disinfectants (ozone, chlorine dioxide, chloramines, etc.) with dissolved organic matter. Some of the commonly found DBPs are listed in this method, but many cannot be determined because well-defined analytical methods are not yet available.

The procedures by which formation potentials and behavior in distribution systems of other DBPs can be measured are exactly as described in 5710A through C, except that different quenching agents may be required for different compounds.

In general, by-product concentrations increase with reaction time, but exceptions exist and different variables may produce different results. For example, at high pH values, THM concentrations increase with time, but if the pH is high enough, trichloroacetic acid will not form at all; at high pH, however, the concentrations of dihaloacetonitriles (DHANs) quickly reach a maximum value in a relatively short time and then decrease because of hydrolysis reactions. Further, some compounds, such as brominated haloacetic acids, are not stable and can degrade during storage—either during a long reaction time (7 d may be too long for some compounds) or after the reaction has been quenched (even if stored at 4°C).

Small concentrations of bromide ion may have significant effects. If a compound contains more than one halogen atom in its molecular formula, compounds containing all the possible combinations of mixed chloro-/bromo- formulas might also be formed.

The most common other disinfection by-products are: trihaloacetic acids (THAAs), including trichloroacetic acid (TCAA), bromodichloroacetic acid (BDCAA), dibromochloroacetic acid (DBCAA), and tribromoacetic acid (TBAA); dihaloacetic acids (DHAAs), including dichloroacetic acid (DCAA), bromochloroacetic acid (BCAA), and dibromoacetic acid (DBAA); monohaloacetic acids (MHAAs), including monochloroacetic acid (MCAA) and monobromoacetic acid (MBAA); chloral hydrate (CH); dihaloacetonitriles (DHANs), including dichloroacetonitrile (DCAN), bromochloroacetonitrile (BCAN), and dibromoacetonitrile (DBAN); 1,1,1-trichloropropanone (111-TCP); chloropicrin (CP), which may be produced either with free or combined chlorine; cyanogen chloride, formed either with free chlorine or monochloramine (and is more stable in solutions containing monochloramine); and dissolved organic halide (DOX) (see Section 5320). Total organic halide (TOX) also may be determined if the sample is not filtered.

The present method differs from those described in 5710B and C, in the means by which reactions are quenched: the different by-products formed require different quenching agents to stop reactions. Sodium sulfite is used to quench the THM reaction, but it may react with and degrade other compounds formed, such as

DHANs. The procedure below lists the recommended quenching steps. Several portions of the same sample may require different quenching agents, depending upon the by-products to be determined. Use appropriate analytical methods to determine the different types of by-products formed.

2. Apparatus

See 5710B.2.
Vials: 40-mL glass vials with TFE-lined screw caps.

3. Reagents

In addition to the reagents listed in 5710B.3, the following reagents also may be needed, depending upon the by-products to be measured:

a. Ammonium chloride solution: Weigh 5 g NH_4Cl and dissolve in 100 mL organic-free water.

b. Nitric acid solution (approx. 3.5N): Dilute 80 mL conc HNO_3 (CAUTION: *strong oxidant*) to 250 mL with organic-free water.

c. Methyl tert-*butyl ether* (*MTBE*), highest purity.

d. Other reagents: Those required by analytical methods for determination of specific by-product concentrations.

4. Procedure

Procedures for the formation of by-products during reactions between disinfectants and dissolved organic matter have been described in 5710A through C, but the quenching agents needed to stop these reactions depend on the specific compounds to be analyzed. For THMs, the quenching agent is sodium sulfite, and its use has already been described (5710B.4.*c*). For the other by-products listed above, substitute as described below. Store all samples headspace-free and sealed with TFE-lined screw caps.

a. Chloral hydrate (CH): This compound may be analyzed with THMs. Use sulfite reducing solution to quench. Adjust sample pH 6 to 7 with (1.0N or 0.1N) HCl. Determine amount of acid to be added with a separate portion containing the same amount of sulfite reducing solution added to sample. Analyze by liquid-liquid extraction, capillary column, GC/electron capture analysis.[1]

b. DHANs, CP, and 111-TCP: Add 4 drops (0.2 mL) ammonium chloride solution to a 40-mL vial and nearly fill with sample. Add a predetermined amount of HCl that will adjust sample pH to 6 to 7 (see ¶ *a* above) and fill completely. (Add 3 drops of phosphate buffer solution for more control of sample pH, if desired, before determining amount of acid solution needed for pH adjustment.) NH_4Cl quenches the reaction by converting free chlorine to monochloramine. Analyze by liquid-liquid extraction, capillary column, GC/electron capture analysis.[1]

c. Haloacetic acids (HAAs), mono-, di-, and trihaloacetic acids: Add 4 drops (0.2 mL) NH_4Cl solution to a 250-mL bottle, and fill bottle completely with sample. Before acidifying and extracting sample, add 1 mL sodium sulfite solution to the 250-mL sample, mix well, and analyze by liquid-liquid extraction, capillary column, GC/electron capture analysis (see Section 6251 or equivalent methods[2]). Sodium sulfite can slowly degrade some of the brominated haloacetic acids during storage; do not add until just before acidification. If a GC/MS method is used, remove water in the sample by drying with sodium sulfate crystals before methylation.

d. Cyanogen chloride: Analyze by GC/MS purge and trap method,[3] which uses ascorbic acid to dechlorinate samples. Also see Section 4500-CN.J. Hydrolysis of cyanogen chloride to cyanate occurs rapidly in the pH range of 8.5 to 9.0 (within 30 min), but the reaction is much slower at pH values of 7.0 and below.

e. DOX: Add 1 mL sodium sulfite solution to a 250-mL bottle and nearly fill with sample. Add sufficient $3.5N$ HNO_3 to lower pH to 2.0 (approximately 12 drops, or 0.6 mL) and fill completely. Analyze according to Section 5320. Total organic halogen (TOX) also may be determined by the same method if the sample is not filtered.

f. Other disinfection by-products not mentioned above: Disinfectants, such as chlorine, ozone, monochloramine, chlorine dioxide, etc., may form other disinfection by-products. Formation potentials or SDS-type data also can be determined for these additional compounds.

5. Calculation

Report concentrations of each compound separately in μg/L. Compare concentrations of different compounds on a micromolar basis (micromoles/L, or μM):

$$\text{Compound, } \mu M = \frac{\text{compound concentration, } \mu g/L}{MW}$$

where:

MW = molecular weight of the compound.

Concentrations of a series of compounds that can be grouped together may sometimes be reported as one value. For example, the dihaloacetic acid group (DHAAs) includes DCAA, BCAA and DBAA and may be reported in terms of a group concentration, obtained by adding the molar concentrations of the separate compounds within the class:*

$$\text{DHAA, } \mu M = \frac{\text{DCAA, } \mu g/L}{129} + \frac{\text{BCAA, } \mu g/L}{173} + \frac{\text{DBAA, } \mu g/L}{219}$$

Alternatively, report in terms of μg/L as DCAA by multiplying the molar concentration by the molecular weight of DCAA:

$$\text{DHAA, } \mu g/L \text{ as DCAA} = \text{DHAA, } \mu M \times 129$$

or, for SDS testing:

$$\text{SDS-DHAA, } \mu g/L \text{ as DCAA} = \text{SDS-DHAA, } \mu M \times 129$$

---

* Federal regulations may require a simple sum in terms of mass units/L.

The definitions given in 5710A through C also are valid. For example, if the initial concentration of disinfectant by-products (DHAAs, for instance) is zero or insignificant, then:

$$\text{DHAAFP} = \text{DHAA}_7.$$

Alternatively, if there is a significant concentration of initial by-product, then:

$$\Delta\text{DHAAFP} = \text{DHAA}_7 - \text{DHAA}_0.$$

6. Quality Control

See 5710B.6 for check on reagent purity or as a check on analytical precision and control by using a reaction with a pure, organic compound under more stable, standardized conditions. The test detailed in Section 5710B.6 applies only to THM formation using excess free chlorination conditions.

7. Precision and Bias

Precision and bias measurements depend, in part, on the analytical procedure used to measure each specific disinfectant by-product concentration. These measurements also depend upon compound properties such as stability toward oxidation and biodegradation. In general, however, formation potential reactions should be reproducible to the extent indicated in 5710B for chlorination reactions. SDS-type reactions (5710C) would not, however, be expected to be as accurate or as precise, although such reactions should predict distribution system concentrations reasonably well.

8. References

1. U.S. ENVIRONMENTAL PROTECTION AGENCY. 1990. Method 551. Methods for the Determination of Organic Compounds in Drinking Water, Supplement I. Off. Research & Development, Washington, D.C.
2. U.S. ENVIRONMENTAL PROTECTION AGENCY. 1990. Methods 552 and 552.1. Methods for the Determination of Organic Compounds in Drinking Water, Supplements I and II. Off. Research & Development, Washington, D.C.
3. FLESCH, J. & P. FAIR. 1988. The analysis of cyanogen chloride in drinking water. Proc. 16th Annu. AWWA Water Quality Technology Conf., Nov. 13–17, 1988, St. Louis, Mo., p. 465. American Water Works Assoc., Denver, Colo.

5910 UV-ABSORBING ORGANIC CONSTITUENTS*

5910 A. Introduction

1. Applications

Some organic compounds commonly found in water and wastewater, such as lignin, tannin, humic substances, and various aromatic compounds, strongly absorb ultraviolet (UV) radiation. UV absorption is a useful surrogate measure of selected organic constituents in fresh waters,[1-3] salt waters,[4-6] and wastewater.[7,8] Strong correlations may exist between UV absorption and organic carbon content, color, and precursors of trihalomethanes (THMs) and other disinfection by-products.[9,10] UV absorption also has been used to monitor industrial wastewater effluents[11] and to evaluate organic removal by coagulation,[10] carbon adsorption,[12,13,14] and other water treatment processes.[10] Specific absorption, the ratio of UV absorption to organic carbon concentration, has been used to characterize natural organic matter.[10,15,16]

Although UV absorption can be used to detect certain individual organic contaminants after separation (e.g., by HPLC), as described in Part 6000, the method described here is not suitable for detection of trace concentrations of individual chemicals. It is intended to be used to provide an *indication* of the aggregate concentration of UV-absorbing organic constituents.

2. References

1. DOBBS, R.A., R.H. WISE & R.B. DEAN. 1972. The use of ultra-violet absorbance for monitoring the total organic carbon of water and wastewater. *Water Res.* 6:1173.
2. WILSON, A.L. 1959. Determination of fulvic acids in water. *J. Appl. Chem.* 9:501.
3. COOPER, W.J. & J.C. YOUNG. 1984. Chemical non-specific organic analysis. *In* R.A. Minear & L. Keith, eds. Water Analysis, Vol. 3. Academic Press, New York, N.Y.

4. OGURA, N. & T. HANYA. 1968. Ultra-violet absorbance of the seawater, in relation to organic and inorganic matters. *Int. J. Oceanol. Limnol.* 1:91.
5. OGURA, N. & T. HANYA. 1968. Ultra-violet absorbance as an index of pollution of seawater. *J. Water Pollut. Control Fed.* 40:464.
6. FOSTER, P. & A.W. MORRIS. 1971. The use of ultra-violet absorption measurements for the estimation of organic pollution in inshore sea waters. *Water Res.* 5:19.
7. BUNCH, R.L., E.F. BARTH & M.B. ETTINGER. 1961. Organic materials in secondary effluents. *J. Water Pollut. Control Fed.* 33:122.
8. MRKVA, M. 1971. Use of ultra-violet spectrophotometry in the determination of organic impurities in sewage. *Wasserwirtsch. Wassertech.* 21:280.
9. SINGER, P.C., J.J. BARRY, III, G.M. PALEN & A.E. SCRIVNER. 1981. Trihalomethane formation in North Carolina drinking waters. *J. Amer. Water Works Assoc.* 73:392.
10. EDZWALD, J.K., W.C. BECKER & K.L. WATTIER. 1985. Surrogate parameters for monitoring organic matter and THM precursors. *J. Amer. Water Works Assoc.* 77(4):122.
11. BRAMER, H.C., M.J. WALSH & S.C. CARUSO. 1966. Instrument for monitoring trace organic compounds in water. *Water Sewage Works* 113:275.
12. BISHOP, D.F., L.S. MARSHALL, T.P. O'FARRELL, R.B. DEAN, B. O'CONNOR, R.A. DOBBS, S.H. GRIGGS & R.V. VILLIERS. 1967. Studies on activated carbon treatment. *J. Water Pollut. Control Fed.* 39:188.
13. SONTHEIMER, H.N., E. HEILKER, M.R. JEKEL, H. NOLTE & F.H. VOLIMER. 1978. The Mülheim process. *J. Amer. Water Works Assoc.* 70(7):393.
14. SUMMERS, R.S., L. CUMMINGS, J. DEMARCO, D.J. HARTMAN, D. METZ, E.W. HOWE, B. MACLEOD & M. SIMPSON. 1992. Standardized Protocol for the Evaluation of GAC. No. 90615, AWWA Research Foundation & American Water Works Assoc., Denver, Colo.
15. THURMAN, E.M. 1985. Organic Geochemistry of Natural Waters. Martinus Nijhoff/Dr. W. Junk Publishers, Dordrecht, Netherlands.
16. OWEN, D.M., G.L. AMY & Z.K. CHOWDHURY. 1993. Characterization of Natural Organic Matter and Its Relationship to Treatability, No. 90631, AWWA Research Foundation & American Water Works Assoc., Denver, Colo.

* Approved by Standard Methods Committee, 1994.

5910 B. Ultraviolet Absorption Method

1. General Discussion

a. Principle: UV-absorbing organic constituents in a sample absorb UV light in proportion to their concentration. Samples are filtered to control variations in UV absorption caused by particles. Adjustment of pH before filtration is optional.

UV absorption is measured at 253.7 nm (often rounded off to 254 nm). The choice of wavelength is arbitrary. Historically, 253.7 nm has been used as the standard wavelength; however, experienced analysts may choose a wavelength that minimizes interferences from compounds other than those of interest while maximizing absorption by the compound(s) of interest. If a wave-

length other than 253.7 nm is used, state that wavelength when reporting results.

b. Interferences: The primary interferences in UV-absorption measurements are from colloidal particles, UV-absorbing organics other than those of interest, and UV-absorbing inorganics, notably ferrous iron, nitrate, nitrite, and bromide. Certain oxidants and reducing agents, such as ozone, chlorate, chlorite, chloramines, and thiosulfate, also will absorb ultraviolet light at 253.7 nm. Many natural waters and waters processed in drinking water treatment plants have been shown to be free of these interferences.

Evaluate and correct for UV absorption contributed by specific interfering substances. If cumulative corrections exceed 10% of the total absorption, select an alternate wavelength and/or use another method. Because UV absorption by organic matter may vary at pH values below 4 or above 10, avoid these values.[1]

A UV absorption scan from 200 to 400 nm can be used to determine presence of interferences. Typical absorption scans of natural organic matter are featureless curves of increasing absorption with decreasing wavelength. Sharp peaks or irregularities in the absorption scan may be indicative of inorganic interferences or unexpected organic contaminants. Because many organic compounds in water and wastewater (e.g., carboxylic acids and carbohydrates) do not absorb significantly in the UV wavelengths, correlate UV absorption to dissolved organic carbon (DOC) or soluble chemical oxygen demand (COD). However, use such correlations with care because they may vary from water to water, seasonally on the same water, and between raw and treated waters. In addition, chemical oxidation (e.g., ozonation, chlorination) of the organic material may reduce UV absorption without removing the organics and thus may change correlations. Because UV absorption and correlations with UV absorption are site-specific, they may not be comparable from one water source to another.

c. *Minimum detectable concentration:* The minimum detectable concentration cannot be determined rigorously because this is a nonspecific measurement. For precise measurement, select cell path length to provide an absorbance of approximately 0.005 to 0.900. Alternatively, dilute high-strength samples. The minimum detectable concentration of a particular constituent depends on the relationship between UV absorption, the desired characteristic (e.g., trihalomethane formation potential or DOC), and any interfering substances.

2. Apparatus

a. *Spectrophotometer,* for use between 200 and 400 nm with matched quartz cells providing a light path of 1 cm. For low-absorbance samples use a path length of 5 or 10 cm. A scanning spectrophotometer is useful.

b. *Filter:* Use a glass-filter* without organic binder. Other filters that neither sorb UV-absorbing organics of interest nor leach interfering substances (e.g., nitrate or organics) into the water may be used, especially if colloidal matter must be removed. Alternatively use filters of TFE, polycarbonate, or silver. Prerinse filter with sample of organic-free water to remove soluble impurities. If alternate separation techniques, filters or filter preparations are used, demonstrate that equivalent results are produced. Filter pore size will influence test results, especially in raw waters.

c. *Filter assembly,* glass, TFE, or stainless steel, capable of holding the selected filters.

3. Reagents

a. *Organic-free water:* Reagent water (see Section 1080) or equivalent water containing less than 0.05 mg DOC/L.

b. *Hydrochloric acid* (optional), HCl, 0.1*N*.

c. *Sodium hydroxide* (optional), NaOH, 0.1*N*.

d. *Phosphate buffer* (optional): Dissolve 4.08 g dried anhydrous KH_2PO_4 and 2.84 g dried anhydrous Na_2HPO_4 in 800 mL or-

ganic-free water. Verify that pH is 7.0 and dilute to 1 L with organic-free water. Store in brown glass bottle at 4°C. Prepare fresh weekly or more frequently if microbial growth is observed.

4. Procedure

a. *Sample volume:* Select sample volume on basis of the cell path length or dilution required to produce a UV absorbance between 0.005 and 0.900. For most applications a 50-mL sample is adequate. Use 100 mL sample if a 10-cm cell path length is required.

b. *Sample preparation:* Wash filter and filter assembly by passing at least 50 mL organic-free water through the filter. For specific applications and correlations, sample pH may be adjusted with HCl or NaOH. In poorly buffered samples an appropriate non-UV-absorbing buffer system such as a phosphate buffer may be used. Take care to avoid precipitate formation during pH adjustment. UV absorbance of fulvic acid solutions apparently remains constant between pH 4 and 10.[1] Report sample pH value used with recorded absorbance. Once sample pH has been adjusted and/or measured, filter sample. Prepare an organic-free water blank and the sample in an identical manner.

c. *Spectrophotometric measurement:* Let spectrophotometer equilibrate according to manufacturer's instructions. Set wavelength to 253.7 nm and adjust spectrophotometer to read zero absorbance with the organic-free water blank. Measure UV absorbance at 253.7 nm of at least two filtered portions of sample at room temperature.

5. Calculation

Report mean UV absorption in units of cm^{-1} using the following notation. To report units in m^{-1} multiply the equation by one hundred.

$$UV_\lambda^{pH} = \left[\frac{\overline{A}}{b}\right] D$$

where:

UV_λ^{pH} = mean UV absorption, cm^{-1} (subscript denotes wavelength used, nm, and superscript denotes pH used if other than 7.0),
b = cell path length, cm,
$\overline{A}$ = mean absorbance measured, and
D = dilution factor resulting from pH adjustment and/or dilution with organic-free water.

$$D = \frac{\text{final sample volume}}{\text{initial sample volume}}$$

Correct results for absorption contributed by known interfering substances. If UV absorption contributed by interfering substances exceeds 10% of the total UV absorption do not use UV absorption at 253.7 nm as an indicator of organics.

6. Quality Control

a. *Replicate measurements:* Use at least two portions of filtered sample.

b. *Duplicate analyses:* Analyze every tenth sample in duplicate (i.e., duplicating the entire procedure) to assess method precision.

---

* Whatman grade 934AH; Gelman type A/E; Millipore type AP40; ED Scientific Specialties grade 161; or other products that give demonstrably equivalent results. Practical filter diameters are 2.2 to 4.7 cm.

TABLE 5910:I. PRECISION OF UV ANALYSES AND CORRELATION TO KHP SAMPLES

Analysis	UV$_{254}$ Result for Given KHP Sample Concentration*cm^{-1}							
	0.54	0.93	1.79	4.87	9.61	25.0	50.0	100.0
Laboratory 1	0.008	0.015	0.034	0.079	0.158	0.323	0.638	1.282
Laboratory 2	0.009	0.016	0.026	0.070	0.134	0.401	0.803	1.612
Laboratory 3	0.010	0.017	0.027	0.081	0.161	0.353	0.695	1.343
Laboratory 4	0.007	0.020	0.033	0.070	0.132	0.319	0.750	1.590
Laboratory 5	0.009	0.018	0.030	0.087	0.140	0.394	0.643	1.447
Mean	0.0086	0.0142	0.0300	0.0774	0.1450	0.3580	0.7058	1.4548
Standard deviation	0.0011	0.0019	0.0035	0.0074	0.0136	0.0384	0.0708	0.1461
% Relative standard deviation†	12.8	11.1	11.7	9.56	9.38	10.7	10.0	10.0

* KHP sample concentration mg/L as C, measured as in Section 5310C.

† The percent relative standard deviation is given by:

$$\% \ RSD = \left[\frac{\text{standard deviation } (S)}{\text{mean } (\overline{X})} \right] \times 100$$

c. Baseline absorbance: Check system baseline UV absorbance at least after every 10 samples by measuring the absorbance of an organic-free water blank. A non-zero absorbance reading for the blank may indicate need for cell cleaning, problems with the reference cell if a dual-beam instrument is being used, or variation in the spectrophotometer response caused by heating or power fluctuations over time.

d. Spectrophotometer check: Difficulties in comparing UV absorption data from different spectrophotometers have been reported. Potassium hydrogen phthalate (KHP), also known as potassium biphthalate, standards were prepared in pH 7, phosphate-buffered (3*d*) reagent water without acidification (see Section 5310B.3*c*) and analyzed in five laboratories. The results are shown in Table 5910:I; these data suggest acceptable precision. These data also are useful for checking spectrophotometer results with KHP standards commonly used for TOC and/or COD analysis. A correlation equation for this 40-sample data set is:

$$UV_{254} = 0.0144 \ KHP \ + \ 0.0018$$

with correlation coefficient (r^2) = 0.987, UV$_{254}$ expressed in cm^{-1}, and KHP expressed as mg/L as C.

This equation can assist in verifying spectrophotometer performance. For example, if a set of UV$_{254}$ analyses is performed and the results are in the 0.010 range, prepare a KHP standard of 0.5 mg/L as C. The projected UV$_{254}$ of this KHP standard would be 0.009 cm^{-1}. If the measured UV$_{254}$ is outside 13% relative standard deviation (RSD) of 0.009 cm^{-1}, the spectrophotometer

TABLE 5910:II. SINGLE-OPERATOR PRECISION FOR UV ABSORPTION MEASUREMENTS OF FULVIC ACID SOLUTIONS

Replicate No.	Result cm^{-1}		
	DOC = 2.5 mg/L	DOC = 4.9 mg/L	DOC = 10.0 mg/L
1	0.110	0.240	0.480
2	0.120	0.230	0.480
3	0.110	0.240	0.470
4	0.100	0.230	0.480
5	0.110	0.240	0.480
6	0.100	0.240	0.470
7	0.110	0.240	0.480
8	0.110	0.230	0.480
9	0.120	0.240	0.480
10	0.110	0.240	0.480
Mean	0.110	0.237	0.478
Standard deviation	0.00667	0.00483	0.00422
% Relative standard deviation	6.06	2.05	0.882

may be suspect and require maintenance. The correlation between UV_{254} and KHP standards is presented solely as a useful means of verifying spectrophotometer performance.

7. Precision and Bias

Table 5910:I shows interlaboratory precision data for 40 KHP samples. The percent relative standard deviations (% RSD) ranged from 9.38 to 12.8.

Single-operator precision data are presented in Table 5910:II for fulvic acid solutions.[2] The % RSD ranged from 0.9 to 6%. Because UV absorption is an aggregate measure of organic carbon, true standards are not available and bias cannot be determined.

8. References

1. EDZWALD, J.K., W.C. BECKER & K.L. WATTIER. 1985. Surrogate parameters for monitoring organic matter and THM precursors. *J. Amer. Water Works Assoc.* 77(4):122.
2. MALLEY, J.P., JR. 1988. A Fundamental Study of Dissolved Air Flotation for Treatment of Low Turbidity Waters Containing Natural Organic Matter. Ph.D. thesis (unpublished), Univ. Massachusetts, Amherst.

PART 6000

INDIVIDUAL ORGANIC COMPOUNDS

6010   INTRODUCTION

6010 A.   General Discussion

The methods presented in Part 6000 are intended for the determination of individual organic compounds. Methods for determination of aggregate concentrations of groups of organic compounds are presented in Part 5000.

Most of the methods presented herein are highly sophisticated instrumental methods for determining very low concentrations of the organic constituents. Stringent quality control requirements are given with each method and require careful attention.

Many compounds are determinable by two or more of the methods presented in Part 6000. Table 6010:I shows the specific analytical methods applicable to each compound. Guidance on selection of method is provided in the introduction to each section.

TABLE 6010:I.   ANALYSIS METHODS FOR SPECIFIC ORGANIC COMPOUNDS*

Compound	Analysis Methods (section number)	Compound	Analysis Methods (section number)
Acenaphthene	6040B; 6410B; 6440B	Chloroisopropyl ether	6410B
Acenaphthylene	6410B; 6440B	Chloromethane	6200B,C
Acetaldehyde	6252B	Chloromethyl benzene	6040B
Aldicarb	6610B	Chloromethylphenol	6410B; 6420B
Aldicarb sulfone	6610B	Chloronaphthalene(s)	6040B; 6410B
Aldicarb sulfoxide	6610B	Chlorophenol(s)	6401B; 6420B
Aldrin	6410B; 6630B,C	Chlorophenoxy benzene	6040B
Aminomethylphosphonic acid (AMPA)	6651B	Chlorophenyl phenyl ether	6410B
		Chlorotoluene	6200B,C
Anthracene	6040B; 6410B; 6440B	Chrysene	6040B; 6410B; 6440B
Baygon	6610B	2,4-D (dichlorophenoxyacetic acid)	6640B
Bentazon	6640B	Dalapon	6640B
Benzaldehyde	6252B	DDD	6410B; 6630B,C
Benzene	6200B,C	DDE	6410B; 6630B,C
Benzidine	6410B	DDT	6410B; 6630B,C
Benzo(a)anthracene	6040B; 6410B; 6440B	Dibenzo(a,h) anthracene	6410B; 6440B
Benzo(a)pyrene	6410B, 6440B	Dibromoacetic acid (DBAA)	6251B
Benzo(b)fluoranthene	6410B, 6440B	Dibromochloromethane	6040B; 6200B,C; 6232B
Benzo(ghi)perylene	6410B, 6440B	Dibromochloropropane	6200B,C; 6231B
Benzo(k)fluoranthene	6410B, 6440B	Dibromoethane	6040B; 6200B,C; 6231B
BHC(s)	6410B; 6630C	Dibromomethane	6200B,C
Bromobenzene	6040B; 6200B,C	Dibutyl phthalate	6410B
Bromochloroacetic acid	6251B	Dicamba	6640B
Bromochloromethane	6200B,C	Dichloran	6630B
Bromodichloromethane	6040B; 6200B,C; 6232B	Dichloroacetic acid (DCAA)	6251B
Bromoform	6040B; 6200B,C; 6232B	Dichlorobenzene(s)	6040B; 6200B,C; 6410B
Bromomethane	6200B,C	Dichlorobenzidine	6410B
Bromophenoxybenzene	6040B	Dichlorodifluoromethane	6200B,C
Bromophenyl phenyl ether	6410B	Dichloroethane	6200B,C
Butyl benzyl phthalate	6410B	Dichloroethene(s)	6200B,C
Butylbenzene(s)	6200B,C	Dichlorophenol(s)	6410B; 6420B
Captan	6630B	Dichloropropane(s)	6200B,C
Carbaryl	6610B	Dichloropropene	6040B; 6200B,C
Carbofuran	6610B	Dieldrin	6410B; 6630B,C
Carbon tetrachloride	6200B,C	Diethyl phthalate	6040B; 6410B
Chlordane	6410B; 6630B,C	Dimethyl phthalate	6410B
Chlorobenzene	6040B; 6200B,C	Dimethylphenol(s)	6410B; 6420B
Chloroethane	6200B,C	Dinitrophenol(s)	6410B; 6420B
Chloroethoxy methane	6040B; 6410B	Dinitrotoluene(s)	6410B
Chloroethyl ether	6040B; 6410B	Dinoseb	6640B
Chloroethylvinyl ether	6200B,C	Di-n-octyl phthalate	6410B
Chloroform	6200B, C; 6232B	Diphenyl hydrazine	6040B

TABLE 6010:I. CONT.

Compound	Analysis Methods (section number)	Compound	Analysis Methods (section number)
Endosulfan	6410B; 6630B,C	Nitrobenzene	6410B
Endosulfan sulfate	6410B; 6630C	Nitrophenol(s)	6410B; 6420B
Endrin	6410B; 6630B,C	Nitrosodi-n-propylamine	6410B
Endrin aldehyde	6410B; 6630C	Nitrosodimethylamine	6410B
Ethenyl benzene (styrene)	6040B	Nitrosodiphenylamine	6410B
Ethylbenzene	6040B; 6200B, C	Oxamyl	6610B
Ethylhexyl phthalate	6410B	Parathion	6630B
Fluoranthene	6040B; 6410B; 6440B	PCB-1016, 1221, 1232, 1242, 1248, 1254, 1260	6410B, 6630C
Fluorene	6040B; 6410B; 6440B	Pentachloronitrobenzene	6630B
Formaldehyde	6252B	Pentachlorophenol	6410B; 6420B; 6640B
Geosmin	6040B	Phenanthrene	6040B; 6410B; 6440B
Glyoxal	6252B	Phenol	6410B; 6420B
Glyphosate	6651B	Phenylbenzamine	6040B
Heptachlor	6410B; 6630B,C	Picloram	6640B
Heptachlor epoxide	6410B; 6630B,C	Propylbenzene	6040B; 6200B,C
Heptaldehyde	6252B	Pyrene	6040B; 6410B; 6440B
Hexachlorobenzene	6040B; 6410B	Silvex (trichlorophenoxy propionic acid)	6640B
Hexachlorobutadiene	6040B; 6200B, C; 6410B		
Hexachlorocyclopentadiene	6410B	Strobane	6630B
Hexachloroethane	6040B; 6410B	Styrene (ethenyl benzene)	6200B,C
3-Hydroxycarbofuran	6610B	2,4,5-T (trichlorophenoxy acetic acid)	6640B
Indeno(1,2,3-cd)pyrene	6410B; 6440B		
Isobutylmethoxy pyrazine	6040B	2,4,5-TP	6640B
Isophorone	6410B	Tetrachloroethane(s)	6040B; 6200B,C
Isopropylbenzene	6200B, C	Tetrachloroethene	6040B; 6200B,C
Isopropyl methoxy pyrazine	6040B	Toluene	6200B,C
Isopropyltoluene	6200B, C	Toxaphene	6410B; 6630B,C
Lindane (γ-BHC)	6630B	Trichloroanisole	6040B
Malathion	6630B	Trichloroacetic acid (TCAA)	6251B
Methane	6211	Trichlorobenzene(s)	6040B; 6200B,C; 6410B
Methiocarb	6610B	Trichloroethane(s)	6040B; 6200B,C
Methoxychlor	6630B	Trichloroethene	6040B; 6200B,C
Methyldinitrophenol(s)	6410B; 6420B	Trichlorofluoromethane	6200B,C
Methylene chloride	6200B, C	Trichlorophenol	6251B; 6410B; 6420B
Methyl glyoxal	6252B	Trichloropropane	6200B,C
Methylisoborneol	6040B	Trifluralin	6630B
Methyl parathion	6630B	Trimethylbenzene(s)	6200B,C
Methomyl	6610B	Vinyl chloride	6200B,C
Mirex	6630B	Xylene(s)	6040B; 6200B,C
Monobromoacetic acid (MBAA)	6251B		
Monochloroacetic acid (MCAA)	6251B		
Naphthalene	6040B; 6200B, C; 6410B; 6440B		

* Compounds are listed under the names by which they are most commonly known and called in specific methods.

6010 B.   Sample Collection and Preservation

1. Volatile Organic Compounds

Use 25- or 40-mL vial equipped with a screw cap with a hole in the center* and TFE-faced silicone septum.† Wash vials, caps, and septa with detergent, rinse with tap and distilled water, and dry at 105°C for 1 h before use in an area free of organic vapors. NOTE—Do not heat seals for extended periods of time (> 1 h) because the silicone layer slowly degrades at 105°C. When bottles

are cool, seal with TFE seals. Alternatively purchase precleaned vials free from volatile organics.

Collect all samples in duplicate and prepare replicate field reagent blanks with each sample set. A sample set is all samples collected from the same general sampling site at approximately the same time. Prepare field reagent blanks in the laboratory by filling a minimum of two sample bottles with reagent water, sealing, and shipping to the sampling site along with empty sample bottles.

Fill sample bottle just to overflowing without passing air bubbles through sample or trapping air bubbles in sealed bottle. When sampling from a water tap, open tap and flush until water tem-

* Pierce 13075 or equivalent.
† Pierce 12722 or equivalent.

perature has stabilized (usually about 10 min). Adjust flow rate to about 500 mL/min and collect duplicate samples from flowing stream. When sampling from an open body of water, fill a 1-L, wide-mouth bottle or breaker with a representative sample and carefully fill duplicate sample bottles from the container.

Preservation of samples is highly dependent on target constituents and sample matrix. Ongoing research indicates the following areas of concern: rapid biodegradation of aromatic compounds, even at low temperatures;[1] dehydrohalogenation reactions such as conversion of pentachloroethane to tetrachloroethane;[2] reactions of alkylbenzenes in chlorinated samples, even after acidification; and possible interactions among preservatives and reductants when dechlorination is used to prevent artifact formation, especially in samples potentially containing many target compounds.

There is as yet no single preservative that can be recommended. Ideally, maintain samples chilled (preferably at 4°C) and analyze immediately. In practice, delays between sampling and analysis often necessitate preservation. The recommended preservation techniques are summarized in Table 6010:II.

1) For samples and field blanks that contain volatile constituents but do not contain residual chlorine, add HCl (4 drops 6N HCl/40 mL) to prevent biodegradation and dehydrohalogenation. NOTE: HCl may contain traces of organic solvents. Verify freedom from contamination before using a specific lot for preservation.

2) For samples and field blanks that contain residual chlorine, also add a reducing agent. Ascorbic acid (25 mg/40 mL) appears to be optimal, but demonstrate that this reductant is appropriate for the specific sample matrix. Sodium thiosulfate (3 mg /40 mL) or sodium sulfite (3 mg /40 mL) also may be appropriate reducing agents, but when either of these is added in the presence of HCl, SO$_2$ formation may interfere with certain packed-column gas chromatographic or GC/MS techniques.

In all cases, run reagent blanks to insure absence of interferences. Add either ascorbic acid or HCl to the sample bottle immediately before shipping it to the sample site or immediately before filling sample bottle. When both preservatives are needed, add only one before filling the sample bottle, to prevent interactions between the acid and the reductant. Add the second preservative once the bottle is almost full. However, if there is evidence that interactions of acid and reducing agent will not create analytical or preservation problems, they may be added simultaneously.

Tightly seal sample bottles, TFE face down. After sampling and preservation invert several times to mix. Chill samples to 4°C immediately after collection and hold chilled in an atmosphere free of organic solvent vapors until analysis. Normally analyze all samples within 14 d of collection. Shorter or longer holding times may be appropriate, depending on constituents and sample matrix. Develop data to show that alternate holding times are appropriate.

2. Other Organic Compounds

See individual methods for sampling and preservation requirements.

3. References

1. BELLAR, T. & J. LICHTENBERG. 1978. Semi-automated headspace analysis of drinking waters and industrial waters for purgeable volatile organic compounds. *In* C. E. Van Hall, ed. Measurement of Organic Pollutants in Water and Wastewater. STP 686, American Soc. Testing & Materials. Philadelphia, Pa.
2. BELLAR, T. & J. LICHTENBERG. 1985. The Determination of Synthetic Organic Compounds in Water by Purge and Sequential Trapping Capillary Column Gas Chromatography. U.S. Environmental Protection Agency, Cincinnati, Ohio.

4. Bibliography

KEITH, L. H., ed. 1988. Principles of Environmental Testing. American Chemical Soc., Washington, D.C.

TABLE 6010:II. RECOMMENDED PRESERVATION FOR VOLATILES

Constituents	Chlorinated Matrix	Non-Chlorinated Matrix
Halocarbons	HCl + reducing agent	HCl
Aromatics	HCl + reducing agent	HCl
THMs	Reducing agent (HCl optional)*	None required
EDB/DBCP	None required	None required

* See 6232B.2.

6010 C. Analytical Methods

1. General Discussion

The methods presented in Part 6000 for identification and quantitation of trace organic constituents in water generally involve isolation and concentration of the organics from a sample by solvent or gas extraction (see Section 6040 and individual methods), separation of the components, and identification and quantitation of the compounds with a detector.

2. Gas Chromatographic Methods

Gas chromatographic (GC) methods are highly sophisticated microanalytical procedures. They should be used only by analysts experienced in the techniques required and competent to evaluate and interpret the data.

a. Gas chromatograph:

1) Principle—In gas chromatography a mobile phase (a carrier gas) and a stationary phase (column packing or capillary column coating) are used to separate individual compounds. The carrier gas is nitrogen, argon-methane, helium, or hydrogen. For packed columns, the stationary phase is a liquid that has been coated on an inert granular solid, called the column packing, that is held in borosilicate glass tubing. The column is installed in an oven with the inlet attached to a heated injector block and the outlet attached to a detector. Precise and constant temperature control of the in-

jector block, oven, and detector is maintained. Stationary-phase material and concentration, column length and diameter, oven temperature, carrier-gas flow, and detector type are the controlled variables.

When the sample solution is introduced into the column, the organic compounds are vaporized and moved through the column by the carrier gas. They travel through the column at different rates, depending on differences in partition coefficients between the mobile and stationary phases.

2) Interferences—Some interferences in GC analyses occur as a result of sample, solvent, or carrier gas contamination, or because large amounts of a compound may be injected into the GC and linger in the detector. Methylene chloride, chloroform, and other halocarbon and hydrocarbon solvents are ubiquitous contaminants in environmental laboratories. Make strenuous efforts to isolate the analytical system from laboratory areas where these or other solvents are in use. An important sample contaminant is sulfur, which is encountered generally only in base/neutral extracts of water, although anaerobic groundwaters and certain wastewaters and sediment/sludge extracts may contain reduced sulfur compounds, elemental sulfur, or polymeric sulfur. Eliminate this interference by adding a small amount of mercury or copper filings to precipitate the sulfur as metallic sulfide. Sources of interference originating in the chromatograph, and countermeasures, are as follows:

• *Septum bleed*—This occurs when compounds used to make the septum on the injection port of the GC bleed from the heated septum. These high-molecular-weight silicon compounds are distinguished readily from compounds normally encountered in environmental samples. Nevertheless, minimize septum bleed by using septum sweep, in which clean carrier gas passes over the septum to flush out the "bleed" compounds.

• *Column bleed*—This term refers to loss of column coating or breakdown products when the column is heated. This interference is more prevalent in packed columns, but also occurs to a much lesser extent in capillary columns. It occurs when the column temperature is high or when water or oxygen are introduced into the system. Solvent injection can damage the stationary phase by displacing it. Certain organic compounds acting as powerful solvents, acids, or bases can degrade the column coating. Injection of large amounts of certain surface-active agents may destroy GC columns.

• *Ghost peaks*—These peaks occur when an injected sample contains either a large amount of a given compound, or a compound that adsorbs to the column coating or injector parts (e.g., septum). When a subsequent sample is injected, peaks can appear as a result of the previous injection. Eliminate ghost peaks by injecting a more dilute sample, by producing less reactive derivatives of a compound that may interact strongly with the column material, by selecting a column coating that precludes these interactions, or by injecting solvent blanks between samples.

b. Detectors: Various detectors are available for use with gas chromatographic systems. See individual methods for recommendations on appropriate detectors.

1) Electrolytic conductivity detector—The electrolytic conductivity detector is a sensitive and element-specific detector that has gained considerable attention because of its applicability to the gas chromatographic analysis of environmentally significant compounds. It is utilized in the analysis of purgeable halocarbons, pesticides, herbicides, pharmaceuticals, and nitrosamines. This detector is capable of operation in each of four specific modes:

halogen (X), nitrogen (N), sulfur (S), and nitrosamine (NO). Only organic compounds containing these elements will be detected.

Compounds eluting from a gas chromatographic column enter a reactor tube heated to 800°C. They are mixed with a reaction gas, hydrogen for X, N, or NO modes, and air for the S mode. The hydrogen catalytically reduces the compounds while the air oxidizes them. The gaseous products are transferred to the detector through a conditioned ion exchange resin or scrubber. In the halogen mode, only HX is detected, while NH_3 or H_2S are eliminated on the resin. In the nitrogen or nitrosamine mode, the NH_3 formed is ionized while HX and H_2S, if present, are eliminated with a KOH/quality wool scrubber. The sulfur mode produces SO_2 or SO_3, which is ionized while HX is removed with a silver wire scrubber. All other products either are not ionizable or are produced in such low yield that they are not detectable.

The electrolytic conductivity detector contains reference and analytical electrodes, a gas-liquid contactor, and a gas-liquid separator. The conductivity solvent enters the cell and flows by the reference electrode. It combines with the gaseous reaction products in the gas-liquid contactor. This heterogeneous mixture is separated into gas and liquid phases in the gas-liquid separator, with the liquid phase flowing past the analytical electrode. The electrometer monitors the difference in conductivity at the reference electrode (solvent) and the analytical electrode (solvent + carrier + reaction products).

2) Electron capture detector—The electron capture detector (ECD) usually is used for the analysis of compounds that have high electron affinities, such as chlorinated pesticides, drugs, and their metabolites. This detector is somewhat selective in its response, being highly sensitive toward molecules containing electronegative groups: halogens, peroxides, quinones, and nitro groups. It is insensitive toward functional groups, such as amines, alcohols, and hydrocarbons.

The detector is operated by passing the effluent from the gas chromatographic column over a radioactive beta particle emitter, usually nickel-63 or tritium adsorbed on platinum or titanium foil. An electron from the emitter ionizes the carrier gas, preferably nitrogen, and produces a burst of electrons. About 100 secondary electrons are produced for each initial beta particle. After further collisions, the energy of these electrons is reduced to the thermal level and they can be captured by electrophilic sample molecules.

The electron population in the ECD cell is collected periodically by applying a short voltage pulse to the cell electrodes and the resulting current is compared with a reference current. The pulse interval is adjusted automatically to keep the cell current constant, even when some of the electrons are being captured by the sample. The change in the pulse rate when a sample enters the ECD is then related to the sample concentration. The ECD offers linearity in the range of 10^4 and subpicogram detection limits for compounds with high electron affinities.

3) Flame ionization detector—The flame ionization detector (FID) is widely used because of its high sensitivity to organic carbon-containing compounds. The detector consists of a small hydrogen/air diffusion flame burning at the end of a jet. When organic compounds enter the flame from the column, electrically charged intermediates are formed. These are collected by applying a voltage across the flame. The resulting current is amplified by an electrometer and measured. The response of the detector is directly proportional to the total mass entering the detector per unit time and is independent of the concentration in the carrier gas.

The FID is perhaps the most widely used detector for gas chromatography because of several advantages: (*a*) it responds to virtually all organic carbon-containing compounds with high sensitivity (approximately 10^{-13} g/mL); (*b*) it does not respond to common carrier gas impurities such as water and carbon dioxide; (*c*) it has a large linear response range (approximately 10^7) and excellent baseline stability; (*d*) it is relatively insensitive to small column flow-rate changes during temperature programming; (*e*) it is highly reliable, rugged, and easy to use; and (*f*) it has low detector dead volume effects and fast response. Its limitations include: (*a*) it gives little or no response to noncombustible gases and all noble gases; and (*b*) it is a destructive detector that changes the physical and chemical properties of the sample irreversibly.

4) *Photoionization detector*—Photoionization occurs when a molecular species absorbs a photon of light energy and dissociates into a parent ion and an electron. The photoionization detector (PID) detects organic and some inorganic species in the effluent of a gas chromatograph with detection limits as low as the picogram range. The PID is equipped with a sealed ultraviolet light source that emits photons which pass through an optically transparent window (made of LiF, MgF_2, NaF, or sapphire) into an ionization chamber where photons are absorbed by the eluted species. Compounds having ionization potential less than the UV source energy are ionized. A positively biased high-voltage electrode accelerates the resulting ions to a collecting electrode and the resulting current is measured by an electrometer. This current is proportional to the concentration.

The PID has high sensitivity, low noise (approximately 10^{-14} A), and excellent linearity (10^7), is nondestructive, and can be used in series with a second detector for more selective detection. The PID can be operated as a universal detector or a selective detector by simply changing the photon energy of the ionization source. Tables of ionization potentials are used to select the appropriate UV source for a given measurement.

5) *Mass spectrometer*—The mass spectrometer (MS) has the ability to detect a wide variety of compounds, coupled with a capacity to deduce compound structures from fragmentation patterns. Among the different types of mass spectrometers, the quadrupole has become the most widely used in water and wastewater analysis.

The mass spectrometer detects compounds by ionizing molecules into charged species with a 70-eV beam. The ions are accelerated toward the quadrupole mass filter through a series of lenses held at 0 to 200 V. The differently sized, charged fragments are separated according to mass-to-charge ratio (related to molecular weight) by means of the quadrupole, which uses varying electric and radiofrequency (rf) fields. The quadrupole is connected to a computer, which varies these fields so that only fragments of one particular mass-to-charge ratio (± 0.5) can traverse the quadrupole at any one time. As the ions leave the quadrupole they are attracted to the electron multiplier through an electrical potential of several thousand volts. The charge fragments, in turn, are detected by the electron multiplier. Because the electric and the rf fields are cycled every few seconds, a fragmentation pattern is obtained. Each cycle is called a mass scan. Most chemicals have unique fragmentation patterns, called mass spectra. The computer contains, and can search, a library of known mass spectra to identify tentatively an unknown compound exhibiting a particular spectrum. Use authentic compounds for confirmation after tentative identifications are made.

Background mass interference can result from the ability of the mass spectrometer to detect any ions created in its ion volume (up to a specified mass). Any compounds continuously present in the source will be detected. Some mass ions always present are due to air components that leak into the system, such as oxygen (masses 16 and 32), nitrogen (masses 14 and 28), carbon dioxide (mass 44), argon (mass 40), and water (mass 18), or to helium carrier gas (masses 4 and 8), or to diffusion pump oil vapors.

3. High-Performance Liquid Chromatographic (HPLC) Methods

a. Principle: HPLC is an analytical technique in which a liquid mobile phase transports a sample through a column containing a liquid stationary phase. The interaction of the sample with the stationary phase selectively retains individual compounds and permits separation of sample components. Detection of the separated sample compounds is achieved mainly through the use of absorbance detectors for organic compounds and through conductivity and electrochemical detectors for metal and inorganic components.

b. Detectors:

1) *Photodiode array detector (PDAD)*—The PDAD measures the absorbance of a sample from an incident light source (UV-VIS). After passing through the sample cell, the light is directed through a holographic grating that separates the beam into its component wavelengths reflected on a linear array of photodiodes. This permits the complete absorbance spectrum to be obtained in 1 s or less and simultaneous multiwavelength analysis.

The PDAD is subject to the interference encountered with all absorbance detectors. Of special concern for HPLC is the masking of the absorbance region of the HPLC mobile phase and its additives. This may reduce the range and sensitivity of the detector to the sample components. Most interferences occur in monitoring the shorter wavelengths (200–230 nm). In this region, many organic compounds absorb light energy and can be sources of interference.

2) *Post column reactor (PCR)*—The PCR consists of in-line sample derivatizing/reacting equipment that permits chemical alteration of certain organic compounds. This equipment is used to enhance detection by attaching a chromophore to the compound(s) of interest. Sensitivity and selectivity of compounds that were initially undetectable are altered to make them detectable.

Interferences from this technique usually arise from the impurities in the reagents used in the reaction. When this technique is coupled with a selective detector such as fluorescence, these interferences are minimized. Generally, only compounds of the same class as the compounds of interest will cause interference.

3) *Fluorescence detector*—The fluorescence detector is an absorbance detector in which the sample is energized by a monochromatic light source. Compounds capable of absorbing the light energy do so and release it as fluorescence emission. Filters permit the detector to respond only to the fluorescent energy. The fluorescence detector is the most sensitive of the current HPLC detectors available and often is used in conjunction with a post column reactor.

Because of instrument sensitivity, minute quantities of contaminants can cause interferences to fluorescence detectors. Contamination can happen from glassware, mobile phase solvents, postcolumn reagents, etc. These sources will raise the background signal and thus narrow the range of the detector. Interference from individual compounds is minimal because of detector specificity (i.e., all interferences must fluoresce).

6020 QUALITY ASSURANCE/QUALITY CONTROL FOR ORGANIC COMPOUNDS

6020 A. Introduction

Quality assurance (QA) and quality control (QC) for organic compound analysis include the operating principles stated in Sections 1020 Quality Assurance, 1030 Data Quality, and 1040 Method Development and Evaluation. This section consolidates the additional requirements common to the methods in Part 6000. The requirements are recommended minimum QA/QC activities; they should be followed unless the individual method gives different, but comparable, specifications. Some methods may have additional QA/QC requirements. Others may have broader acceptance criteria because of the unique difficulties associated with the determination of a constituent, e.g., the extraction efficiency for phenols.

6020 B. Quality Control Practices

This section describes the elements of a quality control program deemed necessary to maintain proper control for organic analyses. The related data quality objectives (DQOs), i.e., the rationale for sampling and analyses, should define which elements are necessary for each individual analytical design.

1. Calibration

a. Initial calibration: Perform initial calibration with a minimum of five concentrations of analytical standards for the analyte(s) of interest (for example: 1, 5, 10, 20, 40). The lowest concentration should be at the minimum reporting level; for example, if the minimum reporting level is 1, the lowest concentration should be 1. The highest concentration should be near the upper end of the calibration range; for example, if the upper end of the calibration range is 40, the high standard should be 40. Concentration ranges should reflect concentrations in actual samples. Choose calibration concentrations with no more than one order of magnitude between concentrations; for example, for a calibration range of 1 to 1000, choose concentrations of 1, 10, 100, and 1000.

Use any of the following calibration functions, as appropriate: response factor for internal standard calibration, calibration factor for external standard calibration, or calibration curve. Calibration curves may be linear or quadratic, and may or may not pass through the origin. Use the following recommended acceptance criteria for the various calibration functions.

If response factors or calibration factors are used, relative standard deviation (RSD) for each analyte should be less than 20%. If RSD is not less than 20% for any analyte, then identify and correct source of lack of linearity before sample quantitation. When using response factors (i.e., for GC/MS analysis), check performance or sensitivity of instrument for the analyte against minimum acceptance values for response factors. Refer to specific analytical method for the acceptance criteria for response factors for each analyte.

If a linear regression is used, the correlation coefficient should be >0.995. If a calibration curve has been constructed, recalculate each calibration point compared to curve. Values should be within ±20%. If any of the recalculated values are not within ±20%, identify and correct source of outlier(s) before sample quantitation.

Use initial calibration with any of the above functions (response factor, calibration factor, or calibration curve) for quantitation of analytes in samples. Use continuing calibration, ¶ *b* below, only for checks on initial calibration and not for sample quantitation. Perform initial calibration when instrument is set up and whenever continuing calibration criteria are not met.

b. Continuing calibration: Continuing calibration is the periodic verification, by analysis of a calibration standard, that instrument performance has not changed significantly from initial calibration. Perform continuing calibration every 10 samples for GC analysis, every 20 samples for GC/MS analysis, or every 12 h, whichever is more frequent. Perform continuing calibration with one or more of the concentrations of analytical standards in the initial calibration, varying actual concentration of continuing calibration standard over the calibration range. The acceptance criteria for continuing calibration should be ±20% (80 to 120% recovery) compared to the known or expected value of the calibration standard. If the acceptance criteria are not met, reanalyze continuing calibration standard or repeat initial calibration. When using response factors (i.e., GC/MS analysis), check performance or sensitivity of instrument for analytes against minimum acceptance values for the response factors. Refer to the specific analytical method for acceptance criteria for response factors for each analyte.

c. Closing standard: Finish all runs with a laboratory-fortified blank (LFB) for VOC analyses or closing standard (for methods with procedural standards) to demonstrate that performance was still acceptable for last sample analyzed. A LFB is a reagent blank to which a known concentration of analytes has been added. See ¶ 3*b* below. All samples must be bracketed by acceptable continuing calibrations.

2. Initial Quality Control

a. Initial demonstration of capability: Before analysis of any sample, require each analyst to demonstrate proficiency with the method of choice. Include at least the analysis of a laboratory reagent blank (LRB) and four laboratory-fortified blanks (LFBs) that have added concentrations between 5 times the minimum reporting level and the midpoint of the calibration curve. The blank should not contain any analyte at a concentration greater than one-fourth the minimum quantitation level. The precision and percent recovery calculated from the four LFBs should be at least as good as the values listed in the method of choice.

b. Method detection level: Determine method detection level (MDL) before any samples are analyzed, using procedure described in Section 1030C or other specified method[1] required for the type of samples the laboratory is intending to analyze. As a starting point for determining concentration to use in performing the laboratory's MDL calculation, try about five times the estimated instrument detection level or refer to the selected method. Determine MDL as an iterative process. Repeat determinations if calculated MDL is not within a factor of 10 of the fortified value. Determine MDL at least annually. Analyze samples for MDL determination over a 3- to 5-d period to generate a more realistic value. Include all applicable sample preparatory techniques in MDL determinations.

c. Minimum quantitation level (MQL): The MQL is the lowest level that can be quantitated accurately. MQL is defined as four times the MDL. Report samples containing compounds of interest at a level less than the MQL as <MQL. Report samples containing compounds of interest at a level less than the MDL as ND (not detected).

e. Sample (batch) set: A sample or batch set is defined as those samples extracted in a single day, not to exceed 20 samples.

f. Analytical day: An analytical day is defined as a 12-h analytical period.

3. Batch Quality Control

a. Reagent blank: A reagent blank consists of all reagents and preservatives that normally contact a sample when it is carried through the entire analytical procedure. Use a reagent blank to determine contribution of reagents and preparative analytical steps to error in the observed value. No analyte of interest should be present in a reagent blank at a level greater than one fourth the MQL. Include a minimum of one reagent blank with each sample set or batch.

b. Laboratory-fortified blank (LFB): An LFB, also known as spiked blank, is a reagent blank containing all the same reagents and preservatives as samples and to which a known concentration of analytes has been added. Use LFB to evaluate laboratory performance and analyte recovery in a blank matrix. Make addition concentration at least 5 times the MQL or the midpoint of the calibration curve, and use to calculate recovery limits and to plot control charts as in Section 1020B. Prepare known-addition solution for blanks and samples from a different primary mix than that used to develop working standard mix. Include a minimum of one LFB with each sample set or batch. Ensure that LFB meets performance criteria in the method of choice.

c. Internal standard (IS): An internal standard is a compound of known concentration added to each standard and sample extract just before sample analysis. This compound should have chromatographic characteristics similar to those of the analytes of interest. Use IS to monitor retention time, relative response, and quantity of analytes in each extract. When quantifying by the internal standard method, measure all analyte responses relative to this standard. Internal standard response should be in the range of ± 30% compared to calibration curve response. The retention time of this compound should separate from all analytes of interest and elute in a representative area of the chromatogram. If a single compound cannot be found to meet these criteria, use additional compounds.

d. Surrogate standard: A surrogate standard is a compound of a known concentration added to each environmental and blank sample before extraction. Use compound(s) that have characteristics similar to those of the analytes of interest and that are unlikely to be found in environmental samples. Carry surrogate standard through entire sample extraction and analytical process to monitor extraction efficiency of the method for each sample. Refer to method of choice for specific surrogates and acceptance criteria.

e. Quality control sample: Analyze an externally generated quality control sample of known quantity as a laboratory-fortified blank at least quarterly or whenever new stock solutions are prepared. This sample is used to validate the laboratory's standards both qualitatively and quantitatively.

f. Laboratory-fortified sample (LFS): A laboratory-fortified sample, also known as laboratory-fortified matrix or matrix spike, is another portion of a sample fortified with the analytes of interest at a concentration at least 5 times the MQL or around the midpoint of the calibration range. Include a minimum of one LFS with each sample set (batch). Make LFSs of sufficient concentrations that sample background levels do not adversely affect the recovery calculations. (Adjust addition concentrations if this is a known sample to be about five times background level.) Base sample batch acceptance on results of LFBs rather than on LFSs, because the sample matrix may interfere with method performance. Prepare addition solution for blanks and samples from a different primary mix than that used to develop working standard mix.

g. Laboratory-fortified sample duplicates (LFSDs): A laboratory-fortified sample duplicate, also known as laboratory-fortified matrix duplicate, spiked sample duplicate, or matrix spike duplicate, is a second portion of the sample to which a known amount of analyte is added. If sufficient sample volume is collected, add to a second portion of fortified sample and compare to first. If sufficient sample volume is not collected, use a second sample to obtain results on two separate LFSs rather than LFSDs. Include a minimum of one LFSD with each sample set (batch). Compare precision and bias to those listed in the method. Base sample batch acceptance on results of reagent blank additions rather than laboratory-fortified sample duplicates.

4. Reference

1. U.S. ENVIRONMENTAL PROTECTION AGENCY. 1995. Definition and procedure for the determination of the method detection limit, revision 1.11. 40 CFR Part 136, Appendix B. *Federal Register* 51:23703.

6040   CONSTITUENT CONCENTRATION BY GAS EXTRACTION*

6040 A.   Introduction

The ability to analyze ultratrace levels of organic pollutants in water has been limited, in part, by the concentration technique. With the development of closed-loop stripping analysis (CLSA) (Method B), organic compounds of intermediate volatility and molecular weight, i.e., from the heavier volatiles to the lighter polynuclear aromatic hydrocarbons, can be extracted from water and concentrated to allow quantitative and semiquantitative analysis (depending on the compound) at nanograms-per-liter levels. This extract can be analyzed on a gas chromatograph (GC) con-

nected to one of several detectors. A CLSA technique coupled with gas chromatographic/mass spectrometric (GC/MS) analysis for the determination of trace organic compounds is presented here. It is applicable to both treated and natural waters.

The purge and trap technique (Method C) is a valuable concentration method applicable to volatile organic compounds. The compounds are concentrated by bubbling an inert gas through the sample followed by collection in, and desorption from, a sorbent trap. This extract may be analyzed by GC or GC/MS methods. The technique is applicable to both water and wastewater.

---

* Approved by Standard Methods Committee, 1997.

6040 B.   Closed-Loop Stripping, Gas Chromatographic/Mass Spectrometric Analysis

1. General Discussion

a. Principle: This CLSA-GC/MS procedure is suitable for the analysis of a broad spectrum of organic compounds in water. It can be used for the identification and quantitation of specific compounds, such as earthy-musty-smelling compounds [e.g., 2-methylisoborneol (MIB) and geosmin][1-3] or U.S. Environmental Protection Agency (EPA) priority pollutants.[4,5] The method is suitable for other taste- and odor-causing compounds (e.g., geranyl acetone 6-methylhept-5-en-2-one, β-ionone, and β-cyclocitral).[6]

In closed-loop stripping, volatile organic compounds of intermediate molecular weight are stripped from water by a recirculating stream of air. The organics are removed from the gas phase by an activated carbon filter. They are extracted from the filter with carbon disulfide (CS_2) or methylene chloride. A portion of the extract is injected into a capillary-column GC/MS for identification of the organic compounds by retention time and spectrum matching; quantification is done by single-ion current integration. Alternatively, analysis may be made by capillary-column GC equipped with a flame ionization detector with identification by retention times on primary and secondary capillary columns.

b. Interference: Organic compounds that are stripped during this procedure may coelute with the compounds of interest. The uniqueness of the mass spectrum of each target compound makes it possible to confirm compound identity with a high probability when coeluting components are present. Problems may arise if several isomers of a compound are present that are not resolvable chromatographically.

c. Detection levels: Trace organics can be detected at low nanogram-per-liter levels. The CLSA-GC/MS detection limits are affected by many factors; especially important are the stripping efficiency and the condition of the GC/MS. Stripping efficiencies can be improved by using an elevated stripping temperature and/or the salting-out technique. The stripping and extraction portion of the method can be evaluated independently of the instrumen-

tation portion. As an option, add internal standard after stripping and extraction, and transfer extract quantitatively.

The method detection levels for five earthy-musty-smelling compounds are shown in Table 6040:I. Detection levels for the salted CLSA method are less than half those for the unsalted method for each compound. Using the elevated stripping temperature rather than the salting-out technique produces comparable recoveries[7] and similar detection levels. Detection levels for various organic compounds of interest, obtained with an elevated stripping temperature/salting-out technique, ranged from 0.1 to 100 ng/L (see Table 6040:II).[8]

2. Apparatus

Use clean glassware in sample collection and calibration standard preparation. Wash with soapy water, rinse with tap water, with demineralized water, and finally with reagent-grade acetone. As an alternative to acetone rinse, bake glassware in an annealing oven for 1 h at 400°C. Air-dry and bake at 180°C for 6 to 12 h. Do not bake sample bottle caps or volumetric ware. After drying and baking, store inverted or cover mouths with aluminum foil to prevent accumulation of dust or other contaminants.

TABLE 6040:I. METHOD DETECTION LEVELS FOR EARTHY-MUSTY SMELLING COMPOUNDS BY CLSA-GC/MS

	Detection Level ng/L*	
	Unsalted Method[1]	Salting-Out Technique[3]
Geosmin	2	0.8
2-Methylisoborneol	2	0.8
2-Isopropyl-3-methoxy pyrazine	2	0.8
2-Isobutyl-3-methoxy pyrazine	2	0.8
2,3,6-Trichloroanisole	5	0.8

* Stripping at 25°C. Selective ion monitoring.

TABLE 6040:II. METHOD DETECTION LEVELS FOR SELECTED ORGANIC COMPOUNDS BY CLSA-GC/MS[8]

Compound	Detection Level ng/L*	Compound	Detection Level ng/L†
1,1,1-Trichloroethane	2.0	1,3,5-Trichlorobenzene	0.1
Trichloroethene	100	1,2,4-Trichlorobenzene	10
Dichlorobromomethane	5.0	1,2,3-Trichlorobenzene	2.0
1,3-Dichloropropene	2.0	bis(2-Chloro-ethoxy)methane	10
1,1,2-Trichloroethane	2.0	Methylisoborneol (MIB)	0.5
Chlorodibromomethane	1.0	Geosmin	0.2
1,2-Dibromoethane	2.0	Naphthalene	100
Tetrachloroethene	100	1,1,2,3,4,4-Hexachloro-1,3-butadiene	2.0
Chlorobenzene	10	1-Chloronaphthalene	0.5
Ethylbenzene	50	2-Chloronaphthalene	0.5
m,p-Xylene	100	Acenaphthene	0.5
Bromoform	1.0	Fluorene	2.0
Ethylbenzene	5.0	Diethylphthalate	100
o-Xylene	50	1-Chloro-4-phenoxybenzene	0.5
1,1,2,2-Tetrachloroethane	50	N-Phenylbenzamine	20
Bromobenzene	0.5	1,2-Diphenylhydrazine (as azobenzene)	1.0
Propylbenzene	0.5	1-Bromo-4-phenoxybenzene	0.5
1-Chloro-3-methylbenzene	0.5	Hexachlorobenzene	1.0
bis (2-Chloroethyl)ether	1.0	Phenanthrene	10
o-Dichlorobenzene	0.1	Anthracene	50
m-Dichlorobenzene	10	Fluoranthene	20
p-Dichlorobenzene	10	Pyrene	20
Hexachloroethane	20	Chrysene	50
N-Nitrosodi-n-propylamine	5.0	Benzo(a)anthracene	50

* Elevated stripping temperature and salting-out both utilized.

† Instrument detection limit based on a 2:1 signal:noise ratio (where a background interference existed, the target compound was required to be at least twice background.)

a. *Sample bottles,* 1-L capacity or larger, glass, with TFE-lined screw caps.

b. *CLSA apparatus,* equipped with the following components (Figure 6040:1) or their equivalents.*

1) *Stripping bottle,* with mark at 1-L level and stainless-steel quick-connect stems (Figure 6040:2) or unpolished spherical glass joints sealed with TFE-covered silicone rubber O-rings and secured with metal clamps.† Immediately after use, clean stripping bottle by rinsing twice with demineralized water and once with organic-free water. For particularly adherent impurities, clean with acetone and bake at 180°C for at least 2 h. Turbid samples may cause a film to deposit on the stripping bottle and frit and may require washing with acid detergent.

2) *Gas heater,* with aluminum heating cylinder and soldering iron (25 W) controlled by a variable transformer (Figure 6040:3). Alternatively, use a temperature-controlled heater block to maintain a fixed temperature at the filter that is 10 to 20°C above temperature of the thermostatic water bath.

3) *Filter holder,* stainless steel or glass.‡ If glass is used, also use an auxiliary heating device, e.g., an infrared light, to maintain proper filter temperature.

4) *Pump,* with stainless-steel bellows,§ providing air flow in the range of 1 to 1.5 L/min. When using the salting-out technique, periodically disassemble the pump and clean. Salt will leave deposits on the pump bellows. If there is a noticeable drop in pump performance, clean valve assembly with acetone or replace.

5) *Automatic timer* (optional), connected to pump.

6) *Circuit,* with stainless-steel parts: 1/8-in. (0.3-cm)-OD tubing, 4-in. (10.2-cm) × 1/4-in. (0.6-cm)-OD flexible tubing, tube fittings, and quick-connect bodies;‖ or glass joints described in ¶ 1) above. Glass sample lines can be used except where circuit enters and exits pump. Use TFE ferrules in making connections to glass and flexible metal tubing. Whenever sample carryover is observed, clean circuit and pump as follows: Connect fittings to the quick-connect bodies on both ends of the circuit to open system. Turn on pump and flush with approximately 100 mL each organic-free water, acetone, and methanol. After last rinse, dry with a heat gun with pump still running (or flush with nitrogen) until there is no residual methanol. (NOTE: Overheating quick-connect units can cause deterioration of internal O-rings.)

7) *Thermostatic water bath,* with 222-mm-OD × 457-mm chromatography jar and thermoregulating system accurate to at least ±0.5°C. When the ambient temperature of the laboratory is greater than 25°C, maintain water bath at 25°C by inserting a coil of copper tubing connected to a cold water tap to recirculate cold water.

8) *Filters,* with 1.5 mg activated carbon# (Figure 6040:4). Use a set of filters matched in solvent flow resistance and performance for each group of samples and calibration standards. Determine filter resistance by measuring solvent flow rate through a cleaned, solvent-wetted filter. First fill the longer glass tube above the charcoal with organic-free water and let flow by gravity through filter disk. Rinse once with acetone, rinse twice with elution solvent, rinse twice with acetone, but on the final rinse, measure time necessary to empty the solvent (0.3-mL volume) from top of filter tube to surface of

---

* Model CLS 1, Tekmar, Cincinnati, OH; Brechbühler AG, 8952 Schlieren ZH, Switzerland, available from Chromapon, Whittier, CA; or equivalent.
† Rotulex Sovirel, Brechbühler AG or equivalent.
‡ Brechbühler AG or equivalent.
§ Metal Bellows Model MB-21, Sharon, MA, or equivalent.

‖ Swagelok fittings or equivalent.
Brechbühler AG, Chromapon, Inc., or equivalent.

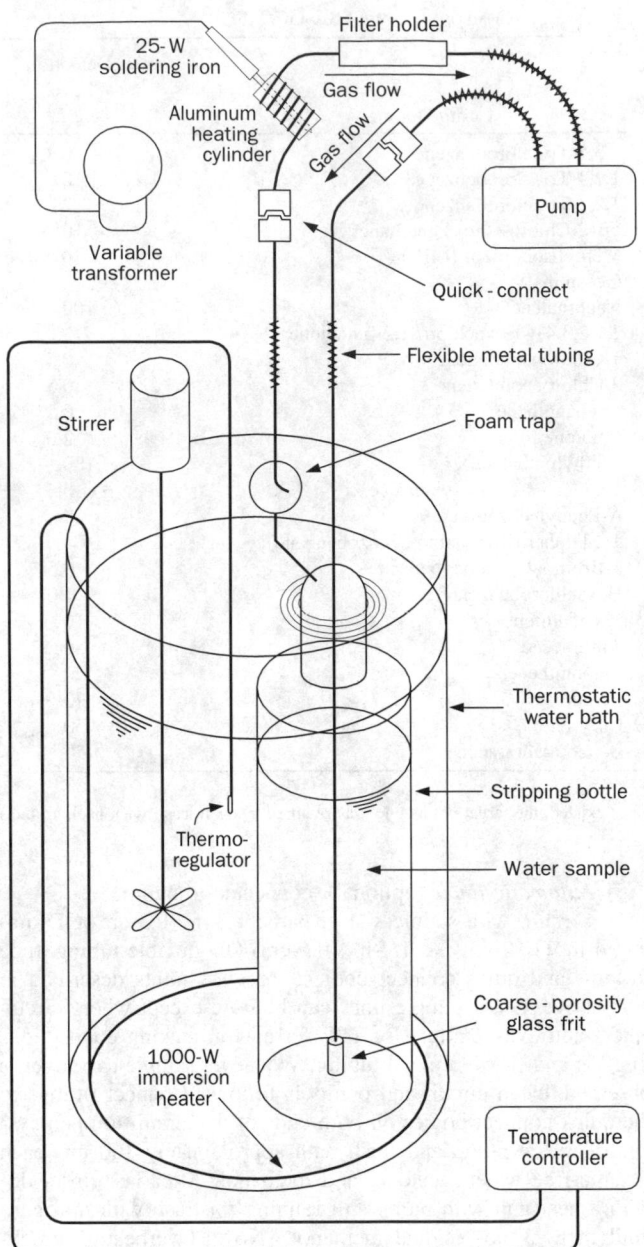

Figure 6040:1. Schematic of closed-loop stripping apparatus (not to scale). Source: KRASNER, S.W., C.J. HWANG & M.J. McGUIRE. 1983. A standard method for quantification of earthy-musty odorants in water. *Water Sci. Technol.* 15(6/7):127.

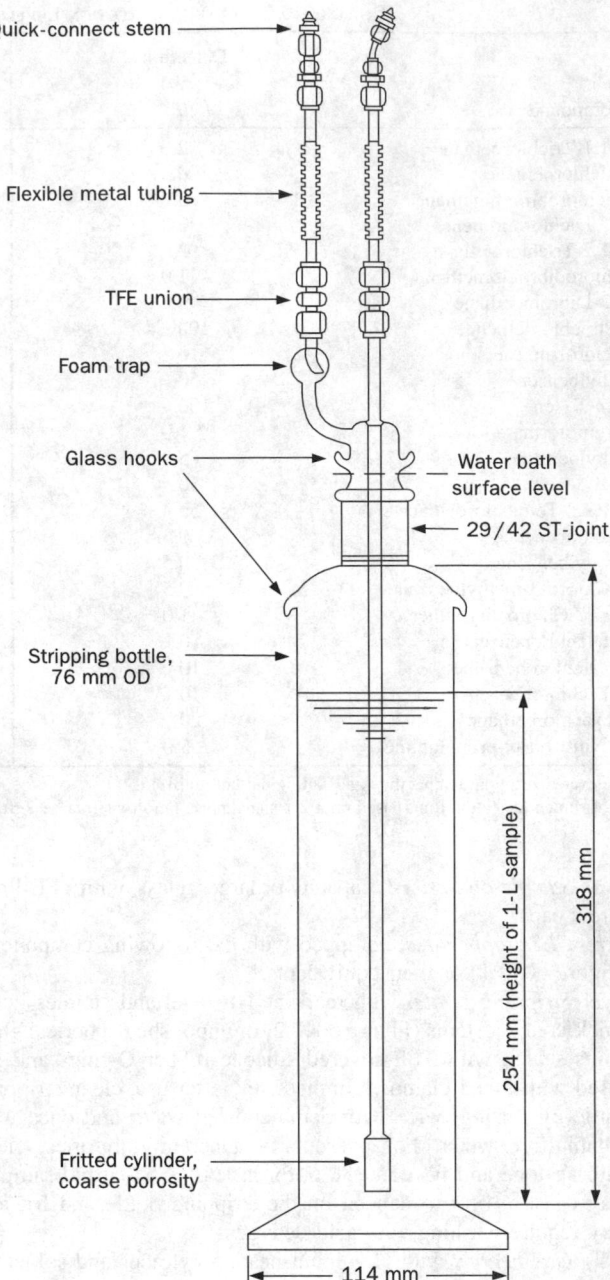

Figure 6040:2. One-liter "tall form" stripping bottle. Source: KRASNER, S.W., C.J. HWANG & M.J. McGUIRE. 1983. A standard method for quantification of earthy-musty odorants in water. *Water Sci. Technol.* 15(6/7):127. (To obtain dimensions in inches, divide dimensions in millimeters by 25.4.)

carbon. Rates for new, commercially prepared filters vary significantly, and decrease with use. Flow rates also depend on the solvent used. Determine optimal flow rate range from analyte recoveries. Preferably, verify filter performance by preparing check standards. Figure 6040:5 shows the reduction in air flow caused by using a "slow" filter. Figure 6040:6 shows the effect of filter resistance on recovery of earthy-musty odorants and one of the internal standards. Clean filter as soon as possible after use. Fill glass tube with organic-free water and let flow through filter. Repeat once with acetone,

twice with elution solvent, and twice with acetone. Measure flow on final acetone rinse. If solvent flow is slow because of salt deposits, pull 1N HNO$_3$ through filter, using a vacuum connection. After acid washing, rinse with distilled water and acetone and continue with cleaning as above. After final rinse, remove residual solvent by connecting filter to a vacuum for approximately 5 min. Clean auxiliary filter after 40 uses or 2 weeks, whichever comes first.

If the salting-out technique is used, Na$_2$SO$_4$ may be carried over and ultimately may clog the filter. The initial water rinse is

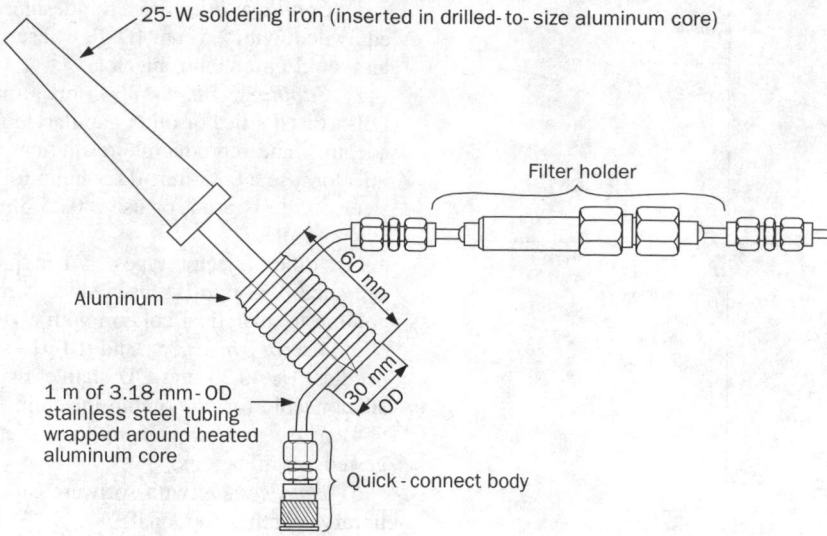

Figure 6040:3. Gas heater.

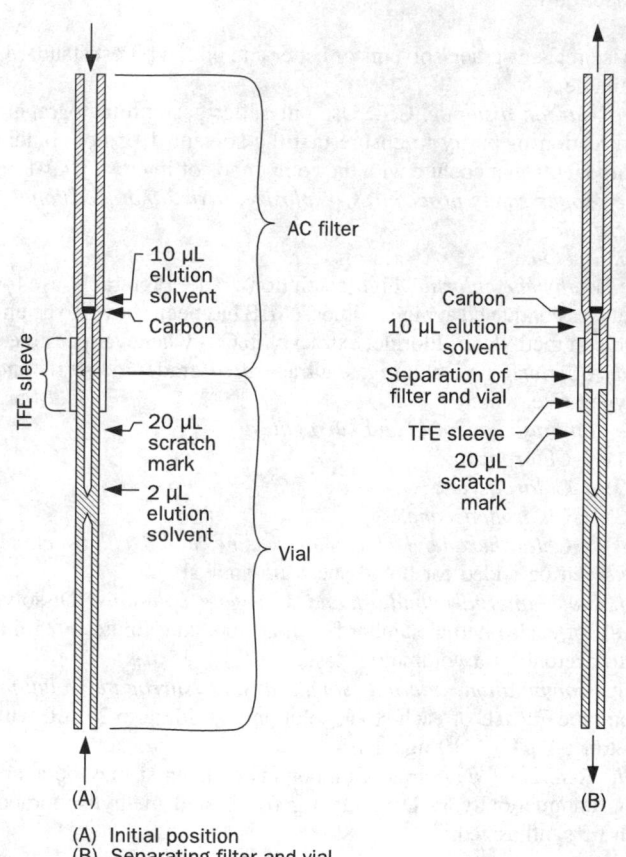

(A) Initial position
(B) Separating filter and vial
(A) Reconnecting

Figure 6040:4. Extraction of filter. Source: KRASNER, S.W., C.J. HWANG & M.J. McGUIRE. 1981. Development of a closed-loop stripping technique for the analysis of taste- and odor-causing substances in drinking water. *In* L.H. Keith, ed. Advances in the Identification and Analysis of Organic Pollutants in Water, Vol. 2. Ann Arbor Science Publishers, Ann Arbor, Mich.

necessary to remove deposited salts and may be avoided if salt is not used.

c. Stirrer (optional), with 5-cm-long TFE stirring bar.

d. Microsyringes, 5-, 10-, 25-, and 50-μL capacity. Use a 25-μL gastight syringe with electrotapered (blunt-end) tip** for transferring extract.

e. Receivers, 50-μL capacity (Figure 6040:4). Receivers can be produced by a custom glass-blowing company: use a 1.6-mm-ID precision-bore capillary glass and grind to 5 mm OD, then

---

** Hamilton Model 1805N electrotapered tip, Reno, NV.

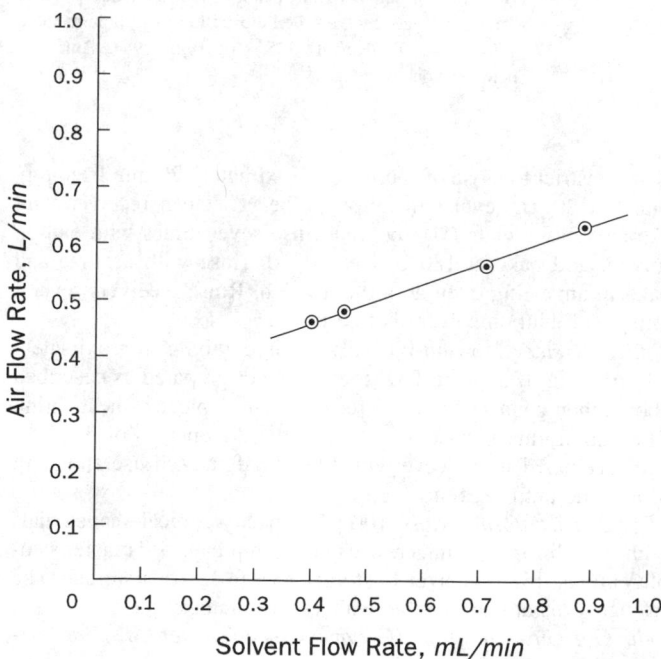

Figure 6040:5. Flow rate through 1.5-mg carbon filter. Air flow rate with no filter is 0.86 L/min.

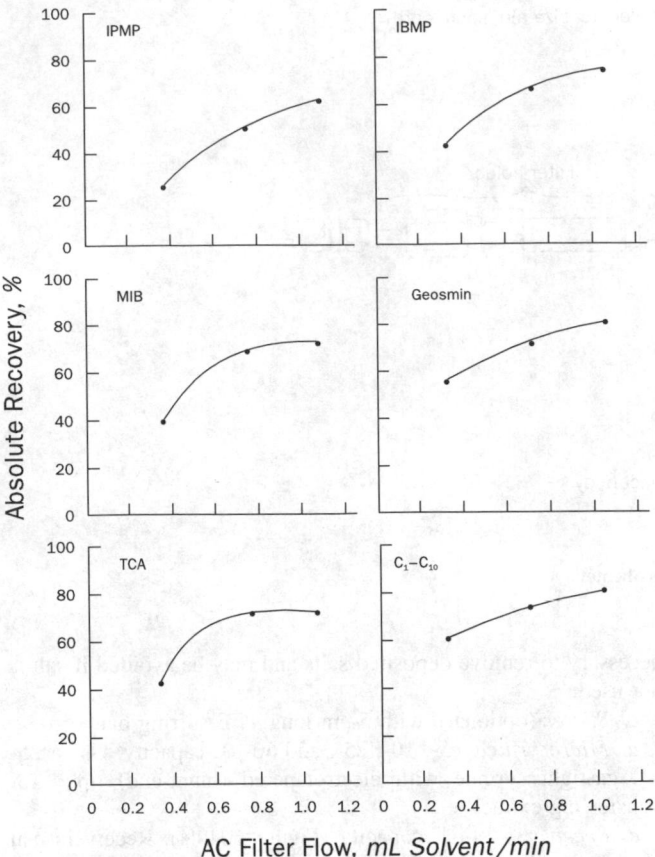

Figure 6040:6. Effect of filter resistance, measured as flow, on recovery of earthy-musty odorants and C₁–C₁₀ internal standard. Reprinted with permission from: HWANG, C.J., S.W. KRASNER, M.J. McGUIRE, M.S. MOYLAN & M.S. DALE. 1984. Determination of subnanogram per liter levels of earthy-musty odorants in water by the salted closed-loop stripping method. *Environ. Sci. Technol.* 18:535. Copyright 1984, American Chemical Society.

heat-constrict to close off bore at approximately 29 mm from top. Mark at 20-μL level with glass scribe. To clean receivers and extract storage vials (¶ *g*, below), rinse seven times with elution solvent and bake at 180°C overnight or rinse with acetone and bake in annealing oven at 400°C for 1 h. Rinse receivers several times with elution solvent before using.

f. TFE sleeve, 5-mm-ID TFE flexible tubing approximately 19 mm long. If a 5-mm-OD receiver is not prepared as described above, then connect filter and receiver with a piece of heat-shrink TFE tubing that is custom-shrunk to the dimensions of the filter and receiver. Rinse sleeve with acetone after each use and store in acetone until ready to use.

g. Extract storage vials, 100-μL capacity conical-shaped vials with TFE-lined septum screw cap or crimp cap.†† Transfer sample extracts from receiver to storage vial for extract storage. The storage vials are compatible with various autosamplers.

h. Gas chromatograph (GC)/mass spectrometer (MS)/data system, equipped with:

†† Pierce Chemical Company #13100 or equivalent.

1) *Capillary injector,* Grob-designed split-splitless injector or equivalent with 2.5-mm-ID glass insert or nonvaporizing, septumless, cold on-column injector.

2) *Capillary column,* 30-m or 60-m × 0.25-mm-ID DB-1 or DB-5 fused silica or other capillary column capable of producing adequate and reproducible resolution. If using a Grob on-column injector, use a 0.32-mm-ID column for injection when a stainless-steel needle is used, or use a 0.25-mm-ID column with a fused-silica needle.

For other injector types, a 1-m precolumn of uncoated, deactivated 0.53-mm-ID fused silica is recommended. Connect precolumn to analytical column with a zero-dead-volume union.

3) *Microsyringes,* 5- and 10-μL capacity, with 75-mm-long needles. Use 0.23-mm-OD stainless-steel or 0.17-mm-OD fused-silica needle for on-column injection.

4) *Mass spectrometer analyzer:* See Section 6200B.2 for suggested specifications.

5) *Data system,* with software capable of performing reverse-library searches (optional).

6) *Autosampler injector* (recommended for improved precision of analysis). When methylene chloride is the elution solvent, manual injection is recommended.

3. Reagents

Use reagent-grade solvents or better and obtain purest standards available.

a. Carbon disulfide, CS₂: Use only after gas chromatographic verification of purity to ensure that the solvent does not contain components that coelute with the compounds of interest. CAUTION: *Use proper safety procedures; explosive, toxic, and occasionally allergenic.*

b. Acetone.

c. Methylene chloride, high-resolution grade. Preferably use for analyses conducted at high altitudes. MIB has been shown to be unstable in methylene chloride; extract with CS₂ whenever possible.[9]

d. Carrier gas: Helium gas, ultrapurified grade, moisture- and oxygen-free.

e. Internal standards and surrogates:

1) *1-Chlorooctane.*

2) *1-Chlorodecane.*

3) *1-Chlorododecane.*

4) *1-Chlorohexane, 1-chlorohexadecane,* and *1-chlorooctadecane* can be added for broad-spectrum analysis.

f. Stock internal standard and surrogate solutions: Dissolve 1 mL of each internal standard‡‡ in acetone and dilute to 25 mL with acetone in a volumetric flask; 1 μL = 35 μg.

g. Combination internal standard and surrogate solution: Combine 7.2 μL of each stock solution and dilute to 25 mL with acetone; 1 μL = 10 ng each.

h. Reference standards: Compounds of interest may be available commercially.§§ Deuterated geosmin and methylisoborneol can be synthesized.[10]

‡‡ 1-Chlorohexadecane and 1-chlorooctadecane solidify upon refrigeration. Warm before removing a portion.
§§ Geosmin and 2-methylisoborneol are available from Wako Bioproducts, Wako Chemicals USA, Inc., 1600 Bellwood Rd., Richmond, VA 23237, or Dalton Chemical Lab, Inc., 4700 Keele St., Suite 119, FARQ, North York, Ontario, Canada M3J 1P3. NOTE: This synthetic geosmin is racemic and includes (+)-geosmin, which has an odor intensity different from that of the natural (−) compound. This precludes its use in quantitative sensory analysis; however, its GC/MS characteristics (i.e., retention time and spectrum) are the same as those of natural geosmin.

i. Stock reference solutions: Dissolve 20 mg of each target compound in acetone and dilute to 10 mL with acetone in a volumetric flask; 1 μL = 2 μg.

j. Combination reference standards solution: Combine 10 μL of each stock reference solution and dilute to 5 mL with acetone; 1 μL = 4 ng each.

k. Organic-free water: Prepare water by treating with activated carbon, mixed-bed deionization, and filtration through a membrane filter.‖‖ Irradiate under ultraviolet light (185 and 254 nm) for 1 h (optional) and prestrip in the CLSA apparatus for 1 h (optional if laboratory blanks are consistently free of interferences), using a clean activated carbon filter or, alternatively, prestrip large quantities of water with nitrogen (ultra-high-purity grade) just before use. Store in a closed bottle tightly capped with TFE-lined screw cap, under nitrogen (optional), in a refrigerator and away from solvent contamination for not longer than 1 week.

l. Sodium sulfate, Na_2SO_4 (optional), granular, anhydrous. Bake at 400°C for 2 h before use; store at room temperature in desiccator.

4. Procedure

a. Sample collection and storage: Rinse sample bottle with sample, fill to overflowing (with no air bubbles), and cap tightly. Collect duplicate samples and in the field keep in an insulated container stocked with ice. In the laboratory store at 4°C, but analyze as soon as possible, preferably within 3 d. For holding longer than 3 d, add 40 mg $HgCl_2$/L to inhibit biological activity. Adding a dechlorinating agent is optional, because disinfection by-products may be affected. CAUTION: *$HgCl_2$-containing samples must be disposed of as hazardous waste. See Section 1090 for precautions.*

b. Treatment of samples:

1) Stripping—Rinse stripping bottle with sample and fill to the 1-L mark, wetting the glass joint with sample. Fill stripping bottle slowly, with minimal aeration, to prevent loss of volatile compounds. Add 10 μL combination internal standard and surrogate solution with the syringe needle tip immersed. Stopper tightly and attach springs. Place in thermostatic water bath at 25°C with glass joint below water level and connect bottle to the circuit. Operate gas heater at 45 to 50°C. Put an "auxiliary" carbon filter in the holder and prestrip for 10 s to flush air contaminants from system. If air quality in the room is demonstrated to be free of interference by analysis of method blank, eliminate prestrip step. Exchange auxiliary filter for a clean one and strip for 2 h (pump flow rate of 1 to 1.5 L/min). Auxiliary filter may be reused many times before cleaning. If stripping bottle has a smaller height-to-diameter ratio than shown in Figure 6040:2, more than 2 h may be required for stripping. Optionally use an automatic timer to terminate each stripping run. Strip time is adjustable, but strip calibration standards and samples under the same conditions.

If sample contains a large amount of algae or turbidity or foaming agents, use only 900 mL sample and 9 μL combination internal standard solution. Because this additional headspace can result in different stripping efficiencies, comparably analyze a calibration standard. Alternatively, dilute concentrated or foaming samples with organic-free water.

2) Alternate stripping techniques—To improve stripping recovery, use a combination of a), b), and c) below to reduce stripping time. Optimize combination, depending on compounds to be analyzed.

a) Elevated stripping temperature—Increase temperature of thermostatic water bath to 45°C to increase recovery of many organic compounds.[5,7] Raise temperature of gas heater to at least 55°C (for a 45°C stripping temperature) to avoid condensation of water vapor on the activated carbon filter. Further increases in stripping temperature reduce recovery.[5]

b) Salting-out technique—Raising the sample ionic strength with Na_2SO_4 before stripping increases the stripping rate of many organic compounds.[3] Bring sample to room temperature before analysis by immersing it in a water bath at 25°C for approximately 15 min. Transfer 800 mL to the stripping bottle and add stirring bar. Using a glass funnel and with the stirring bar (at intermediate setting) in motion, add 72 g Na_2SO_4. Remove funnel and replace with a glass stopper. Continue stirring until Na_2SO_4 has dissolved (not more than 1 min). Remove stopper and stirring bar, then add remaining 100 mL of sample,## rinsing and wetting inside neck of bottle. Add 9 μL combination internal standard solution and strip at 25°C as described in standard stripping procedure above. If analysis is only for MIB and geosmin, a 1-h strip time is adequate. If additional target compounds are being analyzed, verify strip time needed for adequate recoveries of each target compound. Alternatively, combine salt and sample by pouring salt directly into 900 mL sample, stopper tightly, shake vigorously, let stand for several minutes, and add internal standards.

c) Alternative analysis–Use deuterium-labeled geosmin and MIB as internal standards for the determination of geosmin and MIB.[11] Table 6040:III shows a comparison of quantitation and monitoring ions between the 1-chlorodecane internal standard and the MIB-d_3 and geosmin-d_3 internal standards. Variations in stripping or water temperature do not affect the accuracy of the analysis when labeled internal standards are used because MIB and geosmin will strip at the same rate as MIB-d_3 and geosmin-d_3.[11] In addition, labeled standards, if added in the field, will document degradation of the target compound within a 3- to 4-week period. The labeled standards compensate for losses of analyte by physical, chemical, and biological processes during*** sample storage.[11]

3) Extracting the filter—Remove activated carbon filter from holder. In a fume hood, extract with CS_2 as indicated in Figure

---

A total sample volume of 900 mL is preferred to minimize foaming-over due to salt addition.
*** NOTE: Because synthetic labeled geosmin is racemic, best results for the compensation of prolonged biological processes are achieved by monitoring the degradation of natural (−)-geosmin using (−)-geosmin-d_3.

TABLE 6040:III. COMPARISON OF MONITORING AND QUANTITATION IONS FOR CHLORODECANE AND DEUTERATED MIB AND GEOSMIN INTERNAL STANDARDS[11]

Compound	Ions (m/z) Used for Quantitation		Ions (m/z) Monitored	
	A	B	A	B
1-Chlorodecane	91	—	43, 91, 93	—
MIB	95	168	93, 107, 135	168
Geosmin	112	112, 182	111, 112, 125	112, 182
MIB-d_3	—	171	—	171
Geosmin-d_3	—	115, 185	—	115, 185

A = 1-Chlorodecane as internal standard.
B = MIB-d_3 and geosmin-d_3 as internal standard.

‖‖ Millipore Milli-QUV Plus or equivalent.

6040:4. Keep solvents well within the hood to avoid inhalation by analyst or contamination of stripping apparatus. Add 2 μL elution solvent to a clean receiver and connect filter and receiver with a TFE sleeve so as to leave no dead space between glass parts. Place 10 μL elution solvent above the carbon, taking care not to touch carbon with the needle. Warm receiver with hands and alternately pass elution solvent across carbon 10 times. Cool receiver with ice, taking care not to freeze the elution solvent, in order to draw the elution solvent below the carbon. Tap the filter/vial assembly gently on a hard surface to complete transfer of elution solvent to bottom of vial.

Repeat filter extraction with a 10-μL and a 5-μL portion of elution solvent. Separate vial from filter, adjust volume precisely to the 20-μL mark,††† and, using a gastight electrotapered tip syringe, transfer extract to a clean, conical-shaped storage vial. Label and store at −20°C until analysis.

Filter may be extracted while maintaining tight seal between filter and vial during the procedure. Using ice chips, cool closed volume in the vial; solvent accumulates on lower side of the filter disk. Push solvent back to upper side by warming the closed vial between two fingers. Repeat and then extract with more solvent as above.

c. Gas chromatography/mass spectrometry:

1) "Hot-needle" injection technique—To reduce discrimination against higher-boiling compounds by distillation from the needle, use a hot-needle injection technique when the injector is a hot vaporizing type. (Do not use the following procedure for cold on-column injection.) Wet syringe needle and barrel with solvent and expel as much as possible. Pull syringe plunger back, leaving an air gap. Pull up approximately 1.5 μL sample and pull sample totally into syringe barrel. Close the split on the GC injector, wait 10 s, insert syringe needle into the injector, and let needle warm up for 1 to 2 s (optimize time by experience). Rapidly push plunger to bottom of syringe barrel to inject sample. Remove syringe and rinse well with solvent. Open split valve consistently at same time (suggested time 30 s) after the injection.

2) On-column injection technique—To more fully reduce discrimination against higher-boiling compounds, use an on-column injector. A cold on-column injector also can be used to avoid decomposition of thermally labile compounds, e.g., dimethyl polysulfides.[12] Determine thermally labile compounds quantitatively by using a cold on-column injector or an inactive, vaporizing injector.

With an on-column injector, increase sensitivity by injecting large sample volumes (up to 8 μL). To prevent problems from a heavy condensation of solvent with such large-volume injections, use a 2-m retention gap (an empty, deactivated piece of 0.53-mm ID fused silica tubing connected to the head of the column with a zero-dead-volume connector).[13] To preclude backpressure from large-volume injections, inject slowly at about 1 μL/5 s. Keep initial column temperature at 10°C above boiling point of solvent for a full solvent effect and to produce sharp peaks (narrow peak widths).[13] Because the entire injection is deposited directly into the head of the column, the column can develop active sites after as few as 50 to 80 injections. Check activity by injecting a polarity test mixture at least weekly. Breaking approximately 30 cm off the head of the column can restore inertness.

3) Operating conditions for GC/MS—After initial installation of the capillary column, condition it according to the manufac-

turer's instruction. Daily, make a conditioning run with a CS_2 injection or method blank extraction injection before injecting any samples (optional). Typical instrument conditions are given in Table 6040:IV.

d. Calibration standard: The method is semiquantitative for a large number of compounds, but has been shown to be quantitative for many of the compounds listed in this section. Prepare a 20-ng/L target-compound calibration standard by dosing 1 L organic-free water in the stripping bottle with 10 μL combination internal standard solution plus 5 μL combination reference standards solution. (Internal standards concentration is 100 ng/L each.) If the salting-out technique is used, add 72 g Na_2SO_4 to a total volume of 900 mL organic-free water before dosing with 9 μL combination internal standard solution plus 4.5 μL combination reference standards solution. Analyze as directed above. Inject the calibration standard extract, preferably daily, to determine GC/MS response factors and verify spectra.

Verify working linear range by analyzing standards and representative samples with added organics at different concentrations. Calibrate at least every two weeks. Use calibration levels that bracket the levels found in samples. At minimum, use a three-level calibration curve.

e. Blanks: Run a procedural blank daily to assess contamination from reagents, apparatus, and other sources. Run a blank immediately after analyzing any very high-level sample or after installing new parts in the system. Analyze organic-free water with internal standards under the same conditions as samples.

5. Calculations

a. Identification: Identify a compound by matching both retention time and spectra of sample and standard. If they are available, use both a reverse-search computer program with a target-compound library and a forward-search program with the National Institute of Standards and Technology library for tentative identification of other compounds present.

1) Retention times—Use each internal standard to calculate relative retention times for all compounds in the same part of the chromatogram (Table 6040:V). For compounds eluting on the solvent tail, use an early-eluting internal standard (e.g., 1-chloro-

††† Calibration mark needs to be verified periodically.

TABLE 6040:IV. TYPICAL OPERATING CONDITIONS FOR GC/MS ANALYSIS OF CLSA EXTRACTS

Variable	Description or Value
Column	30- or 60-m × 0.25-mm-ID DB-1 or DB-5 fused silica capillary column*
Column temperature program	35°C, 1 min; 35 to 130°C @ 4°C/min; 130 to 220°C @ 10°C/min; 1 min
Carrier gas	Helium
Carrier gas flow rate	1 mL/min
Sample size	About 1.5 μL (splitless injection)
Injector temperature	Cold, on-column
Transfer line temperature	280°C
Ionizer temperature	280°C
Source pressure	About 7×10^4 Pa
Electron energy	70 eV
Mass range scanned	41 to 453 amu
Scan time	1 s

* J&W Scientific, Inc., or equivalent.

TABLE 6040:V. GC/MS DATA FOR THREE INTERNAL STANDARDS AND TWO EARTHY-MUSTY SMELLING COMPOUNDS

Compound	Retention Time* min	Quantification Mass amu	Characteristic Ions (with relative intensities)
1-Chlorooctane	30.8	91	43 (100), 91 (86), 93 (27)
2-Methylisoborneol	36.4	95, 107†	95 (100), 107 (26), 135 (9)
1-Chlorodecane	39.8	91	43 (100), 91 (87), 93 (28)
Geosmin	45.1	112	112 (100), 111 (28), 125 (18)
1-Chlorododecane	47.2	91	43 (100), 91 (61), 93 (19), 85 (12)

* See Table 6040:IV for GC conditions. Data accumulated using 30-m DB-5 capillary column.
† Quantify using two different masses and obtain an average value.

hexane). Sample retention times should match predicted retention times within ± 15 s.

$$\text{Predicted } T_{z,x} = \frac{T_{z,s}}{T_{l,s}} \times T_{l,x}$$

where:

$T_{z,x}$ = retention time of target compound in sample analysis,
$T_{z,s}$ = retention time of target compound in calibration standard analysis,
$T_{l,s}$ = retention time of internal standard in calibration standard analysis, and
$T_{l,x}$ = retention time of internal standard in sample analysis.

2) Spectra—Peaks of at least three characteristic ions should all maximize at the same retention time and have standard intensity ratios (spectra) within ± 20% of those of the calibration-standard compounds. Characteristic ions and their typical relative intensities for three of the internal standards and two earthy-musty-smelling compounds are given in Table 6040:V. Preferably, use reference spectra of 10 to 14 key masses. Determine reference spectra by analysis of standards; verify these frequently. The spectra of MIB are particularly dependent on instrument condition; both 107 and 95 amu have been reported as base peaks (Figure 6040:7). Figure 6040:8 shows the mass spectrum for geosmin.

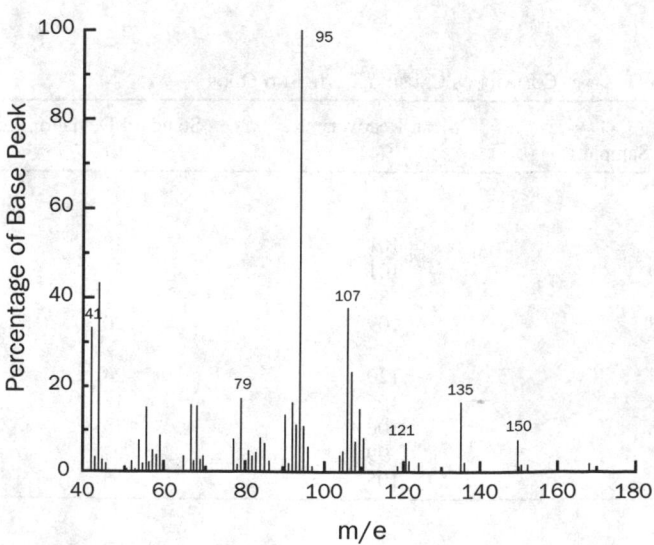

Figure 6040:7. Mass spectrum of 2-methylisoborneol.

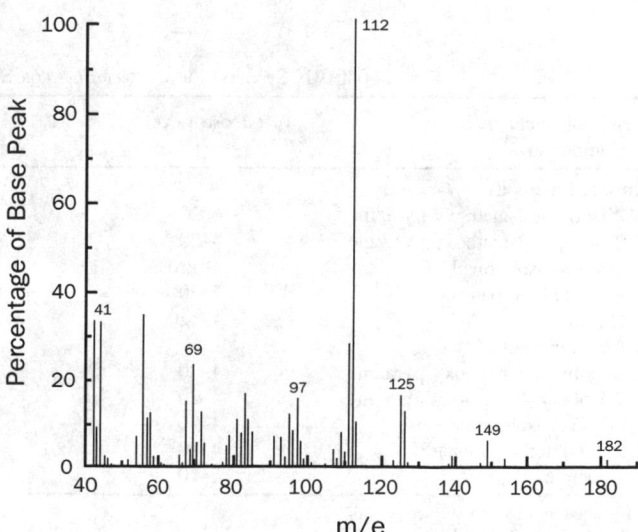

Figure 6040:8. Mass spectrum of geosmin.

b. Quantitation: Determine concentrations by comparison of peak areas of specific quantitation ions. A quantitation ion should be relatively intense in the mass spectrum, yet be free from interference problems caused by closely eluting compounds (see Table 6040:V). Calculate a response factor for each compound from CLSA of a calibration standard as follows:

$$R_z = \frac{A_z \times C_I}{C_z \times A_I}$$

where:

R_z = response factor for target compound z,
A_z = peak area of target compound z,
A_I = peak area of internal standard,
C_I = concentration of internal standard, and
C_z = concentration of target compound z.

Alternatively, a calibration fit can be used.
Compound concentration in the sample (x) is:

$$C_{z,x}, \text{ng/L} = \frac{A_{z,x} \times C_{I,x}}{R_z \times A_{I,x}}$$

where:

$C_{z,x}$ = concentration of target compound in sample,
$A_{z,x}$ = peak area of target compound in sample,
$C_{I,x}$ = concentration of internal standard in sample, and
$A_{I,x}$ = peak area of internal standard in sample.

Use the internal standard 1-chlorodecane for determining response factors. Use other internal standards as a check on the system; calculated values should be to ±20%. Computerized reverse-search spectral matching and automatic quantitation are recommended to improve identification in complex matrices and to facilitate data processing.

Where calibration standards are unavailable, estimate concentrations by comparing the total ion current of the compounds to that of the internal standard 1-chlorodecane.

6. Quality Assurance/Quality Control

The CLSA method is semiquantitative for some compounds because of the variability of stripping efficiencies. However, quantitative data are obtainable for compounds that are reproducibly stripped (e.g., MIB and geosmin).[1-3,7] Follow general quality assurance/quality control requirements (e.g., calibration, initial quality control, and batch quality control) described in Section 6020.

Analyze a replicate sample at least once per 10 samples to check precision. If unusual or unexpected results are obtained, analyze a replicate to confirm. Typically, single-analyst determinations for a relatively simple matrix have a coefficient of variation less than or equal to 10%. Otherwise, precision is usually within 20%. For compounds that are poorly stripped, a higher coefficient of variation may be obtained.

Analyze a sample with a known addition at least once per 10 samples to check accuracy and recovery. If matrix problems exist, this will confirm the accuracy of results. Adjust these recoveries against the calibration standards results. Even when absolute recoveries are less than 50%, standard adjusted recoveries, which correct for stripping efficiencies, are usually between 80 and 120%.

Internal standard response should equal ± 40% of the daily standard. An unacceptable internal standard response requires extract reinjection. If the reinjection is still unacceptable, investigate cause, restrip sample, and reanalyze. If consecutive samples fail the internal standard acceptance criterion, immediately analyze a calibration standard. If the calibration standard internal standard response also is unacceptable, recalibrate the instrument.

Ideally prepare and analyze an intralaboratory check sample monthly. Prepare from an independent stock solution of the standards.

7. Precision and Bias

Precision and bias data are given in Tables 6040:VI and VII for the analysis of earthy-musty-smelling compounds. Table 6040:VIII shows recovery and precision data for selected pollutants.

TABLE 6040:VI. SINGLE-LABORATORY BIAS FOR SELECTED ORGANIC COMPOUNDS CAUSING TASTE AND ODOR

Stripping Technique Compound	Dose Level ng/L	Number of Samples*	Mean Recovery† %	Standard Deviation %
Unsalted method‡				
2-Isobutyl-3-methoxy pyrazine	4–20	23	89	21
2-Isobutyl-3-methoxy pyrazine	4–20	22	101	28
2-Methylisoborneol	4–120	30	101	21
2,3,6-Trichloroanisole	4–20	22	88	27
Geosmin	4–120	28	109	20
Salting-out method‡				
2-Isobutyl-3-methoxy pyrazine	4–20	44	120	24
2-Isobutyl-3-methoxy pyrazine	4–20	48	106	18
2-Methylisoborneol	4–20	48	106	15
2,3,6-Trichloroanisole	4–20	45	99	22
Geosmin	4–20	48	105	15

* Finished and natural surface waters.
† Standard-adjusted recovery.
‡ Stripping at 25°C.
Reprinted with permission from *Environmental Science and Technology*, 18:535.[3] Copyright 1984. American Chemical Society.

TABLE 6040:VII. PRECISION DATA FOR SELECTED ORGANIC COMPOUNDS CAUSING TASTE AND ODOR*

Compound	Dose Level ng/L	Multiple Laboratories†			Single Laboratory‡		
		Mean ng/L	Standard Deviation ng/L	Coefficient of Variation %	Mean ng/L	Standard Deviation ng/L	Coefficient of Variation %
Sample A§							
2-Isopropyl-3-methoxy pyrazine	5.9	5.6	1.6	28	6.6	0.6	9
2-Isobutyl-3-methoxy pyrazine	3.0	3.0	0.7	24	3.2	0.3	11
2-Methylisoborneol	4.3	4.8	1.0	20	5.1	0.1	2
2,3,6-Trichloroanisole	8.7	7.3	3.1	43	9.3	2.0	22
Geosmin	2.9	3.3	0.9	27	3.6	0.5	14
Sample B§							
2-Isopropyl-3-methoxy pyrazine	25	22	8.0	36	27	2.8	10
2-Isobutyl-3-methoxy pyrazine	15	14	4.2	30	17	2.5	14
2-Methylisoborneol	20	18	6.2	34	23	2.0	9
2,3,6-Trichloroanisole	35	32	9.7	30	37	3.3	9
Geosmin	16	16	5.9	37	20	2.8	14

* Stripping at 25°C, unsalted method.
† Five analysts at three laboratories.
‡ Three analysts at one laboratory.
§ Organic-free water dosed with taste and odor compounds.

TABLE 6040:VIII. RECOVERY AND PRECISION DATA FOR SELECTED PRIORITY POLLUTANTS*

Compound	Amount ng	Mean Recovered Amount† ng	Range	Recovery Efficiency %	RSD
Thiophene	25	9	7–11	35	16
Dibromochloromethane	29	17	13–21	57	13
Styrene	22	17	16–20	80	7
Isopropylbenzene	24	26	24–29	107	8
2-Chlorotoluene	26	23	22–27	90	8
bis(2-Chloroethyl)ether	24	3	3–4	12	11
α-Methylstyrene	22	19	18–22	90	6
1,4-Dichlorobenzene	20	18	16–21	93	8
2-Ethyl-1,3-dimethylbenzene	24	22	21–25	92	6
4-Chloro-o-xylene	26	22	20–26	85	9
1,1-Dimethylindan	22	24	22–26	110	7
p-Methylphenol	27	ND‡			
Tetrahydronaphthalene	23	23	20–26	99	10
1,2,4-Trichlorobenzene	23	19	18–21	83	7
Hexachloro-1,3-butadiene	27	31	28–34	114	9
2-Methylbiphenyl	24	25	22–27	101	8
1,6-Dimethylnaphthalene	24	10	8–11	39	12
2-Isopropylnaphthalene	23	21	19–22	91	6
Pentachlorobenzene	26	12	11–14	48	11
Hexachlorobenzene	20	6	5–7	31	13
2,2',4,4'6,6'-Hexachlorobiphenyl	27	28	26–32	104	7
2,2',4,5,5'-Pentachlorobiphenyl	28	23	18–28	82	14

* Stripping at 40°C, unsalted method.
† Based on six purging analyses using single ion quantification.
‡ ND = not detected.
Reprinted with permission from *Environmental Science and Technology*, 17:571.[4] Copyright 1983. American Chemical Society.

8. References

1. KRASNER, S.W., C.J. HWANG & M.J. McGUIRE. 1981. Development of a closed-loop stripping technique for the analysis of taste- and odor-causing substances in drinking water. *In* L.H. Keith, ed. Advances in the Identification and Analysis of Organic Pollutants in Water, Vol. 2. Ann Arbor Science Publishers, Ann Arbor, Mich.
2. KRASNER, S.W., C.J. HWANG & M.J. McGUIRE. 1983. A standard method for quantification of earthy-musty odorants in water. *Water Sci. Technol.* 15(6/7):127.
3. HWANG, C.J., S.W. KRASNER, M.J. McGUIRE, M.S. MOYLAN & M.S. DALE. 1984. Determination of subnanogram per liter levels of earthy-musty odorants in water by the salted closed-loop stripping method. *Environ. Sci Technol.* 18:535.
4. COLEMAN, W.E., J.W. MUNCH, R.W. SLATER, R.G. MELTON & F.C. KOPFLER. 1983. Optimization of purging efficiency and quantification of organic contaminants from water using a 1-L closed-loop-stripping apparatus and computerized capillary column GC/MS. *Environ. Sci. Technol.* 17:571.
5. THOMASON, M.M. & W. BERTSCH. 1983. Evaluation of sampling methods for the determination of trace organics in water. *J. Chromatogr.* 279:383.
6. JONES, G.J. & W. KORTH. 1995. In situ production of volatile odor compounds by river and reservoir phytoplankton populations in Australia. *Water Sci. Technol.* 31(11):145.
7. MALLEVIALLE, J. & I.H. SUFFET, eds. 1987. Identification and Treatment of Tastes and Odors in Drinking Water. American Water Works Association & American Water Works Association Research Foundation, Denver, Colo.
8. Report of Analysis for Semivolatile Organics by Closed-Loop Stripping GC/MS. 1982. Environmental Research Lab., James M. Montgomery, Consulting Engineers, Inc., Pasadena, Calif.
9. KORTH, W., J. ELLIS & K. BOWMER. 1992. The stability of geosmin and MIB and their deuterated analogues in surface waters and organic solvents. *Water Sci. Technol.* 25(2):115.
10. KORTH, W., J. ELLIS & K. BOWMER. 1991. Synthesis of deuterium labelled geosmin and methylisoborneol. *J. Labelled Compounds Radiopharm.* 29:823.
11. KORTH, W., K. BOWMER & J. ELLIS. 1991. New standards for the determination of geosmin and methylisoborneol in water by gas chromatography/mass spectrometry. *Water Res.* 25:319.
12. WAJON, J.E., R. ALEXANDER & R.I. KAGI. 1985. Determination of trace levels of dimethyl polysulphides by capillary gas chromatography. *J. Chromatogr.* 319:187.
13. GROB, K. 1978. On-column injection onto capillary columns. Part 2: Study of sampling conditions; practical recommendations. *J. High Resolution Chromatogr. Chromatogr. Commun.* 1:263.

9. Bibliography

BUDDE, W.L. & J.W. EICHELBERGER. 1979. Organic Analysis Using Gas Chromatography- Mass Spectrometry. Ann Arbor Science Publishers, Ann Arbor, Mich.

GROB, K. & K. GROB, JR. 1978. Splitless injection and the solvent effect. *J. High Resolution Chromatogr. Chromatogr. Commun.* 1:57.

GROB, K. & G. GROB. 1979. Practical gas chromatography—a systematic approach. *J. High Resolution Chromatogr. Chromatogr. Commun.* 2:109.

REINHARD, M., J.E. SCHREINER, T. EVERHART & J. GRAYDON. 1987. Specific analysis of trace organics in water using gas chromatography and mass spectroscopy. *J. Environ. Pathol., Toxicol. Oncol.* 7:417.

6040 C.   Purge and Trap Technique

For applications of this technique to analyses for volatile organics, volatile aromatic organics, and volatile halocarbons, see Section 6200.

6200   VOLATILE ORGANIC COMPOUNDS*

6200 A.   Introduction

1. Source and Significance

Many organic compounds have been detected in ground and surface waters. While most groundwater contamination episodes are traceable to leaking underground fuel or solvent storage vessels, landfills, agriculture practices, and wastewater disposal, the most probable cause for contamination of some aquifers and surface waters has never been firmly established. Contamination may be due to past practices of on-site (leach field) disposal of domestic and industrial wastes or to illegal discharges. Organohalides, particularly the trihalomethanes, are present in most chlorinated water systems, especially those using surface waters as a source of supply. Toxicological studies on animal models have shown that some of these organics have the potential for terato-

* Approved by Standard Methods Committee, 1997.

genesis or carcinogenesis in human beings. To minimize these health risks, sensitive detection and accurate and reproducible quantitation of organics is of paramount importance.

2. Selection of Method

Two capillary gas chromatographic methods for purgeable organic compounds are presented. The scope of analytes is detector-dependent. Method B is a gas chromatographic/ mass spectrometric (GC/MS) technique. Method C combines GC with photoionization detection (PID)/electrolytic conductivity detection (ELCD) in series. Methods B and C are applicable to a wide range of purgeable organics. Both methods can be applied to finished drinking water, drinking water in any stage of treatment, source water, or wastewater.

The methods presented are highly sophisticated micro-analytical procedures that should be used only by analysts experienced in chromatography and data evaluation and interpretation. While the methods are similar, they are not interchangeable from a regulatory point of view.[1,2]

3. Scope

Table 6200:I lists the compounds that can be determined by these methods. All are determinable by both Method B and Method C. Other compounds may be amenable to these methods.

4. Sampling and Storage

See Section 6010B.1

5. Method-Required Quality Control Criteria

a. Initial quality control:

1) Initial demonstration of capability—Conduct initial demonstration of capability study at least once, before analysis of any sample, by each analyst, to demonstrate proficiency with the method of choice. Include at least analysis of a reagent blank and four reagent blank samples fortified at a concentration between 10 times the minimum reporting level and the midpoint of the calibration curve. The blank should not contain any compound of interest at a concentration greater than minimum reporting level. Mean percent recovery for each compound calculated from the four fortified samples should be 80% to 120%, and the relative standard deviation (RSD) should be <20%.

2) Method detection level (MDL)—The MDL is a statistical determination of the minimum concentration that can be measured by the method with a confidence level of 99% that the analyte concentration is greater than zero. Determine MDL before any samples are analyzed, using the procedure described in Section 1030 or other appropriate procedure[3] as required for each matrix to be analyzed. For MDL calculation, start with a concentration about five times the estimated instrument detection level. Perform

TABLE 6200:I. COMPOUNDS DETERMINABLE BY GAS CHROMATOGRAPHIC METHODS FOR PURGEABLE ORGANIC COMPOUNDS

Analyte	Chemical Abstract Services Registry Number	Analyte	Chemical Abstract Services Registry Number
Benzene	71-43-2	2,2-Dichloropropane	590-20-7
Bromobenzene*	108-86-1	1,1-Dichloropropene*	563-58-6
Bromochloromethane	74-97-5	cis-1,3-Dichloropropene*	10061-01-5
Bromodichloromethane	75-27-4	trans-1,3-Dichloropropene*	10061-02-6
Bromoform	75-25-2	Ethylbenzene*	100-41-4
Bromomethane	74-83-9	Hexachlorobutadiene*	87-68-3
n-Butylbenzene*	104-51-8	Isopropylbenzene*	98-82-8
sec-Butylbenzene*	135-98-8	p-Isopropyltoluene*	99-87-6
tert-Butylbenzene*	98-06-6	Methyl t-butyl ether*	1634-04-4
Carbon tetrachloride	56-23-5	Methylene chloride	75-09-2
Chlorobenzene*	108-90-7	Naphthalene*	91-20-3
Chloroethane	75-00-3	n-Propylbenzene*	103-65-1
Chloroform	67-66-3	Styrene*	100-42-5
Chloromethane	74-87-3	1,1,1,2-Tetrachloroethane	630-20-6
2-Chlorotoluene*	95-49-8	1,1,2,2-Tetrachloroethane	79-34-5
4-Chlorotoluene*	106-43-4	Tetrachloroethene*	127-18-4
Dibromochloromethane	124-48-1	Toluene*	108-88-3
1,2-Dibromo-3-chloropropane	96-12-8	1,2,3-Trichlorobenzene*	87-61-6
1,2-Dibromoethane	106-93-4	1,2,4-Trichlorobenzene*	120-82-1
Dibromomethane	74-95-3	1,1,1-Trichloroethane	71-55-6
1,2-Dichlorobenzene*	95-50-1	1,1,2-Trichloroethane	79-00-5
1,3-Dichlorobenzene*	541-73-1	Trichloroethene*	79-01-6
1,4-Dichlorobenzene*	106-46-7	Trichlorofluoromethane	75-69-4
Dichlorodifluoromethane	75-71-8	1,2,3-Trichloropropane	96-18-4
1,1-Dichloroethane	75-34-3	1,2,4-Trimethylbenzene*	95-63-6
1,2-Dichloroethane	107-06-2	1,3,5-Trimethylbenzene*	108-67-8
1,1-Dichloroethene*	75-35-4	Vinyl chloride*	75-01-4
cis-1,2-Dichloroethene*	156-59-4	o-Xylene*	95-47-6
trans-1,2-Dichloroethene*	156-60-5	m-Xylene*	108-38-3
1,2-Dichloropropane	78-87-5	p-Xylene*	106-42-3
1,3-Dichloropropane	142-28-9		

* Compound can be determined using Method 6200C with PID only.

MDL determination as an iterative process. The values listed in Table 6200:III were generated using a concentration of 0.5 µg/L. Conduct MDL determination at least annually. Analyze samples for MDL determination over a 3- to 5-d period to generate a more realistic value.

3) *Quality-control sample*—Analyze an externally generated quality-control sample as a laboratory fortified blank at least quarterly or whenever new stock solutions are generated. Obtain this sample from sources external to the immediate laboratory, and use it to validate the laboratory's standards both qualitatively and quantitatively. Acceptance criteria are supplied by the manufacturer. If all criteria are not met, determine cause of error, and correct it before continuing.

4) *Minimum quantitation level (MQL)*—The MQL is the lowest level that can be quantified accurately. The MQL is defined as four times the MDL.

b. Calibration:

1) *Initial calibration*—Perform initial calibration with a minimum of five concentrations of analytical calibration standards (CALs) for the compound(s) of interest. The lowest concentration should be at the working reporting level; the highest concentration should be at the upper end of the calibration range. Do not report values that are outside of the defined calibration range. For the calibration concentrations, there should be no more than one order of magnitude between concentrations.

Use any of the following calibration functions, as appropriate: response factor for internal standard calibration, calibration factor for external standard calibration, or calibration curve. Calibration curves may be linear through the origin, linear not through the origin, or quadratic through or not through the origin. Use the following recommended acceptance criteria for the various calibration functions.

If using response factors or calibration factors, relative standard deviation (RSD) for each compound of interest should be less than 20%. If the RSD is not less than 20% for any compound of interest, then identify and correct source of lack of linearity before sample quantitation. When using response factors (i.e., for GC/MS analysis), check performance or sensitivity of the instrument for the compound of interest against minimum acceptance values for the response factors. See specific analytical method for the acceptance criteria for the response factors for each compound.

For a linear regression, the correlation coefficient should be >0.994. Recalculate each calibration point compared to curve. Resulting values should be within ±20%. If any of the recalculated values are not within ±20%, identify and correct source of outlier(s) before sample quantitation.

Use initial calibration, with any of the above functions (response factor, calibration factor, or calibration curve) for quantitation of the analytes of interest in samples. Use continuing calibration, described in ¶ 2) below, only for checks on initial calibration and not for sample quantitation. Perform initial calibration when instrument is set up and whenever continuing calibration criteria are not met.

2) *Continuing calibration*—Continuing calibration (CCAL) is the periodic analysis of a calibration standard used to verify that the instrument response has not changed significantly from the initial calibration. Perform continuing calibration every 10 samples for GC analysis, every 20 samples for GC/MS analysis, or every 12 h, whichever is more frequent. Perform continuing calibration with one or more of the concentrations of analytical standards in the initial calibration. Vary actual concentration of continuing calibration standard over calibration range, with a minimum concentration greater than two times the reporting limit. The acceptance criterion for continuing calibration is 70% to 130% recovery compared to the known or expected value of the calibration standard (at the analyst's discretion, the acceptance criterion for the gases may be extended to 60% to 140% recovery). If the acceptance criteria are not met, re-analyze continuing calibration standard or repeat initial calibration. When using response factors, check performance or sensitivity of instrument for analytes of interest against minimum acceptance values for response factors.

3) *Closing standard*—Finish all sample sets with a closing standard to demonstrate that performance was still acceptable for the last sample analyzed. Use acceptance criteria as for the CCAL.

c. Batch quality control:

1) *Analytical day*—An analytical day is defined as a 12-h analytical period.

2) *Sample set (batch)*—A sample set (batch) is defined as those samples extracted in an analytical day, not to exceed 20 samples.

3) *Laboratory reagent blank (LRB)*—A LRB is a blank sample consisting of all reagents that normally contact a sample when carried through the entire analytical procedure. Use reagent blank to determine contribution of reagents and preparative analytical steps to observed value. No compound of interest should be present in reagent blank at a level greater than the MQL. Include a minimum of one reagent blank with each sample set (batch).

4) *Laboratory-fortified blank (LFB)*—See ¶ *b*2) above. NOTE: For this method the LFB and CCAL are the same.

5) *Internal standard (IS)*—An internal standard is a compound of known concentration added to each standard and sample just before sample analysis. Because of the nature of purge and trap analysis, the IS is taken through the entire analytical process, just as is the surrogate standard [see ¶ 6) below]. However, the IS is used for quantitation, whereas the surrogate standard is used to monitor ongoing purge recovery. Use IS to monitor retention time, relative response, and concentration of analytes in each sample. When quantifying by the internal standard method, measure all compound responses relative to this standard. Internal standard area response should be in the range of ±30% compared to the mean calibration curve area response. The IS compound should mimic the chromatographic conditions of the analytes of interest. The retention time of this compound should separate from all analytes of interest and elute in a representative area of the chromatogram. If a specific compound cannot be found to meet these criteria, use additional compounds to satisfy analytical needs.

6) *Surrogate standard (SS)*—A surrogate standard is a compound added to each standard and sample at a known concentration before extraction. Choose a compound(s) that is chemically similar to the analytes and that is unlikely to be found in environmental samples. Carry surrogate standard through entire sample extraction and analytical process to monitor extraction recovery for each sample. Surrogate recovery should remain reasonably constant over time. Recovery should not vary more than 30% from the known value. Refer to method of choice for specific surrogates.

7) *Laboratory-fortified sample (LFS)*—A LFS is an additional portion of a sample to which the analytes of interest have been added at a concentration at least two times the MRL or around the middle of the calibration range. Include a minimum of one LFS with each sample set (batch). Make LFSs at sufficient concentrations that sample background levels do not adversely affect

recovery calculations. (If this is a known sample, adjust addition concentrations to be about five times background level). Base sample batch acceptance on results of CCALs and LFBs rather than on LFSs, because the matrix of the sample may interfere with method performance. Prepare fortifying solution for blanks and samples from a different primary mix than that used to develop working standard mix.

8) *LFS duplicates*—A LFS duplicate is a second LFS used to evaluate the precision of the method in a matrix sample. If sufficient sample volume is collected, fortify a large enough volume to yield two sample portions for analysis. If sufficient sample volume is not collected, use a second bottle of the same sample fortified to the same concentration as the first. Include a minimum of one LFS duplicate with each sample set (batch). Compare precision and bias to those listed in the method. Base sample batch acceptance on results of CCAL and LFB additions rather than LFS duplicates.

6. References

1. U.S. ENVIRONMENTAL PROTECTION AGENCY. 1987. National Primary Drinking Water Regulations—synthetic organic chemicals; monitoring for unregulated contaminants; final rule. 40 CFR 141 & 142; *Federal Register* 52, No. 130.
2. U.S. ENVIRONMENTAL PROTECTION AGENCY. 1986. Guidelines establishing test procedures for the analysis of pollutants under the Clean Water Act. 40 CFR Part 136; *Federal Register* 51, No. 125.
3. U.S. ENVIRONMENTAL PROTECTION AGENCY. 1995. Definition and procedure for the determination of the method detection limit, revision 1.11. 40 CFR Part 136, Appendix B.

6200 B.   Purge and Trap Capillary-Column Gas Chromatographic/Mass Spectrometric Method

This method[1] is applicable to the determination of a wide range of purgeable organic compounds (see Table 6200:I). The method can be extended to include other volatile organic compounds, provided that all performance criteria are met. It should be used only by analysts experienced in the operation of GC/MS systems and in evaluation and interpretation of mass spectra.

1. General Discussion

a. Principle: Volatile organic compounds are transferred efficiently from the aqueous to the gaseous phase by bubbling an inert gas (e.g., helium) through a water sample contained in a specially designed purging chamber at ambient temperature. The vapor is swept through a sorbent trap that adsorbs the analytes of interest. After purging is complete, the trap is heated and backflushed with the same inert gas to desorb the compounds onto a gas chromatographic column. The gas chromatograph is temperature-programmed to separate the compounds. The detector is a mass spectrometer. See Section 6010C for discussion of gas chromatographic and mass spectrometric principles.

b. Interferences: Impurities in the purge gas and organic compounds outgassing from the plumbing upstream of the trap account for most contamination problems. Demonstrate that the system is free from contamination under operational conditions by analyzing laboratory reagent blanks daily. (NOTE: Use blanks for monitoring only; corrections for blank values are unacceptable.) Avoid using non-TFE plastic tubing, non-TFE thread sealants, or flow controllers with rubber components in the purge and trap system. Ensure that the analytical area is not subject to contamination from laboratory solvents, particularly methylene chloride and methyl *tert*-butyl ether (MtBE).

Samples can be contaminated by diffusion of volatile organics (particularly fluorocarbons and methylene chloride) through the septum seal during shipment and storage. Use a field reagent blank prepared from reagent water and carried through the sampling, handling, and shipping procedures as a check on such contamination.

Contamination by carryover can occur whenever high-level and low-level concentration samples are analyzed sequentially. To reduce carryover, rinse purging device and sample syringe with reagent water between samples. Follow analysis of an unusually high concentration sample with a LRB to check for carryover contamination. For samples containing large amounts of water-soluble materials, suspended solids, high boiling compounds, or high levels of volatile compounds, wash purging device with a detergent solution, rinse it with distilled water, and dry it in an oven at 105°C between analyses. The trap and other parts of the system also are subject to contamination; therefore, frequently bake and purge entire system.

c. Detection levels: Method detection levels (MDLs) are compound-dependent and vary with purging efficiency and instrument response. In a single laboratory using reagent water and known-addition concentrations of 0.5 μg/L, observed MDLs were in the range of 0.025 to 0.450 μg/L. The applicable calibration range of this method is compound- and instrument-dependent, but is approximately 0.2 to 200 μg/L. Compounds that are inefficiently purged from water will not be detected when present at low concentrations, but they can be measured with acceptable bias and precision when present at sufficient concentration. Determination of some geometrical isomers (e.g., xylenes) may be hampered by co-elution.

d. Safety: The toxicity or carcinogenicity of each analyte has not been precisely defined. Benzene, carbon tetrachloride, bis(1-chloroisopropyl)ether, 1,4-dichlorobenzene, 1,2-dichloroethane, hexachlorobutadiene, 1,1,2,2-tetrachloroethane, 1,1,2-trichloroethane, chloroform, 1,2-dibromoethane, tetrachloroethene, trichloroethene, and vinyl chloride have been classified tentatively as known or suspect carcinogens. Handle pure standard materials and stock standard solutions of these compounds in a hood and wear a NIOSH/MESA-approved toxic gas respirator when handling high concentrations.

2. Apparatus

a. Purge and trap system: The purge and trap system consists of purging device, trap, and desorber. Several complete systems are available commercially.

1) *Purging device,* designed to accept 25-mL samples with a water column at least 5 cm deep. A smaller 5-mL purging device is acceptable if required method detection levels and performance criteria are met. Keep gaseous headspace between water column and trap to a total volume of less than 15 mL. Pass purge gas through water column as finely divided bubbles with a diameter of less than 3 mm at the origin. Introduce purge gas no more than 5 mm from base of water column. The purging device illustrated in Figure 6200:1 meets these criteria.

Needle spargers may be used instead of the glass frit shown in Figure 6200:1; however, in either case, introduce purge gas at a point <5 mm from base of water column.

2) *Trap,* at least 25 cm long and with an inside diameter of at least 3 mm, packed with the following minimum lengths of adsorbents: 1.0 cm methyl silicone coated packing, 7.7 cm 2,6-diphenylene oxide polymer, 7.7 cm silica gel, and 7.7 cm coconut charcoal. If analysis is not to be made for dichlorodifluoromethane, the charcoal may be eliminated and the polymer section lengthened to 15 cm. Alternative sorbents may be used provided that all quality control criteria are met. Various sorbent traps are available commercially**; ensure that trap keeps total purge gas volume and purge time constant (i.e., 40 mL/min for 11 min) and that performance will meet all quality control criteria. The minimum specifications for the trap are illustrated in Figure 6200:2.

Methyl silicone coated packing is recommended, but not mandatory. The packing protects the diphenylene oxide polymer adsorbent from aerosols, re-coating any active site that may develop during the heating process, and ensures that the polymer is fully enclosed within the heated zone of the trap, thus eliminating potential cold spots. Alternatively, silanized glass wool may be used as a spacer in the trap inlet.

Before initial use, condition trap overnight following manufacturer's instructions. Vent trap effluent to the room, not to analyt-

---

** Tekmar VOCARB 4000 or equivalent.

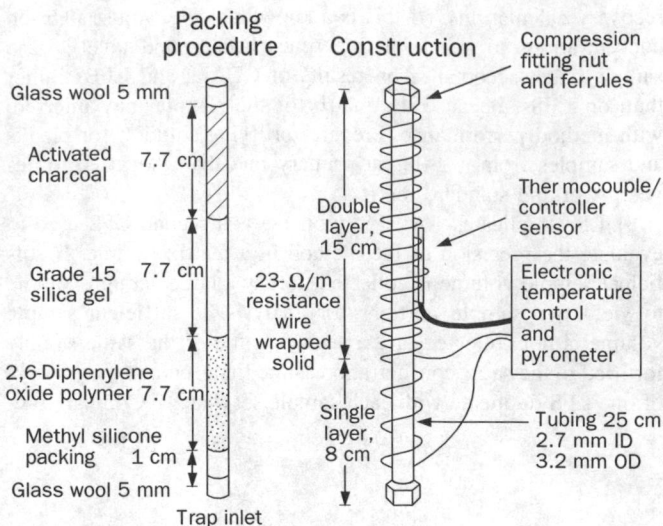

Figure 6200:2. **Trap packings and construction to include desorb capability.**

ical column. Before daily use, condition trap for 10 min with back-flushing. Optimally, vent trap to analytical column during daily conditioning; however, run column through temperature program before sample analysis.

b. Gas chromatograph (GC)†: Use a temperature-programmable GC, suitable for on-column injection. Deactivate all glass components (e.g., injector liners) with a silanizing agent.

c. Capillary GC columns: Use any capillary GC column that meets all performance criteria. Ensure that desorb flow rate is compatible with the column of choice. Four examples of acceptable columns are listed below.

1) *Column 1:* 60-m-long × 0.75-mm-ID VOCOL‡ wide-bore capillary column with 1.5-μm film thickness.

2) *Column 2:* 30-m-long × 0.53-mm-ID DB-624§ mega-bore capillary column with 3-μm film thickness.

3) *Column 3:* 30-m-long × 0.32-mm-ID DB-5§ capillary column with 1-μm film thickness.

4) *Column 4:* 30-m-long × 0.25-mm-ID DB-624§ capillary column with 1.4-μm film thickness.

d. Mass spectrometer, capable of scanning from 35 to 300 amu every 2 s or less, utilizing 70 eV (nominal) electron energy in the electron impact ionization mode, and producing a mass spectrum that meets all criteria in Table 6200:II when 25 ng or less of 4-bromofluorobenzene is introduced into GC. To ensure sufficient precision, the desired scan rate permits acquisition of at least five spectra while a sample component elutes from the GC.

e. Purge and trap — GC/MS interface: Use an open-split or direct-split interface, depending on which column is used. Alternatively, if the narrow-bore column (4) is used, a capillary concentrator preceding the GC may be necessary. This interface condenses desorbed materials onto an uncoated fused silica pre-column and when flash-heated transfers compounds onto the

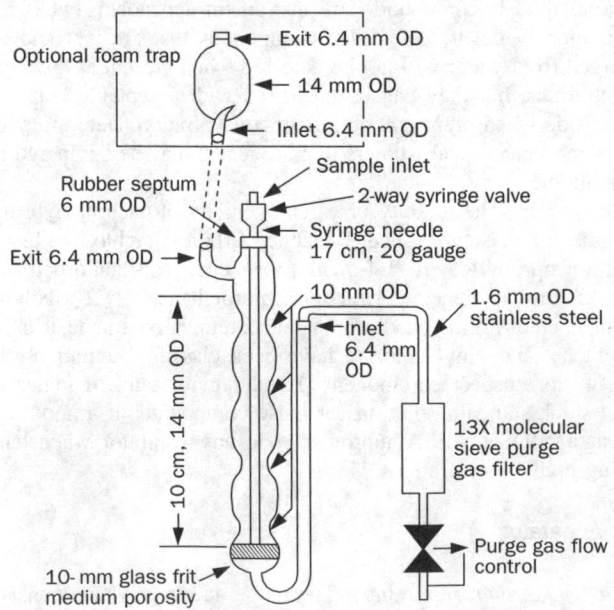

Figure 6200:1. **Purging device.**

---

† Gas chromatographic methods are extremely sensitive to the materials used. Mention of trade names by *Standard Methods* does not preclude the use of other existing or as-yet- undeveloped products that give *demonstrably* equivalent results.
‡ Supelco, Inc. or equivalent.
§ J&W or equivalent.

TABLE 6200:II. BFB KEY *m/z* ABUNDANCE CRITERIA

Mass	*m/z* Abundance Criteria
50	15 to 40% of mass 95
75	30 to 60% of mass 95
95	Base peak, 100% relative abundance
96	5 to 9% of mass 95
173	<2% of mass 174
174	>50% of mass 95
175	5 to 9% of mass 174
176	95 to 101% of mass 174
177	5 to 9% of mass 176

capillary column. The uncoated section of column is cooled to −150°C during desorption and heated to 250°C to transfer condensed materials.

f. Data system: To the mass spectrometer attach a computer that allows continuous acquisition and storage of all mass spectra obtained throughout the chromatographic program. Computer software should allow for a search of all acquired spectra for specific *m/z* (masses) and the plot of such *m/z* abundances versus time or scan number. This type of plot is an extracted ion current profile (EICP). Software also should allow the integration of the abundances in any EICP over a specified time or scan limit.

g. Syringes, 0.5-, 1.0-, 5-, and 25-mL glass hypodermic with detachable tip.‖

h. Syringe valves, two-way, with detachable tip.‖

i. Microsyringes, 10-, 25-, and 100-μL with a 5-cm × 0.15-mm-ID and 220 bevel needle.#

j. Bottles, 40-mL with TFE-lined screw cap.

3. Reagents

a. Reagent water, in which no interferent is observed at or above the MDL of the constituents of interest. Prepare by passing tap water through a carbon filter bed containing about 0.5 kg activated carbon, by distillation, or by using a water purification system.**

b. Trap packing materials:

1) *2,6-Diphenylene oxide polymer,* 60/80 mesh, chromatographic grade.

2) *Methyl silicone packing,* 3 OV-1.

3) *Silica gel,* 35/60 mesh.

c. Methanol, purge-and-trap grade.

d. Hydrochloric acid: HCl, 1 + 1.

e. Vinyl chloride, 99.9% pure.

f. Ascorbic acid.

g. Stock standard solutions: Prepare from pure standard materials or purchase as certified solutions. Prepare stock standard solutions in methanol using assayed liquids or gases as appropriate. CAUTION: *Toxic substances.* See ¶ 1*d*.

Place about 9.8 mL methanol in a 10-mL ground-glass-stoppered volumetric flask. Let stand unstoppered for about 10 min or until all alcohol-wetted surfaces have dried. Weigh flask to nearest 0.1 mg.

Add assayed reference materials as follows: For liquids, using a 100-μL syringe or disposable capillary-tip glass pipet, immediately add two or more drops of assayed reference material to

flask, then reweigh. Ensure that the drops fall directly into the alcohol without contacting flask neck. For halocarbon gases that boil below 30°C (bromomethane, chloroethane, chloromethane, dichlorofluoromethane, trichlorofluoromethane, vinyl chloride), attach a vinyl plastic†† tube to port of gas bottle containing reference material, with open end bubbling into a beaker of methanol showing flow through the tubing; insert needle of 5-mL valved gastight syringe into tube and pull gas into syringe slowly to 5.0-mL mark. Lower syringe needle to within 5 mm of methanol surface and slowly force gas onto surface. The gas will dissolve into the methanol and will be seen as a vortex as it dissolves into the solvent. Reweigh flask (difference is amount of gas dissolved into methanol), dilute to volume, stopper, and mix by inverting several times. Calculate concentration in micrograms per microliter from net gain in weight. When compound purity is assayed to be 96% or greater, calculate concentration of stock standard from uncorrected weight. Preferably use commercially prepared stock standards at any concentration if they are certified by the manufacturer or an independent source. Transfer stock standard solution into a TFE-sealed screw-cap bottle. Store with minimum headspace at −10 to −20°C away from light.

h. Secondary dilution standards: Using stock standard solutions, prepare in methanol secondary dilution standards that contain the compounds of interest, either singly or mixed together. Prepare secondary dilution standards at concentrations that will permit aqueous calibration standards (¶ *j* below) to bracket working range of the analytical system. Store secondary dilution standards with minimal headspace in a freezer and check frequently for signs of evaporation (which would indicate need for regeneration). Always bring to room temperature before preparing calibration standards. Prepare standards fresh weekly for gases. Replace all other standards monthly, or sooner if comparison with check standards indicates a problem.

i. Internal standard/surrogate standard known addition: Prepare a solution containing fluorobenzene (internal standard) and 1,2-dichlorobenzene-d$_4$ (surrogate) in methanol. Alternate internal standard and surrogate compounds may be used, provided that they meet method criteria and do not interfere with any method analyte(s). Prepare secondary dilution standard at a concentration of 5 μg/mL of each compound. Adding 5.0 μL standard to 25.0 mL sample or calibration standard yields a concentration equivalent to 1.0 μg/L. Alternate secondary standard concentrations can be used if addition volume is adjusted accordingly and all internal standard criteria are met. Add this mixture to each sample, standard, and blank.

j. Calibration standards: Prepare at least five concentration levels for each compound by adding appropriate amounts of secondary standard solution to reagent water and inverting water sample twice. Prepare one standard at a concentration near, but above, the MDL (i.e., 4 × MDL for potable-water-type samples) or a level that defines the low end of the working range and the others to correspond to the expected range of sample concentrations or to define the detector working range. Aqueous calibration standards can be stored up to 24 h if held in sealed vials with zero headspace. Otherwise, discard within 1 h. Alternatively, prepare calibration standards by injecting, with a solvent flush, an appropriate amount of a standard mix dilution and internal standard/surrogate mix, directly into a 25-mL syringe filled with reagent water; immediately inject water standard into purge vessel.

‖ Luerlok or equivalent.
Hamilton # 702 or equivalent.
** Millipore Super Q or equivalent.

†† Tygon or equivalent.

TABLE 6200:III. PRIMARY QUANTITATION ION, RETENTION TIMES AND
METHOD DETECTION LEVELS

Analyte	Retention Time min	MDL µg/L	Primary m/z
Dichlorodifluoromethane	1.49	0.190	85
Chloromethane	1.71	0.150	50
Vinyl chloride	1.79	0.120	62
Bromomethane	2.16	0.220	94
Chloroethane	2.28	0.230	64
Trichlorofluoromethane	2.57	0.059	101
1,1-Dichloroethene	3.22	0.130	96
1,1,2-Trichloro-1,2,2-trifluoroethane	3.25	0.065	101
Methylene chloride	3.96	0.099	49
trans-1,2-Dichloroethene	4.40	0.200	96
Methyl t-butyl ether	4.45	0.450	73
1,1-Dichloroethane	5.14	0.047	63
cis-1,2-Dichloroethene	6.30	0.130	96
2,2-Dichloropropane	6.24	0.041	77
Bromochloromethane	6.77	0.032	128
Chloroform	7.00	0.126	83
1,1,1-Trichloroethane	7.24	0.043	97
1,1-Dichloropropene	7.67	0.040	75
Carbon tetrachloride	7.65	0.042	117
Benzene	8.07	0.036	78
1,2-Dichloroethane	8.14	0.055	62
Trichlorethene	9.44	0.045	95
1,2-Dichloropropane	9.85	0.053	63
Dibromomethane	10.07	0.035	93
Bromodichloromethane	10.47	0.112	83
cis-1,3-Dichloropropene	11.29	0.048	75
Toluene	11.81	0.047	91
trans-1,3-Dichloropropene	12.27	0.051	75
1,1,2-Trichloroethane	12.56	0.043	83
1,3-Dichloropropane	12.83	0.090	76
Tetrachloroethene	12.77	0.047	166
Dibromochloromethane	13.24	0.133	129
1,2-Dibromoethane	13.35	0.133	107
Chlorobenzene	14.21	0.052	112
1,1,1,2-Tetrachloroethane	14.37	0.048	131
Ethylbenzene	14.42	0.032	91
m,p-Xylene	14.63	0.038	91
o-Xylene	15.27	0.038	91
Styrene	15.30	0.031	104
Bromoform	15.60	0.131	173
Isopropylbenzene	15.90	0.074	105
Bromobenzene	16.34	0.140	156
1,1,2,2-Tetrachloroethane	16.41	0.066	83
1,2,3-Trichloropropane	16.44	0.072	75
n-Propylbenzene	16.57	0.260	91
2-Chlorotoluene	16.68	0.042	126
4-Chlorotoluene	16.86	0.040	126
1,3,5-Trimethylbenzene	16.88	0.035	105
tert-Butylbenzene	17.38	0.100	119
sec-Butylbenzene	17.46	0.025	105
1,2,4-Trimethylbenzene	17.74	0.046	105
4-Isopropyltoluene	17.99	0.037	119
1,3-Dichlorobenzene	17.89	0.045	146
1,4-Dichlorobenzene	18.04	0.033	146
1,2-Dichlorobenzene	18.64	0.031	146
n-Butylbenzene	18.65	0.028	91
Hexachlorobutadiene	21.69	0.033	225
1,2,4-Trichlorobenzene	21.34	0.043	180
Naphthalene	21.80	0.049	128
1,2,3-Trichlorobenzene	22.32	0.047	180

GC conditions: Column-J&W DB-624, 30 m, 0.25 mm ID, 1.4 µm film; Temperature program-35°C, 4 min; 4°C/min; 50°C, 0 min; 10°C/min; 175°C, 4 min.

4. Procedure

a. Operating conditions: Table 6200:III summarizes recommended operating conditions for the gas chromatograph and gives estimated retention times and MDLs that can be achieved under these conditions. An example of the separations obtained with the specified column is shown in Figure 6200:3. Other chromatographic columns or conditions can be used if the quality control criteria are met.

b. GC/MS performance tests: At beginning of each day on which analyses are to be performed, check GC/MS system by a performance test with BFB before any samples, blanks, or standards are analyzed. Performance tests require the following instrument parameters:

Electron energy: 70 eV (nominal)
Mass range: 35 to 300 amu
Scan time: at least 5 scans/peak but not more than 2 s/scan

Inject 25 ng BFB directly on GC column. If direct injection is not easily performed, add 1 µL 25-µg/mL BFB solution to 25 mL reagent water in syringe used for sample transfer to purge device and analyze as a sample. Obtain a background-corrected mass spectrum of BFB and confirm that all key m/z criteria in Table 6200:II are achieved. If all criteria are not achieved, retune mass spectrometer and repeat test until all criteria are met.

c. Calibration: Calibrate system as follows:

1) System setup—Condition trap initially overnight at 180°C by back-flushing with inert gas at 20 mL/min. Condition trap daily for 10 min at manufacturer's suggested temperature. Connect purge and trap system to GC using recommended temperature program and flow-rate conditions. Calibrate system by either the internal or the external standard technique.

2) Internal standard calibration technique—Select one or more internal standards similar in analytical behavior to the compounds of interest. Fluorobenzene is a recommended internal standard compound. Demonstrate that measurement of internal standard is not affected by method or matrix interference. Because of such

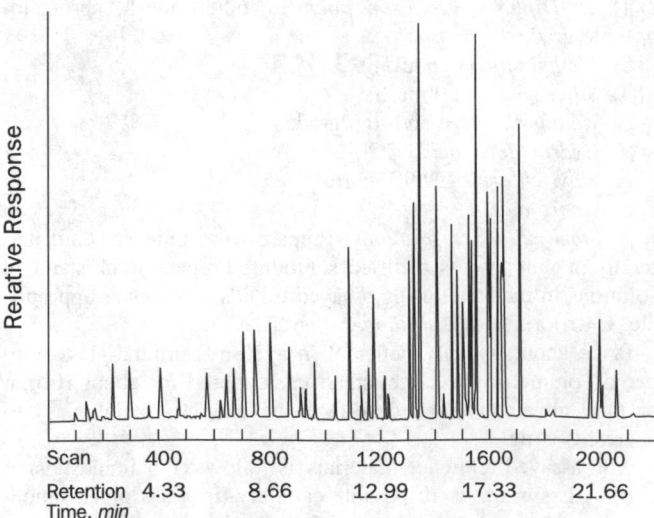

Scan	400	800	1200	1600	2000
Retention Time, min	4.33	8.66	12.99	17.33	21.66

Figure 6200:3. GC/MS chromatogram. Column: J&W DB-624, 30 m, 0.25 mm ID, 1.4 µm film; temperature program: 35°C for 4 min; 4°C/min; 50°C, 0 min; 10°C/min; 175°C, 4 min.:

limitations, no one internal standard may be applicable to all samples. The compounds used as surrogates (e.g., 1,2-dichlorobenzene-d$_4$) for quality control also can be used successfully as internal standards. Prepare calibration standards at a minimum of five concentration levels for each compound as described in ¶ 3*j* above. Prepare a secondary dilution standard containing each of the internal standards (¶ 3*i* above). Analyze each calibration standard according to procedure for samples, adding internal standard solution directly to syringe. Tabulate peak height or area responses against concentration for each compound and internal standard, and calculate response factors (RF) for each compound as follows:

$$RF = \frac{(A_s)(C_{is})}{(A_{is})(C_s)}$$

where:

   A_s = response for compound to be measured,
   A_{is} = response for internal standard,
   C_{is} = concentration for internal standard, and
   C_s = concentration of compound to be measured.

Average RF can be used if RSD is less than 20%.

3) *External standard calibration technique*—Prepare standards as directed in ¶ 3*j*. Analyze each calibration standard and tabulate peak area responses versus concentration. Prepare calibration curve for each compound. Alternatively, if ratio of response to concentration (calibration factor) is a constant over the working range (<20% RSD), assume linearity through the origin and use average calibration factor in place of a calibration curve.

4) *Calibration check*—See ¶ A.5*b*2).

d. Sample analysis: Bring sample to ambient temperature. Remove plunger from 25-mL syringe and close attached valve. Open sample bottle and carefully pour sample into syringe barrel to just short of overflowing. Replace syringe plunger, invert syringe, and open valve. Vent any air and adjust sample volume to 25.0 mL, in duplicate if sufficient sample is available (once sample cap has been removed, sample cannot be stored, because of headspace). Add an appropriate amount of surrogate/internal standard through valve bore, and close valve. Attach to purge device, open valves, and inject sample into purge vessel. Close valves and purge sample for 11.0 min at ambient temperature at a flow rate of 40 mL/min (helium or nitrogen). If water vapor causes problems in the mass spectrometer, use a 3-min dry purge and/or a moisture control module.

Desorb trapped materials onto head of chromatographic column at 180°C while back-flushing trap for 4 min with inert gas at a flow rate compatible with the column of choice, and begin GC temperature program.

Set system auto-drain to empty purge chamber while trap is being desorbed into GC, or alternatively, use sample syringe to empty vessel. Washing chamber with two 25-mL flushes of reagent water is useful if highly contaminated samples are being analyzed. Be sure all areas wetted during purging are also wetted during rinsing to maximize flushing.

Recondition trap by baking at conditioning temperature for 5 to 7 min. Let trap cool to ambient before introduction of next sample into purge vessel. When all sample compounds have been eluted from chromatographic column, end data acquisition and store data files. Use data system software to display full range mass spectra and appropriate extracted ion current profiles

TABLE 6200:IV. SINGLE-LABORATORY BIAS AND PRECISION DATA IN REAGENT WATER*

Analyte	Recovery %	Standard Deviation	Relative Standard Deviation %
Benzene	107	0.046	9
Bromobenzene	111	0.034	6
Bromochloromethane	88	0.052	12
Bromodichloromethane	104	0.036	7
Bromoform	107	0.042	8
Bromomethane	89	0.049	11
n-Butylbenzene	115	0.048	8
sec-Butylbenzene	113	0.043	8
tert-Butylbenzene	116	0.057	10
Carbon tetrachloride	119	0.048	8
Chlorobenzene	108	0.033	6
Chloroethane	115	0.073	13
Chloroform	108	0.043	8
Chloromethane	74	0.036	10
2-Chlorotoluene	111	0.045	8
4-Chlorotoluene	112	0.049	9
Dibromochloromethane	108	0.042	8
1,2-Dibromoethane	102	0.042	8
Dibromomethane	132	0.113	17
1,2-Dichlorobenzene	106	0.043	8
1,3-Dichlorobenzene	108	0.052	10
1,4-Dichlorobenzene	106	0.045	8
Dichlorodifluoromethane	80	0.058	15
1,1-Dichloroethane	109	0.049	9
1,2-Dichloroethane	102	0.031	6
1,1-Dichloroethene	99	0.059	12
cis-1,2-Dichloroethene	103	0.062	12
trans-1,2-Dichloroethene	113	0.045	8
1,2-Dichloropropane	129	0.064	10
1,3-Dichloropropane	107	0.046	9
2,2-Dichloropropane	106	0.049	9
1,1-Dichloropropene	110	0.044	8
cis-1,3-Dichloropropene	99	0.044	9
trans-1,3-Dichloropropene	101	0.038	7
Ethylbenzene	109	0.049	9
Hexachlorobutadiene	112	0.053	9
Isopropylbenzene	112	0.044	8
4-Isopropyltoluene	117	0.046	8
Methylene chloride	85	0.050	12
Methyl *t*-butyl ether	81	0.017	11
Naphthalene	121	0.068	11
n-Propylbenzene	107	0.048	9
Styrene	101	0.039	8
1,1,1,2-Tetrachloroethane	113	0.037	7
1,1,2,2-Tetrachloroethane	104	0.053	10
Tetrachloroethene	106	0.046	9
Toluene	106	0.045	8
1,2,3-Trichlorobenzene	118	0.054	9
1,2,4-Trichlorobenzene	109	0.049	9
1,1,1-Trichloroethane	106	0.040	8
1,1,2-Trichloroethane	97	0.041	9
Trichloroethene	105	0.041	8
Trichlorofluoromethane	105	0.045	9
1,2,3-Trichloropropane	104	0.034	6
1,1,2-Trichloro-1,2,2-trifluoroethane	113	0.042	7
1,2,4-Trimethylbenzene	116	0.044	8
1,3,5-Trimethylbenzene	110	0.051	9
Vinyl chloride	85	0.037	9
m,p-Xylene	110	0.057	10
o-Xylene	106	0.044	8

* For all analytes, seven samples, each of 0.5 µg/L concentration, were analyzed.

(EICP). If any ion abundances exceed system working range, dilute sample in second syringe with reagent water and analyze. NOTE: *Take care with sample because compounds can be very volatile and can be lost if sample is reopened.* Estimate amount of dilution needed and expel excess sample from second syringe, inject that portion into purge vessel, and with a second syringe, add necessary reagent water to a total of 25.0 mL in purge vessel.

5. Calculation

When compounds have been identified, base quantitation on integrated area abundance from the EICP of the primary characteristic *m/z* given in Table 6200:III. If sample produces an interference for the primary *m/z*, calculate a response factor or calibration curve using a secondary characteristic *m/z*, and use secondary *m/z* to quantitate. Report results in micrograms per liter. Report all quality control data with sample results.

6. Quality Control

See Section 6200A.5.

7. Precision and Bias

Typical single-laboratory precision and bias data are shown in Table 6200:IV.

8. Reference

1. U.S. ENVIRONMENTAL PROTECTION AGENCY. 1992. Methods for the Determination of Organic Compounds in Finished Drinking Water and Raw Source Water. U.S. Environmental Protection Agency, Environmental Monitoring & Support Lab., Cincinnati, Ohio.

6200 C.   Purge and Trap Capillary-Column Gas Chromatographic Method

This method[1] is applicable to the determination of purgeable halocarbons and aromatic organic compounds (Table 6200:I) in finished drinking water, raw source water, drinking water in any treatment stage, and wastewater.

1. General Discussion

a. Principle: See Section 6200B.1*a*.
b. Interferences: See Section 6200B.1*b*.
c. Detection levels: In a single laboratory using reagent water and known additions of 0.5 µg/L, calculated method detection levels (MDLs) for these compounds were in the range of 0.01 to 0.05 µg/L, depending on the compound. Some laboratories may not be able to achieve these detection levels because results depend on instrument sensitivity and matrix effects. Analysis of complex mixtures containing partially resolved compounds may be hampered by concentration differences larger than a factor of 10. This problem commonly occurs in analyses of finished drinking waters because of the relatively high trihalomethane content.
d. Safety: The toxicity or carcinogenicity of each reagent has not been defined precisely. Carbon tetrachloride, 1,2-dichloroethane, 1,1,2,2-tetrachloroethane, 1,1,2-trichloroethane, chloroform, 1,2-dibromoethane, tetrachloroethene, trichloroethene, and vinyl chloride have been classified tentatively as known or suspected human or mammalian carcinogens. Prepare primary standards of these compounds in a hood and wear a NIOSH/MESA-approved toxic gas respirator when handling high concentrations.

2. Apparatus

a. Purge and trap system: The purge and trap system consists of three separate pieces of equipment: purging device, trap, and desorber. Several complete systems are commercially available.

1) *Purging device*—See Section 6200B.2*a*1).
2) *Trap*—See Section 6200B.2*a*2). If only compounds boiling above 35°C are to be analyzed, both silica gel and charcoal can be eliminated and polymer increased to fill entire trap. Trap failure is characterized by a pressure drop above 21 kPa across trap during purging or by poor bromoform sensitivities.
3) *Assembly*—See Figures 6200:1 and 2.
b. Gas chromatograph: See Section 6200B.2*b*.
1) *Column*—See Section 6200B.2*c*.
2) *Electrolytic conductivity or microcoulometric detector*—Halogen-specific systems eliminate misidentifications due to non-organohalides that may be coextracted during purging.
3) *Photoionization detector*—A high-temperature detector equipped with a 10.2-eV (nominal) lamp.* Insert between analytical column and halide detector to analyze simultaneously for aromatic and unsaturated volatile organic compounds (see Table 6200:I).
c. Syringes, 5-mL glass hypodermic with detachable tip.†
d. Other equipment: See Section 6200B.2*g* through *j*.

3. Reagents

See Section 6200B.3*a* through *h*.

4. Procedure

a. Operating conditions: Table 6200:V summarizes recommended operating conditions for the gas chromatograph, esti-

---
* Tracor Model 703 or equivalent.
† Luerlok or equivalent.

TABLE 6200:V. RETENTION TIMES AND METHOD DETECTION LEVELS

Analyte	Retention Time min	Method Detection Level	
		Electrolytic Conductivity Detector μg/L	Photo-ionization Detector μg/L
Dichlorodifluoromethane	6.22	0.037	—
Chloromethane	7.09	0.041	—
Vinyl chloride	7.68	0.025	0.088
Bromomethane	9.45	0.103	—
Chloroethane	9.76	0.025	—
Trichlorofluoromethane	11.04	0.042	—
1,1-Dichloroethene	13.59	0.018	0.035
1,1,2-Trichloro-1,2,2-trifluoroethane	13.07	0.047	—
Methylene chloride	15.83	0.068	—
Methyl t-butyl ether	16.49	—	0.411
trans-1,2-Dichloroethene	16.78	0.015	0.015
1,1-Dichloroethane	18.49	0.015	—
2,2-Dichloropropane	20.27	0.220	—
cis-1,2-Dichloroethene	20.54	0.012	0.032
Chloroform	21.04	0.017	—
Bromochloromethane	21.53	0.025	—
1,1,1-Trichloroethane	22.14	0.014	—
1,1-Dichloropropene	22.57	0.019	0.008
Carbon tetrachloride	22.80	0.022	—
Benzene	23.38	—	0.017
1,2-Dichloroethane	23.62	0.074	—
Trichloroethene	25.30	0.012	0.014
1,2-Dichloropropane	25.92	0.021	—
Bromodichloromethane	26.63	0.040	—
cis-1,3-Dichloropropene	28.38	0.067	0.041
Dibromomethane	28.40	0.057	—
Toluene	29.16	—	0.023
trans-1,3-Dichloropropene	30.00	0.029	0.046
1,1,2-Trichloroethane	30.39	0.042	—
Tetrachloroethene	31.04	0.013	0.014
1,3-Dichloropropane	31.18	0.020	—
Dibromochloromethane	31.86	0.039	—
1,2-Dibromoethane	32.36	0.070	—
Chlorobenzene	33.67	0.029	0.027
Ethylbenzene	33.89	—	0.028
1,1,1,2-Tetrachloroethane	33.91	0.020	—
m,p-Xylene	34.08	—	0.021
o-Xylene	35.44	—	0.024
Styrene	35.67	—	0.027
Isopropylbenzene	36.64	—	0.018
Bromoform	36.72	0.023	—
1,1,2,2-Tetrachloroethane	37.43	0.034	—
1,2,3-Trichloropropane	37.88	0.048	—
n-Propylbenzene	37.94	—	0.023
Bromobenzene	37.98	0.026	0.026
1,3,5-Trimethylbenzene	38.44	—	0.019
2-Chlorotoluene	38.48	0.017	0.017
4-Chlorotoluene	38.63	0.026	0.028
1,2,4-Trimethylbenzene	39.61	—	0.030
tert-Butylbenzene	39.76	—	0.018
sec-Butylbenzene	40.34	—	0.018
4-Isopropyltoluene	40.80	—	0.019
1,3-Dichlorobenzene	41.01	0.017	0.028
1,4-Dichlorobenzene	41.40	0.059	0.061
n-Butylbenzene	42.21	—	0.028
1,2-Dichlorobenzene	42.62	0.023	0.031
1,2,4-Trichlorobenzene	48.21	0.019	0.028
Hexachlorobutadiene	48.75	0.026	0.019
Naphthalene	49.05	—	0.043
1,2,3-Trichlorobenzene	49.92	0.018	0.032

GC conditions: Column-Supelco VOCOL, 60 m, 0.75 mm ID, 1.5 μm film; Temperature program-0°C, 8 min; 4°C/min; 185°C, 1.5 min.

mated retention times, and method detection levels. Examples of separations obtained with the specified column are shown in Figures 6200:4 and 5.

b. *Calibration:* See Section 6200A.5b. Use either internal or external calibration technique. If using internal standard technique, prepare a dilution standard as described in Section 6200B.3i.

c. *Instrument performance:* See Section 6200A.5. Ensure that all peaks in standard chromatograms are sharp and symmetrical. Correct any peak tailing significantly in excess of that shown in method chromatograms. Tailing problems generally are traceable to active sites on the GC column or to detector operation. If only compounds eluting before chloroform give random responses or unusually wide peak widths, are poorly resolved, or are missing, the problem usually is traceable to the trap/desorber. If only brominated compounds show poor peak geometry or do not respond properly at low concentrations, replace trap. Excessive detector reactor temperatures also can cause low bromoform response. If negative peaks appear in the chromatogram, replace both ion-exchange column and electrolyte in detector. Check precision between replicate analyses. A properly operating system shows an average relative standard deviation of less than 10%. Poor precision generally is traceable to pneumatic leaks, especially around sample purger and detector reactor inlet and exit, electronic problems, or sampling and storage problems. Monitor retention times for each compound using calibration standards and laboratory control standard. If individual retention times vary by more than 10% over an 8-h period or do not fall within 10% of an established norm, locate and correct source of retention data variance.

d. *Sample analysis:* See Section 6200B.5.

5. Calculation

Identify each organohalide in sample chromatogram by comparing retention time of suspect peak to retention times generated by calibration standards and laboratory control standard. Determine concentrations of individual compounds. If external standard calibration procedure is used, calculate concentration of compound being measured from peak response using calibration curve or calibration factor previously determined.

If internal standard calibration procedure is used, calculate concentration using response factor [*RF*, ¶ B.4c2)] by the following equation:

$$\text{Concentration, } \mu g/L = \frac{A_s \times C_{is}}{A_{is} \times RF}$$

where:

A_s = response for compound to be measured,
A_{is} = response for internal standard, and
C_{is} = concentration of internal standard.

Report results in micrograms per liter without correction for recovery. Report quality control data with sample results.

6. Quality Control

See Section 6200A.5.

7. Precision and Bias

See Table 6200:VI.

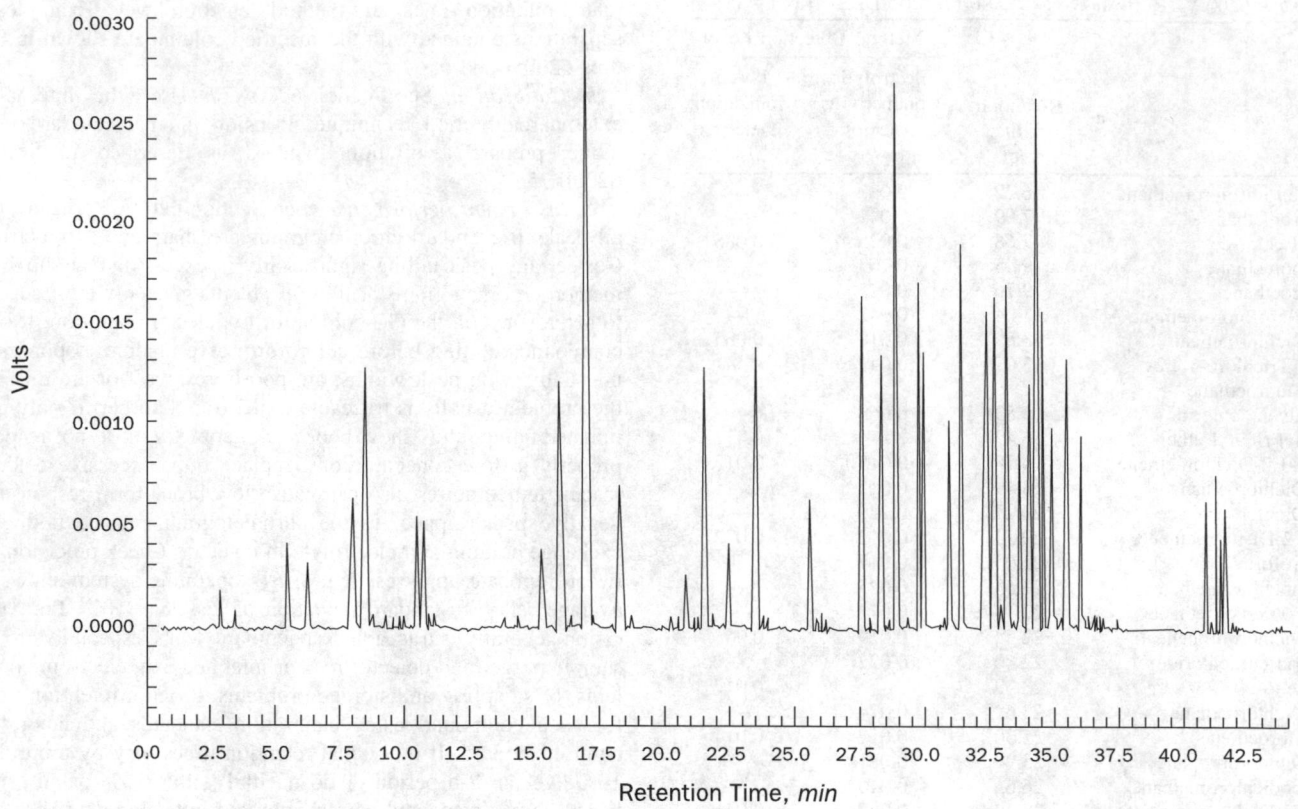

Figure 6200:4. PID chromatogram. GC conditions: Column: Supelco VOCOL, 60 m, 0.75 mm ID, 1.5 μm film; temperature program: 0° C, 8 min: 4° C/min; 185° C, 1.5 min.

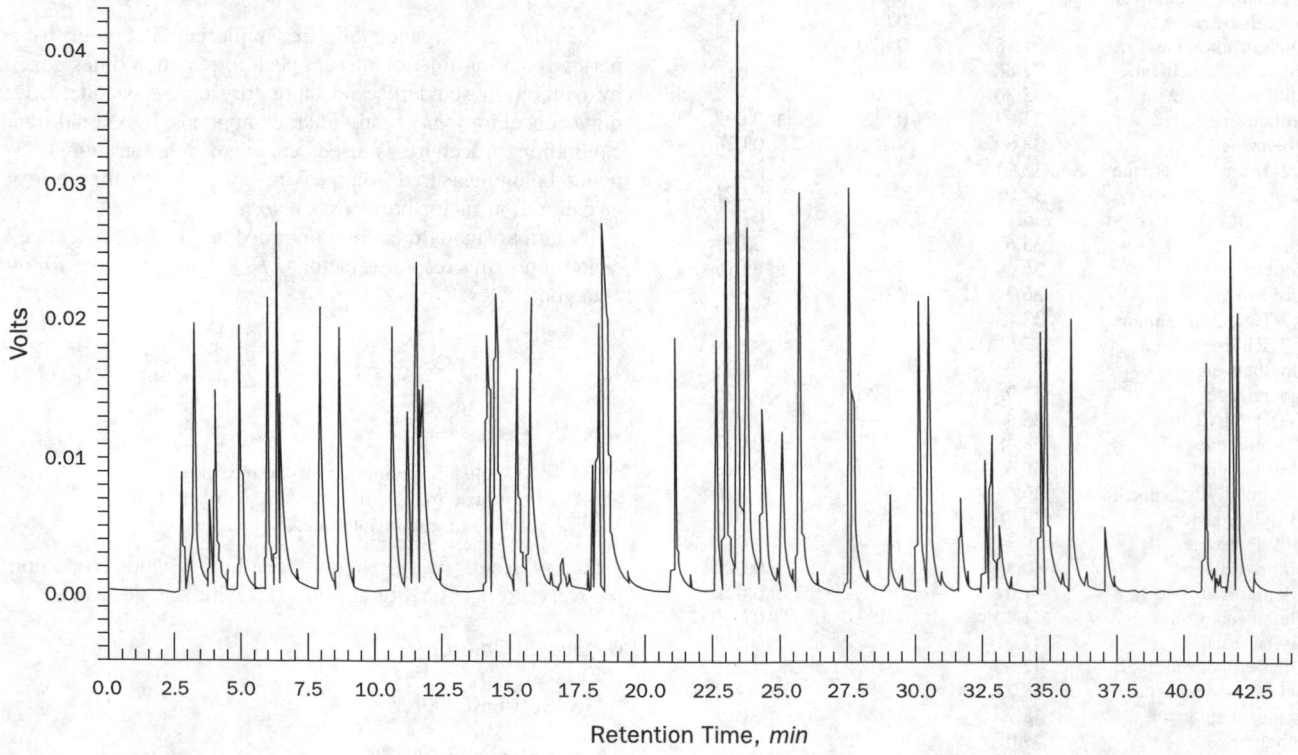

Figure 6200:5. ELCD chromatogram. GC conditions: Column: Supelco VOCOL, 60 m, 0.75 mm ID, 1.5 μm film; temperature program: 0° C, 8 min; 4° C/min; 185° C, 1.5 min.

TABLE 6200:VI. SINGLE-LABORATORY BIAS AND PRECISION DATA IN REAGENT WATER*

Analyte	Photoionization Detector			Electrolytic Conductivity Detector		
	Recovery %	Standard Deviation	Relative Standard Deviation %	Recovery %	Standard Deviation	Relative Standard Deviation %
Benzene	70	0.006	2	—	—	—
Bromobenzene	—	—	—	89	0.008	2
Bromochloromethane	—	—	—	83	0.008	2
Bromodichloromethane	—	—	—	135	0.021	3
Bromoform	—	—	—	81	0.007	2
Bromomethane	—	—	—	73	0.033	9
n-Butylbenzene	63	0.009	3	—	—	—
sec-Butylbenzene	65	0.009	3	—	—	—
tert-Butylbenzene	72	0.006	2	—	—	—
Carbon tetrachloride	—	—	—	79	0.007	2
Chlorobenzene	70	0.009	2	97	0.009	2
Chloroethane	—	—	—	64	0.008	2
Chloroform	—	—	—	83	0.006	1
Chloromethane	—	—	—	96	0.063	13
2-Chlorotoluene	—	—	—	91	0.005	1
4-Chlorotoluene	73	0.009	2	81	0.008	2
Dibromochloromethane	—	—	—	88	0.013	3
1,2-Dibromoethane	—	—	—	139	0.022	3
Dibromomethane	—	—	—	79	0.018	5
1,2-Dichlorobenzene	67	0.010	3	93	0.007	2
1,3-Dichlorobenzene	70	0.009	3	95	0.005	1
1,4-Dichlorobenzene	70	0.019	6	91	0.019	4
Dichlorodifluoromethane	—	—	—	71	0.027	8
1,1-Dichloroethane	—	—	—	82	0.005	1
1,2-Dichloroethane	—	—	—	78	0.024	6
1,1-Dichloroethene	61	0.011	4	81	0.006	1
cis-1,2-Dichloroethene	61	0.010	3	76	0.004	1
trans-1,2-Dichloroethene	79	0.005	1	77	0.005	1
1,2-Dichloropropane	—	—	—	85	0.007	2
1,3-Dichloropropane	—	—	—	148	0.018	2
2,2-Dichloropropane	—	—	—	74	0.045	12
1,1-Dichloropropene	54	0.003	1	74	0.006	2
cis-1,3-Dichloropropene	57	0.013	5	78	0.021	5
trans-1,3-Dichloropropene	63	0.015	5	78	0.009	2
Ethylbenzene	70	0.009	3	—	—	—
Hexachlorobutadiene	55	0.006	2	76	0.008	2
Isopropylbenzene	67	0.006	2	—	—	—
4-Isopropyltoluene	65	0.006	2	—	—	—
Methylene chloride	—	—	—	83	0.022	5
Methyl t-butyl ether†	75	0.130	3	—	—	—
Naphthalene	73	0.014	4	—	—	—
n-Propylbenzene	70	0.007	2	—	—	—
Styrene	70	0.009	3	—	—	—
1,1,1,2-Tetrachloroethane	—	—	—	83	0.001	0
1,1,2,2-Tetrachloroethane	—	—	—	88	0.011	2
Tetrachloroethene	54	0.005	2	79	0.004	1
Toluene	69	0.007	2	—	—	—
1,2,3-Trichlorobenzene	72	0.010	3	84	0.006	1
1,2,4-Trichlorobenzene	70	0.009	3	94	0.006	1
1,1,1-Trichloroethane	—	—	—	79	0.005	1
1,1,2-Trichloroethane	—	—	—	118	0.014	2
Trichloroethene	57	0.004	2	80	0.004	1
Trichlorofluoromethane	—	—	—	70	0.013	4
1,2,3-Trichloropropane	—	—	—	87	0.015	3
1,1,2-Trichloro-1,2,2-trifluoroethane	—	—	—	79	0.029	7
1,2,4-Trimethylbenzene	65	0.010	3	—	—	—
1,3,5-Trimethylbenzene	70	0.006	2	—	—	—
Vinyl chloride	73	0.022	6	67	0.008	2
m,p-Xylene	73	0.007	2	—	—	—
o-Xylene	68	0.008	2	—	—	—

* For all analytes, seven samples, each at a concentration of 0.5 µg/L (unless otherwise noted), were analyzed.
† Sample concentration 5 µg/L.

8. Reference

1. U.S. ENVIRONMENTAL PROTECTION AGENCY. 1991. Volatile organic compounds in water by purge and trap capillary column gas chromatography with photoionization and electrolytic conductivity detectors in series. Method 502.2 *in* Methods for the Determination of Organic Compounds in Finished Drinking Water and Raw Source Water. U.S. Environmental Protection Agency, Environmental Monitoring & Support Lab., Cincinnati, Ohio.

6211   METHANE*

6211 A.   Introduction

1. Occurrence and Significance

Methane (CH_4) is a colorless, odorless, tasteless combustible gas occasionally found in groundwaters. Escape of this gas from water may cause an explosive atmosphere not only in a utility's tanks, pumphouses, and other facilities, but also on the consumer's property, particularly where water is sprayed through poorly ventilated spaces such as public showers.

The explosive limits of CH_4 in air are 5 to 15% by volume. At sea level, a 3.95% CH_4 concentration in air theoretically could be reached in a poorly ventilated space sprayed with hot (68°C) water having a CH_4 concentration of only 0.7 mg/L. At higher water temperatures, the vapor pressure of water is so great that no explosive mixture can form. At lower barometric pressures, the theoretical hazardous concentration of methane in water will be reduced proportionally. In an atmosphere of N_2 or other inert gas, at least 12.8% O_2 must be present for there to be an explosion hazard.

Methane also is produced from wastewater and may be present in sewers and wastewater treatment plants (see Section 2720).

2. Selection of Method

The combustible-gas indicator method (B) offers the advantages of simplicity, speed, and great sensitivity. The volumetric method (C) can be made more accurate for concentrations of 4 to 5 mg/L and higher, but will not be satisfactory for very low concentrations. The volumetric method also can be applied to differentiate between CH_4 and other gases, as when a water supply is contaminated by liquid petroleum gas or other volatile combustible materials.

Methane also may be determined with the gas chromatograph as described in Sludge Digester Gas, Section 2720. This method permits differentiation between H_2 and CH_4, and/or its higher homologs.

---

* Approved by Standard Methods Committee, 1996.

6211 B.   Combustible-Gas Indicator Method

1. General Discussion

a. Principle: An equilibrium according to Henry's law is established between CH_4 in solution and the partial pressure of CH_4 in the gas phase above the solution. The partial pressure of CH_4 can be determined with a combustible-gas indicator. The operation of the instrument is based on the catalytic oxidation of a combustible gas on a heated platinum filament that is made a part of a Wheatstone bridge. The heat generated by the oxidation of the gas increases the electrical resistance of the filament. The resulting imbalance of the electrical circuit causes deflection of a milliammeter that may be calibrated in terms of percentage of CH_4 or percentage of the lower explosive limit of the gas sampled.

b. Interference: Small amounts of ethane usually are associated with CH_4 in natural gas and presumably would be present in water that contains methane. Hydrogen gas has been observed in well waters and would behave similarly to CH_4 in this procedure. Hydrogen sulfide may interfere if the pH of the water is low enough for an appreciable fraction of the total sulfide to exist in the unionized form. The vapors of combustible oils also may interfere. In general, these interferences are of no practical importance because primary interest is in calculating the explosion hazard to which all combustible gases and vapors contribute.

Interference due to H_2S can be reduced by the addition of solid NaOH to the container before sampling.

c. Minimum detectable concentration: The limit of sensitivity of the test is approximately 0.2 mg/L.

d. Sampling: If the water is supersaturated with CH_4, a representative sample cannot be obtained unless the water is under

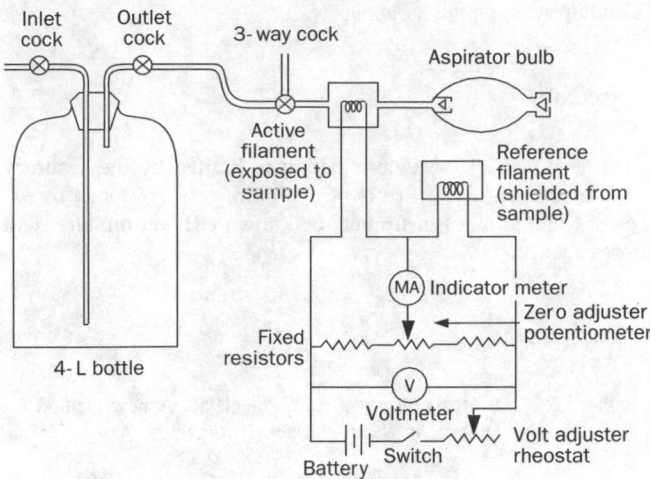

Figure 6211:1. Combustible gas indicator circuit and flow diagram.

sufficient pressure to keep all of the gas dissolved. Operate wells long enough to insure sampling water coming directly from the aquifer. Representative samples can be expected only when the well is equipped with a pump operating at sufficient submergence to assure that no gas escapes from the water.

2. Apparatus

*a. Combustible-gas indicator:** Connect a three-way stopcock to the inlet to zero instrument on atmospheric air immediately before obtaining sample reading. For laboratory use, replace the suction bulb with a filter pump throttled to draw gas through the instrument at a rate of approximately 600 mL/min. See Figure 6211:1.

b. Laboratory filter pump.

c. Glass bottle, 4-L, fitted with a two-hole rubber stopper. Extend inlet tube to within 1 cm of bottom. End outlet tube approximately 1 cm below stopper. Use metal or glass tubes, each fitted with stopcocks or with short (approximately 5-cm) lengths of rubber tubing and pinchcocks. The entire assembly should be capable of holding a low vacuum for several hours. Determine volume of assembly by filling with water and measuring volume, or weight, of water contained.

3. Reagent

Sodium hydroxide, NaOH, pellets.

4. Procedure

a. Rough estimation of CH₄ concentration: Fill bottle about half full of water, using a rubber tube connecting sampling tap and inlet tube, with outlet tube open. With both inlet and outlet tubes closed, shake bottle vigorously for approximately 15 s and let stand for approximately 1 min. Sample gas phase by withdrawing gas from the outlet, leaving inlet open to admit air. If the needle

---

* Marketed under the following trade names: ''Explosimeter,'' ''Methane Gas Detector,'' and ''Methane Tester,'' all manufactured by Mine Safety Appliance Co., Pittsburgh, PA 15235, and ''J-W Combustible Gas Indicator,'' manufactured by Bacharach Instrument Co., Mountain View, CA 94043, or equivalent.

swings rapidly to a high level on the meter and then drops to zero, the CH_4-air mixture is too rich to burn; take a smaller sample for the final test. If needle deflection is too small to be read accurately, take a larger volume of water.

b. Accurate determination: If the water contains H_2S, add approximately 0.5 g NaOH pellets to empty bottle to suppress interference. Evacuate bottle, using filter pump. Fill bottle not more than three-quarters full by connecting inlet tube to sampling cock, with outlet tube closed. After collecting desired volume of water, let bottle fill with air through inlet tube. Close inlet cock, shake bottle vigorously for 60 s, and let stand for at least 2 h. Sample gas phase through outlet tube with inlet cock open. Take reading as rapidly as possible before the entering air has diluted sample appreciably. Measure volume of water sampled.

5. Calculation

The weight of CH_4 (w), in mg, in the sample is given by the equation:

$$w = P \left(\frac{1.928 \, V_g}{T + 273} + \frac{890 \, V_l}{H} \right)$$

where:

P = partial pressure of CH_4, kPa,
T = temperature, °C,
V_g = volume of gas phase, mL,
V_l = volume of liquid phase, mL, and
H = Henry's law constant, kPa/mole CH_4/mole of water.

Values for Henry's constant are as follows:

Temperature °C	Henry's Constant H*	Temperature °C	Henry's Constant H*
0	2.265	40	5.261
5	2.625	45	5.577
10	3.010	50	5.846
15	3.413	60	6.342
20	3.804	70	6.749
25	4.181	80	6.911
30	4.544	90	7.013
35	4.926	100	7.106

*Multiply given values by 10⁶.

For most determinations, it may be assumed that atmospheric pressure is 100 kPa, and that the temperature is 20°C. The concentration of CH_4 in the sample is then given by:

$$\text{mg } CH_4/L = Rf \left(6.7 \frac{V_0 - V_1}{V_1} + 0.24 \right)$$

where:

R = scale reading,
V_0 = total volume of sample bottle, mL,
V_1 = volume of water sampled, mL, and
f = factor depending on instrument used.

If the instrument reads directly in percentage of methane,

$f = 1.00$. If the instrument reads in percentage of the lower explosive limit of CH_4, $f = 0.05$. For instruments that require additional factors, consult the manufacturer. For example, one commercial instrument with a scale that reads in percentage of the lower explosive limit of combustible gases requires an additional factor of 0.77 for CH_4. Hence, the value of f in the above equation would be 0.77×0.05, or 0.0385.

For more accurate work, or in locations where normal barometric pressure is significantly lower than 100 kPa, use the equation:

$$\text{mg } CH_4/L = RBf \left(19.277 \frac{V_0 - V_1}{TV_1} + \frac{8900}{H} \right)$$

where:

B = barometric pressure, kPa,

and other symbols are as above.

6. Accuracy

The accuracy of the determination is limited by the accuracy of the instrument used. Errors of approximately 10% may be expected. Calibration of instrument on known CH_4-air mixtures will improve accuracy.

7. Bibliography

ROSSUM, J.R., P.A. VILLARRUZ & J.A. WADE. 1950. A new method for determining methane in water. *J. Amer. Water Works Assoc.* 42:413.

6211 C.   Volumetric Method

1. General Discussion

a. Principle: If CH_4 is slowly mixed with an excess of O_2 in the presence of a platinum coil heated to yellow incandescence, most of the CH_4 will be converted to CO_2 and H_2O in a smooth reaction. Several passes of the mixed gases may be needed to burn substantially all the CH_4. An excess of O_2 is mixed with the sample before passage through the assembly. By differential absorption and volumetric changes the product CO_2 is measured.

b. Interference: Low-boiling hydrocarbons other than ethane and vapors from combustible oils interfere. These substances, however, are not likely to be present in water in sufficiently high concentration to affect the results significantly.

c. Minimum detectable concentration: This method is not satisfactory for determining CH_4 in water where the concentration is less than 2 mg/L.

d. Sampling: Collect sample as directed in Method B and observe the same precautions to obtain representative samples (Section 6211B.1*d*). Omit NaOH pellets and fill sample bottle with water up to 90% of capacity.

2. Methane Determination

See Section 2720B for a description of apparatus, reagents, procedure, calculation, and precision and bias.

Use percentage of CH_4 found by this method with Henry's law to obtain the CH_4 concentration in original sample. Substitute CH_4 percentage for R (scale reading) and $f = 1$ in the calculation given under Section 6211B.5 preceding.

3. Bibliography

DENNIS, L.M. & M.L. NICHOLS. 1929. Gas Analysis. Macmillan Co., New York, N.Y.

HALDANE, J.S. & J.I. GRAHAM. 1935. Methods of Air Analysis. Charles Griffin & Co., London.

BUSWELL, A.M. & T.E. LARSON. 1937. Methane in ground waters. *J. Amer. Water Works Assoc.* 29:1978.

BERGER, L.B. & H.H. SCHRENK. 1938. Bureau of Mines Haldane gas analysis apparatus. U.S. Bur. Mines Information Circ. No. 7017.

LARSON, T.E. 1938. Properties and determination of methane in ground waters. *J. Amer. Water Works Assoc.* 30:1828.

6231 1,2-DIBROMOETHANE (EDB) AND 1,2-DIBROMO-3-CHLOROPROPANE (DBCP)*

6231 A. Introduction

1. Sources and Significance

Dibromoethane and dibromochloropropane have been found in groundwater supplies in many areas of the United States; typically they are found in agricultural areas where these compounds have been applied in the past as fumigants. Toxicological studies suggest that they may have detrimental effects on human health, and therefore many states have established maximum contaminant levels for them.

_____
* Approved by Standard Methods Committee, 1993.

2. Selection of Method

The liquid-liquid extraction gas chromatographic (GC) method (6231B) uses a microextraction and capillary columns and is the preferred method. In addition, these compounds can be detected by the purge and trap gas chromatographic/mass spectrometric (GC/MS) and GC methods (6200B and C), and dibromoethane by closed-loop stripping analysis (see Section 6040). For additional information on applicability, sensitivity, precision, and bias, see specific methods.

6231 B. Liquid-Liquid Extraction Gas Chromatographic Method

This method[1-3] is applicable to the determination of 1,2-dibromoethane (EDB) and 1,2-dibromo-3-chloropropane (DBCP) in drinking water and untreated groundwater.

1. General Discussion

a. Principle: The sample is extracted with hexane and injected into a gas chromatograph equipped with a linearized electron capture detector for separation and analysis. Identification is confirmed by analyzing the sample with a dissimilar column. See Section 6010C for discussion of gas chromatographic principles.

b. Interferences: Impurities in the extracting solvent usually account for most analytical problems. Analyze solvent blanks on each new bottle of solvent before use. Obtain indirect daily checks on the extracting solvent by monitoring sample blanks; whenever an interference is noted, reanalyze the extracting solvent. If necessary, remove interference by distillation or column chromatography[3] or, more simply, obtain a new source solvent. Interference-free solvent contains less than 0.1 μg/L individual compound interference. Store solvents in an area free of organochlorine solvents.

Accidental sample contamination can occur through diffusion of volatile organics through the septum seal into the sample bottle during shipment and storage. Sample blanks monitor this.

EDB at low concentrations may be masked by very high levels of dibromochloromethane (DBCM) when the confirmation column is used.

For further information on interferences in gas chromatographic methods, see Section 6010C.

c. Detection levels: The method detection levels (MDL)[4] for EDB and DBCP are 0.01 μg/L. The method is useful over a concentration range from approximately 0.03 to 200 μg/L. Actual detection limits are highly dependent on the characteristics of the gas chromatographic system used.

d. Safety: The toxicity or carcinogenicity of each reagent has not been defined precisely. EDB and DBCP have been classified tentatively as known or suspected human or mammalian carcinogens. Handle pure standard materials and stock standard solutions in a hood or glovebox and wear a NIOSH/MESA-approved toxic gas respirator when handling high concentrations.

2. Sampling and Storage

Collect all samples in duplicate and prepare replicate field blanks with each sample set. A sample set is all of the samples collected from the same general sampling site at approximately the same time. Prepare the field reagent blanks in the laboratory by filling a minimum of two sample bottles with reagent water, sealing, and shipping to the sampling site along with sample bottles.

Fill sample bottle to overflowing without air bubbles. When sampling from a water tap, open tap and flush until water temperature has stabilized (usually about 10 min). Adjust flow rate to about 500 mL/min and collect duplicate samples from the flowing stream. When sampling from a well, fill a wide-mouth bottle or beaker with sample, and carefully fill duplicate 40-mL sample bottles.

Keep samples chilled in an atmosphere free of organic solvent vapors, from day of collection until analysis. Do not add sodium thiosulfate as a dechlorinating agent nor acidify.

Analyze all samples within 28 d of collection.

3. Apparatus

a. Sample containers, 40-mL screw-cap vials* each with a TFE-faced silicone septum.† Wash vials and septa with detergent

_____
* Pierce #13075 or equivalent.
† Pierce #12722 or equivalent.

TABLE 6231:I. CHROMATOGRAPHIC CONDITIONS
FOR 1,2-DIBROMOETHANE (EDB) AND
1,2-DIBROMO-3-CHLOROPROPANE (DBCP)

| Compound | Retention Time *min* | |
	Column 1	Column 2
EDB	9.5	8.9
DBCP	17.3	15.0

Column 1 conditions: Durawax-DX 3 (0.25-μm film thickness) in a 30 m long × 0.32-mm ID fused silica capillary column with helium carrier gas at linear velocity of 25 cm/s. Column temperature held isothermal at 40°C for 4 min, then programmed at 8°C/min to 180°C for final hold.

Column 2 conditions: DB-1 (0.25 μm film thickness) in a 30 m long × 0.32-mm ID fused silica capillary column with helium carrier gas at linear velocity of 25 cm/s. Column temperature held isothermal at 40°C for 4 min, then programmed at 10°C/min to 270°C for final hold.

and rinse with tap and distilled water before using. Let vials and septa air dry at room temperature, place in a 105°C oven for 1 h, then remove and let cool in an area free of organics.

b. Vials, auto sampler, screw cap with septa, 1.8 mL.‡

c. Microsyringes, 10- and 100-μL.

d. Microsyringe, 25-μL with a 51- by 0.15-mm needle.§

e. Pipets, 2.0- and 5.0-mL transfer.

f. Volumetric flasks, 10- and 100-mL, glass stoppered.

g. Standard solution storage containers, 15-mL bottles with TFE-lined screw caps.

h. Gas chromatograph:‖ See Section 6200B.2*b*. The system is equipped with a linearized electron capture detector and a capillary column splitless injector.

Two gas chromatography columns are recommended. Column 1 is a highly efficient column that provides separations for EDB and DBCP without interferences from trihalomethanes. Use Column 1 as the primary analytical column unless routinely occurring compounds are not adequately resolved. Use Column 2 as a confirmatory column when GC/MS confirmation is not available.

1) *Column 1,* 30 m long × 0.32-mm ID fused silica capillary with dimethyl silicone mixed phase.# See Table 6231:I. Injector temperature: 200°C; detector temperature: 290°C. See Figure 6231:1 for a sample chromatogram.

2) *Column 2* (confirmation column), 30 m long × 0.32-mm ID fused silica capillary with methyl polysiloxane phase.** See Table 6231:I. Injector temperature: 200°C; detector temperature: 290°C.

4. Reagents

a. Reagent water: See Section 6200B.3*a*.

b. Hexane extraction solvent, UV grade.††

c. Methanol, pesticide quality or equivalent.

d. Sodium chloride, NaCl: Before using, pulverize and place in a muffle furnace at room temperature. Increase temperature to 400°C for 30 min. Store in capped bottle.

‡ Varian #96-000099-00 or equivalent.
§ Hamilton 702N or equivalent.
‖ Gas chromatographic methods are extremely sensitive to the materials used. Mention of trade names by *Standard Methods* does not preclude the use of other existing or as-yet-undeveloped products that give *demonstrably* equivalent results.
Durawax-DX3, 0.25-μm film, or equivalent.
** DB-1, 0.25-μm film, or equivalent.
†† Burdick and Jackson #216 or equivalent.

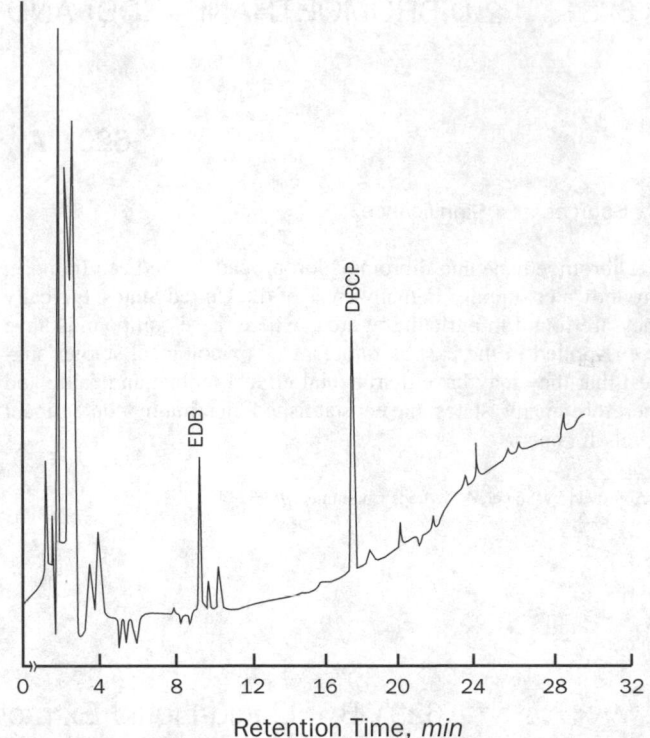

Retention Time, *min*

Figure 6231:1. Extract of reagent water with 0.114 μg/L added EDB and DBCP. Column: fused silica capillary; liquid phase: Durawax-DX3; film thickness: 0.25 μm; column dimensions: 30 m × 0.317 mm ID.

e. 1,2-Dibromoethane, 99%.‡‡

f. 1,2-Dibromo-3-chloropropane, 99.4%.§§

g. Standard stock solutions: See Section 6200B.3*g*. Store in 15-mL bottles with TFE-lined screw caps. Methanol solutions prepared from liquid standard materials are stable for at least 4 weeks when stored at 4°C.

h. Secondary dilution standards: See Section 6200B.3*h*. Dilution standards are as stable as stock solutions.

5. Procedure

a. Operating conditions: Table 6231:I summarizes recommended operating conditions for the gas chromatograph and estimated retention times.

b. Calibration: Prepare calibration standards as directed in Section 6200B.3*j* and analyze according to ¶ 5*d* below. Follow rest of calibration procedure in Section 6200A.5*b*), but limit variations from predicted response to ± 15% rather than ± 20%.

c. Instrument performance: See Section 6200C.4*c*.

d. Sample analysis: Let samples and standards come to room temperature. For samples and field blanks, open bottle, discard 5 mL using a 5-mL transfer pipet, and replace container cap. Weigh to nearest 0.1 g; record weight for subsequent volume determination. For calibration standards, QC check standards, and reagent

‡‡ Such as that available from Aldrich Chemical Company.
§§ Such as that available from AMVAC Chemical Corporation, Los Angeles, CA.

blank, measure 35 mL using a 50-mL graduated cylinder and transfer to a 40-mL sample container.

Remove container cap and add 7 g NaCl. Add 2.0 mL hexane with a transfer pipet. Recap and shake vigorously by hand for 1 min. Let water and hexane phases separate. (If sample is stored at this stage, keep container upside down). Carefully transfer 0.5 mL of hexane layer into an autosampler vial using a disposable glass pipet. Transfer remaining hexane phase, but not any of the water phase, into a second autosampler vial. Hold second vial at 4°C for reanalysis if necessary. Transfer first sample vial to an autosampler set up to inject 2.0-μL portions into the gas chromatograph. Alternatively, manually inject 2-μL portions.

To determine sample volume for samples and field blanks, remove cap and discard remaining sample/hexane mixture. Shake off remaining drops using short, brisk wrist movements. Reweigh empty container with original cap and calculate net weight of sample by difference to the nearest 0.1 g. This net weight is equivalent to the volume of water (in mL) extracted. Alternatively, weigh vial before collection and reweigh full vial. Sample volume then equals gross weight (g) − [tare weight (g) + 5 g].

6. Calculation

Identify EDB and DBCP in sample chromatogram by comparing retention time of suspect peak to retention times generated by calibration and laboratory control standards. Retention times of samples should be within ± 0.1 min of standard for positive identification.

Use calibration curve or calibration factor to calculate uncorrected concentration (C_i) of each compound (e.g., calibration factor × response). Calculate sample volume (V_s) as equal to the net sample weight:

$$V_s = \text{gross weight} - \text{bottle tare}$$

The corrected sample concentration is:

$$\text{Concentration, μg/L} = C_i \times \frac{35}{V_s}$$

Round off results to the nearest 0.1 μg/L or two significant figures.

TABLE 6231:II. SINGLE-LABORATORY PRECISION AND BIAS FOR EDB AND DBCP IN TAP WATER

Compound	Number of Samples	Addition μg/L	Average Bias %	Relative Standard Deviation %
1,2-Dibromoethane	7	0.03	114	9.5
	7	0.24	98	11.8
	7	50.0	95	4.7
1,2-Dibromo-3-chloropropane	7	0.03	90	11.4
	7	0.24	102	8.3
	7	50.0	94	4.8

7. Quality Control

Follow procedures given in Section 6200A.5.

8. Precision and Bias

Single-laboratory precision and bias at several concentrations in tap water are presented in Table 6231:II[5].

9. References

1. GLAZE, W.H. & C.C. LIN. 1984. Optimization of Liquid-Liquid Extraction Methods for Analysis of Organics in Water. EPA-600/S4-83-052, U.S. Environmental Protection Agency.
2. HENDERSON, J.E., G.R. PEYTON & W.H. GLAZE. 1976. A convenient liquid-liquid extraction method for the determination of halomethanes in water at the parts-per-billion level. In L. H. Keith, ed. Identification and Analysis of Organic Pollutants in Water. Ann Arbor Science Publ., Ann Arbor, Mich.
3. RICHARD, J.J. & G.A. JUNK. 1977. Liquid extraction for rapid determination of halomethanes in water. J. Amer. Water Works Assoc. 69: 62.
4. GLASER, J.A., D.L. FOERST, G.D. MCKEE, S. A. QUAVE & W.L. BUDDE. 1981. Trace analyses for wastewater. Environ. Sci. Technol. 15:1426.
5. U.S. ENVIRONMENTAL PROTECTION AGENCY. 1991. 1,2-Dibromoethane (EDB) and 1,2-dibromo-3-chloropropane (DBCP) in water by microextraction and gas chromatography. Method 504 in Methods for the Determination of Organic Compounds in Finished Drinking Water and Raw Source Water. U.S. Environmental Protection Agency, Environmental Monitoring & Support Lab., Cincinnati, Ohio.

6231 C.   Purge and Trap Gas Chromatographic/Mass Spectrometric Method

See Section 6200B for capillary-column method.

6231 D.   Purge and Trap Gas Chromatographic Method

See Section 6200C for capillary-column method.

6232 TRIHALOMETHANES AND CHLORINATED ORGANIC SOLVENTS*

6232 A. Introduction

1. Sources and Significance

The trihalomethane (THM) compounds have been found in most chlorinated water supplies in the United States; typically they are produced in the treatment process as a result of chlorination. The formation of these compounds is a function of precursor concentration, contact time, chlorine dose, and pH. Toxicological studies suggest that chloroform is a potential human carcinogen. Consequently, total trihalomethanes are being regulated in potable waters. Chlorinated organic solvents are found in many raw waters because of industrial contamination.

2. Selection of Method

Several methods are available for measurement of the trihalomethanes and chlorinated organic solvents. Some of these are

specific for these compounds and others have a much broader spectrum. Method 6232B is a simple liquid-liquid extraction gas chromatographic (GC) method that is highly sensitive and very precise for these compounds and certain other chlorinated solvents. Method C refers to purge and trap gas chromatographic/ mass spectrometric (GC/MS) methods that can detect not only THMs but also a wide variety of other compounds. Method D refers to purge and trap GC methods with similar target compounds. All of these methods have approximately the same sensitivity for the trihalomethanes; method choice depends on availability of equipment, operator choice, and the list of desired target compounds. In addition, closed-loop stripping analysis can be used for several of these compounds (see Section 6040).

---

* Approved by Standard Methods Committee, 1993.

6232 B. Liquid-Liquid Extraction Gas Chromatographic Method

This method[1-3] is applicable to the determination of four trihalomethanes (THMs), i.e., chloroform, bromodichloromethane, dibromochloromethane, and bromoform, and the selected chlorinated solvents in finished drinking water, drinking water during intermediate stages of treatment, and in both surface and ground water. For other compounds or sample matrices, collect precision and bias data on actual samples[4] and provide qualitative confirmation of results by gas chromatography/mass spectrometry (GC/MS) to demonstrate the usefulness of the method. Retain documentation to demonstrate method performance. This method is particularly useful when only a few compounds are being monitored.

1. General Discussion

a. Principle: Sample is extracted once with pentane and the extract is injected into a gas chromatograph equipped with a linearized electron capture detector (ECD) for separation and analysis. Extraction and analysis time is 10 to 30 min per sample, depending on analytical conditions.

Confirmatory evidence, where necessary, is obtained by using dissimilar columns, other selective detectors, or mass spectrometry. Component concentrations must be sufficiently high (i.e., >50 μg/L) for confirmatory analyses using a mass spectrometer. See other methods in this section for alternative means of confirming positive results.

Standards added to organic-free water and samples are extracted and analyzed in the same manner, under identical conditions. This step is essential to adjust for the less-than-100% ex-

traction efficiency of the simplified extraction technique. Extreme differences in ionic strength or organic content between standards and samples can result in different equilibria of sample constituents with the extracting solvent and a method bias may result. Monitor known additions recoveries on various matrices for bias.

Where required, sum the concentrations of the four trihalomethanes and report as total trihalomethanes in micrograms per liter.

See Section 6010 for discussion of gas chromatographic principles.

b. Interferences: Impurities contained in the extracting solvent account for many analytical problems. Maintain records of the reagent's manufacturer, lot number, purity, date bottle was opened, and expiration date. Analyze solvent blanks before using a new bottle of solvent. Make indirect daily checks on the extracting solvent by monitoring the sample blanks. Whenever an interference is noted in the method blank, analyze a solvent blank. Discard (or use for another purpose) extraction solvent if a high level of interference is traced to it. Low-level interferences can be removed by distillation or column chromatography;[5] however, it usually is more economical to obtain new solvent or select an approved alternative solvent. Interference-free solvent is defined as a solvent containing less than the laboratory determined detection limit of interference for each constituent. Protect interference-free solvents by storing in an area known to be free of organohalogen solvents. *Do not subtract blank values from sample analysis results as a correction for contamination.*

Sample contamination has been attributed to diffusion of volatile organics through the septum seal on a sample bottle during shipment and storage. Use the trip blank to monitor for this problem.

Contamination also may occur whenever equipment and materials used to store, extract, or analyze samples are inadequately cleaned, prepared, tested, or stored. There are many sources of contamination including contamination of reagents during storage and contamination of equipment reused in the sequential extraction of samples and standards. Maintain records of dates of preparation and cleaning and inclusive dates of use of reagents, standards, bottles, and equipment. Test all reagents and standards before initial use. To reduce possibility of carryover contamination, always clean equipment thoroughly after each use. Where equipment contamination is a concern, processing and analysis of additional method blanks beyond the minimum requirements of this method may be useful. Start by placing reagent water in a sample vial of the same lot that was used for samples and add preservative as was done for samples. Process this method blank in conjunction with samples using the same reagents, materials, and equipment. Where analysis of method blank indicates contamination, investigate possible sources and isolate the cause. Take and document corrective action. Following analysis of a sample containing late-eluting interferences, or containing over-range concentrations of constituents of interest, analyze a solvent blank or method blank to demonstrate freedom from carry-over.

This liquid-liquid extraction technique efficiently extracts a wide boiling range of nonpolar organic compounds and also extracts polar organic components with varying efficiencies. To analyze rapidly for trihalomethanes and chlorinated solvents with sensitivities in the low microgram-per-liter range in the presence of these other organic compounds use the semi-specific electron capture detector. Trihalomethanes are primarily products of the chlorination process and seldom appear in raw unchlorinated source water. The absence of peaks with retention times similar to the trihalomethanes in raw source water analysis is supporting evidence of an interference-free finished drinking water analysis. Because of possible interferences, analysis of a representative raw source water when analyzing finished drinking water provides evidence of freedom from this interference source. When potential interferences are noted in the raw source water, use the alternate chromatographic columns to reanalyze the sample set. If interferences still are noted, make confirmatory qualitative identifications as directed in ¶ 1a. If the peaks are determined to be other than the constituents of interest and they add significantly to the constituents' value in the finished drinking water, analyze sample set by the purge and trap method.[6]

Where chlorinated solvents are present in finished drinking water the most likely source is the raw water. Analyze individual raw water samples to isolate the source of contamination. Always consider the possibility of coeluting interferences. Analysis using capillary chromatography minimizes this possibility. Analysis using dissimilar columns may confirm the presence of interferences through differences in retention time between the constituent of interest and the unknown compound in the sample; however, the most definitive confirmation routinely available is GC/MS.[7]

Because the simplified extraction technique depends on equilibria between solvent and water, because extraction is not 100% efficient, and because efficiency is dependent on concentration, it is important to: extract samples and standards in the same manner; monitor matrix recoveries to assess differences in equilibria; and, where the concentration is found to be above the linear range of the method, to either dilute samples carefully before extraction, or prepare standards in water at the estimated sample concentration and carefully dilute *both* sample and standard extracts. Always process standards and samples together and hold constant such variables as water temperature, solvent temperature, room temperature, extraction time, and separation times.

For further information on interferences in gas chromatographic methods, see Section 6010C.

c. Detection levels: The method is useful for trihalomethane and selected chlorinated solvents at concentrations from approximately 0.1 to 200 µg/L. Actual detection levels are highly dependent on the characteristics of the gas chromatographic system used, the ratio of solvent to water, and interferences present in the solvent. See Section 1030.

2. Sampling and Storage

See Sections 6010B.1 and 5710.

If trihalomethane formation potential is to be measured, do not add any preservatives at the time of sample collection. If chemical stabilization is not used at time of sampling, add the reducing agent just before extracting the sample or add it at the time specified in the formation potential method for quenching the sample.[8,9]

The raw source water sample history should resemble that of the finished drinking water. Take into account the average retention time of the finished drinking water within the water plant when sampling the raw source water.

Store blanks and samples, collected at a given site (sample set), together in a protected area known to be free from contamination. At a water treatment plant, duplicate raw source water, duplicate finished water, and duplicate sample blanks comprise the minimum sample set. When samples are collected and stored under conditions specified in 6010B.1, no measurable loss has been detected over extended periods of time.[8] Analyze samples within 14 d of collection.

For samples collected soon after chlorination, quenching with reducing agent may not be sufficient to prevent further formation of THMs completely, because of hydrolysis of intermediates. In that case, acidification is necessary and consistent with the recommended preservation techniques.

3. Apparatus

a. Sample storage vials: Clean, baked 40-mL glass open screwtop TFE-faced septum VOA vials or equivalent. See Section 6010B for additional information on cleaning, storage, and preparation.

b. Microsyringes, 10, 25, and 100-µL. Microsyringes with extended barrels are suggested for proper injection of methanolic standards when preparing aqueous standards in volumetric flasks.

c. Volumetric flasks, glass-stoppered, 10, 25, 50, 100, 250, 500, and 1000 mL, Class A. Choose size according to final volume of aqueous standard required and concentration of methanolic standards.

d. Extraction vessels: Use sample storage vials. Optionally, where samples are transferred to smaller extraction vessels, select an extraction container on the basis of specific requirements for final extract volume, solvent-to-water volume ratio, and availability. If a separate vessel is used for extraction, place standards along with samples in sample storage vials and follow identical procedures for preservation and transfer to the extraction vessel.

Use the same lots of vessels for sample and standard extraction. Use of clean, oven-baked glass vessels and TFE-faced septa is critical.

e. Extract storage vials: 1.8-mL autosampler vials with open screw-top caps and TFE septa, or equivalent.

f. Gas chromatograph, preferably temperature-programmable with linearized electron-capture detector.

*g. Chromatographic columns:**

1) 0.32-mm ID × 30 m fused silica capillary, 1 μm DB-5,† or equivalent, at linear velocity of 20 cm/s. Temperature program: 35°C for 5 min, ramp 10°C/min to 70°C, then 20°C/min to 200°C. See Figure 6232:1 for a typical standard chromatogram.

2) 0.53-mm ID × 30 m, 1.5 μm DB-5† or equivalent, at 25 cm/s. Starting at 30°C for 1 min, ramp 6°C/min to 150°C.

3) 2-mm ID × 2 m long glass packed with 1% SP-1000‡ on Carbopack B (60/80) operated at 50°C with 60 mL/min flow, or, if temperature-programmable GC is available, 45°C for 1 min, ramp 8°C/min to 240°C.

4) 2-mm ID × 2 m long glass packed with 10% squalene on Chromosorb WAW (80/100 mesh) operated at 67°C with 25 mL/min flow.

5) 2-mm ID × 3 m long glass packed with 6% OV-11/4% SP-2100 on Supelcoport (100/120 mesh); temperature program 45°C for 12 min, then 1°C to 70°C with 25-mL/min flow rate.

h. Mechanical shaker: Optionally, a rotary platform shaker.

i. Solvent pipetor: 2-mL transfer pipet, pipetor, or all-glass and TFE repipetor that attaches to the pentane storage bottle.

j. Transfer pipets, 5 mL. Preferably use a pipetor with disposable tips, cleaned and dried as recommended for TFE septa.

k. Analytical balance, capable of measuring to ± 0.01 g.

4. Reagents

a. Extraction solvent: See ¶ 1*b*. For capillary column split injection technique, preferably use only pentane. For other techniques, recommended solvent is pentane; alternatively, use hexane, methylcyclohexane, methyl-*tert*-butyl ether (MtBE) or 2,2,4-trimethylpentane. For alternative solvents, collect and document precision and bias data, evaluate extraction efficiency and effect of constituent concentration on efficiency, and maintain documents demonstrating applicability. Demonstrate that solvent is free of compounds for which the analysis is being performed.

b. Methyl alcohol, demonstrated to be free of interferences.

c. Neat standard materials: Use materials of 96% purity or greater for:

1) *Calibration standards:* bromoform (CHBr₃), bromodichloromethane (BDCM), dibromochloromethane (DBCM), chloroform (CHCl₃), 1,1,1-trichloroethane (TCA), tetrachloroethene (PCE), trichloroethene (TCE), and carbon tetrachloride (CCl₄).

2) *Internal standard:* 1,2 dibromopropane. A compound selected as an internal standard should have baseline resolution to separate it from constituents of interest and any interferences. Because this requirement is highly dependent on the samples and the analytical conditions and equipment used, no single internal standard is universally applicable. 1,2-dibromopropane has been found to be generally useful.

---

* Chromatographic methods are extremely sensitive to the materials used. Mention of trade names by *Standard Methods* does not preclude the use of other existing or as-yet-undeveloped products that give *demonstrably* equivalent results.
† J&W Scientific.
‡ Supelco.

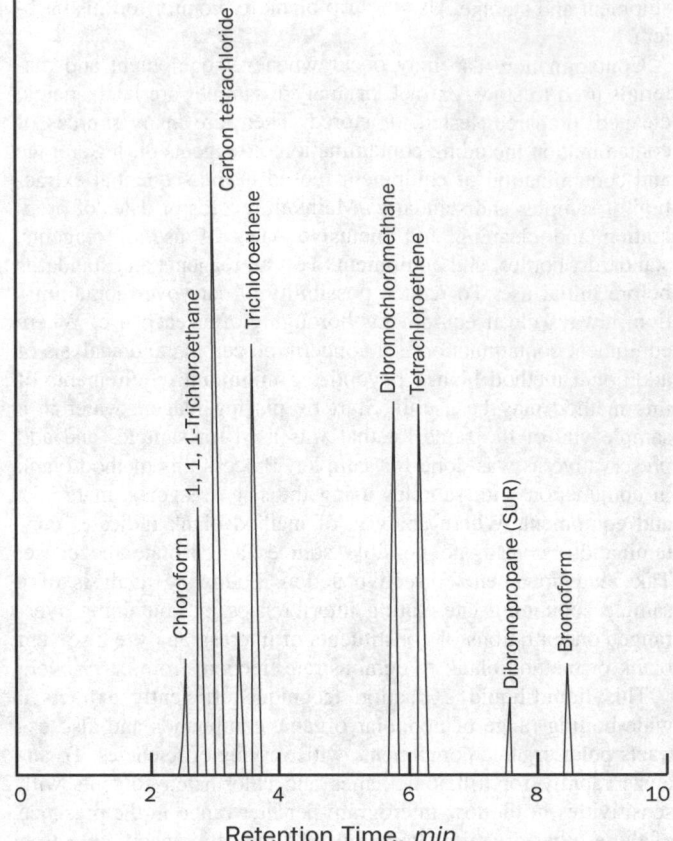

Figure 6232:1. Chromatogram for THMs and chlorinated organic solvents. Concentration was 50 μg/L for each compound; primary column DB-5. Retention times are shown in parentheses.

d. Reagent water: Generate VOC-free water, defined as water free of interference when used in the procedure described herein, by passing tap water through a carbon filter. Alternatively, prepare VOC-free water as follows: boil water for 15 min, then maintain at 90°C while bubbling a contaminant-free inert gas through water at 100 mL/min for 1 h. While water is still hot, transfer to a narrow-mouth screw-cap bottle with a TFE seal. Test VOC-free water each day before use by analyzing a method blank for constituents of interest. If any chlorine residual remains after such treatment, destroy it. See Section 1080 for additional information and general discussion of reagent-grade water.

e. Stock standard solutions: See Section 6200B.3*g*. Alternatively, purchase prepared standard solutions in methanol.
CAUTION: *Trihalomethanes and chlorinated solvents are toxic and may be carcinogenic: prepare primary stock solutions in a hood and wear appropriate personal protective equipment.*

f. Secondary dilution standards: From standard stock solutions, prepare multi-component secondary standards in methyl alcohol so that standards over the working range of the instrument can be prepared using no more than 20 μL methanolic standard solution /100 mL reagent water. See Method 6200B.3*h*.

g. Internal standard solution: Prepare stock solution from neat material in hexane. Make secondary dilution directly into storage container of pentane extracting solvent to produce a concentration of 30 μg internal standard /L pentane.

h. Aqueous calibration standards: Construct a calibration curve for each constituent using a minimum of three different concentrations, but preferably use five to seven concentrations. Bracket each sample with two of the concentrations. Use one concentration near, but above, the laboratory-determined detection limit. Where a sample component exceeds the range bracketed by standards, dilute a fresh volume of sample and re-extract, or prepare new standards in reagent water to bracket the concentration and dilute sample and standard extracts to bring them into the linear range of the detector. To prepare calibration standards, rapidly inject the required volume of alcoholic standard into the expanded area of a reagent-water-filled volumetric flask. Using an extended barrel syringe, inject the methanol *well below* the reagent water surface. Preferably incline the volumetric flask at an approximately 45-deg angle while injecting the standard. Remove syringe and stopper flask. Mix aqueous standards by gently inverting flask three times only. Discard to waste the contents in the neck of flask before transferring standards to sample vials. Add any preservatives to both samples and standards before extraction. Process standards through extraction in conjunction with sample sets. Aqueous standards, when stored with a headspace, are not stable; discard after 1 h. When stored in headspace-free sample storage vials, aqueous standards may be used for 24 h.

Avoid standard preparation procedures that require delivery of less than 10 μL of methanolic standards into volumetric flasks. Instead, use a larger volumetric flask and a larger volume of methanolic standard.

i. Quality control (QC) check standards: Obtain concentrate in methanol from USEPA or NIST for each compound, or if not available, from a second source vendor. If no second source is available, prepare stock standards separately from neat materials used for calibration standards. Prepare a mixed secondary dilution standard containing each compound and then an aqueous QC check standard at a concentration approximating the midlevel calibration standard.

5. Procedure

a. Extraction: Let samples and standards come to room temperature. Open each sample vial and remove 5 mL of sample and discard to waste, preferably using a transfer pipetor with disposable tips. Replace cap, weigh vial to nearest 0.1 g, and record weight.

Using a clean, dedicated volumetric measurement device (¶ 3*i*) carefully measure 2.00 mL pentane and add to sample vial. *Vigorously* shake by hand for 1 min or use a rotary platform shaker set at 60 to 100 rpm.

Let phases separate for at least 2 min. Where emulsions do not separate on standing, centrifuge or transfer entire emulsion to a separate vial and cool extract below 4°C to promote separation. Using a disposable glass pipet, transfer at least 1 mL of upper pentane extract to extract storage vials. Optionally, transfer half of the pentane extract to each of two vials to provide for reanalysis where necessary. Protect pentane extracts from warm temperatures and minimize extract holding time at room temperature. Store extracts at 4°C.

Empty sample/extraction vial to waste, rinse, and shake dry. Reweigh empty container with original cap to nearest 0.1 g and record weight. Calculate weight of sample extracted to the nearest 0.1 g by subtraction of vial-only weight from sample-plus-vial weight. For an assumed density of 1 g /mL, weight of sample extracted is equal to volume of sample extracted, in milliliters. Convert volume in milliliters to liters and record.

b. Sample and standard analysis: Before extraction of samples or standards prepare and analyze a method blank to verify freedom from interferences. Once extracts have been prepared, analyze standards and calculate a calibration curve or calibration factor as outlined in 6. Calculations. Inject 1 to 5 μL of standard extract depending on the configuration of the instrument and the required sensitivity. Inject exactly the same volume of extract each time, preferably using an autosampler. To test that injection volumes are repeatable, inject replicates of a single standard extract, and determine the standard deviation. Percent relative standard deviation (%RSD) should not be more than 5%. If this precision is not routinely achievable, use the internal standard calibration procedure.

After standardization, analyze the method blank, samples, and quality control samples. Extract and analyze a quality-control check standard each twentieth analysis and at the end of the analytical sequence. The percent recovery for the QC check standard should be between 80 and 120%. Develop historical mean control charts of QC check standard recovery for each compound and use the 99% confidence about historical data as the control criteria for rejection of QC check standards validity. Where criteria are failed, repeat analysis of any samples analyzed since the last QC check standard was in control.

c. Internal standard analysis procedure: Add the internal standard to the pentane solvent in the storage container at the concentration specified, and proceed with extraction and analysis of samples and standards as outlined above.

d. Compound identification: Identification of compounds in samples is based on comparison of retention times (RT) of suspect peaks to the confidence limits for RT of the authentic compounds in standards. Using the retention times of the standards analyzed, determine the average retention time for each compound and the standard deviation of the retention time. Tentatively identify peaks in sample chromatograms as compounds on basis of the 99% confidence interval around the calculated mean value using the calculated standard deviation. Nominally, the retention time window would be expected to be no wider than 0.25 min (packed column) and 0.05 min (capillary) before and after the average retention time calculated for the standards. When the 99% confidence limits for the data set are wider than the nominal value, institute corrective action.

Additional evidence of compound identity may be obtained by adding standard material to the suspect extract (standard addition) and reanalyzing. Presence of separate peaks in the extract with the known addition confirms that the suspect peak is not the compound of interest.

If chromatographic data systems are used to identify compounds, follow manufacturer's specifications. If the RT windows calculated by a computerized system are wider than the nominal values, investigate sources of retention time variability and take corrective action.

6. Calculations

a. External standard procedure: Use this procedure only if the volume of the injection can be held constant. Calculate individual response factors (RFs) for each standard analyzed as follows:

$$RF = \frac{\text{Nominal amount compound extracted, } \mu g}{\text{Response (peak area or peak height)}}$$

Calculate the amount of compound for each standard as:

$$W_s = V_s \times C_s$$

where:

W_s = amount of compound, μg,
V_s = volume of standard extracted, L, and
C_s = concentration of prepared standard, $\mu g/L$.

For each compound determine average RF and standard deviation of the RFs using all calibration standards analyzed. If the percent relative standard deviation [%RSD = (SD/mean RF) $\times$ 100] is less than 10% use average RF to calculate sample concentration.

If the %RSD is greater than 10%, plot a calibration curve of amount injected versus response. Use the graph to determine the amount of compound present in each sample. Then determine the concentration by dividing amount, μg, by the volume, L, of sample extracted. Optionally use a data system to prepare a linear regression and use the linear regression equation to calculate compound amounts in samples from response values.

Where average RF is used, determine sample concentration as follows:

$$C_x = \frac{RF \times R_x}{V_x}$$

where:

C_x = compound concentration, $\mu g/L$,
R_x = sample response (mm, area, etc.), and
V_x = volume of sample extracted, L.

Round all final sample results to two significant figures.

b. *Internal standard procedure:* For all analyses made in a given analytical sequence, determine average internal standard response and standard deviation of the internal standard response. Calculate percent relative standard deviation. If the %RSD is greater than 25% take corrective action to improve method precision. Establish the 99% confidence interval for the internal standard response using the calculated mean and standard deviation for the sample set. Reject analyses where the internal standard response is outside these confidence limits, and reanalyze. After analysis of calibration standards, calculate individual relative response factors (RRF) for each compound in each standard as follows:

$$RRF = \frac{R_s \times C_i}{R_i \times C_s}$$

where:

R_s, R_i = responses for calibration standard and internal standards, respectively, and
C_s, C_i = compound concentrations in calibration and internal standards, respectively.

Calculate average RRF for each compound, standard deviation of the RRFs, and %RSD. If %RSD is less than 10% use the average RRF; if it is greater, develop a calibration curve or a linear regression equation as outlined in the external standard procedure.

When using the average internal standard RRF, calculate concentration in samples as follows:

$$C_x = \frac{R_x \times C_i}{R_i \times RRF}$$

where:

C_x = compound concentration in sample, $\mu g/L$, and
R_x = sample response.

c. *Total trihalomethane concentration:* Calculate total trihalomethane concentration by summing the concentration of the four individual trihalomethanes in each sample. This is required for USEPA reporting purposes but it is preferable to report only individual THMs.

7. Quality Control

A minimum program of quality control consists of an initial demonstration of proficiency for each analyst and each instrument system and an ongoing program of quality control analysis. Record initial quality by documenting initial performance relative to published performance criteria. Maintain records of performance by comparing ongoing quality control checks to performance criteria and objectives for data quality. Document this performance as outlined in Sections 1010, 1030, and 6020.

a. *Analyst proficiency:* The analyst should be experienced in the operation of a GC/ECD and produce an initial demonstration of proficiency in accordance with the procedure outlined in 6200A.5a1).

b. *Method blanks:* Prepare and analyze method blanks. Concentrations of compounds in the method blank should not exceed the experimentally determined method detection limit. If the method blank is out of control isolate the source of contamination, apply corrective action, and process a new method blank. Under no circumstances subtract method blank values from the sample result.

c. *Quality control (QC) check standards:* Preferably obtain QC standards from a separate source and prepare independently from calibration standards. Analyze QC check standards as though they were samples. Compare results to known concentration of the check standard and calculate percent recovery. Percent recovery nominally should be between 80 and 120%. Develop mean recovery control charts of QC check standards results and use historical 99% confidence limits to accept or reject the ongoing calibration. Where historical confidence limits are wider than the nominal limits, investigate standard materials, preparation and storage procedures, and other potential sources of error. Take and document corrective actions.

d. *Detector sensitivity:* Maintain a log of detector response, in area counts or peak height, using one standard that is analyzed each day, to monitor changes in detector sensitivity. Optionally, plot these data to observe trends in detector sensitivity. Note the sensitivity at which method detection limit studies were performed and replace or repair detectors where minimum detectable

TABLE 6232:I. PRECISION AND BIAS DATA FOR THM-CHLORINATED ORGANIC SOLVENT METHOD, DB-5 COLUMN

Compound	Added Amount	Amount Recovered $\mu g/L$								Bias % Recovery	Precision % RSD
		A	B	C	D	E	F	G	H		
Chloroform (CHCl$_3$)	20.0	18.7	18.6	19.4	19.5	19.2	18.5	19.5	19.7	95.6	2.39
Bromodichloromethane (BDCM)	20.0	19.8	20.2	20.7	20.8	20.3	19.7	20.6	20.5	101.6	2.10
Dibromochloromethane (DBCM)	20.0	18.7	19.4	20.0	20.2	19.7	18.7	20.1	20.6	98.3	3.58
Bromoform (CHBr$_3$)	20.0	17.4	18.5	18.7	19.2	19.3	17.9	18.8	19.8	93.5	4.08
Trichloroethane (TCA)	20.0	18.5	18.8	19.7	19.9	19.8	18.5	20.1	20.4	97.3	3.83
Carbon tetrachloride (CCl$_4$)	20.0	20.1	20.4	20.0	20.2	20.6	20.0	20.1	20.0	100.8	1.19
Trichloroethene (TCE)	20.0	17.9	18.3	18.9	19.2	19.1	17.9	19.2	19.6	93.8	3.43
Tetrachloroethene (PCE)	20.0	19.8	20.4	20.6	20.9	20.7	19.7	20.7	20.7	102.2	2.24
Internal standard	100.0	99.0	95.0	100.0	102.0	101.0	99.0	105.0	105.0	100.8	3.30

quantities are significantly affected by declining detector sensitivity.

e. Laboratory-fortified samples with known additions: In a laboratory analyzing more than 10 samples daily, extract and analyze a known addition on each tenth sample. Be sure this is representative of different sample types because there is some evidence of matrix effects with liquid-liquid extraction methods. See Section 1020B.6. In a laboratory analyzing fewer than 10 samples daily, each time sample extractions are performed, extract and analyze at least one laboratory-generated known-addition sample. Chart percent recovery as outlined in Section 1020B.12 using a means chart. To evaluate method bias see Section 1030.

f. Duplicate analysis: Randomly select, then extract and analyze in duplicate, 10% of all samples. Maintain an up-to-date log on bias and precision data collected on known-addition samples and duplicate samples. Evaluate results as outlined in Section 1030. If results are significantly different from those cited in ¶ 8 below, check entire analytical scheme to determine why the laboratory's precision and bias limits are excessive.

g. Laboratory control standards (performance evaluation standards): Quarterly, add an external reference laboratory evaluation standard to organic-free water, extract, and analyze. Preferably obtain this standard from a governing agency or other authoritative source. The results from this sample should agree within 20% of the true value for each compound. If not, check each step in preparation and analysis to isolate the problem. Document external reference standard results and any corrective action taken.

8. Precision and Bias

The single-laboratory precision and bias data in Table 6232:I were generated by adding known amounts of trihalomethanes and chlorinated organic solvents to organic-free water. The mixtures were analyzed as true unknowns.

9. References

1. U.S. ENVIRONMENTAL PROTECTION AGENCY. 1980. Analysis of trihalomethanes in drinking water by liquid/liquid extraction. 40 CFR Part 141, Appendix C. Part II.
2. MIEURE, J.P. 1977. A rapid and sensitive method for determining volatile organohalides in water. *J. Amer. Water Works Assoc.* 69:60.
3. REDING, R., W. KOLLMAN, M. WEISNER & H. BRASS. 1978. THM's in drinking water: Analysis by LLE and comparison to purge and trap. *In* C.E. Van Hall, ed. Measurement of Organic Pollutants in Water and Wastewater. STP 686, American Soc. Testing & Materials, Philadelphia, Pa.
4. Handbook for Analytical Quality Control in Water and Waste Water Laboratories. 1972. Analytical Quality Control Lab., National Environmental Research Center, Cincinnati, Ohio.
5. RICHARD, J.J. & G.A. JUNK. 1977. Liquid extraction for rapid determination of halomethanes in water. *J. Amer. Water Works Assoc.* 69: 62.
6. U.S. ENVIRONMENTAL PROTECTION AGENCY. 1979. The Analysis of Trihalomethanes in Finished Water by the Purge and Trap Method. Environmental Monitoring & Support Lab., Environmental Research Center, Cincinnati, Ohio.
7. BUDDE, W.L. & J.W. EICHELBERGER. 1979. Organic Analysis Using Gas Chromatography-Mass Spectrometry. Ann Arbor Science Publ., Ann Arbor, Mich.
8. BRASS, H.J., et al. 1977. *In* R.B. Pojasek, ed. National Organic Monitoring Survey: Sampling and Purgeable Organic Compounds. Drinking Water Quality Through Source Protection, p. 398. Ann Arbor Science Publ., Ann Arbor, Mich.
9. KOPFLER, F.C., et al. 1976. GC/MS determination of volatiles for the National Organics Reconnaissance Survey (NORS) of Drinking Water. *In* L.H. Keith, ed. Identification and Analysis of Organic Pollutants in Water. Ann Arbor Science Publ., Ann Arbor, Mich.
10. WHITE, L.D., et al. 1970. Convenient optimized method for the analysis of selected solvent vapors in industrial atmosphere. *Amer. Ind. Hyg. Assoc. J.* 31:225.

6232 C.   Purge and Trap Gas Chromatographic/Mass Spectrometric Method

See Section 6200B for capillary-column method.

6232 D. Purge and Trap Gas Chromatographic Method

See Section 6200C for capillary column method.

6251   DISINFECTION BY-PRODUCTS: HALOACETIC ACIDS AND TRICHLOROPHENOL*

6251 A.   Introduction

1. Sources and Significance

The haloacetic acids (HAAs) are formed by the chlorination of natural organic (humic and fulvic) matter. Utilities using chlorine as a water disinfectant generate haloacetic acids, usually as the second most prevalent group of known disinfection by-products[1]; the primary group formed is usually the trihalomethanes. Toxicological studies indicate that dichloroacetic acid and trichloroacetic acid are animal carcinogens.[2] The USEPA has proposed a maximum level for the sum of six haloacetic acids,[3] and requires large utilities to monitor drinking water for specified haloacetic acids.[4]

* Approved by Standard Methods Committee, 1994.

2. References

1. U.S. ENVIRONMENTAL PROTECTION AGENCY & ASSOCIATION OF METROPOLITAN WATER AGENCIES. 1989. Disinfection By-Products in U.S. Drinking Water. Vol. 1 Report, James M. Montgomery Consulting Engineers, Pasadena, Calif.
2. ALCEON CORP. 1993. An Overview of Available Information on the Toxicity of Drinking Water Disinfectants and Their By-Products. Cambridge, Mass.
3. U.S. ENVIRONMENTAL PROTECTION AGENCY. 1994. Disinfectant Disinfection Byproduct Rule. Proposed. (June 13, 1994).
4. U.S. ENVIRONMENTAL PROTECTION AGENCY. 1994. Information Collection Rule. Federal Register 59(28):6332.

6251 B.   Micro Liquid-Liquid Extraction Gas Chromatographic Method

This method[1] was developed to analyze simultaneously for monochloroacetic acid (MCAA), monobromoacetic acid (MBAA), dichloroacetic acid (DCAA), trichloroacetic acid (TCAA), bromochloroacetic acid (BCAA), dibromoacetic acid (DBAA), and 2,4,6-trichlorophenol (TCPh) in treated and untreated drinking water. Additional haloacetic acids may be present and analyzed by this method; however, standards for these compounds are not readily available and are less stable.

1. General Discussion

a. Principle: The sample is extracted with methyl *tertiary*-butyl ether (MtBE) at an acidic pH to extract the nondissociated acidic compounds to be determined. A salting agent is added to increase extraction efficiency. The extracted compounds are methylated with diazomethane solution to produce methyl ester or ether derivatives that can be separated chromatographically. A temperature-programmable gas chromatograph using a fused silica capillary column and an electron capture detector (ECD) is used for analysis. Simultaneous analysis and confirmation using a single injection can be effected by setting up both the analytical column and the confirmation column to share a common injection port. Alternatively, use separate analytical and confirmation columns. Alternative detectors may be used if QC criteria can be met. Aqueous calibration standards are extracted, methylated, and an-

alyzed in the same manner to compensate for less than 100% extraction and derivatization efficiencies.

b. Interferences: Impurities in extraction solvent and on glassware and other equipment can interfere. Follow specifications and cleaning procedures carefully to minimize interference. As shown in Figure 6251:1, the analysis separates peaks of haloacetic acids from those of other common disinfection by-products. Use of two columns is recommended because for waters with high carbonate contents, false-positive MCAA peaks have been observed on the column described in ¶ 3*h*4).

c. Detection levels: Method detection levels (MDL) are given in Table 6251:I.[2] The method has been shown to be useful for haloacetic acids over a working range of 0.5 to 30 μg/L (1.0 to 30 μg/L for MCAA) and 0.25 to 15 μg/L for TCPh. The calibration range can be extended, depending on the compound and detector characteristics.

d. Safety: The toxicity and carcinogenicity of each reagent has not been defined precisely. Minimize exposure to these chemicals and use them only in a properly operating ventilation hood.

Avoid exposure to DCAA and TCAA because they are carcinogens.[3] Avoid contact with the other haloacetic acids and their solutions.

MNNG (1-methyl-3-nitro-1-nitrosoguanidine) is carcinogenic. Keep in properly labelled plastic containers, containing activated carbon, with tight-fitting lids and store in a refrigerator used only for chemical storage. Store spatulas and glassware for the han-

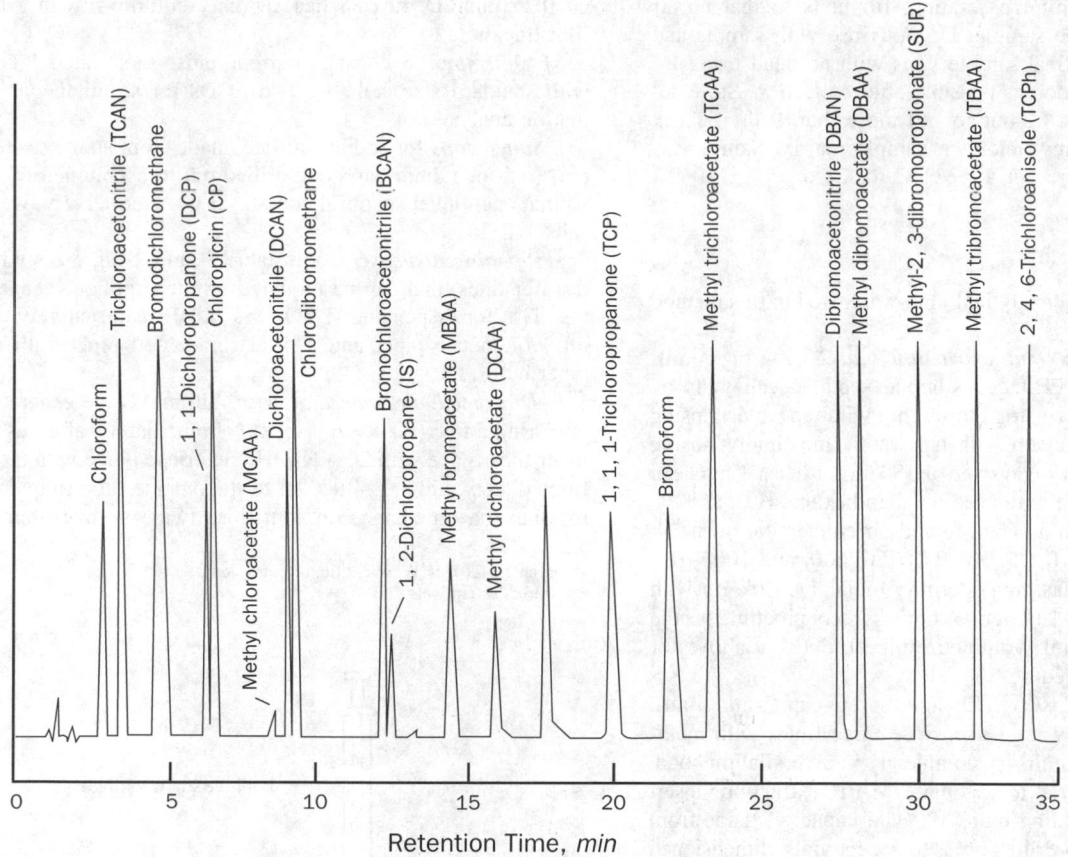

Figure 6251:1. Haloacetic acids separation from other commonly produced disinfection by-products on a DB-1701 column. Chromatogram produced by mixing the methyl esters of each haloacetic acid with other disinfection by-products in MtBE (disinfection by-products tentatively identified).

dling of MNNG in specially labelled plastic containers and use only for MNNG.

Diazomethane is toxic, carcinogenic, and an explosion hazard.[4] Follow special precautions whenever handling this material. Use only in a properly operating fume hood; CAUTION: *Do not breathe vapors.* To avoid explosions, do not heat above 90°C and do not use glassware with ground-glass surfaces (e.g., ground-glass joints, sleeve bearings) or glass stirrers. Special glassware for dia-

zomethane generation and handling, as well as screw-cap volumetric flasks, are available commercially. Always use a safety shield when generating diazomethane. Always quench excess diazomethane with silica gel. Do not store diazomethane/ether solutions; they are extremely hazardous and tend to become contaminated.

Store ether in tightly-closed amber bottles in an explosion-safe or -proof refrigerator. Store only with compatible chemicals. Eliminate all sources of ignition; keep away from heat, sparks, and flames. Handle ether only in a hood and avoid direct physical contact. Do not breathe vapors. If ether is spilled or leaks, evacuate area, ventilate, and absorb on vermiculite or similar material. Wear appropriate OSHA equipment before entering spill area. Also see Section 1090.

2. Sampling and Storage

See Section 6010B.1.

Preferably collect grab samples in quadruplicate to allow sufficient volume for replicates and known additions. Flush sampling tap until water temperature stabilizes and stagnant lines are cleared. Collect samples in nominal 40- or 60-mL vials containing approximately 65 mg crystalline NH₄Cl (bake overnight at >100°C to eliminate contaminants), which converts free chlorine to a combined chlorine residual, and sealed with TFE-faced septa

TABLE 6251:I. METHOD DETECTION LEVELS AND PRECISION DATA*

Compound	Added Conc. μg/L	Found Conc. μg/L	Standard Deviation μg/L	Relative Standard Deviation %	Method Detection Level μg/L
Monochloroacetic acid	0.50	0.54	0.026	4.8	0.082
Monobromoacetic acid	0.50	0.80	0.028	3.4	0.087
Dichloroacetic acid	0.50	0.5	0.017	3.5	0.054
Trichloroacetic acid	0.50	0.5	0.017	3.4	0.054
Bromochloroacetic acid	0.50	0.49	0.015	3.1	0.04
Dibromoacetic acid	0.50	0.47	0.021	4.4	0.065
2,4,6-Trichlorophenol	0.25	0.27	0.011	4.1	0.034

* Based on the analysis of seven portions of reagent water with known additions.[2]

and screw caps. To minimize aeration, fill vials so that no air bubbles pass through the sample. Do not rinse with sample and do not let vial overfill. Seal sample vials with no headspace.

Analyze samples as soon as possible after collection. Store dechlorinated samples at 4°C, but for no more than 9 d;[1,5] check compound stability in any unknown sample matrix. Sample extracts can be held in a freezer at −11°C for 21 d.

3. Apparatus

Preferably dedicate all analytical glassware used in this method to this procedure.

*a. Sample containers and extraction vials,** 40- or 60-mL screw-cap vials with TFE-faced silicone septa. Clean vials by washing with detergent, rinsing thoroughly with tap water, rinsing with 1:10 HCl, rinsing again with tap water, and finally rinsing with reagent water. Heat in an oven at 180° for at least 1 h. Clean caps and septa by rinsing with acetone, then hexane. Heat at 80°C for not more than 1 h in a clean, forced-air convection oven.

b. Microsyringes, 5, 10, 25, 50, 100, 250, 500, and 1000 μL.

c. Syringe, 30-mL glass hypodermic, metal luer-lok tip with 8.9-cm- (3.5-in.-) long × 17 gauge stainless steel pipetting needle (alternatively use a 30-mL volumetric pipet). See ¶ *a* above for glassware cleaning procedure.

d. Micro volumetric flasks,† TFE-lined screw-cap: 2-mL, 5-mL, and 10-mL. Immediately after use, rinse three times with methanol. Invert to drain. Let air-dry completely in a ventilation hood.

e. Mechanical shaker,‡ to automate MtBE extraction. Insert vials into a wooden holding block (20 vial capacity) made from laminated plywood with drilled holes to accept vials, dimensioned to fit snugly onto the shaker table.

f. Extract and standard solution storage container, 1.8-mL clear glass, 7- and 14-mL amber glass screw-cap vials with TFE-lined silicone septa. For cleaning procedures, see ¶ *a* above.

g. Transfer pipets, 14.6- and 23-cm (5.75- and 9-in.) disposable glass pasteur pipets. See ¶ *a* above for glassware cleaning procedure.

h. Gas chromatograph, temperature-programmable (preferably with multiple ramp capability) with injector. Optimally use an autosampler for sample injection and a computer data system for peak integration and quantitation. (A detector base that can mount two electron capture detectors is ideal.)

1) Gas handling equipment: Use carrier (helium) and makeup (nitrogen) gases of high purity (99.999%) grade that pass through indicating calcium sulfate, molecular sieve 5A, activated charcoal, and an oxygen purifying cartridge. Use two-stage metal diaphragm high-purity regulators at the compressed gas sources. Use flow controllers to regulate carrier gas flow. Make all gas lines 0.3-cm (1/8-in.) copper (or stainless steel) tubing; rinse with acetone and bake before use.

2) Injector, split/splitless (using straight open bore insert).

3) Analytical column,§ 30 m long × 0.25 mm ID, fused silica capillary column with a 0.25-μm film thickness or equivalent.

4) Confirmation column, 30 m long × 0.25 mm ID, fused silica capillary column‖ with a 0.25-μm film thickness or a 30 m long

× 0.25 mm ID, fused silica capillary column# with a 0.5-μm film thickness.

5) Detectors, a constant-current pulse-modulated [63]Ni ECD with standard size cell (use two ECDs for simultaneous confirmation analysis).

i. Salt scoops for sodium sulfate, made from stainless steel 1.3-cm- (0.5-in.-) diam bar stock drilled out to a volume of 1.73 mL so that each level scoopful contains 3 g. Alternatively, weigh the salt.

j. Pipetting dispensers, adjustable 5- and 2-mL sizes with TFE transfer lines, that can be mounted on the supplier's reagent bottles. Use for dispensing H_2SO_4 and MtBE. Alternatively use a 3-mL volumetric pipet and a 5-mL graduated pipet with manual pipet bulbs.

k. Diazomethane generator: Use millimole-size generator with "o"-ring joint (Figure 6251:2).** Immediately after use, rinse inner tube twice with 20% NaOH, then rinse twice with tap water. Immediately add 1 g silica gel to the outside tube to quench any residual diazomethane solution, rinse twice with methanol, and

---

\# Durabond-210, J&W Scientific, or equivalent.
** Aldrich or equivalent.

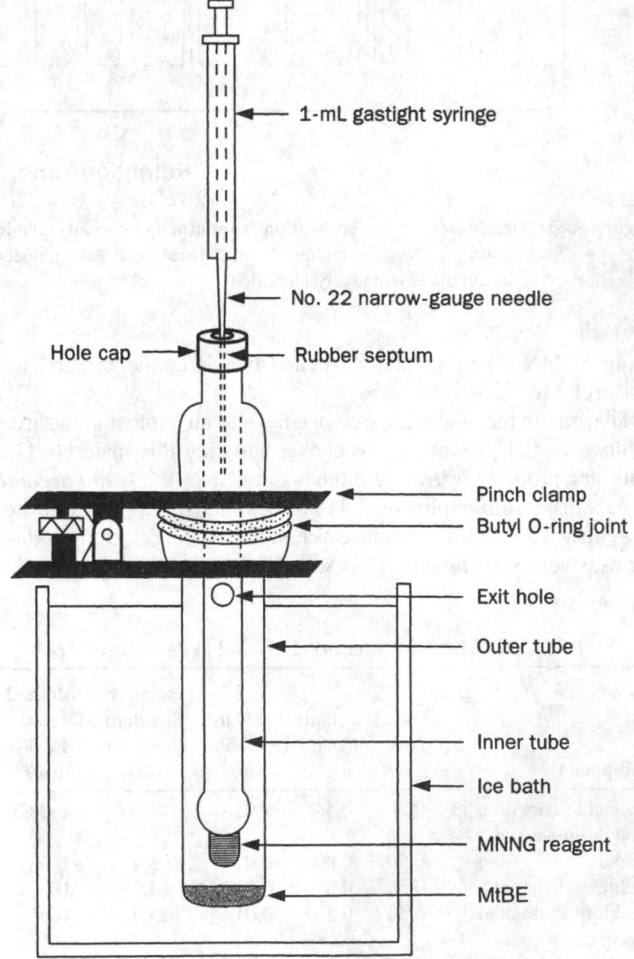

Figure 6251:2. Easy-to-use diazomethane generator apparatus for preparing small amounts of diazomethane in methyl *tertiary*-butyl ether (MtBE).

Labels on figure:
- 1-mL gastight syringe
- No. 22 narrow-gauge needle
- Hole cap
- Rubber septum
- Pinch clamp
- Butyl O-ring joint
- Exit hole
- Outer tube
- Inner tube
- Ice bath
- MNNG reagent
- MtBE

---

* Wheaton: Industrial Glassware, Millville, NJ; or equivalent.
† Kontes or equivalent.
‡ Eberbach or equivalent.
§ Durabond-1701, J&W Scientific, or equivalent.
‖ Durabond-5, J&W Scientific, or equivalent.

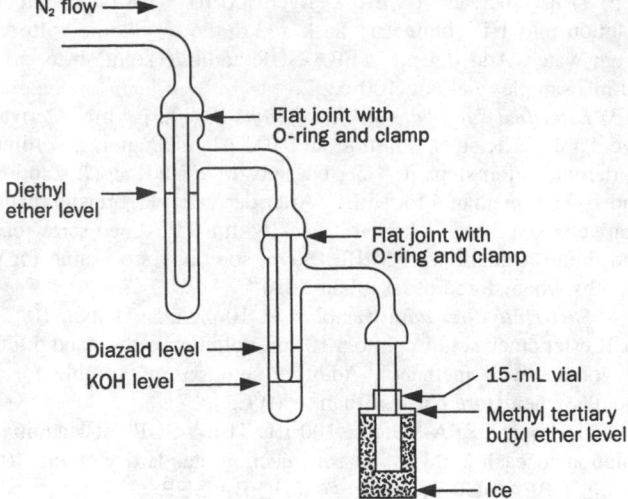

Figure 6251:3. Easy-to-use alternative diazomethane generator for preparing small amounts of diazomethane in MtBE.

TABLE 6251:II. ANALYTICAL STANDARDS

Compound*	Purity %	Molecular Weight	Boiling Point °C @ mm†
MCAA	99	94.5	183
MBAA	99+	138.95	208
DCAA	99	128.94	194
TCAA	98	163.39	198
BCAA	97	173.39	215‡
DBAA	99	217.86	195 @ 250
TCPh	95	197.45	246
IS-DBP	95	201.9	140–142
IS-TCP§	99	147.43	156
SUR-DBPA	99	231.88	160 @ 20
SUR-TFBA	99	194.09	—
MeCA	99+	108.52	130 @ 740
MeBA	98	152.98	144
MeDCA	99+	142.97	143
MeTCA	99	177.42	152–153
MeBCAA	98	187.42	155
MeDBA	—	—	182
TCAn	99	211.48	132 @ 28

* Sources: BCAA, Radian Corp., Austin, TX; DBAA, Fluka Chemika-BioChemika, Switzerland; TCPh obtainable from Chem Service, Inc., Westchester, PA; MeDBA and SUR ester, derivatized acid at Metropolitan Water District of Southern California laboratory; other compounds from Aldrich Chemical Company, Inc., Milwaukee, WI.
† °C at reduced pressure in mm Hg.
‡ Decomposes.
§ Ensure that TCP is not a contaminant when it is used as IS.

twice with tap water. Rinse both inner and outer tubes with reagent water three times. Bake at 180°C until dry in a clean, forced-air convection oven.

Alternatively use the millimole-size diazomethane generator shown in Figure 6251:3.†† To clean, rinse with reagent water and then with methanol, invert, and let air-dry.

l. pH-strips, pH indicating strips, 0 to 2.5 range.

4. Reagents

a. Extraction solvent, 99+ % MtBE, preservative-free.‡‡
b. Sodium sulfate reagents:
1) *Granular sodium sulfate,* Na₂SO₄, reagent grade, suitable for pesticide analysis. Heat at 400°C overnight in a shallow stainless steel pan covered with aluminum foil. Store in a 1-L glass bottle with TFE-lined polypropylene cap.
2) *Acidified sodium sulfate:* To 100 g anhydrous sodium sulfate, heated as above and cooled, add diethyl ether to just cover the solid; make a slurry. Add 0.1 mL conc sulfuric acid and mix thoroughly. Remove ether under low vacuum. Mix 1 g acidified sodium sulfate with 5 mL reagent water and check that pH is less than 4. Store remainder of reagent at 130°C.
c. Methanol, pesticide grade or equivalent.
d. Ammonium chloride, NH₄Cl.
e. Reagents for diazomethane generation with the apparatus shown in Figure 6251:2:
1) *Sodium hydroxide solution,* NaOH, 20%: Dissolve 200 g ACS low-carbonate-grade pellets in 800 mL reagent water.
2) *1-methyl-3-nitro-1-nitrosoguanidine,* MNNG.§§
f. Reagents for diazomethane generation with apparatus shown in Figure 6251:3:
1) *Diethylene glycol monoethyl ether.*‖‖
2) *N-methyl-N-nitroso-p-toluene sulfonamide.*##

3) *Ethyl ether,* absolute.
4) *N-methyl-N-nitroso-p-toluene sulfonamide solution:* Mix 10 g reagent *f* 2) in 100 mL 1:1 (v:v) solution of ethyl ether and reagent *f* 1). Solution is stable for 1 month when stored at 4°C in an amber-colored bottle with a TFE-lined screw cap.
5) *Potassium hydroxide solution,* KOH: Dissolve 37 g in 100 mL reagent water.
g. Silica gel, 35/60 mesh activated at 180°C and stored in a desiccator.
h. Sulfuric acid, H₂SO₄, conc.
i. Copper (II) sulfate pentahydrate, CuSO₄·5H₂O.
j. Standard material, see Table 6251:II for source and physical information.
1) *Individual haloacetic acid standard stock solutions:* Prepare individual haloacetic acid and the trichlorophenol stock solutions as follows: Weigh 0.150 g of each acid. Dilute each standard in MtBE to 10 mL in a screw-top volumetric flask. (NOTE: Do not use methanol for dilution, because spontaneous methylation of the haloacetic acid may occur in methanolic solution.[7] Transfer each to a separate clean 14-mL amber vial and store in a freezer at −11°C. Stock standards are usable for 6 months.
2) *Multicomponent haloacetic acid additive solution:* Prepare a six-component additive solution using individual haloacetic acid stock solutions. Dilute 16.7 μL of each stock standard into a 10-mL volumetric flask containing 9 mL MtBE, but use 8.4 μL of 2,4,6-trichlorophenol solution. After adding all stock solutions, dilute to volume with MtBE. This gives 25 μg/mL of each HAA and 12.5 μg/mL for 2,4,6-trichlorophenol. The additive solution is usable for 3 months when stored at less than −10°C. Alternatively, prepare known-addition solution monthly in methanol and store at 4°C.

†† Paxton Woods Glass, Cincinnati, OH, or equivalent.
‡‡ Omnisolv, manufactured by EM Science, Gibbstown, NJ, or equivalent.
§§ Aldrich or equivalent.
‖‖ Carbitol (Aldrich), or equivalent.
Diazald (Aldrich), or equivalent.

Measure microliter volumes with a gastight syringe using the solvent flush delivery technique. Do solvent flush with a 25-μL syringe by first drawing up 2.5 μL of solvent and then drawing the syringe plunger to the 5-μL mark with air. From the 5-μL mark measure amount of stock solution desired and then deliver the entire contents to the volumetric flask.

3) *Individual haloester standard stock solutions:* Prepare individual methyl ester stock solutions as follows: Weigh 0.1 × (molecular weight of ester/molecular weight of acid) g of each methyl ester in a 10-mL volumetric flask and dilute to mark with MtBE. Prepare methyl ester for dibromoacetic acid by derivatizing 1 mL of a 20 000-μg/mL acid solution with 100 μL methanol (follow derivatization steps in 5e below, but substitute dibromoacetic acid stock solution as the solution added to the outer tube for collection of the diazomethane). After derivatizing, transfer ester quantitatively to a 2-mL volumetric flask with a TFE-lined screw cap and dilute to mark with MtBE. Stock standards are usable for 6 months when stored at less than − 10°C.

4) *Multicomponent haloester additive solution:* Prepare a multicomponent additive solution by diluting 10 μL of each haloester stock standard, but use 5 μL of 2,4,6-trichloroanisole (methyl ether of the phenol), in a 10-mL volumetric flask and bring to volume with MtBE. This will yield a mixture containing approximately 10 μg/mL each, except for 2,4,6-trichloroanisole, which will be approximately 5 μg/mL. Additive solution is usable for 3 months when stored at less than − 10°C.

5) *Direct injection haloester standards:* Prepare direct injection standards using 10 μg/mL multicomponent haloester additive solution, a 30-μg/mL internal standard additive solution [see ¶ 4k2)], and a 10-μg/mL methanol solution of methyl-2,3-dibromopropionate [surrogate ester, see ¶ 4l4)]. Prepare direct injection standards by diluting appropriate volumes of multicomponent haloester additive mix, internal standard additive solution, and surrogate ester solution with enough MtBE to give a final volume of 1.0 mL.

k. Internal standard, 1,2,3-trichloropropane (IS-TCP), 98% pure (alternatively use 1,2-dibromopropane (IS-DBP).

1) *Internal standard stock solutions:* Weigh 50 mg into a 10-mL volumetric flask and bring to volume with methanol. This will yield a 5000-μg/mL stock solution. Stock standards are usable for 6 months when stored at less than − 10°C.

2) *Internal standard additive solution,* 30 μg/mL: Deliver 60 μL internal standard stock solution into a 10-mL volumetric flask and dilute to volume with methanol. Divide evenly among six 1.8-mL vials and store at −11°C. Additive solution is usable for 3 months.

Add 20 μL of internal standard additive solution to each 2 mL extract, yielding internal standard of 300 ng/mL.

l. Surrogate, (DBPA-SUR) 2,3-dibromopropionic acid, 99% pure or 2,3,5,6-tetrafluorobenzoic acid (TFBA-SUR).

1) *Surrogate stock solution,* 20 000 μg/mL: Weigh 0.2000 g SUR acid into a 10-mL screw-cap volumetric flask and dilute to mark with MtBE. Stock solutions are usable for 6 months when stored at less than − 10°C.

2) *Surrogate additive solution:*

a) *DBPA-SUR,* 10 μg/mL: Deliver 5 μL DBPA-SUR stock solution into a 10-mL volumetric flask and dilute to volume with methanol. Divide evenly among six 1.8-mL vials and store at −11°C. Additive solutions are usable for 3 months.

Add 30 μL DBPA-SUR additive solution to each 30-mL sample portion, yielding DBPA-SUR 10 μg/L.

b) *TFBA-SUR,* 20 μg/mL: Deliver 300 μL TFBA-SUR stock solution into 1-L volumetric flask and dilute to volume with reagent water. Add 0.5 mL TFBA-SUR additive solution to each 30-mL sample, yielding 100 μg/L.

3) *Esterified surrogate stock solution,* 10 000 μg/mL: Derivatize 1 mL SUR stock solution and 100 μL methanol, according to derivatization steps in ¶ 5e, but substitute SUR stock solution and 100 μL methanol for MtBE. After derivatizing, transfer quantitatively to a 2-mL volumetric flask with a TFE-lined screw cap and dilute to mark with MtBE. Stock solutions are usable for 6 months when stored at less than − 10°C.

4) *Surrogate ester additive solution,* 10 μg/mL: Deliver 10 μL SUR ester stock solution into a 10-mL volumetric flask and dilute to volume with methanol. Additive solutions are usable for 3 months when stored at less than − 10°C.

Add 10 μL DBPA-SUR or 100 μL TFBA-SUR ester additive solution to each 1 mL of direct injection standard yielding 100 ng/mL DBPA-SUR or 1000 ng/mL TFBA-SUR.

m. Reagent water: See Section 1080.

n. Calibration standards: Prepare aqueous calibration standards in reagent water by injecting a measured amount of the multicomponent haloacetic acid solution directly into water using the solvent flush technique. Prepare five different concentration levels from 0.5 to 30 μg/L for the HAAs and 0.25 to 15 μg/L for 2,4,6-trichlorophenol in 40-mL TFE-lined screw-top bottles containing 30 mL reagent water. Extract these standards and process the same way as samples, using the procedure given below.

5. Procedure

a. Sample preparation: Remove samples and standards from storage and let equilibrate to room temperature. Each time a new sample matrix is analyzed check that the amount of H_2SO_4 added will reduce pH to less than 0.5 before beginning microextraction. Test a separate 10-mL portion by adding 1 g $CuSO_4$, 4 g granular Na_2SO_4 [¶4b1)], and 0.5 mL conc H_2SO_4, mix until salt dissolves, then test using pH indicating strips.

b. Microextraction: Transfer 30 mL from the sample container to a 40- or 60-mL vial with TFE-faced septum and screw cap.

Add surrogate additive solution as indicated in ¶ 4l2) to each sample, including standards and blanks. Add haloacetic acid additive solution at this step for known additions.

Take one vial at a time and add the following in sequence: 1.5 mL conc H_2SO_4, 3 g $CuSO_4$, 12 g baked Na_2SO_4 [¶4b1)], and 3 mL MtBE. Immediately cap and shake briefly by hand to break up any salt clumps.

When using automated extraction, place vials in mechanical shaker wooden holding block. Shake vials at fast speed (approximately 300 cycles/min) for 9 min; alternatively shake manually for 2 min until salt is dissolved.

Remove vials, place upright, and let stand for at least 3 min until the phases separate.

c. Preparation of diazomethane[6]: Using the apparatus in Figure 6251:2, add approximately 130 mg MNNG to the inside tube of the generator. Add 0.5 mL reagent water to the MNNG and secure cap and septum. Add 2 mL MtBE to the outside tube. Place butyl o-ring in glass joint, place inside tube firmly on top of o-ring, and clamp securely with a screw-type pinch clamp.

Place generator in an ice water bath containing enough ice to keep diazomethane MtBE solution at 0°C until used.

Add 600 µL 20% NaOH (1 drop/5 s) using a 1-mL gastight syringe (22-gauge needle) through the generator septum (check that the syringe needle is on the opposite side of the vapor exit hole). Let derivatization continue for 30 min after adding NaOH; use product as soon as possible. Add more ice if necessary to maintain temperature.

If more diazomethane is needed, prepare two or more batches and combine before use.

Alternatively use the apparatus in Figure 6251:3 to prepare diazomethane. Add enough ethyl ether to tube 1 to cover the first impinger. Add 10 mL MtBE to 15-mL collection vial. Set nitrogen flow at 5 to 10 mL/min. Add 4 mL sulfonamide solution and 3 mL 37% KOH solution to the second impinger. Connect tubing as shown and let nitrogen flow purge diazomethane from the reaction vessel into the collection vial for 30 min. Cap vial when collection is complete and hold at 0°C. When stored at 0°C diazomethane solution may be used over a period of 48 h.

d. Separation and concentration: NOTE: Ensure that all items that come into contact with the sample prior to methylation have been washed with a dilute solution of sulfuric acid.

The drying step included here may be used if excess diazomethane is required to maintain the persistent yellow color of the sample (*5e*). It is not necessary in every case and may be used at the discretion of the analyst. Plug a small disposable pipet with a small amount of acid-washed glass wool. Add approximately 1 g acidified Na_2SO_4 [¶4*b*2)] to the pipet and pass exactly 2 mL of the top MtBE sample extract through the salt, being careful not to transfer any of the aqueous phase. Rinse the salt in the pipet with two 250-µL volumes of solvent and collect together in a receiver ampule (a 2-mL volumetric flask with TFE-lined screw cap) for subsequent concentration and methylation.

Concentrate MtBE extract to approximately 1.7 mL using a moderate stream of nitrogen blowing on the surface of the extract.

e. Derivatization: Add 20 µL of 30 µg/mL internal standard additive solution to each concentrated extract. (The internal standard is added at this time to minimize manipulation in the presence of diazomethane.) Cool in an explosion-safe freezer or in an ice bath for 7 min and add diazomethane (cooling extracts is unnecessary if diazomethane is generated by apparatus shown in Figure 6251:3).

Uncap one volumetric flask and add 250 µL of cold diazomethane/MtBE solution. Cap immediately with TFE-lined screw cap; mix gently by inverting once. Repeat for remaining extracts. A persistent yellow color after addition of diazomethane indicates that an excess is available for esterification. If necessary add more diazomethane solution.

Hold for 15 min at 4°C in an explosion-safe or explosion-proof refrigerator. Alternatively, keep extracts in an ice bath (cooling is unnecessary if diazomethane was generated by the apparatus shown in Figure 6251:3).

After holding 15 min, place extracts in a hood and let stand another 15 min until they reach room temperature. Dilute to mark with MtBE and invert flask to mix. If using an autosampler, transfer each extract evenly between two labeled autosampler vials containing approximately 0.01 g silica gel with a 23-cm (9-in.) disposable pasteur pipet to quench excess diazomethane. Keep each extract in contact with diazomethane for approximately the same amount of time before quenching. Store extra autosampler vial in freezer at −10°C as a backup extract. Alternatively, add silica gel to volumetric flask after derivatization is complete.

f. Gas chromatography: Typical operating conditions for the chromatograph are as follows:

Injector temperature 160°C; split valve opened at 0.5 min.

Temperature program: 37°C for 21 min, rising 11°C/min to 136°C, holding 3 min at 136°C, rising 20°C/min to 236°C, holding 3 min at 236°C.

Detector temperature: 300°C.

Carrier gas flow: 30 cm/s at 37°C.

Makeup gas flow: 23 mL/min.

At the beginning of each analytical run, inject two MtBE solvent blanks to condition the GC and to verify that interferences are absent. A 2-µL extract is injected in splitless mode. Always inject same sample volume and use sample dilution to obtain response in the calibration range. If levels greater than the highest standard are obtained, reanalyze diluted sample extract and readjust internal standard concentration. Calculate concentration only for those compounds that were at levels higher than the calibration curve; for other compounds use values obtained from the undiluted sample extract. See Figure 6251:4 for a chromatogram of an extracted and derivatized 30-µg/L standard on the analytical column. See Table 6251:III for retention times. A direct standard may be injected after the MtBE solvent blanks to verify continued system performance.

g. Calibration: Use five levels of calibration standards to define the quantitation range. The lowest standard should be near the limit of quantitation (LOQ) (see Table 6251:IV) for each compound. Use other standards to bracket the expected range of sample concentrations; do not exceed linear range of detector. Prepare standards by adding haloacetic acids and trichlorophenol to reagent water and then extract with the same solvent and derivatize with the same batch of diazomethane as used for the samples. Use the same extraction/esterification procedure for both standards and samples to correct for recovery characteristics. Analyze calibration standards under the same chromatographic conditions as samples.

Analyze calibration standard at one or more levels with each sample set to verify the working calibration curve. If the calibration standard is within ± 15% of the expected value, a new five-level calibration curve is not needed.

6. Calculation

A 2-µL injection of each calibration level will provide peak area (A_a) data for each compound and an internal standard peak area (A_i) for each level; use these peak areas to calculate relative response for each compound.

$$\text{Relative response} = A_a/A_i$$

A calibration curve passing through zero is generated from the plotted points for each compound using the relative response versus standard concentration. Use the internal standard quantitation method to determine unknown concentrations by a linear, quadratic, or point-to-point curve fit.

7. Quality Control

a. Quality control program: Because sample preparation requires many manipulations, chances for errors are increased. Consequently, at least follow minimum quality control requirements to monitor and maintain method performance. Include method

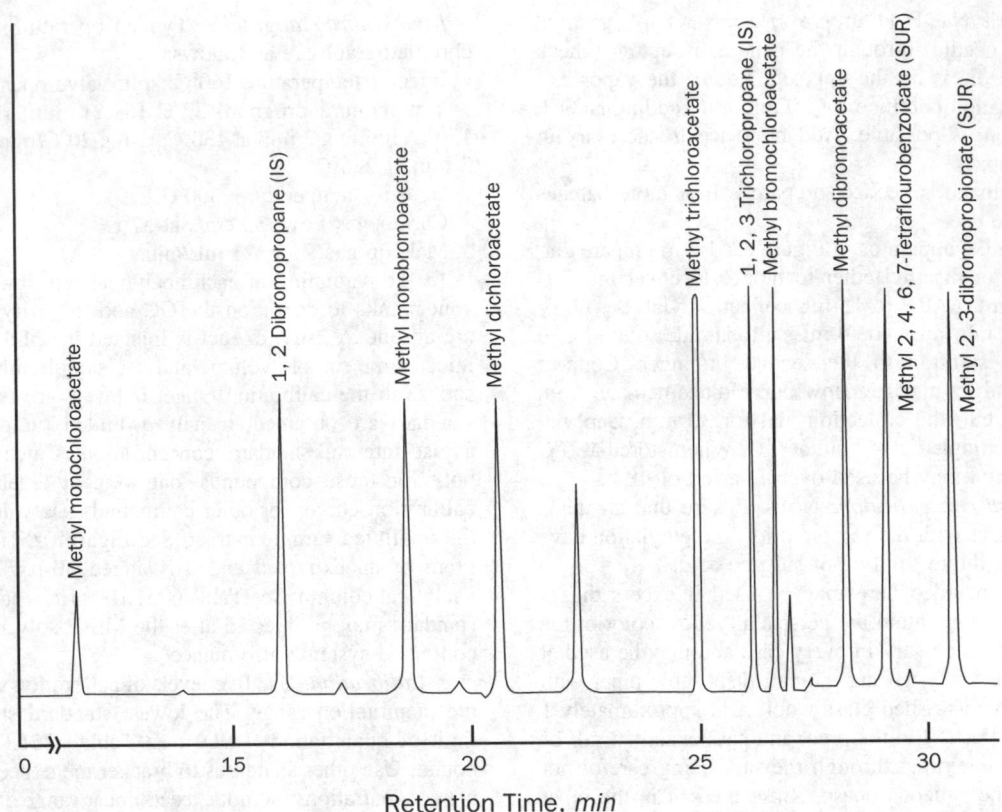

Figure 6251:4. Chromatogram produced by reagent water with known additions: 30 µg/L extracted standard on DB-1701 column.

blanks, an initial demonstration of laboratory capability and detection limits, assessment of the internal standard recovery, determination of surrogate compound recoveries, evaluation of calibration data and curves, sample matrix additions, and precision of replicate sample analysis. Additional quality control measures may be used.

b. Method blanks: Process a method blank (30 mL reagent water) with each set of samples. If the blank produces any peak within the retention time window of a compound that would prevent its determination, seek out and eliminate the source of contamination and reanalyze samples.

c. Initial demonstration of capability: To demonstrate an adequate level of performance, conduct the following operations before analyzing samples and whenever any major analytical change, such as new analyst or switch in type of column, is made.

1) Accuracy as percent recovery—Establish a calibration curve as in ¶ 5*g* and select a representative additive concentration (5 µg/L is convenient) for each target compound. Using a syringe,

TABLE 6251:III. RETENTION TIMES

Compound	Retention Time min		
	DB-1701 Column	DB-5 Column	DB-210 Column
Methyl chloroacetate (MeCA)	11.11	5.80	10.97
Methyl bromoacetate (MeBA)	18.80	9.23	13.03
Methyl dichloroacetate (MeDCA)	20.53	10.11	12.72
Methyl trichloroacetate (MeTCA)	24.78	18.94	14.37
Methyl bromochloroacetate	26.23	19.19	15.11
Methyl dibromoacetate (MeDBA)	28.64	25.51	16.83
2, 4, 6-Trichloroanisole (TCAn)	~ 34.7	33.74	22.08
1,2-Dibromopropane (internal standard)	15.97	10.78	—
1,2,3-Trichloropropane (internal standard)	25.87	18.43	13.87
Methyl-2,3-dibromopropionate (surrogate ester)	30.74	28.60	—
Methyl-2,3,6,7- tetrafluoro-benzoicate (surrogate)	29.19	26.84	—

TABLE 6251:IV. RECOMMENDED QUANTITATION LIMITS

Compound	Recommended Quantitation Limit µg/L
Monochloroacetic acid (MCAA)	1.0
Monobromoacetic acid (MBAA)	0.5
Dichloroacetic acid (DCAA)	0.6
Trichloroacetic acid (TCAA)	0.6
Bromochloroacetic acid (BCAA)	0.8
Dibromoacetic acid (DBAA)	0.6
2,4,6-Trichlorophenol (TCPh)	0.4

TABLE 6251:V. ADDITIVE RECOVERY IN REAGENT WATER*

Compound	Conc. Added µg/L	Mean Conc. Recovered µg/L	Standard Deviation µg/L	Relative Standard Deviation %	Mean Recovery %
Monochloroacetic acid	5.00	4.90	0.19	3.88	98.0
Monobromoacetic acid	5.00	4.95	0.13	2.67	99.0
Dichloroacetic acid	5.00	4.95	0.15	3.11	99.0
Trichloroacetic acid	5.00	5.06	0.16	3.06	101
Dibromoacetic acid	5.00	4.98	0.16	3.11	99.6
2,4,6-Trichlorophenol	2.50	2.51	0.075	2.99	100

* Based on the analysis of seven portions of reagent water with known additions in a single laboratory.

add the appropriate amount of stock standard mix to each of a minimum of seven 30-mL portions of reagent water and analyze.

Calculate average percent recovery (P) and standard deviation of the recovery (S_r). Compare results to the single-laboratory recovery and precision data in Table 6251:V. Compare precision at similar concentrations, that is $P \pm 30\%$ of the additive level. Acceptable mean recovery values are within the interval $P \pm 30\%$. For compounds not meeting this criterion, repeat with another seven samples until satisfactory performance has been demonstrated.

2) *Absolute recovery*—Use direct injection haloester standards, ¶ 4*j*5), to check absolute recoveries of extracted and derivatized haloacetic acids. Calculate absolute recoveries by comparing the ratio (area for compound/area for internal standard) for samples in ¶ 7*c*1) with the area ratio for direct injection standard at a tenfold concentration to account for the extraction (expressed as the haloacetic acid). Typical absolute recoveries are given in Table 6251:VI; acceptable absolute recoveries are within $\pm 30\%$. Recoveries outside this range may indicate insufficient shaking or poor methylation, possibly due to water in the extracts. Correct problem and produce acceptable absolute recoveries before analyzing samples.

Because of rapid advances in chromatography, columns and conditions may be modified to improve separation or to lower cost. Repeat procedure in 7*c* for each modification.

d. Internal standard assessment: The internal standard corrects for any deviation in sample volume injected. A sample injection is acceptable if the area counts of the internal standard peak do not vary more than $\pm 20\%$ from the daily calibration standard(s) IS response.

Reinject an extract exceeding the $\pm 20\%$ range. If reanalysis does not produce acceptable results, reextract and reanalyze. If the reextracted sample results are not acceptable or if samples have exceeded holding time, resample and reanalyze or record results as suspect and out-of-control.

e. Evaluating surrogate recovery: The surrogate is added directly to all samples before acidification and extraction. If the surrogate area is low or absent, it is likely that there has been a derivatization problem (e.g., water in extract) or extraction problem (e.g., water insufficiently acidified).

An extract is acceptable if the area counts of the surrogate standard recovery are $\pm 30\%$ from the surrogate standard recovery for the daily calibration standard(s).

When surrogate recovery is not acceptable, check the following: locate possible errors in calculations or procedure, degradation of standard solution, contamination sources, and instrument performance. If these steps do not reveal the problem, reanalyze the extract. If reanalysis does not produce acceptable results, reextract and reanalyze samples. If the reextracted sample results are not acceptable or if samples have exceeded holding time, record results as suspect and out-of-control.

f. Extracted standard calibration: Quantitation is done by internal standard referencing with relative areas. Produce a minimum five-level extracted standard calibration curve for sample quantitation.

If the response for any compound falls outside the predicted response by more than 15% from a previous calibration, make a new calibration standard and analyze it until an acceptable curve is obtained.

Analyze calibration standards with each sample set after an acceptable five-level calibration curve is generated. If the continuing calibration standards are not within $\pm 15\%$, check for errors or degradation of standards and construct a new calibration curve.

TABLE 6251:VI. ABSOLUTE RECOVERY DATA FOR REAGENT WATER WITH KNOWN ADDITIONS

Compound	Conc. Added µg/L	Mean Conc. Recovered µg/L	Standard Deviation µg/L	Relative Standard Deviation %	Mean Recovery %
Monochloroacetic acid	1.00	0.789	0.047	5.92	78.9
Monobromoacetic acid	1.00	0.706	0.034	4.76	70.6
Dichloroacetic acid	1.00	1.10	0.048	4.38	110
Trichloroacetic acid	1.00	0.927	0.051	5.49	92.7
Bromochloroacetic acid	0.50	0.49	0.015	3.07	98
Dibromoacetic acid	1.00	1.16	0.032	2.75	116
2,4,6-Trichlorophenol	0.50	0.523	0.030	5.89	105

g. Matrix additions: Add each target compound into one sample per sample set (a sample set is all samples extracted within a 24-h period) or 10% of the samples, whichever is greater.

The added concentration should be near to or greater than the background. Take care, particularly with dichloroacetic acid, to ensure that the addition plus background concentration does not exceed calibration range (extract dilution may be needed).

Analyze one sample portion to determine the background concentration (B) of each compound. Add working standard mix to a second sample portion and analyze to determine the concentration of each compound (A). Calculate percent recovery (P_i) as $100 (A - B)/T$, where T is the known concentration of the material added to the sample.

Compare percent recovery (P_i) for each compound with established QC acceptance criteria. Establish QC criteria by initially analyzing seven samples with additions and calculating the average percent recovery (P) and the standard deviation of the percent recovery (S_r).

Calculate QC acceptance criteria as follows:

$$\text{Upper control limit (UCL)} = P + 3S_r$$
$$\text{Lower control limit (LCL)} = P - 3S_r$$

The data generated during the initial demonstration of capability, ¶ 7c1), may be used to set the initial upper and lower control limits.

Monitor all data from dosed samples. Compound recoveries must fall within established control limits. After 10 new recovery measurements, recalculate P and S_r using all the data, and construct new control limits. When the total number of data points reaches 20, update control limits by calculating P and S_r using only the most recent 20 data points.

Compare percent recovery (P_i) for each compound with the QC acceptance criteria established by the control limits. If recovery of any compound falls outside the designated range, performance is judged to be out of control. Seek source of problem immediately and resolve before continuing the analysis.

However, if recovery of a compound meets calibration, blank, internal standard, surrogate, and replicate quality control, laboratory performance is in control, and the recovery problem is matrix-related. Label result for that compound in the sample as suspect/matrix.

h. Replicate analysis: Analyze sample duplicates to monitor precision. Analyze duplicates on at least 10% of all samples randomly selected.

Determine control limits by calculating the range as a function of the relative standard deviation. The range, R, is the absolute difference of the duplicate values, X_1 and X_2, as follows:

$$R = |X_1 - X_2|$$

The normalized range (R_n) is calculated by dividing the range by the average of the duplicate values (X_m):

$$R_n = \frac{R}{X_m}$$
$$X_m = \frac{X_1 + X_2}{2}$$

Calculate mean normalized range (R_m) for 20 pairs of duplicate data points initially and 20 pairs of points quarterly:

$$R_m = \frac{\sum R_n}{n}$$

where:

n = number of duplicate pairs.

$$\text{Variance} = S^2 = \frac{\sum (R_n - R_m)^2}{n - 1}$$

The standard deviation (s) is the square root of the variance. Upper and lower control limits are $R_m + 3s$ and zero, respectively. Acceptable duplicates fall within the control limits. The upper warning limit is $R_m + 2s$. If an R_n value is outside the warning limit, a potential problem is indicated and investigated before the analysis is out of control. Recalculate control limits quarterly using the most recent 20 points not including any data points that are out of control. Recalculate control limits when any major analytical changes are made and after at least 20 points have been collected.

Analyze any problem and correct it. If the duplicate is not acceptable, reextract only for those compounds out of control. If the duplicates are still unacceptable or the sample holding time has been exceeded, resample and reanalyze. If this is not possible record results as suspect and out of control. Do not use such data in range calculations.

i. Additional quality control: Each quarter, analyze QC check standards from an external source. Independent confirmation may include interlaboratory split sampling for comparison. Analyze performance evaluation samples, preferably from USEPA or appropriate state agency, at least once a year. Results for each target compound must be within established acceptance limits.

Shipping blanks are containers filled with reagent water containing appropriate amount of NH_4Cl (see ¶ 2), shipped to all sample locations with sample bottles, and returned with the samples. They are used to assess contamination during sampling and transit. Analyze a shipping blank with each sample set. If the shipping blank contains reportable levels, compare with the laboratory reagent blank. If contamination is not detected in the laboratory reagent blank, the sampling or transportation practices may have caused the contamination. Discard all samples in the set and resample.

Direct ester standards (concentration based on corresponding acid) may be injected at the start of each set to verify the sensitivity, chromatography, and retention times on the gas chromatograph.

Make an instrument check of the entire analytical system daily using data gathered from analyses of reagent blanks, standards, and replicate samples.

8. Precision and Bias

Single-laboratory method detection limits (MDL) and extracted recovery data from reagent water are presented in Table 6251:I. Data for absolute recoveries in reagent water are given in Table 6251:VI. Laboratory data from two different laboratories showing duplicate precision and matrix additions recoveries are presented in Tables 6251:VII and 6251:VIII.

TABLE 6251:VII. SAMPLE DUPLICATE DATA FROM TWO LABORATORIES

Laboratory	Compound	Numer of Pairs of Replicates	Average Difference Between Duplicates %	Standard Deviation of Difference Between Duplicates %
A	MCAA	5	7.6	10.6
	MBAA	3	1.9	1.5
	DCAA	7	1.5	0.8
	TCAA	6	1.4	1.0
	DBAA	5	6.0	6.0
	BCAA	11	1.8	1.3
B	MCAA	10	16.7	14.8
	MBAA	3	8.9	8.4
	DCAA	11	8.5	10.6
	TCAA	11	5.5	3.6
	DBAA	5	5.4	4.0
	BCAA	10	5.3	4.3

TABLE 6251:VIII. FIELD SAMPLE RECOVERY WITH KNOWN ADDITIONS TO DRINKING WATER, IN TWO LABORATORIES

Laboratory	Compound	Added Conc. $\mu g/L$	Number of Samples	Mean Recovery %	Relative Standard Deviation %
A	MCAA	5.0	7	99	4
	MBAA	5.0	7	101	4
	DCAA	5.0	7	96	4
	TCAA	5.0	7	100	3
	BCAA	10.0	14	96	5
	DBAA	5.0	7	102	5
	TCPh	2.5	7	100	6
B	MCAA	5.0	13	101	8
	DCAA	4.0	14	103	7
	TCAA	4.0	14	103	6
	MBAA	5.0	14	97	8
	BCAA	2.0	14	106	8
	DBAA	4.0	14	102	7
	TCPh	0.4	14	104	15

9. References

1. CHINN, R. & S.W. KRASNER. 1989. A simplified technique for the measurement of halogenated organic acids in drinking water by electron capture gas chromatography. Presented at the American Chemical Soc. Pacific Conf. on Chemistry and Spectroscopy, Pasadena, Calif., Oct. 18–21, 1989.
2. U.S. ENVIRONMENTAL PROTECTION AGENCY. 1984. Definition and procedure for the determination of the method detection limit. 40 CFR Part 136, Appendix B. *Federal Register* 49, No. 209.
3. ALCEON CORP. 1993. An Overview of Available Information on the Toxicity of Drinking Water Disinfectants and Their By-Products. Cambridge, Mass.
4. DE BOER, T.J. & H.J. BACKER. 1963. Organic Synthesis, Collective, Vol. IV. John Wiley & Sons, New York, N.Y.
5. U.S. ENVIRONMENTAL PROTECTION AGENCY & ASSOCIATION OF METROPOLITAN WATER AGENCIES. 1989. Disinfection By-Products in U.S. Drinking Water. Vol. 1 Report. James M. Montgomery Consulting Engineers, Pasadena, Calif.
6. FALES, H.M., T.M. JAOUNI & J.F. BABASHAK. 1973. Simple device for preparing ethereal diazomethane without resorting to codistillation. *Anal. Chem.* 45:13.
7. XIE, Y., D.A. RECKOW & R.V. RAJAN. 1993. Spontaneous methylation of haloacetic acids in methanolic stock solutions. *Environ. Sci. Technol.* 27:1232.
8. LIDE, D.R., ed. 1991. CRC Handbook of Chemistry and Physics, 71st ed. CRC Press, Boca Raton, Fla.

10. Bibliography

U.S. ENVIRONMENTAL PROTECTION AGENCY. 1990. Methods for the Determination of Organic Compounds in Drinking Water. Supplement 1. EPA-600/4-90-020.
BARTH, R.C. & P.S. FAIR. 1992. Comparison of the microextraction procedure and Method 552 for the analysis of HAAS and chlorophenols. *J. Amer. Water Works Assoc.* 84:94.

6252 DISINFECTION BY-PRODUCTS: ALDEHYDES (PROPOSED)*

6252 A. Introduction

1. Sources and Significance

Ozone reactions during water treatment are complex and often produce a wide range of unstable oxidation by-products, usually oxygenated and polar. Among the intermediate products formed, when ozone attacks the organic matter present in raw waters, are low-molecular-weight by-products such as aldehydes. If oxidized further, these aldehydes can produce aldo-acids and carboxylic acids. Formaldehyde, a ubiquitous component of the environment, may be introduced into drinking water by ozone treatment, natural metabolism, and commercial processes.

There are two postulated mechanisms for aldehyde formation during ozone treatment. The first involves a two-step Criegee attack at unsaturated C-C bonds by molecular ozone with ozonides or epoxides formed as intermediates.[1] The second involves an indirect reaction of OH radicals.[2] Although the levels of aldehyde

---

* Approved by Standard Methods Committee, 1994.

formation are usually a function of ozone dose, their concentrations are often controlled in water treatment by increasing the pH and thus the alkalinity of the water.

Aldehydes are unlikely to pose a serious health hazard to the consumer at microgram-per-liter concentrations as usually encountered in drinking water treatment. However, they react with nucleophiles even at these low levels and can therefore be a potential threat.[3] Thus, for example, formaldehyde, acetaldehyde, and crotonaldehyde are known animal carcinogens. Formaldehyde is a known human carcinogen.[4] Aldehydes also may serve as important components of assimilable organic carbon in promoting undesirable bioactivity.

2. Selection of Method

The most effective method for the determination of aldehydes in aqueous solutions involves the use of o-(2,3,4,5,6-pentafluorobenzyl)-hydroxylamine (PFBHA)† as a derivatizing agent.

---

† This reagent is known under various synonyms. The more common are o-(2,3,4,5,6-pentafluorophenyl)methylhydroxylamine hydrochloride with CAS RN 57981-02-9 and pentafluorobenzyloxylamine hydrochloride (PFBOA). It also has appeared with the acronym PFBHOX.

PFBHA reacts with low-molecular-weight carbonyl compounds, including aldehydes, to form the corresponding oximes. Unless the carbonyl compound is a symmetrical ketone or formaldehyde, two geometric isomers of the oxime derivatives are formed. These derivatives are extractable with organic solvents and are highly sensitive to analysis by gas chromatography with electron capture detection (GC/ECD) and gas chromatography with selective ion mass spectrometric detection (GC/SIM-MS).

3. References

1. GLAZE, W.H., M. KOGA & D. CANCILLA. 1989. Ozonation by-products. 2. Improvement of an aqueous-phase derivatization method for the detection of formaldehyde and other carbonyl compounds formed by the ozonation of drinking water. *Environ. Sci. Technol.* 23:838.
2. BAILEY, P.S. 1978. Ozonation in Organic Chemistry, Vol. I, Olefinic Compounds. Chap. 4, Academic Press, New York, N.Y.
3. NATIONAL ACADEMY OF SCIENCES. 1987. Drinking Water and Health. National Acad. Sciences, Washington, D.C.
4. AMERICAN CONFERENCE OF GOVERNMENTAL INDUSTRIAL HYGIENISTS. 1993. Documentation of the Threshold Limit Values and Biological Exposure Indices, 6th ed. American Conf. Governmental Industrial Hygienists, Cincinnati, Ohio.

6252 B.   PFBHA Liquid-Liquid Extraction Gas Chromatographic Method

This method measures straight-chain, low-molecular-weight aldehydes in raw and treated drinking water and simultaneously can analyze for C_1-C_{10} mono-carbonyl saturated aliphatic aldehydes, benzaldehyde, the dialdehyde glyoxal, and the keto-aldehyde methyl glyoxal.[1] The effectiveness of the derivatizing agent (PFBHA) in its reactions with these carbonyl compounds has been reviewed.[2]

1. General Discussion

a. Principle: Samples at room temperature are buffered to pH 6, PFBHA is added, and the samples are placed in a constant-temperature water bath. The carbonyl compounds are converted to their corresponding oximes during reaction with PFBHA. Sulfuric acid is used to quench excess PFBHA and the oxime derivatives are extracted with hexane. After H_2SO_4 cleanup, the organic extract is analyzed by gas chromatography where the volatile derivatives are easily separated in a temperature-programmable gas chromatograph equipped with a fused-silica capillary column and either an electron capture detector or selective ion mass spectrometer. Simultaneous analysis and confirmation with a single injection can be effected by setting up both the analytical column and the confirmation column to share a common injection port. Alternatively, use separate analytical and confirmation columns. Aqueous calibration standards similarly are derivatized, extracted, and analyzed. A surrogate recovery standard is added to the samples before derivatization to indicate any variation in derivatization and extraction efficiency.

With the exception of symmetrical ketones and formaldehyde, most carbonyl compounds form two geometrical isomers of oxime derivative. Methyl glyoxal, however, produces only one prominent isomer.

The method described may be used if appropriate quality control can be demonstrated for quantification of all C_1-C_{10} mono-carbonyl, saturated aliphatic aldehydes, benzaldehyde, glyoxal, and methyl glyoxal, but precision and quality control data are presented only for the most commonly found ozonation by-products, namely, formaldehyde, acetaldehyde, heptanal, benzaldehyde, glyoxal, and methyl glyoxal.

b. Interferences: Dissolved ozone, residual chlorine, and other oxidizing substances interfere with the PFBHA reaction. Quantitative addition of sodium thiosulfate as a reducing agent before derivatization or the addition of ammonium chloride or sulfate at the time of sampling (KI if ozone is present) prevents this interference. Ketones and quinones or large quantities of aldehydes may deplete the PFBHA reagent excess necessary to ensure complete reaction. Waters with high sulfide content inhibit the derivatization of carbonyl compounds. The occurrence of artifacts by aldehyde formation from thermal decomposition of water components is a potential positive interference. Because formaldehyde is used as a preservative for membranes, purified water produced by reverse osmosis is also a potential positive interference. In addition, formaldehyde and acetaldehyde are air pollutants and some formaldehyde in the air can be traced to certain insulation materials.

c. Detection levels: The method detection levels (MDL) and precision data for those aldehydes most commonly found in ozonated waters are given in Table 6252:I. These levels were evaluated for the extracted oximes in hexane from aldehyde-free water. The minimum reporting levels (MRL) for these aldehydes are usually set at five times the MDL. In effect, the MRL for all aldehydes analyzed by this method is 0.5 μg/L except for acetaldehyde and glyoxal, where the value is 1 μg/L. Formaldehyde is a ubiquitous contaminant. Method blanks may contain formaldehyde at trace levels. If it cannot be eliminated, raise the MRL. The precision

TABLE 6252:I. METHOD DETECTION LEVELS (MDL) AND PRECISION DATA*

Compound	Added Conc. μg/L	Found Conc. μg/L	Standard Deviation μg/L	Relative Standard Deviation %	Method Detection Level μg/L
Formaldehyde	0.550	0.518	0.026	5.04	0.082
Heptanal	0.255	0.191	0.025	13.1	0.079
Benzaldehyde	0.245	0.227	0.022	9.9	0.079
Glyoxal	0.445	0.315	0.073	23	0.228
Methyl glyoxal	0.420	0.355	0.033	9.4	0.105

* Based on the analysis of seven portions of organic-free water with known additions.

data presented may be matrix-sensitive. Use known standard additions to the matrix if oxime standards are not available.

This method is useful for detecting carbonyl compounds such as short-chain aldehydes (C_1–C_{10}), benzaldehyde, glyoxal, and methyl glyoxal in the range of 1 to 100 μg/L. A clean laboratory reagent water blank, free of these contaminants, is essential.

d. Safety: The toxicity or carcinogenicity of each reagent used in this method has not been defined precisely. Acetaldehyde contains material that can cause cancer in animals, formaldehyde contains material that can cause cancer in humans, and crotonaldehyde causes respiratory tract and eye burns if inhaled or swallowed. Observe proper ventilation and handling procedures. Wear quantitatively-fitted negative-pressure respirators with charcoal air-purifying filter canisters, gloves (such as butyl but not natural rubber, latex, or nitrile[2]), and protective garments resistant to the degrading effects and permeation of these chemicals. Glyoxal and methyl glyoxal are mutagenic in in-vitro tests and the former has subchronic oral toxicity. Take care when handling high concentrations of aldehydes during preparation of primary standards. When handling hexane solutions of the oxime derivatives, wear nitrile gloves (not butyl or latex).

2. Sampling and Storage

See Section 6010B.1 and note the following additional requirements:

Seal sample vials with TFE-lined polypropylene screw caps.* Do not use bakelite black caps made from a formulation containing phenol and formaldehyde. Do not add HCl for this method. Reduce residual free chlorine according to the following method. Additional reagents are required for ozonated samples as described below. If free chlorine is present in the samples, add ammonium chloride or sulfate (0.1 mL of a 20% solution /40 mL sample) before sample collection. Monochloramine may form but will not change the aldehyde concentration of samples subsequently stored at 4°C. If residual ozone is present, the levels of aldehydes may change as the ozone-natural organic matter reaction continues. To prevent this, quantitatively add sodium thiosulfate to the empty vials. Alternatively quench residual ozone by adding 0.1 mL of a 3 g KI/L solution to each 40-mL vial.

Prepare field reagent blanks from organic-free reagent water (¶ 4e).

Ideally, derivatize samples for aldehyde analysis immediately after collection. If this is not feasible, complete derivatization and extraction within 48 h of collection.

3. Apparatus

a. Sample containers and extraction vials: 40-mL screw-top, glass sample vials with aldehyde-free caps. (NOTE: Do not use the thermoset phenol-formaldehyde or urea-formaldehyde.) Prepare these, together with the 14-mL amber vials for storing stock solutions, as follows: Wash with detergent, rinse with tap water, soak in 10% HNO_3 for at least 30 min, rinse with tap water, rinse with laboratory organic-free water (¶ 4e below), and oven dry at 180°C for at least 1 h.

Clean caps and septa by rinsing with methanol, then with hexane, and dry at 80°C for no more than 1 h in a clean, forced-air convection oven.

b. Microsyringes or Eppendorf micro-pipets with glass tips, to measure the following volumes: 5, 10, 25, 50, 100, 250, 500, and 1000 μL.

c. Volumetric flasks, 5, 10, and 25 mL, borosilicate glass. Prepare initially by the method of ¶ 3a, except that after rinsing with organic-free water, rinse with methanol and invert to drain. Air-dry only. *Do not dry in oven.*

d. Syringe: 20-mL glass hypodermic, metal luer lock tip with 8.9-cm- (3.5-in.-) long × 17 gauge stainless steel pipetting needle (alternatively, use a 20-mL volumetric pipet). Clean as above.

e. Automatic pipet dispensers: To simplify batch processing, add reagent by use of these dispensers. Preferably use adjustable 1-mL and 4-mL sizes with PTFE transfer lines that can be mounted on the suppliers' reagent bottles. If these are not available use 1-mL and 4-mL volumetric pipets.

f. Constant-temperature water bath or incubator, capable of holding multiple 40-mL sample vials and maintaining 45°C ± 0.5°C.

g. Pasteur pipets: Have a selection of short-tipped (14.6-cm or 5.75-in.) and long-tipped (23-cm or 9-in.) pipets.

h. Mechanical shaker,† to automate hexane extraction (see Section 6251B.3e). Alternatively, use a vortex mixer or manual shaking for 1 min.

i. Storage vials: 7-mL glass, screw-cap vials with PTFE-lined silicone septa cleaned as described above.

j. Gas chromatograph, with capillary columns, temperature programmable, and supplied with a temperature-controlled injector and electron-capture detector.

1) *Gas handling equipment:* Use carrier (helium) and make-up (nitrogen or 95% argon/5% methane) gases of high purity (99.999%) that pass through indicating calcium sulfate, molecular sieve 5A, activated charcoal, and an oxygen-purifying cartridge.

---

* I-Chem Research, Hayward, CA, or equivalent.

† Eberbach Corp., or equivalent.

Use two-stage metal diaphragm high-purity regulators at the compressed gas sources. Use flow controllers to regulate carrier gas flow. Ensure that all gas lines use 0.3-cm (0.125-in.) copper (or stainless steel) tubing; rinse with high-purity acetone, and bake before use.

2) *Injector*, split/splitless (using straight open-bore insert).

3) *Analytical column*,‡ 30 m long × 0.25 mm ID, fused silica capillary column with a 0.25-μm film thickness.

4) *Confirmation column*,§ 30 m long × 0.25 mm ID, fused silica capillary column with a 0.25-μm film thickness.

5) *Detectors*, a constant-current pulse-modulated ^{63}Ni ECD with standard size cell (use two ECDs for simultaneous confirmation analysis).

4. Reagents

a. *Extraction solvent*, UV-grade, glass-distilled hexane.‖

b. *Solvent for standard preparation*, reagent-grade methanol, free of the target aldehydes.#

c. *Preservation agents*, ammonium chloride, NH_4Cl, or sulfate, $(NH_4)_2SO_4$, or sodium thiosulfate, $Na_2S_2O_3$.

d. *Sulfuric acid*, H_2SO_4, conc and 0.2N.

e. *Organic-free reagent water:* Treat water prepared in commercially available water systems to remove all traces of aldehydes. Two methods have been demonstrated: Either expose reagent water produced by a laboratory purification system** to UV irradiation for 1 h or distill reagent water from acidified potassium permanganate (500 mL water with 64 mg potassium permanganate and 1 mL conc H_2SO_4).

Alternative purification techniques, such as addition of another granular activated carbon filtration step, may be used if they can be shown to effectively eliminate background levels of aldehydes.

---

‡ Durabond-5, J&W Scientific, or equivalent.
§ Durabond-1701, J&W Scientific, or equivalent.
‖ Burdick & Jackson, Muskegon, MI, or equivalent.
Sigma Chemical Co., St. Louis, MO, or equivalent.
** Milli-Q, Millipore Corp., Bedford, MA, or equivalent.

Do not use a reagent water with formaldehyde contamination to quantify formaldehyde in aqueous samples.

f. *Buffer pH 6 reagent:* Prepare in a 200-mL volumetric flask by mixing 100 mL 0.5M potassium hydrogen phthalate solution in aldehyde-free water with 43.6 mL of 1M NaOH and bringing to volume. Add 1 mL of this solution/20 mL aqueous sample before derivatization.

Prepare an alternative buffer solution by dissolving 0.2 g proprietary buffer salt.††

g. *Derivatizing agent, PFBHA:*‡‡ Weigh o-(2,3,4,5,6-pentafluorobenzyl)-hydroxylamine hydrochloride into organic-free water to give a solution concentration of 15 mg/mL. Prepare fresh daily. Prepare enough to add 1 mL/20 mL sample.

h. *Standard materials:* See Table 6252:II for source and physical characteristics of the standards. Obtain purity assay of each purchased standard before use.

1) *Individual aldehyde standard stock solutions:* Prepare by weighing aldehyde standards in methanol. Weigh between 20 and 70 mg of each standard into a 10-mL volumetric flask. Weigh solid standards directly into the empty flask, then fill with methanol. Add liquid standards to the flask, which has been filled to the neck previously with methanol. Place on a weighing balance and stabilize the weight. Inject liquid standard with a microsyringe directly into the bulk of the methanol and determine exact weight after the addition. Because some aldehyde standards are supplied as aqueous solutions, evaluate weight of actual standard component and make an approximate determination of the required volume to be added from either the density (for pure liquids) or the percentage by weight (for solutions). Due to the high volatility of acetaldehyde, keep in the refrigerator at all times and place measurement syringe in a freezer for 10 min before preparing the stock solution. After diluting to 10 mL with methanol, cap flask and invert three times to mix. Transfer stock solutions to separate 14-mL amber vials with screw caps and PTFE liners

---

†† Hydrion, Micro Essential Lab., Inc., Brooklyn, NY.
‡‡ Aldrich Chemical Co., or equivalent.

TABLE 6252:II. ANALYTICAL STANDARDS OF CARBONYL COMPOUNDS USED IN THE PFBHA METHOD

Compound	Purity %	Molecular Weight mg/mmol	Boiling Point °C	Density g/mL
Formaldehyde*	†	33.03	96	1.083
Acetaldehyde‡	99.9	44.05	21	0.788
Propanal (propionaldehyde)*	97	58.08	46–50	0.805
Butanal (n-butyraldehyde)‡	99.9	72.11	75	0.800
Pentanal(n-valeraldehyde)*	99	86.13	103	0.810
Hexanal (caproaldehyde)*	98	100.16	131	0.834
Heptanal (heptaldehyde)*	95	114.19	153	0.818
Octanal (caprylic aldehyde)*	99	128.22	171	0.821
Nonanal (nonyl aldehyde)‡	99.9	142.24	93	0.827
Decanal (decyl aldehyde)‡	99.9	156.27	207–209	0.830
Benzaldehyde‡	99.9	106.12	178–179	1.044
Glyoxal (ethanedial)‡	§	58.04	50	1.14
Methyl glyoxal (pyruvic aldehyde or 2-oxopropionaldehyde)*	§	72.06	72	1.045

* Aldrich Chemical Company, Inc., Milwaukee, WI, or equivalent.

† Available in 37% solution (by weight) in water.

‡ ChemService, Inc., West Chester, PA, or equivalent.

§ Available in 40% solutions (by weight) in water.

and store at 4°C bound with self-adhesive film.§§ Stock solutions (except formaldehyde), are usable for up to 3 months. Let them come to room temperature before pipetting. Overcome presence of turbidity or a precipitate by ultrasonication in warm water. If a precipitate persists, make a new stock solution. Prepare formaldehyde stock solutions each month.

Verify the aldehyde concentrations of the aqueous solutions after filtering through 0.45-μm PTFE filters by the sodium bisulfite-iodine titration method.[3]

2) *Multicomponent aldehyde additive standards*: Prepare additive standards solution using individual stock solutions of those aldehydes of interest. Make mixture weekly. The concentration of each component in this additive standard solution should be about 10 mg/L when added to 20-mL aqueous samples. When preparing a calibration curve in 100 mL of organic-free water, prepare two or more multicomponent additive standards from which a volume in the range 10 to 100 μL can be injected directly into the water. For example, if the stock solution concentration is 50 mg in 10 mL or 5 g/L, 20 μL in 10 mL methanol is required to produce a 10-mg/L additive standard. This is best achieved by first filling the 10-mL volumetric flask to just above the neck with methanol. Inject required volume of each of the stock solutions, using a clean microsyringe for each component, into the bulk of the methanol. After adding all stock solutions, fill to the mark with methanol. Cap and invert three times to mix.

i. Standards derivatives: To determine reaction and extraction efficiency of each aldehyde in different matrices compare the chromatographic response of the derivatized standard in the matrix to that of authentic standards of the oximes. The surrogate standard used to establish optimum conditions for derivatization and laboratory-synthesized oximes for six of the aldehydes have been used to verify recovery of derivatized aldehydes from organic-free water. See Table 6252:III. Unforeseen matrix effects can occur, and because PFBHA-derivatized aldehyde standards (oximes) are not available commercially, some representative syntheses of these derivatives are available.[4]

j. Internal standard, 1,2-dibromopropane and decafluorobiphenyl, 98% purity.‖‖

1) *Internal standard stock solution:* Weigh 50 mg into a 10-mL volumetric flask containing methanol up to the neck. Fill to mark with methanol. This 5-g/L stock solution can be used for up to 6 months when stored as described in ¶ *h*1) above.

§§ Parafilm®, American Can Co., Greenwich, CT, or equivalent.
‖‖ Aldrich Chemical Company, Inc., Milwaukee, WI, or equivalent.

2) *Internal standard working solvent*, 100 μg/L in hexane: Deliver 20 μL internal standard stock solution directly into 1 L hexane in the solvent bottle to be used in the extraction. Cap bottle and invert three times to ensure thorough mixing. This dilution can be used for 4 weeks. To ensure suitability for extraction, run a sample of this working solvent on the GC before extraction of aqueous samples. Before processing samples, provide enough working solvent to extract all calibration and aqueous samples to be analyzed. *Never make up fresh working solvent for use during sample processing.*

k. Surrogate (SUR), 2,3,5,6-tetrafluorobenzaldehyde, 98% pure.‖‖

1) *Surrogate stock solution*, 20 g/L: Weigh 0.2 g SUR into a 10-mL volumetric flask containing methanol up to the neck. After determining the weight difference, fill to mark with methanol. Stock solutions can be used for up to 6 months if stored as described in ¶ *h*1) above.

2) *Surrogate additive solution*, 20 mg/L: Deliver 10 μL SUR stock solution into a 10-mL volumetric flask and dilute to volume with methanol. This solution can be used for up to 3 months when stored at 4°C. At the beginning of sample processing, add 10-μL surrogate additive solution to each 20-mL sample portion, yielding a surrogate concentration of 10 μg/L.

l. Calibration standards: Prepare aqueous calibration standards in 100 mL organic-free water by injecting a measured amount of the multicomponent aldehyde additive standard solution directly into the water using the solvent flush technique. Prepare five different concentration levels within the expected sample range. These would normally be in the range 0.5 to 30 μg/L.

5. Procedure

a. Sample preparation: Remove samples and standard solutions from storage and let reach room temperature.

b. Derivatization: Withdraw 20 mL sample from sample vial using a 20-mL glass syringe or glass pipet. Discard remaining sample, shake the vial dry by hand, and return the syringe contents to the vial. Add 10 μL surrogate additive solution using either a microsyringe or automatic pipettor, to all samples and standards. Add 1 mL pH 6 KHP/NaOH buffer to each portion with an automatic pipettor and swirl to mix. Add 1 mL freshly prepared 15-mg/mL PFBHA solution to each vial by automatic pipet, secure cap, and swirl to mix gently. Place all samples in a constant-temperature water bath set at 45 ± 0.5°C for 1 h and 45 min. Remove vials and cool to room temperature for 15 min.

TABLE 6252:III. RECOVERY OF TRIPLICATE IN-SITU DERIVATIZED ALDEHYDES COMPARED TO THE RECOVERY OF PURE OXIME DERIVATIVES FROM ORGANIC-FREE WATER

Derivatized Aldehyde	Conc. of Oxime Added* μg/L	Mean Conc. Recovered† μg/L	Standard Deviation μg/L	Relative Standard Deviation %	Mean Recovery‡ %
Formaldehyde	19.2	21.2	0.118	0.56	90.4
Acetaldehyde	19.3	19.3	0.095	0.49	100
Heptaldehyde	19.9	23.6	0.104	0.44	84.3
Benzaldehyde	19.8	18.6	0.138	0.74	107
Glyoxal	19.4	23.7	0.235	0.99	82.0
Methyl glyoxal	20.5	22.1	0.194	0.88	92.5

* Amount of standard oxime added to OPW.

† Evaluated by comparing the relative response of the extracted standard oximes to the calibration curve using in-situ derivatized aldehydes.

‡ Calculated recovery of in-situ derivatized aldehydes compared to standard oximes.

c. Microextraction: To each vial add 0.05 mL (approximately 2 drops) conc H_2SO_4 to quench the derivatization reaction and then add 4 mL hexane working solvent containing the internal standard. When using automated extraction, place vials in a mechanical wooden shaker box. Shake vials on fast setting (approximately 300 cycles) for 3.5 min. Alternatively, shake manually for approximately 3 min. Remove vials and place upright. Let stand for approximately 5 min to permit phases to separate.

d. Extraction cleanup: Draw off top hexane layer using a clean 14.6-cm (5.75-in.) disposable pasteur pipet for each sample into a smaller 7-mL clear vial containing 3 mL $0.2N$ H_2SO_4. Shake for 30 s by hand and let stand for approximately 5 min for phase separation. Draw off top hexane layer using another clean 14.6-cm (5.75-in.) disposable pasteur pipet for each sample and place in two 1.8-mL autosampler vials per sample. Store extra autosampler vials in a refrigerator at 4°C as a backup extract.

e. Gas chromatography: Use the following operating conditions for the gas chromatograph: injector temperature 180°C; split valve open at 0.5 min; split flow at 50 mL/min; temperature program: 50°C for 1 min, rising at 4°C/min to 220°C and then at 20°C/min to 250°C; detector temperature: 300°C; carrier gas flow: 1.5 mL/min at 100°C; make-up gas flow: 27 mL/min.

At the beginning of each analysis, inject one hexane solvent blank to condition the GC and to verify that there are no interferences present. Inject 1 µL onto the splitless injector. See Figures 6252:1 and 6252:2 for examples of chromatograms obtained with the above GC conditions for both the analytical and confirmation columns. If dual-column analysis is unavailable, use the column specified in ¶ 3*j*3) above, but be aware of possible interferences. Table 6252:IV lists retention times for both columns.

f. Calibration: The use of five levels of calibration defines the quantitation range. The lowest standard is based on the lowest level of quantitation for each component. Prepare standard levels by adding the aldehydes to reagent-grade water and derivatize with the same PFBHA solution as the samples. This corrects for any recovery characteristics inherent in the method. Analyze the calibration standards under the same GC conditions as the samples.

6. Calculation

a. Standards procedure: Use this procedure when evaluating external, internal, surrogate, and calibration standards. Calculate individual response factor (*RF*) for each standard as follows:

$$RF = \frac{A}{W}$$

where:

 A = response (peak area), and
 W = amount of compound, µg.

For each compound, determine the average *RF* and standard deviation of the *RF* values for all the standards. If the percent relative standard deviation (*%RSD*) is greater than 10%, take corrective action to improve method precision. When *%RSD* is less than 10%, then the mean *RF* is acceptable for use in calculating the relative response ratio of the surrogate/calibration standard *RF* to the internal standard $\overline{RF}$.

b. Determination of constituent concentration: Construct calibration curve for each target constituent from the total relative response factor of all isomers (compared to one of the internal

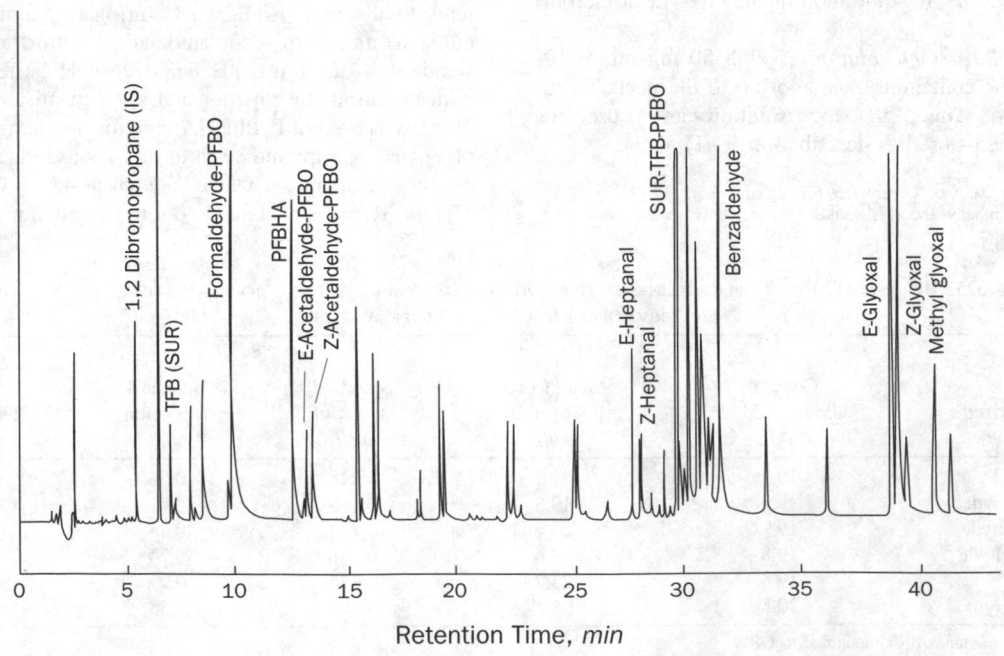

Figure 6252:1. Results for analytical column.

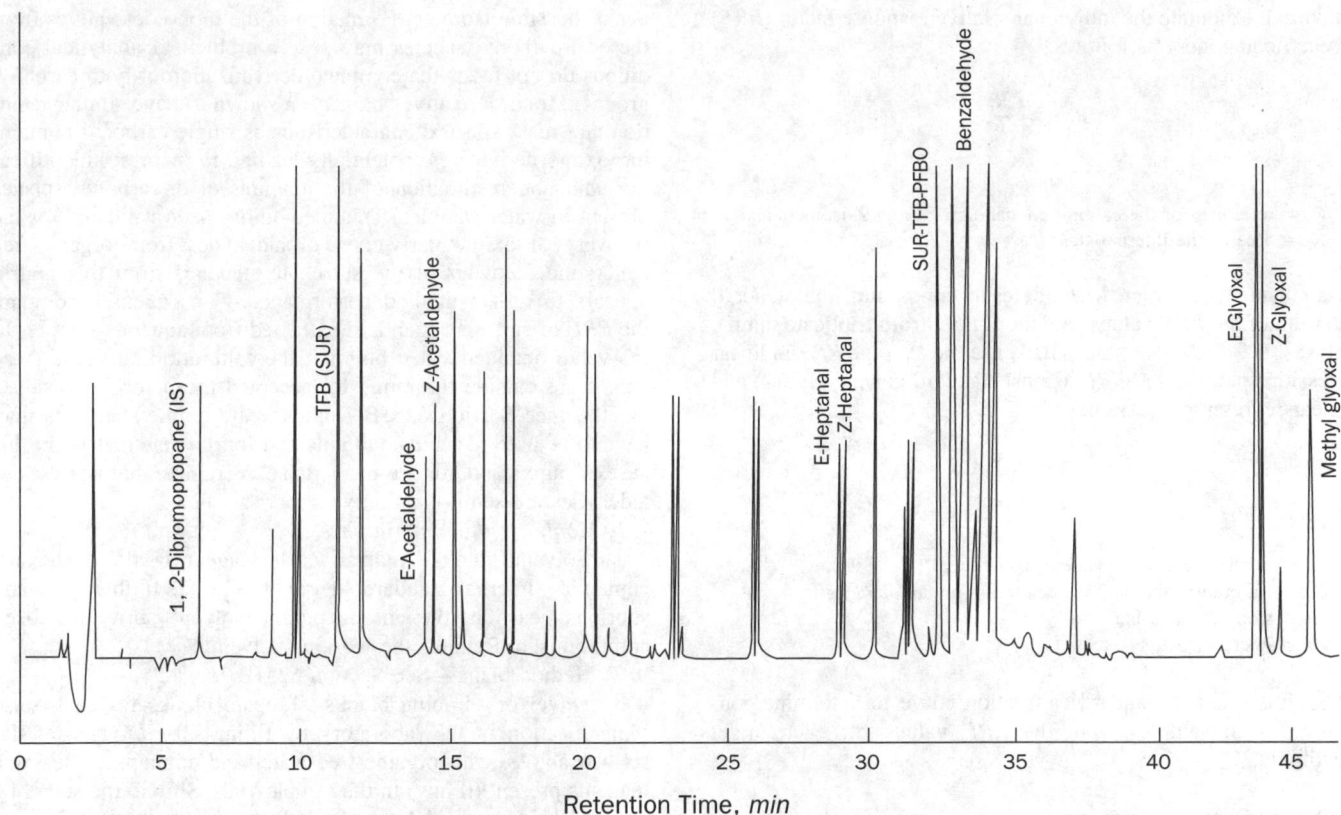

Figure 6252:2. Results for confirmation column.

TABLE 6252:IV. RETENTION TIMES (RTs) FOR DERIVATIZED ALDEHYDES, SURROGATE AND INTERNAL STANDARDS ON ELECTRON-CAPTURE DETECTOR

Compound	Retention Time on Column of ¶ 3*j*3) *min*	Retention Time on Column of ¶ 3*j*4) *min*
1,2-Dibromopropane*	5.21	6.25
SUR - TFB†	7.27	10.88
Formaldehyde (HCHO-PFBO)	9.61	11.21
E-Acetaldehyde (E-CH₃CHO-PFBO)‡	12.88	14.42
Z-Acetaldehyde (Z-CH₃CHO-PFBO)‡	13.15	14.70
E-Heptanal (E-C₆H₁₃CHO-PFBO)‡	27.82	28.95
Z-Heptanal (Z-C₆H₁₃CHO-PFBO)‡	27.93	29.09
SUR - TFB-PFBO§	32.37	32.77
Benzaldehyde (C₆H₅CHO-PFBO)‡	31.41	33.78
E-Glyoxal (E-OHCCHO-PFBO)‡	39.09	43.87
Z-Glyoxal (Z-OHCCHO-PFBO)‡	39.48	44.09
Methyl glyoxal (OCH₃CCHO-PFBO)	41.09	45.72

* Internal standard.

† Underivatized surrogate aldehyde TFB (2,3,5,6-tetrafluorobenzaldehyde).

‡ These aldehydes form E- and Z-PFBO isomers in assumed order of elution but has not been confirmed (-PFBO = pentafluorobenzyloxime).

§ Derivatized surrogate aldehyde.

standards). Calculate the individual relative response factor (*RRF*) for each component as follows:

$$RRF = \frac{R_s}{R_i}$$

where:

R_s = total area of the calibration standard E- and Z-isomers and
R_i = area of the internal standard.

Calculate the average *RRF* for each compound, the standard deviation of the *RRF* values, and the %*RSD* from triplicate sample analyses. If %*RSD* is less than 10%, use the average *RRF* in linear regression (plotting the *RRF* against standard concentrations) and a linear regression equation:

$$RRF_a = m(C_x) + b$$

where:

RRF_a = relative response factor of constituent,
C_x = calibration standard concentration in samples, μg/L,
m = slope of line, and
b = intercept of the y axis.

Use this internal standard calibration curve to determine concentrations in samples from the *RRF* values for each target constituent.

7. Quality Control

a. General considerations: PFBHA is a highly reactive *o*-substituted hydroxylamine. Like hydroxylamine, PFBHA reacts readily with a variety of carbonyl functional groups to produce corresponding oximes. The ease with which PFBHA reacts with carbonyl-containing compounds makes the potential contamination of samples a serious concern. Lower-molecular-weight aldehydes are commonly found in laboratory and outside air and can ultimately contaminate water samples, leading to incorrect calculation of aldehyde concentrations. As a further concern, PFBHA, especially in moist laboratory environments, can react to form oximes when directly exposed to aldehydes in air. For these reasons, exercise care to reduce the sources and exposure of samples, standard solutions, and PFBHA reagents to aldehyde contaminants. Consider storing PFBHA in a desiccator under an inert atmosphere, drying laboratory solvents with molecular sieves, using purified water, and making fresh derivatizing stocks on a regular basis. If, after analysis of appropriate sample blanks, contamination remains a problem, the source of the problem may be in the PFBHA reagent and solutions. Recrystallization of PFBHA may be necessary to remove oximes formed as a result of reagent contamination.

The effects of chromatographic and analytical conditions on E/Z ratios of the oximes have not been fully explored. The possibility of changing E/Z ratios under differing analytical conditions, such as injection temperature, requires that analytical conditions be carefully controlled. E/Z ratios may change as a function of time; therefore analyze samples as soon as possible after preparation and within groups. Use the sum of the isomer peak areas for each constituent for both calibration and quantification. With dicarbonyl species such as glyoxal, E/Z isomerism occurs from oxime formation with both carbonyl groups, increasing the number of possible isomers. Formation of the mono-derivatives from these di-carbonyl species may pose a problem if analytical conditions do not favor the complete derivatization of both carbonyl groups. Mono-derivatives have been shown to have similar retention and mass spectral characteristics as single carbonyl-containing oxime derivatives, potentially leading to incorrect identification and underestimation of the amounts of di-carbonyl species present in water samples. This method has been validated for the recovery of oxime derivatives of aldehydes from organic-free water; the recovery of the surrogate standard from this matrix appears to reflect method performance. Consequently, compare the *RRF* of surrogate standard extracted from aqueous samples to the value obtained when building the calibration curve. If these values are outside the range for accepted mean recovery values of 30% (see Section 6251B.7*c*), authentic oxime standards may have to be used to validate the method for the new matrix. In this case, if pure standards are unavailable, recognize that analyses of aldehydes are semi-quantitative; report as such.

b. Monitoring for interferences:

1) Solvent blank—Analyze each reagent bottle of hexane containing internal standard before it is used. If there are any spurious peaks in the chromatogram, solvent purity has been compromised. Remake the working solvent.

2) Method blank—See Section 6251B.7*b*.

3) Travel or shipping blanks—Prepare blanks for each sampling location in the laboratory by filling 40-mL vials, as described above, with organic-free water and containing the same reagents present (if any) in the sample vials. Ship to the sampling site and back to the laboratory with the sample bottles. Do not open these bottles in the field.

c. Internal standard assessment: Injections of the hexane extracts are acceptable if the area counts of the internal standard peak do not vary more than ± 20% from the mean of all the samples analyzed with the same batch of PFBHA. Reanalyze samples that do not meet this precision. If, after reinjection, criteria are still not met, the sample holding time has not been exceeded, and the same working solvent used for constructing the calibration curve is still available, the second vial may be analyzed.

d. Surrogate standard recovery: Add the surrogate (2,3,5,6-tetrafluorobenzaldehyde) directly to the 20-mL aqueous sample portions before reagent addition to monitor constituent recovery from the sample matrix. If the surrogate area is low or absent, there is likely to be a problem with derivatization or extraction that needs to be resolved before quantification can be undertaken (see Section 6251B.7*e*). A sample extract is acceptable if the area counts of the surrogate peak (or the *RRF* values compared to an acceptable internal standard area) do not vary more than ± 30% from other samples analyzed with the same batch of PFBHA.

e. Sample quantification: See Section 6251B.7*f*.

f. Matrix additions: See Section 6251B.7*g*.

g. Replicate analysis: See Section 6251B.7*h*.

8. References

1. SCLIMENTI, M.J., S.W. KRASNER, W.H. GLAZE & H.S. WEINBERG. 1990. Ozone disinfection by-products: Optimization of the PFBHA derivatization method for the analysis of aldehydes. *In* Advances in Water Analysis and Treatment, Proc. 18th Annu. AWWA Water Quality Technology Conf., Nov. 11–15, 1990, San Diego, Calif., p. 477. American Water Works Assoc., Denver, Colo.

2. FORSBERG, K. 1989. Chemical Protective Clothing Performance Index Book. John Wiley & Sons, New York, N.Y.

3. AMERICAN SOCIETY FOR TESTING AND MATERIALS. 1989. Annual Book of ASTM Standards: Water and Environmental Technology. Vol. 11.01. American Soc. Testing & Materials, Philadelphia, Pa.

4. CANCILLA, D.A., C.-C. CHOU, R. BARTHEL & S.S. QUE HEE. 1992. Characterization of the *o*-(2,3,4,5,6-pentafluorobenzyl)-hydroxylamine hydrochloride (PFBOA) derivatives of some aliphatic mono- and di-aldehydes and quantitative water analysis of these aldehydes. *J. Assoc. Offic. Anal. Chem. Int.* 75:842.

6410   EXTRACTABLE BASE/NEUTRALS AND ACIDS*

6410 A.   Introduction

1. Sources and Significance

The semivolatile compounds covered by this section include many classes of compounds, each characterized by different sources. The compounds include polynuclear aromatic hydrocarbons, often as by-products of petroleum processing or combustion; phthalates, used as plasticizers; phenolics, found most often in wood preservatives; organochlorine pesticides, found most often in agricultural runoff or in wastewaters draining such areas; and PCBs (also see Section 6431A). Many of the listed compounds are toxic or carcinogenic. However, they generally are relatively insoluble in water so they do not occur frequently in potable waters or most wastewaters.

* Approved by Standard Methods Committee, 1993.

2. Selection of Method

Method 6410B is a broad-spectrum gas chromatographic/mass spectrometric (GC/MS) packed- or capillary-column method for detection of these compounds following liquid-liquid extraction. Although this method can be used to determine all the listed compounds, it is not the most sensitive method for individual classes of compounds, which are detected at lower concentrations by GC methods such as those listed in 6420C (phenols), 6440B and C (polynuclear aromatic hydrocarbons), and 6630C and D (organochlorine pesticides and PCBs). In some cases, notably the pesticides, the GC method is substantially more sensitive than the GC/MS method. In other cases, such as the phenols, there is less difference between the methods.

6410 B.   Liquid-Liquid Extraction Gas Chromatographic/Mass Spectrometric Method

This method[1] is applicable to the determination of organic compounds that are partitioned into an organic solvent and are amenable to gas chromatography,* in municipal and industrial discharges.

* *Base/neutral extractables*: acenaphthene, acenaphthylene, anthracene, aldrin, benzo(a)anthracene, benzo(b)fluoranthene, benzo(k)fluoranthene, benzo(a)pyrene, benzo(ghi)perylene, benzyl butyl phthalate, β-BHC, δ-BHC, bis(2-chloroethyl) ether, bis(2-chloroethoxy) methane, bis(2-ethylhexyl) phthalate, bis(2-chloroisopropyl) ether more correctly known as 2,2-oxybis (1-chloropropane), 4-bromophenyl phenyl ether, chlordane, 2-chloronaphthalene, 4-chlorophenyl phenyl ether, chrysene, 4,4'-DDD, 4,4'-DDE, 4,4'-DDT, dibenzo(a,h)anthracene, di-*n*-butylphthalate, 1,3-dichlorobenzene, 1,2-dichlorobenzene, 1,4-dichlorobenzene, 3,3'-dichlorobenzidine, dieldrin, diethyl phthalate, dimethyl phthalate, 2,4-dinitrotoluene, 2,6-dinitrotoluene, di-*n*-octylphthalate, endosulfan sulfate, endrin aldehyde, fluoranthene, fluorene, heptachlor, heptchlor epoxide, hexachlorobenzene, hexachlorobutadiene, hexachloroethane, indeno(1,2,3-cd)pyrene, isophorone, naphthalene, nitrobenzene, *N*-nitrosodi-*n*-propylamine, PCB-1016, PCB-1221, PCB-1232, PCB-1242, PCB-1248, PCB-1254, PCB-1260, phenanthrene, pyrene, toxaphene, 1,2,4-trichlorobenzene.

Acid extractables: 4-chloro-3-methylphenol, 2-chlorophenol, 2,4-dichlorophenol, 2,4-dimethylphenol, 2,4-dinitrophenol, 2-methyl-4,6-dinitrophenol, 2-nitrophenol, 4-nitrophenol, pentachlorophenol, phenol, 2,4,6-trichlorophenol.

The method may be extended to include the following compounds: benzidine, α-BHC, γ-BHC, endosulfan I, endosulfan II, endrin, hexachlorocyclopentadiene, *N*-nitrosodimethylamine, *N*-nitrosodiphenylamine.

1. General Discussion

a. Principle: A measured volume of sample is extracted serially with methylene chloride at a pH above 11 and again at a pH below 2. The extract is dried, concentrated, and analyzed by GC/MS.[2,3] Qualitative compound identification is based on retention time and relative abundance of three characteristic masses (*m/z*). Quantitative analysis uses internal-standard techniques with a single characteristic *m/z*.

b. Interferences:

1) General precautions—See Section 6010C. Method interferences may be caused by contaminants in solvents, reagents, glassware, and other sample-processing hardware that lead to discrete artifacts and/or elevated base lines in detector output. Routinely demonstrate that all materials are free from interferences under the conditions of the analysis by running laboratory reagent blanks as described in Section 6200A.5*c*3).

Clean all glassware thoroughly[4] as soon as possible after use by rinsing with the last solvent used in it, followed by detergent washing with hot water and rinsing with tap water and distilled water. Drain glassware dry and heat in a muffle furnace at 400°C for 15 to 30 min. Some thermally stable materials, such as PCBs, may not be eliminated by this treatment. Solvent rinses with acetone and pesticide-quality hexane may be substituted for the baking. Thorough rinsing with such solvents usually eliminates PCB interference. Do not heat volumetric ware in a muffle furnace. After drying and cooling, seal and store glassware in a clean environment to prevent accumulation of dust or other contaminants. Store inverted or capped with aluminum foil.

Use high-purity reagents and solvents to minimize interference. Purification of solvents by distillation in all-glass systems may be required.

Matrix interferences may be caused by coextracted contaminants. The extent of matrix interferences will vary considerably depending on the sample.

2) *Special precautions*—Benzidine can be lost by oxidation during solvent concentration. Under the alkaline conditions of the extraction step, α-BHC, γ-BHC, endosulfan I and II, and endrin are subject to decomposition. Hexachlorocyclopentadiene is subject to thermal decomposition in the inlet of the gas chromatograph, chemical reaction in acetone solution, and photochemical decomposition. *N*-nitrosodimethylamine is difficult to separate from the solvent under the chromatographic conditions described.

N-nitrosodiphenylamine decomposes in the gas chromatographic inlet and cannot be separated from diphenylamine. Other methods may be preferred for these compounds.[1]

The base-neutral extraction may cause significantly reduced recovery of phenol, 2-methylphenol, and 2,4-dimethylphenol. Results obtained under these conditions are minimum concentrations.

The packed gas chromatographic columns recommended for the basic fraction may not be able to resolve certain isomeric pairs including the following: anthracene and phenanthrene; chrysene and benzo(a)anthracene; and benzo(b)fluoranthene and benzo(k)fluoranthene because retention time and mass spectra for these pairs are not sufficiently different to make unambiguous identification possible. Use alternative techniques, such as the method for polynuclear aromatic hydrocarbons (Section 6440B), to identify and quantify these compounds.

In samples containing many interferences, use chemical ionization (CI) mass spectrometry to make identification easier. Tables 6410:I and II give characteristic CI ions for most compounds covered by this method. Use of CI mass spectrometry to support

TABLE 6410:I. CHROMATOGRAPHIC CONDITIONS, METHOD DETECTION LIMITS, AND CHARACTERISTIC MASSES FOR BASE/NEUTRAL EXTRACTABLES

Compound	Retention Time *min*	Method Detection Limit *µg/L*	Characteristic Masses					
			Electron Impact			Chemical Ionization		
			Primary	Secondary	Secondary	Methane	Methane	Methane
1,3-Dichlorobenzene	7.4	1.9	146	148	113	146	148	150
1,4-Dichlorobenzene	7.8	4.4	146	148	113	146	148	150
Hexachloroethane	8.4	1.6	117	201	199	199	201	203
bis(2-Chloroethyl) ether	8.4	5.7	93	63	95	63	107	109
1,2-Dichlorobenzene	8.4	1.9	146	148	113	146	148	150
bis(2-Chloroisopropyl) ether*	9.3	5.7	45	77	79	77	135	137
N-Nitrosodi-*n*-propylamine			130	42	101			
Nitrobenzene	11.1	1.9	77	123	65	124	152	164
Hexachlorobutadiene	11.4	0.9	225	223	227	223	225	227
1,2,4-Trichlorobenzene	11.6	1.9	180	182	145	181	183	209
Isophorone	11.9	2.2	82	95	138	139	167	178
Naphthalene	12.1	1.6	128	129	127	129	157	169
bis(2-Chloroethoxy) methane	12.2	5.3	93	95	123	65	107	137
Hexachlorocyclopentadiene†	13.9		237	235	272	235	237	239
2-Chloronaphthalene	15.9	1.9	162	164	127	163	191	203
Acenaphthylene	17.4	3.5	152	151	153	152	153	181
Acenaphthene	17.8	1.9	154	153	152	154	155	183
Dimethyl phthalate	18.3	1.6	163	194	164	151	163	164
2,6-Dinitrotoluene	18.7	1.9	165	89	121	183	211	223
Fluorene	19.5	1.9	166	165	167	166	167	195
4-Chlorophenyl phenyl ether	19.5	4.2	204	206	141			
2,4-Dinitrotoluene	19.8	5.7	165	63	182	183	211	223
Diethyl phthalate	20.1	1.9	149	177	150	177	223	251
N-Nitrosodiphenylamine†	20.5	1.9	169	168	167	169	170	198
Hexachlorobenzene	21.0	1.9	284	142	249	284	286	288
α-BHC†	21.1		183	181	109			
4-Bromophenyl phenyl ether	21.2	1.9	248	250	141	249	251	277
γ-BHC†	22.4		183	181	109			
Phenanthrene	22.8	5.4	178	179	176	178	179	207
Anthracene	22.8	1.9	178	179	176	178	179	207
β-BHC	23.4	4.2	181	183	109			
Heptachlor	23.4	1.9	100	272	274			
δ-BHC	23.7	3.1	183	109	181			
Aldrin	24.0	1.9	66	263	220			
Dibutyl phthalate	24.7	2.5	149	150	104	149	205	279

TABLE 6410:I, CONT.

| Compound | Retention Time min | Method Detection Limit μg/L | Characteristic Masses | | | | | |
| | | | Electron Impact | | | Chemical Ionization | | |
			Primary	Secondary	Secondary	Methane	Methane	Methane
Heptachlor epoxide	25.6	2.2	353	355	351			
Endosulfan I†	26.4		237	338	341			
Fluoranthene	26.5	2.2	202	101	100	203	231	243
Dieldrin	27.2	2.5	79	263	279			
4,4'-DDE	27.2	5.6	246	248	176			
Pyrene	27.3	1.9	202	101	100	203	231	243
Endrin†	27.9		81	263	82			
Endosulfan II†	28.6		237	339	341			
4,4'-DDD	28.6	2.8	235	237	165			
Benzidine†	28.8	44	184	92	185	185	213	225
4,4'-DDT	29.3	4.7	235	237	165			
Endosulfan sulfate	29.8	5.6	272	387	422			
Endrin aldehyde			67	345	250			
Butyl benzyl phthalate	29.9	2.5	149	91	206	149	299	327
bis(2-Ethylhexyl) phthalate	30.6	2.5	149	167	279	149		
Chrysene	31.5	2.5	228	226	229	228	229	257
Benzo(a)anthracene	31.5	7.8	228	229	226	228	229	257
3,3'-Dichlorobenzidine	32.2	16.5	252	254	126			
Di-n-octyl phthalate	32.5	2.5	149					
Benzo(b)fluoranthene	34.9	4.8	252	253	125	252	253	281
Benzo(k)fluoranthene	34.9	2.5	252	253	125	252	253	281
Benzo(a)pyrene	36.4	2.5	252	253	125	252	253	281
Indeno(1,2,3-cd)pyrene	42.7	3.7	276	138	277	276	277	305
Dibenzo(a,h)anthracene	43.2	2.5	278	139	279	278	279	307
Benzo(ghi)perylene	45.1	4.1	276	138	277	276	277	305
N-Nitrosodimethylamine†			42	74	44			
Chlordane‡	19–30		373	375	377			
Toxaphene‡	25–34		159	231	233			
PCB 1016‡	18–30		224	260	294			
PCB 1221‡	15–30	30	190	224	260			
PCB 1232‡	15–32		190	224	260			
PCB 1242‡	15–32		224	260	294			
PCB 1248‡	12–34		294	330	262			
PCB 1254‡	22–34	36	294	330	362			
PCB 1260‡	23–32		330	362	394			

* The proper chemical name is 2,2'-oxybis(1-chloropropane).
† See introductory section of text.
‡ These compounds are mixtures of various isomers. (See Figures 6410:2 through 12.)
   Column conditions: Supelcoport (100/120 mesh) coated with 3% SP-2250 packed in a 1.8 m long × 2 mm ID glass column with helium carrier gas at 30 mL/min flow rate. Column temperature held isothermal at 50°C for 4 min, then programmed at 8°C/min to 270°C and held for 30 min.

electron ionization (EI) mass spectrometry is encouraged but not required.

c. *Detection levels:* The method detection level (MDL) is the minimum concentration of a substance that can be measured and reported with 99% confidence that the value is above zero.[5] The MDL concentrations listed in Tables 6410.I and II were obtained with reagent water.[6] The MDL actually obtained in a given analysis will vary, depending on instrument sensitivity and matrix effects.

d. *Safety:* The toxicity or carcinogenicity of each reagent has not been defined precisely. Benzo(a)anthracene, benzidine, 3,3'-dichlorobenzidine, benzo(a)pyrene, α-BHC, β-BHC, δ-BHC, γ-BHC, dibenzo(a,h)anthracene, n-nitrosodimethylamine, 4,4'-DDT, and polychlorinated biphenyls (PCBs) have been tentatively classified as known or suspected, human or mammalian carcinogens. Prepare primary standards of these compounds in a hood

and wear a NIOSH/MESA-approved toxic gas respirator when handling high concentrations.

2. Sampling and Storage

Collect grab samples in 1-L amber glass bottles fitted with a screw cap lined with TFE. Foil may be substituted for TFE if the sample is not corrosive. If amber bottles are not available, protect samples from light. Wash and rinse bottle and cap liner with acetone or methylene chloride, and dry before use. Follow conventional sampling practices[7] but do not rinse bottle with sample. Collect composite samples in refrigerated glass containers. Optionally, use automatic sampling equipment as free as possible of plastic tubing and other potential sources of contamination; incorporate glass sample containers for collecting a minimum of 250 mL. Refrigerate sample containers at 4°C and protect from

TABLE 6410:II. CHROMATOGRAPHIC CONDITIONS, METHOD DETECTION LIMITS, AND CHARACTERISTIC MASSES FOR ACID EXTRACTABLES

Compound	Retention Time min	Method Detection Limit µg/L	Characteristic Masses					
			Electron Impact			Chemical Ionization		
			Primary	Secondary	Secondary	Methane	Methane	Methane
2-Chlorophenol	5.9	3.3	128	64	130	129	131	157
2-Nitrophenol	6.5	3.6	139	65	109	140	168	122
Phenol	8.0	1.5	94	65	66	95	123	135
2,4-Dimethylphenol	9.4	2.7	122	107	121	123	151	163
2,4-Dichlorophenol	9.8	2.7	162	164	98	163	165	167
2,4,6-Trichlorophenol	11.8	2.7	196	198	200	197	199	201
4-Chloro-3-methylphenol	13.2	3.0	142	107	144	143	171	183
2,4-Dinitrophenol	15.9	42	184	63	154	185	213	225
2-Methyl-4,6-dinitrophenol	16.2	24	198	182	77	199	227	239
Pentachlorophenol	17.5	3.6	266	264	268	267	265	269
4-Nitrophenol	20.3	2.4	65	139	109	140	168	122

Column conditions: Supelcoport (100/120 mesh) coated with 1% SP-1240DA packed in a 1.8 m long × 2 mm ID glass column with helium carrier gas at 30 mL/min flow rate. Column temperature held isothermal at 70°C for 2 min then programmed at 8°C/min to 200°C.

light during compositing. If the sampler includes a peristaltic pump, use a minimum length of compressible silicone rubber tubing, but before use, thoroughly rinse it with methanol and rinse repeatedly with distilled water to minimize contamination. Use an integrating flow meter to collect flow-proportional composites.

Fill sample bottles and, if residual chlorine is present, add 80 mg sodium thiosulfate per liter of sample and mix well. Ice all samples or refrigerate at 4°C from time of collection until extraction.

Extract samples within 7 d of collection and analyze completely within 40 d of extraction.

3. Apparatus

a. Separatory funnel, 2-L, with TFE stopcock.

b. Drying column, chromatographic, 400 mm long × 19 mm ID, with coarse frit filter disk.

c. Concentrator tube, Kuderna-Danish, 10-mL, graduated.† Check calibration at volumes used. Use ground-glass stopper to prevent evaporation.

d. Evaporative flask, Kuderna-Danish, 500-mL.‡ Attach to concentrator tube with springs.

e. Snyder column, Kuderna-Danish, three-ball macro.§

f. Snyder column, Kuderna-Danish, two-ball micro.||

g. Vials, 10- to 15-mL, amber glass, with TFE-lined screw cap.

h. Continuous liquid-liquid extractor, equipped with TFE or glass connecting joints and stopcocks requiring no lubrication.#

i. Boiling chips, approximately 10/40 mesh. Heat to 400°C for 30 min or extract in a Soxhlet extractor with methylene chloride.

j. Water bath, heated, with concentric ring cover and temperature control to ±2°C. Use bath in a hood.

k. Balance, analytical, capable of accurately weighing 0.0001 g.

*l. Gas chromatograph:*** An analytical system complete with a temperature-programmable gas chromatograph and all required

accessories including syringes, analytical columns, and gases. Use chromatograph with the injection port designed for on-column injection when packed columns are used and for splitless injection when capillary columns are used.

1) *Column for base/neutrals*, 1.8 m long × 2 mm ID glass, packed with 3% SP-2250 on Supelcoport (100/200 mesh) or equivalent. This column was used to develop the detection limit and precision and bias data presented herein. Guidelines for the use of alternate columns (e.g., DB-5 fused silica capillary) are provided in ¶ 5*b*.

2) *Column for acids*, 1.8 m long × 2 mm ID glass, packed with 1% SP-1240DA on Supelcoport (100/120 mesh) or equivalent. The detection limit and precision and bias data presented herein were developed with this column. For guidelines for the use of alternate columns (e.g., DB-5 fused silica capillary) see ¶ 5*b*.

m. Mass spectrometer, capable of scanning from 35 to 450 amu every 7 s or less, utilizing 70-V (nominal) electron energy in the electron impact ionization mode, and producing a mass spectrum that meets all the criteria in Table 6410:III when 50 ng of decafluorotriphenyl phosphine [DFTPP; bis(perfluorophenyl) phenyl phosphine] is injected through the GC inlet.

n. GC/MS interface: Any GC to MS interface that gives acceptable calibration points at 50 ng or less per injection for each

---

† Kontes K-570050-1025 or equivalent.
‡ Kontes K-570001-0500 or equivalent.
§ Kontes K-503000-0121 or equivalent.
|| Kontes K-569001-0219 or equivalent.
Hershberg-Wolf Extractor, Ace Glass Co., Vineland, NJ, P/N 6841-10, or equivalent.
** Gas chromatographic methods are extremely sensitive to the materials used. Mention of trade names by *Standard Methods* does not preclude the use of other existing or as-yet-undeveloped products that give *demonstrably* equivalent results.

TABLE 6410:III. DFTPP KEY MASSES AND ABUNDANCE CRITERIA

Mass	*m/z* Abundance Criteria
51	30–60% of mass 198
68	Less than 2% of mass 69
70	Less than 2% of mass 69
127	40–60% of mass 198
197	Less than 1% of mass 198
198	Base peak, 100% relative abundance
199	5–9% of mass 198
275	10–30% of mass 198
365	Greater than 1% of mass 198
441	Present but less than mass 443
442	Greater than 40% of mass 198
443	17–23% of mass 442

of the compounds of interest and achieves all acceptable performance criteria may be used. GC to MS interfaces constructed of all glass or glass-lined materials are recommended. Glass can be deactivated by silanizing with dichlorodimethylsilane.

o. Data system: See Section 6200B.2*f.*

4. Reagents

a. Reagent water: See Section 6200B.3*a.*

b. Sodium hydroxide solution, NaOH, 10*N:* Dissolve 40 g NaOH in reagent water and dilute to 100 mL.

c. Sodium sulfate, Na_2SO_4, granular, anhydrous. Purify by heating at 400°C for 4 h in a shallow tray.

d. Sodium thiosulfate, $Na_2S_2O_3 \cdot 5H_2O$, granular.

e. Sulfuric acid, H_2SO_4, 1 + 1: Slowly add 50 mL conc H_2SO_4 to 50 mL reagent water.

f. Acetone, methanol, methylene chloride, pesticide quality or equivalent.

g. Stock standard solutions: Prepare from pure standard materials or purchase as certified solutions. Prepare by accurately weighing about 0.0100 g of pure material, dissolve in pesticide-quality acetone or other suitable solvent, and dilute to volume in a 10-mL volumetric flask; 1 μL = 1.00 μg compound. When compound purity is assayed to be 96% or greater, use the weight without correction to calculate concentration of the stock standard. Use commercially prepared stock standards at any concentration if certified by the manufacturer or by an independent source.

Transfer stock standard solutions into TFE-sealed screw-cap bottles. Store at 4°C and protect from light. Check stock standard solutions frequently for signs of degradation or evaporation, especially just before preparing calibration standards. Replace stock standard solutions after 6 months, or sooner if comparison with check standards indicates a problem.

h. Surrogate standard known-addition solution: Select a minimum of three surrogate compounds from Table 6410:IV. Prepare a surrogate standard solution containing each selected surrogate compound at a concentration of 100 μg/mL in acetone. Adding 1.00 mL to 1000 mL sample is equivalent to a concentration of 100 μg/L of each surrogate standard. Store at 4°C in TFE-sealed glass container. Check solution frequently for stability. Replace solution after 6 months, or sooner if comparison with quality-control check standards indicates a problem.

i. DFTPP standard: Prepare a 25-μg/mL solution of DFTPP in acetone.

j. Calibration standards: Prepare calibration standards at a minimum of three concentration levels for each compound by adding appropriate volumes of one or more stock standards to a volumetric flask. To each calibration standard or standard mixture, add a known constant amount of one or more internal standards (such as those listed in Table 6410:IV), and dilute to volume with acetone. Prepare one calibration standard at a concentration near, but above, the MDL and others corresponding to the expected range of sample concentrations or defining the working range of the GC/MS system.

k. Quality control (QC) check sample concentrate: Obtain a check sample concentrate containing each compound at a concentration of 100 μg/mL in acetone. Multiple solutions may be required. PCBs and multicomponent pesticides may be omitted. If such a sample is not available from an external source, prepare using stock standards prepared independently from those used for calibration.

5. Procedure

a. Extraction: Extraction by means of a separatory funnel, ¶ 1), is most common, but if emulsions will prevent acceptable solvent recovery, use continuous extraction, ¶ 2).

1) Separatory funnel extraction—Normally use a sample volume of 1 L. For sample volumes of 2 L, use 250-, 100-, and 100-mL volumes of methylene chloride for the serial extraction of the base/neutrals and 200-, 100-, and 100-mL volumes of methylene chloride for the acids.

Mark water meniscus on side of sample bottle for later determination of sample volume. Pour entire sample into a 2-L separatory funnel. Pipet 1.00 mL surrogate standard solution into separatory funnel and mix well. Check pH with wide-range pH paper and adjust to pH > 11 with NaOH solution.

Add 60 mL methylene chloride to sample bottle, seal, and shake for 30 s to rinse inner surface. Transfer solvent to separatory funnel and extract sample by shaking for 2 min with periodic venting to release excess pressure. Let organic layer separate from water phase for a minimum of 10 min. If emulsion interface between layers is more than one-third the volume of the solvent layer, use mechanical techniques to complete phase separation. The optimum technique depends on the sample, but may include stirring, filtering emulsion through glass wool, centrifuging, or other physical methods. Collect methylene chloride extract in a 250-mL erlenmeyer flask.

If the emulsion cannot be broken (recovery of less than 80% of the methylene chloride, corrected for water solubility of methylene chloride) in the first extraction, transfer sample, solvent, and emulsion into extraction chamber of a continuous extractor and proceed as described in ¶ 2) below.

Add a second 60-mL volume of methylene chloride to sample bottle and repeat extraction procedure, combining extracts in the erlenmeyer flask. Perform a third extraction in the same manner.

After the third extraction, adjust pH of aqueous phase to <2 using H_2SO_4. Serially extract acidified aqueous phase three times

TABLE 6410:IV. SUGGESTED INTERNAL AND SURROGATE STANDARDS

Base/Neutral Fraction	Acid Fraction
Aniline-d$_5$	2-Fluorophenol
Anthracene-d$_{10}$	Pentafluorophenol
Benzo(a)anthracene-d$_{12}$	Phenol-d$_5$
4,4'-Dibromobiphenyl	2-Perfluoromethyl phenol
4,4'-Dibromooctafluorobiphenyl	
Decafluorobiphenyl	
2,2'-Difluorobiphenyl	
4-Fluoroaniline	
1-Fluoronaphthylene	
2-Fluoronaphthylene	
Naphthalene-d$_8$	
Nitrobenzene-d$_5$	
2,3,4,5,6-Pentafluorobiphenyl	
Phenanthrene-d$_{10}$	
Pyridine-d$_5$	

with 60-mL portions of methylene chloride. Collect and combine extracts in a 250-mL erlenmeyer flask and label combined extracts as the acid fraction.

For each fraction, assemble a Kuderna-Danish (K-D) concentrator by attaching a 10-mL concentrator tube to a 500-mL evaporative flask. Other concentration devices or techniques may be used if the requirements of ¶ 7 are met.

Pour combined extract through a solvent-rinsed drying column containing at least 10 cm anhydrous Na_2SO_4 or more and collect extract in concentrator. Rinse erlenmeyer flask and column with 20 to 30 mL methylene chloride to complete transfer.

Add one or two clean boiling chips to the evaporative flask and attach a three-ball Snyder column. Prewet Snyder column by adding about 1 mL methylene chloride to the top. Place K-D apparatus on a hot water bath (60 to 65°C) in a hood so that concentrator tube is partially immersed in the hot water and entire lower rounded surface of flask is bathed with hot vapor. Adjust vertical position of apparatus and water temperature as required to complete concentration in 15 to 20 min. At proper rate of distillation the column balls actively chatter but the chambers are not flooded with condensed solvent. When the apparent volume of liquid reaches 1 mL, remove K-D apparatus and let drain and cool for at least 10 min.

Remove Snyder column and rinse flask and its lower joint into the concentrator tube with 1 to 2 mL methylene chloride, preferably using a 5-mL syringe.

Add another one or two clean boiling chips to concentrator tube for each fraction and attach a two-ball micro-Snyder column. Prewet Snyder column by adding about 0.5 mL of methylene chloride to the top. Place K-D apparatus on a hot-water bath (60 to 65°C) so that concentrator tube is partially immersed in hot water and continue concentrating as directed above without further solvent addition until apparent volume of liquid reaches about 0.5 mL. After cooling, remove Snyder column and rinse flask and its lower joint into the concentrator tube with approximately 0.2 mL acetone or methylene chloride. Adjust final volume to 1.0 mL with solvent. Stopper concentrator tube and store refrigerated if further processing will not be done immediately. If extract is to be stored longer than 2 d, transfer to a TFE-sealed screw-cap vial and label base/neutral or acid fraction as appropriate.

Determine original sample volume by refilling sample bottle to mark and transferring liquid to a 1000-mL graduated cylinder. Record sample volume to nearest 5 mL.

2) *Continuous extraction*—Mark water meniscus on side of sample bottle, and determine sample volume later as described in ¶ 1). Check pH with wide-range pH paper and adjust to pH > 11 with NaOH solution. Transfer sample to continuous extractor and, using a pipet, add 1.00 mL surrogate standard solution and mix well. Add 60 mL methylene chloride to sample bottle, seal, and shake for 30 s to rinse inner surface. Transfer solvent to extractor. Repeat rinse with an additional 50- to 100-mL portion methylene chloride and add rinse to extractor.

Add 200 to 500 mL methylene chloride to distilling flask, add sufficient reagent water to ensure proper operation, and extract for 24 h. Let cool and detach distilling flask. Dry, concentrate, and seal extract as in ¶ 1) above.

Charge a clean distilling flask with 500 mL methylene chloride and attach it to continuous extractor. Carefully, while stirring, adjust pH of aqueous phase to less than 2 with H_2SO_4. Extract for 24 h. Dry, concentrate, and seal extract as in ¶ 1) above.

b. GC/MS operating conditions: Table 6410:I summarizes the recommended gas chromatographic operating conditions for the base/neutral fraction and Table 6410:II for the acid fraction. Included in these tables are retention times and MDLs that can be achieved under these conditions. Examples of the separations obtained with these columns are shown in Figures 6410:1 through 12. Other packed or capillary (open-tubular) columns or chromatographic conditions may be used if the requirements of ¶ 7 are met.

c. GC/MS performance tests: At the beginning of each day on which analyses are to be performed, check GC/MS system to see if acceptable performance criteria are achieved for DFTPP.[8] Each day that benzidine is to be determined, the tailing factor criterion described in ¶ 2) must be achieved. Each day that the acids are to be determined, the tailing factor criterion described in ¶ 3) must be achieved.

These performance tests have the requirements given in Section 6200B.4*b*, but use following conditions:

Electron energy: 70 V (nominal)
Mass range: 35 to 450 amu
Scan time: To give at least 5 scans per peak but not to exceed 7 s per scan.

1) *DFTPP performance test*—At beginning of each day, inject 2 µL (50 ng) DFTPP standard solution. Obtain a background-corrected mass spectrum of DFTPP and confirm that all the key *m/z* criteria in Table 6410:III are achieved. If not, retune mass spectrometer and repeat test until all criteria are achieved. Meet performance criteria before any samples, blanks, or standards are analyzed. The tailing factor tests in ¶s 2) and 3) may be performed simultaneously with the DFTPP test.

2) *Column performance test for base/neutrals*—At beginning of each day that base/neutral fraction is to be analyzed for benzidine, calculate benzidine tailing factor. Inject 100 ng benzidine either separately or as a part of a standard mixture that may contain DFTPP, and calculate tailing factor, which must be less than 3.0. Calculation of the tailing factor is illustrated in Figure 6410:13.[9] Replace column packing if tailing factor criterion cannot be met.

3) *Column performance test for acids*—At beginning of each day that acids are to be determined, inject 50 ng pentachlorophenol either separately or as a part of a standard mix that may contain DFTPP. The tailing factor for pentachlorophenol must be less than 5. Calculation of the tailing factor is illustrated in Figure 6410:13.[9] Replace column packing if tailing factor criterion cannot be met.

d. Calibration of GC/MS system: Calibrate system daily after performance tests.

Select three or more internal standards similar in analytical behavior to the compounds of interest. Demonstrate that the measurement of the internal standards is not affected by method or matrix interferences. Some recommended internal standards are listed in Table 6410:IV. Use base peak *m/z* as the primary *m/z* for quantification. If interferences are noted, use one of the next two most intense *m/z* quantities for quantification. Using injections of 2 to 5 µL, analyze each calibration standard according to ¶ *e* below and tabulate area of primary characteristic *m/z* (Tables 6410:I and II) against concentration for each compound and internal standard. Calculate response factors (RF) for each compound by the equation given in Section 6200B.4*c*2). If the RF value over the working range is a constant (<35% RSD), it can

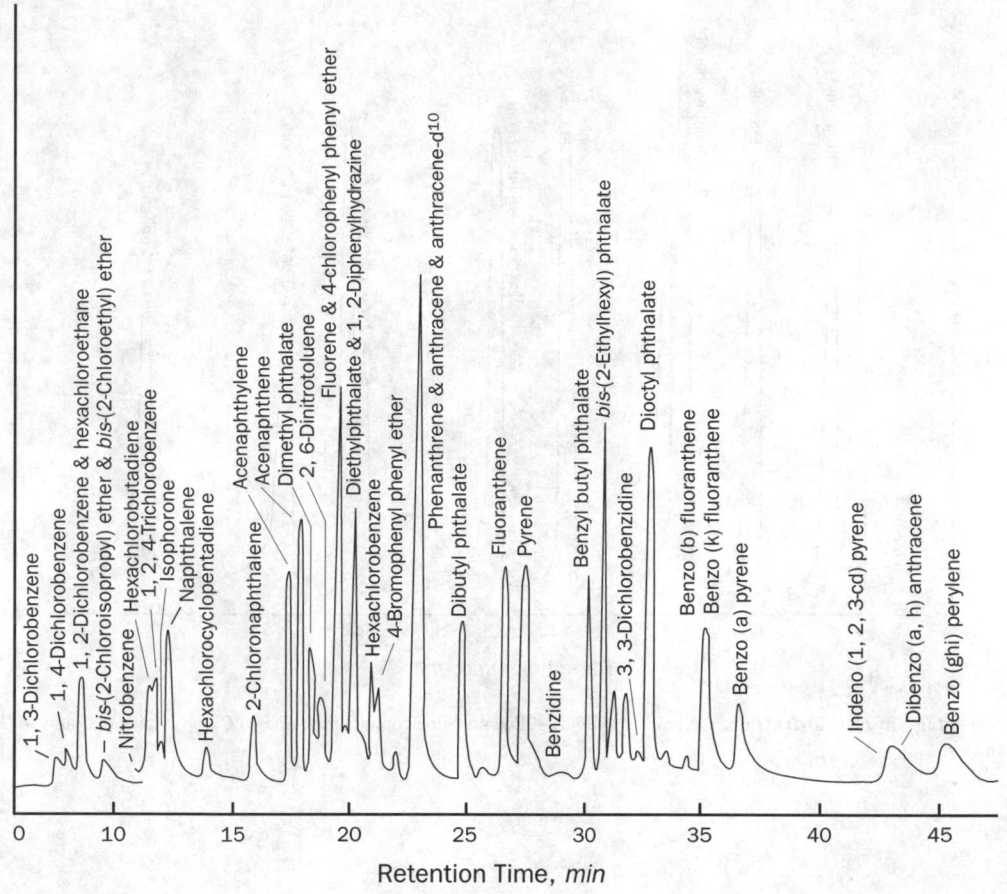

Figure 6410:1. Gas chromatogram of base/neutral fraction. Column: 3% SP-2250 on Supelcoport; program: 50°C for 4 min, 8°C/min to 270°C; detector: mass spectrometer.

be assumed to be invariant; use the average RF for calculations. Alternatively, use the results to plot a calibration curve of response ratios, A_s/A_{is} vs. RF.

Verify working calibration curve or RF on each working day by measuring one or more calibration standards. If the response for any compound varies from the predicted response by more than 20%, repeat test using a fresh calibration standard. Alternatively, prepare a new calibration curve for that compound.

e. Sample analysis: Add internal standard to sample extract, mix thoroughly, and immediately inject 2 to 5 μL of sample extract or standard into GC/MS system using solvent-flush technique[10] to minimize losses due to adsorption, chemical reaction, or evaporation. Smaller (1.0-μL) volumes may be injected if automatic devices are used. Record volume injected to nearest 0.05 μL. If response for any *m/z* exceeds the working range of the GC/MS system, dilute extract and reanalyze. Make all qualitative and quantitative measurements as described below and in ¶ 6. When extract is not being used, store at 4°C, protected from light, in screw-cap vial equipped with unpierced TFE-lined septum.

Obtain EICPs for the primary *m/z* and the two other masses listed in Tables 6410:I and II. See ¶ *d* for masses to be used with internal and surrogate standards. Use the following criteria to make a qualitative identification:

• The characteristic masses of each compound maximize in the same or within one scan of each other.

• The retention time falls within ± 30 s of the retention time of the authentic compound.

• The relative peak heights of the three characteristic masses in the EICPs fall within ± 20% of the relative intensities of these masses in a reference mass spectrum obtained from a standard analyzed in the GC/MS system or from a reference library.

Structural isomers that have very similar mass spectra and less than 30 s difference in retention time can be identified explicitly only if the resolution between authentic isomers in a standard mix is acceptable. Acceptable resolution is achieved if the baseline to valley height between the isomers is less than 25% of the sum of the two peak heights. Otherwise, structural isomers are identified as isomeric pairs.

f. Screening procedure for 2,3,7,8-tetrachlorodibenzo-p-dioxin (2,3,7,8-TCDD): CAUTION: *In screening a sample for 2,3,7,8-TCDD, do not handle reference material without taking extensive safety precautions.* It is sufficient to analyze the base/neutral extract by selected ion monitoring (SIM) GC/MS techniques, as follows:

Concentrate base/neutral extract to a final volume of 0.2 mL. Adjust temperature of base/neutral column to 220°C. Operate

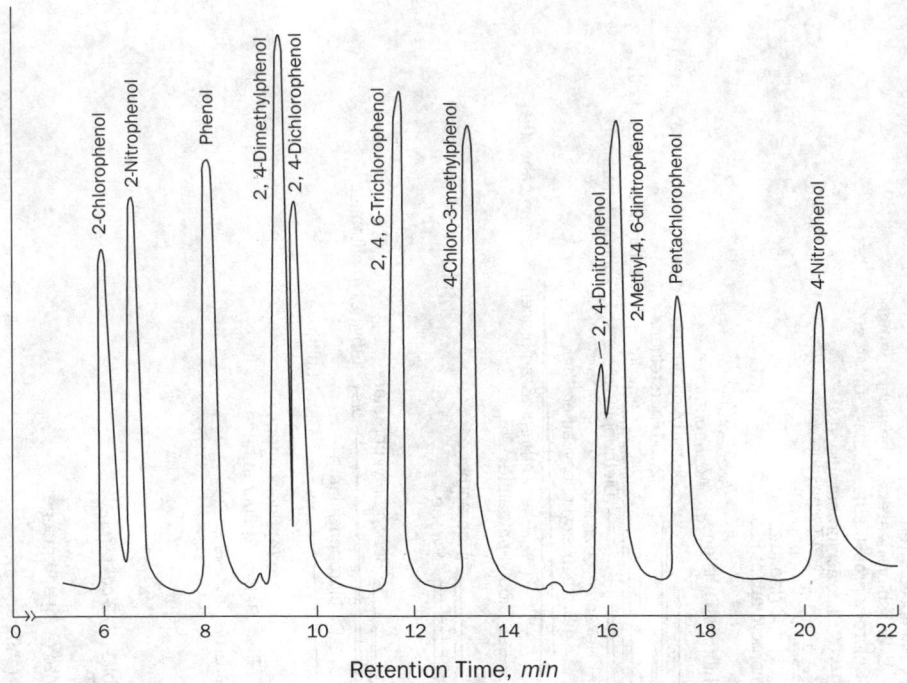

Figure 6410:2. Gas chromatogram of acid fraction. Column: 1% SP-1240DA on Supelcoport; program: 70°C for 2 min, 8°C/min to 200°C; detector: mass spectrometer.

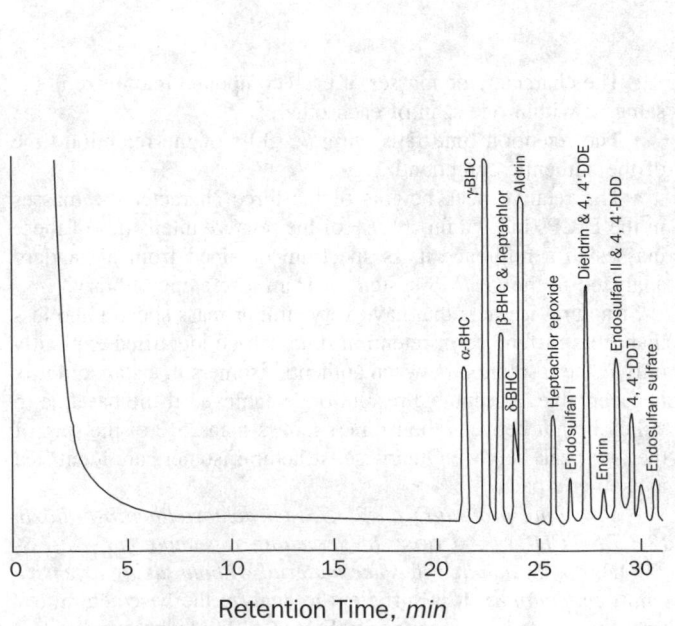

Figure 6410:3. Gas chromatogram of pesticide fraction. Column: 3% SP-2250 on Supelcoport; program: 50°C for 4 min, 8°C/min to 270°C; detector: mass spectrometer.

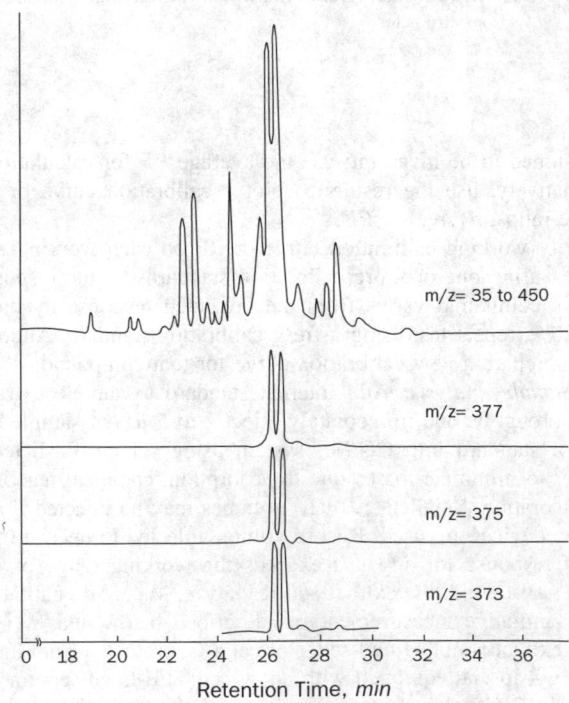

Figure 6410:4. Gas chromatogram of chlordane. Column: 3% SP-2250 on Supelcoport; program: 50°C for 4 min, 8°C/min to 270°C; detector: mass spectrometer.

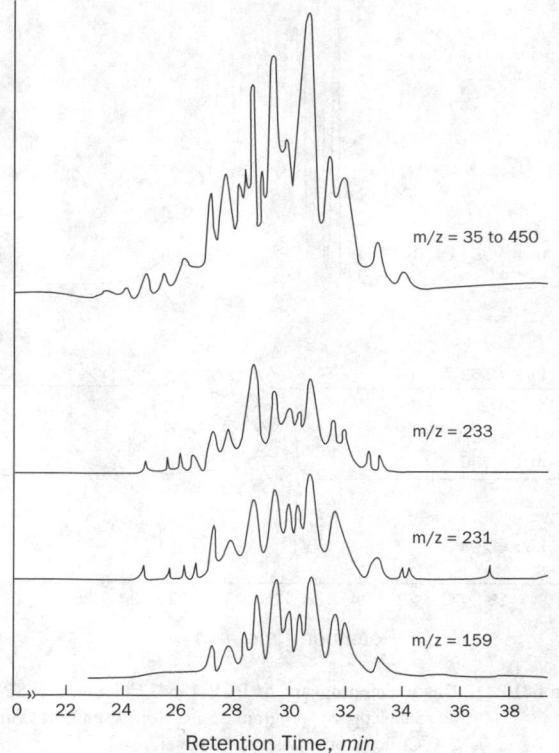

Figure 6410:5. Gas chromatogram of toxaphene. Column: 3% SP-2250 on Supelcoport; program: 50°C for 4 min, 8°C/min to 270°C; detector: mass spectrometer.

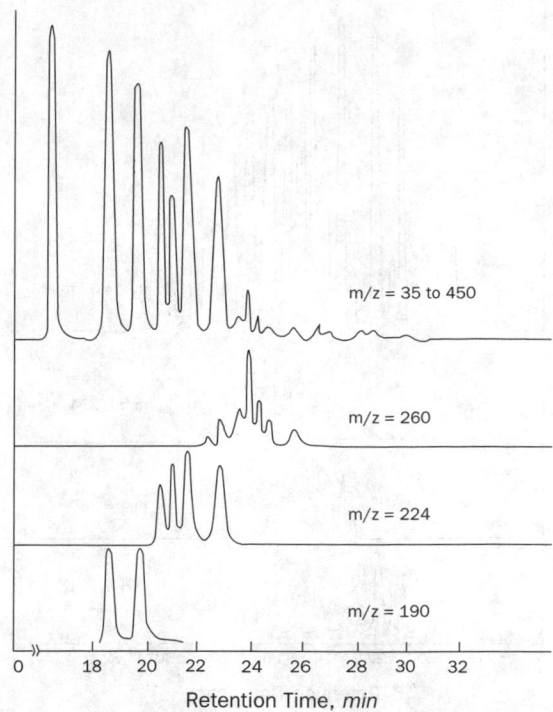

Figure 6410:7. Gas chromatogram of PCB-1221. Column: 3% SP-2250 on Supelcoport; program: 50°C for 4 min, 8°C/min to 270°C; detector: mass spectrometer.

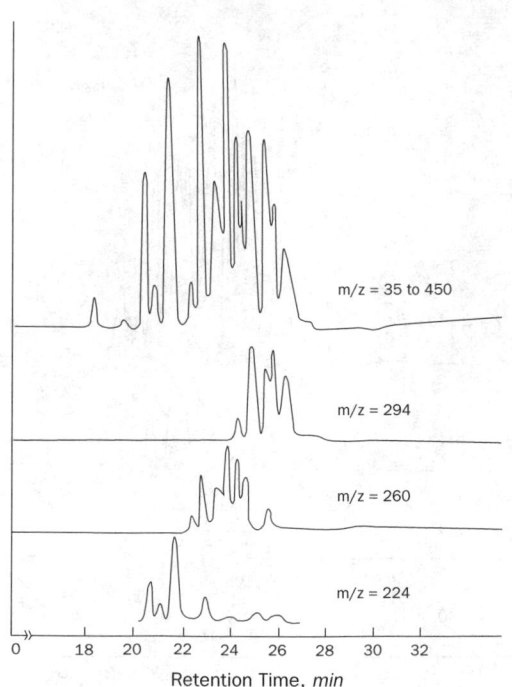

Figure 6410:6. Gas chromatogram of PCB-1016. Column: 3% SP-2250 on Supelcoport; program: 50°C for 4 min, 8°C/min to 270°C; detector: mass spectrometer.

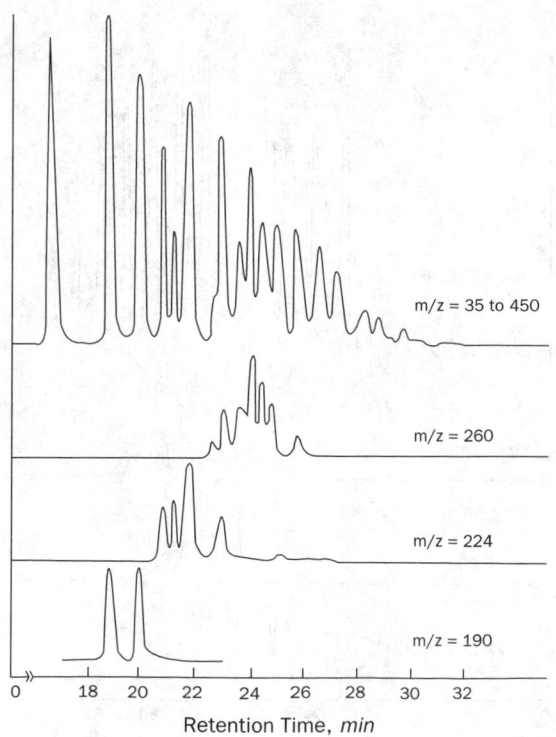

Figure 6410:8. Gas chromatogram of PCB-1232. Column: 3% SP-2250 on Supelcoport; program: 50°C for 4 min, 8°C/min to 270°C; detector: mass spectrometer.

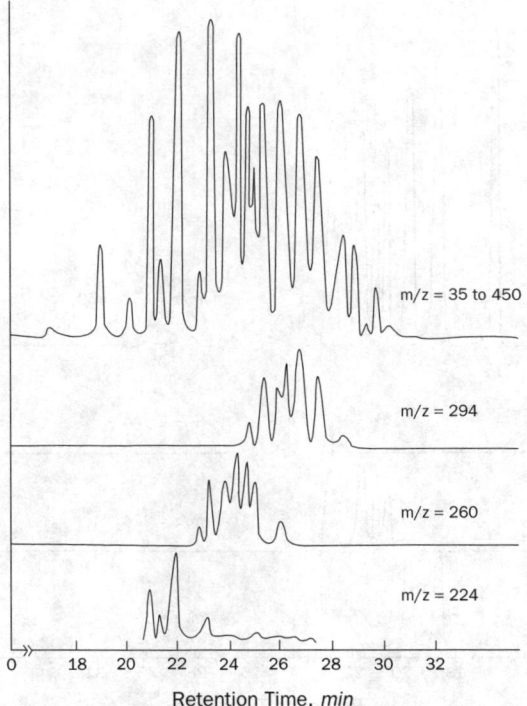

Figure 6410:9. Gas chromatogram of PCB-1242. Column: 3% SP-2250 on Supelcoport; program: 50°C for 4 min, 8°C/min to 270°C; detector: mass spectrometer.

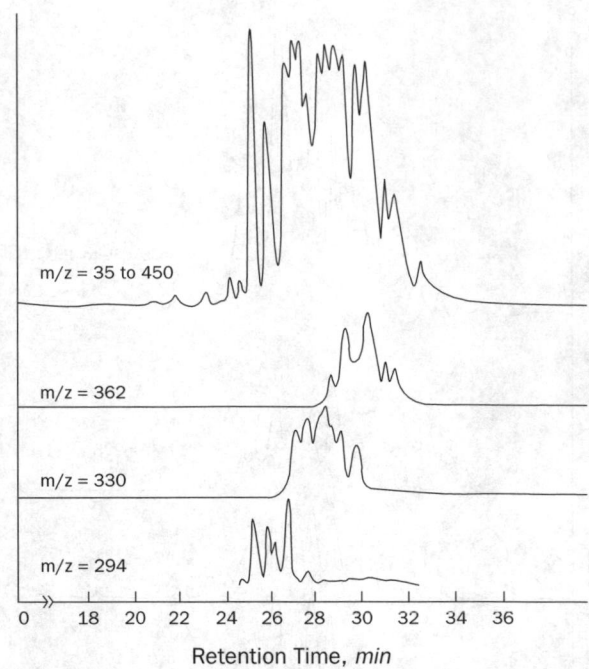

Figure 6410:11. Gas chromatogram of PCB-1254. Column: 3% SP-2250 on Supelcoport; program: 50°C for 4 min, 8°C/min to 270°C; detector: mass spectrometer.

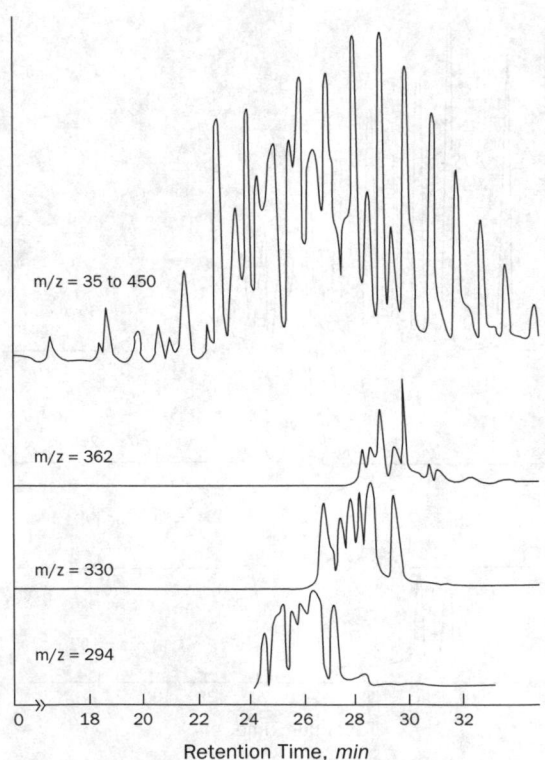

Figure 6410:10. Gas chromatogram of PCB-1248. Column: 3% SP-2250 on Supelcoport; program: 50°C for 4 min, 8°C/min to 270°C; detector: mass spectrometer.

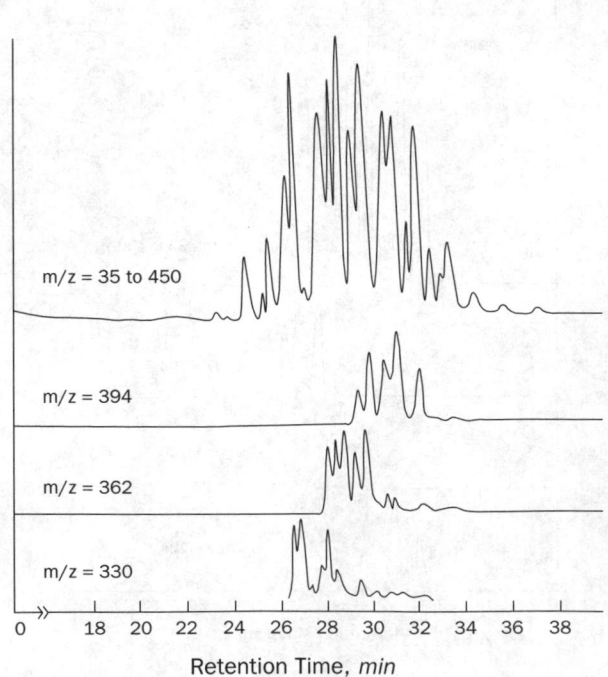

Figure 6410:12. Gas chromatogram of PCB-1260. Column: 3% SP-2250 on Supelcoport; program: 50°C for 4 min, 8°C/min to 270°C; detector: mass spectrometer.

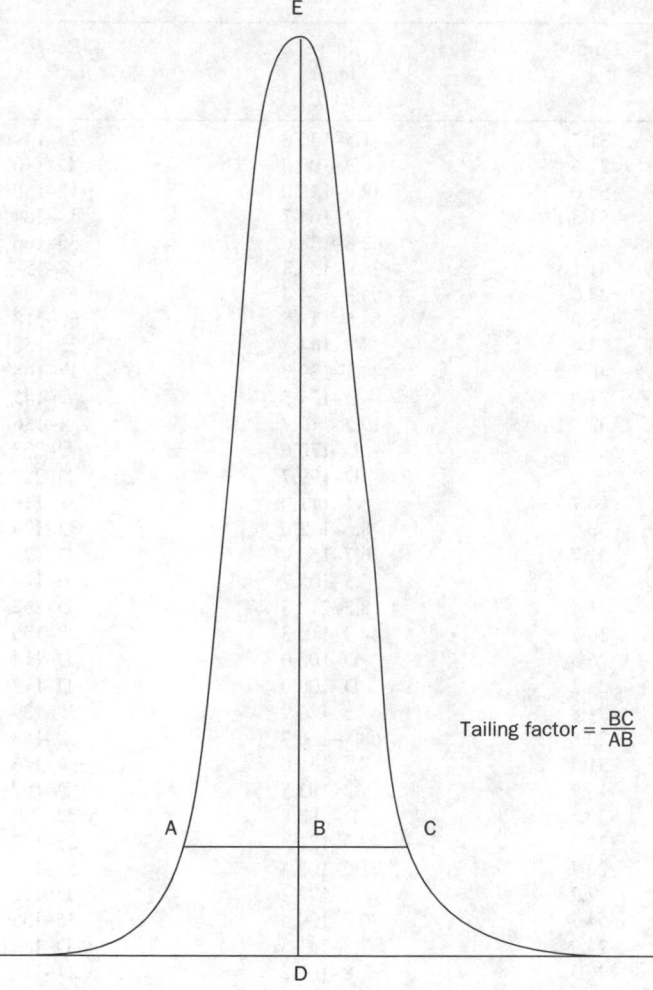

Figure 6410:13. Tailing factor calculation. Example calculation:
Peak height = DE = 100 mm   10% peak height = BD = 10 mm
Peak width at 10% peak height = AC = 23 mm
AB = 11 mm   BC = 12 mm
$$\text{Tailing factor} = \frac{12}{11} = 1.1$$

mass spectrometer to acquire data in the SIM mode using the ions at m/z 257, 320, and 322 and a dwell time no greater than 333 ms/mass. Inject 5 to 7 μL of base/neutral extract. Collect SIM data for a total of 10 min. The possible presence of 2,3,7,8-TCDD is indicated if all three masses exhibit simultaneous peaks at any point in the selected ion current profiles. For each occurrence where the possible presence of 2,3,7,8-TCDD is indicated, calculate and retain the relative abundances of each of the three masses. False positives may be caused by the presence of single or coeluting combinations of compounds whose mass spectra contain all of these masses. Conclusive results of the presence and concentration level of 2,3,7,8-TCDD can be obtained only from a properly equipped laboratory using a specialized test method.[11]

6. Calculation

When a compound has been identified, base quantitation on the integrated abundance from the EICP of the primary characteristic m/z given in Tables 6410:I and II. Use base peak m/z for internal and surrogate standards. If sample produces an interference for the primary m/z, use a secondary characteristic m/z to quantitate.

Calculate sample concentration using the response factor (RF) determined in ¶ 5d and the equation:

$$\text{Concentration, } \mu\text{g/L} = \frac{(A_s)\,(I_s)}{(A_{is})\,(RF)\,(V_e)}$$

where:

A_s = area of characteristic m/z for compound or surrogate standard to be measured,
A_{is} = area of characteristic m/z for internal standard,
I_s = amount of internal standard added to each extract, μg, and
V_e = volume of water extracted, L.

Report results in μg/L without correction for recovery. Report all QC data with sample results.

7. Quality Control

a. Quality control program: See Section 6200A.5.

b. Initial quality control: Proceed according to Section 6200A.5a1) and 2). Use Table 6410:V for acceptance criteria.

c. Analyses of samples with known additions: Use quality acceptance criteria given in Table 6410:V.

TABLE 6410:V. QC ACCEPTANCE CRITERIA*

Compound	Test Concentration μg/L	Limits for s μg/L	Range for $\overline{X}$ μg/L	Range for P, P_s %
Acenaphthene	100	27.6	60.1–132.3	47–145
Acenaphthylene	100	40.2	53.5–126.0	33–145
Aldrin	100	39.0	7.2–152.2	D–166
Anthracene	100	32.0	43.4–118.0	27–133
Benzo(a)anthracene	100	27.6	41.8–133.0	33–143
Benzo(b)fluoranthene	100	38.8	42.0–140.4	24–159
Benzo(k)fluoranthene	100	32.3	25.2–145.7	11–162
Benzo(a)pyrene	100	39.0	31.7–148.0	17–163
Benzo(ghi)perylene	100	58.9	D–195.0	D–219
Benzyl butyl phthalate	100	23.4	D–139.9	D–152

TABLE 6410:V, CONT.

Compound	Test Concentration μg/L	Limits for s μg/L	Range for $\overline{X}$ μg/L	Range for P, P_s %
δ-BHC	100	31.5	41.5–130.6	24–149
β-BHC	100	21.6	D–100.0	D–110
bis(2-Chloroethyl) ether	100	55.0	42.9–126.0	12–158
bis(2-Chloroethoxy) methane	100	34.5	49.2–164.7	33–184
bis(2-Chloroisopropyl) ether†	100	46.3	62.8–138.6	36–166
bis(2-Ethylhexyl) phthalate	100	41.1	28.9–136.8	8–158
4-Bromophenyl phenyl ether	100	23.0	64.9–114.4	53–127
2-Chloronaphthalene	100	13.0	64.5–113.5	60–118
4-Chlorophenyl phenyl ether	100	33.4	38.4–144.7	25–158
Chrysene	100	48.3	44.1–139.9	17–168
4,4'-DDD	100	31.0	D–134.5	D–145
4,4'-DDE	100	32.0	19.2–119.7	4–136
4,4'-DDT	100	61.6	D–170.6	D–203
Dibenzo(a,h)anthracene	100	70.0	D–199.7	D–227
Di-n-butyl phthalate	100	16.7	8.4–111.0	1–118
1,2-Dichlorobenzene	100	30.9	48.6–112.0	32–129
1,3-Dichlorobenzene	100	41.7	16.7–153.9	D–172
1,4-Dichlorobenzene	100	32.1	37.3–105.7	20–124
3,3'-Dichlorobenzidine	100	71.4	8.2–212.5	D–262
Dieldrin	100	30.7	44.3–119.3	29–136
Diethyl phthalate	100	26.5	D–100.0	D–114
Dimethyl phthalate	100	23.2	D–100.0	D–112
2,4-Dinitrotoluene	100	21.8	47.5–126.9	39–139
2,6-Dinitrotoluene	100	29.6	68.1–136.7	50–158
Di-n-octylphthalate	100	31.4	18.6–131.8	4–146
Endosulfan sulfate	100	16.7	D–103.5	D–107
Endrin aldehyde	100	32.5	D–188.8	D–209
Fluoranthene	100	32.8	42.9–121.3	26–137
Fluorene	100	20.7	71.6–108.4	59–121
Heptachlor	100	37.2	D–172.2	D–192
Heptachlor epoxide	100	54.7	70.9–109.4	26–155
Hexachlorobenzene	100	24.9	7.8–141.5	D–152
Hexachlorobutadiene	100	26.3	37.8–102.2	24–116
Hexachloroethane	100	24.5	55.2–100.0	40–113
Indeno(1,2,3-cd)pyrene	100	44.6	D–150.9	D–171
Isophorone	100	63.3	46.6–180.2	21–196
Naphthalene	100	30.1	35.6–119.6	21–133
Nitrobenzene	100	39.3	54.3–157.6	35–180
N-Nitrosodi-n-propylamine	100	55.4	13.6–197.9	D–230
PCB-1260	100	54.2	19.3–121.0	D–164
Phenanthrene	100	20.6	65.2–108.7	54–120
Pyrene	100	25.2	69.6–100.0	52–115
1,2,4-Trichlorobenzene	100	28.1	57.3–129.2	44–142
4-Chloro-3-methylphenol	100	37.2	40.8–127.9	22–147
2-Chlorophenol	100	28.7	36.2–120.4	23–134
2,4-Dichlorophenol	100	26.4	52.5–121.7	39–135
2,4-Dimethylphenol	100	26.1	41.8–109.0	32–119
2,4-Dinitrophenol	100	49.8	D–172.9	D–191
2-Methyl-4,6-dinitrophenol	100	93.2	53.0–100.0	D–181
2-Nitrophenol	100	35.2	45.0–166.7	29–182
4-Nitrophenol	100	47.2	13.0–106.5	D–132
Pentachlorophenol	100	48.9	38.1–151.8	14–176
Phenol	100	22.6	16.6–100.0	5–112
2,4,6-Trichlorophenol	100	31.7	52.4–129.2	37–144

*   s = standard deviation for four recovery measurements,
   $\overline{X}$ = average recovery for four recovery measurements,
  P, P_s = percent recovery measured, and
    D = detected; results must be greater than zero.
† The proper chemical name is 2,2'-oxybis(1-chloropropane).
NOTE: These criteria are based directly upon the method performance data in Table 6410:VI. Where necessary, the limits for recovery were broadened to assure applicability of the limits to concentrations below those used to develop Table 6410:VI.

TABLE 6410:VI. METHOD BIAS AND PRECISION AS FUNCTIONS OF CONCENTRATION*

Compound	Bias as Recovery, X' $\mu g/L$	Single-Analyst Precision, s_r' $\mu g/L$	Overall Precision, S' $\mu g/L$
Acenaphthene	$0.96C + 0.19$	$0.15\overline{X} - 0.12$	$0.21\overline{X} - 0.67$
Acenaphthylene	$0.89C + 0.74$	$0.24\overline{X} - 1.06$	$0.26\overline{X} - 0.54$
Aldrin	$0.78C + 1.66$	$0.27\overline{X} - 1.28$	$0.43\overline{X} + 1.13$
Anthracene	$0.80C + 0.68$	$0.21\overline{X} - 0.32$	$0.27\overline{X} - 0.64$
Benzo(a)anthracene	$0.88C - 0.60$	$0.15\overline{X} + 0.93$	$0.26\overline{X} - 0.28$
Benzo(b)fluoranthene	$0.93C - 1.80$	$0.22\overline{X} + 0.43$	$0.29\overline{X} + 0.96$
Benzo(k)fluoranthene	$0.87C - 1.56$	$0.19\overline{X} + 1.03$	$0.35\overline{X} + 0.40$
Benzo(a)pyrene	$0.90C - 0.13$	$0.22\overline{X} + 0.48$	$0.32\overline{X} + 1.35$
Benzo(ghi)perylene	$0.98C - 0.86$	$0.29\overline{X} + 2.40$	$0.51\overline{X} - 0.44$
Benzyl butyl phthalate	$0.66C - 1.68$	$0.18\overline{X} + 0.94$	$0.53\overline{X} + 0.92$
β-BHC	$0.87C - 0.94$	$0.20\overline{X} - 0.58$	$0.30\overline{X} - 1.94$
δ-BHC	$0.29C - 1.09$	$0.34\overline{X} + 0.86$	$0.93\overline{X} - 0.17$
bis(2-Chloroethyl) ether	$0.86C - 1.54$	$0.35\overline{X} - 0.99$	$0.35\overline{X} + 0.10$
bis(2-Chloroethoxy) methane	$1.12C - 5.04$	$0.16\overline{X} + 1.34$	$0.26\overline{X} + 2.01$
bis(2-Chloroisopropyl) ether†	$1.03C - 2.31$	$0.24\overline{X} + 0.28$	$0.25\overline{X} + 1.04$
bis(2-Ethylhexyl) phthalate	$0.84C - 1.18$	$0.26\overline{X} + 0.73$	$0.36\overline{X} + 0.67$
4-Bromophenyl phenyl ether	$0.91C - 1.34$	$0.13\overline{X} + 0.66$	$0.16\overline{X} + 0.66$
2-Chloronaphthalene	$0.89C + 0.01$	$0.07\overline{X} + 0.52$	$0.13\overline{X} + 0.34$
4-Chlorophenyl phenyl ether	$0.91C + 0.53$	$0.20\overline{X} - 0.94$	$0.30\overline{X} - 0.46$
Chrysene	$0.93C - 1.00$	$0.28\overline{X} + 0.13$	$0.33\overline{X} - 0.09$
4,4'-DDD	$0.56C - 0.40$	$0.29\overline{X} - 0.32$	$0.66\overline{X} - 0.96$
4,4'-DDE	$0.70C - 0.54$	$0.26\overline{X} - 1.17$	$0.39\overline{X} - 1.04$
4,4'-DDT	$0.79C - 3.28$	$0.42\overline{X} + 0.19$	$0.65\overline{X} - 0.58$
Dibenzo(a,h)anthracene	$0.88C + 4.72$	$0.30\overline{X} + 8.51$	$0.59\overline{X} + 0.25$
Di-n-butyl phthalate	$0.59C + 0.71$	$0.13\overline{X} + 1.16$	$0.39\overline{X} + 0.60$
1,2-Dichlorobenzene	$0.80C + 0.28$	$0.20\overline{X} + 0.47$	$0.24\overline{X} + 0.39$
1,3-Dichlorobenzene	$0.86C - 0.70$	$0.25\overline{X} + 0.68$	$0.41\overline{X} + 0.11$
1,4-Dichlorobenzene	$0.73C - 1.47$	$0.24\overline{X} + 0.23$	$0.29\overline{X} + 0.36$
3,3'-Dichlorobenzidine	$1.23C - 12.65$	$0.28\overline{X} + 7.33$	$0.47\overline{X} + 3.45$
Dieldrin	$0.82C - 0.16$	$0.20\overline{X} - 0.16$	$0.26\overline{X} - 0.07$
Diethyl phthalate	$0.43C + 1.00$	$0.28\overline{X} + 1.44$	$0.52\overline{X} + 0.22$
Dimethyl phthalate	$0.20C + 1.03$	$0.54\overline{X} + 0.19$	$1.05\overline{X} - 0.92$
2,4-Dinitrotoluene	$0.92C - 4.81$	$0.12\overline{X} + 1.06$	$0.21\overline{X} + 1.50$
2,6-Dinitrotoluene	$1.06C - 3.60$	$0.14\overline{X} + 1.26$	$0.19\overline{X} + 0.35$
Di-n-octylphthalate	$0.76C - 0.79$	$0.21\overline{X} + 1.19$	$0.37\overline{X} + 1.19$
Endosulfan sulfate	$0.39C + 0.41$	$0.12\overline{X} + 2.47$	$0.63\overline{X} - 1.03$
Endrin aldehyde	$0.76C - 3.86$	$0.18\overline{X} + 3.91$	$0.73\overline{X} - 0.62$
Fluoranthene	$0.81C + 1.10$	$0.22\overline{X} - 0.73$	$0.28\overline{X} - 0.60$
Fluorene	$0.90C - 0.00$	$0.12\overline{X} + 0.26$	$0.13\overline{X} + 0.61$
Heptachlor	$0.87C - 2.97$	$0.24\overline{X} - 0.56$	$0.50\overline{X} - 0.23$
Heptachlor epoxide	$0.92C - 1.87$	$0.33\overline{X} - 0.46$	$0.28\overline{X} + 0.64$
Hexachlorobenzene	$0.74C + 0.66$	$0.18\overline{X} - 0.10$	$0.43\overline{X} - 0.52$
Hexachlorobutadiene	$0.71C - 1.01$	$0.19\overline{X} + 0.92$	$0.26\overline{X} + 0.49$
Hexachloroethane	$0.73C - 0.83$	$0.17\overline{X} + 0.67$	$0.17\overline{X} + 0.80$
Indeno(1,2,3-cd)pyrene	$0.78C - 3.10$	$0.29\overline{X} + 1.46$	$0.50\overline{X} + 0.44$
Isophorone	$1.12C - 1.41$	$0.27\overline{X} + 0.77$	$0.33\overline{X} + 0.26$
Naphthalene	$0.76C + 1.58$	$0.21\overline{X} - 0.41$	$0.30\overline{X} - 0.68$
Nitrobenzene	$1.09C - 3.05$	$0.19\overline{X} + 0.92$	$0.27\overline{X} + 0.21$
N-Nitrosodi-n-propylamine	$1.12C - 6.22$	$0.27\overline{X} + 0.68$	$0.44\overline{X} + 0.47$
PCB-1260	$0.81C - 10.86$	$0.35\overline{X} + 3.61$	$0.43\overline{X} + 1.82$
Phenanthrene	$0.87C - 0.06$	$0.12\overline{X} + 0.57$	$0.15\overline{X} + 0.25$
Pyrene	$0.84C - 0.16$	$0.16\overline{X} + 0.06$	$0.15\overline{X} + 0.31$
1,2,4-Trichlorobenzene	$0.94C - 0.79$	$0.15\overline{X} + 0.85$	$0.21\overline{X} + 0.39$
4-Chloro-3-methylphenol	$0.84C + 0.35$	$0.23\overline{X} + 0.75$	$0.29\overline{X} + 1.31$
2-Chlorophenol	$0.78C + 0.29$	$0.18\overline{X} + 1.46$	$0.28\overline{X} + 0.97$
2,4-Dichlorophenol	$0.87C + 0.13$	$0.15\overline{X} + 1.25$	$0.21\overline{X} + 1.28$
2,4-Dimethylphenol	$0.71C + 4.41$	$0.16\overline{X} + 1.21$	$0.22\overline{X} + 1.31$
2,4-Dinitrophenol	$0.81C - 18.04$	$0.38\overline{X} + 2.36$	$0.42\overline{X} + 26.29$
2-Methyl-4,6-dinitrophenol	$1.04C - 28.04$	$0.10\overline{X} + 42.29$	$0.26\overline{X} + 23.10$

TABLE 6410:VI. CONT.

Compound	Bias as Recovery, X' $\mu g/L$	Single-Analyst Precision, s_r' $\mu g/L$	Overall Precision, S' $\mu g/L$
2-Nitrophenol	$1.07C - 1.15$	$0.16\overline{X} + 1.94$	$0.27\overline{X} + 2.60$
4-Nitrophenol	$0.61C - 1.22$	$0.38\overline{X} + 2.57$	$0.44\overline{X} + 3.24$
Pentachlorophenol	$0.93C + 1.99$	$0.24\overline{X} + 3.03$	$0.30\overline{X} + 4.33$
Phenol	$0.43C + 1.26$	$0.26\overline{X} + 0.73$	$0.35\overline{X} + 0.58$
2,4,6-Trichlorophenol	$0.91C - 0.18$	$0.16\overline{X} + 2.22$	$0.22\overline{X} + 1.81$

* X' = expected recovery for one or more measurements of a sample containing a concentration of C,
s_r' = expected single-analyst standard deviation of measurements at an average concentration found of $\overline{X}$,
S' = expected interlaboratory standard deviation of measurements at an average concentration found of X,
C = true value for the concentration, and
$\overline{X}$ = average recovery found for measurements of samples containing a concentration of C.
† The proper chemical name is 2,2'-oxybis(1-chloropropane).

d. Quality-control check standard analysis: Proceed as in Section 6200A.5a3); prepare QC check standard with 1.0 mL QC check standard concentrate and 1 L reagent water.

e. Bias assessment and records: Assess method bias and maintain records. For example, after the analysis of five wastewater samples, calculate the average percent recovery ($\overline{P}$) and the standard deviation of the percent recovery (s_p). Express bias assessment as a percent recovery interval from $\overline{P} - 2s_p$ to $\overline{P} + 2s_p$. If $\overline{P} = 90\%$ and $s_p = 10\%$, the recovery interval is expressed as 70–110%. Update bias assessment for each compound regularly, (e.g., after each five to ten new accuracy measurements).

f. Use of surrogate compounds: As a quality control check, make known additions to all samples of surrogate standard solution as described in ¶ 5a1), and calculate percent recovery of each surrogate compound.

g. Additional quality-assurance practices: Other desirable practices depend on the needs of the laboratory and the nature of the samples. Analyze field duplicates to assess precision of environmental measurements. Whenever possible, analyze standard reference materials and participate in relevant performance evaluation studies. Certain compounds, such as phthalates, are common laboratory contaminants. When these are measured above the detection limits in sample blanks, locate their source and repeat the analysis after taking corrective action.

8. Precision and Bias

This method was tested by 15 laboratories using reagent water, drinking water, surface water, and industrial wastewaters with additions at six concentrations over the range 5 to 1300 $\mu g/L$.[3] Single-operator precision, overall precision, and method bias were found to be related directly to the compound concentration and essentially independent of the sample matrix. Linear equations describing these relationships are presented in Table 6410:VI.

9. References

1. U.S. ENVIRONMENTAL PROTECTION AGENCY. 1984. Method 625—Base/neutrals and acids. 40 CFR Part 136, 43385; *Federal Register* 49, No. 209.
2. U.S. ENVIRONMENTAL PROTECTION AGENCY. 1977. Sampling and Analysis Procedures for Screening of Industrial Effluents for Priority Pollutants. Environmental Monitoring and Support Lab., Cincinnati, Ohio.
3. U.S. ENVIRONMENTAL PROTECTION AGENCY. 1984. EPA Method Study 30, Method 625—Base/Neutrals, Acids, and Pesticides. EPA-600/4-84-053, National Technical Information Serv., PB84-206572, Springfield, Va.
4. AMERICAN SOCIETY FOR TESTING AND MATERIALS. 1978. Standard practices for preparation of sample containers and for preservation of organic constituents. ASTM Annual Book of Standards, Part 31, D3694-78. Philadelphia, Pa.
5. U.S. ENVIRONMENTAL PROTECTION AGENCY. 1984. Definition and procedure for the determination of the method detection limit. 40 CFR Part 136, Appendix B. *Federal Register* 49, No. 209.
6. OLYNYK, P., W.L. BUDDE & J.W. EICHELBERGER. 1980. Method Detection Limit for Methods 624 and 625. Unpublished report.
7. AMERICAN SOCIETY FOR TESTING AND MATERIALS. 1976. Standard practices for sampling water. ASTM Annual Book of Standards, Part 31, D3370-76. Philadelphia, Pa.
8. EICHELBERGER, J.W., L.E. HARRIS & W.L. BUDDE. 1975. Reference compound to calibrate ion abundance measurement in gas chromatography-mass spectrometry. *Anal. Chem.* 47:995.
9. MCNAIR, N.M. & E.J. BONELLI. 1969. Basic Chromatography. Consolidated Printing, Berkeley, Calif.
10. BURKE, J.A. 1965. Gas chromatography for pesticide residue analysis; some practical aspects. *J. Assoc. Offic. Anal. Chem.* 48:1037.
11. U.S. ENVIRONMENTAL PROTECTION AGENCY. 1984. Method 613—2,3,7,8-Tetrachlorodibenzo-p-dioxin. 40 CFR Part 136, 43368; *Federal Register* 49, No. 209.

6420 PHENOLS*

6420 A. Introduction

1. Sources and Significance

Phenols are found in many wastewaters and some raw source waters in the United States. They generally are traceable to industrial effluents or landfills. These compounds have a low taste threshold in potable waters and also may have a detrimental effect on human health at higher levels.

---

* Approved by Standard Methods Committee, 1993.

2. Selection of Method

For methods of determining total phenols in water and wastewater, see Section 5530.

The methods presented in this section are intended for the determination of individual phenolic compounds. For specific compounds covered, see each method. Method 6420B is a gas chromatographic (GC) method using liquid-liquid extraction and either flame ionization detection (FID) or derivatization and electron capture detection (ECD) to determine a wide variety of phenols at relatively low concentrations. In addition, Method 6420C, a liquid-liquid extraction gas chromatographic/mass spectrometric (GC/MS) method, can be used to determine the phenols at slightly higher concentrations.

6420 B. Liquid-Liquid Extraction Gas Chromatographic Method

This method[1] is applicable to the determination of phenol and certain substituted phenols* in municipal and industrial discharges. When analyzing unfamiliar samples for any or all of these compounds, support the identifications by at least one additional qualitative technique. Alternatively, use the derivatization, cleanup, and electron capture detector gas chromatography (ECD/GC) procedure to confirm measurements made by the flame ionization detector gas chromatographic (FID/GC) procedure. The method for base/neutrals and acids (Section 6410B) provides gas chromatograph/mass spectrometer (GC/MS) conditions appropriate for qualitative and quantitative confirmation of results using the extract produced.

1. General Discussion

a. Principle: See Section 6010C for discussion of gas chromatographic principles. A measured volume of sample is acidified and extracted with methylene chloride. The extract is dried and exchanged to 2-propanol during concentration. The extract is separated by gas chromatography and phenols are measured with a flame ionization detector.[2]

The method provides for a derivatization and column chromatography cleanup procedure to aid in the elimination of interferences.[2,3] Derivatives are analyzed by an electron capture detector.

b. Interferences:

1) General precautions—See Section 6410B.1b.

2) Other countermeasures—The cleanup procedure in ¶ 5c can be used to overcome many of these inteferences, but unique sam-

---

* 4-Chloro-3-methylphenol, 2-chlorophenol, 2,4-dichlorophenol, 2,4-dimethylphenol, 2,4-dinitrophenol, 2-methyl-4,6-dinitrophenol, 2-nitrophenol, 4-nitrophenol, pentachlorophenol, phenol, 2,4,6-trichlorophenol.

ples may require additional cleanup to achieve the method detection limits.

The basic sample wash (¶ 5a) may cause low recovery of phenol and 2,4-dimethylphenol. Results obtained under these conditions are minimum concentrations.

c. Detection levels: The method detection level (MDL) is the minimum concentration of a substance that can be measured and reported with 99% confidence that the value is above zero.[4] The MDL concentrations listed in Tables 6420:I and II were obtained by using reagent water.[5] Similar results were achieved with rep-

TABLE 6420:I. CHROMATOGRAPHIC CONDITIONS AND METHOD DETECTION LIMITS

Compound	Retention Time min	Method Detection Limit µg/L
2-Chlorophenol	1.70	0.31
2-Nitrophenol	2.00	0.45
Phenol	3.01	0.14
2,4-Dimethylphenol	4.03	0.32
2,4-Dichlorophenol	4.30	0.39
2,4,6-Trichlorophenol	6.05	0.64
4-Chloro-3-methylphenol	7.50	0.36
2,4-Dinitrophenol	10.00	13.0
2-Methyl-4,6-dinitrophenol	10.24	16.0
Pentachlorophenol	12.42	7.4
4-Nitrophenol	24.25	2.8

Column conditions: Supelcoport (80/100 mesh) coated with 1% SP-1240DA packed in a 1.8 m long × 2 mm ID glass column with nitrogen carrier gas at 30 mL/min flow rate. Column temperature was 80°C at injection, programmed immediately at 8°C/min to 150°C final temperature. MDLs determined with an FID.

TABLE 6420:II. SILICA GEL FRACTIONATION AND ELECTRON CAPTURE GAS CHROMATOGRAPHY OF PFBB DERIVATIVES

Parent Compound	Percent Recovery By Fraction*				Retention Time *min*	Method Detection Limit $\mu g/L$
	1	2	3	4		
2-Chlorophenol	—	90	1	—	3.3	0.58
2-Nitrophenol	—	—	9	90	9.1	0.77
Phenol	—	90	10	—	1.8	2.2
2,4-Dimethylphenol	—	95	7	—	2.9	0.63
2,4-Dichlorophenol	—	95	1	—	5.8	0.68
2,4,6-Trichlorophenol	50	50	—	—	7.0	0.58
4-Chloro-3-methylphenol	—	84	14	—	4.8	1.8
Pentachlorophenol	75	20	—	—	28.8	0.59
4-Nitrophenol	—	—	1	90	14.0	0.70

Column conditions: Chromosorb W-AW-DMCS (80/100 mesh) coated with 5% OV-17 packed in a 1.8 m long × 2.0 mm ID glass column with 5% methane/95% argon carrier gas at 30 mL/min flow rate. Column temperature held isothermal at 200°C. MDLs determined with an ECD.

* Eluent composition:
  Fraction 1 - 15% toluene in hexane.
  Fraction 2 - 40% toluene in hexane.
  Fraction 3 - 75% toluene in hexane.
  Fraction 4 - 15% 2-propanol in toluene.

resentative wastewaters. The MDL actually obtained in a given analysis will vary, depending on instrument sensitivity and matrix effects.

d. Safety: The toxicity or carcinogenicity of each reagent used in this method has not been defined precisely. Take special care in handling pentafluorobenzyl bromide, which is a lachrymator, and 18-crown-6-ether, which is highly toxic.

2. Sampling and Storage

See Section 6410B.2.

3. Apparatus

Use all the apparatus specified in Section 6410B.3*a–g* and *i–k*, and in addition:

a. Chromatographic column, 100 mm long × 10 mm ID, with TFE stopcock.

b. Reaction flask, 15- to 25-mL round-bottom, with standard tapered joint, fitted with a water-cooled condenser and U-shaped drying tube containing granular calcium chloride.

c. Gas chromatograph:† An analytical system complete with a temperature-programmable gas chromatograph suitable for on-column injection and all required accessories including syringes, analytical columns, gases, detector, and strip-chart recorder. Preferably use a data system for measuring peak areas.

1) *Column for underivatized phenols,* 1.8 m long × 2 mm ID glass, packed with 1% SP1240DA on Supelcoport (80/100 mesh) or equivalent. The detection limit and precision and bias data presented herein were developed with this column. For guidelines for the use of alternate columns (e.g., capillary or megabore) see ¶ 5*b*1).

2) *Column for derivatized phenols,* 1.8 m long × 2 mm ID, glass, packed with 5% OV-17 on Chromosorb W-AW-DMCS (80/100 mesh) or equivalent. This column was used to develop the detection limit and precision and bias data presented herein. For guidelines for the use of alternate columns (e.g., capillary or megabore) see¶ 5*b*1).

3) *Detectors,* flame ionization (FID) and electron capture (ECD). Use the FID to determine parent phenols. Use the ECD when determining derivatized phenols. For guidelines for use of alternative detectors see ¶ 5*b*1).

4. Reagents

Use reagents listed in Section 6410B.4*a–f,* and in addition:

a. Sodium hydroxide solution, NaOH, 1*N:* Dissolve 4 g NaOH in reagent water and dilute to 100 mL.

b. Sulfuric acid, H_2SO_4, 1*N:* Slowly add 58 mL conc H_2SO_4 to 500 mL reagent water and dilute to 1 L.

c. Potassium carbonate, K_2CO_3, powdered.

d. Pentafluorobenzyl bromide (α-bromopentafluorotoluene), 97% minimum purity. (CAUTION: *This chemical is a lachrymator.*)

e. 18-Crown-6-ether (1,4,7,10,13,16-hexaoxacyclooctadecane), 98% minimum purity. (CAUTION: *This chemical is highly toxic.*)

f. Derivatization reagent: Add 1 mL pentafluorobenzyl bromide and 1 g 18-crown-6-ether to a 50-mL volumetric flask and dilute to volume with 2-propanol. Prepare fresh weekly. Prepare in a hood. Store at 4°C and protect from light.

g. Acetone, hexane, methanol, methylene chloride, 2-propanol, toluene, pesticide quality or equivalent.

h. Silica gel, 100/200 mesh.‡ Activate at 130°C overnight and store in a desiccator.

i. Stock standard solutions: Prepare from pure standard materials or purchase as certified solutions. Prepare as directed in Section 6410B.4*g,* but dissolve material in 2-propanol.

j. Calibration standards: Prepare standards appropriate to chosen means of calibration.

1) *External standards:* Prepare at a minimum of three concentration levels for each compound by adding volumes of one or more stock standards to a volumetric flask and diluting to volume with 2-propanol. Prepare one standard at a concentration near, but above, the MDL (see Table 6420:I or II) and the others to correspond to the expected range of sample concentrations or to define the working range of the detector.

2) *Internal standards:* Prepare at a minimum of three concentration levels for each compound by adding volumes of one or more stock standards to a volumetric flask. To each calibration standard, add a known constant amount of one or more internal standards, and dilute to volume with 2-propanol. Prepare one standard at a concentration near, but above, the MDL and the others to correspond to the expected range of sample concentrations or to define the working range of the detector.

k. Quality control (QC) check sample concentrate: Obtain a check sample concentrate§ containing each compound at a concentration of 100 μg/mL in 2-propanol. If such a sample is not available from an external source, prepare using stock standards prepared independently from those used for calibration.

---

† Gas chromatographic methods are extremely sensitive to the materials used. Mention of trade names by *Standard Methods* does not preclude the use of other existing or as-yet-undeveloped products that give *demonstrably* equivalent results.

‡ Davison grade 923 or equivalent.
§ For U.S. federal permit-related analyses, use samples obtainable from U.S. EPA Environmental Monitoring and Support Laboratory, Cincinnati, OH.

5. Procedure

a. Extraction: Mark water meniscus on side of sample bottle for later determination of volume. Pour entire sample into a 2-L separatory funnel. For samples high in organic content, solvent wash sample at basic pH as prescribed in next paragraph, to remove potential interferences. During wash, avoid prolonged or exhaustive contact with solvent, which may result in low recovery of some phenols, notably phenol and 2,4-dimethylphenol. For relatively clean samples, omit wash and extract directly.

To wash, adjust pH to 12.0 or greater with NaOH solution. Add 60 mL methylene chloride and shake the funnel for 1 min with periodic venting to release excess pressure. Discard solvent layer. Repeat wash up to two additional times if significant color is being removed.

Before extraction, adjust to pH of 1 to 2 with H_2SO_4. Extract three times with methylene chloride as directed in Section 6410B.5*a*1). Assemble Kuderna-Danish apparatus, concentrate extract to 1 mL, and remove, drain, and cool K-D apparatus as directed in Section 6410B.5*a*1).

Increase temperature of hot water bath to 100°C. Remove Snyder column and rinse flask and its lower joint into concentrator tube with 1 to 2 mL 2-propanol. Preferably use a 5-mL syringe for this operation. Attach a two-ball micro-Snyder column to concentrator tube and prewet column by adding about 0.5 mL 2-propanol to the top. Place micro-K-D apparatus on water bath so that concentrator tube is partially immersed in hot water. Adjust vertical position of apparatus and water temperature so as to complete concentration in 5 to 10 min. (CAUTION: If temperature is raised too quickly the sample may be blown out of the K-D apparatus). At proper rate of distillation the column balls actively chatter but the chambers are not flooded. When the apparent volume of liquid reaches 2.5 mL, remove K-D apparatus and let drain and cool for at least 10 min. Add 2 mL 2-propanol through top of micro-Snyder column and resume concentrating as before. When the apparent volume of liquid reaches 0.5 mL, remove K-D apparatus and let drain and cool for at least 10 min.

Remove micro-Snyder column and rinse lower joint into concentrator tube with a minimum amount of 2-propanol. Adjust extract volume to 1.0 mL. Stopper concentrator tube and store at 4°C if further processing will not be done immediately. If extract is to be stored longer than 2 d, transfer to a TFE-sealed screwcap vial. If sample extract requires no further cleanup, proceed with chromatographic analysis (¶ *b*). If sample requires further cleanup, proceed to ¶ *c*.

Determine original sample volume by refilling sample bottle to mark and transferring liquid to a 1000-mL graduated cylinder. Record sample volume to nearest 5 mL.

b. Flame ionization detector gas chromatography (FID/GC):

1) Operating conditions—Table 6420:I summarizes the recommended operating conditions for the gas chromatograph and gives retention times and MDLs that can be achieved under these conditions. An example of the separations obtained with this column is shown in Figure 6420:1. Other packed or capillary (opentubular) columns, chromatographic conditions, or detectors may be used if the requirements of ¶ 7 are met.

2) Calibration—To calibrate the system for underivatized phenols, establish gas chromatographic operating conditions equivalent to those given in Table 6420:I. Calibrate using the external or the internal standard technique as follows:

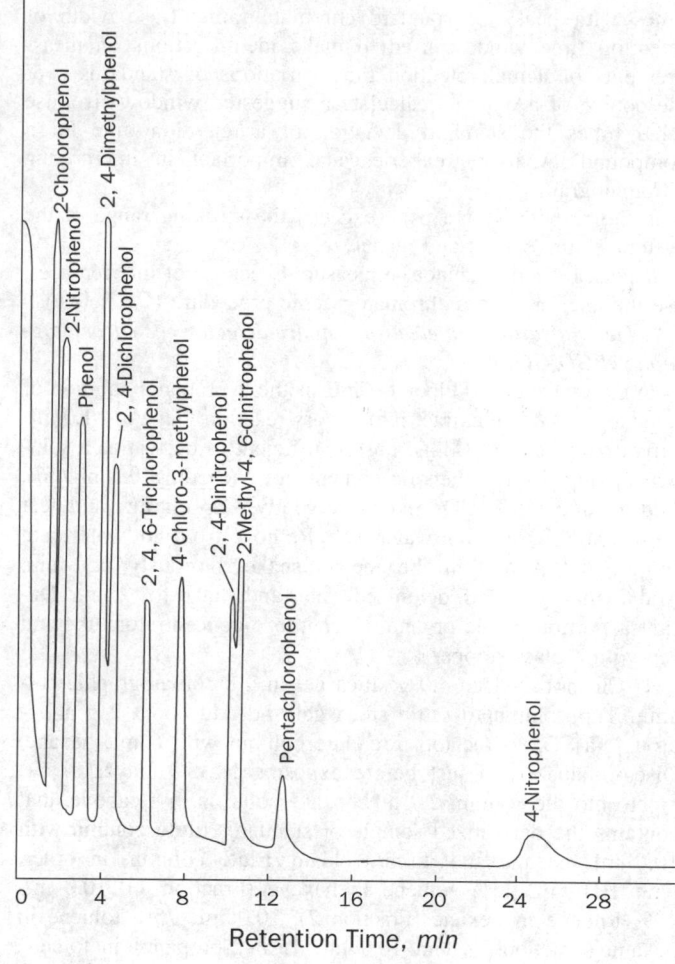

Figure 6420:1. Gas chromatogram of phenols. Column: 1% SP-1240DA on Supelcoport; program: 80°C at injection, immediate 8°C/min to 150°C; detector: flame ionization.

a) External standard calibration procedure—Prepare standards as directed in ¶ 4*j*1) and follow the procedure of ¶ 3) below. Tabulate data and obtain calibration curve or calibration factor as directed in Section 6200B.4*c*3).

b) Internal standard calibration procedure—Prepare samples as directed in ¶ 4*j*2) and follow the procedure of ¶ 3) below. Tabulate data and calculate response factors as directed in Section 6200B.4*c*2).

Verify working calibration curve, calibration factor, or RF on each working day by measuring one or more calibration standards. If the response for any compound varies from the predicted response by more than ±15%, prepare a new calibration curve for that compound.

3) Sample analysis—If the internal standard calibration procedure is used, add internal standard to sample extract and mix thoroughly immediately before injecting 2 to 5 μL sample extract or standard into gas chromatograph using the solvent-flush technique.[6] Smaller (1.0-μL) volumes may be injected if automatic devices are used. Record volume injected to nearest 0.05 μL and resulting peak size in area or peak height units.

Identify compounds in sample by comparing peak retention times with peaks of standard chromatograms. Base width of retention time window used to make identifications on measurements of actual retention time variations of standards over the course of a day. To calculate a suggested window size, use three times the standard deviation of a retention time for a compound. Analyst's experience is important in interpreting chromatograms.

If the response for a peak exceeds the working range of the system, dilute extract and reanalyze.

If peak response cannot be measured because of interferences, use the alternative gas chromatographic procedure (¶ *c* below).

c. Derivatization and electron capture detector gas chromatography (ECD/GC):

1) Derivatization—Pipet 1.0 mL of the 2-propanol solution of standard or sample extract into a glass reaction vial. Add 1.0 mL derivatizing reagent (¶ 4*f*); this is sufficient to derivatize a solution having a total phenolic content not exceeding 0.3 mg/mL. Add about 3 mg K_2CO_3 and shake gently. Cap mixture and heat for 4 h at 80°C in a hot water bath. Remove from hot water bath and let cool. Add 10 mL hexane and shake vigorously for 1 min. Add 3.0 mL distilled, deionized water and shake for 2 min. Decant a portion of the organic layer into a concentrator tube and cap with a glass stopper.

2) Cleanup—Place 4.0 g silica gel in a chromatographic column. Tap column to settle silica gel and add about 2 g anhydrous Na_2SO_4 to the top. Pre-elute column with 6 mL hexane. Discard eluate and just before exposing Na_2SO_4 layer to air, pipet onto the column 2.0 mL hexane solution, ¶ 1) above, that contains the derivatized sample or standard. Elute column with 10.0 mL hexane and discard eluate. Elute column, in order, with 10.0 mL 15% toluene in hexane (Fraction 1), 10.0 mL 40% toluene in hexane (Fraction 2), 10.0 mL 75% toluene in hexane (Fraction 3), and 10.0 mL 15% 2-propanol in toluene (Fraction 4). Prepare all elution mixtures on a volume:volume basis. Elution patterns for the phenolic derivatives are shown in Table 6420:II. Fractions may be combined as desired, depending on the specific phenols of interest or level of interferences.

3) Operating conditions—Table 6420:II summarizes the recommended operating conditions for the gas chromatograph and gives retention times and MDLs that can be achieved under these conditions. An example of the separations obtained with this column is shown in Figure 6420:2.

4) Calibration—Calibrate system daily by preparing a minimum of three 1-mL portions of calibration standards, ¶ 4*j*1), containing each of the phenols of interest and derivatized as above. Analyze 2 to 5 µL of each column eluate collected as in ¶ 5) below and tabulate peak height or area responses against calculated equivalent mass of underivatized phenol injected. Prepare a calibration curve for each compound.

Before using any cleanup procedure, process a series of calibration standards through the procedure to validate elution patterns and to assure absence of interferences from the reagents.

5) Sample analysis—Inject 2 to 5 µL column fractions into the gas chromatograph using the solvent-flush technique. Smaller (1.0-µL) volumes can be injected if automatic devices are used. Record volume injected to nearest 0.05 µL and resulting peak size in area or peak height units. If peak response exceeds linear range of system, dilute extract and reanalyze.

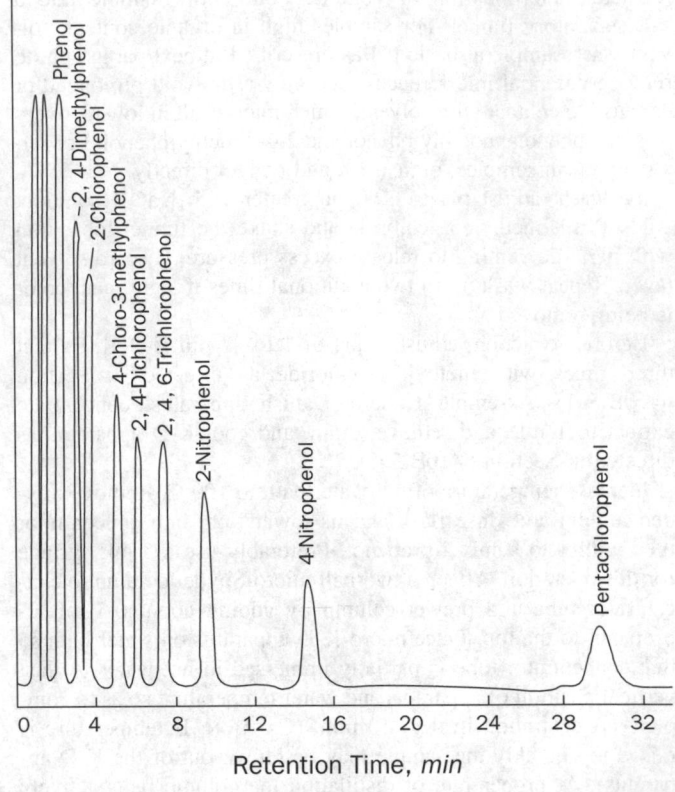

Figure 6420:2. Gas chromatogram of PFB derivatives of phenols. Column: 5% OV-17 on Chromosorb W-AW-DMCS; temperature: 200°C; detector: electron capture.

6. Calculation

a. FID/GC analysis: Determine concentration of individual compounds. If the external standard calibration procedure is used, calculate amount of material injected from peak response using calibration curve or calibration factor determined previously. Calculate sample concentration from the equation:

$$\text{Concentration, µg/L} = \frac{(A)\,(V_t)}{(V_i)\,(V_s)}$$

where:

A = amount of material injected, ng,
V_i = volume of extract injected, µL,
V_t = volume of total extract, µL, and
V_s = volume of water extracted, mL.

If the internal standard calibration procedure is used, calculate concentration in sample using the response factor (RF) determined above and the equation:

$$\text{Concentration, µg/L} = \frac{(A_s)\,(I_s)}{(A_{is})\,(RF)\,(V_o)}$$

where:

A_s = response for compound to be measured,
A_{is} = response for internal standard,
I_s = amount of internal standard added to each extract, μg, and
V_o = volume of water extracted, L.

b. Derivatization and ECD/GC analysis: To determine concentration of individual compounds in the sample, use the equation:

$$\text{Concentration, μg/L} = \frac{(A)\ (V_t)\ (B)\ (D)}{(V_i)\ (V_s)\ (C)\ (E)}$$

where:

A = mass of underivatized phenol represented by area of peak in sample chromatogram, determined from calibration curve in ¶ 5c4), ng,
V_i = volume of eluate injected, μL,
V_t = total volume of column eluate or combined fractions from which V_i was taken, μL,
V_s = volume of water extracted in ¶ 5a, mL,
B = total volume of hexane added in ¶ 5c1), mL,
C = volume of hexane sample solution added to cleanup column in ¶ 5c2), mL,
D = total volume of 2-propanol extract before derivatization, mL, and
E = volume of 2-propanol extract carried through derivatization in ¶ 5c1), mL.

Report results in μg/L without correction for recovery. Report QC data with sample results.

7. Quality Control

a. Quality control program: See Section 6200A.5.

b. Initial quality control: To establish the ability to generate data with acceptable bias and precision, perform the following operations:

Using a pipet, prepare QC check samples at a concentration of 100 μg/L by adding 1.00 mL of 100 μg/mL QC check sample concentrate to each of four 1-L portions reagent water. Analyze check samples according to the method of ¶ 5 and proceed with the check described in Section 6200A.5a1) and 2). Use acceptance criteria given in Table 6420:III.

c. Analyses of laboratory-fortified samples: On an ongoing basis, make known additions to at least 10% of the samples from each sample site being monitored. For laboratories analyzing one to ten samples per month, analyze at least one such sample with a known addition per month. Use the procedure detailed in Section 6200A.5c7) and 8), but use an addition of 100 μg/L rather than 20 μg/L and compare percent recovery for each compound with the corresponding QC acceptance criteria found in Table 6420:III. If the known addition was at a concentration lower than 100 μg/L, use either the QC acceptance criteria in Table 6420:III or optional QC acceptance criteria calculated for the specific addition concentration based on the equations in Table 6420:IV.

d. Quality-control check standard analysis: If analysis of any compound fails to meet the acceptance criteria for recovery, prepare and analyze a QC check standard containing each compound that failed. NOTE: The frequency for the required analysis of a QC check standard will depend on the number of compounds being tested for simultaneously, the complexity of the sample matrix, and the performance of the laboratory.

Prepare the QC check standard by adding 1.0 mL of QC check sample concentrate to 1 L reagent water and proceed as in Section 6200A.5a3) using Table 6420:III.

e. Bias assessment and records: Assess method bias and maintain records as directed in Section 6410B.7e.

8. Precision and Bias

This method was tested by 20 laboratories using reagent water, drinking water, surface water, and three industrial wastewaters with known additions at six concentrations over the range 12 to 450 μg/L.[7] Single-operator precision, overall precision, and method bias were found to be related directly to compound concentration and essentially independent of sample matrix. Linear equations describing these relationships for a flame ionization detector are presented in Table 6420:IV.

TABLE 6420:III. QC ACCEPTANCE CRITERIA*

Compound	Test Conc. μg/L	Limit for s μg/L	Range for $\overline{X}$ μg/L	Range for P, P_s %
4-Chloro-3-methylphenol	100	16.6	56.7–113.4	49–122
2-Chlorophenol	100	27.0	54.1–110.2	38–126
2,4-Dichlorophenol	100	25.1	59.7–103.3	44–119
2,4-Dimethylphenol	100	33.3	50.4–100.0	24–118
4,6-Dinitro-2-methylphenol	100	25.0	42.4 123.6	30–136
2,4-Dinitrophenol	100	36.0	31.7–125.1	12–145
2-Nitrophenol	100	22.5	56.6–103.8	43–117
4-Nitrophenol	100	19.0	22.7–100.0	13–110
Pentachlorophenol	100	32.4	56.7–113.5	36–134
Phenol	100	14.1	32.4–100.0	23–108
2,4,6-Trichlorophenol	100	16.6	60.8–110.4	53–119

* s = standard deviation of four recovery measurements,
  $\overline{X}$ = average recovery for four recovery measurements, and
  P, P_s = percent recovery measured.

NOTE: These criteria are based directly upon the method performance data in Table 6420:IV. Where necessary, the limits for recovery were broadened to assure applicability of the limits to concentrations below those used to develop Table 6420:IV.

TABLE 6420:IV. METHOD BIAS AND PRECISION AS FUNCTIONS OF CONCENTRATION*

Compound	Bias, as Recovery, X' $\mu g/L$	Single-Analyst Precision, s_r' $\mu g/L$	Overall Precision, S' $\mu g/L$
4-Chloro-3-methylphenol	$0.87C - 1.97$	$0.11\overline{X} - 0.21$	$0.16\overline{X} + 1.41$
2-Chlorophenol	$0.83C - 0.84$	$0.18\overline{X} + 0.20$	$0.21\overline{X} + 0.75$
2,4-Dichlorophenol	$0.81C + 0.48$	$0.17\overline{X} - 0.02$	$0.18\overline{X} + 0.62$
2,4-Dimethylphenol	$0.62C - 1.64$	$0.30\overline{X} - 0.89$	$0.25\overline{X} + 0.48$
4,6-Dinitro-2-methylphenol	$0.84C - 1.01$	$0.15\overline{X} + 1.25$	$0.19\overline{X} + 5.85$
2,4-Dinitrophenol	$0.80C - 1.58$	$0.27\overline{X} - 1.15$	$0.29\overline{X} + 4.51$
2-Nitrophenol	$0.81C - 0.76$	$0.15\overline{X} + 0.44$	$0.14\overline{X} + 3.84$
4-Nitrophenol	$0.46C + 0.18$	$0.17\overline{X} + 2.43$	$0.19\overline{X} + 4.79$
Pentachlorophenol	$0.83C + 2.07$	$0.22\overline{X} - 0.58$	$0.23\overline{X} + 0.57$
Phenol	$0.43C + 0.11$	$0.20\overline{X} - 0.88$	$0.17\overline{X} + 0.77$
2,4,6-Trichlorophenol	$0.86C - 0.40$	$0.10\overline{X} + 0.53$	$0.13\overline{X} + 2.40$

* X' = expected recovery for one or more measurements of a sample containing a concentration of C,
  s_r' = expected single-analyst standard deviation of measurements at an average concentration found of $\overline{X}$,
  S' = expected interlaboratory standard deviation of measurements at an average concentration found of $\overline{X}$,
  C = true value for the concentration, and,
  $\overline{X}$ = average recovery found for measurements of samples containing a concentration of C.

9. References

1. U.S. ENVIRONMENTAL PROTECTION AGENCY. 1984. Method 604—Phenols. 40 CFR Part 136, 43290; *Federal Register* 49, No. 209.
2. U.S. ENVIRONMENTAL PROTECTION AGENCY. 1984. Determination of phenols in industrial and municipal wastewaters. Final rep. EPA Contract 68-03-2625, Environmental Monitoring and Support Lab., Cincinnati, Ohio.
3. KAWAHARA, F.K. 1968. Microdetermination of derivatives of phenols and mercaptans by means of electron capture gas chromatography. *Anal. Chem.* 40:100.
4. U.S. ENVIRONMENTAL PROTECTION AGENCY. 1984. Definition and pro-
cedure for the determination of the method detection limit. 40 CFR Part 136, Appendix B. *Federal Register* 49, No. 209.
5. U.S. ENVIRONMENTAL PROTECTION AGENCY. Development of detection limits, EPA Method 604, Phenols. Special letter report for EPA Contract 68-03-2625, Environmental Monitoring and Support Lab., Cincinnati, Ohio.
6. BURKE, J.A. 1965. Gas chromatography for pesticide residue analysis; some practical aspects. *J. Assoc. Offic. Anal. Chem.* 48:1037.
7. U.S. ENVIRONMENTAL PROTECTION AGENCY. 1984. EPA Method Study 14, Method 604—Phenols. EPA-600/4-84-044, National Technical Information Serv., PB84-196211, Springfield, Va.

6420 C.   Liquid-Liquid Extraction Gas Chromatographic/Mass Spectrometric Method

See Section 6410B.

6431   POLYCHLORINATED BIPHENYLS (PCBs)*

6431 A.   Introduction

1. Sources and Significance

The polychlorinated biphenyls (PCBs) are found principally in water supplies contaminated by transformer oils in which PCBs were originally used as a heat-exchange medium. Although the use of these compounds has been banned, there are still numerous transformers in existence that contain PCBs, which results in their occasional discharge into potable water or wastewater. These compounds are toxic, bioaccumulative, and extremely stable, and thus there is a need to monitor them in wastewaters.

2. Selection of Method

The liquid-liquid extraction (LLE) gas chromatographic (GC) method is used to monitor both the PCBs and the organochlorine pesticides simultaneously. This method has excellent sensitivity.

* Approved by Standard Methods Committee, 1993.

The LLE gas chromatographic/mass spectrometric (GC/MS) method also can be used to detect PCBs, but with substantially less sensitivity.

PCBs usually are measured as commercial mixtures of isomers rather than as individual isomers (congeners).

6431 B.   Liquid-Liquid Extraction Gas Chromatographic Method

See Sections 6630B and C.

6431 C.   Liquid-Liquid Extraction Gas Chromatographic/Mass Spectrometric Method

See Section 6410B.

6440   POLYNUCLEAR AROMATIC HYDROCARBONS*

6440 A.   Introduction

1. Sources and Significance

The polynuclear aromatic hydrocarbons (PAHs) often are by-products of petroleum processing or combustion. Many of these compounds are highly carcinogenic at relatively low levels. Although they are relatively insoluble in water, their highly hazardous nature merits their monitoring in potable waters and wastewaters.

* Approved by Standard Methods Committee, 1993.

2. Selection of Method

Method 6440B encompasses both a high-performance liquid chromatographic (HPLC) method with UV and fluorescence detection and a gas chromatographic (GC) method using flame ionization detection. Method 6440C is a gas chromatographic/mass spectrometric (GC/MS) method that also can detect these compounds at somewhat higher concentrations. Certain of these compounds may also be measured by closed-loop stripping analysis, Section 6040.

6440 B.   Liquid-Liquid Extraction Chromatographic Method

This method[1] is applicable to the determination of certain polynuclear aromatic hydrocarbons (PAH)* in municipal and industrial discharges. When analyzing unfamiliar samples for any or all of these compounds, support the identifications by at least one additional qualitative technique. The method for base/neutrals and acids (Section 6410B) provides gas chromatograph/mass spectrometer (GC/MS) conditions appropriate for qualitative and quantitative confirmation of results using the extract produced.

* Acenaphthene, acenaphthylene, anthracene, benzo-(a)anthracene, benzo(a)pyrene, benzo(b)fluoranthene, benzo(ghi)perylene, benzo(k)fluoranthene, chrysene, dibenzo(a,h)anthracene, fluoranthene, fluorene, indeno(1,2,3-cd)pyrene, naphthalene, phenanthrene, and pyrene.

1. General Discussion

a. Principle: A measured volume of sample is extracted with methylene chloride. The extract is dried, concentrated, and separated by the high-performance liquid chromatographic (HPLC) or gas chromatographic (GC) method. If other analyses having essentially the same extraction and concentration steps are to be performed, extraction of a single sample will be sufficient for all the determinations. Ultraviolet (UV) and fluorescence detectors are used with HPLC to identify and measure the PAHs. A flame ionization detector is used with GC.[2]

The method provides a silica gel column cleanup to aid in eliminating interferences. When cleanup is required, sample concentration levels must be high enough to permit separate treatment of subsamples before the solvent-exchange steps.

Chromatographic conditions (¶ 5d) appropriate for the simultaneous measurement of combinations of these compounds may be selected but they do not adequately resolve the following four pairs of compounds: anthracene and phenanthrene; chrysene and benzo(a)anthracene; benzo(b)fluoranthene and benzo(k)fluoranthene; and dibenzo(a,h)anthracene and indeno(1,2,3-cd)pyrene. Unless reporting the sum of an unresolved pair is acceptable, use the liquid chromatographic method, which does resolve all 16 listed PAHs.

b. Interferences: See Section 6410B.1b for precautions concerning glassware, reagent purity, and matrix interferences. Interferences in liquid chromatographic techniques have not been assessed fully. Although HPLC conditions described allow for unique resolution of specific PAHs, other PAH compounds may interfere.

c. Detection levels: The method detection level (MDL) is the minimum concentration of a substance that can be measured and reported with 99% confidence that the value is above zero.[3] The MDL concentrations listed in Table 6440:I were obtained with reagent water.[4] Similar results were achieved with representative wastewaters. MDLs for the GC method were not determined. The MDL actually obtained in a given analysis will vary, depending on instrument sensitivity and matrix effects. This method has been tested for linearity of known-addition recovery from reagent water and has been demonstrated to be applicable over the concentration range from 8 × MDL to 800 × MDL,[4] with the following exception: benzo(ghi)perylene recovery at 80 × and 800 × MDL were low (35% and 45%, respectively).

TABLE 6440:I. HIGH-PERFORMANCE LIQUID CHROMATOGRAPHY
CONDITIONS AND METHOD DETECTION LIMITS

Compound	Retention Time min	Column Capacity Factor k'	Method Detection Limit µg/L*
Naphthalene	16.6	12.2	1.8
Acenaphthylene	18.5	13.7	2.3
Acenaphthene	20.5	15.2	1.8
Fluorene	21.2	15.8	0.21
Phenanthrene	22.1	16.6	0.64
Anthracene	23.4	17.6	0.66
Fluoranthene	24.5	18.5	0.21
Pyrene	25.4	19.1	0.27
Benzo(a)anthracene	28.5	21.6	0.013
Chrysene	29.3	22.2	0.15
Benzo(b)fluoranthene	31.6	24.0	0.018
Benzo(k)fluoranthene	32.9	25.1	0.017
Benzo(a)pyrene	33.9	25.9	0.023
Dibenzo(a,h)anthracene	35.7	27.4	0.030
Benzo(ghi)perylene	36.3	27.8	0.076
Indeno(1,2,3-cd)pyrene	37.4	28.7	0.043

HPLC column conditions: Reverse phase HC-ODS Sil-X, 5 µm particle size, in a 25 cm × 2.6 mm ID stainless steel column. Isocratic elution for 5 min using acetonitrile/water (4 + 6), then linear gradient to 100% acetonitrile over 25 min at 0.5 mL/min flow rate. If columns having other internal diameters are used, adjust flow rate to maintain a linear velocity of 2 mm/s.
* The MDL for naphthalene, acenaphthylene, acenaphthene, and fluorene were determined using a UV detector. All others were determined using a fluorescence detector.

d. Safety: The toxicity or carcinogenicity of each reagent has not been defined precisely. The following compounds have been classified tentatively as known or suspected, human or mammalian carcinogens: benzo(a)anthracene, benzo(a)pyrene, and dibenzo(a,h)anthracene. Prepare primary standards of these compounds in a hood and wear NIOSH/MESA-approved toxic gas respirator when handling high concentrations.

2. Sampling and Storage

For collection and general storage requirements, see Section 6410B.2. Because PAHs are light-sensitive, store samples, extracts, and standards in amber or foil-wrapped bottles to minimize photolytic decomposition.

3. Apparatus

Use all the apparatus specified in Section 6410B.3a–g and i–k, and in addition:

a. Chromatographic column, 250 mm long × 10 mm ID with coarse frit filter disk at bottom and TFE stopcock.

b. High-performance liquid chromatograph (HPLC): An analytical system complete with column supplies, high-pressure syringes, detectors, and compatible strip-chart recorder. Preferably use a data system for measuring peak areas and retention times.

1) Gradient pumping system, constant flow.

2) Reverse phase column, HC-ODS Sil-X, 5-µm particle diam, in a 25-cm × 2.6-mm ID stainless steel column.† This column was used to develop MDL and precision and bias data presented herein. For guidelines for the use of alternate column packings see ¶ 5d1).

3) Detectors, fluorescence and/or UV. Use the fluorescence detector for excitation at 280 mm and emission greater than 389 nm cutoff.‡ Use fluorometers with dispersive optics for excitation utilizing either filter or dispersive optics at the emission detector. Operate the UV detector at 254 nm and couple it to the fluorescence detector. These detectors were used to develop MDL and precision and bias data presented herein. For guidelines for the use of alternate detectors see ¶ 5d1).

c. Gas chromatograph:§ An analytical system complete with temperature-programmable gas chromatograph suitable for on-column or splitless injection and all required accessories including syringes, analytical columns, gases, detector, and strip-chart recorder. Preferably use a data system for measuring peak areas.

1) Column, 1.8 m long × 2 mm ID glass, packed with 3% OV-17 on Chromosorb W-AW-DCMS (100/120 mesh) or equivalent. This column was used to develop the retention time data in Table 6440:II. For guidelines for the use of alternate columns (e.g. capillary or megabore) see ¶ 5d2).

2) Detector, flame ionization. This detector is effective except for resolving the four pairs of compounds listed in ¶ 1a. With the use of capillary columns, these pairs may be resolved with GC. For guidelines for the use of alternate detectors see ¶ 5d2).

4. Reagents

a. Reagent water: See Section 6200B.3a.

b. Sodium thiosulfate, Na₂S₂O₃·5H₂O, granular.

† Perkin Elmer No. 089-0716 or equivalent.
‡ Corning 3-75 or equivalent.
§ Gas chromatographic methods are extremely sensitive to the materials used. Mention of trade names by Standard Methods does not preclude the use of other existing or as-yet-undeveloped products that give demonstrably equivalent results.

TABLE 6440:II. GAS CHROMATOGRAPHIC CONDITIONS AND RETENTION TIMES

Compound	Retention Time min
Naphthalene	4.5
Acenaphthylene	10.4
Acenaphthane	10.8
Fluorene	12.6
Phenanthrene	15.9
Anthracene	15.9
Fluoranthene	19.8
Pyrene	20.6
Benzo(a)anthracene	24.7
Chrysene	24.7
Benzo(b)fluoranthene	28.0
Benzo(k)fluoranthene	28.0
Benzo(a)pyrene	29.4
Dibenzo(a,h)anthracene	36.2
Indeno(1,2,3-cd)pyrene	36.2
Benzo(ghi)perylene	38.6

GC column conditions: Chromosorb W-AW-DCMS (100/120 mesh) coated with 3% OV-17 packed in a 1.8 × 2 mm ID glass column with nitrogen carrier gas at 40 mL/min flow rate. Column temperature held at 100°C for 4 min, then programmed at 8°C/min to a final hold at 280°C.

c. *Cyclohexane, methanol, acetone, methylene chloride, pentane*, pesticide quality or equivalent.

d. *Acetonitrile*, HPLC quality, distilled in glass.

e. *Sodium sulfate*, Na_2SO_4, granular, anhydrous. Purify by heating at 400°C for 4 h in a shallow tray.

f. *Silica gel*, 100/200 mesh, desiccant.‖ Before use, activate for at least 16 h at 130°C in a shallow glass tray, loosely covered with foil.

g. *Stock standard solutions*: Prepare as directed in Section 6410B.4g, using acetonitrile as the solvent.

h. *Calibration standards:* Prepare standards appropriate to chosen means of calibration following directions in Section 6420B.4j, except that acetonitrile is the diluent instead of 2-propanol. See Table 6440:I for MDLs.

i. *Quality control (QC) check sample concentrate:* Obtain a check sample concentrate# containing each compound at the following concentrations in acetonitrile: 100 μg/mL of any of the six early-eluting PAHs (naphthalene, acenaphthylene, acenaphthene, fluorene, phenanthrene, and anthracene); 5 μg/mL of benzo(k)fluoranthene; and 10 μg/mL of any other PAH. If such a sample is not available from an external source, prepare using stock standards prepared independently from those used for calibration.

5. Procedure

a. *Extraction:* Mark water meniscus on side of sample bottle for later determination of volume. Pour entire sample into a 2-L separatory funnel and extract as directed in Section 6410B.5a1) without any pH adjustment.

After extraction, concentrate by adding one or two clean boiling chips to the evaporative flask and attach a three-ball Snyder

‖ Davison, grade 923 or equivalent.
For U.S. federal permit-related analyses, use samples obtainable from U.S. EPA Environmental Monitoring and Support Laboratory, Cincinnati, OH.

column. Prewet Snyder column by adding about 1 mL methylene chloride to the top. Place K-D apparatus on a hot water bath (60 to 65°C) in a hood so that the concentrator tube is partially immersed in the hot water, and the entire lower rounded surface of flask is bathed with hot vapor. Adjust vertical position of apparatus and water temperature as required to complete the concentration in 15 to 20 min. At proper rate of distillation the column balls actively chatter but the chambers are not flooded with condensed solvent. When the apparent volume of liquid reaches 1 mL, remove K-D apparatus and let drain and cool for at least 10 min.

Remove Snyder column and rinse flask and its lower joint into concentrator tube with 1 to 2 mL methylene chloride. Preferably use a 5-mL syringe for this operation. Stopper concentrator tube and store refrigerated if further processing will not be done immediately. If extract is to be stored longer than 2 d, transfer to a TFE-sealed screw-cap vial and protect from light. If sample extract requires no further cleanup, proceed with gas or liquid chromatographic analysis (¶s c through f below). If sample requires further cleanup, first follow procedure of ¶ b before chromatographic analysis.

Determine original sample volume by refilling sample bottle to mark and transferring liquid to a 1000-mL graduated cylinder. Record sample volume to nearest 5 mL.

b. *Cleanup and separation:* Use procedure below or any other appropriate procedure; however, first demonstrate that the requirements of ¶ 7 can be met.

Before using silica-gel cleanup technique, exchange extract solvent to cyclohexane. Add 1 to 10 mL sample extract (in methylene chloride) and a boiling chip to a clean K-D concentrator tube. Add 4 mL cyclohexane and attach a two-ball micro-Snyder column. Prewet column by adding 0.5 mL methylene chloride to the top. Place micro-K-D apparatus on a boiling (100°C) water bath so that concentrator tube is partially immersed in hot water. Adjust vertical position of apparatus and water temperature so as to complete concentration in 5 to 10 min. At proper rate of distillation the column balls actively chatter but the chambers are not flooded. When apparent volume of liquid reaches 0.5 mL, remove K-D apparatus and let drain and cool for at least 10 min. Remove micro-Snyder column and rinse its lower joint into concentrator tube with a minimum amount of cyclohexane. Adjust extract volume to about 2 mL.

To perform silica-gel column cleanup, make a slurry of 10 g activated silica gel in methylene chloride and place in a 10-mm-ID chromatographic column. Tap column to settle silica gel and elute with methylene chloride. Add 1 to 2 cm anhydrous Na_2SO_4 to top of silica gel. Pre-elute with 40 mL pentane. Elute at rate of about 2 mL/min. Discard eluate and just before exposure of Na_2SO_4 layer to the air, transfer all the cyclohexane sample extract onto column using an additional 2 mL cyclohexane. Just before exposure of Na_2SO_4 layer to air, add 25 mL pentane and continue elution. Discard this pentane eluate. Next, elute column with 25 mL methylene chloride/pentane (4 + 6) (v/v) into a 500-mL K-D flask equipped with a 10-mL concentrator tube. Concentrate collected fraction to less than 10 mL as in ¶ 5a. After cooling, remove Snyder column and rinse flask and its lower joint with pentane.

c. *Reconcentration:* Concentrate further as follows:

1) For high-performance liquid chromatography—To extract in a concentrator tube, add 4 mL acetonitrile and a new boiling chip. Attach a two-ball micro-Snyder column and concentrate solvent

as in ¶ 5a (but set water bath at 95 to 100°C.) After cooling, remove micro-Snyder column and rinse its lower joint into the concentrator tube with about 0.2 mL acetonitrile. Adjust extract volume to 1.0 mL.

2) For gas chromatography—To achieve maximum sensitivity with this method, concentrate extract to 1.0 mL. Add a clean boiling chip to methylene chloride extract in concentrator tube. Attach a two-ball micro-Snyder column. Prewet column by adding about 0.5 mL methylene chloride to the top. Place micro-K-D apparatus on a hot water bath (60 to 65°C) and continue concentration as in ¶ 5b. Remove micro-Snyder column and rinse its lower joint into concentrator tube with a minimum amount of methylene chloride. Adjust final volume to 1.0 mL and stopper concentrator tube.

d. Operating conditions:

1) High-performance liquid chromatography—Table 6440:I summarizes the recommended operating conditions for HPLC and gives retention times, capacity factors, and MDLs that can be achieved under these conditions. Preferably use the UV detector for determining naphthalene, acenaphthylene, acenapthene, and fluorene and the fluorescence detector for the remaining PAHs. Examples of separations obtained with this HPLC column are shown in Figures 6440:1 and 2. Other HPLC columns, chromatographic conditions, or detectors may be used if the requirements of ¶ 7 are met.

2) Gas chromatography—Table 6440:II summarizes the recommended operating conditions for the gas chromatograph and gives retention times that were obtained under these conditions.

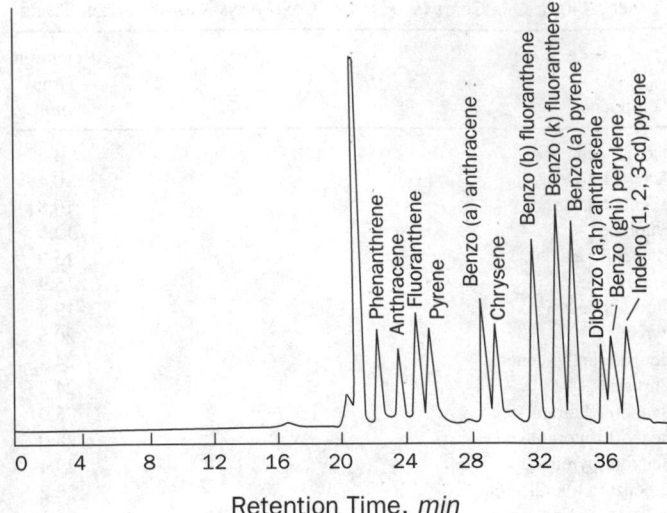

Figure 6440:2. Liquid chromatogram of polynuclear aromatic hydrocarbons. Column: HC—ODS SIL-X; mobile phase: 40% to 100% acetonitrile in water; detector: fluorescence.

An example of the separations is shown in Figure 6440:3. Other packed or capillary (open-tubular) columns, chromatographic conditions, or detectors may be used if the requirements of ¶ 7 are met.

e. Calibration: Calibrate system daily using either external or internal standard procedure.

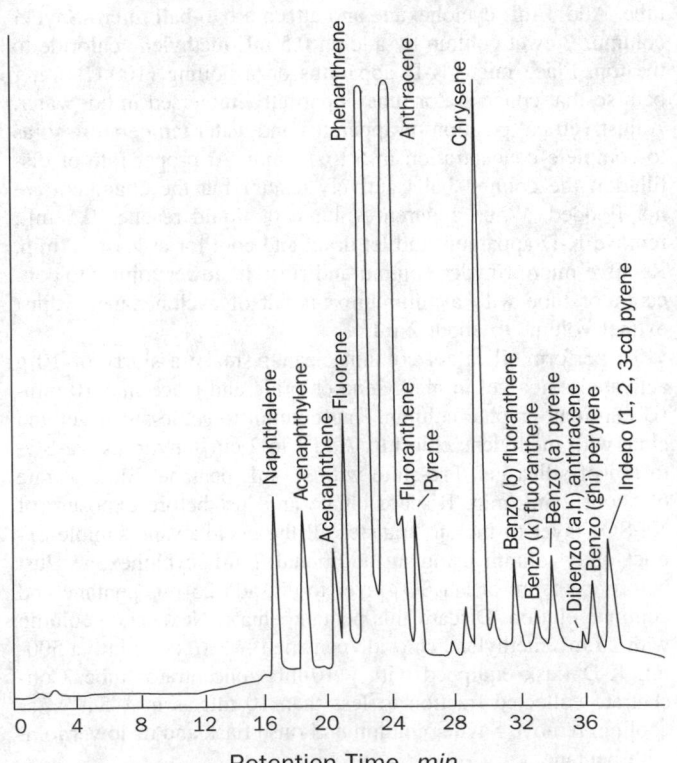

Figure 6440:1. Liquid chromatogram of polynuclear aromatic hydrocarbons. Column: HC—ODS SIL-X; mobile phase: 40% to 100% acetonitrile in water; detector: ultraviolet at 254 nm.

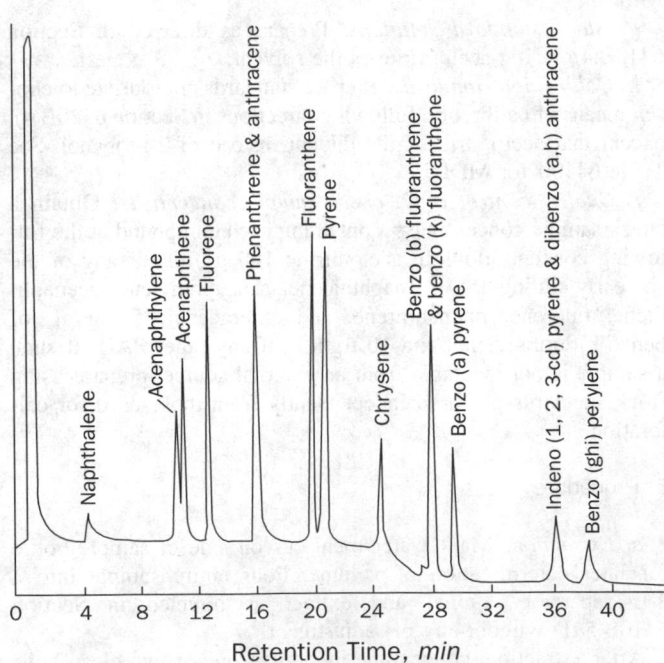

Figure 6440:3. Gas chromatogram of polynuclear aromatic hydrocarbons. Column: 3% OV-17 on Chromosorb W-AW-DCMS; program: 100°C for 4 min, 8°C/min to 280°C; detector: flame ionization.

1) External standard calibration procedure—Prepare standards as directed in ¶ 4h and follow either procedure of ¶ f below. Tabulate data and obtain calibration curve or calibration factor as directed in Section 6200B.4c3).

2) Internal standard calibration procedure—Prepare standards as directed in ¶ 4h and follow either procedure of ¶ f below. Tabulate data and calculate response factors as directed in Section 6200B.4c2).

Verify working calibration curve, calibration factor, or RF on each working day by measuring one or more calibration standards. If the response for any compound varies from the predicted response by more than ± 15%, repeat test using a fresh calibration standard. Alternatively, prepare a new calibration curve for that compound.

Before using any cleanup procedure, process a series of calibration standards through the procedure to validate elution patterns and the absence of interferences from the reagents.

f. Sample analysis:

1) High-performance liquid chromatography—If the internal standard calibration procedure is being used, add internal standard to sample extract and mix thoroughly. Immediately inject 5 to 25 μL sample extract or standard into HPLC using a high-pressure syringe or a constant-volume sample injection loop. Record volume injected to nearest 0.1 μL and resulting peak size in area or peak height units. Re-equilibrate HPLC column at initial gradient conditions for at least 10 min between injections.

Identify compounds in sample by comparing peak retention times with peaks of standard chromatograms. Base width of retention time window used to make identifications on measurements of actual retention time variations of standards over the course of a day. To calculate a suggested window size use three times the standard deviation of a retention time for a compound. Analyst's experience is important in interpreting chromatograms.

If the response for a peak exceeds the working range of the system, dilute extract with acetonitrile and reanalyze.

If peak response cannot be measured because of interferences, further cleanup is required.

2) Gas chromatography—See Section 6420B.5b3). If peak response cannot be measured because of interferences, further cleanup is required.

6. Calculation

Determine concentration of individual compounds using the procedures given in Section 6420B.6a. Report results in μg/L without correction for recovery. Report QC data with sample results.

7. Quality Control

a. Quality-control program: See Section 6200A.5.

b. Initial quality control: To establish the ability to generate data with acceptable precision and bias, proceed as follows: Using a pipet, prepare QC check samples at test concentrations shown in Table 6440:III by adding 1.00 mL of QC check sample concentrate (¶ 4i) to each of four 1-L portions of reagent water. Analyze QC check samples according to the method of ¶ 5. Calculate average recovery and standard deviation of the recovery, compare with acceptance criteria, and evaluate and correct system performance as directed in Section 6200A.5a1) and 2), using acceptance criteria given in Table 6440:III.

c. Analyses of samples with known additions: See Section 6420B.7c. Prepare QC check sample concentrate according to ¶ 4i and use Tables 6440:III and IV. On an ongoing basis, make known additions to at least 10% of the samples from each sample site being monitored. For laboratories analyzing one to ten sam-

TABLE 6440:III. QC ACCEPTANCE CRITERIA*

Compound	Test Conc. μg/L	Limit for s μg/L	Range for $\overline{X}$ μg/L	Range for P, P_s %
Acenaphthene	100	40.3	D–105.7	D–124
Acenaphthylene	100	45.1	22.1–112.1	D–139
Anthracene	100	28.7	11.2–112.3	D–126
Benzo(a)anthracene	10	4.0	3.1– 11.6	12–135
Benzo(a)pyrene	10	4.0	0.2– 11.0	D–128
Benzo(b)fluoranthene	10	3.1	1.8– 13.8	6–150
Benzo(ghi)perylene	10	2.3	D– 10.7	D–116
Benzo(k)fluoranthene	5	2.5	D– 7.0	D–159
Chrysene	10	4.2	D– 17.5	D–199
Dibenzo(a,h)anthracene	10	2.0	0.3– 10.0	D–110
Fluoranthene	10	3.0	2.7– 11.1	14–123
Fluorene	100	43.0	D–119	D–142
Indeno(1,2,3-cd) pyrene	10	3.0	1.2– 10.0	D–116
Naphthalene	100	40.7	21.5–100.0	D–122
Phenanthrene	100	37.7	8.4–133.7	D–155
Pyrene	10	3.4	1.4– 12.1	D–140

* s = standard deviation of four recovery measurements,
  $\overline{X}$ = average recovery for four recovery measurements,
  P, P_s = percent recovery measured, and
  D = detected; result must be greater than zero.

NOTE: These criteria are based directly upon the method performance data in Table 6440:IV. Where necessary, the limits for recovery were broadened to assure applicability of the limits to concentrations below those used to develop Table 6440:IV.

TABLE 6440:IV. METHOD BIAS AND PRECISION AS FUNCTIONS OF CONCENTRATION*

Compound	Bias as Recovery, X' $\mu g/L$	Single-Analyst Precision, s_r $\mu g/L$	Overall Precision, S' $\mu g/L$
Acenaphthene	$0.52C + 0.54$	$0.39\overline{X} + 0.76$	$0.53\overline{X} + 1.32$
Acenaphthylene	$0.69C - 1.89$	$0.36\overline{X} + 0.29$	$0.42\overline{X} + 0.52$
Anthracene	$0.63C - 1.26$	$0.23\overline{X} + 1.16$	$0.41\overline{X} + 0.45$
Benzo(a)anthracene	$0.73C + 0.05$	$0.28\overline{X} + 0.04$	$0.34\overline{X} + 0.02$
Benzo(a)pyrene	$0.56C + 0.01$	$0.38\overline{X} + 0.01$	$0.53\overline{X} - 0.01$
Benzo(b)fluoranthene	$0.78C + 0.01$	$0.21\overline{X} + 0.01$	$0.38\overline{X} - 0.00$
Benzo(ghi)perylene	$0.44C + 0.30$	$0.25\overline{X} + 0.04$	$0.58\overline{X} + 0.10$
Benzo(k)fluoranthene	$0.59C + 0.00$	$0.44\overline{X} - 0.00$	$0.69\overline{X} + 0.01$
Chrysene	$0.77C - 0.18$	$0.32\overline{X} - 0.18$	$0.66\overline{X} - 0.22$
Dibenzo(a,h)anthracene	$0.41C + 0.11$	$0.24\overline{X} + 0.02$	$0.45\overline{X} + 0.03$
Fluoranthene	$0.68C + 0.07$	$0.22\overline{X} + 0.06$	$0.32\overline{X} + 0.03$
Fluorene	$0.56C - 0.52$	$0.44\overline{X} - 1.12$	$0.63\overline{X} - 0.65$
Indeno(1,2,3-cd)pyrene	$0.54C + 0.06$	$0.29\overline{X} + 0.02$	$0.42\overline{X} + 0.01$
Naphthalene	$0.57C - 0.70$	$0.39\overline{X} - 0.18$	$0.41\overline{X} + 0.74$
Phenanthrene	$0.72C - 0.95$	$0.29\overline{X} + 0.05$	$0.47\overline{X} - 0.25$
Pyrene	$0.69C - 0.12$	$0.25\overline{X} + 0.14$	$0.42\overline{X} - 0.00$

* X' = expected recovery for one or more measurements of a sample containing a concentration of C,
  s_r = expected single-analyst standard deviation of measurements at an average concentration found of $\overline{X}$,
  S' = expected interlaboratory standard deviation of measurements at an average concentration found of $\overline{X}$,
  C = true value for concentration, and
  $\overline{X}$ = average recovery found for measurements of samples containing a concentration of C.

ples per month, analyze at least one such sample with a known addition per month. Use the procedure described in Section 6200A.5c7) and 8).

d. Quality-control check standard analysis: See Section 6420B.7d. Prepare QC check standard according to ¶ 4i and use Table 6440:III. If all compounds in Table 6440:III are to be measured in the sample in ¶ c above, it is probable that the analysis of a QC check standard will be required; therefore, routinely analyze the QC check standard with the known-addition sample.

e. Bias assessment and records: See Section 6410B.7e.

8. Precision and Bias

This method was tested by 16 laboratories using reagent water, drinking water, surface water, and three industrial wastewaters with known additions at six concentrations over the range 0.1 to 425 $\mu g/L$.[5] Single-operator precision, overall precision, and method bias were found to be related directly to compound concentration and essentially independent of sample matrix. Linear equations describing these relationships are presented in Table 6440:IV.

9. References

1. U.S. ENVIRONMENTAL PROTECTION AGENCY. 1984. Method 610— Polynuclear aromatic hydrocarbons. 40 CFR Part 136, 43344; *Federal Register* 49, No. 209.

2. U.S. ENVIRONMENTAL PROTECTION AGENCY. 1982. Determination of polynuclear aromatic hydrocarbons in industrial and municipal wastewaters. EPA-600/4-82-025, National Technical Information Serv., PB82-258799, Springfield, Va.

3. U.S. ENVIRONMENTAL PROTECTION AGENCY. 1984. Definition and procedure for the determination of the method detection limit. 40 CFR Part 136, Appendix B. *Federal Register* 49, No. 209.

4. COLE, T., R. RIGGIN & J. GLASER. 1980. Evaluation of method detection limits and analytical curve for EPA Method 610 PNA's. International Symp. Polynuclear Aromatic Hydrocarbons, 5th, Battelle's Columbus Lab., Columbus, Ohio.

5. U.S. ENVIRONMENTAL PROTECTION AGENCY. 1984. EPA Method Study 20, Method 610—PNA's. EPA-600/4-84-063, National Technical Information Serv., PB84-211614, Springfield, Va.

6440 C. Liquid-Liquid Extraction Gas Chromatographic/Mass Spectrometric Method

See Section 6410B.

6610 CARBAMATE PESTICIDES*

6610 A. Introduction

1. Selection of Method

This method is appropriate for the determination of certain N-methylcarbamoyloxime and N-methylcarbamate pesticides in natural ground and surface waters. The procedure is based on high-performance liquid chromatography (HPLC) in conjunction with a post-column derivatization system and a fluorescence detector.

Carbamate pesticides are heat-sensitive and labile, and hence not amenable to analysis by gas chromatography. HPLC is the method of choice, but without preconcentration the usual ultraviolet detector is not adequate because of low sensitivity. A procedure to separate seven carbamate pesticides by HPLC followed by post-column alkaline hydrolysis has been reported.[1] A fluorescent adduct[2-4] is produced and is measured with a fluorescence detector.[5-8]

2. References

1. MOYE, H.A., S.J. SCHERRER & P.A. ST. JOHN. 1977. Dynamic labeling of pesticides for high performance liquid chromatography: Detection of N-methylcarbamates and o-phthalaldehyde. Anal. Lett. 10:1049.
2. SIMONS, S.S., JR. & D.D. JOHNSON. 1977. Communication to the editor. J. Amer. Chem. Soc. 98:7098.

3. KRAUSE, R.T. 1978. Further characterization and refinement of an HPLC post-column fluorometric labeling technique for the determination of carbamate insecticides. J. Chromatogr. Sci. 16:261.
4. KRAUSE, R.T. 1979. Resolution, sensitivity and selectivity of a high-performance liquid chromatographic post-column fluorometric labeling technique for determination of carbamate insecticides. J. Chromatogr. 185:615.
5. HILL, K.M., R.H. HOLLOWELL & L.A. DALCORTIVO. 1984. Determination of N-methylcarbamate pesticides in well water by liquid chromatography and post-column fluorescence derivatization. Anal. Chem. 56:2465.
6. FOERST, D.C. & H.A. MOYE. 1985. Aldicarb and related compounds in drinking water via direct aqueous injection HPLC with post-column derivatization. In Advances in Water Analysis and Treatment, Proc. 12th Annu. AWWA Water Quality Technology Conf., Dec. 2–5, 1984, Denver, Colo., p. 189. American Water Works Assoc., Denver, Colo.
7. U.S. ENVIRONMENTAL PROTECTION AGENCY. 1991. Method 531.1— Measurement of N-methylcarbamoyloximes and N-methylcarbamates in water by direct aqueous injection HPLC with post-column derivatization. In Methods for the Determination of Organic Compounds in Drinking Water. EPA-600/4-88-039, rev. July 1991, U.S. Environmental Protection Agency, Cincinnati, Ohio.
8. U.S. ENVIRONMENTAL PROTECTION AGENCY. 1988. National Pesticide Survey Analytical Methods. Method 5. Measure N-methylcarbamoyloximes and N-methylcarbamates in groundwater by direct aqueous injection HPLC with post-column derivatization. U.S. Environmental Protection Agency, Cincinnati, Ohio.

---

* Approved by Standard Methods Committee, 1996.

6610 B. High-Performance Liquid Chromatographic Method

1. General Discussion

a. Principle: This HPLC method is applicable to the determination of certain carbamate pesticides in water.* The sample is filtered and injected into a reverse-phase HPLC column. The constituents are separated by gradient elution chromatography. After elution the N-methyl compounds are hydrolyzed with sodium hydroxide. The resulting methylamine is reacted with o-phthalaldehyde (OPA) and 2-mercaptoethanol (MERC) to form a highly fluorescent isoindole product that is detected with a fluorescence detector.

b. Interferences: The post-column reaction generally is sensitive to primary amines and will form fluorescent adducts that may cause interference depending on their elution time and fluorescence intensity. Interferences also may be caused by contaminants in solvents, reagents, glassware, and sample-processing equipment. Specific sources of contamination have not been identified. Demonstrate that reagents and apparatus are free from interfer-

ences under the conditions of the analysis by analyzing laboratory method blanks. Use only high-purity reagents and solvents.

High chlorine concentrations, which may be encountered near a chlorine injection point, cause interference and loss of some constituents. Collect samples before chlorination or as far as practical from an injection point. If a raw water source is to be evaluated, collect sample before chlorination.

Matrix interferences may be caused by contaminants in the sample. They vary considerably from source to source. Severe interferences are not expected from most ground and surface waters, but the method probably is not suitable for waste streams, landfill leachate, or wastewater effluents. Confirmatory analysis may be used to increase the confidence of identification and quantitation for compounds determined by the primary analysis.

Interfering contamination in the analysis may occur when a low-level sample is analyzed immediately after a sample containing relatively high concentrations of test compounds. To minimize contamination use disposable syringes, filters, and sample vials. Analyze a reagent water blank after analyzing a sample containing high concentrations of carbamate pesticides, to demonstrate that

---

* Aldicarb sulfoxide, aldicarb sulfone, oxamyl, methomyl, 3-hydroxycarbofuran, aldicarb, baygon, carbofuran, carbaryl, and methiocarb.

TABLE 6610:I. RECOVERY OF KNOWN ADDITIONS OF SELECTED CARBAMATES FROM WATER AND ESTIMATED
DETECTION LIMITS (EDL)

Constituent	Addition $\mu g/L$	No. Data Points	Recovery Average %	Standard Deviation $\mu g/L$	Relative Standard Deviation %	EDL* $\mu g/L$
Aldicarb	1.0	8	107	0.07	7	1.0
Aldicarb sulfoxide	2.0	8	47	0.20	21	2.0
Aldicarb sulfone	2.0	8	83	0.34	20	2.0
Baygon	1.0	7	101	0.32	32	1.0
Carbaryl	2.0	8	97	0.44	23	2.0
Carbofuran	1.5	7	90	0.17	12	1.5
3-Hydroxycarbofuran	2.0	8	108	0.63	29	2.0
Methiocarb	4.0	8	82	0.64	19	4.0
Methomyl	0.50	7	102	0.09	18	0.50
Oxamyl	2.0	8	82	0.29	7	2.0

*EDL = estimated detection limit in sample; calculate by multiplying standard deviation times the Student's t value appropriate for a 99% confidence level and a standard deviation estimate with $n - 1$ degrees of freedom, or a level of compound in sample yielding a peak with a signal-to-noise ratio of approximately 5, whichever value is higher. There were no detectable apparent residues of any constituent in the blank water.

no carryover contamination is present. As an alternative check for carryover, reanalyze any positive samples that immediately followed a high-concentration sample.

c. Detection levels: This method has been validated in a multiple laboratory test. Estimated detection levels (EDLs) are listed in Table 6610:I. Observed detection levels may vary between laboratories and samples, depending on interferences and specific instrumentation.

2. Sampling and Storage

Fill bottle only to shoulder with water to be sampled, leaving space for expansion on freezing. If residual chlorine is present, add 8 mg sodium thiosulfate/100 mL sample before collecting the sample. Keep samples cold (4°C) from time of collection until receipt in the laboratory. Oxamyl, 3-hydrocarbofuran, and carbaryl can degrade quickly in water at room temperature. This short-term degradation is of concern during sample processing and holding at room temperature in autosampler trays. Preserve samples targeted for the analysis of these three compounds by adjusting to pH 3 with monochloroacetic acid buffer solution (1.5 mL buffer/50 mL sample). For maximum protection, add buffer to sample bottle before taking sample. In the laboratory, store samples at 4°C until analysis. Samples are stable for at least 28 d when adjusted to pH 3 and stored at 4°C (see Table 6610:II).

3. Apparatus

a. High-performance liquid chromatograph (HPLC): An analytical system complete with column supplies, high-pressure syringes, detectors, and compatible strip-chart recorder. Preferably use a data system for measuring peak areas and retention times. Use system capable of injecting 200- to 500-μL portions and of performing binary linear gradients at a constant flow rate. See Figure 6610:1.

1) *Primary column:* 150 mm long × 3.9 mm ID stainless steel packed with 4 μm Novapak C18.† If other columns, conditions,

or detectors are used, demonstrate that acceptable results are obtained.

2) *Alternate column:* 250 mm long × 4.6 mm ID stainless steel packed with 5 μm Beckman Ultrasphere ODS.† Data presented herein were obtained with this column. However, newer manufactured columns have not been able to resolve aldicarb sulfone from oxamyl.

3) *Confirmatory column:* 250 mm long × 4.6 mm ID stainless steel packed with 5 μm Supelco LC-1.† This is a trimethylsilyl bonded silica column.

4) *Post-column reactor:* Use a post-column reactor capable of mixing reagents into the mobile phase and equipped with pumps to deliver 0.1 to 1.0 mL/min of each reagent. Use a delivery rate of 0.5 mL/min for both sodium hydroxide and OPA solution. Use PTFE tubing‡ for coils and other post-column lines. Stainless steel, polyetheretherketone (PEEK), and/or nickel lines have been used successfully.

5) *Detector:* Use fluorescence detector capable of excitation at 230 nm and detection of emission energies greater than 418 nm.

b. Filters: For macrofiltration of derivatization solutions and mobile phases, use 47-mm filters.§ For microfiltration of samples before HPLC analysis, use 13-mm filter holder‖ and 13-mm-diam 0.2-μm polyester filters.# If disposable filters** and syringes are to be used, test and verify that comparable results are obtained.

c. Analytical balance, capable of weighing to the nearest 0.0001 g.

d. Sample bottles, 120-mL (or other convenient size) screw-cap polypropylene bottles. Less preferably, use polyethylene bottles or glass containers.

---

† HPLC methods are extremely sensitive to the materials used. Mention of trade names by *Standard Methods* does not preclude the use of other existing or as-yet-undeveloped products that give *demonstrably* equivalent results.

‡ Kratos URS 051 and URA 100, or equivalent.
§ Millipore Type HA, 0.45 μm, for water, and Millipore Type FH, 0.5 μm, for organics, or equivalent.
‖ Millipore stainless steel XX300/200, or equivalent.
Nuclepore 180406, 7035 Commerce Circle, Pleasanton, CA 94566–3294, or equivalent.
** Gelman Sciences Acro LC 13, 0.45-μm disposable filter assembly, or equivalent, for aqueous samples.

TABLE 6610:II. STORAGE STABILITY OF SAMPLES WITH KNOWN ADDITIONS

Constituent	Added Concentration μg/L	Day	Storage Temperature °C	Average Recovery %	Relative Standard Deviation %
Aldicarb	5.0	0	—	100	9
		14	−10	100	4
			+4	110	2
		28	−10	100	6
			+4	83	1
Aldicarb sulfone	10	0	—	99	9
		14	−10	93	3
			+4	99	3
		28	−10	97	0
			+4	86	8
Aldicarb sulfoxide	10	0	—	100	9
		14	−10	91	6
			+4	100	2
		28	−10	98	2
			+4	91	5
Baygon	5.0	0	—	98	10
		14	−10	91	2
			+4	100	4
		28	−10	88	2
			+4	93	9
Carbaryl	10	0	—	100	6
		14	−10	92	4
			+4	95	3
		28	−10	99	18
			+4	89	5
Carbofuran	7.5	0	—	100	9
		14	−10	95	3
			+4	110	7
		28	−10	95	2
			+4	93	3
3-Hydroxycarbofuran	10	0	—	95	9
		14	−10	89	6
			+4	100	3
		28	−10	100	8
			+4	95	11
Methiocarb	20	0	—	110	4
		14	−10	100	1
			+4	98	1
		28	−10	99	0
			+4	94	6
Methomyl	2.5	0	—	110	12
		14	−10	90	6
			+4	96	5
		28	−10	93	2
			+4	96	5
Oxamyl	10	0	—	98	6
		14	−10	85	4
			+4	95	5
		28	−10	100	11
			+4	94	9

Conditions: Samples preserved by adjustment to pH 3 with monochloroacetic acid buffer, and protected from light.

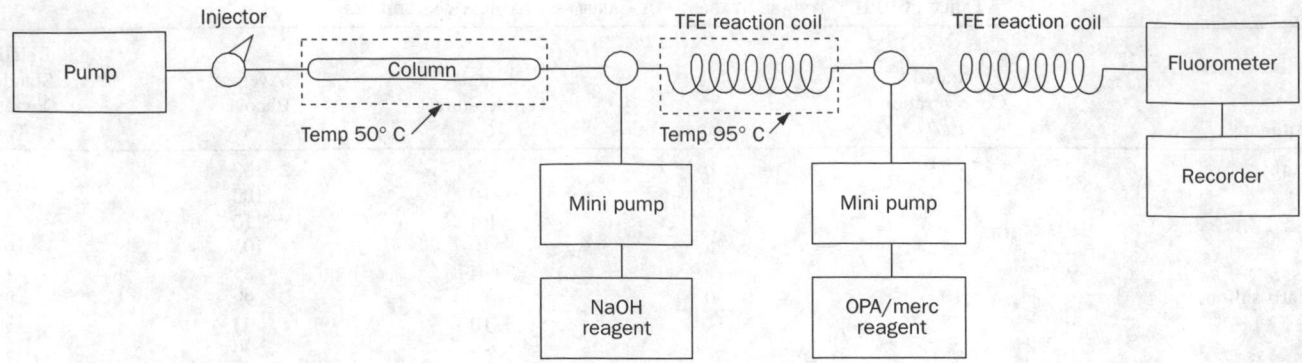

Figure 6610:1. Schematic of post-column reaction HPLC system.

4. Reagents

Use reagent-grade chemicals of high purity and HPLC-grade (tested on HPLC and verified to give no impurity peaks) solvents or equivalent.

a. Methanol, CH₃OH, HPLC grade, or equivalent.

b. Reagent water: Generate by using a water purification system.†† Alternatively, purchase HPLC-grade water commercially. For additional alternatives, see Section 1080.

c. Sodium hydroxide, NaOH, 0.05N: Dissolve 2.0 g NaOH in 1.0 L reagent water. Filter and degas with helium before use.

d. 2-Mercaptoethanol solution: Mix 10.0 mL 2-mercapto-ethanol and 10.0 mL acetonitrile. Cap. Store in refrigerator. (CAUTION: *Stench.*)

e. Sodium borate solution: Dissolve 19.1 g Na₂B₄O₇ · 10H₂O in reagent water. Dilute to 1 L. Prepare a day before use to insure complete dissolution.

f. OPA reaction solution: Dissolve 0.100 g *o*-phthalaldehyde in 10 mL methanol. Add to 1.0 µL sodium borate solution. Mix, filter, and degas with helium. Add 100 µL 2-mercaptoethanol solution and mix. If protected from oxygen, solution is stable for at least 3 d; otherwise, prepare daily.

g. Monochloroacetic acid, 2.5M: Dissolve 236 g monochlo-roacetic acid in 1 L reagent water.

h. Potassium acetate solution, 2.5M: Dissolve 245 g potassium acetate in 1 L reagent water.

i. Monochloroacetic acid buffer: Mix 156 mL 2.5M mono-chloroacetic acid with 100 mL 2.5M potassium acetate solution. CAUTION: *Handle with care; long-term health effects from exposure have not been determined.*

j. Stock pesticide solutions: Prepare stock standard solutions (1.00 µg/µL) by accurately weighing approximately 0.0100 g of each compound into separate 10-mL volumetric flasks. Dissolve in about 5 mL methanol and dilute to mark with methanol. Larger volumes may be used at the convenience of the analyst. If compound purity is certified at 96% or greater, use uncorrected weight in calculations. These solutions are stable for several months when stored in a freezer at −10°C.

k. Calibration standards:

1) *Internal standard:* Prepare an internal standard solution by accurately weighing approximately 0.0010 g 4-bromo-3,5-dime-

thylphenyl *N*-methyl carbamate (BDMC), dissolving in methanol, and diluting to volume in a 10-mL volumetric flask. Add 5 µL of internal standard solution to 50 mL sample to give an internal standard concentration of 10 µg/L.

2) *Stock calibration standard:* Prepare a mixed concentrated standard of the compounds listed in Table 6610:I by adding appropriate amounts of each stock pesticide solution to methanol to yield a concentration of about 1.0 µg/mL each.

3) *Calibration standard:* Prepare working calibration standards by additional dilutions of the stock calibration standard with pH 3 buffered water to yield solutions containing approximately 100, 40, 10, and 2 µg/L of each compound. Any similar series of concentrations is satisfactory. The lowest standard concentration should be near (but above) the EDL (see Table 6610:I). Add internal standard to these working calibration standards to give a final concentration of 10 µg BDMC/L in each.

l. Sodium thiosulfate, Na₂S₂O₃.

5. Procedure

a. Liquid chromatography: Table 6610:III summarizes the recommended operating conditions for the liquid chromatograph, including retention times. An example of the separations achieved under these conditions is shown in Figure 6610:2.

Completely thaw frozen samples before analysis. If it has not already been done, adjust standards and samples to pH 3 by adding 1.5 mL monochloroacetic acid buffer/50 mL sample or standard. Fill a 50-mL volumetric flask to the mark with sample. Add internal standard (BDMC) and mix. Rinse the syringe and filter with 5 mL reagent water and then with 5 mL of prepared sample; discard the filtrate. Filter 5 to 10 mL of prepared sample and inject up to 500 µL of filtrate. Record peak sizes in area units. If the peak response exceeds the working range of the system, dilute sample with pH 3 buffered water and reinject.

b. Calibration:

1) *Internal standard calibration:* Using prepared working calibration standards, inject 500 µL of each standard. The lowest calibration standard should represent compound concentrations near, but above, their respective EDLs. The remaining calibration standards should bracket expected concentration range of samples or should define the working range of the detector. Tabulate peak height or area responses against concentration for each compound

---

†† Millipore, Super-Q, or equivalent.

TABLE 6610:III. PRIMARY AND CONFIRMATORY CHROMATOGRAPHIC CONDITIONS AND RETENTION TIMES FOR SELECTED CARBAMATE PESTICIDES

	Retention Time		
	Primary System	Confirmatory System	Alternate System
Constituent	*min*	*min*	*min*
Aldicarb sulfoxide	6.80	8.5	14.94
Aldicarb sulfone	7.77	8.57	15.23
Oxamyl	8.20	10.03	17.36
Methomyl	8.94	10.39	18.39
3-Hydroxycarbofuran	13.65	12.51	23.32
Aldicarb	16.35	14.11	27.01
Baygon	18.86	16.02	29.25
Carbofuran	19.17	16.33	29.61
Carbaryl	20.29	17.17	30.78
Methiocarb	24.74	20.45	34.93
BDMC (internal standard)	25.28	20.58	35.50

Primary conditions:

Column	150-mm-long × 3.9-mm-ID Waters NovaPak C18 (4 μm)
Mobile phase	Linear gradient from 10:90 methanol:water, hold 2 min, then linear gradient to 80:20 methanol:water in 25 min
Flow rate	1.0 mL/min
Injection volume	500 μL
Detector	Fluorescence; excitation 230 nm; emission >418 nm

Confirmatory conditions:

Column	250-mm-long × 4.6-mm-ID Supelco LC-1 (5 μm)
Mobile phase	Linear gradient from 15:85 methanol:water to 100% methanol in 32 min
Flow rate	1.0 mL/min
Injection volume	500 μL
Detector	Fluorescence; excitation 230 nm; emission >418 nm

Alternate conditions:

Column	250-mm-long × 4.6-mm-ID Beckman Ultrasphere ODS (5 μm)
Mobile phase	Linear gradient from 15:85 methanol:water to 100% methanol in 32 min
Flow rate	1.0 mL/min
Injection volume	500 μL
Detector	Fluorescence; excitation 230 nm; emission >418 nm

Post-column reactor condition for all detectors:

Hydrolysis	NaOH (0.05N), flow rate 0.5 mL/min, 1.0-mL reaction coil at 95°C
Derivatization	OPA solution, flow rate 0.5 mL/min, 1.0-mL reaction coil at ambient temperature

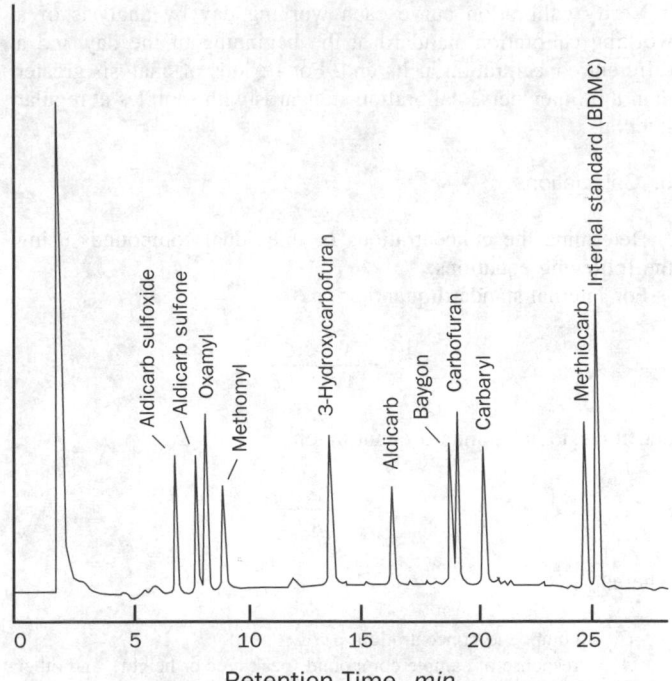

Figure 6610:2. Schematic HPLC-PCD chromatogram of carbamate mix indicating relative response, separations, and retention.

Compound	Concentration
	μg/L
Aldicarb sulfoxide	8.6
Aldicarb sulfone	7.6
Oxamyl	12.6
Methomyl	4.7
3-Hydroxycarbofuran	13.8
Aldicarb	6.5
Baygon	11.6
Carbofuran	11.3
Carbaryl	6.1
Methiocarb	20.4

and the internal standard. Calculate response factors (*RF*) for each compound, using the equation:

$$RF = \frac{A_s \times C_{is}}{A_{is} \times C_s}$$

where:

A_s = peak area or peak height for compound in working standard,
A_{is} = peak area or peak height of internal standard,
C_{is} = concentration of internal standard, μg/L, and
C_s = concentration of compound in working standard, μg/L.

If the *RF* value over the working range is constant (≤ 20% RSD), use average *RF* for calculations.

Alternately, plot a calibration curve using (A_s/A_{is}) vs. C_s.

Verify working calibration curve or RF for each work session by measuring one or more calibration standards. If the response for any compound varies from the predicted response by more than ± 20%, repeat the test using a fresh calibration standard and, if necessary, prepare a new calibration curve.

Verify calibration standards at least quarterly by analyzing a standard prepared using reference material obtained from an independent source.

2) *External standard calibration:* Use working calibration standards (¶ *k*3). Starting with the lowest concentration, analyze each working calibration standard, using 500-μL injections. Tabulate responses, peak height or area versus compound concentration in the standard. Prepare a calibration curve for each compound. Alternately, if the ratio of response to concentration (calibration factor) is constant (≤ 20%) over the working range of the curve, use a calibration factor in place of the calibration curve.

Verify calibration curve each working day by analysis of a working calibration standard at the beginning of the day and a different concentration at its end. For periods of analysis greater than 8 h, intersperse calibration standards with samples at regular intervals.

6. Calculations

Determine the concentrations of individual compounds using the following equations:

For internal standard quantitation,

$$C_x = \frac{A_x \times C_{is}}{A_{is} \times RF}$$

and for external standard quantitation,

$$C_x = \frac{A_x \times C_s}{A_s}$$

where:

C_x = compound concentration, μg/L,
A_x = response of sample compound (peak area or height), and other terms are as defined previously.

7. Quality Control

A minimum quality control program should include the following quality assurance elements: initial demonstration of laboratory capability, laboratory control standards, monitoring of systems performance, blank samples, quality control check samples, duplicate analyses, demonstration of adequate recoveries, assessment of matrix effects, and demonstration of storage stability. Also see Part 1000 for general quality control requirements.

8. Precision and Bias

Single-laboratory recovery data are summarized in Table 6610: II. Comparable results should be obtained in other laboratories. Results are comparable if the calculated percent relative standard deviation (RSD) does not exceed 3 times the single-laboratory RSD or 30%, whichever is greater, and the mean recovery lies within the interval R ± 3SD or R ± 30%, whichever is greater.

Results of an eight-laboratory collaborative test are shown in Table 6610:IV. Test involved six concentrations for each compound (overall range 1 to 109 μg/L).

TABLE 6610:IV. MEAN RECOVERY, SINGLE-ANALYST STANDARD DEVIATION, AND OVERALL STANDARD DEVIATION FOR COLLABORATIVE STUDY DATA*

	Reagent Water			Drinking Water		
Compound	Mean Recovery μg/L	Single-Analyst Standard Deviation μg/L	Overall Standard Deviation μg/L	Mean Recovery μg/L	Single-Analyst Standard Deviation μg/L	Overall Standard Deviation μg/L
Aldicarb	$0.926C + 0.202$	0.32†	$0.022\overline{X} + 0.370$‡	$1.032C + 0.031$	$0.040\overline{X} + 0.046$	$0.101\overline{X} - 0.042$‡
Aldicarb sulfone	$0.942C + 0.446$	$0.025\overline{X} + 0.382$	$0.062\overline{X} + 0.132$	$0.968C - 0.097$	$0.008\overline{X} + 0.276$	$0.039\overline{X} + 0.119$‡
Aldicarb sulfoxide	$0.941C + 0.876$§	$0.040\overline{X} + 0.103$	$0.058\overline{X} + 0.211$	$0.952C + 0.460$§	$0.024\overline{X} + 0.050$	$0.021\overline{X} + 0.440$
Baygon	$0.916C + 0.360$	$0.040\overline{X} + 0.092$	$0.058\overline{X} + 0.230$	$0.994C + 0.101$	$0.046\overline{X} - 0.005$	$0.086\overline{X} - 0.114$
Carbaryl	$0.949C + 0.542$	$0.016\overline{X} + 0.480$	$0.058\overline{X} + 0.219$	$0.958C + 0.439$	$0.039\overline{X} + 0.167$	$0.068\overline{X} + 0.015$
Carbofuran	$0.923C + 0.636$	$0.022\overline{X} + 0.322$	$0.006\overline{X} + 0.564$‡	$0.970C + 0.220$	$0.008\overline{X} + 0.316$	$0.042\overline{X} + 0.178$
3-Hydroxycarbofuran	$0.940C + 0.438$	$0.013\overline{X} + 0.697$‡	$0.038\overline{X} + 0.578$	$0.979C + 0.153$	$0.044\overline{X} + 0.114$	$0.085\overline{X} + 0.045$‡
Methiocarb	$0.923C + 0.887$	$0.005\overline{X} + 1.839$	$0.035\overline{X} + 2.286$	$0.958C + 0.474$	$0.034\overline{X} + 0.046$	$0.057\overline{X} + 0.322$
Methomyl	$0.976C + 0.043$	$0.053\overline{X} + 0.069$	$0.048\overline{X} + 0.133$	$0.988C + 0.000$	0.14†	$0.040\overline{X} + 0.000$
Oxamyl	$0.936C + 0.659$	1.04†	$0.038\overline{X} + 0.699$	$0.998C + 0.045$	$0.025\overline{X} + 0.048$	$0.023\overline{X} + 0.672$‡

*C = true concentration, μg/L; $\overline{X}$ = mean recovery, μg/L.
†Weighted linear regression equation had negative slope; average precision is reported.
‡Coefficient of determination of weighted equation was weak (<0.5).
§Lowest addition recovery not used for this regression.
Source: EDGELL, K.W., L.A. BIEDERMAN & J.E. LONGBOTTOM. 1991. Measurement of *n*-Methylcarbamoyloximes and *n*-Methylcarbamates in Water by Direct Aqueous Injection HPLC with Post Column Derivatization: Collaborative Study. *J. Assoc. Offic. Anal. Chem.* 74:309.

6630 ORGANOCHLORINE PESTICIDES*

6630 A. Introduction

1. Sources and Significance

The organochlorine pesticides commonly occur in waters that have been affected by agricultural discharges. Some of the listed compounds are degradation products of other pesticides detected by this method. Several of the pesticides are bioaccumulative and relatively stable, as well as toxic or carcinogenic; thus they require close monitoring.

_____

* Approved by Standard Methods Committee, 1993.

2. Selection of Method

Methods 6630B and C consist of gas chromatographic (GC) procedures following liquid-liquid extraction of water samples. They are relatively sensitive methods that can be used to detect numerous pesticides. Differences between the methods are minimal after extraction. Method 6630D is a gas chromatographic/mass spectrometric (GC/MS) method that can detect all of the target compounds, but at much higher concentrations. All these methods also are useful for determination of polychlorinated biphenyls (PCBs) (also see Section 6431A).

6630 B. Liquid-Liquid Extraction Gas Chromatographic Method I

1. General Discussion

a. _Application:_ This gas chromatographic procedure is suitable for quantitative determination of the following specific compounds: BHC, lindane (γ-BHC), heptachlor, aldrin, heptachlor epoxide, dieldrin, endrin, captan, DDE, DDD, DDT, methoxychlor, endosulfan, dichloran, mirex, and pentachloronitrobenzene. Under favorable circumstances, strobane, toxaphene, chlordane (tech.), and others also may be determined when relatively high concentrations of these complex mixtures are present and the chromatographic fingerprint is recognizable in packed or capillary column analysis. Trifluralin and certain organophosphorus pesticides, such as parathion, methylparathion, and malathion, which respond to the electron-capture detector, also may be measured. However, the usefulness of the method for organophosphorus or other specific pesticides must be demonstrated before it is applied to sample analysis.

b. _Principle:_ In this procedure the pesticides are extracted with a mixed solvent, diethyl ether/hexane or methylene chloride/hexane. The extract is concentrated by evaporation and, if necessary, is cleaned up by column adsorption-chromatography. The individual pesticides then are determined by gas chromatography. Although procedures detailed below refer primarily to packed columns, capillary column chromatography also may be used. See Section 6010C.2a1) for discussion of gas chromatographic principles and 6010C.2b2) for discussion of electron-capture detector.

As each component passes through the detector a quantitatively proportional change in electrical signal is measured on a strip-chart recorder. Each component is observed as a peak on the recorder chart. The retention time is indicative of the particular pesticide and peak height/peak area is proportional to its concentration.

Variables may be manipulated to obtain important confirmatory data. For example, the detector system may be selected on the basis of the specificity and sensitivity needed. The detector used in this method is an electron-capture detector that is very sensitive to chlorinated compounds. Additional confirmatory identification can be made from retention data on two or more columns where the stationary phases are of different polarities. A two-column procedure that has been found particularly useful is specified. If sufficient pesticide is available for detection and measurement, confirmation by a more definitive technique, such as mass spectrometry, is desirable.

c. _Interference:_ See Sections 6010C.2a2) and 6010C.2b2). Some compounds other than chlorinated compounds respond to the electron-capture detector. Among these are oxygenated and unsaturated compounds. Sometimes plant or animal extractives obscure pesticide peaks. These interfering substances often can be removed by auxiliary cleanup techniques. A magnesia-silica gel column cleanup and separation procedure is used for this purpose. Such cleanup usually is not required for potable waters.

1) Polychlorinated biphenyls (PCBs)—Industrial plasticizers, hydraulic fluids, and old transformer fluids that contain PCBs are a potential source of interference in pesticide analysis. The presence of PCBs is suggested by a large number of partially resolved or unresolved peaks that may occur throughout the entire chromatogram. Particularly severe PCB interference will require special separation procedures.

2) Phthalate esters—These compounds, widely used as plasticizers, cause electron-capture detector response and are a source of interferences. Water leaches these esters from plastics, such as polyethylene bottles and plastic tubing. Phthalate esters can be separated from many important pesticides by the magnesia-silica gel column cleanup. They do not cause response to halogen-specific detectors such as microcoulometric or electrolytic conductivity detectors.

d. _Detection limits:_ The ultimate detection limit of a substance is affected by many factors, for example, detector sensitivity, extraction and cleanup efficiency, concentrations, and detector signal-to-noise level. Lindane (γ-BHC) usually can be determined at 10 ng/L in a sample of relatively unpolluted water; the DDT detection limit is somewhat higher, 20 to 25 ng/L. Increased sensitivity is likely to increase interference with all pesticides.

e. Sample preservation: Some pesticides are unstable. Transport under iced conditions, store at 4°C until extraction, and do not hold more than 7 d. When possible, extract upon receipt in the laboratory and store extracts at 4°C until analyzed. Analyze extracts within 40 d.

2. Apparatus

Clean thoroughly all glassware used in sample collection and pesticide residue analyses. Clean glassware as soon as possible after use. Rinse with water or the solvent that was last used in it, wash with soapy water, rinse with tap water, distilled water, redistilled acetone, and finally with pesticide-quality hexane. As a precaution, glassware may be rinsed with the extracting solvent just before use. Heat heavily contaminated glassware in a muffle furnace at 400°C for 15 to 30 min. High-boiling-point materials, such as PCBs, may require overnight heating at 500°C, but no borosilicate glassware can exceed this temperature without risk. Do not heat volumetric ware. Clean volumetric glassware with special reagents.* Rinse with water and pesticide-quality hexane. After drying, store glassware to prevent accumulation of dust or other contaminants. Store inverted or cover mouth with aluminum foil.

a. Sample bottles: 1-L capacity, glass, with TFE-lined screw cap. Bottle may be calibrated to minimize transfers and potential for contamination.

b. Evaporative concentrator, Kuderna-Danish, 500-mL flask and 10-mL graduated lower tube fitted with a 3-ball Snyder column, or equivalent.

c. Separatory funnels, 2-L capacity, with TFE stopcock.

d. Graduated cylinders, 1-L capacity.

e. Funnels, 125-mL.

f. Glass wool, filter grade.

g. Chromatographic column, 20 mm in diam and 400 mm long, with coarse fritted disk at bottom.

h. Microsyringes, 10- and 25-μL capacity.

i. Hot water bath.

j. Gas chromatograph, equipped with:

1) *Glass-lined injection port.*

2) *Electron-capture detector.*

3) *Recorder:* Potentiometric strip chart, 25-cm, compatible with detector and associated electronics.

4) *Borosilicate glass column,* 1.8 m × 4-mm ID or 2-mm ID.

Variations in available gas chromatographic instrumentation necessitate different operating procedures for each. Therefore, refer to the manufacturer's operating manual as well as gas chromatography catalogs and other references (see Bibliography). In general, use equipment with the following features:

• Carrier-gas line with a molecular sieve drying cartridge and a trap for removal of oxygen from the carrier gas. A special purifier† may be used. Use only dry carrier gas and insure that there are no gas leaks.

• Oven temperature stable to ±0.5°C or better at desired setting.

---

* No Chromix, Godax, 6 Varick Place, New York, NY, or equivalent.
† Hydrox, Matheson Gas Products, P. O. Box E, Lyndhurst, NJ, or equivalent.

• Chromatographic columns—A well-prepared column is essential to an acceptable gas chromatographic analysis. Obtain column packings and pre-packed columns from commercial sources or prepare column packing in the laboratory.

It is inappropriate to give rigid specifications on size or composition to be used because some instruments perform better with certain columns than do others. Columns with 4-mm ID are used most commonly. The carrier-gas flow is approximately 60 mL/min. When 2-mm-ID columns are used, reduce carrier-gas flow to about 25 mL/min. Adequate separations have been obtained by using 5% OV-210 on 100/120 mesh dimethyl-dichlorosilane-treated diatomaceous earth‡ in a 2-m column. The 1.5% OV-17 and 1.95% QF-1 column is recommended for confirmatory analysis. Two additional column options are included: 3% OV-1 and mixed-phase 6% QF-1 + 4% SE-30, each on dimethyl-dichlorosilane-treated diatomaceous earth, 100-120 mesh. OV-210, which is a refined form of QF-1, may be substituted for QF-1. A column is suitable when it effects adequate and reproducible resolution. Sample chromatograms are shown in Figures 6630:1 through 6630:4.

Alternately, use fused silica capillary§ columns, 30 m long with a 0.32-mm ID and 0.25-μm film thickness, or equivalent. See Figure 6630:5. To confirm identification use a column of different polarity.‖

---

‡ Gas Chrom Q, Applied Science Labs., Inc., P. O. Box 440, State College, PA, or equivalent.
§ J & W Scientific, DB-5, DB-1701, or equivalent.
‖ J & W Scientific, DB-1, or equivalent.

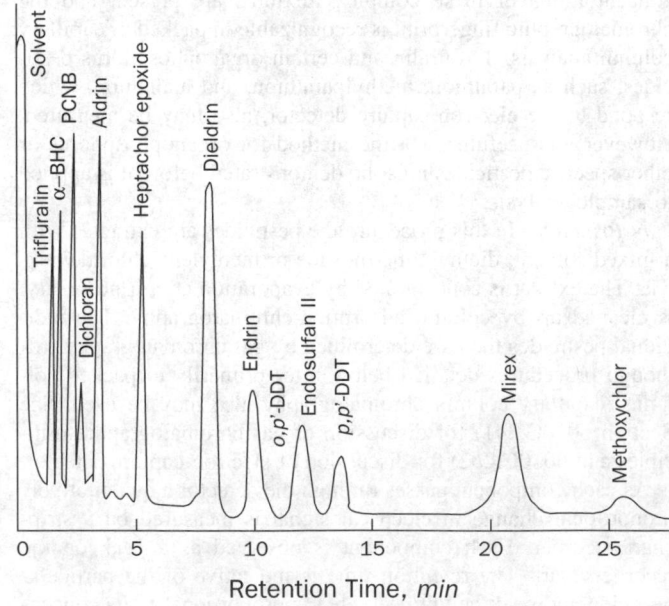

Figure 6630:1. Results of gas chromatographic procedure for organochlorine pesticides. Column packing: 1.5% OV-17 + 1.95% QF-1; carrier gas: argon/methane at 60 mL/min; column temperature: 200°C; detector: electron capture in pulse mode.

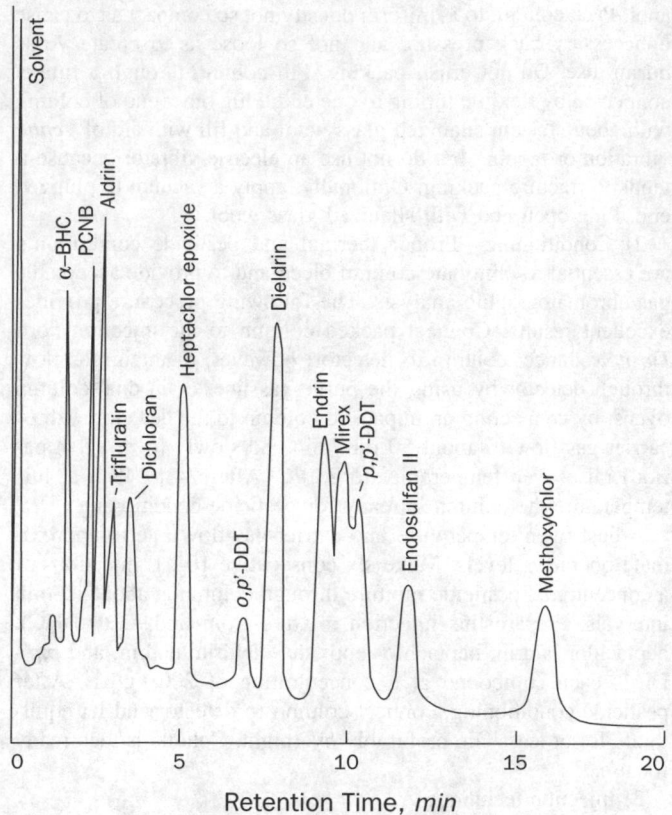

Figure 6630:2. Results of gas chromatographic procedure for organochlorine pesticides. Column packing: 5% OV-210; carrier gas: argon/methane at 70 mL /min; column temperature: 180°C; detector: electron capture.

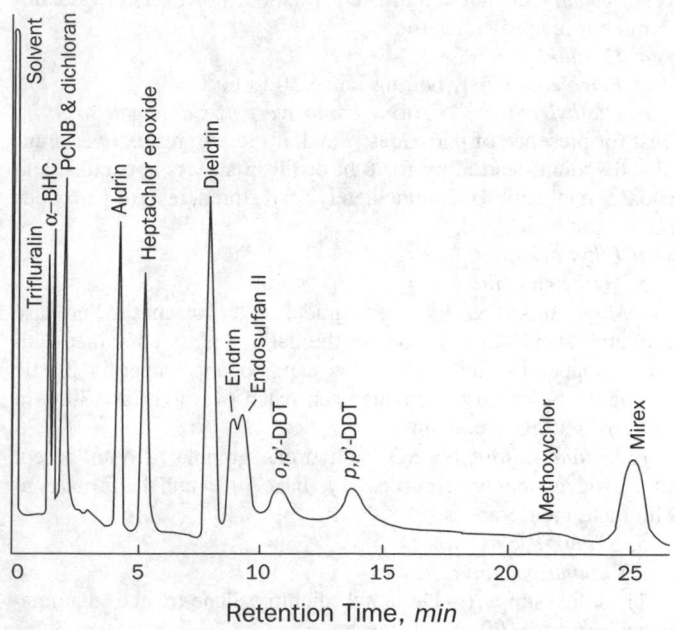

Figure 6630:4. Chromatogram of pesticide mixture. Column packing: 3% OV-1; carrier gas: argon/methane at 70 mL /min; column temperature: 180°C; detector: electron capture.

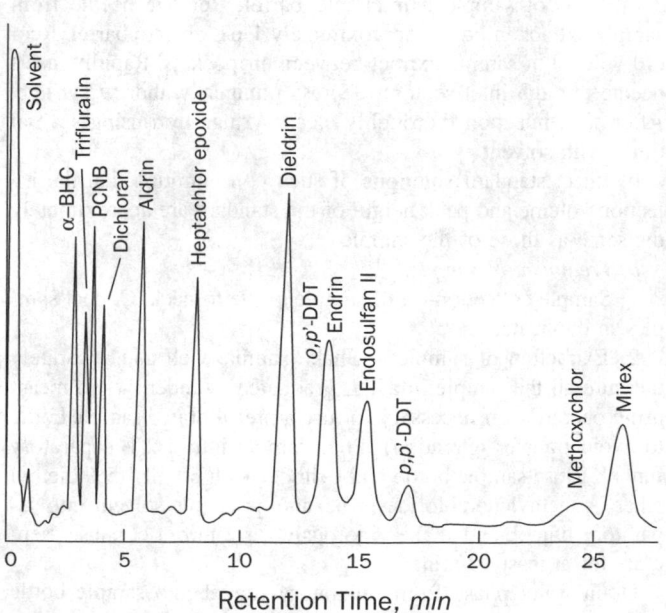

Figure 6630:3. Chromatogram of pesticide mixture. Column packing: 6% QF-1 + 4% SE-30; carrier gas; argon/methane at 60 mL / min; column temperature: 200°C; detector: electron capture.

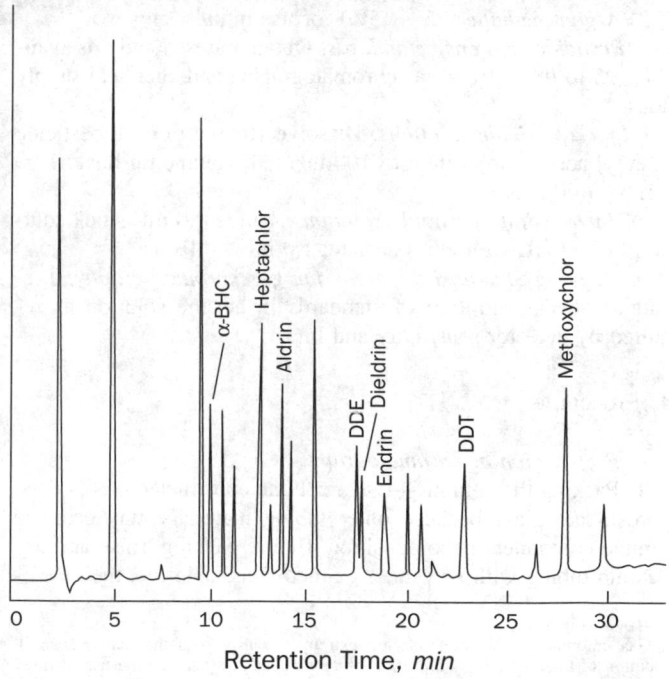

Figure 6630:5. Chromatogram of pesticide mixture. Column DB-5, 30 m long, multilevel program temperature, electron-capture detector.

3. Reagents#

Use solvents, reagents, and other materials for pesticide analysis that are free from interferences under the condition of the analysis. Specific selection of reagents and distillation of solvents in an all-glass system may be required. "Pesticide quality" solvents usually do not require redistillation; however, always determine a blank before use.

a. Hexane.

b. Petroleum ether, boiling range 30 to 60°C.

c. Diethyl ether: CAUTION: *Explosive peroxides tend to form.* Test for presence of peroxides** and, if present, reflux over granulated sodium-lead alloy for 8 h, distill in a glass apparatus, and add 2% methanol. Use immediately or, if stored, test for peroxides before use.

d. Ethyl acetate.

e. Methylene chloride.

f. Magnesia-silica gel,†† PR grade, 60 to 100 mesh. Purchase activated at 676°C and store in the dark in glass container with glass stopper or foil-lined screw cap; do not accept in plastic container. Before use, activate each batch overnight at 130°C in foil-covered glass container.

g. Sodium sulfate, Na_2SO_4, anhydrous, granular: Do not accept in plastic container. If necessary, bake in a muffle furnace to eliminate interferences.

h. Silanized glass wool.

i. Column packing:

1) Solid support—Dimethyl dichlorosilane-treated diatomaceous earth,‡‡ 100 to 120 mesh.

2) Liquid phases—OV-1, OV-210, 1.5% OV-17 (SP 2250) + 1.95% QF-1 (SP 2401), and 6% QF-1 + 4% SE-30, or equivalent.

j. Carrier gas: One of the following is required:

1) *Nitrogen gas,* purified grade, moisture- and oxygen-free.

2) *Argon-methane* (95 + 5%) for use in pulse mode.

k. Pesticide reference standards: Obtain purest standards available (95 to 98%) from gas chromatographic and chemical supply houses.

l. Stock pesticide solutions: Dissolve 100 mg of each pesticide in ethyl acetate and dilute to 100 mL in a volumetric flask; 1.00 mL = 1.00 mg.

m. Intermediate pesticide solutions: Dilute 1.0 mL stock solution to 100 mL with ethyl acetate; 1.0 mL = 10 µg.

n. Working standard solutions for gas chromatography: Prepare final concentration of standards in hexane solution as required by detector sensitivity and linearity.

4. Procedure

a. Preparation of chromatograph:

1) Packing the column—Use a column constructed of silanized borosilicate glass because other tubing materials may catalyze sample component decomposition. Before packing, rinse and dry column tubing with solvent, e.g., methylene chloride, then meth-

anol. Pack column to a uniform density not so compact as to cause unnecessary back pressure and not so loose as to create voids during use. Do not crush packing. Fill column through a funnel connected by flexible tubing to one end. Plug other end of column with about 1.3 cm silanized glass wool and fill with aid of *gentle* vibration or tapping but do not use an electric vibrator because it tends to fracture packing. Optionally, apply a vacuum to plugged end. Plug open end with silanized glass wool.

2) Conditioning—Proper thermal and pesticide conditioning are essential to eliminate column bleed and to provide acceptable gas chromatographic analysis. The following procedure provides excellent results: Connect packed column to the injection port. *Do not* connect column to detector; however, maintain gas flow through detector by using the purge-gas line, or in dual-column ovens, by connecting an unpacked column to the detector. Adjust carrier-gas flow to about 50 mL/min and slowly (over a 1-h period) raise oven temperature to 230°C. After 24 to 48 h at this temperature the column is ready for pesticide conditioning.

Adjust oven temperature and carrier-gas flow rate to approximate operating levels. Make six consecutive 10-µL injections of a concentrated pesticide mixture through column at about 15-min intervals. Prepare this injection mixture from lindane (γ-BHC), heptachlor, aldrin, heptachlor epoxide, dieldrin, endrin, and p,p'-DDT, each compound at a concentration of 200 ng/µL. After pesticide conditioning, connect column to detector and let equilibrate for at least 1 h, preferably overnight. Column is then ready for use.

3) Injection technique

a) Develop an injection technique with constant rhythm and timing. The "solvent flush" technique described below has been used successfully and is recommended to prevent sample blowback or distillation within the syringe needle. Flush syringe with solvent, then draw a small volume of clean solvent into syringe barrel (e.g., 1 µL in a 10-µL syringe). Remove needle from solvent and draw 1 µL of air into barrel. For packed columns, draw 3 to 4 µL of sample extract into barrel. Remove needle from sample extract and draw approximately 1 µL air into barrel. Record volume of sample extract between air pockets. Rapidly insert needle through inlet septum, depress plunger, withdraw syringe. After each injection thoroughly clean syringe by rinsing several times with solvent.

b) Inject standard solutions of such concentration that the injection volume and peak height of the standard are approximately the same as those of the sample.

b. Treatment of samples:

1) Sample collection—Fill sample bottle to neck. Collect samples in duplicate.

2) Extraction of samples—Shake sample well and accurately measure all the sample in a 1-L graduated cylinder in two measuring operations if necessary (or use a precalibrated sample bottle to avoid transfer operation). Pour sample into a 2-L separatory funnel. Rinse sample bottle and cylinder with 60 mL 15% diethyl ether or methylene chloride in hexane, pour this solvent into separatory funnel, and shake vigorously for 2 min. Let phases separate for at least 10 min.

Drain water phase from separatory funnel into sample bottle and carefully pour organic phase through a 2-cm-OD column containing 8 to 10 cm of Na_2SO_4 into a Kuderna-Danish apparatus fitted with a 10-mL concentrator tube. Pour sample back into separatory funnel.

Gas chromatographic methods are extremely sensitive to the materials used. Mention of trade names by *Standard Methods* does not preclude the use of other existing or as-yet-undeveloped products that give *demonstrably* equal results.

** Use E. M. Quant™, MCB Manufacturing Chemists, Inc., 2909 Highland Ave., Cincinnati, OH, or equivalent.

†† Florisil™ or equivalent.

‡‡ Gas-Chrom Q™, Supelcoport, or equivalent.

Rinse sample bottle with 60 mL mixed solvent, use solvent to repeat sample extraction, and pass organic phase through Na_2SO_4. Complete a third extraction with 60 mL of mixed solvent that was used to rinse sample bottle again, and pass organic phase through Na_2SO_4. Wash Na_2SO_4 with several portions of hexane and drain well. Fit Kuderna-Danish apparatus with a three-ball Snyder column and reduce volume to about 7 mL in a hot water bath (90 to 95°C). At this point all methylene chloride present in the initial extracting solvent has been distilled off. Cool, remove concentrator tube from Kuderna-Danish apparatus, rinse ground-glass joint, and dilute to 10 mL with hexane. Make initial gas chromatographic analysis at this dilution.

3) Gas chromatography—Inject 3 to 4 μL of extract solution into a packed column. Always inject the same volume. Inspect resulting chromatogram for peaks corresponding to pesticides of concern and for presence of interferences.

a) If there are presumptive pesticide peaks and no significant interference, rechromatograph the extract solution on an alternate column.

b) Inject standards frequently to insure optimum operating conditions. If necessary, concentrate or dilute (*do not use methylene chloride*) the extract so that peak size of pesticide is very close to that of corresponding peaks in standard. (See dilution factor, ¶ 5a).

c) If significant interference is present, separate interfering substances from pesticide materials by using the cleanup procedure described in the following paragraph.

4) Magnesia-silica gel cleanup—Adjust sample extract volume to 10 mL with hexane. Place a charge of activated magnesia-silica gel§§ (weight determined by lauric-acid value, see Appendix) in a chromatographic column. After settling gel by tapping column, add about 1.3 cm anhydrous granular Na_2SO_4 to the top. Preelute column, after cooling, with 50 to 60 mL petroleum ether. Discard eluate and just before exposing sulfate layer to air, quantitatively transfer sample extract into column by careful decantation and with subsequent petroleum ether washings (5 mL maximum). Adjust elution rate to about 5 mL/min and, separately, collect the eluates in 500-mL Kuderna-Danish flasks equipped with 10-mL receivers.

Make first elution with 200 mL 6% ethyl ether in petroleum ether, and the second with 200 mL 15% ethyl ether in petroleum ether. Make third elution with 200 mL 50% ethyl ether-petroleum ether and the fourth with 200 mL 100% ethyl ether. Follow with 50 to 100 mL petroleum ether to insure removal of all ethyl ether from the column. Alternatively, to separate PCBs elute initially with 0% ethyl ether in petroleum ether and proceed as above to yield four fractions.

Concentrate eluates in Kuderna-Danish evaporator in a hot water bath as in ¶ 4b2) preceding, dilute to appropriate volume, and analyze by gas chromatography.

Eluate composition—By use of an equivalent quantity of any batch of magnesia-silica gel as determined by its lauric acid value (see Appendix) the pesticides will be separated into the eluates indicated below:

6% Ethyl Ether Eluate

Aldrin	Heptachlor	Pentachloro-
BHC	Heptachlor epoxide	nitrobenzene
Chlordane	Lindane (γ-BHC)	Strobane

§§ Florisil™ or equivalent.

DDD	Methoxychlor	Toxaphene
DDE	Mirex	Trifluralin
DDT		PCBs

15% Ethyl Ether Eluate	*50% Ethyl Ether Eluate*
Endosulfan I	Endosulfan II
Endrin	Captan
Dieldrin	
Dichloran	
Phthalate esters	

If present, certain thiophosphate pesticides will occur in each of the above fractions as well as in the 100% ether fraction. For additional information regarding eluate composition and the procedure for determining the lauric acid value, refer to the FDA Pesticide Analytical Manual (see Bibliography). For elution pattern test procedure see Appendix, Section 4.

5) Determination of extraction efficiency—Add known amounts (at concentrations similar to those expected in samples) of pesticides in ethyl acetate solution to 1 L water sample and carry through the same procedure as for samples. Dilute an equal amount of intermediate pesticide solution (¶ 3m above) to the same final volume. Call peak height from standard "a" and peak height from sample to which pesticide was added "b," whereupon the extraction efficiency equals b/a. Periodically determine extraction efficiency and a control blank to test the procedure.

TABLE 6630:I. RETENTION RATIOS OF VARIOUS ORGANOCHLORINE PESTICIDES RELATIVE TO ALDRIN

Liquid phase*	1.5% OV-17 + 1.95% QF-1	5% OV-210	3% OV-1	6% QF-1 + 4% SE-30
Column Temperature	200°C	180°C	180°C	200°C
Argon/methane carrier flow	60 mL/min	70 mL/min	70 mL/min	60 mL/min
Pesticide	RR	RR	RR	RR
α-BHC	0.54	0.64	0.35	0.49
PCNB	0.68	0.85	0.49	0.63
Lindane (γ-BHC)	0.69	0.81	0.44	0.60
Dichloran	0.77	1.29	0.49	0.70
Heptachlor	0.82	0.87	0.78	0.83
Aldrin	1.00	1.00	1.00	1.00
Heptachlor epoxide	1.54	1.93	1.28	1.43
Endosulfan I	1.95	2.48	1.62	1.79
p,p'-DDE	2.23	2.10	2.00	1.82
Dieldrin	2.40	3.00	1.93	2.12
Captan	2.59	4.09	1.22	1.94
Endrin	2.93	3.56	2.18	2.42
o,p'-DDT	3.16	2.70	2.69	2.39
p,p'-DDD	3.48	3.75	2.61	2.55
Endosulfan II	3.59	4.59	2.25	2.72
p,p'-DDT	4.18	4.07	3.50	3.12
Mirex	6.1	3.78	6.6	4.79
Methoxychlor	7.6	6.5	5.7	4.60
Aldrin (Min absolute)	3.5	2.6	4.0	5.6

* All columns glass, 180 cm × 4 mm ID, solid support Gas-Chrom Q (100/200 mesh).

TABLE 6630:II. PRECISION AND BIAS DATA FOR SELECTED ORGANOCHLORINE PESTICIDES

Pesticide	Level Added ng/L	Pretreatment	Mean Recovery ng/L	Recovery %	Precision* ng/L S_T	S_o
Aldrin	15	No cleanup	10.42	69	4.86	2.59
	110		79.00	72	32.06	20.19
	25	Cleanup†	17.00	68	9.13	3.48‡
	100		64.54	65	27.16	8.02‡
Lindane (γ-BHC)	10	No cleanup	9.67	97	5.28	3.47
	100		72.91	73	26.23	11.49‡
	15	Cleanup†	14.04	94	8.73	5.20
	85		59.08	70	27.49	7.75‡
Dieldrin	20	No cleanup	21.54	108	18.16	17.92
	125		105.83	85	30.41	21.84
	25	Cleanup	17.52	70	10.44	5.10‡
	130		84.29	65	34.45	16.79‡
DDT	40	No cleanup	40.30	101	15.96	13.42
	200		154.87	77	38.80	24.02
	30	Cleanup†	35.54	118	22.62	22.50
	185		132.08	71	49.83	25.31

* S_T = overall precision and S_o = single-operator precision.
† Use of magnesia-silica gel column cleanup before analysis.
‡ $S_o < S_T/2$.

Also analyze one set of duplicates with each series of samples as a quality-control check.

5. Calculation

a. Dilution factor: If a portion of the extract solution was concentrated, the dilution factor, D, is a decimal; if it was diluted, the dilution factor exceeds 1.

b. Determine pesticide concentrations by direct comparison to a single standard when the injection volume and response are within 10% of those of the sample pesticide of interest (Table 6630:I). Calculate concentration of pesticide:

$$\mu g/L = \frac{A \times B \times C \times D}{E \times F \times G}$$

where:

A = ng standard pesticide,
B = peak height of sample, mm, or area count,
C = extract volume, μL,
D = dilution factor,
E = peak height of standard, mm, or area count,
F = volume of extract injected, μL, and
G = volume of sample extracted, mL.

Typical chromatograms of representative pesticide mixtures are shown in Figures 6630:1 through 6630:5.

Report results in micrograms per liter without correction for efficiency.

6. Precision and Bias

Ten laboratories in an interlaboratory study selected their own water samples and added four representative pesticides to replicate samples, at two concentrations in acetone. The added pesticides came from a single source. Samples were analyzed with and without magnesia-silica gel cleanup. Precision and recovery data are given in Table 6630:II.

7. Bibliography

MILL, P.A. 1968. Variation of Florisil activity: simple method for measuring absorbent capacity and its use in standardizing Florisil columns. *J. Assoc. Offic. Anal. Chem.* 51:29.

FOOD AND DRUG ADMINISTRATION. 1968 (revised 1978). Pesticide Analytical Manual, 2nd ed. U.S. Dep. Health, Education & Welfare, Washington, D.C.

MONSANTO CHEMICAL COMPANY. 1970. Monsanto Methodology for Arochlors—Analysis of Environmental Materials for Biphenyls, Analytical Chemistry Method 71-35. St. Louis, Mo.

U.S. ENVIRONMENTAL PROTECTION AGENCY. 1971. Method for Organic Pesticides in Water and Wastewater. National Environmental Research Center, Cincinnati, Ohio.

STEERE, N.V., ed. 1971. Handbook of Laboratory Safety. Chemical Rubber Company, Cleveland, Ohio.

GOERLITZ, D.F. & E. BROWN. 1972. Methods for analysis of organic substances in water. *In* Techniques of Water Resources Investigations of the United States Geological Survey, Book 5, Chapter A3, p. 24. U.S. Dep. Interior, Geological Survey, Washington, D.C.

U.S. ENVIRONMENTAL PROTECTION AGENCY. 1978. Method for Organochlorine Pesticides in Industrial Effluents. National Environmental Research Center, Cincinnati, Ohio.

U.S. ENVIRONMENTAL PROTECTION AGENCY. 1978. Method for Polychlorinated Biphenyls in Industrial Effluents. National Environmental Research Center, Cincinnati, Ohio.

U.S. ENVIRONMENTAL PROTECTION AGENCY. 1979. Handbook for Analytical Quality Control in Water and Wastewater Laboratories. National Environmental Research Center, Analytical Quality Control Laboratory, Cincinnati, Ohio.

U.S. ENVIRONMENTAL PROTECTION AGENCY. 1980. Analysis of Pesticide Residues in Human and Environmental Samples. Environmental Toxicology Div., Health Effects Laboratory, Research Triangle Park, N.C.

Appendix—Standardization of Magnesia-Silica Gel* Column by Weight Adjustment Based on Adsorption of Lauric Acid

A rapid method for determining adsorptive capacity of magnesia-silica gel is based on adsorption of lauric acid from hexane solution. An excess of lauric acid is used and the amount not adsorbed is measured by alkali titration. The weight of lauric acid adsorbed is used to calculate, by simple proportion, equivalent quantities of gel for batches having different adsorptive capacities.

1. Reagents

a. Ethyl alcohol, USP or absolute, neutralized to phenolphthalein.

b. Hexane, distilled from all-glass apparatus.

c. Lauric acid solution: Transfer 10.000 g lauric acid to a 500-mL volumetric flask, dissolve in hexane, and dilute to 500 mL; 1.00 mL = 20 mg.

d. Phenolphthalein indicator: Dissolve 1 g in alcohol and dilute to 100 mL.

e. Sodium hydroxide, 0.05N: Dilute 25 mL 1N NaOH to 500 mL with distilled water. Standardize as follows: Weigh 100 to 200 mg lauric acid into 125-mL erlenmeyer flask; add 50 mL neutralized ethyl alcohol and 3 drops phenolphthalein indicator; titrate to permanent end point; and calculate milligrams lauric acid per milliliter NaOH (about 10 mg/mL).

2. Procedure

Transfer 2.000 g magnesia-silica gel to a 25-mL glass-stoppered erlenmeyer flask. Cover loosely with aluminum foil and

_____
* Florisil™ or equivalent.

heat overnight at 130°C. Stopper, cool to room temperature, add 20.0 mL lauric acid solution (400 mg), stopper, and shake occasionally during 15 min. Let adsorbent settle and pipet 10.0 mL supernatant into a 125-mL erlenmeyer flask. Avoid including any gel. Add 50 mL neutral alcohol and 3 drops phenolphthalein indicator solution; titrate with 0.05N NaOH to a permanent end point.

3. Calculation of Lauric Acid Value and Adjustment of Column Weight

Calculate amount of lauric acid adsorbed on gel as follows:

$$\text{Lauric acid value} = \text{mg lauric acid/g gel} = 200 - (\text{mL required for titration} \times \text{mg lauric acid/mL } 0.05N \text{ NaOH}).$$

To obtain an equivalent quantity of any batch of gel, divide 110 by lauric acid value for that batch and multiply by 20 g. Verify proper elution of pesticides by the procedure given below.

4. Test for Proper Elution Pattern and Recovery of Pesticides

Prepare a test mixture containing aldrin, heptachlor epoxide, *p,p'*-DDE, dieldrin, parathion, and malathion. Dieldrin and parathion should elute in the 15% eluate; all but a trace of malathion in the 50% eluate, and the others in the 6% eluate.

6630 C.   Liquid-Liquid Extraction Gas Chromatographic Method II

This method[1] is applicable to the determination of organochlorine pesticides and PCBs*† in municipal and industrial discharges. When analyzing unfamiliar samples for any or all of these compounds, support the identifications by at least one additional qualitative technique. This method includes analytical conditions for a second, confirmatory gas chromatographic column. Alternatively, analyze by a gas chromatographic/mass spectrometric (GC/MS) method for base/neutrals and acids using the extract produced by this method.

Additional PCB congeners can be determined if standards are included.

1. General Discussion

a. Principle: A measured volume of sample is extracted with methylene chloride. The extract is dried and exchanged to hexane

during concentration. If other determinations having essentially the same extraction and concentration steps are to be performed, a single sample extraction is sufficient. The extract is separated by gas chromatography and the compounds are measured with an electron capture detector.[2] See Section 6010C for discussion of gas chromatographic principles.

The method provides procedures for magnesia-silica gel column cleanup and elemental sulfur removal to aid in the elimination of interferences. When cleanup is required, sample concentration levels must be high enough to permit separate treatment of subsamples. Chromatographic conditions appropriate for the simultaneous measurement of combinations of compounds may be selected.

b. Interferences: See Section 6410B.1*b*1) for precautions concerning glassware, reagent purity, and matrix interferences.

Phthalate esters may interfere in pesticide analysis with an electron capture detector. These compounds generally appear in the chromatogram as large, late-eluting peaks, especially in the 15 and 50% fractions from magnesia-silica gel. Common flexible plastics contain phthalates that are easily extracted during laboratory operations. Cross-contamination of clean glassware routinely occurs when plastics are handled during extraction steps,

_____
* Aldrin, α-BHC, β-BHC, δ-BHC, γ-BHC, chlordane, 4,4′-DDD, 4,4′-DDE, 4,4′-DDT, dieldrin, endosulfan I, endosulfan II, endosulfan sulfate, endrin, endrin aldehyde, heptachlor, heptachlor epoxide, toxaphene, PCB-1016, PCB-1221, PCB-1232, PCB-1242, PCB-1248, PCB-1254, PCB-1260.
† The PCBs constitute a class of 209 compounds. This procedure is designed to determine nine commercial formulations known as the Aroclors, each of which is a mixture of PCBs.

especially when solvent-wetted surfaces are handled. Minimize interferences from phthalates by avoiding use of plastics. Exhaustive cleanup of reagents and glassware may be required to eliminate phthalate contamination.[3,4] Phthalate ester interference can be avoided by using a microcoulometric or electrolytic conductivity detector.

c. Detection levels: The MDL is the minimum concentration of a substance that can be measured and reported with 99% confidence that the value is above zero.[5] The MDL concentrations listed in Table 6630:III were obtained by using reagent water.[6] Similar results were achieved with representative wastewaters. The MDL actually obtained in a given analysis will vary, depending on instrument sensitivity and matrix effects. This method has been tested for linearity of known-addition recovery from reagent water and is applicable over the concentration range from $4 \times MDL$ to $1000 \times MDL$ with the following exceptions: Chlordane recovery at $4 \times MDL$ was low (60%); toxaphene recovery was linear over the range of $10 \times MDL$ to $1000 \times MDL$.[6] It is difficult to determine MDLs for mixtures such as these. To calculate the MDLs given, a few of the GC peaks in each mixture were used. Depending on the particular peaks selected, these results may or may not be reproducible in other laboratories.

TABLE 6630:III. CHROMATOGRAPHIC CONDITIONS AND METHOD DETECTION LIMITS*

Compound	Retention Time min		Method Detection Limit μg/L
	Column 1	Column 2	
α-BHC	1.35	1.82	0.003
γ-BHC	1.70	2.13	nd
β-BHC	1.90	1.97	nd
Heptachlor	2.00	3.35	0.003
δ-BHC	2.15	2.20	0.009
Aldrin	2.40	4.10	0.004
Heptachlor epoxide	3.50	5.00	0.083
Endosulfan I	4.50	6.20	0.014
4,4'-DDE	5.13	7.15	0.004
Dieldrin	5.45	7.23	0.002
Endrin	6.55	8.10	0.006
4,4'-DDD	7.83	9.08	0.011
Endosulfan II	8.00	8.28	0.004
4,4'-DDT	9.40	11.75	0.012
Endrin aldehyde	11.82	9.30	0.023
Endosulfan sulfate	14.22	10.70	0.066
Chlordane	mr	mr	0.014
Toxaphene	mr	mr	0.24
PCB-1016	mr	mr	nd
PCB-1221	mr	mr	nd
PCB-1232	mr	mr	nd
PCB-1242	mr	mr	0.065
PCB-1248	mr	mr	nd
PCB-1254	mr	mr	nd
PCB-1260	mr	mr	nd

Column 1 conditions: Supelcoport (100/120 mesh) coated with 1.5% SP-2250/1.95% SP-2401 packed in a 1.8 m long × 4 mm ID glass column with 5% methane/95% argon carrier gas at 60 mL/min flow rate. Column temperature held isothermal at 200°C, except for PCB-1016 through PCB-1248 at 160°C.

Column 2 conditions: Supelcoport (100/120 mesh) coated with 3% OV-1 packed in a 1.8 m long × 4 mm ID glass column with 5% methane/95% argon carrier gas at 60 mL/min flow rate. Column temperature held isothermal at 200°C for the pesticides; at 140°C for PCB-1221 and 1232; and at 170°C for PCB-1016 and 1242 to 1268.

*mr = multiple peak response. See Figures 6630:2 through 10.

nd = not determined.

d. Safety: The toxicity or carcinogenicity of each reagent has not been defined precisely. The following compounds have been classified tentatively as known or suspected, human or mammalian carcinogens: 4,4'-DDT, 4,4'-DDD, the BHCs, and the PCBs. Prepare primary standards of these compounds in a hood and wear a NIOSH/MESA-approved toxic gas respirator when handling high concentrations. Treat and dispose of Hg used for sulfur removal as a hazardous waste.

2. Sampling and Storage

For collection and storage requirements, see Section 6410B.2. If samples will not be extracted within 72 h of collection, adjust pH to the range 5.0 to 9.0 with NaOH or H_2SO_4. Record volume of acid or base used. If aldrin is to be determined, add sodium thiosulfate when residual chlorine is present.

3. Apparatus

Use apparatus specified in Section 6410B.3*a–e*, *g*, and *i–k*. In addition:

a. Chromatographic column, 400 mm long × 22 mm ID, with TFE stopcock and coarse frit filter disk.‡

b. Gas chromatograph:§ An analytical system complete with gas chromatograph suitable for on-column injection and all required accessories including syringes, analytical columns, gases, and strip-chart recorder. Preferably use a data system for measuring peak areas.

1) *Column 1,* 1.8 m long × 4 mm ID, glass, packed with 1.5% SP-2250/1.95% SP-2401 on Supelcoport (100/120 mesh) or equivalent. This column was used to develop the detection limit and precision and bias data presented herein. For guidelines for the use of alternate column packings see ¶ 5*c.*

Although procedures detailed below refer primarily to packed columns, capillary columns may be used if equivalent results can be demonstrated.

2) *Column 2,* 1.8 m long × 4 mm ID, glass, packed with 3% OV-1 on Supelcoport (100/120 mesh) or equivalent.

3) *Detector,* electron-capture. This detector was used to develop the detection limit and precision and bias data presented herein. For use of alternate detectors see ¶ 5*c.*

4. Reagents

This method requires reagents described in Section 6410B.4*a–e*, and in addition:

a. Acetone, hexane, isooctane, methylene chloride, pesticide quality or equivalent.

b. Ethyl ether, nanograde redistilled in glass if necessary. Demonstrate before use freedom from peroxides by means of test strips.‖ Remove peroxides by procedures provided with the test strips. After cleanup, add 20 mL ethyl alcohol preservative per liter of ether.

‡ Kontes K-42054 or equivalent.

§ Gas chromatographic methods are extremely sensitive to the materials used. Mention of trade names by *Standard Methods* does not preclude the use of other existing or as-yet-undeveloped products that give *demonstrably* equivalent results.

‖ E. Merck, EM Science Quant or equivalent.

c. Magnesia-silica gel,# 60/100 mesh. Purchase activated at 1250°F and store in the dark in glass containers with ground-glass stoppers or foil-lined screw caps. Before use, activate each batch for at least 16 h at 130°C in a foil-covered glass container; let cool.

d. Mercury, triple-distilled.

e. Copper powder, activated.

f. Stock standard solutions: Prepare as directed in Section 6410B.4*g,* using isooctane as the solvent.

g. Calibration standards: See Section 6420B.4*j.* Dilute with isooctane and use MDL values from Table 6630:III.

h. Quality control (QC) check sample concentrate: Obtain a check sample concentrate** containing each compound at the following concentrations in acetone: 4,4′-DDD, 10 μg /mL; 4,4′-DDT, 10 μg /mL; endosulfan II, 10 μg /mL; endosulfan sulfate, 10 μg/mL; endrin, 10 μg/mL; any other single-component pesticide, 2 μg/mL. If this method will be used only to analyze for PCBs, chlordane, or toxaphene, the QC check sample concentrate should contain the most representative multicomponent compound at a concentration of 50 μg/mL in acetone. If such a sample is not available from an external source, prepare using stock standards prepared independently from those used for calibration.

5. Procedure

a. Extraction: Mark water meniscus on side of sample bottle for later determination of volume. Pour entire sample into a 2-L separatory funnel and extract with methylene chloride as directed in Section 6410B.5*a*1), without any pH adjustment or solvent wash.

After extracting and concentrating with a three-ball Snyder column, increase temperature of hot water bath to about 80°C. Momentarily remove Snyder column, add 50 mL hexane and a new boiling chip, and reattach Snyder column. Concentrate extract as before but use hexane to prewet column. Complete concentration in 5 to 10 min.

Remove Snyder column and rinse flask and its lower joint into the concentrator tube with 1 to 2 mL hexane. Preferably use a 5-mL syringe for this operation. Stopper concentrator tube and store refrigerated if further processing will not be done immediately. If extract is to be stored longer than 2 d, transfer to a TFE-sealed screw-cap vial. If extract requires no further cleanup, proceed with gas chromatographic analysis. If further cleanup is required, follow procedure of ¶ *b* before chromatographic analysis.

Determine original sample volume by refilling sample bottle to mark and transferring liquid to a 1000-mL graduated cylinder. Record sample volume to nearest 5 mL.

b. Cleanup and separation: Use either procedure below or any other appropriate procedure; however, first demonstrate that the requirements of ¶ 7 can be met. The magnesia-silica gel column allows for a select fractionation of compounds and eliminates polar interferences. Elemental sulfur, which interferes with the electron-capture gas chromatography of certain pesticides, can be removed by the technique described below.

1) *Magnesia-silica gel column cleanup*—Place a weight of magnesia-silica gel (nominally 20 g) predetermined by calibration, ¶ *d*3), into a chromatographic column. Tap column to settle

gel and add 1 to 2 cm anhydrous Na₂SO₄ to the top. Add 60 mL hexane to wet and rinse. Just before exposure of the Na₂SO₄ layer to air, stop elution of hexane by closing stopcock on column. Discard eluate. Adjust sample extract volume to 10 mL with hexane and transfer it from K-D concentrator tube onto column. Rinse tube twice with 1 to 2 mL hexane, adding each rinse to the column. Place a 500-mL K-D flask and clean concentrator tube under chromatographic column. Drain column into flask until Na₂SO₄ layer is nearly exposed. Elute column with 200 mL 6% ethyl ether in hexane (v/v) (Fraction 1) at a rate of about 5 mL /min. Remove K-D flask and set aside. Elute column again, using 200 mL 15% ethyl ether in hexane (v/v) (Fraction 2), into a second K-D flask. Elute a third time using 200 mL 50% ethyl ether in hexane (v/v) (Fraction 3). The elution patterns for the pesticides and PCBs are shown in Table 6630:IV. Concentrate fractions for 15 to 20 min as in ¶ *a,* using hexane to prewet the column, and set water bath temperature at about 85°C. After cooling, remove Snyder column and rinse flask and its lower joint into concentrator tube with hexane. Adjust volume of each fraction to 10 mL with hexane and analyze by gas chromatography, ¶s *c* through *e* below.

2) *Sulfur interference removal*—Elemental sulfur usually will elute entirely in Fraction 1 of the magnesia-silica gel column cleanup. To remove sulfur interference from this fraction or the original extract, pipet 1.00 mL concentrated extract into a clean

TABLE 6630:IV. DISTRIBUTION OF CHLORINATED PESTICIDES AND PCBs INTO MAGNESIA-SILICA GEL COLUMN FRACTIONS[5]

Compound	Recovery by Fraction* %		
	1	2	3
Aldrin	100	—	—
α-BHC	100	—	—
β-BHC	97	—	—
δ-BHC	98	—	—
γ-BHC	100	—	—
Chlordane	100	—	—
4,4′-DDD	99	—	—
4,4′-DDE	98	—	—
4,4′-DDT	100	—	—
Dieldrin	0	100	—
Endosulfan I	37	64	—
Endosulfan II	0	7	91
Endosulfan sulfate	0	0	106
Endrin	4	96	—
Endrin aldehyde	0	68	26
Heptachlor	100	—	—
Heptachlor epoxide	100	—	—
Toxaphene	96	—	—
PCB-1016	97	—	—
PCB-1221	97	—	—
PCB-1232	95	4	—
PCB-1242	97	—	—
PCB-1248	103	—	—
PCB-1254	90	—	—
PCB-1260	95	—	—

* Eluent composition:
    Fraction 1–6% ethyl ether in hexane
    Fraction 2–15% ethyl ether in hexane
    Fraction 3–50% ethyl ether in hexane

#Florisil or equivalent.
** For U.S. federal permit-related analyses, use samples obtainable from U.S. EPA Environmental Monitoring and Support Laboratory, Cincinnati, Ohio.

concentrator tube or TFE-sealed vial. Add 1 to 3 drops of mercury and seal.[7] Mix for 15 to 30 s. If prolonged shaking (2 h) is required, use a reciprocal shaker. Alternatively, use activated copper powder for sulfur removal.[8] Analyze by gas chromatography.

c. Gas chromatography operating conditions: Table 6630:III summarizes the recommended operating conditions for the gas chromatograph and gives retention times and MDLs that can be achieved under these conditions. Examples of separations obtained with Column 1 are shown in Figures 6630:6 to 15. Other packed or capillary (open-tubular) columns,[9] chromatographic conditions, or detectors may be used if the requirements of ¶ 7 are met.

d. Calibration: Calibrate system daily by either external or internal procedure. NOTE: For quantification and identification of mixtures such as PCBs, chlordane, and toxaphene, take extra precautions.[9-11]

1) *External standard calibration procedure*—Prepare standards as directed in ¶ 4*g* and follow procedure of Section 6420B.5*b*3). Tabulate data and obtain calibration curve or calibration factor as directed in Section 6200B.4*c*3).

2) *Internal standard calibration procedure*—Prepare standards as directed in ¶ 4*g* and follow procedure of Section 6420B.5*b*3).

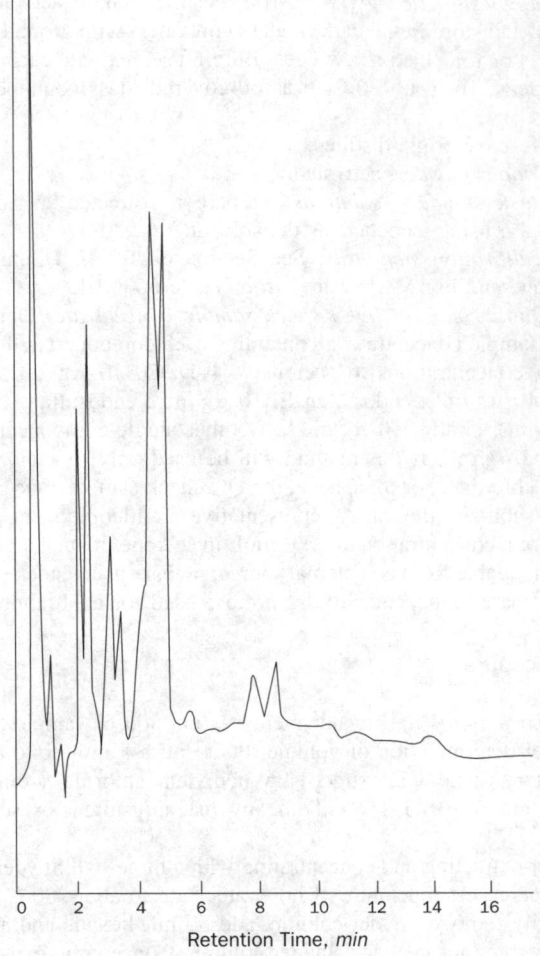

Figure 6630:7. Gas chromatogram of chlordane. Column: 1.5% SP-2250/ 1.95% SP-2401 on Supelcoport; temperature: 200°C; detector: electron capture.

Tabulate data and calculate response factors as directed in Section 6200B.4*c*2).

Verify working calibration curve, calibration factor, or RF on each working day by measuring one or more calibration standards. If the response for any compound varies from the predicted response by more than ± 15%, repeat test using a fresh calibration standard. Alternatively, prepare a new calibration curve for that compound.

3) *Magnesia-silica gel standardization*—Gel from different batches or sources may vary in adsorptive capacity. To standardize the amount used, use the lauric acid value[12], which measures the adsorption from a hexane solution of lauric acid (mg/g gel). Determine the amount to be used for each column by dividing 110 by this ratio and multiplying the quotient by 20 g.

Before using any cleanup procedure, process a series of calibration standards through the procedure to validate elution patterns and the absence of interferences from the reagents.

e. Sample analysis: See Section 6420B.5*b*3). If peak response cannot be measured because of interferences, further cleanup is required.

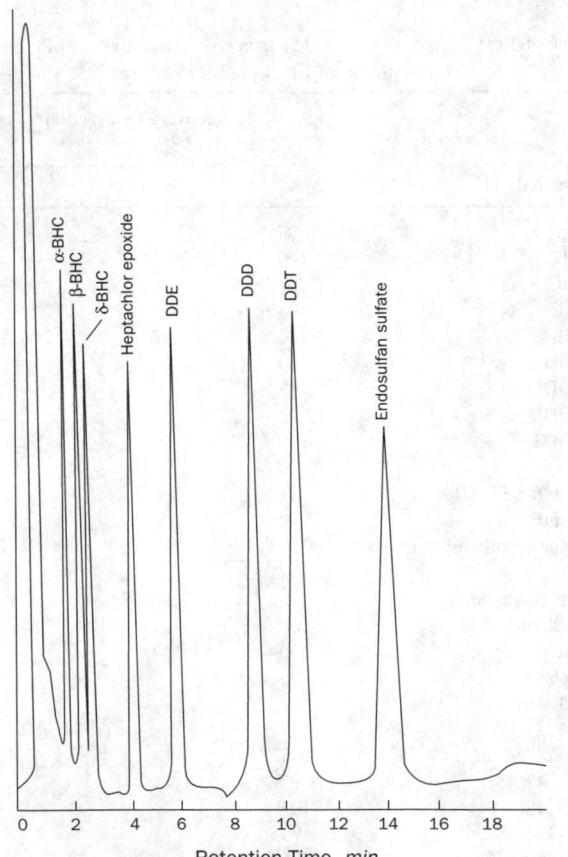

Figure 6630:6. Gas chromatogram of pesticides. Column: 1.5% SP-2250/ 1.95% SP-2401 on Supelcoport; temperature: 200°C; detector: electron capture.

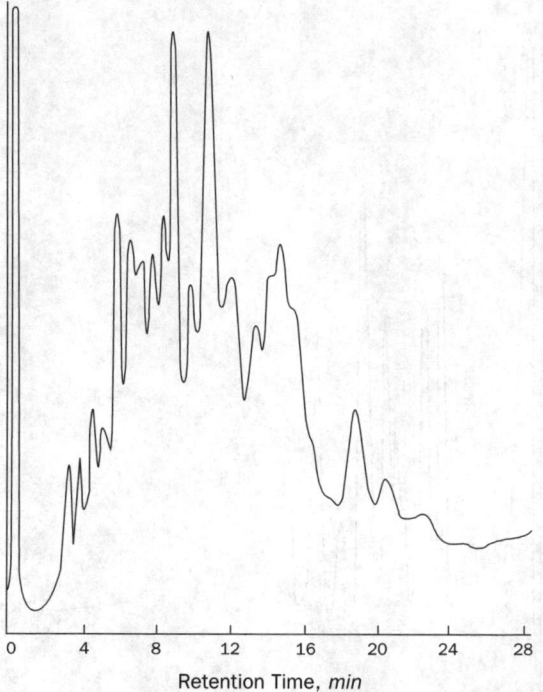

Figure 6630:8. Gas chromatogram of toxaphene. Column: 1.5% SP-2250/ 1.95% SP-2401 on Supelcoport; temperature: 200°C; detector: electron capture.

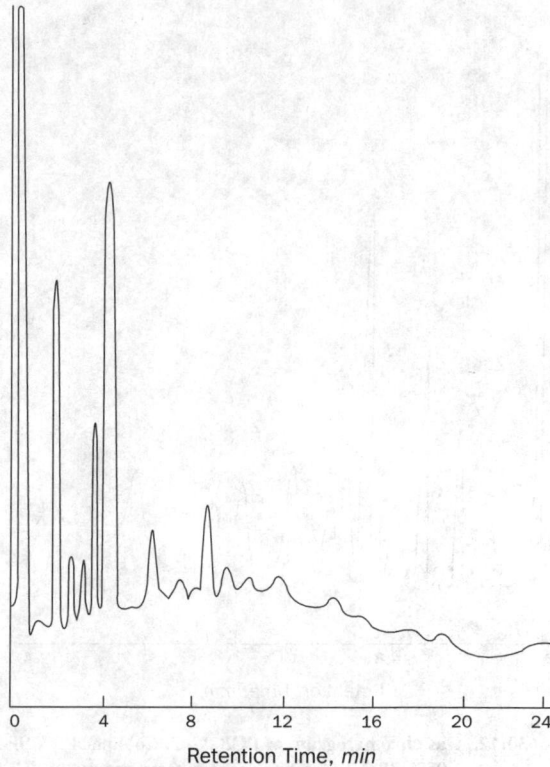

Figure 6630:10. Gas chromatogram of PCB-1221. Column: 1.5% SP-2250/ 1.95% SP-2401 on Supelcoport; temperature: 160°C; detector: electron capture.

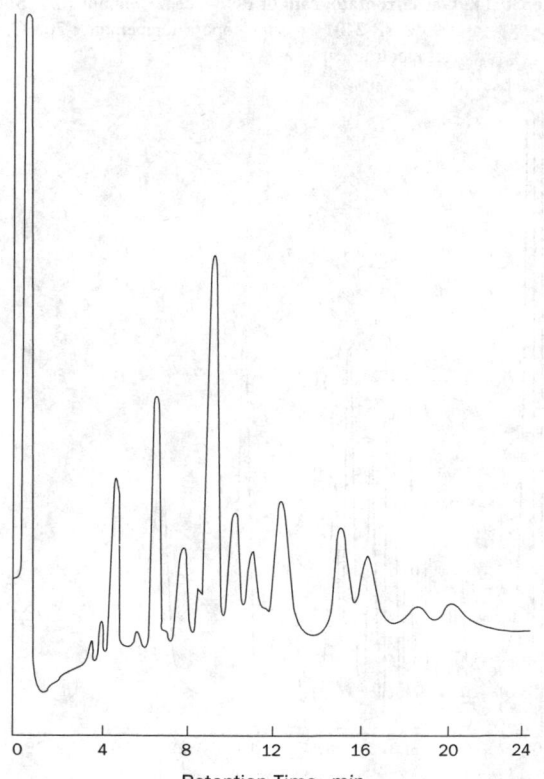

Figure 6630:9. Gas chromatogram of PCB-1016. Column: 1.5% SP-2250/ 1.95% SP-2401 on Supelcoport; temperature: 160°C; detector: electron capture.

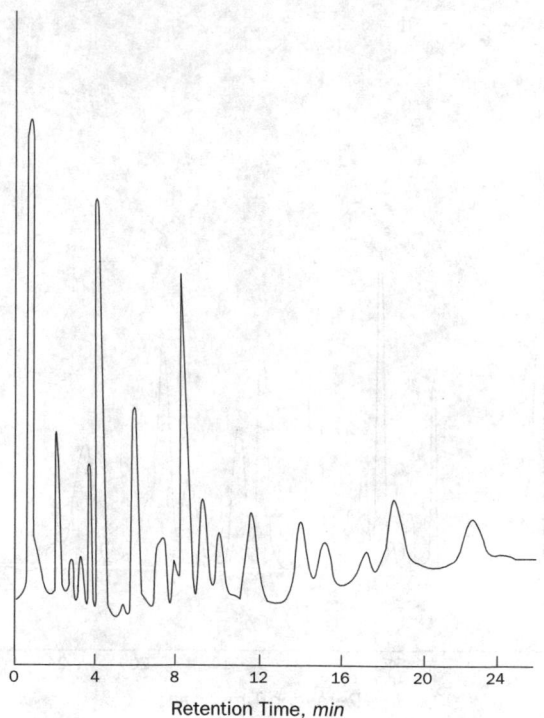

Figure 6630:11. Gas chromatogram of PCB-1232. Column: 1.5% SP-2250/ 1.95% SP-2401 on Supelcoport; temperature: 160°C; detector: electron capture.

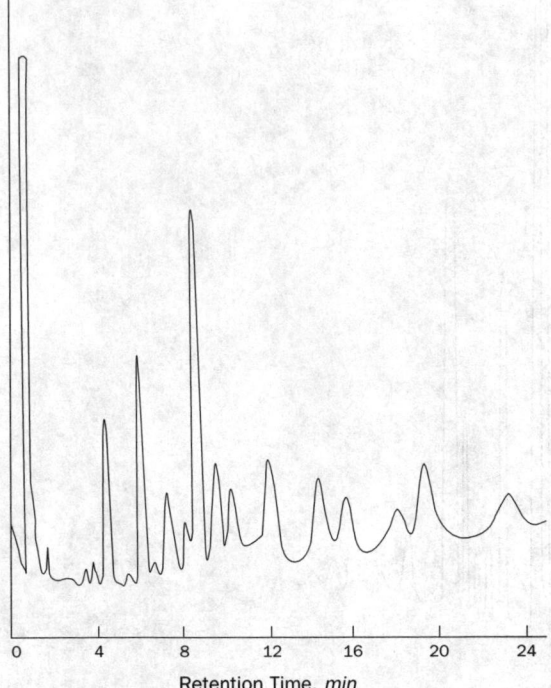

Figure 6630:12. Gas chromatogram of PCB-1242. Column: 1.5% SP-2250/
1.95% SP-2401 on Supelcoport; temperature: 160°C; detector: electron capture.

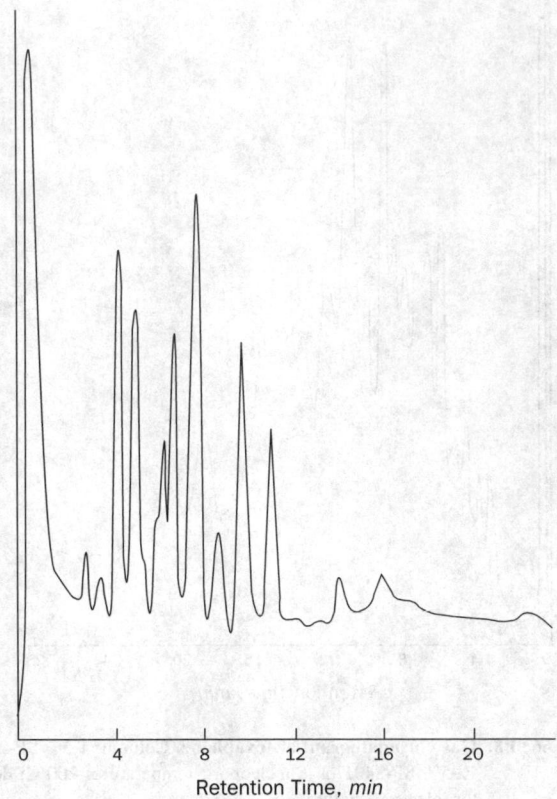

Figure 6630:14. Gas chromatogram of PCB-1254. Column: 1.5% SP-2550/
1.95% SP-2401 on Supelcoport; temperature: 200°C; detector: electron capture.

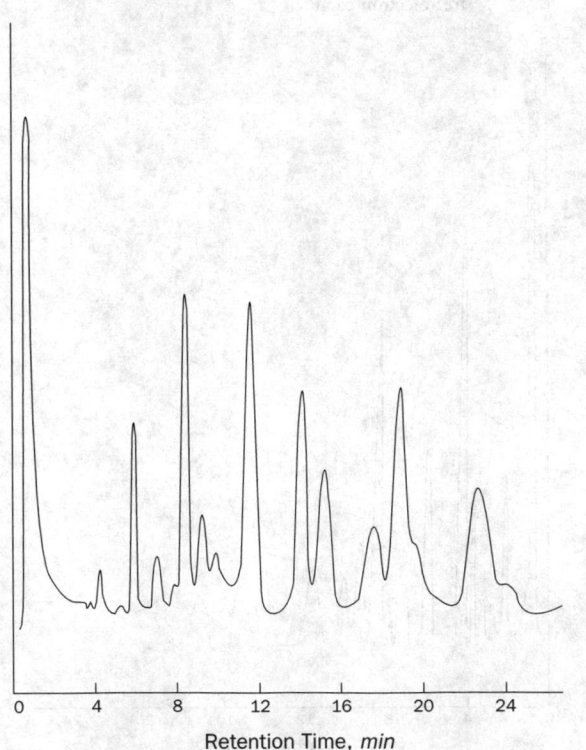

Figure 6630:13. Gas chromatogram of PCB-1248. Column: 1.5% SP-2250/
1.95% SP-2401 on Supelcoport; temperature: 160°C; detector: electron capture.

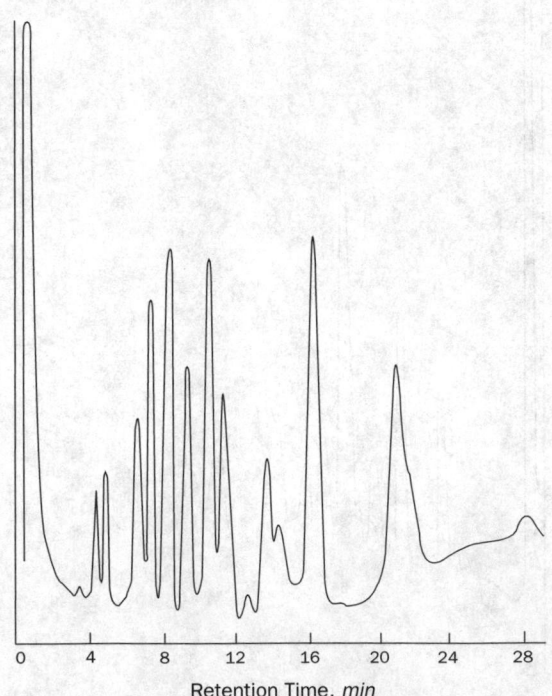

Figure 6630:15. Gas chromatogram of PCB-1260. Column: 1.5% SP-2250/
1.95% SP-2401 on Supelcoport; temperature: 200°C; detector: electron capture.

TABLE 6630:V. QC ACCEPTANCE CRITERIA*

Compound	Test Conc. μg/L	Limit for s μg/L	Range for $\overline{X}$ μg/L	Range for $\overline{P}$, P_s %
Aldrin	2.0	0.42	1.08–2.24	42–122
α-BHC	2.0	0.48	0.98–2.44	37–134
β-BHC	2.0	0.64	0.78–2.60	17–147
δ-BHC	2.0	0.72	1.01–2.37	19–140
γ-BHC	2.0	0.46	0.86–2.32	32–127
Chlordane	50	10.0	27.6–54.3	45–119
4,4'-DDD	10	2.8	4.8–12.6	31–141
4,4'-DDE	2.0	0.55	1.08–2.60	30–145
4,4'-DDT	10	3.6	4.6–13.7	25–160
Dieldrin	2.0	0.76	1.15–2.49	36–146
Endosulfan I	2.0	0.49	1.14–2.82	45–153
Endosulfan II	10	6.1	2.2–17.1	D–202
Endosulfan sulfate	10	2.7	3.8–13.2	26–144
Endrin	10	3.7	5.1–12.6	30–147
Heptachlor	2.0	0.40	0.86–2.00	34–111
Heptachlor epoxide	2.0	0.41	1.13–2.63	37–142
Toxaphene	50	12.7	27.8–55.6	41–126
PCB-1016	50	10.0	30.5–51.5	50–114
PCB-1221	50	24.4	22.1–75.2	15–178
PCB-1232	50	17.9	14.0–98.5	10–215
PCB-1242	50	12.2	24.8–69.6	39–150
PCB-1248	50	15.9	29.0–70.2	38–158
PCB-1254	50	13.8	22.2–57.9	29–131
PCB-1260	50	10.4	18.7–54.9	8–127

* s = standard deviation of four recovery measurements,
    $\overline{X}$ = average recovery for four recovery measurements,
  P, P_s = percent recovery measured, and
    D = detected; result must be greater than zero.

NOTE: These criteria are based directly on the method performance data in Table 6630: VI. Where necessary, the limits for recovery were broadened to assure applicability of the limits to concentrations below those used to develop Table 6630:VI.

6. Calculation

Determine concentration of individual compounds using procedures given in Section 6420B.6a.

If it is apparent that two or more PCB (Aroclor) mixtures are present, the Webb and McCall procedure[13] may be used to identify and quantify the Aroclors, depending on the Aroclors present. Other techniques also are available.

For multicomponent mixtures (chlordane, toxaphene, and PCBs) match peak retention times in standards with peaks in sample. Quantitate every identifiable peak unless interference with individual peaks persists after cleanup. Add peak height or peak area of each identified peak in chromatogram. Calculate as total response in sample versus total response in standard. Environmental degradation of these compounds may make identification difficult. This method is suitable only for intact mixtures such as the original Aroclor or pesticide formulation and it is not suitable for the other altered mixtures that are sometimes found in the environment. In these instances the GC peak pattern would not match the standard.

Report results in micrograms per liter without correction for recovery data. Report QC data with the sample results.

7. Quality Control

a. *Quality control program:* See Section 6200A.5.

b. *Initial quality control:* To establish the ability to generate data with acceptable precision and bias, proceed as follows: Using a pipet, prepare QC check samples at test concentrations shown in Table 6630:V by adding 1.00 mL of QC check sample concentrate (¶ 4h) to each of four 1-L portions of reagent water.

TABLE 6630:VI. METHOD PRECISION AND BIAS AS FUNCTIONS OF CONCENTRATION*

Compound	Bias, as Recovery, X' μg/L	Single-Analyst Precision, s_r μg/L	Overall Precision, S' μg/L
Aldrin	0.81C + 0.04	0.16$\overline{X}$ − 0.04	0.20$\overline{X}$ − 0.01
α-BHC	0.84C + 0.03	0.13$\overline{X}$ + 0.04	0.23$\overline{X}$ − 0.00
β-BHC	0.81C + 0.07	0.22$\overline{X}$ − 0.02	0.33$\overline{X}$ − 0.05
δ-BHC	0.81C + 0.07	0.18$\overline{X}$ + 0.09	0.25$\overline{X}$ + 0.03
γ-BHC	0.82C − 0.05	0.12$\overline{X}$ + 0.06	0.22$\overline{X}$ + 0.04
Chlordane	0.82C − 0.04	0.13$\overline{X}$ + 0.13	0.18$\overline{X}$ + 0.18
4,4'-DDD	0.84C + 0.30	0.20$\overline{X}$ − 0.18	0.27$\overline{X}$ − 0.14
4,4'-DDE	0.85C + 0.14	0.13$\overline{X}$ + 0.06	0.28$\overline{X}$ − 0.09
4,4'-DDT	0.93C − 0.13	0.17$\overline{X}$ + 0.39	0.31$\overline{X}$ − 0.09
Dieldrin	0.90C + 0.02	0.12$\overline{X}$ + 0.19	0.16$\overline{X}$ + 0.16
Endosulfan I	0.97C + 0.04	0.10$\overline{X}$ + 0.07	0.18$\overline{X}$ + 0.08
Endosulfan II	0.93C + 0.34	0.41$\overline{X}$ − 0.65	0.47$\overline{X}$ − 0.20
Endosulfan sulfate	0.89C − 0.37	0.13$\overline{X}$ + 0.33	0.24$\overline{X}$ + 0.35
Endrin	0.89C − 0.04	0.20$\overline{X}$ + 0.25	0.24$\overline{X}$ + 0.25
Heptachlor	0.69C + 0.04	0.06$\overline{X}$ + 0.13	0.16$\overline{X}$ + 0.08
Heptachlor epoxide	0.89C + 0.10	0.18$\overline{X}$ − 0.11	0.25$\overline{X}$ − 0.08
Toxaphene	0.80C + 1.74	0.09$\overline{X}$ + 3.20	0.20$\overline{X}$ + 0.22
PCB-1016	0.81C + 0.50	0.13$\overline{X}$ + 0.15	0.15$\overline{X}$ + 0.45
PCB-1221	0.96C + 0.65	0.29$\overline{X}$ − 0.76	0.35$\overline{X}$ − 0.62
PCB-1232	0.91C + 10.79	0.21$\overline{X}$ − 1.93	0.31$\overline{X}$ + 3.50
PCB-1242	0.93C + 0.70	0.11$\overline{X}$ + 1.40	0.21$\overline{X}$ + 1.52
PCB-1248	0.97C + 1.06	0.17$\overline{X}$ + 0.41	0.25$\overline{X}$ − 0.37
PCB-1254	0.76C + 2.07	0.15$\overline{X}$ + 1.66	0.17$\overline{X}$ + 3.62
PCB-1260	0.66C + 3.76	0.22$\overline{X}$ − 2.37	0.37$\overline{X}$ − 4.86

* X' = expected recovery for one or more measurements of a sample containing a concentration of C,
  s_r = expected single-analyst standard deviation of measurements at an average concentration found of $\overline{X}$,
  S' = expected interlaboratory standard deviation of measurements at an average concentration found of $\overline{X}$,
  C = true value for the concentration, and
  $\overline{X}$ = average recovery found for measurements of samples containing a concentration of C.

Analyze QC check samples according to the method beginning in ¶ 5a. Calculate average recovery and standard deviation of the recovery, compare with acceptance criteria and evaluate and correct system performance as directed in Section 6200A.5a1) and 2).

c. Analyses of samples with known additions: See Section 6420B.7c. Prepare QC check sample concentrates according to ¶ 4h and use Tables 6630:III and IV.

d. Quality-control check standard analysis: See Section 6420B.7d. Prepare QC check standard according to ¶ 4h and use Table 6630:V. If all compounds in Table 6630:V are to be measured in the sample in ¶ c above, it is probable that the analysis of a QC check will be required; therefore, routinely analyze the QC check standard with the known-addition sample.

e. Bias assessment and records: See Section 6410B.7e.

8. Precision and Bias

This method was tested by 20 laboratories using reagent water, drinking water, surface water, and industrial wastewaters with known additions at six concentrations over the range 0.5 to 30 µg/L for single-component pesticides and 8.5 to 400 µg/L for multicomponent samples.[14] Single-operator precision, overall precision, and method bias were found to be related directly to the compound concentration and essentially independent of sample matrix. Linear equations describing these relationships are presented in Table 6630:VI.

9. References

1. U.S. ENVIRONMENTAL PROTECTION AGENCY. 1984. Method 608—Organochlorine pesticides and PCBs. 40 CFR Part 136, 43321; *Federal Register* 49, No. 209.
2. U.S. ENVIRONMENTAL PROTECTION AGENCY. 1982. Determination of pesticides and PCBs in industrial and municipal wastewaters. EPA-600/4-82-023, National Technical Information Serv., PB82-214222, Springfield, Va.
3. GIAM, C.S., H.S. CHAN & G.S. NEF. 1975. Sensitive method for determination of phthalate ester plasticizers in open-ocean biota samples. *Anal. Chem.* 47:2225.
4. GIAM, C.S. & H.S. CHAN. 1976. Control of Blanks in the Analysis of Phthalates in Air and Ocean Biota Samples. U.S. National Bur. Standards, Spec. Publ. 442.
5. U.S. ENVIRONMENTAL PROTECTION AGENCY. 1984. Definition and procedure for the determination of the method detection limit. 40 CFR Part 136, Appendix B. *Federal Register* 49, No. 209.
6. U.S. ENVIRONMENTAL PROTECTION AGENCY. 1980. Method Detection Limit and Analytical Curve Studies, EPA Methods 606, 607, and 608. Special letter rep. for EPA Contract 68-03-2606, Environmental Monitoring and Support Lab., Cincinnati, Ohio.
7. GOERLITZ, D.F. & L.M. LAW. 1971. Note on removal of sulfur interferences from sediment extracts for pesticide analysis. *Bull. Environ. Contam. Toxicol.* 6:9.
8. U.S. ENVIRONMENTAL PROTECTION AGENCY. 1980. Manual of Analytical Methods for the Analysis of Pesticides in Human and Environmental Samples. EPA-600/8-80-038, Health Effects Research Lab., Research Triangle Park, N.C.
9. ALFORD-STEVENS, A. et al. 1986. Characterization of commercial Aroclors by automated mass spectrometric determination of polychlorinated biphenyls by level of chlorination. *Anal. Chem.* 58:2014.
10. ALFORD-STEVENS, A. 1986. Analyzing PCB's. *Environ. Sci. Technol.* 20:1194.
11. ALFORD-STEVENS, A. 1987. Mixture analytes. *Environ. Sci. Technol.* 21, 137.
12. MILLS, P.A. 1968. Variation of florisil activity: Simple method for measuring absorbent capacity and its use in standardizing florisil columns. *J. Assoc. Offic. Anal. Chem.* 51:29.
13. WEBB, R.G. & A.C. MCCALL. 1973. Quantitative PCB standards for election capture gas chromatography. *J. Chromatog. Sci.* 11:366.
14. U.S. ENVIRONMENTAL PROTECTION AGENCY. 1984. EPA Method Study 18, Method 608—Organochlorine Pesticides and PCBs. EPA-600/4-84-061, National Technical Information Serv., PB84-211358, Springfield, Va.

6630 D.   Liquid-Liquid Extraction Gas Chromatographic/Mass Spectrometric Method

See Section 6410B.

6640 ACIDIC HERBICIDE COMPOUNDS*

6640 A. Introduction

1. Sources and Significance

Agricultural chemicals used for weed control can be found as contaminants in various aquatic systems.[1] Although formerly only two chlorinated phenoxy acid herbicides have been among the substances regulated by the U.S. Environmental Protection Agency in drinking water, and thus were routinely measured, several other types of carboxylic acid compounds used for their toxic effects now are under regulatory control.[2]

2. Selection of Method

The micro liquid-liquid extraction gas chromatographic method presented here does not require drying and concentration of large volumes of solvent extract, requires less equipment, and has fewer analytical steps than other methods.[3] A simple procedure for producing diazomethane has been included. The method also provides for simultaneous dual-column confirmation using two electron-capture detectors and dual-channel data acquisition. An optional simplified procedure for alkaline solvent wash of difficult sample matrices allows for additional cleanup of extraneous organics and hydrolysis of the sample.

3. Sampling and Storage

Collect grab samples in quadruplicate. Flush sampling tap until water temperature stabilizes and stagnant lines are cleared. Collect

---

* Approved by Standard Methods Committee, 1994.

samples in nominal 40-mL vials containing 3.2 mg sodium thiosulfate (if residual chlorine is present) and seal with a TFE-faced septum and screw cap. Fill vials so that no air bubbles pass through the sample. Do not rinse vial before filling and do not overfill. Seal vials with no headspace.

Add diluted hydrochloric acid at the sampling site to adjust to pH ≤ 2; check pH using short-range (0 to 3) pH paper. Samples can be held in ice away from light or refrigerated at 4°C for 14 d; sample extracts can be held at 4°C for 14 d.[3] Verify stability of each constituent in any unknown sample matrix.

4. References

1. U.S. ENVIRONMENTAL PROTECTION AGENCY. 1990. National Pesticide Survey: Summary Results of EPA's National Survey of Pesticides in Drinking Water Wells. EPA 570/9-90-015. U.S. Environmental Protection Agency, Cincinnati, Ohio.
2. U.S. ENVIRONMENTAL PROTECTION AGENCY. 1992. National Primary Drinking Water Regulations; Synthetic Organic Chemicals and Inorganic Chemicals; Final Rule. 40 CFR Parts 141 and 142, Part 111; Federal Register 57, No. 138.
3. U.S. ENVIRONMENTAL PROTECTION AGENCY. 1989. Method 515.1, Revision 4.0, Determination of Chlorinated Acids by Gas Chromatography with an Electron Capture Detector. Environmental Monitoring Systems Lab., Off. Research & Development, Cincinnati, Ohio.

6640 B. Micro Liquid-Liquid Extraction Gas Chromatographic Method

1. General Discussion

a. Application: This method can be used to measure various acidic organic compounds and their corresponding acid salts although the form of each acid is not differentiated and the calculated amount of each is expressed as free acid. The method can be applied to groundwater, finished drinking water, raw source waters, and other sample matrices provided that it can meet the necessary quality control objectives. This method measures a wide range of carboxylic acid compounds including, but not limited to, acidic agricultural chemicals such as the acid herbicides (halogenated and nonhalogenated), haloacetic acids, and trichlorophenol. Method performance data have been developed for nine herbicide compounds. The U.S. Environmental Protection Agency (USEPA) cur-

rently requires monitoring of seven of these compounds, namely, 2,2-dichloropropanoic acid (dalapon), 3,6-dichloro-2-methoxybenzoic acid (dicamba), (2,4-dichlorophenoxy)-acetic acid (2,4-D), pentachlorophenol (pcp), (2,4,5-trichlorophenoxy)-propionic acid (2,4,5-TP/silvex), 2-(1-methylpropyl)-4,6-dinitrophenol (dinoseb), and 4-amino-3,5,6-trichloropicolinic acid (picloram). The other two herbicides are 3-(methylethyl)-1H-2,2,3-benzothiadiazin-4(3H)-one 2,3-dioxide (bentazon), which must be monitored in the State of California, and (2,4,5-trichlorophenoxy)-acetic acid (2,4,5-T), which is used in the EPA performance evaluation sample as a laboratory performance check compound.

b. Principle: A 30-mL sample portion is extracted with 3 mL methyl *tertiary*-butyl ether (MtBE), at an acidic pH to extract the nondissociated acid, and with a salting agent to increase extraction efficiency. Extracted compounds are esterified with diazomethane solution to produce methyl ester derivatives that can be chromatographed. Analyze on a temperature-programmable gas chromatograph (GC) using a fused silica capillary column and an electron capture detector (ECD). Analysis and confirmation are simultaneous; the analytical column and the confirmation column share a common injection port. Alternatively, use separate injections for analysis and confirmation.

c. Interferences: Use HPLC-grade MtBE to minimize interferences from any impurities in the extraction solvent. No further distillation or cleanup of the solvent is necessary.

Clean glassware is essential to eliminate contamination and background interferences. Preferably dedicate all analytical glassware used in this procedure. Clean all glassware, except volumetric flasks and diazomethane generators, as follows: Wash with detergent, rinse three times with tap water, and rinse twice with reagent-grade water. Heat at 180°C for at least 1 h in a clean convection oven. Rinse all caps and septa with methanol immediately after use. Heat at 80°C for not more than 1 h in a clean convection oven or air-dry overnight.

Clean millimole diazomethane generators immediately after use. Rinse the inside tube three times with methanol and twice with tap water. Rinse the outer tube three times with methanol, into a waste container containing 11 g silica gel, then twice with tap water. Rinse both inner and outer tubes with reagent water three times. Bake at 180°C in a clean convection oven for at least 1 h until dry.

Immediately after use of volumetric flasks, rinse three to five times with methanol. Invert flasks to drain on a rack and let air-dry in a ventilation hood.

An optional procedure for simplified alkaline solvent wash of difficult matrices is given in ¶ 4b2), below.

CAUTION: *Verification of identity is essential. Be aware of possible interferents, and always confirm with second column and also with GC/MS whenever possible.*

d. Minimum detectable concentration: The method detection levels (Table 6640:I) were calculated from the analysis of seven laboratory fortified blanks.[1] The actual method working range will be matrix dependent; however, as a general guideline the acid herbicides can be measured over a practical working range from approximately five to ten times the MDL for the lower limit to

an upper limit of at least 100 times the MDL. The calibration range can be extended, depending on the compound characteristics and instrument response. Level 1 calibration standard in Table 6640:IV is a suggested starting point for a practical low-level quantitation concentration.

e. Safety: Toxicity and carcinogenicity of all chemicals used in this method have not been fully evaluated; therefore, treat each as a potential health hazard. Minimize exposure and use only in a properly operating ventilation hood. Make material safety data sheets (MSDS) available. Establish a laboratory safety program to meet Occupational Safety and Health Act (OSHA) regulations.

For precautions in handling diazomethane and ether, see 6251B.1*d.*

2. Apparatus

Use apparatus listed in 6251B.3, with the following exceptions: Only 2- and 10-mL micro volumetric flasks are needed; only 14-mL amber glass vials are needed. NOTE: In using the generator shown in Figure 6251:2, inside tube will contain *N*-methyl-*N*-nitroso-*p*-toluene sulfonamide and methanol, not MNNG.

3. Reagents

All reagents listed in 6251B.4*a, b, c, e*1), *f*2), *g, h,* and *i,* and in addition:

a. Sodium thiosulfate, Na$_2$S$_2$O$_4$, anhydrous.

b. Standard material: See Table 6640:II for source and physical information.

1) *Individual herbicide stock solutions:* Weigh 0.0500 g of each compound on an analytical balance. Dilute each in MtBE to the 10-mL mark in a screw-top volumetric flask. Transfer each solution to a separate clean 14-mL amber vial and store in a freezer at −11°C. Alternatively purchase solutions as certified standards. The 5000-µg/mL stock standards are usable for 6 months.

2) *Stock herbicide mix:* Dilute an appropriate amount of each stock standard into a 10-mL volumetric flask containing approximately 7 mL methanol. After all stock solutions have been added,

Table 6640:I. METHOD DETECTION LEVELS*

Compound	Amount Added µg/L	Average Recovery %	Mean Amount Found µg/L	Standard Deviation µg/L	Relative Standard Deviation %	Method Detection Level µg/L
Dalapon	2.0	100	2.00	0.031	1.56	0.1
Dicamba	0.4	122	0.48	0.007	1.55	0.02
2,4-D	2.0	98.1	1.9	0.036	1.85	0.1
Pentachlorophenol	0.2	101	0.20	0.004	2.40	0.02
2,4,5-TP	0.2	119	0.24	0.003	1.58	0.01
2,4,5-T	0.2	97.8	0.20	0.005	2.73	0.02
Dinoseb	0.4	111	0.44	0.012	2.86	0.04
Bentazon	0.4	94.6	0.38	0.032	8.68	0.01
Picloram	0.4	123	0.49	0.011	2.29	0.04
Surrogate				11.3	5.92	
Internal standard				3.74	1.18	

* Based on the analysis of seven portions with known additions.[1]

TABLE 6640:II. ANALYTICAL STANDARDS

Compound*	Molecular Weight	Melting Point
Dalapon	142.97	90–92/14 mm†
Dicamba	221.04	114–116
2,4-D	221.04	136–140
Pentachlorophenol (PCP)	266.34	188–191
2,4,5-TP (silvex)	269.51	175–177
2,4,5-T	255.49	154–158
Dinoseb	240.22	38–42
Bentazon	240.28	137–139
Picloram	241.48	218–219
Internal standard (TCP) 1,2,3-trichloropropane	147.43	−14
Surrogate (TFBA) 2,3,5,6-tetrafluorobenzoic acid	194.09	150–152

* AccuStandard, Inc., New Haven, Conn.
† Boiling point/pressure.

dilute to 10-mL mark with methanol. Solution is usable for 3 months. See Table 6640:III for example solution concentrations.

3) *Herbicide known-addition mix:* Add 200 µL stock herbicide mix to a 2-mL volumetric flask containing about 1.5 mL methanol; dilute to 2-mL mark with methanol (see Table 6640:III).

c. *Internal standard (IS):* 1,2,3-trichloropropane, 99% pure.

1) *Internal standard stock solution:* Weigh 0.1000 g IS into a 10-mL volumetric flask and bring to volume with methanol. This will yield a 10-mg/mL stock solution. Stock standards are usable for 1 year.

2) *MtBE plus internal standard solution,* (*MtBE + IS*), MtBE + 3 µg IS/mL: Mix 240 µL IS stock solution in 800 mL MtBE. Screw 5-mL pump-type dispenser directly on top of the vendor's 1-L bottle and set it to deliver 3 mL. The sample extraction solvent is the same as the internal standard known addition solution.

d. *Surrogate (SUR):* 2,3,5,6-tetrafluorobenzoic acid, 99% pure.

1) *Surrogate stock solution,* 10 mg SUR/mL: Weigh 0.1000 g SUR acid into a 10-mL screw-cap volumetric flask and dilute to mark with methanol. Stock solutions are usable for 1 year.

2) *Surrogate known-addition solution,* 30 µg SUR/mL: Add 30 µL SUR stock solution to a 10-mL volumetric flask and dilute to mark with methanol.

e. *Reagent water:* Purify the water by several stages of cartridge-type purification to filter, demineralize, and trap organic compounds, then distill with an all-glass distillation system.

f. *Calibration standards:* Using a 2-mL screw-cap volumetric flask, add 1.7 mL MtBE for each external standard to be prepared. When using the internal standard method add MtBE + IS to the 2-mL mark and concentrate solvent to approximately 1.7 mL with a gentle stream of nitrogen. Inject a measured amount of herbicide known-addition mix solution and 20 µL SUR known-addition solution to each volumetric flask and place in a freezer at −11°C for 11 min before adding 0.25 mL cold diazomethane/MtBE solution to each standard. Let stand at least 23 min to react and dilute to 2 mL with MtBE. The solution should remain yellow after diazomethane/MtBE addition. Make a minimum of three calibration levels, one near the quantitation limit and the others bracketing actual sample concentrations to be measured. Three low-level standards are suggested in Table 6640:IV to make a short-range calibration curve.

g. *Trimethylsilyl-diazomethane (TMSD)** (optional), 2M in hexane.

4. Procedure

a. *Sample preparation:* See 6251B.5a.

b. *Microextraction:* Transfer 30 mL from sample container to a 40-mL vial with TFE-faced septum and screw cap.

Add 30 µL SUR known additions solution to each sample, including standards and blanks. Add herbicide known additions mix at this step.

Simplified alkaline solvent wash and extraction—To the 30-mL sample, add 9 g Na₂SO₄ and immediately shake vigorously until salt dissolves, then add enough (approximately 100 µL) 20% NaOH to produce pH ≥12. Hold at room temperature for 1 h; shake periodically. Add 3 mL MtBE and shake vigorously for 1 min, then decant as much of the MtBE as possible with a disposable pasteur pipet (taking a little water is acceptable) and discard. Add 1.5 mL conc H_2SO_4, 3 g $CuSO_4 \cdot 5H_2O$, and 3 g

* Aldrich Chemical or equivalent.

TABLE 6640:III. SUGGESTED HERBICIDE SOLUTION CONCENTRATIONS

Compound	Stock* Conc µL/mL	Stock† Amount Added µL	Stock‡ Herbicide Mix µg/mL	Herbicide§ Addition Mix µg/mL
Dalapon	5000	200	100	10
Dicamba	5000	40	20	2
2,4-D	5000	200	100	10
Pentachlorophenol	5000	20	10	1
2,4,5-TP	5000	20	10	1
2,4,5-T	5000	20	10	1
Dinoseb	5000	40	20	2
Bentazon	5000	40	20	2
Picloram	5000	40	20	2

* Individual stock solutions.
† Amount of individual stock added to 10 mL of methanol to make the stock herbicide mix.
‡ Concentration of each compound in stock herbicide mix.
§ Herbicide addition mix is made by adding 200 µL stock herbicide mix to 2 mL methanol.

TABLE 6640:IV. SHORT-RANGE CALIBRATION STANDARDS*

Compound	Level 1 μg/L	Level 2 μg/L	Level 3 μg/L
Dalapon	1.0	2.0	4.0
Dicamba	0.2	0.4	0.8
2,4-D	1.0	2.0	4.0
Pentachlorophenol	0.1	0.2	0.4
2,4,5-TP	0.1	0.2	0.4
2,4,5-T	0.1	0.2	0.4
Dinoseb	0.2	0.4	0.8
Bentazon	0.2	0.4	0.8
Picloram	0.2	0.4	0.8

* Levels 1, 2, and 3 are prepared by adding 2 μL, 4 μL, and 8 μL of herbicide addition mix, respectively, into 2 mL MtBE.

Na_2SO_4, then shake until salt dissolves. Add 3 mL MtBE (for external standard calibration) or MtBE + IS (for internal standard calibration). Immediately cap vial and shake briefly by hand to break up any salt clumps. Lay vial on its side before continuing to next sample. The internal standard can be added later if other quality-control measures are desired.

Place each vial in the wooden holding block. Once all the vials have been inserted, place block immediately on the mechanical shaker. Shake vials at fast speed (approximately 300 cycles/s) for 9 min. (Alternatively, shake manually for about 3 min.)

Remove vials from holder, place upright, and let stand for at least 3 min until the phases separate. If an emulsion forms it can be broken up by briefly sonicating for 10 s in an ultrasonic water bath, then letting the phases separate for at least 3 min.

c. Preparation of diazomethane:[2] Also see ¶ 4d below. With the generator specified, use at least 0.37 g N-methyl-N-nitroso-p-toluene sulfonamide (diazald) (this is enough for derivatizing 10 sample extracts) to initiate a good reaction; if more is needed add about 37 mg more diazald per sample to inside tube of generator. Add enough methanol to diazald cover the yellow crystals, then add 0.3 mL excess methanol before securing cap and septum.

Add up to 5.5 mL MtBE to outside tube of generator, allowing 0.25 mL MtBE per sample extract to be derivatized, then add 0.5 mL MtBE in excess, thereby preparing enough diazomethane/MtBE solution to derivatize at least 20 sample extracts. For example, if derivatizing 10 sample extracts, add 3 mL MtBE to outside tube.

Place butyl-o-ring in the glass joint, and twist inside tube firmly on top of the o-ring, then clamp securely with a screw-type pinch clamp.

Place generator and its contents in an ice-water bath containing enough ice to keep solution at 0°C.

Measure at least 0.7 mL 20% NaOH solution to initiate a good reaction, using a 1-mL gastight syringe with a 22-gauge needle. If a larger batch of diazomethane is being prepared add about 50 μL extra 20% NaOH solution per sample. Working behind a safety shield, slowly add NaOH through the top generator septum (check that syringe needle is on opposite side of vapor exit hole). Add NaOH at rate of 1 mL/3 s. Swirl briefly to mix. Allow derivatization to continue for 30 min. If more diazomethane is needed, prepare two or more batches and combine just before use.

A strong yellow-colored solution in the outer tube indicates that sufficient diazomethane gas has been generated and has dissolved into the MtBE. If the solution remains clear or only faintly yellow, generate another batch of diazomethane until a strong yellow

color is obtained. If poor generation is due to old or decomposed diazald, replace with fresh reagent. Other causes are poor sealing of the diazomethane generator tubes or water present in the MtBE. Alternatively, use the procedure described in 6251B.5c.

d. Separation and concentration: See 6251B.5d. After concentration, internal standard can be added if desired. If not analyzing for dalapon, consider using TMSD (¶ 3g above) in lieu of diazomethane for derivatization because it is less hazardous and is stable in storage.

If using TMSD, after hydrolysis, extraction, and drying of the sample concentrate the 2-mL extract to 1.5 mL so that 0.5 mL of methanol (20% of the sample) may be added. (TMSD produces the most effective methylation of the acid herbicides in a 20% methanol, 80% MtBE solution.) Add approximately 30 μL of 2M TMSD solution to the 2-mL sample extract. Place tube containing extract in a heating block at 50°C and maintain for 1 h. Let extract cool to room temperature. Add 60 μL of 2M acetic acid in methanol to react any excess TMSD. Internal standard may be added at this time. Proceed with analysis and measurement according to ¶ 4f. Verify all quality control requirements as outlined in ¶ 6.

e. Derivatization: Cool extracts in an explosion-safe freezer at approximately −11°C for 11 min and add diazomethane. Uncap one volumetric flask and add 250 μL cold diazomethane/MtBE solution. Cap immediately with a TFE-lined screw cap; mix gently by inverting once. Repeat for the remaining extracts. A persistent faint yellow color should be present after addition of the diazomethane solution, indicating an excess available for esterification.

If using an autosampler, prepare two 1.8-mL vials for each sample. Label vials and add approximately 0.01 g silica gel to quench excess diazomethane.

Let stand in a laboratory fume hood for 23 min to reach room temperature. Dilute to 2-mL mark with MtBE and invert flask to mix. Keep each extract in contact with diazomethane for approximately the same amount of time before quenching. If using an autosampler, transfer each extract evenly between two labeled autosampler vials containing silica gel using a 23-cm (9-in.) disposable pasteur pipet. Store MtBE extract in the freezer at −11°C until analyzed.

f. Gas chromatography: Typical operating conditions are as follows:

Injector temperature 225°C; split valve opened at 0.75 min; split flow 70 mL/min; injection volume 4.0 μL; slow injection rate 1.3 μL/s.

Temperature program: 40°C for 7 min, rising 20°C/min to 160°C, rising 2°C/min to 195°C, and rising 10°C/min to 225°C, holding 6 min at 225°C.

Detector temperature: 300°C; range setting 10.

Carrier gas (helium) flow: 33 cm/s at 37°C.

Makeup gas (nitrogen) flow: 20 mL/min.

At beginning of each analytical run, inject two MtBE solvent blanks to condition GC and to verify that no solvent interferences are present. If available, analyze an extra calibration standard to ensure retention times and response of GC system. Always inject the same amount of sample (3.7 μL), and use sample dilution if necessary to obtain response in the calibration range. See Figures 6640:1 and 2 for sample chromatograms of a single injection onto both the analytical and confirmation columns (gas chromatograph

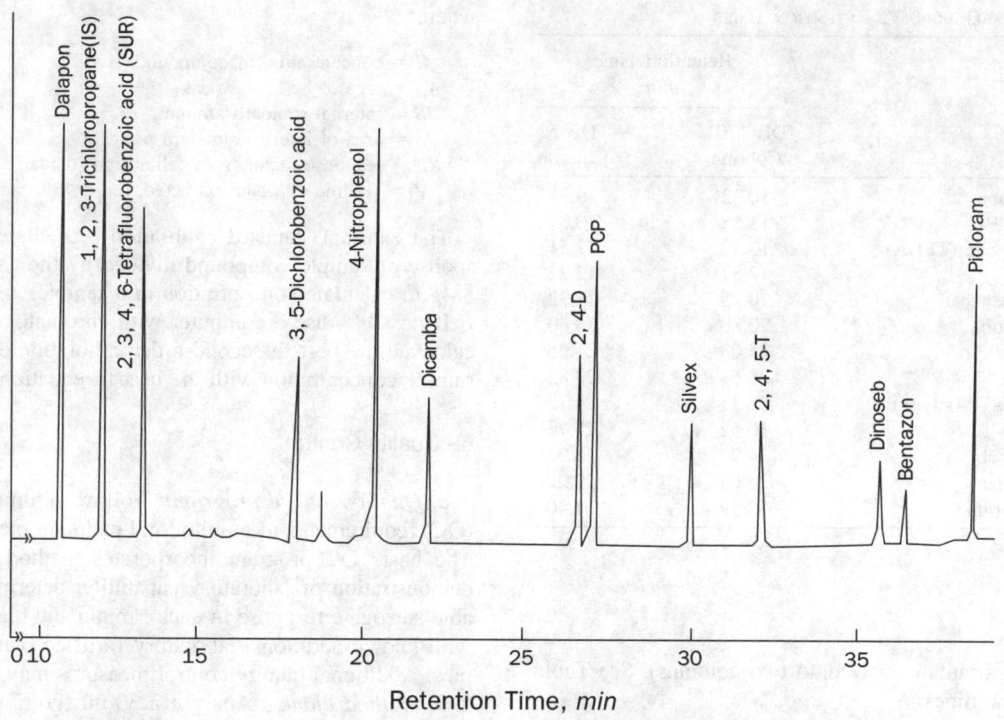

Figure 6640:1. Herbicide standard chromatogram on analytical capillary column. Chromatograms in Figures 6640:1 and 2 generated simultaneously with a single injection using specified analytical conditions; all acidic compounds analyzed as methyl esters.

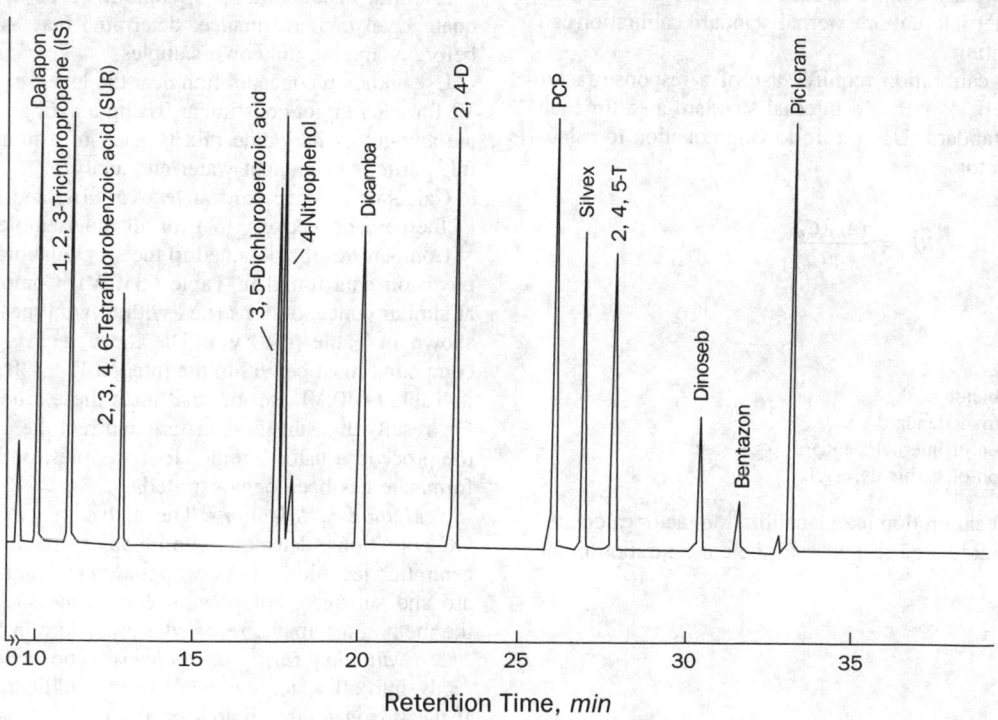

Figure 6640:2. Herbicide standard chromatogram on confirmatory capillary column. Chromatograms in Figures 6640:1 and 2 generated simultaneously with a single injection using specified analytical conditions; all acidic compounds analyzed as methyl esters.

TABLE 6640:V. RETENTION TIMES

Compound	Retention Time min	
	DB-1701 Column	DB-5 Column
Methyl ester of dalapon	10.32	9.34
Internal standard (TCP)	11.58	10.37
Methyl ester of surrogate (TFBA)	13.20	12.51
4-Nitroanisole	17.87	17.22
Methyl-3,5-dichlorobenzoate	20.25	17.32
Methyl ester of dicamba	20.80	18.60
Methyl ester of 2,4-D	25.08	21.26
Pentachloroanisole	25.64	23.82
Methyl ester of 2,4,5-TP (silvex)	28.33	24.73
Methyl ester of 2,4,5-T	30.52	25.72
Methyl ester of dinoseb	34.27	28.05
Methyl ester of bentazon	35.04	29.2
Methyl ester of picloram	38.39	31.40

is set up to inject simultaneously onto two columns). See Table 6640:V for retention times.

g. *Calibration:* Use at least three levels of calibration standards to define the quantitation range of analysis (see Table 6640:IV). Use standards that bracket the expected range of sample concentrations and do not exceed the linear range of the detector. Prepare standards as described in ¶ 3f. Analyze each standard by injecting the same volume as each sample extract, under the same analytical conditions. Either internal or external standard calibration can be used for quantitation.

Internal standard calibration requires use of a response factor calculated from the peak area and internal standard area for each constituent in the standard. Use the following equation to calculate the response factor:

$$RF = \frac{(A_H)(C_I)}{(A_I)(C_H)}$$

where:

RF = response factor,
A_H = area of herbicide peak,
A_I = area of internal standard peak,
C_I = concentration of internal standard, µg/L, and
C_H = concentration of herbicide, µg/L.

External standard calibration uses a calibration factor calculated as the ratio of herbicide area response to the concentration in the standard.

5. Calculation

When using internal standard calibration, calculate compound concentration in a sample with the following equation:

$$C = \frac{(A_H)(IS)}{(A_I)(RF)(V)}$$

where:

C = concentration of compound, µg/L,
A_H = area of herbicide peak,
IS = internal standard amount, µg,
A_I = area of internal standard peak,
RF = response factor from calibration standards in ¶ 4g, and
V = volume of water extracted, L.

The external standard calibration procedure uses the area response of sample compound divided by the calibration factor in ¶ 4g to calculate concentration in a sample.

If possible, use a computer with chromatographic software to calculate the best-fit second-order calibration equation. Calculate sample concentration with the best-fit equation.

6. Quality Control

a. *Quality control program:*[3] Follow minimum quality control (QC) requirements to monitor and maintain method performance. The basic QC program incorporates method blanks, an initial demonstration of laboratory capability, determination of acceptable surrogate response in each sample and blank, sample matrix with known additions, laboratory fortified blanks, and QC samples. Additional quality control measures may be used.

b. *Method blanks:* Analyze a 30-mL portion of reagent water with the same procedure as an actual sample. Each time a set of samples is extracted or reagents are changed, process a laboratory method blank. If the blank produces any peak within the retention time window of a constituent that would prevent its determination, eliminate source of contamination and reanalyze affected samples.

c. *Initial demonstration of capability:* To demonstrate an adequate level of performance, determine bias as percent recovery before analyzing unknown samples.

Use a known concentration near the levels given in Table 6640: VI for each target constituent. Using a syringe, add the appropriate amount of herbicide mix to each of a minimum of four 30-mL portions of reagent water and analyze.

Calculate average percent recovery (R) and standard deviation of the percent recovery (S_r) for all four samples.

Compare results obtained to the single-laboratory recovery and precision data found in Table 6640:VI. Compare precision data at similar concentrations (i.e., within two times the addition level shown in Table 6640:VI). The mean recovery values for each compound must be within the interval R ± 30% using the values in Table 6640:VI. Results that meet these criteria are acceptable. For results that fail these criteria, correct the problem and repeat the procedure using another four samples, until satisfactory performance has been demonstrated.

d. *Method modifications:* The analyst may modify GC columns, GC conditions, detectors, continuous extraction techniques, concentration techniques (i.e., evaporation techniques), internal standard and surrogate compounds. Each time such modifications to the method are made, repeat the procedure in ¶ c above.

e. *Evaluating surrogate recovery:* The surrogate is added directly into all water samples before acidification and extraction. If the surrogate area is low or absent, there has been a derivatization problem (e.g., water in extract) or extraction problem (e.g., water insufficiently acidified).

An extract is acceptable if the surrogate recovery does not exceed 100 ± 30% compared to the average daily calibration standard. If surrogate recovery is unacceptable, check the following

TABLE 6640:VI. SINGLE-LABORATORY BIAS AND PRECISION DATA,* LEVEL 1 ADDITION REAGENT WATER

Compound	Added Amount μg/L	Mean Recovery %	Mean Amount Found μg/L	Standard Deviation μg/L	Relative Standard Deviation %
Dalapon	10	114	11	0.72	6.4
Dicamba	1.0	84	0.83	0.02	2.6
2,4-D	5.0	75	4.0	0.22	5.5
Pentachlorophenol	2.0	107	2.1	0.04	1.7
2,4,5-TP	0.5	107	0.5	0.02	3.0
2,4,5-T	0.5	104	0.50	0.02	3.7
Dinoseb	2.0	90	1.8	0.1	5.4
Bentazon	1.0	84	0.85	0.020	2.3
Picloram	2.0	80	1.60	0.11	7.0

* Based on analysis of four portions of reagent water with known additions.

and correct: possible errors in calculations or procedure, degradation of standard solution, contamination sources, and instrument performance. If the cause of the problem cannot be determined, reanalyze the extract. If surrogate recovery is acceptable, report only data for the reanalyzed extract. If sample extract continues to fail, report all data for that sample as suspect.

If a blank extract reanalysis fails the recovery criterion, identify the problem and correct it before continuing.

f. Internal standard assessment: If the IS procedure is used, monitor the IS area of each sample during each analysis day. The IS area should not deviate from the average daily calibration standards by more than 30%; if it does, optimize instrument performance and inject a second portion of that extract.

If the reinjection is acceptable report the results. If the IS area is still over 30%, re-extract sample if more is available and the holding period has not expired; otherwise report results as suspect.

If consecutive samples fail the IS response immediately analyze a calibration standard. If the standard response factor is within 20% of the predicted value, begin with optimizing instrument performance as described above for each sample failing the IS response. If the standard is above the 20% predicted value, recalibrate as specified in ¶ 4g.

g. Laboratory fortified blank: Analyze at least one laboratory fortified blank (LFB) with every 20 samples or at least one per sample set (all samples extracted within a 24-h period). The concentration of each compound in the LFB should be within two times the Table 6640:VI addition amounts. Calculate accuracy as percent recovery. If recovery of any compound falls outside the control limits, identify the source of the problem and correct it before continuing analyses.

Initially, assess the LFB with the results from the initial demonstration of capability in (¶ c above) and Table 6640:VII. When a minimum of 20 LFB have been analyzed, use these data to develop control charts with the mean percent recovery (X), and the standard deviation (s) of the percent recovery. The upper control limit is $X + 3s$ and the lower control limit is $X - 3s$. After 5 to 10 new measurements, calculate new control limits using only the most recent 20 to 30 data points.

Periodically determine and document the laboratory's detection limit capabilities.[3]

At least quarterly, analyze a QC sample from an outside source.

h. Matrix additions: Add each target constituent into a minimum of 10% of the routine samples at a concentration near or greater than the background concentration. Ensure that the addition plus background concentration does not exceed the calibration range. Eventually, make additions to all routine samples.

Analyze one sample portion to determine the background concentration (*b*) of each constituent. Add herbicide mix to a second

TABLE 6640:VII. SINGLE-LABORATORY BIAS AND PRECISION DATA,* LEVEL 2 ADDITION REAGENT WATER

Compound	Added Amount μg/L	Mean Recovery %	Mean Amount Found μg/L	Standard Deviation μg/L	Relative Standard Deviation %
Dalapon	30	99.3	29.8	0.21	0.71
Dicamba	3.0	92.2	2.77	0.06	2.2
2,4-D	15	90.5	13.6	0.15	1.10
Pentachlorophenol	6.0	95.8	5.75	0.09	1.5
2,4,5-TP	1.5	98.5	1.48	0.02	1.3
2,4,5-T	1.5	86.4	1.30	0.02	1.3
Dinoseb	6.0	90.0	5.40	0.03	5.1
Bentazon	3.0	103	3.09	0.05	1.6
Picloram	6.0	86.0	5.16	0.18	3.5

* Based on analysis of four portions of reagent water with known additions.

TABLE 6640:VIII. LABORATORY PERFORMANCE CHECK SOLUTION

Test	Test Compound	Concentration µg/mL	Requirements
Sensitivity	Dinoseb	0.004	Detection of compound S/N > 3
Chromatographic performance	4-Nitrophenol	1.6	0.70 <PGF<1.05*
Column performance	3,5-Dichlorobenzoic acid	0.6	Resolution >0.40†
	4-Nitrophenol	1.6	

* PGF = peak Gaussian factor. Calculate using the equation:

$$PGF = \frac{1.83 \times W(1/2)}{W(1/10)}$$

where W(1/2) is the peak width at half height and W(1/10) is the peak width at tenth height.

† Resolution between the two peaks as defined by the equation:

$$R = \frac{t}{W}$$

where t is the difference in elution times between the two peaks and W is the average peak width, at the base line, of the two peaks.

Source: U.S. ENVIRONMENTAL PROTECTION AGENCY. 1989. Determination of Chlorinated Acids by Gas Chromatography with an Electron Capture Detector. Method 515.1, rev.4.0. Environmental Monitoring Systems Lab., Off. Research & Development, Cincinnati, Ohio.

TABLE 6640:IX. SAMPLE MATRIX ADDITION RECOVERIES*

Compound	Added Amount µg/L	Source 1 Amount Found µg/L	Source 1 % Found	Source 2 Amount Found µg/L	Source 2 % Found	Treated Source Amount Found µg/L	Treated Source % Found
Dalapon	10	11.6	116	12.1	121	12.5	125
Dicamba	1.0	0.9	91	0.9	94	0.8	83
2,4-D	5.0	4.0	80	4.4	88	4.3	86
Pentachlorophenol	2.0	2.2	110	2.3	113	2.1	105
2,4,5-TP	0.5	0.5	106	0.6	114	0.6	114
2,4,5-T	0.5	0.5	104	0.6	120	0.6	128
Dinoseb	2.0	2.1	105	2.2	110	1.9	94
Bentazon	1.0	0.9	86	0.9	94	0.9	87
Picloram	2.0	1.5	77	1.8	88	2.3	115

* Source 1 is Colorado River water, Source 2 is California State Project Water, and the Treated Source water is the chlorinated effluent from the Weymouth Filtration Plant, La Verne, Calif.

TABLE 6640:X. INTERLABORATORY QUALITY CONTROL COMPARISON DATA*

Sample No.	Compound	Microextraction Results µg/L	True Value µg/L	Acceptance Limits† µg/L
1	2,4-D	32.8	46.3	15.1 –59.1
	Silvex	20.4	18.1	7.47–24.4
2	Dalapon	10.58	12.3	DL–18.4
	Pentachlorophenol	2.28	2.35	DL– 3.92
	Dinoseb	10.9	8.63	DL–11.9
	Picloram	11.0	6.88	DL–12.5

* Source: U.S. ENVIRONMENTAL PROTECTION AGENCY, Water Supply Study Performance Evaluation Study 27 (WS027), February 4, 1991.

† DL = detection limit.

TABLE 6640:XI. RESULTS FOR LABORATORY QUALITY CONTROL WITH LABORATORY FORTIFIED BLANK

Compound	Addition Amount μg/L	Amount Found μg/L	Recovery %
Dalapon	10	9.6	96
Dicamba	1.0	0.82	82
2,4-D	5.0	3.4	68
Pentachlorophenol	2.0	2.1	105
2,4,5-TP	0.5	0.51	102
2,4,5-T	0.5	0.38	76
Dinoseb	2.0	1.70	85
Bentazon	1.0	0.82	82
Picloram	2.0	1.51	76

sample portion and analyze to determine the total matrix concentration (t) for each constituent. Calculate percent recovery of each constituent (R_i) as $100 (t - b)/A$, where A is known concentration of the additions solution.

Compare percent recovery (R_i) for each constituent with established QC acceptance criteria. QC criteria for samples containing no background are established by initially analyzing four reagent water samples with known additions and calculating average percent recovery (R) and standard deviation of the percent recovery (s_r).

If the untreated sample has no background concentrations and the additions are the same as an LFB then the appropriate control limits are those established for the LFB.

If sample does contain background concentrations, determine mean concentration and standard deviation (s_b) for that background level by performing replicate analyses. With replicate analyses measure total matrix addition concentration (background + added amount) and calculate mean concentration (C_t) and standard deviation (s_t). Calculate mean percent recovery (R_t) of the total matrix addition concentration as follows:

$$ R_t = \frac{100(C_t)}{b + A} $$

The control limits are defined as $R_t \pm 3s_s$, where:

$$ s_s = (s_t^2 + s_b^2)^{1/2}/A $$

When recovery of any compound falls outside the control limits and the LFB is in control, the recovery problem is matrix-related; label the result for that compound "suspect/matrix."

i. Laboratory performance check sample: Monitor instrument performance daily by analyzing the laboratory performance check (LPC) sample. The LPC sample contains compounds that indicate instrument sensitivity and column performance. LPC components and performance criteria are given in Table 6640:VIII. If performance criteria cannot be met, reevaluate the instrument system. If the laboratory MDLs differ from those in Table 6640:I, adjust the concentrations of the LPC components to be compatible with the laboratory MDLs.

j. Calibration standards: When using internal-standard calibration, if *RF* values are less than or equal to 20% RSD, either the average *RF* or a calibration curve can be used for quantitation. Verify the calibration or *RF* each working shift by measuring one or more calibration standards. If the response is more than ±20%, repeat the test using a fresh calibration standard. If the repeat test fails, make a new calibration curve with freshly prepared standards. Single-point calibration can be used if the sample extract response is within 20% of the standard.

When using external standard calibration and the calibration factor is less than or equal to 20% RSD, either the average calibration factor or a calibration curve can be used for quantitation. Verify the calibration curve or calibration factor on each working day by measuring at least two calibration standards, one at the beginning and one at the end of the analysis day. Use two standards with different concentrations to verify the calibration curve. If the response is more than ±20%, analyze a fresh calibration standard. If the test fails, make a new calibration curve with freshly prepared standards. Single-point calibration can be used if the sample extract response is within 20% of the standard.

At least quarterly, check the calibration standards against reference material obtained from an independent source; the results must be within the routine calibration criteria.

7. Precision and Bias

Single-laboratory method detection levels[3] and extracted recovery data from reagent water are presented in Table 6640:I. Laboratory data showing single-laboratory accuracy and precision data at two different concentration levels are presented in Tables 6640:VI and VII. Matrix addition recoveries are presented in Table 6640:IX for three different sample matrix types. The interlaboratory comparison data for an Environmental Protection Agency performance evaluation sample with acceptance limits calculated from other participating laboratories[3] are shown in Table 6640:X. Additional quality-control data for a typical laboratory fortified blank are shown in Table 6640:XI.

8. References

1. U.S. ENVIRONMENTAL PROTECTION AGENCY. 1984. Definition and Procedure for the Determination of the Method Detection Limit—Rev. 1.11. 40 CFR Part 136, Appendix B.
2. NGAN, F. & M. TOOFAN. 1991. Modification of preparation of diazomethane for methyl esterification of environmental samples analysis by gas chromatography. *J. Chromatogr. Sci.* 29:8.
3. U.S. ENVIRONMENTAL PROTECTION AGENCY. 1989. Determination of Chlorinated Acids by Gas Chromatography with an Electron Capture Detector. Method 515.1, revision 4.0, Environmental Monitoring Systems Lab., Off. Research & Development, Cincinnati, Ohio.

6651   GLYPHOSATE HERBICIDE*

6651 A.   Introduction

1. Sources and Significance

Glyphosate[N-(phosphonomethyl)glycine] is a broad-spectrum, nonselective, postemergence herbicide that has found widespread agricultural and domestic use. It is sold as a terrestrial and aquatic herbicide under the trade names Roundup® and Rodeo®. Because of low mammalian toxicity (LD50 = 1568 mg/kg rats; oral) there is less concern about water and food contamination than with other pesticides, but the nonselectivity of the herbicide can make nontarget phytotoxicity a problem. Glyphosate's (GLYPH) major metabolite is aminomethylphosphonic acid (AMPA). Contamination of water can occur through runoff and spray drift.

2. Selection of Method

Several methods for determination of GLYPH and AMPA in environmental samples have been developed; those using liquid chromatography are the most precise and accurate. GLYPH and AMPA are not good chromophores or fluorophores and their electrochemical or conductometric detection have not been demonstrated. Sensitive and selective detection has been achieved with the post-column reaction/fluorometric method.[1–3] The absence of a sensitive liquid chromatography technique for confirmation necessitates the use of two different stationary phases.

The liquid chromatographic method presented in 6651B is accurate and precise[3] and includes confirmation by using two col-

umns. Gas chromatography/mass spectrometry confirmation[4] has been used when structural confirmation is required, but the method has not been tested on residues in natural waters.

3. References

1. MOYE, H.A., C.J. MILES & S.J. SCHERER. 1983. A simplified high-performance liquid chromatographic procedure for the determination of glyphosate herbicide and (aminomethyl)phosphonic acid in fruits and vegetables employing postcolumn fluorogenic labeling. *J. Agric. Food Chem.* 31:69.
2. COWELL, J.E., J.L. KUNSTMAN, P.J. NORD, J.R. STEINMETZ & G.R. WILSON. 1986. Validation of an analytical residue method for analysis of glyphosate and metabolite: An interlaboratory study. *J. Agric. Food Chem.* 34:955.
3. OPPENHUIZEN, M.E. & J.E. COWELL. 1991. Liquid chromatographic determination of glyphosate and (aminomethyl)phosphonic acid in environmental water. *J. Assoc. Offic. Anal. Chem.* 74:317.
4. DEYRUP, C.L., S.M. CHANG, R.A. WEINTRAUB & H.A. MOYE. 1985. Simultaneous esterification and acylation of pesticides for analysis by gas chromatography. 1. Derivatization of glyphosphate and (aminomethyl) phosphonic acid with fluorinated alcohol-perfluorinated anhydrides. *J. Agric. Food Chem.* 33:944.

4. Bibliography

BARDALAYE, P.C., W.B. WHEELER & H.A. MOYE. 1985. Analytical techniques of glyphosate residue analysis. *In* E. Grossbard & D. Atkinson, eds. The Herbicide Glyphosate. Butterworths, Woburn, Mass.

_____
* Approved by Standard Methods Committee, 1996.

6651 B.   Liquid Chromatographic Post-Column Fluorescence Method

1. General Discussion

a. Principle: GLYPH and AMPA are separated by anion- or cation-exchange chromatography and measured by post-column fluorescence derivatization. The post-column reactions consist of oxidation of GLYPH (a secondary amine) to glycine (a primary amine) by hypochlorite solution. Glycine then reacts with an *o*-phthalaldehyde (OPA) and mercaptoethanol (MERC) mixed reagent to form an isoindole that is measured fluorometrically. AMPA (a primary amine) reacts directly with the OPA/MERC reagent and is detected (with decreased sensitivity) in the presence of hypochlorite.

b. Interferences: No matrix interferences in water are known. GLYPH degrades in chlorinated water. GLYPH also is known to sorb strongly to minerals and glass surfaces.

c. Minimum detectable concentration: Minimum detection using this method is 25 μg/L for GLYPH and AMPA by direct injection and 0.5 μg/L with the concentration step.

2. Sampling and Storage

Collect a 500-mL representative sample in a polypropylene container. Treatment of sample to remove residual chlorine will prevent glyphosate losses during storage. Destroy chlorine by adding 100 mg/L sodium thiosulfate. Store samples at 4°C away from light and analyze within 2 weeks.

3. Apparatus

a. High-performance liquid chromatograph (HPLC): An analytical system with pump, injector, detectors, and compatible strip chart recorder. Preferably use a data system for measuring peak areas and retention times. Use system capable of injecting 200-μL portions. See Figure 6651:1.

1) *Analytical columns:* Use either a cation exchange resin[1] or an anion-exchange resin[2] packed in a 4.6-mm × 25- to 30-cm

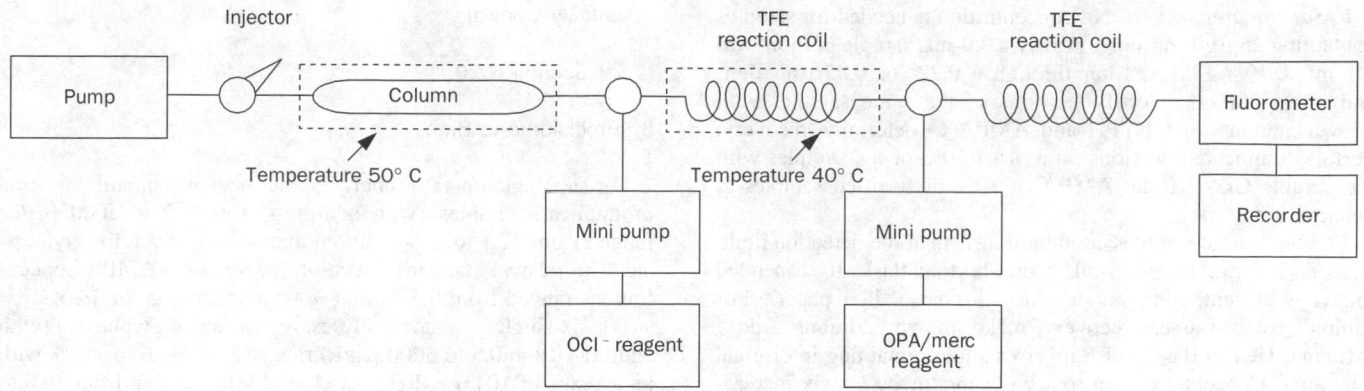

Figure 6651:1. Schematic of post-column reaction HPLC system.

column.* Heat columns to between 50 and 60°C to obtain maximum efficiency.

2) *Post-column reactor:* Use system consisting of two separate pumps capable of delivering reaction solutions at 0.1 to 0.5 mL/min and able to withstand pressures of up to 2000 kPa. Include two woven 1-mL TFE reaction coils[3] (0.5-mm ID × 1.4-mm OD × 5 m) with one maintained at 40°C. Turnkey post-column reactor systems are available commercially.

3) *Fluorescence detector:* Use filter or grating fluorimeter capable of sensitively and selectively measuring the isoindole derivative, with excitation wavelength 230 nm (deuterium), 340 nm (quartz halogen or xenon), and emission wavelength 420 to 455 nm.

4. Reagents

a. Reagent water: See Section 1080.4*b*.

b. Phosphoric acid, H_3PO_4, conc.

c. Sulfuric acid, H_2SO_4, conc. Prepare anion-exchange mobile phase by adding 26 mL conc H_3PO_4 and 2.7 mL conc H_2SO_4 to 5 L water.

d. Hydrochloric acid, HCl, conc.

e. Methanol, CH_3OH, tested on HPLC and verified to give no impurity peaks.

f. Potassium dihydrogen phosphate, KH_2PO_4. Prepare cation-exchange mobile phase by dissolving 0.68 g KH_2PO_4 in 1 L methanol-water (4:96). Adjust to pH 2.1 with conc H_3PO_4. Filter through a 0.22- or 0.45-m membrane filter and degas.

g. Disodium ethylenediamine tetraacetate dihydrate, EDTA sodium salt solutions: Prepare a 0.001*M* solution by dissolving 0.37 g EDTA dihydrate in 1.0 L water and filter through a 0.22- or 0.45-μm filter. Prepare a 0.03*M* solution by dissolving 11.2 g EDTA dihydrate in 1.0 L water and filtering through a 0.33- or 0.45-μm filter.

h. Sodium chloride, NaCl.

i. Sodium hydroxide, NaOH.

j. Calcium hypochlorite, $Ca(OCl)_2$, 70.9% available chlorine.

k. Oxidation reagent: Dissolve 0.5 g $Ca(OCl)_2$ in 500 mL water with rapid magnetic stirring for 45 min. In a 1.0-L volumetric flask, dissolve 1.74 g $K H_2PO_4$, 11.6 g NaCl, 0.4 g NaOH, and 10 mL stock $Ca(OCl)_2$ solution. Dilute to volume, mix well, and filter through a 0.22 or 0.45 μm filter.

l. o-phthalaldehyde, $C_6H_4(CHO)_2$, OPA.

m. 2-mercaptoethanol, $HSCH_2CH_2OH$, MERC.

n. Boric acid powder, H_3BO_3.

o. Potassium hydroxide, KOH.

p. Fluorogenic labeling reagent: Dissolve 100 g boric acid and 72 g KOH in about 700 mL water in a 1.0-L flask. This takes 1 to 2 h. Add 0.8 g OPA dissolved in 5 mL methanol. Add 2.0 mL MERC. Mix well.

q. Glyphosate analytical standard, *N*-(phosphonomethyl) glycine, 99% or greater.

r. Aminomethylphosphonic acid analytical standard, 99% or greater.

s. Glyphosate and AMPA fortification standards: Prepare a solution containing both 0.1 mg GLYPH/mL and 0.1 mg AMPA/mL in water. Make working solutions of 10.0 and 1.0 μg/mL by serial dilution of this stock solution. Store in a refrigerator, in a polypropylene bottle. Prepare fresh monthly.

t. Glyphosate and AMPA HPLC calibration standards: Prepare a solution containing both 0.1 mg GLYPH/mL and 0.1 mg AMPA/mL in 0.001*M* disodium EDTA solution. Make working solutions of 1.00, 0.50, 0.10, 0.05, and 0.025 μg/mL by serial dilution. Store in a refrigerator, in polypropylene bottle. Prepare fresh monthly.

5. Procedure

a. HPLC operation: Equilibrate column at 50°C with mobile-phase flow rate of 0.5 mL/min (see Figure 6651:1). Use an approximate flow rate of 0.5 mL/min for the oxidant and 0.3 mL/min for the OPA-MERC reagent but adjust rates to obtain maximum response. While GLYPH reaches a maximum response at some flow rate of oxidative solution, the AMPA response decreases with any addition of this reagent. Thus, an oxidative reagent flow rate that gives an equal response for both GLYPH and AMPA simultaneously is considered optimum for simultaneous measurements. Reagent flow rates will differ for different mobile phases. Establish a standard curve by injecting calibration standards.

Approximate retention times for GLYPH and AMPA are 13.5 and 10.0 min on the anion-exchange column and 21.5 and 30.0 min on the cation-exchange column.

---

* Aminex, BioRad Labs, A-9 cation exchange and A-27 anion exchange resins, or equivalent.

b. Sample preparation: No concentration is needed for samples containing 25 μg/L or more. Fortify a 9.9-mL sample portion with 0.1 mL 0.10*M* EDTA, filter through a 0.22- or 0.45-μm filter, and inject 200 μL. Fortify portions of the same samples with known amounts of GLYPH and AMPA to determine recovery. Perform duplicate injections on at least 10% of the samples with measurable GLYPH and AMPA or 10% of fortified samples to determine precision.

To concentrate samples containing less than the detection limit, transfer 250 mL to a 500-mL round-bottom flask. If suspended matter is present, filter sample through coarse filter paper.† For samples used to assess recovery, make known additions. Add 5 mL conc HCl to flask and 5 mL to sample remaining in original container. Concentrate on a rotary evaporator by slowly increasing temperature from 20 to 60°C. Before the first portion is completely evaporated, add remaining sample and two 5-mL rinses of the sample bottle. Evaporate to dryness, and if necessary, remove final traces of water with a stream of dry nitrogen. Dissolve residue in 2.9 mL of mobile phase (adjust pH to 2 if necessary) and 0.10 mL 0.03*M* EDTA solution. Filter through 0.45-μm filter to a test tube and inject into the HPLC system.

6. Calculations

Determine concentration of GLYPH and AMPA by regression analysis of the standard curve. Multiply results for samples that were concentrated by the concentration factor, 166.7 (500 mL original sample/3.0 mL), to determine the original water concentration. Report results in milligrams per liter. Report percent recovery but do not correct for recovery.

---

† Whatman #1 or equivalent.

7. Quality Control

See Section 6020.

8. Precision and Bias

For six single-operator analyses, the relative standard deviation of duplicate samples (with additions from 0.5 to 5000 μg/L) ranged from 12.1 to 20% with an average of 14.9% for glyphosate. The relative standard deviation for identical AMPA concentrations ranged from 6.5 to 28.8% with an average of 14.5%.[4]

For six single-operator analyses, recoveries of glyphosate (with additions from 0.5 to 5000 μg/L) ranged from 94.6 to 120% with an average of 104.0%. Recoveries of AMPA ranged from 86.0 to 100% with an average of 93.1%.[4]

9. References

1. COWELL, J.E., J.L. KUNSTMAN, P.J. NORD, J.R. STEINMETZ & G.R. WILSON. 1986. Validation of an analytical residue method for analysis of glyphosate and metabolite: An interlaboratory study. *J. Agric. Food Chem.* 34:955.
2. MOYE, H.A., C.J. MILES & S.J. SCHERER. 1983. A simplified high-performance liquid chromatographic procedure for the determination of glyphosate herbicide and (aminomethyl)phosphonic acid in fruits and vegetables employing postcolumn fluorogenic labeling. *J. Agric. Food Chem.* 31:69.
3. SELVAKA, C.M., K.S. JAIO & I.S. KRULL. 1987. Construction and comparison of open tubular reactors for postcolumn reaction detection in liquid chromatography. *Anal. Chem.* 59:2221.
4. OPPENHUIZEN, M.E. & J.E. COWELL. 1991. Liquid chromatographic determination of glyphosate and (aminomethyl)phosphonic acid in environmental water. *J. Assoc. Offic. Anal. Chem.* 74:317.

PART 7000

RADIOACTIVITY

7010 INTRODUCTION*

7010 A.   General Discussion

1. Occurrence and Monitoring

The radioactivity in water and wastewater originates from both natural sources and human activities. The latter include operations concerned with the nuclear fuel cycle, from mining to reprocessing; medical uses of radioisotopes; industrial uses of radioisotopes; worldwide fallout from atmospheric testing of nuclear devices; and enhancement of the concentration of naturally occurring radionuclides. Monitoring programs for water and wastewater should be designed to assess realistically the degree of environmental radioactive contamination. In some cases, for example, compliance monitoring for drinking water, the conditions are spelled out.[1] In others, it may be necessary to examine the individual situation[2] for consideration of the critical radionuclide(s), the critical pathway by which the critical radionuclide moves through the environment, and a critical population group that is exposed to the particular radionuclide(s) moving along this particular pathway. Use of the critical nuclide-pathway-population approach will help narrow the list of possible radionuclides to monitor.

A list of the most hazardous radionuclides can be selected by examining the radioactivity concentration standards given by the International Committee on Radiation Protection (ICRP)[3], the Federal Radiation Council (FRC)[4], the National Committee on Radiation Protection and Measurement (NCRP)[2], the U.S. Environmental Protection Agency[1], and also agencies in other countries. Individual states within the United States may have their own radioactivity concentration standards if they are Nuclear Regulatory Commission (NRC) agreement states. With few exceptions, these numerical values for radioactivity concentrations in air and water are comparable if certain qualifying assumptions are applied.

Monitoring programs should provide adequate warning of unsafe environmental conditions so that proper precautions can be taken, and of course, to assure that conditions are safe when they are indeed safe. In either circumstance, it is necessary to establish base lines for the kinds and quantities of radionuclides present naturally and to measure additions to this natural background. In this way, measurements may be made to provide information for sound judgments regarding the hazardous or nonhazardous nature of increased concentrations.

2. Types of Measurement

Meaningful measurements require careful application of good scientific techniques. The types of measurements to be made are determined by the objectives of the testing. Gross alpha and gross beta measurements are relatively inexpensive, can be completed quickly, and are useful for screening to determine whether further analysis for specific radionuclides is merited. However, gross measurements give no information about the isotopic composition of the sample, cannot be used to estimate radiation dose, and have poor sensitivity if the concentration of dissolved solids is high. Accurate gross beta and especially gross alpha measurements require careful preparation of standards to determine self-absorption and the ability to prepare samples in a similar manner.

Specific radionuclide measurements are required if dose estimates are to be made, results of gross analyses exceed a certain level, or long-term trends are being monitored. Specific analyses usually are more expensive and time-consuming than a gross analysis. Specific measurements identify radionuclides by the energy of emitted radiation, chemical techniques, half-life, or a combination of these characteristics. Gamma-emitting radionuclides can be measured rapidly and with a minimum of sample preparation by using gamma spectrometry. Measurements requiring chemical separations make it possible to increase the sensitivity by increasing the sample quantity measured.

Knowledge of the chemical and radiochemical characteristics of the radionuclide being measured is critical for satisfactory results. Gross alpha and gross beta results will not provide accurate information about radionuclides having energies significantly different from the energy of the calibration standard. During concentration of water samples by evaporation, radionuclides present in elemental form (e.g., radioiodine, polonium) or as compounds (e.g., tritium, carbon-14) may be lost by volatilization. If the sample is ignited, the chance of volatilization loss is even greater. Groundwater generally contains nuclides of the uranium and thorium series. Use special care in sampling and analyzing such samples because members of these series often are not in secular equilibrium.

3. References

1. U.S. ENVIRONMENTAL PROTECTION AGENCY. 1991. National Primary Drinking Water Regulations; Radionuclides; Proposed Rule. 40 CFR, Part 12. Federal Register 56, No. 38, Part II, USEPA, July 18, 1991.
2. NATIONAL COMMITTEE ON RADIATION PROTECTION AND MEASUREMENTS. 1959. Maximum Permissible Body Burdens and Maximum Permissible Concentrations of Radionuclides in Air and Water for Occupational Exposure. NBS Handbook No. 69, pp. 1, 17, 37, 38, & 93.
3. INTERNATIONAL COMMISSION ON RADIATION PROTECTION. 1979. Limits for Intakes of Radionuclides by Workers. ICRP Publ. 30, Pergamon Press, New York, N.Y.
4. FEDERAL RADIATION COUNCIL. 1961. Background Material for the Development of Radiation Protection Standards. Rep. No. 2 (Sept.), U.S. Government Printing Off., Washington, D.C.

---

* Approved by Standard Methods Committee, 1994.

4. Bibliography

CORYELL, C.D. & N. SUGARMAN, eds. 1951. Radiochemical Studies: The Fission Products. McGraw-Hill Book Co., New York, N.Y.

COMAR, C.I. 1955. Radioisotopes in Biology and Agriculture. McGraw-Hill Book Co., New York, N.Y.

FRIEDLANDER, G., J.W. KENNEDY & J.M. MILLER. 1964. Nuclear and Radiochemistry, 2nd ed. John Wiley & Sons, New York, N.Y.

LEDERER, C.M., J.M. HOLLANDER & I. PERLMANN. 1967. Table of Isotopes, 6th ed. John Wiley & Sons, New York, N.Y.

INTERNATIONAL ATOMIC ENERGY AGENCY. 1971. Disposal of Radioactive Wastes into Rivers, Lakes, and Estuaries. IAEA Safety Ser. No. 36, St. 1/PUB 283.

NATIONAL COUNCIL ON RADIATION PROTECTION AND MEASUREMENTS. 1976. Environmental Radiation Measurements. NCRP Rep. No. 50, Washington, D.C.

7010 B. Sample Collection and Preservation

1. Collection

The principles of representative sampling of water and wastewater apply to sampling for radioactivity testing (see Section 1060).

Because a radioactive element often is present in submicrogram quantities, a significant fraction may be lost by adsorption on the surface of containers used in the examination. Similarly, a radionuclide may be largely or wholly adsorbed on the surface of suspended particles.

Sample containers vary in size from 0.5 L to 18 L, depending on required analyses. Use containers of plastic (polyethylene or equivalent) or glass, except for tritium samples (use glass only). When radioactive industrial wastes or similar materials are sampled, consider the possibility of deposition of radioactivity on surfaces of glassware, plastic containers, and equipment that may cause a loss of radioactivity and possible contamination of subsequent samples collected in inadequately cleansed containers.

2. Preservation

For general information on sample preservation see Section 1060. Table 7010:I gives guidance for sample handling, preservation, and holding times for radionuclides in drinking water. Add preservative at time of collection unless sample will be separated into suspended and dissolved fractions, but do not delay acid addition beyond 5 d. Use conc hydrochloric (HCl) or nitric (HNO_3) acid to obtain a pH <2, except for radiocesium (use only conc HCl) and radioiodine, radon, and tritium (use no preservative). Hold acidified sample at least 16 h before analysis. For further details see references.[1-3]

Test preservatives and reagents for radioactive content.

3. Wastewater Samples

Wastewater often contains larger amounts of nonradioactive suspended and dissolved solids than does water and often most

TABLE 7010:I. SAMPLE HANDLING, PRESERVATION, AND HOLDING TIMES

Constituent	Preservative*	Container†	Maximum Holding Time‡§
Gross alpha	Conc HCl or HNO_3 to pH <2‖	P or G	1 year
Gross beta	Conc HCl or HNO_3 to pH <2‖	P or G	1 year
Radium-226	Conc HCl or HNO_3 to pH <2	P or G	1 year
Radium-228	Conc HCl or HNO_3 to pH <2	P or G	1 year
Radon-222	Cool 4°C#	G with TFE-lined septum	8 d**
Uranium, natural	Conc HCl or HNO_3 to pH <2	P or G	1 year
Radioactive cesium	Conc HCl to pH <2	P or G	1 year
Radioactive strontium	Conc HCl or HNO_3 to pH <2	P or G	1 year
Radioactive iodine	None	P or G	14 d
Tritium	None	G	1 year
Photon-emitters	Conc HCl or HNO_3 to pH <2	P or G	1 year

* (All except radon-222 samples). Add preservative at time of sample collection unless suspended solids activity is to be measured. If sample must be shipped to a laboratory or storage area, acidification (in original sample container) may be delayed for a period not to exceed 5 d. A minimum of 16 h must elapse between acidification and analysis.

† P = plastic, hard or soft; G = glass, hard or soft.

‡ Holding time is time from sampling to analysis. In all cases, analyze samples as soon after collection as possible.

§ A 1-year holding time allows for compositing four quarterly samples.

‖ If HCl is used to acidify samples to be analyzed for gross alpha or gross beta activities, convert the acid salts to nitrate salts before transfer of samples to planchets.

Cooling at 4°C is recommended. Large temperature changes will cause dissolved radon to outgas from sample. Sampling procedures are given elsewhere.[4,5]

** Holding should not exceed regulatory maximum (4 d), if applicable.

TABLE 7010:II. USUAL DISTRIBUTION OF COMMON RADIOELEMENTS BETWEEN THE SOLID AND LIQUID PHASES OF WASTEWATER

In Solution	In Suspension
HCO_3	Ce
Co	Cs
Cr	Mn
Cs	Nb
H	P
I	Pm
K	Pu
Ra	Ra
Rn	Sc
Ru	Th
Sb	U
Sr	Y
	Zn
	Zr

of the radioactivity is in the solid phase. Generally, the use of carriers in the analysis is ineffective without prior conversion of the solid phase to the soluble phase; even then high fixed solids may interfere with radioanalytical procedures. Table 7010:II shows the usual solubility characteristics of common radioelements in wastewater.

The radioelements may exhibit unusual chemical characteristics because of the presence of complexing agents or the method of waste production. For example, tritium may be combined in an organic compound when used in the manufacture of luminous articles; radioiodine from hospitals may occur as complex organic compounds, compared to elemental and iodide forms found in fission products from the processing of spent nuclear fuels; uranium and thorium progeny often exist as inorganic complexes rather than oxides after processing in uranium mills; the strontium-90 titanate waste from a radioisotope heat source is quite insoluble compared to most other strontium wastes. Valuable information on the chemical composition of wastes, the behavior of radioelements, and the quantity of radioisotopes in use appears in the literature.[6,7]

4. References

1. U.S. GEOLOGICAL SURVEY. 1977. Methods for Determination of Radioactive Substances in Water and Fluvial Sediments. U.S. Government Printing Off., Washington, D.C.
2. EML PROCEDURES MANUAL. 1990 (rev. 1992). HASL-300, 27th ed., Vol. 1. Environmental Measurements Lab., U.S. Dep. Energy, New York, N.Y.
3. U.S. ENVIRONMENTAL PROTECTION AGENCY. 1990. Manual for the Certification of Laboratories Involved in Analyzing Public Drinking Water Supplies. EPA-570/9-90-008, Washington, D.C.
4. U.S. ENVIRONMENTAL PROTECTION AGENCY. 1993. Method 913.0. Determination of Radon in Drinking Water by Liquid Scintillation Counting. Environmental Monitoring Systems Lab., Las Vegas, Nev.
5. U.S. ENVIRONMENTAL PROTECTION AGENCY. 1987. Appendix C, NIRS sampling instructions—Radon In Two Test Procedures for Radon in Drinking Water, Intercollaborative Study. EPA-600/2-87-082, ORD Environmental Monitoring Systems Lab., Las Vegas, Nev.
6. INTERNATIONAL ATOMIC ENERGY AGENCY. 1960. Disposal of Radioactive Wastes. International Atomic Energy Agency, Vienna, Austria.
7. NEMEROW, N.L. 1963. Industrial Waste Treatment. Addison-Wesley, Reading, Mass.

7020 QUALITY ASSURANCE/QUALITY CONTROL*

7020 A. Basic Quality Control Program

1. Introduction

Every laboratory that performs radionuclide analyses for environmental water and wastewater samples should have a written and operating quality assurance (QA) plan. This plan can be a separate document or can, by reference, use parts of existing standard operating procedures (SOPs). The quality-control (QC) portion of the QA plan addresses instrumental, background, accuracy, and precision QC. Also essential is a manual of analytical methods, or at least copies of approved methods, available to the analysts.

2. Performance Criteria

For a successful QC program, select acceptable and attainable performance criteria for precision and accuracy. These criteria must reflect the capabilities of the laboratory and the purposes for which the data are to be used.

Performance criteria can be drawn up initially from experience with the analytical method or from criteria set by other laboratories using the same procedure. A tabulation of allowable deviations used by the EPA[1] is given in Table 7020:I. The criteria are a function of the particular analysis under study. A laboratory might use these or other published values until enough data can be compiled to set its own criteria from experience.

3. Minimum Quality Control Program

A radiochemistry QC program is composed of a number of integrated functions, including instrumental, background, precision, and accuracy QC.

A useful way of keeping track of instrument and background performance is through control charts. Certain aspects of radio-

*Approved by Standard Methods Committee, 1996.

chemical instrument and background QC are instrument- or method-specific and are dealt with in the individual methods.

The instrumental control charts[2] are prepared by plotting counts of a reference source on a graph showing time as the abscissa and count rate or total counts as the ordinate. Lines are drawn parallel to the time axis at values (corrected for decay if necessary) for the "true" count rate and for values of ± 2 and ± 3 standard deviations. The true count rate is determined by averaging at least 20 counts with acceptable individual statistics.

Interpret quality control chart data objectively. When a point goes outside the limits, determine whether instrument service is necessary or the result is simply a random occurrence by running a series of repeated measurements and applying another statistical test, such as a Chi-square test, to determine whether the variation was nonstatistical.

Trends from control charts may show other information. For example, if regular measurements of the check source show a movement in one direction, one can infer that some system variable is changing. This variation may not always require instrument service; instead the values of the limits or the "true" value may need to be reevaluated.

a. Instrumental QC: For alpha and beta counters, count check source daily (or before each use) for a predetermined time. Record count rate and plot it on the control chart for the specific system. Compare this value with the $\pm 2\,\sigma$ (warning) limits and the $\pm 3\,\sigma$ (out-of-control) limits, and repeat the procedure if the $\pm 2\,\sigma$ boundary is exceeded. Take appropriate action if repeated values are above the warning levels.

For instruments that produce spectra, such as gamma or alpha spectrometers, many parameters can be tracked, including efficiency response, i.e., count rate or total counts in a given spectral area; peak channel location of one or more spectral photopeaks; difference in channels between two specified peaks; resolution, i.e., width of peak in channels at specified peak height; and certain ratios such as "peak to compton." It is not necessary to track all of these routinely; see specific recommendations in Section 7120. If these basic parameters are outside the limits, other parameters may need to be evaluated.

b. Background QC: At defined frequencies, such as daily or before each use for proportional or liquid scintillation counters, count background for each system for the standard counting time. Make background measurements with each batch of samples. Spectrometers may require specific treatment because background is determined from the sample spectrum in the analysis of activity. In this specific case, longer background counts at less frequent intervals (for example, weekly) may be needed to produce reasonable counts in specific regions of interest.

Record background counts and plot them on control charts. Obtain background "true value" by averaging at least 20 background counts. Calculate 2 and 3 σ counting errors based on the average value and the background counting time. Take appropriate actions in the event of background or standard reference count problems.[2]

Background QC is specific to the type of counting instrument. For instance, gamma spectroscopy background QC represents a specific challenge depending on whether the detector is NaI(Tl) or germanium; see Section 7120.

Analyze reagent blanks often enough to ensure the absence of major interference that would bias the reported results. Ideally, results for a reagent blank will be identical to the background, e.g., for an analysis using a proportional counter, identical to the counts from a blank planchet.

c. Precision QC: Evaluate laboratory's internal precision in performing an analytical procedure by analysis of blind duplicate samples, that is, duplicate portions of randomly selected samples submitted as new samples. It is important that the analyst be unaware either that a given sample will be resubmitted or that it is a reanalysis. This may be difficult in a small laboratory. It is

TABLE 7020:I. LABORATORY PRECISION—ONE STANDARD DEVIATION VALUES FOR VARIOUS ANALYSES IN WATER SAMPLES

Analysis	Activity Level pCi/L	One Standard Deviation for Single Determination	Control Limits*
Gamma-emitters	5 to 100	5 pCi/L	$\mu \pm 8.7$ pCi/L
	>100	5% of known value	$\mu \pm 0.087\,\mu$
Zinc-65, barium-133, ruthenium-106	5 to 50	5 pCi/L	$\mu \pm 8.7$ pCi/L
	>50	10% of known value	$\mu \pm 0.17\,\mu$
Strontium-89, strontium-90	5 to 100	5 pCi/L	$\mu \pm 8.7$ pCi/L
	>100	5% of known value	$\mu \pm 0.087\,\mu$
Gross alpha	≤20	5 pCi/L	$\mu \pm 8.7$ pCi/L
	>20	25% of known value	$\mu \pm 0.43\,\mu$
Gross beta	≤50	5 pCi/L	$\mu \pm 8.7$ pCi/L
	>50 to 100	10 pCi/L	$\mu \pm 17.3$ pCi/L
	>100	15% of known value	$\mu \pm 0.26\,\mu$
Tritium	<4000	$170 \times (known)^{0.0933}$	$\mu \pm 294 \times (\mu)^{0.0933}$
	≥4000	10% of known value	$\mu \pm 0.17\,\mu$
Radium-226	≥0.1	15% of known value	$\mu \pm 0.26\,\mu$
Radium-228	≥0.1	25% of known value	$\mu \pm 0.43\,\mu$
Iodine-131	≤55	6 pCi/L	$\mu \pm 10.4$ pCi/L
	>55	10% of known value	$\mu \pm 0.17\,\mu$
Uranium	≤35	3 pCi/L	$\mu \pm 5.2$ pCi/L
	>35	10% of known value	$\mu \pm 0.17\,\mu$

*Average of three determinations. μ = known value.

important also that samples submitted for duplicate analysis have detectable amounts of radioactivity so that statistical treatment of data does not consist of comparing zeros or "less than" values. In some cases, it may be necessary to substitute samples with known additions for routine workload samples.

Preferably analyze 10 percent duplicate samples to verify internal laboratory precision for a specific analysis; this is a general guideline and may be varied to fit the situation. For example, for a laboratory with a heavy workload and a well-established duplicate analysis program, 5 percent duplicate samples or 20 samples per month may be sufficient to determine whether the data meet established criteria. A discussion of the statistical treatment of duplicate analysis data is given elsewhere.[2,3] One criterion for acceptability of duplicate measurements is given in EPA's Drinking Water Certification Manual. The difference between duplicate measurements should be less than two times the standard deviation of the specific analysis as described in Table 7020:I.[4] If the difference between duplicates exceeds two standard deviations, prior measurements are suspect; examine calculations and procedures and reanalyze samples when necessary.

d. Accuracy QC: Analytical methods are said to be in control when they produce results that are both precise and unbiased. Evaluate accuracy by preparing or obtaining water samples of known radionuclide content, analyzing them, and comparing the results to the known values. Plot data from these standards or known-addition samples on means or individual results control charts,[2,3] and determine whether or not results from these samples are within the control limits and/or warning limits set on the control charts. Laboratories also can set accuracy limits for samples with known additions based on percent recovery of the known value. Preferably run a known-addition or standard sample with each batch of samples. If the known-addition value is outside the limits, investigate method bias. All sample results are suspect until the known-addition results are back within the prescribed limits.

To be certified to perform drinking water analyses under the Safe Drinking Water Act, a laboratory must participate successfully at least three times each year in those EPA laboratory performance evaluation studies that include each of the analyses for which the laboratory is, or wants to be, certified. Two of these studies must be constituent-specific and one must be a mixed blind performance evaluation sample. Analytical results must be within control limits established by EPA for each analysis for which the laboratory is, or wants to be, certified.[1]

Laboratories are encouraged to participate in intercomparison programs such as those sponsored by USEPA,[1] International Atomic Energy Agency (IAEA),† and World Health Organization (WHO).‡

e. Selection of radionuclide standards and sources: Calibrated radionuclide standard solutions are prepared for storage and shipment by the supplier in flame-sealed glass ampules. Preferably perform all dilutions and storage of radionuclides in glass containers and avoid the use of polyethylene.

Standard sources are radioactive sources having adequate activity and an accurately known radionuclide content and radioactive decay rate or rate of particle or photon emission. Each radionuclide standard should have a calibration certificate containing the following information:

Description:	Purity:
Principal radionuclide	Identification of impurities
Chemical form	Activity
Solvent	Included or not included in
Carrier and content	principal activity
Mass and specific gravity or volume	Assumptions:
Standardization:	Decay scheme
Activity per mass or volume	Half-life
Date and time	Equilibrium ratios
Activity of daughter	Production:
Method of standardization	Production method
Accuracy:	Date of separation
Repeatability error	Usable lifetime
Systematic error	
Overall error	
Confidence levels	

Confirm that radionuclide standard sources are traceable to the National Institute of Standards and Technology (NIST). Use such standard sources, or dilutions thereof, for initial calibration. These sources may be purchased from suppliers listed in annually published buyers' guides. Before purchasing standards from commercial suppliers, inquire as to the traceability of the particular radionuclide of interest. A good discussion of NIST traceability is given elsewhere.[5] Example of radionuclide calibration certificates and NIST traceability certificates have been published.[1] At present, standardized radioisotope solutions can be purchased from the IAEA† and NIST.§ Participants in USEPA's Performance Evaluation Studies Program can obtain standardized radionuclide solutions on request.

Use check sources for determining changes in counting rate, counting efficiency, and/or energy calibration. These sources should be of sufficient radiochemical purity and activity to permit correction for decay, but they need not have an accurately known disintegration rate (i.e., need not be a standard source).

Standard reference materials (SRMs) are radioactive materials having adequate activity and accurately known radionuclide content and radioactive decay rate or rate of particle or photon emission. They may be used as internal laboratory control samples, internal tracers, or matrix and blind known additions. They should have calibrations traceable to NIST or equivalent.

4. References

1. DILBECK, G. & P. HONSA. 1994. Environmental Radioactivity Performance Evaluation Studies Program and Radioactive Standards Distribution Program. ORD EMSL-LV, U.S. Environmental Protection Agency, Las Vegas, Nev.

2. KANIPE, L.G. 1977. Handbook for Analytical Quality Control in Radioanalytical Laboratories. E-EP/77-4,TVA,EPA-600/7-77-038, Section 3, U.S. Environmental Protection Agency, Washington, D.C.

3. ROSENSTEIN, M. & A.S. GOLDIN. 1964. Statistical Techniques for Quality Control of Environmental Radioassay. AQCS Rep. Stat-1, U.S. Dep. Health, Education & Welfare, PHS, DRH, Winchester, Mass.

---

† International Atomic Energy Agency, Analytical Quality Control Services, Seibersdorf, P.O. Box 100, A-1400, Vienna, Austria.
‡ World Health Organization, Geneva, Switzerland.

§ U.S. Department of Commerce, Technology Administration, National Institute of Standards and Technology, Standard Reference Materials Program, Building 202, Room 204, Gaithersburg, MD 20899.

4. U.S. ENVIRONMENTAL PROTECTION AGENCY. 1990. Manual for the Certification of Laboratories Analyzing Drinking Water. EPA/570/9-90/008, OGWDW (WH-550D), Office of Ground Water and Drinking Water, Washington, D.C.

5. NATIONAL COUNCIL ON RADIATION PROTECTION AND MEASUREMENTS. 1985. A Handbook of Radioactivity Measurements Procedures. NCRP Rep. No. 58, National Council on Radiation Protection and Measurements, Bethesda, Md.

7020 B.   Quality Control for Wastewater Samples

Generally it is not feasible to perform collaborative (interlaboratory) analyses of wastewater samples because of the variable composition of elements and solids from one facility to the next. The methods included herein have been evaluated by use of homogeneous samples and are useful for nonhomogeneous samples after sample preparation (wet or dry oxidation and/or fusion and solution) resulting in homogeneity. Reference samples used for collaborative testing may be deficient in radioelements exhibiting interferences because of decay during shipment of short-half-life-radionuclides. Generally, however, analytical steps incorporated into the methods eliminate these interferences, even though they may not be necessary for the reference samples.

7020 C.   Statistics

Section 1030 discusses statistics as applied to analysis of chemical constituents. It is applicable also to radioactivity examinations; however, certain statistical concepts peculiar to radioactivity measurements are discussed below.

1. Propagation of Errors

Often it is necessary to calculate the uncertainty of a quantity that is not measured directly, but is derived, by means of a mathematical formula, from directly measured quantities. The uncertainties of the latter are known or can be computed and the uncertainty of the calculated quantity derived from them. Statistically, this is known as propagation of errors. One of the more common applications of propagation of errors is in combining all the sources of error in determining pollutant concentrations in environmental samples such as soil, air, milk, or water. These data include errors from sampling, analysis, and other variables, all of which must be considered in determining the total variability or variance.

The formula for propagation of errors states that when quantities are added or subtracted, the combined error (σ_T) is equal to the square root of the sum of the squares of the individual errors.

$$\sigma_T = (\sigma_1^2 + \sigma_2^2 + \sigma_3^2 + \ldots\ldots + \sigma_n^2)^{1/2}$$

The propagation of errors law application is possible because variance is an additive property:

$$\sigma_T^2 = \sigma_1^2 + \sigma_2^2 + \sigma_3^2 + \ldots\ldots + \sigma_n^2$$

A number of propagation-of-error formulas have possible application to the determination of radionuclide concentrations in water.[1] The most common of these are given in Table 7020:II. The one most widely used in nuclear counting statistics is the first formula, where X represents the activity (counts) of sample + background, and Y represents the activity of the background.

TABLE 7020:II. PROPAGATION-OF-ERROR FORMULAS

Function	Error Formula
$Q = X \pm Y$	$\sigma_Q = (\sigma_x^2 + \sigma_y^2)^{1/2}$
$Q = aX + bY$	$\sigma_Q = (a^2\sigma_x^2 + b^2\sigma_y^2)^{1/2}$
$Q = XY$	$\sigma_Q + XY\left(\dfrac{\sigma_x^2}{X^2} + \dfrac{\sigma_y^2}{Y^2}\right)^{1/2}$
$Q = \dfrac{X}{Y}$	$\sigma_A = \dfrac{X}{Y}\left(\dfrac{\sigma_x^2}{X^2} + \dfrac{\sigma_y^2}{Y^2}\right)^{1/2}$

2. Standard Deviation and Counting Error

The variability of any measurement is described by the standard deviation, which can be obtained from replicate determinations. There is an inherent variability in radioactivity measurements, due to the random nature of radioactive decay, which is described by the Poisson distribution. This distribution is characterized by the standard deviation of a large number of events, N, that equals its square root, or:

$$\sigma(N) = N^{1/2}$$

When $N > 20$, the Poisson distribution approximates the normal (Gaussian) distribution. This approximation simplifies the computation of confidence intervals and permits reasonably accurate estimates of the mean and variance of the distribution of results without performance of replicate counts.

More often, the variable of concern is the standard deviation in the counting rate (number of counts per unit time):

$$R' = \frac{N}{t}$$

where:

t = duration of counting.

The standard deviation of the counting rate, when the appropriate substitutions are made, is:

$$\sigma(R') = \frac{N^{1/2}}{t} = \frac{(R't)^{1/2}}{t} = \left(\frac{R'}{t}\right)^{1/2}$$

In practice, all counting instruments have a background counting rate, B, when no sample is present. With a sample, the counting rate increases to R_o. The net counting rate, R, due to the sample is:

$$R = R_o - B$$

By propagation-of-error methods, the standard deviation of R, the net counting rate, is calculated as follows:

$$\sigma(R) = \left(\frac{R_o}{t_1} + \frac{B}{t_2}\right)^{1/2}$$

where:

R_o = gross counting rate,
B = background counting rate, and
t_1, t_2 = elapsed counting times at which gross sample and background counting rates were measured, respectively.

Counting duration for a given set of conditions depends on the limit of detection required (see below). Preferably divide the counting time into equal periods to check constancy of observed counting rate. For low-level counting, where net count rate is of the same order of magnitude as the background, use $t_1 = t_2$. The error thus calculated includes only uncertainty caused by inherent variability of the radioactive disintegration process and is not the standard deviation of the total analysis. The counting uncertainty is the major portion of the total uncertainty at or near the limit of detection. As concentration levels increase, the percent counting error decreases and systematic errors become the major portion of total error.

Use a confidence level of 95%, or 1.96 standard deviations, as the counting error. Report radioactivity concentration results with the counting error as $X \pm 1.96\sigma$, both in pCi/L.

3. Limit of Detection

Various conventions have been used to estimate the lower limit of detection (LLD) or the minimum detectable activity (MDA). The procedure recommended here, used at the Environmental Measurements Laboratory,[2] is based on hypothesis testing.[3]

The LLD is defined as the smallest quantity of sample radioactivity that will yield a net count for which there is a predetermined level of confidence that radioactivity will be measured. Two errors may occur: Type I, in which a false conclusion is reached that radioactivity is present, and Type II, with a false conclusion that radioactivity is absent.

The LLD may be approximated as follows:

$$LLD = (K_\alpha + K_\beta)S_o$$

where:

K_α = value for upper percentile of standardized normal variate corresponding to preselected risk of concluding falsely that activity is present (α),
K_β = corresponding value for predetermined degree of confidence for detecting presence of activity (1-β), and
S_o = estimated standard error of net sample counting rate.

For sample and background counting rates that are similar (as is expected at or near the LLD) and for α and β equal to 0.05, the smallest amount of radioactivity that has a 95% probability of being detected is,

$$LLD_{95} = 4.66\ S_b$$

where:

S_b = standard deviation of background counting rate, cpm.

To convert LLD to concentration, use the appropriate factors of sample volume, counting efficiency, etc. Note that the approximation $LLD = 4.66\ S_b$ can be used only for determinations where S_b is known so that $S_o = 2^{1/2}S_b$ and there are no counting interferences. Examples of appropriate determinations are tritium, gross alpha or beta, or any single nuclide determination.

Where tracers are added to determine yield or more than one radionuclide is counted in a sample, use the general form of the above equation, for which the 95% confidence level would be:

$$LLD_{95} = 3.29S_o$$

4. References

1. OVERMAN, R.T. & H. CLARK. 1960. Radioisotope Techniques. McGraw Hill Book Co., New York, N.Y.
2. U.S. DEPARTMENT OF ENERGY. 1992. EML Procedures Manual, 27th ed. (rev.). HASL 300, p.4.5.29, Environmental Measurements Lab., U.S. Dep. Energy, New York, N.Y.
3. CURRIE, L.A. 1968. Limits for qualitative detection and quantitative determination—application to radiochemistry. Anal. Chem. 68:586.

7020 D. Calculation and Expression of Results

The results of radioactivity analyses usually are reported in terms of "activity" per unit volume or mass at 20°C. The recognized unit for activity is the Becquerel, Bq, equal to one disintegration per second. A commonly used unit for reporting environmental level concentrations is the picocurie (pCi) = 2.22 disintegrations per minute (dpm), or 1×10^{-12} Curie (Ci) or 1×10^{-3} nanocuries (nCi) or 1×10^{-6} microcuries (μCi).

Specific formulas for the calculation of activity per volume or mass are presented in the individual methods and use the general formula:

$$C = \frac{R_{net}}{e\,y\,i\,v\,d\,u}$$

where:

C = activity per unit volume, in units or activity/mass or volume,
R_{net} = net counting rate, cpm
e = counting efficiency, cpm/dpm,
y = chemical yield,
i = ingrowth correction factor,
v = volume or mass or portion,
d = decay factor, and
u = units correction factor.

Values for variables are method-dependent. Report results in a manner that does not imply greater or lesser precision than that obtained by the method (see Section 1030).

In reporting radiochemical data, include associated random and systematic errors, as well as minimum detectable activity. For some intended data usage, it is necessary to report the result as calculated without regard to the sign of the absolute value, i.e., no "less than" values. The objective of data use, e.g., compliance monitoring, research, dose calculation, or trend monitoring, will dictate the data reporting format.

The following formula illustrates calculation of counting uncertainty at the 95% confidence level:

$$E = \frac{1.96 \left(\dfrac{R_o}{t_1} + \dfrac{B}{t_2} \right)^{1/2}}{e\,y\,i\,v\,d\,u}$$

where:

E = counting error,
R_o = gross sample counting rate, cpm,
t_1 = sample count duration, min,
B = background counting rate, cpm,
t_2 = background count duration, min, and
e, y, i, v, d, and u are as previously defined.

7030 COUNTING INSTRUMENTS*

7030 A. Introduction

Radiochemical analytical instruments operate on the principle that the expenditure of energy by a radiation event is detected and recorded by an instrument suitable for the type of emitted radiation. A description of typical counting instrumentation is presented but is not intended to be all-inclusive. Instrument background and fractional counting efficiency must be measured and integrated into the sample calculations. These characteristics can be compared with historical data and used to evaluate instrument stability.

---

* Approved by Standard Methods Committee, 1994.

7030 B. Description and Operation of Instruments

1. Gas-Flow Proportional Counters

Gas-flow proportional counters operate on the principle that radioactive particles, e.g., alpha and beta particles, cause ionization in the gas with multiplication of resulting electrons and collection at the anode of the counter. The voltage pulse is proportional to the impressed voltage and the number of initial ion pairs formed, hence the term "proportional counter." Alpha particles produce a much higher pulse than beta particles; this provides a means to distinguish between the two.

a. Windowless internal proportional counter:

1) Principle and uses—The internal gas-flow proportional counter accepts planchets within the counting chamber and, at the operating voltage, records the current pulses produced by the radioactive particles or photons emitted into the counting gas. Internal proportional counters are suitable for determining activity at the alpha operating plateau and alpha-plus-beta activity at the beta operating plateau. The alpha or beta activity, or both, may refer to the type of radiation produced by the radionuclides present in the sample. The energy of the activity present in the sample has an effect on the counter detection efficiency and the counting process. These counting effects usually are accounted for during the calibration of the instrument.

Theoretically, half the radiation is emitted in the direction of the planchet and half in the direction of the detector. Depending on the energy or energies of the isotope in the sample, a percentage of the alpha or beta radiation is back-scattered into the counting gas by the sample solids, the planchet, or the walls of the counting chamber. Some of the beta radiation, but only 1 to 2% of the alpha radiation, is back-scattered into the counting gas by sample solids, the planchet, or the walls of the counting chamber, yielding in some cases greater than 50% counting efficiency. For nearly weightless samples, considerably more than 60% of the beta radiation is counted. Take considerable care in sample preparation to prevent the sample or planchet from distorting the electrical field of the counter and depressing the count rate. Avoid nonconducting surfaces, airborne dusts, and vapor from moisture or solvents, because these can damage the counter detector.

2) *Components*—The instrument consists of a counting chamber, a preamplifier, amplifier, scaler, high-voltage power supply, timer, and register. Use the specified counting gas and accessories, make adjustments for sensitivity, and operate in accordance with manufacturer's instruction.

3) *Performance verification*

a) *Radioactive sources*—See Section 7020A.3*e*.

b) *Plateau (alpha or beta)*—Find the operating voltage where the counting rate is consistent, i.e., varies less than 5% over a 150-V change in anode voltage. Check this plateau or, as a minimum, make a response check after each change of counting gas. Determine the plateau by placing a source in the counting chamber and varying the detector voltage until the count rate remains constant. Further increases in high voltage result in little change in the overall detector response until the plateau region is exceeded. CAUTION: *Instrument damage will result from continuous discharge at too high a voltage.* Plot relative counting rate (ordinate) against anode voltage (abscissa). Select an anode voltage near the center of the plateau as the detector operating voltage (see Figure 7030:1).

c) *Background (alpha or beta)*—Characteristic of most detectors is a background or instrument count rate usually due to cosmic radiation, radioactive contaminants in instrument parts, counting room construction material, electrical/electronic noise, and/or the proximity of radioactive sources. The background is roughly proportional to the size or mass of the counting chamber or de-

tector, but it can be reduced by shielding or anti-coincidence guard circuitry. The instrument background count rate should not be confused with the count rate of the laboratory or reagent blanks. However, under some circumstances these count rates might be approximately equal.

Determine instrument background with an empty planchet in the counting chamber. Use a background counting duration as long as the longest sample-counting duration.

d) *Initial calibration (alpha or beta)*—The purpose of initial calibration is to adjust counting efficiency according to the sample thickness (which causes self-absorption of some of the alpha or beta particles within the sample matrix). Initial calibration is procedure-dependent. See individual procedures for specific initial calibration instructions. Absorption curve values do not need to be reestablished after initial calibration if check source response is monitored regularly and indicates instrument stability.

e) *Continuing calibration (source check)*—Verify instrument stability at the operating voltage by counting the check source routinely (see Section 7020A.3*e*). If source count is within two standard deviations of the previously determined average count rate, instrument reliability/stability is established. If the source count rate is not so reproduced, repeat the test. If stability is not attained, service the problem equipment.

f) *Sample counting*—Place prepared sample on a planchet in the counting chamber and assure electrical contact between planchet and chamber. Flush with counting gas and count for a preset duration or a preset count, to give the desired counting precision (see Section 7020D). Count samples in a geometry consistent with initial calibration.

b. *Thin-window proportional counter:*

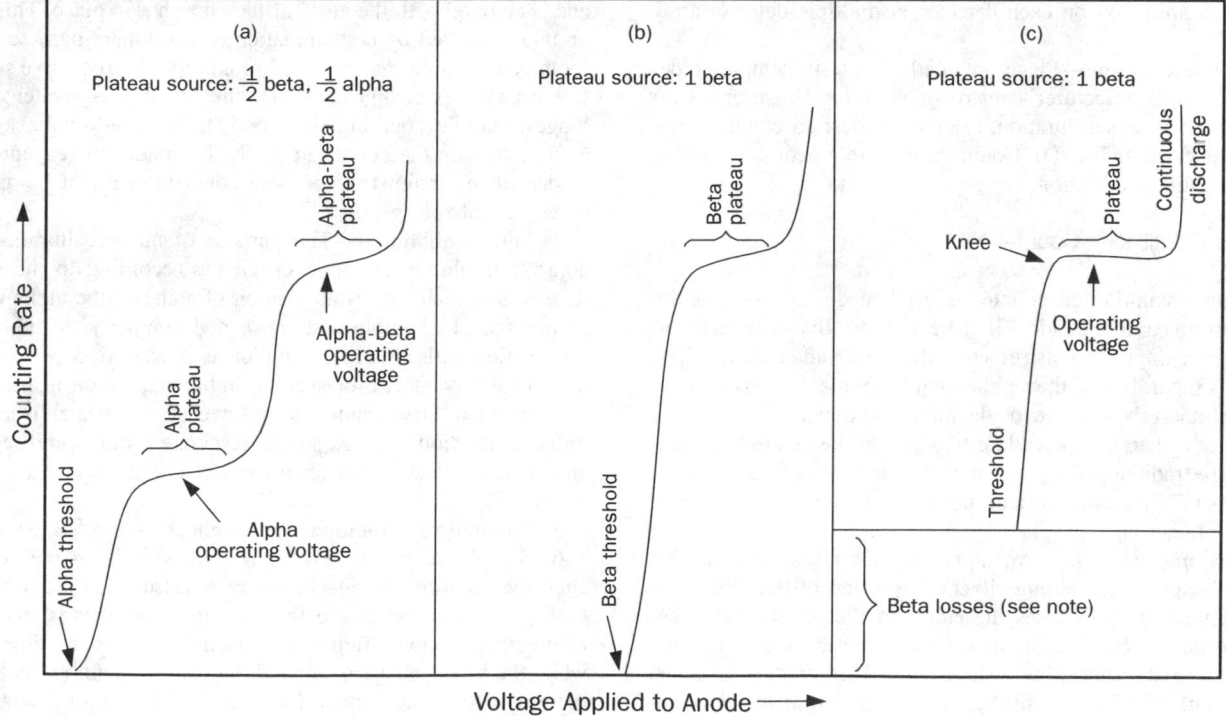

Figure 7030:1. Shape of counting rate–anode voltage curves. Key: (a) and (b) are for internal proportional counter with P-10 gas; (c) is for end-window Geiger-Mueller counter with Geiger gas. (NOTE: Beta losses are dependent on energy of radiation and thickness of window and air path.)

1) *Principle and uses*—The thin-window proportional counter, with or without heavy shielding and/or anti-coincidence guard circuitry, has application for counting various levels of alpha and beta activity. It is approximately 50% as sensitive as an internal proportional counter because the geometry of counting is not as favorable and absorption losses (air path and window) are greater. Because the sample is outside the counting gas, this detector is less affected than the internal proportional counter by contamination from loose residues, losses due to residual moisture, and poor electrical conductance.

Most modern radiochemistry laboratories use shielded counters equipped with anti-coincidence guard circuitry (low-background alpha/beta counter) that will easily give high sensitivity.

Some thin-window detector system manufacturers also offer simultaneous alpha/beta counting. Options of single sample, automatic sample changer, multiple detector arrays, or simultaneous alpha/beta counting can be chosen.

2) *Components*—See ¶ *a*2) above.

3) *Performance verification*

a) *Radioactive sources*—See Section 7020A.3*e*.

b) *Plateau*—See ¶ *a*3)b) above. For multiple detector array systems, determine the plateau on each detector. To select a plateau point for simultaneous counting systems, follow manufacturer's recommendations. Adjust "cross talk" (defined as the percentage of alpha counts represented in the beta channel or the percentage of beta counts in the alpha channel) and monitor it on all systems set up for simultaneous alpha/beta counting. Determine cross-talk values during detector calibration and adjust detector counts to meet needs for a given test.

c) *Background*—See ¶ *a*3)c) above. Measure background on each detector when using multiple-detector array systems.

d) *Initial calibration*—See ¶ *a*3)d) above.

e) *Continuing calibration*—See ¶ *a*3)e) above. Make a check source determination on each detector in multiple-detector array systems.

f) *Sample counting*—Place prepared sample in sample holder according to manufacturer's instructions and set instrument for preset count or preset duration, to give the desired counting precision (see Section 7020D). Count samples in a geometry consistent with initial calibration.

2. Alpha Scintillation Counter

An alpha scintillation counter is used to detect and quantify alpha-emitting radionuclides. If there is more than one radionuclide in the sample, the instrument will detect and count all alpha emissions regardless of their radionuclide source or energy. However, the analysis may be made more radionuclide-specific by using a radiochemical procedure to separate the desired radionuclides. The radionuclide (e.g., naturally-occurring ^{226}Ra, Th, U, etc.) is usually precipitated and mounted as a thin layer (≤ 5 mg/cm^2) on planchets.

a. Principle and uses: An alpha particle interacts with zinc sulfide phosphor (containing silver). A portion of the alpha particle's kinetic energy causes the atoms of the scintillator to become excited. When the atoms return to the ground state, they release the extra energy as visible light. This process is called scintillation. The light is further transformed into an electrical current by an attached photomultiplier tube that amplifies the electrical current into a measurable pulse. These pulses trigger a scaler and the pulse is registered as a "count." Depending on the amount of radionuclide present and required statistics, count a sample long enough to obtain required sensitivity. The counter is calibrated with a thin-layered precipitate of a radionuclide or an electrodeposited radionuclide.

b. Components: The alpha scintillation counter consists of a light-tight sample chamber with a phosphor detector coupled to a photomultiplier tube (PMT) and sample holder, a high-voltage supply, an amplifier-discriminator, a scaler, and readout capability. Generally, the photomultiplier tube has a window diameter greater than that of the samples. Locate the phosphor between sample and photomultiplier tube. Distance between sample surface and face of the PMT usually is about 3 to 5 mm. Arrange phosphor so that the photomultiplier tube window is in optical contact with the scintillator. Under these conditions, the counting efficiency can be 35 to 40%. Radon-222 (radium-226 by radon emanation) also can be counted with an alpha scintillation cell (modified Lucas cell) and a modified sample chamber (see Section 7500-Ra.C). For details on operation and calibration of an alpha scintillation counter, see manufacturer's instructions.

c. Performance verification:

1) *Radioactive source*—See Section 7020A.3*e*.

2) *Plateau*—Find the operating voltage using a radioactive source where the count rate is consistent over some specified voltage range according to the manufacturer's recommendations. See ¶ 1*a*3)b) for additional discussion.

3) *Background*—Two backgrounds need to be considered when dealing with alpha scintillation counters.

a) The electronic instrument background characteristically measures the electronic noise of the system. It is determined with an empty counting chamber and should vary from 0 to 1 cpm. The electronic background data are used to evaluate the performance of the instrument.

b) The second type of background is the chamber background and is counted with the zinc sulfide phosphor in place. This background is caused by contamination of instrument parts, counting room construction material, and proximity of radioactive sources. Use a chamber background counting duration equivalent to the longest sample counting duration. Do not confuse background count rate with the count rate of the laboratory or reagent blank. Under some circumstances, these count rates might be approximately equal.

4) *Initial calibration*—The purpose of initial calibration is to adjust scintillation counting efficiency according to the sample thickness (which causes absorption of some of the alpha or beta particles in the sample matrix) or cell geometry in the case of scintillation cells. Initial calibration is procedure-dependent. See individual procedures for specific initial calibration instructions.

Absorption curve values do not need to be reestablished after initial calibration if check source response is monitored regularly and indicates instrument stability. Recalibrate scintillation cells periodically.

5) *Continuing calibration (source check)*—See ¶ 1*a*3)e).

6) *Sample counting*—Place prepared sample in the counting chamber according to manufacturer's instructions. Take the following precautions: assure that counting chamber is light-tight; assure that photomultiplier tube is not exposed to direct light while the high voltage is applied; let sample chamber dark adapt before starting count; count for a preset duration, or preset count, to give the desired counting precision; assure that the sample is in contact with the phosphor; and count samples in a geometry consistent with initial calibration.

3. Liquid Scintillation Counters

A liquid scintillation spectrometer system is used to detect the photons of light emitted from a scintillation solution, by one or more photomultiplier tubes. Scintillation spectrometers count the number of scintillations (photons) emitted from the sample vial. To a first approximation, the number of photons produced in the scintillator is proportional to the initial energy of the particle. This information is an aid to determining the specific radionuclide present.

a. Principle and uses: A sample containing beta-(or alpha-) emitting radionuclides is mixed within a liquid scintillator. The beta (or alpha) particle is detected by the interchange of energy with the detection medium. The particle excites the solvent molecules, which pass on their excitation energy to the solvent by a collision process. The excited solute molecule rapidly returns to its ground state by the emission of a photon. The intensity of the scintillation depends on the number of solute molecules excited and the initial energy of the particle. Liquid scintillation is commonly used to determine activities of some alpha (i.e., ^{222}Rn and daughters) and most low-energy beta-emitters such a ^{3}H and ^{14}C. Counting efficiencies approaching 100% can be obtained from some radionuclides.

b. Components: The liquid scintillation counting system consists of the liquid scintillator (organic scintillator diluted with an appropriate solvent), a polyethylene or glass vial, and a liquid scintillation counter (with one or more photomultiplier tubes coupled to a single or multichannel analyzer), and readout device.

c. Performance verification:

1) Radioactive sources—See Section 7020A.3*e.*

2) Background—Consider two backgrounds when dealing with liquid scintillation spectrometers.

a) Electronic or instrument background—See ¶ 2*c*3)a)above. Determine background with an empty counting chamber and a dark vial. Some manufacturers supply a dark or "black" vial (a counting vial filled with black material such as graphite) for this purpose.

b) Chamber background—See ¶ 2*c*3)b) above. Determine background using a vial (sometimes supplied by the manufacturer) containing liquid scintillator (cocktail) and an appropriate volume of low-background water (water with no detectable activity). Use a chamber background counting duration equivalent to the longest sample counting duration.

3) Initial calibration—Before sample analysis, optimize the counting conditions for the radionuclide of interest (e.g., some analysts maximize the figure of merit, E^2/B where E = efficiency and B = background count rate) for each radionuclide. Optimum counting conditions do not need to be reestablished after initial calibration if check source response is monitored regularly and indicates instrument stability. Once optimum conditions have been selected by using a pure source, determine sample counting efficiency by one of several methods (see Section 7500-^{3}H and 7500-Rn): standard additions, quench curve, or prepared laboratory standard.

4) Continuing calibration (source check)—See ¶ 1*a*3)e). See Section 7500-Rn for specific calibration procedures for radon-222 by liquid scintillation counting.

5) Sample counting—Place prepared samples in the counting chamber, let samples dark adapt, and count for a preset count or preset duration necessary to obtain the desired sensitivity.

As with other counting techniques, deal with the following interferences: quenching (photon, chemical, or color), chemiluminescence/photoluminescence, static electricity, scintillation volume variations, homogeneity of sample, background, multiple radionuclides, and phase separation.

4. Alpha Spectrometers

a. Principle and uses: Alpha emissions from a sample interact with atoms of a solid-state detector to produce a current proportional to the deposited alpha particle energy. The current is collected, amplified, sorted according to deposited alpha energy, and displayed on a multichannel analyzer. An alpha spectrometer is used to detect, identify, and quantify specific alpha-emitting radionuclides. These radionuclides may be chemically separated from samples and deposited as a thin layer on filter papers or electrodeposited on metal disks. Ideally, use internal tracers on an individual sample basis to ensure accurate quantitation.

b. Components: An alpha spectrometer consists of a sample chamber (with detector, detector/sample holder, and vacuum chamber), a mechanical vacuum pump, a detector bias voltage supply, a preamplifier, a linear amplifier, a multichannel analyzer (with ADC and memory storage), and data readout capability. For details on operation and calibration, see the operation manual provided by the manufacturer.

The detector may be a semiconductor particle detector, that is, a silicon surface detector used for alpha spectrometry. Other charged particle detectors are available from various manufacturers.

c. Performance verification:

1) Radioactive source—See Section 7020A.3*e.*

2) Detector voltage—Set detector operating voltage in accordance with the manufacturer's recommendations.

3) Background—The chamber background consists of counting a clean filter or metal disk. Use a chamber background counting duration equivalent to the longest sample counting duration. This background is caused by contamination of the sample chamber. Do not confuse the background count rate with the count rate of the laboratory or reagent blank. However, under some circumstances, these count rates might be approximately equal.

4) Initial calibration—The purpose of initial calibration is to establish a detector energy calibration and counting efficiency. Use a standard source as described in Section 7020A.3*e.* Before counting, for initial/continuing calibration purposes, ensure that the counting vacuum chamber is well sealed and has an adequate as well as consistent vacuum source. Inadequate vacuum will result in a distorted alpha spectrum. Inconsistent application of vacuum can cause significant drift of the alpha spectra.

5) Continuing calibration (source check)—Verify detector efficiency and energy calibration stability. A mixed alpha-emitting isotopic source (see Section 7020A.3*e*) allows confirmation of the energy calibration by permitting identification of specific alpha-emitting radionuclides. For standard deviation criteria, see ¶ 1*a*3)e).

6) Sample counting—Place prepared sample in the counting chamber according to manufacturer's instruction. Take following precautions: assure that the air in the counting chamber is slowly evacuated to at least 500 μm Hg or less; count for a preset duration, or preset count to give the desired counting precision; assure that the sample is properly positioned on the sample holder;

and, after counting, release vacuum *slowly* to prevent contamination of chamber and detector.

5. Gamma Spectrometers

Gamma spectrometry identifies and quantifies specific energy photons (gamma-rays), thereby quantitating specific radionuclides.

a. Principle and uses: Gamma rays from a sample enter the sensitive volume of the detector and interact with the detector atoms. The interactions are converted into a voltage pulse proportional to the photon energy. Pulses are stored in sequence in finite energy-equivalent increments over the desired spectrum range. After sample counting, the accumulated pulses over a certain area may result in a peak that can be identified and quantified as a specific radionuclide by its location and peak area.

b. Components: A gamma spectrometer consists of a detector, preamplifier and detector bias supply, pulse-height analyzer system, data readout capability, and shielded sample enclosure. The pulse-height analyzer system consists of a linear amplifier, an analog-to-digital converter (ADC), memory storage, and a logic control mechanism. The logic control capabilities allow data storage in various modes and display or recall of data. For details on operation and calibration of a gamma spectrum analyzer, see manufacturer's instructions.

Gamma detectors commonly used consist of intrinsic or high-purity germanium (Ge) or lithium drifted germanium [Ge(Li)], sodium iodide [NaI(Tl)], and silicon [Si(Li)]. Ge(Li) detectors offer good resolution and combine poor to good efficiency (depending on the application and cost). NaI(Tl) detectors offer poor resolution and good efficiency at reasonable cost. Si(Li) detectors offer good efficiency and resolution for low-energy x-rays (10 to 200 KeV). Other detectors are available.

c. Performance verification:

1) Radioactive sources—See Section 7020A.3e.

2) Detector voltage—Set detector operating voltage according to manufacturer's recommendations.

3) Background—Detector background consists of counting an appropriate volume of reagent-grade water in the desired geometry. Use detector background counting duration equivalent to the longest sample counting duration. The background is caused by contamination of the sample chamber and by cosmic, natural, and worldwide fallout in the detector shielding. Do not confuse the background count rate with the count rate of the laboratory or reagent blank. Under some circumstances, these count rates may be approximately equal.

4) Initial calibration—Establish a detector energy calibration and counting efficiency per sample geometry. Use a standard source as described in Section 7020A.3e.

5) Continuing calibration (source check)—Verify detector efficiency and energy calibration stability (see Section 7020A.3e). A mixed gamma-emitting isotopic source allows confirmation of the energy calibration by permitting identification of specific gamma-emitting radionuclides. For standard deviation criteria, see ¶ 1a3)e).

6) Sample counting—Place prepared sample in the counting chamber and proceed according to instrument manufacturer's instructions.

7) Resolution—Semiconductor detectors greatly improve energy resolution over other types of detectors. Energy resolution, R, refers to the ability of the detector to discriminate between two radiations of different energies. Not all detectors are capable of giving energy information; for such devices this characteristic is not relevant. Resolution is defined with reference to a plot of the number of radiations detected against the radiation energy, as shown in Figure 7030:2:

$$R = \frac{\Delta E}{E_o}$$

where E_o is the energy corresponding to the centroid of the peak and ΔE refers to the width of the peak halfway between the base line and the top. ΔE also is called the full width at half maximum (FWHM). Resolution often is expressed as a percentage (e.g., $R = 0.10$ or 10%). For example, if a photopeak with a maximum at 80 V had a width of 8 V at the midpoint of its amplitude, the resolution of the system would be $100 \times (8/80) = 10\%$. The dispersion of activities about the maximum of a peak is approximately Gaussian.

The smaller the value, the better the detector will be at separating two radiations of similar energy. Resolution is never perfect because of electronic noise and the statistical nature of the interactions of radiation with matter. Resolution varies greatly for different types of detectors (see Table 7030:I).

d. Gamma scintillation: A common gamma spectroscopy system is the sodium iodide, thallium-activated [NaI(Tl)] crystal system using scintillation phenomenon. The high atomic number of iodine in NaI gives good counting efficiency for a gamma-ray detector. A small amount of Tl is added to activate the crystal.

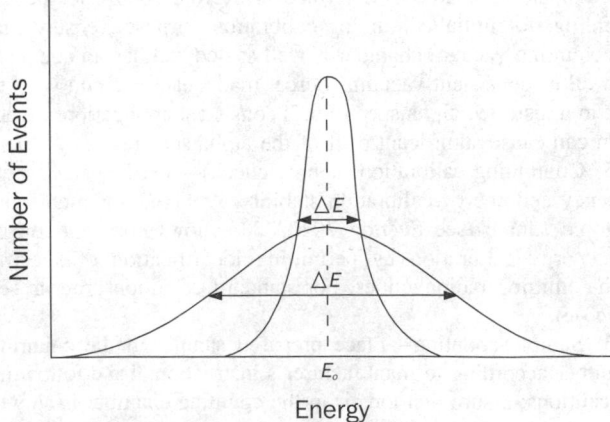

Figure 7030:2. Calculation of the energy resolution of a pulse-type detector.

TABLE 7030:I. ENERGY RESOLUTION FOR VARIOUS DETECTOR TYPES

Detector	Resolution at Given Energy		
	5.9 KeV	122 KeV	1332 KeV
Proportional counter	1.2	—	—
X-ray NaI(Tl)	3.0	12.0	—
3 × 3 NaI(Tl)	—	12.0	60
Si(Li)	0.16	—	—
Planar Ge	0.18	0.5	—
Coaxial Ge	—	0.8	1.8

The best achievable resolution is about 7% with the [137]Cs 662 KeV gamma ray [in a 3-in.- (7.6-cm) diam by 3-in.-long crystal] and slightly worse for smaller or larger detectors.

The light decay time constant in NaI is about 0.25 μs. Typical charged sensitive pre-amplifiers translate this into an output pulse rise time of about 0.5 μs. Fast coincidence measurements cannot achieve the very short resolving times that are possible with plastic, especially at low gamma-ray energies. Many configurations of NaI detectors are available commercially, ranging from very thin crystals for x-ray measurements to large crystals with multiple phototubes. Crystals built with a well to allow nearly spherical (4π) geometry counting of weak samples also are a standard configuration.

e. Solid state detectors: A common high-resolution germanium detector consists of a diode of over 30 cm^3 sensitive volume encased in a 7.6-cm-diam sensitive volume cylinder, with a dipstick immersed in liquid nitrogen in a large cryostat, a pre-amplifier, and a detector bias voltage supply. The detector is cooled with liquid nitrogen to protect the diode and to reduce electronic noise generation. Intrinsic germanium detectors are operated at the temperature of liquid nitrogen. Maintain the detector at the temperature of liquid nitrogen. Use a linear amplifier that will maintain the pulse resolution provided by the detector.

In Ge diode detector systems, the interaction of gamma photons with the detector causes ionization of the detector atoms. A bias voltage applied to the detector allows collection of freed electrons that are proportional to the deposited photon energy with a resolving time of 10^{-9} to 10^{-13}s. Such a system has exceptionally high resolution.

f. Data system: The data systems contain one or more of the following: a visual display for the gamma spectrometer readout indicator, a digital printer, and a computer terminal with associated capabilities. An oscilloscope is helpful in aligning the instrument with standards such as ^{60}Co, ^{137}Cs, [207]Bi, and [154]Eu. Computer capability is essential in data reduction and in complex sample analysis.

6. Bibliography

NADER, J.S., G.R. HAGEE & L.R. SETTER. 1954. Evaluating the performance of the internal counter. *Nucleonics* 12(6):29.

TAYLOR, J.M. 1963. Semiconductor Particle Detectors. Butterworths, Washington, D.C.

GOULDING, F.S. 1964. Semiconductor detectors—their properties and applications. *Nucleonics* 22(5):54.

HEATH, R.L. 1964. Scintillation Spectrometry, Gamma Ray Spectrum. IDO-16880, Vols. 1 & 2. Technical Information Div., U.S. Atomic Energy Comm., Washington, D.C.

SIEGBAHN, K., ed. 1965. Alpha, Beta, and Gamma Ray Spectroscopy, Vol. 1. North Holland Publishing Co., Amsterdam, Netherlands.

DEARNALEY, G. & D.C. NORTHROP. 1966. Semiconductor Counters for Nuclear Radiations, 2nd ed. John Wiley & Sons, New York, N.Y.

DEARNALEY, G. 1966. Nuclear detection by solid state devices. *J. Sci. Instrum.* 43:869.

BOLOGNA, J.A. & S.B. HELMICH. 1969. An Atlas of Gamma Ray Spectra, Los Alamos 4312.

FRENCH, W.R., JR., R.L. LaSHURE & J.L. CURRAN. 1969. Lithium drifted germanium detectors. *Amer. J. Phys.* 37:11.

FRENCH, W.R., JR., W.M. WEHRBEIN & S.E. MOORE. 1969. Measurement of photoelectric, Compton, and pair-production cross sections in germanium. *Amer. J. Phys.* 37:391.

McKENZIE, J.M. 1969. Index to the Literature of Semiconductor Detectors. U.S. Government Printing Off., Washington, D.C.

GOULDING, F.S. & Y. STONE. 1970. Semiconductor radiation detectors. *Science* 170:281.

HEATH, R.L. 1974. Gamma Ray Spectrum Catalogue, Ge(Li) and Si(Li) Spectrometry. ANCR-1000-2, National Technical Information Serv., Springfield, Va.

NATIONAL COUNCIL ON RADIATION PROTECTION AND MEASUREMENTS. 1985. A Handbook of Radioactivity Measurements. NCRP Rep. No. 58, Bethesda, Md.

KNOLL, G. F. 1989. Radiation Detection and Measurement, 2nd ed. John Wiley & Sons, New York, N.Y.

7040 FACILITIES*

7040 A. Counting Room

The design and construction of a counting room may vary widely from one laboratory to another. Provide a room free of dust and fumes to protect the electrical stability of the instrumentation; however, a "clean room" with specific controlled access is not required. Stabilize and reduce background radiation as much as possible by careful choice of building materials for walls, floor, and ceiling as well as by assuring that samples containing appreciable activity are located distant from the instrument area. Construct floors and counter tops of a material that is easy to clean in the event of contamination.

Provide air-conditioning and/or humidity control as necessary to avoid instrument instability. Follow as closely as possible the instrument manufacturer's recommendations for operating temperature and humidity. Most counting instruments are supplied with voltage-regulating circuitry suitable for controlling minor fluctuations in line voltage. For unusual fluctuations use an auxiliary voltage regulator/transformer. Finally, locate the counting room in an area of minimal traffic flow.

A modern chemical laboratory can be used to process routine environmental samples for radiochemical analyses. Preferably segregate nonenvironmental levels of radioactivity to preclude cross-contamination and health concerns. Provide adequate space between instruments to allow necessary access for required periodic maintenance.

---

* Approved by Standard Methods Committee, 1994.

7040 B. Radiochemistry Laboratory

The prime consideration in the design of a radiochemistry laboratory is contamination control. Radioactivity concentration levels found in environmental samples would not normally produce cross-contamination problems if good laboratory practices are followed. In areas where radioactive standards are being prepared, take care to minimize contamination of surfaces, other samples, and personnel. Either bench surfaces of an impervious material covered with adsorbent paper or trays (stainless steel, plastic, or fiberglass) lined with adsorbent paper are acceptable.

1. Chemical Reagents and Reagent-Grade Water

Most reagents contain some radioactivity and other impurities that may result in a systematic error if not accounted for. Quantify the contribution of reagents by analyzing reagent blanks. A reagent blank is a sample having all of the constituents of the unknown except those being determined. Do not use reagents having a radioactivity level high enough to interfere. It is recommended that about 5% of every analysis batch be a reagent blank. In most phases of radiochemistry, it is necessary to use high-purity reagents. In such cases, use reagent-grade chemicals. For example, if barium is to be used as a carrier for radium, it is necessary to determine the radium content of the barium salt used for the analysis. Rare-earth carriers, such as yttrium or cerium, can be contaminated with thorium from the original rare-earth ores.

Distilled or deionized water is used in analytical laboratories for dilution, preparation of reagents, and final rinsing of glassware. Ordinary distilled water usually is not of sufficient purity to be used for certain applications in the environmental radiochemistry laboratory. These applications include background blanks for gamma spectroscopy and liquid scintillation counting. Impurities commonly present in distilled water include radon and tritium. Remove radon by aerating with aged (30 d) air or other inert gases, or aging the distilled water for 30 d to allow the radon and daughters to decay. Tritium presence, which is normal in surface water, can be avoided by using water from deep wells. Deep well water is normally old enough so that the tritium has decayed completely or at least to an acceptable level.

2. Apparatus and Glassware

The same considerations apply to the use and care of glassware in the radiochemistry laboratory as in any analytical laboratory. An excellent discussion is given elsewhere of the kinds of glassware available, the use of volumetric ware, and various cleaning requirements.[1] Certain aspects of glassware usage are peculiar to radiochemistry. Consider glassware and apparatus used in the preparation of standards or for higher-level samples as contaminated; discard or segregate for further use with samples of comparable activity. (A useful rule of thumb is to not analyze samples side by side if it is known that they differ in activity level by three orders of magnitude, i.e., pCi vs. nCi.) As available, preferably use single planchets and auxiliary supplies.

3. Reference

1. U.S. ENVIRONMENTAL PROTECTION AGENCY. 1979. Handbook for Analytical Quality Control in Water and Wastewater Laboratories. Chap. 4. EPA-600/4-79-019, U.S. Environmental Protection Agency, Cincinnati, Ohio.

7040 C. Laboratory Safety

While specific safety criteria are beyond the scope of this discussion, apply general and customary safety practices as a part of good laboratory procedure. Each laboratory should have a safety plan as part of standard operating procedure. Where safety practices are included in an approved method, follow them strictly.

Regard each chemical used as a potential health hazard and maintain exposure as low as reasonably achievable. Each laboratory is responsible for maintaining a current awareness file of applicable regulations regarding the safe handling of the chemicals specified in the methods for radiochemical analysis. Make available a reference file of material data handling sheets to all personnel involved. Use fume hoods when necessary, wear safety glasses or a shield for eye protection, and wear protective clothing at all times.

7040 D. Pollution Prevention

1. Management Techniques

Pollution prevention encompasses any technique that reduces or eliminates the quantity or toxicity of waste at the point of generation. Numerous opportunities for pollution prevention exist. The EPA has established a preferred hierarchy of environmental management techniques that places pollution prevention as the management option of first choice. Whenever feasible, use pollution prevention techniques to address waste generation. When wastes cannot be reduced at the source, the next recommended option is recycling. Further information about pollution prevention that may be applicable to laboratories and research institutions is available.[1]

2. Reference

1. AMERICAN CHEMICAL SOCIETY. 1985. Less is Better: Laboratory Chemical Management for Waste Reduction. American Chemical Soc., Washington, D.C.

7040 E. Waste Management

1. Management Techniques

Establish laboratory waste management practices consistent with all applicable rules and regulations. Protect air, water, and land by minimizing and controlling all releases from hoods and bench operations, complying with the letter and spirit of any sewer discharge permits and regulations, and by complying with all solid and hazardous waste regulations, particularly the hazard-ous waste identification rules and land disposal restrictions. Further information on waste management is available.[1]

2. Reference

1. AMERICAN CHEMICAL SOCIETY. 1990. The Waste Management Manual for Laboratory Personnel. American Chemical Soc., Washington, D.C.

7110 GROSS ALPHA AND GROSS BETA RADIOACTIVITY (TOTAL, SUSPENDED, AND DISSOLVED)*

7110 A. Introduction

1. Occurrence

a. Natural radioactivity: Uranium, thorium, and radium are naturally occurring radioactive elements that have a long series of radioactive daughters that emit alpha, beta, and/or gamma radiations until a stable end-element is produced. These naturally occurring elements, through their radioactive daughter gases, radon and thoron, cause airborne activity and contribute to the radioactivity of rain and groundwaters. Additional naturally radioactive elements include potassium-40, rubidium-87, samarium-147, lutetium-176, and rhenium-187.

b. Artificial radioactivity: With the development and operation of nuclear reactors and radionuclide-generating devices, large quantities of radioactive elements are being produced. These include almost all the elements in the periodic table.

_____
* Approved by Standard Methods Committee, 1996.

2. Significance

Regular measurements of gross alpha and gross beta activity in water may be invaluable for early detection of radioactive contamination and indicate the need for supplemental data on concentrations of more hazardous radionuclides.

With the simpler techniques for routine measurement of gross alpha and beta activity, the presence of contamination may be determined in a matter of hours, whereas days may be required to make the radiochemical analyses necessary to identify radionuclides present.

3. Bibliography

U.S. ENVIRONMENTAL PROTECTION AGENCY. 1976. Drinking Water Regulations. Radionuclides. *Fed. Reg.* 41:28402.

7110 B. Evaporation Method for Gross Alpha-Beta

1. General Discussion

a. Selection of counting instrument: The thin-window, heavily shielded, gas-flow, anticoincidence-circuitry proportional counter is the recommended instrument for counting gross alpha and beta radioactivity because of its superior operating characteristics. These include a very low background and a high sensitivity to detect and count an alpha and beta radiation range that is reasonable but not so wide as that of internal proportional counters. Calibrate the instrument by adding standard nuclide portions to media comparable to the samples and preparing, mounting, and counting the standards exactly as the samples.

An internal proportional or Geiger counter also may be used; however, the internal proportional counter has a higher background for beta counting than the thin-window counter and alpha activity cannot be determined separately with a Geiger counter. Alpha activity can be measured with either a thin-window or internal proportional counter; counting efficiency is higher for the internal counter.

b. Calibration standard: When gross beta activity is assayed in samples containing mixtures of naturally radioactive elements and fission products, the choice of a calibration standard may influence the beta results significantly because self-absorption factors and counting chamber characteristics are beta-energy-dependent.

A standard solution of cesium-137, certified by the National Institute of Standards and Technology (NIST) or traceable to a certified source, is recommended for calibration of counter efficiency and self-absorption for gross beta determinations. The half-life of cesium-137 is about 30 years. The daughter products after beta decay of cesium-137 are stable barium-137 and metastable barium-137, which in turn disintegrates by gamma emission. For this reason, the standardization of cesium-137 solutions may be stated in terms of the gamma emission rate per milliliter or per gram. To convert gamma rate to equivalent beta disintegration rate, multiply the calibrated gamma emission rate by 1.29.

Strontium-90 in equilibrium with its daughter yttrium-90 also is a suitable gross beta standard; its use is recommended by EPA.[1] For gross alpha activity, the recommended standards are natural uranium and thorium-230. Plutonium-239 and americium-241 also are widely used.

Note that gross alpha and beta results are meaningless unless the calibration standard also is reported.

c. Radiation lost by self-absorption: The radiation from alpha-emitters having an energy of 8 MeV and from beta-emitters having an energy of 60 KeV will not escape from the sample if the emitters are covered by a sample thickness of 5.5 mg/cm². The radiation from a weak alpha-emitter will be stopped if covered by only 4 mg/cm² of sample solids. Consequently, for low-level counting it is imperative to evaporate all moisture and preferable to destroy organic matter before depositing a thin film of sample solids from which radiation may enter the counter. In counting water samples for gross beta radioactivity, a solids thickness of 10 mg/cm² or less on the bottom area of the counting pan is recommended. For the most accurate results, determine the self-absorption factor as outlined below.

d. Calibration of overall counter efficiency: Correct observed counting rate for geometry, back-scatter, and self-absorption (sample absorption).

Although it is useful to know the variation in these individual factors, determine overall efficiency by preparing standard sample sources and unknowns.

1) For measuring mixed fission products or beta radioactivity of unknown composition, use a standard solution of cesium-137 or strontium-90 in equilibrium with its daughter yttrium-90.

Prepare a standard (known disintegration rate) in an aqueous solution of sample solids similar in composition to that present in samples. Dispense increments of solution in tared pans and evaporate. Make a series of samples having a solids thickness of 1 to 10 mg/cm² of bottom area in the counting pan. Evaporate carefully to obtain uniform solids deposition. Dry (103 to 105°C), weigh, and count. Calculate the ratio of counts per minute to disintegrations per minute (efficiency) for different weights of sample solids. Plot efficiency as a function of sample thickness and use the resulting calibration curve to convert counts per minute (cpm) to disintegrations per minute (dpm).

2) If other radionuclides are to be tested, repeat the above procedure, using certified solutions of each radionuclide. Avoid unequal distribution of sample solids, particularly in the 0- to 3-mg/cm² range, in both calibration and sample preparation.

3) For alpha calibration, proceed as above, using a standard solution of natural uranium salt in secular equilibrium (not depleted uranium), thorium-230, plutonium-239, or americium-241. Recount alpha standard at the beta operating voltage and determine alpha amplification factor (¶ 5b below). Report calibration standard used with results.

2. Apparatus

a. Counting pans, of metal resistant to corrosion from sample solids or reagents, about 50 mm diam, 6 to 10 mm in height, and thick enough to be serviceable for one-time use. Stainless steel planchets are recommended for acidified samples.

b. Thin end-window proportional counter, capable of accommodating a counting pan.

c. Alternate counters: Other beta counters are internal proportional and Geiger counters.

*d. Membrane filter,** 0.45-μm pore diam.

e. Gooch crucibles.

f. Counting gas, as recommended by the instrument manufacturer.

3. Reagents

a. Methyl orange indicator solution.

b. Nitric acid, HNO₃, 1N.

c. Clear acrylic solution: Dissolve 50 mg clear acrylic† in 100 mL acetone.

d. Ethyl alcohol, 95%.

e. Conducting fluid:‡ Prepare according to manufacturer's directions (for internal counters).

f. Standard certified thorium-230, cesium-137, or strontium-90–yttrium-90 solution.§

g. Standard certified americium-241,§ plutonium-239,§ or natural uranium solution.§ For natural uranium, use material in secular equilibrium.

h. Reagents for wet-combustion procedure:

1) *Nitric acid,* HNO₃, 6N.

2) *Hydrogen peroxide solution:* Dilute 30% H₂O₂ with an equal volume of water.

4. Procedure

a. Total sample activity:

1) For each 20 cm² of counting pan area, take a volume of sample containing not more than 200 mg residue for beta examination and not more than 100 mg residue for alpha examination. The specific conductance test on a nonpreserved sample helps to select the appropriate sample volume.

2) Evaporate by either of the following techniques:

a) Add sample directly to a tared counting pan in small increments, with evaporation at just below boiling temperature. This procedure is not recommended for large samples.

b) Place sample in a borosilicate glass beaker or evaporating dish, add a few drops of methyl orange indicator solution, add 1N HNO₃ dropwise to pH 4 to 6, and evaporate on a hot plate or steam bath to near dryness. Avoid baking solids on evaporation vessel. Transfer to a tared counting pan with the aid of a rubber policeman and distilled water from a wash bottle. Using a rubber policeman, thoroughly wet walls of evaporating vessel with a few drops of acid and transfer washings to counting pan. (Excess alkalinity or mineral acidity is corrosive to aluminum counting pans.)

---

* Type HA, Millipore Filter Corp., Bedford, MA, or equivalent.
† Lucite or equivalent.
‡ Anstac 2M, Chemical Development Corporation, Danvers, MA, or equivalent.
§ U.S. Environmental Protection Agency, Environmental Monitoring Systems Laboratory, Radioanalysis Branch, P.O. Box 93478, Las Vegas, NV 89193.

3) Complete drying in an oven at 103 to 105°C, cool in a desiccator, weigh, and keep dry until counted.

4) Treat sample residues having particles that tend to be airborne with a few drops of clear acrylic solution, then air- and oven-dry and weigh.

5) With a thin end-window counter count alpha and/or beta activity.

6) Store sample in a desiccator and count for decay if necessary. Radionuclides that are volatile under the sample preparation conditions of this method will not be measured. In some geographic areas nitrated water solids (sample evaporated with nitric acid present) will not remain at a constant weight after being dried at 105°C for 2 h and then exposed to the atmosphere before and during counting. Other radioactive substances (such as some chemical forms of radioiodine) also may be lost during sample evaporation and drying. Heat such samples to a dull red heat for a few minutes to convert the salts to oxides. Sample weights then usually are sufficiently stable to give consistent counting rates and a correct counting efficiency can be assigned. Radioisotopes such as those of cesium may be lost when samples are heated to dull red color. Such losses are limitations of the test method.

b. Activity of dissolved matter: Proceed as in ¶ 4a1) above, using a sample filtered through a 0.45-μm membrane filter.

c. Activity of suspended matter:

1) For each 10 cm² of membrane filter area, take a volume of sample not to exceed 50 mg suspended matter for alpha assay and not to exceed 100 mg for beta assay.

2) Filter sample through membrane filter with suction; then wash sides of filter funnel with a few milliliters of distilled water.

3) Transfer filter to a tared counting pan and oven-dry.

4) If sample is to be counted in an internal counter, saturate membrane with alcohol and ignite. (When beta or alpha activity is counted with another type of counter, ignition is not necessary provided that the sample is dry and flat.) When burning has stopped, direct flame of a Meker burner down on the partially ignited sample to fix sample to pan.

5) Cool, weigh, and count alpha and beta activities.

6) If sample particles tend to be airborne, treat sample with a few drops of clear acrylic solution, air-dry, and count.

7) Alternatively, prepare membrane filters for counting in internal counters by wetting filters with conducting fluid, drying, weighing, and counting. (Include weight of membrane filter in the tare.)

d. Activity of suspended matter(alternate): If it is impossible to filter sewage, highly polluted waters, or industrial wastes through membrane filters in a reasonable time, proceed as follows:

1) Determine total and dissolved activity by the procedures given in ¶s 4a and 4b and estimate suspended activity by difference.

2) Filter sample through an ashless mat or filter paper of stated porosity. Dry, ignite, and weigh suspended fixed residue. Transfer and fix a thin uniform layer of sample residue to a tared counting pan with a few drops of clear acrylic solution. Dry, weigh, and count in a thin end-window counter for alpha and beta activity.

e. Activity of nonfatty semisolid samples: Use the following procedure for samples of sludge, vegetation, soil, etc.:

1) Determine total and fixed solids of representative samples according to Section 2540.

2) Reduce fixed solids of a granular nature to a fine powder with pestle and mortar.

3) Transfer a maximum of 100 mg fixed solids for alpha assay and 200 mg fixed solids for beta assay for each 20 cm² of counting pan area (see NOTE below).

4) Distribute solids to uniform thickness in a tared counting pan by (*a*) spreading a thick aqueous cream of solids that is weighed after oven-drying, or (*b*) dispensing dry solids of known weight and spreading with acetone and a few drops of clear acrylic solution.

5) Oven-dry at 103 to 105°C, weigh, and count.

NOTE: The fixed residue of vegetation and similar samples usually is corrosive to aluminum counting pans. To avoid difficulty, use stainless steel pans or treat a weighed amount of fixed residue with HCl or HNO_3 in the presence of methyl orange indicator to pH 4 to 6, transfer to an aluminum counting pan, dry at 103 to 105°C, reweigh, and count.

f. Alternate wet-combustion procedure for biological samples: Some samples, such as fatty animal tissues, are difficult to process according to ¶ 4e above. An alternate procedure consists of acid digestion. Because a highly acid and oxidizing state is created, volatile radionuclides may be lost under these conditions.

1) To a 2- to 10-g sample in a tared silica dish or equivalent, add 20 to 50 mL 6*N* HNO_3 and 1 mL 15% H_2O_2 and digest at room temperature for a few hours or overnight. Heat gently and, when frothing subsides, heat more vigorously but without spattering, until nearly dry. Add two more 6*N* HNO_3 portions of 10 to 20 mL each, heat to near boiling, and continue gentle treatment until dry.

2) Ignite in a muffle furnace for 30 min at 600°C, cool in a desiccator, and weigh.

3) Continue the test as described in ¶s 4e3)-5) above.

5. Calculation and Reporting

a. Alpha activity: Calculate alpha activity, in picocuries per liter, by the equation

$$\text{Alpha} = \frac{\text{net cpm} \times 1000}{2.22e\,\nu}$$

where:

e = calibrated overall counter efficiency (see ¶ 1d, and
ν = volume of sample counted, mL.

Express the counting error as described in ¶ *c* below. Similarly, calculate and report alpha activity in picocuries per kilogram of moist biological material or per kilogram of moist and per kilogram of dry silt.

b. Beta activity: Calculate and report gross beta activity and counting error in picocuries per liter of fluid, per kilogram of moist (live weight) biological material, or per kilogram of moist and per kilogram of dry silt, according to ¶s *a*, above, and *c*, below.

To calculate picocuries of beta activity per liter, determine the value of *e* in the above equation as described in ¶ 1d, above.

When counting beta activity in the presence of alpha activity by gas-flow proportional counting systems (at the beta plateau) alpha particles also are counted. Because alpha particles are more readily absorbed by increased sample thickness than beta particles, alpha/beta count ratios vary. Therefore, prepare a calibration curve by counting standards (americium-241, thorium-230, or plutonium-239) with increasing solids thickness, first on the alpha plateau, then on the beta plateau. Plot the ratios of the two counts against mg/cm^2 thickness, determine the alpha amplification factor (M), and correct the amplified alpha count on the beta plateau for the sample.

$$M = \frac{\text{net cpm on beta plateau}}{\text{net cpm on alpha plateau}}$$

If significant alpha activity is indicated by the sample alpha plateau count, determine beta activity by counting the sample at the beta plateau and calculating:

$$\text{Beta, pCi/L} = \frac{B - AM}{2.22 \times D \times V}$$

where:

B = net beta counts at the beta plateau,
A = net alpha counts at the alpha plateau,
M = alpha amplification factor (from ratio plot),
2.22 = dpm/pCi
D = beta counting efficiency, cpm/dpm and
V = sample volume, L.

Some gas-flow proportional counters have electronic discrimination to eliminate alpha counts at the beta operating voltage. For these instruments the alpha amplification factor will be less than 1.

Where greater precision is desired, for example, when the count of alpha activity at the beta plateau is a substantial fraction of the net counts per minute of gross beta activity, the beta counting error equals $(E_a^2 + E_b^2)^{1/2}$, where E_a is the alpha counting error and E_b the gross beta counting error.

c. Counting error: Determine the counting error, E (in picocuries per sample), at the 95% confidence level from:

$$E = \frac{1.96 \; \sigma(R)}{2.22e}$$

where $\sigma(R)$ is calculated as shown in Section 7020C.2, using $t_1 = t_2$ (in minutes); and e, the counter efficiency, is defined and calculated as in ¶ 1*d*.

d. Miscellaneous information to be reported: In reporting radioactivity data, identify adequately the sample, sampling station, date of collection, volume of sample, type of test, type of activity, type of counting equipment, standard calibration solutions used (particularly when counting standards other than those recommended in ¶ 1*d* are used), time of counting (particularly if short-lived isotopes are involved), weight of sample solids, and kind and amount of radioactivity. So far as possible, tabulate the data for ease of interpretation and incorporate repetitious items in the table heading or in footnotes. Unless especially inconvenient, do not change quantity units within a given table. Always report the counting error to assist in interpretation of results.

6. Precision and Bias

In a collaborative study of two sets of paired water samples containing known additions of radionuclides, 15 laboratories determined the gross alpha activity and 16 analyzed gross beta activity. The samples contained simulated water minerals of approximately 350 mg fixed solids/L. The alpha results of one laboratory were rejected as outliers.

The average recoveries of added gross alpha activity were 86, 87, 84, and 82%. The precision (random error) at the 95% confidence level was 20 and 24% for the two sets of paired samples. The method was biased low, but not seriously.

The average recoveries of added gross beta activity were 99, 100, 100, and 100%. The precision (random error) at the 95% confidence level was 12 and 18% for the two sets of paired samples. The method showed no bias.

7. Reference

1. U.S. ENVIRONMENTAL PROTECTION AGENCY. 1980. Prescribed Procedures of Measurement of Radioactivity in Drinking Water. EPA-600/4-80-032.

8. Bibliography

BURTT, B.P. 1949. Absolute beta counting. *Nucleonics* 5:8, 28.
GOLDIN, A.S., J.S. NADER & L.R. SETTER. 1953. The detectability of low-level radioactivity in water. *J. Amer. Water Works Assoc.* 45:73.
SETTER, L.R., A.S. GOLDIN & J.S. NADER. 1954. Radioactivity assay of water and industrial wastes with internal proportional counter. *Anal. Chem.* 26:1304.
SETTER, L.R. 1964. Reliability of measurements of gross beta radioactivity in water. *J. Amer. Water Works Assoc.* 56:228.
THATCHER, L.L., V.J. JANZER & K.W. EDWARDS. 1977. Techniques of water resources investigations of the US Geological Survey. Chap. A5 *in* Methods for Determination of Radioactive Substances in Water and Fluvial Sediments. Stock No. 024-001-02928-6, U.S. Government Printing Off., Washington, D.C.
JOHNS, F.B., et al. 1979. Radiochemical Analytical Procedures for Analysis of Environmental Samples. EMLS-LV-0539-17, Environmental Monitoring Systems Lab., Off. Research & Development, U.S. Environmental Protection Agency, Las Vegas, Nev.

7110 C. Coprecipitation Method for Gross Alpha Radioactivity in Drinking Water

1. General Discussion

The Evaporation Method for Gross Alpha-Beta, 7110B, does not separate alpha-(or beta-) emitting radionuclides from the sample's dissolved solids. For drinking water samples with high dissolved solids content, e.g., 500 mg/L or higher, Method 7110B is severely limited because of the small sample size possible and the very long counting times necessary to meet required sensitivity (3 pCi/L). The coprecipitation procedure eliminates the problem of high dissolved solids and gives increased sensitivity.

a. Principle: All alpha-emitting radionuclides of interest (mainly radium, uranium, and thorium isotopes) are coprecipitated with barium sulfate and iron hydroxide as carriers, thereby separating alpha-emitting radionuclides from other sample dissolved solids. The combined precipitates are filtered and counted for alpha activity. Relatively large samples can be analyzed so that sensitivity is improved and counting time is minimized.

b. Interferences: Allow at least 3 h for decay of radon progeny before beginning the alpha count.

Soluble ions that coprecipitate and add to the mixed barium sulfate and iron hydroxide precipitate weights result in counting efficiencies that are biased low.

Iron hydroxide precipitates collected on membrane filters without a holding agent flake when dried and are easily lost from the filter. Add 5 mg paper pulp fiber to the sample to help secure the iron hydroxide to the filter. Preferably use glass fiber filters because the surface glass fibers help to secure the precipitate.

c. Calibration: Add at least 100 pCi standard alpha-emitter activity to 500-mL portions of tap water in separate beakers. Add 2.5 mL conc HNO_3 to each beaker. Determine counting efficiency (cpm/pCi) for the alpha-emitter by taking these known additions through the procedure. Make at least six replicate determinations to determine counting efficiency.[1]

Use 500-mL portions with no addition for blank corrections of alpha activity in the tap water and reagents.

$$\text{Efficiency, cpm/pCi} = \frac{C_a - C_b}{\text{pCi}}$$

where:

C_a = sample with added activity, mean cpm,
C_b = mean blank cpm, and
pCi = activity added.

Preferably use thorium 230 (a pure alpha-emitter) for gross alpha efficiency calibration.* As noted in 7110B.1*b*, other alpha-counting standards may be used.

2. Apparatus

a. Hot plate/magnetic stirrer and stirring bars.
b. Filter membranes, 47-mm diam, 0.45-μm pore size, or glass fiber filters.†
c. Drying lamp.
d. Planchets, stainless steel, 2-in. diam.
e. Alpha scintillation counter or low-background proportional counter.

3. Reagents

a. Ammonium hydroxide, NH_4OH, 6N.
b. Barium carrier, 5 mg Ba^{2+}/mL: Dissolve 4.4 g $BaCl_2\cdot2H_2O$ in 500 mL distilled water.
c. Bromocresol purple, 0.1%: Dissolve 100 mg water-soluble reagent in 100 mL distilled water.

d. Iron carrier, 5 mg Fe^{3+}/mL: Dissolve 17.5 g $Fe(NO_3)_3\cdot9H_2O$ in 200 mL distilled water containing 2 mL 16N HNO_3. Dilute to 500 mL.
e. Sulfuric acid, H_2SO_4, 2N: Dilute 55 mL conc H_2SO_4 to 1 L with distilled water.
f. Paper pulp/water mixture: Add a 0.5-g paper pulp pellet to 500 mL distilled water in a plastic bottle. Add 5 drops diluted (1 + 4) detergent. Cap bottle and stir vigorously for 3 h before use. Stir mixture whenever a portion is taken.
g. Detergent: 1 part detergent + 4 parts distilled water.‡

4. Procedure

To a sample of 500 mL to 1 L, or sample diluted to 500 mL, add 5 drops of diluted detergent. Place sample on magnetic stirrer/hot plate and, while stirring, gently add 20 mL 2N H_2SO_4. Boil for 10 min to flush CO_2 (from carbonates and bicarbonates) from sample. Radon also is flushed. Reduce temperature to below boiling, continue stirring, and add 1 mL barium carrier solution. Continue stirring for 30 min. Add 1 mL bromocresol purple indicator solution, 1 mL iron carrier solution, and 5 mL stirred paper pulp/water reagent.

Continue stirring and add 6N NH_4OH dropwise until there is a distinct color change (yellow to purple). Continue warming and stirring for 30 min. Filter sample through a glass fiber filter (or membrane filter if further analysis is to be done). Quantitatively transfer all precipitate to the filter. Wash precipitate with 25 mL distilled water. Hold filter for 3 h for collected radon progeny to decay. Dry filter at 105°C or under a mild heat lamp.

Count filter for gross alpha activity.

Prepare a reagent blank precipitate to determine reagent alpha activity background.

5. Calculations

$$\text{Gross alpha activity, pCi/L} = \frac{C_a - C_b}{E\ V}$$

where:

E = counter efficiency, cpm/pCi,
V = volume analyzed, L,
C_a = sample counts per minute, cpm, and
C_b = reagent blank, cpm.

6. Precision and Bias

In collaborative test with 18 laboratories participating,[2] gross alpha activities for four different samples were calculated with four different alpha-emitting radionuclide standard counting efficiencies. Thorium-230, a pure alpha-emitter, appeared to be the best standard for gross alpha counting efficiency.

Water samples A, B, C, and D contained gross alpha concentrations of 74.0, 52.6, 4.8, and 10.0 pCi/L, respectively, at 3 h after separation of alpha-emitting radionuclides by coprecipitation with iron hydroxide and barium sulfate. Test results using the thorium-230 counting efficiency showed coefficients of variation for repeatability (within laboratory precision) of 7.9, 7.8, 8.7, and

---

* Available from U.S. Environmental Protection Agency, Environmental Monitoring Systems Laboratory, Nuclear Radiation Assessment Division, P.O. Box 93478, Las Vegas, NV 89193-3478.
† Gelman Type A/E, Millipore Type AP, or equivalent.

‡ Rohm and Haas Triton N101 or Triton X100, or equivalent.

8.8%, respectively, for an average of 8.3%. Coefficients of variation for reproducibility (combined within and between laboratory precision) of 20.4, 16.8, 18.7, and 18.5%, respectively, were obtained for an average of 18.6%.

A comparison of the 18 laboratory grand average results (calculated with the ^{230}Th counting efficiency) and known gross alpha particle concentrations showed accuracy indexes of 91.9, 99.4, 122, and 94.5%, respectively, for an average accuracy index of 102%. The *t*-test for bias showed a significant positive bias for Sample C but no significant bias for the other three samples.

7. References

1. U.S. ENVIRONMENTAL PROTECTION AGENCY. 1984. EERF Radiochemistry Procedures Manual. 00-02 Radiochemical Determination of Gross Alpha Activity in Drinking Water by Coprecipitation. EPA-520/5-84-006, USEPA ORP-EERF, Montgomery, Ala.
2. WHITTAKER, E.L. 1986. Test Procedure for Gross Alpha Particle Activity in Drinking Water, Interlaboratory Collaborative Study. EPA-600/4-86/027 July 1986, Pre-issue copy. USEPA EMSL-Las Vegas, Las Vegas, Nev.

7120 GAMMA-EMITTING RADIONUCLIDES*

7120 A. Introduction

For information about occurrence of natural and artificial radioactivity, see Section 7110A.1.

---

* Approved by Standard Methods Committee, 1997.

7120 B. Gamma Spectroscopic Method

1. General Discussion

a. Application: This method describes the use of gamma spectroscopy, using either germanium (Ge) diodes or thallium-activated sodium iodide [NaI(Tl)] crystals, for the measurement of gamma photons emitted from radionuclides present in water. The method is applicable to samples that contain radionuclides emitting gamma photons with energies ranging from about 60 to 2000 KeV.

The method can be used for qualitative and quantitative determinations with Ge detectors or for screening and semi-quantitative and semi-qualitative determinations with [NaI(Tl)] detectors. Exact quantitation using NaI is possible for single nuclides or when the gamma emissions are limited to a few well-separated energies. Detection limits for typical counting systems range from a few picocuries (pCi) of gamma activity for a 100-min count to approximately 100 pCi for a 5-min count, depending on counting geometry and gamma ray energy and abundance.

Determine energy and efficiency calibrations for each detector at several energies between 50 and 2000 KeV for the geometries of interest. Gamma ray libraries[1,2] for Ge spectrometry should contain the nuclides and gamma ray lines most likely to be found in water samples. Have computer software available to list the contents of the library and to add more nuclides and gamma ray lines to the library for peak search routines.

b. Principle: Because gamma spectroscopy is nondestructive, it is possible to analyze for gamma-emitting radionuclides without separating them from the sample matrix. This technique makes it possible to identify and quantitate gamma-emitting radionuclides when the gross beta screen has been exceeded or it is otherwise necessary to define the contribution of gamma-emitters to the total radioactivity present.

A homogeneous water sample is put into a standard geometry for gamma counting. The counting efficiency for this geometry must have been determined with a mixed energy gamma standard containing known radionuclide activities. Sample portions are counted long enough to meet the required sensitivity of measurement.

The gamma spectrum is printed out and/or stored in the appropriate computer-compatible device for data processing (calculation of sample radionuclide concentrations).

Consult a good text on gamma ray spectrometry[3] for a more detailed discussion.

c. Sampling and storage: See Table 7010:I.

d. Interferences: Significant interference occurs when a sample is counted with a NaI(Tl) detector and the sample radionuclides emit gamma photons of nearly identical energies. Such interference is greatly reduced by counting the sample with a Ge detector. Higher-energy gammas that predominate may completely mask minor, less energetic photopeaks for both Ge and NaI detectors by increasing the baseline or Compton continuum.

Interferences can occur with Ge detectors from cascade peak summing, which results when two or more gamma rays are emitted in one disintegration, e.g., with cobalt-60, where 1172 and 1333 KeV gamma rays are emitted in cascade. These can be detected together to produce a sum peak at 2505 KeV or a count in the continuum between the individual peaks and the sum peak, thus causing the loss of counts from one or both of the other two peaks. Cascade summing (as distinct from random summing) is geometry/counting efficiency dependent (but not count rate dependent) with the effect, and hence error, increasing as tighter geometries and more efficient detectors are used. This problem has become more commonplace with the availability of larger, more affordable, and more efficient Ge detectors.

Sample homogeneity is important to gamma count reproducibility and counting efficiency validity. When sample radionuclides are adsorbed on the walls of the counting container, the sample is no longer homogeneous. This problem can be lessened by adding 15 mL 1N HNO$_3$/L sample at collection.

Sample density and composition can affect data quality. Prepare efficiency calibration standards in the same geometry and density as the samples. Ensure reproducible sample geometry to limit bias.[4] Plexiglass spacers may be useful in producing consistent sample positions. Random noise produced by vibration or by improper grounding can increase peak width and introduce additional uncertainty.

e. *Safety:* No unusual hazards are associated with the reagents used in this procedure. Follow routine safety precautions, i.e., wear laboratory coat, plastic gloves, and safety glasses and use a hood, when transferring samples and standards and preparing standards when solutions of gamma-emitting radionuclides are used. Cool germanium diodes with liquid nitrogen when they are being used to count samples. Take care when transferring liquid nitrogen from the storage dewar to the dewar used to supply coolant for the germanium diode. Use cryogenic gloves, protective clothing, and eye protection.

2. Apparatus

a. *Detector,* large-volume (> 50 cm^3) germanium-diode detector or 10.2-cm $\times$ 10.2-cm (4-in. $\times$ 4-in.) thallium-activated sodium iodide crystal [NaI(Tl)] detector. Smaller detectors may be acceptable if inherent limitations, such as reduced counting efficiency, are taken into account. Be sure that large detectors can accommodate re-entrant (Marinelli) beakers. Germanium (Ge) detectors are preferred because of better photon energy resolution. Despite the possibly higher counting efficiencies of NaI(Tl) detectors, the considerably narrower peak shape from a Ge detector leads to fewer baseline counts, thereby improving peak sensitivity. Preferably do not use NaI(Tl) detectors to make both qualitative and quantitative analyses for samples containing multiple gamma-emitting radioisotopes; however, they are preferred if a single nuclide is being quantitated. Ge detectors may be of either intrinsic (pure) germanium type or lithium-drifted germanium [Ge(Li)] type. Both require use of liquid nitrogen for cooling when bias voltage is applied to the detector. Intrinsic Ge detectors can be stored or shipped at ambient temperatures; Ge(Li) detectors must be cooled with liquid nitrogen at all times to avoid damage to detector.

b. *Gamma-ray spectrometer plus analyzer* with at least 2048 channels for Ge or 256 to 512 for NaI(Tl). See Section 7030B.5.

c. *Counting container,* standard geometry for either detector, e.g., 0.5-L cylindrical container, 0.45-L or 4-L re-entrant (Marinelli)[4] polyethylene beaker. Counting containers of other sizes often are used.

d. *Computer:* Use a data acquisition system including a computer (PC, networked-PCs, or larger) supplied with software to automate the processing of raw spectral data as outlined in Sections 4 and 5, below. Software should contain algorithms to: perform energy and efficiency calibrations; locate peaks (deconvoluting multiplets as needed; that may require a separate routine[s] for low-resolution NaI counting data); perform peak integrations; perform nuclide searches; do activity calculations (result, uncertainty, and MDA); and manage the library reference information needed to support these actions.

e. *High-voltage power supply.*

f. *Amplifier,* suitable for spectroscopy with gain and shaping time adjustments and baseline restoration.

g. *Analog-to-digital converter and spectrum storage device.*

3. Reagents

a. *Distilled or deionized water,* radon-free, for standard preparation and sample dilution.

b. *Nitric acid,* HNO$_3$, 1N.

4. Procedure

a. *Energy calibration:* Use NIST or NIST-traceable, or equivalent standards. For a Ge system, adjust analyzer amplifier gain and analog-to-digital converter zero offset to locate each photopeak in its appropriate channel. A 0.5- or 1.0-KeV per channel calibration is recommended. If the system is calibrated for 1 KeV per channel with channel zero representing 0 KeV, the energy will be equal to the channel number. Check and adjust the pole zero cancellation of the amplifier output if required.[5]

Use a standard containing a mixture of gamma energies from about 100 to 2000 KeV for energy calibration. Multiline gamma standards can be obtained commercially or can be prepared by the user. Some laboratories use radium-226 and daughters in equilibrium or europium-152 for this purpose. NIST SRM 4275 is a solid source that is useful for energy calibration and routine monitoring of instrument performance. Solid sources prepared on plastic mounts are stable and are recommended. Count energy calibration standards long enough to minimize uncertainty due to counting statistics; as a rule of thumb accumulate 10 000 counts in each photopeak area resulting in a counting error of 1%.

For a NaI(Tl) system, a 10- or 20-KeV per channel calibration is adequate. A solid multipeak standard source, e.g., bismuth-207, is satisfactory for energy calibration.

b. *Efficiency calibration:* Use NIST, NIST-traceable, or equivalent standards with minimal cascade summing concerns. Use a known amount of a multipeak standard or various radionuclides that emit gamma photons with energies well spaced and distributed over the normal range of analysis; put these into each container geometry and gamma count for a photopeak spectrum accumulation. Count efficiency calibration standards long enough to minimize uncertainty due to counting statistics; as a rule of thumb accumulate 10 000 counts in each photopeak area, resulting in a counting error of 1%.

Determine counting efficiencies for the various gamma energies (photopeaks) from the activity counts of the known-value samples as follows:

$$E = \frac{C}{A \times B}$$

where:

E = efficiency (expressed as counts per minute/gamma rays emitted per minute),

C = net count rate, cpm (integrated counts in the photopeak above the baseline continuum divided by the counting duration),

A = activity of radionuclide added to the given geometry container, dpm (corrected for decay, if necessary), and

B = gamma-ray abundance of the radionuclide being measured, gammas/disintegration.

Plot counting efficiency against gamma energy for each container geometry and for each detector that is to be used. For NaI systems, prepare a library of radionuclide spectra from counts of known radionuclide-water sample concentrations at standard sample geometries.

c. Sample measurement: Measure sample portion in a standard-geometry container calibrated as directed in ¶s *a* and *b* above. Place container and sample on a shielded Ge or NaI(Tl) detector and gamma count for a period of time that will meet the required sensitivity. Print gamma spectrum and/or store the spectrum on the appropriate computer-compatible device.

5. Calculations

The equations (Sections 4 and 5) describe the fundamental relationships between the defined variables and could be used if the calculations were to be performed manually. NaI spectral data, with their high probability for peak overlap due to low resolution, often require complex peak unfolding routines. Modern gamma spectroscopy systems rely on vendor-supplied computer software to process the 256 to 500-plus (for NaI systems) and up to 8000 (for Ge systems) data points. The supplied software should be accompanied by documentation describing the algorithms, which must incorporate the fundamental relationships presented here.

Determine isotopes indicated by the gamma spectrum as follows: Identify all photopeak energies, integrate photopeak regions of the spectrum and subtract the area under the baseline continuum to determine the true photopeak area, and identify isotopes by their appropriate photopeaks, and ratios to each other when more than one gamma photon is emitted by an isotope.

Calculate the sample radionuclide concentrations as follows:

$$A' = \frac{C}{2.22 \times B \times E \times V \times e^{-\lambda t}}$$

where:

A' = sample radionuclide concentration, pCi/L,

V = sample volume, L,

$e^{-\lambda t}$ = decay factor (corrected to sample collection time), with λ = decay constant for the gamma-emitting radionuclide being analyzed, and t = duration of time from sample collection to counting,

2.22 = conversion factor from dpm to pCi,

and B, C, and E are as defined in ¶ *4a* above. Calculate the 2σ counting error term for gamma-emitters as follows:

$$2\sigma, \text{pCi/L} = \frac{2\sqrt{\dfrac{C + 2G}{t_c}}}{2.22 \times B \times E \times V \times e^{-\lambda t}}$$

where:

G = photopeak area below continuum, cpm,

t_c = counting duration, min,

and other terms are as defined above.

Report the result and counting error together in the form:

$$X \pm 2\sigma, \text{pCi/L}$$

Vendor-supplied software usually can calculate a Total Propagated Uncertainty (TPU). If so, the 2σ value reflects the total uncertainty and not just the counting error. The vendor-supplied software also should calculate a Minimum Detectable Concentration (MDC) (see Section 7020C.3 for a general discussion). Report concentration, uncertainty, and MDC for each sample.

6. Quality Control

See Section 7020.

a. Duplicates: Make duplicate analyses for one out of every ten samples. (See Section 7020A.3*c*.) If it is known or strongly suspected that the sample(s) contain no detectable gamma-emitting radionuclides, do not use the sample(s) for duplicate analysis. In such a case, rely on the results of known-addition samples as described below. If desired, analyze known-addition samples in duplicate.

b. Known-addition sample: Analyze a known-addition sample for one out of every ten samples. This may be freshly prepared, or it may be a previously analyzed standard sample such as a performance evaluation sample from EPA-Las Vegas. See Section 7020A.3*d*.

c. Background: See Section 7030B.5*c*3) and 7020A.3*b*. Find and identify any background lines that may be present. If present, subtract the background counts line by line from the sample photopeaks. Even if background does not significantly affect results, monitor it to ensure system integrity. Weekly background counts for durations longer than normal sample counting durations may be needed to quantify low-level activity from nuclides such as cobalt-60.

d. Energy calibration: Check the energy calibration daily or before each use with a multi-line source as in ¶ *4a* above.

e. Efficiency check: Check detector efficiency daily or before each use with a stable multiline source in a reproducible geometry.

f. Records: Collect and maintain results from duplicate pairs and check standards. Include date, results, analyst's name, and any comments relevant to the evaluation of these data.

7. Precision and Bias

The precision of an individual measurement by gamma spectrometry can be improved by increasing sample counting duration. It may be necessary to gamma count for as much as 1000 min to reach desired precision. Other ways of increasing precision of an individual measurement are to increase sample volume, use a more efficient detector, or concentrate the sample. To obtain accurate results, calibrate carefully and use standardized radionuclides at the proper activity and purity levels.

Collaborative test data for a closely defined technique or procedure were not available for gamma-emitters in water. However, data from USEPA's Environmental Radioactivity Performance Evaluation Studies Program are presented here. Table 7120:I is a summary of the recovery, within-laboratory variance, S_r, and total-error variance, S_R, regression line equations for each gamma-emitter studied. These data are from the analysis of gamma-emit-

TABLE 7120:I. GAMMA-EMITTERS RECOVERY AND PRECISION ESTIMATE REGRESSION LINE EQUATIONS

Nuclide	Recovery	S_r pCi/L	S_R pCi/L
Iodine-131*	$y = 1.013x - 0.64$	$y = 0.033x + 1.82$	$y = 0.073x + 2.66$
Cesium-137	$y = 1.004x + 0.79$	$y = 0.024x + 1.40$	$y = 0.049x + 1.93$
Cesium-134	$y = 0.919x + 0.60$	$y = 0.028x + 1.25$	$y = 0.058x + 1.67$
Barium-133	$y = 0.938x + 2.82$	$y = 0.024x + 1.49$	$y = 0.064x + 1.78$
Ruthenium-106	$y = 0.928x + 2.33$	$y = 0.051x + 3.67$	$y = 0.072x + 7.27$
Zinc-65	$y = 1.016x + 0.49$	$y = 0.029x + 2.61$	$y = 0.054x + 3.74$
Cobalt-60	$y = 0.986x + 0.79$	$y = 0.023x + 1.35$	$y = 0.045x + 1.94$
Chromium-51†	$y = 0.997x + 0.30$	$y = 0.058x + 3.99$	$y = 0.081x + 8.67$

* Analyzed singly as a separate study from the other gamma-emitters.
† No longer used.

ting radionuclides in standard samples by participants in the program from 1981 to 1995. It is not possible to say how many investigators used NaI(Tl) detectors in the early years of the data collection period. It is believed that most, if not all, are now using germanium detectors, and that the data are comparable among all participating laboratories. The gamma spectral data from the Gamma Performance Evaluation and Blind Samples from February 1981 through November 1995 (April 1981 through October 1995 for iodine-131) were summarized by study and the data were arrayed for analyses. See Table 7120:II.

Regression equations were generated for: recovery, grand average of each study; estimate of precision, standard deviation (1 σ) of the mean value of each study, $S_{\bar{x}}$; within-laboratory standard deviation (1 σ), also known as the repeatability or random error, S_r; between-laboratory standard deviation (1 σ), also known as the systematic error, S_L; and total error from within and between labs, also known as reproducibility, S_R. S_R equals the square root of the sum of the variance of the within-laboratory error and the between-laboratory error, i.e., the reproducibility variance is equal to the sum of the random variance and the systematic variance.

$$S_R^2 = S_r^2 + S_L^2$$
$$S_R = \sqrt{S_r^2 + S_L^2}$$

8. References

1. U.S. DEPARTMENT OF ENERGY. 1992. EML Procedures Manual, 27th ed. (rev.). HASL 300, 4.5.2.3, Environmental Measurements Lab., U.S. Dep. Energy, New York, N.Y.

2. AMERICAN SOCIETY FOR TESTING AND MATERIALS. 1992. Standard Method for High-Resolution Gamma-Ray Spectrometry of Water. Annual Book of ASTM Standards - Part 11, Water and Environmental Technology, 11.02 D 3649-91. American Soc. Testing & Materials, Philadelphia, Pa.

3. GILMORE, G. & J.D. HEMINGWAY. 1995. Practical Gamma Ray Spectrometry. John Wiley & Sons Ltd., Chichester, England.

4. INSTITUTE OF ELECTRICAL AND ELECTRONIC ENGINEERS. 1978. IEEE Standard Techniques for Determination of Germanium Semiconductor Gamma Ray Efficiency using a Standard (Re-entrant) Beaker Geometry. IEEE Standard 680-1978. Inst. Electrical & Electronic Engineers, Piscataway, N.J.

5. AMERICAN NATIONAL STANDARDS INSTITUTE. 1991. Calibration and Use of Germanium Detectors for Measurements of Gamma-Ray Emission of Radionuclides. ANSI N42.14-1991 (revision of ANSI N42.14-1978). American National Standards Inst., New York, N.Y.

9. Bibliography

AMERICAN NATIONAL STANDARDS INSTITUTE. 1971 (reaffirmed 1983). Test Procedures for Germanium Gamma-Ray Detectors. ANSI/IEEE 325, American National Standards Inst., New York, N.Y.

ERDTMANN, G. & W. SOYKA. 1979. The gamma rays of the radionuclides. Verlag Chemie ICRP, 1983 radionuclide transformations: Energy and intensity of emissions. Ann. International Counc. Radiation Protection, ICRP Pub. 30, Pergamon Press, Elmsford, N.Y.

KOCHER, P.C. 1981. Radioactive Decay Data Tables: A Handbook of Decay Data for Application to Radiation Dosimetry and Radiologi-

TABLE 7120:II. GAMMA-EMITTERS STUDY: SUMMARY OF PARTICIPANTS

Nuclide	No. of Studies*	Concentration Range pCi/L	No. of Participants†	No. of Acceptables‡	No. where σ ≠ 0§
Iodine-131	31	14–148	42–119	39–115	34–105
Cesium-137	72	4–94	45–197	43–180	30–158
Cesium-134	72	2–64	17–196	15–183	8–165
Barium-133	17	49–110	118–192	107–181	94–156
Ruthenium-106	35	15–252	20–193	17–179	11–135
Zinc-65	41	10–165	61–195	55–182	46–163
Cobalt-60	62	8–69	75–196	70–183	48–165
Chromium-51	19	21–302	34–124	32–121	16–112

* Samples with concentrations equal to 0 pCi/L were not included.
† Total number of participants in study, even though all of the data may not have been used.
‡ Number of participants used in calculating grand average and standard deviation of grand average.
§ Participants reporting within laboratory variance equal to zero were not used in calculating the study within-laboratory, between-laboratory, and total error variance.

cal Assessments. DOE/TIC-11026. U.S. Dep. Energy, National Technical Information Center, Springfield, Va.

NATIONAL COUNCIL ON RADIATION PROTECTION AND MEASUREMENTS. 1985. A Handbook of Radioactivity Measurements Procedures. NCRP

Rep. No. 58, National Counc. Radiation Protection & Measurements, Bethesda, Md.

KNOLL, G.F. 1989. Radiation Detection and Measurement. John Wiley & Sons, New York, N.Y.

7500-Cs   RADIOACTIVE CESIUM*

7500-Cs A.   Introduction

Radioactive cesium has been considered one of the more hazardous radioactive nuclides produced in nuclear fission. Upon in-

gestion, like potassium, cesium distributes itself throughout the soft tissue and has a relatively short residence time in the body. Half-lives of ^{134}Cs and ^{137}Cs are 2 and 30 years, respectively, both being beta- and gamma-emitters.

---

* Approved by Standard Methods Committee, 1993.

7500-Cs B.   Precipitation Method

1. General Discussion

Principle: If the activity of cesium is high, radioactive cesium can be determined directly by gamma-counting a large liquid sample (4 L) or the sample can be evaporated to dryness and counted. For lower-level environmental samples, add cesium carrier to an acidified sample and collect the cesium as phosphomolybdate. This is purified and precipitated as Cs_2PtCl_6 for counting. If total radiocesium determined by beta-counting exceeds 30 pCi/L, determine ^{134}Cs and ^{137}Cs by gamma spectrometry.

2. Apparatus

a. *Magnetic stirrer* with TFE-coated magnet bar.

b. *Centrifuge,* bench-size clinical, and centrifuge tubes.

c. *Filter papers* and glass fiber filter,* 2.4 cm diam.

d. *pH paper,* wide range, 1 to 11 pH.

e. *Filtering apparatus:* See Section 7500-Sr.B.2c.

f. *Counting instruments:* Use either a low-background beta counter (see Section 7030B.1) or a gamma spectrometer (see Section 7030B.5).

3. Reagents

a. *Ammonium phosphomolybdate reagent,* $H_{12}Mo_{12}N_3O_{40}P$: Dissolve 100 g molybdic acid (85% MoO_3) in a mixture of 240 mL distilled water and 140 mL conc ammonium hydroxide (NH_4OH). When solution is complete, filter and add 60 mL conc nitric acid (HNO_3). Separately mix 400 mL conc HNO_3 and 960 mL distilled water. After both solutions cool to room temperature,

add, with constant stirring, the $(NH_4)_6Mo_7O_{24}$ solution to the HNO_3 solution. Let stand for 24 h. Filter† and discard insoluble material.

Collect filtrate in a 3-L beaker and heat to 50 to 55°C (never above 55°C). Remove from heating unit. Add 25 g sodium dihydrogen phosphate (NaH_2PO_4) dissolved in 100 mL distilled water, stir occasionally for 15 min, and let settle (approximately 30 min). Filter and wash precipitate with 1% potassium nitrate (KNO_3) and finally with distilled water. Dry precipitate and paper at 100°C for 3 to 4 h. Transfer solid $(NH_4)_3PMo_{12}O_{40}$ to a weighing bottle and store in a desiccator.

b. *Chloroplatinic acid,* 0.1M: Dissolve 51.8 g $H_2PtCl_6·6H_2O$ in distilled water and dilute to 1000 mL.

c. *Cesium carrier:* Dissolve 1.267 g cesium chloride (CsCl) in distilled water and dilute to 100 mL; 1 mL = 10 mg Cs.

d. *Calcium chloride,* 3M: Dissolve 330 g $CaCl_2$ in distilled water and dilute to 1000 mL.

e. *Ethanol,* 95%.

f. *Hydrochloric acid,* HCl, conc, 6N, 1N.

g. *Sodium hydroxide,* NaOH, 6N.

4. Procedure

a. To a 1-L sample, add 1.0 mL cesium carrier and enough conc HCl to make the solution about 0.1N HCl (about 8.6 mL). Slowly add 1 g $(NH_4)_3PMo_{12}O_{40}$ and stir for 30 min using a magnetic stirrer at 800 rpm. Let precipitate settle for at least 4 h and discard supernatant by decanting or using suction (provided by an inverted glass funnel connected to a vacuum source). Using a stream of 1N HCl, quantitatively transfer precipitate to a cen-

---

* Whatman No. 41, 9 cm diam; Whatman No. 42, 2.4 cm diam; or equivalent.

† Whatman No. 42 filter paper or equivalent.

trifuge tube. Centrifuge and discard supernatant. Wash precipitate with 20 mL $1N$ HCl and discard wash solution.

b. Dissolve precipitate by dropwise addition of 3 to 5 mL $6N$ NaOH. Heat over a flame for several minutes to remove ammonium ions. (Moist pH paper turns green as long as NH_3 vapors are evolved.) Dilute to 20 mL with distilled water. Add 10 mL $3M$ $CaCl_2$ and adjust to pH 7 with $6N$ HCl to precipitate $CaMoO_4$. Stir, centrifuge, and filter‡ supernatant into a 50-mL centrifuge tube. Wash precipitate remaining in the original centrifuge tube with 10 mL distilled water, filter through the same filter paper, and combine the wash with filtrate. Discard precipitate and filter paper.

c. Add 2 mL $0.1M$ H_2PtCl_6 and 5 mL ethanol. Cool and stir in ice bath for 10 min. Using distilled water transfer to a tared glass-fiber filter. Wash with successive portions of distilled water, $1N$ HCl, and ethanol.

d. Dry at 110°C for 30 min, cool, weigh, mount on a nylon disk and ring with polyester plastic§ cover, and beta-count or gamma-scan for ^{134}Cs and ^{137}Cs.

---

‡ Whatman No. 41 filter paper or equivalent.
§ Mylar or equivalent.

5. Calculation

Calculate the concentration of radiocesium as follows:

$$Cs,\ pCi/L = \frac{C}{2.22 \times EVR}$$

where:

C = net count rate, cpm,
E = counter efficiency,
V = volume of sample, L, and
R = fractional chemical yield
  $= \dfrac{\text{recovered } Cs_2PtCl_6,\ mg \times 0.3945}{\text{added Cs carrier, mg}}$

6. Bibliography

FINSTON, H.L. & M.T. KINSLEY. 1961. The Radiochemistry of Cesium. Rep. NAS-NS-3035, U.S. Atomic Energy Comm.

KRIEGER, H.L. 1976. Interim Radiochemical Methodology for Drinking Water. EPA-600/4-75-008 (revised), U.S. Environmental Protection Agency, Environmental Monitoring and Support Lab., Cincinnati, Ohio.

7500-I   RADIOACTIVE IODINE*

7500-I A.   Introduction

1. Occurrence and Significance

Radioiodine that results from testing nuclear devices or is released during use and processing of reactor fuels is a major concern in radioactivity monitoring. Fission products may contain iodine-129 through iodine-135. Iodine-129 has a half-life of 1.6×10^7 years but a relatively low specific activity (1.73×10^{-4} Ci/g for ^{129}I as compared to 1.24×10^5 Ci/g for ^{131}I). The half-life of ^{131}I is 8 d while for the other isotopes it is shorter (35 min to 21 h). At present, only ^{131}I is likely to be found in water. When ingested or inhaled, it concentrates in the thyroid gland and may cause thyroid cancer.

2. Selection of Method

Of the three methods, the precipitation method (B) is preferred because it is simple and involves the least time. Method C, in which iodide is concentrated by absorption on an anion resin, purified, and counted in a beta-gamma coincidence system, is sensitive and accurate. Method D uses distillation. With each method it is possible to reach the EPA recommended detection limit of 1 pCi ^{131}I/L.

3. Bibliography

KLEINBERG, J. & G.A. COWAN. 1960. The Radiochemistry of Fluorine, Chlorine, Bromine and Iodine. Rep. NAS-NS-3005, U.S. Atomic Energy Comm.

BRAUER, F.P., J.H. KAYE & R.E. CONNALY. 1970. X-ray and β-γ. Coincidence Spectrometry Applied to Radiochemical Analysis of Environmental Samples. Advances in Chemistry Ser., No. 93, Radionuclides in the Environment, pp. 231–253. American Chemical Soc.

---

* Approved by Standard Methods Committee, 1993.

7500-I B. Precipitation Method

1. General Discussion

Principle: Iodate carrier is added to an acidified sample and, after reduction with Na_2SO_3 to iodide, the ^{131}I is precipitated with $AgNO_3$. The precipitate is dissolved and purified with zinc powder and H_2SO_4 and the solution is reprecipitated as PdI_2 for counting.

2. Apparatus

a. *Counting instrument:* Low-background beta counter (see Section 7030B.1) or gamma spectrometer (7030B.5).

b. *Fine-fritted glass funnel.*

c. *Filter apparatus:* Two-piece filter funnel with filtering equipment.*

d. *Filter materials:* Filter paper;† glass-fiber filter, 2.4 cm diam; or 0.8-μm pore-diam membrane filter, 4.7 cm diam.

3. Reagents

a. *Ammonium hydroxide,* NH_4OH, 6N.

b. *Ethanol,* 95%.

c. *Hydrochloric acid,* HCl, 6N.

d. *Iodate carrier:* Dissolve 1.685 g KIO_3 in distilled water and dilute to 100 mL. Store in dark flask; 1 mL = 10 mg I.

e. *Nitric acid,* HNO_3, conc.

f. *Palladium chloride,* $PdCl_2$: Dissolve 3.3 g $PdCl_2$ in 100 mL 6N HCl; 1 mL = 20 mg Pd.

g. *Silver nitrate,* $AgNO_3$, 0.1M: Dissolve 17 g $AgNO_3$ in distilled water and dilute to 1000 mL. Store in dark flask.

h. *Sodium sulfite,* Na_2SO_3, 1M (freshly prepared): Dissolve 6.3 g Na_2SO_3 in distilled water and dilute to 50 mL.

i. *Sulfuric acid,* H_2SO_4, 2N.

j. *Zinc,* powder, reagent grade.

4. Procedure

a. To a 2000-mL sample, add 15 mL conc HNO_3 and 1.0 mL iodate carrier. Mix well. Add 4 mL freshly prepared 1M Na_2SO_3 and stir for 30 min. Add 20 mL 0.1M $AgNO_3$, stir for 1 h, and let settle for 1 h. Decant and discard as much of the supernatant as possible. Filter remainder through a glass-fiber filter and discard filtrate.

b. Transfer filter to a centrifuge tube and slurry with 10 mL distilled water. Add 1 g zinc powder and 2 mL 2N H_2SO_4 and stir frequently for at least 30 min. Filter, with vacuum, through a fine-fritted glass funnel and collect filtrate in an erlenmeyer flask. Wash both residue and filter with a minimum quantity of distilled water and add wash water to filtrate. Discard residue.

c. Add 2 mL 6N HCl and heat in water bath at 80°C for 10 min. Add 1 mL 0.2M $PdCl_2$ and digest for at least 5 min. Centrifuge and discard supernatant.

d. Dissolve precipitate in 5 mL 6N NH_4OH and heat in boiling water bath for 5 min. Filter through a glass-fiber filter and collect filtrate in a centrifuge tube. Discard filter and residue.

e. Neutralize filtrate with 6N HCl, add 2 mL in excess, and heat in a water bath. Add 1 mL 0.2M $PdCl_2$ to reprecipitate PdI_2 and digest for 10 min. Cool slightly and transfer to a tared filter with distilled water. Wash successively with 5-mL portions of distilled water and 95% ethanol. Dry in a vacuum oven at 60°C for 1 h, weigh precipitate, mount, and beta-count.

f. If final PdI_2 precipitate on a glass-fiber filter is counted in a low-background beta counter, the background counting rate is relatively high (about 1.3 cpm). If precipitate is collected on a 0.8-μm membrane filter and dried for 30 min at 70°C it may be counted in a beta-gamma coincidence scintillation system with a background rate of less than 0.1 cpm.

If a low-background counter is used, confirm identity of ^{131}I by recounting precipitate after about 1 week to check the half-life.

5. Calculation

Calculate concentration of radioiodine as follows:

$$^{131}I, \text{ pCi/L} = \frac{C}{2.22 \times EVR \times A}$$

where:

C = net count rate, cpm,
E = counting efficiency of ^{131}I as function of mass of PdI_2 precipitate,
V = volume of sample, L,
R = fractional chemical yield
$ = \dfrac{\text{recovered } PdI_2 \times 0.0704}{\text{added iodine carrier}}$, and
A = ^{131}I decay factor for the time interval between sample collection and measurement.

6. Bibliography

U.S. ENVIRONMENTAL PROTECTION AGENCY. 1980. Prescribed Procedures for Measurement of Radioactivity in Drinking Water. EPA-600/4-80-032, Environmental Monitoring and Support Lab., Cincinnati, Ohio.

---

* Fisher Filtrator or equivalent.
† Whatman No. 42 or equivalent.

7500-I C. Ion-Exchange Method

1. General Discussion

Principle: A known amount of inactive iodine in the form of KI is added as a carrier and the sample is taken through an oxidation-reduction step using hydroxylamine and sodium bisulfite to convert all iodine to iodide. Iodine, as the iodide, is concentrated by absorption on an anion-exchange column. Following an NaCl wash, iodine is eluted with sodium hypochlorite. Iodine in the iodate form is reduced to I_2, extracted into CCl_4, and back-extracted as iodide into water. The iodine finally is precipitated as PdI_2.

2. Apparatus

a. Counting instrument: Low-background beta counter (Section 7030B.1) or gamma spectrometer (7030B.5).

b. Chromatographic column, 2 cm × 15 cm.

c. Vacuum filter holder, 2.5 cm² filter area.

*d. Filter paper,** 2.4 cm diam.

e. Vacuum oven.

3. Reagents

a. Iodine carrier: Weigh approximately 13 g dried KI to the nearest 0.1 mg. Dissolve in a 1-L volumetric flask containing 100 mL distilled water. Add 10 mL $1M$ NaHSO₃ and dilute to mark with distilled water. Concentration of carrier I, mg/L = g KI × 0.7644.

b. Ethanol, absolute.

c. Hydroxylamine hydrochloride, 1M: Dissolve 6.95 g $NH_2OH \cdot HCl$ in distilled water and dilute to 100 mL.

d. Nitric acid, HNO₃, conc, $8N$, $1.6N$.

e. Sodium bisulfite, 1M: Dissolve 1.04 g NaHSO₃ in distilled water and dilute to 10 mL.

f. Sodium hydroxide, 12N: Dissolve 480 g NaOH in distilled water and dilute to 1 L.

g. Sodium hypochlorite, NaOCl, 5%: Use available household bleach.

h. Anion-exchange resin.†

i. Carbon tetrachloride, CCl₄, reagent grade.

j. Hydrochloric acid, HCl, $3N$, $1N$.

k. Palladium chloride: Dissolve 3.3 g PdCl₂ in 100 mL $6N$ HCl; 1 mL = 20 mg Pd.

l. Sodium chloride, NaCl, $2M$: Dissolve 117 g NaCl in distilled water and dilute to 1 L.

m. Hydroxylamine hydrochloride wash solution: Add 20 mL conc HNO₃ and 20 mL $1M$ $NH_2OH \cdot HCl$ to 100 mL distilled water.

4. Procedure

a. To 1 L sample in a beaker add, while stirring, 2.0 mL iodine carrier and 5 mL 5% NaOCl, and heat for 2 to 3 min to complete

oxidation. After the interchange reaction (2 to 3 min), slowly add 5 mL conc HNO₃. Add 25 mL $1M$ $NH_2OH \cdot HCl$ and stir. Let reaction go on for a few seconds, add 10 mL $1M$ NaHSO₃, and adjust pH to 6.5 with $12N$ NaOH or $1.6N$ HNO₃. Stir thoroughly for a few minutes. (Stir samples containing a large amount of organic material, such as muddy water, for 45 min.) Filter through a glass-fiber filter to remove suspended matter. Discard residue.

b. Pour 20 mL anion-exchange resin into a column and wash sides down with distilled water. Pass sample through ion-exchange column at a flow rate of 20 mL/min. Discard effluent. Wash column with 200 mL distilled water and then with 100 mL $2M$ NaCl at a flow rate of 4 mL/min. Discard wash solutions.

c. Add 50 mL 5% NaOCl in 10- to 20-mL increments, stirring the resin as needed to eliminate gas bubbles, and maintain a flow rate of 2 mL/min. To the eluted volume of 50 to 60 mL, collected in a beaker, carefully add 10 mL conc HNO₃ to make sample 2 to $3N$ in HNO₃ and transfer to a separatory funnel. (Add acid slowly with stirring until vigorous reaction subsides.)

d. Add 50 mL CCl₄ and 10 mL $1M$ $NH_2OH \cdot HCl$. Extract iodine into organic phase by shaking for about 2 min. Let phases separate and transfer organic phase to another separatory funnel. Add 25 mL CCl₄ and 5 mL $1M$ $NH_2OH \cdot HCl$ to the first separatory funnel and shake for 2 min. Combine organic phase with the one obtained from the first extraction. Discard aqueous phase. Add 20 mL $NH_2OH \cdot HCl$ wash solution to the organic phase and shake for 2 min. Let phases separate and transfer organic phase to a clean separatory funnel. Discard wash solution.

e. Add 25 mL distilled water and 10 drops $1M$ NaHSO₃ to organic phase. Shake for 2 min, let phases separate, and discard organic phase. Transfer aqueous phase to a beaker. Add 10 mL $3N$ HCl. Using a stirrer-hot plate, boil and stir the sample until it evaporates to 10 to 15 mL or begins to turn yellow.

f. Add 1.0 mL PdCl₂ solution dropwise. Rinse sides of beaker with $1N$ HCl and add sufficient $1N$ HCl to make a volume of 30 mL. Continue stirring until cool. Place beaker in a stainless steel tray and store at about 4°C overnight.

g. Filter through a tared filter mounted in a filter holder. Wash residue with $1N$ HCl and then with absolute alcohol. Dry in a vacuum oven at 60°C for 1 h. Cool in a desiccator, weigh precipitate, then seal it between polyester tape and polyester plastic film,‡ with the film over the precipitate. Count with a beta-gamma coincidence system.

5. Calculation

Calculate ^{131}I, pCi/L, as in B.5.

6. Bibliography

American Society for Testing and Materials. 1972 Book of ASTM Standards. Part 23. D 2334-68, Philadelphia, Pa.

Gabay, J.J., C.J. Paperiello, S. Goodyear, J.C. Daly & J.M. Matuszek. 1974. A method of determining [129]I in milk and water. *Health Phys.* 26:89.

---

* Whatman No. 42 or equivalent.
† Dowex 1 × 8, 50-100 mesh, chloride form, or equivalent.

---

‡ Mylar or equivalent.

7500-I D.   Distillation Method

1. General Discussion

Principle: Iodine carrier is added to an acidified sample and iodine is distilled into a caustic solution. The distillate is acidified and the iodine is extracted into CCl_4. After back-extraction as iodide, the iodine is purified as PdI_2 for counting.

2. Apparatus

a. Distillation apparatus and 3-L round-bottom flask.

b. Separatory funnel, 60 mL.

c. Filter apparatus: Two-piece filter funnel with filtering equipment.*

d. Filter paper: See B.2*d*.

3. Reagents

a. Ammonium hydroxide, NH_4OH, conc.

b. Carbon tetrachloride, CCl_4.

c. Ethanol, 95%.

d. Hydrochloric acid, HCl, 6*N*, 1*N*.

e. Iodide carrier: Dissolve 2.616 g KI in distilled water, add 2 drops $NaHSO_3$, and dilute to 100 mL. Store in dark flask. 1 mL = 20 mg I.

f. Nitric acid, HNO_3, conc.

g. Palladium chloride: Dissolve 3.3 g $PdCl_2$ in 100 mL 6*N* HCl; 1 mL = 20 mg Pd.

h. Sodium bisulfite, $NaHSO_3$, 1*M:* Dissolve 5.2 g $NaHSO_3$ in distilled water and dilute to 50 mL. Prepare only in small quantities.

i. Sodium hydroxide, NaOH, 0.5*N*.

j. Sodium nitrite, $NaNO_2$, 1*M:* Dissolve 69 g $NaNO_2$ in distilled water and dilute to 1 L.

k. Sulfuric acid, H_2SO_4, 12*N*.

l. Tartaric acid, $C_4H_6O_6$, 50%: Dissolve 50 g $C_4H_6O_6$ in distilled water and dilute to 100 mL.

4. Procedure

a. To a 2000-mL sample in a 3-L round-bottom flask, add 15 mL 50% $C_4H_6O_6$ and 1.0 mL iodide carrier. Mix well, cautiously add 25 mL cold conc HNO_3, and close distillation apparatus (Figure 7500-I:1).

b. Connect an air line to still inlet, adjust flow rate to about 2 bubbles/s, and distill for at least 15 min into 15 mL 0.5*N* NaOH. Cool and transfer NaOH solution to a 60-mL separatory funnel. Discard still residue.

c. Adjust distillate to slightly acid with 1 mL 12*N* H_2SO_4 and oxidize with 1 mL 1*M* $NaNO_2$. Add 10 mL CCl_4 and shake for

---

* Fisher Filtrator or equivalent.

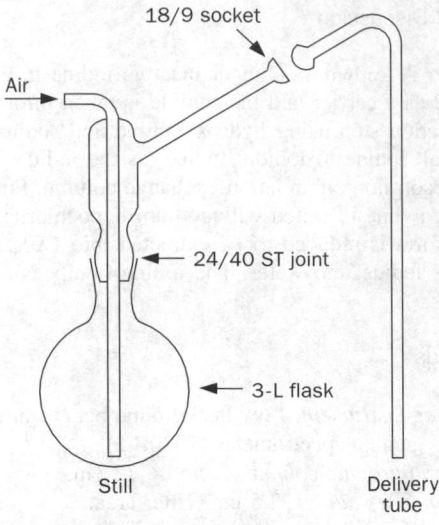

Air

18/9 socket

24/40 ST joint

3-L flask

Still

Delivery tube

Figure 7500-I:1. Distillation apparatus for iodine analysis (not to scale).

1 to 2 min. Transfer organic layer to a clean 60-mL separatory funnel containing 2 mL 1*M* $NaHSO_3$.

d. Add 5 mL CCl_4 and 1 mL 1*M* $NaNO_2$ to original separatory funnel containing the aqueous layer and shake for 2 min. Combine organic fractions. Repeat and discard aqueous layer.

e. Shake separatory funnel thoroughly until CCl_4 layer is decolorized; let phases separate and transfer aqueous layer to a centrifuge tube. Add 2 mL 1*M* $NaHSO_3$ to the separatory funnel containing CCl_4 and shake for several minutes. When phases separate, add aqueous layer to centrifuge tube. Add 1 mL distilled water to separatory funnel and shake for several minutes. When the phases separate, add aqueous layer to centrifuge tube. Discard organic layer.

f. To combined aqueous fractions, add 2 mL 6*N* HCl and heat in water bath at 80°C for 10 min. Add 1.0 mL $PdCl_2$ solution dropwise, with stirring, and digest for 15 min.

g. Cool, stir precipitate, and transfer to a tared filter mounted in a two-piece funnel. Let precipitate settle by gravity for uniform deposition, then apply suction. Wash residue with 10 mL 1*N* HCl, 10 mL distilled water, and then with 10 mL 95% ethanol. Dry in a vacuum oven at 60°C for 1 h. Cool in desiccator, weigh, mount, and make beta count.

5. Calculation

Calculate the concentration of radioiodine as given in B.5.

6. Bibliography

AMERICAN SOCIETY FOR TESTING AND MATERIALS. 1972 Book of ASTM Standards. Part 23. D 2334-68, Philadelphia, Pa.

7500-Ra   RADIUM*

7500-Ra A.   Introduction

1. Occurrence

Radium is a radioactive member of the alkaline earth family and is widely disseminated throughout the earth's crust. It has four naturally occurring isotopes—11.6-d radium-223, 3.6-d radium-224, 1600-year radium-226, and 5.75-year radium-228. Radium-223 is a member of the uranium-235 series, radium-224 and radium-228 are members of the thorium series, and radium-226 is a member of the uranium-238 series. The contribution of radium-228 (a beta-emitter) to the total radium alpha activity is negligible because of the 1.9-year half-life of its first alpha-emitting daughter product, thorium-228. The other three radium isotopes are alpha-emitters; each gives rise to a series of relatively short-lived daughter products, including three more alpha-emitters.

Because of their longer half-lives and health significance, radium-226 and radium-228 are the most important radium isotopes found in water. Radium is a bone-seeker and high concentrations can lead to malignancies.

* Approved by Standard Methods Committee, 1993.

2. Selection of Method

The principles of the two common methods for measuring radium are (a) alpha-counting a barium-radium sulfate precipitate that has been purified, and (b) measurement of radon-222 produced from the radium-226 in a sample or in a soluble concentrate isolated from the sample.

The determination of radium by precipitation (Method B) includes all alpha-emitting radium isotopes; it is a screening technique particularly applicable to drinking water. As long as the concentration of radium is less than the ^{226}Ra plus ^{228}Ra drinking water standard, examination by a more specific method is seldom needed. This method also is applicable to sewage and industrial wastes, provided that steps are taken to destroy organic matter and eliminate other interfering ions (see Gross Alpha and Gross Beta Radioactivity, Section 7110). However, avoid igniting sample ash or a fusion will be necessary.

The emanation technique (Method C), based on measurement of radon-222, is nearly, but not absolutely, specific for radium-226. Procedures for soluble, suspended, and total radium-226 are given.

The sequential precipitation method (Method D) can be used to measure either radium-228 alone or radium-228 and radium-226.

7500-Ra B.   Precipitation Method

1. General Discussion

a. *Application:* This method is suitable for determination of the alpha-emitting isotopes of radium.

b. *Principle:* Because of the difference in half-lives of the nuclides in the series including the alpha-emitting Ra isotopes, these isotopes can be identified by the rate of ingrowth and decay of their daughters in a barium sulfate precipitate.[1-3] The ingrowth of alpha activity from radium-226 increases at a rate governed primarily by the 3.8-d half-life radon-222. The ingrowth of alpha activity in radium-223 is complete by the time a radium-barium precipitate can be prepared for counting. The ingrowth of the first two alpha-emitting daughters of radium-224 is complete within a few minutes and the third alpha daughter activity increases at a rate governed by the 10.6-h half-life of lead-212. The activity of the radium-224 itself, with a 3.6-d half-life, also is decreasing, leading to a rather complicated ingrowth and decay curve.

Lead and barium carriers are added to the sample containing alkaline citrate, then sulfuric acid (H_2SO_4) is added to precipitate radium, barium, and lead as sulfates. The precipitate is purified by washing with nitric acid (HNO_3), dissolving in alkaline EDTA, and reprecipitating as radium-barium sulfate after pH adjustment

to 4.5. This slightly acidic EDTA keeps other naturally occurring alpha-emitters and the lead carrier in solution.

2. Apparatus

a. *Counting instruments:* One of the following is required:

1) *Internal proportional counter,* gas-flow, with scaler and register.

2) *Alpha scintillation counter,* silver-activated zinc sulfide phosphor deposited on thin polyester plastic, with photomultiplier tube, scaler, timer, and register; or

3) *Proportional counter,* thin end-window, gas-flow, with scaler and register.

b. *Membrane filter holder,* or stainless steel or TFE filter funnels, with vacuum source.*

c. *Membrane filters*† or *glass fiber filters.*‡

3. Reagents

a. *Citric acid, 1M:* Dissolve 210 g $H_3C_6H_5O_7 \cdot H_2O$ in distilled water and dilute to 1 L.

* Fisher Filtrator or equivalent.
† Millipore Type HAWP or equivalent.
‡ No. 934-AH, diameter 2.4 cm, H. Reeve Angel and Co., or equivalent.

b. Ammonium hydroxide, conc and 5*N:* Verify strength of old 5*N* NH$_4$OH solution before use.

c. Lead nitrate carrier: Dissolve 160 g Pb(NO$_3$)$_2$ in distilled water and dilute to 1 L; 1 mL = 100 mg Pb.

d. Stock barium chloride solution: Dissolve 17.79 g BaCl$_2$·2H$_2$O in distilled water and dilute to 1 L in a volumetric flask; 1 mL = 10 mg Ba.

e. Barium chloride carrier: To a 100-mL volumetric flask, add 20.00 mL stock BaCl$_2$ solution using a transfer pipet, dilute to 100 mL with distilled water, and mix; 1 mL = 2.00 mg Ba.

f. Methyl orange indicator solution.

g. Phenolphthalein indicator solution.

h. Bromcresol green indicator solution: Dissolve 0.1 g bromcresol green sodium salt in 100 mL distilled water.

i. Sulfuric acid, H$_2$SO$_4$, 18*N.*

j. Nitric acid, HNO$_3$, conc.

k. EDTA reagent, 0.25*M:* Add 93 g disodium ethylenediaminetetraacetate dihydrate to distilled water, dilute to 1 L, and mix.

l. Acetic acid, conc.

m. Ethyl alcohol, 95%.

n. Acetone.

o. Clear acrylic solution:§ Dissolve 50 mg clear acrylic in 100 mL acetone.

p. Standard radium-226 solution: Prepare as directed in Method C, ¶s 3*d-f,* except that in ¶ *f* (standard radium-226 solution), add 0.50 mL BaCl$_2$ stock solution (¶ 3*a*) before adding the ^{226}Ra solution; 1 mL final standard radium solution so prepared contains 2.00 mg Ba/mL and approximately 3 pCi ^{226}Ra/mL after the necessary correcting factors are applied.

4. Procedure for Radium in Drinking Water and for Dissolved Radium

a. To 1 L sample in a 1500-mL beaker, add 5 mL 1*M* citric acid, 2.5 mL conc NH$_4$OH, 2 mL Pb(NO$_3$)$_2$ carrier, and 3.00 mL BaCl$_2$ carrier. In each batch of samples include a distilled water blank.

b. Heat to boiling and add 10 drops methyl orange indicator.

c. While stirring, slowly add 18*N* H$_2$SO$_4$ to obtain a permanent pink color; then add 0.25 mL acid in excess.

d. Boil gently 5 to 10 min.

e. Set beaker aside and let stand until precipitate has settled (3 to 5 h or more).‖

f. Decant and discard clear supernate. Transfer precipitate to a 40-mL or larger centrifuge tube, centrifuge, decant, and discard supernate.

g. Rinse wall of centrifuge tube with a 10-mL portion of conc HNO$_3$, stir precipitate with a glass rod, centrifuge, and discard supernate. Repeat rinsing and washing two more times.

h. To precipitate, add 10 mL distilled water and 1 to 2 drops phenolphthalein indicator solution. Stir and loosen precipitate from bottom of tube (using a glass rod if necessary) and add 5*N* NH$_4$OH, dropwise, until solution is definitely alkaline (red). Add

10 mL EDTA reagent and 3 mL 5*N* NH$_4$OH. Stir occasionally for 2 min. Most of the precipitate should dissolve, but a slight turbidity may remain.

i. Warm in a steam bath to clear solution (about 10 min), but do not heat for an unnecessarily long period.# Add conc acetic acid dropwise until red color disappears; add 2 or 3 drops bromcresol green indicator solution and continue to add conc acetic acid dropwise, while stirring with a glass rod, until indicator turns green (aqua).** BaSO$_4$ will precipitate. Note date and time of precipitation as zero time for ingrowth of alpha activity. Digest in a steam bath for 5 to 10 min, cool, and centrifuge. Discard supernate. The final pH should be about 4.5, which is sufficiently low to destroy the Ba-EDTA complex, but not Pb-EDTA. A pH much below 4.5 will precipitate PbSO$_4$.

j. Wash Ba-Ra sulfate precipitate with distilled water and mount in a manner suitable for counting as given in ¶s *k, l,* or *m* following.

k. Transfer Ba-Ra sulfate precipitate to a tared stainless steel planchet with a minimum of 95% ethyl alcohol and evaporate under an infrared lamp. Add 2 mL acetone and 2 drops clear acrylic solution, disperse precipitate evenly, and evaporate under an infrared lamp. Dry in oven at 110°C, weigh, and determine alpha activity, preferably with an internal proportional counter. Calculate net counts per minute and weight of precipitate.

l. Weigh a membrane filter, a counting dish, and a weight (glass ring) as a unit. Transfer precipitate to tared membrane filter in a holder and wash with 15 to 25 mL distilled water. Place membrane filter in dish, add glass ring, and dry at 110°C. Weigh and count in one of the counters mentioned under ¶ 2*a* above. Calculate net counts per minute and weight of precipitate.

m. Add 20 mL distilled water to the Ba-Ra sulfate precipitate, let settle in a steam bath, cool, and filter through a special funnel with a tared glass fiber filter. Dry precipitate at 110°C to constant weight, cool, and weigh. Mount precipitate on a nylon disk and ring with an alpha phosphor on polyester plastic film,[4] and count in an alpha scintillation counter. Calculate net counts per minute and weight of precipitate.

n. If the isotopic composition of the precipitate is to be estimated, perform additional counting as mentioned in the calculation below.

o. Determination of combined efficiency and self-absorption factor: Prepare standards from 1 L distilled water and the standard radium-226 solution (¶ 3*p* preceding). Include at least one blank. The barium content will impose an upper limit of 3.0 mL on the volume of the standard radium-226 solution that can be used. If *x* is volume of standard radium-226 solution added, then add (3.00 − *x*) mL BaCl$_2$ carrier(¶ 3*e* above). Analyze standards as samples, beginning with ¶ 4*a,* but omit 3.00-mL BaCl$_2$ carrier.

From the observed net count rate, calculate the combined factor, *bc,* from the formula:

$$bc = \frac{\text{net cpm}}{ad \times 2.22 \times \text{pCi radium-226}} \quad \dagger\dagger$$

where:

ad = ingrowth factor (see below) multiplied by chemical yield.

---

§ Lucite or equivalent.

‖ If original concentrations of isotopes of radium other than ^{226}Ra are of interest, note date and time of this original precipitation as the separation of the isotopes from their parents; use a minimal settling time and complete procedure through ¶ *j* without delay. Assuming the presence of and separation of parents, decay of ^{223}Ra and ^{224}Ra begins at the time of the first precipitation, but ingrowth of decay products is timed from the second precipitation (¶ *i*). The time of the first precipitation is not needed if the objective is to check the final precipitate for its ^{226}Ra content only.

If solution does not clear in 10 min, cool, add another mL 5*N* NH$_4$OH, let stand 2 min, and heat for another 10-min period.

** The end point is most easily determined by comparison with a solution of similar composition that has been adjusted to pH 4.5 using a pH meter.

†† See calculation that follows.

If all chemical yields on samples and standards are not essentially equal, the factor bc will not be a constant. In this event, construct a curve relating the factor bc to varying weights of recovered $BaSO_4$.

5. Calculation

$$\text{Radium, pCi/L} = \frac{\text{net cpm}}{a\,b\,c\,d\,e \times 2.22}$$

where:

a = ingrowth factor (as shown in the following tabulation):

Ingrowth h	Alpha Activity from ^{226}Ra
0	1.000
1	1.016
2	1.036
3	1.058
4	1.080
5	1.102
6	1.124
24	1.489
48	1.905
72	2.253

b = efficiency factor for alpha counting,
c = self-absorption factor,
d = chemical yield, and
e = sample volume, L.

The calculations are based on the assumption that the radium is radium-226. If the observed concentration approaches 3 pCi/L, it may be desirable to follow the rate of ingrowth and estimate the isotopic content[2,3] or, preferably, to determine radium-226 by radon-222.

The optimum ingrowth periods can be selected only if the ratios and identities of the radium isotopes are known. The number of observed count rates at different ages must be equal to or greater than the number of radium isotopes present in a mixture. In the general case, suitable ages for counting are 3 to 18 h for the first count; for isotopic analysis, additional counting at 7, 14, or 28 d is suggested, depending on the number of isotopes in mixture. The amounts of the various radium isotopes can be determined by solving a set of simultaneous equations.[3] This approach is most satisfactory when radium-226 is the predominant isotope; in other situations, the approach suffers from statistical counting errors.

6. Precision and Bias

In a collaborative study, 20 laboratories analyzed four water samples for total (dissolved) radium. The radionuclide composition of these reference samples is shown in Table 7500-Ra:I. Note that Samples C and D had a ^{224}Ra concentration equal to that of ^{226}Ra.

TABLE 7500-Ra:I. CHEMICAL AND RADIOCHEMICAL COMPOSITION OF SAMPLES USED TO DETERMINE BIAS AND PRECISION OF RADIUM-226 METHOD

Radionuclide Composition	Samples			
	Pair 1		Pair 2	
	A	B	C	D
Radium-226,* pCi/L	12.12	8.96	25.53	18.84
Thorium-228,* pCi/L	none	none	25.90	19.12
Uranium, natural, pCi/L	105	77.9	27.7	20.5
Lead-210,* pCi/L	11.5	8.5	23.7	17.5
Strontium-90,* pCi/L	49.1	36.3	13.9	10.2
Cesium-137, pCi/L	50.3	37.2	12.7	9.5
NaCl, mg/L	60	60	300	300
$CaSO_4$, mg/L	30	30	150	150
$MgCl_2 \cdot 6H_2O$, mg/L	30	30	150	150
KCl, mg/L	5	5	10	10

* Daughter products were in substantial secular equilibrium.

The four results from each of two laboratories and two results from a third laboratory were rejected as outliers. The average recoveries of radium-226 from the remaining A, B, C, and D samples were 97.5, 98.7, 94.9, and 99.4%, respectively. At the 95% confidence level, the precision (random error) was 28% and 30% for the two sets of paired samples. The method is biased low for radium-226, but not seriously. The method appears satisfactory for radium-226 alone or in the presence of an equal activity of radium-224 when correction for radium-224 interference is made from a second count.

For the determination of ^{224}Ra in Samples C and D, the results of two laboratories were excluded. Hence the average recoveries were 51 and 45% for Samples C and D, respectively. At the 95% confidence level, the precision was 46% for this pair of samples. The results indicated that the method for ^{224}Ra is seriously biased low. When the recoveries for radium-224 did not agree with those for radium-226, this may have been due, in part, to incomplete instructions given in the method to account for the transitory nature of ^{224}Ra activity. The method as given here contains footnotes calling attention to the importance of the time of counting. Still uncertain is the degree of separation of radium-224 from its parent, thorium-228, in ¶s 4a through g above.

Radium-223 and radium-224 analysis by this method may be satisfactory, but special refinements and further investigations are required.

7. References

1. KIRBY, H.W. 1954. Decay and growth tables for naturally occurring radioactive series. *Anal. Chem.* 26:1063.
2. SILL, C. 1960. Determination of radium-226, thorium-230, and thorium-232. Rep. No. TID 7616 (Oct.), U.S. Atomic Energy Comm., Washington, D.C.
3. GOLDIN, A.S. 1961. Determination of dissolved radium. *Anal. Chem.* 33:406.
4. HALLDEN, N.A. & J.H. HARLEY. 1960. An improved alpha-counting technique. *Anal. Chem.* 32:1961.

7500-Ra C. Emanation Method

1. General Discussion

a. Application: This method is suitable for the determination of soluble, suspended, and total radium-226 in water. In this method, *total* radium-226 means the sum of suspended and dissolved radium-226. Radon means radon-222 unless otherwise specified.

b. Principle: Radium in water is concentrated and separated from sample solids by coprecipitation with a relatively large amount of barium as the sulfate. The precipitate is treated to remove silicates, if present, and to decompose insoluble radium compounds, fumed with phosphoric acid to remove sulfite (SO_3^{2-}), and dissolved in hydrochloric acid (HCl). The completely dissolved radium is placed in a bubbler, which is then closed and stored for a period of several days to 4 weeks for ingrowth of radon. The bubbler is connected to an evacuated system and the radon gas is removed from the liquid by aeration, dried with a desiccant, and collected in a counting chamber. The counting chamber consists of a dome-topped scintillation cell coated inside with silver-activated zinc sulfide phosphor; a transparent window forms the bottom (Figure 7500-Ra:1). The chamber rests on a photomultiplier tube during counting. About 4 h after radon collection, the alpha-counting rate of radon and decay products is at equilibrium, and a count is obtained and related to radium-226 standards similarly treated.

The counting gas used to purge radon from the liquid to the counting chamber may be helium, nitrogen, or aged air.

Some radon (emanation) techniques employ a minimum of chemistry but require high dilution of the sample and large chambers for counting the radon-222.[1] Others involve more chemical separation, concentration, and purification of radium-226 before de-emanation into counting cells of either the ionization or alpha scintillation types. The method[2] given here requires a moderate amount of chemistry coupled with a sensitive alpha scintillation count of radon-222 plus daughter products in a small chamber.[3]

c. Concentration techniques: The chemical properties of barium and radium are similar; therefore, because barium does not interfere with de-emanation, as much as 100 mg may be used to aid in coprecipitating radium from a sample to be placed in a single radon bubbler. However, because some radium-226 is present in barium salts, reagent tests are necessary to account for radium-226 introduced in this way.

d. Interferences: Only the gaseous alpha-emitting radionuclides, radon-219 (actinon) and radon-220 (thoron), can interfere. Interference from these radionuclides would be expected to be very rare in water not contaminated by such industrial wastes as uranium mill elements.[2] The half-lives of these nuclides are only 3.92 and 54.5 s, respectively, so only their alpha-emitting decay products interfere.

Interference from stable chemicals is limited. Small amounts of lead, calcium, and strontium, collected by the barium sulfate, do not interfere. However, lead may cause deterioration of platinum ware. Calcium at a concentration of 300 mg/L and other dissolved solids (in brines) at 269 000 mg/L cause no difficulty.[4]

The formation of precipitates in excess of a few milligrams during the radon-222 ingrowth period is a warning that modifications[2] may be necessary because radon-222 recovery may be impaired.

e. Minimum detectable concentration: The minimum detectable concentration depends on counter characteristics, background-counting rate of scintillation cell, length of counting period, and contamination of apparatus and environment by radium-226. Without reagent purification, the overall reagent blank (excluding background) should be between 0.03 and 0.05 pCi radium-226, which may be considered the minimum detectable amount under routine conditions.

2. Apparatus

The de-emanation assembly is shown in Figure 7500-Ra:1, and its major components are described in ¶s *b-e*, below.

a. Scintillation counter assembly with a photomultiplier (PM) tube 5 cm or more in diameter, normally mounted, face up, in a light-tight housing. The photomultiplier tube, preamplifier, high-voltage supply, and scaler may be contained in one chassis; or the PM tube and preamplifier may be used as an accessory with

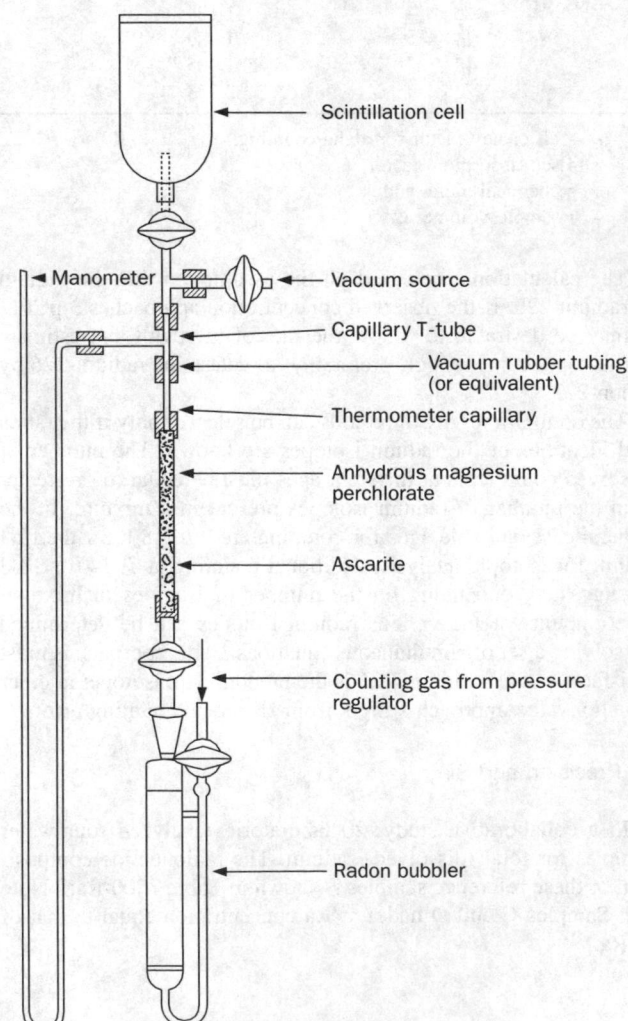

Figure 7500-Ra:1. De-emanation assembly.

a proportional counter or a separate scaler. A high-voltage safety switch should open automatically when the light cover is removed, to avoid damage to the photomultiplier tube.

Use a preamplifier with a variable gain adjustment. Equip counter with a flexible ground wire attached to the chassis and to the neck of the scintillation cell by an alligator clip or similar device. Ascertain operating voltage by determining a plateau using ^{222}Rn in the scintillation cell as the alpha source; the slope should not exceed 2%/100 V. Calibrate and use counter and scintillation cell as a unit when more than one counter is available. The background-counting rate for the counter assembly without the scintillation cell in place should be 0.00 to 0.03 cpm.

b. *Scintillation cells,*[2,3] Lucas-type, preferably having a volume of 95 to 140 mL, made in the laboratory, or commercially available.*

c. *Radon bubblers,* capacity 18 to 25 mL.† Use gas-tight glass stopcocks and a fritted glass disk of medium porosity.‡ Use one bubbler for a standard ^{226}Ra solution and one for each sample and blank in a batch.[2]

d. *Manometer,* open-end capillary tube or vacuum gauge having volume that is small compared to volume of scintillation cell, 0 to 760 mm Hg.

e. *Gas purification tube,* 7 to 8 mm OD standard-wall glass tubing, 100 to 120 mm long, constricted at lower end to hold glass wool plug; thermometer capillary tubing.

f. *Sample bottles,* polyethylene, 2- to 4-L capacity.

g. *Membrane filters.*§

h. *Gas supply:* Helium, nitrogen, or air aged in high-pressure cylinder with two-stage pressure regulator and needle valve. Helium is preferred.

i. *Silicone grease,* high-vacuum.

j. *Sealing wax,* low-melting.‖

k. *Laboratory glassware:* Excepting bubblers, decontaminate all glassware before and between uses by heating for 1 h in EDTA decontaminating solution at 90 to 100°C, then rinse in water, 1N HCl, and again in distilled water to dissolve barium (radium) sulfate, Ba(Ra)SO$_4$.

Removal of previous samples from bubblers and rinsing is described in ¶ 5a17). More extensive cleaning of bubblers requires removal of wax from joints, silicone grease from stopcocks, and the last traces of barium-radium compounds.

l. *Platinum ware:* Crucibles (20 to 30 mL) or dishes (50 to 75 mL), large dish (for flux preparation), and platinum-tipped tongs (preferably Blair type). Clean platinum ware by immersion and rotation in a molten bath of potassium pyrosulfate, remove, cool, rinse in hot tap water, digest in hot 6N HCl, rinse in distilled water, and finally flame over a burner.

3. Reagents

a. *Stock barium chloride solution:* Dissolve 17.79 g BaCl$_2$·2H$_2$O in distilled water and dilute to 1 L; 1 mL = 10 mg Ba.

b. *Dilute barium chloride solution:* Dilute 200.0 mL stock barium chloride solution to 1000 mL as needed; 1 mL = 2.00 mg Ba. Let stand 24 h and filter through a membrane filter.

---

* William H. Johnston Laboratories, 3617 Woodland Ave., Baltimore, MD 21215.
† Available from Corning Glass Works, Special Sales Section, Corning, NY 14830.
‡ Corning or equivalent.
§ Type HAWP, Millipore Filter Corp., Bedford, MA, or equivalent.
‖ Pyseal, Fisher Scientific Co., Pittsburgh, PA, or equivalent.

Optionally, add approximately 40 000 dpm of ^{133}Ba to this solution before dilution. Take account of the stable barium carrier added with the ^{133}Ba and with the diluting solution, so that the final barium concentration is near 2 mg/L. The use of ^{133}Ba provides a convenient means of checking on the recovery of ^{226}Ra from the sample; see 7, below. Use the BaCl$_2$ solution containing ^{133}Ba in steps described in ¶s 5a3), b8), and c3). Do *not* use in d below; instead, use a separate dilution of stock BaCl$_2$ solution for preparing ^{226}Ra standard solutions.

c. *Acid barium chloride solution:* To 20 mL conc HCl in a 1-L volumetric flask, add dilute BaCl$_2$ solution to the mark and mix.

d. *Stock radium-226 solution:* Take every precaution to avoid unnecessary contamination of working area, equipment, and glassware, preferably by preparing ^{226}Ra standards in a separate area or room reserved for this purpose. Obtain a National Institute of Standards and Technology (NIST) gamma ray standard containing 0.1 μg ^{226}Ra as of date of standardization. Using a heavy glass rod, cautiously break neck of ampule, which is submerged in 300 mL acid BaCl$_2$ solution in a 600-mL beaker. Chip ampule unit until it is thoroughly broken or until hole is large enough to give complete mixing. Transfer solution to a 1-L volumetric flask, rinse beaker with acid BaCl$_2$ solution, dilute to mark with same solution, and mix; 1 mL = approximately 100 pg ^{226}Ra.

Determine the time in years, t, since the NIST standardization of the original ^{226}Ra solution. Calculate pCi ^{226}Ra/mL as:

$$\text{pCi } ^{226}\text{Ra} = [1 - (4.3 \times 10^{-4})(t)] \, [100] \, [0.990]$$

e. *Intermediate radium-226 solution:* Dilute 100 mL stock radium-226 solution to 1000 mL with acid BaCl$_2$ solution; 1 mL = approximately 10 pCi ^{226}Ra.

f. *Standard radium-226 solution:* Add 30.0 mL intermediate radium-226 solution to a 100-mL volumetric flask and dilute to mark with acid BaCl$_2$ solution; 1 mL = approximately 3 pCi ^{226}Ra and contains about 2 mg Ba. See ¶ d et seq. above for correction factors.

g. *Hydrochloric acid,* HCl, conc, 6N, 1N, and 0.1N.

h. *Sulfuric acid,* H$_2$SO$_4$, conc and 0.1N.

i. *Hydrofluoric acid,* HF, 48%, in a plastic dropping bottle. (CAUTION.)

j. *Ammonium sulfate solution:* Dissolve 10 g (NH$_4$)$_2$SO$_4$ in distilled water and dilute to 100 mL in a graduated cylinder.

k. *Phosphoric acid,* H$_3$PO$_4$, 85%.

l. *Ascarite,* 8 to 20 mesh.

m. *Magnesium perchlorate,* anhydrous desiccant.

n. *EDTA decontaminating solution:* Dissolve 10 g disodium ethylenediaminetetraacetate dihydrate and 10 g Na$_2$CO$_3$ in distilled water and dilute to 1 L in a graduated cylinder.

o. *Special reagents for total and suspended radium:*

1) *Flux:* Add 30 mg BaSO$_4$, 65.8 g K$_2$CO$_3$, 50.5 g Na$_2$CO$_3$, and 33.7 g Na$_2$B$_4$O$_7$·10H$_2$O, to a 500-mL platinum dish. Mix thoroughly and heat cautiously to expel water, then fuse and mix thoroughly by swirling. Cool flux, grind in a porcelain mortar to pass a 10- to 12-mesh (or finer) screen, and store in an airtight bottle.

2) *Dilute hydrogen peroxide solution:* Dilute 10 mL 30% H$_2$O$_2$ to 100 mL in a graduated cylinder. Prepare daily.

4. Calibration of Scintillation Counter Assembly

a. Test bubblers by adding about 10 mL distilled water and passing air through them at the rate of 3 to 5 mL (free volume)/

min. Air should form many fine bubbles rather than a few large ones; the latter condition indicates nonuniform pores. Do not use bubblers requiring excessive pressure to initiate bubbling. Fritted-glass disks of medium porosity (¶ 2c) usually are satisfactory.

b. Apply silicone grease to stopcocks of a bubbler and, with gas inlet stopcock closed, add 1 mL stock $BaCl_2$ solution and 10 mL (30 pCi) standard radium-226 solution, and fill bubbler two-thirds to three-fourths full with additional acid $BaCl_2$ solution.

c. With bubbler in a clamp or rack, dry joint with lint-free paper or cloth, warm separate parts of the joint, apply sealing wax sparingly to the male part, and make the connection with a twisting motion to spread the wax uniformly in the ground joint. Let cool. Establish zero ingrowth time by purging liquid with counting gas for 15 to 20 min according to ¶ 4j below and adjust inlet pressure to produce a froth a few millimeters thick. Close stopcocks, record date and time, and store bubbler, preferably for 3 weeks or more (with most samples) before collecting and counting ^{222}Rn. A much shorter ingrowth period of 16 to 24 h is convenient for a standard bubbler. Obtain an estimate of ^{222}Rn present at any time from the B columns in Table 7500-Ra:II.

d. Attach scintillation cell as shown in Figure 7500-Ra:1;# substitute a glass tube with a stopcock for bubbler so that the compressed gas can be turned on or off conveniently. Open stopcock on scintillation cell, close stopcock to gas, and gradually open stopcock to vacuum source to evacuate cell. Close stopcock to vacuum source and check manometer reading for 2 min to test system, especially the scintillation cell, for leaks.

e. Open stopcock to counting gas and cautiously admit gas to scintillation cell until atmospheric pressure is reached.

f. Center scintillation cell on photomultiplier tube, cover with light-tight hood and, after 10 min, obtain a background counting rate (preferably over a 100- to 1000-min period, depending on concentration of ^{226}Ra in samples). *Do not expose phototube to external light with the high voltage applied.*

g. Repeat Steps d through f above for each scintillation cell.

h. If the leakage test and background are satisfactory, continue calibration.

i. With scintillation cell and standard bubbler (¶ 4c) on vacuum train, open stopcock on scintillation cell and evacuate scintillation cell and purification system (Figure 7500-Ra:1) by opening stopcock to vacuum source. Close stopcock to vacuum source. Check system for leaks as in Step d above.

j. Adjust gas regulator (diaphragm) valve so that a very slow stream of gas will flow with the needle valve open. Attach gas supply to inlet of bubbler.

k. Note time as beginning of an approximately 20-min de-emanation period. Very cautiously open bubbler outlet stopcock to equalize pressure and transfer all or most of the fluid in the inlet side arm to bubbler chamber.

l. Close outlet stopcock and very cautiously open inlet stopcock to flush remaining fluid from side arm and fritted disk. Close inlet stopcock.

m. Repeat Steps h and l above, four or five times, to obtain more nearly equal pressures on the two sides of bubbler.

n. With outlet stopcock fully open, cautiously open inlet stopcock so that gas flow produces a froth a few millimeters thick at surface of bubbler solution. Maintain flow rate by gradually increasing pressure with regulator valve and continue deemanation until pressure in cell reaches atmospheric pressure. Total elapsed time for the de-emanation should be 15 to 25 min.

o. Close stopcocks to scintillation cell, close bubbler inlet and outlet, shut off and disconnect gas supply, and record date and time as the ends of the ^{222}Rn ingrowth and de-emanation periods and as the beginnings of decay of ^{222}Rn and ingrowth of decay products.

p. Store bubbler for another ^{222}Rn ingrowth in the event a subsequent de-emanation is desired (Table 7500-Ra:II). The standard bubbler may be kept indefinitely.

q. Four hours after de-emanation, when daughter products are in virtual transient equilibrium with ^{222}Rn, place scintillation cell on photomultiplier tube, cover with light-tight hood, let stand for at least 10 min, then begin counting. Record date and time counting was started and finished.

r. Correct net counting rate for ^{222}Rn decay (Table 7500-Ra: II) and relate it to picocuries ^{226}Ra in standard bubbler (see ¶ 6a). Unless the scintillation cell is physically damaged, the calibration will remain essentially unchanged for years. Occasional calibration is recommended.

s. Repeat Steps h through r above on each scintillation cell.

t. To remove ^{222}Rn and prepare scintillation cell for reuse, evacuate and cautiously refill with counting gas. Routinely, repeat evacuation and refilling twice, and repeat process more times if the cells have contained a high ^{222}Rn activity. (Decay products with a half-life of approximately 30 min will remain in the cell. Do not check background on cells until activity of decay products has had time to decay to insignificance.)

5. Procedure

a. *Soluble radium-226:*

1) Using a membrane filter, filter at least 1 L sample or a volume containing up to 30 pCi ^{226}Ra and transfer to a polyethylene bottle as soon after sampling as possible. Save the suspended matter for determination by the procedure described in 5b, below. Record sample volume filtered if suspended solids are to be analyzed as in the procedure for ^{226}Ra in suspended matter.

2) Add 20 mL conc HCl/L of filtrate and continue analysis when convenient.

3) Add 50 mL dilute $BaCl_2$ solution, with vigorous stirring, to 1020 mL acidified filtrate [¶ 2) preceding] in a 1.5-L beaker. In each batch of samples include a reagent blank consisting of distilled water plus 20 mL conc HCl.

4) Cautiously, with vigorous stirring, add 20 mL conc H_2SO_4. Cover beaker and let precipitate overnight.

5) Filter supernate through a membrane filter, using $0.1N$ H_2SO_4 to transfer Ba-Ra precipitate to filter, and wash precipitate twice with $0.1N$ H_2SO_4.

6) Place filter in a platinum crucible or dish, add 0.5 mL HF and 3 drops (0.15 mL) $(NH_4)_2SO_4$ solution, and evaporate to dryness.

7) Carefully ignite over a small flame until carbon is burned off; cool. (After filter is charred a Meker burner may be used.)

8) Add 1 mL H_3PO_4 with a calibrated dropper and heat on hot plate at about 200°C. Gradually raise temperature and maintain at about 300 to 400°C for 30 min.

9) Swirl vessel over a low Bunsen flame, adjusted to avoid spattering, while covering the walls with hot H_3PO_4. Continue to

---

The system as described and shown in Figure 7500-Ra:1 is considered minimal. In routine work, use manifold systems and additional, more precise needle valves. An occasional drop of solution will escape from the bubbler; provide enough free space beyond the outlet stopcock to accommodate this liquid, preventing its entrance into the gas-purifying train.

TABLE 7500-Ra:II. FACTORS FOR DECAY OF RADON-222, GROWTH OF RADON-222 FROM RADIUM-226, AND CORRECTION OF RADON-222 ACTIVITY FOR DECAY DURING COUNTING

Time	Factor for Decay of Radon-222 $A = e^{-\lambda t}$		Factor for Growth of Radon-222 from Radium-226 $B = 1 - e^{-\lambda t}$		Factor for Correction of Radon-222 Activity for Decay during Counting $C = \lambda t/(1 - e^{-\lambda t})$	Time	Factor for Decay of Radon-222 $A = e^{-\lambda t}$		Factor for Growth of Radon-222 from Radium-226 $B = 1 - e^{-\lambda t}$		Factor for Correction of Radon-222 Activity for Decay during Counting $C = \lambda t/(1 - e^{-\lambda t})$
	Hours	Days	Hours	Days	Hours		Hours	Days	Hours	Days	Hours
0.0	1.0000		0.000 00		1.000	29	0.8034	0.0052	0.1966	0.9948	1.113
0.2	0.9985		0.001 51		1.001	30	0.7973	0.0044	0.2027	0.9956	1.118
0.4	0.9970		0.003 01		1.001						
0.6	0.9955		0.004 52		1.002	31	0.7913	0.0036	0.2087	0.9964	1.122
0.8	0.9940		0.006 02		1.003	32	0.7854	0.0030	0.2146	0.9970	1.126
						33	0.7795	0.0025	0.2205	0.9975	1.130
1	0.9925	0.8343	0.007 52	0.1657	1.004	34	0.7736	0.0021	0.2264	0.9979	1.134
2	0.9850	0.6960	0.014 99	0.3040	1.008	35	0.7678	0.0018	0.2322	0.9982	1.138
3	0.9776	0.5807	0.022 40	0.4193	1.011						
4	0.9703	0.4844	0.029 75	0.5156	1.015	36	0.7620	0.0015	0.2380	0.9985	1.142
5	0.9630	0.4041	0.037 05	0.5959	1.019	37	0.7563	0.0012	0.2437	0.9988	1.146
						38	0.7506	0.0010	0.2494	0.9990	1.150
6	0.9557	0.3372	0.044 29	0.6628	1.023	39	0.7449	0.0009	0.2551	0.9991	1.154
7	0.9485	0.2813	0.051 48	0.7187	1.027	40	0.7393	0.0007	0.2607	0.9993	1.159
8	0.9414	0.2347	0.058 61	0.7653	1.031						
9	0.9343	0.1958	0.065 69	0.8042	1.034	41	0.7338	0.0006	0.2662	0.9994	1.163
10	0.9273	0.1633	0.072 72	0.8367	1.038	42	0.7283	0.0005	0.2717	0.9995	1.167
						43	0.7228	0.0004	0.2772	0.9996	1.171
11	0.9203	0.1363	0.079 69	0.8637	1.042	44	0.7173	0.0003	0.2827	0.9997	1.175
12	0.9134	0.1137	0.086 62	0.8863	1.046	45	0.7120	0.0003	0.2880	0.9997	1.179
13	0.9065	0.0948	0.093 49	0.9052	1.050						
14	0.8997	0.0791	0.100 31	0.9209	1.054	46	0.7066	0.0002	0.2934	0.9998	1.184
15	0.8929	0.0660	0.107 07	0.9340	1.058	47	0.7013	0.0002	0.2987	0.9998	1.188
						48	0.6960	0.0002	0.3040	0.9998	1.192
16	0.8862	0.0551	0.1138	0.9449	1.062	49	0.6908	0.0001	0.3092	0.9999	1.196
17	0.8795	0.0459	0.1205	0.9541	1.066	50	0.6856	0.0001	0.3144	0.9999	1.201
18	0.8729	0.0383	0.1271	0.9617	1.069						
19	0.8664	0.0320	0.1336	0.9680	1.073	51	0.6804	0.0001	0.3196	0.9999	1.205
20	0.8598	0.0267	0.1402	0.9733	1.077	52	0.6753	0.0001	0.3247	0.9999	1.209
						53	0.6702	0.0001	0.3298	0.9999	1.213
21	0.8534	0.0223	0.1466	0.9777	1.081	54	0.6652	0.0001	0.3348	0.9999	1.218
22	0.8470	0.0186	0.1530	0.9814	1.085	55	0.6602	0.0000	0.3398	1.0000	1.222
23	0.8406	0.0155	0.1594	0.9845	1.089						
24	0.8343	0.0129	0.1657	0.9871	1.093	56	0.6552	0.0000	0.3448	1.0000	1.226
25	0.8280	0.0108	0.1720	0.9892	1.097	57	0.6503	0.0000	0.3497	1.0000	1.231
						58	0.6454	0.0000	0.3546	1.0000	1.235
26	0.8218	0.0090	0.1782	0.9910	1.101	59	0.6405	0.0000	0.3595	1.0000	1.239
27	0.8156	0.0075	0.1844	0.9925	1.105	60	0.6357	0.0000	0.3643	1.0000	1.244
28	0.8095	0.0063	0.1905	0.9937	1.109						

heat for a minute after precipitate fuses into a clear melt (just below redness) to insure complete removal of SO_3.

10) Fill cooled vessel one-half full with $6N$ HCl, heat on steam bath, then gradually add distilled water to within 2 mm of top of vessel.

11) Evaporate on boiling steam bath until there are no more vapors of HCl.

12) Add 6 mL $1N$ HCl, swirl, and warm to dissolve $BaCl_2$ crystals.

13) Close gas inlet stopcock, add a drop of water to the fritted disk of the fully greased and tested radon bubbler, and transfer sample from platinum vessel to bubbler with a medicine dropper. Use dropper to rinse vessel with at least three 2-mL portions of distilled water. Add distilled water until bubbler is two-thirds to three-fourths full.

14) Dry, wax if necessary, and seal joint. Establish zero ingrowth time as instructed in ¶ 4c preceding.

15) Close stopcocks, record date and time, and store bubbler for ^{222}Rn ingrowth, preferably for 3 weeks for low concentrations of radium-226.

16) De-emanate and count ^{222}Rn as instructed for calibrations in ¶s 4i through r, with sample replacing standard bubbler.

17) The sample in the bubbler may be stored for a second ingrowth or it may be discarded and the bubbler cleaned for reuse.

(A bubbler is readily cleaned while in an inverted position by attaching a tube from a beaker containing 100 mL 0.1N HCl to the inlet and attaching another tube from outlet to a suction flask. Alternately open and close outlet and inlet stopcocks to pass the acid rinse water sequentially through the fritted disk, accumulate in the bubbler, and flush into the suction flask. Drain bubbler with the aid of vacuum, heat ground joint gently to melt wax, and separate joint. More extensive cleaning, as indicated in ¶ 2k above, may be necessary if the bubbler contained more than 10 pCi ^{226}Ra.)

b. Radium-226 in suspended matter:

1) Suspended matter in water usually contains siliceous materials that require fusion with an alkaline flux to insure recovery of radium. Dry suspended matter (up to 1000 mg inorganic material) retained on the membrane filter specified in ¶ 5a1) above in a tared platinum crucible and ignite as in ¶ 5a7).

2) Weigh crucible to estimate residue.

3) Add 8 g flux/g residue, but not less than 2 g flux, and mix with a glass rod.

4) Heat over a Meker burner until melting begins, being careful to prevent spattering. Continue heating for 20 min after bubbling stops, with an occasional swirl of the crucible to mix contents and achieve a uniform melt. A clear melt usually is obtained only when the suspended solids are present in small amount or have a high silica content.

5) Remove crucible from burner and rotate as melt cools to distribute it in a thin layer on crucible wall.

6) When cool, place crucible in a covered beaker containing 120 mL distilled water, 20 mL conc H_2SO_4, and 5 mL dilute H_2O_2 solution for each 8 g flux. (Reduce acid and H_2O_2 in proportion to flux used.) Rotate crucible to dissolve melt if necessary.

7) When melt is dissolved, remove and rinse crucible into beaker. Save crucible for Step 10) below.

8) Heat solution and slowly add 50 mL dilute $BaCl_2$ solution with vigorous stirring. Cover beaker and let stand overnight for precipitation. (Precipitation with cool sample solution also is satisfactory.)

9) Add about 1 mL dilute H_2O_2 and, if yellow color (from titanium) deepens, add more H_2O_2 until there is no further color change.

10) Continue analysis according to ¶s 5a5) through 16).

11) Calculate result as directed in ¶s 6a and b, taking into account that the suspended solids possibly were contained in a sample volume other than 1 L [see ¶ 5a1)].

c. Total radium-226:

1) Total ^{226}Ra in water is the sum of soluble and suspended ^{226}Ra as determined in 5a and b preceding, or it may be determined directly by examining the original water sample that has been acidified with 20 mL conc HCl/L sample and stored in a polyethylene bottle.

2) Thoroughly mix acidified sample and take 1020 mL or a measured volume containing not more than 1000 mg inorganic suspended solids.

3) Add 50 mL dilute $BaCl_2$ solution and slowly, with vigorous stirring, add 20 mL conc H_2SO_4/L sample. Cover and let precipitate overnight.

4) Filter supernate through membrane filter and transfer solids to filter as in ¶ 5a5) preceding.

5) Place filter and precipitate in tared platinum crucible and proceed as in ¶s 5b2) through 10) above but with the following changes in the procedure given in ¶ 5b8): Omit adding dilute

$BaCl_2$ solution, digest for 1 h on a steam bath, and filter immediately after digestion without stirring up $BaSO_4$. (If these changes are not made, filtration will be very slow.)

6) Calculate total radium-226 concentration as directed in ¶s 6a and b.

6. Calculations

a. Calculate the ^{226}Ra in a bubbler, including reagent blank, as follows:

$$^{226}\text{Ra, pCi} = \frac{R_s - R_b}{R_c} \times \frac{1}{1 - e^{-\lambda t_1}} \times \frac{1}{e^{-\lambda t_2}} \times \frac{\lambda t_3}{1 - e^{-\lambda t_3}}$$

where:

λ = decay constant for ^{222}Rn, 0.007 55/h,
t_1 = time interval allowed for ingrowth of ^{222}Rn, h,
t_2 = time interval between de-emanation and counting, h,
t_3 = time interval of counting, h,
R_s = observed counting rate of sample in scintillation cell, cph,
R_b = (previously) observed background counting rate of scintillation cell with counting gas, cph,
R_c = calibration constant for scintillation cell [i.e., observed net counts per hour, corrected by use of ingrowth and decay factors (C/AB from below) per picocurie of Ra in standard],

or:

$$^{226}\text{Ra, pCi} = \frac{(R_s - R_b)}{R_c} \times \frac{C}{AB}$$

where:

A = factor for decay of ^{222}Rn (see Table 7500-Ra:II),
B = factor for growth of ^{222}Rn from ^{226}Ra (see Table 7500-Ra:II), and
C = factor for correction of ^{222}Rn activity for decay during counting (see Table 7500-Ra:II).

For nontabulated times, obtain decay factors for ^{222}Rn by multiplying together the appropriate tabulated "day" and "hour" decay factors, interpolating for less than 0.2 h if indicated by the precision desired. Obtain radon-222 growth factors for nontabulated times most accurately, especially for short periods (e.g., in calibrations), by calculation from ^{222}Rn decay factors given in Column A and using formula given in heading for Column B (of Table 7500-Ra:II). Linear interpolations are satisfactory for routine samples. Obtain the decay-during-counting factors by linear interpolation for all nontabulated times.

In calculating cell calibration constants, use the same equation, but picocuries of ^{226}Ra is known and R_c is unknown.

b. Convert the activity into picocuries per liter of soluble, suspended, or total ^{226}Ra by the following equation:

$$^{226}\text{Ra, pCi/L} = \frac{(D - E) \times 1000}{\text{mL sample}}$$

where:

D = pCi ^{226}Ra found in sample, and
E = pCi ^{226}Ra found in reagent blank.

7. Recovery of Barium (Radium-226) (Optional)

If [133]Ba was added in reagent *b*, check recovery of Ba by removing sample from bubbler, adjusting its volume appropriately, gamma-counting it under standardized conditions, and comparing the result with the count obtained from a 50-mL portion (evaporated if necessary to reduce volume) of dilute barium solution also counted under standardized conditions; add 1 mL H_3PO_4 to the latter portion before counting. The assumption that the Ba and [226]Ra are recovered to the same extent is valid in the method described.

Note that [226]Ra and its decay products interfere slightly even if a gamma spectrometer is used. The technique works best when the ratio of [133]Ba to [226]Ra is high.

Determinations of recovery are particularly helpful with irreplaceable samples, both in gaining experience with the method and in applying the general method to unfamiliar media.

8. Precision and Bias

In a collaborative study, seven laboratories analyzed four water samples for dissolved radium-226 by this method. No result was rejected as an outlier. The average recoveries of added radium-226 from Samples A, B, C, and D (below) were 97.1, 97.3, 97.6, and 98.0%, respectively. At the 95% confidence level, the precision (random error) was 6% and 8% for the two sets of paired samples. Because of the small number of participating laboratories and the low values for random and total errors, there was no evidence of laboratory systematic errors. Neither radium-224 at an activity equal to that of the radium-226 nor dissolved solids up to 610 mg/L produced a detectable error in the results.

Test samples consisted of two pairs of simulated moderately hard and hard water samples containing known amounts of added radium-226 and other radionuclides. The composition of the samples with respect to nonradioactive substances was the same for a pair of samples but varied for the two pairs. The radiochemical composition of the samples is given in Table 7500-Ra:I.

9. References

1. HURSH, J.B. 1954. Radium-226 in water supplies of the U.S. *J. Amer. Water Works Assoc.* 46:43.
2. RUSHING, D.E., W.J. GARCIA & D.A. CLARK. 1964. The analysis of effluents and environmental samples from uranium mills and of biological samples for radium, polonium, and uranium. *In* Radiological Health and Safety in Mining and Milling of Nuclear Materials. International Atomic Energy Agency, Vienna, Austria, Vol. 11, p. 187.
3. LUCAS, H.F. 1957. Improved low-level alpha scintillation counter for radon. *Rev. Sci. Instrum.* 28:680.
4. RUSHING, D.E. 1967. Determination of dissolved radium-226 in water. *J. Amer. Water Works Assoc.* 59:593.

7500-Ra D. Sequential Precipitation Method

1. General Discussion

a. Application: This method can be used to determine soluble radium-228 alone or soluble radium-228 plus radium-226.

b. Principle: Radium-228 and radium-226 in water are concentrated and separated by coprecipitation with barium and lead as sulfates and purified by EDTA chelation. After 36-h ingrowth of actinium-228 from radium-228, actinium-228 is carried on yttrium oxalate, purified, and beta-counted. Radium-226 in the supernatant is precipitated as the sulfate, purified, and alpha-counted (Method B) or it is transferred to a radon bubbler and determined by the emanation procedure (Method C), which is the preferred method.

If analysis of radium-226 is not required, the procedure for radium-228 may be terminated by beta-counting the yttrium oxalate precipitate with a follow-up precipitation of barium sulfate for yield determination. If it is determined that radium-228 is absent, the radium-226 fraction may be alpha-counted directly. If radium-228 is present, radium-226 must be determined by radon emanation.

c. Sampling and storage: To drinking water or a filtered sample of turbid water, add 2 mL conc nitric acid (HNO_3)/L sample at the time of collection or immediately after filtration.

2. Apparatus

a. Counting instruments: One of the following is required:
1) *Internal proportional counter,* gas flow, with scaler, timer, and register; or a thin end-window (polyester plastic)* proportional counting chamber with scaler, timer, register amplifier, and preferably having an anticoincident system (low background).

2) *Scintillation counter assembly:* See ¶ C.2a. This equipment is necessary only if radium-226 is determined sequentially with radium-228 and is analyzed by emanation of radon.

b. Centrifuge, bench-size clinical, with polypropylene tubes.

c. Filter funnels, for 2.4-cm filter paper.

d. Stainless steel pans, 5.1 cm.

e. Infrared drying lamp assembly.

f. Magnetic stirrer hot plate.

g. Membrane filters, 47-mm diam, 0.45-μm pore diam.†

3. Reagents

a. Acetic acid, conc.

b. Acetone, anhydrous.

c. Ammonium hydroxide, NH_4OH, conc.

d. Ammonium oxalate solution: Dissolve 25 g $(NH_4)_2C_2O_4$ in distilled water and dilute to 500 mL.

e. Ammonium sulfate solution: Dissolve 20 g $(NH_4)_2SO_4$ in a minimum of distilled water and dilute to 100 mL.

f. Ammonium sulfide solution: Dilute 10 mL $(NH_4)_2S$ (20 to 24%) to 100 mL with distilled water.

g. Barium carrier standardized: Dissolve 2.846 g $BaCl_2 \cdot 2H_2O$ in distilled water, add 0.5 mL conc HNO_3, and dilute to 100 mL; 1 mL = 16 mg Ba.

---

* Mylar or equivalent.
† Gelman Ga-6 or equivalent.

h. Citric acid, 1M: See Section B.3*a.*

i. EDTA reagent, 0.25M: See Section B.3*k.*

j. Ethanol, 95%.

k. Lead carrier: Solution A: Dissolve 2.397 g $Pb(NO_3)_2$ in distilled water, add 0.5 mL conc HNO_3, and dilute to 100 mL; 1 mL = 15 mg Pb. *Solution B:* Dilute 10 mL Solution A to 100 mL with distilled water; 1 mL = 1.5 mg Pb.

l. Methyl orange indicator solution: Dissolve 0.1 g methyl orange powder in 100 mL distilled water.

m. Nitric acid, HNO_3, conc, 6*N*, and 1*N.*

n. Sodium hydroxide, 18N: Dissolve 720 g NaOH in 500 mL distilled water and dilute to 1 L.

o. Sodium hydroxide, 10N: Dissolve 400 g NaOH in 500 mL distilled water and dilute to 1 L.

p. Sodium hydroxide, NaOH, 1N.

q. Strontium-yttrium mixed carrier: Solution A: Dilute 10.0 mL yttrium carrier to 100 mL. *Solution B:* Dissolve 0.4348 g $Sr(NO_3)_2$ in distilled water and dilute to 100 mL. Combine equal volumes of Solutions A and B; 1 mL = 0.9 mg Sr and 0.9 mg Y.

r. Sulfuric acid, H_2SO_4, 18*N.*

s. Yttrium carrier: Add 12.7 g Y_2O_3 (Section 7500-Sr.B.3*d*) to an erlenmeyer flask containing 20 mL distilled water. Heat to boiling and, while stirring with a magnetic stirring hot plate, add small portions of conc HNO_3. (About 30 mL is necessary to dissolve the Y_2O_3. Small additions of distilled water also may be needed to replace water lost by evaporation.) After total dissolution, add 70 mL conc HNO_3 and dilute to 1 L with distilled water; 1 mL = 10 mg Y.

4. Procedure

a. Radium-228:

1) For 1 L sample add 5 mL 1*M* citric acid and a few drops methyl orange indicator. The solution should be red. Add 10 mL lead carrier (Solution A), 2.0 mL barium carrier, and 2 mL yttrium carrier; stir well. Heat to incipient boiling and maintain at this temperature for 30 min.

2) Add conc NH_4OH until a definite yellow color is obtained; add a few drops excess. Precipitate lead and barium sulfates by adding 18*N* H_2SO_4 until the red color reappears; add 0.25 mL excess. Add 5 mL $(NH_4)_2SO_4$ solution/L sample. Stir frequently and hold at about 90°C for 30 min.

3) Cool and filter with suction through a membrane filter. Quantitatively transfer precipitate to filter. Carefully place filter in a 250-mL beaker. Add about 10 mL conc HNO_3 and heat gently until the filter dissolves completely. Using conc HNO_3 transfer precipitate to a centrifuge tube. Centrifuge and discard supernatant.

4) Wash precipitate with 15 mL conc HNO_3, centrifuge, and discard supernatant. Repeat wash and centrifuge again. Add 25 mL EDTA reagent, heat in a hot water bath, and stir well. Add a few drops 10*N* NaOH if the precipitate does not dissolve readily.

5) Add 1 mL strontium-yttrium mixed carrier and stir thoroughly. Add a few drops 10*N* NaOH if any precipitate forms. Add 1 mL $(NH_4)_2SO_4$ solution and stir thoroughly. Add conc acetic acid until $BaSO_4$ precipitates; add 2 mL excess. The pH should be about 4.5. Digest in a hot water bath (80°C) until precipitate settles. Centrifuge and discard supernatant.

6) Add 20 mL EDTA reagent, heat in a hot water bath, and stir until precipitate dissolves. Repeat Step 5. Note time of last

$BaSO_4$ precipitation as zero time for ingrowth of ^{228}Ac. Dissolve precipitate in 20 mL EDTA reagent, add 0.5 mL yttrium carrier and 1 mL lead carrier (Solution B). If any precipitate forms, dissolve by adding a few drops 10*N* NaOH. Mix well, cap tube, and age at least 36 h.

7) Add 0.3 mL $(NH_4)_2S$ solution and mix well. Add 10*N* NaOH dropwise with vigorous stirring until PbS precipitates; add 10 drops excess. Stir intermittently for about 10 min. Centrifuge and decant supernatant into a clean tube.

8) Add 1 mL lead carrier (Solution B), 0.1 mL $(NH_4)_2S$ solution, and a few drops 10*N* NaOH. Repeat precipitation of PbS. Centrifuge and filter supernatant through filter paper‡ into a clean tube. Wash filter with a few milliliters of distilled water. Discard residue.

9) Add 5 mL 18*N* NaOH (make at least 2*N* in OH^-). Because of the short half-life of ^{228}Ac (6.13 h) complete the following procedure without delay. Mix well and digest in a hot water bath until $Y(OH)_3$ coagulates. Centrifuge and decant supernatant into a beaker. Cover beaker and save supernatant for ^{226}Ra analysis, ¶s *b* or *c* below. Note time of $Y(OH)_3$ precipitation; this is the end of ^{228}Ac ingrowth and beginning of ^{228}Ac decay. (t_3 = time in minutes between last $BaSO_4$ and first $Y(OH)_3$ precipitations.) Dissolve precipitate in 2 mL 6*N* HNO_3. Heat and stir in a hot water bath about 5 min. Add 5 mL distilled water and reprecipitate $Y(OH)_3$ with 3 mL 10*N* NaOH. Heat and stir in a hot water bath until precipitate coagulates. Centrifuge and discard supernatant.

10) Dissolve precipitate with 1 mL 1*N* HNO_3 and heat in hot water bath for several minutes. Dilute to 5 mL with distilled water and add 2 mL ammonium oxalate solution. Heat to coagulate, centrifuge, and discard supernatant. Add 10 mL distilled water, 6 drops 1*N* HNO_3, and 6 drops ammonium oxalate solution. Heat and stir in a hot water bath for several minutes. Centrifuge and discard supernatant. Transfer quantitatively to a tared stainless-steel planchet using a minimum quantity of distilled water. Dry under an infrared lamp to constant weight and count in a low-background beta counter. (t_1 = time in minutes between first $Y(OH)_3$ precipitation and counting.)

If analysis of radium-226 is not required, complete Steps *b*1) and 3) below to obtain the fractional barium yield to be used in calculating ^{228}Ra activity.

b. Radium by precipitation:

1) To the supernatant saved in ¶ *a*9) above add 4 mL conc HNO_3 and 2 mL $(NH_4)_2SO_4$ solution, mixing well after each addition. Add conc acetic acid until $BaSO_4$ precipitates; add 2 mL excess. Digest on a hot plate until precipitate settles. Centrifuge and discard supernatant.

2) Add 20 mL EDTA reagent, heat in a hot water bath, and stir until precipitate dissolves. Add a few drops 10*N* NaOH if precipitate does not dissolve readily. Add 1 mL strontium-yttrium mixed carrier and 1 mL lead carrier (Solution B), and stir thoroughly. Add a few drops 10*N* NaOH if any precipitate forms. Add 1 mL $(NH_4)_2SO_4$ solution and stir thoroughly. Add conc acetic acid until $BaSO_4$ precipitates; add 2 mL excess. Digest in a hot water bath until precipitate settles. Centrifuge, discard supernatant, and note time.

3) Wash precipitate with 10 mL distilled water. Centrifuge and discard supernatant. Transfer quantitatively to a tared stainless-steel planchet using a minimum quantity distilled water. Dry un-

---

‡ Whatman No. 42 or equivalent.

der an infrared lamp to constant weight. If after sufficient beta decay of the actinium fraction ^{228}Ra is found to be absent, make a direct alpha count for ^{226}Ra. If ^{228}Ra is present, determine ^{226}Ra by radon emanation, ¶ c below.

4) Count immediately in an alpha proportional counter.

c. *Radium-226 by radon:* Transfer the final precipitate obtained in b above to a small beaker using a rubber policeman and 14 mL EDTA reagent. Add a few drops 10N NaOH and heat on a hot plate to dissolve. Cool and transfer to a radon bubbler (Figure 7500-Ra:1) rinsing beaker with 1 mL EDTA reagent. Proceed as in Method C beginning with 5a14).

5. Calculation

a. *Calculation of ^{228}Ra concentration:*

$$^{228}\text{Ra,pCi/L} = \frac{C}{2.22 \times EVR} \times \frac{\lambda t_2}{(1 - e^{-\lambda t_2})} \times \frac{1}{(1 - e^{-\lambda t_3})} \times \frac{1}{e^{-\lambda t_1}}$$

where:

C = average net count rate, cpm,
E = counter efficiency, for ^{228}Ac,
V = sample volume, L,
R = fractional chemical yield of yttrium carrier, ¶ 4a10), multiplied by fractional chemical yield of barium carrier, ¶ b3),

λ = decay constant of ^{228}Ac, 0.001 884/min,
t_1 = time between first Y(OH)$_3$ precipitation and start of counting, min,
t_2 = counting time, min, and
t_3 = ingrowth time of ^{228}Ac between last BaSO$_4$ precipitation and first Y(OH)$_3$ precipitation, min.

The factor $\lambda t_2/(1 - e^{-\lambda t_2})$ corrects average count rate to count rate at beginning of counting time.

b. *Calculation of ^{226}Ra (plus any ^{224}Ra and ^{223}Ra) concentration:* See Section B.5.

c. *Calculation of ^{226}Ra (emanation) concentration:* See Section C.6.

6. Bibliography

JOHNSON, J.O. 1971. Determination of radium 228 in natural waters. Radiochemical Analysis of Water. U.S. Geol. Surv. Water Supply Paper 1696-G, U.S. Government Printing Off., Washington, D.C.

KRIEGER, H.L. 1976. Interim Radiochemical Methodology for Drinking Water. EPA-600/4-75-008 (revised), U.S. Environmental Protection Agency, Environmental Monitoring and Support Lab., Cincinnati, Ohio.

KRIEGER, H.L. & E.L. WHITTAKER. 1980. Prescribed Procedures for Measurement of Radioactivity in Drinking Water. EPA-600/4-80-032, U.S. Environmental Protection Agency, Environmental Monitoring and Support Lab., Cincinnati, Ohio.

7500-Rn   RADON*

7500-Rn A.   Introduction

1. Occurrence and Significance

Radon-222 is a gaseous decay product of naturally occurring radium-226. It is an alpha-emitter with a 3.82-d half-life, and normally is of concern only in groundwater. It is considered to be carcinogenic, as are its short-lived daughters. In household air, radon may originate from radium in building materials and the surrounding soil. Where radon concentration in the water supply is high, the water also can be a major source of radon in household air. While radon dissolves readily in water and other solvents, it is easily displaced from water by air; thus, aeration of radon-bearing water in normal household uses can release a significant fraction of the dissolved radon to the air.[1-3]

The average ^{222}Rn concentration in community groundwater systems in the U.S. is estimated to range from 200 to 600 pCi/L,[2-6] with some individual wells having much higher concentrations.

___
*Approved by Standard Methods Committee, 1996.

2. References

1. PARTRIDGE, J.E., T.R. HORTON & E.L. SENSINTAFFAR. 1979. A Study of Radon-222 Released from Water During Typical Household Activities. ORP/EERF-79-1, Eastern Environmental Radiation Facility, Montgomery, Ala.

2. SMITH, B.M., W.N. GRUNE, F.B. HIGGINS, JR. & J.G. TERRILL, JR. 1961. Natural radioactivity in ground water supplies in Maine and New Hampshire. *J. Amer. Water Works Assoc.* 53:75.

3. ALDRICH, L.K., III, M.K. SASSER & D.A. CONNERS, IV. 1975. Evaluation of Radon Concentrations in North Carolina Ground Water Supplies. Dept. Human Resources, Div. Facility Services, Radiation Protection Branch, Raleigh, N.C.

4. O'CONNELL, M.F. & R.F. KAUFMAN. 1976. Radioactivity Associated with Geothermal Waters in the Western United States. U.S. Environmental Protection Agency Technical Note ORP/LV-75-8A.

5. DUNCAN, D.L., T.F. GESELL & R.H. JOHNSON, JR. 1976. Radon-222 in potable water. *In* Proc. Health Physics Soc. 10th Midyear Topical Symposium: Natural Radioactivity in Man's Environment.

6. COTHERN, C.R. & P.A. REBERS. 1990. Radon, Radium, and Uranium in Drinking Water. Lewis Publishers, Inc., Chelsea, Mich.

7500-Rn B.   Liquid Scintillation Method

1. General Discussion

a. Principle: This method is specific for radon-222 (radon) in drinking water. Radon is partitioned selectively into a mineral-oil scintillation cocktail immiscible with the water sample. The sample is dark-adapted and equilibrated, and then counted in a liquid scintillation counter using a region or window of the energy spectrum optimal for radon alpha particles. Results are reported as pCi/L.

The procedure has been developed for the analysis of radon in drinking water supplies from groundwater and surface-water sources. Applications of this analytical procedure to matrices other than drinking water have not been studied; use caution in analyzing any such samples.

b. Interferences: There are no known chemical interferences from species found in drinking water nor from the dilute concentration of acid that may be present in the calibration standards. Uranium, radium, or other radioactive elements would cause a positive bias, if present in quantities significantly greater than the radon.

Diffusion of radon is affected by temperature and pressure. Let samples equilibrate to room temperature before processing.

Precision and accuracy of the method are affected by the background in the energy window used for analysis. A procedure is provided for selection of the analytical window to minimize the background contribution to the measurement.

Some cocktails will become progressively quenched by atmospheric oxygen after opening. This problem has not been noted for the mineral-oil-based cocktail. For other than mineral-oil-based cocktails, check weekly for quenching.

Radon has an affinity for some plastics used in sample containers. Use only glass sample containers or glass scintillation vials with TFE or foil-lined caps.

c. Sample preservation, storage, and holding time: Collect samples from a nonaerated faucet that has been allowed to flow for sufficient time so that the sample is representative of the water in the distribution system or well. The following procedure will minimize the loss of radon from the sample during collection:

Place a glass sample vial in a 300- to 600-mL beaker or other suitable container; attach delivery tube to faucet, and start the flow. Make sure that delivery tube does not let bubbles enter the sample. Fill vial to prevent its floating, then fill beaker until vial is submerged. Place tip of delivery tube about two thirds of the way into vial and fill until approximately two or more vial volumes (50 to 100 mL) have been displaced. Carefully remove vial by hand or with a pair of 25-cm (10 in.) tweezers and cap vial with a TFE or foil-lined cap. Cap sample vials underwater, if possible. Invert sample and check for air bubbles. If any bubbles are present, discard sample and repeat sampling procedure.

Alternatively, collect samples in containers other than scintillation vials, with similar precautions.

Record date and time of sample collection and store sample in a cooler. Transport samples to laboratory in a cooler or other suitable insulated package to avoid large temperature changes and outgassing of radon. Begin counting within 4 d or applicable regulatory specified holding time.

d. Minimum detectable concentration: 18 pCi/L for a 50-min count time, 6 cpm background, 2.7 cpm/dpm efficiency, and energy region optimized by the procedure in ¶ 4b.

2. Apparatus

a. Pipet: Precision 5-mL mechanical pipet or syringe.

b. Scintillation cocktail dispenser adjustable to deliver 5 mL.

c. Liquid scintillation counter: Preferably use a system permitting automatic spectral analysis.

d. Faucet connector or universal faucet adapter.

e. Plastic tubing for connector or adapter.

f. Scintillation vials: 23-mL glass vials with caps, TFE or foil-lined for sampling and plastic* for counting.

g. Volumetric glassware.

h. Sample storage and shipping containers, insulated.

3. Reagents

a. Scintillation cocktail: Water-immiscible high-efficiency mineral oil cocktail or other commercial equivalent.

b. Hydrochloric acid, HCl, conc.

c. Water, radon-free demineralized or equivalent.

d. Radium solution: Use two dilutions for calibration and check standards. Use NIST-traceable (explicit or implicit) radium-226 standard solution.

4. Procedure

a. Calibration: Prepare 100 mL radium-226 in water standard such that the final activity will be approximately 8000 pCi/L by the procedure suggested below. Transfer standard to a scintillation vial or other suitable container, seal, and record initial mass to nearest 0.0001 g. To a 100-mL volumetric flask add 20 mL water and 0.5 mL conc HCl; stopper. Transfer with a pipet, or suitable dropper, the required mass of radium solution into flask; re-weigh vial. Obtain actual mass of radium solution added by difference of final and initial weights. Fill to mark and mix.

Transfer 15 mL diluted standard into scintillation vial, to which has been added 5 mL mineral oil cocktail. Prepare at least three standards and three backgrounds using distilled or deionized water.

Set standards and background samples aside for at least 25 d (99% ingrowth) to allow radon progeny to attain secular equilibrium with radium-226. Determine optimal analytical window as outlined in ¶ 4b below. After ingrowth period, let sample dark-adapt for 3 h if necessary and count for 50 min. Repeat counting two additional times. From pooled results calculate a system calibration factor by the following expression:

$$CF = \frac{S - B}{C \times V}$$

where:

CF = calibration factor, cpm/pCi,
S = standard counting rate, cpm,
B = background counting rate, cpm,
C = concentration of radium-226 standard, pCi/L, and
V = volume of standard used, 0.015 L.

b. Selecting optimal window: Count a radon standard for 5 min

---

*Polyseal core liner, or equivalent.

or sufficient time to acquire several thousand counts or more in the alpha region and generate a sample spectrum. For greater clarity use a log scale for the channel number or energy axis if possible.

The alpha activity region of interest will be obvious as one or two large peaks at the higher end of the energy spectrum. The lower peak is the doublet of radon-222 and polonium-218 and the higher peak is that of polonium-214. The optimal window is formed by extending the region by approximately 10 channels on each side of the alpha peaks. Use this window for subsequent calibration and analysis. Calibration factor should be at least 6 cpm/pCi with the background not exceeding 6 cpm.

For counters not having a spectrum display, set window initially wide-open and count for sufficient time to obtain several thousand counts. Adjust energy window to a width of 5% of full scale at upper end of scale (95 to 100%) and determine count rate in the region. Repeat counts at successively lower regions using the same 5% interval (90 to 95, 85 to 90, 80 to 85, etc.). Plot count rate versus midpoint of interval and choose region of interest, which will be evident by one or two prominent peaks in the upper half of the energy scale. Background should be 6 cpm or less and conversion factor should be approximately 6 cpm/pCi.

c. Analysis of samples: Carefully remove by pipet 8.5 mL sample from the scintillation vial used for collection and add 5 mL water-immiscible scintillation cocktail. Alternatively, pipet, without turbulence, a 15-mL portion to a scintillation vial containing 5 mL cocktail if sample was collected in a different container.

Cap and shake sample for 30 s and set aside in the dark for a minimum of 3 h to equilibrate radon progeny and dark-adapt sample. Count all samples within the regulation specified holding time. The time of sample collection is the initial time for decay correction.

Count a standard for 5 min or longer if required and either examine spectrum or compare results to previous standards to determine if there has been any shift or quench due to changes in the cocktail or instrument drift. Count samples for 50 min or to a percent 2σ counting error of 10% or for a period of time to achieve an uncertainty in the net counting rate corresponding to program data quality objectives using optimized window settings for alpha counting. Make sure the expression used includes the background in the uncertainty computation. This may have to be done manually because most instruments calculate the uncertainty only for the gross counting rate.

5. Calculations

Calculate concentration of radon-222 in pCi/L from the following equation:

$$\text{Rn, pCi/L} = \frac{G - B}{CF \times D \times V}$$

where:

G = gross counting rate of sample, cpm,
B = background counting rate, cpm,
CF = calibration factor (see 4a),
V = volume of sample (~0.015 L), and
D = decay factor for Rn-222 between time of collection and mid-point of counting period for that sample.

Calculate 2σ (95% confidence level) counting uncertainty, as:

$$2\sigma, \text{pCi/L} = \frac{2 \times \sqrt{\dfrac{G}{T_G} + \dfrac{B}{T_B}}}{CF \times D \times V}$$

where:

T_G = duration of sample count, and
T_B = duration of background count.

Report 2σ uncertainty with each drinking water radioactivity concentration result. This term represents the uncertainty due to the random nature of radioactive decay; it is related to count time and can be used to determine whether sample been counted long enough to satisfy any required precision criteria. If percent counting error (2σ counting error divided by activity concentration) exceeds precision requirements, count sample longer or reduce the holding time.

Report result and counting error together in the form:

$$X \pm 2\sigma \text{ pCi/L, } 2\sigma \text{ counting error}$$

For example, for a water sample with calculated radon-222 concentration of 285 pCi/L and 2σ counting error of 27 pCi/L, report result as:

$$^{222}\text{Rn: } 285 \pm 27 \text{ pCi/L, } 2\sigma \text{ counting error}$$

6. Quality Control

a. Background samples: Include a minimum of two background samples with each batch of 20 samples. Place backgrounds as first and next-to-last samples of batch. Use average of these backgrounds to calculate results for batch. The background should be ≤6 cpm.

For a suitable background sample, use laboratory deionized water, or prepare by boiling 2 L laboratory radium- and uranium-free tap water to remove residual radon if present. Store the cooled tap water in a capped 2-L bottle.

b. Duplicate samples: Collect duplicate field samples for one out of every ten samples. Preferably collect all samples in duplicate if the number of samples from an individual client represents a single source. Ensure that at least 10% of the samples analyzed daily are duplicates, and that duplicate analyses have a relative percent difference (RPD) less than or equal to the percent 2σ counting error or 10% of the decay-corrected radon concentration, whichever is greater. Relative percent difference is calculated by the following expression:

$$RPD = \frac{|\text{Analysis 1} - \text{Analysis 2}| \times 200}{\text{Analysis 1} + \text{Analysis 2}}$$

Record the RPD and note acceptability of the duplicate analysis. If RPD exceeds the limits, recount duplicates. If results still exceed limits but RPD for the quality control check standard is acceptable, a problem with the sampling procedure may exist. Resolve problem before collecting and analyzing additional samples.

c. Quality control check standard (QCCS): QCCSs are prepared from a dilution of radium different from that used to prepare standards and should have a nominal activity of ~8000 pCi/L. Place first QCCS immediately after first background and before first sample. Place additional QCCS after every tenth sample in batch, and final QCCS as last sample of the batch.

The relative percent difference (RPD) between sequential pairs of QCCS samples must be less than or equal to the 2σ counting error or 10% of the known value of the QCCS sample, whichever is greater. If RPD exceeds this value, recount the pair of QCCS samples. If RPD is still unacceptable, standards and/or instrument are suspect. Resolve problem and rerun samples between suspect QCCS.

d. Records: Collect and maintain results from backgrounds, duplicate pairs, and QCCS standards in a bound notebook; include date, results, name of analyst, and comments relevant to data evaluation.

Plot averages of backgrounds and QCCS standards on a control chart for the counter.

7. Precision and Accuracy

A collaborative study of this method composed of 36 participants[1] produced the results shown below:

Sample Conc. pCi/L	Accuracy %	Repeatability pCi/L	Reproducibility pCi/L	Bias %
111	101–102	9	12	0.7–2.3
153	102–103	10	16–18	2.3–3.4

8. Reference

1. PIA, S.H. & P.B. HAHN. 1992. Radiation Research and Methods Validation Annual Report 1992. Environmental Monitoring Systems Laboratory—Las Vegas, Off. Research and Development, U.S. Environmental Protection Agency, Las Vegas, Nev.

9. Bibliography

WHITTAKER, E.L., J.D. AKRIDGE & J. GIOVINO. 1987. Two Test Procedures for Radon in Drinking Water. EPA/600/2–87/082, U.S. Environmental Protection Agency, Las Vegas, Nev.

VITZ, E. 1991. Toward a standard method for determining waterborne radon. *Health Phys.* 60:817.

LOWRY, J.D. 1991. Measuring low radon levels in drinking water supplies. *J. Amer. Water Works Assoc.* 1991(4):149.

PRICHARD, H.M. & T.F. GESELL. 1977. Rapid measurements of Rn-222 concentrations in water with a common liquid scintillation counter. *Health Phys. J.* 33:577.

YOUDEN, W.J. & E.H. STEINER. 1975. Statistical Manual of the Association of Official Analytical Chemists. Assoc. of Official Analytical Chemists, McLean, Va.

7500-Sr   TOTAL RADIOACTIVE STRONTIUM AND STRONTIUM-90*

7500-Sr A.   Introduction

The important radioactive nuclides of strontium produced in nuclear fission are ^{89}Sr and ^{90}Sr. Strontium-90 is one of the most hazardous of all fission products. It decays slowly, with a half-life of 28 years. Upon ingestion, strontium is concentrated in the bone.

The method presented in this section is designed to measure total radioactive strontium (^{89}Sr and ^{90}Sr) or ^{90}Sr alone in drinking water or in filtered raw water. It is applicable to sewage and industrial wastes provided that steps are taken to destroy organic matter and eliminate other interfering ions.

* Approved by Standard Methods Committee, 1993.

7500-Sr B.   Precipitation Method

1. General Discussion

a. Principle: A known amount of inactive strontium ions, in the form of strontium nitrate, $Sr(NO_3)_2$, is added as a "carrier." The carrier, alkaline earths, and rare earths are precipitated as the carbonate to concentrate the radiostrontium. The carrier, along with the radionuclides of strontium, is separated from other radioactive elements and inactive sample solids by precipitation as

$Sr(NO_3)_2$ from fuming nitric acid solution. The strontium carrier, together with the radionuclides of strontium, finally is precipitated as strontium carbonate, $SrCO_3$, which is dried, weighed to determine recovery of carrier, and measured for radioactivity. The activity in the final precipitate is due to radioactive strontium only, because all other radioactive elements have been removed. A correction is applied to compensate for losses of carrier and activity during the various purification steps. A delay in the count will give an increased counting rate due to the ingrowth of ^{90}Y.

b. Concentration techniques: Because of the very low amount of radioactivity, a large sample must be taken and the activity concentrated by precipitation. $Sr(NO_3)_2$ and barium nitrate, $Ba(NO_3)_2$, carriers are added to the sample. Sodium carbonate is then added to concentrate radiostrontium by precipitation of alkaline earth carbonates along with other radioactive elements. The supernate is discarded. The precipitate is dissolved and reprecipitated to remove interfering radionuclides.

c. Interference: Radioactive barium (^{140}Ba, ^{140}La) interferes in the determination of radioactive strontium inasmuch as it precipitates with the radioactive strontium. Eliminate this interference by adding inactive $Ba(NO_3)_2$ carrier and separating this from the strontium by precipitating barium chromate in acetate buffer solution. Radium isotopes also are eliminated by this treatment.

In hard water, some calcium nitrate may be coprecipitated with $Sr(NO_3)_2$ and can cause errors in recovery of the final precipitate and in measuring its activity. Eliminate this interference by repeated precipitations of strontium as the nitrate followed by leaching the $Sr(NO_3)_2$ with acetone (CAUTION).

For total radiostrontium, count the precipitate within 3 to 4 h after the final separation and before ingrowth of ^{90}Y.

d. Determination of ^{90}Sr: Because it is impossible to separate the isotopes ^{89}Sr and ^{90}Sr by any chemical procedure, the amount of ^{90}Sr is determined by separating and measuring the activity of ^{90}Y, its daughter. After equilibrium is reached, the activity of ^{90}Y is exactly equal to the activity of ^{90}Sr. Two alternate procedures are given for the separation of ^{90}Y. In the first method, ^{90}Y is separated by extraction into tributyl phosphate from concentrated nitric acid (HNO_3) solution. It is back-extracted into dilute HNO_3 and evaporated to dryness for beta counting. The second method consists of adding yttrium carrier, separating by precipitation as yttrium hydroxide, $Y(OH)_3$, and finally precipitating yttrium oxalate for counting.

2. Apparatus

a. Counting instruments: Use either an internal proportional counter, gas-flow, with scaler, timer, and register; or a thin end-window (polyester plastic film*) proportional or G-M counting chamber with scaler, timer, register amplifier, and preferably having an anticoincident system (low background).

b. Filter paper,† 2.4 cm diam; or glass fiber filters, 2.4 cm diam.

c. Two-piece filtering apparatus for 2.4-cm filters such as TFE filter holder,‡ stainless steel filter holder, or equivalent.

d. Stainless steel pans, about 50 mm diam and 7 mm deep, for counting solids deposited on pan bottom. For counting precipitates on 2.4-cm filters, use nylon disk with ring§ on which the filter samples are mounted and covered by 0.25 mil film.

3. Reagents

a. Strontium carrier, 10 mg Sr^{2+}/mL, standardized: Carefully add 24.16 g $Sr(NO_3)_2$ to a 1-L volumetric flask and dilute with distilled water to the mark. For standardization, pipet three 10.0-mL portions of strontium carrier solution into 40-mL centrifuge tubes and add 15 mL $2N$ Na_2CO_3 solution. Stir, heat in a boiling

water bath for 15 min, and cool. Filter $SrCO_3$ precipitate through a tared fine-porosity sintered-glass crucible of 15-mL size. Wash precipitate with three 5-mL portions of water and then with three 5-mL portions of absolute ethanol (or acetone). Wipe crucible with absorbent tissue and dry to constant weight in an oven at 110°C (20 min). Cool in a desiccator and weigh.

$$Sr, \text{ mg/mL} = \frac{(\text{mg } SrCO_3)\,(0.5935)}{10}$$

b. Barium carrier, 10 mg Ba^{2+}/mL: Dissolve 19.0 g $Ba(NO_3)_2$ in distilled water and dilute to 1 L.

c. Rare earth carrier, mixed: Dissolve 12.8 g cerous nitrate hexahydrate, $Ce(NO_3)_3 \cdot 6H_2O$, 14 g zirconyl chloride octahydrate, $ZrOCl_2 \cdot 8H_2O$, and 25 g ferric chloride hexahydrate, $FeCl_3 \cdot 6H_2O$, in 600 mL distilled water containing 10 mL conc HCl, and dilute to 1 L.

d. Yttrium carrier: Dissolve 12.7 g yttrium oxide,‖ Y_2O_3, in 30 mL conc HNO_3 by stirring and warming. Add an additional 20 mL conc HNO_3 and dilute to 1 L with distilled water; 1 mL is equivalent to 10 mg Y, or approximately 34 mg $Y_2(C_2O_4)_3 \cdot 9H_2O$. Determine exact equivalence by precipitating yttrium carrier in acid solution according to ¶s 4c2)-8), below or by extracting yttrium carrier in acid solution according to ¶s 4b3)-11), below.

e. Acetate buffer solution: Dissolve 154 g $NH_4C_2H_3O_2$ in 700 mL distilled water, add 57 mL conc acetic acid, adjust pH to 5.5 by dropwise addition of conc acetic acid or $6N$ NH_4OH as necessary, and dilute to 1 L.

f. Acetic acid, 6N.

g. Acetone, anhydrous.

h. Ammonium hydroxide, NH_4OH, 6N.

i. Hydrochloric acid, HCl, 6N.

j. Methyl red indicator, 0.1%: Dissolve 0.1 g methyl red in 100 mL distilled water.

k. Nitric acid, HNO_3, fuming (90%), conc, $14N$, $6N$, and $0.1N$.

l. Oxalic acid, saturated solution: Dissolve approximately 11 g $H_2C_2O_4 \cdot 2H_2O$ in 100 mL distilled water.

m. Sodium carbonate solution, $1M$: Dissolve 124 g $Na_2CO_3 \cdot H_2O$ in distilled water and dilute to 1 L.

n. Sodium chromate solution, $0.5M$: Dissolve 117 g $Na_2CrO_4 \cdot 4H_2O$ in distilled water and dilute to 1 L.

o. Sodium hydroxide, $6N$: Dissolve 240 g NaOH in distilled water and dilute to 1 L.

p. Tributyl phosphate, reagent grade: Shake with an equal volume of $14N$ HNO_3 to equilibrate. Separate and discard the HNO_3 washings.

4. Procedure

a. Total radiostrontium:

1) To 1 L of drinking water, or a filtered sample of raw water in a beaker, add 2.0 mL conc HNO_3 and mix. Add 2.0 mL each of strontium and barium carriers and mix well. (A precipitate of $BaSO_4$ may form if the water is high in sulfate ion, but this will cause no difficulties.) A smaller sample may be used if it contains at least 25 pCi strontium. The suspended matter that has been

---

* Mylar, E.I. du Pont de Nemours, Wilmington, DE, or equivalent.
† Whatman No. 42 or equivalent.
‡ Flurolon Laboratory, Box 305, Caldwell, NJ.
§ Control Molding Corp., Staten Island, NY, or equivalent.

‖ Yttrium oxide, Code 1118, American Potash and Chemical Corp., West Chicago, IL, or equivalent. Yttrium oxide of purity less than Code 1118 may require purification because of radioactivity contamination.

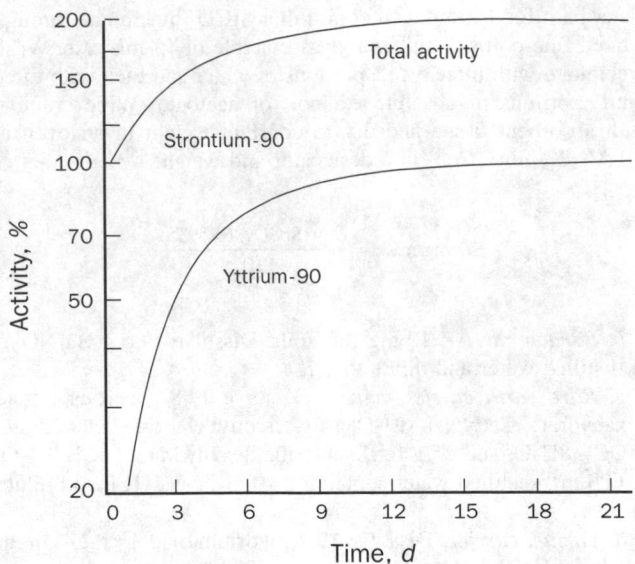

Figure 7500-Sr:1. Yttrium-90 vs. strontium-90 activity as a function of time.

filtered off may be digested [see Gross Alpha and Gross Beta Radioactivity, 7110B.4f1)], diluted, and analyzed separately.

2) Heat to boiling, then add 20 mL 6N NaOH and 20 mL 1M Na$_2$CO$_3$. Stir and let simmer at 90 to 95°C for about 1 h.

3) Set beaker aside until precipitate has settled (about 1 to 3 h).

4) Decant and discard clear supernate. Transfer precipitate to a 40-mL centrifuge tube and centrifuge. Discard supernate.

5) Add, dropwise (CAUTION—*effervescence*), 4 mL conc HNO$_3$. Heat to boiling, stir, then cool under running water.

6) Add 20 mL fuming HNO$_3$, cool 5 to 10 min in ice bath, stir, and centrifuge. Discard supernate.

7) Add 4 mL distilled water, stir, and heat to boiling to dissolve the strontium. Centrifuge while hot to remove remaining insolubles and decant supernate to a clean centrifuge tube. Add 2 mL 6N HNO$_3$, heat to boiling, centrifuge while hot, and combine supernate with aqueous supernate. Discard insoluble residue of SiO$_2$, BaSO$_4$, etc.

8) Cool combined supernates, then add 20 mL fuming HNO$_3$, cool 5 to 10 min in ice bath, stir, centrifuge, and discard supernate.

9) Add 4 mL distilled water and dissolve by heating. Repeat Step 8) preceding.

10) Repeat Step 9) preceding if more than 200 mg Ca were present in the sample.

11) After last HNO$_3$ precipitation, invert tube in a beaker for about 10 min to drain off most excess HNO$_3$. Add 20 mL anhydrous acetone, stir thoroughly, cool, and centrifuge. Discard supernate (CAUTION).

12) Dissolve precipitate of Sr(NO$_3$)$_2$ + Ba(NO$_3$)$_2$ in 10 mL distilled water and boil for 30 s to remove any remaining acetone.

13) Add 0.25 mL (5 drops) mixed rare earth carrier and precipitate rare earth hydroxides by making solution basic with 6N NH$_4$OH. Digest in a boiling water bath for 10 min. Cool, centrifuge, and decant supernate to a clean tube. Discard precipitate.

14) Repeat Step 13) preceding.

Note the time of rare earth precipitation, which marks the beginning of the ^{90}Y ingrowth period. Do not delay procedure more than a few hours after the separation; otherwise, false results will be obtained because of ingrowth of [90]Y.

15) Add 2 drops methyl red indicator and then add 6N acetic acid dropwise with stirring until indicator changes from yellow to red.

16) Add 5 mL acetate buffer solution, heat to boiling, and add dropwise, with stirring, 2 mL Na$_2$CrO$_4$ solution. Digest in a boiling water bath for 5 min. Cool, centrifuge, and decant supernate to a clean tube. Discard residue.

17) Add 2 mL 6N NaOH, add 5 mL 1M Na$_2$CO$_3$ solution, and heat to boiling. Cool in an ice bath (about 5 min) and centrifuge. Discard supernate.

18) Add 15 mL distilled water, stir, centrifuge, and discard wash water.

19) Repeat Step 18), and proceed either as in Step 20)a) or 20)b), below. *Save this precipitate if a determination of* [90]*Sr is required.*

20) Either

a) Slurry precipitate with a small volume of distilled water and transfer to a tared stainless steel pan; dry under an infrared lamp, cool, weigh, and count# the precipitate of SrCO$_3$;** or

b) Transfer precipitate to a tared paper or glass filter mounted in a two-piece funnel. Allow gravity settling for uniform deposition and then apply suction. Wash precipitate with three 5-mL portions of water, three 5-mL portions of 95% alcohol, and three 5-mL portions of ethyl ether or acetone. Dry in an oven at 110 to 125°C for 15 to 30 min, cool, weigh,** mount on a nylon disk and ring with polyester plastic film cover, and count.

21) Calculation

$$\text{Total Sr activity, pCi/L} = \frac{b}{adf \times 2.22}$$

where:

a = beta counter efficiency [see Step 22) below],

$d = \dfrac{\text{mg final SrCO}_3 \text{ precipitate}}{\text{mg SrCO}_3 \text{ in 2 mL of carrier}}$

   = correction for carrier recovery [see Step 23) below],

f = sample volume, L,

b = beta activity, net cpm = $(i/t) - k$,

i = total counts accumulated,

t = time of counting, min, and

k = background, cpm.

22) Counting efficiency—As a first estimate, when mounting sample according to Step 20)a), convert counts per minute to disintegrations per minute, based on the beta activity of cesium-137 standard solutions having a sample thickness equivalent to that of the SrCO$_3$ precipitate. More precise measurements may follow a second count after substantial ingrowth of ^{90}Y from ^{90}Sr, but this precision is not warranted for the usual total radio-

---

Strontium-90 in thick samples is counted with low efficiency; hence, a first count within hours favors ^{89}Sr counting, and a recount after 3 to 6 d that exceeds the first count provides a rough estimate of the ^{90}Y ingrowth—see Figure 7500-Sr:1 and R.J. Velten (1966) below.
** When a determination of total strontium is not required, weigh precipitate [Step 20)a) or 20)b)] for carrier recovery but do not count. Then proceed with ^{90}Sr determination according to ¶ 4b following.

strontium determination. When mounting samples according to Step 20)b), determine self-absorption curves by separately precipitating standard solutions of ^{89}Sr and ^{90}Sr as the carbonate (see gross beta in Section 7110).

23) Correction for carrier recovery—20 mg Sr are equivalent to 33.7 mg SrCO$_3$. Should more than traces of stable strontium be present in the sample, it would act as carrier; hence its determination by flame photometric or atomic absorption spectrometric method would be required.

b. Strontium-90 by extraction of yttrium-90:†† Store SrCO$_3$ precipitate, as in ¶ 4*a*20), for at least 2 weeks to allow ingrowth of ^{90}Y and then proceed as directed here or in an alternate procedure in ¶ 4*c* following.

1) Transfer of precipitate to separatory funnel—Either

a) Place a small funnel upright into mouth of a 60-mL separatory funnel; then place pan with precipitate, as in Step 20)a), in funnel and add, dropwise, 1 mL 6*N* HNO$_3$ (CAUTION—*effervescence*); tilt pan to empty into funnel and rinse pan twice with 2-mL portions of 6*N* HNO$_3$; or

b) Uncover precipitate from filter, as in Step 20)b), and transfer filter with forceps to upright funnel in mouth of 60-mL separatory funnel as in ¶ a) above. Dislodge bulk of precipitate into funnel stem. Dropwise, add with caution 1 mL 6*N* HNO$_3$ to filter, removing residual precipitate and dissolving bulk precipitate. Rinse filter and funnel twice with 2-mL portions 6*N* HNO$_3$.

2) Remove filter or pan and add 10 mL fuming HNO$_3$ to separatory funnel through upright funnel.

3) Remove upright funnel and add 1 mL yttrium carrier in a separatory funnel.

4) Add 5.0 mL tributyl phosphate reagent, shake thoroughly for 3 to 5 min, allow phases to separate, and transfer aqueous layer to a second 60-mL separatory funnel.

5) Add 5.0 mL tributyl phosphate reagent, shake 5 min, allow phases to separate, and transfer aqueous layer to a third 60-mL separatory funnel.

6) Combine organic extractants in the first and second funnels into one funnel and wash organic phase twice with 5-mL portions 14*N* HNO$_3$. Record time as the beginning of ^{90}Y decay (combine acid washings with aqueous phase in third funnel if a second ingrowth of ^{90}Y is desired).

7) Back-extract ^{90}Y from combined organic phases with 10 mL 0.1*N* HNO$_3$ for 5 min.

8) Continue as in ¶s 4*c*6)-8) below or transfer aqueous phase from Step 7) immediately above into a 50-mL beaker and evaporate on a hot plate to 5 to 10 mL.

9) Repeat Step 7) above and transfer aqueous phase to beaker in Step 8) preceding; evaporate to 5 to 10 mL.

10) Transfer residual solution in beaker to a tared stainless steel counting pan and evaporate.

11) Rinse beaker twice with 2-mL portions of 0.1*N* HNO$_3$; add rinsings to counting pan, evaporate to dryness, and weigh.

12) Count in an internal proportional or end-window counter and calculate ^{90}Sr as given in ¶ 4*c*9) following.

c. Strontium-90 by oxalate precipitation of yttrium-90:††

1) Quantitatively transfer SrCO$_3$ precipitate to a 40-mL centrifuge tube with 2 mL 6*N* HNO$_3$. Add acid dropwise during dissolution (CAUTION—*effervescence*). Use 0.1*N* HNO$_3$ for rinsing.

---

†† See footnote to Step 20a) when a determination for only ^{90}Sr is required.

2) Add 1 mL yttrium carrier, 2 drops methyl red indicator and, *dropwise,* add conc NH$_4$OH to the methyl red end point.

3) Add 5 mL more conc NH$_4$OH and *record the time,* which is the end of ^{90}Y ingrowth and the beginning of decay; centrifuge and decant supernate to a beaker (save supernate and washings for a second ingrowth if desired).

4) Wash precipitate twice with 20-mL portions hot distilled water.

5) Add 5 to 10 drops of 6*N* HNO$_3$, stir to dissolve precipitate, add 25 mL distilled water, and heat in a water bath at 90°C.

6) Gradually add 15 to 20 drops saturated oxalic acid reagent with stirring and adjust to pH 1.5 to 2.0 (pH meter or indicator paper) by adding conc NH$_4$OH dropwise. Digest precipitate for 5 min and cool in an ice bath with occasional stirring.

7) Transfer precipitate to a tared glass fiber filter in a two-piece funnel. Let precipitate settle by gravity (for uniform deposition) and apply suction. Wash precipitate in sequence with 10 to 15 mL hot distilled water and then three times with 95% ethyl alcohol and three times with diethyl ether.

8) Air-dry precipitate with suction for 2 min, weigh, mount on a nylon disk and ring with polyester plastic film cover, count, and calculate ^{90}Sr as follows.

9) Calculation

$$^{90}\text{Sr, pCi/L} = \frac{\text{net cpm}}{a\,b\,c\,d\,f\,g \,\times\, 2.22}$$

where:

a = counting efficiency for ^{90}Y,
b = chemical yield of extracting or precipitating ^{90}Y,
c = ingrowth correction factor if not in secular equilibrium,
d = chemical yield of strontium determined gravimetrically or by flame photometry,
f = volume of original sample, L,
g = ^{90}Y decay factor, $e^{-\lambda t}$, and
e = base of natural logarithms,
λ = 0.693/T$_{1/2}$, where T$_{1/2}$ for ^{90}Y is 64.2 h, and
t = time between separation and counting, h.

5. Precision and Bias

In a collaborative study of two sets of paired, moderately hard water samples containing known additions of radionuclides, 12 laboratories determined the total radiostrontium and 10 laboratories determined ^{90}Sr. The results of one sample from one laboratory were rejected as outliers.

The average recoveries of added total radiostrontium from the four samples were 99, 99, 96, and 93%. The precision (random error) at the 95% confidence level was 10 and 12% for the two sets of paired samples. The method was slightly biased on the low side.

6. Bibliography

HAHN, R.B. & C.P. STRAUB. 1955. Determination of radioactive strontium and barium in water. *J. Amer. Water Works Assoc.* 47:335.

GOLDIN, A.S., R.J. VELTEN & G.W. FRISHKORN. 1959. Determination of radioactive strontium. *Anal. Chem.* 31:1490.

GOLDIN, A.S. & R.J. VELTEN. 1961. Application of tributyl phosphate extraction to the determination of strontium 90. *Anal. Chem.* 33:149.

VELTEN, R.J. 1966. Resolution of Sr-89 and Sr-90 in environmental media by an instrumental technique. *Nucl. Instrum. Methods* 42:169.

7500-³H TRITIUM*

7500-³H A. Introduction

Tritium exists fairly uniformly in the environment as a result of natural production by cosmic radiation and residual fallout from nuclear weapons tests. This background level gradually is being increased by the use of nuclear reactors to generate electricity, although tritium from this source is only a small proportion of environmental tritium. Nuclear reactors and fuel-processing plants are localized sources of tritium because of discharges during normal operation. This industry is expected to become the major source of environmental tritium contamination in the fu-

ture. Tritium is produced in light-water nuclear reactors by ternary fission, neutron capture in coolant additives, control rods and plates, and activation of deuterium. About 1% of the tritium in the primary coolant is released in gaseous form to the atmosphere; the remainder eventually is released in liquid waste discharges. Most tritium produced in reactors remains in the fuel and is released when fuel is reprocessed.

Naturally occurring tritium is most abundant in precipitation and lowest in aged water because of its physical decay by beta emission to helium. The maximum beta energy of tritium is 0.018 MeV and its half-life is 12.26 years.

* Approved by Standard Methods Committee, 1993.

7500-³H B. Liquid Scintillation Spectrometric Method

1. General Discussion

a. Principle: A sample is treated by alkaline permanganate distillation to hold back most quenching materials, as well as radioiodine and radiocarbon. Complete transfer of tritiated water is assured by distillation to near dryness. A subsample of distillate is mixed with scintillation solution and the beta activity is counted on a coincidence-type liquid scintillation spectrometer. The scintillation solution consists of 1,4-dioxane, naphthalene, POPOP, and PPO.* The spectrometer is calibrated with standard solutions of tritiated water; then background and unknown samples are prepared and counted alternately, thus nullifying errors that could result from instrument drift or from aging of the scintillation solution.

b. Interferences: Sample distillation effectively removes non-volatile radioactivity and the usual quenching materials. For waters containing volatile organic or radioactive materials, use wet oxidation (Section 4500-N$_{org}$) to remove interference from quenching due to volatile organic material. Distillation at about pH 8.5 holds back volatile radionuclides such as iodides and bicarbonates. Double distillation with an appropriate delay (10 half-lives) between distillations may be required to eliminate interference from volatile daughters of radium isotopes. Some clear-water samples collected near nuclear facilities may be monitored satisfactorily without distillation, especially when the monitoring instrument is capable of discriminating against beta radiation energies higher than those in the tritium range.

2. Apparatus

a. Liquid scintillation spectrometer, coincidence-type.

b. Liquid scintillation vial: 20-mL; polyethylene, low-K glass, or equivalent bottles.

c. Distillation apparatus: 250-mL round-bottom distillation flask, connecting side-arm adapter, condenser, and heating mantle.

3. Reagents

a. Scintillation solution: Thoroughly mix 4 g PPO, 0.05 g POPOP, and 120 g solid naphthalene in 1 L spectroquality 1,4-dioxane. Store in dark bottle. Solution is stable for 2 months. Alternatively, use a commercially prepared scintillation solution available from suppliers of liquid scintillation materials.

b. Low-background water: Use water with no detectable tritium activity (most deep well waters are low in tritium).

c. Standard tritium solution: Dilute available tritium standard solution to approximately 1000 dpm/mL with low-background water.

d. Sodium hydroxide, NaOH, pellets.

e. Potassium permanganate, KMnO$_4$.

4. Procedure

Add three pellets NaOH and 0.1 g KMnO$_4$ to 100 mL sample in 250-mL distillation flask. Distill at 100 to 105°C, discard first 10 mL distillate, and collect next 50 mL. Thoroughly mix 4 mL distillate with 16 mL scintillation solution in tightly capped vial.

Prepare low-background water and standard tritium solution in same manner as samples.

Hold samples, background, and standards in the dark for 3 h. Count samples containing less than 200 pCi/mL for 100 min and samples containing more than 200 pCi/mL for 50 min.

5. Calculations and Reporting

a. Calculate and report tritium, ³H, in picocuries per milliliter (pCi/mL) or its equivalent, nanocuries per liter (nCi/L) as follows:

$$^3H = \frac{(C - B)}{(E \times 4 \times 2.22)}$$

* POPOP = 1,4-di-2-(5-phenyloxazolyl) benzene; PPO = (2,5- diphenyloxazole).

where:

- C = gross counting rate for sample, cpm,
- B = background counting rate, cpm,
- E = counting efficiency, $(S - B)/D$,
- S = gross counting rate for standard solution, cpm, and
- D = tritium activity in standard sample, dpm, corrected for decay to time of counting.

b. Calculate the counting error at the 95% confidence level based on the equation for $\sigma(R)$ given in Section 7010G. A total count of 40 000 within 1 h for a background count rate of about 50 cpm gives a counting error slightly in excess of 1% at the 95% confidence level.

6. Precision and Bias

Samples with tritium activity above 200 pCi/mL can be analyzed with precision of less than ±6% at the 95% confidence level and those with 1 pCi/mL can be analyzed with a precision of less than ±10%.

7. Bibliography

LIBBY, W.F. 1946. Atmospheric helium-3 and radiocarbon from cosmic radiation. *Phys. Rev.* 69:671.

NATIONAL COUNCIL ON RADIATION PROTECTION, SUBCOMMITTEE ON PERMISSIBLE INTERNAL DOSE. 1959. Maximum Permissible Body Burdens and Maximum Permissible Concentrations of Radionuclides in Air and in Water for Occupational Exposure. NBS Handbook 69 (June), National Bur. Standards, Washington, D.C.

INTERNATIONAL COMMISSION ON RADIATION PROTECTION. 1960. Report of Committee II on permissible dose for internal radiation, 1959. *Health Phys.* 3:41.

BUTLER, F.E. 1961. Determination of tritium in water and urine. *Anal. Chem.* 33:409.

FOOD AND AGRICULTURE ORGANIZATION, INTERNATIONAL ATOMIC ENERGY AGENCY & WORLD HEALTH ORGANIZATION. 1966. Methods of Radiochemical Analysis. World Health Org., Geneva.

SMITH, J.M. 1967. The Significance of Tritium in Water Reactors. General Electric Co., San Jose, Calif.

YOUDEN, W.J. 1967. Statistical Techniques for Collaborative Tests. Assoc. Official Analytical Chemists, Washington, D.C.

PETERSON, H.T.J., J.E. MARTIN, C.L. WEAVER & E.D. HARWARD. 1969. Environmental tritium contamination from increasing utilization of nuclear energy sources. Seminar on Agricultural and Public Health Aspects of Environmental Contamination by Radioactive Materials, International Atomic Energy Assoc., Vienna, pp. 35–60.

SODD, V.J. & K.L. SCHOLZ. 1969. Analysis of tritium in water; a collaborative study. *J. Assoc. Offic. Anal. Chem.* 52:1.

WEAVER, C.L., E.D. HARWARD & H.T. PETERSON. 1969. Tritium in the environment from nuclear power plants. *Pub. Health Rep.* 84, 363.

U.S. ENVIRONMENTAL PROTECTION AGENCY. 1975. Tentative Reference Method for Measurement of Tritium in Environmental Waters. EPA 600/4-75-013, Environmental Monitoring and Support Lab., U.S. Environmental Protection Agency, Las Vegas, Nev.

7500-U   URANIUM*

7500-U A.   Introduction

1. Occurrence

Uranium, the heaviest naturally occurring element, is a mixture of three radioactive isotopes: uranium-238 (99.275%), uranium-235 (0.72%), and uranium-234 (0.005%). Most drinking-water sources, especially ground waters, contain soluble carbonates and bicarbonates that complex and keep uranium in solution.

2. Selection of Method

Method B, a radiochemical procedure, determines total uranium alpha activity without making an isotopic uranium analysis. Method C is a radiochemical procedure that determines the isotopic content of the uranium alpha activity; it is consistent with determining the differences among naturally occurring, depleted, and enriched uranium.

---

* Approved by Standard Methods Committee, 1996.

3. Bibliography

GRIMALDI, F. S. et al. 1954. Collected Papers on Methods of Analysis for Uranium and Thorium. Bull. 1006, U.S. Geological Survey.

BLANCHARD, R. 1963. Uranium Decay Series Disequilibrium in Age Determination of Marine Calcium Carbonates. Ph.D. Thesis, Washington Univ., St. Louis, Mo.

BARKER, F. B. et al. 1965. Determination of uranium in natural waters. U.S. Geological Survey, Water Supply Paper 1696-C, U.S. Government Printing Off., Washington, D.C.

EDWARD, K. W. 1968. Isotopic analysis of uranium in natural waters by alpha spectroscopy. U.S. Geological Survey, Water Supply Paper 1696-F, U.S. Government Printing Off., Washington, D.C.

THATCHER, L. L., V. J. JANZER & K. W. EDWARDS. 1977. Methods for Determination of Radioactive Substances in Water and Fluvial Sediments. Book 5, Chapter A5. Techniques of Water-Resources Investigations of the United States Geological Survey. U.S. Government Printing Off., Washington, D.C.

KRIEGER, H. L. & E. L. WHITTAKER. 1980. Prescribed procedures for measurement of radioactivity in drinking water. EPA-600/4-80-032, U.S. Environmental Protection Agency.

7500-U B. Radiochemical Method

1. General Discussion

a. Principle: The sample is acidified with hydrochloric or nitric acid and boiled to eliminate carbonate and bicarbonate ions. Uranium is coprecipitated with ferric hydroxide and subsequently separated. The ferric hydroxide is dissolved, passed through an anion-exchange column, and washed with acid, and the uranium is eluted with dilute hydrochloric acid. The acid eluate is evaporated to near dryness, the residual salt is converted to nitrate, and the alpha activity is counted.

b. Interference: The only alpha-emitting radionuclide that may be carried through this procedure is protactinium-231. However, this isotope, which is a decay product of uranium-235, causes very little interference. Check reagents for uranium contamination by analyzing a complete reagent blank.

c. Sampling: Preserve sample by adjusting its pH to <2 with HCl or HNO_3 at time of collection.

2. Apparatus

a. Counting instrument, gas-flow proportional or alpha scintillation counting system.

b. Ion-exchange column, approximately 13 mm ID $\times$ 150 mm long with 100-mL reservoir.

c. Membrane filter apparatus, 47-mm diam.

3. Reagents

a. Ammonium hydroxide, NH_4OH, 5N, 1%.

b. Anion-exchange resin. *

c. Ferric chloride carrier: Dissolve 9.6 g $FeCl_3 \cdot 6H_2O$ in 100 mL 0.5N HCl; 1 mL = 20 mg Fe^{3+}.

d. Hydriodic acid, HI, 47%.

e. Hydrochloric acid, HCl, conc, 8N, 6N, 0.1N.

f. Iodic acid, 1 mg/mL: Dissolve 100 mg HIO_3 in 100 mL 4N HNO_3.

g. Nitric acid, HNO_3, conc, 4N.

h. Sodium hydrogen sulfite, 1%: Dissolve 1 g $NaHSO_3$ in 100 mL 6N HCl.

i. Uranium standard solution:† Dissolve 177.3 mg natural undepleted uranyl acetate, $UO_2(C_2H_3O_2)_2 \cdot 2H_2O$, in 1000 mL 0.2$N$ HNO_3; 1 mL = 100 µg U = 150 dpm U = 67.6 pCi U. NOTE: Commonly available uranyl salts may be formed from depleted uranium; verify isotopic composition before use.

4. Calibration

Determine counting efficiency, E, for a known amount of uranium standard solution (about 750 dpm) evaporated from 6 to 8 mL of 1 mg/mL HIO_3 solution in a 50-mm-diam stainless steel planchet. After flaming planchet, count for at least 50 min. Run a reagent blank with the standard portions and count.

---

* Dowex 1×4, 100-200 mesh, chloride form, or equivalent.
† Standard radioactive solutions with uranium isotopes in equilibrium are available for participants in the performance evaluation studies program from the U.S. Environmental Protection Agency, NRA/STD, P.O. Box 93478, Las Vegas, NV 89193. A uranium oxide assay standard, CRM 129, is available for purchase from U.S. Department of Energy, Chicago Operations Office, New Brunswick Laboratory, D-350, 9800 South Cass Avenue, Argonne, IL 60439.

$$\text{Counting efficiency, } E = \frac{C - B}{D}$$

where:

C = gross alpha count rate of standard, cpm,
B = alpha background count rate, cpm, and
D = disintegration rate of uranium standard, dpm.

Determine uranium recovery factor by adding a measured amount of uranium standard to the same volume of sample and taking it through the entire procedure. Alpha count the separated, evaporated, and flamed uranium planchet. Determine the recovery factor on at least 10% of all drinking water samples. For non-drinking water samples, it may be necessary to determine the recovery factor in every sample.

$$\text{Recovery factor, } R = \frac{C' - B'}{DE}$$

where:

C' = gross count rate of sample with added uranium, cpm,
B' = count of reagent blank, cpm,
D = disintegration rate of uranium standard, dpm, and
E = counting efficiency.

5. Procedure

a. If the sample has not been acidified, add 5 mL conc HCl or HNO_3 to 1 L sample in a 1500-mL beaker. Add 1 mL $FeCl_3$ carrier. In each batch of samples include a distilled-water blank. Cover with watch glass and heat to boiling for 20 min. If pH is greater than 1, add conc HCl or HNO_3 dropwise to bring pH to 1. While sample is boiling, gently add 5N NH_4OH from a polyethylene squeeze bottle with the delivery tube inserted between the watch glass and the beaker lip. Add 5N NH_4OH until turbidity persists while boiling continues; then add 10 mL more. Continue boiling for 10 min more, then set aside for 30 min to cool and settle. After sufficient settling, decant and filter supernate through a 47-mm, 0.45-µm membrane filter using a large filtering apparatus. Slurry the remaining precipitate, transfer to the filtering apparatus, and filter with suction. Complete transfer using 1% solution of NH_4OH delivered from a polyethylene squeeze bottle. Place filtering apparatus over a clean 250-mL filtering flask, add 25 mL 8N HCl to dissolve precipitate, and filter. Wash filter with an additional 25 mL 8N HCl. (Alternatively, use centrifugation in place of filtration as in 7500-U.C.4a.)

b. Prepare an ion-exchange column by slurrying the anion-exchange resin with 8N HCl and pouring it into a 13-mm-ID column to give a resin bed height of about 80 mm. Transfer solution to the 100-mL reservoir of the ion-exchange column. Rinse side-arm filtering flask twice with 25-mL portions of 8N HCl. Combine in the ion-exchange reservoir. Pass sample solution through the anion-exchange column at a flow rate of not more than 5 mL/min. After sample has passed through column, elute the iron (and plutonium if present) with six column volumes of freshly prepared 8N HCl containing 1 mL 47% HI /9 mL 8N HCl. Wash column with two additional column volumes of 8N HCl.

Discard all washes. Elute uranium into a 100-mL beaker with six column volumes of 0.1N HCl. Evaporate acid eluate to near dryness and convert residue to the nitrate form by three successive treatments with 5-mL portions of conc HNO₃, evaporating to near dryness each time. *Do not bake.* Dissolve residue (of which there may be very little visible) in 2 mL 4N HNO₃. Using a transfer pipet, transfer to a marked planchet. Complete transfer by rinsing beaker three times with 2-mL portions of 4N HNO₃. Evaporate planchet contents to dryness under a heat lamp, flame to remove traces of HIO₃, cool, and count for alpha activity.

c. To regenerate anion-exchange resin column, pass three column volumes of 1% NaHSO₃ in 6N HCl through the column, follow with six column volumes of 6N HCl, and then three column volumes of distilled water. Do not let resin become dry. When ready for the next set of samples, equilibrate by passing six column volumes of 8N HCl through the column.

6. Calculations

$$\text{Uranium alpha activity, pCi/L} = \frac{C'' - B'}{2.22 \times ERV}$$

where:

C'' = gross count rate of sample, cpm,
V = volume of sample, L, and other factors are as defined above.

7. Precision and Bias

In a collaborative study, three sets of triplicate samples with known additions of uranium were analyzed by 18 laboratories. The average recovery was 91.5%. The estimated average 95% repeatability interval was 29.3% of the uranium concentration over the range of 8 to 75 pCi/L. The estimated average 95% reproducibility interval was 37.2% over the same range.

7500-U C.   Isotopic Method

1. General Discussion

a. Principle: The sample is acidified with hydrochloric or nitric acid and uranium-232 is added as an isotopic tracer. Uranium is separated as in the radiochemical method (see Section 7500-U.B) and is electrodeposited onto a stainless steel disk for counting by alpha pulse height analysis using a silicon surface barrier detector.

b. Interferences: The only alpha-emitting radionuclide that may be carried through the procedure is protactinium-231. The presence of this radionuclide can be determined from the alpha spectrum and the interference subtracted. Check reagents for uranium contamination by analyzing a complete reagent blank.

c. Sampling: Preserve sample by adjusting its pH to <2 with HCl or HNO₃ at the time of collection.

2. Apparatus

a. Counting instrument, alpha spectrometer (see 7030B.4), giving a resolution of 50 keV (FWHM) or better and having a counting efficiency greater than 15%.

b. Ion-exchange column, 13 mm ID × 150 mm long with 100-mL reservoir.

c. Electrodeposition apparatus as shown in Figure 7500-U:1. Although the electrodeposition cell is surrounded by water the water is not circulated because cooling is unnecessary. The cathode slide has mirror finish, is 0.05 cm thick, and has an exposed electrodeposition area of 2 cm². The anode is a 1-mm-diam platinum wire with an 8-mm-diam loop at the end above the cathode.

d. DC power supply, 0 to 12 V at 0 to 2 amp, for electrodeposition.

e. Centrifuge, capable of handling 100-mL or larger centrifuge bottles.

3. Reagents

In addition to reagents *d* through *g* from Section 7500U.B, the following are needed:

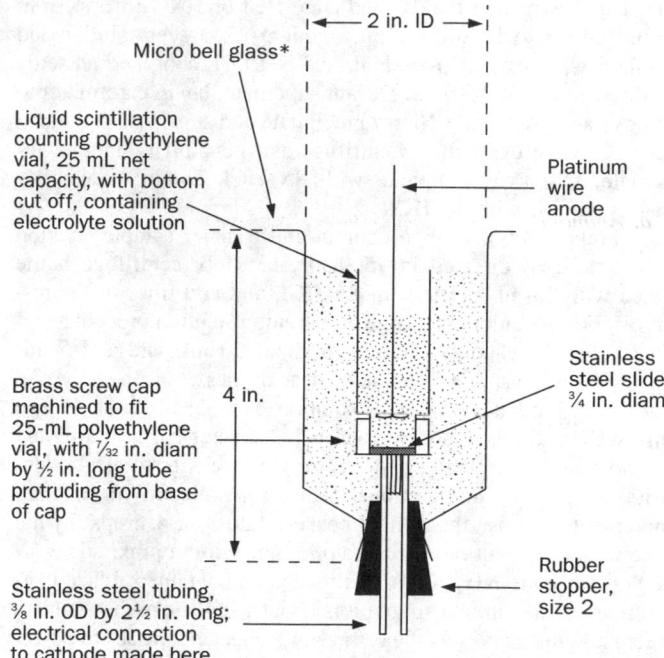

Figure 7500-U:1. Electrodeposition apparatus. To obtain dimensions in centimeters, multiply dimensions in inches by 2.54.

a. Ammonium hydroxide, NH₄OH, 5N, 1.5N, and 0.15N.
*b. Anion-exchange resin.**
c. Ethyl alcohol, made slightly basic with a few drops of conc NH₄OH/100 mL.
d. Preadjusted electrolyte, (NH₄)₂SO₄, 1M, adjusted to pH 3.5 with conc NH₄OH and conc H₂SO₄.
e. Sulfuric acid, H₂SO₄, conc, 3.6N.

---

* Bio Rad AGl-X4, 100-200 mesh, chloride form, or equivalent.

f. Sodium hydrogen sulfate, about 5% in $18N$ H_2SO_4. Dissolve 10 g $NaHSO_4 \cdot H_2O$ in 100 mL water and carefully add 100 mL conc H_2SO_4.

g. Thymol blue indicator, sodium salt, 0.04% solution.

h. Uranium-232 tracer solution, 10 dpm/mL in $1N$ HNO_3: If possible use a ^{232}U standard solution from, or traceable to, the National Institute for Standards and Technology (NIST). Standardize a freshly purified solution of ^{232}U by thoroughly mixing a known amount with a known amount of another uranium standard such as ^{236}U or natural uranium, and electroplating the mixture. Determine specific activity of the ^{232}U solution from an alpha pulse height analysis of the electroplated mixture. Alternatively, evaporate weighed portions of a freshly purified ^{232}U solution (free of HCl) on stainless steel slides and count with a 2π proportional counter. Determine efficiency of the 2π counter accurately with a NIST alpha-particle standard. When using this standard, correct for resolving time and backscattering if necessary.

4. Procedure

a. If the sample has not been acidified, add 5 mL conc HCl or conc HNO_3 to 1 L sample in a 1500-mL beaker. Mix and check pH. If pH is greater than 1, add conc HCl or HNO_3 dropwise to bring the pH to 1. Add 1.0 mL uranium-232 tracer solution and 1 mL $FeCl_3$ carrier. Cover, boil, add NH_4OH, cool, and let settle as directed in 7500U.B.5*a*. Decant supernate, being careful not to remove any precipitate. Slurry precipitate and supernate and transfer to a centrifuge bottle. Centrifuge and pour off remaining supernate. Dissolve precipitate with $8N$ HCl. Dilute to approximately 50 mL with $8N$ HCl.

b. Prepare ion-exchange column and transfer sample solution to reservoir as directed in 7500U.B.5*b*. Rinse centrifuge bottle twice with 25-mL portions of $8N$ HCl, and add rinse to the reservoir. Follow anion-exchange and uranium-elution procedures of 7500U.B.5*b*. Evaporate sample to about 20 mL and add 5 mL conc HNO_3. Evaporate sample to near dryness.

c. Add 2 mL 5% $NaHSO_4$ solution. Add 5 mL conc HNO_3, mix well, and evaporate to dryness but do not bake. Warm and dissolve in 5 mL preadjusted electrolyte. Transfer to electrodeposition cell using an additional 5 to 10 mL electrolyte in small increments to rinse the sample beaker. Add 3 or 4 drops thymol blue indicator solution. If the color is not salmon pink, add $3.6N$ H_2SO_4 (or conc NH_4OH) until this color is obtained. Place platinum anode in solution so that it is about 1 cm above the stainless steel slide that serves as the cathode. Connect electrodes to power supply and adjust to give a current of 1.2 amp (constant current power supplies will not require further adjustments during electrodeposition). Continue electrodeposition for 1 h. When electrodeposition is to be ended, add 1 mL conc NH_4OH and continue for 1 min. Remove anode from cell and then turn off power. Discard solution in cell and rinse two or three times with $0.15N$ NH_4OH. Disassemble cell and wash slide with ethyl alcohol that has been made basic with NH_4OH. Dry slide over a hot plate. Measure activity of the uranium isotopes using an alpha spectrometer (see 7030B.4) within a week of preparation.†

† Electrodeposition was the recommended technique for alpha spectroscopy source preparation in the collaborative test of this method. A number of laboratories are currently using a rare earth fluoride co-precipitation technique for alpha spectroscopy source preparation. If adequate resolution can be obtained, the rare earth fluoride source preparation should be adequate as an alternative to electrodeposition.

5. Calculations

a. Determine total counts for each uranium isotope by summing the counts in the peak at the energy corresponding to the isotope. If two isotopes are close in energy, complete resolution may not be possible. Subtract background from each peak. Make a blank correction for each peak, if necessary.

b. Calculate concentration of each uranium isotope as follows:

$$U_i, \text{pCi/L} = \frac{C_i \times A_t}{2.22 \times C_t V}$$

where:

U_i = concentration of uranium isotope being determined,
C_i = net sample counts in the energy region corresponding to uranium isotope being measured,
A_t = activity of added uranium-232 tracer, dpm,
C_t = net sample counts in the energy region corresponding to uranium-232 tracer, and
V = sample volume, L.

6. Calibration

To calculate uranium recovery, determine absolute counting efficiency (E) of the alpha spectrometer. To determine efficiency count a standard source of a known alpha activity having the same active area as the samples.

$$E = \frac{C_s - B}{D}$$

where:

C_s = gross count rate in the energy region corresponding to the energy of the standard, cpm,
B = background count rate in the energy region corresponding to the energy of the standard, cpm, and
D = disintegration rate of standard, dpm.

$$\text{Recovery factor, } R = \frac{C_t}{t \times A_t E}$$

where:

t = sample counting time, min.

7. Precision and Bias

In a collaborative study, four sets of duplicate samples with known additions of uranium isotopes were analyzed by eight laboratories. Results agreed within 5% of the reference values, except for very low concentrations of uranium (concentrations approaching MDL due to background). Levels less than 0.1 pCi/L can be detected by this method.

8. Bibliography

KRAUS, K.A. & F. NELSON. 1956. Anion Exchange Studies of the Fission Products. Proc. International Conf. on the Peaceful Uses of Atomic Energy, Geneva, 1955, 7, 113, Session 9 Bl, p. 837, United Nations.

GINDLER, J.E. 1962. The Radiochemistry of Uranium. NAS-NS-3050, National Academy of Sciences–National Research Council, Washington, D.C.

BARKER, F.B. 1965. Determination of Uranium in Natural Waters. Radiochemical Analysis of Water. Geological Survey Water Supply Paper 1669-C, U.S. Government Printing Office, Washington, D.C.

EDWARDS, K.W. 1968. Isotopic Analysis of Uranium in Natural Waters by Alpha Spectrometry. Radiochemical Analysis of Water, Geological Survey Water Supply Paper 1696-F, U.S. Government Printing Off., Washington, D.C.

KORKISCH, J. 1969. Modern Methods for the Separation of Rarer Metal Ions. Pergamon Press, New York, N.Y.

BALTAKMENS, F. 1975. Simple method for the determination of uranium in soils by two stage ion exchange. *Anal. Chem.* 47:1147.

ESSINGTON, E.H. & E.B. FOWLER. 1976. Nevada Applied Ecology Group. Soils element activities for period July 1, 1974 to May 1, 1975. *In* M.G. White & P.B. Dunaway, eds. Studies of Environmental Plutonium and Other Transuranics in Desert Ecosystems. Nevada Applied Ecology Group Progress Rep. MVO-159, p. 17.

U.S. ENVIRONMENTAL PROTECTION AGENCY. 1979. Radiometric Method for the Determination of Uranium in Water. EPA-600/7-79-093, Environmental Monitoring and Support Lab., Las Vegas, Nev.

U.S. ENVIRONMENTAL PROTECTION AGENCY. 1984. Radiochemical Determination of Thorium and Uranium in Water. 00-07. Radiochemical Procedures Manual, EPA-520/5-84-006, Eastern Environmental Radiation Facility, Montgomery, Ala.

PART 8000

TOXICITY

8010 INTRODUCTION*

8010 A. General Discussion

1. Uses of Toxicity Tests

Toxicity tests are desirable in water quality evaluations because chemical and physical tests alone are not sufficient to assess potential effects on aquatic biota.[1-3] For example, the effects of chemical interactions and the influence of complex matrices on toxicity cannot be determined from chemical tests alone. Different species of aquatic organisms are not equally susceptible to the same toxic substances nor are organisms equally susceptible throughout the life cycle. Even previous exposure to toxicants can alter susceptibility. In addition, organisms of the same species can respond differently to the same level of a toxicant from time to time, even when all other variables are held constant.

Toxicity tests are useful for a variety of purposes that include determining: *(a)* suitability of environmental conditions for aquatic life, *(b)* favorable and unfavorable environmental factors, such as DO, pH, temperature, salinity, or turbidity, *(c)* effect of environmental factors on waste toxicity, *(d)* toxicity of wastes to a test species, *(e)* relative sensitivity of aquatic organisms to an effluent or toxicant, *(f)* amount and type of waste treatment needed to meet water pollution control requirements, *(g)* effectiveness of waste treatment methods, *(h)* permissible effluent discharge rates, and *(i)* compliance with water quality standards, effluent requirements, and discharge permits. In such regulatory assessments, use toxicity test data in conjunction with receiving-water and site-specific discharge data on volumes, dilution rates, and exposure times and concentrations.

2. Test Procedures

There is a need to use correct terminology (see Section 8010B, Terminology), and environmentally relevant test procedures to meet regulatory, legal, and research objectives.[3-8]

The procedures given below allow measurement of biological responses to known and unknown concentrations of materials in both fresh and saline waters. These toxicity tests are applicable to routine monitoring requirements as well as research needs. Refer to Part 9000 for microbiological methods and Part 10000 for field and other types of biological laboratory methods for water quality evaluations. Refer to Section 10900 for identification aids for aquatic organisms.

Reasonable uniformity of procedures and of data presentation is essential. The use of standardized methods described below will ensure adequate uniformity, reproducibility, and general usefulness of results without interfering unduly with the adaptability of the tests to local circumstances.

Quality assurance practices for toxicity test methods include all aspects of the test that affect the quality of the data. These include sampling and handling, source and condition of test organisms, performance of reference toxicant tests, and the test procedures themselves. Quality assurance/quality control guidelines are available for single compound testing and general laboratory practices[9] and for effluent evaluations in technical guidance manuals for conducting acute and short-term chronic toxicity tests with effluents.[10-12]

3. References

1. U.S. ENVIRONMENTAL PROTECTION AGENCY. 1991. Technical Support Document for Water Quality-Based Control. EPA-505/2-90-001 (PB91-127415), Off. Water, U.S. Environmental Protection Agency, Washington, D.C.
2. U.S. ENVIRONMENTAL PROTECTION AGENCY. 1987. Permit Writer's Guide to Water Quality-Based Permitting for Toxic Pollutants. Off. Water, U.S. Environmental Protection Agency, Washington, D.C.
3. GROTHE, D.R., K.L. DICKSON & D.K. REED-JUDKINS, eds. 1996. Whole Effluent Toxicity Testing: An Evaluation of Methods and Prediction of Receiving System Impacts. SETAC Pellston Workshop on Whole Effluent Toxicity, Sept. 16-25, 1995, Pellston, Mich. SETAC Press, Pensacola, Fla.
4. AMERICAN SOCIETY FOR TESTING AND MATERIALS. 1996. 1996 Annual Book of ASTM Standards, Section 11, Water and Environment Technology. Volume 11.04 Pesticides; Resources Recovery; Hazardous Substances and Oil Spill Responses; Waste Disposal; Biological Effects. American Soc. Testing & Materials, W. Conshohocken, Pa.
5. ORGANIZATION FOR ECONOMIC COOPERATION AND DEVELOPMENT. 1981. OECD Guidelines for Testing of Chemicals. Organization for Economic Cooperation and Development, Paris, France.
6. BERGMAN, H., R. KIMERLE & A.W. MAKI, eds. 1985. Environmental Hazard Assessment of Effluents. Pergamon Press, Inc., Elmsford, N.Y.
7. AMERICAN SOCIETY FOR TESTING AND MATERIALS. 1997. Standard Guide for Conducting Acute Toxicity Tests on Aqueous Effluents with Fishes, Macroinvertebrates, and Amphibians. ASTM -1192-97, American Soc. Testing & Materials, W. Conshohocken, Pa.
8. AMERICAN SOCIETY FOR TESTING AND MATERIALS. 1997. Standard Guide for Conducting Acute Toxicity Tests with Fishes, Macroinvertebrates, and Amphibians. ASTM E 729-96, American Soc. Testing & Materials, W. Conshohocken, Pa.
9. U.S. ENVIRONMENTAL PROTECTION AGENCY. 1987. Federal Insecticide, Fungicide and Rodenticide Act (FIFRA); Good Laboratory Practice Standards. Proposed Rule. 40 CFR Part 160; *Federal Register* 52:48920.
10. WEBER, C.I., ed. 1993. Methods for Measuring the Acute Toxicity of Effluents and Receiving Water to Freshwater and Marine Organisms, 4th ed. EPA-600/4-90-027F, Environmental Monitoring Systems Lab., U.S. Environmental Protection Agency, Cincinnati, Ohio.
11. KLEMM, D.J., G.E. MORRISON, T.J. NORBERG-KING, W.H. PELTIER & M.A. HEBER, eds. 1994. Short-Term Methods for Estimating the Chronic Toxicity of Effluents and Receiving Waters to Marine and Estuarine Organisms, 2nd ed. EPA-600/4-91-003, Environmental Monitoring and Support Lab., U.S. Environmental Protection Agency, Cincinnati, Ohio.

---

* Approved by Standard Methods Committee, 1997.

12. LEWIS, P.A., D.J. KLEMM, J.M. LAZORCHAK, T.J. NORBERG-KING, W.H. PELTIER & M.A. HEBER, eds. 1994. Short-Term Methods for Estimating the Chronic Toxicity of Effluents and Receiving Waters to Freshwaters Organisms, 3rd ed. EPA-600/4-91-002, Environmental Monitoring Systems Lab., U.S. Environmental Protection Agency, Cincinnati, Ohio.

4. Bibliography

RAND, G.M. & S.R. PETROCELLI, eds. 1985. Fundamentals of Aquatic Toxicology. Methods and Applications. Hemisphere, New York, N.Y.
KLAASON, C.D., M.O. AMDUR & J. DOULL, eds. 1986. Casarett and Doull's Toxicology, 3rd ed. Macmillan, New York, N.Y.

8010 B. Terminology

An aquatic toxicity test is a procedure in which the responses of aquatic organisms are used to detect or measure the presence or effect of one or more substances, wastes, or environmental factors, alone or in combination.

1. General Terms

Acclimate—to accustom test organisms to different environmental conditions, such as temperature, light, and water quality.

Response—the measured biological effect of the variable tested. In acute toxicity tests the response usually is death or immobilization. In plant toxicity tests, the response can be death, growth inhibition, or reproductive inhibition. In biostimulation tests, the response is biomass increase.

Control—treatment in a toxicity test that duplicates all the conditions of the exposure treatment but contains no test material.

Range-finding test—preliminary test designed to establish approximate toxicity of a solution. Test design incorporates multiple, widely spaced, concentrations with single replicates; exposure is usually 8 to 24 h.

Screening test—toxicity test to determine if an impact is likely to be observed; test design incorporates one concentration, multiple replicates, exposure 24 to 96 h.

Definitive test—toxicity test designed to establish concentration at which a particular end point occurs. Exposures for these tests are longer than for screening or range-finding tests, incorporating multiple concentrations at closer intervals and multiple replicates.

2. Toxicity Terms

Dose—amount of toxicant that enters the organism. Dose and concentration are not interchangeable.

Toxicity—potential or capacity of a test material to cause adverse effects on living organisms, generally a poison or mixture of poisons. Toxicity is a result of dose or exposure concentration and exposure time, modified by variables such as temperature, chemical form, and availability.

Exposure time—time of exposure of test organism to test solution.

Acute toxicity—relatively short-term lethal or other effect, usually defined as occurring within 4 d for fish and macroinvertebrates and shorter times (2 d) for organisms with shorter life spans.

Chronic toxicity—toxicity involving a stimulus that lingers or continues for a relatively long period of time, often one-tenth of the life span or more. "Chronic" should be considered a relative term depending on the life span of an organism. A chronic toxic effect can be measured in terms of reduced growth, reduced reproduction, etc., in addition to lethality.

Lethal concentration (LCP)—toxicant concentration estimated to produce death in a specified proportion (P) of test organisms. Usually defined as median (50%) lethal concentration, LC50, i.e., concentration killing 50% of exposed organisms at a specific time of observation, for example, 96-h LC50.

Effective concentration (ECP)—toxicant concentration estimated to cause a specified effect in a designated proportion (P) of test organisms. The effect is usually sublethal, such as a change in respiration rate or loss of equilibrium. The exposure time also is specified; for example, the 96 h EC50 for loss of equilibrium is the effective concentration for 50% of the test organisms in 96 h, for this kind of effect.

Inhibition concentration (ICP)—toxicant concentration estimated to cause a specified percentage (P) inhibition or impairment in a qualitative biological function. For example, an IC25 could be the concentration estimated to cause a 25% reduction in growth of larval fish, relative to the control. Use this term with any toxicological test that measures a change in rate, such as respiration, number of progeny, decrease in number of algal cells, etc.

Asymptotic LC50—toxicant concentration at which LC50 approaches a constant for a prolonged exposure time.

Median tolerance limit (TLm)—test material concentration at which 50% of test organisms survive for a specified exposure time. This term has been superseded by median lethal concentration (LC50) and median effective concentration (EC50).

No-observed-effect concentration (NOEC)—in a full- or partial-life-cycle test, the highest toxicant concentration in which the values for the measured response are not statistically significantly different from those in the control.

Lowest-observed-effect concentration (LOEC)—in a full- or partial-life-cycle test, the lowest toxicant concentration in which the values for the measured response are statistically significantly different from those in the control.

3. Biostimulation Terms

Limiting nutrient—nutrient, among those required, which is inadequate in quantity for growth while others remain sufficient.

Nutrient—specific substance required for organism growth.

Maximum standing crop—maximum weight of organisms during a test, specified as wet or dry weight.

4. Solution Renewal Terms

Static test—test in which solutions and test organisms are placed in test chambers and kept there for the duration of the test.

Renewal test—tests in which organisms are exposed to solutions of the same composition that are renewed periodically during the test period (with renewals usually at 24-h intervals). This is accomplished by transferring test organisms or replacing test solution.

Flow-through test—test in which solution is replaced continuously in test chambers throughout the test duration.

5. Evaluation of Results

Maximum allowable toxicant concentration (MATC)—toxicant concentration that may be present in a receiving water without causing significant harm to productivity or other uses. MATC is determined by long-term tests of either partial life cycle with sensitive life stages or a full life cycle of the test organism.

Chronic value (ChV)—geometric mean of the NOEC and LOEC from partial-and full-life-cycle tests and early-life-stage tests.

Acute-to-chronic ratio—numerical relationship between acute and chronic toxicity that is applied to acute toxicity test values to estimate toxicant concentration that is safe for chronic or long-term exposure of a test organism.

6. Bibliography

STEPHAN, C.E., D.I. MOUNT, D.J. HANSEN, J.H. GENTILE, G.A. CHAPMAN & W.A. BRUNGS. 1985. Guidelines for deriving numerical national water quality criteria for the protection of aquatic organisms and their uses. NTIS PB85-227049, U.S. Environmental Research Laboratories, Duluth, Minn.; Narragansett, R.I.; and Corvallis, Ore.

U.S. ENVIRONMENTAL PROTECTION AGENCY. 1991. Technical Support Document for Water Quality-Based Control. EPA-505/2-90-001 (PB91-127415), Off. Water, U.S. Environmental Protection Agency, Washington, D.C.

KLEMM, D.J., G.E. MORRISON, T.J. NORBERG-KING, W.H. PELTIER & M.A. HEBER, eds. 1994. Short-Term Methods for Estimating the Chronic Toxicity of Effluents and Receiving Waters to Marine and Estuarine Organisms, 2nd ed. EPA-600/4-91-003, Environmental Monitoring & Support Lab., U.S. Environmental Protection Agency, Cincinnati, Ohio.

LEWIS, P.A., D.J. KLEMM, J.M. LAZORCHAK, T.J. NORBERG-KING, W.H. PELTIER & M.A. HEBER, eds. 1994. Short-term Methods for Estimating the Chronic Toxicity of Effluents and Receiving Waters to Freshwater Organisms, 3rd ed. EPA-600/4-91-002, Environmental Monitoring Systems Lab., U.S. Environmental Protection Agency, Cincinnati, Ohio.

RAND, G.M., ed. 1995. Fundamentals of Aquatic Toxicology, 2nd ed. Taylor and Francis, Washington, D.C.

8010 C.   Basic Requirements for Toxicity Tests

1. General Requirements

The basic requirements and desirable conditions for toxicity tests are: *(a)* an abundant supply of water of desired quality (see 8010E.4*b*), *(b)* an adequate and effective flowing water system constructed of nonpolluting or absorbing materials (see 8010F.1*a*), *(c)* adequate space and well-planned holding, culturing, and testing equipment and facilities (see 8010E.3), *(d)* an adequate source of healthy experimental organisms (see 8010E.4), and *(e)* appropriate lighting facilities for plant toxicity tests. Much valuable information and advice regarding general requirements and desirable conditions for toxicity testing are available.[1-9]

2. Requirements for Specific Test Purposes

The facilities, equipment, and water supplies needed for effective tests depend on the type of tests and their objectives.[6] For effluent and monitoring compliance tests requiring receiving water as the dilution water, use water immediately upstream and outside the zone of influence of the waste. When studies require the use of laboratory-grade water, use a water supply free from pollution and one that provides for acceptable survival, growth, and reproduction of the aquatic test organisms to be studied. The most important requirements for designing a toxicity testing program are defining objectives of the study and establishing quality control practices, to ensure that the data are of sufficient quality to address the objectives and to ensure credibility.

3. References

1. U.S. DEPARTMENT OF COMMERCE. 1970. Aquarium Design Criteria, special ed. National Fisheries Center Aquarium.
2. CLARK, J.R. & R.L. CLARK, eds. 1964. Sea Water Systems for Experimental Aquariums. U.S. Fish & Wildl. Serv. Bur. Sports Fish & Wildl. Res. Rep. 63, U.S. Government Printing Off., Washington, D.C.
3. SPOTTE, S. 1973. Marine Aquarium Keeping—The Science, the Animals, the Art. Wiley Interscience Publ., New York, N.Y.
4. LASKER, R. & L.L. VLYMER. 1969. Experimental Seawater Aquarium. U.S. Fish & Wildl. Serv. Bur. Commercial Fish. Circ. 334, U.S. Government Printing Off., Washington, D.C.
5. TARZWELL, C.M. 1962. Development of water quality criteria for aquatic life. *J. Water Pollut. Control Fed.* 34:1178.
6. WEBER, C.I., ed. 1993. Methods for Measuring the Acute Toxicity of Effluents and Receiving Waters to Freshwater and Marine Organisms, 4th ed. EPA-600/4-90-027F, Environmental Monitoring and Support Lab., U.S. Environmental Protection Agency, Cincinnati, Ohio.
7. DENNY, J.S. 1987. Guidelines for The Culture of Fathead Minnows *Pimephales promelas* for Use in Toxicity Tests. EPA-600/3-87-001, Environmental Research Lab., U.S. Environmental Protection Agency, Duluth, Minn.
8. STURGIS, T.C. 1990. Guidance for Contracting Biological and Chemical Evaluations of Dredge Material. Tech. Rep. D-90, Dredging Operations Technical Support Program, U.S. Army Corps of Engineers, Vicksburg, Miss.
9. AMERICAN SOCIETY FOR TESTING AND MATERIALS. 1993. 1993 Annual Book of ASTM Standards, Section 11 Water and Environment Technology. Volume 11.04 Pesticides; Resources Recovery; Hazardous Substances and Oil Spill Responses; Waste Disposal; Biological Effects. American Soc. Testing & Materials, Philadelphia, Pa.

8010 D. Conducting Toxicity Tests

1. Types of Toxicity Tests: Their Uses, Advantages, and Disadvantage

Toxicity tests are classified according to (a) duration—short-term, intermediate, and/or long-term, (b) method of adding test solutions—static, renewal, or flow-through, and (c) purpose—effluent quality monitoring, single compound testing, relative toxicity, relative sensitivity, taste or odor, or growth rate, etc.

Short-term toxicity tests are used for routine monitoring suitable for effluent discharge permit requirements and for exploratory tests. They may use end points other than mortality. Acute definitive tests typically use mortality as an end point or other discrete observations to determine effects due to the toxicant (i.e., LC50 or EC50 values). These tests also may be used to indicate a suitable range of toxicant concentrations for intermediate and long-term tests. Short-term tests, rather than longer-duration tests, are used to obtain toxicity data as rapidly and inexpensively as possible. They are valuable for estimation of overall toxicity, for screening test solutions or materials for which toxicity data do not exist, for assessing relative toxicity of different toxicants or wastes to selected test organisms, or for relative sensitivity of different organisms to different conditions of such variables as temperature and pH. The results of these tests can be used to calculate acceptable concentrations for very short exposures, such as those that might occur as organisms pass through an effluent zone of initial dilution or a mixing zone.

Toxicity tests of intermediate duration typically are used when longer exposure durations are necessary to determine the effect of the toxicant on various life stages of long-life-cycle organisms, and to indicate toxicant concentrations for life-cycle tests.

Long-term toxicity tests are generally used for estimating chronic toxicity. Long-term testing may include early-life-stage, partial-life-cycle, or full-life-cycle testing. Exposures may be as short as 7 d to expose specific portions of an organism's life cycle, 21 to 28 d to several months or longer for traditional partial-life-cycle and full-life-cycle tests with fish.

To establish a successful testing program, consider the following: Use caution when static tests are used for evaluation of solutions with high BOD and/or COD levels or high bacterial populations. These tests can be conducted successfully with incorporation of rigorous dissolved oxygen monitoring and acceptable aeration. Volatile or unstable toxicants may decrease in concentration during the test, resulting in an underestimation of the exposure concentration causing an effect on the test organisms. Metabolic products, such as ammonia, may increase to undesirably high concentrations resulting in stress or death of test organisms and overestimation of the concentration that causes a toxic response. Toxicant concentration may be reduced by sorption on sediments, test chamber walls, by the food provided for the test organisms, or by combination with the mucus or metabolic products of the test organisms and in their bodies.

Flow-through toxicity tests are desirable for high-BOD or COD samples and for those that contain unstable or volatile substances. Organisms with high metabolic rates are difficult to maintain under static exposure conditions, whereas flow-through tests provide well-oxygenated test solutions and continuous removal of metabolic wastes. Use flow-through toxicity tests whenever there is evidence or expectation of rapid degradation of the test solution.

Such a change is indicated when the survival time of test animals in a fresh solution is significantly shorter than in a corresponding 2-d-old solution (provided that adequate DO is present throughout both tests). Flow-through toxicity tests are also desirable for industrial effluents and chemicals that are removed appreciably from solution by precipitation, by test organisms, or by other means.

The LC50 values may be useful measures of acute toxicity but they do not represent concentrations that are safe or harmless in aquatic habitats. Concentrations of wastes that are not demonstrably toxic in 96 h may be toxic at longer exposure periods in a receiving water. Thus the 96-h LC50 may represent only a fraction of long-term toxicity. When estimating safe discharge rates or dilution ratios for effluents or other pollutants on the basis of acute toxicity evaluations, use acute-to-chronic ratios determined primarily from life-cycle tests; however, NOEC values determined from shorter-duration chronic toxicity tests can be used. Even the provision of an apparently ample margin of safety can fail to accomplish its purpose when there is cumulative toxicity that cannot be predicted from acute toxicity results.

No single, simple acute-to-chronic ratio is valid for all wastes or toxicants. However, research on effluents has shown that acute-to-chronic ratios for whole effluents often are around 10. An acute-to-chronic ratio of 20 commonly has been used for non-persistent chemicals while a factor of 100 has been used for persistent chemicals. The constituents of a complex waste responsible for acute toxicity may be, but are not necessarily, the constituents responsible for chronic or cumulative toxicity demonstrable in diluted waste that is no longer acutely toxic. The chronic toxicity may be lethal after a long exposure period or it may only cause impairment of function. Knowledge of acute toxicity of a waste often can be very helpful in predicting and preventing acute damage to aquatic life in receiving waters as well as in regulating toxic waste discharges.

2. Short-Term Toxicity Tests

a. Range-finding toxicity tests: For effluents or materials of unknown toxicity conduct short-term (usually 24-h or 48-h), small-scale range-finding or exploratory tests to determine approximate concentration range to be included in definitive short-term tests. For effluents with low or slow-acting toxicity, 48- or 96-h tests may be necessary. Expose test organisms to a wide range of concentrations of the test substance, usually in a logarithmic ratio, such as 0.01, 0.1, 1, 10, and 100% of the sample. Attempt to include concentrations that will kill all organisms and others that will kill very few or no organisms. For short-term, definitive tests, select a geometrically spaced series of concentrations between the highest concentration that killed no, or only a few, test organisms and the lowest concentration that killed most or all test organisms.

Prepare test concentrations as described in Section 8010F.2b.

b. Short-term definitive tests: Because death is an important, easily detected adverse effect, the most commonly used tests are for acute lethality. These tests are most appropriate for routine monitoring and checking conformity with NPDES requirements.[1] If it is not possible to perform a range-finding toxicity test before a definitive acute toxicity test, using a concentration series with a 0.5 (100, 50, 25, 12.5, 6.25%) or 0.3 (100, 30, 10, 3, 1%) dilution factor may be appropriate. Short-term tests may be static, renewal, or flow-through. Exposure periods for these tests usually are 48 h or 96 h. Static or renewal tests often are used when the test organisms are phyto- or zooplankton because these organisms are easily washed out in flow-through tests. Static and renewal tests are considerably less expensive to perform than flow-through tests. Overnight express mail shipments of samples often make static and renewal tests the method of choice for regulatory compliance testing.

Test solutions may be renewed daily if required because of oxygen demand, if the toxicant is unstable or volatile, or in the case of whole effluents, daily variation in the composition of the effluent. Renewals also may be less frequent. If the test material has high BOD and/or COD level or is relatively unstable, use test vessels with maximum surface area-to-volume ratio, or use the renewal or flow-through technique.

Test duration is determined by the toxicant and the test objectives and usually is the same for different groups of organisms. For short-life-cycle organisms such as phytoplankton, the usual exposure time can cover many generations. Determine test duration, in part, by the length of the life cycle. Generally, expose fish and large invertebrates in static and static renewal tests for 96 h and in flow-through tests for an equal period unless composition of the toxicant is variable. In this case longer exposure may be useful to assess impacts of toxicant variability. Expose *Daphnia* and *Ceriodaphnia* for 48 h. Short-term tests have been limited arbitrarily to 96 h, but longer tests sometimes are desirable because death does not always occur within the 48-h or 96-h period. When some test animals, though still alive, are dying or evidently affected after 96-h exposure, prolong the test or express the results of the test as a 48-h or 96-h EC50, defining the observed effect. If tests are continued for longer periods, the test organisms may need to be fed.

Feed test organisms as directed in specific sections of Part 8000. Record feeding and ensure that it is equivalent in each container.

Special tests may be conducted on altered or treated samples of effluent to obtain additional toxicity information. For example, effluent dilution water mixtures may be aged 24 to 48 h before adding the test organisms, to determine changes in toxicity. When special tests are conducted, describe methods in detail.

3. Intermediate-Term Toxicity Tests

No sharp time separation exists between short- and intermediate- or between intermediate- and long-term tests. Usually tests lasting 10 d or less are considered short-term while intermediate tests may last from 11 to 90 d. The length of the test organism's life cycle helps to determine what is short-term, intermediate, or long-term for that species. Intermediate-length tests may be static, renewal, or flow-through, but flow-through tests are recommended for most situations. For conduct of tests see Section 8010F.3*a*.

4. Long-Term, Partial- or Complete-Life-Cycle Toxicity Tests

With few exceptions, use flow-through tests with exposure extending over as much of the life cycle as possible. Continue tests from egg to egg or beyond, or for several life cycles for smaller forms. Determine the maximum concentrations of toxicant not producing harmful effects with continuous exposure. The overall objective of this type of test is to determine NOECs or chronic value (ChV) of effluents, toxicants, or wastes. Use life-cycle tests whenever possible to determine acute-to-chronic ratios and the effects on growth, reproduction, development of sex products, maturation, spawning, success of spawning and hatching, survival of larvae or fry, growth and survival of different life stages, deformities, behavior, and bioaccumulation, although bioaccumulation (or bioconcentration) often is determined with more mature animals in specially designed tests.[2]

In life-cycle or partial-life-cycle tests, ensure that water quality factors such as temperature, pH, salinity, and DO follow the natural seasonal cycle unless the test objective is to study one of these factors. It may be essential that the natural annual cycle be duplicated if the development of sex products, spawning, and development of eggs and larvae are to be normal. Whenever possible, do not let toxicant concentrations vary by more than ±15% from the selected concentration because of uptake by test organisms, absorption, precipitation, or other factors.

In these tests, select five or more concentrations on the basis of short- or intermediate-term tests and set up the exposure chambers at least in duplicate. Vary exposure chambers, spawning chambers, and other equipment to meet the needs of the different organisms. (See Sections 8111 through 8910.) Other apparatus, water supplies, and analytical determinations are listed in Section 8010E.

5. Short-Term Tests for Estimating Chronic Toxicity

Tests are available to estimate long-term effects of a toxicant or effluent after a relatively short (7 d) exposure. End points for the tests, called chronic estimator or rapid bioassessment tests, include lethality, reproductive potential, and growth. Tests estimating chronic toxicity frequently are being included as biomonitoring requirements in discharge permits. The long duration of life-cycle or early-life-stage chronic tests increases the cost and reduces ability of laboratories to conduct long-term tests successfully as the demand for testing increases. The EPA has published a number of short-term chronic estimation toxicity test methods for fresh- and saltwater invertebrates and fishes.[3,4] These tests were designed to evaluate effluent toxicity and may not be appropriate for other testing requirements such as pre-manufacturing testing, development of water quality criteria, etc.

6. Special-Purpose Toxicity Tests

a. Relative sensitivity to a toxicant: To rank the sensitivity of different species to a toxicant, use a standard water and standard exposure conditions. Select exposure conditions (e.g., temperature, DO, pH, CO_2, light, and salinity) in a favorable range for the test species and keep conditions constant throughout the test.

b. Relative sensitivity of various toxicants to selected species: These tests resemble sensitivity tests because the selected test conditions, dilution waters, and test species are kept constant and standard. Prevent any change in sensitivity of test organisms during the tests. If possible, select species from several different groups: an alga, microcrustacean, macrocrustacean, insect, mollusk, or fish.

c. Toxicity reduction evaluation: Use acute and chronic toxicity tests to determine the toxicant in the effluent. A Toxicity Reduction Evaluation (TRE) is a phased approach that first, characterizes the acute or chronic toxicity of an effluent, second, identifies the toxicant(s) of concern (this phase often is termed Toxicity Identification Evaluation, TIE), and third, confirms toxicity. This approach is then used to evaluate the removal of the toxicant(s) by pretreatment or changes at the wastewater treatment plant. EPA guidance manuals for performance of TIEs[5-8] and generalized TRE protocols for municipal[9] and industrial[10] facilities are available. Other TRE protocols are available.[11-15]

d. Flesh tainting tests: CAUTION: *Perform such tests only when there is assurance that the intake of potentially tainting substances through consumption of the organism is safe. In cases where sufficient information on the substances is not available, replace consumption with smell.* Use these tests to determine the maximum concentrations of wastes and materials that do not taint the flesh of edible aquatic organisms. Expose organisms that are large enough to supply portions for a taste panel. Set up exposure tanks as for other flow-through tests. Perform range-finding tests over a wide concentration range to determine the concentrations for a more definitive series of tests.

After exposure, prepare test organisms for taste testing. Clean, prepare for cooking (without seasoning), wrap in aluminum foil, and bake in an oven. When organisms are cooked, divide them into portions, wrap in aluminum foil, assign a code number, and distribute to a taste panel while still warm, along with samples of unexposed organisms similarly cooked, wrapped, and coded. Record the observations of the panel on a prepared form and determine the highest concentration of test material not causing detectable tainting based on either taste or smell. Several tests may be necessary.

e. Growth-rate determinations: Growth rate is an important response of both algae and fish to toxicants and environmental factors. This section discusses the topic with respect to fish.[21] For a discussion related to algae see Section 8111G.3c. Always report details of the method of feeding fish in growth studies. Three techniques are available:

Unrestricted food supply—Provide attractive and palatable food (usually live food such as *Daphnia*, tubificid worms, or brine shrimp) continuously in greater quantities than fish can consume. It is desirable to make a mass balance on food consumed by weighing food introduced and uneaten food removed.

Intermittent satiated food supply—Provide all the attractive food that fish can consume at time of feeding once or twice daily. After fish cease to feed, remove all uneaten food.

Uniformly restricted food supply—Once or twice per day, provide all fish with an amount of food that they will consume completely and without exception. Ideally, hold fish separately in individual aquariums or compartments. For fish held together, feed so that all fish have an equal opportunity to consume food. Uniformity of temperature and DO helps to ensure equal feeding of a group of fish.

While growth studies usually have been conducted with unrestricted and intermittent satiated feeding techniques, it is recommended that each study include at least one test series using uniformly restricted food supply. Only this technique can reveal whether growth rate differences are not the result of the effect of the toxicant on appetite or food consumption rate. The presence of an abundant food supply can obscure toxic effects. For example, fish exposed to toxicants such as cyanide or pentachlorophenol increase food consumption rate to compensate partially for loss of efficiency of food utilization caused by the toxicant. This may not be possible in natural conditions where food supply may be limited.

Ideally, include a series of tests with different, uniformly restricted food rations with the lowest ration near that which results in no growth (or loss of weight) in the control. This is the maintenance level. Determine the effect of the variable under study at any level of food availability and consumption by relating observed growth rates to, for example, toxicant concentration, at each feeding level.

Juvenile fish may gain enough weight in 1 to 3 weeks to determine growth rate satisfactorily. Longer exposures with weighings at intervals of approximately 10 d are needed to determine long-term effects such as acclimation or accumulative toxicity.

Report results as specific growth rates computed as follows:

$$\text{Growth rate} = \frac{\text{Weight gain, g}}{\text{time interval, d}} \times \frac{1}{\text{mean weight, g}}$$

where:

mean weight = [weight at start of time interval (g) ÷ weight at end of time interval (g)] ÷ 2

Determine dry weight, wet weight, and fat (lipid) content of fish at the beginning and end of a test. Weight gain due to increased fat content is not universally considered true growth; some investigators consider that true growth occurs only when there is an increase of protein. However, fat storage is important ecologically and bioenergetically because fat can be used as an energy source during periods of malnutrition, reproduction, and overwintering survival. Fat content also is important in the dynamics of toxicant uptake, storage, and depuration. Fat and water content should typify that of the target species.

7. References

1. WEBER, C.I., ed. 1993. Methods for Measuring the Acute Toxicity of Effluents and Receiving Water to Freshwater and Marine Organisms, 4th ed. EPA-600/4-90-027F, Environmental Monitoring Systems Lab., U.S. Environmental Protection Agency, Cincinnati, Ohio.

2. AMERICAN SOCIETY FOR TESTING AND MATERIALS. 1996. Standard practice for conducting bioconcentration tests with fishes and saltwater bivalve molluscs. E-1022-94, Annual Book of ASTM Standards, Vol. 11.05. American Soc. Testing & Materials, W. Conshohocken, Pa.

3. LEWIS, P.A., D.J. KLEMM, J.M. LAZORCHAK, T.J. NORBERG-KING, W.H. PELTIER & M.A. HEBER, eds. 1994. Short-Term Methods for Estimating the Chronic Toxicity of Effluents and Receiving Waters to Freshwater Organisms, 3rd ed. EPA-600/4-91-002, Environmental Monitoring Systems Lab., U.S. Environmental Protection Agency, Cincinnati, Ohio.

4. KLEMM, D.J., G.E. MORRISON, T.J. NORBERG-KING, W.H. PELTIER & M.A. HEBER, eds. 1994. Short-term Methods for Estimating the Chronic Toxicity of Effluents and Receiving Waters to Marine and Estuarine Organisms, 2nd ed. EPA-600/4-91-003, Environmental Monitoring and Support Lab., U.S. Environmental Protection Agency, Cincinnati, Ohio.

5. NORBERG-KING, T.J., D.I. MOUNT, E. DURHAN, G. ANKLEY, & L. BURKHARD. 1991. Methods for Aquatic Toxicity Identifications: Phase I Toxicity Characterization Procedures. EPA-600/6-91-003. Environmental Research Lab., U.S. Environmental Protection Agency, Duluth, Minn.

6. MOUNT, D.I. & L. ANDERSON-CARNAHAN. 1989. Methods for Aquatic Toxicity Identification Evaluations: Phase II Toxicity Identification Procedures. EPA-600/3-88-035, Environmental Research Lab., Off.

Research and Development, U.S. Environmental Protection Agency, Duluth, Minn.

7. MOUNT, D.I. 1989. Methods for Aquatic Toxicity Identification Evaluations: Phase III Toxicity Confirmation Procedures. EPA-600/3-88-036, Environmental Research Lab., Off. Research and Development, U.S. Environmental Protection Agency, Duluth, Minn.

8. NORBERG-KING, T.J., D.I. MOUNT, J. AMATO, D. JENSEN & J. THOMPSON. 1991. Toxicity Identification Evaluation: Characterization of Chronically Toxic Effluents, Phase I. EPA-600/6-91-005. Environmental Research Lab., U.S. Environmental Protection Agency, Duluth, Minn.

9. BOTTS, J.A., J.W. BRASWELL, J. ZYMAN, W.L. GOODFELLOW & S.B. MOORE. 1989. Toxicity Reduction Evaluation Protocol for Municipal Wastewater Treatment Plants. EPA-600/2-88-062, Risk Reduction Engineering Lab., Off. Research and Development, U.S. Environmental Protection Agency, Cincinnati, Ohio.

10. FAVA, J.A., D. LINDSAY, W.H. CLEMENT, G.M. DeGRAEVE, J.D. COONEY, S. HANSEN, W. RUE, S. MOORE & P. LANNFORD. 1989. Generalized Methodology for Conducting Industrial Toxicity Reduction Evaluations. EPA-600/2-88-070, Risk Reduction Engineering Lab., Off. Research and Development, U.S. Environmental Protection Agency, Cincinnati, Ohio.

11. WALSH, G.E. & R.L. GARNAS. 1983. Determination of bioactivity of chemical fractionations of liquid wastes using freshwater and saltwater algae and crustaceans. Environ. Sci. Technol. 17:180.

12. DOI, J. & D.R. GROTHE. 1989. Use of fractionation/chemical analysis schemes for plant effluent toxicity evaluations. In G.W. Suter II & M.A. Lewis, eds. Aquatic Toxicology and Environmental Fate: Eleventh Volume. ASTM STP 1007, American Soc. Testing & Materials, Philadelphia, Pa., p. 123.

13. PARKHURST, B.R., C.W. GEHRS & I.B. RUBIN. 1979. Value of chemical fractionation for identifying the toxic components of complex aqueous effluents. In L.L. Marking & R.A. Kimerle, eds. Aquatic Toxicology. ASTM STP 667, American Soc. Testing & Materials, Philadelphia, Pa., p. 122.

14. GOODFELLOW, W.C., JR., W.C. McCULLOCH, J.A. BOTTS, A.G. McDEARMON & D.F. BISHOP. 1989. Long-term multispecies toxicity and effluent fractionation study at a municipal wastewater treatment plant. In G.W. Suter II & M.A. Lewis, eds. Aquatic Toxicology and Environmental Fate; Eleventh Volume. ASTM STP 1007, American Soc. Testing & Materials, Philadelphia, Pa., p. 139.

15. SAMOLLOFF, M.R., et al. 1983. Combined bioassay-chemical fractionation scheme for the determination of toxic chemicals in sediments. Environ. Sci. Technol. 17:329.

8010 E. Preparing Organisms for Toxicity Tests

1. Selecting Test Organisms

The prime considerations in selecting test organisms are: their sensitivity to the factors under consideration; their geographical distribution, abundance, and availability within a practical size range throughout the year; their recreational, economic, and ecological importance and relevance to the purpose of the study; their abiotic requirements and whether these requirements approach the conditions normally found at the study site; the availability of culture methods for rearing them in the laboratory and a knowledge of their physiological and nutritional requirements; and their general physical condition and freedom from parasites and disease. To select a best species consider available information on sensitivity, consult with local authorities in pollution control or fish and wildlife agencies, or determine sensitivity with short-term tests. Select the test species based on the considerations listed above as well as organism size and life-cycle length. Sections 8110 through 8910 list plant, invertebrate, and fish species that are commonly used in aquatic toxicity testing. For testing of early life stages of organisms, species having a short life cycle are most cost-effective, but some tests require larger organisms with long life cycles (e.g., bioaccumulation or in situ biomonitoring studies).

For studies to determine effluent effects, select species representative in the area impacted. In most cases, use of laboratory-cultured species is preferable to use of those collected from the field. Laboratory-cultured organisms, either from in-house cultures or purchased from commercial bioassay organism suppliers, are of known age and quality (those taken from the field may already have been selected or biased for the more resistant members of the population). This allows for use of the most sensitive life stages throughout the year. Their use also may be more cost-effective and allows for better quality assurance and control. For each series of tests, use organisms from a single source. Choose organisms that are nearly uniform in size, and for fish, with the largest individual not more than 50% longer than the shortest. Use organisms of the same age group or life stage. Optimally, conduct reference toxicant tests on cultured stocks and on lots of acquired or collected organisms.[1] Report time, place, source, and culture history for cultured organisms and method of collection, transportation, handling, and acclimation of acquired or collected organisms, and their response to reference toxicants.

Knowledge of their environmental requirements and food habits is important in selecting test organisms. Methods for laboratory holding and culturing are well described for a number of standard test species. When the purpose of the testing is site-specific, it may be necessary to collect certain life stages of selected organisms from the field for testing.

2. Collecting Test Organisms

Preferably, use standard laboratory-reared organisms. In special instances, such as when it is important to incorporate genetic variability from wild populations, use species indigenous to the receiving water. This is particularly important for organisms with recreational, commercial, or ecological significance.

In designing a test, consider any unusual past conditions to which the organisms may have been exposed (pesticides, effluents from industries, waste treatment plants, return flows, etc.). Interactive effects of a new toxicant mixed with those presently being discharged to the receiving water may be important. Do not collect test organisms from polluted areas where they are in poor condition, diseased, parasitized, or deformed, or where they have unusually high body burdens of chemicals. Avoid testing organisms with questionable histories.

Many smaller invertebrates and fish can be collected along the shore in dip nets, in coarse plankton nets, or by hand. Catch larger species that occur near shore in seines. Traps, fyke nets, and trawls are valuable tools for collecting organisms, but may be selective for some species. Otter trawls are effective for collecting benthic species and midwater trawls for pelagic species. Various dredges are available to collect benthic species from different types of bottoms or to collect different sizes of organisms. Commercially important species such as lobster, blue crab, and dungeness crab may be taken by the researcher in traps or deep-water trawls or may be purchased from commercial suppliers if proper care is taken before purchase. Species that colonize surfaces, such as barnacles, may be harvested from hard surfaces submerged in the water. Insure that organisms are not damaged during collection, transfer, and transport.

When seining or using trawls, make short hauls. Avoid collecting significant amounts of plant materials, debris, mud, sand, or gravel in net or in bag of seine because these will injure the animals. Always leave seine bag in the water at end of haul, stretch out wings of seine, open bag entrance, dip out organisms with a bucket or hand net, and transfer directly to prepared holding tanks. Do not expose delicate, easily damaged species to air. Take out larger, more hardy species with soft mesh dip nets. Do not collect too many animals at one time. After bringing a trawl up to the boat, bring it over the side without delay and avoid letting the catch hit the boat. Immerse that portion of net containing specimens in a tank of water. Open trawl and remove desired animals by dipping with a bucket or a hand net with small soft mesh. Have adequate quantities of clean water available in tanks before beginning a haul. Transfer organisms to tanks as rapidly and carefully as possible.

If organisms are to be transported any distance by boat, hold in aerated live boxes. If they are transported by truck, put them in large baffled and insulated tanks filled with water from area in which they were collected. Aerate the water and maintain at temperature of collection. Determine water temperature, salinity, DO, and pH at the collecting site. Do not handle organisms more than necessary. Make transfers with suitable containers or hand nets, or for small organisms, by large-bore pipets. Use hand nets made of soft material with several layers around the net rim and free from sharp points or projections. Clean and sterilize all equipment before use. Avoid overcrowding of organisms during transport. Aeration, oxygenation, water exchange, and cooling may reduce distress. Avoid cold shock as much as overheating.

Observe collected animals for possible injury resulting from transport to the laboratory. Examine smaller forms under a dissecting microscope. Criteria for assessing injury depend on the species and are more difficult for sluggish ones. Useful criteria include lost or damaged appendages, inability to maintain a normal body posture (e.g., dorsal side uppermost), abnormal locomotion, refusal to feed, discoloration, or uncoordinated movements of the mouth or other body parts.

For additional information on collecting aquatic organisms, see Part 10000.

3. Handling, Holding, and Conditioning Test Organisms

During transport to the laboratory, organisms sometimes are crowded, bruised, and otherwise stressed, thereby increasing their susceptibility to disease. To avoid outbreaks of disease in stock tanks, treat organisms during transit or on arrival in accordance with procedures in ¶ 5 below and as suggested for each of the different groups (Sections 8210 through 8910). Hold field-collected fish in quarantine for at least 7 d to observe for parasites and disease, and to recover from collection and transport stress; observe invertebrates for at least 2 d. If more than 10% of the collected animals die after the second day or if they are parasitized or diseased beyond control, do not use them. Clean and sterilize all contacted containers and equipment and collect another supply from a different area if possible.

Because it is not always possible to collect from unpolluted areas and the collector cannot always be sure that a particular organism has not been exposed to a toxicant, for certain types of tests it may be necessary to sample collected individuals to determine if they have accumulated pesticides, heavy metals, and toxic materials to be studied. Check animals or materials collected as food for test organisms for disease and content of pesticides, heavy metals, and toxic materials to be studied. Feed test organisms daily during quarantine.

After quarantine period, transfer disease-free animals to regular stock tanks. Discard organisms that touch dry surfaces, are dropped, or are injured during handling. To avoid unnecessary stress, do not subject organisms to rapid temperature or water-quality changes. In general, change water temperature less than 3°C in any 24-h period. For stenothermal, deep-water species use an even smaller rate of temperature change. Preferably keep DO concentrations at or near saturation but never at less than 60% or greater than 100% of saturation. After transfer to stock holding tanks, begin a slow acclimation to laboratory conditions such as temperature, salinity, and hardness. The period of acclimation will be governed by type of organism and extent of changes in water quality. For forms with a life cycle of several months or more, use an acclimation period of at least 2 to 3 weeks.

Inspect organisms closely and frequently to determine stress, unusual behavior, parasites or disease, changes in color, or failure to eat. Avoid crowding. Provide adequate flow-through water so that characteristics such as DO, pH, CO_2, salinity, hardness, and NH_3 are favorable. Check temperature and DO frequently. Do not let metabolic products accumulate. Generally, use a flow-through rate of 6 to 10 tank volumes/d. Usually, greater amounts of flow-through water are required for smaller organisms on a weight-volume basis. For small organisms, use a water flow of at least 3 L/d/g. When brood stock are being held, periodic or continuous treatment for parasite and disease control may be required.[2-10]

Clean tanks and equipment thoroughly and often, removing or flushing out all growths and wastes, preferably daily but at least twice per week. Remove all uneaten food within 24 h. Use different sets of nets and other equipment for different groups of organisms and clean and sterilize them between uses. When handling is necessary, clean hands and nets before touching organisms. Cover tanks and containers to prevent organisms from jumping out. Shield tanks with curtains or by some other means to protect organisms from unnecessary disturbances and noise. Provide photoperiods and light intensities favorable to the organisms (see Section 8010F.3f). Begin acclimation to test conditions at a suitable interval in advance of testing.

It is of utmost importance that animals be kept in excellent condition before the tests. Make no abrupt changes in environmental conditions; preferably follow natural seasonal variations in environmental conditions such as temperature and daylight patterns. Many water supplies are supersaturated with gases, especially in winter when very cold water is brought into the laboratory and warmed. Because there is a danger of gas bubble disease, keep incoming water in an open system and let it cascade over baffles or otherwise aerate it to bring dissolved gases into equilibrium with the air or strip supersaturated nitrogen with pure oxygen.[11]

Acclimate freshwater arthropods by rearing them in the dilution water at the test temperatures, unless temperature is one of the factors being studied. Acclimate other organisms to the dilution water and test temperatures by gradually changing the water from 100% holding water to 100% dilution water over a period of several days. Keep all organisms in 100% dilution water for at least 2 d before use. Do not use a group of organisms if more than 10% die during the 48 h immediately before the beginning of the test.[1] If a group fails to meet these criteria discard or re-treat, hold, and reacclimate if necessary.

Make necessary provisions for organisms that require a special substrate, cover, or materials to use for clinging, support, the building of cases, or hiding.

Hold cold-water, freshwater organisms between 5 and 15°C. Hold warm-water organisms between 10 and 25°C, depending on season and test objectives.

4. Culturing Test Organisms

The advantage of cultured test organisms over field-collected animals is that the age, life history, and existing conditions are documented and thus the responses of these organisms are more consistent between tests lots. For organisms used extensively in effluent biomonitoring programs, EPA has developed a series of test methods[12,13] that are adaptable to most laboratories. Culturing test organisms requires strict adherence to standard protocol, 7-d/week monitoring, and an adequate facility.

a. Facilities, construction materials, and equipment: Do not use construction materials in contact with dilution water that contain leachable substances or adsorb significant amounts of substances from the water. Use tempered glass, fiberglass, or stainless steel (No. 316) and silicone sealant as construction material for fresh-water systems. Do not use rubber or plastics containing toxic fill-ers, additives, stabilizers, plasticizers, etc. Fluorocarbon plastic, nylon, and their equivalents usually are acceptable. Test the tox-icity of all materials before purchasing large quantities. Clean, soak, and flush all new tanks, troughs, and similar equipment with dilution water for several days before use. Use a glass or titanium interface between the water and heating elements for marine wa-ters and glass or stainless steel for fresh waters.

Provide adequate space for test organisms, holding facilities, water storage reservoirs, and water supply systems. Provide dis-tribution of hot and cold water and mixing facilities to obtain any desired temperature. Aerate or vigorously mix to prevent gas su-persaturation caused by heating dilution water.[4] Use oil-free pumps if possible. If air pumps are not oil-free, they should have water seals and filters to prevent oil from entering air lines and contaminating tanks. When large volumes of air are needed, use low-pressure blowers. Do not locate air intakes in shops or fur-nace rooms or near outlets from chemical exhaust hoods, chemical laboratories, or vehicle exhausts.

Provide acclimation and culturing tanks with temperature con-trol and aeration. Design holding facilities for ease of cleaning and prevention of bacterial growths. For holding and culturing fish and many macroinvertebrates, preferably use round or oval tanks of at least 1 to 3 m diam (Figure 8010:1). Provide a stand-pipe drain in the center, threaded below the tank floor so that, when the standpipe is removed, the opening is flush with the tank bottom. Slope tank bottom gently to center. Use tanks with

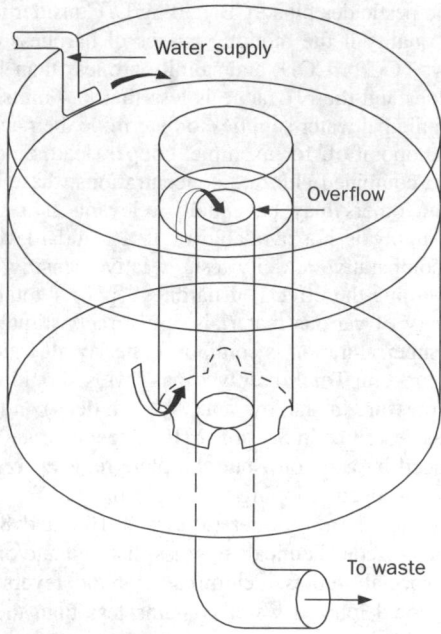

Figure 8010:1. Holding tank design for fish and macroinvertebrates.

smooth surfaces to facilitate cleaning, to prevent injuries to or-ganisms, and to insure that no material will collect in corners, cracks, and crevices. Introduce water into a circular tank as a jet along the edge and above the surface to create a circular move-ment of water around the central standpipe. Fit another pipe, with half-moon cutouts at its base, over the standpipe and screen, so that the outflowing water passes up through the outside pipe, then down the standpipe. This results in a circular current and a certain amount of self-cleaning. Square or rectangular tanks may be used for special purposes or when space is scarce. Provide standpipes at one end for draining, with threads for securing the pipe on the underside. Ensure that tank corners are rounded and that surfaces are smooth.

b. Water supply: Provide a flowing water system for holding, spawning, and rearing a variety of aquatic organisms. In general, reconstituted fresh water and artificial seawater are not cost-ef-fective for large-scale rearing or for flow-through tests. If avail-able, use natural, unpolluted fresh or salt water supplies that have low turbidity, high DO, low BOD, and an annual temperature cycle that approximates that of the test organisms. If a flow-through water supply is not available, use of reconstituted fresh water or artificial salt water in recirculating culture systems with biological filters may be acceptable.

1) *Freshwater supplies*—A good freshwater supply is constant in quality, provides for adequate survival, growth, and reproduc-tion of test organisms, and does not contain more than the des-ignated amounts of the following: suspended solids, 20 mg/L; total organic carbon (TOC), 2 mg/L; chemical oxygen demand (COD), 5 mg/L; un-ionized NH_3, 20 μg/L; total residual chlorine, 0.01 mg/L; total organophosphorus pesticides, 50 ng/L;* total or-

---

* No individual pesticide should exceed the allowable concentration limit set in the National Water Quality Guidelines, EPA, as set in accordance with the Federal Water Pollution Control Act 92–500 as amended 1972.

ganochlorine pesticides plus PCBs, 50 ng/L. Consider water to be of constant quality if the monthly ranges of hardness, alkalinity, conductivity, TOC or COD, and salinity are less than 10% of the average values and the pH range is less than 0.4 units.

Check municipal water supplies to determine their acceptability from the standpoint of, for example, copper, lead, zinc, fluoride, and free and combined chlorine concentrations; these are typical concerns, but others may be equally relevant. If a satisfactory freshwater supply is not available or if a standard water is required for comparative toxicity tests, relative sensitivity tests, or tests to determine the effects of hardness, pH, or total alkalinity on the toxicity of various materials, use a reconstituted standard water. Gas supersaturation is common in nearly all water supplies and merits concern. Total dissolved gas levels should not exceed barometric pressure in shallow tanks ($\leq$1 m deep) and should be monitored as described in Section 2810. Supersaturated gas levels can be reduced by aeration[4], but complete removal requires pretreatment by nitrogen stripping with pure oxygen.

Prepare standard fresh water (Tables 8010:I and 8010:II) by adding reagent-grade chemicals to glass-distilled and/or deionized water. For special studies, determine that the reverse osmosis, distilled and/or deionized water contains less than the indicated constituents:

Conductivity	1 μS/cm
Total organic carbon (TOC)	2 mg/L
or chemical oxygen demand (COD)	5 mg/L
Boron, fluoride	100 μg/L each
Un-ionized ammonia	20 μg/L
Aluminum, arsenic, chromium, cobalt, copper, iron, lead, nickel, zinc	1 μg/L each
Total residual chlorine	< 10 μg/L
Total organophosphorus pesticides	50 ng/L
Total organochlorine pesticides plus polychlorinated biphenyls (PCBs)	50 ng/L

Carbon-filtered deionized water usually is acceptable. Determine conductivity of distilled and/or deionized water for each batch of reconstituted water. Check other constituents periodically. If the water is prepared from a dechlorinated water, test the reconstituted water to determine that first instar daphnids survive for 48 h and can reproduce successfully when mature (Section 8711).[1]

The pH, alkalinity, and hardness of a receiving water influence toxicity of some materials, especially metals. Therefore, it is desirable to have a supply of both hard and soft waters with suitable pH and alkalinity.

It is advantageous to have water with temperatures between 3 and 12°C during the winter and between 12 and 25°C at peak summer temperatures. For general use, the pH should be in the range of 7 to 8.2 and dissolved CO_2 should be 1 mg/L or less.

2) Marine water supplies—Use unpolluted marine water with low turbidity and settleable solids and a pH and salinity favorable for the test organism. Ensure that annual salinity variations are not so wide as to be harmful to the organisms. In general, it is preferable to have a source of higher-salinity water (e.g., ocean water) from which brackish water can be prepared by dilution.

If a suitable marine water supply is not available, use artificial seawater for limited culturing and toxicity testing.[15–17] Prepare artificial seawater by adding the compounds listed in Table 8010:III to 800 mL glass-distilled or deionized water, in the order listed. Make sure that each salt is dissolved before adding the next. Make up to 1 L with distilled or deionized water. The salinity should be 34 ± 0.5 g/kg and pH 8.0 ± 0.2. Obtain desired salinity at time of use by dilution with deionized water. Alternatively, prepare marine water from commercially available salt mixes that have been shown to provide for adequate survival, growth, and reproduction of test organisms.

To increase salinity of a natural water, use a strong natural brine prepared by freezing and then partially thawing seawater.

TABLE 8010:II. QUANTITIES OF REAGENT-GRADE CHEMICALS TO BE ADDED TO AERATED SOFT RECONSTITUTED FRESH WATER FOR BUFFERING pH[5,8]

Desired pH*	Quantity of Chemical to Be Added mL/L water		
	1.0N NaOH	1.0M KH_2PO_4	0.5M H_3BO_3
6.0	1.3	80.0	—
6.5	5.0	30.0	—
7.0	19.0	30.0	—
7.5	—	—	—
8.0	19.0	20.0	—
8.5	6.5	—	40.0
9.0	8.8	—	30.0
9.5	11.0	—	20.0
10.0	16.0	—	18.0

* Approximate equilibrium pH. Do not aerate after adding these chemicals.

TABLE 8010:I. RECOMMENDED COMPOSITION FOR RECONSTITUTED FRESH WATER

Water Type	Salts Required mg/L				Water Quality		
	$NaHCO_3$	$CaSO_4 \cdot 2H_2O$	$MgSO_4$	KCl	pH*	Hardness mg $CaCO_3$/L	Alkalinity mg $CaCO_3$/L
Very soft	12	7.5	7.5	0.5	6.4–6.8	10–13	10–13
Soft	48	30	30	2.0	7.2–7.6	40–48	30–35
Moderately hard	96	60	60	4.0	7.4–7.8	80–100	60–70
Hard	192	120	120	8.0	7.6–8.0	160–180	110–120
Very hard	384	240	240	16.0	8.0–8.4	280–320	225–245

* Approximate equilibrium pH after aeration with fish in water.

TABLE 8010:III. PROCEDURE FOR PREPARING
RECONSTITUTED SEAWATER*[15,17]

Compound in Order of Addition	Final Concentration mg/L
NaF	3
SrCl$_2$·6H$_2$O	20
H$_3$BO$_3$	30
KBr	100
KCl	700
CaCl$_2$·2H$_2$O	1 470
Na$_2$SO$_4$	4 000
MgCl$_2$·6H$_2$O	10 780
NaCl	23 500
Na$_2$SiO$_3$·9H$_2$O	20
Na$_4$EDTA*	1
NaHCO$_3$	200

* Tetrasodium ethylenediaminetetraacetate. Omit when toxicity tests are conducted with metals. Omit when tests are conducted with plankton or larvae. Strip the medium of trace metals.[11,12]

TABLE 8010:IV.A. MACRONUTRIENT STOCK SOLUTION

Compound	Concentration mg/L	Element	Resulting Concentration mg/L
NaNO$_3$	25.5	N	4.20
		Na	11.0
NaHCO$_3$	15.0	C	2.14
K$_2$HPO$_4$	1.04	K	0.469
		P	0.186
MgSO$_4$·7H$_2$O	14.7	S	1.91
		Mg	2.90
MgCl$_2$	5.70		
CaCl$_2$·2H$_2$O	4.41	Ca	1.20

TABLE 8010:IV.B. MICRONUTRIENT STOCK SOLUTION

Compound	Concentration μg/L	Element	Resulting Concentration μg/L
H$_3$BO$_3$	186	B	32.5
MnCl$_2$	264	Mn	115
ZnCl$_2$	3.27	Zn	1.57
CoCl$_2$	0.780	Co	0.354
CuCl$_2$	0.009	Cu	0.004
Na$_2$MoO$_4$·2H$_2$O	7.26	Mo	2.88
FeCl$_3$	96.0	Fe	33.0
Na$_2$EDTA·2H$_2$O	300	—	—

Natural brine also may be prepared (not to exceed 100‰ by evaporation of natural seawater with aeration and low heat (maximum of 35°C).[13] This is satisfactory if only limited amounts of water are needed; for larger volumes, use commercial sea salts or a stronger solution of artificial seawater.

In the preparation of artificial seawater, be sure that an undesirable concentration of metals does not occur. Even reagent chemicals contain traces of several metals and their extensive use can result in a buildup of metals. If large volumes of artificial seawater are not required, remove the metals by passing the seawater through a column containing a cation-exchange resin in the sodium form.

The suitability of any artificial saltwater is enhanced by aging, with aeration, and by the introduction of nitrogen-fixing bacteria from water in which healthy aquatic organisms have been maintained or by addition of marine algae.

c. Food and feeding:

1) Culture of microorganisms—Phytoplankton and zooplankton may be cultured as test organisms for studying biostimulation, toxicity, etc. They may be cultured also as food for other organisms such as copepoda, Daphnia and other microcrustaceans, the larvae and adults of mollusks, and young and adult fish.

a) Culture medium for freshwater algae—Prepare reconstituted fresh water by adding reagent-grade macro- and micronutrients to glass-distilled and/or deionized water in the concentrations given in Tables 8010:IV.A and B.

Prepare a separate stock solution of each micronutrient salt in 1000 times the specified final concentration in glass-distilled or deionized water. Combine trace metals and EDTA in a single micronutrient stock solution in glass-distilled or deionized water at 1000 times the final concentration of each. Note that in toxicity testing of metals the effect of EDTA must be established and reported as part of the test results.

To prepare algal culture medium, add 1 mL of each micronutrient stock solution to 900 mL glass-distilled or deionized water, then add 1 mL trace metal EDTA mixture and make up to 1 L with glass-distilled or deionized water. As an alternative to algal culture medium preparation, commercially supplied nutrient media† are available.

† Fritz Chemical Co., Dallas, TX, or equivalent.

To prepare duckweed culture medium, add 10 mL of each micronutrient stock solution to 900 mL glass-distilled or deionized water, then add 1 mL trace metal EDTA mixture and make up to 1 L. Alternatively, mix 10 mL each of three stock solutions (A, B, and C) to 1 L (8211B.2).

Whenever axenic algal cultures are used, or if bacterial growth interferes with the test, prepare media aseptically: To 800 mL glass-distilled or deionized water add 1 mL of each micronutrient stock solution in the order listed, mixing after each addition. Filter-sterilize by passing through a sterile 0.2-μm-porosity membrane filter (pre-rinsed with 100 mL double-distilled water) into an autoclave-sterilized container. Add 1 mL filter-sterilized micronutrient solution, and make up to 1 L with sterile distilled or deionized water.

Store uninoculated sterile reference medium in the dark to avoid photochemical changes.

When sterility is desired in algal tests, check sterility periodically by adding 1 mL inoculated test culture to tubes of sterile nutrient test medium and incubate in the dark at the test temperature for 2 weeks. The appearance of opalescence in the test medium indicates contamination.

Prepare sterile bacterial nutrient test medium by adding the following quantities of chemicals to 1 L glass-distilled water:

Sodium glutamate	250 mg
Sodium acetate	250 mg
Glycine	250 mg
Sucrose	250 mg
Sodium lactate	250 mg
DL alanine	250 mg
Nutrient agar	50 mg

Bring medium to a boil, dispense to test tubes, and sterilize by autoclaving.

b) *Culture medium for marine algae*—To artificial seawater (Table 8010:III) add nutrients listed in Table 8010:V to give the indicated concentrations in the algal culture medium.

When an unpolluted seawater is used, prepare the medium by enriching filter-sterilized seawater with micronutrients at one-half the indicated concentrations. If sterile techniques are required, follow procedures for fresh water. When sterilization is performed by autoclaving, add vitamins after autoclaving. When filter-sterilization is used, use a positive pressure of 72 kPa.

2) *Mass production of algae as food for other organisms*—Rearing zooplankton, various filter-feeders, the larvae of crustaceans, and fish requires large quantities of phytoplankters. These needs must be met with an apparatus capable of producing continuous amounts of desired organisms at high densities. Such an apparatus (Figure 8010:2) permits easy assembly, cleaning, and sterilization and efficient utilization of light energy, and is constructed for continuous use. The main body of the unit is a 60-cm section of 15-cm-diam borosilicate glass drainage pipe. The top section is a 15-cm to 5-cm concentric reducer and the bottom section is a 15-cm to 5-cm ell. Each section accommodates a No. 12 silicone rubber stopper held in place by a carboy or similar type clamp. Hold the three sections together by aluminum ring clamps and seal adjoining surfaces by silicone "O" rings made from small-diameter tubing. Use material that is autoclavable and nontoxic.

Use this or similar device to supply cells on a periodic or continuous basis. As the cells are withdrawn, add more medium. The following species have been grown at the indicated concentrations: *Skeletonema costatum*, 4.3×10^6 cells/mL; *Dunaliella tertiolecta*, 4.4×10^6 cells/mL; *Isochrysis galbana*, 7.0×10^6 cells/mL; *Monochrysis lutheri*, 5.0×10^6 cells/mL.

3) *Food for macroinvertebrates and fish*—A suitable food is essential for rearing various macroinvertebrate and early life stages of some fishes. Distinguish carnivores from herbivores to supply the correct type of food. Organisms taken as food differ for different life stages of a species. As organisms grow they require progressively larger food organisms. Many feed on pelagic organisms whose movements should be sufficient to attract the predator but slow enough so they can be caught readily. Use food organisms that are nutritious, easily digested, uncontaminated, and readily obtainable. The nutritional requirements of salt- and freshwater organisms can differ greatly; check specific requirements of particular species. Distribute zooplankton food in rearing tanks to match distribution of organisms using it. Provide an adequate amount of food with a ratio of number of prey to predator varying from 50:1 to 200:1, depending on the feeding efficiency of the cultured species. If a small number of organisms is reared in a large tank, provide an acceptable food organism density to insure that enough are captured. Some algae and diatoms used for food have a tendency to settle to the bottom. Circular movement of the water in rearing tanks, provided as described in Section 8010E.4*a*, keeps food materials in suspension.

When using cultured microorganisms as food, be aware of possible environmental changes they may cause. In addition to the possible presence of toxic metabolites, algal blooms may produce excess oxygen and result in supersaturation and gas bubble disease.[11] Use live food whenever possible. Analyze food for toxicants, especially pesticides and heavy metals. Supplement natural

TABLE 8010:V. NUTRIENTS FOR ALGAL CULTURE MEDIUM IN SEAWATER

Compound	Concentration		Concentration of Nutrient
$NaNO_3$	25.0	mg/L	4.2 mg N/L
K_2HPO_4	1.05	mg/L	0.19 mg P/L
$FeCl_3$	72.6	μg/L	
$MnCl_2$	2.30	μg/L	
$ZnCl_2$	2.10	μg/L	
$Na_2MoO_4 \cdot 2H_2O$	2.50	μg/L	
$CuCl_2$	0.20	μg/L	
Na_2EDTA	300	μg/L	
Vitamins:			
Thiamine	0.100	mg/L	
Biotin	0.50	μg/L	
B_{12}	0.50	μg/L	

foods with commercially available dried and pelleted foods.‡ (See Section 8910 for fish feeding.) These foods should be attractive to the organisms, supply necessary nutrients and trace elements, and contain binders to insure pellet stability.[20] Nutritionally deficient diets may cause significant differences in sensitivity of test organisms to toxicants and will affect reproductive performance.

Methods for rearing freshwater organisms have been described.[21] Methods for rearing larvae of marine animals with special reference to their food organisms have been summarized.[22] The literature on laboratory feeding larvae of marine fish[23] and freshwater fish[24] has been reviewed. Standard fish hatchery culture facilities and operations and other useful freshwater and marine aquaculture references are available.[25–27]

d. *Cleaning containers and equipment:*

1) *Cleaning holding, acclimation, testing, and dilution water tanks*—Clean test containers and toxicant delivery systems before use. Soak new solvent and acid resistant containers overnight in tap or dilution water, then wash with laboratory detergent, rinse with 100% acetone, water, acid (such as 5% HNO_3 or HCl), and rinse twice with tap water. After each test, wash the system appropriately, e.g., acid to remove metals and bases; detergent, sodium hypochlorite§ (NaOCl) solution (200 mg/L), organic solvent, or activated carbon to remove organic compounds, etc. Immediately before testing, rinse again with dilution water.[1]

2) *Removal of unused food and wastes*—Do not let unused food or fecal material accumulate. Whenever possible, build holding and testing containers with sloping bottoms so food and feces can be removed easily with a siphon. The amount and frequency of cleaning depends on the organism, ratio of dilution water to weight and volume of organisms, and feeding schedule. Clean holding containers at least once every other day. If growths occur on sides of containers, dislodge and let settle for removal.

5. Parasites and Disease

a. *Stress in relation to parasites and infectious disease:* Unexpected and often unexplained mortalities in experimental and

‡ Some foods that have been used widely include Glenco Trout Food, Glenco, MN 53336; Biorell and TetraMin, available from local pet shops; Oregon Moist, Warrenton, OR 97146; and Cerophyl®, Agri-Tech, Inc., Kansas City, MO 64112. The latter has been used for the small forms and as a food for organisms providing food for the higher species.

§ Do not use acid and hypochlorite together.

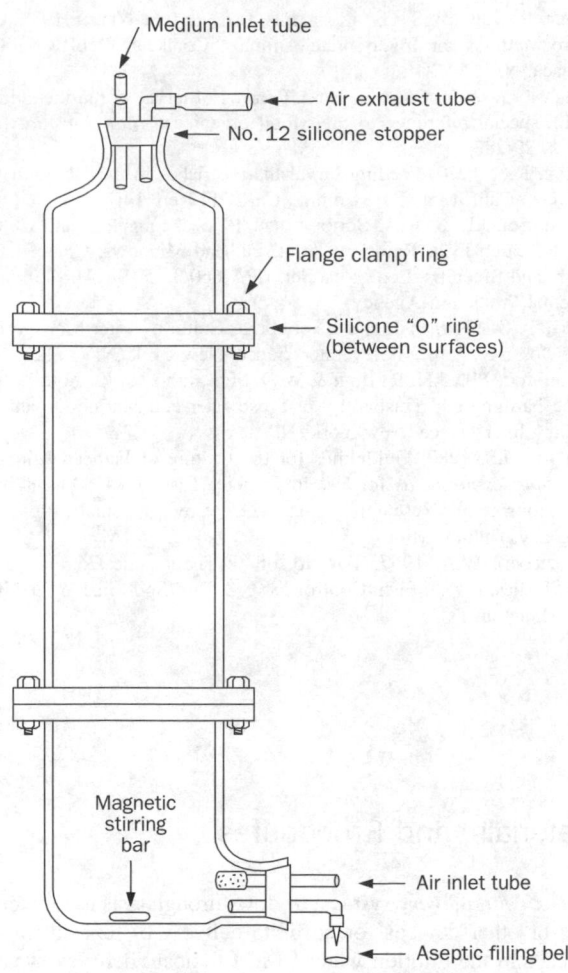

Figure 8010:2. Algal culture units. Cells are withdrawn from unit through the aseptic filling bell.

control animals interfere with acute or chronic test results. While many factors may be responsible for the death of an animal, diseases due to specific pathogens are among the most significant. In general obtain fish and other animals from specific pathogen-free stocks (commercial bioassay organism supplier, specific hatcheries, etc.) rather than stressing populations by parasite and disease controls. Also optimize laboratory conditions for the individual species to prevent the fostering of disease conditions.

When large numbers of organisms are retained in a relatively small space, undesirable growths, infectious diseases, and parasites may become a problem. Pathogens and parasites that might be very rare in natural waters become potential and ever-present dangers in intensive culture.

Filtration and/or sterilization of water, regular cleaning of holding vessels, strict sanitation practices, and sterilization of equipment are essential for healthy animals. Uniform food distribution, limiting the amount of unused food, and expeditious removal of unused food and waste materials are also important.

Organisms exposed to toxicants become stressed and weakened, and may become more susceptible to parasites and disease. Because other environmental factors contribute to reduced resis-

tance, pay careful attention to nutrition, oxygen supply, and water quality.

b. Control methods: UV light and ozonation have been used successfully to control disease and parasites. Antibiotics used in holding tanks reduce bacterial populations. To reduce mortality and to avoid introduction of disease into stock tanks, treat with a wide-spectrum antibiotic immediately after collection, during transport, during egg production and hatching, or on arrival at the laboratory. Holding in a tetracycline-based antibiotic,‖ 15 mg/L for 24 to 48 h, may be helpful. Other chemotherapeutic agents are available, but use care in their application because some are toxic at low concentrations.[23] Do not use treated organisms for tests for at least 10 d after treatment unless eggs are treated and less than 48-h-old organisms are required for the test. If contamination is suspected, disinfect tanks and containers with 200 mg NaOCl/L for 1 h.

For larval tests, use strict sanitary measures including sterilization of utensils and containers, filtration and UV sterilization of water, and removal of metabolic products. If disease signs appear in larval cultures, discard the entire culture.

For tests using adult fish and shellfish, early diagnosis and prompt treatment, when available, can prevent losses.

6. References

1. WEBER, C.I., ed. 1993. Methods for Measuring the Acute Toxicity of Effluents and Receiving Water to Freshwater and Marine Organisms, 4th ed. EPA-600/4-90-027F, Environmental Monitoring Systems Lab., U.S. Environmental Protection Agency, Cincinnati, Ohio.
2. MOUNT, D.I. & W.A. BRUNGS. 1967. A device for continuous treatment of fish in holding chambers. *Trans. Amer. Fish. Soc.* 96:55.
3. CLINE, T.F. & G. POST. 1972. Therapy for trout eggs infected with *Saprolegnia. Progr. Fish-Cult.* 34:148.
4. DAVIS, H.S. 1953. Culture and diseases of game fishes. Univ. California Press, Berkeley.
5. HERWIG, N. 1979. Handbook of Drugs and Chemicals Used in the Treatment of Fish Diseases. Charles C. Thomas, Pub., Springfield, Ill.
6. HOFFMAN, G.L. & F.P. MEYER. 1974. Parasites of Freshwater Fishes. THF Publ., Inc., Neptune City, N.J.
7. HOFFMAN, G.L. & A.J. MITCHELL. 1980. Some Chemicals That Have Been Used for Fish Diseases and Pests. Fish Farming Experimental Sta., U.S. Fish & Wildlife Serv., Stuttgart, Ariz.
8. REICHENBACH-KLINKE, H. & E. ELKAN. 1965. The Principal Diseases of Lower Vertebrates. Academic Press, New York, N.Y.
9. SNIEWZKO, S.F., ed. 1970. A Symposium on Diseases of Fishes and Shellfishes. Spec. Publ. No. 5, American Fisheries Soc., Washington, D.C.
10. VAN DUIJN, C., JR. 1973. Diseases of Fishes, 3rd ed. Charles C. Thomas Publ., Springfield, Ill.
11. BOUCK, G.R., R.E. KING & G. SCHMIDT. 1984. Comparative removal of gas supersaturation by plunges, screens and packed columns. *Aquacult. Eng.* 3:159.
12. LEWIS, P.A., D.J. KLEMM, J.M. LAZORCHAK, T.J. NORBERG-KING, W.H. PELTIER & M.A. HEBER, eds. 1994. Short-Term Methods for Estimating the Chronic Toxicity of Effluents and Receiving Waters to Freshwater Organisms, 3rd ed. EPA-600/4-91-002, Environmental Monitoring Systems Lab., U.S. Environmental Protection Agency, Cincinnati, Ohio.
13. KLEMM, D.J., G.E. MORRISON, T.J. NORBERG-KING, W.H. PELTIER & M.A. HEBER, eds. 1994. Short-Term Methods for Estimating the Chronic Toxicity of Effluents and Receiving Waters to Marine and

‖ Terramycin or equivalent.

Estuarine Organisms, 2nd ed. EPA-600/4-91-003, Environmental Monitoring and Support Lab., U.S. Environmental Protection Agency, Cincinnati, Ohio.

14. MARKING, L.L. & V.K. DAWSON. 1973. Toxicity of Quinaldine Sulfate to Fish. Invest. Fish Control 48, U.S. Bur. Sport Fish & Wildlife, Washington, D.C.

15. KESTER, E., I. DREDALL, D. CONNERS & R. PYTOWICZ. 1967. Preparation of artificial seawater. Limnol. Oceanogr. 12:176.

16. SPOTTE, S., G. ADAMS & P.M. BUBUCIS. 1984. GP2 as an artificial seawater for culture or maintenance of marine organisms. Zool. Biol. 3:229.

17. ZAROOGIAN, G.E., G. PESCH & G. MORRISON. 1969. Formulation of an artificial sea water media suitable for oyster larvae development. Amer. Zool. 9:1141.

18. ZILLIOUS, E.J., H.R. FOUCK, J.C. PRAGER & J.A. CARDIN. 1973. Using Artemia to assay oil dispersant toxicities. J. Water Pollut. Control Fed. 45:2389.

19. DAVEY, E.W., J.H. GENTILE, S.J. ERICKSON & P. BETZER. 1970. Removal of trace metals from marine culture medium. Limnol. Oceanogr. 15:486.

20. MEYERS, S.P. & Z.P. ZEIN-ELDIN. 1972. Binders and pellet stability in development of crustacean diets. Proc. 3rd Annu. Workshop World Maricult. Soc., p. 351.

21. NEEDHAM, J.G., P.S. GALTSOFF, F.E. LUTZ & P.S. WELSH. 1937. Culture Methods for Invertebrate Animals. Comstock Publ. Co., Inc., Ithaca, N.Y. XXXII.

22. HIRANO, R. & Y. OSHIMA. 1963. Rearing of larvae of marine animals with special reference to their food organisms. Bull. Jap. Soc. Sci. Fish. 29:282.

23. MAY, R.C. 1970. Feeding larval marine fishes in the laboratory, A review. Calif. Mar. Res. Comm., CalCOFl Rep. 14:76.

24. BRAUGHN, J.L. & R.A. SCHOETTGER. 1975. Acquisition and Culture of Research Fish: Rainbow Trout, Fathead Minnows, Channel Catfish, and Bluegills. Ecol. Res. Ser. EPA-660/3-75-011, U.S. Environmental Protection Agency.

25. SPOTTE, S. 1979. Fish and Invertebrate Culture: Water Management in Closed Systems. Wiley Interscience, New York, N.Y.

26. BARDACH, J.E., J.H. RYTHER & W.O. MCLARNEY. 1982. Aquaculture: The Farming and Husbandry of Freshwater and Marine Organisms. Wiley Interscience, New York, N.Y.

27. DENNY, J.S. 1987. Guidelines for the Culture of Fathead Minnows Pimephales promelas for Use in Toxicity Tests. EPA-600/3-87-001. Environmental Research Lab., U.S. Environmental Protection Agency, Duluth, Minn.

28. WILLFORD, W.A. 1967. Toxicity of 22 Therapeutic Compounds to Six Fishes. Invest. Fish. Control 18, U.S. Bur. Sport Fish & Wildlife, Washington, D.C.

8010 F. Toxicity Test Systems, Materials, and Procedures

1. Water Supply Systems and Testing Equipment

a. Composition of materials used: Construct all components of a test system, including water heating and cooling units, constant-level troughs and head boxes, valves and fittings, diluters, pumps, mixing equipment, tanks, and exposure chambers, from inert materials. Acceptable materials include lead-free glass, perfluorocarbon plastics, silicone sealant and tubing, polyvinyl clear flexible plastic,* PVC, nylon, fiberglass, and No. 316 stainless steel. Unplasticized plastics, such as polyethylene or polypropylene, may be used in the dilution portion including holding and acclimation tanks. Avoid contact with brass, copper, lead, or rubber. Some neoprene formulations have been acutely toxic while others have been proven safe for toxicity testing. Cure materials in dilution water at least 48 h before use.

b. Temperature regulation: Obtain dilution water of the desired temperature by mixing hot and cold water of constant temperatures in the correct proportions, by heat exchangers, water baths, or by heaters or coolers in constant-level troughs and head boxes. A heated room or high-low incubator with thermostatic controls usually is suitable for static tests on warm-water organisms. Hold dilution water in tanks until it reaches ambient temperature for conducting static tests. For cold-water species use a specially insulated constant-temperature room or large water bath equipped with temperature controls and adequate circulating water. A satisfactory design for a small laboratory to conduct short-term static tests has been described.[1-3] Special facilities required for different groups of organisms are described in Sections 8111 through 8910.

c. Toxicant delivery system: In flow-through tests use metering pumps or other devices for accurate delivery of toxicant or test material into the dilution water.[1] Most toxicant delivery systems have been designed for fresh water and may not be applicable to all test substances. Deliver dilution water from constant-head troughs or head boxes by siphons, constricted tubing, nozzles, or pumps. Deliver toxicants by siphons from constant-head reservoirs, pumps, calibrated glass nozzles, solenoid valves, or Mariotte bottles.[4] Mix dilution water and toxicant in tanks with baffles or stirrers or in mixing troughs.[5] Since the introduction of the serial diluter,[6] various methods and types of diluters have been described.[1,7-25]

The choice of a toxicant-delivery system for flow-through toxicity tests depends on such factors as dilution, flow rate, quantity of toxicant available, and the presence of suspended solids. The proportional diluter[17] was designed to handle a dilution factor (i.e., the factor by which a concentration is multiplied to calculate the next lower concentration) between 0.50 and 0.90. The serial diluter has been modified to provide a narrow range of concentrations.[18] Other diluters operate well with a dilution factor of 0.50.[1,23] Flow rates through test chambers may vary from 400 mL/min for the proportional diluter[6] to 6000 mL/min.[23] Some stock solutions[1,20,22-24] and other systems have been modified to handle larger volumes of toxicant and high suspended solids concentrations.[1,20] These diluters are most suitable for effluent toxicity tests. Several diluters have been designed for flow-through toxicity tests with embryo-larval stages of aquatic organisms.[1,22] One of these systems was designed to eliminate the air/water interface for tests using volatile compounds and compounds with very low solubility.[22]

---

* Tygon® or equivalent.

The basic components of a flow-through system are shown in Figure 8010:3. The diluent water reservoir is large enough to provide water for at least 5 d. If dilution water is added to this reservoir continuously, a smaller capacity is preferred. Dilution water flows at a constant rate by gravity from this reservoir to a constant-head diluent-supply head box through a nonmetallic float-controlled valve or other device and then to the diluter. Provide head box with heating or cooling equipment and a thermostat to maintain constant temperature. Equip test containers with an overflow system designed to prevent organisms from entering outlets. Clean test containers daily as described in 8010E.4d.[25]

The constant-head toxicant supply is a constant-level tank, a Mariotte bottle, or other device. If toxicant is added at greater than a drip rate, adjust toxicant temperature to that of the diluter via a heat exchange; otherwise, the toxicant heat exchanger in Figure 8010:3 is not necessary. If the toxicant stock solution is unstable, renew it before it degrades. If metering pumps are used, the toxicant supply system need not maintain a constant head.

A simple valve control system for regulation of flow rates of dilution water and toxicant solution has been described.[26] For more toxic materials, less toxicant is required and a Mariotte bottle or syringe pump that delivers a very slow but constant flow is useful.[27,28]

A diluter meters dilution water from the constant-head box and toxicant from the constant-head tank or other containers and mixes them in the proper proportions for each of the test chambers. After proper calibration of the diluter, make toxicant stock solutions to the proper concentration. Shield the toxicant supply reservoir from light when necessary.

Provide a mixing chamber between diluter and test container for each concentration. If duplicate test containers are used, run separate delivery tubes from the mixing chamber to each duplicate or use specialized glassware that will split the mixed solution equally. Use flow rates through test containers of at least 6 tank water volumes/24 h. Do not let rates through test containers vary temporally or between containers by more than ± 10%. Calibrate the toxicant and dilution water volumes used in each portion of the toxicant delivery system and the flow rate through each test container. Check operation of toxicant delivery system daily during test.

2. Preparing Test Materials

a. Dilution water: Whenever possible, test toxicity of effluents on site where ample supplies of toxicant and dilution water are available. Consistent overnight delivery service is available virtually anywhere in the United States and elsewhere and the cost and logistic problems in performing on-site studies may make it more advantageous to perform the tests under more stable conditions in permanent laboratories. On-site testing does permit temperature, DO, pH, hardness, salinity, turbidity, and other qualities of the dilution water to vary normally with those of the receiving water.

Convey the effluent sample to testing chambers with as little modification as possible. Do not unnecessarily aerate, heat, cool, or agitate. In cases where the testing facility is remote from the effluent discharge site, artificial or reconstituted water may be used as the diluent. If the diluted effluent is low in DO, adjust flow-through and loading in the test chambers so that DO is not reduced significantly; aerate as a last resort. Hold temperature at or near that of the receiving water.

If the receiving water is deficient in DO and has temperatures above the locally applicable water quality standards, bring these into compliance so that allowable levels of a specific waste can be assessed meaningfully. Determine toxicity of test waste in conjunction with other contaminants present in receiving water by taking dilution water from receiving water just outside the area of effect of test waste. This is especially necessary when effluents contain metal salts, cyanide complexes, ammonium compounds, or other materials, the toxicity of which is greatly influenced by changes in pH, hardness, temperature, etc. If there are wide variations in receiving water quality characteristics, determine waste toxicity at the upper and lower limits of the range.

Evaluate receiving water effects on aquatic biota by using two controls, one with the receiving water and another (either natural or synthetic) with an unpolluted water of similar quality. If appropriate to the study, adjust calcium, magnesium, sulfate, alkalinity, pH, and DO for freshwater controls to those of the receiving water before adding wastes to determine if dilution water itself is unfavorable for the more sensitive aquatic species in the area. Conduct chronic tests to detect the most subtle sources of stress.

Turbidity of dilution water is important in determining harmful concentrations of potential toxicants because some toxicants are sorbed on particles. Turbid dilution water may limit visual inspection and photosynthesis of algae, affect the response of test organisms to tested material, form deposits, and clog water systems. When large amounts of settleable solids significantly remove toxicants from the water, determine concentrations of toxicants in bottom sediments and their toxicity by appropriate tests with benthic organisms.

When the purpose of the test is other than to determine toxicity of an effluent, use for dilution water only a nonpolluted natural or synthetic dilution water of constant and reproducible quality that is favorable for aquatic life. Warm or cool dilution water to test temperature and bring to equilibrium with atmospheric gases (i.e., DO, CO_2) before use.

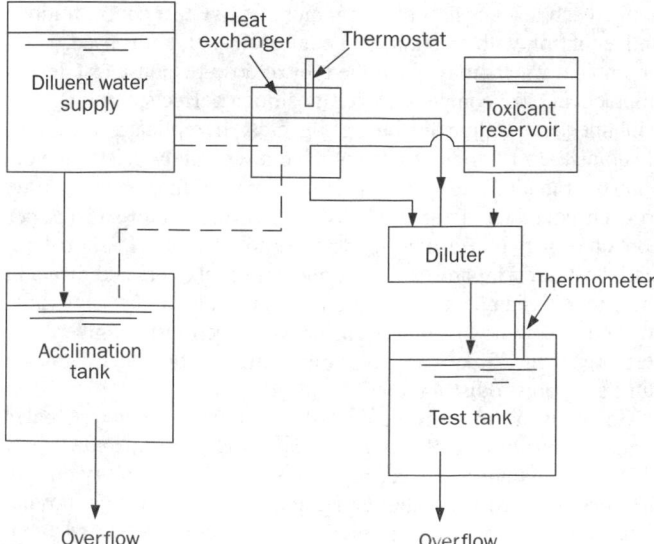

Figure 8010:3. Basic components of flow-through system.

Use standard water conditions and organisms for comparative toxicity and sensitivity tests. Use reconstituted fresh or marine water [Sections 8010E.4*b*1) and 2)] if natural supply is not suitable. Because of the effects of water quality on toxicity, it may be desirable to use both hard and soft water for tests on freshwater organisms.

Many marine organisms spend a portion of their life cycles in estuaries. In life-cycle tests, change dilution water in accordance with their requirements at different life stages. If the effects of temperature are not being studied, keep it within a favorable range.

For warm-water ($\geq$ 20°C) species, keep DO greater than 40% of saturation; for cold-water species, keep DO greater than 60% of saturation. Some larval forms, such as those of marine crustaceans, may require higher DO concentrations.

Determine pH requirements for test organisms. In long-term studies, other than effluent and monitoring tests, keep pH within 0.4 units of the desired value. Avoid rapid changes in temperature, pH, or CO_2 content. A rapid increase in the CO_2 content of marine waters indicates that some significant change has occurred that should be investigated at once. Freshwater organisms are more tolerant of pH changes and accommodate to much wider variations than strictly marine forms. Changes in pH drastically alter toxicity of many materials, for example, cyanide and NH_3.

In working with estuarine and marine organisms and different life stages that may be marine or estuarine, salinity is of prime importance. Use the natural salinity for each test species and its different developmental stages.[29]

Keep acidity, total alkalinity, and hardness of dilution water constant. Alkalinity and hardness influence toxicity of some metals and total alkalinity is an important factor in photosynthesis and algal growth.

b. Toxicant solution: Prepare toxicant solution in advance and add immediately to the dilution water for static tests. If a toxicant is unstable, determine its stability and replace as necessary. If possible, measure toxicant concentrations during the test. Prepare all solutions for each series of tests from the same source sample. Disperse undissolved material uniformly by shaking or gently mixing.

If solvents are necessary, use acetone, dimethylformamide (DMF), ethanol, methanol, isopropanol, acetonitrile, dimethylacetamide, ethylene glycol, or triethylene glycol to prepare stock solutions. Certain surfactants may be useful. Use only the minimal amount of solvent necessary to disperse the toxicant. Do not exceed 0.5 mg/L in static and 0.1 mg/L in flow-through test solutions. A solvent control is also necessary when a carrier is used.

If a solvent is used, use two sets of controls, one containing no solvent and the other containing the highest concentration of solvent used during the test.

Some effluents, especially oily wastes, are difficult to distribute evenly. The nature of the test material (e.g., single chemical or solid versus effluent) governs the preparation of test concentrations and frequency of test medium replacement. Common problems include insolubility, adsorption on exposed surfaces, decomposition, photolysis, loss of volatiles, high BOD and/or COD levels, and bacterial growth. These can change the apparent test concentration and lead to erroneous results. Effluent samples that vary in composition with time may require a series of tests to characterize toxicity.

Store effluent samples, in containers from which all the air has been expelled, at 4 $\pm$ 1°C. Do not store samples longer than absolutely necessary because toxicity may change with time. (Choose sample containers to minimize changes in concentration of constituents in sample.)

Thoroughly mix test materials before use. Use material directly as a stock solution of toxicant or prepare a stock solution using filtered dilution water. Make stock solutions with dilution water on a volume-to-volume basis. If the effluent is liquid, designate the percentage waste in each test concentration. If the waste is a solid, dilute on a weight-to-volume basis, e.g., milligrams per liter.

If the waste contains both solids and liquids, mix thoroughly to disperse before using as a stock toxicant, and provide agitation in the stock reservoir and test containers. If larger organisms are tested, a propeller placed under a screen or perforated false bottom may be used to maintain consistency of the solution. If the solids settle out rapidly and do not contact pelagic organisms, test only the liquid portion. After thorough mixing, let settle and decant or drain off the liquid for use as test toxicant. If the solid waste portion is toxic, set up test chambers having a certain weight-to-volume ratio of bottom material and expose benthic and burrowing organisms of the receiving water area. Mix wastes and let settle before adding organisms. If waste contains sparingly water-soluble materials check solubility and if below or at very low toxic concentrations, solvents, emulsifying agents, or water-miscible solvents may be used to disperse.

c. Test organisms: Select test organisms as described in Section 8010E.1 and handle as indicated in Sections 8010E.3 and 4. For long-term tests, use only the highest quality test organisms available. At the end of the test, control organisms still should be in good condition.

3. Test Procedures

a. Experimental design: Expose test organisms in at least duplicate containers of each experimental concentration. By using more organisms and replicate test containers for each toxicant concentration, intratest variability can be evaluated. Individual test methods may specify the number of replicates. Statistical methods also may dictate numbers of replicates. Use only true replicates with no water connection between test containers. Typically, each test consists of a minimum of five test concentrations and a control, with an additional control if a solvent is used. Tests of ambient waters may make the typical dose-response test design impractical. To compensate for positional effects, arrange test containers at random in the testing area. If replicates are used, randomize each series of test containers separately. Distribute organisms randomly to test containers either adding one at a time to each container if there are to be less than 11 organisms per container or two at a time if there are to be more. In short-term static tests, add organisms to intermediate containers and then add them to test chambers containing the toxicant all at the same time. Minimize volume of liquid transferred to each test chamber with test organisms. Take care to avoid contamination of treatments during organism distribution.

Generally, short-term, acute tests with fish and invertebrates require control survival of 90% or more to be considered acceptable. If this is not achieved, repeat the test. Acceptability criteria also may be used with other end points such as spawning, normal development, growth, reproduction, or general apparent health, but the various levels of control performance are dependent on the specific test procedures. Lower levels of survival may be acceptable in longer exposure assays or in tests that use life stages where survival, even under ideal conditions, is limited. Examples of acceptable reduced survival include short-term tests for estimating the chronic toxicity of effluent to freshwater and marine organisms.[2,3] In these tests acceptable survival for *Mysidopsis bahia, Cyprinodon variegatus, Ceriodaphnia dubia,* and *Pimephales promelas* is 80%.[2,3]

In short-term static or renewal tests with fish, use 20 or more test organisms in each toxicant concentration. Use of larger numbers of organisms per test concentration for smaller organisms is desirable. The number of organisms exposed in each test concentration is governed by size of the organism; expected normal mortality; extent of cannibalism; availability of dilution water, toxicant, and test organisms; and desired test precision. Test precision depends on variability of organism response, number of organisms exposed to each concentration, number of replications, differences between concentrations and range of concentrations tested, and toxicant concentration and its variability.

For a given test increasing the number of test organisms increases confidence in estimates of effect and precision. It is recommended that the 95% confidence interval be less than ±30% of the mean. With test organisms for which culture methods are not available, this precision may be difficult or impossible to attain. Reference toxicant tests are used to evaluate the condition of the test organisms and may be used to evaluate intertest performance of the method.[1-3] Make specifically designed intra- or inter-laboratory tests to evaluate precision.

Determine length and weight of representative organisms before the test to establish loading rates and acceptability with respect to size variation. After acclimation has begun, handle test organisms as little as possible. Increases in weight or growth may be determined by adding more animals than required initially so that some may be removed to make necessary measurements. At the end of the test, measure weight and length to determine sublethal impacts for computation of statistical end points (see 8010.G).

b. Selecting test concentrations: Express liquid waste concentrations as percent on a volume-to-volume basis. Express concentrations of nonaqueous wastes and of individual chemicals as milligrams or micrograms per liter. Clearly indicate what the weight represents as inclusion of water of hydration is part of the weight of the solute (e.g., $CuSO_4 \cdot 5H_2O$). It often is more appropriate to express weight as weight of toxicant (e.g., μg Cu/L). When an impure chemical is tested, especially in a formulation containing added inert ingredients, indicate the chemical composition by weight and whether the EC value is based on concentration of total material or active ingredient.

Although test end points such as EC or LOEC may be determined by using any appropriate series of test concentrations, the geometric series of concentration values is simplest to use when the approximate toxicity range is unknown. Multiply the highest and succeeding concentrations by a constant factor (0.3 to 0.5) to obtain concentrations that are evenly spaced on a logarithmic scale. Range-finding tests can help determine the dilution factor to use in subsequent tests.

The magnitude of concentration intervals to establish an EC by interpolation depends on the required degree of confidence in the point estimate and on the experimental data. Intervals spaced more tightly around the expected EC will give a more precise estimate of the true EC.

c. Loading: For static tests do not exceed an organism loading of 0.8 g/L in the test container. In tests with small organisms and tropical forms, decrease loading to as low as 0.1 g/L and accommodate large test organisms by using larger or duplicate test containers. Limit the number of test organisms per volume of test solution so that during the test *(a)* DO remains greater than 60% of saturation for cold-water species and greater than 40% of saturation for warm-water species; *(b)* toxicant concentration is not lowered significantly; *(c)* concentrations of metabolic products (e.g., NH_3, CO_2) do not become too high; and *(d)* organisms are not stressed by crowding or increased potential for cannibalism. Do not let concentration of un-ionized ammonia exceed 20 μg NH_3-N/L (Table 8010:VI).

For flow-through studies, use a flow rate of at least 6 tank volumes/24 h to maintain desirable temperature and DO and safe concentrations of metabolites. If the DO concentration drops below the desired level, increase the turnover rate within the diluter. If this is inadequate, aeration of the test chambers may be permissible.

d. Physical and chemical determinations:

1) Dilution water analysis—For fresh water, measure hardness, alkalinity, pH, TOC (or COD), and suspended solids at least once every 30 d and at the beginning and end of the test. If water quality is variable, test more frequently. Tap water can be dechlorinated by active aeration (using air stones) for 24 h, filtration through activated carbon, or use of sodium thiosulfate. If a treated tap water is used, measure residual chlorine by one of the methods given in Section 4500-Cl.[31]

Analyze weekly for pH, alkalinity, and hardness to define test-water variability. If characteristics are affected by the toxicant, test samples from each toxic concentration at least once every other week. For brackish or marine dilution water, measure salinity, pH, DO, and temperature two or three times daily; and suspended solids and TOC at least once every 30 d and at the beginning and end of each test.

2) Toxicant analysis—For flow-through life-cycle tests it may not be necessary to make routine detailed analyses, but make periodic tests to insure that the correct ratio of effluent to dilution water is maintained in exposure tanks.

For studies to determine chemical water quality criteria, it is desirable to measure concentration of toxicants in each container at beginning and at least once during the test or weekly in longer tests. Chemical measurements should be made in at least one container at the next-to-lowest toxicant concentration (or within the calibration range of analytical method). Whenever a malfunction

TABLE 8010:VI. PERCENTAGE OF AMMONIA UN-IONIZED IN DISTILLED WATER*

Temperature °C	Percentage Un-ionized at Given pH								
	6.0	6.5	7.0	7.5	8.0	8.5	9.0	9.5	10.0
5	0.01	0.04	0.11	0.40	1.1	3.6	10	27	54
10	0.02	0.06	0.18	0.57	1.8	5.4	15	36	64
15	0.03	0.08	0.26	0.83	2.6	7.7	21	45	72
20	0.04	0.12	0.37	1.2	3.7	11	28	55	80
25	0.05	0.17	0.51	1.7	5.1	14	35	63	84
30	0.07	0.23	0.70	2.3	7.0	19	43	70	88

* Prepared from data given in Sillen and Martell.[30]

is detected in any part of toxicant delivery system, check toxicant concentration in at least one container. For replicate test containers use a ratio of highest measured concentration to lowest measured concentration of less than 1.15; if this is exceeded, check toxicant delivery system and analyze additional samples from test containers to determine if the sampling or analytical method is sufficiently precise. Do not accept measured toxicant concentrations differing by more than ±15% from the calculated concentration unless specific reasons justify a greater difference. Precision of chemical test methods is important for assessing measurements of nominal toxicant concentrations, and test methods ideally should be calibrated in a range that includes the nominal toxicant concentration.

Record temperature at least hourly throughout the test (24 h/d) in at least one test container and make additional measurements on dilution water and other test solutions. If test is performed in a laboratory, continuous monitoring of temperature of the testing area (i.e., water bath, environmental chamber, etc.) may be appropriate. Measure DO, pH, and salinity at the beginning of the test and daily thereafter in the control, high, medium, and low toxicant concentrations. Generally, variation should not exceed ±1.0°C.

Take water samples for chemical analysis at the center of the exposure tank; do not include surface scum or material from tank bottom or sides. If analytical results are not affected by storage, collect daily, equal-volume grab samples and composite for a week. Analyze sufficient samples throughout the test to determine whether the concentration of toxicant is reasonably constant. If it is not, analyze enough samples weekly to show the variability of toxicant concentration. If methods are available, determine in the next-to-lowest concentration the loss of toxicant. If the loss is more than 10%, attempt to alleviate by using either a faster flow rate or a lower loading.

When necessary, analyze mature and immature test organisms for toxicant residues. For larger organisms analyze muscle and liver and possibly gills, blood, brain, bone, kidney, GI tract, gonads, and skin. For large organisms, analysis of whole specimens can be used but does not replace analysis of individual tissues, especially muscle (edible fillet).

e. Biological data and observations: In short-term tests with macroinvertebrates and fish, count number of dead or affected organisms in each container at least daily throughout the test. With certain fast-acting biocides it may be useful to count number of dead or affected organisms in each container at 1.5, 3, 6, 12, and 24 h after beginning the test. Remove dead organisms as soon as observed. Often it is more important to obtain data that will define the shape of the toxicity curve than to obtain data at pre-specified times.

Death is the adverse effect most often used to reflect acute toxicity. The usual criterion for death is no movement, especially no gill movement in fish, and no reaction to gentle prodding. Death is not easily determined for some invertebrates. Cessation of movement of antennae, mouth parts, or other organs may be used. When death cannot be determined, use EC50 rather than LC50. The effect usually used for determining EC50 with daphnids, midge larvae, copepods, and other organisms is immobilization, defined as inability to move, except for minor activity of appendages. Other effects can be used to determine EC50, but always report the effect and its definition. Consistency in defining the effect directly influences toxicity test precision and repeatability. Also report such effects as erratic swimming, loss of reflex,

discoloration, changes in behavior, excessive mucus production, hyperventilation, opaque eyes, curved spine, hemorrhaging, molting, and cannibalism.

In short-term tests, organism reactions during the first few hours may indicate the nature of the toxicant and serve as a guide for further tests.

In long-term partial- or full-life-cycle tests a photographic method for counting and measuring small test organisms has proven useful.[32] This is rapid and accurate, and does not entail handling the organism. With this method, use exposure tanks with glass bottoms and drains that allow the water level to be drawn down. To count and measure test organisms, draw water down to a depth of 2 to 3 cm and transfer tank to a light box having fluorescent lights under a square millimeter grid of adequate size. Photograph aquarium bottom; this shows organisms over the grid. On an enlargement of the picture, count and measure organisms.

f. Photoperiod and artificial light: In long-term studies to determine water quality requirements for those species requiring annual-light-cycle photoperiods, simulate natural seasonal daylight and darkness periods at the locality or some central location.[33] (See Section 8910 for more information on light cycles for fish.)

Use cool white fluorescent tubes or a wide-spectrum lamp as a light source.† Some organisms require subdued light, others need a place to hide, and some, such as lake trout eggs, require darkness during certain life stages. Base exposure to light on what is normal to, and required by, the species. Measure light intensity at the water surface. In short-term tests a standard photoperiod of 16 h light, 8 h dark is suggested.

g. Exposure chambers: For organisms weighing more than 0.5 g, use a test solution between 15 and 30 cm deep. In short-term tests, these organisms often are exposed in about 15 L solution in 20-L wide-mouth, soft-glass bottles. Fabricate test containers of other sizes by welding (not soldering) stainless steel, by gluing double-strength or stronger window glass with clear silicone adhesive formulated for aquarium use, or by modifying glass bottles, battery jars, or beakers to provide screened overflow holes or V-notches. Because silicone adhesives absorb some organochlorine and organophosphorus pesticides, expose as little of the adhesive as possible to the water. Place extra beads of adhesive for added strength only on the outside of containers. Expose smaller organisms in 30-mL to 2-L beakers that contain 15 to 1500 mL solution. Expose daphnids, midge larvae, copepods, and other small organisms in loosely covered beakers or other containers. Disposable plastic containers may be used for tests involving compounds that do not react with the plastic. Disposable plastic containers are recommended for such tests as the 7-d *Ceriodaphnia* survival and reproduction assay[3] and *Champia parvula* sexual reproduction assay.[2]

With flow-through tests keep liquid surface area/volume ratio small to reduce loss of volatiles. For various exposure chamber designs see Sections 8210 through 8910.

4. References

1. WEBER, C.I., ed. 1993. Methods for Measuring the Acute Toxicity of Effluents and Receiving Water to Freshwater and Marine Organisms, 4th ed. EPA-600/4-90-027F, Environmental Monitoring Systems Lab., U.S. Environmental Protection Agency, Cincinnati, Ohio.

† For example, OPTIMA 50, Duro-Test Corp., North Bergen, NJ 07047, or equivalent.

2. KLEMM, D.J., G.E. MORRISON, T.J. NORBERG-KING, W.H. PELTIER & M.A. HEBER, eds. 1994. Short-Term Methods for Estimating the Chronic Toxicity of Effluents and Receiving Waters to Marine and Estuarine Organisms, 2nd ed. EPA-600/4-91-003, Environmental Monitoring and Support Lab., U.S. Environmental Protection Agency, Cincinnati, Ohio.

3. LEWIS, P.A., D.J. KLEMM, J.M. LAZORCHAK, T.J. NORBERG-KING, W.H. PELTIER & M.A. HEBER, eds. 1994. Short-Term Methods for Estimating the Chronic Toxicity of Effluents and Receiving Waters to Freshwater Organisms, 3rd ed. EPA-600/4-91-002, Environmental Monitoring Systems Lab., U.S. Environmental Protection Agency, Cincinnati, Ohio.

4. MCALLISTER, W.A., JR., W.L. MAUCH & F.L. MAYER, JR. 1972. A simplified device for metering chemicals in intermittent-flow bioassays. Trans. Amer. Fish. Soc. 101:555.

5. LOWE, J.I. 1964. Chronic exposure of spot, Leiostomus xanthurus, to sublethal concentrations of toxaphene in seawater. Trans. Amer. Fish. Soc. 93:396.

6. MOUNT, D.I. & R.E. WARNER. 1965. A Serial Dilution Apparatus for Continuous Delivery of Various Concentrations of Material in Water. PHS Publ. No. 999-WP-23, Environ. Health Ser., U.S. Dep. Health, Education & Welfare, Washington, D.C.

7. MOUNT, D.I. & C. STEPHAN. 1967. A method for establishing acceptable toxicant limits for fish—Malathion and the butoxyethanol ester of 2,4-D. Trans. Amer. Fish. Soc. 96:185.

8. CLINE, T.F. & G. POST. 1972. Therapy for trout eggs infected with Saprolegnia. Progr. Fish-Cult. 34:148.

9. CHANDLER, J.H., H.O. SANDERS & D.F. WALSH. 1974. An improved chemical delivery apparatus for use in intermittent-flow bioassays. Bull. Environ. Contam. Toxicol. 12:123.

10. SCHIMMEL, S.C., D.J. HANSEN & J. FORESTER. 1974. Effects of aroclor 1254 on laboratory-reared embryos and fry of sheepshead minnows (Cyprinodon variegatus). Trans. Amer. Fish. Soc. 103:582.

11. FREEMAN, R.A. 1971. A constant flow delivery device for chronic bioassay. Trans. Amer. Fish. Soc. 100:135.

12. BENGTSSON, B.E. 1972. A simple principle for dosing apparatus in aquatic systems. Arch. Hydrobiol. 70:413.

13. GRANMO, A. & S.C. KOLLBERG. 1972. A new simple water flow system for accurate continuous flow tests. Water Res. 6:1597.

14. BENOIT, D.A. & F.A. PUGLISI. 1973. A simplified flow-splitting chamber and siphon for proportional diluters. Water Res. 7:1915.

15. LICHATOWICH, J.A., P.W. O'KEEFE, J.A. STRAND & W.L. TEMPLETON. 1973. Development of methodology and apparatus for the bioassay of oil. In Proc. Joint Conf. Prevention and Control of Oil Spills, p. 659. American Petroleum Inst., U.S. Environmental Protection Agency & U.S. Coast Guard, Washington, D.C.

16. ABRAM, F.S.H. 1973. Apparatus for control of poison concentration in toxicity studies with fish. Water Res. 7:1875.

17. MOUNT, D.I. & W.A. BRUNGS. 1967. A simplified dosing apparatus for fish toxicology studies. Water Res. 1:21.

18. THATCHER, T.O. & J.F. SANTNER. 1966. Acute toxicity of LAS to various fish species. Proc. 21st Ind. Waste Conf., Purdue Univ., Eng. Ext. Bull. No. 121:996.

19. SHUMWAY, D.L. & J.R. PALENSKY. 1973. Impairment of the Flavor of Fish by Water Pollutants. Ecological Res. Ser. No. EPA-R3-73-101, U.S. Environmental Protection Agency, Washington, D.C.

20. DEFOE, D.L. 1975. Multichannel toxicant injection system for flow-through bioassays. J. Fish. Res. Board Can. 32:544.

21. RILEY, C.W. 1975. Proportional diluter for effluent bioassays. J. Water Pollut. Control Fed. 47:2620.

22. BIRGE, W.J., J.A. BLACK, J.E. HUDSON & D.M. BRUSER. 1979. Embryo-larval toxicity tests with organic compounds. In L.L. Marking & R.A. Kimerle, eds. Aquatic Toxicology. ASTM STP 667, American Soc. Testing & Materials, Philadelphia, Pa., p. 313.

23. GARTON, R.R. 1980. A simple continuous-flow toxicant delivery system. Water Res. 14:227.

24. BENOIT, D.A., V.R. MATTSON & D.L. OLSON. 1982. A continuous-flow mini-diluter system for toxicity testing. Water Res. 16:457.

25. LEMKE, A.E. 1964. A new device for constant-flow test chambers. Progr. Fish-Cult. 26:136.

26. JACKSON, H.W. & W.A. BRUNGS. 1966. Biomonitoring of industrial effluents. Proc. 21st Ind. Waste Conf., Purdue Univ., Eng. Ext. Bull. 121:117.

27. SURBER, E.W. & T.O. THATCHER. 1963. Laboratory studies of the effects of alkyl benzene sulfonate (ABS) on aquatic invertebrates. Trans. Amer. Fish. Soc. 92:152.

28. BURROWS, R.E. 1949. Prophylactic treatment for control of fungus, Saprolegnia parasitica. Progr. Fish-Cult. 11:97.

29. DAVEY, E.W., J.H. GENTILE, S.J. ERICKSON & P. BETZER. 1970. Removal of trace metals from marine culture medium. Limnol. Oceanogr. 15:486.

30. SILLEN, L.C. & A.E. MARTELL. 1964. Stability Constants of Metal Ion Complexes. Spec. Publ. 17, Chemical Soc., London, England.

31. ANDREW, R.W. & G.E. GLASS. 1974. Amperometric methods for determining residual chlorine, ozone and sulfite. U.S. Environmental Protection Agency, National Water Quality Lab., Duluth, Minn.

32. MCKIM, J.M. & D.A. BENOIT. 1971. Effect of long-term exposures to copper on survival, reproduction and growth of brook trout Salvelinus fontinalis (Mitchill). J. Fish. Res. Board Can. 28:655.

33. DRUMMOND, R.A. & W.F. DAWSON. 1970. An inexpensive method for simulating a diel pattern of lighting in the laboratory. Trans. Amer. Fish. Soc. 99:434.

5. Bibliography

COMMITTEE ON METHODS FOR TOXICITY TESTS WITH AQUATIC ORGANISMS. 1975. Methods for Acute Toxicity Tests with Fish, Macroinvertebrates, and Amphibians. EPA-660/3-75-009, U.S. Environmental Protection Agency, Corvallis, Ore.

LEE, D.R. 1980. Reference toxicants in quality control of aquatic bioassays. In A.L. Buikema, Jr. & J. Cairns, Jr., eds. Aquatic Invertebrate Bioassays. American Soc. Testing & Materials, Philadelphia, Pa.

AMERICAN SOCIETY FOR TESTING AND MATERIALS. 1996. Standard practice for conducting acute toxicity tests with fishes, macroinvertebrates, and amphibians. E 729-88, Annual Book of ASTM Standards. American Soc. Testing & Materials, W. Conshohocken, Pa.

AMERICAN SOCIETY FOR TESTING AND MATERIALS. 1996. Standard Guide for Conducting Acute Toxicity Tests on Aqueous Effluents with Fishes, Macroinvertebrates, and Amphibians, ASTM E 1192-88, Annual Book of ASTM Standards. American Soc. Testing & Materials, W. Conshohocken, Pa.

U.S. ENVIRONMENTAL PROTECTION AGENCY. 1980. Proposed good laboratory practice guidelines for toxicity testing. Federal Register 45:26377.

8010 G.   Calculating, Analyzing, and Reporting Results of Toxicity Tests

This section identifies statistical methods used in analyzing data from acute and chronic toxicity tests, evaluates methods appropriate for each type of data, and discusses advantages and disadvantages of the statistical methods that are used most often. If unfamiliar with basic statistical concepts and methods, consult a statistician when questions concerning data analysis arise.

The precision of a biological test is limited by a number of factors including the normal biological variation among individuals of a species, ruggedness of test protocols, and analyst proficiency. Calculation of test end points from experimental data therefore will reflect the net effect of these test variables.[1] In general, the statistical methods presented in this section are applicable to both acute and chronic methods. Traditionally, however, some techniques have been used specifically for acute test results, and others for chronic test results.

1. Toxicity Data Analysis

Data generated in acute toxicity tests are quantal, that is, responses are measured with yes/no-type observations (e.g., did exposure cause immobilization, death, or not?). Continuous measurements that are measured in quantitative or graded tests, such as length, weight, or number of young produced, usually are not utilized as end points in an acute toxicity test. Data generated in chronic toxicity tests may be quantal (e.g., long-term survival), continuous (e.g., growth), or count (e.g., number of young produced).

Results from chronic toxicity tests may be analyzed by one or more statistical methods, depending on the purpose of the test. Regression methods, both parametric and nonparametric, may be used to generate ECs, evaluate conventional concentration-response test designs, and provide confidence intervals for point estimates (e.g., EC50 or LC50) interpolated from the appropriate model. Hypothesis test methods (both null and alternate) can be used to generate estimates of effect threshold (e.g., NOEC, LOEC) or estimates of equivalence (e.g., "bioequivalence"). The advantages and drawbacks of these different statistical applications have been reviewed in detail.[2,3] All of the methods are computationally intense, but availability of computer software applications and descriptive material facilitate their use in most instances. Although more than one statistical method can be used to treat a given data set, the experimental design chosen for the bioassay may affect the robustness of the statistical results. For example, point estimates and confidence intervals derived from regression models benefit from designs with more doses near the effect level of interest, while hypothesis test methods improve (in terms of statistical "power") with an increase in number of replicates at tested doses. In order to estimate an effect (or the test concentration below a specified effect) regression methods interpolate points on the modelled concentration-response curve, whereas the hypothesis test results are constrained to be one of the doses tested. In this sense, both regression and hypothesis test methods benefit from having the selected test concentrations bracket close to the "true" effect level of interest. If the purpose of testing is to compare sources or evaluate changes in toxicity over time, then effect levels from appropriate regression methods are preferred because the results from multiple tests are amenable to normal population statistics, whereas multiple NOECs are not.

ECP values can be calculated from parametric models such as probit, logit, and GLiM.[4-6] Choose the model appropriately for the type of data generated in the test. For example, probit analysis is only suitable for quantal data,[7] whereas the GLiM models require the appropriate data link.[5] The nonparametric ICP or linear interpolation model is also available for evaluation of nonquantal data.[8] Care must be given to evaluating goodness of fit and biological plausibility of the resultant toxicity curves and confidence interval data with any model results. The model should be able to account for apparent thresholds or low dose enhancement (nutritive or hormetic effects) in test data. Models that "force" data to fit by averaging adjacent dose response data ("smoothing") or extrapolating to zero may not be appropriate for use on a given data set.

The NOEC (no-observed-effect concentration) or LOEC (lowest-observed-effect concentration) values calculated from hypothesis test statistics are based on statistical significance (usually at the $P = 0.05$ level): the NOEC is the highest dose tested with a response not statistically different from control response, and the LOEC is the lowest dose tested having a statistically significant difference from control response. The NOEC can be used for analysis of any biological end point and data type commonly encountered in aquatic toxicity testing. The Chronic Value, ChV (sometimes referred to as MATC) is a point estimate determined as the geometric mean of the LOEC and NOEC doses. Hypothesis test results are sensitive to intratest variability which controls statistical power, and results must be examined for homogeneity in variance among test groups as well. Data transformations usually are required (see ¶ 2a below).

Recently, the alternative hypothesis method (sometimes termed "bioequivalence") has been proposed and discussed[3,9] as an adjunct method for comparing the toxic response of a specified sample concentration to a threshold or acceptable level of response. This method has application where it is desirable to determine statistically that a specific sample response is equivalent to, rather than different from, a targeted value. It is useful in pharmacokinetic applications or other experimental designs where limited numbers of doses can be tested, and when emphasis on the impacts of false-negative rather than false-positive statistical errors is important. In simplest form it is applied in a two-concentration test design (control and one dose, or two different doses), and is subject to the same concerns over variance and power as the null test.

Either hypothesis test method can be used to compare statistical significance between point estimates derived from regression models. This may be useful for comparing different species' response to toxicants, or different temporal responses.

a. Calculation of LC50: Acute toxicity test results generally are characterized by the median lethal concentration, LC50, when mortality is the end point or median effective concentration, EC50, when a sublethal effect is the end point. The LC50 is an estimate of the true median lethal concentration of the test material for the entire test species. Therefore, also provide a measure of statistical confidence in the point estimate, such as the 95% confidence interval of the LC50; values other than 50% can be used to characterize toxicity; however, the precision of test results for a typical sigmoid (or "s"-shaped) cumulative distribution dose-response curve may be best in the vicinity of the 50% effect level because this is the straightest part of the curve. LC values near the tails of this curve (e.g., LC10 or LC90) have wider confidence intervals.

Numerous procedures are available for analyzing quantal toxicity data. LC50 calculations include parametric procedures such as probit analysis,[7,10] logit,[11] and generalized linear models (GLiM).[4,5] The most commonly used nonparametric procedures are the Spearman-Karber method and the trimmed Spearman-Karber method,[12] while numerical interpolation techniques include graphical interpolation, moving average interpolation, and the binomial distribution. No single method is most appropriate for all data sets, but graphical interpolation and binomial distribution methods are simple to use. Personal computers facilitate the use of more sophisticated statistical methods and models, which may provide better fit of the experimental data for many data sets. The statistical procedure cannot always be selected before the test is conducted because the data generated must meet certain model assumptions or minimum criteria before the applicability of various methods can be determined.

Parametric procedures transform dose-response data to a known or expected functional form before the LC50 determination (parametric methods). The probit method[10] probably is the most widely used LC50 calculation procedure and uses the probit transformation of mortality data in combination with a standard curvefitting technique. A second parametric procedure utilizes a logit transformation[11] of mortality data. Common disadvantages of the parametric computational methods are that the distribution properties data must meet the model assumptions or the LC50 produced will not be considered appropriate. The probit and logit methods yield symmetric dose-response curves; they are not valid if the true curve is asymmetric. Models (GLiM) have been used to avoid such drawbacks.[4,5] Additionally, unless 0 and 100% mortality data are adjusted, the parametric methods can be utilized only when at least two partial kills are present in the data set.

The nonparametric Spearman-Karber method and the trimmed Spearman-Karber method[12] do not require that data meet model assumptions and do not depend on the presence of partial kills. These methods estimate the LC50 if the dose-response curve is symmetric. However, if the true curve is asymmetric, the Spearman-Karber method estimates the mean of tolerances and the trimmed Spearman-Karber method estimates the trimmed mean of tolerances; the LC50 is the median of tolerances. One drawback of the Spearman-Karber method is that the test data must cover the range from 0 to 100% mortality.

The moving average and moving average angle (transformation of mortality to angular values) methods also do not assume that the data distribution fits a predetermined model nor require partial kills for the LC50 calculation.[13] Shortcomings of these two methods are that the concentration series of the test must be equally spaced, and the methods cannot be used to calculate an LC value other than the LC50. Median effective time (ET50) for mortality at each concentration is estimated by plotting percentage mortality on a probit scale against time on a logarithmic scale and then using probit analysis techniques similar to those given above.[10,14–16] Procedures are the same, although more frequent observations of mortality may be required.

Single chemical toxicity tests often use preliminary range-finding tests with a wide range of test concentrations to obtain an estimate of toxicity. For subsequent definitive tests, a dilution series with more tightly spaced concentrations is used. For effluent toxicity tests, range-finding tests may not be feasible, because of stipulated limitations on effluent sample holding times. In these instances, effluent tests often produce results in which all organisms live at one test concentration, while all organisms die at the next higher concentration (i.e., no partial kills). For this type of data, the binomial test provides an approximate LC50 and specifies the concentration that can serve as statistically sound 95% confidence limits. However, the binomial test does not provide a true LC50 estimate and should be used only when the other available methods (e.g., probit) cannot be used. A preferred alternative to the binomial distribution method is the Williams test.[17]

Manual techniques for graphical determination of LC50 and confidence limits are available, and have been described in past editions of *Standard Methods*. A number of user-friendly computer software packages,* which can be used for parametric and nonparametric methods for treatment of data and generation of confidence limits, also are available. Because each of these methods has limitations with respect to certain data characteristics that can be analyzed (i.e., concentration series, number of partial kills, data type, data distribution, etc.) take care in using the appropriate method with the experimental data.

b. Control mortality: Generally, control mortality should not be greater than 10%. More than this is usually unsatisfactory and requires repetition of the test. Sometimes with long-term tests or with some invertebrates that have considerable mortality under the best possible conditions, it is necessary to use Abbott's formula:[18]

$$P = \frac{P^* - C}{1 - C}$$

where:

P and P^* = corrected and observed proportions responding to the experimental stimulus and
C = proportion responding in the control.

This approach does not solve the problem of possible interaction effects of the toxicant with whatever is causing mortality in the control.

Stephan[19] reviewed the methods available for calculating LC50s and recommended use of the moving average method and log concentration when one or more partial kills are present in a data set, and use of either binomial or moving average when no partial mortality is observed. Data sets of the latter type are not desirable when accurate estimates of toxicity are required.

A review of LC50 calculation methods[20] concluded that if the effect data are normally distributed, then probit was the most efficient method to use for quantal data, whereas the trimmed Spearman-Karber method was preferred when normality did not hold. While statistical models other than these have been recommended and used[5,8] for continuous and count data, similar evaluations of how the data fit the model assumptions must be made for any regression model. Because there may not be any single LC50 technique that is always best in all situations, goodness-of-fit estimates of the model outputs, as well as confidence limits, must be evaluated.

c. Hypothesis test methods for analyzing toxicity data: Traditionally, these methods have been used for chronic toxicity test results rather than for acute test results. However, with develop-

---

* Such as TOXSTAT®, University of Wyoming, Laramie; TOXCAL®, Tidepool Scientific Software, P.O. Box 2203, McKinleyville, CA; Toxdat, Statistical Support Staff, Biological Methods Branch, Environmental Monitoring and Support Laboratory, U.S. EPA, Cincinnati, OH.

ment of short-term critical life-stage test methods, many of which have lethality end points, the distinction between acute and chronic is sometimes specious; thus these statistical methods need not be limited to one type of toxicity test. Determine NOECs and LOECs with available computerized statistical methods for hypothesis testing (Dunnett's Test, Bonferroni's T-Test, Steel's Many-One Ranked Test, or Wilcoxon Rank Sum Test.[21] Parametric statistical methods require that individual test data meet normality assumptions, and that the variances of different treatment groups within a test are homogeneous. Therefore, before using these methods check the data to ensure that model assumptions are met. Use the Chi-square or Shapiro-Wilk's tests to test for normality, and Bartlett's test for homogeneity of variance. Statistical software packages (¶ 1a above) are suitable for these procedures.

Often it is necessary to transform data to conduct desired statistical analyses. Perform certain transformations routinely before any analysis. One example is percentage data (e.g., percent fertilized, hatch, survival). Adjust percent or proportion data by using the arcsine square-root transformation before using Dunnett's Test. This transformation corrects for the bias in treatment group variances that is expected for percent data. Transformations (e.g., log, square root) of quantitative data such as number of young or larval weight sometimes is useful in helping the data meet the assumptions of normality and homogeneity of variance.

d. Regression models for point estimates: ECP values other than median lethal concentrations are frequently desirable, particularly in chronic tests. Point estimation techniques for LC, EC, and IC determinations include probit, logit, GLiM, and linear interpolation. Linear interpolation and probit methods are available in the software packages referenced in ¶ 1a. GLiM models also are available in software products.† Probit usually is considered to be effective near median effect levels, and should be used only for quantal data; check to see that normality assumptions are met. Nonmonotonic dose-response curves, especially those with low-dose enhancement or thresholds, may best be analyzed by the appropriate GLiM model with data link,[4,5] or the nonparametric linear interpolation method.[8] The latter is often the most efficient method for calculating point estimates from nonquantal test results. One of the advantages of using an appropriate regression model is that nominal confidence intervals can be calculated for any given point estimate. This information can be used to help evaluate fit from alternative models, and, more importantly, provides an expression of statistical confidence in the test result. Confidence intervals can be constructed from the entire data set, or just the means of the experimental replicates. Typically, expect better coverage of the nominal confidence interval range from the former. "Bootstrapping" techniques typically are used for constructing confidence intervals, but the user should refer to the specific software provider or available literature for more information on specific methods.

e. Plotting the data: Many anomalies and trends in test results are not obvious unless the data are plotted. Plot chronic toxicity test data before making statistical analysis to assist in proper interpretation of results. By plotting the data, answers can be obtained to important questions, including the following: *(a)* Is the variability among replicates homogeneous across all concentrations? *(b)* What is the pattern of response vs. concentration— is the dose-response curve monotonic? *(c)* Do some concentrations

have enhanced response relative to controls? *(d)* Do some data appear to be highly unusual (outliers)? *(e)* Is the effect at the LOEC consistent with the concentrations above and below it, or could it be different just by chance? and *(f)* Do the confidence limits look plausible for the shape of the curve?

f. Outliers: Outliers are data points that are inconsistent with the trends exhibited by the majority of the data. They are detected most often by visual examination of plots of the data. Standard procedures for statistically detecting outliers are available[21] but should be used with extreme caution. In many instances, outliers are a result of measurement error, transcription, or data entry error. Because apparent outliers are usually mean values computed from replicates, inspection may show that the mean was affected by just one replicate, or that the variance among replicates was quite large. If the error can be traced and be corrected justifiably, then it is acceptable to do so. If not, make the analysis with and without the data point in question and report the results of both analyses. Regression methods usually are more robust to the effects of an outlier than are hypothesis test methods, and thus are a more critical concern in the latter.

2. Reporting Results

Report results from toxicity tests as completely as possible so that any conclusions can be evaluated independently. Include all of the following that are applicable: *(a)* test organisms used, including species, age, life stage, food used in cultures, acclimation, mean length, and weight, reference toxicant test data for the test population, diseases and treatment, source, and observations on behavior during test; *(b)* tested material: its source, storage, physical and chemical characteristics, and collection method and time; *(c)* dilution water: its source, storage, physical and chemical characteristics, collection method and time, pretreatments, additives, preparation (if applicable) and known contaminants; *(d)* test solution: its physical and chemical properties, especially toxicant concentrations (if applicable), and temperature; *(e)* test method, end point(s) of test, deviations from referenced procedures, data and time of initiation and termination, type and volume of test chambers, volume of test solutions, number of replicate test chambers per treatment, number of organisms per replicate, toxicant delivery, system, and flow rate or frequency of renewal; and *(f)* quality assurance methods used to ensure data integrity.

Present raw data for individual biological end points (e.g., mortality) and water quality measurements. Reference statistical methods and provide tabular summaries on toxic end points (e.g., LC50 with confidence limits, LOEC, NOEC, chronic value) and physical and chemical data. If applicable (e.g., effluent testing), include quality assurance data, such as results of reference toxicant tests, in tabular form.

3. References

1. BURTON, G.A., JR., W.R. ARNOLD, L.W. AUSLEY, J.A. BLACK, G.M. DEGRAEVE, F.A. FULK, J.F. HELTSHE, W.H. PELTIER, J.J. PLETL & J.H. RODGERS, JR. 1996. Effluent toxicity test variability. *In* Whole Effluent Toxicity Testing. D.R. Grothe, K.L. Dickson & D.K. Reed-Judkins, eds. SETAC Press, Pensacola, Fla.
2. PACK, S. 1993. A Review of Statistical Data Analysis and Experimental Design in OECD Aquatic Toxicology Test Guidelines. Report to OECD. Paris.
3. CHAPMAN, G.A., B.S. ANDERSON, A.J. BAILER, R.B. BAIRD, R. BERGER, D.T. BURTON, D.L. DENTON, W.L. GOODFELLOW, M.A. HERBER,

---

† Such as SAS®, SAS Institute, Cary, NC; The GLiM System, Release 3.77, Numerical Algorithms Group, Oxford, UK; or S-Plus®, Statistical Sciences, Seattle, WA.

L.L. McDonald, T.J. Norberg-King & P.J. Ruffier. 1996. Methods and appropriate endpoints. *In* Whole Effluent Toxicity Testing. D.R. Grothe, K.L. Dickson & D.K. Reed-Judkins, eds. SETAC Press, Pensacola, Fla.

4. Kerr, D.R. & J.P. Meador. 1996. Modelling dose response using generalized linear models. *Environ. Toxicol. Chem.* 15:395.

5. Bailer, A.J. & J.T. Oris. 1997. Estimating inhibition concentrations for different response scales using generalized linear models. *Environ. Toxicol. Chem.* 16:1554.

6. Dobson, A. 1990. An Introduction to Generalized Linear Models. Chapman & Hall, London, UK.

7. Bliss, C.I. 1934. The method of probits. *Science* 79:38.

8. Lewis, P.A., D.J. Klemm, J.M. Lazorchak, T.J. Norberg-King, W.H. Peltier & M.A. Heber, eds. 1994. Short-Term Methods for Estimating the Chronic Toxicity of Effluents and Receiving Waters to Freshwater Organisms, 3rd ed. EPA-600/4-91-002, Environmental Monitoring Systems Lab., U.S. Environmental Protection Agency, Cincinnati, Ohio.

9. Erickson, W.P. & L.L. McDonald. 1995. Tests for bioequivalence of control media and test media in studies of toxicity. *Environ. Toxicol. Chem.* 14:1247.

10. Finney, D.J. 1971. Probit Analysis, 3rd ed. Cambridge Univ. Press, London & New York.

11. Berkson, J. 1953. A statistically precise and relatively simple method of estimating the bioassay with quantal response based on the logistic function. *J. Amer. Statist. Assoc.* 48:565.

12. Hamilton, M.A., R. Russo & R.V. Thurston. 1977. Trimmed Spearman-Karber method for estimating median lethal concentrations in toxicity bioassays. *Environ. Sci. Technol.* 11:714.

13. Pickering, O.H. & W.N. Vigor. 1965. The acute toxicity of zinc to eggs and fry of the fathead minnow. *Progr. Fish-Cult.* 27:153.

14. Litchfield, J.T. 1949. A method for rapid graphic solution of time-percent effect curves. *Pharmacol. Exp. Ther.* 97:399.

15. Shepard, M.P. 1955. Resistance and tolerance of young speckled trout (*Salvelinus fontinalis*) to oxygen lack, with special reference to low oxygen acclimation. *J. Fish. Res. Board Can.* 12:387.

16. Sprague, J.B. 1973. The ABC's of pollutant bioassay using fish. *In* J. Cairns & K.L. Dickson, eds. Biological Methods for the Assessment of Water Quality. ASTM STP 528, p. 6. American Soc. Testing & Materials, Philadelphia, Pa.

17. Williams, D.A. 1986. Interval estimation of the median lethal dose. *Biometrics* 42:641; correction: *Biometrics* 43:1035.

18. Abbott, W.S. 1925. A method of computing the effectiveness of an insecticide. *J. Econ. Entomol.* 18:265.

19. Stephan, C.E. 1977. Methods for calculating an LC50. *In* F.L. Mayer & J.L. Hamelink, Aquatic Toxicology and Hazard Evaluation. ASTM STP 634, American Soc. Testing & Materials, Philadelphia, Pa.

20. Gelber, R.D., P.T. Lavin, C.R. Mehta & D.A. Schoenfeld. 1984. Statistical analysis. *In* G.M. Rand & S.R. Petrocelli, eds. Fundamentals of Aquatic Toxicology. Method and Applications. Hemisphere, New York, N.Y.

21. Snedecor, G.W. & W.G. Cochran. 1980. Statistical Methods, 7th ed. Iowa State Univ. Press, Ames.

4. Bibliography

Dunnett, C.W. 1955. A multiple comparison procedure for comparing several treatments with a control. *J. Amer. Statist. Assoc.* 50:1096.

Steel, R.G.D. & J.H. Torrin. 1960. Principles and Procedures of Statistics with Special Reference to Biological Sciences. McGraw-Hill Publ., New York, N.Y.

Jensen, A.L. 1972. Standard error of LC50 and sample size of fish bioassays. *Water Res.* 6:85.

Finney, D.J. 1978. Statistical Methods in Biological Assay, 3rd ed. Griffin Press, London, England.

Conover, W.J. 1980. Practical Nonparametric Statistics, 2nd ed. John Wiley & Sons, New York, N.Y.

Draper, N.R. & J.A. John. 1981. Influential observations and outliers in regression. *Technometrics* 23:21.

Miller, R.G. 1981. Simultaneous Statistical Inference. Springer-Verlag, New York, N.Y.

Dixon, W.J. & F.J. Massey, Jr. 1983. Introduction to Statistical Analysis, 4th ed. McGraw Hill, New York, N.Y.

Finney, D.J. 1985. The median lethal dose and its estimation. *Arch. Toxicol.* 56:215.

8010 H. Interpreting and Applying Results of Toxicity Tests

1. Interpretation of Results

The 48- and 96-h LC50 values produced from standard acute toxicity tests are useful estimates of relative acute lethal toxicity to test organisms under specified conditions. Without additional data, however, these values do not necessarily have any direct meaning in terms of "safe" or "hazardous" conditions in natural water (e.g., exposure analysis). Long-term exposure to much lower concentrations may be lethal to fish and other organisms and/or may cause nonlethal impairment of their function. Similarly, short-term exposure to these or higher values of total contaminants may cause no discernible effect. Numerous site-specific factors may influence the effect of the test material on the biota of a receiving water body.[1,2]

2. Influence of Test Conditions

The results obtained in a toxicity test, in large part, depend on the conditions and nature of exposure; they are the product of operationally defined procedures. Therefore, it is important to select the testing procedures carefully to provide appropriate conditions and ensure that the results are applicable to the water quality problem at hand. At the outset define the problem carefully and succinctly and establish how the results of the toxicity test will assist in the problem solution. Selection of type and species of test organism, life stage, test response, duration of test, and physical and chemical conditions of test are key factors in obtaining useful, interpretable toxicity test results.

Although not always possible, it usually is desirable to prescribe toxicity test conditions that are as close as possible to the natural environmental conditions. For some variables such as pH, temperature, and contaminants, even small differences between the laboratory test conditions and those in the natural environment can affect substantially, and therefore reduce, the utility of the results. Some situations dictate that "standard" toxicity tests (i.e., incorporating a standard set of test conditions such as 96-h exposure, standard chemical and physical conditions, and standard test organisms) be run in addition to those incorporating conditions more similar to those of the environment of concern. This practice provides results that could be compared with those reported in the literature as well as a potential route for adapting or interpreting results of other standard toxicity tests. In some instances, in-stream toxicity tests with caged organisms or laboratory tests using receiving-stream waters are a more reliable means by which to evaluate ambient toxicity. Such procedures can test for the impact of a complex variety of contaminants.

In the interpretation of toxicity test results to predict in-stream effects,[3] consider such factors as expected rate of dilution, aquatic chemistry, bioavailability, and duration and pattern of organism exposure in the environment. In addition, consider the function and sensitivity of the test organisms, compared to resident species of concern. Apply hazard assessment approach[4] to interpret more accurately the impact on aquatic organisms or designated beneficial uses of the water.

In recent years, aquatic toxicity testing has been applied to a variety of different regulatory and scientific purposes, including toxicity testing of municipal and industrial effluents as part of monitoring/permit compliance,[2-3,5-7] the derivation of national and site-specific water quality criteria for individual chemicals,[1,8,9] product safety evaluations,[10] chemical persistence studies,[5,11] effluent interaction studies, testing of leachates and sediments, and studies included in Toxicity Reduction Evaluation (TRE) programs to identify constituents causing toxicity in effluents.[2] These diverse applications have broadened the utility of toxicity testing, and made more important the judicious interpretation of their results.

3. Statistical Interpretation

It is crucial in conducting or interpreting the results of a toxicity test to understand clearly that statistically significant differences between control and test organisms, upstream and downstream populations, seasonal variations, etc., are not necessarily changes or differences that have ecological impact. Conversely, trends or other changes that appear to have biological/beneficial use significance may not be statistically demonstrable because of sample size or other limitations.

While the application and interpretation of toxicity tests in water quality management programs may appear to be more difficult than the more frequently used chemical test, they offer certain advantages: they directly address bioavailability and the complex interaction of multiple chemicals, they yield a single integrated measurement of organism response to a chemically complex sample, and they may be less expensive and easier to interpret than a series of chemical measurements.

4. References

1. U.S. ENVIRONMENTAL PROTECTION AGENCY. 1984. Water Quality Standards Handbook. Off. Water Regulations and Standards (WH-585), Washington, D.C.
2. U.S. ENVIRONMENTAL PROTECTION AGENCY. 1991. Technical Support Document for Water Quality-Based Control. EPA-505/2-90-001 (PB91-127415), Off. Water, U.S. Environmental Protection Agency, Washington, D.C.
3. GROTHE, D.R., K.L. DICKSON & D.K. REED-JUDKINS, eds. 1996. Whole Effluent Toxicity Testing: An Evaluation of Methods and Prediction of Receiving System Impacts. SETAC Pellston Workshop on Whole Effluent Toxicity, Sep. 16–25, 1995, Pellston, Mich. SETAC Press, Pensacola, Fla.
4. BERGMAN, H.E. KIMERLE & A.W. MAKI, eds. 1985. Environmental Hazard Assessment of Effluents. Pergamon Press, Inc., Elmsford, N.Y.
5. WEBER, C.I., ed. 1993. Methods for Measuring the Acute Toxicity of Effluents and Receiving Water to Freshwater and Marine Organisms, 4th ed. EPA-600/4-90-027F, Environmental Monitoring Systems Lab., U.S. Environmental Protection Agency, Cincinnati, Ohio.
6. KLEMM, D.J., G.E. MORRISON, T.J. NORBERG-KING, W.H. PELTIER & M.A. HEBER, eds. 1994. Short-Term Methods for Estimating the Chronic Toxicity of Effluents and Receiving Waters to Marine and Estuarine Organisms, 2nd ed. EPA-600/4-91-003, Environmental Monitoring and Support Lab., U.S. Environmental Protection Agency, Cincinnati, Ohio.
7. LEWIS, P.A., D.J. KLEMM, J.M. LAZORCHAK, T.J. NORBERG-KING, W.H. PELTIER & M.A. HEBER, eds. 1994. Short-Term Methods for Estimating the Chronic Toxicity of Effluents and Receiving Waters to Freshwater Organisms, 3rd ed. EPA-600/4-91-002, Environmental Monitoring Systems Lab., U.S. Environmental Protection Agency, Cincinnati, Ohio.
8. U.S. ENVIRONMENTAL PROTECTION AGENCY. 1985. Guidelines for Deriving Numerical National Water Quality Criteria for the Protection of Aquatic Organisms and Their Uses. NTIS-PB85-227049, National Technical Information Services, Springfield, Va.
9. U.S. ENVIRONMENTAL PROTECTION AGENCY. 1994. Interim Guidance on Determination and Use of Water Effect Ratios for Metals. EPA/823-B-94-001, U.S. EPA Off. Water, Washington, D.C.
10. U.S. ENVIRONMENTAL PROTECTION AGENCY. 1985. Environmental Effects Testing Guidelines. 40 CFR Part 797; *Federal Register* 50:39321.
11. U.S. ENVIRONMENTAL PROTECTION AGENCY. 1989. Method for Conducting Laboratory Toxicity Degradation Evaluations with Complex Effluents. Battelle Rep., March 1989.
12. BOTTS, J.A., J.W. BRASWELL, J. ZYMAN, W.L. GOODFELLOW & S.B. MOORE. 1989. Toxicity Reduction Evaluation Protocol for Municipal Wastewater Treatment Plants. EPA-600/2-88-062, Risk Reduction Engineering Lab., Off. Research and Development, U.S. Environmental Protection Agency, Cincinnati, Ohio.
13. FAVA, J.A., D. LINDSAY, W.H. CLEMENT, G.M. DeGRAEVE, J.D. COONEY, S.R. HANSEN, W. RUE, S. MOORE & P. LANKFORD. 1989. Generalized Methodology for Conducting Industrial Toxicity Reduction Evaluations. EPA-600/2-88-070, Risk and Reduction Engineering Lab., Off. Research and Development, U.S. Environmental Protection Agency, Cincinnati, Ohio.

8010 I.   Selected Toxicological Literature

Toxicity testing has become an integral part of the evaluation of the effects of waste discharges on the aquatic environment. Listed below are the principal scientific and engineering journals publishing results of original toxicological research. Textbooks and manuals that summarize and synthesize the results of original research also are listed.

1. Journals

Aquatic Toxicology
Archives of Experimental Contamination and Toxicology
Archives of Toxicology
Bulletin of Environmental Contamination and Toxicology

Bulletin of Marine Pollution
Chemosphere
Ecotoxicology and Environmental Health
Environmental Pollution
Environmental Toxicology and Chemistry
Marine Environmental Research

2. Textbooks and Manuals

BURTON, G.A., JR. 1992. Sediment Toxicity Assessment. Lewis Publ., Boca Raton, Fla.

ECOBICHON, D.J., ed. 1992. The Basis of Toxicity Testing. Lewis Publ., Boca Raton, Fla.

DALLINGER, R. & P.S. RAINBOW, eds. 1993. Ecotoxicology of Metals in Invertebrates. Lewis Publ., Boca Raton, Fla.

CALOW, P., ed. 1993-1994. Handbook of Toxicology. Vols. 1 and 2. Blackwell Scientific, Cambridge, Mass.

CAIRNS, J., JR. & B.R. NIEDEFLEHNER, eds. 1994. Ecological Toxicity Testing, Scale, Complexity, and Relevance. Lewis Publ., Boca Raton, Fla.

COCKERHAM, L.G. & B.S. SHANE. 1994. Basic Environmental Toxicology. Lewis Publ., Boca Raton, Fla.

HOFFMAN, D.J., ed. 1994. Handbook of Ecotoxicology. Lewis Publ., Boca Raton, Fla.

MALINS, D.C. & G.K. OSTRANDER, eds. 1994. Aquatic Toxicology. Lewis Publ., Boca Raton, Fla.

BREYER, W.N., ed. 1995. Interpreting Environmental Contamination in Animal Tissue. Lewis Publ., Boca Raton, Fla.

LANDIS, W.G. & M.H. YU. 1995. Introductions of Environmental Toxicology. Lewis Publ., Boca Raton, Fla.

NEWMAN, M.C. 1995. Quantitative Methods in Aquatic Toxicology. Lewis Publ., Boca Raton, Fla.

RAND, G.M., ed. 1995. Fundamentals of Aquatic Toxicology, 2nd ed. Taylor & Francis, Washington, D.C.

RÖMBKE, J. & J.F. MOLTMANN, eds. 1995. Applied Toxicology. Lewis Publ., Boca Raton, Fla.

TIMBRELL, J.A. 1995. Introduction to Toxicology, 2nd ed. Taylor & Francis, Bristol, Mass.

AMERICAN SOCIETY FOR TESTING AND MATERIALS. 1996. Annual Book of ASTM Standards, 11.05. American Soc. Testing & Materials, W. Conshohocken, Pa.

CHANG, L.W., ed. 1996. Toxicology of Metals. Lewis Publ., Boca Raton, Fla.

NEWMAN, M.C. & C.H. JAGOE, eds. 1996. Ecotoxicology: A Hierarchical Treatment. Lewis Publ., Boca Raton, Fla.

OSTRANDER, G.K., ed. 1996. Techniques in Aquatic Toxicity. Lewis Publ., Boca Raton, Fla.

STINE, K.E. & T.M. BROWN. 1996. Principles of Toxicology. Lewis Publ., Boca Raton, Fla.

8020 QUALITY ASSURANCE AND QUALITY CONTROL IN LABORATORY TOXICITY TESTS

8020 A. General Discussion

Quality assurance and quality control (QA/QC) are essential elements of laboratory bioassay procedures. A good QA/QC program provides framework and criteria for assessing data quality, including a well-defined chain of responsibility, explicit data quality objectives, procedures and protocols for testing, and a mechanism for identifying and correcting potential problems. Elements to be included in a quality assurance plan (QAP) are outlined in Section 1020A; other resources for developing a comprehensive QAP for laboratory toxicity testing programs are available.[1-5] As a minimum, QAPs for laboratories performing aquatic toxicity testing should provide specific guidance on data quality objectives, test procedures, sample handling, data management, internal quality control, and corrective action.

References

1. AMERICAN NATIONAL STANDARDS INSTITUTE & AMERICAN SOCIETY FOR QUALITY CONTROL. 1993. Quality Systems Requirements for Environmental Programs. ANSI/ASQC E4-1993, Milwaukee, Wisc.

2. RATLIFF, T.A., JR. 1990. The Laboratory Quality Assurance System—A Manual of Quality Procedures with Related Forms. Van Nostrand Reinhold, New York, N.Y.

3. U.S. ENVIRONMENTAL PROTECTION AGENCY. 1993. EPA Requirements for Quality Management Plans. EPA QA/R-2, Off. Research & Development, Quality Assurance Management Staff, U.S. Environmental Protection Agency, Washington, D.C.

4. U.S. ENVIRONMENTAL PROTECTION AGENCY. 1993. Guidance for Data Collection in Support of Environmental Decision-Making using the Data Quality Objective Process. EPA QA/G-4, Off. Research & Development, Quality Assurance Management Staff, U.S. Environmental Protection Agency, Washington, D.C.

5. U.S. ENVIRONMENTAL PROTECTION AGENCY. 1993. EPA Requirements for Quality Assurance Project Plans. EPA QA/R-5, Off. Research & Development, Quality Assurance Management Staff, U.S. Environmental Protection Agency, Washington, D.C.

8020 B. Elements of QA/QC

1. Data Quality Objectives

Data quality objectives are either qualitative or quantitative statements describing the overall acceptable uncertainty in results or decisions derived from environmental data. Such objectives for evaluating toxicity must ensure that information obtained will provide an accurate and precise estimate of environmental effects. They identify the types of measurements to be made, the allowable bias, and desired precision of measurements.

Accuracy is defined as the degree of agreement between an observed value and the true value or an accepted reference value. For water quality parameters, a measurement of accuracy might include calibration against a known standard. For toxicity testing, a reference toxicant test (i.e., exposing the test organism to a contaminated matrix of known toxicity) can be used as a measurement of accuracy of organism response). Precision is defined as the degree of agreement among repeated measurements collected under identical conditions and usually is described by a measure of variance (e.g., variance, standard deviation, coefficient of variation). For toxicity testing, precision of organism response can be described in a control chart of responses to a reference toxicant.

If the response (e.g., survival) of test organisms exposed to a sediment/water sample is significantly different from the response to a reference or control, then the organism has been affected by the sample. Traditionally, decisions of statistical significance are made at $\alpha = 0.05$. This means that the probability of a false positive result (detecting a difference when in fact none exists) must remain below 5%. Data quality objectives must control levels of bias (i.e., the difference between the measured value and true value) and precision to ensure that statistical significance is not affected by measurement error.

Minimum data quality objectives should be provided for: water quality in the test chamber (e.g., temperature, salinity, alkalinity, hardness, dissolved oxygen, pH, and ammonia); frequency and acceptable limits; minimum control survival; sensitivity of test organisms (e.g., reference toxicant testing); and frequency and number of observations.

Limits for bias and desired levels of precision generally are not stipulated in standardized test protocols described herein, but should be specified in the laboratory's own manual of Standard Operating Procedures (SOPs). Performance criteria (e.g., acceptable levels for control survival or water quality measurements) for most of these categories may be found in the test protocols for the organism of interest.

2. Test Procedures

Test procedures describe how to make all routine measurements associated with toxicity testing and related QA/QC activities. Follow these procedures to ensure integrity and quality of data.

Use SOPs and standardized data forms to ensure quality and consistency of toxicological testing and reporting. Write SOPs for all routine laboratory activities and periodically review and update them. Examples of quality control checklists, project schedule lists, procedural checklists, and test and reference toxicant procedures are available.[1-5]

Steps taken in the laboratory to reduce the potential for bias include blind testing, random assignment of organisms to test chambers, statistical designs (e.g., the randomized block) procedures to prevent cross-contamination, confirmation and witnessing of recorded observations, use of reference toxicant tests, and control charting.

Blind testing, where the experimental treatment is unknown to the analyst, prevents the analyst from applying biases upon the treatments based upon any preconceived expectations.

Use randomized designs to eliminate bias due to test chamber position within the test array. The completely randomized block, in which treatments are allocated to the experimental units at random, is the simplest form of the design. Each unit has an equal chance of receiving a particular treatment. In addition, process the units in a random order at all subsequent stages of a test where order has the potential to affect results. For example, position test containers maintained in a water bath under a light source randomly within the testing area. When replicates receiving a single treatment arc placed together, observed differences cannot be attributed solely to treatment; differences may have resulted from placement as well as treatment. Discussions of randomized block design, completely randomized block design, and other statistical aspects of experiment design are available.[6-10]

During setup and conduct of toxicity tests, prevent contamination from an external source and cross-contamination between treatments. Preventive measures include cleaning of equipment between contact with treatments, proper conditioning of laboratory test apparatus to minimize leaching, and covering test chambers to minimize loss of volatiles or extraneous contamination. Preferably, also analyze food, dilution water, and control water/sediment periodically for background contamination.

Periodic double checks of observations and calculations and witnessing of all raw data sheets (i.e., having a coworker review and sign each raw data sheet) are good preventative steps for early identification and correction of errors. Important preventative procedures should include counting animals twice to assure accuracy before adding them to the test chamber and periodic confirmation of calibration and measurements, particularly if environmental factors seem to be out of range.

Use reference toxicant tests to assess sensitivity of test organisms. Plot results from reference toxicant tests on control charts (Section 1020B) to determine whether the sensitivity of test organisms to a given reference toxicant is within a predetermined range of acceptability. Construct control charts by plotting successive values, for example, LC50s, for a reference toxicant, and evaluating temporal changes in sensitivity. Recalculate the mean and standard deviation with each plot until the statistics stabilize. Evaluate individual values in relation to the mean and standard deviation. Procedures for developing and using control charts are described in detail in Section 1020B and elsewhere.[11]

3. Sample Handling, Storage, and Shipment

Consistency in sample handling and tracking is most important for the testing of samples where legal ramifications are possible. To make technically sound decisions that withstand potential litigation, it is essential that samples be handled appropriately and be traceable to their source. Key components of this QA/QC element include established chain-of-custody procedures as well as procedures for sample sieving, subdividing, homogenization, compositing, storage, and monitoring.

Chain-of-custody procedures require an unbroken record of possession of a sample from its collection through analysis or testing, and possibly up to and during a court proceeding.[12] The goals of chain-of-custody are twofold: to ensure that the sample collected was the sample tested and to ensure that the sample has not been tampered with or altered in any way. Chain-of-custody can be accomplished through use of custody seals and sample tracking forms. Examples of such forms are available.[1,12]

a. Water and wastewater: Guidance for handling effluent samples under the National Pollutant Discharge Elimination System (NPDES) program dictates that samples are to be stored at 4°C and that the lapsed time from collection to initiation of testing should not exceed 36 h.[2] However, holding times may be adjusted depending on study objectives and other specific logistical considerations (e.g., shipment of samples from remote areas). If water samples are to be stored, keep headspace to a minimum. Before storage, floating debris may be removed, if necessary, by pouring water samples through 2- to 4-mm mesh sieve. If there is a possibility of interference due to the presence of indigenous organisms that show predation, competition, etc., pass samples through a 60-μm mesh sieve.[2] However, if volatile contaminants are of concern, take care to minimize aeration during collection, handling, storage, and testing.

b. Sediment: Sediment samples may require sieving before testing. Decisions regarding sieving are driven by presence of debris, such as twigs or leaves, that may impact recovery of test animals at test termination and/or the presence of indigenous species in the sample that may serve as food for, compete with, or actually prey upon the test organism. In any case, test results can be biased. If sieving is required, press-sieve all sediments without adding water (including reference and control sediments) before testing. In most cases, a 0.5-mm screen size is sufficient for removing predators while larger sieves may be used for removing debris. Recommendations regarding sieving of test material usually are found in specific standardized test protocols.

Depending on test objectives, samples may be composited, homogenized, and/or subdivided before testing. Use clean, noncontaminating containers and implements to handle and store samples. Suggested materials are stainless steel, TFE, Lexan®, high-density polyethylene, and glass. Other appropriate materials may be specified. Homogenize sediments to a consistent color and texture. Samples may be homogenized by hand with a spatula made of noncontaminating materials, or by mechanical mixing. Verify efficiency of homogenization by chemical analysis.

Sediments frequently are stored before testing. Current guidance for dredged material evaluations permits pre-test storage of sediment samples for up to 8 weeks from time of collection.[13] Preferably store sediment samples at 4°C with zero headspace or under an inert gas such as argon. Rehomogenize samples just before testing.

Maximum time limits for storage of sediments prior to testing are of concern; test samples as soon after collection as possible.

4. Data Recording, Reduction, Validation, and Reporting

Quality control of recording, reducing, validating, and reporting data is necessary for production of complete and scientifically defensible reports. Issues to be considered include maintenance of laboratory notebooks, data management, reporting and validation procedures, identification and handling of unacceptable data and outliers, measurements of completeness and comparability, and procedures for data archival.

Standardization of data recording facilitates electronic transfer and manipulation of data. At a minimum, standardize procedures for intralaboratory data entry. Identify no-data entries with a mark, "−," to indicate that data were not omitted. Use abbreviations for names of personnel and routine laboratory observations to reduce data recording and entry time; standardize these whenever possible. Attach a list of definitions and code descriptions to data sheets and project files. Record data in indelible ink; make corrections by drawing a single line through the mistake, correcting the mistake, dating and initialing the correction, and giving an initialed explanation for the lined-out data in a footnote at the bottom of the data sheet. More detailed guidance on maintaining laboratory notebooks can be found elsewhere.[14]

Validate all original data at each level of transcription (e.g., entering data from bound laboratory notebooks into computer databases). Arrange for an independent QA/QC review on a minimum of 10% of the data. Review laboratory record daily for outlier or unusual observations so that any necessary corrective action can be taken.

Criteria for establishing outlier values are program-specific. Toxicity endpoint outliers such as survival, growth, or reproduction may be more important than water quality outliers. Depending on program requirements, identify outliers and either accept them as "real" or reject and selectively remove. If outliers are removed from a data set, note this and clearly justify the reason. For example, an outlier for mortality in a given replicate might be reasonably excluded from a data set when it is clearly related to spurious low dissolved oxygen. If there is no rational explanation for the outlier, it must be assumed that the value is real and representative of the variability of the test system.

Completeness and comparability are two ways in which data quality can be assessed. Completeness is a measure of the amount of data obtained versus the amount of data originally intended for collection. Generally 80 to 90% is an acceptable level of completeness for water quality data. However, endpoint data, such as survival or reproduction, should be 100% complete; otherwise the statistical power of the test may be compromised. If data are less than 80% complete, use professional judgment to assess the data's usefulness for decision-making. Comparability is defined as the confidence with which one data set can be compared against another. Comparability and confidence can be enhanced through interlaboratory calibration including use of reference toxicants and control charts.

5. Internal Quality Control Checks

Internal quality control checks are "in-house" procedures implemented by the laboratory to ensure high-quality data. Internal quality control checks include review of documentation to determine that all samples are tested, sample holding times are not exceeded, holding conditions are acceptable, test protocols followed, instruments calibrated and maintained, and control survival and water quality conditions are within acceptable ranges. Other important issues are verification of the taxonomy and viability of test organisms.

Document source and culture history of test organisms. If possible, preserve a subsample of the test organisms for future identification in the event of aberrant toxicity. The age, size, and/or maturity of the test organisms usually are specified in the test protocol; verify these. Specify appropriate holding time and acclimation procedures either in the test protocols or the laboratory's SOPs; ensure that resulting documentation is available for audit.

Two widely accepted ways to assess test organism viability are the use of test-validation controls and reference toxicant tests. A test validation control is a group of organisms that, with the exception of the treatment factor, are handled in a manner identical to the other organisms in the test. Acceptable levels of mortality in the test validation controls for most acute lethality tests are limited to ≤10% (i.e., survival ≥90%). If less than 90% survival is achieved in the test validation control, the test is considered invalid and must be repeated. For chronic sublethal tests, the test validation control also may include acceptable limits for other endpoint data such as growth and reproduction. Reference toxicant tests are designed to assess sensitivity to a specific contaminant. In a reference toxicant test, organisms are exposed to a range of concentrations of a single contaminant or contaminant mixture in water-only exposures and an LC50 (usually 96 h) is calculated. Evaluate results of reference toxicant tests in a laboratory control chart (see Section 8020B.2).

Before testing, develop guidance for defining deviations, deficiencies, and appropriate corrective action. Corrective action may be required when a deficiency or deviation from planning documents or procedures is discovered or when there are deviations from established data quality objectives.

Deviations are defined as data outside the range specified in data quality objectives. Out-of-compliance data may be due to deviations in test protocols or deficiencies associated with toxicological tests. Examples of deviations from the DQO in toxicity tests include excessive control mortality, out-of-range water quality conditions, lack of randomization, lack of required reference, control, and/or out-of-range reference toxicant results.

Poor control survival, loss of control over exposure conditions, major mechanical errors, or mishandling of test organisms may result in a decision to retest. However, brief episodes of out-of-range water quality conditions or incomplete test monitoring information may require only that data be flagged and qualified. A number of typical test deviations and suggested corrective actions are summarized in Table 8020:I.

Corrective actions may include, but are not limited to, reviewing the data and calculations, identifying and qualifying suspicious data, and retesting. Review all "out-of-limit" events as soon as data are tabulated and validated.

6. References

1. MOORE, D.W., T.M. DILLON, J.Q. WORD & J.A. WARD. 1994. Quality Assurance/Quality Control (QA/QC) Guidance for Laboratory Dredged Material Bioassays—Results of QA/QC Workshop Held May 26-27, 1993, in Seattle, Washington. Misc. Paper D-94-3, U.S. Army Corps of Engineers Waterways Experiment Sta., Vicksburg, Miss.

2. U.S. ENVIRONMENTAL PROTECTION AGENCY. 1991. Methods for Measuring the Acute Toxicity of Effluents and Receiving Waters to Freshwater and Marine Organisms, 4th ed. EPA-600/4-90-027, Off. Research & Development, Washington, D.C.

3. U.S. ENVIRONMENTAL PROTECTION AGENCY. 1994. Methods for Measuring the Toxicity and Bioaccumulation of Sediment-Associated Contaminants with Freshwater Invertebrates. EPA-600/R-94-024. Off. Research & Development, Washington, D.C.

4. U.S. ENVIRONMENTAL PROTECTION AGENCY. 1994. Methods for Assessing the Toxicity of Sediment-Associated Contaminants with Estuarine and Marine Amphipods. EPA-600/R-94-025. Off. Research & Development, Washington, D.C.

5. U.S. ENVIRONMENTAL PROTECTION AGENCY. 1991. Manual for the Evaluation of Laboratories Performing Acute Toxicity Tests. EPA/600-4-90/031, U.S. Environmental Protection Agency, Cincinnati, Ohio.

6. SOKAL, R.R. & F.J. ROHLF. 1981. Biometry—The Principles and Practices of Statistics in Biological Research, 2nd ed. W.H. Freeman & Co., New York, N.Y.

7. COCHRAN, W.G. & G.M. COX. 1957. Experimental Designs, 2nd ed. John Wiley & Sons, New York, N.Y.

8. GAD, S.C. & C.S. WEIL. 1988. Statistics and Experimental Design for Toxicologists. Telford, Caldwell, N.J.

9. HICKS, C.R. 1982. Fundamental Concepts in the Design of Experiments. Holt, Rinehart & Winston, New York, N.Y.

10. HURLBERT, S.H. 1984. Pseudo replication and the design of ecological field experiments. *Ecol. Monogr.* 54(2):187.

11. AMERICAN SOCIETY FOR TESTING AND MATERIALS. 1992. Manual on Presentation of Data and Control Chart Analysis, 6th ed. ASTM Manual Ser. MNL 7, Revision of Spec. Tech. Publ. (STP) 15D. American Soc. Testing & Materials, Philadelphia, Pa.

12. RATLIFF, T.A., JR. 1990. The Laboratory Quality Assurance System—A Manual of Quality Procedures with Related Forms. Van Nostrand Reinhold, New York, N.Y.

13. U.S. ENVIRONMENTAL PROTECTION AGENCY & U.S. ARMY CORPS OF ENGINEERS. 1994. Evaluation of Dredged Material Proposed for Discharge in Inland and Near Coastal Waters—Testing Manual. EPA-823-B-94-002, U.S. Environmental Protection Agency, Off. Water and Dep. of the Army, U.S. Army Corps of Engineers, Washington, D.C.

14. KANARE, H.M. 1985. Writing the Laboratory Notebook. American Chemical Soc., Washington, D.C.

TABLE 8020:I. SUMMARY OF TYPICAL TEST DEVIATIONS AND NEED FOR RETESTING

Deviation	Need for Retesting	
	Required	Possible*
Lack of test array randomization		√
Testing not blind		√
Required references, controls not tested	√	
Test chambers not identical		√
Test containers broken or misplaced		√
Mean control mortality exceeds acceptable limits	√	
Excessive control mortality in a single replicate		√
Test organisms not randomly assigned to test chambers		√
Test organisms not from the same population		√
Test organisms not all the same species or species complex	√	
Test organism holding time exceeded		√
Water quality parameters consistently out of range	√	
Brief episodes of out-of-range water quality parameters		√
Test monitoring documentation incomplete		√
Sample holding times exceeded	√†	
Sample storage conditions outside acceptable ranges		√†

* If not retested, data may have to be qualified depending on study objectives.
† Unless evidence provided to clearly show that sample quality (physico-chemistry and contaminant levels) has not been affected.

8030 MUTAGENESIS*

8030 A. Introduction

1. Significance

Mutagenesis may be defined as the induction of a heritable change in an organism's genetic material. Studies of human cancer have established that a mutagenic event is very likely the initiating factor in some kinds of cancers. Most known carcinogenic chemicals and radiations also are mutagenic. Therefore, demonstration of mutagenic activity suggests that the substance may be (but need not be) carcinogenic. The common association between mutagenic activity and carcinogenicity is the basis for using short-term mutagenesis tests with bacteria or cultured cells to detect potential carcinogens.

It is also known that mutagens of natural origin are ubiquitous in the environment.[1] Therefore, the relationship between mutagenic activity of an environmental sample and chemical pollutants in the sample must be examined very carefully in order to draw conclusions about the source of mutagenic activity.

2. Selection of Method

Many tests to detect mutagenicity exist.[2] Tests using bacteria, particularly the *Salmonella* microsomal mutagenicity (Ames) test, are most common.[3,4] The latter test uses nonvirulent "tester strains" of *Salmonella typhimurium*. Because many mutagenic chemicals require some metabolic processing by enzymes that are lacking in the bacteria, mammalian enzyme preparations can be added. The test is simple, inexpensive, and sensitive, gives results in 2 d, and shows good correlation between mutagenicity/nonmutagenicity and carcinogenicity/noncarcinogenicity in rodents. Disadvantages of the test are that some mutagens active in mammalian cells will not be detected, carcinogens that are not also mutagens (e.g., asbestos) will not be detected, and the relative potency of mutagens in the test does not necessarily correlate with the carcinogenic potency in mammals. Although the test provides *qualitative* information regarding mutagens, it shows great variability among laboratories in *quantitative* terms. However, it is of great value as a preliminary screening test.

Because environmental samples are complex mixtures, no standard application of the *Salmonella* microsomal mutagenicity test is possible.[5,6] The design of an individual test will depend on the sample and circumstances and the desired information regarding mutagenic activity. For example, to determine whether a particular mutagen is present, a specific chemical extraction procedure may be necessary. If a concentration step is used, it must be considered in reporting results. To determine whether a particular treatment process (e.g., chlorination or ozonation) leads to mutagen production, samples should be assayed before and after treatment. Sample preparation depends on the nature of materials in the raw water.[6,7] The methods described here are for the Ames plate incorporation test. Generally they are applicable to samples soluble in aqueous or organic solvents. Within limits, they will provide preliminary data on the presence of mutagenic materials.

* Approved by Standard Methods Committee, 1996.

For definitive information, tailor the Ames test and sample preparation to the specific situation. When reporting results, state the sample preparation used.

3. Sample Collection, Storage, and Preparation

Because of variability among samples, a single procedure for sample preparation cannot be provided, but general principles apply:[5-9]

Make tests as soon as possible after sample collection. Delay in testing will be accompanied by a progressive loss in mutagenic activity, no matter how the sample is stored. To minimize loss of mutagenic activity, store samples at or below $-20°C$, under an inert atmosphere (N_2, argon), and protected from light. Many mutagens, particularly polycyclic aromatic hydrocarbons (PAHs) are readily photooxidized to nonmutagenic, but still toxic, compounds. Add no preservatives.

Exposure of the *Salmonella* tester strains requires the sample to be in a solvent compatible with the aqueous suspension of the bacteria. Dimethylsulfoxide (DMSO) is used frequently. If the sample is in a volatile solvent such as dichloromethane (DCM) or hexane, add DMSO and remove the organic solvent with a stream of N_2, leaving behind the relatively less volatile DMSO. This is called solvent-exchanging. After the sample has been transferred to DMSO analyze immediately.[10]

4. References

1. AMES, B.N. 1989. What are the major carcinogens in the etiology of human cancer? Environmental pollution, natural carcinogens, and the causes of human cancer: Six errors. *In* V.T. DeVita, Jr., S. Hellman & S.A. Rosenberg, eds. Important Advances in Oncology. J.B. Lippincott, Philadelphia, Pa.

2. HOLLSTEIN, M. & J. McCANN. 1979. Short-term tests for carcinogens and mutagens. *Mutation Res.* 65:133.

3. AMES, B.N., J. McCANN & E. YAMASAKI. 1975. Methods for detecting carcinogens and mutagens with the *Salmonella*/mammalian microsome mutagenicity test. *Mutation Res.* 31:347.

4. MARON, D. & B.N. AMES. 1983. Revised methods for the *Salmonella* mutagenicity test. *Mutation Res.* 113:173.

5. MEIER, J.R. 1988. Genotoxic activity of organic chemicals in drinking water. *Mutation Res.* 196:211.

6. WANG, Y.Y., C.P. FLESSEL, L.R. WILLIAMS, K. CHANG, M.J. DIBARTOLOMEIS, B. SIMMONS, H. SINGER & S. SUN. 1987. Evaluation of guidelines for preparing wastewater samples for Ames testing. *In* S.S. Sandhu, D.M. DeMarini, M.J. Mass, M.M. Moore & J.L. Mumford, eds. Short-Term Bioassays in the Analysis of Complex Environmental Mixtures V. Plenum Publishing Corp., New York, N.Y.

7. WANG, Y.Y., C.P. FLESSEL, K.-I. CHANG, D.A. HOLLANDER, P.J. MARSDEN & L.R. WILLIAMS. 1989. Evaluation of a protocol for preparing drinking water samples for Ames mutagenicity testing. *In* R.L. Jolley et al., eds. Water Chlorination: Chemistry, Environmental Impact and Health Effects, Vol. 6. Lewis Publishers, Inc., Chelsea, Mich.

8. ICAIR. 1985. Guidelines for preparing environmental and waste samples for mutagenicity (Ames) testing: Interim procedures and panel

meeting proceedings. EPA-600/4-85-058. Environmental Monitoring Systems Lab., U.S. Environmental Protection Agency, Las Vegas, Nev.

9. MARSDEN, P.J., D.F. GURKA, L.R. WILLIAMS, J.S. HEATON & J.P. HELLERSTEIN. 1987. Interim procedures for preparing environmental samples for mutagenicity (Ames) testing. In I.H. Suffet & M. Ma-

laigandi, eds. Organic Pollutants in Water: Sampling, Analysis and Toxicity Testing. ACS Advances in Chemistry Ser. No. 214. American Chemical Soc., Washington, D.C.

10. MARON, D., J. KATZENELLENBOGEN & B.N. AMES. 1981. Compatibility of organic solvents with the Salmonella/microsome test. Mutation Res. 88:343.

8030 B.   *Salmonella* Microsomal Mutagenicity Test

1. General Discussion

a. Principle: Tester strains of *S. typhimurium* require the amino acid histidine for growth. Reversion of the tester strains from histidine dependence to independence, i.e., reversion of the histidine requirement, is evidence of mutagenicity. The bacteria can be cultivated on simple media and show reproducible responses to test mutagens. The bacteria are exposed to the sample, with or without additional activating enzymes, and are plated on minimal agar containing a trace amount of histidine. Mutants (i.e., revertants to histidine independence) are able to grow and form macroscopic colonies. The dose-response can be quantified by varying sample concentration and counting revertant colonies per plate at each concentration. The number of revertants per unit dose of sample is calculated with statistical methods.

b. Tester strains: The tester strains currently most widely used in testing environmental samples are TA98 and TA100. Strains TA97a and TA102 can be used but are found to give high and variable rates of background mutation. Because the different strains are reverted by different classes of mutagens, using multiple strains provides information on the nature of the mutagenic chemical(s) present. These strains also contain the R-factor plasmid pKM101, which confers resistance to ampicillin. Other commonly used strains include TA1535 and TA1538, which do not contain the R-factor plasmid. Details of the tester strain mutations and other available strains have been published.[1-4] Quality assurance requirements and procedures,[5,6] as well as data production and analysis methods,[7,8] are available.

Tester strains presently are available without cost by written request to Dr. Bruce N. Ames, Department of Biochemistry, University of California, Berkeley, California 94720.

2. Apparatus

a. Autoclave: See Section 9030B.3.

b. Water bath, reciprocating, for use at 37°C.

c. Water bath for use at 45°C to 50°C.

d. Incubator for use at 37°C.

e. Refrigerator or cold room.

f. Freezer, −80°C (or liquid nitrogen refrigerator).

g. pH meter.

h. Centrifuge, capable of $10\,000 \times g$.

i. Vortex mixer.

j. Magnetic stir-plate and stirring bars.

k. Hot plate.

l. Colony counter, manual and automatic (optional).

m. Microscope, dissecting or light.

n. Micropipetor, 10 μL, 100 μL, 500 μL, and 5 mL.

3. Media and Reagents

a. Nutrient broth: Use dehydrated nutrient broth prepared in accordance with manufacturer's directions.*

b. Nutrient agar plates: These are used to test for sensitivity to crystal violet. Add 15 g agar/L nutrient broth, before sterilizing.† Mix well and autoclave at 121°C for 15 min with slow exhaust. Remove from autoclave and let cool to about 50°C. Pour 25 to 30 mL into 100-mm petri plates and lct harden on a level surface. To evaporate excess moisture, hold covered plates in a clean, draft-free environment overnight. Store prepared plates in a tightly covered container in the refrigerator.

c. Vogel-Bonner medium E (50X): This is an inorganic salt medium[2] used in the preparation of minimal agar. To prepare 1 L of 50X concentrate, heat 670 mL water to 45°C and add in order (making sure each salt is completely dissolved before adding the next) 10 g magnesium sulfate heptahydrate, $MgSO_4 \cdot 7H_2O$; 100 g citric acid monohydrate; 500 g potassium phosphate, dibasic K_2HPO_4 (anhydrous); and 175 g sodium ammonium phosphate, $NaNH_4HPO_4 \cdot 4H_2O$. Make up to 1 L with water in a loosely capped 2-L flask or bottle and sterilize by autoclaving for 20 min at 121°C. After cooling, tighten cap and store at room temperature.

d. Glucose solution, 40%: To 600 mL water add 400 g D-glucose. Stir and make up to 1 L with water. Mix well and sterilize in a loosely capped flask by autoclaving for 20 min at 121°C (slow exhaust). Alternatively, dispense to 250-mL or 500-mL bottles with rubber-lined screw caps before autoclaving. Leave caps loose during autoclaving. Tighten after solutions have cooled to room temperature.

e. Minimal agar plates for use in the mutagenicity test. To 930 mL water in a 2-L flask add 15 g agar and a magnetic stirring bar. Mix, cap loosely, and autoclave for 20 min at 121°C, with slow exhaust. Remove from autoclave, cool slightly, and add the following sterile solutions slowly with continuous stirring: 20 mL 50X Vogel-Bonner medium E and 50 mL 40% glucose. Mixing is facilitated if the salts and glucose solutions are first warmed to 45°C.

Place agar in a 45°C water bath and pour approximately 25 to 30 mL into 15-mm × 100-mm petri plates. Let agar harden on a level surface and cool to room temperature. To let excess moisture evaporate, hold plates covered in a clean, draft-free area overnight or up to 2 d. A convenient method of plate storage is to return them to the plastic bags in which they were originally packaged and to seal the bags securely with tape. Long-term storage

---

* Oxoid #2, Oxoid Ltd., Basingstoke, Hants, England, available in U.S.A. from KC Biological, Inc., Lenexa, KS, or equivalent.
† Bacteriological agar, BBL Select, Oxoid #L28, or equivalent.

at room temperature is acceptable. If plates are stored under refrigeration let them come to room temperature before use. Autoclave and discard any plates showing contamination.

f. Histidine-biotin solution, 0.5 mM: This solution is added to the top agar in the proportion of 10 mL to 100 mL top agar. (It provides a necessary trace of histidine to permit bacteria to undergo a few cell divisions. The tester strains also are biotin-dependent, but as this requirement is the result of a gene deletion, it cannot be reverted.) Add 12.4 mg D-biotin to 100 mL water. Dissolve by heating to the boiling point and add 9.6 mg L-histidine·HCl. Sterilize by filtering through a 0.22-μm-pore-diam filter or by autoclaving for 20 min at 121°C, with slow exhaust.

g. Top agar:

Agar	6 g
Sodium chloride, NaCl	5 g
Distilled water	1 L

Add magnetic stirring bar, mix and autoclave for 20 min at 121°C, with slow exhaust. While agar is still melted, mix thoroughly and dispense 100-mL portions into sterile screw-capped bottles of a convenient size (100 to 250 mL). Alternatively, make top agar in 100-mL amounts (0.6 g agar, 0.5 g NaCl, 100 mL H_2O) and autoclave in loosely capped bottles. Cool to room temperature, tighten caps, and store at 4°C. Before use remelt top agar in a boiling water bath or microwave oven and add 10 mL sterile 0.5 mM histidine-biotin solution. Hold top agar in a 45°C water bath or dry heat device.

h. Phosphate-buffered saline (PBS),[9] for washing bacteria in the micro suspension mutagenicity test method.[10] To 900 mL distilled water add 8.0 g sodium chloride, NaCl; 0.2 g potassium chloride, KCl; 0.2 g potassium phosphate, monobasic, KH_2PO_4; 0.1 g magnesium chloride hexahydrate, $MgCl_2 \cdot 6H_2O$; and 1.15 g sodium phosphate, dibasic, Na_2HPO_4.

Dissolve completely and add 0.10 g calcium chloride, $CaCl_2$, dissolved in a little water. Adjust to pH 7.4 with either HCl or NaOH as appropriate. Make up to 1 L. Sterilize by filtration through a 0.22-μm-pore-diam filter or equivalent.

i. Sodium phosphate buffer, pH 7.4, used in S9 mix (see ¶ *m* below). Prepare stock solutions:

1) *Sodium phosphate,* monobasic, monohydrate, $NaH_2PO_4 \cdot H_2O$, 13.8 g/500 mL water.

2) *Sodium phosphate,* dibasic, anhydrous, Na_2HPO_4, 14.2 g/500 mL water.

Mix 60 mL Solution 1) with 440 mL Solution 2). Check pH and adjust if necessary to pH 7.4 by adding more of one of the stock solutions. To lower pH add Solution 1); to raise it add Solution 2). Sterilize by autoclaving for 20 min at 121°C with slow exhaust.

j. Ampicillin stock solution, 8 mg/mL: Make a solution of 0.80 g ampicillin trihydrate in 100 mL 0.02N NaOH. Sterilize by filtering through a 0.22-μm-pore-diam membrane filter. Store in capped glass bottle at 4°C. Add 3.15 mL/L master plate agar solution and pour plates as usual.

k. Master plate agar: To a 2-L flask add 15 g agar, 914 mL water, and a magnetic stirring bar. Mix and autoclave for 20 min at 121°C with slow exhaust. Remove from autoclave and add, with stirring, 20 mL 50X Vogel-Bonner medium E salts; 50 mL 40% glucose; 10 mL sterile L-histidine·HCl solution (0.5 g/100 mL); 10 mL sterile D-biotin solution (12.2 mg/100 mL); and 3.15 mL ampicillin stock solution. (NOTE: The final concentration of

histidine in the master plates is approximately fivefold greater than in top agar. Clearly label plates to distinguish them from minimal agar plates. For strains without the R-factor plasmid, omit ampicillin.)

l. Crystal violet solution, 0.1%: Dissolve 0.1 g in 100 mL water. Mix well and store at 4°C in screw-cap glass bottle in the dark. Use to confirm presence of the *rfa* mutation.

m. S9 mix: S9, a cell-free fraction prepared by homogenization and centrifugation of rat liver (or other tissue) at 9000 × *g* for 10 min, is added when metabolic activation is required. Prepare S9 from the liver of rats pretreated with polychlorinated biphenyls (PCBs, Aroclor 1254) to increase activity of liver enzymes.[2] Unless animal facilities are available, preferably obtain S9 commercially.‡

Because S9 contains temperature-sensitive enzymes, store frozen at − 80°C or below and thaw only for immediate use. Each mutagen may have an optimum concentration of S9 for maximal mutagenic activity. Because this optimum cannot be specified in advance, and the amount of sample often is limited, standardize on an S9 concentration. Between 20 and 40 mg protein/mL S9 is common, but consistency is essential.

Standardize S9 according to protein content. Determine protein content of a small portion of undiluted S9.[11] Freshly prepared S9 has a protein content of about 40 mg/mL.[2] Adjust protein concentration immediately before use to the desired concentration with 0.1*M* sodium phosphate buffer, pH 7.2–7.4, or with 0.15*M* KCl. Check for sterility by spreading 0.1 mL on a minimal agar plate containing histidine and biotin and incubate for 2 d at 37°C. Discard contaminated S9 (more than 10 colonies/0.1 mL).

In an assay add cofactors to provide necessary reducing activity for the cytochrome enzymes. Add reduced nicotinamide adenine dinucleotide phosphate (NADPH) or a NADPH generating system consisting of nicotinamide adenine dinucleotide phosphate (NADP), glucose-6-phosphate (G-6-P), and $MgCl_2$. The combination of S9 and cofactors is termed ''S9 mix.'' Ideally, select the amount of S9 in the mix to give optimum mutagenic response with the sample. Practically, use either 4 mL or 10 mL S9/100 mL S9 mix (i.e., 4% or 10% S9, equivalent to approximately 1.6 mg or 4.0 mg protein/mL S9 mix, respectively).[2] For consistency adjust the protein concentration in S9 mix to these concentrations. Once prepared, keep S9 mix on ice and use immediately; do not refreeze.

Use the following stock solutions for preparing S9 mix:

1) *Potassium chloride,* KCl, 1.65*M*: Dissolve 12.3 g KCl in 80 mL water and make up to 100 mL. Autoclave for 20 min at 121°C and store at room temperature.

2) *Magnesium chloride hexahydrate*, $MgCl_2 \cdot 6H_2O$, 0.4*M*: Dissolve 8.13 g $MgCl_2 \cdot 6H_2O$ in 80 mL distilled water and make up to 100 mL. Autoclave for 20 min at 121°C and store at room temperature.

3) *Sodium phosphate buffer:* See ¶ 3*i* above.

4) *NADP (nicotine adenine dinucleotide phosphate),* 0.1*M*: Dissolve 743 mg in 80 mL sterile water and make up to 100 mL with sterile water. Sterilize by filtering through a 0.22-μm-pore-diam membrane filter. Store 5- or 10-mL portions at − 20°C for

‡ AMC Cancer Research Center and Hospital, c/o Dr. Elias Balbinder, 6401 W. Colfax Ave., Lakewood, CO 80214; Litron Laboratories, 1351 Mt. Hope Ave., Suite 207, Rochester, NY 14620; Microbiological Associates, c/o Dr. Steve Haworth, 5221 River Road, Bethesda, MD 20816; MolTox, 335 Paint Branch Drive, College Park, MD 20742; or Organon Teknika, 1 Technology Court, Malvern, PA 19355.

up to 6 months. (NOTE: The weight given here is the formula weight of anhydrous free acid. The sodium salt typically is used and it may have associated water. Thus the amount required will vary from lot to lot. Most suppliers provide a specification sheet with the calculated formula weight. Use this value in preparing the solution. Typical values range from 750 to 825.)

5) *Glucose-6-phosphate, 1M:* Dissolve 6.5 g in 20 mL sterile water and make up to 25 mL. Sterilize by filtering through a 0.22-μm-pore-diam membrane filter. Store 5- or 10-mL portions at $-20°C$ for up to 6 months.

To prepare 50 mL S9 mix, add in order, to a vessel in an ice bath: 18.75 mL or 15.75 mL sterile water; 25 mL 0.2M sodium phosphate buffer; 2.0 mL 0.1M NADP; 0.25 mL 1.0M glucose-6-phosphate; 1.0 mL 1.65M KCl; 1.0 mL 0.4M MgCl$_2$·6H$_2$O; and 2.0 or 5.0 mL rat liver S9. (NOTE: The amount of S9 can be varied to obtain desired protein concentration. Preferably use either 2 mL (4%) or 5 mL (10%) of rat liver S9/50 mL S9 mix.[2] Adjust water volume to maintain concentrations of other reagents.)

Prepare S9 mix immediately before use. Discard any unused portion. Do not refreeze.

4. Tester Strain Stock Cultures

a. Preparation of stock cultures: Tester bacteria are available (see ¶ 1b) as small paper disks saturated with the *Salmonella* culture, sealed in small sterile plastic bags with a little agar. On receipt, aseptically remove disks and make subcultures by wiping the disk across an agar master plate and then placing disk in sterile nutrient broth. Adjust volume of nutrient broth, depending on how many frozen stock vials are to be produced. To an overnight culture in nutrient broth add dimethylsulfoxide (DMSO spectrophotometric grade, 0.09 mL DMSO/mL culture). Mix well and dispense aseptically into sterile 1.5-mL cryotubes, filling each tube almost to the top. Label, freeze in crushed dry ice, and store at $-80°C$ or in liquid nitrogen refrigerator. If frozen cultures are to be used repeatedly do not allow them to thaw, because this tends to increase the spontaneous background mutation rate and the chance of contamination. If freezer facilities are not available, less preferably maintain strains by repeated subculture and confirm strain characteristics at each mutagenicity test.[2] To prepare bacteria for a test remove a small amount of the frozen culture with a sterile spatula and inoculate a broth culture. Cultures also can be preserved by lyophilization.[2]

As an alternative to repeatedly sampling the frozen stocks use the "master plate" method.[2] Streak a drop of a broth culture on a minimal agar plate and incubate for 48 h at 37°C. The plate may be stored in a refrigerator for up to 2 weeks. Use well-separated colonies from this plate to initiate broth cultures. Prepare new master plates from broth cultures initiated from the frozen stocks.

b. Characterization of tester strains: Characterize tester strains immediately on receipt and preferably as a part of each mutagenesis assay. The procedures described here in are for strains TA98 and TA100. Procedures for other strains are available.[2–4] Strains TA98 and TA100 require histidine and biotin for growth. In addition, they contain the *rfa* mutation affecting the cell wall, the *uvrB* deletion leading to reduced DNA repair capacity, and the R-factor plasmid pKM101 conferring resistance to ampicillin.

Confirm the histidine requirement by streaking a sample of broth culture on a minimal agar plate containing biotin, but lacking histidine. The biotin requirement is the result of a deletion and cannot revert. No bacterial growth should be seen on the histidine-deficient plate.

The *rfa* mutation renders the cell wall permeable to large molecules; growth of the tester strains is inhibited by crystal violet. To test for this inhibition, add 0.1 mL of broth culture to 2 mL melted top agar, and spread the mixture on a nutrient agar plate. Place a sterile 6.4-mm-(¼-in.-)diam disk of filter paper§ in center of plate and add 10 μL 0.1% crystal violet solution to the filter paper. The crystal violet will diffuse into the agar; sensitive strains show a clear zone around the paper disk, indicating growth inhibition. The *uvrB* deletion confers increased sensitivity to ultraviolet light, which can be demonstrated by comparison with the wild-type strain.[12] The presence of the R-factor plasmid allows tester strains to grow on agar containing ampicillin.

The strains are reverted by different mechanisms and their response to chemical mutagens depends on the mode of interaction of the chemical with the bacterial DNA. Strain TA98 is reverted by frame-shift mutagens that generally are large molecules such as polycyclic aromatic hydrocarbons. The reading frame of the DNA is shifted by the deletion or addition of a base pair. Strain TA100 is reverted by mutagens that cause base-pair substitutions in the bacterial DNA. In this mutation, an adenine-thymine base pair is replaced by a guanine-cytosine pair, or vice-versa. For strains TA98 and TA100 diagnostic mutagens are recommended;[2] these are relatively more mutagenic for one strain than for the other. See Table 8030:I. In using diagnostic mutagens, relative mutagenic activity is important; absolute activity varies between laboratories. When a test is conducted with the diagnostic mutagens, a dose-response curve is generated in the same manner as with samples.

c. Safety precautions: The tester strains were derived from *S. typhimurium* type LT2, which is of low virulence. They have been modified further to increase their sensitivity to mutagens but incidentally to decrease their virulence. Aside from deliberate ingestion, the tester strains do not present a health hazard to workers. Nevertheless, good laboratory practice dictates certain precautions. Never pipet by mouth. Do not eat, drink, or smoke in the laboratory. Autoclave all live cultures and culture plates with colonies before disposal.

Handle plates or cultures or materials containing positive control mutagens or carcinogens as hazardous materials. Treat all test plates

---

§ Schleicher & Schuell No. 740-E or equivalent.

TABLE 8030:I. DIAGNOSTIC MUTAGENS FOR TESTER STRAINS TA98 AND TA100

Chemical	Amount per Plate	S9	Histidine Revertants per Plate TA98	Histidine Revertants per Plate TA100	Animal Toxicity
Sodium azide	1.5 μg	−	3.0	3000	Highly toxic
Daunomycin	6.0 μg	−	3123	47	Toxic
Methyl methanesulfonate	1.0 μL	−	23	2730	Carcinogen
2-Aminofluorene	10 μg	+	6194	3026	Carcinogen
Benzo(a)pyrene	1.0 μg	+	143	937	Carcinogen

(except for negative controls) and other materials prepared for testing as potentially hazardous, and dispose of them accordingly.

5. Procedure

a. Without metabolic activation: Conduct all operations in subdued light or under dim incandescent, nonactinic lighting. Yellow ''bug-lights'' or ''gold'' fluorescent tubes that emit very little ultraviolet radiation are satisfactory.

1) Culture preparation—Prepare overnight broth cultures of tester strains by inoculation with either frozen stock or master plate colonies. Each test plate requires 0.1 mL of broth culture (approximately 1 to 2×10^8 bacteria). Prepare the volume of broth culture accordingly. Incubate cultures overnight (10 to 12 h) at 37°C, with shaking at approximately 210 rpm to ensure adequate aeration.

2) Test material preparation—Most mutagens are soluble in the recommended DMSO. Other relatively nontoxic, nonmutagenic solvents such as ethanol or acetone may be used provided that excessive toxicity is absent.[13] Because many compounds are water-soluble use water whenever possible. When using a volatile solvent take precautions to prevent solvent evaporation before it is added to the top agar or incubation mixture. Keep solutions with volatile solvents tightly capped and on ice. The final concentration of solvent in the top agar should not exceed 5%, except when the solvent is water. Dissolve test material at the highest concentration to be tested and prepare appropriate dilutions. Use a concentration range of at least three logs, with test concentrations at half-log intervals, e.g., 1, 3, 10, 30, 100, 300, and 1000 µg/plate. It may be more convenient to specify dose levels for water samples in terms of original water volume equivalents per plate. In the latter case, mutagenic activity, if present, has been detected in the range of 0.05 to 2.0 liter-equivalents.[14] If negative results are obtained at the highest concentration and no toxicity to the bacteria is apparent, increase concentration of test material to 5 or even 10 mg/plate,[5,7] unless limited by toxicity or precipitation. Prepare sample solutions immediately before use; do not store.

3) Plate incorporation test—Prepare at least three plates for each concentration of test material with minimal base agar. Add 0.1 mL bacterial culture, 2 mL melted top agar at 43 to 45°C, and the test material in 0.1 mL or less of DMSO (or other compatible solvent) to a 13- × 100-mm or 12- × 75-mm sterile glass test tube. It may be convenient to place the required number of tubes containing melted top agar in the 45°C water bath in advance and to add bacteria and test material just before pouring test plates. Briefly mix with a vortex mixer, pour on minimal agar plates, and immediately tilt and rotate plates so that the top agar forms a uniform layer before hardening. Place plates on a level surface and let top agar solidify. (Use a carpenter's level to confirm that the surface is level.) Work quickly. Negative controls receive solvent (e.g., DMSO) or no addition to the mixture of bacteria and top agar. When the top agar has solidified, invert and incubate plates in the dark at 37°C.

4) Scoring plates—Count revertant (histidine-independent) colonies after 48 to 72 h. Longer incubation will result in a very slight increase in the number of revertant colonies but also will increase the chance of contamination. Hold incubation time constant in any test series. Because *Salmonella* strains are motile, the size of colonies depends on whether their bacteria were at the surface or embedded in the top agar. Surface colonies tend to be large (up to 2 mm), circular, and flat while embedded colonies are small (<1 mm) and dense. All colonies are a uniform light cream color, slightly translucent, with a smooth surface and a regular smooth edge to the surface colonies. The appearance of colonies with differing morphology indicates contamination; discard plate and check stock culture for purity. Colony size is inversely proportional to the number of revertants per plate because of exhaustion of nutrients. A trace of histidine in the top agar enables the bacteria to undergo several replications before the histidine is exhausted; it effectively increases the number of bacteria at risk for mutagenesis. This replication also serves as an indicator of toxicity. Replication of nonrevertant bacteria gives a hazy appearance to the top agar between the revertant colonies. If this background lawn replication is absent, it indicates excessive toxicity to the bacteria, and an unreliable mutation test. The variability of the numbers of revertant colonies between replicate plates is usually about 10 to 15%. It tends to become less with increased experience of the analyst.

Plates with bacteria only show a characteristic number of spontaneous revertant colonies. Among a group of eight laboratories the spontaneous revertant range for strain TA98 was 15 to 75 and for strain TA100 between 60 and 220 colonies/plate.[7] Within a given laboratory, these values should remain relatively constant. If marked variation occurs investigate immediately by recharacterizing the strains and confirming media formulation. These numbers may be slightly different on plates with S9 mix. Plates that have received bacteria exposed to the solvent only may show slightly higher numbers of spontaneous revertants.

b. With metabolic activation:

1) Without preincubation—Except for the addition of S9 mix to the standard plate test to convert certain mutagenic chemicals to their active forms, the general procedure is the same as that described above.

Add 0.5 mL S9 mix to the top agar, test material, and bacteria as described above. Incubate and score plates as above. Include a positive control mutagen when using S9 to demonstrate that the preparation is active and to detect possible variation between batches of S9. Recommended control compounds requiring metabolic activation include 2-aminofluorene,[2] 7,12-dimethylbenz(a)-anthracene, and 2-acetylaminofluorene.[7]

2) With preincubation—To enhance test sensitivity for materials or mixtures containing low levels of mutagenic activity, preincubate the bacteria with the test material, with or without S9 mix.[15] To a 13- × 100-mm tube containing the test material in 0.1 mL sodium phosphate buffer add 0.5 mL S9 mix and 0.1 mL of desired tester strain (approximately 1 to 2×10^8 cells). If S9 mix is not used, replace it with 0.1 mL buffer. Incubate at room temperature or at 37°C for 20 min. Add 2 mL melted top agar and pour mixture on a minimal agar base plate as above. The temperature and length of preincubation affect the yield of revertant colonies and should be optimized and standardized for the material being tested. The S9 mix remains metabolically active for 30 to 45 min; avoid prolonged incubation.

3) Microsuspension modification—When the amount of material available for testing is small or the mutagenic activity is very weak, the microsuspension assay may be used.[16] Prepare overnight bacterial cultures as described above. Harvest the bacteria by centrifugation (10 000 × *g* for 10 min) and resuspend in the same volume of phosphate-buffered saline (PBS) or sodium phosphate buffer. Centrifuge again and resuspend the bacterial pellet in a volume of PBS equivalent to *one-tenth* that of the

original broth culture. The incubation mixture contains 0.1 mL bacterial suspension, 0.1 mL S9 mix or phosphate buffer, and the test material dissolved in 10 µL or less of a compatible solvent. Following incubation for up to 90 min, add top agar, mix, and pour plates. Sensitivity is increased up to tenfold or more because the bacteria are in contact with higher concentrations of the test material or mutagenic metabolites during preincubation. However, the toxic effects of the material also may be increased. Make standard assay to provide a reference. Optimize incubation conditions for the particular material being tested.

6. Data Presentation and Analysis

a. Data presentation: Preferably report raw data (i.e., revertant colonies per plate, including control plates) as well as interpreted data.[7,8] If this is not feasible, indicate means of replicate plates, number of replicates, number of experiments, and a measure of the variability (e.g., standard deviation). In cases with a clearly positive result, present data graphically to facilitate comparisons. Direct comparisons of different samples, such as extracts from different waters, usually are done in a single experiment that subsequently is replicated. Presentation of data transformed to revertants per weight of test material or water volume equivalents or to ratios of treated to control cultures is acceptable only if the original data also are available. Report any pretreatment used on the sample.

b. Data analysis: Usually it will be apparent if the test material is mutagenic, so involved statistical analysis of the results is not required. However, when dealing with weakly mutagenic materials, preferably establish objective criteria for deciding whether a material is mutagenic or contains mutagenic substances. An important first criterion is that a reproducible dose-response relationship can be demonstrated, i.e., the number of revertants per plate is proportional to the amount of test material added per plate over some part of the range of amounts tested. As a guide, use the *modified twofold increase rule,*[1,5] which states that a test is considered positive if two consecutive dose levels, or the highest nontoxic dose level, produced a response at least twice that of the solvent control, *and* at least two of these consecutive doses showed a dose-response relationship. If results are in doubt, further testing is indicated, perhaps with test modifications.

It may be desirable, for quantitative comparison purposes, to express mutagenic activity in terms of *Salmonella* revertants per microgram of material or per water volume equivalent, but often the dose-response curve is nonlinear. With low amounts of test material, the number of revertants per plate may increase with increasing amount of material but beyond a certain amount, the number of revertants per plate no longer increases, and even may decline. The reasons for this nonlinearity include toxicity of the test material to the bacteria and limited solubility of the test material in the aqueous medium. Therefore, mutagenic activity of an unknown compound or mixture often is described in terms of the slope of the linear portion of the dose-response curve with low amounts of test material. Statistical methods have been described for deciding which points on these nonlinear curves to use in defining specific mutagenic activity of the sample.[9,17]

Failure to demonstrate mutagenic activity with one or more *Salmonella* tester strains does not amount to proof that the sample contains no mutagenic material. The individual strains have certain mutagens to which they are sensitive, the concentration of S9 may be critical, and some mutagens and carcinogens are not mutagenic for any commonly used strain. Therefore, qualify conclusions regarding mutagenic activity on the basis of the characteristics of the tester strains used as well as the test conditions.[2-4,8]

7. References

1. AMES, B.N., J. McCANN & E. YAMASAKI. 1975. Methods for detecting carcinogens and mutagens with the *Salmonella*/mammalian-microsome mutagenicity test. *Mutation Res.* 31:347.

2. MARON, D. & B.N. AMES. 1983. Revised methods for the *Salmonella* mutagenicity test. *Mutation Res.* 113:173.

3. LEVIN, D.E., E. YAMASAKI & B.N. AMES. 1982. A new Salmonella tester strain (TA97) for the detection of frame shift mutagens. A run of cytosines as a mutational hot-spot. *Mutation Res.* 94:315.

4. LEVIN, D.E., M. HOLLSTEIN, M.F. CHRISTMAN, E.A. SCHWIERS & B.N. AMES. 1982. A new Salmonella tester strain (TA102) with A-T base pairs at the site of mutation detects oxidative mutagens. *Proc. Nat. Acad. Sci. (U.S.A.)* 79:7445.

5. WILLIAMS, L.R. & J.E. PRESTON. 1983. Interim Procedures for Conducting the *Salmonella*/Microsomal Mutagenicity Assay (Ames Test). EPA-600/4-82-068, Environmental Monitoring Systems Lab., U.S. Environmental Protection Agency, Las Vegas, Nev.

6. WILLIAMS, L.R. 1985. Quality assurance considerations in conducting the Ames test. *In* J.K. Taylor & T.W. Stanley, eds. Quality Assurance for Environmental Measurements. ASTM STP 867, American Soc. Testing & Materials, Philadelphia, Pa.

7. DESERRES, F.J. & M.D. SHELBY. 1979. Recommendations on data production and analysis using the *Salmonella*/microsome mutagenicity assay. *Mutation Res.* 64:159.

8. CLAXTON, L.D., J. ALLEN, A. AULETTA, K. MORTELMANS, E. NESTMANN & E. ZEIGER. 1987. Guide for the *Salmonella typhimurium*/ mammalian microsome tests for bacterial mutagenicity. *Mutation Res.* 189:83.

9. BERNSTEIN, L., J. KALDOR, J. McCANN & M.C. PIKE. 1982. An empirical approach to the statistical analysis of mutagenesis data from the *Salmonella* test. *Mutation Res.* 97:267.

10. DULBECCO, R. & M. VOGT. 1954. Plaque formation and isolation of pure cell lines with poliomyelitis viruses. *J. Exp. Med.* 98:167.

11. LOWRY, O.H., N.J. ROSEBROUGH, A.L. FARR & R.J. RANDALL. 1951. Protein measurement with Folin phenol reagent. *J. Biol. Chem.* 193:265.

12. AMES, B.N., F.D. LEE & W.E. DURSTON. 1973. An improved bacterial test system for the detection and classification of mutagens and carcinogens. *Proc. Nat. Acad. Sci. (U.S.A.)* 70:782.

13. MARON, D., J. KATZENELLENBOGEN & B.N. AMES. 1981. Compatibility of organic solvents with the *Salmonella*/microsome test. *Mutation Res.* 88:343.

14. MEIER, J. 1988. Genotoxic activity of organic chemicals in drinking water. *Mutation Res.* 196:211.

15. YAHAGI, T., M. NAGAO, Y. SEINO, T. MATSUSHIMA, T. SUGIMURA & M. OKADA. 1977. Mutagenicities of N-nitrosamines on *Salmonella*. *Mutation Res.* 48:121.

16. KADO, N.Y., D. LANGLEY & E. EISENSTADT. 1983. A simple modification of the *Salmonella* liquid incubation assay. Increased sensitivity for detecting mutagens in human urine. *Mutation Res.* 121:25.

17. MARGOLIN, G., N. KAPLAN & E. ZEIGER. 1981. Statistical analysis of the Ames *Salmonella*/microsome test. *Proc. Nat. Acad. Sci. (U.S.A.)* 78:3779.

8050   BACTERIAL BIOLUMINESCENCE

8050 A.   Introduction

The bacterial bioluminescence test (BBT) is a metabolic inhibition test that uses a standardized suspension of luminescent bacteria as test organisms under standardized conditions. This test method provides a rapid, reliable, and convenient means of determining the toxicity of waste material. The BBT has been validated for a variety of environmental applications including effluent monitoring, groundwater testing, sediment testing, hazardous wastes testing, assessing the efficiency of bioremediation processes, and general biomonitoring.

Luminescent bacteria possess several attributes that support their use for toxicity testing. Their small size (less than 1 μ in diameter) provides a very high surface-to-volume ratio. This feature, as well as the relatively simple morphology and lack of membrane-sided compartmentalization of internal functions, provides many target sites at or near the cytoplasmic membrane.

In bacteria, many metabolic pathways that function in respiration, oxidative phosphorylation, osmotic stabilization, and transport of chemicals and protons into and out of the cell are located within, or very near, the cytoplasmic membrane. The luciferase pathway, which functions as a shunt for electrons directly to oxygen at the level of reduced flavin mononucleotide, also is located within the cell membrane complex.

These cellular characterizations, coupled with the fact that bacterial respiration is 10 to 100 times greater than that of mammalian cells, provide for a dynamic metabolic system that can be quantitated easily and accurately by measuring the rate of light output from a bacterial suspension. Typically, the suspension contains approximately 10^6 individual organisms. The light intensity is substantial (well within the operating range of common light sensors), and the number of organisms compensates for variations among individuals that may influence results of tests on statistically limited numbers of individuals.

8050 B.   Bacterial Bioluminescence Test

1.   General Discussion

Principle: Certain strains of luminescent bacteria divert up to 10% of their respiratory energy into a specific metabolic pathway that converts chemical energy into visible light. This pathway is intrinsically tied to respiration; any change in cellular respiration or disruption of cell structures results in a change in respiration with a concurrent change in the rate of bioluminescence.

In the BBT, the light output of test organisms is measured under standard conditions, the organisms are exposed to the test sample for a specified time, and their light output is measured again. Reduction in light output between the first and second measurements is essentially proportional to the toxicity of the test sample.

2.   Test Procedures

Several milliliters of sample are required. Standard organisms and equipment are available commercially.* When serial dilutions of the sample are tested, suitable controls are used, and appropriate data reduction methods are applied, a dose/response curve is produced, which allows identification of an inhibition concentration (i.e., 20%, 50%) with good precision.

Standardization is achieved by providing the test cells in lyophilized form, designed to capture and maintain their optimum physical state. This method of cell preservation assures consistent sensitivity and stability of the test cells. Data reported in the literature were produced in tests using standard, commercially available, freeze-dried strains of the bacterium *Photobacterium phosphoreum*,[1,2] allowing meaningful comparison of results from tests from different laboratories.[3-5]

3.   References

1. BULICH, A.A. 1982. Microtox® System Operating Manual. Beckman Publ. No. 015-555879, Beckman Instruments, Inc., Carlsbad, Calif.
2. MICROBICS CORPORATION. 1992. Microtox^R Manual, Vol. 1–5. Microbics Corp., Carlsbad, Calif.
3. CORDINA, J.C., A. PÉREZ-GARCIA, P. ROMERO & A. DE VINCENTE. 1993. A comparison of microbial bioassays for the detection of metal toxicity. *Arch. Environ. Contam. Toxicol.* 25:250.
4. RAMAIAH, N. & D. CHANDRAMOHAN. 1993. Ecological and laboratory studies on the role of luminous bacteria and their luminescence in coastal pollution. *Mar. Pollut. Bull.* 26:190.
5. SCHIEWE, M.H., E.G. HAWK, D.I. ACTOR & M.M. KRAHN. 1985. Use of bacterial bioluminescence assay to assess toxicity of contaminated marine sediments. *Can. J. Fish. Aquat. Sci.* 42:1244.

4.   Bibliography

JOHNSON, B.T. Potential geotoxicity of sediments from the Great Lakes. *Environ. Toxicol. Water Quality* 7:373.
HO, K.T.Y. & J.G. QUINN. 1993. Bioassay-directed fractionation of organic contaminants in an estuarine sediment using the new mutagenic bioassay Microtox®. *Environ. Toxicol. Chem.* 12:823.
MASSAY, J., M.D. AITKEN, L.M. BALL & P.E. HECK. 1994. Mutagenicity screening of reaction products from the enzyme-catalyzed oxidation of phenolic pollutants. *Environ. Toxicol. Chem.* 13:1743.

* Microtox®, Microbics Corp., Carlsbad, CA, or equivalent.

8070 P450 REPORTER GENE RESPONSE TO DIOXIN-LIKE ORGANIC COMPOUNDS

8070 A. Introduction

A cell culture of human liver cancer cells can be used to detect the presence of toxic and/or carcinogenic organic compounds in environmental samples. A sample of water, soil, aquatic sediment, or tissue is extracted with a solvent to remove semivolatile organic compounds, and a small amount of the extract is applied to a culture well containing cells attached to the bottom of the well in medium. After exposure, the cells are rinsed and lysed, the cell fragments are removed by centrifugation, and the extract is tested for luminescence. A luminescent enzyme (luciferase) is produced by the cells if dioxin-like compounds were present in the extract, because a reporter gene (plasmid) from the firefly has been attached to the human chromosome at the site induced by dioxin and other planar compounds (at the CYP1A1 gene). The amount of light produced, which is quantified by a luminometer, is a function of the concentrations and induction potency of the organic compounds in the extract. Dioxin (2,3,7,8-tetrachlorodi-benzo-*p*-dioxin, TCDD) has the strongest affinity for the receptor on the cell membrane (Ah-receptor), and therefore will be detected at the lowest concentration.

This assay is both a detection system and a meaningful biological response to the toxicants in environmental samples. It can be used to screen environmental samples for the presence of some of the most toxic and carcinogenic compounds. Only those compounds that are dioxin-like and that attach to the Ah-receptor will induce the CYP1A1 gene and result in the production of luciferase. Such induction would occur in humans or wildlife, including aquatic species, if these compounds came in contact with their tissues. Induction of the CYP1A1 gene is one of the key factors used in designating a compound a carcinogen. High levels of such induction in fish (P450 measurements) have been shown to correlate with histological damage and reduced reproductive capacity.

8070 B. The P450 RGS Test

1. Principle and Significance

The P450 Reporter Gene System (RGS) is a biomarker test for the detection of toxic and/or carcinogenic organic compounds, using a transgenic cell line (101L)* derived from the human hepatoma cell line (HepG2). Under appropriate test conditions, induction of the CYP1A1 gene in mammalian cells normally results in the production of the enzyme P4501A1. This response is evidence that the cells have been exposed to one or more xenobiotic organic compounds, including dioxins, furans, coplanar polychlorinated biphenyls (PCBs), and several polycyclic aromatic hydrocarbons (PAHs). Detection of induction has been made rapid and inexpensive by the stable integration of a firefly plasmid, such that Ah-receptor binding and subsequent transcription results in the production of the luminescent enzyme, luciferase. This RGS test has shown concentration-response relationships using dilutions of TCDD, 2,3,7,8-TCDF, five coplanar PCBs, and eight PAHs, and has responded to application of extracts from environmental samples.[1] Extracts of environmental samples (water, tissue, soil, or aquatic sediments) may be tested for the presence of toxic and carcinogenic organic compounds by methods available elsewhere.[2]

The organic compounds that induce CYP1A1 site on the chromosome are toxic, often carcinogenic, and several have been shown to bioconcentrate and biomagnify. Various birds, mammals, and fish exposed to these compounds have exhibited physiological, reproductive, and histopathological effects.[3-5] Use of a screening tool such as the RGS will permit selection of the most contaminated samples and exclusion of those not requiring further chemical characterization.

2. Test Summary

Details of the culture and testing methods have been published.[6,7] Dichloromethane (DCM) extracts of environmental samples are added to individual wells (six-well plates) containing approximately one million cells. Exposure time is 6 to 18 h, usually the latter. Application of 2 to 20 μL solvent produces a low background (blank) induction when applied to cells in 2 mL of culture medium. The luminescence (in relative light units, RLU) of the combined cytoplasm from the cells in each well is compared to that of other replicate wells, the solvent control, and two reference toxicants [benzo(a)pyrene and TCDD], using a 96-well luminometer. The mean RLU of the control wells is set to unity. Mean RLUs of samples are reported as fold induction (times background), which is derived by dividing by the mean RLU of the solvent (background control). The RGS response represents the integrated CYP1A1 induction from all dioxin-like compounds present in the extract. Results may be expressed as fold induction or, by use of the reference toxicant, as TCDD or benzo(a)pyrene [B(a)P] equivalents. The initial dry weight (determined on a separate subsample) of the extracted sample, the final volume of solvent containing the extracted material (1 or 2 mL), and the amount applied to the cells are recorded. Induction may be expressed as ng TCDD or μg B(a)P equivalents per gram dry weight or per liter.

3. References

1. ANDERSON, J.W., S.S. ROSSI, R.H. TUKEY, T. VU & L.C. QUATTROCHI. 1995. A biomarker, P450 RGS, for assessing the induction potential of environmental samples. *Environ. Toxicol. Chem.* 14:1159.
2. U.S. ENVIRONMENTAL PROTECTION AGENCY. 1995. Methods 3540 and 3550 *in* EPA Test Methods for Evaluating Solid Waste, Physical-

*Cells are available from Columbia Analytical Services, Carlsbad, CA 92009.

Chemical Methods, SW-846, 3rd ed., Update 2B, March, 1995. U.S.. Environmental Protection Agency, Washington, D.C.

3. SAFE, S. 1994. Polychlorinated biphenyls (PCBs): Environmental impact, biochemical and toxic responses, and implications for risk assessment. *Crit. Rev. Toxicol.* 24:87.

4. HAHN, M.E., A. POLAND, E. GLOVER & J.J. STEGEMAN. 1992. The Ah-receptor in marine animals: phylogenetic distribution and relationship to cytochrome P450 A inducibility. *Mar. Environ. Res.* 34:87.

5. STEGEMAN, J., M. BROUWER, R.T. DI GIULIO, L. FORLIN, B.A. FOWLER, B.M. SANDERS & P.A. VAN VELD. 1992. Molecular responses to environmental contamination: enzyme and proteins systems as indicators

of chemical exposure and effect. *In* R.J. Huggett, R.A. Kimerle, P.M. Mehrle, Jr. & H.L. Bergman, eds., Biomarkers: Biochemical, Physiological, and Histological Markers of Anthropogenic Stress, p. 235. Lewis Publishers, Boca Raton, Fla.

6. POSTLIND, H., T.P. VU, R.H. TUKEY & L.C. QUATTROCHI. 1993. Response of human CYPl-luciferase plasmids to 2,3,7,8-tetrachlorodibenzo-p-dioxin and polycyclic aromatic hydrocarbons. *Toxicol. Appl. Pharmacol.* 118:255.

7. ANDERSON, J.W., K. BOTHNER, T. VU & R.H. TUKEY. 1996. Using a biomarker (P-450 RGS) test method on environmental samples. *In* G.K. Ostrander, ed., Techniques in Aquatic Toxicology, p. 277. Lewis Publishers/CRC Press, Boca Raton, Fla.

8080 SEDIMENT POREWATER TESTING* (PROPOSED)

8080 A. Introduction

1. Applications

The standard approach for assessing the quality or potential toxicity of marine or estuarine sediments has been to expose macrobenthic organisms directly to whole sediments for a specified time, after which the survival of the test species was determined. Whole-sediment methods[1,2] have limitations, including use of adult macrobenthic organisms and the use of mortality as the primary end point. In addition, the standard amphipod test protocol[1,2] underestimates the potential toxicity of contaminated sediments because the pore water is flushed out and replaced with fresh overlying water before the start of the exposure.

The porewater toxicity test approach offers several advantages over the standard whole-sediment method. Sensitive life stages of sensitive species can be used in tests utilizing sublethal end points. There are no artifacts produced by sediment texture. A dilution series test design can be used easily for better differentiation among highly toxic samples. Whereas whole sediment preferably should be tested within 2 weeks of collection, studies indicate that pore water can be stored in the frozen state for extended periods of time without any change in tox-

icity.[3] Most of the studies on porewater testing have focused on marine and estuarine species.

2. References

1. U.S. ENVIRONMENTAL PROTECTION AGENCY & U.S. ARMY CORPS OF ENGINEERS. 1991. Evaluation of Dredged Material Proposed for Ocean Disposal. EPA-503/8-91/001, U.S. Environmental Protection Agency Off. Research & Development & U.S. Army Corps of Engineers, Washington, D.C.

2. AMERICAN SOCIETY FOR TESTING AND MATERIALS. 1995. Standard Guide for Conducting 10-Day Static Sediment Toxicity Tests with Marine and Estuarine Amphipods. E1367-92, American Soc. Testing & Materials, Philadelphia, Pa.

3. CARR, R.S. & D.C. CHAPMAN. 1995. Comparison of methods for conducting marine and estuarine sediment pore water toxicity tests. I. Extraction, storage and handling techniques. *Arch. Environ. Contam. Toxicol.* 28:69.

3. Bibliography

ADAMS, D.D. 1991. Sediment pore water sampling. Chapter 7 *in* A. Mudroch & S.D. MacKnight, eds., Handbook of Techniques for Aquatic Sediments Sampling. CRC Press, Inc., Boca Raton, Fla.

BURTON, G.A., JR., ed. 1992. Sediment Toxicity Assessment. Lewis Publishers, Inc., Boca Raton, Fla.

---

*Approved by Standard Methods Committee, 1997.

8080 B. Sediment Collection and Storage

Sediment collection methods vary considerably with specific study objectives. Remove as much overlying water as possible from the sample before placing it in the sample container or adding it to a composite sample. Nearly fill sample container to minimize head space, allowing some room for sample to be rehom-

ogenized in its original container. Extract pore water as soon as possible after sample collection. If the sediment sample cannot be processed immediately, store on ice or refrigerate at 4°C. The toxicity of pore water extracted from refrigerated sediments can change considerably after a period of weeks or even days.

8080 C.   Extraction of Sediment Pore Water

Methods that have been used for obtaining sediment pore (interstitial) water including centrifugation,[1-3] pressurized (pneumatic or mechanical "squeezing") extraction,[4-8] vacuum (suction) methods,[9,10] and equilibration methods using dialysis membranes or fritted glass samplers.[4,11,12] Studies comparing recovery efficiencies of trace metals and organics for different extraction methods indicate that substantial losses of nonpolar contaminants (e.g., fluoranthene and p,p'-DDE) can occur with all methods.[13] Toxicity tests with echinoderm gametes and embryos have been conducted to compare the toxicity of pore water obtained by various extraction techniques.[14] These studies suggest that centrifugation minimizes loss of nonpolar contaminants. Loss of metals is comparable among the various extraction methods. Centrifugation is preferable to filtration for removal of particulates because it minimizes adsorptive loss of contaminants.

Sandy sediments do not compact appreciably during centrifugation, making pore water recovery difficult, whereas the pneumatic extraction method (Figures 8080:1 and 2) is particularly effective.[11] The vacuum method is least expensive for small-scale projects.

Regardless of the method used for initial extraction, centrifuge extracted pore water to remove suspended particulates for fertilization and embryo development assays with echinoderms and mollusks.

1. Centrifugation

Use a centrifuge equipped with a swinging bucket-type rotor capable of spinning 100- to 1000-mL bottles at 10 000 × g. Use

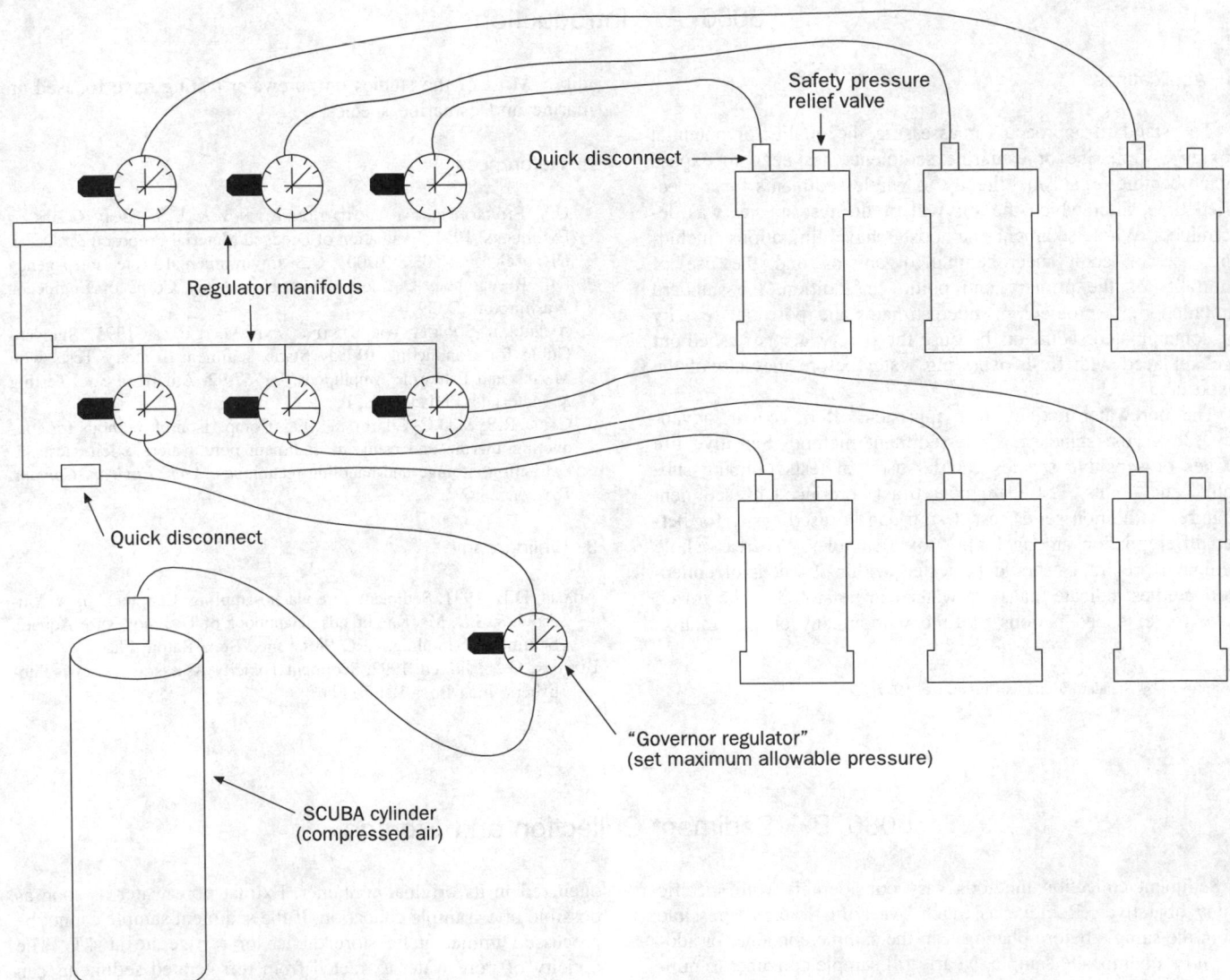

Figure 8080:1. Pneumatic system for porewater extraction. Source: CARR, R.S. & D.C. CHAPMAN. 1995. Comparison of methods for conducting marine and estuarine sediment porewater toxicity tests. I. Extraction, storage and handling techniques. *Arch. Environ. Contam. Toxicol.* 28:29.

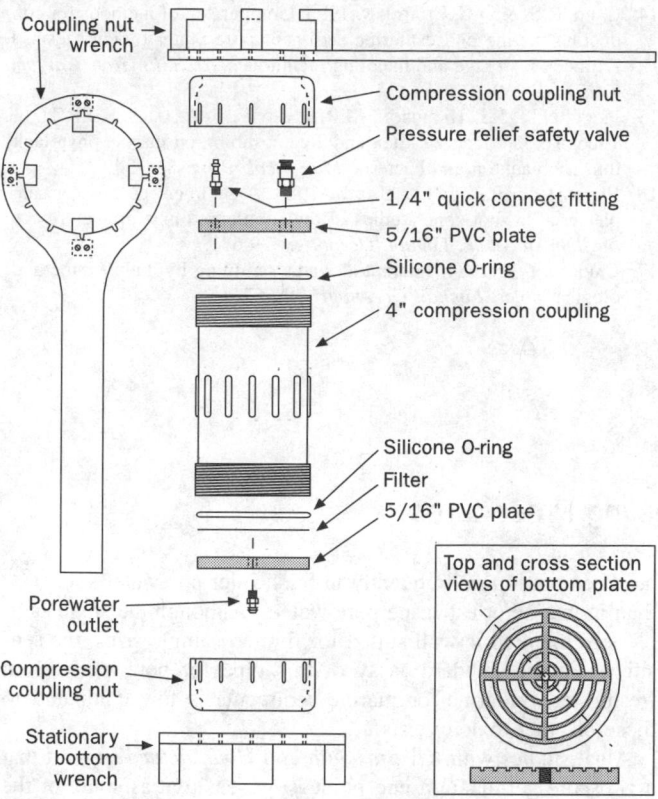

Figure 8080:2. Detail of porewater extraction cylinder. (For dimensions in centimeters, multiply dimensions in inches by 2.54.) Source: CARR, R.S. & D.C. CHAPMAN. 1995. Comparison of methods for conducting marine and estuarine sediment porewater toxicity tests. I. Extraction, storage and handling techniques. *Arch. Environ. Contam. Toxicol.* 28:29.

tubes or bottles composed of glass or polycarbonate to minimize adsorption of soluble contaminants on container wall. For some sediments it may be possible to decant the supernatant without disturbing the pellet, but for most sediments, use a pipet to transfer the supernatant to a separate container.

2. Pressurized Squeeze Extraction

The most common squeeze extraction devices use compressed air (or nitrogen) to pressurize a cylinder containing the sediment. Normally use a filter in the bottom of the cylinder to minimize introduction of sediment into the porewater sample. Some filters (e.g., glass fiber filters) can adsorb a high percentage of nonpolar contaminants from solution.[13,14] Other filter materials (e.g., polyester and nylon) are preferable. Test any part of the extraction device that contacts the pore water during the extraction process for toxicity before use. Fill extraction device with a small volume of test dilution water and, after a minimum of 8 h, test dilution water for toxicity. Soak new filters in deionized water or test dilution water with several exchanges for at least 24 h to remove any residual contaminants before use. Between samples, acid-wash the parts of the extraction devices that come in contact with the sample.

3. Vacuum Extraction

The simplest vacuum extraction system is a fused-glass air stone attached with aquarium air-line tubing to a polypropylene syringe. Apply vacuum by bracing the syringe plunger or using a vacuum pump. Modify the system with TFE* tubing and a glass syringe when loss of contaminants due to adsorption is a concern. This method is inexpensive and may retain more volatile compounds than the centrifugation or pressurized extraction methods. Vacuum methods may be more time-consuming than other extraction methods when large (>1 L) volumes are needed, particularly for fine-grained sediments. Thoroughly rinse all system components before use to remove residual toxicants.[15] Determine effectiveness of the rinsing procedure by testing the toxicity of test dilution water held in the vacuum extraction system for a minimum of 8 h. Pore water extracted by the vacuum methods from sandy sediments has a higher particulate content than pore water obtained by the other methods; if the suspended particulates are not removed before testing, they may produce a response in fertilization and embryo development toxicity tests.

4. Equilibration Methods

A small-volume vessel with a membrane placed in the sediment and allowed to equilibrate with the surrounding interstitial water is the most commonly used equilibration technique for collecting pore water.[11,16,17] The limitations to this technique are that only milliliter volumes can be obtained within a reasonable time (days). Test toxicity of components used to construct the equilibration device by soaking device in clean test dilution water or clean sediment for at least the same length of time as the longest equilibration period to be used.

5. References

1. EDMUNDS, W.M. & A.H. BATH. 1976. Centrifuge extraction and chemical analysis of interstitial waters. *Environ. Sci. Technol.* 10: 467.
2. GIESY, J.P., R.L. GRANEY, J.L. NEWSTED, C.J. ROSIU, A. BENDA, R.G. KREIS & F.J. HORVATH. 1988. Comparison of three sediment bioassay methods using Detroit River sediments. *Environ. Toxicol. Chem.* 7: 483.
3. LANDRUM, P.F., S.R. NIHART, B.J. EADIE & L.R. HERCHE. 1987. Reduction in bioavailability of organic contaminants to the amphipod *Pontoporeia hoyi* by dissolved organic matter of sediment interstitial waters. *Environ. Toxicol. Chem.* 6:11.
4. BENDER, M., W. MARTIN, J. HESS, F. SAYLES, L. BALL & C. LAMBERT. 1987. A whole-core squeezer for interfacial pore-water sampling. *Limnol. Oceanogr.* 32:1214.
5. CARR, R.S., D.C. CHAPMAN, C.L. HOWARD & J. BIEDENBACH. 1996. Sediment quality triad assessment survey in the Galveston Bay, Texas system. *Ecotoxicol.* 5:341.
6. CARR, R.S. & D.C. CHAPMAN. 1992. Comparison of whole sediment and pore-water toxicity tests for assessing the quality of estuarine sediments. *Chem. Ecol.* 7:19.
7. JAHNKE, R. A. 1988. A simple, reliable, and inexpensive pore-water sampler. *Limnol. Oceanogr.* 33:483.
8. REEBURGH, W.S. 1967. An improved interstitial water sampler. *Limnol. Oceanogr.* 12:163.
9. KNEZOVICH, J.P. & F.L. HARRISON. 1987. A new method for determining the concentrations of volatile organic compounds in sediment interstitial water. *Bull. Environ. Contam. Toxicol.* 38:937.

*Teflon® or equivalent.

10. WINGER, P.V. & P.J. LASIER. 1991. A vacuum-operated pore-water extractor for estuarine and freshwater sediments. *Arch. Environ. Contam. Toxicol.* 21:321.

11. DI TORO, D.M., J.D. MAHONY, D.J. HANSEN, K.J. SCOTT, M.B. HICKS, S.M. MAYR & M.S. REDMOND. 1990. Toxicity of cadmium in sediments: the role of acid volatile sulfide. *Environ. Toxicol. Chem.* 9:1487.

12. HESSLIN, R.H. 1976. An *in situ* sampler for close interval pore water studies. *Limnol. Oceanogr.* 21:912.

13. SCHULTS, D.W., S.P. FERRARO, L.M. SMITH, F.A. ROBERTS & C.K. POINDEXTER. 1992. A comparison of methods for collecting interstitial water for trace organic compounds and metals analyses. *Water Res.* 26:989.

14. CARR, R.S. & D.C. CHAPMAN. 1995. Comparison of methods for conducting marine and estuarine sediment pore water toxicity tests. I. Extraction, storage and handling techniques. *Arch. Environ. Contam. Toxicol.* 28:69.

15. PRICE, N.M, P.J. HARRISON, M.R. LANDRY, F. AZAM & K.J.F. HALL. 1986. Toxic effects of latex and Tygon tubing on marine phytolankton, zooplankton and bacteria. *Mar. Ecol. Prog. Ser.* 34:41.

16. BOTTOMLEY, E.Z. & I.L. BAYLY. 1984. A sediment pore water sampler used in root zone studies of the submerged macrophyte, *Myriophyllum spicatum. Limnol. Oceanogr.* 29:671.

17. CARIGNAN, R. 1984. Interstitial water sampling by dialysis: methodological notes. *Limnol. Oceanogr.* 29:667.

8080 D.   Toxicity Testing Procedures

1. General Procedures

Because of the difficulty in obtaining large volumes of pore water, organisms and life stages that require only small volumes are most amenable to testing with pore water. For tests requiring more than 7 d to complete, preferably use a static renewal test design to ensure acceptable water quality. Short-term toxicity tests have been used most frequently with pore water. Much of the general guidance provided in Section 8010 is applicable to testing with pore water. More specific guidance can be found in sections for particular species or groups of organisms (e.g., Sections 8510, 8610, 8710, 8720).

2. Exposure Chambers

The type of exposure chamber used depends on the test. Most porewater tests are conducted in relatively small volumes, i.e., ≤ 10 mL. Preferably cover test chambers to minimize evaporation and resulting salinity increases during the exposure period. Scintillation vials (20 mL) with polyethylene or polypropylene cap liners are ideal inexpensive disposable test chambers for many species. Avoid caps with urea-formaldehyde liners because these can be toxic. Stender dishes with ground-glass lids (20-mL capacity with 10 mL of exposure media) make excellent exposure chambers for tests that require microscopic examination of the test organisms without transferring them to another container (e.g., the *Dinophilus gyrociliatus* life-cycle test).[1]

3. Organisms

Many types of organisms have been used in porewater tests. Minute species or larval forms are preferable not only for their small volume requirements, but also because they tend to be the most sensitive. Most of the studies on porewater testing have focused on marine and estuarine species.

a. Marine and estuarine species: A commercially available test system,* which detects changes in the photoluminescence of the marine bacterium *Photobacterium phosphoreum* as an end point, has been used more frequently in freshwater porewater studies[2-4] than in marine or estuarine pore waters. Although the small sample size required is well suited for limited sample sizes, the sensitivity of the standard assay of this type for pore water from freshwater, estuarine, or marine sediments is low compared to those of other toxicity tests.

Algal studies with *Ulva fasciata* and *Ulva lactuca* suggest that a zoospore germination end point is as sensitive as some of the most sensitive embryological development assays used in porewater testing. This test appears to be particularly resistant to ammonia toxicity. Many algal species used in microplate procedures could easily be adapted for use with porewater samples.

Porewater toxicity testing has been conducted with the polychaete *Dinophilus gyrociliatus*.[1,5,6] Other minute polychaetes such as *Ctenodrilus serratus* or *Ophryotocha* spp.[7,8] can be tested in small volumes.

The mollusk tests used most successfully with pore water are fertilization and embryological development tests with the abalone *Haliotes refugens*. Other more common embryological development tests with oysters[9] and clams[10] could be adapted for use with porewater samples.

Most of the toxicity testing with marine and estuarine pore water has been conducted with sea urchin gametes and embryos.[6,11,12] The species most commonly used is the sea urchin, *Arbacia punctulata*, but other species of sea urchin (e.g., *Strongylocentrotus* spp. and *Lytechinus* spp.) as well as the sand dollar (e.g., *Dendraster* spp.) also have been used successfully. Types of tests include fertilization tests, embryological development tests, and cytogenetic assay.[13]

Fish embryos and larvae of red drum *Sciaenops ocellatus* also have been used successfully in porewater testing.[14]

b. Freshwater species: Only a limited number of species have been used in porewater studies with fresh water. A number of studies with a commercially available system* have been reported.[2,3,15] The freshwater amphipod *Hyalella azteca* has been used to test the toxicity of pore water from freshwater sediments.[3,15] *Ceriodaphnia dubia* also has been used in life-cycle tests with pore water.

---

*Microtox® or equivalent.

*Microtox® or equivalent.

4. References

1. CARR, R.S., J.W. WILLIAMS & C.T.B. FRAGATA. 1989. Development and evaluation of a novel marine sediment pore water toxicity test with the polychaete *Dinophilus gyrociliatus*. *Environ. Toxicol. Chem.* 8:533.

2. GIESY, J.P., R.L. GRANEY, J.L. NEWSTED, C.J. ROSIU, A. BENDA, R.G. KREIS & F.J. HORVATH. 1988. Comparison of three sediment bioassay methods using Detroit River sediments. *Environ. Toxicol. Chem.* 7:483.

3. GIESY, J.P., C.J. ROSIU, R.L. GRANEY & M.G. HENRY. 1990. Benthic invertebrate bioassays with toxic sediment and pore water. *Environ. Toxicol. Chem.* 9:233.

4. ANKLEY, G.T., K. LODGE, D.J. CALL, M.D. BALCER, L.T. BROOKE, P.M. COOK, R.J. KREIS, JR., A.R. CARLSON, R.D. JOHNSON, G.J. NIEMI, R.A. HOKE, C.W. WEST, J.P. GIESY, P.D. JONES & Z.C. FUYING. 1992. Integrated assessment of contaminated sediments in the lower Fox River and Green Bay, Wisconsin. *Ecotoxicol. Environ. Safety* 23:46.

5. CARR, R. S., M.D. CURRAN & M. MAZURKIEWICZ. 1986. Evaluation of the archiannelid *Dinophilus gyrociliatus* for use in short-term life-cycle toxicity tests. *Environ. Toxicol. Chem.* 5:703.

6. CARR, R.S. & D.C. CHAPMAN. 1992. Comparison of whole sediment and pore-water toxicity tests for assessing the quality of estuarine sediments. *Chem. Ecol.* 7:19.

7. REISH, D.J. & R.S. CARR. 1978. The effect of heavy metals on the survival, reproduction, development and life cycles for two species of polychaetous annelids. *Mar. Pollut. Bull.* 9:24.

8. CARR, R.S. & D.J. REISH. 1977. The effect of petroleum hydrocarbons on the survival and life history of polychaetous annelids. *In* D.A. Wolfe, ed., Fate and Effects of Petroleum Hydrocarbons in Marine Ecosystems and Organisms. Pergamon Press, New York, N.Y.

9. LONG, E.R., M.R. BUCHMAN, S.M. BAY, R.J. BRETELER, R.S. CARR, P.M. CHAPMAN, J.E. HOSE, A.L. LISSNER, J. SCOTT & D.A. WOLFE. 1990. Comparative evaluation of five toxicity tests with sediments from San Francisco Bay and Tomales Bay, California. *Environ. Toxicol. Chem.* 9:1193.

10. LAUGHLIN, R.B., JR., R.G. GUSTAFSON & P. PENDOLEY. 1989. Acute toxicity of tributyltin (TBT) to early life history stages of the hard shell clam, *Mercenaria mercenaria. Bull. Environ. Contam. Toxicol.* 42:352.

11. CARR, R.S. & D.C. CHAPMAN. 1995. Comparison of methods for conducting marine and estuarine sediment pore water toxicity tests. I. Extraction, storage and handling techniques. *Arch. Environ. Contam. Toxicol.* 28:69.

12. LONG, E.R., R.S. CARR, G.A. THURSBY & D.A. WOLFE. 1995. Sediment toxicity in Tampa Bay: Incidence, severity, and spatial extent. *Fla. Sci.* 58:163.

13. HOSE, J.E., H.W. PUFFER, P.S. OSHIDA & S.M. BAY. 1983. Developmental and cytogenetic abnormalities induced in the purple sea urchin by environmental levels of benzo(a)pyrene, *Arch. Environ. Contam. Toxicol.* 12:319.

14. ROACH, R.W., R.S. CARR, C.L. HOWARD & B.W. CAIN. 1992. An assessment of produced water impacts in Galveston Bay system. U.S. Fish Wild. Serv. Rep.

15. WINGER, P.V., R.J. LASIER & H. GEITNER. 1993. Toxicity of sediments and pore water from Brunswick Estuary, Georgia. *Arch. Environ. Contam. Toxicol.*

8110 ALGAE

Algae are unicellular to multicellular plants that occur in fresh water, marine water, and damp terrestrial environments. All algae possess chlorophyll, the green pigment essential for photosynthesis. Algae may contain additional pigments such as fucoxanthin (brown) or phycoerythrin (red), which can mask the green color of chlorophyll. The life cycle of algae may be simple, involving cell division, or complex, involving alternation of generations. Algae are primary producers of organic matter upon which animals depend either directly or indirectly through the food chain.

Test procedures using algae are valuable for determining the primary productivity of a water and for testing the toxicity of chemicals present in a water. Section 8111, Biostimulation (algal productivity), measures the response of a cultured species of algae to the nutritional condition of the water.

Section 8112, Phytoplankton, measures the response of an algal species to materials that interfere with its normal metabolism.

Together, the tests allow the assessment of the effects of point or nonpoint discharges in fresh and marine waters.

8111 BIOSTIMULATION (ALGAL PRODUCTIVITY)*

8111 A. General Principles

Algal assays consist of three steps: (a) selection and measurement of appropriate factors or conditions during the assay (e.g., biomass indicators such as measured or calculated dry weight); (b) presentation and statistical analysis of measurements; and (c) interpretation of results.

Interpretation of results involves assessment of receiving water to determine its nutritional status and its potential sensitivity to change, effects of chemical constituents on algal growth in receiving waters, effects of changes in waste treatment processes on algal growth in receiving waters, impact of nutrients in tributary waters on algal growth in lakes and confluent receiving waters, and effects of measures such as those used in lake restoration and advanced waste treatment.

The maximum standing crop is a response that can be estimated from growth measurements. It is proportional to the initial amount of limiting nutrient available.

_____
* Approved by Standard Methods Committee, 1997.

The algal test procedure for determining primary productivity of a water sample is based on Liebig's "Law of the Minimum," which states that growth is limited by the substance that is present in minimal quantity in respect to the need of the organism. Biostimulants are substances that increase algal growth or the potential for algal growth.

Algal species used in biostimulation tests are selected to allow for a standardized test of growth response using a well-characterized organism under standard laboratory conditions. See Sections 10010, 10200, and 10300 for methods appropriate to field studies.

Effects of various substances on maximum crop of selected algal species cultured under specified conditions are measured in this text. Results are assessed by comparing growth in the presence of selected nutrient and chelator additions to growth in controls. Experimental designs must incorporate sufficient replication to permit statistical evaluation of results.

8111 B. Planning and Evaluating Algal Assays

1. Sampling

Because water quality may vary greatly with time and point of collection, establish sampling programs to obtain representative and comparable data.

Consider all pertinent environmental factors in planning an assay, to insure that valid results and conclusions are obtained. In a stratified lake or impoundment, collect only depth-integrated (composite) euphotic zone samples. In most cases, the euphotic zone is defined as the depth to which at least 1% of the surface light is available. For euphotic zone depths of more than 8 m, subsample at least at the surface and at each 3-m depth interval. Likewise, for euphotic zones of less than 8 m, sample at least at the surface and at 2-m intervals. Composite equal-volume depth samples in a suitable nonmetallic container, mix thoroughly, and subsample for algal assay and chemical and biological analysis, including indigenous algal biomass and identification.[1]

Transect lines are helpful in sampling. Samples from a transect can be taken from predetermined euphotic zones. Representative river samples can be identified by specific conductance measurements that show the homogeneity of the sampling transect. In rivers and streams, useful information may be obtained by taking samples upstream and downstream from suspected pollutant sources or confluent tributaries.[1]

The nutrient content of natural waters and wastewaters often varies greatly with time; variation may be seasonal, or even hourly in wastewaters. When sampling, consider and minimize effects of these variations.

2. Test Variables

Deficiency of any essential nutrient may limit algal growth, but tests are made for those few nutrients most likely to be growth-limiting (nitrogen, phosphorus, trace elements). Measurement of the algal growth potential of water distinguishes between the nutrients in the sample (as determined by chemical analysis) and nutrient forms that are actually available for algal growth.[1]

To evaluate the potential effect of a substance on receiving waters, consider the following factors: amount and distribution, chemical and/or physical nature, fate and persistence, pathways by which it will reach the receiving water, dilution by the receiving body, and selection of appropriate test water.[1,2]

When the algal assay is used to measure stimulation of growth by a given effluent, include the following in the overall evaluation: effluent quality, growth measurements and test organisms, concentration of growth-limiting nutrient, and potential nutrient concentration and changes in availability.

3. References

1. MILLER, W.E., J.C. GREENE & T. SHIROYAMA. 1978. The *Selenastrum capricornutum* Printz Algal Assay Bottle Test: Experimental Design, Application, and Data Interpretation Protocol. EPA-600/9-78-018, U.S. Environmental Protection Agency, Environmental Research Lab. Corvallis, Ore.
2. NATIONAL EUTROPHICATION RESEARCH PROGRAM. 1971. Algal Assay Procedure: Bottle Test. U.S. Environmental Protection Agency, Pacific Northwest Environmental Research Lab., Corvallis, Ore.

4. Bibliography

McGAUHEY, P.H., D.B. PORCELLA & G.L. DUGAN. 1970. Eutrophication of surface waters—Indian Creek reservoir. First Progress Rep., FWQA Grant No. 16010 DNY. U.S. Environmental Protection Agency, Pacific Northwest Environmental Research Lab., Corvallis, Ore.

MALONEY, T.E., W.E., MILLER & T. SHIROYAMA. 1971. Algal responses to nutrient additions in natural waters. Spec. Symp., American Soc. Limnology & Oceanography. Special Symposium on Nutrients and Eutrophication: Limiting-Nutrient Controversy 1:134.

MILLER, W.E. & T.E. MALONEY. 1971. Effects of secondary and tertiary wastewater effluents on algal growth in a lake-river system. *J. Water Pollut. Control Fed.* 43:2361.

MALONEY, T.E., W.E. MILLER & N.L. BLIND. 1972. Use of algal assays in studying eutrophication problems. *Proc. Int. Conf. Water Pollut. Res. 6th*, p. 205. Pergamon Press, Oxford, England & New York, N.Y.

SCHERFIG, J., P.S. DIXON, R. APPLEMAN & C.A. JUSTICE. 1973. Effect of Phosphorus Removal on Algal Growth. Ecol. Res. Ser. 660/3-75-015, U.S. Environmental Protection Agency.

MILLER, W.E., T.E. MALONEY & J.C. GREENE. 1974. Algal productivity in 49 lakes as determined by algal assays. *Water Res.* 8:667.

SPECHT, D.T. 1975. Seasonal variation of algal biomass production potential and nutrient limitation in Yaquina Bay, Oregon. *In* E.J. Middlebrooks, D.H. Falkenborg, and T.E. Maloney, eds. Proceedings Workshop on Biostimulation and Nutrient Assessment, Utah State Univ., Logan, Sept. 10–12, 1975. PRWG 168-1; also published as Biostimulation and Nutrient Assessment. Ann Arbor Science Publ., Ann Arbor, Mich.

DAVIS, J. & J. DeCOSTA. 1980. The use of algal assays and chlorophyll concentrations to determine fertility of water in small impoundments in West Virginia. *Hydrobiologia* 71:19.

McCOY, G.A. 1983. Nutrient limitation in two arctic lakes, Alaska. *Can. J. Fish. Aquat. Sci.* 40:1195.

NOVAK, J.T. & D.E. BRUNE. 1985. Inorganic carbon limited growth kinetics of some freshwater algae. *Water Res.* 19:215.

GOPHEN, M. & M. GOPHEN. 1986. Trophic relations between two agents of sewage purification systems: Algae and mosquito larvae. *Agr. Wastes* 15:159.

GREENE, J.C., W.E. MILLER & E. MERWIN. 1986. Effects of secondary effluents on eutrophication in Las Vegas Bay, Lake Mead, Nevada. *Water, Air, Soil Pollut.* 29:391.

LANGIS, R., P. COUTURE, J. DE LA NOUE & N. METHOT. 1986. Induced responses of algal growth and phosphate removal by three molecular weight DOM fractions from a secondary effluent. *J. Water Pollut. Control Fed.* 58:1073.

YUSOFF, F.M. & C.D. McNABB. 1989. Effects of nutrient availability on primary productivity and fish production in fertilized tropical ponds. *Aquaculture* 78:303.

8111 C. Apparatus

1. Sampling and Sample Preparation

a. Sampler, nonmetallic.

b. Sample bottles, borosilicate glass, linear polyethylene, polycarbonate, or polypropylene, capable of being autoclaved.

c. Membrane filter apparatus, for use with 47-mm or 104-mm prefilters (e.g., glass fiber filter) and 0.45-μm-porosity filters.

d. Autoclave or pressure cooker, capable of producing 108 kPa at 121°C.

2. Culturing and Incubation

a. Culture vessels: Use erlenmeyer flasks of good-quality borosilicate glass. When trace nutrients are being studied, use special glassware made of high-silica glass or polycarbonate. While flask size is not critical, the surface-to-volume ratios of the growth medium are, because of CO_2 limitation. Use the following:

25 mL sample in 125-mL flask
50 mL sample in 250-mL flask
100 mL sample in 500-mL flask

b. Culture closures: Use demonstrably nontoxic foam plugs,* loose-fitting aluminum foil, or inverted beakers to permit some gas exchange and prevent contamination. Determine for each batch of closures whether that batch has any significant effect on maximum specific growth rate and/or maximum standing crop.

c. Constant-temperature room: Provide constant-temperature room, or equivalent incubator, capable of maintaining temperature of 18 ± 2°C (marine) to 24 ± 2°C (freshwater).

d. Illumination: Use "cool-white" fluorescent lighting to provide 4304 lux ± 10% or 2152 lux ± 10% measured adjacent to the flask at the liquid level with closure in place.

e. Light measurement device: Calibrate device against a standard light source or light meter.

3. Other Apparatus[1]

a. Analytical balance capable of weighing 100 g with a precision of ± 0.1 mg.

---

* Gaymar white, polyurethane foam plugs, VWR Scientific or Gaymar Industries, Inc., 701 Seneca St., Buffalo, NY 14210, or *demonstrably nontoxic* equivalent.

b. *Electronic particle (cell) counter.*

c. *Fluorometer,* suitable for chlorophyll *a.*

d. *Microscope and illuminator,* good quality, general purpose.

e. *Hemocytometer or plankton counting slide.*

f. *Shaker table,* capable of 100 oscillations/min.

g. *pH meter* to measure to ± 0.1 pH unit.

h. *Dry-heat oven* capable of operating at up to 120°C.

i. *Centrifuge* capable of a relative centrifugal force of at least 1000 × *g.*

j. *Desiccator.*

4. Reference

1. MILLER, W.E., J.C. GREENE & T. SHIROYAMA. 1978. The *Selenastrum capricornutum* Printz Algal Assay Bottle Test: Experimental Design, Application, and Data Interpretation Protocol. EPA-600/9-78-018, U.S. Environmental Protection Agency, Environmental Research Lab, Corvallis, Ore.

8111 D.   Sample Handling

1. Sampling Procedure

Use a nonmetallic water sampler and autoclavable storage container. Leave a minimum of air space in the transport container and keep it in the dark at 0 to 4°C.

2. Removal of Indigenous Algae

To use unialgal test species, "remove" indigenous algae before assay by autoclaving and filtering. Always prepare sample as soon as possible (within 24 h) after collection.

Use autoclaving followed by filtration to determine amount of algal biomass that can be grown from all bioavailable nutrients in the water, including those contained in filterable organisms. Autoclave freshwater samples at 108 kPa and 121°C for 30 min or 10 min/L of sample, whichever is longer. Pasteurize marine or estuarine samples for 4 h at 60°C. After autoclaving and cooling to room temperature, equilibrate sample by bubbling with a 1% mixture of carbon dioxide in air for at least 2 min/L. This will restore carbon dioxide lost during autoclaving and lower pH to its original level (usually it will rise on autoclaving). In some instances, waters with total hardness greater than 150 mg/L will lose calcium and phosphorus during autoclaving. The precipitate may be resistant to resolubilization by the addition of carbon dioxide and air. In waters containing high levels of hardness and alkalinity the pH may not increase during autoclaving. Filter carbon-dioxide-equilibrated sample through pre-filter, if necessary, followed by a 0.45-μm membrane filter[1].

3. Storage

Changes occur in water samples during storage regardless of storage conditions. The extent and nature of these changes is not well known. Therefore, keep storage duration to a minimum after sample preparation. Store samples in full containers with no air space. Before sample preparation, store samples in the dark at 0 to 4°C. If prolonged storage is anticipated, prepare sample first and then store in the dark at 0 to 4°C.

4. Reference

1. MILLER, W.E., J.C. GREENE & T. SHIROYAMA. 1978. The *Selenastrum capricornutum* Printz Algal Assay Bottle Test: Experimental Design, Application, and Data Interpretation Protocol. EPA-600/9-78-018, U.S. Environmental Protection Agency, Environmental Research Lab, Corvallis, Ore.

8111 E.   Synthetic Algal Culture Medium

See Section 8010E.4*c*1).

8111 F. Inoculum

1. Recommended Test Algae

The following selected species are used primarily in the United States, Canada, and northern Europe. The tests are probably valid for other species worldwide but would require validation testing. If diatoms are the selected test species, silica must be added to the synthetic algal culture medium.

a. Freshwater algae:

Selenastrum capricornutum Printz (see Section 10900, Plate 1A:G).

b. Marine algae:

1) *Dunaliella tertiolecta* Butcher (DUN Clone) Woods Hole Oceanographic Institution.

2) *Thalassiosira pseudonana* (Hasle and Heimdal) (CN Clone) (old *Cyclotella nana*) Univ. Rhode Island. Do not shake. (See Section 10900, Plate 1B:T; Plates 29, 31.)

3) *Skeletonema costatum* (Greville) Cleve. (See Section 10900, Plate 1B:W; Plate 35.)

2. Sources of Test Algae

Obtain algal cultures from recognized sources.* After receipt of cultures, check identity and purity.

3. Maintaining Stock Cultures

a. Medium: See Section 8010E.4*c*1).

b. Incubation conditions:

1) Freshwater species—Temperature 24 ± 2°C under continuous cool-white fluorescent lighting at 4304 lux ± 10% for *S. capricornutum;* shake at 110 oscillations/min.

2) Marine species—Temperature 18 ± 2°C under continuous cool-white fluorescent lighting at 4304 lux ± 10% for *D. tertiolecta* (shake at 110 oscillations/min) and for *T. pseudonana* (do

not shake but swirl daily). Higher temperatures (up to 24°C) may be justified for appropriate test species used in the Gulf of Mexico and other warm-water marine systems. If other species are used, always relate growth of those species to *D. tertiolecta* to insure comparability.

c. First stock transfer: Upon receipt of inoculum species, transfer a portion to the algal culture medium. (Example: 1 mL of inoculum in 50 mL in a 125-mL erlenmeyer flask).

d. Subsequent stock transfers: Make a new stock transfer, using aseptic technique, as the first operation on opening a stock culture. The volume transferred is not critical so long as enough cells are included to overcome significant growth lag. Make weekly stock transfers to provide a continuing supply of "healthy" cells. Check algal cultures microscopically to insure that the stock cultures remain unialgal.

e. Age of inoculum: Use cultures 1 to 3 weeks old as a source of inoculum. For *Selenastrum* and *Dunaliella,* a 5- to 7-d incubation often is sufficient to provide enough cells.

4. Preparing Inoculum

Centrifuge stock culture and discard supernatant. Resuspend sedimented cells in an appropriate volume of glass-distilled water containing 15 mg NaHCO$_3$/L for freshwater species and artificial seawater minus nutrients for marine species [Section 8010E.4*c*1); Table 8010:II] diluted to appropriate salinity, and again centrifuge. Resuspend sedimented algae in the proper solution and use as the inoculum.

5. Amount of Inoculum

Count cells suspended in the prepared inoculum and pipet into the test water to give a starting cell concentration as follows:

S. capricornutum	10^3 cells/mL
D. tertiolecta	10^3 cells/mL

Calculate volume of transfer to result in the above concentrations in the test flasks (e.g., for *S. capricornutum,* if there are 5×10^5 cells/mL in stock culture, transfer 0.2 mL/100 mL test water).

---

* American Type Culture Collection, 12301 Parklawn Drive, Rockville, MD 20852 (phone: 800-638-6597; e-mail: sales@atcc.org); UTEX Culture Collection of Algae, Department of Botany, University of Texas at Austin, TX 78713-7640 (internet: www.botany.utexas.edu); Provasoli-Guillard National Center for Culture of Marine Phytoplankton (CCMP), McKown Point, West Boothbay Harbor, ME 04575 (phone: 207-633-9630; e-mail: ccmp@bigelow.org).

8111 G. Test Conditions and Procedures

1. Temperature

Keep temperature at 18 ± 2°C for marine species and 24 ± 2°C for freshwater species.

2. Illumination

See Section 8111F.3*b*. Measure light intensity adjacent to the flask at the liquid level.

3. Procedure

a. Preparation of glassware: Wash all glassware with detergent (nonphosphate or sodium carbonate) and rinse thoroughly with tap water. Then rinse with a warm 10% (v/v) solution of reagent-grade HCl. Fill vials and centrifuge tubes with 10% HCl. Fill all containers to about one-tenth capacity with HCl solution and swirl to bathe entire inner surface. After HCl rinse, rinse glassware five times with tap water, then five times with deionized water. An automatic laboratory glassware washer may be used and is the preferred method. The acid-washed glassware should be neutralized with a saturated solution of Na_2CO_3 prior to washing in an automatic washer.

If an electronic particle counter will be used, add a final rinse of deionized water that has been filtered through a $0.22\text{-}\mu m$ filter.

Dry clean glassware at $105°C$ in an oven and store either in closed cabinets or on open shelves with tops covered with aluminum foil.

Before use, autoclave culture flasks covered with aluminum foil at 108 kPa for 15 min. After autoclaving, prerinse flasks with culture medium and invert on absorbent paper for 20 to 30 min to drain. Close culture flasks with foam plugs.

Use disposable pipets to minimize possibility of contamination.

b. pH control: To insure the availability of CO_2, keep the pH below 8.5 by using optimum surface-to-volume ratios (Section 8111C.2), continuously shaking the flask (approximately 100 oscillations/min).

c. Growth measurement: Describe the growth of a test alga in the bottle test[1] by maximum standing crop.[1] Generally, these measurements can be made on Days 3, 5, 7, 10, 12, and 14 if a growth curve is desired. If determination of maximum standing crop is the goal, count only on Days 12 and 14.

Maximum standing crop—Maximum standing crop in any flask is defined as the maximum algal biomass achieved during incubation. For practical purposes, it may be assumed that the maximum standing crop has been achieved when the increase in biomass is less than 5%/d. The maximum standing crop usually is achieved in the algal assay test after 12 to 14 d of incubation.

4. Biomass Monitoring

Several methods may be used, but the selected measurements should be related to dry weight.

a. Dry weight: Use either the aluminum dish or membrane filter method. To use the first, centrifuge a suitable portion of algal suspension, wash sedimented cells three times in distilled water, transfer to tared crucibles or aluminum cups, dry overnight in a hot-air oven at $105°C$, cool to room temperature in a desiccator jar, and weigh.

For the membrane filter method, rinse each filter with 50 mL deionized water and place in folded sheets of paper or on an aluminum weighing dish on which identification codes have been written. Dry overnight in a hot-air oven at $60°C$, cool to room temperature in a desiccator jar, and determine tare weight. Filter a suitable measured portion of algal suspension through a tared $0.45\text{-}\mu m$-pore-diameter membrane filter under a vacuum of 51 kPa. Use ≤ 50 mL as the cell density dictates. Rinse filter funnel with 50 mL deionized water using a wash bottle and let rinsings pass through filter. Dry in an oven for several hours at $60°C$, cool in a desiccator, and weigh.

b. Electronic particle counting: Suspend *S. capricornutum* cells in a 1% NaCl electrolyte solution in a ratio of 1.0 mL cell suspension to 9 mL of $0.22\text{-}\mu m$-filtered saline (10:1 dilution). Pass the resulting suspension through a $100\text{-}\mu m$-diam aperture. Each cell that passes through the aperture causes a voltage drop proportional to its displaced electrolyte volume, which is recorded as a count. A knowledge of both the number of particles (cells) counted per unit volume of sample (usually 0.5 mL) and the mean particle (cell) volume displaced allows changes in cell biomass (in microliters per liter) to be calculated. Equations that can accurately relate volume to dry weight must be developed by each laboratory.

c. Chlorophyll: All algae contain chlorophyll and measuring this pigment can yield some insight into the relative amount of algal biomass present. To measure chlorophyll by fluorescence, swirl test flask to suspend cells. Pipet a portion of cell suspension (5 to 6 mL minimum) into a cuvette and read fluorescence. Zero fluorometer with a distilled water blank before each sample reading.

d. Direct microscopic counting: Use a hemocytometer or plankton counting cell (Section 10200F.2). For filamentous algae break up the algal filaments by using a syringe, an ultrasonic bath, a high-speed blender, or vigorous stirring with glass beads. Each of these techniques has drawbacks, but expelling the sample forcefully through a syringe against the inside of the flask is most satisfactory. Other methods of biomass measurement such as dry weight, absorbance, or chlorophyll fluorescence are more precise than cell counts for growth assessment of filamentous algae.

5. Bibliography

WEISS, C.M. & R.W. HELMS. 1971. Interlaboratory Precision Test—An Eight Laboratory Evaluation of the Provisional Algal Assay Procedure: Bottle Test. Dep. Environmental Science & Engineering, School Public Health, Univ. North Carolina, Chapel Hill.

JORDAN, C. & P. DINSMORE. 1985. Determination of biologically available phosphorus using a radiobioassay technique. *Freshwater Biol.* 15: 597.

8111 H. Effect of Additions

1. Procedures

The quantity of cells produced in a given medium is limited by the concentration of nutrient present in the lowest relative quantity with respect to that required by the organism. If a quantity of the limiting nutrient is added to the medium, cell production increases until this additional supply is depleted or until some other nutrient becomes limiting. Additions of substances other than that which is limiting would yield no increase in cell production. Nutrient and chelator additions may be made singly or in combination and the growth response compared to that of untreated controls to identify those substances that limit growth. The selection of additives, e.g., nitrogen, phosphorus, iron, EDTA, wastewater effluents, will depend on the requirements of the test.

In all cases, keep volume of added nutrient or chelator solution as small as possible, but make it large enough to yield a potentially measurable response. Relate the concentrations of additions to nutrient levels in the sample. To assess the effect of nutrient and chelator additions, compare treated sample to an untreated control. For highly productive controls, flask-to-flask variations may be high and might mask the effect of small additions of the limiting nutrient.

It is sometimes necessary to check the test water for the presence of toxic constituents. To do this, treat the sample with an appropriate dilution of the complete synthetic medium. If no growth, or less than expected growth, occurs, toxic materials are suspected. In some situations, sample dilution or addition of a chelating agent will eliminate toxic effects.

2. Bibliography

KLOTZ, R.L. 1985. Influence of light on the alkaline phosphatase activity of *Selenastrum capricornutum* (Chlorophyceae) in streams. *Can. J. Fish. Aquat. Sci.* 42:384.

KUWABARA, J.S., J.A. DAVIS & C.C.Y. CHANG. 1986. Algal growth response to particle-bound orthophosphate and zinc. *Limnol. Oceanogr.* 31:503.

STORCH, T.A. & V.L. DUNHAM. 1986. Iron-mediated changes in the growth of Lake Erie phytoplankton and axenic algal cultures. *J. Phycol.* 22:109.

WILLIAMS, T.G., & D.H. TURPIN. 1987. Photosynthetic kinetics determine the outcome of competition for dissolved inorganic carbon by freshwater microalgae: Implications for acidified lakes. *Oecologia* 73:307.

ELLIS, B.K. & J.A. STANFORD. 1988. Phosphorus bioavailability of fluvial sediments determined by algal assays. *Hydrobiologia* 160:9.

ANKLEY, G.T., A. KATKO & J.W. ARTHUR. 1990. Identification of ammonia as an important sediment-associated toxicant in the lower Fox River and Green Bay, Wisconsin. *Environ. Tox. Chem.* 9:313.

BRABAND, A., B.A. FAAFENC & J.P.M. NILSSEN. 1990. Relative importance of phosphorus supply to phytoplankton production: Fish excretion versus external loading. *Can. J. Fish. Aquat. Sci.* 47: 364.

CARR, O.J. & R. GOULDER. 1990. Fish-farm effluents in rivers: II. Effects on inorganic nutrients, algae and the macrophyte *Ranunculus penicillatus. Water Res.* 24:639.

ENGLER, D.L. & O. SARNELLE. 1990. Algal use of sedimentary phosphorus from an Amazon floodplain lake: Implications for total phosphorus analysis in turbid waters. *Limnol. Oceanogr.* 35:483.

IRELAND, F.A., B.M. JUDY, W.R. LOWER, M.W. THOMAS & G.F. KRAUSE. 1991. Characterization of eight soil types using the *Selenastrum capricornutum* bioassay. *In* Plants For Toxicity Assessment: Second Volume, p. 217. STP 1115, American Soc. Testing & Materials, Philadelphia, Pa.

DIERBERG, F.E. 1993. Decomposition of desiccated submersed aquatic vegetation and bioavailability of released phosphorus. *Lake Reserv. Manage.* 8:31.

DJOMO, J.E., A. DAUTA & F. MOREAU. 1993. Bioassessment of the eutrophication potential in relation to phosphorus leaching from petroleum drilling muds. *Ann. Limnol.* 29:103.

8111 I. Data Analysis and Interpretation

1. Reporting Requirements

The fundamental measure used in the algal assay to determine biostimulation is the amount of suspended solids (dry weight) produced and determined gravimetrically. Other biomass indicators may be used, but all results must include experimentally determined conversion factors and the dry weight of suspended solids. Use several biomass indicators whenever possible, because biomass indicators respond differently to any given nutrient-limiting condition.

Report results of addition assays with results from two types of reference samples: the assay reference medium and untreated water samples. Report results of assays as maximum standing crop (with time at which it was reached).

Determine the available concentration of growth-limiting nutrient by comparing maximum standing crop in an untreated sample with a maximum standing crop in reference medium.

To determine the nutrients that limit standing crop by single-nutrient additions, treat a number of replicate flasks with single nutrients, determine the maximum standing crop for each flask, and compare the averages by Student's *t* test or other appropriate statistical tests.

To identify standing-crop-limiting nutrients by multiple-nutrient tests, make analysis-of-variance calculations. Account for possible interaction between difference nutrients by using factorial analysis.[1]

Report maximum standing crop with confidence interval. Base the calculation of confidence interval for the average values on at least five samples. Consequently, make a minimum of five replications when an unfamiliar source water is first analyzed. Use these results to calculate the standard deviation. For subsequent samples from the same source use only three replicates and report with the confidence interval established for that source water.

The overall evaluation of assay results consists of first determining whether a result is significant when considered as a laboratory measurement. Several methods are available, such as the Student's *t* test and analysis-of-variance techniques.

The second part of the evaluation is the correlation of laboratory assay results to effects observed or predicted in the field. Specific guidelines are not yet available, but note the general considerations in Section 8111B.

2. Reference

1. EUTROPHICATION AND LAKE RESTORATION BRANCH. 1974. Marine Algal Assay Procedure: Bottle Test. U.S. Environmental Protection Agency, Pacific Northwest Environmental Research Lab., Corvallis, Ore.

8112 PHYTOPLANKTON*

8112 A. Introduction

The phytoplankton are primary producers in the aquatic community and, as such, are at the base of aquatic food chains. Because of this, they must be tested in bioassays that predict and determine the potential effects of a substance on the aquatic environment. The

_____
* Approved by Standard Methods Committee, 1993.

same general principles and techniques used in determining biostimulation (Section 8111) are used to determine toxicity to phytoplankton. The procedure applies to fresh water, estuarine, and marine phytoplankton. See Sections 10200D through H and 10300D for additional information on phytoplankton.

8112 B. Inoculum

In addition to the marine or estuarine algae listed in 8111F.1*b*, *Monochrysis lutheri* Droop may be used. Maintain the test species in full-strength media [Section 8010E.4*c*1) and Tables 8010:III

and 8010:IV.A and B]. Test species must be in the logarithmic growth phase; therefore, transfer them to fresh culture medium every 4 to 5 d.

8112 C. Test Conditions and Procedures

1. Maximum Specific Growth Rate

Add test material to test vessels to give desired concentrations. Prepare triplicate vessels for each concentration. Use dilutions of culture medium to simulate chemical conditions of specific receiving waters. For optimum surface-to-volume ratios, see Section 8111C.2*a*.

The maximum specific growth rate (μ_{max}) occurs during the logarithmic phase of growth, usually between Day 0 and Day 5. Therefore, measure biomass at least daily during the first 5 d of incubation. Indirect measurements of biomass, such as chlorophyll *a* or cell numbers, usually will be required because accurate gravimetric measurements at low cell densities are difficult. See Section 8111G.3*c* and 8111G.4 for methods.

Test a geometric series of concentrations initially (see Section 8010F.3*b*). After this preliminary test, progressively bisect intervals on a logarithmic scale. Narrow the range of test concentrations to determine the concentration that reduces the maximum specific growth rate (μ_{max}) to 50% that of the control. This requires that two of the concentrations tested fall on each side of the concentration that inhibited (μ_{max}) to 50% (see Section 8010G).

Compare the maximum specific growth rate (μ_{max})to that obtained in the synthetic freshwater or artificial seawater culture medium. Regional and seasonal variations in quality make natural waters unsuitable as standard test media for comparative toxicity tests. Therefore, use a synthetic freshwater medium and/or artificial seawater. Add various concentrations of toxicants to the culture medium in triplicate and inoculate with test species.

2. Other Tests

For other types of tests, such as those to determine effluent requirements or compliance with water quality standards, take dilution water from the receiving body near the outfall but outside its influence. Remove undesirable organisms before making growth rate tests with selected sensitive species (Section 8111D). Determine maximum specific growth rates (μ_{max}) in test vessels and compare with controls and EC50s based on percent of growth reduction. An alternative approach that provides a number that may relate to natural conditions should be reviewed.[1]

3. Reference

1. MILLER, W.E., J.C. GREENE & T. SHIROYAMA. 1978. The *Selenastrum capricornutum* Printz Algal Assay Bottle Test. U.S. Environmental Protection Agency Rep. EPA-600/9-78-018, National Technical Information Serv., U.S. Dep. Commerce, Springfield, Va.

4. Bibliography

ERICKSON, S.J., N. LACKIE & T.E. MALONEY. 1970. A screening technique for estimating copper toxicity to estuarine phytoplankton. *J. Water Pollut. Control Fed.* 42:R270.

WALSH, G.E. 1972. Effects of herbicides on photosynthesis and growth of marine unicellular algae. *Hyacinth Control J.* 10:45.

GREEN, J.C., W.E. MILLER, T. SHIROYAMA & E. MERWIN. 1975. Toxicity of Zinc to the Green Alga *Selenastrum capricornutum* as a Function of Phosphorus or Ionic Strength. U.S. Environmental Protection

Agency Rep. EPA-660/3-75-034, National Technical Information Serv., U.S. Dep. Commerce, Springfield, Va.

GREEN, J.C., W.E. MILLER, T. SHIROYAMA, R.A. SOLTERO & K. PUTNAM. 1976. Use of algal assays to assess the effects of municipal and smelter wastes upon phytoplankton production. *In* Terrestrial and Aquatic Ecological Studies of the Northwest. Eastern Washington State College Press, Cheney.

KOELMANS, A.A., C.S. JIMÉNEZ & L. LUKLEMA. 1993. Sorption of chlorobenzenes to mineralization of phytoplankton. *Environ. Toxicol.Chem.* 12:1425.

STRANGE, K. & D.L. SWACKHAMER. 1994. Factors affecting phytoplankton species-specific differences in accumulation of 40 polychlorinated biphenyls (PCBs). *Environ. Toxicol. Chem.* 13:1849.

BROWN, L.S. & D.U.S. LEAN. 1995. Toxicity of selected pesticide to lake phytoplankton measured using photosynthetic inhibition compared to maximal uptake rates of phosphate and ammonium. *Environ. Toxicol. Chem.* 14:93.

8200 AQUATIC FLOWERING PLANTS

Aquatic flowering plants belong to phylum Spermatophyta, characterized by possession of true roots, stems, and leaves and production of seeds from flowers. The phylum contains most of the conspicuous land plants of the world. Aquatic flowering plants are almost exclusively fresh-water inhabitants. Test procedures are described for duckweed (8210), which floats on the surface of the water, and rooted plants (8220), which have roots extending into the sediment.

8211 DUCKWEED*

8211 A. Introduction

1. Organism Characteristics

Lemna minor L. (Figure 8211:1) (also known as common duckweed) is a small flowering aquatic macrophyte widely distributed in quiescent fresh water and estuaries ranging from tropical to temperate zones. It is the most common species of the family Lemnaceae in the United States and many other parts of the world. It is morphologically simple, consisting only of frond and root. The frond size is approximately 2 to 4 mm and root length is up to 50 mm. The plant is colonial, multiplies sexually and asexually, and has a growth rate far exceeding those of other flowering plants.[1] Duckweed is a food for waterfowl and small animals, and provides food, shelter, and shade for fish and other aquatic organisms. Furthermore, it serves as a habitat for various invertebrates.

2. Test Applications

Common duckweed is an ideal organism for testing aquatic phytotoxicity of herbicides, industrial and municipal wastewaters, and other contaminants.[2] Because many ambient waters and effluents are colored and/or turbid, they are difficult to test for toxicity by using algal testing without filtering, which decreases sam-

* Approved by Standard Methods Committee, 1997.

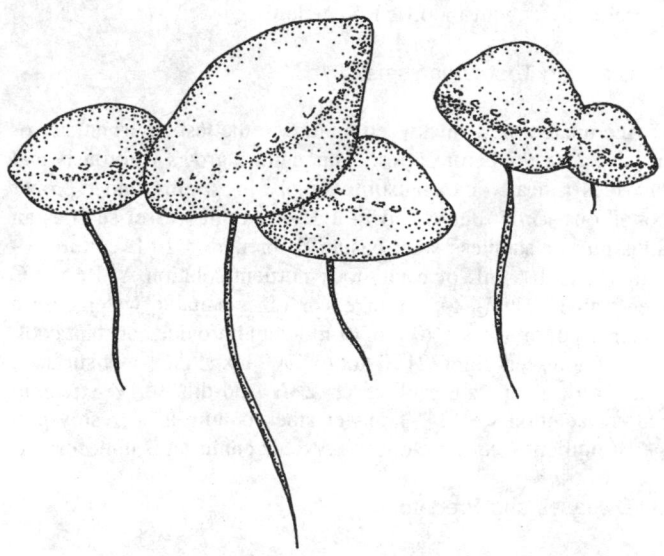

Figure 8211:1. Common duckweed: *Lemna minor.*

ple integrity. In addition, some samples contain labile constituents and require either renewal or flow-through methods. Algal testing may be inappropriate for these samples, whereas the duckweed toxicity test can be modified easily by either method.

The duckweed toxicity test is useful, especially for determining phytotoxicity at the air-water interface where surface-active substances, oil and grease, and toxic organic compounds may be concentrated. The test also is useful for determining toxicity of metals,[3] organic compounds,[4,5] and industrial and municipal effluents.[6–9] It is generally described as a simple, sensitive, and cost-effective test.[2]

Common duckweed, being a floating plant, may underestimate the toxicity of a substance (such as bromacil) that adsorbs on particular matter and precipitates during a static test. Also, substances (such as atrazine) that concentrate at the air-water interface tend to affect duckweed more than other aquatic plants, such as algae or submerged plants. Gentle shaking or stirring test vessels to increase mixing may overcome these problems.

3. References

1. HILLMAN, W.S. & D.D. CULLEY. 1978. The use of duckweed. *Amer. Scientist* 66:442.
2. WANG, W. 1990. Literature review on duckweed toxicity testing. *Environ. Res.* 51:7.
3. WANG, W. 1986. Phytotoxicity tests of aquatic pollutants by using common duckweed. *Environ. Pollut.* (Ser. B) 11:1.
4. KING, J.M. & K.S. COLEY. 1985. Toxicity of aqueous extracts of natural and synthetic oils to three species of *Lemna*. Spec. Tech. Publ. 891, American Soc. Testing & Materials, Philadelphia, Pa.
5. HUGHES, J.S., M.M. ALEXANDER & K. BALU. 1988. An evaluation of appropriate expressions of toxicity in aquatic plant bioassays as demonstrated by the effects of atrazine on algae and duckweed. Spec. Tech. Publ. 921, American Soc. Testing & Materials, Philadelphia, Pa.
6. ROWE, E.L., R.J. ZIOBRO, C.J.K. WANG & C.W. DENCE. 1982. The use of an alga *Chlorella pyrenoidosa* and a duckweed *Lemna perpusilla* as test organisms for toxicity bioassays of spent bleaching liquors and their compounds. *Environ. Pollut.* (Ser. A) 27:289.
7. WANG, W. & J. WILLIAMS. 1988. Screening and biomonitoring of industrial effluents using phytotoxicity tests. *Environ. Toxicol. Chem.* 7: 645.
8. LOCKHART, W.L. & A.P. BLOUW. 1979. Phytotoxicity tests using the duckweed *Lemna minor*. *In* Toxicity Tests Freshwater Organisms. Canadian Spec. Publ. Fish. Aquat. Sci. 44:112.
9. TARALDSEN, T.E. & T.J. NORBERG-KING. 1990. New method for determining toxicity using duckweed (*Lemna minor*). *Environ. Toxicol. Chem.* 9:761.

4. Bibliography

HILLMAN, W.S. 1961. The Lemnaceae, or duckweed, a review of the descriptive and experimental literature. *Bot. Rev.* 27:221.
HOLST, R.W. & T.C. ELLWANGER. 1982. Pesticide Assessment Guidelines, Subdivision J Hazard Evaluation: Nontarget Plants. EPA 540/9-82-020, U.S. Environmental Protection Agency, Washington, D.C.

8211 B. Selecting and Preparing Test Organisms

1. Test Species

The procedure is designed for use with *Lemna minor*. The organism can be obtained from commercial sources, testing laboratories, or the field. It must be identified and confirmed taxonomically before use.[1] Other duckweed species, such as *L. gibba, L. perpusilla, L. pancicostata,* and *L. polyrrhiza*, have been used successfully[2,3] with modified procedures.

2. Culturing Test Organisms

Acclimate a new duckweed culture to the test environment for at least 2 weeks before a test. This culture grows vigorously and provides a nearly inexhaustible supply for testing under proper conditions. Grow duckweed in a 15-L culture vessel such as an aquarium or stainless steel basin. To prepare a 10-L culture solution, add 100 mL of each stock nutrient solution A, B, and C (see Table 8211:I), to deionized or other suitable water. Use a water depth of at least 40 mm or more and provide constant cool-white fluorescent light (2150 lux to 4300 lux at the water surface). Maintain a temperature of 25 ± 2°C. Add diluted (¼ strength) culture solution weekly. Transfer stock culture to a freshly prepared nutrient solution monthly. Axenic culturing is unnecessary.

3. Diseases and Predators

Diseases, phytophagous insects, or other pests usually do not pose any problem to a duckweed culture. If the culture appears

TABLE 8211:I. DUCKWEED NUTRIENT SOLUTION

Solution	Stock Solution Concentration		Element	Final Concentration	
A:					
NaNO$_3$	25.5	g/L	N	42.0	mg/L
			Na	110.0	mg/L
NaHCO$_3$	15.0	g/L	C	21.4	mg/L
K$_2$HPO$_4$	1.04	g/L	K	4.69	mg/L
			P	1.86	mg/L
B:					
CaCl$_2$·2H$_2$O	4.41	g/L	Ca	12.0	mg/L
MgCl$_2$	5.7	g/L	Mg	29.0	mg/L
FeCl$_3$	0.096	g/L	Fe	0.33	mg/L
Na$_2$EDTA·2H$_2$O	0.3	g/L			
MnCl$_2$	0.264	g/L	Mn	1.15	mg/L
C:					
MgSO$_4$·7H$_2$O	14.7	g/L	S	19.1	mg/L
H$_3$BO$_3$	0.186	g/L	B	325	µg/L
Na$_2$MoO$_4$·2H$_2$O	7.26	mg/L	Mo	28.8	µg/L
ZnCl$_2$	3.27	mg/L	Zn	15.7	µg/L
CoCl$_2$	0.78	mg/L	Co	3.54	µg/L
CuCl$_2$	0.009	mg/L	Cu	0.04	µg/L

NOTE:

1. Omit Na$_2$EDTA·2H$_2$O in Solution B if test samples contain toxic metals. In that case, acidify Solution B to pH 2 to prevent precipitation.
2. To prepare the duckweed nutrient solution, add 1 mL of each stock solution to 100 mL deionized water. Adjust to pH 7.5 − 8.0.

unhealthy, destroy the old culture and start a new one. It is good practice to maintain several cultures isolated from each other.

4. References

1. CORRELL, D.S. & H.B. CORRELL. 1972. Aquatic and Wetland Plants of Southwestern United States. U.S. Environmental Protection Agency, Washington, D.C.

2. KING, J.M. & K.S. COLEY. 1985. Toxicity of aqueous extracts of natural and synthetic oils to three species of *Lemna*. Spec. Tech. Publ. 891, American Soc. Testing & Materials, Philadelphia, Pa.

3. HUGHES, J.S., M.M. ALEXANDER & K. BALU. 1988. An evaluation of appropriate expressions of toxicity in aquatic plant bioassays as demonstrated by the effects of atrazine on algae and duckweed. Spec. Tech. Publ. 921, American Soc. Testing & Materials, Philadelphia, Pa.

8211 C. Toxicity Test Procedure

1. General Considerations

Use static, renewal,[1] or flow-through methods.[1,2] Usually, if a test solution is stable (e.g., a solution with low microbial population, high toxic metal concentration, or low volatility), use a static test. If samples are unstable, use renewal (daily renewal) or flow-through methods (see Section 8010D).

2. Preparing Test Materials

As used herein, dilution water and control water are identical to duckweed nutrient solution. For preparation of this solution, see Table 8211:I.

For preparation of toxicant solutions, see Section 8010F.2*b*.

3. Test Procedures and Conditions

Use these procedures in screening, range-finding, or definitive tests.

In a screening test, use a predetermined concentration (e.g., 100% effluent) to determine if a sample is toxic in comparison with the control water. If the sample is toxic, test it further using range-finding and definitive tests. In the range-finding test, examine a series of concentrations, usually at ratios of 10, e.g., 10%, 1%, 0.1%, etc.

Devise a definitive test on the basis of range-finding test results. Use five concentrations of sample in a ratio of 0.5, e.g., 10%, 5%, 2.5%, etc. Ideally, prepare a series of solutions in which the midpoint concentration produces approximately 50% inhibitory effect, and the highest and lowest concentrations produce approximately 90 and 10% inhibitory effects.

Use four replicates of each test treatment and control, including a negative control containing only duckweed nutrient solution or modified solution[3] as well as a positive control containing 20 mg potassium chromate (as Cr)/L. Preferably use 60×15-mm glass petri dishes as test vessels. (Although the dish depth may be less than root length, duckweed growth is not adversely affected.[3]) Plastic vessels can be used, but make sure duckweed specimens do not adhere to the vessel wall. Add the same amount of nutrients to all control and test samples, i.e., 1 mL of each nutrient stock solution (A, B, and C) to make 100 mL sample. Prepare a 15-mL portion of test solution (or control sample). Select duckweed specimens from stock cultures that have been grown under the same conditions. Cut all roots to reduce algal contamination, if desired. Use only unblemished colonies containing two fronds of approximately equal size per colony. Alternatively, four three-frond plants, three four-frond plants, or other combinations can be used. Place 12 duckweed fronds in each vessel and cover. Illuminate with continuous cool-white fluorescent light at 2150 lux to 4300 lux at water surface and incubate at $25 \pm 2°C$.

In testing effluent toxicity in a receiving water, use daily renewal with fresh effluent and use receiving water as diluent. Otherwise, use a standard water as diluent. Complete frond counts daily to determine any intermediate toxic effects. In some cases, duckweed may exhibit delayed effects; it takes about 2 to 3 d for the growth rate to change when duckweed is transferred to a new solution. Stimulatory effects may occur in some instances. Test duration is 96 h.

4. Test Results

Observe duckweed plants under a lighted magnifying glass (2 × or higher) for symptoms, including chlorosis (loss of pigment/yellowing), necrosis (localized dead tissue), colony breakup, root destruction, loss of buoyancy, and gibbosity (humpback or swelling). Compare affected fronds with duckweed specimens in the control. These observations are of use in establishing a "no-observed-effect concentration" (NOEC).

The most commonly used and seemingly reliable method of evaluation is frond increase, a quantal value directly reflecting duckweed growth. To measure frond increase, count every visible, protruding bud. This nondestructive method allows repeated observations of the same solution.

Other methods that have been used include ^{14}C uptake, chlorophyll content, biomass, frond area, plant colony counts, total root number and root length.[4,5] Mortality alone is of limited value.[5]

Some substances, at low concentrations, stimulate, rather than inhibit, duckweed growth. Report this effect if it is observed.

5. Statistical Analysis

Follow general procedures described in Section 8010G.

In screening tests, the key question is whether a sample is toxic or stimulatory as compared with the water control. Express toxicity (or stimulation) in percent inhibition (or stimulation) relative to the control

$$\% I = 100 (C - T)/C$$

where C and T are average increases in number of fronds in control and test samples, respectively.

Results of definitive tests may be graphed using linear, semilog, or log-log plots. Typically, the concentration-effect relationship is sigmoidal. Determine IC10, IC50, and IC90 values (the concentration causing 10, 50, and 90% inhibitory effects) and SC20 value (the concentration causing 20% stimulatory effect) by graphical or statistical methods. The slope of dose-response relationship is toxicant-specific and thus can be valuable information.

6. Quality Control

The negative control sample is necessary for quality control. A test is not acceptable if more than 10% of the control specimens die, or show adverse symptoms. Under normal conditions, the doubling time for duckweed is less than 2 d. If the control sample yields less than a twofold increase in fronds in 96 h, the test is not acceptable.

Another measure of quality control is the use of reference chemicals at a single specified concentration as a positive control. The chromate ion has been found to be an ideal reference toxicant. Its toxicity is not affected by water quality and thus it can be used in natural waters, wastewaters, leachates, and the like; 20 mg/L Cr(VI) concentration causes approximately 60% inhibitory effect of duckweed growth.[6,7] Include this positive control in every test.

7. References

1. DAVIS, J.A. 1981. Comparison of Static-Replacement and Flow-Through Bioassays using Duckweed, *Lemna gibba* G-3. EPA-560/6-81-003, U.S. Environmental Protection Agency, Washington, D.C.
2. WALBRIDGE, C.T. 1977. A Flow-Through Testing Procedure with Duckweed (*Lemna minor* L.). EPA-600/3-77-108, U.S. Environmental Protection Agency, Duluth, Minn.
3. WANG, W. 1992. Toxicity reduction of photo processing wastewaters. *J. Environ. Sci. Health* A27:1313.
4. TARALDSEN, J.E. & T.J. NORBERG-KING. 1990. New method for determining toxicity using duckweed (*Lemna minor*). *Environ. Toxicol. Chem.* 9:761.
5. WANG, W. & J. WILLIAMS. 1988. Screening and biomonitoring of industrial effluents using phytotoxicity tests. *Environ. Toxicol. Chem.* 7:645.
6. WANG, W. 1986. The effect of river water on phytotoxicity of barium, cadmium, and chromium ions. *Environ. Poll.* (Sec. B) 11:193.
7. WANG, W. 1987. Chromate ion as a reference toxicant in aquatic phytotoxicity tests. *Environ. Toxicol. Chem.* 6:953.

8. Bibliography

U.S. ENVIRONMENTAL PROTECTION AGENCY. 1983. Lemna Acute Toxicity Test. EPA-560/6-82-002, National Technical Information Serv., Springfield, Va.

8220   AQUATIC EMERGENT PLANTS*

8220 A.   Introduction

1. Organism Characteristics

Emergent plants are important components of aquatic and wetland ecosystems. They are among the primary producers, providing oxygen, food, and habitat for many life forms, including invertebrates, fish, amphibians, birds, and mammals. These plants also are important in nutrient cycling and in the stabilization of sediments of near-shore environments.

2. Test Applications

This method is designed for evaluating the effects of water contaminants and general water quality on germination and seedling growth of emergent plants. The method provides for a group of indicators as measures of toxic response, including seed germination, root elongation, root dry weight, and dry weight of the seedling shoot. Seed germination and seedling growth represent the first phase of plant development. Significant inhibition of this developmental phase will affect the ability of plants to compete and survive in their environment.

The seed germination and seedling growth tests are simple, versatile, and useful for screening toxicity in water.[1–3] The tests

* Approved by Standard Methods Committee, 1997.

are useful for evaluating toxicity of metals,[4–6] organic compounds,[5,7] and complex effluents.[2,3] They can be conducted in fresh, sea, or brackish water, with the use of appropriate plant species. They are applicable to turbid and/or discolored aqueous samples.[3,] In addition, the test sediments can be screened for toxicity by testing aqueous extracts, pore water, or whole sediment.

One advantage of seed germination and seedling growth tests is that seeds can be obtained in bulk quantity and stored for extended periods with minimal maintenance costs. Stored plant seeds are dormant and resistant to environmental stress. Under favorable conditions for germination, seeds undergo rapid changes and become highly sensitive to the environment.[1] Another advantage is that plant seeds are self-sufficient and require only water, diatomic oxygen, and appropriate temperature and light regimes to germinate, although some species require special treatment.[1] Standard water (Section 8010) can be used as dilution water and control solution. Because nutrients and adjuvants (e.g., chelating agents) are not required in germination and early plant growth, there is no potential interference and interaction of these substances. The tests are highly desirable to complement other tests where the potential interaction may exist.

The tests can be conducted in darkness or light with the species suggested. The tests are especially useful for evaluating photodegradable compounds or samples. When both dark- and light-phase experiments are conducted, the tests can be used to evaluate the effects of photodegradation of toxicants and effects of toxicants on the light reaction of photosynthesis.[2]

The tests can be performed with a relatively small volume of test solution (30 mL/vessel or less) compared to aquatic animal testing (100 to 200 mL/vessel). A small volume is desirable because it will minimize the expense of sample collection and disposal. Perform the tests using static, renewal, or flow-through procedures.

3. References

1. MAYER, A.M. & A. POLJAKOFF-MYBER. 1982. The Germination of Seeds, 3rd ed. Pergamon Press, Oxford, England.
2. WALSH, G.E., D.E. WEBER, T.L. SIMON & L.K. BRASHERS. 1991. Toxicity tests of effluents with marsh plants in water and sediments. *Environ. Toxicol. Chem.* 10:517.
3. WANG, W. & J. WILLIAMS. 1988. Screening and biomonitoring of industrial effluents using phytotoxicity tests. *Environ. Toxicol. Chem.* 7: 645.
4. WONG, M.H. & A.D. BRADSHAW. 1982. A comparison of the toxicity of heavy metals, using root elongation of rye grass, *Lolium perenne. New Phytol.* 91:255.
5. WANG, W. 1987. Root elongation method for toxicity testing of organic and inorganic pollutants. *Environ. Toxicol. Chem.* 6:409.
6. GORSUCH, J.W., R.O. KRINGLE & K.A. ROBILLARD. 1990. Chemical effects on the germination and early growth of terrestrial plants. ASTM STP 1091, American Soc. Testing & Materials, Philadelphia, Pa.

7. RATSCH, H.C. 1983. Interlaboratory root elongation testing of toxic substances on selected plant species. EPA-600/3-83-051, U.S. Environmental Protection Agency, Corvallis Environmental Research Lab., Corvallis, Ore.
8. WALSH, G.E., D.E. WEBER, L.K. BRASHERS & T.L. SIMON. 1990. Artificial sediments for use in tests with wetland plants. *Environ. Exper. Bot.* 30:341.

4. Bibliography

HOLST, R.W. & T.C. ELLWANGER. 1982. Pesticide Assessment Guidelines. Subdivision J. Hazard Evaluation: Non-Target Plants. Off. Pesticides and Toxic Substances, U.S. Environmental Protection Agency, Washington, D.C.

U.S. ENVIRONMENTAL PROTECTION AGENCY. 1985. Toxic Substances Control Act Test Guidelines: Environmental Effects Testing Guidelines. 40 CFR Part 797, *Federal Register* 50 (188): 39389.

U.S. FOOD AND DRUG ADMINISTRATION. 1987. Seed germination and root elongation. *In* Environmental Assessment Technical Handbook, 4.06. Center for Food Safety and Applied Nutrition, Center for Veterinary Medicine, Washington, D.C.

BOUTIN, C., K.E. FREEMARK & C.J. KEDDY. 1993. Proposed guidelines for registration of chemical pesticides: Nontarget plant testing and evaluation. Environment Canada Tech. Rep. Ser. No. 145, Hull, Quebec, Canada.

AMERICAN SOCIETY FOR TESTING AND MATERIALS. 1995. Practice for Conducting Early Seedling Growth Tests. E1598-94, American Soc. Testing & Materials, Philadelphia, Pa.

8220 B. Selecting and Preparing Test Organisms

1. Test Species

The criteria for selecting test species are economic and ecological relevancy, seed availability, consistent performance, and high germination percentage. There are many emergent species[1] but only a few are readily available.[2,3]

The following freshwater plant species are recommended and typically are available from commercial vendors:*

Echinochloa crusgalli (Figure 8220:1)	Japanese (duck) millet
Leersia oryzoides	Rice cutgrass
Nelumbo lutea	American lotus
Oryza sativa	Rice
Rorippa nasturtium-aquaticum	Watercress
Zizania aquatica	Wild rice

* For example, Wildlife Nurseries, P.O. Box 2724, Oshkosh, WI 54903; Environmental Concern, P.O. Box P, St. Michaels, MD 21663; Kester's Wild Game Food Nurseries, Inc., Omro, WI 54963; or Mangelsdorf Seed Co., 1415 13th Street, St. Louis, MO 63106.

The germination of American lotus seeds requires scarification that may affect toxicity test results.

Field-collected seeds can be used. Obtain species variety (cultivar), and any certification information such as germination percentage and date collected. Remove weed seed if present. Do not use plant seeds treated with fungicides, repellents, or micronutrients (e.g., boron, manganese), etc.

2. Preparing Test Organisms

Obtain sufficient seeds for one year of testing. Store fresh seeds at −10°C (or 4°C if seeds cannot tolerate lower temperature). Test seeds regularly during storage for percent germination. Some species can be stored longer than one year without a decrease in percent germination.

Use seeds from the same storage lot and year or season of collection. Before beginning toxicity testing, separate seeds into size classes using standard seed dockage sieves. Discard damaged seeds. Use the size class containing most seeds exclusively for the test.

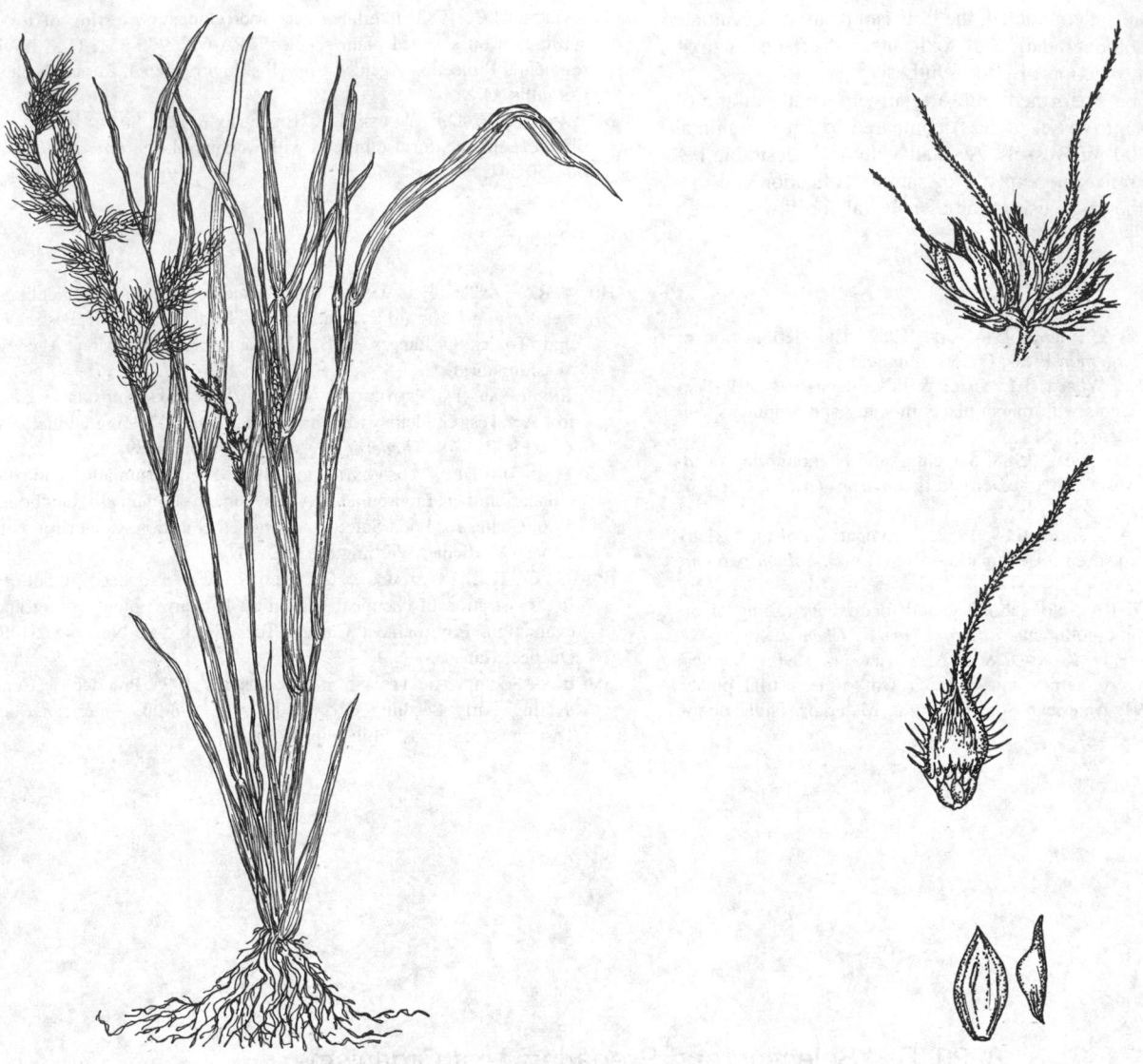

Figure 8220:1. *Echinochloa crusgalli* (**Japanese millet or duck millet).** Left: entire plant; right, top to bottom: spikelet cluster, spikelet, and floret seed.

3. References

1. CORRELL, D.S. & H.B. CORRELL. 1972. Aquatic and Wetland Plants of Southwestern United States. U.S. Environmental Protection Agency, Washington, D.C.
2. FREEMARK, K., P. MACQUARRIE, S. SWANSON & H. PETERSON. 1990.

Development of Guidelines for Testing Pesticide Toxicity to Nontarget Plants for Canada. ASTM STP 1091, American Soc. Testing & Materials, Philadelphia, Pa.
3. SWANSON, S.M., C.P. RICKARD, K.E. FREEMARK & P. MACQUARRIE. 1991. Testing for Pesticide Toxicity to Aquatic Plants. ASTM STP 1115, American Soc. Testing & Materials, Philadelphia, Pa.

8220 C. Toxicity Test Procedure

1. General Considerations

Seed germination and seedling growth tests may be conducted using either static, renewal, or flow-through methods. As a general rule, if test samples are not highly volatile or degradable, use a static test. Otherwise, use a renewal (daily renewal) or flow-through procedure with fresh sample.

2. Preparing Test Materials

Use reconstituted freshwater, Table 8010:I, as dilution water and control water. Alternatively, use deionized and distilled water.

For preparation of toxicant solutions, see Section 8010F.2b.

3. Test Procedures and Conditions

Use these procedures in screening, range-finding, or definitive tests.

In a screening test, use a predetermined concentration (e.g., 100% effluent) to determine if a sample is inhibitory in comparison with the control. If a sample is inhibitory by more than 10%, test it further using range-finding and definitive tests. In the range-finding test, examine at least three concentrations, usually at ratios of 0.1, e.g., 10%, 1%, and 0.1%, plus the control water.

Devise a definitive test on the basis of range-finding test results. Use at least five concentrations of toxicant solutions in a ratio of 0.5, e.g., 10%, 5%, 2.5%, etc. Ideally, prepare a series of solutions in which the midpoint concentration produces an inhibitory effect of approximately 50% and the highest and lowest concentrations produce approximately 90 and 10% inhibition. Stimulatory effects may occur in some instances.

Perform all tests in at least quadruplicate. Include control water in each test. Use a 100- × 15-mm or 47-mm culture dish, test tube, or equivalent, as the test vessel. Use only one kind of vessel in one test. Do not use filter paper or seed pack growth pouch, because it may cause erratic results.[1-3] Pipet 10 mL test solution into each test vessel. Seed trays may be used, requiring 30 mL test solution.[3]

Place 10 to 15 seeds in each test vessel. Seeds should not be in contact with each other or with sides of culture dishes. Place test vessels in a seed germinator or other growth facility. Use a randomized complete block design with blocks delineated within the growth facility; if blocking is not feasible, totally randomize vessels in the growth facility.

Test duration, temperature, and light regimes will vary depending on experimental design and test species. Typically, incubate seeds until the expected percentage of control seed germination has been attained (See Section 8220D) and control seed root development has reached at least 20 mm.[4]

Table 8220:I provides an example of appropriate test conditions.

If fungi or other microorganisms interfere with germination, pretreat test seeds with a reagent-grade sodium hypochlorite solution (3.33 g OCl^-/L) or 10% household bleach for 20 min. Rinse seeds eight to ten times with deionized or distilled water. Remove excess solution with tissue paper. Omit pretreatment step if this affects germination of the control by more than 10%.

Conduct the renewal test, if necessary, by transferring seeds or seedlings into clean vessels containing fresh toxicant or control water daily. Modify a proportional diluter for flow-through testing.

4. Test Results

A seed is counted as having germinated if the radicle reaches a length of 5 mm or longer. Record all data on root and shoot elongation of germinated seeds in each dish.

Determine root elongation by measuring the length of each primary root from the transition point of the hypocotyl to the tip of the root. Use a digitizer interfaced with a computer, if possible.

Alternatively, cut primary roots in each dish, combine, dry to constant weights at 70°C, and weigh. Likewise determine shoot biomass. Report abnormal appearance such as discoloration, stunted growth, and chlorosis.

Some substances, at low concentration, stimulate, rather than inhibit, seed germination or root elongation. Report this effect if it is observed.

5. Statistical Analysis

Follow general procedure described in Section 8010G.

Express sample toxicity in percent inhibition relative to the control.[5,6]

$$\% I = 100 \ (C - T)/C$$

where C and T are mean seed germination percentages in control water and test solutions, respectively, if seed germination is used as the test indicator. If root elongation is used as the test indicator, C and T are root length (in mm) in control and test solutions, respectively.

Determine IC10, IC50, IC90 values (the concentrations causing 10, 50, and 90% inhibitory effects) and SC20 value (the concentration causing 20% stimulatory effect) by statistical curve-fitting methods. Report concentration-inhibition relationship and confidence limit of test results.

TABLE 8220:I. EXAMPLE OF SEED GERMINATION AND SEEDLING GROWTH TEST CONDITIONS

Test Variable	Condition or Value in Example Test
Test species	*Echinochloa crusgalli* (Japanese millet)
Pretreatment	20 min, hypochlorite solution (3.33 g OCl^-/L)
Test type	Static or renewal
Temperature	25 ± 1°C
Light quality	Dark or light
Test vessel	100 × 15-mm culture dish
Test solution	8 mL/vessel
Specimens	15 seeds/vessel
Replicates	4
Control solution and dilution water	Standard water
Test duration	120 h
Indicators	Seed germination (radicle 5 mm or longer)
	Root elongation
	Root dry biomass
	Shoot dry biomass
	Abnormal appearance

6. Quality Control

The negative control sample is needed for quality control. Derive empirical performance criteria (e.g., germination percent, mean and standard deviation of root elongation of the negative control sample, and test duration for primary root reaching 20 mm) and statistical confidence limits for each species for 3 months. Any time the performance of a test in the next 3 months falls below the confidence limits, repeat tests; if two consecutive tests fall below the criterion, replace the seed lot and discard the data.

7. References

1. GORSUCH, J.W., R.O. KRINGLE & K.A. ROBILLARD. 1990. Chemical Effects on the Germination and Early Growth of Terrestrial Plants. ASTM STP 1091, American Soc. Testing & Materials, Philadelphia, Pa.

2. WANG, W. & J. WILLIAMS. 1990. The use of phytotoxicity tests (common duckweed, cabbage, and millet) for determining effluent toxicity. *Environ. Monit. Assess.* 14:45.

3. WANG, W. 1993. Comparative rice seed toxicity tests using filter paper, Growth Pouch-TM, and seed tray methods. *Environ. Monit. Assess.* 24:257.

4. U.S ENVIRONMENTAL PROTECTION AGENCY. 1985. Toxic Substances Control Act Test Guidelines: Environmental Effects Testing Guidelines. CFR 40 Part 797, *Federal Register* 50 (188): 39389.

5. RATSCH, H.C. & D. JOHNDRO. 1986. Comparative toxicity of six test chemicals to lettuce using two root elongation test methods. *Environ. Monit. Assess.* 6:267.

6. WALSH, G.E., D.E. WEBER, T.L. SIMON & L.K. BRASHERS. 1991. Toxicity tests of effluents with marsh plants in water and sediments. *Environ. Toxicol. Chem.* 10:517.

8310   CILIATED PROTOZOA (PROPOSED)*

8310 A.   Introduction

1. General Discussion

Ciliated protozoans (Kingdom Protista, Phylum Ciliophora) are ubiquitous unicellular eukaryotes. They inhabit freshwater and marine aquatic environments, soils, and sediments. Ciliates are important organisms in the transfer and transformation of nutrients in ecological food chains.[1,2] In the aquatic environment, ciliates are an integral part of the zooplankton community, feed predominantly on bacteria and small phytoplankton,[3-5] and mediate the transfer of energy from the microbial food web to larger metazoan zooplankton.[1] In soil and sediment, ciliates feed primarily on bacteria and organic detritus.[6] The prevalence of this group and their importance in trophic processes make them particularly appropriate as organisms used for the assessment of water quality.[7-9]

Recent advances in the assessment of environmental toxicity have focused on microscale testing, more rapid bioassessment techniques, and more sensitive indicators of water quality (i.e., sublethal versus lethal effects).[10] The potential for using ciliates to evaluate water quality was recognized some time ago.[11,12] More recently, investigators have focused increasingly on ciliates as test and/or indicator organisms for the assessment of eutrophic and contaminated media, because they represent a neglected trophic level in most bioassay batteries, and are sensitive to a broad range of toxicants in the natural environment.[13] A review of this field can be found elsewhere.[14]

2. Selection of Method

This section includes three standardized toxicity test methods using ciliated protozoa as test organisms. The first two utilize ciliates common in fresh water, and can be used for whole water testing with effluents and pure chemicals, while the third utilizes a soil ciliate; this test is most appropriate as an elutriate test, where contaminated soils and mine tailings are potentially implicated.

3. References

1. FENCHEL, T. 1987. Ecology of Protozoa—The Biology of Free-Living Phagotrophic Protists. Science Tech Publishers, Madison, Wisc.

2. PORTER, K.G., E.B. SHERR, B.F. SHERR, M. PACE & R.W. SANDERS. 1985. Protozoa in planktonic food webs. *J. Protozool.* 32:409.

3. PACE, M.L. & J.D. ORCUTT, JR. 1981. The relative importance of protozoans, rotifers, and crustaceans in a freshwater zooplankton community. *Limnol. Oceanogr.* 26:822.

4. SHERR, E.B. & B.F. SHERR. 1987. High rates of consumption of bacteria by pelagic ciliates. *Nature* 325:710.

5. PRATT, J.R. & J. CAIRNS, JR. 1985. Functional groups in the protozoa. *J. Protozool.* 32:415.

6. CLARHOLM, M. 1985. Interactions of bacteria, protozoa and plants leading to mineralization of soil nitrogen. *Soil Biol. Biochem.* 17:181.

7. BICK, H. 1968. Autokologische und saprobiologische Untersuchungen an Susswasserciliaten. *Hydrobiologie* 31:17.

8. FOISSNER, W., H. BLATTERER, H. BERGER & F. KOHMANN. 1991. Taxonomische und ökologisches Revision der Ciliaten des Saprobien-

---

* Approved by Standard Methods Committee, 1997.

system, Band I: Cyrtophorida, Oligotrichida, Hydrotrichida, Colpodea. Informationsnberichte der Bayerische Landesamt für Wasserwirtschaft, Munich, Germany.

9. SLADECEK, V. 1973. System of water quality from the biological point of view. *Arch. Hydrobiol. Beih. Ergebn. Limnol.* 7:1.

10. BLAISE, C. 1991. Microbiotests in aquatic ecotoxicology: Characteristics, utility, and prospects. *Environ. Toxicol. Water Qual.* 6:145.

11. ANTIPA, G.A. 1977. Use of commensal protozoa as biological indicators of water quality and pollution. *Trans. Amer. Microscop. Soc.* 96:482.

12. CAIRNS, J., JR. 1974. Protozoans (Protozoa). Pollution Ecology of Freshwater Invertebrates. Academic Press Inc., New York, N.Y.

13. LYNN, D.H. & G.L. GILRON. 1992. A brief review of approaches using ciliated protists to assess aquatic ecosystem health. *J. Aquat. Ecosys. Health* 1:263.

14. GILRON, G.L. & D.H. LYNN. 1997. Ciliated protozoa as test organisms in toxicity assessments. *In* P. Wells, C. Blaise & K. Lee, eds. Microscale Aquatic Toxicology—Advances, Techniques and Practice, p. 323. CRC Press, Boca Raton, Fla.

8310 B. Growth Inhibition Test with Freshwater Ciliate *Colpidium campylum*

1. Background

The short generation time of ciliates and the usefulness of growth as a sensitive biological characteristic have made growth inhibition a widely used end point for bioassays. Such bioassays measure population growth rate of the test ciliate species in response to a gradient of test concentrations.

This method, a short-term bioassay,[1-4] is based on a change in the population growth of *Colpidium campylum* (Figure 8310:1) over a 24-h period. The number of cells produced during 24 h in the presence of the toxicant is compared to the growth in a control culture. The bioassay has a broad range of application for single toxicants and contaminant mixtures such as effluents. Intercalibration data and technical review[5] support its usefulness as a standard method.

Other examples of test methods that have used growth inhibition as a test end point for aquatic assessments are described elsewhere.[6,7]

2. Source of Test Organisms

Cultures of *Colpidium campylum* (ATCC 50414) can be obtained from the American Type Culture Collection.*

3. Holding and Culturing Test Organisms

a. Culture maintenance: Culture *Colpidium campylum* Stokes axenically in Proteose Peptone Yeast Extract and Serum (PPYS) medium[8] enriched with bovine serium albumin.[9]† Incubate cultures at 28°C in the dark; subculture each week.

b. Preparing organisms for testing: Acclimate organisms to monoxenic cultivation. Grow them with commercially available lyophilized *Escherichia coli*, strain ATCC 11303,‡ strain ATCC 9637,§ or strain K12.‖

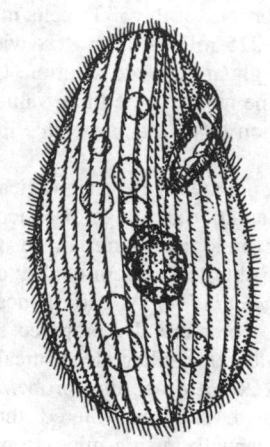

Figure 8310:1. *Colpidium campylum.*

Prepare Minimal Medium (MM) used for the test as follows:

CaCl$_2$·2H$_2$O	107 mg
NaCl	14.5 mg
NaNO$_3$	4.5 mg
MgSO$_4$·7H$_2$O	75.7 mg
Na$_2$SO$_4$	39.5 mg
NaHCO$_3$	135 mg
Reagent water	1 L

Mix thoroughly. pH should be 8.15 ± 0.02. Filter through a 0.45-μm membrane filter and store at 4°C.

Inoculate two 125-mL sterile borosilicate erlenmeyer or sterile cell culture bottles# containing 10 mL MM medium and 0.4 mL *E. coli* suspension (2.5 mg/mL in MM medium) with 2 drops of axenic *C. campylum* culture.

After 48 h incubation at 28°C, count the cells. Prepare the definitive inoculum in 500-mL sterile borosilicate erlenmeyer or cell culture bottles** by inoculating 50 mL MM medium and 2 mL

---

* ATCC, Rockville, MD.
† A4503, Sigma Chemical Co., St. Louis, MO, or equivalent.
‡ Sigma EC11303.
§ Sigma EC9637.
‖ Sigma EC1.

Nunclon, 50 mL, 25 cm^2, or equivalent.
** Corning, 270 mL, 75 cm^2, or equivalent.

E. coli (2.5 mg/mL in MM medium) with 1000 cells/mL. After 48 h growth at 28°C, the inoculum can be used for the bioassay.

4. Test Conditions and Procedures

a. Test vessels: Perform test in 30-mL crystal polystyrene screw-capped vials. Alternatively, if using electronic Coulter counting, use counter cuvettes directly. If products tested can adsorb on plastic or alter it, use borosilicate glass or TFE vials.

b. Test initiation: To each vial add in the given order:

Toxicant solution in MM (1.25 final concentration in the vial)	4	mL
E. coli, suspension 2.5 mg/mL (in MM medium)	0.25	mL
C. campylum dilution (3333 cells/mL)	0.75	mL

Start test timing when the ciliates are added. The final volume is 5 mL with an initial cell concentration of 500 cells/mL. For each test on a substance, use one vial per concentration and three control vials (without toxicant).

To verify true value of the inoculum (500 cells/mL in theory), distribute 0.75-mL portions of the 3333 cells/mL dilution in three separate vials, add 0.225 mL MM, and fix with 1 mL commercially prepared 2.5% glutaraldehyde solution. Count samples and calculate densities. The mean of the three values is considered as the initial concentration (N_0). Incubate vials in the dark at 28°C for 24 h.

At the same time, initiate a reference toxicant test (potassium dichromate) with an appropriate range of concentrations to verify the sensitivity of the biological material. An EC50 of 10 to 15 mg/L indicates acceptable test system quality control.[1]

c. Counting and calculation: At end of incubation period, fix each vial with 1 mL commercially prepared 5% glutaraldehyde solution. Count ciliate cells either electronically with a Coulter counter fitted with a 200-μm aperture probe, after dilution with a 1% NaCl electrolyte solution filtered through a 0.45-μm membrane filter or manually, using microscopy and the aid of a counting chamber (such as a hemocytometer, Palmer cell, or Sedgwick-Rafter cell). Compute the number of cells produced (*CP*) by:

$$CP = N - N_o$$

where:

N = final counted population, and
N_o = initial concentration, ¶ *b* above.

5. Evaluating and Reporting Test Results

The statistical end point of the test is the EC50. The cells produced in each concentration of toxicant are estimated as percentage of the control (mean of the three vials). Determine the EC50 by using a computer program, e.g., Stephan LC50 program[10]; TOXDAT and TOXSTAT (see Section 8010G).

6. References

1. Dive, D., S. Robert, E. Angrand, C. Bel, H. Bonnemain, L. Brun, Y. Demarque, A. Le Du, R. El Bouhouti, M.N. Fourmaux, L. Guery, O. Hanssens & M. Murat. 1989. A bioassay using the measurement of the growth inhibition of a ciliate protozoan: *Colpidium campylum* Stokes. *Hydrobiologia* 188/189:181.
2. Dive, D., C. Blaise & A. Le Du. 1991. Standard protocol proposal for undertaking the *Colpidium campylum* ciliate protozoan growth inhibition test. *Angewandte Zool.* 1:79.
3. Dive, D. & H. Leclerc. 1977. Utilisation du protozoaire *Colpidium campylum* pour le mesure de la toxicité et de l'accumulation des micropollutants: Analyse critique et applications. *Environ. Pollut.* 14:169.
4. Dive, D. & H. Leclerc. 1975. Standardized test method using protozoa for measuring water pollutant toxicity. *Prog. Water Technol.* 7(2):67.
5. Dive, D., C. Blaise, S. Robert, A. Le Du, N. Bermingham, R. Cardin, A. Kwan, R. Legault, L. MacCarthy, D. Moul & L. Veilleux. 1990. Canadian workshop on the *Colpidium campylum* ciliate protozoan growth inhibition test. *Angewandte Zool.* 1:49.
6. Forge, T.A., M.L. Berrow, J.F. Darbyshire & A. Warren. 1993. Protozoan bioassays of soil amended with sewage sludge and heavy metals, using the common soil ciliate *Colpoda steinii. Biol. Fertil. Soils* 16:282.
7. Janssen, M.P.M., C. Oosterhoff, G.J.S.M. Heijmans & H. Van der Voet. 1995. The toxicity of metal salts and the population growth of the ciliate *Colpoda cucculus. Bull. Environ. Contamin. Toxicol.* 54:597.
8. Plesner, P., L. Rasmussen & E. Zeuthen. 1964. Techniques used in the study of synchronous *Tetrahymena. In* E. Zeuthen, ed. Synchrony in Cell Division and Growth, p. 543. John Wiley & Sons, New York, N.Y.
9. Dive, D.G. & L. Rasmussen. 1978. Growth studies on *Colpidium campylum* under axenic conditions. *J. Protozool.* 25(3):42A.
10. Stephan, C.E. 1977. Methods for calculating an LC50. *In* F.L. Mayer & J.L. Hamelink, eds. Aquatic Toxicology and Hazard Evaluation, p. 65. ASTM STP 634, American Soc. Testing & Materials, Philadelphia, Pa.

8310 C. Chemotactic Test with Freshwater Ciliate *Tetrahymena thermophila*

1. Background

The movement of ciliates toward or away from chemicals (i.e., chemosensory behavior) is a well-studied physiological response in ciliates.[1,2] These types of tests measure chemosensory behavior or the inhibition of chemosensory behavior as the biological end point.[3–5]

Chemotaxis inhibition bioassays have a broad range of application for single toxicants and contaminant mixtures such as effluents. The T-maze toxitactic assay (TMTA)[6], upon which this method is based, has undergone species comparison validation, technical refinement, and interlaboratory calibration.

2. Source of Test Organisms

Cultures of *Tetrahymena thermophila* (Figure 8310:2) [ATCC 30382 Strain B—18684 (1975) or ATCC 30383 Strain B—18686 (1975)] can be obtained from the American Type Culture Collection.* Other tetrahymenine species (e.g., *T. vorax*, ATCC 30421) can be obtained from the same source.

3. Holding and Culturing Test Organisms

a. Culture medium preparation: Prepare Proteose Peptone Yeast Extract (PPYE) medium as follows (depending on culture medium requirements):

Dextrose	0.5 g
Proteose peptone†	2.0 g
Yeast extract†	2.0 g
Distilled water	400.0 mL

Heat distilled water in a beaker over a Bunsen burner. Add dextrose and stir. Add proteose peptone and mix. Add yeast extract, but *do not* stir. Heat solution until yeast extract is dissolved, but do not let solution boil. Dispense 10-mL portions into culture (test) tubes (20 × 150 mm or 15 × 150 mm). Cap tubes and autoclave for 20 min at 103 kP. The shelf life of the culture medium is 1 month, provided that it is refrigerated and covered with plastic film.‡

b. Culture transfer and maintenance: Using sterile technique, transfer culture every 2 weeks. Keep cultures at room temperature with regular ambient lighting.

c. Preparation of cultures: Ensure that all solutions to be used in the bioassay are at room temperature (20° ± 2°C).

Inoculate, using sterile technique, 10 mL sterile PPYE with about 1 mL stock *Tetrahymena thermophila* culture. After 48 h, aseptically transfer the 10-mL culture to 50 mL sterile PPYE in a 250-mL erlenmeyer flask. At this time, soak the corks for the mazes in dilution water (spring water).

After 24 h, harvest 50 mL PPYE culture by centrifuging in centrifuge tubes (preferably use conical 12- to 15-mL tubes) at 1200 rpm for 3 min. With a Pasteur pipet, remove the supernatant, ensuring that the pellet of cells is not disturbed.

Gently and completely resuspend the pellets into a centrifuge tube and add 12 to 15 mL dilution water. Centrifuge at 1200 rpm

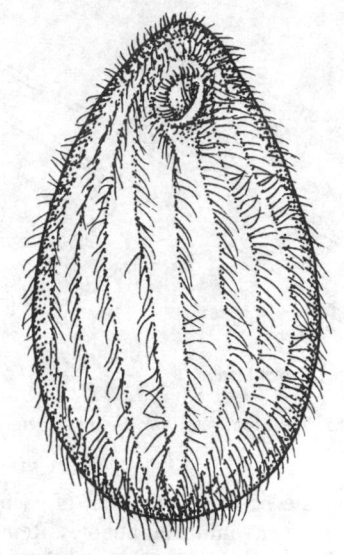

Figure 8310:2. *Tetrahymena thermophila.*

for 3 min. Repeat resuspension, centrifugation, and dilution twice more, removing supernatant each time. Gently and completely resuspend the pellet in a small amount of dilution water. Transfer cells with a Pasteur pipet to 50 mL dilution water in a 250-mL erlenmeyer flask. Leave for 18 h under conditions of ambient temperature and lighting conditions to starve the culture.

Harvest cells by centrifuging at 1200 rpm for 3 min and carefully removing supernatant with Pasteur pipet. Gently and completely resuspend the pellets into one centrifuge tube. Using dilution water, adjust cell density to approximately 400 000 cells/mL (± 10%).

4. Test Conditions and Procedures

a. Test apparatus and design: A schematic diagram outlining the apparatus used in the TMTA procedure is presented in Figure 8310:3. The test design comprises at least three replicate glass T-mazes, 30 cm in longest dimension, for each concentration, with five concentrations in a serial dilution and a control. Before running the definitive test, perform a preliminary motility test to ensure that cells are motile in the test medium. For full test, set up five concentrations and a control, each comprising three replicate T-mazes at each concentration (total of 18 mazes).

b. Test exposure: Turn stopcock for each maze so that the bore is in line with the third (upright) arm. Label maze arms "test" and "control" with tape and/or marker. Using Pasteur pipets (14.6-cm/5.75-in.), fill arms of each T-maze apparatus, one at a time, with the respective solutions [one test (toxicant solution), one control, in that order]. Stop each arm with a rubber cork. Ensure that no air bubbles are caught in the arms, particularly around the stopcocks. Holding one arm upwards at a 45-deg angle, shake out all air bubbles by firmly hitting the T-maze apparatus on the palm of the hand; repeat for second arm. Recork, if necessary, to release any air bubbles.

Using a 23-cm Pasteur pipet, transfer cells from a homogeneous suspension into each stopcock barrel (the solution is filled

* ATCC, Rockville, MD.
† Difco or equivalent.
‡ Parafilm™ or equivalent.

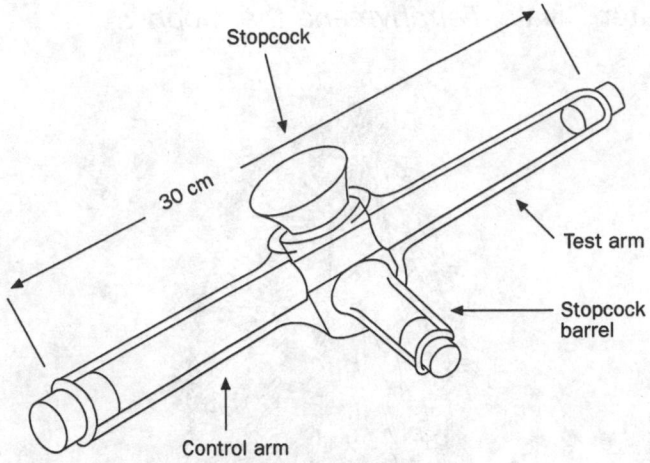

Figure 8310:3. Test apparatus for T-maze chemotactic test.

above the level of the bore). Tap bottom of each T-maze apparatus gently to remove any initial air bubbles. Remove air bubbles from all stopcock barrels before commencing the test for all mazes. NOTE: *Removal of air bubbles is crucial to conducting the assay properly, because bubbles will prevent organisms from migrating into the arms.*

After all T-mazes are filled completely, begin 20-min exposure period. Turn stopcock so that cells are able to migrate freely through the arms. Grease stopcocks sparingly with high- vacuum grease before use. Ensure that the stopcock barrel is completely aligned with the stopcock arms. After 20-min exposure period, turn stopcocks again to a closed position to terminate the test.

In parallel with each run of the T-maze tests, perform a standard reference toxicant test (using sodium chloride) with an appropriate range of concentrations to verify the sensitivity of the biological material. A lowest-observed-effect concentration (LOEC) of 2000 to 3000 mg/L indicates acceptable test system quality control.[6]

c. Test termination and enumeration: Immediately after the test is completed, empty arms of T-maze into counting tubes (e.g., test tubes, Coulter counter cuvettes). Using a 14.6-cm Pasteur pipet, rinse each arm with the test solution from that arm to ensure that all cells have been removed. Enumerate cells under 400 × magnification. Evenly disperse cells in counting tubes by inverting tubes or using a vortex mixer. Take five 10-μL samples from each counting tube and add this to five wells of a polystyrene 96-well microplate with flat wells.§ Add 20 μL dilution water and 10 μL Lugol's iodine solution (see Section 10200B.2*a*) to each of the five wells.§ Count no fewer than three of the five wells per arm. If necessary, count a smaller or larger portion, depending on cell density. As a guideline, the densest arm (where accumulation/attraction has occurred) should have at least 100 cells/well or 10 000 cells/mL. Record replicate counts.

§ Corning No. 25880-96 or equivalent.

5. Evaluating and Reporting Test Results

Follow general procedures described in Section 8010G.

The statistical end points of the test are the lowest-observed-effect concentration (LOEC) and the EC50. They are determined by calculating the I_{tox} values defined below for the concentration series, plotting them graphically, and applying statistical analysis.

Calculate a "toxitactic" index (I_{tox}) for each T-maze as follows:

$$I_{tox} = \frac{T}{T + C}$$

where:

T = mean number of cells in test arm, and
C = mean number of cells in control arm.

To determine LOEC value, plot the I_{tox} values for the concentrations tested and a control (*y*-axis) against concentration of the toxicant (*x*-axis). When there is a response, an increase in I_{tox} with increasing concentration denotes attraction, and a decrease in I_{tox} with increasing concentration denotes repulsion.

Conduct an analysis of variance (ANOVA) and a multivariate test (e.g., William's or Dunnett's tests)[7] on all data for a given test, to determine the lowest concentration at which the I_{tox} value is statistically, significantly different from the control I_{tox}. That concentration is the LOEC.

Calculate the IC50 by the linear interpolation method[8] with a software package.||

6. References

1. HELLUNG-LARSEN, P., V. LEICK, N. TOMMERUP & D. KRONBORG. 1990. Chemotaxis in *Tetrahymena. Europ. J. Protistol.* 25:229.
2. VAN HOUTEN, J., E. MARTEL & T. KASCH. 1982. Kinetic analysis of chemokinesis of *Paramecium. J. Protozool.* 29:226.
3. BERK, S.G., J.H. GUNDERSON & L.A. DERK. 1985. Effects of cadmium and copper on chemotaxis of marine and freshwater ciliates. *Bull. Environ. Contam. Toxicol.* 34:897.
4. ROBERTS, R.O. & S.G. BERK. 1990. Development of a protozoan chemoattraction bioassay for evaluating toxicity of aquatic pollutant. *Toxic. Assess.* 5:279.
5. BERK, S.G., B.A. MILLS, K.C. STEWART, R.S. TING & R.O. ROBERTS. 1990. Reversal of phenol and naphthalene effects on ciliate chemoattraction. *Bull. Environ. Contam. Toxicol.* 44:181.
6. GILRON, G.L., D.H. LYNN & J. BROADFOOT. 1996. Further Development of a Sublethal Bioassay for Pulp and Paper Mill Effluents Using Ciliated Protozoans. Final Report. Environment Canada, Ottawa, Ont., Canada.
7. SNEDECOR, G.W. & W.G. COCHRAN. 1980. Statistical Methods, 7th ed. Iowa State Univ. Press, Ames.
8. NORBERG-KING, T.J. 1993. An Interpolation Estimate for Chronic Toxicity: The ICP Approach. U.S. Environmental Protection Agency, Environmental Research Lab. Duluth, Minn.

|| BOOTSTRP, available from U.S. Environmental Protection Agency, Cincinnati, Ohio.

8310 D. Growth Inhibition Test with the Soil Ciliate *Colpoda inflata*

1. Background

This bioassay measures the population growth rate of the test ciliate species in response to a gradient of test concentrations. It is similar to that used for algal growth inhibition tests.[1,2] Growth, in this case, may be inhibited by direct effects on the ciliate cell maintenance or by effects that suppress energy intake (i.e., feeding).

The method is based on a test method[3] for the evaluation of solid-phase media (i.e., soils and soil elutriates) using the soil ciliate, *Colpoda inflata* (Section 10900, Plate 6,D). The number of cells produced during a 24-h period in the presence of toxicant is compared to the value obtained in a control culture. The method also has been applied successfully to mining effluents.[4]

2. Source of Test Organisms

Dry cysts of *Colpoda inflata* (ATCC 30917) can be obtained from the American Type Culture Collection.* Other colpodid species (*C. steinii*, ATCC 30920, and *C. cucullus*, ATCC 30916) can be obtained from the same source.

3. Holding and Culturing Test Organisms

a. Holding organisms: Dry cysts can be held at room temperature on filter paper for extended periods (1 to 2 years). Cysts in spent cultures can be stored wet for periods up to months without loss of viability. Grow these cultures or dry cysts as needed by adding culture medium as described below.

b. Maintaining cultures: Maintain cultures developed from stored cysts for 2 to 7 d in 10% Sonneborn's *Paramecium* medium, prepared as follows: Boil 2.5 g cereal grass leaves† for 5 min in 1 L distilled, deionized water; filter,‡ adjust volume to 1 L with distilled water, and add 0.5 g Na$_2$HPO$_4$. Dilute full-strength medium before use with distilled, deionized water and autoclave in 50-mL portions. Add to cultures a food bacterium such as nonpathogenic *Klebsiella pneumoniae* (ATCC 27889) as recommended by ATCC.

4. Test Conditions and Procedures

a. Test vessels: Conduct tests in sterile, 24-well, polystyrene tissue culture plates,§ using either 10% Sonneborn's medium (¶ 3*b* above) or a minimal salts medium consisting of 6 mg KCl, 4 mg CaHPO$_4$, and 2 mg MgSO$_4$ in 1 L sterile distilled, deionized water. Test volume is 2 mL per well.

b. Test initiation: Dispense sterile medium into wells and amend with toxicant from stock solutions of reagent-grade chemicals prepared in sterile distilled, deionized water. Alternatively, conduct tests using percentage dilutions of complex mixtures

(e.g., wastewater effluents, soil extracts[4]). For mixture tests, prepare minimal salts medium in concentrated form (10 ×) and dilute with sterile distilled water.

After medium and toxicant are dispensed into test wells, add ciliates from log-phase cultures (48 to 96 h old) along with food bacteria. The volume of well-mixed culture added should ensure that equal numbers of ciliates (approximately 100 cells/mL) are added to each well.

At the same time, perform a reference toxicant test (using copper sulfate) with an appropriate range of concentrations to verify the sensitivity of the biological material. A toxicant concentration corresponding to a 50% inhibition of growth relative to controls (IG50) between 25 and 70 µg/L indicates acceptable test system quality control.

c. Counting and calculation: After 24 h, remove subsamples from each test well and enumerate using a direct counting technique as follows: Thoroughly mix each well with a micropipettor and transfer a 20-µL subsample to a clean microscope slide as 3 or 4 drops. Scan all drops immediately at low magnification (40×) on a stereomicroscope to search for active cells. Active cells are always moving and can be distinguished easily from bacterial aggregates or cysts. For a given well, repeat subsampling at least three times to assure that an accurate estimate of the population in the well is obtained. If repeat counts of subsamples vary by more than 30%, continue subsampling until population estimates stabilize. Repeat this procedure for each well, and compute mean of subsample estimates for a given replicate. Alternatively, use direct particle counting techniques (i.e., Coulter counter) to enumerate cells.

5. Evaluating and Reporting Test Results

Follow general procedures described in Section 8010G.

Estimate IG50 by regressing cell number on log dose and then using inverse prediction[5] to estimate IG50 from the control response (mean = 100%).

6. References

1. GREENE, J.C., C.L. BARTELS, W.J. WARREN-HICKS, B.R. PARKHURST, G.L. LINDER, S.A. PETERSON & W.E. MILLER. 1989. Protocol for short term toxicity screening of hazardous waste sites. EPA-600/3-88-029, U.S. Environmental Protection Agency, Environmental Research Lab, Corvallis, Ore.
2. ENVIRONMENT CANADA. 1992. Biological Test Method: Growth Inhibition Test Using the Freshwater Alga, *Selenastrum capricornutum*. ON. EPS Rep. Conservation and Protection, Ottawa, Ont., Canada.
3. PRATT, J.R., D. MOCHAN & Z. XU. 1997. Rapid toxicity estimation using soil ciliates: sensitivity and bioavailability. *Bull. Environ. Contam. Toxicol.* 58:387.
4. BOWERS, N., J.R. PRATT, D. BEESON & M. LEWIS. 1997. Comparative evaluation of soil toxicity using lettuce seeds and soil ciliates. *Environ. Toxicol. Chem.* 16:207.
5. SNEDECOR, G.W. & W.G. COCHRAN. 1980. Statistical Methods, 7th ed. Iowa State Univ. Press, Ames.

---

* ATCC, Rockville, MD.
† Cerophyl®, Agri-Tech, Inc., Kansas City, MO 64112, or equivalent.
‡ Whatman No. 1 filter paper or equivalent.
§ Costar, Inc. or equivalent.

8420 ROTIFERS* (PROPOSED)

8420 A. Introduction

1. Ecological Significance

Rotifers are classified in the Phylum Rotifera, one of several phyla of lower invertebrates. There are approximately 2000 rotifer species named; they are divided into two classes, Digononta and Monogononta.[1,2] The use of rotifer cysts for toxicity testing has been discussed in the literature.[3] Most rotifer species inhabit fresh water,[4] but there are some genera, like *Synchaeta*, in which most species are marine.[5] In coastal marine habitats, rotifers sometimes are the dominant portion of the biomass.[6] They also are abundant in marine interstitial habitats, interstitial water of soils,[7] and water clinging to mosses, liverworts, and lichens.[8] In freshwater lake plankton[9] and in river sediments,[10] rotifers often are abundant, with high species diversity.

Rotifers play an important role in the ecological processes of many aquatic communities.[11] As suspension feeders, planktonic rotifers influence algal species composition through selective grazing.[12–15] Rotifers often compete with cladocera and copepods for phytoplankton in the 2- to 18-μm size range. Along with crustaceans, rotifers contribute substantially to nutrient recycling.[16]

2. Types of Toxicity Tests

The procedures in Section 8420 serve as guidelines for using rotifers to estimate sublethal toxicity with population growth rate as endpoint. These procedures have been adapted for examining surface water and effluents as well as sediment pore water. Several other types of rotifer tests described are based on endpoints such as mortality,[17] ingestion,[18] swimming,[19] enzyme activity,[20] and stress protein gene expression.[21]

3. References

1. NOGRADY, T., R.I. WALLACE & T.W. SNELL. 1993. Rotifera, Vol. 1. Biology, Ecology and Systematics. SPB Academica Publishing bv., The Hague, The Netherlands.
2. BUIKEMA, A.L., JR., J. CAIRNS, JR. & G.W. SULLIVAN. 1974. Evaluation of *Philodina acuticornis* (Rotifera) as a bioassay organism for heavy metals. *Water Resour. Bull.* 10:648.
3. SNELL, T.W. & C.R. JANSSEN. 1995. Rotifers in Ecotoxicology: A review. *Hydrobiologia* 313/314:231.

4. WALLACE, R.L. & T.W. SNELL. 1991. Rotifera. *In* J.H. Thorp & A.P. Covich, eds. Ecology and Classification of North American Freshwater Invertebrates. Academic Press, New York, N.Y.
5. NOGRADY, T. 1982. Rotifera. *In* S.P. Parker, ed. Synopsis and Classification of Living Organisms. McGraw-Hill, New York, N.Y.
6. EGLOFF, D.A. 1988. Food and growth relations of the marine zooplankter, *Synchaeta cecelia* (Rotifera). *Hydrobiologia* 157:129.
7. POURRIOT, R. 1979. Rotiferes du sol. *Rev. Ecol. Biol. Sol* 16:279.
8. RICCI, C. 1983. Life histories of some species of Rotifera Bdelloidea. *Hydrobiologia* 104:175.
9. STEMBERGER, R.S. 1990. An inventory of rotifer species diversity of northern Michigan inland lakes. *Arch. Hydrobiol.* 118:283.
10. SCHMID-ARAYA, J.M. 1995. Disturbance and population dynamics of rotifers in bed sediments. *Hydrobiologia* 313/314:279.
11. PACE, M.L. & J.D. ORCUTT. 1981. The relative importance of protozoans, rotifers and crustaceans in freshwater zooplankton communities. *Limnol. Oceanogr.* 26:822.
12. BOGDAN, K.G. & J.J. GILBERT. 1987. Quantitative comparison of food niches in some freshwater zooplankton. *Oecologia* 72:331.
13. STARKWEATHER, P.L. 1987. Rotifera. *In* T.J. Pandian & F.J. Vernberg, eds. Animal Energetics Vol. 1: Protozoa through Insecta. Academic Press, Orlando, Fla.
14. WILLIAMSON, C.E. 1983. Invertebrate predation on planktonic rotifers. *Hydrobiologia* 104:385.
15. ARNDT, H. 1993. Rotifers as predators on components of the microbial web (bacteria, heterotrophic flagellates, ciliates)—a review. *Hydrobiologia* 255/256:231.
16. EJSMONT-KARABIN, J. 1983. Ammonia nitrogen and inorganic phosphorus excretion by the planktonic rotifers. *Hydrobiologia* 104:231.
17. AMERICAN SOCIETY FOR TESTING AND MATERIALS. 1996. Standard Guide for Acute Toxicity Test with the Rotifer *Brachionus*. ASTM 11.05, E1440-91, American Soc. Testing & Materials, W. Conshohocken, Pa.
18. JUCHELKA, C.M. & T.W. SNELL. 1994. Rapid toxicity assessment using rotifer ingestion rate. *Arch. Environ. Contam. Toxicol.* 26:549.
19. CHAROY, C.P., C.R. JANSSEN, G. PERSOONE & P. CLEMENT. 1995. The swimming behavior of *Brachinous calyciflorus* (Rotifera) under toxic stress: I. The use of automated trajectometry for determining sublethal effects of chemicals. *Aquat. Toxicol.* 32:271.
20. BURBANK, S.E., & T.W. SNELL. 1994. Rapid toxicity assessment using esterase biomarkers in *Brachionus calyciflorus* (Rotifera). *Environ. Toxicol. Water Qual.* 9:171.
21. COCHRANE, B.J., Y.D. DE LAMA & T.W. SNELL. 1994. Polymerase chain reaction as a tool for developing stress protein probes. *Environ. Toxicol. Chem.* 13:1221.

* Approved by Standard Methods Committee, 1997.

8420 B. Selecting and Preparing Testing Organisms

1. Selecting Test Organisms

The rotifers recommended for use were chosen because of the existence of published reports describing protocols, a database of

responses to pure toxicants, and the availability of cysts (resting eggs). In *Brachionus calyciflorus* and *B. plicatilis*, for which standardized tests exist, the recommended strain also is indicated. A summary of ecological and test conditions to be considered in tests with these organisms is given in Table 8420:I. Substantial differences in sensitivity to toxicants have been reported among rotifer strains of different geographic origin.[1] In accord with the criteria listed in Section 8010E.1, the recommended test species include (but are not restricted to) the following:

a. *Freshwater rotifers:*
Class: Monogononta
  Brachionus calyciflorus (Gainesville strain)[2]
  Brachionus rubens[3]
  Brachionus patulus[4]
  Asplanchna brightwelli[5]
Class: Digononta
  Philodina roseola[6]
  Philodina acutiocornis[7]

See Section 10900, Plate 8 for drawings of several freshwater rotifer species.

b. *Marine rotifers:*
  Brachionus plicatilis (Russian strain)[1]

2. Obtaining Test Organisms

a. *Rotifer cysts: B. calyciflorus* in fresh water and *B. plicatilis* in marine waters are hatched from cysts. Rotifer cysts hatch synchronously, providing test animals of similar age in uniform physiological condition. Detailed descriptions of rotifer cyst hatching are available.[1,2]

b. *Cyst hatching:* Incubate *B. calyciflorus* cysts in standard synthetic fresh water for 15 to 16 h before the start of a test. To initiate hatching, place about 40 mL standard fresh water in a glass petri dish or tissue-culture-grade polystyrene dish. Incubate rotifer cysts at 25°C in light of 3000 to 4000 lux. Hatching should start after about 15 to 16 h; within 2 h remove dish from incubator to transfer rotifers to test tubes. Cooler temperatures, low or high pH, and elevated hardness and alkalinity can delay hatching. If hatching is delayed, check cysts hourly to ensure collecting test animals within 2 h of hatching.

Hatch *B. plicatilis* cysts by a similar procedure in standard synthetic seawater. *B. plicatilis* cysts usually initiate hatching after 24 to 26 h at 25°C in light of 3000 to 4000 lux.

c. *Food and feeding:* See ¶ C.3 below.

d. *Rotifer reproduction:* Rotifers reproduce asexually via ameiotic parthenogenesis.[8] Monogononts also can reproduce sexually, but this capacity usually is not utilized in toxicity tests (with certain exceptions[9]). Asexual rotifer reproduction allows simple sublethal toxicity tests to be conducted using population growth as an end point.

3. Parasites and Diseases

Fungal parasites on rotifers have been reported in a few natural populations, but never in laboratory populations used for toxicity

TABLE 8420:I. SUMMARY OF ECOLOGICAL AND TEST CONDITIONS THAT SHOULD BE CONSIDERED WHEN CONDUCTING TOXICITY TESTS WITH *B. CALYCIFLORUS* (BC) OR *B. PLICATILIS* (BP) ROTIFERS

Condition	Comment
Geographical	Pan-global distribution
Habitat	Pelagic zooplankter
Life cycle	Parthenogenetic and sexual reproduction
Lifespan	5–7 d at 25°C
Temperature	10–32°C
Salinity	BC − 0–5, BP − 1–60
Nutrition	Suspension feeders on microalgae
Photoperiod	No special requirements
Control mortality	Not to exceed 10%
Life-cycle test	Water, pore water[1]

testing. No known diseases affect the use of brachionid rotifers in toxicity tests.

4. References

1. SNELL, T.W., B.D. MOFFAT, C.R. JANSSEN & G. PERSOONE. 1991. Acute toxicity tests using rotifers: III. Effects of temperature, strain and exposure time on the sensitivity of *Brachionus plicatilis*. *Environ. Toxicol. Water Qual.* 6:63.

2. SNELL, T. W., B.D. MOFFAT, C.R. JANSSEN & G. PERSOONE. 1991. Acute toxicity tests using rotifers: IV. Effects of cyst age, temperature, and salinity on the sensitivity of *Brachionus calyciflorus*. *Ecotoxicol. Environ. Safety* 21:308.

3. HALBACH, U., M. WIEBERT, M. WESTMAYER & C. WISSEL. 1983. Population ecology of rotifers as a bioassay tool for ecotoxicological tests in aquatic environments. *Ecotoxicol. Environ. Safety* 7:484.

4. RAO, T. & S.S.S. SARMA. 1986. Demographic parameters of *Brachionus patulus* Muller (Rotifera) exposed to sublethal DDT concentrations at low and high food levels. *Hydrobiologia* 139:193.

5. ROGERSON, A., J. BERGER & C.M. GROSSO. 1982. Acute toxicity of ten crude oils on the survival of the rotifer *Asplanchna sieboldi* and sublethal effects on rates of prey consumption and neonate production. *Environ. Pollut.* 29:179.

6. SCHAEFER, E.D. & W.O. PIPES. 1973. Temperature and toxicity of chromate and arsenate to the rotifer, *Philodina roseola. Water Res.* 7: 1781.

7. BUIKEMA, A.L., JR., J. CAIRNS, JR. & G.W. SULLIVAN. 1974. Evaluation of *Philodina acuticornis* (Rotifera) as a bioassay organism for heavy metals. *Water Resour. Bull.* 10:648.

8. WALLACE, R.L. & T.W. SNELL. 1991. Rotifera. *In* J.H. Thorp & A.P. Covich, eds. Ecology and Classification of North American Freshwater Invertebrates. Academic Press, New York, N.Y.

9. SNELL, T.W. & M.J. CARMONA. 1995. Comparative toxicant sensitivity of sexual and asexual reproduction in the rotifer *Brachionus calyciflorus. Environ. Toxicol. Chem.* 14:415.

8420 C.   Aquatic Toxicity Test Procedures

1. General Procedures

Use exploratory tests to determine the toxicant concentrations for short-term tests (see Section 8010D). Prepare control and test solutions in standard synthetic fresh water or seawater and introduce them into test containers as described in Section 8010F.

2. Water Supplies

a. Artificial fresh water: See Section 8010E.4*b*1) and Table 8010:I for preparation of a moderately hard water. Adjust to pH 7.5 with 10*M* KOH or HCl.

b. Artificial seawater: Prepare standard synthetic seawater[2] with a salinity of 15 by adding 11.31 g NaCl, 0.36 g KCl, 0.54 g CaCl$_2$, 1.97 g MgCl$_2$·6H$_2$O, 2.39 g MgSO$_4$·7H$_2$O, and 0.17 g NaHCO$_3$ to 1 L deionized or distilled water. Mix well on a magnetic stirrer and adjust pH to 8.0 with 10*M* KOH or HCl.

c. Deionized water: Prepare all media with high-quality deionized or distilled water (see Section 1080). Water from certain commercially available systems* is suitable.

3. Food and Feeding

a. Nannochloris oculata (food for Brachionus calyciflorus): Maintain unialgal stock cultures of the green alga *Nannochloris* cells† in sterile test tube cultures containing 20 mL Bold's Basal Medium (BBM), prepared as follows: To 1 L water (¶ 2*c*), add 250 mg NaNO$_3$, 75 mg MgSO$_4$·7H$_2$O, 175 mg KH$_2$PO$_4$, 25 mg CaCl$_2$·2H$_2$O, 75 mg K$_2$HPO$_4$, 25 mg NaCl, 2 mL vitamin stock (see below), and 2 mL trace metal stock (see below).

To prepare 500 mL trace metal stock water, add 2.5 g Na-FeEDTA, 11 mg ZnSO$_4$·7H$_2$O, 90 mg MnCl$_2$·4H$_2$O, 5 mg CoCl$_2$·6H$_2$O, 5 mg CuSO$_4$·5H$_2$O, and 3.2 mg NaMoO$_4$·2H$_2$O to water (¶ 2*c*).

To prepare 500 mL vitamin stock water, add 100 mg thiamine, 5 mg biotin, and 5 mg B$_{12}$ to water (¶ 2*c*).

Propagate cultures by serial transfer using sterile technique. To inoculate a large *Nannochloris* culture, pour contents of a dark green 20-mL test tube culture into 2 L BBM; this will yield an initial density of about 10^3 cells/mL. Aerate this culture with filtered air and maintain at 25°C in light for 3 to 6 d until the cells reach log-phase growth, at which point they have the highest nutritional quality. Harvest and concentrate algal cells by centrifugation at 5000 × *g* for 10 min. Concentrated algal cells can be stored in the refrigerator for 3 to 4 d without loss of nutritional quality. Quantify algal cell density using a Neubauer slide hemacytometer according to the manufacturer's protocol. Add smallest volume of algae stock necessary to each test solution to make a suspension of 10^6 cells/mL. Pour 100 mL of each test solution into a 150-mL beaker and stir gently on a magnetic stirrer (approximately 120 rpm) using a small stir bar (about 1.5 cm). Use test solutions promptly; do not stir for more than 30 min. Pipet 12 mL control solution into each of seven replicate test tubes. Repeat for each test solution.

b. Nannochloropsis sp. (food for Brachionus plicatilis): Maintain unialgal stock cultures of the Eustigmatophycean alga *Nannochloropsis*‡ cells in test tubes containing 20 mL of sterile ASPM algal growth medium prepared as follows: To 1 L synthetic seawater (¶ 2*b*), add 150 mg NaNO$_3$, 10 mg K$_2$HPO$_4$, 2 mL trace metal stock (¶ 3*a*), and 2 mL vitamin stock (¶ 3*a*). Propagate cultures by serial transfer using sterile technique. Follow procedures given in ¶ 3*a* for culturing, concentrating, and dispensing to test tubes.

4. Exposure Chambers

Use standard, disposable 16- ×150-mm borosilicate glass test tubes as exposure chambers.

5. Conducting the Test

An overview of the test is shown in Figure 8420:1.

a. Adding test animals: To begin test, transfer six newly hatched rotifers (neonates) into each test tube. *B. calyciflorus* rotifers are approximately 250 μm in length, about 1/4 the size of newborn *Daphnia*. Their small size and slow swimming speed have some advantages for capturing and manipulation. Newly hatched rotifers are white, so they are most visible on a dark background at about 10 × magnification. The best type of illumination is a darkfield setting. Because they are moderately phototactic, rotifers tend to congregate around the edges of the hatching dish. Squeeze the transfer micropipet gently to provide the right amount of suction. Practice to develop a feel for the right pressure. Confirm that each tube receives exactly six rotifers by

‡ University of Texas at Austin, Culture Collection of Algae, LB 2164.

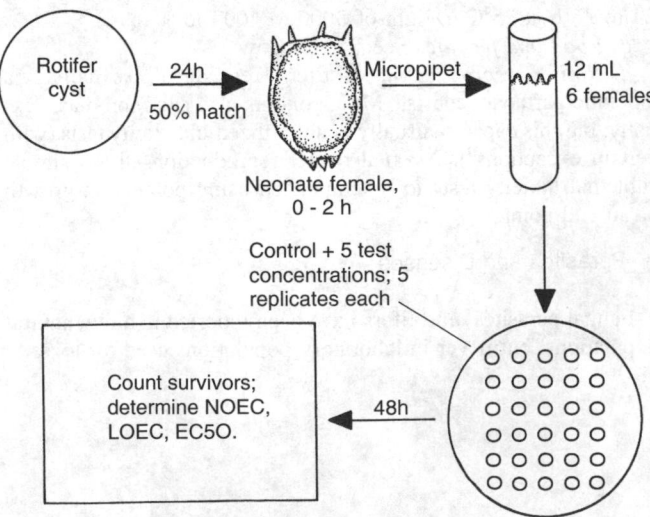

Figure 8420:1. Schematic diagram of rotifer static life-cycle toxicity tests. Test conditions: Generations – 2; end point – reproductive rate $r = (\ln N_t - \ln N_o)/T$; temperature – 25°C; photoperiod – darkness; medium – synthetic freshwater; food – *Nannochloris*.

---

* Nanopure II system with one pretreatment, one high-capacity, and two ultrapure cartridges, or equivalent.
† University of Texas at Austin, Culture Collection of Algae, LB 1998.

watching their entry into the tube under the microscope. Cap and immediately place tubes on a wheel rotator in a 25°C incubator in darkness. Rotation rate should be 10 to 120 revolutions per hour to maintain the algae in suspension. Do not use a shaker because it will damage the animals. Repeat until all remaining test solutions have been inoculated with rotifers. Record time at which neonates are placed in control treatment as the beginning of the 48-h incubation period.

b. Duration and type of test: The test uses rotifer asexual reproduction to estimate sublethal toxicity. A typical schedule of reproduction is presented in Figure 8420:2. The 48-h population growth rate is calculated and its decline with increasing toxicity quantified.

c. Scoring the test: Remove test tubes from rotator after 48 h. Empty contents of one tube into a petri dish and count number of animals per tube, discriminating between live and dead individuals. Repeat until all tubes have been counted. Calculate *r*, the population growth rate for each tube as:

$$r = \frac{\ln N_t - \ln N_o}{T}$$

where:

N_t = number of live rotifers in tube after 2 d,
N_o = initial number of rotifers in tube (6), and
T = incubation period (2 d).

Typically, *r* values range from 0.7 to 1.2 offspring per female per day. An analysis of variance and Dunnett's test can be calculated to compare each toxicant concentration to the control. This is a one-way analysis of variance with five treatments, each with seven replicates. From these data, no-observed-effect concentration (NOEC), lowest-observed-effect concentration (LOEC), and chronic values can be calculated. An example of results is shown in Table 8420:II. An inhibiting concentration (IC50) can be calculated by linear regression of log toxicant concentration on *r*. The regression equation, if significant, can be used to calculate the toxicant concentration yielding 50% reduction in *r* (IC50) as compared to controls.

d. Reference toxicant test: Perform a reference toxicant test, or positive control, with every fifth test. This verifies that the animals will respond to toxicity if it is present. Perform reference tests according to the protocol described above. Cadmium chloride, expressed as cadmium, is commonly used as a reference

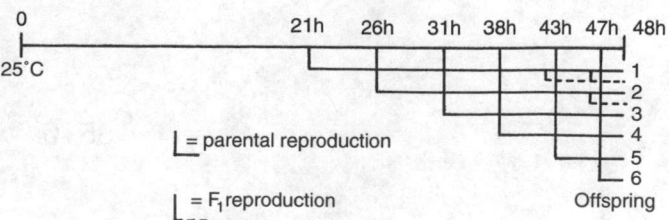

Figure 8420:2. Schedule of reproduction.

toxicant. Other metal chlorides or organic compounds may be used.

6. Data Evaluation

The data obtained from this test can be considered valid if the *r* value in the control is at least 0.70. This represents the minimum acceptable population growth rate. Lower values suggest that there is an unidentified problem with the dilution water, algae, or rotifers. Often when low population growth rates are observed, there is a problem with algae quality. For additional guidance in data analysis and other statistical considerations, see Sections 8010G and H.

7. References

1. U.S. ENVIRONMENTAL PROTECTION AGENCY. 1993. Methods for Measuring the Acute Toxicity of Effluents and Receiving Waters to Freshwater and Marine Organisms. C.I. Weber, ed. EPA-600/4-90-027F, U.S. Environmental Protection Agency, Washington, D.C.,
2. GUILLARD, R.R.L. 1983. Culture of phytoplankton for feeding marine invertebrates. *In* C.J. Berg, Jr., ed. Culture of Marine Invertebrates. Hutchinson-Ross, Stroudsberg, Pa.
3. STARR, R.C. & J.A. ZEIKUS. 1993. UTEX—The culture collection of algae at the University of Texas at Austin. *J. Phycol.* 29:1.

8. Bibliography

JANSSEN, C.R., G. PERSOONE & T.W. SNELL. 1994. Cyst-based toxicity tests. VIII. Short-chronic toxicity tests with the freshwater rotifer *Brachionus calyciflorus. Aquat. Toxicol.* 28:243.
SNELL, T.W. & B.D. MOFFAT. 1992. A two day life cycle test with the rotifer *Brachionus calyciflorus. Environ. Toxicol. Chem.* 11:1249.

TABLE 8420:II. SAMPLE TEST RESULTS

Species	Toxicant	mg/L 24-h LC50	48-h Reproductive Rate NOEC	48-h IC50	CV %
B. calyciflorus	Copper	0.03	0.02	0.03	44
	Cadmium	1.3	0.04	0.07	2
	Pentachlorophenol	1.2	0.11	0.27	29
B. plicatilis	Copper	0.06	0.01	—	—
	Cadmium	39	1.0	—	—
	Pentachlorophenol	1.9	0.5	—	—

8510 ANNELIDS*

8510 A. Introduction

The phylum Annelida includes three classes: Polychaeta, Oligochaeta, and Hirudinea. Polychaetes are an important, often predominant, component of marine and estuarine biota. In subtidal benthic environments, they comprise about 30 to 75% of the macroinvertebrate species and individuals. They include a variety of feeding types with the majority being either filter or detritus feeders. Deposit-feeding polychaetes affect surface sediments by their burrowing and irrigating habits. They are important food for snails, large crustaceans, fish, and birds. Many species have short life cycles.

Oligochaetes are among the most common benthic invertebrates in all types of aquatic environments. Particular species assemblages are recognized indicators of environmental quality. In grossly polluted freshwater habitats, oligochaetes dominate the benthic fauna, whereas in estuarine areas they and polychaete worms are often the most common benthic organisms. They feed mainly on bacteria, although other feeding types occur. They affect surface sediments as do the polychaetes. They are an important primary or alternate food for leeches, crustaceans, fish, and birds.

Hirudinea are leeches, either free-living or parasitic; they have not been used for toxicity tests.

The following procedures are intended to serve as guidelines for the use of polychaetes and oligochaetes in various toxicity tests. These procedures also can be, and have been, adapted to testing sediments.

* Approved by Standard Methods Committee, 1997.

8510 B. Selecting and Preparing Test Organisms

1. Selecting Test Organisms

In accord with the criteria listed in Section 8010E.1, the recommended test species include (but are not restricted to) the following:

a. Marine polychaetes:

1) Family Nereidae

Neanthes arenaceodentata (*Neanthes acuminata* and *Neanthes caudata* of some authors) (New England, Florida, California coasts, Europe) (Figure 8510:1A).
Neanthes succinea (all of U.S. coasts).
Neanthes virens (east coast of U.S.).

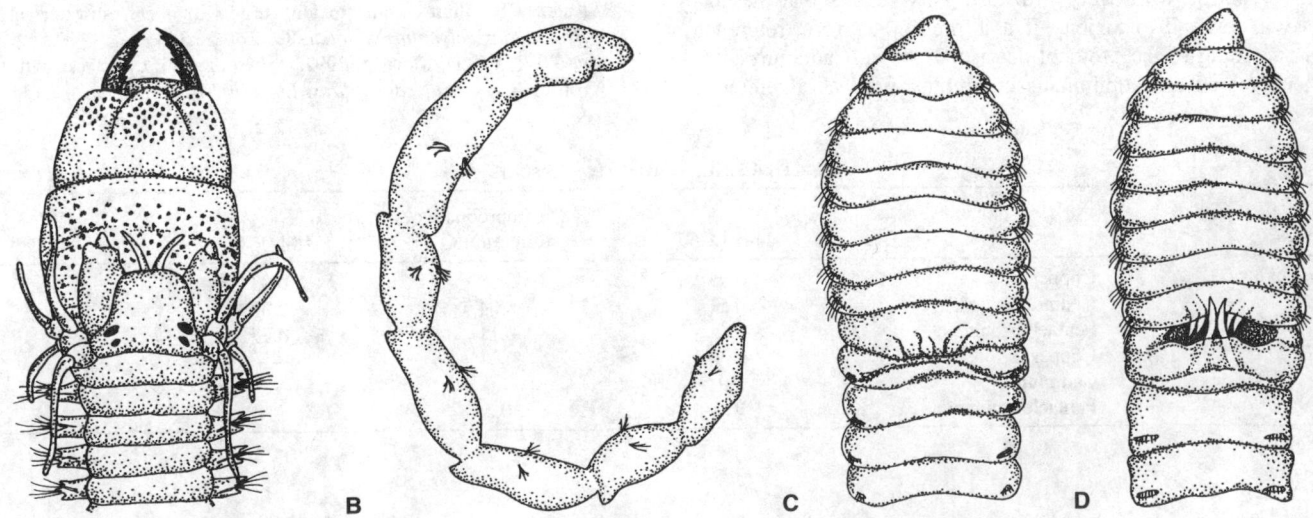

Figure 8510:1. Marine polychaetes. A—*Neanthes arenaceodentata*, anterior end, dorsal view; B—*Ctenodrilus serratus*, adult; C—*Capitella capitata*, female, dorsal view; D—*Capitella capitata*, male, dorsal view.

2) Family Capitellidae

Capitella capitata (cosmopolitan) (Figures 8510:1C and D).

3) Family Ctenodrilidae

Ctenodrilus serratus (cosmopolitan) (Figure 8510:1B).

4) Family Dinophilidae

Dinophilus gyrociliatus (Figure 8510:2A).

5) Family Dorvilleidae

Ophryotrocha diadema (west coast of U.S.) (Figure 8510: 2B).

6) Other species used in toxicity tests but not discussed include: *Nereis diversicolor, Arenicola cristata,* and *Abarenicola pacifica.*

b. *Freshwater oligochaetes:*

1) Family Tubificidae

Limnodrilus hoffmeisteri (cosmopolitan).

Tubifex tubifex (cosmopolitan) (Figure 8510:3A).

Branchiura sowerbyi (cosmopolitan) (Figure 8510:3B).

2) Family Lumbriculidae

Stylodrilus heringianus (holarctic) (Figure 8510:3C).

Lumbriculus variegatus (holarctic).

c. *Marine oligochaetes:*

1) Family Tubificidae

Monopylephorus cuticulatus (N.E. Pacific).

Tubificoides "fraseri" (all North American coasts).

Tectichilus verrucosus (all North American coasts).

d. *Other freshwater and marine oligochaetes* used in toxicity tests but not discussed include: *Quistadrilus multisetosus, Spirosperma ferox, Spirosperma nikolskyi, Rhyacodrilus montana, Varichaetadrilus pacifica, Ilyodrilus frantzi, Nais communis, Paranais frici,* and *Paranais litoralis.*

2. Collecting Test Organisms

a. *Marine polychaetes:*

1) *Neanthes arenaceodentata, Neanthes succinea,* and *Capitella capitata* inhabit intertidal and subtidal mud flats in estuarine

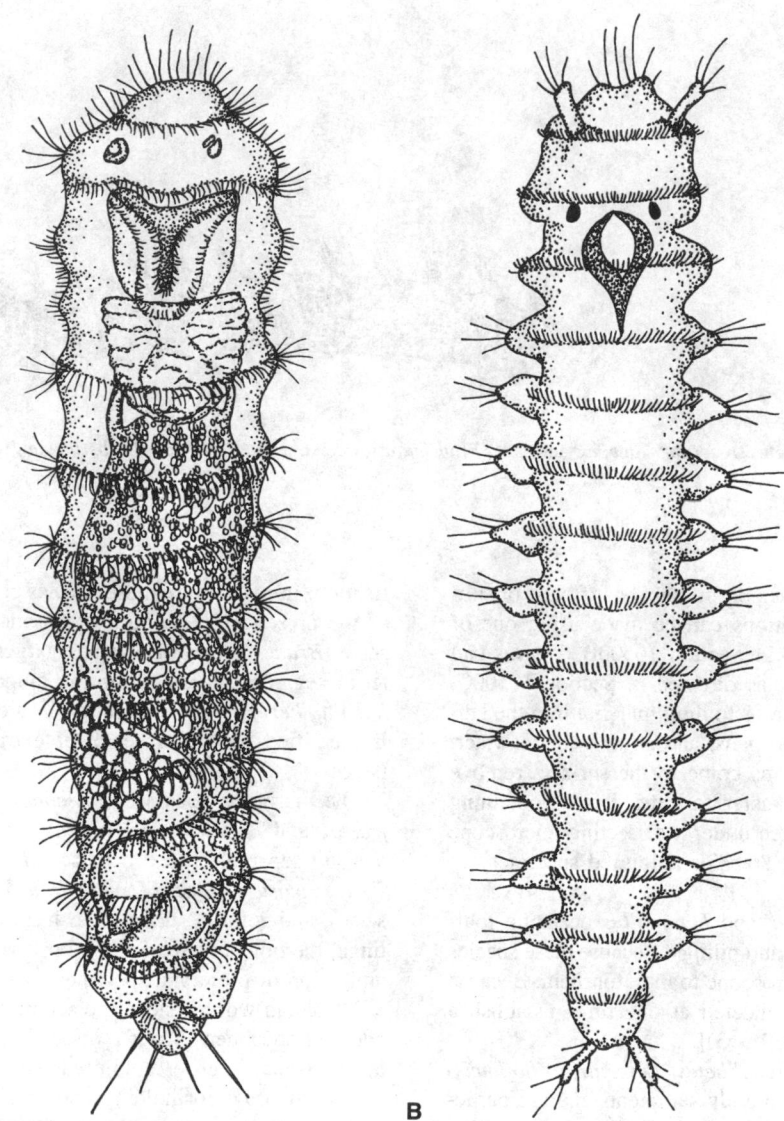

A   B

Figure 8510:2. Marine polychaetes. A—*Dinophilus gyrociliatus,* adult; B—*Ophryotrocha diadema,* adult.

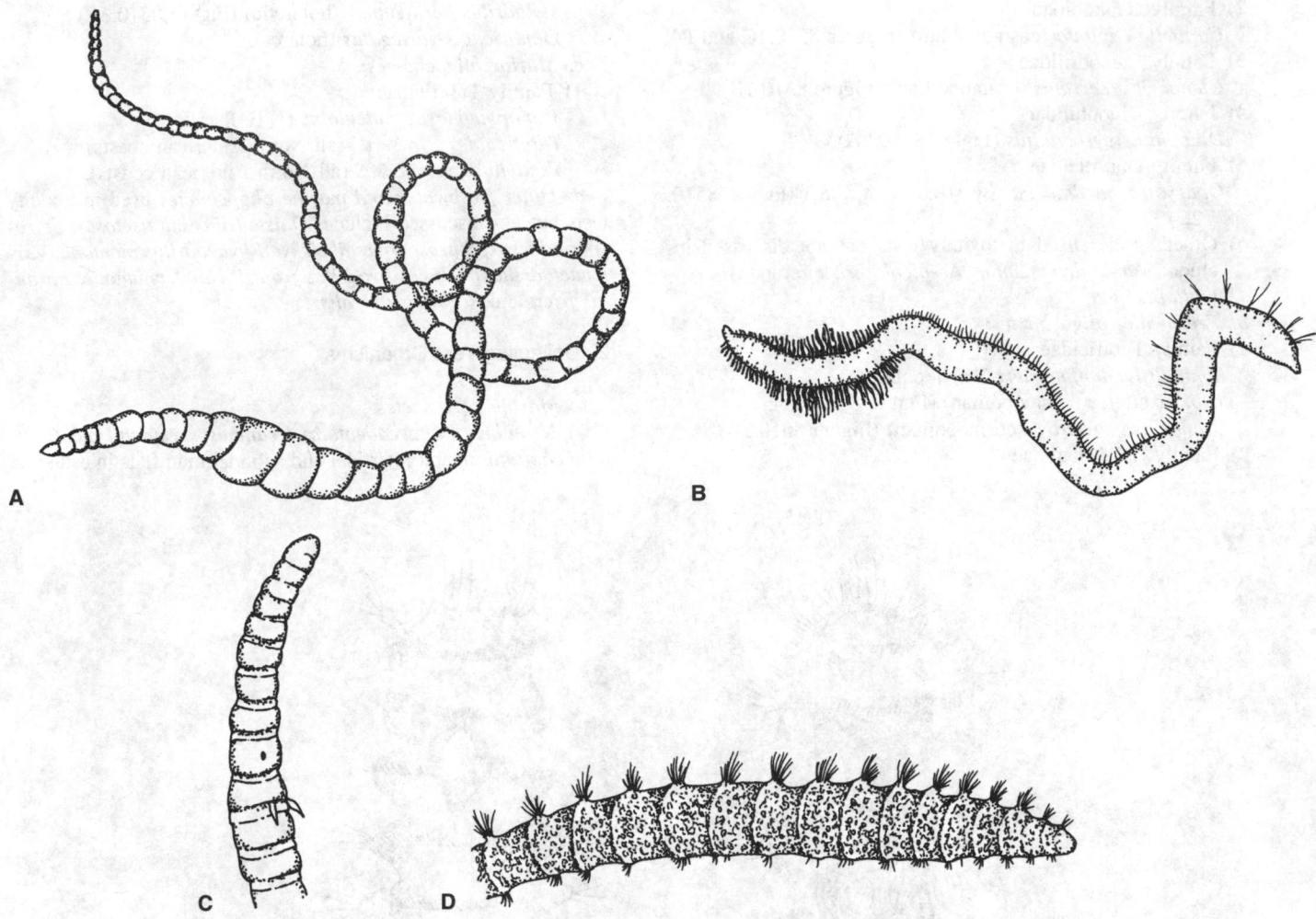

Figure 8510:3. Freshwater oligochaetes. A—*Tubifex tubifex,* adult; B—*Branchiura sowerbyi,* adult; C—*Stylodrilus heringianus,* adult; D—*Quistradrilus multisetosus.*

areas and the fouling communities on pilings, boat floats, or submerged objects. Subtidal collections can be made using one of the sampling devices described in Section 10500B. Worms can be separated from the sediment as directed in Section 10500C. To obtain worms, bring substrate or fouling material into the laboratory, place in white enameled pans, and cover with seawater. After a period of time, the worms come to the surface; remove them with a fine brush and transfer to petri dishes containing seawater. Examine each specimen under a dissecting microscope and discard all injured worms. Transfer uninjured specimens to 4-L aquariums or shallow trays.

2) *Ctenodrilus, Ophryotrocha,* and *Dinophilus* occur on fouling organisms attached to floats and pilings. Because these species are minute, use a dissecting microscope to look for them. Because only a small number can be collected at one time, establish a laboratory colony [Section 8510B.3a3)].

b. *Freshwater and marine oligochaetes: Limnodrilus hoffmeisteri* and *Tubifex tubifex* inhabit muddy sediments and are particularly common in areas of gross organic pollution. Examination of preserved specimens usually is necessary for positive identi-

fication. In live culture these species can be separated on the basis of the presence (*T. tubifex*) or absence (*L. hoffmeisteri*) of hair setae. *Branchiura sowerbyi* is a larger worm with gills on posterior segments; it is common in muddy, warm-water areas. *Stylodrilus heringianus* is common in areas of clean, fine sand and is identified by the presence of extra rings and nonretractable penes.

The marine species *Monopylephorus cuticulatus, Tubificoides fraseri,* and *Tectidrilus verrucosus,* are found in muddy sediments and are separable based on size (*M. cuticulatus* > *T. fraseri* > *T. verrucosus*), setae (*M. cuticulatus* has occasional irregular hair setae), and color (*T. verrucosus* has a papillate skin and greenish tinge, the others are dark red). Positive identification requires examination of preserved specimens.

To obtain worms, sieve the sediments through a 0.5-mm sieve and sort specimens under a dissecting microscope. Discard damaged worms. Transfer uninjured specimens to aerated aquariums or shallow trays for holding and feeding. Ensure that worms are collected from an uncontaminated area because rapid resistance to some toxicants can occur.[1]

3. Culturing

a. Marine polychaetes:

1) Condition of animals—Discard animals injured during collection. Some species such as *Neanthes arenaceodentata* can regenerate a tail; thus it is not always necessary to discard worms missing tails when establishing a culture. Save worms with gametes in the coelom for starting cultures, but normally do not use them for toxicity tests.

2) Food and feeding—Cultures of the polychaete species mentioned here can be maintained without sediment; therefore, the worms must be fed, as should worms in long-term experiments [see 8510C.4*a*3)]. Cultures of the larger species (i.e., *N. arenaceodentata*) have better survival, growth, and tube production if fed a mixed diet that consists of a macroalga for tube construction and a commercially prepared invertebrate or fish diet for nutrition. The green alga, *Enteromorpha* sp., is convenient because it grows abundantly in most estuarine areas of North America. Collect in quantity, wash with seawater, dry, and store indefinitely. Before use, soak the alga in seawater and knead to separate individual filaments. Other macroalgae, for example, cultured brown *Ectocarpus siliculosus*, produce excellent results in polychaete cultures but are not as convenient to use. Place sufficient macroalga in culture containers to allow worms to construct tubes. Add commercially prepared diet* to the worm cultures three times weekly. Vigorously mix flakes with a small

---

* Prawn Flakes, Plankton Flakes, TetraMarin, or equivalent.

amount of seawater to moisten and break them up before adding them to the cultures. To minimize overfeeding, examine each culture container before adding the commercial diet. If most of the diet material is uneaten, do not add more and add less food at subsequent feedings.

A powdered diet is suitable for small species (*Ctenodrilus serratus*, *Capitella capitata*, *Ophryotrocha diadema*, and *Dinophilus gyrociliatus*) and the larvae of *N. arenaceodentata*. Prepare a fine powder from dried *Enteromorpha* sp. or one of the commercial diets by grinding the dry material in a blender and sieving it to smaller than 0.061 mm. *C. serratus* can be fed at a rate of 0.1 mL/worm/week of a mixture of 1.0 g powder/100 mL seawater.

Feed living *Dunaliella* sp. to larval *N. succinea* until the larvae settle. For culturing instructions see Section 8010E.4*c*1)*b*). Feed *Dunaliella* sp. at a minimum of 2 000 000 cells/L of worm culture or at a rate great enough to maintain a green color in the seawater. After the larvae settle, feed *Enteromorpha* sp. until the swimming reproductive epitoke stage is reached.

3) Producing test organisms

a) Capitella capitata—Laboratory-cultured specimens begin to mature in about 15 to 25 d after hatching. A mature female develops white masses of eggs in the coelom from about segment 10 posteriorly and a mature male develops specialized setae on the dorsal surface of segments 8 and 9. The female lays fertilized eggs along the inside lining of her tube where larval development continues until the trochophore larvae emerge 4 to 6 d later (Figure 8510:4). To obtain free-swimming trochophore larvae, examine tubes under a dissecting microscope to detect those con-

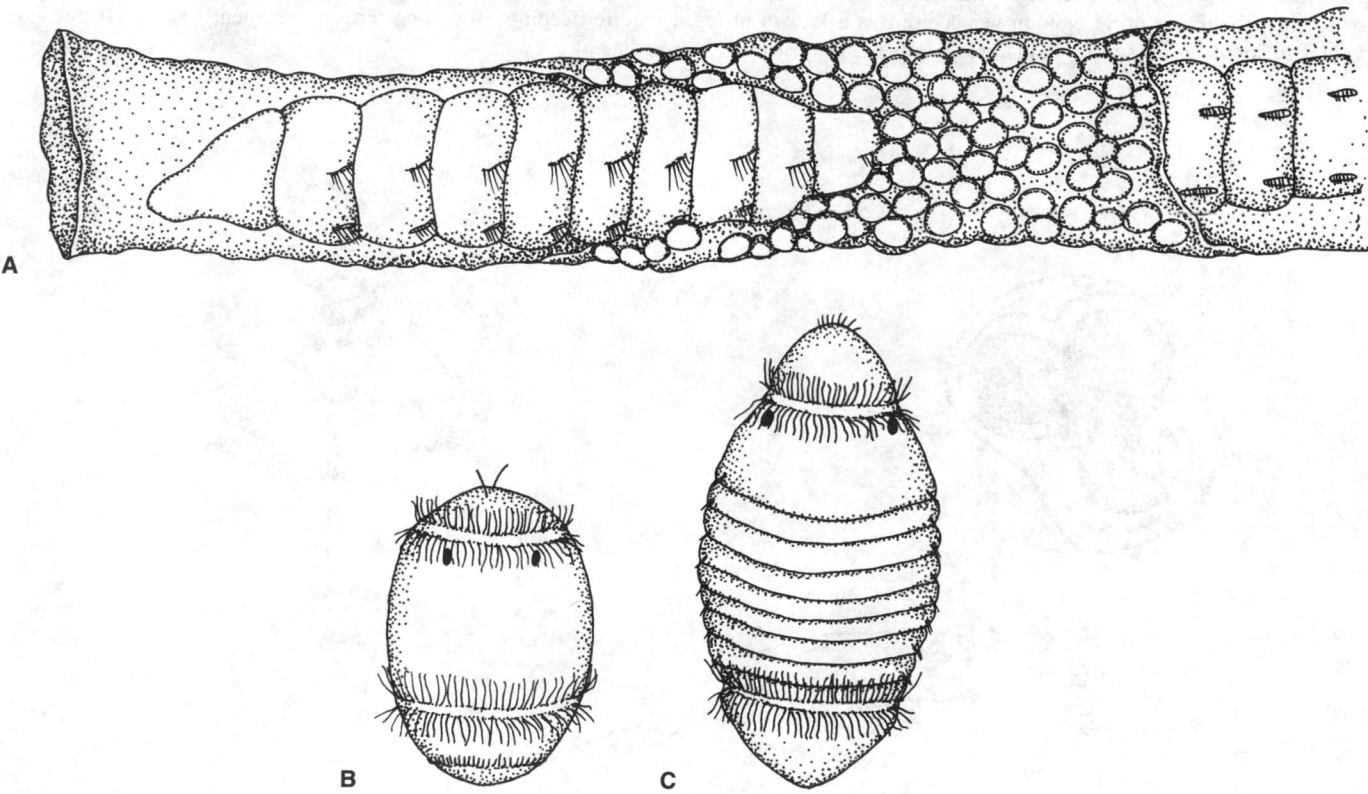

Figure 8510:4. Life stages of *Capitella capitata*. A—Female incubating developing embryos; B—Recently hatched trochophore larva; C—Metatrochophore stage, ready to settle.

taining eggs or larvae. Recently fertilized eggs appear white, but as they mature they become grey-green and can be seen moving about. Place tubes containing larvae in a petri dish. Under a dissecting microscope open the tubes to free the trochophores. One female provides 200 to 300 trochophores. Remove and discard the female and the tube containing any larvae that did not swim free. Use the free-swimming larvae in tests or let them develop for later use.

Sibling species of *Capitella capitata* have been described;[2] however, the taxonomic status of this species complex is still in question.

b) *Neanthes succinea*—Take nearly mature epitokes from the field or laboratory colony and hold until they complete sexual metamorphosis. Mature epitokes swim to the water surface and release gametes. If fertilization is successful, separate the zygotes into several 4-L jars containing aerated water and let them develop to the three-setiger stage (about 1 week). These larvae are ready for use in tests. One fertilization provides more than 2000 larvae.

c) *Neanthes arenaceodentata*—Before spawning, either the male or female enters the tube or burrow of another worm. If the worms are of different sex, they remain together and spawn within the tube. The female dies within 1 d after spawning and the male incubates the eggs for about 3 weeks, at which time they have 18 to 21 setigerous segments. At that time the young worms leave the tube, begin feeding, and construct their own tubes. Feed them *Enteromorpha* as indicated in Section 8510B.3a2). Under laboratory conditions (20°C) sexual maturity is reached in 3 to 4 months. It is impossible to distinguish the sex of immature forms morphologically. Distinguish by observing whether or not they fight when placed together. Like sexes fight; unlike sexes do not. Use a female with maturing eggs in her coelom as a known individual to identify the sex of immature worms. The most con-

venient time to obtain young juveniles is shortly after they have left the parent's tube and have begun to feed.

d) *Ctenodrilus serratus*—This species reproduces asexually about every 14 d at 20°C by transverse division. Each individual produces about five to eight new specimens. Large colonies can be maintained with minimum care.

e) *Dinophilus gyrociliatus* completes its life cycle in 7 to 10 d at 20°C. The female lays two to five eggs in a capsule. The larger eggs develop into females and the smaller ones into males. The male mates with the female before hatching from the capsule (Figure 8510:5A), then dies. Large colonies can be maintained with a minimum of care.

f) *Ophryotrocha diadema* (Figure 8510:2B) completes its life cycle in 30 to 40 d at 20°C. This species is hermaphroditic with Segments 3 and 4 containing the male elements. The remaining segments posterior to Segment 4 are female. Eggs are laid in a loose jelly capsule (Figure 8510:5B) and number 10 to 14. They hatch from the capsule in 8 d as a four-segmented larva (Figure 8510:5C). They begin feeding the next day. Large colonies can be maintained with a minimum amount of care; however, subcultures should be established every 5 to 6 weeks in clean containers.

b. *Freshwater and marine oligochaetes:*

1) Condition of animals—Oligochaetes show great regenerative abilities; hence it is not always necessary to discard injured specimens. Mature individuals with well-developed clitellar regions are particularly important for culture establishment. Keep cultures in the dark or under natural light/dark regimes.

2) Food and feeding—Oligochaetes feed mainly on bacteria in sediments; therefore, in experiments with natural sediments additional feeding is unnecessary. Short-term experiments do not require feeding; for long-term experiments (>10 d) provide sediment.

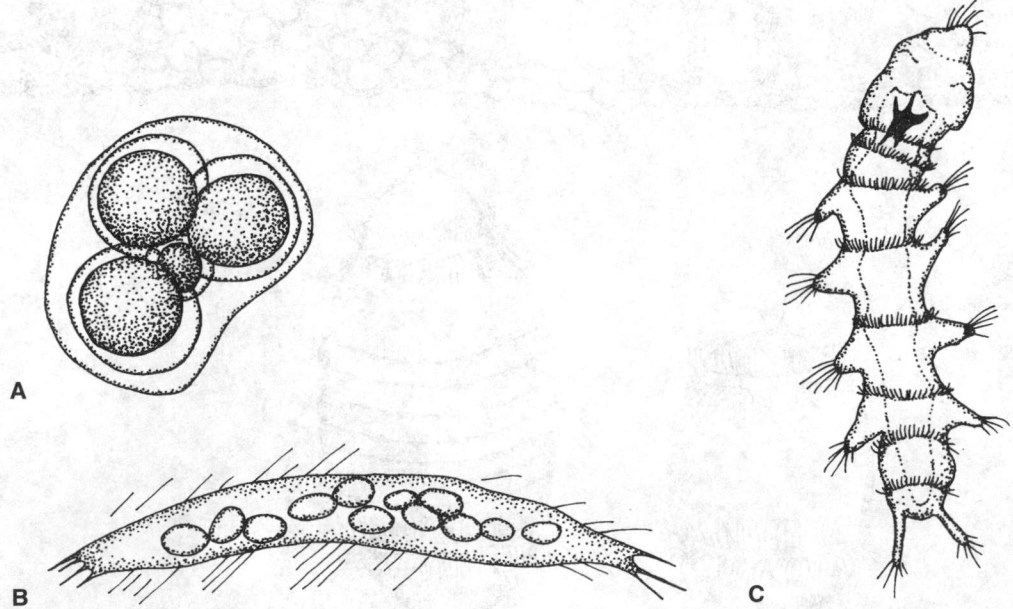

Figure 8510:5. Life stages of selected marine polychaetes. A—*Dinophilus gyrociliatus,* three female and one male (small) embryos in a developing capsule; B—*Ophryotrocha diadema,* developing embryos in egg capsule; C—*Ophryotrocha diadema,* larva recently emerged from egg capsule.

Condition sterile sediments by preparing an inoculum of *Enteromorpha* (for marine worms) or lettuce (for freshwater worms) consisting of the aqueous material remaining after decay of the plant fibers in diluent water. Add inoculum directly to culture containers in a volume not to exceed 10% of the total. Preferably use sediments of fine sand with some silt content, rather than more muddy sediments in which the worms are difficult to find. Check cultures periodically for spoilage; if this occurs, clean cultures and restart. Oligochaetes have no larval stage. No separate feeding regime is required for juveniles.

3) *Producing test organisms*—Gravid worms in culture lay eggs that hatch in 3 to 14 d, depending on species and temperature. Newly hatched worms lack the full component of adult setae but rapidly develop these.

Freshwater species generally grow better in mixed culture. The following combinations are recommended: *L. hoffmeisteri* and *T. tubifex, L. hoffmeisteri* and *B. sowerbyi, T. tubifex* and *B. sowerbyi, S. heringianus* and *L. hoffmeisteri. T. fraseri* is a parthenogenic species and is particularly amenable to culturing; *T. verrucosus, L. variegatus,* and *M. cuticulatus* can be cultured as pure species.

4. Parasites and Diseases

Microbial growth can result from overfeeding, improper conditioning of food, or insufficient DO. Prevent fungal growths by providing sufficient aeration. To minimize overfeeding, examine each aquarium before feeding. If most of the diet is uneaten, do not add more and add less food at subsequent feedings. Generally there is adequate DO in 4-L aquariums; however, aeration can be increased to correct for any deficiency. The internal protozoan parasitic gregarines have been observed to reduce the vitality of some species of laboratory populations of polychaetes. Gregarines are common in polychaetes and oligochaetes but it is not known if they cause similar problems in these species.

5. References

1. KLERBS, P.L. & J.S. LEVINTON. 1989. Rapid evolution of metal resistance in a benthic oligochaete inhabiting a metal-polluted site. *Biol. Bull.* 176:135.
2. GRASSLE, J.F. & J.P. GRASSLE. 1976. Sibling species in the marine pollution indicator *Capitella* (Polychaeta). *Science* 192:567.

6. Bibliography

BAILY, H.C. & D.H.W. LIU. 1980. *Lumbriculus variegatus,* a benthic oligochaete, as a bioassay organism. *In* J.C. Eaton, P.R. Parrish & A.C. Hendricks, eds. Aquatic Toxicology. ASTM STP 707, American Soc. Testing & Materials, Philadelphia, Pa.
CHAPMAN, P.M. & R.O. BRINKHURST. 1984. Lethal and sublethal tolerances of aquatic oligochaetes with reference to their use as a biotic index of pollution. *Hydrobiologia* 115:139.
REISH, D.J. 1985. The use of *Neanthes arenaceodentata* as a laboratory experimental animal. *Tethys* 11:335.
CARR, R.S., M.D. CURRAN & M. MAZURKIEWICZ. 1986. Evaluation of the archiannelid *Dinophilus gyrociliatus* for use in short-term life cycle toxicity tests. *Environ. Toxicol. Chem.* 5:703.
PESCH, C.E. & P.S. SCHAUER. 1988. Flow-through culture techniques for *Neanthes arenaceodentata* (Annelida: Polychaeta). *Environ. Toxicol. Chem.* 7:961.
JENNER, H.A. & T. BOWER. 1992. The accumulation of metals and toxic effects in *Nereis virens. Environ. Monit. Assess.* 21:85.
MILBUNK, G. 1987. Biological characterization of sediment by standardized tubificid bioassays. *Hydrobiologia* 155:267.
REISH, D.J. & T.V. GERLINGER. 1997. A review of the toxicological studies with polychaetous annelids. *Bull. Mar. Sci.* 60:584.
SMITH, D.P., J.H. KENNEDY & K.L. DICKSON. 1991. An evaluation of a naiadid oligochate as a toxicity test organism. *Environ. Toxicol. Chem.* 10:1459.

8510 C.   Toxicity Test Procedures

1. General Procedures

Use exploratory tests (see Section 8010D) to determine toxicant concentrations for short-term tests. Prepare dilution water and toxicant solutions and introduce them into test containers as described in Section 8010F.

2. Water Supply

a. Artificial seawater: See Section 8010E.4*b*2). Use a salinity of approximately 35.5 g/kg and a pH of about 7.8 for marine populations; use lower salinity for estuarine worms.

b. Natural seawater: Determine and report quality routinely. Maintain dilution water salinity at or near selected or normal concentration. During a test, do not allow salinity to vary by more than ±3 g/kg. Filter seawater through a 0.45-μm membrane filter.

c. Distilled or tap water: Determine quality (hardness, alkalinity, chemical constituents) and report routinely. Use a near-neutral (pH 7.0) water.

3. Exposure Chambers

a. Marine polychaetes: Use 4-L aquariums or glass jars for short-term and intermediate static and renewal tests and for long-term tests where flow-through facilities are not appropriate. Cover aquariums to prevent entrance of foreign materials. Do not add more than 2.5 L test solution to each 4-L aquarium. Use 500-mL erlenmeyer flasks, containing 100 mL seawater, for either short-term or long-term experiments when only one organism is placed in each flask. Close the flask with a No. 7 TFE stopper fitted with a glass tube for aeration. Use small stender dishes (30 mL) for larval tests. For flow-through tests, use exposure chambers described in Section 8010F.1*c*). In the case of cannibalistic species such as *Neanthes arenaceodentata* (Figure 8510:6), isolate individuals during testing. Container size depends on biomass; maintain loading densities below 0.5 g/L for static conditions and below 0.5 g/L/d for flow-through tests at 20°C. For tests with sediment, use glass crystalling dishes of the appropriate size for species tested and number of individuals per dish. Fill dishes with sediment 1 to 4 cm deep. Let clean seawater flow over top of sediment.

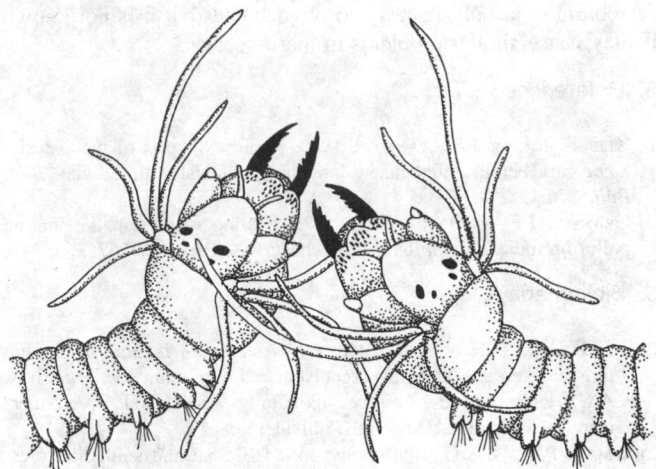

Figure 8510:6. *Neanthes arenaceodentata.* Adults of same sex in a fighting position.

b. Freshwater and marine oligochaetes: Conduct test in a manner similar to that described for the polychaete larval tests. Use shallow disposable polyethylene petri dishes with covers for static or replacement tests. Container size depends on biomass; maintain loading densities below 0.5 g/L, and preferably below 0.2 g/ L. Place 10 worms in each container per test concentration plus controls. Run duplicate tests. Worms can be tested individually, with 20 individuals per test concentration. For flow-through tests, consider using or adapting exposure chambers described in Section 8010F.1*c*).

4. Conducting the Toxicity Tests

a. Setting up the test chambers: For static and renewal tests, set up as described in Section 8010D. In short-term tests, do not clean exposure containers. In long-term tests in which the organisms are fed, remove unused food and other materials as described in Section 8010E.4*d*. It is unnecessary to provide a bottom substrate for any but long-term oligochaete toxicity tests. Photoperiod and light intensity do not appear to be factors in polychaete tests; however, test oligochaetes either in the dark or with a natural light/dark simulation.[1] Keep temperatures within ±2°C of the natural habitat unless the effect of temperature is being tested.

1) Marine polychaetes—Use a minimum of 20 worms for each test concentration. For cannibalistic species, such as *N. arenaceodentata*, use one worm per container. For other species, place 2, 5, or 10 worms in each container (depending on biomass and container size, see Section 8010F.3*c*). For tests with sediments, use a minimum of three replicate dishes per sediment concentration and a minimum of five worms per dish. Individuals of cannibalistic species do not have to be separated with sediment in the dishes.

2) Freshwater and marine oligochaetes—Use a minimum of 20 worms for each test concentration, preferably in two replicates of 10 worms each. Although the worms will intertwine when healthy, toxified individuals remain separate and toxic effects (e.g., the progressive disintegration of posterior segments) will be manifest.

b. Duration and type of test:

1) Short-term tests—The length of short-term or acute tests depends on the length of the organisms's life cycle (see Section 8010F.3*a*). Short-term tests may be conducted under static conditions (recommended for tests of shorter duration, i.e., <4 d) or with periodic renewals for tests of longer duration (i.e., >4 d).

2) Intermediate-length tests—Use tests of intermediate length for determining adult polychaete survival. For most species conduct these renewal or flow-through tests for 20 to 28 d.

3) Long-term tests—Long-term tests are either partial or full life-cycle tests. Partial life-cycle tests begin with the polychaete trochophore larval or oligochaete egg-case stage and end with sexual maturity. Full-life-cycle tests also begin with larvae or egg-case stage but continue through reproduction and subsequent egg production or larval settlement of the offspring. Because of the long duration, conduct these tests using either periodic renewal or flow-through conditions.

Select and prepare test concentrations as described in Section 8010F.2*b*. Measure and mix dilution water and stock toxicant solutions by proportional diluters and deliver to exposure chambers as described in Section 8010F.1. Make tests in flow-through exposure chambers similar to those described in Sections 8740C.3 and 8010F.1). Renewal tests using up to 4-L exposure chambers may be necessary if flowing water is unavailable.

The duration of long-term tests depends on length of life-cycle of the organism. For example, for polychaetes it varies from about 1 month with *C. capitata* to 3 or more months with *N. succinea, N. virens,* and *N. arenaceodentata.*

c. Test organisms: See Section 8510B.

d. Performing tests:

1) Short-term tests—Set up and conduct renewal tests as described in Section 8010D.2. Determine survival of adults by checking exposure chambers at 1, 2, 4, 8, 18, and 24 h, then once or twice daily thereafter. Dead specimens generally are pale and swollen and lie on the bottom; live specimens usually respond to physical stimulation. If the tests are more than 4 d long, renew solutions, preferably daily but at least every fourth day. In short-term tests with polychaete larvae, determine survival after 96 h by microscopic examination. The absence of larvae generally indicates death because decomposition of small larvae is rapid.

2) Intermediate-length tests—Set up test chambers described in ¶ 4*a* above to determine adult lethality (LC50 or incipient LC50). Examine test containers daily to determine survival. If no organisms are killed after a certain length of exposure, report the period beyond which there is no further kill and the percentage killed in each test concentration. For contaminated sediment tests, sieve contents of each replicate dish and count number of survivors. If a graded series of mixed sediments has been used, calculate LC50 based on percentage of contaminated sediment.

3) Polychaete life-cycle tests beginning with the trochophore larval stages—Set up as described previously in this section. Conduct test through sexual maturity with test periods varying from 3 to 4 weeks with *C. capitata*, and from 2 to 3 months or longer for *N. arenaceodentata, N. succinea*, and *N. virens*. Feed larvae as described in Section 8510B.3*a*2). Determine survival at least twice weekly for *C. capitata* and once a week for *Neanthes*. During the early part of the study, count organisms on bottom of exposure chambers. If a renewal test is being conducted, decant supernatant fluid, examine under a dissecting microscope, and replace fluid with fresh test solution. For flow-through tests, re-

move chambers from exposure box, count organisms, and replace chamber. If no organisms are observed by the third examination, terminate that test chamber.

When *C. capitata* is the test organism, remove test chambers after about 15 to 16 d and every 2 d thereafter to check with a dissecting microscope for presence of eggs in the coelom and later for the presence of zygotes along the sides of the tube. Remove females when developing eggs are in the trochophore stage and count the larvae. Discard females and larvae after counting larvae and record the number of dead and deformed larvae. Continue to examine each exposure chamber every 2 d for detection of females incubating larvae until all females have been removed and the total number of larvae recorded.

Abnormal larvae of *C. capitata* (Figure 8510:7) have been observed during life-cycle tests when exposed to sublethal concentrations of chromium, zinc, or detergents.[2]

For *N. succinea*, set up exposure chambers as described in ¶ 4a above with 25 larvae in each 1-L exposure chamber or 10 larvae in each flow-through exposure chamber. Use 10 chambers per concentration tested. Because these worms fight and are cannibalistic when crowded, prepare additional exposure chambers or reduce numbers in each test chamber to five individuals after the first month. Continue tests until animals reach the epitoke stage; then determine individual lengths and total weights and compare with those in the control.

4) Polychaete life-cycle tests beginning with the newly settled larval stage—These tests will vary in duration from about 1 month for *C. capitata* to 3 months or more for *N. arenaceodentata*. Set up tests as described previously with newly settled larvae. Use a minimum of two specimens per flask and 10 flasks per concentration. As tests progress, count organisms as above. Examine for survival once or twice per week as in ¶ d3).

For *N. arenaceodentata*, use recently emerged juveniles having approximately 18 to 21 setigerous segments. If a renewal test is conducted, place four worms in each of five 4-L exposure chambers with 2.5 L test solution. Set up five containers for each test concentration and control. For flow-through tests, place two larvae in each of 10 exposure chambers for each test concentration and the controls. At 25 d, examine worms by viewing from outside the container for the presence of eggs in the coelom. If necessary, move mature worms among the replicates of a given treatment to pair males with females. Mature eggs reach 450 μm diam and are yellowish-orange. Examine at 5-d intervals until eggs are

noted and then at 2- to 3-d intervals to determine whether eggs are being laid. The females die within 1 d after laying eggs and the males incubate them for about 3 weeks. The life cycle is complete when the juvenile worms emerge from the parental tube. Remove males and count larvae.

5) Oligochaete life-cycle tests—Set up as described above. Test duration depends on test conditions and end points chosen. Use procedures similar to those for polychaetes.

5. References

1. REISH, D.J. 1980. The effect of different pollutants on ecologically important polychaete worms. EPA 600/3-80-053, U.S. Environmental Protection Agency, U.S. Government Printing Off., Washington, D.C.
2. REISH, D.J., F.M. PILTZ, J.M. MARTIN & J.Q. WORD. 1974. The induction of abnormal polychaete larvae by heavy metals. *Mar. Pollut. Bull.* 5:125.

6. Bibliography

BELLAN, G., D.J. REISH & J.P. FORET. 1972. The sublethal effects of a detergent on the reproduction, development and settlement in the polychaetous annelid *Capitella capitata*. *Mar. Biol.* 14:183.

REISH, D.J., C.E. PESCH, J.H. GENTILE, G. BELLAN & D. BELLAN-SANTINI. 1978. Interlaboratory calibration experiments using the polychaetous annelid *Capitella capitata*. *Mar. Environ. Res.* 1:109.

CHAPMAN, P.M., M.A. FARRELL & R.O. BRINKHURST. 1982. Relative tolerances of selected aquatic oligochaetes to individual pollutants and environmental factors. *Aquat. Toxicol.* 2:47.

CHAPMAN, P.M., M.A. FARRELL & R.O. BRINKHURST. 1982. Effects of species interactions on the survival and respiration of *Limnodrilus hoffmeisteri* and *Tubifex tubifex* (Oligochaeta, Tubificidae) exposed to various pollutants and environmental factors. *Water Res.* 16:1405.

McLEESE, D.W., L.E. BURRIDGE & J. VAN DINTER. 1982. Toxicities of five organochlorine compounds in water and sediment to *Nereis virens*. *Bull. Environ. Contam. Toxicol.* 28:216.

PESCH, C.E. & G.L. HOFFMAN. 1983. Interlaboratory comparison of a 28-day toxicity test with the polychaete *Neanthes arenaceodentata*. *In* W.E. Bishop, R.D. Cardwell & B.B. Heidolph, eds. Aquatic Toxicology and Hazard Assessment, 6th Symp., American Soc. Testing & Materials, Philadelphia, Pa.

REISH, D.J. & T.V. GERLINGER. 1984. The effects of cadmium, lead, and zinc on survival and reproduction in polychaetous annelid *Neanthes arenaceodentata* (F. Nereididae). *Linn. Soc. New South Wales*, p. 383.

PESCH, C.E., P.S. SCHAVER & M.A. BALBONI. 1986. Effect of diet on copper toxicity to *Neanthes arenaceodentata* (Annelida: Polychaeta). *In* T.M. Poston & R. Purdy, eds. Aquatic Toxicology and Environmental Fate, 9th Symp., American Soc. Testing & Materials, Philadelphia, Pa.

CHAPMAN, P.M. & D.G. MITCHELL. 1986. Acute tolerance tests with the oligochaete *Nais commonis* (Naididae) and *Ilyodrilus frontzi* (Tubificidae). *Hydrobiologia* 137:61.

WIEDERHOLM, T., A.-M. WIEDERHOLM & G. MILBUNK. 1987. Bulk sediment bioassays with five species of fresh-water Oligochaete. *Water, Air, Soil Pollut.* 36:131.

CHAPMAN, P.M. 1987. Oligochaete respiration as a measure of sediment toxicity in Puget Sound, Washington. *Hydrobiologia* 155:249.

MOORE, D.W., T.M. DILLON & B.S. SUEDAL. 1991. Chronic toxicity of tributyltin on the marine polychaete worm *Nereis (Neanthes) arenaceodentata*. *Aquat. Toxicol.* 21:181.

REISH, D.J. & J.A. LEMAY. 1991. Toxicity and bioconcentration of metals and organic compounds by polychaetous annelids. *In* M.E. Petersen & J.B. Kirkegaard, eds. Systematics, Biology and Morphology of World Polychaeta. *Ophelia*, suppl. 5:653.

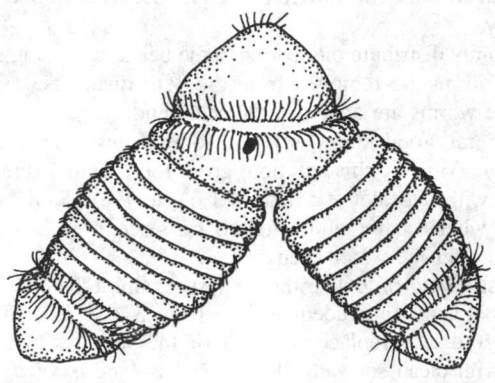

Figure 8510:7. *Capitella capitata.* Abnormal larva.

8510 D. Sediment Test Procedures Using the Marine Polychaete
Neanthes arenaceodentata (PROPOSED)

1. General Procedures

Before conducting a definitive sediment test with the polychaete *Neanthes arenaceodentata*, conduct preliminary tests to become familiar with the test procedures given below.

2. Water Supply

a. Artificial seawater: See Section 8010E.4*b*2). Use a salinity of approximately 28 to 35 g/kg and a pH of approximately 7.8.

b. Natural seawater: Determine and report salinity routinely. During the test, do not let the salinity vary more than ±3 g/kg.

3. Sediment

a. Collection: Collect sediment with a benthic grab such as a van Veen bottom sampler [see Section 10500B.3*a*4)]. Collect multiple samples at a site to obtain the required 25 L of sediment. Press sediments through a 1.0-mm sieve, using a minimum amount of water, to remove macroscopic invertebrates. Place all sieved sediments in a plastic bucket and mix later in the laboratory. Hold sediments at 4°C until start of test; use within 2 weeks of collection, if possible. Collect uncontaminated or reference sediments from an area with similar-sized particles free from contaminants.

b. Sediment chemistry: Analyze sediments for grain size, total organic carbon, metals, and organic compounds.

4. Exposure Chambers

Use as a test chamber a 1-L glass beaker with an internal diameter of 10 cm. Cover beakers with a watch glass to minimize evaporation and to reduce contamination. Aerate each chamber through a 1-mm-opening glass pipet that extends between the beaker spout and the watch glass cover to a depth of not closer than 2 cm from the sediment surface. See Figure 8510:8.

5. Conducting the Sediment Toxicity Test

a. Setting up the test chambers: Use five replicates with each sediment sample. Add sufficient sediment to each beaker to make a 2.0-cm layer. Carefully add approximately 750 mL sea water to each beaker with minimal physical disturbance of the sediment. Aerate at the rate of about 1 to 2 bubbles/s. Prepare sediments and overlying water the day before starting the test to provide time for sediment and seawater to adjust to test temperature.

b. Environmental conditions: Although photoperiod is not critical, usually make sediment tests with *Neanthes arenaceodentata* at either 16 h light/8 h dark or 12 h light and dark. Make tests at 20 ± 2°C.

c. Test animals: The male *N. arenaceodentata* incubates the developing embryos in his tube for approximately 3 weeks after fertilization. Embryos do not feed during this time but subsist on the yolk within the embryo. At 21 to 24 d the juvenile worms leave the tube of the male parent and begin to feed. In the sediment test, use juvenile worms (Figure 8510:9) that are 2 to 3 weeks post-emergent or are approximately 5 to 6 weeks of age

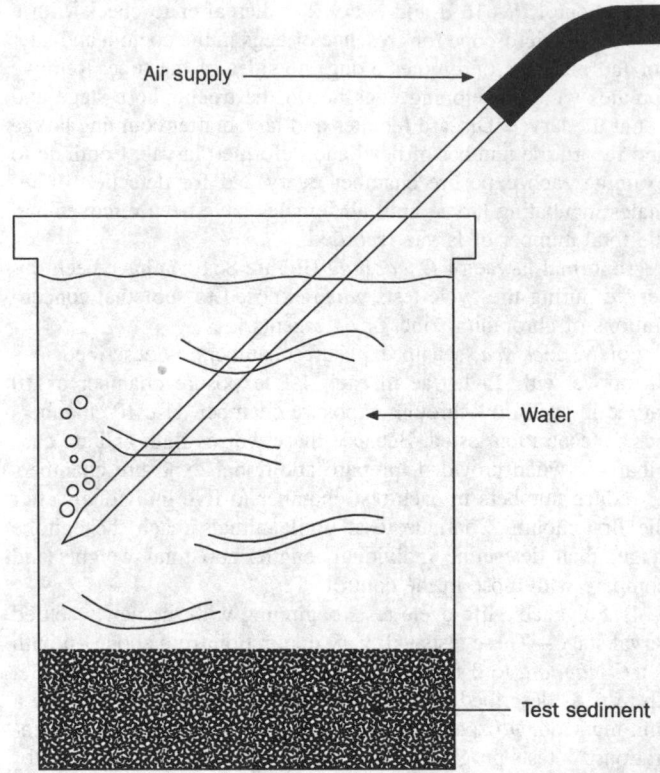

Figure 8510:8. Experimental setup for sediment testing.

Air supply
Water
Test sediment

from the time of fertilization. See Section 8510B.3*a*3)c). Place all juvenile worms to be used in a test together in a white, enamel pan and select uniform-sized worms. Generally place 20 to 30% more worms than needed in the pan and discard the larger and smaller ones. Place five worms in a petri dish with seawater with the number of petri dishes equaling the total number of replicates to be used in the test plus five additional dishes each with five worms. Choose five dishes at random and weigh the five worms together to obtain initial dry weight. See Section 8510D.4*e*. Select additional worms for the reference toxicant test. See Section 8510D.4*f*.

Randomly distribute the worms to the beakers, making sure that all five worms are removed from the petri dish. The test begins when the worms are added to the sediment.

Add 5 mL food slurry solution to each test container every other day. To make this solution, grind food* into a fine powder and mix with seawater at a ratio of 1.0 g food to 25 mL seawater. On days when feeding and water changes occur on the same day, add food after the water change.

d. Test monitoring: During the test, examine beakers daily to ensure that aeration is adequate. On Days 3, 6, 9, 12, 15, and 18 for the 20-d test, replace one half of the seawater within each beaker with clean seawater. Measure dissolved oxygen, temper-

---

* TetraMarin® or equivalent, widely available in aquatic pet supply stores.

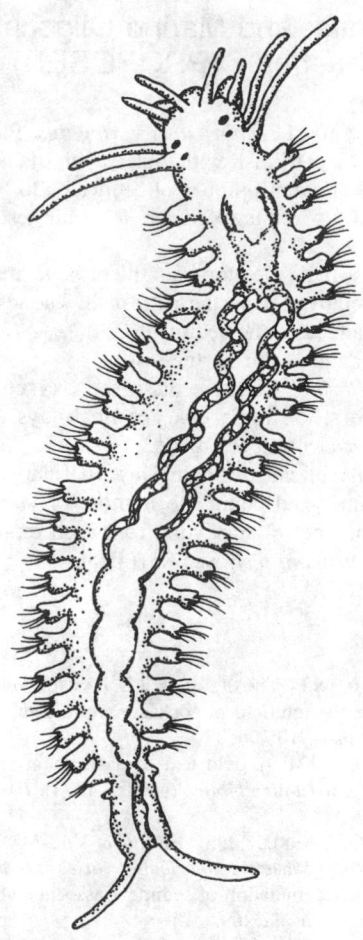

Figure 8510:9. *Neanthes arenaceodentata.* Larva recently emerged from male parent's tube.

ature, and salinity on the initial and terminal days and the days specified above before the water change.

e. Termination of the test: On Day 20 remove worms from each test container, count, place in a clean small petri dish, and wash in distilled water. Place worms from each replicate in a preweighed aluminum pan and dry at 50°C until constant weight (overnight); record dry weight of worms in each pan. Record as the dry weight per worm per day as calculated using the formula:

$$G = (W_t - W_i) / T$$

where:

G = estimated individual growth rate, mg dry weight/d,
W_t = mean estimated individual dry weight, mg, at termination of test,
W_i = mean estimated individual dry weight, mg, at start of test, and
T = exposure time, d.

The use of the mean individual growth rate per worm per day facilitates comparison of results between different sediments tested and with other experiments. Because each test is not started with exactly the same size worm (weight), expressing growth as a rate function rather than absolute mass per worm normalizes results.

f. Reference toxicant test: Make a reference toxicant test, or positive control, concurrently with the sediment test and use it to check the health of the test animals. Use reference test animals from the same batch as those used in the sediment test. Cadmium chloride, expressed as cadmium, is commonly used as the reference toxicant; however, other chlorides of metals or organic compounds may be used. The reference toxicant test is a standard 96-h test without sediment.

6. Bibliography

CARR, R.S., J.W. WILLIAMS & C.T.B. FREGABOR. 1989. Development and valuation of a novel marine sediment pore water test with the polychaetous annelid *Dinophilus gyrociliatus. Environ. Toxicol. Chem.* 8:533.

JOHNS, D.M., R.A. PASTOROK & T.C. GINN. 1991. A sublethal sediment toxicity test using juvenile *Neanthes* sp. (Polychaeta: Annelida). *In* Aquatic Toxicity and Hazard Assessment, 14th Symp., p. 280. ASTM STP No. 1124, American Soc. Testing & Materials, Philadelphia, Pa.

DILLON, T.M., D.W. MOORE & A.B. GIBSON. 1993. Development of a chronic sublethal sediment bioassay with the marine polychaete worm, *Nereis (Neanthes) arenaceodentata. Environ. Toxicol. Chem.* 12:589.

MOORE, D.W. & T.M. DILLON. 1993. The relationship between growth and reproduction in the marine polychaete *Nereis (Neanthes) arenaceodentata* (Moore): Implications for chronic sublethal sediment bioassays. *J. Exp. Mar. Biol. Ecol.* 173:231.

PESCH, C.E., W.R. MUNNS, JR. & R. GUTJAR-GODBELL. 1991. Effects of contaminated sediments on the life history traits and population growth rate of *Neanthes arenaceodentata* (Polychaeta, Nereidae) in the laboratory. *Environ. Toxicol. Chem.* 10:805.

DILLON, T.M., D.W. MOORE & D.J. REISH. 1995. A 28 day bioassay with the marine polychaete *Nereis (Neanthes) arenaceodentata. In* J.S. Hughes, G.R. Biddinger & E. Monis, eds. Environmental Toxicology and Risk Assessment. ASTM STP No. 1218, p. 201. American Soc. Testing & Materials, Philadelphia, Pa.

GERLINGER, T.V., D.J. REISH & M. FANIZZA. 1995. Survival and growth of juvenile *Neanthes arenaceodentata* (Annelida: Polychaeta) in marine sediment taken from the vicinity of an ocean outfall. *Bull. South. Calif. Acad. Sci.* 94:65.

8510 E. Sediment Test Procedures Using the Freshwater and Marine Oligochaetes *Pristina leidyi, Tubifex tubifex,* and *Lumbriculus variegatus* (PROPOSED)

1. General Procedure

The following procedures comprise short-term acute tests. Oligochaetes also can be used to measure reproductive effects[1,2] and in combined toxicity and bioaccumulation tests.[3]

Before conducting a definitive sediment test with the oligochaetes, *Pristina leidyi, Tubifex tubifex,* or *Lumbriculus variegatus,* conduct preliminary tests to become familiar with the test procedures given below.

2. Water Supply

Use fresh water for all tests. It may be moderately hard synthetic water prepared with a commercially available system,* deionized water and reagent-grade chemicals, receiving water, or synthetic water modified to reflect receiving water hardness.

3. Sediment

Collect sediment with a benthic grab such as an Ekman or van Veen sampler. See Sections 10500B.3a4) and 7) and 8510D.3a and b.

4. Exposure Chambers

Use as a test chamber a 250-mL beaker with an internal diameter of 6 cm. Cover beakers with a watch glass to minimize evaporation and to reduce contamination. Aeration is unnecessary.

5. Conducting the Sediment Toxicity Test

a. Setting up the test chamber: See 8510D.5, but add only about 100 mL water to each beaker.

b. Environmental conditions: Set lighting for a 16 h light/8 h dark photoperiod at an intensity of 550 to 1050 lux (50 to 100 ft-c). Make tests with *Pristina leidyi* at 24 ± 1°C and at 20 to 25°C with *Tubifex tubifex* and *Lumbriculus variegatus.*

c. Test animals: Use specimens of mixed age in these tests. Place worms in a white, enamel pan and select healthy-appearing specimens. Generally place 20 to 30% more worms than needed in the pan. Use five worms of *Pristina leidyi* and *Tubifex tubifex*

---

* Millipore Milli or equivalent.

per replicate and 10 of *Lumbriculus variegatus.* Place worms for each replicate in a petri dish with water, with the number of petri dishes equaling the total number of replicates to be used. Select additional worms for the reference toxicant test. See Section 8510E.5f.

Randomly distribute the worms to the beakers, making sure that all worms are removed from the petri dish. The test begins when the worms are added to the sediment.

Do not feed animals during the test.

d. Test monitoring: Measure dissolved oxygen concentration and temperature at the initial and terminal days as well as any days when water changes are made.

e. Termination of the test: On Day 10 remove worms from each test container and count the number of survivors. Record separately the number of survivors from each replicate.

f. Reference toxicant test: See 8510D.5f.

6. References

1. REYNOLDS, T.B., S.P. THOMPSON & J.L BAMSEY. 1991. A sediment bioassay using the tubificid oligochaete worm *Tubifex tubifex. Environ. Toxicol. Chem.* 10:1061.
2. REYNOLDS, T.B. 1994. A field text of a sediment bioassay with the oligochaete worm *Tubifex tubifex* (Muller, 1774). *Hydrobiologia* 278:223.
3. PHIPPS, G.T., G.T. ANKLEY, D.A. BENOIT & V.R. MATTSON. 1993. Use of the aquatic oligochaete *Lumbriculus variegatus* for assessing the toxicity and bioaccumulation of sediment associated components. *Environ. Toxicol. Chem.* 12:269.

7. Bibliography

BAILEY, N.C. & D.N.W. LUI. 1980. *Lumbriculus variegatus,* a benthic oligochaete, as a bioassay organism. *In* J.C. Eaton, P.R. Parrish & A.C. Hendricks, eds. Aquatic Toxicology, ASTM STP No. 507, p. 202. American Soc. Testing & Materials, Philadelphia, Pa.
ANKLEY, G.T., R.A. HOKE, D.A. BENOIT, E.N. LEONARD, C.W. WEST, G.L. PHIPPE, V.R. MATTSON & L.A. ANDERSON. 1993. Development and evaluation of test methods for benthic invertebrates and sediments: effects of flow rate and feeding on water quality and exposure conditions. *Arch. Environ. Contam. Toxicol.* 25:12.
U.S. ENVIRONMENTAL PROTECTION AGENCY & U.S. ARMY CORPS OF ENGINEERS. 1995. Evaluation of Dredged Material Proposed for Discharge in Waters of the United States. Testing Manual, U.S. EPA, 823-B-94-002, Washington, D.C.

8510 F. Data Evaluation

1. Short-Term and Intermediate Adult Survival Studies

Determine the LC50 values for each exposure period as described in Section 8010G. A useful supplementary measure is to determine LT50 (time to 50% mortality) in comparative exposures to single toxicant, effluent, water, or sediment concentrations. LT50s also provide useful ancillary information for LC50 studies.

2. Polychaete Life-Cycle Studies Beginning with the Trochophore and Settled Larval Stages

The number of females forming and laying eggs and the number of offspring produced are inversely related to sublethal toxicant concentrations at levels below the LC50. They provide a more subtle measure of effects than the LC50. Record life-cycle data for each concentration of toxicant as follows: number of females forming eggs, number of females laying eggs, and number of eggs and live offspring produced. Compare these data, expressed on a percentage basis, for all test concentrations with those obtained from the controls. Use statistical and reporting techniques described in Sections 8010G and H.

8610 MOLLUSKS*

8610 A. Introduction

1. Characteristics and Ecology

The Phylum Mollusca, the second largest phylum of the animal kingdom, is made up of such forms as clams, mussels, oysters, snails, slugs, octopuses, squids, as well as others. Approximately 80% of all mollusks are smaller than 5 cm (maximum shell size). The life cycle of mollusks varies from about 1 to about 10 years. In numbers of species (about 100 000 living and 35 000 fossil), mollusks are second only to the arthropods. Their ecology also is very diversified: they have been able to live successfully in nearly all terrestrial, freshwater, and marine habitats, from the greatest depths of the ocean to the highest altitudes recorded for animal life.[1,2] Mollusks are important ecologically and are a source of food for human beings around the world.

2. Types of Tests

Mollusks, particularly marine gastropods (snails and slugs) and bivalves (clams, mussels, oysters, etc.), have been used extensively as bioassay test species throughout the United States. Two types of larval development toxicity tests are described in this section. The marine bivalve larval development test is suitable for water toxicity testing for all coasts and uses oysters (*Crassostrea* spp.), mussels (*Mytilus* spp.), or quahog clams (*Mercenaria mercenaria*). Larvae of the red abalone, an edible marine gastropod (*Haliotus rufescens*), are used in bioassays on the Pacific Coast.

A second type of test exposes adult bivalves, such as bent-nose clams (*Macoma nasuta*), to sediments in the laboratory. At test termination, tissues can be excised and analyzed chemically for pollutants, such as pesticides and heavy metals, to determine whether animals have accumulated toxic substances above background levels.

3. References

1. BARNES, R.D. 1987. Invertebrate Zoology, 5th ed. Sanders College Publishing, Philadelphia, Pa.
2. HICKMAN, C.P. & C.P. HICKMAN, JR. 1979. Integrated Principles of Zoology. C.V. Mosby Co., St. Louis, Mo.

* Approved by Standard Methods Committee, 1997.

8610 B. Selecting and Preparing Test Organisms

1. Selecting Test Organisms

In accordance with the criteria listed in Section 8010E.1, the recommended test species include (but are not restricted to) the following:

    a. Marine gastropods:

    Family Haliotidae:

    Haliotus rufescens (west coast of U.S.)

    b. Marine bivalves:

    Family Ostreidae (Section 10900, Plate 21N):

        Crassostrea gigas (west coast of U.S.)

        Crassostrea virginica (east coast of U.S.)

    Family Veneridae:

        Mercenaria mercenaria (all of U.S. coasts) (Section 10900, Plate 21J)

    Family Tellinidae:

        Macoma nasuta (west coast of U.S.)

    Family Mytilidae:

        Mytilus spp. (cosmopolitan) (Section 10900, Plate 21L)

    c. Other freshwater and marine mollusks: Also used in toxicity tests, but not discussed, are the freshwater *Anodonta imbecillis* (Section 10900, Plate 21E), and the marine species *Ostrea lurida*, *Argopecten irradians irradians*, *Spisula solidissima*, *Mulinia lateralis*, *Macoma balthica*, and *Rangia cuneata* (Section 10900, Plate 21M).

2. Collecting and Conditioning Test Organisms

    a. Marine gastropods: Collect *Haliotus rufescens* (red abalone) adults from the field or purchase from commercial dealers. Mature red abalone can be collected on rocky substrates from the intertidal zone to depths exceeding 30 m. They are most commonly found in crevices where there is an abundance of macroalgae. State collection permits are always required for collecting abalone. While abalone captured in the wild can be induced to spawn, those grown or conditioned in culturing facilities have been more dependable. Commercial mariculture facilities can supply ripe organisms (sexually mature abalones are usually about 70 mm or more in shell length). In any case, obtain brood stock from sources free of contamination by toxic substances to avoid genetic or physiological preadaptation to pollutants. In the laboratory, identify the sex of the organism by inspecting the gonads, located under the right posterior edge of the shell. An abalone placed upside down on a flat surface will soon relax and begin moving the foot trying to right itself. During this movement, bend the foot away from the gonad area and determine the sex. The ovary is jade green, the testes are cream-colored. When the gonad fully envelopes the dark blue-gray conical digestive gland and is bulky along its entire length, the abalone is ready for spawning.[1] Ripe spawners have a distinct color difference between the gray digestive gland and the green or cream-colored gonad. Less developed gonads appear gray (in females) or brown (in males).

    b. Marine bivalves:

    1) *Crassostrea gigas* (Pacific oysters), *Crassostrea virginica* (Eastern oysters), *Mercenaria mercenaria* (quahog clams), and *Mytilus* spp. (mussels) may be collected in the field, but those purchased from commercial suppliers who deal in bioassay organisms usually will spawn more consistently. Identify all field-caught bivalves to species.[2]

    2) *Macoma nasuta* (bent-nose clams) commonly inhabit intertidal areas such as bays and harbors. They reside in the upper few millimeters of fine sediment, which may be collected by hand and sieved to acquire specimens. Identify field-caught organisms to species before their use in testing procedures. Animals also are commercially available. Hold clams as briefly as possible and introduce into test sediments soon after collection.

3. Culturing

    a. Marine gastropods: Keep abalone separated in aerated tanks with flowing seawater. Ideal maintenance temperature is $15 \pm 1°C$, the toxicity test temperature. If brood stock are to be held for longer than 5 d at the testing facility, feed brood stock with blades of the giant kelp, *Macrocystis* (Section 10900, Plate 21A,E). Feed to slight excess; large amounts of uneaten algae will foul culture water. If *Macrocystis* is unavailable, substitute other brown algae (*Nerocystis* [Section 10900, Plate 21H,G], *Egregia, Eisenia*) or any fleshy red alga.[1] For brood stock, preferably use abalone 7 to 10 cm in shell length. They are easier to handle than larger ones, and can be spawned more often (about every 4 months under suitable culture conditions).[3]

    b. Marine bivalves: Maintain adult bivalves in glass aquaria or fiberglass tanks and continuously supply with high-quality, fresh seawater (salinity 18 to 34 g/kg, depending on location and species). If a continuous flow-through system is not available, use natural seawater or commercially available sea-salt preparations made with deionized water in a recirculating system. Water may require efficient activated charcoal or other type of filtration to maintain high water quality. Observe animals daily, and discard any obviously unhealthy animals. Before spawning or direct use in a toxicity test, brush or gently scrape animals to remove barnacles or other encrusting organisms. If conditioning bivalves to spawn, hold for 1 to 8 weeks at 20°C for oysters and quahogs and 15 to 18°C for mussels. During extended holding and conditioning, provide animals with an adequate supply of natural or cultivated phytoplankton to ensure adequate nutrition. Alternatively, some suppliers will precondition animals before shipping. Animals then may be received on the same day as the test, and the maintenance of an elaborate seawater holding system can be avoided.

4. References

1. HAHN, K.O. 1989. Handbook of Culture of Abalone and Other Marine Gastropods. CRC Press, Inc., Boca Raton, Fla.
2. RANSOM, J.E. 1981. Harper and Row's Complete Field Guide to North American Wildlife. Harper and Row, New York, N.Y.
3. AULT, J. 1985. Some quantitative aspects of reproduction and growth of the red abalone, *Haliotis rufescens* Swainson. *J. World Maricult. Soc.* 16:398.

5. Bibliography

HUNT, J.W., B.S. ANDERSON, S.L. TURPIN, H.R. BARBER, M. MARTIN, D.L. DENTON & F.H. PALMER. 1991. Marine Bioassay Project Sixth Report: Interlaboratory Comparison and Protocol Development with

Four Marine Species. Rep. No. 91-21-WQ, State Water Resource Control Board, Sacramento, Calif.

U.S. Environmental Protection Agency & U.S. Army Corps of Engineers. 1991. Evaluation of Dredged Material Proposed for Ocean Disposal. Testing Manual. EPA-503/8-91-001, U.S. Environmental Protection Agency, Off. Research & Development, Washington, D.C..

U.S. Environmental Protection Agency & U.S. Army Corps of Engineers. 1994. Evaluation of Dredged Material Proposed for Discharge in Waters of the United States. Testing Manual. EPA-832/B-94-002, U.S. Environmental Protection Agency, Off. Research & Development, Washington, D.C.

U.S. Environmental Protection Agency. 1995. Short-term Methods for Estimating the Chronic Toxicity of Effluents and Receiving Waters to West Coast Marine and Estuarine Organisms. EPA-600/R-95-136, U.S. Environmental Protection Agency, Washington, D.C.

American Society for Testing and Materials. 1996. Standard Guide for Conducting Bioconcentration Tests with Fishes and Saltwater Bivalve Mollusks. E724-94, Annual Book of ASTM Standards, Vol. 11.05, American Soc. Testing & Materials, W. Conshohocken, Pa.

8610 C. Toxicity Test Procedures

1. General Procedures

Use exploratory tests (see Section 8010D) to determine toxicant concentrations for short-term tests. Prepare dilution water and toxicant solutions and introduce them into test containers as described in Section 8010F.

2. Water Supply

a. Marine gastropod larvae: Tests require a marine laboratory with a supply of clean, 20-μm-filtered seawater, with salinity 34 ± 2 g/kg.

b. Marine bivalve larvae: Use high-quality natural seawater, preferably filtered to 20 μm. Test salinity (18 to 34 g/kg) is dependent on the location and species selected for the test, but regardless of selected salinity, do not vary the salinity by more than ±1 among treatments.

3. Exposure Chambers

Use 30- to 600-mL borosilicate glass beakers or nontoxic disposable containers, five chambers per concentration. Cover the chambers during the test with glass plates or individual caps to avoid contamination from air and excessive evaporation of test solutions. Soak new test chambers in dilution water overnight.

4. Conducting the Toxicity Tests

a. Spawning and fertilization: If mature brood stock are shipped by a supplier, allow 2 d or more for laboratory acclimation before spawning induction; this should increase the probability of achieving a successful spawn of viable gametes. Always bring brood stock up to acclimation temperature slowly to avoid premature spawning. Before beginning the spawning induction process, be sure that test solutions will be mixed, sampled, and temperature-equilibrated in time to receive the newly fertilized eggs. Spawning induction generally takes about 3 h, but if embryos are ready before the test solutions are at the proper temperature, the delay may allow embryos to develop past the one-cell stage before transfer to the toxicant. Transfer can then damage the embryos, leading to unacceptable tests results.

1) Marine gastropod larvae—Ripe abalone can be induced to spawn by stimulating the synthesis of prostaglandin-endoperoxide in the reproductive tissues.[1] This can be done by addition of hydrogen peroxide to seawater buffered with trishydroxymethylaminomethane (Tris)[1] or by irradiation of seawater with ultraviolet light.[2] Select four ripe male and four ripe female abalone. Place organisms of each sex in separate, clean polyethylene buckets filled with 6 L of 0.22-μm-filtered seawater. Aerate the water and keep at 15°C. Prepare buffer by placing 12.1 g Tris in a 125-mL erlenmeyer glass flask and adding 50 mL deionized water. Mix solution thoroughly until the buffer reagent is completely dissolved. Then, into a flask, mix 12 mL of 30% H_2O_2 with 38 mL deionized water. Use hydrogen peroxide less than 1 year old from purchase date, refrigerated, and opened for no more than 30 d.

Add 25 mL buffer solution to each bucket followed by 25 mL H_2O_2 solution; mix thoroughly. Keep buckets covered and undisturbed for next 2.5 h of exposure. After exposure, empty, rinse, and refill buckets with 6 L 0.22-μm-filtered seawater. Continue moderately heavy aeration until spawning begins. Siphon eggs into 1-L glass beaker fertilization container. Fertilize eggs within 1 h of release by adding 100 to 200 mL sperm-laden water. Use a gentle flow of filtered seawater poured along the edge of the fertilization container to roil the eggs, mixing them with sperm. Let fertilized eggs settle for 15 min before siphoning off the sperm-laden water. Refill container with filtered seawater and settle eggs again. After 15 min, siphon fertilized eggs into a 1-L beaker for counting. Evenly mix the embryos in the 1-L beaker by gentle vertical stirring with a clean perforated plunger.[3] Never let embryos settle densely in the bottom of the beaker, and take care not to crush embryos while stirring. Using a 1-mL wide-bore graduated pipet, take five samples of evenly suspended embryos. Empty contents of pipet onto a Sedgwick-Rafter slide and count embryos. Take the mean of five samples to estimate number of embryos. Number of embryos in the beaker should be between 200 and 300 embryos/mL. Dilute if concentration is too high, let embryos settle and pour off excess water if concentration is too low.

2) Marine bivalve larvae—Remove at least a dozen bivalves from conditioning chamber and place them in a container filled with seawater at about 20°C for oysters and quahogs and 15°C for mussels. They should resume pumping within about 30 min. Over a 15- to 20-min period, slowly increase the temperature; do not exceed 32°C for oysters and quahogs and 20°C for mussels. If animals do not spawn within 30 min, return them to water at the original temperature, and, after about 15 min, raise temperature again. Other methods have been used successfully to induce bivalves to spawn. The addition of algae into the water may work for all bivalves. The injection of the posterior adductor muscle

with 0.5M potassium chloride has been used successfully for mussels, and the addition of heat-killed sperm has worked for oysters.

When individuals are observed to be shedding gametes, place each spawning bivalve in an individual chamber containing seawater at 20°C for oysters and clams and 15°C for mussels. Examine a small sample of gametes from each spawner. Although it is highly desirable to pool populations of sperm and eggs, it is more important that they are of high quality. Use only sperm that are highly motile and eggs that are not vacuolated, small, or abnormally shaped. Combine the best sperm and filter through a 36-μm screen* to remove extraneous material or clumped gametes. Filter the eggs through a 75-μm nylon screen into a 1-L beaker and dilute to a density of 5000 to 8000 eggs/mL. Use a Sedgwick-Rafter slide to verify density. While using a perforated plunger to continually suspend eggs, remove three 100-mL portions of the egg suspension and fertilize these with 1, 3, and 10 mL of the sperm suspension. After 1.5 to 2.5 h, inspect each of these portions microscopically. Use the zygote suspension with the lowest amount of sperm giving normal embryo development (i.e., ≥90%). Each suspension contains sufficient numbers of embryos to perform multiple tests. The number of tests, however, will depend on the initial density of the egg solution, the test chamber volume, and the final density of embryos in the test chambers.

b. Setting up test chambers:

1) Marine gastropod larvae—Conduct assay at 15°C and then maintain within ±1°C of that temperature. Maintain photoperiod at 16 h light and 8 h dark; ambient laboratory lighting between 550 and 1050 lux is sufficient. Prepare test chambers as described in Section 8010F. Prepare sufficient water volume to initiate five replicate chambers per test concentration and control with 10 mL of water in each replicate. Measure dissolved oxygen in one replicate from each concentration. Do not aerate chambers unless dissolved oxygen falls below 4.0 mg/L. If aeration is required, do not exceed that necessary to maintain an acceptable minimum oxygen level. Inoculate each test chamber to provide a minimum of 300 embryos from the adjusted stock per 10 mL.

2) Marine bivalve larvae—Conduct assay at 20 ± 1°C for oysters and clams and 15 ± 1°C for mussels. For photoperiod, lighting, test chamber preparation, and dissolved oxygen control, see ¶ 1) above. Prepare sufficient water volume for four replicate chambers per dilution and control, with 10 mL water in each. Calculate volume of fertilized egg suspension necessary to provide 15 to 30 larvae/mL and add to each replicate.

c. Performing tests:

1) Marine gastropod larvae—Incubate for 48 h at 15 ± 1°C. At the end of the 48-h period, pour entire test solution with larvae through a 37-μm-mesh screen if transferring to a container with smaller volume. Wash larvae from screen into 25-mL vials. Add buffered formalin to preserve the larvae in a 5% solution (alternatively, use glutaraldehyde in a 0.5% solution). Pipet a sample from each vial onto a Sedgwick-Rafter counting slide and examine 100 larvae. Return larvae to vials for future reference.

_____
* Nitex or equivalent.

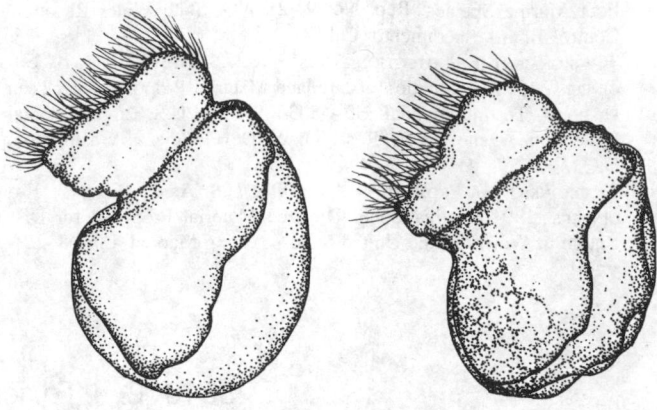

Figure 8610:1. Abalone: (left) normal veliger; (right) abnormal veliger.

Count number of normal and abnormal larvae in each replicate container. Use larval shell development as the test end point (Figure 8610:1). Further details and test variability are available.[4]

2) Marine bivalve larvae—Incubate for 48 h at 20 ± 1°C for oysters and clams and 15 ± 1°C for mussels. After incubation period, proceed as directed in ¶ 1) above. Further details and test variability are available.[3,5]

5. Statistical Analysis

Calculate and report results in accordance with Section 8010G. Computer methods for data processing and analysis are available.[3–5]

6. References

1. MORSE, D.E., H. DUNCAN, N. HOOKER & A. MORSE. 1977. Hydrogen peroxide induces spawning in molluscs, with activation of prostaglandin endoperoxide synthetase. *Science* 196:298.

2. KICUCHI, S. & N. UKI. 1974. Technical study on artificial spawning of abalone, genus *Haliotis* II. Effect of irradiated seawater with ultraviolet rays on inducing to spawn. *Bull. Tohoku Reg. Fish. Res. Lab.* 33:79.

3. AMERICAN SOCIETY FOR TESTING AND MATERIALS. 1996. Standard Guide for Conducting Static Acute Toxicity Tests Starting with Embryos of Four Species of Saltwater Bivalve Molluscs. E 724-94, Annual Book of ASTM Standards, Vol. 11.05. American Soc. Testing & Materials, W. Conshohocken, Pa.

4. HUNT, J.W., B.S. ANDERSON, S.L. TURPEN, H.R. BARBER, M. MARTIN, D.L. DENTON & F.H. PALMER. 1991. Marine Bioassay Project Sixth Report: Interlaboratory Comparison and Protocol Development with Four Marine Species. Rep. No. 91-21-WQ, State Water Resource Control Board, Sacramento, Calif.

5. U.S. ENVIRONMENTAL PROTECTION AGENCY. 1995. Short-Term Methods for Estimating the Chronic Toxicity of Effluents and Receiving Waters to West Coast Marine and Estuarine Organisms. EPA/600/R-95-136, U.S. Environmental Protection Agency, Washington, D.C.

8610 D. Sediment Test Procedures Using Marine Bivalves

1. General Procedures

Before conducting a definitive sediment test with the bivalve *Macoma nasuta*, conduct preliminary tests to become familiar with the test procedures given below.

2. Water Supply

Use seawater with a salinity of approximately 35 g/kg and a pH of about 7.8. A minimum salinity of 25 g/kg is required for the test but any salinity within the range 25 to 35 that reflects site-specific conditions is appropriate. Filter seawater to 20 μm or less in quantities sufficient to support this testing scheme for its duration. Routinely analyze and document source seawater for metals, especially heavy metals, and other potential contaminants.

3. Sediment

Collect sediment from the site with a benthic sampling device (see Section 10500B). Multiple samples may be required from each individual site to acquire sufficient material. Collect control sediment at the same time and in the same general location as the test organisms. Collect reference sediment from a site known to be generally free of contaminants. Sieve test, control, and reference sediments through a 1.0-mm screen to remove resident organisms. Store sediment at 4°C until required for test initiation; recommended holding times for sediment range from 2 to 6 weeks.[1] Begin tests within 6 weeks of sediment collection.

4. Exposure Chambers

Use 20- to 40-L aquaria for long-term flow-through or static-renewal exposures of *Macoma*; these will generate sufficient quantities of tissue for the chemical analyses necessary to determine bioaccumulated tissue concentrations of virtually any analyte. The duration of bioaccumulation tests (30 d including a depuration period of 2 d) requires the use of relatively large sediment volumes. A 40-L aquarium should contain a 5- to 8-cm depth of sediment with overlying water reaching 4 to 6 cm below aquarium height. Maintain ratio of sediment to seawater for smaller aquaria to ensure that there is enough material to support the animals throughout the 28-d exposure and 2-d depuration period. Approximately 50 g wet weight sediment per gram of wet flesh (without shell) is an appropriate minimum; however, consider the final volume of tissue needed for the analytes of concern when choosing the number of animals per replicate and ultimately the amount of sediment and water required to support them. Preferably use flow-through facilities if they are available; the clean seawater should flow over the surface of the sediment at the bottom of the tanks at a rate of at least 1 drop/s. If static-renewal procedures are used, siphon 80% of the overlying water in each aquarium and replace with clean seawater every other day at minimum. Cover aquaria to prevent introduction of foreign materials.

5. Conducting the Toxicity Tests

a. Setting up test chambers: Conduct assay at a chosen temperature between 12°C and 16°C and maintain within ±2°C of that temperature. Although photoperiod does not appear to be a critical parameter for sediment testing with clams, maintain and report a 16:8, 14:10, or 12:12 light:dark period. Ambient laboratory lighting is sufficient. Prepare test chambers as described in Section 8010F. Prepare sufficient aquaria to initiate five replicate chambers per site, as well as five replicates each for control and reference sediments. Each site is an individual test material; there is no dilution series or sediment concentration associated with bioaccumulation testing with clams (although such a design is possible). Label each chamber with an identifier. Place an appropriate volume of sediment (as described in C.3) at bottom of each test chamber. A total volume of about 60 L sediment per site (before sieving) is sufficient for testing in 40-L aquaria. Adjust volume to reflect test chamber size and number of animals used per replicate. Choose a test temperature between 12°C and 16°C, but maintain within ±2°C of the chosen temperature.

b. Performing tests: Set up test chambers as described above. After equilibration, add mollusks to each tank. Thirty clams will be well supported by 5 to 8 cm of sediment in a 40-L aquarium with overlying seawater. This number of animals is sufficient to ensure adequate tissue for the chemical analyses typically associated with this type of testing. This assay is often used to expose more than one organism type simultaneously. Should additional animals be included in the analysis, ensure adequate sediment and water for their health. After adding animals, record initial measurements of temperature, dissolved oxygen, pH, and salinity, and confirm that values are within the appropriate range for testing. Remove a water sample from each tank and hold for ammonia analysis. The duration of exposure for determining bioaccumulation is 28 d with an additional 2-d exposure to clean sediment for tissue depuration. See Section 8610B.2*b*2) for organism information.

Examine each replicate daily and monitor for mortality. Remove dead organisms immediately to preserve water quality. All dead clams will not necessarily be accounted for because of tissue degradation and scavenging by remaining clams. Note removal of dead organisms on water quality observation sheet. Record daily monitoring of salinity, dissolved oxygen, pH, and temperature on a water quality data sheet. On the day of initiation, every fifth day during the test period, and at test termination, remove a subsample from one replicate of each sediment type (i.e., site, control, or reference) for ammonia analysis. Adjust flow rate of the water if dissolved oxygen falls below 4 mg/L.

On Day 28 of the assay, transfer the animals to clean seawater to depurate for 48 h. Label equivalent size test chambers with the same identifiers and place control sediment in the bottom to the same level as test sediment. Sieve contents of each test chamber with clean at the test seawater temperature and transfer surviving organisms into control sediment. In preparation for test termination, label food-grade resealable plastic bags with species name and replicate identifiers. After depuration period, sieve contents of each tank at the test seawater temperature and transfer animals to the corresponding labeled bag. Freeze tissue samples for transport to site of chemical analyses.

6. Statistical Analysis

Assemble, analyze, evaluate, and report data as described in Section 8010G.

7. Reference

1. U.S. Environmental Protection Agency & U.S. Army Corps of Engineers. 1991. Evaluation of Dredged Material Proposed for Ocean Disposal. Testing Manual. EPA-503/8-91-001, U.S. Environmental Protection Agency, Off. Research & Development, Washington, D.C.

8. Bibliography

U.S. Environmental Protection Agency & U.S. Army Corps of Engineers. 1994. Evaluation of Dredged Material Proposed for Discharge in Waters of the United States. Testing Manual. EPA-832/B-94-002, U.S. Environmental Protection Agency, Off. Research & Development, Washington, D.C..

American Society for Testing and Materials. 1996. Standard Guide for Conducting Bioconcentration Tests with Fishes and Saltwater Bivalve Mollusks. E 724-94, Annual Book of ASTM Standards, Vol. 11.05. American Soc. Testing & Materials, W. Conshohocken, Pa.

8710 ARTHROPODS

Plylum Arthropoda is the largest group of animals; it comprises more than one million species, the majority of which are insects. Other arthropods include crustaceans, spiders, ticks, mites, and other less-known species. Arthropods are found in all environments, including both fresh and marine waters. Two classes of arthropods are used extensively in toxicity testing, the crustaceans and the insects. Test procedures are described for several different crustacean groups including *Daphnia* (8711), *Ceriodaphnia* (8712), mysids (8714), and decapods (8740). Representatives of the insect orders belonging to stoneflies, mayflies, caddisflies, and dipterans are the most commonly used groups in aquatic testing (8750).

8711 *DAPHNIA**

8711 A. Introduction

Daphnia sp. (Figure 8711:1) are small freshwater crustaceans. They have been used for many years to assess the acute and chronic effects of single chemicals and complex mixtures.[1]

Daphnia are valuable as test organisms because of their sensitivity to toxic substances, ease of identification and handling, ubiquitous distribution, and extensive use in toxicity testing. *Daphnia* are fecund and reproduce parthenogenically, which allows for the establishment of clones with little genetic variability and with reproducible testing results.

1. Life History

D. pulex attains a maximum length of approximately 3.5 mm, whereas *D. magna* is much larger and attains a length of 5 to 6 mm. These species are differentiated with certainty only by determining the size and number of spines on the postabdominal claws when using a dissecting or compound microscope (see Figures 8711:2 and 3).[2]

The life span of *Daphnia*, from the release of the egg into the brood chamber until adult death, is highly variable and depends

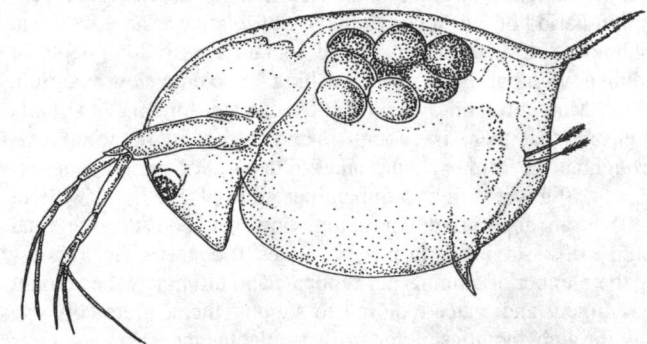

Figure 8711:1. *Daphnia* sp., adult female.

on species and environmental conditions.[2] Generally, it increases as temperature decreases. The average life span of *D. magna* is about 40 d at 25°C and about 56 d at 20°C. The average life span of *D. pulex* at 20°C is approximately 50 d. Four distinct life-cycle periods are recognized: egg, juvenile, adolescent, and adult. The adolescent period is a single instar between the last juvenile instar

_____
* Approved by Standard Methods Committee, 1997.

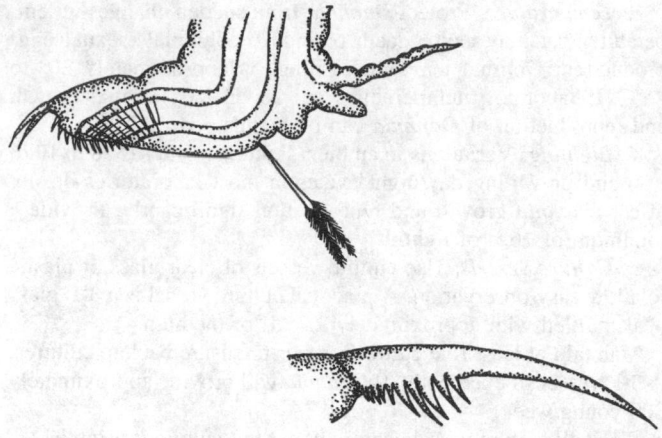

Figure 8711.2. *Daphnia pulex:* (above) postabdomen; (below) postabdominal claw.

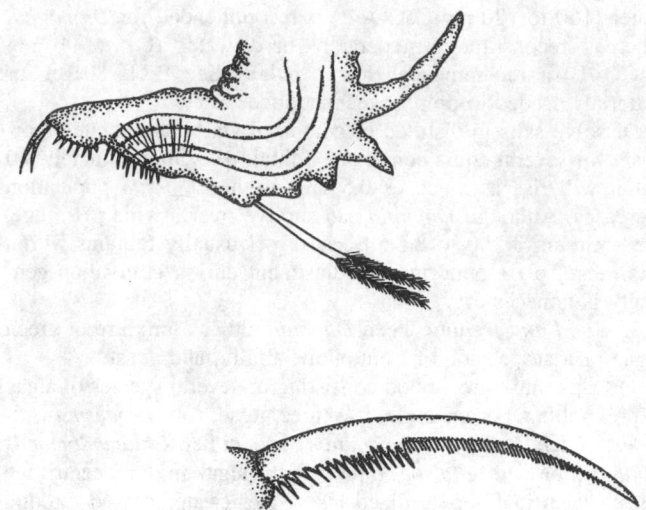

Figure 8711.3. *Daphnia magna:* (above) postabdomen; (below) postabdominal claw.

and the first adult instar; during this instar the first clutch of eggs reaches full development in the ovary. Under laboratory conditions, a clutch of 6 to 10 eggs (15 to 20 eggs in older animals) typically is released into the brood chamber. The eggs hatch and the juveniles, already similar in form to the adults, are released in approximately 2 d when the female molts. The time required to reach sexual maturity varies from 6 to 10 d and appears to depend on temperature. The growth rate is greatest during juvenile stages (early instars); body size may double during each of these stages. *D. pulex* has three to four juvenile instars, whereas *D. magna* has three to five juvenile instars. Each instar stage is terminated by a molt. Growth occurs immediately after each molt while the new carapace is still elastic.

Populations of *Daphnia* consist almost exclusively of females during most of the year; males are abundant only in spring or autumn. For most of the year reproduction is parthenogenic, and only females produce young. Males are distinguished from females by their smaller size, larger antennules, modified postabdomen, and first legs having a stout hook used in clasping. Production of males appears to be induced principally by high population densities and subsequent accumulations of excretory products and/or a decrease in available food. These conditions, along with exposure to temperature extremes, may induce the appearance of sexual (resting) eggs in cases (ephippia) that are cast off during the next molt. The shift towards male and sexual egg production appears related to the metabolic rate of the parent.

As a rule, males and ephippia will not be observed unless stock cultures are neglected or the culture experiences stress.

2. References

1. WEBER, C.I., ed. 1991. Methods for Measuring the Acute Toxicity of Effluents to Freshwater and Marine Organisms, 4th ed. EPA-600/4-90-027, Environmental Monitoring Systems Lab., U.S. Environmental Protection Agency, Cincinnati, Ohio.
2. PENNAK, R.W. 1989. Freshwater Invertebrates of the United States, 3rd ed. John Wiley & Sons, New York, N.Y.

3. Bibliography

ANDERSON, B.G. & L.J. ZUPANCIC, JR. 1937. Growth and variability in *Daphnia pulex. Biol. Bull.* 73:444.
ANDERSON, B.G. & J.C. JENKINS. 1942. A time study of the events in the life span of *Daphnia pulex. Biol. Bull.* 83:260.
ADEMA, D.M.M. 1978. *Daphnia magna* as a test animal in acute and chronic toxicity tests. *Hydrobiologia* 59:125.
DOMA, S. 1979. Ephippia of *Daphnia magna* Straus—A technique for their mass production and quick revival. *Hydrobiologia* 67:183.
CARVALHO, G.R. & R.N. HUGHES. 1983. The effect of food availability, female culture-density and photoperiod on ephippia *Daphnia magna* Straus (Crustacea:Cladocera). *Freshwater Biol.* 13:37.

8711 B.   Selecting and Preparing Test Organisms

1. Obtaining and Selecting Test Species

Daphnia are widely available from many laboratories and commercial biological supply houses. Only 20 to 30 organisms are needed to start a culture. Some biologists prefer *D. pulex* to *D. magna* because it is more widely distributed and easier to culture. However, *D. magna* neonates (first instar) are larger and somewhat easier to use. Verify species used.

2. Culturing Organisms

a. Water supply: Although *Daphnia* cultures can be maintained successfully in some natural waters, preferably use a synthetic (reconstituted) water medium. Reconstituted water is easily prepared, is of known standardized quality, produces predictable results, and permits adequate growth and reproduction. Because daphniads are very sensitive to media hardness, reconstituted hard water (160 to 180 mg CaCO₃/L) is recommended for *D. magna*, whereas reconstituted moderately hard water (80 to 90 mg CaCO₃/L) is recommended for *D. pulex*.[1] See Table 8010:I for materials needed to prepare reconstituted water.

Dissolve salts in distilled or deionized water and aerate vigorously for several hours before use. Initial pH is approximately 8.0 but it will rise as much as 0.5 unit as the *Daphnia* population increases. Although *Daphnia* can survive over a wide pH range, the optimum is 7.0 to 8.6.[2] Because pH usually remains within this range, pH monitoring or adjustment during cultivation generally is unnecessary.

b. Food and feeding: Feed *Daphnia* either a mixture of green algae or a suspension of trout chow, alfalfa, and yeast.

1) Algae mixture—Food consisting of several species of algae is preferable.[3] For example, use three algae, *Ankistrodesmus falcatus, Selenastrum capricornutum,* and either *Chlamydomonas reinhardi* or *Chlorella* sp. To prepare the algal mixture, centrifuge algae, wash in filter-sterilized lake water (water passed through 0.22-μm filter), and centrifuge again. Transfer *Daphnia* to fresh culture water and feed using a sterile pasteur pipet by adding to *Daphnia* ≤ 9 to 10 d old, 2 drops of each alga per *Daphnia* culture beaker or to *Daphnia* 9 to 10 d old, 1 drop of each alga per 2 adults, rounding up when there is an odd number of adults.

At the end of a work week (e.g., Friday) add 1 extra drop of each alga per *Daphnia* culture beaker. Adjust algae feed so that the algae are almost cleared before *Daphnia* are transferred to fresh culture beakers. If only 2 of the 3 algae are available, add proportionately more of the two algae.

2) Trout chow suspension—Place 6.3 g trout chow pellets,* 2.6 g dried yeast,† and 0.5 g dried alfalfa‡ in a blender jar. Add 500 mL deionized water and mix at high speed for 5 min. Let settle in a refrigerator for 1 h. Decant and save top 300 mL; discard remainder. Freeze 30- to 50-mL portions in small (50- to 100-mL) polyethylene bottles with screw caps. Thaw portions as needed. After thawing, refrigerate but do not hold for longer than 1 week.

Feed 1.5 mL prepared food per 1000 mL of medium, three times per week. There may be excess food at this rate of feeding, but if the medium is aerated continuously and replaced each week, no problems should result.

3) *Selenastrum capricornutum*—The green alga *Selenastrum capricornutum* (Printz) can be used as a *Daphnia* food source.[4] Combinations of other green algae are also suitable [see three-algae mixture, ¶ 1) above]. The *Selenastrum capricornutum* culture procedure produces 7-d-old cultures containing four to five million algal cells per mL and 2- to 4-d-old cultures containing one to three million cells per mL. Prepare algal food and feed it three times per week to *Daphnia* as follows:

Combine volumes of 7-d-old and 3-d-old algal cultures in a ratio of two volumes to one, respectively. Centrifuge algal cells and resuspend in a volume of reconstituted moderately hard or hard water calculated to yield a combined algal culture containing approximately ten million cells/mL. Add sufficient volume of cell suspension to stock cultures daily to provide approximately 300 000 algal cells/mL of culture, e.g., add approximately 30 mL cell suspension to 1 L *Daphnia* stock culture.

c. Temperature: Protect *Daphnia* from sudden changes in temperature that may cause death or induce ephippial (sexual egg) production. Optimal temperature range is approximately 20° to 25°C. If laboratory temperatures are 20 ± 2°C, normal growth and reproduction of *Daphnia* can be maintained.

d. Lighting: Variations in ambient light intensities (538 to 1076 lux) and prevailing day/night cycles in most laboratories do not affect *Daphnia* growth and reproduction significantly. Provide a minimum of 16 h of light/d.

e. Culture vessels: Use culture vessels of clear glass or plastic to allow easy observation. A practical culture vessel is a 3-L glass beaker filled with approximately 2.75 L of medium.

Maintain at least five culture vessels to ensure backup cultures. A 3-L vessel stocked with 30 *Daphnia* will provide approximately 300 young/week.

Wash all culture vessels before use. After culture is established, clean each chamber weekly with distilled or deionized water and wipe with a clean sponge to remove accumulated food and dead *Daphnia*. Monthly, wash each vessel with detergent during medium replacement. After washing, rinse three times with tap water and then with culture medium to remove all traces of detergent.

f. Air supply: *Daphnia* can survive when the dissolved oxygen concentration is as low as 3 mg/L but grow better when the concentration is above 6 mg/L. Gently but continuously aerate each culture vessel using an aquarium air pump or a general laboratory compressed air supply (oil-free).

g. Culture maintenance: Replace medium in each stock culture vessel weekly. If large tanks (100 L) are used, weekly replacement may be unnecessary.

Cull *Daphnia* populations weekly to about 30 adults per stock vessel to prevent overcrowding, preferably during medium replacement. Transfer *Daphnia* with large-bore (15-mm-diam) glass pipet (with fire-polished end) or disposable plastic pipet.

3. Selecting Test Organisms

Use *D. magna* or *D. pulex* neonates (first instar ≤ 24 h old), preferably from the second or third brood, to initiate tests. To obtain young for a test, remove females bearing embryos from the stock cultures 24 h before starting the test and place them in 400-mL beakers containing 300 mL medium and either 0.5 mL trout chow-yeast-alfalfa suspension (see B.2*b*) or 10 mL cultured algae. Use the young found in the beakers within 24 h. Five beakers, each containing 10 adults, usually will supply enough first instars for one toxicity test.

Because the appearance of ephippia is indicative of unfavorable conditions, do not use *Daphnia* from cultures producing ephippia.

4. References

1. WEBER, C.I., ed. 1991. Methods for Measuring the Acute Toxicity of Effluents to Freshwater and Marine Organisms, 4th ed. EPA-600/4-90-027, Environmental Monitoring Systems Lab., U.S. Environmental Protection Agency, Cincinnati, Ohio.
2. LEWIS, P.A. & C.I. WEBER. 1985. A study of the reliability of *Daphnia* acute toxicity tests. *In* R.D. Cardwell, R. Purdy & R.C. Bahner, eds., Seventh Symposium on Aquatic Toxicology and Hazard Assessment. ASTM STP 854, American Soc. Testing & Materials, Philadelphia, Pa.
3. GOULDEN, C.E., R.M. COMOTTO, J.A. HENDRICKSON, JR., L.L. HORNIG & K.L. JOHNSON. 1982. Procedure and recommendations for the culture and use of *Daphnia* in bioassay studies. *In* J.G. Pearson, R.B.

---

* Conforming to U.S. Fish and Wildlife Service Specification PR(11)-78; obtainable at livestock feed stores.
† Fleishmann's or equivalent.
‡ Obtainable at most health food stores.

Foster & W.E. Bishop, eds. Aquatic Toxicology and Hazard Assessment. Fifth Symposium on Aquatic Toxicology. ASTM STP 766, American Soc. Testing & Materials, Philadelphia, Pa.

4. MILLER, W.E., J.C. GREENE & T. SHIROYAMA. 1978. The *Selenastrum capricornutum* Printz, Algal Assay Bottle Test. EPA-600/9-78-018, U.S. Environmental Protection Agency, Environmental Research Lab., Corvallis, Ore.

5. Bibliography

DAVIS, P. & G.W. OZBURN. 1969. The pH tolerance of *Daphnia pulex* (Leydig, emend., Richard). *Can. J. Zool.* 47:1173.

BIESINGER, K.E. & G.M. CHRISTENSEN. 1972. Effects of various metals on survival, growth, reproduction, and metabolism of *Daphnia magna. J. Fish. Res. Board Can.* 29:1691.

WINNER, R.W., T. KEELING, R. YEAGER & M.P. FARRELL. 1977. Effect of food type on the acute chronic toxicity of copper to *Daphnia magna. Freshwater Biol.* 7:343.

TENBERGE, W.F. 1978. Breeding *Daphnia magna. Hydrobiologia* 59:121.

PARENT, S. & R.D. CHEETHAM. 1980. Effects of acid precipitation on *Daphnia magna. Bull. Environ. Contam. Toxicol.* 25:298.

SCHULTZ, T.W., S.R. FREEMAN & N.N. DUMONT. 1980. Uptake, depuration and distribution of selenium in Daphnia and its effects on survival and ultrastructure. *Arch. Environ. Contam. Toxicol.* 9:23.

PUCKE, S.C. 1981. Development and standardization of *Daphnia* culturing and bioassays. M.S. thesis, Univ. Cincinnati, Cincinnati, Ohio.

LEONHARD, S.L. & S.C. LAWRENCE. 1981. *Daphnia magna* (Straus), *Daphnia pulex* (Leydig) Richard. *In* S.G. Lawrence, ed. Manual for the Culture of Selected Freshwater Invertebrates. *Can. Spec. Publ. Fish. Aquat. Sci.* 54:31.

GOPHEN, M. & B. GOLD. 1981. The use of inorganic substances to stimulate gut evacuation in *Daphnia magna. Hydrobiologia* 80:43.

HAVAS, M. 1981. Physiological response of aquatic animals to low pH. *In* R. Singer, ed. Effects of Acidic Precipitation on Benthos. North American Benthological Soc., Springfield, Ill.

ALIBONE, M.R. & P. FAIR. 1981. The effects of low pH on the respiration of *Daphnia magna* Straus. *Hydrobiologia* 85:185.

FRANCE, R.L. 1982. Comment on *Daphnia* respiration in low pH water. *Hydrobiologia* 94:195.

WALTON, W.E., S.M. COMPTON, J.D. ALLAN & R.E. DANIELS. 1982. The effect of acid stress on survivorship and reproduction of *Daphnia pulex* (Crustacea:Cladocera). *Can. J. Zool.* 60:573.

8711 C. Procedures

1. Short-Term Tests

a. Preparation of test materials and medium: Prepare test materials and concentrations, dilution water, and toxicant solutions as described in Section 8010F. Make up test solutions and controls in 100-mL quantities in 125-mL wide-mouth flint-glass bottles or equivalent vessels.

b. Performing tests: After preparing test solutions, segregate neonates that have been released from the mothers' brood chambers during the preceding 24 h at 20°C or 25°C and collect in one vessel (use neonates cultured at the test temperature). Introduce the same number of neonates (at least 10) into each test vessel and control. Use a plastic, disposable pipet with a 5-mm bore for collecting and transferring neonates. Alternatively, use a glass bulb pipet.

Introduce neonates to test solutions by releasing them below the surface of the solution. Observe animals regularly, ideally after 1 h and 4 h and daily thereafter. A 48-h exposure is generally accepted for a *Daphnia* acute toxicity test.[1] Record number of motile animals in each test vessel. Consider an animal nonmotile if it shows no independent movement even after gentle squirting with test solution from a pipet (nonmotile animals are not necessarily dead). At threshold concentrations of such substances as ethanol, acetone, and chlorobutanol, animals may show no movement and the heart may have ceased to beat but on transfer to dilution water they will recover. However, such animals maintained in the test medium will die. In addition to immobilization, note behaviors and features such as the number of *Daphnia* that are on bottom, lethargic, swimming, caught on the bottom or on debris, floating on surface, swimming erratically, or have a flared carapace.

Record conditions of the medium such as whether it is cloudy or if any particulate matter, precipitate, undissolved material, or film is present. Continue observations for a minimum of 48 h or as long as there is no more than 10% control mortality. Run tests in replicates of at least three.

Do not feed animals during tests. Longer-term tests require modifications of standard conditions.

c. Criteria for test acceptability: An acceptable test will have no more than 10% mortality in the negative control. A clear dose-response must be apparent from a sample plot of the data. Ideally one concentration will have no effect (no more mortality than control) and the EC50 will be bracketed by concentrations producing mortality. Use a scatterplot of the 48 h data to identify outliers. Examine records for clerical or experimental errors when outliers exist.

2. Long-Term Tests

a. Determination of toxicity effect(s) on survival, growth, and reproduction: Sublethal effects may occur at lower toxicant concentrations than those causing acute toxicity. Precede long-term tests by acute (48-h) toxicity tests to establish the maximum concentration to be used.

b. Preparation of test medium: Prepare test medium as regular culture medium, but use water representative of that receiving the effluent discharge, or the dilution water used to culture the daphnids when testing chemicals. Prepare a series of 6 to 10 1-L quantities of medium to which graded amounts of effluents, mixtures, or chemicals have been added. Use as the highest concentration of chemical or effluent the equivalent to the 48-h LC50 or EC50 values. Reduce each successive concentration in a consistent progression (e.g., geometric). Use test dilution water as a negative control. Dispense each liter of test medium in 100-mL quantities to each of 10 glass or plastic chambers. Run tests in replicates of at least six (the minimum needed to detect statistical significance).

c. Performing tests: Preferably conduct test according to Good Laboratory Practice standards/regulations.[2–4] Segregate and collect 24-h-old neonate *Daphnia* that have been cultured at the test temperature. Introduce one neonate into each chamber randomly. On the following day and on alternate days thereafter, add an appropriate amount of food (see Section 8711B.2b). First-generation Daphnia (those animals used to begin the test) may be transferred to new media as necessary, but at least three times weekly. Make daily observations and note dead or immobilized animals. As animals grow and reproduce, remove young and record their number. Cover all test chambers loosely with plate glass or equivalent to minimize evaporation. Continue test for 21 d at 25°C. If desired, continue observations until a set number of broods, e.g., six, are reached in control animals; this may take 30 d at 20°C. Handle animals as in the individual cultures of stock animals. This test design may not be appropriate for highly volatile chemicals because of possible evaporation.

At the end of the exposure period, analyze results and test for statistically significant differences in the number of young produced, first-generation Daphnia survival, and, if appropriate, the dry weights of surviving animals in each treatment. Note appearance of first broods and number of broods.

d. Criteria for test acceptability: A test is invalid if control mortality exceeds 20% during exposure. The average brood production of the controls should be 60 ± 10 young over the duration of the test. If ephippia are produced in the controls, the test is invalid.

3. Statistical Analysis

Assemble, analyze, evaluate, and report data as described in Section 8010G.

4. Quality Assurance/Quality Control

Quality assurance (QA) practices for hazardous-waste toxicity tests consist of all aspects of the test that affect data quality,[5-7] including sampling and handling, source and condition of test organisms, condition of equipment, test conditions, instrument calibration, replication, use of reference toxicants, record keeping, and data evaluation.

Prepare a control chart for the reference toxicant. Plot and examine successive toxicity values (LC50) to determine whether the results are within prescribed limits. In this technique, a running plot is maintained for the toxicity values from successive tests with reference toxicant.

Run reference toxicant tests periodically. Suggested reference toxicants are $CdCl_2$ and sodium dodecyl sulfate. If the LC50 from a given test with the reference toxicant does not fall in the expected range for Daphnia, the sensitivity of the organisms and the overall credibility of the test system are suspect. In this case, examine test procedure for defects and repeat with a different batch of Daphnia.

5. References

1. WEBER, C.I., ed. 1991. Methods for Measuring the Acute Toxicity of Effluents to Freshwater and Marine Organisms, 4th ed. EPA-60/4-90-027, Environmental Monitoring Systems Lab., U.S. Environmental Protection Agency, Cincinnati, Ohio.
2. U.S. FOOD AND DRUG ADMINISTRATION. 1987. Good Laboratory Practice (GLP) Regulations for Nonclinical Laboratory Studies. 21 CFR, Part 58.
3. ORGANIZATION FOR ECONOMIC COOPERATION AND DEVELOPMENT. 1989. Principles of good laboratory practice, Annex 2, OECD guidelines for testing of chemicals C(81) 30(Final). *Offic. J. Europ. Commun.* 32(315):1.
4. GARNER, W.Y. & M.S. BARGE, eds. 1988. Good Laboratory Practices—An Agrochemical Perspective. ACS Symp. Ser. 369, American Chemical Soc., Washington, D.C.
5. GREENE, J.C., C.L. BARTELS, W.J. WARREN-HICKS, B.R. PARKHURST, G.L. LINDER, S.A. PETERSON & W.E. MILLER. 1988. Protocols for short term toxicity screening of hazardous waste sites. EPA-600/3-88-029. Environmental Research Lab., Corvallis, Ore.
6. U.S. ENVIRONMENTAL PROTECTION AGENCY. 1979. Good laboratory practice standards for health effects. Paragraph 772.1110-1, Part 772. Standards for development of test data. *Federal Register* 44:27362.
7. U.S. ENVIRONMENTAL PROTECTION AGENCY. 1996. Good laboratory practice standards. 40 CFR 792.

6. Bibliography

AMERICAN SOCIETY FOR TESTING AND MATERIALS. 1984. Guide for Conducting Acute Toxicity Tests with Fish, Macroinvertebrates and Amphibians. E729. American Soc. Testing & Materials, Philadelphia, Pa.

BIESINGER, K.E., L.R. WILLIAMS, H.W. VAN DER SCHALIE. 1987. Procedures for Conducting *Daphnia pulex* toxicity bioassays. EPA-600/8-87-011. Environmental Monitoring Systems Lab., U.S. Environmental Protection Agency, Las Vegas, Nev.

DEWOSKIN, R.S. 1984. Good laboratory practice regulations: a comparison. Research Triangle Inst., Research Triangle Park, N.C.

GERSICH, F.M. & D.P. MILLAZO. 1990. Evaluations of a 14-day static renewal toxicity test with *Daphnia magna* Straus. *Arch. Environ. Contam. Toxicol.* 19:72.

GOULDEN, C.E. & L.L. HENRY. 1990. *Ceriodaphnia* and *Daphnia* Bioassay Workshop Manual. Acad. Natural Sciences of Philadelphia, Philadelphia, Pa.

MASTERS, J.A., M.A. LEWIS & D.H. DAVIDSON. 1991. Validation of a four-day *Ceriodaphnia* toxicity test and statistical considerations in data analysis. *Environ. Toxicol. Chem.* 10:47.

PORCELLA, D.B. 1983. Protocol for bioassessment of hazardous waste sites. EPA-600/2-83-054, U.S. Environmental Protection Agency, Corvallis, Ore.

WEBER, C.I., W.H. PELTIER, T.J. NORBERG-KING, W.B. HORNING, F.A. KESSLER, J.R. MENKEDICK, T.W. NEIHEISEL, P.A. LEWIS, D.J. KLEMM, Q.H. PICKERING, E.L. ROBINSON, J.M. LAZORCHAK, L.J. WYMER & R.W. FREYBERG. 1989. Short-Term Methods for Estimating the Chronic Toxicity of Effluents and Receiving Waters for Freshwater Organisms, 2nd ed. EPA-600/4-89-001, Environmental Monitoring Systems Lab., U.S. Environmental Protection Agency, Cincinnati, Ohio.

WEBER, C.I., ed. 1993. Methods for Measuring the Acute Toxicity of Effluents and Receiving Waters to Freshwater and Marine Organisms, 4th ed. EPA-600/4-90-027F, Environmental Monitoring Systems Lab, U.S. Environmental Protection Agency, Cincinnati, Ohio.

8712   *CERIODAPHNIA**

8712 A.   Introduction

Ceriodaphnia is a genus of cladocerans that are smaller in size than their closely related and morphologically similar counterpart, *Daphnia*. *Ceriodaphnia* produce three to four broods per week under optimal conditions, whereas *Daphnia*, because of their larger size, do not reproduce until the fourth to sixth instar stage after hatching.[1,2] *Ceriodaphnia* were first used in effluent toxicity evaluations in 1984.[3]

1. Life History

Ceriodaphnia have a life history similar to those of other daphnids and are believed to occur in limnetic areas all over the world.[2] *Ceriodaphnia* are pond- and lake-dwelling species that are usually common among the vegetation in littoral areas. The life span of *Ceriodaphnia*, from release of the egg into the brood chamber until death, is variable and depends on temperature as well as other environmental conditions. As with *Daphnia*, *Ceriodaphnia* life span usually is related to temperature; at 25°C and 20°C, the average life span for *Ceriodaphnia dubia* is 30 d and 50 d, respectively. The increase in life span at lower water temperatures is attributed to a lower metabolic activity.

Currently, no distinct developmental stages are recognized in the life cycle of *Ceriodaphnia*. The organism is referred to as a neonate during its first instar stage (when it is still less than 24 h old). The time to sexual maturity for *Ceriodaphnia dubia* varies from 3 to 5 d and probably depends on body size and environmental conditions, particularly temperature, water quality, and food availability. Typically, a clutch of 4 to 10 eggs is released into the brood chamber, but clutches with as many as 20 eggs occur. The eggs hatch in the brood chamber and the neonates are released in about 38 h, just before the adult female molts. The growth rate of the organism is greatest during the early instar stages and body size may double during each instar. Each instar stage is terminated by a molt. Growth occurs immediately after each molt while the new carapace is still elastic.

---

* Approved by Standard Methods Committee, 1997.

Ceriodaphnia populations consist almost exclusively of parthenogenic females during most of the year. Males appear primarily in the autumn and in the late spring. The factors responsible for the appearance of males are not fully understood.[1-2] Production of male eggs has been attributed partially to overcrowding of females, a decrease in available food, and a decrease in water temperature.[1-2] If continued, these same conditions appear to induce the production of sexual eggs. Gametogenic females are morphologically similar to parthenogenic females; however, sexual females produce only a few "resting eggs" and can copulate with males. When the fertilized eggs of these females pass into the brood chamber, the walls of the brood chamber become dark and thick and form an ephippium. Ephippia are embryos encased in a tough covering and are resistant to drying. The development of ephippia among cladocerans is an adaptation to adverse environmental conditions and allows populations to survive both drought and freezing conditions.

2. References

1. WEBER, C.I., ed. 1991. Methods for Measuring the Acute Toxicity of Effluents to Freshwater and Marine Organisms, 4th ed. EPA-600/4-90-027, Environmental Systems Lab., U.S. Environmental Protection Agency, Cincinnati, Ohio.
2. PENNAK, R.W. 1989. Freshwater Invertebrates of the United States, 3rd ed. John Wiley & Sons, New York, N.Y.
3. MOUNT, D.I. & T.J. NORBERG. 1984. A seven-day life-cycle cladoceran toxicity test. *Environ. Toxicol. Chem.* 3:425.

3. Bibliography

DEGRAEVE, G.M. & J.D. COONEY. 1987. *Ceriodaphnia:* An update on effluent toxicity testing and research needs. *Environ. Toxicol. Chem.* 6:331.
U.S. ENVIRONMENTAL PROTECTION AGENCY. 1984. Development of water-quality based permit limitations for toxic pollutants: National policy. *Federal Register* 49:9016.

8712 B.   Selecting and Preparing Test Organisms

1. Obtaining and Selecting Test Species

Ceriodaphnia are available from many laboratories and commercial biological supply houses. A culture can be started with 10 to 20 organisms. Start cultures of test organisms at least 2 weeks before neonates will be needed for testing to ensure an adequate supply.

Only *Ceriodaphnia dubia* is now used in effluent toxicity tests. *Ceriodaphnia dubia*, toothed-pecten variety, is considered a morphological variant of *Ceriodaphnia dubia*. The distinguishing

morphological characteristics between *Ceriodaphnia dubia* and *Ceriodaphnia dubia*, toothed-pecten variety, are shown in Figures 8712.1 through 3. Verify species of *Ceriodaphnia* before using.[1]

2. Culturing Organisms

a. Water supply and renewal: Culture water for *Ceriodaphnia* can be either an acceptable surface or well water source or a synthetic (reconstituted) water medium. Synthetic water usually is recommended because it is of known standardized quality, is

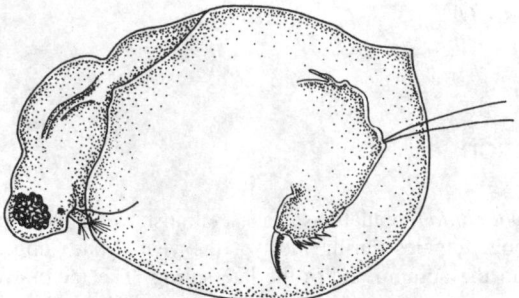

Figure 8712:1. *Ceriodaphnia dubia.*

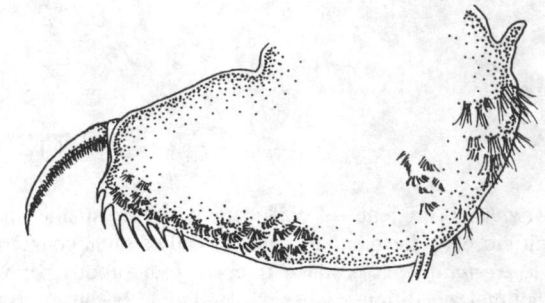

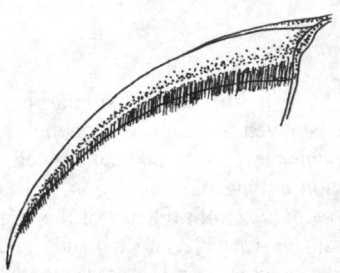

Figure 8712:2. *Ceriodaphnia dubia:* (above) postabdomen; (below) postabdominal claw.

easily prepared, and yields reproducible results. Ceriodaphnids are believed to perform better in moderately hard (80 to 100 mg $CaCO_3$/L total hardness) synthetic waters than in soft synthetic water (30 to 50 mg $CaCO_3$/L).

Prepare soft reconstituted water with 50-g/L stock solutions of $NaHCO_3$, $MgSO_4$, and KCl to provide 48.0 mg $NaHCO_3$/L, 30.0 mg $MgSO_4$/L, and 2.0 mg KCl/L. Add $CaSO_4 \cdot 2H_2O$ in powdered form to provide a concentration of 30.0 mg/L. Double these concentrations for preparing moderately hard reconstituted water. Use reagent-grade chemicals. See Section 8710 for further details. Aerate water for 24 to 48 h before use. Pass air through a water filter containing granular activated charcoal. After confirmation of water quality parameters (hardness, pH, etc.), add selenium, as sodium selenite, and vitamin B-12 to the water to obtain a concentration of 3 μg/L of each.[1]

Reconstituted waters have been associated with culture and bioassay performance problems due to unidentified and toxic components in the deionized water used to prepare the media[2] or due to a nutritional deficiency caused by the absence of a full complement of trace elements (synthetic waters contain four salts).[3] Carbon-treated distilled water is best for preparing synthetic culture waters.

Renewal frequency also has been shown to have an effect on the survival and reproductive success of *C. dubia*. Daily renewals of culture medium produced the best survival and reproduction results, while triweekly renewal conditions gave poor results and relatively higher coefficient of variations for reproduction.[4]

b. Food and feeding: Cladocerans are filter feeders that are most abundant in eutrophic lakes during summer phytoplankton blooms. Examination of *Ceriodaphnia* gut contents reveals algae, bacteria, and detritus.[1] *Ceriodaphnia* can be fed a combination food (known as YCT) consisting of yeast, dried cereal grass,* and commercially available flaked fish food† or trout chow. Supplement the diet with a 50:50 mixture of a green algal suspension of *Selenastrum capricornutum* and YCT.[5,6] *Selenastrum capricornutum* cultures can be established and maintained by following the algal culture procedures in 8111F.

Prepare food (YCT) for *Ceriodaphnia* as follows: Place 5.0 g of fish food flakes† or trout chow pellets in a blender containing 1 L synthetic water. Blend at high speed for 5 min. Transfer contents of blender to a 1-L separatory funnel and aerate continuously through spigot opening for 1 week at ambient laboratory temperature. Place mixture in a refrigerator and let settle for 1 h. Pour supernatant into a clean bottle and discard remainder.

Place 5.0 g dry yeast in 1 L synthetic water. Place in a blender at low speed for 5 min. Transfer to a bottle and refrigerate for 4 h. Mix well before combining for YCT.

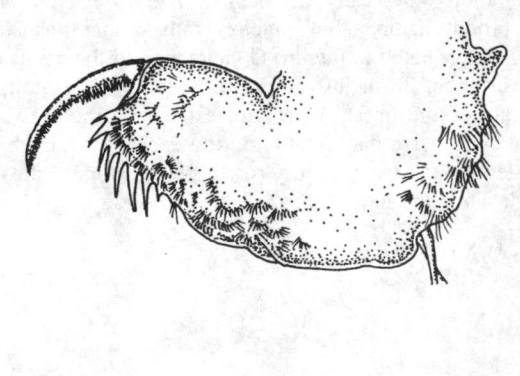

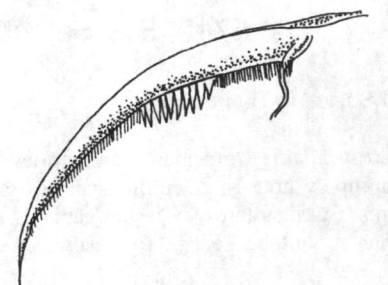

Figure 8712:3. *Ceriodaphnia dubia,* **toothed-pecten variety:** (above) postabdomen; (below) postabdominal claw.

---

* Such as Cerophyl®, Agri-Tech, Inc., Kansas City, MO 64112, or equivalent.
† TetraMin® or equivalent.

Place 5.0 g dried cereal grass‡ in 1 L synthetic water in a blender. Mix at high speed for 5 min. Transfer to a bottle, refrigerate, and let settle for 4 h. Pour supernatant into clean bottle and discard the remaining solids.

Mix 300-mL volumes of each of the three components above. Filter mixture through a nylon§ 110-μm-mesh filter. Determine dry weight (i.e., total solids) concentration on each batch of YCT mixture. The food should contain 1700 to 1900 mg total solids/L. Adjust level of total solids by dilution with synthetic water if YCT dry weight is greater than 1900 mg/L. Place 30- to 50-mL portions in small polyethylene bottles with screw caps and freeze. Thaw portions as needed. Keep refrigerated and discard unused portions after 2 weeks.

Feed *Ceriodaphnia* mass cultures daily at the rate of 4 mL food/L medium. Feed individual cultures or test chambers daily at the rate of 0.1 mL YCT/d and 0.1 mL algae/d. The quality of each new batch of *Ceriodaphnia* food can be determined in a 7-d reproduction test with control water. Culture grid records also can be used to evaluate food quality.[1]

c. Temperature: Sudden changes of several degrees in temperature may cause death of *Ceriodaphnia*. Optimal temperature range is approximately 25 ± 2°C. Maintain cultures within this temperature range.

d. Lighting: Variations in ambient light intensities and prevailing day/night cycles in most laboratories do not affect *Ceriodaphnia* reproduction significantly. A light intensity of 550 to 1050 lux and a photoperiod of 16 h light and 8 h dark is recommended.

e. Culture vessels and maintenance: Culture individual adults in 30-mL plastic cups containing 15 to 20 mL culture medium. Feed organisms daily and transfer them to new culture media in new plastic cups at least three times per week; more frequent transfers are desirable. Construct styrofoam (or similar rigid material) culture boards consisting of slots that hold the 30-mL plastic cups. Maintain a minimum of four boards, containing 20 individuals per board, to ensure backup cultures. Individual cultures of *Ceriodaphnia* are used as the immediate source of neonates for toxicity tests. Do not use culture boards exceeding 20% adult mortality to supply neonates used in toxicity tests. A healthy culture board will supply neonates for approximately 14 d.

f. Air supply: *Ceriodaphnia* can survive when the dissolved oxygen concentration is as low as 3 mg/L but reproduce better when the concentration is above 6 mg/L. As long as the culture water source is properly aerated, individual culture boards of *Ceriodaphnia* do not require aeration. Maintain DO levels within criteria for warmwater aquatic life.

3. Selecting Test Organisms

Use first instar neonate *Ceriodaphnia* less than 24 h old from a brood of eight or more from a female on her third or fourth brood. To obtain neonates for a test, transfer adult females to a new culture vessel containing fresh culture solution with a disposable, wide-mouth (approximately 4 mm) pipet. Keep tip of pipet under water surface when the *Ceriodaphnia* are released to prevent air from being trapped under the organism's carapace. Group cups of neonates together until enough are available to initiate a test.

4. References

1. WEBER, C.I., ed. 1991. Methods for Measuring the Acute Toxicity of Effluents to Freshwater and Marine Organisms, 4th ed. EPA-600/4-90-027, Environmental Systems Lab., U.S. Environmental Protection Agency, Cincinnati, Ohio.
2. DEGRAEVE, G.M. & J.D. COONEY. 1987. *Ceriodaphnia:* An update on effluent toxicity testing and research needs. *Environ. Toxicol. Chem.* 6:331.
3. KEATING, K.I., P.B. CAFFREY & K.A. SCHULTZ. 1989. Inherent problems in reconstituted water. *In* U.M. Cowgill & L.R. Williams, eds. Aquatic Toxicology and Hazard Assessment: 12th Volume. ASTM STP 1027, American Soc. Testing & Materials, Philadelphia, Pa.
4. FERRARI, B. & J.F. FERARD. 1996. Effects of nutritional renewal frequency on survival and reproduction of *Ceriodaphnia dubia. Environ. Toxicol. Chem.* 15:765.
5. COONEY, J.D., G.M. DEGRAEVE, E.L. MOORE, W.D. PALMER & T.L. POLLOCK. 1992. Effects of food and water quality on culturing of *Ceriodaphnia dubia. Environ. Toxicol. Chem.* 11:823.
6. PATTERSON, P.W., K.L. DICKSON, W.T. WALLER & J.H. RODGERS, JR. 1992. The effect of nine diet and water combinations on the culture health of *Ceriodaphnia dubia. Environ. Toxicol. Chem.* 11:1023.

5. Bibliography

KEATING, K.I. & B.C. DAGBUSAN. 1984. Effect of selenium deficiency on cuticle integrity in the cladocera (Crustacea). *Proc. Nat. Acad. Sci.* 81:3433.
KNIGHT, J.T. & W.T. WALLER. 1992. Influence of the addition of Cerophyl® on the *Selenastrum capricornutum* diet of the cladoceran *Ceriodaphnia dubia. Environ. Toxicol. Chem.* 11:521.

---

‡ Cerophyl® or equivalent.
§ Nitex® or equivalent.

8712 C. Procedures

1. Acute Toxicity Tests

Prepare test materials and concentrations, dilution water, and toxicant solutions as described in Section 8010F. Make up test solutions and controls in 100-mL quantities in 125-mL wide-mouth flint-glass bottles or equivalent vessels.

After preparing test solutions, segregate neonates that have been released from the mothers' brood chambers during the preceding 24 h at 25°C ± 2°C, and collect in one vessel. Use neonates cultured at the test temperature. At test initiation, organisms should be less than 24 h old, although in some circumstances, organisms less than 48 h old are acceptable. Introduce five neonates per replicate into each test solution vessel and control exposure.[1] Use a minimum of four replicates with each test concentration. Use a plastic, disposable pipet with a 4-mm bore for collecting and transferring neonates.

Introduce neonates to test solutions by releasing them below the surface of the solution. Observe animals regularly, initially after 1 h and 4 h and daily thereafter. A 48-h exposure is generally accepted for a *Ceriodaphnia* acute toxicity test.[1] However, depending on the study objectives, tests may be conducted for 24, 48, or 96 h. If the test is conducted for 96 h, renew test solutions and feed organisms after 48 h. At each 24-h interval during the test, record either the number of surviving animals or the number of motile animals in each test vessel. Consider an animal nonmotile if it shows no independent movement even after gentle squirting with test solution from a pipet (nonmotile animals are not necessarily dead).

If lethality is selected as the test endpoint, an LC50 value can be generated. If immobility is selected as the test endpoint, an EC50 value can be generated. The LC50 value represents the concentration of test solution or chemical at which 50% of the animals have died. Similarly, the EC50 value represents the concentration of test solution or chemical at which 50% of the animals exposed are immobilized. Test results usually are considered acceptable as long as there is no more than 10% control mortality. Do not feed *Ceriodaphnia* during most short-term tests; however, feed animals a minimum of 2 h before test initiation.

2. Short-Term Chronic Toxicity Tests

a. Determination of toxicity effect(s) on survival and reproduction: Sublethal effects may occur at lower toxicant concentrations than those causing acute toxicity. Precede long-term tests by acute (48-h) toxicity tests to establish the maximum concentration to be used.

b. Preparation of test medium: Prepare by methods used for preparing regular culture medium, but use water representative of that receiving the waste effluent, or the dilution water used to culture *C. dubia* when testing chemicals. Prepare a series of six to ten 250-mL quantities of medium to which log concentrations of effluents, mixtures, or chemicals have been added. Use as the highest concentration of chemical or effluent, the equivalent to the 48-h LC50 or EC50 values. Reduce each successive concentration in a consistent progression (e.g., geometric). Tests conducted with effluent generally use a 0.5 serial dilution such that the effluent concentrations tested are 6.25%, 12.5%, 25%, 50%, and 100% effluent. Use test dilution water as a negative control (i.e., 0% effluent). Dispense each 250 mL test medium in 15-mL quantities to each of ten glass or plastic 30-mL chambers. Preferably run tests in replicates of ten; a minimum of six is required for statistical significance.

c. Performing tests: Preferably conduct test according to Good Laboratory Practice standards/regulations.[2-5] Segregate and collect <24-h-old neonate *Ceriodaphnia* that have been cultured at the test temperature. Introduce one neonate into each chamber of the test and array randomly. At 24-h intervals, transfer the first-generation organisms (i.e., *Ceriodaphnia* used to begin the test) into fresh test solutions and add an appropriate amount of food to each test chamber (see Section 8711B.2b). Under some chronic test protocols, *Ceriodaphnia* may be transferred to new test media less frequently; however, feed organisms daily. Make daily observations and note number of dead animals. As animals grow and reproduce, remove young and record their number. Cover all test chambers loosely with plate glass or equivalent to minimize evaporation but not affect oxygen transfer. Continue test for 7 to 8 d at 25°C. Handle test animals as in the individual cultures of stock animals.

At end of exposure period, analyze results and test for statistically significant differences in the number of young produced, first-generation *Ceriodaphnia* survival, and reproduction (i.e., number of neonates produced) in each treatment. Note appearance of first broods and number of broods. Generally, chronic toxicity tests are considered acceptable if control mortality is less than 20%, surviving control females produce an average of 15 neonates per adult, and 60% of control females produce three broods or more.[6] The coefficient of variation (CV) for reproduction between replicates in the test control should be equal to or less than 20%. Determine CV by dividing the standard deviation for reproduction by the average number of young per surviving female and multiplying by 100. A CV greater than 20% indicates that conditions between replicates were substantially different. Substantial differences among control replicates may interfere with statistical detection of substantial differences between treatment exposures.

3. Criteria for Test Acceptability

An acceptable test will have no more than 10% mortality in the negative control. A clear dose-response must be apparent from a sample plot of the data. Ideally one concentration will have no effect (no more mortality than control) and the EC50 will be bracketed by concentrations producing mortality. Use a scatter plot of the 48-h data to identify outliers. Examine records for clerical or experimental error when outliers exist.

4. Statistical Analysis

Assemble, analyze, evaluate, and report data as described in Section 8010G.

5. References

1. WEBER, C.I., ed. 1991. Methods for Measuring the Acute Toxicity of Effluents for Freshwater and Marine Organisms, 4th ed. EPA-600/4-90-027, Environmental Monitoring Systems Lab., U.S. Environmental Protection Agency, Cincinnati, Ohio.

2. Food and Drug Administration. 1987. Good Laboratory Practice (GLP) Regulations for Nonclinical Laboratory Studies. 21 CFR, Part 58.

3. U.S. Environmental Protection Agency. 1989. Good Laboratory Practice Standards. 40 CFR, Parts 160 and 792.

4. Organization for Economic Cooperation and Development. 1989. Principles of good laboratory practice, Annex 2, OECD guidelines for testing of chemicals C(81) 30(Final). *Offic. J. Europ. Commun.* 32(315):1.

5. Garner, W.Y. & M.S. Barge, eds. 1988. Good Laboratory Practice—An Agrochemical Perspective. ACS Symp. Ser. 369, American Chemical Soc., Washington, D.C.

6. U.S. Environmental Protection Agency. 1994. Short-Term Methods for Estimating the Chronic Toxicity of Effluents and Receiving Waters to Freshwater Organisms, 3rd ed. EPA-600/4-91-002, Environmental Monitoring Systems Lab., U.S. Environmental Protection Agency, Cincinnati, Ohio.

6. Bibliography

Stephan, C.E. & J.W. Rogers. 1985. Advantages of using regression analysis to calculate results of chronic toxicity tests. *In* R.C. Bahner & D.J. Hansen, eds. Aquatic Toxicity and Hazard Assessment. 8th Symp. Aquatic Toxicology. ASTM STP 891, American Soc. Testing & Materials, Philadelphia, Pa.

Stephen, C.E. 1989. Topics on expressing and predicting results of life cycle tests. *In* G.W. Sutter II & M.A. Lewis, eds. Aquatic Toxicology and Environmental Fate. 11th Symp. Aquatic Toxicology. ASTM STP 1007, American Soc. Testing & Materials, Philadelphia, Pa.

Masters, J.A., M.A. Lewis & D.H. Davidson. 1991. Validation of a four-day *Ceriodaphnia* toxicity test and statistical considerations in data analysis. *Environ. Toxicol. Chem.* 10:47.

DeGraeve, G.M., J.D. Cooney, B.H. Marsh, T.L. Pollock & N.G. Reichenbach. 1992. Variability in the performance of the 7-d *Ceriodaphnia dubia* survival and reproduction test: An intra- and interlaboratory study. *Environ. Toxicol. Chem.* 11:851.

Cooney, J.D., G.M. DeGraeve, E.L. Moore, B.J. Lenoble, T.L. Pollock & G.J. Smith. 1992. Effects of environmental and experimental design factors on culturing and toxicity testing of *Ceriodaphnia dubia*. *Environ. Toxicol. Chem.* 11:839.

8714 MYSIDS*

8714 A. Introduction

1. Suitability for Toxicity Tests

Mysids are an important component of both the pelagic and epibenthic communities. They are preyed upon by many species of fish, birds, and larger invertebrate species, and they are predators of smaller crustaceans and larval stages of invertebrates. In some cases, they feed on algae. Mysids are sensitive to both organic and inorganic toxicants. The ecological importance of mysids, their wide geographical distribution, ability to be cultured in the laboratory, and sensitivity to contaminants make them appropriate toxicity test organisms.[1-8] Juvenile mysids used in these tests are taken from cultures shortly after release from the brood and exposed to varying concentrations of a toxicant in static or flow-through conditions. These procedures will be useful for conducting toxicity tests with other species of mysids, although modifications may be necessary. The tests are applicable to most chemicals, either individually or in formulations, commercial products, and known or unknown mixtures, and with appropriate modifications, can be used to conduct tests on factors such as temperature, salinity, and dissolved oxygen. These methods also can be used to assess the toxicity of potentially toxic discharges such as municipal wastes, oil drilling fluids, produced water from oil well production, and other types of industrial wastes.

___
* Approved by Standard Methods Committee, 1997.

2. References

1. Brandt, O., R. Fujimura & B. Finlayson. 1993. Use of *Neomysis mercedis* (Crustacea:Mysidacea) for estuarine toxicity tests. *Trans. Amer. Fish. Soc.* 122:279.

2. Gentile, S., J. Gentile, J. Walter & J. Heltshe. 1982. Chronic effects of cadmium on two species of mysid shrimp: *Mysidopsis bahia* and *Mysidopsis bigelowi*. *Hydrobiologia* 93:195.

3. Nimmo, D. & E. Iley, Jr. 1982. Culturing and chronic toxicity of *Mysidopsis bahia* using artificial seawater. Pub. PA902, U.S. Environmental Protection Agency, Gulf Breeze, Fla.

4. Lussier, S., J. Gentile & J. Walker. 1985. Acute and chronic effects of heavy metals and cyanide on *Mysidopsis bahia* (Crustacea:Mysidacea). *Aquat. Toxicol.* 7:25.

5. Davidson, B., A. Valkira & P. Seligman. 1986. Acute and chronic effects of tributyltin on the mysid *Acanthomysis sculpta* (Crustacea: Mysidacea). *Proc. Oceans 86* 4:1218.

6. American Society for Testing and Materials Committee E47. 1992. Standard Guide for Static and Flow-Through Acute Toxicity Tests with Mysids from the West Coast of the United States. ASTM E-1463-92, American Soc. Testing & Materials, Philadelphia, Pa.

7. American Society for Testing and Materials Committee E47. 1990. Standard Guide for Conducting Life-Cycle Toxicity Tests with Saltwater Mysids. ASTM E-1191-90, American Soc. Testing & Materials, Philadelphia, Pa.

8. American Society for Testing and Materials Committee E47. 1988. Standard Guide for Conducting Acute Toxicity Tests with Fishes, Macroinvertebrates, and Amphibians. ASTM E-729-88a, American Soc. Testing & Materials, Philadelphia, Pa.

8714 B.   Selecting and Preparing Test Organisms

1. Selection of Test Species

Test species may be designated by a particular regulation (e.g., Federal Insecticide, Fungicide and Rodenticide Act; Toxic Substances Control Act). If the toxicity test is for nonregulatory purposes, any number of species can be tested if the test conditions are suitable for culturing the species. If it is desirable to use mysids that are not cultured routinely, it may be necessary to collect them from a single field source.

Select test species that meet the following criteria: *(a)* The species preferably occurs, or is closely related to a species that occurs, in the receiving water being studied; *(b)* the species is available in unbiased (i.e., not prescreened for resistant individuals by prior exposure to adverse conditions) numbers sufficient for the tests; *(c)* the species can be held in the laboratory in a healthy condition (i.e., active, feeding, free of lesion, etc.); and *(d)* the species represents an important trophic link or economic resource in habitats similar to that of the receiving water. If the data are available when selecting species, consider relative sensitivities of different species and life stages.

In accordance with the criteria listed in Section 8010E.1, the recommended test species include (but are not restricted to) the following:

a. Estuarine and freshwater mysids:
  Neomysis mercedis (Figure 8714:l)
  Mysidopsis almyra[1] [= *Americamysis almyra*]* (Figure 8714:2)
b. Marine mysids:
  Holmesimysis costata[2] [= *Acanthomysis sculpta*]† (Figure 8714:3)

---

* *Mysidopsis almyra, M. bahia,* and *M. bigelowi* are three of six species forming the group *Americamysis,* confined to the northwestern Atlantic and endemic to estuarine waters along the east coasts from New England to Colombia, and now distinguished from the group *Mysidopsis.* To avoid confusion with previously published methods, they are referred to here by genus *Mysidopsis.*
† *Holmesimysis costata* is one of five species of the genus *Holmesimysis* which is present in the North Pacific Ocean. Confusion has existed about genus in which to place the mysid used in toxicity tests in California. Up to 1988 all authors referred to this mysid species as *Acanthomysis sculpta* (Tattersall). All known species of the genus *Acanthomysis* from the Pacific Coast of North America were placed in the new genus *Holmesimysis.*

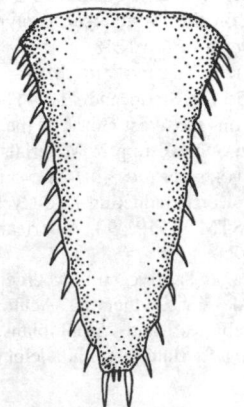

Figure 8714:1. *Neomysis mercedes:* telson.

Mysidopsis bahia[1] [= *Americamysis bahia*]* (Figure 8714:4)

Mysidopsis bigelowi[1] [= *Americamysis bigelowi*]* (Figure 8714:5)

2. Collecting and Handling Test Organisms

Collecting equipment and methods are described in Sections 8010E.2, 10200B, and 10500B. Handling and holding are discussed in Section 8010E.3 and 4. NOTE: Avoid subjecting mysids to unnecessary stress such as inappropriate capture and transport, temperature shock, or water quality change.

a. Estuarine and freshwater mysids: Neomysis mercedis (Figure 8714:1) is a Pacific coast species living in fresh and estuarine water to 18 g/kg salinity estuarine water; its temperature range is 6 to 22°C.[3] *N. mercedis* ranges from Prince William Sound, Alaska, to south of Point Conception, California.

Collect *N. mercedis* by hand dip nets or plankton tows in rivers and estuaries. Collection with a dip net (0.5- to 1.0-mm mesh) at night results in minimal mechanical damage[4] and yields many specimens in good condition and with little accompanying debris. Transfer specimens to a 100-L (30-gal) plastic container filled with site water and transport to the laboratory. Aerate with a portable air pump. Separate *N. mercedis* from other organisms and discard any specimen that is injured or does not appear to be in good condition. Specimens can be picked up by using a bulb pipet with a 5-mm bore or with a plastic spoon. Alternatively, collect organisms by towing a plankton net (0.5-mm mesh) from a boat in open water. This technique can result in high mysid mortality and much accompanying detritus. *N. mercedis* are abundant between February and July but scarce during the remainder of the year.[5]

Mysidopsis almyra [= *Americamysis almyra*][1] (Figure 8714:2) is an East Coast species often living sympatrically with *M. bahia,* but preferring lower salinities ranging from 10 to 20 g/kg at temperatures over 20°C.[6] It is found in inshore waters along the entire coast of the Gulf of Mexico and northward along the Atlantic coast to Patapsco River, Maryland.[1] Collect *M. almyra* with hand dip nets (350-μm mesh) or a 1.5-m beam trawl with a 0.9-mm mesh size pulled by hand in shallow areas of estuaries.[7] The dip net method yields many specimens in good condition and with little accompanying debris. Remove possible predator species, such as ctenophores, immediately. Transfer specimens to an insulated 4-L (1-gal) or larger plastic container filled with site water and transport to the laboratory. Aerate with a portable air pump. Separate *M. almyra* from other organisms and discard any specimen injured or not appearing to be in good condition. Specimens can be picked up by using a bulb pipet with a 5-mm bore. Alternatively, collect organisms by towing a plankton net (350-μm mesh) from a boat in open water at night. However, this technique can result in high mysid mortality and much accompanying detritus. *M. almyra* are abundant throughout the year in the southern latitudes.

b. Marine mysids: Holmesimysis costata [= *Acanthomysis sculpta*][2]† (Figure 8714:3) is the principal species of the genus in

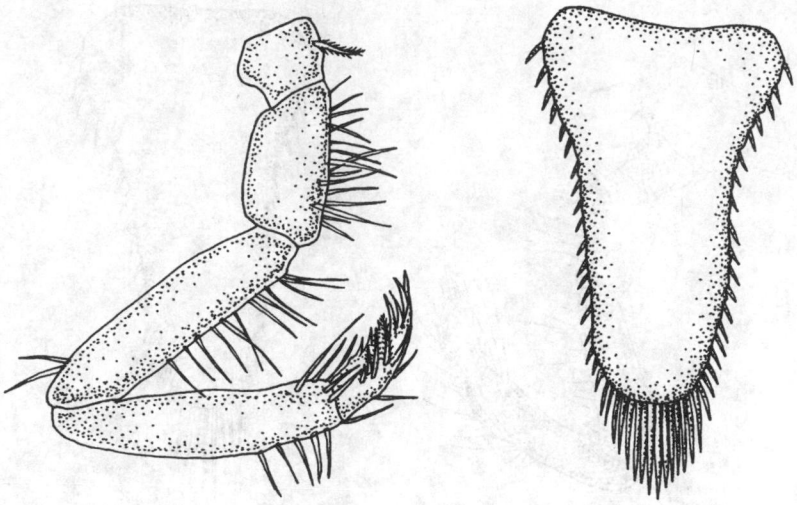

Figure 8714:2. *Mysidopsis almyra:* (left) endopod of thoracic leg 2; (right) telson.

California marine waters. *H. costata* occurs abundantly offshore among the fronds of the giant kelp especially during the summer months.[8] Collect *H. costata* from a boat by passing a hand net (0.5- to 1.0-mm mesh) through the kelp canopy. Transfer specimens to a 20-L (5-gal) bucket filled with seawater and transport to the laboratory. Pour contents of the bucket into one or more pans and separate *H. costata* from the other organisms. Discard any specimen that is injured or does not appear to be in good condition. Some specimens might be parasitized externally by a marine leech; do not use these specimens or place them in the

laboratory stock colony. Mysids can be picked up by using a bulb pipet with a 5-mm diam.

Mysidopsis bahia [= *Americamysis bahia*][1] (Figure 8714:4), often sympatric with *M. bigelowi* and *M. almyra*, at temperatures over 20°C,[6] but preferring higher salinities ranging from 10 to 30 g/kg,[9] is found in inshore waters along the coast of the Gulf of Mexico, and northward along the Atlantic coast to Narragansett, R.I.[1] *M. bigelowi* [= *Americamysis bigelowi*][1] (Figure 8714:5), is found on the Atlantic coast from Massachusetts (Georges Bank) southward to Florida. It often occurs sympatrically with *M. bahia*,

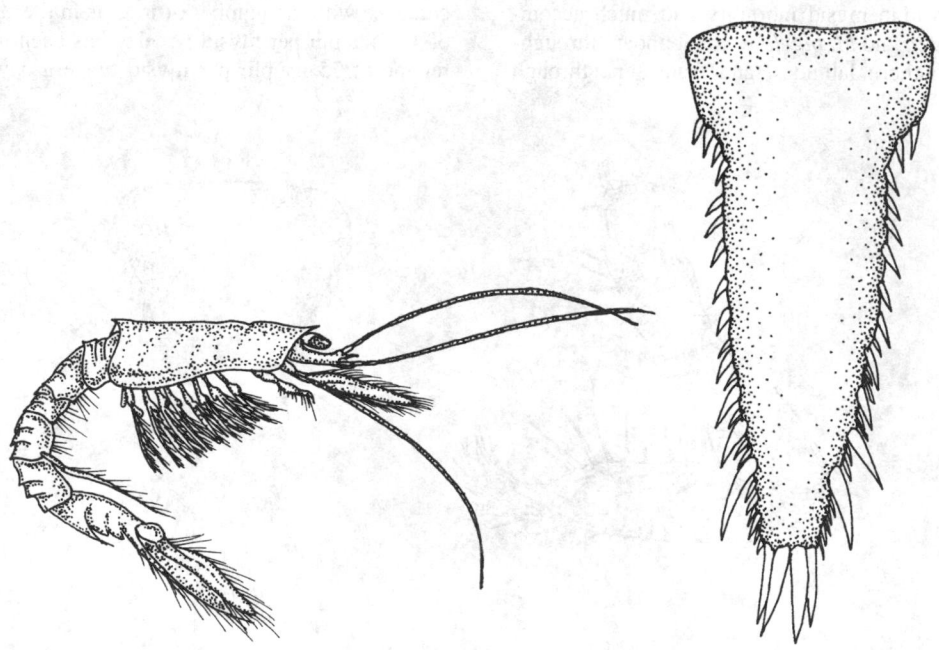

Figure 8714:3. *Holmesimysis costata:* (left) entire animal; (right) telson.

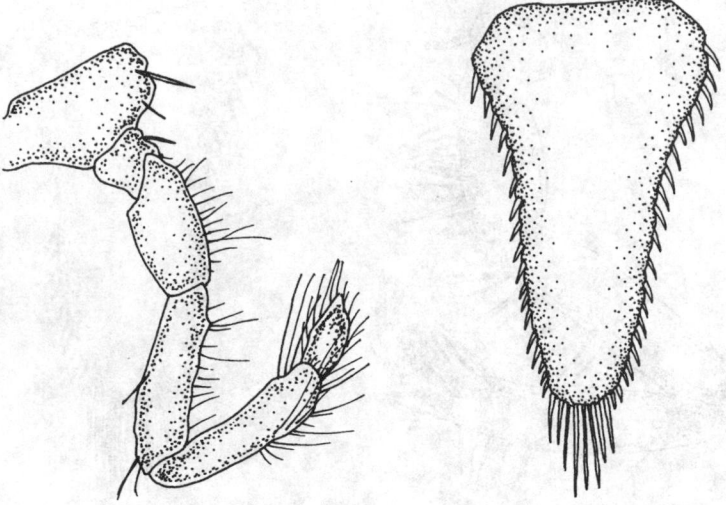

Figure 8714:4. *Mysidopsis bahia:* (left) endopod of thoracic leg 2; (right) telson.

with a salinity range from 30 to 35 g/kg[9] and in water temperatures from 2 to 30°C.

Collect *M. bahia* and *M. bigelowi* by using hand dip nets (350-μm mesh) in shallow areas of salt ponds and estuaries.[10] This method yields many specimens in good condition and with little accompanying debris. Remove possible predator species, such as ctenophores, immediately. Transfer specimens to an insulated 4-L (1-gal) or larger plastic container filled with site water and transport to the laboratory. Aerate with a portable air pump. Separate *M. bahia* and *M. bigelowi* from the other organisms and discard any specimen injured or not appearing to be in good condition. Specimens can be picked up by using a bulb pipet with a 5-mm bore. Alternatively, collect by towing a plankton net (350-μm mesh) from a boat in open water at night. However, this technique can result in high mysid mortality and much accompanying detritus. *M. bahia* and *M. bigelowi* are abundant throughout the year in the southern latitudes, and from June through September in temperate latitudes in shallow water with a temperature above 20°C.

3. Holding, Acclimating, and Culturing Organisms

See Sections 8010E.3 and 4.

Keep mysids in tanks, aquariums, or screened enclosure depending on size and number. Use good quality dilution water (see Section 8010E.4*b*) for acclimation. Feed mysids live brine shrimp nauplii daily during acclimation. At least once daily, feed live brine shrimp nauplii in excess to mysids in brood stock tanks and in test chambers, to maintain live nauplii in the chambers at all times to prevent cannibalism and to support adequate survival, growth, and reproduction in the brood stock. Adjust ration in accordance with the number of mysids in the stock colony. A ration of 150 nauplii per mysid per day has been used successfully.[8] A regime of 75 nauplii per mysid twice a day or 50 nauplii three

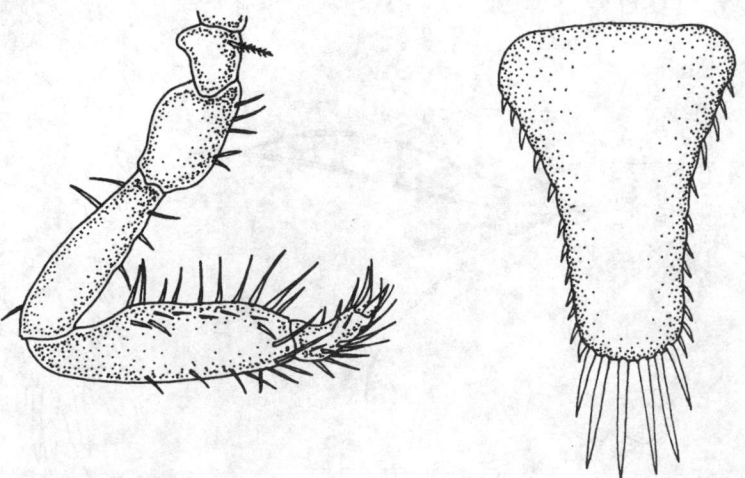

Figure 8714:5. *Mysidopsis bigelowi:* (left) endopod of thoracic leg 2; (right) telson.

times a day might improve growth and reproduction in the brood stock. Use diets that are certified toxicant-free or test for toxic substances before use.

a. Estuarine and freshwater mysids:

1) *N. mercedis*—Culture in either a static or a flow-through system. For static system, use 75- to 114-L aquariums supplied with aeration and a subsurface filter of dolomite 3 to 5 cm in thickness. *N. mercedis* is extremely sensitive to nitrogenous wastes; clean aquariums daily to remove excess food. For a flow-through system, supply sufficient water for a minimum of two tank volumes per day. Successful cultures have been maintained at a temperature of between 15 and 19°C (optimum 17°C), hard fresh water (150 to 200 mg CaCO₃/L, hardness and alkalinity), and additional natural seawater or reconstituted seawater (see Section 8010E.4b) to salinity of 1 to 3 g/kg (optimum 2 g/kg).[4,8] Feed *Artemia salina* nauplii[11] to mysids three times a day at the rate of 50 nauplii/mysid/feeding (a total of 150 nauplii/ mysid/d) and add an artificial food supplement containing vitamins and minerals (0.02 to 0.06 mg/mysid) every other day. Preferably supplement diet with commercially available food‡ having micronutrients and vitamins; supplements of rotifers and algae also may be beneficial.

Under laboratory conditions and water temperatures of 15 to 19°C, a life cycle for *N. mercedis* is completed in approximately 3 months. A gravid female will carry an average of 20 embryos in a brood, of which an average of seven will be released.[4]

To collect young *N. mercedis* for acute static or flow-through toxicity testing, place females carrying embryos, which are in the eye-development stage, in brood chambers, 7 to 14 d before starting the test. For brood chambers, use cages covered with 0.25- to 0.50-mm nylon mesh, which allows the neonates to escape into the main body of the aquarium but retains the adults. Remove neonates each day from the aquarium with a fine (0.5-mm) mesh dip net and transfer to a dish. Remove healthy specimens for testing. Pool the young released over a 2- to 3-d period and transfer them to a holding vessel until a sufficient number are obtained for a test.

2) *M. almyra*—Preferably culture in a flow-through system supplied with sufficient water for a minimum of two tank volumes per day for best control of salinity and nitrogenous wastes.[9] Alternatively, culture in static 76-L (20-gal) or larger aquariums supplied with aeration and a subsurface filter of dolomite 3 to 5 cm deep. Successful cultures have been maintained at a temperature of 25 to 27°C (optimum 26°C), in natural seawater or reconstituted seawater (see Section 8010E.4b) at salinity of 10 to 20 g/kg. Feed *Artemia* nauplii[11] to mysids twice a day at the rate of 75 nauplii/mysid/feeding (a total of 150 nauplii/mysid/d). Preferably supplement *Artemia* cultures with an enhancement product§ to ensure the amino acid content of the nauplii.[11] Supplementing the diet of *M. almyra* with rotifers and algae also may be beneficial. There is very little information regarding the life cycle and brood size for *M. almyra*, but growth, respiration, and energetics studies have been conducted.[7]

Juveniles for acute static or flow-through toxicity testing can be collected in several ways, depending on the frequency of tests and number of animals needed. If juveniles of the same age are required intermittently, place females carrying embryos, which are in the eye-development stage, in brood chambers or aerated finger bowls, 1 d before starting the test. Use as brood chambers either 4-L beakers

with netted (1-mm) bottoms placed within wide-mouth separatory funnels, or cages covered with 1-mm mesh‖ placed in an aquarium allowing the neonates to escape into the main body of the aquarium but retaining the adults. Remove neonates the next day from the separatory funnel, aquarium, or bowl with a fine (0.5-mm) mesh dip net and transfer to a dish. Remove healthy specimens for testing.[9] When age-standardized mysid juveniles are not required, place adults collected from cultures into large mesh containers within separate aquariums, allowing adults to be lifted out into new aquariums every few days and leaving juveniles to be siphoned or netted out as needed. Pool young released over a 2- to 3-d period and transfer to a holding vessel until sufficient number are obtained for a test. When juveniles are needed almost daily for testing, use a siphon entrapment system, or "mysid generator." In this system, juveniles are continuously siphoned out of an aquarium (adults are excluded by a 750-µm screen) into a collection vessel. Juveniles are collected daily and can be used either for testing or for starting up new culture tanks.[12]

b. Marine mysids:

1) *H. costata*—The organisms can be picked up with a bulb pipet having a 5-mm diam. For acclimation, place *H. costata* in an aquarium with aeration at a density of approximately 10 to 20 specimens/L seawater. Change the water if it becomes cloudy. Animals can be cultured in the laboratory on a diet of *Artemia* nauplii larvae, powdered fish flake food, and fresh fronds of the giant kelp (*Macrocystis*). Preferably add a fresh, carefully washed frond of the giant kelp *Macrocystis* to the brood stock to provide additional substrate and food for mysids.

H. costata can complete three or four life cycles a year under laboratory conditions. Females will produce more than one brood set under laboratory conditions. To obtain young mysids, place adult *H. costata* in a cage within an aquarium. Use a cage covered with nylon screening with a 0.25-mm mesh, which allows the newborn to escape into the main body of the aquarium but retains the adults. Remove newborn from the aquarium with a fine dip net or glass pipet and transfer to a dish where specimens can be observed and removed for testing.

2) *M. bahia* and *M. bigelowi*—Follow culturing instructions for *M. almyra* [¶ 3a2) above], but use salinity of 25 g/kg.

Under laboratory conditions with water temperatures of 25°C a life cycle of *M. bahia* and *M. bigelowi* is completed in 1 month or less, at which time the first eggs are laid. A gravid female will carry an average of eight embryos in a brood, all of which are normally released as healthy postlarvae; the first brood is released after 14 to 18 d and succeeding broods are released every 7 d. Productivity gradually declines in the last third of the 3- to 5-month life span.[9]

Collect juveniles for acute static or flow-through toxicity testing as directed in ¶ 3a2) above.

4. Parasites and Diseases

For general problems and control procedures see Sections 8010E.5.

Unexpected and often unexplained mortalities in experimental and control animals interfere with test results and interpretations. Optimize laboratory conditions for each species to prevent the development of disease. Maintain salinity and temperature appro-

---

‡ TetraMarin® or equivalent.
§ SELCO or equivalent.

‖ Nytex® or equivalent.

priate for particular species and consistent with specified test conditions.

Reproduction will be depressed when culture density is high. This phenomenon has not occurred when cultures are maintained at densities of 10 mysids/L or less. Therefore, when cultures are not being used for supplying test organisms, remove enough adults at least every 2 weeks to stimulate reproduction. Preferably keep neonates, juveniles, and adults of mysids in separate tanks. Keep brood stock tanks free of other animals, such as amphipods, hydroids, and worms. If an outbreak of these animals or others occurs, remove all mysids and clean tank thoroughly. Examine mysids thoroughly before replacement, and discard any having hydroids attached. Clean tanks with hot water and a 5% solution of hydrochloric or nitric acid.[9] Wash, dry, and autoclave substrate (i.e., dolomite or oyster shells), or discard it.

Handle mysids as little as possible. When handling is necessary, proceed gently, carefully, and quickly to reduce stress. Dip nets are best for removing gravid female mysids from brood-stock tanks. Such nets are commercially available or can be made from 350-μm mesh nylon netting, silk bolting cloth, plankton netting, or similar knotless material. Discard mysids that touch dry surfaces or are dropped or injured. Sterilize equipment used to handle mysids between uses by autoclaving. Wash new equipment with detergent and rinse with water, a water-miscible organic solvent, water, acid (such as 10% conc HCl), and at least twice with deionized, distilled, or dilution water. At the end of a test, clean all equipment by a procedure appropriate for removing the test material (e.g., acid to remove metals and bases; detergent, organic solvent, or activated carbon to remove organic chemicals), and rinse at least twice with deionized, distilled, or dilution water.[13]

5. References

1. PRICE, W., R. HEARD & L. STUCK. 1994. Observations on the genus *Mysidopsis* Sars, 1864 with the designation of a new genus, *Americamysis*, and the descriptions of *Americamysis alleni* and *A. stucki* (Peracarida:Mysidacea:Mysidae), from the Gulf of Mexico. *Proc. Biol. Soc. Washington* 107:680.
2. HOLMQUIST, C. 1979. *Mysis costata* Holmes, 1900, and its relations (Crustacea:Mysidacea). *Zool. Jahrb. Abteil. System., Oekol. Geogr. Tierre* 106:471.
3. SIMMONS, M. & A. KNIGHT. 1975. Respiratory response of *Neomysis intermideia* (Crustacea:Mysidacea) to changes in salinity, temperature, and season. *Comp. Biochem. Physiol.* 50A:181.
4. BRANDT, O., R. FUJIMURA & B. FINLAYSON. 1993. Use of *Neomysis mercedis* (Crustacea:Mysidacea) for estuarine toxicity tests. *Trans. Amer. Fish. Soc.* 122:279.
5. ORSI, J. & A. KNUTSON. 1979. An extension of the known range of *Neomysis mercedis*, the opossum shrimp. *Calif. Fish Game* 65:127.
6. PRICE, W. 1976. The Abundance and Distribution of Mysidacea in the Shallow Waters of Galveston Island, Texas. Ph.D. thesis, Texas A&M Univ., College Station.
7. REITSEMA, L. 1981. The Growth, Respiration and Energetics of *Mysidopsis almyra* (Crustacea:Mysidacea) in Relation to Temperature, Salinity, and Hydrocarbon Exposure. Ph.D. thesis, Texas A&M Univ., College Station.
8. AMERICAN SOCIETY FOR TESTING AND MATERIALS COMMITTEE E47. 1992. Standard Guide for Static and Flow-Through Acute Toxicity Tests with Mysids from the West Coast of the United States. ASTM E-1463-92, American Soc. Testing & Materials, Philadelphia, Pa.
9. LUSSIER, S., A. KUHN, M. CHAMMAS & J. SEWALL. 1988. Techniques for the laboratory culture of *Mysidopsis* species (Crustacea:Mysidacea). *Environ. Toxicol. Chem.* 7:969.
10. LUSSIER, S., A. KUHN, M. CHAMMAS & AND J. SEWALL. 1991. Life history and toxicological comparisons of temperate and subtropical mysids. *Amer. Fish. Soc. Symp.* 9:169.
11. AMERICAN SOCIETY FOR TESTING AND MATERIALS COMMITTEE E47. 1992. Practice for Using Brine Shrimp Nauplii as Food for Test Animals in Aquatic Toxicity Tests. ASTM E-1203-92, American Soc. Testing & Materials, Philadelphia, Pa.
12. REITSEMA, L. & J. NEFF. 1980. A recirculating artificial seawater system for the laboratory culture of *Mysidopsis almyra* (Crustacea: Pericaridea). *Estuaries* 3:321.
13. AMERICAN SOCIETY FOR TESTING AND MATERIALS COMMITTEE E47. 1990. Standard Guide for Conducting Life-Cycle Toxicity Tests with Saltwater Mysids. ASTM E-1191-90, American Soc. Testing & Materials, Philadelphia, Pa.

6. Bibliography

BANNER, A.H. 1948. A taxonomic study of the mysidacea and the euphausiacea of the Northwestern Pacific. II. Mysidacea from the Tribe Mysini through Subfamily Mysidellinae. *Trans. Royal Can. Inst.* 27: 47.

TATTERSALL, W.M. 1951. A Review of the Mysidacea of the United States National Museum. Bull. U.S. Nat. Mus. 201.

RUCKER, R.R. & K. HODGEBOOM. 1953. Observations on gas-bubble disease of fish. *Progr. Fish-Cult.* 15:24.

BANNER, A.H. 1954. A Supplement to W.M. Tattersall's Review of the Mysidacea of the United States National Museum. *Proc. U.S. Nat. Mus.* 103:525.

BOWMAN, T.E. 1964. *Mysidopsis almyra*, a new estuarine mysid crustacean from Louisiana and Florida. *Tulane Stud. Zool.* 12:15.

HEUBACH, W. 1969. *Neomysis awatschensis* in the Sacramento-San Joaquin River Estuary. *Limnol. Oceanogr.* 14:533.

MOLENOCK, J. 1969. *Mysidopsis bahia*, A new species of mysid (Crustacea:Mysidacea) from Galveston Bay, Texas. *Tulane Stud. Zool. Bot.* 15:113.

WIGLEY, R. & B. BURNS. 1971. Distribution and biology of mysids (Crustacea:Mysidacea) from the Atlantic coast of the United States in the NMFS Woods Hole collection. *Fish. Bull.* 69:717.

MAUCHLINE, J. 1980. Advances in Marine Biology, Vol. 18. Academic Press, New York, N.Y.

HOLMQUIST, C. 1982. Mysidacea (Crustacea) secured during investigations along the West Coast of North America by the National Museums of Canada, 1955-1956, and some inferences drawn from the results. *Zool. Jahrb. Abteil. System., Oekol. Geogr. Tierre* 109:469.

PRICE, W. 1982. Key to the shallow water mysidacea of the Texas coast with notes on their ecology. *Hydrobiologia* 93:9.

ORSI, J.J. & A.C. KNUTSON, JR. 1983. The role of mysid shrimp in the Sacramento-San Joaquin Estuary and factors affecting their abundance and distribution. *In* T.J. Conomos, ed. San Francisco Bay: The Urbanized Estuary. Pacific Div., American Assoc. for the Advancement of Science, San Francisco, Calif.

KATHMAN, R., W. AUSTIN, J. SALTMAN & J. FULTON. 1986. Identification Manual to the Mysidacea and Euphausiacea of the Northeast Pacific. Can. Fish. & Aquat. Sci. Spec. Publ. 93.

8714 C.   Toxicity Test Procedures

1. Short-Term Procedures

a. General test procedures: Short-term testing can be used to determine relative toxicity of substances. Tests are made to determine LC50 or EC50 values and to estimate toxicant concentrations for intermediate- and long-term tests. Basic requirements for toxicity tests are described in Section 8010C.

Short-term tests may be static, static with renewal, or flow-through, depending on the objective of the test and the character of the toxicant or effluent (see Section 8010D). Acute static, static with renewal, or flow-through toxicity tests are conducted preferably with young mysids in accordance with other studies with this group.[1–3]

Collect young mysids of nearly uniform size in accordance with instructions given in Section 8714B.3. *H. costata* used in acute toxicity tests should be 3 to 7 d post-release from the brood sac, and *N. mercedis*, 1 to 5 d post-release. *Mysidopsis* species should be less than 24 h post-release from the brood sac. Use 10 to 20 mysids per toxicant concentration. Transfer mysids to each test chamber with a glass pipet. Feed mysids with brine shrimp larvae three times a day at the rate of 30 nauplii/mysid (a total of 90 nauplii/mysid/d) during the test period. Examine test chambers daily, record mortality, and remove all dead specimens and debris. Generally it is not necessary to consider mysid weight/L test solution given the low weight of mysids, but in flow-through tests use less than 10 g mysids/L test solution for tests at temperatures at or below 17°C and 5 g mysids/L test solution for tests at higher temperatures. For static testing, do not load above 0.8 g mysids/L at 17°C or less and 0.5 g mysids/L at temperatures above 20°C. Limit loading to ensure that concentrations of dissolved oxygen and test material do not fall below acceptable limits, concentrations of metabolic products do not exceed acceptable levels, and test mysids are not stressed because of cannibalism, aggression, or crowding.

b. Specific test procedures:

1) Freshwater and estuarine mysids

a) Equipment and physical conditions—Ensure that equipment and facilities that contact stock solutions, test solutions, or any water into which test organisms will be placed do not contain substances that can be leached or dissolved by aqueous solutions in amounts that adversely affect mysids. General requirements for toxicity test systems and materials are described in Section 8010F.1. In addition, choose equipment and facilities that contact stock or test solutions to minimize sorption of test materials from water. Use glass, Type 316 stainless steel, nylon, and fluorocarbon plastics to minimize dissolution, leaching, and sorption, but do not use stainless steel in tests on metals with salt water. Do not use cast iron pipe with salt water and preferably avoid its use in a fresh water-supply filter system because colloidal iron will be added to the dilution water and strainers will be needed to remove rust particles. Do not let brass, copper, lead, galvanized metal, or natural rubber contact dilution water, stock solutions, or test solutions before or during the test. Avoid items made of neoprene rubber or other materials not mentioned previously unless it has been shown that their use will not adversely affect survival, growth, or reproduction of mysids.[1,2]

Use test material of reagent grade or better unless a test of formulation, commercial product, or technical-grade or use-grade material is specifically needed. Before a test is begun, note the following about the test material: identities and concentrations of major ingredients and major impurities; solubility and stability in dilution water; precision and bias of the analytical method at the planned concentration(s) of the test material; and estimate of toxicity to humans.

Select temperature appropriate for the species being tested, and hold test temperature within ±2°C of mean test temperature during a 96-h test, or ±1°C for any 48-h period. Conduct tests with *Mysidopsis almyra* in a temperature range of 26 to 28°C,[2] and tests with *Neomysis mercedis* at 15 to 19°C.[2] Keep salinity within the tolerance range of the selected species. The optimum salinity for *M. almyra* is 10 to 20 g/kg and for *N. mercedis* is 1 to 3 g/kg.[2] If a test salinity other than the optimum is used, set up an additional control at the optimum salinity. Use ambient laboratory lighting with a photoperiod of 16 h light/8 h dark, preferably with 15- to 30-min dusk/dawn transition period to acclimate mysids to the test photoperiod.

Use dilution water from a surface source, well, or spring, or use reconstituted water (see Section 8010E.4). *N. mercedis* cultures have not been reported for media of reconstituted fresh water. Do not use chlorinated water as, or in the preparation of, dilution water, because chlorine-produced oxidants are toxic to mysids.[3] Establish a supply of dilution water that is available in adequate quantities, acceptable to test organisms, uniform in quality, and not likely to affect test results unnecessarily. An acceptable dilution water is one in which the test species will survive, grow, and reproduce satisfactorily. Maintain uniform quality of the dilution water so that the test organisms are cultured or acclimated, and the test conducted in water of the same quality.[1,2,4]

Use at least two test chambers (in which containers may be placed) for each concentration; these can consist of standard 57-L aquariums or can be constructed by gluing strong window glass with clear silicone adhesive. Because adhesives can sorb some organochlorine or organophosphorus pesticides, apply as little adhesive as possible. Finger bowls can be used for static acute toxicity test containers; for a 2-L bowl, use 20 animals; for a 350-mL bowl, 10 animals; each bowl constitutes a replicate. Flow-through toxicity tests can be conducted in a 2.5-L wide-mouth glass jar with a central standpipe. The test solution enters the compartment directly and flows through the standpipe into a drain. Cover standpipe with a 200- to 235-µm mesh nylon screen to avoid escape of the young mysids.[1,2,4]

For information pertaining to species selection, collection, holding, acclimation, disease control, and culturing see Sections 8010E and 8714B.

b) Test procedure

Range-finding test: If the approximate toxicity of test material is unknown, conduct an abbreviated range-finding test to determine the concentrations to use in definitive tests. Use three to five widely spaced toxicant concentrations (for example, a decade test having concentrations a factor of ten from each other). Static tests may be acceptable as would use of fewer mysids, e.g., five per container. Run this test for 24 to 96 h.

Definitive test: To determine LC50 or EC50 values, use a 96-h test period with a minimum of five toxicant concentrations (according to the results of the range-finding test) and a control. In some cases the test solution can be added directly to the dilution water, but usually it is dissolved in a solvent to form a stock solution that is then added to dilution water. If a stock solution is used, determine the concentration and stability of the test material in it and dilution water before beginning the test. If the test material is subject to photolysis, shield stock solution from light. Use a solvent control if dosing solutions are prepared in an organic solvent. Acceptable solvents are dimethylformamide, ethanol, methanol, acetone, and triethylene glycol (see Section 8010F.2). Limit concentrations of solvent to 0.1 mL/L of test solution. If a solvent other than water is used and the concentration of solvent is the same in all test solutions that contain test material, include at least one solvent control containing the same concentration of solvent and using solvent from the same batch used to make the stock solution, and also include a dilution water control. If a solvent other than water is used and the concentration of solvent is not the same in all test solutions that contain test material, include both a solvent control, containing the highest concentration of solvent present in any other treatment and using solvent from the same batch used to make the stock solution, and a dilution water control. The percentage of organisms that show signs of disease or stress, such as discoloration, unusual behavior, or death, must be 10% or less in the solvent control and in the dilution water control.

To establish definitive test concentrations, prepare solutions using a dilution ratio of 1.5 to 2 between successive concentrations (see Section 8010F.3).

c) *Test initiation*—On day of test, remove a sufficient number of mysids from the holding facility at one time to provide about one-third more animals than are needed. Select a set of test chambers [one test chamber for each test concentration plus control(s)] to be processed together to avoid possible selective bias during loading. Transfer mysids with a wide-bore (larger than the largest mysid) glass pipet with a smooth tip. Begin static tests by placing test organisms in the chambers within 30 min after the test material was added to the dilution water. Begin flow-through tests by placing test organisms in the chambers after the test solutions have been flowing through the chambers long enough for the concentrations of test material to have reached steady state.

d) *Biological observations*—Monitor survival by daily inspection (see Section 8010F.3). The criteria for death of mysids are opaque white coloration, immobility (especially absence of movement of respiratory and feeding appendages), and lack of reaction to gentle prodding. Count, record, and remove dead mysids daily. Count live animals at the beginning of the experiment and daily to account for cannibalism or death resulting from impingement on the sides of test compartments. Record missing or dead impinged animals. Do not stress live test organisms in an attempt to determine whether they are dead, immobilized, or otherwise affected. Prodding of organisms and movement of test chambers during test should be done very gently. Some organisms exposed to some organophosphorus compounds seem to be very sensitive to sudden changes in light intensity.

e) *Chemical data recording*—Analyze water in control and test chambers daily for pH, dissolved oxygen, salinity (for marine or estuarine species), and temperature (see Section 8010F.3d). Maintain DO concentrations at ≥60% saturation. When testing volatile substances, do not aerate test solution. However, take care that chemical substances that create a dissolved oxygen demand do not result in conditions inconsistent with the dissolved oxygen criterion of the test as well as of mysid health. As a last resort to maintain dissolved oxygen above the criterion, use aeration. If aeration is used, make frequent measurements to confirm test chemical concentrations and aerate all test chambers, including controls.

f) *Verification of exposure*—Before and during the test verify exposure concentrations of the test chemical (see Section 8010F.3d). In static and renewal tests, measure the concentration of test material, if possible, in at least the control and high, medium, and low concentrations of test material at the beginning and end of test. Measurement of degradation products may be desirable. Measure concentration of test material in flow-through test chambers as often as practical during test. Measure in all chambers concurrently at least once during test, preferably near the beginning; except for the control treatment, measure each test chamber (especially for those concentrations closest to the LC50) at least one additional time during the test on a schedule designed to give reasonable confidence in the concentration of the material in the test chambers during the entire exposure period, taking into account the flow rate and the number of independent metering devices; and measure at least one appropriate chamber whenever a malfunction is detected in any part of the metering system.

2) *Marine mysids*—The test procedures for marine mysids are essentially identical to those for freshwater and estuarine mysids except for the noted differences in biological and environmental requirements. Conduct tests with *H. costata*[2] at temperatures of 17 ± 2°C for south of Point Conception, California, and 15 ± 2°C for north of Point Conception, and with salinity of 30 to 35 g/kg. *H. costata* cultures have not been reported for media of reconstituted seawater. Conduct tests with *M. bahia* and *M. bigelowi*[1] at temperatures of 27 ± 1°C and salinity of 20 to 30 g/kg.

2. Life-Cycle Test Procedures for Marine and Estuarine Mysids

a. *General test procedures:* Life-cycle testing can be used to determine relative long-term toxicity of substances. Tests are conducted to determine changes in numbers and weights of individuals resulting from effects of the test material on survival, growth, and reproduction. Results may be used to predict long-term effects in field situations, compare chronic sensitivities of different species, and chronic toxicities of different materials.

Life-cycle toxicity tests are flow-through (see Sections 8010D.1 and 4) with a flow rate through each test container of at least 5 to 10 volume additions per 24 h. Start tests with young mysids in accordance with other studies with this group.[1,2,4] Mysids of *Mysidopsis* species used in life-cycle toxicity tests should be less than 24 h post-release from the brood sac; collect in accordance with instructions given in Section 8714B.3. Start test with 2 containers of 15 or 3 containers of 10 mysids each, in at least 4 to 8 true replicate chambers per concentration. Transfer mysids to each test chamber with a glass pipet. Feed mysids with brine shrimp larvae three times a day at the rate of 50 nauplii/mysid (a total of 150 nauplii/mysid/d) during test period.[5] Examine test chambers daily, record mortality, and remove all dead specimens and debris. Generally it is not necessary to consider mysid weight/L test solution given the low weight of mysids, but in flow-through tests use less than 5 g of mysids/L of test solution at temperatures above 20°C. Limit loading to ensure that the concentrations of dissolved oxygen and test material do not fall below acceptable limits, concentrations of metabolic products do not exceed acceptable levels, and the test mysids are not stressed because of cannibalism, aggression, or crowding.[6]

b. Specific test procedures: Conduct *Mysidopsis bahia, Mysidopsis bigelowi,* and *Mysidopsis almyra* tests at a temperature of 27°C for approximately 28 d. Conduct life-cycle toxicity tests by two general methods, pairing and non-pairing.[4] The method using pairing may make it easier to collect population data for life table analysis. If the method with pairing is to be used, start the test with at least two containers of 15 randomly selected mysids each, in at least two true replicate chambers per concentration [see ¶ 3) below]. Pair mysids at sexual maturity (Day 12 to 14), with one female and one male in each test container. Preferably use at least 20 randomly selected pairs per treatment; transfer them between containers within a test chamber, but not from one test chamber to another, to create as many pairs as possible. Pair all mysids on the same day of the test. If the method without pairing is to be used, start test with at least three containers of ten randomly selected mysids each, in at least four true replicate chambers per concentration[6] [see ¶ 3) below]. Keep mysids in these containers throughout the test.

1) *Equipment and physical conditions*—For general guidance on equipment and materials, see ¶ 1*b*1)a). For the test material, in addition to the items noted in ¶ 1*b*1)a), ascertain acute toxicity to the test species, measurement or estimate of chronic toxicity to the test species, and recommended handling procedures.

Select temperature appropriate for the species being tested, and hold test temperature within ±1°C of mean test temperature. Conduct tests with *Mysidopsis* species in a temperature range of 26 to 28°C. Keep salinity within the tolerance range of the selected species. The optimum salinity for *M. bahia* and *M. bigelowi* is 20 to 30 g/kg and for *M. almyra*, 10 to 20 g/kg (see Section 8714B.2). If a test salinity other than the optimum is used, set up an additional control at the optimum salinity. Use ambient laboratory lighting with a photoperiod of 16 h light/8 h dark, preferably with 15- to 30-min dusk/dawn transition period to acclimate mysids to the test photoperiod.

For dilution water requirements, see ¶ 1*b*1)a).

Calculate minimum number of test chambers, test containers, and pairs of mysids per treatment, expected variance within test chambers, expected variance between test chambers within a treatment, and either the maximum acceptable confidence interval on a point estimate or the minimum detectable difference using hypothesis testing.[4,6]

Test solution can flow from one container to another but not from one chamber to another. Test chambers can be constructed by gluing strong window glass with clear silicone adhesive. Because adhesives can sorb some organochlorine or organophosphorus pesticides, apply as little adhesive as possible. Cover chambers to prevent contamination and reduce evaporation. Test solution may enter the container directly or containers may be oscillated in the test solution, or the water level in the test chamber may be varied by means of a self-starting siphon. Test containers used successfully include 250-mL glass beakers with holes drilled in the sides and covered with 250-μm mesh*, 90- or 140-mm-ID glass petri dish bottoms with collars made of 210- or 250-μm screen,[7] and 110- by 180- by 200-mm deep glass rectangular chambers partitioned into six containers with a 65-mm-high, 330-μm mesh nylon collar. Provide metering system that will accommodate type and concentration(s) of test material and necessary

flow rates of the test solutions, mix test material with dilution water immediately before entrance to test chambers, and supply selected concentration(s) of test material reproducibly.[1] Ensure that mysids remain submerged and are not stressed by crowding or turbulence in exposure system. Use test containers that provide a surface area of at least 25 cm²/mysid and a solution depth of at least 25 mm at all times.[4,6]

For information pertaining to species selection, collection, holding, acclimation, disease control, and culturing, see Sections 8010E and 8714B.

2) *Test procedure*—For general information on test procedures, see Section 8010F.3. If a life-cycle test is intended to allow calculation of an endpoint, include one or more control treatments and a geometric series of at least five concentrations of test material, each of which is at least 50% of the next higher one. Use results from range-finder or definitive tests to determine the appropriate range of concentrations. To determine whether a specific concentration reduces survival, growth, or reproduction, only that concentration and control(s) are necessary; however, two additional concentrations at about one-half and twice the specific concentration of concern are preferable.

While test solution sometimes can be added directly to dilution water, preferably dissolve it in a solvent (reagent-grade or better) to form a stock solution and add stock solution to dilution water in the metering system. If a stock solution is used, determine the concentration and stability of the test material in it before beginning the test. If test material is subject to photolysis, shield stock solution from light.

Use a solvent control if dosing solutions are prepared in an organic solvent. Acceptable solvents are triethylene glycol, methanol, ethanol, and acetone.[4] Limit concentrations of solvent to 0.1 mL/L test solution. Do not use surfactant in preparation of a stock solution. If a solvent other than water is used and solvent concentration is the same in all test solutions that contain test material, include at least one solvent control containing the same concentration of solvent and using solvent from the same batch used to make the stock solution, as well as a dilution water control. If a solvent other than water is used and the concentration of solvent is not the same in all test solutions that contain test material, conduct a solvent test to determine whether survival, growth, or reproduction of the test species is related to the concentration of solvent over the range used in the toxicity test if such a solvent test with the same dilution water and test species has not already been conducted. A life-cycle test is unacceptable if any treatment contains a concentration of solvent in the range of effect. If no effect on the test species using the same dilution water is found at the test concentration of solvent, a life-cycle test may be conducted using solvent concentrations within the tested range, but include in such tests a solvent control containing the highest concentration of solvent present in any other treatment using solvent from the same batch used to make the stock solution, and a dilution water control.

The percentage of organisms that show signs of disease or stress, such as discoloration, unusual behavior, or death, must be 30% or less in the solvent control and in the dilution water control.

To establish test concentrations, set highest concentration of life-cycle test to be equal to the lowest concentration that caused adverse effects in a comparable definitive acute test. Use a dilution ratio of 1.5 to 2 between successive concentrations.[1] For more information on experimental design, see Section 8010F.3*a*.

---

* Nytex® or equivalent.

3) *Test initiation*—On the day that the toxicity test is initiated, remove a sufficient number of mysids from the holding facility at one time to provide about one-third more animals than are needed. Transfer mysids gently with a wide-bore (larger than the largest mysid) glass pipet with a smooth tip. Begin flow-through life-cycle tests by placing test organisms into randomly selected containers after test solutions have been flowing through the test chambers long enough for the concentrations of test material to have reached steady state.

4) *Biological observations*—See ¶ 1*b*1)d). Also count, record, and remove live young in each container daily. Record day of brood release.

Determine dry weight (dried at 60°C for 72 to 96 h or to constant weight) of each individual first-generation mysid alive at end of test to nearest microgram. Rinse mysids with deionized water to remove salt before drying. Weigh males and females separately to determine sex-specific effects.[8] Remove brine shrimp nauplii caught in female brood sacs before drying. Total body length (total midline body length from anterior tip of carapace to posterior margin of endopod of uropod, excluding setae) may be determined for mysids alive at the end of the test, but not for preserved mysids because of body curvature. Note any abnormal development or aberrant behavior for first- and second-generation mysids.

5) *Chemical data*—Analyze water in control and test chambers daily for salinity and temperature. Measure pH and dissolved oxygen (DO) at beginning, end, and at least weekly during the test in the control and pH at least once in the highest test concentration. Also measure DO whenever there is an interruption of the flow of test solution. For DO maintenance requirements, see ¶ 1*b*1)e).

6) *Verification of exposure*—Before and during the test verify exposure concentrations of test chemical. Measure concentration of test material in the flow-through test chambers at least weekly in each treatment, including the control(s) during the test. Measurement of degradation products may be desirable. If a malfunction that could alter the concentration of test material is detected in the metering system, take water samples immediately from affected test chambers and analyze as soon as possible.

3. Statistical Analysis

Assemble, analyze, evaluate, and report data as described in Section 8010G.

4. References

1. AMERICAN SOCIETY FOR TESTING AND MATERIALS COMMITTEE E47. 1988. Standard Guide for Conducting Acute Toxicity Tests with Fishes, Macroinvertebrates, and Amphibians. ASTM E-729-88a, American Soc. Testing & Materials, Philadelphia, Pa.
2. AMERICAN SOCIETY FOR TESTING AND MATERIALS COMMITTEE E47. 1992. Standard Guide for Conducting Static and Flow-Through Acute Toxicity Tests with Mysids from the West Coast of the United States. ASTM E-1463-92, American Soc. Testing & Materials, Philadelphia, Pa.
3. U.S. ENVIRONMENTAL PROTECTION AGENCY. 1985. Ambient Water Quality Criteria for Chlorine-1984. EPA 440/5-80-030, National Technical Information Serv., Springfield, Va.
4. AMERICAN SOCIETY FOR TESTING AND MATERIALS COMMITTEE E47. 1990. Standard Guide for Conducting Life-Cycle Toxicity Tests with Saltwater Mysids. ASTM E-1191-90, American Soc. Testing & Materials, Philadelphia, Pa.
5. AMERICAN SOCIETY FOR TESTING AND MATERIALS COMMITTEE E47. 1992. Practice for Using Brine Shrimp Nauplii as Food for Test Animals in Aquatic Toxicity Tests. ASTM E-1203-92, American Soc. Testing & Materials, Philadelphia, Pa.
6. LUSSIER, S., D. CHAMPLIN, A. KUHN & J. HELTSHE. 1997. Mysid (*Mysidopsis bahia*) Life-Cycle Test: Design Comparisons and Assessment. *In* D. Bengtson & D. Henshel, eds. Environmental Toxicology and Risk Assessment: Biomarkers and Risk Assessment, Vol. 5. ASTM STP 1306, American Soc. Testing & Materials, Philadelphia, Pa.
7. GENTILE, S., J. GENTILE, J. WALKER & J. HELTSHE. 1982. Chronic effects of cadmium on two species of mysid shrimp: *Mysidopsis bahia* and *Mysidopsis bigelowi*. *Hydrobiologia* 93:195.
8. BRETELER, R., J. WILLIAMS & R. BUHL. 1982. Measurement of chronic toxicity using the opossum shrimp *Mysidopsis bahia*. *Hydrobiologia* 93:189.

5. Bibliography

NIMMO, D.R., L.H. BAHNER, R.A. RIGBY, J.M. SHEPPARD & A.J. WILSON, JR. 1977. *Mysidopsis bahia*: An estuarine species suitable for life-cycle toxicity tests to determine the effects of a pollutant. *In* F.L. Mayer & J.L. Hamelink, eds. Aquatic Toxicology and Hazard Evaluation, First Annual Symposium. Amer. Soc. Testing & Materials STP 634, p. 109. American Soc. Testing & Materials, Philadelphia, Pa.

CRIPE, G.M., D.R. NIMMO & T.L. HAMAKER. 1981. Effects of two organophosphate pesticides on swimming stamina of the Mysid *Mysidopsis bahia*. *In* F.J. Vernberg, F. Thurberg, A. Calabrese & W. Vernberg, eds. Biological Monitoring of Marine Pollutants, p. 21. Academic Press, New York, N.Y.

GENTILE, J.H., S.M. GENTILE, G. HOFFMAN, J.F. HELTSHE & N. HAIRSTON, JR. 1983. The effects of a chronic mercury exposure on survival, reproduction and population dynamics of *Mysidopsis bahia*. *Environ. Toxicol. Chem.* 2:61.

DAVIDSON, B.M., A.O. VALKIRA & P.F. SELIGMAN. 1986. Acute and chronic effects of tributyltin on the mysid *Acanthomysis sculpta* (Crustacea, Mysidacea). *Proc. Oceans 86* 4:1218.

REISH, D.J. & J.A. LEMAY. 1988. Bioassay Manual for Dredged Sediments. U.S. Army Corps of Engineers, Los Angeles Dist., Calif.

FISHER, D., D. BURTON, L. HALLIG, R. PAULSON & C. HERSH. 1988. Standard Operating Procedures for Short-Term Chronic Effluent Toxicity Test with Freshwater and Saltwater Organisms, Section 17—Mysid (*Neomysis americana*) Survival, Growth, and Reproduction Test Method. Johns Hopkins Univ., Shady Side, Md.

MARTIN, M., J.W. HUNT, B.S. ANDERSIB, S.L. TURPEN & F.H. PALMER. 1989. Experimental evaluation of the mysid *Holmesimysis costata* as a test organism for effluent toxicity testing. *Environ. Toxicol. Chem.* 8:1003.

ASATO, S.L. & D.J. REISH. 1989. The effects of heavy metals on the survival and feeding of *Holmesimysis costata* (Crustacea:Mysidacea). *In* Biologia Marina, Mem. del VII Simposium, La Paz, Baja California Sur, Mexico, p. 113.

BRANDT, O., R. FUJIMURA & B. FINLAYSON. 1993. Use of *Neomysis mercedis* (Crustacea:Mysidacea) for estuarine toxicity tests. *Trans. Amer. Fish. Soc.* 122:279.

FINLAYSON, B.J., J.A. HARRINGTON, R. FUJIMURA & G. ISSAC. 1993. Identification of methyl parathion toxicity in Colusa Basin Drain water. *Environ. Toxicol. Chem.* 12:291.

U.S. ENVIRONMENTAL PROTECTION AGENCY. 1995. Short-Term Methods for Estimating the Chronic Toxicity of Effluents and Receiving Waters to West Coast Marine and Estuarine Organisms. EPA-600/R-95-136, U.S. Environmental Protection Agency, National Exposure Research Lab., Cincinnati, Ohio.

8740 DECAPODS*

8740 A. Introduction

Decapod crustaceans are among the most commercially important invertebrates.

Larval, postlarval, or adult stages of several species of decapods may be found in large numbers in estuaries and rocky intertidal habitats near the shore, where they are vulnerable to various types of discharges. Because of their phylogenetic relationship to insects, and the fact that pesticides often are applied in watersheds draining to estuaries, use of decapods in the testing of pesticide toxicity is particularly relevant. Postlarvae of penaeid shrimp use the estuaries as nursery grounds until they are

---

* Approved by Standard Methods Committee, 1997.

large enough to migrate offshore. Early life stages are particularly vulnerable.

There is considerable diversity among the decapods, with the basic separation of adult animals into the more active swimmers (Natantia), encompassing the shrimp and lobsters, and the more sedentary crabs (Reptantia). While there are exceptions and variations, shrimp are generally planktivores, while the lobsters and crabs are predators and scavengers. There is much greater similarity within the decapods at the larval and postlarval stages, when zooplankton is the primary food. Laboratory holding and testing is easier at these early stages because brine shrimp (*Artemia salina*) nauplii usually are an appropriate food.

8740 B. Selecting and Preparing Test Organisms

Some species used frequently in toxicity studies are penaeid shrimp (*Penaeus* sp.) larvae, postlarvae, and juveniles,[1-3] which have high commercial value on the Gulf of Mexico coast, and larvae of the American lobster,[4] an equally important commercial species along the northeast coast of the U.S. The grass shrimp, *Palaemonetes* sp., is another species that has been used in many toxicity investigations,[5-12] because of the ease of collecting and holding, abundance in marshes and estuaries along the Gulf of Mexico and southeastern coast of the U.S., and sensitivity, especially to pesticides.

1. Selecting Test Organisms

Many of the species used previously in toxicity testing are listed below. Consult specific references to determine which life stages were tested and the methods used for collecting, holding, feeding, and exposing the animals. For information on selecting test organisms, as well as the handling, holding, and conditioning of the animals, see Section 8010E.1-3. Regional references to location and identification of decapods are given in Section 10900.

a. *Marine and estuarine decapods:*
1) Suborder Natantia
a) Section Penaeidea—Species of *Penaeus* are prominent commercial shrimp, harvested in the Gulf of Mexico.

 Penaeus setiferus
 Penaeus aztecus
 Penaeus duorarum
b) Section Caridea
Crangon—cosmopolitan
Palaemonetes pugio—southeast coast of U.S., and Gulf of Mexico.
Palaemonetes vulgaris
Palaemonetes intermedius
Pandalus danae—Pacific Northwest, including Alaska.

Pandalus hypsinotus—Pacific Northwest, including Alaska.
2) Suborder Reptantia
a) Section Macrura
Panulirus (spiny lobster)—Point Conception south along coasts of southern California and Baja California.
Homarus americanus—northeast coast of U.S.
Petrolisthes—U.S. coast of Gulf of Mexico.
Rithropanopeus harrisii—U.S. coast of Gulf of Mexico.
Panopeus herbstii—U.S. coast of Gulf of Mexico.
Menippe mercenaria—rock or reef areas of Florida.
Cancer productus—Pacific coast of U.S.
Cancer oregonensis—Pacific coast of U.S.
Cancer magister (see Section 10900, Plate 12, M)—Pacific coast of U.S.
Callinectes sapidus (see Section 10900, Plate 12, N)—southeast and Gulf of Mexico coasts of U.S.
Uca pugilator—southeast coast of U.S.
b. *Freshwater decapods:*
Cambarus—43 species, between Blue Ridge Mountains and Mississippi River (see Section 10900, Plate 12, H).
Procambarus—97 species, New England and Great Lakes to Mexico.
Orconectes—59 species, Maine to Texas, most in Central Basin.
Macrobrachium ohione—Atlantic coastal plain from North Carolina to Georgia, along Mississippi River from St. Louis southward, and Texas.
Palaemonetes kadiakensis—west of the Alleghenies from southern Ontario and Great Lakes to Gulf coast of northeastern Mexico.

2. Collecting Test Organisms

a. *Marine and estuarine decapods:* In shallow estuarine environments and tidal flats, collect juvenile or adult decapods with seine nets. Hold the collection within the net in very shallow water while gently dipping individuals from the water to a water-filled ice chest. Cool the water to be used for transporting the animals to slow activity and to increase the initial levels of dissolved oxygen. Air pumps with battery packs may be necessary to provide sufficient oxygen during transport, particularly if animal density is high. This method of collection is also applicable to gravid adults of some species (blue crabs, *Palaemonetes*, etc.), to hatch the larvae in the laboratory and conduct larval toxicity testing.

Lobsters and blue crabs (*Callinectes*) are best obtained from traps, but a scientific collecting permit may be required to capture and retain gravid females. Penaeid shrimp, *Pandalus* species, and other decapods are best obtained by using short-duration (10-min) otter trawls, so that the animals are subjected to less stress from crowding in the cod end. Rapidly transfer the animals into cool, well-aerated water in an ice chest. Avoid excessive crowding during holding on shipboard and during transportation to the laboratory. Provide aeration during transport.

Some crab species (*Rithropanopeus, Petrolisthes, Menippe, Cancer,* etc.) are best collected by hand at low tide, but it will be difficult to obtain a sufficient number of individuals to conduct tests with adults. These species are best suited for tests with larvae.

b. Freshwater decapods: While adults may be tested when a sufficient supply of individuals is obtained, preferably collect gravid females and use the offspring in testing.

3. Holding, Acclimating, and Culturing Test Organisms

Guidelines for the culturing of test organisms, including decapods, are found in Section 8010E.4. The following sections provide guidelines for collecting, handling, and obtaining larvae from crayfish, crabs (e.g., *Cancer magister*), American lobster (*Homarus americanus*), and shrimp (*Penaeus* and *Palaemonetes*).

a. Water supply: See Section 8010E.4b.

b. Acclimating, holding, and maintaining stock cultures: See Sections 8010E.3 and 4. Risks in handling most adult crustaceans usually are not great because of their rigid exoskeleton and general durability. Both larval and adult forms of many species are cannibalistic and readily attack each other in the soft-shell stage. Hold juveniles and adults in individual compartments in long troughs or divided tanks. Form the compartments with perforated separators that slide into slots on the sides. Use stainless steel for freshwater forms and glass, acrylic, plastic, or plywood covered with fiberglass for marine forms. Provide rigid, transparent covers to prevent loss of the highly motile specimens. Use perforated separators to ensure a flow of water through each compartment to remove metabolic products and provide DO. The crustacean growth process, which involves a periodic ecdysis or sloughing of the rigid exoskeleton, imposes a lack of uniformity in test animals that is not readily detectable in advance. In the pre-ecdysis stage and during ecdysis animals are heavily stressed and more sensitive to unsatisfactory environmental conditions and toxicants.

1) *Crayfish*—Collect specimens from their natural habitat by trapping, seining, or by hand (see Section 8010E.2). General procedures for holding and acclimating are described in Sections 8010E.3 and 4. Because crayfish are cannibalistic, hold all but the young stages in separate compartments. Suitable holding, acclimating, and culturing chambers are stainless steel, glass, fiberglass-covered wood, or plastic troughs, 180 cm long, 30 cm wide, and 20 cm deep, with a divider down the center to make two long troughs. Make shallow channels on the sides and central divider every 15 cm into which separators can be slipped to make 12 compartments on each side, each approximately 15 × 15 cm square and 20 cm deep. This size is suitable for crayfish. The number and size of compartments depend on the size and number of organisms to be tested. To hold a large number of small crayfish, remove the separators to make a tank of the desired length. Provide separators with a large number of perforations so they operate as screens. Control water depth in test chambers by a standpipe in the last compartment of the trough. When cleaning the separators, temporarily raise them a short distance from the bottom to allow excess food and wastes to be washed out and remove the standpipe in the last compartment to insure strong flows. Clean routinely with a siphon and a brush to loosen materials from compartments, screens, walls, and bottoms. Supply water adjusted to the desired temperature and DO to the two head compartments by a siphon from a constant-head box. Use a minimum flow of 10 trough volumes/d. Adjust volume to maintain favorable water quality in each compartment. Required water depth depends on size of organisms; 15 cm is preferred. Provide each set of troughs with a transparent lid. For life-cycle studies beginning with eggs or newly hatched young, collect ovigerous females and place in flow-through troughs under natural water conditions. Begin acclimation to different conditions after 2 d. Hold animals in troughs until young hatch. Remove compartment dividers to provide freedom of movement of young. Clean as described in Section 8010E.4d. Use macerated fish food for juveniles and adults. Alternatively use prepared dry fish food. Use very finely divided pieces of fish and commercial fish food pellets as food for the newly hatched.

2) *Crabs*—Static culture of brachyuran crab larvae has been achieved for several species of Atlantic coast crabs.[13–16] Long-term static or renewal bioassays with these species have been performed. Culture of dungeness crab, *Cancer magister*, larvae has been reported.[17] Culturing crab larvae requires a favorable water supply and control of competitors, predators, and disease. Filter water and disinfect it by UV light treatment. For unpolluted open ocean water, little or no treatment is required. If the supply is from an estuary receiving organic wastes, purify before use. Filter seawater for the flow-through system by gravity flow through a coarse, quartz sand filter and adjust to the desired salinity, approximately 25 to 30 g/kg, by adding fresh water. To remove other organisms, refilter under pressure through sequential layers of 40/60-mesh garnet, 20/30-mesh silica sand, and 0.3-cm hard charcoal. Follow by filtration through a polishing filter and treat with UV light. Use constant-level head boxes equipped with heating, cooling, and stirring devices, to deliver constant measured flows by siphons, selected nozzles, or constant and accurate delivery pumps.

Collect ovigerous females or purchase from fishermen and place in holding tanks or in flow-through troughs similar to, but larger than, those described for crayfish. Acclimate and condition as described in Section 8010E.3. When eggs are ready to hatch, transfer females to static tanks provided with aerated and UV-disinfected water at 30 g/kg and 13°C. As eggs hatch, dip out swimming first-stage larvae with beakers and transfer to culture beakers with large-bore pipets. Many crab larvae are positively phototactic. Collection of crab larvae may be simplified by applying an intense light source to one side of the hatching container.

Dungeness crab larvae have long delicate spines that make their culture in flowing systems difficult. Culture larvae to the fourth and fifth stage in 250-mL beakers that have a hole 15 mm in diameter blown through their sides near the bottom. Using silicone cement, fasten a plastic screen having 360-μm openings over this hole on the inside of the beaker and plastic screen with 210-μm openings over the hole on the outside. Because of the lip created by blowing the glass, the two screens are 3 to 4 mm apart. The larger-mesh screen on the inside is less likely to catch and damage spines of larval crabs while the smaller-mesh screen on the outside does not come in contact with larvae but does retain food organisms, brine shrimp nauplii. Set the 250-mL beakers in glass trays or aquariums large enough to accommodate 10 beakers and provide a depth of at least 10 cm. Supply trays with a constant flow of water by a tube that discharges near the tray bottom. Provide an automatic siphon at the outlet so there is continual filling and drawdown (Figure 8740:1). Construct the automatic siphon so that when the water reaches the high point and the siphon is activated, the beakers contain approximately 200 mL, and when the siphon is broken the beakers contain about 150 mL. The automatic siphon consists of a silicone rubber stopper drilled to receive an 8-mm-ID right-angle glass tube on one end and a 5-mm-ID right-angle glass tube on the other end. In a 1.3-cm-diam hole blown through side of tray, insert stopper with 8-mm hole on inside of tray. Insert tubes into stopper as shown in Figure 8740:1. Placement of hole inside of tray controls water level in beakers at 200 mL. The distance between the top of the inside hole in the stopper and the bottom of the inside siphon leg is equal to the difference in depth between 200 mL and 150 mL. Make siphon intake perfectly flat and smooth to prevent air from being drawn into the siphon. Adjust tube diameters to give a 15-min cycle, 10-min filling and 5-min drawdown.

When culture chambers are set up and functioning, place 10 first-stage larvae in each beaker with a smooth large-bore pipet. The larvae can be fed nonliving food but preferably feed first-stage brine shrimp nauplii at the rate of 70 for each crab larva three times/week through the third stage, then 100 brine shrimp for each crab larva. The nutritional quality of brine shrimp will vary depending on source. This will affect the sensitivity of the larvae and the results of the test. Keep density of crab larvae low and that of food organisms high to minimize crab larvae contacts that may result in cannibalism. Before feeding, transfer larvae to clean chlorine-disinfected and rinsed beakers. Maintain a temperature of 12 to 13°C, pH 8, and a salinity of 25 to 30 g/kg. Adjust photoperiod to correspond with natural conditions or, if the cycle is off-season, to correspond to the normal annual cycle of light and dark. Exclude natural light and use fluorescent light (see Section 8010F.3f). Under these conditions, survival of 80 to 90% through the fourth zoeal stage has been attained. Larvae usually begin molting into the fifth zoeal stage by the 45th day. Mortalities then increase.

Juvenile and adult dungeness crabs are much less susceptible to disease than larvae. Older life stages are much less sensitive than larvae and will be more tolerant of conditions. With strict sanitation and unpolluted open-sea water, sand filtration alone provides sufficient water quality control. Hold juvenile and adult crabs in trough compartments similar to, but larger than, those used for crayfish. To allow sufficient space for each juvenile crab, use compartments 15 × 15 cm and 15 to 20 cm deep. For adult crabs use 30- × 30-cm or 40- × 40-cm compartments with a depth of about 30 cm. For large specimens use deeper water. For ease of supplying water, arrange troughs on stands having three shelves with space on each for two troughs. Feed cut-up or macerated fresh fish, clams, or mussels, or commercial dried fish

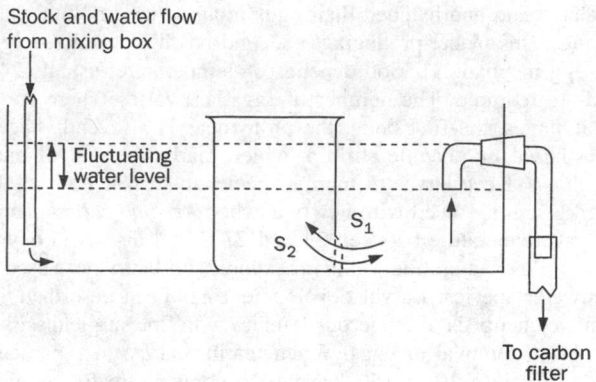

Figure 8740:1. Rearing and exposure beaker and automatic siphon for dungeness crab larvae. After BUCHANAN, D.V., M.J. MYERS & R.S. CALDWELL. 1975. An improved flowing water apparatus for culture of brachyuran crab larvae (unpublished).

foods to juveniles and adults. Remove unused food within 24 h to reduce fouling. Routinely clean sides and bottoms of compartments and remove wastes with vacuum or siphon cleaners. Raise screen separators a few millimeters and flush as suggested for crayfish troughs.

3) *American lobster, Homarus americanus*—Culture procedures for the American lobster are being revised, but are not yet completed. The culture procedures for the American lobster in the 19th edition of *Standard Methods* should not be used. Consult the literature for lobster seed stock.

4) *Shrimp (Natantia)*—Obtain by collection or purchase from bait dealers. Seine shrimp of the genera *Penaeus, Palaemonetes,* and *Crangon* from estuaries. Check animals for parasites, disease, and general condition. For general instructions on collecting, handling, transferring, holding, acclimating, and culturing, see Section 8010E.

a) Palaemonetes—Several marine and freshwater species of the genus *Palaemonetes* have been reared through metamorphosis.[19-25] They are suitable for life-cycle studies and can be brought from the field for direct use or for laboratory rearing. Place field-collected adult shrimp in suitable flow-through aquarium water. Feed freshwater species macerated parts of local fishes; feed marine forms macerated mollusks or fish. Examine shrimp periodically to detect ovigerous females. When eggs are nearly ready to hatch, remove desired number of females from tank and put into individual containers. Keep females in these containers, preferably with flow-through water, until eggs begin to hatch. During this period, feed macerated fish or other suitable food. After eggs hatch, remove female and feed prelarvae or prezoeae 1-d-old *Artemia salina* nauplii. The rearing procedure for larvae is similar for all six species. Use equipment and procedures similar to those for the dungeness crab, but with rearing chambers with a capacity of 1 L set in a deeper tray. Place 10 larvae in each beaker and feed with newly hatched brine shrimp nauplii and ideally maintain at 25°C. Filter and disinfect water. During the larval period, provide 14 h light and 10 h dark cycle (see Section 8010F.3f). Inspect larvae and feed daily. If sediments or wastes tend to collect, remove them daily with a siphon. At 25°C the larval period lasts 16 to 24 d. The average length of larval life is between 19 and 20 d. To rear through entire life cycle, immediately place females that have laid and hatched their eggs in an aquarium with males. Mating takes place, producing a second batch of fertilized eggs. The egg incubation period depends on temperature; usually 24 to 28 d are required. The number of eggs laid varies. There are six larval stages, the first being the protozoea. The seventh stage is a postlarval or juvenile shrimp, which marks the end of metamorphosis. Keep larvae of marine species in seawater adjusted to 25 g/kg salinity. Feed with newly hatched *Artemia salina* nauplii. Rear at temperature between 23 and 27°C and dissolved oxygen above 60% of saturation, using procedures similar to those used for freshwater species. Larval development has been described.[19-24] Remove chelipeds of ovigerous females with fine surgical scissors to prevent removal of eggs. When rearing larvae to a particular age, maintain a 10 to 15% surplus to compensate for mortality and to provide for other uses.

b) Penaeus—The rearing and culturing of larvae of this genus have been described in several studies.[26-32] Hold shrimp in glass tanks of at least 30-L capacity. Provide each tank with flow-through water, 2 to 3 cm of sand over the bottom, and a screen over the top to prevent the shrimp from jumping out. Avoid overloading. Keep no more than 22 to 24 animals in a 30-L tank. For *Penaeus* spp., use a minimum flow of 7.5 L/g/d. Flows up to 22 L/g/d may be desirable to insure DO above 60% of saturation and the removal of metabolic products. Acclimate to laboratory test conditions for about 2 weeks. For short-term or medium-length tests with adults and juveniles, shrimp can be field-collected. Cut-up fish is a satisfactory food. Cut a fillet from mullet, grouper, or other abundant species into 1- to 2-cm pieces. Feed one piece per shrimp each 2 to 3 d, depending on the size of the shrimp. Remove uneaten food daily. For larval tests or life-cycle studies, collect gravid females offshore, let them spawn, and rear larvae at least to the postlarval stage. Penaeid shrimp can be reared from egg to postlarvae in the laboratory. Use the diatom *Skeletonema* as food for protozoeal stages. When diatoms are used as food, use air-lift pumps to prevent accumulation of the diatoms. Feed freshly hatched brine shrimp, *Artemia*, to the mysis and postlarval stages.

The protozoeal stages, I through III, of the penaeid shrimp require algae as food. Because larval shrimp are pelagic and unable to search for food during the early part of their life cycle, maintain the required density of the phytoplankton *Skeletonema costatum* and *Tetraselmis* sp. Add these to larval culture chambers according to stage of development, number of larvae present, and volume of water:

Protozoeal I—*Skeletonema*, 50 000 cells/mL
Protozoeal II—*Skeletonema*, 150 000 cells/mL
Protozoeal III—*Tetraselmis*, 20 000 cells/mL
Mysis I—*Artemia* nauplii, 3/mL
Mysis II—*Artemia* nauplii, 3/mL
Mysis III—*Artemia* nauplii, 3/mL
Postlarvae I-IV—*Artemia* nauplii, 3/mL

Maintain phytoplankton in continuous culture or harvest and freeze to use later. Algal culture production units shown in Figure 8010:2, and described in Section 8010E.4c2), will produce daily 7.5 L of culture containing 4.3×10^6 *Skeletonema costatum*/mL or 7.0×10^6 *Isochrysis galbana*/mL and several other species of algae at similar concentrations. Add algae as food for the larval shrimp as either a fresh or a frozen concentrate. Concentrate algae by centrifuging and discard growth medium. Use a temperature of 28 to 30°C and a salinity of 27 to 35 g/kg. Omit antibiotics

from the larval culture medium if EDTA (disodium salt) is substituted at a concentration of 10 mg/L seawater. Feed juvenile shrimp with fresh pieces of fish, clams, or mussels or with prepared dried foods.

4. Parasites, Diseases, and Harmful Growths

Consult Section 8010E.5 for general problems of parasites, infectious diseases, and control procedures. Parasites and diseases of crustaceans, including specific infections known to occur in decapods, include bacteria, phycomycetes, and fungi.[24-39] Parasites have been found in many species of crustaceans and their presence can influence results. In *Uca*, an ectoparasitic isopod is found on the gills, nematodes in the gut, and metacercaria in the green glands. Species of *Lagenidium* similar to the one that occurs in shrimp occur in other marine crustaceans. *L. callinectes* occurs in eggs and larvae of the blue crab. The blue crab has a barnacle (*Octolasmus lowei*) living in association with its gills and gill chamber, metacercariae in various organs, and the sacculinid *Loxothylacus taxanas* beneath its abdomen. *Saprolegia parasitica* attacks larvae of the shrimp *Palaemonetes kadiakensis*.

Adult shellfish in recirculated or flow-through systems are susceptible to biotoxins and pathogens. Remove metabolites and dead individuals from recirculating systems. Juvenile and adult lobsters, crabs, and shrimps are subject to bacterial and fungal infections. *Gaffkya*, a bacterial pathogen, is particularly prevalent in tank-held lobsters, while *Vibrio* disease occurs in tank-held postlarval adult shrimp. Most captive crustaceans are subject to "shell disease," produced by chitin-destroying bacteria. A systemic fungal disease has been described in European prawns and several fungal infections occur in wild shrimp populations. The larval stages of the lobster and several other crustaceans are prone to infections of the ubiquitous marine bacterium *Leucothrix mucor*, which has produced mortalities of over 90% in larval cultures. The exuvia and the new exoskeleton after molting become entangled in the long dense filaments of the bacteria and the larvae are unable to swim or feed adequately. This organism also can produce high mortalities by causing pelagic eggs to sink and by interfering with the filtering apparatus of larval forms and the functioning of gills. In some instances, it may be necessary to culture larvae in artificial seawater to avoid *L. mucor* infection. Place ovigerous females in a bath of malachite green (5 mg/L) for 1 min only or rinse them several times in artificial seawater of the correct salinity that contains streptomycin, 2 mL/L, from a stock solution containing 2 g antibiotic/L.

Maintaining a 1-mg/L concentration of antibiotic throughout the larval culture period prevents infections. Twice daily cleaning also is a good preventive method. Seawater, filtered and exposed to UV radiations, should be nearly bacteria-free. A disease of lobster larvae tentatively has been associated with the phycomycete *Haliphtorus*. It appears as a scab on the first segment of the thoracic appendages up to and surrounding the first row of gills. Thorough cleaning and UV treatment of the water supply is the only known treatment. In most cases, these scabs adhere to both old and new carapaces and thus cause a mechanical impediment to molting. Mortality appears to be restricted to larvae and young juveniles. No deaths of specimens with a carapace length over 27 mm have been observed. The fungus *Lagenidium* sp. causes serious problems in rearing larval shrimp. The disease first becomes apparent in the second protozoeal stage and disappears as the shrimp reach the first mysis stage. Shrimp become immobilized by replacement of muscle tissue by fungal mycelium.

5. References

1. ANDERSON, J.W., J.M. NEFF, B.A. COX, H.E. TATEM & G.M. HIGHTOWER. 1974. Characteristics of dispersions and water-soluble extracts of crude and refined oils and their toxicity to estuarine crustaceans and fish. *Mar. Biol.* 27:75.

2. CRIPE, G.M. 1994. Comparative acute toxicities of several pesticides and metals to *Mysidopsis bahia* and post-larval *Penaeus duorarum*. *Environ. Toxicol. Chem.* 13:1867.

3. BAMBANG, Y., G. CHARMANTIER, P. THUET & J-P. TRILLES. 1994. Effect of cadmium on survival and osmoregulation of various developmental stages of the shrimp *Penaeus japonicus* (Crustacea:Decapoda). *Mar. Biol.* 123:443.

4. YOUNG-LAI, W.W., M. CHARMANTIER-DAURES & G. CHARMANTIER. 1991. Effect of ammonia on survival and osmoregulation in different stages of the lobster *Homarus americanus*. *Mar. Biol.* 110: 293.

5. WILSON, J.H., P.A. CUNNINGHAM, D. EVANS & J.D. COSTLOW, JR. 1995. Using grass shrimp embryos to determine the effects of sediment on the toxicity and persistence of diflubenzuron in laboratory microcosms. *In* J.S. Hughes, G.R. Biddinger & E. Mones, eds. Environmental Toxicology and Risk Assessment, 3rd Vol. ASTM STP 1218, p. 267. American Soc. Testing & Materials, Philadelphia, Pa.

6. ANDERSON, J.W., J.M. NEFF, B.A. COX, H.E. TATEM & G.M. HIGHTOWER. 1974. The effects of oil on estuarine animals: Toxicity, uptake and depuration, respiration. *In* F.J. & W.B. Vernberg, eds. Pollution and Physiology and Marine Organisms, p. 285. Academic Press, New York, N.Y.

7. BRECKEN-FOLSE, J.A., F.L. MAYER, L.E. PEDIGO & L.L. MARKING. 1994. Acute toxicity of 4-nitrophenol, 2,4-dinitrophenol, terbufos and trichlorfon to grass shrimp (*Palaemonetes* spp.) and sheephead minnows (*Cyprinodon variegatus*) as affected by salinity and temperature. *Environ. Toxicol. Chem.* 13:67.

8. ROESIJADI, G., J.W. ANDERSON, S.R. PETROCELLI & C.S. GIAM. 1976. Osmoregulation of the grass shrimp *Palaemonetes pugio* exposed to polychlorinated biphenyls (PCBs) I. Effect on chloride and osmotic concentrations and chloride and water exchange kinetics. *Mar. Biol.* 38:343.

9. ROESIJADI, G., S.R. PETROCELLI, J.W. ANDERSON, C.S. GIAM & G.E. NEFF. 1976. Toxicity of polychlorinated biphenyls (Aroclor 1254) to adult, juvenile and larval stages of the shrimp *Palaemonetes pugio*. *Bull. Environ. Contam. Toxicol.* 156:297.

10. ROESIJADI, G., J.W. ANDERSON & C.S. GIAM. 1976. Osmoregulation of the grass shrimp *Palaemonetes pugio* exposed to polychlorinated biphenyls (PCBs). II. Effect on free amino acids of muscle tissue. *Mar. Biol.* 38:357.

11. TATEM, H.E., J.W. ANDERSON & J.M. NEFF. 1976. Seasonal and laboratory variations in the health of grass shrimp *Palaemonetes pugio*: Dodecyl sodium sulfate bioassay. *Bull. Environ. Contam. Toxicol.* 16:368.

12. TATEM, H.E., B.A. COX & J.W. ANDERSON. 1978. The toxicity of oils and petroleum hydrocarbons to estuarine crustaceans. *Estuarine Coastal Mar. Sci.* 6:365.

13. SPOTTE, S. 1979. Fish and Invertebrate Culture: Water Management in Closed Systems. Wiley Interscience, New York, N.Y.

14. COSTLOW, J.D. & C.G. BOOKHOUT. 1960. A method of developing brachyuran crab eggs in vitro. *Limnol. Oceanogr.* 5:212.

15. RICE, A.L. & D.I. WILLIAMSON. 1970. Methods for rearing larval decapod crustacea. *Helgolander wiss. Meeresunters* 20.

16. SASTRY, A.N. 1970. Culture of brachyuran crab larvae using a recirculating sea water system in the laboratory. *Helgolander wiss. Meeresunters* 20:406.

17. REED, P.H. 1969. Culture methods and effects of temperature and salinity on survival and growth of dungeness crab, *Cancer magister* larvae in the laboratory. *J. Fish. Res. Board Can.* 26:389.

18. AIKEN, D.F. & S.L. WADDY. 1984. Production of seed stock for lobster culture. *Aquaculture* 44:103.

19. BROAD, A.C. & J.H. HUBSCHMAN. 1963. The larval development of *Palaemonetes kadiakensis*, M.J. Rathbun, in the laboratory. *Trans. Amer. Microsc. Soc.* 82:185.

20. HUBSCHMAN, J.H. & A.C. BROAD. 1974. The larval development of *Palaemonetes intermedius* Holthuis 1949 (Decapoda:Palaemonidae) reared in the laboratory. *Crustaceana* 26:89.

21. DOBKIN, S. 1963. The larval development of *Palaemonetes paludosus* (Gibbes 1850) (Decapoda:Palaemonidae) reared in the laboratory. *Crustaceana* 6:41.

22. HUBSCHMAN, J.H. & J.A. ROSE. 1969. *Palaemonetes kadiakensis* Rathbun: Post embryonic growth in the laboratory (Decapoda:Palaemonidae). *Crustaceana* 16:81.

23. FAXON, W. 1879. On the development of *Palaemonetes vulgaris*. *Bull. Mus. Comp. Zool.* (Harvard) 5:303.

24. BROAD, A.C. 1957. Larval development of *Palaemonetes pugio* Holthuis. *Biol. Bull.* 112:144.

25. BROAD, A.C. 1957. The relationship between diet and larval development of *Palaemonetes*. *Biol. Bull.* 112:162.

26. COOK, H.L. & M.A. MURPHY. 1966. Rearing penaeid shrimp from eggs to postlarvae. *Proc. 19th Annu. Conf. S.E. Assoc. Game Fish. Comm.* 19:283.

27. COOK, H.L. & M.A. MURPHY. 1969. The culture of larval penaeid shrimp. *Trans. Amer. Fish. Soc.* 98:751.

28. COOK, H.L. 1967. A method of rearing penaeid shrimp larva for experimental studies. *FAO (Food Agr. Organ. U.N.) Fish. Rep.* 3:709.

29. MOCK, C.R. & M.A. MURPHY. 1970. Techniques for raising penaeid shrimp from egg to postlarvae. *Proc. 1st Annu. Workshop World Maricult. Soc.* 1:143.

30. MOCK, C.R. 1974. Larval Culture of Penaeid Shrimp at the Galveston Biological Laboratory. NOAA (Nat. Ocean. Atmos. Admin.) Tech. Rep., *NMFS (Nat. Mar. Fish. Serv.) Circ.* 388:33.

31. MOCK, C.R., R.A. NEAL & B.R. SALSER. 1973. A closed raceway for the culture of shrimp. *Proc. 4th Annu. Workshop World Maricult. Soc.* 4:247.

32. ZEIN-ELDIN, Z.P. & G.W. GRIFFITH. 1969. An appraisal of the effects of salinity and temperature on growth and survival of post-larval penaeids. *FAO (Food Agr. Organ. U.N.) Fish. Rep.* 3:1015.

33. ANDERSON, J.I.W. & D.A. CONROY. 1968. The significance of disease in preliminary attempts to raise crustacea in sea water. *Bull. Off. Inform. Epizoot.* 69:1239.

34. BROCK, T.D. 1966. The habitat of *Leucothrix mucor*, a widespread marine organism. *Limnol. Oceanogr.* 11:303.

35. JOHNSON, P.W., J.M. SIEBURTH, A. SASTRY, C.R. ARNOLD & M.S. DOTY. 1971. *Leucothrix mucor* infestation of benthic crustacea, fish eggs and tropical algae. *Limnol. Oceanogr.* 16:962.

36. LIGHTNER, D.V. & C.T. FONTAIN. 1973. A new fungus disease of the white shrimp *Penaeus setiferus*. *J. Invertebr. Pathol.* 22:94.

37. COUCH, J.H. 1942. A new fungus on crab eggs. *J. Elisha Mitchell Sci. Soc.* 58:158.

38. ROGERS-TALBERT, R. 1948. The fungus *Lagenidium callinectes* Couch on eggs of the blue crab in Chesapeake Bay. *Biol. Bull.* 95:214.

39. HUBSCHMAN, J.H. & J.A. SCHMITT. 1969. Primary mycosis in shrimp larvae. *J. Invertebr. Pathol.* 13:351.

6. Bibliography

COSTLOW, J.D., C.G. BOOKHOUT & R. MONROE. 1962. Salinity-temperature effects on the larval development of the crab *Panopeus herbstii* Milne-Edwards reared in the laboratory. *Physiol. Zool.* 35:78.

COSTLOW, J.D. & C.G. BOOKHOUT. 1962. The larval development of *Sesarma recticulatum* Say reared in the laboratory. *Crustaceana* 4:281.

WILLIAM, A.B. 1965. Marine decapod crustaceans of the Carolinas. *Fish Bull.* 65:10298.

SAILA, S., J. FLOWERS & J.T. HUGHES. 1968. Fecundity of the American lobster *Homarus americanus. Trans. Amer. Fish. Soc.* 98:537.

NIMMO, D.R., A.J. WILSON, JR. & R.R. BLACKMAN. 1970. Localization of DDT in the body organs of pink and white shrimp. *Bull. Environ. Contam. Toxicol.* 5:333.

PERKINS, H.C. 1972. Developmental rates at various temperatures of embryos of the northern lobster (*Homarus americanus* Milne-Edwards). *Fish. Bull.* 70:96.

SALSER, B.R. & C.R. MOCK. 1973. An airlift circulator for algal culture tanks. *Proc. 4th Annu. Workshop World Maricult. Soc.* 4:295.

VERNBERG, W.B., P. DeCOURSEY & W.J. PADGETT. 1973. Synergistic effects of environmental variables on larvae of *Uca pugilator. Mar. Biol.* 22:307.

BARDACH, J.E., J.H. RYTHER & W.O. MCLARNEY. 1982. Aquaculture: The Farming and Husbandry of Freshwater and Marine Organisms. Wiley Interscience, New York, N.Y.

AMERICAN SOCIETY FOR TESTING AND MATERIALS. 1993. Standard Guide for Conducting Sediment Toxicity Tests with Freshwater Invertebrates. ASTM E1383-93, American Soc. Testing & Materials, Philadelphia, Pa.

8740 C.   Toxicity Test Procedures

Refer to Section 8010 for a basic discussion of toxicity testing, including terminology and basic procedures. Specific procedures for conducting toxicity tests with crayfish, crab larvae, adult, juvenile or larval lobster, and adult or larval shrimp are described below.

1. Toxicity Test Procedures Using Larvae or Postlarvae of Crabs or Shrimp

a. General procedures: Read and understand the basic procedures and concepts described in Section 8010 before initiating tests. Conduct preliminary tests to become familiar with the specific organisms and test procedures given below.

1) Collecting adult decapods—Depending on the species, gravid adult decapods may be collected either by trap, otter trawl, seine net, or by hand at low tide (see Section 8740B.2). Small crab species may be found under rocks, while larger crabs may hide in dense marine grasses. Females with eggs (sponge) may be observed with yellow to brown egg masses in a brood pouch (shrimp), or extending from the undersurface (crabs and lobsters). Because there are hundreds to millions of eggs per female, only a few gravid animals will be sufficient to conduct a toxicity test. Careful handling and decreased temperature will decrease the possibility of the females releasing the eggs during transport, before the exposures are ready.

2) Collecting larval or postlarval stages—Before attempting to raise larval decapods to postlarval stages, determine if this or a related species is being reared at a commercial or government facility for mariculture/aquaculture. Staff members at these facilities have the appropriate expertise to answer questions, and it is often possible to purchase or obtain the needed test organisms from a facility normally dealing with mass cultures. See Section 8740B.4 for procedures related to the acclimating, holding, and maintaining of stock cultures.

2. Static, Short-Term, Early-Life-Stage Test

As an example of a test system that may be used with several different species, the following is a toxicity test procedure that has been used with embryos of the blue crab, *Callinectes sapidus.*[1]

a. General procedures: Conduct preliminary tests to become familiar with the test procedures.

1) Collecting adult female *Callinectes sapidus* with sponge (embryos)—Collect *Callinectes sapidus* with sponge in crab traps or buy from local fishermen. They are available from March through October in the southeastern United States. Sponges that are bright yellow are preferred (Stage 3 embryos). Later stages include orange (Stage 6) and red-brown (Stage 7). Each sponge has 2 to 3 million embryos.

2) Collecting embryos from the sponge of *Callinectes sapidus*—Using forceps, remove pieces of sponge from a sponge-carrying female and shake pieces gently in a beaker of seawater (salinity between 18 and 33 parts per thousand). Take up embryos (Figure 8740:2) shaken from the sponge with a pipet and transfer

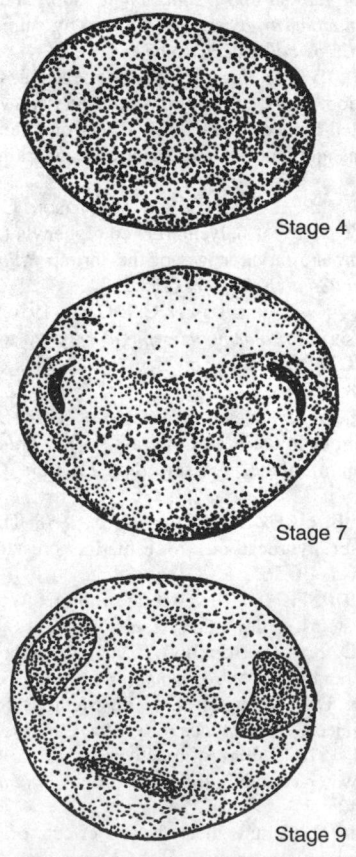

Stage 4

Stage 7

Stage 9

Figure 8740:2. Crustacean embryos.

to culture plates containing natural seawater. Embryos that are Stage 6 or younger can be kept in seawater at 3°C in the refrigerator for up to 1 month. When needed for toxicity tests, bring embryos to room temperature (20°C).

3) *Exposure chambers*—Use sterile 24-well polystyrene culture plates (well diameter 16 mm).

b. Conducting the toxicity tests: Add 10 embryos to each well. Add toxicants dissolved in water or 1 µL solvent (ethanol or acetone). For solvent controls use 1 µL solvent. Use five replicates for each concentration. Add 2 mL seawater to each well. Incubate plates at 28°C in the dark and examine each day until zoea (hatching stage) emerge from the egg sacs. Determine hatching by checking embryos each day under a dissecting microscope. Stage 3 embryos take approximately 7 d at 28°C to hatch. Stage 6 embryos take about 4 d to hatch. The zoea (Figure 8740:3) emerge over a 12-h period from Stage 9.

c. Interpreting results: Calculate EC50 values using probit analysis procedures, after counting the number of emerged zoea in control and toxicant treated wells. Hypothesis testing procedures (Dunnett's Test, etc.—see Section 8010G) may be used to estimate the NOEC and LOEC. An additional end point that may be of value is the concentration that produces a 25% inhibition in normal survival (IC25).

3. Long-Term Tests in Flowing Exposure Systems

a. General procedures: Become familiar with the test species and procedures described below before conducting a test with an effluent or specific toxicant.

b. Collecting and holding animals: See Sections 8740B.2 through 4 for information on collecting and holding the organism (crab, shrimp, lobster) selected for testing. Static systems (closed aquariums) can be used to hold decapods either before testing or to obtain larvae, but the water must be changed frequently, depending on the number of individuals and size of animals in each aquarium. Monitor ammonia levels in the water to determine when to renew the water. Avoid excess feeding, which leads to water cloudiness from bacterial growth. A flow-

ing clean seawater system, where the water is completely replaced in 24 h (about 7 tank volumes/d), is ideal for holding decapods.

c. Preparation of exposure system: Figure 8740:1 shows apparatus that is also appropriate for use in the flowing-water exposure of fresh water or marine decapod larvae or postlarvae. These trays, which fill and empty as a function of the flow rate and the high and low levels of the automatic siphon, may be constructed to hold several partitions or several beakers with nylon mesh over drilled holes. One tray can hold all the replicate exposure chambers of a given exposure concentration, or if space permits there can be two trays per concentration, representing true replicates of a specific concentration. Include five toxicant concentrations plus a control in a test as well as a solvent control if a carrier solvent is used. The flow to a specific tray therefore must contain a given concentration of the test substance produced by mixing clean seawater and the high concentration of the toxicant (or full- strength effluent). Each tray (or set of replicate trays) must receive one of the six concentrations of test material (including control), and the same total flow rate of water.

Two types of exposure systems have been used frequently. In a simple system, a Mariotte bottle of toxicant is placed above a head-box receiving a flow of clean water.[2] The head-box has an overflow tube to keep the head of water and flow rate stable, and the main flow from the box goes to the delivery box. As Figure 8740:4 shows, the slow drip from the bottle enters the flow of clean water so that mixing will occur before, and as, this water enters the delivery box. This long, narrow delivery box or tube receiving the high concentration of toxicant can be fitted with stoppers and glass tubing that exit the stopper and then make a 90° angle, paralleling the tube. The outer portion of the tubing then is bent to curve downward. When all stoppers and tubes are in place and the delivery box is receiving test solution, the tubes can be bent downward, such that the water begins to flow out the end. A funnel held in place over the exposure tray receives this water at a given rate from the test contaminated water supply, and a similar system (with larger tubing and higher flow rates) can be used to in-

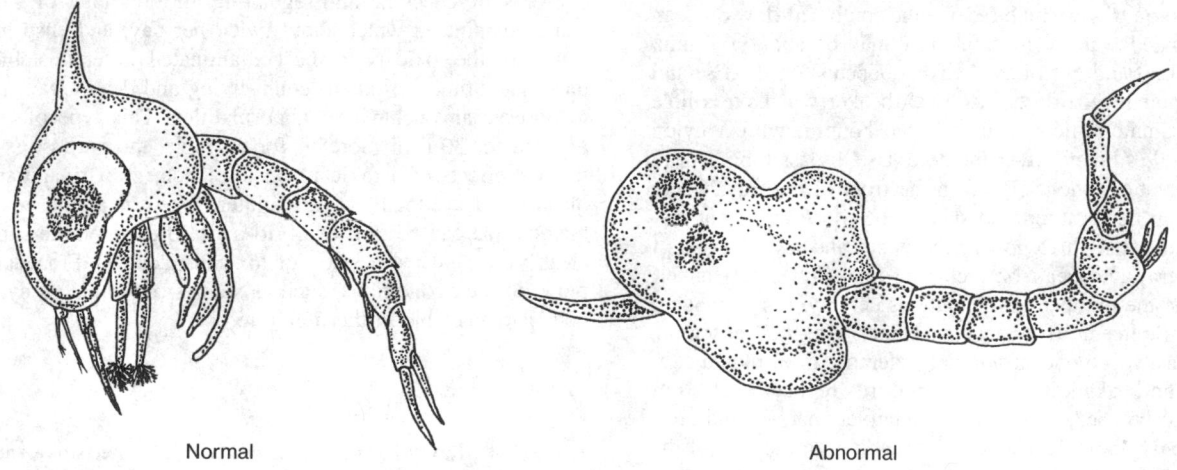

Normal   Abnormal

Figure 8740:3. Crustacean larvae.

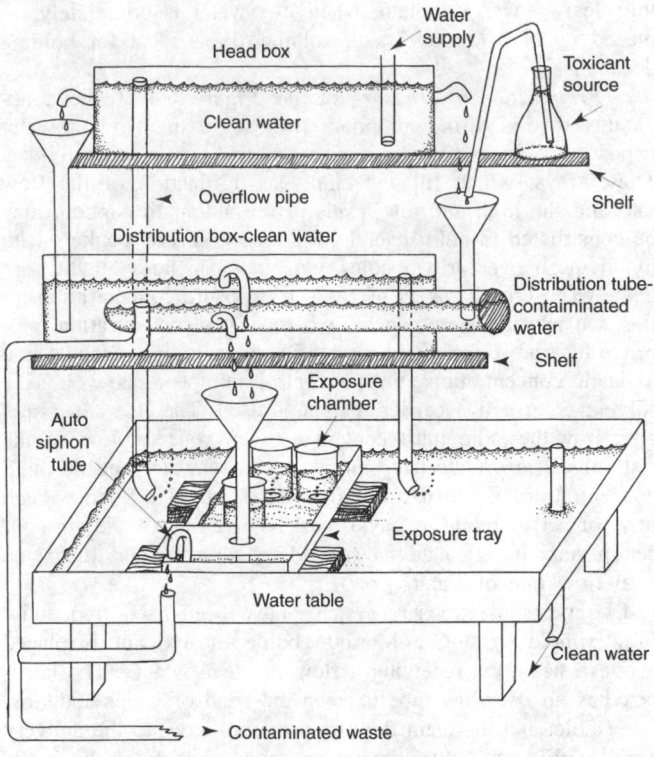

Figure 8740:4. Water table.

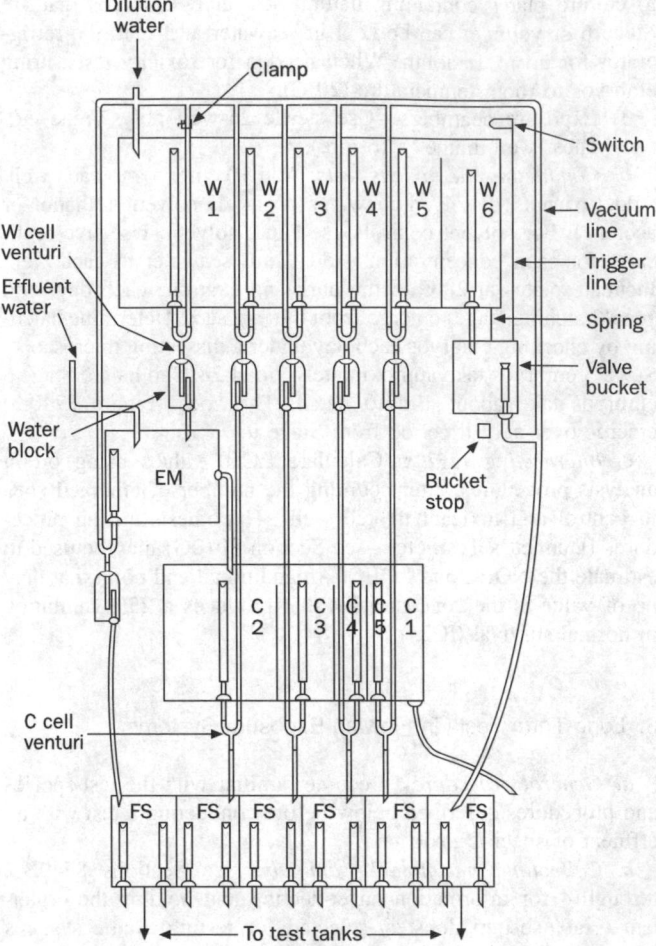

Figure 8740:5. Proportional diluter. Source: LEMKE, A.E., W.R. BRUNGS & B.J. HALLIGAN. 1978. Manual for Construction and Operation of Toxicity-Testing Proportional Diluters. EPA-600/3-78-072, U. S. Environmental Protection Agency, Duluth, Minn.

troduce the clean water to the same funnel. The flow rates of both systems can be regulated by the size of the tubing and the direction downward of the end of the tubing. The total flow rate in each tray must be the same (±15%), and the concentrations of the exposure dilutions should vary by a factor of approximately 0.5 (100%, 50%, 25%, etc.).

A second exposure system is the Brungs-Mount or Mount-Brungs diluter (Figure 8740:5). Another description and diagram of the system are found elsewhere.[3] Once constructed, this system will deliver six concentrations of contaminant to two replicate tanks, as long as needed.

d. Conducting the test: Expose postlarval decapods in beakers that are placed on trays, which receive the combined flow of clean and contaminated water. The total flow may be about 1 L/min, but the flow into and out of the beakers depends on the rise and fall of the water level in the tray. Each beaker will have a hole drilled by a diamond-hole drill near the bottom, with a nylon mesh cemented (silicone) over the hole. Use a glass tube bent to form an automatic siphon, placed in the front of the tray, where the effluent can be captured and sent to the waste treatment system. Adjust tube length to govern the upper and lower levels of water in the tray and beakers; incoming total flow rate (funnel) will determine the number of fluctuations per day. For a standard test use two replicates (trays) of each inflowing concentration, and use at least four concentrations (preferably five) plus a control. NOTE: The beakers in a tray are not true replicates. If 10 or 12 trays are to be used, they must be relatively narrow and preferably hold only about 10 beakers. Depending on the size of the eggs, larvae, postlarvae, or juvenile, there may be from 1 to 5 animals per beaker. Use at least 10 animals per replicate to con-

duct a toxicity test. With this type of exposure system, the primary concerns are checking and regulating the flow rates of the clean and contaminated water about twice per day, and ensuring that inflow of the toxicant to the contaminated water is stable. Use these same time periods to count living and dead organisms and to observe any behavioral abnormalities. This type of test can proceed for 30 d or more if, for example, the purpose is to determine effects of a toxicant on the hatch, growth, and survival of larvae. The Mount-Brungs diluter system is another exposure method that can be run for 10 to over 30 d, given the supply of clean water and the volume of toxicant required. Procedures for more difficult conditions, such as studies of petroleum hydrocarbons, are available in the literature.[4-7]

4. References

1. LEE, R.F., K. O'MALLEY & Y. OSHIMA. 1996. Effects of toxicants on developing oocytes and embryos of the blue crab, *Callinectes sapidus*. *Mar. Environ. Res.* 42:125.

2. VANDERHORST, J.R., C.I. GIBSON, L.J. MOORE & P. WILKERSON. 1977. Continuous-flow apparatus for use in petroleum bioassay. *Bull. Environ. Contam. Toxicol.* 17:577.

3. LEMKE, A.E., W.A. BRUNGS & B.J. HALLIGAN. 1978. Manual for construction and operation of toxicity-testing proportional diluters. EPA-600/3-78-072, U.S. Environmental Protection Agency, Duluth, Minn.

4. ANDERSON. J.W., S.L. KIESSER & J.W. BLAYLOCK. 1980. The cumulative effect of petroleum hydrocarbons on marine crustaceans during constant exposure. *Rapp. P.-v. Reun. Cons. Int. Explor. Mer.* 179:62.

5. ANDERSON, J.W., S.L. KIESSER, D.L. McQUERRY, R.G. RILEY & M.L. FLEISCHMANN. 1984. Toxicity testing with constant or diluting concentrations of chemically dispersed oil. *In* E.A. Thomas, ed., Oil Spill Chemical Dispersants: Research, Experience, and Recommendations. ASTM STP 840, p. 14. American Soc. Testing & Materials, Philadelphia, Pa.

6. ANDERSON, J.W. 1986. Predicting the effects of complex mixtures on marine invertebrates by use of a toxicity index. *In* C.H. Ward & B.T. Walton, eds. Environmental Hazard Assessment of Effluents, Spec. Publ. Soc. Environmental Toxicology & Chemistry, p. 115. Pergamon Press, New York, N.Y.

7. WELLS, P.G.K., J.W. ANDERSON & D. MACKAY. 1984. Uniform methods for exposure regimes in aquatic toxicity experiments with chemically dispersed oils. *In* T.E. Allen, ed. Oil Spill Chemical Dispersants: Research, Experience, and Recommendations. ASTM STP 840, p. 23. American Soc. Testing & Materials, Philadelphia, Pa.

5. Bibliography

GREEN, F.A., JR., J.W. ANDERSON, S.R. PETROCELLI, B.J. PRESLEY & R. SIMS. 1976. Effect of mercury on survival, respiration and growth of postlarval white shrimp *Penaeus setiferus*. *Mar. Biol.* 37:443.

CHEM, J.-C. 1990. Lethal effect of ammonia and nitrite on *Penaeus chilensis* juveniles. *Mar. Biol.* 107:427.

LOMBARDO, R.J., et al. 1991. Effects of lindane and acetone on the development of larvae of the southern king crab (*Lithodes antarcticus* Jaquinot). *Bull. Environ. Contam. Toxicol.* 46:185.

REDDY, M.S., et al. 1991. Toxic impact of aldrin on acid and alkaline phosphatase activity of penaeid prawn, *Metapenaeus monoceros: In vitro* study. *Bull. Environ. Contam. Toxicol.* 46:479.

RODRIQUEZ, E.M. & O.A. AMIN. 1991. Acute toxicity of parathion and 2,4-D to larval and juvenile stages of *Chasmagnathus granulata* (Decapoda, Brachyura). *Bull. Environ. Contam. Toxicol.* 46:634.

JAMES, M.O., A.H. ALTMAN, C-L. J. LI & S.M. BOYLE. 1992. Dose- and time-dependent formation of benzo(a)pyrene metabolite DNA adducts in the spiny lobster, *Panularis argus*. *Mar. Environ. Res.* 34:299.

STEELE, C.W., S. STRICKLER-SHAW & D.H. TAYLOR. 1992. Attraction of crayfishes *Procambarus clarkii, Orconectes rusticus* and *Cambarus bartoni* to a feeding stimulant and its suppression by a blend of metals. *Environ. Toxicol. Chem.* 11:1323.

ABDULLAH, A.R., A. KUMAR & J.C. CHAPMAN. 1994. Inhibition of acetylcholinesterase in the Australian freshwater shrimp (*Paratya australiensis*) by profenofos. *Environ. Toxicol. Chem.* 13:1861.

BARRE, J.S., et al. 1994. Toxicity of the water-soluble fraction of diesel fuel to stop settlement of juvenile crabs (*Menippe mercenaria*). *Bull. Mar. Sci.* 55:235.

BLECKMANN, C.A., B. RABE, S.J. EDGMON & D. FILLINGAME 1995. Aquatic toxicity variability for fresh- and saltwater-species in refinery wastewater effluent. *Environ. Toxicol. Chem.* 14:1219.

CANLI, M. & R.W. FURNESS. 1995. Mercury and cadmium uptake from seawater and from food by the Norway lobster *Nephrops norvegicus*. *Environ. Toxicol. Chem.* 14:819.

HALL, L.W., M.C. ZIEGENFUSS, R.D. ANDERSON & W.D. KILLEN, JR. 1995. Use of estuarine water column tests for detecting toxic conditions in ambient areas of the Chesapeake Bay watershed. *Environ. Toxicol. Chem.* 14:267.

HORST, M.N. & A.N. WALKER. 1995. Biochemical effects of diflubenzuron on chitin synthesis in the postmolt blue crab *Callinectes sapidus*. *J. Crust. Biol.* 15:401.

WOLFE, D.A., et al. 1995. Comparative toxicities of polar and non-polar organic fractions from sediments affected by the *Exxon Valdez* oil spill in Prince William Sound, AK. *Chem. Ecol.* 10:137.

8740 D. Data Evaluation

1. Calculating the Results

Section 8010G describes methods for calculating, analyzing, and reporting results of toxicity tests. In acute toxicity tests, the LC50 or EC50 values may be determined by probit, linear interpolation, or even graphical methods, which also may provide the 25% effects concentration (EC25). Computer programs* not only provide these values, but present the 95% confidence limits. Chronic tests may produce EC25 and EC50 values; also use hypothesis testing (Dunnett's, etc.) to produce NOEC and LOEC values. Most available computer programs will first test the data for assumptions of normality, and also test that the variances of the different treatment groups are homogenous. These criteria should be met before conducting hypothesis testing to determine NOEC and LOEC values. Transform percentage data by arcsine square-root transformation before using Dunnett's test. Transformations of quantitative data (log, square-root, etc.) as number of larvae, or weight of larvae often are useful in helping the data meet assumptions of normality and homogeneity of variance.

2. Reporting the Results

See Sections 8010G.3 and 8010H.

---

* Such as ToxCalc, TOXSTAT, and EPA programs.

8750 AQUATIC INSECTS*

8750 A. Introduction

1. Ecological Importance

Aquatic insects are important components of lake and stream biota.[1-3] In trout streams, they comprise 50 to 90% of the macroinvertebrate species. Such groups as mayflies, stoneflies, caddisflies, and midges are major food items for many species of fish.[1,4] Aquatic insects may be more sensitive to certain pollutants than are fish.[5,6]

2. Suitability for Toxicity Tests

The wide variety of aquatic insects, their abundance in unpolluted streams, their sensitivity to low concentrations of pollutants, and the ease of maintaining many species under laboratory conditions make them useful test animals. Procedures using aquatic insects have been developed for determining acceptable environmental conditions or concentrations of toxicants.[2,7] Most studies have been short-term, but procedures are available for long-term tests.

Toxicants may interfere with survival, growth, reproduction, emergence, and metabolism of aquatic insects. Because effects of long-term exposure to sublethal concentrations of toxicants may be more relevant than effects of infrequent short-term exposure to higher concentrations, flow-through, long-term tests are recommended for these applications.

_____
* Approved by Standard Methods Committee, 1993.

3. References

1. MERRITT, R.W. & K.W. CUMMINS. 1996. An Introduction to the Aquatic Insects of North America, 3rd ed. Kendall/Hunt Publishing Co., Dubuque, Iowa.
2. RESH, V.H. & D.M. ROSENBURG. 1984. The Ecology of Aquatic Insects. Praeger Scientific, New York, N.Y.
3. WILLIAMS, D.D. & B.W. FELTMATE. 1992. Aquatic Insects. CAB International, Wallingford, U.K.
4. HYNES, H.B. 1970. Ecology of Running Waters. Univ. Toronto Press, Buffalo, N.Y. & Toronto, Ont., Canada.
5. HART, C.W., JR. & S.C.H. FULLER. 1974. Pollution Ecology of Freshwater Invertebrates. Academic Press, New York, N.Y.
6. MAYER, F.L., JR. & M.R. ELLERSIECK. 1986. Manual of Acute Toxicity: Interpretation and Data Base for 410 Chemicals and 66 Species of Freshwater Animals. U.S. Dep. Interior, Fish & Wildlife Serv., Resource Publ. 160, Washington, D.C.
7. SURBER, E.W. & T.O. THATCHER. 1963. Laboratory studies of the effects of alkyl benzene sulfonate on aquatic invertebrates. _Trans. Amer. Fish. Soc._ 92:152.

4. Bibliography

EDMUNDSON, W.T. 1959. Freshwater Biology, 2nd ed. John Wiley & Sons, Wiley Interscience, New York, N.Y.
MACAN, T.T. 1963. Freshwater Ecology. John Wiley & Sons, Wiley Interscience, New York, N.Y.
HYNES, H.B. 1970. Biology of Polluted Waters. Univ. Toronto Press, Buffalo, N.Y. & Toronto, Ont., Canada.
GAUFIN, A.R. 1972. Water Quality Requirements of Aquatic Insects. Final Rep. Contract 14-12-438, Water Quality Off., U.S. Environmental Protection Agency.

8750 B. Selecting and Preparing Test Organisms

1. Species Selection

Use insects that are important food for fishes, readily available and abundant, relatively easy to keep and culture in the laboratory, and most sensitive to the materials under investigation.

a. _Suggested test organisms:_
1) Stoneflies (See Section 10900, Plate 13, A–D.)
   Pteronarcys dorsata
   Pteronarcys californica
   Hesperoperla lycorias
   Hesperoperla pacifica
2) Mayflies (See Section 10900, Plate 13, E–I.)
   Hexagenia bilineata
   Hexagenia limbata
   Hexagenia rigida
   Ephemerella subvaria
3) Caddisflies (See Section 10900, Plate 15, H–K.)
   Brachycentrus americanus
   Brachycentrus occidentalis
   Clistoronia magnifica
b. _Other species that have been used:_
1) Stoneflies
   Isogenus frontalis
   Perlesta placida
   Paragnetina media
   Phasganophora capitata
   Acroneuria californica
2) Mayflies
   Ephemerella cornuta
   Ephemerella grandis
   Ephemerella doddsi
   Ephemerella needhami
   Ephemerella tuberculata
   Stenonema ithaca
3) Caddisflies
   Hydropsyche betteni

Macronemum zebratum
Arctopsyche grandis
Hydropsyche bifida
4) Diptera (See Section 10900, Plate 16, A–B.)
Chironomus plumosus
Chironomus attenuatus
Chironomus tentans
Chironomus californicus
Glyptochironomus labiferus
Goeldichironomus holoprasinus
Tanypus grodhausi
Tanytarsus (paratanytarsus) dissimilis

For each test, use early instar larvae or nymphs when possible, especially for growth studies. Of the listed species, only chironomids complete a generation in one summer. Use late instars for adult emergence tests. For all tests, insects that are cultured are preferable because their source and history are known.

2. Collecting Test Animals

When cultured animals are not available, collect all test specimens from clean, natural waters rich in aquatic insects (see Section 10500, Benthic Macroinvertebrates). Collect larger stream species from riffle areas of clean, well-aerated gravel rubble streams with hand screens or bottom samplers. Stir bottom and let current carry dislodged insects downstream into net.

Immediately after collection, gently place net contents in a 15- to 20-L insulated container partly filled with stream water. Transport to laboratory. Remove and discard larger rocks after it has been determined that they are free of insects. If transportation time exceeds 30 min, provide for aeration and temperature control. In laboratory, swirl water in containers and dip it out. Pour through a screen-bottom container (of a mesh that will retain insects required), held partly submerged in a tank of water. Wash screenings into a holding tank. If it is desired to separate insects, wash into a large white enamel pan containing 3 to 5 cm of water. Remove desired species with a large-bore pipet or small spoon-shaped screen and place in holding tanks. For riffle insects use oval or round flow-through tanks[1] provided with rocks for cover and paddle wheels to provide a current in dilution water.[1] Alternatively collect insects by gently picking up rocks, rubble, or gravel, and carefully washing or picking, then placing desired insects in insulated containers for transport to laboratory.

To obtain benthic insects, sample bottom materials with Eckman, Petersen, or Ponar dredges. Empty dredge into a large pail, add water, and swirl by hand. Partly submerge an appropriate mesh washing screen, pour a portion of swirling sample into it, and wash by moving up and down in the water. Place washed insects in an insulated container and continue until enough insects have been collected.

Chironomids probably will be the dominant insect species in silt bottom material. However, other important immature insects such as dragonflies, damselflies, several species of Diptera, beetles, and mayflies may be found in and on silt bottoms. The mayfly, *Hexagenia limbata*, is a large species often occurring in great abundance in soft, unpolluted muds rich in organic matter that occur in deep pools, ponds, lakes, and reservoirs. Obtain these by collecting top 8 cm of mud and washing as described previously.

3. Holding, Acclimating, and Culturing

a. General considerations: As soon after collection as possible, examine insects for injury. Place all uninjured specimens in holding chambers, supply them with food, and hold for at least 1 week for observation and acclimation to desired temperature. Acclimate stream species in flowing water. Keep in oval troughs that have a current of water or in stainless steel wire cages in running water.[1] In these troughs include flat stones covered with attached algae as cover and food for herbivorous species. Supply insects with materials to build larval and pupal cases. For caddisflies, use sand grains, small pieces of wood, and plant materials retained by a 16-mesh screen. Permit insects that construct tubes or cases to do so. Hold benthic species in aquariums provided with a 3- to 5-cm layer of unsterilized mud from the site where they were collected. *Hexagenia* require a substrate in which to burrow.[2] For chironomids use the highly organic ooze that overlies the bottom where they were collected. Alternatively, either silica sand or shredded paper toweling may be used as a substrate for chironomid larva.[3,4]

Provide water, DO, and other conditions as described in Section 8010E and F. Maintain final holding temperature within 3°C of temperature at which organisms were collected. For long holding periods, maintain natural seasonal temperatures. When aquatic insects are collected in winter at water temperatures of 1°C or lower, acclimate them to higher temperatures if they are to be used in short-term tests (Section 8010E.3).

Different species require different light intensities. Stoneflies require stones under which they can hide from direct light. Fix light cycle at a certain day length, or vary it seasonally to correspond with natural annual photoperiod. For *Chironomus plumosus*, use a 16-h photoperiod. Lamps and fixtures are described in Section 8010F.3*f*.

b. Food and feeding: *Acroneuria, Isogenus,* and *Paragnetina* are predators requiring live food. Feed to excess with small midges, blackfly larvae, mosquitoes, or small caddisfly larvae from an unpolluted environment.[2] Feed *Pteronarcys* and *Ephemerella* to excess with coarse, chopped maple, birch, or aspen leaves that have fallen naturally and have been dried and then soaked in test water for at least 2 weeks before feeding. Feed *Hexagenia, Hydropsyche,* and *Arctopsyche* finely ground leaves and fish-food pellets. If the substrate is rich in organic matter, additional food may not be required for *Hexagenia*. Avoid overfeeding with fish food because it causes DO depletion. The larvae of some Hydropsychidae are highly carnivorous and cannibalistic; keep them well-fed with plankton, microcrustacea, blackfly larvae, and other organisms, collected from fish hatcheries, ponds, lakes, and streams with a net of No. 20 bolting silk.

Feed chironomids twice per week. Keep in jars supplied with algal culture medium [Section 8010E.4c1)a)] inoculated with algae including diatoms. Alternatively use a mixture of 5 g fish food plus 1 g powdered dried cereal grass* blended in 1 L of water. Add about 100 mL of this suspension to each culture per feeding. If there is no flow-through, remove 100 mL of test solution before feeding. Use 10-L culture jars containing 8 L or less of medium with a screen cover to retain adults.[5,6] Keep in a constant-temperature room at 21 to 24°C. For long-term studies follow natural temperature cycle of water from which chironomids were taken.

---

* Cerophyl®, Agri-Tech, Inc., Kansas City, MO 64112, or equivalent.

Because the jars have a mud substrate, do not clean them or overfeed the organisms. Collect emerging adults for breeding in wire screen cylinders placed over the culture jars.[7,8]

4. References

1. SURBER, E.W. & T.O. THATCHER. 1963. Laboratory studies of the effects of alkyl benzene sulfonate on aquatic invertebrates. *Trans. Amer. Fish. Soc.* 92:152.

2. FREMLING, C.R. & G.L. SCHOENING. 1973. Artificial substrates for *Hexagenia* may-fly nymphs. *In* Proc. 1st Int. Conf. Ephemeroptera, p. 209.

3. GREER, I.E. 1993. Standard Operating Procedures for Culture of Chironomids (SOP B5.25) and *Hyallella azteca* (SOP B5.38). National Biological Serv., Columbia, Mo.

4. U.S. ENVIRONMENTAL PROTECTION AGENCY. 1994. Methods for Measuring the Toxicity and Bioaccumulation of Sediment-Asssociated Contaminants with Freshwater Invertebrates. EPA/600/R-94/024, U.S. Environmental Protection Agency, Washington, D.C.

5. NEBEKER, A.V. 1972. Effect of high winter water temperatures on adult emergence of aquatic insects. *Water Res.* 5:777.

6. BAY, E.C. 1967. An inexpensive filter-aquarium for rearing and experimenting with aquatic invertebrates. *Turtox News* 45:146.

7. BREVER, K.D. 1965. A rearing technique for the colonization of chironomid midges. *Ann. Entomol. Soc. Amer.* 58:135.

8. NEBEKER, A.V. & A.E. LEMKE. 1968. Preliminary studies on the tolerance of aquatic insects to heated waters. *J. Kans. Entomol. Soc.* 41: 413.

5. Bibliography

USINGER, R.L., ed. 1956. Aquatic Insects of California—With Keys to North American Genera and California Species. Univ. California Press, Berkeley.

ROBACK, S.S. 1957. The Immature Tendipedids of the Philadelphia Area. Monogr. Acad. Natural Sci. Philadelphia No. 9, Philadelphia, Pa.

FREMLING, C.R. 1967. Methods for mass-rearing *Hexagenia* (Ephemeroptera: Ephemeridae). *Trans. Amer. Fish. Soc.* 96:407.

NEBEKER, A.V. 1972. Effect of low oxygen concentration on survival and emergence of aquatic insects. *Trans. Amer. Fish. Soc.* 101:675.

GAUFIN, A.R., R. CLUBB & R. NEWELL. 1974. Studies on the tolerance of aquatic insects to low oxygen concentrations. *Great Basin Natur.* 31:45.

ANDERSON, N.H. 1977. Continuous rearing of the Limnephilid caddisfly, *Clistoronia magnifica* (Banks). Proc. 2nd Symp. on Trichoptera, 1977. Junk, The Hague, Netherlands.

MERRITT, R.W. & K.W. CUMMINS. 1984. An Introduction to the Aquatic Insects of North America, 2nd ed. Kendell/Hunt Publishing Co., Dubuque, Iowa.

PENNAK, R. 1989. Fresh Water Invertebrates of the United States, 3rd ed. John Wiley & Sons Inc., New York, N.Y.

8750 C. Toxicity Test Procedures

1. General Procedures

Conduct tests as described in Section 8010D. If possible, use a minimum of 20 specimens for each toxicant concentration with an additional 40 animals for growth studies. Two species may be tested in the same tank if precautions are taken to avoid predation.

Do not use static testing with stream insects unless air stones or water movement can simulate natural water conditions. Use static tests with certain lake or reservoir species if required DO levels are maintained. For long-term tests, see Section 8010D.

a. Test tanks: Use glass and stainless steel aquariums of either 8-L or 20-L size for quiet-water species. For stream species, use round or oval, stainless steel or epoxy-painted troughs [1,2,3] (90 cm long, 15 cm wide, and 15 cm deep) in which natural stream flow is simulated. Set tanks side by side so paddle wheels on one long shaft can be used to circulate water in them all.[1] Jetted incoming water from the diluter also can maintain adequate flow.

b. Flow rate: Use flows to each tank of no less than 6 to 10 tank volumes/24 h. In aquariums without water-circulating devices use much higher flows for stream species to simulate stream flow. In oval test tanks use velocities near 0.5 cm/s. For quiet-water forms, such as *Hexagenia* and *Chironomus*, do not disturb mud substrate with water flow.

c. Aeration: Aeration is unnecessary; however, use if desired with nonvolatile toxicants to increase or control water movement, especially for tank tests with lake and reservoir species or if DO levels drop.

d. Cleaning: See Section 8010E.4d. Siphon out detritus on tank bottom weekly during long-term testing. If a mud substrate is used, no cleaning is necessary. Avoid overfeeding.

e. Substrate: For all stream riffle species use fine-mesh stainless steel screens formed into cylinders or cubes, which provide 10 to 15 cm²/insect. Place cages in oval troughs or in glass cylinders.[1] For 30- to 90-d adult emergence tests, obtain clean rocks, 5 to 10 cm in diameter (one for every three insects) from collection site for a substrate. Provide fine screen or sticks that protrude above water surface for adult emergence tests.

f. Light and photoperiod: See Section 8010F.3f. Use natural photoperiod at time of testing for locality in which test is conducted. Increase day length during adult emergence tests by 0.5 h every 2 weeks.

g. Temperature: See Section 8010F.1b. Use 10°C as a winter temperature. For trout stream insects, use summer temperatures near 15°C. Increase temperature during adult emergence tests by 1°C each week up to a maximum of 5°C above initial temperature. When using warm-water stream or lake insects, follow natural temperature cycle.

h. Time of year: Under natural conditions, most species emerge as adults in spring. Therefore start adult emergence tests no later than March 1st. *Hexagenia limbata* and most midge species are exceptions, emerging throughout summer in most localities.

2. Toxicant Preparation

See Sections 8010F.1 and 2b.

3. Test Procedures for *Hexagenia*

Use *Hexagenia* for short-term survival (96 to 168 h), survival for 5 to 60 d, adult emergence, or full-life-cycle tests (90 to 120 d). Use a minimum of 20 organisms per aquarium of not less than 8 L capacity. Use a water depth of 8 to 20 cm. Provide a fine organic ooze substrate 4 to 5 cm deep and as similar as possible to that where naiads occur naturally. When using newly hatched *Hexagenia* to start a test, use 50/tank. When *Hexagenia* eggs are used as a source of larvae, pipet them into petri dishes (about 200/dish) with 200 mL test water at about 20°C, and let hatch.

When substrate is mud, determine survival by counting number of dead animals that have left their burrows and/or by counting number of new burrows formed after disturbing mud surface sufficiently to destroy entrances to old burrows. If counts do not agree, use the latter. For acute toxicity tests, alternatively use an artificial substrate of epoxy resin to facilitate observation and monitoring of test animals. For growth or emergence tests, set up an additional set of containers so that naiads can be removed periodically for measurement. Remove 10 naiads from their burrows after 20 to 60 d to determine growth. Do not remove more than 50% of surviving animals before conclusion of these tests. Keep a record of total number removed. Use these animals to provide additional data on growth and emergence. Record body length, head capsule width, and live weight.

In acute toxicity tests, determine survival after 1.5, 3, 6, and 12 h and twice daily thereafter. As a sign of death, use failure of specimens to respond by movement to gentle probing or flash light illumination. In longer-term studies, check tanks daily to remove and record dead animals and cast naiad skins, which indicate successful molting.

For growth studies, determine initial range and mean of total length, head capsule width, and weight from specimens in holding tank. Kill all animals in warm water (40 to 50°C) before measuring. Take measurements twice during testing, using animals that are to be discarded. Obtain final measurements for all survivors. Make two counts: number of adults and cast skins; if different, use cast skins because some adults may have escaped.

Determine and record percentage of adults that emerge, sex, incidence of incomplete emergence (i.e., half-out of nymphal skin, wings unsuccessfully unfolded, etc.), adult length, weight, and head capsule width, and number of mature eggs.

4. Test Procedures for *Chironomus*

Follow procedures described in Section 8010F. For each concentration, use duplicate 20-L aquariums with mud or powdered dried cereal grass* substrate and screen covers. Maintain flow to each test container at about 2 L/h. Use a mud substrate similar to that for *Hexagenia*. Use lighting and photoperiod as described in Section 8010F.3f. Do not feed animals during short-term tests. Feed during 30-d and emergence tests as in Section 8750B.3b. If prepared food is used, add powdered dried cereal grass* or about 100 mL food suspension to each container twice per week.

For long-term tests, place 50 first-instar larvae (about 1.5 mm long and less than 24 h old) in each test aquarium. Transfer larvae

_____
* Cerophyl® or equivalent.

with an eyedropper. Determine number of emerging adult males and females. Count both adults and pupal cases. If counts differ, use pupal case count. At 25 ± 1°C emergence takes about 1 month. To determine success of fertilization of eggs, take 50 eggs and determine the percent hatchability. If it is impossible to separate and count eggs, hatch fertilized egg masses in beakers with same test water from which adults emerged. Count 60 larvae into a petri dish and examine for injured larvae. Transfer often will injure early instar larvae; if a correction for this is not made in the count, errors in percent survival result. After examination, count 50 larvae and return to test chamber and rear them to adult stage. End points for taking and analyzing data are emergence of adults, egg production, and hatching of young. Repeat complete test at least once.

5. References

1. SURBER, E.W. & T.O. THATCHER. 1963. Laboratory studies of the effects of alkyl benzene sulfonate on aquatic invertebrates. *Trans. Amer. Fish. Soc.* 92:152.
2. NEBEKER, A.V. 1972. Effect of high winter water temperatures on adult emergence of aquatic insects. *Water Res.* 5:777.
3. NEBEKER, A.V. & A.E. LEMKE. 1968. Preliminary studies on the tolerance of aquatic insects to heated waters. *J. Kans. Entomol. Soc.* 41:413.
4. FREMLING, C.R. & G.L. SCHOENING. 1973. Artificial substrates for *Hexagenia* may-fly nymphs. *In* Proc. 1st Int. Conf. Ephemeroptera, p. 209.

6. Bibliography

SANDERS, H.O. & O.B. COPE. 1968. The relative toxicities of several pesticides to naiads of three species of stone-flies. *Limnol. Oceanogr.* 13:112.

GAUFIN, A.R. & S. HERN. 1971. Laboratory studies on tolerance of aquatic insects to heated waters. *J. Kans. Entomol. Soc.* 44:240.

ROUSSEL, J.S. 1972. Standard methods for detection of insecticide resistance of *Diabrotica* and *Hypera* beetles. *Bull. Entomol. Soc. Amer.* 18:179.

FRIESEN, M.K. 1979. Use of eggs of the burrowing mayfly *Hexagenia rigida* in toxicity testing. *In* E. Scherer, ed. Toxicity Tests for Freshwater Organisms. *Can. Spec. Publ. Fish. Aquat. Sci.* 44:27.

AMERICAN SOCIETY FOR TESTING AND MATERIALS. 1993. Standard Practice for Conducting Acute Toxicity Tests with Fishes, Macroinvertebrates, and Amphibians. ASTM Standard E 729-88, Philadelphia, Pa.

FREMLING, C.R. & W.L. MAUCK. 1980. Methods for using nymphs of burrowing mayflies (Ephemeroptera, *Hexagenia*) as toxicity test organisms. *In* A.L. Buikema & J. Cairns, eds. Aquatic Invertebrate Bioassays. ASTM STP 715, American Soc. Testing & Materials, Philadelphia, Pa.

SAOUTER, E., L. HARE, P.G.C. CAMPBELL, A. BOUDOU & F. RIBEYRE. 1993. Mercury accumulation in the burrowing mayfly, *Hexagenia rigida* (Ephemeroptera) exposed to CH₃HgCl or HgCl₂ in water and sediment. *Water Res.* 27:1041.

AMERICAN SOCIETY FOR TESTING AND MATERIALS. 1993. ASTM Standards on Aquatic Toxicology and Hazard Evaluation. ASTM Committee E-47, ASTM Publ. 03-547093-16, American Soc. Testing & Materials, Philadelphia, Pa.

8750 D.   Data Evaluation

Analyze, evaluate, and report data from various tests as described in Section 8010G.

8810   ECHINODERM FERTILIZATION AND DEVELOPMENT (PROPOSED)*

8810 A.   Introduction

1. Background

The Phylum Echinodermata encompasses a widely distributed and diverse group of marine animals. The class Echinoidea includes sand dollars and sea urchins, organisms that are common inhabitants of rocky shores and ocean bottoms over all depth ranges. Many echinoid species are maintained easily in the laboratory and are responsive to simple methods of spawning inducement. Gametes and embryos are easily obtained and reared in the laboratory and have been the subject of scientific research for over 100 years.[1] Numerous species have been used and laboratory techniques have been developed that utilize various stages of the organism's life cycle.[2–4]

Toxicity tests utilizing the short-term exposure of gametes or embryos are of comparable or greater sensitivity to many contaminants than tests with other marine species and life stages.[5–9] Echinoid toxicity tests can be performed on small volumes (≥ 2 mL) over short time periods (1 to 96 h), and under static conditions without feeding. These tests have been used successfully to evaluate the toxicity of effluents, receiving waters, chemicals, and sediments, provided the salinity of the test samples is near typical ocean levels (28 to 34 g/kg). Recent adaptations of these test methods have expanded applications to include evaluation of genotoxic effects,[10] interstitial water,[11] and toxicity identification evaluation (TIE) studies.[12] Methods similar to these have been proposed or recommended as components of regulation programs.[13]

2. References

1. HINEGARDNER, R.T. 1969. Growth and development of the laboratory cultured sea urchin. *Biol. Bull.* 137:465.
2. DINNEL, P.A., G.G. PAGANO & P.S. OSHIDA. 1988. A sea urchin test system for environmental monitoring. *In* R.D. Burke, P.V. Mladenov, P. Lambert & R.L. Parsley, eds. Echinoderm Biology. A.A. Balkema, Rotterdam, The Netherlands.
3. BAY, S., R. BURGESS & D. NACCI. 1993. Status and applications of echinoid (phylum echinodermata) toxicity test methods. *In* W.G. Landis, J.S. Hughes & M.A. Lewis, eds. Environmental Toxicology and Risk Assessment. ASTM STP 1179, American Soc. Testing & Materials, Philadelphia, Pa., p. 281.
4. KOBAYASHI, N. 1995. Bioassay data for marine pollution using echinoderms. *In* P.N. Chermisinoff, ed. Encyclopedia of Environmental Control Technology. Gulf Publ. Co., Houston, Tex, p. 539.
5. OKUBO, K. & T. OKUBO. 1962. Study of the bio-assay method for the evaluation of water pollution—II. Use of the fertilized eggs of sea urchins and bivalves. *Bull. Tokai Reg. Fish. Res. Lab.* 32:131.
6. NACCI, D., E. JACKIM & R. WALSH. 1986. Comparative evaluation of three rapid marine toxicity tests: Sea urchin early embryo growth test, sea urchin sperm cell toxicity test and Microtox. *Environ. Toxicol. Chem.* 5:521.
7. DINNEL, P.A., J.M. LINK, Q.J. STOBER, M.W. LETOURNEAU & W.E. ROBERTS. 1989. Comparative sensitivity of sea urchin sperm bioassays to metals and pesticides. *Archiv. Environ. Contam. Toxicol.* 18: 748.
8. MORRISON, G., E. TORELLO, R. COMELEO, R. WALSH, A. KUHN, R. BURGESS, M. TAGLIABUE & W. GREENE. 1989. Intralaboratory precision of saltwater short-term chronic toxicity tests. *Res. J. Water Pollut. Control Fed.* 61:1707.
9. SCHIMMEL, S.C., G.E. MORRISON & M.A. HEBER. 1989. Marine complex effluent toxicity program: Test sensitivity, repeatability and relevance to receiving water toxicity. *Environ. Toxicol. Chem.* 8:739.
10. HOSE, J.E. 1985. Potential uses of sea urchin embryos for identifying toxic chemicals: Description of a bioassay incorporating cytologic, cytogenetic and embryologic endpoints. *J. Appl. Toxicol.* 5:245.
11. CARR, R.S. & D.C. CHAPMAN. 1995. Comparison of methods for conducting marine and estuarine sediment porewater toxicity tests—extraction, storage, and handling techniques. *Arch. Environ. Contam. Toxicol.* 28:69.
12. BAILEY, H.C., J.L. MILLER, M.J. MILLER & B.S. DHALIWAL. 1995. Application of toxicity identification procedures to the echinoderm fertilization assay to identify toxicity in a municipal effluent. *Environ. Toxicol. Chem.* 14:2181.
13. PASTOROK, R.A., J.W. ANDERSON, M.K. BUTCHER & J.E. SEXTON. 1994. West Coast marine species chronic protocol variability study. Final Rep. for Washington Dep. Ecology, Olympia.

3. Bibliography

PAGANO, G., G. CORSALE, A. ESPOSITO, P.A. DINNEL & L.A. ROMANA. 1989. Use of sea urchin sperm and embryo bioassay for testing the sublethal toxicity of realistic pollutant levels. *Advan. Appl. Biotechnol. Ser.* 5:153.

---

* Approved by Standard Methods Committee, 1997.

8810 B.   Selecting and Preparing Test Organisms

1. Selecting Test Organisms

In accord with the criteria listed in Section 8010E.1, the recommended test species include (but are not restricted to) the following:

Scientific Name	Common Name	Location	Approximate Spawning Season
Arbacia punctulata	Atlantic sea urchin	Atlantic coast Gulf coast	Summer Winter
Strongylocentrotus droebachiensis	Green sea urchin	Northern Atlantic and Pacific coasts	Winter
Stronglyocentrotus purpuratus	Pacific purple sea urchin	Pacific coast	Fall-spring
Dendraster excentricus	Pacific eccentric sand dollar	Pacific coast	Spring-summer

The use of the above species is encouraged to increase the comparability of results from different laboratories. Successful toxicity tests can be conducted with other species, such as the Hawaiian sea urchins *Echinometra* spp. and *Tripneustes gratilla*[1] or *Lytechinus* spp.[2] Use of alternate species may be advantageous in certain regions, but modifications to the test method may be necessary and the results may not be comparable.

2. Collecting Broodstock

Obtain test organisms (gametes or embryos) from broodstock collected from the field during their natural spawning season and held in the laboratory until needed. Collect in areas away from obvious sources of pollution and having water quality similar to that used for holding and testing. Organisms obtained from a commercial supplier may be used. *Dendraster excentricus* forms dense aggregations in intertidal or subtidal sandy areas; collect individuals by hand at low tide, by diving, or by dredge. Regular sea urchin species inhabit rocky or sandy areas of the intertidal and subtidal zones. Collect individuals by hand, either at low tide or by diving.

Discard any individuals damaged during collection or subsequent handling. Avoid sudden or extreme variations in temperature, salinity, or other environmental factors during collection and transport, because they may induce premature spawning. Animals may be shipped by overnight mail service in insulated containers containing an ice substitute. Do not ship animals submerged in water because the oxygen will be depleted rapidly and premature spawning may occur. Instead, wrap animals in seaweed or towels soaked in seawater to maintain high humidity.

3. Culture Techniques

Hold sea urchins and sand dollars in aquariums with either a flow-through seawater supply or recirculating filter system.[3] Feed sea urchins *ad libitum* brown macroalgae, such as *Macrocystis* spp. or *Egregia* spp. Substitute romaine lettuce[4] or commercial fish feed if fresh seaweed is unavailable. Rehydrated seaweed purchased from food markets also has been used successfully. Aquariums containing sand dollars should contain several centimeters of sand if animals are to be held more than a few days. Sand dollars feed on suspended or benthic materials (e.g., detritus or plankton); provide a source of unfiltered natural seawater or prepared food (e.g., powdered fish feed) if the animals will be held for long periods.

Holding temperature varies with the species and should be similar to that at the collection site. Recommended temperatures are 15 to 18°C for *A. punctulata*, 12 to 16°C for *D. excentricus*, 8 to 15°C for *S. purpuratus*, and 8 to 12°C for *S. droebachiensis*. Hold animals at 28 to 34 g/kg salinity.

4. Parasites and Diseases

A variety of commensal organisms (e.g., annelids and crustaceans) are often associated with sea urchins and sand dollars. These organisms are not harmful and are not essential to the survival of the broodstock.

Excessive microbial growth can result from the accumulation of feces in aquariums. These growths produce metabolites (e.g., hydrogen sulfide) that may be toxic or cause stress to the animals. Clean aquariums several times per week.

5. Gamete Preparation

Induce sea urchins or sand dollars to spawn just before the beginning of a test. Pool gametes from at least three individuals of each sex to provide a representative sample for testing. Females and males of most echinoid species usually can be induced to spawn by injection of $0.5M$ potassium chloride (KCl). Use a hypodermic syringe (20-gauge needle) to pierce the peristomial membrane surrounding the mouth and inject approximately 1.0 mL (0.5 mL for sand dollars) into the coelomic cavity. Use sterile needles and KCl to guard against disease if the animals will be returned to laboratory aquariums. Usually, two injections of 0.5 mL each are made on opposite sides of the mouth. Place sea urchins upright (oral side down) and observe for evidence of gamete release through the genital pores, located around the anus on the aboral surface. Sperm are milky white in color while eggs are orange to red, depending on the species.

Electrical stimulation is an alternate spawning method that has been used with success on some species (primarily *A. punctulata*). Place electrodes from a 12-V DC power source on either side of the anal pore of the urchin, and spawning occurs until the electrodes are removed.[4] This method has the advantage of permitting a check of gamete type and quality by applying the electrodes briefly. Neither spawning method kills the animal, which may be respawned in 30 to 60 d if held under appropriate conditions.

Invert females releasing eggs (oral side up) and place on a beaker filled to the brim with seawater of the appropriate temperature. The eggs will fall to the bottom of the beaker after extrusion. Collect sperm in the "dry" condition, without contact with seawater that activates the sperm. Remove sea urchin sperm from the gonopore area with a glass transfer pipet or automatic pipet (with enlarged tip) and place in a small conical test tube. Collect sand dollar sperm by inverting males over 5- to 10-mL beakers of seawater. The sand dollar sperm fall to the bottom with little dilution and can be removed easily by pipet. Use care to avoid transferring fecal material with the gametes.

Gametes for most species may be stored for several hours in an ice bath or refrigerator. Keep eggs from each female separate until evaluated for quality. Store *A. punctulata* eggs at the culture temperature. Examine subsamples of sperm and eggs from each animal under a compound microscope to evaluate their quality. Eggs should be round and a germinal vesicle (clear spot) should not be visible. The presence of a germinal vesicle indicates immature eggs. Viability also can be checked by removing a subsample, adding sperm and determining fertilization. Use eggs from a more mature female if more than a few percent of immature eggs are present. Evaluate sperm quality by diluting a subsample in seawater and checking for motility.

Pool equal volumes (at least 0.025 mL for sea urchins) of sperm from at least three males and store in a conical tube in an ice bath. Avoid additional dilution until just before test. Activate sperm by dilution in seawater and use within 30 min. Sperm have limited energy reserves and their viability declines exponentially within minutes of activation; carefully monitor holding times of activitated sperm and preferably standardize within the laboratory. Longer holding times (e.g., 60 min) can be used successfully, provided sperm are chilled on ice and higher sperm densities are used in the test. The final density of sperm needed for the test varies according to test type, species, and maturation stage. Determine this value from trial fertilization tests (see 8810C.4*d*) or previous experience.

Gently wash eggs once by centrifugation or let them settle in the spawning beaker and decant off excess seawater. Gently resuspend eggs in fresh seawater, pool, and let settle again. Use care when washing sand dollar eggs; they are surrounded by thick jelly coat that is easily disturbed (with adverse effects on fertilization).

Prepare a working stock solution of known egg density (dependent on test volume and test type). Check density by mixing stock solution well, removing a small subsample, diluting it 1:100 × with seawater, and counting number of eggs in a known volume on a Sedgwick-Rafter cell. Use of a perforated plunger (plastic disk containing numerous holes, attached to a plastic rod) is strongly recommended to provide a homogeneous suspension of gametes and embryos for this and other steps of the procedure. Adjust density by adding or removing seawater. Store the working stock at or below the exposure temperature (depending on species) and use within 2 h if possible.

6. References

1. DINNEL, P.A. 1988. Adaptation of the sperm/fertilization bioassay protocol to Hawaiian sea urchin species. Final Rep. to State of Hawaii Dep. Health, Honolulu.
2. ENVIRONMENT CANADA. 1992. Biological Test Method: Fertilization Assay Using Echinoids (Sea Urchins and Sand Dollars). EPS 1/RM/27, Environmental Protection, Environment Canada, Ottawa, Ont.
3. LEAHY, P.S., T.C. TUTSCHELTE, R. J. BRITTEN & E.H. DAVIDSON. 1978. A large-scale laboratory maintenance system for gravid purple sea urchins (*Strongylocentrotus purpuratus*). *J. Exper. Zool.* 204:369.
4. WEBER, C.I., W.B. HORNING, II, D.J. KLEMM, T.W. NEIHEISEL, P.A. LEWIS, E.L. ROBINSON, J. MENKEDICK & F. KESSLER, eds. 1988. Short-term methods for estimating the chronic toxicity of effluents and receiving waters to marine and estuarine organisms. EPA-600/4-87-028, Environmental Monitoring and Support Lab., U.S. Environmental Protection Agency, Cincinnati, Ohio.

8810 C.   Echinoderm Fertilization Test

1. General Procedures

Conduct exploratory tests (see Section 8010D) first if the concentration range to be tested is not known. Prepare dilution water and toxicant solutions and introduce them into test containers as described in Section 8010F.

Observe the following general precautions in procedures that involve handling of gametes. First, take stringent measures to avoid cross-contamination of egg and sperm solutions, e.g., use separate pipets and glassware for each sex. Seemingly minute amounts of sperm are sufficient to fertilize an egg stock prematurely and invalidate an entire test. Second, enlarge the opening of disposable pipet tips used for dispensing gametes. Trim tip with a razor blade to produce an opening of 1 to 2 mm when transferring eggs. Preferably also enlarge pipet tips used for concentrated sperm solutions to facilitate transfer of this highly viscous suspension. Modified pipet tips are not required for sand dollar sperm shed into seawater. Finally, avoid inadvertently warming gamete or embryo solutions during preparation steps, because this may greatly hasten degradation during storage. Use a temperature-controlled room or chilled water baths to ensure that all preparatory steps are conducted at or below test temperature.

2. Water Supplies

Maintain salinity of dilution water within 2 g/kg of the holding salinity. Adjust pH to 7.7 to 8.3, unless altered pH is an important factor in the experimental design. Dilution water quality should be sufficient to produce ≥70% fertilization in control samples.

a. Artificial seawater: See Section 8010E.4*b*2). Avoid commercial sea salt mixes because they are often toxic to echinoid gametes and embryos. However, some seawater formulations based on reagent-grade chemicals have been used successfully.[1,2] Conduct preliminary tests to determine suitability of each batch of artificial salts before use.

b. Natural seawater: Choose a source of natural seawater free of contamination and of uniform quality. Pass the water through a filter with an effective pore size ≤1.0 μm to remove parasites and predators. Additional treatment (e.g., aeration, additional filtration, sterilization, or activated carbon treatment) may be needed to obtain acceptable water quality, especially during storage. Avoid prolonged storage of seawater if possible, because aging (>24 h) natural seawater can produce potentially toxic metabolites.

c. Salinity adjustment: Echinoids have limited osmoregulation ability. Adjust salinity of test samples that deviate by more than

2 g/kg from the culture environment to eliminate potential interferences. Use hypersaline brine (HSB) or an artificial sea salt mixture for salinity adjustment. Exercise caution in selecting the material used to adjust salinity so that the toxicity of the sample is not altered by the introduction of chemicals such as chelators (e.g., EDTA) or toxic contaminants (e.g., heavy metals).

Partial freezing and thawing of seawater is a convenient method of preparing HSB in sufficient quantities for fertilization or embryo development tests.[3] Freezing (-10 to $-20°C$) one or two 4-L containers (glass or plastic) overnight for 6 to 12 h will provide sufficient 80- to 100-g/kg brine for most tests. Evaporation also is an effective method of HSB preparation.[3] Salinity of the HSB should not exceed 100 g/kg.

Use the following formula to determine the volume of brine (V_B) to be added to the sample:

$$V_B = V_S \times (S_T - S_S)/(S_B - S_T)$$

where:

V_S = volume of test sample to be added, mL,
S_T = desired test salinity after adjustment,
S_S = initial salinity of sample, and
S_B = salinity of brine.

Check pH of adjusted samples. Add dilute hydrochloric acid or sodium hydroxide to adjust pH, if necessary.

3. Exposure Chambers

Conduct tests in glass culture tubes or vials of 10- to 20-mL capacity. Cover chambers loosely to prevent contamination during the test. Sealed chambers may be used provided that acceptable control performance is obtained. Disposable glass tubes, or scintillation or shell vials make convenient exposure chambers that can be discarded after the test. Ensure that all test chambers and equipment used to prepare test solutions are clean and noncontaminating. Disposable test chambers usually can be used straight from the box, although it is a good precaution to prerinse or soak them in distilled water or seawater. Avoid use of detergent and hypochlorite solutions in cleaning other equipment because of potential toxicity to test organisms. In multipurpose laboratories, use glassware dedicated solely to use in toxicity tests.

4. Conducting the Test

a. Setting up test chambers: Set up the test as described in Section 8010D. Prepare all solutions and equilibrate to test temperature before beginning to spawn animals. Prepare at least four replicates of each solution if hypothesis tests (e.g., Dunnett's test) are to be used to determine the NOEC or LOEC (see 8010B). Two or three replicates are adequate if point estimation techniques are to be used to determine values such as the EC50.

b. Duration and type of test: In the fertilization test, add a predetermined number of sperm to test solution and expose for 20 min (other times ranging from 5 to 120 min have been used). Then add eggs to produce a specific ratio of sperm to eggs and allow 20 min for fertilization to occur. Preserve samples by addition of formalin and examine under a compound microscope. Toxic effects are manifested by an impaired ability of the sperm to fertilize eggs, indicated by lack of an obvious fertilization membrane around the egg.

c. Test organisms: Fertilization tests can be conducted with all the recommended species.

d. Performing the test:

1) Preparation—Arrange replicate 5-mL samples of test solution in random order by assigning random numbers to individual test chamber numbers (i.e., replicate chambers of the first treatment group will have unrelated numbers such as 16, 31, and 4, instead of sequential numbers); then arrange the chambers in a rack in numerical order.

Other test volumes may be used if preferred (e.g., 2 mL or 10 mL), but adjustments to the following instructions will be necessary to maintain the desired sperm-to-egg ratios in the test chambers. Measure water quality of each test substance concentration on additional samples of test material. A single initial measurement is sufficient for most parameters unless the test material is highly unstable. Use of the proper sperm-to-egg ratio is critical to obtaining good test sensitivity and precision. Because a fixed number of eggs is used in the test, sperm-to-egg ratios are altered by varying the number of sperm added to the test chambers. The proper number of sperm is the least amount that produces >80% fertilization. Use of excessive (>2 × the optimal number) amounts of sperm may reduce test sensitivity.

2) Sperm density measurement—Use a portion of the concentrated pooled sperm to determine the density (be sure to reserve sufficient sperm to conduct the test and a possible trial fertilization). Use the following procedures to aid in accurately pipetting the highly viscous concentrated sperm: Enlarge pipet tip opening to about 2 mm, wipe off any sperm adhering to the outside of the pipet tip before delivery of the sample (take care not to wick away sperm from inside the tip), and repeatedly rinse pipet tip with dilution water after sample delivery until all sperm inside has been removed.

Add 0.025 mL sperm to approximately 180 mL seawater in a graduated cylinder. Bring mixture up to 200 mL using 10% acetic acid (kills sperm) to produce an 8000 × dilution. Cover the cylinder, mix well by inversion, and let bubbles dissipate. Add a sample of the mixture to each side of a hemocytometer. Alternate dilution volumes (e.g., 1 mL sperm solution in 100 mL) may be more suitable if a less dense sperm solution is used.

Let sperm settle for about 10 min. Examine a sufficient number of small squares on the slide so that about 100 sperm are counted. Examine the same number of squares on the opposite side of the hemocytometer. If the two counts are within 20%, use the mean to calculate the density according to the equation below. Reload hemocytometer and repeat counts if variability exceeds 20%.

$$\text{sperm/mL} = \frac{(\text{dilution factor}) \ (\text{count}) \ (4000 \ \text{squares/mm}^3) \ (1000 \ \text{mm}^3/\text{mL})}{\text{No. squares counted}}$$

Alternatively, use a ratio turbidimeter with a 1-cm-diam cuvette to determine sperm density rapidly.[4] The relationship between turbidity and sperm density is linear, but may vary with species, individual, season, or type of instrument. Use hemocytometer counts for initial calibration of turbidimeter and verification of method suitability.

3) Sperm-to-egg ratio selection—The sperm-to-egg ratio for the test usually is selected on the basis of the control performance of previous tests. The correct value may vary, depending on species and time of year. Conduct a trial fertilization test (just before the actual test) if adequate data are not available to determine the sperm-to-egg ratio [see ¶ 5 below]. For purple sea urchins, use sperm-to-egg ratios ≤500:1 if fertilization in controls is accept-

able. Higher ratios may reduce test sensitivity and should only be used when prior experiments or trial fertilization results indicate the ratio is needed to obtain acceptable control fertilization ($\geq 70\%$). Sperm-to-egg ratios above 3000:1 indicate unacceptable quality of purple sea urchin sperm; use additional animals or a different species (in better spawning condition). Optimum sperm-to-egg ratios are species-specific.[1,5]

4) *Egg stock preparation*—Add a sufficient volume of washed eggs to seawater to make 100 to 500 mL of a stock solution containing 2000 to 2500 eggs/mL. Determine density of eggs as directed in Section 8810B.5. Adjust density to proper range by adding or removing seawater. Verify that stock volume is sufficient for number and size of test chambers.

5) *Trial fertilization*—Determine density of pooled sperm as directed in ¶ 2) above. Prepare egg stock solution as directed in ¶ 4) above. Calculate volume of seawater needed to dilute 0.025 mL pooled sperm and produce a trial stock solution such that a sperm-to-egg ratio of 3000:1 will result when 0.1 mL of stock is added to test chamber (e.g., if 1000 eggs will be in each chamber and 0.1 mL sperm stock is added to the test sample, then a sperm stock density of 3.0×10^7 sperm/mL is needed). Prepare duplicate test chambers for each sperm-to-egg ratio to be tested (including 3000:1), each containing 5 mL seawater at correct temperature. Prepare trial sperm stock solution. Prepare several dilutions of stock that will produce desired range of sperm-to-egg ratios in test chambers.

A suggested dilution series is as follows:

Sperm-to-Egg Ratio	Trial Stock *mL*	Seawater *mL*
3000:1	No dilution	—
1304:1	5	6.5
545:1	2	9.0
231:1	1	12.0
100:1	0.5	14.5

Add 0.1 mL trial stock or dilution to appropriate test chambers. After 20 min, add 1000 eggs. Add formalin preservative after 20 additional min. Determine percent fertilized in a subsample of 100 eggs from each replicate. Select lowest sperm-to-egg ratio producing $\geq 90\%$ fertilization. It may be necessary to determine the ratio by interpolation. Verify sperm density in stock dilution corresponding to the chosen sperm-to-egg ratio by a hemocytometer count.

6) *Sperm stock preparation*—Calculate sperm stock solution density required to produce desired sperm-to-egg ratio in the test chambers (e.g., a stock containing 2.5×10^6 sperm/mL is needed to produce a sperm-to-egg ratio of 250:1 when 1000 eggs are present). Remove 0.025 mL pooled sperm and dilute with seawater to produce a stock solution of the desired density. Mix the solution well and use within 30 min.

7) *Sperm addition*—Use an automatic pipet to add 0.1 mL sperm stock to each test chamber. Mix stock periodically during inoculations. Add sperm in a steady rhythm, with each addition at intervals of about 5 s.

8) *Egg addition*—add eggs after a 20-min exposure period (other sperm exposure times of 5 to 60 min can be used for comparability with other laboratories). Use egg stock to add 1000 eggs (e.g., 0.5 mL stock) to each chamber using the same order and rhythm as for sperm. Use a perforated plunger or equivalent device to gently mix egg stock thoroughly during additions.

9) *Test termination*—Stop test by adding 0.25 mL concentrated formalin to each tube 20 min after egg addition (5% final concentration of formalin). Glutaraldehyde may be used as a preservative instead of formalin. Lugol's solution (Section 10200B) also is an effective preservative and has the advantage of being less toxic to the analyst. Cap test chambers securely, store at room temperature, and determine fertilization within 48 h, or as soon as practical. The samples can be stored indefinitely, but appearance of egg or fertilization membrane may change upon storage, making detection of the endpoint more difficult. Be extremely careful not to contaminate test equipment or laboratory furniture with preservative.

10) *Sample evaluation*—Examine eggs in exposure vial using an inverted compound microscope or transfer a representative subsample to a Sedgwick-Rafter cell for use with a conventional microscope. It is often convenient to concentrate the eggs before transfer by removing most of the overlying water with a pipet. Mix remaining sample well before transfer to counting chamber. Discard formalin-contaminated exposure chambers promptly (do not reuse chambers).

Examine at least 100 eggs (40 to 100 × magnification) from each replicate and score for presence or absence of an elevated fertilization membrane. Avoid bias by counting all eggs in subsamples transferred to counting chambers. Newly fertilized eggs usually have a completely elevated membrane around the egg (Figure 8810:1, A). The fertilization membrane may change in appearance with prolonged storage, partially collapsing or touching a portion of the egg. Consequently, count eggs showing any elevation of the fertilization membrane as fertilized. Exclude unusually small, immature, or abnormally-shaped eggs from counts.

Calculate percentage of fertilized eggs in each sample.

5. Statistical Analysis

Assemble, analyze, evaluate, and report data as described in Section 8010G.

6. Quality Assurance

Continued success in conducting this toxicity test depends on an overall effort to maintain and improve laboratory techniques and equipment.[3] Accurate background information regarding sample characteristics and test organism condition is necessary to enable correct interpretation of test results. Measure basic water quality parameters (pH, dissolved oxygen, salinity, temperature) on representative samples of controls and test materials. Because ammonia can be highly toxic to marine organisms, measure total ammonia with a sensitive method (e.g., 4500-NH$_3$.E) and report concentration of un-ionized ammonia (NH$_3$). Include additional controls or blanks in the experimental design to verify that special treatments (e.g., storage, centrifugation, pH adjustment, carrier solvent addition) do not produce unanticipated effects.

Use reference toxicant tests to provide a measure of test precision and possible organism condition.[3] The reference toxicant test usually consists of replicate exposures to three to five concentrations of a stable chemical (e.g., Cu, Cd, sodium dodecyl sulfate) that are sufficient to calculate a point estimate of effect (e.g., EC50). Preferably include a reference toxicant test with each experiment, or test at least monthly. Plot cumulative mean and confidence limits on a control chart to identify outlier values. Outliers indicate potential problems with the technique or test organisms.

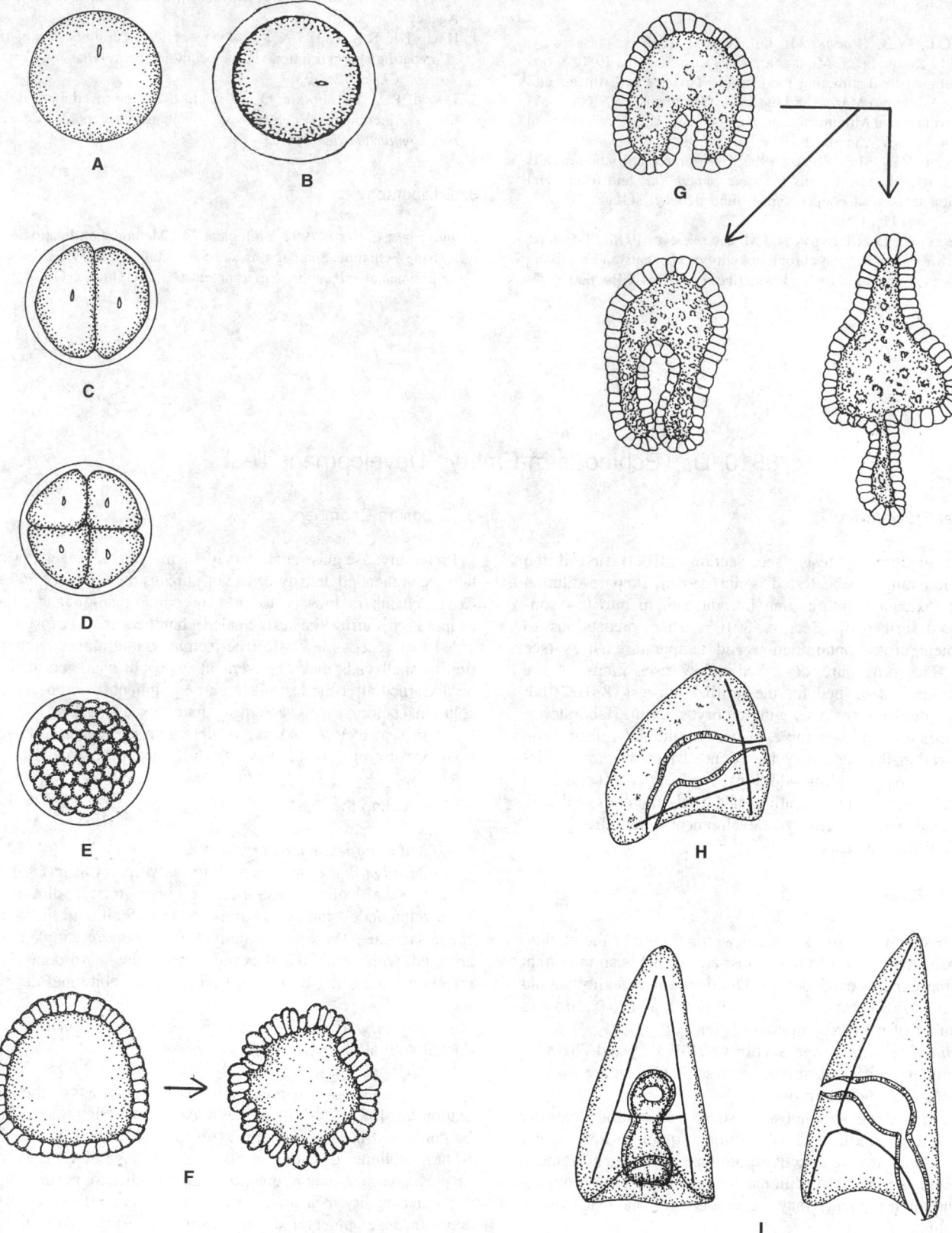

Figure 8810:1. Early developmental stages of sea urchins and sand dollars. A—unfertilized egg; B—fertilized egg; C, D, E—early cleavage; F—blastula with arrow indicating abnormal example; G—gastrula with arrows indicating abnormal examples; H—prism; I—frontal and lateral views of normal pluteus.

7. References

1. WEBER, C.I., W.B. HORNING, II, D.J. KLEMM, T.W. NEIHEISEL, P.A. LEWIS, E.L. ROBINSON, J. MENKEDICK & F. KESSLER, eds. 1988. Short-Term Methods for Estimating the Chronic Toxicity of Effluents and Receiving Waters to Marine and Estuarine Organisms. EPA-600/4-87-028, Environmental Monitoring and Support Lab., U.S. Environmental Protection Agency, Cincinnati, Ohio.
2. NEIHEISEL, T.W. & M.E. YOUNG. 1992. Use of three artificial sea salts to maintain fertile sea urchins (*Arbacia punctulata*) and to conduct fertilization tests with copper and sodium dodecyl sulfate. *Environ. Toxicol. Chem.* 11:1179.
3. CHAPMAN, G.A., D.L. DENTON & J.M. LAZORCHAK. 1995. Short-term methods for estimating the chronic toxicity of effluents and receiving waters to west coast marine and estuarine organisms. EPA-600/R-95-136, National Exposure Research Lab., U.S. Environmental Protection Agency, Cincinnati, Ohio.
4. HALL, T.J., R.K. HALEY & K.J. BATTAN. 1993. Turbidity as a method of preparing sperm dilutions in the echinoid sperm bioassay. *Environ. Toxicol. Chem.* 12:2133.
5. DINNEL, P.A., J.M. LINK & Q.J. STOBER. 1987. Improved methodology for a sea urchin sperm cell bioassay for marine waters. *Archiv. Environ. Contam. Toxicol.* 16:23.

8. Bibliography

ENVIRONMENT CANADA. 1992. Biological Test Method: Fertilization Assay Using Echinoids Sea Urchins and Sand Dollars). EPS 1/RM/27, Environmental Protection, Environment Canada, Ottawa, Ont.

8810 D.   Echinoderm Embryo Development Test

1. General Procedures

Conduct exploratory tests (see Section 8010D) first if the concentration range to be tested is not known. Prepare dilution water and toxicant solutions and introduce them into test containers as described in Section 8010F. Take precautions to avoid gamete cross-contamination and temperature stress (see 8810C.1). The procedure described below uses many of the same techniques described for the fertilization test (8810C) but has been optimized for use with embryos. Some laboratories conduct both tests on a sample to gain additional information. The fertilization test also may be extended into an embryo development test by including additional replicate chambers. Such an approach requires modification of the test methods and may reduce precision of the embryo development results because of variable fertilization rates.

2. Water Supplies

Maintain salinity of dilution water within 2 g/kg of the holding salinity. Adjust pH to 7.7 to 8.3, unless altered pH is an important factor in the experimental design. Dilution water quality should be sufficient to produce ≥70% normal development (relative to initial number of embryos) in control samples.

a. Artificial seawater: See Sections 8010E.4*b*2) and 8810C.2. Avoid commercial sea salt mixes because they often are toxic to echinoid gametes and embryos.

b. Natural seawater: Choose a source of natural seawater free of contamination and of uniform quality. Pass water through a filter with an effective pore size ≤1.0 μm to remove parasites and predators. Additional treatment (e.g., aeration, activated carbon treatment) may be needed to obtain acceptable water quality.

c. Salinity adjustment: The sea urchin development test usually is more sensitive to deviations in salinity than is the fertilization test. Use the methods described in Section 8810C.2 to adjust salinity of samples that deviate by more than 2 g/kg.

3. Exposure Chambers

Preferably, use glass chambers of 10-mL to 1-L capacity. Maintain recommended density of test organisms regardless of volume. Cover chambers loosely to prevent contamination and reduce evaporation during the test. Sealed chambers may be used provided that acceptable control performance is obtained. Scintillation or shell vials make convenient exposure chambers that can be discarded after the test. Clean all equipment for preparing test solutions before use. Clean test chambers by soaking in fresh water or seawater. Avoid use of detergent or hypochlorite solutions because of potential toxicity to the test organisms.

4. Conducting the Test

a. Setting up test chambers: See 8810C.4*a*.

b. Duration and type of test: Various volumes of test solution (5 to 1000 mL) may be used. Add embryos to test solution and let develop under static conditions for 48 to 96 h until the pluteus stage is reached. Preserve subsamples (or the entire sample if vials are used) with formalin and examine under the microscope. Toxic effects are indicated by embryo mortality or abnormal development.

c. Test organisms: Embryo development tests can be conducted with all the recommended species.

d. Performing tests:

1) Preparation—Test preparation is the same as for the fertilization test [8810C.4*d*1)] with two exceptions. First, the test may be conducted in larger volumes (up to 1000 mL) if desired. Use of large volumes provides no distinct advantage in test sensitivity or precision, but does allow exposure chamber to be subsampled for water quality measurement or to determine effects at various times or developmental stages. Second, measure both initial and final water quality for each treatment group. Include one additional replicate test chamber in each treatment group for final water quality measurements when tests are conducted in small volumes (e.g., 10 mL).

Quantify toxic response by either complete count or relative count. For complete count, calculate percentage of normal pluteus larvae at the end of exposure, based on counts of all preserved test organisms and number of embryos added at the start. This method provides the most comprehensive assessment of effects because all instances of embryo mortality and aberrant development are included in the percentage. A relative count requires counts of only a subsample of organisms at the end of the test. Both embryo mortality and aberrant development are reflected in this method as well, but toxic effects may be underestimated if the test solution causes rapid decomposition of dead embryos and consequent failure to detect them during microscopic examination. Choose the evaluation method before the test is started so that all required information will be obtained.

2) *Egg density adjustment*—Add a sufficient volume of washed eggs to seawater to make 100 to 500 mL of a stock solution containing 1000 eggs/mL. It may be more convenient to prepare a more concentrated solution (e.g., 10 000 eggs/mL) if test volumes larger than 50 mL are used. Determine density of eggs as directed in Section 8810B.5. Adjust density to desired value by adding or removing seawater. Verify that stock volume is more than sufficient (approximately 50% greater) for the number and size of test chambers used.

3) *Sperm stock preparation*— Prepare a sperm stock by adding about 0.025 mL dry sperm to 50 mL seawater. If volume of stock solution needed to fertilize the eggs is not known from prior experience, determine density of the sperm stock with a hemocytometer (see 8810C.4*d*).

4) *Egg fertilization*—Add a sufficient volume of sperm stock to egg stock to produce a sperm-to-egg ratio of 200 to 1000:1. Mix well and examine a subsample after about 10 min to assess fertilization percentage. Add more sperm if less than 90% of the eggs are fertilized. A fertilization rate of less than 90% after the second addition of sperm indicates that the gametes are of poor quality. Spawn additional animals to obtain better gametes if possible. Add the embryos to the test containers as soon as possible (generally within 2 h but no later than 4 h after fertilization).

5) *Embryo addition*—Use an automatic pipet to add sufficient embryo stock solution to each test chamber to result in about 25 embryos/mL (e.g., 0.25 mL/10 mL of sample). It is important that the same number of embryos is added to each test chamber. Use a perforated plunger to mix the stock solution thoroughly during embryo addition.

If the complete count method will be used to evaluate toxic effects, add embryos to at least five additional test chambers containing control water. Intersperse these chambers throughout the experimental array and add embryos in the same manner used for the other chambers. Examine these additional chambers promptly to estimate actual number of embryos added.

6) *Exposure*—Loosely cover the test chambers and leave undisturbed for 48 to 96 h under static conditions. The optimum exposure length varies with species and test temperature. The exposure time should be long enough to allow the embryos to develop to the pluteus stage, yet short enough (≤96 h) that internal food reserves are not exhausted. The following exposure conditions are recommended to provide consistency with results from other laboratories: *A. punctulata*, 48 h at 20°C; *D. excentricus*, 72 h at 15°C; *S. purpuratus*, 72 h at 15°C or 96 h at 12°C; and *S. droebachiensis*, 96 h at 12°C. These are target times; a few extra hours may be allowed to help assure that most (>90%) control

larvae have attained the normal pluteus stage. Ambient laboratory light levels and photoperiods are adequate for all species.

7) *Test termination*—Preserve organisms for later microscopic examination by adding sufficient borax-buffered formalin (pH>7.0, see 10200B.2*a*) to produce a 5% concentration. Unbuffered formalin may be used provided that the samples are examined rapidly (within a few days), before skeletal components dissolve. Add formalin directly to test chamber if disposable vials or culture tubes are used. Otherwise, thoroughly mix test chamber contents, transfer a 10-mL subsample to a vial, and add formalin.

8) *Test evaluation*—Examine preserved embryos and larvae with a compound microscope at a magnification of 100 ×. Concentrate test organisms by removing overlying water from the storage vial and transfer to a Sedgwick-Rafter counting chamber for examination with a conventional microscope. Alternatively, use an inverted microscope to examine the organisms in the storage vial and eliminate losses due to transfer.

Normally developing embryos develop synchronously through a series of characteristic stages including early cleavage, blastula, gastrula, prism, and pluteus (Figure 8810:1, B-I). The appearance of pluteus larvae varies with species, but all normal plutei should have the following features: a pyramid shape supported by a framework of skeletal rods, an internal gut that is attached to the body wall at both ends and consists of three distinctive regions, and at least one pair of post-oral arms (Figure 8810:1, I). The length of the post-oral arms varies with species. Count as abnormal all grossly deformed pluteus larvae, deformed embryos, mostly normal-appearing embryos that have not attained the pluteus stage (inhibited development), and uncleaved fertilized eggs. Do not count unfertilized eggs.

Determine percentage of normal pluteus larvae for each replicate using either the complete or relative count method (below).

a) *Complete count method*—Count all embryos in preserved sample. Preferably use an inverted microscope to minimize counting errors. If a conventional compound microscope is used, use a consistent and efficient method to transfer embryos to counting chamber because lost embryos (remaining in vial or stuck to transfer pipet) are assumed to have died. Variability in recovery of larvae from the storage vial may introduce experimental error that reduces the ability to detect statistically significant effects. Calculate percentage of embryos developing to normal pluteus larvae, P_n, as follows:

$$P_n = 100(E_n/E_i)$$

where:

E_n = number of normal larvae at end of test, and
E_i = number of embryos at start of test.

b) *Relative count method*—It is easier and usually just as effective to determine percentage of normal development in a representative sample of at least 100 embryos and larvae at the end of the test. Calculate this value as follows:

$$P_n = 100[E_n/(E_n + E_a)]$$

where:

E_a = number of abnormal embryos/larvae, and other terms are as defined above.

5. Statistical Analysis

See Section 8010G.2 for general information.

Control performance may vary between tests because of factors such as variations in test temperature and gamete condition. Normalize response data to the control performance before statistical evaluation (e.g., EC50) and to facilitate comparisons between tests as follows:

$$P_{adj} = 100(P_n/M)$$

where:

P_{adj} = normalized value,
P_n = percent normal or fertilized for the sample, and
M = mean percent normal or fertilized for controls.

6. Quality Assurance

See 8810C.6.

7. Bibliography

AMERICAN SOCIETY FOR TESTING AND MATERIALS. 1995. Standard guide for conducting static acute toxicity tests with echinoid embryos. ASTM E-1563, Annual Book of ASTM Standards, Vol. 11.04. American Soc. Testing & Materials, Philadelphia, Pa.

8910 FISH*

8910 A. Introduction

Fish have been regarded as good test species for the assessment of aquatic toxicity because of their ecological and economic importance. While many different fish species may be used in tox-

* Approved by Standard Methods Committee, 1997.

icity studies, the selection of test species will depend on the objective of the test, the availability of the species, and the ease of culturing and handling individuals. Section 8910 provides guidance for species selection, culturing, and testing procedures for fish in aquatic toxicity studies.

8910 B. Fish Selection and Culture Procedures

1. Selection of Test Species

General guidelines for selecting test organisms are outlined in Section 8010E.1. Of prime consideration in the selection of a fish species is the purpose of the test. For example, test species may be designated by a particular regulation (e.g., FIFRA, TSCA), or, if the toxicity test is for non-regulatory purposes, any number of species may be tested provided that the environmental requirements of the species can be duplicated for the test. If it is desirable to use a fish that is not routinely cultured, it may be necessary to make collections from a single field source.

Select test species according to the following criteria: (a) the species should be available in unbiased (i.e., not pre-screened for resistant individuals by prior exposure to adverse conditions) numbers sufficient for the tests; and (b) the species should be capable of being held in the laboratory in a healthy condition (i.e., active, feeding, free of lesions, etc.) for at least 1 month. Consider relative sensitivities of different species and life stages if the data are available when selecting species.

2. Collecting and Handling Test Fish

Collecting equipment and methods are described in Sections 8010E.2 and 10600. Handling and holding are discussed in Section 8010E.3. It is extremely important to avoid subjecting fish

to unnecessary stress such as inappropriate capture and transport, temperature shock, or water quality change.

a. Freshwater fish: Whenever possible, obtain routinely cultured species. Salmonid fish usually are available from private, state, and federal hatcheries. Obtain trout certified pathogen-free, if possible. When fish cannot be obtained from hatcheries, appropriate field collection is acceptable. Collecting permits usually are required by state agencies. Avoid fish from bait dealers or fishermen because information on source, handling, holding time, etc., usually is not available.

b. Marine and estuarine fish: Various life stages of marine fish may be collected from the field for laboratory tests. Vertical movement of early larval stages may necessitate nighttime collection. Many marine fish and most marine fish larvae are extremely fragile; handle carefully during collection, sorting, and transfer. For sorting and transferring larvae during and after collection, use a pipet appropriate to the size of the larval fish. Whenever possible, transfer larger larvae, juvenile, and adult fish by dipping or gently pouring. Fine-mesh dip nets also are suitable if transfers are made gently.

3. Holding and Acclimating

See Sections 8010E and F for additional discussion.
Keep fish stocks in tanks, small ponds, live boxes, or screen

pens, depending on fish size and number. Use good-quality dilution water (See Section 8010E.4b) for acclimation. Feed fish natural or commercially available prepared foods daily during acclimation. Detailed information on handling, holding, care, and feeding of fish is available.[1-4] Because food requirements vary with the species and size of fish, select an appropriate diet for the species. Fish obtained from a hatchery should be provided initially with food to which they are accustomed. Many fish can be maintained for long periods on dried food but live food supplements may be desirable. Do not overfeed. Diets should be certified toxicant-free or tested for toxic substances before use.

While maintaining fish during holding and acclimation, watch carefully for signs of disease, stress, physical damage, and mortality. Remove dead and abnormal individuals immediately. If mortality rate exceeds 10%, discard the entire stock.

Handle fish carefully and as quickly as possible.[1,2] For extensive handling such as weighing, measuring, or taking other data, anesthetize fish.[3]

For short-term tests, use fish of similar size and source. The length of the longest fish should be not more than 1.5 times the length of the shortest fish. Acclimate fish to laboratory conditions before the test. Standard acclimation periods range from 48 h to 14 d. However, when using early-life-stage fish in short-term tests, this may not be possible.

4. Parasites and Disease

a. Stress in relation to parasites and disease: Unexpected and often unexplained mortalities in experimental and control animals interfere with test results and interpretations. Optimize laboratory conditions for each particular species to prevent the development of disease.

When large numbers of organisms are retained in a relatively small space, microbial diseases or parasites may become epidemic. If the water is unpolluted and poor in nutrients, disease often can be prevented by strict sanitation. Disease may arise if the water is enriched with organic materials or if toxic substances are present. Pathogens and parasites that might be very rare in natural waters may become epidemic in intensive culture. Uneaten food and fecal material provide a potential source of bacteria, parasites, and toxic products. Filtration and/or sterilization of water, adequate feeding, regular cleaning of holding vessels, sterilization of equipment, and securing disease-free fish are the first lines of defense.

Organisms exposed to toxicants may become more susceptible to parasites and disease. Because various environmental factors may contribute to reduced resistance, pay careful attention to nutrition, oxygen supply, and water quality. To minimize accumulation of fecal material and hence dissolved oxygen demand, do not feed fish during the 2 d before initiating short-term tests for cold-water species and 1 d for warm-water fishes. Siphon daily from the holding tank any fecal material or uneaten food. In testing of young fish, feeding live brine shrimp nauplii may be desirable if starvation is a possibility during the test.

b. Control methods:

1) General—Ultraviolet light and ozonation have been used to control disease and parasites present in dilution and/or culture water. Antibiotics used in holding tanks reduce bacterial populations. To reduce mortality and to avoid introduction of disease into stock tanks, treat fish with a broad-spectrum antibiotic immediately after collection, during transport, or on arrival at the laboratory. Do not place treated organisms into holding tanks for 4 d after treatment or use treated organisms for tests until at least 14 d after treatment. Clean and disinfect tanks and containers with 200 mg sodium hypochlorite (NaOCl)/L for 1 h after removal of

TABLE 8910:I. RECOMMENDED PROPHYLACTIC AND THERAPEUTIC TREATMENTS FOR FRESHWATER FISH TO BE USED FOR EXPERIMENTAL PURPOSES

Disease	Chemical	Concentration mg/L	Application
External bacteria	Hyamine 1662® or 3500®*	1–2 AI†	30–60 min in flow-through system‡
	Nitrofurazone (water mix)	3–5 AI	30–60 min in flow-through system‡
	Neomycin sulfate	25	30–60 min in flow-through system‡
	Oxytetracycline hydrochloride (water-soluble)	25 AI	30–60 min in flow-through system‡
Monogenetic trematodes, fungi, and external protozoa§	Formalin *plus* zinc-free malachite green oxalate	25 ± 0.1	1–2 h in static systems, 30–60 min in flow-through system‡
	Formalin	150–250	
	KMnO$_4$	2–6	1–2 h in static system, 30–60 min in flow-through system‡
	NaCl	14 000–30 000 2000–4000	5–10 min dip 24 h minimum, but may be continued indefinitely
	Para-dimethyl aminobenzene-diazo sodium sulfonate (35% AI)‖	20	30–60 min in flow-through system‡
Parasitic copepods	Trichlorfon#	0.25 AI	Weekly for up to 4 weeks if necessary in static or flow-through systems. Do not use at >27°C.

* Benzalkonium chloride.
† AI = active ingredient.
‡ Add concentrated stock solution to the inflowing water by a drip system or by the technique of Brungs and Mount.[5]
§ One treatment usually is sufficient except for *Ichtyophthirius*, which must be treated daily or every other day until no sign of the protozoans remains. This may take 4 to 5 weeks at 10°C and 11 to 13 d at 15 to 21°C. A temperature of 32°C is lethal to *Ichthyophthirius* in 1 week.
‖ Dexon® or equivalent.
Masoten® or equivalent.

diseased fish. Dechlorinate with sodium thiosulfate and rinse with clear water before reusing tanks.

2) *Recommended disease therapy*—Treat freshwater fish to cure or prevent disease by the methods in ¶ *b*1) above and Table 8910:I. These methods have been found dependable, but their efficacy may be altered by temperature or water quality. If fish are severely diseased, destroy the entire stock. A number of good reviews of fish diseases and parasites and methods for their control have been published.[5-11]

Published information on related topics includes a summary of problems in marine fish larval culture,[12] a description of a larval culture system,[13] and a discussion of disinfection of water supplies.[14]

5. Culturing Test Fish

a. Freshwater fish: More than 30 species of freshwater fish have been reared for stocking fresh waters. The culture methods can be adapted to laboratory scale to produce various life stages of fish.[15-25] Methods are given below for three freshwater species commonly used in toxicity experiments: the rainbow trout, *Oncorhynchus mykiss*; the bluegill sunfish, *Lepomis macrochirus*; and the channel catfish, *Ictalurus punctatus*. For specifics regarding fathead minnow, *Pimephales promelas*, see Section 8921. These species are representative of test organisms that can be found in various freshwater habitats, ranging from free-flowing streams and rivers to ponds and lakes.

1) *Rainbow trout, Oncorhynchus mykiss*—Rainbow trout of various ages can be purchased from certified specific pathogen-free hatcheries. Life stages range from unfertilized eggs and sperm (fertilization can be performed in the testing laboratory) to juvenile fish. Preferably obtain eyed embryos, and grow these to testing size unless spawning, fertilization and early embryonic stages are important for the test.

Overnight courier shipment of hatchery-raised eyed trout embryos in special insulated cartons is standard. These cartons are adequate to maintain cold temperatures during shipping. Upon receipt, measure the ambient temperature of the egg mass and if necessary slowly temper the eggs ($\pm$ 3°C/h) to the testing temperature or culture conditions. Maintain embryos at temperatures $\pm$ 2°C in a range from 8 to 12°C.[26]

To perform egg fertilization in the laboratory, obtain gametes in plastic bags from the supplier within 24 h of removal from the adult. Hold gametes in unopened bags and slowly acclimate to test or culture temperature. When eggs and milt are at the desired temperature, mix together with either ovarian fluid or a small volume of 0.75% saline solution. Gently stir and let stand for about 1 min while fertilization occurs. Pour off excess water and milt and replace with fresh dilution water. Repeat several times until the embryos are in clear water. It may be useful to let eggs rest for 1.5 to 2 h to harden before transferring to incubation cups or culture systems.

Place fertilized eggs (embryos) in incubation chambers until hatching. For small-scale cultures, such as those that may be manipulated experimentally during an early-life-stage toxicity test (Section 8910C.2), incubation cups may be constructed from 8-cm sections of 5-cm-OD plastic or glass tubing with nylon or stainless steel screening cemented over one end to retain the embryos. During incubation supply a flow of fresh, high-quality, well oxygenated water over the embryos. Oscillate incubation cups (containing less than 250 embryos) in holding tank (or test)

water by a rocker arm using a 2-rpm electric motor to supply the necessary flow. Incubation chambers for mass culture of trout generally consist of wire-screen trays with rectangular or oblong mesh (15- $\times$ 3.5-mm openings) stacked vertically in deep flow-through troughs; or spaced along horizontal troughs.[23,24] Agitate embryos (shocking) periodically. Every 2 d remove dead embryos that turn white. Maintain incubation chambers in the dark because exposure to light may result in premature hatching or death.[23] The rate of hatching is determined by the temperature regime; the optimum range, 7 to 10°C, produces hatching in 44 to 68 d.

After development to the free-swimming fry stage, provide at least one complete exchange of water per hour and use a 20-gal (75-L) aquarium per 100 fry. Maintain water pH between 6.5 and 8.5, dissolved oxygen above 5 mg/L, dissolved solids above 50 mg/L, and insure that the water is free of pollutants. Feed fry slightly to excess, as often as 10 times/d.[23,24] Daily rations generally average 7 to 9% body weight. Commercial dry food has proven successful in hatchery production; for laboratory culture, live brine shrimp nauplii are preferred for early stages.

When trout reach the fingerling stage, reduce quantity of fish in a tank to approximately 1 g fish/L flow /d.[23] Feed fish at a rate equivalent to 4 to 5% of body weight.[23,24] As fish grow, grade and sort them into separate tanks according to size to reduce size-dependent adverse hierarchical feeding dominance.

2) *Bluegill, Lepomis macrochirus*—Breeding and cultivating of bluegills may be carried out in a variety of ponds or tanks.[23-25] Adult bluegills average 100 to 150 g in weight (12 to 18 cm in length) and generally spawn when 1 year old. The breeding period is May to August and a given individual will spawn more than once during the season.

Stock spawning ponds with a 1:1 or, preferably, a 2:3 ratio of male to female adults. Although bluegills do not require the highly controlled environment required by trout, maintain adequate DO and water quality conditions (see Section 8010E.4*b*). Bluegills adapt readily to a wide variety of commercial feeds; feed to satiation.[25]

Provide spawning ponds with small piles of pea gravel in the shallow water (0.5 to 1.0 m deep) around the edges.[24] Male bluegills will use this material to build nesting areas. Space gravel piles at least 1 m apart to reduce aggression between males guarding territories.

If dense spawning for mass fry-production is desired, place spawning stalls side by side around the perimeter of the ponds.[24] Make stalls 1 m long and enclose on three sides by wood or concrete, place gravel on the bottom, and orient the open side toward the pond center. Hatching should take place within 5 d. By slipping a screen over the open end of the stall after the females leave the nest, fry can be captured after hatching but before they have dispersed. Stock fry in growing ponds at densities of 40 fry/m².

It is not recommended to rear fry in spawning ponds with adults, but if this is necessary reduce stocking densities of brood fish to limit losses due to predation by adults. For fairly high densities of fry (40/m²) maintain only two or three pairs of spawning adults per 4000 m². If large fry are desired for harvest, use a lower stock density: one breeding pair per 4000 m² to produce 10 fry/m². Under intensive culture, periodically sort juvenile bluegills according to size to facilitate uniformity of specimens for toxicity tests.

3) *Channel catfish, Ictalurus punctatus*—To establish breeding stock, collect adult catfish 3 or more years old, preferred size 1

to 4.5 kg. Segregate adults by sex in holding ponds until spawning is desired. Feed adult catfish commercially available pellets supplemented with fresh or frozen cut fish and live minnows. Daily rations of dry feed should equal approximately 3% of the weight of the stock.[23,24]

When spawning season begins (April or May), increase the daily ration to 4% of body weight. Methods have been developed for spawning in ponds and pens,[23] but aquarium spawning is most efficient in terms of space and rates of successful spawning.[23] In this method, pair catfish according to size in troughs (23- to 240-L provided with flowing water and a spawning compartment (e.g., stainless steel milk can or similar structure). Inject females intraperitoneally with hormones: three doses of 10 mg acetone-dried fish pituitary material/kg female at intervals from 6 h in warm water to 24 h in cold climates or a single dose of 2200 IU/kg body weight of human chorionic gonadotropin.[23] Most injected fish will spawn within 16 to 24 h after the last injection. Fertilized eggs adhere to each other in oval gelatinous masses. New eggs are golden, later turning pink as the embryos develop. After spawning, remove spawners and eggs. Use the troughs for additional spawning.

Incubate embryos either in hatching troughs or in open-mouth hatchery jars. Hatching troughs may be of any convenient size but at least 25 cm deep and supplied with running water. Retain the egg mass in a wire-mesh basket suspended in the hatching trough and place a paddlelike agitator driven by an electric motor alongside each basket to insure mixing of the eggs. If hatchery jars are used, place each egg mass in a separate 6- to 8-L jar. Introduce a gentle flow of water just above the mass by a rubber tube to simulate the agitation provided in nature by the fanning activities of the male.

Catfish embryos hatch in 8 to 10 d at 24°C. The fry have light-colored bodies with pink yolk sacs. Remove fry from hatching troughs or hatchery jars near the time when the yolk sac disappears, but before complete absorption, approximately 3 to 5 d after hatching. Transfer fry by siphoning through a large-bore glass tube into rearing troughs.

Catfish fry can be reared according to the methods established for trout except that warmer temperatures (24 to 28°C) are necessary. For the first 4 to 5 d in the rearing troughs, feed fry sparingly with finely ground fish food 10 times/d. Siphon off uneaten food after 2 h. Increase daily rations until they equal approximately 4 to 5% of total body weight.

b. Marine and estuarine fish: Culture methods for marine and estuarine fish species are less developed than those for freshwater species. General information about marine fish culture methods can be found in several sources.[13,14,26–38] For any species, maintain adult brood stock, eggs, and larvae under conditions approximating those in the natural environment.

A method is described below for culture of the sheepshead minnow, *Cyprinodon variegatus*, which is routinely cultured and used in egg-to-embryo or embryo-larval toxicity tests.

The sheepshead minnow thrives over a wide range of salinities and temperatures. Acclimate adult fish ≥ 27 m standard length to laboratory conditions for at least 2 weeks at a salinity of at least 10 to 20 g/kg (‰, parts per thousand) and a temperature of 30°C. Hold photoperiod at 12 h light, 12 h dark. During this period feed liberally on fresh or frozen adult brine shrimp. Eggs from natural spawning may be obtained by placing a pair of adult fish in a spawning chamber about $12 \times 18 \times 10$ cm high. Place spawning trays (2 cm deep, formed by 0.5-mm nylon screen attached to a frame and covered with 2-mm nylon screen) in the spawning chambers. Larger spawning tanks with several spawning pairs may be used, provided that each male has space to establish a territory. As embryos are deposited they fall through the screen into the trays, thus preventing predation by adult fish and allowing easy removal. Each pair may spawn a maximum of 10 to 30 embryos/d but average production is about 8/pair/d.

Alternatively, sheepshead minnows may be induced to produce eggs by hormone injection.[32] Inject each female intraperitoneally with 50 IU human chorionic gonadotropic hormone. Repeat after 2 d. On the third day most females can be readily stripped to obtain ripe eggs. Strip or dissect eggs into filtered seawater in a beaker and add macerated testes. The number of eggs produced per female by this method is 100 to 200, depending on fish size. This method has the advantage of producing eggs at specified times; however, the fish typically are sacrificed to obtain eggs and sperm, thus reducing brood stock.

Fertilized eggs may be hatched in flowing or static water systems. For a flowing water system, place embryos in a hatching chamber formed by gluing a 9-cm-high collar of 0.5-mm mesh nylon screen around a petri dish. Suspend hatching chambers in flow-through seawater aquariums with self-starting siphons. As the water level in the aquarium changes, water in the hatching baskets is exchanged gently. Alternatively, place embryos in separator funnels and aerate gently.[37,38] Sheepshead minnow fry hatch after 5 d at 30°C at a salinity between 15 and 20 g/kg. As embryos hatch, transfer to a rearing aquarium and immediately feed newly hatched brine shrimp (live or frozen) or a dry food. Supplement a dry-food diet occasionally with live organisms. Juveniles become sexually distinguishable when about 24 mm long and females may produce eggs within 3 months after hatching.

6. References

1. U.S. Environmental Protection Agency. 1991. Methods of Measuring the Acute Toxicity of Effluents and Receiving Waters of Freshwater and Marine Organisms. EPA-600/4-90-027, Off. Research & Development, EMSL, Cincinnati, Ohio.
2. American Society for Testing & Materials Committee E-35. 1988. Standard Practices for Conducting Acute Toxicity Tests with Fishes, Macroinvertebrates and Amphibians. ASTM Des. No. E-729, American Soc. Testing & Materials, Philadelphia, Pa.
3. Committee on Methods for Toxicity Tests with Aquatic Organisms. 1975. Methods for Acute Toxicity Tests with Fish, Macroinvertebrates, and Amphibians. Ecol. Res. Ser. EPA-660/3-75-009, U.S. Environmental Protection Agency, Duluth, Minn.
4. Hunn, J.B., R.A. Schoettger & E.W. Whealdon. 1968. Observations on the handling and maintenance of bioassay fish. *Progr. Fish-Cult.* 30:164.
5. Brungs, W.A. & D.I. Mount. 1967. A device for continuous treatment of fish in holding chambers. *Trans. Amer. Fish. Soc.* 96:55.
6. Snieszko, S.F. 1970. A Symposium on Diseases of Fishes and Shell fishes. Spec. Publ. 5, American Fisheries Soc., Washington, D.C.
7. Hoffman, G.L. 1967. Parasites of North American Freshwater Fishes. Univ. California Press, Berkeley & Los Angeles.
8. Van Duijn, D., Jr. 1973. Diseases of Fishes, 3rd ed. Charles C. Thomas Co., Springfield, Ill.
9. Davis, H.S. 1953. Culture and Diseases of Game Fishes. Univ. California Press, Berkeley & Los Angeles.
10. Hoffman, G.L. & F.P. Meyer. 1974. Parasites of Freshwater Fishes: A Review of Their Control and Treatment. T.F.H. Publications Inc., Ltd., Neptune City, N.J.
11. Sindermann, C.J. 1970. Principal Diseases of Marine Fish and Shellfish. Academic Press, New York, N.Y.

12. HOUDE, E.D. 1973. Some recent advances and unresolved problems in the culture of marine fish larvae. *Proc. World Maricult. Soc.* 3: 83.

13. HOUDE, E.D. & A.J. RAMSEY. 1971. A culture system for marine fish larvae. *Progr. Fish-Cult.* 33:156.

14. HOFFMAN, G.L. 1974. Disinfection of contaminated water by ultraviolet irradiation with emphasis on whirling disease (*Myxosoma cerebralis*), and its effects on fish. *Trans. Amer. Fish. Soc.* 103:541.

15. NATIONAL ACADEMY OF SCIENCES. 1973. Nutrient requirements of trout, salmon and catfish. Publ. Off. Nat. Acad. Sci., Washington, D.C. 11:1.

16. STALNAKER, C.B. & R.E. GRESSWELL. 1974. Early life history and feeding of young mountain white fish. EPA-660/3-73-019, Off. Research & Development, U.S. Environmental Protection Agency. U.S. Government Printing Off., Washington, D.C.

17. CARLSON, A.R. & J.G. HALE. 1972. Successful spawning of largemouth bass *Micropterus salmoides* (Lacepede) under laboratory conditions. *Trans. Amer. Fish. Soc.* 101:539.

18. HOKANSON, K.E.F., J.H. McCORMICK, B.R. JONES & J.H. TUCKER. 1973. Thermal requirements for maturation, spawning, and embryo survival of the brook trout *Salvelinus fontinalis. J. Fish. Res. Board Can.* 30:975.

19. HOKANSON, K.E.F., J.H. McCORMICK & B.R. JONES. 1973. Temperature requirements for embryos and larvae of the northern pike, *Esox lucius* (Linnaeus). *Trans. Amer. Fish. Soc.* 102:89.

20. SIEFERT, R.E. 1972. First food of larval yellow perch, white sucker, bluegill, emerald shiner and rainbow smelt. *Trans. Amer. Fish Soc.* 101:219.

21. SMITH, W.E. 1973. A cyprinodontid fish, *Jordanella floridae*, as a laboratory animal for rapid chronic bioassays. *J. Fish. Res. Board Can.* 30:329.

22. STEVENS, R.E. 1966. Hormone-induced spawning of striped bass for reservoir stocking. *Progr. Fish-Cult.* 28:19.

23. BARDACH, J.E., J.R. RYTHER & W.O. McLARNEY. 1972. Aquaculture: The Farming and Husbandry of Freshwater and Marine Organisms. Wiley Interscience, John Wiley & Sons, Inc., New York, N.Y.

24. HUET, M. 1970. Textbook of Fish Culture: Breeding and Cultivation of Fish. Thanet Press, Margate, England.

25. SMITH, W.E. 1976. Larval feeding and rapid maturation of bluegill in the laboratory. *Progr. Fish-Cult.* 38:95.

26. SHELBOURNE, J.E. 1964. The artificial propagation of marine fish. *Adv. Mar. Biol.* 2:1.

27. MAY, R.C. 1970. Feeding larval marine fishes in the laboratory: A review. Calif. Mar. Res. Comm., CalCOFI Rep. 14:76.

28. MIDDAUGH, D.P., M.J. HEMMER & L.R. GOODMAN. 1987. Methods for Spawning, Culturing and Conducting Toxicity Tests with Early Life Stages of Four Atherinid Fishes: The Inland Silverside (*Menidia beryllina*), Atlantic Silverside (*M. menidia*), Tidewater Silverside (*M. peninsulae*) and California Grunion (*Lenresthes tenuis*). EPA-600/8-87-004, Off. Research & Development, U.S. Environmental Protection Agency, Washington, D.C.

29. HOUDE, E.D. & B.J. PALKO. 1970. Laboratory rearing of the clupeid fish, *Harengula pensacolae*, from fertilized eggs. *Mar. Biol.* 5:354.

30. LASKER, R. 1964. An experimental study of the effect of temperature on the incubation time, development and growth of Pacific sardine embryos and larvae. *Copeia* 1964:399.

31. BOYD, J.F. & R.C. SIMMONS. 1974. Continuous laboratory production of fertile *Fundulus heteroclitus*, Walbaum eggs lacking chorionic fibrils. *J. Fish. Biol.* 6:389.

32. HANSEN, D.J. & P.R. PARRISH. 1975. Suitability of sheepshead minnows (*Cyprinodon variegatus*) for life cycle toxicity tests. *In* F.L. Mayer & J.L. Hamelink, Aquatic Toxicity and Hazard Identification. ASTM STP 634, American Soc. Testing & Materials, Philadelphia, Pa.

33. KUO, C., Z.H. SHEHADEH & K.K. MILISEN. 1973. A preliminary report on the development, growth and survival of laboratory reared larvae of the grey mullet, *Mugilcephalus L. J. Fish. Biol.* 5:459.

34. MIDDAUGH, D.P. & R.L. YOAKUM. 1974. The use of chorionic gonadotropin to induce laboratory spawning of the Atlantic Croaker, *Micropogon undulatus*, with notes on subsequent embryonic development. *Cheasapeake Sci.* 15:110.

35. HOFF, F.H. 1972. Artificial spawning of black seabass, *Centropristis striata*, aided by chorionic gonadotropin hormones. Florida Dep. Natural Resources Marine Research Lab. (mimeograph).

36. KRAMER, D. & J.R. ZWEIEL. 1970. Growth of anchovy larvae, *Engraulis mordax*, Girard in the laboratory as influenced by temperature. *Rep. Calif. Coop. Oceanic Fish. Invest.* 14:84.

37. HANSEN, D.J. 1978. Laboratory culture of sheepshead minnows (*Cyprinodon variegatus*). *In* Bioassay Procedures for the Ocean Disposal Permit Program. EPA-600/9-78-010, U.S. Environmental Protection Agency.

38. HANSEN, D.J., P.R. PARRISH, S.C. SCHIMMEL & L.R. GOODMAN. 1978. Life-cycle toxicity test using sheepshead minnows (*Cyprinodon variegatus*). *In* Bioassay Procedures for the Ocean Disposal Permit Program. EPA-600/9-78-010, U.S. Environmental Protection Agency.

8910 C. Test Procedures

1. Short-Term Tests

a. General test procedures: Short-term testing can be used to determine relative toxicity of substances. Tests are made to determine LC50 or EC50, and to estimate toxicant concentrations for intermediate- and long-term tests.

Short-term tests may be static, static with renewal, flow-through, or recirculating, depending on the objective of the test, the life stage being tested, and the character of the toxicant or effluent (see Section 8010D.1 and 2 and Section 8921).

Although any life stage may be used, short-term tests are performed most frequently with small species (less than 5 g body weight) or juvenile forms of large fish species (see Section 8910B.1). The life stage selected depends on test purpose, availability, and laboratory facilities (see Section 8910B.1). If culturing is necessary to obtain a particular life stage, see Section 8910B.5.

Select fish of near uniform size, with the longest no more than 1.5 times the length of the shortest. Use 10 to 20 fish per toxicant concentration. Additional replicates may be used to increase the number of test fish at each concentration. For juvenile and adult fish, terminate feeding 48 h before initiating tests. For all tests, limit fish weight/L test solution. This practice minimizes oxygen depletion, metabolic waste accumulation, and crowding-induced stress. In flow-through tests, use less than 10 g of fish/L of test solution for tests at or below 17°C or 5 g of fish/L at higher temperatures. For static testing, do not load above 0.8 g/L at 17°C or less and 0.5 g/L above 20°C at higher temperatures.

b. Specific test procedures:

1) Freshwater fish

a) Equipment and physical conditions—Use test equipment made of glass, No. 316 stainless steel, or perfluorocarbon plastics.* Use unplasticized plastics such as polyethylene, polypropylene, and polyvinylchloride† or silastic in the water delivery system. In static and flow-through systems with low rates of exchange, avoid certain types of TFE stoppers; pretest stoppers to insure absence of toxicity.

Select temperature appropriate for the species being tested, hold test temperature between ± 2°C of mean test temperature during a 96-h test, or ± 1°C during any 48 h. The photoperiod at the test site can be ambient laboratory lighting with a photoperiod of 16 h light/8 h dark. A 15- to 30-min dusk/dawn transition period is desirable to acclimate test fish to the photoperiod.[1,2]

Use dilution water from a surface source, well, or spring, or use reconstituted water.[3] If the source potentially is contaminated with pathogens, irradiate with UV before using. Analyze water in control and assay chambers daily for pH, dissolved oxygen, and temperature. Maintain DO concentration at ≥60% saturation. When testing volatile substances, do not aerate test solutions. However, take care that chemical substances that create a dissolved oxygen demand do not result in conditions inconsistent with the dissolved oxygen criterion of the test as well as of fish health. As a last resort to maintain dissolved oxygen above the criterion, use aeration. If aeration is used, it may be desirable to make frequent analytical measurements to confirm test chemical concentrations.

The length of a short-term test varies, generally from 24 to 96 h. Longer test periods may be used, depending upon characteristics and variability of the waste and the purpose of the test. Dilution factors in the receiving stream provide guidance in determining effluent concentrations for range-finding tests.

For information pertaining to species selection, collection, holding, acclimation, disease control, and culturing see Sections 8010E and 8910B.

b) Test procedure

Range-finding test: If the approximate toxicity of test material is unknown, conduct an abbreviated range-finding test to determine the concentrations that should be used in the definitive tests. Do this in tests at three to five widely spaced toxicant concentrations (for example, a decade test having concentrations a factor of ten from each other). For these tests, static tests may be acceptable as would use of fewer fish, e.g., five per chamber. Run this test for 24 to 96 h.

Definitive test: To determine LC50 or EC50, use a 96-h test with a minimum of five toxicant concentrations and a control according to the results of the range-finding test. Use a carrier control if dosing solutions are prepared in an organic solvent. Acceptable carriers are dimethylformamide, ethanol, methanol, acetone, and triethylene glycol. Limit concentrations of carrier to 0.1 mL/L of test solution. Carrier controls should not result in increased mortality. If more than 10% of the fish in a control system die, repeat the test. To establish definitive test concentrations, prepare solutions using a dilution ratio of 1.5 to 2 between successive concentrations.

2) Marine fish—The test procedure for marine species is essentially identical to that for freshwater fish. Additionally, determine salinity daily. Conduct toxicity tests requiring saline dilution water using natural or reconstituted seawater having a salinity appropriate to requirements of the test fish. Run stenohaline species with seawater having a salinity of 30 to 34‰ and euryhaline species at 10 to 25‰ (± 1 to 2‰) salinity.

Be aware that effluent testing often requires exposing fish to 50 to 100% effluent. This exposure may create a salinity-induced stress on stenohaline organisms that could invalidate test results unless salinity is controlled by adding appropriate amounts of sea salt or through addition of saline brine produced by evaporating natural seawater.

2. Early-Life-Stage Toxicity Tests

a. General test procedures: Start fish early-life-stage toxicity tests with newly fertilized eggs and expose them through their developmental stages to an early juvenile age.[4] Testing procedures for the freshwater rainbow trout and marine/estuarine sheepshead minnow are discussed below. For fathead minnow, see Section 8921. Other species can be used if the specific environmental requirements of the fish can be approximated in the laboratory.

Common to all these tests are end points such as time-to-hatch, survival during the different life stages, and growth. Also observe behavior to determine behavioral effects of the test compound. Histological, physiological, or biochemical end points relevant to the study objectives also can be measured.

1) Equipment and physical conditions—For a description of suitable diluter systems see Section 8010F. Set up a minimum of two replicate exposure chambers per test concentration. Construct egg hatching cups of glass tubing (8 cm diam × 10 cm long) with 40 mesh stainless steel or equivalent nylon‡ screen glued at one end with clear silicone sealant. Suspend egg cups from a rocker arm assembly and low-speed motor designed to oscillate the egg cups slowly up and down approximately 2 to 3 cm. Use a flow rate through the exposure chambers sufficient to replace 90% of the water in 8 to 12 h. A self-starting siphon tube in each exposure chamber can be substituted for the rocker-arm system to insure test solution exchange.

Provide light intensities over the chambers of approximately 400 to 800 lux and establish a 16-h light and 8-h dark photoperiod. Preferably provide a 15- to 30-min dusk/dawn transition period.

2) Chemical data recording—Periodically measure dissolved oxygen, pH, conductivity, temperature, hardness, and alkalinity. Typically measure hardness, alkalinity, and conductivity at least weekly. Record temperature at least daily or continuously in one chamber that is centrally located in the row of exposure chambers. Measure dissolved oxygen and pH at least weekly in each exposure chamber containing surviving fish.

3) Verification of exposure concentrations—Before and during the test verify exposure concentrations of the test chemical. This will confirm actual versus nominal concentrations.

Initially, analyze concentration in each replicate test chamber. If consistent concentrations are observed among replicates at each concentration, make subsequent measurements on composite samples from each replicate. However, make other checks of diluter function to ensure proper operating conditions.

It is not necessary that measured concentrations be within any specific percentage of the nominal concentrations. Indeed, such

---

* Teflon® or equivalent.
† Tygon® or equivalent.

‡ Nitex or equivalent.

characteristics as hydrolysis rate and volatility may make it impossible. It is important to maintain consistent concentrations during exposure and to confirm concentrations by chemical measurements. Pre-exposure monitoring will confirm that the diluter system is operating properly and that fish exposure may begin. Fluctuating exposure concentrations may indicate an improperly operating diluter system, while slowing rising concentrations may indicate that system equilibration (e.g., volatility, adsorption to chamber surfaces, hydrolysis, etc.) has not been achieved.

Once concentrations have stabilized, start exposure. Measure exposure concentrations at least weekly until the test ends.

Dichotomous data have end points that fall within two categories such as mortality (alive-dead) and hatch (hatched-not hatched). These data can be analyzed with 2×2 contingency tables, logit, probit, and the chi-square statistic.[7]

b. Specific test procedures:

1) Rainbow trout—The rainbow trout early-life-stage test is conducted for 60 d post-hatch at $12 \pm 2°C$.

a) General considerations—Run the rainbow trout early-life-stage test beginning with newly fertilized embryos or eyed embryos. Commercial suppliers of eggs and sperm make it convenient to begin with male and female gametes and fertilize the eggs in the laboratory. Study designs may expose gametes before and/or during fertilization, as well as immediately after fertilization. Selection may depend on available time because use of newly fertilized eggs adds approximately 1 month to the duration of the test.

b) Equipment and physical condition—For a description of suitable diluter systems see Section 8010F. Use an exposure system similar to that described in 8910C.2a1) with the following changes. On bottom of the egg cup use 16 mesh stainless steel or nylon screen.‡ Incubate under little or no light. When eggs hatch, provide light intensities over the chambers of approximately 400 to 800 lux on a 16-h light and 8-h dark photoperiod. Preferably include 15- to 30-min dusk/dawn transition. Hold at $12 \pm 2°C$.

c) Test initiation—The test begins with distribution of newly fertilized eggs or eyed eggs to each incubation cup. Select number of embryos based on the desired discriminating power for the test and the number of replicates. At a minimum, apportion 60 embryos per experimental group among four incubation cups and place two cups in each of two replicate test chambers.

d) Biological observations—Monitor survival by daily inspecting embryos or hatchlings. Record observations on fish behavior, noting abnormal and normal individuals along with characteristics of any abnormalities.

During embryo incubation remove dead eggs to prevent fungal infections. Dead eggs can be distinguished from living eggs by their white color.

When hatching is greater than 95% complete, count post-hatch alevins and release them to their respective growth chambers. For approximately 2 weeks following hatching, alevins will feed from their yolk sacs. When the first few alevins begin to swim up, feed fry with brine shrimp nauplii in combination with a standard commercial fish food at least three times per day as needed.

Because hatching may occur over a 3- to 6-d period, use the time to obtain at least 95% hatch of control to establish the 60-d post-hatch growth period. Determine growth at 30 d post-hatch (midpoint of growth period) and 60 d post-hatch (end of test). Use a method of measurement that least stresses the fish. The photographic method[5] has been used successfully to estimate weight and lengths when fish are not to be sacrificed.

e) Chemical data recording—See Section 8910C.2a2).

f) Verification of exposure concentrations—See Section 8910C.2a3).

2) Sheepshead minnow—Run this test for 35 d at $25 \pm 2°C$.

a) Equipment and physical conditions—For a description of suitable diluter systems see Section 8010F. Use an exposure system similar to that described in 8910C.2a1) except with seawater salinity of 10 to 20‰.

b) Test initiation—See Section 8910B.5b for egg collection. The test begins with distribution of newly fertilized eggs to each embryo cup. Select a number of embryos based upon the desired discriminating power for the test and the number of replicates. At a minimum, distribute 60 embryos per experimental group among four incubation cups and place two cups in each of two replicate test chambers. Sheepshead minnows hatch in about 7 d at 25°C.[6,7] Release larvae from hatching cup to the test chambers and start feeding. Initially feed a combination of live brine shrimp nauplii, then shift to brine shrimp and commercial food after 7 to 10 d.[6]

c) Biological observations—See Section 8910C.2b1)d).

d) Chemical data recording—In addition to those water quality factors described in Section 8910C.2b1)e), measure dilution water salinity at least weekly.

e) Verification of exposure concentrations—See Section 8910C.2a3).

3. Reproductive Toxicity Tests

a. General test procedures: Use newly spawned eggs, newly hatched larvae, juveniles, or sexually immature fish to start a test. The life stage selected depends on species, laboratory space and facilities, availability of the life stages, and test objective.

Expose enough fish to each concentration of toxicant to insure adequate numbers of each sex at maturity but low enough to prevent stress due to crowding. Additional fish may be introduced to tanks with similar treatment to provide specimens for histological examination, residue analysis, or selected physiological measures of condition. The exact number of fish required depends on life stage at start of test and test species (see Section 8910C.3b).

At start of test, measure total length and weight of all fish. Repeat measurement for any fish that die during the test. To prevent injury, anesthetize§ large fish before handling. Larvae and small fish may be measured by a photographic method.[5] At the end of a test, record length, weight, and, if possible, sex and gonadal condition of each fish.

For viability and hatchability tests, incubate eggs from each spawning at an optimum temperature in control water. Count live and dead eggs and remove dead eggs daily. Evaluate egg viability for all spawnings by incubating eggs until development clearly is observed (some defined stage of embryogenesis, e.g., eyeing, is reached). Determine hatchability for all spawnings in all exposure chambers or from a predetermined number of spawnings when the species tested is one that spawns continuously (many times per season). Count number of dead, deformed, and normal larvae hatched daily, using a dissecting microscope if necessary.

To evaluate larval growth and survival at each toxicant concentration, collect a uniform number of larvae (usually 20 to 50) at random from two or more successful hatches and place in

---

‡ Nitex or equivalent.

§ 3-aminobenzoic acid ethyl ester or equivalent.

chambers for that toxicant concentration. Determine length and number of larvae upon transfer to growth chambers, preferably by the photographic method. Determine total length of larvae at selected intervals and at end of test. Count, measure, and remove dead larvae daily.

For methods of toxicant mixing and delivery see Section 8010F.1c.

Use spawning tanks, exposure tanks, and growth chambers appropriate for the test species. Design each growth chamber so that test solutions can be drained down to 2.5 to 3 cm and the chamber transferred to a fluorescent light box provided with a millimeter grid for photographing fish.[5]

Monitor fish and embryos maintained for physiological, biochemical, and histological tests carefully. As a minimum, report all pertinent data for each test container at the beginning, about a third of the way through, and at end of test. Include number and weight of individuals, number of spawnings, number of embryos, and total lengths of normal, deformed, and injured mature and immature males and females. Count and record all survivors and mortalities. Calculate mean incubation time for median spawning and hatch dates if known. The hatchability, fry survival, growth, and percent deformities also may be determined.

Measure toxicant concentration in all tanks at each concentration weekly. Composites of equal-volume daily grab samples for 1 week may be used if it has been shown that analytical results for the test compound are unaffected by storage. Include samples for assessing recovery (i.e., known additions) and blanks. Analyze enough samples throughout the test to determine whether toxicant concentrations are constant. If this is not possible, analyze enough samples weekly to establish variability of toxicant concentration (see Section 8010F.3d).

Record temperature continuously in a centrally-located tank. Measure oxygen levels periodically in each tank. Analyze water from the control and one exposure tank at least weekly for pH, hardness, alkalinity, and conductivity in freshwater systems and pH and salinity in marine systems. If any characteristic is affected by the toxicant, analyze that characteristic at least 5 d/week, rotating among tanks so that each is analyzed once every other week.

When possible, analyze mature fish and/or eggs, larvae, and juveniles for toxicant residues.

b. Specific test procedures: Procedures used for reproduction tests are described below for a freshwater and a marine species.

1) Brook trout, Salvelinus fontinalis (freshwater)—This test procedure extends over only a part of the life cycle because of the longevity of brook trout. It follows the life cycle from the yearling stage through spawning, egg hatching, and development for 90 d.

a) Equipment and physical conditions—For description of suitable diluter systems see Section 8010F. Set up duplicate tanks for each test concentration and control. Premix each concentration before delivery to duplicate spawning tanks and growth chambers.

Construct alevin-to-juvenile growth chambers with dimensions of $18 \times 15 \times 18$ cm of glass or stainless steel with a glass bottom. Maintain water depth at about 13 cm. Design each chamber so that the water can be drained down to a depth of 2 to 3 cm to allow the chamber to be placed over a millimeter grid on a fluorescent light box for photographing fish for measurements of length.

Construct spawning tanks of No. 316 stainless steel, with dimensions of $80 \times 30 \times 40$ cm. Use a 30-cm water depth. Place a spawning substrate or nest[6-8] in spawning tanks at the appropriate time. Use spawning nest, $28 \times 33 \times 7.5$ cm, made of double-strength glass or stainless steel. Large fish may require a larger nest. Drill three 2.5-cm holes in each end, 2.5 cm from the bottom, and cover with 10-mesh stainless steel wire to allow water in the box to drain to a depth of 2.5 cm when the box is removed from the spawning chamber. Place a bottomless screen egg-retainer ($27 \times 32 \times 1.3$ cm with 2.5-cm square compartments, constructed from 1.3-cm-wide strips of 7-mesh stainless steel screen) in the spawning box. Place 2-mesh stainless steel screen, 27×32 cm, to which 1.3- to 2.5-cm gravel is attached with silicone adhesive, on top of the screen egg retainer.[8] Use smooth gravel to prevent injury to active, spawning fish. This spawning box is readily removed from the spawning tanks to collect eggs for transfer to incubation cups. For spawning stocks, select yearling fish that will not grow too large for the spawning box. Fish weighing not more than 50 to 70 g at time of selection and 150 g at spawning are appropriate. If fish weigh more than 150 g, use a larger spawning box.

Provide a daylight period conducive to the spawning of brook trout. Ideally use 15- to 30-min dawn/dusk transition times and provide conditions described above in Section 8910C.1b1)a).

Provide water flow of 6 to 10 tank volumes/d to maintain oxygen levels above 60% saturation. Remove uneaten food and wastes from growth chambers daily. Brush interior surfaces to remove attached growths as needed.

b) Exposure procedures—To begin the test, collect juveniles from the field no later than March 1 and acclimate for at least 1 month or use cultured stock of equivalent age. Judge suitability of fish for testing on the basis of acceptance of food, apparent lack of disease, and occurrence of less than 2% mortality during acclimation and no mortality during the 2 weeks before the test.

Begin exposure by placing at least 12 acclimated yearling brook trout in each duplicate tank at each test concentration and suitable controls using a stratified random assignment (see 8010F.3a). This allows about a 4-month exposure to toxicant before the onset of secondary or rapid-growth phase of the gonads. Extra test animals may be included at the beginning so that fish can be removed periodically for special examination or for chemical analysis.

Use a particulate or pelleted trout food. Feed fish the largest particle or pellet they will take, at least twice daily. Base amount on a reliable hatchery feeding schedule.[9] Analyze each batch of food for pesticides.

Record mortalities daily and measure total length and weight of fish directly at initiation of tests and every 3 months thereafter. Do not feed fish for 24 h before weighing. Lightly anaesthetize them to facilitate measuring.

When secondary sexual characteristics are well-developed (approximately 2 weeks before spawning), separate males, females, and undeveloped fish in each tank and randomly reduce number of sexually mature fish to two males and four females per tank. Record number of mature, immature, deformed, and injured males and females in each tank and number from each category to be discarded. Thoroughly clean, sterilize, and rinse the spawning substrates and place one for each male in each spawning tank. As soon as spawning begins, set up incubation cups (as described in 8910B.5a1) or a suitable alternate system to receive embryos for hatching. Remove embryos from the substrate at a fixed time each day, preferably so fish are not disturbed during early part of the light period.

Randomly select 50 embryos from the first eight spawnings of 50 embryos or more in each duplicate spawning chamber and place in an embryo incubator cup. Count remaining embryos from the first eight spawnings and all embryos from subsequent spawnings and place them in separate incubator cups to determine viability as evidenced by development to a specific stage, e.g., formation of neural keel after 11 to 12 d at 9°C or eyeing. Remove and record number of dead embryos from each spawn. Never place more than 250 embryos in one incubator cup. Incubate all embryos to determine viability and discard after reaching some clearly distinguishable stage (development of neural keel or eyeing). Discarded embryos may be analyzed chemically or used for other measurements.

Obtain additional information on hatchability and alevin survival by transferring embryos from control tanks immediately after spawning (a) to tanks having test concentrations where spawning is reduced or absent and (b) to tanks where an effect is seen on survival of embryos or alevins, and by transferring embryos from those test concentrations to control tanks. Always reserve two growth chambers in each duplicate spawning tank for embryos produced in that tank.

Remove dead embryos daily from incubator cups. When hatching begins, record number of alevins hatching daily in each cup. On completion of hatching in any cup, transfer fish to a culture dish and randomly sample 25 alevins. Count dead or deformed alevins. Transfer 25 selected alevins to a growth chamber and place it over the light box to measure by the photographic method. After photographing, return alevins to incubator cup. Never net alevins, but transfer by gentle pouring or by large-bore pipets. Transport in growth chambers containing enough test solution to limit harsh contact with the bottom screening. Preserve unused alevins in formalin for subsequent histological examination. Record length and weight of discarded alevins separately from the data for fish kept for continued exposure.

For 90-d growth and survival exposures randomly select 20 alevins from each duplicate incubator cup for each test concentration and control. Because embryos from one spawn may hatch over a 3- to 6-d period, use the median hatch date to establish the start time of the 90-d growth and survival period. For growth tests select two groups of 20 alevins that are less than 3 weeks apart in age. Use any remaining groups only for hatchability testing. After photographing to determine length, preserve for weight determination. To equalize effects of incubator cups on growth, keep all groups selected for 90-d exposure in the incubator cups for 3 weeks after the median hatch date, then release into growth chambers. Begin feeding immediately. Keep the two groups from the same exposure chambers separate for replication of each test concentration. Record mortalities daily, total lengths at 30 to 60 d after hatching (by the photographic method), and total length and weight at 90 d after hatching. At the end of the test cease feeding juveniles for 24 h and then weigh. Terminate survival and growth studies after 3 months, at which time fish may be used for chemical analysis of tissue and physiological measurements of toxicant-related effects.

End exposure of all parental fish after 3 weeks in which no spawning occurs in any tank. Record mortality and weight, measure total length of parental fish, and check sex and condition of gonads (e.g., reabsorption, degree of maturation, spent ovaries).

Report, for each tank of a partial-life-cycle test, number and individual weights and total lengths of immature males and females at initiation of test, after 3 months, at reduction in numbers, and at end of test. Report individual weights and total lengths of normal, deformed, and injured fish, number maturing, number dying during test, number of spawnings and eggs, hatchability and fry survival, growth, and deformities. Calculate a mean incubation time based on date of spawning and median hatch dates.

For additional information on the life cycle of brook trout consult other sources.[8-17]

2) Fathead minnow, *Pimephales promelas* (freshwater)—See Section 8921.

3) Reproduction tests with other freshwater species—Partial-life-cycle toxicity tests have been performed with the bluegill, *Lepomis macrochirus*, *Oryzias latipes*, and the flagfish, *Jordanella floridae*.[18-23] When culture techniques are available, other native freshwater species may be used as appropriate.

4) Sheepshead minnow, *Cyprinodon variegatus* (euryhaline)

a) Equipment and physical system—Use test apparatus similar to that used for freshwater fish with spawning aquariums at least 30 × 18 × 20 cm. Place a spawning chamber in each aquarium as described in 8910B.5b.

b) Exposure procedures—Begin with adult fish or preferably with embryos. Secure embryos either by natural spawning or by hormone-induced spawning (as described in 8910B.5b). Keep water temperature above 22°C, preferably at 30°C, with salinities above 15 mg/kg. When starting with adults, set up five or six spawning aquariums for each test concentration and controls. Use breeding fish, all from the same stock, that have been kept in holding tanks for at least 2 weeks, during which less than 2% mortality occurred. Feed fish a combination of frozen adult brine shrimp and dry trout food. Maintain water flow through spawning aquariums at 6 to 10 tank volumes/d. Use natural seawater filtered to remove planktonic larvae 15 μm and larger.

When embryos are produced, remove them from spawning chambers and place in hatching chambers for each concentration being tested as well as the controls. Start toxicant dosing in exposure chambers before the hatching chambers are placed inside them. Construct hatching chambers by cementing a 9-cm-wide strip of 500-μm nylon screen around a petri dish. Place the hatching chambers in 90- × 30- × 30-cm exposure chambers in 7 cm of water with flow-through of the toxicant. As embryos hatch, feed fry with newly hatched brine shrimp nauplii. Clean screens on incubation cups and chambers daily. Check and record daily survival of embryos and fry, which constitute the first filial (F_1) generation.

On the first day after hatching remove each chamber and count and measure fry photographically. During the first 2 weeks feed with newly hatched brine shrimp nauplii. During the following 2 weeks supplement this diet with dry trout pellets or dry mollie flakes. After 4 weeks count and measure fish by the photographic method and reduce the number to 50 for each test concentration and controls. Record length, weight, condition, and number of living, deformed, and dead fish remaining. Determine percent mortality and abnormality in each test concentration and controls. Preserve specimens for future tests or discard. Place the 50 selected fish, 25 each, in duplicate growth chambers having a glass bottom and provisions for drawing the water level down to 1 to 2 cm. Feed a mixed diet of brine shrimp and dry food twice daily and examine daily for dead specimens. At 8 weeks measure again by the photographic method. Twice daily, feed dry food supplemented with frozen adult brine shrimp until maturity. Check each batch of food for pesticides, PCBs, or other contaminants of concern. Clean all exposure aquariums and spawning and hatching chambers two to three times per week. Siphon out all wastes.

As fish approach sexual maturity, place separate pairs in spawning chambers, five pairs from each duplicate exposure chamber; i.e., 10 pair for each test concentration and controls, and continue exposure. Count, measure, and weigh all unused fish from each duplicate exposure chamber. Record number deformed and dead in each test concentration and controls, condition of fish, and other pertinent data. Preserve some fish for whole-body tissue residue analysis. As fertilized eggs are produced, remove at a specified time daily, count, and place 25 F_2 fish in a hatching chamber as for the F_1 generation. Record the total number of embryos produced in each chamber, time required to hatch, hatching success, and survival of embryos. Test those not placed in hatching chambers for fertility and record percent of fertile females.

Keep pairs in each spawning chamber until all needed embryos have been obtained. At termination, measure and weigh spawning pairs and record all other pertinent data. Preserve for toxicant analyses, if desired.

Expose embryos of F_2 generation in hatching chambers in their respective duplicate exposure chambers for each test concentration and controls as before. Count and measure by photographic method[5] as for the F_1 generation. Feed fish and record results. At the end of 4 weeks terminate the test. Weigh and measure all fish; record number of deformed fish and determine number that died. Preserve for histological examination and tissue analyses. Determine effects of each test concentration and calculate safe levels. During tests, record temperature daily, and oxygen concentration, pH, and salinity at least weekly. If possible, chemically analyze test water for toxicant at the beginning, at regular intervals (e.g., weekly) during exposure, and on completion of tests. Analyze lots of 10 fish from highest and lowest exposure concentrations and from controls for toxicant accumulation. Analyze dilution water for toxicant at beginning and end of test.

For additional information about the life cycle of *Cyprinodon variegatus*, and testing procedures, consult other sources.[22–26]

5) *Life-cycle tests with other marine fishes*—Life-cycle tests may be performed with other marine fishes such as *Fundulus heteroclitus* and *Menidia menidia*. For information on life cycle and culture of these species, consult other sources.[27–29]

4. Statistical Analysis

Analyze, handle, and report data as in Section 8010G.

5. References

1. U.S. ENVIRONMENTAL PROTECTION AGENCY. 1985. Standard Evaluation Procedure: Acute Toxicity Test for Freshwater Fish. EPA-540/9-85-006, Off. Pesticide Programs, Hazard Evaluation Div., Washington, D.C.
2. DRUMMOND, R.A. & W.F. DAWSON. 1970. An impressive method for simulating diel patterns of lighting in the laboratory. *Trans. Amer. Fish. Soc.* 99:434.
3. COMMITTEE ON METHODS FOR TOXICITY TESTS WITH AQUATIC ORGANISMS. 1975. Methods for Acute Toxicity Tests with Fish, Macroinvertebrates and Amphibians. EPA 660/3-75-009, U.S. Environmental Protection Agency, Washington, D.C.
4. GOODMAN, L.R. 1985. Comparative Toxicological Relationships Demonstrated in Early Life Stage Tests with Marine Fish. EPA-600/9-95-135. U.S. Environmental Protection Agency, Washington, D.C.
5. MARTIN, J.W. 1967. A method of measuring lengths of juvenile salmon from photographs. *Progr. Fish-Cult.* 29:238.
6. HANSEN, D.J. & P.R. PARRISH. 1975. Suitability of sheepshead minnows (*Cyprinodon variegatus*) for life cycle toxicity tests. In F.L. Mayer & J.L. Hamelink, Aquatic Toxicity and Hazard Identification. ASTM STP634, American Soc. Testing & Materials, Philadelphia, Pa.
7. U.S. ENVIRONMENTAL PROTECTION AGENCY. 1986. Standard Evaluation Procedure. Fish Early-Life-Stage Test. EPA-540/9-86-139, Off. Pesticide Programs, Hazard Evaluation Div., Washington, D.C.
8. BENOIT, D.A. 1974. Artificial laboratory spawning substrate for brook trout (*Salvelinus fontinalis* Mitchell). *Trans. Amer. Fish. Soc.* 103: 144.
9. PIPER, R.G., I.B. MCELWAIN, L.E. ORME, J.P. MCCRAREN, L.G. FOWLER & J.R. LEONARD. 1982. Fish Hatchery Management. U.S. Dept. Interior, Fish & Wildlife Serv., Washington, D.C.
10. ATZ, J.W. & G.E. PICKFORD. 1959. The use of pituitary hormones in fish culture. *Endeavor* 18:125.
11. MCKIM, J.M. & D.A. BENOIT. 1971. Effect of long-term exposures to copper on survival, reproduction and growth of brook trout, *Salvelinus fontinalis* (Mitchell). *J. Fish. Res. Board Can.* 28:655.
12. ALLISON, L.N. 1951. Delay of spawning in eastern brook trout by means of artificially prolonged light intervals. *Progr. Fish-Cult.* 13: 111.
13. CARSON, B.W. 1955. Four years progress in the use of artificially controlled light to induce early spawning of brook trout. *Progr. Fish-Cult.* 17:99.
14. HALE, J.G. 1968. Observations on brook trout, *Salvelinus fontinalis* spawning in 10-gallon aquaria. *Trans. Amer. Fish. Soc.* 97:299.
15. HENDERSON, N.E. 1962. The annual cycle in the testes of the eastern brook trout. *Salvelinus fontinalis* (Mitchell). *Can. J. Zool.* 40:631.
16. HENDERSON, N.E. 1963. Influence of light and temperature on the reproductive cycle of the eastern brook trout *Salvelinus fontinalis* (Mitchell). *J. Fish Res. Board Can.* 20:859.
17. WYDOSKI, R.S. & E.L. COOPER. 1966. Maturation and fecundity of brook trout from infertile streams. *J. Fish. Res. Board Can.* 23:623.
18. EATON, J.G. 1974. Chronic cadmium toxicity to the bluegill (*Lepomis macrochirus* Rafinesque). *Trans. Amer. Fish. Soc.* 103:729.
19. SMITH, W.E. 1973. A cyprinodontid fish, *Jordanella floridae*, as a laboratory animal for rapid chronic bioassays. *J. Fish. Res. Board Can.* 30:329.
20. EATON, J.G. 1970. Chronic malathion toxicity to the bluegill, *Lepomis macrochirus. Water Res.* 4:673.
21. MCCOMISH, T.S. 1968. Sexual differentiation of bluegills by the urogenital opening. *Progr. Fish-Cult.* 30:28.
22. FOSTER, N.R., J. CAIRNS, JR. & R.L. KAESLER. 1969. The flagfish *Jordanella floridae*, as a laboratory animal for behavioral bioassay studies. *Proc. Acad. Natur. Sci. Philadelphia* 121:129.
23. BENOIT, D.A. 1975. Chronic effects of copper on survival, growth and reproduction of the bluegill (*Lepomis macrochirus*). *Trans. Amer. Fish. Soc.* 104:353.
24. SCHIMMEL, S.C., D.J. HANSEN & J. FORESTER. 1974. Effects of Aroclor 1254 on laboratory-reared embryos and fry of sheepshead minnows (*Cyprinodon variegatus*). *Trans. Amer. Fish. Soc.* 103:582.
25. HANSEN, D.J., S.C. SCHIMMEL & J. FORESTER. 1973. Aroclor 1254 in eggs of sheepshead minnows: Effects on fertilization success and survival of embryos and fry. *Proc. 27th Annu. Conf. S.E. Assoc. Game Fish Comm.*: 420.
26. HANSEN, D.J. 1978. Laboratory culture of sheepshead minnows (*Cyprinodon variegatus*). In Bioassay Procedures for the Ocean Disposal Permit Program. EPA-600/9-78-010, U.S. Environmental Protection Agency.
27. BOYD, J.F. & R.C. SIMMONS. 1974. Continuous laboratory production of fertile *Fundulus heteroclitus*, Walbaum eggs lacking chorionic fibrils. *J. Fish. Biol.* 6:389.
28. RUBINOFF, I. 1958. Raising the atherinid fish *Menidia menidia* in the laboratory. *Copeia* 1958:146.
29. MIDDAUGH, D.P., M.J. HEMMER & L.R. GOODMAN. 1987. Methods for Spawning, Culturing and Conducting Toxicity Tests with Early Life Stages of Four Atherinid Fishes: The Inland Silverside (*Menidia beryllina*), Atlantic Silverside (*M. menidia*), Tidewater Silverside (*M. peninsulae*) and California Grunion (*Leuresthes tenuis*). EPA-600/8-87-004. Off. Research & Development, U.S. Environmental Protection Agency, Washington, D.C.

8921   FATHEAD MINNOW*

8921 A.   Introduction

The fathead minnow, *Pimephales promelas* rafinesque, is a small, common, and widely distributed freshwater fish of the family Cyprinidae. This minnow is maintained easily in the laboratory and can be spawned year-round. These attributes have led to its widespread use in aquatic toxicology studies, particularly those utilizing early life stages (i.e., embryos and larvae), such as the short-term tests for measuring the chronic toxicity of effluents.

1. Description

Adult fathead minnows typically range in size from 43 to 102 mm, averaging about 51 mm (2 in.) in total length.[1,2] Young and nonbreeding adults are light in color with a distinct lateral band from caudal peduncle to head; males and females at this stage are difficult to differentiate, except that males are typically larger. In breeding condition (Figure 8921:1), males are distinguished from females by the presence of nuptial tubercles on the snout and by coloration. Mature males are dark in overall coloration with a saddle-like pattern behind the head, whereas females are quite drab.

2. Distribution, Biology, and Life History

The fathead minnow is tolerant of adverse conditions including high temperature and turbidity and low oxygen concentrations.[1] Because of this tolerance, the fathead minnow is found in a diversity of habitats and is widely distributed throughout central North America from Canada to northern Mexico.[1,2] It is most abundant in muddy streams, brooks, ponds, and small lakes.[1] This

---

* Approved by Standard Methods Committee, 1997.

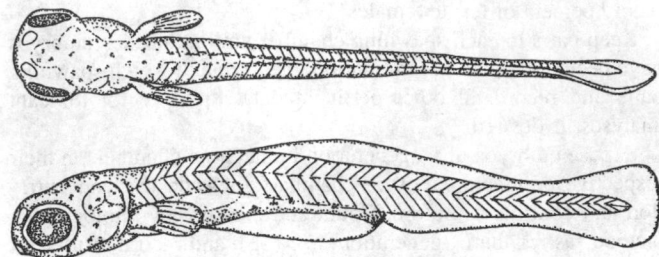

Figure 8921:2. Newly hatched fathead minnow larvae: (above) top view; (below) lateral view.

species is a popular bait fish and has been introduced to areas both within and outside of its native range because of the relative ease with which it is maintained and propagated.

The fathead minnow rarely lives beyond an age of 2 years. In warm, food-rich waters, it grows rapidly and may reach adult size and begin spawning in as little as 3 months. In cold waters, it make take a year to reach maturity.[2] It is omnivorous, with diet consisting at times of algae, organic detritus, aquatic insects, worms, small crustaceans, and planktonic organisms. Because it is highly prolific, can utilize many foods, and is widely preyed upon by other fish and fish-eating birds, the fathead minnow is considered an ideal forage fish and important bait species.

In the wild, fathead minnows begin spawning in the spring and often continue to spawn throughout the summer. Spawning typically begins when the water temperature reaches about 16 to 18°C, although this temperature may vary with the population and latitude.[1] Spawning usually occurs in the early morning in shallow water less than 1 m deep. The male selects a suitable substrate (e.g., underside of a log, branch, root, large rock, board) and herds a receptive female into position. Using her ovipositor, the female deposits her adhesive eggs (from 100 to 500 per spawn) on the substrate and is then driven off by the territorial male. The male aggressively guards the nest and often seeks out additional females to spawn in the nest. Time to egg hatching depends on water temperature. For example, eggs hatch in about 1 week at 22°C and between 4 and 5 d at 25°C. Newly hatched larvae are approximately 5 mm long and opaque white in color, and have large black eyes (Figure 8921:2).[1,3]

3. References

1. WEBER, C.I., ed. 1993. Appendix A, Distribution, life cycle, taxonomy, and culture methods A.5. (Fathead minnow *Pimephales promelas*). *In* Methods for Measuring the Acute Toxicity of Effluents to Freshwater and Marine Organisms, 4th ed. EPA-600/4-90-027F, Environmental Monitoring Systems Lab., U.S. Environmental Protection Agency, Cincinnati, Ohio.

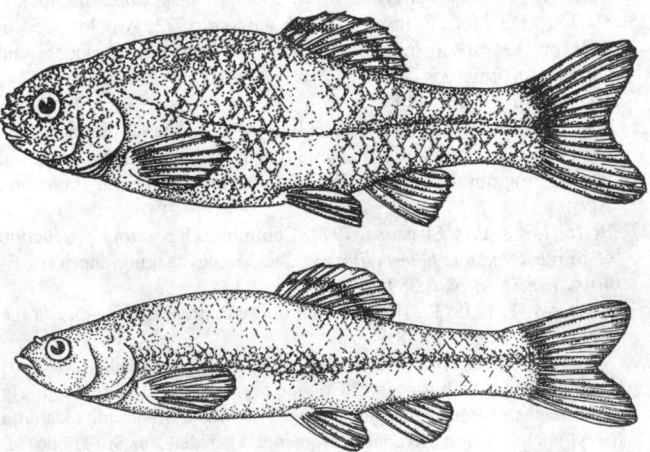

Figure 8921:1. Adult fathead minnows in breeding condition: (above) male; (below) female.

2. SCOTT, W. & E. CROSSMAN. 1973. Freshwater Fishes of Canada. Bull. 184, Fisheries Research Board of Canada, Ottawa, Ont., Canada.
3. SYNDER, D.E., M.B. MULLHALL SYNDER & S.C. DOUGLAS. 1977. Identification of golden shiner, *Notemigonus crysoleucas*, spotfin shiner, *Notropis spilopterus*, and fathead minnow, *Pimephales promelas*, larvae. *J. Fish. Res. Board Can.* 34:1397.

4. Bibliography

MARKUS, H. 1934. Life history of the blackhead minnow (*Pimephales promelas*). *Copeia* 1934:115.

8921 B. Culture and Maintenance of Test Organisms

1. Obtaining Test Organisms

Organisms of various ages may be obtained for toxicity testing from commercial breeders, biological supply houses, or an in-house culture facility. An in-house breeding facility is recommended, and may be required, if testing is to be conducted with early life stages or fish of a specific age. Ensure that fish, particularly those from outside sources, are certified as to their identity, age, and freedom from disease. Preferably use sources that supply reference toxicant data with their shipments. In general, except as source of "new" genes, avoid using organisms from bait shops, hatcheries, or field populations because their suitability for use and disease status cannot be assured.

2. Culturing and Care of Test Organisms

Fathead minnows, including all life stages from egg to adult, may be successfully cultured in the laboratory using static, recirculating, or flow-through systems. Basic information on establishing and maintaining a culture facility is presented below. For additional information, see Sections 8010E.4 and 8910B, and other sources.[1-3]

a. Water supply and culture system: Supply cultures with good-quality water; reconstituted (synthetic) water, dechlorinated municipal water, and natural water (see Section 8010E.4b) are all acceptable. Natural water is preferred provided that its quality is relatively constant and it meets minimum acceptability criteria. Reconstituted water usually is recommended to be of moderate hardness (see Table 8910:I). Analyze all water supplies periodically for chlorine (free and combined), ammonia, toxic metals (e.g., Cr, Cd, Cu, Pb), organic compounds (e.g., pesticides, PCBs), and basic water quality factors (e.g., pH, DO, conductivity, hardness, alkalinity).

Choose a spawning unit designed to simulate natural spawning conditions (e.g., light, temperature). Typically, fish are bred in small (60- to 120-L) tanks or aquaria. Flow-through culture systems are recommended; however, static or recirculating systems may be used provided that adequate water quality is maintained through use of activated carbon filtration and/or other treatments. See Section 8910E.4 for a discussion of appropriate construction materials and equipment. For spawning, hold water temperature at $25 \pm 2°C$. Aerate water as necessary to maintain DO concentration at or near saturation. Establish a controlled photoperiod of 16 h light (e.g., 5:00 AM to 9:00 PM) and 8 h dark.

b. Establishment of breeding: Establish breeding units with 15 to 20 mature (>6-month-old) adults and two or three spawning substrates per tank. Construct spawning substrates from inverted halves (semicircular sections) of Schedule 40 PVC pipe (7.5 cm ID × 7.5 cm length) or other nontoxic material (e.g., glass, stainless steel). If the inverted surface is smooth, roughen it to aid embryo adhesion. At first it may be difficult to differentiate the sex of fish, but this process becomes easier over the next week as organisms develop their secondary sexual characteristics (Figure 8921:1). Remove excess males to maintain a sex ratio of approximately six females per male and no more than two males per tank.[1,2] As an alternative, pair females and males in divided tanks.

c. Embryo collection and incubation: Check spawning substrates daily in the late morning or early afternoon. Record the number of eggs per substrate and incubate embryos using one of the techniques described below.[1,2] Examine embryos daily and remove all that are dead (milky and opaque) or show fungal growth. Embryos maintained at 22 to 25°C will hatch in about 4 to 7 d.

1) Incubation of embryos on substrate—Place several substrates on end in a circular pattern (embryos on the inside) around a source of gentle aeration. Maintain a constant temperature and sufficient water depth to cover substrates.

2) Incubation of embryos in a separatory funnel—Remove embryos from substrates with a gentle rolling action of the index finger.[4] Incubate embryos in a 2-L separatory funnel containing approximately 1.5 L water. Hold separatory funnel in a constant-temperature bath and maintain constant gentle aeration from bottom of funnel.

3) Incubation of embryos in incubation cups—Remove embryos from substrates with a gentle rolling action of the index finger.[4] Place embryos in incubation cups attached to a rocker-arm assembly[5] that maintains constant water movement over the embryos.

d. Rearing of larvae and juveniles: Each day, transfer newly hatched larvae to small (10- to 60-L) rearing tanks. Use large-bore pipets or other methods that do not require direct handling of larvae. Do not use nets until fish are approximately 30 d old. Establish initial density at or below 15 larvae/L. Reduce density proportionally as fish grow larger by thinning fish or moving them to larger tanks. Hold juveniles at a density of ≤1 fish/L, typically in tanks of 200 L or more. Maintain aeration and a relatively constant temperature (20 to 25°C). Keep tanks holding replacement spawners (brood stock) at or near spawning temperature ($25 \pm 2°C$).

e. Food and feeding: Feed larvae up to 30 d old two to three times a day with newly hatched brine shrimp (*Artemia salina*). Culturing of brine shrimp is described elsewhere.[6] Supplement live food once or twice a day with commercial fish starter food. Feed older fish with frozen adult brine shrimp, commercial fish starter, and tropical fish flake food.[1,2] Fish may be fed *ad libitum* but avoid overfeeding because it increases tank maintenance, decreases water quality, and may increase stress and susceptibility to disease.

f. Parasite and disease control: Observe fish daily for disease and abnormal behavior. Parasites and disease will rarely be a problem if proper water quality and aeration are maintained in rearing tanks. If necessary, provide treatment as described in Section 8910B.4. Clean and disinfect tanks and related equipment on a regular basis and, in particular, before new fish are added or after any disease outbreaks. Avoid spread of disease by disinfecting dip nets before use.

3. Acclimating and Holding Test Organisms

When possible, quarantine and acclimate (preferably for at least 48 h) test organisms obtained from outside sources; however, no acclimation is necessary for tests initiated with fish early-life stages. During acclimation, protect fish from large changes in temperature or water quality (e.g., pH, hardness) and minimize handling. Avoid overcrowding and maintain sufficient DO concentrations. During acclimation, change water from 100% holding water to 100% dilution water. Keep all organisms in 100% dilution water for at least 48 h before use.

Observe fish daily for signs of stress and disease; remove dead or abnormal organisms promptly. Mortality of 5 to 10% is not unusual during the first 48 h of acclimation; however, do not use organisms in tests if mortality exceeds 5% in the 48-h period preceding test initiation. See Sections 8010E.3 and 8910B.3 for additional information.

4. References

1. WEBER, C.I., ed. 1993. Appendix A, Distribution, life cycle, taxonomy, and culture methods A.5. (Fathead minnow *Pimephales promelas*). *In* Methods for Measuring the Acute Toxicity of Effluents to Freshwater and Marine Organisms, 4th ed. EPA-600/4-90-027F, Environmental Monitoring Systems Lab., U.S. Environmental Protection Agency, Cincinnati, Ohio.
2. LEWIS, P.A., D.J. KLEMM, J.M. LAZORCHAK, T.J. NORBERG-KING, W.H. PELTIER & M.A. HEBER, eds. 1994. Short-Term Methods for Measuring the Chronic Toxicity of Effluents and Receiving Waters to Freshwater Organisms, 3rd ed. EPA-600/4-91-002, Environmental Monitoring Systems Lab., U.S. Environmental Protection Agency, Cincinnati, Ohio.
3. DENNY, J.S. 1987. Guidelines for the Culture of Fathead Minnows *Pimephales promelas* for Use in Toxicity Tests. EPA-600/3-87-001, Environmental Research Lab., U.S. Environmental Protection Agency, Duluth, Minn.
4. GAST, M.H. & W.A. BRUNGS. 1973. A procedure for separating eggs of the fathead minnow. *Prog. Fish. Cult.* 35:54.
5. MOUNT, D.I. Chronic toxicity of copper to fathead minnows (*Pimephales promelas*). *Water Res.* 2:214.
6. WEBER, C.I., ed. 1993. Appendix A, Distribution, life cycle, taxonomy, and culture methods A.4. (Brine shrimp *Artemia salina*). *In* Methods for Measuring the Acute Toxicity of Effluents to Freshwater and Marine Organisms, 4th ed. EPA-600/4-90-027F, Environmental Monitoring Systems Lab., U.S. Environmental Protection Agency, Cincinnati, Ohio.

8921 C. Procedures

1. Short-Term (Acute) Test

Test procedures and conditions common to short-term tests on fathead minnows are summarized in Table 8921:I; additional procedures and conditions for specific short-term tests are shown in Table 8921:II.

a. Scope and application: Short-term tests are conducted with the fathead minnow to determine toxicity of pure compounds, formulations and mixtures, effluents, and receiving waters. Test populations are usually mixed sex. Mortality is the primary test endpoint (see Table 8910:II); other endpoints such as loss of equilibrium should be noted and an EC50 (median effect concentration) value determined. Test results may be used to compare toxicity among chemicals to determine sensitivity of different species, for regulatory purposes, or for ecological risk assessments.

b. General test procedures: Short-term (acute) test procedures applicable to the fathead minnow are described in several sources.[1–5] For test duration and types, see Table 8910:II. Choose type on basis of availability of test compound, mixture or effluent, toxicant characteristics such as volatility and solubility, and age/size of minnows used for testing (see Sections 8010D.1 and 2). Early life stages (i.e., larvae) or juveniles are preferred over adults for use as test organisms because they are typically more sensitive and require smaller test solution volumes. Selection of larvae or juveniles depends on test purpose, availability, and laboratory facilities (see Section 8910B.1). Testing of effluents and receiving waters requires use of early life stages[1] (see Section 8921B.2).

TABLE 8921:I. TEST CONDITIONS COMMON TO VARIOUS FATHEAD MINNOW SHORT-TERM TESTS

Test Condition	Type or Value
Light quality and intensity	Ambient laboratory levels; 550–1050 lux (50–100 ft-c)
Photoperiod	16 h light; 8 h dark
pH	6.0–9.0 (if outside this range, adjust to pH 7.0 and perform a parallel test without pH adjustment - see text)
Dilution water	High-quality fresh water. May consist of natural water, receiving water, moderately hard reconstituted fresh water, or dechlorinated municipal water (see Section 8010E.4*b*1) and 8010F.2*a*.
Test concentrations	≥5 plus a control; factor of ≥0.5 between concentrations preferred
Carrier solvent	For testing low-solubility compounds, use ≤0.1 mL/L of a suitable solvent (acetone, dimethylformamide, ethanol, methanol, isopropanol, acetonitrile, or ethylene glycol) (see Section 8010F.2*b*).
Test solution aeration	Not needed unless DO concentration drops below 4.0 mg/L; avoid supersaturation.

Initially determine approximate toxicity of the test material in a range-finding test of 24 to 96 h. Use three to five widely spaced concentrations (e.g., dilution factor of 10) and, optionally, fewer

TABLE 8921:II. TEST CONDITIONS SPECIFIC TO VARIOUS FATHEAD MINNOW SHORT-TERM TESTS

Test Condition	Acute Test	Survival and Growth Test	Embryo-Larval Survival and Teratogenicity Test
Test type	Static, static-renewal, recirculating, flow-through	Static-renewal	Static-renewal
Duration	24, 48, or 96 h (typically)	7 d	7 or 8 d
Temperature	20°C ± 1°C or 25°C ± 1°C	25°C ± 1°C	25°C ± 1°C
Test chamber size	≥250 mL	≥500 mL	≥150 mL; ≥250 mL (preferred)
Test solution volume	≥200 mL	≥250 mL	≥70 mL; ≥200 mL (preferred)
Test solution renewal	After 48 h (minimum); after 24 h (preferred)	Daily	Daily
Age of test organisms			
Effluents and receiving waters	1–14 d (post hatching), ≤24-h range in age	<24-h-old larvae (post hatch); if larvae are not obtained from in-house cultures they should be <48 h old (<24-h range in age)	≤36-h-old embryos, ≤24-h range in age (maximum of 48 h if shipped)
Pure compounds and mixtures	1–14 d (post hatching), ≤24-h range in age or 30–60 d (post hatching); ≤24-h range in age		
Organisms per test chamber	≥10; 20–25 (preferred)	≥10; 15–25 (preferred)	≥10; 15–25 (preferred)
Replicate chambers per concentrations	≥2; 4 (preferred)	≥3; 4 (preferred)	≥3; 4 (preferred)
Organisms per concentration	≥20; 80–100 (preferred)	≥30; 60–100 (preferred)	≥30; 60–100 (preferred)
Feeding:			
Larvae (>2 d old)	0.2 mL *Artemia* (brine shrimp) nauplii concentrate before test initiation and 2 h before test solution renewal at 48 h	0.1 g newly hatched *Artemia* nauplii (<24 h old)	Not required
Juveniles	Feed before test initiation; do not feed during test.	Three times daily at 4-h intervals, or at a minimum, 0.15 g twice daily with 6 h between feedings. No feeding during final 12 h.	
Cleaning of test chambers	As required. Generally not required if test is conducted flow-through or solutions are renewed after 24 or 48 h.	Siphon daily, immediately before test solution renewal	Not required
Test endpoints	Mortality (normally LC50 and NOEC), behavior (activity, swimming, buoyancy, feeding, etc.)	Mortality and growth (weight)	Mortality and teratogenicity (deformed larvae)
Test acceptability	Mortality of control organisms ≤10%	Mortality of control organisms ≤20%; average dry weight per surviving organism in control chambers of ≥0.25 mg	Mortality of control organisms ≤20%

organisms per concentration than normally recommended for definitive tests. Use results of the range-finding test to determine appropriate concentrations for use in the subsequent definitive test.

c. Specific test procedures:

1) Equipment and physical conditions—Always use materials that minimize sorption and leaching of toxic substances, such as tempered glass, perfluorocarbon plastics,* and No. 316 stainless steel (see Section 8010F.1). Clean and rinse all equipment, including new glassware, before use (see Section 8010E.4*d*). Flow-through tests may require use of a diluter system (see Section 8010F.1*c*) or continuous-flow-low-volume (e.g., peristaltic) pumps. Conduct testing in well-ventilated, temperature-controlled facility. Maintain test temperature and light conditions as directed

in Tables 8910:I and II. Cover test vessels with clear plastic or glass allowing air circulation.

For dilution water requirements, see Table 8921:I. Prepare test solutions as described in Section 8010F.2*b*. Use a dilution factor of 0.5 or greater for determining test concentrations in the definitive test. For testing compounds of low solubility, follow carrier-solvent recommendation in Table 8921:I and use a solvent control. Analyze water in control and test chambers daily for pH, DO, and temperature. For renewal tests, make these measurements in new solutions and before renewing solutions, to see the range of conditions. Avoid aeration, particularly for volatile compounds, except to maintain DO concentrations at a minimum of 4 mg/L.

Outside pH range 6.0 to 9.0, the toxicity of metals and organics may be masked by the toxic effects of low or high pH. In such cases, preferably make two parallel tests, one with the pH adjusted

_____
* Teflon® or equivalent.

to 7.0 and one without an adjusted pH. Adjust sample pH by adding 1N NaOH or 1N HCl dropwise, as required, being careful to avoid overadjustment.

2) *Test initiation*—Obtain suitable fathead minnow larvae (1 to 14 d old) or juveniles (30 to 60 d old) from an in-house culture or outside supplier. Larvae are often the most sensitive fish life stage and usually are required for compliance testing of effluents.[1] Distribute test organisms randomly between replicate test chambers containing test solutions. Use a minimum of 2 replicates with 10 organisms each. If sufficient numbers of test organisms are available, preferably test 80 to 100 organisms at each concentration (preferably 4 replicates each with 20 to 25 organisms). For static and static-renewal tests, use live weight loading in the test solutions below 0.65 g/L (20°C) or 0.40 g/L (25°C). For flow-through tests, use live weight loading below 5.0 g/L (20°C) or 2.5 g/L (25°C).

3) *Solution renewal*—Unless the supply of test material or effluent is limited, renew test solutions daily or every other day. Periodic renewal is especially important in tests of volatile compounds, chemicals with low solubilities, and compounds that degrade rapidly.

4) *Feeding*—See Table 8921:II. Avoid overfeeding because it may reduce toxicant concentrations and DO levels. Culturing *Artemia* is discussed elsewhere.[1,6]

5) *Biological data and observations*—Monitor and record mortality daily starting approximately 24 h after test initiation; remove dead organisms. Criteria for establishing death include lack of movement and no reaction to gentle prodding. Record general observations of fish appearance (coloration, deformities) and behavior (lethargy, lack of schooling, loss of equilibrium).

6) *Chemical data recording*—Measure conductivity, hardness, and alkalinity in freshly prepared solutions at test initiation, each test solution renewal, and at test termination in the highest concentration of test solution and in the dilution water. Measure pH, DO, and temperature at test initiation and daily thereafter in all test concentrations. In static-renewal tests, make measurements in both freshly prepared and 24-h-old solutions.

7) *Verification of exposures*—If resources are available for analyses, verify exposures by measuring concentrations of the test chemical in exposure solutions at test initiation and termination. Base statistical analyses and results (LC50, NOEC) on measured concentrations rather than nominal concentrations.

8) *Test termination*—End test after 24, 48, or 96 h. Before termination, record number of dead and abnormal fish in each test chamber.

2. Short-Term Methods for Estimating Chronic Toxicity

Several test methods are available for estimating the long-term (chronic) effects of a toxicant or effluent after a relatively short (7-d) period of exposure (see Section 8010D.5). Two of these tests, the larval survival and growth test and the embryo-larval survival and teratogenicity test, use fathead minnows as test organisms and are described below. These short-term tests were developed as cost-effective alternatives to long-term early-life-stage and life-cycle tests. They were designed primarily to evaluate effluent toxicity[7–10] and are often included as biomonitoring requirements in discharge permits. They also have been used successfully to estimate the potential chronic toxicity of pure compounds.[9,11–13] In addition, because embryos and larvae often are the most sensitive stages in a fish's life cycle,[14,15] studies with embryo-larval stages have been used for investigating teratogenesis and identifying developmental toxicants.[16–19]

a. Larval survival and growth test: This method estimates the chronic toxicity of effluents and chemicals using newly hatched fathead minnow larvae in a 7-d test. Test results are based on survival and weight of larvae. Test conditions and procedures are given in Table 8921:I and II. For testing effluents and receiving waters, see available literature.[7]

1) *Equipment and physical conditions*—For effluents and most chemicals, static-renewal exposures are typically used. Other types of exposure (e.g., static, flow-through) may be used if supplies of test materials are limited or if test compounds are volatile or degrade rapidly. Suitable test chambers include 500-mL or 1-L beakers made of borosilicate glass or nontoxic disposable plastic. For other requirements, see ¶ 1*c* above.

2) *Test initiation*—Obtain organisms and set up test chambers according to requirements of Table 8921:II. Begin tests with effluents as soon as possible, preferably within 24 h of sample collection. From a pool of larvae, randomly select and distribute one or two larvae at a time until each test chamber contains a minimum of 10, but preferably 15, larvae. Use a large-bore pipet or similar device for transferring larvae; do not use a dip net. During transfer, avoid adding excess water to test chambers because this will dilute exposure concentrations.

3) *Solution renewal*—Unless the supply of test material or effluent is limited or flow-through procedures are used, renew test solutions daily. Before renewal, remove uneaten and dead *Artemia*, dead larvae, and other debris by siphon. Take care not to accidentally remove or injure larvae. Use of a light box will enhance larval visibility and simplify this task. Add new test solutions to test chambers after removal of approximately 80 to 90% of old solutions. Add solutions slowly down the side of the test chamber to avoid injury to larvae.

4) *Feeding*—See Table 8921:II. Rinse *Artemia* with fresh water before feeding. Avoid overfeeding because this may reduce toxicant concentrations and DO levels. Do not feed during final 12 h of test. Refer to the literature for information on culturing of *Artemia*.[1,6]

5) *Biological data and observations*—See ¶ 1*c*5) above.

6) *Chemical data recording*—See ¶ 1*c*6) above.

7) *Verification of exposures*—See ¶ 1*c* 7) above.

8) *Test termination*—Terminate test after 7 d. Before termination, record number of dead and abnormal fish in each test chamber (replicate). Prepare fish in each replicate for dry-weight determination. If necessary, preserve larvae in 70% ethanol or 4% formalin for up to 7 d before drying and weighing. Rinse each group of larvae with deionized water, transfer to a labeled and tared weighing boat, and dry at 60°C for 24 h (or at 100°C for 6 h). Let cool in a desiccator, weigh to nearest 0.01 mg, and record. For each replicate, determine mean individual dry weight per fish (using the number of original larvae) to the nearest 0.001 mg. For controls, calculate mean weight per surviving fish.

9) *Test acceptability criteria*—The test is considered acceptable if control survival is ≥80% and average dry weight per surviving larvae of fish in the control replicates is ≥0.25 mg.

10) *Statistical analysis*—Assemble, analyze, evaluate, and report data as described in Section 8010G and other references.[1,7] Calculate LC50 by a point estimation technique such as regression analysis. Obtain lowest-observed-effect concentration (LOEC) and no-observed-effect concentration (NOEC) values for survival

and growth by using hypothesis-testing techniques such as Dunnett's Procedure or Steel's Many-one Rank Test.

b. Embryo-larval survival and teratogenicity test: This method estimates the chronic toxicity of effluents and chemicals by exposing fathead minnow embryo-larval stages in a 7- or 8-d test. Tests are initiated with fertilized embryos. Exposure is continued for several days after hatching of larvae (approximately 4 to 5 d after fertilization at 25°C), depending on age of embryos at test initiation. Test results are based on total frequency of both mortality and gross morphological deformities (terata). Examples of abnormally developed fathead minnow larvae are shown in Figure 8921:3. The test is useful in screening for teratogens (agents that produce terata) because organisms are exposed during early embryonic development when they are most susceptible. General test conditions and procedures are given in Tables 8921:I and II. For testing of effluents and receiving waters, see available literature.[7]

1) *Equipment and physical conditions*—For effluents and most chemicals, static-renewal exposures are typically used. Other types of exposure (e.g., static, flow-through) may be used if supplies of test materials are limited or if test compounds are volatile or degrade rapidly. Suitable test chambers include 250-mL or 500-mL beakers made of borosilicate glass or nontoxic disposable plastic. Chambers as small as 150 mL may be used when small solution volumes are used. For other requirements, see ¶ 1*c* above.

2) *Test initiation*—Use embryos, preferably less than 36 h old, to initiate tests. Obtain embryos from an in-house culture (see Section 8921B) or commercial supplier. Remove embryos from spawning substrates within 12 h of spawning. If organisms must be shipped from a supplier, embryos up to 48-h old may be used, provided all embryos are of the same approximate age. Set up test chambers according to requirements of Table 8921:II.

Preferably use larger volumes for testing compounds that are rapidly degraded, volatile, or have a low water solubility. Begin tests with effluents as soon as possible, preferably within 24 h of sample collection.

From a pool of embryos from three or more spawns, randomly select and distribute several at a time until each test chamber contains a minimum of 10, but preferably 15 to 25, embryos. Exclude abnormal and nonviable (milky-colored and opaque) embryos as well as any showing signs of fungal infection. A light box and stereoscopic microscope are recommended for examining and counting embryos. Use a large-bore pipet or similar device for transferring embryos. During transfer, avoid adding excess water to test chambers because this will dilute exposure concentrations.

3) *Solution renewal*—Unless the supply of test material or effluent is limited, or flow-through procedures are used, renew test solutions daily. Before renewal, remove dead embryos or larvae. Take care not to accidentally remove or injure embryos or larvae. A light table enhances fish visibility and simplifies this task. Add new test solutions to test chambers after removal of approximately 80 to 90% of old solutions. Add solutions slowly down the side of the test chamber to avoid injury to embryos and larvae.

4) *Feeding*—No feeding is required.

5) *Biological data and observations*—Each day, approximately 24 h after test initiation and before solution change, record number of dead (milky and opaque) and live embryos in each test chamber. After hatching begins, record each day the number of hatched, dead, live, and deformed larvae. Deformed larvae are those with gross morphological abnormalitites (Figure 8921:3) or

other characteristics that preclude survival (Figure 8921:2 shows normal larvae). See other sources for detailed information on the embryology and development of the fathead minnow and identification of abnormalities.[14-20] Note that larvae typically do not become active for 1 to 2 d after hatching and may remain relatively immobile on the bottom of the test chamber during this time.

6) *Chemical data recording*—See ¶ 1*c*6) above.

7) *Verification of exposures*—See ¶ 1*c*7) above.

8) *Test termination*—Terminate test after 7 or 8 d. Before termination, record number of surviving, dead, and abnormal larvae in each test chamber (replicate).

9) *Test acceptability criteria*—The test is considered acceptable if control survival is ≥80%.

3. Early-Life-Stage Test

In this test, fathead minnow early life-stages, beginning with newly fertilized eggs, are exposed through embryo-larval development to an early juvenile age.[20-23] This test is run at 25°C for approximately 33 d (28 d post hatch) under flow-through conditions. Test endpoints include time-to-hatch, percent hatch, survival during different life stages, and growth. The intent is to determine the lowest effect and highest no-effect concentration of the test substance.

a. Equipment and physical condition: For a description of suitable diluter systems, see Section 8010F. Construct egg hatching cups (8 cm diam × 10 cm long) of glass tubing or other acceptable materials such as 316 stainless steel or TFE. Glue a 40 mesh nylon or stainless steel screen to one end of the cup using clear silicone sealant. Suspend egg cups from a rocker-arm assembly and low-speed motor designed to oscillate the egg cups slowly up and down approximately 2 to 3 cm. Use a low rate through the exposure chambers sufficient to replace 90% of the water in 8 to 12 h. A self-starting siphon tube in each exposure

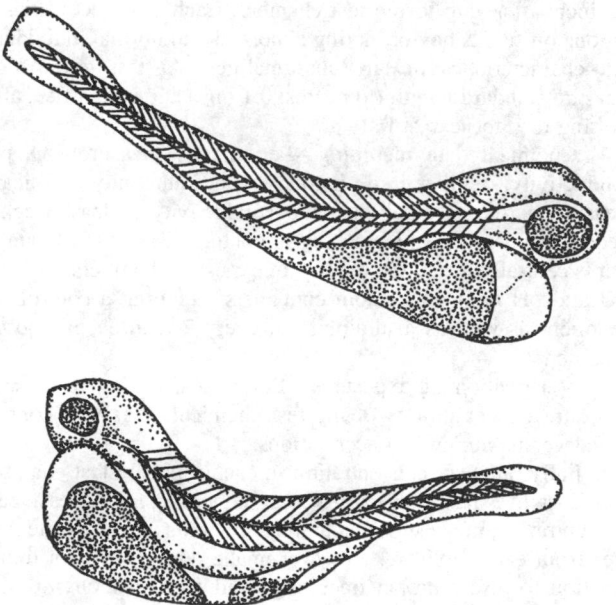

Figure 8921:3. Examples of abnormal fathead minnow larvae. Compare to Figure 8921:2.

chamber can be substituted for the rocker-arm system to ensure test solution exchange.

Provide light intensities over the chambers of approximately 400 to 800 lux and establish a 16-h light and 8-h dark photoperiod, preferably with a 15- to 30-min dusk-dawn transition period. Hold temperature at 25 ± 2°C.

For dilution water requirements, see Table 8921:I. Prepare test solutions as described in Section 8010F.2b. For testing of compounds of low solubility, follow a carrier-solvent recommendation in Table 8921:I and use solvent control.

b. Test organisms: Use embryos, preferably less than 48-h old, to initiate tests. Obtain embryos from an in-house culture (see Section 8921B) or commercial supplier.

c. Test procedures: Use a minimum of five exposure concentrations and one control, and a dilution factor for determining test concentrations of ≥0.5. Set up a minimum of two (preferably four) replicate exposure chambers per test concentration. Use a range-finding test of 4 to 10 d conducted with juveniles to determine the test concentrations for the definitive study.

1) Test initiation—From a pool of embryos from three or more spawns, randomly select and distribute several at a time until 60 embryos are distributed per test concentration, divided among the replicate embryo incubation cups suspended in each test chamber. Exclude abnormal and nonviable (milky-colored and opaque) embryos as well as any showing signs of fungal infection. A light box and stereoscopic microscope are recommended for examining and counting embryos. Use a large-bore pipet or similar device for transferring embryos.

2) Feeding—Provide live brine shrimp (≤24-h old) three times daily during the first 5 d following hatching.[6] At about Day 7, supplement brine shrimp diet with a fine grade of commercial fish meal (fish starter). A slightly larger grade may be substituted as the fish grow.

3) Biological data and observations—Monitor survival by daily inspection. After hatching is complete, record number of live larvae, live embryos, dead embryos, and unaccounted-for embryos for each incubation cup, then release the larval fish from the incubation cup to the test chamber. Each day, record observations on fish behavior, noting abnormal and normal individuals and characteristics of any abnormalities. At test termination, measure standard length (to nearest 0.1 mm) and weigh fish, after blotting dry, to nearest 0.01 g.

4) Chemical data recording—Periodically measure DO, pH, conductivity, temperature, hardness, and alkalinity. Typically measure hardness, alkalinity, and conductivity at least weekly. Record temperature at least daily or continuously in one chamber that is centrally located in the row of exposure chambers. Measure DO and pH daily in random containers including a control and treatment as well as a minimum of every 7 d in each exposure chamber.

5) Verification of exposures—Before and during test, verify exposure concentrations of the test chemical. This will confirm actual versus nominal concentrations.

Initially, analyze concentration in each replicate test chamber. If consistent concentrations are observed among replicates at each concentration, make subsequent measurements on composite samples from each replicate. However, make other checks of diluter function to insure proper operating conditions. Although desirable, it is not necessary that measured concentrations be within any specific percentage of the nominal concentrations (e.g., ±20%). Such characteristics as hydrolysis rate and volatility may make it

impossible. Maintain consistent concentrations by chemical measurements. Once concentrations have stabilized, start exposure. Measure exposure at least weekly until test ends. Base statistical analyses and results (LOECs, NOECs) on measured concentrations rather than nominal concentrations.

6) Test termination—Terminate test 28 d post-hatching.

7) Test acceptability—The test is acceptable if the survival of all control fish at the end of the test is ≥80% (based on the initial egg count) and survival is not less than 70% in any one control replicate.

4. Life-Cycle Reproductive Toxicity Test

This type of test uses newly spawned eggs or newly hatched larvae to start a test, continues through fish maturation and reproduction, and ends not less than 28 d after the hatching of the second generation. See Section 8910C.3 for a description of general test procedures for this test and other sources for additional details.[24–26]

a. Equipment and physical conditions: See Sections 8010F and 8910C.3. The physical systems are similar to those described for testing the brook trout (Section 8910C.3b).

Use one of the following two arrangements of test tanks (made of glass or stainless steel with viewing windows): The first consists of duplicate spawning tanks for each of the five or more test concentrations and controls, measuring 30 × 30 × 90 cm with a 30-cm-square portion at one end, screened off and divided in half to form two larval chambers for the progeny; deliver test water separately to the larval and spawning chambers of each tank, with about one-third of the water volume going to each larval chamber. Alternatively, use duplicate progeny tanks measuring 30 × 30 × 60 cm plus duplicate progeny tanks for each spawning tank. Use a larval tank with minimum dimensions of 30 × 30 × 30 cm, divided to form two separate larval chambers with separate standpipes, or separate 30 × 15 × 30 cm tanks. Supply test solutions and water for controls as in Section 8010F.1. Maintain a water depth of 15 cm in all tanks.

Flow rate, oxygen requirements, aeration, cleaning and operation are as described for the brook trout in Section 8010C.3b1).

Fathead minnows deposit eggs on the underside of submerged objects. For spawning substrates, use inverted semicircular sections of ceramic drain tile or PVC pipe (7.5 cm ID, 7 to 10 cm long), or equivalent (see Section 8921B.2b). If the inverted surface of the substrate is smooth, roughen it to aid embryo adhesion. Place substrate parallel to the long axis of the spawning tank so that each end is readily accessible to the fish. Fasten incubation cups, such as those described in 8921C.3, to a rocker arm with a vertical travel distance of 3 to 5 cm. For illumination, see 8010F.3f.

Use a 16-h light/8-h dark photoperiod, preferably with a 15- to 30-min dawn-dusk transition time.

Maintain temperature at 25 ± 2°C and record continuously.

b. Test initiation: Initiate tests with embryos or larvae from at least three females. Begin the life-cycle test by randomly selecting and distributing embryos or 1- to 5-d old larvae to each duplicate spawning tank for each test concentration. Extra fish may be added at the beginning so that some can be removed periodically for special examinations.

Exclude abnormal and nonviable (milky-colored and opaque) embryos as well as any showing signs of fungal infection. A light box and stereoscopic microscope are recommended for examining and counting embryos. Use a large-bore pipet or similar device for transferring embryos.

c. Feeding: Feed newly hatched larvae minimal amounts of live brine shrimp nauplii.[6] Avoid overfeeding. Continue feeding larval and juvenile fish twice daily with live brine shrimp nauplii for 30 to 60 d. Thereafter, frozen adult brine shrimp may be supplemented by pelleted and/or flake food. Feed quantitatively among all test groups.

d. Thinning and preparation for spawning: When test fish are 60 ± 2 d old, discard injured or deformed individuals and randomly reduce the number in each tank to 15. Record number, length, and weight of discarded and deformed fish. To obtain 15 fish per tank, it may be necessary to transfer or combine fish from duplicate tanks. Continue routine feeding and cleaning until fish mature and are almost ready to spawn. Place five spawning tiles in each duplicate spawning tank, separated fairly widely to reduce fighting among the territorial male fish. Place tiles so that their undersides and guard males can be seen from the tank end. When fish are fully mature (i.e., have well-defined secondary sexual characteristics–see Section 8921A and Figure 8921:1) and spawning is imminent, reduce the number of males to no more than four per tank. Reserve the fifth tile as cover for females. Do not remove males having established territories under tiles where a recent spawn has occurred.

e. Spawning and embryo incubation: Each day, check spawning tiles and remove those with newly deposited embryos, beginning about 6 h after start of the light period. Loosen embryos from spawning tiles and at the same time separate them from one another by lightly placing a finger on the egg mass and moving it in a circular pattern with increasing pressure until the embryos begin to roll. Wash groups of embryos into separate containers and return to spawning tanks. Count embryos, select those needed for incubation, and discard remainder. Check all embryos for different stages of development.[27] If more than one distinct stage is present, consider each stage as one spawning and handle separately as described below.

Each day, randomly select 50 unbroken embryos from a single spawn and place in an incubator cup to determine viability and hatchability. Count, record, and discard remaining embryos. Determine viability and hatchability on each spawn of at least 50 embryos until number of spawns (≥ 50 embryos) in each tank equals number of females in that tank. Subsequently, test for hatchability only on subsamples from every third spawning of at least 50 embryos. Remove spawns from tiles, count and record embryos, and discard.

If no spawning occurs for a week, cease testing of parental fish. Record total length and weight, sex, and gonadal condition of parental (F_0) fish, then discard.

Each day, record live and dead embryos in incubator cups, remove dead embryos, and clean cup screens. After larvae begin to hatch (about 4 to 5 d), cease handling or removing them from cups until all have hatched. At that time, if enough larvae are still alive, select 40 at random and transfer immediately to a larval growth chamber to determine survival and growth of the second (F_1) generation. Count and discard incubation groups not used for survival and growth studies.

f. F_1 generation larval-juvenile survival and growth: Select larvae for 30- and 60-d growth and survival exposures from early spawned embryos in each duplicate tank. Plan their distribution for hatchability tests so that a new group of larvae is ready to be tested as soon as possible after the previously tested group is removed from the larval chambers. Record mortality and larval lengths at 30 and 60 d after hatching. Weigh juveniles when exposures are terminated (60 d). Do not feed fish (larvae, juveniles, or adults) for 24 h before weighing.

g. Extended testing: Normally testing is concluded with the F_1 generation larval-juvenile exposures; however, an extended life-cycle test may be conducted through an additional generation, if desired. In this case, transfer 50 of the 60-d post-hatch F_1 fish from each growth chamber to the corresponding spawning chamber. Follow procedures used for the F_1 generation to determine survival of embryos, larvae, and juveniles of the F_2 generation. Cease testing adult fish on completion of spawning. Continue post-hatch study to 60 d.

h. Biological data and observations: Record the following data for each tank and the controls: total number and length of normal and deformed individuals at the end of 30 and 60 d for each generation; total length, weight, and number of each sex, both normal and deformed, at the end of the tests; mortality during tests; number of spawns and embryos produced in each and total embryo production by each generation; and percentage of larvae surviving and growth of juveniles as well as deformities produced.

Use fish and embryos obtained from the test for physiological, biochemical, histological, and other tests for toxicant-produced effects, as necessary.

i. Chemical data recording: Periodically measure dissolved oxygen, pH, conductivity, temperature, hardness, and alkalinity. See Section 8921C.3c4) for additional information.

j. Verification of exposures: Before and during the test, verify exposure concentrations of the test chemical. See Section 8921C.3c5). Base statistical analyses and results (LOECs, NOECs) on measured rather than nominal concentrations.

5. Statistical Analysis

Assemble, analyze, evaluate, and report data as described in Section 8010G.

6. References

1. WEBER, C.I., ed. 1993. Methods for Measuring the Acute Toxicity of Effluents and Receiving Waters to Freshwater and Marine Organisms, 4th ed. EPA-600/4-90-027F, Environmental Monitoring Systems Lab., U.S. Environmental Protection Agency, Cincinnati, Ohio.
2. U.S. ENVIRONMENTAL PROTECTION AGENCY. 1985. Standard Evaluation Procedure: Acute Toxicity Test for Freshwater Fish. EPA-540/9-85-006. Off. Pesticide Programs, Hazard Evaluation Div., Washington, D.C.
3. AMERICAN SOCIETY FOR TESTING AND MATERIALS. 1995. Standard guide for conducting acute toxicity tests with fishes, macroinvertebrates, and amphibians. E729-88a, Annual Book of ASTM Standards, Vol. 11.04. American Soc. Testing & Materials, Philadelphia, Pa.
4. AMERICAN SOCIETY FOR TESTING AND MATERIALS. 1995. Standard guide for conducting acute toxicity tests on aqueous effluents with fishes, macroinvertebrates, and amphibians. E1192-88, Annual Book of ASTM Standards, Vol. 11.04. American Soc. Testing & Materials, Philadelphia, Pa.
5. PARRISH, P.R. 1995. Acute toxicity tests. Appendix A *in* G.M. Rand, ed., Fundamentals of Aquatic Toxicology: Effects, Environmental Fate, and Risk Assessment, 2nd ed. Taylor & Francis, Washington, D.C.
6. AMERICAN SOCIETY FOR TESTING AND MATERIALS. 1995. Standard practice for using brine shrimp nauplii as food for test animals in

aquatic toxicology. E1203-92, Annual Book of ASTM Standards, Vol. 11.04. American Soc. Testing & Materials, Philadelphia, Pa.

7. LEWIS, P.A., D.J. KLEMM, J.M. LAZORCHAK, T.J. NORBERG-KING, W.H. PELTIER & M.A. HEBER, eds. 1994. Short-Term Methods for Estimating the Chronic Toxicity of Effluents and Receiving Waters to Freshwater Organisms, 3rd ed. EPA-600/4-91-002, Environmental Monitoring Systems Lab., U.S. Environmental Protection Agency, Cincinnati, Ohio.

8. NORBERG, T.J. & D.I. MOUNT. 1985. A new fathead minnow (*Pimephales promelas*) subchronic toxicity test. *Environ. Toxicol. Chem.* 4:711.

9. BIRGE, W.J., J.A. BLACK & A.G. WESTERMAN. 1985. Short-term fish and amphibian embryo-larval test for determining the effects of toxicant stress on early-life stages and estimating chronic values for single compounds and complex effluents. *Environ. Toxicol. Chem.* 4:807.

10. BIRGE, W.J., J.A. BLACK, T.M. SHORT & A.G. WESTERMAN. 1989. A comparative ecological and toxicological investigation of a secondary wastewater treatment plant effluent and its receiving stream. *Environ. Toxicol. Chem.* 8:437.

11. NORBERG-KING, T.J. 1989. An evaluation of the fathead minnow seven day subchronic test for estimating chronic toxicity. *Environ. Toxicol. Chem.* 8:1075.

12. PICKERING, Q.H. 1988. Evaluation and comparison of two short-term fathead minnow tests for estimating chronic toxicity. *Water Res.* 22: 883.

13. PICKERING, Q.H. & J.M. LAZORCHAK. Evaluation of the robustness of the Fathead Minnow, *Pimephales promelas*, Larval Survival and Growth Test, U.S. EPA Method 1000.0. *Environ. Toxicol. Chem.* 14: 653.

14. WEIS, J.S. & P. WEIS. 1989. Effects of environmental pollutants on early fish development. *CRC Crit. Rev. Aquat. Sci.* 1:45.

15. McKIM, J.M. 1977. Evaluation of tests with early life stages of fish for predicting long-term toxicity. *J. Fish. Res. Board Can.* 34:1148.

16. BIRGE, W.J., J.A. BLACK, A.G. WESTERMAN & B.A. RAMEY. 1983. Fish and amphibian embryos—A model system for evaluating teratogenicity. *Fund. Appl. Toxicol.* 3:237.

17. WEIS, J.S. & P. WEIS. 1987. Pollutants as developmental toxicants in aquatic organisms. *Environ. Health Perspect.* 71:77.

18. SILBERHORN, E.M. 1992. Evaluation and use of a fish embryo-larval assay for the detection of development toxicants. Ph.D. dissertation, Univ. Kentucky, Lexington.

19. BIRGE, W.J., J.A. BLACK, J.E. HUDSON & D.M. BRUSER. 1979. Embryo-larval toxicity tests with organic compounds. *In* L.L. Marking & R.A. Kimerle, eds. Aquatic Toxicology. ASTM STP 667, American Soc. Testing & Materials, Philadelphia, Pa.

20. McKIM, J.M. 1995. Early life stage toxicity tests. Appendix B *in* G.M. Rand, ed. Fundamentals of Aquatic Toxicology: Effects, Environmental Fate, and Risk Assessment, 2nd ed. Taylor & Francis, Washington, D.C.

21. U.S. ENVIRONMENTAL PROTECTION AGENCY. 1995. Fish early life stage toxicity test. 40 CFR §797.1600 40. (First published in Toxic Substances Control Act Guidelines; Final Rules. *Federal Register* 50: 39252-39516, 1985).

22. U.S. ENVIRONMENTAL PROTECTION AGENCY. 1986. Standard Evaluation Procedure: Fish Early-Life Stage Test. EPA-540/9-86-138, Off. Pesticide Programs, Hazard Evaluation Div., Washington, D.C.

23. AMERICAN SOCIETY FOR TESTING AND MATERIALS. 1995. Standard guide for conducting early life-stage toxicity tests with fishes. E1241-92, Annual Book of ASTM Standards, Vol. 11.04. American Soc. Testing & Materials, Philadelphia, Pa.

24. U.S. ENVIRONMENTAL PROTECTION AGENCY. 1986. Standard Evaluation Procedure: Fish Life-Cycle Tests. EPA-540/9-86-137, Off. Pesticide Programs, Hazard Evaluation Div., Washington, D.C.

25. U.S. ENVIRONMENTAL PROTECTION AGENCY. 1982. User's Guide for Conducting Life-Cycle Chronic Toxicity Tests with Fathead Minnows (*Pimephales promelas*). EPA-600/8-81-011, Environmental Research Lab., Duluth, Minn.

26. MOUNT, D.I. 1968. Chronic toxicity of copper to fathead minnow (*Pimephales promelas* Rafinesque). *Water Res.* 2:215.

27. MANNER, H.W. & C.M. DEWESE. 1974. Early embryology of the fathead minnow *Pimephales promelas* Rafinesque. *Anat. Rec.* 180:99.

7. Bibliography

COONEY, J.E. 1995. Freshwater tests. *In* G.M. Rand, ed. Fundamentals of Aquatic Toxicology: Effects, Environmental Fate, and Risk Assessment, 2nd ed. Taylor & Francis, Washington, D.C.

PART 9000

MICROBIOLOGICAL EXAMINATION

9010 INTRODUCTION*

The following sections describe procedures for making microbiological examinations of water samples to determine sanitary quality. The methods are intended to indicate the degree of contamination with wastes. They are the best techniques currently available; however, their limitations must be understood thoroughly.

Tests for detection and enumeration of indicator organisms, rather than of pathogens, are used. The coliform group of bacteria, as herein defined, is the principal indicator of suitability of a water for domestic, industrial, or other uses. The cultural reactions and characteristics of this group of bacteria have been studied extensively.

Experience has established the significance of coliform group density as a criterion of the degree of pollution and thus of sanitary quality. The significance of the tests and the interpretation of results are well authenticated and have been used as a basis for standards of bacteriological quality of water supplies.

The membrane filter technique, which involves a direct plating for detection and estimation of coliform densities, is as effective as the multiple-tube fermentation test for detecting bacteria of the coliform group. Modification of procedural details, particularly of the culture medium, has made the results comparable with those given by the multiple-tube fermentation procedure. Although there are limitations in the application of the membrane filter technique, it is equivalent when used with strict adherence to these limitations and to the specified technical details. Thus, two standard methods are presented for the detection and enumeration of bacteria of the coliform group.

It is customary to report results of the coliform test by the multiple-tube fermentation procedure as a Most Probable Number (MPN) index. This is an index of the number of coliform bacteria that, more probably than any other number, would give the results shown by the laboratory examination; it is not an actual enumeration. By contrast, direct plating methods such as the membrane filter procedure permit a direct count of coliform colonies. In both procedures coliform density is reported conventionally as the MPN or membrane filter count per 100 mL. Use of either procedure permits appraising the sanitary quality of water and the effectiveness of treatment processes. Because it is not necessary to provide a quantitative assessment of coliform bacteria for all samples, a qualitative, presence-absence test is included.

Fecal streptococci and enterococci also are indicators of fecal pollution and methods for their detection and enumeration are given. A multiple-tube dilution and a membrane filter procedure are included.

Methods for the differentiation of the coliform group are included. Such differentiation generally is considered of limited value in assessing drinking water quality because the presence of any coliform bacteria renders the water potentially unsatisfactory and unsafe. Speciation may provide information on colonization of a distribution system and further confirm the validity of coliform results.

Coliform group bacteria present in the gut and feces of warm-blooded animals generally include organisms capable of producing gas from lactose in a suitable culture medium at $44.5 \pm 0.2°C$. Inasmuch as coliform organisms from other sources often cannot produce gas under these conditions, this criterion is used to define the fecal component of the coliform group. Both the multiple-tube dilution technique and the membrane filter procedure have been modified to incorporate incubation in confirmatory tests at $44.5°C$ to provide estimates of the density of fecal organisms, as defined. Procedures for fecal coliforms and *Escherichia coli* include a 24-h multiple-tube test using A-1 medium, a 7-h rapid method, and chromogenic substrate coliform tests. This differentiation yields valuable information concerning the possible source of pollution in water, and especially its remoteness, because the *nonfecal* members of the coliform group may be expected to survive longer than the *fecal* members in the unfavorable environment provided by the water.

The heterotrophic plate count may be determined by pour plate, spread plate, or membrane filter method. It provides an approximate enumeration of total numbers of viable bacteria that may yield useful information about water quality and may provide supporting data on the significance of coliform test results. The heterotrophic plate count is useful in judging the efficiency of various treatment processes and may have significant application as an in-plant control test. It also is valuable for checking quality of finished water in a distribution system as an indicator of microbial regrowth and sediment buildup in slow-flow sections and dead ends.

Experience in the shipment of un-iced samples by mail indicates that noticeable changes may occur in type or numbers of bacteria during such shipment for even limited periods of time. Therefore, refrigeration during transportation is recommended to minimize changes, particularly when ambient air temperature exceeds 13°C.

Procedures for the isolation of certain pathogenic bacteria and protozoa are presented. These procedures are tedious and complicated and are not recommended for routine use. Likewise, tentative procedures for enteric viruses are included but their routine use is not advocated.

Examination of routine bacteriological samples cannot be regarded as providing complete information concerning water quality. Always consider bacteriological results in the light of information available concerning the sanitary conditions surrounding the sample source. For a water supply, precise evaluation of quality can be made only when the results of laboratory examinations are interpreted in the light of sanitary survey data. Consider inadequate the results of the examination of a single sample from a given source. When possible, base evaluation of water quality on the examination of a series of samples collected over a known and protracted period of time.

Pollution problems of tidal estuaries and other bodies of saline water have focused attention on necessary modification of existing bacteriological techniques so that they may be used effectively. In the following sections, applications of specific techniques to saline water are not discussed because the methods used for fresh waters generally can be used satisfactorily with saline waters.

Methods for examination of the waters of swimming pools and other bathing places are included. The standard procedures for the plate count, fecal coliforms, and fecal streptococci are identical with those used for other waters. Procedures for *Staphylococcus* and *Pseudomonas aeruginosa,* organisms commonly associated with the upper respiratory tract or the skin, are included.

---

* Approved by Standard Methods Committee, 1993.

Procedures for aquatic fungi and actinomycetes are included.

Sections on rapid methods for coliform testing and on the recovery of stressed organisms are included. Because of increased interest and concern with analytical quality control, this section continues to be expanded.

The bacteriological methods in Part 9000, developed primarily to permit prompt and rapid examination of water samples, have been considered frequently to apply only to routine examinations. However, these same methods are basic to, and equally valuable in, research investigations in sanitary bacteriology and water treatment. Similarly, all techniques should be the subject of investigations to establish their specificity, improve their procedural details, and expand their application to the measurement of the sanitary quality of water supplies or polluted waters.

9020 QUALITY ASSURANCE/QUALITY CONTROL*

9020 A. Introduction

1. General Considerations

The growing emphasis on microorganisms in water quality standards and enforcement activities and their continuing role in research, process control, and compliance monitoring require the establishment and effective operation of a quality assurance (QA) program to substantiate the validity of analytical data.

A laboratory quality assurance program is the integration of intralaboratory and interlaboratory quality control (QC), standardization, and management practices into a formal, documented program with clearly defined responsibilities and duties to ensure that the data are of the type, quality, and quantity required.

The program must be practical and require only a reasonable amount of time or it will be bypassed. Generally, about 15% of overall laboratory time should be spent on different aspects of a quality assurance program. However, more time may be needed for more important analytical data, e.g., data for enforcement actions. When properly administered, a balanced, conscientiously applied QA program will optimize data quality without adversely affecting laboratory productivity.

Because microbiological analyses measure constantly changing living organisms, they are inherently variable. Some quality control tools used by chemists, such as reference standards, instrument calibration, and quality control charts, may not be available to the microbiologist.

Because QA programs vary among laboratories as a result of differences in organizational mission, responsibilities, and objectives; laboratory size, capabilities, and facilities; and staff skills and training, this provides only general guidance. Each laboratory should determine the appropriate QA level for its purpose.

2. Guidelines for a Quality Assurance Program

Develop a QA program to meet the laboratory's specific needs and the planned use of the data. Emphasis on the use of data is particularly important where significant and costly decisions depend on analytical results. An effective QA program will confirm the quality of results and increase confidence in the data.

a. Management responsibilities: Management must recognize the need for quality assurance, commit monetary and personnel resources, assume a leadership role, and involve staff in development and operation of the QA program. Management should meet with the laboratory supervisor and staff to develop and maintain a comprehensive program and establish specific responsibility for management, supervisors, and analysts.

b. Quality assurance officer: In large laboratories, a QA officer has the authority and responsibility for application of the QA program. Ideally, this person should have a staff position reporting directly to upper management, not a line position. The QA officer should have a technical education, be acquainted with all aspects of laboratory work, and be familiar with statistical techniques for data evaluation. The QA officer is responsible for initiating the program, convincing staff of its value, and providing necessary information and training to the staff. Once the QA program is functioning, the coordinator conducts frequent (weekly to monthly) reviews with the laboratory supervisor and staff to determine the current status and accomplishments of the program and to identify and resolve problems. The QA officer also reports periodically to management to secure backing in actions necessary to correct problems that threaten data quality.

c. Staff: Laboratory and field staffs participate with management in planning the QA program, preparing standard operating procedures, and most importantly, implementing the QC program in their daily tasks of collecting samples, conducting analyses, performing quality control checks, and calculating and reporting results. Because the staffs are the first to see potential problems, they should identify them and work with the supervisor to correct and avoid them. It is critical to the success of the QA program that staff understand and actively support it.

3. Quality Assurance Program Objectives

The objectives of a QA program include providing data of known quality, ensuring a high quality of laboratory performance, maintaining continuing assessment of laboratory operations, identifying weaknesses in laboratory operations, detecting training needs, and improving documentation and recordkeeping.

4. Elements of a Quality Assurance Program

Each laboratory should develop and implement a written QA plan describing the QA program and QC activities of the labo-

---

* Approved by Standard Methods Committee, 1997.

ratory. The plan should address the following basic common aspects:

a. Statement of objectives, describing the specific goals of the laboratory.

b. Sampling procedures, including selection of representative sites and specified holding time and temperature conditions. If data may be subjected to litigation, use chain-of-custody procedures.

c. Personnel policies, describing specific qualification and training requirements for supervisors and analysts.

d. Equipment and instrument requirements, providing calibration procedures and frequency and maintenance requirements.

e. Specifications for supplies, to ensure that reagents and supplies are of high quality and are tested for acceptability.

f. Analytical methods, i.e., standardized methods established by a standards-setting organization and validated. Ideally, these laboratory methods have documented precision, bias, sensitivity, selectivity, and specificity.

g. Analytical quality control measures, including such analytical checks as duplicate analyses, positive and negative controls, sterility checks, and verification tests.

h. Standard operating procedures (SOPs), i.e., written statement and documentation of all routine laboratory operations.

i. Documentation requirements, concerning data acquisition, recordkeeping, traceability, and accountability.

j. Assessment requirements:

1) Internal audits of the laboratory operations, performed by the QA officer and supervisor.

2) On-site evaluations by outside experts to ensure that the laboratory and its personnel are following an acceptable QA program.

3) Performance evaluation studies, in which the QA officer works with the supervisor to incorporate unknown challenge samples into routine analytical runs and laboratories are encouraged to participate in state and national proficiency testing and accreditation programs. The collaborative studies confirm the abilities of a laboratory to generate acceptable data comparable to those of other laboratories and identify potential problems.

k. Corrective actions: When problems are identified by the staff, supervisor, and/or QA coordinator, use standard stepwise procedures to determine the causes and correct them. Nonconformances identified by external laboratory evaluation are corrected, recorded, and signed off by the laboratory manager and QA officer.

Detailed descriptions of quality assurance programs are available.[1-4]

The QC guidelines discussed in 9020B and 9020C are recommended as useful source material, but all elements need to be addressed in developing a QA program.

5. References

1. GASKIN, J.E. 1992. Quality Assurance in Water Quality Monitoring. Inland Water Directorate, Conservation & Protection, Ottawa, Ont., Canada.
2. RATLIFF, T.A., JR. 1990. The Laboratory Quality Assurance System. A Manual of Quality Procedures with Related Forms. Van Nostrand Reinhold, New York, N.Y.
3. GARFIELD, F.M. 1984. Quality Assurance Principles of Analytical Laboratories. Assoc. Official Analytical Chemists, Arlington, Va.
4. DUX, J.P. 1983. Quality assurance in the analytical laboratory. *Amer. Lab.* 26:54.

9020 B. Intralaboratory Quality Control Guidelines

All laboratories have some intralaboratory QC practices that have evolved from common sense and the principles of controlled experimentation. A QC program applies practices necessary to minimize systematic and random errors resulting from personnel, instrumentation, equipment, reagents, supplies, sampling and analytical methods, data handling, and data reporting. It is especially important that laboratories performing only a limited amount of microbiological testing exercise strict QC. A listing of key QC practices is given in Table 9020:I. Other sources of QC practices are available.[1-3] These practices and guidelines will assist laboratories in establishing and improving QC programs. Laboratories should address all of the QC guidelines discussed herein, but the depth and details may differ for each laboratory.

1. Personnel

Microbiological testing should be performed by a professional microbiologist or technician trained in environmental microbiology whenever possible. If not, a professional microbiologist should be available for guidance. Train and evaluate the analyst in basic laboratory procedures. The supervisor periodically should review procedures of sample collecting and handling, media and glassware preparation, sterilization, routine analytical testing,

counting, data handling, and QC techniques to identify and eliminate problems. Management should assist laboratory personnel in obtaining additional training and course work to advance their skills and career.

2. Facilities

a. Ventilation: Plan well-ventilated laboratories that can be maintained free of dust, drafts, and extreme temperature changes. Whenever possible, laboratories should have air conditioning to reduce contamination, permit more stable operation of incubators, and decrease moisture problems with media and instrumentation.

b. Space utilization: Design and operate the laboratory to minimize through traffic and visitors, with a separate area for preparing and sterilizing media, glassware, and equipment. Use a vented laminar-flow hood for dispensing and preparing sterile media, transferring microbial cultures, or working with pathogenic materials. In smaller laboratories it may be necessary, although undesirable, to carry out these activities in the same room.

c. Laboratory bench areas: Provide at least 2 m of linear bench space per analyst and additional areas for preparation and support activities. For stand-up work, typical bench dimensions are 90 to 97 cm high and 70 to 76 cm deep. For sit-down activities such

TABLE 9020:I. KEY QUALITY CONTROL PRACTICES

Item	Action	Frequency	Further Information in Section 9020B, ¶
Reagent water	Monitor quality		See Table 9020:II
Bench surface	Monitor for contamination	Weekly	2e
Air in workplace	Monitor bacterial density	Monthly	2e
Thermometers	Check accuracy	Semiannually	3a
Balances and weights	Check accuracy	Monthly	3b
Balances	Service and recalibrate	Annually	3b
pH meter	Standardize	Each use	3c
	Check against another meter	Monthly	3c
Media-dispensing apparatus	Check volume accuracy	Each use	3f
Hot-air oven	Check performance	Monthly	3g
Autoclave	Check performance	Each use	3h
Refrigerator	Check temperature	Daily	3i
Freezer	Check temperature	Daily	3j
	Defrost	Semiannually	3j
Membrane filtration equipment	Check for leaks and surface scratches	Each use	3k
UV lamps	Test with UV meter	Quarterly	3l
Biohazard hood	Monitor air and UV lamps	Monthly	3m
	Inspect for airflow	Quarterly	3m
Incubator	Check temperature	Twice daily	3n and o
Microscope	Clean optics and stage	Each use	3p
Glassware	Inspect for cleanliness, chips, and etching	Each use	4a
	Check pH	Each batch	4a1)
	Conduct inhibitory residue test	Annually	4a2)
Dilution water bottles	Check pH and volume	Each use	4c
Media	Check pH and appearance	Each use	4i1)
Autoclave	Check performance	Weekly	4i2)
Plate counts	Perform duplicate analyses	Weekly	8a4)
	Repeat counts	Monthly	8a2)

as microscopy and plate counting, benches are 75 to 80 cm high. Specify bench tops of stainless steel, epoxy plastic, or other smooth, impervious surface that is inert and corrosion-resistant, has a minimum number of seams, and has adequate sealing of any crevices. Install even, glare-free lighting with about 1000 lux (100 ft-candles) intensity at the working surface.

d. Walls and floors: Assure that walls are covered with a smooth finish that is easily cleaned and disinfected. Specify floors of smooth concrete, vinyl, asphalt tile, or other impervious, sealed washable surfaces.

e. Work-area monitoring: Maintain high standards of cleanliness in work areas. Monitor air, at least monthly, with air density plates. The number of colonies on the air density plate test should not exceed 160/m²/15 min exposure (15 colonies/plate/15 min).

Plate or the swab method[1] can be used weekly or more frequently to monitor bench surface contamination. Although uniform limits for bacterial density have not been set, each laboratory can use these tests to establish a base line and take action on a significant increase.

f. Laboratory cleanliness: Regularly clean laboratory rooms and wash benches, shelves, floors, and windows. Wet-mop floors and treat with a disinfectant solution; do not sweep or dry-mop. Wipe bench tops and treat with a disinfectant before and after use. Do not permit laboratory to become cluttered.

3. Laboratory Equipment and Instrumentation

Verify that each item of equipment meets the user's needs for precision and minimization of bias. Perform equipment maintenance on a regular basis as recommended by the manufacturer or obtain preventive maintenance contracts on autoclave, balances, microscopes, and other equipment. Directly record all quality control checks in a permanent log book.

Use the following quality control procedures:

a. Thermometer/temperature-recording instruments: Check accuracy of thermometers or temperature-recording instruments semiannually against a certified National Institute of Standards and Technology (NIST) thermometer or one traceable to NIST and conforming to NIST specifications. For general purposes use thermometers graduated in increments of 0.5°C or less. Maintain in water or glycerol for air incubators and refrigerators and glycerol for freezers and seal in a flask. For a 44.5°C water bath, use a submersible thermometer graduated to 0.2°C or less. Record temperature check data in a quality control log. Mark the necessary NIST calibration corrections on each thermometer and incubator, refrigerator, or freezer. When possible, equip incubators and water baths with temperature-recording instruments that provide a continuous record of operating temperature.

b. Balances: Follow manufacturer's instructions in operation and routine maintenance of analytical and top-loading balances. Balances should be serviced and recalibrated by a manufacturer technician annually or more often as conditions change or problems occur. In weighing 2 g or less, use an analytical balance with a sensitivity less than 1 mg at a 10-g load. For larger quantities use a pan balance with sensitivity of 0.1 g at a 150-g load.

Wipe balance before use with a soft brush. Clean balance pans after use and wipe spills up immediately with a laboratory tissue. Inspect weights with each use and replace if corroded. Use only a plastic-tip forceps to handle weights. Check balance and working weights monthly against a set of reference weights (ANSI/

ASTM Class 1 or NIST Class S) for accuracy, precision, and linearity.[4] Record results.

c. pH meter: Use a meter graduated in 0.1 pH units or less, that includes temperature compensation. Preferably use digital meters and commercial buffer solutions. With each use, standardize meter with two buffers that bracket the pH of interest and record. Date buffer solutions when opened and check monthly against another pH meter. Discard solution after each use and replace buffer supply before expiration date. For full details of pH meter use and maintenance, see Section 4500-H$^+$.

d. Water purification system: Commercial systems are available that include some combination of prefiltration, activated carbon, mixed-bed resins, and reverse-osmosis with final filtration to produce a reagent-grade water. The life of such systems can be extended greatly if the source water is pretreated by distillation or by reverse osmosis to remove dissolved solids. Such systems tend to produce the same quality water until resins or activated carbon are near exhaustion and quality abruptly becomes unacceptable. Some deionization components are available now that automatically regenerate the ion exchange resins. Do not store reagent water unless a commercial UV irradiation device is installed and is confirmed to maintain sterility.

Monitor reagent water continuously or daily with a calibrated conductivity meter and analyze at least annually for trace metals. Replace cartridges at intervals recommended by the manufacturer based on the estimated usage and source water quality. Do not wait for column failure. If bacteria-free water is desired, include aseptic final filtration with a 0.22-μm-pore membrane filter and collect in a sterile container. Monitor treated water for contamination and replace the filter as necessary.

e. Water still: Stills produce water of a good grade that characteristically deteriorates slowly over time as corrosion, leaching, and fouling occur. These conditions can be controlled with proper maintenance and cleaning. Stills efficiently remove dissolved substances but not dissolved gases or volatile organic chemicals. Freshly distilled water may contain chlorine and ammonia (NH_3). On storage, additional NH_3 and CO_2 are absorbed from the air. Use softened water as the source water to reduce frequency of cleaning the still. Drain and clean still and reservoir according to manufacturer's instructions and usage.

f. Media dispensing apparatus: Check accuracy of volumes dispensed with a graduated cylinder at start of each volume change and periodically throughout extended runs. If the unit is used more than once per day, pump a large volume of hot reagent water through the unit to rinse between runs. Correct leaks, loose connections, or malfunctions immediately. At the end of the work day, break apparatus down into parts, wash, rinse with reagent water, and dry. Lubricate parts according to manufacturer's instructions or at least once per month.

g. Hot-air oven: Test performance monthly with commercially available *Bacillus subtilis* spore strips or spore suspensions. Monitor temperature with a thermometer accurate in the 160 to 180°C range and record results. Use heat-indicating tape to identify supplies and materials that have been exposed to sterilization temperatures.

h. Autoclave: Record items sterilized, temperature, pressure, and time for each run. Optimally use a recording thermometer. Check and record operating temperature weekly with a minimum/maximum thermometer. Test performance with *Bacillus stearothermophilus* spore strips, suspensions, or capsules monthly. Use heat-indicating tape to identify supplies and materials that have been sterilized.

i. Refrigerator: Maintain temperature at 1 to 4°C. Check and record temperature daily and clean monthly. Identify and date materials stored. Defrost as required and discard outdated materials quarterly.

j. Freezer: Maintain temperature at −20°C to −30°C. Check and record temperature daily. A recording thermometer and alarm system are highly desirable. Identify and date materials stored. Defrost and clean semiannually; discard outdated materials.

k. Membrane filtration equipment: Before use, assemble filtration units and check for leaks. Discard units if inside surfaces are scratched. Wash and rinse filtration assemblies thoroughly after use, wrap in nontoxic paper or foil, and sterilize.

l. Ultraviolet lamps: Disconnect lamps monthly and clean bulbs with a soft cloth moistened with ethanol. Test lamps quarterly with an appropriate (short- or long-wave) UV light meter* and replace bulbs if output is less than 70% of the original. For short-wave lamps used in disinfecting work areas, expose plate count agar spread plates containing 200 to 300 organisms of interest, for 2 min. Incubate plates at 35°C for 48 h and count colonies. Replace bulb if count is not reduced 99%.

CAUTION: *Although short-wave (254-nm) UV light is known to be more dangerous than long-wave UV (365-nm), both types of UV light can damage eyes and skin and potentially are carcinogenic.[5] Protect eyes and skin from exposure to UV light. (See Section 1090B.)*

m. Biohazard hood: Once per month expose plate count agar plates to air flow for 1 h. Incubate plates at 35°C for 48 h and examine for contamination. A properly operating biohazard hood should produce no growth on the plates. Disconnect UV lamps and clean monthly by wiping with a soft cloth moistened with ethanol. Check lamps' efficiency as specified above. Inspect cabinet for leaks and rate of air flow quarterly. Use a pressure monitoring device to measure efficiency of hood performance. Have laminar-flow safety cabinets containing HEPA filters serviced by the manufacturer. Maintain hoods as directed by the manufacturer.

n. Water bath incubator: Verify that incubators maintain test temperature, such as 35 ± 0.5°C or 44.5 ± 0.2°C. Keep an appropriate thermometer (¶ 3*a*, above) immersed in the water bath; monitor and record temperature twice daily (morning and afternoon). For optimum operation, equip water bath with a gable cover. Use only stainless steel, plastic-coated, or other corrosion-proof racks. Clean bath as needed.

o. Incubator (air, water jacketed, or aluminum block): Verify that incubators maintain appropriate test temperatures. Also, verify that cold samples are incubated at the test temperature for the required time. Check and record temperature twice daily (morning and afternoon) on the shelves in use. If a glass thermometer is used, submerge bulb and stem in water or glycerine to the stem mark. For best results use a recording thermometer and alarm system. Place incubator in an area where room temperature is maintained between 16 and 27°C (60 to 80°F).

p. Microscopes: Use lens paper to clean optics and stage after each use. Cover microscope when not in use.

Permit only trained technicians to use fluorescence microscope and light source. Monitor fluorescence lamp with a light meter

---

* Fisher Scientific, short wave meter (Cat. No. 11-924-54) and long wave meter (Cat. No. 11-984-53), Pittsburgh, PA 15219-4785, or equivalent.

and replace when a significant loss in fluorescence is observed. Log lamp operation time, efficiency, and alignment. Periodically check lamp alignment, particularly when the bulb has been changed; realign if necessary. Use known positive 4 + fluorescence slides as controls.

4. Laboratory Supplies

a. Glassware: Before each use, examine glassware and discard items with chipped edges or etched inner surfaces. Particularly examine screw-capped dilution bottles and flasks for chipped edges that could leak and contaminate the analyst and the area. Inspect glassware after washing for excessive water beading and rewash if necessary. Make the following tests for clean glassware as necessary:

1) pH check—Because some cleaning solutions are difficult to remove completely, spot check batches of clean glassware for pH reaction, especially if soaked in alkali or acid. To test clean glassware for an alkaline or acid residue add a few drops of 0.04% bromthymol blue (BTB) or other pH indicator and observe the color reaction. BTB should be blue-green (in the neutral range).

To prepare 0.04% bromthymol blue indicator solution, add 16 mL 0.01N NaOH to 0.1 g BTB and dilute to 250 mL with reagent water.

2) Test for inhibitory residues on glassware and plasticware— Certain wetting agents or detergents used in washing glassware may contain bacteriostatic or inhibiting substances that require 6 to 12 rinsings to remove all traces and insure freedom from residual bacteriostatic action. Perform this test annually and before using a new supply of detergent. If prewashed, presterilized plasticware is used, test it for inhibitory residues. Although the following procedure describes testing of petri dishes for inhibitory residue, it is applicable to other glass or plasticware.

a) Procedure—Wash and rinse six petri dishes according to usual laboratory practice and designate as Group A.

Wash six petri dishes as above, rinse 12 times with successive portions of reagent water, and designate as Group B.

Rinse six petri dishes with detergent wash water (in use concentration), and air-dry without further rinsing, and designate as Group C.

Sterilize dishes in Groups A, B, and C by the usual procedure.

For presterilized plasticware, set up six plastic petri dishes and designate them as Group D.

Prepare and sterilize 200 mL plate count agar and hold in a 44 to 46°C water bath.

Prepare a culture of *E. aerogenes* known to contain 50 to 150 colony-forming units/mL. Preliminary testing may be necessary to achieve this count range. Inoculate three dishes from each test group with 0.1 mL and the other three dishes from each group with 1 mL culture.

Analyze the four sets of six plates each, following heterotrophic plate count method (Section 9215B), and incubate at 35°C for 48 h. Count plates with 30 to 300 colonies and record results as CFU/mL.

b) Interpretation of results—Difference in averaged counts on plates in Groups A through D should be less than 15% if there are no toxic or inhibitory effects.

Differences in averaged counts of less than 15% between Groups A and B and greater than 15% between Groups A and C indicate that the cleaning detergent has inhibitory properties that are eliminated during routine washing. Differences between B and D greater than 15% indicate an inhibitory residue.

b. Utensils and containers for media preparation: Use utensils and containers of borosilicate glass, stainless steel, aluminum, or other corrosion-resistant material (see Section 9030). Do not use copper utensils.

c. Dilution water bottles: Use scribed bottles made of nonreactive borosilicate glass or plastic with screwcaps containing inert liners. Clean before use. Disposable plastic bottles prefilled with dilution water are available commercially and are acceptable. Before use of each lot, check pH and volume and examine sterile bottles of dilution water for a precipitate; discard if present. Reclean bottles with acid if necessary, and remake the dilution water. If precipitate repeats, procure a different source of bottles.

d. Reagent-grade water quality: The quality of water obtainable from a water purification system differs with the system used and its maintenance. See 3d and e above. Recommended limits for reagent water quality are given in Table 9020:II. If these limits are not met, investigate and correct or change water source. Although pH measurement of reagent water is characterized by drift, extreme readings are indicative of chemical contamination.

e. Use test for evaluation of reagent water, media, and membranes: When a new lot of culture medium, membrane filters, or a new source of reagent-grade water is to be used make comparison tests, at least quarterly, of the current lot in use (reference lot) against the new lot (test lot).

1) Procedure—Use a single batch of control water (redistilled or distilled water polished by deionization), glassware, membrane filters, or other needed materials to control all variables except the one factor under study. Make parallel pour or spread plate or

TABLE 9020:II. QUALITY OF REAGENT WATER USED IN MICROBIOLOGY TESTING

Test	Monitoring Frequency	Maximum Acceptable Limit
Chemical tests:		
Conductivity	Continuously or with each use	>0.5 megohms resistance or <2 μmhos/cm at 25°C
pH	With each use	5.5–7.5
Total organic carbon	Monthly	<1.0 mg/L
Heavy metals, single (Cd, Cr, Cu, Ni, Pb, and Zn)	Annually*	<0.05 mg/L
Heavy metals, total	Annually*	<0.10 mg/L
Ammonia/organic nitrogen	Monthly	<0.10 mg/L
Total chlorine residual	Monthly or with each use	<0.01 mg/L
Bacteriological tests:		
Heterotrophic plate count (See Section 9215)	Monthly	< 1000 CFU/mL
Use test (see 4e)	Quarterly and for a new source	Student's $t \leq 2.78$

* Or more frequently if there is a problem.

membrane filter plate tests on reference lot and test lot, according to procedures in Sections 9215 and 9222. As a minimum, make single analyses on five different water samples positive for the target organism. Replicate analyses and additional samples can be tested to increase the sensitivity of detecting differences between reference and test lots.

When conducting the use test on reagent water, perform the quantitative bacterial tests in parallel using a known high-quality water as a control water. Prepare dilution/rinse water and media with new source of reagent and control water. Test water for all uses (dilution, rinse, media preparation, etc.).

2) *Counting and calculations*—After incubation, compare bacterial colonies from the two lots for size and appearance. If colonies on the test lot plates are atypical or noticeably smaller than colonies on the reference lot plates, record the evidence of inhibition or other problem, regardless of count differences. Count plates and calculate the individual count per 1 mL or per 100 mL. Transform the count to logarithms and enter the log-transformed results for the two lots in parallel columns. Calculate the difference, d, between the two transformed results for each sample, including the $+$ or $-$ sign, the mean, $\overline{d}$ and the standard deviation s_d of these differences (see Section 1010B).

Calculate Student's t statistic, using the number of samples as n:

$$t = \frac{\overline{d}}{s_d/\sqrt{n}}$$

These calculations may be made with various statistical software packages available for personal computers.

3) *Interpretation*—Use the critical t value, from a Student's t table for comparison against the calculated value. At the 0.05 significance level this value is 2.78 for five samples (four degrees of freedom). If the calculated t value does not exceed 2.78, the lots do not produce significantly different results and the test lot is acceptable. If the calculated t value exceeds 2.78, the lots produce significantly different results and the test lot is unacceptable.

If the colonies are atypical or noticeably smaller on the test lot or the Student's t exceeds 2.78, review test conditions, repeat the test, and/or reject the test lot and obtain another one.

f. Reagents: Because reagents are an integral part of microbiological analyses, their quality must be assured. Use only chemicals of ACS or equivalent grade because impurities can inhibit bacterial growth, provide nutrients, or fail to produce the desired reaction. Date chemicals and reagents when received and when first opened for use. Make reagents to volume in volumetric flasks and transfer for storage to good-quality inert plastic or borosilicate glass bottles with borosilicate, polyethylene, or other plastic stoppers or caps. Label prepared reagents with name and concentration, date prepared, and initials of preparer. Include positive and negative control cultures with each series of cultural or biochemical tests.

g. Dyes and stains: In microbiological analyses, organic chemicals are used as selective agents (e.g., brilliant green), as indicators (e.g., phenol red), and as microbiological stains (e.g., Gram stain). Dyes from commercial suppliers vary from lot to lot in percent dye, dye complex, insolubles, and inert materials. Because dyes for microbiology must be of proper strength and stability to produce correct reactions, use only dyes certified by the Biological Stain Commission. Check bacteriological stains before use with at least one positive and one negative control culture and record results.

h. Membrane filters and pads: The quality and performance of membrane filters vary with the manufacturer, type, brand, and lot. These variations result from differences in manufacturing methods, materials, quality control, storage conditions, and application.

1) Membrane filters and pads for water analyses should meet the following specifications:

a) Filter diam 47 mm, mean pore diam 0.45 μm. Alternate filter and pore sizes may be used if the manufacturer provides data verifying performance equal to or better than that of 47-mm-diam, 0.45-μm-pore size filter. At least 70% of filter area must be pores.

b) When filters are floated on reagent water, the water diffuses uniformly through the filters in 15 s with no dry spots on the filters.

c) Flow rates are at least 55 mL/min/cm² at 25°C and a differential pressure of 93 kPa.

d) Filters are nontoxic, free of bacterial-growth-inhibiting or stimulating substances, and free of materials that directly or indirectly interfere with bacterial indicator systems in the medium; ink grid is nontoxic. The arithmetic mean of five counts on filters must be at least 90% of the arithmetic mean of the counts on five agar spread plates using the same sample volumes and agar media.

e) Filters retain the organisms from a 100-mL suspension of *Serratia marcescens* containing 1×10^3 cells.

f) Water-extractables in filter do not exceed 2.5% after the membrane is boiled in 100 mL reagent water for 20 min, dried, cooled, and brought to constant weight.

g) Absorbent pad has diam 47 mm, thickness 0.8 mm, and is capable of absorbing 2.0 ± 0.2 mL Endo broth.

h) Pads release less than 1 mg total acidity calculated as $CaCO_3$ when titrated to the phenolphthalein end point with 0.02N NaOH.

i) If filter and absorbent pad are not sterile, they should not be degraded by sterilization at 121°C for 10 min. Confirm sterility by absence of growth when a membrane filter is placed on a pad saturated with tryptone glucose extract broth or tryptone glucose extract agar and incubated at 35 ± 0.5°C for 24 h.

j) Some lots of membrane filters yield low recoveries, poor differentiation, or malformation of colonies due to toxicity, chemical composition, or structural defects.[6] Perform the use test (¶ 4e) on new lots of filters.

2) Standardized tests:

Standardized tests are available for evaluating retention, recovery, extractables, and flow rate characteristics of membrane filters.[7]

Some manufacturers provide information beyond that required by specifications and certify that their membranes are satisfactory for water analysis. They report retention, pore size, flow rate, sterility, pH, percent recovery, and limits for specific inorganic and organic chemical extractables. Although the standard membrane filter evaluation tests were developed for the manufacturers, a laboratory can conduct its own tests.

To maintain quality control inspect each lot of membranes before use and during testing to insure they are round and pliable, with undistorted gridlines after autoclaving. After incubation, colonies should be well-developed with well-defined color and shape as defined by the test procedure. The gridline ink should not channel growth along the ink line nor restrict colony development. Colonies should be distributed evenly across the membrane surface.

i. Culture media: Because cultural methods depend on properly prepared media, use the best available materials and techniques in media preparation, storage, and application. For control of quality, use commercially prepared media whenever available but note that such media may vary in quality among manufacturers and even from lot to lot from the same manufacturer.

Order media in quantities to last no longer than 1 year. Use media on a first-in, first-out basis. When practical, order media in quarter pound (114 g) multiples rather than one pound (454 g) bottles, to keep the supply sealed as long as possible. Record kind, amount, and appearance of media received, lot number, expiration date, and dates received and opened. Check inventory quarterly for reordering.

Store dehydrated media at an even temperature in a cool dry place, away from direct sunlight. Discard media that cake, discolor, or show other signs of deterioration. If expiration date is given by manufacturer, discard unused media after that date. A conservative time limit for unopened bottles is 2 years at room temperature. Compare recovery of newly purchased lots of media against proven lots, using recent pure-culture isolates and natural samples.

Use opened bottles of media within 6 months. Dehydrated media are hygroscopic. Protect opened bottles from moisture. Close bottles as tightly as possible, immediately after use. If caking or discoloration of media occurs, discard media. Store opened bottles in a dessicator.

1) *Preparation of media*—Prepare media in containers that are at least twice the volume of the medium being prepared. Stir media, particularly agars, while heating. Avoid scorching or boil-over by using a boiling water bath for small batches of media and by continually attending to larger volumes heated on a hot plate or gas burner. Preferably use hot plate-magnetic stirrer combinations. Label and date prepared media. Prepare media in reagent water. Measure water volumes and media with graduates or pipets conforming to NIST and APHA standards, respectively. Do not use blow-out pipets. After preparation and storage, remelt agar media in boiling water or flowing steam.

Check and record pH of a portion of each medium after sterilization and cooling. Check pH of solid medium with a surface probe. Record results. Make minor adjustments in pH (<0.5 pH units) with $1N$ NaOH or HCl solution to the pH specified in formulation. If the pH difference is larger than 0.5 units, discard the batch and check preparation instructions and pH of reagent water to resolve the problem. Incorrect pH values may be due to reagent water quality, medium deterioration, or improper preparation. Review instructions for preparation and check water pH. If water pH is unsatisfactory, prepare a new batch of medium using water from a new source (see 9020B.3*d* and *e*). If water is satisfactory, remake medium and check; if pH is again incorrect, prepare medium from another bottle.

Record pH problems in the media record book and inform the manufacturer if the medium is indicated as the source of error. Examine prepared media for unusual color, darkening, or precipitation and record observations. Consider variations of sterilization time and temperature as possible causes for problems. If any of the above occur, discard the medium.

2) *Sterilization*—Sterilize media at 121 to 124°C for the minimum time specified. A double-walled autoclave permits maintenance of full pressure and temperature in the jacket between loads and reduces chance for heat damage. Follow manufacturer's directions for sterilization of specific media. The required expo-

TABLE 9020:III. TIME AND TEMPERATURE FOR AUTOCLAVE STERILIZATION

Material	Time at 121°C*
Membrane filters and pads	10 min
Carbohydrate-containing media (lauryl tryptose, BGB broth, etc.)	12–15 min
Contaminated materials and discarded cultures	30 min
Membrane filter assemblies (wrapped), sample collection bottles (empty)	15 min
Buffered dilution water, 99 mL in screw-cap bottle	15 min
Rinse water, volume > 100 mL	Adjust for volume

* Except for media, times are guidelines; check for sterility.

sure time varies with form and type of material, type of medium, presence of carbohydrates, and volume. Table 9020:III gives guidelines for typical items. Do not expose media containing carbohydrates to the elevated temperatures for more than 45 min. Exposure time is defined as the period from initial exposure to removal from the autoclave.

Some currently available autoclave models are automatic and include features such as vertical sliding, self-sealing and opening doors, programmable sterilization cycles, and continuous multipoint monitoring of chamber temperature and pressure. These units also may incorporate solution cooling and vapor removal features. When sterilizer design includes heat exchangers and solution cooling features as part of a factory-programmed liquid cycle, strict adherence to the 45-min total elapsed time in the autoclave is not necessary provided that printout records verify normal cycle operation and chamber cooling during exhaust and vapor removal.

Remove sterilized media from autoclave as soon as chamber pressure reaches zero, or, if a fully automatic model is used, as soon as the door opens. Do not reautoclave media.

Check effectiveness of sterilization weekly by placing *Bacillus stearothermophilus* spore suspensions or strips (commercially available) inside glassware. Sterilize at 121°C for 15 min. Place in trypticase soy broth tubes and incubate at 55°C for 48 h. If growth of the autoclaved spores occurs after incubation at 55°C, sterilization was inadequate. A small, relatively inexpensive 55°C incubator is available commercially.†

Sterilize heat-sensitive solutions or media by filtration through a 0.22-μm-pore-diam filter in a sterile filtration and receiving apparatus. Filter and dispense medium in a safety cabinet or biohazard hood if available. Sterilize glassware (pipets, petri dishes, sample bottles) in an autoclave or an oven at 170°C for 2 h. Sterilize equipment, supplies, and other solid or dry materials that are heat-sensitive, by exposing to ethylene oxide in a gas sterilizer. Use commercially available spore strips or suspensions to check dry heat and ethylene oxide sterilization.

3) *Use of agars and broths*—Temper melted agars in a water bath at 44 to 46°C until used but do not hold longer than 3 h. To monitor agar temperature, expose a bottle of water or medium to the same heating and cooling conditions as the agar. Insert a thermometer in the monitoring bottle to determine when the temperature is 45 to 46°C and suitable for use in pour plates. If possible, prepare media on the day of use. After pouring agar plates for streaking, dry agar surfaces by keeping dish slightly open for at

† 3M Health Care, St. Paul, MN 55144, or equivalent.

least 15 min in a bacteriological hood to avoid contamination. Discard unused liquid agar; do not let harden or remelt for later use.

Handle tubes of sterile fermentation media carefully to avoid entrapping air in inner tubes, thereby producing false positive reactions. Examine freshly prepared tubes to determine that gas bubbles are absent.

4) Storage of media—Prepare media in amounts that will be used within holding time limits given in Table 9020:IV. Protect media containing dyes from light; if color changes occur, discard the media. Refrigerate poured agar plates not used on the day of preparation. Seal agar plates with loose-fitting lids in plastic bags if held more than 2 d. Prepare broth media that will be stored for more than 2 weeks in screw-cap tubes, other tightly sealed tubes, or in loose-capped tubes placed in a sealed plastic bag or other tightly sealed container to prevent evaporation.

Mark liquid level in several tubes and monitor for loss of liquid. If loss is 10% or more, discard the batch. If media are refrigerated, incubate overnight at test temperature before use and reject the batch if false positive responses occur. Prepared sterile broths and agars available from commercial sources may offer advantages when analyses are done intermittently, when staff is not available for preparation work, or when cost can be balanced against other factors of laboratory operation. Check performance of these media as described in ¶ 5 below.

5) Quality control of prepared media—Maintain in a bound book a complete record of each prepared batch of medium with name of preparer and date, name and lot number of medium, amount of medium weighed, volume of medium prepared, sterilization time and temperature, pH measurements and adjustments, and preparations of labile components. Compare quantitative recoveries of new lots with previously acceptable ones. Include sterility and positive and negative control culture checks on all media as described below.

5. Standard Operating Procedures (SOPs)

SOPs are the operational backbone of an analytical laboratory. SOPs describe in detail all laboratory operations such as preparation of reagents, reagent water, standards, culture media, proper use of balances, sterilization practices, and dishwashing procedures, as well as methods of sampling, analysis, and quality control. The SOPs are unique to the laboratory. They describe the tasks as performed on a day-to-day basis, tailored to the laboratory's own equipment, instrumentation, and sample types. The SOPs guide routine operations by each analyst, help to assure uniform operations, and provide a solid training tool.

TABLE 9020:IV. HOLDING TIMES FOR PREPARED MEDIA

Medium	Holding Time
Membrane filter (MF) broth in screw-cap flasks at 4°C	96 h
MF agar in plates with tight-fitting covers at 4°C	2 weeks
Agar or broth in loose-cap tubes at 4°C	2 weeks
Agar or broth in tightly closed screw-cap tubes or other sealed containers	3 months
Poured agar plates with loose-fitting covers in sealed plastic bags at 4°C	2 weeks
Large volume of agar in tightly closed screw-cap flask or bottle at 4°C	3 months

6. Sampling

a. Planning: Microbiologists should participate in the planning of monitoring programs that will include microbial analyses. They can provide valuable expertise on the selection of sampling sites, number of samples and analyses needed, workload, and equipment and supply needs. For natural waters, knowledge of the probable microbial densities, and the impact of season, weather, tide and wind patterns, known sources of pollution, and other variables, are needed to formulate the most effective sampling plan.

b. Methods: Sampling plans must be specific for each sampling site. Prior sampling guidance can be only general in nature, addressing the factors that must be considered for each site. Sampling SOPs describe sampling equipment, techniques, frequency, holding times and conditions, safety rules, etc., that will be used under different conditions for different sites. From the information in these SOPs sampling plans will be drawn up.

7. Analytical Methods

a. Method selection: Because minor variations in technique can cause significant changes in results, microbiological methods must be standardized so that uniform data result from multiple laboratories. Select analytical methods appropriate for the sample type from *Standard Methods* or other source of standardized methods and ensure that methods have been validated in a multilaboratory study with the sample types of interest.

b. Data objectives: Review available methods and determine which produce data to meet the program's needs for precision, bias, specificity, selectivity, and detection limit. Ensure that the methods have been demonstrated to perform within the above specifications for the samples of interest.

c. Internal QC: The written analytical methods should contain required QC checks of positive and negative control cultures, sterile blank, replicate analyses (precision), and a known quantitative culture, if available.

d. Method SOPs: As part of the series of SOPs, provide each analyst with a copy of the analytical methods written in step-wise fashion exactly as they are to be performed and specific to the sample type, equipment, and instrumentation used in the laboratory.

8. Analytical Quality Control Procedures

a. General quality control procedures:

1) New methods—Conduct parallel tests with the standard procedure and a new method to determine applicability and comparability. Perform at least 100 parallel tests across seasons of the year before replacement with the new method for routine use.

2) Comparison of plate counts—For routine performance evaluation, repeat counts on one or more positive samples at least monthly and compare the counts with those of other analysts testing the same samples. Replicate counts for the same analyst should agree within 5% and those between analysts should agree within 10%. See 9020B.10*b* for a statistical calculation of data precision.

3) Control cultures—For each lot of medium check analytical procedures by testing with known positive and negative control cultures for the organism(s) under test. See Table 9020:V for examples of test cultures.

TABLE 9020:V. CONTROL CULTURES FOR MICROBIOLOGICAL TESTS

Group	Control Culture	
	Positive	Negative
Total coliforms	*Escherichia coli*	*Staphylococcus aureus*
	Enterobacter aerogenes	*Pseudomonas* sp.
Fecal coliforms	*E. coli*	*E. aerogenes*
		Streptococcus faecalis
Escherichia coli	*E. coli*	*E. aerogenes*
Fecal streptococci	*Enterococcus faecalis*	*Staphylococcus aureus*
		E. coli
Enterococci	*S. faecalis*	*S. mitis/salivarius*

TABLE 9020:VI. CALCULATION OF PRECISION CRITERION

Sample No.	Duplicate Analyses		Logarithms of Counts		Range of Logarithms ($R_{\log}$) ($L_1 - L_2$)
	D_1	D_2	L_1	L_2	
1	89	71	1.9494	1.8513	0.0981
2	38	34	1.5798	1.5315	0.0483
3	58	67	1.7634	1.8261	0.0627
.	.	.	.	.	.
.	.	.	.	.	.
.	.	.	.	.	.
14	7	6	0.8451	0.7782	0.0669
15	110	121	2.0414	2.0828	0.0414

Calculations:

1) Σ of $R_{\log} = 0.0981 + 0.0483 + 0.0627 + \ldots + 0.0669 + 0.0414$

     $= 0.718\ 89$

2) $\overline{R} = \dfrac{\Sigma R_{\log}}{n} = \dfrac{0.718\ 89}{15} = 0.0479$

3) Precision criterion $= 3.27\ \overline{R} = 3.27\ (0.0479) = 0.1566$

4) *Duplicate analyses*—Perform duplicate analyses on 10% of samples and on at least one sample per test run. A test run is defined as an uninterrupted series of analyses. If the laboratory conducts less than 10 tests/week, make duplicate analyses on at least one sample each week.

5) *Sterility checks*—For membrane filter tests, check sterility of media, membrane filters, buffered dilution and rinse water, pipets, flasks and dishes, and equipment as a minimum at the end of each series of samples, using sterile reagent water as the sample. If contaminated, check for the source. For multiple-tube and presence-absence procedures, check sterility of media, dilution water, and glassware. To test sterility of media, incubate a representative portion of each batch at an appropriate temperature for 24 to 48 h and observe for growth. Check each batch of buffered dilution water for sterility by adding 20 mL water to 100 mL of a nonselective broth. Alternatively, aseptically pass 100 mL or more dilution water through a membrane filter and place filter on growth medium suitable for heterotrophic bacteria. Incubate at 35 ± 0.5°C for 24 h and observe for growth. If any contamination is indicated, determine the cause and reject analytical data from samples tested with these materials. Request immediate resampling and reanalyze.

b. Precision of quantitative methods: Calculate precision of duplicate analyses for each different type of sample examined, for example, drinking water, ambient water, wastewater, etc., according to the following procedure:

1) Perform duplicate analyses on first 15 positive samples of each type, with each set of duplicates analyzed by a single analyst. If there is more than one analyst, include all analysts regularly running the tests, with each analyst performing approximately an equal number of tests. Record duplicate analyses as D_1 and D_2.

2) Calculate the logarithm of each result. If either of a set of duplicate results is <1, add 1 to both values before calculating the logarithms.

3) Calculate the range (R) for each pair of transformed duplicates as the mean ($\overline{R}$) of these ranges.

See sample calculation in Table 9020:VI.

4) Thereafter, analyze 10% of routine samples in duplicate. Transform the duplicates as in ¶ 2) and calculate their range. If the range is greater than 3.27 $\overline{R}$, there is greater than 99% probability that the laboratory variability is excessive. Determine if increased imprecision is acceptable; if not, discard all analytical results since the last precision check (see Table 9020:VII). Identify and resolve the analytical problem before making further analyses.

5) Update the criterion used in ¶ 4) by periodically repeating the procedures of ¶s 1) through 3) using the most recent sets of 15 duplicate results.

9. Verification

For the most part, the confirmation/verification procedures for drinking water differ from those for other waters because of specific regulatory requirements.

a. Multiple-tube fermentation (MTF) methods:

1) Total coliform procedure (9221B)

a) Drinking water—Carry samples through confirmed phase only. Verification is not required. For QC purposes, if normally there are no positive results, analyze at least one positive source water quarterly to confirm that the media produce appropriate responses. For samples with a history of heavy growth without gas in presumptive-phase tubes, carry the tubes through the confirmed phase to check for false negative responses for coliform bacteria. Verify any positives for fecal coliforms or *E. coli*.

b) Other water types—Verify by performing the completed MTF Test on 10% of samples positive through the confirmed phase.

2) Enzyme substrate coliform test (total coliform/*E. coli*) (9223B)

a) Drinking water—Verify at least 5% of total coliform positive results from enzyme substrate coliform tests by inoculating growth from a known positive sample and testing for lactose fermentation or for β-D-galactopyranosidase by the *o*-nitrophenyl-β-D-galactopyranoside (ONPG) test and indophenol by the cyto-

TABLE 9020:VII. DAILY CHECKS ON PRECISION OF DUPLICATE COUNTS*

Date of Analysis	Duplicate Analyses		Logarithms of Counts		Range of Logarithms	Acceptance of Range†
	D_1	D_2	L_1	L_2		
8/29	71	65	1.8513	1.8129	0.0383	A
8/30	110	121	2.0414	2.0828	0.0414	A
8/31	73	50	1.8633	1.6990	0.1643	U

* Precision criterion $= (3.27\overline{R}) = 0.1566$.

† A = acceptable; U = unacceptable.

chrome oxidase (CO) test. See 9225D for these tests. Coliforms are ONPG-positive and cytochrome-oxidase-negative. Verify *E. coli* using the EC MUG test (see 9221F).

b) Other water types—Verify at least 10% of total coliform positive samples as in ¶ 2a above.

3) Fecal streptococci procedure—Verify as in 9230C.5. Growth of catalase-negative, gram-positive cocci on bile esculin agar at 35°C and in brain-heart infusion broth at 45°C verifies the organisms as fecal streptococci. Growth at 45°C and in 6.5% NaCl broth indicates the streptococci are members of the enterococcus group.

4) Include known positive and negative pure cultures as a QC check.

b. Membrane filter methods:

1) Total coliform procedures

a) Drinking water—Pick all, up to 5 typical and 5 atypical (nonsheen) colonies from positive samples on M-Endo medium and verify as in 9222B.5g. Also verify any positives for fecal coliforms or *E. coli*. If there are no positive samples, test at least one known positive source water quarterly.

b) Other water types—Verify positives monthly by picking at least 10 sheen colonies from a positive water sample as in 9222B.5g. Adjust counts based on percent verification.

c) To determine false negatives, pick representative atypical colonies of different morphological types and verify as in 9222B.5g.

2) Fecal coliform procedure

a) Verify positives monthly by picking at least 10 blue colonies from one positive sample. Verify in lauryl tryptose broth and EC broth as in 9221B.3 and 9221E. Adjust counts based on percent verification.

b) To determine false negatives, pick representative atypical colonies of different morphological types and verify as in 9221B.3 and 9221E.

3) *Escherichia coli* procedure

a) Drinking water—Verify at least 5% of MUG-positive and MUG-negative results. Pick from well-isolated sheen colonies that fluoresce on nutrient agar with MUG (NA MUG), taking care not to pick up medium, which can cause a false positive response. Also verify nonsheen colonies that fluoresce. Verify by performing the citrate test and the indole test as described in 9225D, but incubate indole test at 44.5°C. *E. coli* are indole-positive and yield no growth on citrate.

b) Other water types—Verify one positive sample monthly as in ¶ a) above. Adjust counts based on percentage of verification.

4) Fecal streptococci procedure—Pick to verify monthly at least 10 isolated esculin-positive red colonies from m-Enterococcus agar to brain heart infusion (BHI) media. Verify as described in 9230C. Adjust counts based on percentage of verification.

5) *Enterococci* procedures—Pick to verify monthly at least 10 well-isolated pink to red colonies with black or reddish-brown precipitate from EIA agar. Transfer to BHI media as described in 9230C. Adjust counts based on percentage of verification.

6) Include known positive and negative pure cultures as a quality control check.

10. Documentation and Recordkeeping

a. QA plan: The QA program documents management's commitment to a QA policy and sets forth the requirements needed to support program objectives. The program describes overall policies, organization, objectives, and functional responsibilities for achieving the quality goals. In addition, the program should develop a project plan that specifies the QC requirements for each project. The plan specifies the QC activities required to achieve the data representativeness, completeness, comparability, and compatibility. Also, the QA plan should include a program implementation plan that ensures maximum coordination and integration of QC activities within the overall program (sampling, analyses, and data handling).

b. Sampling records: A written SOP for sample handling records sample collection, transfer, storage, analyses, and disposal. The record is most easily kept on a series of printed forms that prompt the user to provide all the necessary information. It is especially critical that this record be exact and complete if there is any chance that litigation may occur. Such record systems are called chain of custody. Because laboratories do not always know whether analytical results will be used in future litigation, some maintain chain-of-custody on all samples. Details on chain of custody are available in Section 1060B and elsewhere.[1]

c. Recordkeeping: An acceptable recordkeeping system provides needed information on sample collection and preservation, analytical methods, raw data, calculations through reported results, and a record of persons responsible for sampling and analyses. Choose a format agreeable to both the laboratory and the customer (the data user). Ensure that all data sheets are signed and dated by the analyst and the supervisor. The preferable record form is a bound and page-numbered notebook, with entries in ink and a single line drawn through any change with the correction entered next to it.

Keep records of microbiological analyses for at least 5 years. Actual laboratory reports may be kept, or data may be transferred to tabular summaries, provided that the following information is included: date, place, and time of sampling, name of sample collector; identification of sample; date of receipt of sample and analysis; person(s) responsible for performing analysis; analytical method used; the raw data and the calculated results of analysis. Verify that each result was entered correctly from the bench sheet and initialed by the analyst. If an information storage and retrieval system is used, double check data on the printouts.

11. Data Handling

a. Distribution of bacterial populations: In most chemical analyses the distribution of analytical results follows the Gaussian curve, which has symmetrical distribution of values about the mean (see Section 1010B). Microbial distributions are not necessarily symmetrical. Bacterial counts often are characterized as having a skewed distribution because of many low values and a few high ones. These characteristics lead to an arithmetic mean that is considerably larger than the median. The frequency curve of this distribution has a long right tail, such as that shown in Figure 9020:1, and is said to display positive skewness.

Application of the most rigorous statistical techniques requires the assumption of symmetrical distributions such as the normal curve. Therefore it usually is necessary to convert skewed data so that a symmetrical distribution resembling the normal distribution results. An approximately normal distribution can be obtained from positively skewed data by converting numbers to their logarithms, as shown in Table 9020:VIII. Comparison of the fre-

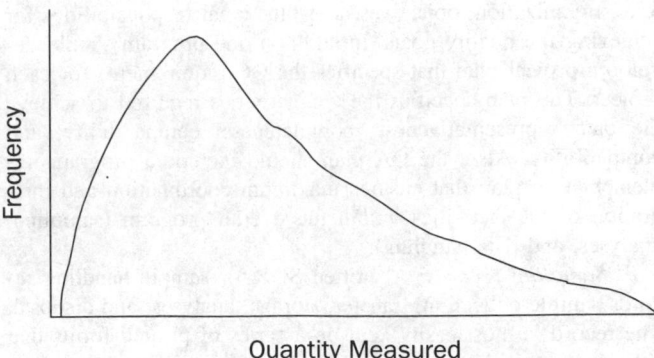

Figure 9020:1. Frequency curve (positively skewed distribution).

quency tables for the original data (Table 9020:IX) and their logarithms (Table 9020:X) shows that the logarithms approximate a symmetrical distribution.

b. Central tendency measures of skewed distribution: If the logarithms of numbers from a positively skewed distribution are approximately normally distributed, the original data have a lognormal distribution. The best estimate of central tendency of lognormal data is the geometric mean, defined as:

$$\bar{x}_g = {}^n\sqrt{(x_1)(x_2) \cdots (x_n)}$$

and

$$\log \bar{x}_g = \frac{\sum (\log x_i)}{n}$$

that is, the geometric mean is equal to the antilog of the arithmetic mean of the logarithms. For example, the following means

TABLE 9020:VIII. COLIFORM COUNTS AND THEIR LOGARITHMS

MPN Coliform Count No./100 mL	log MPN
11	1.041
27	1.431
36	1.556
48	1.681
80	1.903
85	1.929
120	2.079
130	2.114
136	2.134
161	2.207
317	2.501
601	2.779
760	2.881
1020	3.009
3100	3.491

$\bar{x} = 442$ $\bar{x}_g = $ antilog $2.1825 = 152$

TABLE 9020:IX. COMPARISON OF FREQUENCY OF MPN DATA

Class Interval	Frequency (MPN)
0 to 400	11
400 to 800	2
800 to 1200	1
1200 to 1600	0
1600 to 2000	0
2000 to 2400	0
2400 to 2800	0
2800 to 3200	0

calculated from the data in Table 9020:VIII are drastically different.

$$\log \bar{x}_g = \frac{\sum (\log x_i)}{n} = \frac{32.737}{15} = 2.1825$$

geometric mean

$$\bar{x}_g = \text{antilog} (2.1825) = 152$$

and arithmetic mean

$$\bar{x} = \frac{\sum x_i}{n} = \frac{6632}{15} = 442$$

Therefore, although regulations or tradition may require or cause microbiological data to be reported as the arithmetic mean or median, the preferred statistic for summarizing microbiological monitoring data is the geometric mean. An exception may be in the evaluation of data for risk assessment. The arithmetic mean may be a better measure for this purpose because it may generate a higher central tendency value and possibly provide a greater safety factor.[8]

c. "Less than" (<) values: There has always been uncertainty as to the proper way to include "less than" values in calculation and evaluation of microbiological data because such values cannot be treated statistically without modification. Proposed modifications involve changing such numbers to zero, choosing values halfway between zero and the "less than" value, or assigning the "less than" value itself, i.e., changing <1 values to 1, ½, or 0.

There are valid reasons for not including < values, whether modified or not. If the database is fairly large with just a few < values, the influence of these uncertain values will be minimal

TABLE 9020:X. COMPARISON OF FREQUENCY OF LOG MPN DATA

Class Interval	Frequency (log MPN)
1.000 to 1.300	1
1.300 to 1.600	2
1.600 to 1.900	1
1.900 to 2.200	5
2.200 to 2.500	1
2.500 to 2.800	2
2.800 to 3.100	2
3.100 to 3.400	0
3.400 to 3.700	1

and of no benefit. If the database is small or has a relatively large number of < values, inclusion of modified < values would exert an undue influence on the final results and could result in an artificial negative or positive bias. Including < values is particularly inappropriate if the < values are <100, <1000, or higher because the unknown true values could be anywhere from 0 to 99, 0 to 999, etc. When < values are first noted, adjust or expand test volumes. The only exception to this caution would be regulatory testing with defined compliance limits, such as the <1/100 mL values reported for drinking water systems where the 100-mL volume is required.

12. References

1. BORDNER, R.H., J.A. WINTER & P.V. SCARPINO, eds. 1978. Microbiological Methods for Monitoring the Environment, Water and Wastes. EPA-600/8-78-017, Environmental Monitoring & Support Lab., U.S. Environmental Protection Agency, Cincinnati, Ohio.
2. AMERICAN SOCIETY FOR TESTING AND MATERIALS. 1995. Standard guide for good laboratory practices in laboratories engaged in sampling and analysis of water. D-3856-95, Annual Book of ASTM Standards, Vol. 11.01, American Soc. Testing & Materials, Philadelphia, Pa.
3. AMERICAN SOCIETY FOR TESTING AND MATERIALS. 1996. Standard practice for writing quality control specifications for standard test methods for water analysis. D-5847-96, Annual Book of ASTM Standards, Vol. 11.01, American Soc. Testing & Materials, West Conshohocken, Pa.
4. AMERICAN SOCIETY FOR TESTING AND MATERIALS. 1993. Annual Book of ASTM Standards, Vol. 14.02, General Methods and Instrumentation. E-319-86 (reapproved 1993), Standard Practice for Evaluation of Single-Pan Mechanical Balances, and E-898-88 (reapproved 1993), Standard Method of Testing Top-Loading, Direct-Reading Laboratory Scales and Balances. American Soc. Testing & Materials, Philadelphia, Pa.
5. SCHMITZ, S., C. GARBE, B. TEBBE & C. ORFANOS. 1994. Long wave ultraviolet radiation (UVA) and skin cancer. Hautarzt 45:517.
6. BRENNER, K. & C.C. RANKIN. 1990. New screening test to determine the acceptability of 0.45 μm membrane filters for analysis of water. Appl. Environ. Bacteriol. 56:54.
7. AMERICAN SOCIETY FOR TESTING AND MATERIALS. 1977. Annual Book of ASTM Standards. Part 31, Water. American Soc. Testing & Materials, Philadelphia, Pa.
8. HAAS, C.N. 1996. How to average microbial densities to characterize risk. Water Res. 30:1036.

9020 C.   Interlaboratory Quality Control

1. Background

Interlaboratory QC programs are a means of establishing an agreed-upon, common performance criteria system that will assure an acceptable level of data quality and comparability among laboratories with similar interests and/or needs.

These systems may be volunteer, such as that for the cities in the Ohio River Valley Water Sanitation Commission (ORSANCO), or regulatory, such as the Federal Drinking Water Laboratory Certification Program (see below). Often, the term "accreditation" is used interchangeably with certification. Usually, interlaboratory quality control programs have three elements: uniform criteria for laboratory operations, external review of the program, and external proficiency testing.

2. Uniform Criteria

Interlaboratory quality control programs begin as a volunteer or mandatory means of establishing uniform laboratory standards for a specific purpose. The participants may be from one organization or a group of organizations having common interests or falling under common regulations. Often one group or person may agree to draft the criteria. If under regulation, the regulating authority may set the criteria for compliance-monitoring analyses.

Uniform sampling and analytical methods and quality control criteria for personnel, facilities, equipment, instrumentation, supplies, and data handling and reporting are proposed, discussed, reviewed, modified if necessary, and approved by the group for common use. Criteria identified as necessary for acceptable data quality should be mandatory. A formal document is prepared and provided to all participants.

The QA/QC responsibilities of management, supervisors, and technical staff are described in 9020A. In large laboratories, a QA officer is assigned as a staff position but may be the supervisor or other senior person in smaller laboratories.

After incorporation into laboratory operations and confirmation that the QA program has been adapted and is in routine use, the laboratory supervisor and the QA officer conduct an internal program review of all operations and records for acceptability, to identify possible problems and assist in their resolution. If this is done properly, there should be little concern that subsequent external reviews will find major problems.

3. External Program Review

Once a laboratory has a QA program in place, management informs the organization and a qualified external QA person or team arranges an on-site visit to evaluate the QA program for acceptability and to work with the laboratory to solve any problems. An acceptable rating confirms that the laboratory's QA program is operating properly and that the laboratory has the capability of generating valid defensible data. Such on-site evaluations are repeated and may be announced or unannounced.

4. External Proficiency Testing

Whenever practical, the external organization conducts formal performance evaluation studies among all participant laboratories. Challenge samples are prepared and sent as unknowns on a set schedule for analyses and reporting of results. The reported data are coded for confidentiality and evaluated according to an agreed-upon scheme. The results are summarized for all laboratories and individual laboratory reports are sent to participants. Results of such studies indicate the quality of routine analyses of each laboratory as compared to group performance. Also, results of the group as a whole characterize the performance that can be expected for the analytical methods tested.

5. Example Program

In the Federal Drinking Water Laboratory Certification Program, public water supply laboratories must be certified according to minimal criteria and procedures and quality assurance described in the EPA manual on certification:[1] criteria are established for laboratory operations and methodology; on-site inspections are required by the certifying state agency or its surrogate to verify minimal standards; annually, laboratories are required to perform acceptably on unknown samples in formal studies, as samples are available; the responsible authority follows up on problems identified in the on-site inspection or performance evaluation and requires corrections within a set period of time. Individual state programs may exceed the federal criteria.

On-site inspections of laboratories in the present certification program show that primary causes for discrepancies in drinking water laboratories have been inadequate equipment, improperly prepared media, incorrect analytical procedures, and insufficiently trained personnel.

6. References

1. U.S. ENVIRONMENTAL PROTECTION AGENCY. 1997. Manual for the Certification of Laboratories Analyzing Drinking Water, 4th ed. EPA-814B-92-002, U.S. Environmental Protection Agency, Cincinnati, Ohio.

9030   LABORATORY APPARATUS*

9030 A.   Introduction

This section contains specifications for microbiological laboratory equipment. For testing and maintenance procedures related to quality control, see Section 9020.

---

* Approved by Standard Methods Committee, 1993.

9030 B.   Equipment Specifications

1. Incubators

Incubators must maintain a uniform and constant temperature at all times in all areas, that is, they must not vary more than $\pm 0.5°C$ in the areas used. Obtain such accuracy by using a water-jacketed or anhydric-type incubator with thermostatically controlled low-temperature electric heating units properly insulated and located in or adjacent to the walls or floor of the chamber and preferably equipped with mechanical means of circulating air.

Incubators equipped with high-temperature heating units are unsatisfactory, because such sources of heat, when improperly placed, frequently cause localized overheating and excessive drying of media, with consequent inhibition of bacterial growth. Incubators so heated may be operated satisfactorily by replacing high-temperature units with suitable wiring arranged to operate at a lower temperature and by installing mechanical air-circulation devices. It is desirable, where ordinary room temperatures vary excessively, to keep laboratory incubators in special rooms maintained at a few degrees below the recommended incubator temperature.

Alternatively, use special incubating rooms well insulated and equipped with properly distributed heating units, forced air circulation, and air exchange ports, provided that they conform to desired temperature limits. When such rooms are used, record the daily temperature range in areas where plates or tubes are incubated. Provide incubators with open metal wire or perforated sheet shelves so spaced as to assure temperature uniformity throughout the chamber. Leave a 2.5-cm space between walls and stacks of dishes or baskets of tubes.

Maintain an accurate thermometer, traceable to the National Institute of Standards and Technology (NIST), with the bulb immersed in liquid (glycerine, water, or mineral oil) on each shelf in use within the incubator and record daily temperature readings (preferably morning and afternoon). It is desirable, in addition, to maintain a maximum and minimum registering thermometer within the incubator on the middle shelf to record the gross temperature range over a 24-h period. At intervals, determine temperature variations within the incubator when filled to maximum capacity. Install a recording thermometer whenever possible, to maintain a continuous and permanent record of temperature.

Ordinarily, a water bath with a gabled cover to reduce water and heat loss, or a solid heat sink incubator, is required to maintain a temperature of $44.5 \pm 0.2°C$. If satisfactory temperature control is not achieved, provide water recirculation. Keep water depth in the incubator sufficient to immerse tubes to upper level of media.

2. Hot-Air Sterilizing Ovens

Use hot-air sterilizing ovens of sufficient size to prevent internal crowding; constructed to give uniform and adequate sterilizing temperatures of 170 ± 10°C; and equipped with suitable thermometers. Optionally use a temperature-recording instrument.

3. Autoclaves

Use autoclaves of sufficient size to prevent internal crowding; constructed to provide uniform temperatures within the chambers (up to and including the sterilizing temperature of 121°C); equipped with an accurate thermometer the bulb of which is located properly on the exhaust line so as to register minimum temperature within the sterilizing chambers (temperature-recording instrument is optional); equipped with pressure gauge and properly adjusted safety valves connected directly with saturated-steam supply lines equipped with appropriate filters to remove particulates and oil droplets or directly to a suitable special steam generator (do not use steam from a boiler treated with amines for corrosion control); and capable of reaching the desired temperature within 30 min. Confirm, by chemical or toxicity tests, that the steam supply has not been treated with amines or other corrosion-control chemicals that will impart toxicity.

Use of a vertical autoclave or pressure cooker is not recommended because of difficulty in adjusting and maintaining sterilization temperature and the potential hazard. If a pressure cooker is used in emergency or special circumstances, equip it with an efficient pressure gauge and a thermometer the bulb of which is 2.5 cm above the water level.

4. Gas Sterilizers

Use a sterilizer equipped with automatic controls capable of carrying out a complete sterilization cycle. As a sterilizing gas use ethylene oxide (CAUTION: *Ethylene oxide is toxic—avoid inhalation, ingestion, and contact with the skin. Also, ethylene oxide forms an explosive mixture with air at 3-80% proportion.*) diluted to 10 to 12% with an inert gas. Provide an automatic control cycle to evacuate sterilizing chamber to at least 0.06 kPa, to hold the vacuum for 30 min, to adjust humidity and temperature, to charge with the ethylene oxide mixture to a pressure dependent on mixture used, to hold such pressure for at least 4 h, to vent gas, to evacuate to 0.06 kPa, and finally, to bring to atmospheric pressure with sterile air. The humidity, temperature, pressure, and time of sterilizing cycle depend on the gas mixture used.

Store overnight sample bottles with loosened caps that were sterilized by gas, to allow last traces of gas mixture to dissipate. Incubate overnight media sterilized by gas, to insure dissipation of gas.

In general, mixtures of ethylene oxide with chlorinated hydrocarbons such as freon are harmful to plastics, although at temperatures below 55°C, gas pressure not over 35 kPa, and time of sterilization less than 6 h, the effect is minimal. If carbon dioxide is used as a diluent of ethylene oxide, increase exposure time and pressure, depending on temperature and humidity that can be used.

Determine proper cycle and gas mixture for objects to be sterilized and confirm by sterility tests.

5. Optical Counting Equipment

a. Pour and spread plates: Use Quebec-type colony counter, dark-field model preferred, or one providing equivalent magnification (1.5 diameters) and satisfactory visibility.

b. Membrane filters: Use a binocular microscope with magnification of 10 to 15×. Provide daylight fluorescent light source at angle of 60 to 80° above the colonies; use low-angle lighting for nonpigmented colonies.

c. Mechanical tally.

6. pH Equipment

Use electrometric pH meters, accurate to at least 0.1 pH units, for determining pH values of media.

7. Balances

Use balances providing a sensitivity of at least 0.1 g at a load of 150 g, with appropriate weights. Use an analytical balance having a sensitivity of 1 mg under a load of 10 g for weighing small quantities (less than 2 g) of materials. Single-pan rapid-weigh balances are most convenient.

8. Media Preparation Utensils

Use borosilicate glass or other suitable noncorrosive equipment such as stainless steel. Use glassware that is clean and free of residues, dried agar, or other foreign materials that may contaminate media.

9. Pipets and Graduated Cylinders

Use pipets of any convenient size, provided that they deliver the required volume accurately and quickly. The error of calibration for a given manufacturer's lot must not exceed 2.5%. Use pipets having graduations distinctly marked and with unbroken tips. Bacteriological transfer pipets or pipets conforming to APHA standards may be used. *Do not pipet by mouth; use a pipet aid.*

Use graduated cylinders meeting ASTM Standards (D-86 and D-216) and with accuracy limits established by NIST where appropriate.

10. Pipet Containers

Use boxes of aluminum or stainless steel, end measurement 5 to 7.5 cm, cylindrical or rectangular, and length about 40 cm. When these are not available, paper wrappings for individual pipets may be substituted. To avoid excessive charring during sterilization, use best-quality sulfate pulp (kraft) paper. *Do not use copper or copper alloy cans or boxes as pipet containers.*

11. Refrigerator

Use a refrigerator maintaining a temperature of 1 to 4.4°C to store samples, media, reagents, etc. Do not store volatile solvents, food, or beverages in a refrigerator with media. Frost-free refrigerators may cause excessive media dehydration on storage longer than 1 week.

12. Temperature-Monitoring Devices

Use glass or metal thermometers graduated to 0.5°C to monitor most incubators and refrigerators. Use thermometers graduated to 0.1°C for incubators operated above 40°C. Use continuous recording devices that are equally sensitive. Verify accuracy by comparison with a NIST-certified thermometer, or equivalent.

13. Dilution Bottles or Tubes

Use bottles or tubes of resistant glass, preferably borosilicate glass, closed with glass stoppers or screw caps equipped with liners that do not produce toxic or bacteriostatic compounds on sterilization. Do not use cotton plugs as closures. Mark graduation levels indelibly on side of dilution bottle or tube. Plastic bottles of nontoxic material and acceptable size may be substituted for glass provided that they can be sterilized properly.

14. Petri Dishes

For the plate count, use glass or plastic petri dishes about 100 × 15 mm. Use dishes the bottoms of which are free from bubbles and scratches and flat so that the medium will be of uniform thickness throughout the plate. For the membrane filter technique use loose-lid glass or plastic dishes, 60 × 15 mm, or tight-lid dishes, 50 × 12 mm. Sterilize petri dishes and store in metal cans (aluminum or stainless steel, but not copper), or wrap in paper—preferably best-quality sulfate pulp (kraft)—before sterilizing. Presterilized petri dishes are commercially available.

15. Membrane Filtration Equipment

Use filter funnel and membrane holder made of seamless stainless steel, glass, or autoclavable plastic that does not leak and is not subject to corrosion. Field laboratory kits are acceptable but standard laboratory filtration equipment and procedures are required.

16. Fermentation Tubes and Vials

Use fermentation tubes of any type, if their design permits conforming to medium and volume requirements for concentration of nutritive ingredients as described subsequently. Where tubes are used for a test of gas production, enclose a shell vial, inverted. Use tube and vial of such size that the vial will be filled completely with medium, at least partly submerged in the tube, and large enough to make gas bubbles easily visible.

17. Inoculating Equipment

Use wire loops made of 22- or 24-gauge nickel alloy* or platinum-iridium for flame sterilization. Use loops at least 3 mm in

diameter. Sterilize by dry heat or steam. Single-service hardwood or plastic applicators also may be used. Make these 0.2 to 0.3 cm in diameter and at least 2.5 cm longer than the fermentation tube; sterilize by dry heat and store in glass or other nontoxic containers.

18. Sample Bottles

For bacteriological samples, use sterilizable bottles of glass or plastic of any suitable size and shape. Use bottles capable of holding a sufficient volume of sample for all required tests and an adequate air space, permitting proper washing, and maintaining samples uncontaminated until examinations are completed. Ground-glass-stoppered bottles, preferably wide-mouthed and of resistant glass, are recommended. Plastic bottles of suitable size, wide-mouthed, and made of nontoxic materials such as polypropylene that can be sterilized repeatedly are satisfactory as sample containers. Presterilized plastic bags, with or without dechlorinating agent, are available commercially and may be used. Plastic containers eliminate the possibility of breakage during shipment and reduce shipping weight.

Metal or plastic screw-cap closures with liners may be used on sample bottles provided that no toxic compounds are produced on sterilization.

Before sterilization, cover tops and necks of sample bottles having glass closures with aluminium foil or heavy kraft paper.

19. Bibliography

COLLINS, W.D. & H.B. RIFFENBURG. 1923. Contamination of water samples with material dissolved from glass containers. *Ind. Eng. Chem.* 15:48.

CLARK, W.M. 1928. The Determination of Hydrogen Ion Concentration, 3rd ed. Williams & Wilkins, Baltimore, Md.

ARCHAMBAULT, J., J. CUROT & M.H. McCRADY. 1937. The need of uniformity of conditions for counting plates (with suggestions for a standard colony counter). *Amer. J. Pub. Health* 27:809.

BARKWORTH, H. & J.O. IRWIN. 1941. The effect of the shape of the container and size of gas tube in the presumptive coliform test. *J. Hyg.* 41:180.

RICHARDS, O.W. & P.C. HEIJN. 1945. An improved dark-field Quebec colony counter. *J. Milk Technol.* 8:253.

COHEN, B. 1957. The measurement of pH, titratable acidity, and oxidation-reduction potentials. *In* Manual of Microbiological Methods. Society of American Bacteriologists. McGraw-Hill Book Co., New York, N.Y.

MORTON, H.E. 1957. Stainless-steel closures for replacement of cotton plugs in culture tubes. *Science.* 126:1248.

McGUIRE, O.E. 1964. Wood applicators for the confirmatory test in the bacteriological analysis of water. *Pub. Health Rep.* 79:812.

BORDNER, R.H., J.A. WINTER & P.V. SCARPINO, eds. 1978. Microbiological Methods for Monitoring the Environment, Water and Wastes. EPA-600/8-78-017, Environmental Monitoring & Support Lab., U.S. Environmental Protection Agency, Cincinnati, Ohio.

AMERICAN PUBLIC HEALTH ASSOCIATION. 1993. Standard Methods for the Examination of Dairy Products, 16th ed. American Public Health Assoc., Washington, D.C.

---

* Chromel, nichrome, or equivalent.

9040 WASHING AND STERILIZATION*

Cleanse all glassware thoroughly with a suitable detergent and hot water, rinse with hot water to remove all traces of residual washing compound, and finally rinse with laboratory-pure water. If mechanical glassware washers are used, equip them with influent plumbing of stainless steel or other nontoxic material. Do not use copper piping to distribute water. Use stainless steel or other nontoxic material for the rinse water system.

Sterilize glassware, except when in metal containers, for not less than 60 min at a temperature of 170°C, unless it is known from recording thermometers that oven temperatures are uniform, under which exceptional condition use 160°C. Heat glassware in metal containers to 170°C for not less than 2 h.

Sterilize sample bottles not made of plastic as above or in an autoclave at 121°C for 15 min.

For plastic bottles loosen caps before autoclaving to prevent distortion.

---

* Approved by Standard Methods Committee, 1993.

9050 PREPARATION OF CULTURE MEDIA*

9050 A. General Procedures

1. Storage of Culture Media

Store dehydrated media (powders) in tightly closed bottles in the dark at less than 30°C in an atmosphere of low humidity. Do not use them if they discolor or become caked and lose the character of a free-flowing powder. Purchase dehydrated media in small quantities that will be used within 6 months after opening. Additionally, use stocks of dehydrated media containing selective agents such as sodium azide, bile salts or derivatives, antibiotics, sulfur-containing amino acids, etc., of relatively current lot number (within a year of purchase) so as to maintain optimum selectivity. See also Section 9020.

Prepare culture media in batches that will be used in less than 1 week. However, if the media are contained in screw-capped tubes they may be stored for up to 3 months. See Table 9020:IV for specific details. Store media out of direct sun and avoid contamination and excessive evaporation.

Liquid media in fermentation tubes, if stored at refrigeration or even moderately low temperatures, may dissolve sufficient air to produce, upon incubation at 35°C, a bubble of air in the tube. Incubate fermentation tubes that have been stored at a low temperature overnight before use and discard tubes containing air.

Fermentation tubes may be stored at approximately 25°C; but because evaporation may proceed rapidly under these conditions—resulting in marked changes in concentration of the ingredients—do not store at this temperature for more than 1 week. Discard tubes with an evaporation loss exceeding 1 mL.

2. Adjustment of Reaction

State reaction of culture media in terms of hydrogen ion concentration, expressed as pH.

The decrease in pH during sterilization will vary slightly with the individual sterilizer in use, and the initial reaction required to obtain the correct final reaction will have to be determined. The decrease in pH usually will be 0.1 to 0.2 but occasionally may

be as great as 0.3 in double-strength media. When buffering salts such as phosphates are present in the media, the decrease in pH value will be negligible.

Make tests to control adjustment to required pH with a pH meter. Measure pH of prepared medium as directed in Section 4500-H$^+$. Titrate a known volume of medium with a solution of NaOH to the desired pH. Calculate amount of NaOH solution that must be added to the bulk medium to reach this reaction. After adding and mixing thoroughly, check reaction and adjust if necessary. The required final pH is given in the directions for preparing each medium. If a specific pH is not prescribed, adjustment is unnecessary.

The pH of reconstituted dehydrated media seldom will require adjustment if made according to directions. Such factors as errors in weighing dehydrated medium or overheating reconstituted medium may produce an unacceptable final pH. Measure pH, especially of rehydrated selective media, regularly to insure quality control and media specifications.

3. Sterilization

After rehydrating a medium, dispense promptly to culture vessels and sterilize within 2 h. Do not store nonsterile media.

Sterilize all media, except sugar broths or broths with other specifications, in an autoclave at 121°C for 15 min after the temperature has reached 121°C. When the pressure reaches zero, remove medium from autoclave and cool quickly to avoid decomposition of sugars by prolonged exposure to heat. To permit uniform heating and rapid cooling, pack materials loosely and in small containers. Sterilize sugar broths at 121°C for 12 to 15 min. The maximum elapsed time for exposure of sugar broths to any heat (from time of closing loaded autoclave to unloading) is 45 min. Preferably use a double-walled autoclave to permit preheating before loading to reduce total needed heating time to within the 45-min limit. Presterilized media may be available commercially.

---

* Approved by Standard Methods Committee, 1993.

4. Bibliography

BUNKER, G.C. & H. SCHUBER. 1922. The reaction of culture media. *J. Amer. Water Works Assoc.* 9:63.

RICHARDSON, G.H., ed. 1985. Standard Methods for the Examination of Dairy Products, 15th ed. American Public Health Assoc., Washington, D.C.

BALOWS, A., W.J. HAUSLER, JR., K.L. HERRMANN, H.D. ISENBERG & H.J. SHADOMY, eds. 1991. Manual of Clinical Microbiology, 5th ed. American Soc. Microbiology, Washington, D.C.

9050 B. Water

1. Specifications

To prepare culture media and reagents, use only distilled or demineralized reagent-grade water that has been tested and found free from traces of dissolved metals and bactericidal or inhibitory compounds. Toxicity in distilled water may be derived from fluoridated water high in silica. Other sources of toxicity are silver, lead, and various unidentified organic complexes. Where condensate return is used as feed for a still, toxic amines or other boiler compounds may be present in distilled water. Residual chlorine or chloramines also may be found in distilled water prepared from chlorinated water supplies. If chlorine compounds are found in distilled water, neutralize them by adding an equivalent amount of sodium thiosulfate or sodium sulfite.

Distilled water also should be free of contaminating nutrients. Such contamination may be derived from flashover of organics during distillation, continued use of exhausted carbon filter beds, deionizing columns in need of recharging, solder flux residues in new piping, dust and chemical fumes, and storage of water in unclean bottles. Store distilled water out of direct sunlight to prevent growth of algae and turn supplies over as rapidly as possible. Aged distilled water may contain toxic volatile organic compounds absorbed from the atmosphere if stored for prolonged periods in unsealed containers. Good housekeeping practices usually will eliminate nutrient contamination.

See Section 9020.

2. Bibliography

STRAKA, R.P. & J.L. STOKES. 1957. Rapid destruction of bacteria in commonly used diluents and its elimination. *Appl. Microbiol.* 5:21.

GELDREICH, E.E. & H.F. CLARK. 1965. Distilled water suitability for microbiological applications. *J. Milk Food Technol.* 28:351.

MACLEOD, R.A., S.C. KUO & R. GELINAS. 1967. Metabolic injury to bacteria. II. Metabolic injury induced by distilled water or Cu^{++} in the plating diluent. *J. Bacteriol.* 93:961.

9050 C. Media Specifications

The need for uniformity dictates the use of dehydrated media. Never prepare media from basic ingredients when suitable dehydrated media are available. Follow manufacturer's directions for rehydration and sterilization. Commercially prepared media in liquid form (sterile ampule or other) also may be used if known to give equivalent results. See Section 9020 for quality-control specifications.

The terms used for protein source in most media, for example, peptone, tryptone, tryptose, were coined by the developers of the media and may reflect commercial products rather than clearly defined entities. It is not intended to preclude the use of alternative materials provided that they produce equivalent results.

NOTE—The term "percent solution" as used in these directions is to be understood to mean "grams of solute per 100 mL solution."

1. Dilution Water

a. Buffered water: To prepare stock phosphate buffer solution, dissolve 34.0 g potassium dihydrogen phosphate (KH_2PO_4), in 500 mL reagent-grade water, adjust to pH 7.2 ± 0.5 with $1N$ sodium hydroxide (NaOH), and dilute to 1 L with reagent-grade water.

Add 1.25 mL stock phosphate buffer solution and 5.0 mL magnesium chloride solution (81.1 g $MgCl_2 \cdot 6H_2O$/L reagent-grade water) to 1 L reagent-grade water. Dispense in amounts that will provide 99 ± 2.0 mL or 9 ± 0.2 mL after autoclaving for 15 min.

b. Peptone water: Prepare a 10% solution of peptone in distilled water. Dilute a measured volume to provide a final 0.1% solution. Final pH should be 6.8.

Dispense in amounts to provide 99 ± 2.0 mL or 9 ± 0.2 mL after autoclaving for 15 min.

Do not suspend bacteria in any dilution water for more than 30 min at room temperature because death or multiplication may occur.

2. Culture Media

Specifications for individual media are included in subsequent sections. Details are provided where use of a medium first is described.

9060 SAMPLES*

9060 A. Collection

1. Containers

Collect samples for microbiological examination in nonreactive borosilicate glass or plastic bottles that have been cleansed and rinsed carefully, given a final rinse with deionized or distilled water, and sterilized as directed in Sections 9030 and 9040. For some applications samples may be collected in presterilized plastic bags.

2. Dechlorination

Add a reducing agent to containers intended for the collection of water having residual chlorine or other halogen unless they contain broth for direct planting of sample. Sodium thiosulfate ($Na_2S_2O_3$) is a satisfactory dechlorinating agent that neutralizes any residual halogen and prevents continuation of bactericidal action during sample transit. The examination then will indicate more accurately the true microbial content of the water at the time of sampling.

For sampling chlorinated wastewater effluents add sufficient $Na_2S_2O_3$ to a clean sterile sample bottle to give a concentration of 100 mg/L in the sample. In a 120-mL bottle 0.1 mL of a 10% solution of $Na_2S_2O_3$ will neutralize a sample containing about 15 mg/L residual chlorine. For drinking water samples, the concentration of dechlorinating agent may be reduced: 0.1 mL of a 3% solution of $Na_2S_2O_3$ in a 120-mL bottle will neutralize up to 5 mg/L residual chlorine.

Cap bottle and sterilize by either dry or moist heat, as directed (Section 9040). Presterilized plastic bags or bottles containing $Na_2S_2O_3$ are available commercially.

Collect water samples high in metals, including copper or zinc (>1.0 mg/L), and wastewater samples high in heavy metals in sample bottles containing a chelating agent that will reduce metal toxicity. This is particularly significant when such samples are in transit for 4 h or more. Use 372 mg/L of the disodium salt of ethylenediaminetetraacetic acid (EDTA). Adjust EDTA solution to pH 6.5 before use. Add EDTA separately to sample bottle before bottle sterilization (0.3 mL 15% solution in a 120-mL bottle) or combine it with the $Na_2S_2O_3$ solution before addition.

3. Sampling Procedures

When the sample is collected, leave ample air space in the bottle (at least 2.5 cm) to facilitate mixing by shaking, before examination. Collect samples that are representative of the water being tested, flush or disinfect sample ports, and use aseptic techniques to avoid sample contamination.

Keep sampling bottle closed until it is to be filled. Remove stopper and cap as a unit; do not contaminate inner surface of stopper or cap and neck of bottle. Fill container without rinsing, replace stopper or cap immediately, and if used, secure hood around neck of bottle.

a. Potable water: If the water sample is to be taken from a distribution-system tap without attachments, select a tap that is supplying water from a service pipe directly connected with the main, and is not, for example, served from a cistern or storage tank. Open tap fully and let water run to waste for 2 or 3 min, or for a time sufficient to permit clearing the service line. Reduce water flow to permit filling bottle without splashing. If tap cleanliness is questionable, choose another tap. If a questionable tap is required for special sampling purposes, disinfect the faucet (inside and outside) by applying a solution of sodium hypochlorite (100 mg NaOCl/L) to faucet before sampling; let water run for additional 2 to 3 min after treatment. Do not sample from leaking taps that allow water to flow over the outside of the tap. In sampling from a mixing faucet remove faucet attachments such as screen or splash guard, run hot water for 2 min, then cold water for 2 to 3 min, and collect sample as indicated above.

If the sample is to be taken from a well fitted with a hand pump, pump water to waste for about 5 to 10 min or until water temperature has stabilized before collecting sample. If an outdoor sampling location must be used, avoid collecting samples from frost-proof hydrants. If there is no pumping machinery, collect a sample directly from the well by means of a sterilized bottle fitted with a weight at the base; take care to avoid contaminating samples by any surface scum. Other sterile sampling devices, such as a trip bailer, also may be used.

In drinking water evaluation, collect samples of finished water from distribution sites selected to assure systematic coverage during each month. Carefully choose distribution system sample locations to include dead-end sections to demonstrate bacteriological quality throughout the network and to ensure that localized contamination does not occur through cross-connections, breaks in the distribution lines, or reduction in positive pressure. Sample locations may be public sites (police and fire stations, government office buildings, schools, bus and train stations, airports, community parks), commercial establishments (restaurants, gas stations, office buildings, industrial plants), private residences (single residences, apartment buildings, and townhouse complexes), and special sampling stations built into the distribution network. Preferably avoid outdoor taps, fire hydrants, water treatment units, and backflow prevention devices. Establish sampling program in consultation with state and local health authorities.

b. Raw water supply: In collecting samples directly from a river, stream, lake, reservoir, spring, or shallow well, obtain samples representative of the water that is the source of supply to consumers. It is undesirable to take samples too near the bank or too far from the point of drawoff, or at a depth above or below the point of drawoff.

c. Surface waters: Stream studies may be short-term, high-intensity efforts. Select bacteriological sampling locations to include a baseline location upstream from the study area, industrial and municipal waste outfalls into the main stream study area, tributaries except those with a flow less than 10% of the main stream, intake points for municipal or industrial water facilities, down-

---

* Approved by Standard Methods Committee, 1997.

stream samples based on stream flow time, and downstream recreational areas. Dispersion of wastewaters into the receiving stream may necessitate preliminary cross-section studies to determine completeness of mixing. Where a tributary stream is involved, select the sampling point near the confluence with the main stream. Samples may be collected from a boat or from bridges near critical study points. Choose sampling frequency to be reflective of changing stream or water body conditions. For example, to evaluate waste discharges, sample every 4 to 6 h and advance the time over a 7- to 10-d period.

To monitor stream and lake water quality establish sampling locations at critical sites. Sampling frequency may be seasonal for recreational waters, daily for water supply intakes, hourly where waste treatment control is erratic and effluents are discharged into shellfish harvesting areas, or even continuous.

d. Bathing beaches: Sampling locations for recreational areas should reflect water quality within the entire recreational zone. Include sites from upstream peripheral areas and locations adjacent to drains or natural contours that would discharge stormwater collections or septic wastes. Collect samples in the swimming area from a uniform depth of approximately 1 m. Consider sediment sampling of the water-beach (soil) interface because of exposure of young children at the water's edge.

To obtain baseline data on marine and estuarine bathing water quality include sampling at low, high, and ebb tides.

Relate sampling frequency directly to the peak bathing period, which generally occurs in the afternoon. Preferably, collect daily samples during the recognized bathing season; as a minimum include Friday, Saturday, Sunday, and holidays. When limiting sampling to days of peak recreational use, preferably collect a sample in the morning and the afternoon. Correlate bacteriological data with turbidity levels or rainfall over the watershed to make rapid assessment of water quality changes.

e. Sediments and biosolids: The bacteriology of bottom sediments is important in water supply reservoirs, in lakes, rivers, and coastal waters used for recreational purposes, and in shellfish-growing waters. Sediments may provide a stable index of the general quality of the overlying water, particularly where there is great variability in its bacteriological quality.

Sampling frequency in reservoirs and lakes may be related more to seasonal changes in water temperatures and stormwater runoff. Bottom sediment changes in river and estuarine waters may be more erratic, being influenced by stormwater runoff, increased flow velocities, and sudden changes in the quality of effluent discharges.

Microbiological examination of biosolids from water and wastewater treatment processes is desirable to determine the impact of their disposal into receiving waters, ocean dumping, land application, or burial in landfill operations.

Collect and handle biosolids with less than 7% total solids using the procedures discussed for other water samples. Biosolids with more than 7% solids and exhibiting a ''plastic'' consistency or ''semisolid'' state typical of thickened sludges require a finite shear stress to cause them to flow. This resistance to flow results in heterogeneous distribution of biosolids in tanks and lagoons. Use cross-section sampling of accumulated biosolids to determine distribution of organisms within these impoundments. Establish a length-width grid across the top of the impoundment, and sample at intercepts. A thief sampler that samples only the solids layer may be useful. Alternatively use weighted bottle samplers that can be opened up at a desired depth to collect samples at specific locations.

Processed biosolids having no free liquids are best sampled when they are being transferred. Collect grab samples across the entire width of the conveyor and combine into a composite sample. If solids are stored in piles, classification occurs. Exteriors of uncovered piles are subject to various environmental stresses such as precipitation, wind, fugitive dusts, and fecal contamination from scavengers. Consequently, surface samples may not reflect the microbiological quality of the pile. Therefore, use cross-section sampling of these piles to determine the degree of heterogeneity within the pile. Establish a length-width grid across the top of the pile, and sample intercepts. Sample augers and corers may prove to be ineffective for sampling piles of variable composition. In such cases use hand shovels to remove overburden.

f. Nonpotable samples (manual sampling): Take samples from a river, stream, lake, or reservoir by holding the bottle near its base in the hand and plunging it, neck downward, below the surface. Turn bottle until neck points slightly upward and mouth is directed toward the current. If there is no current, as in the case of a reservoir, create a current artificially by pushing bottle forward horizontally in a direction away from the hand. When sampling from a boat, obtain samples from upstream side of boat. If it is not possible to collect samples from these situations in this way, attach a weight to base of bottle and lower it into the water. In any case, take care to avoid contact with bank or stream bed; otherwise, water fouling may occur.

g. Sampling apparatus: Special apparatus that permits mechanical removal of bottle stopper below water surface is required to collect samples from depths of a lake or reservoir. Various types of deep sampling devices are available. The most common is the ZoBell J-Z sampler,[1] which uses a sterile 350-mL bottle and a rubber stopper through which a piece of glass tubing has been passed. This tubing is connected to another piece of glass tubing by a rubber connecting hose. The unit is mounted on a metal frame containing a cable and a messenger. When the messenger is released, it strikes the glass tubing at a point that has been slightly weakened by a file mark. The glass tube is broken by the messenger and the tension set up by the rubber connecting hose is released and the tubing swings to the side. Water is sucked into the bottle as a consequence of the partial vacuum created by sealing the unit at time of autoclaving. Commercial adaptations of this sampler and others are available.

Bottom sediment sampling also requires special apparatus. The sampler described by Van Donsel and Geldreich[2] has been found effective for a variety of bottom materials for remote (deep water) or hand (shallow water) sampling. Construct this sampler preferably of stainless steel and fit with a sterile plastic bag. A nylon cord closes the bag after the sampler penetrates the sediment. A slide bar keeps the bag closed during descent and is opened, thereby opening the bag, during sediment sampling.

For sampling wastewaters or effluents the techniques described above generally are adequate; in addition see Section 1060.

4. Size of Sample

The volume of sample should be sufficient to carry out all tests required, preferably not less than 100 mL.

5. Identifying Data

Accompany samples by complete and accurate identifying and descriptive data. Do not accept for examination inadequately identified samples.

6. References

1. ZOBELL, C.E. 1941. Apparatus for collecting water samples from different depths for bacteriological analysis. *J. Mar. Res.* 4:173.
2. VAN DONSEL, D.J. & E.E. GELDREICH. 1971. Relationships of Salmonellae to fecal coliforms in bottom sediments. *Water Res.* 5:1079.

7. Bibliography

PUBLIC HEALTH LABORATORY SERVICE WATER SUB-COMMITTEE. 1953. The effect of sodium thiosulphate on the coliform and *Bacterium coli* counts of non-chlorinated water samples. *J. Hyg.* 51:572.

SHIPE, E.L. & A. FIELDS. 1956. Chelation as a method for maintaining the coliform index in water samples. *Pub. Health Rep.* 71:974.
HOATHER, R.C. 1961. The bacteriological examination of water. *J. Inst. Water Eng.* 61:426.
COLES, H.G. 1964. Ethylenediamine tetra-acetic acid and sodium thiosulphate as protective agents for coliform organisms in water samples stored for one day at atmospheric temperature. *Proc. Soc. Water Treat. Exam.* 13:350.
DAHLING, D.R. & B.A. WRIGHT. 1984. Processing and transport of environmental virus samples. *Appl. Environ. Microbiol.* 47:1272.
U.S. ENVIRONMENTAL PROTECTION AGENCY. 1992. Environmental Regulations and Technology Control of Pathogens and Vector Attraction in Sewage Sludge. EPA-625/R-92-013. Washington, D.C.

9060 B.   Preservation and Storage

1. Holding Time and Temperature

a. General: Start microbiological analysis of water samples as soon as possible after collection to avoid unpredictable changes in the microbial population. For most accurate results, ice samples during transport to the laboratory, if they cannot be processed within 1 h after collection. If the results may be used in legal action, employ special means (rapid transport, express mail, courier service, etc.) to deliver the samples to the laboratory within the specified time limits and maintain chain of custody. Follow the guidelines and requirements given below for specific water types.

b. Drinking water for compliance purposes: Preferably hold samples at <10°C during transit to the laboratory. Analyze samples on day of receipt whenever possible and refrigerate overnight if arrival is too late for processing on same day. Do not exceed 30 h holding time from collection to analysis for coliform bacteria. Do not exceed 8 h holding time for heterotrophic plate counts.

c. Nonpotable water for compliance purposes: Hold source water, stream pollution, recreational water, and wastewater samples below 10°C during a maximum transport time of 6 h. Refrigerate these samples upon receipt in the laboratory and process within 2 h. When transport conditions necessitate delays in delivery of samples longer than 6 h, consider using either field laboratory facilities located at the site of collection or delayed incubation procedures.

d. Other water types for noncompliance purposes: Hold samples below 10°C during transport and until time of analysis. Do not exceed 24 h holding time.

2. Bibliography

CALDWELL, E.L. & L.W. PARR. 1933. Present status of handling water samples—Comparison of bacteriological analyses under varying temperatures and holding conditions, with special reference to the direct method. *Amer. J. Pub. Health* 23:467.
COX, K.E. & F.B. CLAIBORNE. 1949. Effect of age and storage temperature on bacteriological water samples. *J. Amer. Water Works Assoc.* 41: 948.
PUBLIC HEALTH LABORATORY SERVICE WATER SUB-COMMITTEE. 1952. The effect of storage on the coliform and *Bacterium coli* counts of water samples. Overnight storage at room and refrigerator temperatures. *J. Hyg.* 50:107.
PUBLIC HEALTH LABORATORY SERVICE WATER SUB-COMMITTEE. 1953. The effect of storage on the coliform and *Bacterium coli* counts of water samples. Storage for six hours at room and refrigerator temperatures. *J. Hyg.* 51:559.
MCCARTHY, J.A. 1957. Storage of water sample for bacteriological examinations. *Amer. J. Pub. Health* 47:971.
LONSANE, B.K., N.M. PARHAD & N.U. RAO. 1967. Effect of storage temperature and time on the coliform in water samples. *Water Res.* 1: 309.
LUCKING, H.E. 1967. Death rate of coliform bacteria in stored Montana water samples. *J. Environ. Health* 29:576.
MCDANIELS, A.E. & R.H. BORDNER. 1983. Effect of holding time and temperature on coliform numbers in drinking water. *J. Amer. Water Works Assoc.* 75:458.
MCDANIELS, A.E. et al. 1985. Holding effects on coliform enumeration in drinking water samples. *Appl. Environ. Microbiol.* 50:755.

9211 RAPID DETECTION METHODS*

9211 A. Introduction

There is a generally recognized need for methods that permit rapid estimation of the bacteriological quality of water. Applications of rapid methods may range from analysis of wastewater to potable water quality assessment. In the latter case, during emergencies involving water treatment plant failure, line breaks in a distribution network, or other disruptions to water supply caused

___
* Approved by Standard Methods Committee, 1994.

by disasters, there is urgent need for rapid assessment of the sanitary quality of water.

Ideally, rapid procedures would be reliable and have sensitivity levels equal to those of the standard tests routinely used. However, sensitivity of a rapid test may be compromised because the bacterial limit sought may be below the minimum bacterial concentration essential to rapid detection. Rapid tests fall into two categories, those involving modified conventional procedures and those requiring special instrumentation and materials.

9211 B. Seven-Hour Fecal Coliform Test (SPECIALIZED)

This method[1,2] is similar to the fecal coliform membrane filter procedure (see Section 9222D) but uses a different medium and incubation temperature to yield results in 7 h that generally are comparable to those obtained by the standard fecal coliform method.

1. Medium

M-7 h FC agar: This medium may not be available in dehydrated form and may require preparation from the basic ingredients.

Proteose peptone No. 3 or polypeptone	5.0	g
Yeast extract	3.0	g
Lactose	10.0	g
d-Mannitol	5.0	g
Sodium chloride, NaCl	7.5	g
Sodium lauryl sulfate	0.2	g
Sodium desoxycholate	0.1	g
Bromcresol purple	0.35	g
Phenol red	0.3	g
Agar	15.0	g
Reagent-grade water	1	L

Heat in boiling water bath. After ingredients are dissolved heat additional 5 min. Cool to 55 to 60°C and adjust pH to 7.3 ± 0.1 with 0.1N NaOH (0.35 mL/L usually required). Cool to about 45°C and dispense in 4- to 5-mL quantities to petri plates with tight-fitting covers. Store at 2 to 10°C. Discard after 30 d.

2. Procedure

Filter an appropriate sample volume through a membrane filter, place filter on the surface of a plate containing M-7 h FC agar medium, and incubate at 41.5°C for 7 h. Fecal coliform colonies are yellow (indicative of lactose fermentation).

3. References

1. VAN DONSEL, D.J., R.M. TWEDT & E.E. GELDREICH. 1969. Optimum temperature for quantitation of fecal coliforms in seven hours on the membrane filter. *Bacteriol. Proc. Abs.* No. G46, p. 25.
2. REASONER, D.J., J.C. BLANNON & E.E. GELDREICH. 1979. Rapid seven hour fecal coliform test. *Appl. Environ. Microbiol.* 38:229.

9211 C. Special Techniques (SPECIALIZED)

Special rapid techniques are summarized in Table 9211:I. Most are not sensitive enough for potable water quality measurement or are not specific. They may be useful in monitoring wastewater effluents and natural waters but require reagents not generally available, are tedious, or require special handling or incubation schemes incompatible with most water laboratory schedules. Except for the colorimetric test, none are suitable for routine use but they may be used as research tools. The user should refer to the literature citations for the technique listed in the table for proce-

dural details, conditions for use, and method limitations. Only the adenosine triphosphate (ATP) procedure (the firefly bioluminescence system), the colorimetric test to estimate total microbial density, and a radiometric fecal coliform procedure that uses a ^{14}C-labeled substrate can be recommended.

Correlate initial concentration of bacteria with ATP concentration by extracting ATP from serial dilutions of a bacterial suspension, or for the ^{14}C radiometric method, standardize by determining the $^{14}CO_2$ released by known concentrations of fecal

TABLE 9211:I. SPECIAL RAPID TECHNIQUES

Microbial Group	Rapid Method	Test Time h	Sensitivity cells/mL	Reference
Nonspecific microflora	Bioluminescence	1	100 000	1–3
	Chemiluminescence	1	500 000	3–5
	Impedance	3–12	100 000	6–9
	Colorimetric	0.02	10 000	10
	Epifluorescence/fluorometric	<1–several	—	11–13
Fecal coliforms	Radiometric	4–5	2–20	14
	Glutamate decarboxylase	10–13	0.01–500 000	15–17
	Electrochemical	1–7	1 000 000	18–20
	Impedance	6–12	200–100 000	6–9
	Gas chromatographic assay	9–12	>50	21
	Colorimetric	8–20	5–130 000	22
	Potentiometric	3.5–15	0.1->10 000 000	23
Gram-negative bacteria	Limulus assay	2	500–3000	24–27
	Fluorescent antibody	2–3	—	28–30

coliform organisms in natural samples, not pure cultures. In using any rapid procedure, determine the initial bacterial density by using an appropriate procedure such as heterotrophic plate count (Section 9215) or total (9221) or fecal (9222) coliforms, and correlate with results from the special rapid technique.

1. Bioluminescence Test (Total Viable Microbial Measurement)

The firefly luciferase test for ATP in living cells is based on the reaction between the luciferase enzyme, luciferin (enzyme substrate), magnesium ions, and ATP. Light is emitted during the reaction and can be measured quantitatively and correlated with the quantity of ATP extracted from known numbers of bacteria. When all reactants except ATP are in excess, ATP is the limiting factor. Addition of ATP drives the reactions, producing a pulse of light that is proportional to the ATP concentration.

The assay is completed in less than 1 h.[1–3] For monitoring microbial populations in water, the ATP assay is limited primarily by the need to concentrate bacteria from the sample to achieve the minimum ATP sensitivity level, which is 10^5 cells/mL. When combined with membrane filtration of a 1-L sample, ATP assay can provide the sensitivity level needed.

2. Radiometric Detection (Fecal Coliforms)

In this test, $^{14}CO_2$ is released from a ^{14}C-labeled substrate.[14] The technique permits presumptive detection of as few as 2 to 20 fecal coliform bacteria in 4.5 h. The test uses M-FC broth, uniformly labeled ^{14}C-mannitol, and two-temperature incubation; 2 h at 35°C followed by 2.5 h at 44.5°C for fecal coliform specificity. Add labeled substrate at start of 44.5°C incubation. Use membrane filtration to concentrate organisms from sample and place membrane filter in M-FC broth in a sealable container. The $^{14}CO_2$ released is trapped by exposure to $Ba(OH)_2$-saturated filter paper disk. ^{14}C activity is assayed by liquid scintillation spectrometry. Except for the use of the ^{14}C-mannitol substrate and liquid scintillation spectrometry to count the activity of the $^{14}CO_2$

released by the fecal coliforms, this procedure is similar to those given in Section 9222.

3. References

1. CHAPPELLE, E.W. & G.L. PICCIOLO. 1975. Laboratory Procedures Manual for the Firefly Luciferase Assay for Adenosine Triphosphate (ATP). NASA GSFC Doc. X-726-75-1, National Aeronautics & Space Admin., Washington, D.C.
2. PICCIOLO, G.L., E.W. CHAPPELLE, J.W. DEMING, R.R. THOMAS, D.A. NIBLE & H. OKREND. 1981. Firefly luciferase ATP assay development for monitoring bacterial concentration in water supplies. EPA-600/S2:81-014, U.S. Environmental Protection Agency, Cincinnati, Ohio; NTIS No. PB 88-103809/AS, National Technical Information Serv., Springfield, Va.
3. NELSON, W.H., ed. 1985. Instrumental Methods for Rapid Microbiological Analysis. VCH Publishers, Inc., Deerfield Beach, Fla.
4. SEITZ, W.R. & M.P. NEARY. 1974. Chemiluminescence and bioluminescence. Anal. Chem. 46:188A.
5. OLENIAZ, W.S., M.A. PISANO, M.H. ROSENFELD & R.L. ELGART. 1968. Chemiluminescent method for detecting microorganisms in water. Environ. Sci. Technol. 2:1030.
6. WHEELER, T.G. & M.C. GOLDSCHMIDT. 1975. Determination of bacterial cell concentrations by electrical measurements. J. Clin. Microbiol. 1:25.
7. SILVERMAN, M.P. & E.F. MUNOZ. 1979. Automated electrical impedance technique for rapid enumeration of fecal coliforms in effluents from sewage treatment plants. Appl. Environ. Microbiol. 37:521.
8. MUNOZ, E.F. & M.P. SILVERMAN. 1979. Rapid, single-step most-probable-number method for enumerating fecal coliforms in effluents from sewage treatment plants. Appl. Environ. Microbiol. 37:527.
9. FIRSTENBERG-EDEN, R. & G. EDEN. 1984. Impedance Microbiology. John Wiley & Sons, Inc., New York, N.Y.
10. WALLIS, C. & J.L. MELNICK. 1985. An instrument for the immediate quantification of bacteria in potable waters. Appl. Environ. Microbiol. 49:1251.
11. BITTON, G., R.J. DUTTON & J.A. FORAN. 1984. A new rapid technique for counting microorganisms directly on membrane filters. Stain Technol. 58:343.
12. SIERACKI, M.E., P.W. JOHNSON & J.M. SIEBURTH. 1985. Detection, enumeration, and sizing of planktonic bacteria by image-analyzed epifluoresence microscopy. Appl. Environ. Microbiol. 49:799.

13. McCoy, W.F. & B.H. Olson. 1985. Fluorometric determination of the DNA concentration in municipal drinking water. *Appl. Environ. Microbiol.* 49:811.

14. Reasoner, D.J. & E.E. Geldreich. 1978. Rapid detection of waterborne fecal coliforms by $^{14}CO_2$ release. *In* A.N. Sharpe & D.S. Clark, eds. Mechanizing Microbiology. Charles C. Thomas, Publisher, Springfield, Ill.

15. Moran, J.W. & L.D. Witter. 1976. An automated rapid test for *Escherichia coli* in milk. *J. Food Sci.* 41:165.

16. Moran, J.W. & L.D. Witter. 1976. An automated rapid method for measuring fecal pollution. *Water Sewage Works* 123:66.

17. Trinel, P.A., N. Hanoune & H. LeClerc. 1980. Automation of water bacteriological analysis: running test of an experimental prototype. *Appl. Environ. Microbiol.* 39:976.

18. Wilkins, J.R., G.E. Stoner & E.H. Boykin. 1974. Microbial detection method based on sensing molecular hydrogen. *Appl. Microbiol.* 27:947.

19. Wilkins, J.R. & E.H. Boykin. 1976. Analytical notes—electrochemical method for early detection of monitoring of coliforms. *J. Amer. Water Works Assoc.* 68:257.

20. Grana, D.C. & J.R. Wilkins. 1979. Description and field test results of an *in situ* coliform monitoring system. NASA Tech. Paper 1334, National Aeronautics & Space Admin., Washington, D.C.

21. Newman, J.S. & R.T. O'Brien. 1975. Gas chromatographic presumptive test for coliform bacteria in water. *Appl. Environ. Microbiol.* 30:584.

22. Warren, L.S., R.E. Benoit & J.A. Jessee. 1978. Rapid enumeration of faecal coliforms in water by a colorimetric β-galactosidase assay. *Appl. Environ. Microbiol.* 35:136.

23. Jouenne, T., G.-A. Junter & G. Carriere. 1985. Selective detection and enumeration of fecal coliforms in water by potentiometric measurement of lipoic acid reduction. *Appl. Environ. Microbiol.* 50:1208.

24. Tencate, J.W., H.R. Buler, A. Sturk & J. Levin. 1985. Bacterial Endotoxins. Structure, Biomedical Significance, and Detection with the *Limulus* Amebocyte Lysate Test. Alan R. Liss, Inc., New York, N.Y.

25. Jorgensen, J.H., J.C. Lee, G.A. Alexander & H.W. Wolf. 1979. Comparison of *Limulus* assay, standard plate count, and total coliform count for microbiological assessment of renovated wastewater. *Appl. Environ. Microbiol.* 37:928.

26. Jorgensen, J.H. & G.A. Alexander. 1981. Automation of the *Limulus* amebocyte lysate test by using the Abbott MS-2 microbiology system. *Appl. Environ. Microbiol.* 41:1316.

27. Tsugi, K., P.A. Martin & D.M. Bussey. 1984. Automation of chromogenic substrate *Limulus* amebocyte lysate assay method for endotoxin by robotic system. *Appl. Environ. Microbiol.* 48:550.

28. Abshire, R.L. 1976. Detection of enteropathogenic *Escherichia coli* strains in wastewater by fluorescent antibody. *Can. J. Microbiol.* 22:365.

29. Abshire, R.L. & R.K. Guthrie. 1973. Fluorescent antibody techniques as a method for the detection of fecal pollution. *Can. J. Microbiol.* 19:201.

30. Thomason, B.M. 1981. Current status of immunofluorescent methodology. *J. Food Protect.* 44:381.

9211 D.   Coliphage Detection

Coliphages are bacteriophages that infect and replicate in coliform bacteria and appear to be present wherever total and fecal coliforms are found. Correlations between coliphages and coliform bacteria in fresh water generally show that coliphages may be used to indicate the sanitary quality of water.[1-5] Because coliphages are more resistant to chlorine disinfection than total or fecal coliforms, they may be a better indicator of disinfection efficiency than coliform bacteria.[4] The quantitative relationship between coliphages and coliform bacteria in disinfected waters is different from that in natural fresh waters because of differences in their survival rates.

1. Materials and Culture Media

a. *Host culture:* *Escherichia coli* C, ATCC No. 13706.

b. *Media:*

1) *Tryptic(ase) soy agar (TSA),* to maintain *E. coli* C host stock cultures:

Tryptone (pancreatic digest of casein) or equivalent	15.0	g
Soytone (soybean peptone) or equivalent	5.0	g
Sodium chloride, NaCl	5.0	g
Agar	15.0	g
Reagent-grade water	1	L

pH should be 7.3 ± 0.1 at 25°C; if necessary, adjust pH with 0.1 or 1.0*N* NaOH or HCl. Heat to boiling to dissolve, then autoclave for 15 min at 121°C. For agar slants, dispense 5 to 8 mL in 16- × 125-mm screw-capped tubes before sterilizing; for plates, dispense 20 to 25 mL per petri dish after autoclaving and cooling to about 45°C.

2) *Tryptic(ase) soy broth (TSB):*

Tryptone (pancreatic digest of casein), or equivalent	17.0	g
Soytone (soybean peptone), or equivalent	3.0	g
Dextrose	2.5	g
Sodium chloride, NaCl	5.0	g
Dipotassium hydrogen phosphate, K_2HPO_4	2.5	g
Reagent-grade water	1	L

pH should be 7.3 ± 0.1 at 25°C; adjust with 0.1 or 1.0*N* NaOH or HCl, if necessary. Warm and agitate to dissolve completely. Dispense in appropriate volumes as needed; sterilize in autoclave for 15 min at 121°C.

3) *Modified tryptic(ase) soy agar (MTSA):* To the ingredients of TSB, add ammonium nitrate, NH_4NO_3, 1.60 g; strontium nitrate, $Sr(NO_3)_2$, 0.21 g; and agar, 15 g. pH should be 7.3 ± 0.1 at 25°C; if necessary, adjust pH with 0.1 or 1.0*N* NaOH or HCl. Heat to boiling to dissolve, dispense 5.5 mL in 16- × 25-mm screw-capped tubes, and autoclave for 15 min at 121°C.

4) *Glycerine:* Add 10% (w/v) to tryptic(ase) soy broth before autoclave sterilization.

5) *2,3,5-triphenyl tetrazolium chloride (TPTZ),* 1% (w/v) in ethanol. Add to MTSA tempered at 45 to 46°C to enhance plaque visibility. Prepare fresh weekly.

2. Procedure

a. *Frozen host preparation:* Inoculate *E. coli* C from a stock agar slant (on TSA) into a tube(s) containing 10 mL TSB and

10% glycerine (w/v) and incubate overnight at 35°C. Then inoculate each tube into a flask containing 25 mL TSB plus 10% glycerine and incubate at 35 ± 0.5°C until an optical density of 0.5 at 520 nm is obtained (equivalent to about 1×10^9 *E. coli* C cells/mL). Measure optical density using a spectrometer. Zero spectrometer with sterile TSB plus 10% glycerine.

Aseptically dispense 4.5-mL portions of cell suspension in sterile plastic test tubes, cap, chill to 9°C, and freeze at −20°C. Store for no longer than 6 weeks in non-frost-free freezer to reduce loss of frozen host culture viability.

b. Assay procedure: The procedure is directly applicable to samples containing more than 5 coliphage/100 mL; if sample contains more than 1000 coliphage/100 mL, dilute sample 1:5 or 1:10 with sterile distilled water before proceeding.

Thaw tube(s) of frozen host *E. coli* C in 44.5°C water bath. Use one tube of host culture per sample. Add 1.0 mL of host *E. coli* C culture, 5 mL sample or dilution, and 0.08 mL TPTZ[6] to each of four tubes of MTSA (melted and held at about 45°C).

Mix thoroughly and pour into separate 100- × 15-mm labeled petri dishes, cover, and let agar gel. Incubate inverted plates at 35°C. Count plaques after incubating for 4 to 6 h.

3. Interpreting and Reporting Results

Bacteriophage infect and multiply in sensitive bacteria. This results in lysis of the bacterial cells and a release of phage particles to infect adjacent cells. As the infected coliform bacteria are lysed, visible clear areas known as plaques develop in the lawn of confluent bacterial growth. Count plaques on each plate and record. Obtain the number of plaques/100 mL of sample by summing the plaques on the four plates and multiplying by 5. If a diluted sample has been used, additionally multiply by the reciprocal of the dilution factor.

Based on coliphage counts, estimate total and fecal coliform numbers as shown below.[4] Independently verify equations for specific types of samples and locations.

Total coliforms:

$$\log y = 0.627 (\log x) + 1.864$$

where:

> y = total coliforms/100 mL and
> x = coliphages/100 mL.

Fecal coliforms:

$$\log y = 0.805 (\log x) + 0.895$$

where:

> y = fecal coliforms/100 mL and
> x = coliphages/100 mL.

4. References

1. WENTZEL, R.S., P.E. O'NEILL & J.F. KITCHENS. 1982. Evaluation of coliphage detection as a rapid indicator of water quality. *Appl. Environ. Microbiol.* 43:430.
2. ISBISTER, J.D. & J.L. ALM. 1982. Rapid coliphage procedure for water treatment processes. *In* Proc. Amer. Water Works Assoc. Water Quality Technol. Conf., Seattle, Wash., Dec. 6–9, 1981.
3. ISBISTER, J.D., J.A. SIMMONS, W.M. SCOTT & J.F. KITCHENS. 1983. A simplified method for coliphage detection in natural waters. *Acta Microbiol. Polonica* 32:197.
4. KOTT, Y., N. ROZE, S. SPERBER & N. BETZER. 1974. Bacteriophages as viral pollution indicators. *Water Res.* 8:165.
5. KENNEDY, J.D., JR., G. BITTON & J.L. OBLINGER. 1985. Comparison of selective media for assay of coliphages in sewage effluent and lake water. *Appl. Environ. Microbiol.* 49:33.
6. HURST, C.J., J.C. BLANNON, R.L. HARDAWAY & W.C. JACKSON. 1994. Differential effect of tetrazolium dyes upon bacteriophage plaque assay titers. *Appl. Environ. Microbiol.* 60:3462.

9212 STRESSED ORGANISMS*

9212 A. Introduction

1. General Discussion

Indicator bacteria, including total coliforms, fecal coliforms, and fecal streptococci, may become stressed or injured in waters and wastewaters. These injured bacteria are incapable of growth and colony formation under standard conditions because of structural or metabolic damage. As a result, a substantial portion of the indicator bacteria present, i.e., 10 to greater than 90%, may not be detected.[1,2] These false negative bacteriological findings could result in an inaccurate definition of water quality and lead to the acceptance of a potentially hazardous condition resulting

from contamination by resistant pathogens[3] or the penetration of undetected indicator bacteria through treatment barriers.[4]

Stressed organisms are present under ordinary circumstances in treated drinking water and wastewater effluents, saline waters, polluted natural waters, and relatively clean surface waters. High numbers of injured indicator bacteria may be associated with partial or inadequate disinfection and the presence of metal ions or other toxic substances. These and other factors, including extremes of temperature and pH and solar radiation, may lead collectively to significant underestimations of the number of viable indicator bacteria.

Publications support the health significance of injured coliform bacteria.[2,5−7] These reports show that enteropathogenic bacteria are less susceptible than coliforms to injury under conditions sim-

---

* Approved by Standard Methods Committee, 1994.

ilar to those in treated drinking water and wastewater, that injured pathogens retain the potential for virulence, and that they recover after being ingested. Hence, methods allowing for the enumeration of injured coliform bacteria yield more sensitive determinations of potential health risks. This conclusion is further supported by the observation that viruses and waterborne pathogens that form cysts also are more resistant than indicator bacteria to environmental stressors.

2. Sample Handling and Collection

Certain laboratory manipulations following sample collection also may produce injury or act as a secondary stress to the organisms.[2,8] These include excessive sample storage time, prolonged holding time (more than 30 min) of diluted samples before inoculation into growth media and of inoculated samples before incubation at the proper temperature, incorrect media formulations, incomplete mixing of sample with concentrated medium, and exposure to untempered liquefied agar media. Excessive numbers of nonindicator bacteria also interfere with detection of indicators by causing injury.[9]

3. References

1. McFeters, G.A., J.S. Kippin & M.W. LeChevallier. 1986. Injured coliforms in drinking water. *Appl. Environ. Microbiol.* 51:1.
2. McFeters, G.A. 1990. Enumeration, occurrence, and significance of injured indicator bacteria in drinking water. *In* G.A. McFeters, ed. Drinking Water Microbiology: Progress and Recent Developments, p. 478. Springer-Verlag, New York.
3. LeChevallier, M.W. & G.A. McFeters. 1985. Enumerating injured coliforms in drinking water. *J. Amer. Water Works Assoc.* 77:81.
4. Bucklin, K.E., G.A. McFeters & A. Amirtharajah. 1991. Penetration of coliforms through municipal drinking water filters. *Water Res.* 25:1013.
5. LeChevallier, M.W., A. Singh, D.A. Schiemann & G.A. McFeters. 1985. Changes in virulence of waterborne enteropathogens with chlorine injury. *Appl. Environ. Microbiol.* 50:412.
6. Singh, A. & G.A. McFeters. 1986. Repair, growth and production of heat-stable enterotoxin by *E. coli* following copper injury. *Appl. Environ. Microbiol.* 51:738.
7. Singh, A., R. Yeager & G.A. McFeters. 1986. Assessment of in vivo revival, growth, and pathogenicity of *Escherichia coli* strains after copper- and chlorine-induced injury. *Appl. Environ. Microbiol.* 52:832.
8. McFeters, G.A., S.C. Cameron & M.W. LeChevallier. 1982. Influence of diluents, media and membrane filters on the detection of injured waterborne coliform bacteria. *Appl. Environ. Microbiol.* 43:97.
9. LeChevallier, M.W. & G.A. McFeters. 1985. Interactions between heterotrophic plate count bacteria and coliform organisms. *Appl. Environ. Microbiol.* 49:1338.

4. Bibliography

Clark, H.F., E.E. Geldreich, H.L. Jeter & P.W. Kabler. 1951. The membrane filter in sanitary bacteriology. *Pub. Health Rep.* 66:951.
McKee, J.E., R.T. McLaughlin & P. Lesgourgues. 1958. Application of molecular filter techniques to the bacterial assay of sewage. III. Effects of physical and chemical disinfection. *Sewage Ind. Wastes* 30:245.
Rose, R.E. & W. Litsky. 1965. Enrichment procedure for use with the membrane filter for the isolation and enumeration of fecal streptococci from water. *Appl. Microbiol.* 13:106.
Maxcy, R.B. 1970. Non-lethal injury and limitations of recovery of coliform organisms on selective media. *J. Milk Food Technol.* 33:445.
Lin, S.D. 1973. Evaluation of coliform tests for chlorinated secondary effluents. *J. Water Pollut. Control Fed.* 45:498.
Braswell, J.R. & A.W. Hoadley. 1974. Recovery of *Escherichia coli* from chlorinated secondary sewage. *Appl. Microbiol.* 28:328.
Stevens, A.P., R.J. Grasso & J.E. Delaney. 1974. Measurements of fecal coliform in estuarine water. *In* D.D. Wilt, ed., Proceedings of the 8th National Shellfish Sanitation Workshop, U.S. Dep. Health, Education, & Welfare, Washington, D.C.
Bissonnette, G.K., J.J. Jezeski, G.A. McFeters & D.S. Stuart. 1975. Influence of environmental stress on enumeration of indicator bacteria from natural waters. *Appl. Microbiol.* 29:186.
Bissonnette, G.K., J.J. Jezeski, G.A. McFeters & D.S. Stuart. 1977. Evaluation of recovery methods to detect coliforms in water. *Appl. Environ. Microbiol.* 33:590.

9212 B.   Recovery Enhancement

This section describes some general procedures and considerations regarding recovery of stressed indicator organisms.

For chlorinated samples, insure that sufficient dechlorinating agent is present in the sample bottle (see Section 9060A.2).[1] Collect water samples with elevated concentrations of heavy-metal ions in a sample bottle containing a chelating agent[2] (see Section 9060A.2) and minimize sample storage time (see Section 9060B). Use buffered peptone dilution water rather than buffered water (see Section 9050C.1) when preparing dilutions of samples containing heavy-metal ions. After making dilutions, inoculate test media within 30 min.

Resuscitation of stressed or injured organisms is enhanced by inoculating samples and initially culturing organisms in an enriched, noninhibitory medium at a moderate temperature.

Although no simple test is available to establish the presence of injured bacteria in a given sample, bacteria in water known to contain stressors such as disinfectants or heavy metals frequently will be injured.[1,3] When multiple-tube fermentation test results consistently are higher than those obtained from parallel membrane filter tests, or there is other indication of suboptimal recovery, consider injury probable and use one or more of the following procedures.

1. Recovery of Injured Total Coliform Bacteria Using Membrane Filtration

a. m-T 7 agar: Use m-T 7 agar[4] in the procedure described for the membrane filter test (see Section 9222B).

Proteose peptone No. 3	5.0 g
Yeast extract	3.0 g
Lactose	20.0 g
Tergitol 7	0.4 mL
Polyoxyethylene ether W1	5.0 g

Bromthymol blue	0.1	g
Bromcresol purple	0.1	g
Agar	15.0	g
Reagent-grade water	1	L

Adjust to pH 7.4 with 0.1*N* NaOH after sterilization at 121°C for 15 min. Aseptically add 1.0 μg penicillin G/mL when medium has cooled to about 45°C.

After filtering sample place filter on m-T 7 agar and incubate at 35°C for 22 to 24 h. Coliform colonies are yellow. Verify not less than 10% of coliform colonies by the procedure in Section 9222B.5*f*. With some drinking water samples containing many non-coliform bacteria, confluent growth occurs. To obtain reliable results, carefully distinguish target yellow colonies from background growth.

b. Addition of sodium sulfite: The addition of sodium sulfite to some media (0.05 to 0.1%) can improve the detection of coliform bacteria following exposure to chloramine but not chlorine.[5] Such modified medium is applicable to clean-water systems using chloramination or to chlorinated discharges such as wastewater effluent containing high levels of organic compounds.

2. Recovery of Injured Fecal Coliform Bacteria Using Membrane Filtration

a. Enrichment-temperature acclimation: Use two-layer agar (M-FC agar with a nonselective overlay medium that does not contain glucose, i.e., tryptic soy agar) with a 2-h incubation at 35°C followed by 22 h at 44.5°C.[6] Prepare the M-FC agar plate in advance but do not add the overlay agar more than 1 h before use.

Alternatively, use a pre-enrichment in phenol red lactose broth incubated at 35°C for 4 h followed by M-FC agar at 44.5°C for 22 h.[7]

As a third option, prepare enrichment two-layer medium containing specific additives and incubate for 1.5 h at room temperature (22 to 26°C) followed by 35°C for 4.5 h and 44.5°C for 18 h.[8]

b. Temperature acclimation:[9] Modify elevated temperature procedure by preincubation of M-FC cultures for 5 h at 35°C, followed by 18 ± 1 h at 44.5°C. Use a commercially available temperature-programmed incubator to make the change from 35 to 44.5°C after the 5 h preincubation period to eliminate inconvenience and provide a practical method of analysis.

c. Deletion of suppressive agent:[10] Eliminate rosolic acid from M-FC medium and incubate cultures at 44.5°C ± 0.2°C for 24 h. Fecal coliform colonies are intense blue on the modified medium and are distinguished from the cream, gray, and pale-green colonies typically produced by nonfecal coliforms.

d. Alternative medium-temperature acclimation: Use m-T 7 medium with an 8 h incubation at 37°C followed by 12 h at 44.5°C.[11]

e. Verification of stressed fecal coliform bacteria: Modifications of media and procedures may decrease selectivity and dif-

ferentiation of fecal coliform colonies. Therefore, if any procedural modifications are used, verify not less than 10% of the blue colonies from a variety of samples. Use lauryl tryptose broth (Section 9221B) (35°C for 48 h) with transfer of gas-producing cultures to EC broth (Section 9221E) (44.5°C for 24 h). Gas production at 44.5°C confirms the presence of fecal coliforms.

3. Recovery of Stressed Fecal Streptococci Using Membrane Filtration

Using bile broth medium yields fecal streptococcus recoveries comparable with multiple-tube fermentation tests.[12] Preincubate membrane filters on an enrichment medium for 2 h at 35°C and follow by plating on m-Enterococcus agar (Section 9230) for 48 ± 2 h at 35°C.

Verification of stressed fecal streptococci—Verify not less than 10% of the colonies from a variety of samples using the confirmed test procedure given in Section 9230B.3.

4. References

1. McFeters, G.A. & A.K. Camper. 1983. Enumeration of coliform bacteria exposed to chlorine. *In* A.I. Laskin, ed. Advances in Applied Microbiology, Vol. 29, p. 177.
2. Domek, M.J., M.W. LeChevallier, S.C. Cameron & G.A. McFeters. 1984. Evidence for the role of metals in the injury process of coliforms in drinking water. *Appl. Environ. Microbiol.* 48:289.
3. LeChevallier, M.W. & G.A. McFeters. 1985. Interactions between heterotrophic plate count bacteria and coliform organisms. *Appl. Environ. Microbiol.* 49:1338.
4. LeChevallier, M.W., S.C. Cameron & G.A. McFeters. 1983. New medium for the improved recovery of coliform bacteria from drinking water. *Appl. Environ. Microbiol.* 45:484.
5. Watters, S.K., B.H. Pyle, M.W. LeChevallier & G.A. McFeters. 1989. Enumeration of *E. cloacae* after chlorine exposure. *Appl. Environ. Microbiol.* 55:3226.
6. Rose, R.E., E.E. Geldreich & W. Litsky. 1975. Improved membrane filter method for fecal coliform analysis. *Appl. Microbiol.* 29:532.
7. Lin, S.D. 1976. Membrane filter method for recovery of fecal coliforms in chlorinated sewage effluents. *Appl. Environ. Microbiol.* 32:547.
8. Stuart, D.S, G.A. McFeters & J.E. Schillinger. 1977. Membrane filter technique for quantification of stressed fecal coliforms in the aquatic environment. *Appl. Environ. Microbiol.* 34:42.
9. Green, B.L., E.M. Clausen & W. Litsky. 1977. Two-temperature membrane filter method for enumerating fecal coliform bacteria from chlorinated effluents. *Appl. Environ. Microbiol.* 33:1259.
10. Presswood, W.G. & D. Strong. 1977. Modification of M-FC medium by eliminating rosolic acid. *Amer. Soc. Microbiol. Abs. Annu. Meeting.* ISSN-0067-2777:272.
11. LeChevallier, M.W., P.E. Jakanoski, A.K. Camper & G.A. McFeters. 1984. Evaluation of m-T7 agar as a fecal coliform medium. *Appl. Environ. Microbiol.* 48:371.
12. Lin, S.D. 1974. Evaluation of fecal streptococci tests for chlorinated secondary effluents. *J. Environ. Eng. Div., Proc. Amer. Soc. Civil Engr.* 100:253.

9213 RECREATIONAL WATERS*

9213 A. Introduction

1. Microbiological Indicators

Recreational waters include freshwater swimming pools, whirlpools, and naturally occurring fresh and marine waters. Many local and state health departments require microbiological monitoring of recreational waters. Historically, the most common microbiological tests to assess sanitary quality have been heterotrophic counts and total and fecal coliform tests. Total coliform tests and heterotrophic counts usually are performed on treated waters and fecal coliform tests performed on untreated waters. Although detection of coliform bacteria in water indicates that it may be unsafe to drink, other bacteria have been isolated from recreational waters that may suggest health risks through body contact, ingestion, or inhalation. Other bacteria suggested as indicators of recreational water quality include *Pseudomonas aeruginosa,* fecal streptococci, enterococci, and staphylococci. Ideally, recreational water quality indicators are microorganisms for which densities in the water can be related quantitatively to potential health hazards resulting from recreational use, particularly where upper body orifices are exposed to water. The ideal indicator is the one with the best correlation between density and the health hazards associated with a given type of pollution. The most common potential sources of infectious agents in recreational waters include untreated or poorly treated municipal and industrial effluents or sludge, sanitary wastes from seaside residences, fecal wastes from pleasure craft, drainage from sanitary landfills, stormwater runoff, and excretions from animals. In addition, the source of infectious agents may be the aquatic environment itself. The potential health hazards from each of these sources are not equal. Exposure to untreated or inadequately treated human fecal wastes is considered the greatest health hazard. The presence of microbiological indicators in treated swimming pools or whirlpools indicate possible insufficient water exchange, disinfection, and maintenance. Bather density is a major factor in determining the probability of swimmer-associated illnesses with swimming pools, particularly when there is insufficient disinfection and water circulation. The bathers themselves may be the source of pollution by shedding organisms associated with the mouth, nose, and skin.

2. Infectious Diseases from Water Exposure

In general, infections or disease associated with recreational water contact fall into two categories. The first group is gastroenteritis resulting from unintentional ingestion of water contaminated with fecal wastes. Enteric microorganisms that have been shown to cause gastroenteritis from recreational water contact include *Giardia, Cryptosporidium, Shigella, Salmonella, E. coli* 0157:H7, Hepatitis A, Coxsackie A and B, and Norwalk virus. Leptospirosis is not an enteric infection but also is transmitted through contact with waters contaminated with human or animal wastes. The second group or category of infections or disease is associated mainly with microorganisms that are indigenous to the environment, which include the following: *Pseudomonas aeruginosa, Staphylococcus* sp., *Legionella* sp., *Naegleria fowleri, Mycobacterium* sp., and *Vibrio* sp. The illnesses or waterborne diseases caused by these organisms include dermatitis or folliculitis, otitis externa, Pontiac fever, granulomas, primary amebic meningoencephalitis (PAM), and conjunctivitis. Commonly occurring illnesses or infections associated with recreational water contact are dermatitis caused by *Pseudomonas aeruginosa* and otitis externa, "swimmer's ear," frequently caused by *Pseudomonas aeruginosa* and *Staphylococcus aureus.*

3. Microbiological Monitoring Limitations

Routine examination for pathogenic microorganisms is not recommended except for investigations of water-related illness and special studies; in such cases, focus microbiological analyses on the known or suspected pathogen. Methods for several of these pathogens are given in Section 9260, Detection of Pathogenic Bacteria, Section 9510, Detection of Enteric Viruses, and Section 9711, Pathogenic Protozoa. Because some pathogenic organisms such as *Giardia, Cryptosporidium, Mycobacterium,* and *Naegleria* are more resistant to changes in environmental conditions than indicator bacteria, routine monitoring may not always reflect the risk of infection from these organisms. Described below are recommended methods for microbial indicators of recreational water quality. Consider the type(s) of water examined in selecting the microbiological method(s) or indicator(s) to be used. No single procedure is adequate to isolate all microorganisms from contaminated water. While bacterial indicators may not adequately reflect risk of viral, fungal, or parasitic infection from recreational waters, available technology limits monitoring for such organisms in routine laboratory operations.

4. Bibliography

CABELLI, V.J. 1977. Indicators of recreational water quality. *In* Bacterial Indicators/Health Hazards Associated with Waters. STP 635, American Soc. Testing & Materials, Philadelphia, Pa.

DUFOUR, A.P. 1986. Diseases caused by water contact. *In* Waterborne Diseases in the United States. CRC Press Inc., Boca Raton, Fla.

MOE, C.L. 1996. Waterborne transmission of infectious agents. *In* Manual of Environmental Microbiology. American Soc. Microbiology, ASM Press, Washington, D.C.

---

* Approved by Standard Methods Committee, 1997.

9213 B. Swimming Pools

1. General Discussion

a. Characteristics: A swimming pool is a body of water of limited size contained in a holding structure.[1] The pool water generally is potable and treated with additional disinfectant but also may come from thermal springs or salt water. Modern pools have a recirculating system for filtration and disinfection.

b. Monitoring requirements:

1) General—Monitor water quality in pools for changes in chemical and physical characteristics that may result in irritation to the bather's skin, eyes, and mucosal barriers or may adversely affect disinfection. Microorganisms of concern typically are those from the bather's body and its orifices and include those causing infections of the eye, ear, upper respiratory tract, skin, and intestinal or genitourinary tracts. Water quality depends on the efficacy of disinfection, sanitary conditions, the number of bathers in the pool at any one time, and the total number of bathers per day.

2) Disinfected indoor pools—Swimming pools should be disinfected continuously when in use. Test swimming pool water for residual chlorine and pH when the pool is initially opened and at least three times/d. Collect samples from at least two locations for these determinations. Evaluate clarity of the swimming pool water before opening for the day and during periods of heavy usage.[2] The heterotrophic plate count is the primary indicator of disinfection efficacy. Indicators of health risk include normal skin flora that are shed, such as *Pseudomonas* and *Staphylococcus*.[3-6] These organisms account for a large percentage of swimming-pool-associated illnesses. In special circumstances *Mycobacterium*, *Legionella*, or *Candida albicans* may be associated with health risks related to recreational waters. Take samples for microbiological examination while the pool is in use. APHA recommends for public swimming pools that not more than 15% of the samples collected during any 30-d period shall have a heterotrophic plate count of 200/mL or show a positive confirmed total coliform test in any of five 10-mL portions of sample examined with the multiple-tube fermentation test or more than 1 total coliform/50 mL when the membrane filter test is used. Whenever swimming pool samples are examined for total staphylococci or *Staphylococcus aureus*, not more than 50 organisms/100 mL should be present.[2]

3) Disinfected outdoor pools—Fecal coliform bacteria and *Pseudomonas* species are the primary indicators of contamination from animal pets, rodents, stormwater runoff, and human sources. Supporting indicators include coliform bacteria, the heterotrophic plate count, and staphylococci.

4) Untreated pools—The primary indicator may be fecal coliform bacteria. Supporting indicators are those described for disinfected pools. Untreated pools are not recommended for recreational use due to increased health risks.

2. Samples

a. Containers: Collect samples for bacteriological examination of swimming pool water as directed in Section 9060A. Use containers with capacities of 120 to 480 mL, depending on analyses to be made. Add sufficient sodium thiosulfate, $Na_2S_2O_3$, to the sample to provide a concentration of approximately 100 mg/L in the sample. Do this by adding 0.1 mL of 10% solution of $Na_2S_2O_3$ to a 120-mL bottle or 0.4 mL to a 480-mL bottle. After adding $Na_2S_2O_3$, stopper or cap and sterilize container.

b. Sampling procedure: Collect samples during periods of maximum bather load. Information on number of bathers may be helpful in subsequent interpretation of laboratory results. Use sampling frequency consistent with state and local health regulations.

Collect samples by carefully removing cap of a sterile sample bottle and holding bottle near the base at an angle of 45 deg. Fill in one slow sweep down through the water, with the mouth of the bottle always ahead of the hand. Avoid contamination of the sample by floating debris. Replace cap. Do not rinse bottle (i.e., retain sodium thiosulfate). For pools equipped with a filter, samples may be collected from sampling cocks provided in the return and discharge lines from the filter.

Most bacteria shed by bathers are in body oils, saliva, and mucus discharges that occur near the surface; collect additional samples of the surface microlayer from the area in 1-m-deep water. Collect microlayer samples by plunging a sterile glass plate (approximately 20 cm by 20 cm) vertically through the water surface and withdrawing it upward at a rate of approximately 6 cm/s. Remove surface film and water layer adhering to both sides of plate with a sterile silicone rubber scraper and collect in a sterile glass bottle. Repeat until desired volume is obtained. To minimize microbial contamination, wrap glass plate and scraper in metal foil and sterilize by autoclaving before use. Wear sterile rubber or plastic gloves during sampling or hold glass plate with forceps, clips, or tongs.

Determine residual chlorine or other disinfectant at poolside at the time of sample collection (see Section 4500-Cl.G). Residual disinfectant levels, chemical, and physical quality of pool water should be consistent with local, state, or APHA standards. The permissible bathing load should adhere to local, state, or APHA-recommended regulations.

c. Sample storage: Analyze microbiological samples as soon as possible after collection (see Section 9060B).

d. Sample volume: See Section 9222B.5.

e. Sample dilution: If sample dilutions are required, use 0.1% peptone water or buffered dilution water as diluent to optimize recovery of stressed organisms (see Section 9222 for suggested sample volume). Because peptone water has a tendency to foam, avoid including air bubbles when pipetting to assure accurate measure.

3. Heterotrophic Plate Count

Determine the heterotrophic plate count as directed in Section 9215. Use at least two plates per dilution.

4. Tests for Total Coliforms

Determine total coliform bacteria as directed in Sections 9221, 9222, or 9223.

5. Tests for Fecal Coliforms

Test for fecal coliforms according to the multiple-tube fermentation technique (Section 9221), the membrane filter technique (Section 9222), or rapid methods (Section 9211).

6. Test for Staphylococci or *Staphylococcus aureus*

a. Baird-Parker agar base:

Tryptone	10.0 g
Beef extract	5.0 g
Yeast extract	1.0 g
Glycine	12.0 g
Sodium pyruvate	10.0 g
Lithium chloride	5.0 g
Agar	20.0 g
Reagent-grade water	1 L

Sterilize by autoclaving. Cool to 50°C and aseptically add 50 mL commercial egg yolk tellurite enrichment/L. Mix well. Final pH should be 7.0 ± 0.2.

b. Procedure: Use membrane filter technique to prepare samples. Place membrane filter on Baird-Parker agar and incubate at 35 ± 0.5°C for 48 ± 4 h. Staphylococci typically form slate gray to jet black, smooth, entire colonies. If *S. aureus* is present egg yolk clearing may be observed if the membrane filter is raised from the medium. Verify some differentiated colonies with a commercial multi-test system or on the basis of such key characteristics as catalase reaction, coagulase production, aerobic and anaerobic acid production from certain carbohydrates, and typical microscopic morphology.

7. Test for *Staphylococcus aureus*

Use a modified multiple-tube procedure.

a. Media:

1) *M-staphylococcus broth:*

Tryptone	10.0	g
Yeast extract	2.0	g
Lactose	2.0	g
Mannitol	10.0	g
Dipotassium hydrogen phosphate, K_2HPO_4	5.0	g
Sodium chloride, NaCl	75.0	g
Sodium azide, NaN_3	0.049	g
Reagent-grade water	1	L

Sterilize by boiling for 4 min; pH should be 7.0 ± 0.2. For 10-mL inocula prepare and use double-strength medium.

2) *Lipovitellenin-salt-mannitol agar:* This medium may not be available in dehydrated form and may require preparation from the basic ingredients or by addition of egg yolk to a dehydrated base.

Beef extract	1.0	g
Polypeptone	10.0	g
Sodium chloride, NaCl	75.0	g
d-Mannitol	10.0	g
Agar	15.0	g
Phenol red	0.025	g
Egg yolk	20.0	g
Reagent-grade water	1	L

Sterilize by autoclaving; pH should be 7.4 ± 0.2.

b. Procedure: Inoculate tubes of M-staphylococcus broth as directed in Section 9221. Incubate at 35 ± 1°C for 24 h. Hold original enrichment sample but streak from positive (turbid) tubes on plates of lipovitellenin-salt-mannitol agar and incubate at 35 ± 1°C for 48 h. Opaque (24 h), yellow (48 h) zones around the colonies are positive evidence of lipovitellenin-lipase activity (opaque) and mannitol fermentation (yellow).

If the plate is negative, streak another plate from the original enrichment tube before discarding. Lipovitellenin-lipase activity has a 95% positive correlation with coagulase production. If necessary, confirm positive isolates as catalase-positive, coagulase-positive, fermenting mannitol, fermenting glucose anaerobically, yielding typical microscopic morphology, and gram-positive.

8. Tests for *Pseudomonas aeruginosa*

Tests for *P. aeruginosa* are presented in Sections 9213E and F and include a membrane filter procedure and a multiple-tube technique.

9. Test for Streptococci or Enterococci

Determine fecal streptococci or enterococci as described in Section 9230, and if necessary, perform additional biochemical tests to identify species.

10. References

1. CENTERS FOR DISEASE CONTROL. 1983. Swimming Pools—Safety and Disease Control through Proper Design and Operation. DHHS—CDC No. 83-8319, Centers for Disease Control, Atlanta, Ga.
2. AMERICAN PUBLIC HEALTH ASSOCIATION. 1981. Public Swimming Pools. Recommended Regulations for Design and Construction, Operation and Maintenance. American Public Health Assoc., Washington, D.C.
3. SEYFRIED, P.L., R.S. TOBIN, N.E. BROWN & P.F. NESS. 1985. A prospective study of swimming-related illness. II. Morbidity and the microbiological quality of water. *Amer. J. Pub. Health* 75:1071.
4. KLAPES, N.A. & D. VESLEY. 1988. Rapid assay for in situ identification of coagulase-positive staphylococci recovered by membrane filtration from swimming pool water. *Appl. Environ. Microbiol.* 52:589.
5. COVERT, T.C. & P.V. SCARPINO. 1987. Comparison of Baird-Parker agar, Vogel-Johnson agar, and M-Staphylococcus broth for the isolation and enumeration of *Staphylococcus aureus* in swimming pool waters. Abstr. Annu. Meeting American Soc. Microbiology, Atlanta, Ga., American Soc. Microbiology, Washington, D.C.
6. CHAROENCA, N. & R.S. FUJIOKA. 1995. Association of staphylococcal skin infections and swimming. *Water Sci. Technol.* 32:11.

11. Bibliography

WORKING PARTY OF THE PUBLIC HEALTH LABORATORY SERVICE. 1965. A bacteriological survey of swimming baths in primary schools. *Monthly Bull. Min. Health & Pub. Health Lab. Serv.* 24:116.

GUNN, B.A., W.E. DUNKELBERG, JR. & J.R. CRUTZ. 1972. Clinical evaluation of 2% LSM medium for primary isolation and identification of staphylococci. *Amer. J. Clin. Pathol.* 57:236.

HATCHER, R.F. & B.C. PARKER. 1974. Investigations of Freshwater Surface Microlayers. VPI-SRRC-BULL 64. Virginia Polytechnic Inst. and State Univ., Blacksburg.

U.S. ENVIRONMENTAL PROTECTION AGENCY. 1985. Test Methods for *Escherichia coli* and Enterococci in Water by the Membrane Filter Procedure. EPA-600/4-85/076.

HURST, C.J. 1991. Disinfection of drinking water, swimming-pool-water and treated sewage effluent. *In* S.S. Block. Disinfection, Sterilization and Preservation, 4th ed. Lea & Febiger, Philadelphia, Pa.

9213 C. Whirlpools

1. General Discussion

a. Characteristics: A whirlpool is a shallow pool with a maximum water depth of 1.2 m; it has a closed-cycle water system, a heated water supply, and usually a hydrojet recirculation system. It may be constructed of plastic, fiberglass, redwood, or epoxy-lined surfaces. Whirlpools are designed for recreational as well as therapeutic use and may accommodate one or more bathers. These pools usually are not cleaned, drained, and refilled after each use. They are located in homes, apartments, hotels, athletic facilities, rehabilitation centers, and hospitals.

b. Monitoring requirements: Whirlpool-associated infections are common because of the inherent design and characteristics of whirlpools, which include high temperature, reduced disinfection efficacy, and increased organic material. All these factors contribute to favorable conditions for growth of microorganisms, especially *Pseudomonas aeruginosa.* Studies have shown that whirlpools can serve as a reservoir of *Legionella pneumophila.* Therefore, frequent testing for residual disinfectant levels and pH, along with scheduled maintenance, is necessary for safe whirlpool water quality.[1-5]

c. Microbiological indicators: The primary indicator of disinfection efficacy is *P. aeruginosa,* with total coliforms, heterotrophic plate count, and staphylococci as supporting indicators of water quality. The standard index of water quality, i.e., total coliforms, may be insufficient to judge the microbiological quality of whirlpool water. *Pseudomonas aeruginosa* is frequently isolated from whirlpool water that is coliform-negative.[6] In the event of a whirlpool-associated outbreak, collect samples as close as possible to the time of the outbreak. Analyze for the suspected pathogen and *P. aeruginosa.* Methods for *P. aeruginosa* are described in Sections 9213E and 9213F.

d. Sample preservation: Examine samples as soon as possible after collection. See Section 9060B.

2. References

1. CENTERS FOR DISEASE CONTROL. 1981. Suggested Health and Safety Guidelines for Public Spas and Hot Tubs. DHHS-CDC #99-960. United States Government Printing Off., Washington, D.C.
2. SOLOMON, S.L. 1985. Host factors in whirlpool-associated *Pseudomonas aeruginosa* skin disease. *Infect. Control* 6:402.
3. HIGHSMITH, A.K., P.N. LEE, R.F. KHABBAZ & V.P. MUNN. 1985. Characteristics of *Pseudomonas aeruginosa* isolated from whirlpools and bathers. *Infect. Control* 6:407.
4. GROOTHUIS, D.G., A.H. HAVELAAR & H.R. VEENENDAAL. 1985. A note on legionellas in whirlpools. *J. Appl. Bacteriol.* 58:479.
5. HIGHSMITH, A.K. & M.S. FAVERO. 1985. Microbiological aspects of public whirlpools. *Clin. Microbiol. Newsletter* 7:9.
6. HALL, N. 1984. Whirlpools and *Pseudomonas aeruginosa.* UHL Lab Hotline 21:9.

3. Bibliography

GELDREICH, E.E., A.K. HIGHSMITH & W.J. MARTONE. 1985. Public whirlpools—the epidemiology and microbiology of disease. *Infect. Control* 6:392.

9213 D. Natural Bathing Beaches

1. General Discussion

a. Characteristics: A natural bathing beach is any area of a stream, lake, ocean, impoundment, or hot spring that is used for recreation. A wide variety of pathogenic microorganisms can be transmitted to humans through use of natural fresh and marine recreational waters contaminated by wastewater.[1,2] These include enteric pathogens such as *Salmonella, Shigella,* enteroviruses, protozoa, multicellular parasites, and "opportunists" such as *P. aeruginosa, Klebsiella* sp., *Vibrio* sp., and *Aeromonas hydrophila,* which can multiply in recreational waters with sufficient nutrients. Other organisms of concern are those associated with the skin, mouth, or nose of bathers, such as *Staphylococcus aureus* and other organisms, e.g., nontuberculous mycobacteria and leptospira, and *Naegleria* sp..[3-9]

b. Monitoring requirements: Historically, fecal coliforms have been recommended as the indicator of choice for evaluating the microbiological quality of recreational waters. Many states have adopted use of this indicator in their water quality standards. Recent studies have demonstrated that *E. coli* and enterococci showed a stronger correlation with swimming-associated gastroenteritis than do fecal coliforms, and that both indicators were equally acceptable for monitoring fresh-water quality. For marine water, enterococci showed the strongest relationship of density to gastroenteritis. The recommended densities of these indicator organisms were calculated to approximate the degree of protection previously accepted for fecal coliforms. EPA-recommended water quality criteria are based on these findings.[10] While the primary indicators of water quality are *E. coli* and enterococci, the enumeration of *P. aeruginosa, Aeromonas hydrophila,* and *Klebsiella* sp. in recreational waters may be useful in cases of discharge of pulp and paper wastes and effluents from textile finishing plants into receiving waters.

2. Samples

a. Containers: Collect samples as directed in Section 9060A. The size of the container varies with the number and variety of tests to be performed. Adding $Na_2S_2O_3$ to the bottle is unnecessary.

b. Sampling procedure: Collect samples 0.3 m below the water surface in the areas of greatest bather load. Take samples over the range of environmental and climatic conditions, especially during times when maximal pollution can be expected, i.e., periods of tidal, current, and wind influences, stormwater runoff, wastewater treatment bypasses. See Section 9213B.2*b* for methods of sample collection and Section 9222 for suggested sample volumes.

c. Sample storage: See Section 9060B.

3. Tests for *Escherichia coli*

a. Media:
1) mTEC agar:*

Proteose peptone	5.0	g
Yeast extract	3.0	g
Lactose	10.0	g
Sodium chloride, NaCl	7.5	g
Dipotassium phosphate, K_2HPO_4	3.3	g
Monopotassium phosphate KH_2PO_4	1.0	g
Sodium lauryl sulfate	0.2	g
Sodium desoxycholate	0.1	g
Bromcresol purple	0.08	g
Bromphenol red	0.08	g
Agar	15.0	g
Reagent-grade water	1	L

Sterilize by autoclaving; pH should be 7.3 ± 0.2. Pour 4 to 5 mL liquefied agar into culture dishes (50 × 10 mm). Store in refrigerator.

2) Urea substrate:*

Urea	2.0	g
Phenol red	10	mg
Reagent-grade water	100	mL

Adjust pH to between 3 and 4. Store at 2 to 8°C. Use within 1 week.

b. Procedure: Filter sample through a membrane filter (see Section 9222), place membrane on mTEC agar, incubate at 35 ± 0.5°C for 2 h to rejuvenate injured or stressed bacteria, and then incubate at 44.5 ± 0.2°C for 22 h. Transfer filter to a filter pad saturated with urea substrate. After 15 min, count yellow or yellow-brown colonies, using a fluorescent lamp and a magnifying

---

* Difco or equivalent.

lens. *E. coli* produces yellow or yellow-brown colonies. Verify a portion of these differentiated colonies with a commercial multi-test system [see Section 9222B.5*f*2)*b*)].

4. Tests for Enterococci

Perform tests for enterococci by the multiple-tube technique (Section 9230B) or membrane filter technique (Section 9230C).

5. Tests for *Pseudomonas aeruginosa*

Perform tests for *P. aeruginosa* as directed in Sections 9213E and F. Use the multiple-tube test with samples but note that the procedures may not be applicable to marine samples.

6. Tests for *Salmonella/Shigella*

See Section 9260.

7. References

1. CABELLI, V.J. 1980. Health Effects Criteria for Marine Recreational Waters. EPA-600/1-80-031, U.S. Environmental Protection Agency, Research Triangle Park, N.C.
2. DUFOUR, A.P. 1984. Health Effects Criteria for Fresh Recreational Waters. EPA-600/1-84-004, U.S. Environmental Protection Agency, Research Triangle Park, N.C.
3. KESWICK, B.H., C.P. GERBA & S.M. GOYAL. 1981. Occurrence of enteroviruses in community swimming pools. *Amer. J. Pub. Health* 71:1026.
4. DUTKA, B.J. & K.K. KWAN. 1978. Health indicator bacteria in water surface microlayers. *Can. J. Microbiol.* 24:187.
5. CABELLI, V.J., H. KENNEDY & M.A. LEVIN. 1976. *Pseudomonas aeruginosa* and fresh recreational waters. *J. Water Pollut. Control Fed.* 48:367.
6. SHERRY, J.P., S.R. KUCHMA & B.J. DUTKA. 1979. The occurrence of *Candida albicans* in Lake Ontario bathing beaches. *Can. J. Microbiol.* 25:1036.
7. STEVENS, A.R., R.L. TYNDALL, C.C. COUTANT & E. WILLAERT. 1977. Isolation of the etiological agent of primary amoebic meningoencephalitis from artificially heated waters. *Appl. Environ. Microbiol.* 34:701.
8. WELLINGS, F.M., P.T. AMUSO, S.L. CHANG & A.L. LEWIS. 1977. Isolation and identification of pathogenic *Naegleria* from Florida lakes. *Appl. Environ. Microbiol.* 34:661.
9. N'DIAYE, A., P. GEORGES, A. N'GO & B. FESTY. 1985. Soil amoebas as biological markers to estimate the quality of swimming pool waters. *Appl. Environ. Microbiol.* 49:1072.
10. U.S. ENVIRONMENTAL PROTECTION AGENCY. 1986. Ambient Water Quality Criteria for Bacteria—1986. EPA-440/5-84-002, U.S. Environmental Protection Agency, Washington, D.C.

8. Bibliography

OLIVIERI, V.P., C.W. DRUSE & K. KAWATA. 1977. Microorganisms in Urban Stormwater. EPA-600/2-77-087, U.S. Environmental Protection Agency, Cincinnati, Ohio.
RICE, E.W., T.C. COVERT, D.K. WILD, D. BERMAN, S.A. JOHNSON & C.H. JOHNSON. 1993. Comparative resistance of *Escherichia coli* and *Enterococci* to chlorination. *J. Environ. Health.* A28:89.

9213 E. Membrane Filter Technique for *Pseudomonas aeruginosa*

1. Laboratory Apparatus

See Section 9222B.1.

2. Culture Media

a. M-PA agar: This agar may not be available in dehydrated form and may require preparation from the basic ingredients.

L-lysine HCl	5.0 g
Sodium chloride, NaCl	5.0 g
Yeast extract	2.0 g
Xylose	2.5 g
Sucrose	1.25 g
Lactose	1.25 g
Phenol red	0.08 g
Ferric ammonium citrate	0.8 g
Sodium thiosulfate, $Na_2S_2O_3$	6.8 g
Agar	15.0 g
Reagent-grade water	1 L

Adjust to pH 6.5 $\pm$ 0.1 and sterilize by autoclaving. Cool to 55 to 60°C; readjust to pH 7.1 $\pm$ 0.2 and add the following dry antibiotics per liter of agar base: sulfapyridine,* 176 mg; kanamycin,* 8.5 mg; nalidixic acid,* 37.0 mg; and cycloheximide,* 150 mg. After mixing dispense in 3-mL quantities in 50- $\times$ 12-mm petri plates. Store poured plates at 2 to 8°C. Discard unused medium after 1 month.

b. Modified M-PA agar.†
c. Milk agar (Brown and Scott Foster Modification):

Mixture A:

Instant nonfat milk‡	100 g
Reagent-grade water	500 mL

Mixture B:

Nutrient broth	12.5 g
Sodium chloride, NaCl	2.5 g
Agar	15.0 g
Reagent-grade water	500 mL

---

* Sigma Chemical Co., St. Louis, MO, or equivalent.
† Commercially available as M-PA-C agar. Contains magnesium, sulfate, kanamycin, and nalidixic acid.
‡ Carnation or equivalent.

Separately prepare and sterilize Mixtures A and B; cool rapidly to 55°C; aseptically combine mixtures and dispense 20 to 25 mL per petri dish.

3. Procedure

a. Presumptive tests: Filter 200 mL or less of natural waters or up to 500 mL of swimming pool waters through sterile membrane filters. Place each membrane on a poured plate of modified M-PA agar so that there is no air space between the membrane and the agar surface. Invert plates and incubate at 41.5 $\pm$ 0.5°C for 72 h.

Typically, *P. aeruginosa* colonies are 0.8 to 2.2 mm in diameter and flat in appearance with light outer rims and brownish to greenish-black centers. Count typical colonies, preferably from filters containing 20 to 80 colonies. Use a 10- to 15-power magnifier as an aid in colony counting.

b. Confirmation tests: Use milk agar to confirm a number of typical and atypical colonies. Make a single streak (2 to 4 cm long) from an isolated colony on a milk agar plate and incubate at 35 $\pm$ 1.0 °C for 24 h. *P. aeruginosa* hydrolyzes casein and produces a yellowish to green diffusible pigment.

4. Interpretation and Calculation of Density

Confirmation is not routinely required. In the absence of confirmation, report results as the number of presumptive *P. aeruginosa*/100 mL.

5. Bibliography

DRAKE, C.H. 1966. Evaluation of culture media for the isolation and enumeration of *Pseudomonas aeruginosa. Health Lab. Sci.* 3:10.

BROWN, M.R.W. & J.H. SCOTT FOSTER. 1970. A simple diagnostic milk medium for *Pseudomonas aeruginosa. J. Clin. Pathol.* 23:172.

LEVIN, M.A. & V.J. CABELLI. 1972. Membrane filter technique for enumeration of *Pseudomonas aeruginosa. Appl. Microbiol.* 24:864.

DUTKA, B.J. & K.K. KWAN. 1977. Confirmation of the single-step membrane filter procedure for estimating *Pseudomonas aeruginosa* densities in water. *Appl. Environ. Microbiol.* 33:240.

BRODSKY, M.H. & B.W. CIEBIN. 1978. Improved medium for recovery and enumeration of *Pseudomonas aeruginosa* from water using membrane filters. *Appl. Environ. Microbiol.* 36:26.

9213 F. Multiple-Tube Technique for *Pseudomonas aeruginosa*

1. Laboratory Apparatus

See Section 9221.

2. Culture Media

a. Asparagine broth: This medium may not be available in dehydrated form and may require preparation from the basic ingredients.

Asparagine, DL	3.0 g
Anhydrous dipotassium hydrogen phosphate, K_2HPO_4	1.0 g
Magnesium sulfate, $MgSO_4 \cdot 7H_2O$	0.5 g
Reagent-grade water	1 L

Adjust pH to 6.9 to 7.2 before sterilization.

b. Acetamide broth: This medium may not be available in dehydrated form and may require preparation from the basic ingredients.

Acetamide .... 10.0 g
Sodium chloride, NaCl .... 5.0 g
Anhydrous dipotassium hydrogen phosphate, K_2HPO_4 .. 1.39 g
Anhydrous potassium dihydrogen phosphate, KH_2PO_4 .. 0.73 g
Magnesium sulfate, $MgSO_4 \cdot 7H_2O$ .... 0.5 g

Dissolve 1.2 g phenol red in 100 mL 0.01N NaOH and add 1 mL/L of acetamide broth. Use phenol red stock solution within 1 year. Adjust pH to 7.1 to 7.3 before sterilization. Final pH should be 7.0 ± 0.2. Prepare acetamide broth as described above. If agar slants are preferred, prepare as described above but add 15 g agar/L, heat to dissolve agar, and dispense 8-mL quantities in 16-mm tubes. After autoclaving, incline tubes while cooling to provide a large slant surface.

3. Procedure

a. Presumptive test: Perform a five-tube multiple-tube test. Use 10 mL single-strength asparagine broth for inocula of 1 mL or less and 10 mL double-strength asparagine broth for 10-mL inocula. For swimming pools, higher dilutions may be necessary. Incubate inoculated tubes at 35 to 37°C. After 24 h and again after 48 h of incubation, examine tubes under long-wave ultraviolet light (black light) in a darkened room. Production of a green fluorescent pigment constitutes a positive presumptive test.

b. Confirmed test: Confirm positive tubes by inoculating 0.1 mL of culture into acetamide broth or onto the surface of acetamide agar slants. Development of purple color (alkaline pH) within 24 to 36 h of incubation at 35 to 37°C is a positive confirmed test for *Pseudomonas aeruginosa.*

c. Computing and reporting results: Refer to Table 9221:V and to Section 9221D.

9215   HETEROTROPHIC PLATE COUNT*

9215 A.   Introduction

1. Applications

The heterotrophic plate count (HPC), formerly known as the standard plate count, is a procedure for estimating the number of live heterotrophic bacteria in water and measuring changes during water treatment and distribution or in swimming pools. Colonies may arise from pairs, chains, clusters, or single cells, all of which are included in the term "colony-forming units" (CFU). The final count also depends on interaction among the developing colonies; choose that combination of procedure and medium that produces the greatest number of colonies within the designated incubation time. To compare data, use the same procedure and medium. Three different methods and four different media are described.

2. Selection of Method

a. Pour plate method: The pour plate method (9215B) is simple to perform and can accommodate volumes of sample or diluted sample ranging from 0.1 to 2.0 mL. The colonies produced are relatively small and compact, showing less tendency to encroach on each other than those produced by surface growth. On the other hand, submerged colonies often are slower growing and are difficult to transfer. A thermostatically controlled water bath is essential for tempering the agar, but even so, significant heat shock to bacteria from the transient exposure of the sample to 45 to 46°C agar may occur.

b. Spread plate method: The spread plate method (9215C) causes no heat shock and all colonies are on the agar surface where they can be distinguished readily from particles and bubbles. Colonies can be transferred quickly, and colony morphology easily can be discerned and compared to published descriptions. However, this method is limited by the small volume of sample or diluted sample that can be absorbed by the agar: 0.1 to 0.5 mL, depending on the degree to which the prepoured plates have been dried. To use this procedure, maintain a supply of suitable predried, absorbent agar plates.

c. Membrane filter method: The membrane filter method (9215D) permits testing large volumes of low-turbidity water and is the method of choice for low-count waters (< 1 to 10 CFU/mL). This method produces no heat shock but adds the expense of the membrane filter. Further disadvantages include the smaller display area, the need to detect colonies by reflected light against a white background if colored filters or contrast stains are not used, possible damage to cells by excessive filtration pressures, and possible variations in membrane filter quality (see Section 9020B.4*h*).

3. Work Area

Provide a level table or bench top with ample area in a clean, draft-free, well-lighted room or within a horizontal-flow laminar hood. Use table and bench tops having nonporous surfaces and disinfect before any analysis is made.

4. Samples

Collect water as directed in Section 9060A. Initiate analysis as soon as possible after collection to minimize changes in bacterial

---

* Approved by Standard Methods Committee, 1994.

population. The recommended maximum elapsed time between collection and analysis of samples is 8 h (maximum transit time 6 h, maximum processing time 2 h). When analysis cannot begin within 8 h, maintain sample at a temperature below 4°C but do not freeze. Maximum elapsed time between collection and analysis must not exceed 24 h.

5. Sample Preparation

Mark each plate with sample number, dilution, date, and any other necessary information before examination. Prepare at least duplicate plates for each volume of sample or dilution examined. For the pour or spread plate methods use sterile glass (65 cm^2) or presterilized disposable plastic (57 cm^2) petri dishes.

Thoroughly mix all samples or dilutions by rapidly making about 25 complete up-and-down (or back-and-forth) movements. Optionally, use a mechanical shaker to shake samples or dilutions for 15 s.

6. Media

Compare new lots of media with current lot in use according to Section 9020B.4i.

a. Plate count agar (tryptone glucose yeast agar): Use for pour and spread plate methods. This high-nutrient agar, widely used in the past, gives lower counts than R2A or NWRI agar. It is included for laboratories wishing to make comparisons of media or to extend the continuity of old data.

Tryptone	5.0	g
Yeast extract	2.5	g
Glucose	1.0	g
Agar	15.0	g
Reagent-grade water	1	L

pH should be 7.0 ± 0.2 after autoclaving at 121°C for 15 min.

b. m-HPC agar:† Use this high-nutrient medium only for the membrane filter method.

Peptone	20.0	g
Gelatin	25.0	g
Glycerol	10.0	mL
Agar	15.0	g
Reagent-grade water	1	L

Mix all ingredients except glycerol. Adjust pH to 7.1, if necessary, with 1N NaOH, heat slowly to boiling to dissolve thoroughly, add glycerol, and autoclave at 121°C for 5 min.‡

c. R2A agar: Use for pour, spread plate, and membrane filter methods. This low-nutrient agar gives higher counts than high-nutrient formulations.

Yeast extract	0.5	g
Proteose peptone No. 3 or polypeptone	0.5	g
Casamino acids	0.5	g
Glucose	0.5	g
Soluble starch	0.5	g
Dipotassium hydrogen phosphate, K$_2$HPO$_4$	0.3	g
Magnesium sulfate heptahydrate, MgSO$_4$·7H$_2$O	0.05	g
Sodium pyruvate	0.3	g

† Formerly called m-SPC agar.
‡ This medium may not be sterile; use with care to avoid contamination.

Agar	15.0	g
Reagent-grade water	1	L

Adjust pH to 7.2 with solid K$_2$HPO$_4$ or KH$_2$PO$_4$ before adding agar. Heat to dissolve agar and sterilize at 121°C for 15 min.

d. NWRI agar (HPCA): Use for pour, spread plate, and membrane filter methods. This low-nutrient medium is likely to produce higher colony counts than high-nutrient media. It is not currently available in dehydrated form and requires preparation from the basic ingredients; this makes its usage less desirable.

Peptone	3.0	g
Soluble casein	0.5	g
K$_2$HPO$_4$	0.2	g
MgSO$_4$	0.05	g
FeCl$_3$	0.001	g
Agar	15.0	g
Reagent-grade water	1	L

Adjust pH to 7.2 before autoclaving for 15 min at 121°C.

7. Incubation

For compliance monitoring purposes under U.S. EPA's Surface Water Treatment Rule (40 CFR 141.74), provision on heterotrophic bacteria, incubate pour plates at 35°C for 48 h. Otherwise, select from among recommended times and temperatures for monitoring changes in water quality. The highest counts typically will be obtained from 5- to 7-d incubation at a temperature of 20 to 28°C.

During incubation maintain humidity within the incubator so that plates will have no moisture weight loss greater than 15%. This is especially important if prolonged incubation is used. A pan of water placed at the bottom of the incubator may be sufficient but note that to prevent rusting or oxidation the inside walls and shelving should be of high-grade stainless steel or anodized aluminum. For long incubation in nonhumidified incubators, seal plates in plastic bags.

8. Counting and Recording

a. Pour and spread plates: Count all colonies on selected plates promptly after incubation. If counting must be delayed temporarily, store plates at 5 to 10°C for no more than 24 h, but avoid this as routine practice. Record results of sterility controls on the report for each lot of samples.

Use an approved counting aid, such as the Quebec colony counter, for manual counting. If such equipment is not available, count with any other counter provided that it gives equivalent magnification and illumination. Automatic plate counting instruments are available. These generally use a television scanner coupled to a magnifying lens and an electronics package. Their use is acceptable if evaluation in parallel with manual counting gives comparable results.

In preparing plates, pipet sample volumes that will yield from 30 to 300 colonies/plate. The aim is to have at least one dilution giving colony counts between these limits, except as provided below.

Ordinarily, do not pipet more than 2.0 mL of sample; however, when the total number of colonies developing from 2.0 mL is less than 30, disregard the rule above and record result observed. With this exception, consider only plates having 30 to 300 colonies in determining the plate count. Compute bacterial count per milliliter by the following equation:

$$\text{CFU/mL} = \frac{\text{colonies counted}}{\text{actual volume of sample in dish, mL}}$$

If there is no plate with 30 to 300 colonies, and one or more plates have more than 300 colonies, use the plate(s) having a count nearest 300 colonies. Compute the count as above and report as estimated CFU per milliliter.

If plates from all dilutions of any sample have no colonies, report the count as less than one (< 1) divided by the corresponding largest sample volume used. For example, if no colonies develop from the 0.01-mL sample volume, report the count as less than 100 (< 100) estimated CFU/mL.

If the number of colonies per plate far exceeds 300, do not report result as "too numerous to count" (TNTC). If there are fewer than 10 colonies/cm^2, count colonies in 13 squares (of the colony counter) having representative colony distribution. If possible, select seven consecutive squares horizontally across the plate and six consecutive squares vertically, being careful not to count a square more than once. Multiply sum of the number of colonies in 13 representative square centimeters by 5 to compute estimated colonies per plate when the plate area is 65 cm^2. When there are more than 10 colonies/cm^2, count four representative squares, take average count per square centimeter, and multiply by the appropriate factor to estimate colonies per plate. The factor is 57 for disposable plastic plates and 65 for glass plates. When bacterial counts on crowded plates are greater than 100 colonies/cm^2, report result as greater than (>) 6500 divided by the smallest sample volume plated for glass plates or greater than (>) 5700 divided by the smallest sample volume plated for plastic plates. Report as estimated colony-forming units per milliliter.

If spreading colonies (spreaders) are encountered on the plate(s) selected, count colonies on representative portions only when colonies are well distributed in spreader-free areas and the area covered by the spreader(s) does not exceed one-half the plate area.

When spreading colonies must be counted, count each of the following types as one: a chain of colonies that appears to be caused by disintegration of a bacterial clump as agar and sample were mixed; a spreader that develops as a film of growth between the agar and bottom of petri dish; and a colony that forms in a film of water at the edge or over the agar surface. The last two types largely develop because of an accumulation of moisture at the point from which the spreader originates. They frequently cover more than half the plate and interfere with obtaining a reliable plate count.

Count as individual colonies similar-appearing colonies growing in close proximity but not touching, provided that the distance between them is at least equal to the diameter of the smallest colony. Count impinging colonies that differ in appearance, such as morphology or color, as individual colonies.

If plates have excessive spreader growth, report as "spreaders" (Spr). When plates are uncountable because of missed dilution, accidental dropping, and contamination, or the control plates indicate that the medium or other material or labware was contaminated, report as "laboratory accident" (LA).

b. Membrane filter method: Count colonies on membrane filters using a stereoscopic microscope at 10 to 15 × magnification. Preferably slant petri dish at 45° angle on microscope stage and adjust light source vertical to the colonies. Optimal colony density per filter is 20 to 200. If colonies are small and there is no crowding, a higher limit is acceptable.

Count all colonies on the membrane when there are ≤ 2 colonies per square. For 3 to 10 colonies per square count 10 squares and obtain average count per square. For 10 to 20 colonies per square count 5 squares and obtain average count per square. Multiply average count per square by 100 and divide by the sample volume to give colonies per milliliter. If there are more than 20 colonies per square, record count as > 2000 divided by the sample volume. Report averaged counts as estimated colony-forming units. Make estimated counts only when there are discrete, separated colonies without spreaders.

9. Computing and Reporting Counts

The term "colony-forming units" (CFU) is descriptive of the methods used; therefore, report all counts as colony-forming units. Include in the report the method used, the incubation temperature and time, and the medium. For example: CFU/mL, pour plate method, 35°C/48 h, plate count agar.

To compute the heterotrophic plate count, CFU/mL, divide total number of colonies or average number (if duplicate plates of the same dilution) per plate by the sample volume. Record sample volumes used and number of colonies on each plate counted or estimated.

When colonies on duplicate plates and/or consecutive dilutions are counted and results are averaged before being recorded, round off counts to two significant figures only when converting to colony-forming units.

Avoid creating fictitious precision and accuracy when computing colony-forming units by recording only the first two left-hand digits. Raise the second digit to the next higher number when the third digit from the left is 5, 6, 7, 8, or 9; use zeros for each successive digit toward the right from the second digit. For example, report a count of 142 as 140 and a count of 155 as 160, but report a count of 35 as 35.

10. Analytical Bias

Avoid inaccuracies in counting due to carelessness, damaged or dirty optics that impair vision, or failure to recognize colonies. Laboratory workers who cannot duplicate their own counts on the same plate within 5% and the counts of other analysts within 10% should discover the cause and correct such disagreements.

9215 B.   Pour Plate Method

1. Samples and Sample Preparation

See 9215A.4 and 9215A.5.

2. Sample Dilution

Prepare water used for dilution blanks as directed in Section 9050C.

a. Selecting dilutions: Select the dilution(s) so that the total number of colonies on a plate will be between 30 and 300 (Figure 9215:1). For example, where a heterotrophic plate count as high as 3000 is suspected, prepare plates with 10^{-2} dilution.

For most potable water samples, plates suitable for counting will be obtained by plating 1 mL and 0.1 mL undiluted sample and 1 mL of the 10^{-2} dilution.

b. Measuring sample portions: Use a sterile pipet for initial and subsequent transfers from each container. If pipet becomes contaminated before transfers are completed, replace with a sterile pipet. Use a separate sterile pipet for transfers from each different dilution. Do not prepare dilutions and pour plates in direct sunlight. Use caution when removing sterile pipets from the container; to avoid contamination, do not drag pipet tip across exposed ends of pipets in the pipet container or across lips and necks of dilution bottles. When removing sample, do not insert pipets more than 2.5 cm below the surface of sample or dilution.

c. Measuring dilutions: When discharging sample portions, hold pipet at an angle of about 45° with tip touching bottom of petri dish or inside neck of dilution bottle. Lift cover of petri dish just high enough to insert pipet. Allow 2 to 4 s for liquid to drain from 1-mL graduation mark to tip of pipet. If pipet is not a blow-out type, touch tip of pipet *once* against a dry spot on petri dish bottom. Less preferably, use a cotton-plugged blow-out-type pipet and gently blow out remaining volume of sample dilution. When 0.1-mL quantities are measured, let diluted sample drain from

chosen reference graduation until 0.1 mL has been delivered. Remove pipet without retouching it to dish. Pipet 1 mL, 0.1 mL, or other suitable volume into sterile petri dish before adding melted culture medium. Use decimal dilutions in preparing sample volumes of less than 0.1 mL; in examining sewage or turbid water, do not measure a 0.1-mL inoculum of original sample, but prepare an appropriate dilution. Prepare at least two replicate plates for each sample dilution used. After depositing test portions for each series of plates, pour culture medium and mix carefully. Do not let more than 20 min elapse between starting pipetting and pouring plates.

3. Plating

a. Melting medium: Melt sterile solid agar medium in boiling water or by exposure to flowing steam in a partially closed container, but avoid prolonged exposure to unnecessarily high temperatures during and after melting. Do not resterilize plating medium. If the medium is melted in two or more batches, use all of each batch in order of melting, provided that the contents remain fully melted. Discard melted agar that contains precipitate.

Maintain melted medium in a water bath between 44 and 46°C until used, preferably no longer than 3 h. In a separate container place a thermometer in water or medium that has been exposed to the same heating and cooling as the plating medium. Do not depend on the sense of touch to indicate proper medium temperature when pouring agar.

Use plate count agar, R2A agar, or NWRI agar as specified in Section 9215A.6. Before using a new lot of medium test its suitability.

b. Pouring plates: Limit the number of samples to be plated in any one series so that no more than 20 min (preferably 10 min) elapse between dilution of the first sample and pouring of the last plate in the series. Pour at least 10 to 12 mL liquefied medium

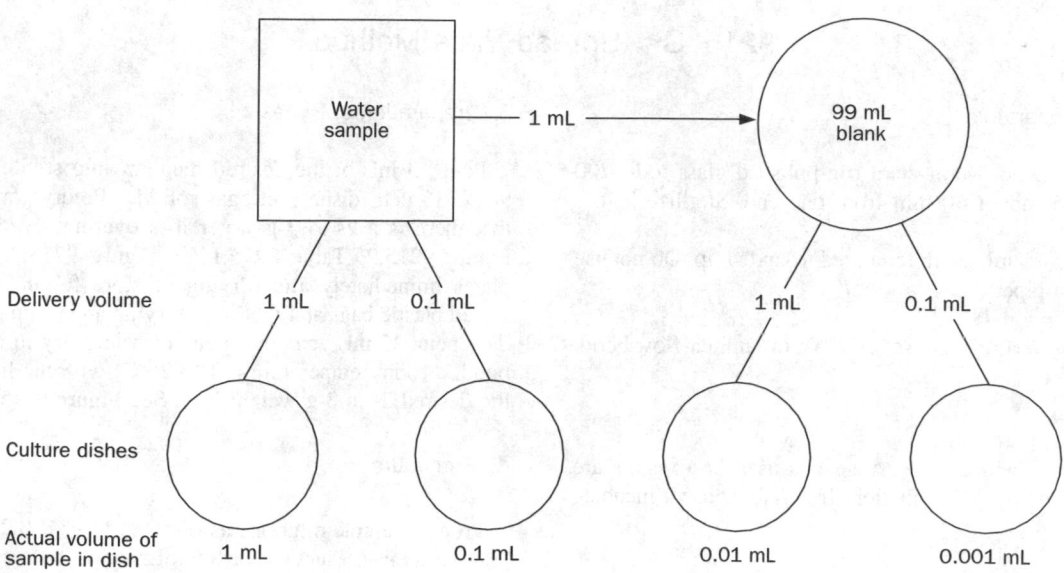

Figure 9215:1. Preparation of dilutions.

maintained at 44 to 46°C into each dish by gently lifting cover just high enough to pour. Carefully avoid spilling medium on outside of container or on inside of dish lid when pouring. When pouring agar from flasks or tubes that have been held in a water bath, wipe with clean paper towel and flame the neck before pouring. As each plate is poured mix melted medium thoroughly with test portions in petri dish, taking care not to splash mixture over the edge, by rotating the dish first in one direction and then in the opposite direction, or by rotating and tilting. Let plates solidify (within 10 min) on a level surface. After medium solidifies, invert plates and place in incubator.

c. Sterility controls: Check sterility of medium and dilution water blanks by pouring control plates for each series of samples. Prepare additional controls to determine contamination of plates, pipets, and room air.

4. Incubation

See Section 9215A.7.

5. Counting, Recording, Computing, and Reporting

See Sections 9215A.8 and 9215A.9.

6. Bibliography

BREED, R.S. & W.D. DOTTERER. 1916. The number of colonies allowable on satisfactory agar plates. Tech. Bull. 53, New York Agricultural Experiment Sta.

BUTTERFIELD, C.T. 1933. The selection of a dilution water for bacteriological examinations. *J. Bacteriol.* 23:355; *Pub. Health Rep.* 48: 681.

ARCHAMBAULT, J., J. CUROT & M.H. MCCRADY. 1937. The need of uniformity of conditions for counting plates (with suggestions for a standard colony counter). *Amer. J. Pub. Health* 27:809.

RICHARDS, O.W. & P.C. HEIJN. 1945. An improved dark-field Quebec colony counter. *J. Milk Technol.* 8:253.

BERRY, J.M., D.A. MCNEILL & L.D. WITTER. 1969. Effect of delays in pour plating on bacterial counts. *J. Dairy Sci.* 52:1456.

GELDREICH, E.E., H.D. NASH, D.J. REASONER & R.H. TAYLOR. 1972. The necessity of controlling bacterial populations in potable waters: Community water supply. *J. Amer. Water Works Assoc.* 64:596.

GELDREICH, E.E. 1973. Is the total count necessary? Proc. 1st Annu. AWWA Water Quality Technol. Conf., Dec. 3-4, 1973. Cincinnati, Ohio, p. VII-1. American Water Works Assoc., Denver, Colo.

GINSBURG, W. 1973. Improved total count techniques. Proc. 1st Annu. AWWA Water Quality Technol. Conf., Dec. 3-4, 1973. Cincinnati, Ohio, p. VIII-1. American Water Works Assoc., Denver, Colo.

DUTKA, B.J., A.S.Y. CHAU & J. COBURN. 1974. Relationship of heterotrophic bacterial indicators of water pollution and fecal sterols. *Water Res.* 8:1047.

KLEIN, D.A. & S. WU. 1974. Stress: a factor to be considered in heterotrophic microorganism enumeration from aquatic environments. *Appl. Microbiol.* 37:429.

GELDREICH, E.E., H.D. NASH, D.J. REASONER & R.H. TAYLOR. 1975. The necessity for controlling bacterial populations in potable waters: Bottled water and emergency water supplies. *J. Amer. Water Works Assoc.* 67:117.

BELL, C.R., M.A. HOLDER-FRANKLIN & M. FRANKLIN. 1980. Heterotrophic bacteria in two Canadian rivers.—I. Seasonal variation in the predominant bacterial populations. *Water Res.* 14:449.

MEANS, E.G., L. HANAMI, G.F. RIDGWAY & B.H. OLSON. 1981. Evaluating mediums and plating techniques for enumerating bacteria in water distribution systems. *J. Amer. Water Works Assoc.* 73: 585.

AMERICAN PUBLIC HEALTH ASSOCIATION. 1993. Standard Methods for the Examination of Dairy Products, 16th ed. American Public Health Assoc., Washington, D.C.

REASONER, D.J. & E.E. GELDREICH. 1985. A new medium for the enumeration and subculture of bacteria from potable water. *Appl. Environ. Microbiol.* 49:1.

9215 C.   Spread Plate Method

1. Laboratory Apparatus

a. Glass rods: Bend 4-mm-diam fire-polished glass rods, 200 mm in length, 45° about 40 mm from one end. Sterilize before using.

b. Pipet, glass, 1.1 mL, with tempered, rounded tip. Do not use disposable plastic pipets.

c. Turntable (optional).*

d. Incubator or drying oven, set at 42°C, or laminar-flow hood.

2. Media

See 9215A.6*a, c,* and *d*. If R2A agar is used best results are obtained at 28°C with 7 d incubation; if NWRI is used, incubate at 20°C for 7 d.

3. Preparation of Plates

Pour 15 mL of the desired medium into sterile 100 × 15 or 90 × 15 petri dishes; let agar solidify. Predry plates inverted so that there is a 2- to 3-g water loss overnight with lids on. See Figure 9215:2, Table 9215:I, or Figure 9215:3. Use predried plates immediately after drying or store for up to 2 weeks in sealed plastic bags at 4°C. For predrying and using plates the same day, pour 25 mL agar into petri dish and dry in a laminar-flow hood at room temperature (24 to 26°C) with the lid off to obtain the desired 2- to 3-g weight loss. See Figure 9215:3.

4. Procedure

Prepare sample dilutions as directed in 9215B.2.

a. Glass rod: Pipet 0.1 or 0.5 mL sample onto surface of predried agar plate. Using a sterile bent glass rod, distribute inoculum over surface of the medium by rotating the dish by hand or on a

---

* Fisher Scientific, hand operated, No. 08-758 or Lab-Line, motor driven, No. 1580, or equivalent.

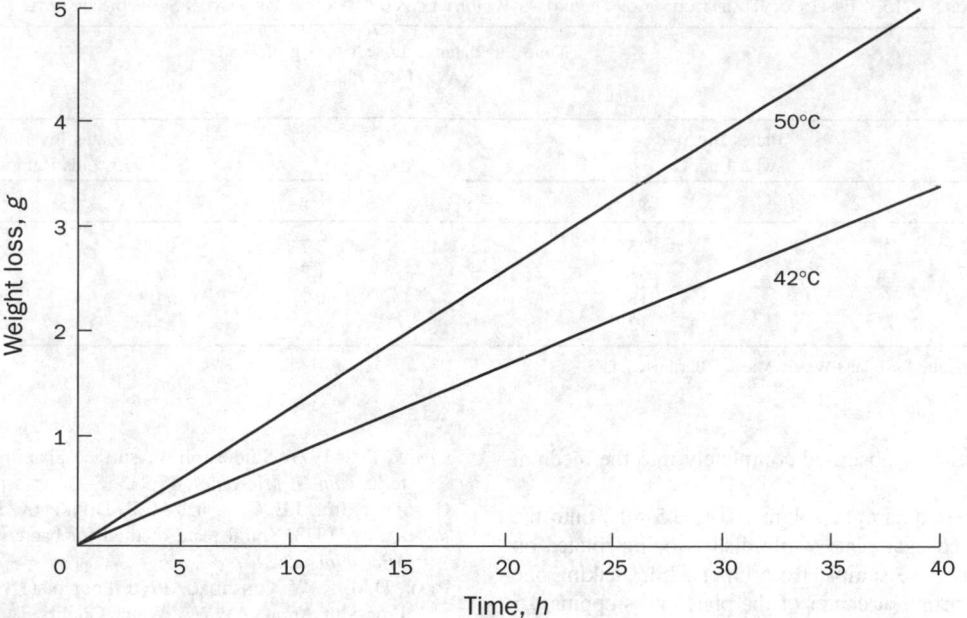

Figure 9215:2. Drying weight loss of 15-mL agar plates stored separately, inverted with lids on. Source: Unpublished data. Water Purification Lab., Chicago Dep. Water.

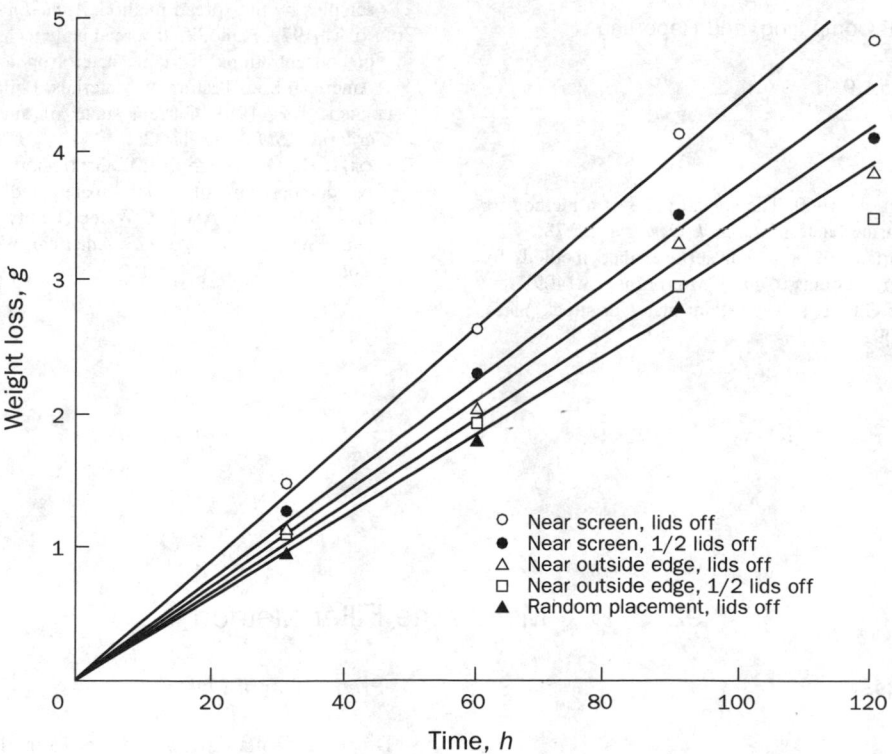

- ○ Near screen, lids off
- ● Near screen, 1/2 lids off
- △ Near outside edge, lids off
- □ Near outside edge, 1/2 lids off
- ▲ Random placement, lids off

Figure 9215:3. Weight loss of 25-mL agar plates (100 × 15 mm) dried separately in a laminar-flow hood at room temperature (24 to 26°C), relative humidity (30 to 33%), and air velocity 0.6 m/s. Source: Unpublished data. Alberta Environmental Centre, Vegreville, Alta.

TABLE 9215:I. EFFECT OF TEMPERATURE OF DRYING ON WEIGHT LOSS OF 15-mL AGAR PLATES STORED SEPARATELY*

Temp. °C	Time for Plates to Lose 1 to 4 g of Water (Avg. for 5 Plates) h							
	Plates Inverted with Lids On				Plates Inverted with Lids Removed			
	1 g	2 g	3 g	4 g	1 g	2 g	3 g	4 g
24	32	64	95	125	3.7	7.0	10.5	14.0
37	17	35	51	67	1.7	3.5	5.3	7.0
50	6	12	18	24	0.7	1.3	1.9	2.7
60	4	8	12	16	—	—	—	—

* Referenced in Canada Centre for Inland Waters Manual, Burlington, Ont.

turntable. Let inoculum be absorbed completely into the medium before incubating.

b. Pipet: Pipet desired sample volume (0.1, 0.5 mL) onto the surface of the predried agar plate while dish is being rotated on a turntable. Slowly release sample from pipet while making one to-and-fro motion, starting at center of the plate and stopping 0.5 cm from the plate edge before returning to the center. Lightly touch the pipet to the plate surface. Let inoculum be absorbed completely by the medium before incubating.

5. Incubation

See 9215A.7.

6. Counting, Recording, Computing, and Reporting

See 9215A.8 and 9215A.9.

7. Bibliography

BUCK, J.D. & R.C. CLEVERDON. 1960. The spread plate as a method for the enumeration of marine bacteria. *Limnol. Oceanogr.* 5:78.

CLARK, D.S. 1967. Comparison of pour and surface plate methods for determination of bacterial counts. *Can. J. Microbiol.* 13:1409.

VAN SOESTBERGAN, A.A. & C.H. LEE. 1969. Pour plates or streak plates. *Appl. Microbiol.* 18:1092.

CLARK, D.S. 1971. Studies on the surface plate method of counting bacteria. *Can. J. Microbiol.* 17:943.

GILCHRIST, J.E., J.E. CAMPBELL, C.B. DONNELLY, J.T. PELLER & J.M. DELANEY. 1973. Spiral plate method for bacterial determination. *Appl. Microbiol.* 25:244.

PTAK, D.M. & W. GINSBURG. 1976. Pour plate vs. streak plate method. Proc. 4th Annu. AWWA Water Quality Technol. Conf., Dec. 6-7, 1976. San Diego, Cal., p. 2B-5. American Water Works Assoc., Denver, Colo.

DUTKA, B.J., ed. 1978. Methods for Microbiological Analysis of Waters, Wastewaters and Sediments. Inland Waters Directorate, Scientific Operation Div., Canada Centre for Inland Waters, Burlington, Ont.

KAPER, J.B., A.L. MILLS & R.R. COLWELL. 1978. Evaluation of the accuracy and precision of enumerating aerobic heterotrophs in water samples by the spread method. *Appl. Environ. Microbiol.* 35:756.

YOUNG, M. 1979. A modified spread plate technique for the determination of concentrations of viable heterotrophic bacteria. STP 673:41-51, American Soc. Testing & Materials, Philadelphia, Pa.

GELDREICH, E.E. 1981. Current status of microbiological water quality criteria. *ASM News* 47:23.

TAYLOR, R.H., M.J. ALLEN & E.E. GELDREICH. 1981. Standard plate count: A comparison of pour plate and spread plate methods. Proc. 9th Annu. AWWA Water Quality Technol. Conf., Dec. 6-9, 1981. Seattle, Wash., p. 223. American Water Works Assoc. Denver, Colo.

9215 D.   Membrane Filter Method

1. Laboratory Apparatus

See Section 9222B.1.

2. Media

See 9215A.6. Use m-HPC agar, or alternatively R2A or NWRI agar.

3. Preparation of Plates

Dispense 5-mL portions of sterile medium* into 50- × 9-mm petri dishes. Let solidify at room temperature. Prepared plates may be stored inverted in a plastic bag or tight container in a refrigerator, for no longer than 2 weeks.

* m-HPC agar may not be sterile.

4. Sample Size

The volume to be filtered will vary with the sample. Select a maximum sample size to give 20 to 200 CFU per filter.

5. Procedure

Filter appropriate volume through a sterile 47-mm, 0.45-μm, gridded membrane filter, under partial vacuum. Rinse funnel with three 20- to 30-mL portions of sterile dilution water. Place filter on agar in petri dish.

6. Incubation

Place dishes in close-fitting box or plastic bag containing moistened paper towels. Incubate at $35 \pm 0.5°C$ for 48 h if using m-HPC agar, or longer if using R2A medium, or at 20 to 28°C for 5 to 7 d if using NWRI or R2A agar. Duplicate plates may be incubated for other time and temperature conditions as desired.

7. Counting, Recording, Computing, and Reporting

See 9215A.8 and 9215A.9. Report as CFU/mL, membrane filter method, time, medium.

8. Bibliography

CLARK, H.F., E.E. GELDREICH, H.L. JETER & P.W. KABLER. 1951. The membrane filter in sanitary bacteriology. *Pub. Health Rep.* 66:951.

STOPERT, E.M., W.T. SOKOSKI & J.T. NORTHAM. 1962. The factor of temperature in the better recovery of bacteria from water by filtration. *Can. J. Microbiol.* 8:809.

TAYLOR, R.H. & E.E. GELDREICH. 1979. A new membrane filter procedure for bacterial counts in potable water and swimming pool samples. *J. Amer. Water Works Assoc.* 71:402.

CLARK, J.A. 1980. The influence of increasing numbers of non-indicator organisms upon the detection of indicator organisms by the membrane filter and presence-absence tests. *Can. J. Microbiol.* 20: 827.

DUTKA, B.J., ed. 1981. Membrane Filtration, Applications, Techniques, and Problems. Marcel Dekker, Inc., New York, N.Y. and Basel, Switzerland.

HOADLEY, A.W. 1981. Effect of injury on the recovery of bacteria on membrane filters. *In* B. J. Dutka, ed. Membrane Filtration, Applications, Techniques, and Problems, p. 413. Marcel Dekker, Inc., New York, N.Y. and Basel, Switzerland.

9216   DIRECT TOTAL MICROBIAL COUNT*

9216 A.   Introduction

Direct total cell counts of bacteria in water or wastewater usually exceed counts obtained from heterotrophic plate counts and

* Approved by Standard Methods Committee, 1993.

most probable number methods because, unlike those procedures, direct counts preclude errors caused by viability-related phenomena such as selectivity of growth media, cell clumping, and slow growth rates.

9216 B.   Epifluorescence Microscopic Method

1. General Discussion

The epifluorescence microscopic method produces direct total cell counts with relative speed (20 to 30 min from time of sampling) and sensitivity. It does not permit differentiation of bacterial cells on the basis of taxonomy, metabolic activity, or viability, and it cannot be used to estimate the microbial biomass because of considerable variation in the volume of individual cells. The method requires an experienced technician who can distinguish microbial cells from debris on the basis of morphology.

The method consists of sample fixation for storage, staining with a chemical fluorochrome, vacuum filtration onto a nonfluorescing polycarbonate membrane, and enumeration by counting with an epifluorescence microscope.

2. Apparatus

a. Microscope, vertical UV illuminator for epifluorescence with flat field $100\times$ oil immersion objective lens, to give total magnification of at least $1000\times$.

b. Counting graticule, ocular lens micrometer* calibrated with stage micrometer.*

c. Filters,† including excitation filters (KP 490 and LP 455), beam splitter (LP 510), and barrier filter (LP 520 using mercury lamp, HBO 50).

d. Blender or vortex mixer.

e. Filtration unit, suitable for use with 25-mm-diam membrane filters.

* Bausch & Lomb No. 31-16-13.
† K. Zeiss or equivalent.

f. Membrane filters, polycarbonate,‡ 25-mm-diam, 0.2-μm pore size (purchase nonfluorescent or prepare by soaking membrane in Irgalan black [2 g/L in 2% acetic acid] for 24 h, then rinse in water and air dry); cellulosic§ 25-mm-diam, 5-μm pore size.

g. Syringes, 3-mL, disposable, with disposable syringe filters, 0.2-μm pore size.

h. Test tubes, glass, screw-capped, 13- × 125-mm.

3. Reagents

a. Phosphate buffer: Dissolve 13.6 g KH_2PO_4 in water and dilute to 1 L. Adjust to pH 7.2 if necessary; filter through 0.2-μm membrane filter.

b. Fixative, 5.0% (w/v) glutaraldehyde in phosphate buffer. Prepare fresh daily.

c. Fluorochrome, 0.1% (w/v) acridine orange‖ in phosphate buffer.

d. Immersion oil, low fluorescing.#

4. Procedure

Collect water samples as directed in Section 9060. Add 9.0 mL sample to test tube containing 1.0 mL fixative. Fixed samples can be stored at 4°C for up to 3 weeks without significant decrease in cell numbers.

Disperse and dilute samples from mesotrophic or eutrophic sources to obtain reproducible results. Mix sample using blender or vortex mixer, then make tenfold dilutions in phosphate buffer as necessary. Clean water samples may not require dilution but larger sample volumes (>100 mL) may be required to obtain reliable counts.

Place 1 mL sample or dilution on a nonfluorescent polycarbonate filter supported by a cellulosic membrane filter in filter holder. Using disposable sterile syringe filters, add 1 mL fluoro-

---

‡ Nuclepore Corp. or equivalent.
§ Millipore Corp. or equivalent.
‖ Sigma Chemical Co., biological grade, or equivalent.
Cargille Laboratories, Inc., Type B, or equivalent.

chrome and wait 2 min, then add about 3 mL filtered phosphate buffer to promote more even cell distribution. Alternatively, combine fluorochrome with sample in a small clean vial, let react, and add mixture to filter holder. Filter with vacuum (about 13 kPa). Wash with 2 mL phosphate buffer and filter. Remove polycarbonate filter with forceps and air dry for 1 to 2 min. The filter can be cut into quarter sections and saved if needed. Place dried filter on a drop of immersion oil on a clean glass microscope slide. Add a small drop of immersion oil to filter surface. Gently cover filter with a clean glass cover slip. Samples can be stored in the dark for several months without significant loss of fluorescence.

Examine at least 10 randomly selected fields on the filter using the 100× oil immersion lens to establish that distribution of microbial cells is uniform and that individual cells can be enumerated (if not, dilute sample and repeat). Preferably count 10 to 50 cells per field. Count number of cells in at least 20 squares using the calibrated counting graticule.

5. Calculations

Calculate the average number of cells per filter. Obtain effective filter area from specifications of filtration unit. Extrapolate to determine number of cells per milliliter of sample:

Total cells/mL = (avg cells/square) × (squares/filter)
$$\times \text{ (dilution factor) / sample volume, mL.}$$

6. Bibliography

HOBBIE, J.E., R.J. DALEY & S. JASPER. 1977. Use of nuclepore filters for counting bacteria by fluorescence microscopy. *Appl. Environ. Microbiol.* 33:1225.

SIERACK, M.E., P.W. JOHNSON & J.McH. SIEBURTH. 1985. Detection, enumeration, and sizing of planktonic bacteria by image-analyzed epifluorescence microscopy. *Appl. Environ. Microbiol.* 49:799.

AMERICAN SOCIETY FOR TESTING AND MATERIALS. 1987. Standard test method for enumeration of aquatic bacteria by epifluorescence microscopy counting procedure. ASTM D4455-85, Annual Book of ASTM Standards, Vol. 11.02, Water. American Soc. Testing & Materials, Philadelphia, Pa.

9217 ASSIMILABLE ORGANIC CARBON*

9217 A. Introduction

1. Significance

Growth of bacteria in drinking water distribution and storage systems can lead to the deterioration of water quality, violation of water quality standards, and increased operating costs. Growth or regrowth results from viable bacteria surviving the disinfection process and utilizing nutrients in the water and biofilm to

---

* Approved by Standard Methods Committee, 1997.

sustain growth.[1] Factors other than nutrients that influence regrowth include temperature,[2] residence time in mains and storage units,[3] and the efficacy of disinfection.[4] Tests to determine the potential for bacterial regrowth focus on the concentration of nutrients.[5–7]

Not all organic compounds are equally susceptible to microbial decomposition; the fraction that provides energy and carbon for bacterial growth has been called labile dissolved organic carbon,[8,9] biodegradable organic carbon (BDOC),[7] or assimilable organic carbon (AOC).[5] Easily measured chemical surrogates for AOC are not available now.[10,11] As alternatives to chemical methods, bioassays have been proposed.[5-7,12-14]

In a bioassay, the growth of a bacterial inoculum to maximum density can be used to estimate the concentrations of limiting nutrients; the underlying assumptions of the AOC bioassay are that nitrogen and phosphorus are present in excess, i.e., that organic carbon is limiting, and that the bioassay organism(s) represent the physiological capabilities of the distribution system microflora. Various bioassay procedures use an inoculum of one to four species of bacteria[5,12,13,15,16] growing in log phase or present in late stationary phase, or may use undefined bacteria attached to a sand substratum,[7] suspended in the sample,[6] or filtered from the sample and then resuspended.[14] Incubation vessels vary as to material,[17] size,[18,19] closure,[18] and cleaning procedure.[5,18,19] Water to be tested for nutrient concentrations has been variously prepared.[5,7,14] The AOC bioassay is an indirect or surrogate method, wherein nutrient concentrations are not measured directly, but colony-forming units (CFU) of the bioassay organism(s) are the test variable. Nutrient concentrations have been estimated directly from changes in dissolved organic carbon concentrations within the test vessel[7] or indirectly from epifluorescence microscopic counts of the maximum number of bacterial cells grown,[13,14] turbidity,[14] or incorporation of tritiated thymidine into bacterial DNA.[6,20] CFU densities, total cell densities, or bacterial production are converted to nutrient concentration by the growth yield of bacteria, defined as either the ratio between CFU or cells produced and organic carbon used, or biomass produced and organic carbon used.[5,6]

2. Selection of Method

The method described below is a two-species bioassay using *Pseudomonas fluorescens* strain P-17 and *Spirillum* strain NOX (van der Kooij)[10] that has been modified to reduce problems of bacterial and carbon contamination.[18,19] It uses a defined inoculum and miniaturized incubation vessels, requires no specialized equipment, and has been related to the presence of coliforms in a drinking water distribution system.[22] The two-species inoculum probably underestimates the total quantity of AOC, is consistently lower than BDOC estimates, and does not provide an estimate of refractory organic carbon.[23] Critical aspects of the proposed method, including the preparation of the incubation vessel, test water, and inoculum, and enumeration of the test organisms, are transferable to alternate AOC assays that use a different defined inoculum.

With an undefined bacterial inoculum, enumeration by the spread plate technique is not applicable; alternate response variables, such as changes in dissolved organic carbon (DOC) concentration, turbidity, epifluorescence microscopic counts, bacterial mortality, or bacterial growth, have been used.[6,7,14]

3. Sampling and Storage

Follow precautions outlined in Section 9060A and B for collecting and storing samples. Pasteurized and dechlorinated water samples probably can be held for several days without deterioration if properly sealed. Initiate the AOC assay as quickly as possible after pasteurization (see ¶ B.4c).

4. References

1. CHARACKLIS, W.G. 1988. Bacterial Regrowth in Distribution Systems. American Water Works Assoc. Research Foundation Research Rep., American Water Works Assoc., Denver, Colo.
2. FRANSOLET, G., G. VILLERS & W.J. MASSCHELEIN. 1985. Influence of temperature on bacterial development in waters. *Ozone Sci. Eng.* 7: 205.
3. MAUL, A., A.H. EL-SHAARAWI & J.C. BLOCK. 1985. Heterotrophic bacteria in water distribution systems. I. Spatial and temporal variation. *Sci. Total Environ.* 44:201.
4. LECHEVALLIER, M.W., C.D. CAWTHON & R.G. LEE. 1988. Factors promoting survival of bacteria in chlorinated water supplies. *Appl. Environ. Microbiol.* 54:649.
5. VAN DER KOOIJ, D., A. VISSER & W.A.M. HIJNEN. 1982. Determining the concentration of easily assimilable organic carbon in drinking water. *J. Amer. Water Works Assoc.* 74:540.
6. SERVAIS, P., G. BILLEN & M.C. HASCOET. 1987. Determination of the biodegradable fraction of dissolved organic matter in waters. *Water Res.* 21:445.
7. JORET, J.C., Y. LEVI, T. DUPIN & M. GILBERT. 1988. Rapid method for estimating bioeliminable organic carbon in water. *In* Proc. Annu. Conf. American Water Works Association, June 19–23, 1988, Orlando, Fla., p. 1715. American Water Works Assoc., Denver, Colo.
8. WETZEL, R.G. & B.A. MANNY. 1972. Decomposition of dissolved organic carbon and nitrogen compounds from leaves in an experimental hard-water stream. *Limnol. Oceanogr.* 17:927.
9. OGURA, N. 1975. Further studies on decomposition of dissolved organic matter in coastal seawater. *Mar. Biol.* 31:101.
10. VAN DER KOOIJ, D. 1988. Assimilable Organic Carbon (AOC) in Water. *In* The Search for a Surrogate. AWWA Research Foundation/KIWA Cooperative Research Rep. p. 311. American Water Works Assoc. Research Foundation, Denver, Colo.
11. KAPLAN, L.A. & T.L. BOTT. 1990. Nutrients for Bacterial Growth in Drinking Water: Bioassay Evaluation. EPA Project Summary, EPA-600/S2-89-030: 1-7. U.S. Environmental Protection Agency, Washington, D.C.
12. KENNY, F.A., J.C. FRY & R.A. BREACH. 1988. Development and Operational Implementation of Modified and Simplified Method for Determination of Assimilable Organic Carbon (AOC) in Drinking Water. International Assoc. Water Pollution Research & Control, Brighton, U.K., pp. 1–5.
13. NEDWELL, D.B. 1987. Distribution and pool sizes of microbially available carbon in sediment measured by a microbiological assay. *Microbiol. Ecol.* 45:47.
14. WERNER, P. 1984. Investigations on the substrate character of organic substances in connection with drinking water treatment. *Zentralbl. Bakt. Hyg.* 180:46.
15. VAN DER KOOIJ, D. & W.A.M. HIJNEN. 1983. Nutritional versatility of a starch utilizing *Flavobacterium* at low substrate concentrations. *Appl. Environ. Microbiol.* 45:804.
16. VAN DER KOOIJ, D. & W.A.M. HIJNEN. 1984. Substrate utilization of an oxalate-consuming *Spirillum* species in relation to its growth in ozonated water. *Appl. Environ. Microbiol.* 47:551.
17. COLBOURNE, J.S., R.M TREW & P.J. DENNIS. 1988. Treatment of water for aquatic bacterial growth studies. *J. Appl. Bacteriol.* 65:79.
18. KAPLAN, L.A. & T.L. BOTT. 1989. Measurement of assimilable organic carbon in water distribution systems by a simplified bioassay technique. *In* Advances in Water Analysis and Treatment, Proc. 16th Annu. AWWA Water Quality Technology Conf., Nov. 13–17, 1988, St. Louis, Mo., p. 475. American Water Works Assoc., Denver, Colo.
19. KAPLAN, L.A., T.L. BOTT & D.J. REASONER. 1993. Evaluation and simplification of the assimilable organic carbon nutrient bioassay for bacterial growth in drinking water. *Appl. Environ. Microbiol.* 59: 1532.
20. MORIARTY, D.J.W. 1986. Measurement of bacterial growth rates in aquatic systems from rates of nucleic acid synthesis. *In* K.C. Marshall, ed. *Advan. Microb. Ecol.* 9:245.

21. VAN DER KOOIJ, D., W.A.M. HIJNEN & J.C. KRUITHOF. 1989. The effects of ozonation, biological filtration and distribution on the concentration of easily assimilable organic carbon (AOC) in drinking water. *Ozone Sci. Eng.* 11:297.

22. LeCHEVALLIER, M.W., W.H. SHULZ & R.G. LEE. 1989. Bacterial nutrients in drinking water. *In* M.W. LeChevallier, B.H. Olson & G.A. McFeters, eds. Assessing and Controlling Bacterial Regrowth in Distribution Systems. American Water Works Assoc. Research Foundation Research Rep., American Water Works Assoc., Denver, Colo.

23. PREVOST, M., D. DUCHESNE, J. COALLIER, R. DESJARDINS & P. LAFRANCE. 1990. Full-scale evaluation of biological activated carbon filtration for the treatment of drinking water. *In* Advances in Water Analysis and Treatment, Proc. 17th Annu. AWWA Water Quality Technology Conf., Nov. 12–16, 1989, Philadelphia, Pa., p. 147. American Water Works Assoc., Denver, Colo.

5. Bibliography

VAN DER KOOIJ, D. 1979. Characterization and classification of *fluorescent pseudomonads* isolated from tap water and surface water. *Antonie van Leeuwenhoek* 45:225.

VAN DER KOOIJ, D., A. VISSER & W.A.M. HIJNEN. 1980. Growth of *Aeromonas hydrophila* at low concentrations of substrates added to tap water. *Appl. Environ. Microbiol.* 39:1198.

WERNER, P. 1981. Microbial studies on the chemical and biological treatment of ground water containing humic acid. *Vom Wasser* 57:157.

OLSON, B.H. 1982. Assessment and implications of bacterial regrowth in water distribution systems. EPA Project Summary, EPA-600/S2-82-072:1-10. U.S. Environmental Protection Agency, Washington, D.C.

RIZET, M., F. FIESSINGER & N. HOUEL. 1982. Bacterial regrowth in a distribution system and its relationship with the quality of the feed water: case studies. *In* Proc. Annu. Conf. American Water Works Association, May 16–20, 1982, Miami Beach, Fla., p. 1199. American Water Works Assoc., Denver, Colo.

VAN DER KOOIJ, D., J.P. ORANJE & W.A.M. HIJNEN. 1982. Growth of *Pseudomonas aeruginosa* in tap water in relation to utilization of substrates at concentrations of a few micrograms per liter. *Appl. Environ. Microbiol.* 44:1086.

CAMPER, A.K., M.W. LeChevallier, S.C. BROADAWAY & G.A. McFETERS. 1986. Bacteria associated with granular activated carbon particles in drinking water. *Appl. Environ. Microbiol.* 52:434.

WENG, C., D.L. HOVEN & B.J. SCHWARTZ. 1986. Ozonation: An economic choice for water treatment. *J. Amer. Water Works Assoc.* 78(11):83.

CARLUCCI, A.F., S.L. SHIMP & D.B. CRAVEN. 1987. Bacterial response to labile dissolved organic matter increases associated with marine discontinuities. Fed. European Microbiological Societies, *Microbiol. Ecol.* 45:211.

LeCHEVALLIER, M.W., T.M. BABCOCK & R.G. LEE. 1987. Examination and characterization of distribution system biofilms. *Appl. Environ. Microbiol.* 53:2714.

THINGSTAD, T.F. 1987. Utilization of N, P, and organic C by heterotrophic bacteria. I. Outline of a chemostat theory with a consistent concept of maintenance metabolism. *Marine Ecol. Progr. Ser.* 35:99.

ANSELME, C., I.H. SUFFET & J. MALLEVIALLE. 1988. Effects of ozonation on tastes and odors. *J. Amer. Water Works Assoc.* 80(10):45.

FRANSOLET, G., A. DEPELCHIN, G. VILLERS, R. GOOSSENS & W.J. MASSCHELEIN. 1988. The role of bicarbonate in bacterial growth in oligotrophic waters. *J. Amer. Water Works Assoc.* 80(11):57.

9217 B.   *Pseudomonas fluorescens* Strain P-17, *Spirillum* Strain NOX Method

1. General Discussion

a. Principle: The AOC bioassay using *Pseudomonas fluorescens* strain P-17 and *Spirillum* strain NOX involves growth to a maximum density of a small inoculum in a batch culture of pasteurized test water. Pasteurization inactivates native microflora. The test organisms are enumerated by the spread plate method for heterotrophic plate counts (Section 9215C) and the density of viable cells is converted to AOC concentrations by an empirically derived yield factor for the growth of P-17 on acetate-carbon and NOX on oxalate-carbon as standards. The number of organisms at stationary phase is assumed to be the maximum number of organisms that can be supported by the nutrients in the sample and the yields on acetate carbon and oxalate carbon are assumed to equal the yield on naturally occurring AOC.[1,2]

b. Interferences: Untreated surface waters, especially those with high concentrations of suspended solids or high turbidity, can contain large numbers of spore-forming bacteria that may survive pasteurization, grow, and interfere with the enumeration of P-17 and NOX on spread plates. Such waters generally have high AOC concentrations and can be diluted with organic-free water amended with mineral salts or prefiltered through carbon-free filters. Potable waters that have been disinfected and carry a disinfectant residual will inhibit growth of the test organism unless the disinfectant is neutralized. Surface waters from reservoirs treated with copper sulfate also may be inhibitory unless a chelating agent is added to the sample,[3] and lime-softened waters

with elevated pH values may require pH adjustment. Any amendment to a sample requires a control for AOC contamination.

c. Minimum detectable concentration: In theory, concentrations of less than 1 µg C/L can be detected. In practice, organic carbon contamination during glassware preparation and sample handling imposes a limit of detection of approximately 5 to 10 µg AOC/L.

2. Apparatus

a. Incubation vessels: Organic-carbon-free borosilicate glass vials (45 mL capacity) with TFE-lined silicone septa.

b. Incubator, set at 15 ± 0.5°C.

c. Hot water bath capable of achieving and holding 70°C.

d. Continuously adjustable pipet* capable of delivering between 10 and 100 µL.

e. Erlenmeyer flask, 125-mL, with ground-glass stopper.

f. Apparatus for preparing dilution water and making heterotrophic plate counts: See Sections 9050C and 9215C.

3. Reagents

a. Sodium acetate stock solution, 400 mg acetate-C/L: Dissolve 2.267 g $CH_3COONa \cdot 3H_2O$ in 1 L organic-carbon-free, deionized water. Transfer to 45-mL vials, fill to shoulder, cap tightly, and autoclave. Although standard autoclave practice is to loosen caps,

---

* Eppendorf or equivalent.

keep vials with septa capped tightly for autoclaving. Store at 5°C in tightly capped vials. Solution may be held for up to 6 months.

b. Sodium thiosulfate solution: Dissolve 30 g Na₂S₂O₃ in 1 L deionized water. Transfer to 45-mL vials and autoclave as directed in ¶ 3*a*.

c. Buffered water: See Section 9050C.

d. R2A agar: See Section 9215A.6*c*.

e. Sodium persulfate solution, 10% (w/v): Dissolve 100 g Na₂S₂O₈ in 1 L deionized water.

f. Organic-free water: See Section 5710B.3*e*. Alternatively, use HPLC-grade bottled water.

g. Mineral salts solution: Dissolve 171 mg K_2HPO_4, 767 mg NH_4Cl, and 1.444 g KNO_3 in 1 L carbon-free water. Transfer to 45-mL vials and autoclave as directed in ¶ 3*a*.

h. Cultures of strains P-17 (ATCC 49642) and NOX (ATCC 49643).†

4. Procedure

a. Preparation of incubation vessels: Wash 45-mL vials with detergent, rinse with hot water, 0.1*N* HCl two times, and deionized water three times, dry, cap with foil, and heat to 550°C for 6 h. Soak TFE-lined silicone septa in a 10% sodium persulfate solution for 1 h at 60°C; rinse three times with carbon-free deionized water. Alternatively, use pre-cleaned water sampling vials[4] or an equivalent AOC-free vial.‡ Use same cleaning procedure for all glassware.

b. Preparation of stock inoculum: Prepare individual turbid suspensions of P-17 and NOX by transferring growth from a slant culture on R2A agar into 2 to 3 mL filtered (0.2 μm), autoclaved sample. Use slant not older than 6 months. The autoclaved sample can be any water that supports growth of P-17 and NOX and is organic-carbon-limited. Neutralize chlorinated samples with sodium thiosulfate (42 μL/50 mL). Transfer 100 μL of suspension to 50 mL filtered, autoclaved sample in a sterile 125-mL ground-glass-stoppered erlenmeyer flask. Add 125 μL sodium acetate solution (suspension contains 1 mg acetate-C/L). Incubate at room temperature (≤ 25°C) until the viable cell count reaches the stationary phase. Organic-carbon limitation will insure complete utilization of acetate-C so that no AOC is transferred with the inoculum. The stationary phase is reached when the viable cell count, as measured by spread plates, reaches maximum value. Store stock cultures for not more than 6 months at 5°C. Before inoculating a bioassay vessel, make a viable count of the culture (spread plate) to determine the appropriate volume of inoculum to be added to each bioassay vessel.

c. Preparation of incubation water: Collect samples directly into 10 45-mL vials. Use 9 vials for AOC measurement and 1 for growth control. Fill each vial to the neck (40 mL) within as short a time as possible. Place septa on the vials, TFE side down, and secure with open-topped screw caps. Alternatively, collect 500 mL sample in an organic-carbon-free vessel and pour into each vial. Neutralize samples containing disinfectant residuals with 33 μL sodium thiosulfate solution added to each vial or 0.5 mL per 500-mL sample. Preferably, collect an extra vial to check for residual chlorine after neutralization. In the laboratory, cap vials tightly and pasteurize in 70°C water bath for 30 min.

d. Inoculation and incubation: Cool, inoculate with 500 colony-forming units (CFU)/mL each of P-17 and NOX, either by injecting through the septum or by removing cap and using a carbon-free pipet. Plastic, sterile tips for continuously adjustable pipets are suitable. Use the following equation to calculate volume of inoculum:

$$\text{Volume of inoculum} = \frac{(500 \text{ CFU/mL}) \times (40 \text{ mL/vial})}{\text{CFU/mL stock inoculum}}$$

Hold vials at 15°C in the dark for 1 week. If a 15°C incubator is unavailable, incubate at room temperature not to exceed 25°C. Because incubation temperature influences growth yield, record and report temperature. Determine yields as directed below if an alternative temperature is used.

e. Enumeration of test bacterium: On incubation days 7, 8, and 9 remove three vials from the incubator. Sample an individual vial on only 1 d. Shake vials vigorously for 1 min, remove 1 mL with a sterile pipet, and prepare a dilution series (see Section 9215B). Plate three dilutions (10^{-2}, 10^{-3}, and 10^{-4}) in duplicate. Incubate plates at 25°C for 3 to 5 d and score the number of colonies of each strain. P-17 colonies appear on plates first; they are 3 to 4 mm in diameter with diffuse yellow pigmentation. NOX colonies are small (1- to 2-mm diam) white dots. It may be necessary to count P-17 and NOX colonies at different dilutions. Sample vials on three separate days to check whether maximum density has been reached. Day-to-day variations of between 11 and 16% of the mean for batch cultures of P-17 in stationary phase are typical.[1] A consistent increase in cell densities of 20% or more over the 3-d period indicates that the cultures are not in stationary phase; repeat assay with longer incubation period. Alternatively, collect more samples (three for each additional sampling day) and prepare as in ¶ *c* above so that extended incubation can be used. A sharp population decrease of approximately 0.5 log over the 3 d is unusual, but may occur. If this happens repeat the assay.

f. Determination of yield of P-17 and NOX: The yields of P-17 and NOX on model carbon compounds should be constant if organic carbon is limiting and the incubation temperature is kept constant. It is acceptable to use the previously derived empirical yield values of 4.1×10^6 CFU P-17/μg acetate-C, 1.2×10^7 CFU-NOX/μg acetate-C, and 2.9×10^6 CFU-NOX/μg oxalate-C at 15°C.[5] However, the determination of a yield control provides an important check on both the bioassay (see also 6. Quality Control, below) and carbon limitation in the sample.

5. Calculation

a. AOC concentration: Average viable count results for the 3 d and calculate concentration of AOC as the product of the mean of the viable counts and the inverse of the yield:

μg AOC/L = [(mean P-17 CFU/mL)(1/yield)
   + (mean NOX CFU/mL)(1/yield)](1000 mL/L)

When the empirical yield factors[5] are used, the equation becomes:

μg AOC/L = [(mean P-17 CFU/mL)(μg acetate-C/4.1 × 10⁶ CFU)
   + (mean NOX CFU/mL)(μg oxalate-C/2.9 × 10⁶ CFU)]
   (1000 mL/L)

---

† Available from the American Type Culture Collection.
‡ Pierce Vari-Clean.

or

$$\mu g\ AOC/L = [(\text{mean P-17 CFU/mL})(2.44 \times 10^{-7}\ \mu g\ \text{acetate-C/CFU}) \\ + (\text{mean NOX CFU/mL})(3.45 \times 10^{-7}\ \mu g\ \text{oxalate-C/CFU})] \\ (1000\ \text{mL/L})$$

In practice, the densities of organisms vary during the stationary phase. Using average density over 3-d period provides a more accurate estimate of the real maximum density.

Reporting AOC as μg C/L assumes that the yields on acetate and oxalate are equal to the yields on naturally occurring AOC. To permit data comparisons report incubation temperature, contribution of each species to AOC, and yield factors used.

6. Quality Control

See Section 9020B for general quality control procedures. Quality control specific to the AOC bioassay includes testing the inoculum for purity and viability by plating a portion on R2A agar, testing the incubation vessel, inoculum, thiosulfate solution, and any supplemental procedure such as filtration or dilution for organic carbon contamination, testing the P-17 and NOX inocula for yield, and testing the sample for carbon limitation or inhibition of assay organisms. Test all deviations in procedure (see ¶ 6f).

To make these tests, use separate controls for blank, yield, and growth. The controls outlined below use a single vial and are meant as a trouble-shooting guide. Definitive determination, for example, that the yield is different from a published value or that a sample is inhibitory, requires replication and statistical analysis.

a. Blank control: Dilute mineral salts solution 10:1 with carbon-free water. Follow procedures outlined above: Fill a vial to the shoulder with organic-carbon-free water, add 100 μL mineral salts and 100 μL sodium thiosulfate, pasteurize, inoculate with P-17/NOX, incubate, and enumerate growth.

b. Yield control: Dilute sodium acetate or sodium oxalate solution 10:1 with carbon-free water, preparing 40 mg C/L working concentrations. Follow procedures outlined above: Fill a vial to the shoulder with carbon-free water, add 100 μL mineral salts, 100 μL sodium thiosulfate, and 100 μL sodium acetate or sodium oxalate working solution, pasteurize, inoculate with P-17/NOX, incubate, and enumerate growth. P-17, unlike NOX, will not grow with oxalate as sole carbon source (oxalate is considered a major by-product of ozonation). NOX growth in HPLC-grade water presumed to be organic carbon-free is to be expected. The yield control is a quality control measurement and is not intended to provide a conversion factor for the calculation of AOC.

c. Growth control: Use additional sample of test water collected with the nine AOC vials, (¶ 4c above) but amend with 100 μL diluted mineral salts and 100 μL of diluted acetate or oxalate solution per vial before pasteurization. As with other controls, inoculate with P-17/NOX, incubate, and enumerate growth.

d. Yield calculations: If previously derived empirical yield values (see ¶ 4f above) are not used, a conversion factor can be derived empirically by using pure cultures of P-17 and NOX. Mixed cultures of the organisms cannot be used and a separate blank control for each species is required. Convert density units to CFU/L by multiplying CFU/mL by 1000, and divide by 100 μg acetate or oxalate-C/L. Express yield as CFU P-17 or NOX/

μg acetate-C or oxalate-C. For P-17 and acetate-C, the equation is:

$$\text{Yield} = \frac{[(\text{P-17 CFU/mL:yield control}) - (\text{P-17 CFU/mL:blank control})] \times (1000\ \text{mL/L})}{100\ \mu g\ \text{acetate-C/L}}$$

$$= \text{P-17 CFU}/\mu g\ \text{acetate-C}$$

e. Interpretation of growth control: Subtract densities of P-17 and NOX that grew in the sample amended with only thiosulfate from the densities of P-17 and NOX that grew in the growth control. Compare difference to the difference between yield and blank controls.

If: (growth control − sample) = (yield control − blank control)

Then: sample is carbon-limited and not inhibitory

If: (growth control − sample) < (yield control − blank control)

Then: sample is inhibitory to bioassay organism

If: (growth control − sample) > (yield control − blank control)

Then: sample is not carbon-limited

f. Supplemental procedure check: When using such supplemental procedures as filtration, dilution, or chemical amendment check for carbon contribution to the AOC values. To test a procedure, use carbon-free water and blank control as a base line. Perform the supplemental procedure on additional carbon-free water and compare to densities of P-17 and NOX that grow in the blank control.

7. Precision and Bias

The P-17 bioassay performed in a single laboratory using 45-mL vials had a precision of ± 17.5% based on a total of 58 assays with 14 different samples.[6]

8. References

1. KAPLAN, L.A. & T.L. BOTT. 1989. Measurement of assimilable organic carbon in water distribution systems by a simplified bioassay technique. *In* Advances in Water Analysis and Treatment, Proc. 16th Annu. AWWA Water Quality Technology Conf., Nov. 13–17, 1988, St. Louis, Mo., p. 475. American Water Works Assoc., Denver, Colo.
2. VAN DER KOOIJ, D., A. VISSER & J.P. ORANJE. 1982. Multiplication of fluorescent pseudomonads at low substrate concentrations in tap water. *Antonie van Leeuwenhoek* 48:229.
3. LECHEVALLIER, M.W., W.H. SHULZ & R.G. LEE. 1989. Bacterial nutrients in drinking water. *In* M.W. LeChevallier, B.H. Olson & G.A. McFeters, eds. Assessing and Controlling Bacterial Regrowth in Distribution Systems. American Water Works Assoc. Research Foundation Research Rep., American Water Works Assoc., Denver, Colo.
4. KAPLAN, L.A. & T.L. BOTT. 1990. Modifications to simplify an AOC bioassay for routine use by utilities monitoring bacterial regrowth potential in water distribution systems. *In* Advances in Water Analysis and Treatment, Proc. 17th Annu. AWWA Water Quality Technology Conf., Nov. 12–16, 1989, Philadelphia, Pa., p. 1031. American Water Works Assoc., Denver, Colo.

5. VAN DER KOOIJ, D., W.A.M. HIJNEN & J.C. KRUITHOF. 1989. The effects of ozonation, biological filtration and distribution on the concentration of easily assimilable organic carbon (AOC) in drinking water. *Ozone Sci. Eng.* 11:297.
6. KAPLAN, L.A. & T.L. BOTT. 1990. Nutrients for bacterial growth in drinking water. Bioassay evaluation. EPA Project Summary, EPA-600/S2-89-030: 1-7. U.S. Environmental Protection Agency, Washington, D.C.

9. Bibliography

KING, E.O., M.K. WARD & D.E. RANEY. 1954. Two simple media for the demonstration of pyocyanin and fluorescin. *J. Lab. Clin. Med.* 44: 301.
MASON, J. & D.P. KELLY. 1988. Thiosulfate oxidation by obligately heterotrophic bacteria. *Microbial Ecol.* 15:123.

9221 MULTIPLE-TUBE FERMENTATION TECHNIQUE FOR MEMBERS OF THE COLIFORM GROUP*

9221 A. Introduction

The coliform group consists of several genera of bacteria belonging to the family Enterobacteriaceae. The historical definition of this group has been based on the method used for detection (lactose fermentation) rather than on the tenets of systematic bacteriology. Accordingly, when the fermentation technique is used, this group is defined as all facultative anaerobic, gram-negative, non-spore-forming, rod-shaped bacteria that ferment lactose with gas and acid formation within 48 h at 35°C.

The standard test for the coliform group may be carried out either by the multiple-tube fermentation technique or presence-absence procedure (through the presumptive-confirmed phases or completed test) described herein, by the membrane filter (MF) technique (Section 9222) or by the enzymatic substrate coliform test (Section 9223). Each technique is applicable within the limitations specified and with due consideration of the purpose of the examination. Production of valid results requires strict adherence to quality control procedures. Quality control guidelines are outlined in Section 9020.

When multiple tubes are used in the fermentation technique, results of the examination of replicate tubes and dilutions are reported in terms of the Most Probable Number (MPN) of organisms present. This number, based on certain probability formulas, is an estimate of the mean density of coliforms in the sample. Coliform density, together with other information obtained by engineering or sanitary surveys, provides the best assessment of water treatment effectiveness and the sanitary quality of source water.

The precision of each test depends on the number of tubes used. The most satisfactory information will be obtained when the largest sample inoculum examined shows gas in some or all of the tubes and the smallest sample inoculum shows no gas in all or a majority of the tubes. Bacterial density can be estimated by the formula given or from the table using the number of positive tubes in the multiple dilutions (9221C.2). The number of sample portions selected will be governed by the desired precision of the result. MPN tables are based on the assumption of a Poisson distribution (random dispersion). However, if the sample is not adequately shaken before the portions are removed or if clumping

of bacterial cells occurs, the MPN value will be an underestimate of the actual bacterial density.

1. Water of Drinking Water Quality

When drinking water is analyzed to determine if the quality meets the standards of the U.S. Environmental Protection Agency (EPA), use the fermentation technique with 10 replicate tubes each containing 10 mL, 5 replicate tubes each containing 20 mL, or a single bottle containing a 100-mL sample portion. When examining drinking water by the fermentation technique, process all tubes or bottles demonstrating growth with or without a positive acid or gas reaction to the confirmed phase (9221B.2). Apply the completed test (9221B.3) to not less than 10% of all coliform-positive samples per quarter. Obtain at least one positive sample per quarter. A positive EC broth (9221E) or a positive EC MUG broth (9221F) test result is considered an alternative to the positive completed test phase.

For the routine examination of public water supplies the object of the total coliform test is to determine the efficiency of treatment plant operation and the integrity of the distribution system. It is also used as a screen for the presence of fecal contamination. A high proportion of coliform occurrences in a distribution system may be attributed not to treatment failure at the plant or the well source, but to bacterial regrowth in the mains. Because it is difficult to distinguish between coliform regrowth and new contamination, assume all coliform occurrences to be new contamination unless otherwise demonstrated.

2. Water of Other than Drinking Water Quality

In the examination of nonpotable waters inoculate a series of tubes with appropriate decimal dilutions of the water (multiples and submultiples of 10 mL), based on the probable coliform density. Use the presumptive-confirmed phase of the multiple-tube procedure. Use the more labor-intensive completed test (9221B.3) as a quality control measure on at least 10% of coliform-positive nonpotable water samples on a seasonal basis. The object of the examination of nonpotable water generally is to estimate the density of bacterial contamination, determine a source of pollution, enforce water quality standards, or trace the survival of micro-

---

* Approved by Standard Methods Committee, 1994.

organisms. The multiple-tube fermentation technique may be used to obtain statistically valid MPN estimates of coliform density. Examine a sufficient number of samples to yield representative results for the sampling station. Generally, the geometric mean or median value of the results of a number of samples will yield a value in which the effect of sample-to-sample variation is minimized.

3. Other Samples

The multiple-tube fermentation technique is applicable to the analysis of salt or brackish waters as well as muds, sediments, and sludges. Follow the precautions given above on portion sizes and numbers of tubes per dilution.

To prepare solid or semisolid samples weigh the sample and add diluent to make a 10^{-1} dilution. For example, place 50 g sample in sterile blender jar, add 450 mL sterile phosphate buffer or 0.1% peptone dilution water, and blend for 1 to 2 min at low speed (8000 rpm). Prepare the appropriate decimal dilutions of the homogenized slurry as quickly as possible to minimize settling.

9221 B. Standard Total Coliform Fermentation Technique

1. Presumptive Phase

Use lauryl tryptose broth in the presumptive portion of the multiple-tube test. If the medium has been refrigerated after sterilization, incubate overnight at room temperature (20°C) before use. Discard tubes showing growth and/or bubbles.

a. Reagents and culture medium:

1) *Lauryl tryptose broth:*

Tryptose	20.0 g
Lactose	5.0 g
Dipotassium hydrogen phosphate, K_2HPO_4	2.75 g
Potassium dihydrogen phosphate, KH_2PO_4	2.75 g
Sodium chloride, NaCl	5.0 g
Sodium lauryl sulfate	0.1 g
Reagent-grade water	1 L

Add dehydrated ingredients to water, mix thoroughly, and heat to dissolve. pH should be 6.8 ± 0.2 after sterilization. Before sterilization, dispense sufficient medium, in fermentation tubes with an inverted vial, to cover inverted vial at least one-half to two-thirds after sterilization. Alternatively, omit inverted vial and add 0.01 g/L bromcresol purple to presumptive medium to determine acid production, the indicator of a positive result in this part of the coliform test. Close tubes with metal or heat-resistant plastic caps.

Make lauryl tryptose broth of such strength that adding 100-mL, 20-mL, or 10-mL portions of sample to medium will not reduce ingredient concentrations below those of the standard medium. Prepare in accordance with Table 9221:I.

b. Procedure:

1) Arrange fermentation tubes in rows of five or ten tubes each in a test tube rack. The number of rows and the sample volumes selected depend upon the quality and character of the water to be examined. For potable water use five 20-mL portions, ten 10-mL portions, or a single bottle of 100 mL portion; for nonpotable water use five tubes per dilution (of 10, 1, 0.1 mL, etc.).

In making dilutions and measuring diluted sample volumes, follow the precautions given in Section 9215B.2. Use Figure 9215:1 as a guide to preparing dilutions. Shake sample and dilutions vigorously about 25 times. Inoculate each tube in a set of five with replicate sample volumes (in increasing decimal dilutions, if decimal quantities of the sample are used). Mix test portions in the medium by gentle agitation.

2) Incubate inoculated tubes or bottles at 35 ± 0.5C. After 24 ± 2 h swirl each tube or bottle gently and examine it for growth, gas, and acidic reaction (shades of yellow color) and, if no gas or acidic reaction is evident, reincubate and reexamine at the end of 48 ± 3 h. Record presence or absence of growth, gas, and acid production. If the inner vial is omitted, growth with acidity signifies a positive presumptive reaction.

c. Interpretation: Production of an acidic reaction or gas in the tubes or bottles within 48 ± 3 h constitutes a positive presumptive reaction. Submit tubes with a positive presumptive reaction to the confirmed phase (9221B.2).

9221:I. Preparation of Lauryl Tryptose Broth

Inoculum mL	Amount of Medium in Tube mL	Volume of Medium + Inoculum mL	Dehydrated Lauryl Tryptose Broth Required g/L
1	10 or more	11 or more	35.6
10	10	20	71.2
10	20	30	53.4
20	10	30	106.8
100	50	150	106.8
100	35	135	137.1
100	20	120	213.6

The absence of acidic reaction or gas formation at the end of 48 ± 3 h of incubation constitutes a negative test. Submit drinking water samples demonstrating growth without a positive gas or acid reaction to the confirmed phase (9221B.2). An arbitrary 48-h limit for observation doubtless excludes occasional members of the coliform group that grow very slowly (see Section 9212).

2. Confirmed Phase

a. Culture medium: Use brilliant green lactose bile broth fermentation tubes for the confirmed phase.

Brilliant green lactose bile broth:

Peptone	10.0	g
Lactose	10.0	g
Oxgall	20.0	g
Brilliant green	0.0133	g
Reagent-grade water	1	L

Add dehydrated ingredients to water, mix thoroughly, and heat to dissolve. pH should be 7.2 ± 0.2 after sterilization. Before sterilization, dispense, in fermentation tubes with an inverted vial, sufficient medium to cover inverted vial at least one-half to two-thirds after sterilization. Close tubes with metal or heat-resistant plastic caps.

b. Procedure: Submit all presumptive tubes or bottles showing growth, any amount of gas, or acidic reaction within 24 ± 2 h of incubation to the confirmed phase. If active fermentation or acidic reaction appears in the presumptive tube earlier than 24 ± 2 h, transfer to the confirmatory medium; preferably examine tubes at 18 ± 1 h. If additional presumptive tubes or bottles show active fermentation or acidic reaction at the end of a 48 ± 3- h incubation period, submit these to the confirmed phase.

Gently shake or rotate presumptive tubes or bottles showing gas or acidic growth to resuspend the organisms. With a sterile loop 3.0 to 3.5 mm in diameter, transfer one or more loopfuls of culture to a fermentation tube containing brilliant green lactose bile broth or insert a sterile wooden applicator at least 2.5 cm into the culture, promptly remove, and plunge applicator to bottom of fermentation tube containing brilliant green lactose bile broth. Remove and discard applicator. Repeat for all other positive presumptive tubes.

Incubate the inoculated brilliant green lactose bile broth tube at 35 ± 0.5°C. Formation of gas in any amount in the inverted vial of the brilliant green lactose bile broth fermentation tube at any time (e.g., 6 ± 1 h, 24 ± 2 h) within 48 ± 3 h constitutes a positive confirmed phase. Calculate the MPN value from the number of positive brilliant green lactose bile tubes as described in Section 9221C.

c. Alternative procedure: Use this alternative only for polluted water or wastewater known to produce positive results consistently.

If all presumptive tubes are positive in two or more consecutive dilutions within 24 h, submit to the confirmed phase only the tubes of the highest dilution (smallest sample inoculum) in which all tubes are positive and any positive tubes in still higher dilutions. Submit to the confirmed phase all tubes in which gas or acidic growth is produced only after 48 h.

3. Completed Phase

To establish the presence of coliform bacteria and to provide quality control data, use the completed test on at least 10% of positive confirmed tubes (see Figure 9221:1). Simultaneous inoculation into brilliant green lactose bile broth for total coliforms and EC broth for fecal coliforms (see Section 9221E below) or EC-MUG broth for *Escherichia coli* may be used. Consider positive EC and EC-MUG broths elevated temperature (44.5°C) results as a positive completed test response. Parallel positive brilliant green lactose bile broth cultures with negative EC or EC-MUG broth cultures indicate the presence of nonfecal coliforms.

a. Culture media and reagents:

1) *LES Endo agar:* See Section 9222B. Use 100- × 15-mm petri plates.

2) *MacConkey agar:*

Peptone	17	g
Proteose peptone	3	g
Lactose	10	g
Bile salts	1.5	g
Sodium chloride, NaCl	5	g
Agar	13.5	g
Neutral red	0.03	g
Crystal violet	0.001	g
Reagent-grade water	1	L

Add ingredients to water, mix thoroughly, and heat to boiling to dissolve. Sterilize by autoclaving for 15 min at 121°C. Temper agar after sterilization and pour into petri plates (100 × 15 mm). pH should be 7.1 ± 0.2 after sterilization.

3) *Nutrient agar:*

Peptone	5.0 g
Beef extract	3.0 g
Agar	15.0 g
Reagent-grade water	1 L

Add ingredients to water, mix thoroughly, and heat to dissolve. pH should be 6.8 ± 0.2 after sterilization. Before sterilization, dispense in screw-capped tubes. After sterilization, immediately place tubes in an inclined position so that the agar will solidify with a sloped surface. Tighten screw caps after cooling and store in a protected, cool storage area.

4) *Gram-stain reagents:*

a) Ammonium oxalate-crystal violet (Hucker's): Dissolve 2 g crystal violet (90% dye content) in 20 mL 95% ethyl alcohol; dissolve 0.8 g $(NH_4)_2C_2O_4 \cdot H_2O$ in 80 mL reagent-grade water; mix the two solutions and age for 24 h before use; filter through paper into a staining bottle.

b) Lugol's solution, Gram's modification: Grind 1 g iodine crystals and 2 g KI in a mortar. Add reagent-grade water, a few milliliters at a time, and grind thoroughly after each addition until solution is complete. Rinse solution into an amber glass bottle with the remaining water (using a total of 300 mL).

c) Counterstain: Dissolve 2.5 g safranin dye in 100 mL 95% ethyl alcohol. Add 10 mL to 100 mL reagent-grade water.

d) Acetone alcohol: Mix equal volumes of ethyl alcohol (95%) with acetone.

b. Procedure:

1) Using aseptic technique, streak one LES Endo agar (Section 9222B.2) or MacConkey agar plate from each tube of brilliant green lactose bile broth showing gas, as soon as possible after the observation of gas. Streak plates in a manner to insure presence of some discrete colonies separated by at least 0.5 cm. Observe the following precautions when streaking plates to obtain a high

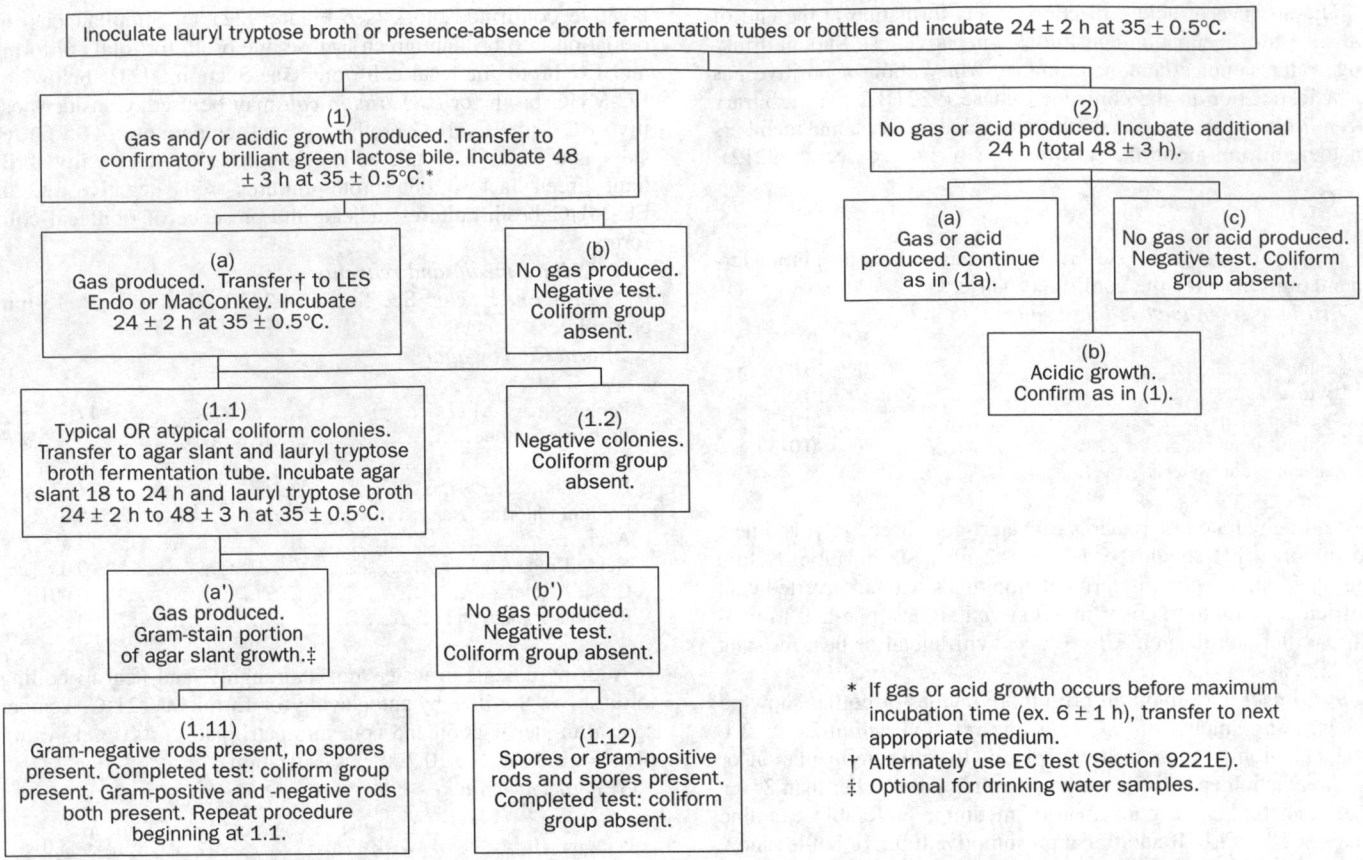

Figure 9221:1. Schematic outline of presumptive, confirmed, and completed phases for total coliform detection.

proportion of successful isolations if coliform organisms are present: (*a*) Use a sterile 3-mm-diam loop or an inoculating needle slightly curved at the tip; (*b*) tap and incline the fermentation tube to avoid picking up any membrane or scum on the needle; (*c*) insert end of loop or needle into the liquid in the tube to a depth of approximately 0.5 cm; and (*d*) streak plate for isolation with curved section of the needle in contact with the agar to avoid a scratched or torn surface. Flame loop between second and third quadrants to improve colony isolation.

Incubate plates (inverted) at $35 \pm 0.5°C$ for 24 ± 2 h.

2) The colonies developing on LES Endo agar are defined as *typical* (pink to dark red with a green metallic surface sheen) or *atypical* (pink, red, white, or colorless colonies without sheen) after 24 h incubation. Typical lactose-fermenting colonies developing on MacConkey agar are red and may be surrounded by an opaque zone of precipitated bile. From each plate pick one or more typical, well-isolated coliform colonies or, if no typical colonies are present, pick two or more colonies considered most likely to consist of organisms of the coliform group, and transfer growth from each isolate to a single-strength lauryl tryptose broth fermentation tube and onto a nutrient agar slant. (The latter is unnecessary for drinking water samples.)

If needed, use a colony magnifying device to provide optimum magnification when colonies are picked from the LES Endo or MacConkey agar plates. When transferring colonies, choose well-isolated ones and barely touch the surface of the colony with a

flame-sterilized, air-cooled transfer needle to minimize the danger of transferring a mixed culture.

Incubate secondary broth tubes (lauryl tryptose broth with inverted fermentation vials inserted) at $35 \pm 0.5°C$ for 24 ± 2 h; if gas is not produced within 24 ± 2 h reincubate and examine again at 48 ± 3 h. Microscopically examine Gram-stained preparations from those 24-h nutrient agar slant cultures corresponding to the secondary tubes that show gas.

3) *Gram-stain technique*—The Gram stain may be omitted from the completed test for potable water samples only because the occurrences of gram-positive bacteria and spore-forming organisms surviving this selective screening procedure are infrequent in drinking water.

Various modifications of the Gram stain technique exist. Use the following modification by Hucker for staining smears of pure culture; include a gram-positive and a gram-negative culture as controls.

Prepare separate light emulsions of the test bacterial growth and positive and negative control cultures on the same slide using drops of distilled water on the slide. Air-dry and fix by passing slide through a flame and stain for 1 min with ammonium oxalate-crystal violet solution. Rinse slide in tap water and drain off excess; apply Lugol's solution for 1 min.

Rinse stained slide in tap water. Decolorize for approximately 15 to 30 s with acetone alcohol by holding slide between the fingers and letting acetone alcohol flow across the stained smear until the solvent flows colorlessly from the slide. Do not over-decolorize. Counterstain with safranin for 15 s, rinse with tap water, blot dry with absorbent paper or air dry, and examine microscopically. Gram-positive organisms are blue; gram-negative organisms are red. Results are acceptable only when controls have given proper reactions.

c. Interpretation: Formation of gas in the secondary tube of lauryl tryptose broth within 48 ± 3 h and demonstration of gram-negative, nonspore-forming, rod-shaped bacteria from the agar culture constitute a positive result for the completed test, demonstrating the presence of a member of the coliform group.

4. Bibliography

MEYER, E.M. 1918. An aerobic spore-forming bacillus giving gas in lactose broth isolated in routine water examination. *J. Bacteriol.* 3:9.

HUCKER, G.J. & H.J. CONN. 1923. Methods of Gram Staining. N.Y. State Agr. Exp. Sta. Tech. Bull. No. 93.

NORTON, J.F. & J.J. WEIGHT. 1924. Aerobic spore-forming lactose fermenting organisms and their significance in water analysis. *Amer. J. Pub. Health* 14:1019.

HUCKER, G.J. & H.J. CONN. 1927. Further Studies on the Methods of Gram Staining. N.Y. State Agr. Exp. Sta. Tech. Bull. No. 128.

PORTER, R., C.S. MCCLESKEY & M. LEVINE. 1937. The facultative sporulating bacteria producing gas from lactose. *J. Bacteriol.* 33:163.

COWLES, P.B. 1939. A modified fermentation tube. *J. Bacteriol.* 38:677.

SHERMAN, V.B.D. 1967. A Guide to the Identification of the Genera of Bacteria. Williams & Wilkins, Baltimore, Md.

GELDREICH, E.E. 1975. Handbook for Evaluating Water Bacteriological Laboratories, 2nd ed. EPA-670/9-75-006, U.S. Environmental Protection Agency, Cincinnati, Ohio.

EVANS, T.M., C.E. WAARVICK, R.J. SEIDLER & M.W. LECHEVALLIER. 1981. Failure of the most-probable number technique to detect coliforms in drinking water and raw water supplies. *Appl. Environ. Microbiol.* 41:130.

SEIDLER, R.J., T.M. EVANS, J.R. KAUFMAN, C.E. WAARVICK & M.W. LECHEVALLIER. 1981. Limitations of standard coliform enumeration techniques. *J. Amer. Water Works Assoc.* 73:538.

GERHARDS, P., ed. 1981. Manual of Methods for General Bacteriology. American Soc. Microbiology, Washington, D.C.

KRIEG, N.R. & J.G. HOLT, eds. 1984. Bergey's Manual of Systematic Bacteriology, Vol 1. Williams & Wilkins, Baltimore, Md.

GREENBERG, A.E. & D.A. HUNT, eds. 1985. Laboratory Procedures for the Examination of Seawater and Shellfish, 5th ed. American Public Health Assoc., Washington, D.C.

U.S. ENVIRONMENTAL PROTECTION AGENCY. 1989. National primary drinking water regulations: analytical techniques; coliform bacteria; final rule. *Federal Register* 54(135):29998 (July 17, 1989).

9221 C. Estimation of Bacterial Density

1. Precision of Fermentation Tube Test

Unless a large number of sample portions is examined, the precision of the fermentation tube test is rather low. For example, if only 1 mL is examined in a sample containing 1 coliform organism/mL, about 37% of 1-mL tubes may be expected to yield negative results because of random distribution of the bacteria in the sample. When five tubes, each with 1 mL sample, are used under these conditions, a completely negative result may be expected less than 1% of the time.

Consequently, exercise great caution when interpreting the sanitary significance of coliform results obtained from the use of a few tubes with each sample dilution, especially when the number of samples from a given sampling point is limited.

2. Computing and Recording of MPN

To calculate coliform density, compute in terms of the Most Probable Number (MPN). The MPN values, for a variety of planting series and results, are given in Tables 9221:II, III, and IV. Included in these tables are the 95% confidence limits for each MPN value determined. If the sample volumes used are those found in the tables, report the value corresponding to the number of positive and negative results in the series as the MPN/100 mL or report as total or fecal coliform presence or absence.

The sample volumes indicated in Tables 9221:II and III relate more specifically to finished waters. Table 9221:IV illustrates MPN values for combinations of positive and negative results when five 10-mL, five 1.0-mL, and five 0.1-mL volumes of samples are tested. When the series of decimal dilutions is different from that in the table, select the MPN value from Table 9221:IV

for the combination of positive tubes and calculate according to the following formula:

MPN value (from table)

$$\times \frac{10}{\substack{\text{largest volume tested in dilution} \\ \text{series used for MPN determination}}} = \text{MPN/100 mL}$$

When more than three dilutions are used in a decimal series of dilutions, use the results from only three of these in computing the MPN. To select the three dilutions to be used in determining the MPN index, choose the highest dilution that gives positive results in all five portions tested (no lower dilution giving any negative results) and the two next succeeding higher dilutions. Use the results at these three volumes in computing the MPN index. In the examples given below, the significant dilution results are shown in boldface. The number in the numerator represents positive tubes; that in the denominator, the total tubes planted; the combination of positives simply represents the total number of positive tubes per dilution:

Example	1 mL	0.1 mL	0.01 mL	0.001 mL	Combination of positives	MPN Index /100 mL
a	5/5	**5/5**	**2/5**	**0/5**	5-2-0	5000
b	5/5	**4/5**	**2/5**	0/5	5-4-2	2200
c	**0/5**	**1/5**	**0/5**	0/5	0-1-0	20

In *c*, select the first three dilutions so as to include the positive result in the middle dilution.

TABLE 9221:II. MPN INDEX AND 95% CONFIDENCE LIMITS FOR VARIOUS COMBINATIONS OF POSITIVE AND NEGATIVE RESULTS WHEN FIVE 20-mL PORTIONS ARE USED

No. of Tubes Giving Positive Reaction Out of 5 of 20 mL Each	MPN Index/ 100 mL	95% Confidence Limits (Approximate)	
		Lower	Upper
0	<1.1	0	3.0
1	1.1	0.05	6.3
2	2.6	0.3	9.6
3	4.6	0.8	14.7
4	8.0	1.7	26.4
5	>8.0	4.0	Infinite

TABLE 9221.III. MPN INDEX AND 95% CONFIDENCE LIMITS FOR VARIOUS COMBINATIONS OF POSITIVE AND NEGATIVE RESULTS WHEN TEN 10-mL PORTIONS ARE USED

No. of Tubes Giving Positive Reaction Out of 10 of 10 mL Each	MPN Index/ 100 mL	95% Confidence Limits (Approximate)	
		Lower	Upper
0	<1.1	0	3.0
1	1.1	0.03	5.9
2	2.2	0.26	8.1
3	3.6	0.69	10.6
4	5.1	1.3	13.4
5	6.9	2.1	16.8
6	9.2	3.1	21.1
7	12.0	4.3	27.1
8	16.1	5.9	36.8
9	23.0	8.1	59.5
10	>23.0	13.5	Infinite

TABLE 9221.IV. MPN INDEX AND 95% CONFIDENCE LIMITS FOR VARIOUS COMBINATIONS OF POSITIVE RESULTS WHEN FIVE TUBES ARE USED PER DILUTION (10 mL, 1.0 mL, 0.1 mL)

Combination of Positives	MPN Index/ 100 mL	95% Confidence Limits		Combination of Positives	MPN Index/ 100 mL	95% Confidence Limits	
		Lower	Upper			Lower	Upper
				4-2-0	22	9.0	56
0-0-0	<2	—	—	4-2-1	26	12	65
0-0-1	2	1.0	10	4-3-0	27	12	67
0-1-0	2	1.0	10	4-3-1	33	15	77
0-2-0	4	1.0	13	4-4-0	34	16	80
1-0-0	2	1.0	11	5-0-0	23	9.0	86
1-0-1	4	1.0	15	5-0-1	30	10	110
1-1-0	4	1.0	15	5-0-2	40	20	140
1-1-1	6	2.0	18	5-1-0	30	10	120
1-2-0	6	2.0	18	5-1-1	50	20	150
				5-1-2	60	30	180
2-0-0	4	1.0	17				
2-0-1	7	2.0	20	5-2-0	50	20	170
2-1-0	7	2.0	21	5-2-1	70	30	210
2-1-1	9	3.0	24	5-2-2	90	40	250
2-2-0	9	3.0	25	5-3-0	80	30	250
2-3-0	12	5.0	29	5-3-1	110	40	300
				5-3-2	140	60	360
3-0-0	8	3.0	24				
3-0-1	11	4.0	29	5-3-3	170	80	410
3-1-0	11	4.0	29	5-4-0	130	50	390
3-1-1	14	6.0	35	5-4-1	170	70	480
3-2-0	14	6.0	35	5-4-2	220	100	580
3-2-1	17	7.0	40	5-4-3	280	120	690
				5-4-4	350	160	820
4-0-0	13	5.0	38	5-5-0	240	100	940
4-0-1	17	7.0	45	5-5-1	300	100	1300
4-1-0	17	7.0	46	5-5-2	500	200	2000
4-1-1	21	9.0	55	5-5-3	900	300	2900
4-1-2	26	12	63	5-5-4	1600	600	5300
				5-5-5	≥1600	—	—

When a case such as that shown below in line *d* arises, where a positive occurs in a dilution higher than the three chosen according to the rule, incorporate it in the result for the highest chosen dilution, as in *e*:

Example	1 mL	0.1 mL	0.01 mL	0.001 mL	Combination of positives	MPN Index /100 mL
d	**5/5**	**3/5**	**1/5**	1/5	5-3-2	1400
e	**5/5**	**3/5**	**2/5**	0/5	5-3-2	1400

When it is desired to summarize with a single MPN value the results from a series of samples, use the geometric mean or the median.

Table 9221:IV shows the most likely positive tube combinations. If unlikely combinations occur with a frequency greater than 1% it is an indication that the technique is faulty or that the statistical assumptions underlying the MPN estimate are not being fulfilled. The MPN for combinations not appearing in the table, or for other combinations of tubes or dilutions, may be estimated by Thomas' simple formula:

$$\text{MPN/100 mL} = \frac{\text{no. of positive tubes} \times 100}{\sqrt{\left(\begin{array}{c}\text{mL sample in} \\ \text{negative tubes}\end{array} \times \begin{array}{c}\text{mL sample in} \\ \text{all tubes}\end{array}\right)}}$$

While the MPN tables and calculations are described for use in the coliform test, they are equally applicable to determining the MPN of any other organisms provided that suitable test media are available.

3. Bibliography

McCRADY, M.H. 1915. The numerical interpretation of fermentation tube results. *J. Infect. Dis.* 12:183.

McCRADY, M.H. 1918. Tables for rapid interpretation of fermentation-tube results. *Pub. Health J.* 9:201.

HOSKINS, J.K. 1933. The most probable numbers of *B. coli* in water analysis. *J. Amer. Water Works Assoc.* 25:867.

HOSKINS, J.K. 1934. Most Probable Numbers for evaluation of *coli-aerogenes* tests by fermentation tube method. *Pub. Health Rep.* 49:393.

HOSKINS, J.K. & C.T. BUTTERFIELD. 1935. Determining the bacteriological quality of drinking water. *J. Amer. Water Works Assoc.* 27:1101.

HALVORSON, H.O. & N.R. ZIEGLER. 1933–35. Application of statistics to problems in bacteriology. *J. Bacteriol.* 25:101; 26:331,559; 29:609.

SWAROOP, S. 1938. Numerical estimation of *B. coli* by dilution method. *Indian J. Med. Res.* 26:353.

DALLA VALLE, J.M. 1941. Notes on the most probable number index as used in bacteriology. *Pub. Health Rep.* 56:229.

THOMAS, H.A., JR. 1942. Bacterial densities from fermentation tube tests. *J. Amer. Water Works Assoc.* 34:572.

WOODWARD, R.L. 1957. How probable is the Most Probable Number? *J. Amer. Water Works Assoc.* 49:1060.

McCARTHY, J.A., H.A. THOMAS, JR. & J.E. DELANEY. 1958. Evaluation of the reliability of coliform density tests. *Amer. J. Pub. Health* 48:1628.

U.S. ENVIRONMENTAL PROTECTION AGENCY. 1989. National primary drinking water regulations: analytical techniques; coliform bacteria; final rule. *Federal Register* 54(135):29998 (July 17, 1989).

DE MAN, J.C. 1977. MPN tables for more than one test. *European J. Appl. Microbiol.* 4:307.

9221 D. Presence-Absence (P-A) Coliform Test

The presence-absence (P-A) test for the coliform group is a simple modification of the multiple-tube procedure. Simplification, by use of one large test portion (100 mL) in a single culture bottle to obtain qualitative information on the presence or absence of coliforms, is justified on the theory that no coliforms should be present in 100 mL of a drinking water sample. The P-A test also provides the optional opportunity for further screening of the culture to isolate other indicators (fecal coliform, *Aeromonas, Staphylococcus, Pseudomonas,* fecal streptococcus, and *Clostridium*) on the same qualitative basis. Additional advantages include the possibility of examining a larger number of samples per unit of time. Comparative studies with the membrane filter procedure indicate that the P-A test may maximize coliform detection in samples containing many organisms that could overgrow coliform colonies and cause problems in detection.

The P-A test is intended for use on routine samples collected from distribution systems or water treatment plants. When sample locations produce a positive P-A result for coliforms, it may be advisable to determine coliform densities in repeat samples. Quantitative information may indicate the magnitude of a contaminating event.

1. Presumptive Phase

a. Culture media:

1) *P-A broth:* This medium is commercially available in dehydrated and in sterile concentrated form.

Beef extract	3.0	g
Peptone	5.0	g
Lactose	7.46	g
Tryptose	9.83	g
Dipotassium hydrogen phosphate, K_2HPO_4	1.35	g
Potassium dihydrogen phosphate, KH_2PO_4	1.35	g
Sodium chloride, NaCl	2.46	g
Sodium lauryl sulfate	0.05	g
Bromcresol purple	0.0085	g
Reagent-grade water	1	L

Make this formulation triple (3×) strength when examining 100-mL samples. Dissolve the P-A broth medium in water without heating, using a stirring device. Dispense 50 mL prepared medium into a screw-cap 250-mL milk dilution bottle. A fermentation tube insert is not necessary. Autoclave for 12 min at 121°C with the total time in the autoclave limited to 30 min or less. pH should be 6.8 ± 0.2 after sterilization. When the PA medium is sterilized by filtration a 6× strength medium may be used. Aseptically dispense 20 mL of the 6× medium into a sterile 250-mL dilution bottle or equivalent container.

2) *Lauryl tryptose broth:* See Section 9221B.1.

b. Procedure: Shake sample vigorously for 5 s (approximately 25 times) and inoculate 100 mL into a P-A culture bottle. Mix thoroughly by inverting bottle once or twice to achieve even distribution of the triple-strength medium throughout the sample. Incubate at 35 ± 0.5°C and inspect after 24 and 48 h for acid reactions.

c. Interpretation: A distinct yellow color forms in the medium when acid conditions exist following lactose fermentation. If gas also is being produced, gently shaking the bottle will result in a foaming reaction. Any amount of gas and/or acid constitutes a positive presumptive test requiring confirmation.

2. Confirmed Phase

The confirmed phase is outlined in Figure 9221:1.

a. Culture medium: Use brilliant green lactose bile fermentation tubes (see 9221B.2).

b. Procedure: Transfer all cultures that show acid reaction or acid and gas reaction to brilliant green lactose bile (BGLB) broth for incubation at 35 ± 0.5°C (see Section 9221B.2).

c. Interpretation: Gas production in the BGLB broth culture within 48 ± 3 h confirms the presence of coliform bacteria. Re-port result as presence-absence test positive or negative for total coliforms in 100 mL of sample.

3. Completed Phase

The completed phase is outlined in Section 9221B.3 and Figure 9221:1.

4. Bibliography

WEISS, J.E. & C.A. HUNTER. 1939. Simplified bacteriological examination of water. *J. Amer. Water Works Assoc.* 31:707.

CLARK, J.A. 1969. The detection of various bacteria indicative of water pollution by a presence-absence (P-A) procedure. *Can. J. Microbiol.* 15:771.

CLARK, J.A. & L.T. VLASSOFF. 1973. Relationships among pollution indicator bacteria isolated from raw water and distribution systems by the presence-absence (P-A) test. *Health Lab. Sci.* 10:163.

CLARK, J.A. 1980. The influence of increasing numbers of nonindicator organisms upon the detection of indicator organisms by the membrane filter and presence-absence tests. *Can. J. Microbiol.* 26:827.

CLARK, J.A., C.A. BURGER & L.E. SABATINOS. 1982. Characterization of indicator bacteria in municipal raw water, drinking water and new main water samples. *Can. J. Microbiol.* 28:1002.

JACOBS, N.J., W.L. ZEIGLER, F.C. REED, T.A. STUKEL & E.W. RICE. 1986. Comparison of membrane filter, multiple-fermentation-tube, and presence-absence techniques for detecting total coliforms in small community water systems. *Appl. Environ. Microbiol.* 51:1007.

RICE, E.W., E.E. GELDREICH & E.J. READ. 1989. The presence-absence coliform test for monitoring drinking water quality. *Pub. Health Rep.* 104:54.

9221 E.　Fecal Coliform Procedure

Elevated-temperature tests for distinguishing organisms of the total coliform group that also belong to the fecal coliform group are described herein. Modifications in technical procedures, standardization of methods, and detailed studies of the fecal coliform group have established the value of this procedure. The test can be performed by one of the multiple-tube procedures described here or by membrane filter methods as described in Section 9222. The procedure using A-1 broth is a single-step method.

The fecal coliform test (using EC medium) is applicable to investigations of drinking water, stream pollution, raw water sources, wastewater treatment systems, bathing waters, seawaters, and general water-quality monitoring. Prior enrichment in presumptive media is required for optimum recovery of fecal coliforms when using EC medium. The test using A-1 medium is applicable to source water, seawater, and treated wastewater.

1. Fecal Coliform Test (EC Medium)

The fecal coliform test is used to distinguish those total coliform organisms that are fecal coliforms. Use EC medium or, for a more rapid test of the quality of shellfish waters, treated wastewaters, or source waters, use A-1 medium in a direct test.

a. EC medium:

Tryptose or trypticase	20.0 g
Lactose	5.0 g
Bile salts mixture or bile salts No. 3	1.5 g
Dipotassium hydrogen phosphate, K_2HPO_4	4.0 g
Potassium dihydrogen phosphate, KH_2PO_4	1.5 g
Sodium chloride, NaCl	5.0 g
Reagent-grade water	1 L

Add dehydrated ingredients to water, mix thoroughly, and heat to dissolve. pH should be 6.9 ± 0.2 after sterilization. Before sterilization, dispense in fermentation tubes, each with an inverted vial, sufficient medium to cover the inverted vial at least partially after sterilization. Close tubes with metal or heat-resistant plastic caps.

b. Procedure: Submit all presumptive fermentation tubes or bottles showing any amount of gas, growth, or acidity within 48 h of incubation to the fecal coliform test.

1) Gently shake or rotate presumptive fermentation tubes or bottles showing gas, growth, or acidity. Using a sterile 3- or 3.5-mm-diam loop or sterile wooden applicator stick, transfer growth from each presumptive fermentation tube or bottle to EC broth (see Section 9221B.2).

2) Incubate inoculated EC broth tubes in a water bath at 44.5 ± 0.2°C for 24 ± 2 h.

Place all EC tubes in water bath within 30 min after inoculation. Maintain a sufficient water depth in water bath incubator to immerse tubes to upper level of the medium.

c. Interpretation: Gas production with growth in an EC broth culture within 24 ± 2 h or less is considered a positive fecal coliform reaction. Failure to produce gas (with little or no growth) constitutes a negative reaction. If multiple tubes are used, calculate MPN from the number of positive EC broth tubes as described in Section 9221C. When using only one tube for subculturing from a single presumptive bottle, report as presence or absence of fecal coliforms.

2. Fecal Coliform Direct Test (A-1 Medium)

a. A-1 broth: This medium may be used for the direct isolation of fecal coliforms from water. Prior enrichment in a presumptive medium is not required.

Lactose	5.0 g
Tryptone	20.0 g
Sodium chloride, NaCl	5.0 g
Salicin	0.5 g
Polyethylene glycol *p*-isooctylphenyl ether*	1.0 mL
Reagent-grade water	1 L

Heat to dissolve solid ingredients, add polyethylene glycol *p*-isooctylphenyl ether, and adjust to pH 6.9 ± 0.1. Before sterilization dispense in fermentation tubes with an inverted vial sufficient medium to cover the inverted vial at least partially after sterilization. Close with metal or heat-resistant plastic caps. Sterilize by autoclaving at 121°C for 10 min. Store in dark at room temperature for not longer than 7 d. Ignore formation of precipitate.

---

* Triton X-100, Rohm and Haas Co., or equivalent.

Make A-1 broth of such strength that adding 10-mL sample portions to medium will not reduce ingredient concentrations below those of the standard medium. For 10-mL samples prepare double-strength medium.

b. Procedure: Inoculate tubes of A-1 broth as directed in Section 9221B.1*b*1). Incubate for 3 h at 35 ± 0.5°C. Transfer tubes to a water bath at 44.5 ± 0.2°C and incubate for an additional 21 ± 2 h.

c. Interpretation: Gas production in any A-1 broth culture within 24 h or less is a positive reaction indicating the presence of fecal coliforms. Calculate MPN from the number of positive A-1 broth tubes as described in Section 9221C.

3. Bibliography

PERRY, C.A. & A.A. HAJNA. 1933. A modified Eijkman medium. *J. Bacteriol.* 26:419.

PERRY, C.A. & A.A. HAJNA. 1944. Further evaluation of EC medium for the isolation of coliform bacteria and *Escherichia coli*. *Amer. J. Pub. Health* 34:735.

GELDREICH, E.E., H.F. CLARK, P.W. KABLER, C.B. HUFF & R.H. BORDNER. 1958. The coliform group. II. Reactions in EC medium at 45°C. *Appl. Microbiol.* 6:347.

GELDREICH, E.E., R.H. BORDNER, C.B. HUFF, H.F. CLARK & P.W. KABLER. 1962. Type distribution of coliform bacteria in the feces of warm-blooded animals. *J. Water Pollut. Control Fed.* 34:295.

GELDREICH, E.E. 1966. Sanitary significance of fecal coliforms in the environment. FWPCA Publ. WP-20-3 (Nov.). U.S. Dep. Interior, Washington, D.C.

ANDREWS, W.H. & M.W. PRESNELL. 1972. Rapid recovery of *Escherichia coli* from estuarine water. *Appl. Microbiol.* 23:521.

OLSON, B.H. 1978. Enhanced accuracy of coliform testing in seawater by a modification of the most-probable-number method. *Appl. Microbiol.* 36:438.

STRANDRIDGE, J.H. & J.J. DELFINO. 1981. A-1 Medium: Alternative technique for fecal coliform organism enumeration in chlorinated wastewaters. *Appl. Environ. Microbiol.* 42:918.

9221 F. *Escherichia coli* Procedure (PROPOSED)

Escherichia coli is a member of the fecal coliform group of bacteria. This organism in water indicates fecal contamination. Enzymatic assays have been developed that allow for the identification of this organism. In this method *E. coli* are defined as coliform bacteria that possess the enzyme β-glucuronidase and are capable of cleaving the fluorogenic substrate 4-methylumbelliferyl-β-D-glucuronide (MUG) with the corresponding release of the fluorogen when grown in EC-MUG medium at 44.5°C within 24 ± 2 h or less. The procedure is used as a confirmatory test after prior enrichment in a presumptive medium for total coliform bacteria. This test is performed as a tube procedure as described here or by the membrane filter method as described in Section 9222. The chromogenic substrate procedure (Section 9223) can be used for direct detection of *E. coli*.

Tests for *E. coli* (using EC-MUG medium) are applicable for

the analysis of drinking water, surface and ground water, and wastewater. *E. coli* is a member of the indigenous fecal flora of warm-blooded animals. The occurrence of *E. coli* is considered a specific indicator of fecal contamination and the possible presence of enteric pathogens.

1. *Escherichia coli* Test (EC-MUG medium)

Use EC-MUG medium for the confirmation of *E. coli*.
a. EC-MUG medium:

Tryptose or trypticase	20.0 g
Lactose	5.0 g
Bile salts mixture or bile salts No. 3	1.5 g
Dipotassium hydrogen phosphate, K_2HPO_4	4.0 g

Potassium dihydrogen phosphate, KH_2PO_4	1.5	g
Sodium chloride, NaCl	5.0	g
4-methylumbelliferyl-β-D-glucuronide (MUG)	0.05	g
Reagent-grade water	1	L

Add dehydrated ingredients to water, mix thoroughly, and heat to dissolve. pH should be 6.9 ± 0.2 after sterilization. Before sterilization, dispense in tubes that do not fluoresce under long-wavelength (366 nm) ultraviolet (UV) light. An inverted tube is not necessary. Close tubes with metal or heat-resistant plastic caps.

b. Procedure: Submit all presumptive fermentation tubes or bottles showing growth, gas, or acidity within 48 ± 3 h of incubation to the *E. coli* test.

1) Gently shake or rotate presumptive fermentation tubes or bottles showing growth, gas, or acidity. Using a sterile 3- or 3.5-mm-diam metal loop or sterile wooden applicator stick, transfer growth from presumptive fermentation tube or bottle to EC-MUG broth.

2) Incubate inoculated EC-MUG tubes in a water bath or incubator maintained at $44.5 \pm 0.2°C$ for 24 ± 2 h. Place all EC-MUG tubes in water bath within 30 min after inoculation. Maintain a sufficient water depth in the water-bath incubator to immerse tubes to upper level of medium.

c. Interpretation: Examine all tubes exhibiting growth for fluorescence using a long-wavelength UV lamp (preferably 6 W). The presence of bright blue fluorescence is considered a positive response for *E. coli*. A positive control consisting of a known *E. coli* (MUG-positive) culture, a negative control consisting of a thermotolerant *Klebsiella pneumoniae* (MUG-negative) culture, and an uninoculated medium control may be necessary to interpret the results and to avoid confusion of weak auto-fluorescence of the medium as a positive response. If multiple tubes are used, calculate MPN from the number of positive EC-MUG broth tubes as described in Section 9221C. When using only one tube or subculturing from a single presumptive bottle, report as presence or absence of *E. coli*.

2. Bibliography

FENG, P.C.S. & P.A. HARTMAN. 1982. Fluorogenic assays for immediate confirmation of *Escherichia coli*. *Appl. Environ. Microbiol.* 43:1320.

HARTMAN, P.A. 1989. The MUG (glucuronidase) test for *E. coli* in food and water. *In* A. Balows et al., eds., Rapid Methods and Automation in Microbiology and Immunology. Proc. 5th Intl. Symp. on Rapid Methods and Automation in Microbiology & Immunology, Florence, Italy, Nov. 4–6, 1987.

SHADIX, L.C. & E.W. RICE. 1991. Evaluation of β-glucuronidase assay for the detection of *Escherichia coli* from environmental waters. *Can. J. Microbiol.* 37:908.

9222   MEMBRANE FILTER TECHNIQUE FOR MEMBERS OF THE COLIFORM GROUP*

9222 A.   Introduction

The membrane filter (MF) technique is highly reproducible, can be used to test relatively large sample volumes, and usually yields numerical results more rapidly than the multiple-tube fermentation procedure. The MF technique is extremely useful in monitoring drinking water and a variety of natural waters. However, the MF technique has limitations, particularly when testing waters with high turbidity or large numbers of noncoliform (background) bacteria. When the MF technique has not been used previously, it is desirable to conduct parallel tests with the method the laboratory is using currently to demonstrate applicability and comparability.

1. Definition

As related to the MF technique, the coliform group is defined as those facultative anaerobic, gram-negative, non-spore-forming, rod-shaped bacteria that develop red colonies with a metallic (golden) sheen within 24 h at 35°C on an Endo-type medium containing lactose. Some members of the total coliform group may produce dark red, mucoid, or nucleated colonies without a metallic sheen. When verified these are classified as atypical coliform colonies. When purified cultures of coliform bacteria are tested, they produce negative cytochrome oxidase and positive β-galactosidase test reactions.† Generally, pink (non-mucoid), blue, white, or colorless colonies lacking sheen are considered noncoliforms by this technique.

2. Applications

Turbidity caused by the presence of algae, particulates, or other interfering material may not permit testing of a sample volume sufficient to yield significant results. Low coliform estimates may be caused by the presence of high numbers of noncoliforms or of toxic substances. The MF technique is applicable to the examination of saline waters, but not wastewaters that have received only primary treatment followed by chlorination because of turbidity in high volume samples or wastewaters containing toxic metals or toxic organic compounds such as phenols. For the detection of stressed total coliforms in treated drinking water and chlorinated secondary or tertiary wastewater effluents, use a method designed for stressed organism recovery (see Section 9212B.1). A modified MF technique for fecal coliforms (Section 9212) in chlorinated wastewater may be used if parallel testing over a 3-month period with the multiple-tube fermentation technique shows comparability for each site-specific type of sample.

---

* Approved by Standard Methods Committee, 1997.

† ONPG is a substrate for the β-galactosidase test.

The standard volume to be filtered for drinking water samples is 100 mL. This may be distributed among multiple membranes if necessary. However, for special monitoring purposes, such as troubleshooting water quality problems or identification of coliform breakthrough in low concentrations from treatment barriers, it may be desirable to test 1-L samples. If particulates prevent filtering a 1-L sample through a single filter, divide sample into four portions of 250 mL for analysis. Total the coliform counts on each membrane to report the number of coliforms per liter. Smaller sample volumes will be necessary for source or recreational waters and wastewater effluents that have much higher coliform densities.

Statistical comparisons of results obtained by the multiple-tube method and the MF technique show that the MF is more precise (compare Tables 9221:II and III with Table 9222:II). Data from each test yield approximately the same water quality information, although numerical results are not identical (see Section 9010B for drinking water).

3. Bibliography

CLARK, H.F., E.E. GELDREICH, H.L. JETER & P.W. KABLER. 1951. The membrane filter in sanitary bacteriology. *Pub. Health Rep.* 66:951.

KABLER, P.W. 1954. Water examinations by membrane filter and MPN procedures. *Amer. J. Pub. Health* 44:379.

THOMAS, H.A. & R.L. WOODWARD. 1956. Use of molecular filter membranes for water potability control. *J. Amer. Water Works Assoc.* 48: 1391.

MCCARTHY, J.A., J.E. DELANEY & R.J. GRASSO. 1961. Measuring coliforms in water. *Water Sewage Works* 108:238.

LIN, S. 1973. Evaluation of coliform test for chlorinated secondary effluents. *J. Water Pollut. Control Fed.* 45:498.

MANDEL, J. & L.F. NANNI. 1978. Measurement evaluation. *In* S.L. Inhorn, ed. Quality Assurance Practices for Health Laboratories, p. 209. American Public Health Assoc., Washington, D.C.

9222 B.   Standard Total Coliform Membrane Filter Procedure

1. Laboratory Apparatus

For MF analyses use glassware and other apparatus composed of material free from agents that may affect bacterial growth.

a. Sample bottles: See Section 9030B.18.

b. Dilution bottles: See Section 9030B.13.

c. Pipets and graduated cylinders: See Section 9030B.9. Before sterilization, loosely cover opening of graduated cylinders with metal foil or a suitable heavy wrapping-paper substitute. Immediately after sterilization secure cover to prevent contamination.

d. Containers for culture medium: Use clean borosilicate glass flasks. Any size or shape of flask may be used, but erlenmeyer flasks with metal caps, metal foil covers, or screw caps provide for adequate mixing of the medium contained and are convenient for storage.

e. Culture dishes: Use sterile borosilicate glass or disposable, presterilized plastic petri dishes, 60 × 15 mm, 50 × 9 mm, or other appropriate size. Wrap convenient numbers of clean, glass culture dishes in metal foil if sterilized by dry heat, or suitable heavy wrapping paper when autoclaved. Incubate loose-lidded glass and disposable plastic culture dishes in tightly closed containers with wet paper or cloth to prevent moisture evaporation with resultant drying of medium and to maintain a humid environment for optimum colony development.

Presterilized disposable plastic dishes with tight-fitting lids that meet the specifications above are available commercially and are used widely. Reseal opened packages of disposable dish supplies for storage.

f. Filtration units: The filter-holding assembly (constructed of glass, autoclavable plastic, porcelain, or stainless steel) consists of a seamless funnel fastened to a base by a locking device or by magnetic force. The design should permit the membrane filter to be held securely on the porous plate of the receptacle without mechanical damage and allow all fluid to pass through the membrane during filtration. Discard plastic funnels with deep scratches on inner surface or glass funnels with chipped surfaces.

Wrap the assembly (as a whole or separate parts) in heavy wrapping paper or aluminum foil, sterilize by autoclaving, and store until use. Alternatively expose all surfaces of the previously cleaned assembly to ultraviolet radiation (2 min exposure) for the initial sanitization before use in the test procedure, or before re-using units between successive filtration series. Field units may be sanitized by dipping or spraying with alcohol and then igniting or immersing in boiling water for 2 min. After submerging unit in boiling water, cool it to room temperature before reuse. Do not ignite plastic parts. Sterile, disposable field units may be used.

For filtration, mount receptacle of filter-holding assembly on a 1-L filtering flask with a side tube or other suitable device (manifold to hold three to six filter assemblies) such that a pressure differential (34 to 51 kPa) can be exerted on the filter membrane. Connect flask to a vacuum line, an electric vacuum pump, a filter pump operating on water pressure, a hand aspirator, or other means of securing a pressure differential (138 to 207 kPa). Connect a flask of approximately the same capacity between filtering flask and vacuum source to trap carry-over water.

g. Membrane filter: Use membrane filters (for additional specifications, see Section 9020) with a rated pore diameter such that there is complete retention of coliform bacteria. Use only those filter membranes that have been found, through adequate quality control testing and *certification by the manufacturer*, to exhibit: full retention of the organisms to be cultivated, stability in use, freedom from chemical extractables that may inhibit bacterial growth and development, a satisfactory speed of filtration (within 5 min), no significant influence on medium pH (beyond ± 0.2 units), and no increase in number of confluent colonies or spreaders compared to control membrane filters. Use membranes grid-marked in such a manner that bacterial growth is neither inhibited nor stimulated along the grid lines when the membranes with entrapped bacteria are incubated on a suitable medium. Preferably use fresh stocks of membrane filters and if necessary store them in an environment without extremes of temperature and humidity. Obtain no more than a year's supply at any one time.

Preferably use presterilized membrane filters for which the manufacturer has certified that the sterilization technique has neither induced toxicity nor altered the chemical or physical properties of the membrane. If membranes are sterilized in the laboratory, autoclave for 10 min at 121°C. At the end of the sterilization period, let the steam escape rapidly to minimize accumulation of water of condensation on filters.

h. Absorbent pads consist of disks of filter paper or other material certified for each lot by the manufacturer to be of high quality and free of sulfites or other substances of a concentration that could inhibit bacterial growth. Use pads approximately 48 mm in diameter and of sufficient thickness to absorb 1.8 to 2.2 mL of medium. Presterilized absorbent pads or pads subsequently sterilized in the laboratory should release less than 1 mg total acidity (calculated as $CaCO_3$) when titrated to the phenolphthalein end point, pH 8.3, using 0.02N NaOH and produce pH levels of 7 ± 0.2. Sterilize pads simultaneously with membrane filters available in resealable kraft envelopes, or separately in other suitable containers. Dry pads so they are free of visible moisture before use. See sterilization procedure described for membrane filters above and Section 9020 for additional specifications on absorbent pads.

i. Forceps: Smooth flat forceps, without corrugations on the inner sides of the tips. Sterilize before use by dipping in 95% ethyl or absolute methyl alcohol and flaming.

j. Incubators: Use incubators to provide a temperature of 35 ± 0.5°C and to maintain a humid environment (60% relative humidity).

k. Microscope and light source: To determine colony counts on membrane filters, use a magnification of 10 to 15 diameters and a cool white fluorescent light source adjusted to give maximum sheen discernment. Optimally use a binocular wide-field dissecting microscope. Do not use a microscope illuminator with optical system for light concentration from an incandescent light source for discerning coliform colonies on Endo-type media.

2. Materials and Culture Media

The need for uniformity dictates the use of commercial dehydrated media. Never prepare media from basic ingredients when suitable dehydrated media are available. Follow manufacturer's directions for rehydration. Store opened supplies of dehydrated media in a desiccator. Commercially prepared media in liquid form (sterile ampule or other) may be used if known to give equivalent results. See Section 9020 for media quality control specifications.

Test each new medium lot against a previously acceptable lot for satisfactory performance as described in Section 9020B. With each new lot of Endo-type medium, verify a minimum 10% of coliform colonies, obtained from natural samples or samples with known additions, to establish the comparative recovery of the medium lot.

Before use, test each batch of laboratory-prepared MF medium for performance with positive and negative culture controls. Check for coliform contamination at the beginning and end of each filtration series by filtering 20 to 30 mL of dilution or rinse water through the filter. If controls indicate contamination, reject all data from affected samples and request resample.

a. LES Endo agar:*

Yeast extract	1.2	g
Casitone or trypticase	3.7	g
Thiopeptone or thiotone	3.7	g
Tryptose	7.5	g
Lactose	9.4	g
Dipotassium hydrogen phosphate, K_2HPO_4	3.3	g
Potassium dihydrogen phosphate, KH_2PO_4	1.0	g
Sodium chloride, NaCl	3.7	g
Sodium desoxycholate	0.1	g
Sodium lauryl sulfate	0.05	g
Sodium sulfite, Na_2SO_3	1.6	g
Basic fuchsin	0.8	g
Agar	15.0	g
Reagent-grade water	1	L

Rehydrate product in 1 L water containing 20 mL 95% ethanol. Do not use denatured ethanol, which reduces background growth and coliform colony size. Bring to a near boil to dissolve agar, then promptly remove from heat and cool to 45 to 50°C. Do not sterilize by autoclaving. Final pH 7.2 ± 0.2. Dispense in 5- to 7-mL quantities into lower section of 60-mm glass or plastic petri dishes. If dishes of any other size are used, adjust quantity to give an equivalent depth of 4 to 5 m. Do not expose poured plates to direct sunlight; refrigerate in the dark, preferably in sealed plastic bags or other containers to reduce moisture loss. Discard unused medium after 2 weeks or sooner if there is evidence of moisture loss, medium contamination, or medium deterioration (darkening of the medium).

b. M-Endo medium:†

Tryptose or polypeptone	10.0	g
Thiopeptone or thiotone	5.0	g
Casitone or trypticase	5.0	g
Yeast extract	1.5	g
Lactose	12.5	g
Sodium chloride, NaCl	5.0	g
Dipotassium hydrogen phosphate, K_2HPO_4	4.375	g
Potassium dihydrogen phosphate, KH_2PO_4	1.375	g
Sodium lauryl sulfate	0.05	g
Sodium desoxycholate	0.10	g
Sodium sulfite, Na_2SO_3	2.10	g
Basic fuchsin	1.05	g
Agar (optional)	15.0	g
Reagent-grade water	1	L

1) Agar preparation—Rehydrate product in 1 L water containing 20 mL 95% ethanol. Heat to near boiling to dissolve agar, then promptly remove from heat and cool to between 45 and 50°C. Dispense 5- to 7-mL quantities into 60-mm sterile glass or plastic petri dishes. If dishes of any other size are used, adjust quantity to give an equivalent depth. Do not sterilize by autoclaving. Final pH should be 7.2 ± 0.2. A precipitate is normal in Endo-type media.

Refrigerate finished medium in the dark and discard unused agar after 2 weeks.

2) Broth preparation—Prepare as above, omitting agar. Dispense liquid medium (at least 2.0 mL per plate) onto absorbent pads (see absorbent pad specifications, Section 9222B.1) and carefully remove excess medium by decanting the plate. The broth may have a precipitate but this does not interfere with medium performance if pads are certified free of sulfite or other toxic

---

* Dehydrated Difco M-Endo Agar LES (No. 0736), dehydrated BBL M-Endo Agar LES (No. 11203), or equivalent.

† Dehydrated Difco M-Endo Broth MF (No. 0749), dehydrated BBL *m*-Coliform Broth (No. 11119), or equivalent may be used if absorbent pads are used.

agents at a concentration that could inhibit bacterial growth. Refrigerated broth may be stored for up to 4 d.

c. Buffered dilution rinse water: See Section 9050C.1.

3. Samples

Collect samples as directed in Sections 9060A and B.

4. Coliform Definition

Bacteria that produce a red colony with a metallic (golden) sheen within 24 h incubation at 35°C on an Endo-type medium are considered members of the coliform group. The sheen may cover the entire colony or may appear only in a central area or on the periphery. The coliform group thus defined is based on the production of aldehydes from fermentation of lactose. While this biochemical characteristic is part of the metabolic pathway of gas production in the multiple-tube test, some variations in degree of metallic sheen development may be observed among coliform strains. However, this slight difference in indicator definition is not considered critical to change its public health significance, particularly if suitable studies have been conducted to establish the relationship between results obtained by the MF and those obtained by the standard multiple-tube fermentation procedure.

5. Procedures

a. Selection of sample size: Size of sample will be governed by expected bacterial density. In drinking water analyses, sample size will be limited only by the degree of turbidity or by the noncoliform growth on the medium (Table 9222:I). For regulation purposes, 100 mL is the official sample size.

An ideal sample volume will yield 20 to 80 coliform colonies and not more than 200 colonies of all types on a membrane-filter surface. Analyze drinking waters by filtering 100 to 1000 mL, or by filtering replicate smaller sample volumes such as duplicate 50-mL or four replicates of 25-mL portions. Analyze other waters by filtering three different volumes (diluted or undiluted), depending on the expected bacterial density. See Section 9215B.2 for preparation of dilutions. When less than 10 mL of sample (diluted or undiluted) is to be filtered, add approximately 10 mL sterile dilution water to the funnel before filtration or pipet the sample volume into a sterile dilution bottle, then filter the entire dilution. This increase in water volume aids in uniform dispersion of the bacterial suspension over the entire effective filtering surface.

b. Sterile filtration units: Use sterile filtration units at the beginning of each filtration series as a minimum precaution to avoid accidental contamination. A filtration series is considered to be interrupted when an interval of 30 min or longer elapses between sample filtrations. After such interruption, treat any further sample filtration as a new filtration series and sterilize all membrane filter holders in use. See Section 9222B.1f for sterilization procedures and Section 9020B.2m and n for UV cleaning and safety guidelines.

c. Filtration of sample: Using sterile forceps, place a sterile membrane filter (grid side up) over porous plate of receptacle. Carefully place matched funnel unit over receptacle and lock it in place. Filter sample under partial vacuum. With filter still in place, rinse the interior surface of the funnel by filtering three 20- to 30-mL portions of sterile dilution water. Alternatively, rinse funnel by a flow of sterile dilution water from a squeeze bottle. This is satisfactory only if the squeeze bottle and its contents do not become contaminated during use. Rinsing between samples prevents carryover contamination. Upon completion of final rinse and the filtration process disengage vacuum, unlock and remove funnel, immediately remove membrane filter with sterile forceps, and place it on selected medium with a rolling motion to avoid entrapment of air. If the agar-based medium is used, place prepared filter directly on agar, invert dish, and incubate for 22 to 24 h at 35 ± 0.5°C.

If liquid medium is used, place a pad in the culture dish and saturate with at least 2.0 mL M-Endo medium and carefully remove excess medium by decanting the plate. Place prepared filter directly on pad, invert dish, and incubate for 22 to 24 h at 35 ± 0.5°C.

Differentiation of some colonies from either agar or liquid medium substrates may be lost if cultures are incubated beyond 24 h.

Insert a sterile rinse water sample (100 mL) after filtration of a series of 10 samples to check for possible cross-contamination or contaminated rinse water. Incubate the rinse water control membrane culture under the same conditions as the sample.

For nonpotable water samples, preferably decontaminate filter unit after each sample (as described above) because of the high number of coliform bacteria present in these samples. Alternatively, use an additional buffer rinse of the filter unit after the filter is removed to prevent carryover between samples.

TABLE 9222:I. SUGGESTED SAMPLE VOLUMES FOR MEMBRANE FILTER TOTAL COLIFORM TEST

Water Source	Volume (X) To Be Filtered mL							
	100	50	10	1	0.1	0.01	0.001	0.0001
Drinking water	X							
Swimming pools	X							
Wells, springs	X	X	X					
Lakes, reservoirs	X	X	X					
Water supply intake			X	X	X			
Bathing beaches			X	X	X			
River water				X	X	X	X	
Chlorinated sewage				X	X	X		
Raw sewage					X	X	X	X

d. Alternative enrichment technique: Place a sterile absorbent pad in the lid of a sterile culture dish and pipet at least 2.0 mL lauryl tryptose broth, prepared as directed in 9221B.1.*a*1), to saturate pad. Carefully remove any excess liquid from absorbent pad by decanting plate. Aseptically place filter through which the sample has been passed on pad. Incubate filter, without inverting dish, for 1.5 to 2 h at 35 ± 0.5°C in an atmosphere of at least 60% relative humidity.

If the agar-based Endo-type medium is used, remove enrichment culture from incubator, lift filter from enrichment pad, and roll it onto the agar surface, which has been allowed to equilibrate to room temperature. Incorrect filter placement is at once obvious, because patches of unstained membrane indicate entrapment of air. Where such patches occur, carefully reseat filter on agar surface. If the liquid medium is used, prepare final culture by removing enrichment culture from incubator and separating the dish halves. Place a fresh sterile pad in bottom half of dish and saturate with at least 2.0 mL of M-Endo medium and carefully remove excess liquid from absorbent pad by decanting plate. Transfer filter, with same precautions as above, to new pad. Discard used enrichment pad.

With either the agar or the liquid medium, invert dish and incubate for 20 to 22 h at 35 ± 0.5°C. Proceed to ¶ *e* below.

e. Counting: To determine colony counts on membrane filters, use a low-power (10 to 15 magnifications) binocular wide-field dissecting microscope or other optical device, with a cool white fluorescent light source directed to provide optimal viewing of sheen. The typical coliform colony has a pink to dark-red color with a metallic surface sheen. Count both typical and atypical coliform colonies. The sheen area may vary in size from a small pinhead to complete coverage of the colony surface. Atypical coliform colonies can be dark red, mucoid, or nucleated without sheen. Generally pink, blue, white, or colorless colonies lacking sheen are considered noncoliforms. The total count of colonies (coliform and noncoliform) on Endo-type medium has no consistent relationship to the total number of bacteria present in the original sample. A high count of noncoliform colonies may interfere with the maximum development of coliforms. Refrigerating cultures (after 22 h incubation) with high densities of noncoliform colonies for 0.5 to 1 h before counting may deter spread of confluence while aiding sheen discernment.

Samples of disinfected water or wastewater effluent may include stressed organisms that grow relatively slowly and produce maximum sheen in 22 to 24 h. Organisms from undisinfected sources may produce sheen at 16 to 18 h, and the sheen subsequently may fade after 24 to 30 h.

f. Coliform verification: Occasionally, typical sheen colonies may be produced by noncoliform organisms and atypical colonies (dark red or nucleated colonies without sheen) may be coliforms. Preferably verify all typical and atypical colony types. For drinking water, verify all suspect colonies by swabbing the entire membrane or pick at least five typical colonies and five atypical colonies from a given membrane filter culture. For waters other than drinking water, at a minimum, verify at least 10 sheen colonies (and representative atypical colonies of different morphological types) from a positive water sample monthly. See Section 9020B.8. Based on need and sample type, laboratories may incorporate more stringent quality control measures (e.g., verify at least one colony from each typical or atypical colony type from a given membrane filter culture, verify 10% of the positive sam-

ples). Adjust counts on the basis of verification results. Verification tests are listed below.

1) *Lactose fermentation*—Transfer growth from each colony or swab the entire membrane with a sterile cotton swab (for presence-absence results in drinking water samples) and place in lauryl tryptose broth; incubate the lauryl tryptose broth at 35 ± 0.5°C for 48 h. Gas formed in lauryl tryptose broth and confirmed in brilliant green lactose broth (Section 9221B.2 for medium preparation) within 48 h verifies the colony as a coliform. Simultaneous inoculation of both media for gas production is acceptable. Inclusion of EC broth inoculation for 44.5 ± 0.2°C incubation will provide information on the presence of fecal coliforms. Use of EC-MUG with incubation at 44.5 ± 0.2°C for 24 h will provide information on presence of *E. coli*. See Section 9222G for MF partition procedures.

2) *Alternative coliform verifications*—Apply this alternative coliform verification procedure to isolated colonies on the membrane filter culture. If a mixed culture is suspected or if colony separation is less than 2 mm, streak the growth to M-Endo medium or MacConkey agar to assure culture purity or submit the mixed growth to the fermentation tube method.

a) *Rapid test*—A rapid verification of colonies utilizes test reactions for cytochrome oxidase (CO) and β-galactosidase. Coliform reactions are CO negative and β-galactosidase positive within 4 h incubation of tube culture or micro (spot) test procedure.

b) *Commercial multi-test systems*—Verify the colony by streaking it for purification, selecting a well-isolated colony, and inoculating into a multi-test identification system for Enterobacteriaceae that includes lactose fermentation and/or β-galactosidase and CO test reactions.

6. Calculation of Coliform Density

Compute the count, using membrane filters with 20 to 80 coliform colonies and not more than 200 colonies of all types per membrane, by the following equation:

$$(\text{Total}) \text{ coliforms}/100 \text{ mL} = \frac{\text{coliform colonies counted} \times 100}{\text{mL sample filtered}}$$

If no coliform colonies are observed, report the coliform colonies counted as "<1 coliform/100 mL."

For verified coliform counts, adjust the initial count based upon the positive verification percentage and report as "verified coliform count/100 mL."

Percentage verified coliforms

$$= \frac{\text{number of verified colonies}}{\text{total number of coliform colonies subjected to verification}} \times 100$$

a. Water of drinking water quality: While the EPA Total Coliform Rule for public water supply samples requires only a record of coliform presence or absence in 100-mL samples, it may be advisable to determine coliform densities in repeat sampling situations. This is of particular importance when a coliform biofilm problem is suspected in the distribution system. Quantitative information may provide an indication of the magnitude of a contaminating event.

With water of good quality, the occurrence of coliforms generally will be minimal. Therefore, count all coliform colonies (disregarding the lower limit of 20 cited above) and use the formula given above to obtain coliform density.

If confluent growth occurs, covering either the entire filtration area of the membrane or a portion thereof, and colonies are not discrete, report results as "confluent growth with (or without) coliforms." If the total number of bacterial colonies, coliforms plus noncoliforms, exceeds 200 per membrane, or if the colonies are not distinct enough for accurate counting, report results as "too numerous to count" (TNTC) or "confluent," respectively. For drinking water, the presence of coliforms in such cultures showing no sheen may be confirmed by either transferring a few colonies or placing the entire membrane filter culture into a sterile tube of brilliant green lactose bile broth. As an alternative, brush the entire filter surface with a sterile loop, applicator stick, or cotton swab and inoculate this growth to the tube of brilliant green lactose bile broth. If gas is produced from the brilliant green bile broth tube within 48 h at 35 ± 0.5°C, coliforms are present. For compliance with the EPA Total Coliform Rule, report confluent growth or TNTC with at least one detectable coliform colony (which is verified) as a total coliform positive sample. Report confluent growth or TNTC without detectable coliforms as invalid. For invalid samples, request a new sample from the same location within 24 h and select more appropriate volumes to be filtered per membrane, observing the requirement that the standard drinking water portion is 100 mL, or choose another coliform method that is less subject to heterotrophic bacterial interferences. Thus, to reduce interference from overcrowding, instead of filtering 100 mL per membrane, filter 50-mL portions through two separate membranes, 25-mL portions through each of four membranes, etc. Total the coliform counts observed on all membranes and report as number per 100 mL.

b. Water of other than drinking water quality: As with potable water samples, if no filter has a coliform count falling in the ideal range, total the coliform counts on all filters and report as number per 100 mL. For example, if duplicate 50-mL portions were examined and the two membranes had five and three coliform colonies, respectively, report the count as eight coliform colonies per 100 mL, i.e.,

$$\frac{[(5 + 3) \times 100]}{(50 + 50)} = 8 \text{ coliforms/100 mL}$$

Similarly, if 50-, 25-, and 10-mL portions were examined and the counts were 15, 6, and <1 coliform colonies, respectively, report the count as 25/100 mL, i.e.,

$$\frac{[(15 + 6 + 0) \times 100]}{(50 + 25 + 10)} = 25 \text{ coliforms/100 mL}$$

On the other hand, if 10-, 1.0-, and 0.1-mL portions were examined with counts of 40, 9, and <1 coliform colonies, respectively, select the 10-mL portion only for calculating the coliform density because this filter had a coliform count falling in the ideal range. The result is 400/100 mL, i.e.,

$$\frac{(40 \times 100)}{10} = 400 \text{ coliforms/100 mL}$$

In this last example, if the membrane with 40 coliform colonies also had a total bacterial colony count greater than 200, report the coliform count as ≥400/100 mL.

TABLE 9222:II. CONFIDENCE LIMITS FOR MEMBRANE FILTER COLIFORM RESULTS USING 100-mL SAMPLE

Number of Coliform Colonies Counted	95% Confidence Limits	
	Lower	Upper
0	0.0	3.7
1	0.1	5.6
2	0.2	7.2
3	0.6	8.8
4	1.0	10.2
5	1.6	11.7
6	2.2	13.1
7	2.8	14.4
8	3.4	15.8
9	4.0	17.1
10	4.7	18.4
11	5.4	19.7
12	6.2	21.0
13	6.9	22.3
14	7.7	23.5
15	8.4	24.8
16	9.2	26.0
17	9.9	27.2
18	10.7	28.4
19	11.5	29.6
20	12.2	30.8

Report confluent growth or membranes with colonies too numerous to count as described in *a* above. Request a new sample and select more appropriate volumes for filtration or utilize the multiple-tube fermentation technique.

c. Statistical reliability of membrane filter results: Although the precision of the MF technique is greater than that of the MPN procedure, membrane counts may underestimate the number of viable coliform bacteria. Table 9222:II illustrates some 95% confidence limits. These values are based on the assumption that bacteria are distributed randomly and follow a Poisson distribution. For results with counts, c, greater than 20 organisms, calculate the approximate 95% confidence limits using the following normal distribution equations:

$$\text{Upper limit} = c + 2\sqrt{c} \quad \text{Lower limit} = c - 2\sqrt{c}$$

7. Bibliography

FIFIELD, C.W. & C.P. SCHAUFUS. 1958. Improved membrane filter medium for the detection of coliform organisms. *J. Amer. Water Works Assoc.* 50:193.

McCARTHY, J.A. & J.E. DELANEY. 1958. Membrane filter media studies. *Water Sewage Works* 105:292.

RHINES, C.E. & W.P. CHEEVERS. 1965. Decontamination of membrane filter holders by ultraviolet light. *J. Amer. Water Works Assoc.* 57:500.

GELDREICH, E.E., H.L. JETER & J.A. WINTER. 1967. Technical considerations in applying the membrane filter procedure. *Health Lab. Sci.* 4:113.

WATLING, H.R. & R.J. WATLING. 1975. Note on the trace metal content of membrane filters. *Water SA* 1:28.

LIN, S.D. 1976. Evaluation of Millipore HA and HC membrane filters for the enumeration of indicator bacteria. *Appl. Environ. Microbiol.* 32:300.

STANDRIDGE, J.H. 1976. Comparison of surface pore morphology of two brands of membrane filters. *Appl. Environ. Microbiol.* 31:316.

GELDREICH, E.E. 1976. Performance variability of membrane filter procedure. *Pub. Health Lab.* 34:100.

GRABOW, W.O.K. & M. DU PREEZ. 1979. Comparison of m-Endo LES, MacConkey and Teepol media for membrane filtration counting of total coliform bacteria in water. *Appl. Environ. Microbiol.* 38:351.

DUTKA, B.D., ed. 1981. Membrane Filtration Applications, Techniques and Problems. Marcel Dekker, Inc., New York, N.Y.

EVANS, T.M., R.J. SEIDLER & M.W. LECHEVALLIER. 1981. Impact of verification media and resuscitation on accuracy of the membrane filter

total coliform enumeration technique. *Appl. Environ. Microbiol.* 41: 1144.

FRANZBLAU, S.G., B.J. HINNEBUSCH, T.M. KELLEY & N.A. SINCLAIR. 1984. Effect of noncoliforms on coliform detection in potable ground-water: improved recovery with an anaerobic membrane filter technique. *Appl. Environ. Microbiol.* 48:142.

MCFETERS, G.A., J.S. KIPPIN & M.W. LECHEVALLIER. 1986. Injured coliforms in drinking water. *Appl. Environ. Microbiol.* 51:1.

9222 C.   Delayed-Incubation Total Coliform Procedure

Modification of the standard MF technique permits membrane shipment or transport after filtration to a distant laboratory for transfer to another substrate, incubation, and completion of the test. This delayed-incubation test may be used where it is impractical to apply conventional procedures. It also may be used: (*a*) where it is not possible to maintain the desired sample temperature during transport; (*b*) when the elapsed time between sample collection and analysis would exceed the approved time limit; or (*c*) where the sampling location is remote from laboratory services.

Independent studies using both fresh- and salt-water samples have shown consistent results between the delayed incubation and standard direct test. Determine the applicability of the delayed-incubation test for a specific water source by comparing with results of conventional MF methods.

To conduct the delayed-incubation test, filter sample in the field immediately after collection, place filter on the transport medium, and ship to the laboratory. Complete the coliform determination in the laboratory by transferring the membrane to standard M-Endo or LES Endo medium, incubating at 35 ± 0.5°C for 20 to 22 h, and counting typical and atypical coliform colonies that develop. For drinking water samples collected for compliance with the EPA Total Coliform Rule, report the presence or absence of verified coliforms in 100-mL samples. Verify colonies as outlined previously in Section 9222B.5*f*.

Transport media are designed to keep coliform organisms viable and generally do not permit visible growth during transit time. Bacteriostatic agents in holding/preservative media suppress growth of microorganisms en route but allow normal coliform growth after transfer to a fresh medium.

The delayed-incubation test follows the methods outlined for the total coliform MF procedure, except as indicated below. Two alternative methods are given, one using the M-Endo preservative medium and the other the M-ST holding medium.

1. Apparatus

a. Culture dishes: Use disposable, sterile, plastic petri dishes (50 × 12 mm) with tight-fitting lids. Such containers are light in weight and are less likely to break in transit. In an emergency or when plastic dishes are unavailable, use sterile glass petri dishes wrapped in plastic film or similar material. See Section 9222B.1*e* for specifications.

b. Field filtration units: See Section 9222B.1*f* for specifications. Disinfect by adding methyl alcohol to the filtering chamber, igniting the alcohol, and covering unit to produce formaldehyde.

Ultraviolet light disinfection also may be used in the field if an appropriate power source is available (115 V, 60 Hz). Glass or metal filtration units may be sterilized by immersing in boiling water for 2 min. Use a hand aspirator to obtain necessary vacuum.

2. Materials and Transport Media

a. M-Endo methods:

1) *M-Endo preservative medium:* Prepare M-Endo medium as described in Section 9222B.2*b*. After cooling to below 45°C, aseptically add 3.84 g sodium benzoate (USP grade)/L or 3.2 mL 12% sodium benzoate solution to 100 mL medium. Mix ingredients and dispense in 5- to 7-mL quantities to 50- × 9-mm petri plates. Refrigerate poured plates. Discard unused medium after 96 h.

2) *Sodium benzoate solution:* Dissolve 12 g $NaC_7H_5O_2$ in sufficient reagent water to make 100 mL. Sterilize by autoclaving or by filtering through a 0.22-μm pore size membrane filter. Discard after 6 months.

3) *Cycloheximide:** Optionally add cycloheximide to M-Endo preservative medium. It may be used for samples that previously have shown overgrowth by fungi, including yeasts. Prepare by aseptically adding 50 mg cycloheximide/100 mL to M-Endo preservative medium. Store cycloheximide solution in refrigerator and discard after 6 months. Cycloheximide is a powerful skin irritant; handle with caution according to the manufacturer's directions.

b. M-ST method:

M-ST holding medium:

Sodium phosphate, monobasic, $NaH_2PO_4 \cdot H_2O$	0.1 g
Dipotassium hydrogen phosphate, KH_2PO_4	3.0 g
Sulfanilamide	1.5 g
Ethanol (95%)	10 mL
Tris (hydroxymethyl) aminomethane	3.0 g
Reagent-grade water	1 L

Dissolve ingredients by rehydrating in water. Sterilize by autoclaving at 121°C for 15 min. Final pH should be 8.6 ± 0.2. Dispense at least 2.0 mL to tight-lidded plastic culture dishes containing an absorbent pad and carefully remove excess liquid from pad by decanting plate. Store in refrigerator for use within 96 h.

---

* Actidione®, manufactured by the Upjohn Company, Kalamazoo, MI, or equivalent.

3. Procedure

a. Sample preservation and shipment: Place absorbent pad in bottom of sterile petri dish and saturate with selected coliform holding medium (see Section 9222C.2 above). Remove membrane filter from filtration unit with sterile forceps and roll it, grid side up, onto surface of medium-saturated pad. Protect membrane from moisture loss by tightly closing plastic petri dish. Seal loose-fitting dishes with an appropriate sealing tape to prevent membrane dehydration during transit. Place culture dish containing membrane in an appropriate shipping container and send to the laboratory for test completion. The sample can be held without visible growth for a maximum of 72 h on the holding/preservative medium. This usually allows use of the mail or a common carrier. Visible growth occasionally begins on the transport medium when high temperatures are encountered during transit.

b. Transfer and incubation: At the laboratory, transfer membrane from holding medium on which it was shipped to a second sterile petri dish containing M-Endo or LES Endo medium and incubate at 35 ± 0.5°C for 20 to 22 h.

4. Estimation of Coliform Density

Proceed as in Section 9222B.6 above. Record times of collection, filtration, and laboratory examination, and calculate the elapsed time. Report elapsed time with coliform results.

5. Bibliography

GELDREICH, E.E., P.W. KABLER, H.L. JETER & H.F. CLARK. 1955. A delayed incubation membrane filter test for coliform bacteria in water. *Amer. J. Pub. Health* 45:1462.

PANEZAI, A.K., T.J. MACKLIN & H.G. COLES. 1965. *Coli-aerogenes* and *Escherichia coli* counts on water samples by means of transported membranes. *Proc. Soc. Water Treat. Exam.* 14:179.

BREZENSKI, F.T. & J.A. WINTER. 1969. Use of the delayed incubation membrane filter test for determining coliform bacteria in sea water. *Water Res.* 3:583.

CHEN, M. & P.J. HICKEY. 1986. Elimination of overgrowth in delayed-incubation membrane filter test for total coliforms by M-ST holding medium. *Appl. Environ. Microbiol.* 52:778.

9222 D. Fecal Coliform Membrane Filter Procedure

Fecal coliform bacterial densities may be determined either by the multiple-tube procedure or by the MF technique. See Section 9225 for differentiation of *Escherichia coli*, the predominant fecal coliform. If the MF procedure is used for chlorinated effluents, demonstrate that it gives comparable information to that obtainable by the multiple-tube test before accepting it as an alternative. The fecal coliform MF procedure uses an enriched lactose medium and incubation temperature of 44.5 ± 0.2°C for selectivity. Because incubation temperature is critical, submerge waterproofed (plastic bag enclosures) MF cultures in a water bath for incubation at the elevated temperature or use an appropriate solid heat sink incubator or other incubator that is documented to hold the 44.5°C temperature within 0.2°C throughout the chamber, over a 24-h period. Areas of application for the fecal coliform method in general are stated in the introduction to the multiple-tube fecal coliform procedures, Section 9221E.

1. Materials and Culture Medium

a. M-FC medium: The need for uniformity dictates the use of dehydrated media. Never prepare media from basic ingredients when suitable dehydrated media are available. Follow manufacturer's directions for rehydration. Commercially prepared media in liquid form (sterile ampule or other) also may be used if known to give equivalent results. See Section 9020 for quality control specifications.

M-FC medium:

Tryptose or biosate	10.0 g
Proteose peptone No. 3 or polypeptone	5.0 g
Yeast extract	3.0 g
Sodium chloride, NaCl	5.0 g
Lactose	12.5 g
Bile salts No. 3 or bile salts mixture	1.5 g
Aniline blue	0.1 g
Agar (optional)	15.0 g
Reagent-grade water	1 L

Rehydrate product in 1 L water containing 10 mL 1% rosolic acid in 0.2N NaOH.* Heat to near boiling, promptly remove from heat, and cool to below 50°C. Do not sterilize by autoclaving. If agar is used, dispense 5- to 7-mL quantities to 50- × 12-mm petri plates and let solidify. Final pH should be 7.4 ± 0.2. Refrigerate finished medium, preferably in sealed plastic bags or other containers to reduce moisture loss, and discard unused broth after 96 h or unused agar after 2 weeks.

Test each medium lot against a previously acceptable lot for satisfactory performance as described in Section 9020B, by making dilutions of a culture of *E. coli* (Section 9020) and filtering appropriate volumes to give 20 to 60 colonies per filter. With each new lot of medium verify 10 or more colonies obtained from several natural samples, to establish the absence of false positives. For most samples M-FC medium may be used without the 1% rosolic acid addition, provided there is no interference with background growth. Such interference may be expected in stormwater samples collected during the first runoff (initial flushing) after a long dry period.

Before use, test each batch of laboratory-prepared MF medium for performance with positive and negative culture controls. Check for coliform contamination at the beginning and end of each filtration series by filtering 20 to 30 mL of dilution or rinse

---

* Rosolic acid reagent will decompose if sterilized by autoclaving. Refrigerate stock solution in the dark and discard after 2 weeks or sooner if its color changes from dark red to muddy brown.

water through filter. If controls indicate contamination, reject all data from affected samples and request resample.

b. Culture dishes: Tight-fitting plastic dishes are preferred because the membrane filter cultures are submerged in a water bath during incubation. Place fecal coliform cultures in plastic bags or seal individual dishes with waterproof (freezer) tape to prevent leakage during submersion. Specifications for plastic culture dishes are given in Section 9222B.1*e*.

c. Incubator: The specificity of the fecal coliform test is related directly to the incubation temperature. Static air incubation may be a problem in some types of incubators because of potential heat layering within the chamber, slower heat transfer from air to the medium, and the slow recovery of temperature each time the incubator is opened during daily operations. To meet the need for greater temperature control use a water bath, a heat-sink incubator, or a properly designed and constructed incubator shown to give equivalent results. A temperature tolerance of 44.5 ± 0.2°C can be obtained with most types of water baths that also are equipped with a gable top for the reduction of water and heat losses.

2. Procedure

a. Selection of sample size: Select volume of water sample to be examined in accordance with the information in Table 9222: III. Use sample volumes that will yield counts between 20 and 60 fecal coliform colonies per membrane.

When the bacterial density of the sample is unknown, filter several volumes or dilutions to achieve a countable density. Estimate volume and/or dilution expected to yield a countable membrane and select two additional quantities representing one-tenth and ten times this volume, respectively.

b. Filtration of sample: Follow the same procedure and precautions as prescribed under Section 9222B.5*b* above.

c. Preparation of culture dish: Place a sterile absorbent pad in each culture dish and pipet at least 2.0 mL M-FC medium, prepared as directed above, to saturate pad. Carefully remove any excess liquid from culture dish by decanting the plate. Aseptically, place prepared filter on medium-impregnated pad as described in Section 9222B above.

As a substrate substitution for the nutrient-saturated absorbent pad, add 1.5% agar to M-FC broth as described in Section 9222B above.

d. Incubation: Place prepared dishes in waterproof plastic bags or seal, invert, and submerge petri dishes in water bath, and incubate for 24 ± 2 h at 44.5 ± 0.2°C. Anchor dishes below water surface to maintain critical temperature requirements. Place all prepared cultures in the water bath within 30 min after filtration. Alternatively, use an appropriate, accurate solid heat sink or equivalent incubator.

e. Counting: Colonies produced by fecal coliform bacteria on M-FC medium are various shades of blue. Nonfecal coliform colonies are gray to cream-colored. Normally, few nonfecal coliform colonies will be observed on M-FC medium because of selective action of the elevated temperature and addition of rosolic acid salt reagent. Count colonies with a low-power (10 to 15 magnifications) binocular wide-field dissecting microscope or other optical device.

f. Verification: Verify typical blue colonies and any atypical grey to green colonies as described in Section 9020 for fecal coliform analysis. Simultaneous inoculation at both temperatures is acceptable.

3. Calculation of Fecal Coliform Density

a. General: Compute the density from the sample quantities that produced MF counts within the desired range of 20 to 60 fecal coliform colonies. This colony density range is more restrictive than the 20 to 80 total coliform range because of larger colony size on M-FC medium. Calculate fecal coliform density as directed in Section 9222B.6 above. Record densities as fecal coliforms per 100 mL.

b. Sediment and biosolid samples: For total solid (dry weight basis) see Section 2540G.

Calculate fecal coliforms per gram dry weight for biosolid analysis as follows:

Fecal coliforms per gram dry weight

$$= \frac{\text{colonies counted}}{(\text{dilution chosen}) \times (\% \text{ dry solids})}$$

where dilution and % dry solids are expressed in decimal form.

Example 1: There were 22 colonies observed on the 1:10 000 dilution plate of a biosolid with 4% dry solids.

$$\frac{22}{(0.0001)(0.04)} = 5.5 \times 10^6 \text{ fecal coliform/g dry weight}$$

If no filter has a coliform count falling in the ideal range (20

TABLE 9222:III. SUGGESTED SAMPLES VOLUMES FOR MEMBRANE FILTER FECAL COLIFORM TEST

Water Source	Volume (X) To Be Filtered mL							
	100	50	10	1	0.1	0.01	0.001	0.0001
Lakes, reservoirs	X	X						
Wells, springs	X	X						
Water supply intake		X	X	X				
Natural bathing waters		X	X	X				
Sewage treatment plant			X	X	X			
Farm ponds, rivers				X	X	X		
Stormwater runoff				X	X	X		
Raw municipal sewage				X	X	X		
Feedlot runoff				X	X	X		
Sewage sludge					X	X	X	

to 60), total the coliform counts on all countable filters and report as fecal coliforms per gram dry weight.

Example 2: There were 18 colonies observed on the 1:10 000 dilution plate and 2 colonies observed on the 1:100 000 dilution plate of a biosolid sample with 4% dry solids.

$$\frac{(18 + 2)}{(0.0001 + 0.00001)(0.04)} = 4.5 \times 10^6$$

To compute a geometric mean of samples, convert coliform densities of each sample to $\log_{10}$ values. Determine the geometric mean for the given number of samples (usually seven) by averaging the $\log_{10}$ values of the coliform densities and taking the antilog of that value.

4. Bibliography

GELDREICH, E.E., H.F. CLARK, C.B. HUFF & L.C. BEST. 1965. Fecal-coliform-organism medium for the membrane filter technique. *J. Amer. Water Works Assoc.* 57:208.

ROSE, R.E., E.E. GELDREICH & W. LITSKY. 1975. Improved membrane filter method for fecal coliform analysis. *Appl. Microbiol.* 29:532.

LIN, S.D. 1976. Membrane filter method for recovery of fecal coliforms in chlorinated sewage effluents. *Appl. Environ. Microbiol.* 32:547.

PRESSWOOD, W.G. & D.K. STRONG. 1978. Modification of M-FC medium by eliminating rosolic acid. *Appl. Environ. Microbiol.* 36:90.

GREEN, B.L., W. LITSKY & K.J. SLADEK. 1980. Evaluation of membrane filter methods for enumeration of faecal coliforms from marine waters. *Mar. Environ. Res.* 67:267.

SARTORY, D.P. 1980. Membrane filtration faecal coliform determinations with unmodified and modified M-FC medium. *Water SA* 6:113.

GRABOW, W.O.K., C.A. HILNER & P. COUBROUGH. 1981. Evaluation of standard and modified M-FC, MacConkey, and Teepol media for membrane filter counting of fecal coliform in water. *Appl. Environ. Microbiol.* 42:192.

RYCHERT, R.C. & G.R. STEPHENSON. 1981. Atypical *Escherichia coli* in streams. *Appl. Environ. Microbiol.* 41:1276.

PAGEL, J.E., A.A. QURESHI, D.M. YOUNG & L.T. VLASSOFF. 1982. Comparison of four membrane filter methods for fecal coliform enumeration. *Appl. Environ. Microbiol.* 43:787.

U.S. ENVIRONMENTAL PROTECTION AGENCY. 1992. Environmental Regulations and Technology. Control of Pathogens and Vector Attraction in Sewage Sludge. EPA-626/R-92-013, Washington, D.C.

U.S. ENVIRONMENTAL PROTECTION AGENCY. 1993. Standards for the Use or Disposal of Sewage Sludge: Final Rule. 40 CFR Part 257; *Federal Register* 58:9248, Feb. 19, 1993.

9222 E.   Delayed-Incubation Fecal Coliform Procedure

This delayed-incubation procedure is similar to the delayed-incubation total coliform procedure (Section 9222C). Use the delayed-incubation test only when the standard immediate fecal coliform test cannot be performed (i.e., where the appropriate field incubator is not available, or where, under certain circumstances, a specialized laboratory service is advisable to examine, confirm, or speciate the suspect colonies).

Results obtained by this delayed method have been consistent with results from the standard fecal coliform MF test under various laboratory and field use conditions. However, determine test applicability for a specific water source by comparison with the standard MF test, especially for saline waters, chlorinated wastewaters, and waters containing toxic substances.

To conduct the delayed-incubation test filter sample in the field immediately after collection, place filter on M-ST holding medium (see Section 9222C.2b below), and ship to the laboratory. Complete fecal coliform test by transferring filter to M-FC medium, incubating at 44.5°C for 24 ± 2 h, and counting fecal coliform colonies.

The M-ST medium keeps fecal coliform organisms viable but prevents visible growth during transit. Membrane filters can be held for up to 3 d on M-ST holding medium with little effect on the fecal coliform counts.

1. Apparatus

a. *Culture dishes:* See Section 9222C.1a for specifications.

b. *Field filtration units:* See Section 9222C.1b.

2. Materials and Transport Medium

a. *M-ST medium:* Prepare as described in Section 9222C.2b.

b. *M-FC medium:* Prepare as described in Section 9222D.1a.

3. Procedure

a. *Membrane filter transport:* Place an absorbent pad in a tight-lid plastic petri dish and saturate with M-ST holding medium. After filtering sample remove membrane filter from filtration unit and place it on medium-saturated pad. Use only tight-lid dishes to prevent moisture loss; however, avoid having excess liquid in the dish. Place culture dish containing membrane in an appropriate shipping container and send to laboratory. Membranes can be held on the transport medium at ambient temperature for a maximum of 72 h with little effect on fecal coliform counts.

b. *Transfer:* At the laboratory remove membrane from holding medium and place it in another dish containing M-FC medium.

c. *Incubation:* After transfer of filter to M-FC medium, place tight-lid dishes in waterproof plastic bags, invert, and submerge in a water bath at 44.5°C ± 0.2°C for 24 ± 2 h or use a solid heat sink or equivalent incubator.

d. *Counting:* Colonies produced by fecal coliform bacteria are various shades of blue. Nonfecal coliform colonies are gray to cream-colored. Count colonies with a binocular wide-field dissecting microscope at 10 to 15 magnifications.

e. *Verification:* Verify typical blue colonies and any atypical (grey to green) colonies as described in Section 9020 for fecal coliform analysis.

4. Estimation of Fecal Coliform Density

Count as directed in Section 9222D.2e above and compute fecal coliform density as described in Section 9222D.3. Record time of collection, filtration, and laboratory examination, and calculate and report elapsed time.

5. Bibliography

CHEN, M. & P.J. HICKEY. 1983. Modification of delayed-incubation procedure for detection of fecal coliforms in water. *Appl. Environ. Microbiol.* 46:889.

9222 F.   *Klebsiella* Membrane Filter Procedure

Klebsiella bacteria belong to the family Enterobacteriaceae and are included in the total coliform group. The outermost layer of *Klebsiella* bacteria consists of a large polysaccharide capsule, a characteristic that distinguishes this genus from most other bacteria in this family; this capsule provides some measure of protection from disinfectants. *Klebsiella* bacteria are commonly associated with coliform regrowth in large water supply distribution systems.

Klebsiellae may be opportunistic pathogens that can give rise to bacteremia, pneumonia, urinary tract, and several other types of human infection. Approximately 60 to 80% of all *Klebsiella* from feces and from clinical specimens are positive in the fecal coliform test and are *Klebsiella pneumoniae*.

Klebsiella bacteria also are widely distributed in nature, occurring in soil, water, grain, vegetation, etc. Wood pulp, paper mills, textile finishing plants, and sugar-cane processing operations contain large numbers of klebsiellae in their effluents (10^4 to 10^6), and *Klebsiella* sp. are often the predominant coliform in such effluents.

Rapid quantitation may be achieved in the MF procedure by modifying M-FC agar base through substitution of inositol for lactose and adding carbenicillin or by using M-Kleb agar. These methods reduce the necessity for biochemical testing of pure strains. Preliminary verification of differentiated colonies is recommended.

1. Apparatus

a. Culture dishes: See Section 9222B.1e for specifications.
b. Filtration units: See Section 9222B.1f.

2. Materials and Culture Medium

a. Modified M-FC agar (M-FCIC agar): This medium may not be available in dehydrated form and may require preparation from the basic ingredients:

Tryptose or biosate	10.0 g
Proteose peptone No. 3 or polypeptone	5.0 g
Yeast extract	3.0 g
Sodium chloride, NaCl	5.0 g
Inositol	10.0 g
Bile salts No. 3 or bile salts mixture	1.5 g
Aniline blue	0.1 g
Agar	15.0 g
Reagent-grade water	1 L

Heat medium to boiling and add 10 mL 1% rosolic acid dissolved in 0.2N NaOH. Cool to below 45°C and add 50 mg carbenicillin.* Dispense aseptically in 5- to 7-mL quantities into 50- × 9-mm plastic petri dishes. Refrigerate until needed. Discard unused agar medium after 2 weeks. Do not sterilize by autoclaving. Final pH should be 7.4 ± 0.2.

b. M-Kleb agar:

Phenol red agar	31.0 g
Adonitol	5.0 g
Aniline blue	0.1 g
Sodium lauryl sulfate	0.1 g
Reagent-grade water	1 L

Sterilize by autoclaving for 15 min at 121°C. After autoclaving, cool to 50°C in a water bath; add 20 mL 95% ethyl alcohol (not denatured) and 0.05 g filter sterilized carbenicillin/L. Shake thoroughly and dispense aseptically into 50- × 9-mm plastic culture plates. The final pH should be 7.4 ± 0.2. Refrigerated medium can be held for 20 d at 4 to 8°C.

3. Procedure

a. See Section 9222B.5 for selection of sample size and filtration procedure. Select sample volumes that will yield counts between 20 and 60 *Klebsiella* colonies per membrane. Place membrane filter on agar surface; incubate for 24 ± 2 h at 35 ± 0.5°C. *Klebsiella* colonies on M-FCIC agar are blue or bluish-gray. Most atypical colonies are brown or brownish. Occasional false positive occurrences are caused by *Enterobacter* species. *Klebsiella* colonies on M-Kleb agar are deep blue to blue gray, whereas other colonies most often are pink or occasionally pale yellow. Count colonies with a low-power (10 to 15 magnifications) binocular wide field dissecting microscope or other optical device.

b. Verification: Verify *Klebsiella* colonies from the first set of samples from ambient waters and effluents and when *Klebsiella* is suspect in water supply distribution systems. Verify a minimum of five typical colonies by transferring growth from a colony or pure culture to a commercial multi-test system for gram-negative speciation. Key tests for *Klebsiella* are citrate (positive), motility (negative), lysine decarboxylase (positive), ornithine decarboxylase (negative), and urease (positive). A *Klebsiella* strain that is indole-positive, liquefies pectin, and demonstrates a negative fecal coliform response is most likely of nonfecal origin.

4. Bibliography

DUNCAN, D.W. & W.E. RAZELL. 1972. *Klebsiella* biotypes among coliforms isolated from forest environments and farm produce. *Appl. Microbiol.* 24:933.

* Available from Geopen, Roerig-Pfizer, Inc. New York, NY.

STRAMER, S.L. 1976. Presumptive identification of *Klebsiella pneumoniae* on M-FC medium. *Can. J. Microbiol.* 22:1774.

BAGLEY, S.T. & R.J. SEIDLER. 1977. Significance of fecal coliform-positive *Klebsiella. Appl. Environ. Microbiol.* 33:1141.

KNITTEL, M.D., R.J. SEIDLER, C. EBY & L.M. CABE. 1977. Colonization of the botanical environment by *Klebsiella* isolates of pathogenic origin. *Appl. Environ. Microbiol.* 34:557.

EDMONSON, A.S., E.M. COOK, A.P.D. WILCOCK & R. SHINEBAUM. 1980. A comparison of the properties of *Klebsiella* isolated from different sources. *J. Med. Microbiol.* (U.K.) 13:541.

SMITH, R.B. 1981. A Critical Evaluation of Media for the Selective Iden-

tification and Enumeration of *Klebsiella*. M.S. thesis, Dep. Civil & Environmental Engineering, Univ. Cincinnati, Ohio.

NIEMELA, S.I. & P. VAATANEN. 1982. Survival in lake water of *Klebsiella pneumoniae* discharged by a paper mill. *Appl. Environ. Microbiol.* 44:264.

GELDREICH, E.E. & E.W. RICE. 1987. Occurrence, significance, and detection of *Klebsiella* in water systems. *J. Amer. Water Works Assoc.* 79:74.

DUNCAN, I.B.R. 1988. Waterborne *Klebsiella* and human disease. *Toxicity Assess.* 3:581.

9222 G.   MF Partition Procedures

1. *Escherichia coli* Partition Methods

a. Applications: Escherichia coli is a member of the fecal coliform group of bacteria; its presence is indicative of fecal contamination. Rapid quantitation and verification may be achieved with the MF procedure by transferring the membrane from a total-coliform- or fecal-coliform-positive sample to a nutrient agar substrate containing 4-methylumbelliferyl-β-D-glucuronide (MUG). In this method *E. coli* is defined as any coliform that produces the enzyme β-glucuronidase and hydrolyzes the MUG substrate to produce a blue fluorescence around the periphery of the colony.

In the examination of drinking water samples, use this method to verify the presence of *E. coli* from a total-coliform-positive MF on Endo-type media. In the examination of wastewater and other nonpotable water samples, use this procedure to verify positive filters from mFC medium used in the fecal coliform MF procedure.

b. Apparatus:

1) *Culture dishes*: See Section 9222B.1*e*.
2) *Filtration units*: See Section 9222B.1*f*.
3) *Forceps*: See Section 9222B.1*i*.
4) *Incubator*: See Section 9222B.1*j*.
5) *Ultraviolet lamp*, long wave (366 nm), preferably 6 W.
6) *Microscope and light source:* See Section 9222B.1 *k*.

c. Materials and culture medium:

1) *Nutrient agar with MUG (NA-MUG):*

Peptone	5.0 g
Beef extract	3.0 g
Agar	15.0 g
4-methylumbelliferyl-β-D-glucuronide	0.1 g
Reagent-grade water	1 L

Add dehydrated ingredients to reagent-grade water, mix thoroughly, and heat to dissolve. Sterilize by autoclaving for 15 min at 121°C. Dispense aseptically into 50-mm plastic culture plates. The final pH should be 6.8 ± 0.2. Refrigerated prepared medium may be held for 2 weeks.

2) *EC broth with MUG (EC-MUG):*

Tryptose or trypticase	20.0 g
Lactose	5.0 g

Bile salts mixture or bile salts No. 3	1.5 g
Dipotassium hydrogen phosphate, K$_2$HPO$_4$	4.0 g
Potassium dihydrogen phosphate, KH$_2$PO$_4$	1.4 g
Sodium chloride, NaCl	5.0 g
4-methylumbelliferyl-β-D-glucuronide	0.1 g
Reagent-grade water	1 L

Add dehydrated ingredients to reagent-grade water, mix thoroughly and heat to dissolve. pH should be 6.9 ± 0.2 after sterilization. Before sterilization, dispense into culture tubes and cap with metal or heat-resistant plastic caps.

d. Procedure: See Section 9222B.5 for selection of sample size and filtration procedure. For drinking water samples using Endo-type medium, count and record the metallic golden sheen colonies. Before transfer of the membrane, transfer a small portion of each target colony to the appropriate total coliform verification medium, using a sterile needle. See Section 9222B.5 for total coliform verification procedures.

Alternatively, after transfer and incubation on NA-MUG, swab the surface growth on the filter and transfer to the appropriate total coliform verification medium. Aseptically transfer the membrane from the Endo-type medium to NA-MUG or EC-MUG medium. If differentiation of the total coliforms is desired using NA-MUG medium, mark each sheen colony with a fine-tipped marker or by puncturing a hole in the membrane adjacent to the colony with a sterile needle. Incubate NA-MUG at 35 ± 0.5°C for 4 h or EC-MUG at 44.5 ± 0.2 for 24 ±2 h. Observe individual colonies or tubes using a long-wave-length (366-nm) ultraviolet light source, preferably containing a 6-W bulb. The presence of a blue fluorescence in the tube, on the periphery (outer edge) of a colony, or observed from the back of the plate is considered a positive response for *E. coli*. Count and record the number of target colonies, if quantification is desired, or just record presence or absence of fluorescence.

For nonpotable water samples, use mFC medium for initial isolation before transfer to NA-MUG or EC-MUG medium. The procedure is the same as the above, with the exception of the total coliform verification process.

For the EC-MUG method, a positive control consisting of a known *E. coli* (MUG-positive) culture, a negative control consisting of a thermotolerant *Klebsiella pneumoniae* (MUG-negative) culture, and an uninoculated medium control may be necessary to interpret the results and to avoid confusion of weak autofluorescence of the medium as a positive response. See Section 9221F.

2. Fecal Coliform Partition Method

a. Applications: Further partitioning of total coliforms from the original MF coliform-positive culture in a presence/absence search for fecal coliform in a drinking water sample may be achieved within 24 h. This procedure provides additional information from the original sample.

b. Materials and culture medium: EC broth. See Section 9221E.1a.

c. Procedure: See Section 9222B.5 for selection of sample size and filtration procedure. For drinking water samples using Endo-type media, count and record the metallic (golden) sheen colonies. Before transfer of membrane or swabbing of plate, transfer a small portion of each target colony to the appropriate total coliform verification media using a sterile needle (see Section 9222B.5f). Use a sterile cotton swab to collect bacteria from the membrane surface, or pick discrete colonies with a 3-mm loop or sterile applicator stick, or transfer the entire membrane to inoculate a tube of EC medium. Incubate inoculated EC broth in a water bath at 44.5 ± 0.2°C for 24 ± 2 h. Place all EC tubes in water bath within 30 min after inoculation. Maintain a sufficient water depth in water bath incubator to immerse tubes to upper level of the medium. Gas production in an EC broth culture in 24 h or less is considered a positive response for fecal coliform bacteria.

3. Bibliography

U.S. ENVIRONMENTAL PROTECTION AGENCY. 1989. Drinking Water; National Primary Drinking Water Regulations; Total Coliforms (Including Fecal Coliforms and *E. coli*); Final Rule. 40 CFR Parts 141 and 142. *Federal Register* 54:27544, June 29, 1989.

MATES, A. & M. SHAFFER. 1989. Membrane filtration differentiation of *E. coli* from coliforms in the examination of water. *J. Appl. Bacteriol.* 67:343.

U.S. ENVIRONMENTAL PROTECTION AGENCY. 1991. National Primary Drinking Water Regulations; Analytical Techniques; Coliform Bacteria. 40 CFR Part 141, *Federal Register* 56:636, Jan. 8, 1991.

MATES, A. & M. SHAFFER. 1992. Quantitative determination of *Escherichia coli* from coliforms and fecal coliforms in sea water. *Microbios* 71:27.

SARTORY, D. & L. HOWARD. 1992. A medium detecting beta-glucuronidase for the simultaneous membrane filtration enumeration of *Escherichia coli* and coliforms from drinking water. *Lett. Appl. Microbiol.* 15:273.

SHADIX, L.C., M.E. DUNNIGAN & E.W. RICE. 1993. Detection of *Escherichia coli* by the nutrient agar plus 4-methylumbelliferyl-β-D-glucuronide (MUG) membrane filter method. *Can. J. Microbiol.* 39:1066.

9223   ENZYME SUBSTRATE COLIFORM TEST*

9223 A.   Introduction

The enzyme substrate test utilizes hydrolyzable substrates for the simultaneous detection of total coliform bacteria and *Escherichia coli* enzymes. When the enzyme technique is used, the total coliform group is defined as all bacteria possessing the enzyme β-D-galactosidase, which cleaves the chromogenic substrate, resulting in release of the chromogen. *Escherichia coli* are defined as bacteria giving a positive total coliform response and possessing the enzyme β-glucuronidase, which cleaves a fluorogenic substrate, resulting in the release of the fluorogen. The test can be used in either a multiple-tube, multi-well, or a presence-absence (single 100-mL sample) format.

1. Principle

a. Total coliform bacteria: Chromogenic substrates, such as ortho-nitrophenyl-β-D-galactopyranoside (ONPG) or chlorophenol red-β-D-galactopyranoside (CPRG), are used to detect the enzyme β-D-galactosidase, which is produced by total coliform bacteria. The β-D-galactosidase enzyme hydrolyzes the substrate and produces a color change, which indicates a positive test for total coliforms at 24 h (ONPG) or 28 h (CPRG) without additional procedures. Noncoliform bacteria, such as *Aeromonas* and *Pseu-*

domonas species, may produce small amounts of the enzyme β-D-galactosidase, but are suppressed and generally will not produce a positive response within the incubation time unless more than 10^4 colony-forming units (CFU)/mL (10^6 CFU/100 mL) are present.

b. Escherichia coli: A fluorogenic substrate, such as 4-methylumbelliferyl-β-D-glucuronide (MUG), is used to detect the enzyme β-glucuronidase, which is produced by *E. coli*. The β-glucuronidase enzyme hydrolyzes the substrate and produces a fluorescent product when viewed under long-wavelength (366-nm) ultraviolet (UV) light. The presence of fluorescence indicates a positive test for *E. coli*. Some strains of *Shigella* spp. also may produce a positive fluorescence response. Because *Shigella* spp. are overt human pathogens, this is not considered a detriment for testing the sanitary quality of water.

2. Applications

The enzyme substrate coliform test is recommended for the analysis of drinking and source water samples. Formulations also are available for the analysis of marine waters. Initially, laboratories planning to use this procedure should conduct parallel quantitative testing (including seasonal variations) with one of the standard coliform tests to assess the effectiveness of the test for

---

* Approved by Standard Methods Committee, 1997.

the specific water type being analyzed and to determine the comparability of the two techniques. This is particularly important when testing source waters.

Water samples containing humic or other material may be colored. If there is background color, compare inoculated tubes to a control tube containing only water sample. In certain waters, high calcium salt content can cause precipitation but this should not affect the reaction.

Do not use the enzyme substrate test to verify presumptive coliform cultures or membrane filter colonies, because the substrate may be overloaded by the heavy inoculum of weak β-D-galactosidase-producing noncoliforms, causing false-positive results.

9223 B.   Enzyme Substrate Test

1. Substrate Media

Formulations are available commercially* in disposable tubes for the multiple-tube procedure, in disposable multi-wells† for the multi-well procedure, or in containers that will hold 100-mL samples for the presence-absence approach.* Appropriate preweighed portions of the reagent for mixing and dispensing into multiple tubes for 10-mL test portions or other containers for 100-mL samples also are available. The need for good quality assurance and uniformity requires the use of a commercial substrate medium. Avoid prolonged exposure of the substrate to direct sunlight. Store media according to directions and use before expiration date. Discard discolored media.

2. Procedure

a. Multiple-tube procedure: Select the appropriate number of tubes per sample with predispensed media for the multiple-tube test and label. Follow manufacturer's instructions for preparing serial dilutions for various formulations. Aseptically add 10 mL sample to each tube, cap tightly, and mix vigorously to dissolve. The mixture remains colorless with ONPG-based tests and turns yellow with the CPRG format. Some particles may remain undissolved throughout the test; this will not affect test performance. Incubate at 35 ± 0.5°C for period specified by substrate manufacturer.

The procedure also can be performed by adding appropriate amounts of the substrate media to the sample, mixing thoroughly, and dispensing into five or ten sterile tubes. Incubate as stated for multiple-tube procedure.

b. Multi-well procedure: The multi-well procedure is performed with sterilized disposable packets. Add sample to 100-mL container with substrate, shake vigorously, and pour into tray. The tray sealer dispenses the sample into the wells and seals the package. Incubate at 35 ± 0.5°C for period specified by substrate manufacturer. The MPN value is obtained from the table provided by the manufacturer.

c. Presence-absence procedure (P/A): Aseptically add preweighed enzymatic medium to 100-mL sample in a sterile, transparent, nonfluorescent borosilicate glass or equivalent bottle or container. Optionally, add 100-mL sample to the enzymatic substrate in a sterile container provided by the manufacturer. Asep-

tically cap and mix thoroughly to dissolve. Incubate as specified in manufacturer's instructions.

3. Interpretation

a. Total coliform bacteria: After the minimum proper incubation period, examine tubes or containers for the appropriate color change (Table 9223:I). ONPG is hydrolyzed by the bacterial enzyme to yield a yellow color. CPRG is hydrolyzed by the bacterial enzyme to yield a red or magenta color. If the color response is not uniform throughout the tube, mix by inversion before reading. Read manufacturer's instructions for interpretation guidelines. Some manufacturers suggest comparing sample tubes against a color comparator available through the manufacturer. Samples are negative for total coliforms if no color is observed in ONPG tests or if the tube is yellow when CPRG is used. If a chromogenic response is questionable after 18 or 24 h for ONPG, incubate up to an additional 4 h. If response is negative after 28 h for CPRG, incubate up to an additional 20 h. If the chromogen intensifies, the sample is total-coliform positive; if it does not, the sample is negative.

b. Escherichia coli: Examine positive total coliform tubes or containers for fluorescence using a long-wavelength (366-nm) ultraviolet lamp (preferably 6-W bulb). Compare each tube against the reference comparator available from a commercial source of the substrate. The presence of fluorescence is a positive test for *E. coli*. If fluorescence is questionable, incubate for an additional 4 h for ONPG tests and up to an additional 20 h for CPRG tests; intensified fluorescence is a positive test result.

4. Reporting

If performing an MPN procedure, calculate the MPN value for total coliforms and *E. coli* from the number of positive tubes as described in Section 9221C. If using the presence-absence procedure, report results as total coliform and *E. coli* present or absent in 100-mL sample.

---

* Colilert® and Colilert 18® for multi-tube, P/A, and tray formats, Colilert MW® for multi-tube format, and Colisure™ for multi-tube and P/A formats, available from IDEXX Laboratories, Inc., Westbrook, ME.
† Quanti-Tray® or Quanti-Tray®/2000, available from IDEXX Laboratories, Inc., Westbrook, ME.

TABLE 9223:I. COLOR CHANGES FOR VARIOUS MEDIA

Substrate	Total Coliform Positive	*E. coli* Positive	Negative Result
ONPG-MUG	Yellow	Blue fluorescence	Colorless/no fluorescence
CPRG-MUG	Red or magenta	Blue fluorescence	Yellow/no fluorescence

5. Quality Control

Test each lot of media purchased for performance by inoculation with three control bacteria: *Escherichia coli*, a total coliform other than *E. coli* (e.g., *Enterobacter cloacae*), and a noncoliform. Also add a sterile water control. If the sterile water control exhibits faint fluorescence or faint positive coliform result, discard and use a new batch of substrate. Avoid using a heavy inoculum. If *Pseudomonas* is used as the representative noncoliform, select a nonfluorescent species. Incubate these controls at $35 \pm 0.5°C$ as indicated above. Read and record results. Other quality-control guidelines are included in Section 9020.

6. Bibliography

EDBERG, S.C., M.J. ALLEN, D.B. SMITH & THE NATIONAL COLLABORATIVE STUDY. 1988. National field evaluation of a defined substrate method for the simultaneous enumeration of total coliforms and *Escherichia coli* from drinking water: Comparison with the standard multiple tube fermentation method. *Appl. Environ. Microbiol.* 54:1595.

EDBERG, S.C. & M.M. EDBERG. 1988. A defined substrate technology for the enumeration of microbial indicators of environmental pollution. *Yale J. Biol. Med.* 61:389.

COVERT, T.C., L.C. SHADIX, E.W. RICE, J.R. HAINES & R.W. FREYBERG. 1989. Evaluation of the Autoanalysis Colilert test for detection and enumeration of total coliforms. *Appl. Environ. Microbiol.* 55:2443.

EDBERG, S.C. & D.B. SMITH. 1989. Absence of association between total heterotrophic and total coliform bacteria from a public water supply. *Appl. Environ. Microbiol.* 55:380.

EDBERG, S.C., M.J. ALLEN, D.B. SMITH & THE NATIONAL COLLABORATIVE STUDY. 1989. National field evaluation of a defined substrate method for the simultaneous detection of total coliforms and *Escherichia coli* from drinking water: Comparison with presence- absence techniques. *Appl. Environ. Microbiol.* 55:1003.

EDBERG, S.C., M.J. ALLEN, D.B. SMITH & N.J. KRIZ. 1990. Enumeration of total coliforms and *Escherichia coli* from source water by the defined substrate technology. *Appl. Environ. Microbiol.* 56:366.

RICE, E.W., M.J. ALLEN & S.C. EDBERG. 1990. Efficacy of β-glucuronidase assay for identification of *Escherichia coli* by the defined-substrate technology. *Appl. Environ. Microbiol.* 56:1203.

RICE, E.W., M.J. ALLEN, D.J. BRENNER & S.C. EDBERG. 1991. Assay for β-glucuronidase in species of the genus *Escherichia* and its application for drinking water analysis. *Appl. Environ. Microbiol.* 57:592.

SHADIX, L.C. & E.W. RICE. 1991. Evaluation of β-glucuronidase assay for the detection of *Escherichia coli* from environmental waters. *Can. J. Microbiol.* 37:908.

EDBERG, S.C., M.J. ALLEN & D.B. SMITH. 1991. Defined substrate technology method for rapid and simultaneous enumeration of total coliforms and *Escherichia coli* from water: Collaborative study. *J. Assoc. Offic. Anal. Chem.* 74:526.

EDBERG, S.C., F. LUDWIG & D.B. SMITH. 1991. The Colilert® System for Total Coliforms and *Escherichia coli*. American Water Works Association Research Foundation, Denver, Colo.

COVERT, T.C., E.W. RICE, S.A. JOHNSON, D. BERMAN, C.H. JOHNSON & P.M. MASON. 1992. Comparing defined-substrate coliform tests for the detection of *Escherichia coli* in water. *J. Amer. Water Works Assoc.* 84(5):98.

MCCARTY, S.C., J.H. STANDRIDGE & M.C. STASIAK. 1992. Evaluating a commercially available defined-substrate test for recovery of chlorine-treated *Escherichia coli*. *J. Amer. Water Works Assoc.* 84(5): 91.

PALMER, C.J., Y. TSAI, A.L. LANG & L.R. SANGERMANO. 1993. Evaluation of Colilert-marine water for detection of total coliforms and *Escherichia coli* in the marine environment. *Appl. Environ. Microbiol.* 59:786.

CLARK, J.A. & A.H. SHAARAWI. 1993. Evaluation of commercial presence-absence test kits for detection of total coliforms, *Escherichia coli*, and other indicator bacteria. *Appl. Environ. Microbiol.* 59:380.

U.S. ENVIRONMENTAL PROTECTION AGENCY. 1994. National Primary and Secondary Drinking Water Regulation: Analytical methods for regulated drinking water contaminants; Final Rule. 40 CFR Parts 141 & 143; *Federal Register* 59:62456.

MCFETERS, G.A., S.C. BROADWAY, B.H. PYLE, M. PICKETT & Y. EGOZY. 1995. Comparative performance of Colisure™ and accepted methods in the detection of chlorine-injured total coliforms and *E. coli*. *Water Sci. Technol.* 31:259.

9225 DIFFERENTIATION OF THE COLIFORM BACTERIA*

9225 A. Introduction

Identification of bacteria that constitute the coliform group sometimes is necessary to determine the nature of pollution. It is of particular importance in reference to distinguishing the presence of *Escherichia coli*. Special procedures for detection of *E. coli* are given in Sections 9221F, 9222G, and 9223. Differential tests for identification must be used with the knowledge that all strains taxonomically assigned to the coliform group do not conform necessarily to the coliform definition stated in this manual

because they may not ferment lactose, or if they do, they may not produce gas. Furthermore, gram-negative bacteria other than coliforms ferment lactose and produce sheen (e.g., *Aeromonas* spp.) and not all strains of a species will react uniformly in media. Unusual strains (such as *E. coli*, inactive, Table 9225:I), mutants, and injured organisms may not give classical responses. The traditional "IMViC" tests (i.e., indole, methyl red, Voges-Proskauer, and citrate utilization) are useful for coliform differentiation, but do not provide complete identification. Additional biochemical tests often are necessary. Commercial kits for identification are available and may serve as economical alternatives

---

* Approved by Standard Methods Committee, 1994.

to traditional differential media. Automated systems of identifying large numbers of isolates also are available.

The significance of various coliform organisms in water has been and is a subject of considerable study. Collectively, the co-liforms are referred to as indicator organisms. The genera *Enterobacter, Klebsiella, Citrobacter,* and *Escherichia* usually are represented in the majority of isolations made from raw and treated municipal water supplies.

9225 B. Culture Purification

1. Procedure

A pure culture is essential for accurate identification. Obtain a pure culture by carefully picking a well-isolated colony that gives typical responses on an appropriate solid medium or membrane filter, and streaking on a tryptic soy or nutrient agar plate. Better distribution of colonies in the subculture is obtained if a portion of the picked colony is emulsified in peptone broth or physiological saline (0.85% w/v) and then streaked. When picking a colony from a primary culture on a selective medium, be aware that viable cells, which have not formed colonies themselves, may surround the picked colony. Incubate the subculture at 35 ± 0.5°C for 24 h and test a single well-isolated colony by the Gram stain to confirm the sole presence of gram-negative, non-spore-forming rods (Section 9221B). Also determine that the culture is oxidase-negative (Section 9225E). Oxidase-positive, gram-negative, non-spore-forming rods are not coliform bacteria, but may be organisms such as *Aeromonas*, which is not regarded as an indicator of fecal pollution.

Variation in organisms of the coliform group occurs occasionally and mixed reactions in differential media may indicate a pure culture undergoing variation. Persistent variations of reactions in differential media indicate a mixed culture caused by inadequate purification.

2. Bibliography

Ptak, D.J., W. Ginsburg & B.F. Willey. 1974. Aeromonas, the great masquerader. Proc. AWWA Water Quality Technology Conf., Dallas, Tex., p. V-1. American Water Works Assoc., Denver, Colo.

Van Der Kooj, D. 1988. Properties of aeromonads and their occurrence and hygienic significance in drinking water. *Zentralbl. Bacteriol. Hyg. B* 187:1.

Hartman, P.A., B. Swaminathan, M.S. Curiale, R. Firstenberg-Eden, A.N. Sharpe, N.A. Cox, D.Y.C. Fung & M.C. Goldschmidt. 1992. Rapid methods and automation. *In:* C. Vanderzant & D.F. Splittstoesser, eds., Compendium of Methods for the Microbiological Examination of Foods, 3rd. ed. p.665. American Public Health Assoc., Washington, D.C.

Stager, C.E. & J.R. Davis. 1992. Automated systems for identification of microorganisms. *Clin. Microbiol. Rev.* 5:302.

Rice, E.W., M.J. Allen, T.C. Covert, J. Langewis & J. Standridge. 1993. Identifying Escherichia species with biochemical test kits and standard bacteriological tests. *J. Amer. Water Works Assoc.* 85(2): 74.

9225 C. Identification

1. Definition

Coliforms are defined here as facultative anaerobic, gram-negative non-spore-forming rods that ferment lactose with gas formation within 48 h at 35°C or, as applied to the membrane filter method, produce a dark red colony with a metallic sheen within 24 h on an Endo-type medium containing lactose. However, anaerogenic (non-gas-producing) lactose-fermenting strains of *Escherichia coli* and coliforms that do not produce metallic sheen on Endo medium may be encountered. These organisms, as well as typical coliforms, can be considered indicator organisms, but they are excluded from the current definition of coliforms. More extensive testing may be required for proper identification.

2. Characteristics and Tests

Coliforms belong to the bacterial taxonomic family Enterobacteriaceae. Table 9225:1 provides data on some of the biochemical reactions used for differentiating these organisms.

Preparing differential media and reagents may not be as economical for many laboratories as using commercially prepared and prepackaged multiple-test kits, which reduce quality-control work. These commercial kits are simple to store and use, and give reproducible and generally accurate results. Periodically test reactions with known stock cultures of bacteria to assure accuracy and reproducibility of results. Make further tests if the kit provides equivocal results.

3. Bibliography

Krieg, N.R., ed. 1984. Bergey's Manual of Systematic Bacteriology, Vol. I. Williams & Wilkins Co., Baltimore, Md.

Edwards, P.R. & W.H. Ewing. 1986. Identification of Enterobacteriaceae, 4th ed. Burgess Publ. Co., Minneapolis, Minn.

TABLE 9225:I. BIOCHEMICAL REACTIONS OF KEY SPECIES OF THE FAMILY ENTEROBACTERIACEAE*

Species	Lactose Fermentation	ONPG Hydrolysis	Indole Production	Methyl Red	Voges-Proskauer	Simmons' Citrate	Ornithine Decarboxylase	Lysine Decarboxylase	Sorbitol Fermentation	Cellobiose Fermentation	Motility, 35–37°C	Yellow Pigment
Citrobacter diversus	35	96	99	100	0	99	99	0	99	99	95	0
Citrobacter freundii	50	95	5	100	0	95	20	0	98	55	95	0
Enterobacter aerogenes	95	100	0	5	98	95	98	98	100	100	97	0
Enterobacter agglomerans	40	90	20	50	70	50	0	0	30	55	85	75
Enterobacter cloacae	93	99	0	5	100	100	96	0	95	99	95	0
Escherichia coli	95	95	98	99	0	1	65	90	94	2	95	0
Escherichia coli, inactive	25	45	80	95	0	1	20	40	75	2	5	0
Escherichia fergusonii	0	83	98	100	0	17	100	95	0	96	93	0
Escherichia hermannii	45	98	99	100	0	1	100	6	0	97	99	98
Escherichia vulneris	15	100	0	100	0	0	0	85	1	100	100	50
Hafnia alvei	5	90	0	40	85	10	98	100	0	15	85	0
Klebsiella oxytoca	100	100	99	20	95	95	0	99	99	100	0	1
Klebsiella ozaenae	30	80	0	98	0	30	3	40	65	92	0	0
Klebsiella pneumoniae	98	99	0	10	98	98	0	98	99	98	0	0
Klebsiella rhinoscleromatis	0	0	0	100	0	0	0	0	100	100	0	0
Serratia fonticola	97	100	0	100	9	91	97	100	100	6	91	0
Serratia marcescens	2	95	1	20	98	98	99	99	99	5	97	0

Percentage Isolates Positive in 1 to 2 Days†

* Modified after Farmer, J.J. III, 1985. Clinical identification of new species and biogroups of Enterobacteriaceae. *J. Clin. Microbiol.* 21:46.
† Reactions that become positive after 2 d are not considered.

9225 D.   Media, Reagents, and Procedures

Commercially available media and reagents can reduce work and cost; however, include negative and positive controls with known stock cultures to assure accuracy and reliability. Detailed methods are available. Expected test results are shown in Table 9225:I.

1. Lactose, Sorbitol, and Cellobiose Fermentation Tests

Suspend 16 g phenol red broth base and 5 g of the desired carbohydrate in 1 L reagent-grade water and stir to dissolve completely. Dispense in tubes to a depth of one-third tube length. To determine gas production place a small inverted vial (Durham tube) in the tubes of media at the time of preparation. Close tubes and sterilize at 121°C for 15 min. Store tubes in the dark (refrigeration preferred) and discard if evaporation exceeds 10% of the volume.

To conduct a test, inoculate with a loopful of growth from a well-isolated colony or slant and incubate for 24 to 48 h at 35 ± 0.5°C. Carbohydrate fermentation (acid production) is indicated by a decrease in pH, resulting in a change in color of the pH indicator, phenol red, from red-orange to yellow (pH <6.6). Alternatively, for lactose fermentation, lauryl tryptose broth (Section 9221B) may be used.

2. ONPG Hydrolysis

Numerous commercial test kits and disks for determining ONPG hydrolysis are available, or an ONPG-containing medium (Section 9222) can be used. Alternatively, prepare peptone water by dissolving 1 g peptone and 0.5 g NaCl in 100 mL reagent-grade water. Sterilize at 121°C for 15 min. Also prepare ONPG solution by dissolving 0.6 g *o*-nitrophenyl-β-D-galactopyranoside (ONPG) in 100 mL 0.01M Na_2HPO_4, sterilize by filtration, and store in the dark at 4 to 10°C. To prepare ONPG broth, aseptically combine 25 mL ONPG solution and 75 mL peptone water, dispense aseptically in 2.5-mL amounts in sterile 13- × 100-mm tubes, and store in the dark for up to 1 month at 4 to 10°C. Do not use the ONPG solution if it becomes yellow.

To conduct the test, inoculate 0.5 mL ONPG broth with a heavy loopful of growth from a slant and incubate at 35 ± 0.5°C for up to 24 h. A yellow color, compared with an uninoculated tube or (preferably) a tube inoculated with an ONPG-negative culture, is a positive test. Interpret tests of yellow-pigmented organisms with caution. Do not use the enzyme substrate method (Section 9223) to test ONPG hydrolysis.

3. Indole Test

Indole is a product of the metabolism of tryptophane.
a. Reagents:
1) *Medium:* Use tryptophane broth. Dissolve 10.0 g tryptone or trypticase/L reagent-grade water. Dispense in 5-mL portions in test tubes and sterilize.
2) *Test reagent:* Dissolve 5 g *p*-dimethylaminobenzaldehyde in 75 mL isoamyl (or normal amyl) alcohol, ACS grade, and add 25 mL conc HCl. The reagent should be yellow. Some brands of *p*-dimethylaminobenzaldehyde are not satisfactory and some good brands become unsatisfactory on aging.

The amyl alcohol solution should have a pH value of less than 6.0. Purchase both amyl alcohol and benzaldehyde in as small amounts as will be consistent with the volume of work to be done.

b. Procedure: Inoculate 5-mL portions of medium from a pure culture and incubate at 35 ± 0.5°C for 24 ± 2 h. Add 0.2 to 0.3 mL test reagent and gently shake. Let stand for about 10 min and observe results.

A dark red color in the amyl alcohol surface layer constitutes a positive indole test; the original color of the reagent, a negative test. An orange color probably indicates the presence of skatole, a breakdown product of indole.

4. Methyl Red Test

The methyl red test measures the ability of organisms to produce stable acid end products from glucose fermentation.

a. Reagents:

1) *Medium:* Use buffered glucose broth. Dissolve 7.0 g proteose peptone or equivalent peptone, 5.0 g glucose, and 5.0 g dipotassium hydrogen phosphate (K_2HPO_4) in 1 L reagent-grade water. Dispense in 5-mL portions in test tubes and sterilize by autoclaving at 121°C for 12 to 15 min, making sure that total time of exposure to heat is not longer than 30 min.

2) *Indicator solution:* Dissolve 0.1 g methyl red in 300 mL 95% ethyl alcohol and dilute to 500 mL with reagent-grade water.

b. Procedure: Inoculate 10-mL portions of medium from a pure culture. Incubate at 35 ± 0.5°C for 5 d. To 5 mL of the culture add 5 drops methyl red indicator solution.

Incubation for 48 h is adequate for most cultures, but do not incubate for less than 48 h. If test results are equivocal at 48 h repeat with cultures incubated for 4 or 5 d. In such cases incubate duplicate cultures at 22 to 25°C. Testing of culture portions at 2, 3, 4, and 5 d may provide positive results sooner.

Record a distinct red color as methyl-red-positive and a distinct yellow color as methyl-red-negative. Record a mixed shade as questionable and possibly indicative of incomplete culture purification.

5. Voges-Proskauer Test

The Voges-Proskauer test measures the ability of organisms to produce a neutral end product (acetoin) from glucose fermentation.

a. Reagents:

1) *Medium:* See ¶ 4a1) above.

2) *Naphthol solution:* Dissolve 5 g purified α-naphthol (melting point 92.5°C or higher) in 100 mL absolute ethyl alcohol. When stored at 5 to 10°C, this solution is stable for 2 weeks.

3) *Potassium hydroxide*, 7N: Dissolve 40 g KOH in 100 mL reagent-grade water.

b. Procedure: Inoculate 5 mL medium and incubate for 48 h at 35 ± 0.5°C. To 1 mL of culture add 0.6 mL naphthol solution and 0.2 mL KOH solution. Shake well after the addition of each reagent. Development of a pink to crimson color at the surface within 5 min constitutes a positive test. Do not read after 10 min. Disregard tubes developing a copper color.

6. Simmons' Citrate Test

The citrate test measures the ability of bacteria to utilize citrate as the sole source of carbon.

a. Medium: Use Simmons' citrate agar. To make Simmons' citrate agar, add 0.2 g $MgSO_4 \cdot 7H_2O$, 1.0 g ammonium dihydrogen phosphate ($NH_4H_2PO_4$), 1.0 g K_2HPO_4, 2.0 g sodium citrate dihydrate, 5.0 g NaCl, 15.0 g agar, and 0.08 g bromthymol blue to 1 L reagent-grade water. Tube for long slants.

b. Procedure: Inoculate agar medium by the streak technique using a light inoculum.

Incubate 48 h at 35 ± 0.5°C. Record growth on the medium with a blue color as a positive reaction; record absence of growth or color change as negative.

7. Motility Test

The motility test measures whether an organism is motile in a semi-solid medium.

a. Medium: Use motility test medium made by adding 3.0 g beef extract, 10.0 g peptone, 5.0 g NaCl, and 4.0 g agar to 1 L reagent-grade water. Adjust pH to 7.4, dispense in 3-mL portions in 13- × 100-mm tubes or 8-mL portions in 16- × 125-mm tubes, and sterilize.

b. Procedure: Inoculate by stabbing into the center of the medium, using an inoculating needle, to a depth of 5 mm. Incubate for 1 to 2 d at 35°C. If negative, incubate an additional 5 d at 22 to 25°C.

Diffuse growth through the medium from the point of inoculation is positive. In a negative test, growth is visible only along the stab line and the surrounding medium stays clear. Alternatively, prepare the medium without agar and examine a young culture using the hanging drop slide technique for motile organisms.

8. Lysine and Ornithine Decarboxylase Tests

This procedure tests the ability of bacteria to metabolize the amino acids lysine and ornithine.

a. Reagents:

1) *Media:* Use a basal medium made according to the Moeller or Falkow methods. For the Moeller method, dissolve 5.0 g peptone (Orthana special, thiotone, or equivalent), 5.0 g beef extract, 0.625 mL bromcresol purple (1.6%), 2.5 mL cresol red (0.2%), 0.5 g glucose, and 5.0 mg pyridoxal in 1 L reagent-grade water and adjust to pH 6.0 to 6.5. For the Falkow method, dissolve 5.0 g peptone, 3.0 g yeast extract, 1.0 g glucose, and 1.0 mL bromcresol purple (1.6%) in 1 L reagent-grade water and adjust to pH 6.7 to 6.8. For either decarboxylase test divide into three portions: make no addition to the first portion, add enough L-lysine dihydrochloride to the second portion to make a 1% solution, and add L-ornithine dihydrochloride to the third to make 1% (for the Falkow method, add only 0.5% of the L-amino acid). After adding ornithine readjust pH of the medium to 6.0 ± 0.2. Dispense in 3- to 4-mL portions in screw-capped test tubes and sterilize by autoclaving at 121°C for 10 min. A floccular precipitate in the ornithine medium does not interfere with its use.

2) *Mineral oil:* Use mineral oil sterilized by autoclaving at 121°C for 30 to 60 min depending on the size of the container.

b. Procedure: Lightly inoculate each of the three media, add a layer of about 10 mm thickness of mineral oil, and incubate at 37°C for up to 4 d. Examine tubes daily. A color change from yellow to violet or reddish-violet constitutes a positive decarboxylase test; a change to bluish gray indicates a weak positive; no color change or a yellow color represents a negative test. See Table 9225:I.

9. Oxidase Test

The oxidase test determines the presence of oxidase enzymes. Coliform bacteria are oxidase-negative.

a. Reagents:

1) *Media:* Use either nutrient agar or tryptic soy agar plates to streak cultures and produce isolated colonies. From these obtain the inoculum for oxidase testing on impregnated filter paper. Do not use any medium that includes a carbohydrate in its formulation. Use only tryptic soy agar if reagent is dropped on colonies.

Tryptic soy agar:

Tryptone	15.0 g
Soytone	5.0 g
Sodium chloride, NaCl	5.0 g
Agar	15.0 g
Reagent-grade water	1.0 L

pH should be 7.3 ± 0.2 after sterilization.

2) *Tetramethyl p-phenylenediamine dihydrochloride*, 1% aqueous solution, freshly prepared or refrigerated for no longer than 1 week. Impregnate a filter paper strip* with this solution. Alternatively, prepare a 1% solution of dimethyl *p*-phenylenediamine hydrochloride. Single-use reagent ampules, commercially available, are convenient and economical, but use them with caution. When the reagent is to be dropped directly on colonies, use tryptic soy agar plates because nutrient agar plates give inconsistent results; when smearing a portion of a picked colony on reagent-impregnated filter paper, do not transfer any medium with the culture material.

---

* Whatman No. 1 or equivalent.

b. Procedure: Remove some of a colony from agar plate with a platinum wire, a wooden or plastic applicator stick, or a glass rod and smear on the test strip. Do not use iron or other reactive wire because it will cause false positive reactions. A dark purple color that develops within 10 s is a positive oxidase test. Test positive and negative cultures concurrently. If the liquid reagent is used, drop it on colonies on the culture plate. Oxidase-positive colonies develop a pink color that successively becomes maroon, dark red, and finally, black.

10. Yellow Pigment

Observe isolated colonies on nutrient agar slants and plates or plates of tryptic soy agar incubated at 35 ± 0.5°C for up to 48 h. Pigmentation often intensifies as time of incubation proceeds.

11. Bibliography

MacFaddin, J.F. 1976. Biochemical Tests for Identification of Medical Bacteria. Williams & Wilkins Co., Baltimore, Md.

Farmer, J.J., III. 1985. Biochemical identification of new species and biogroups of *Enterobacteriaceae. J. Clin. Microbiol.* 21:46.

Washington, J.A., ed. 1985. Laboratory Procedures in Clinical Microbiology, 2nd ed. Springer-Verlag, New York, N.Y.

Balows, A.H., W.J. Hausler, Jr., K.L. Hermann, H.P. Isenberg & H.J. Shadomy, eds. 1991. Manual of Clinical Microbiology, 5th ed. American Soc. for Microbiology, Washington, D.C.

9230   FECAL STREPTOCOCCUS AND ENTEROCOCCUS GROUPS*

9230 A.   Introduction

1. Fecal Streptococcus Group

The fecal streptococcus group consists of a number of species of the genus *Streptococcus,* such as *S. faecalis, S. faecium, S. avium, S. bovis, S. equinus,* and *S. gallinarum.* They all give a positive reaction with Lancefield's Group D antisera[1] and have been isolated from the feces of warm-blooded animals. In addition, *S. avium* sometimes reacts with Lancefield's Group Q antisera. *S. faecalis* subsp. *liquefaciens* and *S. faecalis* subsp. *zymogenes* are differentiated based on the ability of these strains to liquefy gelatin and hemolyze red cells. However, the validity of these subspecies is questionable.[2,3]

The normal habitat of fecal streptococci is the gastrointestinal tract of warm-blooded animals. *S. faecalis* and *S. faecium* once were thought to be more human-specific than other *Streptococcus* species. Other species have been observed in human feces but less frequently.[4] Similarly, *S. bovis, S. equinus,* and *S. avium* are not exclusive to animals, although they usually occur at higher densities in animal feces.[5] Certain streptococcal species predom-

inate in some animal species and not in others, but it is not possible to differentiate the source of fecal contamination based on speciation of fecal streptococci.

The fecal streptococci have been used with fecal coliforms to differentiate human fecal contamination from that of other warm-blooded animals. Editions of *Standard Methods* previous to the 17th suggested that the ratio of fecal coliforms (FC) to fecal streptococci (FS) could provide information about the source of contamination. A ratio greater than four was considered indicative of human fecal contamination, whereas a ratio of less than 0.7 was suggestive of contamination by nonhuman sources. The value of this ratio has been questioned because of variable survival rates of fecal streptococcus group species. *S. bovis* and *S. equinus* die off rapidly, once exposed to aquatic environments, whereas *S. faecalis* and *S. faecium* tend to survive longer.[6] Furthermore, disinfection of wastewaters appears to have a significant effect on the ratio of these indicators, which may result in misleading conclusions regarding the source of contaminants.[7] The ratio is affected also by the methods for enumerating fecal streptococci. The KF membrane filter procedure has a false-positive rate ranging from

---

* Approved by Standard Methods Committee, 1993.

10 to 90% in marine and fresh waters.[8–10] For these reasons, the FC/FS ratio cannot be recommended, and should not be used as a means of differentiating human and animal sources of pollution.

2. Enterococcus Group

The enterococcus group is a subgroup of the fecal streptococci that includes *S. faecalis*, *S. faecium*, *S. gallinarum*, and *S. avium*. The enterococci are differentiated from other streptococci by their ability to grow in 6.5% sodium chloride, at pH 9.6, and at 10°C and 45°C.

The enterococci portion of the fecal streptococcus group is a valuable bacterial indicator for determining the extent of fecal contamination of recreational surface waters. Studies at marine and fresh water bathing beaches indicated that swimming-associated gastroenteritis is related directly to the quality of the bathing water and that enterococci are the most efficient bacterial indicator of water quality.[11,12] Water quality guidelines based on enterococcal density have been proposed for recreational waters.[13] For recreational fresh waters the guideline is 33 enterococci/100 mL while for marine waters it is 35/100 mL. Each guideline is based on the geometric mean of at least five samples per 30-d period during the swimming season.

3. Selection of Method

The multiple-tube technique is applicable primarily to raw and chlorinated wastewater and sediments, and can be used for fresh and marine waters. The membrane filter technique also may be used for fresh and saline water samples, but is unsuitable for highly turbid waters.

4. References

1. SNEATH, P.H.A., N.S. MAIR, M.E. SHARPE & J.G. HOLT. eds. 1986. Bergey's Manual of Systematic Bacteriology. Vol. 2, Williams & Wilkins, Baltimore, Md.

2. JACOB, A.E., G.J. DOUGLAS & S.J. HOBBS. 1973. Self-transferable plasmids determining the haemolysin and bacteriocin of *Streptococcus faecalis* var. *zymogenes*. *J. Bacteriol.* 121:863.

3. OLIVER, D.R., B.L. BROWN & D.B. CLEWELL. 1977. Characterization of plasmids determining haemolysin and bacteriocin production in *Streptococcus faecalis* 5952. *J. Bacteriol.* 130:948.

4. WATANABE, T., H. SHIMOHASHI, Y. KAWAI, & M. MUTAI. 1981. Studies on streptococci. I. Distribution of fecal streptococci in man. *Microbiol. Immunol.* 25:257.

5. THOMAS, C.D. & M.A. LEVIN. 1978. Quantitative analysis of group D streptococci. Abs. Annual Meeting, American Soc. Microbiology, p. 210.

6. FEACHAM, R. 1975. An improved role for faecal coliform to faecal streptococcus ratios in the differentiation between human and non-human pollution sources. *Water Res.* 9:689.

7. ROSSER, P.A.E. & D.P. SARTORY. 1982. A note on the effect of chlorination of sewage effluents on faecal coliform to faecal streptococcus ratios in the differentiation of faecal pollution sources. *Water S.A.* 8: 66.

8. FUJIOKA, R.S., A.A. UENO & O.T. NARIKAWA. 1984. Recovery of False Positive Fecal Streptococcus on KF Agar from Marine Recreational Waters. Tech. Rep. No. 168, Water Resources Research Center, Univ. Hawaii at Manoa, Honolulu.

9. OLIVIERI, V.P., C.W. KRUSE, K. KAWATA & J.E. SMITH. 1977. Microorganisms in Urban Stormwater. EPA-600/2-77-087, U.S. Environmental Protection Agency, Edison, N.J.

10. ERICKSEN, T.H., C. THOMAS & A. DUFOUR. 1983. Comparison of two selective membrane filter methods for enumerating fecal streptococci in freshwater samples. Abs. Annual Meeting, American Soc. Microbiology, p. 279.

11. CABELLI, V.J. 1983. Health Effects Criteria for Marine Waters. EPA-600/1-80-031, U.S. Environmental Protection Agency, Cincinnati, Ohio.

12. DUFOUR, A.P. 1984. Health Effects Criteria for Fresh Recreational Waters. EPA-600/1-84-004, U.S. Environmental Protection Agency, Cincinnati, Ohio.

13. U.S. ENVIRONMENTAL PROTECTION AGENCY. 1986. Ambient Water Quality Criteria for Bacteria—1986. EPA-440/5-84-002, U.S. Environmental Protection Agency, Washington, D.C.

9230 B.   Multiple-Tube Technique

1. Materials and Culture Media

a. Azide dextrose broth:[1]

Beef extract	4.5	g
Tryptone or polypeptone	15.0	g
Glucose	7.5	g
Sodium chloride, NaCl	7.5	g
Sodium azide, NaN₃	0.2	g
Reagent-grade water	1	L

pH should be 7.2 ± 0.2 at 25°C after sterilization.

b. Pfizer selective enterococcus (PSE) agar:[2]

Peptone C	17.0	g
Peptone B	3.0	g
Yeast extract	5.0	g
Bacteriological bile	10.0	g
Sodium chloride, NaCl	5.0	g
Sodium citrate	1.0	g
Esculin	1.0	g
Ferric ammonium citrate	0.5	g
Sodium azide, NaN₃	0.25	g
Agar	15.0	g
Reagent-grade water	1	L

pH should be 7.1 ± 0.2 after sterilization. Hold medium for not more than 4 h at 45 to 50°C before plates are poured.

2. Presumptive Test Procedure

Inoculate a series of tubes of azide dextrose broth with appropriate graduated quantities of sample. Use sample of 10 mL portions or less. Use double-strength broth for 10-mL inocula. The portions used will vary in size and number with the sample character. Use only decimal multiples of 1 mL (see Section 9221 for suggested sample sizes).

Incubate inoculated tubes at 35 ± 0.5°C. Examine each tube for turbidity at the end of 24 ± 2 h. If no definite turbidity is present, reincubate, and read again at the end of 48 ± 3 h.

3. Confirmed Test Procedure

Subject all azide dextrose broth tubes showing turbidity after 24- or 48-h incubation to the confirmed test.

Streak a portion of growth from each positive azide dextrose broth tube on PSE agar. Incubate the inverted dish at 35 ± 0.5°C for 24 ± 2 h. Brownish-black colonies with brown halos confirm the presence of fecal streptococci.

Brownish-black colonies with brown halos may be transferred to a tube of brain-heart infusion broth containing 6.5% NaCl. Growth in 6.5% NaCl broth and at 45°C indicates that the colony belongs to the enterococcus group.

4. Computing and Recording of MPN

Estimate fecal streptococci densities from the number of tubes in each dilution series that are positive on PSE agar. Similarly, estimate enterococci densities from the number of tubes in each dilution series containing streptococci that can grow in 6.5% NaCl broth. Compute the combination of positives and record as the most probable number (MPN). Refer to Section 9221D.

5. References

1. MALLMAN, W.L. & E.B. SELIGMANN. 1950. A comparative study of media for the detection of streptococci in water and sewage. *Amer. J. Pub. Health* 40:286.
2. ISENBERG, H.D., D. GOLDBERT & J. SAMPSON. 1972. Laboratory studies with a selective enterococcus medium. *Health Lab. Sci.* 9:289.

9230 C.   Membrane Filter Techniques

1. Laboratory Apparatus

See Section 9222B.1.

2. Materials and Culture Media

a. mE agar for enterococci:[1]

Peptone	10.0	g
Sodium chloride, NaCl	15.0	g
Yeast extract	30.0	g
Esculin	1.0	g
Actidione (cycloheximide)	0.05	g
Sodium azide, NaN$_3$	0.15	g
Agar	15.0	g
Reagent-grade water	1	L

Heat to dissolve ingredients, sterilize, and cool in a water bath at 44 to 46°C. Mix 0.25 g nalidixic acid in 5 mL reagent-grade water, add a few drops of 0.1N NaOH to dissolve the antibiotic, and add to the basal medium. Add 0.15 g 2,3,5-triphenyl tetrazolium chloride and mix well to dissolve. Pour the agar into 9- × 50-mm petri dishes to a depth of 4 to 5 mm (approximately 4 to 6 mL), and let solidify. The final pH should be 7.1 ± 0.2. Store poured plates in the dark at 2 to 10°C. Discard after 30 d. (NOTE: This medium is recommended for culturing enterococci in fresh and marine recreational waters.)

b. EIA substrate:[1]

Esculin	1.0	g
Ferric citrate	0.5	g
Agar	15.0	g
Reagent-grade water	1	L

The pH should be 7.1 ± 0.2 before autoclaving. Heat to dissolve ingredients, sterilize, and cool in a water bath at 44 to 46°. Pour medium into 50-mm petri dishes to a depth of 4 to 5 mm (approximately 4 to 6 mL) and let solidify. Store poured plates in the dark at 2 to 10°C. Discard after 30 d.

c. m Enterococcus agar for fecal streptococci:[2]

Tryptose	20.0	g
Yeast extract	5.0	g
Glucose	2.0	g
Dipotassium phosphate, K$_2$HPO$_4$	4.0	g
Sodium azide, NaN$_3$	0.4	g
2,3,5-Triphenyl tetrazolium chloride	0.1	g
Agar	10.0	g
Reagent-grade water	1	L

Heat to dissolve ingredients. Do not autoclave. Dispense into 9- × 50-mm petri plates to a depth of 4 to 5 mm (approximately 4 to 6 mL), and let solidify. Prepare fresh medium for each set of samples. (NOTE: This medium is recommended for Group D streptococci in fresh and marine waters.)

d. Brain-heart infusion broth:

Infusion of calf brain	200	g
Infusion of beef heart	250	g
Proteose peptone	10.0	g
Glucose	2.0	g
Sodium chloride, NaCl	5.0	g
Disodium hydrogen phosphate, Na$_2$HPO$_4$	2.5	g
Reagent-grade water	1	L

The pH should be 7.4 after sterilization.

e. Brain-heart infusion agar: Add 15.0 g agar to the ingredients for brain-heart infusion broth. The pH should be 7.4 after sterilization. Tube for slants.

f. Bile esculin agar:[3]

Beef extract	3.0	g
Peptone	5.0	g
Oxgall	40.0	g
Esculin	1.0	g
Ferric citrate	0.5	g
Agar	15.0	g
Reagent-grade water	1	L

Heat to dissolve ingredients. Dispense 8 to 10 mL into tubes for slants or an appropriate volume into a flask for subsequent pouring into plates. Autoclave at 121°C for 15 min. Do not overheat because this may cause darkening of the medium. Cool to 44 to 46°C and slant the tubes or dispense 15 mL into 15- × 100-mm petri dishes. The final pH should be 6.6 ± 0.2 after sterilization. Store at 4 to 10°C.

3. Procedures

a. mE Method:[1]

1) Selection of sample size and filtration—Filter appropriate sample volumes through a 0.45-μm, gridded, sterile membrane to give 20 to 60 colonies on the membrane surface. Transfer filter to agar medium in petri dish, avoiding air bubbles beneath the membrane.

2) Incubation—Invert culture plates and incubate at 41°C ± 0.5°C for 48 h.

3) Substrate test—After 48 h incubation, carefully transfer filter to EIA medium. Incubate at 41°C ± 0.5°C for 20 min.

4) Counting—Pink to red enterococci colonies develop a black or reddish-brown precipitate on the underside of the filter. Count colonies using a fluorescent lamp and a magnifying lens.

b. m Enterococcus method:[2]

1) Selection of sample size and filtration—See ¶ 3a.

2) Incubation—Let plates stand for 30 min, then invert and incubate at 35 ± 0.5°C for 48 h.

3) Counting—Count all light and dark red colonies as enterococci. Count colonies using a fluorescent lamp and a magnifying lens.

4. Calculation of Fecal Streptococci or Enterococci Density

Compute density from sample quantities producing membrane filter counts within the desired 20- to 60-fecal streptococcus or enterococcus colony range. Calculate as in Section 9222B.6. Record densities as fecal streptococci or enterococci per 100 mL.

5. Verification Tests

Pick selected typical colonies from a membrane and streak for isolation onto the surface of a brain-heart infusion agar plate. Incubate at 35°C ± 0.5°C for 24 to 48 h.

Transfer a loopful of growth from a well-isolated colony on brain-heart infusion agar into a brain-heart infusion broth tube and to each of two clean glass slides. Incubate the brain-heart infusion broth at 35 ± 0.5°C for 24 h. Add a few drops of freshly prepared 3% hydrogen peroxide to the smear on a slide. The appearance of bubbles constitutes a positive catalase test and indicates that the colony is not a member of the fecal streptococcus group. If the catalase test is negative, i.e., no bubbles, make a Gram stain of the second slide. Fecal streptococci and enterococci are gram-positive, ovoid cells, 0.5 to 1.0 μm in diameter, mostly in pairs or short chains.

Transfer a loopful of growth from the brain-heart infusion broth to each of the following media: bile esculin agar (incubate at 35 ± 0.5°C for 48 h); brain-heart infusion broth (incubate at 45 ± 0.5°C for 48 h); brain-heart infusion broth with 6.5% NaCl (incubate at 35 ± 0.5°C for 48 h).

Growth of catalase-negative, gram-positive cocci on bile esculin agar and at 45°C in brain-heart infusion broth verifies that the colony is of the fecal streptococcus group. Growth at 45°C and in 6.5% NaCl broth indicates that the colony belongs to the enterococcus group.

6. Serological Verification of Group D Fecal Streptococci

An alternate verification test for Group D streptococci can be performed using the precipitin method of Lancefield.[4] This test is highly specific for S. faecalis, S. faecium, S. avium, S. gallinarum, S. bovis, and S. equinus.

a. Antigen preparation: Pick typical single colonies from the membrane filter and streak for isolation on brain-heart infusion agar or blood agar plates. Pick a well-isolated colony and inoculate into 30 to 50 mL of Todd-Hewitt broth.[5] Incubate at 35°C for 24 h under aerobic conditions. Concentrate bacterial suspension by centrifuging (3000 × g for 5 min). Draw supernatant off and resuspend cells in 0.5 mL saline solution. Autoclave resuspended cells for 15 min at 121°C. Centrifuge the bacteria and decant clear supernatant fluid containing the group antigen.

b. Capillary precipitin test: Antisera for this test may be obtained from commercial sources.

Dip a 1.2- to 1.5-mm-OD capillary tube into antiserum and draw up about 1 cm of serum. Place a finger over upper end of tube so that no air will be drawn up and carefully wipe off excess antiserum. Dip tube into streptococcal antigen extract solution and draw up an equal volume of antigen. Carefully wipe off excess extract. Place a finger over upper end of tube and force lower end into plasticine to plug lower opening. Invert tube and place it in plasticine groove of a capillary holding rack.

TABLE 9230:I. SELECTED KEY BIOCHEMICAL CHARACTERISTICS OF THE STREPTOCOCCUS SPECIES WITHIN THE FECAL STREPTOCOCCUS AND ENTEROCOCCUS GROUPS*

| | Fecal Streptococcus Group | | | | | |
| | Enterococcus Group | | | | | |
Test	*S. faecalis*	*S. faecium*	*S. avium*	*S. gallinarum*	*S. bovis*	*S. equinus*
Catalase	−		−	−		−
40% Bile	+	+	+	+	+	+
Esculin[6]	+	+	+	+	+	+
Growth at 45°C	+	+	+	+	+	+
Growth in 6.5% NaCl	+	+	+	+	−	−
Growth at 10°C	+	+	+	+	−	−
Pyruvate utilization[7]	+	−	−	−	−	−
Phosphatase activity[8]	+	−	+	+	−	−
Arginine hydrolysis[8]	+	+	−d	−	−	−
L-Sorbose fermentation[9]	−	−	+	−	−	−
Lactose fermentation[8]	+	+	+	+	+	+
n-Acetyl-β-glucoseaminidase activity[9]	−d		−	+	−	−
Starch[6]	−	−	−	−	+	−
Arabinose[6]	−	+	+	−	−	−

* + = 90% or more of strains are positive
− = 90% or more of strains are negative
d = reactions variable

A positive test for Group D antigen is characterized by a white precipitate that appears at the antigen-antiserum interface within 15 min and usually by 5 min. If no reaction has occurred by 30 min, the test is negative. Examination of the tubes is more effective if they are read in a bright light against a dark background.

Serological verification of Group D streptococci also can be done using commercially available agglutination tests.* The slide agglutination tests are simple and appear to be reliable. Group D streptococci are verified directly from isolated colonies on membrane filter plates or from broth culture tubes. To verify presumptive enterococci cultures, test also for salt tolerance (growth in 6.5% NaCl broth).

7. Identification of Individual Species within Fecal Streptococcus and Enterococcus Groups

Table 9230:I shows some of the key biochemical reactions for identifying fecal streptococci, enterococci, and species within these two groups.

---

* Phadebact Strep D test, Pharmacia Diagnostics, Piscataway, NJ and Streptex Test, Burroughs-Wellcome Co., Research Triangle Park, NC.

8. References

1. LEVIN, M.A., J.R. FISCHER & V.J. CABELLI. 1975. Membrane filter technique for enumeration of enterococci in marine waters. *Appl. Microbiol* 30:66.
2. SLANETZ, L.W. & C.H. BARTLEY. 1957. Numbers of enterococci in water, sewage and feces determined by the membrane filter technique with an improved medium. *J. Bacteriol.* 74:591.
3. PHILLIPS, E. & P. NASH. 1985. Culture media. *In* E.H. Lennette, A. Ballows, W.J. Hausler, Jr. & H.J. Shadomy, eds. Manual of Clinical Microbiology, 5th ed. American Soc. Microbiology, Washington, D.C.
4. LANCEFIELD, R.C. 1933. A serological differentiation of human and other groups of hemolytic streptococci. *J. Exp. Med.* 57:571.
5. DIFCO LABORATORIES. 1984. Difco Manual, 10th ed.
6. FACKLAM, R.R. & R.B. CAREY. 1985. Streptococci and aerococci. *In* E.H. Lennette, A. Ballows, W.J.Hausler, Jr. & H.J. Shadomy, eds. Manual of Clinical Microbiology, 5th ed. American Soc. Microbiology, Washington, D.C.
7. GROSS, J.C., M.P. HOUGHTON & L.B. SENTERFIT. 1975. Presumptive speciation of *Streptococcus bovis* and other Group D streptococci from human sources by using arginine and pyruvate tests. *J. Clin. Microbiol.* 1:54.
8. COWAN, S.T. & K.J. STEEL. 1965. Manual for the Identification of Medical Bacteria. Cambridge Univ. Press, Cambridge, England.
9. BRIDGE, P.D. & P.H.A. SNEATH. 1983. Numerical taxonomy of streptococcus. *J. Gen. Microbiol.* 129:565.

9240　IRON AND SULFUR BACTERIA*

9240 A.　Introduction

The group of nuisance organisms collectively designated "iron and sulfur bacteria" is morphologically and physiologically heterogeneous, having in common the ability to transform or deposit significant amounts of iron or sulfur, usually in the form of objectionable slimes. However, iron and sulfur bacteria are not the sole producers of bacterial slimes and in some cases may be associated with slimes of other bacteria.

The iron and sulfur bacteria may be filamentous or single-celled, autotrophic or heterotrophic, aerobic or anaerobic. The taxonomic position of these bacteria is very diverse. They are studied as iron and sulfur bacteria, because these elements and their transformations may be important in water treatment and distribution systems and may be especially bothersome in waters for industrial use, such as cooling and boiler waters. Iron bacteria may cause, or be associated with, fouling and plugging of wells. Also, the growth of these bacteria may result in consumer com-

---

* Approved by Standard Methods Committee, 1996.

plaints of red water in distribution systems for potable water and sulfate-reducing bacteria may cause rusty water and tuberculation of pipes. These organisms also may cause odor, taste, frothing, color, and increases in turbidity in waters.

The nutrient supply for iron and sulfur bacteria may be wholly or partly inorganic. Many of these bacteria can grow under oligotrophic conditions when attached to a substrate in flowing water. This seems quite important in the case of certain sulfur bacteria utilizing small amounts of hydrogen sulfide or in the case of organisms such as *Gallionella*, which obtain their energy from the oxidation of ferrous iron. *Thiobacillus ferrooxidans* contributes to the problem of acid mine drainage and can be identified by tests for transformation of ferrous to ferric iron or oxidation of reduced sulfur compounds under conditions of low pH. Temperature, light, pH, and oxygen supply are critical to the growth of iron and sulfur bacteria. Under different environmental conditions some bacteria may appear either as iron or as sulfur bacteria.

9240 B. Iron Bacteria

1. General Characteristics

"Iron bacteria" are considered to be capable of metabolizing reduced iron present in their aqueous habitat and of depositing it in the form of hydrated ferric oxide on or in their mucilaginous secretions. A somewhat similar mechanism is used by bacteria that utilize manganese. The large amount of brown slime so produced will impart a reddish tinge and an unpleasant odor to drinking water and may render the supply unsuitable for domestic or industrial purposes. These bacteria obtain energy by the oxidation of iron from the ferrous to the ferric state; the ferric form is precipitated as ferric hydroxide [$Fe(OH)_3$]. Iron may be obtained from the pipe itself if it is made of iron or from the water within it. Because of the energetics involved, the amount of ferric hydroxide deposited is very large in comparison with the enclosed cells.

Some bacteria that do not oxidize ferrous iron may cause it to be dissolved or deposited indirectly. In their growth, either they liberate iron by utilizing organic radicals to which the iron is attached or they alter environmental conditions to permit the solution or deposition of iron. Under these conditions, less ferric hydroxide may be produced, but taste, odor, and fouling may be engendered.

2. Collection of Samples and Identification

Examples of iron bacteria are shown in Figures 9240:1 through 9240:5. Identification of nuisance iron bacteria usually has been made on the basis of microscopic examination of the suspected material. Directly examine bulked activated sludge, masses of microbial growth in lakes, rivers, and streams, and slime growths in cooling-tower waters. Suspected development of iron bacteria in

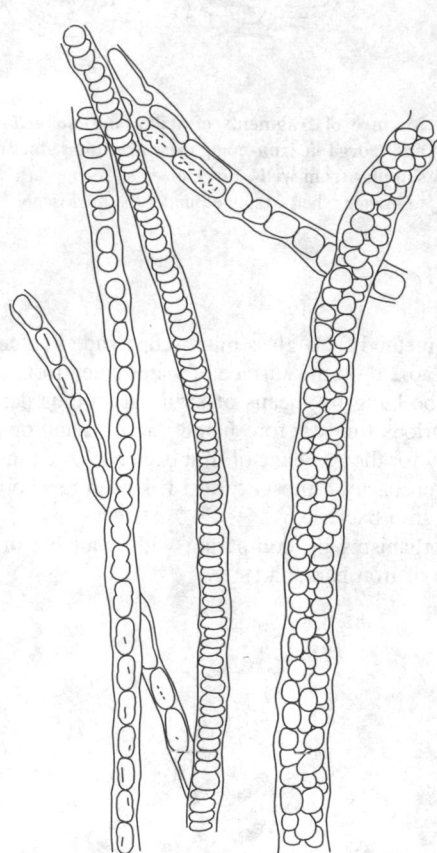

Figure 9240:1. Filaments of *Crenothrix polyspora* showing variation of size and shape of cells within the sheath. Note especially the multiple small round cells, or "conidia," found in one of the filaments. This distinctive feature is the reason for the name *polyspora*. Young growing colonies usually are not encrusted with iron or manganese. Older colonies often exhibit empty sheaths that are heavily encrusted. Cells may vary considerably in size: Rod-shaped cells average 1.2 to 2.0 μm in width by 2.4 to 5.6 μm in length; coccoid cells of "conidia" average 0.6 μm in diameter.

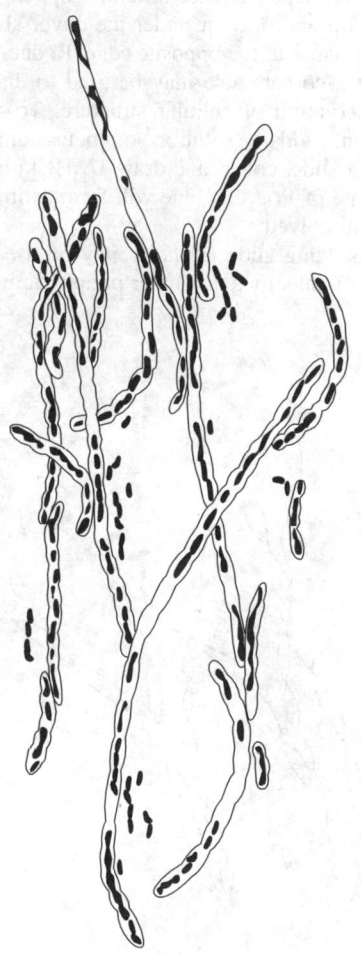

Figure 9240:2. Filaments of *Sphaerotilus natans*, showing cells within the filaments and some free "swarmer" cells. Filaments show false branching and areas devoid of cells. Individual cells within the sheath may vary in size, averaging 0.6 to 2.4 μm in width by 1.0 to 12.0 μm in length; most strains are 1.1 to 1.6 μm wide by 2.0 to 4.0 μm long.

water wells or in distribution systems may require special efforts to secure samples useful for identification.

Settle or centrifuge samples drawn directly from wells and examine sediment microscopically. Place a portion of sediment on a microscope slide, cover with a cover slip, and examine under a low-power microscope for filaments and iron-encrusted filaments. The material trapped by filters placed in front of back-surge valves often has yielded excellent specimens of iron bacteria. Water pumped from wells may be passed through a 0.45-μm membrane filter and the filter examined microscopically after drying and clearing with immersion oil applied directly to the membrane. Phase-contrast microscopes have made possible the examination of unstained culture material. Use india ink or lactophenol blue for staining when conventional light microscopy is used. Also, iron bacteria have been observed when the epifluorescence microscopic method (Section 9216B) was used.

Continued heavy deposition of iron caused by the oxidation of ferrous iron by air or by other environmental changes often hides the sheaths or stalks of iron bacteria. The cells within the filaments often die and disintegrate and the filaments tend to be fragmented or crushed by the mass of the iron precipitate.

To dissolve iron deposits place several drops of 1N HCl at one edge of cover slip and draw it under the cover slip by applying filter or blotting paper to the opposite edge. Reducing compounds such as sodium ascorbate also may be used to dissolve deposits and permit observation of cellular structure. To verify that the material is iron, add a solution of potassium ferrocyanide to a sample on a slide, cover, and draw 1N HCl under cover slip. A blue precipitate of Prussian blue will form as iron around cells or filaments is dissolved.

Floating or hanging slide methods may be used to determine the presence of filamentous and other periphytic iron bacteria. In

Figure 9240:4. Mixture of fragments of stalks of *Gallionella ferruginea* and inorganic iron-manganese precipitate found in natural samples from wells. Fragmented stalks appear golden yellow to orange when examined under the microscope.

the floating method fix a glass microscope slide in a cork and let it float for 1 to 2 d on the surface of water taken from the source. Slides may be hung by means of string at various depths in the water for various times before being removed and observed microscopically for the presence of iron bacteria.[1] A cultural method called the iron bacteria presence test has been developed for the detection of iron bacteria.[2]

Identify organisms by comparing with available drawings or photographs of iron bacteria.[1–17]

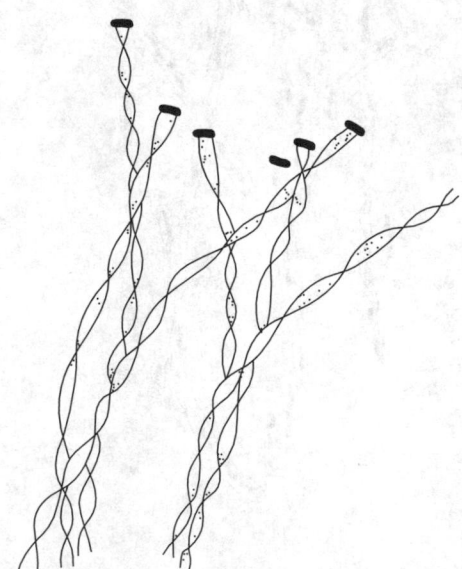

Figure 9240:3. Laboratory culture of *Gallionella ferruginea*, showing cells, stalks excreted by cells, and branching of stalks where cells have divided. A precipitate of inorganic iron on and around the stalks often blurs the outlines. Cells at tip of stalk average 0.4 to 0.6 μm in width by 0.7 to 1.1 μm in length.

Figure 9240:5. Single-celled iron bacterium *Siderocapsa treubii*. Cells are surrounded by a deposit of ferric hydrate. Individual cells average 0.4 to 1.5 μm in width by 0.8 to 2.5 μm in length.

3. References

1. WOJCIK, W. & M. WOJCIK. 1986. Monitoring biofouling. *In* D.R. Cullimore, ed. Proc. International Symposium on Biofouled Aquifers: Prevention and Restoration. American Water Resources Assoc., Bethesda, Md.

2. CULLIMORE, D.R. & A.E. McCANN. 1977. The identification, cultivation and control of iron bacteria in ground water. *In* F.A. Skinner & J.M. Shewan, eds. Aquatic Microbiology. Academic Press, New York, N.Y.

3. LUESCHOW, L.A. & K.M. MACKENTHUN. 1962. Detection and enumeration of iron bacteria in municipal water supplies. *J. Amer. Water Works Assoc.* 54:751.

4. STARKEY, R.L. 1945. Transformations of iron by bacteria in water. *J. Amer. Water Works Assoc.* 37:963.

5. STOKES, J.L. 1954. Studies on the filamentous sheathed iron bacterium *Sphaerotilus natans*. *J. Bacteriol.* 67:278.

6. KUCERA, S. & R.S. WOLFE. 1957. A selective enrichment method for *Gallionella ferruginea*. *J. Bacteriol.* 74:344.

7. WAITZ, S. & J.B. LACKEY. 1958. Morphological and biochemical studies on the organism *Sphaerotilus natans*. *Quart. J. Fla. Acad. Sci.* 21:335.

8. WOLFE, R.S. 1958. Cultivation, morphology, and classification of the iron bacteria. *J. Amer. Water Works Assoc.* 50:1241.

9. WOLFE, R.S. 1960. Observations and studies of *Crenothrix polyspora*. *J. Amer. Water Works Assoc.* 52:915.

10. WOLFE, R.S. 1960. Microbial concentration of iron and manganese in water with low concentrations of these elements. *J. Amer. Water Works Assoc.* 52:1335.

11. DONDERO, N.C., R.A. PHILIPS & H. HEUKELEKIAN. 1961. Isolation and preservation of cultures of *Sphaerotilus*. *Appl. Microbiol.* 9:219.

12. MULDER, E.G. 1964. Iron bacteria, particularly those of the *Sphaerotilus-Leptothrix* group, and industrial problems. *J. Appl. Bacteriol.* 27:151.

13. DRAKE, C.H. 1965. Occurrence of *Siderocapsa treubii* in certain waters of the Niederrhein. *Gewasser Abwasser* 39/40:41.

14. STALEY, J.T., M.P. BRYANT, N. PFENNING & J.G. HOLT, eds. 1989. Bergey's Manual of Systematic Bacteriology, Volume 3. Williams & Wilkins, Baltimore, Md.

15. EDMONDSON, W.T., ed. 1959. Ward & Whipple's Fresh Water Biology, 2nd ed. John Wiley & Sons, New York, N.Y.

16. SKERMAN, V.B.D. 1967. A Guide to the Identification of the Genera of Bacteria, 2nd ed. Williams & Wilkins, Baltimore, Md.

17. GHIORSE, W.C. 1984. Biology of iron and manganese depositing bacteria. *Annu. Rev. Microbiol.* 38:515.

9240 C.   Sulfur Bacteria

1. General Characteristics

The bacteria that oxidize or reduce significant amounts of inorganic sulfur compounds exhibit a wide diversity of morphological and biochemical characteristics. One group, the sulfate-reducing bacteria, consists mainly of single-celled forms that grow anaerobically and reduce sulfate, SO_4^{2-}, to hydrogen sulfide, H_2S. One member of this group, *Desulfonema*, is multicellular and exhibits gliding motility. A second group, the photosynthetic green and purple sulfur bacteria, grows anaerobically in the light and uses H_2S as a hydrogen donor for photosynthesis. Members of a third colorless filamentous group are myxotrophic and utilize organic sources of carbon but may get their energy from the oxidation of reduced sulfur compounds. The sulfide is oxidized to sulfur or sulfate. A fourth group, the aerobic sulfur-oxidizers, oxidizes reduced sulfur compounds aerobically to obtain energy for chemoautotrophic growth.

The sulfur bacteria of most importance in the water and wastewater field are the sulfate-reducing bacteria, which include *Desulfovibrio*, and the single-celled aerobic sulfur-oxidizers of the genus *Thiobacillus*. The sulfate-reducing bacteria contribute greatly to tuberculations and galvanic corrosion of water mains and to taste and odor problems in water. *Thiobacillus*, by its production of sulfuric acid, has contributed to the destruction of concrete sewers and the acid corrosion of metals.

2. Collection of Samples and Identification

Identification of nuisance sulfur bacteria usually has been made on the basis of microscopic examination of the suspected material.

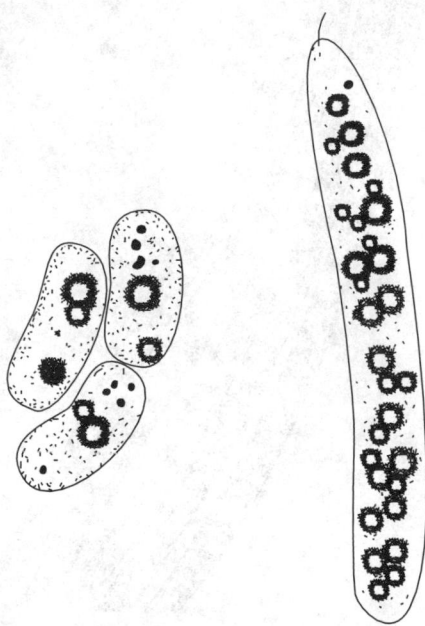

Figure 9240:6. Photosynthetic purple sulfur bacteria. Large masses of cells have brown-orange to purple color—may appear chalky if there is a large amount of sulfur within the cells. Left: cells of *Chromatium okenii* (5.0 to 6.5 μm wide by 8 to 15 μm long) containing sulfur globules. Right: *Thiospirillum jenense* (3.5 to 4.5 μm wide by 30 to 40 μm long); cell contains sulfur globules and polar flagellum is visible.

Examine samples of slimes suspended in waters, scrapings from exposed surfaces, or sediments directly.

Three groups of sulfur bacteria may be recognized microscopically; green and purple sulfur bacteria; large, colorless filamentous sulfur bacteria; and large, colorless, nonfilamentous sulfur bacteria. The fourth group, consisting of sulfate-reducing bacteria and sulfur-oxidizing bacteria of the genus *Thiobacillus*, cannot be identified by appearance alone.

a. Green and purple sulfur bacteria:

1) Green sulfur bacteria most frequently occur in waters containing H$_2$S. They are small, ovoid to rod-shaped, nonmotile organisms, generally less than 1 µm in diameter, and with a yellowish-green color in masses. Sulfur globules are seldom if ever deposited within the cells.

2) Purple sulfur bacteria (Figure 9240:6) occur in waters containing H$_2$S. They are large, generally stuffed with sulfur globules, and often so intensely pigmented as to make individual cells appear red. Large, dense, highly colored masses are detected easily by the naked eye. The presence of photosynthetic bacteria in concentrated masses can be confirmed by extracting the mass with ether and scanning the extract absorbance in the infrared region. Bacterial chlorophyll will absorb strongly in the range of 660 to 870 nm.

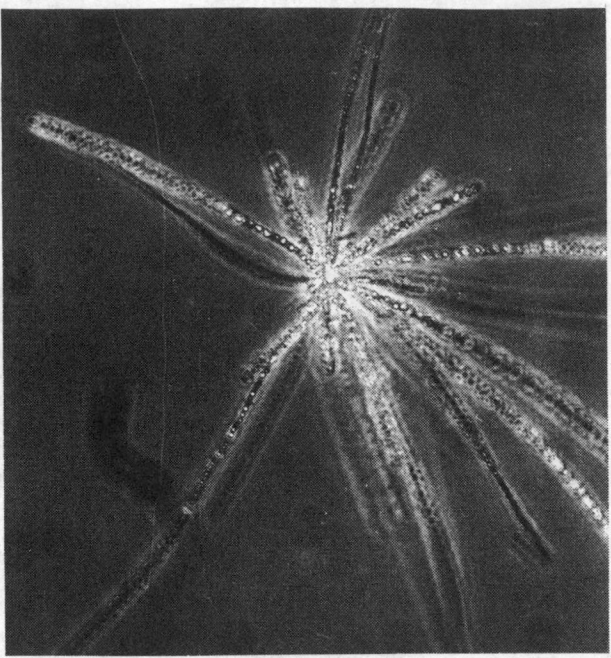

Figure 9240:8. ***Thiothrix nivea*** **in a rosette formation.** Filaments are 1 to 1.5 µm in diameter and of varying length. Individual cells are 1 to 1.5 µm wide and 2 to 4 µm long.

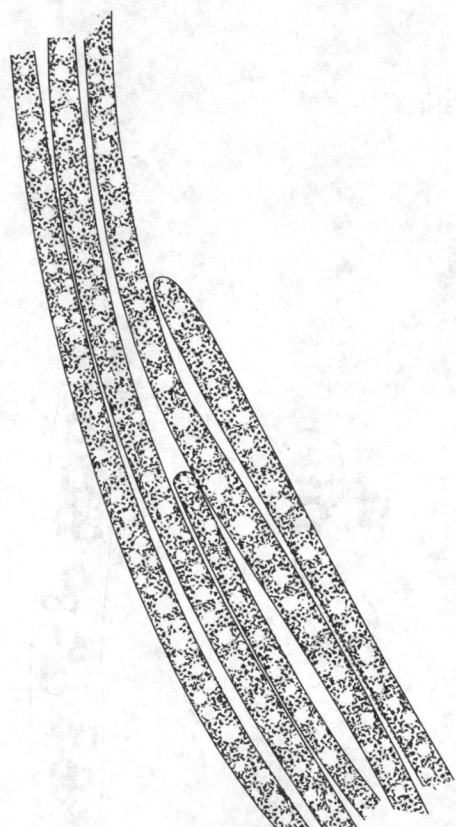

Figure 9240:7. Colorless filamentous sulfur bacteria: ***Beggiatoa alba*** **trichomes, containing globules of sulfur.** Filaments are composed of a linear series of individual rod-shaped cells that may be visible when not obscured by light reflecting from sulfur granules. Trichomes are 2 to 15 µm in diameter and may be up to 1500 µm long; individual cells, if visible, are 4.0 to 16.0 µm long.

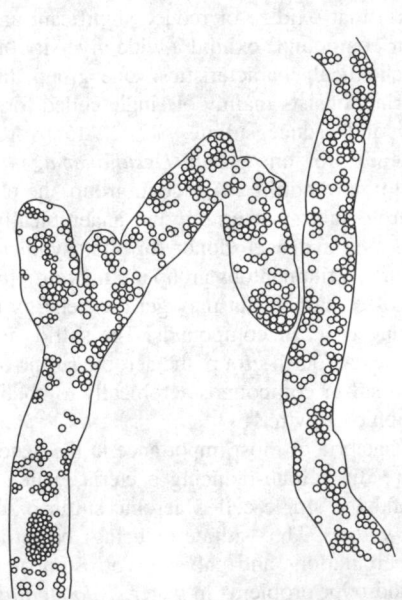

Figure 9240:9. Colorless filamentous sulfur bacteria: portion of a colony, showing branching of the mucoid filament, identified as ***Thiodendron mucosum.***[1] Because the name *Thiodendron* previously had been used in bacterial taxonomy, its use here is illegitimate and this organism remains unnamed. Individual cells (1.0 to 2.5 µm wide by 3 to 9 µm long) have been found within the jelly-like material of the filaments. The long axis of the cells runs parallel to the long axis of the filaments.

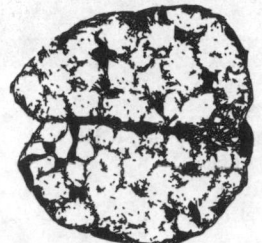

Figure 9240:10. Colorless nonfilamentous sulfur bacteria: dividing cell of *Thiovolum majus*, containing sulfur globules. Cells may measure 9 to 17 μm in width by 11 to 18 μm in length and are generally found in nature in a marine littoral zone rich in organic matter and hydrogen sulfide.

b. Colorless filamentous sulfur bacteria: Colorless filamentous sulfur bacteria (Figures 9240:7, 9240:8, and 9240:9) occur in waters where both oxygen and H_2S are present. They may form mats with a slightly yellowish-white appearance due to deposition of internal sulfur globules. They generally are large and may be motile with a characteristic gliding movement. Identify by comparing organisms with available photographs.[1-4]

c. Colorless nonfilamentous sulfur bacteria: Colorless, nonfilamentous sulfur bacteria (see Figure 9240:10, for example) usually are associated with decaying algae. They are extremely motile, ovoid to rod-shaped with sulfur globules and possible calcium carbonate deposits. They generally are very large.

d. Colorless small sulfur bacteria and sulfate-reducing bacteria: The small single-celled bacteria, *Thiobacillus* spp., and the sulfate-reducing bacteria, such as *Desulfovibrio,* cannot be identified by direct microscopic examination. *Thiobacillus* types are small, colorless, motile, and rod-shaped and are found in an environment containing H_2S. Sulfur globules are absent. Identify *Thiobacillus* types, *Desulfovibrio,* or other sulfate-reducing bacteria physiologically.

3. References

1. LACKEY, J.B. & E.W. LACKEY. 1961. The habitat and description of a new genus of sulfur bacterium. *J. Gen. Microbiol.* 26:28.
2. FAUST, L. & R.S. WOLFE. 1961. Enrichment and cultivation of *Beggiatoa alba. J. Bacteriol.* 81:99.
3. MORGAN, G.B. & J.B. LACKEY. 1965. Ecology of a sulfuretum in a semitropical environment. *Z. Allg. Mikrobiol.* 5:237.
4. STALEY, J.T., M.P. BRYANT, N. PFENNING & J.G. HOLT, eds. 1989. Bergey's Manual of Systematic Bacteriology, Volume 3. Williams & Wilkins, Baltimore, Md.

9240 D. Enumeration, Enrichment, and Isolation of Iron and Sulfur Bacteria

There are no good means of enumerating iron and sulfur bacteria other than the sulfate-reducing bacteria and the thiobacilli. Laboratory cultivation and isolation of pure cultures is difficult and successful isolation is uncertain. This is especially true of attempts to isolate filamentous bacteria from activated sludge or other sources where many different bacterial types are present.

1. Media

a. Casitone-glycerol-yeast autolysate broth (CGY), for the Sphaerotilus group: This medium may not be available in dehydrated form and may require preparation from the basic ingredients. It may be solidified by adding 1.5% agar.

Casitone	5.0	g
Glycerol	10.0	mL
Yeast autolysate	1.0	g
Reagent-grade water	1	L

b. Isolation medium (iron bacteria): This medium may not be available in dehydrated form and may require preparation from the basic ingredients.

Glucose	0.15	g
Ammonium sulfate, $(NH_4)_2SO_4$	0.5	g
Calcium nitrate, $Ca(NO_3)_2$	0.01	g
Dipotassium hydrogen phosphate, K_2HPO_4	0.05	g
Magnesium sulfate, $MgSO_4 \cdot 7H_2O$	0.05	g
Potassium chloride, KCl	0.05	g
Calcium carbonate, $CaCO_3$	0.1	g
Agar	10.0	g

Cyanocobalamin	0.01	mg
Thiamine	0.4	mg
Reagent-grade water	1	L

c. Maintenance (SCY) medium (iron bacteria): This medium may not be available in dehydrated form and may require preparation from the basic ingredients.

Sucrose	1.0	g
Casitone	0.75	g
Yeast extract	0.25	g
Trypticase soy broth without dextrose	0.25	g
Agar	10.0	g
Cyanocobalamin	0.01	mg
Thiamine	0.4	mg
Reagent-grade water	1	L

d. Mn agar No. 1: This medium may not be available in dehydrated form and may require preparation from the basic ingredients.

Manganous carbonate, $MnCO_3$	2.0	g
Beef extract	1.0	g
Ferrous ammonium sulfate, $Fe(NH_4)_2(SO_4)_2$	150	mg
Sodium citrate	150	mg
Yeast extract	75	mg
Cyanocobalamin	0.005	mg
Agar	10.0	g
Reagent-grade water	1	L

Prepare and sterilize the medium without cyanocobalamin. Separately sterilize cobalamin by filtration and add aseptically just before medium solidifies.

e. Mn agar No. 2:[1] Prepare fresh each time from basic ingredients:

Manganous sulfate, $MnSO_4 \cdot H_2O$	10	mg
Agar	15.0	g
Natural water	1	L

f. Iron oxidizing medium (Thiobacillus ferrooxidans): This medium may not be available in dehydrated form and may require preparation from the basic ingredients.

Basal salts:

Ammonium sulfate, $(NH_4)_2SO_4$	3.0	g
Potassium chloride, KCl	0.10	g
Dipotassium hydrogen phosphate, K_2HPO_4	0.50	g
Magnesium sulfate, $MgSO_4 \cdot 7H_2O$	0.50	g
Calcium nitrate $Ca(NO_3)_2$	0.01	g
H_2SO_4, 10N	1.0	mL
Reagent-grade water	700	mL

Energy source:

Ferrous sulfate, $FeSO_4 \cdot 7H_2O$, 14.74% solution (w/v)	300	mL

Separately sterilize basal salts and energy source and combine when cool. Store in the refrigerator and discard after 2 weeks. A precipitate will form and the medium will be opalescent and green. The pH should be 3.0 to 3.6.

g. Ferrous sulfide agar (Gallionella ferruginea): This medium may not be available in dehydrated form and may require preparation from the basic ingredients.

Agar layer:

Ferrous sulfide, FeS (washed precipitate and liquid)	500	mL
Sodium sulfide, Na_2S	15.6	g
Ferrous ammonium sulfate, $Fe(NH_4)_2(SO_4)_2 \cdot 6H_2O$	78.4	g
Boiling reagent-grade water	1	L
Agar (liquid) (30 g/L)	500	mL

Liquid overlay:

Ammonium chloride, NH_2Cl	1.0	g
Dipotassium hydrogen phosphate, K_2HPO_4	0.5	g
Magnesium sulfate, $MgSO_4 \cdot 7H_2O$	0.2	g
Calcium chloride, $CaCl_2$	0.1	g
Reagent-grade water	1	L

Prepare FeS by reacting equal molar quantities of Na_2S and $Fe(NH_4)_2(SO_4)_2$ in boiling reagent-grade water. Let precipitate settle from the hot solution in a completely filled and stoppered bottle. Wash precipitate four times by decanting supernatant and replacing with boiling water. Store FeS in a glass stoppered bottle completely filled with additional boiling water.

Add equal volumes of FeS and 3% agar at 45°C. Prepare slants in screw-capped tubes. Prepare liquid overlay, bubble CO_2 through it for 10 to 15 s, and add several milliliters to agar slant. Modifications to this medium and procedure are available.[2–4]

A variation of the basic medium requires adding 0.5 mL formalin (40% formaldehyde solution) to a screw-capped dilution bottle containing 10 mL FeS agar and 100 mL liquid overlay. This variation is said not to work in all cases.[3,4] Add 0.001% bromthymol blue and 0.004% bromcresol purple to liquid overlay.

h. Sulfate-reducing medium: This medium may not be available in dehydrated form and may require preparation from the basic ingredients.

Sodium lactate	3.5	g
Beef extract	1.0	g
Peptone	2.0	g
Magnesium sulfate, $MgSO_4 \cdot 7H_2O$	2.0	g
Sodium sulfate, Na_2SO_4	1.5	g
Dipotassium hydrogen phosphate, K_2HPO_4	0.5	g
Ferrous ammonium sulfate, $Fe(NH_4)_2(SO_4)_2 \cdot 6H_2O$	0.392	g
Calcium chloride, $CaCl_2$	0.10	g
Sodium ascorbate	0.10	g
Reagent-grade water	1	L

pH should be 7.5 ± 0.3 after sterilization. Prepare medium excluding ferrous ammonium sulfate and sodium ascorbate, dispense in screw-capped test tubes, and sterilize. For use, the tubes must be completely filled; therefore, in a flask sterilize extra medium to be added to tubes for filling. On day of use, prepare separate solutions of ferrous ammonium sulfate (3.92 g/100 mL) and sodium ascorbate (1.00 g/100 mL), sterilize by filtration through a 0.45-μm membrane filter, and aseptically add 0.1 mL each solution/10 mL basal medium.

i. Thiosulfate oxidizing medium (Thiobacillus thioparus): This medium may not be available in dehydrated form and may require preparation from the basic ingredients.

Sodium thiosulfate, $Na_2S_2O_3 \cdot 5H_2O$	10.0	g
Dipotassium hydrogen phosphate, K_2HPO_4	2.0	g
Magnesium sulfate, $MgSO_4 \cdot 7H_2O$	0.1	g
Calcium chloride, $CaCl_2 \cdot 2H_2O$	0.1	g
Ammonium sulfate, $(NH_4)_2SO_4$	0.1	g
Ferric chloride, $FeCl_3 \cdot 6H_2O$	0.02	g
Reagent-grade water	1	L

pH should be 7.8 after sterilization. Separately sterilize $Na_2S_2O_3$ and $(NH_4)_2SO_4$ and add before use. If this medium is used to isolate *Thiobacillus thioparus*, check isolates to ensure they are autotrophs.

j. Sulfur medium (Thiobacillus thiooxidans): This medium may not be available in dehydrated form and may require preparation from the basic ingredients.

Sulfur, elemental	10.0	g
Potassium dihydrogen phosphate, KH_2PO_4	3.0	g
Magnesium sulfate, $MgSO_4 \cdot 7H_2O$	0.5	g
Ammonium sulfate, $(NH_4)_2SO_4$	0.3	g
Calcium chloride, $CaCl_2 \cdot 2H_2O$	0.25	g
Ferric chloride, $FeCl_3 \cdot 6H_2O$	0.02	g
Reagent-grade water	1	L

pH should be 4.8 after sterilization. Weigh sulfur into 250-mL flasks using 1 g/flask. Add 100 mL medium to each flask and sterilize with intermittent steam (30 min for each of 3 consecutive d).

k. Media for Beggiatoa *and myxotrophic strains of* Thiothrix:

1) *Extracted hay:* Extract hay or grass at least five times by boiling in water for 30 min, with two rinses in cold water between each extraction. Dry the extracted hay, place 10 to 30 blades in a large test tube, and sterilize by autoclaving.

2) *Basal medium*

Ammonium chloride, 4% solution	5	mL
Dipotassium hydrogen phosphate, K_2HPO_4, 1% solution	1	mL

Magnesium sulfate, $MgSO_4 \cdot 7H_2O$, 1% solution	1	mL
Calcium sulfate, $CaSO_4 \cdot 2H_2O$, saturated solution	20	mL
Trace elements [see ¶ 3) below]	5	mL
Reagent-grade water	968	mL

3) Trace elements

Reagent-grade water	920	mL
Zinc sulfate, $ZnSO_4 \cdot 7H_2O$, 0.1% solution	10	mL
Manganous sulfate, $MnSO_4 \cdot 4H_2O$, 0.02% solution ...	10	mL
Copper sulfate, $CuSO_4 \cdot 5H_2O$, 0.00005% solution	10	mL
Boric acid, H_3BO_3, 0.1% solution	10	mL
Cobalt nitrate, $Co(NO_3)_2$ or $CoCl_2 \cdot 6H_2O$, 0.01% solution	10	mL
Sodium molybdate, $Na_2MoO_4 \cdot 2H_2O$, 0.01% solution	10	mL
EDTA solution, 2% EDTA with 7% ferrous sulfate, $FeSO_4 \cdot 7H_2O$, with 1 mL conc HCl/100 mL	20	mL

4) *MP agar:* This medium may not be available in dehydrated form and should be made fresh before use.

Basal medium	1	L
Sodium acetate	0.1	g
Sodium sulfide solution [see ¶ 5) below]	3	mL
Agar	15	g

Adjust pH to between 7.0 and 7.5 and heat to dissolve agar. Sterilize by autoclaving for no more than 15 min and cool in a 45 to 50°C water bath. Add sodium sulfide solution immediately before pouring plates. If the medium is to be used for screw-capped tubes the sulfide may be added before autoclaving.

5) *Sodium sulfide solution:* Make up and separately autoclave a 10% solution of $Na_2S \cdot 9H_2O$.

6) *MY agar:* This medium may not be available in dehydrated form and should be made fresh before use.

Basal medium	1	L
Sodium acetate	0.1	g
Nutrient broth powder (Difco)	0.1	g
Yeast extract (Difco)	0.1	g
Sodium sulfide solution [see ¶ 5) above]	3	mL
Agar	15	g

Adjust pH to between 7.0 and 7.5 and heat to dissolve agar. Sterilize by autoclaving for no more than 15 min and cool in a 45 to 50°C water bath. Add sodium sulfide solution immediately before pouring plates. If the medium is to be used for screw-capped tubes the sulfide may be added before autoclaving.

l. Media for heterotrophic strains of Thiothrix: Make the media 3) through 9) with solutions 1) and 2), plus indicated additives. After adding all ingredients, adjust pH to 7.2 to 7.5. For solid media add 12 g agar/L.

1) MSV

Ammonium sulfate, $(NH_4)_2SO_4$	0.5	g
Dipotassium hydrogen phosphate, K_2HPO_4	0.11	g
Potassium dihydrogen phosphate, KH_2PO_4	0.085	g
Magnesium sulfate, $MgSO_4 \cdot 7H_2O$	0.1	g
Calcium chloride, $CaCl_2 \cdot 2H_2O$	0.05	g
Ferric chloride, $FeCl_3 \cdot H_2O$	0.002	g
EDTA	0.003	g

Vitamin mix	1	mL
Reagent-grade water	1	L

2) Vitamin mix

Calcium pantothenate	0.01	g
Niacin	0.01	g
Biotin	0.0005	g
Cyanocobalamin	0.0005	g
Folic acid	0.0005	g
Pyridoxine	0.01	g
p-aminobenzoic acid	0.01	g
Cocarboxylase	0.01	g
Inositol	0.01	g
Thiamine	0.01	g
Riboflavin	0.01	g
Reagent-grade water	100	mL

3) AcS

Sodium acetate	0.15	g
Sodium sulfide, $Na_2S \cdot 9H_2O$	0.187	g
MSV	1	L

4) SS

Sucrose	0.15	g
Sodium sulfide, $Na_2S \cdot 9H_2O$	0.187	g
MSV	1	L

5) GS

Glucose	0.15	g
Sodium sulfide, $Na_2S \cdot 9H_2O$	0.187	g
MSV	1	L

6) SUC

Sodium succinate	0.15	g
MSV	1	L

7) I

Glucose	0.15	g
MSV	1	L

8) S

Sodium sulfide, $Na_2S \cdot 9H_2O$	0.187	g
MSV	1	L

9) LT

Sodium lactate	0.5	g
Sodium thiosulfate, $Na_2S_2O_3$	0.5	g
MSV	1	L

2. Iron Bacteria

a. Sphaerotilus-Leptothrix:

1) Iron bacteria, especially those belonging to the *Sphaerotilus-Leptothrix* group, thrive in media too dilute to support the proliferation of more rapidly growing organisms. One medium[5] is par-

tially selective for *Sphaerotilus* (BOD dilution water, Section 5210B, supplemented with 100 mg/L sodium lactate). Dispense 50 mL of this medium into French square bottles and autoclave at 69 kPa for 15 min. To inoculate sample add 25-mL portions of stream water or 1-, 5-, and 10-mL portions of settled wastewater or process liquor to duplicate bottles of medium. Incubate at 22 to 25°C for 5 d and observe for filamentous growth. Isolate pure cultures by picking a filament from the BOD-lactate broth and streaking on 0.05% meat extract agar. After incubating for 24 h at 25°C, pick typical curling filaments with the aid of a dissecting microscope and transfer to casitone-glycerol-yeast autolysate (CGY) broth. If a pellicle with no underlying turbidity develops in 2 to 3 d, transfer a filament to a CGY agar slant, incubate at 25°C until growth is visible, and store in a refrigerator. In addition, extracted alfalfa straw or pea straw may be used for enrichments.[4]

A detailed key for identifying filamentous microorganisms in complex mixtures such as wastewater and activated sludge is available.[6]

2) Isolation and maintenance media have proven quite successful for identifying various groups of filamentous organisms, including iron bacteria.[7] Prepare agar slants of these media and aseptically pipet 3 mL sterile, dechlorinated tap water onto surface of slants. Inoculate tubes and incubate at room temperature until turbid growth has developed in liquid layer. The cells will remain viable for 3 months in the refrigerator.

3) Another good maintenance medium for cultivating the *Sphaerotilus* group is CGY.[8]

4) *Leptothrix (Sphaerotilus discophorous)* can be distinguished from *Sphaerotilus natans* by its ability to oxidize manganous ion. Use Mn agar No. 1 as the differential medium.[9] Alternatively, *Leptothrix* may be grown by direct plating on Mn agar No. 2.[1]

b. Thiobacillus ferrooxidans: Although this organism also is a sulfur-oxidizing bacterium,[10,11] its main importance has been in acid mine drainage. A medium suitable for enumeration of the MPN is available.[12] Some oxidation of iron occurs during sterilization but the loss of ferrous iron is not appreciable. The medium has a precipitate (probably ferrous and ferric phosphates), is opalescent and green, has a pH of 3.0 to 3.6, and contains 9000 mg/L ferrous iron. Growth of the organism is manifested by a decrease in pH and an increase in concentration of oxidized iron. With practice and use of uninoculated controls, an increase of deep orange-brown color can be seen in positive enrichment tubes or flasks as compared to negative ones. Shake test-tube dilutions daily because these organisms are highly aerobic.

c. Gallionella ferruginea: For cultivation of this organism use ferrous sulfide agar.[13,14] Inoculate tubes with a drop of suspension of a suspected *Gallionella* deposit. Growth at room temperature usually occurs in 18 to 36 h and appears as a white deposit on sides of test tube. The ring of colonies occurring at a certain level reflects a balance between upward diffusion of ferrous ions and downward diffusion of oxygen molecules. Supplementation of ferrous sulfide agar with formalin to isolate pure cultures may be successful.[1,4,15]

d. Other iron bacteria: An acid-tolerant (pH 3.5 to 5.0) filamentous iron-oxidizing *Metallogenium* has been isolated with a medium[16] containing $(NH_4)_2SO_4$, 0.1%; $CaCO_3$, 0.01%; $MgSO_4$, 0.02%; K_2HPO_4, 0.001%; potassium acid phthalate, 0.4%; and 250 mg/L ferrous iron from an acidified $FeSO_4 \cdot 7H_2O$, solution. Add 0.4% formalin to 100 mL of the isolating medium in a 250-mL erlenmeyer flask.

For heterotrophic iron-precipitating bacteria[17] use a ferric ammonium citrate medium consisting of: $(NH_4)_2SO_4$, 0.5 g/L; $NaNO_3$, 0.5 g/L; K_2HPO_4, 0.5 g/L; $MgSO_2 \cdot 7H_2O$, 0.5 g/L; and ferric ammonium citrate, 10.0 g/L. Adjust pH to 6.6 to 6.8 and sterilize. To make the medium solid add 15 g agar/L.

Alternatively grow iron bacteria by combining 20 mL liquid broth medium, 10 mL raw water or inoculum, and 3 g iron oxide. Incubate 48 to 72 h at 25°C on a wrist-action shaker to produce good, visible growth.

3. Sulfur Bacteria

a. Sulfate-reducing bacteria:

1) To enumerate sulfate-reducing bacteria such as *Desulfovibrio*, use a sulfate-reducing medium.[18] Inoculate tubes and fill completely with sterile medium to create anaerobic conditions. For comparative purposes, incubate one or two uninoculated controls with each set of inoculated tubes. To sample volumes greater than 10 mL, pass sample through a 0.45-μm membrane filter and transfer filter to screw-cap test tube with medium. If sulfate-reducing bacteria are present, tubes will show blackening within 4 to 21 d of incubation at 20 to 30°C.

2) An agar medium suitable for growth and enumeration of sulfate-reducing bacteria also is available.[19] The medium consists of trypticase soy agar, 4.0%, fortified with additional agar, 0.5%, to which is added 60% sodium lactate (0.4% v/v), hydrated magnesium sulfate, 0.2%, and ferrous ammonium sulfate, 0.2%. Adjust pH to 7.2 to 7.4 and sterilize. Medium should be clear and free from precipitate. Inoculate all plates within 1 or at most 4 h after agar hardens to prevent saturation with oxygen. To prevent moisture condensation on petri dish covers, replace covers with sterile absorbent tops until 10 to 15 min after agar hardens. Place uninverted plates in desiccator or Brewer jars and replace atmosphere with tank hydrogen or nitrogen by successive evacuation and gas replacement. Alternatively, use a disposable anaerobic generating system.* Incubate at room temperature (21 to 24°C) or at 28 to 30°C, the optimum temperature for these organisms. Growth and blackening around the colonies is typical of sulfate-reducing bacteria and may occur at any time between 2 and 21 d, although the usual time is 2 to 7 d.

3) Media suitable for enumeration or isolation of various species of sulfate-reducing bacteria are available.[20–22]

b. Photosynthetic purple and green sulfur bacteria: Because these organisms are so specialized and rarely cause problems in water and wastewater treatment processes, methods for their isolation and enumeration are not included here. In certain instances they can be beneficial because of their ability to oxidize hydrogen sulfide and thus reduce odor. An excellent review is available.[23] Also, media formulations and methods for cultivating specific members of this group of bacteria are available.[23]

c. Thiobacillus spp.: The growth and physiology of different species of the single-celled sulfur-oxidizing bacteria of the genus *Thiobacillus* have been evaluated carefully.[24,25] Media[26] suitable for enumeration of *Thiobacillus thioparus* and *Thiobacillus thiooxidans* by an MPN technique are listed in 9240D.1. Inoculate medium and incubate for 4 to 5 d at 25 to 30°C. Growth of thiobacilli produces elemental sulfur, which sinks to the bottom with a coincident decrease in pH and turbidity of the medium.

---

* GasPak, BBL, or equivalent.

Chemical tests for formation of sulfate are necessary to confirm presence of *Thiobacillus*.

d. Filamentous sulfur-oxidizing bacteria:

1) *Beggiatoa*—*Beggiatoa* exist in most aquatic habitats where sulfide and oxygen are both present,[27] including fresh and marine water, sediments, and wastewater systems. Because *Beggiatoa* is a multicellular bacterium there is no accurate method of determining the number of viable cells in a sample. To determine the population of *Beggiatoa* make a direct microscopic count of the number of filaments.

Marine beggiatoas may be quite large, up to 100 μm or more in diameter.[28] They have not been grown in the laboratory. However, small marine species (up to 5 μm diam) have been isolated.[29] The media for isolating freshwater and marine beggiatoas differ slightly.

a) Enrichment—Inoculate a tube of extracted hay, ¶ 1*k*1) above, with enough water and a little mud from the sample site to fill the tube to a depth of at least 8 cm. Incubate for at least 1 week and examine for the presence of "tufts" or "puff-balls" consisting of tangled filaments of *Beggiatoa*. Examine by phase-contrast microscopy for the presence of individual filaments.[30] If no "puff-balls" are found, continue incubation for another week and repeat the examination. Continue examining the enrichment for up to 4 weeks before discarding it.

b) Isolation—With sterilized fine-tipped forceps and using a dissecting microscope, transfer tufts of *Beggiatoa* from the enrichments to a small petri dish containing sterile basal medium,[30] ¶ 1*k*2) above. Shake the tufts with the forceps to remove adherent bacteria and transfer tufts to a new petri dish with sterile basal medium. Continue until tufts have been washed at least five times.

Transfer tufts to a "drying plate" containing basal medium and 1.6% agar for about 1 min to remove excess fluid, then transfer the tufts to the center of separate plates of either MY or MP medium,[30,31] ¶s 1*k*3) and 4) above. Incubate plates at room temperature or below and examine with a dissecting microscope every 5 to 10 h for the presence of gliding filaments of *Beggiatoa*. Select filaments that have glided well away from the other filaments and appear to be uncontaminated and transfer them to separate plates of the same medium with a sterile, flattened wire, inoculating needle, or toothpick. Take a little agar with the filament to avoid drying the filament during transfer. Examine the first transfer plates every 5 to 10 h as before and transfer pure filaments to fresh media.

2) *Thiothrix*—Obligately myxotrophic *Thiothrix* have been isolated and characterized from sulfur springs and other bodies of flowing water that contain sulfide.[31,32] Heterotrophic strains have been isolated and characterized from activated sludge wastewater treatment plants.[33–35] While the techniques for the isolation of the myxotrophic and heterotrophic strains are similar, the isolation media differ.

a) Isolation of myxotrophic *Thiothrix*—Collect tufts of the bacterium from rocks, water pipes, or other substrates. Using a dissecting microscope, pick up filaments with fine-tipped forceps, shake to remove contaminating bacteria, and transfer to sterile basal medium, ¶ 1*k*2 above. Repeat at least five times to try to obtain filaments with little or no contamination.

With a sterile Pasteur pipet transfer separate drops to the edge of either MP or MY agar [¶s 1*k*3) and 4), above] petri dishes and tip the dishes so that the drops run from one side of the dish to the other. Draw off excess moisture with the pipet and incubate plates at room temperature or below for about 48 h. Examine plates under a dissecting microscope for the appearance of typical filamentous colonies. With sterile toothpicks, pick colonies that are widely separated and transfer individually to fresh plates of the same medium. Streak transferred material with a wire loop. After about 48 h incubation, examine plates and restreak colonies that appear to be pure.

b) Isolation of heterotrophic *Thiothrix*—There are two different procedures to use, depending on the concentration and size of the *Thiothrix* filaments in the sample.[34]

In one procedure, wash individual filaments or rosettes of large strains of *Thiothrix* several times in MSV broth, ¶ 1*l* 1) above, by transfer with a Pasteur pipet while observing with a dissecting microscope. After several washings transfer filaments to a small amount of MSV (1 to 3 mL) and plate on one or more of the following solid media, ¶s 1*l* 3)–9) above.

If the filaments are small or scarce concentrate the sample by centrifuging. Then dilute 1:5 in MSV, sonicate at 30 W for 10 s, and wash three times by centrifugation at 1900 × *g* for 2 to 5 min. The supernatant contains free filaments that are used to inoculate one or more of the solid media, ¶s 1*l* 3)–9) above.

4. References

1. GHIORSE, W.C. 1984. Biology of iron- and manganese-depositing bacteria. *Annu. Rev. Microbiol.* 38:515.

2. HALLBRECK, E.L. & K. PEDERSON. 1986. The biology of *Gallionella*. *In* D.R. Cullimore, ed. Proc. International Symposium on Biofouled Aquifers: Prevention and Restoration. American Water Works Assoc., Denver, Colo.

3. STARR, M.P., H. STOLP, H.G. TRUPER, A. BALOWS & H.C. SCHLEGE. 1981. The Prokaryotes. A Handbook on Habitats, Isolation and Identification of Bacteria, Volume 1. Springer-Verlag, New York, N.Y.

4. STALEY, J.T., M.P. BRYANT, N. PFENNIG & J.G. HOLT, eds. 1989. Bergey's Manual of Systematic Bacteriology, Volume 3. Williams & Wilkins, Baltimore, Md.

5. ARMBRUSTER, E.H. 1969. Improved technique for isolation and identification of *Sphaerotilus*. *Appl. Microbiol.* 17:320.

6. FARQUHAR, G.J. & W.C. BOYLE. 1971. Identification of filamentous microorganisms in activated sludge. *J. Water Pollut. Control Fed.* 43:604.

7. VANVEEN, W.L. 1973. Bacteriology of activated sludge, in particular the filamentous bacteria. *Antonie van Leeuwenhoek* (Holland) 39:189.

8. DONDERO, N.C., R.A. PHILIPS & H. HEUKELEKIAN. 1961. Isolation and preservation of cultures of *Sphaerotilus*. *Appl. Microbiol.* 9:219.

9. MULDER, E.G. & W.L. VANVEEN. 1963. Investigations on the *Sphaerotilus-Leptothrix* group. *Antonie van Leeuwenhoek* (Holland) 29:121.

10. UNZ, R.F. & D.G. LUNDGREN. 1961. A comparative nutritional study of three chemoautotrophic bacteria: *Ferrobacillus ferrooxidans*, *Thiobacillus ferrooxidans*, and *Thiobacillus thiooxidans*. *Soil Sci.* 92:302.

11. MCGORAN, C.J.M., D.W. DUNCAN & C.C. WALDEN. 1969. Growth of *Thiobacillus ferrooxidans* on various substrates. *Can. J. Microbiol.* 15:135.

12. SILVERMAN, M.P. & D.C. LUNDGREN. 1959. Studies on the chemoautotrophic iron bacterium *Ferrobacillus ferrooxidans*. *J. Bacteriol.* 77:642.

13. KUCERA, S. & R.S. WOLFE. 1957. A selective enrichment method for *Gallionella ferruginea*. *J. Bacteriol.* 74:344.

14. WOLFE, R.S. 1958. Cultivation, morphology, and classification of the iron bacteria. *J. Amer. Water Works Assoc.* 50:1241.

15. NUNLEY, J.W. & N.R. KRIEG. 1968. Isolation of *Gallionella ferruginea* by use of formalin. *Can. J. Microbiol.* 14:385.

16. WALSH, F. & R. MITCHELL. 1972. A pH dependent succession of iron bacteria. *Environ. Sci. Technol.* 6:809.

17. Clark, F.M., R.M. Scott & E. Bone. 1967. Heterotrophic, iron-precipitating bacteria. *J. Amer. Water Works Assoc.* 59:1036.

18. Lewis, R.F. 1965. Control of sulfate-reducing bacteria. *J. Amer. Water Works Assoc.* 57:1011.

19. Iverson, W.P. 1966. Growth of *Desulfovibrio* on the surface of agar media. *Appl. Microbiol.* 14:529.

20. Lechevalier, H.A. & D. Pramer. 1970. The Microbes, 1st ed. J.B. Lippincott Co., Philadelphia, Pa.

21. Mara, D.D. & D.J.A. Williams. 1970. The evaluation of media used to enumerate sulphate reducing bacteria. *J. Appl. Bacteriol.* 33:543.

22. Kreig, N.R. & J.G. Holt. 1986. Bergey's Manual of Systematic Bacteriology, Volume 1. Williams & Wilkins. Baltimore, Md.

23. Pfennig, N. 1967. Photosynthetic bacteria. *Annu. Rev. Microbiol.* 21:285.

24. Hutchinson, M., K.I. Johnstone & D. White. 1965. The taxonomy of certain thiobacilli. *J. Gen. Microbiol.* 41:357.

25. Hutchinson, M., K.I. Johnstone & D. White. 1966. Taxonomy of the acidophilic thiobacilli. *J. Gen. Microbiol.* 44:373.

26. Starkey, R.L. 1937. Formation of sulfide by some sulfur bacteria. *J. Bacteriol.* 33:545.

27. Lackey, J.B., W.W. Lackey & G.B. Morgan. 1965. Taxonomy and ecology of the sulfur bacteria. *Eng. Prog.*, Univ. Fla. Bull. Ser. 119, 19:3.

28. Nelson, D.C., C.O. Wirsen & H.W. Jannasch. 1989. Characterization of large, autotrophic *Beggiatoa* spp. abundant at hydrothermal vents of the Guaymas Basin. *Appl. Environ. Microbiol.* 55:2909.

29. Nelson, D.C., J.B. Waterbury & H.W. Jannasch. 1982. Nitrogen fixation and nitrate utilization by marine and freshwater *Beggiatoa.* *Arch. Microbiol.* 133:172.

30. Strohl, W.R. & J.M. Larkin. 1978. Enumeration, isolation, and characterization of *Beggiatoa* from freshwater sediments. *Appl. Environ. Microbiol.* 36:755.

31. Larkin, J.M. 1980. Isolation of *Thiothrix* in pure culture and observation of a filamentous epiphyte on *Thiothrix.* *Curr. Microbiol.* 4: 155.

32. Larkin, J.M. & D.L. Shinabarger. 1983. Characterization of *Thiothrix nivea.* *Int. J. System. Bacteriol.* 33:841.

33. Eikelboom, D.H. 1975. Filamentous organisms observed in activated sludge. *Water Res.* 9:365.

34. Williams, T.M. & R.F. Unz. 1985. Isolation and characterization of filamentous bacteria present in bulking activated sludge. *Appl. Microbiol. Technol.* 22:273.

35. Williams, T.M. & R.F. Unz. 1985. Filamentous sulfur bacteria of activated sludge: characterization of *Thiothrix, Beggiatoa,* and Eikelboom type 021N strains. *Appl. Environ. Microbiol.* 49:887.

9250 DETECTION OF ACTINOMYCETES*

9250 A. Introduction

1. General Discussion

Earthy-musty odors affect the quality and public acceptance of municipal water supplies in many parts of the world. They are among the naturally occurring odors that plant operators find most difficult to remove by conventional treatment. As early as 1929, it was assumed that these odors could be attributed to volatile metabolites formed during normal actinomycete development.[1] Two such compounds, geosmin and 2-methylisoborneol, have been isolated[2-8] and identified as the agents responsible for earthy-musty odor problems in surface water.[8-10] Both, however, are produced also by some filamentous blue-green algae.[11-15] Geosmin and 2-methylisoborneol have threshold odor concentrations well below the microgram-per-liter level. Thus, traces of these products are sufficient to impart a disagreeable odor to water or a muddy flavor to fish. In areas periodically plagued by this problem, it is prudent to enumerate actinomycetes. Identification of their relative abundance in a drinking water source can provide yet another means to assess water quality. The methods described are well-established techniques that have been used with success in the isolation and enumeration of actinomycetes related to public water supplies.[16,17] Actinomycetes also have been recognized as a cause of disruptions in wastewater treatment. Massive growths are capable of producing thick foam in the activated sludge process.[18,19]

Of the general properties of actinomycetes, the most striking is their fungal-type morphology. Although actinomycetes were looked upon initially as fungi, later research revealed that they were filamentous, branching bacteria.[20] The actinomycetes are represented most commonly by saprophytic forms that have an extensive impact on the environment by decomposing and transforming a wide variety of complex organic residues. Widely distributed in nature, actinomycetes constitute a considerable proportion of the population of soil and lake and river muds. Most actinomycetes from which geosmin and 2-methylisoborneol have been identified are members of the genus *Streptomyces*, which is considered the most likely to be significant in water supply problems.

2. Samples

a. Collection: Collect samples as directed in Section 9060A.

b. Storage: Analyze samples as promptly after collection as possible. Store water samples below 10°C if they cannot be processed promptly.

---

* Approved by Standard Methods Committee, 1993.

3. References

1. ADAMS, B.A. 1929. *Cladothrix dichotoma* and allied organisms as a cause of an "indeterminate" taste in chlorinated water. *Water & Water Eng.* 31:327.
2. GERBER, N.N. & H.A. LECHEVALIER. 1965. Geosmin, an earthy-smelling substance isolated from actinomycetes. *Appl. Microbiol.* 13:935.
3. GERBER. N.N. 1968. Geosmin, from microorganisms, is trans-1,10-dimethyl-trans-9-decalol. *Tetrahedron Lett.* 25:2971.
4. MARSHALL, J.A. & A.R. HOCHSTETLER. 1968. The synthesis of ($\pm$)-geosmin and the other 1,10-dimethyl-9-decalol isomers. *J. Org. Chem.* 33:2593.
5. ROSEN, A.A., R.S. SAFFERMAN, C.I. MASHNI & A.H. ROMANO. 1968. Identity of odorous substances produced by *Streptomyces griseoluteus. Appl. Microbiol.* 16:178.
6. MEDSKER, L.L., D. JENKINS & J.F. THOMAS. 1969. Odorous compounds in natural waters: 2-exo-hydroxy-2-methylbornane, the major odorous compound produced by several actinomycetes. *Environ. Sci. Technol.* 3:476.
7. GERBER, N.N. 1969. A volatile metabolite of actinomycetes, 2-methylisoborneol. *J. Antibiot.* 22:508.
8. ROSEN, A.A., C.I. MASHNI & R.S. SAFFERMAN. 1970. Recent developments in the chemistry of odour in water: The cause of earthy/musty odour. *Water Treat. Exam.* 19:106.
9. PIET, G.J., B.C.J. ZOETEMAN & A.J.A. KRAAYEVELD. 1972. Earthy-smelling substances in surface waters of the Netherlands. *Water Treat. Exam.* 21:281.
10. YURKOWSKI, M. & J.A.L. TABACHEK. 1974. Identification, analysis, and removal of geosmin from muddy-flavored trout. *J. Fish. Res. Board Can.* 31:1851.
11. SAFFERMAN, R.S., A.A. ROSEN, C.I. MASHNI & M.E. MORRIS. 1967. Earthy-smelling substances from a blue-green alga. *Environ. Sci. Technol.* 1:429.
12. MEDSKER, L.L., D. JENKINS & J.F. THOMAS. 1968. Odorous compounds in natural waters. An earthy-smelling compound associated with blue-green algae and actinomycetes. *Environ. Sci. Technol.* 2: 461.
13. KIKUCHI, T., T. MIMURA, K. HARIMAYA, H. YANO, M. ARIMOTO, Y. MASADA & T. INOUE. 1973. Odorous metabolites of blue-green alga *Schizothrix muelleri* Nageli collected in the southern basin of Lake Biwa. Identification of geosmin. *Chem. Pharm. Bull.* 21:2342.
14. TABACHEK, J.L. & M. YURKOWSKI. 1976. Isolation and identification of blue-green algae producing muddy odor metabolites, geosmin and 2-methylisoborneol, in saline lakes in Manitoba. *J. Fish. Res. Board Can.* 33:25.
15. IZAGUIRRE, G., C.J. HWANG, S.W. KRASNER & M.J. McGUIRE. 1982. Geosmin and 2-methylisoborneol from cyanobacteria in three water supply systems. *Appl. Environ. Microbiol.* 43:708.
16. SAFFERMAN, R.S. & M.E. MORRIS. 1962. A method for the isolation and enumeration of actinomycetes related to water supplies. Robert A. Taft Sanitary Engineering Center Tech. Rep. W62-10, U.S. Public Health Serv., Cincinnati, Ohio.
17. KUSTER, E. & S.T. WILLIAMS. 1964. Selection of media for isolation of *Streptomyces. Nature* 202:928.
18. LECHEVALIER, H.A. 1975. Actinomycetes of sewage-treatment plants. Environ. Protection Technol. Ser., EPA-600/2-75-031, U.S. Environmental Protection Agency, Cincinnati, Ohio.
19. LECHEVALIER, M.P. & H.A. LECHEVALIER. 1974. *Nocardia amarae*, sp. nov., an actinomycete common in foaming activated sludge. *Int. J. Syst. Bacteriol.* 24:278.
20. LECHEVALIER, H.A. & M.P. LECHEVALIER. 1967. Biology of actinomycetes. *Annu. Rev. Microbiol.* 21:71.

9250 B. Actinomycete Plate Count

1. General Discussion

A plating method using a double-layer agar technique has been adapted for determining actinomycete density. Because only the thin top layer of the medium is inoculated with sample, surface colonies predominate and identification and counting of colonies is facilitated.

2. Preparation and Dilution

Prepare and dilute samples as directed in Section 9215 or 9610. Dilutions up to 1:1000 (10^{-3}) usually are suitable for raw water, while treated waters may be examined directly. For soil samples, use dilutions from 1:1000 (10^{-3}) to 1:1 000 000 (10^{-6}).

3. Medium

Starch-casein agar:

Soluble starch	10.0	g
Casein	0.3	g
Potassium nitrate, KNO$_3$	2.0	g
Sodium chloride, NaCl	2.0	g
Dipotassium hydrogen phosphate, K$_2$HPO$_4$	2.0	g
Magnesium sulfate, hydrate, MgSO$_4$·7H$_2$O	0.05	g
Calcium carbonate, CaCO$_3$	0.02	g
Ferrous sulfate, hydrate, FeSO$_4$·7H$_2$O	0.01	g
Agar	15.0	g
Reagent-grade water	1	L

No pH adjustment is required. Medium is used to prepare double-layer plates. Store medium for bottom layer in bulk or in tubes in 15-mL amounts. Store medium for surface layer in tubes in 17.0-mL amounts.

4. Procedure

a. Plating: Prepare three plates for each dilution to be examined. Aseptically transfer 15 mL of sterile starch-casein agar to a petri dish and let agar solidify, thus forming the bottom layer. To a test tube containing 17.0 mL liquefied starch-casein agar at 45 to 48°C, add 2 mL of appropriately diluted sample and 1 mL of

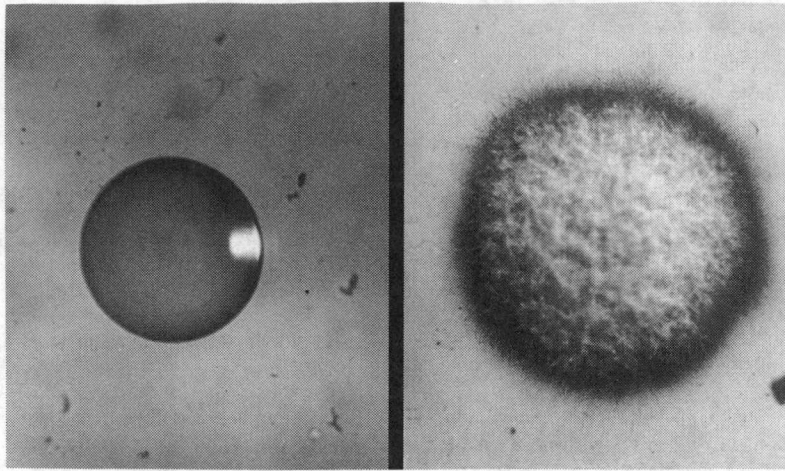

Figure 9250:1. Bacterial colonies—typical colony type vs. actinomycete colony type, 50×. Left: A typical bacterial colony characterized by a smooth mucoid appearance and a relatively distinct smooth border. Right: An actinomycete colony characterized by the mass of branching filaments that result in the fuzzy appearance of its border and by the dull powdery appearance of the spore-laden, aerial hyphae.

the antifungal antibiotic, cycloheximide,* prepared in reagent-grade water (1 mg/mL) and sterilized by autoclaving for 15 min at 121°C. Pipet 5 mL of inoculated agar over the hardened bottom layer with gentle swirling to obtain even distribution of the surface layer.

b. Incubation: Invert and incubate at 28°C until no new colonies appear. Usually this requires 6 to 7 d.

c. Counting: Plates suitable for counting contain 30 to 300 colonies. Identify actinomycetes by gross colony appearance. If necessary, verify by microscopic examination at a magnification of 50 to 100×, as shown in Figure 9250:1. Actinomycete colonies, because of filamentous growth, typically have a fuzzy colonial border. Table 9250:I lists the distinguishing characteristics commonly used to differentiate actinomycete from other bacterial

_____
* Actidione®, Upjohn and Company, Kalamazoo, MI, or equivalent.

colonies. Cycloheximide generally suppresses fungal growth; however, fungal colonies, if present, can be recognized by their wooly appearance. Microscopically, fungi reveal a considerably larger cell diameter than actinomycetes.

5. Calculation

Report actinomycetes per milliliter of water or gram (dry weight) of soil. If three plates are used per sample, the average number of colonies on all plates (total number of colonies/3), times 2, times the reciprocal of the dilution (10/1, 100/1, 1000/1, etc.) equals the actinomycete colony count per milliliter of original sample. For solid or semisolid samples, correct for water content and report actinomycete colonies per gram, dry weight, of sample.

TABLE 9250:I. GENERAL MACROSCOPIC PROPERTIES OF BACTERIAL COLONIES ON SOLID MEDIUM

Characteristic	Typical Colony Type	Actinomycete Colony Type*
Appearance	Shiny or opalescent	When young it is composed of hyphae, but in some species these may later fragment. Substrate and surface hyphae have no distinctive color. As the colony matures, fluffy aerial hyphae that carry spores form and give to colonies of different species various colors and sometimes a chalky appearance. Soluble pigments, either melanin or brightly colored type, that diffuse into the medium, also are common.
Texture	Soft	Strong and leathery
Degree of adherence to solid medium	Weak	Strong
Edge of colony	Regular, continuous, and not different from colony as a whole	Irregular, intermittent, slightly less dense than colony as a whole, and of hyphal appearance

* Actinomycetes are authentic bacteria by all modern criteria, except for their hyphal character and mode of spore formation.

9260 DETECTION OF PATHOGENIC BACTERIA*

9260 A. Introduction

1. General Discussion

One purpose of drinking water and wastewater treatment is to reduce the numbers of viable organisms to acceptable levels, and to remove or inactivate all pathogens capable of causing human disease. Despite the remarkable success of water treatment and sanitation programs in improving public health, sporadic cases and point-source outbreaks of waterborne diseases continue to occur. Water and wastewater may contain a wide variety of bacteria that are opportunistic or overt pathogens of animals and humans. Waterborne pathogens enter human hosts through intact or compromised skin, inhalation, ingestion, aspiration, and direct contact with the mucous membranes of the eye, ear, nose, mouth, and genitals. This section provides an introduction to the etiologic agents responsible for diseases transmitted by drinking and recreational waters in the U.S.

Over 80 genera of bacteria that are nonpathogenic for humans have their natural habitat in water. In addition, some opportunistically pathogenic bacteria (*Pseudomonas, Serratia, Acinetobacter, Chromobacterium, Achromobacter, Aeromonas,* etc.) occur naturally in water. Other opportunists (*Bacillus, Enterobacter, Klebsiella, Actinomyces, Streptomyces,* etc.) are sometimes washed into water from their natural habitat in soil or on vegetative matter. Opportunistic pathogens also may be seeded from regrowth and biofilms in water treatment plants and distribution systems.

Water contamination and disease transmission may result from conditions generated at overloaded and/or malfunctioning sanitary waste disposal and potable water treatment systems. In addition, common outdoor recreational activities such as swimming (including pools and hot tubs), boating, camping, and hiking, all place humans at risk of waterborne diseases from ingestion or direct contact with contaminated water.[1] Outbreaks of gastroenteritis, pharyngoconjunctivitis, folliculitis, otitis, and pneumonia are associated with these recreational activities. Overcrowded parks and recreational areas contribute to the contamination of surface and groundwater.

National statistics on outbreaks of waterborne diseases have been compiled in the U.S. since 1920.[2,3] Since 1971, the Centers for Disease Control and Prevention, the U.S. Environmental Protection Agency, and the Council of State and Territorial Epidemiologists have maintained a collaborative surveillance program on waterborne disease outbreaks of drinking water and recreational water origin.[4] A summary of waterborne diseases in the U.S. has been published.[5] Summary data from outbreaks reported through the national waterborne disease surveillance system for drinking water and recreation from 1985 to 1994 are shown in Table 9260:I.

Laboratory diagnosis of infectious disease depends on isolation of the etiologic agent or demonstration of antibody response in the patient. Environmental microbiological examinations are conducted for compliance monitoring of the environment, to trouble-

TABLE 9260:I. SUMMARY DATA FROM WATERBORNE BACTERIAL DISEASE OUTBREAKS, 1985–94

Type of Water	Variable	Number
Drinking water	Total outbreaks	21
	Agent:	
	Shigella	12
	Campylobacter	6
	Salmonella	2
	E. coli O157:H7	1
	System:	
	Noncommunity	10
	Community	8
	Individual	3
	Source:	
	Well	17
	Lake	2
	Spring	1
	Cistern	1
	Cause:	
	Untreated groundwater	9
	Distribution system deficiency	7
	Treatment deficiency	4
	Unknown	1
Recreational water	Total outbreaks	71
	Agent:	
	Pseudomonas	44
	Shigella	17
	Legionella	6
	Leptospira	2
	E. coli O157:H7	2
	Location:	
	Hotel/motel	23
	Outdoor recreation area (surface water)	21
	Home	14
	Spa or public swimming pool	5
	Resort	4
	Apartment complex/condominum	4
	Source:	
	Whirlpool/hot tub	47
	Lake/pond	20
	Swimming pool	3
	Stream	1

shoot problems in treatment plants and distribution systems, and in support of epidemiological investigations of disease outbreaks. Ideally, the public health microbiologist can contribute expertise in both clinical and environmental microbiology, thereby facilitating epidemiological investigations.

When testing for pathogens in environmental samples, it usually is advisable to include analyses for indicator organisms. Besides coliform indicators (total coliform, fecal coliform, and *E. coli*), fecal streptococci, enterococci, *Clostridium perfringens,* and *Aeromonas* have been proposed as indicators of water quality. No single indicator provides assurance that water is pathogen-

---

* Approved by Standard Methods Committee, 1997.

pfpfree. The choice of monitoring indicator(s) presupposes an understanding of the parameters to be measured and the relationship of the indicator(s) to the pathogen(s). Some bacterial pathogens, such as *Pseudomonas, Aeromonas, Plesiomonas, Yersinia, Vibrio, Legionella,* and *Mycobacterium,* may not correlate with coliform indicators. Traditional bacterial indicators also may not correlate with viruses or parasites in pristine waters or groundwaters, and they may be of limited utility in estuarine and marine waters. Nevertheless, tests for total and fecal bacteria and *E. coli* are useful, because it is rare to isolate bacterial enteric pathogens in the absence of fecal contamination.

Other more general indicators also may be of value for assessing the potential for pathogen contamination and interpreting culture results. Heterotrophic plate count provides information about the total numbers of aerobic organotrophic bacteria and an indication of the total organic composition of the aquatic environment. Physicochemical factors, such as turbidity, pH, salinity, temperature, assimilable organic carbon, dissolved oxygen, biochemical oxygen demand, and ammonia may provide useful information about contamination or the potential of water to support bacterial growth. For treated waters, chlorine residual should be measured at the sample collection point.

This section contains methods for *Salmonella, Shigella,* pathogenic *E. coli, Campylobacter, Vibrio cholerae, Leptospira, Legionella, Yersinia entercolitica, Aeromonas,* and *Mycobacterium.* Methods for isolation and enumeration of *P. aeruginosa* are found in Section 9213E and F. Methods for other pathogens are found elsewhere.[6]

The methods outlined below may be used to analyze samples associated with disease outbreaks, or in other studies on the occurrence of pathogens in water and wastewater. Methods for recovery of bacterial pathogens from water and wastewater have not changed significantly in the past 30 years. The methods presented below are not standardized, and the procedures may need modification to fit a particular set of circumstances. No single procedure is available for reliable detection of any pathogen or group of pathogens. Because the presence of pathogens is intermittent and the survival times in the environment are variable, routine examination of water and wastewater for pathogenic bacteria is not recommended. Even in outbreak situations, the recovery of pathogens from water and wastewater may be limited by lack of facilities, untrained personnel, inadequate methods, and high costs.

2. References

1. PITLIK, S., S.A. BERGER & D. HUMINER. 1987. Nonenteric infections acquired through contact with water. *Rev. Infect. Dis.* 9:54.
2. CRAUN, G.F., ed. 1986. Waterborne Diseases in the United States. CRC Press, Inc., Boca Raton, Fla.
3. LIPPY, E.C. & S.C. WALTRIP. 1984. Waterborne disease outbreaks—1946–1980: A thirty-five year perspective. *J. Amer. Water Works Assoc.* 76:60.
4. KRAMER, M.H., B.L. HERWALDT, G.F. CRAUN, R.L. CALDERON & D.D. JURANEK. 1996. Waterborne diseases: 1993 and 1994. *J. Amer. Water Works Assoc.* 88:66.
5. KRAMER, M.H., B.L. HERWALDT, G.F. CRAUN, R.L. CALDERON & D.D. JURANEK. 1996. Surveillance for waterborne-disease outbreaks—United States, 1993–1994. *Morbid. Mortal. Week. Rep.* 45(SS-1):1.
6. MURRAY, P.R., E.J. BARON, M.A. PFALLER, F.C. TENOVER & R.H. YOLDEN, eds. 1995. Manual of Clinical Microbiology, 6th ed. American Soc. Microbiology Press, Washington, D.C.

9260 B. General Qualitative Isolation and Identification Procedures for *Salmonella*

Rather than a specific protocol for *Salmonella* detection in water, a brief summary of methods suitable for recovery of these organisms is given. Methods currently available have been used in numerous field investigations to demonstrate *Salmonella* in both fresh and marine water environments. The occurrence of *Salmonella* in water is highly variable; there are limitations and variations in both the sensitivity and selectivity of accepted *Salmonella* isolation procedures for the detection of the more than 2300 *Salmonella* serotypes currently recognized. Thus, a negative result by any of these methods does not imply the absence of salmonellae, nor does it imply the absence of other pathogens.

1. Concentration Techniques

Salmonella are ubiquitous in the environment and can be detected at low concentrations in most surface waters. These organisms are usually present in small numbers compared to coliforms; therefore, it is necessary to examine a relatively large sample to isolate the organisms.[1]

a. Swab technique: Prepare swabs from cheesecloth 23 cm wide, folded five times at 36-cm lengths, and cut lengthwise to within 10 cm from the head into strips approximately 4.5 cm wide. Securely wrap the uncut or folded end of each swab with 16-gauge wire for use in suspending the swab in water. Place the swabs in kraft-type bags and sterilize at 121°C for 15 min. Place swab just below the surface of the sampling location for 1 to 3 d.[2,3] (Longer swab exposure will not increase entrapment of pathogens.) Gauze pads of similar thickness may be substituted. During sampling, particulate matter and microorganisms are concentrated from the water passing through or over the swab. After exposure, retrieve the swab, place it in a sterile plastic bag, ice, and send to the laboratory. Maximum storage-transit time allowable is 6 h. Do not transport swabs in enrichment media; ambient transport temperature may cause sufficient proliferation of competitive organisms to mask salmonellae. In the laboratory, place pad or portions of it in enrichment media. When flasks of enrichment medium containing iced swabs are to be incubated at 40 to 41°C, place flasks in a 44.5°C water bath for 5 min before incubation in an air incubator.

b. Diatomaceous earth technique: Place an absorbent pad (not a membrane filter) on a membrane filter funnel receptacle, assemble funnel, and add 2.5 g sterile diatomaceous earth* to pack the funnel neck loosely. Apply vacuum and filter 2 L of sample. After filtration, disassemble funnel, divide resulting "plug" of diatomaceous earth and absorbent pad in half aseptically with a knife-

---

* Celite, World Minerals, Inc., Lompoc, CA or equivalent.

edged, sterile spatula, and add to suitable enrichment media. Alternatively, place entire plug in enrichment medium.

c. Large-volume sampler: Use a filter composed of borosilicate glass microfibers bonded with epoxy resin to examine several liters or more of sample, provided that sample turbidity does not limit filtration.[4] The filter apparatus consists of a 2.5- × 6.4-cm cartridge filter and a filter holder.† Sterilize by autoclaving at 121°C for 15 min. Place sterile filter apparatus (connected in series with tubing to a 20-L water bottle reservoir and vacuum pump) in the 20-L sample container appropriately calibrated to measure volume of sample filtered. Apply vacuum and filter an appropriate volume. When filtration is complete, remove filter and place in a selective enrichment medium.

d. Membrane filter technique: To examine low-turbidity water, filter several liters through a sterile 142-mm-diam membrane of 0.45-μm pore size.[5] For turbid waters, precoat the filter: make 1 L of sterile diatomaceous earth suspension (5 g/L reagent-grade water) and filter about 500 mL. Without interrupting filtration, quickly add sample (1 L or more) to remaining suspension and filter. After filtration, place membrane in a sterile blender jar containing 100 mL sterile 0.1% (w/v) peptone water and homogenize at high speed for 1 min. Add entire homogenate to 100 mL double-strength selective enrichment medium. Alternatively, use multiple 47-mm-diam membrane filters to filter the sample. Immerse each membrane aseptically in 50 mL single-strength selective enrichment medium and incubate.

Qualitative detection of *Salmonella* in suspect potable water also may be achieved successfully by further analysis of selected M-Endo MF cultures (from 100 mL sample volume) that contain significant background growth and total coliforms.[6] After completing routine coliform count, place entire filter with mixed growth into 10 mL tetrathionate broth (containing 1:50 000 brilliant green dye) for *Salmonella* enrichment before differential colony isolation on brilliant green agar. This unique approach requires no special large sample collections and can be an extension of the routine total coliform analysis.

2. Enrichment

Selectively enrich the concentrated sample in a growth medium that suppresses growth of coliform bacteria. Sample enrichment is essential, because the pathogens usually are present in low numbers and solid selective media for colony isolation are somewhat toxic, even to pathogens. No single enrichment medium can be recommended that allows optimum growth of all *Salmonella* serotypes. Use two or more selective enrichment media in parallel for optimum detection. Elevated incubation temperatures including 40°, 41.5°, and 43°C and the addition of brilliant green dye to media help suppress background growth and may improve detection of *Salmonella*, but these modifications also suppress growth of some serotypes, including *Salmonella typhi*.

a. Selenite cystine broth inhibits gram-positive and nonpathogenic enterobacteria while allowing for recovery of most species of *Salmonella*, including *Salmonella typhi*. Optimum incubation time for maximum recovery of *Salmonella* is 48 h at 35 to 37°C. Repeat streaking from tubes with turbidity several times during first day, and daily up to 5 d to increase potential recovery of all serotypes that may be present. Transfer 1 mL selenite broth culture to a fresh tube of same medium for continued incubation to enrich further *Salmonella* growth and enhance recovery of streak plates.

b. Selenite-F broth allows for optimum recovery of most *Salmonella* species, including *Salmonella typhi*, after 24 h at 35 to 37°C. This increased recovery of *Salmonella* is accompanied by a slight decrease in selectivity when compared to selenite cystine. Most significantly, *E. coli* growth is not inhibited. Repeat streaking from tubes with turbidity several times during first day, and daily up to 5 d to increase potential recovery of all serotypes that may be present. Transfer 1 mL selenite broth culture to a fresh tube of same medium for continued incubation to enrich further *Salmonella* growth and enhance recovery of streak plates.

c. Tetrathionate broth, incubated at 35°C, inhibits coliforms and Gram-positive bacteria, permitting selective enrichment of most *Salmonella* species, including *S. typhi*. It has been reported that tetrathionate broth is more selective for *Salmonella* than selenite-based media when incubated for 48 h at 43°C. While this formulation is highly selective, it is unable to inhibit *Proteus mirabilis*, which shows optimum growth. Growth of *Proteus* and *Citrobacter* can be inhibited with addition of brilliant green (see Section 9260B.3*a*). Incubation at 43°C and addition of brilliant green also will inhibit some species of *Salmonella*, including *S. typhi*.

3. Selective Growth

Further separation of pathogens from the remaining nonpathogenic bacterial population is facilitated by proper choice of incubation temperature for primary enrichment followed by secondary differentiation on selective solid media.[7] These factors, incubation temperature, enrichment medium, and isolation medium, are interrelated. No one combination is optimum for recovery of all *Salmonella* serotypes. Method comparisons are encouraged to determine the best combination for a given circumstance.

Solid media commonly used for enteric pathogen detection may be classed into three groups: (*a*) differential media with little or no inhibition toward nonpathogenic bacteria, such as EMB (containing sucrose); (*b*) selective media containing bile salts or sodium desoxycholate as inhibitors,[8] such as MacConkey's agar, desoxycholate agar, or xylose lysine desoxycholate (XLD) agar; and (*c*) selective media containing brilliant green dye, such as brilliant green agar or bismuth sulfite agar. Any medium selected must provide optimum suppression of coliforms while permitting good recovery of the pathogenic group. Great skill at screening for these pathogens is necessary because of the competing growth of various nonpathogens. Streaking duplicate plates, one heavily and one lightly, often aids in recognition of enteric pathogens in the presence of large numbers of interfering organisms.

a. Brilliant green agar: Typical well-isolated *Salmonella* colonies grown on this medium are pinkish white with a red background. *S. typhi* and a few other species of *Salmonella* grow poorly because of the brilliant green dye content. Lactose-fermenters not subject to growth suppression will form greenish colonies or may produce other colorations. Occasionally, slow lactose-fermenters (*Proteus, Citrobacter*, and *Pseudomonas*) will produce colonies resembling those of a pathogen. Suppress spreading effect of pseudomonads by increasing agar concentration to 2%. In some instances, *Proteus* has been observed to "swarm"; reduce this tendency by using agar plates dried to remove surface moisture. If suspect *Salmonella* colonies are not observed after 24 h incubation, reincubate for an additional 24 h

---

† Balston Type AA filter with Type 90 holder, or equivalent.

to permit slow-growing or partially inhibited organisms to develop visible colonies. If typical colonies are not observed or if the streak plate is crowded, isolate in pure culture a few colonies for biochemical characterization. Non-lactose-fermenting colonies in close proximity to lactose-fermenting colonies may be masked.

b. Bismuth sulfite agar (Wilson and Blair medium[9]): Luxuriant growth of many *Salmonella* species (including *S. typhi*) can be expected on this medium. Examine bismuth sulfite plates after 24 h incubation for suspect colonies; reincubate for 24 h to detect slow-growing strains. Typical colonies usually develop a black color, with or without a metallic sheen, and frequently this blackening extends beyond the colony to give a "halo" effect. A few species of *Salmonella* develop a green coloration; therefore, isolate some of these colony types when typical colonies are absent. As with brilliant green agar, typical colony coloration may be masked by numerous bordering colonies after 48 h incubation. A black color also is developed by other H$_2$S-producing colonies, for example, *Proteus* and certain coliforms.

c. Xylose lysine desoxycholate agar: Compared to brilliant green dye, sodium desoxycholate is only slightly toxic to fastidious *Salmonella*. *Salmonella* and *Arizona* organisms produce black-centered red colonies. Coliform bacteria, *Proteus*, and many *Enterobacter* produce yellow colonies. Optimum incubation time is 24 h. If plates are incubated longer, an alkaline reversion and subsequent blackening occur with H$_2$S-positive nonpathogens (*Citrobacter, P. vulgaris,* and *P. mirabilis*).

d. Xylose lysine brilliant green agar: This medium is especially good for *Salmonella* from marine samples. The brilliant green inhibits many *Proteus, Enterobacter,* and *Citrobacter* species.

4. Biochemical Reactions

Many enteric organisms of little or no pathogenicity share certain major biochemical characteristics with *Salmonella*. The identification of pathogens by colony characteristics on selective solid media has limitations inherent in the biological variations of certain organisms and cannot be relied on for even tentative identification. Suspected colonies grown on selective solid media must be purified and further characterized by biochemical reactions; final verification is based on serological identification. Usually a large number of cultures will be obtained from the screening procedure.

Commercially available differential media kits (see Section 9225) may be used as an alternative to Phases 1, 2, and 3 described below, before serological confirmation. These kits give 95 to 98% agreement with conventional tests, although more significant tests will be necessary to achieve further differentiation among strains of *Enterobacteriaceae*.

When such kits are not used, follow a sequential pattern of biochemical testing that will result in a greater saving of media and time for laboratory personnel.[10]

Phase 1—Preliminary screening, phenylalanine deaminase activity: Discard phenylalanine deaminase-positive cultures immediately as indicative of the *Proteus* group. In this test, spot isolates on phenylalanine agar and incubate for 24 h at either 35 or 37°C. Phenylalanine deaminase activity is indicated by a green zone that develops around the colony after flooding of the plate with a 0.5M FeCl$_3$ solution. Subject phenylalanine deaminase-negative cultures to the biochemical tests of Phase 2.

Phase 2—Biochemical tests: The tests used are:

Medium	Purpose of Test
TSI	Fermentation pattern: H$_2$S production
LIA	Lysine decarboxylase activity, H$_2$S production
Urea broth	Urease production

Conformance to the typical biochemical patterns of the *Salmonella* determines whether to process cultures further (Phase 3). Aberrant cultures may be encountered that do not conform to all the classical reactions attributed to each pathogenic group. In all cases, therefore, review reactions as a whole and do not discard cultures on the basis of a small number of apparent anomalies.

Phase 3—Fermentation reactions: Test fermentation reactions in dextrose, mannitol, maltose, dulcitol, xylose, rhamnose, and inositol broths to characterize further the biochemical capabilities of the isolates. This additional sorting reduces the possible number of positive cultures to be processed for serological confirmation. If the testing laboratory is equipped for serological confirmation (see 9260B.5), this series of biochemical tests may be eliminated.

5. Genus Identification by Serological Techniques

Upon completion of the recommended biochemical tests, inoculate the suspected *Salmonella* pure culture onto a brain-heart infusion agar slant and incubate for 18 to 24 h at 35 to 37°C. With wax pencil (china marker), divide an alcohol-cleaned glass slide into four sections. Prepare a dense suspension of test organism by suspending growth from an 18- to 24-h agar slant in 0.5 mL 0.85% NaCl solution. Place a drop of *Salmonella* "O" polyvalent antiserum in the first section and antiserum plus 0.85% NaCl in the second section. Using a clean inoculating loop, transfer a loopful of bacterial suspension to the third section containing 0.85% NaCl solution and to the fourth section containing 0.85% NaCl solution plus antiserum. Gently rock slide back and forth. If agglutination is not apparent in the fourth section at the end of 1 min, the test is negative. All other sections should remain clear.

When biochemical reactions are characteristic of *S. typhi* and the culture reacts with "O" polyvalent antiserum, check other colonies from the same plate for Vi antigen reaction. If there is no agglutination with *Salmonella* Vi antiserum, the culture is not *S. typhi*. Identification of *Salmonella* serotypes requires determination of H antigens and phase of the organism as described by Edwards and Ewing.[10] Isolates yielding biochemical reactions consistent for *Salmonella* and positive with polyvalent "O" antiserum may be identified as "*Salmonella* sp., serotype or bioserotype undetermined." If species identification is necessary, send isolates confirmed as *Salmonella* by biochemical tests and polyvalent "O" antisera to reference laboratories for further analysis.

6. References

1. CHERRY, W.B., J.B. HANKS, B.M. THOMASON, A.M. MURLIN, J.W. BIDDLE & J.M. GROOM. 1972. Salmonellae as an index of pollution of surface waters. *Appl. Microbiol.* 24:334.
2. MOORE, B. 1948. The detection of paratyphoid carriers in towns by means of sewage examination. *Mon. Bull. Mist. Health Pub. Health Lab. Serv.* 7:241.

3. MOORE, B., E.L. PERRY & S.T. CHARD. 1952. A survey by the sewage swab method of latent enteric infection in an urban area. *J. Hygiene* 50:137.

4. LEVIN, M.A., J.R. FISCHER & V.J. CABELLI. 1974. Quantitative large-volume sampling technique. *Appl. Microbiol.* 28:515.

5. PRESNELL, M.W. & W.H. ANDREWS. 1976. Use of the membrane filter and a filter aid for concentrating and enumerating indicator bacteria and *Salmonella* from estuarine waters. *Water Res.* 10:549.

6. CANLAS, L. 1975. Personal communication. Guam Environmental Protection Agency, Agana, Guam.

7. CHEN, H., A.D.E. FRASER & H. YAMAZAKI. 1993. Evaluation of the toxicity of *Salmonella* selective media for shortening the enrichment period. *Int. J. Food Microbiol.* 18:151.

8. LEIFSON, E. 1935. New culture media based on sodium desoxycholate for the isolation of intestinal pathogens and for enumeration of colon bacilli in milk and water. *J. Pathol. Bacteriol.* 40:581.

9. WILSON, W.J. & E.M. McV. BLAIR. 1926. Combination of bismuth and sodium sulfite affording enrichment and selective medium for typhoid and paratyphoid groups of bacteria. *J. Pathol. Bacteriol.* 29:310.

10. EDWARDS, P.R. & W.H. EWING. 1986. Identification of Enterobacteriaeceae, 4th ed. Elsevier Science Publ. Co., Inc., New York, N.Y.

7. Bibliography

MÜLLER, G. 1947. Der Nachweis von Keimer der Typhus-Paratyphus-gruppe in Wasser. H.H. Nolke Verlag, Hamburg, Germany.

GREENBERG, A.E., R.W. WICKENDEN & T.W. LEE. 1957. Tracing typhoid carriers by means of sewage. *Sewage Ind. Wastes* 29:1237.

McCOY, J.H. 1964. *Salmonella* in crude sewage, sewage effluent, and sewage polluted natural waters. *In* Int. Conf. Water Pollut. Res., 1st, London, 1962. Vol. 1:205. MacMillan, New York, N.Y.

BREZENSKI, F.T., R. RUSSOMANNO & P. DEFALCO, JR. 1965. The occurrence of *Salmonella* and *Shigella* in post-chlorinated and nonchlorinated sewage effluents and receiving waters. *Health Lab. Sci.* 2:40.

SPINO, D.E. 1966. Elevated temperature technique for the isolation of *Salmonella* from streams. *Appl. Microbiol.* 14:591.

GALTON, M.M., G.K. MORRIS & W.T. MARTIN. 1968. *Salmonella* in foods and feeds. Review of isolation methods and recommended procedures. Public Health Serv. Bur. Disease Prevention & Environmental Control, National Center for Disease Control, Atlanta, Ga.

BREZENSKI, F.T. & R. RUSSOMANNO. 1969. The detection and use of *Salmonella* in studying polluted tidal estuaries. *J. Water Pollut. Control Fed.* 41:725.

MORINIGO, M.A., M.A. MUNOZ, E. MARTINEZ-MANZANARES, J.M. SANCHEZ & J.J. BORREGO. 1993. Laboratory study of several enrichment broths for the detection of *Salmonella* spp. particularly in relation to water samples. *J. Appl. Bacteriol.* 74:330.

U.S. FOOD AND DRUG ADMINISTRATION. 1995. Bacteriological and Analytical Manual, 8th ed. Assoc. Official Analytical Chemists International, Gaithersburg, Md.

9260 C. Immunofluorescence Identification Procedure for *Salmonella*

The direct fluorescent antibody (FA) technique is a rapid and effective means of detecting salmonellae in fresh- and seawater samples. It may be used as a screening technique to provide rapid results for large numbers of samples, such as those from recreational or shellfish-harvesting waters. Positive FA tests are presumptive evidence for the presence of *Salmonella*. Because of potential cross-reactivity of antibodies, positive FA results should be confirmed by other methods. Sample volumes used depend on the degree of contamination. Where gross pollution is present, use smaller samples. When background information is absent, analyze a 2-L sample, using the diatomaceous earth concentration technique.

1. Apparatus for Fluorescence Microscopy

Standard fluorescent antibody microscopy equipment may be obtained separately or in a package containing the essential instrumentation (*a-f*):

a. Light microscope with microscope stand.

b. Light source, providing energy in the short-wavelength region of the spectrum. A high-pressure mercury 200-W arc enclosed in a quartz envelope, a 75- to 150-W xenon high-pressure lamp, or a low-voltage 100-W quartz halogen lamp may satisfy this requirement. A significant portion of the energy should be emitted in the ultraviolet and blue region of the spectrum.

c. Power pack to provide constant voltage and wattage output for the selected lamp.

d. Basic filters including heat-absorbing filter (KG-1 or KG-2, or equivalent): red-absorbing filter (BG-38 or equivalent); exciter filter (BG-12 or equivalent, BG-12 being also a blue filter); and barrier filter (OG-1 or blue-absorbing filter). New interference excitation filters (KP500 or equivalent) having very high transmission in the blue portion of the spectrum (490 nm) are available. Barrier or suppression filters used with these have a sharp cutoff at 500 to 510 nm.

e. Optics: The fluorescence microscope must have high-quality optics. A 100 × objective with an iris diaphragm to reduce the numerical aperture (N.A.) for dark-field work is essential. Because the N.A. is similar for all 100 × objectives (1.25 to 1.30), base selection on desire for a flat-field (plano) lens.

f. Cardioid dark-field condenser for illuminating specimen: A 95 × oil immersion objective with build-in iris diaphragm is desirable. True dark-field illumination can be achieved only if the objective N.A. is smaller than the condenser N.A., i.e., of the illuminating cone of light. (Difference in N.A. between objective and condenser should be at least 0.05.) Reduce N.A. of an oil immersion objective by using the built-in diaphragm or by putting a funnel stop onto the objective.

g. FA pre-cleaned micro slides, 7.6- × 2.5-cm, 0.8- to 1.0-mm thickness.

h. Cover glass for FA slides, No. 1 1/2, 0.16- to 0.19-mm thickness.

i. Staining assembly consisting of dish, cover, and slide rack with handle. Five dishes are required; for Kirkpatrick's fixative, 95% ethanol, first PBS rinse, second PBS rinse, and reagent water.

j. Moist chamber used to incubate slides containing smears with added conjugate. A simple chamber consists of water-saturated toweling with a culture dish bottom (150 by 20 mm) placed over the wet toweling.

2. Reagents

a. Nondrying immersion oil, Type A (low fluorescence, PCB-free).*

b. FA Kirkpatrick fixative, consisting of 60 mL absolute ethanol, 30 mL chloroform, and 10 mL formaldehyde.†

c. Phosphate-buffered saline (PBS): Add 10 g buffer‡ to 1000 mL freshly prepared distilled water. Stir until the powder dissolves completely. Adjust with 0.2N NaOH to pH 8.0.

d. FA mounting fluid: Use standardized reagent-grade glycerine adjusted to pH 9.0 with 0.2N NaOH and intended for mounting slides to be viewed with the FA microscope.

e. Reagent (laboratory pure) water: Use double-distilled water from an all-glass still or other high-quality analytical-grade laboratory water.

f. FA Salmonella panvalent conjugate is a fluorescein-conjugated anti-*Salmonella* globulin.§ To rehydrate, add 5 mL reagent water to a vial or conjugate. Determine working dilution (see ¶ 5e). Store unused rehydrated conjugate in a freezer, preferably at −60°C. Avoid repeated freezing and thawing.

g. Zn-CdS: Ag phosphor particle.‖

3. Concentration Technique

Place an absorbent pad on a membrane filter funnel and add sufficient sterile diatomaceous earth# to pack funnel neck loosely. Filter 2 L of sample. Rinse funnel with 50 to 100 mL sterile phosphate-buffered dilution water or 0.1% peptone water. Disassemble funnel and remove resulting "plug" of diatomaceous earth and the absorbent pad. Repeat with a second 2-L sample.

4. Enrichment

Immerse one plug and absorbent pad in a flask containing 300 mL selenite cystine broth. Immerse second plug and absorbent pad in a flask containing 300 mL tetrathionate broth supplemented with 3 mL 1:1000 aqueous solution of brilliant green dye and 3 mg *l*-cystine. Incubate at either 35 or 37°C for 24 h.

5. Fluorescent Antibody Reaction and Analysis

a. Prepare spot plates of brilliant green agar (BGA) and xylose lysine brilliant green (XLBG) agar by placing 1 drop (about 0.01 mL, delivered with a wire or sterile plastic loop) of the enrichment medium (selenite cystine or tetrathionate broth) at each of four separate points on the agar surface.[1] Space drops on agar plate so that FA microscope slide will cover two inoculation points. This

is essential because glass slide impression smears of the inoculated points will be made after incubation of plates.

b. Incubate BGA and XLBG plates at 37 ± 0.5°C for 2.5 to 3 h. After incubation, micro CFUs will develop. Make impression smears by taking a *clean* FA microscope glass slide and placing it over two inoculated points on the medium. Press down lightly, being careful not to move glass slide horizontally. Do not apply too much pressure, because it will cause movement of the slide and collection of additional agar. Repeat this process for the other two inoculation points and for inoculation points on second agar medium. Prepare a total of four FA slides in this manner.

c. Air-dry smears and fix for 2 min in Kirkpatrick's fixative. Rinse slides briefly in 95% ethanol and let air dry. *Do not blot.*

d. Cover fixed smears with 1 drop of *Salmonella* panvalent conjugate. Before use, dilute commercial conjugate and determine appropriate working dilution. Most batches are effective at a 1:4 dilution but this will vary with the type of fluorescence equipment used, light source, alignment, magnification, cultures, etc. Determine working dilution (titer) of each lot of conjugate.

e. To determine conjugate titer use a known 18- to 24-h *Salmonella* culture grown in veal infusion broth and make smears on FA glass slide. Dilute conjugate and treat as outlined in *c* and *d* above. For example, if the following results are obtained:

Dilution of Conjugate	Fluorescence
1:2	4+
1:4	4+
1:6	4+
1:8	2+
1:10	1+

use the second highest dilution giving 4+ fluorescence. In the above example use a 1:4 dilution of conjugate. Diluting conjugate insures minimum cross-reactivity. Prepare fresh diluted conjugate daily.

f. After covering each smear with 1 drop of appropriate dilution of conjugate, place slides in a moist chamber to prevent evaporation of staining reagent. After 30 min wash away excess reagent by dipping slides into phosphate-buffered saline (pH 8.0). Place slides in second bath of buffered saline for 10 min. Remove, rinse in distilled water, and drain dry. *Do not blot.*

g. Place a small drop of mounting fluid (pH 9.0) on the smear and cover with a No. 1 1/2 cover slip. Seal edges of cover slip with clear fingernail polish. Examine sealed slides within a few hours while fluorescence is of optimum intensity. Examine under a fluorescence microscope unit fitted with appropriate filters.

h. Include a positive control slide with each set of samples. This checks conjugate reactivity and FA equipment generally.

6. Recording and Interpreting Results

The intensity of organisms fluorescing in any given field is important in assessing positive *Salmonella* smears. If the majority of cells present fluoresce (4+ or 3+) the smear is positive. Carefully scrutinize smears showing only a few scattered fluorescing cells. Critical examination of cellular morphology may distinguish between these cells and salmonellae. The degree of fluorescence is the criterion on which positivity is based. Consider weakly

---

* R.P. Cargille Laboratories, Inc., Cedar Grove, NJ, or equivalent.
† Difco No. 3188 or equivalent.
‡ Difco Bacto-FA Buffer, dried, or equivalent.
§ Difco or equivalent.
‖ General Electric or equivalent.
Celite, World Minerals Inc., Lompoc, CA, or equivalent.

fluorescing cells (2+ and 1+) negative. Confirm all positive FA results by cultural techniques (see Section 9260B).

Reaction	Description	Fluorescence Intensity
Positive	Brilliant yellow-green fluorescence, cells sharply outlined	4+
Positive	Bright yellow-green fluorescence, cells sharply outlined with dark center	3+
Negative	Dull yellow-green fluorescence, cells not sharply outlined	2+
Negative	Faint green fluorescence discernible in dense areas, cells not outlined	1+
Negative	No fluorescence	0

7. Quantitative Immunofluorescence Microspectrofluorometric Microscopy

To make such analyses use a system consisting of analyzing and illumination sections. The analyzing section includes an eyepiece monochromator assembly and a photomultiplier-photometer. The eyepiece uses a beam splitter that reflects to the monochromator and the observer's eye, allowing for simultaneous visual observation and quantitative analysis of the yellow-green fluorescence intensity. The photometer package provides meter readout in milliamperes so that visual observation of fluorescence can be correlated with objective reading. Microspectrofluorometry can be done with a conventional fluorescence microscope.

8. Reference

1. KATZ, I.J. & F.T. BREZENSKI. 1973. Detection of *Salmonella* by fluorescent antibody. U.S. Environmental Protection Agency, Edison, N.J.

9. Bibliography

SCHULTE, S.J., J.S. WITZEMAN & W.M. HALL. 1968. Immunofluorescent screening for *Salmonella* in foods: comparison with culture methods. *J. Amer. Org. Agr. Chem.* 51:1334.

THOMASON, B.M. & J.G. WALLS. 1971. Preparation and testing of polyvalent conjugates for F.A. detection of Salmonellae. *Appl. Microbiol.* 22:876.

THOMASON, B.M. 1971. Rapid detection of *Salmonella* microcolonies by fluorescent antibody. *Appl. Microbiol.* 22:1064.

9260 D. Quantitative *Salmonella* Procedures

This procedure describes one approach for estimating *Salmonella* density in water samples. Other methods have been described in the literature and a comparative study is recommended to select the best quantitative method for any given application. The following procedure must be modified for use with solid or semisolid samples.

Because of the high ratio of coliform bacteria to pathogens, large samples (1 L or more) are required. Any concentration method in Section 9260B.1 may be used but preferably concentrate the sample by the membrane filter technique (Section 9260B.1*d*). After blending the membrane with 100 mL sterile 0.1% (w/v) peptone water, use a quantitative MPN procedure by proportioning homogenate into a five-tube, three-dilution multiple-tube procedure using either selenite cystine, selenite-F, or tetrathionate broth as the selective enrichment medium (See Section 9260B.3). Incubate for 24 h as specified or required for the enrichment medium used and streak from each tube to plates of brilliant green and xylose lysine desoxycholate agars. Incubate for 24 h at 35°C. Select from each plate at least one, and preferably two to three, colonies suspected of being *Salmonella*, inoculate a slant each of triple sugar iron (TSI) and lysine iron (LIA) agars, and incubate for 24 h at 35°C. Test cultures giving a positive reaction for *Salmonella* by serological techniques (see Section 9260B.5). From the combination of *Salmonella* negative and positive tubes, calculate the MPN/1.0 L of original sample (see Section 9221E).

9260 E. *Shigella*

Shigellosis is an acute gastrointestinal disease of humans, caused by four species or serogroups of the genus *Shigella*, *S. dysentariae* (Group A), *S. flexneri* (Group B), *S. boydii* (Group C), and *S. sonnei* (Group D). Shigellae invade the intestinal mucosa, producing dysentery characterized by abdominal pain, fever, and diarrhea. The infectious dose for *Shigella* spp. is low, and most cases result from person-to-person transmission. When outbreaks occur, they are usually associated with fecal contamination of foods and, less frequently, water. The shigellosis case rate has gradually risen in the U.S. over the past 30 years from 6 cases/100 000 population in 1965 to 12 cases/100 000 population in 1995.[1] In the U.S., *S. sonnei* (66.5%) is the most common cause of shigellosis, followed by *S. flexneri* (16.4%), *S. boydii* (1.1%), and *S. dysentariae* (0.5%). The serogroup is not reported for 15.5% of cases.

Shigellosis is most common among children. Outbreaks from direct transmission have been reported in schools, day-care centers, and institutions providing custodial care. Waterborne outbreaks are associated with fecal contamination together with inadequate chlorination of private or noncommunity water supplies, as the result of cross-connections between wastewater and potable water lines, and from exposure to fecally contaminated recreational waters.

Shigellae are sensitive to chlorination at normal levels, and they do not compete favorably with other organisms in the environment. Their survival time is measured in hours and days, and is a function of the extent of pollution, as well as physical conditions such as temperature and pH. Shigellae survive up to 4 d in river water. However, the time required to establish a laboratory diagnosis by culture of patient specimens (1 to 2 d) makes it improbable that shigellae can be recovered from an environmental source unless there is a continuous source of contamination such as wastewater seepage. Shigellae can survive in a viable but nonculturable state after 21 d.[2] The public health significance of nonculturable shigellae in the environment is unknown.

Methods for the reliable quantitative recovery of shigellae from the environment are not yet available. Culture of shigellae is usually either not attempted or unsuccessful. Methods that have resulted in isolation of *Shigella* include membrane filtration[3,4] and centrifugation[5,6] with or without subsequent broth enrichment. Recently, the polymerase chain reaction (PCR) has shown promise for detection of shigellae in environmental samples.[7-9]

1. Sampling and Storage

Collect a water sample in a sterile 1-L container. Collect soil, sediment, sludge, or other samples in plastic bags* or glass or plastic bottles. Hold samples at 2 to 8°C until they are processed. Process samples as soon as possible after collection.

2. Enrichment

Choose a selective enrichment medium to minimize accumulation of volatile acid by-products derived from growth of potentially antagonistic bacteria. Selenite F broth has been used successfully to recover shigellae from water and sand.[5,6] While GN broth facilitates better recovery of shigellae from stools than Selenite F broth, the only reported attempt to use GN broth as an enrichment for membrane filters for isolation of shigellae failed to recover the organism.[10]

Alternatively, use reduced-strength nutrient medium adjusted to pH 8.0 (0.15 g tryptic soy broth, added directly to the sample). During outbreak investigations, the enrichment medium may be made selective by incorporation of antibiotics to which the clinical isolates have shown resistance, such as tetracycline and streptomycin at concentrations of 150 μg/mL.[11]

3. Membrane Filter Procedure

This procedure is suitable for low-turbidity potable and surface waters with low concentrations of coliform bacteria. Filter 100-mL to 1-L samples through 0.45-μm pore size membranes and place filters face up on the surface of XLD or MacConkey agar plates; incubate plates at 35°C overnight. Where growth is confluent, sweep growth from plate and inoculate GN or Selenite F broth enrichments; incubate for 6 h and streak onto MacConkey and XLD plates for colony isolation. Pick colorless colonies (lactose nonfermenters) from membrane or plates to TSI and LIA

---

* WhirlPak™, Ziploc™, or equivalent.

TABLE 9260:II. REACTIONS OF COMMON BACTERIA ON TSI AND LIA MEDIA

Organism	TSI*	LIA*
Shigella	K/A−	K/A−
Salmonella	K/Ag+	K/A+
Escherichia	A/Ag−	K/K−
Proteus	A/Ag+ or K/Ag+	R/A+
Citrobacter	A/Ag+	K/A+
Enterobacter	A/Ag−	K/A−
Aeromonas	A/A−	K/A−
Yersinia	A/A− or K/A−	K/A−
Plesiomonas	K/A−	K/A−

* Fermentation reactions = slant/butt; H₂S production = + or −; K = alkaline, A = acid, R = red (deaminase reaction); g = gas produced.

slants; incubate overnight at 35°C. For biochemical reactions and serological grouping, see ¶ 5 below.

4. Centrifugation Procedure

This procedure is suitable for surface waters, wastewater, and sediments. Centrifuge 200- to 250-mL water samples at 1520 × g for 15 min and pour off all but last 2 mL of supernatant. Resuspend pellet and add 8 mL Selenite F or GN broth. Incubate suspension for 24 h at 35°C. Mix suspension and inoculate one loopful to each of several MacConkey and XLD plates. Streak plates for isolation and incubate overnight at 35°C. Examine plates for colorless colonies, and pick suspect colonies to TSI and LIA slants; incubate at 35°C overnight. For biochemical reactions and serological grouping, see ¶ 5 below.

For solid samples (sediments, soil, sludge, etc.) suspend 10 g sample in 100 mL Selenite F or GN broth and mix thoroughly. Incubate suspension overnight at 35°C. Resuspend sediment and streak one loopful onto each of several MacConkey and XLD agar plates; incubate overnight at 35°C. Pick colorless colonies to TSI and LIA slants, and proceed as above. For biochemical reactions and serological grouping, see ¶ 5 below.

5. Biochemical Identification and Serological Grouping

Examine the TSI and LIA slants for the reactions shown in Table 9260:II. Cultures that are presumptively identified as *Shigella* spp. are serogrouped by a slide agglutination test using polyvalent and group specific antisera. Refer cultures to a public health reference laboratory if molecular typing is desirable for outbreak-related strains.

6. References

1. CENTERS FOR DISEASE CONTROL AND PREVENTION. 1996. Summary of notifiable diseases, United States 1995. *Morbid. Mortal. Week. Rep.* 44:1.
2. COLWELL, R.R., P.R. BRAYTON, D.J. GRIMES, D.B. ROSZAK, S.A. HUQ & L.M. PALMER. 1985. Viable but non-culturable *Vibrio cholerae* and related pathogens in the environment: implications for release of genetically engineered microorganisms. *Bio/Technology* 3:817.
3. DANIELSSON, D. & G. LAURELL. 1968. A membrane filter method for the demonstration of bacteria by the fluorescent antibody technique. *Acta. Path. Microbiol. Scand.* 72:251.

4. LINDELL, S.S. & P. QUINN. 1973. *Shigella sonnei* isolated from well water. *Appl. Microbiol.* 26:424.

5. CODY, R.M. & R.G. TISCHER. 1965. Isolation and frequency of occurrence of *Salmonella* and *Shigella* in stabilization ponds. *J. Water Pollut. Control Fed.* 37:1399.

6. DABROWSKI, J. 1982. Isolation of the *Shigella* genus bacteria from the beach sand and water of the bay of Gdansk. *Biul. Inst. Med. Morskiej.* 33:49.

7. BEJ, A.K., J.L. DiCESARE, L. HAFF & R.M. ATLAS. 1991. Detection of *Escherichia coli* and *Shigella* spp. in water by using the polymerase chain reaction and gene probes for *uid*. *Appl. Environ. Microbiol.* 57:1013.

8. ISLAM, M.S., M.K. HASAN, M.A. MIAH, G.C. SUR, A. FELSENSTEIN, M. VENKATESAN, R.B. SACK & M.J. ALGERT. 1993. Use of the polymerase chain reaction and fluorescent-antibody methods for detecting viable but nonculturable *Shigella dysenteriae* Type 1 in laboratory microcosms. *Appl. Environ. Microbiol.* 59:536.

9. SETHABUTR, O., P. ECHEVERRIA, C.W. HOGE, L. BODHIDATTA & C. PITARANGSI. 1994. Detection of *Shigella* and enteroinvasive *Escherichia coli* by PCR in the stools of patients with dysentery in Thailand. *J. Diarrh. Dis. Res.* 12:265.

10. MAKINTUBEE, S., J. MALLONEE & G. ISTRE. 1987. Shigellosis outbreak associated with swimming. *Amer. J. Pub. Health* 77:166.

11. ROSENBERG, M.L., K.K. HAZLET, J. SCHAEFER, J.G. WELLS & R.C. PRUNEDA. 1976. Shigellosis from swimming. *J. Amer. Water Works Assoc.* 236:1849.

9260 F.   Pathogenic *Escherichia coli*

Escherichia coli is a normal inhabitant of the human digestive tract; however, some *E. coli* cause diarrheal diseases in humans.[1] These pathogenic *E. coli* are classed into five groups: enterotoxigenic (ETEC), enterohemorrhagic (EHEC), enteroinvasive (EIEC), enteropathogenic (EPEC), and the newly recognized group called enteroadherent-aggregative *E. coli* (EA-AggEC) for its aggregative or "stacked-brick"-like adherence to cultured mammalian cells.[2] Pathogenic *E. coli* can be grouped on the basis of serology but, because they are classed on the basis of distinct pathogenic factors, definitive identification requires the determination of the characteristic virulence properties associated with each group. These include: plasmid-mediated cell invasion, plasmid-mediated colonization and enteroadherence factors, production of several potent cytotoxins, hemolysins, as well as heat-labile and stable enterotoxins.[3]

Although pathogenic *E. coli* have most often been implicated in foodborne illness, several major waterborne outbreaks have been reported.[4] Outbreaks have involved both water supplies[5–7] and recreational waters.[8,9] Some *E. coli* pathogens have a low infectious dose.

1. Examination Procedures

The pathogenic *E. coli* groups are phenotypically diverse; hence, no standard microbiological methods have been developed for these pathogens. Unlike typical *E. coli*, some pathogenic groups like EIEC do not ferment lactose[3]; hence, coliform methods based on lactose fermentation are not suitable for detection of EIEC. Also, many fecal coliform confirmation or enrichment procedures use elevated incubation temperature, which is inhibitory to the growth of EHEC.[10] Elevated temperatures and sodium lauryl sulfate used in lauryl tryptose broth (LTB) for MPN analysis also have been found to cause the loss of plasmid, which encodes many of the virulence-associated factors.[11]

Pathogenic *E. coli* that ferment lactose and are not affected by elevated temperatures still can be presumptively distinguished from non-*E. coli* by the MPN fecal coliform procedure (9221E) or the fecal coliform membrane filter method (9222D) followed by serotyping and virulence analysis. These methods, as well as methods from other sources,[12] also have been modified to detect specific pathogenic groups. Regardless of the method, however, when testing for pathogenic *E. coli*, first identify isolates as *E. coli* either by conventional biochemical testing or by using commercially available biochemical identification kits (see Section 9260B.4) before serotyping and assaying for the virulence factors associated with the respective pathogenic groups.

a. EHEC O157:H7: The following procedure is a modification of the standard total coliform fermentation technique (9221B) for detecting *E. coli* O157:H7 in water.[13] Inoculate a 100-mL sample into 50 mL $3 \times$ lauryl tryptose broth (LTB) and incubate at 35°C for 24 h. Serially dilute the sample, spread plate (0.1 mL) onto sorbitol MacConkey agar (SMAC)* and incubate at 35°C for 18 to 24 h. EHEC O157:H7 form colorless colonies because they do not ferment, or are slow fermenters of, sorbitol. Pick ten sorbitol-negative colonies, transfer individually into LTB-MUG (4-methylumbelliferone glucuronide; 0.1 g/L)[14] and incubate at 35°C for 18 to 24 h. EHEC O157:H7 ferment lactose, but do not have β-glucuronidase activity to hydrolyze MUG, so cultures will appear gas-positive and nonfluorescent. Assay these for positive glutamate decarboxylase activity,[13] then identify biochemically as *E. coli*.

Larger volumes of sample also may be examined by the following procedure modified from a procedure for detecting O157:H7 in food.[15] This procedure has not been tested for use in water analysis; however, it has been used extensively to detect O157:H7 bacteria in apple juice. Centrifuge 200 mL sample at 10 000 × *g* for 10 min. Resuspend pellet in 225 mL EHEC enrichment broth (EEB) and incubate at 35°C for 6 h. Spread plate 0.1 mL from EEB and a 1:10 dilution of EEB onto tellurite cefixime SMAC (TC SMAC). Both EEB and TC SMAC contain antibiotics to reduce growth of normal flora bacteria; therefore, they are best suited for highly contaminated samples. Incubate EEB sample and TC SMAC at 35°C for 18 to 24 h. Observe TC SMAC plates for isolated, colorless colonies. If none are evident, serially dilute the overnight EEB sample and replate onto TC SMAC. Test colorless colonies for positive indole reaction and identify biochemically as *E. coli* before serotyping and virulence analysis for the Shiga toxin genes.

b. EPEC, ETEC, EIEC: With the exception of EIEC, use either the MPN fecal coliform procedure (9221E) or the fecal coliform membrane filter method (9222D) for presumptive isolation of these pathogenic *E. coli* groups from water. Alternatively plate

---

* Oxoid USA, Columbia, MD; Difco, Detroit, MI.

presumptive positive samples onto selective media, such as LES Endo and MacConkey (MAC) agars (see Section 9221B.3, Completed Phase). In food analysis, L-EMB agar also has been used. For EIEC, which ferment lactose slowly or not at all, the MPN method is not useful; however, the membrane filter method (9222D) can be used. In food testing for EIEC, Hektoen agar (HE), *Salmonella-Shigella* (SS) agar and MAC are used for selective plating, but HE and MAC appear less inhibitory and are best suited for the isolation of EIEC[10]. In the analysis of each pathogenic *E. coli* group, preferably pick 10 typical (lactose-positive) and 10 atypical (lactose-negative) colonies for biochemical identification. Identify all isolates as *E. coli* before serological typing and analysis for the group-specific virulence factors.

2. Serotyping

For definitive identification, serotype for the O:H antigens any isolates presumptively identified as pathogenic *E. coli* by microbiological methods. Polyvalent antisera are available commercially, but only for the common serotypes. Several anti-O157 and anti-H7 latex agglutination kits are available for typing O157:H7 isolates. Serotype information also is essential for epidemiological investigations.

3. Virulence Analysis

The pathogenic potential of an *E. coli* isolate can be determined only by testing for its distinctive virulence properties. A simple antibody-bound latex agglutination kit and several enzyme linked immunosorbent assay kits are available for testing Shiga cytotoxins of EHEC†. An agglutination kit also is available for testing labile and stable enterotoxins of ETEC,‡ but analysis of other virulence factors may require bioassays using animal models, tissue cultures, or other antibody and nucleic-acid-based molecular methods. A partial listing of commercially available assays and media for pathogenic *E. coli* is available.[12] Most of the assays are specific for EHEC O157:H7 and introduced only recently for food analysis; hence, few have been evaluated by collaborative studies.

4. References

1. ORSKOV, F. & ORSKOV, I. 1992. *Escherichia coli* serotyping and disease in man and animals. *Can. J. Microbiol.* 38:699.

---

† VEROTEST, MicroCarb; Premier EHEC, Meridian; Verotox-F, Denka Seiken.
‡ VET-RPLA, Unipath.

2. VIAL, P.A., R. ROBINS-BROWNE, H. LIOR, V. PRADO, J.B. KAPER, J.P. NATARO, D. MENEVAL, A.-E.-D. ELSAYED & M.M. LEVINE. 1988. Characterization of enteroadherent-aggregative *Escherichia coli*, a putative agent of diarrheal disease. *J. Infect. Dis.* 158:70.
3. LEVINE, M.M. 1987. *Escherichia coli* that cause diarrhea: enterotoxigenic, enteropathogenic, enteroinvasive, enterohemorrhagic and enteroadherent. *J. Infect. Dis.* 155:377.
4. FENG, P. 1995. *Escherichia coli* serotype O157:H7: novel vehicles of infection and emergence of phenotypic variants. *Emerging Infec. Dis.* 2:47.
5. SCHROEDER, S.A., J.R. CALDWELL, T.M. VERNON, P.C. WHITE, S.I. GRANGER & J.V. BENNETT. 1968. A waterborne outbreak of gastroenteritis in adults associated with *Escherichia coli*. *Lancet* 1:737.
6. ROSENBERG, M.L., J.P. KOPLAN, I.K. WACHSMUTH, J.G. WELLS, E.J. GANGAROSA, R.L. GUERRANT & D.A. SACK. 1977. Epidemic diarrhea at Crater Lake from enterotoxigenic *Escherichia coli*. *Ann. Intern. Med.* 86:714.
7. SWERDLOW, D.L., B.A. WOODRUFF, R.C. BRADY, P.M. GRIFFIN, S. TIPPEN, H.D. DONNELL, E. GELDREICH, B.J. PAYNE, A. MEYER, J.G. WELLS, K.D. GREENE, M. BRIGHT, N.H. BEAN & P.A. BLAKE. 1992. A waterborne outbreak in Missouri of *Escherichia coli* O157:H7 associated with bloody diarrhea and death. *Ann. Intern. Med.* 117:812.
8. KEENE, W.E., J.M. MCANULTY, F.C. HOESLY, L.P. WILLIAMS, K. HEDBERG, G.L. OXMAN, T.J. BARRETT, M.A. PFALLER & D.W. FLEMING. 1994. A swimming-associated outbreak of hemorrhagic colitis caused by *Escherichia coli* O157:H7 and *Shigella sonnei*. *N. England J. Med.* 331:579.
9. BREWSTER, D.H., M.I. BROWNE, D. ROBERTSON, G.L. HOUGHTON, J. BIMSON & J.C.M. SHARP. 1994. An outbreak of *Escherichia coli* O157 associated with a children's paddling pool. *Epidemiol. Infect.* 112:441.
10. DOYLE, M.P. & V.V. PADHYE. 1989. *Escherichia coli*. *In* M.P. Doyle, ed. Foodborne Bacterial Pathogens. Marcel Dekker, Inc., N.Y.
11. HILL, W.E. & C.L. CARLISLE. 1981. Loss of plasmids during enrichment for *Escherichia coli*. *Appl. Environ. Microbiol.* 41:1046.
12. U.S. FOOD AND DRUG ADMINISTRATION. 1995. Bacteriological Analytical Manual, 8th ed. Assoc. Official Analytical Chemists International, Gaithersburg, Md.
13. RICE, E.W., C.H. JOHNSON & D.J. REASONER. 1996. Detection of *Escherichia coli* O157:H7 in water from coliform enrichment cultures. *Lett. Appl. Microbiol.* 23:179.
14. FENG, P. & P.A. HARTMAN. 1982. Fluorogenic assay for immediate confirmation of *Escherichia coli*. *Appl. Environ. Microbiol.* 43:1320.
15. HITCHINS, A.D., P. FENG, W.D. WATKINS, S.R. RIPPEY & L.A. CHANDLER. 1995. *Escherichia coli* and the coliform bacteria. *In* Bacteriological Analytical Manual, 8th ed. Assoc. Official Analytical Chemists International, Gaithersburg, Md.

9260 G.   *Campylobacter jejuni*

Campylobacters are commonly found in the normal gastrointestinal and genitourinary flora of wild animals, birds, and domestic animals including sheep, cattle, swine, goats, and chickens.[1] *Campylobacter* infections often are acquired by the fecal oral route, often as zoonoses through exposure to infected animals. Large outbreaks have resulted from contaminated milk, uncooked meat or fowl, and contaminated water systems.[2] *Campylobacter* has been reported to be the most common cause of bacterial enteritis worldwide.[3]

Waterborne transmission of *Campylobacter* has resulted from drinking untreated surface water, contamination of groundwater with surface water, faulty disinfection, and contamination by wild bird feces.[4] In remote mountain areas, the infection has been associated with drinking surface water from cold mountain streams.[5] Occurrence of campylobacters in surface water is variable and appears to be seasonally dependent, with lowest levels occurring in summer. Survival in surface water is affected by both temperature and sunlight.[6] Between 1978 and 1986, 57 outbreaks of campylobacteriosis were reported, including 11 waterborne outbreaks, 7 of which occurred in community water supplies.

1. Water Collection and Filtration Method

Collect large-volume water samples in sterile 10-L plastic containers. Process samples immediately after collection or store at 4°C and process as soon as possible. Filter one to several liters of the water through a 0.45- or 0.22-µm-pore-size, 47-mm-diam, cellulose nitrate membrane filter. Remove filter and place face down on selective medium (see isolation section). Incubate microaerophilically at 42°C for 24 h. Remove filter from the plate and place it face down on another selective plate. Incubate both plates at 42°C for up to 5 d.[7]

For turbid water pre-filtration is necessary. Use a stainless steel filtration device with a 1.5-L reservoir.* Assemble with the following filter sequence: Place a 142-mm, 3.0-µm filter on the screen inside reservoir with a 124-mm prefilter on top. In the bottom tubing adapter place a 47-mm, 1.2-µm filter. Then place Swinnex filter holders in parallel with a 47-mm, 0.65-µm filter in the upstream filter holder and a 47-mm, 0.45-µm filter in the downstream holder. Add 1 L sample to the reservoir, seal, and apply pressure of about 350 kPa. After filtration, remove the 0.45-µm pore-size filter and culture on selective plate medium as described above.

2. Isolation

Campylobacter isolation requires use of selective media containing antimicrobial agents, microaerophilic atmosphere (5% O_2, 10% CO_2, and 85% N_2), and 42°C incubation temperature, to suppress the growth of most common bacteria.[8] The thermophilic campylobacters (*C. jejuni, C. coli, C. lari*, and *C. upsaliensis*) grow well at 42°C. However, other campylobacters (*C. jejuni* subsp. *doylei* and *C. fetus*) do not grow well at 42°C; incubate plates at both 37°C and 42°C for optimal isolation of these bacteria.[9] Microaerophilic conditions can be provided by using commercially available systems and equipment.†

Several selective media for plating campylobacters are commercially available. Skirrow's medium contains blood agar base with lysed horse blood, trimethoprim, vancomycin, and polymixin B. Campy-BAP contains *Brucella* agar base with sheep blood, trimethoprim, vancomycin, polymixin B, amphotericin B, and cephalothin (to which some campylobacters are sensitive). Butzler's medium contains thioglycollate agar with sheep blood, bacitracin, novobiocin, cycloheximide, and cefazolin. Preston's medium contains *Campylobacter* agar base with horse blood, cycloheximide, rifampicin, trimethoprim, and polymyxin B. Other media, such as *Campylobacter* blood-free selective medium and *Campylobacter* charcoal differential agar, can be used to isolate campylobacters.[10] Use of enrichment broth will improve recovery of campylobacters.

Several enrichment media, such as *Campylobacter* broth, Campy-thio broth, Gifu anaerobe-modified semisolid medium, and Preston medium, are used to enhance recovery of campylobacters.[9] Add 10 mL water sample to 10 mL *Campylobacter* enrichment broth tubes in duplicate, and incubate cultures at 37°C

and 42°C for 8 h or overnight. Pre-enrichment of water sample in a selective enrichment broth for 4 h at 37°C may be important for recovery of stressed cells of *C. jejuni* that show less tolerance to elevated growth temperatures. For pre-enrichment of water sample, add 10 mL water to 10 mL enrichment medium and incubate culture for 4 h at 37°C, then transfer the cultures to another incubator at 42°C for overnight incubation.[11,12]

C. jejuni may be induced to a nonculturable state in water, and it is not clear whether pre-enrichment or enrichment will facilitate isolation of these bacteria.[13] Use of a decreased substrate concentration enhances metabolic activity in nonculturable campylobacters from water.[14]

3. Identification

a. Culture examination: Examine *Campylobacter* plates at 24 and 48 h for characteristic colonies, which can range from flat, spreading colonies that cover the entire surface of the plate, to very small, convex, translucent colonies. Colony colors range from gray to yellowish or pinkish.

b. Microscopy identification: *Campylobacter* spp. do not stain well by the conventional Gram stain. If safranin is used as a counterstain, apply it for 2 to 3 min; carbol fuchsin is a better alternative. Even 24-h cultures of campylobacters appear pleomorphic in stained smears, and cells range from small Gram-negative rods and coccoid forms to longer rods that may show an "S" or seagull shape, and long spirals, particularly from older cultures.[15]

c. Motility test: *Campylobacter* normally are motile by a single polar flagellum at one or both ends. Suspend cells in Mueller-Hinton or nutrient broth, and observe motility using phase microscopy or brightfield microscopy with reduced illumination. Do not use saline or distilled water because they may inhibit motility.[8] Young cells are 0.2 to 0.8 µm wide by 1.5 to 5 µm long, curved or spiral, and motile with darting or corkscrew-like motion.[16]

d. Biochemical tests: Despite numerous studies, campylobacters remain relatively difficult to rapidly identify, classify, and type biochemically.[17] Campylobacters do not ferment or oxidize carbohydrates, and they are inert in most biochemical media used to characterize bacterial isolates.[18] Although no standard methods for the characterization of campylobacters have been published, oxidase, catalase, nitrite and nitrate reduction, H_2S production, hippurate hydrolysis, resistance to various agents, temperature tolerances, and growth requirements are among the common phenotypic tests used to characterize campylobacters.[3]

4. Serological Identification Tests

Commercially available kits‡ for serotyping campylobacters are available. These kits use latex particles coated with polyvalent immunoglobulins for several *Campylobacter* species. They are designed for rapid presumptive identification of the thermophilic, enteropathogenic *Campylobacter* species (*C. jejunei, C. coli*, and *C. lari*); use in accordance with manufacturer's instructions.[19]

Other techniques that are not widely available in all laboratories include lectin agglutination, cellular fatty acid profiles, nucleic acid probes, polymerase chain reaction, and other genomic meth-

---

* Millipore No. 316 or equivalent.
† Campy Pak II, BioBag Environmental Chamber or BioBag Type Cfj, Becton Dickenson; Gas Generating Kit System BR56 or Campy Gen, Oxoid; Poly Bag System, Fisher Scientific; or equivalents.

---

‡ Such as Campyslide, BBL Microbiology Systems; Meritec-Campy, Meridian Diagnostics; and Microscreen, Mercia Diagnostics.

ods that can be used in reference and research laboratories for detection and identification of campylobacters.[3]

5. References

1. RYAN, K.J. 1990. *Vibrio* and *Campylobacter*. *In* J.C. Sherris, ed. Medical Microbiology: An Introduction to Infectious Diseases. Elsevier, New York, N.Y.
2. BARON, E.J., R.S. CHANG, D.H. HOWARD, J.N. MILLER & J.A. TURNER, eds. 1994. Medical Microbiology: A Short Course. Wiley-Liss, New York, N.Y.
3. ON, S.L.W. 1996. Identification methods for campylobacters, helicobacters, and related organisms. *Clin. Microbiol. Rev.* 9:405.
4. TAUXE, R.V. 1992. Epidemiology of *Campylobacter jejuni* infections in the United States and other industrialized nations. *In* I. Nachamkin, M.J. Blaser & L.S. Tompkins, eds. *Campylobacter jejuni*: Current Status and Future Trends. American Soc. Microbiology, Washington, D.C.
5. TAYLOR, D.N., K.T. MCDERMOTT, J.R. LITTLE, J.G. WELLS & M.J. BLASER. 1983. *Campylobacter* enteritis from untreated water in the Rocky Mountains. *Ann. Intern. Med.* 99:38.
6. VOGT, R.L., H.E. SOURS, T. BARRETT, R.A. FELDMAN, R.J. DICKINSON & L. WITHERELL. 1982. *Campylobacter* enteritis associated with contaminated water. *Ann. Intern. Med.* 96:292.
7. PEARSON, A.D., M. GREENWOOD, T.D. HEALING, D. ROLLINS, M. SHAHAMAT, J. DONALSON & R.R. COLWELL. 1993. Colonization of broiler chickens by waterborne *Campylobacter jejuni*. *Appl. Environ. Microbiol.* 59:987.
8. ISENBERG, H.D., ed. 1992. Clinical Microbiology Procedures Handbook. Vol. 1. American Soc. Microbiology, Washington, D.C.
9. GOOSSENS, H. & J.P. BUTZLER. 1992. Isolation and identification of *Campylobacter* spp. *In* I. Nachamkin, M.J. Blaser & L.S. Tompkins, eds. *Campylobacter jejuni*: Current Status and Future Trends. American Soc. Microbiology, Washington, D.C.
10. PARKS, L.C., ed. 1993. Handbook of Microbiological Media. CRC Press, Boca Raton, Fla.
11. HUMPHREY, T.J. 1989. An appraisal of the efficacy of preenrichment for the isolation of *Campylobacter jejuni* from water and food. *J. Appl. Bacteriol.* 66:119.
12. HUMPHREY, T.J. 1986. Techniques for the optimum recovery of cold injured *Campylobacter jejuni* from milk or water. *J. Appl. Bacteriol.* 61:125.
13. ROLLINS, D.M. & R.R. COLWELL. 1986. Viable but nonculturable stage of *Campylobacter jejuni* and its role in survival in the natural aquatic environment. *Appl. Environ. Microbiol.* 52:531.
14. ROLLINS, D.M. 1987. Characterization of Growth, Decline, and the Viable but Nonculturable State of *Campylobacter jejuni*. Ph.D dissertation, Univ. Maryland, College Park.
15. KAPLAN, R.L. & A.S. WEISSFELD. 1994. *Campylobacter, Helicobacter* and related organisms. *In* B.J. Howard et al., eds. Clinical and Pathogenic Microbiology, 2nd ed. Mosby, St. Louis, Mo.
16. BEUCHAT, L.R. 1986. Methods for detecting and enumerating *Campylobacter jejuni* and *Campylobacter coli* in poultry. *Poultry Sci.* 65:2192.
17. DUBREUIL, J.D., M. KOSTRZYNSKA, S.M. LOGAN, L.A. HARRIS, J.W. AUSTIN & T.J. TRUST. 1990. Purification, characterization, and localization of a protein antigen shared by thermophilic campylobacters. *J. Clin. Microbiol.* 28:1321.
18. CARDARELLI-LEITE, P., K. BLOM, C.M. PATTON, M.A. NICHOLSON, A.G. STEIGERWALT, S.B. HUNTER, D.J. BRENNER, T.J. BARRETT & B. SWAMINATHAN. 1996. Rapid identification of *Campylobacter* species by restriction fragment length polymorphism analysis of a PCR-amplified fragment of the gene coding for 16S-rRNA. *J. Clin. Microbiol.* 34:62.
19. HODINKA, R.L. & P.H. GILLIGAN. 1988. Evaluation of the Campyslide agglutination test for confirmatory identification of selected *Campylobacter* species. *J. Clin. Microbiol.* 26:47.

9260 H. *Vibrio cholerae*

Vibrio cholerae is the causative agent of cholera, a waterborne illness with symptoms ranging from mild to severe and potentially fatal diarrheal disease.[1,2] This is a well-defined species on the basis of biochemical tests and DNA studies, but the serotypes within the species can be quite diverse in their ability to produce infection. The O1 serogroup is associated with epidemic and pandemic cholera, especially in developing countries. The current (seventh) pandemic has affected over 100 countries, including the United States, with over one million reported cases and 10 000 deaths.[3] The newly identified O139 Bengal serogroup[4] also is capable of producing epidemic cholera. In contrast, the great majority of non-O1/non-O139 strains, which are more common in the environment, do not produce cholera toxin, and are not associated with epidemic cholera. However, these strains occasionally are associated with potentially fatal extra-intestinal infections. *V. cholerae* occurs as part of the normal microflora in estuarine areas, with non-O1/non-O139 strains being much more common than are O1 strains.

1. Concentration Techniques

Levels of *V. cholerae* in natural waters and sewage usually are quite low. Thus, methods of concentration or enrichment usually are employed. One method for isolating *V. cholerae* O1 from contaminated waters is placement of Moore swabs in flowing wastewater for periods up to 1 week, followed by placement into enrichment media at a 1:1 (weight/volume) ratio.[5]

2. Enrichment Procedures

Samples are enriched in alkaline peptone broth (1% peptone, 1% NaCl, pH 8.4), using appropriate concentration of broth relative to sample volume. Incubate enrichment cultures for 6 to 8 h at 35°C, then streak a loopful of the enrichment broth onto thiosulfate-citrate-bile salts-sucrose (TCBS) agar and incubate these plates at 35°C for 18 to 24 h.[6] Other enrichment and plating media have been reviewed.[7,8]

3. Selective Growth

Suspected *V. cholerae* colonies appear yellow, a result of sucrose fermentation. A variety of other sucrose-fermenting vibrios also appear on TCBS, however, including *V. fluvialis, V. furnissii, V. alginolyticus, V. metschnikovii, V. cincinnatiensis,* and *V. carchariae*.[2]

4. Presumptive Tests to Differentiate *V. cholerae*

The following key tests are used to identify *V. cholerae*:

Test	Reaction
Gram-negative rod	+
Cytochrome oxidase	+
Glucose fermented (no gas)	+
Growth in nutrient broth:	
No NaCl added	+
8% NaCl added	−
Arginine dihydrolase	−
Ornithine decarboxylase	+
ONPG hydrolysis	+

After isolation on TCBS, streak presumptive *V. cholerae* isolates to a nonselective medium, such as trypticase soy agar, containing a minimum of 0.5% NaCl.

5. Classification of Isolates as *V. cholerae*

The tests listed below may be used for a more extensive phenotypic characterization of *V. cholerae*.[7] To determine the serogroup, use agglutination assays.

Test	Reaction
ONPG	+
Nitrate reduction	+
Indole	+
0/129 sensitivity:	
10 mg	+
150 mg	+
Swarming	−
Luminescence	v*
Thornley's arginine dihydrolase	−
Lysine decarboxylase	+
Ornithine decarboxylase	+
Growth at 42°C	+
Growth at % NaCl:	
0%	+
3%	+
6%	v
8%	−
10%	−
Voges-Proskauer reaction	v
Gas from glucose fermentation	−
Fermentation to acid:	
L-Arabinose	−
m-Inositol	−
D-Mannose	v
Sucrose	+
Enzyme production:	
Alginase	−
Amylase	+
Chitinase	+
Gelatinase	+
Lipase	+

Test	Reaction
Utilization as sole source of carbon:	
γ-Aminobutyrate	−
Cellobiose	−
L-Citruline	−
Ethanol	−
D-Gluconate	+
D-Glucuronate	−
L-Leucine	−
Putrescine	−
Sucrose	+
D-Xylose	−

* v = variable, differs for strains within the species.

6. Serological Identification

Slide agglutination with polyvalent antisera can be used to identify the serogroups of *V. cholerae*. Polyvalent antiserum for *V. cholerae* O1 is available commercially.* The O1 serogroup can be further divided into two primary serotypes, Ogawa and Inaba.

7. Biotypes of Serogroup O1 *V. cholerae*

V. cholerae can be divided into two biotypes or biovars, classical and El Tor, which differ in several characteristics. The El Tor biotype currently is the most important biotype.

Test[7]	Biovar	
	Classical	El Tor
Hemolysis of sheep erythrocytes	−	v*
Voges-Proskauer reaction	−	+
Chicken erythrocyte agglutination	−	+
Antibiotic sensitivity:		
Polymyxin B (50 IU)	+	−
Bacteriophage susceptibility:		
Mukerjee classical phage IV	Lysis	No lysis
Mukerjee El Tor phage 5	No lysis	Lysis

* v = different reaction within the serovar.

8. Other Procedures

Environmental samples also may be examined by fluorescent-antibody techniques, but the number of *V. cholerae* cells in aquatic samples is generally quite low.[7] Nucleic acid probes are not routinely used for the identification of *V. cholerae*, although DNA probes are extremely useful in determining which strains of this species contain the cholera toxin gene.[2] This distinction is especially important in examining environmental isolates of *V. cholerae* because the great majority of these strains lack the cholera toxin gene.

9. References

1. KAPER, J.B., J.G. MORRIS, JR. & M.M. LEVINE. 1995. Cholera. *Clin. Microbiol. Rev.* 8:48.

---

* Difco or equivalent.

2. OLIVER, J.D. & J.B. KAPER. 1997. *Vibrio* species. *In* M.P. Doyle, L.R. Beuchat & T.J. Montville, eds. Fundamentals of Food Microbiology. American Soc. Microbiology, Washington, D.C.
3. CENTERS FOR DISEASE CONTROL. 1995. Update: *Vibrio cholerae* O1—Western hemisphere, 1991–1994, and V. cholerae O139—Asia, 1994. *Morbid. Mortal. Week. Rep.* 44:215.
4. ALBERT, M.J. 1994. *Vibrio cholerae* O139 Bengal. *J. Clin. Microbiol.* 32:2345.
5. BARRETT, T.J., A. BLAKE, G.K. MORRIS, N.D. PUHR, H.B. BRADFORD & J.G. WELLS. 1980. Use of Moore swabs for isolating *Vibrio cholerae* from sewage. *J. Clin. Microbiol.* 11:385.
6. SPECK, M.L., ed. 1984. Compendium of Methods for the Microbiological Examination of Foods, 2nd ed. American Public Health Assoc., Washington, D.C.
7. WEST, P.A. & R.R. COLWELL. 1984. Identification and classification of *Vibrionaceae*—an overview. *In* R.R. Colwell, ed. Vibrios in the Environment. John Wiley & Sons, New York, N.Y.
8. KAYSNER, C.A. & W.E. HILL. 1994. Toxigenic *Vibrio cholerae* O1 in food and water. *In* I.K. Wachsmuth, P.A. Blake & O. Olsvik, eds. *Vibrio cholerae* and Cholera: Molecular to Global Perspectives. ASM Press, Washington, D.C.

9260 I.   *Leptospira*

Leptospira spp. are motile, aerobic spirochetes that require fatty acids for growth.[1] Serum or polysorbate enrichments must be incorporated into artificial media, and some pathogenic strains may require CO_2 upon initial isolation. Leptospires are divided into two groups, based on their pathogenicity and growth characteristics. The saprophytic leptospires are assigned to the Biflexa Complex, and the pathogenic leptospires make up the Interrogens Complex. Pathogenic strains have an optimal growth temperature of 28 to 30°C, and they grow over a pH range from 5.2 to 7.7. Saprophytic strains prefer a growth temperature between 5 and 10°C below pathogenic strains. Leptospires prefer alkaline conditions, and they persist longest in warm, moist environments protected from sunlight. Under favorable temperature and pH conditions, leptospires survive for 3 to 5 d in damp soil and up to 10 d in natural waters. They survive for 12 to 14 h in undiluted wastewater, up to 3 d in aerated wastewater, and up to 4 weeks in sterile tapwater at pH 7. Nonpathogenic leptospires are ubiquitous, and they have been isolated from municipal water supplies.[2] Generally, pathogenic leprospires require an animal host and do not survive and propagate in the environment.

Leptospirosis is a worldwide zoonotic disease of wild animals.[3] Reservoirs of leptospires in wildlife include deer, foxes, raccoons, skunks, opossums, muskrats, and rodents. Domestic animals harboring leptospires include horses, cattle, goats, pigs, and sheep. Dogs may become infected but cats are spared. Humans are incidental hosts.

Humans acquire leptospirosis (Weil's disease) directly from animals, and from occupational or recreational exposure to urine-contaminated water[4–6] or environmental surfaces. Swimming and other water sports,[7] travel to tropical areas with occupational or recreational exposure to surface waters,[8] and natural disasters that affect sewer systems and runoff[9,10] increase risk of the disease. Outbreaks of leptospirosis associated with drinking water are extremely unusual, and are invariably caused by contamination of domestic water reservoirs with urine of infected rodents.[11]

Leptospirosis ranges from mild nonspecific febrile illness to severe or fatal renal, hepatic, or meningeal disease.[12,13] Leptospires enter through imperfections in the skin, through mucous membranes, or by ingestion of contaminated water. Urine of infected animals and humans may contain 10^6 to 10^8 organisms/mL. Leptospires may be shed into the environment up to 3 months after clinical recovery from disease.

Diagnosis of disease in animals and humans usually is based upon serology, darkfield examination of urine sediments, examination of histopathological stains, or culture of the organism from urine or tissues. Recently polymerase chain reaction (PCR) methods have been introduced for diagnosis and typing of leptospires.

While leptospirosis remains relatively common in tropical regions of the world, only 40 to 120 cases/year have been reported in the U.S. over the past 30 years. Leptospirosis was dropped from the list of notifiable diseases in 1994.

Leptospires are recovered from environmental sources with great difficulty.[14–17] Because both saprophytic and pathogenic strains of leptospires may be recovered from environmental samples, their presence has no public health significance apart from an epidemiological context.

1. Sample Collection

Collect water samples of 100 mL to 1 L in sterile containers for transport to the laboratory at ambient temperature within 72 h of collection. Multiple samples from each sample site usually are required for successful isolation because finding leptospires in 10 to 20% of samples of surface waters receiving farm runoff is considered a high yield. Leptospires find their ecological niche at the interface between sediment and shallow water. Gently agitate the water to bring some of the sediment to the surface of shallow bodies of water to improve the probability of recovering organisms.[18] For soil samples, collect 10 to 20 g of soil in sterile bottles or plastic bags. Use a small, tightly sealed container to protect sample from drying. A small amount of sterile deionized water may be added to soil samples to prevent drying.

2. Sample Processing

Centrifuge a portion of a water sample at $5000 \times g$ for 10 min and examine sediment by darkfield microscopy for leptospires. Their presence indicates that conditions are favorable for leptospire survival, but does not differentiate saprophytic from pathogenic forms. In the laboratory, thoroughly mix soil samples with three volumes of sterile deionized water and let coarse particulate material settle by gravity. Process remaining suspension as a water sample. *Leptospira* can pass through 0.22-μm membrane filters (¶ *a* below); this ability has been exploited to separate them from other bacteria in environmental samples and in mixed cultures. Similarly, guinea pig inoculation (¶ *b* below) has been used as a biological filter for isolation of leptospires from contaminated samples.

a. Filtration method: Filter surface water samples through filter paper* to remove coarse debris before membrane filtration. Occasionally, samples may have to be passed through a series of prefilters of decreasing pore sizes (8-μm, 4-μm, 1-μm, 0.65-μm, and 0.45-μm) to prevent clogging of the final 0.22-μm filter.

b. Animal inoculation method: Filter water through a 0.45-μm membrane filter and inoculate 1 to 3 mL intraperitoneally into weanling guinea pigs. After 3 to 6 d, inject a small amount of sterile saline and withdraw fluid for darkfield examination. If leptospires are seen, perform a cardiac puncture to obtain blood for inoculation of culture media. If no leptospira are seen by darkfield examination, record rectal temperatures daily until a fever spike indicates infection, then repeat the darkfield examination of peritoneal fluid for leptospires. Exsanguinate guinea pigs at 4 weeks and save serum for serological tests. Culture blood, kidney, and brain of guinea pigs with serological evidence of infection. Details of the method are described elsewhere.[19]

3. Culture

Cultures of environmental samples usually will be contaminated with other bacteria unless the samples are filtered through a 0.22-μm membrane filter before inoculation. Filtration also may be used to isolate leptospires from mixed cultures, by direct filtration or another method.[20] Unless sample filtration is used in conjunction with selective media or animal inoculation, a culture contamination rate of 60 to 80% is not uncommon. The amount of sample cultured will depend on the amount of particulate material in the sample. Generally, culture sample volumes from a few drops to 3.5 mL.

a. Culture media: Pathogenic leptospires have been cultured in liquid, semisolid, and solid media, but not all pathogenic strains will grow on solid media. Optimal pH of culture media is 7.2 to 7.4 and optimal incubation temperature is 30°C. Leptospires are sensitive to detergents, so keep glassware free of detergent residues. When using serum enrichments in culture media, use serum free of antibody to leptospires. Bovine serum albumin shows manufacturer and lot variations; test new batches for their ability to support growth of leptospires.

Modifications of the Ellinghausen-McMullough formulation (EMJH) that incorporate bovine serum albumin fraction V and polysorbates are used as serum replacements.[21–24] EMJH base is available commercially. Neomycin is used in culture media at concentrations between 5 and 25 μg/mL to inhibit competing microflora, but it may be toxic to some strains.[25] 5-fluorouracil is used at 100 or 200 μg/mL in culture media, but it too is toxic for some strains, particularly at concentrations above 100 μg/mL.[26]

b. Culture methods:

1) Direct culture method—To recover leptospires from surface waters, place a few drops of water in EMJH liquid medium and incubate overnight at 30°C. Filter inoculated medium through a 0.22-μm membrane filter into a sterile tube and reincubate at 30°C for up to 6 weeks.

2) Dilution method—When samples may contain reasonable numbers of organisms in the presence of inhibitors or competing microflora, prepare 10-fold dilutions in duplicate, and inoculate

0.1 mL undiluted sample and each dilution into EMHJ medium. One tube of each pair may be made selective by addition of a single 30-μg neomycin antimicrobial susceptibility disk to the media before incubation. Incubate cultures at 20 to 30°C for up to 4 months.

3) Animal inoculation method—Add 1 to 2 drops of heart blood from infected guinea pigs to each of three to five tubes of EMJH medium. Incubate cultures at 20°C for up to 4 months.

c. Culture examination: Leptospires usually are detected in cultures of environmental samples within 7 to 14 d; however, incubate and examine cultures weekly for 6 weeks before discarding them as negative. Observe tubes for a lightly turbid ring of growth just below the surface of the medium. This band of maximum turbidity at the zone of optimal oxygen tension is referred to as Dinger's ring. Remove a drop of the culture weekly for darkfield examination and prepare subcultures if motile leptospires are observed. Generally, saprophytic leptospires grow at lower temperatures, and form rings closer to the surface of culture media than pathogenic serovars. Cultures remain viable in semisolid media for at least 8 weeks at room temperature.

4. Identification

Experience and skill are required to differentiate artifacts from leptospires by darkfield microscopy. The biochemical tests previously thought to differentiate between pathogenic and saprophytic serovars do not reliably predict pathogenicity of leptospires, and they are not recommended. Leptospira are identified to serogroup by the microscopic agglutination test using reference antisera. Identification to serovar requires use of adsorbed antisera that are available only in reference laboratories. Over 200 serotypes of *Leptospira* are known.

5. References

1. FAINE, S. 1992. The genus *Leptospira. In* A. Balows, H.G. Trüper, M. Dworkin, W. Harder & K.H. Schleifer, eds. The Prokaryotes, Vol. IV. Springer-Verlag, New York, N.Y.
2. HENRY, R.A. & R.C. JOHNSON. 1978. Distribution of the genus Leptospira in soil and water. *Appl. Envrion. Microbiol.* 35:492.
3. MICHNA, S.W. 1970. Leptospirosis. *Vet. Record* 86:484.
4. ANDERSON, D.C., D.S. FOLLAND, M.D. FOX, C.M. PATTON & A.F. KAUFMANN. 1978. Leptospirosis: a common-source outbreak due to leptospires of the grippotyphosa serogroup. *Amer. J. Epidemiol.* 107:538.
5. COGGINS, W.J. 1962. Leptospirosis due to *Leptospira pomona*: an outbreak of nine cases. *J. Amer. Med. Assoc.* 181:1077.
6. VENKATARAMAN, K.S. & S. NEDUNCHELLIYAN. 1992. Epidemiology of an outbreak of leptospirosis in man and dog. *Comp. Immun. Microbiol. Infect. Dis.* 15:243.
7. SHAW, R.D. 1992. Kayaking as a risk factor for leptospirosis. *Missouri Med.* 89:354.
8. VAN CREVEL, R., P. SPEELMAN, C. GRAVEKAMP & W.J. TERPSTRA. 1994. Leptospirosis in travelers. *Clin. Infect. Dis.* 19:132.
9. FUORTES, L. & M. NETTLEMAN. 1994. Leptospirosis: a consequence of the Iowa flood. *Iowa Med.* 84:449.
10. KAT, A.R., S. MANEA & D.M. SASAKI. 1991. Leptospirosis on Kauai: investigation of a common source waterborne outbreak. *Amer. J. Pub. Health* 81:1310.
11. CACCIAPUOTI, B., L. CICERONI, C. MAFFEI, F. DI STANISLAO, P. STRUSI, L. CALEGARI, R. LUPIDI, G. SCALISE, G. CAGNONI & G. RENGA. 1987. A waterborne outbreak of leptospirosis. *Amer. J .Epidemiol.* 126:535.

---

* Whatman No. 1 or equivalent.

12. HEATH, C.W., A.D. ALEXANDER & M.M. GALTON. 1965. Leptospirosis in the United States (concluded). Analysis of 483 cases in man, 1949–1961. *N. England J. Med.* 272:915.

13. HEATH, C.W., A.D. ALEXANDER & M.M. GALTON. 1965. Leptospirosis in the United States. Analysis of 483 cases in men, 1949–1961. *N. England J. Med.* 273:857.

14. ALEXANDER, A.D., H.G. STOENNER, G.E. WOOD & R.J. BYRNE. 1962. A new pathogenic *Leptospira*, not readily cultivated. *J. Bacteriol.* 83:754.

15. BAKER, M.F. & H.J. BAKER. 1970. Pathogenic *Leptospira* in Malaysian surface waters I. A method of survey for Leptospira in natural waters and soils. *Amer. J. Trop. Med. Hyg.* 19:485.

16. DIESCH, S.L. & W.F. McCULLOCH. 1966. Isolation of pathogenic leptospires from water used for recreation. *Pub. Health Rep.* 81:299.

17. GILLESPIE, W.H., S.G. KENZY, L.M. RINGEN & F.K. BRACKEN. 1957. Studies on bovine leptospirosis. III. Isolation of *Leptospira pomona* from surface water. *Amer. J. Vet. Res.* 18:76.

18. BRAUN, J.L., S.L. DIESCH & W.F. McCULLOCH. 1968. A method for isolating leptospires from natural surface waters. *Can. J. Microbiol.* 14:1011.

19. FAINE, S. 1982. Guidelines for the control of leptospirosis. WHO offset publ. No. 67. World Health Organization, Geneva, Switzerland.

20. SMIBERT, R.M. 1965. A technique for the isolation of leptospirae from contaminating microorganisms. *Can. J. Microbiol.* 11:743.

21. ELLINGHAUSEN, H.C., JR. & W.G. McCULLOUGH. 1965. Nutrition of *Leptospira pomona* and growth of 13 other serotypes: a serum-free medium employing oleic albumin complex. *Amer. J. Vet. Res.* 26:39.

22. ELLINGHAUSEN, H.C., JR. & W.G. McCULLOUGH. 1965. Nutrition of *Leptospira pomona* and growth of 13 other serotypes: fraction of oleic albumin complex and a medium of bovine albumin and polysorbate 80. *Amer. J. Vet. Res.* 26:45.

23. TURNER, L.H. 1970. Leptospirosis III. *Trans. Roy. Soc. Trop. Med. Hyg.* 64:623.

24. ADLER, B., S. FAINE, W.L. CHRISTOPHER & R.J. CHAPPEL. 1986. Development of an improved selective medium for isolation of leptospires from clinical material. *Vet. Microbiol.* 12:377.

25. MYERS, D.M. & V.M. VARELA-DÍAZ. 1973. Selective isolation of leptospiras from contaminated material by incorporation of neomycin to culture media. *Appl. Microbiol.* 25:781.

26. JOHNSON, R.C. & P. ROGERS. 1964. 5-fluorouracil as a selective agent for growth of leptospirae. *J. Bacteriol.* 87:422.

9260 J. *Legionella*

The Legionellaceae have been implicated in outbreaks of disease occurring since 1947.[1] Two forms of disease are recognized: a pneumonic form called Legionnaires' Disease and a nonpneumonic form called Pontiac fever. The first species was isolated following the historic outbreak associated with the Legionnaires' Convention in Philadelphia, Pa., in 1976. Epidemiological findings and animal studies have shown that the organism is transmitted via the airborne route[2] and is ubiquitous in moist environments. The reservoirs for most outbreaks have been either contaminated air conditioning cooling tower water or contaminated potable water distribution systems.[3,4] *Legionella* species also have been isolated in non-disease-related circumstances from a wide variety of aquatic environments such as lakes, streams, reservoirs, and sewage.[5,6] The organisms are able to survive for prolonged periods in laboratory distilled and tap water.[7]

The Legionellaceae are composed of a single genus, *Legionella,* and more than 35 different species.[8] The organisms are Gram-negative, aerobic, non-spore-forming bacteria. They are 0.5 to 0.7 μm wide and 2 to 20 μm long. They possess polar, subpolar, and/or lateral flagella. With the exception of *L. oakridgensis*, all require cysteine and iron salts for growth.

Although *Legionella* originally were isolated in guinea pigs and embryonated hen's eggs, it has been shown that plating directly on artificial media is more sensitive than animal inoculation for *L. pneumophila*.[9] The most widely used medium is an ACES (*N*-2-acetamideo-2-aminoethanesulfonic acid) buffered (pH 6.9) charcoal yeast extract (BCYE) agar supplemented with cysteine, ferric pyrophosphate, and optimally, alpha-ketoglutarate (BCYE-alpha).[10]

No one medium will be optimal for the recovery of *Legionella* from every environmental site; thus different selective media with various antibiotic combinations in a BCYE base may be necessary.[10-12] Also, pretreating samples with hydrochloric acid-potassium chloride, pH 2.2, is useful for eliminating non-*Legionella* organisms.[13] The two most commonly used selective media are GPVA medium (BCYE-alpha supplemented with glycine anisomycin, vancomycin, and polymyxin B) and CCVC medium (BCYE-alpha supplemented with polymyxin B, cephalothin, vancomycin, and cycloheximide). The GPVA medium is less inhibitory to some *Legionella* species. Use CCVC medium in combination with a less selective medium.

Recovery of legionellae from environmental water samples sometimes is difficult. Legionellae may take up to a week to grow on plate media, and even with acid pretreatment and the addition of antibiotics to the medium, faster-growing organisms may overgrow legionellae. In addition, other organisms, including *Pseudomonas* spp., secrete into surrounding media bacterial products that can inhibit *Legionella* growth.[14]

Rapid methods for detecting *Legionella* utilizing direct fluorescent antibody staining (DFA) or polymerase chain reaction technology (PCR) also are available and may be more sensitive than culture-based assays.[15-17] DFA can be quantitative, but may be subject to interference due to cross-reactivity with other organisms.[18] The PCR method is semiquantitative. Both DFA and PCR may not distinguish between viable and nonviable bacteria.

1. Sample Collection

Collect water samples from the littoral zone or from cooling towers, condenser coils, storage tanks, showers, water taps, etc. In most instances, a 1-L water sample is sufficient. Larger volumes of water (1 to 10 L)[6] may be needed in water having low bacterial counts. In addition to collecting water samples, it may be useful to swab various fixtures (e.g., shower heads) and plate directly on selective media. Transport samples to the laboratory in insulated containers. Refrigerate samples that cannot be processed immediately. Treat chlorinated water with sodium thiosulfate (see Section 9060A.2).

2. Immunofluorescence Procedure

Centrifuge 100 mL at 3500 × *g* for 30 min at room temperature and reconstitute the sedimented material in 6 to 10 mL filter-sterilized (0.2-mm filter) water from sample. Prepare smears for DFA by filling two 1.5-cm circles on a microscope slide with the concentrate. Air-dry sample smears, gently heat-fix, treat with 10% formalin for 10 min, rinse with phosphate-buffered saline (pH 7.6), and react with specific fluorescent antibodies.[6,19] The DFA procedure lacks specificity[20] and cannot determine viability. Some environmental bacteria (i.e., *Pseudomonas* spp. and *Xanthomonas-Flavobacterium* group) cross-react with the *Legionella* DFA reagents.

To determine whether organisms are viable, use secondary staining with a tetrazolium dye.[18] Confirm *Legionella* using direct isolation procedures.

3. Media and Reagents

a. Buffered charcoal yeast extract alpha base:[19]

Norit SG charcoal	2.0	g
Yeast extract	10.0	g
ACES buffer	10.0	g
Ferric pyrophosphate, soluble	0.25	g
L-cysteine, HCl·H$_2$O	0.4	g
Agar	17.0	g
Potassium alpha-ketoglutarate	1.0	g
Reagent-grade water	1.0	L

Dissolve yeast extract, agar, charcoal, glycine, and alphaketoglutarate in approximately 850 mL water; boil. Dissolve 10 g ACES buffer in 100 mL warm water, adjust pH to 6.9 with 1*N* KOH and add. Autoclave 15 min at 121°C. Cool to 50°C. Dissolve 0.4 g cysteine and 0.25 g ferric pyrophosphate in 10 mL of water each and filter sterilize separately (0.22 μm). After base has cooled, add cysteine, ferric pyrophosphate, and dyes in that order. Adjust pH to 6.9 with sterile 1*N* KOH and dispense.

b. GPVA medium:[11,12]*

Glycine	0.3	%
Polymyxin B	100	units/mL
Vancomycin	5	μg/mL
Anisomycin	80	μg/mL

To cooled BCYE-alpha base with glycine, add filter-sterilized antibiotics and mix. Adjust pH to 6.9 with sterile 1*N* KOH and dispense.

c. CCVC medium[11]†

Cephalothin	4	μg/ml
Colistin	16	μg/mL
Vancomycin	0.5	μg/mL
Cycloheximide	80	μg/mL

* Available commercially.
† This medium may not be available in dehydrated form and may require preparation from the basic ingredients.

To cooled BCYE-alpha base add filter-sterilized antibiotics and mix. Adjust to pH 6.9 with sterile 1*N* KOH and dispense.

d. Acid treatment reagent,[11] pH 2.0 (0.2*M* KCl/HCl):
Solution A—0.2*M* KCl (14.9 g/L in distilled water).
Solution B—0.2*M* HCl (16.7 mL/L 10*N* HCl in distilled water).
Mix 18 parts of Solution A with 1 part of Solution B. Check pH against a pH 2.0 standard buffer. Dispense into screw-cap tubes in 1.0-mL volumes and sterilize by autoclaving.

e. Alkaline neutralizer reagent[11] (0.1*N* KOH):
Stock solution—0.1*N* KOH (6.46 g/L in deionized water). Dilute 10.7 mL of stock solution with deionized water to 100 mL. Dispense into screw-cap tubes in convenient volumes and sterilize by autoclaving. The pH of *d* and *e* combined in equal volumes should be 6.9.

4. Sample Preparation

a. Low-bacterial-count water: Concentrate water that has a low total bacterial count either by filtration[11] or continuous-flow centrifugation.[21] Filter samples through sterile 47-mm filter funnel assemblies containing a 0.2-μm porosity polycarbonate filter.‡ After filtration, immediately remove the filter aseptically and place it in a 50-mL centrifuge tube or similar-size vessel containing 10 mL sterile tap water or phosphate buffer. If more than one filter is required to concentrate a sample, combine them.

b. High-bacterial-count water: Process water that has a high total bacterial count directly. Place 10 mL sample in a 50-mL centrifuge tube or similar-size vessel containing 10 mL of sterile tap water or phosphate buffer.

c. Sample dispersion: Disperse organisms from filter or aggregates by mixing with a vortex mixer (3 × 30 s).

d. Plating: Plate acid-treated and non-acid-treated samples on two types of BCYE: plain and selective with antibiotics.

1) No acid treatment—Inoculate three plates each of BCYE-alpha and selective BCYE-alpha (GPVA or CCVC) with 0.1 mL of suspension. Spread with a sterile smooth glass rod. Save remainder of specimen for acid treatment and store at 4°C.

2) Acid treatment—Place 1.0 mL of suspension in a sterile 13 × 100-mm screw-capped tube containing 1.0 mL acid treatment reagent and mix. Final pH of mixture should be approximately 2.2. Let stand for 15 min at room temperature, neutralize by adding 1.0 mL alkaline neutralizer reagent, and mix. Inoculate 0.1 mL onto three plates each of BCYE-alpha and selective BCYE-alpha (GPVA or CCVC) and spread with a sterile smooth glass rod.

3) Incubation—Incubate all plates at 35°C in a humidified atmosphere (>50%) for up to 10 d. A candle jar or humidified CO$_2$ incubator (2 to 5% CO$_2$) is acceptable.

e. Total bacterial count examination: Determine the adequacy of processing for each high-bacterial-count water. Some samples may require dilution, concentration, or animal inoculation. If the total count of the acid-treated sample exceeds 300 colonies on BCYE selective medium, make a further 10-fold dilution of the sample stored at 4°C. Repeat acid-treatment and plating.

If the total count of the non-acid-treated sample is less than 30 colonies on BCYE agar, concentrate and treat the collected water as previously described for low-bacterial-count water.

‡ Nuclepore Corp., 7035 Commerce Circle, Pleasanton, CA, or equivalent.

5. Examination of Cultures of Legionellae

With the aid of a dissecting microscope, examine all cultures daily after 48 h incubation for the presence of opaque bacterial colonies that have a "ground-glass" appearance. Place plates with *Legionella*-like colonies in a biological safety cabinet equipped with a burner, a bacteriological needle, and a loop. Aseptically pick each suspect colony onto BCYE-alpha agar and a BCYE agar plate prepared without L-cysteine. Streak the inoculated portion of each plate with a sterile loop to provide areas of heavy growth and incubate for 24 h.

Reincubate plates without growth an additional 24 h. Plates demonstrating growth on only BCYE-alpha agar are presumptive for *Legionella*. Confirm *Legionella* by slide agglutination or direct immunofluorescence. If these confirmatory techniques are not available, send subcultures of the presumptive legionellae to a reference laboratory for further identification. Because there are many serotypes in some species, especially *L. pneumophila*, investigation of environmental sites as possible reservoirs of epidemic-causing strains may be useful.[22] Effective investigatory techniques include monoclonal antibody subtyping, electrophoretic isoenzyme analysis, restriction endonuclease tests, and plasmid analysis.

6. Polymerase Chain Reaction Procedure

A test kit utilizing the polymerase chain reaction (PCR) is available commercially§ and has been used successfully in an epidemiological investigation of an outbreak of Pontiac fever.[23] Perform tests according to manufacturer's instructions. The kit provides sample processing reagents, PCR primers, detection strips, and positive and negative controls. Specific probes allow for the detection of twenty-five *Legionella* species as well as specific detection of *Legionella pneumophila*. The test is semiquantitative, based on a colorimetric comparison to control strips equivalent to 10^3 cells/mL.

7. References

1. McDade, J.E., C.C. Shepard, D.W. Frasier, T.R. Tsai, M.A. Redus, W.T. Dowdle & the Laboratory Investigation Team. 1977. Legionnaires's Disease: isolation of a bacterium and demonstration of its role in other respiratory disease. *N. England J. Med.* 297:1197.

2. Berendt, R.F., et al. 1980. Dose-response of guinea pigs experimentally infected with aerosols of *Legionella pneumophila*. *J. Infect. Dis.* 141:186.

3. Fliermans, C.B., W.B. Cherry, L.H. Orrison, S.J. Smith, D.L. Tison & D.H. Pope. 1981. Ecological distribution of *Legionella pneumophila*. *Appl. Environ. Microbiol.* 41:9.

4. Tobin, J.O.H., R.A. Swan & C.L.R. Bartlett. 1981. Isolation of *Legionella pneumophila* from water systems: methods and preliminary results. *Brit. Med. J.* 282:515.

§ Enviroamp PCR, Perkin-Elmer Roche, Alameda, CA.

5. Cherry, W.B., G.W. Gorman, L.H. Orrison, C.W. Moss, A.G. Steigerwalt, H.W. Wilkinson, S.E. Johnson, R.M. McKinney & D.J. Brenner. 1982. *Legionella jordanis:* a new species of *Legionella* isolated from water and sewage. *J. Clin. Microbiol.* 15:290.

6. Fliermans, C.B., W.B. Cherry, L.H. Orrison & L. Thacker. 1979. Isolation of *Legionella pneumophila* from nonepidemic related aquatic habitats. *Appl. Environ. Microbiol.* 37:1239.

7. Skaliy, P. & H.V. McEachern. 1979. Survival of the Legionnaires's Disease bacterium in water. *Ann. Intern. Med.* 90:662.

8. Brenner, D.J., A.G. Steigerwalt, G.W. Gorman, H.W. Wilkinson, W.F. Bibb, M. Hackel, R.L. Tyndall, J. Campbell, J.C. Feeley, W.L. Thacker, P. Skaliy, W.T. Martin, B.J. Brake, B.S. Fields, H.W. McEachern & L.K. Corcoran. 1985. Ten new species of *Legionella. Int. J. System. Bacteriol.* 35:50.

9. Feeley, J.C., R.J. Gibson, G.W. Gorman, N.C. Langford, J.K. Rasheed, D.C. Macel & W.B. Baine. 1979. Charcoal-yeast extract agar: primary isolation medium for *Legionella pneumophila. J. Clin. Microbiol.* 10:437.

10. Edelstein, P.H. 1982. Comparative studies of selective media for isolation of *Legionella pneumophila* from potable water. *J. Clin. Microbiol.* 16:697.

11. Gorman, G.W., J.M. Barbaree & J.C. Feeley. 1983. Procedures for the Recovery of *Legionella* from Water. Developmental Manual, Centers for Disease Control, Atlanta, Ga.

12. Wadowsky, R.M. & R.B. Yee. 1981. Glycine-containing selective medium for isolation of *Legionellaceae* from environmental specimens. *Appl. Environ. Microbiol.* 42:768.

13. Bopp, C.A., J.W. Sumner, G.K. Morris & J.G. Wells. 1981. Isolation of *Legionella* spp. from environmental water samples by low-pH treatment and use of selective medium. *J. Clin. Microbiol.* 13:714.

14. Paszko-Kolva, C., P.A. Hacker, M.A. Stewart & R.L. Wolfe. 1993. Inhibitory effect of heterotrophic bacteria on the cultivation of *Legionella dumoffi. In* J.M. Barbaree, R.F. Breiman & A.P. Dufour, eds. *Legionella*: Current Status and Emerging Perspectives. ASM Press, Washington, D.C.

15. Palmer, C.J., Y. Tsai, C. Paszko-Lolva, C. Mayer & L.R. San-Germano. 1993. Detection of *Legionella* species in sewage and ocean water by polymerase chain reaction, direct fluorescent-antibody, and plate culture methods. *Appl. Environ. Microbiol.* 59:3618.

16. Palmer, C.J., G.F. Bonilla, B. Roll, C. Paszko-Kolva, L.R. San-Germano & R.S. Fujioka. 1995. Detection of *Legionella* species in reclaimed water and air with the Enviroamp *Legionella* PCR kit and direct fluorescent antibody staining. *Appl. Environ. Microbiol.* 61:407.

17. Williams, H.N., C. Paszko-Kolva, M. Shahamat, C.J. Palmer, C. Pettis & J. Kelley. 1996. Molecular techniques reveal high prevalence of *Legionella* in dental units. *J. Amer. Dental Assoc.* 127:1188.

18. Fliermans, C.B., R.J. Soracco & D.H. Pope. 1981. Measure of *Legionella pneumophila* activity *in situ. Curr. Microbiol.* 6:89.

19. Jones, G.L. & G.A. Hebert. 1979. Legionnaires—the disease, the bacterium and methodology. U.S. Dep. Health, Education, & Welfare, Centers for Disease Control, Atlanta, Ga.

20. Edelstein, P.H., R.M. McKinney, R.D. Meyer, M.A.C. Edelstein, C.J. Krause & S.M. Finegold. 1980. Immunologic diagnosis of Legionnaires' Disease: cross reactions with anaerobic and microaerophilic organisms and infections caused by them. *J. Infect. Dis.* 141:652.

21. Voss, L., K.S. Button, M.S. Rheins & O.H. Tuovinen. 1984. Sampling methodology for enumeration of Legionella spp. in water distribution systems. *In* C. Thornsberry, A. Balows, J.C. Feeley & W. Jakubowski, eds. *Legionella*, Proc. 2nd International Symposium. American Soc. Microbiology, Washington, D.C.

22. Barbaree, J.M., G.W. Gorman, W.T. Martin, B.S. Fields & W.E. Morrill. 1987. Protocol for sampling environmental sites for Legionellae. *Appl. Environ. Microbiol.* 53:1454.

23. MILLER, L.A., J.I. BEEBE, J.C. BUTLER, W. MARTIN, R. BENSON, R.E. HOFFMAN & B.S. FIELDS. 1993. Use of polymerase chain reaction in epidemiological investigations of Pontiac fever. *J. Infect. Dis.* 168:769.

8. Bibliography

CENTERS FOR DISEASE CONTROL, NATIONAL INSTITUTE OF ALLERGY AND INFECTIOUS DISEASES & WORLD HEALTH ORGANIZATION. 1979. International Symposium on Legionnaire's Disease. *Ann. Intern. Med.* 90:489.

BLACKMAN, J.A., F.W. CHANDLER, W.B. CHERRY, A.C. ENGLAND, J.C. FEELEY, M.D. HICKLIN, R.M. McKINNEY & H.W. WILKINSON. 1981. Legionellosis. *Amer. J. Pathol.* 103:427.

DUFOUR, A. & W. JAKUBOWSKI. 1982. Drinking water and Legionnaire's Disease. *J. Amer. Water Works Assoc.* 74:631.

THORNSBERRY, C., A. BALOWS, J.C. FEELEY & W. JAKUBOWSKI, eds. 1984. *Legionella,* Proc. 2nd International Symposium. American Soc. Microbiology, Washington, D.C.

9260 K.   *Yersinia enterocolitica*

Yersinia enterocolitica is a gram-negative bacterium that can cause acute gastroenteritis and can be found in water in cold or temperate areas of the United States. Many wild, domestic, and farm animals are reservoirs of this organism, including wild animals associated with water habitats (beavers, minks, muskrats, nutrias, otters, and racoons).[1,2] The organism can grow at temperatures as low as 4°C with a generation time of 3.5 to 4.5 h if at least trace amounts of organic nitrogen are present.[3] Most environmental strains of *Y. enterocolitica* and the closely related species, *Y. kristensenii, Y. frederiksenii,* and *Y. intermedia,* generally are considered nonpathogenic, but disease outbreaks have been associated with environmental sources. Some strains lacking classic virulence markers also may be associated with disease.[4] *Y. enterocolitica* has become recognized worldwide as an important human pathogen and in several countries it is nearly as common as *Salmonella* and *Campylobacter* as a leading cause of acute or chronic enteritis.[5] *Y. enterocolitica* usually is associated with sporadic cases of gastroenteritis in the U.S.; however, epidemiologic investigations suggest that the predominant pathogenic serotype isolated in the U.S. has been changing.[4] *Y. enterocolitica* serogroup O:3 has replaced O:8 as the most common species recovered from patients, reflecting the same pattern seen in other parts of the world.[4, 5] Two reported incidents of waterborne gastroenteritis possibly caused by *Yersinia* occurred during the period 1971 to 1978.[3,6,7]

Yersinia has been isolated from untreated surface and ground waters in the Pacific Northwest, New York, and other regions of North America, with highest isolations during the colder months.[8–10] Concentrations have ranged from 3 to 7900 CFU/100 mL. Laboratory tests used to isolate and enumerate yersiniae do not discriminate between pathogenic and nonpathogenic strains. *Yersinia* isolations do not correlate with levels of total and fecal coliforms or total plate count bacteria.[9] There is little information on *Yersinia* survival in natural waters and water treatment processes.

In studies of *Yersinia* in chlorinated-dechlorinated secondary effluent and receiving (river) water, the organism was isolated in 27% of the effluent samples, 9% of the upstream samples, and 36% of the downstream samples.[11] Mean total and fecal coliform reductions in effluent chlorination were 99.93 and 99.95%, respectively. In a survey of untreated and treated (chlorination or filtration plus chlorination) drinking water supplies, *Yersinia* was found in 14.0 and 5.7% of the samples, respectively.[9] Of water samples with less than 2.2 coliforms/100 mL, 15.9% were *Yersinia*-positive. *Yersinia* isolation did not correlate with presence of total or fecal coliforms in this study. Another study confirmed that *E. coli* also is not a good indicator for *Yersinia* in water and that *Y. enterocolitica* O:3 strains harboring a virulence plasmid have enhanced resistance to chlorine compared to non-virulent strains.[12]

Because of the existence of animal reservoirs, widespread occurrence and persistence of *Yersinia* in natural and treated water in at least some geographic areas, the evidence for possible waterborne outbreaks, and the lack of definitive information on its reduction by treatment processes, this pathogen is of potential importance in drinking water.

1. Concentration and Cultivation

A membrane filter method for enumerating and isolating *Yersinia enterocolitica* is available.[13] The method may be used for examining large volumes of low-turbidity water and for presumptively identifying the organism without transferring colonies to multiple confirmatory media.

Filter sample through a membrane filter (see Section 9260B.1*d*). Place membrane filter on a cellulose pad saturated with m-YE recovery broth. Incubate for 48 h at 25°C. Aseptically transfer the membrane to a lysine-arginine agar substrate and incubate anaerobically at 35°C. After 1 h, puncture a hole in the membrane next to each yellow to yellow-orange colony with a needle, transfer the membrane to a urease-saturated absorbent pad, and incubate at 25°C for 5 to 10 min. Immediately count all distinctly green or deep bluish-purple colonies. The green or bluish colonies are sorbitol-positive, lysine- and arginine-negative, and urease-positive. They may be presumptively identified as *Y. enterocolitica* or a closely related *Yersinia* species. Additional biochemical testing is necessary to determine species. Reasonably simple tests have been described to screen isolates for pathogenicity.[14] Comprehensive biochemical and serological characterization or the use of molecular methods is necessary to confirm virulence, but these methods are not generally available.

2. References

1. WETZLER, T.F. & J. ALLARD. 1977. *Yersinia enterocolitica* from trapped animals in Washington State. Paper presented at International Conf. Disease in Nature Communicable to Man. Yellow Bay, Mont.

2. WETZLER, T.F., J.T. REA, G. YUEN & W. TURNBERG. 1978. *Yersinia enterocolitica* in waters and wastewaters. Paper presented at 106th Annual Meeting, American Public Health Assoc., Los Angeles, Calif.

3. HIGHSMITH, A.K., J.C. FEELEY, P. SKALIY, J.G. WELLS & B.T. WOOD. 1977. The isolation and enumeration of *Yersinia enterocolitica* from well water and growth in distilled water. *Appl. Environ. Microbiol.* 34:745.

4. BISSETT, M.J., C. POWERS, S.L. ABBOTT & J.M. JANDA. 1990. Epidemiologic investigations of *Yersinia enterocolitica* and related species: sources, frequency, and serogroup distribution. *J. Clin. Microbiol.* 28:910.

5. FENWICK, S.G. & M.D. McCARTY. 1995. *Yersinia enterocolitica* is a common cause of gastroenteritis in Auckland. *N. Zealand Med. J.* 108:269.

6. EDEN, K.V., M.L. ROSENBERG, M. STOOPLER, B.T. WOOD, A.K. HIGHSMITH, P. SKALIY, J.G. WELLS & J.C. FEELEY. 1977. Waterborne gastrointestinal illness at a ski-resort—isolation of *Yersinia enterocolitica* from drinking water. *Pub. Health Rep.* 92:245.

7. KEET, E. 1974. *Yersinia enterocolitica* septicemia. *N.Y. State J. Med.* 74:2226.

8. HARVEY, S., J.R. GREENWOOD, M.J. PICKETT & R.A. MAH. 1976. Recovery of *Yersinia enterocolitica* from streams and lakes of California. *Appl. Environ. Microbiol.* 32:352.

9. WETZLER, T.F., J.R. REA, G.J. MA & M. GLASS. 1979. Non-association of *Yersinia* with traditional coliform indicators. *In* Proc. Annu. Meeting American Water Works Assoc., American Water Works Assoc., Denver, Colo.

10. SHAYEGANI, M., I. DeFORGE, D.M. McGLYNN & T. ROOT. 1981. Characteristics of *Yersinia enterocolitica* and related species isolated from human, animal, and environmental sources. *J. Clin. Microbiol.* 14:304.

11. TURNBERG, W.L. 1980. Impact of Renton Treatment Plant effluent upon the Green-Duwamish River. Masters Thesis, Univ. Washington, Seattle.

12. LUND, D. 1996. Evaluation of *E. coli* as an indicator for the presence of *Campylobacter jejuni* and *Yersinia enterocolitica* in chlorinated and untreated oligotrophic lake water. *Water Res.* 30:1528.

13. BARTLEY, T.D., T.J. QUAN, M.T. COLLINS & S.M. MORRISON. 1982. Membrane filter technique for the isolation of *Yersinia enterocolitica*. *Appl. Environ. Microbiol.* 43:829.

14. FARMER, J.J., G.P. CARTER, V.L. MILLER, S. FALKOW & I.W. WACHSMUTH. 1992. Pyrazinamidase, CR-MOX agar, salicin fermentation-esculin hydrolysis, and d-xylose fermentation for identifying pathogenic serotypes of *Yersinia enterocolitica*. *J. Clin. Microbiol.* 30:2589.

3. Bibliography

HIGHSMITH, A.K., J.C. FEELEY & G.K. MORRIS. 1977. *Yersinia enterocolitica:* a review of the bacterium and recommended laboratory methodology. *Health Lab. Sci.* 14:253.

BOTTONE, E.J. 1977. *Yersinia enterocolitica:* a panoramic view of a charismatic microorganism. *CRC Crit. Rev. Microbiol.* 5:211.

YANKO, W.A. 1993. Occurrence of Pathogens in Distribution and Marketing Municipal Sludges. National Technical Information Serv. Rep. PB88-154273-AS, Springfield, Va.

9260 L. *Aeromonas* (PROPOSED)

1. Introduction

Aeromonas spp. are natural inhabitants of aquatic environments worldwide. These Gram-negative, facultatively anaerobic, glucose-fermenting organisms have been isolated from groundwater, treated drinking water, surface waters, wastewater, sludge, and sediment. Their populations are seasonal in all natural waters, with the highest numbers present in warmer months. Aeromonads cause serious diseases of aquatic animals and represent an economic threat to the aquaculture industry. The motile aeromonads have emerged as a serious microbial threat to human populations, especially the immunocompromised.[1]

As a result of recent taxonomic studies, *Aeromonas* bacteria have been removed from the family *Vibrionaceae* and established as the sole genus of the new family *Aeromonadaceae*. The genus *Aeromonas* comprises 14 recognized and 2 proposed DNA hybridization groups with 13 named phenospecies and 4 unnamed genospecies. The extreme difficulty of phenotypically differentiating aeromonads and the unavailability of DNA hybridization techniques in most laboratories have lead clinical microbiologists to report aeromonads as *A. hydrophila*, *A. sobria*, or *A. caviae*, according to a published classification scheme.[2] Environmental microbiologists usually combine all motile, mesophilic aeromonads into the *Aeromonas hydrophila* complex, or simply report isolates as *A. hydrophila*. These practices obscure understanding of the medical and public health significance of aeromonads isolated from clinical specimens, environmental samples, and public water supplies; identification of *Aeromonas* isolates according to established taxonomic principles is preferable.[3]

While no waterborne outbreaks of gastroenteritis attributed to aeromonads have implicated public drinking water supplies in the U.S., this does not mean that none have occurred. The epidemiologic association between ingestion of untreated well water and subsequent *Aeromonas* gastrointestinal illness has been widely documented. Numerous cases and outbreak investigations of water- and food-transmitted illnesses caused by aeromonads have been reported.[4] Outbreaks of gastroenteritis caused by aeromonads have occurred in custodial care institutions, nursing homes, and day-care centers. *Aeromonas* contamination of drinking water has been documented as a cause of travelers' diarrhea.[5]

For many years, *Aeromonas* have been considered nuisance organisms by environmental microbiologists because they were reported to interfere with coliform multiple tube fermentation (MTF) methods. While aeromonads comprise 12% of bacteria isolated from drinking water by presence-absence methods, no data have demonstrated inhibition of coliform organisms by aeromonads in drinking water. Slight turbidity of LTB tubes, with or without a small bubble of gas in the inverted tube, is suggestive of aeromonads. When the MTF method is used for drinking water samples, cultures producing turbidity at 35°C that remain clear at 44.5°C are suggestive of aeromonads. The presence of aeromonads can be verified by subculturing a loopful of turbid broth to a MacConkey plate and screening colorless colonies for gelatinase and oxidase production. No data are available to support invalidation of coliform MTF tests based on turbidity of tubes in the absence of gas production.

The ecology of mesophilic aeromonads in aquatic environments, including water treatment plants and distribution systems, has been reviewed.[6] The Netherlands and the Province of Quebec have established drinking water standards for *Aeromonas* at 20 CFU/100 mL for water leaving the treatment plant, and 200 CFU/100 mL for distribution system water. Canada has established an *Aeromonas* MCL of 0 (zero) for bottled water. A resuscitation method for recovery of aeromonads in bottled water has been published.[7]

The ability to isolate, enumerate, and identify acromonads from water and wastewater sources is important because of their role in causing human and animal disease, their ability to colonize treatment plants and distribution systems, and their presence and distribution as alternative indicators of the trophic state of waters. The diversity of aeromonads in drinking water plants and distribution systems was shown by several investigators.[8–10]

Many media and methods have been proposed for the isolation and enumeration of aeromonads.[11,12] The methods presented below represent a compromise, because no single enrichment method, isolation medium, or enumeration method is capable of recovering all aeromonads present in a water sample. The methods were chosen on the basis of reproducibility of results, objectivity of interpretation, availability of materials, and specificity of the method for detection of aeromonads in the presence of other heterotrophic bacteria. Consult the literature for additional methods for use in special circumstances.[13]

2. Sample Collection

Collect water samples in sterile screw-capped glass or plastic bottles or plastic bags.* Sample volumes of 200 mL to 1 L are sufficient for most analyses. For chlorinated waters, add sodium thiosulfate (see Section 9060A.2). The potentially toxic effect of heavy metals is neutralized by adding EDTA (see Section 9060A.2).

Transport samples to the laboratory at 2 to 8°C within 8 h. Samples for presence-absence analyses may be transported at ambient temperatures within 24 h. Grab samples are most common. Moore swabs (see 9260B.1*a*) have been used for wastewater sampling, and Spira bottles have been used for tapwater sampling.[13] Both of these methods are used in conjunction with enrichment in 1% alkaline peptone water (APW), pH 8.6.[13] Place sediment and sludge samples in bottles or bags and submit in same way as water samples.

3. Enrichment Methods

Do not use enrichment methods for ecological studies because the predominant strain(s) will overgrow other organisms. Reserve enrichments for presence-absence tests for aeromonads in drinking water, foods, stools, or for monitoring acromonad populations in wastewater or marine environments, where organisms may be present in low numbers or require resuscitation due to injury from exposure to inimical agents or hostile physical environments. For isolation of aeromonads from clear water samples, filter through 0.45 μm membrane filters, place filters in a bottle with 10 mL APW, incubate overnight at 35°C, and inoculate to plating media for isolation. Optimally, to sample clear water intended for drinking, filter a volume of water through a mini-capsule filter†, decant residual water from inlet, plug ends with sterile rubber stoppers, and fill filter with APW, pH 8.6, through syringe port. Incubate

---

* WhirlPak™, ZipLoc™, or equivalent.
† Gelman 12123 or equivalent.

filter at 35°C for 6 h or overnight and streak loopfuls of broth onto selective and differential plating media.[14]

4. Enumeration Methods

a. Spread plates: Enumerate samples expected to contain predominantly acromonads in high numbers (sludge, sediments, wastewater effluents, polluted surface waters, etc.) directly by spreading 0.1-mL portions of decimal dilutions on ampicillin dextrin agar (ADA)[15–17] plates. Incubate plates at 35°C overnight and count bright yellow colonies 1 to 1.5 mm in diameter. Presumptively identify colonies using the screening methods below.

b. Membrane filtration (MF): Enumerate aeromonads in drinking water samples or other low-turbidity waters by using MF procedures with ADA medium and incubating aerobically overnight at 35°C. Filter sample volumes equivalent to 1 mL, 10 mL, and 100 mL. To achieve a countable plate (1 to 30 colonies), prepare decimal dilutions when aeromonads are present in high numbers. Count bright yellow colonies, 1 to 1.5 mm in diameter, and pick to screening media.

c. Multiple-tube fermentation tests (MTF): Multiple-tube fermentation tests using APW, pH 8.6, or trypticase soy broth (TSB) containing ampicillin at 30 μg/mL (TSB30) have been applied to foods; however, they have not been used for enumeration of aeromonads in water samples. Some aeromonads are sensitive to ampicillin and will not grow in TSB30 medium. ADA without agar has been used to enumerate aeromonads in drinking water.[8] Use MTF methods only for clean samples such as groundwater or treated drinking water samples, because the effect of competing microflora present in surface waters on recovery of aeromonads in broth media has not been studied adequately. Similarly, the correlation between MTF population estimates and other enumeration methods has not been examined adequately for matrices other than foods.

5. Screening Tests

Pick 3 to 10 colonies resembling aeromonads on differential and selective plating media or membrane filters and stab-inoculate into deeps of Kaper's multi-test medium[19] or one tube each of triple sugar iron (TSI) agar and lysine iron agar (LIA). Incubate cultures at 30°C for 24 h. Perform a spot oxidase test on growth taken from the LIA slant. Do not test for oxidase on growth from TSI slants, MacConkey agar, or other selective or differential media, because acid production interferes with the oxidase reaction.

TABLE 9260:III. REACTIONS OF ENTERIC BACTERIA ON TSI AND LIA MEDIA

Organism	TSI Reactions*	LIA Reactions*
Shigella	K/A −	K/A −
Salmonella	K/Ag +	K/K +
Escherichia	A/Ag −	K/A −
Proteus	A/Ag + or K/Ag +	R/A +
Citrobacter	A/Ag +	K/A +
Enterobacter	A/Ag −	K/A −
Aeromonas	A/A −	K/A −
Yersinia	A/A − or K/A −	K/A −
Klebsiella	A/Ag −	K/A −

* Fermentation reactions = slant/butt, H₂S production = + or − , K = alkaline, A = acid, R = red (deaminase reaction), g = gas produced.

TABLE 9260:IV. REACTIONS OF *AEROMONAS* AND ENTERIC BACTERIA ON KAPER'S MEDIUM

Organism	Fermentation Pattern*	Motility	H₂S	Indole
Aeromonas hydrophila	K/A	+	−	+
Klebsiella pneumoniae	A/A	−	−	−
Klebsiella oxytoca	A/A	−	−	+
Escherichia coli	K/K or K/A	+ or −	−	+
Salmonella spp.	K/K, K/A, A/K or A/A	+	+	−
Enterobacter spp.	K/K, K/N or N/N	+	−	−
Proteus spp.	R/K or R/A	+	+ or −	+
Yersinia enterocolitica	K/K, K/N or N/N	−	−	+ or −
Citrobacter spp.	K/K or K/A	+	+	−
Serratia spp.	K/K, K/N, or N/N	+	−	−

* K = alkaline; A = acid; N = neutral; R = red (deamination reaction).

Reactions of enteric bacteria on TSI and LIA media are shown in Table 9260:III. When Kaper's medium is used instead of TSI/LIA slants, colonies may be picked and inoculated onto sheep blood agar plates; incubate at 35°C overnight to provide growth for the oxidase test and to record hemolysin production. Cultures are identified presumptively using Kaper's medium according to the characteristics shown in Tables 9260:IV. If species identification is desirable, submit presumptively identified *Aeromonas* cultures to a reference laboratory. Cultures with potential public health or regulatory significance may be subtyped using various molecular methods to determine clonality for outbreak investigations and trouble-shooting of treatment plant or distribution system problems.[1]

6. References

1. JANDA, J.M. & S.L. ABBOTT. 1996. Human Pathogens. *In* B. Austin, M. Altwegg, P. Gosling & S.W. Joseph, eds. The Genus *Aeromonas*, p. 151. John Wiley & Sons, Chichester, U.K.
2. POPOFF, M. & M. VERON. 1976. A taxonomic study of the *Aeromonas hydrophila-Aeromonas punctata* group. *J. Gen. Microbiol.* 94:11.
3. CARNAHAN, A.M. & M. ALTWEGG. 1996. Taxonomy. *In* B. Austin, M. Altwegg, P. Gosling & S.W. Joseph, eds. The Genus *Aeromonas*, p. 1. John Wiley & Sons, Chichester, U.K.
4. JOSEPH, S.W. 1996. Aeromonas gastrointestinal disease: a case study in causation?. *In* B. Austin, M. Altwegg, P. Gosling & S.W. Joseph, eds. The Genus *Aeromonas*, p. 311. John Wiley & Sons, Chichester, U.K.
5. HANNINEN, M.L., S. SALMI, L. MATTILA, R. TAIPALINEN & A. SIITONEN. 1995. Association of *Aeromonas* spp. with travellers' diarrhoea in Finland. *J. Med. Microbiol.* 42:26.
6. HOLMES, P., L.M. NICCOLLS & D.P. SARTORY. 1996. The ecology of mesophilic *Aeromonas* in the aquatic environment. *In* B. Austin, M. Altwegg, P. Gosling & S.W. Joseph, eds. The Genus *Aeromonas*, p. 127. John Wiley & Sons, Chichester, U.K.
7. WARBURTON, D.W., J.K. MCCORMICK & B. BOWEN. 1993. Survival and recovery of *Aeromonas hydrophila* in water: development of methodology for testing bottled water in Canada. *Can. J. Microbiol.* 40:145.
8. HANNINEN, M.-L. & A. SIITONEN. 1995. Distribution of *Aeromonas* phenospecies and genospecies among strains isolated from water, foods or from human clinical samples. *Epidemiol. Infect.* 115:39.
9. HUYS, G., I. KERSTERS, M. VANCANNEYT, R. COOPMAN, P. JANSSEN & K. KERSTERS. 1995. Diversity of *Aeromonas* sp. in Flemish drinking water production plants as determined by gas-liquid chromatographic analysis of cellular fatty acid methyl esters (FAMEs). *J. Appl. Bacteriol.* 78:445.
10. MOYER, N.P., G.M. LUCCINI, L.A. HOLCOMB, N.H. HALL & M. ALTWEGG. 1992. Application of ribotyping for differentiating aeromonads isolated from clinical and environmental sources. *Appl. Environ. Microbiol.* 58:1940.
11. GAVRIEL, A. & A.J. LAMB. 1995. Assessment of media used for selective isolation of *Aeromonas* spp. *Lett. Appl. Microbiol.* 21:313.
12. JEPPESEN, C. 1995. Media for *Aeromonas* spp., *Plesiomonas shigelloides* and *Pseudomonas* spp. from food and environment. *Int. J. Food Microbiol.* 26:25.
13. MOYER, N.P. 1996. Isolation and enumeration of aeromonads. *In* B. Austin, M. Altwegg, P. Gosling & S.W. Joseph, eds. The Genus *Aeromonas*, p. 39. John Wiley & Sons, Chichester, U.K.
14. MOYER, N.P., G. MARTINETTE, J. LÜTHY-HOTTENSTEIN & M. ALTWEGG. 1992. Value of rRNA gene restriction patterns of *Aeromonas* spp. for epidemiological investigations. *Curr. Microbiol.* 24:15.
15. HANDFIELD, M., P. SIMARD & R. LETARTE. 1996. Differential media for quantitative recovery of waterborne *Aeromonas hydrophila*. *Appl. Environ. Microbiol.* 62:3544.
16. HAVELAAR, A.H., M. DURING & J.F. VERSTEEGH. 1987. Ampicillindextrin agar medium for the enumeration of *Aeromonas* species in water by membrane filtration. *J. Appl. Bacteriol.* 62:279.
17. HAVELAAR, A.H. & M. VONK. 1988. The preparation of ampicillin dextrin agar for the enumeration of *Aeromonas* in water. *Lett. Appl. Microbiol.* 7:169.
18. ALTWEGG, M. 1996. Subtyping methods for *Aeromonas* species. *In* B. Austin, M. Altwegg, P. Gosling & S.W. Joseph, eds. The Genus *Aeromonas*, p. 109. John Wiley & Sons, Chichester, U.K.
19. KAPER, J., R.J. SEIDLER, H. LOCKMAN & R.R. COLWELL. 1979. Medium for the presumptive identification of *Aeromonas hydrophila* and Enterobacteriaceae. *Appl. Environ. Microbiol.* 38:1023.

9260 M. *Mycobacterium* (PROPOSED)

The genus *Mycobacterium* comprises over 70 characterized species that are non-motile spore-forming, aerobic, acid-fast bacilli measuring 0.2 to 0.6 $\times$ 1 to 10 μm. Most organisms in this genus are saprophytes, but some species are capable of causing disease in humans. The primary pathogens in this group include *Mycobacterium tuberculosis* and *Mycobacterium leprae,* the causative agents of tuberculosis and leprosy, respectively. Recently there has been an increase in the incidence of disease caused by nontuberculosis mycobacteria, probably related to the increasing numbers of immunocompromised patients.[1-3] In the genus *Mycobacterium*, the most important opportunistic pathogens include *M. avium-intracellulare, M. kansasii, M. marinum,* and *M. simiae,* which are capable of causing disease when the immune system is compromised. Some of the common hosts and environmental reservoirs of *Mycobacteria* are shown in Table 9260:V.

Because of the complex nature of the cell wall, which is rich in lipids and therefore has a hydrophobic surface, this genus is resistant to many common disinfectants. As a result, several members of this genus are becoming important waterborne pathogens in the immunocompromised population. Mycobacteria also are acid-fast and extremely slow-growing. Some species such as *M. avium-intracellulare* require from 3 to 8 weeks to form colonies on culture media.

Mycobacterium avium and *Mycobacterium intracellulare* exhibit overlapping properties, making speciation extremely difficult. As a result, these two pathogens are grouped together and called *M. avium-intracellulare* or refered to as the MAC complex. Organisms from this group are ubiquitous in the environment and have been isolated from potable water systems, including those in hospitals[4-6] as well as from soil and dairy products. This pathogen causes a chronic pulmonary disease in immunocompetent hosts that is clinically and pathologically indistinguishable from tuberculosis; it also causes disseminated disease in immunocompromised hosts. The primary route of transmission is believed to be through ingestion, but increasing numbers of cases originate in the respiratory tract, indicating an aerosol route of transmission.

1. Sample Collection and Concentration

Mycobacteria typically constitute a minority of the microflora, especially in finished waters, and require sample concentration. Collect water samples in sterile 1-L polypropylene containers. For finished, disinfected waters, add 1 mL 10% sodium thiosulfate solution/L water collected. Transport samples to laboratory immediately after collection. If samples cannot be analyzed immediately, store at 4°C and begin analysis within 24 h of sampling.

2. Screening Water Samples by Direct Fluorescent Assay

Before committing the sample to a lengthy culture incubation, survey for acid-fast bacteria by using a combination solution of Auramine-Rhodamine (A-R) fluorescent dye.[7]* Auramine and Rhodamine nonspecifically bind to mycolic acids and resist decolorization by acid alcohol.[8]

Filter a minimum of 500 mL finished water or 100 mL source water (depending on turbidity), through a sterile 0.45-μm-poros-

TABLE 9260:V. MYCOBACTERIA OF WATERBORNE OR UNKNOWN ORIGIN

Mycobacterium Species	Environmental Contaminant	Reservoir
M. kansasii	Rarely	Water, swine, cattle
M. marinum	Rarely	Fish, water
M. simiae	No	Primates, possibly water
M. scrofulaceum	Possibly	Soil, water, foodstuffs
M. szulgai	No	Unknown
M. avium-intracellulare	Possibly	Soil, water, swine, cattle, birds
M. xenopi	Possibly	Water
M. ulcerans	No	Unknown
M. fortuitum	Yes	Soil, water, animals, marine life
M. chelonae	Yes	Soil, water, animals, marine life

ity, 47-mm-diam black filter. Aseptically transfer filter to a sterile polypropylene 50-mL tube and add 5 mL of buffered dilution water. Resuspend organisms from filter by vortexing for 2 min. Aspirate suspension and aseptically transfer to a sterile 15-mL polypropylene centrifuge tube. Centrifuge suspension at 5000 g for 10 min and discard all but about 0.5 mL of supernatant. Resuspend pellet by vortexing. Transfer 100 mL of the concentrate to a clean glass slide and air-dry and heat-fix at 60 to 70°C for 2 h or overnight. Primary stain the smear with A-R (15 min), decolorize with acid-alcohol† for 2 to 3 min, rinse with deionized water, apply secondary potassium permanganate counterstain (no longer than 2 to 4 min), rinse, and let air-dry. Examine smear at 100 $\times$ and 400 $\times$ with a microscope fitted with a BG-12 or 5113 primary filter with a OG-1 barrier filter. Acid-fast organisms will stain yellow-orange on a black background. To confirm for acid-fastness, apply a traditional acid-fast stain (Ziehl-Nielsen with Kenyon modification) directly to the prepared smear following the A-R stain.

For wastewater or highly turbid source waters, collect a 10-mL subsample and transfer to a sterile polypropylene 15-mL tube. Centrifuge at 5000 g for 10 min and discard all but about 0.5 mL of supernatant. Follow slide preparation procedure and staining as above.

3. Decontamination and Culture Methods

Mycobacteria grow very slowly on laboratory media. Therefore, eliminate from the sample naturally occurring organisms that can out-compete and overgrow the mycobacteria. Various isolation and identification methods have been described for the recovery of mycobacteria, especially in the hospital environment.[4-6] Decontamination of the sample concentrate is required for the selection for mycobacteria before culture. In addition, the matrix may affect the success of the recovery of mycobacteria. Several methods (*a* through *c* below) are detailed for recovering mycobacteria from water samples; determine which method performs best with the matrix to be examined.

a. Filter 500-mL water sample through a sterile 0.45-μm-porosity, 47-mm-diam filter. Aseptically transfer filter to a sterile polypropylene 50-mL tube. Add 5 mL sterile distilled water and

---

* Catalog #40-090, Remel, Lenexa, KS, or equivalent.

† Truant-Moore or equivalent.

TABLE 9260:VI. PHENOTYPIC CHARACTERISTICS OF CLINICALLY SIGNIFICANT ENVIRONMENTAL MYCOBACTERIA*

Mycobacterium Species	Growth Rate	Pigmentation	Urease	Nitrate Reduction	Hydrolysis of Polyoxyethylene Sorbitan Monooleate†
M. kansasii	S	P	±	+	+
M. marinum	S	P	+	−	+
M. simiae	S	P	±	−	−
M. scrofulaceum	S	S	±	−	−
M. szulgai	S	S/P	+	+	±
M. xenopi	S	S	−	−	−
M. avium-intracellulare	S	N	−	−	−
M. ulcerans	S	N	±	−	−
M. fortuitum	R	N	+	+	±
M. chelonae	R	N	+	−	±

* S = slow (3 to 8 weeks), R = rapid (7 d or longer), P = photochromogenic, S = scotochromogenic, N = nonphotochromogenic, S/P = scotochromogenic at 37°C and photochromogenic at 24°C.

† Tween 80®.

resuspend organisms off the filter by shaking with two 5-mm glass beads for 1 h on a mechanical shaker.[9] Add a 3% sodium lauryl sulfate, 1% NaOH solution.[10] Spread portions of this suspension onto a selective agar medium as described in ¶ 4 below.

b. Filter 500-mL water sample through sterile 0.45-μm-porosity, 47-mm-diam filter. Aseptically transfer filter to a sterile polypropylene 50-mL tube. Add 5 mL sterile distilled water and resuspend organisms off the filter by shaking with glass beads for 5 min on a mechanical shaker. Add 10 mL 1M NaOH for 20 min followed by centrifugation at 8600 g at 4°C for 15 min. Discard supernatant and add 5 mL 5% oxalic acid for 20 min. Re-centrifuge, discard supernatant, and add 30 mL sterile distilled water to neutralize. Centrifuge again, and resuspend in 0.7 mL distilled water.[11] Use portions of this material for selective growth (¶ 4 below).

c. Add 20 mL 0.04% (w/v) cetylpridinium chloride (CPC) to 500-mL water sample and leave at room temperature for approximately 24 h. Filter sample and wash filter with 500 mL sterile water.[12] A study of decontamination methods for the isolation of mycobacteria from drinking water samples found a CPC concentration of 0.005% (w/v) to yield the highest isolation rate and lowest contamination rate for the water examined.[13]

4. Selective Growth

Culture all samples in duplicate. After sample decontamination, either spread portions of the concentrates or use sterile forceps to place filters on selective media. One common egg-based medium that successfully isolates mycobacteria from environmental concentrates is Lowenstein-Jensen agar. An agar-based medium containing cycloheximide (7H10) is a general growth medium for mycobacteria as well. Place plates in humid chambers or gas-permeable bags to prevent dehydration, and incubate at 37°C. Additional plates also can be incubated at 30°C in a humidified chamber to detect mycobacteria that grow optimally at lower temperatures. Examine plates periodically during a 3- to 8-week incubation period. Count suspect colonies (acid-fast coccobacilli) and subculture to a tube of 7H9 broth. After 5 d, remove subsamples and stain with Ziehl-Nielsen stain with Kenyon modification. Subculture coccobacillary acid-fast organisms further onto 7H10 plates. Conduct phenotypic tests (Table 9260:VI) as a first step toward identification. If biochemical tests do not allow speciation,

use other methods, such as fatty acid profile by HPLC or GLC, serological typing, and/or molecular tests such as DNA probes, and RFLP, which have been used for rapid detection of a limited number of species.[14]

Although phenotypic tests have been the standard for species identification, there are several inherent problems in this approach. First, because initial identification of mycobacteria can take 3 to 8 weeks, observing biochemical changes entails additional time for the isolates (especially those of nontuberculosis mycobacteria) to metabolize specific substrates or to exhibit certain characteristics. Second, phenotypic traits are not stable; thus some species of mycobacteria are untypable by conventional methods.

One approach that can successfully speciate *Mycobacterium* is sequencing amplified rDNA.[15] The method produces objective results in 2 d, gives reproducible results (due to the stability of the 16S rRNA), and can identify new species. These techniques have been used in clinical diagnostic laboratories, and now are available in some full-service environmental testing laboratories.

5. References

1. GOOD, R.C. Opportunistic pathogens in the genus *Mycobacterium*. 1985. *Annu. Rev. Microbiol.* 39:347.
2. CARSON, L.A., L.A. BLAND, L.B. CUSICK, M.S. FAVERO, G.A. BOLAN, A.L. REINGOLD & R.C. GOOD. 1988. Prevalence of nontuberculous mycobacteria in water samples of hemodialysis centers. *Appl. Environ. Microbiol.* 54:3122.
3. DUMOULIN, G.C. & K.D. STOTTMEIR. 1986. Waterborne mycobacteria: an increasing threat to health. *ASM News* 52: 525.
4. DUMOULIN, G.C., K.D. STOTTMEIR, P.A. PELLETIER, A.Y. TSANG & J. HEDLEY-WHITE. 1988. Concentration of *Mycobacterium avium* by hospital water systems. *J. Amer. Med. Assoc.* 260:1599.
5. POWELL, B.L. & J.E. STEADHAM. 1981. Improved technique for isolation of *Mycobacterium kansasii* from water. *J. Clin. Microbiol.* 13:969.
6. CARSON, L.A., L.B. CUSICK, L.A. BLAND & M.S. FAVERO. 1988. Efficiency of chemical dosing methods for isolating nontuberculous mycobacteria from water supplies of dialysis centers. *Appl. Environ. Microbiol.* 54:1756.
7. NOLTE, F.S. & B. METCHOCK. 1995. *Mycobacterium. In* P.R. Murray, E.J. Baron, M.A. Pfaller, F.C. Tenover & R.H. Yolken, eds. Manual of Clinical Microbiology. American Soc. Microbiology Press, Washington, D.C.

8. CHAPIN, K. 1995. Clinical Miscroscopy. *In* P.R. Murray, E.J. Baron, M.A. Pfaller, F.C. Tenover & R.H. Volken, eds. Manual of Clinical Microbiology. American Soc. Microbiology Press, Washington, D.C.

9. KAMALA, T., C.N. PARAMASIVAN, D. HERBERT, P. VENKATESAN & R. PRABHAKAR. 1994. Evaluation of procedures for isolation of nontuberculous mycobacteria from soil and water. *Appl. Environ. Microbiol.* 60:1021.

10. ENGEL, H.W.B., L.G. BERWALD & A.H. HAVELAAR. 1980. The occurrence of *Mycobacterium kansasii* in tapwater. *Tubercle* 61:21.

11. IIVANANINEN, E.K., P.J. MARTIKAINEN, P.K. VAANANEN & M.-L.KATILA. 1993. Environmental factors affecting the occurrence of mycobacteria in brook waters. *Appl. Environ. Microbiol.* 59:398.

12. DUMOULIN, G.C. & K.D. STOTTMEIR. 1978. Use of cetylpyridinium chloride in the decontamination of water culture of mycobacteria. *Appl. Environ. Microbiol.* 36:771.

13. SCHULZE-ROBBECKE, R., A. WEBER & R. FISCHEDER. 1991. Comparison of decontamination methods for the isolation of mycobacteria from drinking water samples. *J. Microbiol. Methods* 14:177.

14. ANDREW, P.W. & G.J. BOULNOIS. 1990. Early days in the use of DNA probes for *Mycobacterium tuberculosis* and *Mycobacterium avium* complexes. *In* A.J.L. Macario & E.C. de Macario, eds. Gene Probes for Bacteria. Academic Press, San Diego, Calif.

15. ROGALL, T., T. FLOHR & E.C. BOTTGER. 1990. Differentiation of *Mycobacterium* species by direct sequencing of amplified DNA. *J. Gen. Microbiol.* 136:1915.

6. Bibliography

TSUKAMURA, M. 1981. A review of the methods of identification and differentiation of mycobacteria. *Rev. Infect. Dis.* 3:841.

GOOD, R.C. 1985. Opportunistic pathogens in the genus *Mycobacterium*. *Annu. Rev. Microbiol.* 39:347.

ICHIYAMA, S. & K. SHIMOKATA. 1988. The isolation of *Mycobacterium avium* complex from soil, water, and dusts. *Microbiol. Immunol.* 32:733.

BROADLEY, S.J., P.A. JENKINS, J.R. FURR & A.D. RUSSELL. 1991. Antimycobacterial activity of biocides. *Lett. Appl. Microbiol.* 13:118.

FISCHEDER, R., R. SCHULZE-ROBBECKE & A. WEBER. 1991. Occurrence of mycobacteria in drinking water samples. *Zentralbl. Hyg. Umweltmed.* 192:154.

JENKINS, P. A. 1991. Mycobacteria in the environment. *J. Appl. Bacteriol.* 70:137.

SCHULZE-ROBBECKE, R., B. JANNING & R. FISCHEDER. 1992. Occurrence of mycobacteria in biofilm samples. *Tubercle Lung Dis.* 73:141.

COLLINS, J. & M. YATES. 1994. Mycobacteria in water. *J. Appl. Bacteriol.* 1984. 57:193.

JENSEN, P.A. 1997. Airborne *Mycobacterium* spp. *In* C.J. Hurst, G.R. Knudsen, M.J. McInerney, L.D. Stetzenbach & M.V. Walter, eds. Manual of Environmental Microbiology. American Soc. Microbiology Press, Washington, D.C.

9510   DETECTION OF ENTERIC VIRUSES*

9510 A.   Introduction

1. Occurrence

Viruses excreted with feces or urine from any species of animal may pollute water. Especially numerous, and of particular importance to health, are the viruses that infect the gastrointestinal tract of man and are excreted with the feces of infected individuals. These viruses are transmitted most frequently from person to person by the fecal-oral route. However, they also are present in domestic sewage which, after various degrees of treatment, is discharged to either surface waters or the land. Consequently, enteric viruses may be present in sewage-contaminated surface and ground waters that are used as sources of drinking water. The viruses known to be excreted in relatively large numbers with feces include polioviruses, coxsackieviruses, echoviruses, and other enteroviruses, adenoviruses, reoviruses, rotaviruses, the hepatitis A (infectious hepatitis) virus(es), and the Norwalk-type agents that can cause acute infectious nonbacterial gastroenteritis. With the possible exception of hepatitis A, each group or subgroup consists of a number of different serological types; thus more than 100 different human enteric viruses are recognized.[1-4]

In temperate climates enteroviruses occur at peak levels in sewage during the late summer and early fall. However, hepatitis A virus (HAV), Norwalk-type viruses, and rotaviruses may be important exceptions because the incidence of the diseases due to

these viruses increases in the colder months. Quantitative information on seasonal patterns of occurrence in water and wastewater of these latter viruses is lacking because they cannot be assayed readily with conventional cell culture techniques. The Norwalk-type viruses have not been cultivated in any cell cultures, although immunochemical assay methods have been developed to detect them as antigens.[5,6] Human rotaviruses and HAV have been cultivated recently in cell cultures, but the techniques are difficult and require concomitant use of immunoassays such as immunofluorescence to detect virus growth or gene probes.[7-11]

Viruses are not normal flora in the intestinal tract; they are excreted only by infected individuals, mostly infants and young children. Infection rates vary considerably from area to area, depending on sanitary and socioeconomic conditions. Viruses usually are excreted in numbers several orders of magnitude lower than those of coliform bacteria. Because enteric viruses multiply only within living, susceptible cells, their numbers cannot increase in sewage. Sewage treatment, dilution, natural inactivation, and water treatment further reduce viral numbers. Thus, although large outbreaks of waterborne viral disease may occur when massive sewage contamination of a water supply takes place,[12] waterborne transmission of viral infection and disease in technologically advanced nations depends on whether minimal quantities of viruses are capable of producing infections. It has been demonstrated that infection can be produced experimentally by a very few virus units,[13] although the risk of infection increases with

---

* Approved by Standard Methods Committee, 1996.

increasing ingested doses.[14] The risk of infection incurred by an individual in a community with a water supply containing a very few virus units has not been determined. Risk analysis has suggested that significant risk of infection could result from low numbers of enteric viruses present in a drinking water supply.[15] The percentage of individuals who develop clinical illness may be as low as 1% for poliovirus and as great as 97% for hepatitis A.[15]

Most recognized waterborne virus disease outbreaks in the U.S. have been caused by obvious sewage contamination of untreated or inadequately treated private and semipublic water supplies. Virus disease outbreaks in community water supply systems usually are caused by contamination through the distribution system.[16]

2. Testing for Viruses

The routine examination of water and wastewater for enteric viruses is not recommended now. However, in special circumstances such as wastewater reclamation, disease outbreaks, or special research studies, it may be prudent or essential to conduct virus testing. Such testing should be done only by competent and specially trained water virologists having adequate facilities.

Laboratories planning to concentrate viruses from water and wastewater should do so with the clear understanding that the available methodology has important limitations.[17] Even the most current methods for concentrating viruses from water still are being researched and continue to be modified and improved. The efficiency of a virus concentration method may vary widely depending on water quality. Furthermore, none of the available virus detection methods have been tested adequately with representatives from all of the virus groups of public health importance. Most virus concentration methods have achieved adequate virus recoveries with water or wastewater samples that have been contaminated experimentally with known quantities of a few specific enteric viruses. Although method effectiveness in field trials is difficult to evaluate, some virus concentration methods have been used successfully to recover naturally occurring enteric viruses. Some of these methods require large equipment for sample processing and virus assay and identification procedures usually require cell culture and related virology laboratory facilities.

Detecting viruses in water through recovery of infectious virus requires three general steps: (a) collecting a representative sample, (b) concentrating the viruses in the sample, and (c) identifying and estimating quantities of the concentrated viruses. Particular problems associated with the detection of viruses of public health interest in the aquatic environment are: (a) the small size of virus particles (about 20 to 100 nm in diameter), (b) the low virus concentrations in water and the variability in amounts and types that may be present, (c) the inherent instability of viruses as biological entities, (d) the various dissolved and suspended materials in water and wastewater that interfere with virus detection procedures, and (e) the present limitations of virus estimation and identification methods.

3. Selection of Concentration Method

The densities of enteric viruses in water and wastewater usually are so low that virus concentration is necessary, except possibly for raw sewage in certain areas or seasons.[18] Numerous methods for concentrating waterborne enteric viruses have been proposed, tested under laboratory conditions with experimentally contaminated samples, and in some cases used to detect viruses under field conditions.[19,20]

Virus concentration methods often are capable of processing only limited volumes of water of a given quality. In selecting a virus concentration method consider the probable virus density, the volume limitations of the concentration method for that type of water, and the presence of interfering constituents. A sample volume less than 1 L and possibly as small as a few milliliters may suffice for recovery of viruses from raw or primary treated sewage. For drinking water and other relatively nonpolluted waters, the virus levels are likely to be so low that hundreds or perhaps thousands of liters must be sampled to increase the probability of virus detection.

Three different techniques used to concentrate viruses from water are described herein: adsorption to and elution from microporous filters (Methods B and C); aluminum hydroxide adsorption-precipitation (Method D); and polyethylene glycol (PEG) hydroextraction-dialysis (Method E).[19,20] A separate technique (Method F) for recovering viruses from solids in small volumes of water also is described. Virus concentration by adsorption to and elution from microporous filters can be used for both small volumes of wastewater and large volumes of natural and finished waters. The aluminum hydroxide adsorption-precipitation and PEG hydroextraction-dialysis methods are impractical for processing large fluid volumes. However, they are suitable for concentrating viruses from wastewater or other waters having relatively high virus densities and for second-step concentration (reconcentration) of viruses in primary eluates obtained by processing large sample volumes through microporous filters.

4. Recovery Efficiencies

In examining a particular water include a preliminary evaluation of virus recovery efficiency. To do this add a known quantity of one or more test virus types to the required volume of sample, process the sample by the concentration method, and assay the concentrate for test viruses to determine virus recovery efficiency. Ideally, such seeded samples should be used whenever field samples are processed. If seeded samples are used concurrently with field samples, take appropriate steps, including disinfection and sterilization and the use of aseptic technique, to prevent accidental contamination of samples.

5. References

1. RAO, V.C. & J.M. MELNICK. 1986. Environmental Virology. American Soc. Microbiology, Washington, D.C.
2. FEACHEM, R.G., D.J. BRADLEY, H. GARELICK & D.D. MARA. 1983. Sanitation and Disease. Health Aspects of Excreta and Wastewater Management. John Wiley & Sons, New York, N.Y.
3. WILLIAMS, F.P. & E.W. AKIN. 1986. Waterborne gastroenteritis. *J. Amer. Water Works Assoc.* 78:34.
4. FEACHEM, R., H. GARELICK & J. SLADE. 1981. Enteroviruses in the environment. *Trop. Dis. Bull.* 78:185.
5. BLACKLOW, N.R. & G. CUKOR. 1980. Viral gastroenteritis agents, Chap. 90 in E.H. Lennette, A. Balows, W.J. Hausler, Jr. & J.P. Truant, eds. Manual of Clinical Microbiology, 3rd ed. American Soc. Microbiology, Washington, D.C.
6. KAPIKIAN, A.Z., R.H. YOLKEN, H.B. GREENBERG, R.G. WYATT, A.R. KALICA, R.M. CHANOCK & H.W. KIM. 1979. Gastroenteritis viruses. *In* E.H. Lennette & N.J. Schmidt, eds. Diagnostic Procedures for Viral, Rickettsial and Chlamydial Infections. American Public Health Assoc., Washington, D.C.

7. SOBSEY, M.D., S.E. OGLESBEE, D.A. WAIT & A.I. CUENEA. 1984. Detection of hepatitis A in drinking water. *Water Sci. Technol.* 17: 23.

8. SMITH, E.M. & C.P. GERBA. 1984. Development of a method for detection of human rotavirus in water. *Appl. Environ. Microbiol.* 43: 1440.

9. SATO, K., Y. INABA, T. SHINOZAKI, R. FUJII & M. MATUMOTO. 1981. Isolation of human rotavirus in cell cultures. *Arch. Virol.* 69:155.

10. HEJKAL, T.W., E.M. SMITH & C.P. GERBA. 1984. Seasonal occurrence of rotavirus in sewage. *Appl. Environ. Microbiol.* 47:588.

11. JIANG, X., M.K. ESTES & T.G. METCALF. 1987. Detection of hepatitis A virus by hybridization with single-stranded RNA probes. *Appl. Environ. Microbiol.* 53:2487.

12. MELNICK, J.L. 1957. A water-borne urban epidemic of hepatitis. *In* Hepatitis Frontiers. Little, Brown & Co., Boston, Mass.

13. WARD, R.L., D.I. BERNSTEIN & E.C. YOUNG. 1986. Human rotavirus studies in volunteers: Determination of infectious dose and serological response to infection. *J. Infect. Dis.* 154:871.

14. AKIN, E. 1981. A review of infective dose data for enteroviruses and other enteric microorganisms in human subjects. *In* Microbial Health

Considerations of Soil Disposal of Domestic Wastewaters. EPA-600/9-83-017, U.S. Environmental Protection Agency, Washington, D.C.

15. GERBA, C.P. & C.N. HAAS. 1988. Assessment of risks associated with enteric viruses in contaminated drinking water. *In* J.J. Lichtenberg, J.A. Winter, C.I. Weber & L. Frankin, eds. Chemical and Biological Characterization of Sludges, Sediments, Dredge Spoils, and Drilling Muds. ASTM STP 976. American Soc. Testing & Materials, Philadelphia, Pa.

16. CRAUN, G.F. 1986. Waterborne Disease in the United States. CRC Press, Boca Raton, Fla.

17. SOBSEY, M.D. 1982. Quality of currently available methodology for monitoring viruses in the environment. *Environ. Internat.* 7:39.

18. BURAS, N. 1976. Concentration of enteric viruses in wastewater and effluent: A two year survey. *Water Res.* 10:295.

19. SOBSEY, M.D. 1976. Methods for detecting enteric viruses in water and wastewater. *In* G. Berg, H.L. Bodily, E.H. Lennette, J.L. Melnick & T.G. Metcalf, eds. Viruses in Water. American Public Health Assoc., Washington, D.C.

20. GERBA, C.P. & S.M. GOYAL. 1982. Methods in Environmental Virology. Marcel Dekker, New York.

9510 B. Virus Concentration from Small Sample Volumes by Adsorption to and Elution from Microporous Filters

1. General Discussion

Viruses can be concentrated from aqueous samples by reversibly adsorbing them to microporous filters and then eluting them from the filters in a small liquid volume.[1] The virus-containing sample is pressure-filtered through microporous filters having large surface areas to which viruses adsorb, presumably by both electrostatic and hydrophobic interactions.[2] Two general types of adsorbent filters are available: electronegative (negative surface charge) and electropositive (positive surface charge). The former filters are composed of either cellulose esters or fiberglass with organic resin binders. They adsorb viruses most efficiently in the presence of multivalent cations such as Al^{3+} and Mg^{2+} and/or at low pH, usually pH 3.5. The latter filters are composed of either fiberglass or cellulose and a positively charged organic, polymeric resin. They adsorb viruses efficiently over a wide pH range without added polyvalent salts. If the sample is neutral or acidic, it can be processed with these filters without chemical conditioning.

Electropositive filters have given virus recoveries comparable to those with electronegative filters.[3-5] They have been used in field studies,[6,7] and were evaluated with a variety of virus types[8-13] and waters.

Adsorbed viruses usually are eluted from the surfaces of microporous filters by pressure-filtering a small volume of eluent fluid through the filters in situ. The eluent is either a slightly alkaline proteinaceous fluid such as beef extract or a more alkaline buffer such as glycine-NaOH, pH 10.5 to 11.5. If glycine-NaOH is used as eluent, preferably use pH 10.5 because of the greater likelihood of virus inactivation at the higher pH.[14,15]

Microporous filter methods suffer from three main limitations. Sample suspended matter tends to clog the adsorbent filter, thereby limiting the volume that can be processed and possibly interfering with the elution process.[16] Dissolved and colloidal organic matter in some waters can interfere with virus adsorption to filters, presumably by competing with viruses for adsorption

sites,[17-19] and they also can interfere with virus elution. Finally, viruses adsorbed to suspended matter may be removed in any clarification procedure applied before virus adsorption. These solids-associated viruses are lost from the sample unless special efforts are made to recover the solids and process them for viruses.[16] A method for recovering solids-associated viruses from small volumes of water and wastewater is given in Section 9510F. Despite these limitations, virus concentration by adsorption to and elution from microporous filters is a most promising technique for detecting viruses.

2. Equipment and Apparatus

a. Adsorbent filter holder, 47-, 90-, or 142-mm diam, equipped with pressure relief valve.

b. Pressure vessel, 12- or 20-L capacity.

c. Positive pressure source up to about 400 kPa with regulator: laboratory air line, air pump, or cylinder of compressed air or nitrogen gas.

d. Autoclavable vinyl plastic tubing with plastic or metal connectors (quick-disconnect type), for connecting positive pressure source, pressure vessel, and filter holder in series.

e. pH meter.

f. Beakers, 50- to 500-mL.

g. Laboratory balance.

h. Graduated cylinders, 25- to 100-mL.

i. Pipets, 1-, 5-, and 10-mL.

3. Materials

a. Electronegative virus adsorbent filter: Use either:
1) *Cellulose nitrate filter,* 0.45-μm porosity.*

---

* Type HA, Millipore Corp., Bedford, MA, or equivalent.

2) *Fiberglass-acrylic resin filter,* 0.45-μm porosity.† Filter media available commercially only as flat sheets can be cut to the desired disk diameter with scissors.

b. Electropositive virus adsorbent filter: Use either:

1) Surface modified cellulose and filter aid disk depth-filter.‡

2) Surface modified cellulose and filter aid thin-sheet medium, 0.20-μm porosity.§

c. Prefilter: Use one or more cellulose nitrate or fiberglass-acrylic resin filters or equivalent, with porosities greater than 0.45 μm to prevent clogging of the virus adsorbent filter by suspended matter. Place prefilters on top of the 0.45-μm-porosity virus adsorbent filter in the same filter holder.

4. Reagents

a. Hydrochloric acid, HCl, 0.1, 1.0, and 10N.

b. Sodium hydroxide, NaOH, 0.1, 1.0, and 10N.

c. Aluminum chloride, $AlCl_3 \cdot 6H_2O$, 0.15N, or magnesium chloride, $MgCl_2 \cdot 6H_2O$, 5N (necessary only for electronegative filters).

d. Sodium thiosulfate, $Na_2S_2O_3 \cdot 5H_2O$, 0.5% (*w/v*).

e. Sodium chloride, 0.14N, pH 3.5: Dissolve 8.18 g in 1 L reagent-grade water and adjust to pH 3.5 with HCl (necessary only for electronegative filters).

f. Virus eluent: Use either:

1) *Glycine-NaOH,* pH 10.5 or 11.5: Prepare 0.05M glycine solution, autoclave, and adjust to pH 10.5 or 11.5 with 1 to 10N NaOH. Add phenol red, 0.0005%, as a pH indicator.

2) *Beef extract,* 3%, pH 9.0: Dissolve 30 g beef extract paste or 24 g beef extract powder in 1000 mL reagent-grade water, adjust to pH 9.0 with 1 to 10N NaOH, and sterilize by autoclaving.

g. Glycine-HCl, pH 1.5: Prepare 0.05M glycine solution, autoclave, and adjust to pH 1.5 with 1 to 10N HCl. Add phenol red, 0.0005%, as a pH indicator.

h. Nutrient broth, 10X, pH 7.5: Dissolve 8.0 g nutrient broth in 90 mL reagent-grade water, adjust to pH 7.5, dilute to 100 mL with reagent-grade water, and sterilize by autoclaving.

i. Antibiotics: Use either:

1) *Penicillin-streptomycin,* 10X: Contains 5000 IU penicillin/mL and 5000 μg streptomycin/mL. Use commercially available form or prepare by dissolving powdered sodium or potassium penicillin-G and streptomycin sulfate in reagent-grade water and sterilizing by filtration. Store frozen.

2) *Gentamycin-kanamycin,* 100X: Contains 5000 μg/mL each of gentamycin (base) and kanamycin (base). Prepare by combining aseptically equal volumes of commercially available sterile gentamycin and kanamycin solutions, 10 000 μg/mL, respectively, or by dissolving powdered gentamycin sulfate and kanamycin sulfate in reagent-grade water and sterilizing by filtration. Store refrigerated or frozen.

j. Hanks balanced salt solution, 10X: Use commercially available form or prepare following a standard protocol.[20]

k. Sodium hypochlorite, 5.25% available chlorine (household bleach).

5. Procedure

a. Sterilization of apparatus, materials, and reagents: Most reagents, virus adsorbent filters, filter holders, tubing, and labware

† No. 8025-035, Filterite Corp., Timonium, MD, or equivalent.
‡ Zeta-plus 50S or 60S, CUNO, Meriden, CT, or equivalent.
§ 1-MDS Virozorb, CUNO, Meriden, CT, or equivalent.

can be sterilized by autoclaving or made virus-free by streaming steam. To sterilize filters load into their holders; if several filters are to be placed in one holder, place filter with smallest porosity on the bottom with progressively larger filters on top. Do not use an automatic drying cycle when autoclaving virus adsorbent filters. Sterilize apparatus and material that cannot be autoclaved or treated with streaming steam by treating with 10-mg/L free chlorine solution, pH 7.0, for 30 min and rinse or flush with 50-mg/L sterile $Na_2S_2O_3$ solution. Do not treat adsorbent filters with chlorine. Use aseptic technique during all virus concentration operations to prevent extraneous microbial contamination.

b. Sample size and choice of filter size: Sample size and, hence, filter diameter depend partly on water quality and the probable virus concentration. Single-stage microporous filter adsorption-elution methods have been used to recover viruses from 100 mL raw sewage on 47-mm-diam filters[21] and from 3.8 to 4.6 L secondary and tertiary sewage effluent on 90- or 142-mm-diam filters.[18,21,22] Based on the diameter and solids-holding characteristics of the filters, the scale and volume capacity of the apparatus and materials, and the quality of the samples, the practical limits for sample size are 20, 8, and 2 L for 142-, 90-, and 47-mm-diam filters, respectively.

c. Choice of filter type: Virus adsorption to electropositive filters decreases above pH 8 and pH adjustment below this value may be necessary for optimal virus adsorption.[4] Virus recovery from raw sewage may be less than with electronegative filters.[10]

d. Sample collection and storage: Collect samples aseptically in sterile containers. If they contain residual chlorine, immediately add $Na_2S_2O_3$ solution to give a final concentration of 50 mg/L. Process samples as soon as possible after collection; do not hold samples for more than 2 h at up to 25°C or 48 h at 2 to 10°C. Do not freeze samples unless they cannot be processed within 48 h; then freeze and store at −70°C or less.

e. Sample processing of electronegative filters: Adjust sample to pH 3.5 and 0.0015N $AlCl_3$ or to between pH 6.0 and 3.5 and 0.1N $MgCl_2$. Make sample adjustments either in a pressure vessel or in another appropriate container. Mix sample vigorously during addition of 1.0 or 0.1N HCl and $AlCl_3$ solution (1 part solution to 100 parts sample) or $MgCl_2$ solution (1 part solution to 50 parts sample). Because $AlCl_3$ is an acid salt, it may decrease sample pH slightly. Do not let sample pH fall below 3.0.

Place sample in a pressure vessel connected to a source of positive pressure and connect pressure vessel outlet to inlet of virus adsorbent filter holder. With pressure relief valve on filter holder opened, apply a slight positive pressure to purge air from filter holder. When sample just begins to flow from pressure relief valve, quickly close valve and continue filtration at a rate not exceeding 28 mL/min/cm² of filter area (about 130, 250, and 4000 mL/min for 47-, 90-, and 142-mm-diam filters, respectively). After filtering entire sample let positive pressure source purge excess fluid from filter holder.

Wash filters with 0.14N NaCl to remove excess Al^{3+} or Mg^{2+} from virus adsorbent filter. Use about 1.5 mL NaCl solution/cm² filter area (25, 100, and 240 mL for 47-, 90-, and 142-mm-diam filters, respectively). Place wash solution in a pressure vessel connected to filter holder inlet, use positive pressure to filter solution through virus adsorbent filter, discard filtrate, and let positive pressure purge virus adsorbent filter of excess wash solution.

Elute viruses from filters with a recommended eluent. Use about 0.45 mL eluent/cm² filter surface area (about 7.5, 28, and 71 mL for 47-, 90-, and 142-mm-diam filters, respectively). With pressure relief valve on filter holder open, add eluent to filter holder so that it completely covers filter surface. When eluent begins to discharge from pressure relief valve, quickly close valve. If pH 11.5 glycine-NaOH is the eluent, place a sterile beaker under filter outlet and apply positive pressure so that filtrate flows slowly from filter holder outlet. Collect filtrate in sterile beaker and, when filtrate no longer flows, slowly increase pressure to force retained fluid from filters. Quickly check eluate (filtrate) pH. If it is less than 11.0, elute with additional pH 11.5 glycine-NaOH until an eluate with a pH $\geq$ 11.0 is obtained. Immediately after checking pH, adjust eluate to a pH between 9.5 and 7.5 with pH 1.5 glycine-HCl or 0.1N HCl while mixing vigorously. Complete elution and eluate pH adjustment to 7.5 to 9.5 in 5 min or less to avoid the possibility of appreciable virus inactivation.

If pH 10.5 glycine-NaOH is the eluent, proceed as with pH 11.5 glycine-NaOH, but pass the eluate through the filters a total of five times. For each elution, collect the filtrate, readjust to pH 10.5 with 1.0 or 0.1N NaOH, and then pass through the filter. After the fifth elution, adjust filtrate to pH 7.4 with glycine-HCl, pH 1.5, or 0.1N HCl.

If 3% beef extract, pH 9.0, is the eluent, place a sterile beaker under filter outlet, apply a slight positive pressure to eluent-containing filter holder so that filtrate flows slowly from the outlet, and collect filtrate. Slowly increase pressure to force additional retained fluid from filters.

Measure eluate volume and add 1/10 of the measured volume each of penicillin-streptomycin or gentamycin-kanamycin, Hanks balanced salt solution, and 10X nutrient broth (add last item to glycine eluates only). Adjust sample to pH 7.4 with glycine-HCl or 0.1N HCl while mixing vigorously. Store at either 4 or $-70°C$, depending on the time until virus assay. Maximum storage at 4°C is 48 h.

f. Processing of electropositive filters: Processing for electropositive filters is identical to that for electronegative filters except that addition of Al^{3+} and Mg^{2+} and sample pH adjustments are unnecessary; because Al^{3+} and Mg^{2+} are not added, it is not necessary to wash filters with 0.14N NaOH before elution. If sample pH is greater than 8.0, adjust to less than pH 8 by adding 1.0 or 0.1N HCl.

6. References

1. FARRAH, S.R., C.P. GERBA, C. WALLIS & J.L. MELNICK. 1976. Concentration of viruses from large volumes of tapwater using pleated membrane filters. *Appl. Environ. Microbiol.* 31:221.

2. FARRAH, S.R., D.O. SHAH & L.O. INGRAM. 1981. Effects of chaotropic and antichaotropic agents on the elution of poliovirus adsorbed to membrane filters. *Proc. Nat. Acad. Sci. U.S.* 18:1229.

3. SOBSEY, M.D. & B.L. JONES. 1979. Concentration of poliovirus from tap water using positively charged microporous filters. *Appl. Environ. Microbiol.* 37:588.

4. SOBSEY, M.D. & J.S. GLASS. 1980. Poliovirus concentration from tap water with electropositive adsorbent filters. *Appl. Environ. Microbiol.* 40:201.

5. SOBSEY, M.D., R.S. MOORE & J.S. GLASS. 1981. Evaluating adsorbent filter performance for enteric virus concentrations in tap water. *J. Amer. Water Works Assoc.* 73:542.

6. CHANG, L.T., S.R. FARRAH & G. BITTON. 1981. Positively charged filters for virus recovery from wastewater treatment plant effluents. *Appl. Environ. Microbiol.* 42:921.

7. HEJKAL, T.W., B. KESWICK, R.L. LABELLE, C.P. GERBA, Y. SANCHEZ, G. DREESMAN, B. HAFKIN & J.L. MELNICK. 1982. Viruses in a community water supply associated with an outbreak of gastroenteritis and infectious hepatitis. *J. Amer. Water Works Assoc.* 74:318.

8. SCHLAAK, M., E. TISCHER & J.M. LOPEZ. 1983. Evaluation of current procedures for the concentration of viruses in water. *Zentralbl. Bakteriol. Microbiol. Hyg. I. Abt. Orig. B* 177:127.

9. GUTTMAN-BASS, N. & R. ARMON. 1983. Concentration of Simian rotavirus SA-11 from tap water by membrane filtration and organic flocculation. *Appl. Environ. Microbiol.* 45:850.

10. ROSE, J.B., S.N. SINGH, C.P. GERBA & L.M. KELLEY. 1984. Comparison of microporous filters for concentration of viruses from wastewater. *Appl. Environ. Microbiol.* 45:989.

11. RAPHAEL, R.A., S.A. SATTAR & V.S. SPRINGTHROPE. 1985. Rotavirus concentration from raw water using positively charged filters. *J. Virol. Methods* 11:131.

12. NUPEN, E.M. & B.W. BATEMAN. 1985. The recovery of viruses from drinking water by means of an in-line electropositive filter. *Water Sci. Technol.* 17:63.

13. TORANZOS, G.A. & C.P. GERBA. 1989. An improved method for the concentration of rotaviruses from large volumes of water. *J. Virol. Methods* 24:131.

14. SOBSEY, M.D., J.S. GLASS, R.J. CARRICK, R.R. JACOBS & W.A. RUTALA. 1980. Evaluation of the tentative standard method for enteric virus concentration from large volumes of tap water. *J. Amer. Water Works Assoc.* 72:292.

15. SOBSEY, M.D., J.S. GLASS, R.R. JACOBS & W.A. RUTALA. 1980. Modification of the tentative standard method for improved virus recovery efficiency. *J. Amer. Water Works Assoc.* 72:350.

16. WELLINGS, F.M., A.L. LEWIS & C.W. MOUNTAIN. 1976. Demonstration of solids-associated virus in wastewater and sludge. *Appl. Environ. Microbiol.* 31:354.

17. FARRAH, S.R., S.M. GOYAL, C.P. GERBA, C. WALLIS & P.T.B. SHAFFER. 1976. Characteristics of humic acid and organic compounds concentrated from tapwater using the aquella virus concentrator. *Water Res.* 10:897.

18. WALLIS, C. & J.L. MELNICK. 1967. Concentration of viruses from sewage by adsorption on Millipore membranes. *Bull. World Health Org.* 36:219.

19. SOBSEY, M.D., C. WALLIS, M. HENDERSON & J.L. MELNICK. 1973. Concentration of enteroviruses from large volumes of water. *Appl. Microbiol.* 26:529.

20. SCHMIDT, N.J. 1979. Tissue culture technics for diagnostic virology. *In* E.H. Lennette & N.J. Schmidt, eds. Diagnostic Procedures for Viral and Rickettsial Infections, 5th ed. American Public Health Assoc., Washington, D.C.

21. RAO, V.C., U. CHANDORKAR, N.U. RAO, P. KUMARAN & S.B. LAKHE. 1972. A simple method for concentrating and detecting viruses in wastewater. *Water Res.* 6:1565.

22. GERBA, C.P., S.R. FARRAH, S.M. GOYAL, C. WALLIS & J.L. MELNICK. 1978. Concentration of enteroviruses from large volumes of tap water, treated sewage, and seawater. *Appl. Environ. Microbiol.* 35:540.

9510 C.   Virus Concentration from Large Sample Volumes by Adsorption to and Elution from Microporous Filters

1. General Discussion

This section describes a two-stage process for concentrating viruses from large sample volumes. Viruses in eluate volumes too large to be conveniently and economically assayed directly in cell cultures, such as those obtained from processing large volumes of water through cartridge or large disk filters, can be concentrated further (reconcentrated) by several alternative methods. Viruses in proteinaceous eluates can be reconcentrated by either "organic flocculation,"[1,2] aluminum hydroxide adsorption-precipitation (Section 9510D), or polyethylene glycol hydroextraction-dialysis (Section 9510E). These reconcentration techniques can be used for both proteinaceous and organic buffer eluates from all types of water. Organic flocculation, now used widely, involves precipitating viruses by acidifying eluates to pH 3.5, recovering the precipitate by centrifugation, and then resuspending it in a small volume of alkaline buffer.[1]

Additionally, viruses in nonproteinaceous eluates such as glycine-NaOH can be reconcentrated by adsorption to and elution from small microporous filters. The eluate is adjusted to pH and ionic conditions for optimum virus adsorption, filtered through a secondary adsorbent, and adsorbed viruses are eluted with a small volume of eluent. This procedure can be used only for reconcentrating primary eluates obtained from processing drinking water

and other highly finished waters because of potential interfering substances likely to be present in primary eluates from natural and less finished waters.

Figure 9510:1 shows the alternative microporous filter adsorption-elution and reconcentration methods.

For general information on microporous filter techniques, see Section 9510B.1.

2. Equipment and Apparatus

a. Apparatus for first-stage concentration (Figure 9510:2):

1) *First-stage virus adsorbent filter holder.*

2) *Chemical additive system.* Use either:

a) *Fluid proportioner* with four feed pumps (quadraplex) and a mixing chamber.*

b) *Venturi-type proportioning injector*† with plastic or metal connectors (quick-disconnect type) and a length of vinyl tubing for the chemical feed line.[3] To feed two separate additives, attach a "Y" or "T" connector and two lengths of vinyl tubing to the

---

* Johanson and Son Machine Corp., Clifton, NJ, or equivalent.
† Models 202-P, 203-P or 204-P, Dema Engineering Co., St. Louis, MO, or equivalent.

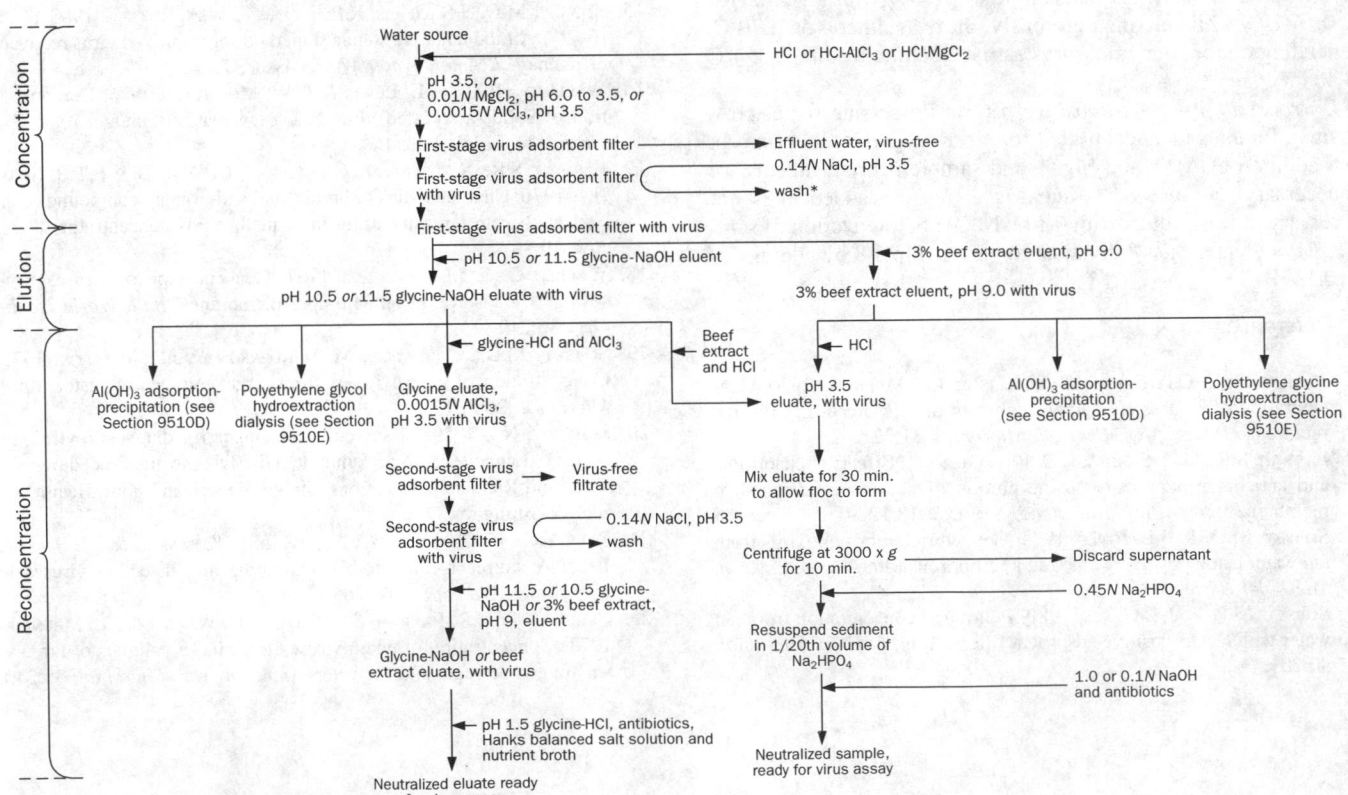

*When using MgCl₂ or AlCl₃ to enhance virus adsorption.

Figure 9510:1. Two-stage microporous filter adsorption-elution method for concentrating viruses from large volumes of water with electronegative filters.

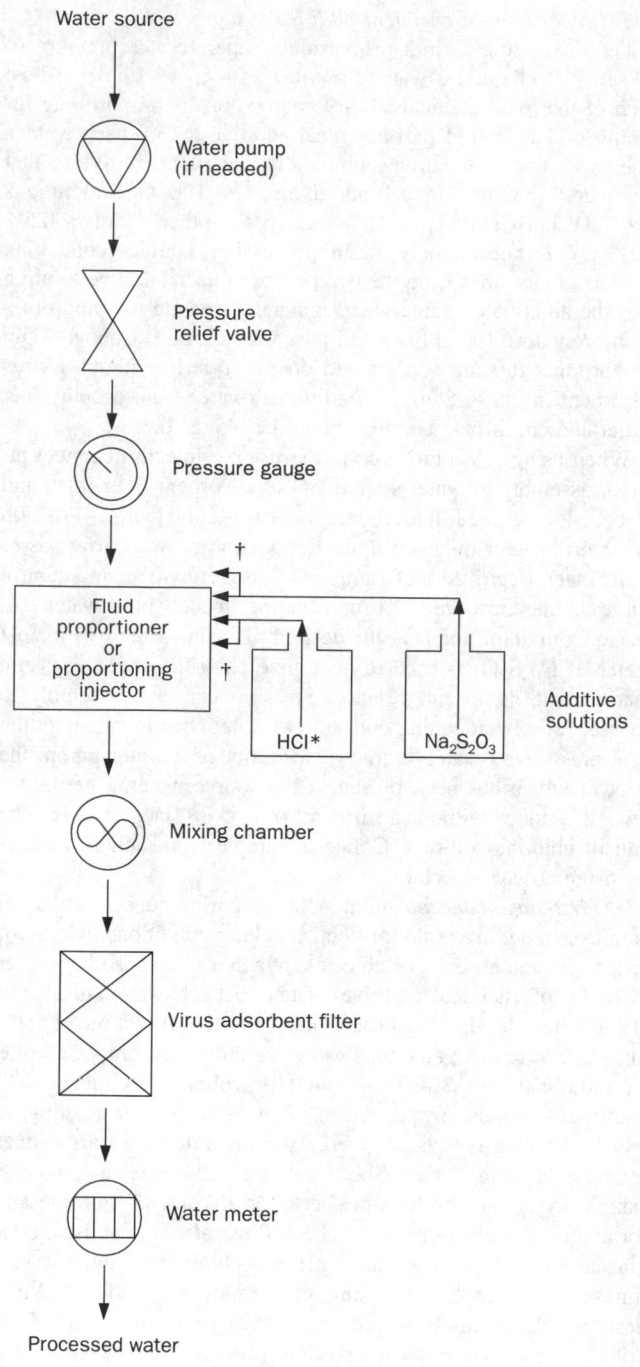

Figure 9510:2. Schematic of apparatus for first-stage concentration with negatively charged filters.

* Plus AlCl₃ or MgCl₂, if used.
† Required for fluid proportioner but not proportioning injector.

chemical feed port, or alternatively, use two separate proportioning injectors. It may be necessary to use a bypass system with the injector to prevent loss of chemical feed due to back pressure from the water line.[4] This bypass system consists of "T" pipe fittings on the injector inlet and outlet ports connected by a length of flexible hose with an in-line shut-off/control valve (see Figure 9510:2).

Proportioning injectors available commercially will process water at flow rates of 3 to 33 L/min with water-to-chemical feed ratios between 10 to 1 and 1110 to 1. Select equipment and operating conditions providing a water-to-chemical feed ratio of 100 to 1.

 3) *Water flow meter.*

 4) *Pressure gauge*, 0 to 400 kPa.

 5) *Vinyl plastic tubing*, autoclavable, with plastic or metal connectors (quick-disconnect type).

 6) *Pressure relief valve* (optional).

 7) *Carboys*, 20- to 50-L, or similar containers.

 8) *Positive pressure source* up to 400 kPa with regulator: laboratory air line, positive pressure pump, or cylinder of compressed air or nitrogen gas.

 9) *Pump* (if source water is not under pressure).

 b. pH meter.

 c. Laboratory balance.

 d. Beakers, 2- or 4-L.

 e. Pressure vessel, 4-L.

 f. Graduated cylinders, 1- and 2-L.

 g. Pipets, 1-, 5-, and 10-mL.

 h. Centrifuge with rotor and buckets for 250- to 500-mL-capacity bottles.‡

 i. Centrifuge bottles, 250- to 500-mL.

3. Materials

 a. First-stage electronegative virus adsorbent filters: Use one of the following:

 1) 293-mm-diam, 8.0- and 1.2-μm-porosity cellulose nitrate filter series.§

 2) 17.8-cm-long, 8.0-μm-porosity fiberglass-epoxy filter tube.‖

 3) 25.4-cm-long, 0.25- or 0.45-μm-porosity fiberglass-acrylic resin pleated filter cartridge.#

 b. Second-stage electronegative virus adsorbent filters: 47-mm-diam, 3.0-, 0.45-, and 0.25-μm-porosity fiberglass-acrylic resin filter series. Use to reconcentrate highly finished water samples only.#

 c. First-stage electropositive adsorbent filters: Use one of the following:

 1) 293-mm-diam surface modified cellulose and filter aid filters.**

 2) 25-cm-long, 0.20-μm-porosity surface modified thin-sheet media pleated filter cartridge.††

4. Reagents

 a. Hydrochloric acid, HCl, 0.06, 1,‡‡ and 6N.

 b. Sodium hydroxide, NaOH, 10N.

 c. Aluminum chloride, AlCl₃·6H₂O, 0.15 and 6N.‡‡

 d. Magnesium chloride, MgCl₂·6H₂O, 10N.‡‡

---

‡ Required for alternative reconcentration procedure using 3% beef extract.
§ Millipore Corp., Bedford, MA, or equivalent.
‖ Balston, Inc., Lexington, MA, or equivalent.
Filterite Corp., Timonium, MD, or equivalent.
** 50S, 60S, or 1-MDS Virozorb, CUNO, Meriden, CT, or equivalent.
†† 1-MDS Virozorb, CUNO, Meriden, CT, or equivalent.
‡‡ Recommended for first-stage virus adsorption with electronegative filters only.

e. Sodium thiosulfate, $Na_2S_2O_3 \cdot 5H_2O$, 0.5% (w/v).

f. Sodium hypochlorite, 5.25% available chlorine (household bleach).

g. Eluent: Use either:

1) *Glycine-NaOH*, pH 10.5 or 11.5: See Section 9510B.4*f*1). Use within 2 h of pH adjustment.

2) *Beef extract*, 3%, pH 9.0.§§ See Section 9510B.4*f*2).

h. Eluate neutralizing solution: Use either:

1) *Glycine-HCl*, pH 1.5: Prepare 0.05*M* glycine solution and adjust to pH 1.5 with 6*N* HCl. Add phenol red, 0.0005%, as a pH indicator. Use within 2 h of pH adjustment.

2) HCl, 1.0*N*.

i. Nutrient broth, 10X, pH 7.5: Dissolve 8.0 g nutrient broth in 90 mL distilled water, adjust to pH 7.5 with 10*N* NaOH, dilute to 100 mL with distilled water, and sterilize by autoclaving.

j. Disodium phosphate, 0.45*N*: Dissolve 40.2 g $Na_2HPO_4 \cdot 7H_2O$ in 1 L distilled water and sterilize by autoclaving.

k. Antibiotics: See Section 9510B.4*i*.

l. Sodium chloride, 0.14*N*: Dissolve 8.18 g NaCl in 1 L distilled water (necessary only with electronegative filters).

m. Hanks balanced salt solution, 10X: See Section 9510B.4*j*.

5. Procedure

When using electronegative filters, follow ¶s *a–f* below for production of primary eluate. When using electropositive filters, first see ¶ *g* for procedural modifications.

a. Sterilization of apparatus, materials, and reagents: See Section 9510B.5*a*.

b. Sample size: For drinking water use a minimum sample of 400 L, although 2000 L or more may have to be processed to detect viruses at a concentration of 1 to 2 infectious units/400 L.

c. Preparation of feed solutions for electronegative filters: Use an HCl additive solution to adjust sample pH to 3.5 for virus adsorption to filters. If acidification to pH 3.5 is inadequate for obtaining maximum virus adsorption, add either $AlCl_3$ or $MgCl_2$ solution.

When only HCl is used, prepare additive solution as follows: Determine concentration of HCl additive solution by titrating a 1-L sample of dechlorinated water to pH 3.5 with 0.06*N* HCl and noting volume required. The volume, in milliliters, of titrant required is equal to the volume of 6*N* HCl needed/L distilled water for making the additive solution. Make at least 5 L additive solution for 400 L of sample.

When $AlCl_3$ is used to enhance virus adsorption use pH 3.5 and a final concentration of added $AlCl_3$ of 0.0015*N*. Because $AlCl_3$ is an acid salt, titrate a 1-L sample to about pH 4.0 with 0.06*N* HCl, add $AlCl_3$ to a concentration of 0.0015*N* and continue titration to pH 3.5, noting volume of titrant used. Prepare additive solution by adding titrant volume (mL) of 6.0*N* HCl/L of 0.15*N* $AlCl_3$.

When $MgCl_2$ is used to enhance virus adsorption, use a pH between 3.5 and 6.0 and a final concentration of added $MgCl_2$ of 0.1*N*. To prepare additive solution titrate a 1-L sample to desired pH with 0.06*N* HCl as previously described and note volume of titrant used. Add the titrant volume of 6.0*N* HCl/L 10*N* $MgCl_2$ to make the additive solution.

---

§§ For alternative reconcentration procedure: organic flocculation.

d. Preparation of chemical additive system:

1) When using a fluid proportioner, operate at a pressure of 100 to 700 kPa and a water flow rate of 4 to 40 L/min. Adjust each of the four chemical additive pumps of the proportioner for a ratio of 1 to 200 (1 part chemical additive to 200 parts water). Use two pumps, operating reciprocally, for each additive so that the overall dilution for each additive is 1 to 100. One additive is either HCl, HCl-$AlCl_3$ or HCl-$MgCl_2$; the other additive, 0.5% $Na_2S_2O_3$, is needed only when processing samples containing chlorine. Place lines from the two pumps of each additive solution into the additive containers and manually operate the pump metering rods to fill feed lines and purge them of air. Connect fluid proportioner to source water and operate briefly without a virus adsorbent in place. Sample conditioned water from proportioner outlet and check pH. The pH should be 3.5 ± 0.3.

When using a Venturi-type proportioning injector, connect injector assembly to water source and to adsorbent filter inlet, and place additive feed line(s) into additive container(s). Position valve on injector outlet to drain line position (away from adsorbent filter). Begin flow of sample. Adjust screw-operated control valve on chemical feed of proportioning injector until water collected from drain line is at the desired pH as measured with a pH meter. If $Na_2S_2O_3$ is used to neutralize chlorine, check to insure that chlorine is absent. Connect virus concentrator assembly to source water by attaching concentrator inlet hose to valved outlet of a pressurized water source or to outlet of a water pump, the inlet of which has been placed in the source water. Operate for several minutes without a virus adsorbent in place to purge the unit of chlorine solution. Collect a sample from outlet of meter to insure absence of chlorine.

e. First-stage concentration: After preparing concentration apparatus and additive solutions and checking conditioned water for proper pH and absence of chlorine, attach a virus adsorbent filter to outlet of chemical additive system. Attach water meter and effluent hose to virus adsorbent outlet. Record initial meter reading and add to this value the desired volume to be processed plus an additional 1 or 2% (to account for volume of either 1 or 2 additive solutions, respectively). This gives meter reading at which sampling is to be stopped. Turn on water and start a timer (or record starting time). Shortly after filtration begins collect a sample from filter outlet and check for absence of chlorine and for appropriate pH value. Also check flow rate. Do not use a flow rate above 40 L/min. Recheck pH and chlorine residual several times during sample processing, or monitor continuously. When desired volume has been processed, turn water off. Purge filter holder of excess water with positive pressure from an air or nitrogen gas source.

f. Washing and virus elution: If $AlCl_3$ or $MgCl_2$ has been used, wash excess Al^{3+} or Mg^{2+} from filter with 4 L 0.14*N* NaCl. Omit washing if only HCl was used. Place wash solution in a 4-L pressure vessel and pass through filter with positive pressure. Purge filter of excess wash solution with positive pressure and discard entire filtrate.

Using aseptic technique, elute virus from filter as soon as possible in the field or after returning to the laboratory. If filter holders with adsorbed viruses must be returned to the laboratory, seal filter holder openings, place filter holder in a sterile plastic bag, and chill.

Use pH 10.5 or 11.5 glycine-NaOH or 3% beef extract, pH 9.0, to elute viruses from first-stage adsorbent filters. Because some viruses are inactivated when pH 11.5 glycine-NaOH is used, alternatively elute with pH 10.5 glycine-NaOH or 3% beef extract, pH 9.0.[1,3,4]

To elute, place eluent in a pressure vessel. Use minimum eluent volumes of 1 L and 300 mL for cartridge and 293-mm-diam disk filters, respectively. To elute with pH 11.5 glycine-NaOH, connect pressure vessel to inlet of filter holder and with pressure relief valve on filter holder open, apply a small positive pressure to the system so that eluent fills void volume of filter holder. When eluent begins to discharge from pressure relief valve, quickly close it. Filter remaining eluent slowly through filter within 1 to 2 min and collect filtrate (eluate) in a sterile 2- or 4-L beaker. When filtrate no longer appears, slowly increase pressure to force additional fluid from filter. If using pH 11.5 glycine-NaOH eluent, immediately check filtrate pH and if it is less than 11.0, elute with 1 L more of pH 11.5 glycine-NaOH. Immediately after checking pH, adjust filtrate to a pH between 7.5 and 9.5 with pH 1.5 glycine-HCl while mixing vigorously. Complete elution and pH adjustment to 7.5 to 9.5 in 5 min or less to avoid possibility of appreciable virus inactivation.

To elute with pH 10.5 glycine-NaOH, use either batch or continuous-flow eluent recirculation. For the batch method, begin elution as with pH 11.5 glycine-NaOH. Collect the filtrate, measure pH, and readjust to pH 10.5 with 1.0 or 0.1N NaOH while mixing vigorously. Then, using this eluate, elute filters four more times, readjusting filtrate to pH 10.5 before each elution. After the fifth elution, adjust filtrate to pH 7.4 with pH 1.5 glycine-HCl or 1.0N HCl while mixing vigorously.

Alternatively, elute with pH 10.5 glycine-NaOH by continuous recirculation. Place eluent in a sterile beaker. Attach short lengths of sterile vinyl or rubber tubing to inlet and outlet openings of filter holder and place free ends of tubing in eluent beaker; slip midsection of filter inlet tubing into a peristaltic or roller pump. Open pressure relief valve on filter holder and operate pump at slow speed so that eluent fills void volume of filter holder. When eluent begins to discharge from pressure relief valve, quickly close it. Increase pump speed so that eluent recirculates through filter assembly and beaker at a minimum flow rate of 100 mL/min. After 5 min recirculation, remove filter inlet tube from beaker and pump remaining fluid from filter assembly. Connect filter inlet to positive pressure source to force additional eluent from filter. Adjust eluate to pH 7.4 with pH 1.5 glycine-HCl or 1.0N HCl while mixing vigorously.

To elute with 3% beef extract, pH 9.0, follow the procedure described above for pH 11.5 glycine-NaOH. Adjust collected filtrate to pH 7.4 with pH 1.5 glycine-HCl or 1N HCl while mixing vigorously. The 5 min time limit to complete elution with pH 11.5 glycine-NaOH is not necessary when beef extract is used.

g. Sample processing of electropositive filters: Processing for electropositive filters is identical to that for electronegative filters except that addition of Al^{3+} and Mg^{2+} and sample pH adjustments are unnecessary; because Al^{3+} and Mg^{2+} are not added, it is not necessary to wash filters with 0.14N NaOH before elution. If the sample pH is greater than 8.0, adjust to less than pH 8 by adding 1.0 or 0.1N HCl.

h. Reconcentration of primary eluates: Further concentrate (reconcentrate) viruses in primary eluates either by organic flocculation, $Al(OH)_3$ adsorption-precipitation (Section 9510D), polyethylene glycol hydroextraction-dialysis (Section 9510E), or

adsorption to and elution from microporous filters. The latter technique can be used only for glycine or other organic buffer eluates.

To concentrate (reconcentrate) viruses in glycine eluates by adsorption to and elution from filters, adjust to pH 3.5 with pH 1.5 glycine-HCl and add $AlCl_3$ to a final concentration of 0.0015N while mixing vigorously. Transfer sample to a 4-L pressure vessel. Filter through a 47-mm-diam 3.0-, 0.45-, and 0.25-μm-porosity fiberglass-acrylic resin filter series at a flow rate of no more than 130 mL/min and discard filtrate. Rinse filters with 25 mL 0.14N NaCl to remove excess Al^{3+}. Pipet NaCl solution directly into filter inlet or place in a small pressure vessel connected to the inlet. Use positive pressure to pass NaCl solution through filter and discard filtrate. Elute adsorbed viruses from filter with 7-mL portions of either pH 10.5 or 11.5 glycine-NaOH or 3% beef extract, pH 9.0. Pipet 7 mL eluent directly into filter holder inlet or into a small pressure vessel connected to filter inlet and connect to a positive pressure source. Carefully apply positive pressure so that eluate flows slowly from filter outlet into a sterile container. When filtrate no longer flows from outlet, increase pressure to force retained fluid from filters. If using pH 11.5 glycine-NaOH, measure eluate pH and immediately adjust to pH between 7.5 and 9.5 with pH 1.5 glycine-HCl. Repeat this elution procedure with another 7-mL portion of pH 11.5 glycine-NaOH. Complete reconcentration within 5 min. If neither eluate portion had a final pH of 11.0 or more, repeat elution procedure with additional 7-mL portions of pH 11.5 glycine-NaOH until an eluate portion has a pH of at least 11.0. Combine all eluates.

If using pH 10.5 glycine-NaOH, elute five successive times with 7-mL volumes of eluent. After each elution, readjust eluate to pH 10.5 with 0.1N NaOH while mixing vigorously. After the fifth elution, adjust eluate to pH 7.4 with pH 1.5 glycine-HCl or 0.1N HCl while mixing vigorously.

If using 3% beef extract, pH 9.0, elute with two 7-mL volumes, combine filtrates, and adjust to pH 7.4 if necessary.

Measure total eluate volume. For glycine eluates, add 1/10th the measured sample volume of 10X Hanks balanced salt solution and 10X nutrient broth. To all eluates add appropriate volumes of antibiotics (1/10th volume penicillin-streptomycin or 1/100th volume gentamycin-kanamycin, or both). Store at 4 or −70°C, depending on time until virus assay.

Further concentrate viruses in beef extract eluates by precipitation at pH 3.5 (organic flocculation). Viruses in glycine eluates also can be reconcentrated by this technique by first supplementing them with beef extract to a final concentration of 1 to 3%. Use sterile beef extract paste (about 80% beef extract) or sterile 20% beef extract solution made from powder to bring glycine eluates to the desired beef extract concentration. While mixing vigorously, adjust eluate to pH 3.5 by adding 1N HCl dropwise. Continue to mix at slow speed for 30 min and centrifuge at 3000 × g for 10 min. Decant and discard supernatant. With vigorous mixing, resuspend sediment in 1/20 the initial sample volume of 0.45N Na_2HPO_4. Add antibiotics (1/10 final sample volume penicillin-streptomycin, 1/100 final sample volume gentamycin-kanamycin, or both) and while mixing vigorously adjust to pH 7.4 with 1.0 or 0.1N NaOH. Check electrical conductivity of sample. If conductivity is > 13 000 μmhos, dialyze sample against Hanks balanced salt solution before assay. Store at 4 or −70°C, depending on time until virus assay.

6. References

1. KATZENELSON, E., B. FATTAL & T. HOSTOVESKY. 1976. Organic floc-culation: an efficient second-step concentration method for the detection of viruses in tapwater. *Appl. Environ. Microbiol.* 32:638.
2. BITTON, G., B.N. FELDBERG & S.R. FARRAH. 1979. Concentration of enteroviruses from seawater and tap water by organic flocculation using non-fat dry milk and casein water. *Air Soil Pollut.* 10:187.
3. PAYMENT, P. & M. TRUDEL. 1980. A simple low cost apparatus for conditioning large volumes of water for virological analysis. *Can. J. Microbiol.* 26:548.
4. PAYMENT, P. & M. TRUDEL. 1981. Improved method for the use of proportioning injectors to condition large volumes of water for virological analysis. *Can. J. Microbiol.* 27:455.

9510 D.   Virus Concentration by Aluminum Hydroxide Adsorption-Precipitation

1. General Discussion

Viruses can be concentrated from small volumes of water, wastewater, and adsorbent filter eluates by precipitation with aluminum hydroxide.[1-4] This process probably involves both electrostatic interactions between the negatively charged virus surface and the positively charged aluminum hydroxide [Al(OH)$_3$] surfaces and coordination of the virus surface by hydroxo-aluminum complexes.[5] Viruses are adsorbed to an Al(OH)$_3$ precipitate that is either added to the sample or formed in the sample from a soluble aluminum salt and a base such as sodium carbonate (Na$_2$CO$_3$) or sodium hydroxide (NaOH). Viruses are allowed to adsorb to the Al(OH)$_3$ precipitate and the virus-containing precipitate is collected by filtration or centrifugation. The recovered precipitate may be inoculated directly into laboratory hosts for virus assay or the viruses are eluted from the precipitate with an alkaline buffer or a proteinaceous solution before virus assay.

The major limitations of this method are that sample size is limited to perhaps a few liters, soluble organic matter can interfere with virus adsorption, and virus recovery from the precipitate may be incomplete. Virus adsorption may be improved by forming the Al(OH)$_3$ precipitate in the sample instead of adding it preformed. Although virus adsorption can be maximized by using large amounts of Al(OH)$_3$, the adsorbed viruses become more difficult to elute. Therefore, some intermediate amount of Al(OH)$_3$ is used to achieve maximum virus recovery. Also, Al(OH)$_3$ is a relatively nonspecific adsorbent so that other substances may be concentrated with viruses. The presence of such impurities may cause the concentrated sample to be toxic for the cell cultures normally used for virus assay.

Several modifications of the Al(OH)$_3$ adsorption-precipitation procedure have been used to concentrate viruses from water, wastewater, and eluates from adsorbent filters. Initially, preformed Al(OH)$_3$ precipitates were made by adding Na$_2$CO$_3$ to AlCl$_3$ solutions and the Al(OH)$_3$ precipitate was resuspended in 0.15N NaCl. This was added to the wastewater and the mixture was stirred gently for 1 h or more to allow viruses to adsorb to the precipitate. The precipitate was recovered by filtration, resuspended in cell culture media, and inoculated into cell cultures.[3,4] More recent procedural modifications include: (*a*) Al(OH)$_3$ precipitate formation within the sample,[1,2,6-8] (*b*) recovery of the Al(OH)$_3$ precipitate by centrifugation followed by elution of viruses from the precipitate with alkaline eluents,[1,2,6,7] and (*c*) a large-volume method in which the precipitate is formed in the sample and collected on a cartridge filter, and viruses are eluted from the precipitate on the filter with alkaline eluent.[9] The method described here is for relatively small sample volumes and uses Al(OH)$_3$ that is either preformed or generated within the sample. The latter modification is preferable because some viruses are not adsorbed efficiently by preformed precipitates.[10]

2. Equipment and Apparatus

a. Centrifuge, with rotor and buckets, capable of operating at about 1900 × *g.*

b. Centrifuge bottles and tubes.

c. Beaker, 100-mL or larger.

d. pH meter.

e. Magnetic stirrer and stirring bars or alternative mixing device.

f. Graduated cylinders, 100-mL or larger.

g. Pipets, 1-, 5-, and 10-mL.

h. Laboratory balance.

*i. Vacuum-type filter holder or Buchner filter funnel,** 47-mm diam or larger.

*j. Filter flask.**

*k. Spatula,** flat blade, metal or autoclavable plastic.

*l. Vacuum source,** vacuum pump or laboratory vacuum line.

3. Materials

*Filter:** Fiberglass-acrylic resin filter† or microporous filter, 0.45-μm porosity,‡ 47-mm diam or larger. To prevent virus adsorption, filter 0.1% polyoxyethylene sorbitan monooleate solution (¶ 4*h*) through the filters, using about 1 mL solution/cm^2 of filter surface area. Rinse filter with distilled water, using about 10 mL/cm^2 of filter surface area. Sterilize treated filters by autoclaving.

4. Reagents

a. Hydrochloric acid, HCl, 0.1 and 1.0N.

b. Sodium hydroxide, NaOH, 0.1 and 1.0N.

c. Sodium carbonate, Na$_2$CO$_3$, 4N§.

d. Aluminum chloride, AlCl$_3$, 0.075N§ or 0.9N.

e. Sodium chloride, NaCl, 0.14N.

f. Beef extract, 3%, pH 7.4: Dissolve 3 g beef extract paste or 2.4 g beef extract powder in 90 mL distilled water, adjust to pH

---

* Required for optional method for collecting Al(OH)$_3$ precipitate from sample.
† Millipore AP20 or equivalent.
‡ Millipore HA or equivalent.
§ For alternative procedure using preformed Al(OH)$_3$ precipitate.

7.4 with 1.0 or 0.1N NaOH, dilute to 100 mL with distilled water, and sterilize by autoclaving.

g. Antibiotics: Use either:

1) *Penicillin-streptomycin,* 10X, containing 5000 IU penicillin/mL and 5000 µg streptomycin/mL. Available commercially or prepare by dissolving powdered sodium or potassium penicillin-G and streptomycin sulfate in distilled water and sterilizing by filtration.

2) *Gentamycin-kanamycin,* 100X, containing 5000 µg/mL each of gentamycin base and kanamycin base (Section 9510B.4*i*).

h. Polyoxyethylene sorbitan monooleate,‖ 0.1% (*v/v*) in distilled water.

5. Procedure

a. Sterilization of apparatus, materials, and reagents: See Section 9510B.5*a*.

b. Preparation of preformed Al(OH)$_3$ precipitate: While mixing 100 mL 0.075N AlCl$_3$ at room temperature, slowly add 4N Na$_2$CO$_3$ solution to form precipitate and adjust to pH 7.2. Continue mixing for 15 min and, if necessary, add more Na$_2$CO$_3$ to maintain pH 7.2. Centrifuge at 1100 $\times$ g for 15 min and discard supernatant. Resuspend sediment in 0.14N NaCl and recentrifuge. Discard supernatant, resuspend sediment in 0.14N NaCl, and sterilize by autoclaving. Cool, centrifuge again, decant supernatant, and resuspend Al(OH)$_3$ sediment in 50 mL sterile 0.14N NaCl. Store at 4°C.

c. Sample size, collection, and storage: Process samples of no more than several liters because the method is too cumbersome and time-consuming for larger volumes. See Section 9510B.5*d* for sample collection and storage procedures.

d. Sample processing: Do not prefilter sample[11,12] because substantial virus losses can occur. Adjust sample to pH 6.0 with 1.0 or 0.1N HCl while mixing vigorously. Form Al(OH)$_3$ precipitate in sample by adding 1 part 0.9N AlCl$_3$ solution to 100 parts sample to give a final 0.009N Al^{3+} concentration. Check sample pH and re-adjust to 6.0 with 1.0 or 0.1N NaOH or HCl, if necessary. Mix slowly for 15 min at room temperature.

Alternatively, use preformed Al(OH)$_3$ precipitate by adding 1 part stock Al(OH)$_3$ suspension/100 parts sample and mix slowly for 2 h at 4 to 10°C to allow for virus adsorption.

Collect virus-containing Al(OH)$_3$ precipitate by centrifugation or filtration. To collect precipitate by centrifugation, centrifuge at 1700 $\times$ g for 15 to 20 min, discard supernatant, and resuspend sediment in 1/1000 to 1/20 original sample volume of 3% beef extract, pH 7.4.

‖ Tween 80®, ICI United States, Inc., Wilmington, DE, or equivalent. Required for optional method for collecting Al(OH)$_3$ precipitate from sample.

To collect precipitate by filtration, vacuum filter sample through a treated filter (¶ 3 above) held in a vacuum-type filter holder or Buchner funnel, using additional filters if filter clogs before entire sample is filtered. Carefully scrape precipitate from filter(s) with a sterile spatula and resuspend in 1/1000 to 1/20 original sample volume of 3% beef extract, pH 7.4.

Regardless of collection method, vigorously mix the Al(OH)$_3$ beef extract suspension and, if necessary, adjust to pH 7.4 with 0.1N HCl or NaOH. Continue mixing for a total of 10 min. Centrifuge at 1900 $\times$ g for 30 min. Decant supernatant, add 1/10 the volume of the concentrate of penicillin-streptomycin solution or 1/100 volume of gentamycin-kanamycin and store at 4 or −70°C.

6. References

1. PAYMENT, P., C.P. GERBA, C. WALLIS & J.L. MELNICK. 1976. Methods for concentrating viruses from large volumes of estuarine water on pleated membranes. *Water Res.* 10:893.
2. FARRAH, S.R., S.M. GOYAL, C.P. GERBA, C. WALLIS & J.L. MELNICK. 1977. Concentration of enteroviruses from estuarine water. *Appl. Environ. Microbiol.* 33:1192.
3. WALLIS, C. & J.L. MELNICK. 1967. Concentration of viruses on aluminum hydroxide precipitates. *In* G. Berg, ed. Transmission of Viruses by the Water Route. Interscience Publ., New York, N.Y.
4. WALLIS, C. & J.L. MELNICK. 1967. Virus concentration on aluminum and calcium salts. *Amer. J. Epidemiol.* 85:459.
5. COOKSON, J.T., JR. 1974. The chemistry of virus concentration by chemical methods. *Develop. Ind. Microbiol.* 15:160.
6. LYDHOLM, B. & A.L. NIELSEN. 1979. Methods for detection of virus in wastewater applied to samples from small scale treatment systems. *Water Res.* 14:169.
7. SELNA, M.W. & R.P. MIELE. 1977. Virus sampling in wastewater-field experiences. *J. Environ. Eng. Div., Proc. Amer. Soc. Civil Eng.* 103:693.
8. DOBBERKAU, H.J., R. WALTER & S. RUDIGER. 1981. Methods for virus concentration from water. *In* M. Goddard & M. Butler, eds. Viruses and Wastewater Treatment. Pergamon Press, New York, N.Y.
9. FARRAH, S.R., C.P. GERBA, C. WALLIS & J.L. MELNICK. 1978. Concentration of poliovirus from tapwater onto membrane filters with aluminum chloride at ambient pH levels. *Appl. Environ. Microbiol.* 35:624.
10. FARRAH, S.R., G.M. GOYAL, C.P. GERBA, R.H. CONKLIN & E.M. SMITH. 1978. Comparison between adsorption of poliovirus and rotavirus by aluminum hydroxide and activated sludge flocs. *Appl. Environ. Microbiol.* 35:360.
11. SOBSEY, M.D., C.P. GERBA, C. WALLIS & J.L. MELNICK. 1977. Concentration of enteroviruses from large volumes of turbid estuary water. *Can. J. Microbiol.* 23:770.
12. HOMMA, A., M.D. SOBSEY, C. WALLIS & J.L. MELNICK. 1973. Virus concentration from sewage. *Water Res.* 7:945.

9510 E. Hydroextraction-Dialysis with Polyethylene Glycol

1. General Discussion

Polyethylene glycol (PEG) hydroextraction is an ultrafiltration process in which the sample is placed in a cellulose dialysis bag and exposed to PEG, a hygroscopic material. Water and microsolutes leave the sample by passing across the semipermeable dialysis membrane into the hygroscopic PEG.[1] Viruses and other macrosolutes, including PEG, cannot cross the dialysis membrane. The sample volume in the dialysis bag is reduced by water loss to the PEG, thereby concentrating viruses and other macrosolutes. The viruses retained in the dialysis bag are recovered by opening the bag, collecting the remaining sample, and eluting any viruses possibly adsorbed to the inner walls of the bag with a small volume of slightly alkaline proteinaceous solution such as 3% beef extract, pH 9.0. The collected concentrate and eluate are combined and assayed for viruses.

The main limitations of this method are that only small samples (less than 1 L) can be processed conveniently, virus elution from the walls of the dialysis bag may be incomplete unless the elution is done painstakingly, and other macrosolutes in the sample that are concentrated with viruses may interfere with virus assays by being cytotoxic.

Initial investigations of this method reported low and highly variable virus recoveries from wastewater.[2,3] The type of dialysis tubing and eluent solution as well as the thoroughness of the elution step have been found to influence virus recovery efficiency. More recently, with modified procedures, efficient and consistent virus recoveries have been obtained from wastewater and from adsorbent filter eluates.[4,5]

2. Equipment and Apparatus

a. *Beakers,* 100-mL or larger.

b. *Graduated cylinders,* 100-mL or larger.

c. *Dialysis tubing clamps.**

d. *Pan,* approximately 30 × 30 × 12 cm, autoclavable.

e. *Magnetic stirrer* and stirring bars or alternative mixing device.

f. *Centrifuge,* with rotor and buckets, capable of operating at about 1900 × g.

g. *pH meter.*

h. *Pipets,* 1-, 5-, and 10-mL.

i. *Tape roller†* or similar device to aid in washing the inside walls of dialysis bags with eluting fluid.

j. *Ultrasonic disruptor-emulsifier,‡* probe type, capable of generating 100 W of acoustical output.

3. Materials

a. *Dialysis tubing,* seamless, regenerated cellulose, 4.8-nm average pore diameter.§

b. *Polyethylene glycol (PEG),‖* dry flakes.

4. Reagents

See Section 9510D.4.

5. Procedure

a. *Sterilization of apparatus, materials, and reagents:* See Section 9510B.5a. Do not sterilize PEG.

b. *Sample size, collection, and storage:* Process samples of no more than a few hundred milliliters. See Section 9510B.5d for sample collection and storage procedures.

c. *Preparation of dialysis tubing:* Cut a length of dialysis tubing long enough to accommodate entire sample. Close one end with a clamp. Do not tie knots to close dialysis tubing. Fill tubing bag with distilled water, sterilize by autoclaving, and let cool.

d. *Sample processing:* Aseptically remove dialysis bag from distilled water and drain. Fill bag with sample and close open end with a second clamp. Place bag in a pan containing a 5-cm layer of PEG, making sure that bag does not touch pan walls. Cover tubing with an additional 5 cm PEG and store at 4°C (for about 18 h) until sample volume has been reduced to no more than a few milliliters. (If PEG 6000 is used the process time is reduced to 4 to 6 h.) Although sample may be allowed to dewater completely, do not let it remain in this state.

Remove dialysis bag from PEG and quickly wash PEG from outside of bag with sterile distilled water. Remove clamp from one end of bag and carefully collect sample concentrate. Add about 1/200 to 1/20 the original sample volume of 3% beef extract, pH 9.0, and clamp closed. Thoroughly wash inside walls of bag with beef extract by rubbing fluid from one end to the other several times using either fingers or a roller device. Remove clamp from one end of bag and collect fluid, kneading or squeezing to recover the last traces. Add recovered fluid to previously collected sample concentrate.

Adjust to pH 7.5 with 1.0 or 0.1N HCl while mixing vigorously. To disperse solids-associated viruses in sample, stir overnight (about 18 h) in the cold (about 4°C) or treat with ultrasonics at 100 W for 1 to 2 min. Prevent sample temperature from rising above 37°C during ultrasonic treatment by chilling in an ice bath. Centrifuge at 1900 × g for 30 min. Decant supernatant, add 1/10 the volume of the concentrate of penicillin-streptomycin solution or 1/100 volume of gentamycin-kanamycin, and store at 4 or −70°C.

6. References

1. SOBSEY, M.D. 1976. Methods for detecting enteric viruses in water and wastewater. *In* G. Berg, H.L. Bodily, E.H. Lennette, J.L. Melnick & T.G. Metcalf, eds. Viruses in Water. American Public Health Assoc., Washington, D.C.
2. CLIVER, D.O. 1967. Detection of enteric viruses by concentration with polyethylene glycol. *In* G. Berg., ed. Transmission of Viruses by the Water Route. Interscience Publ., New York, N.Y.
3. SHUVAL, H.I., S. CYMBALISTA, B. FATTAL & N. GOLDBLUM. 1967. Concentration of enteric viruses in water by hydro-extraction and two-phase separation. *In* G. Berg, ed. Transmission of Viruses by the Water Route. Interscience Publ., New York, N.Y.
4. WELLINGS, F.M., A.L. LEWIS, C.W. MOUNTAIN & L.V. PIERCE. 1975. Demonstration of virus in groundwater after effluent discharge onto soil. *Appl. Microbiol.* 29:751.
5. RAMIA, S. & S.A. SATTAR. 1979. Second-step concentration of viruses in drinking and surface waters using polyethylene glycol hydroextraction. *Can. J. Microbiol.* 25:587.

---

* Fisher Scientific No. 8-670-11A or equivalent.

† Optional, Fisher Scientific No. 14-245-21 or equivalent.

‡ Optional.

§ Made by Union Carbide Corp. and available from many scientific supply companies.

‖ Carbowax 20 000 or 6000 or equivalent.

9510 F.   Recovery of Viruses from Suspended Solids in Water and Wastewater

1. General Discussion

Viruses in the aquatic environment often are associated with solids or particulate matter, either adsorbed to particulate surfaces or embedded within the solid.[1-3] Both freely suspended and solids-associated viruses are concentrated from water by the methods described above. There is evidence that solids-associated viruses are not eluted efficiently from adsorbent filters or from $Al(OH)_3$ precipitates and organic flocs. Recovery of solids-associated viruses by microporous filter methods employing in-situ elution is inconsistent.[2] Solids-associated viruses on adsorbent filters are eluted more efficiently by disrupting filters in elution fluid than by in-situ elution,[2] but this is cumbersome and time-consuming, especially for large-diameter disk filters and cartridge filters.

For small volumes of water and wastewater, solids-associated viruses can be recovered expediently by separating the solids by centrifuging, decanting the supernatant, and eluting viruses from the solids by resuspending in a small volume of eluent.[4] Viruses in the supernatant can be concentrated by one of the procedures described in Sections 9510B, C, D, or E. Viruses eluted from the resuspended solids are separated from the solids by centrifuging and are assayed directly or concentrated further by organic flocculation.[5,6] Major limitations of these methods are incomplete virus elution and poor virus recoveries due to interferences from sample constituents.

2. Equipment and Apparatus

a. *Centrifuge,* with rotor and buckets for 250- to 1000-mL-capacity bottles, capable of operating at about $1250 \times g$.

b. *Centrifuge bottles,* 250- to 1000-mL.

c. *pH meter.*

d. *Laboratory balance.*

e. *Graduated cylinder,* 250-mL or larger.

f. *Beaker,* 250-mL or larger.

g. *Sample bottles,* 250-mL or larger.

h. *Magnetic stirrer and stirring bars,* or alternative mixing device.

i. *Pipets,* 1-, 5-, and 10-mL.

3. Reagents

a. *Hydrochloric acid,* HCl, 0.1 and 1.0*N.*

b. *Sodium hydroxide,* NaOH, 0.1 and 1.0*N.*

c. *Eluent:* Dissolve 10 g beef extract, 1.34 g disodium phosphate heptahydrate, $Na_2HPO_4 \cdot 7H_2O$, and 0.12 g citric acid in 90 mL distilled water, adjust to pH 7.0 with 1*N* HCl or NaOH, dilute to 100 mL with distilled water, and sterilize by autoclaving.

d. *Antibiotics:* See Section 9510B.4*i.*

4. Procedure

a. *Sterilization of apparatus, materials, and reagents:* See Section 9510B.5*a.*

b. *Sample size, collection, and storage:* Collect and process samples of no more than 10 L, depending on capacity of centrifuge. See Section 9510B.5*d* for sample collection and storage procedures.

c. *Sample processing:* Aseptically transfer 250- to 1000-mL sample volumes to centrifuge bottles and centrifuge at $1250 \times g$ for 20 min. Decant and pool supernatants for subsequent processing for viruses by one of the methods for water or wastewater described previously.

Elute viruses from the sedimented solids by resuspending in eluent. Use 40 mL eluent per quantity of sediment from 250 mL of original sample. Pool resuspended sediments from multiple centrifuge bottles in a sterile beaker. Alternatively, keep resuspended sediments from small numbers of centrifuge bottles in the bottles and process them individually. While vigorously mixing with a magnetic stirrer, adjust to pH 7.0 by slowly adding 1*N* NaOH or HCl, if necessary. Reduce mixing speed and continue mixing for 30 min. During this period, check sample pH and readjust to pH 7.0 as necessary. As an alternative to mixing for 30 min, sonicate samples at 100 W for 15 min in a rosette cooling cell maintained at 4°C. Return sample to centrifuge bottles. Centrifuge at $1250 \times g$ and 4°C for 15 min, collect supernatant for subsequent assay or further concentration, and discard the sediment.

If desired, further concentrate viruses from this supernatant by organic flocculation (see Section 9510E). For supernatants that will be assayed directly for viruses with no further concentration, adjust to pH 7.4, add 1/10 the volume of sample of penicillin-streptomycin or 1/100 volume of gentamycin-kanamycin and store at 4 or −70°C.

5. References

1. SCHAUB, S.A. & B.P. SAGIK. 1975. Association of enteroviruses with natural and artificially introduced colloidal solids in water and infectivity of solids-associated virions. *Appl. Microbiol.* 30:212.
2. WELLINGS, F.M., A.L. LEWIS & C.W. MOUNTAIN. 1976. Viral concentration techniques for field sample analysis. *In* L.B. Baldwin, J.M. Davidson & J.F. Gerber, eds. Virus Aspects of Applying Municipal Waste to Land. Univ. Florida, Gainesville.
3. WELLINGS, F.M., A.L. LEWIS & C.W. MOUNTAIN. 1974. Virus survival following wastewater spray irrigation of sandy soils. *In* J.F. Malina, Jr. & B.P. Sagik, eds. Virus Survival in Water and Wastewater Systems. Univ. Texas, Austin.
4. BERG, G. & D.R. DAHLING. 1980. Method for recovering viruses from river water solids. *Appl. Environ. Microbiol.* 39:850.
5. GERBA, C.P. 1982. Detection of viruses in soil and aquatic sediments. *In* C.P. Gerba & S.M. Goyal, eds. Methods in Environmental Virology. Marcel Dekker, Inc., New York, N.Y.
6. FARRAH, S.R. 1982. Isolation of viruses associated with sludge particles. *In* C.P. Gerba & S.M. Goyal, eds. Methods in Environmental Virology. Marcel Dekker, Inc., New York, N.Y.

9510 G. Assay and Identification of Viruses in Sample Concentrates

1. Storage of Sample Concentrates

Because it often is impossible to assay sample concentrates immediately, store them at room temperature (about 25°C) for up to 2 h or at refrigerator temperatures (4 to 10°C) for up to 48 h to minimize virus losses. Freeze samples requiring storage longer than 48 h at −70°C or less. Do not freeze samples at −10 to −20°C because extensive inactivation of some enteric viruses may occur. Store sample concentrates from finished waters in separate freezers or physically separated from other virus-containing material in common freezers.

2. Decontamination of Sample Concentrates

Sample concentrates, especially those from wastewater, are likely to be contaminated with bacteria and fungi that can overgrow cell cultures and interfere with virus detection and assay. Do not decontaminate by centrifugation or filtration because virus losses are likely to occur. For many samples, especially those from finished waters, contamination is controlled adequately by antibiotics such as penicillin-streptomycin or gentamycin-kanamycin that are added immediately after the sample is obtained. To provide additional protection against fungal contamination, add amphotericin B or nystatin at concentrations of 2.5 and 50 μg/mL, respectively.[1] If penicillin-streptomycin or gentamycin-kanamycin are inadequate, use one or more additional antibiotics such as aureomycin, neomycin, or polymyxin B. To maximize the antibiotic effects, incubate samples for 1 to 3 h at 25 to 37°C after adding the antibiotics. Bacterial destruction is further enhanced by freezing at −70°C after incubation with antibiotics. Keep samples frozen until assayed for viruses. To determine if antibiotic treatment has been effective, plate a small subsample on a general-purpose medium such as plate count agar by the spread plate technique and incubate at 37°C for 24 to 48 h.

If extensive bacterial contamination persists after antibiotic treatment, treat with chloroform. Add 1/10 volume of sample of chloroform ($CHCl_3$) and mix vigorously for 30 min at room temperature or homogenize 1 to 2 min at 4 to 10°C. For phase separation, centrifuge at ≥ 1000 × g or store overnight in a refrigerator. Separate sample (upper layer) from $CHCl_3$ (bottom layer) by aspirating with a pipet and bubble with filter-sterilized air for about 15 min to remove dissolved $CHCl_3$. It may be necessary to place sample in a sterile, shallow container and expose it to the atmosphere in a sterile air environment (laminar air flow clean bench or biological safety cabinet) for up to several hours to remove remaining traces of $CHCl_3$. Do not use ether to decontaminate samples because of the hazard of explosion or fire.

3. Laboratory Facilities and Host Systems for Virus Assay

Because viruses are obligate, intracellular parasites, they grow (multiply) only in living host cells. This ability to multiply in, and thereby destroy, their host cells is the basis for virus detection and assay. The two major host cell systems for human enteric viruses are whole animals (usually mice) and mammalian cell cultures of primate origin.

A complete description of facilities, equipment, materials, and methods for conducting virus assays is beyond the scope of this book; see standard handbooks on virology and cell culture.[1-4] Virus assay is beyond the capability of most water and wastewater microbiology laboratories. It should be done only by a trained virologist working in specially equipped virology laboratory facilities. Take particular care to prevent samples or inoculated hosts from becoming contaminated with viruses from other sources and to prevent virus cross-contamination arising from sample concentrates or inoculated hosts. Process and handle samples in a Class II Type I biological safety cabinet[5] or in a "sterile" room or cubicle. The use of such cabinets or facilities is mandatory for testing drinking water or other finished water samples.

There is no single, universal host system for all enteric viruses. Some enteric viruses, notably hepatitis A virus, human rotaviruses, and Norwalk-type gastroenteritis viruses, cannot be assayed routinely in any convenient laboratory host systems. However, most of the known enteric viruses can be detected by using two or more cell culture systems and perhaps suckling mice. The latter previously were considered essential for the detection of group A coxsackie-viruses, but recent studies indicate that the RD cell line may be nearly as sensitive as suckling mice for the isolation of these viruses as well as other enteroviruses.[6,7] In general, the more different host systems used, the greater the enteric virus recovery rate. However, the number of different host systems used is limited by practical and economic considerations.

There have been numerous comparative studies on relative sensitivities of various cell culture systems for enteric virus detection,[6-26] but no systematic, comprehensive study has been reported for enteric virus recoveries from water and wastewater. Primary or secondary human embryonic kidney (HEK) cell cultures appear to be the single most sensitive host system for enteric virus isolations, but they are becoming increasingly more difficult to obtain regularly and, when available from commercial sources, they are expensive. Primary or secondary African green, cynomolgus, or rhesus monkey or baboon kidney cells are sensitive hosts for many enteroviruses and reoviruses, but are not particularly suitable for recovering adenoviruses or group A coxsackieviruses. BGM, a continuous line derived from African green monkey kidney cells, may be comparable in sensitivity to primary monkey kidney cells for enteric virus recovery.[7,18,21,25,26] A number of other continuous cell lines as well as human fetal diploid cell strains have been evaluated for enteric virus recoveries. Some human fetal diploid cell strains give virus isolation rates comparable to primary monkey kidney cells, but plentiful supplies of specific human fetal diploid cell strains are not readily available and many are difficult to maintain. Furthermore, each different cell strain must be characterized for virus susceptibility. Most continuous cell lines generally are less effective than primary cells, but comparable isolation rates for some enteric virus groups have been obtained with Hep-2[11] and HeLa[17,25] cells.

Assay the entire sample concentrate for enteric viruses, using at least two different host systems and dividing entire sample equally among the hosts. Preferably use primary (or secondary) HEK cells with either primary (or secondary) monkey kidney or BGM cells for the recovery of most enteroviruses, adenoviruses, and reoviruses. Additional use of either suckling mice or RD cells provides for enhanced recovery of group A coxsackieviruses. Different host systems may be substituted for these if it is demonstrated that they have equivalent sensitivity.

4. Virus Quantitation Procedures for Sample Concentrates

a. Advantages and disadvantages of different quantitation procedures: Virus assays in suckling mice or other animals are quantal assays and in cell cultures they can be done either by quantal (most probable number or 50% endpoint) or enumerative (plaque) methods. Selection between cell culture assay methods depends on the sample and the choice between achieving either maximum virus sensitivity or maximum precision and accuracy in estimating virus concentration. The plaque technique generally is more precise and accurate than the quantal assay because relatively large numbers of individual infectious units can be counted directly as discrete, localized areas of infection (plaques). Quantal assays are more sensitive than monolayer plaque assays, but are less sensitive than an agar cell suspension plaque assay.[26]

Because virus plaques are discrete areas of infection arising from a single infectious virus unit, it is relatively easy to recover viruses from individual plaques and then to inoculate them into additional cell cultures to obtain a pure virus culture for identification. However, large proportions of so-called "false-positive" plaques that do not confirm as virus-positive when material from these plaques is further passaged in cell cultures have been reported.[27,28] Whether this problem is due to nonviral, plaque-like areas of cytotoxicity from the sample or to technical inability to passage viruses successfully from the initial plaques remains uncertain.[27,28]

The use of specific plaque assay conditions for optimizing the recovery of certain enteric virus groups may preclude efficient recovery of other enteric groups requiring different plaque assay conditions. Furthermore, some viruses, such as adenoviruses, do not form plaques efficiently under any conditions. Cytotoxicity due to water or wastewater constituents in sample concentrates is difficult to control in plaque assay systems because the agar overlay medium is difficult to remove and replace.

A potential limitation of quantal assays is the possibility that two or more different virus types will be inoculated into the same cell culture and thus produce a simple positive culture. This not only results in an underestimation of virus concentration but also requires separation of the individual virus types by further passage in cell culture. Such mixed cultures may go undetected unless virus isolates are identified serologically. Recent results indicate that mixed positive cultures are encountered rarely when samples are divided into small portions for inoculation into a series of replicate cell cultures.[7,25]

Cytotoxicity due to constituents of sample concentrates usually can be controlled in quantal assay cell cultures by replacing the culture medium before the cells die.

b. Cell culture procedures for virus isolation and assay: To assay sample concentrates in cell cultures by quantal or plaque methods, drain the medium from newly confluent cultures. To reduce toxicity, rinsing with buffered saline solution may be helpful. Inoculate with unit volumes of sample. Use no more than 0.06 mL sample/cm^2 of cell layer surface, e.g., maximum volumes of 1.0, 3.0, and 6.0 mL in cell culture flasks with areas of 25, 75, and 150 cm^2, respectively.[29] If samples are expected to contain such large quantities of viruses that it would be difficult to make reliable estimates of concentration, inoculate cell cultures with small sample volumes or dilutions of concentrates. Allow viruses to adsorb to cells for 2 h at 37 $\pm$ 0.5°C. Redistribute inoculum over the cell layer manually every 15 min or keep cultures on a mechanical rocker during the adsorption period. Add liquid maintenance medium to cultures for quantal assays or agar-containing medium for plaque assays. Invert plaque assay cultures so that cell (agar) side of culture faces up and incubate at 37°C.

Microscopically examine quantal assay cultures for the appearance of cytopathic effects (CPE) daily during the first 3 d and then periodically for a total of at least 14 d. Do not change cell culture medium unless cytotoxicity or cell deterioration occurs. Freeze cultures developing CPE at −70°C when more than 75% of the cells become involved. After 14 or more days, freeze at −70°C all remaining cultures, including those remaining negative for CPE as well as controls. Thaw cultures and clarify culture fluid-cell lysate by slow-speed centrifugation or filtration through sterile 0.22- or 0.45-μm porosity filters. Inoculate clarified material from each initial (first-passage) culture into a second (second-passage) culture by transferring 20% of the total initial culture into newly confluent cell cultures of the same type. Microscopically examine second-passage cultures for development of CPE periodically over a period of 14 or more days. Consider second-passage cultures developing CPE as confirmed virus-positive. Freeze and store at −70°C for virus identification. Discard as negative any virus cultures negative for CPE after this second incubation period of 14 or more days.

Periodically examine plaque assay cultures for appearance of plaques over a 14-d period. Mark and tally plaques as they appear. Transfer viruses from each plaque directly to at least two newly confluent, liquid-medium cell cultures of the same type[27] before plaques become too large and grow together or before the entire cell layer deteriorates. Do not store material obtained from plaques before transfer to new cell cultures, as this may result in loss of virus titer and unsuccessful transfers. Microscopically examine these second-passage cultures periodically over 14 d for development of CPE. Freeze cultures developing CPE at −70°C for virus identification.

c. Virus isolation and assay in mice: To detect group A and B coxsackieviruses in mice, inoculate samples into animals no older than 24 h using standard procedures.[2,3,8] Use either the intracerebral or intraperitoneal route, inoculating 0.02 and 0.05 mL, respectively. Observe mice daily over a 14-d period for development of weakness, tremors, and either flaccid (due to group A coxsackieviruses) or spastic (due to group B coxsackieviruses) paralysis. Sacrifice animals developing symptoms, and using sterile technique, prepare 20% tissue suspensions in Hanks balanced salt solution of the entire skinned, eviscerated torso or just the brain and legs. Store suspensions at −70°C until used for further passage and identification. For second passage in mice, follow general procedures used for the initial inoculations. However, making a second passage in cell cultures is preferable to making a second passage in mice because it is easier to do subsequent virus identification by neutralization tests.

d. Estimating virus concentration: Determining the amount of virus in a sample concentrate depends on the assay used. If a sample concentrate is assayed in cell cultures by the plaque technique, count all plaques and calculate the virus concentration, expressed as plaque-forming units (PFU).

If a sample concentrate is assayed by the quantal method, estimate the virus concentration by the most probable number (MPN) method and express as most probable number of infectious units (MPNIU), or by a 50% end point method and express as 50% infectious or lethal dose (ID_{50} or LD_{50}).[2,4,30−32] If the undiluted sample concentrate or a single sample dilution was inoculated into a series of replicate cell cultures (or mice), calculate the MPNIU from the number of confirmed CPE-negative cultures (or mice), q, per total number of cultures (or mice) inoculated, n, according to the formula

$$MPN = -\ln(q/n)$$

If more than one sample dilution was inoculated into cell cultures (or mice), calculate the MPNIU from the formula developed by Thomas:[33]

$$MPN/mL = \frac{P}{\sqrt{NQ}}$$

where:

P = total number of positive cultures (or mice) from all dilutions,
N = total mL sample inoculated for all dilutions, and
Q = total mL sample in all negative cultures (or mice).

In using this formula, exclude from the computation all dilutions containing only positive cultures (or mice).

For MPN values obtained from a single sample dilution, the 95% confidence interval is based on the standard error of the binomial distribution when more than 30 cultures (or mice) are inoculated or from the confidence coefficient table of Crow[32,34] when 30 or fewer cultures (or mice) are inoculated.

Make 50% end-point estimates arithmetically by either the Reed-Muench or Karber method.[2,4] These methods require results from several equally spaced sample dilutions, preferably with about the same number of dilutions above and below the 50% end point, and may not be useful for sample concentrates containing relatively low virus levels.

e. Identification of virus isolates: Identify enteric viruses isolated from sample concentrates by standard serological techniques, although preliminary identification of genus (enterovirus, reovirus, or adenovirus) sometimes can be made on the basis of information obtained from the isolation procedure. Enteric viruses recovered in suckling mice are likely to be either group A or B coxsackieviruses. For enteric viruses isolated in cell cultures, preliminary identification of genus often can be made from the characteristic appearance of cytopathic effects (CPE) in infected cell cultures.

Confirm preliminary identification of suspected adenovirus and reovirus isolates by detecting their respective group specific antigens by complement fixation tests using clarified, second-passage cell-culture lysate as the antigen. Identify specific reovirus serotypes by hemagglutination-inhibition (HI) or neutralization (Nt) tests. Adenovirus serotypes can be separated into four groups on the basis of their ability (or inability) to hemagglutinate rhesus monkey or rat erythrocytes.[2,3,8] Except for type 18, the first 28 numbered adenoviruses can be identified as to specific serotype by HI. Alternatively, identify all adenovirus serotypes by Nt tests using either individual type-specific antisera or intersecting antisera pools. Also identify specific enterovirus serotypes by neutralization tests in cell cultures using intersecting pools of hyperimmune sera.[2,3,8,35] Use mice for Nt tests for group A and B coxsackieviruses only if the virus isolates fail to propagate in cell cultures.[36] Because polioviruses often are the most prevalent enteroviruses in water and wastewater, test enterovirus isolates for neutralization by an antisera pool against the three types of poliovirus before making neutralization tests with intersecting antisera pools.

5. References

1. PAUL, J. 1975. Cell and Tissue Culture, 5th ed. Churchill Livingstone, New York, N.Y.
2. LENNETTE, E.H. & N.J. SCHMIDT, eds. 1979. Diagnostic Procedures for Viral and Rickettsial Infections, 5th ed. American Public Health Assoc., Washington, D.C.
3. LENNETTE, E.H., A. BALOWS, W.J. HAYSLER & J.P. TRUANT, eds. 1980. Manual of Clinical Microbiology, 3rd ed. American Soc. Microbiology, Washington, D.C.
4. FRESHNEY, R.I. 1987. Culture of Animal Cells. A Manual of Basic Technique, 2nd ed. Alan R. Liss, Inc., New York, N.Y.
5. U.S. PUBLIC HEALTH SERVICE. 1976. Guidelines for Research Involving Recombinant DNA Molecules. Appendix D-I, Biological Safety Cabinets. National Inst. Health, Bethesda, Md.
6. SCHMIDT, N.J., H.H. HO & E.H. LENNETTE. 1975. Propagation and isolation of group A coxsackieviruses in RD cells. *J. Clin. Microbiol.* 2:183.
7. SCHMIDT, N.J., H.H. HO, J.L. RIGGS & E.H. LENNETTE. 1978. Comparative sensitivity of various cell culture systems for isolation of viruses from wastewater and fecal samples. *Appl. Environ. Microbiol.* 36:480.
8. HSIUNG, G.D. 1982. Diagnostic Virology, 3rd ed. Yale Univ. Press, New Haven, Conn.
9. KELLY, S., J. WINSSER & W. WINKELSTEIN. 1957. Poliomyelitis and other enteric viruses in sewage. *Amer. J. Pub. Health* 47:72.
10. KELLY, S. & W.W. SANDERSON. 1962. Comparison of various tissue cultures for the isolation of enteroviruses. *Amer. J. Pub. Health* 52:455.
11. PAL, S.R., J. McQUILLIN & P.S. GARDNER. 1963. A comparative study of susceptibility of primary monkey kidney cells, Hep 2 cells and HeLa cells to a variety of faecal viruses. *J. Hyg., Camb.* 61:493.
12. LEE, L.H., C.A. PHILLIPS, M.A. SOUTH, J.L. MELNICK & M.D. YOW. 1965. Enteric virus isolations in different cell cultures. *Bull. World Health Org.* 32:657.
13. SCHMIDT, N.J., H.H. HO & E.H. LENNETTE. 1965. Comparative sensitivity of human fetal diploid kidney cell strains and monkey kidney cell cultures for isolation of certain human viruses. *Amer. J. Clin. Pathol.* 43:297.
14. BERQUIST, K.R. & G.J. LOVE. 1966. Relative efficiency of three tissue culture systems for the primary isolation of viruses from feces. *Health Lab. Sci.* 3:195.
15. HERRMANN, E.C. 1967. The usefulness of human fibroblast cell lines for the isolation of viruses. *Amer. J. Epidemiol.* 85:200.
16. FAULKNER, R.S. & C.E. VAN ROOYEN. 1969. Studies on surveillance and survival of viruses in sewage in Nova Scotia. *Can. J. Pub. Health* 60:345.
17. LUND, E. & C.E. HEDSTROM. 1969. A study on sampling and isolation methods for the detection of virus in sewage. *Water Res.* 3:823.
18. SHUVAL, H., B. FATTAL, S. CYMBALISTA & N. GOLDBLUM. 1969. The phase-separation method for the concentration and detection of viruses in water. *Water Res.* 3:225.
19. SCHMIDT, N.J. 1972. Tissue culture in the laboratory diagnosis of virus infections. *Amer. J. Clin. Pathol.* 57:820.
20. COONEY, M.K. 1973. Relative efficiency of cell cultures for detection of viruses. *Health Lab. Sci.* 4:295.
21. DAHLING, D.R., G. BERG & D. BERMAN. 1974. BGM: A continuous cell line more sensitive than primary rhesus and African green kidney cells for the recovery of viruses from water. *Health Lab. Sci.* 11:275.
22. SCHMIDT, N.J., H.H. HO & E.H. LENNETTE. 1976. Comparative sensitivity of the BGM cell line for the isolation of enteric viruses. *Health Lab. Sci.* 13:115.
23. HATCH, M.H. & G.E. MARCHETTI. 1971. Isolation of echoviruses with human embryonic lung fibroblast cells. *Appl. Microbiol.* 22:736.
24. RUTALA, W.A., D.F. SHELTON & D. ARBITER. 1977. Comparative sensitivities of viruses to cell cultures and transport media. *Amer. J. Clin. Pathol.* 67:397.

25. IRVING, L.G. & F.A. SMITH. 1981. One-year survey of enteroviruses, adenoviruses and reoviruses isolated from effluent at an activated-sludge purification plant. *Appl. Environ. Microbiol.* 41:51.

26. MORRIS, R. & W.M. WAITE. 1980. Evaluation of procedures for recovery of viruses from water—II detection systems. *Water Res.* 14: 795.

27. LEONG, L.Y.C., S.J. BARRETT & R.R. TRUSSELL. 1978. False-positives in testing of secondary sewage for enteric viruses. Abs. Annu. Meeting, American Soc. Microbiology, Washington, D.C.

28. KEDMI, S. & B. FATTAL. 1981. Evaluation of the false-positive enteroviral plaque phenomenon occurring in sewage samples. *Water Res.* 15:73.

29. PAYMENT, P. & M. TRUDEL. 1985. Influence of inoculum size, incubation temperature, and cell culture density on virus detection in environmental samples. *Can. J. Microbiol.* 31:977.

30. CHANG, S.L., G. BERG, K.A. BUSCH, R.E. STEVENSON, N.A. CLARKE & P.W. KABLER. 1958. Application of the Most Probable Number method for estimating concentrations of animal viruses by tissue culture technique. *Virology* 6:27.

31. CHANG, S.L. 1965. Statistics of the infective units of animal viruses. *In* G. Berg, ed. Transmission of Viruses by the Water Route. Interscience Publ., New York, N.Y.

32. SOBSEY, M.D. 1976. Field monitoring techniques and data analysis. *In* L.B. Baldwin, J.M. Davidson & J.F. Gerber, eds. Virus Aspects of Applying Municipal Waste to Land. Univ. Florida, Gainesville.

33. THOMAS, H.A., JR. 1942. Bacterial densities from fermentation tube tests. *J. Amer. Water Works Assoc.* 34:572.

34. CROW, E.L. 1956. Confidence intervals for a proportion. *Biometrika* 43:423.

35. MELNICK, J.L., V. RENNICK, B. HAMPIL, N.J. SCHMIDT & H.H. HO. 1973. Lyophilized combination pools of enterovirus equine antisera: Preparation and test procedures for the identification of field strains of 42 enteroviruses. *Bull. World Health Org.* 48:263.

36. MELNICK, J.L., N.J. SCHMIDT, B. HAMPIL & H.H. HO. 1977. Lyophilized combination pools of enterovirus equine antisera: Preparation and test procedures for the identification of field strains of 19 group A coxsackievirus serotypes. *Intervirology* 8:1720.

9610   DETECTION OF FUNGI*

9610 A.   Introduction

1. Significance

Fungi, including yeasts and filamentous species or molds, are ubiquitously distributed, achlorophyllous, heterotrophic organisms with organized nuclei and usually with rigid walls. They may be found wherever nonliving organic matter occurs, although some species are pathogenic and others are parasitic. In spring water near the source, the number of fungus spores usually is minimal. Unpolluted stream water has relatively large numbers of species representing the true aquatic fungi (species possessing flagellated zoospores and gametes), aquatic Hyphomycetes, and soil fungi. Moderately polluted water may carry cells or spores of the three types; however, it has fewer true aquatic fungi and aquatic Hyphomycetes, and soil fungi are more numerous. Heavily polluted water has large numbers of soil fungi. The group designated as soil fungi includes yeast-like fungi, many species of which have been isolated from polluted waters.

The association between fungal densities and organic loading suggests that fungi may be useful indicators of pollution. Unfortunately, no single species or group of fungi has been identified as important in this role. There may be some exceptional special cases; for example, the principal distinction between the yeasts *Candida lambica* and *C. krusei* is the ability to use pentose sugars. Because the former species grows well on pentoses, it could be used as an indicator of pulp and paper mill wastes, which are high in such sugars. Certain species of yeasts and filamentous fungi are characteristic of warmer waters and may be useful indicators of thermal pollution.

Because fungi possess broad enzymatic capabilities, they can degrade actively most complex natural substances and certain synthetic compounds, including some pesticides. Most fungi are aerobic or microaerophilic, although a few species show limited anaerobic metabolism and a very few are capable of totally anaerobic growth.[1] Some species do not require light.

2. Occurrence and Survival

Fungi are present in, and have been recovered from, diverse, remote, and extreme aquatic habitats including lakes, ponds, rivers, streams, estuaries, marine environments, wastewaters, sludge, rural and urban stormwater runoff, well waters, acid mine drainage, asphalt refineries, jet fuel systems, and aquatic sediments.

a. Fungi in potable water: Fungi have been found in potable water[2-7] and on the inner surface of distribution system pipes.[8] Either they survive water treatment or they enter the system after treatment and remain viable. Tuberculate macroconidia of *Histoplasma capsulatum*[9] can pass through a 0.75-m rapid sand filter. Plain sedimentation or alum flocculation and settling removed 80 to 99% of the spores. If these relatively large (8- to 14-μm) globose, tuberculate macroconidia pass through treatment, it is not surprising that other fungi, typically with smaller spores, are found in treated water.

Having survived treatment or having been introduced after treatment, fungal spores can remain viable for extended periods of time. Pathogenic fungi have been stored effectively in sterile distilled water for relatively long periods.[10] Spores of *H. capsulatum,* stored in raw Ohio River water and sterile tap water remained highly infective for mice after 400 d.[11]

Tastes and odors in potable water are associated with the presence of procaryotic organisms such as bacteria, actinomycetes, and cyanobacteria. However, fungi may be involved.[6,7] Propagules from 19 genera of filamentous fungi have been isolated from a chlorinated surface water system and a nonchlori-

---

* Approved by Standard Methods Committee, 1993.

nated groundwater distribution system.[2] The mean number of colony-forming units (CFU) was 18/100 mL in the groundwater system and 34/100 mL in the surface system.

In Finland,[3] fungi were isolated from rivers, lakes, and ponds supplying nine communities with sand-filtered water, three with artificially recharged groundwater, of which two used chemical coagulation, and three with chemically coagulated and disinfected water. Mesophilic fungi were common in all raw water samples; however, thermotolerant fungi were more abundant in river than in lake water. Chemical coagulation and disinfection proved far more efficient in removing fungi than sand filtration and disinfection. *Aspergillus fumigatus* was the most common fungus.

Five chlorinated groundwater systems[4] in the U.S. yielded an average count per positive sample of about 5.5 CFU/100 mL. In France[5] yeasts were recovered from 50% of 38 samples and filamentous fungi from 81%.

Except for *Aspergillus fumigatus*[3,4], *A. flavus*,[4,5] and *A. niger*,[4] the fungi isolated from potable water usually are not considered to be medically important. Fungus infections may be significant for individuals with compromised immune systems. Most of the fungi are common soil saprobes.

b. Fungi in recreation waters: Some fungi pathogenic to humans may be expected in recreational waters such as pools and beaches and in accompanying washing facilities such as shower stalls.

Trichophyton mentagrophytes, the cause of tinea pedis or athlete's foot, has been isolated[12] from the wooden flooring of a shower stall. Seven species of pathogenic and potentially pathogenic fungi were isolated from 361 samples of beach sand in Hawaii.[13] Beach sand on the German Baltic coast, in Portugal, and on the Adriatic coast yielded *Epidermophyton*.[14]

c. Survival on chlorination: An unidentified yeast was isolated from other organisms that survived chlorination of wastewater effluents.[15] This yeast survived 1 mg free chlorine/L for 20 min in contrast with *E. coli,* which failed to survive 5 min contact with 0.03 mg free chlorine/L. The amount of chlorine for fungus control is known at least for *C. albicans.* It has been shown[16] that cells of *C. albicans* were inactivated effectively with 4 mg chlorine/L in 30 min when the initial cell count was 10^5 cells/mL. In an Illinois study with *C. parapsilosis,*[17,18] a commonly isolated yeast known to cause health problems in the tropics, greater amounts of chlorine were required to inactivate the organism than coliform bacteria. Mechanisms of inactivation by chlorine on assimilative stages of yeasts and other microorganisms have been suggested.[19] Fungal cells, especially conidia, can survive much higher doses of chlorine than coliform bacteria,[20] including a 10-min exposure to 10 mg chlorine/L when the initial spore count was approximately 10^6/mL.

3. Growth Patterns and Identification

In water there are two basic patterns of fungal growth. True aquatic fungi produce zoospores or gametes that are motile by means of flagella, either of the whiplash or tinsel type. Some fungi, particularly the Trichomycetes (fungi that inhabit the hind gut of certain worms, mosquito larvae, etc.), have amoeboid stages. Aquatic fungi typically are collected by exposing suitable baits (solid foodstuffs) in the habitat being examined or in a sample within the laboratory. Relatively little work on these fungi in polluted water has been done in the U.S. They have been studied

more extensively in polluted waters in England, Germany, and Japan.

The second fungal growth form is nonmotile in all stages of the life cycle. Growth and reproduction usually are asexual (anamorphic). Three growth processes have been recognized: (*a*) filamentous growth with blastic spores or spores produced in special structures; (*b*) filamentous growth with the filaments breaking up in an arthric (fragmenting) manner to form separate spores called arthroconidia as in *Geotrichum* and related genera; and (*c*) single-celled growth produced on each parent cell, called budding, typical of the yeasts.

Identification of fungi, which are considerably larger than bacteria, is dependent on colonial morphology on a solid medium, growth and reproduction morphology, and, for yeasts, physiological activity in laboratory cultures. Increasing numbers of fungi usually indicate increasing organic loadings in water or soil. Large numbers of similar fungi suggest excessive organic load while a highly diversified mycobiota indicates populations adjusted to the environmental organics. Despite their wide occurrence, little attention has been given to the presence and ecological significance of fungi in aquatic habitats. The relevance of fungi and their activities in water is emphasized by increasing knowledge of their pathogenicity for humans, animals, and plants; their role as food or energy sources; their activity in natural purification processes; and their function in sediment formation.

A survey of the literature of fungi occurring in water, wastewater, and related organically polluted substrata listed 984 species[21]: 133 species were assigned to the Mastigomycotina (fungi with flagellated zoospores); 79 to the Zygomycotina, mostly mucoraceous fungi; 161 to the Ascomycotina, including perfect (teleomorphic) states of some of those assigned to the Fungi Imperfecti; 18 to the Basidiomycotina, including perfect states of several yeasts; and 593 to the Deuteromycotina or Fungi Imperfecti. Of the total, 133 species were zoosporic, 131 species were yeast-like, and 718 species were filamentous. Most zoosporic species were recovered from mildly polluted or unpolluted waters; of the remaining species fewer than half were recovered in numbers large enough to indicate membership in a population for even a brief period of time. The significance of fungi in both aquatic and terrestrial environments has been discussed in detail.[22-38]

Quantitative enumeration of fungi is not equivalent to that of unicellular bacteria because a fungal colony may develop from a single cell (spore), an aggregate of cells (a cluster of spores or a single multi-celled spore), or from a mycelial or pseudomycelial fragment (containing more than one viable cell). It is assumed that each fungal colony developing in laboratory culture originates from a single colony-forming unit (CFU), which may or may not be a single cell.

4. References

1. Tabak, H. & W.B. Cooke. 1968. Growth and metabolism of fungi in an atmosphere of nitrogen. *Mycologia* 60:115.
2. Nagy, L.A. & B.H. Olson. 1982. The occurrence of filamentous fungi in water distribution systems. *Can. J. Microbiol.* 28:667.
3. Niemi, R.M., S. Kunth & K. Lundstrom. 1982. Actinomycetes and fungi in surface waters and potable water. *Appl. Environ. Microbiol.* 43:378.
4. Rosenzweig, W.D., H. Minnigh & W.O. Pipes. 1986. Fungi in potable water distribution systems. *J. Amer. Water Works Assoc.* 78: 53.

5. HINZELIN F. & J.C. BLOCK. 1985. Yeast and filamentous fungi in drinking water. *Environ. Tech. Letters* 6:101.

6. BURMAN, N.P. 1965. Symposium on consumer complaints. 4. Taste and odour due to stagnation and local warming in long lengths of piping. *Proc. Soc. Water Treat. Exam.* 14:125.

7. BAYS, L.R., N.P. BURMAN & W.M. LEWIS. 1970. Taste and odour in water supplied in Great Britain. A survey of the present position and problems for the future. *Proc. Soc. Water Treat. Exam.* 19:136.

8. NAGY, L.A. & B.H. OLSON. 1985. Occurrence and significance of bacteria, fungi, and yeasts associated with distribution pipe surfaces. Proc. American Water Works Assoc., Water Quality Technology Conf., p. 213.

9. METZLER, D.F., C. RITTER & R.L. CULP. 1956. Combined effect of water purification processes on the removal of *Histoplasma capsulatum* from water. *Amer. J. Pub. Health* 46:1571.

10. CASTELLANI, A. 1963. The cultivation of pathogenic fungi in sterile distilled water. *Commentarii* 1(10):1.

11. COOKE, W.B. & P.W. KABLER. 1953. The survival of *Histoplasma capsulatum* in water. *Lloydia* 16:252.

12. AJELLO, L. & M.E. GETZ. 1954. Recovery of dermatophytes from shoes and shower stalls. *J. Invest. Derm.* 22:17.

13. KISHIMOTO, R.A. & G.E. BAKER. 1969. Pathogenic and potentially pathogenic fungi isolated from beach sands and selected soils of Oahu, Hawaii. *Mycologia* 61:539.

14. MULLER, G. 1973. Occurrence of dermatophytes in the soils of European beaches. *Sci. Total Environ.* 2:116.

15. ENGELBRECHT, R.S., D.H. FOSTER, E.O. GREENING & S.H. LEE. 1974. New microbial indicators of waste water efficiency. Environ. Protect. Technol. Ser. No. 670/2-73-082.

16. JONES, J. & J.A. SCHMITT. 1978. The effect of chlorination on the survival of cells of *Candida albicans. Mycologia* 70:684.

17. ENGELBRECHT, R.S. & C.N. HAAS. 1977. Acid-fast bacteria and yeasts as disinfection indicators: Enumeration methodology. Proc. American Water Works Assoc. Water Quality Technology Conf. 1977, p. 1.

18. HAAS, C.N. & R.S. ENGELBRECHT. 1980. Chlorine dynamics during inactivation of coliforms, acid-fast bacteria, and yeasts. *Water Res.* 14:1749.

19. HAAS, C.N. & R.S. ENGELBRECHT. 1980. Physiological alterations of vegetative microorganisms resulting from chlorination. *J. Water Pollut. Control Fed.* 52:1976.

20. ROSENZWEIG, D.W., H.A. MINNIGH & W.O. PIPES. 1983. Chlorine demand and inactivation of fungal propagules. *Appl. Environ. Microbiol.* 45:182.

21. COOKE, W.B. 1986. The Fungi of "Our Mouldy Earth". *Beihefte zur Nova Hedwigia* 85:1.

22. BARLOCHER, F. & B. KENDRICK. 1976. Hyphomycetes as intermediates of energy flow in streams. *In* E. B. Jones, ed. Recent Advances in Aquatic Mycology. Elek Science, London, England.

23. COOKE, W.B. 1961. Pollution effects on the fungus population of a stream. *Ecology* 42:1.

24. COOKE, W.B. 1965. The enumeration of yeast populations in a sewage treatment plant. *Mycologia* 57:696.

25. COOKE, W.B. 1970. Our Mouldy Earth. FWPCA Res. Contract Ser. Publ. No. CWR, Cincinnati, Ohio.

26. COOKE, W.B. 1971. The role of fungi in waste treatment. *CRC Critical Rev. Environ. Control.* 1:581.

27. COOKE, W.B. 1976. Fungi in sewage. *In* E.B.G. Jones, ed. Recent Advances in Aquatic Mycology. Elek Science, London, England.

28. COOKE, W.B. 1979. The Ecology of Fungi. CRC Press, Boca Raton, Fla.

29. DICK, M.W. 1971. The ecology of Saprolegniales in the lentic and littoral muds with a general theory of fungi in the lake ecosystem. *J. Gen. Microbiol.* 65:325.

30. HARLEY, J.L. 1971. Fungi in ecosystems. *J. Appl. Ecol.* 8:627.

31. MEYERS, S.P., D.G. AHEARN & W.L. COOK. 1970. Mycological studies of Lake Champlain. *Mycologia* 62:504.

32. NOELL, J. 1973. Slime-inhabiting geofungi in a polluted stream (winter/spring). *Mycologia* 65:57.

33. PARK, D. 1972. Methods of detecting fungi in organic detritus in water. *Trans. Brit. Mycol. Soc.* 58:281.

34. QURESHI, A.A. & B.J. DUTKA. 1974. A preliminary study on the occurrence and distribution of geofungi in Lake Ontario near the Niagara River. Proc. 17th Conf. Great Lakes Research, International Soc. Great Lakes Research.

35. SHERRY, J.P. & A.A. QURESHI. 1986. Isolation and enumeration of fungi using membrane filtration. *In* B.J. Dutka, ed. Membrane Filtration Applications, Techniques, and Problems. Marcel Dekker, New York, N.Y.

36. SIMARD, R.E. 1971. Yeasts as an indicator of pollution. *Marine Poll. Bull.* 12:123.

37. SPARROW, F.K. 1968. Ecology of fresh water fungi. *In* G.C. Ainsworth & A.S. Sussman, eds. The Fungi, An Advanced Treatise. Vol. 3. The Fungal Population. Academic Press, New York, N.Y.

38. TOMLINSON, T.G. & I.L. WILLIAMS. 1975. Fungi, *In* C.R. Curds & H.A. Hawkes, eds. Ecological Aspects of Used-Water Treatment. Vol. 1. The Organisms and their Ecology. Academic Press, New York, N.Y.

5. Bibliography

EMERSON, R. 1958. Mycological organization. *Mycologia* 50:589.

COOKE, W.B. 1958. Continuous sampling of trickling filter populations. I. Procedures. *Sewage Ind. Wastes* 30:21.

COOKE, W.B. & A. HIRSCH. 1958. Continuous sampling of trickling filter populations. II. Populations. *Sewage Ind. Wastes* 30:139.

COOKE, W.B. 1959. Trickling filter ecology. *Ecology* 40:273.

SPARROW, F.K. 1959. Fungi. (Ascomycetes, Phycomycetes); including W.W. Scott, Key to genera, Fungi Imperfecti (Aquatic Hyphomycetes only). *In* W.T. Edmondson, ed. Ward & Whipple's Fresh Water Biology, 2nd ed. John Wiley & Sons, New York, N.Y.

COOKE, W.B. 1963. A Laboratory Guide to Fungi in Polluted Waters, Sewage, and Sewage Treatment Systems, Their Identification and Culture. USPHS Publ. 999-WP-1, Cincinnati, Ohio.

FULLER, M.S. & R.O. PAYTON. 1964. A new technique for the isolation of aquatic fungi. *BioScience* 14:45.

WILLOUGHBY, L.C. & V.G. COLLINS. 1966. A study of fungal spores and bacteria in Blelham Tarn and its associated streams. *Nova Hedwigia* 12:150.

COOKE, W.B. & G.S. MATSUURA. 1969. Distribution of fungi in a waste stabilization pond system. *Ecology* 50:689.

BROCK, T.D. 1970. Biology of Microorganisms. Prentice-Hall, Englewood Cliffs, N.J.

COOKE, W.B. 1970. Fungi in the Lebanon sewage treatment plant and in Turtle Creek, Warren Co., Ohio. *Mycopathol. Mycol. Appl.* 42:89.

PATERSON, R.A. 1971. Lacustrine fungal communities. *In* J. Cairns, ed. Structure and Function of Microbial Communities. Symposium, American Microscopical Soc., Burlington, Vt. 1969. Virginia Polytechnic Inst. & State Univ. Res. Div. Mono. 3:209.

JONES, E.B.G. 1971. Aquatic Fungi. *In* C. Booth, ed. Methods in Microbiology 4:335. Academic Press, New York, N.Y.

FARR, D.F. & R.A. PATERSON. 1974. Aquatic fungi in rivers: Their distribution and response to pollutants. VPI-WRRC Bull. 68, Virginia Water Resources Research Center, Virginia Polytechnic Inst. & State Univ., Blacksburg.

GARETH JONES, E.B., ed. 1976. Recent Advances in Aquatic Mycology. Elek Science, London, England.

FULLER, M.S., ed. 1978. Lower Fungi in the Laboratory. *Palfrey Contrib. in Botany* 1:1-212. Athens, Ga.

ALEXOPOULOS, C.J. & C.W. MIMS. 1979. Introductory Mycology, 3rd ed. John Wiley & Sons, New York, N.Y.

9610 B. Pour Plate Technique

1. Samples

a. Containers: Collect samples as directed in Section 9060A. Alternatively, use cylindrical plastic vials with snap-on caps. These vials usually are sterile as received. Transport them in an upright position to minimize the chance of leakage and discard after use.

b. Storage: Hold samples for not more than 24 h. If analysis is not begun promptly after sample collection, refrigerate.

2. Media

For counting, neopeptone-glucose-rose bengal-aureomycin® agar is the usual medium of choice, although experience may indicate that Czapek agar (for *Aspergillus, Penicillium,* and related fungi) and yeast extract-malt extract-glucose agar or Diamalt agar (for yeasts) may be preferable. For inventory, use neopeptone-glucose agar.

a. Neopeptone-glucose-rose bengal aureomycin® agar: Add 5.0 g neopeptone, 10.0 g glucose, 3.5 mL rose bengal solution (1 g/100 mL reagent-grade water), and 20.0 g agar to 1 L reagent-grade water. Because this medium is used for making pour plates, prepare and store basal agar either in bulk, or more conveniently, in tubes in 10-mL amounts. Sterilize by autoclaving; the final pH should be about 6.5.

Separately prepare a solution of chlortetracycline or tetracycline (1.0 g water-soluble antibiotic/150 mL reagent-grade water) and refrigerate. Before use, sterilize by filtration. To complete the medium, add 0.05 mL sterile solution to 10 mL melted basal agar at about 45°C.

This medium may not be available in dehydrated form and may require preparation from the basic ingredients. Dehydrated Cooke's rose bengal agar may be used in place of the agar base. This medium is useful for isolating a broad spectrum of fungal species.

b. Czapek (or Czapek-Dox) agar: Dissolve 30.0 g sucrose, 3.0 g sodium nitrate ($NaNO_3$), 1.0 g dipotassium hydrogen phosphate (K_2HPO_4), 0.5 g magnesium sulfate ($MgSO_4$), 0.5 g potassium chloride (KCl), 0.01 g ferrous sulfate ($FeSO_4$), and 15.0 g agar in 1 L reagent-grade water. The pH should be 7.3 after sterilization.

This medium is useful for isolating species of *Aspergillus, Penicillium, Paecilomyces,* and some other fungi with similar physiological requirements.

c. Yeast extract-malt extract-glucose agar: Dissolve 3.0 g yeast extract, 3.0 g malt extract, 5.0 g neopeptone (or equivalent), 10.0 g glucose, and 20.0 g agar in 1 L reagent-grade water. No adjustment of pH is required.

This medium is useful for isolating yeasts.

d. Diamalt agar: Dissolve 150 g diamalt and 20.0 g agar in 1 L reagent-grade water. No adjustment of pH is required. The medium will be turbid but filtration is unnecessary.

This medium is useful in the purification of yeast isolates and for study of yeast species in various specified tests.

e. Neopeptone-glucose agar: Dissolve 5.0 g neopeptone (or equivalent), 10.0 g glucose, and 20.0 g agar in 1 L reagent-grade water. The pH should be about 6.5 after sterilization. (This medium is known also as Emmons' Sabouraud Agar or Emmons' Sabouraud Dextrose Agar.)

This medium is useful for maintaining stock cultures. It is comparable to neopeptone-glucose-rose bengal aureomycin® agar but contains neither rose bengal nor an antibiotic.

3. Procedure

As many as 40 samples can be analyzed simultaneously by a single analyst by the following procedure; however, 20 samples represents the optimum number.

a. Preparation and dilution: To a sterile 250-mL erlenmeyer flask add 135 mL sterile reagent-grade water and 15 mL sample to obtain a 1:10 sample dilution. Use a sterile measuring device for each sample, or, less preferably, rinse the measure with sterile reagent-grade water between samples. Mix sample well before withdrawing the 15-mL portion. Shake flask on a rotary shaker at about 120 to 150 oscillations/min for about 30 min or transfer flask contents to a blender jar, cover and blend at low speed for 1 min or high speed for 30 s. Preferably use a sterile blender jar and appurtenances for each sample or wash jar thoroughly between samples and rinse with sterile water. Further dilutions may be made by adding 45 mL sterile water to 5 mL of a 1:10 diluted suspension.

For stream water samples a dilution of 1:10 usually is adequate. Dilute samples with large amounts of organic material, such as sediments, to 1:100 or 1:1000. Dilute stream bank or soil samples to 1:1000 or 1:10 000.

b. Plating: Prepare five plates for each dilution to be examined. To use neopeptone-glucose-rose bengal-aureomycin® agar, aseptically transfer 10 mL of medium at 45°C to a 9-cm petri dish. Add 1 mL of appropriate sample dilution and mix thoroughly by tilting and rotating dish (see plating procedure under heterotrophic plate count, Section 9215). Alternatively add to petri dish 1 mL sample, 0.05 mL antibiotic solution, and 10 mL liquefied agar medium at 43 to 45°C. Solidify agar as rapidly as possible. (In arid areas use more medium to prevent dehydration during incubation.)

c. Incubation: Stack plates but do not invert. Incubate at room conditions of temperature (20 to 24°C) and lighting but avoid direct sunlight. Examine and count plates after 3, 5, and 7 d.

d. Counting and inventory: The fungus plate count will provide the basis for rough quantitative comparisons among samples; the inventory will give relative importance of at least the more readily identifiable species or genera.

In preparing plates, use sample portions that will give about 50 to 60 colonies on a plate. Determine this volume by trial and error. When first examining a new habitat plate at least two sample dilutions. Estimates of up to 300 colonies may be made, but discard more crowded plates. The medium containing rose bengal tends to produce discrete colonies and permits slow-growing organisms to develop.

The inventory includes the direct identification of fungi based on colonial morphology and the counting of colonies assignable to various species or genera. When discrete colonies cannot be identified and identification is important, use a nichrome wire with its tip bent in an L-shape to pick from each selected colony and streak on a slant of neopeptone-glucose agar (¶ 2e). If five plates are used per sample, the average number of colonies on all plates (total number of colonies counted/5) times the reciprocal of the dilution (10/1, 100/1, 1000/1, etc.) equals the fungus colony count per milliliter of original sample. For solid or semisolid samples, use a correction for the water content to report fungus colonies per gram dry weight. Determine water content by drying paired 15-mL portions of original sample at 100°C overnight; the difference between wet and dry weights is the amount of water lost from the sample.

9610 C.   Spread Plate Technique

The spread plate technique is an alternative procedure for obtaining quantitative data on colony-forming units.

1. Samples

See Section 9610B.1.

2. Media

Use any of the following media. Aureomycin®-rose bengal-glucose-peptone agar (ARGPA) (e) and streptomycin-terramycin®-malt extract agar (STMEA) (f) are useful in analyzing sewage and polluted waters.[1]

a. Neopeptone-glucose-rose bengal aureomycin® agar: See Section 9610B.2a.

b. Czapek (or Czapek-Dox) agar: See Section 9610B.2b.

c. Yeast extract-malt extract-glucose agar: See Section 9610B.2c.

d. Diamalt agar: See Section 9610B.2d.

e. Aureomycin®-rose bengal-glucose-peptone agar: Dissolve 10.0 g glucose, 5.0 g peptone, 1.0 g potassium dihydrogen phosphate (KH_2PO_4), 0.5 g magnesium sulfate ($MgSO_4 \cdot 7H_2O$), 0.035 g rose bengal, and 20.0 g agar in 800 mL reagent-grade water and sterilize. Dissolve 70.0 mg aureomycin® hydrochloride in 200 mL reagent-grade water, sterilize by filtration, and add to the cooled (42 to 45°C) agar base. The pH should be 5.4. Pour 25-mL portions into sterile petri dishes (100 × 15 mm) and let agar harden. Poured plates may be held up to 4 weeks at 4°C.

f. Streptomycin-terramycin®-malt extract agar: Dissolve 30.0 g malt extract, 5.0 g peptone, and 15.0 g agar in 800 mL reagent-grade water and sterilize. Dissolve 70.0 mg each of streptomycin and terramycin® in separate 100-mL portions reagent-grade water, sterilize by filtration, and add to the cooled (42 to 45°C) agar base. The pH should be 5.4. Pour about 20-mL portions into sterile petri dishes (60 × 15 mm) and let agar harden. Poured plates may be held up to 4 weeks at 4°C.

3. Procedure

a. Preparation and dilution: See Sections 9215A.5 and 9610B.3a. Make dilutions with buffered water (Section 9050C.1) and select dilutions that yield 20 to 150 colonies per plate.

b. Plating: Pre-dry plates separately with lids off in a laminar-flow hood at room temperature and about 30% relative humidity for 1 to 1.5 h. Prepare at least three plates, or five plates if data are to be analyzed statistically, per sample or dilution. Using a sterile pipet, transfer 1.0 mL of sample or dilution onto surface of a pre-dried agar plate. Spread sample over entire agar surface using a sterile L-shaped glass rod or use a mechanical device to rotate plate and ensure proper sample distribution.

c. Incubation: With dish covers on, let plates dry at room temperature, invert plates, and incubate at 15°C for 7 d in an atmosphere of high humidity (90 to 95%). Alternatively incubate at 20°C for 5 to 7 d. Slow-growing fungi may not produce noticeable colonies until 6 or 7 d.

d. Counting and recording: Using a Quebec colony counter, count all colonies on each selected plate. If counting must be delayed temporarily, hold plates at 4°C for not longer than 24 h. Depending on colony size, plates with as many as 150 colonies can be counted but the optimal maximum number is 100 colonies.

Record results as colony-forming units (CFU)/100 mL original sample. For solid or semisolid samples report CFU/g wet or dry, preferably dry. If three or more plates are used per sample, use average number of colonies times the reciprocal of the dilution (see 9610B) to give colony count. If no plates have colonies, record count as < 1 for the highest dilution. If there are more than 150/plate, record as Too Numerous To Count (TNTC) but indicate a count of > 150 for the appropriate dilution. If colonies are crowded and overlapping with spreaders, record as "obscured" (OBSC) and repeat analysis with higher dilution or earlier observations.

4. Reference

1. EL-SHAARAWI, A., A.A. QURESHI & B.J. DUTKA. 1977. Study of microbiological and physical parameters in Lake Ontario adjacent to the Niagara River. *J. Great Lakes Res.* 3:196.

9610 D. Membrane Filter Technique

For general information on the membrane filter technique and apparatus needed see Section 9222. However, except for comparisons of different manufacturers' membranes, no critical tests have been reported for membrane filters for fungal isolation efficiency. Media components, pH levels, and antibiotics have been used in routine plating procedures. It appears that the reported procedures are satisfactory.

1. Samples

See Sections 9610B.1 and 9060A.

2. Media

Use unmodified or modified aureomycin®-rose bengal-glucose-peptone agar (MARGPA) or modified streptomycin-terramycin®-malt extract agar (MSTMEA).[1] These media are prepared identically to the unmodified media described in 9610C.2e and f except that the concentration of each antibiotic is increased from 70 mg/L to 200 mg/L. Dispense media in portions of 5 to 7 mL in glass or plastic petri dishes (60 × 15 mm); plastic dishes with tight-fitting lids are preferred.

3. Procedure

a. *Preparation and dilution:* See Sections 9215A.5 and 9610B.3a. Select dilutions to yield 20 to 100 colonies per membrane.

b. *Filtration:* Filter appropriate volumes of well-shaken sample or dilution, in triplicate, through membrane filters with pore diameter of 0.45 or 0.8 μm. See Section 9222.

c. *Incubation:* Transfer filters to dishes, invert dishes or not, and incubate at 15°C for 5 d in a humid atmosphere. Alternatively incubate at 20°C for 3 d, or longer depending on fungi present.

d. *Counting and recording:* Using a binocular dissecting microscope at a magnification of 10×, count all colonies on each selected plate. If counting must be delayed temporarily, hold plates at 4°C for not longer than 24 h. Ideal plates have 20 to 80 colonies per filter. See Section 9610C.

4. Reference

1. QURESHI, A.A. & B.J. DUTKA. 1978. Comparison of various brands of membrane filter for their ability to recover fungi from water. *Appl. Environ. Microbiol.* 32:445.

9610 E. Technique for Yeasts

Of the total number of fungal colonies obtained from polluted waters, as many as 50% may be yeast colonies. Solid media such as those described above do not permit growth of all yeasts; thus, a quantitative enrichment technique may be useful in addition to the plate count (see also Fungi Pathogenic to Humans, Section 9610H).

1. Media

For enrichment, use yeast nitrogen base-glucose broth; for isolation, use yeast extract-malt extract-glucose agar or diamalt agar.

a. *Yeast nitrogen base-glucose broth:* Dissolve 13.4 g yeast nitrogen base in 1 L reagent-grade water; sterilize by filtration. Prepare 500 mL each of 2% and 40% aqueous glucose solutions and sterilize separately by filtration. To make final medium, aseptically add to a sterile 250-mL erlenmeyer flask 25 mL yeast nitrogen base solution and 25 mL of either 2% or 40% glucose solutions to make 1% or 20% final glucose concentrations. Stopper flask with a gauze-wrapped cotton stopper and store until used.

b. *Yeast extract-malt extract-glucose agar:* See Section 9610B.2c.

c. *Diamalt agar:* See Section 9610B.2d.

2. Procedure

a. *Sample preparation and dilution:* Prepare as directed in Section 9610B.

b. *Enrichment:* In 250-mL erlenmeyer flasks prepare one flask each of yeast nitrogen base medium containing 1% and 20% glucose. Inoculate with 1 mL of appropriate sample dilution and incubate at room temperature on a rotary shaker operating at 120 to 150 oscillations/min for at least 64 h. Shaken cultures are necessary to prevent overgrowth by filamentous fungi.

c. *Isolation:* Remove flasks from shaker and let settle 4 to 5 h. Yeast cells, if present, will settle to the bottom, bacteria and filamentous fungi will remain in suspension, and filamentous fungi will float on the surface or will be attached to the glass surface at or above the meniscus. With a nichrome wire loop remove a loopful of sediment at the sediment-supernatant interface from a tilted flask and smear-streak on yeast extract-malt extract-glucose agar. Use three plates per flask. Incubate at room temperature but out of direct sunlight for 2 to 3 d. It is not necessary to invert dishes. To obtain pure cultures, pick from reasonably isolated colonies and restreak on the same medium or on diamalt agar plates. Obtain pure cultures of as many different colonies as can be recognized.

d. Counting: It is impossible to obtain a meaningful plate count after this type of enrichment isolation. If it is assumed that one cell in the original sample will produce one or more colonies on the plates after enrichment, it can be stated that yeasts, or specific types of yeasts, occur at a minimal number dependent on the highest positive dilution. The reciprocal of this dilution is the indicated number of yeasts in the sample.

3. Bibliography

LODDER, J., ed. 1970. The Yeasts, A Taxonomic Study, 2nd ed. North Holland Publ. Co., Amsterdam.

BUCK, J.D. 1975. Distribution of aquatic yeasts—effect of inoculation temperature and chloramphenicol concentration on isolation. *Mycopathologia* 56:73.

9610 F. Zoosporic Fungi

1. Occurrence and Significance

Most fungi found in lacustrine (lake) and lotic (river) habitats that reproduce asexually by motile, uniflagellate spores and have determinate growth of the fungal body belong to the class Chytridiomycetes. Fungi with indeterminate growth, asexual reproduction by motile, biflagellate spores, and sexual reproduction involving oogonia and antheridia, are members of the class Oomycetes. A reduction in numbers of species of both classes tends to occur in polluted areas of rivers but more species of Oomycetes than of Chytridiomycetes can be found in polluted situations. Species of the Oomycete genera *Saprolegnia* (notably *S. ferax*) and *Leptomitus* appear to be more tolerant than other forms. Bioassay studies indicate that Oomycetes are more tolerant to zinc, cyanide, and mannitol than are Chytridiomycetes. The latter appear to be more tolerant to treatment with surfactants than do the Oomycetes.

Some Chytridiomycetes may parasitize planktonic and other algae. In the case of epidemic fungal infections of phytoplankton species, the activities of fungi may affect the composition of phytoplankton communities by delaying the time of algal maxima and by reducing the population of certain algae so that other phytoplankters will replace the infected algal populations. In the case of nonepidemic infections, fungi may not influence algal populations; instead, they may infect only phytoplankters during periods of decline and thus only hasten decomposition of the algae.

Filamentous Oomycetes, particularly members of the Saprolegniaceae and Pythiaceae, are found in virtually all types of freshwater habitats and damp-to-wet soils. Most of the nearly 250 species involved occur as saprobes on dead and decaying organic matter such as insect exuviae, algae, and submerged vascular plant remains. A few occur as parasites of algae, aquatic invertebrates, fish, and vascular plants; none are associated with human disease.

Rarely do any of these fungi develop in sufficient numbers to be observed or collected directly. Consequently, various techniques have been devised for their collection and isolation.

2. Sampling and Baiting

Collect samples in sterile 35-mL plastic vials, refrigerate, and start analysis within 6 to 8 h. Place each sample in a sterile plate (20 × 100 mm) and dilute with 10 to 15 mL sterile reagent-grade water. Add three to four split hemp seed halves (*Cannabis sativa*), or whole seeds of mustard (*Brassica*) or sesame (*Sesamum*) as bait to each culture. Incubate at 18 to 23°C and examine daily for fungal growth on the bait. As growth becomes evident, usually within 72 h, remove the infected bait, wash it thoroughly with water from a wash bottle, and transfer to a fresh plate of water containing two to three halves of hemp or other seed. Genera may be identified from spore arrangement within the sporangium and the manner in which spores are released. Specific determination requires microscopic examination of the sexual reproductive structures.

To collect the few naturally occurring parasites or pathogens, place the host organisms in a plate containing sterile water and hemp seed.

3. Isolation

Although most filamentous Oomycetes can be cultivated on plain cornmeal agar, selective media for isolating *Saprolegnia* from fresh water have been developed.[1]

Obtain axenic cultures by drawing spores into a micropipet as they emerge from the sporangium. Less preferably, use hyphal tips, but note that several different genera and species frequently occur on a single piece of bait. Transfer the spore suspension or hyphal tip to a plate of cornmeal agar. When growth on the agar has occurred, remove bacteria-free hyphal tips aseptically by cutting out a small block of agar. Transfer to fresh medium or water. If growth is not free from contamination after one transfer, make additional transfers to insure pure cultures. Other methods have been outlined.[2]

4. Dilution Plating

Make serial dilutions with sterile reagent-grade water (1:100 000 to 1:700 000) and spread 1 mL over surface of a *freshly* prepared cornmeal agar plate. Remove each developing colony and transfer to water for identification. This method permits numerical estimation and determination of composition of the Oomycete community but requires at least 10 plates.

5. References

1. HO, H.H. 1975. Selective media for the isolation of *Saprolegnia* spp. from fresh water. *Can. J. Microbiol.* 21:1126.

2. SEYMOUR, R.L. 1970. The Genus *Saprolegnia*. *Nova Hedwigia* Beihefte 19:1.

6. Bibliography

WILLOUGHBY, L.G. 1962. The occurrence and distribution of reproductive spores of Saprolegniales in fresh water. *J. Ecol.* 50:733.

9610 G. Aquatic Hyphomycetes

1. Occurrence and Significance

Freshwater Hyphomycetes are a very specialized group of conidial fungi that usually occur on partially decayed, submerged leaves and occasionally wood of angiosperms. The mycelium, which is branched and septate, ramifies through the leaf tissue, especially in petioles and veins. The conidiophores project into the water and the conidia that usually develop are liberated under water. Mature conidia also can be found in the surface foam of most rivers, streams, and lakes. The conidia of the majority of these fungi are hyaline, thin-walled, and either tetraradiately branched, that is, with four divergent arms, or sigmoid (S-shaped) with the curvature in more than one plane. A special feature of the conidia is that while suspended in water, even for long periods, they do not germinate. However, if they come to rest on a solid surface, germ tubes are produced within a few hours. The size and morphology of these spores make them potentially more prominent in plankton analysis work than the spores of other fungi.

Ecological investigations of freshwater Hyphomycetes have been limited to substrate, habitat, dispersal, and their role in the enhancement of leaf substrates as food for aquatic invertebrates. The most common substrates of these organisms are submerged, decaying leaves of angiosperms such as alder (*Alnus*), oak (*Quercus*), hazelnut (*Corylus*), elm (*Ulmus*), maple (*Acer*), chestnut (*Castanea*), blackberry (*Rubus*), ash (*Fraxinus*), and willow (*Salix*). Submerged gymnosperm leaves usually are free of aquatic Hyphomycetes. The usual habitat of these fungi is well-oxygenated water, such as alpine brooks, mountain streams, and fast-flowing rivers. However, they also have been found in slow-running, often contaminated, rivers, stagnant or temporary pools, melting snow, and soil. There is often an increase in the numbers of species and individuals of aquatic Hyphomycetes from autumn until spring, with a decline between April and June.

2. Sample Collection and Storage

For most freshwater environments, collect foam or partially decayed, submerged, angiosperm leaves in sterile bottles. Refrigerate sample until it is examined.

3. Sample Treatment and Analysis

Wash the leaf samples in sterile distilled water and place one to three leaves in a sterile petri dish about 1 cm deep containing sterile pond, river, or lake water. Incubate at room temperature. Within 1 to 2 d, the mycelium and conidia develop. Conidiophores and conidia can be observed with a dissecting microscope on any portion of a leaf surface, but most frequently are seen on petioles and veins. When released, the conidia either remain suspended in the water or settle to the dish bottom. Using a dissecting microscope, pick up single conidia with a micropipet. Transfer each conidium to a microscope slide in a drop of water for identification. The conidium may be transferred with a sterile needle to a plate of 2% malt extract agar for colony production. Search for conidia in foam samples with a dissecting microscope and isolate single conidia as described above. Submerge mycelial plugs from stock culture isolates of aquatic Hyphomycetes in autoclaved pond water in deep petri dishes; conidiogenesis usually occurs within 2 to 10 d.

Conidia in all stages of development can be preserved on slides with lactophenol mounting medium in which either acid fuchsin or cotton blue is dissolved and sealed with clear fingernail polish. To permit good adherence of the nail polish, avoid excessive amounts of mounting medium.

9610 H. Fungi Pathogenic to Humans

1. Occurrence and Significance

Routine isolations of fungi from polluted streams and wastewater treatment plants usually have yielded relatively few species pathogenic to human and other higher animals. *Geotrichum candidum*, an arthroconidium-producing fungus, for which there is, in the United States, presumptive evidence of an association with disease, is isolated almost universally. When its teleomorphic stage (an ascus) develops, it is known as *Endomyces candidus*. *Rhinocladiella mansonii*, now called *Exophiala mansonii*, a causal agent of one form of chromomycosis, usually in the tropics, is equally widespread. It has been listed also as *Phialophora jeanselmei* and *Trichosporium heteromorphum*. *Aspergillus fumigatus*, a causal agent of pulmonary aspergillosis, is commonly isolated. *Pseudallescheria (Petriellidium, Allescheria) boydii* is a causal agent of eumycotic mycetomas and other eumycotic conditions grouped[1] under the heading "Pseudallescheriasis." Infection may follow from a puncture wound with contaminated materials or breathing such contaminated materials as sprays at wastewater treatment plants or contaminated air. It usually is recovered in its anamorphic state, *Scedosporium (Monosporium) apiospermum*. The presence of these fungi in stream water probably represents soil runoff because virtually all zoopathogenic fungi exist saprobically in soil as their natural reservoir. Other zoopathogenic fungi occasionally are recovered in low frequencies from streams, polluted or not. Another fungus, the yeast *Candida albicans*, can be recovered in varying numbers from wastewater treatment plant effluents, streams receiving such effluents, and recreational waters. In humans this fungus is usually a commensal organism, like *Geotrichum candidum*, coexisting in harmony with its host organism; up to 80% of normal, healthy adults have detectable levels of *C. albicans* in their feces, while about 35% harbor it in their oral cavities in the absence of any overt disease. A very large proportion of the female population has vaginal candidiasis in varying degrees of severity. The presence of *C. albicans* in raw wastewater, wastewater treatment plant effluent, or contaminated water is not surprising. *C. albicans* has been isolated from these habitats on routine media heavily supplemented with antibacterial drugs, but not on media or with techniques described in Sections 9610B or E. It also has been isolated from estuarine and marine habitats on a maltose-yeast nitrogen base-chloramphenicol-cycloheximide medium.

2. Identification of *C. albicans*

C. albicans can be identified among the white and pink yeasts growing on an 0.8-μm black membrane filter on maltose-yeast nitrogen base-chloramphenicol-cycloheximide medium. From each colony, inoculate a 0.5-mL portion of calf or human blood serum, incubate at 37°C for 2 to 3 h, transfer a drop or two to a slide, and examine microscopically for the production of germ tubes from a majority of the cells. Of the white yeasts, only *C.*

albicans produces these short hyphae from the parent cell within 2 to 3 h incubation.[2]

3. References

1. RIPPON, J. 1982. Medical Mycology. W.B. Saunders Co., Philadelphia, Pa.
2. BUCK, J.D. & B.M. BUBACIS. 1978. Membrane filter procedure for enumeration of *Candida albicans* in natural waters. *Appl. Environ. Microbiol.* 35:237.

4. Bibliography

PAGAN, E.F. 1970. Isolation of human pathogenic fungi from river water. Ph.D. dissertation, Botany Dep., Ohio State Univ., Columbus.

9711 PATHOGENIC PROTOZOA*

9711 A. Introduction

1. Significance

Pathogenic intestinal protozoa are significant problems in drinking water supplies. These organisms cause diarrhea or gastroenteritis of varying severity; numerous outbreaks have occurred. During the 1970s, waterborne outbreaks due to *Giardia lamblia* were noted with increasing frequency, especially in communities using unfiltered surface water sources.[1] By the mid 1980s, waterborne outbreaks due to *Cryptosporidium parvum* began to appear, and in 1993, this organism was responsible for the largest waterborne outbreak in U.S. history.[2] Recreational waterborne outbreaks also have been reported.[3] *Cryptosporidium* can cause a severe diarrhea that is self-limiting in immunocompetent individuals but may be prolonged and life-threatening in the immunocompromised.[4]

Microscopic antibody-based methods generally have been used for detecting and quantifying the environmentally resistant cyst stages of the pathogenic intestinal protozoa in water samples. Using these methods, surveys have demonstrated the wide distribution and occurrence of *Giardia* cysts and *Cryptosporidium* oocysts in raw and treated water supplies.[5-7] Recently, another protozoan intestinal pathogen, *Cyclospora cayetanensis*, has been associated with waterborne[8] and foodborne illness.[9] Another group of protozoan organisms, Microsporidia, while not yet associated with a waterborne outbreak, appear to be widely distributed in nature. Microsporidia have caused intestinal illness and conjunctivitis, primarily in the immunocompromised.[10] Although *Cyclospora* have been found in drinking water,[11] the methods for detecting both *Cyclospora* and Microsporidia in environmental samples are developmental.

Giardia and *Cryptosporidium* occur in domestic and feral animals as well as in humans. The environment may become contaminated through direct deposit of human and animal feces or

through sewage and wastewater discharges to receiving water. Ingestion of water containing these organisms may cause disease.

2. References

1. CRAUN, G.F. 1990. Waterborne *giardiasis*. *In* E.A. Meyer, ed. *Giardiasis.* Elsevier, New York, N.Y.
2. MACKENZIE, W.R., N.J. HOXIE, M.E. PROCTOR, M.S. GRADUS, K.A. BLAIR, D.E. PETERSON, J.J. KAZMIERCZAK, D.G. ADDISS, K.R. FOX, J.B. ROSE & J.P. DAVIS. 1994. A massive outbreak in Milwaukee of *Cryptosporidium* infection transmitted through the public water supply. *N. England J. Med.* 331(3):161.
3. ROSE, J.B., J.T. LISLE & M. LECHEVALLIER. 1997. Waterborne cryptosporidiosis: incidence, outbreaks and treatment strategies. *In* R. Fayer, ed. *Cryptosporidium* and Cryptosporidiosis. CRC Press, Boca Raton, Fla.
4. UNGAR, B.L.P. 1994. *Cryptosporidium* and cryptosporidiosis. *In* S. Broder, T.C. Merigan, Jr. & D. Bolognesi, eds. Textbook of AIDS Medicine. Williams & Wilkins, Baltimore, Md.
5. LECHEVALLIER, M.W., W.D. NORTON & R.G. LEE. 1991. Occurrence of *Giardia* and *Cryptosporidium* spp. in surface water supplies. *Appl. Environ. Microbiol.* 57:2610.
6. LECHEVALLIER, M.W., W.D. NORTON & R.G. LEE. 1991. *Giardia* and *Cryptosporidium* spp. in filtered drinking water supplies. *Appl. Environ. Microbiol.* 57:2617.
7. ROSE, J.B., C.P. GERBA & W. JAKUBOWSKI. 1991. Survey of potable water supplies for *Cryptosporidium* and *Giardia*. *Environ. Sci. Technol.* 25:1393.
8. SOAVE, R. & W.D. JOHNSON, JR. 1995. *Cyclospora*: conquest of an emerging pathogen. *Lancet* 345:667.
9. CENTERS FOR DISEASE CONTROL AND PREVENTION. 1996. Update: outbreaks of *Cyclospora cayetanensis* infection—United States and Canada, 1996. *Morbid. Mortal. Weekly Rep.* 45:611.
10. WEBER, R., R.T. BRYAN, D.A. SCHWARTZ & R.L. OWEN. 1994. Human microsporidial infection. *Clin. Microbiol. Rev.* 7:426.
11. RABOLD, J.G., C.W. HOGE, D.R. SHLIM, C. KEFFORD, R. RAJAH & P. ECHEVERRIA. 1994. *Cyclospora* outbreak associated with chlorinated drinking water. *Lancet* 344:1360.

---

* Approved by Standard Methods Committee, 1997.

9711 B.　Detection and Enumeration Methods

1. General Discussion

Methods for the simultaneous detection and enumeration of *Giardia* cysts and *Cryptosporidium* oocysts in water have appeared in previous editions of *Standard Methods*. These methods were developed to assist in the investigation of suspected waterborne disease outbreaks. They were applied subsequently to occurrence and distribution studies and to the determination of drinking water treatment effectiveness. The need for quantitative methods for regulatory purposes[1] resulted in method evaluation studies that underscored the deficiencies of the existing antibody-based immunofluorescence methods.[2] These deficiencies included requiring analysts with a high degree of experience, lengthy analysis time, high expense, lack of specificity, erratic efficiency, low precision, and difficulty in determining viability.[3]

The regulatory requirements for precise, sensitive, and quantitative methods for these organisms have stimulated interest and research on methods for their detection. No method is included in this edition of *Standard Methods* because the methods for *Giardia* and *Cryptosporidium* are evolving rapidly. A regulatory method based on the widely used immunofluorescence assay is available.[4] In addition, a draft method that addresses some of the limitations of the regulatory method has been published.[5] The reader with a need to analyze water samples for pathogenic in-

testinal protozoa is advised to consult the current literature for the latest methodology.

2. References

1. U.S. ENVIRONMENTAL PROTECTION AGENCY. 1996. Monitoring requirements for public drinking water supplies: *Cryptosporidium*, *Giardia*, viruses, disinfection byproducts, water treatment plant data and other information requirements. 40 CFR Part 141; *Federal Register* 61: 24353.
2. Development of Performance Evaluation (PE) Sample Preparation Protocols for *Giardia* Cysts and *Cryptosporidium* Oocysts. 1996. EPA Contract No. 68-C3-0365, available from EPA Water Docket, 202/260-3027.
3. JAKUBOWSKI, W., S. BOUTROS, W. FABER, R. FAYER, W. GHIORSE, M. LECHEVALLIER, J. ROSE, S. SCHAUB, A. SINGH & M. STEWART. 1996. Status of environmental methods for *Cryptosporidium*. *J. Amer. Water Works Assoc.* 88(9):107.
4. FOUT, G.S., F.W. SCHAEFER, J.W. MESSER, D.R. JAHLING & R.E. STETLER. 1996. ICR Microbial Laboratory Manual. EPA 600/R-95-178, U.S. Environmental Protection Agency, Washington, D.C. [available online]
5. U.S. ENVIRONMENTAL PROTECTION AGENCY. 1997. Method 1622: *Cryptosporidium* in Water by Filtration/IMS/FA (draft). EPA-821/R-97-023, U.S. Environmental Protection Agency, Washington, D.C. [available online]

10010 INTRODUCTION*

Physical and chemical characteristics of water bodies affect the abundance, species composition, stability, productivity, and physiological condition of aquatic organism populations. Biological methods used for assessing water quality include the collection, counting, and identification of aquatic organisms; biomass measurements; measurements of metabolic activity rates; measurements of the toxicity, bioconcentration, and bioaccumulation of pollutants; and processing and interpretation of biological data.

Information from these methods may serve one or more of the following purposes:

1. To explain the cause of color, turbidity, odor, taste, or visible particulates in water;

2. To aid in the interpretation of chemical analyses, for example, in relating the presence or absence of certain biological forms to oxygen deficiency or supersaturation in natural waters;

3. To identify the source of a water that is mixing with another water;

4. To explain the clogging of pipes, screens, or filters, and to aid in the design and operation of water and wastewater treatment plants;

5. To determine optimum times for treatment of surface water with algicides and to monitor treatment effectiveness;

6. To determine the effectiveness of drinking water treatment stages and to aid in determining effective chlorine dosage within a water treatment plant;

7. To identify the nature, extent, and biological effects of pollution;

8. To indicate the progress of self-purification in bodies of water;

9. To aid in determining the condition and effectiveness of unit processes and biological wastewater treatment methods in a wastewater treatment plant;

10. To document short- and long-term variability in water quality caused by natural phenomena and/or human activities;

11. To provide data on the status of an aquatic system on a regular basis;

12. To correlate the biological mass or components with water chemistry or conditions.

The specific nature of a problem and the reasons for collecting samples will dictate which communities of aquatic organisms will be examined and which sampling and analytical techniques will be used.

The following communities of aquatic organisms are considered in specific sections that follow:

1. PLANKTON (Section 10200): A community of plants (phytoplankton) and animals (zooplankton), usually drifting or suspended in water, nonmotile or insufficiently motile to overcome transport by currents. In fresh water they generally are small or microscopic in size; in the marine or estuarine environment, larger forms are observed more frequently.

2. PERIPHYTON (Section 10300): A community of microscopic plants and animals associated with the surfaces of submersed objects. Some are attached, some move about. Many of the protozoa and other minute invertebrates and algae found in the plankton also occur in the periphyton.

3. MACROPHYTON (Section 10400): The larger plants of all types. They are sometimes attached to the bottom (benthic), sometimes free-floating, sometimes totally submersed, and sometimes partly emergent. Complex types usually have true roots, stems, and leaves; the macroalgae are simpler but may have stem- and leaf-like structures.

4. MACROINVERTEBRATES (Section 10500): The invertebrates defined here are those retained by the US Standard No. 30 sieve. They are generally bottom-dwelling organisms (benthos).

5. FISH (Section 10600): Vertebrates of diverse morphology, ecology, and behavior, inhabiting (and generally limited to) aquatic systems. They have fins and gills.

6. AMPHIBIANS, AQUATIC REPTILES, BIRDS, AND MAMMALS: These vertebrates also may be affected directly or indirectly by spills or other discharges of pollutants and may be useful in monitoring the presence of toxic substances or long-term changes in water quality. Discussion of these organisms is not included.

Large numbers of bacteria and fungi are present in the plankton and periphyton and constitute an essential element of the total aquatic ecosystem. Although their interactions with living and dead organic matter profoundly affect the larger aquatic organisms, techniques for their investigation are not included herein (see Part 9000).

Field observations are indispensable for meaningful biological interpretations, but many biological factors cannot be evaluated directly in the field. These must be analyzed as field data or field samples within the laboratory. Because the significance of the analytical result depends upon the representativeness of the sample, attention is given to field methods as well as to associated laboratory procedures.

Before sampling begins, clearly define study objectives. For example, the frequency of a repetitive sampling program may vary from hourly, for a detailed study of diel variability, to every third month (quarterly) for a general assessment of seasonal conditions, depending on objectives. The scope of the study must be adjusted to limitations in personnel, time, and budget. Before the development of a study plan, examine historic data for the study area and conduct a literature search of work by previous investigators.

Whenever practicable, biologists should collect their own samples. Much of the value of an experienced biologist lies in personal observations of conditions in the field and in the ability to recognize signs of environmental changes as reflected in the various aquatic communities.

The primary orientation of Part 10000 is toward field collection and associated laboratory analyses to aid in determining the status of aquatic communities under field conditions and to aid in interpreting the influence of past and present environmental conditions. The methods selected are necessary for the appraisal of water quality. Principal emphasis is on methods and equipment, rather than on interpretation or application of results. The complex interrelationships existing in an aquatic environment often require many different field and laboratory procedures; consequently, frequent cross-references between sections have been made.

---

* Approved by Standard Methods Committee, 1994.

Many other types of studies may be, and are being, conducted that are oriented more toward laboratory research. Such laboratory studies will develop further basic knowledge of community and/or organism responses under controlled conditions and will aid in predicting effects of future changes in environmental conditions on the aquatic communities. However, such studies are not within the scope of this book.

10200 PLANKTON*

10200 A. Introduction

The term "plankton" refers to those microscopic aquatic forms having little or no resistance to currents and living free-floating and suspended in natural waters. Planktonic plants, "phytoplankton," and planktonic animals, "zooplankton," are covered in this section. The phytoplankton (microscopic algae) occur as unicellular, colonial, or filamentous forms. Many are photosynthetic and are grazed upon by zooplankton and other aquatic organisms. Other organisms occurring in the same environment are dealt with elsewhere: zoosporic fungi in Section 9610F; aquatic hyphomycetes in Section 9610G; and bacteria in Part 9000. The zooplankton in fresh water comprise principally protozoans, rotifers, cladocerans, and copepods; a greater variety of organisms occurs in marine waters.

1. Significance

Plankton, particularly phytoplankton, long have been used as indicators of water quality.[1-4] Some species flourish in highly eutrophic waters while others are very sensitive to organic and/or chemical wastes. Some species develop noxious blooms, sometimes creating offensive tastes and odors[5] or anoxic or toxic conditions resulting in animal deaths or human illness.[6] The species assemblage of phytoplankton and zooplankton also may be useful in assessing water quality.[7]

Because of their short life cycles, plankters respond quickly to environmental changes, and hence their standing crop and species composition are more likely to indicate the quality of the water mass in which they are found. They strongly influence certain nonbiological aspects of water quality (such as pH, color, taste,

and odor), and in a very practical sense, they are a part of water quality. Certain taxa often are useful in determining the origin or recent history of a given water mass. Because of their transient nature, and often patchy distribution, however, the utility of plankters as water quality indicators may be limited. Information on plankton as indicators is interpreted best in conjunction with concurrently collected, physicochemical and other biological data.

Planktonic organisms predominate in ponds, lakes, and oceans. Potamoplankton develop in large rivers with slow-moving waters that approach lentic conditions. Because their origin can be uncertain and the duration of their exposure to pollutants unknown, plankters generally are less valuable as water quality indicators in lotic than in lentic environments.

2. References

1. PALMER, C.M. 1969. A composite rating of algae tolerating organic pollution. *J. Phycol.* 5:78.
2. PALMER, C.M. 1963. The effect of pollution on river algae. *Bull. N.Y. Acad. Sci.* 108:389.
3. RAWSON, D.S. 1956. Algal indicators of trophic lake types. *Limnol. Oceanogr.* 1:18.
4. STOERMER, E.F. & J.J. YANG. 1969. Plankton Diatom Assemblages in Lake Michigan. Spec. Rep. No. 47, Great Lakes Research Div., Univ. Michigan, Ann Arbor.
5. PRESCOTT, G.W. 1968. The Algae: A Review. Houghton Mifflin Co., Boston, Mass.
6. CARMICHAEL, W., ed. 1981. The Water Environment, Algal Toxins and Health. Plenum Press, New York, N.Y.
7. GANNON, J.E. & R.S. STEMBERGER. 1978. Zooplankton (especially crustaceans and rotifers) as indicators of water quality. *Trans. Amer. Microsc. Soc.* 97:16.

---

* Approved by Standard Methods Committee, 1994.

10200 B. Sample Collection

1. General Considerations

The frequency and location of sampling is dictated by the purpose of the study.[1] Locate sampling stations as near as possible to those selected for chemical and bacteriological sampling to insure maximum correlation of findings. Establish a sufficient number of stations in as many locations as necessary to define adequately the kinds and quantities of plankton in the waters stud-

ied. The physical nature of the water (standing, flowing, or tidal) will influence greatly the selection of sampling stations. The use of sampling sites selected by previous investigators usually will assure the availability of historical data that will lead to a better understanding of current results and provide continuity in the study of an area.

In stream and river work, locate stations upstream and downstream from suspected pollution sources and major tributary streams and at appropriate intervals throughout the reach under investigation. If possible, locate stations on both sides of the river because lateral mixing of river water may not occur for great distances downstream. In a similar manner, investigate tributary streams suspected of being polluted but take care in the interpretation of data from a small stream because much of the plankton may be periphytic in origin, arising from scouring of natural substrates by the flowing water. Plankton contributions from adjacent lakes, reservoirs, and backwater areas, as well as soil organisms carried into the stream by runoff, also can influence data interpretation. The depth from which water is discharged from upstream stratified reservoirs also can affect the nature of the plankton.

Because water of rivers and streams usually is well mixed vertically, subsurface sampling, i.e., the upper meter or a composite of two or more strata, often is adequate for collection of a representative sample. There may be problems caused by stratification due to thermal discharges or mixing of warmer or colder waters from tributaries and reservoirs. Always sample in the main channel of a river and avoid sloughs, inlets, or backwater areas that reflect local habitats rather than river conditions. In rivers that are mixed vertically and horizontally, measure plankton populations by examining periodic samples collected at midstream 0.5 to 1 m below the surface.

If it can be determined or correctly assumed that the plankton distribution is uniform and normal, use a scheme of random sampling to accomodate statistical testing. Include both random selection of sampling sites and transects as well as the random collection of samples at each selected site. On the other hand, if it is known or assumed that plankton distribution is variable or patchy, include additional sampling sites, collect composite samples, and increase sample replication. Use appropriate statistical tests to determine population variability.

In sampling a lake or reservoir use a grid network or transect lines in combination with random procedures. Take a sufficient number of samples to make the data meaningful. Sample a circular lake basin at strategic points along a minimum of two perpendicular transects extending from shore to shore; include the deepest point in the basin. Sample a long, narrow basin at several points along a minimum of three regularly spaced parallel transects that are perpendicular to the long axis of the basin, with the first near the inlet and the last near the outlet. Sample a large bay along several parallel transects originating near shore and extending to the lake proper. Because many samples are required to appraise completely the plankton assemblage, it may be necessary to restrict sampling to strategic points, such as the vicinity of water intakes and discharges, constrictions within the water body, and major bays that may influence the main basin.

In lakes, reservoirs, and estuaries where plankton populations can vary with depth, collect samples from all major depth zones or water masses. The sampling depths will be determined by the water depth at the station, the depth of the thermocline or an isohaline, or other factors. In shallow areas of 2 to 3 m depth, subsurface samples collected at 0.5 to 1 m may be adequate. In deeper areas, collect samples at regular depth intervals. In estuaries sample above and below the pyncocline. Depth intervals for sampling vary for estuaries of different sizes and depths, but use depths representative of the vertical range. Composite sampling above and below the pyncocline often is used. In marine sampling, the intent and scope of the study will determine the collection extent.

Over the continental shelf, take samples at stations approximately equidistant from the shore seaward. Take a vertical series from surface to near bottom at each station, gradually adding more stations across the shelf. It is important to sample the entire vertical range over a continental shelf. Benthic grab samples may be taken to collect dormant resting cells or cysts. Beyond the shelf in pelagic waters, sample in the photic zone from the surface to the thermocline for phytoplankton and to deeper depths for zooplankton. Sampling depths vary, but often are at 10- to 25-m intervals above the thermocline, then at 100- to 200-m intervals below the thermocline to 1000 m, and thereafter at 500- to 1000-m intervals.

Samples usually are referred to as "surface" or "depth" (subsurface) samples. The latter are samples taken from some stated depth, whereas surface samples may be interpreted as samples collected as near the water surface as possible. A "skimmed" sample of the surface film plankton (neuston)[2] can be revealing; however, ordinarily do not include a disproportionate quantity of surface film in a surface sample because a neustonic flora[3] as well as plankton often are trapped on top or at the surface film together with pollen, dust, and other detritus. Various methods have been used for sampling surface organisms.

Sampling frequency depends on the intent of the study as well as the range of seasonal fluctuations, the immediate meteorological conditions, adequacy of equipment, and availability of personnel. Select a sampling frequency at some interval shorter than community turnover time. This requires consideration of life-cycle length, competition, predation, flushing, and current displacement. Frequent plankton sampling is desirable because of normal temporal variability and migratory character of the plankton community. Daily vertical migrations occur in response to sunlight, and random horizontal migrations or drifts are produced by winds, shifting currents, and tides. Ideally, collect daily samples and, when possible, sample at different times during the day and at different depths. When this is not possible, weekly, biweekly, monthly, or even quarterly sampling still may be useful for determining major population changes.

In river, stream, and estuarine regions subject to tidal influence, expect fluctuations in plankton composition over a tidal cycle. A typical sampling pattern at a station within an estuary includes a vertical series of samples taken from the surface, across the pyncocline, to near bottom, collected at 3-h intervals, over at least two complete tidal cycles. Once a characteristic pattern is recognized the sampling routine may be modified.

A useful series of monographs on oceanographic methodology has been published.[4-7] Representative taxonomic references for estuarine and marine phytoplankton include diatoms,[8-11] dinoflagellates,[12-14] coccolithophores,[15] and cyanophyceae[16] (cyanobacteria).

2. Sampling Procedures

Once sampling locations, depths, and frequency have been determined, prepare for field sampling. Label sample containers with sufficient information to avoid confusion or error. On the label indicate date, cruise number, sampling station, study area (river, lake, reservoir), type of sample, and depth. Use waterproof labels. When possible, enclose collection vessels in a protective container to avoid breakage. If samples are to be preserved immediately after collection, add preservative to container before sampling. Sample size depends on type and number of determinations to be made; the number of replicates depends on statistical design of the study and statistical analyses selected for data interpretation. Always design a study around an objective with a statistical approach rather than fit statistical analyses to data already collected.

In a field record book note sample location, depth, type, time, meteorological conditions, turbidity, water temperature, salinity, and other significant observations. Engineer's field notebooks with waterproof paper are very suitable. Field data are invaluable when analytical results are interpreted and often help to explain unusual changes caused by the variable character of the aquatic environment. Collect coincident samples for chemical analyses to help define environmental variations having a potential effect on plankton.

a. Phytoplankton: In oligotrophic waters or where phytoplankton densities are expected to be low collect a sample of up to 6 L. For richer, eutrophic waters collect a sample of 0.5 to 1 L.

Because of their small size, nannoplankton and picoplankton can pass through collection nets, making nets unsuitable for most phytoplankton sampling.

For qualitative and quantitative evaluations collect whole (unfiltered and unstrained) water samples with a water collection bottle consisting of a cylindrical tube with stoppers at each end and a closing device. Lower the open sampler to the desired depth and close by dropping a weight, called a messenger, which slides down the supporting wire or cord and trips the closing mechanism. If possible, obtain composite samples from several depths or pool samples from one depth from several casts. The most commonly used samplers that operate on this principle are the Kemmerer,[17] Van Dorn[18] (Figure 10200:1), Niskin, and Nansen samplers.

Because these samplers collect whole water samples, all size classes of phytoplankton are collected. Different size categories of phytoplankton can be separated by subsequently filtering these whole water samples through netting of the appropriate mesh size. Select appropriate mesh sizes for concentrating the various size categories of phytoplankton typical of the aquatic system under study.[19,20]

The Van Dorn usually is the preferred sampler for standing crop, primary productivity, and other quantitative determinations because its design offers no inhibition to free flow of water through the cylinder. In deep-water situations, the Niskin bottle is preferred. It has the same design as the Van Dorn sampler except that the Niskin sampler can be cast in a series on a single line for simultaneous sampling at multiple depths with the use of auxiliary messengers. Because the triggering devices of these samplers are very sensitive, avoid rough handling. Always lower the sampler into the water; do not drop. Kemmerer and Van Dorn samplers have capacities of 0.5 L or more. Polyethylene or polyvinyl chloride sampling devices are preferred to metal samplers because the latter liberate metallic ions that may contaminate the sample. Use polyethylene or glass sample storage bottles. Metallic ion contamination can lead to significant errors when algal assays or productivity measurements are made.

For shallow waters use the Jenkins surface mud sampler,[21] one of the bottle samplers modified so that it is held horizontally,[22] or an appropriate bacteriological sampler.[23]

For greater speed of collection and to obtain large, accurately measured quantities of organisms, use a pump. Diaphragm and peristaltic pumps are less damaging to organisms than centrifugal pumps.[24] Centrifugal pump impellers can damage organisms as

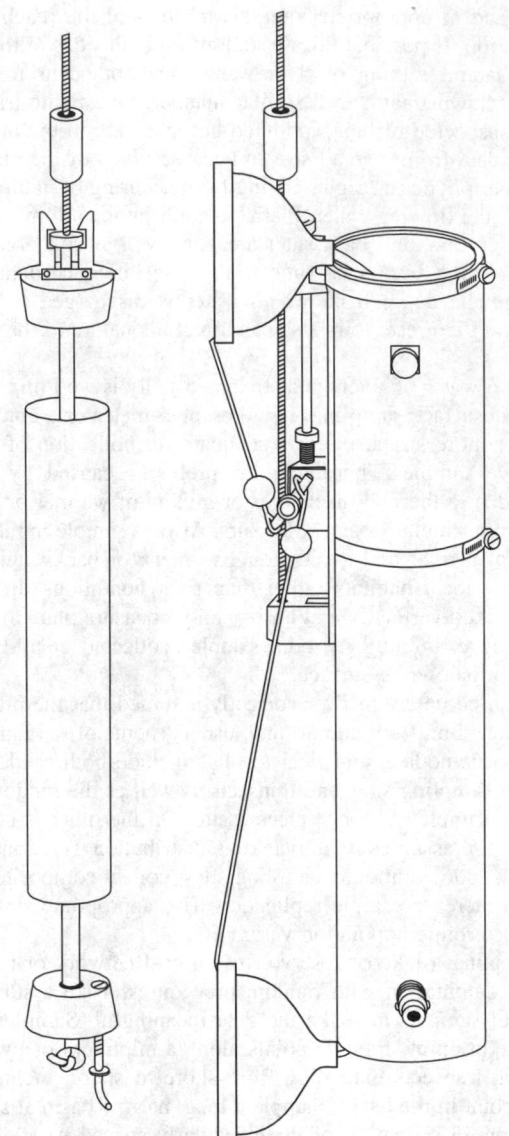

Figure 10200:1. Structural features of common water samplers, Kemmerer (left) and Van Dorn (right).

can passage through the hose.[25] Lower a weighted hose, attached to a suction pump, to the desired depth, and pump water to the surface. The pump is advantageous because it supplies a homogeneous sample from a given depth or an integrated sample from the surface to a particular depth. If a centrifugal pump is used, draw samples from the line before they reach the impeller. For samples to be analyzed for organochlorine compounds use TFE tubing.

To examine live samples fill containers partially and store in a refrigerator or ice chest in the dark, or preferably, hold at ambient temperature. Examine specimens promptly after collection.

If it is impossible to examine living material or if phytoplankton are to be counted later, preserve the sample. For a sample that will be preserved, fill the container completely. The most suitable phytoplankton preservative is Lugol's solution, which can be used for most forms including the naked flagellates. Unfortunately, acidic Lugol's solution (or formalin) dissolves the coccoliths of Coccolithophores, which are common in estuarine and marine waters.

Lugol's solution: To preserve samples with Lugol's solution add 0.3 mL Lugol's solution to 100 mL sample and store in the dark. For long-term storage add 0.7 mL Lugol's solution per 100 mL sample and buffered formaldehyde to a minimum of 2.5% final concentration after 1 h. Prepare Lugol's solution by dissolving 20 g potassium iodide (KI) and 10 g iodine crystals in 200 mL distilled water containing 20 mL glacial acetic acid.[26] Utermohl's[27] modification of Lugol's solution results in a neutral or slightly alkaline solution. Prepare modified Lugol's solution by dissolving 10 g KI and 5 g iodine crystals in 20 mL distilled water, then adding 50 mL distilled water in which 5 g anhydrous sodium acetate has been dissolved. This allows preservation of Coccolithophores, but would be less effective for other flagellates.

Other acceptable preservatives are:

Formalin: To preserve samples with formalin, add 40 mL buffered formalin (20 g sodium borate, $Na_2B_2O_4$, + 1 L 37% formaldehyde) to 1 L of sample immediately after collection. In estuarine and marine collections, adjust pH to at least 7.5 with sodium borate for samples containing Coccolithophores.

Merthiolate: To preserve samples with merthiolate add 36 mL merthiolate solution to 1 L of sample and store in the dark. Prepare merthiolate solution by dissolving 1.0 g merthiolate, 1.5 g sodium borate, and 1.0 mL Lugol's solution in 1 L distilled water. Merthiolate-preserved samples are not sterile, but can be kept effectively for 1 year, after which time formalin must be added.[28]

"M^3" fixative: Prepare by dissolving 5 g KI, 10 g iodine, 50 mL glacial acetic acid, and 250 mL formalin in 1 L distilled water (dissolve the iodide in a small quantity of water to aid in solution of the iodine). Add 20 mL fixative to 1 L sample and store in the dark.

Glutaraldehyde: Preserve samples by adding neutralized glutaraldehyde to yield a final concentration of 1 to 2%.

Other commonly used preservatives include 95% alcohol, and 6-3-1 preservative, (6 parts water, 3 parts 95% alcohol, and 1 part formalin). Use equal volumes of preservative and sample.

To retain color in preserved plankton, store samples in the dark or add 1 mL saturated copper sulfate ($CuSO_4$) solution/L.

Most preservatives distort and disrupt certain cells,[29,30] especially those of delicate forms such as *Euglena, Cryptomonas, Synura, Chromulina,* and *Mallamonas.* Lugol's iodine solution usually is least damaging for these phytoflagellates. To become familiar with live specimens and preservation-caused distortions, use reference collection from biological supply houses or consult experienced co-workers.

b. Zooplankton: The choice of sampler depends on the type of zooplankton, the kind of study (distribution, productivity, etc.) and the body of water being investigated. Zooplankton populations invariably are distributed in a patchy way, making both sampling and data interpretation difficult.

For collecting microzooplankton (20 to 200 μm) such as protozoa, rotifers, and immature microcrustacea, use the bottle samplers described for phytoplankton. The small zooplankters usually are sufficiently abundant to yield adequate samples in 5- to 10-L bottles; however, composite samples over depth and time are recommended. Water bottle samplers are suitable especially for discrete-depth samples. If depth-integrated samples are desired, use pumps or nets. The larger and more robust microzooplankters

(e.g., loricate forms and crustacea) may be concentrated by passing the whole water through a 20-μm mesh net. If quantitative estimates of other nonloricate, delicate forms are required, do not screen. Fix 0.5 to 5 L of whole water for enumeration of these forms.

Bottle samplers usually are unsuitable for collecting larger zooplankton, such as mature microcrustacea, that, unlike the smaller forms, are much less numerous and are sufficiently agile to avoid capture. Although comparatively large water volumes, and consequently adequate numbers of microcrustacea, can be sampled with a pump, avoidance by larger, more agile zooplankters at the pump head can cause sampling error. Consequently, larger trap samplers or nets are the preferred collection methods.

The Juday trap[31] operates on the same principle as the water bottle samplers but is generally larger (10 L). The larger size makes the Juday trap more suitable for collecting zooplankters, especially larger copepods. However, it is awkward to use and its 10-L capacity is inadequate for oligotrophic lakes or other water bodies with few zooplankters. Because it is constructed of metal it is unsuited if heavy metals analyses are required.

The Schindler-Patalas trap[32] (Figure 10200:2) usually is preferred to the Juday trap because it is constructed of clear acrylic plastic and is transparent. It can be lowered into the water with minimal disturbance and is suitable for collecting larger zooplankters. Models of 10- to 12-L capacity are available but the 30-L size is preferred. It has no mechanical closing mechanism and thus is convenient for cold-weather sampling when mechanical devices tend to malfunction. Like the Juday trap, it can be fitted with nets of various mesh sizes, but the No. 20 mesh net is used most often.

Plankton nets are preferred to bottles and traps for sampling where plankters are few or where only qualitative data or a large biomass is needed for analysis. Because they were designed originally for qualitative sampling, modifications are required for quantitative work.

The mesh size, type of material, orifice size, length, hauling method, type of tow, and volume sampled will depend on the particular needs of the study.[33,34] Type of netting and mesh size determine filtration efficiency, clogging tendencies, velocity, drag, and the condition of the sample after collection. Silk, formerly the common mesh material in plankton nets, is not recommended because of shrinkage of mesh openings and rotting with age. Nylon monofilament mesh is preferred because of its mesh size accuracy and durability. Nylon nets of different mesh sizes still are labelled by the silk rating system: characteristics of commonly used nylon plankton nets are listed in Table 10200:I. Finer mesh sizes clog more readily than coarser mesh; a compromise must be made between mesh size small enough to retain desired organisms effectively and a size large enough to preclude a serious clogging problem. If clogging occurs, reduce its effects by decreasing the length of tow.

The maximum volume, V_M, of water that can be filtered through a net during a vertical tow can be estimated with the formula,

$$V_M = \pi r^2 d$$

where:

r = radius of net orifice and
d = depth to which net is lowered.

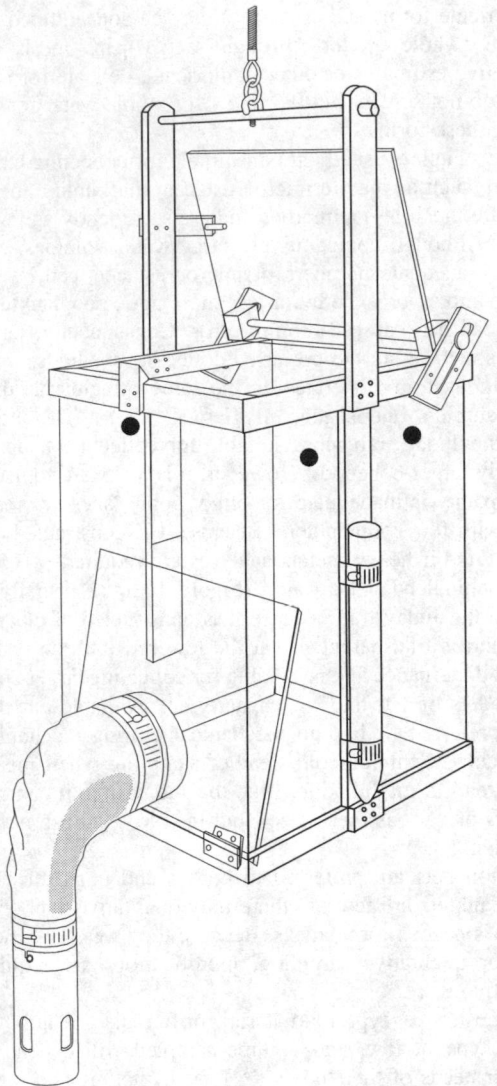

Figure 10200:2. The Schindler-Patalas plankton trap.

This volume is a maximum because clogging of the net's meshes by phytoplankton and other particles and, for fine netting, even the netting itself can cause some water to be diverted from

the net's path.[35,36] Keep net towing distance as short as practical to alleviate clogging. If the net has a pronounced green or brown color after towing, clogging probably has occurred.

To estimate sampling volume, V_A, mount a calibrated flow meter midway between the net rims and mouth center (the meter is mounted off-center to avoid flow reduction associated with the towing bridle).[37] Equip meter with lock mechanisms to prevent it turning in reverse or while in air. Record flow-meter readings before and after collecting sample. Calculate filtration efficiency, E, from:

$$E = V_A/V_M$$

If E is less than about 0.8, substantial clogging has occurred. Take steps to increase efficiency. Clogging not only decreases the volume filtered, but also leads to biased samples because filtration efficiency is nonuniform during the tow.[34]

Various types of plankton nets are shown in Figure 10200:3. Simple conical nets have been used for many years with little modification in design or improvement in accuracy. Their major source of error is that the filtration characteristics of conical nets usually are unknown. Filtration efficiency in No. 20 mesh cone nets ranges from 40 to 77%. To improve efficiency, place a porous cylinder collar or nonporous truncated cone in front of the conical portion of the net. The Juday net exemplifies a commonly used net with a truncated cone. For good filtration characteristics the ratio of filtering area of net to orifice area should be at least 3:1. Bridles attaching the net to the towing line also adversely influence filtration efficiency and increase turbulence in front of the net, thereby increasing the potential for net avoidance by larger zooplankters. The tandem, Bongo net design (Figure 10200:3) reduces these influences and permits duplicate samples to be collected simultaneously.

Three types of tows are used: vertical, horizontal, and oblique. Vertical tows are preferred to obtain an integrated water column sample. To make a vertical tow, lower the weighted net to a given depth, then raise vertically at an even speed of 0.5 m/s.

In small water bodies haul the net hand over hand with a steady, unhurried motion approximating the speed of 0.5 m/s. In large bodies where long net hauls and vessel drifting are expected, use a davit, meter wheel, angle indicator, and winch. Attach a 3- to 5-kg weight to hold the net down. Determine depth of the net by multiplying the length of the extended wire by the cosine of the wire's angle with the vertical direction. Maintain wire angle as close to the vertical as possible by controlling the boat's speed

TABLE 10200:I. CHARACTERISTICS OF COMMONLY USED PLANKTON NETS

Silk No.	Size of Aperture μm	Approximate Open Area %	Classification
000	1024	58	Largest zooplankton and ichthyoplankton
00	752	54	Larger zooplankton and ichthyoplankton
0	569	50	Large zooplankton and ichthyoplankton
2	366	46	Large microcrustacea
6	239	44	Microcrustacea
10	158	45	Microcrustacea and most rotifers
20	76	45	Net phyto- and zooplankton
25	64	33	Nannoplankton

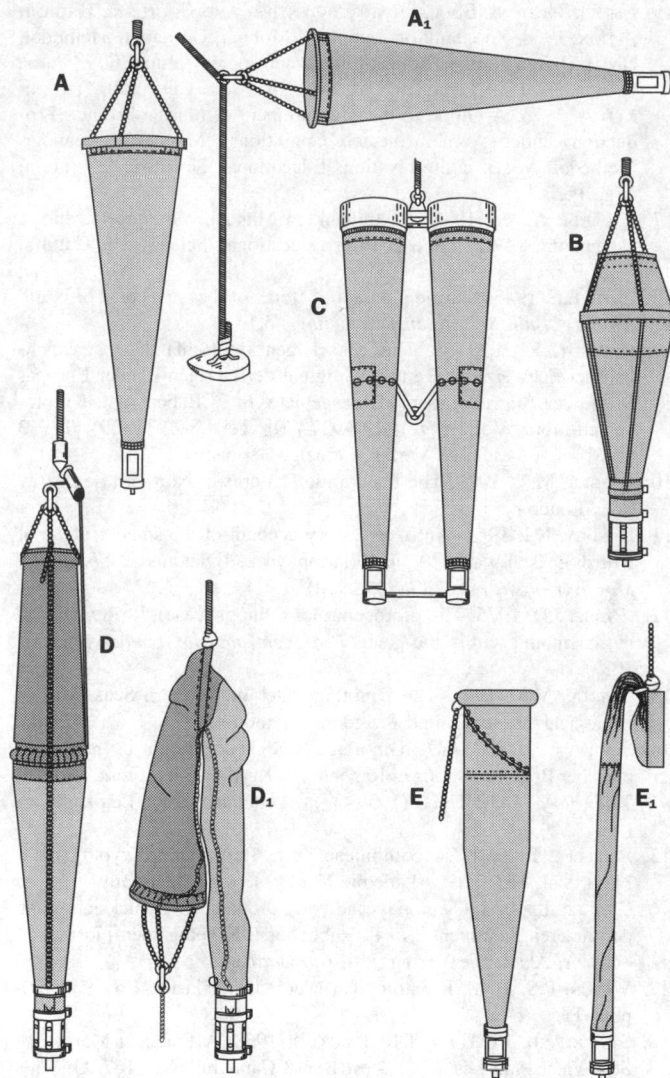

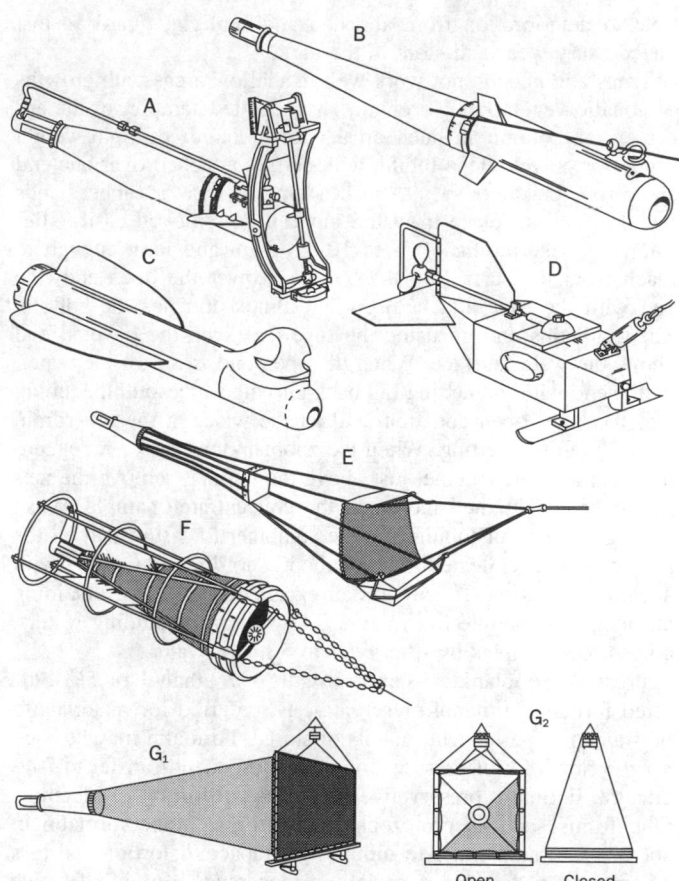

Figure 10200:4. **Examples of commonly used high-speed zooplankton samplers.** (A) Clarke-Bumpus sampler; (B) Miller sampler; (C) Hardy plankton indicator; (D) Hardy continuous plankton recorder; (E) Issacs-Kidd mid-water trawl; (F) Gulf V sampler; (G) Tucker trawl, G_1-side view, G_2-front view open and closed.

Figure 10200:3. **Examples of commonly used plankton sampling nets.** (A) Simple conical tow-net; A—rigged for vertical tows; A_1—for oblique or horizontal tows; (B) Wisconsin (Birge) tow-net with truncated cone to improve filtration efficiency; (C) Bongo net, can be fitted with flow meters and opening/closing mechanisms; (D) Wisconsin net fitted with messenger-activated closing mechanism, D—open, D_1—closed; (E) Free-fall net, E—open, E_1—closed.

Horizontal tows usually are used to obtain depth distribution information on zooplankton. Although a variety of horizontal samplers is available (see Figure 10200:4), use the Clarke-Bumpus sampler[38] for quantitative collection of zooplankton because of its built-in flowmeter and opening-closing device. For horizontal tows use a boat equipped as above and determine sampler depth as above. Lower sampler to preselected depth, open, tow at that depth for 5 to 10 min, then close and raise it.

A variety of zooplankton sampling methods can be used in flowing water. The method of choice depends largely on flow velocity. Properly weighted bottles, traps and pump hoses, and nets can be used in medium- to slow-flowing waters. In turbulent, well-mixed waters, collect surface water by bucket and filter it through the appropriate mesh size. Select sample size based on concentration of zooplankters.

Give plankton nets proper care and maintenance. Do not let particulate matter dry on the net because it can significantly reduce size of mesh apertures and increase frequency of clogging. Wash net thoroughly with water after each use. Periodically clean with a warm soap solution. Because nylon net material is suscep-

null against the wind drift, or wherever feasible, do vertical hauls from an anchored boat.

Vertical and oblique tows collect a composite sample, whereas horizontal tows collect a sample at a discrete depth. Oblique tows usually are preferred over vertical tows in shallow water or wherever a longer net tow is required. For oblique tows, lower the net or sampler to some predetermined depth and then raise at a constant rate as the boat moves forward. Oblique tows do not necessarily sample a true angle from the bottom to the surface. Under best conditions the pattern is somewhat sigmoid due to boat acceleration and slack in the tow line.

tible to deterioration from abrasion and sunlight, guard against unnecessary wear and store in the dark.

Traps and nets do not work well in shallow areas with growths of aquatic vegetation. To obtain an integrated sample for the entire water column in such areas, use a length of light-weight rubber or polyethylene tubing with netting attached over one end and a rope on the other.[39] Attach netting by tape or rubber bands that will stay in place in water, but can be removed easily after sampling. Use tubing of 5- to 10-cm diam and long enough to reach from the surface to the bottom. Lower the open end (the end with the rope attached) until it almost touches the bottom. Then pull this end up using the rope and keep the covered end above the water surface. When the open end is out of the water, let the end with the netting fall back into the water, pull the tubing into the boat, open end first, and let the water in the tube drain out through the netting. When the zooplankton has been concentrated in a small volume, just above the netting, remove the netting over a container and catch the concentrated sample. Wash netting and end of tubing into the container to assure that all the zooplankton is collected. This method is not limited to areas with aquatic vegetation. It provides an excellent method of obtaining an integrated sample from any shallow area. In standing waters, collect tow samples by filtering 1 to 5 m³ of water.

Preserve zooplankton samples with 70% ethanol or 5% buffered formalin. Ethanol preservative is preferred for materials to be stained in permanent mounts or stored. Formalin may be used for the first 48 h of preservation with subsequent transfer to 70% ethanol. Formalin preservative may cause distortion of pleomorphic forms such as protozoans and rotifers. Make formalin in sucrose-saturated water to minimize carapace distortion and loss of eggs in crustaceans, especially cladocerans.[40] Bouin's fixative produces reasonable results for soft-bodied microzooplankton.[41] This fixative is picric acid saturated in calcium carbonate-buffered formaldehyde containing 5% (v/v) acetic acid. Dilute Bouin's fixative 1:19 with the sample. Because rapid fixation is necessary, pour the sample onto the fixative or inject fixative rapidly into the sample.

Use a narcotizing agent such as carbonated water, menthol-saturated water, or neosynephrine to prevent or reduce contraction or distortion of organisms, especially rotifers, cladocerans, and many marine invertebrates.[42,43] Adding a few drops of detergent prevents clumping of preserved organisms. Preserve samples as soon as most animal movement has ceased, usually within a half hour of narcotization. To prevent evaporation, add 5% glycerin to the concentrated sample. In turbid samples, differentiate animal and detrital material by adding 0.04% rose bengal stain, which intensely stains the carapace (shell) of zooplankters and is a good general cytoplasmic stain.

3. References

1. U.S. ENVIRONMENTAL PROTECTION AGENCY. 1982. Handbook for Sampling and Sample Preservation of Water and Wastewater. EPA-600/4-82-029.
2. PARKER, B.C. & R.F. HATCHER. 1974. Enrichment of surface freshwater microlayers with algae. *J. Phycol.* 10:185.
3. TAGUCHI, S. & K. NAKAJIMA. 1971. Plankton and seston in the sea surface of three inlets of Japan. *Bull. Plankton Soc. Japan* 18:20.
4. UNITED NATIONS EDUCATIONAL, SCIENTIFIC AND CULTURAL ORGANIZATION. 1966. Determination of Photosynthetic Pigments in Sea-water. Monogr. Oceanogr. Methodol. No. 1. United Nations Educational, Scientific & Cultural Org., Paris.
5. UNITED NATIONS EDUCATIONAL, SCIENTIFIC AND CULTURAL ORGANIZATION. 1968. Zooplankton Sampling. Monogr. Oceanogr. Methodol. No. 2. United Nations Educational, Scientific & Cultural Org., Paris.
6. UNITED NATIONS EDUCATIONAL, SCIENTIFIC AND CULTURAL ORGANIZATION. 1973. A Guide to the Measurement of Marine Primary Production under Some Special Conditions. Monogr. Oceanogr. Methodol. No. 3. United Nations Educational, Scientific & Cultural Org., Paris.
7. SOURNIA, A., ed. 1978. Phytoplankton Manual. Monogr. Oceanogr. Methodol. No. 6. United Nations Educational, Scientific & Cultural Org., Paris.
8. CUPP, E.E. 1943. Marine plankton diatoms of the west coast of North America. *Bull. Scripps Inst. Oceanogr.* 5:1.
9. HUSTEDT, F. 1927–66. Die Kieselalgen Deutschlands, Österreichs und der Schweiz mit Berucksichtigung der Übrigen Lander Europas Sowie der Angrenzenden Meeresgebiete. *In* L. Rabenhorst, Kryptogamen-Flora. Vol. 7: Teil 1 (1927–30); Teil 2 (1931–59); Teil 3 (1961–66). Akademie Verlag, Leipzig, Germany.
10. LEBOUR, M.V. 1930. The Planktonic Diatoms of Northern Seas. Ray Soc., London.
11. HENDEY, N.I. 1964. An introductory account of the smaller algae of British coastal waters, V. Bacillariophyceae (Diatoms). *Fish. Invest. Min. Agr. Fish. Food (G.B.)*, Ser. IV:1.
12. DODGE, J.D. 1975. The prorocentrales (Dinophyceae), II. Revision of the taxonomy within the genus *Prorocentrum. Bot. Limnol. Soc.* 71:103.
13. LEBOUR, M.V. 1925. The Dinoflagellates of Northern Seas. Marine Biological Assoc. United Kingdom, Plymouth.
14. SCHILLER, J. 1931–37. Dinoflagellatae (Peridineae) in monographischer Behandlung. *In* L. Rabenhorst, Kryptogamen-Flora. Vol. 10; Teil 1 (1931–33); Teil 2 (1935–37). Akademie Verlag, Leipzig, Germany.
15. SCHILLER, J. 1930. Coccolithineae. *In* L. Rabenhorst, Kryptogamen-Flora. Vol. 10, p. 89. Akademie Verlag, Leipzig, Germany.
16. GEITLER, L. 1932. Cyanophyceae von Europa unter Berucksichtigung der anderen Kontinente. *In* L. Rabenhorst, Kryptogamen-Flora. Vol. 14, p. 1. Akademie Verlag, Leipzig, Germany.
17. WELCH, P.S. 1948. Limnological Methods. Blakiston Co., Philadelphia, Pa.
18. STRICKLAND, J.D.H. & T.R. PARSONS. 1968. A Practical Manual of Sea Water Analysis. Fish. Res. Board Can. Bull. No. 167. Queen's Printer, Ottawa, Ont.
19. DUSSART, B.M. 1965. Les différentes catégories de plancton. *Hydrobiologia* 26:72.
20. SIEBURTH, J.McN., V. SMETACEK & J. LENZ. 1978. Pelagic ecosystem structure: Heterotrophic compartments of plankton and their relationship to plankton size fractions. *Limnol. Oceanogr.* 23:1256.
21. MORTIMER, C.H. 1942. The exchange of dissolved substances between mud and water in lakes. *J. Ecol.* 30:147.
22. VOLLENWEIDER, R.A. 1969. A Manual on Methods for Measuring Primary Production in Aquatic Environments. IBP Handbook No. 12. Blackwell Scientific Publ., Oxford, England.
23. GELDREICH, E.E., H.D. NASH, D.F. SPINO & D.J. REASONER. 1980. Bacterial dynamics in a water supply reservoir: a case study. *J. Amer. Water Works Assoc.* 72:31.
24. BEERS, J.R. 1978. Pump sampling. *In* A. Sournia, ed. Phytoplankton Manual. United Nations Educational, Scientific and Cultural Org., Paris.
25. EXTON, R.J., W.M. HOUGHTON, W. ESAIAS, L.W. HAAS & D. HAYWARD. 1983. Spectral differences and temporal stability of phycoerythrin fluorescence in estuaries and coastal waters due to the domination of labile cryptophytes and stable cyanobacteria. *Limnol. Oceanogr.* 28:1225.
26. EDMONDSON, W.T., ed. 1959. Freshwater Biology, 2nd ed. John Wiley & Sons, New York, N.Y.

27. UTERMÖHL, H. 1958. Zur Vervollkommung der quantitativen Phytoplankton-Methodik. *Int. Ver. Theoret. Angewand. Limnol., Commun.* No. 9.

28. WEBER, C.I. 1968. The preservation of phytoplankton grab samples. *Trans. Amer. Microsc. Soc.* 87:70.

29. PAERL, H.W. 1984. An evaluation of freeze fixation as a phytoplankton preservation method for microautoradiography. *Limnol. Oceanogr.* 29:417.

30. SILVER, M.W. & P.J. DAVOLL. 1978. Loss of ^{14}C activity after chemical fixation of phytoplankton: Error source for autoradiography and other productivity measurements. *Limnol. Oceanogr.* 23:362.

31. JUDAY, C. 1916. Limnological apparatus. *Trans. Wis. Acad. Sci.* 18: 566.

32. SCHINDLER, D.W. 1969. Two useful devices for vertical plankton and water sampling. *J. Fish. Res. Board Can.* 26: 1948.

33. SCHWOERBEL, J. 1970. Methods of Hydrobiology. Pergamon Press, Toronto, Ont.

34. TRANTER, D.J., ed. 1980. Reviews on Zooplankton Sampling Methods. United Nations Educational, Scientific & Cultural Org., Switzerland.

35. GANNON, J.E. 1980. Towards improving the use of zooplankton in water quality surveillance of the St. Lawrence Great Lakes. Proc. 1st Biol. Surveillance Symp., 22nd Conf. Great Lakes Research Can. Tech. Rep. Fish. Aquat. Sci. 976, p. 87.

36. ROBERTSON, A. 1968. Abundance, distribution, and biology of plankton in Lake Michigan with the addition of a Research Ships of Opportunity project. Spec. Rep. No. 35, Great Lakes Research Div., Univ. Michigan, Ann Arbor.

37. EVANS, M.S. & D.W. SELL. 1985. Mesh size and collection characteristics of 50-cm diameter conical plankton nets. *Hydrobiologia* 122: 97.

38. CLARKE, G.L. & D.F. BUMPUS. 1940. The Plankton Sampler: An Instrument for Quantitative Plankton Investigations. Spec. Publ. No. 5, Limnological Soc. America.

39. PENNAK, R.W. 1962. Quantitative zooplankton sampling in littoral vegetation areas. *Limnol. Oceanog.* 7:487.

40. HANEY, J.F. & D.J. HALL. 1973. Sugar-coated Daphnia; A preservation technique for Cladocera. *Limnol. Oceanogr.* 18:331.

41. COATS, D.W. & J.F. HEINBOKEL. 1982. A study of reproduction and other life cycle phenomena in plankton protists using an acridine orange fluorescence technique. *Mar. Biol.* 67:71.

42. GANNON, J.E. & S.A. GANNON. 1975. Observations on the narcotization of crustacean zooplankton. *Crustaceana* 28(2):220.

43. STEEDMAN, H.F. 1976. Narcotizing agents and methods. *In* H.F. Steedman, ed. Zooplankton Fixation and Preservation. Monogr. Oceanogr. Methodol. No. 4. United Nations Educational, Scientific & Cultural Org., Paris.

10200 C.   Concentration Techniques

The organisms contained in water samples sometimes must be concentrated in the laboratory before analysis. Three techniques for concentrating phytoplankton, namely, sedimentation, membrane filtration, and centrifugation, are described below. A special technique for zooplankton also is given.

1. Sedimentation

Sedimentation is the preferred method of concentration because it is nonselective (unlike filtration) and nondestructive (unlike filtration or centrifugation), although many of the picoplankton, the smaller nannoplankton, and actively swimming flagellates (in unpreserved samples) may not settle completely. The volume concentrated varies inversely with the abundance of organisms and is related to sample turbidity. It may be as small as 1 mL for use with an inverted microscope or as large as 1 L for general phytoplankton and zooplankton enumeration.

Allow 1 h settling/mm of column depth. For a treated sample (10 mL liquid detergent/L) allow about 0.5 h settling/mm depth.[1] The sample may be concentrated in a series of steps by quantitatively transferring the sediment from the initial container to sequentially smaller ones. Use cylindrical settling chambers with thin, clear glass bottoms. Fill settling chambers without forming a vortex, keep them vibration-free, and move them carefully to avoid nonrandom distribution of settled matter. Carefully siphon or decant the supernatants to obtain the desired final volume (5 mL for diatom mounts). Store the concentrated sample in a closed, labeled glass vial.

2. Membrane Filtration

The filtration method permits use of high magnification for enumerating small plankters including flagellates and cyanobacteria. However, delicate forms such as "naked" flagellates are distorted by even gentle filtration. When populations are dense and the content of detritus is high, the filter clogs quickly and silt may crush the organisms or obscure them from view.

Pour a measured volume of well-mixed sample into a funnel equipped with a membrane filter having a pore diameter of 0.45 μm. Apply a vacuum of less than 50 kPa to the filter until about 0.5 cm of sample remains on filter. Break vacuum, then apply low vacuum (about 12 kPa) to remove remaining water but not to dry the filter.

For samples with a low phytoplankton and silt content the method does not require counting of individual plankters to assemble enumeration data and it increases the probability of observing less abundant forms.[2] Samples also may be concentrated on a filter, inverted onto a microscope slide, and quick-frozen, permitting the removal of the filter and transfer of plankton to the slide.[3,4]

3. Centrifugation

Plankton can be concentrated by batch or continuous centrifugation. Centrifuge batch samples at 1000 g for 20 min. The Foerst continuous centrifuge is no longer recommended as a quantitative device but it may be desirable to continue its use in existing

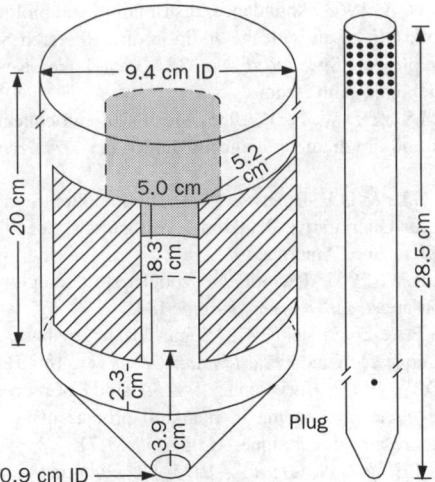

Figure 10200:5. Filter funnel for concentrating zooplankton. This device, originally designed for rotifers, can be modified for other zooplankters by changing the dimensions and mesh size. (After Likens and Gilbert.[5])

programs to assure continuity with previously collected data. Although centrifugation accelerates sedimentation, it may damage fragile organisms.

4. Zooplankton Concentration

Zooplankton samples often need to be concentrated in the field, especially when large water bottles or pump methods of sampling

are used. Moreover, samples obtained by nets or other methods sometimes need to be concentrated further for storage or preparation for examination. When only small volume reductions are needed, pour sample back into the bucket of traps or nets. In processing large volumes of water as with pump sampling, use larger plankton buckets or funnels with greater water volume retention and filtration surface area. Construct a filter funnel similar to that shown in Figure 10200:5 of clear acrylic plastic or other suitable material.[5] The volume of the apparatus and the mesh size depend on volume of water to be filtered and size of organisms to be retained. The mesh size of the filter funnel normally is the same as that of the net or other field sampling device.

5. References

1. Furet, J.E. & K. Benson-Evans. 1982. An evaluation of the time required to obtain sedimentation of fixed algal particles prior to enumeration. *Brit. Phycol. J.* 17:253.
2. McNabb, C.D. 1960. Enumeration of freshwater phytoplankton concentrated on the membrane filter. *Limnol. Oceanogr.* 5:57.
3. Hewes, C.D. & O. Holm-Hansen. 1983. A method for recovering nanoplankton from filters for identification with the microscope: The filter-transfer-freeze (FTF) technique. *Limnol. Oceanogr.* 28:389.
4. Hewes, C.D., F.M.H. Reid & O. Holm-Hansen. 1984. The quantitative analysis of nanoplankton: A study of methods. *J. Plankton Res.* 6:601.
5. Likens, G.E. & J.J. Gilbert. 1970. Notes on quantitative sampling of natural populations of planktonic rotifers. *Limnol. Oceanogr.* 15:816.

10200 D. Preparing Slide Mounts

1. Phytoplankton Semi-Permanent Wet Mounts

Agitate the settled sample concentrate and withdraw a subsample with an accurately calibrated pipet. Clean pipet regularly. To prepare wet mounts transfer 0.1 mL to a glass slide, place a cover slip over the sample, and ring the cover slip with an adhesive such as clear nail polish to prevent evaporation. For semi-permanent mounts, add a few drops of glycerin to the slide. As the sample ages the water evaporates, leaving the organisms imbedded in the glycerin. If the cover slip is ringed with adhesive, the slide can be retained for a few years if stored in the dark.

2. Phytoplankton Permanent Mounts

a. Membrane filter mounts: Place two drops of immersion oil on a labeled slide. Immediately after filtering place the filter on top of the oil with a pair of forceps and add two drops of oil on top of the filter. The oil impregnates the filter and makes it trans-

parent. Impregnation time is 24 to 48 h. This procedure can be completed in 1 to 2 h by applying heat (70°C). Once the filter has cleared, place a few additional drops of oil on it and cover with a cover slip. The mounted filter is now ready for microscopic examination. Alternatively, mount membrane filters in mounting medium.* Immerse filters in 1-propanol to displace residual water and transfer to xylol for several minutes to clear filters. Place a section of filter or entire filter on a microscope slide with the mounting medium, cover with a cover glass, and dry at low temperature.[1]

b. Sedimented slide mounts: Two techniques are available for making permanent, resin mounts of natural phytoplankton that has been deposited by sedimentation on a microscope slide or cover glass and dehydrated by ethanol vapor substitution.[2,3]

---

* Permount, Fisher Scientific Co., or equivalent.

3. Diatom Mounts

Samples concentrated for diatom analysis by settling or centrifugation may contain dissolved materials, such as marine salts, formalin, and detergents, that will leave interfering residues. Wash well with distilled water before slide preparation. Transfer several drops of washed concentrate by means of a large-bore disposable pipet or large-bore dropper to a cover glass on a hot plate warmed enough to increase the evaporation rate but not enough to cause boiling (use a large-bore pipet or dropper to prevent possible selective filtration, thus exclusion, of larger forms or those forming colonies or chains). If the cleaned material is very concentrated, improve distribution of diatoms by adding the drops to a cover glass already flooded with distilled water. Evaporate to dryness. Repeat addition and evaporation until a sufficient quantity of sample has been transferred to the cover glass, but avoid producing a residue so dense that organisms cannot be recognized. If in doubt about the density, examine under a compound microscope. After evaporation, incinerate the residue on the cover glass on a hot plate at 300 to 500°C; alternatively, use a muffle furnace. This usually requires 20 to 45 min. Mount as described below.

Treat samples concentrated for diatom analysis by membrane filtration as described by Patrick and Reimer.[4] Mix equal volumes of conc nitric acid (HNO_3) and sample. CAUTION: *When working with conc HNO_3 wear safety goggles and an acid-resistant apron and gloves, and work under a hood.* Add a few grains of potassium dichromate ($K_2Cr_2O_7$)[5] to facilitate digestion of the filter and cellular organic matter. Add more dichromate if solution color changes from yellow to green. Place sample on a hot plate and boil down to approximately one-third the original volume. Alternatively, let treated sample stand overnight. This cleaning process destroys organic matter and leaves only diatom shells (frustules). Cool, wash with distilled water, and mount as described above. Transfer cleaned frustules to a cover glass and dry as described above.

Place a drop of mounting medium in the center of a labeled slide. Use 25- by 75-mm slides with frosted ends. Using a suitable high-refractive-index microscopic mounting medium assures permanent, easily handled mounts for examination under oil immersion. Heat the slide to near 90°C for 1 to 2 min before applying the heated cover slip with its sample residue to hasten evaporation of solvent in the mounting medium. Remove the slide to a cool surface and, during cooling (5 to 10 s), apply firm but gentle pressure to the cover glass with a broad, flat instrument.

4. Zooplankton Mounts

For zooplankton analyses, withdraw a 5-mL subsample from the concentrate and dilute or concentrate further as necessary. Transfer sample to a counting cell or chamber (see below) for analysis as a wet mount. Use polyvinyl lactyl phenol† for preparing semipermanent zooplankton mounts. The mounts are good for about a year, after which time the clearing agent causes deterioration of organisms. For long-term storage ring cover slip with clear lacquer (fingernail polish) to retard mountant crystallization. For permanent mounting, other mountants are available.‡

For the protozoan portion of the microzooplankton, a protargol staining procedure[6] not only provides a permanent mount but also reveals the cytological details often necessary for identification. This procedure is qualitative and is especially important in taxonomic studies of the ciliated protozoa.

5. References

1. MILLIPORE FILTER CORPORATION. 1966. Biological examination of water, sludge and bottom materials. Millipore Techniques, Water Microbiology, p. 25.
2. SANFORD, G.R., A. SANDS & C.R. GOLDMAN. 1962. A settle-freeze method for concentrating phytoplankton in quantitative studies. *Limnol. Oceanogr.* 14:790.
3. CRUMPTON, W.G. & R.G. WETZEL. 1981. A method for preparing permanent mounts of phytoplankton for critical microscopy and cell counting. *Limnol. Oceanogr.* 26:976.
4. PATRICK, R. & C.W. REIMER. 1967. The Diatoms of the United States. Vol. 1. Monogr. 13, Philadelphia Acad. Natur. Sci.
5. HOHN, M.H. & J. HELLERMAN. 1963. The taxonomy and structure of diatom populations for three eastern North American rivers using three sampling methods. *Trans. Amer. Microsc. Soc.* 62:250.
6. SMALL, E.B. & D.H. LYNN. 1985. Phylum Ciliophora Doflein, 1901. *In* J.J. Lee, S.H. Hunter & E.C. Bovee, eds. An Illustrated Guide to the Protozoa. Soc. Protozoology, Lawrence, Kansas.

† Biomedical Specialists, Box 1687, Santa Monica, CA.
‡ CMC-10, Master's Chemical Co., P.O. Box 2382, Des Plaines, IL; Hydramount, Biomedical Specialists, Box 1687, Santa Monica, CA; or equivalent.

10200 E. Microscopes and Calibrations

1. Compound Microscope

Use either a standard or an inverted compound microscope for algal identification and enumeration. Equip either type with a mechanical stage capable of moving all parts of a counting cell past the objective lens. Standard equipment is a set of $10\times$ or $12.5\times$ oculars and $10\times$, $20\times$, $40\times$, and $100\times$ objectives. Use objectives to provide adequate working distance for the counting chamber. Magnification requirements vary with the plankton fraction being investigated, the type of microscope, counting chamber used, and optics. With standard objectives, the Sedgwick-Rafter chamber limits magnification to approximately $200\times$ and the Palmer-Maloney cell limits magnification to approximately $500\times$. Inverted microscopes are limited in resolution by their optics. The useful upper limit of magnification for any objective is 1000 times the numerical aperture (NA). Above this magnification, no greater detail can be resolved. Use combinations of oculars, intermediate magnifiers, and objectives to obtain the greatest magnification without exceeding the useful limit of magnification. When the limit is exceeded, empty magnification results. Empty magnification occurs where the image is larger but no greater resolution is achieved. Optics providing contrast enhancement such as phase contrast or differential interference contrast are useful.

2. Stereoscopic Microscope

The stereoscopic microscope is essentially two complete microscopes assembled into a binocular instrument to give a stereoscopic view and an erect rather than an inverted image. Use this microscope for the study and counting of large plankters such as mature microcrustacea. Include 10× to 15× paired oculars in combination with 1× to 8× objectives. This combination of optics bridges the gap between the hand lens and the compound microscope and provides magnification ranging from 10× to 120×. Alternatively, use a good-quality zoom-type instrument with comparable magnification.

3. Inverted Compound Microscope

The inverted compound microscope often is used routinely for plankton counting in many laboratories.[1-3] This instrument is unique in that the objectives are below a movable stage and the illumination comes from above, thus permitting viewing of organisms that have settled to the bottom of a chamber. Place samples in a cylindrical settling chamber having a thin, clear glass bottom. Chambers of various capacities are available; the appropriate size depends on the density of organisms. After a suitable period of settling (see Section 10200C.1), count organisms in the settling chamber.

The major advantage of the inverted microscope is that by a simple rotation of the nosepiece a specimen can be examined (or counted) directly in the settling chamber at any desired magnification. Although not recommended, oil immersion objectives have some useful applications. No preparation or manipulation other than settling is required. Generally, examine a preserved sample. Techniques are available for samples with an abundance of organisms that tend to float.[4]

4. Epifluorescence Microscope

An epifluorescence microscope may be either standard or inverted. It uses incident light to excite electrons in intracellular compounds, such as pigments or absorbed stains, with the energy emitted during electron return to the ground state being measured as fluorescent light. The technique has been applied to the microscopic identification of chlorophyll-containing cells (autotrophs) and nonpigmented heterotrophic plankton; fluorescent stains such as primulin or proflavin also have been used to differentiate nannoplanktonic primary and secondary producers.[5-7] Excitation and emission wavelengths are unique for each pigment and stain and require distinct light filter combinations and light sources. Select the filter combinations for the particular application. Epifluorescence microscopy is particularly useful for the enumeration of picoplankton and heterotrophic flagellate populations common to most aquatic systems. Concentrate samples by membrane filtration. Use epifluorescence microscopy as a complementary procedure to standard light microscope counting techniques.

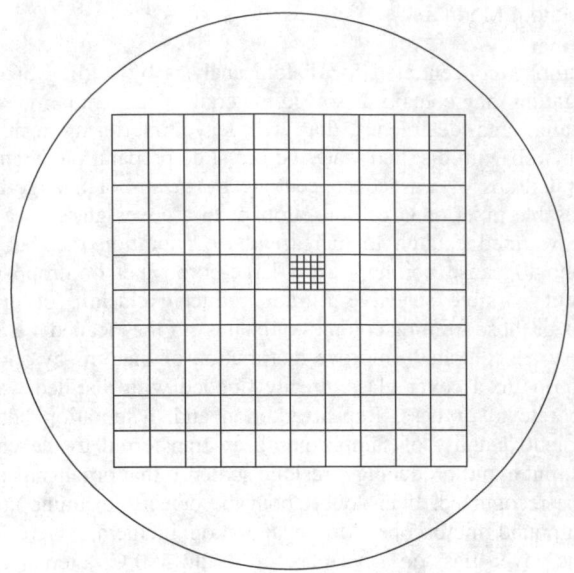

Figure 10200:6. Ocular micrometer ruling. A Whipple micrometer reticule is illustrated.

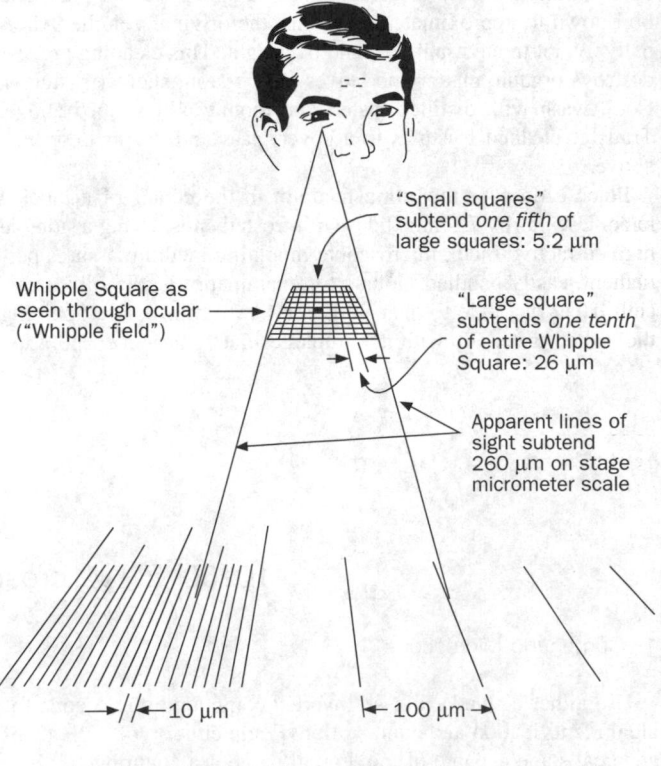

"Small squares" subtend *one fifth* of large squares: 5.2 μm

Whipple Square as seen through ocular ("Whipple field")

"Large square" subtends *one tenth* of entire Whipple Square: 26 μm

Apparent lines of sight subtend 260 μm on stage micrometer scale

10 μm

100 μm

Portion of magnified image of stage micrometer scale

Figure 10200:7. Calibration of Whipple Square, as seen with 10× ocular and 43× objective (approximately 430× total magnification).

5. Microscope Calibration

Microscope calibration is essential. The usual equipment for calibration is a Whipple grid (ocular micrometer, reticle, or reticule) placed in an eyepiece of the microscope and a stage micrometer that has a standardized, accurately ruled scale on a glass slide. The Whipple disk (Figure 10200:6) has an accurately ruled grid subdivided into 100 squares. One square near the center is subdivided further into 25 smaller squares. The outer dimensions of the grid are such that with a $10\times$ objective and a $10\times$ ocular, it delimits an area of approximately 1 mm^2 on the microscope stage. Because this area may differ from one microscope to another, carefully calibrate the Whipple grid for each microscope.

With the ocular and stage micrometers parallel and in part superimposed, match the line at the left edge of the Whipple grid with the zero mark on the stage micrometer scale (Figure 10200:7). Determine the width of the Whipple grid image to the nearest 0.01 mm from the stage micrometer scale. Should the width of the image of the Whipple grid be exactly 1 mm (1000 μm), the larger squares will be 1/10 mm (100 μm) on a side and each of the smaller squares 1/50 mm (20 μm).

When the microscope is calibrated at higher magnifications, the entire scale on the stage micrometer will not be seen; make measurements to the nearest 0.001 mm. Additional details for calibration are available.[8]

6. References

1. WETZEL, R.G. & G.E. LIKENS. 1991. Limnological Analyses, 2nd ed. Springer-Verlag, New York, N.Y.
2. LUND, J.W.G., C. KIPLING & E.D. LeCREN. 1958. The inverted microscope method of estimating algal numbers and the statistical basis of estimations by counting. *Hydrobiologia* 11:143.
3. SICKO-GOAD, L. & E.F. STOERMER. 1984. The need for uniform terminology concerning phytoplankton cell size fractions and examples of picoplankton from the Laurentian Great Lakes. *J. Great Lakes Res.* 10:90.
4. REYNOLDS, C.S. & G.H.M. JAWORSKI. 1978. Enumeration of natural *Microcystis* populations. *Brit. Phycol. J.* 13:269.
5. DAVIS, P.G. & J. McN. SIEBURTH. 1982. Differentiation of phototrophic and heterotrophic nanoplankton populations in marine waters by epifluorescence microscopy. *Ann. Inst. Oceanogr.* 58:249.
6. CARON, D.A. 1983. Techniques for enumeration of heterotrophic and phototrophic nanoplankton, using epifluorescence microscopy, and comparison with other procedures. *Appl. Environ. Microbiol.* 46:491.
7. SHERR, E.B. & B.F. SHERR. 1983. Double-staining epifluorescence techniques to assess frequency of dividing cells and bacteriovory in natural populations of heterotrophic microprotozoa. *Appl. Environ. Microbiol.* 46:1388.
8. JACKSON, H.W. & L.G. WILLIAMS. 1962. Calibration and use of certain plankton counting equipment. *Trans. Amer. Microsc. Soc.* 81:96.

10200 F.   Phytoplankton Counting Techniques

1. Counting Units

Some phytoplankton are unicellular while others are multicellular (colonial). The variety of configurations poses a problem in enumeration. For example, should a four-celled colony of *Scenedesmus* (Plate 41) be reported as one colony or four individual cells? Listed below are suggestions for reporting:

Enumeration Method	Counting Unit	Reporting Unit
Total cell count	One cell	Cells/mL
Natural unit count[1] (clump count)	One organism (any unicellular organism or natural colony)	Units/mL
Areal standard unit count*	400 μm^2	Units/mL

Making a total cell count is time-consuming and tedious, especially when colonies consist of thousands of individual cells. The natural unit or clump is the most easily used system; however, it is not necessarily accurate because sample handling and preserving may dislodge cells from the colony. The unit method also may not be quantitatively accurate nor reflect abundance of biomass or biovolume. Whatever method is chosen, identify it in reporting results.

If the distribution of organisms is random and the population fits a Poisson distribution, the counting error may be estimated.[2] For example, the approximate 95% confidence limits, as a percentage of the number of units counted (N), equals:

$$\frac{2}{\sqrt{N}}\ (100\%)$$

Thus, if 100 units are counted, the 95% confidence limits approximate ± 20%. For a count of 400 units, the limits are about 10%.

2. Counting Procedures

To enumerate plankton use a counting cell or chamber that limits the volume and area for ready calculation of population densities.

When counting with a Whipple grid, establish a convention for tallying organisms lying on an outer boundary line. For example, in counting a "field" (entire Whipple square), designate the top and left boundaries as "no-count" sides, and the bottom and right boundaries as "count" sides. Thus, tally every plankter touching a "count" side from the inside or outside but ignore any touching

---

* Areal standard unit equals area of four small squares in Whipple grid at a magnification of 200.

a ''no-count'' side. If significant numbers of filamentous or other large forms cross two or more boundaries of the grid, count them separately at a lower magnification and include their number in the total count.

To identify organisms use standard bench references (see Section 10900).

Do not count dead cells or broken diatom frustules. Tally empty centric and pennate diatoms separately as ''dead centric diatoms'' or ''dead pennate diatoms'' for use in converting the diatom species proportional count to a count per milliliter.

Magnification is important in phytoplankton identification and enumeration. Although magnifications of 100× to 200× are useful for counting large organisms or colonies, much higher magnifications often are required. It is useful to categorize techniques for phytoplankton counting according to the magnifications provided.

a. Low-magnification (up to 200×) methods: The Sedgwick-Rafter (S-R) cell is a device commonly used for plankton counting because it is easily manipulated and provides reasonably reproducible data when used with a calibrated microscope equipped with an eyepiece measuring device such as the Whipple grid.

The greatest disadvantage associated with the cell is that objectives providing high magnification cannot be used. As a result, the S-R cell is not appropriate for examining nannoplankton. The S-R cell is approximately 50 mm long by 20 mm wide by 1 mm deep. The total area of the bottom is approximately 1000 mm^2 and the total volume is approximately 1000 mm^3 or 1 mL. Carefully check the exact length and depth of the cell with a micrometer and calipers before use.

1) *Filling the cell*—Before filling the S-R cell with sample, place the cover glass diagonally across the cell and transfer sample with a large-bore pipet (Figure 10200:8). Placing cover slip in this manner will help prevent formation of air bubbles in cell corners. The cover slip often will rotate slowly and cover the inner portion of the S-R cell during filling. Do not overfill because this would yield a depth greater than 1 mm and produce an invalid count. Do not permit large air spaces caused by evaporation to develop in the chamber during a lengthy examination. To prevent formation of air spaces, occasionally place a small drop of distilled water on edge of cover glass.

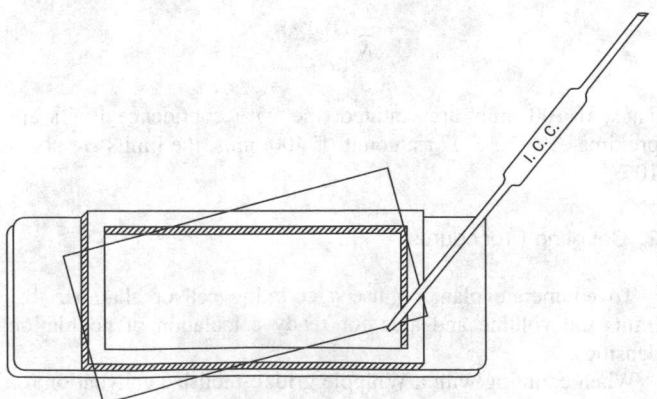

Figure 10200:8. Counting cell (Sedgwick-Rafter), showing method of filling. Source: WHIPPLE, G.C., G.M. FAIR, and M.C. WHIPPLE. 1927. The Microscopy of Drinking Water. John Wiley & Sons, New York, N.Y.

Before counting let the S-R cell stand for at least 15 min to settle plankton. Count plankton on the bottom of the S-R cell. Some phytoplankton, notably some blue-green algae or motile flagellates in unpreserved samples, may not settle but rise to the underside of the cover slip. When this occurs, count these organisms and add to total of those counted on the cell bottom to derive total number of organisms. Count algae in strips or fields.

2) *Strip counting*—A ''strip'' the length of the cell constitutes a volume approximately 50 mm long, 1 mm deep, and the width of the total Whipple grid.

The number of strips to be counted is a function of the precision desired and the number of units (cells, colonies, or filaments) per strip. Derive number of plankton in the S-R cell from the following:

$$\text{No./mL} = \frac{C \times 1000 \text{ mm}^3}{L \times D \times W \times S}$$

where:

C = number of organisms counted,
L = length of each strip (S-R cell length), mm,
D = depth of a strip (S-R cell depth), mm,
W = width of a strip (Whipple grid image width), mm, and
S = number of strips counted.

Multiply or divide number of cells per milliliter by a correction factor to adjust for sample dilution or concentration.

3) *Field counting*—On samples containing many plankton (10 or more plankters per field), make field counts rather than strip counts. Count plankters in random fields each consisting of one Whipple grid. The number of fields counted will depend on plankton density and statistical accuracy desired (see 10200F.1). Calculate the number of plankton per milliliter as follows:

$$\text{No./mL} = \frac{C \times 1000 \text{ mm}^3}{A \times D \times F}$$

where:

C = number of organisms counted,
A = area of a field (Whipple grid image area), mm^2,
D = depth of a field (S-R cell depth), mm, and
F = number of fields counted.

Multiply or divide the number of cells per milliliter by a correction factor to adjust for sample dilution or concentration.

b. Intermediate magnification (low to 500×) methods: The Palmer-Maloney (P-M) nannoplankton cell[3] is designed specifically for nannoplankton enumeration. It has a circular chamber with a 17.9-mm diam, 0.4-mm depth, and 0.1-mL volume. The shallow depth permits use of 40 to 45× objectives with sufficient working distance. The principal disadvantage of the P-M cell is that these magnifications (400 to 450×) often are insufficient for nannoplankton identification and enumeration.

Because a relatively small sample portion is examined in the P-M cell do not use it unless the sample contains a dense population (10 or more plankters per field). Such a small sample portion from a less dense population causes serious underestimation of density.

Introduce sample with a pipet into one of the 2- by 5-mm channels on the side of the chamber with the cover slip in place.

After a 10-min settling period count the plankters in random fields, with the number of fields depending on density and variety of plankton and the statistical accuracy desired. Strips may be counted in this or any other circular cell by measuring the effective diameter and counting two perpendicular strips that cross at the center. Calculate the number per milliliter as follows:

$$\text{No./mL} = \frac{C \times 1000 \text{ mm}^3}{A \times D \times F}$$

where:

    C = number of organisms counted,
    A = area of a field (Whipple grid image), mm^2,
    D = depth of a field (P-M cell depth), mm, and
    F = number of fields counted.

Multiply or divide the number of cells per milliliter by a correction factor to adjust for sample dilution or concentration.

Another readily available chamber is the standard medical hemacytometer used for enumerating blood cells. It has a ruled grid machined into a counting plate and is fitted with a ground-glass cover slip. The grid is divided into 1-mm^2 divisions; the chamber is 0.1 mm deep. Introduce sample by pipet and view under 450× magnification. Count all cells within the grid. The chamber comes from the manufacturer with a detailed instruction sheet containing directions on calculations and proper usage. A disadvantage to these counting cells is that the sample must have a very high plankton density to yield statistically reliable data.

c. High-magnification methods: Examination of phytoplankton at high magnification requires the use of oil immersion objectives. Suitable procedures include using inverted microscope chambers, membrane filter mounts, sedimented slide mounts, the Lackey drop method, and diatom mounts.

1) *Inverted microscope counts*—Prepare a sample for examination by filling the settling chamber. After the desired settling time (see Section 10200C.1), transfer the chamber to the microscope stage. Count perpendicular strips across the center of the bottom cover glass. Strip counts may be made by using a Whipple grid or special counting oculars that have a pair of adjustable parallel hairs and a single cross hair. Determine the width of the strip with a stage micrometer and tally organisms as they pass the single cross hair that functions as a reference point. Hold strip width constant for any series of samples. Alternatively examine random nonoverlapping fields until at least 100 units of the dominant species are counted. For highest accuracy, particularly because algae distribution may be nonuniform, count the entire chamber floor. Alternatively, make a random field-minimum count to attain a precision level of at least 85%.[4]

$$\text{Strip count (No./mL)} = \frac{C \times A_t}{L \times W \times S \times V}$$

where:

    C = number of organisms counted,
    A_t = total area of bottom of settling chamber, mm^2,
    L = length of a strip, mm,
    W = width of a strip (Whipple grid image width), mm,
    S = number of strips counted, and
    V = volume of sample settled, mL.

$$\text{Field count (No./mL)} = \frac{C \times A_t}{A_f \times F \times V}$$

where:

    A_f = area of a field (Whipple grid image area), mm^2,
    F = number of fields counted,

and other terms are as defined above.

2) *Membrane filter mounts*—Concentrate sample as directed in Section 10200C.2 and prepare membrane filter as directed in Section 10200D.2a.

Examine samples, concentrated on unlined membrane filters and mounted in oil as described above. Count enough random fields to ensure desired level of statistical accuracy (see 10200F.1). Select magnification level and size of microscope field (quadrat) such that the most abundant species appear in at least 70% but not more than 90% of microscopic fields examined (80% is optimum). Adjust microscope field size by using part or all of the Whipple grid. Examine 30 random microscope fields and record number of fields in which each species occurred. Report results as organisms per milliliter, calculated as follows:

$$\text{No./mL} = \frac{N \times Q}{V \times D}$$

where:

    N = density (organisms/field) from Table 10200:II,
    Q = number of fields per filter,
    V = milliliters filtered, and
    D = dilution factor (0.96 for 4% formalin preservative).

3) *Sedimented slide mounts*—Examine mounts prepared as directed in Section 10200D.2b.

4) *Lackey drop method*—The Lackey drop (microtransect) method[5] is a simple method of obtaining counts of considerable accuracy with samples containing a dense plankton population. It is similar to the S-R strip count.

Prepare slides as directed in Section 10200D.1. Oil immersion objectives can be used with the semipermanent slides. Count organisms in enough strips to ensure desired level of statistical accuracy (see 10200F.1). Calculate number of organisms per milliliter as follows:

$$\text{No./mL} = \frac{C \times A_t}{A_s \times S \times V}$$

where:

    C = number of organisms counted,
    A_t = area of cover slip, mm^2,
    A_s = area of one strip, mm^2,
    S = number of strips counted, and
    V = volume of sample under the cover slip, mL.

5) *Diatom mounts*—Prepare samples as directed in Section 10200D.3.

For diatom species proportional count, examine diatom samples under oil immersion at a magnification of at least 900×. Scan lateral strips the width of the Whipple grid until at least 250 cells are counted. Available time and accuracy required dictate the number of cells to be counted. Determine percentage abundance of each species from tallied counts and calculate counts per milliliter of each species by multiplying percent abundance by total live and dead diatom count obtained from the plankton counting

TABLE 10200:II. CONVERSION TABLE FOR MEMBRANE FILTER TECHNIQUE
(Based on 30 Scored Fields)

Total Occurrence	F* %	N†
1	3.3	0.03
2	6.7	0.07
3	10.0	0.10
4	13.3	0.14
5	16.7	0.18
6	20.0	0.22
7	23.3	0.26
8	26.7	0.31
9	30.0	0.35
10	33.3	0.40
11	36.7	0.45
12	40.0	0.51
13	43.3	0.57
14	46.7	0.63
15	50.0	0.69
16	53.3	0.76
17	56.7	0.83
18	60.0	0.91
19	63.3	1.00
20	66.7	1.10
21	70.0	1.20
22	73.3	1.32
23	76.7	1.47
24	80.0	1.61
25	83.3	1.79
26	86.7	2.02
27	90.0	2.30
28	93.3	2.71
29	96.7	3.42
30	100.0	?

$$* F = \frac{\text{total number of species occurrences} \times 100}{\text{total number of fields examined}}$$

† N = number of organisms per field.

chamber. For greater accuracy distinguish between living and dead diatoms at the species level.

6) Phytoplankton staining technique—Staining algae permits differentiation between "live" and "dead" diatoms.[6] This permits enumerating total phytoplankton in a single sample without sacrificing detailed diatom taxonomy. It also results in permanent reference slides. The procedure is most useful when diatoms are major components of phytoplankton and it is important to distinguish between living and dead diatoms.

Preferably preserve samples in Lugol's solution or alternatively in formalin (see 10200B.3). For analysis thoroughly mix the sample and filter a portion through a 47-mm-diam membrane filter (pore diam 0.45 or 0.65 μm). Use a vacuum of 16 to 20 kPa and never let sample dry. Add 2 to 5 mL aqueous acid fuchsin solution (dissolve 1 g acid fuchsin in 100 mL distilled water to which 2 mL glacial acetic acid has been added; filter) to the filter and let stand for 20 min. After staining, filter sample, wash briefly with distilled water, and filter again. Administer successive rinses of 50%, 90%, and 100% propanol to the sample while filtering. Soak for 2 min in a second 100% propanol wash, filter, and add xylene. At least two washes are required; let the final one soak 10 min before filtering. Trim the xylene-soaked filter and place on a microscope slide on which there are several drops of mounting medium.† Apply several more drops of medium to top of filter and install a cover glass. Carefully squeeze out excess mounting medium. Make the final mount permanent by lacquering the edges of the cover glass.

Count organisms using the most appropriate magnification. "Live" diatoms typically are red while "dead" ones are unstained. Oil immersion is necessary for species identifications of diatoms and many other algae. Count either strips or random fields and calculate plankton densities per milliliter:

$$\text{No./mL} = \frac{C \times A_t}{A_c \times V}$$

where:

C = number of organisms counted,
A_t = total area of effective filter before trimming and mounting,
A_c = area counted (strips or fields), and
V = volume of sample filtered, mL.

3. References

1. INGRAM, W.M. & C.M. PALMER. 1952. Simplified procedures for collecting, examining, and recording plankton in water. *J. Amer. Water Works Assoc.* 44:617.
2. STRICKLAND, J.D.H. & T.R. PARSONS. 1968. A Practical Manual of Sea Water Analysis. Fish. Res. Board Can. Bull. No. 167. Queen's Printer, Ottawa, Ont.
3. PALMER, C.M. & T.E. MALONEY. 1954. A New Counting Slide for Nannoplankton. Spec. Publ. No. 21, American Soc. Limnology & Oceanography.
4. SOURNIA, A., ed. 1978. Phytoplankton Manual. Monogr. Oceanogr. Methodol. No. 6. United Nations Educational, Scientific & Cultural Org., Paris.
5. LACKEY, J.B. 1938. The manipulation and counting of river plankton and changes in some organisms due to formalin preservation. *Pub. Health Rep.* 53:2080.
6. OWEN, B.B., JR., M. AFZAL & W.R. CODY. 1978. Staining preparations for phytoplankton and periphyton. *Brit. Phycol. J.* 13:155.

† Permount, Fisher Scientific Co., or equivalent.

10200 G. Zooplankton Counting Techniques

1. Subsampling

Count entire samples having low zooplankton numbers (<200 zooplankters) without subsampling. However, most zooplankton samples will contain more organisms than can be enumerated practically; therefore, use a subsampling procedure. Before subsampling, remove and enumerate all large uncommon organisms such as fish larvae in fresh water or coelenterates, decapods, fish larvae, etc., in salt water. Subsample by the pipet or splitting method.

In the pipet method, adjust sample to a convenient volume in a graduated cylinder or Imhoff cone. Concentrating the plankton by using a rubber bulb and clear acrylic plastic tube with fine mesh netting fitted on the end is convenient and accurate (Figure 10200:9). For picoplankton and the smaller microzooplankton, use sedimentation techniques described for concentrating phytoplankton. Transfer sample to a beaker or other wide-mouth vessel for subsampling with a Hensen-Stempel or similar wide-bore pipet. Gently stir sample completely and randomly with the pipet and quickly withdraw 1 to 5 mL. Transfer to a suitable counting chamber.

Alternatively, subsample by splitting with any of a number of devices of which the Folsom plankton splitter[1] is best known (Figure 10200:10). Level splitter before using. Place sample in the splitter and divide into subsplits. Rinse splitter into the subsamples. Repeat until a workable number (200 to 500 individuals) is obtained in a subsample. Exercise care to provide unbiased splits. Even when using the Folsom splitter unbiased subsamples cannot be unquestioningly assumed;[2] therefore, count animals in several subsamples from the same sample to verify that the splitter is unbiased and to determine the sampling error introduced by using it.

Another method permits abundance estimates of more equivalent levels of precision among taxa than obtained with either the Hensen-Stempel pipet or the Folsom splitter.[3] Normal counting procedures tally organisms on the basis of their abundance in a sample. Therefore, in a sample with a dominant organism making up 50% of total numbers, the tally of the dominant taxon will be large and have a small error. However, error about the subdominants will increase as the tally of each taxon decreases. By accepting one level of precision, the technique[3] has been developed to obtain the same error about dominants and subdominants, permitting quantitative comparisons between taxa over successive times or between stations.

2. Enumeration

Using a compound microscope and a magnification of 100×, enumerate small zooplankton (protozoa, rotifers, and nauplii) in a 1- to 5-mL clear acrylic plastic counting cell fitted with a glass cover slip. For larger, mature microcrustacea use a counting chamber holding 5 to 10 mL. A Sedgwick-Rafter cell is not suitable because of size. An open counting chamber 80 by 50 mm and 2 mm deep is desirable; however, an open chamber is difficult to move without jarring and disrupting the count. A mild detergent solution placed on the chamber before counting reduces organism movements or special counting trays with parallel or circular grooves or partitions[4,5] can be used. Count mi-

crocrustacea with a binocular dissecting microscope at 20× to 40× magnification. If identification is questionable, remove organisms with a microbiological transfer loop and examine at a higher magnification under a compound microscope.

Report smaller zooplankton as number per liter and larger forms as number per cubic meter:

$$No./m^3 = \frac{C \times V'}{V'' \times V'''}$$

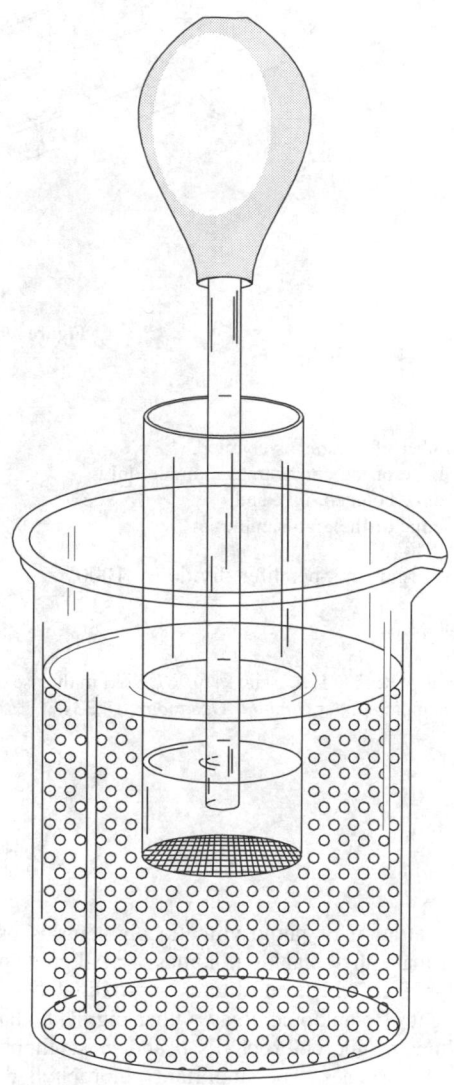

Figure 10200:9. A simple, efficient device for concentrating plankton. The tube is lowered into the beaker containing the sample. Water filtering into the tube is removed with the rubber bulb. The filter is nylon monofilament screen cloth that is glued to the bottom of the tube. The mesh size should be sufficiently small to prevent zooplankters from entering the filtrate (after Dodson and Thomas[5]).

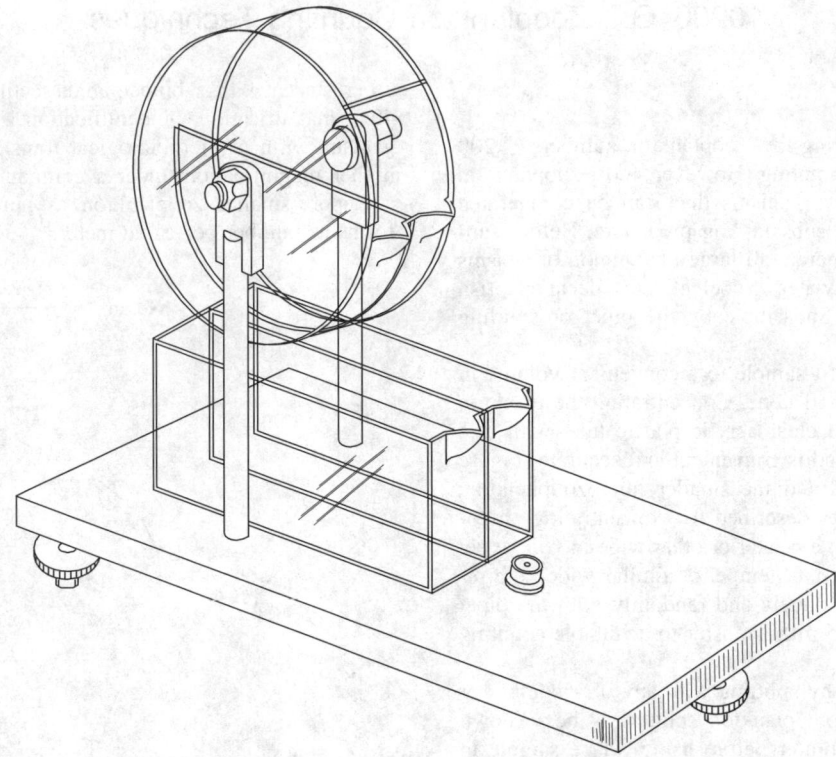

Figure 10200:10. The Folsom plankton splitter.

where:

C = number of organisms counted,
V' = volume of the concentrated sample, mL,
V'' = volume counted, mL, and
V''' = volume of the grab sample, m^3.

To obtain organisms per liter divide by 1000.

3. References

1. LONGHURST, A.R. & D.L.R. SEIBERT. 1967. Skill in the use of Folsom's plankton sample splitter. *Limnol. Oceanogr.* 12:334.

2. MCEWEN, G.F., M.W. JOHNSON & T.R. FOLSOM. 1954. A statistical analysis of the Folsom sample splitter based upon test observations. *Arch. Meteorol. Geophys. Bioklimatol.,* Ser. A, 6:502.

3. ALDEN, R.W., III, R.C. DAHIYA & R.J. YOUNG, JR. 1982. A method for the enumeration of zooplankton samples. *J. Exp. Mar. Biol. Ecol.* 59:185.

4. GANNON, J.E. 1971. Two counting cells for the enumeration of zooplankton micro-crustacea. *Trans. Amer. Microsc. Soc.* 90:486.

5. DODSON, A.N. & W.H. THOMAS. 1964. Concentrating plankton in gentle fashion. *Limnol. Oceanogr.* 9:455.

10200 H.   Chlorophyll

The concentration of photosynthetic pigments is used extensively to estimate phytoplankton biomass.[1,2] All green plants contain chlorophyll *a,* which constitutes approximately 1 to 2% of the dry weight of planktonic algae. Other pigments that occur in phytoplankton include chlorophylls *b* and *c,* xanthophylls, phycobilins, and carotenes. The important chlorophyll degradation products found in the aquatic environment are the chlorophyllides, pheophorbides, and pheophytins. The presence or absence of the various photosynthetic pigments is used, among other features, to separate the major algal groups.

The three methods for determining chlorophyll *a* in phytoplankton are the spectrophotometric,[3–5] the fluorometric,[6–8] and the high-performance liquid chromatographic (HPLC) techniques.[9] Fluorometry is more sensitive than spectrophotometry, requires less sample, and can be used for in-vivo measurements.[10] These optical methods can significantly under- or overestimate chlorophyll *a* concentrations,[11–18] in part because of the overlap of the absorption and fluorescence bands of co-occurring accessory pigments and chlorophyll degradation products.

Pheophorbide *a* and pheophytin *a,* two common degradation products of chlorophyll *a,* can interfere with the determination of chlorophyll *a* because they absorb light and fluoresce in the same region of the spectrum as does chlorophyll *a.* If these pheopigments are present, significant errors in chlorophyll *a* values will result. Pheopigments can be measured either by spectrophotometry or fluorometry, but in marine and freshwater environments the fluorometric method is unreliable when chlorophyll *b* co-occurs. Upon acidification of chlorophyll *b*, the resulting fluorescence emission of pheophytin *b* is coincident with that of pheophytin *a*, thus producing underestimation and overestimation of chlorophyll *a* and pheopigments, respectively.

HPLC is a useful method for quantifying photosynthetic pigments[9,13,15,16,19–21] including chlorophyll *a*, accessory pigments (e.g., chlorophylls *b* and *c*), and chlorophyll degradation products (chlorophyllides, pheophorbides, and pheophytins). Pigment distribution is useful for quantitative assessment of phytoplankton community composition and zooplankton grazing activity.[22]

1. Pigment Extraction

Conduct work with chlorophyll extracts in subdued light to avoid degradation. Use opaque containers or wrap with aluminum foil. The pigments are extracted from the plankton concentrate with aqueous acetone and the optical density (absorbance) of the extract is determined with a spectrophotometer. The ease with which the chlorophylls are removed from the cells varies considerably with different algae. To achieve consistent complete extraction of the pigments, disrupt the cells mechanically with a tissue grinder.

Glass fiber filters are preferred for removing algae from water. The glass fibers assist in breaking the cells during grinding, larger volumes of water can be filtered, and no precipitate forms after acidification. Inert membrane filters such as polyester filters may be used where these factors are irrelevant.

a. Equipment and reagents:

1) *Tissue grinder:** Successfully macerating glass fiber filters in tissue grinders with grinding tube and pestle of conical design may be difficult. Preferably use round-bottom grinding tubes with a matching pestle having grooves in the TFE tip.

2) *Clinical centrifuge.*

3) *Centrifuge tubes,* 15-mL graduated, screw-cap.

4) *Filtration equipment,* filters, glass fiber† or membrane (0.45-μm porosity, 47-mm diam); vacuum pump; solvent-resistant disposable filter assembly, 1.0-μm pore size;‡ 10-mL solvent-resistant syringe.

5) *Saturated magnesium carbonate solution:* Add 1.0 g finely powdered $MgCO_3$ to 100 mL distilled water.

6) *Aqueous acetone solution:* Mix 90 parts acetone (reagent-grade BP 56°C) with 10 parts saturated magnesium carbonate solution. For HPLC pigment analysis, mix 90 parts HPLC-grade acetone with 10 parts distilled water.

b. Extraction procedure:

1) Concentrate sample by centrifuging or filtering as soon as possible after collection. If processing must be delayed, hold samples on ice or at 4°C and protect from exposure to light. Use opaque bottles because even brief exposure to light during storage will alter chlorophyll values. Samples on filters taken from water having pH 7 or higher may be placed in airtight plastic bags and stored frozen for 3 weeks. Process samples from acidic water promptly after filtration to prevent possible chlorophyll degradation from residual acidic water on filter. Use glassware and cuvettes that are clean and acid-free.

2) Place sample in a tissue grinder, cover with the 2 to 3 mL 90% aqueous acetone solution, and macerate at 500 rpm for 1 min. Use TFE/glass grinder for a glass-fiber filter and glass/glass grinder for a membrane filter.

3) Transfer sample to a screw-cap centrifuge tube, rinse grinder with a few milliliters 90% aqueous acetone, and add the rinse to the extraction slurry. Adjust total volume to 10 mL, with 90% aqueous acetone. Use solvent sparingly and avoid excessive dilution of pigments. Steep samples at least 2 h at 4°C in the dark. Glass fiber filters of 25- and 47-mm diam§ have dry displacement volumes of 0.03 and 0.10 mL, respectively, and introduce errors of about 0.3 and 1.0% if a 10-mL extraction volume is used.

4) Clarify by filtering through a solvent-resistant disposable filter (to minimize retention of extract in filter and filter holder, force 1 to 2 mL air through the filter after the extract), or by centrifuging in closed tubes for 20 min at 500 g. Decant clarified extract into a clean, calibrated, 15-mL, screw-cap centrifuge tube and measure total volume. Proceed as in 2, 3, 4, or 5 below.

2. Spectrophotometric Determination of Chlorophyll

a. Equipment and reagents:

1) *Spectrophotometer,* with a narrow band (pass) width (0.5 to 2.0 nm) because the chlorophyll absorption peak is relatively narrow. At a spectral band width of 20 nm the chlorophyll *a* concentration may be underestimated by as much as 40%.

2) *Cuvettes,* with 1-, 4-, and 10-cm path lengths.

3) *Pipets,* 0.1- and 5.0-mL.

4) *Hydrochloric acid,* HCl, 0.1N.

b. Determination of chlorophyll a *in the presence of pheophytin* a: Chlorophyll *a* may be overestimated by including pheopigments that absorb near the same wavelength as chlorophyll *a*. Addition of acid to chlorophyll *a* results in loss of the magnesium atom, converting it to pheophytin *a*. Acidify carefully to a final molarity of not more than $3 \times 10^{-3}M$ to prevent certain accessory pigments from changing to absorb at the same wavelength as pheophytin *a*.[13] When a solution of pure chlorophyll *a* is converted to pheophytin *a* by acidification, the absorption-peak-ratio (OD664/OD665) of 1.70 is used in correcting the apparent chlorophyll *a* concentration for pheophytin *a*.

Samples with an OD664 before/OD665 after acidification ratio ($664_b/665_a$) of 1.70 are considered to contain no pheophytin *a* and to be in excellent physiological condition. Solutions of pure pheophytin show no reduction in OD665 upon acidification and have a $664_b/665_a$ ratio of 1.0. Thus, mixtures of chlorophyll *a* and pheophytin *a* have absorption peak ratios ranging between 1.0 and 1.7. These ratios are based on the use of 90% acetone as solvent. Using 100% acetone as solvent results in a chlorophyll *a* before-to-after acidification ratio of about 2.0.[3]

Spectrophotometric procedure—Transfer 3 mL clarified extract to a 1-cm cuvette and read optical density (OD) at 750 and 664 nm. Acidify extract in the cuvette with 0.1 mL 0.1N HCl. Gently agitate the acidified extract and read OD at 750 and at 665 nm, 90 s after acidification. The volumes of extract and acid and the time after acidification are critical for accurate, consistent results.

The OD664 before acidification should be between 0.1 and 1.0. For very dilute extracts use cuvettes having a longer path length. If a larger cell is used, add a proportionately larger volume of

---

* Kontes Glass Co., Vineland, NJ 08360: Glass/glass grinder, Model No. 8855: Glass/TEE grinder, Model 886000; or equivalent.
† Whatman GF/F (0.7 μm), GFB (1.0 μm), Gelman AE (1 μm),[23] or equivalent.
‡ Gelman Acrodisc or equivalent.

---

§ GF/F or equivalent.

acid. Correct OD obtained with larger cuvettes to 1 cm before making calculations.

Subtract the 750-nm OD value from the readings before (OD 664 nm) and after acidification (OD 665 nm).

Using the corrected values calculate chlorophyll a and pheophytin a per cubic meter as follows:

$$\text{Chlorophyll } a, \text{ mg/m}^3 = \frac{26.7 \, (664_b - 665_a) \times V_1}{V_2 \times L}$$

$$\text{Pheophytin } a, \text{ mg/m}^3 = \frac{26.7 \, [1.7 \, (665_a) - 664_b] \times V_1}{V_2 \times L}$$

where:

V_1 = volume of extract, L,
V_2 = volume of sample, m^3,
L = light path length or width of cuvette, cm, and
$664_b, 665_a$ = optical densities of 90% acetone extract before and after acidification, respectively.

The value 26.7 is the absorbance correction and equals $A \times K$

where:

A = absorbance coefficient for chlorophyll a at 664 nm = 11.0, and
K = ratio expressing correction for acidification.

$$= \frac{\left(\dfrac{664_b}{665_a}\right) \text{ pure chlorophyll a}}{\left(\dfrac{664_b}{665_a}\right) \text{ pure chlorophyll a} - \left(\dfrac{664_b}{665_a}\right) \text{ pure pheophytin a}}$$

$$= \frac{1.7}{1.7 - 1.0} = 2.43$$

c. Determination of chlorophyll a, b, and c (trichromatic method): Spectrophotometric procedure—Transfer extract to a 1-cm cuvette and measure optical density (OD) at 750, 664, 647, and 630 nm. Choose a cell path length or dilution to give OD664 between 0.1 and 1.0.

Use the optical density readings at 664, 647, and 630 nm to determine chlorophyll a, b, and c, respectively. The OD reading at 750 nm is a correction for turbidity. Subtract this reading from each of the pigment OD values of the other wavelengths before using them in the equations below. Because the OD of the extract at 750 nm is very sensitive to changes in the acetone-to-water proportions, adhere closely to the 90 parts acetone:10 parts water (v/v) formula for pigment extraction. Turbidity can be removed easily by filtration through a disposable, solvent-resistant filter attached to a syringe or by centrifuging for 20 min at 500 g.

Calculate the concentrations of chlorophyll a, b, and c in the extract by inserting the corrected optical densities in following equations:[5]

a) C_a = 11.85(OD664) − 1.54(OD647) − 0.08(OD630)
b) C_b = 21.03(OD647) − 5.43(OD664) − 2.66(OD630)
c) C_c = 24.52(OD630) − 7.60(OD647) − 1.67(OD664)

where:

C_a, C_b, and C_c = concentrations of chlorophyll a, b, and c, respectively, mg/L, and
OD664, OD647, and OD630 = corrected optical densities (with a 1-cm light path) at the respective wavelengths.

After determining the concentration of pigment in the extract, calculate the amount of pigment per unit volume as follows:

$$\text{Chlorophyll } a, \text{ mg/m}^3 = \frac{C_a \times \text{extract volume, L}}{\text{volume of sample, m}^3}$$

3. Fluorometric Determination of Chlorophyll a

The fluorometric method for chlorophyll a is more sensitive than the spectrophotometric method and thus smaller samples can be used. Calibrate the fluorometer spectrophotometrically with a sample from the same source to achieve acceptable results. Optimum sensitivity for chlorophyll a extract measurements is obtained at an excitation wavelength of 430 nm and an emission wavelength of 663 nm. A method for continuous measurement of chlorophyll a in vivo is available, but is reported to be less efficient than the in-vitro method given here, yielding about one-tenth as much fluorescence per unit weight as the same amount in solution. Pheophytin a also can be determined fluorometrically.[24]

a. Equipment and reagents: In addition to those listed under 1a and 2a above:

Fluorometer,‖ equipped with a high-intensity F4T.5 blue lamp, photomultiplier tube R-446 (red-sensitive), sliding window orifices 1×, 3×, 10×, and 30×, and filters for light emission (CS-2-64) and excitation (CS-5-60). A high-sensitivity door is preferable.

b. Extraction procedure: Prepare sample as directed in 1b above.

1) Calibrate fluorometer with a chlorophyll solution of known concentration as follows: Prepare chlorophyll extract and analyze spectrophotometrically. Prepare serial dilutions of the extract to provide concentrations of approximately 2, 6, 20, and 60 µg chlorophyll a/L. Make fluorometric readings for each solution at each sensitivity setting (sliding window orifice): 1×, 3×, 10×, and 30×. Using the values obtained, derive calibration factors to convert fluorometric readings in each sensitivity level to concentrations of chlorophyll a, as follows:

$$F_s = \frac{C_a'}{R_s}$$

where:

F_s = calibration factor for sensitivity setting S,
R_s = fluorometer reading for sensitivity setting S, and,
C_a' = concentration of chlorophyll a determined spectrophotometrically, µg/L.

2) Measure sample fluorescence at sensitivity settings that will provide a midscale reading. (Avoid using the 1× window because of quenching effects.) Convert fluorescence readings to

‖ Model 10-005, Turner Designs, Sunnyvale, CA or equivalent.

concentrations of chlorophyll a by multiplying the readings by the appropriate calibration factor.

c. Determination of chlorophyll a *in the presence of pheophytin* a: This method normally is not applicable to freshwater samples. See discussion under 10200H and ¶2*b* above.

1) Equipment and reagents—In addition to those listed under 1*a* and 2*a* above, pure chlorophyll *a*# (or a plankton chlorophyll extract with a spectrophotometric before-and-after acidification ratio of 1.70 containing no chlorophyll *b*).

2) Fluorometric procedure—Calibrate fluorometer as directed in ¶ 3*b*1). Determine extract fluorescence at each sensitivity setting before and after acidification. Calculate calibration factors (F_s) and before-and-after acidification fluorescence ratio by dividing fluorescence reading obtained before acidification by the reading obtained after acidification. Avoid readings on the $1\times$ scale and those outside the range of 20 to 80 fluorometric units.

3) Calculations—Determine the "corrected" chlorophyll *a* and pheophytin *a* in sample extracts with the following equations:[8,24]

$$\text{Chlorophyll } a, \text{ mg/m}^3 = F_s \frac{r}{r-1} (R_b - R_a) \frac{V_e}{V_s}$$

$$\text{Pheophytin } a, \text{ mg/m}^3 = F_s \frac{r}{r-1} (rR_a - R_b) \frac{V_e}{V_s}$$

where:

F_s = conversion factor for sensitivity setting S (see ¶ 2*b*, above),
R_b = fluorescence of extract before acidification,
R_a = fluorescence of extract after acidification,
r = R_b/R_a, as determined with pure chlorophyll *a* for the instrument (redetermine r and F_s if filters or light source are changed),
V_e = volume of extract, and
V_s = volume of sample.

d. Extraction of whole water, nonfiltered samples: Alternatively, to prevent cell lysis during filtration, extract whole water sample.

1) Equipment and reagents—Fluorometer equipped with a high-sensitivity R928 phototube** with output impedance of 36 ma/W at 675 nm and a high-sensitivity door. Place neutral density filter (40–60N) in the rear light path,†† selected to permit reagent blanking on the highest sensitivity scale.

2) Extraction procedure—Decant 1.5 mL sample into screw-cap test tube and add 8.5 mL 100% acetone. Mix with vortex mixer and hold in the dark for 6 h at room temperature. Filter through glass fiber filter‡‡ or centrifuge. Measure fluorescence as described in Section 10200H.3 and estimate concentrations as in ¶ 3*c*. Because humic substances interfere, if they are present filter a sample portion (see 10200H.1*b*) and process filtrate with sample. Subtract filtrate (blank) fluorescence from that of sample.

4. High-Performance Liquid Chromatographic Determination of Algal Chlorophylls and Their Degradation Products

a. Equipment and reagents: In addition to those listed for pigment extraction, ¶ 1*a* above:

1) *High-pressure liquid chromatograph* capable of a flow rate of 2.0 mL/m.

2) *High-pressure injector valve* equipped with a 100-μL sample loop.

3) *Guard column* (4.0 $\times$ 0.5 cm, C_{18} packing material, 3-μm particle size, or equivalent protection system) for extending life of primary column.

4) *Reverse-phase HPLC column.*§§

5) *Fluorescence detector* capable of excitation at 430 ± 30 nm and measuring emission at wavelengths greater than 600 nm.

6) *Data recorder device:* Strip chart recorder or, preferably, an electronic integrator.

7) *Syringe,* glass, 250-μL.

8) *HPLC eluents:* System A (80:15:5; methanol:reagent water: ion-pairing solution) and System B (80:20; methanol:acetone). Use HPLC-grade solvents; measure volumes before mixing. Filter eluents through a solvent-resistant 0.4-μm filter before use and degas with helium. Prepare the ion-pairing (IP) solution from 15 g tetrabutylammonium acetate‖‖ and 77 g ammonium acetate## made up to 1 L with reagent water.[15]

9) *Calibration standards:* Individually dissolve 1 mg each pure chlorophyll *a* and *b*‖‖ in 100 mL 90% acetone. Determine the exact concentrations spectrophotometrically (ε_{664} for chlorophyll *a* in 90% acetone = 87.67 L g^{-1} cm^{-1}; ε_{647} for chlorophyll *b* in 90% acetone = 51.36 L g^{-1} cm^{-1}).[5] Prepare pheophytin *a* + *a'* and *b* + *b'* standards from the primary chlorophyll *a* and *b* standards by acidification with hydrochloric acid; correct respective concentrations for Mg^{2+} loss. Extract chlorophyll *c* with 90% acetone from diatoms, purify by thin-layer chromatography (TLC)[25] and calibrate spectrophotometrically (ε_{631} for a mixture containing equal amounts of chlorophylls c_1 and c_2 in 90% acetone containing 1% pyridine = 42.6 L g^{-1} cm^{-1}; the absence of this small amount of pyridine is presumed to cause only small differences in the absorption properties of chlorophyll *c*.[26] Alternatively, determine the chlorophyll *c* content of a 90% acetone extract made from diatoms, spectrophotometrically (chlorophyll c_1 + c_2, μg/mL = 24.36E_{630} − 3.73E_{664})[5] and use as standard. Prepare chlorophyllide *a* from diatoms,[27] purify by TLC[25] and calibrate spectrophotometrically in 90% acetone (ε_{664} for chlorophyllide *a* = 128 L g^{-1} cm^{-1}).[28] Prepare pheophorbide *a* by acidification of chlorophyllide *a*, purify by TLC[25] and calibrate spectrophotometrically in 90% acetone (ε_{665} for pheophorbide *a* = 69.8 L g^{-1} cm^{-1}).[28] Standards stored under nitrogen in the dark at −20°C are stable for about 1 month.

b. Procedure:

1) Set up and equilibrate the HPLC system with solvent System A at a flow rate of 2 mL/min. Adjust fluorometer sensitivity to provide full-scale reading with the most concentrated chlorophyll *a* standard.

2) Calibrate HPLC system by preparing working standards from the primary standards (on day of use). Once retention times of the standards are determined for a particular system, simplify standardization by preparing serial dilutions from mixed standards. Prepare separately mixed standards for the chlorophylls and chlorophyllide *a* and for the pheophytins and pheophorbide *a*. Mix 1-mL portions of standards with 300 μL ion-pairing solutions and equilibrate for 5 min before injection (use of ion-pairing agents greatly enhances separation of dephytolated pigments,

---

Purified chlorophyll *a,* Sigma Chemical Company, St. Louis, MO, or equivalent.
** Hammamatsu Corp., Middlesex, NJ, or equivalent.
†† If using Model 10-005, Turner Designs, or equivalent.
‡‡ Whatman GF/F or equivalent.

---

§§ Microsorb C_{18} column, 10 cm long, 3-μm particle size, Rainin Co., or equivalent.
‖‖ Fluka Chemical Corp., 980 South Second Street, Ronkonkoma, NY, or equivalent.
Sigma Chemical Company, or equivalent.

chlorophyllide *a*, chlorophyll *c,* and pheophorbide *a*). Prepare blanks by mixing 1 mL 90% acetone with 300 μL IP solution. Rinse syringe twice with 150 μL standard and draw about 250 μL standard into syringe for injection. Place syringe in injector valve, overfilling the 100-μL sample loop. Construct calibration curves by plotting fluorescence peak areas (or heights) against standard pigment concentrations.

3) Prepare samples for injection by mixing a 1-mL portion of the 90% acetone pigment extract with 300 μL IP solution.

4) Use a two-step solvent program to optimize separation of the chorophylls from their degradation products.[15] After injection, change from solvent System A to System B over 5 min and follow with System B for 15 min at a flow rate of 2 mL/min. Re-equilibrate the column with System A for 5 min before the next injection for a total analysis time of approximately 25 min. Degas the solvent systems with helium during analysis. Increase lifetime of HPLC column by storing it in 100% methanol between runs. Periodically flush the HPLC system with reagent water to avoid buildup of ion pairing agents.

5) Calculate individual pigment concentrations using the following formula:

$$C_i = \frac{A_s \, F_i \, V_E}{V_I \, V_S}$$

where:

C_i = individual pigment concentration, mg/L,
A_s = area of individual pigment peak from sample injection,
F_i = standard response factor (mg pigment/0.1 mL standard divided by corresponding peak area).
V_I = injection volume (0.1 mL),
V_E = extraction volume, mL, and
V_S = sample volume, L.

6) This method is designed only for quantification of chlorophylls and their degradation products. Detect carotenoid pigments, which also are present in 90% acetone extracts but do not fluoresce, by absorbance spectroscopy (at about 440 nm).[21]

7) The elution order and approximate retention times for the major chlorophyll pigments and their degradation products are shown in Figure 10200:11. The detection limits (*s/n* = 2) vary with fluorometer configuration and flow rate; however, they range from 10 to 100 pg per injection for most chlorophylls and their degradation products.[15,21,29] The accuracy of the HPLC method depends primarily on purity of pigment standards. Preferably measure absorption spectra (350 to 750 nm) of the standards and compare with published data. Pigment purity also can be assessed by HPLC analysis, providing there are no co-eluting contaminants with absorption and fluorescence bands overlapping those of the standards. HPLC and spectrophotometrically derived pigment concentrations for available EPA standards agree reasonably well (± 20%) if spectrophotometric results are corrected for the presence of pheopigments and the HPLC results are expressed as pigment equivalents (e.g., chlorophyll *a* equivalents = chlorophyllide *a* + chlorophyll *a* + chlorophyll *a'*, provided that the proper molecular weight corrections are applied).[30] Thus, if significant amounts of chlorophyll derivatives are present, pigment concentrations determined spectrophotometrically will be overestimated. The agreement between HPLC and fluorometrically derived results depends on the presence of accessory chlorophylls *b*, *c*, and their derivatives. Triplicate injections of a fivefold dilution of an EPA sample

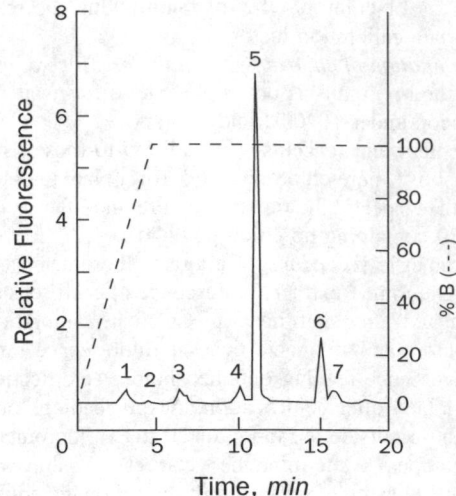

Figure 10200:11. Reverse-phase HPLC chromatogram for a fivefold dilution of EPA sample. Injection volume 100 μL; peaks detected by fluorescence spectroscopy (λ_{ex}: 400–460 nm; λ_{ex}: >600 nm). Peak identities are: 1—chlorophyllide *a;* 2—chlorophyll *c;* 3—pheophorbide *a;* 4—chlorophyll *b;* 5—chlorophyll *a;* 6—pheophytin *a;* and 7—pheophytin *a*. The chlorophyll *b* degradation products, pheophytin *b* and pheophytin *b'*, were below detection limits. Peak identities confirmed by on-line diode array spectroscopy (350–550 nm).

gave coefficients of variation of 7.5% (chlorophyllide *a*), 9.1% (chlorophyll *c*), 13.4% (pheophorbide *a*), 9.6% (chlorophyll *b*), 0.5% (chlorophyll *a*), 6.2% (pheophytin *a*), and 22.9% (pheophytin *a'*), with an average value of 10% for the seven pigments analyzed.

5. High-Performance Liquid Chromatographic Determination of Algal Chlorophyll and Carotenoid Pigments (PROPOSED)

a. Equipment and reagents: In addition to those listed for pigment extraction, ¶ 1*a* above:

1) *High-performance liquid chromatographic pump* capable of gradient delivery of three different solvents at a flow rate of 1 mL/min.

2) *High-pressure injector valve* equipped with a 200-μL sample loop.

3) *Guard column* (50 × 4.6 mm, C_{18} packing material,*** 5-μm particle size) for extending life of primary column.

4) *Reverse-phase HPLC column* with endcapping (250 × 4.6 mm, 5-μm particle size, C_{18} column***).

5) *Variable wavelength or filter absorbance detector* with low-volume flowthrough cell. Detection wavelength is 436 nm.

6) *Data recording device:* Strip chart recorder or, preferably, an electronic integrator or computer equipped with hardware and software for chromatographic data analysis.

7) *Syringe, glass,* 500-μL.

8) *HPLC eluents:* Eluent A (80:20, v:v; methanol:0.5*M* ammonium acetate, pH 7.2); Eluent B (90:10, v:v; acetonitrile:water), and Eluent C, ethyl acetate. Use HPLC-grade solvents. Meas-

---

*** Spherisorb ODS-2, Phase Separations Inc., Norwalk, CT, or equivalent.

TABLE 10200:III. EXTINCTION COEFFICIENTS AND CHROMATOGRAPHIC PROPERTIES OF PIGMENTS SEPARATED BY REVERSE-PHASE HIGH-PERFORMANCE LIQUID CHROMATOGRAPHY (CF. FIGURE 10200:12)

Pigment Identity	Wavelength (solvent) nm	E_{1cm} L g^{-1} cm^{-1}	Ref. No.	Retention Time min	% c.v. ($n = 3$ inj)	Absorption Maxima in Eluent* nm		
Chlorophyllide a	664 (90% acetone)	128.0	28	7.8	5.7	nd†	nd	nd
Chlorophyll c_{1+2}	631 (90% acetone)	42.6	26	8.9	0.6	444	576	630
Peridinin	466 (acetone)	134.0	33	10.0	1.2		472	
Fucoxanthin	449 (acetone)	160.0	44	11.0	0.9		446	(466)
Neoxanthin	439 (ethanol)	224.3	35	11.5	5.9	416	441	470
Violaxanthin	443 (ethanol)	255.0	35	13.2	2.6	416	440	470
Diadinoxanthin	448 (acetone)	223.0	36	14.6	6.0	422	446	476
Lutein	445 (ethanol)	255.0	35	17.5	0.7	(422)	446	476
Zeaxanthin	450 (ethanol)	254.0	35	18.0	2.2	(428)	454	478
Chlorophyll b	647 (90% acetone)	51.36	5	21.1	1.0	456	596	646
Chlorophyll a	664 (90% acetone)	87.67	5	22.3	0.8	431	618	665
β,β-Carotene	453 (90% acetone)‡	262.0	35	25.4	2.0	427	462	480

* All absorption maxima are from Wright et al.[32] except those for chlorophyll c_{1+2} (R.R. Bidigare and M. Latasa, unpublished data).
† Not determined.
‡ Because of a potential insolubility problem of β,β-carotene in ethanol, prepare this standard in 90% acetone, not ethanol. It is assumed that the extinction coefficient of β,β-carotene in 90% acetone is the same as that in ethanol.

ure volumes before mixing. Filter eluents through a solvent-resistant 0.4-μm filter before use and degas with helium.

9) *Calibration standards:* Chlorophylls a and b, and β,β-carotene can be purchased††† as can zeaxanthin and lutein.‡‡‡ Other pigment standards can be purified from plant extracts by thin-layer chromatography (TLC)[25] or preparative-scale HPLC. Determine concentration of all standards using a monochromator-based spectrophotometer in the appropriate solvents before calibration of the HPLC system. The recommended extinction coefficients for most common algal pigments found in freshwater systems are given in Table 10200:III. Measure absorbance in a 1-cm cuvette at the appropriate wavelength (usually at λ_{max}) and 750 nm (to correct for light scattering). Calculate concentrations of standards as follows:

$$C_i = \frac{(A_{\lambda max} - A_{750nm})}{E_{1cm} \times b} \times 1000$$

where:

C_i = individual pigment concentration, mg/L,
A = absorbance at specific wavelength,
E_{1cm} = weight-specific absorption coefficient, L g^{-1} cm^{-1},
b = pathlength of cuvette, cm, and
1000 = conversion factor, g to mg.

Standards stored under nitrogen in the dark at $-20°C$ are stable for about 1 month.

b. *Procedure:*

1) Set up and equilibrate the HPLC system with Eluent A at a flow rate of 1 mL/min.

2) Calibrate the HPLC using working standards (about 0 to 1000 ng/mL) prepared from primary standards on day of use. Mix 1 mL standard with 300 μL distilled water, shake, and equilibrate for 5 min before injection (diluting standards and sample extracts with water increases affinity of pigments for the column in the loading step, resulting in an improved separation of more polar pigments).

††† Sigma Chemical Co., St. Louis, MO, or equivalent.
‡‡‡ Roth Chemical Co., distributed by Atomergic Chemetals Corp., Farmingdale, NY, or equivalent.

Rinse syringe twice with 300 μL standard and draw 500 μL standard into syringe for injection. Place syringe in injector valve, overfilling the 200 μL sample loop 2.5-fold. To check for possible interferences in the extraction solvent and/or filter, prepare a blank by extracting a glass fiber filter in 90% acetone; mixing 1 mL 90% acetone filter extract and 300 μL distilled water; and injecting into the HPLC system. Plot absorbance peak areas (or heights) against standard pigment concentrations. Calculate response factors as the slope of the regression between the weights of the injected standards (ng) and the areas of the parent pigment (plus areas of structurally-related isomers if present). These isomers contribute to the absorption signal of the standards; disregarding them results in over-estimation of pigments in sample extracts.[31]

3) Prepare samples for injection by mixing a 1 mL portion of the 90% acetone pigment extract and 300 μL distilled water, shake, and equilibrate for 5 min before injection.

4) Following sample injection, use a gradient program to op-

TABLE 10200:IV. HPLC SOLVENT SYSTEM PROGRAM

Time min	Flow Rate mL/min	Percentage of Eluent A	B	C	Conditions
Analysis protocol:					
0.0	1.0	100	0	0	Injection
2.0	1.0	0	100	0	Linear gradient
2.6	1.0	0	90	10	Linear gradient
13.6	1.0	0	65	35	Linear gradient
18.0	1.0	0	31	69	Linear gradient
23.0	1.0	0	31	69	Hold
25.0	1.0	0	100	0	Linear gradient
26.0	1.0	100	0	0	Linear gradient
34.0	1.0	100	0	0	Hold
Shutdown protocol:					
0.0	1.0	100	0	0	Analysis complete
3.0	1.0	0	100	0	Linear gradient
6.0	1.0	0	0	100	Linear gradient
16.0	1.0	0	0	100	Washing
17.0	0.0	0	0	100	Shutdown

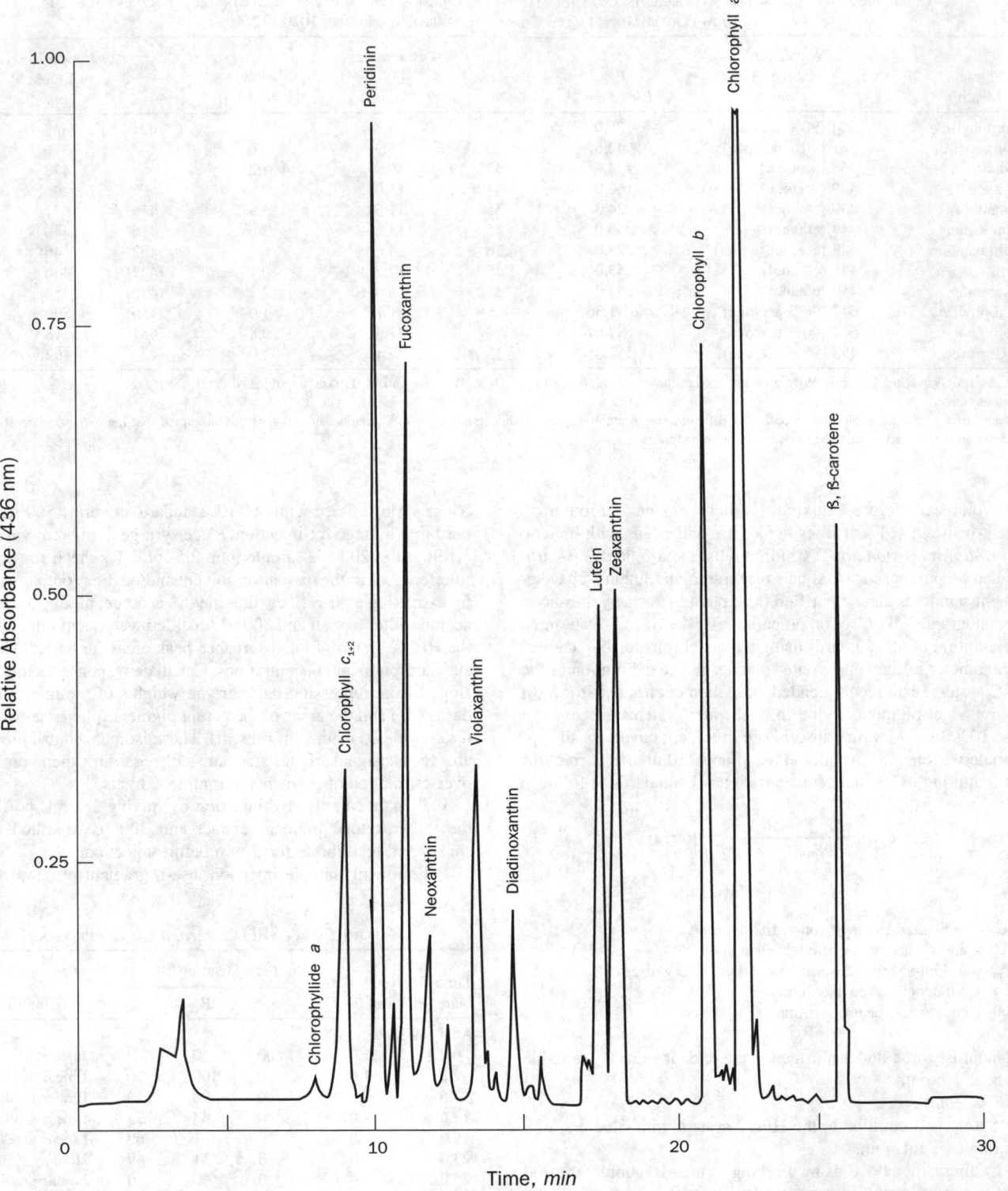

Figure 10200:12. Reverse-phase HPLC pigment chromatogram for a mixture of common algal pigments found in freshwater systems. For further data see Table 10200:III. Sample contained a natural extract with authentic known additions. The small unlabeled peaks are pigment degradation products.

timize separation of chlorophyll and carotenoid pigments. The system described in Table 10200:IV has been developed from the original method[32] to insure elution of most hydrophobic pigments. Degas solvent system with helium during analysis. Periodically flush HPLC system with distilled water to avoid accumulation of ion-pairing reagents.

5) Routinely determine peak identities by comparing retention times of sample peaks with those of pure standards. Confirm peak identities spectrophotometrically by collecting eluting peaks from the column outlet (or directly with an on-line diode array spectrophotometer). Absorption maxima for most common pigments found in freshwater systems are given in Table 10200:III.

6) Calculate individual pigment concentrations using the formula given in ¶ 4b5) preceding.

7) This method is designed for separation of chlorophyll and carotenoid pigments (Figure 10200:12), however, it also separates major chlorophyll breakdown products.

8) Method precision was assessed by making triplicate injections of a mixture of phytoplankton and plant extracts. Coefficients of variation ranged from 0.6 to 6.0% (Table 10200:III). Using an appropriate internal standard increases precision.

6. References

1. ROTT, E. Spectrophotometric and chromatographic chlorophyll analysis: comparison of results and discussion of the trichromatic method. *Ergebn. Limnol.* (Suppl. to *Arch. Hydrobiol.*) 14:37.

2. MARKER, A.F.H., E.A. NUSCH, H. RAI & B. RIEMANN. 1980. The measurement of photosynthetc pigments in freshwaters and standardization of methods: Conclusions and recommendations. *Ergebn. Limnol.* (Suppl. to *Arch. Hydrobiol.*) 14:91.

3. LORENZEN, C.J. 1967. Determination of chlorophyll and pheo-pigments: spectrophotometric equations. *Limnol. Oceanogr.* 12:343.

4. FITZGERALD, G.P. & S.L. FAUST. 1967. A spectrophotometric method for the estimation of percentage degradation of chlorophylls to pheo-pigments in extracts of algae. *Limnol. Oceanogr.* 12:335.

5. JEFFREY, S.W. & G.F. HUMPHREY. 1975. New spectrophotometric equations for determining chlorphylls a, b, and c, in higher plants, algae and natural phytoplankton. *Biochem. Physiol. Pflanzen* 167:191.

6. YENTSCH, C.S. & D.W. MENZEL. 1963. A method for the determination of phytoplankton chlorophyll and phaeophytin by fluorescence. *Deep Sea Res.* 10:221.

7. LOFTUS, M.E. & J.H. CARPENTER. 1971. A fluorometric method for determining chlorophylls a, b, and c. *J. Mar. Res.* 29:319.

8. HOLM-HANSEN, O., C.J. LORENZEN, R.W. HOLMES & J.D.H. STRICKLAND. 1965. Fluorometric determination of chlorophyll. *J. Cons. Cons. Perma. Int. Explor. Mer* 30:3.

9. ABAYCHI, J.K. & J.P. RILEY. 1979. The determination of phytoplankton pigments by high-performance liquid chromatography. *Anal. Chim. Acta* 107:1.

10. LORENZEN, C.J. 1966. A method for the continous measurement of *in vivo* chlorophyll concentration. *Deep Sea Res.* 13:223.

11. JACOBSEN, T.R. 1978. A quantitative method for the separation of chlorophylls a and b from phytoplankton pigments by high-pressure liquid chromatography. *Mar. Sci. Comm.* 4:33.

12. BROWN, L.M., B.T. HARGRAVE & M.D. MACKINNON. 1981. Analysis of chlorophyll a in sediments by high-performance liquid chromatography. *Can. J. Fish. Aquat. Sci.* 38:205.

13. GIESKES, W.W. & G.W. KRAAY. 1983. Unknown chlorophyll a derivatives in the North Sea and the tropical Atlantic Ocean revealed by HPLC analysis. *Limnol. Oceanogr.* 28:757.

14. GOWEN, R.J., P. TETT & B.J.B. WOOD. 1983. Changes in the major dihydroporphyrin plankton pigments during the spring bloom of phytoplankton in two Scoottish sea-lochs. *J. Mar. Biol. Assoc. U.K.* 63:27.

15. MANTOURA, R.F.C. & C.A. LLWEWLLYN. 1983. The rapid determination of algal chlorophyll and caroteniod pigments and their breakdown products in natural waters by reverse-phase high-performance liquid chromatography. *Anal. Chim. Acta* 151:297.

16. GIESKES, W.W.C. & G.W. KRAAY. 1984. Phytoplankton, its pigments, and primary production at a central North Sea station in May, July and September 1981. *Neth. J. Sea Res.* 18.71.

17. HALLEGRAEFF, G.M. & S.E. JEFFREY. 1985. Description of new chlorophyll a alteration products in marine phytoplankton. *Deep Sea Res.* 32:697.

18. TREES, C.C., M.C. KENNICUTT II & J.M. BROOKS. 1985. Errors associated with the standard fluorometric determination of chlorophylls and phaeopigments. *Mar. Chem.* 17:1.

19. ESKINS, K., C.R. SCHOFIELD & H.J. DUTTON. 1977. High-performance liquid chromatography of plant pigments. *J. Chromatogr.* 135:217.

20. WRIGHT, S.W. & J.D. SHEARER. 1984. Rapid extraction and high performance liquid chromatography of chlorophylls and carotenoids from marine phytoplankton. *J. Chromatogr.* 294:281.

21. BIDIGARE, R.R., M.C. KENNICUTT II & J.M. BROOKS. 1985. Rapid determination of chlorophylls and their degradation products by high-performance liquid chromatography. *Limnol. Oceanogr.* 30:432.

22. JEFFRY, S.W. 1974. Profiles of photosynthetic pigments in the ocean using thin-layer chromatography. *Mar. Biol.* 26:101.

23. PHINNEY, D.A. & C.S. YENTSCH. 1985. A novel phytoplankton chlorophyll technique: toward automated analysis. *J. Plankton Res.* 7:633.

24. STRICKLAND, J.D.H. & T.R. PARSONS. 1968. A Practical Manual of Sea Water Analysis. Fish. Res. Board Can. Bull. No. 167. Queen's Printer, Ottawa, Ont.

25. JEFFREY, S.W. 1981. An improved thin-layer chromatographic technique for marine phytoplankton pigments. *Limnol. Oceanogr.* 26:191.

26. JEFFREY, S.W. 1972. Preparation and some properties of crystalline chlorophyll c_1 and c_2 from marine algae. *Biochim. Biophys. Acta* 279:15.

27. BARRETT, J. & S.W. JEFFREY. 1971. A note on the occurrence of chlorophyllase in marine algae. *J. Exp. Mar. Biol. Ecol.* 7:255.

28. LORENZEN, C.J. & J. NEWTON DOWNS. 1986. Specific absorption coefficients of chlorophyllide a and pheophorbide a in 90 percent acetone, and comments on the fluorometric determination of chlorophyll and pheopigments. *Limnol. Oceanogr.* 31:449.

29. SARTORY, D.P. 1985. The determination of algal chlorophyllous pigments by high performance liquid chromatography and spectrophotometry. *Water Res.* 19:605.

30. MURRAY, A.P., C.F. GIBBS & A.R. LONGMORE. 1986. Determination of chlorophyll in marine waters: Intercomparison of a rapid HPLC method with full HPLC, spectrophotometric and fluorometric methods. *Mar. Chem.* 19:211.

31. BIDIGARE, R.R. 1991. Analysis of algal chlorophylls and carotenoids. *In* D.C. Hurd & D.W. Spencer, eds., Marine Particles: Analysis and Characterization. America Geophysical Union, Washington, D.C.

32. WRIGHT, S.W., S.W. JEFFREY, R.F.C. MANTOURA, C.A. LLEWELLYN, T. BJORNLAND, D. REPETA & N. WELSCHMEYER. 1991. Improved HPLC method for the analysis of chlorophylls and carotenoids from marine phytoplankton. *Mar. Ecol. Prog. Ser.* 77:183.

33. JEFFREY, S.W. & F.T. HAXO. 1968. Photosynthetic pigments of symbiotic dinoflagellates (zooxanthallae) from corals and clams. *Biol. Bull.* 135:149.

34. JENSEN, A. 1978. Chlorophylls and carotenoids. *In* J.A. Helleburst & J.S. Craige, eds., Handbook of Phycological Methods: Physiological and Biochemical Methods. Cambridge University Press, Cambridge, England.

35. DAVIS, B.H. 1976. Carotenoids. *In* T.W. Goodwin, ed., Chemistry and Biochemistry of Plant Pigments. Academic Press, New York, N.Y.

36. JOHANSEN, J.E., W.A. SVEC & S. LIAAEN-JENSEN. 1974. Carotenoids of the Dinophyceae. *Phytochem.* 13:2261.

10200 I. Determination of Biomass (Standing Crop)

Biomass is a quantitative estimate of the total mass of living organisms within a given area or volume. It may include the mass of a population (species biomass) or of a community (community biomass) but gives no information on community structure or function. The most accurate methods for estimation of biomass are dry weight, ash-free dry weight, and volume of living organisms. Indirect methods include estimates of total carbon, caloric content, nitrogen, lipids, carbohydrates, silica (diatoms), and chlorophyll (algae). Adenosine triphosphate[1] (ATP) and deoxyribonucleic acid[2,3] (DNA) also have been used as indirect estimates. All estimates of biomass can be affected by the presence of organic and inorganic detritus; ATP and DNA analyses include contributions from the bacterial flora.[4]

1. Chlorophyll a

Chlorophyll a is used as an algal biomass indicator.[5] Assuming that chlorophyll a constitutes, on the average, 1.5% of the dry weight of organic matter (ash-free weight) of algae, estimate the algal biomass by multiplying the chlorophyll a content by a factor of 67.

2. Biovolume (Cell Volume)

Plankton data derived on a volume-per-volume basis often are more useful than numbers per milliliter.[6] Determine cell volume by using the simplest geometric configuration that best fits the shape of the cell being measured (such as sphere, cone, cylinder).[7] Cell sizes of an organism can differ substantially in different waters and from the same waters at different times during the year; therefore, average measurements from 20 individuals of each species for each sampling period. Calculate the total biovolume of any species by multiplying the average cell volume in cubic micrometers by the number per milliliter.

Compute total wet algal volume as:

$$V_t = \sum_{i=1}^{n} (N_i \times V_i)$$

where:

V_t = total plankton cell volume, mm^3/L,
N_i = number of organisms of the ith species/L, and
V_i = average volume of cells of ith species, μm^3.

3. Cell Surface Area

An estimation of cell surface area is valuable in analyzing interactions between the cell and surrounding waters. Compute average surface area in square micrometers and multiply by the number per milliliter of the species being considered.

4. Displacement Volume

This method[8] measures an equivalent volume of liquid that is displaced by the sample. Displacement volume may be determined by several methods; for simple, direct measurement proceed as follows: Place sample in sieve of mesh size equal to or smaller than net used in capture; let sample drain and transfer to a measured volume of water in a graduated cylinder; measure the new volume containing sample plus known volume. The displacement volume equals the new volume minus original measured volume of water.

5. Gravimetric Methods

The biomass of the plankton community can be estimated from gravimetric determinations, although silt and organic detritus interfere. Determine dry weight by placing 100 mg wet concentrated sample in a clean, ignited, and tared porcelain crucible and dry at 105°C for 24 h. Alternatively, filter a known volume of sample through 0.45-μm-pore-diam membrane or a prerinsed, dried, and preweighed glass-fiber filter. (Note that the small sample used in direct filtration may lead to error if not handled properly.) Cool sample in a desiccator and weigh. Obtain ash weight by igniting the dried sample at 500°C for 1 h. Cool, rewet ash with distilled water, and bring to constant weight at 105°C. The ash is rewetted to restore water of hydration of clays and other minerals; this may amount to as much as 10% of weight lost during incineration.[9] The ash-free dry weight is the difference between the dry weight and the weight of the ash residue after ashing. The ash-free dry weight is preferred to dry weight to compare mixed assemblages. The ash content may constitute 50% or more of the dry weight in phytoplankton having inorganic structures, such as the diatoms. In other forms the ash content is only about 5% of dry weight.

6. Adenosine Triphosphate (ATP)

Methods of measuring adenosine triphosphate (ATP) in plankton provide the only means of determining the total viable plankton biomass. ATP occurs in all plants and animals, but only in living cells; it is not associated with nonliving particulate material. The ratio of ATP to biomass varies from species to species, but appears to be constant enough to permit reliable estimates of biomass from ATP measurements.[10] The method is simple and relatively inexpensive and the instrumentation is stable and reliable. The method also has many potential applications in entrainment and bioassay work, especially plankton mortality studies.

a. Equipment and reagents:

1) *Glassware:* clean, sterile, dry borosilicate glass flasks, beakers, and pipets.

2) *Filters:* 47-mm-diam, 0.45-μm-porosity membrane filters.

3) *Filtration equipment.*

4) *Freezer* (-20°C).

5) *Boiling water bath.*

6) *Detection instruments* designed specifically for measuring ATP.*

7) *Microsyringes:* 10-, 25-, 50-, 100-, and 250-μL.

8) *Reaction cuvettes and vials.*

9) *Tris buffer* (0.02M, pH 7.75): Dissolve 7.5 g trishydroxymethylaminomethane in 3 L distilled water and adjust to pH 7.75 with 20% HCl. Autoclave 150-mL portions at 115°C for 15 min.

---

* Beckman, JRB, Turner Designs, or equivalent.

10) *Luciferin-luciferase enzyme preparation:*† Rehydrate frozen (−20°C) lyophilized extracts of firefly lanterns with Tris buffer as directed by the supplier; let stand at room temperature 2 to 3 h, then centrifuge at 300 × *g* for 1 min and decant the supernatant into a clean, dry test tube; let stand at room temperature for 1 h.

11) *Purified ATP standard:* Dissolve 12.3 mg disodium ATP in 1 L distilled water and dilute 1.0 mL to 100 mL with Tris buffer; 0.2 mL = 20 ng ATP.

b. Procedure:

1) Calibration—To determine the calibration factor (*F*), prepare a series of dilutions of purified ATP standard and record the light emission from several portions of each concentration of standard. Correct mean area of standards by subtracting peak reading or mean area of several blanks using 0.2 mL Tris buffer. Calculate calibration factor F_S as:

$$F_S = \frac{C}{A_S}$$

where:

F_S = calibration factor at sensitivity S,
A_S = peak reading or mean area under standard ATP curve corrected for blank, and
C = concentration of ATP in standard solution, ng/mL.

2) Sample analysis—Collect a 1- to 2-L sample in a clean, sterile sampler. Pass through a 250-μm net to remove large zooplankton[10] and filter through a 47-mm 0.45-μm-porosity filter by applying a vacuum of about 30 kPa. (IMPORTANT: Break the suction before the last film of water is pulled through the filter.) Quickly place filter in a small beaker. Immediately cover filter with 3 mL boiling Tris buffer, using an automatic pipet. Place beaker in boiling water bath for 5 min and, with a Pasteur pipet, transfer extract to a clean, dry, calibrated test tube. Rinse filter and beaker with 2 mL boiling Tris buffer; combine extracts, record volume, bring volume up to 5 mL with Tris buffer, cover tubes with parafilm and, if samples cannot be analyzed immediately, freeze at −25°C. Extracts may be stored for many months in a freezer. Prepare at least triplicate extracts of each sample.

The analytical procedure depends on detection equipment used. If a scintillation counter is used, pipet 0.2 mL enzyme preparation into a glass vial. Measure the light emission of the enzyme preparation (blank) for 2 to 3 min at sensitivity settings near that anticipated for the sample. Add 0.2 mL sample extract to the vial, record the time, and swirl. Start recording light output 10 s after combining ATP extract and enzyme preparation; record output for 2 to 3 min, using the same time period for all samples. Determine the mean of areas under the curves obtained and correct by subtracting mean of areas under the

curves obtained from blanks prepared as directed in Strickland and Parsons.[11]

c. Calculations: Calculate concentration of ATP as:

$$ATP, \ ng/L = \frac{A_c \times V_e \times F_s}{V_s}$$

A_c = mean corrected area under extract curves,
V_e = extract volume, mL,
V_s = volume of sample, L, and
F_s = calibration factor.

If an ATP content of 2.4 μg ATP/mg dry weight organic matter is assumed,[12] total living plankton biomass (*B*), as dry weight organic matter, is given as:

$$B, \ mg/L = \frac{ATP}{(2.4)(1000)}$$

7. References

1. HOLM-HANSEN, O. & C.R. BOOTH. 1966. The measurement of adenosine triphosphate in the ocean and its ecological significance. *Limnol. Oceanogr.* 11:510.
2. HOLM-HANSEN, O., W.H. SUTCLIFFE, JR. & J. SHARP. 1968. Measurement of deoxyribonucleic acid in the ocean and its ecological significance. *Limnol. Oceanogr.* 13:507.
3. HOLM-HANSEN, O. 1969. Determination of microbial biomass in ocean profiles. *Limnol. Oceanogr.* 14:740.
4. PAERL, H.W., M.M. TILZER & C.R. GOLDMAN. 1976. Chlorophyll *a* vs. ATP as algal biomass indicators in lakes. *J. Phycol.* 12:242.
5. CREITZ, G.I. & F.A. RICHARDS. 1955. The estimation and characterization of plankton populations by pigment analysis. *J. Mar. Res.* 14:211.
6. KUTKUHN, J.H. 1958. Notes on the precision of numerical and volumetric plankton estimates from small sample concentrations. *Limnol. Oceanogr.* 3:69.
7. VOLLENWEIDER, R.A. 1969. A Manual on Methods for Measuring Primary Production in Aquatic Environments. IBP Handbook No. 12. Blackwell Scientific Publ., Oxford, England.
8. JACOBS, F. & G.C. GRANT. 1978. Guidelines for zooplankton sampling in quantitative baseline and monitoring programs. EPA-600/3-78-026, U.S. Environmental Protection Agency.
9. NELSON, D.J. & D.C. SCOTT. 1962. Role of detritus in the productivity of a rock-outcrop community in a Piedmont stream. *Limnol. Oceanogr.* 7:396.
10. RUDD, J.W.M. & R.D. HAMILTON. 1973. Measurement of adenosine triphosphate (ATP) in two precambrian shield lakes of northwestern Ontario. *J. Fish. Res. Board Can.* 30:1537.
11. STRICKLAND, J.D.H. & T.R. PARSONS. 1968. A Practical Manual of Sea Water Analysts. Fish. Res. Board Can. Bull. No. 167. Queen's Printer, Ottawa, Ont.
12. WEBER, C.I. 1973. Recent developments in the measurement of the response of plankton and periphyton to changes in their environment. *In* G. Glass, ed. Bioassay Techniques and Environmental Chemistry. Ann Arbor Science Publ., Ann Arbor, Mich.

---

† Dupont, Sigma Chemical, or equivalent.

10200 J.   Metabolic Rate Measurements

The physiological condition and the spectrum of biological interactions of the aquatic community must be considered for evaluation of the state of natural waters. Earlier, numbers, species composition, and biomass were the prime considerations. Recognition of the limitations of this approach led to the measurement of rates of metabolic processes such as photosynthesis (productivity), nitrogen fixation, respiration, and electron transport. These provide a better understanding of the complex nature of the aquatic ecosystem. An indication of photosynthetic efficiency can be determined by the productivity index (mg C fixed/unit chlorophyll a).[1]

1. Nitrogen Fixation

The ability of an organism to fix nitrogen is a great competitive advantage and plays a major role in population dynamics. Two reliable methods for estimating nitrogen fixation rates in the laboratory are the ^{15}N isotope tracer method[2,3] and the acetylene reduction method.[4] Because the rate of nitrogen fixation varies greatly with different organisms and with the concentration of combined nitrogen, nitrogen fixation rates cannot be used to estimate biomass of nitrogen-fixing organisms. However, the acetylene reduction method is useful in measuring nitrogen budgets and in algal assay work.[5]

2. Productivity, Oxygen Method

Productivity is defined as the rate at which inorganic carbon is converted to an organic form. Chlorophyll-bearing organisms (phytoplankton, periphyton, macrophytes) serve as primary producers in the aquatic food chain. Photosynthesis ultimately results in the formation of a wide range of organic compounds, release of oxygen, and reduction of carbon dioxide (CO_2) in the surrounding waters. Primary productivity[6] can be determined by measuring the changes in oxygen and CO_2 concentrations.[7] In poorly buffered waters, pH can be a sensitive property for detecting variations in the system. As CO_2 is removed during photosynthesis, the pH rises. This shift can be used to estimate both photosynthesis and respiration.[8] The sea and many fresh waters are too highly buffered to make this useful, but it has been applied successfully to productivity studies in some lake waters.

Two methods of measuring the rate of carbon uptake and net photosynthesis in situ are the oxygen method[9] and the carbon 14 method.[10] In both methods, clear (light) and darkened (dark) bottles are filled with water samples and suspended at regular depth intervals for an incubation period of several hours or samples are incubated under controlled conditions in environmental growth chambers in the laboratory.

The basic reactions in algal photosynthesis involve uptake of inorganic carbon and release of oxygen, summarized by the relationship:

$$CO_2 + H_2O \text{ N } (CH_2O)_x + O_2$$

The chief advantages of the oxygen method are that it provides estimates of gross and net productivity and respiration and that analyses can be performed with inexpensive laboratory equipment and common reagents. The dissolved oxygen (DO) concentration is determined at the beginning and end of the incubation period. Productivity is calculated on the assumption that one atom of carbon is assimilated for each molecule of oxygen released.

a. Equipment:

1) *BOD bottles*, numbered, 300-mL, clear and opaque borosilicate glass, with ground glass stopper and flared mouth, for sample incubation. Acid-clean the bottles, rinse thoroughly with distilled water, and just before use, rinse with water being tested. Do not use phosphorus-containing detergents.

If suitable opaque bottles are not available, make clear BOD bottles opaque by painting them black and wrapping with black waterproof tape. As a further precaution, wrap entire bottle in aluminum foil or place in light-excluding container during incubation.

2) *Supporting line or rack* that does not shade suspended bottles.

3) *Nonmetallic opaque acrylic Van Dorn sampler* or equivalent, of 3- to 5-L capacity.

4) *Equipment and reagents for dissolved oxygen determinations:* See Section 4500-O.

5) *Pyrheliometer.*

6) *Submarine photometer.*

7) *Thermometer.*

b. Procedure:

1) Obtain a profile of the input of solar radiation for the photoperiod with a pyrheliometer.

2) Determine depth of euphotic zone (the region that receives 1% or more of surface illumination) with a submarine photometer. Select depth intervals for bottle placement. The photosynthesis-depth curve will be approximated closely by placing samples at intervals equal to one-tenth the depth of the euphotic zone. Estimate productivity in relatively shallow water with fewer depth intervals.

3) Measure oxygen concentration with probe or by titration and temperature and salinity to determine whether water is supersaturated with respect to oxygen (see Table 4500-O:I). If water is supersaturated, bubble nitrogen gas through sample to lower initial oxygen concentration to less than 80% saturation.

4) Keep samples out of direct sunlight during handling. Introduce samples taken from each preselected depth into duplicate clear, darkened, and initial-analysis bottles. Insert delivery tube of sampler to bottom of sample bottle and fill so that three volumes of water are allowed to overflow. Remove tube slowly and close bottle. Use water from the same grab sample to fill a "set" (one light, one dark, and one initial bottle).

5) Immediately treat (fix) samples taken for the chemical determination of initial dissolved oxygen (see Section 4500-O) with manganous sulfate ($MnSO_4$), alkaline iodide, and sulfuric acid (H_2SO_4) or check with an oxygen probe. Analyses may be delayed several hours if necessary, if samples are fixed or iced and stored in the dark.

6) Suspend duplicate paired clear and darkened bottles at the depth from which the samples were taken and incubate for at least 2 h, but never longer than it takes for oxygen-gas bubbles to form in the clear bottles or DO to be depleted in the dark bottles.

7) At the end of the exposure period, immediately determine DO as described above.

c. Calculations: The increase in oxygen concentration in the light bottle during incubation is a measure of net production which, because of the concurrent use of oxygen in respiration, is somewhat less than the total (or gross) production. The loss of oxygen in the dark bottle is used as an estimate of total plankton respiration. Thus:

Net photosynthesis = light bottle DO − initial DO

Respiration = initial DO − dark bottle DO

Gross photosynthesis = light bottle DO − dark bottle DO

Average results from duplicates.

1) Calculate the gross or net production for each incubation depth and plot:

$$\text{mg carbon fixed/m}^3 = \text{mg oxygen released/L} \times 12/32 \times 1000 \text{ L/m}^3 \times K$$

where K is the photosynthetic quotient (PQ), ranging from 1 to 2, depending on the nitrogen supply.[11,12]

Use the factor 12/32 to convert oxygen to carbon; under ideal conditions 1 mole of O_2 (32 g) is released for each mole of carbon (12 g) fixed.

2) Productivity is defined as the rate of production and generally is reported in grams carbon fixed per square meter per day. Determine the productivity of a vertical column of water 1 m square by plotting productivity for each exposure depth and graphically integrating the area under the curve.

3) Using the solar radiation profile and photosynthesis rate during incubation adjust the data to represent phytoplankton productivity for the entire photoperiod. Because photosynthetic rates vary widely during the daily cycle,[13,14] do not attempt to convert data to other test circumstances.

3. Productivity, Carbon 14 Method

A solution of radioactive carbonate ($^{14}CO_3^{2-}$) is added to light and dark bottles that have been filled with sample as described for the oxygen method. After incubation in situ, collect the plankton on a membrane filter, treat with hydrochloric acid (HCl) fumes to remove inorganic carbon 14, and assay for radioactivity. The quantity of carbon fixed is proportional to the fraction of radioactive carbon assimilated.

This procedure differs from the oxygen method in that it affords a direct measurement of carbon uptake and measures only net photosynthesis.[15] It is basically more sensitive than the oxygen method, but fails to account for organic materials that leach from cells[16,17] during incubation.

a. Equipment and reagents:

1) *Pyrheliometer.*

2) *Submarine photometer.*

3) *BOD bottles and supporting apparatus:* See ¶s 2a1) and 2), above.

4) *Membrane-filtering device and 25-mm filters* with pore diameters of 0.22, 0.30, 0.45, 0.80, and 1.2 μm.

5) *Counting equipment for measuring radioactivity:* Scaler with end-window tube, gas flow meter, or liquid scintillation counter (see Section 7010D). The thin-window tube is the least expensive detector and, when used with a small scaler, provides acceptable data at modest cost.

6) *Fuming chamber:* Use a glass desiccator with a depth of about 1.4 cm conc HCl in desiccant chamber. The fuming chamber is recommended for filter decontamination.[18,19]

7) *Syringe or pipet,* nonmetallic.

8) *Chemical reagents:* See Sections 4500-CO_2 (Carbon Dioxide) and 2320 (Alkalinity).

9) *Radioactive carbonate solutions:*

a) *Sodium chloride dilution solution,* 5% NaCl (w/v): Add 0.3 g sodium carbonate (Na_2CO_3) and one pellet sodium hydroxide (NaOH) per liter. Use for marine studies only.

b) *Carrier-free radioactive carbonate solution,* commercially available in sealed vials having approximately 5 μCi ^{14}C/mL. Confirm absence of suspended and dissolved toxic metals[20] or filter and pass through an ion-exchange column.*

c) *Working solutions* with activities of 1, 5, and 25 μCi ^{14}C/2 mL. For studies of fresh water use carrier-free radioactive carbonate and for studies of marine water prepare by diluting carrier-free radioactive carbonate solution with NaCl dilution solution.

d) *Stock ampules:* Prepare ampules containing 2 mL of required working solution. Fill ampules and autoclave sealed ampules at 121°C for 20 min.[21]

b. Procedure:

1) Obtain a record of incident solar radiation for the photoperiod with a pyrheliometer.

2) Determine depth intervals for sampling and incubation as described above.

3) Use duplicate light and dark bottles at each depth. Also use dark bottles or bottles harvested at time zero. Fill bottles with sample, add 2 mL radioactive carbonate solution (using a nonmetallic pipet) to the bottom of each bottle, and mix thoroughly by repeated inversion. The concentration of carbon 14 should be approximately 10 μCi/L in relatively productive waters, to 100 μCi/L, or higher, in oliogotrophic (open ocean) waters. To obtain statistical significance, have at least 1000 cpm in the filtered sample. Take duplicate samples at each depth to determine initial concentration of inorganic carbon (CO_2, HCO_3^-, and CO_3^{2-}) available for photosynthesis (see Section 4500-CO_2). For estuarine and marine samples, estimate total inorganic carbon concentrations with a simple titration procedure[22] and make initial temperature, salinity, and pH measurements.

4) Incubate samples for up to 4 h. If measurements are required for the entire photoperiod, overlap 4-h periods from dawn until dusk. A 4-h incubation period may be sufficient provided energy input is used as the basis for integrating incubation period to entire photoperiod. For incubation procedure, see ¶ 2b6) above.

5) After incubating remove sample bottles and immediately place in the dark. Filter unpreserved samples without delay. Avoid sample preservation to avoid lysing cells or determine extracellular products.

6) Filter two portions of each sample through a membrane filter, taking care that the largest pore size is consistent with quantitative retention of plankton. Although the 0.45-μm pore filter usually is adequate, determine the efficiency of sample retention immediately before analysis, with a wide range of pore sizes.[23,24] Apply approximately 30 kPa of vacuum during filtration. Excess vacuum may cause extensive cell rupture and loss of radioactivity through the membrane.[25] Use maximum sample volume consistent with rapid filtration (1 to 2 min), but do not clog filter.

7) Place membranes in HCl fumes for 20 min. Count filters as soon as possible, although extended storage in a desiccator is acceptable.

---

* Chelex 100 or equivalent.

8) Determine radioactivity by counting with an end-window tube, windowless gas flow detector, or liquid scintillation counter.

9) Determine counting geometry of thin-window and windowless gas flow detectors.[26] Using three ampules of carbon 14, prepare a series of barium carbonate ($BaCO_3$) precipitates on tared 0.45-μm membrane filters as directed below. The precipitates will contain the same amount of carbon 14 activity but will have different thicknesses ranging from 0.5 to 6.0 mg/cm^2. Dilute each ampule to 500 mL with a solution of 1.36 g Na_2CO_3/L CO_2-free distilled water. Pipet 0.5-mL portions into each of seven conical flasks containing 0, 0.5, 1.5, 2.5, 3.5, 4.5, and 5.5 mL, respectively, of a solution of 1.36 g Na_2CO_3 /L CO_2-free distilled water. Add, respectively, 0.3, 0.6, 1.2, 1.8, 2.4, 3.0, and 3.6 mL 1.04% barium chloride ($BaCl_2$) solution. Let $BaCO_3$ precipitate stand 2 h with gentle swirling every half hour. Collect each precipitate on a filter (using an apparatus with a filtration area comparable to that of the samples). With suction, dry filters without washing; place in a desiccator for 24 h, weigh, and count. The counting rate increases exponentially with decreasing precipitate thickness. Extrapolate graphically (or mathematically) to zero precipitate thickness and multiply the zero-thickness counting rate by 1000 to correct for ampule dilution. This represents the amount of activity added to each sample bottle used to determine fraction of carbon 14 taken up in light and dark bottles.

c. Calculations:

1) Subtract the mean dark-bottle or time-zero sample count from the mean light-bottle counts for each replicate pair.

2) Determine the total dissolved inorganic carbon available for photosynthesis (carbonate, bicarbonate, and free CO_2) from pH and alkalinity measurements; make direct measurement of total CO_2 according to Section 4500-CO_2 or the methods described in the literature.[27-30]

3) Determine quantity of carbon fixed by using the following relationship:

$$\text{mg carbon fixed/L} = \frac{\text{counting rate of filtered sample}}{\text{total activity added to sample}}$$
$$\times \frac{300}{\text{volume filtered}} \times \text{mg/L initial inorganic carbon} \times 1.064\dagger$$

4) Integrate productivity for the entire depth of euphotic zone and express as grams carbon fixed per square meter per day [see ¶ 2*c*2) above].

5) Using the solar radiation records and photosynthesis rates during incubation, adjust data to represent phytoplankton productivity for the entire photoperiod. If samples were incubated for less than the full photoperiod, apply a correction factor.

4. References

1. GUNDERSEN, K. 1973. *In-situ* determination of primary production by means of the new incubator, ISIS. *Helgolander wiss. Meeresunters.* 24:465.
2. BURRIS, R.H., F.J. EPPLING, H.B. WAHLIN & P.W. WILSON. 1942. Studies of biological nitrogen fixation with isotopic nitrogen. *Proc. Soil Sci. Soc. Amer.* 7:258.
3. NEESS, J.C., R.C. DUGDALE, V.A. DUGDALE & J.J. GOERING. 1962. Nitrogen metabolism in lakes. I. Measurement of nitrogen fixation with N^{15}. *Limnol. Oceanogr.* 7:163.
4. STEWART, W.D.P., G.P. FITZGERALD & R.H. BURRIS. 1967. *In situ* studies on N$_2$ fixation using the acetylene reduction technique. *Proc. Nat. Acad. Sci.* 58:2071.
5. STEWART, W.D.P., G.P. FITZGERALD & R.H. BURRIS. 1970. Acetylene reduction assay for determination of phosphorus availability in Wisconsin lakes. *Proc. Nat. Acad. Sci.* 66:1104.
6. GOLDMAN, C.R. 1968. Aquatic primary production. *Amer. Zoologist* 8:31.
7. ODUM, H.T. 1957. Primary production measurements in eleven Florida springs and a marine turtle-grass community. *Limnol. Oceanogr.* 2:85.
8. BEYERS, R.J. & H.T. ODUM. 1959. The use of carbon dioxide to construct pH curves for the measurements of productivity. *Limnol. Oceanogr.* 4:499.
9. GAARDER, T. & H.H. GRAN. 1927. Investigations of the production of plankton in Oslo Fjord. *Rapp. Proces-Verbaux. Reunions Cons. Perma. Int. Explor. Mer* 42:1.
10. STEEMAN-NEILSEN, E. 1952. The use of radioactive carbon (C-14) for measuring organic production in the sea. *J. Cons. Perma. Int. Explor. Mer* 18:117.
11. WILLIAMS, P.J. LEB., R.C.T. RAINE & J.R. BRYAN. 1979. Agreement between the ^{14}C and oxygen methods of measuring phytoplankton production: Reassessment of the photosynthetic quotient. *Oceanol. Acta* 2:411.
12. DAVIES, J.M. & P.J. LEB. WILLIAMS. 1984. Verification of ^{14}C and O$_2$ derived primary organic production using an enclosed system. *J. Plankton Res.* 6:457.
13. RYTHER, J.H. 1956. Photosynthesis in the ocean as a function of light intensity. *Limnol. Oceanogr.* 1:61.
14. FEE, E.J. 1969. A numerical model for the estimation of photosynthetic production, integrated over time and depth, in natural waters. *Limnol. Oceanogr.* 14:906.
15. STEEMAN-NEILSEN, E. 1964. Recent advances in measuring and understanding marine primary production. *J. Ecol.* 52(Suppl.):119.
16. ALLEN, M.B. 1956. Excretion of organic compounds by *Chlamydomonas*. *Arch. Mikrobiol.* 24:163.
17. FOGG, G.E. & W.D. WATT. 1965. The kinetics of release of extracellular products of photosynthesis by phytoplankton. *In* C.R. Goldman, ed. Primary Productivity in Aquatic Environments. Suppl. 18, Univ. California Press, Berkeley.
18. WETZEL, R.G. 1965. Necessity for decontamination of filters in C^{14} measured rates of photosynthesis in fresh waters. *Ecology* 46:540.
19. MCALLISTER, C.D. 1961. Decontamination of filters in the C^{14} method of measuring marine photosynthesis. *Limnol. Oceanogr.* 6:447.
20. CARPENTER, E.J. & J.S. LIVELY. 1980. Review of estimates of algal growth using ^{14}C tracer techniques. *In* P.G. Falkowski, ed. Primary Productivity in the Sea. Brookhaven Symp. Biol. No. 31. Plenum Press, New York, N.Y.
21. STRICKLAND, J.D.H. & T.R. PARSONS. 1968. A Practical Manual of Sea Water Analysis. Fish. Res. Board Can. Bull. No. 167. Queen's Printer, Ottawa, Ont.
22. PARSONS, T.R., Y. MAITA & C.M. LALLI. 1984. A Manual of Chemical and Biological Methods for Seawater Analysis. Pergamon Press, New York, N.Y.
23. LASKER, R. & R.W. HOLMES. 1957. Variability in retention of marine phytoplankton by membrane filters. *Nature* 180:1295.
24. HOLMES, R.W. & C.G. ANDERSON. 1963. Size fractionation of C^{14}-labelled natural phytoplankton communities. *In* C.H. Oppenheimer, ed. Symposium on Marine Microbiology. Charles C. Thomas, Springfield, Ill.
25. ARTHUR, C.R. & F.H. RIGLER. 1967. A possible source of error in the C^{14} method of measuring primary productivity. *Limnol. Oceanogr.* 12:121.
26. JITTS, H.R. & B.D. SCOTT. 1961. The determination of zero-thickness activity in Geiger counting of C^{14} solutions used in marine productivity studies. *Limnol. Oceanogr.* 6:116.

† Correction for isotope effect.

27. SAUNDERS, G.W., F.B. TRAMA & R.W. BACHMANN. 1962. Publ. No. 8, Great Lakes Research Div., Univ. Michigan, Ann Arbor.
28. DYE, J.F. 1944. The calculation of alkalinities and free carbon dioxide in water by use of nomographs. *J. Amer. Water Works Assoc.* 36: 859.

29. MOORE, E.W. 1939. Graphic determination of carbon dioxide and the three forms of alkalinity. *J. Amer. Water Works Assoc.* 31:51.
30. PARK, K., D.W. HOOD & H.T. ODUM. 1958. Diurnal pH variation in Texas bays and its application to primary production estimations. *Publ. Inst. Mar. Sci. Univ. Tex.* 5:47.

10300   PERIPHYTON*

10300 A.   Introduction

1. Definition and Significance

Microorganisms growing on stones, sticks, aquatic macrophytes, and other submerged surfaces are useful in assessing the effects of pollutants on lakes, streams, and estuaries. Included in this group of organisms, here designated periphyton,[1,2] are the zoogleal and filamentous bacteria, attached protozoa, rotifers, and algae, and the free-living microorganisms that swim, creep, or lodge among the attached forms.

Unlike the plankton, which often do not respond fully to the influence of pollution in rivers for a considerable distance downstream, the periphyton show marked responses immediately below pollution sources. Examples are the beds of *Sphaerotilus* and other "slime organisms" commonly observed in streams below discharges of organic wastes. Because the abundance and com-

position of the periphyton at a given location are governed by the water quality at that point, observations of their condition generally are useful in evaluating conditions in bodies of water.

The use of periphyton in assessing water quality often is hindered by the lack of suitable natural substrates at the desired sampling station. Furthermore, it often is difficult to collect quantitative samples from these surfaces. To circumvent these problems artificial substrates have been used to provide a uniform surface type, area, and orientation.[3]

2. References

1. ROLL, H. 1939. Zur Terminologie des Periphytons. *Arch. Hydrobiol.* 35:39.
2. YOUNG, O.W. 1945. A limnological investigation of periphyton in Douglas Lake, Michigan. *Trans. Amer. Microsc. Soc.* 64:1.
3. SLÁDEČKOVÁ, A. 1962. Limnological investigation methods for the periphyton ("Aufwuchs") community. *Bot. Rev.* 28:286.

---

* Approved by Standard Methods Committee, 1996.

10300 B.   Sample Collection

1. Station Selection

In rivers, locate stations a short distance upstream and at one or more points downstream from the suspected pollution source or intended study area in the areas of central mixing. In large rivers, sample both sides of the stream in main flow areas. Because the effects of a pollutant depend on the assimilative capacity of the stream and on the nature of the pollutant, progressive changes in water quality downstream from the pollution source may be caused entirely by dilution and cooling—as in the case of nutrients, toxic industrial wastes, and thermal pollution—or by gradual mineralization of degradable organic compounds. Cursory examination of shoreline and bottom periphyton growths on natural substrates downstream from an outfall may indicate conspicuous zones of biological response to water quality that will be useful in determining appropriate sites for sampling stations. When an intensive sampling program is not feasible, a minimum of three sampling stations, one in a reference area upstream from

a pollution source and the others in the community downstream from the source, where complete mixing with the receiving water has occurred, will provide minimal data on the periphyton community.

In lentic waters (e.g., lakes, reservoirs, ponds) and other standing-water bodies where zones of pollution may be arranged concentrically, locate stations in areas adjacent to a waste outfall and in unaffected areas. Use control stations similar to the affected ones (e.g., similar in water depth and distance from shore).

2. Sample Collection

a. Natural substrates: Collect qualitative samples by scraping submerged stones, sticks, pilings, and other available substrates. Many devices have been developed to collect quantitative samples from irregular surfaces. Appropriate techniques for the removal of periphyton from both living and nonliving surfaces have been described.[1-4]

b. Artificial substrates: The most widely used artificial substrate is the standard, plain, 25- by 75-mm glass microscope slide, but other materials such as clear vinyl plastic also are suitable. Do not change substrate type during a study because colonization varies with substrate. In small, shallow streams and in the littoral regions of lakes and reservoirs where light penetrates to the bottom, place slides or other substrates vertically in frames anchored to the bottom. In large, deep streams or standing-water bodies where turbidity varies widely, place slides vertically with the slide face at right angles to the prevailing current. A floating rack, as shown in Figure 10300:1,* is suitable. Expose several slides (minimally five; three for biomass, one for species, and one backup for each time interval) for each type of analysis to assure collecting sufficient material and to determine variability in results caused by normal differences in colonization of individual slides. In addition to effects of pollutants, length of substrate exposure and seasonal changes in temperature and other natural environmental conditions may have a profound effect on sample composition. No community on an artificial substrate is representative of the natural community.

Place, expose, and handle all artificial substrate samplers in conditions as nearly identical as possible, whether they are replicate samplers at a particular sampling location or samplers at different locations. Sampler type and/or construction cause changes in surrounding physical conditions that in turn affect periphyton growth. Variations of 10 to 25% between sample replicates are common. Therefore, to reduce sampling error and increase interpretive power, reduce the magnitude of all possible test variables and use sufficient replication.

c. Exposure period: Colonization on clean slides proceeds at an exponential rate for the first 1 or 2 weeks and then slows. Because exposures of less than 2 weeks may result in very sparse collections, and exposures of more than 2 weeks may result in loss of material due to sloughing, sample for 2 weeks during the summer. This exposure period precludes collecting sexually mature thalli of larger, slow-growing filamentous algae such as *Cladophora* and *Stigeoclonium*. To obtain optimum growth during the winter, use a longer exposure period. For the most exacting work, determine the optimum exposure period by testing colonization rates over a period of about 6 weeks.

Secondary problems associated with macroinvertebrate infestation and grazing may occur, often within 7 to 14 d. To reduce the confounding influence of grazing, increase substrate sampling area and expose for 7 to 10 d.

3. Sample Preservation

Preserve samples that are taken for counting and identification in 5% neutralized formalin, Lugol's iodine, or merthiolate (see Section 10200B.4).

Preserve slides intact in bottles of suitable size or scrape into containers in the field. Air-dry slides for dry and ash-free dry weight in the field and store in a 3.0- × 7.7-cm glass bottle. Place slides for chlorophyll analyses in acetone or methanol in the field or collect and freeze with trichlorotrifluoroethane† (or alternative) or CO_2 and hold on dry ice until returned to the laboratory. Store all samples in the dark.

---

* Wildlife Supply Co., 301 Cass St., Saginaw, MI 48602, or equivalent.
† Freon or equivalent.

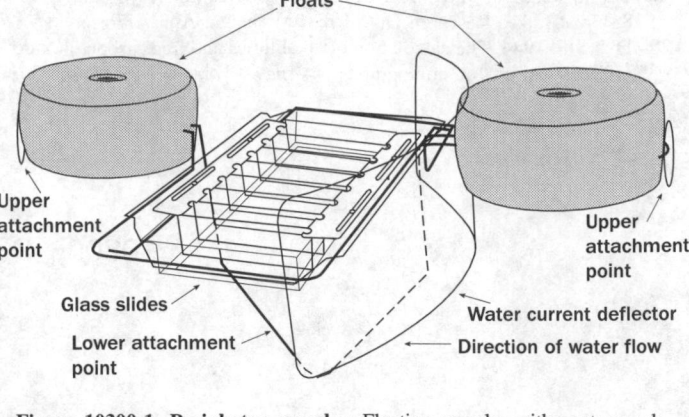

Figure 10300:1. Periphyton sampler. Floating sampler with upstream deflecting baffle and transparent, removable slide rack holding up to eight microscope slides.

4. References

1. SLÁDEČKOVÁ, A. 1962. Limnological investigation methods for the periphyton ("Aufwuchs") community. *Bot. Rev.* 28:286.
2. GOUGH, S.B. & W.J. WOELKERLING. 1976. On the removal and quantification of algal aufwuchs from macrophyte hosts. *Hydrobiologia* 48:203.
3. BOOTH, W.E. 1981. A method for removal of some epiphytic diatoms. *Botanica Marina* 24:603.
4. DELBECQUE, E.J.P. 1985. Periphyton on nymphaeids: An evaluation of methods and separation techniques. *Hydrobiologia* 124:85.

5. Bibliography

COOKE, W.B. 1956. Colonization of artificial bare areas by microorganisms. *Bot. Rev.* 22:613.
HOHN, M.H. 1966. Artificial substrate for benthic diatoms—collection, analysis, and interpretation. *In* K.W. Cummings, C.A. Tryon, Jr., & R.T. Hartman, eds. Organism-Substrate Relationships in Streams. Spec. Publ. No. 4, p. 87. Pymatuning Lab. Ecology, Univ. Pittsburgh, Pittsburgh, Pa.
KEVERN, N.R., J.L. WILHM & G.M. VAN DYNE. 1966. Use of artificial substrata to estimate the productivity of periphyton communities. *Limnol. Oceanogr.* 11:499.
ARTHUR, J.W. & W.B. HORNING. 1969. The use of artificial substrates in pollution surveys. *Amer. Midland Natur.* 82:83.
TIPPETT, R. 1970. Artificial surfaces as a method of studying populations of benthic micro-algae in fresh water. *Brit. Phycol. J.* 5:187.
ERTL, M. 1971. A quantitative method of sampling periphyton from rough substrates. *Limnol. Oceanogr.* 16:576.
ANDERSON, M.A. & S.L. PAULSON. 1972. A simple and inexpensive wood-float periphyton sampler. *Progr. Fish-Cult.* 34:225.
NORTH AMERICAN BENTHOLOGICAL SOCIETY. 1974–1991. (Annual) Current and Select Bibliographies on Benthic Biology. Springfield, Ill.
MARKER, A.F.H., C.A. CROWTHER & R.J.M. GUNN. 1980. Methanol and acetone as solvents for estimating chlorophyll a and phaeopigments by spectrophotometry. *Arch. Hydrobiol. Ergebn. Limnol.* 14:52.
NEROZZI, A. & P. SILVER. 1983. Periphytic community analysis in a small oligotropic lake. *Proc. Penn. Acad. Sci.* 57:138.
WETZEL, R., ed. 1983. Periphyton of Freshwater Ecosystems. Developments in Hydrobiology 17. Dr. W. Junk BV Publishers, The Hague, The Netherlands.
HAMILTON, P.B. & H.C. DUTHIE. 1984. Periphyton colonization of rock surfaces in a boreal forest stream studied by scanning electron microscopy and track autoradiography. *J. Phycol.* 20:525.

Nielsen, T.S., W.H. Funk, H.L. Gibbons & R.M. Duffner. 1984. A comparison of periphyton growth on artificial and natural substrates in the Upper Spokane River, Washington, USA. *Northwest Sci.* 58: 243.

Pip, E. & G.G.C. Robinson. 1984. A comparison of algal periphyton composition on 11 species of submerged macrophytes. *Hydrobiol. Bull.* 18:109.

Poulin, M., L. Berard-Therriault & A. Cardinal. 1984. Benthic diatoms from hard substrates of marine and brackish waters of Quebec Canada 3. Fragilarioideae, Fragilariales, Fragilariaceae. *Nat. Can.* (Que). 111:349.

Stevenson, R.J. 1984. How currents on different sides of substrates in streams affect mechanisms of benthic algal accumulation. *Int. Rev. ges. Hydrobiol.* 69:241.

Vymazal, J. 1984. Short-term uptake of heavy metals by periphytic algae. *Hydrobiologia* 119:171.

Austin, A. & J. Deniseger. 1985. Periphyton community changes along a heavy metals gradient in a long narrow lake. *Environ. Exper. Bot.* 25:41.

Flower, R.J. 1985. An improved epilithon sampler and its evaluation in two acid lakes. *Brit. Phycol. J.* 20:109.

Lamberti, G.A. & V.H. Resh. 1985. Comparability of introduced tiles and natural substrates for sampling lotic bacteria, algae, and macro-invertebrates. *Freshwater Biol.* 15:21.

Piekarczyk, R. & E. McArdle. 1985. Pioneer colonization and interaction of photosynthetic and heterotrophic microorganisms on an artificial substrate of polyurethane foam in E.J. Beck Lake, Illinois, USA. *Trans. Ill. State Acad. Sci.* 78:81.

10300 C.   Sample Analysis

1. Sedgwick-Rafter Counts

Remove periphyton from slides with a razor blade and rubber policeman. Disperse scrapings in 100 mL or other suitable volume of preservative with vigorous shaking, or use a blender. Transfer a 1-mL portion to a Sedgwick-Rafter cell, and make a strip count as described in Section 10200F.2a. If material in the Sedgwick-Rafter cell is too dense to count directly, discard and replace with a diluted sample.

Sedgwick-Rafter cells do not permit examination at magnifications higher than $200\times$. The Palmer cell[1], a thinner version of the S-R cell, permits examination at 400 to $500\times$ with a standard compound microscope.

Express counts as cells or filaments per square millimeter of substrate area, calculated as in ¶ C.2.

2. Inverted Microscope Method Counts

Using an inverted microscope for periphyton counts permits magnifications higher than those possible with the Sedgwick-Rafter cell. Remove periphyton quantitatively from slides with a razor blade and policeman. Transfer a measured portion, after serial dilution if necessary, into a standardized plankton sedimentation chamber. After a suitable period of settling (see Section 10200C.1), count organisms in the settling chamber by counting all organisms within a known number of strips or random fields. Calculate algal density per unit area of substrate as follows:

$$\text{Organisms/mm}^2 = \frac{N \times A_t \times V_t}{A_c \times V_s \times A_s}$$

where:

N = number of organisms counted,
A_t = total area of chamber bottom, mm^2,
V_t = total volume of original sample suspension, mL,
A_c = area counted (strips or fields), mm^2,
V_s = sample volume used in chamber, mL, and
A_s = surface area of slide or substrate, mm^2.

Separation of periphyton from silt and detritus may be enhanced by adding a drop or less of a saturated iodine solution to the counting chamber just before counting. This method is especially useful when Chlorophyta are the predominant organisms because iodine stains starch food reserves blue. Iodine can be added even to preserved samples.

3. Diatom Species Counts

Preparation of permanent diatom mounts from periphyton samples differs from preparation of mounts from plankton samples because of the need to remove extracellular organic matter (such as gelatinous materials). If this organic matter is not removed it will produce a thick brown or black carbonaceous deposit on the cover glass when the sample is incinerated. Clear organic matter by incineration by placing a small, known volume of sample (< 1 mL) directly on a cover slip. Let water evaporate and ash at 525°C (not more) for 6 to 10 min. Mount cover slip for direct examination of diatom frustules. Alternatively, decompose organic substances by oxidation with ammonium persulfate or with HNO$_3$ or 30% H$_2$O$_2$ and K$_2$Cr$_2$O$_7$ (see Section 10200D.3) before mounting sample. To oxidize with persulfate place a measured sample of approximately 5 mL in a disposable 10-mL vial. Let stand 24 h, withdraw supernatant liquid by aspiration, replace with a 5% solution of (NH$_4$)$_2$S$_2$O$_8$, and mix thoroughly. Do not exceed a total volume of 8 mL. Heat vial to approximately 90°C for 30 min. Let stand 24 h, withdraw supernatant liquid, and replace with reagent-grade water. After three changes of reagent-grade water, with a disposable pipet transfer a drop of the diatom suspension to a cover glass, evaporate to dryness, and prepare and count a mount as described for plankton (Section 10200). Count as least 500 frustules and express results as relative numbers or percentage of each species per unit area.

4. Stained Sample Preparation and Counting

Staining periphyton samples permits distinguishing algae from detritus and "live" from "dead" diatoms. This distinction is especially important because periphyton often contains many dead diatoms of planktonic as well as periphytic origin.

In the first method, cells are exposed to a vital stain to evaluate the percentages of live, senescent, and dead algae, particularly diatoms, by estimating relative metabolic activities. The colorless

tetrazolium violet is reduced in the cytochrome system of metabolically active cells to form violet-colored triphenylformazan.
When cells are senescent or dead, the reaction fails.

Make tetrazolium violet solution by adding 2.0 g tetrazolium
violet to 1.0 L water. The solution may be buffered to a pH of
7.5 to 7.7 with tris-hydroxymethyl amine. Add 1 mL tetrazolium
violet solution to 9 mL sample and incubate 2 to 4 h at room
temperature. Count diatom frustules and other cells (at least 300/
sample) and place into the following categories: a) active: violet
precipitate observed within the cell or mitochondria; b) senescent:
chlorophyll present, but no violet precipitate; c) dead: no chlorophyll or violet precipitate present.

In the second method, all algal components of periphyton may
be studied in one preparation, without sacrificing detailed diatom
taxonomy.[2] This method yields permanent slides for reference
collections.

Thoroughly mix preserved samples in the preservative solution.
Prepare acid fuchsin stain by dissolving 1 g acid fuchsin in 100
mL reagent-grade water, adding 2 mL glacial acetic acid, and
filtering. Place a measured sample in a centrifuge tube with 10 to
15 mL acid fuchsin stain. Mix sample and stain several times
during a 20-min staining period; centrifuge at 1000 g for 20 min.

Decant stain, being careful not to disturb sediment, or siphon
off supernatant. Add 10 to 15 mL 90% propanol, mix, centrifuge
for 20 min, and decant supernatant. Repeat using two washes of
100% propanol and one wash of xylene. Centrifuge, decant xylene, and add fresh xylene. At this stage, store sample in well-
sealed vials or prepare slides.

Slides for periphyton examinations require random dispersion
of a known amount of xylene suspension. Use a microstirrer to
break up clumps of algae before removing sample portion from
xylene suspension. Count a number of drops of suspended sample
into a thin ring of mounting medium* on a slide. Mix the xylene
suspension and medium with a spatula until the xylene has evaporated. Warm the slide on a hot plate at 45°C and cover sample
with a cover slip.

Count diatoms on the prepared slides using the magnification
most appropriate to the desired level of taxonomic identification.
Count strips or random fields. Calculate diatom density per unit
area of substrate:

$$\text{Organisms/area sampled} = \frac{N \times A_t \times V_t}{A_c \times V_s \times A_s}$$

where the terms are as defined in 10300C.2.

5. Dry and Ash-Free Weight

Collect at least three replicate slides for weight determinations.[3]
Slides air-dried in the field can be stored indefinitely if protected
from abrasion, moisture, and dust. Use slides expressly designated
for dry and ash-free weight analysis.

a. Equipment:

1) *Analytical balance,* with a sensitivity of 0.1 mg.

2) *Drying oven,* double-wall, thermostatically controlled to
within ±1°C.

3) *Electric muffle furnace* with automatic temperature control.

4) *Crucibles,* porcelain, 30-mL capacity.

---

* Naphrax®, Northern Biological Supply, 3 Betts Avenue, Martlesham Heath, Ips-
wich IP5 7RH, United Kingdom, or equivalent.

5) *Single-edge razor blades or rubber policeman.*

b. Procedure:

1) Dry slides to constant weight at 105°C, and ignite for 1 h
at 500°C. If weights are to be obtained from field-dried material,
re-wet dried material with reagent-grade water and remove from
slides with a razor blade or rubber policeman. Place scrapings
from each slide in a separate prewashed, prefired, tared crucible;
dry to constant weight at 105°C; cool in a desiccator and weigh;
and ignite for 1 h at 500°C.

2) Re-wet ash with reagent-grade water and dry to constant
weight at 105°C. This reintroduces water of hydration of clay and
other minerals, which is not driven off at 105°C but is lost during
ashing. If not corrected for, this water loss will be recorded as
volatile organic matter.[4]

c. Calculations: Calculate mean weight from slides and report
as dry weight [(crucible + sample weight at 105° C) minus (tare
weight of crucible)] per square meter of exposed surface. If 25-
by 75-mm slides are used, then

$$\text{g/m}^2 = \frac{\text{g/slide (average)}}{0.003\ 75}$$

Calculate ash weight for sample [(crucible + sample weight at
500°C) minus (tare weight of crucible)]. Subtract ash weight from
dry weight to obtain ash-free weight, and report as ash-free weight
per square meter of exposed surface.

6. Chlorophyll and Pheophytin

The chlorophyll content of attached algal communities is a useful index of the phytoperiphyton biomass. Quantitative chlorophyll determinations require the collection of periphyton from a
known surface area. Extract the pigments with aqueous acetone
or methanol (see Section 10200G) and use a spectrophotometer
or fluorometer for analysis. If immediate pigment extraction is
not possible, samples may be stored frozen for as long as 30 d if
kept in the dark.[5] The ease with which chlorophylls are removed
from cells varies considerably with different algae; to achieve
complete pigment extraction disrupt the cells mechanically with
a grinder, blender, or sonic disintegrator, or freeze them. Grinding
is the most rigorous and effective of these methods.

The Autotrophic Index (AI) is a means of determining the
trophic nature of the periphyton community (see Section
10200G). It is calculated as follows:

$$\text{AI} = \frac{\text{Biomass (ash-free weight of organic matter), mg/m}^2}{\text{Chlorophyll } a, \text{ mg/m}^2}$$

Normal AI values range from 50 to 200; larger values indicate
heterotrophic associations or poor water quality. Nonviable organic material affects this index. Depending on the community,
its location and growth habit, and method of sample collection,
there may be large amounts of nonliving organic material that
may inflate the numerator and produce disproportionately high AI
values. Nonetheless, the AI is an approximate means of describing
changes in periphyton communities between sampling locations.

a. Equipment and reagents: See Section 10200G.

b. Procedure: In the field, place individual glass microscope
slides used as substrates directly into 100 mL of a mixture of
90% acetone (water with 10% saturated $MgCO_3$ solution). Im-

mediately store on dry ice in the dark. (NOTE: Vinyl plastic is soluble in acetone. If vinyl plastic is used as the substrate, scrape periphyton from it before solvent extraction.) If extraction cannot be carried out immediately, freeze samples in the field and keep frozen until processed.

Rupture cells by grinding in a tissue homogenizer and steep in acetone for 24 h in the dark at or near 4°C.

To determine pigment concentration, follow the procedures given in Section 10200G.

c. Calculation: After determining pigment concentration in the extract, calculate amount of pigment per unit surface area of sample as follows:

$$\text{mg chlorophyll } a/\text{m}^2 = \frac{C_a \times \text{volume of extract, L}}{\text{area of substrate, m}^2}$$

where:

C_a is as defined in Section 10200G.

7. References

1. WETZEL, R.G. & G.E. LIKENS. 1991. Limnological Analyses, 2nd ed. Springer-Verlag, New York, N.Y.
2. OWEN, B.B., JR. 1977. The effect of increased temperatures on algal communities of artificial stream channels. Ph.D. dissertation, Univ. Alberta, Edmonton.
3. NEWCOMBE, C.L. 1950. A quantitative study of attachment materials in Sodon Lake, Michigan. *Ecology* 31:204.
4. NELSON, D.J. & D.C. SCOTT. 1962. Role of detritus in the productivity of a rock outcrop community in a piedmont stream. *Limnol. Oceanogr.* 7:396.
5. GRZENDA, A.R. & M.L. BREHMER. 1960. A quantitative method for the collection and measurement of stream periphyton. *Limnol. Oceanogr.* 5:190.

8. Bibliography

EATON, J.W. & B. MOSS. 1966. The estimation of numbers and pigment content in epipelic algal populations. *Limnol. Oceanogr.* 11:584.
MOSS, B. 1968. The chlorophyll *a* content of some benthic algal communities. *Arch. Hydrobiol.* 65:51.
CRIPPEN, R.R. & J.L. PERRIER. 1974. The use of neutral red and Evans blue for live-dead determinations of marine plankton. *Stain Technol.* 49:97.
OWEN, B.B., M. AFZAL & W.R. CODY. 1978. Staining preparations for phytoplankton and periphyton. *Brit. Phycol. J.* 13:155.
OWEN, B.B., M. AFZAL & W.R. CODY. 1979. Distinguishing between live and dead diatoms in periphyton communities. *In* R.L. Weitzel, ed. Methods and Measurements of Periphyton Communities: A Review. STP 690, American Soc. Testing & Materials, Philadelphia, Pa.
WETZEL, R.G., ed. 1983. Periphyton of Freshwater Ecosystems. Developments in Hydrobiology 17. Dr. W. Junk BV Publishers, The Hague, The Netherlands.
DELBECQUE, E.J.P. 1985. Periphyton on Nymphaeids: An evaluation of methods and separation techniques. *Hydrobiologia* 124:85.
TREES, C.C., M.C. KENNICUTT & J. M. BROOKS. 1985. Errors associated with the standard fluorometric determination of chlorophylls and phaeopigments. *Mar. Chem.* 17:1.

10300 D.   Primary Productivity

The productivity of periphyton communities is a function of water quality, substrate, and seasonal patterns in temperature and solar illumination. Productivity may be estimated from temporal changes in biomass (standing crop) or from the rate of oxygen evolution or carbon uptake.[1]

1. Biomass Accumulation

a. Ash-free dry weight: The accumulation rate of organic matter on artificial substrates by attachment, growth, and reproduction of colonizing organisms has been used widely to estimate the productivity of streams and reservoirs.[2,3] To use this method, expose several replicate clean substrates for a predetermined period, scrape the accumulated material from the slides, and ash as described previously.

$$P = \frac{\text{mg ash-free weight/slide}}{tA}$$

where:

P = net productivity, mg ash-free weight/m²/d,
t = exposure time, d, and
A = area of a slide, m².

Obtain estimates of seasonal changes in biomass of established communities by placing many replicate substrates at a sampling point and then retrieving a few at a time at regular intervals. Replace removed slides with new clean slides. The recommended collection interval ranges from 2 to 4 weeks for a year or longer.[2] Gain in ash-free weight per unit area from one collection period to the next is a measure of net production.

b. ATP estimates: Measurement of adenosine triphosphate (ATP) has been used in recent years to estimate microbial biomass in water. This technique is applicable to periphyton.[4] It provides an additional tool for assessing the magnitude and rate of biomass accumulation on substrates in natural waters. At present, the procedure should be limited to communities colonizing artificial substrates.

1) Equipment and reagents—See Section 10200H.5*a.*

2) Procedure—Either scrape periphyton from an exposed artificial substrate or, if standard glass microscope slides are used, place them in polyethylene slide mailers containing preheated (99°C) Tris buffer. Immerse in a boiling water bath for 10 min to extract ATP. If samples are not assayed immediately, freeze at −25°C; they may be stored in a freezer for up to several months. Complete analysis as directed in Section 10200H.5*b.* Slides exposed in waters containing high turbidity may collect substantial amounts of particulates including clays. ATP sorbs to these materials; the sorption results in a quenching effect.

3) Calculations—See Section 10200H.5*c.*

2. Standing Water Productivity Measured by Oxygen Method

Hourly and daily rates of oxygen evolution and carbon uptake by periphyton growing in standing water can be studied by confining this community briefly in bottles, bell jars, or other chambers. In contrast, the metabolism of organisms in flowing water is highly dependent on current velocity and cannot be determined with precision under static conditions. Productivity estimates for flowing waters and those for standing waters present different problems; therefore, separate procedures are given.

Productivity and respiration of epilithic and epipelic periphyton in littoral regions of lakes and ponds can be determined by inserting transparent and opaque bell jars or open-ended plastic chambers into substrata along transects perpendicular to the shoreline.[5,6] Chambers are left in place for one-half the daily photoperiod. The DO concentration in a chamber is determined at the beginning and end of the exposure period. Gross productivity is the sum of the net gain in DO in the transparent chamber and the oxygen used in respiration. Values obtained are doubled to estimate productivity for the entire photoperiod. Alternatively, determine the proportion of the incubation period of the total insolation during the photoperiod more accurately by measuring the insolation of the incubation period as a percentage of the total daily insolation. Both these methods assume that photosynthesis is proportional to irradiance (i.e., not light saturated and no photoinhibition).

Failure to account for changes in DO in chambers caused by phytoplankton photosynthesis and respiration may cause serious errors in the estimates of periphyton metabolism. It is essential that these values be obtained at the time the periphyton is studied by using the light- and dark-bottle method (see Section 10200I).

a. Equipment and reagents:

1) *Clear and darkened glass or plastic* chambers,* approximately 20 cm in diameter and 30 cm high, with a median lateral port, sealed with a serum bottle stopper for removal of small water samples for DO analyses or for the insertion of an oxygen probe. Fit the chamber with a small, manually operated, propeller-shaped stirring paddle.

2) *Dissolved oxygen probe, or equipment and reagents required for Winkler dissolved oxygen determinations:* See Section 4500-O.

b. Procedure: At each station place both a transparent and an opaque chamber over the substrate at sunrise or mid-daylight and leave in place for one-half the daily photoperiod. In extremely productive environments or to define the hourly primary productivity changes throughout the day, use incubation periods shorter than one-half the photoperiod. The minimum incubation period giving reliable results is 2 h. Determine DO concentration at the beginning of the incubation period.

Include a set of Gaarder-Gran light- and dark-bottle productivity and respiration measurements with each set of chambers to obtain a correction for phytoplankton metabolism. Incubate for the same time period as the chambers. See Section 10200I.

At end of exposure period, carefully mix the water in the chambers and determine DO concentration.

c. Calculations: When the exposure period is one-half of the photoperiod, calculate gross primary productivity of the peri-

---

* Plexiglas or equivalent.

phyton community as:

$$P_G = \frac{2[V_c(C'_{fc} - C'_{ic}) + V_o(C'_{io} - C'_{fo})]}{A}$$

where:

P_G = gross production, mg $O_2/m^2/d_{12h}$,
V_c = volume of clear chamber, L,
C'_{fc} and C'_{ic} = final and initial concentrations, respectively, of DO in the clear chamber, mg/L, corrected for phytoplankton metabolism,
V_o = volume of opaque chamber, L,
C'_{io} and C'_{fo} = initial and final concentrations, respectively, of DO in the opaque chamber, mg/L, corrected for phytoplankton metabolism, and
A = substrate area, m^2.

Correct for the effects of phytoplankton metabolism in the overall oxygen change in the clear chamber by the following equations:

$$C'_{fc} = C_{fc} - C_{flb}$$
$$C'_{ic} = C_{ic} - C_{ilb}$$
$$C'_{fo} = C_{fo} - C_{fdb}$$
$$C'_{io} = C_{io} - C_{idb}$$

where:

C_{fc} = final DO concentration in clear chamber, mg/L,
C_{flb} = final DO concentration in light bottle, mg/L,
C_{ic} = initial DO concentration in clear chamber, mg/L,
C_{ilb} = initial DO concentration in light bottle, mg/L,
C_{fo} = final DO concentration in opaque chamber, mg/L,
C_{fdb} = final DO concentration in dark bottle, mg/L,
C_{io} = initial DO concentration in opaque chamber, mg/L, and
C_{idb} = initial DO concentration in dark bottle, mg/L.

Calculate periphyton community respiration by:

$$R = \frac{24 V_o(C'_{io} - C'_{fo})}{tA}$$

where:

R = community respiration, mg $O_2/m^2/d_{24h}$, and
t = length of exposure, h.

Determine the net periphyton community productivity (P_N) as the difference:

$$P_N = P_G - R$$

If the incubation time is different from one-half the photoperiod, modify the daily gross production calculation as follows:

$$P_G = \frac{t_p[V_c(C'_{fc} - C'_{ic}) + V_o(C'_{io} - C'_{fo})]}{tA}$$

where:

t_p = length of the daily photoperiod, h.

Community respiration and net production calculations for incubation periods other than one-half the photoperiod are not changed.

3. Standing Water Productivity Measured by Carbon-14 Method

The approach is similar to that described above for the oxygen method. Transparent and opaque chambers are placed over the substrate, carbon-14-labeled Na_2CO_3 is injected into the chamber by syringe, mixed well, and allowed to incubate with the periphyton for one-half the photoperiod. The concentration of dissolved inorganic carbon available for photosynthesis is determined by titration. At the end of the incubation period, the periphyton is removed from the substrate and assayed for carbon-14.[5]

a. *Equipment and reagents:*

1) *Incubation chamber:* See Section 10300D.2a.

2) *Special equipment and reagents:* See Section 10200I.

3) *Carbon-14-labeled solution of sodium carbonate,* having a known specific activity of approximately 10 μCi/mL.

4) *Other equipment and reagents:* See Section 4500-CO_2.

b. *Procedure:* At each station place a transparent and opaque chamber over the substrate and add approximately 10 μCi carbon-14/L of chamber volume. Mix water in the chambers well, taking care to avoid disturbing the periphyton. Determine concentration of dissolved inorganic carbon as described in Section 2320. At end of exposure period, remove surface centimeter of periphyton and sediment enclosed in the chamber, freeze, and store frozen in a vacuum desiccator.

Immediately before analysis, expose sample to fumes of HCl for 10 to 15 min to drive off all inorganic carbon-14 retained in the periphyton. Combust sample (or portion) by the Van Slyke method[6] or oxidize by heating in a closed system. Collect all CO_2 for radioassay either by flushing CO_2 into a two-vial train of ethanolamine (2-aminoethanol) or alternative CO_2 absorber, such as methoxyethanol (1:7)[7] or flushing CO_2 produced by combustion into a gas-flow counter or electrometer. Alternatively, extract known amounts of periphyton biomass with a tissue solubilizer,† using, for example, 1.0 mL in closed vials at 60°C for 48 h.[8] Radioassay subsamples (100-μL) by liquid scintillation.

c. *Calculations:*

$$P_N = {}^{12}C \text{ available} \times \frac{{}^{14}C \text{ assimilated} \times \text{conversion factors}}{{}^{14}C \text{ available (added)}}$$

$$P_N = \frac{(a)\ (b)\ (d)\ (e)}{(c)}$$

where:

P_N = net primary productivity per unit area of substrate per unit time, mg C/m²/d,

a = ${}^{12}C$ available = dissolved inorganic carbon, mg ${}^{12}C$/L = (total alkalinity − phenolphthalein alkalinity) × 0.240[6] = mg ${}^{12}C$/L,

b = ${}^{14}C$ assimilated = [(radioactivity of sample in light chamber × k_1) − (background activity of dark chamber × k_2)] × (isotope effect, 1.06). Express radioactivity as disintegrations per second (dps), i.e., counts per second corrected to 100% radioassay counter efficiency.

k_1 = correction factor to convert individually different light-chamber volumes to 1 L,

k_2 = correction factor to convert individually different dark-chamber volumes to 1 L,

1.06 = isotope effect to correct for slightly greater mass of ${}^{14}C$ than of ${}^{12}C$, which results in a 6% slower assimilation rate,

---

† Beckman BTS-450 or equivalent.

c = ${}^{14}C$ available = ${}^{14}C$ activity added = (μCi ${}^{14}C$ added) × (disintegrations of ${}^{14}C$/s/μCi) = 3.7 × 10⁴ μCi ${}^{14}C$ added, mL,

d = a dimensional factor to convert area of substrate sampled to m², and

e = factor to expand incubation period to the total daylight period. After integration by planimetry or electronic digitizer of the total amount of insolation for the day, determine percentage of total represented by the incubation period.

4. Flowing Water Productivity Measured by Oxygen Method

Primary productivity of the periphyton community in a stream or river ecosystem can be related to changes in DO. These changes are the integrated effects of photosynthesis, affected by light levels and turbidity, that occur during the photoperiod by stream phytoplankton, periphyton, and the submerged portions of macrophytes. Respiration results from metabolism of plant communities, aquatic animals, and attached and free-floating microbial heterotrophs. Water depth, turbulence, and water temperature all influence the process of reaeration. Oxygen also can enter by accrual of groundwater and surface waters. Daily fluctuations in photosynthetic production of oxygen are imposed on the relatively steady demand of respiratory activity. However, this latter process may fluctuate greatly in streams receiving a significant load of organic wastes, particularly under intermittent loads such as oxygen demand from urban stormwater runoff. Respiration rates also may vary diurnally under certain conditions, but the factors involved are not well understood.

The rate of change in stream DO (q) in grams per cubic meter per hour is represented by the following function of the photosynthetic rate (p), respiration (r), reaeration (d), and accrual from groundwater inflow and surface runoff (a):[9]

$$q = p - r + d + a$$

If the equation is multiplied through by depth in meters (z), the resulting values are in terms of grams oxygen per square meter per hour. Figure 10300:2 illustrates this conceptual relationship between q, primary productivity, and respiration of the stream plant community.

The procedure measures the time-variable oxygen concentrations in a stream over a 24-h period. Compensations are made for oxygen changes due to physical factors (accrual and reaeration) and the rate of oxygen change due to biological activity that is separated into components due to respiration and primary production. The metabolic rates are the sum of the activity of the entire stream community. Planktonic productivity and respiration can be separated from overall community activity by the use of the light- and dark-bottle oxygen technique (see Section 10200I). However, in most small streams planktonic production is insignificant. The component of production and respiration due to macrophytes is very difficult to separate from periphytic metabolic activity in systems where vascular plants are common.

Because periphyton attach to plant surfaces as well as nonliving substrates, radiotracer techniques are required to separate the component of production due to macrophytes from that due to attached algae.[10] When vascular plants are present use techniques discussed in Section 10400 to estimate their contribution to net primary productivity.

Respiration by fish and benthic fauna also is difficult to quantitate directly and usually is not separated from periphyton res-

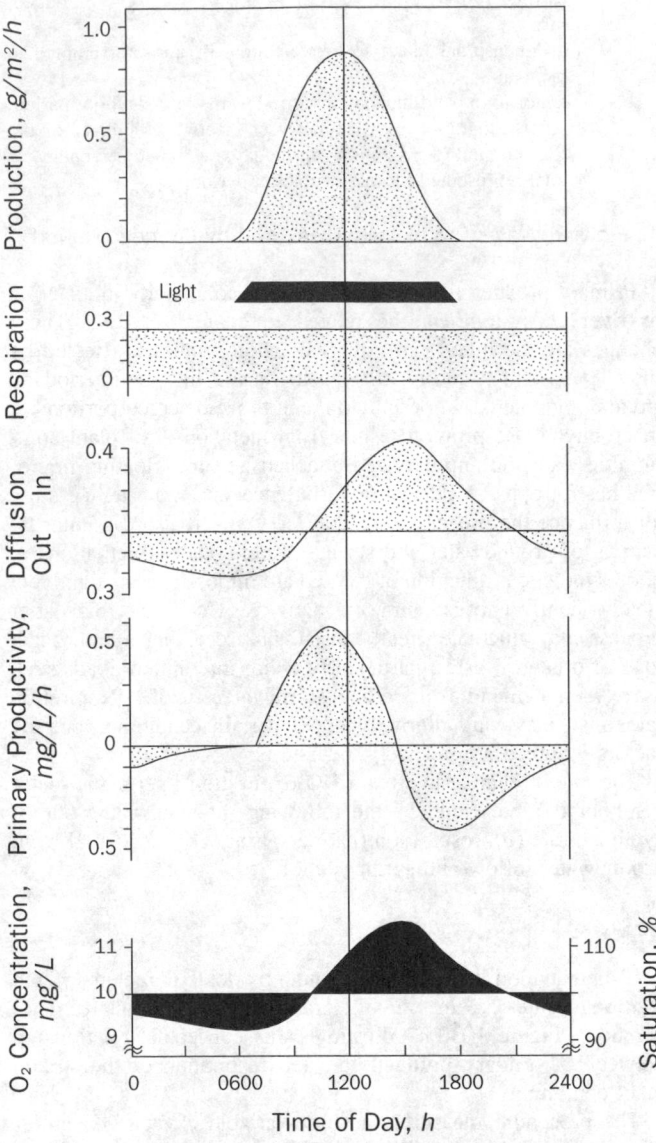

Figure 10300:2. Component processes in the oxygen metabolism of a section of a hypothetical stream during the course of a cloudless day. Production (P), respiration (R), and diffusion (D) are given on an areal basis. The combined effect of these rate processes for a stream 1 m deep is given in mg/L/h (q). The actual oxygen values that would result in a stream with a long homogeneous community are given in the lowermost curve. Source: ODUM, H.T. 1956. Primary production in flowing waters. *Limnol. Oceanogr.* 1:102.

piration. If compartmentalized animal metabolism is required, calculate this contribution from laboratory respiration rates extrapolated to the field situation based on animal population sizes.[11,12]

Estimate primary productivity in flowing water by either the free water demand method or the chamber method.[13,14] The first does not introduce artificiality to the system; however, it is difficult to separate the components of metabolic activity except for

the contribution due to plankton. The chamber method measures periphyton activity alone.[15–19]

Depending on the hydrologic characteristics of the stream system, accrual and reaeration may be significant. Accrual can be accounted for by simple mixing equations if estimates of the accrued flow and its oxygen concentration are known. In practice, select for study reaches that do not incur significant accrual. Measure reaeration rates either directly[15–18] or by estimation from physical and hydrodynamic features of the stream itself.[17,18]

a. Equipment:

1) *BOD bottles,* for light- and dark-bottle measurements. See Section 10200I.

2) *DO meter and probe* for measurement of DO.

3) *Bottom chamber,* 60 × 20 × 10 cm, with 32-cm lengthwise dividing baffle, rheostat-controlled submersible pump, temperature thermistor, and DO probe.[14] Use clear and opaque plastic sleeves for covering chamber and petri dishes or other means of placing periphyton within chambers.

4) *Current meter,* capable of detecting water current velocities ranging from 0.03 to 3 m/s in water depths as shallow as 0.3 m.

5) *Tape measure (30 m) and depth staff,* or similar equipment, as required to measure stream cross sections.

6) *Fluorometer,* capable of detecting fluorescent dye concentration at 0.5 to 100 µg/L (required only if direct measurement of reaeration is made).

7) *Liquid scintillation counter,* capable of sensitive detection of ^{85}Kr and ^{3}H (required only if direct measurement of reaeration is made).

b. Procedure:

1) Light- and dark-chamber method—Grow samples of typical periphyton communities on artificial substrate or collect natural material. Transfer identical portions to both clear and opaque chambers, taking care to use sufficient periphyton to make the ratio of chamber volume to periphyton area equivalent to the ratio of stream volume to periphyton substrate area. Measure current in the stream and match the circulation rate in the clear and opaque chambers to the current. Measure DO concentrations initially in both clear and opaque chambers and after 1 to 3 h to estimate the rate of oxygen increase or decrease. Make concurrent measurements of phytoplankton activity using light- and dark-bottle techniques as described in Section 10200I.2. Incubate light and dark bottles for the same time interval as the chambers.

Make several measurements during the photoperiod to define daily primary productivity. In addition, collect sufficient natural substrate samples of the study reach to estimate periphyton biomass (see Section 10300B). At end of incubation period harvest enclosed periphyton and determine ash-free biomass (see Section 10300B).

2) Free-water diurnal curve methods—Measure, hourly or continuously, DO concentration and water temperature for a 24-h period at one or two stations, depending on stream conditions, precision desired, and availability of equipment. If similar conditions exist for some distance upstream from the reach being studied, diurnal measurements of DO at a single station are sufficient to estimate productivity. Where upstream conditions are significantly different from those in the reach being studied, make measurements at the upstream and downstream limits of the reach.

If the single-station method is used, measure depth at several points along the study reach to define average depth. Map and/or make physical surveys to estimate magnitude of possible sources of accrual via effluents or tributary streams and springs. If the two-station method is used, measure the wetted cross-sectional stream area as well as current velocity at several points to define flow (in cubic meters per second) and average cross-sectional area. Correct for phytoplankton activity by light- and dark-bottle measurements (see Section 10200I.2).

3) *Direct measurement of reaeration*[16]—Under special circumstances it may be desirable to estimate reaeration directly although the results may not be more accurate than those of the empirical formulations usually used. The tracer gas technique is satisfactory, but is difficult and requires sophisticated equipment not routinely available. Use this method with care and with full recognition of its restrictions. Depending on stream flow, release 10 to 250 μCi ^{85}Kr with 5 to 125 μCi ^{3}H at the upstream end of the reach together with sufficient fluorescent dye to produce a concentration of 10 μg/L when completely mixed across the river cross section. Make fluorometric measurements at the downstream end of the reach until the dye peak appears, then collect water samples to measure the ^{85}Kr/^{3}H ratio by liquid scintillation techniques. Record time of travel for the dye peak from the injection point.

c. Calculations:

1) *Chamber method*—Calculation is analogous to that used for the bell jar technique discussed in Section 10300D.2.

$$P_n = \frac{V_c(C'_{fc} - C'_{ic})B}{tW_c}$$

where:

    P_n = hourly rate of net primary production, mg O$_2$/m^2/h,
    V_c = volume of clear chamber, L,
    B = average periphyton biomass estimated for the study reach, mg/m^2,
    t = incubation period, h,
    W_c = total biomass of periphyton contained in clear chamber, mg,
    C'_{fc} = final oxygen concentration in clear chamber, corrected for phytoplankton metabolism, mg/L:
        $C'_{fc} = C_{fc} - C_{flb}$
        C_{fc} = final DO in clear chamber,
        C_{flb} = final DO in light bottle, and
    C'_{ic} = initial oxygen concentration in clear chamber corrected for light-bottle measurement, mg/L:
        $C'_{ic} = C_{ic} - C_{ilb}$
        C_{ic} = initial DO in clear chamber, and
        C_{ilb} = initial DO in light bottle.

$$r = \frac{V_o(C'_{io} - C'_{fo})B}{tW_o}$$

where:

    r = hourly periphyton respiration rate, mg O$_2$/m^2/h,
    V_o = volume of opaque chamber, L,
    B = average periphyton biomass for the study reach, mg/m^2,
    W_o = total biomass of periphyton contained in opaque chamber, mg,
    C'_{io} = initial oxygen concentration in opaque chamber, corrected for phytoplankton respiration, mg/L:
        $C'_{io} = C_{io} - C_{idb}$
        C_{io} = initial DO in opaque chamber, mg/L,
        C_{idb} = initial DO in dark bottle, mg/L, and
    C'_{fo} = final oxygen concentration in opaque chamber, mg/L:
        $C'_{fo} = C_{fo} - C_{fdb}$
        C_{fo} = final DO in opaque chamber, mg/L, and
        C_{fdb} = final DO in dark bottle, mg/L.

For each pair of chamber measurements,

$$P_g = P_n + r$$

where:

    P_g = hourly gross periphytic primary production, mg O$_2$/m^2/h.

P_G is the area under the curve of primary production per hour through the photoperiod, mg O$_2$/m^2/d (see Figure 10300:3). Also,

$$R = \left(\frac{\sum_1^n r_n}{n}\right) \times 24$$

where:

    R = total periphyton community respiration, mg O$_2$/m^2/d, and
    n = number of observations.

Thus,

$$P_N = P_G - R$$

where:

    P_N = net periphytic production, mg O$_2$/m^2/d.

2) *Free water methods*

a) *Calculation of reaeration or diffusion for both the single and upstream-downstream methods*—Calculate k_2 from radio-tracer data as follows:

$$K_{Kr} = \frac{-1}{t}\ln\frac{(C_{Kr}/C_H)_d}{(C_{Kr}/C_H)_u}$$

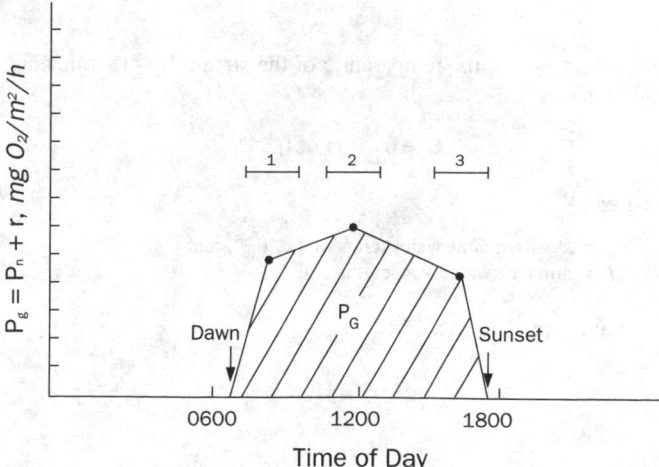

Figure 10300:3. Gross periphytic primary production (P_G) determined by the O'Connell-Thomas Chamber. P_G is the area under the curve obtained by graphical integration planimetry. Each point is the run $P_g = P_n + r$ for incubation periods 1, 2, and 3, which are denoted by the indicated lines.

and

$$k_2 = \frac{K_{kr}}{0.83}$$

where:

k_2 = reaeration coefficient (base e), d^{-1},
K_{Kr} = base e transfer coefficient for ^{85}Kr, d^{-1},
t = time of travel, d,
$(C_{Kr}/C_H)_u$ = ratio of released radioactivities (μCi/mL) ^{85}Kr to ^3H at the upstream station, and
$(C_{Kr}/C_H)_d$ = ratio of radioactivities (μCi/mL) ^{85}Kr to ^3H at the downstream station.

The reaeration coefficient also can be calculated from an equation relating the rate of energy dissipation in a stream to k_2.[16,17]

$$k_2 = K \frac{\Delta h}{t}$$

where:

K = escape coefficient,
Δh = change in water surface elevation in a stream reach, and
t = time of flow through a stream reach.

This can be expressed in terms of hydrodynamic and physical data:

$$k_{2_{20}} = K' \frac{\Delta H}{\Delta X} \times V$$

where:

K' = 28.3×10^3 s/m · d for stream flows between 0.028 and 0.28 m^3/s; 21.3×10^3 s/m · d for stream flows between 0.28 and 0.56 m^3/s; and 15.3×10^3 s/m · d for stream flows above 0.56 m^3/s,
$k_{2_{20}}$ = reaeration coefficient, d^{-1}, at 20°C,
$\frac{\Delta H}{\Delta X}$ = slope, m/km, and
V = velocity, m/s.

Convert $k_{2_{20}}$ to the temperature of the stream by the following equation:

$$k_{2t} = k_{2_{20}} \, (1.024)^{(T-20)}$$

where:

k_{2t} = k_2 at ambient water temperature, d^{-1}, and
T = ambient water temperature, °C.

Convert to D in mg/L/h:

$$D = \frac{k_{2t}C_s}{24}$$

where:

C_s = oxygen concentration at saturation at ambient stream temperatures, mg/L.

b) *Single-station method—Calculation of primary productivity and respiration from diurnal oxygen and temperature measure*

ments at a single station is summarized in Figure 10300:4 and Table 10300:I.

Tabulate hourly DO measurements and temperatures. Determine C_s (DO of air-saturated H_2O at each temperature from Table 4500-O:I) and compute uncorrected DO consumption, milligrams per liter per hour, for each period:

$$\Delta DO_{\substack{hours \\ 1\ to\ 2}} = DO_{hour\ 2} - DO_{hour\ 1}$$

Plot on the half hour, as shown in Figure 10300:4b.

Calculate the net primary production and respiration of phytoplankton as shown in Section 10200I. Determine the 24-h average hourly plankton respiration, $(\sum_1^n r_p)/n$, in milligrams per liter per hour every half hour. Calculate the hourly net phytoplankton production and tabulate for the approximate hours during the photoperiod. Plot as shown on Figure 10300:4c.

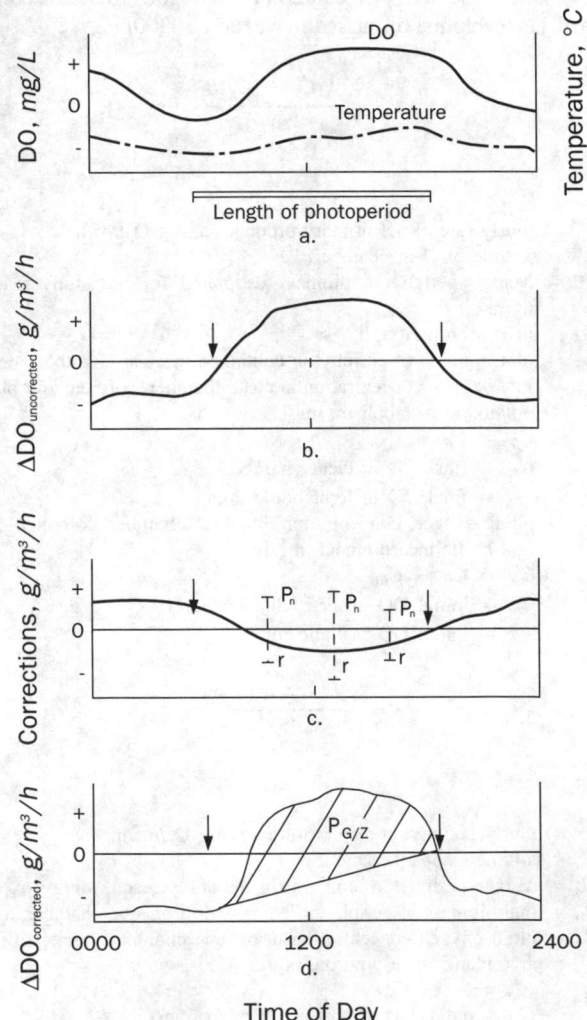

Figure 10300:4. Calculation of gross primary production at a single station. P_g, g O_2/m^2/h = area of corrected rate of change curve integrated for the length of the photoperiod multiplied by average water depth, z, for the reach in meters.

TABLE 10300:I. SAMPLE CALCULATION LEDGER FOR COMPUTATION OF CORRECTED RATE OF OXYGEN CHANGE FROM A SINGLE-STATION DIURNAL CURVE

Time h	DO mg/L	Water Temp. °C	C_s* mg/L	Uncorrected ΔDO† mg/L/h	P_p‡ mg/L/h	R_p§ mg/L/h	k^2 d^{-1}	D mg/L/h	Corrected $\Delta DO\|$ mg/L/h
Midnight									
0030									
0100									
0230									
.									
.									
Noon									
1230									
1300									
.									
.									
Midnight									

* DO concentration at 100% saturation for a given water temperature, from Table 4500-O:I.
† Hourly rate of change of DO. For example, for noon to 1300, $DO_{1200-1300} = DO_{1300} - DO_{1200}$; plot at 1230.
‡ Phytoplankton net production.
§ Phytoplankton respiration rate.
$\|$ $\Delta DO_{corrected} = \Delta DO_{uncorrected} - D - P_p - R_p$

Calculate and tabulate k_{2t} and substitute D for each C_s, as outlined in ¶ a), above. Plot as shown in Figure 10300:4c.

Correct each ΔDO for diffusion and phytoplankton metabolism:

$$\Delta DO_{corrected}, \text{mg/L/H} = \Delta DO_{uncorrected} - D - P_p - R_p$$

Plot each point as shown in Figure 10300:4d.

Gross primary productivity of the benthic and attached algal populations is computed as the area under the curve in Figure 10300:4d from sunrise to sunset. This is primary production in grams per cubic meter per day. Multiply by the average depth for a reach, z meters, to obtain P_G in grams per square meter per day. Calculate community respiration:

$$R = 24 \, z \, F$$

where:

R = community respiration, g/m²/d,
z = depth, m, and
F = average hourly ΔDO for the dark period (without regard to sign), mg/L/h.

Calculate net primary productivity P_N as:

$$P_N = P_G - R$$

c) Upstream-downstream method—Calculation of primary productivity and respiration for a stream reach from upstream and downstream pairs of diurnal curves of oxygen and water temperature is summarized in Figure 10300:5 and Table 10300:II. Alternatively, calculate as below, with oxygen change expressed as

TABLE 10300:II. SAMPLE CALCULATION LEDGER FOR COMPUTATION OF CORRECTED RATES OF OXYGEN CHANGE FROM THE UPSTREAM-DOWNSTREAM DIURNAL CURVES OF OXYGEN CONCENTRATION AND TEMPERATURE

Time h	DO mg/L Upstream	Downstream	Uncorrected ΔDO mg/L	Water Temp. °C	C_s* mg/L	P_p† mg/L	R_p‡ mg/L	k_2 d^{-1}	Corrected $\Delta DO\|$ mg/L
Midnight									
0100									
0200									
.									
.									
Noon									
1300									
.									
.									
Midnight									

* DO concentration at 100% saturation for a given water temperature, from Table 4500-O:I.
† Change in oxygen concentration in the light bottle per hour multiplied by travel time between the upstream and downstream station.
‡ Change in oxygen concentration in the dark bottle multiplied by travel time between the upstream and downstream station.
$\|$ $\Delta DO_{corrected} = \Delta DO_{uncorrected} - D - P_p - R_p$

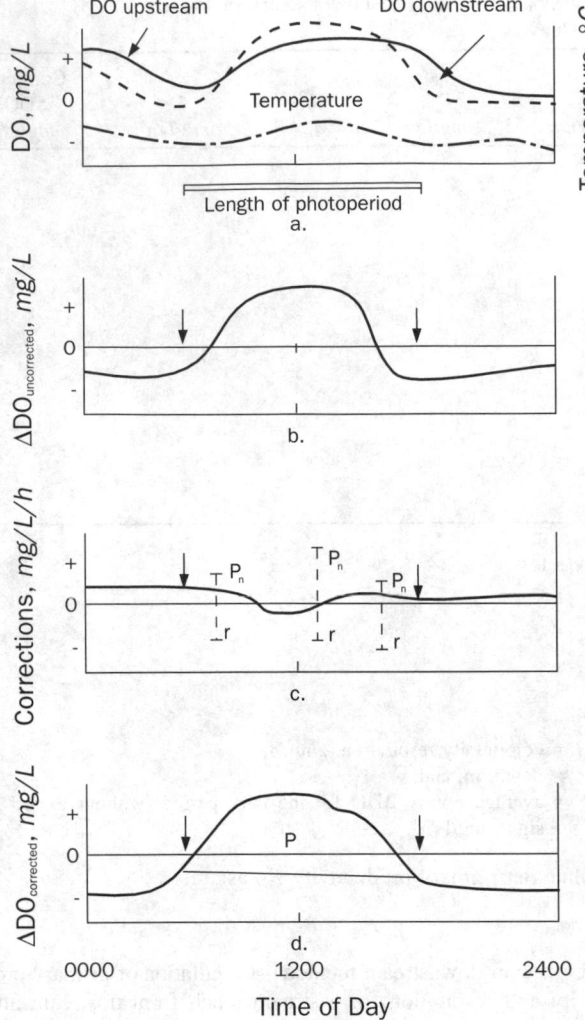

Figure 10300:5. Calculation of gross periphytic primary productivity from upstream-downstream diurnal curves. *P* is the area under the corrected rate of change graph.

the difference between stations rather than as change per hour. The calculations are analogous. Multiply the area under a curve of oxygen change between two stations, corrected for diffusion and plankton metabolism and expressed in milligrams per liter, by the discharge in cubic meters per hour, and divide by the water surface area between the two stations. This, multiplied by 24, yields gross primary productivity in grams per square meter per day.

To compute gross primary productivity by this method, tabulate upstream and downstream DO and average water temperature for the reach at each hour. Calculate ΔDO between upstream and downstream stations for each hour as

$$\Delta DO = DO_{downstream} - DO_{upstream}$$

Tabulate C_s and determine the planktonic activity. Correct for planktonic respiration by relating average hourly dark bottle DO change to the time of travel in the stream reach; correct for plank-

tonic production by the hourly change in DO in the light bottle times the time of travel (see Table 10300:II).

Calculate or tabulate k_2 and convert it to the total oxygen diffusion for the reach. Because diffusion, D, is expressed as milligrams per liter per hour, multiply it by the travel time to obtain the diffusion correction.

Correct each hourly upstream-downstream ΔDO as shown in Table 10300:II. Integrate the area under this ΔDO curve from sunrise to sunset to give P as in Figure 10300:5d.

$$P_G, \text{ g/m}^2/\text{d} = \frac{Q}{A} P$$

where:

Q = flow, m³/h, and
A = reach area, m² (average reach width × reach length).

$$\text{Respiration, } R, \text{ g O}_2/\text{m}^2/\text{d} = \frac{\Delta DO_{dark} \times Q \times 24}{A}$$

and

$$\text{Net production } P_N = P_G - R$$

5. References

1. VOLLENWEIDER, R.A., ed. 1969. A Manual on Methods for Measuring Primary Production in Aquatic Environments. IBP Handbook No. 12. F.A. Davis Co., Philadelphia, Pa.
2. SLÁDEEK, V. & A. SLÁDEČKOVA. 1964. Determination of periphyton production by means of the glass slide method. *Hydrobiologia* 23: 125.
3. KING, D.L. & R.C. BALL. 1966. A qualitative and quantitative measure of *aufwuchs* production. *Trans. Amer. Microsc. Soc.* 82:232.
4. CLARK, J.R., D.I. MESSENGER, K.L. DICKSON & J. CAIRNS, JR. 1978. Extraction of ATP from *aufwuchs* communities. *Limnol. Oceanogr.* 23:1055.
5. WETZEL, R.G. 1963. Primary productivity of periphyton. *Nature* 197: 1026.
6. WETZEL, R.G. 1964. A comparative study of the primary production of higher aquatic plants, periphyton, and phytoplankton in a large shallow lake. *Int. Rev. ges. Hydrobiol.* 49:1.
7. LOEB, S.L. 1981. An in situ method for measuring the primary productivity and standing crop of the epilithic periphyton community in lentic systems. *Limnol. Oceanogr.* 26:394.
8. BEER, S., A.J. STEWART & R.G. WETZEL. 1982. Measuring chlorophyll *a* and ¹⁴C-labeled photosynthate in aquatic angiosperms by use of a tissue solubilizer. *Plant Physiol.* 69:54.
9. ODUM, H.T. 1956. Primary production in flowing waters. *Limnol. Oceanogr.* 1:102.
10. ALLEN, H.L. 1971. Primary productivity, chemo-organotrophy, and nutritional interactions of epiphytic algae and bacteria or macrophytes in the littoral of a lake. *Ecol. Monogr.* 41:97.
11. HALL, C.A.S. 1972. Migration and metabolism in a temperate stream ecosystem. *Ecology* 53:585.
12. NIXON, S.W. & C.A. OVIATT. 1974. Ecology of a New England salt marsh. *Ecol. Monogr.* 43:463.
13. McINTIRE, C.D., R.L. GARRISON, H.K. PHINNEY & C.E. WARREN. 1964. Primary production in laboratory streams. *Limnol. Oceanogr.* 9:92.
14. THOMAS, N.A. & R.L. O'CONNELL. 1966. A method for measuring primary production by stream benthos. *Limnol. Oceanogr.* 11:386.

15. COPELAND, B.J. & W.R. DUFFER. 1964. Use of a clear plastic dome to measure gaseous diffusion rates in natural waters. *Limnol. Oceanogr.* 9:494.

16. TSIVOGLOU, E.C. & L.A. NEAL. 1976. Tracer measurement of reaeration. III. Predicting the capacity of inland streams. *J. Water Pollut. Control Fed.* 48:2669.

17. GRANT, R.S. 1976. Reaeration-coefficient measurements of 10 small streams in Wisconsin. U.S. Geol. Surv. Water Resources Publ. 76-96.

18. ODUM, H.T. & C.M. HOSKIN. 1958. Comparative studies of the metabolism of marine water. *Publ. Inst. Mar. Sci. Univ. Tex.* 4:115.

19. BOTT, T.L., J.T. BROCK, C.E. CUSHING, S.V. GREGORY, D. KING & R.C. PETERSEN. 1978. A comparison of methods for measuring primary productivity and community respiration in streams. *Hydrobiologia* 60:3.

6. Bibliography

POMEROY, L.R. 1959. Algal productivity in salt marshes. *Limnol. Oceanogr.* 4:386.

CASTENHOLZ, R.W. 1961. An evaluation of a submerged glass method of estimating production of attached algae. *Verh. Int. Ver. Limnol.* 14:155.

WHITFORD, L.A. & G.J. SCHUMACHER. 1964. Effect of a current on respiration and mineral uptake in *Spirogyra and Oedogonium*. *Ecology* 45:168.

DUFFER, W.R. & T.C. DORRIS. 1966. Primary productivity in a southern Great Plains stream. *Limnol. Oceanogr.* 11:143.

McINTIRE, C.D. 1966. Some factors affecting respiration of periphyton communities in lotic environments. *Ecology* 47:918.

CUSHING, C.E. 1967. Periphyton productivity and radionuclide accumulation in the Columbia River, Washington, USA. *Hydrobiologia* 29:125.

HANSMANN, E.W., C.B. LANE & J.D. HALL. 1971. A direct method of measuring benthic primary production in streams. *Limnol. Oceanogr.* 16:822.

SCHINDLER, D.W., V.E. FROST & R.V. SCHMIDT. 1973. Production of epilithiphyton in two lakes of the experimental lakes area, northwestern Ontario. *J. Fish. Res. Board Can.* 30:1511.

NORTH AMERICAN BENTHOLOGICAL SOCIETY. 1974–1990 (annual). Current and Select Bibliographics on Benthic Biology. Springfield, Ill.

WETZEL, R.G. & G.E. LIKENS. 1991. Limnological Analyses, 2nd ed. Springer-Verlag, New York, N.Y.

10300 E.   Interpreting and Reporting Results

Although several systems have been developed to organize and interpret periphyton data, no single method is universally accepted. The methods may be qualitative or quantitative. Qualitative methods deal with the taxonomic composition of the communities in zones of pollution, whereas quantitative methods deal with community structure using diversity indices, similarity indices, and numerical indices of saprobity.

1. Qualitative Methods (Indicator Species and Communities)

The saprobity system developed by Kolkwitz and Marsson is a widely used method of interpreting periphyton data. This scheme divides polluted stream reaches into polysaprobic, α and ß mesosaprobic, and oligosaprobic zones, and lists the characteristics of each. The system has been refined[1,2] and enlarged by Fjerdingstad[3,4] and Sládeček.[5,6]

Evaluation of the saprobity system requires microscopic evaluation of living indicator biota, particularly for the sensitive sessile protozoans. Glass slides and other transparent substrates are advantageous because they permit direct microscopic examination and identification. Removal of periphyton from slides and preservation for subsequent examination may be acceptable for diatoms, but observation of preserved material is not acceptable for most flagellated organisms.

2. Quantitative Methods

These methods use cell counts per unit area of substrate and numerical indices of pollution or water quality. Considerable data on cell densities and species composition of periphyton collected on glass slides in polluted rivers in England are available.[7]

Other indices include the Shannon-Weaver,[8] Simpson's,[9] and Pinkham-Pearson.[10] The saprobity system[11] also may be used

where code numbers assigned for the saprobial value and the abundance of individual species are used to calculate a Mean Saprobial Index. Results also may be expressed by the truncated-log normal distribution of diatom species[12,13] as well as the Autotrophic Index (AI).[14]

3. References

1. KOLKWITZ, R. 1950. Oekologie der saprobien. *Ver Wasser-, Boden, Lufthyg. Schriftenreihe* (Berlin) 4:1.

2. LIEBMANN, H. 1951. Handbuch der Frischwasser und Abwasserbiologie. Bd. I. Oldenbourg, Munich, Germany.

3. FJERDINGSTAD, E. 1964. Pollution of streams estimated by benthal phytomicroorganisms. I. A saprobic system based on communities of organisms and ecological factors. *Int. Rev. ges. Hydrobiol.* 49:63.

4. FJERDINGSTAD, E. 1965. Taxonomy and saprobic valency of benthic phytomicroorganisms. *Int. Rev. ges. Hydrobiol.* 50:475.

5. SLÁDEČEK, V. 1966. Water quality system. *Verh. Int. Ver. Limnol.* 16:809.

6. SLÁDEČEK, V. 1973. System of water quality from the biological point of view. *Arch. Hydrobiol. Ergebn. Limnol.* 7:1.

7. BUTCHER, R.W. 1946. Studies in the ecology of rivers. VI. The algal growth in certain highly calcareous streams. *J. Ecol.* 33:268.

8. SHANNON, C.E. 1948. The Mathematical Theory of Communication. Univ. Illinois Press, Urbana.

9. SIMPSON, E.H. 1949. Measurement of diversity. *Nature* 163:688.

10. PINKHAM, C.F.A. & J.G. PEARSON. 1976. Applications of a new coefficient of similarity to pollution surveys. *J. Water Pollut. Control Fed.* 48:717.

11. PANTLE, R. & H. BUCK. 1955. Die biologische überwachung der Gewasser und der Darstellung der Ergebnisse. *Gas-Wasserfach* 96:604.

12. PATRICK, R., M.H. HOHN & J.H. WALLACE. 1954. A new method for determining the pattern of the diatom flora. *Bull. Philadelphia Acad. Natur. Sci.* 259:1.

13. PATRICK, R. 1973. Use of algae, especially diatoms, in the assessment of water quality. *In* J. Cairns, Jr., ed. Biological Methods for the

Assessment of Water Quality. ASTM STP 528, American Soc. Testing & Materials, Philadelphia, Pa.

14. WEBER, C. 1973. Recent developments in the measurement of the response of plankton and periphyton to changes in their environment. *In* G. Glass, ed. Bioassay Techniques and Environmental Chemistry. Ann Arbor Science Publ., Ann Arbor, Mich.

4. Bibliography

FJERDINGSTAD, F. 1950. The microflora of the River Mølleaa, with special reference to the relation of the benthal algae to pollution. *Folia Limnol. Scand.* 5:1.

BLUM, J.L. 1956. The ecology of river algae. *Bot. Rev.* 22:291.

YOUNT, J.L. 1956. Factors that control species numbers in Silver Springs, Florida. *Limnol. Oceanogr.* 1:286.

BUTCHER, R.W. 1959. Biological assessment of river pollution. *Proc. Linnean Soc. London* 170:159.

HOHN, M.H. 1959. The use of diatom populations as a measure of water quality in selected areas of Galveston and Chocolate Bay, Texas. *Publ. Inst. Mar. Sci. Univ. Tex.* 5:206.

HOHN, M.H. 1961. Determining the pattern of the diatom flora. *J. Water Pollut. Control Fed.* 33:48.

PATRICK, R. 1963. The structure of diatom communities under varying ecological conditions. *Ann. N.Y. Acad. Sci.* 108:359.

SCHLICHTING, H.E., JR. & R.A. GEARHEART. 1966. Some effects of sewage effluent upon phyco-periphyton in Lake Murray, Oklahoma. *Proc. Okla. Acad. Sci.* 46:19.

SLÁDEČKOVA, A. & V. SLÁDEČEK. 1966. Periphyton as indicator of reservoir water quality. *Technol. Water* (Czech) 7:507.

TAYLOR, M.P. 1967. Thermal effects on the periphyton community in the Green River. Tennessee Valley Authority, Div. Health & Safety, Water Qual. Br., Biol. Sect., Chattanooga, Tenn.

PATRICK, R. 1968. The structure of diatom communities in similar ecological conditions. *Amer. Natur.* 102:173.

DICKMAN, M. 1969. A quantitative method for assessing the toxic effects of some water soluble substances, based on changes in periphyton community structure. *Water Res.* 3:963.

BESCH, W.K., M. RICARD & R. CANTIN. 1970. Use of benthic diatoms as indicators of mining pollution in the N.W. Miramichi River. *Tech. Rep. Fish. Res. Board Can.* 202:1.

NUSCH, E.A. 1970. Ecological and systematic studies of the Peritricha (Protozoa, Ciliata) in the periphyton community of reservoirs and dammed rivers with different degrees of saprobity. *Arch. Hydrobiol.* (Suppl.) 37:243.

ROSE, F.L. & C.D. McINTIRE. 1970. Accumulation of dieldrin by benthic algae in laboratory streams. *Hydrobiologia* 35:481.

WHITTON, B.A. 1970. Toxicity of zinc, copper and lead to Chlorophyta from flowing waters. *Arch. Mikrobiol.* 72:353.

BURROWS, E.M. 1971. Assessment of pollution effects by the use of algae. *Proc. Roy. Soc. Lond. Ser. B.* 177:295.

PATRICK, R. 1971. The effects of increasing light and temperature on the structure of diatom communities. *Limnol. Oceanogr.* 16:405.

ARCHIBALD, R.E.M. 1972. Diversity of some South African diatom associations and its relation to water quality. *Water Res.* 6:1229.

CAIRNS, J., JR., B.R. LANZA & B.C. PARKER. 1972. Pollution-related structural and functional changes in aquatic communities with emphasis on freshwater algae and protozoa. *Proc. Acad. Natur. Sci. Philadelphia* 124:79.

OLSON, T.A. & T.O. ODLAUG. 1972. Lake Superior Periphyton in Relation to Water Quality. Water Pollut. Control Res. Ser., 18080 DEM 02/72. Univ. Minnesota School Public Health, Minneapolis.

HANSMANN, E.W. 1973. Effects of logging on periphyton in coastal streams of Oregon. *Ecology* 54:194.

RUTHVEN, J.A. & J. CAIRNS, JR. 1973. Response of fresh-water protozoan artificial communities to metals. *J. Protozool.* 20:127.

MIDWEST BENTHOLOGICAL SOCIETY. 1964–1973 (annual). Current and Select Bibliographies on Benthic Biology. Springfield, Ill.

NORTH AMERICAN BENTHOLOGICAL SOCIETY. 1974–1991 (annual). Current and Select Bibliographies on Benthic Biology. Springfield, Ill.

BAXTER, R.M. 1977. Environmental effects of dams and impoundments. *Annu. Rev. Ecol. Systematics* 8:255.

WEITZEL, R.L., ed. 1979. Methods of Measurement of Periphyton Communities: A Review. ASTM Spec. Tech. Publ. 690, American Soc. Testing & Materials, Philadelphia, Pa.

WETZEL, R.G., ed. 1983. Periphyton of Freshwater Ecosystems. Developments in Hydrobiology 17. Dr. W. Junk B.V. Publ., The Hague, The Netherlands.

STEVENSON, R.J. 1984. Epilithic and epipelic diatoms in the Sandusky River USA with emphasis on species diversity and water pollution. *Hydrobiologia* 114:161.

KOSINSKI, R.J. 1984. The effect of terrestrial herbicides on the community structure of stream periphyton. *Environ. Pollut. Ser. A, Ecol. Biol.* 36:165.

LINDSTROM, E.A. & T.S. TRASAN. 1984. Influence of current velocity on periphyton distribution and succession in a Norwegian soft water river. *Verh. Int. Ver. Limnol.* 22:1965.

McGUIRE, M.J., R.M. JONES, E.G. MEANS, G. IZAGUIRRE & A.E. PRESTON. 1984. Controlling attached blue-green algae with copper sulfate. *J. Amer. Water Works Assoc.* 76:60.

PARKER, B.C., G.J. SCHUMACHER & L.A. WHITFORD. 1984. Some rarely reported algae of the Appalachian Mountains, Eastern North America: Why so rare? *Va. J. Sci.* 35:197.

10400   MACROPHYTON*

10400 A.   Introduction

1. General Discussion

The macrophyton consists principally of aquatic vascular flowering plants, but it also includes the aquatic mosses, liverworts, ferns, and the larger macroalgae. Like other primary producers, these plants respond to the quality of the water in which they grow. The use of biota, including macrophyton, is an increasingly important and recognized technique for assessing aquatic habitats.[1] Macrophyton often constitute a dominant factor in the habitat of other aquatic organisms.

Freshwater forms range in size from the tiny watermeal (*Wolffia* spp.), about the size of a pinhead, to plants such as the cattail

* Approved by Standard Methods Committee, 1994.

(*Typha* spp.), up to 4 m high, and finally to cypress trees (*Taxodium* spp.), up to 50 m high. Higher aquatic plants often are clustered in large numbers, many in pure stands, covering extensive areas of shallow lakes, reservoirs, marshes, and canals. A few of the larger freshwater algae (*Chara* spp. and *Nitella* spp.) resembling higher plants in size, form, and habit sometimes are included in the macrophyton. In marine water, the intertidal rockweeds (*Fucus* spp. and *Ascophyllum* spp.) and offshore kelps (*Fucus* spp. and *Macrocystis* spp.) are conspicuous. Vascular marine or estuarine plants, such as the eelgrass (*Zostera* spp.) and the marshgrass (*Spartina* spp.), are essential to the aquatic ecosystem.

Three growth forms of macrophyton generally are recognized: floating, submersed, and emersed. Floating plants may be rooted or free-floating; their principal foliage or crown floats freely on the water surface. All or most of the foliage of submersed plants grows beneath the water surface: nearly all submersed vascular plants have roots. Growing tips of submersed plants may emerge to flower and some species can produce floating leaves. Emersed plants have their principal foliage in the air at or above the water surface; they are attached by roots to the bottom mud. In some cases the same species may grow as a floating or emersed type, or as a submersed or emersed type. Submersed and emersed vascular plants typically are rooted to the bottom but they may be found detached and floating.

The distribution and abundance of higher plants is subject to considerable spatial and temporal variation. Among the many factors that determine their presence, density, and morphology are sediment type, water turbidity, water current, nutrient concentrations, water depth, shoreline disturbance, herbivore grazing, and human activities. Zonation in the littoral zone of lakes and shallow, slow-moving streams is common. Emergent macrophytes generally are found in the most shallow portion of the littoral zone. During periods of low water level they may occupy the terrestrial as well as the aquatic habitat. The depth of inhabitation seldom exceeds 1 m. Floating-leaved plants commonly are found in the shallower littoral areas in water depths between 1 and 3 m.

Submersed plants may occur from the edge of the shore to the interface of the littoral and profundal zones, but rarely extend beyond a depth of 10 m because of limitation of underwater light.

2. Reference

1. NORRIS, R.H., B.T. HART & M. FIULAYSAR, eds. 1995. Use of biota to assess water quality. *Austral. J. Ecol.* 20:1.

3. Bibliography

BUTCHER, R.W. 1933. Studies on the ecology of rivers. I. On the distribution of macrophytic vegetation in the rivers of Britain. *J. Ecol.* 21:58.

WELCH, P.S. 1948. Limnological Methods. Blakiston Co., Philadelphia, Pa.

MILLIGAN, H.F. 1969. Management of aquatic vascular plants and algae. *In* Eutrophication: Causes, Consequences, Correctives. National Acad. Science, Washington, D.C.

BOYD, C.E. 1971. The limnological role of aquatic macrophytes and their relationship to reservoir management. *In* G.E. Hall, ed. Reservoir Fisheries and Limnology. Spec. Publ. Amer. Fish. Soc. No. 8.

HUTCHINSON, G.E. 1975. A Treatise on Limnology. Vol. III. Limnological Botany. John Wiley & Sons, New York, N.Y.

WESTLAKE, D.F. 1975. Macrophytes. *In* B.A. Whitton, ed. River Ecology. Univ. California Press, Berkeley & Los Angeles.

WILE, I. 1975. Lake restoration through mechanical harvesting of aquatic vegetation. *Verh. Int. Ver. Limnol.* 19:660.

WOOD, R.D. 1975. Hydrobotanical Methods. University Park Press, Baltimore, Md.

RASCHKE, R.L. 1978. Macrophyton. *In* W.T. Mason, Jr., ed. Methods for the Assessment and Prediction of Mineral Mining Impacts on Aquatic Communities: A Review and Analysis. U.S. Dep. Interior, Harpers Ferry, W. Va.

WETZEL, R.G. 1983. Limnology, 2nd ed. Saunders College Publishing, Philadelphia, Pa.

DENNIS, W.M. & B.G. ISOM, eds. 1984. Ecological Assessment of Macrophyton: Collection, Use, and Meaning of Data. ASTM STP 843, American Soc. Testing & Materials, Philadelphia, Pa.

10400 B. Preliminary Survey

1. General Considerations

A macrophyte survey includes species identification, location, assessment of health, and quantity. More detailed studies may involve functioning of aquatic plants in nutrient and heavy metal uptake and turnover, use of plants as indicator organisms, and effects of plants on water quality conditions.

Several sampling protocols are required to meet diverse survey needs. The usefulness of a given study and the appropriate types of statistical analyses are determined and fixed initially.

To develop a good sample design, determine what information is desired, prevailing environmental conditions, the life and growth form of the species being sampled, and the methods for obtaining reproducible data that are comparable to other or future studies. In defining reporting requirements, consider matters such as the use of scientific names; the selection of appropriate descriptors such as frequency, density, biomass, cover, diversity, productivity, and outer limit of vegetation growth; and the use of proper statistical techniques.

2. Pre-Field Investigations

During pre-field investigations assemble maps, charts, aerial photographs, taxonomic keys, and past reports. Maps, charts, and aerial photographs help determine access routes, project size, plant community distribution, habitat characteristics that may influence plant distribution, and sampling obstacles or hazards. These items also provide a base for doing field work and reporting results. They may be available locally from municipal engineering departments, zoning boards, planning commissions, drainage districts, and land conservation commissions. At the state level information may come from natural resource agencies, natural history survey organizations, universities, and transportation departments. At the federal level, the Geological Survey, Natural

Resources Conservation Services, Bureau of Land Management, Forest Service, Park Service, Fish and Wildlife Service, Tennessee Valley Authority, Bureau of Indian Affairs, Army Corps of Engineers, Environmental Protection Agency, National Biological Service, and the National Oceanographic and Atmospheric Administration have many maps, charts, and aerial photographs available. A final source is private map companies that publish hydrographic maps for fishermen and recreational boaters.

Past reports provide historical records useful for planning sampling logistics and interpreting results. Comparable studies taken at different times provide a dynamic look at the vegetation. An often-overlooked resource is a herbarium storing pressed and mounted plant specimens. These generally are located in universities and natural history museums.

3. Field Reconnaissance

Sampling efficiency is improved and a sampling scheme can be refined during a field reconnaissance. It provides an opportu-

nity to learn the species of plants present and to sketch their distribution. The species-area curve technique frequently is used to determine the likelihood of finding more species during a preliminary survey. A field reconnaissance allows the investigator to answer logistical questions that are the bane of all sampling efforts.

4. Bibliography

ONDOK, J.P. & J. KVET. 1978. Selection of sampling areas in assessment of production. *In* D. Dykyjova & J. Kvet, eds. Pond Littoral Ecosystems: Structure and Functioning. Ecological Studies 28. Springer-Verlag, Berlin.

MACEINA, M.J., J.V. SHIREMAN, K.A. LANGELAND & D.E. CANFIELD, JR. 1984. Prediction of submersed plant biomass by use of a recording fathometer. *J. Aquat. Plant Manage.* 22:35.

Also see Section 10400A.2.

10400 C.   Vegetation Mapping Methods

Mapping vegetation stands may be necessary. Do this during the preliminary survey.[1]

1. Baseline Method

Vegetation maps constructed using the baseline method or the basepoint-stadia rod-alidade method generally are limited to pure stands of floating or emersed littoral macrophyton in all bodies of water. In clear water, the outline of pure stands of submersed vegetation can be determined by using a viewing box (usually a wooden or plastic box with a watertight glass lens) from the surface or by underwater observation by a diver using a snorkel or SCUBA (self-contained underwater breathing apparatus). The baseline method and the basepoint-stadia rod-alidade method provide accurate maps of vegetation in areas up to 1×10^5 m^2 where most of the vegetation outline is visible. The baseline method uses intercepting lines from each end of a predetermined base line to closely spaced markers (i.e., chaining pins) around the stand. By presetting the map scale, the ratio between the length of the base line and its reduction on the map (drawn on a plane table) can be determined. The basepoint-stadia rod-alidade method is a modification of the baseline method in that the distance between the vegetational outline and the base point is determined with an alidade, range finder, or portable Loran-C unit. One unit on the stadia rod, as viewed between the cross-hairs of the alidade, is equivalent to a distance of 3.05 m between the stadia rod and the alidade. Chaining pins are not required when this method is used. In practice, more readings closely spaced along the vegetational outline usually are taken using the basepoint-stadia rod-alidade method.

It is not necessary to measure all distances and angles; some can be determined trigonometrically. After all angles and dis-

tances are calculated, fill in irregularities through inspection and use of other maps and photographs.

The technology for using global positioning systems (GPS), especially when linked to geographical information systems (GIS), is evolving rapidly.[2] These systems probably will have wide applicability to mapping aquatic vegetation, thereby rendering older surveying techniques obsolete or at least time-consuming and tedious.

2. Line Intercept Method

The line intercept method is preferable for mapping mixed stands and/or large areas. Select sampling points at equal intervals along a base line. Choose interval length by the degree of accuracy desired: the closer the sampling points, the more accurate the map. Run transects perpendicularly from the base line to the boundary of the plant stand. Use an intercept line (transect line) of plastic-coated wire rope to prevent stretching. If line flotation is a problem, use lead weights applied at regular intervals to sink the line and act as interval markers on the rope to designate sampling units. Use 0.5-m segments (in dense vegetation) to 5-m sampling units (in sparse vegetation) for determining plant species that vertically intercept the line at each segment. During underwater surveys, record data with a wax or soft lead pencil on a writing board constructed of plastic overlays. Construct the vegetation map by placing points where plant species are found within an outline map (or aerial photograph) of the sample area. Determine the area that a single species or total vegetation occupies by planimetry or digitization and computer calculation.

Determine frequency from line intercept or quadrat data, or with a set sampler consisting of a 2-cm steel tube, 2 m long, to which five 0.75-cm by 25-cm steel rods are attached on 40-cm centers. Record vegetation touching each of the five points within

2.5 cm of the distal tip. If more than one plant species is touching, record the plant nearest the tip. If no plant is touched, record bare ground.

3. Remote Sensing

a. General considerations: Remote sensing techniques are used to detect, assess, and monitor aquatic macrophytes. These techniques include analog aircraft and satellite serial photography, digital aircraft and satellite multispectral scanning in the visible, infrared, and thermal bands; microwave techniques, primarily side-looking airborne radar (SLAR); and shuttle imaging radar (SIR). They provide a synoptic view of large areas, and allow quick surveys to further delineate areas of interest and repeated viewing at relatively low cost.

In selecting remote sensing system(s) consider expected results, time for project completion, and available resources. The larger the area, the greater the advantage of remote sensing. Remote sensing also lends itself to studies over time, because each image is a historical record.

For determining general associations of a widespread macrophytic growth, ground resolutions of between 30 and 80 m are recommended. Widespread multitemporal coverage is available at scales ranging from 1:100 000 to 1:1 000 000 at a reasonable cost in the form of paper prints, transparencies, and digital formats. Landsat Multispectral Scanner (MSS) and Thematic Mapper (TM) have been used to identify areas with emergent vegetation or topped-out submergent vegetation. Surface roughness is a requisite for an imaging return with SLAR and SIR. Landsat provides limited capability for species discrimination, but availability, cost, and repeat cycles make it useful in determining presence of large populations over time. For a detailed vegetation survey, including discrete species identification, use much larger-scale imagery, (1:10 000 or greater). High-altitude aerial coverage available through the National High Altitude Mapping Program (NHAP) and other sources at original scales of approximately 1:60 000 to 1:120 000 provides both good initial areal coverage and the capacity for enlargement up to five times.[3]

After determining scale, select film/filter or sensor combinations. These include black and white imagery, color infrared, and black and white-infrared imagery; color infrared film used with a yellow filter is widely applicable, but other combinations also are useful.[4–7] For detailed flight planning consider growth stage of plants, water depth and clarity, tidal conditions, cloud cover, and sun angle.[3,8,9] Available resources ultimately determine remote sensing activities.

b. Aerial photography:

1) Equipment—For all-format aerial photography, use a good-quality 35-mm single-lens reflex (SLR) camera with manual or through-the-lens metering or any good 70-mm camera system, preferably with a motor drive. Intervalometers providing for exact exposure intervals for stereo photography are available for both systems. A 28-mm-focal-length lens with a 35-mm SLR camera gives wide coverage at low altitudes; 40-mm and 80-mm lenses are successful with 70-mm camera systems. When photographing with black-and-white or color film, use a skylight or haze filter. When photographing with color infrared film below 1700 m, use a Wratten 12 filter; at altitudes above 1700 m use a Wratten 15 filter. Time of year and condition of vegetation also may influence the choice of yellow/orange filter. In turbid waters, color film*

provides better water penetration and is more useful in mapping submersed vegetation than is color infrared film.[4] Color infrared† yields more detail, is more useful in mapping emergent vegetation or wetlands, and may provide more detail in clear, nonturbid waters.

Preferably mount camera in the belly of a high-wing, single-engine aircraft for low-altitude, small-format photography. Belly mounts require special aircraft modification, but provide a stable platform, protection for the camera, and good access. Alternatively, camera may be mounted on the aircraft door.

2) Procedures—Because sun angle is critical in obtaining high-quality aerial photography, schedule flight for a time when solar altitude is between 40 and 68 degrees.[7]

Set camera at designated ISO reading for the film (assume 100 to 125 for color infrared film) and shoot in the automatic exposure mode. At a typical airspeed of about 190 km/h (approximately 120 mph) a shutter speed of 1/250 or 1/500 is adequate. Determine proper f-stop from an aerial exposure computer; in general, f 5.6 to 11 (f 8 is optimum) gives acceptable color exposure. Proper exposure of color infrared film depends on such factors as time of day and year, altitude, humidity, and type of landscape.

Process film through the manufacturer, a photo laboratory, or aerial photography service. Development of color infrared film is available solely through the manufacturer.

c. Fathometry: Recording fathometers are best applied in water more than 1 m deep where the instruments can determine accurately the height and distribution of subsurface macrophytes.

A recording fathometer can be mounted on most boats and can accurately determine one-dimensional (percent cover) and two-dimensional (percent vertical area) profiles of submersed vegetation. Fathometry is especially useful for determining the outer edge of plant growth. Make linear and planimetric measurements on chart tracings that provide permanent records for ready comparison over time. Calculate percent cover by dividing the linear measurement for a macrophyte species or community by the total chart paper length for any given transect. Dividing the area of the tracing with macrophytes by the total water area gives percent vertical area. Use a fathometer accurate to the nearest 0.1 m.

Mount the transducer for the recording fathometer with brackets on the boat's transom. Keep speed and recorder speed constant to produce tracings of similar length and resolution. Only a few minutes are required to replicate transects several kilometers long. Unless gross morphological differences exist, species discrimination on the chart tracings is difficult or impossible. Mark boundaries of monotypic colonies and community types with a fixed line on the chart tracings. Dense vegetation mats that reach the surface may impede boat movement, prevent the transducer signal from reaching the hydrosoil, and merge tracings of macrophytes with the transducer line. Tracing patterns from water less than 1 m deep may be difficult to interpret.[10] Location of transects or points along a transect also can be determined using GPS techniques.[2]

4. References

1. RASCHKE, R.L. 1984. Mapping—Surface or ground surveys. *In* W.M. Dennis & B.G. Isom, eds. Ecological Assessment of Macrophyton: Collection, Use, and Meaning of Data. Spec. Tech. Publ. 843, American Soc. Testing & Materials, Philadelphia, Pa.

---

* Kodak Ektachrome Professional film type 5036 or equivalent.

† Kodak Aerochrome Infrared film type 2443 or equivalent.

2. KRESS, R. & D. MORGAN. 1995. Application of New Technologies for Aquatic Plant Management. U.S. Army Waterways Experiment Sta., Corps of Engineers, Aquatic Plant Control Research Program. Bulletin, Vol. A-95.

3. AVERY, E.T. & G.L. BERLINE. 1985. Interpretation of Aerial Photographs, 4th ed. Burgess Publishing Co., Minneapolis, Minn.

4. ANDREWS, D.S., D.H. WEBB & A.L. BATES. 1984. The use of aerial remote sensing in quantifying submersed aquatic macrophytes. *In* W.M. Dennis & B.G. Isom, eds. Ecological Assessment of Macrophyton: Collection, Use, and Meaning of Data. Spec. Tech. Publ. 843, American Soc. Testing & Materials, Philadelphia, Pa.

5. LONG, K.S. 1979. Remote Sensing of Aquatic Plants. Tech. Rep., Waterways Experiment Sta., U.S. Army Corps of Engineers, Vicksburg, Miss., Vol. A-79-2.

6. BREEDLOVE, B.W. & W.M. DENNIS. 1984. The use of all-format aerial photography in aquatic macrophyton sampling. *In* W.M. Dennis & B.G. Isom, eds. Ecological Assessment of Macrophyton: Collection, Use, and Meaning of Data. Spec. Tech. Publ. 843, American Soc. Testing & Materials, Philadelphia, Pa.

7. BENTON, A.R., JR. 1976. Monitoring aquatic plants in Texas. Tech. Rep. RSC-76, Texas A & M Remote Sensing Center, College Station, Tex.

8. COLWELL, R.H., ed. 1983. Manual of Remote Sensing, 2nd ed. American Soc. Photogrammetry & Remote Sensing, Falls Church, Va.

9. SLAMA, C., ed. 1980. Manual of Photogrammetry, 4th ed. American Soc. Photogrammetry & Remote Sensing, Falls Church, Va.

10. MACEINA, M.J. & J.V. SHIREMAN. 1980. The use of a recording fathometer for determination of distribution and biomass of hydrilla. *J. Aquat. Plant Manage.* 18:34.

5. Bibliography

SCHMID, W.D. 1965. Distribution of aquatic vegetation as measured by line intercept with SCUBA. *Ecology* 48:816.

OSBORNE, J.A. 1977. Ground Truth Measurements of *Hydrilla verticillata* Royle and Those Factors Influencing Underwater Light Penetration to Coincide with Remote Sensing and Photographic Analysis. Res. Rep. No. 2, Bur. Aquatic Plant Research & Control, Florida Dep. Natural Resources, Tallahassee.

10400 D.   Population Estimates

1. Sampling Design

The design of a sampling program depends on study aims, collection methods, variation and distribution of vegetation, personnel and funds available, and accuracy expected. Variation in space usually is not random; distribution is determined by water depth, shoreline activity, sediment type, or other factors. The parametric statistic for estimating the true population mean assumes that the population being sampled has a normal distribution and that all sample units have the same probability of being selected. Avoid fixed sampling stations in sampling programs to determine population means, unless they are chosen at random at the beginning of the study. Because normally distributed plant populations may not be a characteristic of contiguous plant communities, use parametric statistics with caution.

The simple random sampling design is best applied to homogeneous, noncontiguous plant communities. The number of stations required to obtain an estimate of the true population mean with a predetermined level of confidence and permissible error can be determined by applying the data from a pilot study to the following equation:

$$N = \left(\frac{t \times S}{d \times \bar{x}} \right)^2$$

where:

N = number of sampling stations,
t = Student's t at a given probability level; because N is unknown, set t = 2.0; t is approximately equal to 2.0 for $N > 30$,
S = standard deviation,
$\bar{x}$ = estimator of true population mean usually determined by conducting a pilot study; and
d = permissible error of the final mean; $d = 0.1$ is recommended for vegetation studies ($\pm$ 10%).

An estimate of sampling program cost may be obtained as the sum of initial fixed cost (such as cost of equipment purchase) and variable cost (cost per sample multiplied by number of samples).

Apply stratified random sampling to populations having many homogeneous stands. This design is best applied to populations with obvious gradients and, in practice, to gain precision by the minimized variance within strata. Determine placement of strata by a pilot study. To maximize precision, place stratum boundaries around homogeneous areas; generally, the fewer strata, the greater precision. Allocate sampling in stratified random sampling design according to:

$$\frac{N_i}{N} = \frac{W_i S_i}{\Sigma(W_i S_i)}$$

where:

N_i = number of samples in stratum I,
N = total number of samples,
W_i = a weight reflecting the size (number of quadrats, for example) of stratum i, and
S_i = standard deviation of sampled characteristic within stratum I.

Means for population measurement taken along randomly placed stations on a transect line do not represent large areas of lake populations unless the transect line is placed randomly. Arbitrarily placed transect lines within a sampling area may or may not reflect the true variation of the vegetation within.

2. Collection Methods

a. Field inventory/reconnaissance:
1) Manual collection—If water depth, clarity, temperature, flow, and other circumstances permit, collect specimens by hand. Under ideal conditions, manual collection by wading, snorkeling, or with SCUBA in deeper water habitats permits a detailed and comprehensive evaluation of the macrophyte community.

2) Drag chains—Construct drag chains by welding sharpened U-shaped hooks to a short length (0.6 to 1.0 m) of medium-weight chain. Attach chain to a rope and pull it through the water. Attach a float to the end of the rope to prevent its loss if the chain is snagged and/or dropped. The drag chain can be used readily from a slow-moving or stationary boat and is most efficient in collection of submersed macrophyte species with tall growth forms.

3) Rakes and tongs—Rakes with various handle lengths and oyster tongs may be useful in collecting macrophytes. A rope may be attached to the rake handle for sampling in deep water or to facilitate sampling over a wider radius.

4) Grab samplers—Devices developed for sampling benthic organisms, such as the Ekman, Ponar, and similar grab samplers (see Section 10500B.3), may be used to collect macrophytes. The light weight of the Ekman grab makes it preferable for the rapid and numerous samplings often required for survey inventories.

5) Recording fathometers—Use to determine height and distribution of subsurface macrophytes. Species with similar morphology usually cannot be distinguished from chart tracing; use supplemental methods to identify species.

b. Quantitative sampling: Numerical data collected to describe vegetation commonly include such measures of abundance as density, frequency, cover, and biomass/standing crop.[1-3] Collect these data from plots or quadrats or, less frequently, by plotless sampling techniques. The choice of analytical method depends on vegetation density and types, water depth, flow, height of vegetation in the water column, and nature of the sediment.

1) Line intercepts[4]—This plotless sampling technique entails use of a weighted nylon or lead core line laid along the bottom between two known points or oriented by a compass reading. For dense floating mat vegetation, a floating line may be laid on top of the mat. A surveyor measures the linear distance occupied by various species that underlie the transect line. Express these as a percentage of the total line length for individual species as well as for all species combined. If frequency data are desired, mark the line in increments (e.g., 1 m) and treat species presence/absence in a manner similar to data from quadrat sampling. The line intercept has been used to characterize and map aquatic communities[5] and to correlate distribution of macrophytes with selected environmental factors.[6] Line transects also are useful for determining patterns of plant distribution.[7] In aquatic environments, the line-intercept method is time-consuming and may require a diver equipped with SCUBA. Problems arise in determining whether a plant underlies the transect line.[8]

2) Belt transects—This technique is similar to the line transect and is useful for biomass or density determinations. Data are collected along a fixed line, but from a two-dimensional plot or belt. The belt can be treated as a series of contiguous quadrats or quadrat location may be selected on the basis of a fixed interval or water depth. Use floating or sinking frames.

3) Quadrats—Quadrats can be used for such population and community estimates as frequency, cover, density, and biomass. Quadrats can be any two-dimensional shape but are typically round or rectangular. The sampling area of quadrat samplers can be of any size, but typically varies from 0.1 to 1 m.[2]

With the exception of frames, most sampling devices described have been used to obtain estimates of above-ground biomass (standing crop).[9] Above-ground biomass generally is used because of the difficulty in collecting underground plant parts, such as rhizome and roots. Without the underground parts, however, the data are of limited value for estimates of primary production.

a) Manual samplers—These are relatively simple devices for sampling macrophytes, such as cutting shears. Although they can be used in deep water and manipulated by a diver, they work best in shallow water. They are relatively inexpensive and can be constructed easily or purchased from commercial sources.

Frames are suitable for sampling in shallow water. For sampling short, erect plants, use a square sinking frame constructed of metal. For dense or tangled vegetation, a square assembly frame with pins or wing nuts at the corners or a fixed-corner three-sided frame may be useful. Decide whether to include only macrophytes rooted within the frame or also overlapping plants. In deep water, difficulty and bias may occur in sampling tall submersed vegetation. For macrophytes forming a dense floating mat, use a floating frame constructed of wood or PVC pipe.

Box samplers are useful for sampling where water is shallow and the bottom consists of unconsolidated sediments. The sampler consists of an open-ended box with a metal cutting flange at the bottom and lateral handles; a sampler constructed of 7-mm plexiglass with dimensions 0.5 m × 0.5 m × 0.6 m and aluminum cutting flange and corner reinforcements weighs about 12 kg. With modifications, a box sampler can be used in deep water.[10]

Benthic dome (BeD) samplers[11] may be used for sampling in deep flowing waters. The sampler consists of a plastic dome with a stainless steel circular collar that can be pushed into the substrate. It weighs approximately 11 kg and has a sampling area of 0.25 m.[2]

Various samplers[12,13] developed for macroinvertebrate sampling also may be used to collect macrophytes. These include the Surber sampler, suitable for shallow rivers with moderate current [see Section 10500B.3*b*1)]; the stovepipe (cylindrical) sampler, suitable for wadable waters with unconsolidated sediment bottoms [see Section 10500B.3*c*3)]; and the Ekman grab sampler, best suited for soft sediment bottoms with short, erect vegetation [see Section 10500B.3*a*6)].

b) Mechanically operated samplers—Mechanical sampling devices are costly and complex, and require a floating platform with winches, cables, and booms. The samplers described below are useful in deep water. They may decrease sample collection time, increase accuracy of above-ground and total biomass estimates, and be subject to less bias than many manual methods.

CAUTION: *Use extreme caution for safe operation.*

The Louisiana box sampler is an open-ended 35-cm-high box made of sheet metal or similar material that samples a 61- × 61-cm quadrat (sampling area = 0.37 m²).[14] It can be used from a V-hulled or pontoon boat and is hoisted above the water with a cable and boom. A quick-release mechanism lets the sampler fall free through the water column. Aquatic vegetation is trapped against the bottom and severed by cutting edges along the base of sampler. A nylon net sack over the top retains severed plant fragments. A diver inserts a horizontal cutting blade in a slot at the level of the substrate before the sampler is hoisted to the surface. Manual insertion of the cutting plate by a diver makes use of the Louisiana box sampler comparatively efficient. In soft sediments, the sampler may penetrate too deeply and require lifting before the cutting plate can be inserted. Rocks, stumps, roots, and other debris may prevent complete closure of the cutting door.

The Osborne sampler[15] is a stainless steel box having outside dimensions of 50 cm × 50 cm × 60 cm high and a sampling area of 0.25 m².[2] The sampler weighs 110 kg and is operated by winch and cable from a pontoon boat. After hoisting and suspending the sampler alongside the pontoon boat, a quick-release mechanism allows free-fall through the water column. Tempered steel blades along the bottom edge of the sampler cut vegetation during the descent. A wire mesh screen fastened to the top prevents loss of plant fragments. A hinged slotted door is closed with a lift cable and the sampler is winched to the pontoon boat platform for removal of macrophytes and sediments. Because the sampler penetrates and collects sediments, the sample includes roots and rhizomes and can be used to estimate total biomass as well as above-ground biomass. Efficient operation and accurate biomass estimates require an unconsolidated substrate free of rocks and other debris.

The Waterways Experiment Station (WES)[16] sampler is made of perforated stainless steel and operated from a pontoon boat with an overhead beam that allows it to be hydraulically raised and lowered through a circular opening in the pontoon's platform. Two types are available: one is cylindrical with a sampling area of $0.28 \, m^2$ and the other is square with sampling area of $0.39 \, m^2$. Rotating cutting blades at the base of each sampler sever vegetation as the samplers are lowered. The bottom cutting plate of each is closed hydraulically. A major advantage of the WES sampler is its capability to obtain plant samples from any depth. The Louisiana Box and Osborne samplers, once released, free-fall to the substrate, whereas the hydraulic operation of the WES sampler controls its descent. The size and weight of the trailer and pontoon boat for the WES sampler restrict its use in certain water bodies and require an improved ramp for launching. Although the WES square sampling head is reported to provide a more accurate estimate of above-ground biomass than the circular one, a substantial underestimate of actual above-ground biomass still is reported.[16]

3. Sample Preparation and Analysis

a. Biomass:

1) Fresh weight (wet weight)—Wash samples free of silt and debris, place in a nylon bag (mesh size 0.75 cm) and spin in a garment washer at 560 rpm for 6 to 7 min to remove excess moisture. Weigh sample to nearest 0.1 g.

2) Dry weight—Dry subsample (not less than 10%) in a forced-air oven at 105°C for 48 h or until a constant weight is achieved. The coefficient of variation for a series of subsamples *should not exceed 10%*. Calculate dry weight by dividing dry weight of subsample by fresh weight of subsample times fresh weight of sample.

3) Ash-free dry weight—Transfer dried subsample to a covered and preweighed crucible. Ignite at 550°C for 6 h. The amount of ash is the weight of material remaining after combustion. Calculate ash-free dry weight by determining the ratio between ash and dry weight times dry weight of sample (see Section 10300C.5).

b. Chlorophyll content: Extract fresh plant material with acetone made basic with $MgCO_3$. Grind the plant material and centrifuge at 2500 rpm for 10 to 15 min. Wash residue with acetone and add filtered washings to extract. Dry overnight in a container with anhydrous Na_2SO_4. Dilute with 90% acetone and water. Determine chlorophyll content (see Section 10200G).

c. Carbon content: Most plants (entire) contain 46 to 48% carbon on a dry-weight basis. A factor (46.5%) can be used to calculate carbon content and make comparisons.

d. Caloric content: Determine energy content by bomb calorimetry.

e. Species identification:

1) Sample preparation—Use fresh specimens for identification wherever possible. Avoid immature plants or plants lacking flowers. Because aquatic plants contain from 80 to 95% water and have less supportive tissue than terrestrial forms, a different procedure is required for drying, preserving, and mounting them. Collect plants during peak growth when flowers and/or fruits are present, if practical. Collect the entire plant (stems, rhizomes, leaves, roots, flowers, and fruits).

After collection, either press plants in the field[17-19] or wrap specimen in several layers of paper and submerge in water. Label wrapped specimens with date and location of collection on an index card and place sample and card in a plastic bag. Preferably use an ice chest containing crushed ice for storage in the field. Press plants as soon as practical. They can be kept for several days under refrigeration at 4°C.

Clean plant of all silt and residue. Prepare a mount by centering the plant on 100% rag herbarium paper. Place emergent plants immediately on paper because they take on a natural posture. Place a limp plant in a shallow pan of water and slide the herbarium paper under it; with a slow motion, raise the paper at a 30° angle while keeping the plant centered. Leaves and stems should lie flat on the paper. Drain off excess water, cover with wax paper to prevent plant from sticking to blotters, and place in a plant press between paper and blotters. Place plant press in a dryer. Dry plants at room temperature, but change blotters at least every other day until the plant is sufficiently dry for permanent mounting.

To prepare a wet mount place specimen in an airtight glass vessel filled with 1 part 10% formalin, 3 parts water, and a trace of powdered copper sulfate. Plants will remain lifelike and retain their color for many years in this condition.

2) Identification—A stereomicroscope is needed to identify many plants, especially aquatic grasses and sedges. Observe vegetative and floral structures by dissecting them, under magnification, with forceps and fine needle probes.

Preferably identify to species. Numerous references are available to assist in identifying aquatic macrophytes (see Section 10900G).

3) Plant label—An important part of the species collection is the label that identifies the plant, the collector, the location of the collection, and the date of the collection.[19] Attach label to the sheet with the mounted plant. The mounted plant is a permanent record that is most useful when placed in an herbarium where it can be utilized by others.

4. Data Presentation

Express fresh weight (wet weight), dry weight, and ash-free dry weight as grams or kilograms per square meter. Data are best expressed as ash-free dry weight of total biomass. Determine significant digits for dry weight and ash-free dry weight from the accuracy of the scale used to obtain fresh weight: do not use more significant digits than those used for expressing fresh weight. Report pigment as grams chlorophyll per gram dry plant matter and caloric value as gram calories per gram dry plant matter.

5. References

1. Cox, G.W. 1967. Laboratory Manual of General Ecology. William C. Brown Co., Dubuque, Iowa.

2. KERSHAW, K.A. 1971. Quantitative and Dynamic Ecology. Edward Arnold Co., London, England.

3. MADSEN, J.D. & J.A. BLOOMFIELD. 1993. Aquatic vegetation quantification symposium: an overview. *Lakes Reservoir Manage.* 7:137.

4. TITUS, J.E. 1993. Submersed macrophyte vegetation and distribution within lakes: Line transect sampling. *Lakes Reservoir Manage.* 7:155.

5. LIND, C.T. & G. COTTAM. 1969. The submerged aquatics of University Bay: A study in eutrophication. *Amer. Midland Natur.* 81:353.

6. SCHMID, W.D. 1965. Distribution of aquatic vegetation as measured by line intercept with SCUBA. *Ecology* 46:816.

7. LUDWIG, J.A. & J.F. REYNOLDS. 1985. Statistical Ecology. Wiley Interscience, New York, N.Y.

8. RASCHKE, R.L. & P.C. RUSANOWSKI. 1984. Aquatic macrophyton field collection methods and laboratory analyses. *In* W.M. Dennis & B.G. Isom, eds. Ecological Assessment of Macrophyton: Collection, Use, and Meaning of Data. Spec. Tech. Publ. 843, American Soc. Testing & Materials, Philadelphia, Pa.

9. MADSEN, J.D. 1993. Biomass techniques for monitoring and assessing control of aquatic vegetation. *Lakes Reservoir Manage.* 7:141.

10. PUCKERSON, L.L. & G.E. DAVID. 1975. An *in situ* quantitative epibenthic sampler. U.S. Dep. Interior National Park Serv. Rep., Everglades National Park, Homestead, Fla.

11. RASCHKE, R.L. & P.J. FREY. 1981. Benthic dome (BeD) sampler. *Progr. Fish-Cult* 43:56.

12. EDMONDSON, W.T. & G.G. WINBERG. 1971. A Manual on Methods for the Assessment of Secondary Productivity in Fresh Waters. IBP Handbook No. 17. Blackwell Scientific Publ., Oxford and Edinburgh, U.K.

13. ISOM, B.G. 1978. Benthic macroinvertebrates. *In* W.T. Mason, Jr., ed. Methods for the Assessment and Prediction of Mineral Mining Impacts on Aquatic Communities: A Review and Analysis. U.S. Dep. Interior, Harpers Ferry, W.Va.

14. MANNING, J.H. & R.E. JOHNSON. 1975. Water level fluctuation and herbicide application: An integrated control method for *Hydrilla* in a Louisiana reservoir. *J. Aquat. Plant Manage.* 13:11.

15. OSBORNE, J.A. 1984. The Osborne submersed aquatic plant sampler for obtaining biomass measurements. *In* W.M. Dennis & B.G. Isom, eds. Ecological Assessment of Macrophyton: Collection, Use, and Meaning of Data. Spec. Tech. Publ. 843, American Soc. Testing & Materials, Philadelphia, Pa.

16. SABOL, B.M. 1984. Development and use of the Waterways Experiment Station's hydraulically operated submersed aquatic plant sampler. *In* W.M. Dennis & B.G. Isom, eds. Ecological Assessment of Macrophyton: Collection, Use, and Meaning of Data. Spec. Tech. Publ. 843, American Soc. Testing & Materials, Philadelphia, Pa.

17. HAYNES, R.R. 1984. Techniques for collecting aquatic and marsh plants. *Ann. Missouri Bot. Gard.* 71:229.

18. TSUDA, R.T. & I.A. ABBOTT. 1985. Collection, handling, preservation, and logistics. *In* M.M. Littler & D.S. Littler, eds. Handbook of Phycological Methods. Cambridge University Press, Cambridge, England.

19. HELLQUIST, C.B. 1993. Taxonomic considerations in aquatic vegetation assessments. *Lakes Reservoir Manage.* 7:175.

6. Bibliography

FORSBERG, C. 1959. Quantitative sampling of sub-aquatic vegetation. *Oikos* 10:233.

STEEL, R.G.D. & J.H. TORRIE. 1960. Principles and Procedures of Statistics with Special Reference to the Biological Sciences. McGraw-Hill Book Co., New York, N.Y.

FAGER, E.W., A.O. PLECHSIG, R.F. FORD, R.I. CLUTTER & R.J. GHELARDI. 1966. Equipment for use in ecological studies using SCUBA. *Limnol. Oceanogr.* 11:503.

LIVERMORE, D.F. & W.E. WUNDERLICH. 1969. Mechanical removal of organic production from waterways. *In* Eutrophication: Causes, Consequences, Correctives. National Acad. Sciences, Washington, D.C.

Also see Sections 10400A.2 and 10400C.5.

10400 E. Productivity

1. General Discussion

The complexity and heterogeneity of form, function, phenology, and distribution of aquatic macrophytes have resulted in diverse ways of determining their productivity. These methods can be grouped broadly as biomass methods, based on biomass or on changes in biomass, and metabolic methods, based on estimates of inorganic carbon or oxygen exchange resulting from photosynthesis. The biomass methods generally are simpler than the metabolic methods and require little specialized equipment or expertise and fewer assumptions. Biomass methods integrate responses to environmental conditions and may provide estimates of above-ground production only. They are best used for long-term comparisons (several months to a year) because they are easily confounded by seasonal changes. Biomass methods are insensitive to losses due to fragmentation, herbivory, and secretion or leaching of dissolved organic compounds. In contrast, metabolic methods provide instantaneous measures of photosynthesis and thus reflect responses of plants to different environmental conditions. Metabolic methods for estimating plant productivity also can provide insight into factors controlling distribution and success. The principal drawbacks of the metabolic methods are

that they require specialized equipment and assumptions that may be tenuous and are based on photosynthetic rates (typically measured over a period of minutes to hours) and require extrapolation to net assimilation over longer periods.

In choosing a method consider why macrophyte production is of interest and the use for the data, the habit (growth form and phenology) and habitat of the population, and the cost and effort required to obtain the desired information. The common methods are listed in Table 10400:I and are described below.

2. Biomass Methods

a. Biomass harvest methods: Biomass measurements vary from a simple, one-time sample of maximum biomass to complicated evaluations of seasonal biomass dynamics by methods originally intended for grassland plants.[1-3] These methods are applicable mainly to emergent and submersed macrophytes. Preferably evaluate productivity of floating plants by using growth rates, permanent quadrats, and random samples[4] or by the turnover and metabolic methods discussed in ¶s 2b and 3, below.

1) Above-ground biomass measures—Peak above-ground biomass is measured by the above-ground biomass (usually as ash-

TABLE 10400:I. METHODS USED TO DETERMINE MACROPHYTE PRODUCTION*

| | Plant Habit | | | | | |
| | Emergent | | Floating | | Submersed | |
Method	Deciduous	Evergreen	Deciduous	Evergreen	Deciduous	Evergreen
Biomass harvest:						
Above-ground biomass	+	−	+	−	+	−
Biomass dynamics	+	−	−	−	−	−
Biomass tagging:						
Turnover, growth increment, and						
summed shoot maximum	+	+	+	+	+	+
Cohort	+	−	+	−	+	−
Below-ground biomass	+	+	+	+	+	+
Oxygen measurement:						
Light and dark bottle	−	−	−	−	+	+
Open system	−	−	−	−	+	+
Radiocarbon incorporation	−	−	−	−	+	+
Inorganic carbon exchange:						
Continuous CO_2 exchange	+	+	+	+	+	+
Discrete inorganic C measurement	−	−	−	−	+	+
Potentiometric C flux	−	−	−	−	+	+

* + designates applicable method: − designates method not commonly applied. Evergreen implies retention of substantial above-ground biomass year round: deciduous implies that 10% or less of seasonal maximum biomass is present year round.

free dry weight per unit area) at the time of apparent maximum above-ground biomass (usually time of flowering). This single measurement does not account for biomass carried over from the previous season, losses of material before the peak, or growth after the peak, and therefore generally underestimates net above-ground annual production (NAAP) to an extent depending on the relationship of annual turnover to maximum biomass.[5] The method provides a reasonable estimate of NAAP for many submersed species.

The seasonal biomass accumulation method[6] is a modification of the peak above-ground biomass method that considers only the positive changes in live material. Live above-ground biomass is determined at the beginning of the growing season and at the time of maximum above-ground biomass; the NAAP is calculated as the difference. This method accounts for yearly carryover of living material, but in some cases this will result in a further underestimate of NAAP relative to the peak above-ground biomass.[2]

When recruitment is continuous during the growing season, the biomass peak is less well-defined and greater losses may occur before the seasonal maximum. Under such conditions, above-ground biomass methods yield poor estimates of the NAAP.[3]

2) Biomass dynamics methods—These methods are applicable to emergent plants when dead material remains near the site of decomposition.

The Smalley method[7] estimates net production on the basis of samplings of live and dead material (per unit area) at regular 3- to 6-week intervals. Net production equals the increase in material between samplings: a decrease in live and dead material indicates no net production; an increase in live material and decrease in dead material indicates production equal to the increase in live material; and a decrease in live material and increase in dead biomass with a negative sum indicates no production, while a positive sum indicates production. The method underestimates production if dead material from other areas is present or if new growth is undetected when mortality is high. It is sensitive to sampling frequency[3] and requires a homogeneous area large

enough to accommodate replicate sampling. A modification of this method can be used where above-ground biomass varies little from year to year; in this case net above-ground production is assumed to equal the summed losses of dead material.[8]

More complex procedures based on harvests from a series of paired plots have been proposed.[3,9,10]

b. Biomass tagging methods:

1) Biomass turnover and growth increment methods—Leaf turnover and biomass marking studies involve marking individual plant leaves to follow production, growth, and loss over the year or growing season. For plants with large, long-lived leaves and basal growth, e.g., Vallisneria, Zostera, macroalgae, determine short-term growth of individual leaves. Use these methods for studies of populations (species) rather than communities. Marking methods are particularly useful for evergreen plants, where there may be little seasonal change in biomass, and for plants where the ratio of production to biomass is either very much greater than, or very much less than, one. Turnover measures estimate production and biomass loss and are considered better than harvest techniques, both with respect to accuracy and for the additional phenological information. Tagging methods require major efforts in regular censusing and SCUBA diving for sampling deep, submersed populations.

Where leaves or plant parts are about the same size seasonally, the method is relatively simple, requiring only periodic censusing of plant parts. These methods are not appropriate for species with much branching or different rates of leaf production and loss per branch, and are confounded by intense grazing or by sloughing of newly produced parts.

a) Biomass turnover—To determine turnover time of leaves and shoots, where individuals are easy to distinguish, or where vegetative spread is insignificant or easy to account for and leaf size is fairly constant, choose 10 to 50 individual plants and mark each leaf of each plant. If all new leaves are initiated to the inside of older leaves tag only the newest leaves initially. Mark new leaves or stems with anything that does not interfere with normal growth and is not easily lost, e.g., staples, hole punches, plastic bird rings, fishing line, indelible markers. At regular intervals, revisit plant, tag new leaves, and record the number of new leaves and total number of leaves present. Because new leaves may not be fully expanded at a sampling visit, use a convention concerning the developmental stage of leaves.[11] The sampling interval (weeks to months) depends on research needs and the likelihood of losing the youngest tagged leaf. Compute the annual leaf turnover for each individual plant as the number of new leaves produced annually divided by the maximum number of leaves present at any time in the season. Because the turnover rate is the ratio of production to biomass, calculate NAAP by multiplying the turnover ratio by the maximum above-ground biomass. If vegetative spread is common, modify this method so that all stems or leaves of a species within plots of a given area are tagged.[12,13] Revisit plots at regular intervals, tag new leaves, and record total number of leaves present. Calculate production as above. This method will not account for changes in plant size (increase in weight between years), for mortality, or for recruitment into the population, because only plants initially present that survive the season are included.

b) *Growth increment measurements*—If leaf size changes throughout the growing season or if several types of above-ground parts are present, use methods that account for such differences. Such methods frequently are used for seagrasses.[14–16] Mark each leaf of every plant within a quadrat at a set level above the bottom, relative to a stationary stake or frame. If plants are not buried deeply, use a set distance above the base (i.e., rhizome or root-shoot interface). After a predetermined time, remove all leaves or shoots at the level of the stake or at the set distance. Weigh unmarked leaves produced during the interval. Remove and weigh growth on older leaves (the portion below the marking, but above the ground or base). The combined weights are the net leaf production during the interval. These data also can be used to calculate relative growth rates as [(ln Δ weight)/Δ time]. Make detailed measurements of leaf growth rates by marking plants at the reference level more frequently, e.g., every other day.[17]

A modification of this method[18] permits computation of the growth rate of individual leaves and production of different plant parts.

c) *Summed shoot maximum method*—Where nondestructive measures are required,[3] determine shoot size and number and estimate production.[3] Choose permanent quadrats of a size allowing easy enumeration of plant parts. Label every stem in each quadrat. On regular sampling visits label and count new stems. Develop length:weight regressions using plants collected from areas near but outside the quadrats. This procedure is required only once or twice during the year, depending on the characteristics of study species. Estimate net above-ground (shoot) primary productivity using the number of new parts produced and their weights (based on length:weight regressions). Estimate production per unit area as the mean leaf turnover, multiplied by maximum shoot mass.

For certain species, more complex variations of this method have been used.[19,20]

2) *Cohort methods*—Cohort methods often are used to determine net production of aquatic plants subject to substantial biomass loss before the seasonal maximum biomass is attained. They are useful for species in which groups of individuals, or subunits, initiated at the same time (cohorts), can be identified, that is, for plants where shoots emerge only during one time, or several discrete times, during an annual cycle.

The Allen curve method has been adapted for aquatic macrophytes. It provides an estimate of net production from tables of the numbers and weights of all individuals. It is particularly appropriate for populations for which shoot death and initiation occur throughout the growing season. The method can account for periods of negative production during the year.

Tag all members of a cohort (all leaves or shoots) shortly after initiation or emergence. This sampling is usually the maximum number of individuals present at any time in the cohort, because mortality will decrease the number thereafter.[21] Some new individuals (but still members of the same cohort) may appear by the second sampling visit.[3] Record the number of individuals and their mean weight (dry weight per leaf or shoot). Determine weight from size:weight regressions constructed from data for plants outside the study plot, but using members of the same cohort. Alternatively, harvest adjacent plots of cohort members to estimate the average weight per individual. Repeat for several replicate plots. Revisit plots regularly and record number and mean weight of individuals. Visit frequently enough to minimize potential for loss of young stems or leaves before they have been counted. Plot values (see Figure 10400:1) and determine total area beneath the curve by planimetery, digitally, or gravimetrically, to estimate net annual above-ground production. Repeat if more than one cohort emerges per year; net above-ground annual production then is the sum of the areas beneath several curves.[3,21] If negative production occurs during the year add this loss back to yield net production. Losses usually can be avoided by initiating studies after winter senescence.[21]

An adaptation of the cohort production method[22] recognizes the hierarchical structure of many aquatic macrophytes. This method may be used when differential turnover rates for plant subunits would confound the simple cohort method.

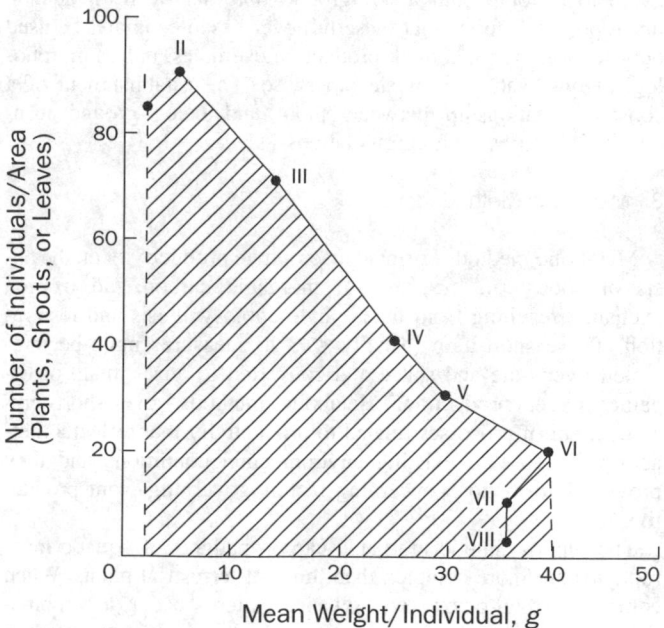

Figure 10400:1. Allen curve for a cohort of a population of aquatic macrophytes. Sample data are indicated by Roman numerals. Net annual above-ground production is proportional to the shaded area.

c. Estimates of below-ground production: The below-ground portions of aquatic macrophytes often comprise a substantial portion of the plant's biomass and net production, and may have important ecosystem implications because of metabolic activity and decay.[5] The amount of below-ground biomass varies dramatically between species, within a species, and seasonally for populations of the same species growing in different habitats.

1) Peak biomass—The maximum below-ground biomass is commonly taken as the net annual below-ground production. This may substantially over- or underestimate actual production, depending on biomass turnover. For many submersed nonevergreens, it provides a reasonable estimate of production, especially where shoot turnover approaches 1 and little below-ground biomass overwinters. For most floating-leaved and emergent macrophytes, it will overestimate biomass.

2) Seasonal biomass accumulation—A more appropriate, but more difficult, method involves repeated sampling and determination of live below-ground biomass at regular intervals during the year. For marsh plants, use sediment cores (25 cm deep, 6- to 10-cm diam) taken at regular intervals (2 to 4 weeks).[23,24] Sieve the cores to remove below-ground biomass and determine weight of live portion. Live material usually can be distinguished by its light color, texture, and turgor. Stains also may be used; chlorazol black stains dead material[25] and tetrazolium stains live material. Use the seasonal change in live biomass as an estimate of production. As for shoots, the loss of plant material before the maximum biomass is attained and any production after the maximum biomass is attained are not accounted for.

3) Root turnover measures—For many applications, the most reasonable way to determine below-ground production is by extrapolation from shoot production or turnover. For many plants, production and turnover of below-ground biomass is directly related to growth and turnover of above-ground plant parts. However, the seasonal translocation of carbohydrate reserves between the below-ground and above-ground portions by most aquatic macrophytes requires that these turnover-based estimates be used only for long-term, annual, production estimates, unless morphological observations indicate otherwise. The establishment of a constant relationship between shoot and below-ground turnover[18,26] justifies such extrapolations.

3. Metabolic Methods

Metabolic methods estimate macrophyte productivity on the basis of short-term measures of inorganic carbon and oxygen exchange resulting from macrophyte photosynthesis and respiration. These short-term (usually < 3 h) measures must be integrated over time and space and converted to an estimate of organic matter production. Metabolic methods give short-term assessment of processes related to productivity that reflect a metabolic response to existing environmental conditions and they provide information concerning factors governing plant productivity.

The photosynthetic inorganic carbon budgets of aquatic macrophytes are more complex than those of terrestrial plants. When complex carbon or oxygen exchange patterns occur, a combination of metabolic methods may be used,[27,28] or a biomass method may be preferable.

a. Oxygen measurement: Determining changes in dissolved oxygen (DO) resulting from photosynthesis and respiration is the most common metabolic method for estimating macrophyte pro-

ductivity. Changes in DO can be determined for plants in chambers (in the field or laboratory) or for the whole system when macrophyte metabolism dominates the oxygen dynamics. Measure DO concentrations with polarographic oxygen electrodes or by titration (Section 4500-O). For the most part, the methodology is similar to that described for phytoplankton and periphyton; however, several additional considerations warrant attention.

Measurements of net oxygen release to water or of changes in the utilization of inorganic carbon by photosynthesis of submersed macrophytes assume acquisition from, or release to, the surrounding water. Many rooted submersed plants do not release dissolved oxygen at rates proportional to photosynthesis.[29–32] Some oxygen diffuses internally from leaves to sites of high respiratory demand in rooting tissues. Additionally, some oxygen is released from leaves and is utilized by epiphytic bacteria and from roots and is utilized by rhizospheric bacteria.[33]

1) Light and dark bottle (chamber) method—This method is similar to those described for other primary producers, where portions of plants or the above-ground portions of several plants are enclosed in either clear or opaque containers. The change in oxygen per unit plant per unit time is an expression of plant photosynthesis (or respiration in the dark bottle) that is used to calculate productivity during the incubation period. Daily or longer-term production is estimated, using additional information on environmental conditions during the incubation period and during the period over which a production estimate is needed, and the relationship between the response of the plant segment incubated and the response of the entire plant (or population) to those environmental conditions.[34–38] See Sections 10200I.2 and 10300D.2 for procedural details. Mixing to simulate natural water movement is desirable.[39–46] Equipment modifications have been discussed,[47–53] as have specific analytical requirements.[34,54–57]

Determining productivity for periods longer than the test hours requires including respirations during the dark (24-h net production = daytime net production minus nighttime respiration or daytime gross production minus 24-h respiration) and extrapolating from the incubation period to these longer periods. Estimate net photosynthesis during the daylight period from productivity measurements in 3- to 4-h intervals from dawn to dusk daily, or by multiplying the production measured during a 3- to 4-h midday incubation by the fraction of total daily light received during the incubation period. For monthly or yearly periods there are several methods for extrapolating.[58,59] Most estimates of macrophyte production by these methods rely on a single midday incubation conducted at weekly-to-monthly intervals. There are three ways to estimate production for longer periods. First, measure productivity at 3- to 4-h intervals from dawn to dusk, on a clear day, several times during the growing season, to establish a relationship between midday photosynthesis and total daily photosynthesis. Estimate daily production on the basis of midday incubation at regular intervals (weekly to monthly), using the relationship between light during the incubation period to daily light and photosynthesis during the incubation period to daily photosynthesis as determined above. Estimate production on intervening days, when photosynthesis is not measured, on the basis of the available light on those days. Second, by using experimental evidence (P-I curves from plants incubated at different depths) or literature values, establish a relationship between light and productivity and a midday incubation (weekly to monthly) as a scaling factor. Carefully estimate production on the basis of available light for any day.[60,61] Third, use more involved modeling that accounts for growth characteristics and environmental conditions to provide estimates of macrophyte production.[62,63]

The epiphyton associated with a macrophyte may influence the determination of photosynthesis and respiration,[64-68] but the respective contributions to photosynthesis cannot be resolved by the oxygen method.[66] Alternatively, use the ^{14}C method for estimating macrophyte photosynthesis.

Short-term measures of macrophyte production using the light-dark bottle method do not account for the loss of organic substances that may be as much as 10% of the recently fixed carbon.[69] Other losses of fixed carbon, as well as sloughing, grazing, and fragmentation, also are unaccounted for. Further, the light-dark bottle oxygen method cannot predict the allocation of photosynthate to below-ground growth or reproduction.[5,70] The inability to account for such factors is a shortcoming of all metabolic methods, but oxygen methods provide reasonable estimates of short-term biomass net accumulation for some plants.[38,57]

2) *Open-system oxygen method*—In flowing-water systems where macrophytes dominate primary production, analysis of diurnal oxygen curves can be used to estimate macrophyte production.[39,71] The change in the oxygen content of a parcel of water is the result of both community metabolism and oxygen diffusion across the water surface (see Section 10300D.4). Entry of groundwater or surface runoff is assumed negligible during measurements. Criteria[72] for determining the suitability of a system for open-system monitoring include extent of uniform area, high biological activity, water depth and residence time (influencing observable changes in water chemistry), turbulence (influencing spatial variation for monitoring and gaseous exchange with the atmosphere), and uniformity of the channel (allowing the calculation of production per unit area and providing uniform residence time). The accuracy and sensitivity of several methods have been compared.[73]

The single-station method (Section 10300D.4) is used most commonly. In addition to the radiotracer method presented, the reaeration coefficient can be determined by direct methods or calculations based on physical parameters.[73-75] To use a single station assume stream homogeneity above the region of measurement.

To use an automated two-station system, calculate a continuous function (Fourier series) to determine an exact solution to the oxygen mass balance equation, rather than the approximate finite difference solution.[76] This system models net productivity so that the oxygen concentrations at the downstream station can be predicted accurately from upstream values. Additionally, it provides detailed information, such as hourly variations in net production and seasonal changes in community photosynthetic characteristics.[77]

For analytical details, see Section 10300D.4*a* and *b*.

Determine production from the change in DO between the two stations:

$$dc/dr = K(c_s - c) + \beta(t)$$

where:

dc/dr = change in DO concentration between stations,
K = reaeration coefficient,
c_s = saturation DO concentration, mg/L,
c = DO concentration, mg/L,
$\beta(t)$ = net productivity, and
t = flow time between stations.

For short time intervals, write t as a Fourier cosine series:

$$\beta(t) = A_o/2 + \sum_{n=1}^{\infty} A_n \cos n\omega t$$

where:

A_n = Fourier coefficients and
ω = $2\pi/48$.

Determine enough Fourier coefficients to solve for $\beta(t)$.[76] After solving the equation plot estimated net photosynthetic rates as a function of incoming solar radiation to evaluate the solution for net production. An anomalous relationship between light and net production, or an unreasonable variation in net photosynthesis over time suggests solution errors. The estimate of K is the weak point of the method; an iterative approach to estimation of K, based on the morning and afternoon relationship between net production and light,[73] has been proposed.

b. Radiocarbon incorporation method: The radiocarbon (^{14}C) tracer method[78] has been used extensively for estimating the productivity of virtually all aquatic primary producers. It is based on the measurement of radiocarbon incorporation when a small but known quantity is made available to a plant. The proportion of tracer added to that incorporated indicates the fraction of stable carbon that is incorporated. This method is more sensitive than oxygen methods and thus can be used where photosynthetic rates are very low or where carbon consumption is excessive.

Although this method directly measures the incorporation of external inorganic carbon, the relationship of this incorporation to net or gross photosynthesis is not without controversy.[79] The general consensus is that the ^{14}C method for macrophytes provides a rate of carbon incorporation between net and gross photosynthesis, but closer to net photosynthesis.[35,38,47,52,57,80] The primary drawback of the ^{14}C method is that it provides no measure of dark-period respiration. If the photoassimilation of ^{14}C is assumed to estimate net photosynthesis for macrophytes (gross photosynthesis minus respiration), then respiration during the dark period must be measured separately to account for carbon lost at night in the calculation of 24-h net production (24-h net primary production = gross photosynthesis − 24-h respiration, or here, 24-h NPP = daytime ^{14}C uptake − nighttime respiration).[57] As noted for the oxygen method, respiration usually is assumed to be identical during the day and night for macrophytes.[38,57]

Portions of CO_2 from photorespiration and mitochondrial respiration are stored internally in gas lacunal spaces and are intensively recycled.[81] Some CO_2 from rooting tissues and rhizosphere bacterial metabolism can diffuse internally in these gas channels to photosynthetic tissues.[82] Thus, rates of incorporation of $^{14}CO_2$ from the water may not be proportional (underestimate) to true rates of carbon fixation.

For procedural details see above for the light and dark bottle oxygen method and Section 10300D.3. Analysis of organic fractions into which carbon 14 may be incorporated has been described[30,64,83] as have procedures for carbon-14.[84-88]

c. Inorganic carbon exchange methods: Changes in the inorganic carbon concentration in air or water surrounding aquatic macrophytes as a result of photosynthesis and respiration can be determined by several methods. These methods provide a highly sensitive, direct measure of photosynthetic carbon uptake, thus requiring few assumptions, and can provide additional information on plant physiology. They require considerable expertise and expensive equipment. Carbon-exchange methods that use closed chambers have the same potential problems as the light-dark chamber oxygen method; however, the formation of bubbles during macrophyte photosynthesis usually does not interfere.

1) *Continuous infrared gas analysis (IRGA)*—The continuous IRGA method is based on measuring the change in CO_2, resulting from photosynthesis and respiration, in an air stream that has passed over the leaf or leaves of a plant. The change in CO_2 concentration is determined from the difference in concentration between a reference and a sample air flow. For submersed plants, enclose portions of the plant in a cuvette filled with a bathing solution (e.g., water from the site of collection), and bubble a gas stream through the solution. The CO_2 in the air stream equilibrates with the CO_2 (all components of the carbonate system) in the water. The change in CO_2 concentration of the air stream reflects plant metabolism and can be measured as the air stream passes through the IR analyzer. Details of equipment and procedures have been discussed.[80,89-95]

2) *Discrete inorganic carbon exchange*—Metabolic inorganic carbon exchange with the air or water also can be measured by determining the change in inorganic carbon in sealed containers after a discrete incubation period, using IRGA,[68,92,96-98] gas chromatography,[99] or a total organic carbon analyzer operating in the inorganic carbon mode (see Section 5310). The change in CO_2 in the air, for emergent and floating plants, or the change in dissolved inorganic carbon (DIC) in the water, for submersed plants, is determined for subsamples of the incubation medium. This method is analogous to the light and dark bottle oxygen method, except that changes in the CO_2 or DIC concentrations are determined, rather than the change in DO concentration.[38] This method requires more expensive equipment than the oxygen method does; however, in low-DIC waters it is much easier to measure the change in DIC resulting from plant metabolism than the corresponding change in DO (due to the relative abundance of DO and DOC and the high sensitivity of the DIC method). See references cited above for procedural details.

The change in CO_2 in the light is net photosynthesis. A darkened chamber can be used to determine respiration. Extrapolation to estimate production and additional methodological considerations are covered in the discussion for the light and dark bottle oxygen method. As for other carbon exchange methods, the exchange of CO_2 with the atmosphere, as well as marl formation, can result in errors in the determination of carbon exchange.

3) *Potentiometric measurement of inorganic carbon flux*—The physical-chemical relationships of the carbonate buffer systems in natural waters dictate a definable relationship between DIC (or alkalinity) and pH. On the basis of such relationships, changes in the pH of the water resulting from plant-mediated changes in DIC concentrations (due to photosynthesis and respiration) can be used to estimate the inorganic carbon exchange for submersed macrophytes. The method presented here is based on the estimation of the DIC or total carbon (C_T) determined by a Gran titration.[100] This approach requires only a good pH meter capable of readings to 0.1 pH unit, an electrode, and careful laboratory procedure. However, it is not suitable for waters of very high or low pH, and has limitations similar to those of the oxygen method. While it has been applied primarily to laboratory studies,[101-103] it is adaptable for field use.

To measure inorganic carbon flux,[104] incubate and collect sample as described for the light and dark bottle method. Use 125-mL gastight bottle or polypropylene syringe for sampling. Determine temperature of a 2.0- to 100-mL water sample and add sample to a titration vessel containing a magnetic stir bar, or add

stir bar to collection bottle for direct titration. Assure that titration vessel is partially sealed around the outlet to provide protection from the atmosphere. Record initial pH. Stir sample and use a 0.5- to 5-mL-capacity piston syringe-buret to titrate stepwise with HCl or other appropriate acid of known normality. Use acid of such normality that the total solution volume is not changed by more than 10% by the end of titration. Record pH and volume of titrant added at three points between pH 7.6 and 6.7 (or lower if necessary) and another three points between pH 4.4 and 3.7.

Calculate net photosynthesis by the change in carbon in the light bottle or container, and respiration by the change in carbon in the dark container as follows: Calculate F_2 from pH readings in the lower pH range as

$$F_2 = [\text{antilog } (a - \text{pH})] \times [(V_s + v)/V_s]$$

where:

> F_2 = antilogarithmic Gran functions for pH change with titrant additions,
> a = any convenient number above the pH range, e.g., 5,
> V_s = sample volume, mL, and
> v = titrant volume, mL.

Plot F_2 against v and fit with a straight line, locating the intersection of the line on the v axis ($v52$). Calculate $F51$ from pH readings in the higher pH range as

$$F_1 - [\text{antilog } (b - \text{pH})] \times (V_s + v) \times (v_2 - v)$$

where:

> F_1 = antilogarithmic Gran functions for pH change with titrant additions and
> b = any convenient number above the pH range, e.g., 8.

Plot F_1 against v, locating the intersection of the best-fitting straight line with the v axis (v_1). Then,

$$V_I = v_I \times \frac{1000}{V_s} \times n$$

$$C_T = V_2 - V_I = \frac{1000}{V_s} \times n \times (v_2 - v_I)$$

$$V_2 = v_2 \times \frac{1000}{V_s} \times n$$

where:

> V_I = acidity, meq/L,
> V_2 = total alkalinity, meq/L,
> C_T = total DIC, mmol/L, and
> n = normality of the acid titrant.

Use of antilog paper simplifies this procedure. The method can be used for water samples with an initial pH of less than 7 if an equal amount of NaOH is added to all samples.[101,104]

A formulation of a relationship of pH to C_T for a water of constant alkalinity and changing pH (e.g., as a result of photosynthesis or respiration) eliminates the need for titrations to determine the change in C_T in an incubation vessel resulting from macrophyte photosynthesis, as only the initial and final pH would be required.[104]

4. Data Presentation

Express seasonal and annual rates of macrophyte production in units of carbon or dry weight per unit area of colonization or littoral region. Occasionally, energy units (kcal) may be used but are not recommended. Net annual above-ground production, expressed as grams C per square meter per year or grams dry weight (or ash-free dry weight) per square meter per year is used often but variously underestimates total net annual production. Express the results of short-term (minutes, hours, days) measurements of production or photosynthesis as carbon fixed per unit shoot dry weight or per unit chlorophyll a. For emergent plants, report as carbon uptake per unit leaf or shoot surface area. The value of data collected depends on clear statements of how the production values are calculated and expressed (e.g., dried at 105°C, dry weight or ash- free dry weight, per unit area colonized or per area lake bottom, etc.) and on the provision of ancillary information to allow the data to be re-expressed and compared with values from other studies (chlorophyll content, ash content, above:below ground weight, etc.).

5. References

1. LINTHURST, R.A. & R.J. REIMOLD. 1978. An evaluation of methods for estimating the net aerial primary productivity of estuarine angiosperms. *J. Appl. Ecol.* 15:919.
2. SHEW, D.M., R.A. LINTHURST & E.D. SENECA. 1981. Comparisons of production computation methods in a southeastern North Carolina salt marsh. *Estuaries* 4:97.
3. DICKERMAN, J.A., A.J. STEWART & R.G. WETZEL. 1986. Estimates of net above ground production: Sensitivity to sampling frequency. *Ecology* 67:650.
4. TUCKER, C.S. & T.A. DEBUSK. 1981. Productivity and nutritive value of *Pistia stratioles* and *Eichhornia crassipes. J. Aquat. Plant Manage.* 19:61.
5. WESTLAKE, D.F. 1982. The primary productivity of water plants. *In* J.J. Symoens & P. Compere, eds. Studies on Aquatic Vascular Plants. Royal Botanical Soc. Belgium, Brussels.
6. MILNER, C. & R.E. HUGHES. 1968. Methods for the Measurement of Primary Production of Grasslands. IBP Handbook No. 6. Blackwell Scientific Publ., Oxford, England.
7. KIRBY, C.J. & J.G. GOSSELINK. 1976. Primary production in a Louisiana Gulf coast *Spartina alterniflora marsh. Ecology* 58:1052.
8. VALIELA, I., J.M. TEAL & W.J. SASS. 1975. Production and dynamics of salt marsh vegetation and the effects of experimental treatment with sewage sludge. *J. Appl. Ecol.* 12:973.
9. WIEGERT, R.G. & F.C. EVANS. 1964. Primary production and the disappearance of dead vegetation in an old field. *Ecology* 45:49.
10. LOMNICKI, R.A., E. BANDOLA & K. JANKOWSKA. 1968. Modification of the Wiegert-Evans method for the estimation of net primary production. *Ecology* 49:147.
11. MOELLER, R.E. 1978. Seasonal changes in biomass, tissue chemistry, and net production of the evergreen hydrophyte, *Lobelia dortmanna. Can. J. Bot.* 56:1425.
12. SAND-JENSEN, K. 1975. Biomass, net production and growth dynamics in an eelgrass (*Zostera marina* L.) population in Vellerup Vig, Denmark. *Ophelia* 14:185.
13. SAND-JENSEN, K. & M. SONDERGAARD. 1978. Growth and production of isoetids in oligotrophic Lake Kalgaard, Denmark. *Verh. Int. Ver. Limnol.* 20:659.
14. ZIEMAN, J.C. & R.G. WETZEL. 1980. Productivity in seagrasses: Methods and rates. *In* R.C. Phillips & C.P. McRoy, eds. Handbook of Seagrass Biology. Garland STPM Press, New York, NY.
15. ZIEMAN, J.C. 1974. Methods for the study of the growth and production of turtle grass, *Thalassia testudinum* Konig. *Aquaculture* 4:139.
16. ZIEMAN, J.C. 1975. Quantitative and dynamic aspects of the ecology of turtle grass *Thalassia testudinum. Estuarine Res.* 1:541.
17. BROUNS, J.J.W.M. & F.M.L. HEIJS. 1986. Production and biomass of the seagrass *Enhalus acoroides* (L.f) Royle and its epiphytes. *Aquat. Bot.* 25:21.
18. JACOBS, R.P.W.M. 1979. Distribution and aspects of the production and biomass of eelgrass, *Zostera marina* L., at Roscoff, France. *Aquat. Bot.* 7:151.
19. JACKSON, D., S.P. LONG & C.F. MASON. 1986. Net primary production, decomposition and export of *Spartina anglica* on a Suffolk saltmarsh. *J. Ecol.* 74:647.
20. DAWSON, F.H. 1976. The annual production of the aquatic macrophyte *Ranunculus penicillatus* var. *calcareus* (R.W. Butcher) C.D.K. Cook. *Aquat. Bot.* 2:51.
21. MATHEWS, C.P. & D.F. WESTLAKE. 1969. Estimation of production by populations of higher plants subject to high mortality. *Oikos* 20: 156.
22. CARPENTER, S.R. 1980. Estimating net shoot production by a hierarchical cohort method of herbaceous plants subject to high mortality. *Amer. Midland Natur.* 140:163.
23. VALIELA, I., J.M. TEAL & N.Y. PERSSON. 1976. Production dynamics of experimentally enriched salt march vegetation: Belowground biomass. *Limnol. Oceanogr.* 21:245.
24. GALLAGHER, J.L. & F.G. PLUMLEY. 1979. Underground biomass profiles and productivity in Atlantic coastal marshes. *Amer. J. Bot.* 66: 156.
25. WILLIAMS, D.D. & N.E. WILLIAMS. 1974. A counterstaining technique for use in sorting benthic samples. *Limnol. Oceanogr.* 19:152.
26. WIUM-ANDERSON, S. & J. BORUM. 1984. Biomass variation and autotrophic production of an epiphyte-macrophyte community in coastal Danish area: I. Eelgrass (*Zostera marina* L.) biomass and net production. *Ophelia* 23:33.
27. KEELEY, J.E. & G. BUSCH. 1984. Carbon assimilation characteristics of the aquatic CAM plant *Isoetes howellii. Plant Physiol.* 76:525.
28. BOSTON, H.L. & M.S. ADAMS. 1986. The contribution of crassulacean acid metabolism to the annual productivity of two aquatic vascular plants. *Oecologia* 68:615.
29. ODUM, H.T. 1956. Primary production in flowing waters. *Limnol. Oceanogr.* 1:103.
30. HOUGH, R.A. 1979. Photosynthesis, respiration, and organic carbon release in *Elodea canadensis* Michx. *Aquat. Bot.* 7:1.
31. SAND-JENSEN, K. & C. PRAHL. 1982. Oxygen exchange with the lacunae and across leaves and roots of the submerged vascular macrophyte, *Lobelia dortmanna* L. *New Phytol.* 91:103.
32. SAND-JENSEN, K., C. PRAHL & H. STOCKHOLM. 1982. Oxygen release from roots of submerged aquatic macrophytes. *Oikos* 38:349.
33. WETZEL, R.G. 1990. Land-water interfaces: Metabolic and limnological regulators. *Verh. Int. Ver. Limnol.* 24:6.
34. HILL, B.H., J.R. WEBSTER & A.E. LINKINS. 1984. Problems in the use of closed chambers for measuring photosynthesis by a lotic macrophyte. *In* W.M. Dennis & B.G. Isom, eds. Ecological Assessment of Macrophyton: Collection, Use, and Meaning of Data. Spec. Tech. Publ. 843, American Soc. Testing & Materials, Philadelphia, Pa.
35. WETZEL, R.G. 1965. Techniques and problems of primary productivity measurements in higher aquatic plants and periphyton. *In* C.R. Goldman, ed. Primary Production in Aquatic Environments. Mem. Ist. Ital. Idrobiol., 18 Suppl., Univ. California Press, Berkeley.
36. WESTLAKE, D.F. 1978. Rapid exchange between plant and water. *Verh. Int. Ver. Limnol.* 20:2363.
37. MOESLUND, B., M.G. KELLY & N. THYSSEN. 1981. Storage of carbon and transport of oxygen in river macrophytes: Mass-balance, and the measurement of primary productivity in rivers. *Arch. Hydrobiol.* 93: 45.
38. KEMP, M.W., M.R. LEWIS & T.W. JONES. 1986. Comparison of methods for measuring production by the submersed macrophyte, *Potamogeton perfoliatus* L. *Limnol. Oceanogr.* 31:1322.

39. RAVEN, J.A. 1970. Exogenous inorganic carbon sources in plant photosynthesis. *Biol. Rev.* 45:167.

40. SMITH, F.A. & N.A. WALKER. 1980. Photosynthesis by aquatic plants: Effects of unstirred layers in relation to assimilation of CO_2 and HCO_3 and carbon isotope discrimination. *New Phytol.* 86:245.

41. BLACK, M.A., S.C. MABERLY & D.H.N. SPENCE. 1981. Resistance to carbon dioxide fixation in four submerged freshwater macrophytes. *New Phytol.* 69:54.

42. WESTLAKE, D.F. 1967. Some effects of low-velocity current on the metabolism of aquatic macrophytes. *J. Exp. Bot.* 18:187.

43. MADSEN, T.V. & M. SONDERGAARD. 1983. The effects of current velocity on the photosynthesis of *Callitrishe stagnalia* Scop. *Aquat. Bot.* 15:187.

44. JENKINS, J.T. & M.C.F. PROCTER. 1985. Water velocity, growth-form and diffusion resistances to photosynthesis CO_2 uptake in aquatic bryophytes. *Plant Cell. Environ.* 8:317.

45. RODGERS, J.H., JR., K.L. DICKSON & J. CAIRNS, JR. 1978. A chamber for *in situ* evaluation of periphyton productivity in lotic systems. *Arch. Hydrobiol.* 84:389.

46. DAWSON, F.H., D.F. WESTLAKE & G.I. WILLIAMS. 1981. An automatic system to study the responses of respiration and photosynthesis by submerged macrophytes to environmental variables. *Hydrobiologia* 77:277.

47. WETZEL, R.G. 1964. A comparative study of the primary productivity of higher aquatic plants, periphyton, and phytoplankton in a large shallow lake. *Int. Rev. Ges. Hydrobiol.* 49:1.

48. LOVE, R.J.R. & G.G.C. ROBINSON. 1977. The primary productivity of submerged macrophytes in West Blue Lake Manitoba. *Can. J. Bot.* 55:118.

49. LITTLER, M.M. & K.E. ARNOLD. 1985. Electrodes and chemicals. *In* M.M. Littler and D.S. Littler, eds. Handbook of Phycological Methods, Ecological Field Methods: Macroalgae. Cambridge University Press, Cambridge, England.

50. HATCHER, B.G. 1977. An apparatus for measuring photosynthesis and respiration of intact large marine algae and comparison of results with those from experiments with tissue segments. *Mar. Biol.* 43:381.

51. LOBBAN, C.S. 1978. Translocation of ^{14}C in *Macrocystis pyrifera* (giant kelp). *Plant Physiol.* 61:585.

52. BITTAKER, H.F. & R.L. IVERSON. 1976. *Thalassia testudinum* productivity: A field comparison of measurement methods. *Mar. Biol.* 37:39.

53. VOOREN, C.M. 1981. Photosynthetic rates of benthic algae from the deep coral reef of Curacao. *Aquat. Bot.* 10:143.

54. ADAMS, M.S., J. TITUS & M. MCCRACKEN. 1974. Depth distribution of photosynthetic activity in a *Myriophyllum spicatum* community in Lake Wingra. *Limnol. Oceanogr.* 19:377.

55. GUILIZZONI, P. 1977. Photosynthesis of the submergent macrophyte *Ceratophyllum demersum* in Lake Wingra. *Trans. Wis. Acad. Sci. Arts Lett.* 65:152.

56. PRIDDLE, J. 1980. The production ecology of benthic plants in some antarctic lakes II. Laboratory physiological studies. *J. Ecol.* 68:155.

57. YIPKIN, Y., S. BEER, E.P.H. BEST, T. KAIRESALO & K. SALONEN . 1986. Primary production of macrophytes: terminology, approaches and a comparison of methods. *Aquat. Bot.* 26:129.

58. WESTLAKE, D.F. 1974. Methods for measuring production rates—Calculation of day rates per unit of lake surface (*b*) periphyton and macrophytes. *In* R.A. Vollenweider, ed. A Manual on Methods for Measuring Primary Production in Aquatic Environments, 2nd ed. IBP Handbook No. 12. Blackwell Scientific Publ., Oxford, England.

59. WETZEL, R.G. & G.E. LIKENS. 1991. Limnological Analyses, 2nd ed. Springer-Verlag, New York, N.Y.

60. GOULDER, R. 1970. Day-time variations in the rates of production by two natural communities of submerged freshwater macrophytes. *J. Ecol.* 58:521.

61. MCCRACKEN, M.D., M.S. ADAMS, J. TITUS & W. STONE. 1975. Diurnal course of photosynthesis in *Myriophyllum spicatum* and *Oedogonium. Oikos* 26:355.

62. TITUS, J.E., R.A. GOLDSTEIN, M.S. ADAMS, J.B. MANKIN, R.V. O'NEILL, P.R. WEILER, JR., H.H. SHUGART & R.S. BOOTH. 1975. A production model for *Myriophyllum spicatum* L. *Ecology* 56:1129.

63. BEST, E.P.H. 1981. A preliminary model for the growth of *Ceratophyllum demersum* L. *Verh. Int. Ver. Limnol.* 21:1484.

64. ALLEN, H.A. 1971. Primary productivity, chemo-organotrophy, and nutritional interactions of epiphytic algae and bacteria on macrophytes in the littoral of a lake. *Ecol. Monogr.* 41:97.

65. HUTCHINSON, G.E. 1975. A Treatise on Limnology. Vol. 3. Limnological Botany. John Wiley & Sons, New York, N.Y.

66. SAND-JENSEN, K. 1977. Effects of epiphytes on eelgrass photosynthesis. *Aquat. Bot.* 3:55.

67. CATTANEO, A. & J. KALFF. 1980. The relative contribution of aquatic macrophytes and their epiphytes to the production of macrophyte beds. *Limnol. Oceanogr.* 25:280.

68. KAIRESALO, T. 1983. Photosynthesis and respiration with an *Equisetum fluviatile* L. stand in Lake Paajarvi, southern Finland. *Arch. Hydrobiol.* 96:317.

69. HOUGH, R.A. & R.G. WETZEL. 1972. A ^{14}C-assay for photorespiration in aquatic plants. *Plant Physiol.* 49:987.

70. WESTLAKE, D.R. 1975. Primary production of freshwater macrophytes. *In* J.P. Cooper, ed. Photosynthesis and Productivity in Different Environments. IBP Handbook No. 3. Cambridge University Press, Cambridge, England.

71. EDWARDS, R.W. & M. OWENS. 1962. The effects of plants on river conditions. IV. The oxygen balance of a chalk stream. *J. Ecol.* 50:207.

72. KINSEY, D.W. 1985. Open-flow systems. *In* M.M. Littler & D.S. Littler, eds. Handbook of Phycological Methods, Ecological Field Methods: Macroalgae. Cambridge University Press, Cambridge, England.

73. KOSINSKI, R.J. 1984. A comparison of the accuracy and precision of several open-water oxygen productivity techniques. *Hydrobiologia* 119:139.

74. OWENS, M. 1974. Methods of measuring production rates in running water. *In* R.A. Vollenweider, ed. A Manual on Methods for Measuring Primary Production in Aquatic Environments. IBP Handbook No. 12. Blackwell Scientific Publ., Oxford, England.

75. WILCOCK, R.J. 1982. Simple predictive equations for calculating stream reaeration coefficients. *N. Zealand Sci.* 25:53.

76. KELLY, M.G., G.M. HORNBERGER & B.J. COSBY. 1974. Continuous automated measurements of rates of photosynthesis and respiration in an undisturbed river community. *Limnol. Oceanogr.* 19:305.

77. KELLY, M.G., N. THYSSEN & B. MOESLAND. 1983. Light and the annual variation of oxygen- and carbon-based measurements of productivity in a macrophyte-dominated river. *Limnol. Oceanogr.* 28:503.

78. STEEMANN NIELSEN, E. 1952. The use of radio-active carbon (C^{14}) for measuring organic production in the sea. *J. Cons. Perm. Int. Explor. Mer* 18:117.

79. PETERSON, B.J. 1980. Aquatic primary productivity and the ^{14}C-CO_2 method: A history of the productivity problem. *Annu. Rev. Ecol. System.* 11:359.

80. TITUS, J.E., M.S. ADAMS, T.D. GUSTAFSON, W.H. STONE & D.F. WESTLAKE. 1979. Evaluation of differential infrared gas analysis for measuring gas exchange by submersed aquatic plants. *Photosynthetica* 13:294.

81. SØNDERGAARD, M. & R.G. WETZEL. 1980. Photorespiration and internal recycling of CO_2 in the submersed angiosperm *Scirpus subterminalis* Torr. *Can. J. Bot.* 58:591.

82. WETZEL, R.G., E.S. BRAMMER, K. LINDSTRÖM & C. FORSBERG . 1985. Photosynthesis of submersed macrophytes in acidified lakes. I. Carbon limitations and utilization of benthic CO_2 sources. *Aquatic Biol.* 22:107.

83. ARNOLD, K.E. & M.M. LITTLER. 1985. The carbon-14 method for measuring primary productivity. *In* M.M. Littler & D.S. Littler, eds. Handbook of Phycological Methods, Ecological Field Methods: Macroalgae. Cambridge University Press, Cambridge, England.

84. BURNISON, B.K. & K.T. PEREZ. 1974. A simple method for the dry combustion of ^{14}C-labeled materials. *Ecology* 55:899.

85. BEER, S., A.J. STEWART & R.G. WETZEL. 1982. Measuring chlorophyll *a* and ^{14}C-labeled photosynthate in aquatic angiosperms by use of a tissue solubilizer. *Plant Physiol.* 69:57.

86. FRANKO, D.A. 1986. Measurement of algal chlorophyll-*a* and carbon assimilation by a tissue solubilizer method: A critical analysis. *Arch. Hydrobiol.* 106:327.

87. HISCOX, J.D. & G.F. ISRAELSTAM. 1979. A method for the extraction of chlorophyll from leaf tissue without maceration. *Can. J. Bot.* 57:1332.

88. FILBIN, G.J. & R.A. HOUGH. 1984. Extraction of ^{14}C-labeled photosynthate from aquatic plants with dimethyl sulfoxide (DMSO). *Limnol. Oceanogr.* 29:426.

89. BROWSE, J.A. 1985. Measurement of photosynthesis by infrared gas analysis. *In* M.M. Littler & D.S. Littler, eds. Handbook of Phycological Methods, Ecological Field Methods: Macroalgae. Cambridge University Press, Cambridge, England.

90. VAN, T.K., W.T. HALLER & G. BOWES. 1976. Comparison of the photosynthetic characteristics of three submersed aquatic plants. *Plant Physiol.* 58:761.

91. CEULEMANS, R., F. KOCKELBERGH & I. IMPENS. 1986. A fast, low cost and low power requiring device for improving closed loop CO_2 measuring systems. *J. Exp. Bot.* 37:1234.

92. BARKO, J.W., P.G. MURPHY & R.G. WETZEL. 1977. An investigation of primary production and ecosystem metabolism in a Lake Michigan dune pond. *Arch. Hydrobiol.* 81:155.

93. SALE, P.J.M., P.T. ORR, G.S. SHELL & D.J.C. ERSKINE. 1985. Photosynthesis and growth rates in *Salivinia molesta* and *Eichhornia crassipes. J. Appl. Ecol.* 22:125.

94. BROWSE, J.A. 1979. An open-circuit infrared gas analysis system for measuring aquatic plant photosynthesis at physiological pH. *Aust. J. Plant Physiol.* 6:493.

95. VINCENT, W.F. & C. HOWARD-WILLIAMS. 1986. Antarctic stream ecosystems: physiological ecology of a blue-green algae epilithon. *Freshwater Biol.* 16:219.

96. TITUS, J.E. & W.H. STONE. 1982. Photosynthetic response of two submersed macrophytes to dissolved inorganic carbon concentrations and pH. *Limnol. Oceanogr.* 27:151.

97. MADSEN, T.V. 1985. A community of submerged aquatic CAM plants in Lake Kalgaard, Denmark. *Aquat. Bot.* 23:97.

98. SALONEN, K. 1981. Rapid and precise determination of total inorganic carbon and some gases in aqueous solutions. *Water Res.* 15:403.

99. STAINTON, M.P. 1973. A syringe gas-stripping procedure for gas-chromatographic determination of dissolved inorganic carbon in fresh water and carbonates in sediments. *J. Fish. Res. Board Can.* 30:1441.

100. GRAN, G. 1952. Determination of the equivalence point in potentiometric titrations. Part II. *Analyst* 77:661.

101. DENNY, P., P.T. ORR & D.J. ERSKINE. 1983. Potentiometric measurements of carbon dioxide flux of submerged aquatic macrophytes in pH-stated natural waters. *Freshwater Biol.* 13:507.

102. ALLEN, E.D. & D.H.N. SPENCE. 1981. The differential ability of aquatic plants to utilize the inorganic carbon supply in freshwaters. *New Phytol.* 87:269.

103. MABERLY, S.C. & D.H.N. SPENCE. 1983. Photosynthetic inorganic carbon use by freshwater plants. *J. Ecol.* 71:705.

104. TALLING, J.F. 1973. The application of some electrochemical methods to the measurement of photosynthesis and respiration in fresh waters. *Freshwater Biol.* 3:335.

6. Bibliography

EDWARDS, R.W. & M. OWENS. 1960. The effects of plants on river conditions. I. Summer crops and estimates of net productivity of macrophytes in a chalk stream. *J. Ecol.* 48:151.

WESTLAKE, D.F. 1963. Comparisons of plant productivity. *Biol. Rev.* 38:385.

WESTLAKE, D.F. 1965. Some basic data for investigations of the productivity of aquatic macrophytes. *In* C.R. Goldman, ed. Primary Production in Aquatic Environments. Mem. Ist. Ital. Idrobiol., 18 Suppl., Univ. California Press, Berkeley.

WESTLAKE, D.F. 1967. Some effects of low-velocity currents on the metabolism of aquatic macrophytes. *J. Exp. Bot.* 18:187.

WETZEL, R.G. 1969. Excretion of dissolved organic compounds of aquatic macrophytes. *Bioscience* 19:539.

VOLLENWEIDER, R.A., ed. 1974. A Manual on Methods for Measuring Primary Production in Aquatic Environments, 2nd ed. IBP Handbook 12. Blackwell Scientific Publ., Oxford, England.

COOPER, J.P., ed. 1975. Photosynthesis and Productivity in Different Environments. IBP Handbook 3. Cambridge University Press, Cambridge, England.

HUTCHINSON, G.E. 1975. A Treatise on Limnology. Vol. III. Limnological Botany. John Wiley & Sons, New York, N.Y.

WETZEL, R.G. 1983. Limnology, 2nd ed. Saunders College Publishing. Philadelphia, Pa.

BOTT, T.L., J.T. BROCK, C.E. CUSHINGS, S.V. GREGORY, D. KING & R.C. PETERSEN. 1978. A comparison of methods for measuring primary productivity and community respiration in streams. *Hydrobiologia* 60:3.

CATTANEO, A. & J. KALFF. 1978. Seasonal changes in the epiphyte community of natural and artificial macrophytes in Lake Memphremagog. (Que.-Vt.). *Hydrobiologia* 60:135.

ONDOK, J.P. & J. KVET. 1978. Selection of sampling areas in assessment of production. *In* D. Dykyjova & J. Kvet, eds. Pond Littoral Ecosystems: Structure and Functioning. Ecological Studies 28. Springer-Verlag, Berlin.

GILLESPIE, D.M. & A.C. BENKE. 1979. Methods for calculating cohort production from field data—some relationships. *Limnol. Oceanogr.* 24:171.

LEWIS, M.R., W.M. KEMP, J.J. CUNNINGHAM & J.C. STEVENSON. 1982. A rapid technique for preparation of aquatic macrophyte samples for measuring ^{14}C incorporation. *Aquat. Bot.* 13:203.

SCHUBAUER, J.P. & C.S. HOPKINSON. 1984. Above- and belowground emergent macrophyte production and turnover in a coastal marsh ecosystem, Georgia. *Limnol. Oceanogr.* 29:1052.

MADSEN, J.D. 1993. Biomass techniques for monitoring and assessing control of aquatic vegetation. *Lakes Reservoir. Manage.* 7:141.

10500   BENTHIC MACROINVERTEBRATES*

10500 A.   Introduction

1. Definition

Benthic macroinvertebrates are animals inhabiting the sediment, or living on or in other available bottom substrates of freshwater, estuarine, and marine ecosystems. During all or part of their life cycles, these organisms may construct attached cases, tubes, or nets that they live on or in; roam freely over rocks, organic debris, and other substrates; or burrow freely in substrates. Although they vary in size from small forms, difficult to see without magnification, to other individuals large enough to see without difficulty, macroinvertebrates are considered historically by definition to be visible to the unaided eye and retained on a U.S. Standard No. 30 sieve (0.595-mm or 0.600-mm openings).[1]

The standard sieve for collecting freshwater, estuarine, and marine benthic macroinvertebrates is the U.S. Standard No. 30 sieve; however, some estuarine and marine programs use the U.S. Standard No. 50 sieve (0.300-mm openings) or the U.S. Standard No. 35 sieve (0.500-mm openings). For all aquatic assessment programs, use of the No. 30 sieve to collect benthic fauna of freshwater, estuarine, and marine habitats or from any water transport system is recommended. To accommodate some old historical data bases and if the data-quality objectives of the study permit, a U.S. Standard No. 28 sieve (1.0-mm openings) might be utilized. To obtain a more representative sample of the benthos that would include smaller forms or early life-stages, and other taxa of macroinvertebrates, a U.S. Standard No. 60 sieve (0.250-mm openings) may be used.

The standardization of bioassessment for species composition, taxa richness, diversity, evenness, trophic levels, and major taxonomic spatial and temporal patterns may be enhanced significantly by the conventional use of a U.S. Standard No. 30 sieve.

The major macroinvertebrates found in freshwater are flatworms, annelids, mollusks, crustaceans, and insects. The major macroinvertebrate groups included in estuarine and marine waters are bryozoans, sponges, annelids, mollusks, roundworms, cnidarians (coelenterates), crustaceans, insects, and echinoderms.

2. Response to Environment

The species composition and population or species density (numbers of individuals per unit area) of macroinvertebrate communities in streams, lakes, estuaries, and marine waters can be uniform from year to year in unperturbed environments. Typically, however, life-cycle dynamics produce variations in species composition and abundance either temporally or spatially.

Most aquatic habitats, particularly free-flowing streams and waters with acceptable water quality and substrate conditions, support diverse macroinvertebrate communities in which there is a reasonably balanced distribution of species among the total number of individuals present. Such communities respond to changing habitats and water quality by alterations in community structure (invertebrate abundance and composition). However, many habitats, especially disturbed ones, are dominated by a few species.

Macroinvertebrate community responses to environmental changes are useful in assessing the impact of municipal, industrial, oil, and agricultural wastes, and impacts from other land uses on surface waters. Four types of environmental changes for which patterns of macroinvertebrate community structure change have been documented are increased inorganic micronutrients, increased organic loading, substrate alteration, and toxic chemical pollution. Inorganic micronutrients and severe organic pollution usually result in a restriction in the variety of macroinvertebrates to only the most tolerant ones and a corresponding increase in density of those tolerating the polluted conditions, usually associated with low dissolved oxygen concentration. In some cases severe organic pollution, siltation, or toxic chemical pollution may reduce or even eliminate the entire macroinvertebrate community from an affected area. Not all cases conform to those described because conditions may be mediated by other environmental (biological, chemical, and physical) conditions.

Assessing the impact of a pollution source generally involves comparing macroinvertebrate communities and their physical habitats at sites influenced by pollution with those collected from adjacent unaffected sites. This can include a gradient away from point sources of contamination. The procedure includes sampling and analyzing types of communities from different sites and subsequently determining whether the presumed pollution-affected community differs from nonaffected ones. The basic information required for most community structure analyses is a count of individuals per species. From the count data, communities can be characterized and compared according to community structure, density, diversity, community metrics, pollution indicators, or other analyses,[1] including various statistical methods (see 10500D). Biomass and productivity estimates also can be determined with the organisms collected.[1-4] Equally desirable is a characterization of the dissolved oxygen concentration, substrate, water depth, type of sediment, grain size of the sediment, total organic carbon (TOC), and other site- and situation-specific characteristics.

While the following macroinvertebrate methods traditionally are used for sampling and quantifying benthic invertebrate communities, other methods also are being evaluated in an effort to develop and implement narrative biological criteria for surface waters.[5] Not discussed here are EPA-developed rapid bioassessment techniques[6] and Environmental Monitoring and Assessment Program (EMAP) protocols for field operations and methods and laboratory methods for sampling macroinvertebrates and assessing the ecological conditions of wadeable streams.

3. References

1. KLEMM, D.J., P.A. LEWIS, F. FULK & J.M. LAZORCHAK. 1990. Macroinvertebrate Field and Laboratory Methods for Evaluating the Biological Integrity of Surface Waters. EPA-600/4-90-030. Environmental

---

* Approved by Standard Methods Committee, 1997.

Monitoring Systems Laboratory, U.S. Environmental Protection Agency, Cincinnati, Ohio.

2. WETZEL, R.G. & G.E. LIKENS. 1991. Limnological Analyses, 2nd ed. Springer-Verlag, New York, N.Y.

3. PLANTE, C. & J.A. DOWNING. 1989. Production of freshwater invertebrate populations in lakes. *Can. J. Fish. Aquat. Sci.* 46:1489.

4. PLANTE, C. & J.A. DOWNING. 1990. Empirical evidence for differences among methods for calculating secondary production. *J. N. Amer. Benthol. Soc.* 9(1):9.

5. U.S. ENVIRONMENTAL PROTECTION AGENCY. 1996. Biological Criteria. Technical Guidance for Streams and Small Rivers. G.R. Gibson, Jr., ed. EPA-822-B-96-001, Off. Water, U.S. Environmental Protection Agency, Washington, D.C.

6. PLAFKIN, J.L., M.T. BARBOUR, K.D. PORTER, S.K. GROSS & R.M. HUGHES. 1989. Rapid Bioassessment Protocols for Use in Streams and Rivers. Benthic Macroinvertebrates and Fish. EPA-440/4-89-001, Off. Water, U.S. Environmental Protection Agency Washington, D.C.

4. Bibliography

NEEDHAM, J.G. & P.R. NEEDHAM. 1941. A Guide to the Study of Freshwater Biology. Comstock Publishing Co., Ithaca, N.Y.

HUTCHINSON, G.E. 1957. A Treatise on Limnology. John Wiley & Sons, New York, N.Y.

HYNES, H.B.N. 1963. The Biology of Polluted Waters. Liverpool Univ. Press, England.

MACAN, T.T. 1963. Freshwater Ecology. John Wiley & Sons, New York, N.Y.

RUTTNER, F. 1966. Fundamentals of Limnology. Univ. Toronto Press, Toronto, Ont.

MACKENTHUN, K.M. & W.M. INGRAM. 1967. Biological Associated Problems in Freshwater Environments. Federal Water Pollution Control Admin., Washington, D.C.

HYNES, H.B.N. 1970. The Ecology of Running Waters. Univ. Toronto Press, Toronto, Ont.

ODUM, E.P. 1971. Fundamentals of Ecology, 3rd ed. Saunders Publishing Co., Philadelphia, Pa.

McNULTY, J.K. 1971. Effects of sewage pollution in Biscayne Bay, Florida: sediments and the distribution of benthic and fouling macroorganisms. *Bull. Mar. Sci. Gulf Carib.* 11:394.

KINNER, P., D. MAURER & V. LEATHEM. 1974. Benthic invertebrates in Delaware Bay: Animal-sediment associations of the dominant species. *Int. Rev. Ges. Hydrobiol.* 59:685.

COULL, B.C., ed. 1977. Ecology of Marine Benthos. University of South Carolina Press, Columbia.

NICHOLS, J.A. 1977. Benthic community structure near the Woods Hole sewage outfall. *Int. Rev. Ges. Hydrobiol.* 52:235.

PEARSON, T.H. & R. ROSENBERG. 1978. Macrobenthic succession in relation to organic enrichment and pollution of the marine environment. *Oceanogr. Mar. Biol. Annu. Rev.* 16:229.

FAUCHALD, K. & P.A. JUMARS. 1979. The diet of worms: A study of polychaete feeding guilds. *Oceanogr. Mar. Biol. Annu. Rev.* 17:193.

EWING, R.M. & D.M. DAUER. 1982. Macrobenthic community structure of the lower Chesapeake Bay. I. Old Plantation Creek, Kings Creek, Cherrystone Inlet and the adjacent offshore area. *Int. Rev. Ges. Hydrobiol.* 67:777.

TOURTELLOTE, G.H. & D.M. DAUER. 1983. Macrobenthic communities of the lower Chesapeake Bay. II. Lynnhaven Roads, Lynnhaven Bay, Broad Bay, and Linkhorn Bay. *Int. Rev. Ges. Hydrobiol.* 68:59.

HAWTHORNE, S.D. & D.M. DAUER. 1983. Macrobenthic communities of the lower Chesapeake Bay. III. Southern Branch of the Elizabeth River. *Int. Rev. Ges. Hydrobiol.* 68:193.

DAUER, D.M., T.L. STOKES, JR., H.R. BARKER, JR., R.M. EWING & J.W. SOURBEER. 1984. Macrobenthic communities of the lower Chesapeake Bay. IV. Baywide transects and inner continental shelf. *Int. Rev. Ges. Hydrobiol.* 69:1.

RESH, V.H. & D.M. ROSENBERG. 1984. The Ecology of Aquatic Insects. Praeger Publ., New York, N.Y.

BROWER, J.R. & J.H. ZAR. 1984. Field and Laboratory Methods for General Ecology. Wm. C. Brown Publ., Dubuque, Iowa.

WARD, J.V. 1984. Ecological perspectives in the management of aquatic insects. *In* V.H. Resh & D.M. Rosenberg, eds. The Ecology of Aquatic Insects. p. 558. Praeger Scientific, New York, N.Y.

WIEDERHOLM, T. 1984. Responses of aquatic insects to environmental pollution. *In* V.H. Resh & D.M. Rosenberg, eds. The Ecology of Aquatic Insects. p. 508. Praeger Scientific, New York, N.Y.

CUMMINS, K.W. & M.A. WILZBACH. 1985. Field Procedures for Analysis of Functional Feeding Groups of Stream Macroinvertebrates. Contribution 1611. Appalachian Environ. Lab., Univ. Maryland, Frostburg.

MINSHALL, G.W., K.W. CUMMINS, R.C. PETERSEN, C.E. CUSHING, D.A. BRUNS, J.R. SEDELL & R.L. VANNOTE. 1985. Developments in stream ecosystem theory. *Can. J. Fish. Aquat. Sci.* 42:1045.

LEONARD, P.M. & D.J. ORTH. 1986. Application and testing of an index of biotic integrity in small, coolwater streams. *Trans. Amer. Fish. Soc.* 115:401.

WARWICK, R.M. 1986. A new method for determining pollution effects on marine macrobenthic communities. *Mar. Biol.* 92:557.

HILSENHOFF, W.L. 1987. An improved biotic index of organic stream pollution. *Great Lakes Entomol.* 20:31.

ROHM, C.M., J.W. GIESE & C.C. BENNETT. 1987. Evaluation of aquatic ecoregion classification of streams in Arkansas. *Freshwater Ecol.* 4: 127.

SHACKELFORD, B. 1988. Rapid Bioassessments of Lotic Macroinvertebrate Communities: Biocriteria Development. Arkansas Dep. Pollution Control & Ecology, Little Rock.

STEEDMAN, R.J. 1988. Modification and assessment of an index of biotic integrity to quantify stream quality in southern Ontario. *Can. J. Fish. Aquat. Sci.* 45:492.

WHITTIER, T.R., R.M. HUGHES & D.P. LARSEN. 1988. Correspondence between ecoregions and spatial patterns in stream ecosystems in Oregon. *Can. J. Fish. Aquat. Sci.* 45:1264.

U.S. ENVIRONMENTAL PROTECTION AGENCY. 1988. Proc. 1st National Workshop on Biological Criteria, Lincolnwood, Ill., Dec. 2–4, 1987. Rep. No. 905/9–89/003, U.S. Environmental Protection Agency, Chicago, Ill.

OHIO ENVIRONMENTAL PROTECTION AGENCY. 1989. Biological Criteria for the Protection of Aquatic Life: Volume III. Standardized Biological Field Sampling and Laboratory Methods for Assessing Fish and Macroinvertebrate Communities. Ohio Environmental Protection Agency, Columbus, Ohio.

REYNOLDSON, T.B., D.W. SCHLOESSER & B.A. MANNY. 1989. Development of a benthic invertebrate objective for mesotrophic Great Lakes waters. *J. Great Lakes Res.* 15:669.

BURD, B.J., A. NEMEC & R.O. BRINKHURST. 1990. The development and application of analytical methods in benthic marine infaunal studies. *Advan. Mar. Biol.* 26:169.

NATIONAL RESEARCH COUNCIL. 1990. Managing Troubled Waters: The Role of Marine Environmental Monitoring. National Academy Press, Washington, D.C.

WESTON, D. 1990. Quantitative examination of macrobenthic changes along an organic enrichment gradient. *Mar. Ecol. Prog. Ser.* 61:233.

YOUNT, J.D. & G.J. NIEMI. 1990. Recovery of lotic communities and ecosystems from disturbance—a narrative review of case studies. *Environ. Manag.* 14:547.

BLAKE, J.A. & A. LISSNER. 1993. Taxonomic atlas of the benthic fauna of the Santa Maria Basin and Western Santa Barbara Channel, Vol. 1. Santa Barbara Mus. Natural History, Santa Barbara, Calif.

ROSENBERG, D.M. & V.H. RESH, eds. 1993. Freshwater Biomonitoring and Benthic Macroinvertebrates. Chapman-Hall, New York, N.Y.

10500 B. Sample Collection

1. General Considerations

Before conducting a benthic survey, determine specific data quality objectives (DQOs) and define clearly the information sought. DQOs are qualitative and quantitative statements developed to specify the quality of data needed to support specific decisions and conclusions about the information sought. Discussion with water chemists, hydrologists, limnologists, and individuals from other disciplines will be helpful. Ultimate selection of a methodology will depend on whether the habitat to be studied is a stream, lake, reservoir, or marine area. For example, to determine whether the macroinvertebrate community downstream from a discharge is damaged, only a few sampling stations upstream and downstream from the discharge are needed. However, if the objective is to delimit the extent of damage from a discharge or series of discharges, it is necessary to have reference stations upstream from all discharges, to bracket each discharge with stations, and to establish stations downstream. In marine waters, it may be necessary to sample a nearby estuary or, for open ocean waters, to sample some distance from the discharge point.

Characterize the physicochemical properties of faunal sampling station substrate and overlying water. Measure such properties as sediment size class distribution (sand, silt, and clay); organic content and toxic pollutant concentrations; temperature, salinity, hardness, alkalinity, dissolved oxygen, total organic carbon (TOC), ammonia, sulfides, and nutrient (total and dissolved) concentrations; and biochemical oxygen demand and water depth.

After gaining a thorough understanding of the factors involved with a particular body of water, select specific areas to be sampled. There is no set number of sampling stations that will be appropriate to monitor all possible waste discharges. No water quality survey is routine, nor can one be conducted totally on a "cookbook" basis. However, if some basic rules such as the following are adhered to, a sound survey can be designed:

1. Always establish a reference station(s) upstream or at a point remote from all wastewater discharges of concern. Because most surveys are made to determine the damage that pollution causes to aquatic life, this will be the basis for comparison of the biota in polluted and unpolluted areas. Preferably have at least two reference stations, one well away from, or upstream from, the discharge and the other directly above, or in the immediate vicinity of, the effluent discharge, but not subject to its influence. Whenever it is feasible, use reference stations having physicochemical characteristics similar to those of the substrate and overlying water of the receiving area.

2. Locate a station immediately downstream or in the affected area in the immediate vicinity of each discharge.

3. If the discharge does not mix completely on entering the body of water, but channels along one side or disperses in a specific direction, locate stations in the left-bank (looking upstream), midchannel, and right-bank sections of the stream, and in concentric arcs in lakes and oceanic waters, or any other configuration that will meet study objectives.

4. Establish stations at various distances downstream from the last discharge of concern to determine the linear extent of damage. In the marine environment, an estuary nearby may be sampled or in open ocean waters samples may be taken in a nearby area comparable with respect to currents, depth, sediment characteristics, and salinity.

5. To permit comparison of macroinvertebrate communities, be sure that all sampling stations are ecologically similar. For example, select stations that are similar with respect to bottom substrate (e.g., sand, gravel, rock, mud, organic content), depth, presence of riffles and pools, stream width, gradient, flow velocity, bank or shore cover, salinity, or hardness, TOC, nutrient and dissolved oxygen concentrations, and wave exposure.

6. Collect samples for physical, toxicological (if applicable), and chemical analyses as close as possible to biological sampling stations to assure correlation of findings; take such samples at the same time and from the same grab when possible. Collect substrate samples for physicochemical analyses from the upper few centimeters where most organisms live.

7. Locate sampling stations for macroinvertebrates in the best physical habitat areas that are not influenced by atypical conditions (e.g., bridges, dams, etc.).

8. Discharges in areas near a coast may be subject to variation in degree of salt water intrusion (salt water wedge). In such areas, macroinvertebrate populations may change drastically; document and/or allow for this effect.

9. When sampling in small, wadeable, first- to third-order streams, initiate sampling at the most downstream station and then proceed upstream to minimize disruptions induced by the sampling itself. This is not necessary for nonwadeable streams and rivers.

For a long-term biological monitoring program collect macroinvertebrates at each station at least once during each of the annual seasons. More frequent sampling may be necessary if the characteristics of the effluents change or if spills occur. Make allowance for collections at night where "drift" or night feeding organisms are of special concern. In general, the most critical period for macroinvertebrates in streams is during periods of high temperature and low flow, whereas in estuarine and marine environments it is the period of maximum stratification and poor vertical mixing. If available time and funds limit sampling frequency, make at least one survey during the critical time.

2. Sampling Design

In biology, the term *population* refers to a group of individuals that are all members of the same species or taxonomic group. In statistics, *population* refers to the entire set of values for the characteristic of interest in a whole sampling universe. For example, researchers interested in determining the mean density of worms in the bottom of a lake might take ten grabs from the lake sediments. The number of worms in each grab would be an *observation*, the density of worms would be the *characteristic of interest*, and the contents within each grab would be the *experimental unit* or *sampling unit*. The entire bottom of the lake would be the *sampling universe* and enough grabs to equal the area of the entire lake bottom would be the *population* (of units).

Similarly, the term *sample* has two, often contradictory, usages. In typical studies, observations usually are not made for all the possible sampling units. Instead, observations are made that make up only a small fraction of the total possible number of observations that could be made. Statistically, this set of observations is referred to as a sample. In the example given above the ten grabs would be a sample. However, in everyday language and as used in this book and most scientific publications, the term "sample" has been used to signify a portion of the real world that has been selected for measurement, such as a water sample, plankton haul, or bottom grab. Therefore, each of the grabs in the example above would be considered a sample, i.e., "ten samples were taken."

Collecting a representative sample is difficult because of variation in successive samples. Without knowledge of sampling variability, the degree to which the data truly represent the population cannot be known. Make replicate observations of a population if definitive statistical inferences about the population are to be made.[1-10]

Standardize sampling design to consider the following requirements:

1. Approximate the set of all samples that can be selected (i.e., separate the sampling universe into all possible samples). For example, if the location (site) containing the population has an area of 1000 m^2 and the sampling device samples an area of 1 m^2, there are 1000 samples that could be collected in the sampling universe.

2. Assign each sample an equal probability of being selected. Using the situation above, divide the area to be sampled into 1000 discrete units.

3. Use a table of random numbers to select sites for sampling, i.e., sample randomly, not haphazardly.

4. The sampling design outlined above is known as simple random sampling. It is often advantageous to determine the number of samples necessary for a certain level of precision while using this type of sampling design. Use the following formula to estimate this number:

$$N = \left(\frac{t \times s}{D \times \bar{x}}\right)^2$$

where:

N = number of samples,
t = tabulated t value at 0.05 level with the degrees of freedom of preliminary survey (generally $t \approx 2.0$ at larger sampling sizes),
s = sampling standard deviation of samples, known from a preliminary survey,
D = required level of precision expressed as a decimal (0.30 to 0.35 usually yields a statistically reliable estimate), and
$\bar{x}$ = sample mean density of preliminary survey.

Specific information (the mean and standard deviation) about the population to be sampled is necessary to estimate the number of samples. Because this information is unknown because sampling has yet to take place, estimate the population's mean and standard deviation by one of three ways: conduct a pilot study; use results from an earlier or a similar study; or make educated estimates of the population mean and standard deviation.[2,11] As an example, in determining the mean chironomid density in relatively homogeneous lake sediments during the summer, and having information that six grabs taken the previous summer produced a mean density of 4230 chironomids/m^2 and a standard deviation of 1628 chironomids/m^2, it may be satisfactory if the final estimate of mean chironomid density is correct within ± 30%, with a 5% probability of error ($\alpha = 0.05$). Using the formula given above,

$$N = \left(\frac{2.5706 \times 1628}{0.30 \times 4230}\right)^2$$

($t = 2.5706$ at a 5% probability of error and 5 degrees of freedom)

$$N = 10.88 \approx 11$$

Thus it is estimated that 11 grabs will be necessary.

5. Simple random sampling is useful in sampling relatively homogeneous areas. However, most taxa are not distributed uniformly over water bottoms. Different habitats (sand, mud, gravel, or organic material) support different densities and species of organisms. In such circumstances, use a stratified random design. In this sampling design a heterogeneous universe (different bottom substrates, current velocities, depths, temperatures, etc.) is divided into more homogeneous strata. Once the strata are defined, use random sampling, as above, within each stratum. Stratified random sampling has two important advantages: it is often valuable to have data on the various subsets of a population (e.g., density of benthic invertebrates in each of the sediment types), and stratified sampling often reduces variability because it deals with more homogeneous subpopulations, allowing for more accurate (closer to the actual value) and precise (less variation among the values) population estimates. Prior information is necessary to divide the population into the various strata. This is usually accomplished through pre-study reconnaissance (a pilot study). Systematic sampling, a third type of sampling design (in addition to simple random and stratified random), often is used in such pilot studies. In a systematic-transect design, conduct sampling at equal intervals along a number of transects within a habitat. This design can be used to identify and locate the existent strata.[2,11]

6. In descriptive studies investigators should take at least three replicate sampling units per station.[2,12] If statistical testing is planned, more replicates probably are necessary.

7. Standardize data acquisition and recording when practical. Use metric units.

3. Sampling Devices, Quantitative

Quantitative and qualitative samplers have been designed to collect organisms from the bottom of different water bodies. The most common quantitative sampling devices are the Ponar, Petersen, and Ekman grabs and the Surber or square-foot stream bottom sampler, all described below.

a. Grab samplers:

Measure each grab-type sampler for actual surface area sampled before it is first used.

1) *The Ponar grab* (Figure 10500:1) is used increasingly in medium to deep rivers, lakes, and reservoirs.[13] It is similar to the Petersen grab in size, weight, lever system, and sample compartment, but has side plates and a screen on top of the sample compartment to prevent sample loss during closure. With one set of weights, the standard 23- by 23-cm sampler weighs 20 kg. A 15- by 15-cm petite Ponar may be used. The large surface disturbance associated with a Ponar grab can be reduced by installing hinged, rather than fixed, screen tops, thereby reducing the pressure wave associated with the sampler's descent. This sampler is best used for sand, gravel, or small rocks with mud, but it can be used in all substrates except bedrock.

2) *The Petersen grab* (Figure 10500:2) is used for sampling hard bottoms such as sand, gravel, marl, and clay in swift currents and deep water. It is an iron, clam-type grab manufactured in various sizes that will sample an area of from 0.06 to 0.09 m.^{2} It

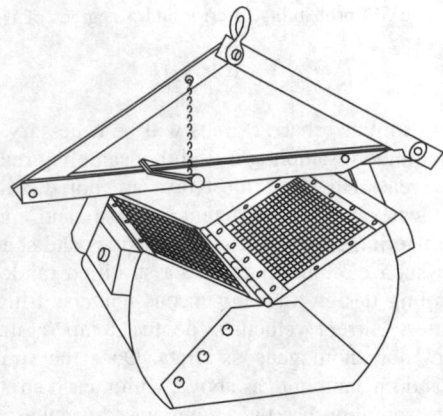

Figure 10500:1. Ponar grab.

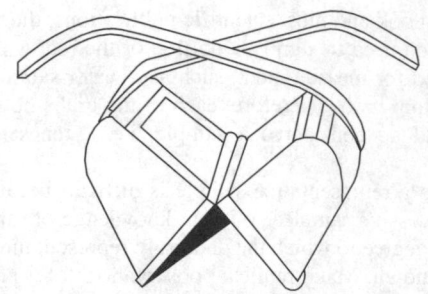

Figure 10500:3. Van Veen grab.

weighs approximately 13.7 kg, but may weigh as much as 31.8 kg when auxiliary weights are bolted to its sides. The primary advantage of the extra weights is to make the grab stable in swift currents and to give additional cutting force in fibrous or firm bottom materials. Modify the sampler by adding end plates, by cutting large strips out at the top of each side, and by adding a hinged 30-mesh screen as in the Ponar grab.[14]

To use the Petersen grab, set the hinged jaws and lower to the bottom slowly to avoid disturbing lighter bottom materials. Ease rope tension to release the catch. As the grab is raised the lever system closes the jaws.

3) *The Van Veen grab* (Figure 10500:3) is used to sample in the open sea and in large lakes. The long arms tend to stabilize the sampler without disturbing the water at the water-substrate interface. It is basically an improved version of the Petersen grab for mud, gravel, pebble, and sand substrates. The sampler is heavy; lower it from a boat or ship platform with mechanical or hydraulic lifts.

4) *The Smith-McIntyre grab* (Figure 10500:4) has the heavy steel construction of the Petersen, but its jaws are closed by strong coil springs.[15] Chief advantages are its stability and easier control in rough water. Its bulk and heavy weight require operation from a large boat equipped with a winch. The 45.4-kg grab can sample an area of 0.2 m,[16-18] but smaller models (0.1 m² or 0.05 to 0.06 m²) are available.

5) *The Shipek grab* (Figure 10500:5) is designed to take a sample 0.04 m² in surface area and approximately 10 cm deep at the center. The sample compartment is composed of two concentric half cylinders. When the grab touches bottom, inertia from a self-contained weight releases a catch and helical springs rotate the inner half cylinder by 180°. The sample bucket may be disengaged from the upper semi-cylinder by releasing two retaining latches. This grab is for special use in marine waters and large inland bodies of water.

6) *The Ekman grab* (Figure 10500:6) is useful only for sampling silt, muck, and sludge in water with little current. It is difficult to use when rocky or sandy bottoms or moderate macrophyte growth are present because small pebbles or grit or macrophyte stems prevent proper jaw closure. The grab weighs approximately 3.2 kg. The box-like part holding the sample has spring-operated jaws on the bottom that must be cocked manually (exercise caution in cocking and handling the grab because of possible injuries if jaws are tripped accidentally). At the top of the grab are two hinged overlapping lids that are held open partially during descent by water passing through the sample compartment. These lids are held shut by water pressure when the sampler is being retrieved. The grab is made in three sizes: 15 × 15 cm, 23 × 23 cm, and 30 × 30 cm, but the smallest size usually is adequate. A taller model of this sampler, either 23 cm or 30.5 cm tall, is available. To prevent sample overflow and loss, place a Standard U.S. No. 30 sieve insert in the top for deep sediments.

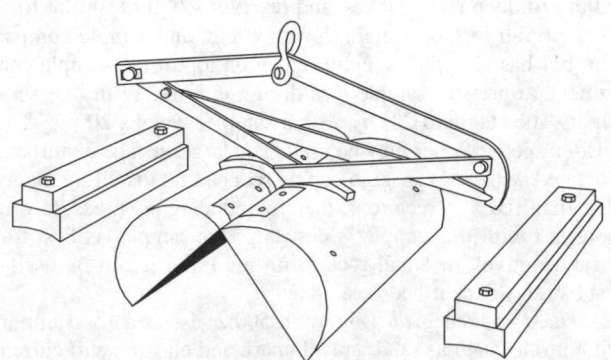

Figure 10500:2. Petersen grab.

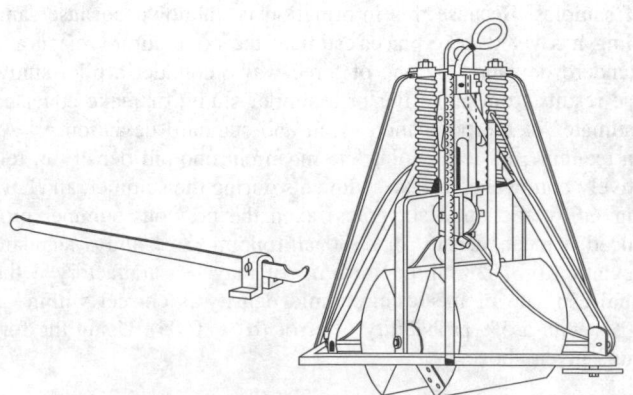

Figure 10500:4. Smith-McIntyre grab.

b. Riffle/run samplers:

1) *Surber-type samplers* (Figure 10500:7)[19] consist of two brass frames, each 30.5 cm (1 ft) square, hinged together along one edge. When in use, the two frames are locked at right angles, one frame marking off the area of substrate to be sampled, and the other supporting a net to collect organisms washed into it from the sample area.

The net usually is 69 cm long with the first few centimeters and the wings constructed of heavier material (canvas, taffeta) to increase durability. Standard mesh size is 9 threads/cm. While a smaller mesh size might increase the number of smaller invertebrates and young instars collected, it also will clog more easily and exert more resistance to the current than a larger mesh. This could result in a loss of organisms due to backwashing from the sample net. This sampler is specific for macrobenthos; many microcomponents of the benthos are not collected.

Use this sampler in shallow (30 cm or less), flowing water. When it is used in deeper water some organisms may be carried over the top of the sampler. Position sampler securely on the stream bottom parallel to water flow with the net portion downstream. Take care not to disturb the substrate upstream from sampler. Leave no gaps under the edges of the frame that would allow water to wash under the net. Fill gaps that may occur along the back edge of the sampler by carefully shifting rocks and gravel along the outside edge. When the sampler is in place (it may be necessary to hold it in place with one hand in a strong current), carefully turn over and lightly hand-rub all rocks and large stones inside the frame to dislodge organisms clinging to them. Examine each stone for organisms, larval or pupal cases, etc., that may be clinging to it before discarding. Scrape attached algae, insect cases, etc., from the stones into the sampler net. Stir remaining gravel and sand with the hands or a stick to a depth of 5 to 10 cm, depending on the substrate, to dislodge bottom-dwelling organisms. It may be necessary to hand-pick some mussels and snails that are not carried into the net by the current.

Remove sample by inverting net into sample container. Carefully examine net for small organisms clinging to it. Remove these, preferably with forceps to avoid damage, and include in sample. Rinse sampler net after each use.

A common problem in using the Surber sampler is that organisms wash under the bottom edge of the sampler. The following modifications have been suggested for different substrates:

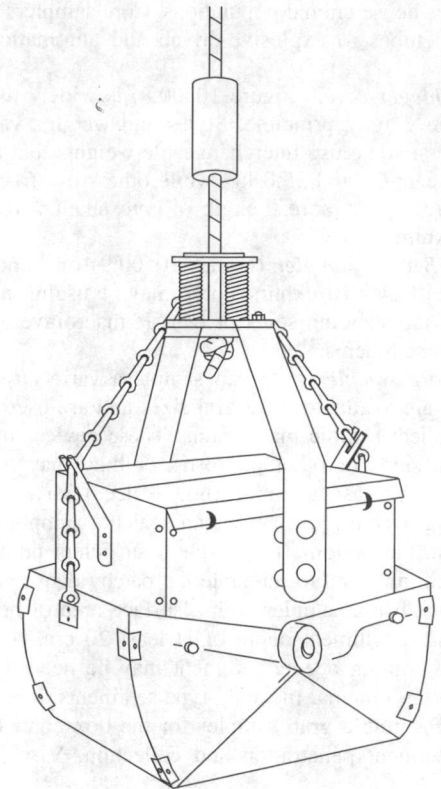

Figure 10500:6. Ekman grab.

For loose gravel—Extend bottom edge of Surber frame to 5 or more cm allowing for insertion of frame into substrate to a greater depth. This method works well in soft substrates such as sand and gravel where the current causes substrate shifting.

For coarse gravel and rock—Add serrated extension to back edge of frame to secure it and reduce washing from under this edge. This method is helpful in hard gravel and rock substrates where sinking the entire frame is impossible.

For gravel and bedrock—Add a 5-cm band of flexible material to bottom edge of sampler to create a seal in rocky, uneven substrates. Make band of foam rubber or fine-textured synthetic sponge. Remove organisms that stick to foam and include in sample.

2) *Hess-type samplers* are cylindrical with enclosed sides and an open top. They function similarly to the Surber-type samplers.[2]

c. Core or cylindrical samplers: Use core or cylindrical samplers to sample sediments in depth. Efficient use as surface sam-

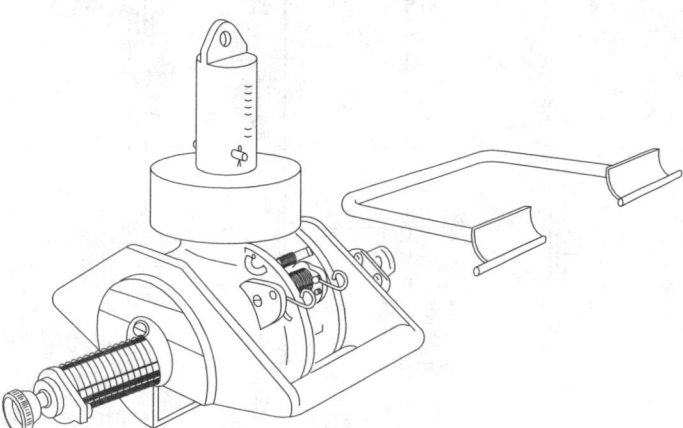

Figure 10500:5. Shipek grab.

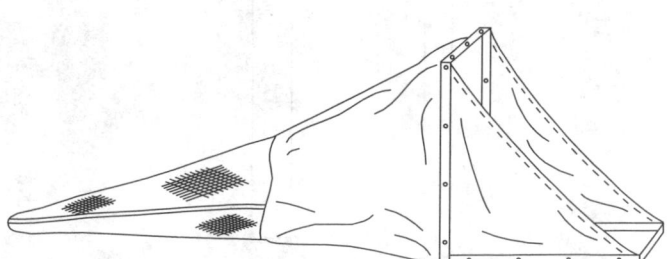

Figure 10500:7. Surber or square-foot sampler.

plers requires dense animal populations. Core samplers vary from hand-pushed tubes to explosive-driven and automatic-surfacing models.[2,20]

1) *The Phleger corer* (Figure 10500:8) is widely used. It operates on the gravity principle. Styles and weights vary among manufacturers; some use interchangeable weights that allow variations between 7.7 and 35.0 kg, while others use fixed weights weighing 41.0 kg or more. Length of core taken will vary with substrate texture.

2) *The KB core sampler* (Figure 10500:9) or a modification known as the Kajak-Brinkhurst corer, may be useful in obtaining estimates of the standing stock of benthic macroinvertebrates inhabiting soft sediments.[21]

3) *Box core samplers*[22-25] can sample a variety of sediment types. They are available in several sizes and are used in marine waters to collect benthic macrofauna. These devices may be deployed from ships or other platforms or they may be used by divers. Preferably use a box coring device with a rectangular corer, having a cutting arm that can seal the sample before retraction from the bottom. To sample a sufficient number of individuals and taxa, and to integrate the patchy distribution of the benthic fauna, use a sampler with a surface area of no less than 100 cm^2 and a sediment depth of at least 20 cm. A box corer capable of sampling deeper sediment may be needed to collect deep burrowing infauna. In sandy-type sediments, it may be necessary to substitute a grab sampler for the box corer to achieve adequate sediment penetration and collection. Visually inspect

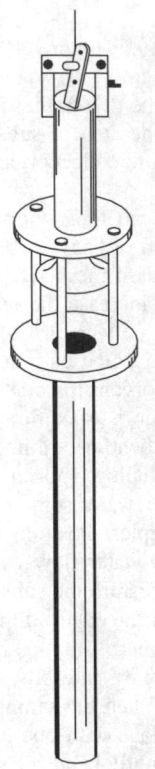

Figure 10500:9. KB corer.

each sample to ensure that an undisturbed and adequate amount of sample is collected.

4) *The Wilding or stovepipe sampler* (Figure 10500:10)[26] is made in various sizes and with many modifications. It is especially useful for quantitatively sampling a bottom with dense, vascular plant growth. It may be used to sample vegetation, mud-water interface sediment, or most shallow stream substrates. However, large volumes of vegetation, when sampled in this way, may require a great deal of time for laboratory processing.

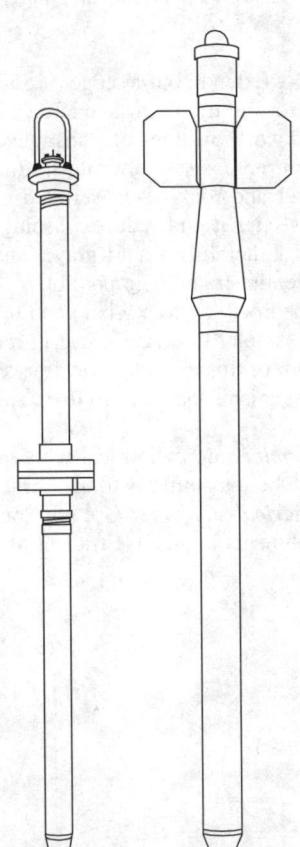

Figure 10500:8. Phleger core sampler.

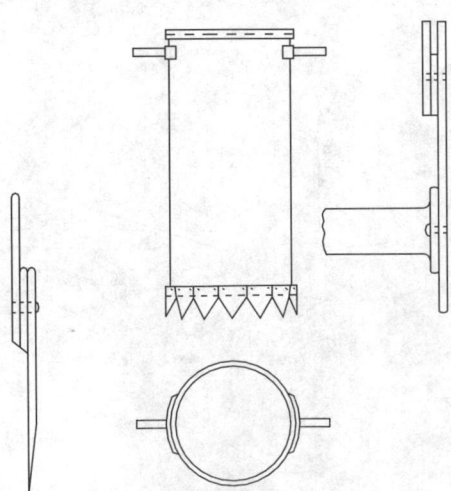

Figure 10500:10. Wilding or stovepipe sampler.

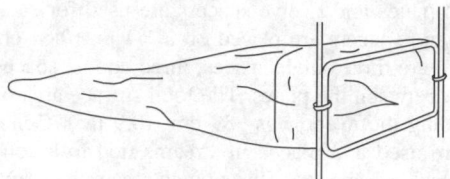

Figure 10500:11. Drift net sampler.

d. Drift samplers: Drift samplers, usually in the form of nets (Figure 10500:11), are anchored in flowing water for capture of macroinvertebrates that have migrated or have been dislodged from the bottom substrates into the current. Drift organisms are important to the stream ecosystem because they are prey for stream fish and should be considered in the study of fish populations. Drift organisms respond to pollutional stresses, including spills, by increased drift from an affected area; therefore, drift is important in water-quality investigations, especially of spills of toxic materials. Drift also is a factor in recolonizing denuded areas and it contributes to recovery of disturbed streams.

Use nets having a 929-cm² upstream opening and mesh equivalent to U.S. Standard No. 30 screen (0.595-mm pore size). After placing the net in the water, frequently remove organisms and debris to prevent clogging and subsequent diversion of water at the net opening. Use replicate samples as appropriate to meet study objectives. Set drift-net samples for any specified time (usually 1 to 3 h) but use the same time for each station. Sampling between dusk and 1 AM is optimum.

The total quantity (numbers or biomass) of organisms drifting past a given station is the best measure of drift intensity. Report data in terms of numbers or biomass/m³/24 h.[27–29]

4. Sampling Devices, Qualitative

When sampling qualitatively, search for as many different organisms as possible. Collect samples by any method that will capture representative species.

a. Dip, kick nets are the most versatile collecting devices for shallow, flowing water, and are useful also for shoreline collecting in lakes. When combined with a standardized kicking technique,[30] these nets are appropriate for quantitatively sampling macroinvertebrates.[31]

b. Tow nets, dredges, or trawls range from simple sled-mounted nets to complicated devices incorporating teeth that dig into the bottom. Some models feature special apparatus to hold the net open during towing and to close it during descent and retrieval. Available styles have been discussed elsewhere.[20,32,33]

5. Sampling Devices, Artificial Substrate Samplers

Artificial substrate samplers are devices of standard composition and configuration placed in the water for a predetermined exposure period for colonization by macroinvertebrate communities. Because many physical variables encountered in bottom sampling are minimized, e.g., depth, light penetration, temperature differences, and species substrate preferences, artificial substrate sampling complements other types of sampling. Like natural submerged substrates such as logs and pilings, artificial substrates are colonized primarily by immature aquatic insects,

crustaceans, coelenterates, bryozoans, and to some extent worms, gastropods, and mollusks. In lotic systems the organisms that colonize artificial substrates are primarily drift organisms, such as immature insects and eggs, carried by water currents. Placement conditions should be similar, so the numbers and kinds of organisms reflect capacity to support aquatic life.

Position artificial substrates in the euphotic zone (0.3 m) for maximum abundance and diversity of macroinvertebrates.[12] Optimum time for substrate colonization is 6 weeks for most waters. For uniformity of depth, suspend sampler from floats on a 3.2-mm steel cable. If vandalism is a problem, use subsurface floats or place sampler on the bottom. Regardless of installation technique, use uniform procedures.

At shallow water stations (less than 1.2 m deep), install samplers so that they are located midway in the water column at low flow. For samplers installed in July when the water depth is about 1.2 m and the August average low flow is 0.6 m, install 0.3 m above the bottom. Take care not to let samplers touch the bottom or they may become covered with silt, thereby increasing the sampling error. In shallow streams with sheet rock bottoms, secure artificial substrates to 0.95-cm steel rods that are driven into the substrate or secure to rods that are mounted on low, flat, rectangular blocks.

Before removing samples from the water, it may be necessary to enclose them in an oversized plastic bag (double wrapping) that is tightly sealed to prevent possible loss of organisms or to use a large dip net (openings equivalent to a U.S. Standard No.

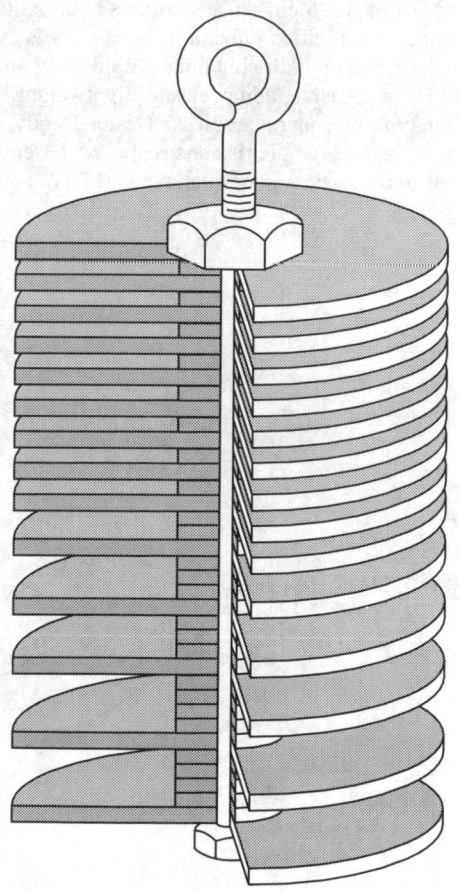

Figure 10500:12. Hester-Dendy artificial substrate unit.

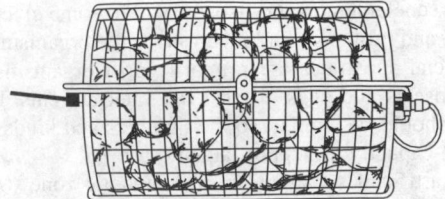

Figure 10500:13. Basket sampler.

30 sieve) when the sample is removed. Disassemble sampler and brush in a pan of water in the field or add preservative to the bag containing the intact sampler, and disassemble and brush later in the laboratory.

Although many different styles of artificial substrate samplers have been tested,[34] the Fullner[35] modification of the Hester-Dendy[36] multiplate and the basket sampler[12] are used widely.

a. Multiple-plate or modified Hester-Dendy sampler (Figure 10500:12) is constructed of 0.3-cm-thick tempered hardboard with 7.5-cm round plates and 2.5-cm round spacers that have center-drilled holes. The plates are separated by spacers on a 0.63-cm-diam eyebolt, held in place by a nut at the top and bottom. In each sampler, 14 large plates and 24 spacers are used. Separate the top 9 plates by a single spacer, Plate 10 by two spacers, Plates 11 and 12 by three spacers, and Plates 13 and 14 by four spacers. The sampler is approximately 14 cm long and 7.5 cm in diameter, has an exposed surface area of approximately 1300 cm², and weighs about 0.45 kg. Do not reuse samplers exposed to oils and chemicals that may inhibit colonization. Because it is cylindrical, the sampler fits a wide-mouth container for shipping and storage. The sampler is inexpensive, compact, and lightweight.[12,35,36]

A different type of square modified Hester-Dendy, multiple-plate artificial substrate sampler is constructed of 0.3-cm tempered hardboard cut into 7.6-cm square plates and 2.5-cm square spac-

ers.[37] Eight plates and twelve spacers are used for each sampler. The plates and spacers are placed on a 1/4-in. (0.64-cm) eyebolt so that there are three single spaces, three double spaces, and one triple space between the plates. The total surface area of the sampler, excluding the eyebolt, is 939 cm² (0.9 m²). Generally, five samplers are used and placed in streams tied to a concrete construction block as anchor. This prevents samplers from coming into contact with the natural substrates.

b. The basket sampler[12] (Figure 10500:13) is a cylindrical "barbecue" basket 28 cm long and 17.8 cm in diameter, filled with approximately 30 5.1-cm-diam rocks or rocklike material weighing 7.7 kg. A hinged side door allows access to the contents. The sampler provides an estimated 0.24 m² of surface area for colonization. The factors governing proper installation and collection are the same as those described for the multiplate sampler. Some investigators prefer using the basket because natural substrate materials are used for colonization.

c. Marsh net sampler (Figure 10500:14) is used for sampling macroinvertebrates in estuarine and marine environments.[38] It can be used in different habitats (e.g., marsh, beach, tidal creek, and tidal flat) of estuarine and marine intertidal zones to depths of 3 m. The metal frame is constructed of No. 22 galvanized sheet metal and 1/4-in. (6-mm) welding rods. A 0.5-m plankton net of nylon monofilament screen is laced to the posterior end of the frame. The net has a bayonet-type cod end for easy removal. The mesh size of the plankton net and cod end is about 1 mm (bar measure). The frame and net weigh 5 kg. The collecting procedures are the same in all habitats of the intertidal zone. The net is placed at one end of the sampling area and 30 m of rope is paid out in an arc to prevent the operator from disturbing the sampling site. The net is then retrieved by hand at a rate of about 0.3 m/s. Advantages are that the sampling distance does not have to be measured before taking the sample, the net can be towed at a constant speed, and samples also can be taken over soft mud bottoms.[38]

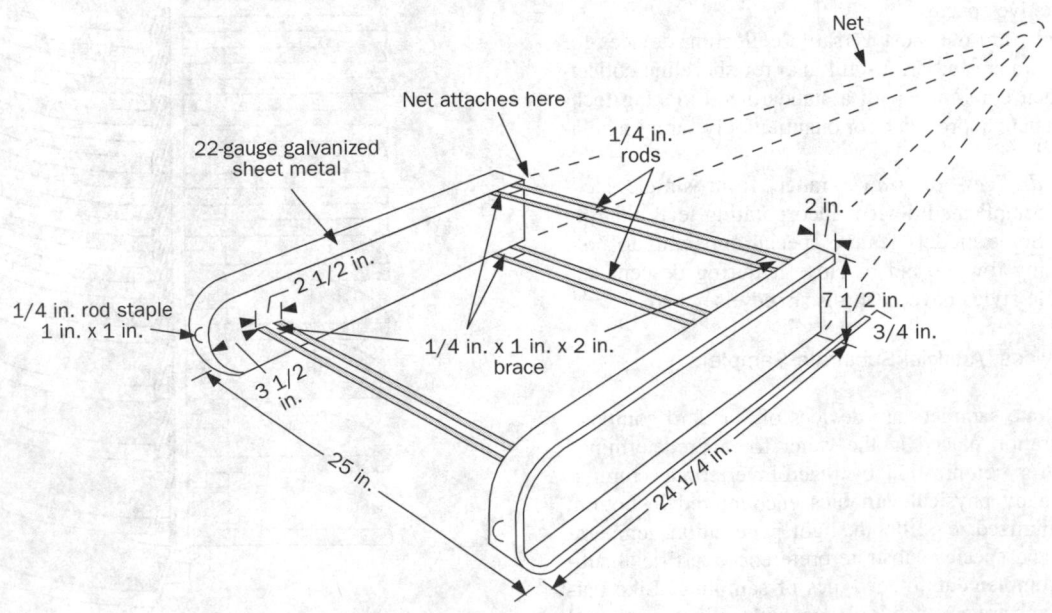

Figure 10500:14. Marsh net sampler.

6. Suction Samplers

Suction samplers are used widely for collecting benthic macroinvertebrate samples.[39,40] These samplers can be placed directly on specific sampling sites, but a scuba diver is required to collect samples.[41] Improved accuracy of locating sampling sites and ability to collect a large number of replicate samples may outweigh the disadvantage of using a diver. Suction samplers have been used widely in sampling marine environments, but they have obvious depth limitations.

7. References

1. SNEDECOR, G.W. & W.G. COCHRAN. 1967. Statistical Methods. Iowa State Univ. Press, Ames.
2. KLEMM, D.J., P.A. LEWIS, F. FULK & J.M. LAZORCHAK. 1990. Macroinvertebrate Field and Laboratory Methods for Evaluating the Biological Integrity of Surface Waters. EPA-600/4-90-030. Environmental Monitoring Systems Laboratory, U.S. Environmental Protection Agency, Cincinnati, Ohio.
3. GREEN, R.H. 1979. Sampling Design and Statistical Methods for Environmental Biologists. John Wiley & Sons, New York, N.Y.
4. ZAR, J.H. 1984. Biostatistical Analysis, 2nd ed. Prentice-Hall, Englewood Cliffs, N.J.
5. UNDERWOOD, A.J. & C.H. PETERSON. 1988. Towards an ecological framework for investigating pollution. *Mar. Ecol. Progr. Ser.* 46:227.
6. MANTEL, N. 1987. The detection of disease clustering and a generalized regression approach. *Cancer Res.* 27:200.
7. LEGENDRE, P. & M.J. FORTIN. 1989. Spatial patterns and ecological analysis. *Vegetation* 80:107.
8. GREEN, R.H., J.M. BOYD & J.S. MACDONALD. 1993. Relating sets of variables in environmental studies: the sediment quality triad as a paradigm. *Environmetrics* 4:439.
9. UNDERWOOD, A.J. 1994. On beyond BACI: sampling designs that might reliably detect environmental disturbances. *Ecol. Appl.* 4:3.
10. LANDIS, W.G., G.B. MATTHEWS, R.A. MATTHEWS & A. SARGENT. 1994. Application of multivariate techniques to endpoint determination, selection and evaluation in ecological risk assessment. *Environ. Toxicol. Chem.* 13:1917.
11. GAUGUSH, R.F. 1987. Sampling Design for Reservoir Water Quality Investigations. Instruction Rep. E-87-1, Waterways Experiment Station, Vicksburg, Miss. National Technical Information Serv., Springfield, Va.
12. MASON, W.T., JR., C.I. WEBER, P.A. LEWIS & E.C. JULIAN. 1973. Factors affecting the performance of basket and multiplate macroinvertebrate samplers. *Freshwater Biol.* 3:409.
13. POWERS, C.F. & A. ROBERTSON. 1967. Design and Evaluation of an All-Purpose Benthos Sampler. Spec. Rep. No. 30, Great Lakes Research Div., Univ. Michigan, Ann Arbor.
14. WEBER, C.I., ed. 1973. Biological Field and Laboratory Methods for Measuring the Quality of Surface Waters and Effluents. EPA-670/4-73-001, U.S. Environmental Protection Agency, Cincinnati, Ohio.
15. SMITH, W. & A.D. MCINTYRE. 1954. A spring-loaded bottom sampler. *J. Mar. Biol. Assoc. U.K.* 33:257.
16. ELLIOTT, J.M. 1971. Some Methods for the Statistical Analysis of Samples of Benthic Invertebrates. Freshwater Biological Assoc., Sci. Publ. No. 25.
17. MCINTYRE, A.D. 1971. Efficiency of Marine Bottom Samplers. *In* N.A. Holme & A.D. McIntyre, eds. Methods for the Study of Marine Benthos. IBP Handbook No. 16, p. 140. Blackwell Scientific Publications, Oxford, England.
18. WIGLEY, R.L. 1967. Comparative efficiency of Van Veen and Smith-McIntyre grab samplers as recorded by motion pictures. *Ecology* 48:168.
19. SURBER, E. 1937. Rainbow trout and bottom fauna production in one mile of stream. *Trans. Amer. Fish. Soc.* 66:193.
20. BARNES, H. 1959. Oceanographic and Marine Biology. George Allen and Unwin, Ltd., London, England.
21. BRINKHURST, R.O., K.E. CHUA & E. BATOOSINGH. 1969. Modifications in sampling procedures as applied to studies on the bacteria and tubificid oligochaetes inhabiting aquatic sediments. *J. Fish Res. Board Can.* 26:2581.
22. HOLME, N.A. & A.D. MCINTYRE, eds. 1984. Methods for the Study of Marine Benthos. IBP Handbook 16, Blackwell Scientific Publ., Oxford, England.
23. HESSLER, R.R. & P.A. JUMARS. 1974. Abyssal community analysis from replicate box cores in the central north Pacific. *Deep-Sea Res.* 21:185.
24. PROBERT, P.K. 1984. A comparison of macrofaunal samples taken by box corer and anchor-box dredge. *New Zealand Oceanogr. Inst. Rec.* 4(13):149.
25. ELEFTERIOU, A. & N.A. HOLME. 1984. Macrofauna techniques. *In* N.A. Holme & A.D. McIntyre, eds. Methods for the Study of Marine Benthos. IBP Handbook 16, Chapter 6, p. 140. Blackwell Scientific Publ., Oxford, England.
26. WILDING, J.L. 1940. A new square-foot aquatic sampler. Limnol. Soc. Amer. Special Pub. No. 4, Ann Arbor, Mich.
27. WATERS, T.F. 1961. Standing crop and drift of stream bottom organisms. *Ecology* 42:532.
28. DIMOND, J.B. 1967. Pesticides and Stream Insects. Bull. No. 2, Maine Forest Serv., Augusta, & Conservation Foundation, Washington, D.C.
29. WATERS, T.F. 1972. The drift of stream insects. *Annu. Rev. Entomol.* 17:253.
30. FROST, S., A. HUN & W. KERSHAW. 1971. Evaluation of a kicking technique for sampling stream bottom fauna. *Can. J. Zool.* 49:167.
31. CROSSMAN, J.S., J. CAIRNS, JR. & R.L. KAESLER. 1973. Aquatic Invertebrate Recovery in the Clinch River Following Hazardous Spills and Floods. Water Resour. Res. Center Bull. 63, Virginia Polytechnic Inst. & State Univ., Blacksburg.
32. WELCH, P.S. 1948. Limnological Methods. Blakiston Co., Philadelphia, Pa.
33. USINGER, R.L. 1956. Aquatic Insects of California, with Keys to North American Genera and California Species. Univ. California Press, Berkeley.
34. BEAK, T.W., T.C. GRIFFING & G. APPLEBY. 1974. Use of artificial substrates to assess water pollution. *In* Proceedings Biological Methods for the Assessment of Water Quality. American Soc. Testing & Materials, Philadelphia, Pa.
35. FULLNER, R.W. 1971. A comparison of macroinvertebrates collected by basket and modified multiple-plate samplers. *J. Water Pollut. Control Fed.* 43:494.
36. HESTER, F.E. & J.B. DENDY. 1962. A multiple-plate sampler for aquatic macroinvertebrates. *Trans. Amer. Fish. Soc.* 91:420.
37. OHIO ENVIRONMENTAL PROTECTION AGENCY. 1989. Biological criteria for the protection of aquatic life: Volume II, Users Manual for Biological Field Assessment of Ohio Surface Waters & Volume III, Standardized Biological Field Sampling and Laboratory Methods for Assessing Fish and Macroinvertebrates Communities. Ohio Environmental Protection Agency, Div. Water Quality Monitoring & Assessment, Surface Water Section, Columbus.
38. PULLEN, E.J., C.R. MOCK & R.D. RINGO. 1968. A net for sampling the intertidal zone of an estuary. *Limnol. Oceanogr.* 13:200.
39. GALE, W. & J. THOMPSON. 1975. A suction sampler for quantitatively sampling benthos on rocky substrates in rivers. *Trans. Amer. Fish. Soc.* 104:398.
40. LARSEN, P.F. 1974. A remotely operated shallow water benthic suction sampler. *Chesapeake Sci.* 15:176.
41. SIMMONS, G.M., JR. 1977. The Use of Underwater Equipment in Freshwater Research. VPI-SG-77–03, Virginia Polytechnic Inst. & State Univ., Blacksburg.

8. Bibliography

MACAN, T.T. 1958. Methods of sampling bottom fauna in stony streams. *Mitt. Int. Ver. Limnol.* 8:1.

DICKSON, K.L., J. CAIRNS, JR. & J.C. ARNOLD. 1971. An evaluation of the use of a basket type artificial substrate for sampling macroinvertebrate organisms. *Trans. Amer. Fish. Soc.* 100:553.

WASHINGTON, H.G. 1984. Diversity, biotic and similarity indices. A review with special relevance to aquatic ecosystems. *Water Res.* 18:653.

WRONA, F.J., P. CALOW, T. FORD, D.J. BAIRD & L. MALTBY. 1986. Estimating the abundance of stone-dwelling organisms: a new method. *Can. J. Fish. Aquat. Sci.* 43:2025.

BRITTAIN, J.E. & T.J. EIKELAND. 1988. Invertebrate drift—A review. *Hydrobiologia* 166:77.

FERRARO, S.P., F.A. COLE, W.A. DEBEN & R.C. SWARTZ. 1989. Power-cost efficiency of eight macroinvertebrate sampling schemes in Puget Sound, Washington, U.S.A. *Can. J. Fish. Aquat. Sci.* 46:2157.

GOLDSTEIN, R. 1989. Power and sample size via MS/PC-DOS computers. *Amer. Statist.* 43:253.

PETERMAN, R.M. 1990. Statistical power analysis can improve fisheries research and management. *Can. J. Fish. Aquat. Sci.* 47:2.

SCRIMGEOUR, G.J., J.M. CULPAND & N.E. GLOZIER. 1993. An improved technique for sampling lotic invertebrates. *Hydrobiologia* 254(2):65.

AMERICAN SOCIETY FOR TESTING AND MATERIALS. 1996. 1996 Annual Book of Standards. Section 11, Vol. 11.05, Biological Effects and Environmental Fate; Biotechnology; Pesticides, Standards D-4342-84, D-4343-84, D-4344-84, D-4345-84, D-4346-84, D-4347-84, D-4348-84, D-4387-84, D-4401-84, D-4407-84, D-4556-85, D-4557-85, D-4558-85. American Soc. Testing & Materials, W. Conshohocken, Pa.

10500 C. Sample Processing and Analysis

1. Sample Processing

After collecting a bottom grab sample, transfer it to either specially designed sieve tables (or hoppers) or a container. If a container (such as small trash can) is used, dilute with ambient water, and swirl. Pour slurry gradually into a sieve bucket. Gently wash slurry over screen to prevent damaging or losing specimens. Slurries that clog the screen require removal of screened material. A series of one or two coarser screens (e.g., 1-cm and 0.5-cm mesh) will hold back larger materials such as leaves, sticks, shells, and gravel while permitting organisms and smaller materials to pass through to the bottom sieve. Carefully check rocks, sticks, shells, and other objects for attached or burrowed organisms before discarding. A soft-bristled toothbrush may be used to remove attached invertebrates from rocks, sticks, and similar objects.

Wash residual on the screen into a container. A cheesecloth bag is very useful because it does not restrict the quantity of wash water. Label containers with a collection code but do not affix labels to lids. Similar labels can be written with pencil or indelible ink on high-rag-content paper and placed in the container. Record label code on a field sheet that describes location, date, type of sample, collector's name, and other pertinent information.

Use laboratory elutriation devices[1,2] as appropriate to reduce time required to sort benthic organisms from samples containing large amounts of silt, mud, or clay. Wash screened material into a container and fix the contents in a solution of 10% buffered formalin or 70% ethanol.[3–6] If ethanol is used, do not fill more than one-half the container with screened material. Preserve and store animals with calcareous shells or exoskeletons, i.e., mussels, snails, crayfish, and ostracods, in 70% ethanol.[6,7]

Some macroinvertebrates (soft-bodied animals) are identified more easily if they are relaxed to prevent constriction during preservation. Common relaxants include carbonated water (soda water) or carbon dioxide added to water. Other relaxants include aqueous solutions of 70% ethyl alcohol, 2% nicotine sulfate, propylene phenoxetol, or 5 to 10% solutions of either chlorotone, chloral hydrate, or magnesium sulfate added gradually to the water containing the soft-bodied animals until the degree of relaxation sought is reached. Narcotize organisms before fixing them. Fix annelid specimens in 5 to 10% formalin before preserving them in 70 to 80% ethanol (note that alcohol is not a satisfactory tissue fixative). Fixation stabilizes tissue proteins to retain characteristics of the soft body (e.g., segmented worms) form.[8]

For qualitative samples place rocks, sticks, and other objects in a white pan partially filled with water. Many animals will float free from these objects and can be removed with forceps.

Assign identification numbers either in the field or at the laboratory and transcribe information from the labels to a permanent ledger. The ledger provides a convenient reference in identifying number of samples collected at various places, time of sampling, and water characteristics.

Preserve and store organisms taken from artificial substrates and sieved with a U.S. Standard No. 30 sieve in 70% ethanol. Fix soft-bodied organisms first with 5 to 10% buffered formalin, and preserve in 5 to 10% buffered formalin or 70% ethanol.

2. Sorting and Identification

Whether organisms are sorted in the field or the laboratory, follow consistent procedures. Before processing a sample, transfer information from the label to a data sheet that provides space for scientific names and number of individuals. Place sample directly in a shallow white tray with water for sorting. To facilitate sorting organisms from detritus, the organisms may be stained with rose bengal (200 mg/L or achieve a light pink color) in the formalin or ethanol preservative for at least 24 h.[9] NOTE: Excessive staining may prevent specific identification of some specimens. Examine entire sample and separate organisms unless they occur in very large numbers. If a subsample is sorted, take care that rare forms are not excluded. As organisms are picked from the sample, sort under a scanning lens or stereoscopic microscope, separate them into different taxonomic categories, identify to the lowest taxonomic level to meet the data quality objectives, and record on the data sheet. Place animals in separate vials according to category and fill vials with 5 to 10% formalin or 70% ethanol. Place inside vials labels containing sample tracking number, date collected, sampling location, and names of organisms.

Identify animals in each vial using a stereoscopic and compound microscope, according to need, and available experience and resources. Identify organisms to species level if possible. Additional sources of information on laboratory techniques and identification of macroinvertebrates are available (see Bibliography and Section 10900).

3. References

1. WORSWICK, J.M. & M.T. BARBOUR. 1974. An elutriation apparatus for macroinvertebrates. *Limnol. Oceanogr.* 19:538.

2. LAUFF, G.H., K.W. CUMMINS, C.H. ERIKSON & M. PARKER. 1961. A method for sorting bottom fauna samples by elutriation. *Limnol. Oceanogr.* 6:462.

3. EDMONDSON, W.T., ed. 1959. Ward and Whipple's Freshwater Biology, 2nd ed. John Wiley & Sons, New York, N.Y.

4. COOK, D.G. & R.O. BRINKHURST. 1973. Marine Flora and Fauna of the Northeastern United States, Annelida: Oligochaeta. NOAA Tech. Rep. NMFS CIRC-374, U.S. Dep. Commerce, National Oceanic Atmospheric Admin., National Marine Fisheries Serv., Seattle, Wash.

5. KLEMM, D.J. 1982. Leeches (Annelida Hirudinea:) of North America. EPA-600/3-82-025, Environmental Monitoring & Support Lab., U.S. Environmental Protection Agency, Cincinnati, Ohio.

6. PENNAK, R.W. 1989. Freshwater Invertebrates of the United States–Protozoa to Mollusca, 3rd ed. John Wiley & Sons, Inc., New York, N.Y.

7. BURCH, J.B. 1972. Freshwater Sphaeriacean Clams (Mollusca: Pelecypoda) of North America. U.S. Environmental Protection Agency, Cincinnati, Ohio.

8. KLEMM, D.J., ed. 1985. A Guide to the Freshwater Annelida (Polychaeta, Naidid and Tubificid Oligochaeta, and Hirudinea) of North America. Kendall/Hunt Publ. Co., Dubuque, Iowa.

9. MASON, W.T., JR. & P.P. YEVICH. 1967. The use of phloxine B and rose bengal stains to facilitate sorting benthic samples. *Trans. Amer. Microsc. Soc.* 86:221.

4. Bibliography

HARTMAN, O. 1941. Polychaetous annelids. Part IV. Pectinariidae, with a review of all species from the western hemisphere. *Allan Hancock Pacific Exp.* 7:325.

HARTMAN, O. 1944. Polychaetous annelids. Part IV. Paraonidae, Magelonidae, Longosomidae, Ctenodrillidae, and Sabellariidae. *Allan Hancock Pacific Exp.* 10:311.

HARTMAN, O. 1945. The Marine Annelids of North Carolina. Duke University Press, Durham, N.C.

HARTMAN, O. 1947. Polychaetous annelids. Part VII. Capitellidae. *Allan Hancock Pacific Exp.* 10:391.

PETTIBONE, M.H. 1963. Marine polychaete worms of the New England region. I. Families Aphroditidae through Trochochaetidae. *U.S. Nat. Mus. Bull.* 227:1.

SMITH, R.I., ed. 1964. Keys to marine invertebrates of the Woods Hole Region. Contrib. No. 11, Systematics-Ecology Program, Marine Biological Lab., Woods Hole, Mass.

McDAIN, J.C. 1968. The Caprellidae (Crustacea: Amphipoda) of the Western North Atlantic. Smithsonian Institute Bull. 278, Washington, D.C.

SCHULTZ, G.A. 1969. How to Know the Marine Isopod Crustaceans. Wm. C. Brown Company Publ., Dubuque, Iowa.

HOLME, N.A. & A.D. McINTYRE. 1971. Methods for the Study of Marine Benthos. IBP Handbook No. 16. Blackwell Scientific Publications, Oxford, England.

FOSTER, N.M. 1971. Spionidae (Polychaete) of the Gulf of Mexico and the Caribbean Sea. Stud. Fauna Curacao Other Caribbean Islands 36.

GOSNER, K.L. 1971. Guide to Identification of Marine and Estuarine Invertebrates. Cape Hatteras to the Bay of Fundy. Wiley-Interscience, New York, N.Y.

LEWIS, P.A. 1972. References for the Identification of Freshwater Macroinvertebrates. EPA-R4-F2-006, U.S. Environmental Protection Agency, Cincinnati, Ohio.

WASS, M.L. et al. 1972. A checklist of the biota of the lower Chesapeake Bay. Special Scientific Rep. No. 65, Virginia Inst. Marine Science, Gloucester Point.

BOUSFIELD, E.L. 1973. Shallow-Water Gammaridean Amphipoda of New England. Cornell University Press, Ithaca, N.Y.

DAY, J.H. 1973. New Polychaeta from Beaufort, with a key to all species recorded from North Carolina. U.S. Circ. No. 375, National Oceanic Atmospheric Admin., National Marine Fisheries Serv., Washington, D.C.

WATLING, L. & D. MAURER. 1973. Guide to the Macroscopic Estuarine and Marine Invertebrates of the Delaware Bay Region. Delaware Bay Rep. Ser. Vol. 5, p. 178. Univ. Delaware, Newark.

WILLIAMS, A.B. 1974. Marine flora and fauna of the northeastern United States. Crustacean: Decapoda. U.S. Circ. No. 389, National Oceanic Atmospheric Admin., National Marine Fisheries Serv., Washington, D.C.

FOX, R.S. & K.H. BYNUM. 1975. The amphipod crustaceans of North Carolina estuarine waters. *Chesapeake Sci.* 16:223.

GARDINER, S.L. 1975. Errant polychaete annelids from North Carolina. *J. Elisha Mitchell Sci. Soc.* 91:77.

MORRIS, P.A. 1975. A Field Guide to Shells of the Atlantic and Gulf Coasts and the West Indies. Houghton Mifflin Co., Boston, Mass.

SMITH, R.I. & J.T. CARLTON, eds. 1975. Light's Manual: Intertidal Invertebrates of the Central California Coast, 3rd ed. University of California Press, Berkeley.

BUTLER, T.H. 1980. Shrimps of the Pacific Coast of Canada. *Can. Bull. Fish. Aquat. Sci.* 202:1.

BLAXTER, J.H.S., S.F.S. RUSSELL & S.M. YONGE. 1980. The species of mysids and key to genera. *Advan. Mar. Biol.* 18:7.

SIEG, J. & R.N. WINN. 1981. The Tanaidae (Crustacea: Tanaidacea) of California, with a key to the world genera. *Proc. Biol. Soc. Wash.* 94(2):315.

EWING, R.M. & D.M. DAUBER. 1981. A new species of Amastigos (Polychaeta: Capitellidae) from the Chesapeake Bay and Atlantic coast of the United States with notes on the Capitellidae of the Chesapeake Bay. *Proc. Biol. Soc. Wash.* 94:163.

HEARD, R.W. 1982. Guide to common tidal marsh invertebrates of the Northeastern Gulf of Mexico. Alabama Sea Grant Consortium. MASGP-79-004.

PRICE, W.W. 1982. Key to the shallow water Mysidacea of the Texas coast with notes on their ecology. *Hydrobiologia* 93:9.

WRONA, F.J., J.M. CULP & R.W. DAVIES. 1982. Macroinvertebrate subsampling: a simplified apparatus and approach. *Can. J. Fish. Aquat. Sci.* 39:1051.

WILLIAMS, A.B. 1984. Shrimp, lobsters, and crabs of the Atlantic Coast of the Eastern United States, Maine to Florida. Smithsonian Institution Press, Washington, D.C.

BRINKHURST, R.O. 1986. Guide to the Freshwater Aquatic Microdrile Oligochaetes of North America. Canadian Spec. Publ. Fisheries & Aquatic Science 84, Dep. Fisheries & Oceans, Ottawa, Ont.

PENNAK, R.W. 1989. Fresh-Water Invertebrates of the United States. Protozoa to Mollusca, 3rd ed. John Wiley & Sons, Inc., New York, N.Y.

VECCHIONE, M., C.F.E. ROPER & M.J. SWEENEY. 1989. Marine Flora and Fauna of the Eastern United States. Mollusca: Cephalopoda. National Marine Fisheries Serv., National Systematics Lab., Washington D.C.

KLEMM, D.J., P.A. LEWIS, F. FULK & J.M. LAZORCHAK. 1990. Macroinvertebrate Field and Laboratory Methods for Evaluating the Biological Integrity of Surface Waters. EPA-600/4-90-030. Environmental Monitoring Systems Lab., U.S. Environmental Protection Agency, Cincinnati, Ohio.

PACKARSKY, B.L., P.R. FRAISSINET, M.A. PENTON & D.J. CONKLIN, JR. 1990. Freshwater Macroinvertebrates of Northeastern North America. Cornell University Press, Ithaca, N.Y.

KLEMM, D.J. 1991. Taxonomy and pollution ecology of the Great Lakes Region leeches (Annelida: Hirudinea). *Mich. Acad.* 24:37.

THORP, J.H. & A.P. COVICH, eds. 1991. Ecology and Classification of North American Freshwater Invertebrates. Academic Press, Inc., New York, N.Y.

MERRITT, R.W. & K.W. CUMMINS, eds. 1996. An Introduction to the Aquatic Insects of North America, 3rd ed. Kendall/Hunt Publishing Co., Dubuque, Iowa.

10500 D.   Data Evaluation, Presentation, and Conclusions

There are two basic approaches used in evaluating effects of pollution on aquatic life. The first is to make a qualitative inventory of the benthic fauna, "above and below" or "before and after" the suspected or known areas of pollution, thereby determining species presence or absence. Then, through an understanding of the responses of various species to certain pollutants and habitat degradation, determine the significance of damage or change. The second approach is to make a quantitative analysis of the numbers of individuals, species, and structure (abundance and composition) of the aquatic community affected by pollution and then to compare with reference information. In most pollution surveys these approaches are integrated because each provides valuable interpretative information.

1. Qualitative Data Evaluation

No two aquatic organisms react identically to a pollutant because of complex relationships between genetic factors and environmental conditions. However, certain taxa are relatively sensitive to certain types of pollution such as siltation and turbidity, organic enrichment, acidity, heavy metals and other industrial toxic wastes, oil production, agricultural products, radioactive wastes, and thermal effects. For example, operculate snails, immature stages of certain mayflies, stoneflies, caddisflies, riffle beetles, hellgrammites, many marine amphipods, mysids, bivalve larvae, and echinoderms are sensitive to many pollutants. Pollution-tolerant macroinvertebrates such as certain sludge worms, midge larvae, leeches, pulmonate snails, and some polychaetes usually increase in number under organically enriched conditions. Facultative organisms, those that tolerate moderate pollution, include most snails, sowbugs, scuds, and blackfly larvae. Tolerant organisms may be found in either clean or polluted situations; thus their presence is not definitive. However, a population of tolerant organisms combined with an absence of intolerant ones is a good indication of the presence of pollution. The same species found in different geographical areas may well react differently or be present in different numbers throughout the year.

2. Quantitative Data Evaluation

Statistical methods of data evaluation and mathematical description of community structure are valuable tools in data analysis. Analysis of biological data commonly begins with the calculation of descriptive statistics (mean, standard deviation, standard error, and confidence intervals). Analysis proceeds by application of robust statistical methods of comparison (Chi-square, Student's *t*, regression, correlation, analyses of variance, or nonparametric equivalents).[1,2]

Mathematical expressions, including numerical indices of community structure, are useful in characterizing and describing aquatic communities. These expressions usually are based on the structural and functional stability of the system.[2]

Diversity indices, although limited, condense considerable biological data into single numerical values.[2–9] Unfortunately, useful information may be lost by condensing biological data.

Select methods for analyzing multivariate benthic community data using two important criteria: the methods should test specific impact-related hypotheses suggested by the data quality objectives and study design, and the methods should objectively identify relationships among variables. Use methods that make a priori assumptions about relationships among variables only secondarily for presentations, not for primary analysis.

More powerful multivariate statistical analyses generally are less subject to criticism and may be more appropriate for some bioassessment studies.[1,10] Recommended data analyses approaches are: regression of species (or taxa) richness on abundance, analysis of variance followed by linear orthogonal contrasts,[11] various other multivariate approaches (e.g., cluster techniques and ordination, analyzing principal components, ANOVA, discriminate analyses), and macroinvertebrate community metrics[2,12] for assessing biomonitoring data and water quality.

For statistical evaluation of the data collected in pollution surveys, it always is beneficial to identify the sources of variability commonly found. Variability in macroinvertebrate data comes from the methods of sampling and the distribution of organisms. Perhaps the major source is sampling error. Organisms generally are clustered in relation to habitat distribution; therefore, random samples often show high variability among replicates. In statistical analyses of quantitative data, large numbers of samples often are required to detect statistically significant differences. Exercise care in using parametric statistical methods because the basic assumption of normal distribution is not always true. Data often have to be transformed before being tested. Do not assume that a statistically significant difference is ecologically significant.

3. Data Presentation

Data presentation may take many forms. The basic techniques include tables, bar graphs (horizontal and vertical), pie diagrams, pictorial charts (ideographs), line graphs, frequency distribution tables and graphs, histograms, frequency polygons, and cumulative frequency polygons. These may be superimposed on maps. Several reports that may be useful in analyzing macroinvertebrate data have been included in the bibliography. Methods for interpreting benthic invertebrate data with measures of contamination and toxicity are available.[13]

4. Conclusions

Despite detailed data quality objectives, field methodology, and laboratory analysis and data presentation, it often requires extensive professional experience and skill and knowledge of the scientific literature to draw defensible conclusions from a data set. Even in the best circumstances, there can be more than one conclusion drawn from a study. When more than one conclusion is possible, it is appropriate to present all options.

5. References

1. GREEN, R.H. 1979. Sampling Design and Statistical Methods for Environmental Biologists. John Wiley & Sons, New York, N.Y.
2. KLEMM, D.J., P.A. LEWIS, F. FULK & J.M. LAZORCHAK. 1990. Macroinvertebrate Field and Laboratory Methods for Evaluating the Biological Integrity of Surface Waters. EPA-600/4-90-030. Environmental Monitoring Systems Lab., U.S. Environmental Protection Agency, Cincinnati, Ohio.
3. WASHINGTON, H.G. 1984. Diversity, biotic and similarity indices: a review with special reference to aquatic ecosystems. Water Res. 18:653.
4. WILHM, J.L. 1967. Comparison of some diversity indices applied to populations of benthic macroinvertebrates in a stream receiving organic wastes. J. Water Pollut. Control Fed. 39:1673.
5. WILHM, J.L. & T.C. DORIS. 1968. Biological parameters for water quality criteria. Bioscience 18:477.
6. WILHM, J.L. 1970. Range of diversity index in benthic macroinvertebrate populations. J. Water Pollut. Control Fed. 42:R221.
7. WILHM, J.L. 1972. Graphic and mathematical analyses of biotic communities in polluted streams. Annu. Rev. Entomol. 17:223.
8. CAIRNS, J., JR., D.W. ALBAUGH, F. BUSEY & M.D. CHANAY. 1968. The sequential comparison index—a simplified method for nonvirologists to estimate relative differences in biological diversity in stream pollution studies. J. Water Pollut. Control Fed. 40:1607.
9. BOESCH, D.F. 1977. Application of Numerical Classification in Ecological Investigations of Water Pollution. Ecol. Res. Ser., EPA-600/3-77-033, U.S. Environmental Protection Agency, Corvallis, Ore.
10. SMITH, W., V.R. GIBSON, L.S. BROWN-LEGER & J.F. GRASSLE. 1979. Diversity as an indicator of pollution. Cautionary results from microcosm experiments. In J.P. Grassle, G.P. Patil, W. Smith & C. Taille, eds. Ecological Diversity in Theory and Practice. International Publ. House, Fairland, Md.
11. HOKE, R.A., J.P. GIESY & J.R. ADAMS. 1990. Use of linear orthogonal contrast in analysis of environmental data. Environ. Toxicol. Chem. 9:875.
12. PLAFKIN, J.L., M.T. BARBOUR, K.D. PORTER, S.K. GROSS & R.M. HUGES. 1989. Rapid Bioassessment Protocols for Use in Streams and Rivers. EPA-440/40-89-001, Off. Water, Assessment and Watershed Protection Div., U.S. Environmental Protection Agency, Washington, D.C.
13. CHAPMAN, P.M. 1996. Presentation and interpretation of sediment quality triad data. Ecotoxicology 5:1.

6. Bibliography

BECK, W.M. 1955. Suggested method for reporting biotic data. Sewage Ind. Wastes 27:1193.
INGRAM, W.M. 1960. Effective methods for collecting and recording data from water pollution surveys. In C.M. Tarzwell, compiler. Biological Problems in Water Pollution, p. 260. U.S. Dep. Health, Education & Welfare, Cincinnati, Ohio.
INGRAM, W.M. & A.F. BARTSCH. 1960. Graphic expression of biological data in water pollution reports. J. Water Pollut. Control Fed. 32:297.
PIELOU, E.C. 1966. The measurement of diversity in different types of biological collections. J. Theor. Biol. 13:131.
LLOYD, M., J.H. ZAR & J.R. KARR. 1968. On the calculation of information—Theoretical measures of diversity. Amer. Midland Natur. 79:257.
CAIRNS, J., JR., K.L. DICKSON, R.E. SPARKS & W.T. WALLER. 1970. A preliminary report on rapid biological information systems for water pollution control. J. Water Pollut. Control Fed. 45:685.
CAIRNS, J., JR. 1971. A simple method for the biological assessment of the effects of waste discharges on aquatic bottom-dwelling organisms. J. Water Pollut. Control Fed. 43:755.
ERMAN, D.C. & W.T. HELM. 1971. Comparison of some species importance values and ordination techniques used to analyze benthic invertebrate communities. Oikos 22:240.
CLIFFORD, H.T. & W. STEPHENSON. 1975. An Introduction to Numerical Classification. Academic Press, New York, N.Y.
BOESCH, D.F. 1977. Application of Numerical Classification in Ecological Investigations of Water Pollution. EPA-600/3-77-033, U.S. Environmental Protection Agency Research Lab., Corvallis, Ore.
ORLOCI, L., C.R. ROA & W.M. STITELLER, eds. 1978. Multivariate Methods in Ecological Work. International Cooperative Publ. House, Fairland, Md.
CONOVER, W.J. 1980. Practical Nonparametric Statistics, 2nd ed. John Wiley & Sons, New York, N.Y.
POLLARD, J.E. 1981. Investigator differences associated with a kicking method for sampling macroinvertebrates. J. Freshwater Ecol. 1:215.
GAUCH, H.G., JR. 1982. Multivariate Analysis in Community Ecology. Cambridge University Press, New York, N.Y.
PLATTS, W.S., W.F. MEGAHAN & G.W. MINSHALL. 1983. Methods for Evaluating Stream, Riparian, and Biotic Conditions. General Technical Rep. INT-138. U.S. Dept. Agriculture, U.S. Forest Serv., Ogden, Utah.
ALLAN, J.D. 1984. Hypothesis testing in ecological studies of aquatic insects. In V.H. Resh & D.M. Rosenberg, eds. The Ecology of Aquatic Insects, p. 484. Praeger Scientific, New York, N.Y.
ZAR, J.H. 1984. Biostatistical Analysis, 2nd ed. Prentice-Hall, Inc., Englewood Cliffs, N.J.
FAUSCH, D.D., J.R. KARR & P.R. YANT. 1984. Regional application of an index of biotic integrity based on stream fish communities. Trans. Amer. Fish. Soc. 113:39.
DAWSON, C.L. & R.A. HELLENTHAL. 1986. A Computerized System for the Evaluation of Aquatic Habitats Based on Environmental Requirements and Pollution Tolerance Association of Resident Organisms. Rep. No. EPA-600/53-85-019, U.S. Environmental Protection Agency, Cincinnati, Ohio.
HUGHES, R.M., D.P. LARSEN & J.M. OMERNIK. 1986. Regional reference sites: A method for assessing stream potentials. Environ. Management 10:629.
LEONARD, P.M. & D.J. ORTH. 1986. Application and testing of an index of biotic integrity in small, coolwater streams. Trans. Amer. Fish. Soc. 115:401.

KARR, J.R., K.D. FAUSCH, P.L. ANGERMEIER, P.R. YANT & I.J. SCHLOSSER. 1986. Assessing Biological Integrity in Running Waters: A Method and Its Rationale. Spec. Publ. 5, Illinois Natural History Surv., Champaign.

STEEDMAN, R.J. 1988. Modification and assessment of an index of biotic integrity to quantify stream quality in southern Ontario. *Can. J. Fish. Aquat. Sci.* 45:492.

PLANTE, C. & J.A. DOWNING. 1989. Production of freshwater invertebrate populations in lakes. *Can. J. Fish. Aquat. Sci.* 46:1489.

PLANTE, C. & J.A. DOWNING. 1990. Empirical evidence for differences among methods for calculating secondary production. *J.N. Amer. Benthol. Soc.* 9:1.

WETZEL, R.G. & G.E. LIKENS. 1991. Limnological Analyses, 2nd. ed. Springer-Verlag, New York, N.Y.

10550   NEMATOLOGICAL EXAMINATION*

10550 A.   Introduction

1. Occurrence and Impact

Nematodes are aquatic animals present in fresh, brackish, and salt waters and soil worldwide. Freshwater nematodes propagate in slow sand filters and aerobic wastewater treatment plants. They appear in large numbers in secondary wastewater effluents.

A freshwater nematode has been defined as "any (nematode) species inhabiting either fresh water or non-brackish swampy soil below the water table; hence a species that will not drown in fresh water; a species fitted to utilize oxygen dissolved in fresh water."[1] Nematodes may affect the freshwater ecosystem by serving as a food source for invertebrates, small vertebrates such as fish, and a variety of fungi. They may have an impact on drinking water. It has been estimated that there are hundreds of millions of nematodes per acre in the top 7.6 cm of a drinking water filter bed.[2] Coexisting with nematodes in the filter beds are rotifers, protozoa, and many other invertebrates.

Predaceous nematodes from the families Aporcelaimidae, Diplogasteridae, Dorylaimidae, and Mononchidae abound in fresh water, devouring other nematodes, oligochaetes, and other small invertebrates. Their role in ingestion of algae and diatoms is unknown; however, dorylaimids occasionally are seen with bright amber, yellow, or green coloration as a consequence of algae in the gut.

Freshwater nematodes ingest bacteria including human enteric pathogens. Bacteria and enteric viruses can survive chlorination inside nematode bodies. Nematodes are able to survive water filtering systems and can emerge alive from domestic water taps. Live nematodes have been detected in both raw and processed water at 16 of 22 water purification sites surveyed in the United States.[3] In Puerto Rico, 13 genera of phytoparasitic nematodes and 13 genera of free-living nematodes were recovered from tap water samples from 11 sites.[4]

2. Nematode Characterization

Aquatic nematodes are well adapted to their habitat. Their long, slender bodies (see Figure 10550:1) are encased by a strong, protective, usually smooth cuticle (skin) inflated by a high-turgor pressure system. Some aquatic species have a long, filamentous, nonmuscular tail. Whipping of the tail, together with the undulating sinuous body movement, propels the nematode through the water at a very rapid rate. Many aquatic nematodes possess glands in the tail that produce a sticky secretion controlled by a spinneret at the tail tip. These secretions temporarily fasten the nematode to a substrate so that anchored nematodes can function without interference by water currents. Oxygen requirements of most freshwater nematodes are low and the metabolism of some species may be nearly anaerobic.

The body is pierced by six to eight distinct openings and few to many minute apertures. First is the oral aperture at the apex of the anterior end, followed by two amphids on the head or neck region, the excretory pore (usually near the esophagus base), the vulva and anus in females, and the cloaca in males, and, on some nematodes, two small pores on the tail called phasmids. Cuticular ornamentation such as engravings, pores, spines, alae, or inflations also may be present.

Internally a stoma may be present, armed with teeth, or unarmed, uncollapsed or collapsed, or sometimes modified to form a hollow spear. An esophagus follows the stoma and terminates in esophageal glands. The intestine extends from the base of the esophageal glands to the rectum, which leads to the anus in females or cloaca in males.

The female gonad is single or paired and consists of an ovary, uterus, and vagina. It exits at the vulva. The male gonad consists of one or two testes, vas efferans, and vas deferens, and exits in the cloaca. Males possess spicules, which are the male copulatory organ, and their guide, the gubernaculum.

The nervous system comprises a nerve ring encircling the esophagus and connected ganglia and nerve cells. Appropriate muscle cell groups are present.

Although nematodes do not possess respiratory and circulatory organs, they tolerate large variations in the levels of salts and other environmental chemicals. Aerobic metabolism is dependent on the diffusion of oxygen into their tissues. Lacking a circulatory system, nematodes rely on diffusion through the tissues for translocation of nutrients, respiratory gases, and waste products.

---

* Approved by Standard Methods Committee, 1997.

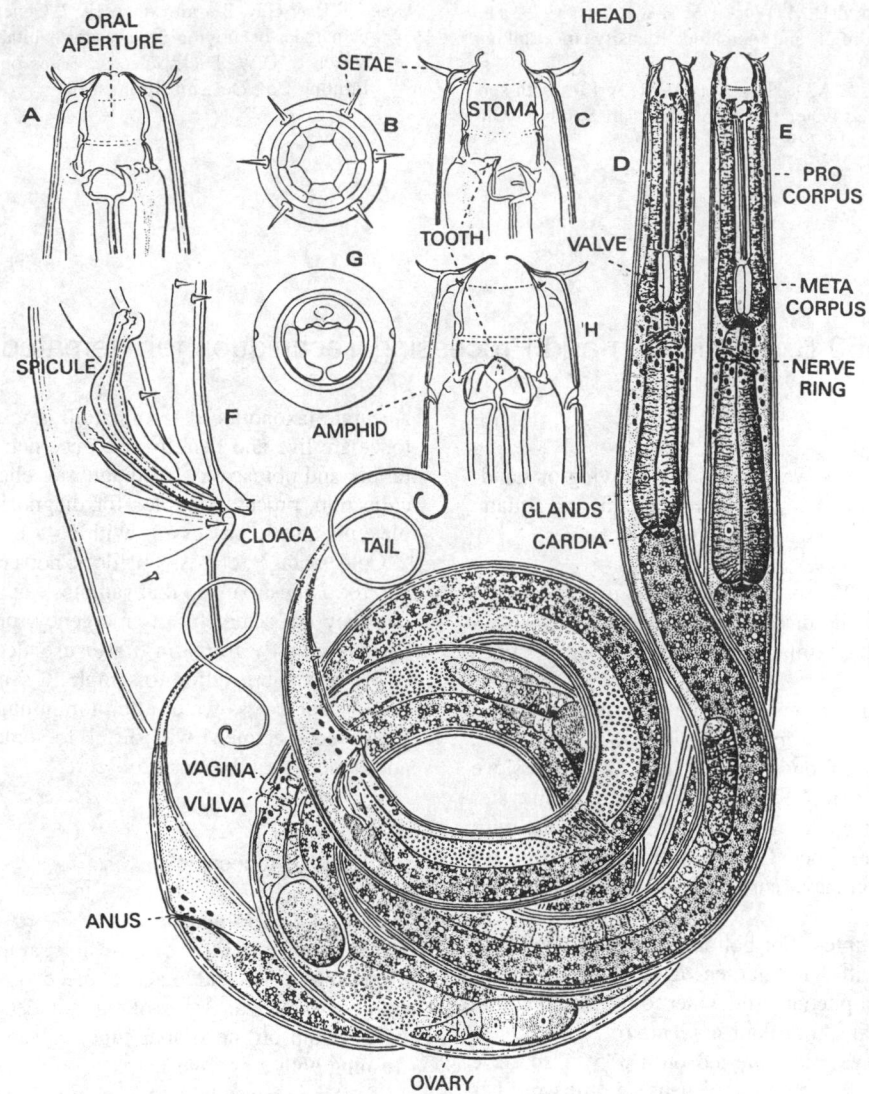

Figure 10550:1. _Butlerius_ sp., a freshwater nematode. A. head; B. _en face_ view of head showing six setae and a central oral aperture; C. head lateral view showing stoma and tooth; D. male; E. female; F. male tail portion showing spicules, gubernaculum, and cloaca; G. head section showing aphids; H. ventral view of head showing aphids and tooth. Russell, C.C., Department of Plant Pathology, Oklahoma State Univ., Stillwater.

3. References

1. Cobb, N.A. 1914. North American fresh-water nematodes. _Trans. Amer. Microscop. Soc._ 33:35.
2. Cobb, N.A. 1918. Filter-bed nemas: Nematodes of the slow sand filter-beds of American cities. _In_ Cobb, N.A. 1935. Contributions to a Science in Nematology. Collected Papers between 1914 and 1935. Williams & Wilkins, Baltimore, Md.
3. Chang, S.L., R.L. Woodward & P.W. Kabler. 1960. Survey of free-living nematodes isolated from water supplies. _J. Amer. Water Works Assoc._ 52:613.
4. Roman, J. & X. Rivas. 1972. Nematodes found in tap water from different localities in Puerto Rico. _J. Agricult. Univ. Puerto Rico_ 56:187.

4. Bibliography

Peters, B.G. 1930. A biological investigation of sewage. _J. Helminth._ 8: 133.

Edmondson, W.T., ed. 1959. Ward & Whipple's Fresh Water Biology, 2nd ed. John Wiley & Sons, New York, N.Y.

Chang, S.L. 1960. Survival and protection against chlorination of human enteric pathogens in free-living nematodes isolated from water supplies. _Amer. J. Trop. Med. Hyg._ 9:136.

Sasser, J.N. & W.R. Jenkins, eds. 1960. Nematology: Fundamental and Recent Advances with Emphasis on Plant Parasitic and Soil Forms. Univ. North Carolina, Chapel Hill.

Chang, S.L. & P.W. Kabler. 1962. Free-living nematodes in aerobic treatment effluent. _J. Water Pollut. Control Fed._ 34:1356.

Calaway, W.T. 1963. Nematodes in wastewater treatment. _J. Water Pollut. Control Fed._ 35:1006.

Walters, J.V. & R.R. Holcomb. 1967. Isolation of an enteric pathogen from sewage borne nematodes. _Nematologica_ 13:155 (abs.).

Chang, S.L. 1970. Interactions between animal viruses and higher forms of microbes. _J. San. Eng. Div., Proc. Amer. Soc. Civil Eng._ 96:151.

Ferris, V.R. & J.M. Ferris. 1979. Thread Worms. _In_ C.W. Hart, Jr. & S.L.H. Fuller, eds. Pollution Ecology of Estuarine Invertebrates. Academic Press, New York, N.Y.

Tombes, A.S., A.R. Abernathy, D.M. Welch & S.A. Lewis. 1979. The relationship between rainfall and nematode density in drinking water. *Water Res.* 13:619.

Mott, J.B., G. Mulamoottil & A.D. Harrison. 1981. A 13-month survey of nematodes at three water treatment plants in Southern Ontario, Canada. *Water Res.* 15:729.

Esser, R.P., & G.R. Buckingham. 1987. Genera and species of free-living nematodes occupying fresh water habitats in North America. *In* J.A. Veech & D.W. Dickson, eds. Vistas on Nematology. E.O. Painter Printing Co., DeLeon Springs, Fla.

10550 B.   Collection and Processing Techniques for Nematodes

1. Samples

Principal samples are: tap or well water, free-flowing or standing water without bottom sediment, bottom sediment, and aquatic plants and coarse detritus such as stones, twigs, or leaves.

a. Sample collection:

1) Tap water—Place a 20-cm, 325 mesh (45-μm pore size) sieve at a 45° angle under the discharge. Adjust water flow to a moderately slow rate, with no splashing, striking upper one-third of sieve. Run for 4 h.

2) Free-flowing or standing water—Take samples from sites where bottom sediments are absent or too deep to be collected. Collect five subsamples as follows: Hold 20-cm, 325 mesh sieve firmly at a 45° angle. Dip 3- to 4-L stainless steel pitcher in water and fill to 1-L mark. Pour contents slowly through top one-third of sieve. Repeat three more times. Collect additional 1 L to wash and concentrate detritus on sieve surface from top to bottom of sieve.

3) Bottom sediment sample—For bottom-to-surface depth less than 20 cm, stir bottom with hand garden rake. Scoop up stirred sediments in stainless steel pitcher. Add water to pitcher to within 5 cm of top. Stir, then wait 30 s. Pour contents of pitcher on to a 20 mesh (1-mm pore size) sieve nested on a 325 mesh sieve with the surface held at a 45° angle until dense detritus reaches pitcher lip. (Usually about 9/10 of the pitcher is poured off.)

For bottom-to-surface depth of 20 to 30 cm, collect duplicate samples by holding a 325 mesh sieve at a 90° angle near the bottom. Using hand rake, stir bottom sediments so that they roil up in a dense cloud in front of sieve. Let cloud settle about 10 s, then move sieve into cloud about 2.5 to 5 cm above the bottom. Bring sieve out of water while holding it at a 45° angle.

4) Aquatic plants, plant or inorganic debris—Randomly collect live floating or submerged plants of one species from target site and place in 1-L jars filled with collection site water. Do not fill more than half of jar with plant material. If several plant species are present, take two or more samples. Place plant and inorganic debris (sticks, leaves, pebbles, etc.) in 1-L jar to about half its volume.

b. Sample concentration: Concentrate detritus present on sieve face by washing tap water across sieve face from top to bottom. Place sieve on lip of a clean, empty 250-mL beaker, bring beaker forward until bottom side is up at a 45° angle. Wash detritus into beaker by flushing tap water from another beaker through the bottom one-third of the bottom side. Pour sample into jar.

c. Sample transport and storage: Regardless of collection mode keep sample jars cool. On very hot days use ice to cool them.

Accurate taxonomic determination is most effective when nematodes are live and healthy. Because nematode mortality, deterioration, and obfuscation of diagnostic characters begins at time of collection, process samples for diagnosis within 24 h and complete diagnostic processing within 48 h.

Cold storage retards, but does not entirely halt, deterioration and rot. Plan survey so that samples can be processed on the same day they are taken. In an emergency, preserve entire sample indefinitely in 4% formalin (*never* use alcohol). Add equal volume of 8% formalin solution to sample. If sample jar is more than half full, decant excess water after a minimum 40-min settling period. Preserved specimens will shrink to some degree and body pores and lumens may be made obscure.

2. Sample Processing

a. Specialized apparatus:

1) *Custom pipet,* for clean-water samples—Take a 29-cm-long disposable pipet and place a piece of 12-cm-long rubber tube snugly over about 3 cm of the conical pickup end. Add a wire buret clamp on the rubber tube. (Clean by removing clamp and flushing with a syringe.)

2) *Baermann funnel,* for samples containing debris—Use a glass funnel with a 15.5-cm top opening and 1.5-cm tube. Fit a rubber tube to the exit tube and close with a buret wire clamp. Place an 8- to 10-cm-diam coarse screen wire disk (3-mm pores) in the funnel opening. Add tap water until it lies just above the wire disk. Insert a facial tissue over the disk.

b. Procedure:

1) Clear or relatively clear water—Shake to obtain homogenous mix, then pour slowly onto the surface of a 7.6-cm, 325 mesh sieve. Concentrate as indicated in ¶ 1*b*, above. Pour concentrated residues into 50-mL conical-bottom centrifuge tube or tubes. Let nematodes settle for 40 min. Insert a custom pipet, ¶ *a*1), above, with rubber tube closed by finger pressure, to tube bottom. Depress rubber tube to take up the ball of nematodes on bottom of cone. Discharge about 0.05 mL (small drop) of pipet contents on to a microscope slide. Cover drop with a 22-mm cover slip. Diagnose nematodes using a *compound* microscope.

2) Samples with much debris—Pour concentrated samples very slowly onto the facial tissue in a Baermann funnel. After 24 h flush funnel into a 250-mL beaker. Process as directed in ¶ 2*b*1), above.

3) Samples containing live plants, plant debris, or inorganic material—Process samples immediately on return to laboratory.

Shake vigorously and pour contents into beaker. Concentrate samples as directed in ¶ 1b, above, and, depending on clarity of sample, proceed according to ¶ 2b1) or 2), above. If the laboratory is equipped to process samples with an excess of debris using the centrifugal flotation technique,[1] preferably use this technique.

3. Reference

1. CAVENESS, F.E. & H.J. JENSEN. 1955. Modification of the centrifugal technique for the concentration of nematodes and their eggs from soil and plant tissue. *Proc. Helminth. Soc. Wash.* 22:87.

10550 C.   Illustrated Key to Freshwater Nematodes

1. General Discussion

The following key was devised so that persons trained in biology, but not necessarily in nematology, could use it. The illustrations include original drawings, photocopies of published drawings, or photocopies on which figures were redrawn. The two most important references were Goodey[1] and Chitwood & Chitwood.[2] Other publications used as references and for illustrative material are listed in the bibliography.

2. Key

Published literature indicates that several genera in this key contain species predominantly associated with terrestrial habitats. Presence of such nematodes suggests runoff from banks or higher ground in which various plant species (often food sources for these nematodes) are growing. These genera are indicated by an asterisk (*).

Refer to Couplet No.

1.   Cephalic setae indistinct or absent 2

        Cephalic setae absent but setae-like head appendages present 64

        Cephalic setae present 69

2.(1)   Stylet present 3

        Stylet absent 38

3.(2)   Base of stylet knobbed or flanged 4

        Stylet knobs or flanges absent 29

Refer to
Couplet No.

4.(3)   Valvate median esophageal bulb present   5

        Valvate median esophageal bulb absent 22

5.(4)   Females eel-like   6

        Females swollen 21

6.(5)   Vulva at mid-body   7

        Vulva on lower third of body 14

7.(6)   Esophagus not overlapping intestine   8

        Esophagus overlapping intestine 11

8.(7)   Stylet length less than 50 µm   9

        Stylet length
        greater than 80 µm . . . . *Dolichodorus*

9.(8)   Tail terminus pointed . . . . *Tetylenchus**

        Tail terminus not pointed 10

10.(9) Tail terminus knobbed . . . . *Psilenchus**

        Tail terminus never knobbed or
        pointed . . . . *Tylenchorhynchus**

11.(7) Labium offset. 12

Labium flattened, amalgamated or nearly so 13

12.(11) Stylet massive, 40–50 μm long . . . . *Hoploaimus**

Stylet long and thin, longer than
90 μm. . . . . *Belonolaimus**

13.(11) Body 0.5–1.0 mm long,
tail tip not mucronate . . . . *Radopholus**

Body 2–3 mm long, tail tip usually
mucronate . . . . *Hirschmanniella*

14.(6) Cuticle heavily annulated, stylet elongate 15

Cuticle not heavily annulated, stylet short 17

15.(14) Cuticular sheath absent 16

Cuticular sheath present . . . . *Hemicycliophora*

16.(15) Annules with cuticular spines or scales . . . . *Criconema**

Annules plain without spines or
scales . . . . *Criconemoides**

17.(14) Body death position straight 18

Body death position spiral. . . . . *Helicotylenchus**

Refer to
Couplet No.

18.(17) Median esophageal bulb
distinct but not pronounced 19

Median esophageal bulb
well developed . . . . *Aphelenchoides*

19.(18) Esophagus overlapping intestine 20

Esophagus not overlapping intestine . . . . *Tetylenchus**

20.(19) Median bulb and valves small,
stylet usually weak . . . . *Ditylenchus**

Median bulb, valves and stylet
well developed, labium flattened . . . .*Pratylenchus**

21.(5) Female body soft, white, with
few or no internal eggs . . . .*Meloidogyne**

Female body a rigid brown cyst
usually with many internal eggs . . . . *Heterodera**

22.(4) Stylet short, less than 100 μm 23

Stylet long, greater than 100 μm . . . . *Xiphinema**

23.(22) Stylet complex 24

Stylet simple 25

24.(23) Stylet with anterior arch-like
portion . . . . *Diphtherophora**

Stylet with dorsal thickening
piece . . . . *Tylencholaimellus**

25.(23) Stylet knobs elongate, flange-like 26

Stylet knobs round 27

26.(25) Filiform tail . . . . *Aulolaimoides*

Round tail . . . . *Enchodelus*

27.(25) Tail rounded 28

Tail pointed . . . . *Nothotylenchus*

28.(27) Esophagus base elongate . . . . *Tylencholoaimus**

Esophagus base oval . . . . *Doryllium*

29.(3) Valvate median esophageal bulb absent .30

Valvate median esophageal bulb present. 37

30.(29) Stomal walls not cuticularized 31

Stomal walls cuticularized
(*Actinolaimus, Metactinolaimus,
Paractinolaimus*) . . . . *Actinolaiminae*

31.(30) Esophagus with basal expansions 32

Esophagus
expanding uniformly . . . . *Oionchus*

32.(31) Terminal fifth or sixth
of esophagus an ovoid bulb 33

Posterior third of esophagus swollen 36

33.(32) Stylet axial, positioned centrally 34

Stylet not axial, originating from tooth in
stoma wall . . . . *Campydora**

34.(33) Gonads paired; vulva usually near mid-body 35

Gonad single, posterior to vulva; vulva anterior
to mid-body . . . . *Tyleptus**

35.(34) Stylet slender . . . . *Leptonchus**

Stylet not slender . . . . *Dorylaimoides**

36.(32) Stylet axial, positioned centrally
(*Dorylaimus, Eudorylaimus,
Labronema, Mesodorylaimus,
Thornia, Laimydorus,
Prodorylaimus*) . . . . Dorylaiminae

Stylet not axial, originating from tooth in
stoma wall . . . . Nygolaimus

37.(29) Tail pointed . . . . *Seinura**

Tail rounded . . . . *Aphelenchus**

38.(2) Teeth present, prominent 39

Teeth absent, minute, or indistinct 50

39.(38) Esophagus without mid-region expansion 40

Esophagus expanded at mid-region 49

40.(39) Tail pointed or tapering 41

Tail rounded 47

41.(40) Male tail without setae 42

Male tail with setae . . . . *Oncholaimus*

42.(41) Stoma with denticles 43

Stoma without denticles 45

43.(42) Denticles scattered or in longitudinal rows 44

Denticles in transverse rows . . . . *Mylonchulus*

44.(43) Denticles situated on longitudinal rib
of stoma . . . . *Prionchulus*

Denticles scattered on stoma wall . . . . *Sporonchulus*

45.(42) Tooth anteriorly directed 46

Tooth retrorse . . . . *Anatonchus*

46.(45) Tooth in basal part of stoma . . . . *Iotonchus*

Tooth in anterior part of stoma . . . . *Mononchus*

47.(40) Stoma with prominent medial or apical tooth 48

Stoma with small basal tooth . . . . *Bathyodontus*

48.(47) Stoma with 3 teeth, without small
basal tooth, caudal glands opening
terminally . . . . *Enoplocheilus*

Stoma with large anterior and small
basal tooth, caudal glands opening
ventrally . . . . *Mononchulus*

49.(39) Lip region with rib-like armature . . . . *Mononchoides*

Lip region without rib-like armature . . . . *Diplogaster*

50.(38) Esophagus with
basal expansions 51

Esophagus uniformly
cylindrical 60

51.(50) Esophagus without
mid-region expansion 52

Esophagus expanded at mid-region 55

52.(51) Amphids distinct 53

Amphids indistinct 54

53.(52) Stoma walls anteriorly inflated with
minute tooth . . . . *Microlaimus*

Stoma walls without tooth and with straight,
tapering sides . . . . *Leptolaimus*

54.(52) Stoma with 3 rod-like
thickenings . . . . *Rhabdolaimus*

Stoma without rod-like
thickenings . . . . *Monochromadora*

55.(51) Gonads paired 56

Gonads single 58

56.(55) Stomal walls straight, amalgamated 57

Stomal walls separated, not straight . . . . *Alloionema*

57.(56) Moderately swollen metacorpus, stoma not
excessively elongate . . . . *Rhabditis*

Elongate, cylindrical metacorpus, stoma
elongate . . . . *Cylindrocorpus*

58.(55) Tail with sharp terminus 59

Tail bluntly conical . . . . *Cephalobus*

59.(58) Anterior part of stoma a broad,
open chamber . . . . *Panagrolaimus*

Stoma narrow, collapsed . . . . *Eucephalobus*

60.(50) Stoma absent or indistinct 61

Stoma distinct 63

61.(60) Lip region narrow, tooth absent 62

Lip region broad, small denticle apparent
in stomal area . . . . *Tripyla*

62.(61) Amphid aperture appearing
as large slit. . . . . *Amphidelus*

Amphid aperture appearing
as minute pores . . . . *Alaimus*

63.(60) Stoma narrow and long . . . . *Cryptonchus*

Stoma wide and shallow . . . . *Bathyonchus*

64.(1) Body symmetrical 65

Body asymmetrical, bearing
series of protuberances on side . . . . *Bunonema**

65.(64) Lip appendages not elaborate 66

Lip appendages elaborate 68

Refer to
Couplet No.

66.(65) Lateral lip appendages thorn-like, directed laterally . . . . *Diploscapter*

Lateral lip appendages not thorn-like or directed laterally 67

67.(66) Papillae or setae horn-like . . . . *Macrolaimus*

Lips flap-like and pointed anteriorly . . . . *Teratocephalus*

68.(65) Lip appendages forked and elaborately fringed . . . . *Acrobeles**

Lip appendages membranous and wing-like . . . . *Wilsonema**

69.(1) Post-cephalic setae absent 70

Post-cephalic setae present (may be very faint ex. *Tobrilus*) 92

70.(69) Stylet absent 71

Stylet present 91

71.(70) Teeth absent, minute or indistinct 72

Teeth usually present, prominent 85

72.(71) Esophagus with basal expansions 73

Esophagus uniformly cylindrical 82

73.(72) Amphids oval, spiral, or stirrup-shaped 74

Amphids circular 80

Refer to Couplet No.

74.(73) Amphids spiral..75

Amphids not spiral...79

75.(74) Cuticular punctations absent...76

Cuticular punctations present..78

76.(75) Esophageal bulb without valves....................................77

Esophageal bulb valvate.................................... *Plectus & Anaplectus*

77.(76) Esophageal-intestinal valve elongate.................................... *Paraplectonema*

Esophageal-intestinal valve shortened.................................... *Paraphanolaimus*

78.(75) Labial region characteristically flap-like.................................... *Euteratocephalus*

Labial region not flap-like, lips bluntly rounded.................................... *Ethmolaimus*

79.(74) Amphids oval.................................... *Greenenema*

Amphids stirrup-shaped.................................... *Chronogaster*

80.(73) Esophageal-intestinal valve shortened..81

Esophageal-intestinal valve elongate.................................... *Desmolaimus*

81.(80) Excretory pore and large
excretory gland present . . . . *Domorganus*

Excretory pore and gland
indistinct or absent . . . . *Monhystera*

82.(72) Stoma wide and shallow, conspicuous,
tail filiform . . . . *Prismatolaimus*

Stoma narrow, elongate,
collapsed or inconspicuous 83

83.(82) Gonad single . . . . *Cylindrolaimus*

Gonads paired 84

84.(83) Amphids inconspicuous . . . . *Tripyla*

Amphids conspicuous . . . . *Aphanolaimus*

85.(71) Terminal fifth or sixth of
esophagus an ovoid bulb 86

Esophagus uniformly cylindrical,
stoma with massive teeth . . . . *Ironus*

86.(85) Cuticular punctations present 87

Cuticular punctations absent 89

87.(86) Amphids not spiral 88

Amphids spiral . . . . *Achromadora*

88.(87) Four longitudinal rows of cuticular
         markings present . . . . *Chromadora*

         No longitudinal rows of cuticular
         markings present . . . . *Prochromadorella*

89.(86) Amphids distinct 90

         Amphids indistinct . . . . *Butlerius*

90.(89) Female gonad double, amphids
         hook-shaped . . . . *Anonchus*

         Female gonad single, amphid
         circular . . . . *Monhystrella*

91.(70) Lip region annulated, not set off . . . . *Atylenchus*

         Lip region smooth, set off . . . . *Eutylenchus*

92.(69) Esophagus with basal expansion 93

         Esophagus uniformly cylindrical 98

93.(92) Cuticular punctation present,
         amphids not circular 94

         Cuticular punctation present, amphids circular 97

94.(93) Ocelli (eye spots) present 95

Ocelli absent 96

95.(94) Stoma with three equal-sized teeth . . . . *Chromadorina*

Stoma with at least one large tooth . . . . *Punctodora*

96.(94) Cuticle with lateral
longitudinal rows
of punctation . . . . *Hypodontolaimus*

Cuticle without lateral
differentiations . . . . *Chromadorita*

97.(93) Esophageal bulb valvate . . . . *Prodesmodora*

Esophageal bulb without valves . . . . *Odontolaimus*

98.(92) Amphid anterior on body 99

Amphid posteriorly located . . . . *Bastiania*

99.(98) Amphid spiral . . . . *Paracyatholaimus*

Amphid cup-shaped or obscure 100

Refer to
Couplet No.

100.(99) Stomal teeth massive . . . . *Oncholaimus*

Stomal teeth small . . . . *Tobrilus*

3. References

1. GOODEY, T. 1963. Soil and Freshwater Nematodes, 2nd ed. (Revised by J.B. Goodey). The Methuen Co., London, & John Wiley & Sons, New York, N.Y.
2. CHITWOOD, B.G. & M.B. CHITWOOD. 1937. An Introduction to Nematology, Section I: Anatomy (rev. ed., 1950). Monumental Printing Co., Baltimore, Md.

4. Bibliography

THORNE, G. 1939. A monograph of the nematodes of the superfamily Dorylaimoidea. *Capita. Zool.* 8:1.
GERLACH, S.A. 1954. Brasilianische Meeres-Nematoden 1. *Bol. Inst. Oceanog.* 5:3.
CHITWOOD, B.G. & A.C. TARJAN. 1957. A redescription of *Atylenchus decalineatus* Cobb, 1913 (Nematoda: Tylenchinae). *Proc. Helminth. Soc. Wash.* 24:48.
ANDRÁSSAY, I. 1959. Nematoden aus dem Psammon des Adige-Flusses, I. *Mem. Mus. Civ. Stor. Nat., Verona* 7:163.
HOPPER, B.E. & E.J. CAIRNS. 1959. Taxonomic Keys to Plant, Soil and Aquatic Nematodes. Alabama Polytechnic Inst., Auburn.
CHITWOOD, B.G. 1960. A preliminary contribution on the marine nemas (Adenophorea) of Northern California. *Trans. Amer. Microsc. Soc.* 79:347.
LUC, M. 1960. *Dolichodorus profundus* n. sp. (Nematoda-Tylenchida). *Nematologica* 5:1.

LOOF, P.A.A. 1961. The nematode collection of Dr. J.G. de Man. *Meded. Lab. Fytopath.* 190:169.
THORNE, G. 1964. Nematodes of Puerto Rico: Belondiroidea new superfamily, Leptonchidae, Thorne, 1935, and Belonenchidae new family (Nemata, Adenophorea, Dorylaimida). Univ. Puerto Rico Agr. Exp. Sta. Tech. Paper 39.
EDWARD, J.C. & S.L. MISRA. 1966. *Criconema vishwanathum* n. sp. and four other hitherto described Criconematinae. *Nematologica* 11:566.
HOPPER, B.E. & S.P. MEYERS. 1967. Folliculous marine nematodes on turtle grass, *Thalassia testudinum* König, in Biscayne Bay, Florida. *Bull. Mar. Sci.* 17:471.
MULVEY, R.H. & H.J. JENSEN. 1967. The Mononchidae of Nigeria. *Can. J. Zool.* 45:667.
ALLEN, M.W. & E.M. NOFFSINGER. 1968. Revision of the genus *Anaplectus* (Nematoda: Plectidae). *Proc. Helminth. Soc. Wash.* 35:77.
ANDRÁSSAY, I. 1968. Fauna Paraguayensis 2. Nematoden asu den Galeriewaldern des Acaray-Flusses. *Opusc. Zool. Boest.* 8:167.
DeGRISSE, A. 1968. Bijdrage tot de morfologie en de systematiek van Criconematidae (Taylor, 1936) Thorne, 1949 (Nematoda). Plantenatlas Sleutel, Gent.
ANDRÁSSAY, I. 1973. Nematoden asu strand- und höhoenbiotopen von Kuba. *Acta Zool. Hung.* 19(3–4):233.
FERRIS, V.R., J.M. FERRIS, & J.P. TJEPKEMA. 1973. Genera of Freshwater Nematodes (Nematoda) of Eastern North America. Biota of Freshwater Ecosystems. Identification Manual No. 10, U.S. Environmental Protection Agency.
TARJAN, A.C. & B.E. HOOPER. 1974. Nomenclatorial Compilation of Plant and Soil Nematodes. Society of Nematologists. O.E. Painter Printing Co., DeLeon Springs, Fla.

10600   FISH*

10600 A.   Introduction

1. Ecological Importance

Fish are a major component of most aquatic habitats, with over 775 species in North American streams, rivers, and lakes. They are the focus of economically important sport and commercial fisheries; licensing fees for both private and commercial sectors provide funds for state and federal agencies. They are an impor-

tant source of food and recreation and are a key unit in many natural food webs. They have an impact on the physicochemical properties of the system in which they occur and affect plankton, macrophytes, and other aquatic organisms. They also can serve as environmental indicators. Changes in the composition of a fish assemblage often indicate a variation in pH, salinity, temperature regime, solutes, flow, clarity, dissolved oxygen, substrate composition, or pollution level. The gain or loss of certain species is a common consequence of environmental change. Because fish

---

* Approved by Standard Methods Committee, 1994.

are conspicuous they often are the primary indicators of the toxification of streams and lakes. In extreme cases the presence of dead or moribund fish may adversely affect potability and recreational use of waters, create foul odors, and corrupt shorelines.

Because fish are ecologically important, there are often intense commercial and recreational interests surrounding their study. These diverse interests translate into a need for the scientist, often supported by public funds, to be aware of the sensitivity of such investigations.

Fish share many physiological properties with mammals and are used in both the laboratory and the field by the environmental manager and health specialist in biological assays.[1]

2. Definitions

A *population* is a group of individuals of any one kind of organism occupying a particular space. Its study includes definition of taxonomic position, habitat and mobility, diet, numbers of individuals by age, size, weight, sex, fecundity, and sources of mortality.

An *assemblage* (association, community) is a group of several populations sharing a common geographical area. The study of their coordinated activity is key to the understanding of the environmental system.

3. Scope of Analysis

An analysis of a target unit commences with review of existing data followed by sampling and preservation, identification, demographic estimation, and pathological examination.

The guidelines provided here are directed to the general practitioner who may need specialists such as the commercial and sport angler, fishery biologist, taxonomist, histopathologist, population statistician, systems ecologist, and toxicologist. Adapting to the particular situation is the key element in the study of fish in their natural habitat.[2-7]

4. References

1. HUGGETT, R.J., R.A. KIMERLE, P.M. MEHRLE, JR. & H.L. BERGMAN. 1992. Biomarkers, Biochemical, Physiological, and Histological Markers of Anthropogenic Stress. Lewis Publ., Boca Raton, Fla.
2. SCHRECK, C.B. & P.B. MOYLE, eds. 1990. Methods for Fish Biology. American Fisheries Soc., Bethesda, Md.
3. ROYCE, W.F. 1984. Introduction to the Practice of Fishery Science. Academic Press, Orlando, Fla.
4. TEMPLETON, R.G. 1984. Freshwater Fisheries Management. Fishing News Books Ltd., Farnham, Surrey, England.
5. NIELSEN, L. & D.L. JOHNSON, eds. 1983. Fisheries Techniques. American Fisheries Soc., Bethesda, Md.
6. EVERHART, W.H. & W.D. YOUNGS. 1981. Principles of Fishery Science. Cornell Univ. Press, Ithaca, N.Y.
7. BAGENAL, T., ed. 1978. Methods for the Assessment of Fish Production in Fresh Waters, 3rd ed. IBP Handbook No. 3. Blackwell Scientific Publ., Oxford, England.

10600 B. Data Acquisition

1. Planning and Organization

a. Objectives/variables: Define study objectives before collecting data. Understand the relative importance of such variables as time of day and season, weather, flow/flood and tidal conditions, method and its selectivity, and competence and experience of the investigators.

b. Regulations: A detailed understanding of licensing and permit requirements for the collection of specimens is essential. Most states have strictly enforced regulations on both the collection and disposition of fish specimens. Effective public relations usually involves the guidance of local residents and waterfront associations on the activities planned.

c. Units: Carefully choose units of measure, giving attention to the conventions and expectations of those sponsoring and using the data. Units may influence methods used or be fixed by the intent to duplicate earlier efforts.

d. Site inspection: Study topographic maps and other relevant data before visiting the site. Make an early and thorough visual examination of the study site. Use glasses with polarized lenses to aid examination of bottom features and detection of fish. Binoculars allow field identification of fish and help in the logging of needed behavioral information. In clear water use face plate and snorkel or SCUBA (self contained underwater breathing apparatus) to define habitat, identify fish, and observe behavior.

Use wet or dry suits during colder periods and even under ice but note that special safety training is essential.

e. Data forms: Print data forms on good-quality bond paper or record data on plastic sheets. Electronic media, e.g., polycorders and computer notebooks, are becoming popular. Data can be transferred directly into computer files to save time and reduce transcription errors. Back up files often and make hard copies to store at a second location. Include on the data form the following information in an order best suited for the study:

• Date and time of collection or observation;

• Exact location using the Universal Mercator System (UTM) or a local variant; township, range and section numbers; county and state; physical features such as a stream confluence, islands, bays, etc.; and station identification number or code;

• Site conditions as required, e.g., presence of ice, flood state, tidal stage, meteorological events such as air temperature, occurrence of storms and rainfall in last 48 h, water temperature, discharge, vegetational cover on nearby shores, etc.;

• Purpose of activity or project;

• Description of collections or observations made including preservatives, photograph numbers, and gear type;

• Personnel and their functions;

• Name of person recording the data;

• Chain of custody signatures and dates;

• Habitat information as required, e.g., dissolved oxygen, turbidity, pH, substrate, aquatic vegetation, etc.; and

• Information on fishes collected or observed (identity, number, mass, presence of disease, etc.).

f. Description: Collected materials may define the species represented, describe the population of a particular species, describe a species assemblage, or characterize impacts of some event such as a chemical spill. The detailed analysis may include:

• Preliminary species assignment for each specimen;

• Number of individuals of each species;

• Standard, fork, or total length for each specimen;

• Sex, if discernible;

• Maturity as indicated by gonadal condition and coloration;

• Weight of each specimen (displacement volume may be used);

• Description of unusual features such as tags, deformities, lesions, tumors, or parasites; and

• Materials taken to determine the age of fish, such as scales, otoliths, spines, or opercular bones, or stomachs for diet analysis.

g. Conduct of field workers: Obtain necessary permits for collecting, holding, transporting, and stocking of specimens. Provide adequate advance notice of activities to conservation officers, wardens, local law-enforcement agencies, and officers of lake or watershed associations. Inform them in detail of the actions planned. Understand trespass law for the study area and request access.

Be sensitive to the use of waterfront areas. Avoid, where possible, damage to amenity plantings and capture of favored specimens, e.g., a pair of large smallmouth bass holding a territory next to a dock. Deal pleasantly and in a well-informed manner with the questions of onlookers. Display the name and address of the study group through such means as name tags, arm patches, or equipment decals.

Demonstrate regard for safety. Become proficient with boat safety, proper operation of sampling equipment, and first aid including cardiopulmonary resuscitation (CPR). Wear life preservers when appropriate. Avoid wearing waders on board vessels when in deep water. Follow local restrictions on boat speed. Handle gear proficiently. Use fail-safe switches on all electrical gear—especially electroshocking equipment.

Identify all gear with name, address, telephone number, and permit or license numbers of the using agency.

Position gear inconspicuously to minimize tampering and vandalism. Avoid navigational channels and other heavily used sites. Whenever appropriate, submerge indicator buoys and mark their location by paired range points such as navigational aids or landmarks. When transferring gear from one locality to another be careful not to translocate organisms. Maintain gear in a professional manner. Avoid use of tattered nets and casually repaired and dirty equipment.

Dispose of processed specimens rationally and legally. Check with regional museums and/or academic institutions that regularly voucher specimens. Obtain accession policies as necessary. Preserved specimens are needed in biology classes in most local school districts. Contact principals and science teachers if suitable material is available. If necessary, bury or burn specimens fixed in formalin at appropriate and legal facilities. Pay attention to local legislation regarding humane care of vertebrates including fishes. Do not overharvest. A 2-h evening set of a gill net may yield a sufficient number of specimens for a particular study while an overnight set would be wasteful. Avoid sampling nontarget species.

Be prepared to sample on short notice (such as after a storm or flood) or at inconvenient times (such as at night or in the winter) to take advantage of all sampling opportunities.

2. Existing Data

Published and unpublished data already exist for most larger lakes and river systems. Natural history museums and academic institutions are primary sources and often can provide preserved materials as well as the names of local specialists. Members of the American Fisheries Society, the American Society of Ichthyologists and Herpetologists, and the Ecological Society of America also can provide information. Private engineering and environmental consulting firms often maintain detailed regional files. State agencies such as departments of health, environment, conservation, fisheries, wildlife management, and planning, and federal agencies such as the U.S. Fish and Wildlife Service, National Biological Survey, U.S. Environmental Protection Agency, Bureau of Reclamation, U.S. Army Corps of Engineers, and U.S. Geological Survey are good resources. Large data sets may be available from power utilities, refineries, food producers, and chemical companies having riparian facilities. Cooperative extension services and public colleges and universities provide expertise and information. Commercial fishermen and master sport anglers of an area are other important sources. Local libraries, newspapers, and local residents may provide useful material, including photographs, dates, and even specimens.

3. Collection and Observation Methods

All methods of collecting and observing fish are selective and, therefore, biased. No one method of collection will completely portray the composition of a fish assemblage. The goals of an individual study dictate the appropriate methods(s). Carefully and completely describe methods so that users of the data may properly assess them.

a. Angling: The use of hook and line is an ancient means and involves relatively simple gear[1] but its effective use is a matter of skill. The services of a competent angler are often valuable. The techniques depend on the resourcefulness and skill of the angler, and may be time-consuming and expensive.

b. Set line: A set line is a heavy line anchored at each end bearing regularly spaced leaders having baited hooks. It is widely used for commercial and private fishing. It usually is fished overnight on the bottom and can be used to great depths.

c. Trolling: Trolling is towing a hook and line behind a vessel. It is effective for larger fish of open and deep waters. Use metal line equipped with metal weights or wing depressors to achieve desired depths. Specialized lures that reflect sonar often are used with acoustical electrical gear to determine depth and to locate target fish. Trolling may be the most economical means of capturing some species.

d. Spear and bow and arrow: The use of barbed and/or hooked (some automated) spears is of limited utility. It often is prohibited. Sport bow-and-arrow fishing may yield large numbers of carp, gars, or other larger fish in shallow water. Spearing through ice often is effective, especially for sturgeon and larger esocids and percids. Efficiency varies seasonally.

e. Nets—General remarks: Netting is used in static gear such as traps and weirs and in active gear such as seines and trawls. Netting may be made of cotton, plastic, or metal. Nets of natural fiber are subject to microbial decomposition and have been supplanted by other materials but are still of value where loss of gear is likely. Plastic netting is exceedingly durable but is weakened by ultraviolet irradiation; avoid prolonged exposure to sunlight. Netting is available in colors that may hold some sampling advantage. Mesh sizes are measured in terms of "bar," i.e., along the edge of the frame, "diagonal," i.e., from opposite angles of the frame, and "stretched," i.e., from opposite angles when the net is under tension. Knotting varies; some knots are abrasive to captured fish. Knotless netting also is available.

f. Hoop, fyke, and trap nets: Elongated, tapered nets supported on hoops and variously divided into chambers with secondary funnel net sections and anchored to the bottom are common. Usually they are used at depths less than 3 m. They may be kept in place for a protracted period but usually are visited daily to remove the catch from the inner chamber or cod. Orient the cod end into the current in slowly moving water. The basic hoop or ring net may be converted into a fyke net by adding panels of netting at the open end. Those added to the sides are called wings; the single panel placed at the center of the mouth may be quite long and is called the lead. Wings and leads usually are equipped with floats and weights and are placed to deflect and trap or confuse the normal movements of fishes along shore into the net. The main body of a typical hoop or fyke net may range from 5 to 10 m in length and up to 2.5 m in diameter. A net hung on rectangular framing is called a trap net (Figure 10600:1). On commercial fishing grounds researchers usually contract for the catches of larger trap nets rather than use their own. Another net type of this general design is the pound net.

Hoop, fyke, and trap nets are effective for larger fishes that are mobile or that seek cover. They are especially useful in capturing live fish for spawning stock or fish to be used in mark-and-recapture population assessment. These nets can be baited to increase their efficiency in capturing some species.

Nets of this type are available through net supply houses and commonly are built to specification. They require a boat to set and inspect. They often are conspicuous and thus attract the public. They require surveillance and the clear posting of their ownership and purpose.

g. Traps: The term "fish trap" usually is reserved for smaller, portable units commonly made of galvanized wire. They are fitted with one or more conical inserts and an opening to remove the catch.

Smaller devices usually called "minnow traps" may be useful. Many types are available, including highly durable plastic units that can be easily stacked, stored, and transported. Minnow traps may be placed in a well, a cave pool, near hydroelectric facilities, or in other awkward localities with good results. They may be made more effective through baiting. A readily degradable element (e.g., wooden lath in a plastic or metal trap) should be a part of the design of all traps used in the field so that lost traps do not continue to capture and kill fish.

h. Weirs: Weirs are stationary traps usually installed along the course of a stream or river. They are of complex design and may be incorporated into a fish ladder and dam. They guide fish into a sampling or capture sector called the "pot."

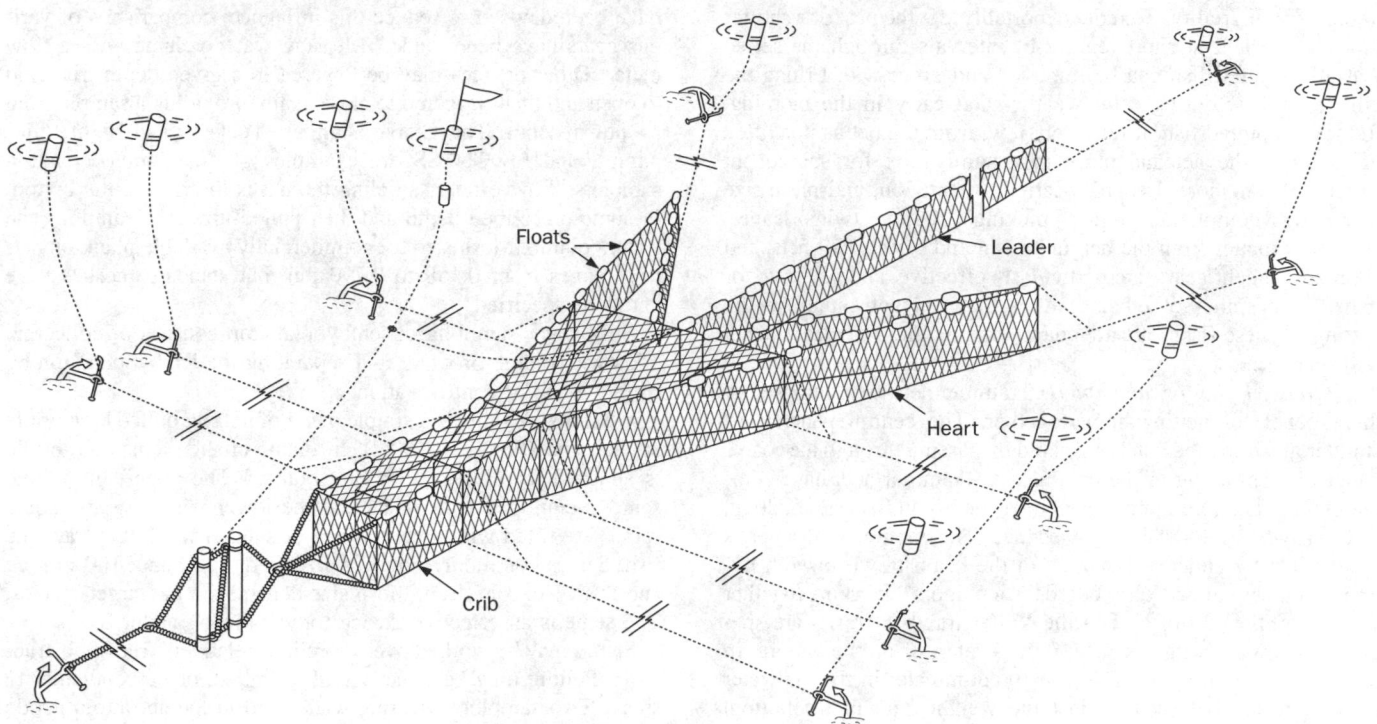

Figure 10600:1. Diagram of a sunken trap net. Source: BAGENAL, T., ed. 1978. Methods for Assessment of Fish Production in Fresh Waters. IBP Handbook No. 3. Blackwell Scientific Publ., Oxford, England.

i. Gill nets: Nets constructed of thin line with mesh large enough for the target species to penetrate just beyond the operculum are called "gill nets." The fish become entrapped while attempting to swim through the net and are harvested. Gill nets are composed of panels of netting of the same or diverse mesh size suspended between a stronger "float line" equipped with flotation devices and attached along the upper working edge and another heavier "lead line" attached along the lower working edge. Adjust weights and floats to position the net on the bottom, surface, or at an intermediate depth. The ends of the gill net are equipped with anchors, tether lines, and buoys.

The gill net may be set under ice through appropriate holes. Gill nets may be used in standing water; they are less successful in flowing water, but can be set parallel to the current.

The setting and recovery of gill nets require special attention. The sampling site must be free of ensnaring objects such as submerged trees. The boat used to set the net must be free of projections that can catch the net during the payout and the net boat must proceed at a speed in harmony with the workers discharging the net. The net must be well pleated and free of snarls to set; the anchors, tethers, and buoys must be ready for quick release as well. At the conclusion of the set pull the tether to extend the net to its full working length. The tether line must be long enough so that the buoy is not pulled out of sight. The marker buoy can be set below the surface to reduce vandalism and paired range points may be used to relocate the site. If the tether line and buoys are lost, attempt retrieval by grappling hook or SCUBA. Failure to recover lost gill netting can be environmentally disastrous because it will continue to catch and kill fish. Report lost gill netting to local fisheries authorities.

Overnight setting is common; however, a test set during dawn or dusk or a short night set may be advisable to avoid the capture of an excessive number of specimens. Use of gill nets can result in high fish mortality. To reduce mortality, set the net for a shorter time or check it at short (e.g., 2-h) intervals through the set so that all captured fish can be removed and processed. Lifting the gill net is best during calm weather and early in the morning. Remove captured fish, if relatively few, from the net as it is lifted or retain in the net and place in a sturdy box for subsequent "picking" on shore. Record location, orientation, and mesh size of net. On completion of the "picking" remove twigs, leaves, and other matter from the net and clean and dry it. Gill nets, and all entanglement gears, are particularly effective, and selective for fishes with spines or other features that can get caught in the netting. If these fishes are abundant, time required to clear the net will increase.

j. Trammel, flag or tangle nets: Trammel nets are composed of three panels of netting hung together. The central panel is of smaller mesh and the fish is ensnared by passing through the coarse panel to form a bag in the central net. Trammel nets have commercial applications but are used infrequently in fisheries biology.

k. Trawls: A trawl is a towed net. The mouth of the net is maintained by either a frame, as in the beam trawl, or with hydraulic planes called "otter boards" or "doors" working together with weights and floats as in the "otter trawl." Trawls are specialized to work at the surface, in midwater, or on the bottom. In the surface trawl, buoying devices predominate; in the midwater trawl they are balanced against the weight; and in the bottom trawl the weights predominate. The bottom trawl usually has abrasion skirting on the lower surfaces, rollers that facilitate movement over obstructions, and special chains (ticklers) that run along the lower leading lip of the mouth. The tickler stimulates fish to rise off the bottom and into the net. Some trawls have one or more conical inserts before the cod or terminal part. The cod is held closed with a cinch line that can be pulled to release the catch onto the sorting deck or tables. The trawl is pulled by a bridle and warp worked from a hydraulic winch. Smaller trawls may be worked strenuously by hand. The length of warp required depends on speed and depth of sampling; however, a 30° angle of warp to the water surface is typical. The speed of towing relates to the gear but is around 2 or 3 knots. Depth and the character of the bottom may be defined by echo sounder. The duration of a tow ranges from a few minutes to several hours. Night trawling may yield larger catches but is more difficult in many inland waters because of navigational aids, anchorage buoys, and other obstructions.

l. Drop samplers: A drop sampler is an active sampler that is particularly useful for sampling small fishes in turbid water or in submerged or emergent vegetation.[2-4] The device is often an open-ended cylinder made of sheet plastic or netting. It usually has a metal skirt on the bottom that penetrates the substrate to assure complete closure. Drop samplers usually are suspended by a bridle from a boom on the bow of a boat, and then quietly dropped to quantitatively sample a small area (1 to 3 m²). The enclosed area can be treated with a piscicide. A quantitative sample can be obtained by collecting the specimens with a pump and plankton net.

m. Ichthyoplankton sampling: Ichthyoplankton consists of the eggs and very young stages of fish (larvae, sacfry, postlarvae).[5,6] Ichthyoplankton usually is collected either by plankton nets or bulk water sampling.[7] Nets having a mouth diameter smaller than the main body of the net may be towed faster than the usual 2 to 3 knots.[8] The towing bridle of the net affects sample collection and a number of designs such as the double net or "bongo net" have been devised to reduce this influence; comparisons of various gears have been made.[9] Measure water volume with a flow meter. Other devices may be lowered to a given depth triggered to open and then triggered to close again, providing a sample from a known depth. The Clarke-Bumpus, Tucker trawl, light traps, pumps, and MOCNESS are examples of quantitative plankton samplers.[10] In vertical sampling the net is lowered to the bottom or some prescribed depth and then pulled upwards, sampling the water column. Mesh size for commercially available plankton netting ranges from 0.158 to 0.790 mm but standard meshes have yet to be specified.

Bulk water sampling of ichthyoplankton consists of collecting a known volume of water and separating the ichthyoplankton by filtration and/or centrifugation.

n. Seines: A seine is a simple panel of netting pulled by a bridle at each end (Figure 10600:2). In many smaller seines the bridle is attached to pulling poles or "brails." The upper line of the seine is equipped with floats and the lower with weights. Some seines are fitted with a central bag of smaller mesh that traps the fish. Seines commonly range in size from 1 to about 100 m long and from 1 to 3 m deep. Mesh size depends on the target species. The seine is an effective device for sampling smaller fishes.

Seines may be worked over shorelines relatively free of obstructions. Pulling may be either parallel, angled, or perpendicular to shore. Two samplers, wearing waders, form the net into a gentle "U" while pulling. After a suitable distance the seine, with the lead line on the bottom, is pulled to and up on the shore. A series of shorter passes may be more productive than one long one.

Figure 10600:2. Bag seine in operation in a stream. Photo courtesy of New York State Museum.

Small seines may be used over cobbled stream beds by placing the net poles firmly in the bottom and then rolling the cobbles upcurrent of the net; this action dislodges hidden fish to drift into the net. Benthic species are especially prone to capture in this manner.

Large seines (beach) usually are set from a boat. Fix a long warp on shore and pay it out by a boat working offshore to a set distance. Swing the boat parallel to shore and pay the net out. Attach a second warp and return the boat to shore. Draw the seine to shore by hand or by draft engine using the two warps.

A block seine may be used to block the mouth of an embayment. This sampling is effective when water elevations change. A self-contained block net is the closing seine consisting of a circular panel of netting strung on a weighted bottom ring and a floating upper ring. Drop the device onto the bottom enclosing the fish and remove them with a dip net. The technique is quantitative to a degree but emplacement of the gear causes fish to leave the study area.

The commercial purse seine is a larger version of the closing net and is used in deeper water for the capture of schooling fish. The lower edge is equipped with a pursing warp or line that allows the net to be drawn together under the surrounded school. After closure the fish are dip netted or hydraulically removed.

o. Lift, dip, and throw nets: These nets may be operated by a single worker. The lift net consists of a square panel of net held open by a pair of diagonal braces and is lifted at the crossing point by a cord. Such nets usually are baited while resting on the bottom and lifted when a sufficient number of fish have been gathered.

Dip or pole nets are conical nets attached to a ring frame which in turn is attached to a pole. They are effective in the capture of salmonids, smelt, and various clupeids (such as blueback herring and alewife).

Throw or cast nets are circular nets often having a pursable and weighted margin operated by a draw string. The net is thrown over a school of fish or a baited sector of bottom, the margin pursed and the closed net removed from the bottom.

p. Goin dredge: The Goin dredge is a wooden box with a net bottom and one wall missing. Handles or hand holes are placed on the three remaining sides and are used to work the device through vegetation.

q. Electrofishing: Electrical devices also may be used for collecting.[11-13] They are particularly useful in areas where uneven bottoms, fast-flowing water, or obstructions make other collecting techniques difficult or impossible. Several factors affect the selectivity of electrofishing operations, including water depth and velocity, stream width, conductivity of the water, fish size and morphology, and fish behavior. An electrical field in the water is produced by passing a current between two submersed electrodes or between one electrode and the ground. Depending upon design, electrical devices produce either alternating current (AC) or direct current (DC). AC stuns fish in its field, allowing them to be dipped from the water, whereas DC induces galvanotaxis so that the fish move toward one of the poles, from where they are recovered. DC devices are particularly effective in turbid water or in waters with numerous obstructions or heavy vegetation. AC devices are more likely to kill fish. Effectiveness of electrofishing is affected by such environmental factors as water hardness and availability of electrolytes. Addition of salt to raise conductivity of very soft water may be necessary.

Electrofishing gear ranges from large, gasoline-motor rigs (Figure 10600:3) mounted in a boat to small, portable back-pack units powered by a gasoline motor or a battery. The electric seine consists of a conventional seine in which electrodes of an AC source run along the lead and float lines, thus greatly increasing collecting efficiency.

Sampling with electricity requires that the collector be aware of the special dangers associated with this technique. Specially train staff, provide and use protective equipment, and carefully monitor the sampling area before and during the operation.

In certain studies, e.g., when assessing stream fish production by habitat type, electrofishing can provide a quantitative sample. In streams of manageable size and flow, sampling areas can be isolated with block nets, allowing capture and processing of all fish in the area. A second advantage is that fish mortality can be controlled carefully. Keep captured fish in live cars (containers, on occasion with aeration and refrigeration capabilities) during sampling and return immediately after processing.

r. Ichthyocides: Ichthyocides (piscicides) are the least selective method of collecting but are the most efficient in terms of percentage of total population sampled.[14] Chemicals used may act as a metabolic poison, as a vasoconstrictor (suffocant), or as an anesthetic. Most have some serious fault in terms of their value as a fish-collecting tool, such as slow reaction time, lack of sufficient killing power except in high doses, deleterious side effects to either the environment or user, or expense. Rigorous time-consuming permitting procedures usually are required. In some places, use of ichthyocides is illegal; review ordnances. Among the chemicals or chemical preparations used in the past are cresol, sodium cyanide, sodium hypochlorite, antimycin A and B, quinaldine, tricaine methanosulfonate (MS-222), urethane, and rote-

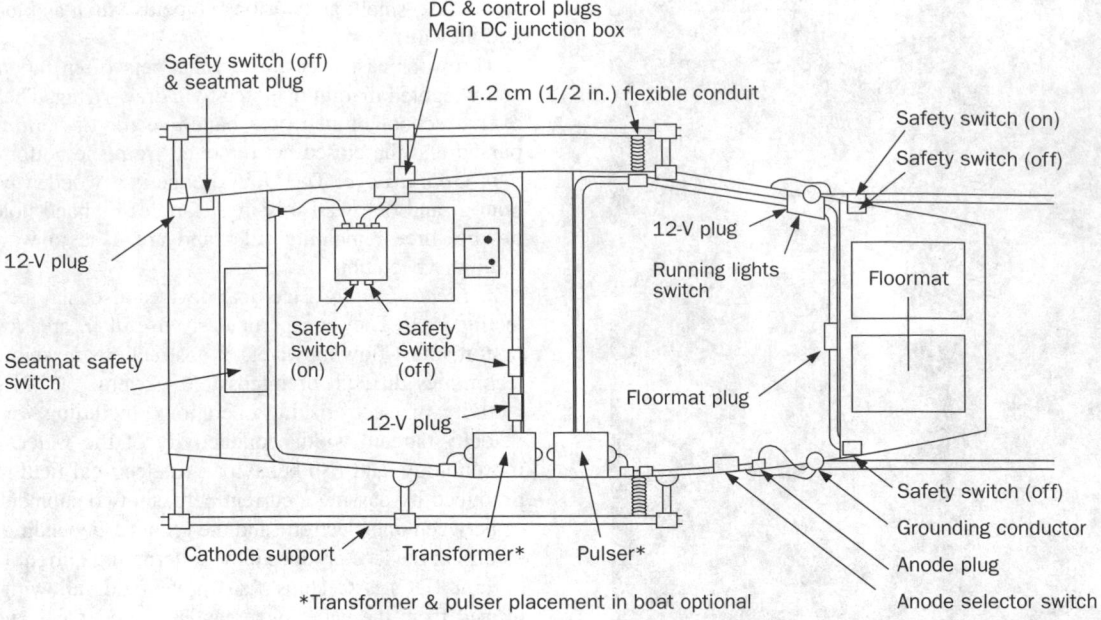

Figure 10600:3. Diagram of electrofishing boat. Source: Novotny, D.W. & G.R. Priegel. 1974. Electrofishing Boats, Improved Designs and Operational Guidelines to Increase the Effectiveness of Boom Shockers. Tech. Bull. No. 73, Dep. Natural Resources, Madison, Wis.

none; of these, only antimycin and rotenone now are approved for use by the U.S. Environmental Protection Agency.

Rotenone has been most widely used as a general collecting tool. It is the commercial name for a crystalline ketone ($C_{23}H_{22}O_6$) found in six genera of leguminous plants, particularly in the genus *Derris*. Rotenone functions as a vasoconstrictor and fish affected by it literally suffocate. Rotenone is selective for gill-breathing organisms; fish collected using rotenone can be eaten. It is rapidly detoxified by potassium permanganate, is adversely affected by light and high temperatures, and rapidly breaks down when subjected to such conditions.

Rotenone is available as powder, resin, crystals, or liquid emulsions. A 5% preparation is the usual strength employed in collecting. The powdered preparations are least stable and the resins and crystalline products the most stable. A ready-to-use liquid emulsion is most convenient, but it is relatively expensive. If weight and space problems exist, use crystals or resins. Mix resins and crystals with solvents immediately before use. For the resins, mix 100 g fragmented rotenone (broken up before mixing) with 1 L commercial-grade acetone and 100 mL emulsifier. For the crystals, mix 20 g with 3.8 L acetone. Crystals are the recommended form. The amount of rotenone to be used per collecting station varies according to volume and current of water, reaction time of the toxicant, water temperature, and other factors. A sample bioassay may be useful in determining dosage. In general, be conservative because good-quality rotenone is surprisingly effective. To use rotenone have properly trained personnel.

s. Concussion: Concussive methods such as explosive devices and substances have been used; however, most states and the federal government prohibit their use. A simple concussive method called ''tunking'' or ''stoning'' consists of striking an emergent boulder with another rock and then turning the boulder over for collection of the fishes underneath it. Sculpins and minnows may be captured in this manner.

t. Creel census: The systematic collection of data from sport anglers is a primary means of analysis. Field examination of fish caught, coupled with a standard interview of the angler, is common. Allocation of fishing diaries among anglers followed by their systematic retrieval is another method. Statistically rigorous sampling methods and well-designed interview protocols are essential for the generation of reliable results.

u. Slurp gun: The slurp gun consists of a valved cylinder fitted with a plunger. It is used with SCUBA or while snorkeling and is especially effective for selective sampling of fish at nesting sites, or within otherwise inaccessible interstices.

v. Stomach and gut examination: The contents of the digestive tracts of fish often include examples of additional fish species not otherwise sampled.

w. Serendipity: Useful specimens can be collected after various kinds of fish mortalities. The release of toxic substances, lethal changes in water temperature, anoxic water, construction of cofferdams, dewatering of power-plant flumes, drying up of natural water bodies, and the stranding of fish after floods exemplify events that can yield study materials.

x. Acoustic methods: Acoustic means of fish detection and quantification involve the generation of a series of sound pulses by an electroacoustic transducer mounted on a vessel, a towed body, or a fixed land feature.[15] Vertical and side or horizontal scanning are widely used. For side scanning, one or more transducers are mounted at some depth and oriented horizontally, as on the piers of a bridge crossing a river or on the walls or floor of a lock chamber. The sound impulse is reflected by suspended objects and the bottom. It is received by the same transducer, which reconverts the sound signal into an electrical signal that is amplified and displayed and/or recorded on a paper tape, VCR tape, liquid display, photographic film, etc. The usual paper tape provides a profile of the water showing the bottom and the objects suspended above it. Objects having high ''acoustic impedance,'' such as fish with air blad-

Key:
- a strap tag
- b plastic arrow
- c bachelor button
- d hydrostatic (Lea) tag
- e plastic flag tag
- f ivorine/silver plate tag
- g spaghetti tags
- h Petersen disc tag
- i barb and trailer tag
- j spring anchor tag
- k dart and capsule tag
- l Carlin tag
- m Sphyrion tag
- n roll and anchor tag
- o subcutaneous tag
- p body cavity tag
- q jaw tag with pennant
- r jaw tag

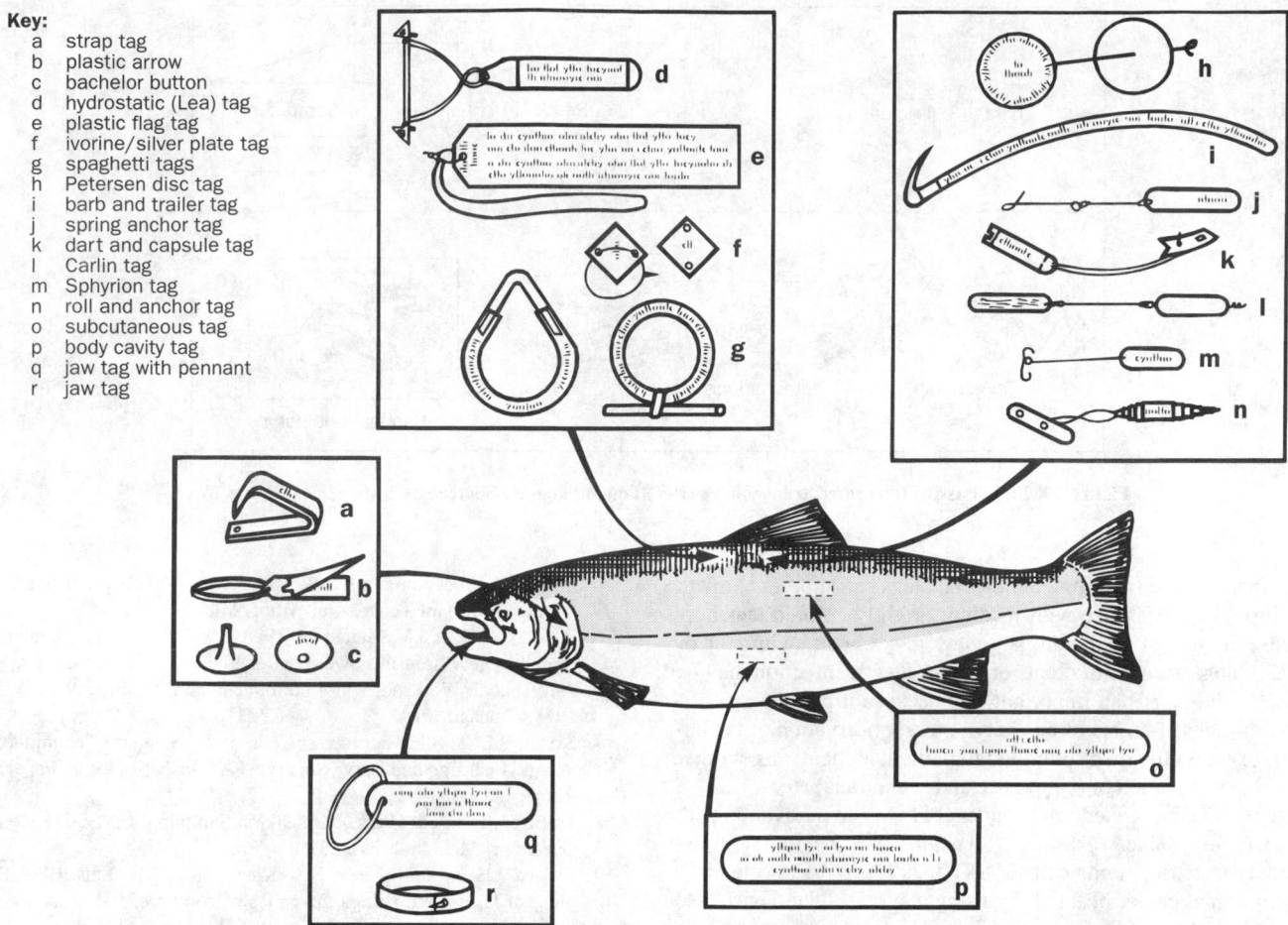

Figure 10600:4. Types of tags commonly used. Source: STOTT B. 1971. Marking and tagging. *In* BAGENAL, T., ed. 1978. Methods for Assessment of Fish Production in Fresh Waters. IBP Handbook No. 3. Blackwell Scientific Publ., Oxford, England.

ders, are especially vivid targets and produce strong signals. Gas bubbles, plankton, particulate matter, and density differences associated with a thermocline, floating leaves, and other objects may produce acoustic traces. In the more typical recording type of sounding apparatus a single, relatively stable fish will produce a chevron-shaped "echogram," with the point of the chevron pointing upwards. Several fish usually can be resolved into a corresponding number of such figures, but a school of fish may produce a cloud-like mark, complicating analysis. Under these circumstances estimation of biomass usually is undertaken.

Verify identity of species being recorded by capture or direct observation (e.g., SCUBA). Weighing captured specimens permits extrapolation to biomass estimation.

Apart from quantification, sonar methods provide high-quality data on the vertical and horizontal location of fish stocks.

y. Snorkeling and SCUBA: Snorkeling is the process of using a face plate to facilitate underwater vision and a "J"-shaped breathing tube, to allow sustained observation of underwater conditions. SCUBA consists of a compressed-air tank, a gas-flow regulator, and a hose and mouth piece. Ancillary but crucial gear includes a weight belt for adjusting buoyancy, fins, an inflatable life vest, and either a wetsuit or a dry suit.

Depths less than 15 m are accessible to those with good training. Greater depths require considerable training and experience.

Underwater study permits precise gear placement, direct observation of gear function, rapid definition of resource use by the species represented, assessment of impact of chemical spills in deeper water (when dead and moribund specimens may be resting on the bottom), the study of behavior, and estimation of population size. A team of divers can swim side-by-side along a defined course noting fish present and making simultaneous records of observation using underwater recording materials. An underwater pad consisting of sanded vinyl plastic can be marked with the usual graphite pencil and erased with a conventional rubber eraser. Aluminum foil taped onto a plastic sheet, again marked with a pencil, is effective and has the added advantage of providing a permanent record.

4. Tags and Tagging

The marking of a fish or group of fish followed by their release and recapture provides information on movement, rates of growth, and population characteristics. The methods include: dyes and stains (including fluorescent materials), finclips, attachment of tags (Figure 10600:4), emplacement of encoded wires, sonic (transmitter) tags, the "PIT" tag (see below), branding, radioisotopes, and marker chemicals that can be detected easily or influence ageable features such as the scales, otoliths, or spines. Chemical marking may be restricted under certain circumstances. Beware the influence

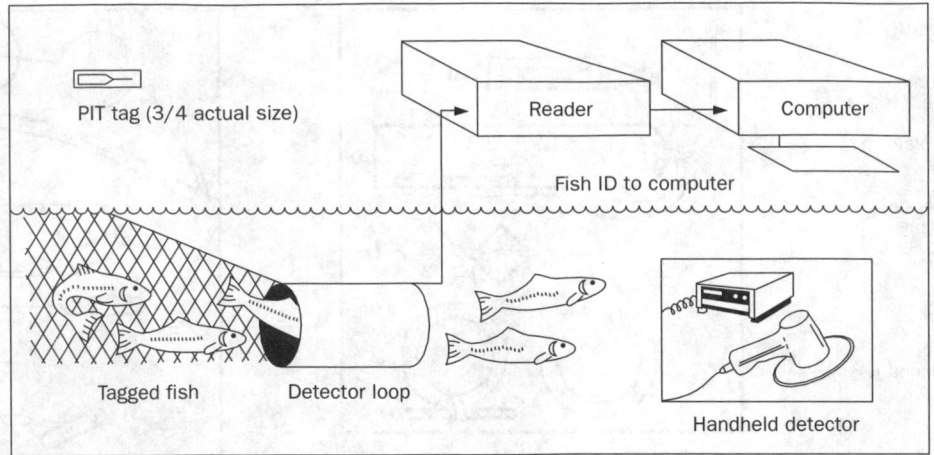

Figure 10600:5. Passive integrated transponder (PIT) tagging system. Source: BioSonics, Inc., Seattle, Wash.

of the mark on the fish: for example, jaw tags can suppress growth rate through interference with feeding; brightly colored tags may increase the role of predation; large tags may impair swimming or cause entanglement with plants or other objects; infection may be induced. Marking is an important tool because it permits identification of genetic strains of hatchery stocks, observation of the behavior of individual fish submitted to particular chemicals or other stressors, observation of dominance and social rank, etc.

The passive integrated transponder (PIT)* is an innovation with considerable value.[16] Small (10-mm × 2.1-mm) glass-imbedded electrical units called ''PIT tags'' may be injected into the abdominal cavity of a fish by using a syringe-like device (see Figure 10600:5). Fish 5 cm and larger may be marked in this fashion. The tag is activated by a hand-held excitatory unit that elicits a distinctive signal permitting recognition of the individual fish and immediate data processing by computer. The specimen may be processed with little or no handling by using a water vessel or a flow-through cylinder (as is done with migratory salmon moving through a weir) during the tag excitation. The tags last for several years. Salmon smolt can be tagged and individually recognized years later as the adult fish return for spawning. Coded wire tags (CWT) also have been used extensively but require excision of the wire for identification.

5. References

1. McClane, A.J., ed. 1974. McClane's New Standard Fishing Encyclopedia and International Angling Guide, 2nd ed. Holt, Rinehart & Winston, New York, N.Y.
2. Zimmerman, R., T.J. Minello & G. Zamora. 1984. Selection of vegetated habitat by brown shrimp, *Penaeus aztecus*, in a Galveston Bay salt marsh. *U.S. Fish. Bull.* 82:325.
3. Zimmerman, R. & T. Minello. 1984. Densities of *Penaeus aztecus*, *Penaeus setiferus*, and other natant macrofauna in a Texas salt marsh. *Estuaries* 7:421.
4. Baltz, D.M., C. Rakocinski & J.W. Fleeger. 1993. Microhabitat use by marsh-edge fishes in a Louisiana estuary. *Environ. Biol. Fishes* 36:109.
5. Snyder, D.E. 1976. Terminology for intervals of larval fish development. *In* J. Boreman, ed. Great Lakes Fish Egg and Larvae Identification: Proceedings of a Workshop, p. 41. U.S. Fish Serv., National Powerplant Team, Ann Arbor, Mich.
6. Bowles, R.R. & J.V. Merriner. 1978. Evaluation of ichthyoplankton sampling gear used in power plant studies. *In* L.D. Jensen, ed. Proc. 4th Annual Workshop on Entrainment and Impingement, 5 Dec. 1978, Chicago, Ill.
7. Smith, P.E. & S.L. Richardson, eds. 1977. Standard Techniques for Pelagic Fish Egg and Larva Surveys. FAO Fisheries Tech. Paper 175, Rome, Italy.
8. Tranter, D.J., ed. 1968. Zooplankton Sampling. UNESCO Monogr. Oceanogr. Methodol. No. 2, Paris, France.
9. Choat, J.H., P.J. Doherty, B.A. Kerrigan & J.M. Leis. 1993. Sampling of larvae and pelagic juveniles of coral reef fishes: A comparison of towed nets, purse seine and light aggregation devices. *Fish. Bull.* 91:195.
10. Wiebe, P.H., K.H. Burt, S.H. Boyd & A.W. Morton. 1976. A multiple opening/closing net and environmental sensing system for sampling zooplankton. *J. Mar. Res.* 34:313.
11. Novotny, D. & G.R. Priegel. 1974. Electrofishing Boats, Improved Designs and Operational Guidelines to Increase the Effectiveness of Boom Shockers. Tech. Bull. No. 73, Dep. Natural Resources, Madison, Wis.
12. Vibert, R., ed. 1967. Fishing with Electricity, Its Application to Biology and Management, Contributions to a Symposium. European Inland Fisheries Advisory Comm., FAO, Fishing News Books Ltd., London, U.K.
13. Webster, D.A., J.L. Forney, R.H. Gibbs, Jr., J.H. Severs & W.F. Van Woert. 1955. A comparison of alternating and direct currents in fishery work. *N.Y. Fish Game J.* 2:106.
14. Schnick, R.A., F.P. Meyer & D.L. Gray. 1986. A Guide to Approved Chemicals in Fish Production and Fishery Resource Management. National Fish Research Lab., La Crosse, Wis.
15. Mitson, R.B. 1983. Fisheries Sonar. Fishing News Books Ltd., Farnham, Surrey, England.
16. Prentice, E.F., C.W. Sims & D.L. Park. 1985. A Study to Determine the Biological Feasibility of a New Fish Tagging System. Bonneville Power Admin. Div. Fish & Wildlife, Portland, Ore.

6. Bibliography

Larval and Immature Fishes
Nansen, F. 1915. Closing nets for vertical hauls and for horizontal towing. *J. Cons. Perma. Int. Explor. Mer* 67:1.

* Available from BioSonics, Inc., 3670 Stone Way North, Seattle, WA 98103.

GIBBONS, S.G. & J.H. FRASER. 1937. The centrifugal pump and suction hose as a method of collecting plankton samples. *J. Cons. Perma. Int. Explor. Mer* 12:155.

ARON, W. 1958. The use of a large capacity portable pump for plankton patchiness. *J. Mar. Res.* 16:158.

BARY, B.M., J.G. DeSTEFANO, M. FORSYTH & J. VON DEN KERKHOF. 1958. A closing high-speed plankton catcher for use in vertical and horizontal towing. *Pacific Sci.* 12:46.

CLARKE, W.D. 1964. The jet net, a new high-speed plankton sampler. *J. Mar. Res.* 22:284.

ARON, W., E.H. AHLSTROM, B.M. BARY, A.W.H. BE & W.D. CLARKE. 1965. Towing characteristics of plankton sampling gear. *Limnol. Oceanogr.* 10:333.

BEERS, J.R., G.L. STEWART & J.D.H. STRICKLAND. 1967. A pumping system for sampling small plankton. *J. Fish. Res. Board Can.* 24:1.

FRASER, J.H. 1968. Zooplankton sampling. *Nature* 211:915.

CROCE, N.D. & A. CHIARABINI. 1971. A suction pipe for sampling midwater and bottom organisms in the sea. *Deep Sea Res.* 18:851.

SHERMAN, K. & K.A. HONEY. 1971. Size selectivity of the Gulf III and bongo zooplankton samplers. *Int. Comm. Northwest Atl. Fish. Res. Bull.* 8:45.

Mature Fishes

HASKELL, D.C., J. MacDOUGAL & D. GEDULDIG. 1954. Reactions and motion of fish in a direct current electric field. *N.Y. Fish & Game J.* 1:47.

LAGLER, K. F. 1956. Freshwater Fishery Biology, 2nd ed. Wm. C. Brown Co., Dubuque, Iowa.

HURLBERT, S.H. 1971. The nonconcept of species diversity: a critique and alternative parameters. *Ecology* 52:577.

LOVE, R.H. 1971. Dorsal-aspect target strength of an individual fish. *J. Acoust. Soc. Amer.* 49:816.

LOVE, R.H. 1971. Measurements of fish target strength: a review. *Fish Bull. NOAA/NMFS* 69:703.

FORBES, S.T. & O. NAKKEN, eds. 1972. Manual of Methods for Fisheries Resource Survey and Appraisal. Part 2: The Use of Acoustic Instruments for Fish Detection and Abundance Estimation. FAO Man. Fisheries Sci. No. 5, Food & Agriculture Org. United Nations, Rome, Italy.

WEATHERLEY, A.H. 1972. Growth of Fish Populations. Academic Press, New York, N.Y.

KUSHLAN, J.A. 1974. Quantitative sampling of fish populations in shallow, freshwater environments. *Trans. Amer. Fish Soc.* 103:348.

URICK, R.J. 1975. Principles of Underwater Sound. McGraw-Hill, New York, N.Y.

GULLAND, J.A., ed. 1977. Fish Population Dynamics, John Wiley & Sons, New York, N.Y.

GRINSTED, B.G., R.M. GENNINGS, G.R. HOOPER, C.A. SCHULTZ & D.A. HORTON. 1978. Estimation of standing crop of fishes in the predator-stocking-evaluation reservoirs. *Proc. Annu. Conf. S.E. Assoc. Game Fish Comm.* 30:120.

BURCZYNSKI, J. 1979. Introduction to the Use of Sonar Systems for Estimating Fish Biomass. FAO Fish. Tech. Pap. No. 191, Food & Agriculture Org. United Nations, Rome, Italy.

SHORROCKS, B. 1979. The Genesis of Diversity. University Park Press, Baltimore, Md.

HOCUTT, C.H. & J.R. STAUFFER, JR., eds. 1980. Biological Monitoring of Fish. Lexington Books, D.C. Heath and Co., Lexington, Mass.

WELCH, H.E. & K.H. MILLS. 1981. Marking fish by scarring soft fin rays. *Can. J. Fish Aquat. Sci.* 38:1168.

NAKKEN, O. & S.C. VENEMA, eds. 1983. Symposium on Fishery Acoustics. Fish. Rep. 300, Food & Agriculture Org. United Nations, Unipub, Ann Arbor, Mich.

BROWNIE, C., D.R. ANDERSON, K.P. BURNHAM & D.S. ROBSON. 1985. Statistical Inference from Band Recovery Data. Colorado Fish & Wildlife Res. Unit, Fort Collins.

10600 C.   Sample Preservation

The decision to preserve specimens depends on study objectives. Preserved material may be necessary to confirm identity of a species, to evaluate certain demographic characteristics of the population, or to estimate incidence of parasitic infection or disease. It also may be essential evidence in legal proceedings.

Do not preserve specimens unless there is clear need for them. Fixatives are toxic, dangerous if used improperly, and covered by regulations on hazardous substances. Preserved specimens require expensive and time-consuming curation.

Fix specimens in 10% formalin (a 9:1 ambient water dilution of 100% formalin). Formaldehyde gas reaches saturation in water at about 37% by weight and by convention this saturated solution is called 100% formalin. Fix fish less than 10 cm in total length without opening the visceral cavity. Larger specimens require injection of preservative into the visceral cavity or slitting of right ventral body wall for about 25% of body length. Specimens larger than 25 cm total length (and especially oily species) usually require injection of concentrated formalin into the dorsal muscle mass.

The placement of fish in the sampler container, i.e., head up or down, depends on intended use. A good ratio of specimen mass to preservative is 1:1 or with the level of preservative submerging the specimen by at least an inch.

Fix ichthyoplankton in formalin. To facilitate sorting, 1 g rose bengal stain/L fixative may be added to stain living tissue. If the sample contains a large amount of biomass (detritus, non-target organisms) split the sample into two or more sample jars rather than increase the concentration of formalin.

Ideally use wide-mouthed containers with a durable, screw-type plastic cover. If using metal caps, add about 1 g sodium borate/L preserved material.

Formaldehyde is highly allergenic and a listed carcinogen; minimize direct contact with skin and avoid breathing fumes. Formaldehyde is best transported in tightly sealed plastic containers.

After several days to two or more weeks in the fixative (depending on the size of the fish), transfer specimens to 70% ethyl or isopropyl alcohol for long-term preservation. Preserve ichthyoplankton samples in a 3 to 5% solution of formalin because alcohol tends to shrink and distort the specimens. If possible, reuse formalin.

It may be harmful and/or illegal to release even small amounts of 10% formalin into wastewater collection systems. Examine ordinances before disposal of fixatives and preservatives.

Isopropyl alcohol is a less expensive and less flammable substitute for ethyl alcohol, but weigh these advantages against the fact that isopropyl is not a good fixative and may damage the specimens for histology. Both alcohols are highly flammable when stored in bulk. Quantities stored in any building may be limited by codes. If otoliths will be used for fish age determination, do not fix specimens in formalin until after the otoliths are removed. If the otoliths cannot be removed before long-term preservation, fix the specimens in alcohol, which is adequate for small fish, or freeze them.

For pathology fix whole fish or organs for at least 24 h in 10 times their volume of neutral buffered formalin before further processing:

37% Formaldehyde (100% formalin)	100 mL
Distilled water	900 mL
Sodium phosphate monobasic, $NaH_2PO_4 \cdot H_2O$	4 g
Sodium phosphate dibasic, Na_2HPO_4	6.5 g

Fix fish less than 8 cm long whole. For larger specimens fix viscera either by injecting the fixative into the body cavity or by cutting the body cavity from the anus to below the head. Fixation is most effective on live fish or as soon after death as possible. A delay results in autolysis. In larger fish that cannot be preserved whole, preserve representative sections of organs about $5 \times 5 \times 2$ mm in size. Obtain such tissue blocks from organ areas having irregular color, size, or consistency, using a scalpel or razor blade and fix immediately. Except in large fish, fix the entire brain. See Figure 10600:6 for morphological and anatomical guidance.

Rigorously document preserved materials. Place labels composed of highly resistant bond paper, bearing the key facts of lot number, collection date, locality, name(s) of collector(s), and other particulars written in graphite or waterproof india ink, inside each container or attach to larger specimens held in plastic bags.

Ship fixed specimens packed in absorbent paper or cloth (cheesecloth or paper towels) lightly moistened with the fixative or preservative and sealed in a plastic bag in turn sealed in a second or third bag. Pad package and place in a box or canister for transport with a copy of the chain-of-custody documents. Shipment in bulk fixative or preservative via the United States mails is prohibited.

If preserved specimens are no longer needed after completion of a study, offer them to a regional or national museum. Planning for eventual disposal of specimens is both economical and provides for the best long-term use of the materials collected.

10600 D. Analysis of Collections

1. Identification

a. General remarks: Identification of fish is based on diagnostic characters such as body form, color and size, shape and position of fins, meristic features such as the number of rays in a fin or the number of scales in a specific series, the presence of distinctive organs such as barbels, or the lateral line and various proportions such as the ratio of the length of the head to the total length of the body (see Figure 10600:6). Diagnostic features may vary with age, sex, reproductive condition, social status, time of year, and habitat of the fish. Diagnostic keys and other descriptive materials are available for all regions of North America and a list of selected taxonomic works appears in the bibliography.

Common names of North American fishes are listed in a special publication of The American Fisheries Society.[1]

Identification may be performed on both fresh and preserved specimens. Fresh materials are essential for color although some preserved specimens retain color if they have been preserved in a color preservative, such as BHT. Fixed specimens are suitable for determining meristic or mensural characteristics. Use a dissecting binocular microscope with illumination and dissecting tools to examine specimens less than 10 cm long.

b. Ichthyoplankton: The identification of fish eggs and larval fish is a special discipline and selected works are cited in the bibliography. Enumeration often involves knowledge of adult populations such as sexual maturity, spawning migration, spawning groups, and presence/absence of the species in the watershed. Intensive studies may involve captive spawning, rearing, and documentation of larval development of selected species. Obtain spe-cialized assistance for identification of ichthyoplankton and the more difficult taxa such as the Cyprinidae (minnows), Clupeidae (herrings), Catostomidae (suckers), Poeciliidae (livebearers), Cyprinodontidae (killifishes), Percidae (perches), and others.

c. Rare and endangered forms: Pay special attention to rare and endangered forms that are protected by law. If a rare or endangered form is present a special permit or memorandum of understanding may be required. Do not intentionally collect rare or endangered forms; if they are accidentally taken and fixed contact the responsible agency and transfer specimens to a designated museum.

2. Diet

The contents of the digestive tract provide information on the amount and kind of foods eaten. Stomach contents may be extracted from some living fish by inserting a smooth and moistened glass or metal cylinder down the esophagus into the stomach. Most analyses, however, involve the sacrifice of the fish. After capture, quickly preserve either the entire fish or the viscera.

To characterize diet, use a dissecting binocular microscope with the organisms suspended in water in a shallow transparent container. Define the contents to the lowest practical taxon and express as frequency of occurrence. Each dietary item by weight, volume, or caloric content may be enumerated in special studies.

3. Structure of Populations and Assemblages

a. General remarks: The actual properties of a population such as number, average size, and weight are estimated statistics.[2] For

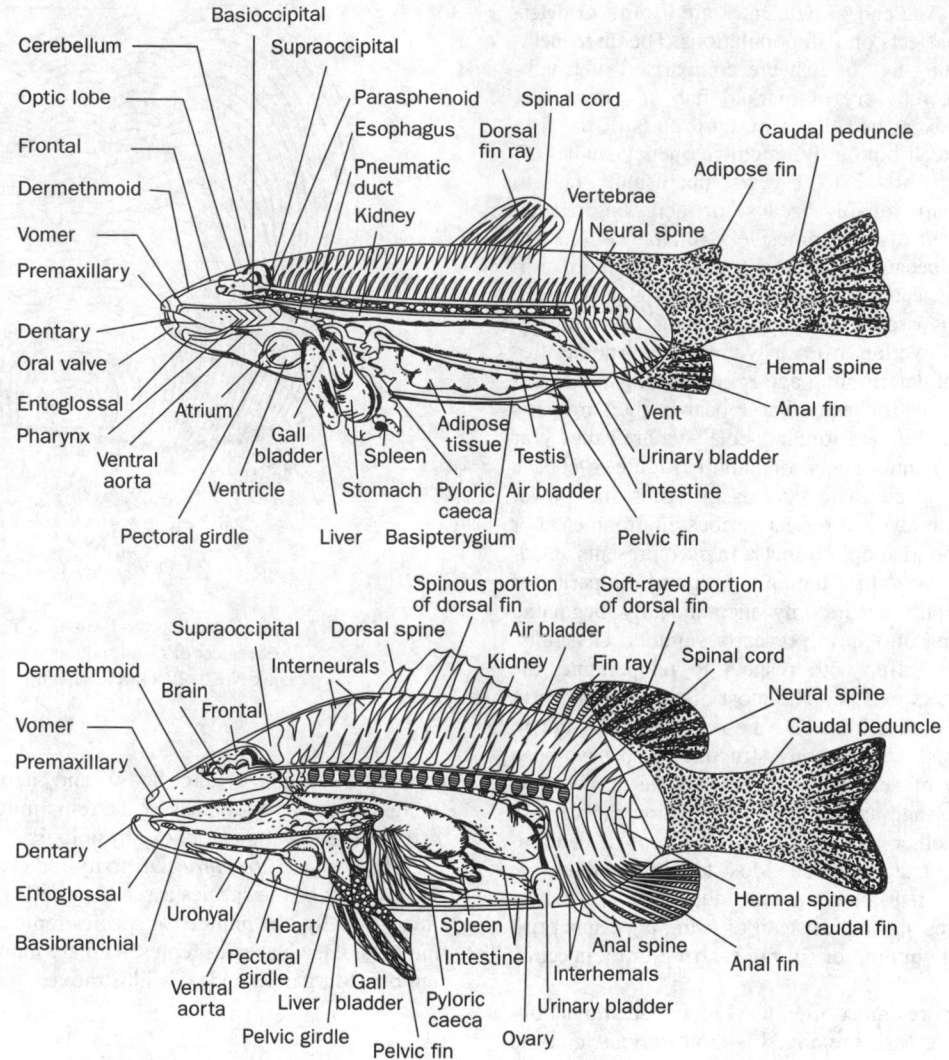

Figure 10600:6. Key organs and external body parts of a soft-rayed (upper) and spiny-rayed (lower) fish. Source: LAGLER, K.F. 1956. Freshwater Fishery Biology, 2nd ed. Wm. C. Brown Co., Dubuque, Iowa.

ecological and management purposes evaluate the numbers of individuals and the biomass (total weight) along with the factors that regulate these. Four key variables define number: natality, i.e., number of individuals being added through reproduction, mortality, immigration, and emigration.

Biomass is the net result of dietary income and its conversion efficiency, respiration, defecation, and loss to other compartments of the ecosystem such as predators, parasites, and pathogens.

The ideal static characterization of a population consists of graphs that show by sex:

Age frequency and maturation
Length frequency
Weight frequency
Age versus length
Age versus weight
Length versus weight

b. Population size: One of the simplest and most practical population estimation methods is the Petersen ratio based on marked fish. Collect a sample of fish, mark, and release. Collect a second or recapture sample including both marked and unmarked fish. The method assumes random distribution and no immigration or emigration. The larger the number of marked fish, the more reliable the estimate. It is particularly important to mark large numbers of the smaller species, if possible, because individuals of such species are less easily recovered. Estimate population size using the mark-recapture method by the formula

$$\hat{N} = MC/R$$

where:

$\hat{N}$ = estimate of population size,
M = number of fish marked and released,
C = recapture sample size, including both marked and unmarked fish, and
R = number of marked fish recaptured.

The estimate is of the population present at the time of the first (marking and release) sample, not at the time of recapture.

c. Age and growth: Age and growth rates are useful for determining environmental effects on fish populations. The three methods used for determining age of fish are comparison of length-frequency distribution, recovery of marked fish of known age, and interpretation of layers laid down on hard parts of the fish. Determination of age distribution by length-frequency studies often is adequate for the first 2 to 4 years, but usually fails to separate older age groups reliably because of increasing overlap in length distribution. It also becomes less reliable as one approaches the equator because breeding seasons are more protracted in warmer areas, causing yearly age groups to be less well defined. In cold-water assemblages, growth rates often are retarded and age groups overlap, even in young fish. Another frequently used method of determining age is interpreting and counting growth zones or growth checks that appear in the hard parts of fish. Those considered to be formed once a year are called year marks, annual marks, annual rings, or annuli (Figure 10600:7). They are formed during alternate periods of faster and slower growth (or no growth at all) and reflect various environmental or internal influences. The assumption that a mark represents an annulus requires validation.[3,4] In a temperate region, the period of little or no growth usually occurs only once annually, beginning in winter and extending into spring or early summer. Generally, the more the seasons differ with respect to temperature, the sharper the annual marks will be. The most distinct annual rings are developed in temperate climates of the northern and southern hemispheres. Scales and several bony structures also have been of value in the study of seasonal growth. Otoliths provide the most reliable record of age and growth, but require that the fish be sacrificed. Still another method of age and growth determination is by marking or tagging fish. Most tagging methods are not applicable to small fish, or if they are, they may cause mortality, so that recaptures are few. Instead of using tags, small fish may be marked by fin clipping or fin-ray scarring or by injecting dye.

Scales are the structures most often used in age determinations of fish because they are easily removed without serious injury to the fish. For scaleless fish, removal of other structures (otoliths, vertebrae, fin spines) may be necessary. Take scales usually from the upper mid-side of the body where they are large and symmetrical. Wherever scales are taken, remove them from the same part of the body in all individuals to be compared. Several scales may be needed for analysis, because an annulus that appears doubtful on one scale may be clear on another. In addition, some scales may have been regenerated, i.e., replaced, and may not show all annuli.

d. Index of condition: The coefficient or index of condition (also condition factor) is the length-weight relationship used to express relative plumpness or robustness of the fish. This, in turn, is related to environmental conditions. The equation usually used is:

$$K = \frac{(W \times 10^5)}{L^3}$$

where:

K = coefficient or index of condition,
W = weight, g, and
L = length, mm.

The gonadosomal coefficient (weight of gonads divided by the

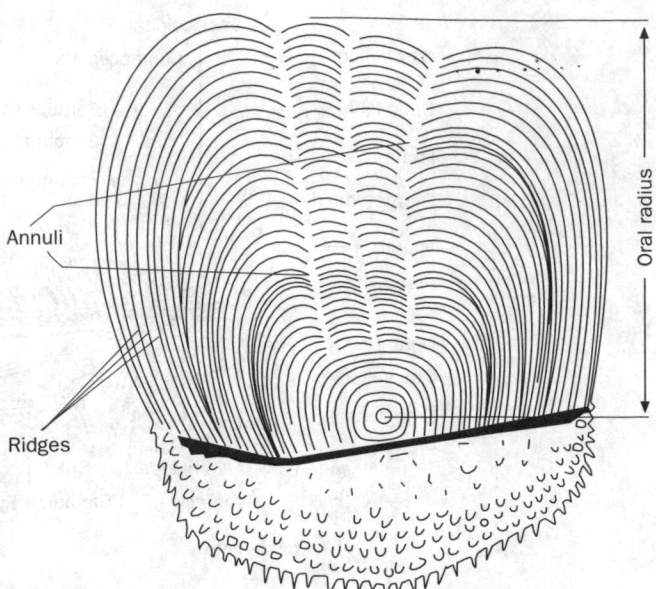

Figure 10600:7. Fish scale. Source: BAGENAL, T., ed. 1978. Methods for Assessment of Fish Production in Fresh Waters. IBP Handbook No. 3. Blackwell Scientific Publ., Oxford, England.

remaining weight of the body) and hepatosomal coefficient (weight of the liver divided by the remaining weight of the body) also are used as indicators of physiological condition.

e. Assemblage structure: Diversity indices are used to quantify the structure of the species assemblage in a particular habitat over time.[5] Usually, the number of species represented and the relative numbers of each species represented are incorporated into a single number. Margalef's index is illustrative.

$$d = \frac{(S - 1)}{\ln N}$$

where:

d = Margalef Index,
S = number of species represented in sample, and
N = total number of individuals in the sample.

The index of biotic integrity (IBI) is another method for assessing community structure. It produces a score for each sample site that allows assessment of environmental conditions at that site by comparing it to scores from other sites. It is a method for comparing changes in environmental conditions at a single site over time or at several sites within a carefully defined geographical area.

The IBI score is generated by using several carefully selected metrics that reflect different aspects of the aquatic system. Typically there are twelve metrics that fall into three broad categories. The first group assesses the composition of the fish assemblage. These metrics explore the relationship between the number of species and the number of individuals within each species or category. The second group examines the trophic composition of the assemblage. To score these metrics, some knowledge of the feeding ecology of the fishes is required. The third group looks at fish abundance and condition. These metrics deal with the number of individuals caught and their health.

Once the metrics are identified, develop the scoring criteria. Scores are based on the deviation from conditions at a relatively undisturbed site. Typically, there are three choices: a high score (5) is assigned if the condition is equal to that found at a similar but undisturbed site (key), a middle score (3) if the condition is worse than that found at the key site, and a low score (1) if conditions are much worse than those at the key site.

The total IBI score is the sum of the metric scores. In the example above, the key site would have an IBI score of 60 (12 metrics, maximum metric score of 5). The scores from all the sampled sites would be compared to this maximum. Sites where environmental conditions are relatively undisturbed would score near the maximum whereas degraded sites would score near the minimum of 12. The greatest advantage to the IBI is that comparisons are easy to make and the results are obvious.

The IBI metrics and scoring criteria are not universal and metrics and scoring protocols have been or must be developed for any geographic region.[6,7]

4. References

1. ROBINS, C.R., R.M. BAILEY, C.E. BOND, J.R. BOOKER, E.A. LACHNER, R.N. LEA & W.B. SCOTT, eds. 1990. A List of Common and Scientific Names of Fishes from the United States and Canada, 5th ed. Spec. Publ. No. 20, American Fisheries Soc., Bethesda, Md.

2. CUSHING, D.H. 1975. Marine Ecology and Fisheries. Cambridge Univ. Press, Cambridge, U.K.

3. BAGENAL, T.B., ed. 1974. Ageing of Fish. Proc. Internat. Symp. Univ. Reading, England, 19–20 July, 1973. The Gresham Press, Surrey, England.

4. STEVENSON, D.K. & S.E. CAMPANA, eds. 1992. Otolith Microstructure Examination and Analysis. Can. Spec. Publ. Fish. Aquat. Sci. 117, Ottawa, Ont.

5. PIELOU, E.C. 1975. Ecological Diversity. Wiley-Interscience Publ., John Wiley & Sons, New York, N.Y.

6. KARR, J.R., K.D. FAUSCH, P.L. ANGERMEIER, P.R. YANT & I.J. SCHLOSSER. 1986. Assessing Biological Integrity in Running Waters: A Method and Its Rationale. Illinois Natural History Surv. Spec. Publ. 5, Champaign.

7. MILLER, D.L., P.L. ANGERMEIER, R.M. HUGHES, J.R. KARR, P.B. MOYLE, L.H. SCHRADER, B.A. THOMPSON, R.A. DANIELS, K.D. FAUSCH, G.A. FITZHUGH, J.R. GAMMON, D.B. HALLIWELL, P.M. LEONARD & D.J. ORTH. 1988. Regional applications of an index of biotic integrity for use in water resource management. *Fisheries* 13(5):12.

5. Bibliography

Larval and Immature Fishes

HUBBS, C.L. 1943. Terminology of early stages of fishes. *Copeia* 1943(4): 160.

MANSUETI, A.J. & J.D. HARDY, JR. 1967. Development of Fishes of the Chesapeake Bay Region: An Atlas of Egg, Larval and Juvenile Stages. Natural Resources Inst., Univ. Maryland, College Park.

LIPPSON, A.J. & R.L. MORAN. 1974. Manual for Identification of Early Developmental Stages of Fishes of the Potomac River Estuary. Martin Marietta Corp. Environ. Tech. Center, Baltimore, Md.

JONES, P.W., F.D. MARTIN & J.D. HARDY, JR. 1978. Development of Fishes of the Mid-Atlantic Bight—An Atlas of Egg, Larval and Juvenile Stages. Vol. I. Acipenseridae through Ictaluridae. U.S. Fish & Wildlife Serv. Biol. Serv. Program FWS/OBS - 78/11.

HARDY, J.D., JR. 1978. Development of Fishes of the Mid-Atlantic Bight—An Atlas of Egg, Larval, and Juvenile Stages. Vol. II. An-

guillidae through Syngnathidae. U.S. Fish & Wildlife Serv. Biol. Serv. Program FWS/OBS - 78/12, Fort Collins, Colo.

HARDY, J.D., JR. 1978. Development of Fishes of the Mid-Atlantic Bight—An Atlas of Egg, Larval and Juvenile Stages. Vol. III. Aphredoderidae through Rachycentridae. U.S. Fish & Wildlife Serv. Biol. Serv. Program FWS/OBS - 78/12, Fort Collins, Colo.

JOHNSON, G.D. 1978. Development of Fishes of the Mid-Atlantic Bight— An Atlas of Egg, Larval and Juvenile Stages. Vol. IV. Carangidae through Ephippidae. U.S. Fish & Wildlife Serv. Biol. Serv. Program FWS/OBS - 78/12, Fort Collins, Colo.

FRITZSCHE, R.A. 1978. Development of Fishes of the Mid-Atlantic Bight—An Atlas of Egg, Larval and Juvenile Stages. Vol. V. Chaetodontidae through Ophidiidae. U.S. Fish & Wildlife Serv. Biol. Serv. Program FWS/OBS - 78/12, Fort Collins, Colo.

MARTIN, F.D. & G.E. DREWRY. 1978. Development of Fishes of the Mid-Atlantic Bight—An Atlas of Egg, Larval and Juvenile Stages. Vol. VI. Stramateidae through Ogcocephalidae. U.S. Fish & Wildlife Serv. Biol. Serv. Program FWS/OBS - 78/12, Fort Collins, Colo.

WANG, J.C.S. & R.J. KERNEHAN. 1979. Fishes of the Delaware Estuaries: A Guide to the Early Life Histories. EA Communications, Ecological Analysts, Inc., Towson, Md.

SNYDER, D.E. 1981. Contributions to a Guide to the Cypriniform Fish Larvae of the Upper Colorado River System in Colorado. Biol. Sci. Ser. No. 3, Bur. Land Management, Colorado.

WANG, J.C.S. 1981. Taxonomy of the Early Life Stages of Fishes—Fishes of the Sacramento - San Joaquin Estuary and Moss Landing Harbor, Elkhorn Slough, Calif. Ecological Analysts, Inc., Concord, Calif.

AUER, N.A., ed. 1982. Identification of Larval Fishes of the Great Lakes Basin with Emphasis on the Lake Michigan Drainage. Great Lakes Fisheries Research Center, Ann Arbor, Mich.

Mature Fishes

General

PERLMUTTER, A. 1961. Guide to Marine Fishes. New York Univ. Press, New York, N.Y.

LEE, D.S., C.R. GILBERT, C.H. HOCUTT, R.E. JENKINS, D.E. McALLISTER & J.R. STAUFFER, JR. 1980 et seq. Atlas of North American Freshwater Fishes. North Carolina State Mus. Natural History, Raleigh.

CASTRO, J.I. 1983. Sharks of the North American Waters. Texas A & M Univ. Press, College Station.

KUEHNE, R.A. & R.W. BARBOUR. 1983. The American Darters. Univ. Press of Kentucky, Lexington.

PAGE, L.M. 1983. Handbook of Darters. TFH Publications, Inc., Neptune, N.J.

BEHNKE R.J. 1992. Native Trouts of Western North America. American Fisheries Soc. Monogr. No. 6, American Fisheries Soc., Bethesda, Md.

Pacific Slope and Northwest

McPHAIL, J.D. & C.C. LINDSEY. 1970. Freshwater Fishes of Northwestern Canada and Alaska. Bull. 173, Fisheries Research Board Canada, Ottawa, Ont.

MILLER, D.J. & R.N. LEA. 1972. Guide to the Coastal Marine Fishes of California. Fish. Bull. 157, California Dep. Fish & Game, Sacramento.

HART, J.L. 1973. Pacific Fishes of Canada. Bull. 180, Fisheries Research Board Canada, Ottawa, Ont.

MOYLE, P.B. 1976. Inland Fishes of California. Univ. California Press, Berkeley.

WYDOSKI, R.S. & R.R. WHITNEY. 1979. Inland Fishes of Washington. Univ. Washington Press, Seattle.

MORROW, J.E. 1980. The Freshwater Fishes of Alaska. Alaska Northwest Publishing Co., Anchorage.

ESCHMEYER, W.N., E.S. HERALD & H. HAMMAN. 1983. A Field Guide to Pacific Coast Fishes. The Peterson Field Guide Series, Houghton Mifflin Co., Boston, Mass.

Southwest and Intermountain

KOSTER, W.J. 1957. Guide to the Fishes of New Mexico. Univ. New Mexico Press, Albuquerque.

LARIVERS, I. 1962. Fish and Fisheries of Nevada. Nevada State Fish & Game Comm., Carson City.

EVERHART, W.H. & W.R. SEAMAN. 1971. Fishes of Colorado. Colorado Game and Parks Div., Denver.

MINCKLEY, W.L. 1973. Fishes of Arizona. Arizona Game & Fish Dep., Phoenix.

SIMPSON, J.C. & R.L. WALLACE. 1982. Fishes of Idaho. Univ. Press of Idaho, Moscow.

Great Plains

KNAPP, F.T. 1953. Fishes Found in the Fresh Waters of Texas. Ragland Studio and Lithograph Printing Co., Brunswick, Ga.

BAILEY, R.M. & M.O. ALLUM. 1962. Fishes of South Dakota. Misc. Publ. Mus. Zool. Univ. Mich. 119, Ann Arbor.

SIGLER, W.F. & R.R. MILLER. 1963. Fishes of Utah. Utah Game & Fish Dep., Salt Lake City.

BAXTER, G.T. & J.R. SIMON. 1970. Wyoming Fishes. Bull. 4, Wyoming Game and Fish Comm., Cheyenne.

EVERHART, W.H. & W.R. SEAMAN. 1971. Fishes of Colorado. Colorado Game & Parks Div., Denver.

MORRIS, J., L. MORRIS & L. WITT. 1972. The Fishes of Nebraska. Nebraska Game and Parks Comm., Lincoln.

MILLER, R.J. & H.W. ROBINSON. 1973. The Fishes of Oklahoma. Oklahoma State Univ. Press, Stillwater.

SCOTT, W.B. & E.J. CROSSMAN. 1973. Freshwater Fishes of Canada. Bull. 184, Fisheries Research Board Canada, Ottawa, Ont.

HOESE, H.D. & R.H. MOORE. 1977. Fishes of the Gulf of Mexico: Texas, Louisiana, and Adjacent Waters. Texas A & M Univ. Press, College Station.

TOMELLERI, J.R. & M.E. EBERLE. 1990. Fishes of the Central United States. Univ. Press of Kansas, Lawrence.

Piedmont and Florida

HILDEBRAND, S.F. & W.C. SCHROEDER. 1928. Fishes of Chesapeake Bay. Fishery Bull. 43, U.S. Bur. Fisheries.

SMITH-VANIZ, W.F. 1968. Freshwater Fishes of Alabama. Agricultural Experiment Sta., Auburn Univ. Press, Auburn, Ala.

DAHLBERG, M.D. & D.C. SCOTT. 1971. The Freshwater Fishes of Georgia. Bull. Georgia Acad. Sci. 29.

DAHLBERG, M.D. 1975. Guide to Coastal Fishes of Georgia and Nearby States. Univ. Georgia Press, Athens.

MENHINICK, E.F. 1992. The Freshwater Fishes of North Carolina. North Carolina Wildlife Resource Comm., Raleigh, N.C.

JENKINS, R.E. & N.M. BURKHEAD. 1994. Freshwater Fishes of Virginia. American Fisheries Soc., Bethesda, Md.

Mississippi-Ohio Valley

COOK, F.A. 1959. Freshwater Fishes of Mississippi. Mississippi Game & Fish Comm., Jackson.

DOUGLAS, N.H. 1974. Freshwater Fishes of Louisiana. Claitors Publ. Div., Baton Rouge, La.

CLAY, W.M. 1975. The Fishes of Kentucky. Kentucky Dep. Fish & Wildlife Resources, Frankfort.

PFLIEGER, W.L. 1975. The Fishes of Missouri. Missouri Dep. Conservation, Columbia.

HOESE, H.D. & R.H. MOORE. 1977. Fishes of the Gulf of Mexico: Texas, Louisiana and Adjacent Waters. Texas A & M Univ. Press, College Station.

SMITH, P.W. 1979. The Fishes of Illinois. Univ. Illinois Press, Urbana.

TRAUTMAN, M.B. 1981. The Fishes of Ohio, 2nd ed. Ohio State Univ. Press, Columbus.

Great Lakes

HUBBS, C.L. & K.F. LAGLER. 1958. Fishes of the Great Lakes Region. Univ. Michigan Press, Ann Arbor.

SCOTT, W.B. & E.J. CROSSMAN. 1973. Freshwater Fishes of Canada, Bull. 184. Fisheries Research Board Canada, Ottawa, Ont.

TRAUTMAN, M.B. 1981. The Fishes of Ohio, 2nd ed. Ohio State Univ. Press, Columbus.

BECKER, G.C. 1983. Fishes of Wisconsin. Univ. Wisconsin Press, Madison.

Northeastern Region

TEE-VAN, J., ed. 1948 et seq. Fishes of the Western North Atlantic. Mem. Sears Foundat. Mar. Res., New Haven, Conn.

BIGELOW, H.B. & W.C. SCHROEDER. 1953. Fishes of the Gulf of Maine. Fish Bull. 74, U.S. Fish & Wildlife Serv.

LEIM, A.H. & W.B. SCOTT. 1966. Fishes of the Atlantic Coast of Canada. Bull. 155, Fisheries Research Board Canada, Ottawa, Ont.

SCOTT, W.B. & E.J. CROSSMAN. 1973. Freshwater Fishes of Canada. Bull. 184, Fisheries Research Board Canada, Ottawa, Ont.

SMITH, C.L. 1985. The Inland Fishes of New York State. New York State Dep. Environmental Conservation, Albany, N.Y.

HARTELL, K.E. 1992. Non-native fishes known from Massachusetts fresh water, with a list of the fishes of the state. Occ. Pap. Mus. Comparative Zoology Fish Dept. No. 2, Cambridge, Mass.

10600 E.   Investigation of Fish Kills

Fish kills vary, from the individual fish that dies of old age to the catastrophic kill, from partial to complete, and from natural occurrence to the results of human activity. No single investigative procedure can be appropriate for all situations. The following brief description may serve as an aid in investigating kills. Getting to the scene promptly before the evidence has decomposed or drifted away is vital. If surveillance of a particular body of water or area is involved, have available preset plans and equipment on a standby basis.[1-3]

1. Causes of Fish Kills

Fish kills may be caused by such natural events as acute temperature change, storms, ice and snow cover, decomposition of natural materials, salinity change, spawning mortalities, parasites, and bacterial and viral epizootics. Human-caused fish kills may be attributed to municipal or industrial wastes, agricultural activities, and water control activities.

2. Classification of Kills

One dead fish in a stream may be called a fish kill; however, in a practical sense, adopt some minimal range in number of dead fish observed, plus additional qualifications, in reporting and classifying fish kills. Any fish kill is significant if it affects fish of sport or commercial value, results from a suspected negligent discharge or malfunctioning waste treatment facility, or causes widespread environmental damage. The following definitions, based on a stream about 60 m wide and 2 m deep, are suggested as guidelines. For other size streams, make proportional adjustments.

MINOR KILL: 1 to 100 dead or dying fish confined to a small area or stream stretch. If recurrent, it could be significant; investigate.

MODERATE KILL: 100 to 1000 dead or dying fish of various species in 1 to 2 km of stream or equivalent area of a lake or estuary.

MAJOR KILL: 1000 or more dead or dying fish of many species in a reach of stream up to 16 km or greater, or equivalent area of a lake or estuary.

3. Investigation Techniques

In preparing for a field investigation, study area maps and determine the zone of fish mortality and the access to it. Identify waste dischargers. Contact participating laboratories to discuss number and size of samples that will be submitted, types of analyses required, dates of sample receipt, method of sample shipment, date by which results are needed, and to whom results are to be reported. Use two information record forms for fish kill investigations, an initial contact form and a field investigation form.

On all fish kill investigations take a thermometer, dissolved oxygen test kit, conductivity and pH meters; or a general chemical kit, biological sampling gear, sample bottles, fixatives, and other specimen containers. Take a camera to provide a photographic record of the event. Include in the investigating team at least one person who is experienced in investigating fish kills.

The field investigation consists of visual observations, sampling of fish, water, and other biota, and physical measurements of the environment. The first local observer of the kill is a useful guide to the area, which should be reconnoitered initially to establish that a fish kill actually has occurred.

If a fish kill has taken place, immediately start fish sampling because collection of dying or recently dead fish is critical. Moribund fish usually are preferred. For purposes of comparison, if possible, collect healthy fish from an unaffected area.

Examine moribund or recently dead animals immediately for external and internal abnormalities. Record all changes in color, size, location and consistency of organs. Record the location of lesions by organ (Figure 10600:6). Colored photographs, with indication of scale, taken at the site are an excellent way of documenting observations and greatly help the consulting pathologist. If photography is not possible describe exactly what was seen, recording the number and size of abnormalities. Provide healthy fish of the same species when possible.

Do not freeze samples for pathology. If fixatives are unavailable place samples in plastic bags on ice and rush to the pathologist.

If virology or bacteriology testing is indicated freeze additional specimens of key organs such as the liver, kidney, spleen, heart, and brain and other parts showing abnormality or lesions, label, and forward for analysis.

Bleed dying fish at collection time to obtain at least 1 g blood. Collect a blood sample in a chemically clean, solvent-washed glass bottle with a TFE-lined screw cap.

Identify and count dead fish. In a large river count dead fish from a fixed station such as a bridge during a fixed period of time. Extrapolate to the total time involved. Alternatively, in a large river or lake, make a shore count and project to entire area of kill. In smaller water bodies traverse entire area to enumerate dead fish.

Collect water samples representative of unpolluted and polluted areas in accordance with the instruction given in Section 10200. As a minimum, measure temperature, pH, dissolved oxygen, and conductivity. Make additional tests depending on suspected causes of the fish kill. Take samples for examination of plankton, periphyton, macrophyton, and macroinvertebrates.

Record observations on water appearance, streamflow, and weather conditions. Color photographs are valuable in recording conditions.

4. References

1. BURDICK, G.E. 1965. Some problems in the determination of the cause of fish kills. *In* Biological Problems in Water Pollution. Publ. No. 999-WP-25, U.S. Public Health Serv., Washington, D.C.
2. SMITH, L.L., JR., et al. 1956. Procedures for Investigation of Fish Kills. A Guide for Field Reconnaissance and Data Collection. Ohio River Sanitary Comm., Cincinnati, Ohio.
3. MEYER, F.P. & L.A. BARCLAY. 1990. Field Manual for the Investigation of Fish Kills. U.S. Fish & Wildlife Serv. Res. Publ. No. 177, Washington, D.C.

5. Bibliography

FEDERAL WATER POLLUTION CONTROL ADMINISTRATION. 1967 & 1968. Pollution Caused Fish Kills. Publ. No. CWA-7, U.S. Dep. Interior.

FEDERAL WATER POLLUTION CONTROL ADMINISTRATION. 1970. Investigating Fish Mortalities. Publ. No. CWT-5, U.S. Dep. Interior (also available as No. 0-380-257, U.S. Government Printing Off., Washington, D.C.).

GRIZZLE, J.M. & W.A. ROGERS. 1976. Anatomy and Histology of the Channel Catfish. Agricultural Experiment Sta., Auburn Univ., Auburn, Ala.

GORMAN, D.B. 1982. Histology of the Striped Bass. AFS Monogr. No. 3, American Fisheries Soc., Bethesda, Md.

ROBERTS, R.J., ed. 1982. Microbial Diseases of Fish. Academic Press, New York, N.Y.

YASUTAKE, W.T. & J.H. WALES. 1983. Microscopic Anatomy of Salmonids: An Atlas. U.S. Fish & Wildlife Serv. Resource Publ. 150, U.S. Dep. Interior, Washington, D.C.

CAIRNS, V.W., P.V. HODSON & J.O. NRIAGU, eds. 1984. Contaminant Effects on Fisheries. John Wiley & Sons, New York, N.Y.

ELLIS, A.E. 1985. Fish and Shellfish Pathology. Academic Press, New York, N.Y.

STOSKOPF, N.K. 1993. Fish Medicine. W.B. Saunders Co., Philadelphia, Pa.

10900 IDENTIFICATION OF AQUATIC ORGANISMS*

Experienced aquatic biologists will be familiar with most organisms illustrated in Plates 1 through 35, and seldom will need the assistance of keys to identify organisms to the level illustrated. Because these plates are not intended for critical identification, specific (species) names are not cited. Organisms most likely to be observed are illustrated. For the convenience of those less familiar with the organisms referred to in preceding sections, a se-

_____
* Approved by Standard Methods Committee, 1997.

ries of short keys is presented to enable them to identify most organisms to the level illustrated by the plates.

In conformity with preceding sections, organisms are arbitrarily divided into microscopic and macroscopic, depending on whether or not they pass through a U.S. Standard No. 30 sieve (0.5 mm). For the study of microscopic forms, use a compound microscope. For examination of the smaller macroscopic organisms and to resolve the finer structures of larger forms, use a wide-field stereoscopic microscope.

10900 A. Procedure in Identification

Critical identification of a specimen often is time-consuming, even for an experienced biologist. Before using any key or other aid to identification, carefully study the specimen for one to several minutes. If necessary, find other examples and compare them with the unknown.

It is important to know where or under what conditions the organism lived before attempting to identify it. For example, did it come from fresh water—a lake or a stream? Is it marine—from the open ocean, shoreline, or estuary? Was it a free swimmer or floater in the water? Was it a bottom organism, attached, crawling, or burrowing? Finally, turn to the following key to major groups.

Only the more common types of aquatic organisms are illus-

trated here, with special attention to those most frequently used in water quality evaluation. When specimens do not fit obviously into one of the types listed, consult a professional biologist, a microbiologist for the bacteria and fungi, or some of the references provided. Descriptions of color and movement refer to freshly collected or living specimens, or, in the case of microscopic forms, to those preserved as described in Section 10200.

Sizes of the organisms illustrated in Plates 1 through 27 are shown in parentheses in the legend. These are intended to represent *common* sizes, not absolute maxima or minima. Exceptional individuals and even whole localized populations may be encountered that are considerably larger or smaller than the sizes cited.

10900 B. Key to Major Groups of Aquatic Organisms
(Plates 1–35)

Beginning with couplet 1a and 1b of the Keys, choose one of the contrasting statements. Proceed to the couplet number indicated at the end of the chosen statement and repeat the process. Continue until the name of an organism or a plate number is cited instead of another couplet number. Additional information is provided in many of the plate legends.

		Refer to couplet No.
1a.	Macroscopic: The organism, mass, or colony is visible to the naked eye	13
1b.	Microscopic: Not readily visible to the naked eye	2

1. Key to Microscopic Organisms

2a.	Specimen a single cell or a mass or colony of relatively independent cells (shapeless, rounded, or threadlike)	3
2b.	Specimen a many-celled, highly organized plant or animal..	7
3a.	Cells contain one or more pigments, including chlorophyll *a* (overall color may range through various shades of green, blue, red, brown, or yellow). ALGAE (for details, see Section 10900D following, "Key for Identification of Freshwater Algae")	4
3b.	Cells typically colorless, lacking chlorophyll *a*	12
4a.	Nuclei present; pigment confined to chloroplasts	5
4b.	Nuclei, plastids, or vacuoles absent (pseudovacuoles may be present in certain filamentous forms). Pigment generally diffused throughout cytoplasm. CYANOBACTERIA, Plate 1A.	
5a.	Cell wall permanently rigid, composed of SiO_2, geometrical in appearance, and with regular patterns of fine markings; composed of two essentially similar halves, one placed over the other as a cover. Golden brown to greenish in color. DIATOMS, Plate 1B.	

Refer to
couplet
No.

5b.   Cell wall, if present, capable of sagging or bending, rigidity depending on internal pressure of cell contents. Cell walls usually of one piece 6

     6a.   Cells or colonies nonmotile. Usually some shade of green. NONMOTILE GREEN ALGAE, Plates 1A, 1B, 28–35 (in part)

     6b.   Cells or colonies move by means of relatively long whiplike flagella. PIGMENTED FLAGELLATES, Plates 4A, 4B, 28–35 (in part).

7a.   Body with cilia (hairlike structures used for locomotion) 8

7b.   Body without cilia 9

     8a.   Body generally covered with cilia, usually somewhat elongate or wormlike, bilaterally symmetrical. Minute FLATWORMS (Platyhelminthes), relatives of *Planaria*, Plate 9.

     8b.   Cilia confined to one or two crowns at anterior end, which often present the illusion of rotating wheels. Internal jaws present. ROTIFERS (Rotifera), Plate 8.

9a.   Long, slender, unsegmented worms that move by sinuous crawling or thrashing motion. ROUNDWORMS (Nematoda), Plate 9.

9b.   Possess external skeleton and jointed appendages 10

     10a.   Crawl about or swim by means of jointed appendages thrust out from between two clamlike shells. All appendages can be withdrawn entirely within shells when disturbed. OSTRACODS (Ostracoda), Plate 11.

     10b.   Swim rapidly by means of a pair of enlarged jointed appendages (antennae) that cannot be withdrawn inside carapace or shell 11

11a.   Locomotor appendages (antennae) branched. Microcrustacea, CLADOCERA (Cladocera), Plate 11.

11b.   Locomotor appendages (antennae) unbranched; body tapers toward rear. Microcrustacea, COPEPODS (Copepoda), Plate 11.

     12a.   Ingest and digest food internally (ingested food of various colors may be visible through body wall). Single-celled or colonial, attached or free-living. PROTOZOANS (Protozoa), Plates 4B, 5A, 5B, and 6.

     12b.   Digest food externally and absorb products through cell wall. Often secrete masses of slime. BACTERIA and FUNGI, Plates 26 and 27.

2. Key to Macroscopic Organisms

13a.   Specimen a mass of filaments or a glob of gelatinous or semisolid material containing many tiny units, requiring microscopic examination to determine details of structure 2

13b.   Specimen a well-organized unit or colony 14

     14a.   Organism plantlike; flowerlike structures, if present, do not respond when touched, generally are colored some shade of green, brown, or red 16

     14b.   Organism animal-like; usually responds rapidly when touched, whether attached or free-living 15

15a.   Internal backbone present (vertebrates) 17

15b.   No internal backbone present (macroinvertebrates)* 18

     16a.   Plant structure relatively simple. Attachment structures may be present, but no true roots or fibrous tissue. Larger ALGAE, Plates 2A, 2B, 28 (*Nitella*), 33 (*Chara* and *Batrachospermum*), 35 (numerous).

     16b.   Plant structure usually includes true roots, stems, and leaves. Fibers or vascular tissue usually present; flowers or seeds may be observed. (One atypical group, "watermeal," consists only of tiny roundish masses, 0.5 to 1 mm in diameter, often misidentified as algae.) HIGHER PLANTS, Plates 3A, 3B, and 3C.

17a.   Side appendages, if present, are flat fins. FISHES, Plate 24.

17b.   Side appendages, if present, are footlike, with separate digits. AMPHIBIANS, Plate 25.

3. Key to Macroinvertebrates

     18a.   Body bilaterally symmetrical (with right and left sides, but may be superficially coiled into a spiral); animal usually not attached but may live inside an attached cocoon or case, or crawl about; usually solitary 23

     18b.   Symmetry not bilateral 19

19a.   Body typically radially symmetrical 21

19b.   Body or colony nonsymmetrical 20

     20a.   Body mass generally porous; not a colony, sometimes finger- or antler-like. Freshwater representatives generally are fragile, colored green or brown; marine forms tougher, various colors. SPONGES (Porifera), Plate 7.

     20b.   Body mass otherwise 22

21a.   Animals with soft smooth bodies and tentacles around a mouth; no anus. Solitary or colonial. Larger colonies usually have rigid limy skeleton of massive, branched, or fan-shaped form. HYDRAS, SEA ANEMONES, JELLYFISHES, CORALS, etc. (Cnidaria), Plate 7.

21b.   Body covering usually spiny, soft or rigid, flattened or elongate, typically having five radii, with or without spines or arms; anus present. Solitary. Marine only. STARFISHES and relatives (Echinodermata), Plate 23.

     22a.   Colony a jellylike mass, a network of branching tubes, a plant-like tuft, or a lacy, limy crust or mass. MOSS ANIMALS (Ectoprocta), Plate 22.

---

* Invertebrates retained on a U.S. Standard No. 30 sieve (0.5 mm).

Refer to
couplet
No.

22b.   Exclusively marine. Surface of body or colony relatively smooth but tough. Solitary forms, sac-like, with two external openings. Exhibit all degrees of colonialism. Compound forms range from thin slimy masses, with organisms arranged in tiny radial patterns, to huge, shapeless masses resembling tough frozen gelatin. SEA SQUIRTS, SEA PORK (Ascidiacea, Urochorda, Chordata), not illustrated.

23a.   Animal living within a hard, limy shell, soft body (Mollusca or Brachiopoda) 29

23b.   Animal without a limy shell 24

    24a.   Jointed legs present (may not be functional). Body may be hard or soft.. 30

    24b.   Jointed legs absent, body covering mostly soft, animal pliable (a hardened head capsule may be present) 25

25a.   Body girded by annulations or creases at regular intervals, dividing it into many small segments much wider than long 26

25b.   Segments present or absent; if present, not much wider than they are long 27

    26a.   Body with suction disk at one or both ends, in length usually less than 10 times its width. LEECHES (Annelida, Hirudinea), Plate 9.

    26b.   Body without suction disks, in length usually more than 10 times its width; hairs or bristles often evident.   SEGMENTED WORMS (Annelida), Plates 9 and 10.

27a.   Body unsegmented 28

27b.   Body segmented, slender. TWO-WINGED FLIES (Diptera), Plate 16.

    28a.   Body long and slender, appearing smooth, evenly tapered to a fine point at one end. ROUNDWORMS (Nematoda), Plate 9.

    28b.   Body flat, elongate, or oblong; pigmented spots on head; head often spade-shaped. FLATWORMS (Platyhelminthes, Turbellaria), Plate 9.

    28c.   Body long and slender with a circle of tentacles at one end; lives in a sandy tube, marine (Phoronidea), Plate 22.

29a.   Shell consisting of two hinged halves. BIVALVES (Pelecypoda) or LAMP SHELLS (Brachiopoda), Plate 21

29b.   Shell entire, usually spiral but may be "coolie hat"-shaped. SNAILS (Gastropoda), Plate 20.

    30a.   Body with functional legs 31

    30b.   Body without functional legs, mummy- or capsule-like, living in a cocoon. PUPAE (Insecta), Plates 15, 16, and 17.

31a.   Body with three pairs of legs. Larvae, nymphs, and some adults (Insecta) 42

31b.   Body with more than three pairs of legs 32

    32a.   Body compact, spider-like, with four conspicuous pairs of legs (two other pairs of appendages present). WATER MITES (Acari) or SEA SPIDERS (Pycnogonidea), Plates 12 and 22.

    32b.   Body with at least five conspicuous pairs of legs. CRUSTACEANS (Crustacea) 33

    32c.   Body covering hard; divided into broad head, truncate body, and sharp tail section (marine). HORSESHOE CRAB (Arachnoidea), Plate 22.

4. Key to Crustacea

33a.   Paired compound eyes on stalks 34

33b.   Paired compound eyes, if present, sessile 36

    34a.   Carapace, if present, does not fuse with more than four thoracic segments.................................... 35

    34b.   Carapace fused with all thoracic segments; pincers present. Decapoda, Plate 12.

35a.   Carapace covers most of the thoracic segments. Mysidacea, Plate 11.

35b.   Carapace extends over the anterior segments of abdominal region. Leptostraca, Plate 12.

    36a.   Body flattened horizontally.. 37

    36b.   Body flattened laterally. SCUDS (Amphipoda), Plate 12.

37a.   First pair thoracic legs with pincers. Tanaidacea, Plate 12.

37b.   First pair of thoracic legs pointed and similar to all thoracic legs. Isopoda, Plate 12.

5. Key to Insect Pupae

    38a.   Back of pupa with small, paired, hook-bearing plates. CADDISFLIES (Trichoptera), Plate 15.

    38b.   Back without paired hook-bearing plates but may have knobs or bristles................................... 39

39a.   Developing wings (pads) held free from body. BEETLES (Coleoptera), Plate 17.

39b.   Wing pads closely appressed to body, mummy-like, or appendages not evident 40

    40a.   With one closely appressed pair of wing pads, but not fused to body; or capsule-like, appendages not evident. TWO-WINGED FLIES (Diptera), Plate 16.

    40b.   Two pairs of wing pads.. 41

41a.   First two or three abdominal segments with spiracles (holes for breathing) on each side; body without numerous projections. AQUATIC MOTHS (Lepidoptera), not illustrated.

41b.   Body differing from above, may have numerous knobs or other projections on back. HELLGRAMMITES (Neuroptera and Megaloptera), Plate 15.

6. Key to Insect Larvae, Nymphs, and Some Adults

    42a.   Animal flea-like, with a bifid projecting appendage on the underside. SPRINGTAILS (Collembola), Plate 22.

    42b.   Animal otherwise.. 43

Refer to couplet No.

43a. Body ending in long segmented filaments 44

43b. Long filaments absent or, if present, not segmented 45

    44a. Two tail filaments, all legs ending in two claws. STONEFLIES (Plecoptera), Plate 13.

    44b. Middle and hind legs ending with one claw, three tail filaments (usually). MAYFLIES (Ephemeroptera), Plate 13.

45a. Back of body covered with two hard wing covers, a pair of membranous wings underneath the covers. ADULT BEETLES (Coleoptera), Plate 17.

45b. Back without hard wing covers 46

    46a. Body with exposed membranous wings or wing pads on back.. 47

    46b. Body without membranous wings or wing pads (larvae)... 49

47a. Membranous wings present; held flat and in a V-shape on back. Mouth parts formed into a long, sharply pointed beak folded underneath body. TRUE BUGS (Hemiptera), Plate 18.

47b. Membranous wings absent, wing pads present. Mouth parts formed into an extendable, scoop-like mask that covers face. (Odonata) 48

    48a. Body ending in three oblong, fan-like plates. DAMSELFLIES (Zygoptera), Plate 14.

    48b. Fan-like plates absent. DRAGONFLIES (Anisoptera), Plate 14.

49a. Mouth parts formed into slender curved rods nearly half as long as body (less than 10 mm). SPONGILLA FLIES (Neuroptera), not illustrated.

49b. Mouth parts adapted for biting or chewing 50

    50a. Body with five paired knobs on underside of abdominal segments, legs on first three segments short and stubby. Often found on lily pads. AQUATIC MOTHS (Lepidoptera), not illustrated.

    50b. Body without paired knobs on underside of abdomen... 51

51a. Sides of each abdominal segment with a slender, tapering process 52

51b. Sides of each abdominal segment without a tapering process, but may have hair-like or tubular processes 53

    52a. Body ending in a pair of hook-bearing fleshy legs or in a single tapering filament. HELLGRAMMITES and relatives (Megaloptera), Plate 15.

    52b. Body otherwise. BEETLES (Coleoptera), Plate 17.

53a. Body covering mostly hard; knobs, hairlike processes, or other special ornamentation may be present on back, or else body is entirely soft except for a hardened head capsule. BEETLES (Coleoptera), Plate 17.

53b. Most of body soft except for a hardened head capsule and with one to three hard plates on the back of first body segments; tubular processes may be present on sides of the body in various arrangements. Body may end in a pair of hook-bearing legs. Most larvae living in portable cases made of bits of sticks, leaves, or sand or in attached fibrous cases. CADDISFLIES (Trichoptera), Plate 15.

ACKNOWLEDGMENTS

Plates 1 through 35, which follow on succeeding pages, present over 300 marine and freshwater organisms commonly found in natural, polluted, and treated waters. These plates were drawn for this work by Eugene Schunk of the Cincinnati Art Services, Inc. and by Lhisa J. Reish of Olympia, Washington. Figure 15E was drawn by Dr. Harold Walker; Figures 16A, K, L, and O also are based on his work. The figures in Plate 27 are based on original drawings by Dr. W. Bridge Cooke. Plates 28 through 35 were drawn by Dr. Harold Walker and Sharon Adams. In a number of instances, it would have been impossible to illustrate a certain organism for the purposes of this manual were it not for the courtesy of other publishers, who permitted illustrations from their publications to be incorporated herein. The following organisms were so reproduced:

Plate

1B S—*Achnanthes*, courtesy of Veb Gustav Fischer Verlag, Jena. Source: Die Susswasser—Flora Mitteleuropas, Heft 10, by F. Hustedt, 1930.

U—*Coscinodiscus*, courtesy of E. Schweizerbart'sche Verlagsbuchhandlung, Stuttgart. Source: Das Phytoplankton des Susswassers, Die Binnengewasser, Band XVI, Teil II, Halfte II, by G. Huber-Pestalozzi and F. Hustedt, 1942. Plates CVIII-CXVI and CXXIII.

W—*Skeletonema*, courtesy of Academische Verlagsgesellschaft, Leipzig. Source: Die Kieselalgen, by F. Hustedt. In: L. Rabenhorst, Kryptogamen-Flora von Deutschland, Osterreich und der Schweiz, Band VII, 1930.

8 I—*Notholca* Robert W. Pennak, Fresh-Water Invertebrates of the United States, Copyright© 1953, The Ronald Press Company, New York. Figure 116*N*, page 190, adapted for Figure 8, courtesy of The Ronald Press.

11 H—*Diaptomus* (copepod)

12 H—*Cambarus* (crayfish, crawdad) courtesy of Holden-Day, Inc., San Francisco, California. Source: Needham & Needham's Guide to the Study of Freshwater Biology, 1951. Figures 1 and 10, Plate 14, page 37; Figures 16, 18, and 20, Plate 24, page 61; and Figure 9, Plate 14, page 37.

19 E—*Tarebia*,

H—*Lymnaea* (pond snail),

I—Orb snail, and

J—*Ferrissia* (limpet), courtesy of R.M. Sinclair, Advisor for Biological Sampling and Analysis (American Public Health Association), 13th ed.

K—*Lanx* (limpet) courtesy of John Wiley & Sons, Inc., New York. Source: Ward & Whipple, Fresh Water Biology (2nd ed.), W.T. Edmondson, Editor, 1959. Figures 43.62B(E), 43.13(H), 43.20(I) and 43.14(K).

22 C—*Limulus* (horseshoe crab) courtesy of Western Publishing Company, Inc., Golden Press Division, Racine, Wisconsin. Source: Seashores, a Golden Nature Guide, 1955. Page 79.

23 A—*Asterias*, and

C—*Thyrone*, courtesy of Connecticut State Geological and Natural History Survey: Echinoderms of Connecticut, by Wesley Roswell Coe, 1912.

25 C—*Ambystoma* (terrestrial adult), courtesy of Dover Publications, Inc., New York. Source: Biology of the Amphibia, by G.K. Noble, 1931. Figure 147C, page 471.

D—*Ambystoma* (aquatic larva), courtesy of the New York State Museum and Science Service, Albany, New York. Source: The Salamanders of New York, by Sherman C. Bishop, 1941. Figure 33b, page 166. [Bulletin 324, New York State Museum, Albany.]

E—*Necturus*, courtesy of Dover Publications, Inc., New York. Source: Biology of the Amphibia, by G.K. Noble, 1931. Figure 35B, page 99.

G—*Siren intermedia* (siren), reprinted from Sherman Bishop: Handbook of Salamanders. Copyright 1943 by Comstock Publishing Company, Inc. Used by permission of Cornell University Press.

26 A—Micrococcus,

B—Streptococcus,

C—Sarcina,

D—Bacillus,

E—Vibrio and

F—Spirillum, courtesy of John Wiley & Sons, Inc., New York. Source: Ward & Whipple, Fresh Water Biology (2nd ed.), W.T. Edmondson, Editor, 1959. Figure 3.1.

K—Actinomycete growth form, Selman A. Waksman, The Actinomycetes. Copyright© 1957, The Ronald Press Company, New York. Figure 2–6, page 18, adapted for Figure 26K, courtesy The Ronald Press.

O—*Tetracladium*, and

P, Q—*Achlya*, courtesy of John Wiley & Sons, Inc., New York. Source: Ward & Whipple, Fresh Water Biology (2nd ed.), W.T. Edmondson, Editor, 1959. Figure 4.119 and 4.79.

28–35 Courtesy of U.S. Environmental Protection Agency. Source: Algae and Water Pollution by C.M. Palmer (R.L. Lewis, ed.), 1977. U.S. Environmental Protection Agency, Off. Research and Development, Cincinnati, Ohio.

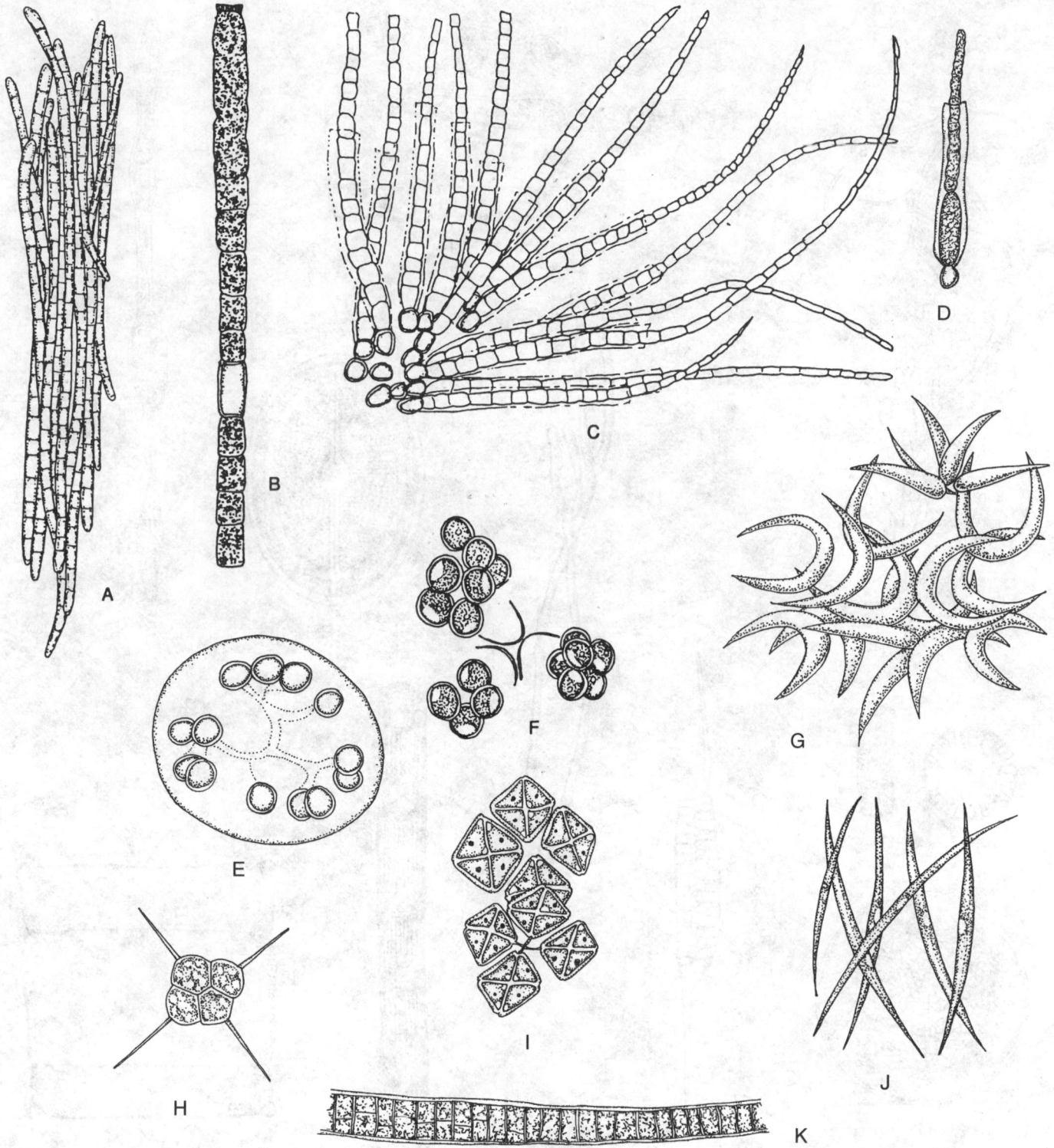

Plate 1A. Cyanobacteria and green algae. Dimensions refer to individual cells or filaments. All organisms inhabit fresh water.

Cyanobacteria (Phylum Cyanophyta):

A—*Aphanizomenon*, aggregate of filaments (3–6 μm)
B—*Aphanizomenon*, detail
C—*Gloeotrichia*, portion of colony (cells 7–9 μm diameter)
D—*Gloeotrichia*, detail

Green algae (Phylum Chlorophyta):

E—*Dictyosphaerium* (8–14 μm)

F—*Westella* (5–7 μm)
G—*Selenastrum* (6–7 μm)
H—*Tetrastrum* (5–9 μm)
I—*Crucigenia* (5–8 μm)
J—*Ankistrodesmus* (2–3 μm)
K—*Schizomeris* (12–18 μm)

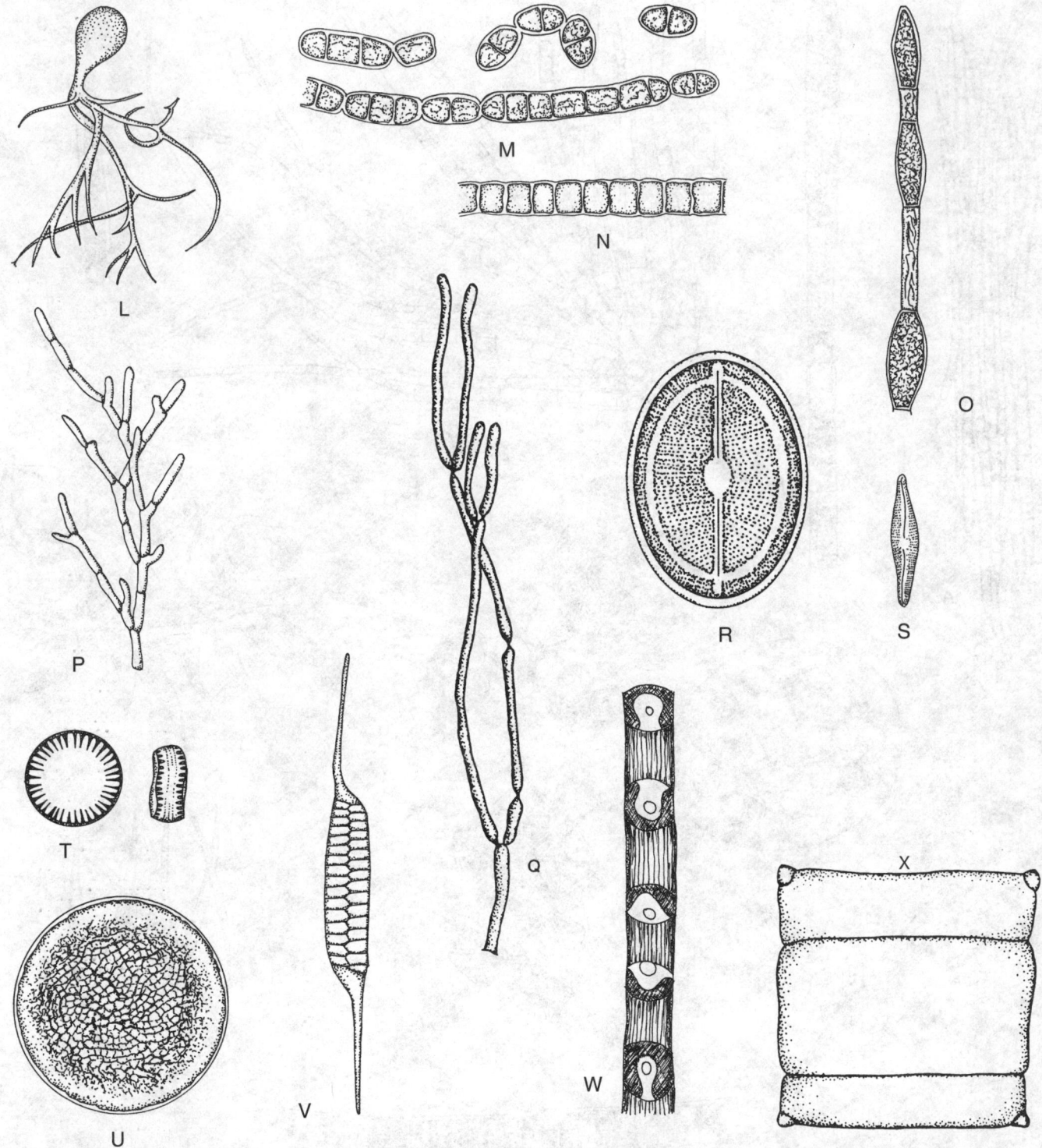

Plate 1B. Yellow-green, green, and golden-brown algae. Dimensions refer to individual cells or filaments. All organisms inhabit fresh water.

Yellow-green algae (Phylum Xanthophyta):		Golden-brown algae (Phylum Chrysophyta) (diatoms):	
L—*Botrydium*	(1–2 μm)	R—*Cocconeis*	(10 μm)
Green algae (Phylum Chlorophyta):		S—*Achnanthes*	(10 μm)
M—*Stichococcus*	(3 μm)	T—*Cyclotella*	(10 μm)
N—*Hyalotheca*	(12–30 μm)	U—*Coscinodiscus*	(120 μm)
O—*Pithopora*	(50–100 μm)	V—*Rhizosolenia*	(5–15 μm)
P—*Microthamnion*	(2–4 μm)	W—*Skeletonema*	(5–15 μm)
Q—*Dichotomosiphon*	(50–100 μm)	X—*Biddulphia*	(100 μm)

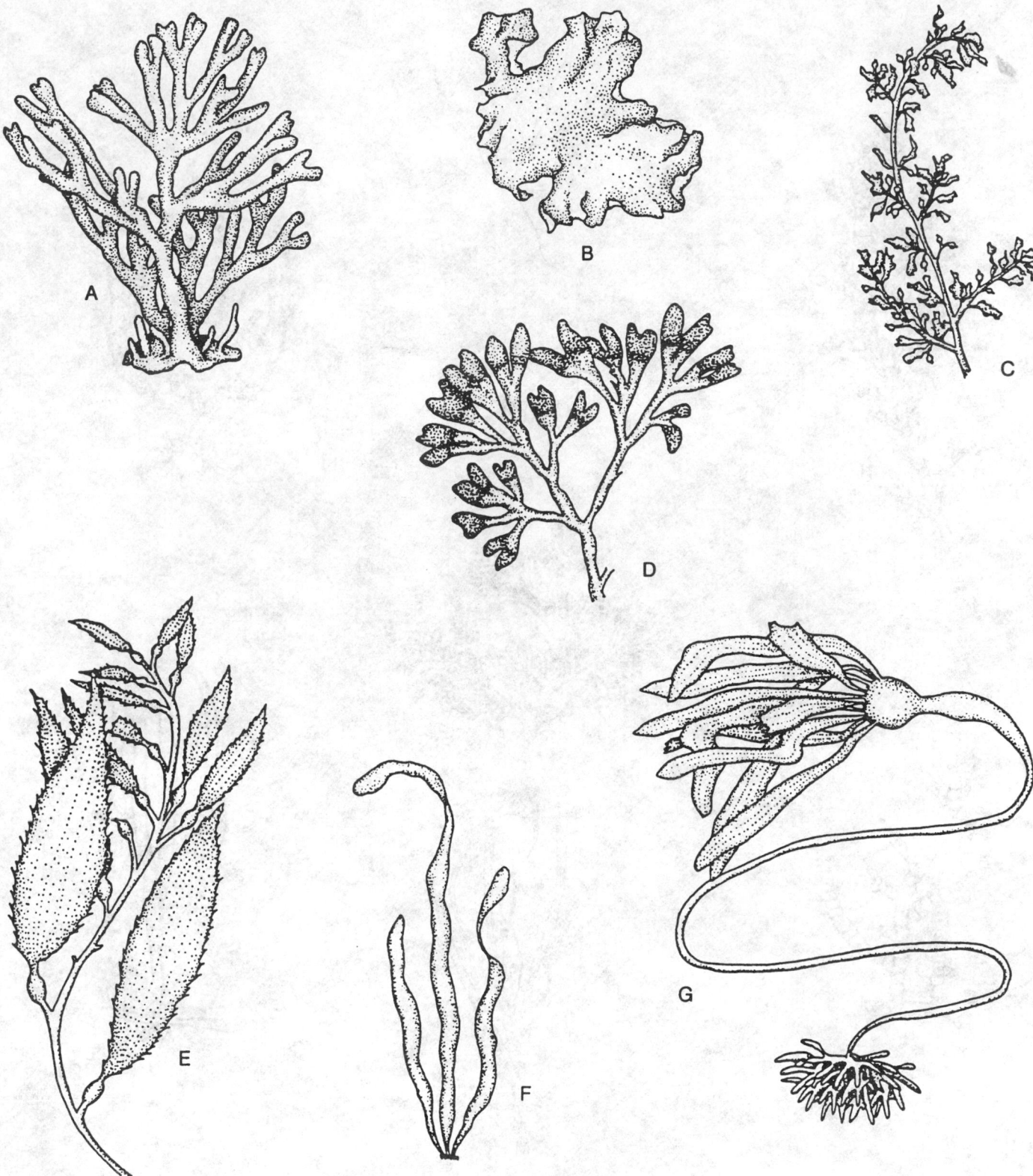

Plate 2A. Types of large marine algae.

Green algae (Phylum Chlorophyta):
    A—Sponge weed, *Codium*   (30–40 cm)
    B—Sea lettuce, *Ulva*   (20 cm)
    C—*Sargassum*   (20–100 cm)

Brown algae (Phylum Phaeophyta):
    D—Rockweed, *Fucus*   (75 cm)
    E—Giant kelp, *Macrocystis*   (30 m)
    F—*Enteromorpha*   (40 cm)
    G—Bull whip, *Nereocystis*   (20 m)

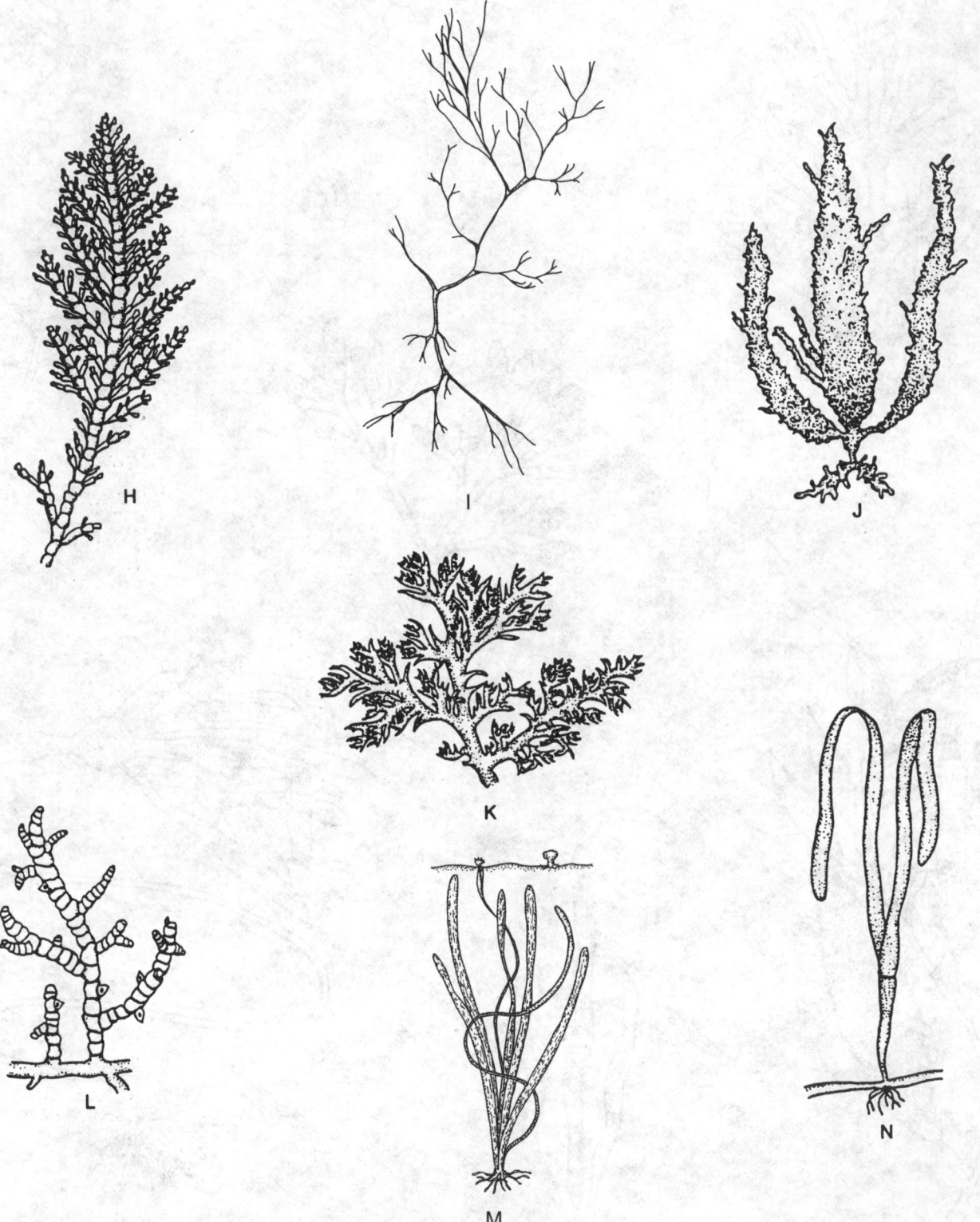

Plate 2B. Types of large marine algae and marine grasses:

Red algae (Phylum Rhodophyta):
   H—*Corallina*   (4 cm)
   I—*Gracilaria*   (50 cm)
   J—*Gigartina*   (30–40 cm)
   K—*Plocamium*   (30 cm)

   L—*Champia*   (1–6 cm)
Marine grasses (Phylum Spermatophyta):
   M—Eelgrass, *Vallisneria*   (50 cm)
   N—Eelgrass, *Zostera*   (45 cm)

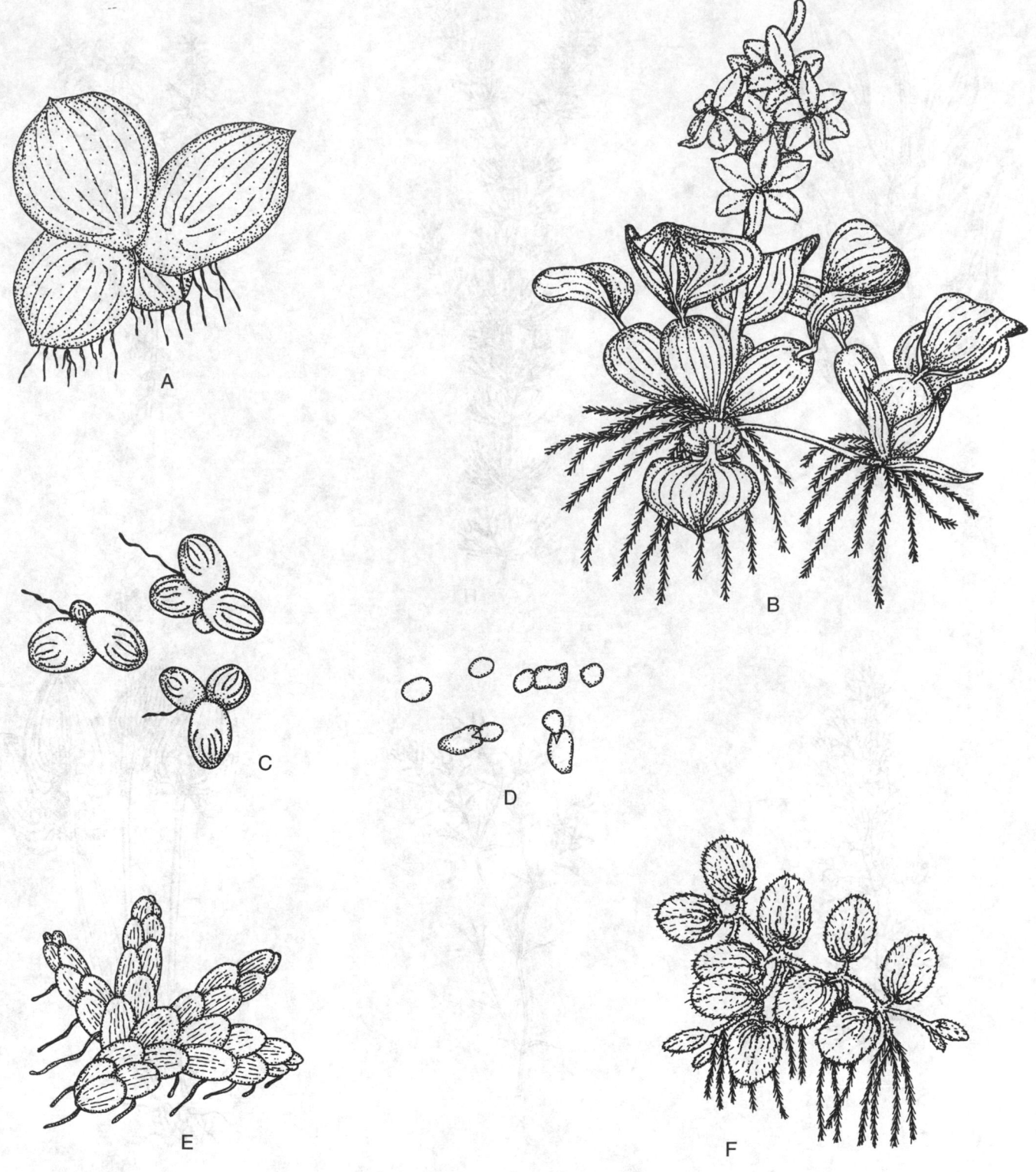

Plate 3A. Higher plants. All floating, freshwater inhabitants.

Phylum Spermatophyta:
  A—Great duckweed, *Spirodela*   (8 mm)
  B—Water hyacinth, *Eichhornia*   (22 cm)
  C—Lesser duckweed, *Lemna*   (5 cm)

D—Watermeal, *Wolffia*   (1–1.5 mm)
Phylum Pteridophyta:
  E—Water velvet, *Azolla*   (1 cm)
  F—Water fern, *Salvinia*   (4 cm)

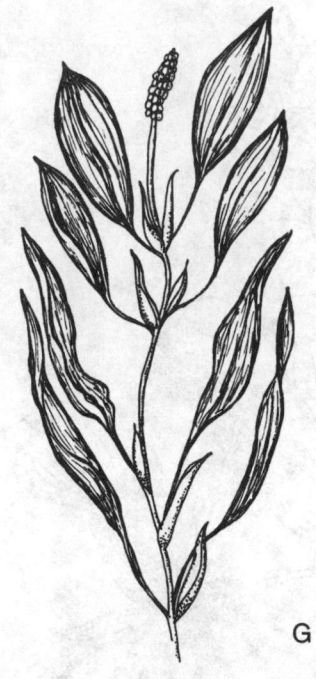

G

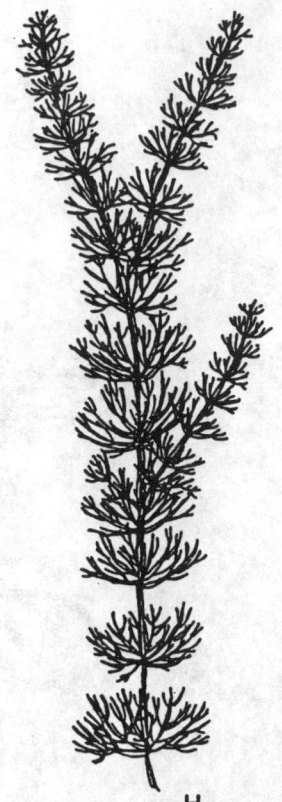

H

I

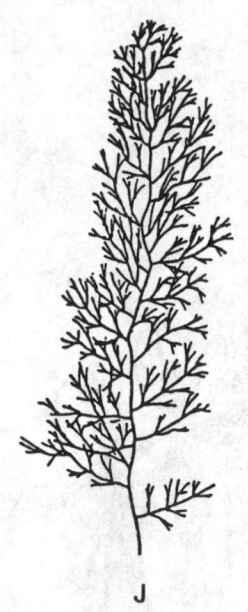

J

K

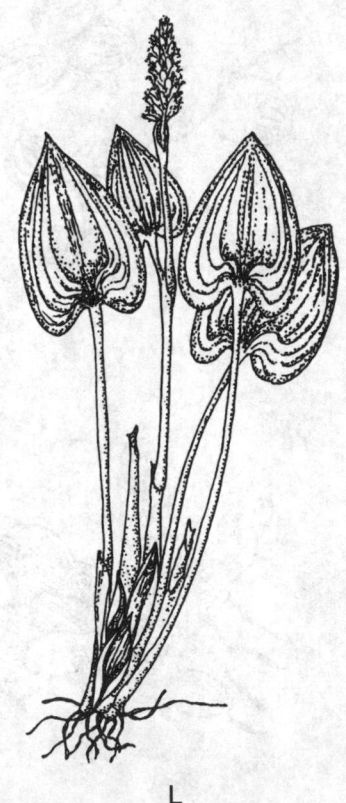

L

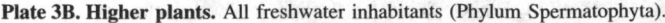

Plate 3B. Higher plants. All freshwater inhabitants (Phylum Spermatophyta).

Submerged:
  G—Pondweed, *Potamogeton*   (30–60 cm)
  H—Coontail, *Ceratophyllum*   (30 cm)
  I—Waterweed, *Elodea*   (15 cm)

  J—Water milfoil, *Myriophyllum*   (30 cm)
  K—Naiad, *Najas*   (60 cm)
Emersed:
  L—Pickerelweed, *Pontederia*   (60 cm)

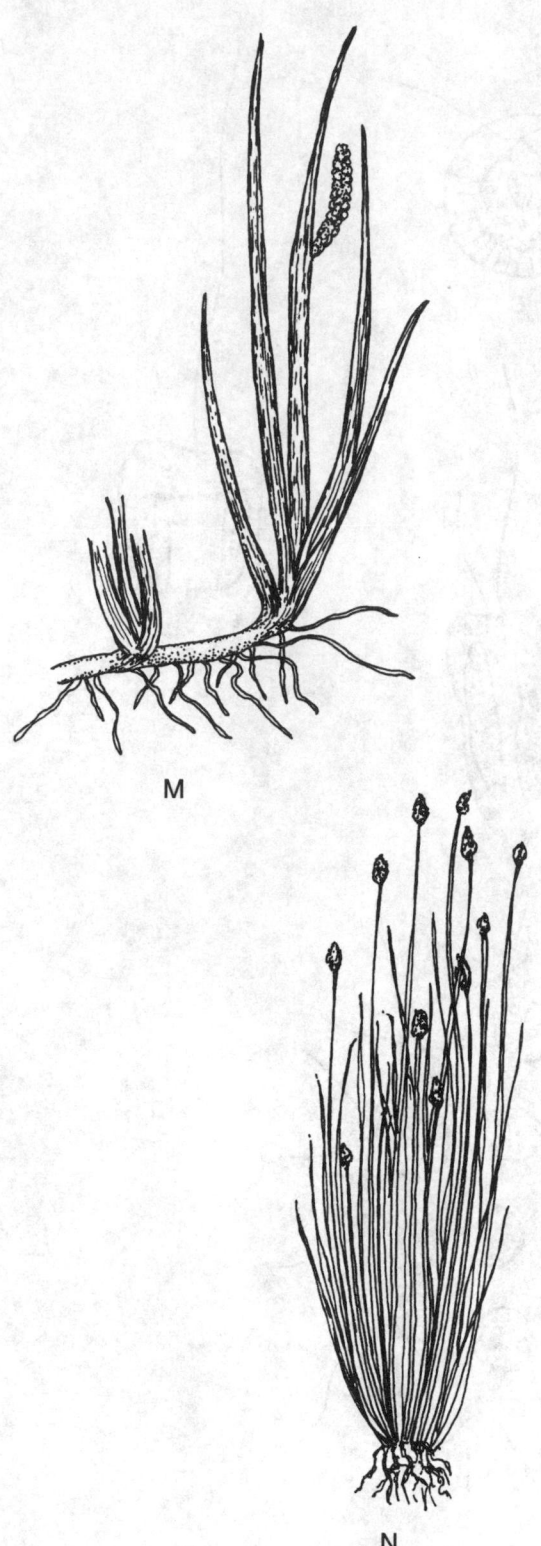

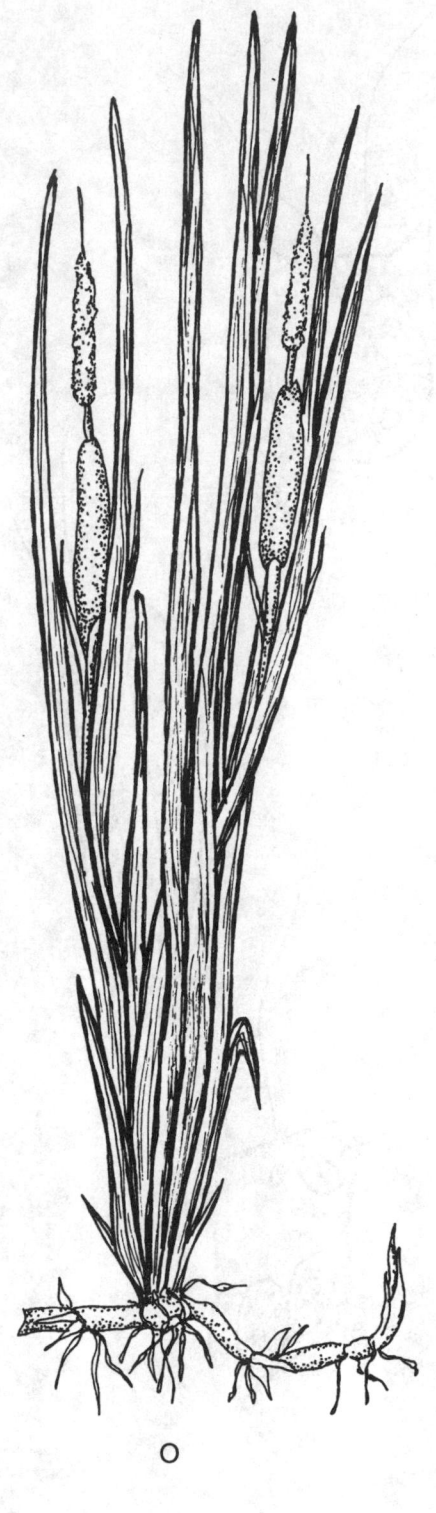

Plate 3C. Higher plants. All freshwater inhabitants (Phylum Spermatophyta), emersed.

M—Sweet flag, *Acorus*   (30 cm)   O—Cattail, *Typha*   (1–2 m)
N—Spike rush, *Eleocharis*   (30 cm)

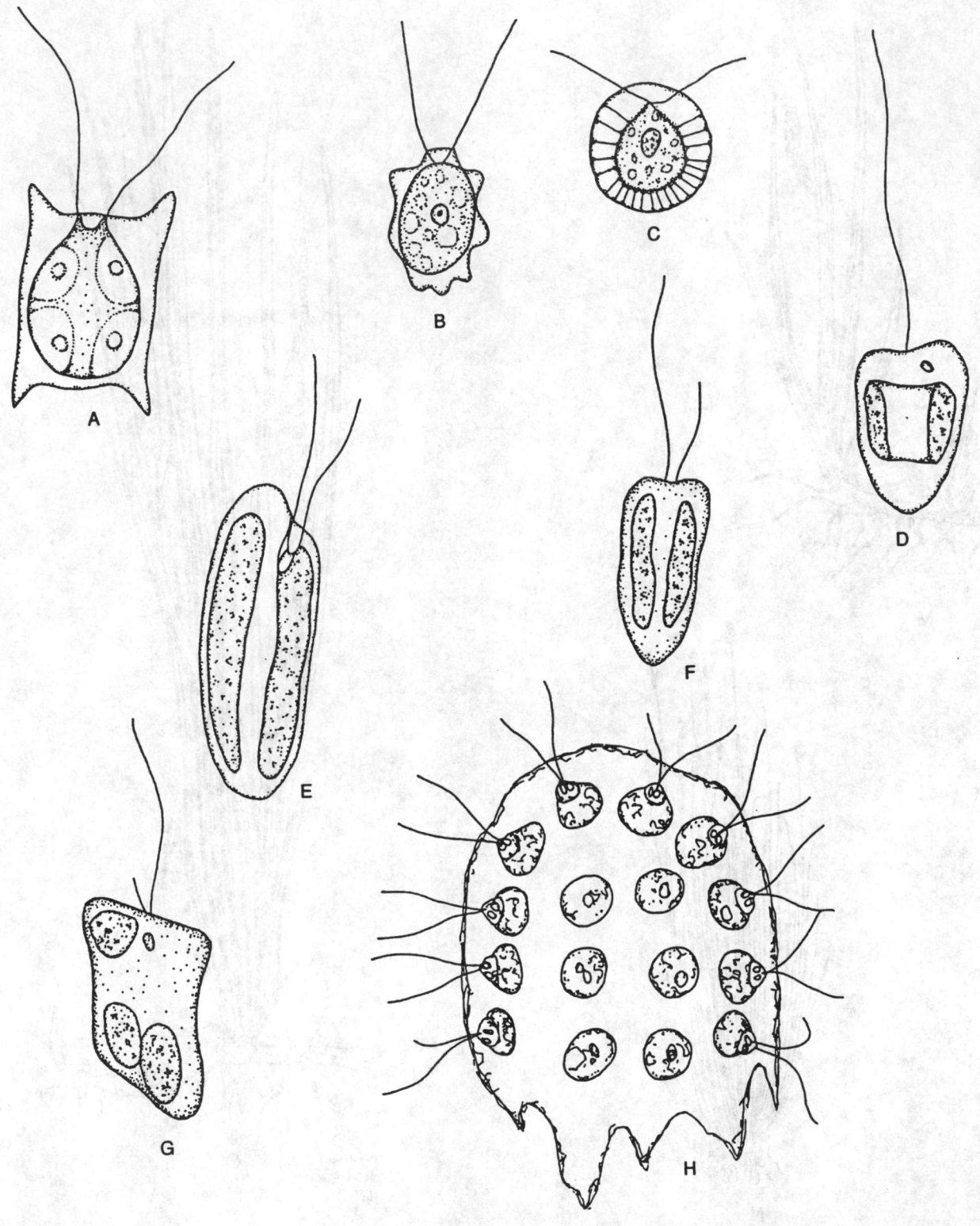

Plate 4A. Flagellates. Pigmented flagellates, various phyla. Freshwater inhabitants.

A—*Pteromonas*	(5–18 μm)	E—*Cryptomonas*	(6–12 μm)
B—*Lobomonas*	(5–14 μm)	F—*Ochromonas*	(7–14 μm)
C—*Haematococcus*	(40–45 μm)	G—*Chloramoeba*	(10–15 μm)
D—*Chromulina*	(4–10 μm)	H—*Platydorina*	(66–70 μm, colony)

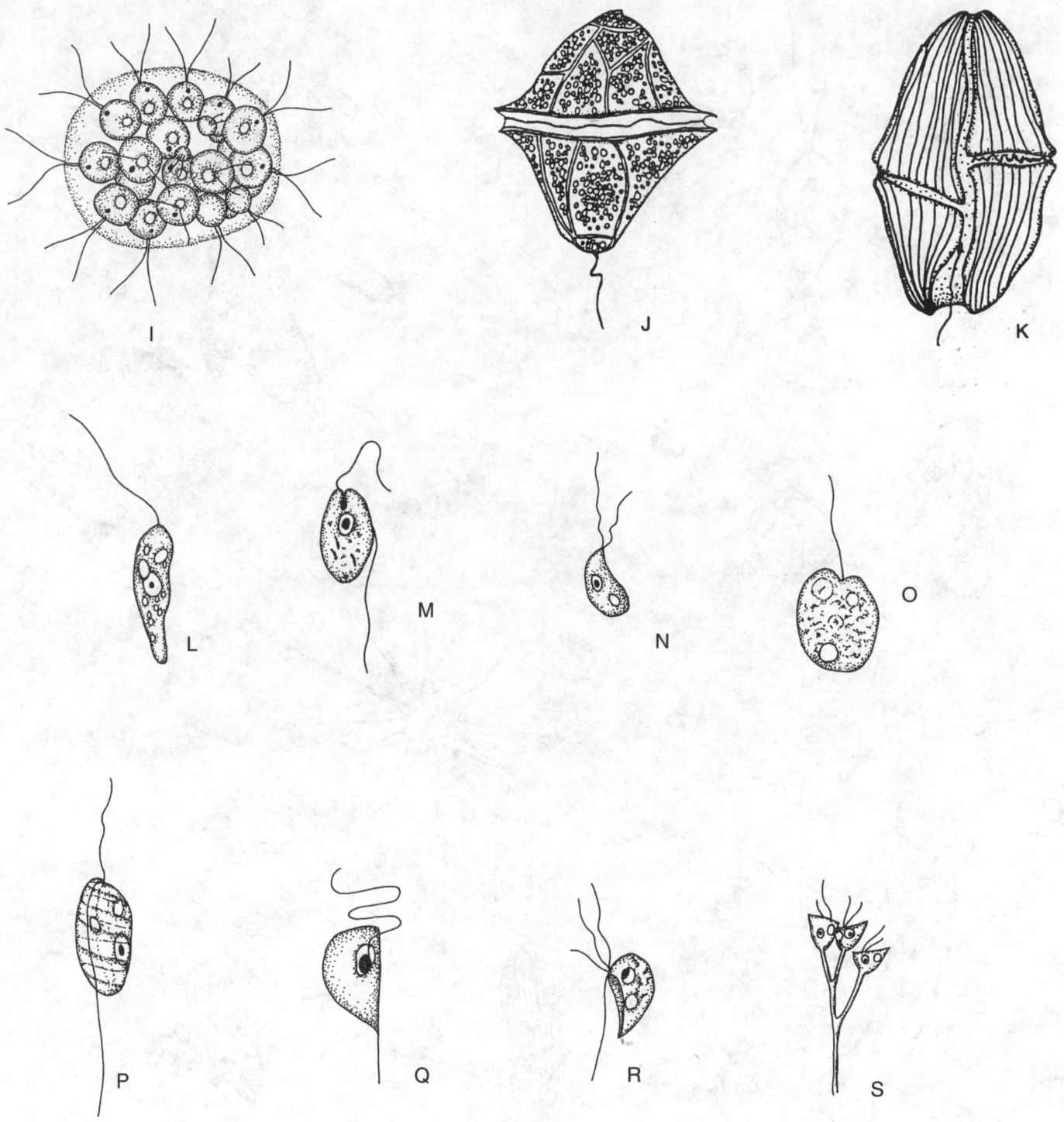

Plate 4B. Flagellates. Freshwater inhabitants except as noted.

Pigmented (various phyla):
  I—*Pleodorina*   (8–10 μm)
  J—*Gonyaulax* (marine) (10 μm)
  K—*Gymnodinium* (marine) (10 μm)
Nonpigmented (Phylum Protozoa):
  L—*Astasia*   (40–50 μm)
  M—*Bodo*   (11–22 μm)

N—*Dinomonas*   (15–16 μm)
O—*Oikomonas*   (5–20 μm)
P—*Anisonema*   (14–60 μm)
Q—*Cercomonas*   (10–36 μm)
R—*Tetramitus*   (11–30 μm)
S—*Dendromonas*   (8 μm)

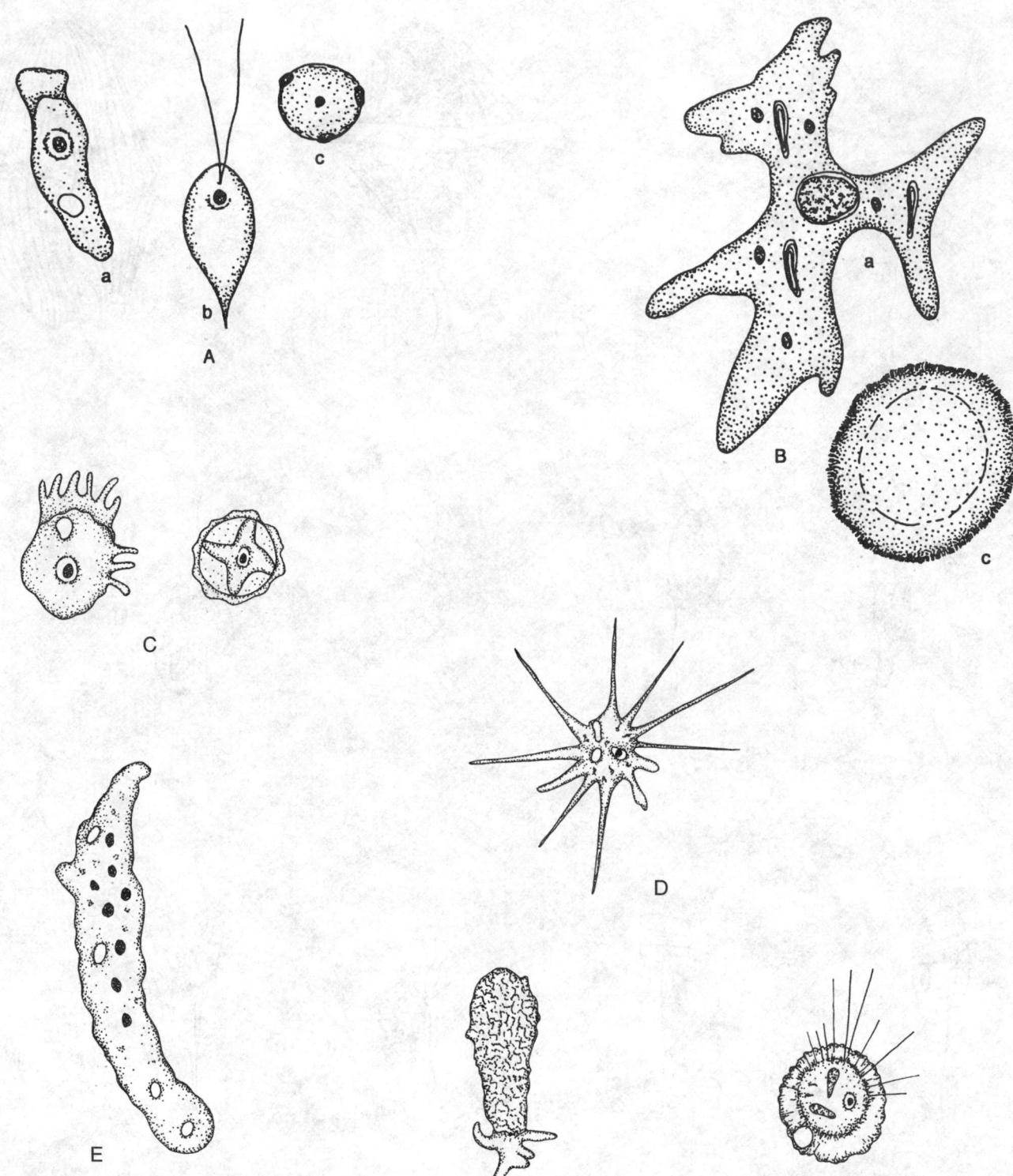

Plate 5A. Amoebas (Phylum Protozoa).

A—*Naegleria*, (a) amoeboid stage, (b) flagellated
　　stage, (c) cyst stage　　　　　　　　　　　　(10–36 μm)
B—*Amoeba*, (a) amoeboid state, (c) cyst stage　　(30–600 μm)
C—*Acanthamoeba* (*Hartmannella*), (a) amoeboid
　　stage, (c) cyst stage　　　　　　　　　　　　(15–25 μm)

D—*Amoeba radiosa*　　　　　　　　　　　　　(30–120 μm)
E—*Pelomyxa*　　　　　　　　　　　　　　　　(0.25–3 mm)
F—*Difflugia*　　　　　　　　　　　　　　　　(40 μm)
G—*Actinophrys*　　　　　　　　　　　　　　　(25–50 μm)

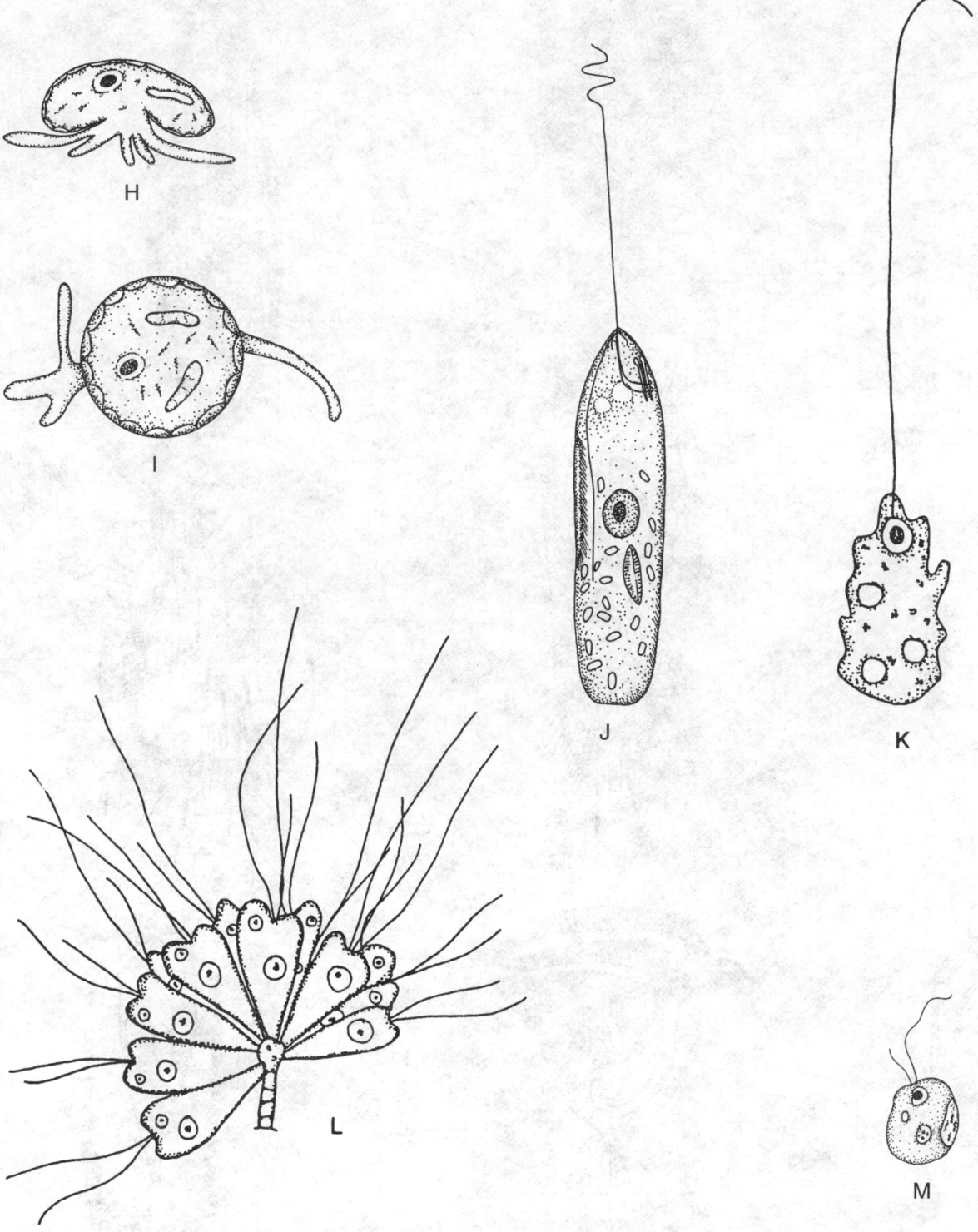

Plate 5B. Amoebas and nonpigmented flagellates (Phylum Protozoa).

Amoebas:

H—*Arcella* (side view)   (30–260 μm)
I—*Arcella* (top view)   (30–260 μm)

Nonpigmented flagellates:

J—*Peranema*   (40–70 μm)
K—*Mastigamoeba*   (28–200 μm)
L *Anthophysa*   (5–6 μm)
M—*Monas*   (5–16 μm)

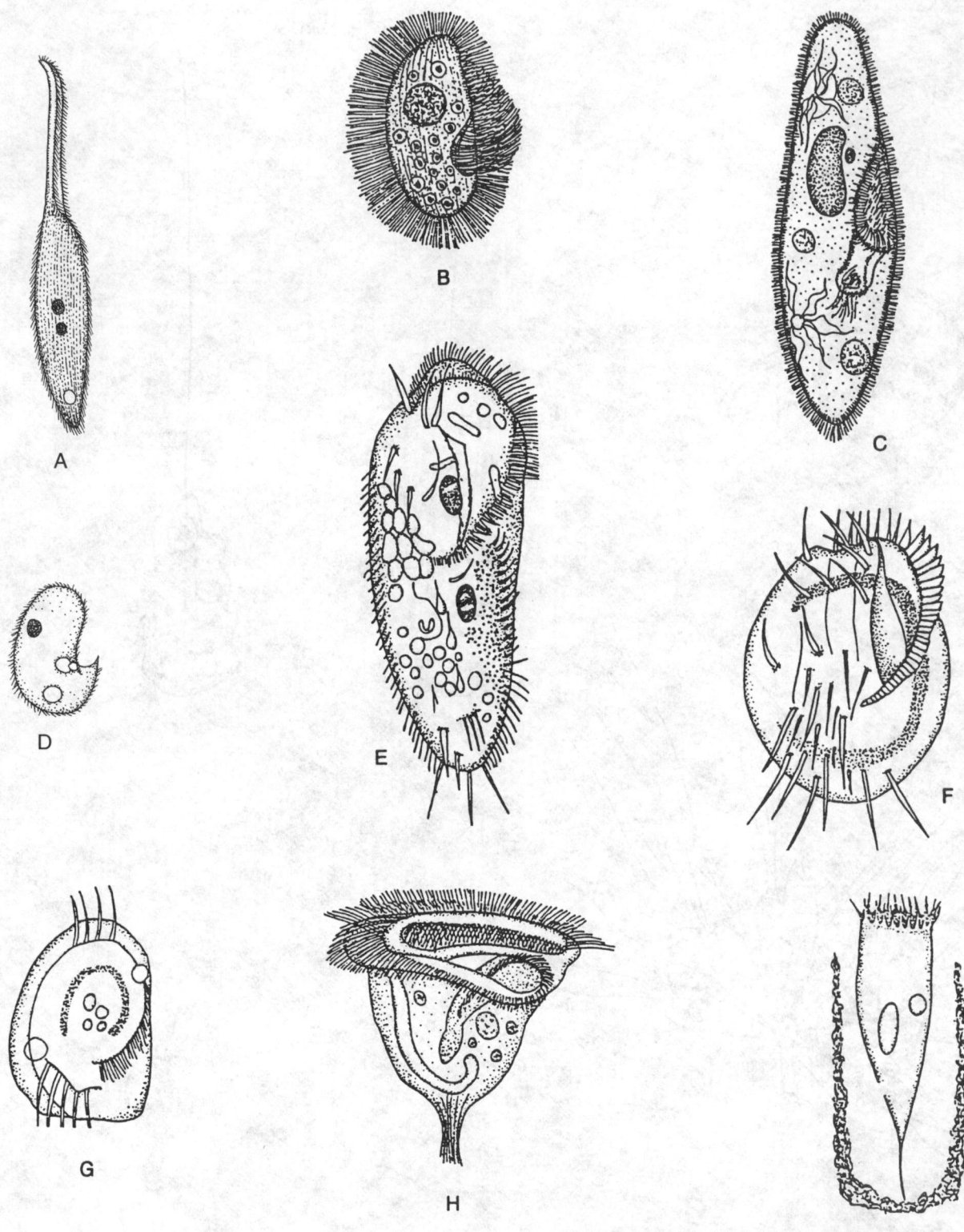

Plate 6. Ciliates (Phylum Protozoa).

A—*Lionotus*	(100 μm)	F—*Euplotes*		(70–195 μm)
B—*Pleuronema*	(38–120 μm)	G—*Aspidisca*		(30–50 μm)
C—*Paramecium*	(50–330 μm)	H—*Vorticella*		(40–175 μm)
D—*Colpoda*	(12–110 μm)	I—*Tintinnidium*		(40–200 μm)
E—*Stylonychia*	(100–300 μm)			

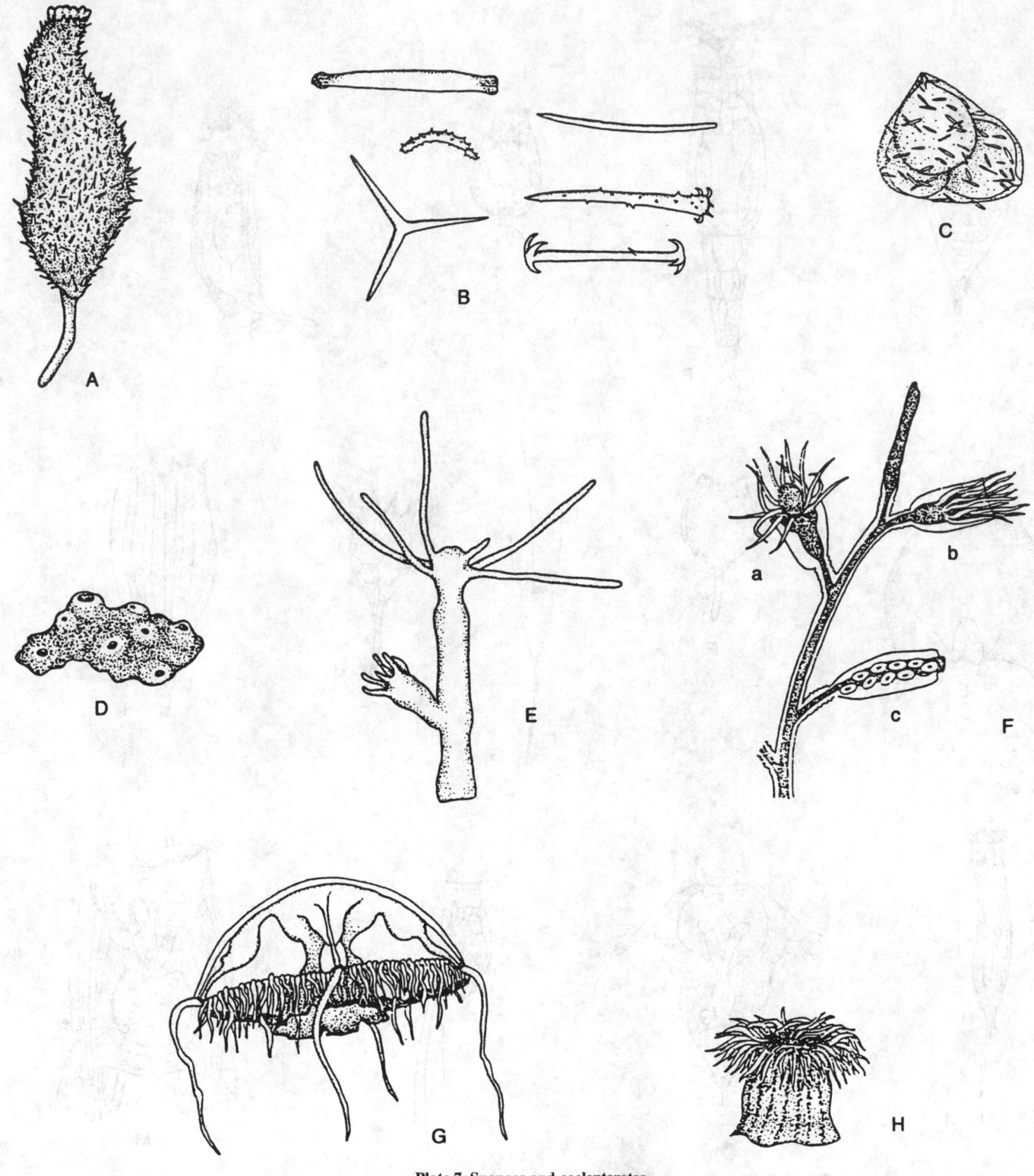

Plate 7. Sponges and coelenterates.

Sponges (Phylum Porifera):

A—Vase sponge, marine   (1 cm)
B—Sponge spicule types   (0.2–0.5 cm)
C—*Trochospongilla*, 3 gemmules, freshwater   (1 mm)
D—Purple sponge, *Haliclona*, marine   (1–10 cm)

Coelenterates (Phylum Cnidaria):

E—*Hydra* extended with bud, freshwater   (2 cm)
F—*Obelia* extended (a) and contracted
   (b) hydranth and gonotheca (c), marine   (2–10 cm)
G—Jellyfish (medusoid stage), *Craspedascusta*,
   freshwater   (10–15 mm)
H—Sea anemone, *Anthopleura*, marine   (5–15 cm)

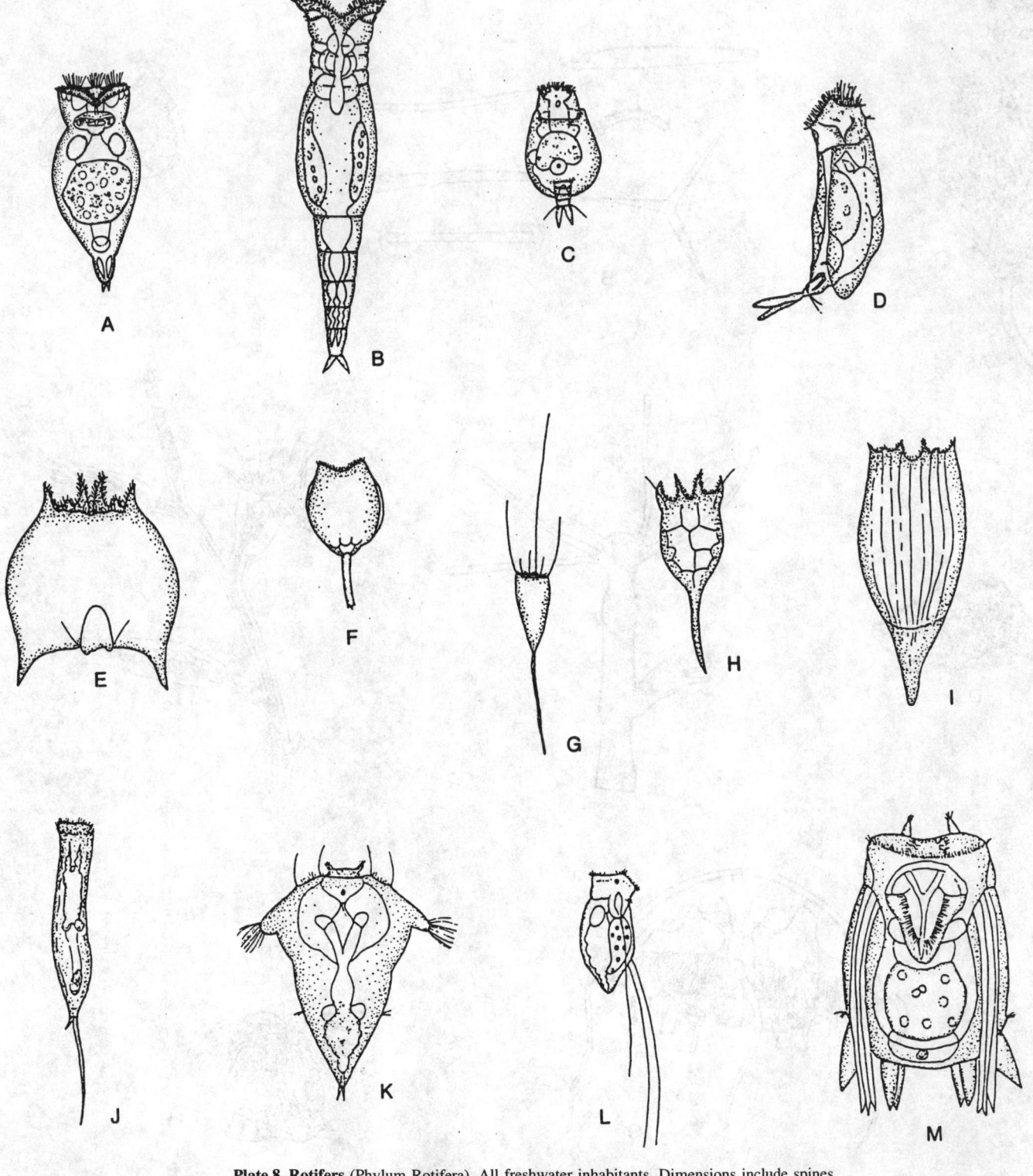

Plate 8. Rotifers (Phylum Rotifera). All freshwater inhabitants. Dimensions include spines.

A—*Epiphanes*   (0.6 mm)
B—*Philodina*   (0.4 mm)
C—*Euchlanis*   (0.25 mm)
D—*Proales*   (0.45 mm)
E—*Brachionus*   (0.2 mm)
F—*Monostyla*   (0.15 mm)
G—*Kellicottia*   (1 mm)

H—*Keratella*   (0.2 mm)
I—*Notholca*   (0.2 mm)
J—*Trichocerca*   (0.6 mm)
K—*Synchaeta*   (0.26 mm)
L—*Filinia*   (0.15 mm)
M—*Polyarthra*   (0.175 mm)

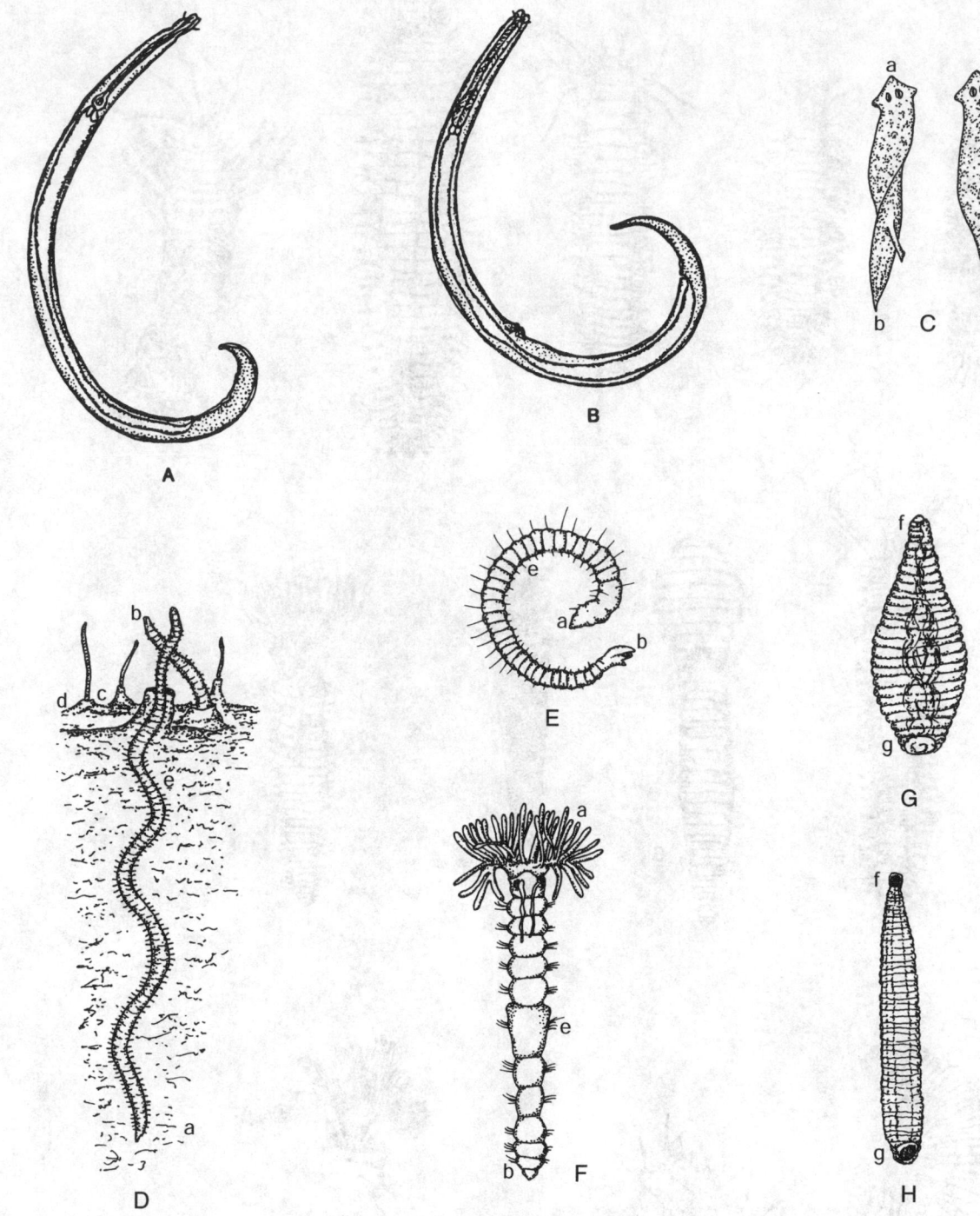

Plate 9. Roundworms, flatworms, and segmented worms. (a) anterior end, (b) posterior end, (c) tubes, (d) mud surface, (e) setae, (f) anterior sucker, (g) posterior sucker. All freshwater inhabitants.

Roundworms (Phylum Nematoda):
  A—*Plectus*   (1.0 mm)
  B—*Tripyla*   (1.0 mm)

Flatworms (Phylum Platyhelminthes):
  C—*Planaria*   (5–13 mm)

Segmented worms (Phylum Annelida, Class
      Oligochaeta):
  D—*Tubifex*, sludgeworm   (2.5–5 cm)

E—*Dero*   (3–7 mm)

Segmented worms (Phylum Annelida, Class
      Polychaeta):
  F—*Manayunkia*, lives in tube   (5 mm)

Segmented worms (Phylum Annelida, Class
Hirudinea):
  G—Leech   (3–6 cm)
  H—Leech   (3–6 cm)

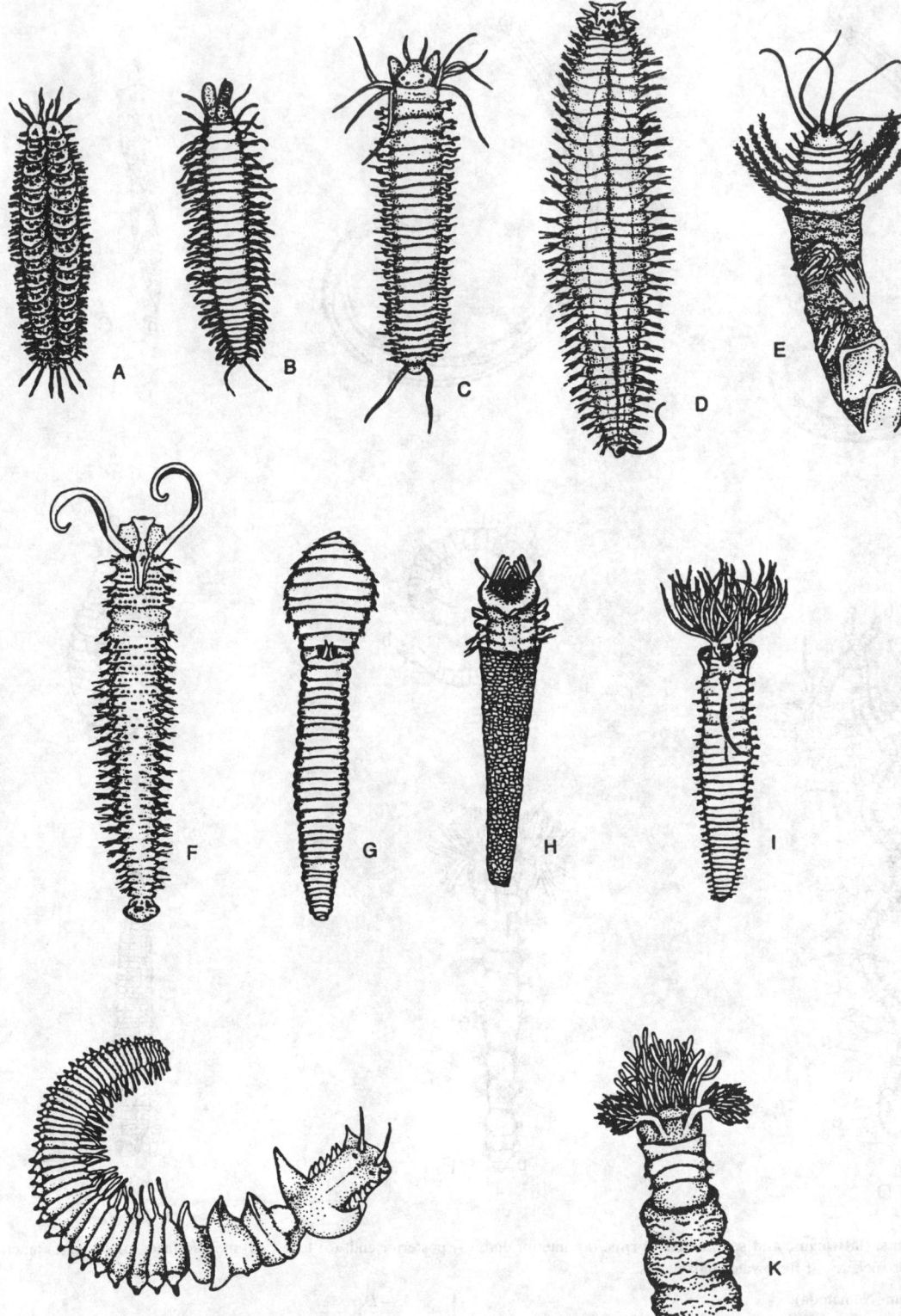

Plate 10. Segmented marine worms (Phylum Annelida, Class Polychaeta).

A—Scale worm, *Halosydna*   (5 cm)
B—Syllids, *Typosyllis*   (2–5 cm)
C—Nereid, *Nereis*   (4–15 cm)
D—Nephtyid, *Nephtys*   (5–25 cm)
E—Onuphid, *Diopatra*, in tube covered with shells   (5–25 cm)
F—Spionid, *Polydora*   (2–4 cm)

G—Capitellid, *Capitella*   (2–4 cm)
H—Ice cream cone worm, *Pectinaria*   (1–3 cm)
I—Feather duster worm, *Sabella*   (10–15 cm)
J—Chaetopterid, *Chaetopterus*, lives in U-shaped tube   (10–20 cm)
K—Terebellid, *Pista*   (10–15 cm)

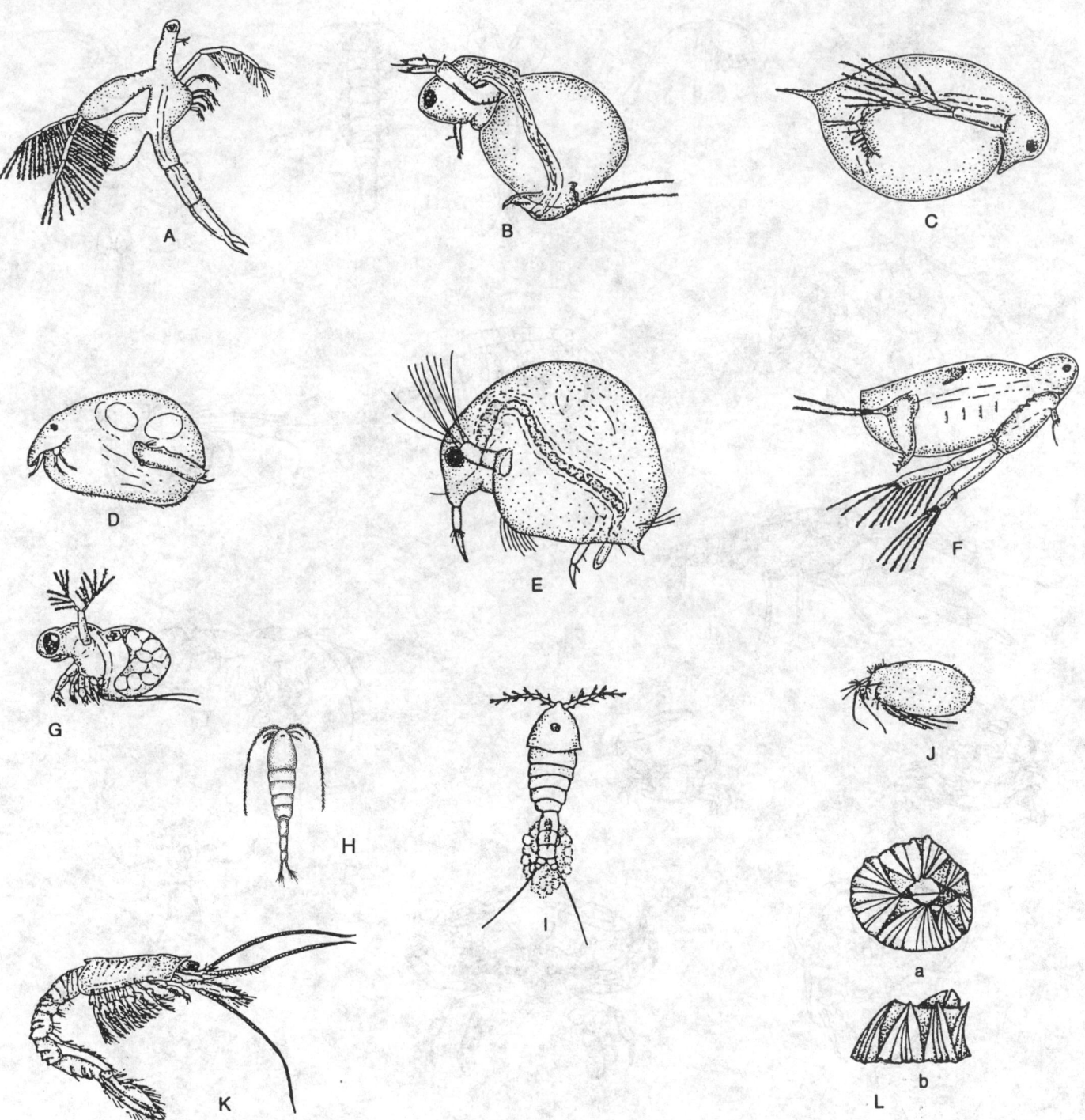

Plate 11. Crustaceans (Phylum Arthropoda)—all freshwater inhabitants, except as noted.

Order Cladocera:
A—*Leptodora* (9 mm)
B—*Moina* (1.5 mm)
C—*Daphnia* (2 mm)
D—*Alona* (0.4 mm)
E—*Bosmina* (0.4 mm)
F—*Diaphanosoma* (1.5 mm)
G—*Polyphemus* (1.5 mm)

Order Copepoda:
H—*Diaptomus*, freshwater (2 mm)
I—*Tigriopus*, marine (2–4 mm)
Order Ostracoda:
J—*Ostracod*, marine (2–3 mm)
Order Mysidacea:
K—*Holmesimysis*, marine (10–20 mm)
Order Cirripedia:
L—Barnacle, *Balanus*, marine, a. top view,
b. side view (10 mm)

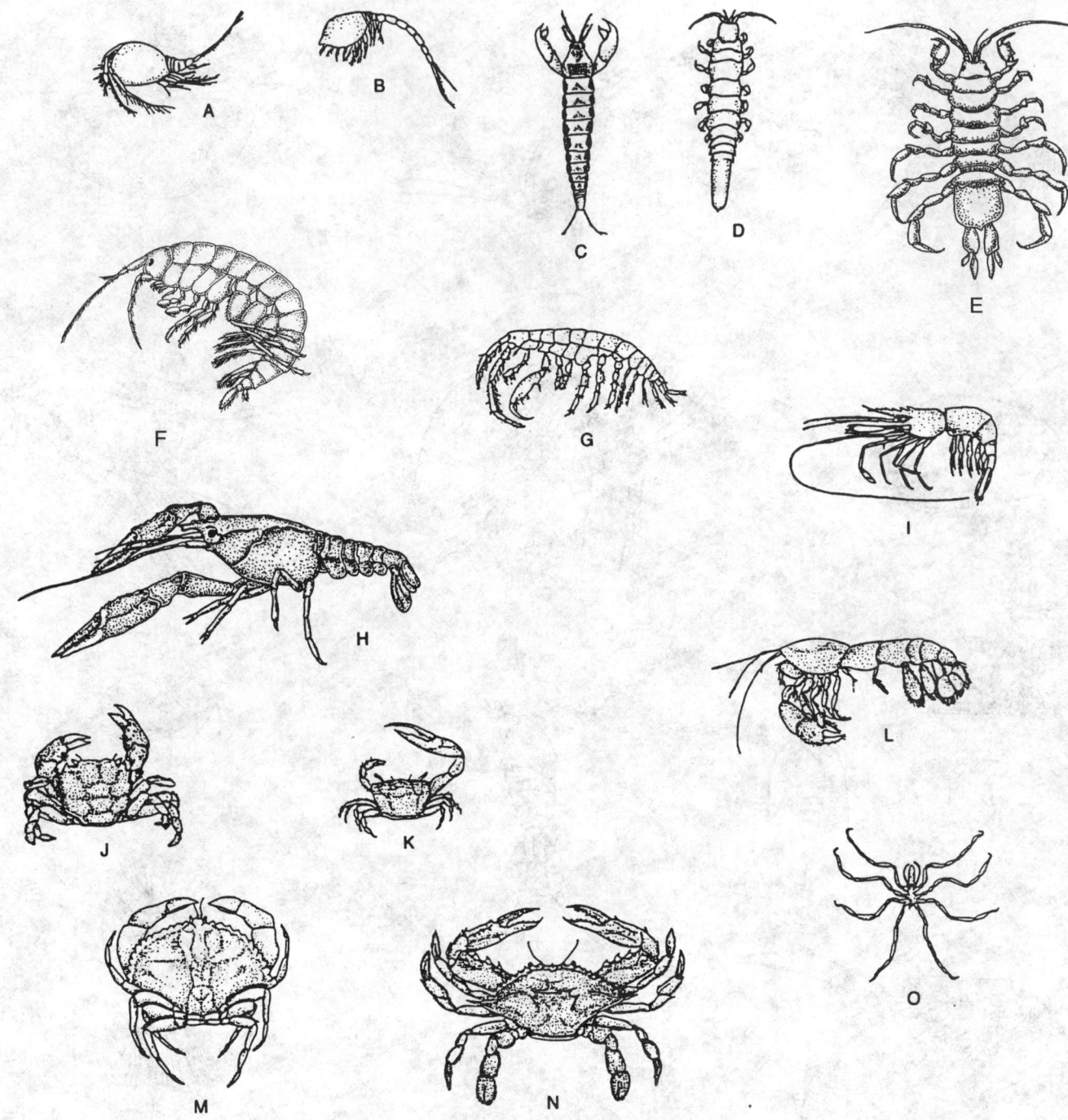

Plate 12. Crustaceans and Pycnogonid (Phylum Arthropoda).

Crustaceans:
Order Leptostraca:
    A—*Epinebalia*, marine   (10–20 mm)
Order Cumacea:
    B—*Oxyurostylis*, marine   (8–12 mm)
Order Tanaidacea:
    C—*Anatanis*, marine   (8–10 mm)
Order Isopoda:
    D—*Idotea*, marine   (1–3 cm)
    E—Sowbug, *Asellus*, freshwater   (2 cm)
Order Amphipoda:
    F—Scud, *Gammarus*, freshwater   (1.5 cm)

G—*Jassa*, marine   (1.5 cm)
Order Decapoda:
    H—Crayfish or crawdad, *Cambarus*, freshwater   (150 cm)
    I—Shrimp, *Spirontocaris*, marine   (3–6 cm)
    J—Mud crab, *Hemigrapsus*, marine   (3–6 cm)
    K—Fiddler crab, *Uca*, marine   (3–6 cm)
    L—Ghost shrimp, *Callianassa*, marine   (5–10 cm)
    M—Cancer crab, *Cancer*, marine   (3–18 cm)
    N—Blue crab, *Callinectes*, marine   (10–20 cm)
Pycnogonid, Class Pycnogonidea:
    O—Sea spider, pycnogonid, marine   (10–20 mm)

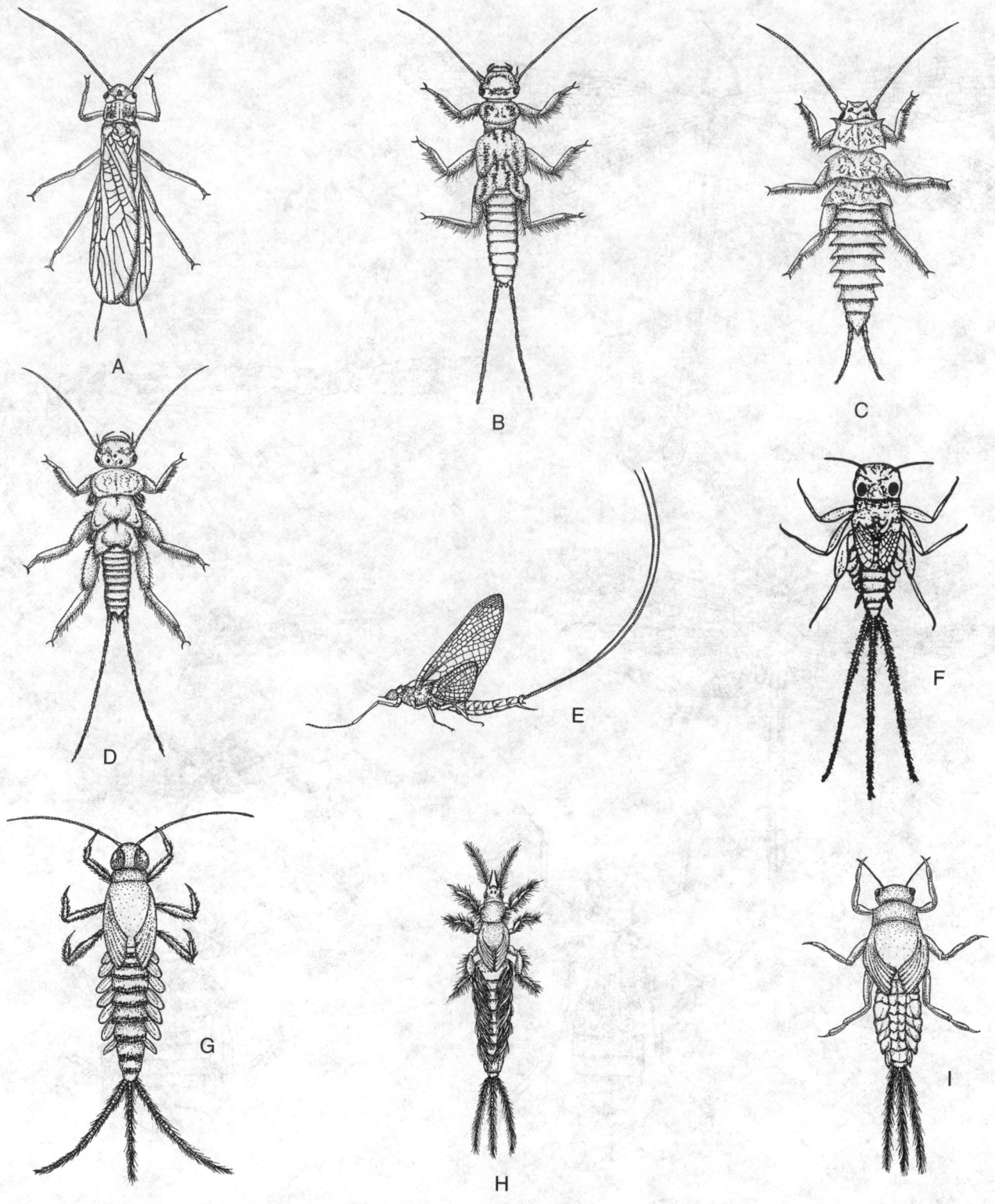

Plate 13. Stoneflies and mayflies. All freshwater inhabitants.

Stoneflies (Order Plecoptera):

A—Adult, *Isoperla*, Family Perlodidae	(14–23 mm)
B—Nymph, *Isoperla*, Family Perlodidae	(10–14 mm)
C—Nymph, *Pteronarcys*, Family Pteronarcidae	(10–40 mm)
D—Nymph, *Acroneuria*, Family Perlidae	(20–30 mm)

Mayfiles (Order Ephemeroptera):

E—Adult, mayfly, Family Heptageniidae	(12–18 mm)
F—Nymph, *Stenonema*, Family Heptageniidae	(10–14 mm)
G—Nymph, *Baetis*, Family Baetidae	(7–14 mm)
H—Nymph, *Hexagenia*, Family Ephemeridae	(20–30 mm)
I—Nymph, *Ephemerella*, Family Ephemerellidae	(8–15 mm)

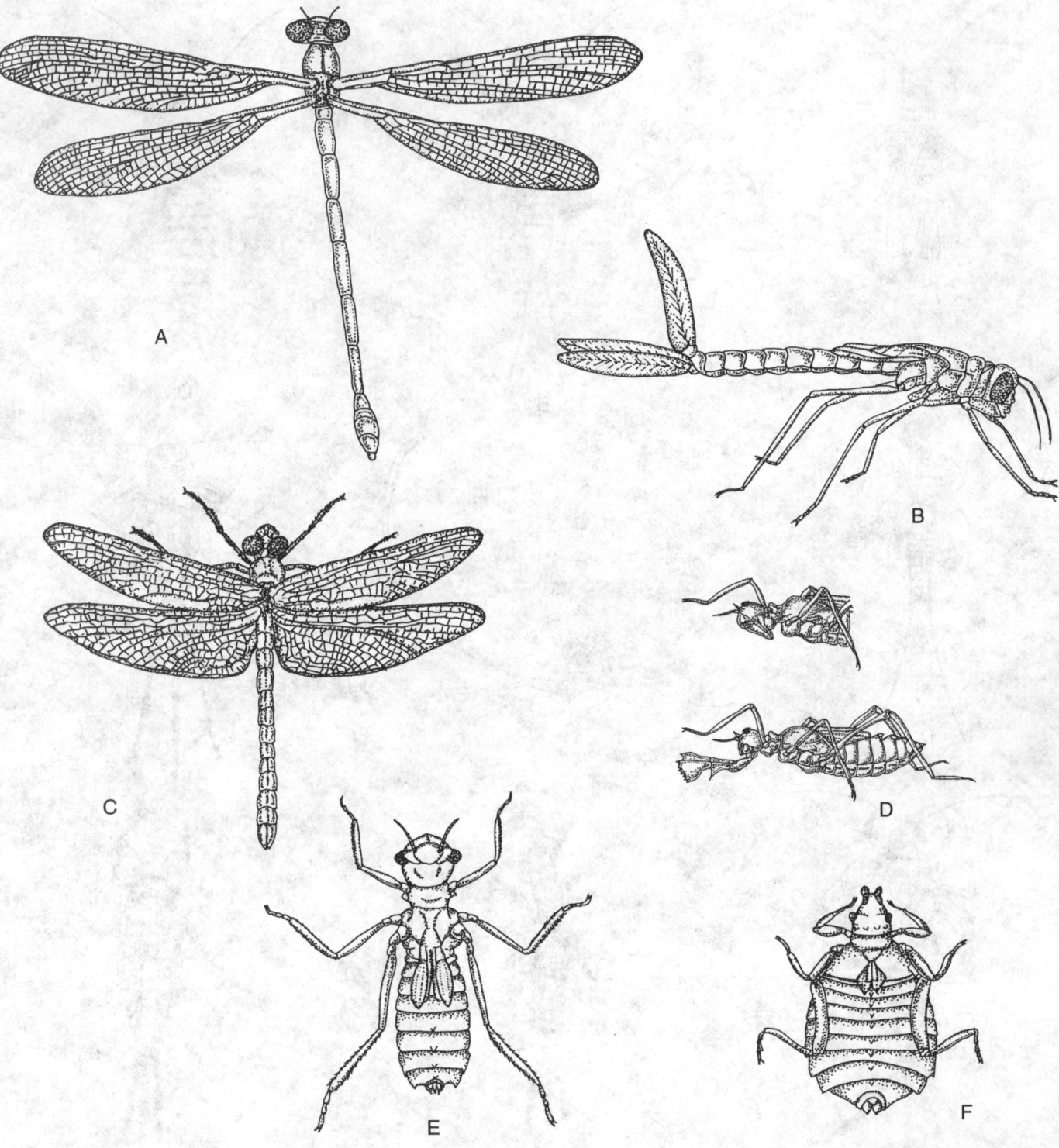

Plate 14. Damselflies, dragonflies (Order Odonata), all freshwater inhabitants.

A—Adult, damselfly	(35–55 mm)	
B—Nymph, damselfly, *Lestes*, Family Lestidae	(20–30 mm)	
C—Adult, dragonfly, *Macromia*, Family Macromiidae	(50–70 mm)	

D—Nymph, dragonfly, *Macromia*, Family Macromiidae, showing "mask" both extended and contracted (15–45 mm)

E—Nymph, dragonfly, *Helocordulia*, Family Libellulidae (15–45 mm)

F—Nymph, dragonfly, *Hagenius*, Family Gomphidae (15–20 mm)

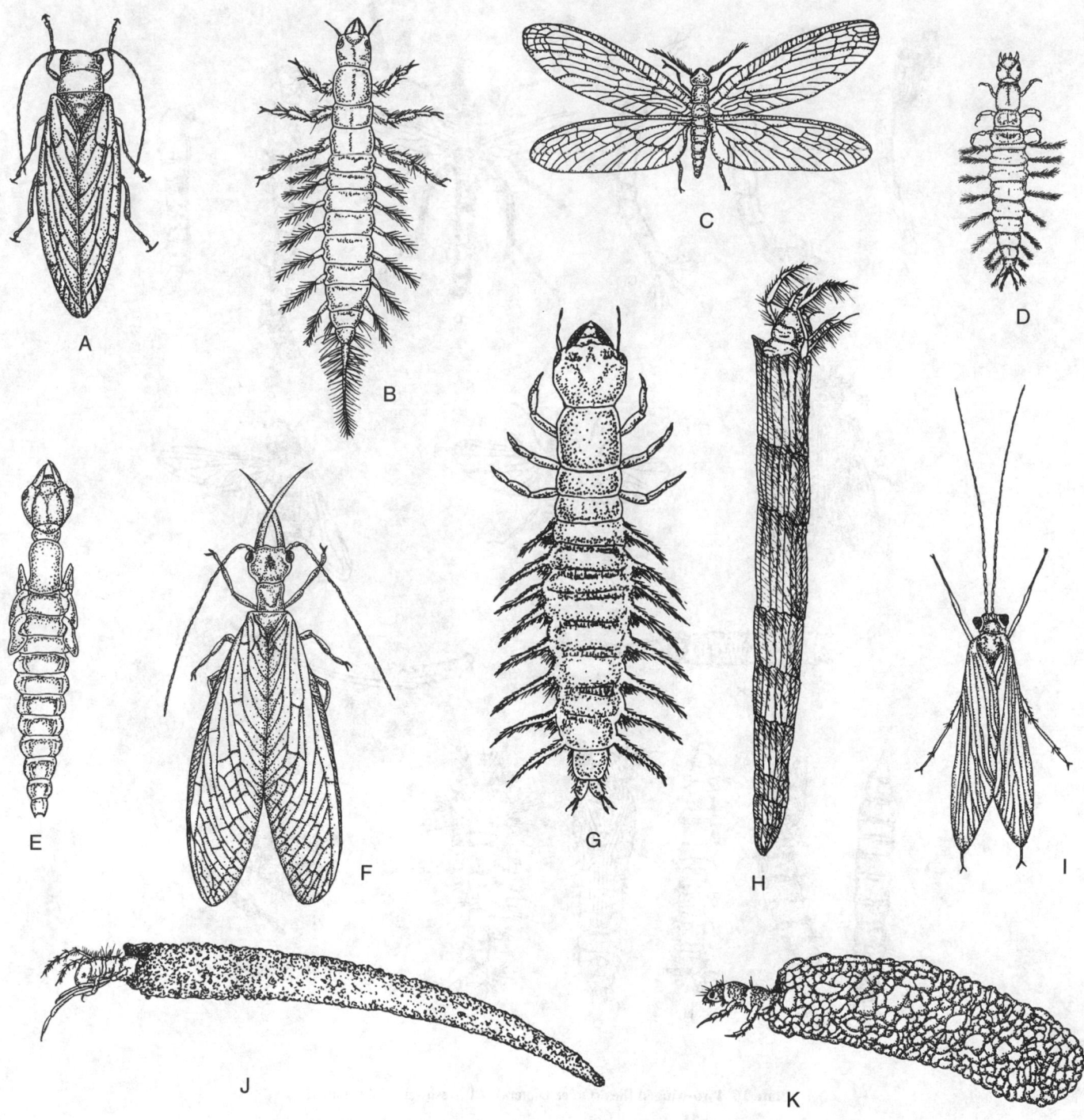

Plate 15. Hellgrammite and relatives, and caddisflies. All freshwater inhabitants.

Hellgrammite and relatives:

A—Adult, alderfly, *Sialis*, Family Sialidae   (9–15 mm)
B—Larva, alderfly, *Sialis*, Family Sialidae   (15–30 mm)
C—Adult, fishfly, *Chauliodes*, Family Corydalidae   (15–30 mm)
D—Larva, fishfly, *Chauliodes* Family Corydalidae   (20–40 mm)
E—Pupa, hellgrammite, *Corydalus*, Family
    Corydalidae
F—Adult, dobsonfly, *Corydalus*, Family
    Corydalidae   (25–70 mm)

G—Larva, dobsonfly, *Corydalus*, Family
    Corydalidae   (25–90 mm)

Caddisflies:

H—Larva and case, *Triaenodes*, Family
    Leptoceridae   (10–14 mm)
I—Adult, *Hydropsyche*, Family Hydropsychidae   (20–30 mm)
J—Larva and case, *Ochrotricha*, Family
    Hydroptilidae   (4–6 mm)
K—Larva and case, *Leptocella*, Family
    Leptoceridae   (14–18 mm)

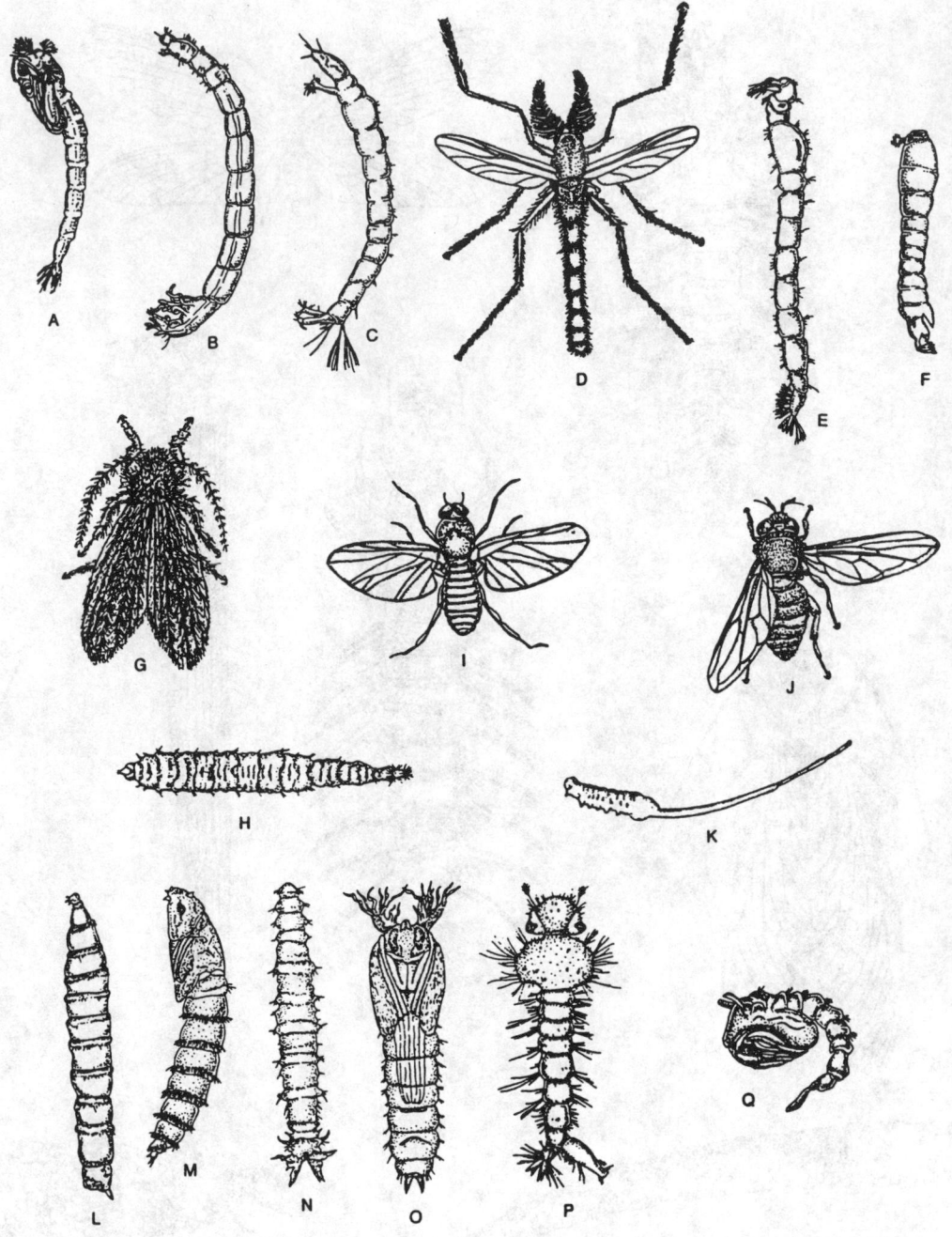

Plate 16. Two-winged flies (Order Diptera). All freshwater inhabitants.

A—Pupa, midge, *Chironomus*, Family
　　Chironomidae

B—Larva, midge, *Chironomus*, Family
　　Chironomidae   (5–30 mm)

C—Larva, midge, *Ablabesmyia*, Family
　　Chironomidae   (5–10 mm)

D—Adult, midge, Chironomidae   (4–12 mm)

E—Larva, phantom midge, *Chaoborus*, Family
　　Culicidae   (8–12 mm)

F—Larva, housefly, *Musca*, Family Muscidae   (3–8 mm)

G—Adult, sewage fly, *Psychoda*, Family
　　Psychodidae   (2–5 mm)

H—Larva, sewage fly, *Psychoda*, Family
　　Psychodidae   (4–6 mm)

I—Adult, blackfly, *Simulium*, Family Simuliidae   (2–6 mm)

J—Adult, drone fly, Family Syrphidae   (10–15 mm)

K—Rat-tailed maggot, *Eristalis*, Family Syrphidae   (15–30 mm)

L—Larva, *Tabanus*, Family Tabanidae   (30–40 mm)

M—Blowfly, *Tabanus*, Family Tabanidae

N—Larva, cranefly, *Tipula*, Family Tipulidae   (30–40 mm)

O—Pupa, cranefly, *Antocha*, Family Tipulidae

P—Larva, mosquito, *Aedes*, Family Culicidae   (10–15 mm)

Q—Pupa, mosquito, *Culex*, Family Culicidae

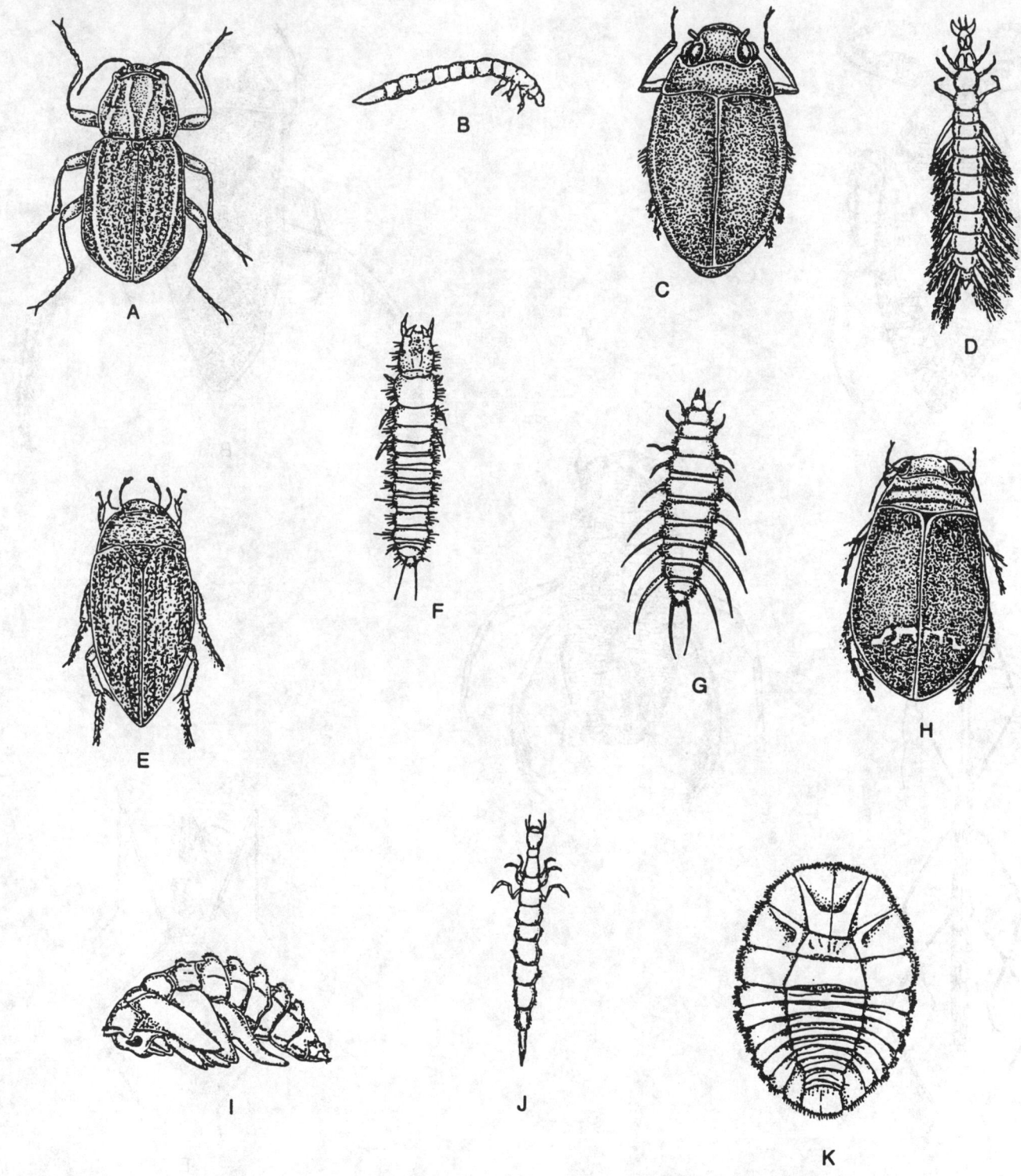

Plate 17. Beetles (Order Coleoptera). All freshwater inhabitants.

A—Adult, riffle beetle, *Stenelmis*, Family Elmidae (2–5 mm)
B—Larva, *Narpus*, Family Elmidae (4–10 mm)
C—Adult, whirligig beetle, *Dineutus*, Family
 Gyrinidae (7–15 mm)
D—Larva, *Dineutus*, Family Gyrinidae (10–30 mm)
E—Adult, water scavenger beetle, *Hydrophilus*,
 Family Hydrophilidae (2–40 mm)
F—Larva, *Berosus*, Family Hydrophilidae (5–20 mm)

G—Larva, *Enochrus*, Family Hydrophilidae (10–25 mm)
H—Adult, predaceous diving beetle, *Dytiscus*, Family
 Dytiscidae (2–40 mm)
I—Pupa, *Cybister*, Family Dytiscidae
J—Larva, *Cybister*, Family Dytiscidae (10–25 mm)
K—Larva, water penny, *Psephenus*, Family
 Psephenidae (3–10 mm)

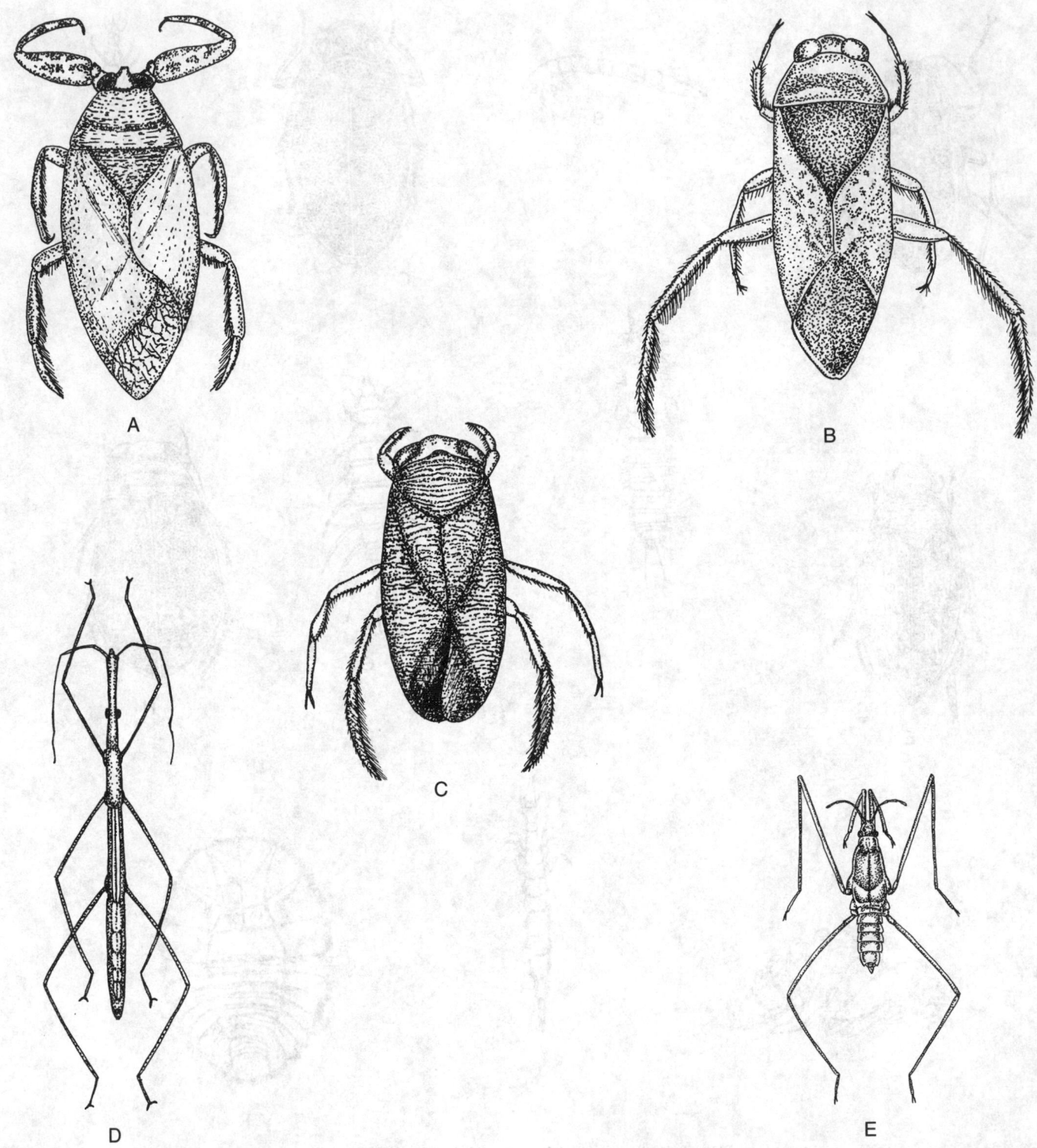

Plate 18. True bugs (Order Hemiptera, all adults). All freshwater inhabitants.

A—Electric light bug, *Lethocerus*, Family
    Belostomidae   (20–70 mm)
B—Backswimmer, *Notonecta*, Family
    Notonectidae   (5–17 mm)
C—Water boatman, *Sigara*, Family Corixidae   (3–12 mm)

D—Marsh treader, *Hydrometra*, Family
    Hydrometridae   (8–11 mm)
E—Water strider, *Gerris*, Family Gerridae   (2–15 mm)

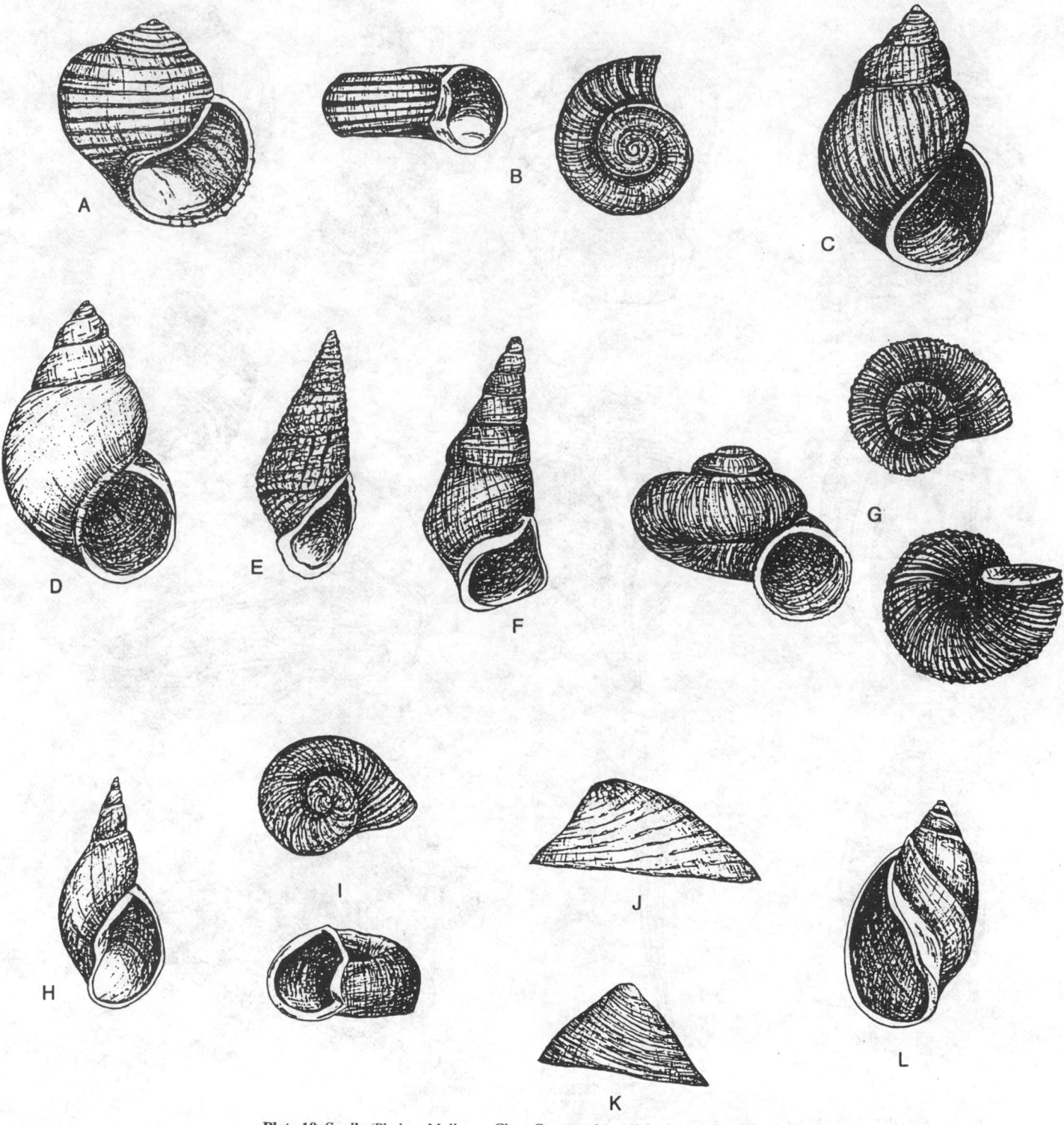

Plate 19. Snails (Phylum Mollusca, Class Gastropoda). All freshwater inhabitants.

A—Apple snail, *Pomacea*, Family Pilidae   (5 cm)
B—*Marisa*, side and top views, Family Pilidae   (15 mm)
C—*Campeloma*, Family Viviparidae   (4 cm)
D—Faucet snail, *Bithynia*, Family Ammicolidae   (2 cm)
E—*Tarebia*, Family Thiaeidae   (15 mm)
F—River snail, *Pleurocera*, Family Pleuroceridae   (3 cm)

G—*Valvata*, side, top and bottom views, Family
     Valvatidae   (1 cm)
H—Pond snail, *Lymnaea*, Family Lymnaeidae   (15 mm)
I—Orb snail, top and side views, *Helisoma*,
     Family Planorbidae   (1 cm)
J—Limpet, *Ferrissia*, Family Ancylidae   (2 mm)
K—Limpet, *Lanx*, Family Lancidae   (10 mm)
L—Pouch snail, *Physa*, Family Physidae   (3–5 mm)

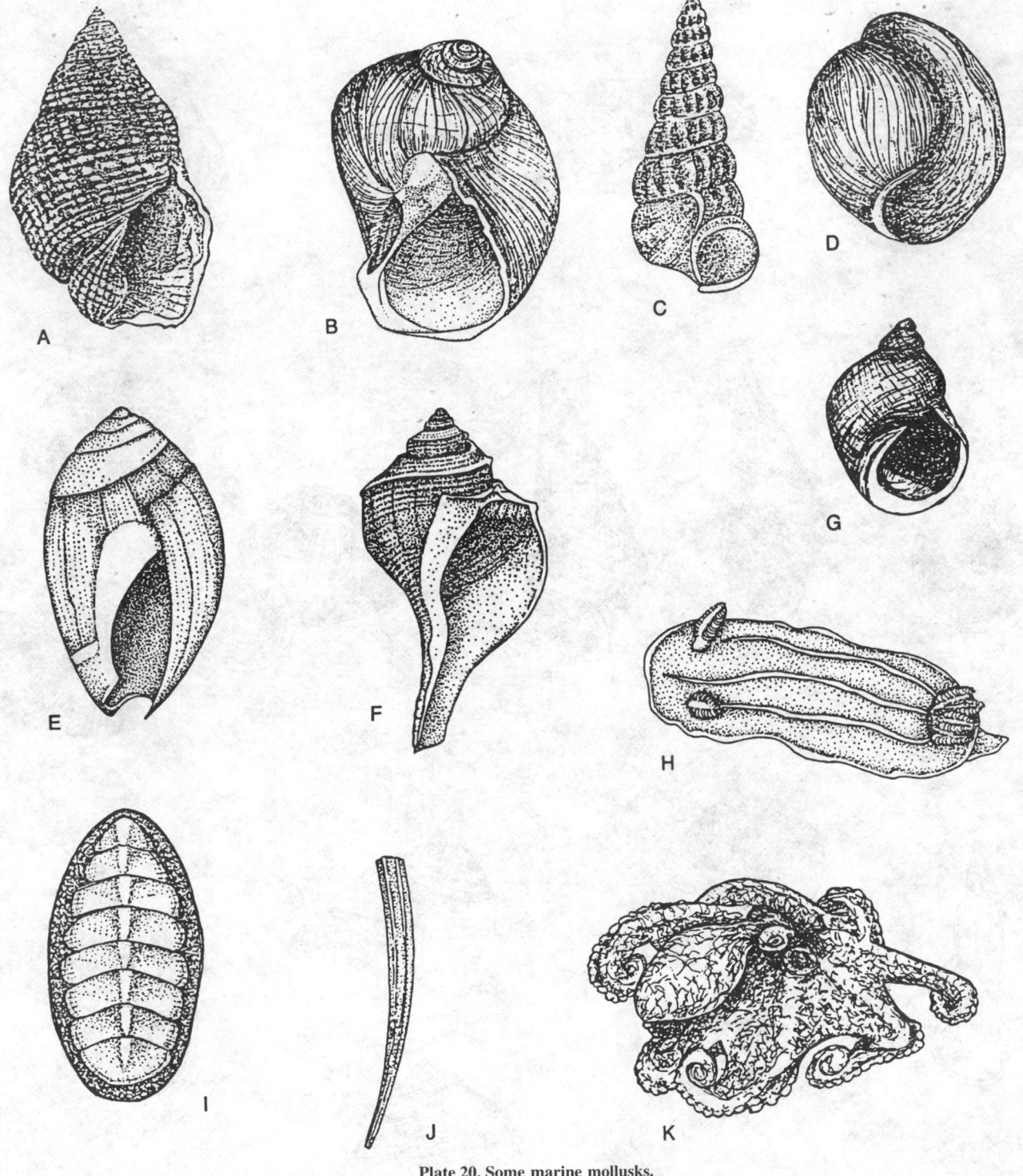

Plate 20. Some marine mollusks.

Snails (Class Gastropoda):
  A—Basket snail, *Nassarius*, Family Nassidae   (2–5 cm)
  B—Moon snail, *Polinices*, Family Naticidae   (5–8 cm)
  C—Horn snail, *Cerithidea*, Family Cerithiidae   (1–3 cm)
  D—Bubble snail, *Bulla*, Family Bullidae   (3–5 cm)
  E—Olive snail, *Olivella*, Family Olividae   (2–3 cm)
  F—*Busycon*, Family Melongenidae   (5–8 cm)
  G—Periwinkle, *Littorina*, Family Littorinidae   (1 cm)
  H—Sea slug, *Chromodoris*

Chitons (Class Polyplacophora):
  I—Conspicious chiton, *Stenoplax*   (3–6 cm)

Tusk shell (Class Scaphopoda):
  J—*Dentalium*   (2–4 cm)

Octopus (Class Cephalopoda):
  K—*Octopus*   (3–10 cm up to
                                              1 m or more)

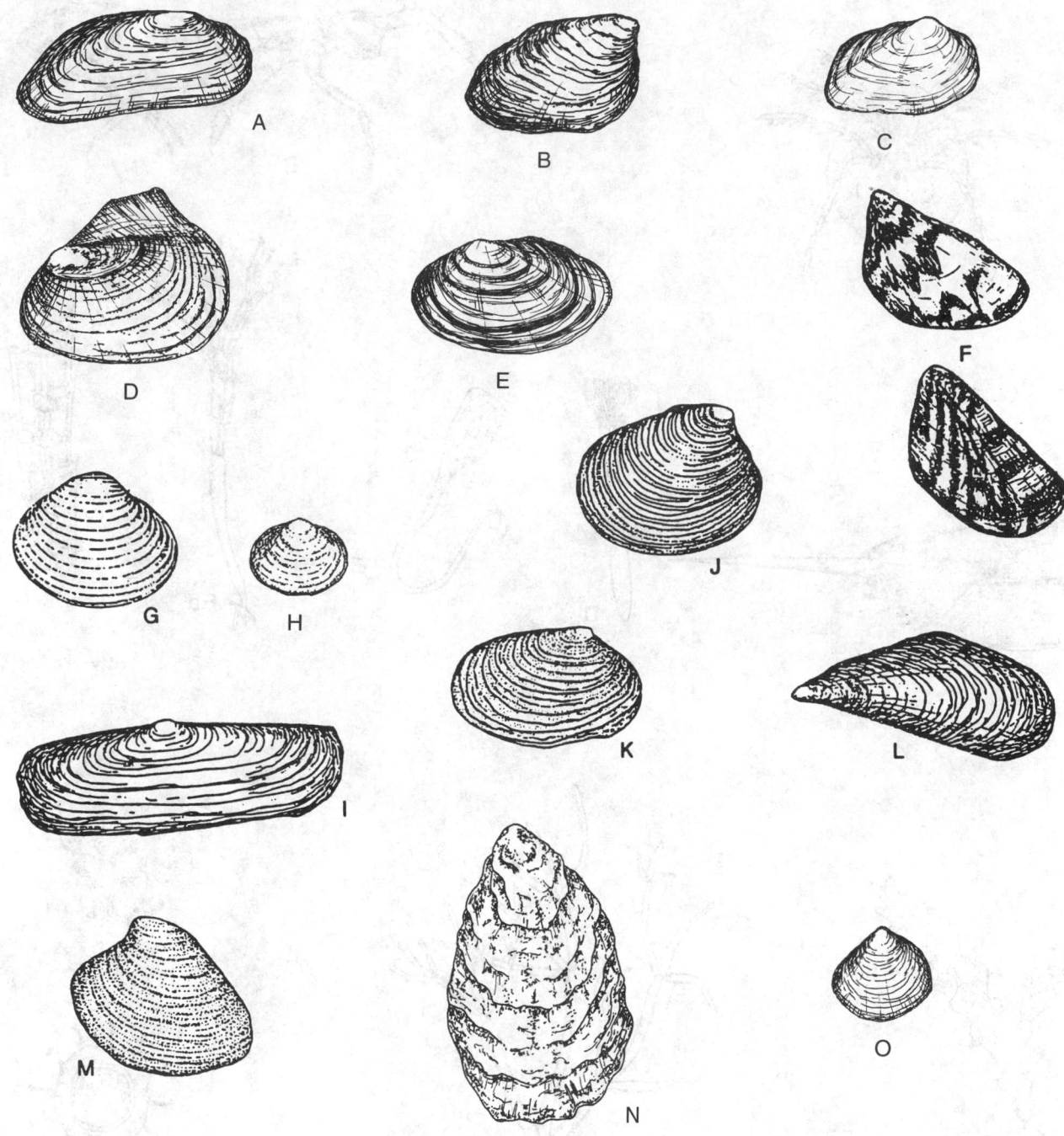

Plate 21. Bivalves (Phylum Mollusca, Class Pelecypoda).

Freshwater species:

A—Spectacle case, *Margaritifera*, Family
    Margaritiferidae   (10 cm)

B—Pearly mussel, *Pleurobema*, Family
    Unionidae   (10 cm)

C—Pearly mussel, *Gonidea*, Family Unionidae   (10 cm)

D—Winged lampshell, *Proptera*, Family
    Lampsilinae   (13 cm)

E—Papershell, *Anodonta*, Family Anodontidae   (14 cm)

F—Zebra mussel (examples of color patterns),   (5 cm)
Dreissena   (4 cm)

G—Asiatic clam, *Corbicula*, Family Corbiculidae

H—Fingernail clam, *Sphaerium*, Family
    Sphaeriidae   (1 cm)

Marine species:

I—Jackknife clam, *Tagelus*, Family
    Psammobiidae   (6–10 cm)

J—Quahog clam, *Mercenaria*, Family Veneridae   (6–10 cm)

K—Soft-shelled clam, *Mya*, Family Myidae   (5–10 cm)

L—Blue musssel, *Mytilus*, Family Mytilidae   (6 cm)

M—Rangia clam, *Rangia*, Family Matricidae   (5 cm)

N—Oyster, *Crassostrea*, Family Ostreidae   (9 cm)

O—Marsh clam, *Polymesoda*, Family
    Corbiculidae   (4 cm)

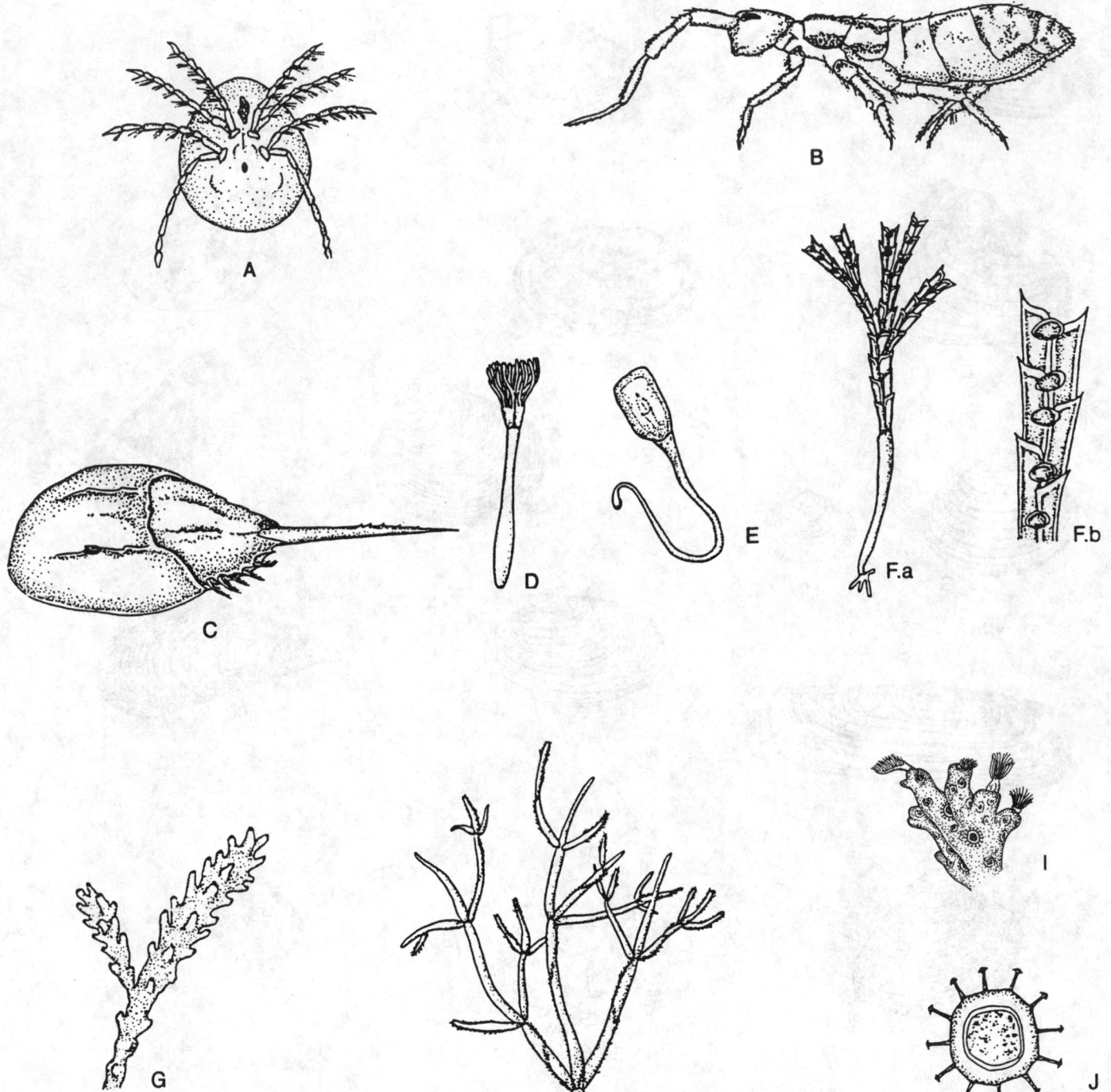

Plate 22. Miscellaneous invertebrates.

Arthropods (Phylum Arthropoda):
  A—Water mite, *Limnochares*, Class
Arachnoidea, freshwater   (3 mm)
  B—Springtail, *Orchesella*, Class Insecta, Order
    Collembola, freshwater   (2 mm)
  C—Horsehoe crab, *Limulus*, Class Arachnoidea,
    marine   (10–30 cm)
  D—*Phoronis*, Phylum Phoronidea, marine   (2–4 cm)
  E—*Glottidia*, Phylum Brachiopoda, marine   (3–5 cm)

Phylum Ectoprocta (Bryozoa):
  F—*Bugula*, a. colony, b. enlarged view, marine   (1–3 cm)
  G—*Crisulipora*, calcareous colony, marine   (1–3 cm)
  H—*Zoobotryon*, gelatinous colony, marine   (20–100 cm)
  I—Jellyball, *Pectinatella*, young colony,
    freshwater   (11 cm)
  J—*Pectinatella*, statoblast, freshwater   (1 mm)

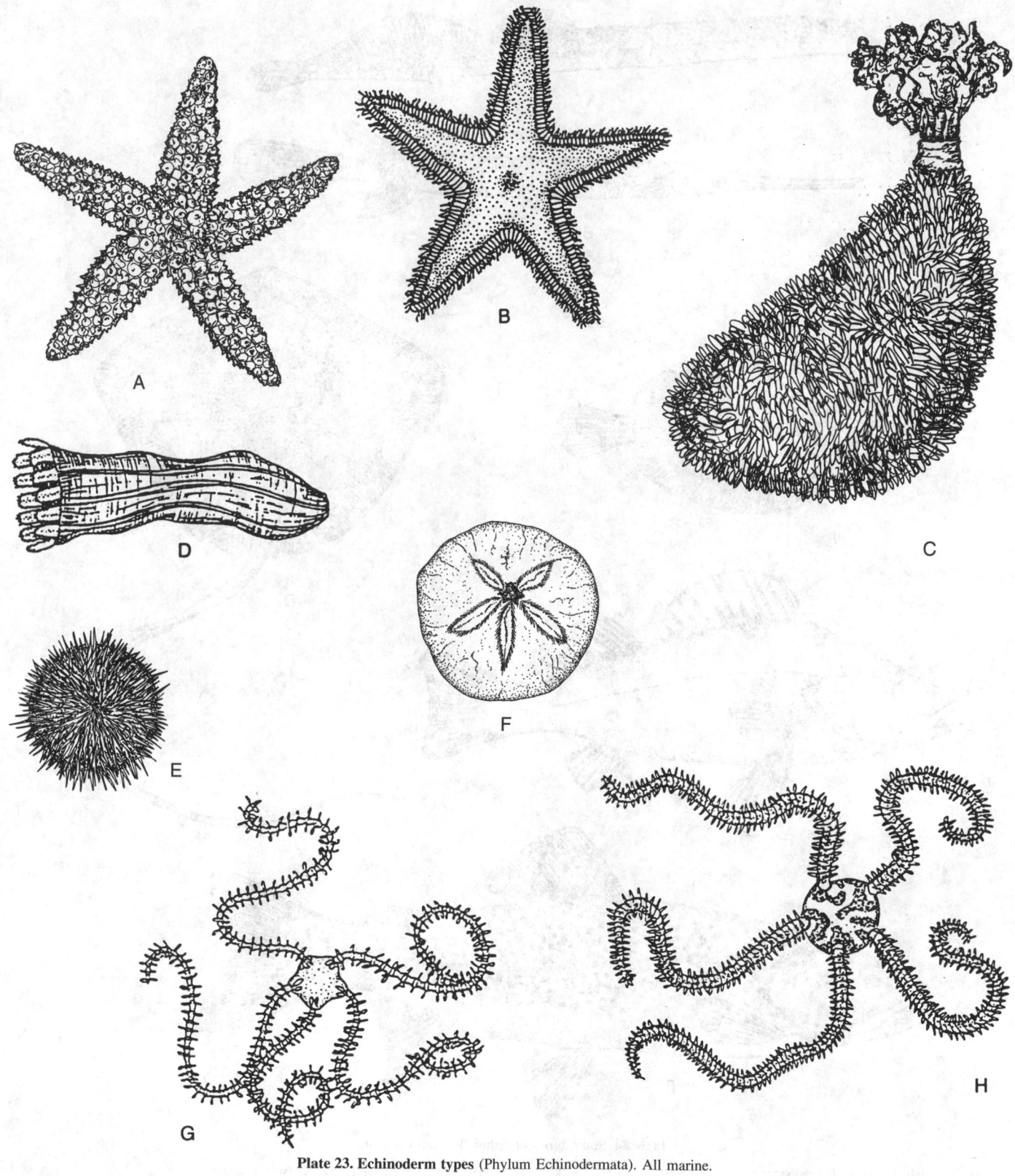

Plate 23. Echinoderm types (Phylum Echinodermata). All marine.

A—Starfish, *Asterias*, Class Asteroidea   (15 cm)
B—Sand starfish, *Astropecten*, Class Asteroidea   (10–20 cm)
C—Sea cucumber, *Thyrone*, Class Holothuroidea   (10 cm)
D—Sea cucumber, *Leptosynpta*, Class
    Holothuroidea   (5–8 cm)

E—Sea urchin, *Strongylocentrotus*, Class
    Echinoidea   (6 cm)
F—Sand dollar, *Echinarachnius*, Class Echinoidea   (7 cm)
G—Brittle star, *Amphiodia*, Class Ophiuroidea   (disk 15 mm)
H—Brittle star, *Ophiopholis*, Class Ophiuroidea   (disk 15 mm)

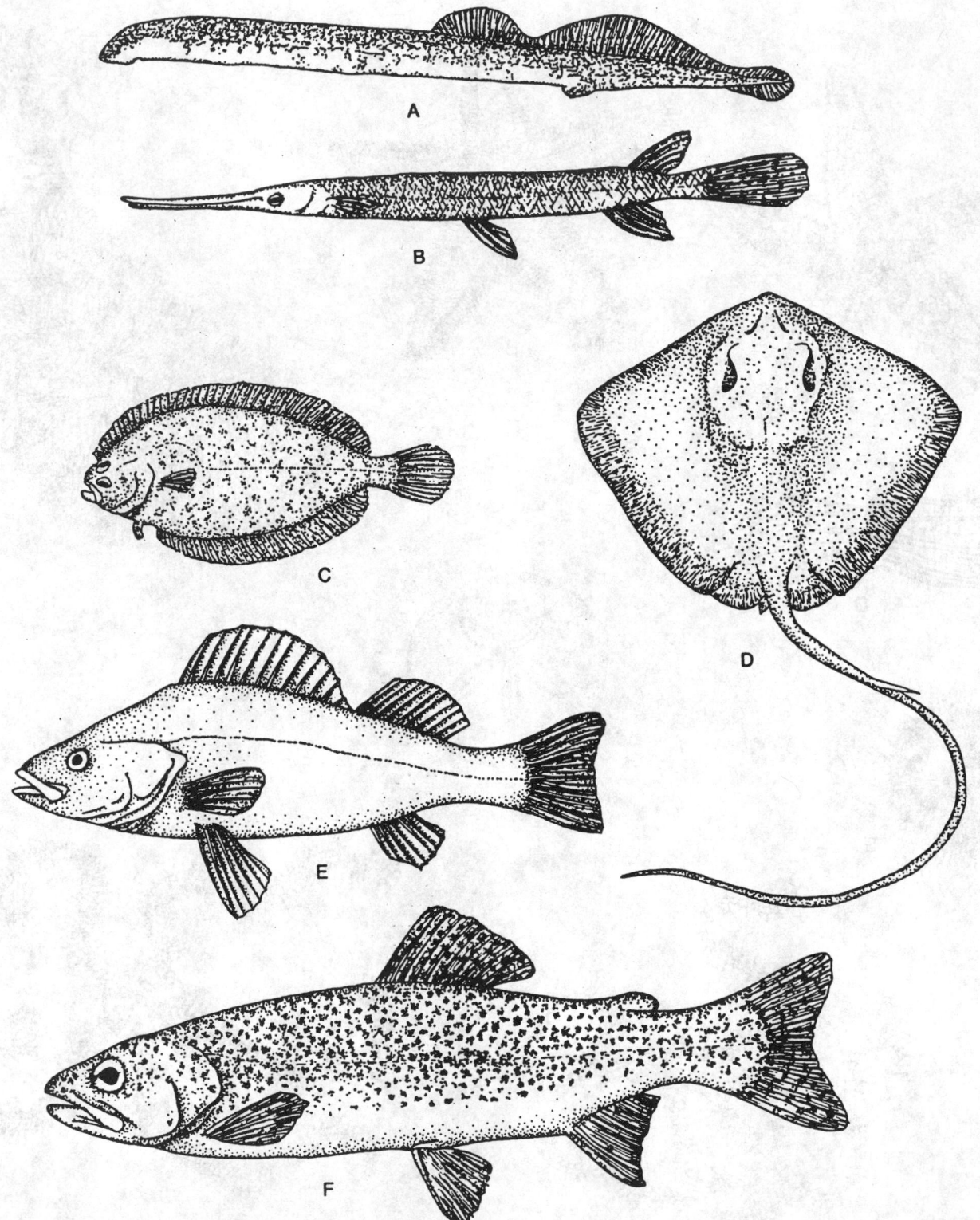

Plate 24. Some types of fishes (Phylum Chordata).

A—Jawless fish, lamprey, *Petromyzon*, Class
    Agnatha, freshwater   (30–45 cm)
B—Ganoid fish, long-nosed gar, *Lepisosteus*, Class
    Osteichthys, freshwater   (2.4 m)
C—Flatfish, flounder, *Paralichthys*, Class
    Osteichthys, marine   (30–60 cm)

D—Cartilage fish, stingray, *Dasyatis*, Class
    Chondrichthys, marine   (2 m)
E—Spiny-rayed fish, perch, *Perca*, Class
    Osteichthys, freshwater   (30 cm)
F—Soft-rayed fish, rainbow trout, *Salmo*, Class
    Osteichthys, freshwater   (30 cm)

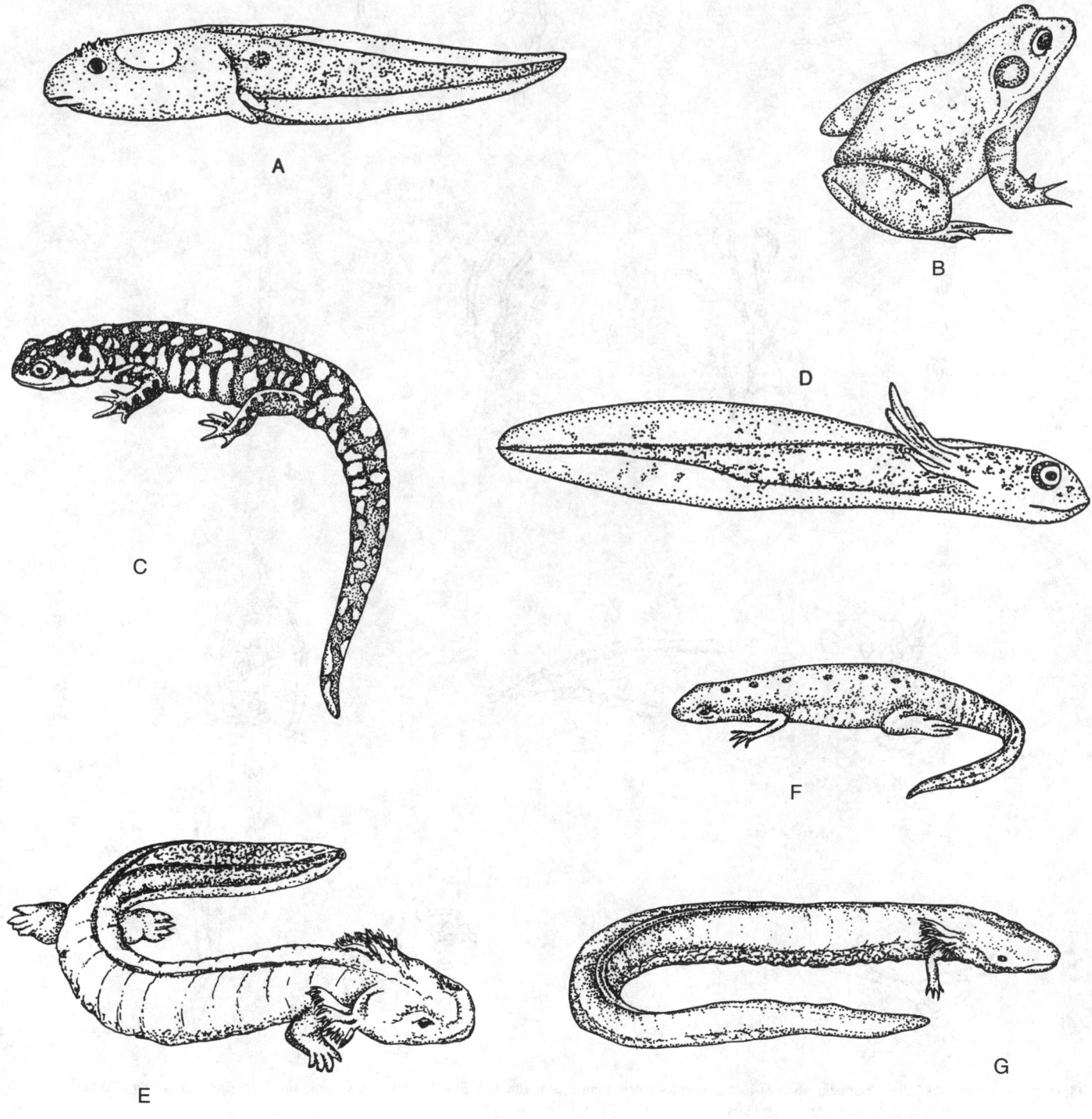

Plate 25. Types of amphibians (Phylum Chordata, Class Amphibia). All freshwater.

Frogs and toads (Order Salientia):
  A—Tadpole larva with hind legs beginning to
    develop from the body of the tadpole
  B—Frog, *Rana*   (30 cm)

Salamanders (Order Caudata):
  C—*Ambystoma*, adult, typically terrestrial   (20 cm)

D—*Ambystoma*, larva, inhabits water   (5–15 cm)
E—Water dog or mud puppy, *Necturus*, gills
    present in adult   (to 30 cm)
F—Salamander with flat tail, *Notophthalmus*   (9 cm)
G—Salamander, *Siren*, with gills and front pair of
    legs present   (1 m)

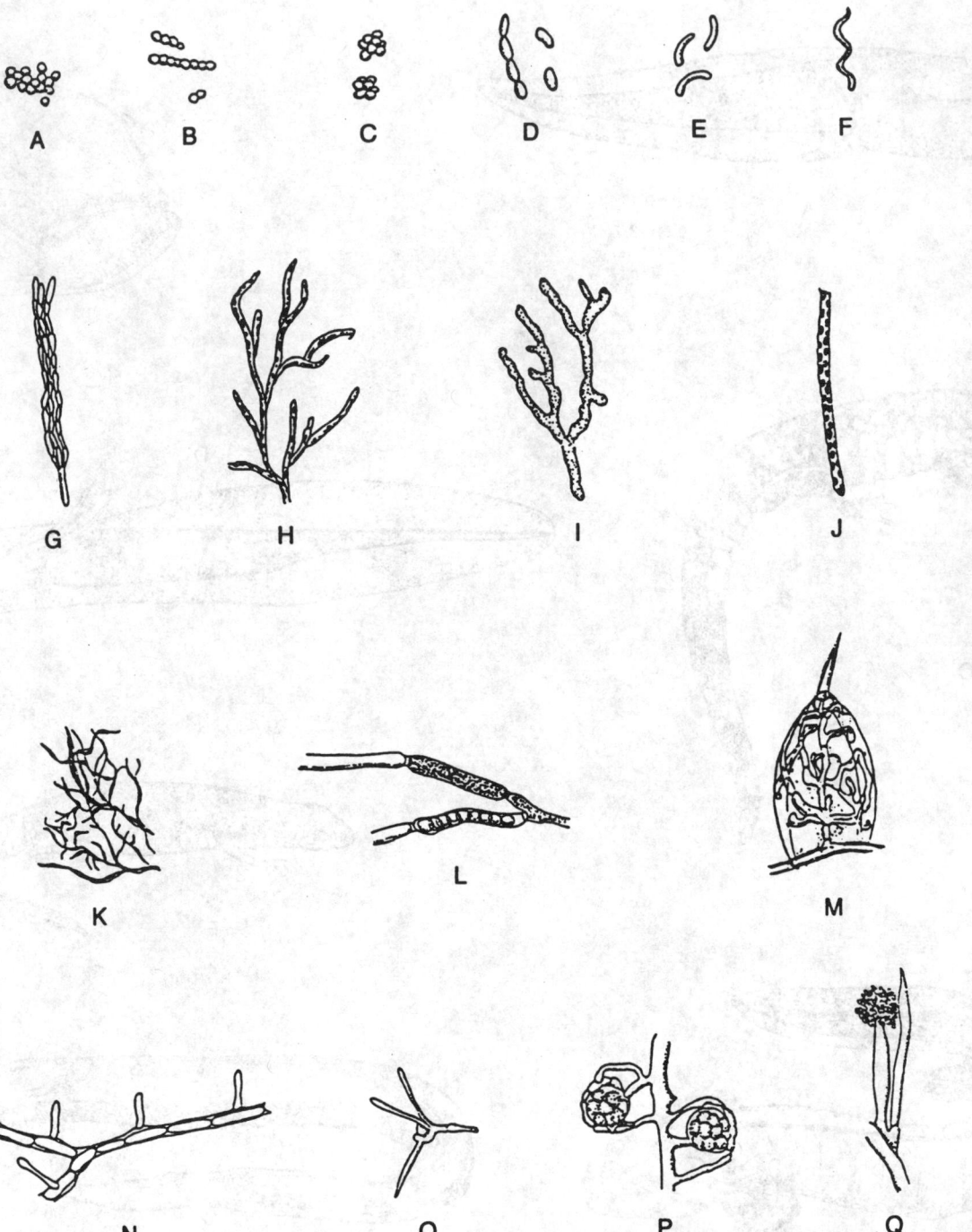

Plate 26. Bacteria and fungi. Diameter of most bacterial cells is less than 2 μm, although *Beggiatoa* (J) may range up to 16 μm in diameter and be of indefinite length. All freshwater inhabitants.

Bacterial cellular forms and arrangements:
    A—*Micrococcus*
    B—*Streptococcus*
    C—*Sarcina*
    D—*Bacillus*
    E—*Vibrio*
    F—*Spirillum*
Sewage organisms:
    G—*Sphaerotilus* cells
    H—*Sphaerotilus* growth form

I—*Zoogloea* growth form
J—*Beggiatoa* (sulfur bacterium)
K—An actinomycete growth form from compost
Fungi:
    L—*Leptomitus*, showing zoospores and cellulin
        plugs
    M—*Zoophagus* with rotifer on mycelial peg   (diam 3 μm)
    N—*Zoophagus*, showing mycelial pegs
    O—*Tetracladium*   (diam 2.5–3.5 μm)
    P—*Achlya*, with oospores
    Q—*Achlya*, with encysted oospores

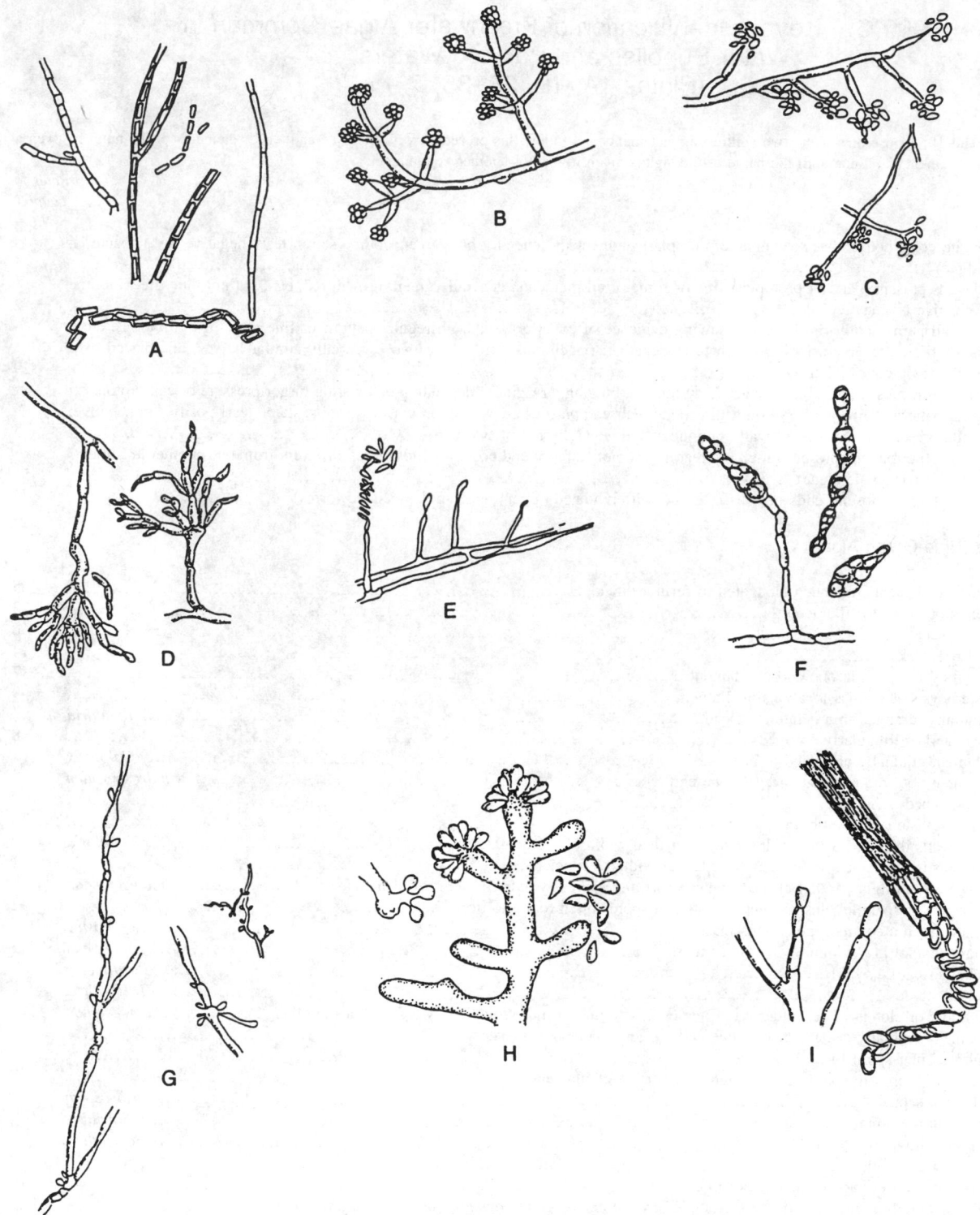

Plate 27. Fungi. All freshwater inhabitants.

A—*Geotrichum candidum*   colonies white
B—*Trichoderma viride*   colonies green
C—*Exophiala jeanselmei*   colonies black
D—*Cladosporium cladosporioides*   colonies black
E—*Fusarium* sp.   colonies orange

F—*Alternaria alternata*   colonies gray
G—*Trichosporon cutaneum*   colonies white
H—*Pseudodallescheria boydii*   colonies white
I—*Klasterskaya splendens*

Note: Color Plates follow 10-163

10900 C.   Key for Identification of Freshwater Algae Common in Water Supplies and Polluted Waters*
(Plates 1A, 1B, 28–35)

Beginning with 1a and 1b, choose one of the two contrasting statements and follow this procedure with the "a" and "b" statements of the number given at the end of the chosen statement. Continue until the name of the alga is given instead of another key number.

Refer to couplet No.

1a.   Plastid (separate color body) absent; complete protoplast pigmented; generally blue-green; iodine starch test† negative (cyanobacteria, blue-green algae) 4

1b.   Plastid or plastids present; parts of protoplast free of some or all pigments; generally green, brown, red, etc., but not blue-green; iodine starch test† positive or negative 2

    2a.   Cell wall permanently rigid (never showing evidence of collapse), and with regular pattern of fine markings (striations, etc.); plastids brown to green; iodine starch test† negative; flagella absent; wall of two essentially similar halves, one placed over the other as a cover (diatoms)... 29

    2b.   Cell wall, if present, capable of sagging, wrinkling, bulging, or rigidity, depending on existing turgor pressure of cell protoplast; regular pattern of fine markings on wall generally absent; plastids green, red, brown, etc.; iodine starch test† positive or negative; flagella present or absent; cell wall continuous and generally not of two parts.................................... 3

3a.   Cell or colony motile; flagella present (often not readily visible); anterior and posterior ends of cell different from one another in contents and often in shape (flagellate algae) 51

3b.   Nonmotile; true flagella absent; ends of cells often not differentiated (green algae and associated forms) 77

1. Cyanobacteria (Blue-Green Algae)

    4a.   Cells in filaments (or much elongated to form a thread)... 5

    4b.   Cells not in (or as) filaments 23

5a.   Heterocysts present 6

5b.   Heterocysts absent 14

    6a.   Heterocyst located at one end of filament 7

    6b.   Heterocysts at various locations in filament 9

7a.   Filaments radially arranged in a gelatinous bead .... *Rivularia*

7b.   Filaments isolated or irregularly grouped 8

    8a.   Filament gradually narrowed to one end .... *Calothrix*

    8b.   Filament not gradually narrowed to one end .... *Cylindrospermum*

9a.   Filament unbranched 10

9b.   Filament with occasional (false) branches 13

    10a.   Crosswalls in filament much closer together than width of filament .... *Nodularia*

    10b.   Crosswalls in filament at least as far apart as width of filament 11

11a.   Filaments normally in tight parallel clusters; heterocysts and spores cylindric to long oval in shape. .... *Aphanizomenon*

11b.   Filaments not in tight parallel clusters; heterocysts and spores often round to oval 12

    12a.   Filaments in a common gelatinous mass .... *Nostoc*

    12b.   Filaments not in a common gelatinous mass .... *Anabaena*

13a.   False branches in pairs .... *Scytonema*

13b.   False branches, single .... *Tolypothrix*

    14a.   Filament or elongated cell attached at one end, with one or more round cells (spores) at the other .... *Chamaesiphon*

    14b.   Filament generally not attached at one end; no terminal spores present.............................. 15

15a.   Filament with regular spiral form throughout 16

15b.   Filament not spiral, or with spiral form limited to a portion of filament 17

    16a.   Filament septate .... *Arthrospira*

    16b.   Filament nonseptate .... *Spirulina*

17a.   Filament very narrow, only 0.5 to 2.0 μm wide .... *Schizothrix*

17b.   Filament 3 to 95 μm wide 18

    18a.   Filaments loosely aggregated or not in clusters 19

    18b.   Filaments tightly aggregated and surrounded by a common gelatinous secretion that may be invisible 22

19a.   Filament surrounded by wall-like sheath that frequently extends beyond the ends of the filament of cells; filament generally without movement 20

19b.   Filament not surrounded by a wall-like sheath; filament may show movement 21

    20a.   Cells separated from one another by a space .... *Johannesbaptistia*

    20b.   Cells in contact with adjacent cells .... *Lyngbya*

21a.   All filaments short, with less than 20 cells; one or both ends of filament sharply pointed .... *Raphidiopsis*

---

* Source: PALMER, C.M. 1977. Algae and Water Pollution. Publ. No. 600/9-77-036, U.S. Environmental Protection Agency, Cincinnati, Ohio.
† Add 1 drop of Lugol's (iodine) solution, diluted 1:1 with distilled water. In about 1 min, if positive, starch is stained blue and later black. Other structures (such as nucleus, plastids, cell wall) also may stain, but turn brown to yellow.

21b.   Filaments long, with more than 20 cells; filaments commonly without sharp-pointed ends .... *Oscillatoria*
     22a.   Filaments arranged in a tight, essentially parallel bundle .... *Microcoleus*
     22b.   Filaments arranged in irregular fashion, often forming a mat .... *Phormidium*
23a.   Cells in a regular pattern of parallel rows, forming a plate .... *Merismopedia*
23b.   Cells not regularly arranged to form a plate 24
     24a.   Cells regularly arranged near surface of a spherical gelatinous bead 25
     24b.   Gelatinous bead, if present, not spherical 26
25a.   Cells ovate to heart-shaped, connected to center of bead by colorless stalks .... *Gomphosphaeria*
25b.   Cells round, without gelatinous stalks .... *Coelosphaerium* type
     26a.   Cells cylindric-oval .... *Aphanothece*
     26b.   Cells spherical 27
27a.   Two or more distinct layers of gelatinous sheath around each cell or cell cluster .... *Gloeocapsa*
27b.   Gelatinous sheath around cells not distinctly layered 28
     28a.   Cells isolated or in colonies of 2 to 32 cells .... *Chroococcus*
     28b.   Cells in colonies composed of many cells .... *Anacystis (Microcystis, Polycystis)*

2. Diatoms

29a.   Front (valve) view circular in outline; markings radial in arrangement; cells may form a filament (centric diatoms) 30
29b.   Front (valve) view elongate, not circular; transverse markings in one or two longitudinal rows; cells, if grouped, not forming a filament (pennate diatoms) 32
     30a.   Cells in persistent filaments with valve faces in contact; therefore, cells commonly seen in side (girdle) view ...... *Melosira*
     30b.   Cells isolated or in fragile filaments, often seen in front (valve) view 31
31a.   Radial markings (striations), in valve view, extending from center to margin; short spines often present around margin (valve view) .... *Stephanodiscus*
31b.   Area of prominent radial markings, in valve view, limited to approximately outer half of circle, marginal spines generally absent .... *Cyclotella*
     32a.   Cell longitudinally symmetrical in valve view 33
     32b.   Cell longitudinally unsymmetrical (two sides unequal in shape), at least in valve view 49
33a.   Raphe at or near the edge of the valve 34
33b.   Raphe or pseudoraphe median or submedian 35
     34a.   Marginal, keeled raphe areas lie opposite one another on the two valves .... *Hantzschia*
     34b.   Marginal, keeled raphe areas lie diagonal to one another on the two valves .... *Nitzschia*
35a.   Cell transversely symmetrical in valve view 36
35b.   Cell transversely unsymmetrical (two ends unequal in shape or size), at least in valve view 44
     36a.   Cell round-oval in valve view, not more than twice as long as it is wide .... *Cocconeis*
     36b.   Cell elongate, more than twice as long as it is wide 37
37a.   Cell flat (girdle face wide, valve face narrow) .... *Tabellaria*
37b.   Girdle and valve faces about equal in width 38
     38a.   Cell with several markings (septa) extending without interruption across the valve face; no marginal line of pores present .... *Diatoma*
     38b.   Cross-markings (striations or costae) on valve surface, interrupted by either longitudinal space (pseudoraphe), or line (raphe), or line of pores (carinal dots) 39
39a.   Cells attached side by side to form a ribbon of several to many cells .... *Fragilaria*
39b.   Cells isolated or in pairs 40
     40a.   Cell narrow, linear, often narrowed to both ends; true raphe absent .... *Synedra*
     40b.   Cell commonly "boat-shape" in valve view; true raphe present 41
41a.   Cell longitudinally unsymmetrical in girdle view; sometimes with attachment stalk .... *Achnanthes*
41b.   Cell symmetrical in girdle as well as valve view; generally not attached 42
     42a.   Area without striations extending as a transverse belt around middle of cell .... *Stauroneis*
     42b.   No continuous clear belt around middle of cell 43
43a.   Cell with coarse transverse markings (costae), which appear as solid lines even under high magnification .... *Pinnularia*
43b.   Cell with fine transverse markings (striae), which appear as lines of dots under high magnification .... *Navicula*
     44a.   Cells attached together at one end only to form radiating colony .... *Asterionella*
     44b.   Cell not forming a loose radiating colony 45
45a.   Cells in fan-shaped colonies .... *Meridion*
45b.   Cells isolated or in pairs 46
     46a.   Prominent wall markings in addition to striations present just below lateral margins on valve surface of cell .... *Surirella*
     46b.   Wall markings along sides of valve limited to striations 47
47a.   Cell elongate, sides almost parallel except for terminal knobs .... *Asterionella*
47b.   Sides of cell converging toward one end 48
     48a.   Cells bent in girdle view .... *Rhoicosphenia*
     48b.   Cells straight in girdle view .... *Gomphonema*
49a.   Valves with transverse septa or costae .... *Epithemia*

*Refer to
couplet
No.*

49b.   Valves with no transverse septa or costae 50
    50a.   Raphe located almost through center of valve .... *Cymbella*
    50b.   Raphe excentric, near concave edge of valve .... *Amphora*

3. Flagellate Algae

51a.   Cell in a loose, rigid conical sac (lorica); isolated or in a branching colony .... *Dinobryon*
51b.   Case or sac, if present, not conical; colony, if present, not branching 52
    52a.   Cells isolated or in pairs.. 53
    52b.   Cells in a colony of four or more cells 71
53a.   Prominent transverse groove encircles cell 54
53b.   Cell without transverse groove 56
    54a.   Cell with prominent rigid projections, one forward and two or three on posterior end .... *Ceratium*
    54b.   Cell without several rigid polar projections.. 55
55a.   Portions above and below transverse groove about equal .... *Peridinium*
55b.   Front portion distinctly larger than posterior portion .... *Massartia*
    56a.   Cell with long bristles extending from surface plates .... *Mallomonas*
    56b.   Cell without bristles and surface plates 57
57a.   Cell protoplast enclosed in loose, rigid covering (lorica) 58
57b.   Ccll with tight membrane or wall but no loose, rigid covering 60
    58a.   Lorica flattened; cell with two flagella .... *Phacotus*
    58b.   Lorica not flattened; cell with one flagellum 59
59a.   Lorica often opaque, generally dark brown to red; plastid green .... *Trachelomonas*
59b.   Lorica often transparent, colorless to light brown; plastid light brown .... *Chrysococcus*
    60a.   Plastids brown to red to olive or blue-green... 61
    60b.   Plastids grass green.. 64
61a.   Plastids blue-green to blue .... *Chroomonas*
61b.   Plastids brown to red to olive green 62
    62a.   Plastid brown; one or two flagella... 63
    62b.   Plastids red, red-brown, or olive green; two flagella .... *Rhodomonas*
63a.   Anterior end of cell oblique; two flagella .... *Cryptomonas*
63b.   Anterior end of cell rounded or pointed; one flagellum .... *Chromulina*
    64a.   Cell with colorless rectangular wing .... *Pteromonas*
    64b.   No wing extending from cell.. 65
65a.   Cells flattened; margin rigid .... *Phacus*
65b.   Cell not flattened; margin rigid or flexible 66
    66a.   Pyrenoid present in the single plastid; no paramylon; margin not flexible; two or more flagella per cell 67
    66b.   Pyrenoid absent; paramylon present; several plastids per cell; margin flexible or rigid; one flagellum per cell.... 70
67a.   Cells fusiform (tapering at each end) .... *Chlorogonium*
67b.   Cells not fusiform, generally almost spherical 68
    68a.   Plastids numerous .... *Vacuolaria*
    68b.   Plastids few, commonly one... 69
69a.   Two flagella per cell .... *Chlamydomonas*
69b.   Four flagella per cell .... *Carteria*
    70a.   Cell flexible in form; paramylon a capsule or disk; cell elongate .... *Euglena*
    70b.   Cell rigid in form; paramylon ring-shaped; cell almost spherical .... *Lepocinclis*
71a.   Plastids brown 72
71b.   Plastids green 73
    72a.   Cells in contact with one another .... *Synura*
    72b.   Cells separated from one another by space .... *Uroglenopsis*
73a.   Colony flat, one cell thick .... *Gonium*
73b.   Colony rounded, more than one cell thick 74
    74a.   Cells in contact with one another 75
    74b.   Cells separated from one another by space 76
75a.   Cells radially arranged .... *Pandorina*
75b.   Cells all facing one direction .... *Pyrobotrys (Chlamydobotrys)*
    76a.   Cells more than 400 per colony .... *Volvox*
    76b.   Cells less than 75 per colony .... *Eudorina*

4. Green Algae and Associated Forms

77a.   Cells jointed together to form a net .... *Hydrodictyon*
77b.   Cells not forming a net 78

78a. Cells attached side by side to form a plate or ribbon one cell wide and thick; number of cells commonly two, four, or eight .... *Scenedesmus*

78b. Cells not attached side by side 79

79a. Cells isolated or in nonfilamentous or nontubular thalli 80

79b. Cells in filaments or other tubular or threadlike thalli 111

80a. Cells isolated and narrowest at the center because of incomplete fissure (desmids)............................ 81

80b. Cells isolated or in clusters but without central fissure... 84

81a. Each half of cell with three spinelike or pointed knobular extensions .... *Staurastrum*

81b. Cell margin with no such extensions 82

82a. Semicells with a median incision or depression... 83

82b. Semicells with no median incision or depression .... *Cosmarium*

83a. Margin with rounded lobes .... *Euastrum*

83b. Margin with sharp-pointed teeth .... *Micrasterias*

84a. Cells elongate.. 85

84b. Cells round to oval or angular 92

85a. Cell radiating from a central point .... *Actinastrum*

85b. Cells isolated or in irregular clusters 86

86a. Cells with terminal spines .... *Schroederia*

86b. Cells without terminal spines.. 87

87a. Cells with colorless attachment area at one end .... *Characium*

87b. No attachment area at one end of cell 88

88a. Plastids two per cell; unpigmented area across center of cell .... *Closterium*

88b. Cell with plastid that continues across the center... 89

89a. Cell 5 to 10 times as long as it is broad 90

89b. Cell 2 to 4 times as long as it is broad 91

90a. Pyrenoid absent, or one per cell .... *Ankistrodesmus*

90b. Pyrenoids several per cell .... *Closteriopsis*

91a. Cells semicircular; cell ends pointed but with no terminal spines .... *Selenastrum*

91b. Cells arcuate but less than semicircular; cell ends pointed and each with a short spine .... *Closteridium*

92a. Cells regularly arranged in a tight, flat colony .... *Pediastrum*

92b. Cells not in a tight, flat regular colony... 93

93a. Cells angular 94

93b. Cells round to oval 95

94a. Two or more spines at each angle .... *Polyedriopsis*

94b. Spines none or less than two at each angle .... *Tetraedron*

95a. Cells with long, sharp spines 96

95b. Long, sharp spines absent 99

96a. Cells round... 97

96b. Cells oval... 98

97a. Cells isolated .... *Golenkinia*

97b. Cells in colonies .... *Micractinium*

98a. Each cell end has one spine .... *Diacanthos*

98b. Each cell end has more than one spine .... *Chodatella*

99a. Colony of definite regular form, round to oval 100

99b. Colony, if present, not a definite oval or sphere; or cells may be isolated 104

100a. Colony a tight sphere of cells... 101

100b. Colony a loose sphere of cells enclosed by a common membrane... 102

101a. Sphere solid, slightly irregular, no connecting processes between cells .... *Planktosphaeria*

101b. Sphere hollow, regular; short connecting processes between cells .... *Coelastrum*

102a. Cells round... 103

102b. Cells oval .... *Oocystis*

103a. Cells connected to center of colony by branching stalk .... *Dictyosphaerium*

103b. No stalk connecting cells .... *Sphaerocystis*

104a. Oval cells, enclosed in a somewhat spherical, often orange-colored matrix .... *Botryococcus*

104b. Cells round, isolated or in colorless matrix... 105

105a. Adjoining cells with straight, flat walls between their protoplasts 106

105b. Adjoining cells with rounded walls between their protoplasts 107

106a. Cells embedded in a common gelatinous matrix .... *Palmella*

106b. No matrix or sheath outside of cell walls .... *Phytoconis (Protococcus)*

107a. Cells loosely arranged in a large gelatinous matrix .... *Tetraspora*

107b. Cells isolated or tightly grouped in a small colony 108

108a. Cells located inside of protozoa .... *Zoochlorella*

108b. Cells not inside of protozoa 109

109a. Plastid filling ⅔ or less of cell 110

109b.   Plastid filling ¾ or more of cell .... *Chlorococcum*

    110a.   Cell diameter 2 μm or less; reproduction by cell division .... *Nannochloris*

    110b.   Cell diameter 2.5 μm or more; reproduction by internal spores .... *Chlorella*

111a.   Cells attached end to end in an unbranched filament 112

111b.   Thallus branched, or more than one cell wide 119

    112a.   Plastids in form of one or more marginal spiral ribbons .... *Spirogyra*

    112b.   Plastids not in form of spiral ribbons... 113

113a.   Filaments, when breaking, separating through middle of cells 114

113b.   Filaments, when breaking, separating irregularly or at ends of cells 115

    114a.   Starch test positive; cell margin straight; one plastid, granular .... *Microspora*

    114b.   Starch test negative; cell margin slightly bulging; several plastids .... *Tribonema*

115a.   Marginal indentations between cells .... *Desmidium*

115b.   No marginal indentations between cells 116

    116a.   Plastids, two per cell .... *Zygnema*

    116b.   Plastid, one per cell (sometimes appearing numerous).. 117

117a.   Some cells with walls having transverse wrinkles near one end; plastid an irregular net .... *Oedogonium*

117b.   No apical wrinkles in wall; plastid not porous 118

    118a.   Plastid a flat or twisted axial ribbon .... *Mougeotia*

    118b.   Plastid an arcuate marginal band .... *Ulothrix*

119a.   Thallus a flat plate of cells .... *Hildenbrandia*

119b.   Thallus otherwise 120

    120a.   Thallus a long tube without crosswalls .... *Vaucheria*

    120b.   Thallus otherwise 121

121a.   Thallus a leathery strand with regularly spaced swellings and a continuous surface membrane of cells .... *Lemanea*

121b.   Thallus otherwise 122

    122a.   Filament unbranched .... *Schizomeris*

    122b.   Filament branched... 123

123a.   Branches in whorls (clusters) 124

123b.   Branches single or in pairs 126

    124a.   Thallus embedded in gelatinous matrix .... *Batrachospermum*

    124b.   Thallus not embedded in gelatinous matrix.. 125

125a.   Main filament one cell thick .... *Nitella*

125b.   Main filament three cells thick .... *Chara*

    126a.   Most of filament surrounded by a layer of cells .... *Compsopogon*

    126b.   Filament not surrounded by a layer of cells 127

127a.   End cell of branches with a rounded or blunt-pointed tip 128

127b.   End cell of branches with a sharp-pointed tip 130

    128a.   Plastids green; starch test positive 129

    128b.   Plastids red; starch test negative .... *Audouinella*

129a.   Some cells dense, swollen, dark green (spores); other cells light green, cylindric .... *Pithophora*

129b.   All cells essentially alike, light to medium green, cylindric .... *Cladophora*

    130a.   Filaments embedded in gelatinous matrix... 131

    130b.   Filaments not embedded in gelatinous matrix 132

131a.   Cells of main filament much wider than even the basal cells of branches .... *Draparnaldia*

131b.   No abrupt change in width of cells from main filament to branches .... *Chaetophora*

    132a.   Branches very short, with no cross-walls .... *Rhizoclonium*

    132b.   Branches long, with cross-walls... 133

133a.   Branches ending in an abrupt spine having a bulbous base .... *Bulbochaete*

133b.   Branches gradually reduced in width, ending in a long pointed cell, with or without color .... *Stigeoclonium*

10900 D. Index to Illustrations

10900 E.   Selected Taxonomic References

The most useful references for the nonspecialist are listed below. These references are primarily regional and will aid the biologist in the identification of both freshwater and marine plants and animals. Each reference is listed once under the broadest classification, i.e., a general reference to the identification of invertebrates is listed under "Invertebrates, General" and is not repeated under each individual phylum. As a rule, more academic and specialized reports on a single genus or family are not listed; however, these often are listed in the bibliography of a cited reference.

1. General, Introductory

MINER, R.W. 1950. Field Book of Seashore Life. G.P. Putnam's Sons, New York, N.Y.

DAVIS, C.C. 1955. The Marine and Fresh-Water Plankton. Michigan State Univ. Press, East Lansing.

EDMONDSON, W.T., ed. 1959. Ward and Whipple's Fresh Water Biology, 2nd ed. John Wiley & Sons, New York, N.Y.

NEEDHAM, J.G. & P.R. NEEDHAM. 1962. A Guide to the Study of Fresh-Water Biology, 5th ed. Holden-Day Inc., San Francisco, Calif.

NEWELL, G.E. & R.C. NEWELL. 1963. Marine Plankton, A Practical Guide. Hutchinson Educational Ltd., London, England.

REID, G.K. 1967. Pond Life. A Guide to Common Plants and Animals of North American Ponds and Lakes. Golden Press, New York, N.Y.

VOSS, G.L. 1976. Seashore Life of Florida and the Caribbean. E.A. Seemann Publ. Inc., Miami, Fla.

STERRER, W., ed. 1986. Marine Fauna and Flora of Bermuda. John Wiley & Sons, New York, N.Y.

RICKETTS, E.F., et al. 1988. Between Pacific Tides, 5th ed. Revised by D.W. Phillips. Stanford Univ. Press, Stanford, Calif.

DAILEY, M.D., D.J. REISH & J.W. ANDERSON, eds. 1994. Ecology of the Southern California Bight. Univ. California Press, Berkeley.

REISH, D.J. 1995. Marine Life of Southern California, 2nd ed. Kendall/Hunt Publ. Co., Dubuque, Iowa.

2. Algae

General

TAYLOR, W.R. 1957. Marine Algae of the Northeastern Coast of North America, 2nd ed. Univ. Michigan Press, Ann Arbor.

PALMER, C.M. 1959. Algae in Water Supplies. U.S. Pub. Health Serv. Publ. No. 657, Washington, D.C.

ABBOTT, I.A. & G.J. HOLLENBERG. 1976. Marine Algae of California. Stanford Univ. Press, Stanford, Calif.

EDWARDS, P. 1976. Illustrated Guide to the Seaweeds and Sea Grasses in the Vicinity of Port Arkansas, Texas. Univ. Texas Press, Austin.

ABBOTT, I.A. & E.Y. DAWSON. 1978. How to Know the Seaweeds, 2nd ed. Wm. C. Brown Co., Dubuque, Iowa.

PRESCOTT, G.W. 1978. How to Know the Fresh Water Algae, 3rd ed. Wm. C. Brown Co., Dubuque, Iowa.

HUMM, H.J. 1979. The Marine Algae of Virginia. Univ. Virginia Press, Charlottesville.

ROUND, F.E. 1981. The Ecology of Algae. Cambridge Univ. Press, New York, N.Y.

ESSER, K. 1982. Cryptograms: Cyanobacteria, Algae, Fungi, Lichens. Cambridge University Press, New York, N.Y.

BOLD, H.C. & M.J. WYNNE. 1985. Introduction to the Algae: Structure and Reproduction, 2nd ed. Prentice-Hall, Inc., Englewood Cliffs, N.J.

ABBOTT, I.A., ed. 1985–1995. Taxonomy of Economic Seaweeds, Vols. 1–5. California Sea Grant College System, Univ. California, La Jolla.

DILLARD, G.E. 1989–1993. Freshwater Algae of the Southeastern United States. Parts 1–6. Cramer-Borntraeger, Stuttgart, Germany.

SCHNEIDER, C.W. & R.B. SEARLES. 1991. Seaweeds of the Southeastern United States: Cape Hatteras to Cape Canaveral. Duke Univ. Press, Durham, N.C.

STEWART, J.G. 1991. Marine Algae and Seagrasses of San Diego County, California. California Sea Grant College System, Univ. California, La Jolla.

CANTER-LUND, H. & J.W.G. LUND. 1995. Freshwater Algae: Their Microscopic World Explored. Biopress Limited, Bristol, U.K.

VAN DEN HOEK, C., D.G. MANN & H.M. JAHNS. 1995. Algae: An Introduction to Phycology. Cambridge University Press, Cambridge, U.K.

TOMAS, C.R., ed. 1997. Identifying Marine Phytoplankton. Academic Press, Orlando, Fla.

Blue-Green Algae

WELCH, H. 1964. An introduction to the bluegreen algae, with a dichotomous key to all the genera. Limnol. Soc. S. Afr. News Letter 1:25.

FOGG, G.E., W.D.P. STEWART, P. FAY & A.E. WASLBY. 1973. The Blue-Green Algae. Academic Press, London, England.

HUMM, H.J. & S.B. WICKS. 1980. Introduction and Guide to the Marine Blue-Green Algae. Wiley-Interscience, Somerset, England.

VANLANDINGHAM, S.L. 1982. Guide to the Identification, Environmental Requirements and Pollution Tolerance of Blue-Green Algae (Cyanophyta). EPA-600/3-82-073, U.S. Environmental Protection Agency.

ANAGNOSTIDIS, K. & J. KOMAREK. 1985–1990. Modern Approach to the Classification System of Cyanophytes. I. Introduction. Arch. Hydrobiol. Suppl. 71:291; 2. Chroococcales. Arch. Hydrobiol. Suppl. 73:157; 3. Oscillatoriales. Arch. Hydrobiol. Suppl. 80:327; 4. Nostocales. Algol. Stud. 56:247; 5. Stigonematales. Algol. Stud. 59:1.

Green Algae

PICKETT-HEAPS, J.D. 1975. Green Algae: Structure, Reproduction and Evolution in Selected Genera. Sinauer Assoc., Inc., Sunderland, Mass.

PRESCOTT, G.W., H.T. CROASDALE, W.C. VINYARD, C.E. BICUDO & M. BICUDO. 1975-1983. A Synopsis of North American Desmids. Part II. Desmidiaceae: Placodermae. Sections 1–6. Univ. Nebraska Press, Lincoln.

DILLARD, G.E. 1989–1993. Freshwater Algae of the Southeastern United States. Parts 1–6. Biblio. Phycol. Bands 81, 83, 85, 89, 90 & 93.

Red Algae

WRAY, J.L. 1977. Calcareous Algae. Elsevier Science Publishing Co., Amsterdam, Netherlands.

KAPRAUN, D.F. 1980. An Illustrated Guide to the Benthic Marine Algae of Coastal North Carolina. I. Rhodophyta. Univ. North Carolina Press, Chapel Hill.

Phytoplankton and Diatoms

HUSTEDT, F. 1955. Marine littoral diatoms, Beaufort, North Carolina. Duke Univ. Mar. Sta. Bull. 6:5.

PATRICK, R. & C.W. REIMER. 1966. The Diatoms of the United States. Vol. 1. Philadelphia Acad. Natur. Sci. Monogr. No. 13, Philadelphia, Pa.

WEBER, C.I. 1971. A Guide to the Common Diatoms at Water Pollution Surveillance System Stations. U.S. Environmental Protection Agency, National Environmental Research Center, Cincinnati, Ohio.

PATRICK, R. & C.W. REIMER. 1975. The Diatoms of the United States Exclusive of Alaska and Hawaii. Vol. 2, Fragilariaceae, Eunotiaceae, Acanthaceae, Naviculaceae. Philadelphia Acad. Natur. Sci. Monogr. No. 13, Philadelphia, Pa.

SOURNIA, A. 1978. Phytoplankton Manual. Monogr. on Oceanographic Methodology No. 6, United Nations Educational, Scientific & Cultural Org., Paris.

BELCHER, H. & E. SWALE. 1979. An Illustrated Guide to River Phytoplankton. Her Majesty's Stationery Off., London, England.

VINYARD, W.C. 1980. Diatoms of North America. Mad River Press, Eureka, Calif.

SPECTOR, D., ed. 1984. Dinoflagellates. Academic Press, Orlando, Fla.

MARSHALL, H.G. 1986. Identification Manual for Phytoplankton of the United States Atlantic Coast. EPA-600/4-86-003, U.S. Environmental Protection Agency, Cincinnati, Ohio.

HARRIS, G.P. 1986. Phytoplankton Ecology, Structure, Function and Fluctuation. Chapman and Hall, New York, N.Y.

KRAMMER, K. & H. LANGE-BERTALOT. 1986–1991. Bacillariophyceae. Teil 1–4. Gustav Fischer Verlag, Stuttgart, Germany.

CHRETIENNOT-DINET, M.J., A. SOURNIA, M. RICARD & C. BILLARD. 1993. A Classification of the Marine Phytoplankton of the World from Class to Genus. Phycologia 32:159.

ROUND, F.E., R.M. CRAWFORD & D.G. MANN. 1990. The Diatoms: Biology and Morphology of the Genera. Cambridge Univ. Press, Cambridge, U.K.

SIMS, P.A., ed. 1996. An Atlas of British Diatoms. Biopress Ltd., Bristol, U.K.

3. Fungi

General

COOKE, W.B. 1986. The Fungi of "Our Mouldy Earth." Beiheft Nova Hedwigia Berlin, Stuttgart 85:1.

MOSS, S.T., ed. 1986. The Biology of Marine Fungi. Cambridge Univ. Press, New York, N.Y.

KOHLMEYER, J. & B. VOLKMANN-KOHLMEYER. 1991. Illustrated key to the filamentous higher fungi. Botanica Marina 34:1.

CARLILE, M.J. & S. WATKINSON. 1995. The Fungi. Academic Press, Orlando, Fla.

Phycomycetes

SPARROW, F.K. 1960. Aquatic Phycomycetes, 2nd ed. Univ. Michigan Press, Ann Arbor.

DICK, M.W. 1969. Morphology and taxonomy of the Oomycetes, with special reference to Saprolegniaceae, Leptomytaceae and Pythiaceae. 1. Sexual reproduction. New Phytol. 68:751.

DICK, M.W. 1973. Saprolegniales. In G.C. Ainsworth, F.K. Sparrow & A.S. Sussman, eds. The Fungi. Vol. IV B. Academic Press, New York, N.Y.

SPARROW, F.K. 1973. Chytridiomycetes, Hyphochytridiomycetes. In G.C. Ainsworth, F.K. Sparrow & A.S. Sussman, eds. The Fungi. Vol. IV B. Academic Press, New York, N.Y.

KARLING, J.S. 1977. Chytridiomycetarum Iconographia. J. Cramer, Germany.

BARR, D.J.S. 1978. Taxonomy and phylogeny of Chytrids. *BioSystems* 10:153.

KARLING, J.S. 1980. The Simple Biflagellate Holocarpic Phycomycetes, 2nd ed. J. Cramer, Germany.

Ascomycetes

INGOLD, C.T. 1954. Aquatic Ascomycetes: Discomycetes from lakes. *Trans. Brit. Mycol. Soc.* 37:1.

INGOLD, C.T. 1955. Aquatic Ascomycetes: Further species from the English Lake District. *Trans. Brit. Mycol. Soc.* 38:157.

INGOLD, C.T. 1976. The morphology and biology of freshwater fungi excluding the Phycomycetes. *In* E.B. Gareth Jones, ed. Recent Advances in Aquatic Mycology. John Wiley & Sons, New York, N.Y.

Fungi Imperfecti

RANZONI, F.V. 1953. The aquatic Hyphomycetes of California. *Farlowia* 4:353.

CRANE, J.L. 1968. Freshwater Hyphomycetes of the northern Appalachian highlands including New England and three coastal plain states. *Amer. J. Bot.* 55:996.

KENDRICK, W.B., ed. 1971. Taxonomy of Fungi Imperfecti. Univ. Toronto Press, Toronto, Ont.

KENDRICK, W.B. & J.W. CARMICHAEL. 1973. Hyphomycetes. *In* G.C. Ainsworth, F.K. Sparrow & A.S. Sussman, eds. The Fungi, An Advanced Treatise. Vol. IV A. Academic Press, New York, N.Y.

INGOLD, C.T. 1975. An Illustrated Guide to Aquatic and Waterborne Hyphomycetes (Fungi Imperfecti) with Notes on their Biology. Freshwater Biol. Assoc. Sci. Publ. 30.

INGOLD, C.T. 1976. The morphology and biology of freshwater fungi excluding the Phycomycetes. *In* E.B. Gareth Jones, ed. Recent Advances in Aquatic Mycology. John Wiley & Sons, New York, N.Y.

CARMICHAEL, J.W., W.B. KENDRICK, I.I. CONNORS & L. SIGLER. 1980. Genera of Hyphomycetes. Univ. Alberta Press, Edmonton.

4. Higher Plants

FERNALD, M.L. 1950. Gray's Manual of Botany, 8th ed. D. Van Nostrand Co., New York, N.Y.

FASSETT, N.C. 1960. A Manual of Aquatic Plants (with a revision appendix by E.C. Ogden). Univ. Wisconsin Press, Madison.

STEWARD, A.N., L.R. DENNIS & H.M. GILKEY. 1963. Aquatic Plants of the Pacific Northwest, with Vegetative Keys, 2nd ed. Oregon State Univ. Press, Corvallis.

WINTERRINGER, G.S. & A. LOPINOT. 1966. Aquatic Plants of Illinois. Vol. VI. Popular Sci. Ser. Illinois State Mus., Springfield, Ill.

MASON, H.L. 1969. Flora of the Marshes of California. Univ. California Press, Berkeley.

CORRELL, D.S. & H.B. CORRELL. 1972. Aquatic and Wetland Plants of Southwestern United States. 16030 DNL, U.S. Environmental Protection Agency.

VOSS, E.G. 1972. Michigan Flora. Part I, Gymnosperms and Monocots. Cranbrook Inst. Science, Bloomfield Hills, Mich.

MOUL, E.T. 1973. Higher Plants of the Marine Fringe. Marine Flora and Fauna of the Northeastern U.S. Circ. 384, National Oceanic Atmospheric Admin., National Marine Fisheries Serv., U.S. Government Printing Off., Washington, D.C.

BEAL, E.O. 1977. A Manual of Marsh and Aquatic Vascular Plants of North Carolina. Tech. Bull. No. 247, North Carolina Agriculture Exper. Sta., Raleigh, N.C.

GODFREY, R.K. & J.W. WOOTEN. 1979-1981. Dicotyledons. Aquatic and Wetland Plants of Southeastern United States, Volume 1: Monocotyledons. Volume 2: Univ. Georgia Press, Athens.

HELLQUIST, C.B. & C.E. CROW. 1980-1984. Aquatic Vascular Plants of New England. Part I, Zosteraceae, Potamogetonaceae, Zannichelliaceae, Najadaceae; Part 3, Alismataceae; Part 5, Araceae, Lemnaceae, Xyridaceae, Eriocaulaceae, Pontederiaceae; Part 7, Cambom-

baceae, Nymphaeaceae, Nelumbonaceae, Certophyllaceae. Bull. No. 515, 518, 523, 527, New Hampshire Agricultural Experiment Sta., Durham.

CROW, G.E. & C.B. HELLQUIST. 1981-1985. Aquatic Vascular Plants of New England. Part 2, Typhaceae and Sparganiaceae; Part 4, Juncaginaceae, Scheuchzeriaceae, Butomaceae, Hydrocharitaceae; Part 6, Trapaceae, Haloragaceae, Hippuridaceae; Part 8, Lentibulariaceae. Bull. No. 517, 520, 524, 528, New Hampshire Agricultural Experiment Sta., Durham.

VOSS, E.G. 1985. Michigan Flora. Part II, Dicots (Saururaceae-Cornaceae). Cranbrook Inst. Science, Bloomfield Hills, Mich.

GLEASON, H.A. & A. CRONQUIST. 1991. Manual of Vascular Plants of Northeastern United States and Adjacent Canada, 2nd ed. New York Botanical Garden, New York, N.Y.

5. Invertebrates, General

HYMAN, L.H. 1940-67. The Invertebrates. Vols. 1-6. McGraw-Hill, New York, N.Y.

GOSNER, K.L. 1971. Guide to Identification of Marine and Estuarine Invertebrates; Cape Hatteras to the Bay of Fundy. Wiley-Interscience, New York, N.Y.

HULINGS, N.C. & J.S. GRAY. 1971. A Manual for the Study of Meiofauna. Smithsonian Contrib. Zool. No. 78.

WATLING, L. & D. MAURER. 1973. Guide to the Macroscopic Estuarine and Marine Invertebrates of the Delaware Bay Region. Vol. 5. College Marine Studies, Univ. Delaware, Newark.

SMITH, R.I. & J.T. CARLTON, eds. 1975. Light's Manual: Intertidal Invertebrates of the Central California Coast, 3rd ed. Univ. California Press, Berkeley.

MILNE, L. & M. MILNE. 1976. Invertebrates of North America. Doubleday & Co., New York, N.Y.

BRUSCA, R.C. 1980. Common Invertebrates of the Gulf of California, 2nd ed. Univ. Arizona Press, Tucson.

MORRIS, R.H., D.P. ABBOTT & E.C. HADERLIE. 1980. Intertidal Invertebrates of California. Stanford Univ. Press, Stanford, Calif.

FOX, J.C., P.R. FITZGERALD & C. LUE-HING. 1981. Sewage Organisms: A Color Atlas. Lewis Publ., Chelsea, Mich.

HEARD, R.W. 1982. Guide to Common Tidal Marsh Invertebrates of the Northeastern Gulf of Mexico. Mississippi Alabama Sea Grant Consortium, MASGP-79.

KOZLOFF, E.N. 1987. Marine Invertebrates of the Pacific Northwest. Univ. Washington Press, Seattle.

RUPPERET, E. & R. FOX. 1988. Seashore Animals of the Southeast: A Guide to Common Shallow-Water Invertebrates of the Southeastern Atlantic Coast. Univ. South Carolina Press, Columbia.

PENNAK, R.W. 1989. Freshwater Invertebrates of the United States, 3rd ed. John Wiley & Sons, New York, N.Y.

PECKARSKY, B.L., P.R. FRAISSINET, M.A. PENTON & D.J. CONKLIN, JR., eds. 1990. Freshwater Macroinvertebrates of Northeastern North America. Cornell Univ. Press, Ithaca, N.Y.

THORP, J.H. & A.P. CORRICH. 1991. Ecology and Classification of North American Freshwater Invertebrates. Academic Press, Orlando, Fla.

BARNES, R.S.K. 1994. The Brackish-Water Fauna of Northwestern Europe. An Identification Guide to Brackish-Water Habitats, Ecology and Macrofauna for Field Workers, Naturalists and Students. Cambridge Univ. Press, Cambridge, U.K.

HAYWARD, R. & J.S. RYLAND. 1995. Handbook of the Marine Fauna of North-West Europe. Oxford Univ. Press, Oxford, U.K.

HUBBARD, M.D., M.L. PESCADOR, A.K. RASMUSSEN, I.S. ASKEVOLD, J. JONES, R.W. FLOWERS & J.H. EPLER. 1995. The freshwater macroinvertebrates of Florida: A guide to references for their identification. *Florida Entomol.* 78:161.

GOSLINER, T.M., D.W. BEHRENS & G.C. WILLIAMS. 1996. Coral Reef Animals of the Indo-Pacific—Animal Life from Africa to Hawaii Exclusive of the Vertebrates. Sea Challengers, Inc., Monterey, Calif.

6. Protozoa

CUSHMAN, J.A. 1948. Foraminifera, 4th ed. Harvard Univ. Press, Cambridge, Mass.

JAHN, T.L. & F.F. JAHN. 1949. How to Know the Protozoa. Wm. C. Brown Co., Dubuque, Iowa.

KUDO, R.R. 1966. Protozoology, 5th ed. C.C. Thomas, Springfield, Ill.

CURDS, C.R. 1969. An Illustrated Key to the British Freshwater Ciliated Protozoa Commonly Found in Activated Sludge. Water Pollution Res. Tech. Paper No. 12. Majesty's Stationary Office, London.

LEWIS, K.B. 1970. A key to the recent genera of the foraminiferida. New Zealand Oceanogr. Inst. Mem. No. 45.

BICK, H. 1972. Ciliated Protozoa. An Illustrated Guide to the Species Used as Bacteriological Indicators in Fresh Water Biology. World Health Org., Geneva.

PAGE, F.C. 1976. An Illustrated Key to Freshwater and Soil Amoeba. Freshwater Biol. Assoc. Sci. Pub. 34, Ambleside, U.K.

CORLISS, J.O. 1979. The Cilated Protozoa: Characterization, Classification, and Guide to the Literature, 2nd ed. Pergamon Press, New York, N.Y.

FINLAY, B.J. & C. OCHSENBEIN-GATLLEN. 1982. Ecology of Free-Living Protozoa. Occas. Publ. Freshwater Biol. Assoc., Ambleside, U.K.

MUNSON, D.A. 1992. Marine amoebae from Georgia coastal surface waters. *Trans. Amer. Microsc. Soc.* 111:360.

LAYBOURN-PARRYM, J. 1992. Protozoan Plankton Ecology. Chapman and Hall, London, U.K.

CARY, P.C. 1992. Marine Interstitial Ciliates—An Illustrated Key. Chapman and Hall, New York, N.Y.

PATTERSON, D.J. 1996. Free-living Freshwater Protozoa. Oxford Univ. Press, Oxford, U.K.

7. Porifera

DELAUBENFELS, M.W. 1932. The marine and fresh water sponges of California. *Proc. U.S. Nat. Mus.* 81, Art. 4.

PENNEY, J.T. & A.A. RACEK. 1968. Comprehensive Revision of a Worldwide Collection of Freshwater Sponges (Porifera: Spongillidae). U.S. Nat. Mus. Bull. 272.

ACKERS, R.G., D. MOSS, B.E. PECTON & S.M. STONE. 1985. Sponges of the British Isles. Marine Conservation Soc., Rosson Wye, U.K.

SMITH, D.G. 1990. Keys to the freshwater macroinvertebrates of Massachusetts (No. 5): Porifera: Spongillidae (freshwater sponges). Massachusetts Dep. Environmental Protection, Div. Water Cont., Westborough.

RUETZLER, K., ed. 1990. New Perspectives in Sponge Biology. Smithsonian Inst. Press, Washington, D.C.

DAWSON, E.W. 1993. The Marine Fauna of New Zealand. Index to the Fauna: 2. Porifera. New Zealand Oceanogr. Inst. Mem. No. 100, Wellington, New Zealand.

GREEN, K. & G. BAKUS. 1993. The Porifera. *In* Taxonomic Atlas of the Benthic Fauna of the Santa Maria Basin and the Western Santa Barbara Channel. Santa Barbara Mus. Natural History, Santa Barbara, Calif.

8. Cnidaria

FRASER, C.M. 1937. Hydroids of the Pacific Coast of Canada and the United States. Univ. Toronto Press, Toronto, Ont.

FISHER, W.K. 1938. Hydrocorals of the North Pacific Ocean. *Proc. U.S. Nat. Mus.* 84:493.

CARLGREN, O. 1952. Actiniaria from North America. *Arkiv. Zool.* (ser. 2) 3:373.

FRASER, C.M. 1954. Hydroids of the Atlantic Coast of North America. Univ. Toronto Press, Toronto, Ont.

HAND, C. 1954–1955. The Sea Anemones of Central California. Parts 1–3. *Wasmann J. Biol.* 12:345; 13:37; 13:189.

CALDER, D.R. 1971. Hydroids and Hydromedusae of Southern Chesapeake Bay. Va. Inst. Mar. Sci., Special Papers in Marine Science No. 1.

SMITH, F.G.W. 1971. Atlantic Reef Corals. Univ. Miami Press, Miami, Fla.

CAIRNS, S.D. 1986. A Revision of the Northwest Atlantic Stylasteridae (Coelenterata: Hydrozoa). Smithsonian Contrib. Zool. No. 418.

CAMPBELL, R.D. 1987. A new species of *Hydra* (Cnidaria: Hydrozoa) from North America with comments on species within the genus. *Zool. J. Linn. Soc.* 91:253.

MANUEL, R.L. 1988. British Anthozoa, 2nd ed. Synopsis of British Fauna. Field Council Studies, Shrewsbury, U.K.

LARSON, R.J. 1990. Scyphomedusae and Cubomedusae from the Eastern Pacific. *Bull. Mar. Sci.* 47:546.

9. Rotifera

DONNER, J. 1966. Rotifers. Warne, London & New York.

PONTIN, R.M. 1978. A Key to the Freshwater Planktonic and Semi-Planktonic Rotifera of the British Isles. Freshwater Biol. Soc. Assoc. Publ. 38, Ambleside, U.K.

SLÁDEČEK, V. 1983. Rotifers as indicators of water quality. *Hydrobiologia* 109:169.

STEMBERGER, R.S. 1990. An inventory of rotifer species diversity of Northern Michigan inland lakes. *Arch. Hydrobiol.* 118:283.

10. Platyhelminthes

HYMAN, L.H. 1953. The polyclad flatworms of the Pacific Coast of North America. *Bull. Amer. Mus. Nat. Hist.* 100:269.

BALL, I.R., et al. 1981. The Planarians (Turbellaria) of Temporary Waters in Eastern North America. Life Sci. Contrib. Royal Ont. Mus. No. 127.

CANNON, L.R.G. 1986. Turbellaria of the World: A Guide to Families and Genera. Queensland Mus., Brisbane, Australia.

KOLASA, J., et al. 1987. Microturbellarian from interstitial waters, streams, and springs in Southeastern New York. *J. North Amer. Benthol. Soc.* 6:125.

HILBIG, B. & J.A. BLAKE. 1993. Platyhelminthes (Class Turbellaria). *In* Taxonomic Atlas of the Benthic Fauna of the Santa Maria Basin and Western Santa Barbara Channel, Vol. 1, Part 5. Santa Barbara Mus. Natural History, Santa Barbara, Calif.

11. Nematoda

HOPE, W.D. & D.G. MURPHY. 1972. A taxonomic hierarchy and check list of the genera and higher taxa of marine nematodes. Smithsonian Contrib. Zool. No. 137.

FERRIS, V.R., et al. 1973. Genera of Freshwater Nematodes (Nematoda) of Eastern North America. Biota Freshwater Ecosystems Identification Manual No. 10.

TARJAN, A.C., R.P. ESSER & S.L. CHANG. 1977. An illustrated key to the nematodes found in fresh water. *J. Water Pollut. Control Fed.* 49:2318.

PLATT, H.M. & R.M. WARWICK. 1988. Free-living Marine Nematodes (Part II). Synopsis of British Fauna. Field Studies Council, Shrewsbury, U.K.

12. Nemertea

CORREA, D.D. 1964. Nemerteans from California and Oregon. *Proc. Calif. Acqd. Sci.* 31:19.

GIBSON, R. & J.O. YOUNG. 1976. Freshwater Nemerteans. *J. Linn. Soc. Zool.* 58:177.

BLAKE, J.A. 1993. Phylum Nemertea. *In* Taxonomic Atlas of the Benthic Fauna of the Santa Maria Basin and Western Santa Barbara Channel, Vol. 1, Part 6, Santa Barbara Mus. Natural History, Santa Barbara, Calif.

13. Nematomorpha

CHANDLER, C.M. 1985. Horsehair worms (Nematomorpha, Gordioidea) from Tennessee, with a review of the taxonomy and distribution in the United States. *J. Tenn. Acad. Sci.* 60:50.

14. Gastrotricha

BRONSON, K.B. 1950. An introduction to the taxonomy of the Gastrotricha with a study of eighteen species from Michigan. *Trans. Amer. Microsc. Soc.* 69:325.

15. Annelida

General

KLEMM, D.J., ed. 1985. A Guide to the Freshwater Annelida (Polychaeta, Naidid and Tubificid Oligochaeta, and Hirudinea of North America). Kendall/Hunt Publ. Co., Dubuque, Iowa.

DAVIS, R.W. 1991. Annelida: Leeches, Polychaetes and Acanthobdellids. *In* J.H. Thorp & A.P. Covich, eds. Ecology and Classification of North American Freshwater Invertebrates. Academic Press, San Diego, Calif.

BLAKE, J.A., C. ERSEUS & B. HILBIG. 1994. Annelida (Oligochaeta & Polychaeta). *In* Taxonomic Atlas of the Benthic Fauna of the Santa Maria Basin and the Western Santa Barbara Channel, Vol. 2. Santa Barbara Mus. Natural History, Santa Barbara, Calif.

Polychaeta

PETTIBONE, M.H. 1963. Marine Polychaete Worms of the New England Region. 1. Aphroditae through Trochochaetidae. U.S. Nat. Mus. Bull. 227.

HARTMAN, O. 1968. Atlas of the Errantiate Polychaetous Annelids from California. Allan Hancock Foundation, Univ. Southern California, Los Angeles.

HARTMAN, O. 1969. Atlas of the Sedentariate Polychaetous Annelids from California. Allan Hancock Foundation, Univ. Southern California, Los Angeles.

DAY, J.H. 1973. New Polychaeta from North Carolina, with a Key to All Species Recorded from North Carolina. Circ. 375, National Oceanic Atmospheric Admin., National Marine Fisheries Serv., Seattle, Wash.

BANSE, K. & K.D. HOBSON. 1974. Benthic Errantiate Polychaetes of British Columbia and Washington. Fish. Res. Bd. Can. Bull. No. 185.

FAUCHALD, K. 1977. The Polychaete Worms; Definitions and Keys to the Order, Families, and Genera. Los Angeles Co. Mus. Natur. Hist. Sci. Ser. 28, Los Angeles, Calif.

HOBSON, K.D. & K. BANSE. 1981. Sedentariate and Archiannelid Polychaetes of British Columbia and Washington. Can. Fish. Aquatic Sci. Bull. No. 209.

UEBELACKER, J.M. & B.G. JOHNSON. 1984. Taxonomic Guide to the Polychaetous Annelids of the Northern Gulf of Mexico. Barry A. Vittor & Assoc., Mobile, Ala.

BLAKE, J.A., B. HILBIG & P.H. SCOTT, eds. 1995. Polychaeta. *In* Taxonomic Atlas of the Benthic Fauna of the Santa Maria Basin and Western Santa Barbara Channel, Vols. 5(2), 6(3), and 7(4). Santa Barbara Mus. Natural History, Santa Barbara, Calif.

Oligochaeta and Hirudinea

BRINKHURST, R.O. & B.G.M. JAMIESON. 1971. Aquatic Oligochaeta of the World. Oliver & Boyd, Edinburgh.

HILTUNEN, J.K. & D.J. KLEMM. 1980. A Guide to the Naididae (Annelida: Clitellata: Oligochaeta) of North America. EPA-600/4-80-031, U.S. Environmental Protection Agency, Cincinnati, Ohio.

BRINKHURST, R.O. 1982. British and other Marine and Estuarine Oligochaetes. Linn. Soc. Synopsis Brit. Fauna, New Ser. No. 21.

KLEMM, D.J. 1982. Leeches (Annelida: Hirudinea) of North America.
EPA-600/3-82-025, U.S. Environmental Protection Agency, Cincinnati, Ohio.

STIMPSON, K.S., D.J. KLEMM & J.K. HILTUNEN. 1982. A Guide to the Freshwater Tubificidae (Annelida: Clitellata: Oligochaeta) of North America. EPA-600/3-82-033, U.S. Environmental Protection Agency, Cincinnati, Ohio.

BRINKHURST, R.O. 1986. Guide to the Freshwater Aquatic Microdrile Oligochaeta of North America. Can. Fish. Aquatic Sci. Spec. Bull. No. 87.

SAWYER, R.T. 1986. Leech Biology and Behavior. II. Feeding Biology, Ecology, and Systematics. Oxford Univ. Press, New York, N.Y.

HENDRIX, P.F. 1995. Earthworm Ecology and Biography in North America. CRC Press, Inc., Boca Raton, Fla.

KLEMM, D.J. 1995. Identification Guide to the Freshwater Leeches (Annelida: Hirudinea) of Florida and Other Southern States. Florida Dep. Environmental Protection, Tallahassee.

Branchiobdella

HOLT, P.C. 1986. Newly established families of the order Branchiobdellida (Annelida: Clitellata) with a synopsis of the genera. *Proc. Biol. Soc. Wash.* 99:676.

16. Sipuncula and Echiura

STEPHEN, A.C. & S.J. EDMONDS. 1982. The Phyla Sipuncula and Echiura. British Mus. Natural History, London.

RICE, M.E. 1993. Two new species of *Phascolion* (Sipuncula Phascolionidae) from tropical and subtropical waters of the Central Western Atlantic. *Proc. Biol. Soc. Wash.* 106:591.

CUTLER, E.B. 1995. The Sipuncula. Their Systematics, Biology and Evolution. Cornell Univ. Press, Ithaca, N.Y.

17. Crustaceans

General

McLAUGHLIN, P.A. 1980. Comparative Morphology of Recent Crustacea. W.H. Freeman and Company, San Francisco, Calif.

FITZPATRICK, J.F. 1983. How to Know the Freshwater Crustaceans. W.C. Brown, Dubuque, Iowa.

SCHRAM, F.R. 1986. Crustacea. Oxford Univ. Press, New York, N.Y.

DAHMS, H.U. 1993. Pictorial keys for the identification of crustacean nauplii from the marine meiobenthos. *J. Crustacean Biol.* 23:609.

Branchiopoda

BROOKS, J.L. 1957. The systematics of North American Daphnia. Mem. Conn. Acad. Sci. 13.

BELK, D. 1975. Key to the Anostraca (fairy shrimps) of North America. *Southwest Nat.* 20:91.

FREY, D.G. 1987. The Taxonomy and Biogeography of the Cladocera. *Hydrobiologia* 145:5.

FREYER, G. 1987. A new classification of the branchiopod Crustacea. *Zool. J. Linn. Soc.* 91:357.

Ostracoda

DELORME, L.D. 1970–1971. Freshwater Ostracodes of Canada. Parts I–VI. *Can. J. Zool.* 48:153, 253, 1099; 49:49.

KORNICKER, L.S. 1981. Revision, Distribution, Ecology, and Ontogeny of the Ostracoda Subfamily Cyclasteropinae (Myodocopina: Cylindroleberidae). Smithsonian Contrib. Zool. No. 319.

KORNICKER, L.S. 1986. Sarsiellidae of the Western Atlantic and Northern Gulf of Mexico, and Revision of the Sarsiellinae (Ostracoda: Myodocopina). Smithsonian Contrib. Zool. No. 415.

KORNICKER, L.S. 1986. Cylindroleberididae of the Western North Atlantic, Northern Gulf of Mexico, and Zoogeography of the Myodocopina (Ostracoda). Smithsonian Contrib. Zool. No. 425.

ANGEL, M.V. 1993. Marine Planktonic Ostracoda. Synopsis of British Fauna. Field Council Studies, Shrewsbury, U.K.

Copepoda

HARDING, J.P. & W.A. SMITH. 1974. A Key to the British Freshwater Cyclopid and Calanoid Copepods, with Ecological Notes. Freshwater Biol. Assoc. Sci Publ. 18, Ambleside, U.K.

COULL, B.C. 1977. Copepoda: Harpactioidea. Marine Flora and Fauna of the Northeastern U.S. Circ. 399, National Oceanic Atmospheric Admin., National Marine Fisheries Serv., U.S. Government Printing Off., Washington, D.C.

DAWSON, J.K. & G. KNATZ, 1980. Illustrated key to the planktonic copepods of San Pedro Bay, California. *Tech. Rep. Allan Hancock Foundation* 2:1.

ISHIDA, T. 1987. Freshwater harpacticoid copepods of Hokkaido, northern Japan. *Sci. Rep. Hokkaido Salmon Hatchery* 41:77.

REID, J.W. 1988. Copepoda (Crustacea) from a seasonally flooded marsh in Rock Creek Regional Park, Maryland. *Proc. Biol. Soc. Wash.* 101:31.

Cirripedia

CORNWALL, I.E. 1955. The Barnacles of British Columbia. B.C. Prov. Mus. Handbook No. 7.

NEWMAN, W.A. 1976. Revision of the Balanomorph barnacles; including a catalog of the species. *San Diego Soc. Natur. Hist. Mem.* 9:1.

Leptostraca

MARTIN, J.W., E.W. VETTER & C.E. CASHCLARK. 1996. Description, external morphology and natural history observations of *Nebalia hessleri*; New species (Phyllocarida: Leptostraca), from Southern California, with a key to the extant families of the Leptostraca. *J. Crustacean Biol.* 16:347.

Cumacea

LIE, U.M. 1969. Cumacea from Puget Sound and off the NW coast of Washington, with descriptions of two new species. *Crustaceana* 17:19.

JONES, N.S. 1976. British Cumaceans. Synopsis of the British Fauna (New Series) No. 7, Academic Press, London.

WATLING, L. 1979. Crustacea: Cumacea. Marine Flora and Fauna of the Northeastern U.S. Circ. 423, National Oceanic and Atmospheric Admin., National Marine Fisheries Serv., U.S. Government Printing Off., Washington, D.C.

Mysidacea

TATTERSALL, W.M. 1951. A Review of the Mysidacea of the U.S. National Museum. U.S. Nat. Mus. Bull. 201.

KATHMAN, R.D., W.C. AUSTIN, J.C. SALTMAN & J.D. FULTON. 1986. Identification Manual to the Mysidacea and Euphausiacea of the Northeast Pacific. Can. Fish Aquatic Sci. Spec. Publ. No. 93.

DALY, K.L. & C. HOLMQUIST. 1986. A key to the Mysidacea of the Pacific Northwest. *Can. J. Zool.* 64:1201.

PRICE, W.W., R.H. HEARD & L. STUCK. 1994. Observations on the genus *Mysidopsis* Sars, 1864 with the designation of a new genus *Americamysis, Americamysis alleni* and *A. stucki* (Peracardia Mysidacea: Mysidae), from the Gulf of Mexico. *Proc. Biol. Soc. Wash.* 107:680.

Tanaidacea and Isopoda

VANNAME, W.C. 1936. The American land and freshwater isopod crustaceans. *Bull. Amer. Mus.* 71:1.

SCHULTZ, G.A. 1969. How to Know the Marine Isopod Crustaceans. W.C. Brown, Dubuque, Iowa.

WILLIAMS, W.D. 1972. Freshwater Isopods (Asellidae) of North America. Biota of Freshwater Ecosystems. Ident. Manual No. 7, U.S. Environmental Protection Agency, U.S. Government Printing Off., Washington, D.C.

KENSLEY, B. & M. SCHOTTE. 1989. Guide to the Marine Isopod Crustaceans of the Caribbean. Smithsonian Institution Press, Washington, D.C.

Amphipoda

MCCAIN, J.C. 1968. The Caprellidae (Crustacea: Amphipoda) of the Western North America. U.S. Nat. Mus. Bull. 278.

HOLSINGER, J.R. 1972. The Fresh Water Amphipod Crustaceans (Gammaridae) of North America. Biota of Fresh Water Ecosystems. Ident. Manual No. 5, U.S. Environmental Protection Agency, U.S. Government Printing Off., Washington, D.C.

LAUBITZ, D.R. 1972. The Caprellida (Crustacea, Amphipoda) of Atlantic and Arctic Canada. Publ. Biol. Oceanogr. Nat. Mus., Canada, No. 4.

BOUSFIELD, E.L. 1973. Shallow-Water Gammaridean Amphipoda of New England. Cornell Univ. Press, Ithaca, N.Y.

FOX, R.S. & K.H. BYNUM. 1975. The amphipod crustaceans of North Carolina estuarine waters. *Chesapeake Sci.* 16:223.

LINCOLN, R.J. 1979. British Marine Amphipoda: Gammaridea. British Museum (Natural History), London.

BRUSCA, G.J. 1981. Annotated keys to the Hyperiidea (Crustacea: Amphipoda) of North American coastal waters. *Tech. Rep. Allan Hancock Foundation* 5:1.

BARNARD, J.L. & C.M. BARNARD. 1983. Freshwater Amphipoda of the World. Hayfield Assoc., Mt. Vernon, Va.

BARNARD, J.L. & G.S. KARAMAN. 1991. The families and genera of marine gammaridean Amphipoda (except marine gammaroids), Parts 1 and 2. Rec. Australian Mus. 13, Sydney, Australia.

BLAKE, J.A., L. WATLING & P.H. SCOTT, eds. 1995. The Amphipoda. *In* Taxonomic Atlas of the Benthic Fauna of the Santa Maria Basin and Western Santa Barbara Channel, Vol. 12, Parts 3–5. Santa Barbara Mus. Natural History, Santa Barbara, Calif.

VINOGRADOV, M.W., A.F. VOLKOV & T.N. SEMENOVA, eds. 1996. Hyperiid Amphipods (Amphipoda, Hyperiidea) of the World Oceans. Science Publ., Lebanon, N.H.

Decapoda

GARTH, J.S. 1958. Brachyura of the Pacific coast of America. Oxyrhyncha. *Allan Hancock Pac. Exped.* 21:1.

GARTH, J.S. & W. STEPHENSON. 1966. Brachyura of the Pacific Coast of America. Brachyrhyncha: Portunidae. Allan Hancock Monogr. Mar. Biol. 1.

MANNING, R.B. 1969. Stomatopod crustacea of the western Atlantic. *Stud. Trop. Oceanogr.* 8:1.

PEREZ, F.E. 1969. Western Atlantic shrimps of the genus *Penaeus*. *Fish. Bull.* 67:1.

HOBBS, H.H., JR. 1972. Crayfish (Astacidae) of North and Middle America. Biota of Freshwater Ecosystems. Ident. Manual No. 9, U.S. Environmental Protection Agency, U.S. Government Printing Off., Washington, D.C.

MANNING, R.B. 1972. Stomatopod crustacea. Eastern Pacific expeditions of the New York Zoological Society. *Zoologica* 56:95.

WILLIAMS, A.B. 1974. Crustacea: Decapoda. Marine Flora and Fauna of Northeastern U.S. Circ. 389, National Oceanic Atmospheric Admin., National Marine Fisheries Serv., U.S. Government Printing Off., Washington, D.C.

CRANE, J. 1975. Fiddler Crabs of the World (Ocypodidae: Genus *Uca*). Princeton Univ. Press, Princeton, N.J.

NATIONS, J.D. 1975. The genus *Cancer* (Crustacea: Brachyura): Systematics, biogeography and fossil record. Nat. Hist. Mus., Los Angeles County, Sci. Bull. 23, Los Angeles, Calif.

BUTLER, T.H. 1980. Shrimps of the Pacific Coast of Canada. Can. Fish Aquatic Sci. Bull. No. 202.

HOBBS, H.H., JR. 1981. The crayfishes of Florida. Smithsonian Contrib. Zool. No. 318.

HART, J.F.L. 1982. Crabs and their Relatives of British Columbia. British Columbia Prov. Mus. Handbook No. 40, Victoria.

WILLIAMS, A.B. 1984. Shrimps, Lobsters, and Crabs of the Atlantic Coast of the Eastern United States, Maine to Florida. Smithsonian Inst. Press, Washington, D.C.

HART, C.W., JR. & J. CLARK. 1989. An Interdisciplinary Bibliography of Freshwater Crayfishes (Astacoidea and Parastacoidea) from Aristotle through 1987. Smithsonian Contrib. Zool. No. 455.

WILLIAMS, A.B., L.G. ABELE, D.L. FELDER, H.H. HOBBS, JR., R.B. MAN-

NING, P.A. McLAUGHLIN & I.P. FARFANTE. 1989. Common and Scientific Names of Aquatic Invertebrates from the United States and Canada: Decapod Crustaceans. AFS Spec. Publ. 17, American Fisheries Soc., Bethesda, Md.

SQUIRES, H.J. 1990. Decapod Crustacea of the Atlantic Coast of Canada. Dept. Fisheries & Oceans, Ottawa, Ont., Canada.

WICKSTEN, M.K. 1990. Key to the hippolytid shrimp of the Eastern Pacific Ocean. *U.S. Nat. Mar. Fish Serv. Fish Bull.* 88:587.

JENSEN, G.C. 1995. Pacific Coast Crabs and Shrimps. Sea Challengers, Inc., Monterey, Calif.

18. Insects

General and Introductory

USINGER, R.L., ed. 1956. Aquatic Insects of California. With Keys to North American Genera and California Species. Univ. California Press, Berkeley.

BORROR, D.J. & R.E. WHITE. 1970. A Field Guide to the Insects of America North of Mexico. Peterson Field Guide Ser., Houghton Mifflin Co., Boston, Mass.

BLAND, R.G. & H.E. JAQUES. 1978. How to Know the Insects, 3rd ed. Wm. C. Brown Co., Dubuque, Iowa.

CHU, H.F. 1979. How to Know the Immature Insects. Wm. C. Brown Co., Dubuque, Iowa.

LEHMKUHL, D.M. 1979. How to Know the Aquatic Insects. Wm. C. Brown Co., Dubuque, Iowa.

McCAFFERTY, W.P. 1981. Aquatic Entomology. The Fishermen's and Ecologists' Illustrated Guide to Insects and Their Relatives. Science Books International, Boston, Mass.

BRIGHAM, A.R., W.U. BRIGHAM & A. GNILKA, eds. 1982. Aquatic Insects and Oligochaetes of North and South Carolina. Midwest Aquatic Enterprises, Mahomet, Ill.

MERRITT, R.W. & K.W. CUMMINS, eds. 1996. An Introduction to the Aquatic Insects of North America, 3rd ed. Kendall/Hunt Publ. Co., Dubuque, Iowa.

STEHR, F.W. 1987. Immature Insects. Kendall/Hunt Publ. Co., Dubuque, Iowa.

BORROR, D.J., D.M. DELONG & C.A. TRIPLEHORN. 1989. An Introduction to the Study of Insects, 6th ed. Saunders College Publ., Philadelphia, Pa.

ARNETT, R.H., JR. 1993. American Insects: A Handbook of the Insects of America North of Mexico. Sandhill Crane Press, Inc., Gainesville, Fla.

Mayflies (Ephemeroptera)

BURKS, B.D. 1953. The mayflies, or Ephemeroptera, of Illinois. *Bull. Ill. Natur. Hist. Surv.* 26:1.

LEONARD, J.W. & F.A. LEONARD. 1962. Mayflies of Michigan Trout Streams. Cranbrook Inst. Science, Mich.

EDMUNDS, G.F., JR., S.L. JENSEN & L. BERNER. 1976. The Mayflies of North and Central America. Univ. Minnesota Press, Minneapolis.

BERNER, L. & M.L. PESCADOR. 1988. The Mayflies of Florida. Univ. Florida Press, Gainesville.

Dragonflies and Damselflies (Odonata)

NEEDHAM, J.G. & M.J. WESTFALL, JR. 1955. A Manual of the Dragonflies of North America, Including the Greater Antilles, and the Provinces of the Mexican Border. Univ. California Press, Berkeley.

WALKER, E.M. & P.S. CORBET. 1975. The Odonata of Canada and Alaska, Vol. III. Univ. Toronto Press, Toronto, Ont.

WALKER, E.M. 1983. The Odonata of Canada and Alaska, Vols. I and II. Univ. Toronto Press, Toronto, Ont.

Stoneflies (Plecoptera)

FRISON, T.H. 1935. The stoneflies, or Plecoptera, of Illinois. *Bull. Ill. Natur. Hist. Surv.* 20:281.

FRISON, T.H. 1942. Studies of North American Plecoptera, with special reference to the fauna of Illinois. *Bull. Ill. Natur. Hist. Surv.* 22:235.

JEWETT, S.G. 1960. The stoneflies (Plecoptera) of California. *Bull. Calif. Insect Surv.* 6:125.

HITCHCOCK, S.W. 1974. Guide to the Insects of Connecticut. Part VII. The Plecoptera or Stoneflies of Connecticut. State Geol. Natur. Hist. Surv. Conn. Bull. 107.

STEWART, K.W. & B.P. STARK. 1988. Nymphs of North American Stonefly Genera (Plecoptera). Thomas Say Found. 12.: Entomol. Soc. America.

Megaloptera and Neuroptera

ROSS, H.H. 1937. Studies of nearctic aquatic insects. I. Nearctic alder flies of the genus *Sialis*. *Bull. Ill. Natur. Hist. Surv.* 21:57.

PARFIN, S.I. & A.B. GURNEY. 1956. The Spongilla-flies, with special reference to those of the Western Hemisphere (Sisyridae, Neuroptera). *Proc. U.S. Nat. Mus.* 105:421.

WESTFALL, M.J. & M.L. MAY. 1996. Damselflies of North America. Scientific Publishers, Gainesville, Fla.

Caddisflies (Trichoptera)

ROSS, H.H. 1944. The caddisflies, or Trichoptera of Illinois. *Bull. Ill. Natur. Hist. Surv.* 23:1.

SCHUSTER, G.A. & D.A. ETNIER. 1978. A Manual for the Identification of the Larvae of the Caddisfly genera *Hydropsyche* Pictet and *Symphitopsyche* Ulmer in Eastern and Central North America (Trichoptera: Hydropsychidae). EPA-600/4-78-060, U.S. Environmental Protection Agency, Cincinnati, Ohio.

WALLACE, I.D., B. WALLACE & G.N. PHILIPSON. 1990. Key to the Case-Bearing Caddisfly (Trichoptera) Larvae of Florida. Florida Dep. Environmental Protection, Tallahassee.

WIGGINS, G.B. 1996. Larvae of North American Caddisfly Genera (Trichoptera), 2nd ed. Univ. Toronto Press, Toronto, Ont., Canada.

Diptera

JOHANNSEN, O.A. 1934, 1935, 1937. Aquatic Diptera. Parts I–IV (Pt. V by L.C. Thomsen). Memoirs Cornell Univ. Agr. Exp. Sta. Reproduced in 1969 by Entomological Reprint Specialists, Los Angeles, Calif.

BRYCE, D. & A. HOBART. 1972. The biology and identification of the larvae of the Chironomidae (Diptera). *Entomol. Gaz.* 23:175.

SIMPSON, K.W. & R.W. BODE. 1980. Common Larvae of Chironomidae (Diptera) from New York State Streams and Rivers with Particular Reference to the Fauna of Artificial Substrates. Bull. No. 439, New York State Mus., Albany.

TESKEY, H.J. 1981. Key to Families—Larvae. *In* Manual of Nearctic Diptera, Vol. 1. Monogr. No. 27, Biosystems Research Inst., Ottawa, Ont.

BODE, R.W. 1983. Larvae of North American *Eukiefferiella* and *Tvetenia* (Diptera: Chironomidae). Bull. N.Y. State Mus. No. 452, Albany.

OLIVER, D.R. & M.E. ROUSSEL. 1983. Insects and Arachnids of Canada. Part 11—Genera of Larval Midges of Canada. Publ. 1746, Can. Gov. Pub. Centre, Ottawa, Ont., Canada.

WIEDERHOLM, T., ed. 1983. Chironomidae of the Holarctic Region. Pt. 1. Larvae. Entomol. Scand., Suppl. 19.

WIEDERHOLM, T., ed. 1986. Chironomidae of the Holarctic Region. Pt. 2. Pupae. Entomol. Scand., Suppl. 28.

BODE, R.W. 1990. Chironomidae. *In* B.L. Peckarsky, P.R. Fraissinet, M.A. Penton & D.J. Conklin, Jr., eds. Freshwater Macroinvertebrates of Northeastern North America. Cornell Univ. Press, Ithaca, N.Y.

CROSSKEY, R.W. 1990. The Natural History of Blackflies. John Wiley & Sons, Chichester, U.K.

EPLER, J.H. 1995. Identification Manual for the Larval Chironomidae (Diptera) of Florida. Florida Dep. Environmental Protection, Tallahassee.

Beetles (Coleoptera)

JAQUES, H.E. 1951. How to Know the Beetles. Wm. C. Brown Co., Dubuque, Iowa.

DILLON, E.S. & L.S. DILLON. 1961. Manual of Common Beetles of Eastern North America. Harper and Row, New York, N.Y.

BROWN, H.D. 1972. Aquatic Dryopoid Beetles (Coleoptera) of the United States. Biota of Freshwater Systems. Ident. Manual No. 6, U.S. Environmental Protection Agency, U.S. Government Printing Off., Washington, D.C.

ARNETT, R.H. 1973. The Beetles of the United States. American Entomological Inst., Ann Arbor, Mich.

Hemiptera

HUNGERFORD, H.B. 1919. The biology and ecology of aquatic and semiaquatic Hemiptera. *Univ. Kans. Sci. Bull.* 9:1.

HUNGERFORD, H.B. 1948. The Corixidae of the Western Hemisphere (Hemiptera). *Univ. Kans. Sci. Bull.* 32:1.

SAVAGE, A.A. 1989. Adults of the British Aquatic Hemiptera Heteroptera: A Key with Ecological Notes. Freshwater Biol. Assoc. Sci. Publ. 50, Ambleside, U.K.

19. Tardigrada

MORGAN, C.I. & P.E. KING. 1976. British Tardigradas. Academic Press, London, England.

20. Hydracarina

COOK, D.R. 1974. Water Mite Genera and Subgenera. Mem. Amer. Entomol. Inst., Ann Arbor, Mich.

SMITH, B.P. 1990. Hydrachnidia. *In* B.L. Peckarsky, P.R. Fraissinet, M.A. Penton & D.J. Conklin, Jr., eds. Freshwater Macroinvertebrates of Northeastern North America. Cornell Univ. Press, Ithaca, N.Y.

21. Pycnogonida

KING, P.E. 1973. Pycnogonids. St. Martin's Press, New York, N.Y.

ARNAUD, F. & R.V. BAMBER. 1987. The biology of Pycnogonida. *Advan. Mar. Biol.* 24:1.

22. Mollusks

General

ANDREWS, J. 1971. Sea Shells of the Texas Coast. Univ. Texas Press, Austin.

ABBOTT, R.T. 1974. American Seashells, 2nd ed. Van Nostrand Reinhold Co., New York, N.Y.

KEEN, A.M. & E. COAN. 1974. Marine Mollusca Genera of Western North America: With Illustrated Key. Stanford Univ. Press, Stanford, Calif.

EMERSON, W.K. & M.K. JACOBSON. 1976. The American Museum of Natural History Guide to Shells—Land, Freshwater, and Marine from Nova Scotia to Florida. Alfred Knopf, New York, N.Y.

McLEAN, J.H. 1978. Marine Shells of Southern California. Los Angeles Co. Mus. Natur. Hist., Sci. Ser. 24, Zool. No. 11.

TAYLOR, D.W. 1981. Freshwater mollusks of California: A distributional checklist. *Calif. Fish & Game.* 67:140.

JONES, A.M. & J.M. BAXTER. 1987. Mollusca: Caudofoveata, Solenogastres, Polyplacophora and Scaphopoda. Synopsis of British Fauna. Field Studies Council, Shrewsbury, U.K.

TURGEON, D.A., A.E. BOGAN, E.V. COAN, W.K. EMERSON, W.G. LYONS, W.L. PRATT, C.F.E. ROPER, A. SCHELTEMA, F.G. THOMPSON & J.D. WILLIAMS. 1988. Common and Scientific Names of Aquatic Invertebrates from United States and Canada: Mollusks. Amer. Fish. Soc. Spec. Rep. No. 16, Bethesda, Md.

Gastropods

WALTER, H.J. & J.B. BURCH. 1957. Key to the Genera of Fresh Water Gastropods (Snails and Limpets) Occurring in Michigan. Mus. Zool. Univ. Mich. Circ. No. 3, Ann Arbor.

BURCH, J.B. 1962. How to Know Eastern Land Snails. Wm. C. Brown Co., Dubuque, Iowa.

LIMBERG, D.R. 1981. Acmaeidae, Gastropoda, Mollusca. Boxwood Press.

GRAHAM, A. 1988. Mollusca: Prosobranch and Pyramidellid Gastropods. Synopses of British Fauna. Field Studies Council, Shrewsbury, U.K.

BURCH, J.B. 1989. North American Freshwater Snails. Malacological Publications, Hamburg, Mich.

BEHRENS, D.W. 1991. Pacific Coast Nudibranchs—A Guide to the Opisthobranchs, Alaska to Baja California, 2nd ed., revised. Sea Challengers, Inc., Monterey, Calif.

SCOTT, P.H., J.A. BLAKE & A.L. LISSNER. 1996. The Mollusca, Part 2. The Gastropoda. *In* Taxonomic Atlas of the Benthic Fauna of the Santa Maria Basin and the Western Santa Barbara Channel. Santa Barbara Mus. Natural History, Santa Barbara, Calif.

Bivalves (Pelecypoda)

HEARD, W.H. & J. BURCH. 1966. Keys to the Genera of Freshwater Pelecypods of Michigan. Mus. Zool. Univ. Mich. Circ. No. 4, Ann Arbor.

BURCH, J.B. 1975. Freshwater Sphaeriacean Clams (Mollusca: Pelecypoda) of North America. Malacological Publications, Hamburg, Mich.

BURCH, J.B. 1975. Freshwater Unionacean Clams (Mollusca: Pelecypoda) of North America. Malacological Publications, Hamburg, Mich.

BUCHANAN, A.C. 1980. Mussels (Naiades) of the Meramec River Basin, Missouri. Aquatic Ser. No. 17, Missouri Dep. Conservation, Columbia.

MACKIE, G.L., D.S. WHITE & T.W. ZDEBA. 1980. A Guide to Freshwater Mollusks of the Laurentian Great Lakes with special emphasis on the Genus *Pisidium*. EPA-600/3-80-068, U.S. Environmental Protection Agency Research Lab., Duluth, Minn.

VOKES, H.E. 1981. Genera of the Bivalvia: A Systematic and Bibliographic Catalogue. Paleontological Research Inst., Ithaca, N.Y.

BERNARD, F.R. 1983. Catalogue of the Living Bivalvia of the Eastern Pacific Ocean: Bering Strait to Cape Horn. Can. Fish Aquatic Sci. Spec. Publ. No. 61.

CUMMINGS, K.S. & C.A. MAYER. 1992. Field Guide to Freshwater Mussels of the Midwest. Illinois Nat. History Survey, Champaign, Ill.

WILLIAMS, J.D., M.L. WARREN, JR., K.S. CUMMINGS, J.L. HARRIS & R.J. NEVES. 1993. Conservation status of freshwater mussels of the United States and Canada. *Fisheries* 18(9):9.

CLAUDI, R. & G.L. MACKIE. 1994. Practical Manual for Zebra Mussel Monitoring and Control. Lewis Publ., Boca Raton, Fla.

STRAYER, D.L. & K.J. JIRKA. 1997. The Pearly Mussels of New York State. New York State Museum Memoir 26.

23. Bryozoa

OSBORN, R.C. 1950–1953. Bryozoa of the Pacific Coast of North America. Parts 1–3. *Allan Hancock Foundation Pac. Exped.* 14:1.

BRUSHNELL, J.H. 1965. On the taxonomy and distribution of fresh-water Ectoprocta in Michigan. Parts I, II, IV. *Trans. Amer. Microscop. Soc.* 84:231, 339, 529.

SOULE, J.D. & D.F. SOULE. 1969. Systematics and biogeography of burrowing bryozoans. *Inst. Zool.* 9:791.

WOOLLACOTT, R.M. & W.J. NORTH. 1971. Bryozoans of California and northern Mexico kelp beds. *Nova Hedwigia*, 32:455.

MUNDY, S.F. 1980. A Key to the British and European Freshwater Bryozoans. Sci. Publ. Freshwater Biol. Assoc. U.K. 41.

WOOD, T.S. 1989. Ectoproct Bryozoans of Ohio. Ohio Biological Surv., Columbus.

RYLAND, J.S. & P.J. HAYWARD. 1991. Marine Fauna and Flora of the Northeastern USA: Erect Bryozoa. NOAA Tech Rep. Nat. Mar. Fish. Serv. No. 99, Seattle, Wash.

SOULE, F.F., J.D. SOULE & H.W. CHANEY. 1995. The Bryozoa. *In* Taxonomic Atlas of the Benthic Fauna of the Santa Maria Basin and the Western Santa Barbara Channel, Vol. 13. Santa Barbara Mus. Natural History, Santa Barbara, Calif.

24. Echinoderms

FISHER, W.K. 1911–1930. Asteroidea of the North Pacific and adjacent waters. U.S. Nat. Mus. Bull. 76.

MORTENSEN, T. 1928–1951. Monograph of the Echinoidea. 5 vols. Reitzel, Copenhagen.

CLARK, A.M. 1962. Starfishes and Their Relations. British Mus. (Natural History), London.

PAWSON, D.L. & H.B. FELL. 1965. A revised classification of the dendrochirate holothurians. *Breviora* 214:1.

GRAY, I.E., M.E. DOWNEY & M.J. CERAME-VIVAS. 1968. Seastars of North Carolina. *Fish. Bull.* 67:127.

KYTE, M.A. 1969. A synopsis and key to the recent Ophiuroidea of Washington state and southern British Columbia. *J. Fish Res. Board Can.* 26:1727.

AUSTIN, W.C. & M.P. HAYLOCK. 1973. British Columbia Marine Faunistic Survey Report: Ophiuroidea from the Northeast Pacific. Fish. Res. Board Can. Tech. Rep. 426.

DOWNEY, M.E. 1973. Starfishes from the Caribbean and Gulf of Mexico. Smithsonian Contrib. Zool. No. 126.

PAWSON, D.L. 1977. Echinodermata: Holothuroidea. Marine Flora and Fauna of the Northeastern U.S. Circ. No. 405, National Oceanic Atmospheric Admin., National Marine Fisheries Serv., Washington, D.C.

LAMBERT, P. 1981. The Sea Stars of British Columbia. British Columbia Prov. Mus. Handbook No. 39, Victoria.

HENDLER, G., J.E. MILLER, D.L. PAWSON & P.M. KIER, eds. 1995. Sea Stars, Sea Urchins and Allies. Echinoderms of Florida and the Caribbean. Smithsonian Inst. Press, Washington, D.C.

CUTRESS, B.M. 1996. Changes in dermal ossicles during somatic growth in Caribbean littoral sea cucumbers (Echinodermata: Holothuroidea: Aspidochirotida). *Bull. Mar. Sci.* 58:44.

25. Urochordata (Tunicates)

VAN NAME, W.G. 1945. The north and south American ascidians. *Bull. Amer. Mus. Natur. Hist.* 84:1.

BERRILL, N.J. 1950. The Tunicata. Royal Soc., London.

MILLAR, R.H. 1970. British Ascidians. Academic Press, London.

MILLAR, R.H. 1971. The biology of ascidians. *Advan. Mar. Biol.* 9:1.

26. Fishes

See Section 10600D for references to the identification of fishes.

27. Amphibians

STEBBINS, R.C. 1966. A Field Guide to Western Reptiles and Amphibians. Houghton Mifflin Co., Boston, Mass.

COCHRAN, D.M. & C.J. GOIN. 1970. The New Field Book of Reptiles and Amphibians. Putnam Nature Field Book. G.P. Putnam's Sons, New York, N.Y.

CONANT, R. & J.T. COLLINS. 1991. Field Guide to the Reptiles and Amphibians of Eastern and Central North America, 3rd ed. Houghton Mifflin Co., Boston, Mass.

ALGAE COLOR PLATES

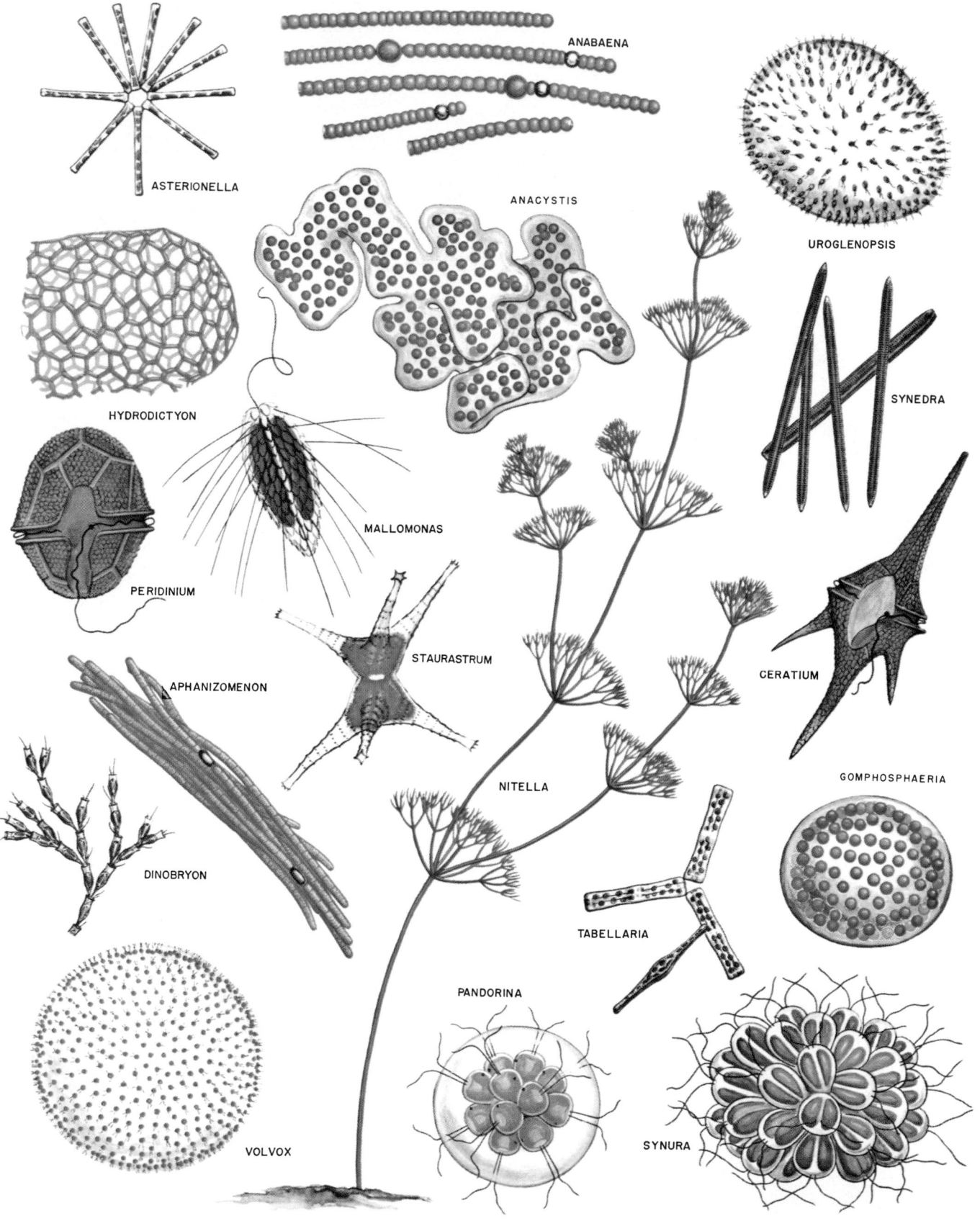

Plate 28. Taste and odor algae.

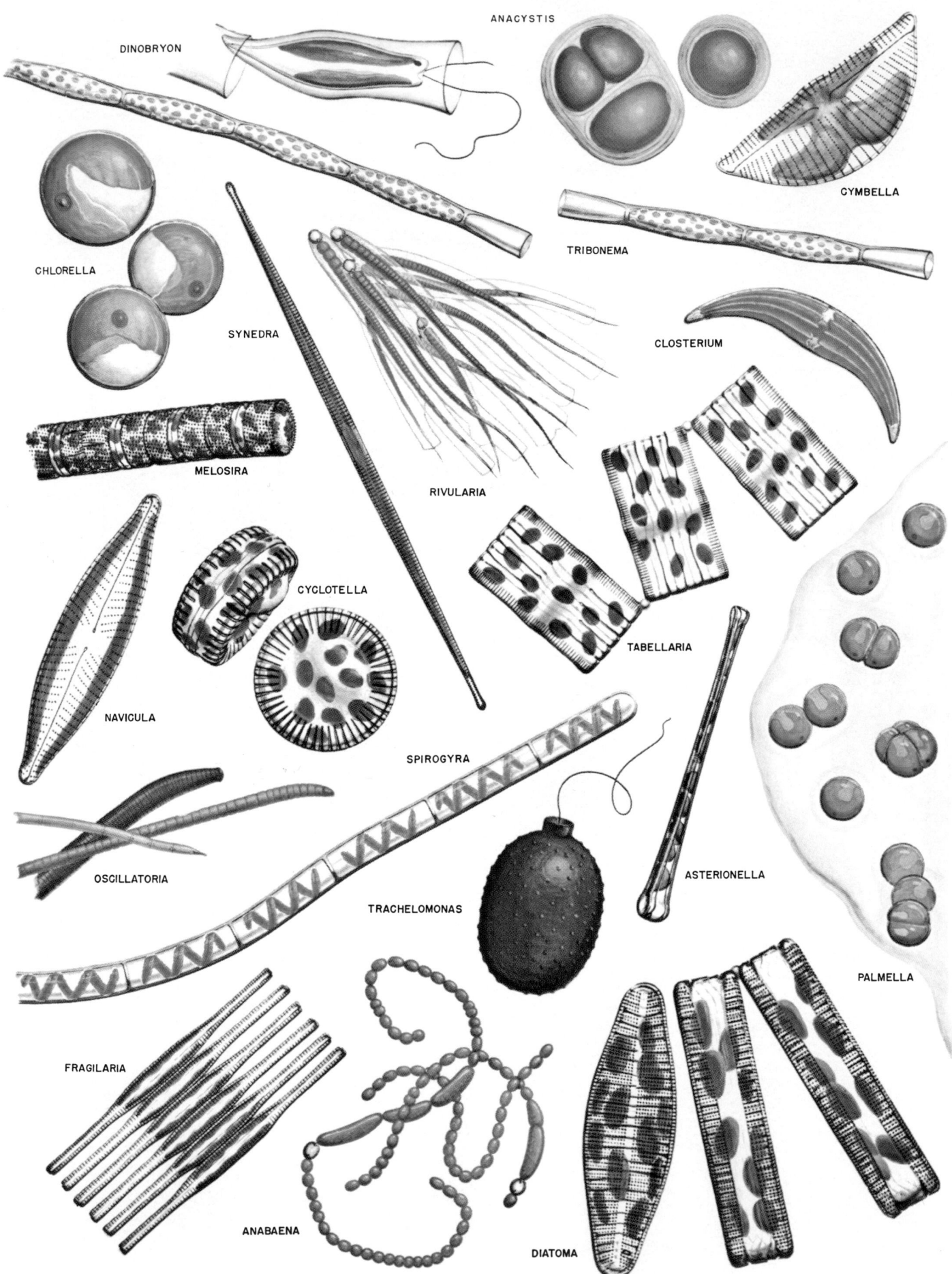

Plate 29. Filter- and screen-clogging algae.

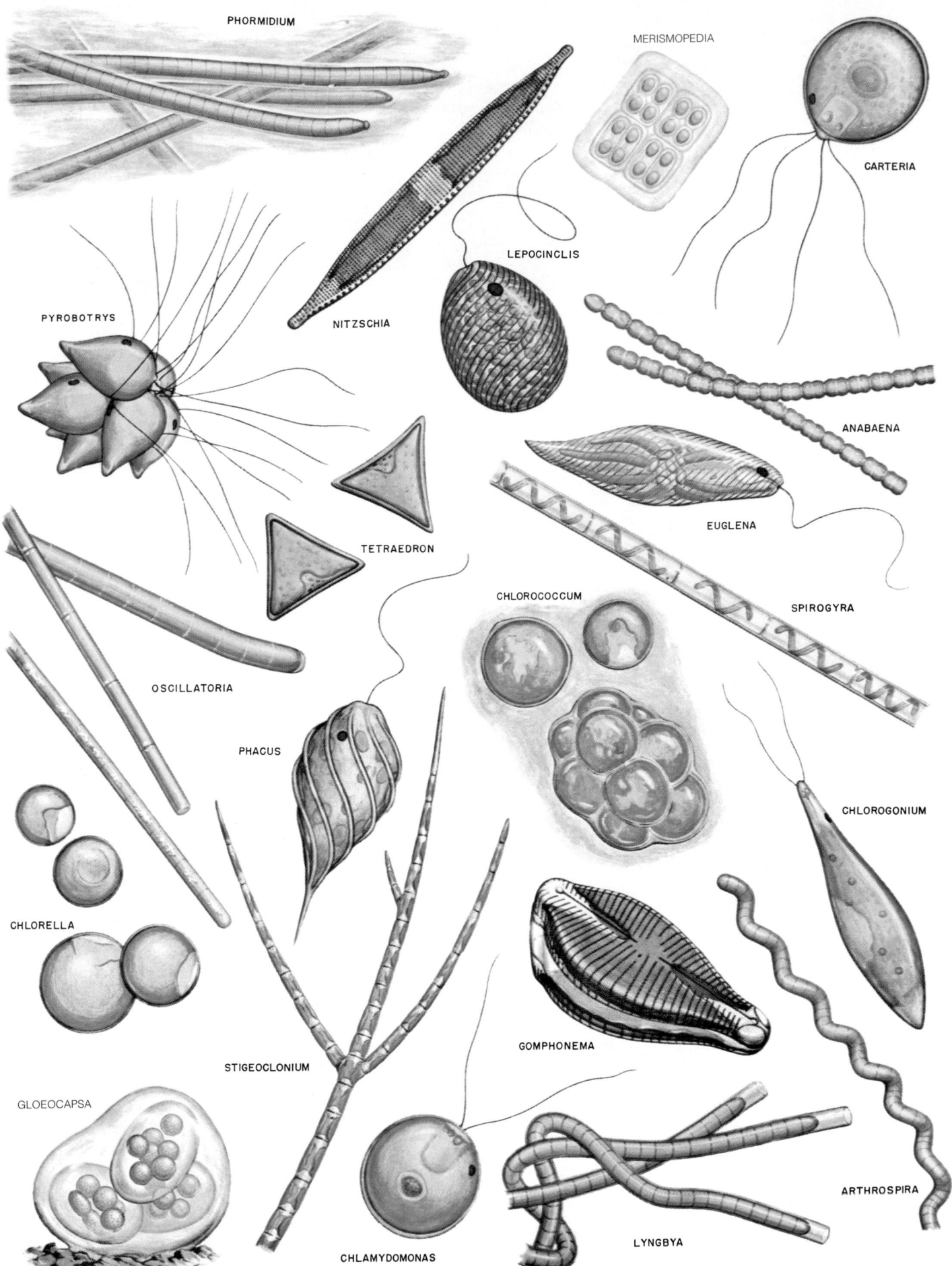

PHORMIDIUM

MERISMOPEDIA

CARTERIA

LEPOCINCLIS

NITZSCHIA

PYROBOTRYS

ANABAENA

TETRAEDRON

EUGLENA

CHLOROCOCCUM

SPIROGYRA

OSCILLATORIA

PHACUS

CHLOROGONIUM

CHLORELLA

STIGEOCLONIUM

GOMPHONEMA

GLOEOCAPSA

CHLAMYDOMONAS

LYNGBYA

ARTHROSPIRA

Plate 30. Fresh-water pollution algae.

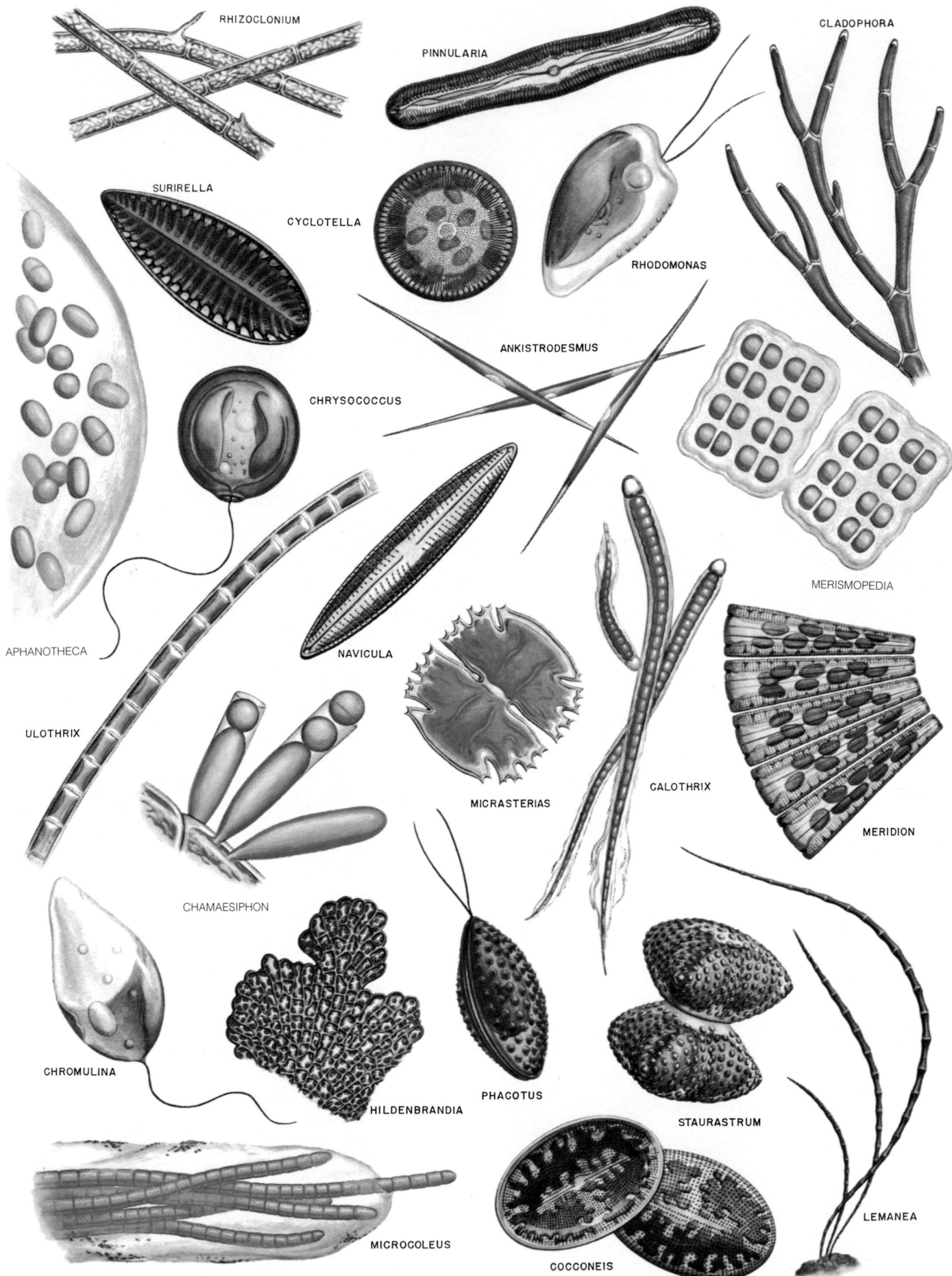

RHIZOCLONIUM

PINNULARIA

CLADOPHORA

SURIRELLA

CYCLOTELLA

RHODOMONAS

CHRYSOCOCCUS

ANKISTRODESMUS

MERISMOPEDIA

APHANOTHECA

NAVICULA

ULOTHRIX

MICRASTERIAS

CALOTHRIX

MERIDION

CHAMAESIPHON

CHROMULINA

HILDENBRANDIA

PHACOTUS

STAURASTRUM

LEMANEA

MICROCOLEUS

COCCONEIS

Plate 31. Clean-water algae.

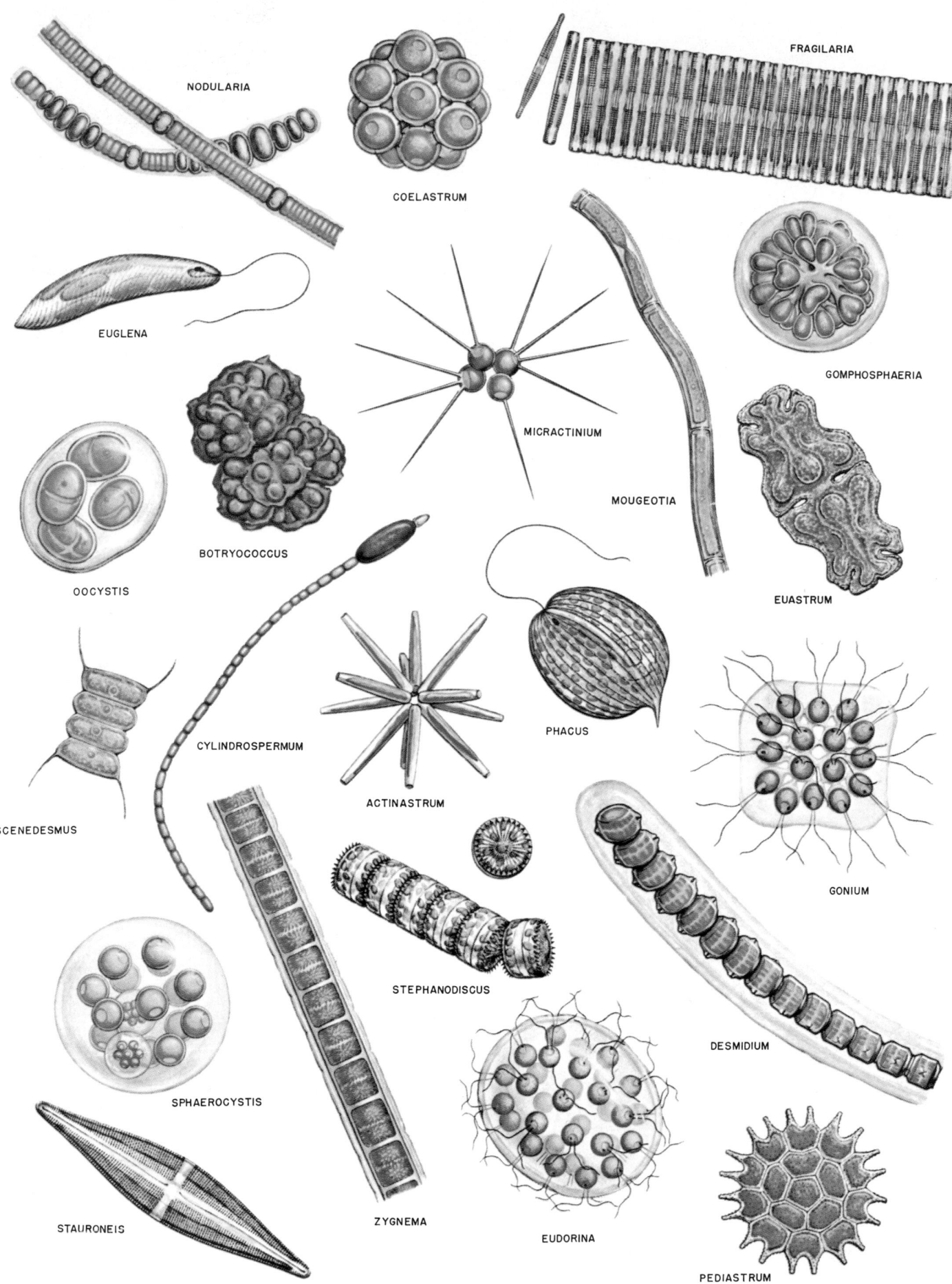

NODULARIA

COELASTRUM

FRAGILARIA

EUGLENA

MICRACTINIUM

GOMPHOSPHAERIA

MOUGEOTIA

BOTRYOCOCCUS

OOCYSTIS

EUASTRUM

SCENEDESMUS

CYLINDROSPERMUM

ACTINASTRUM

PHACUS

GONIUM

STEPHANODISCUS

DESMIDIUM

SPHAEROCYSTIS

STAURONEIS

ZYGNEMA

EUDORINA

PEDIASTRUM

Plate 32. Plankton and other surface-water algae.

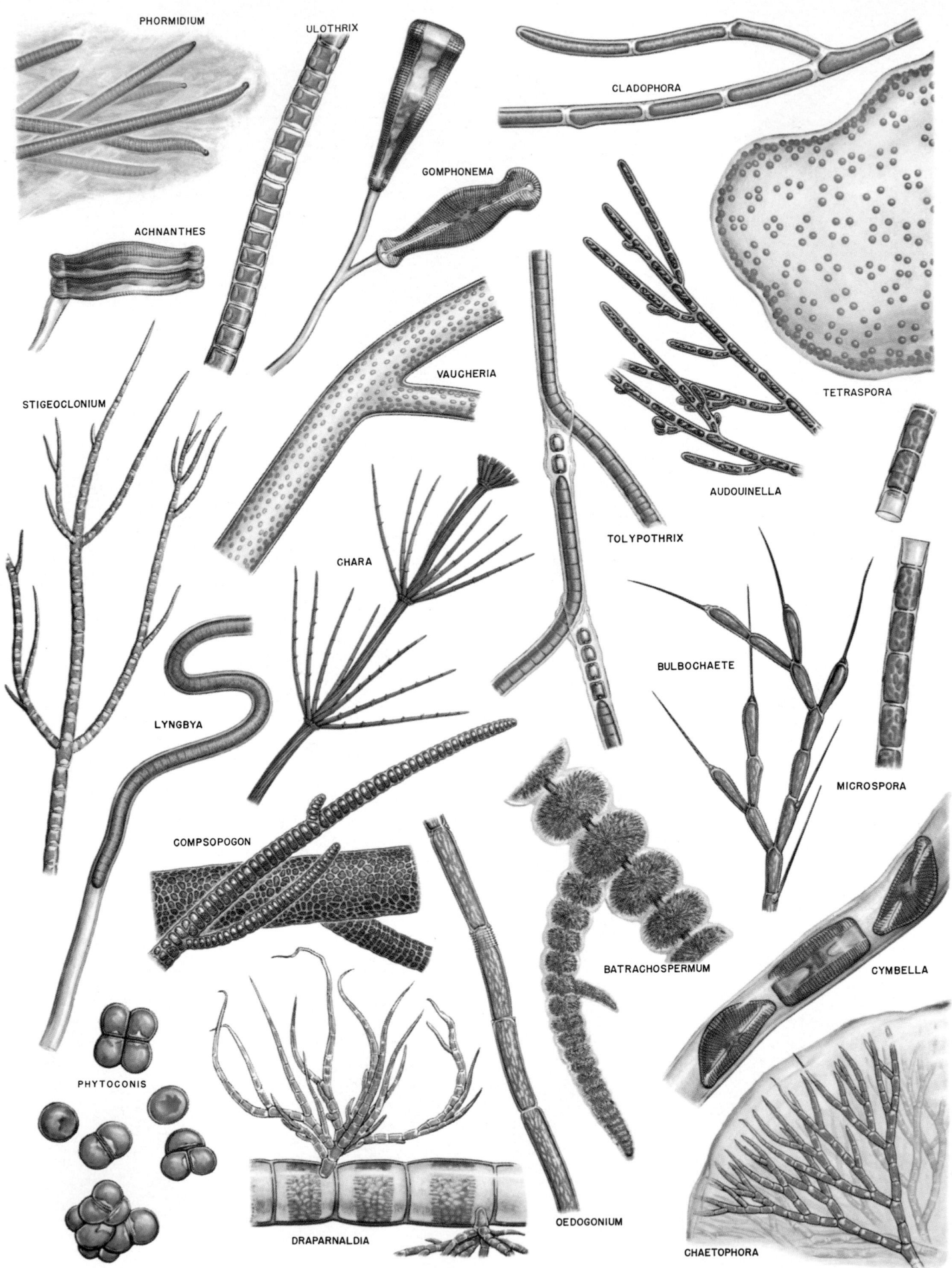

Plate 33. Algae growing on surfaces.

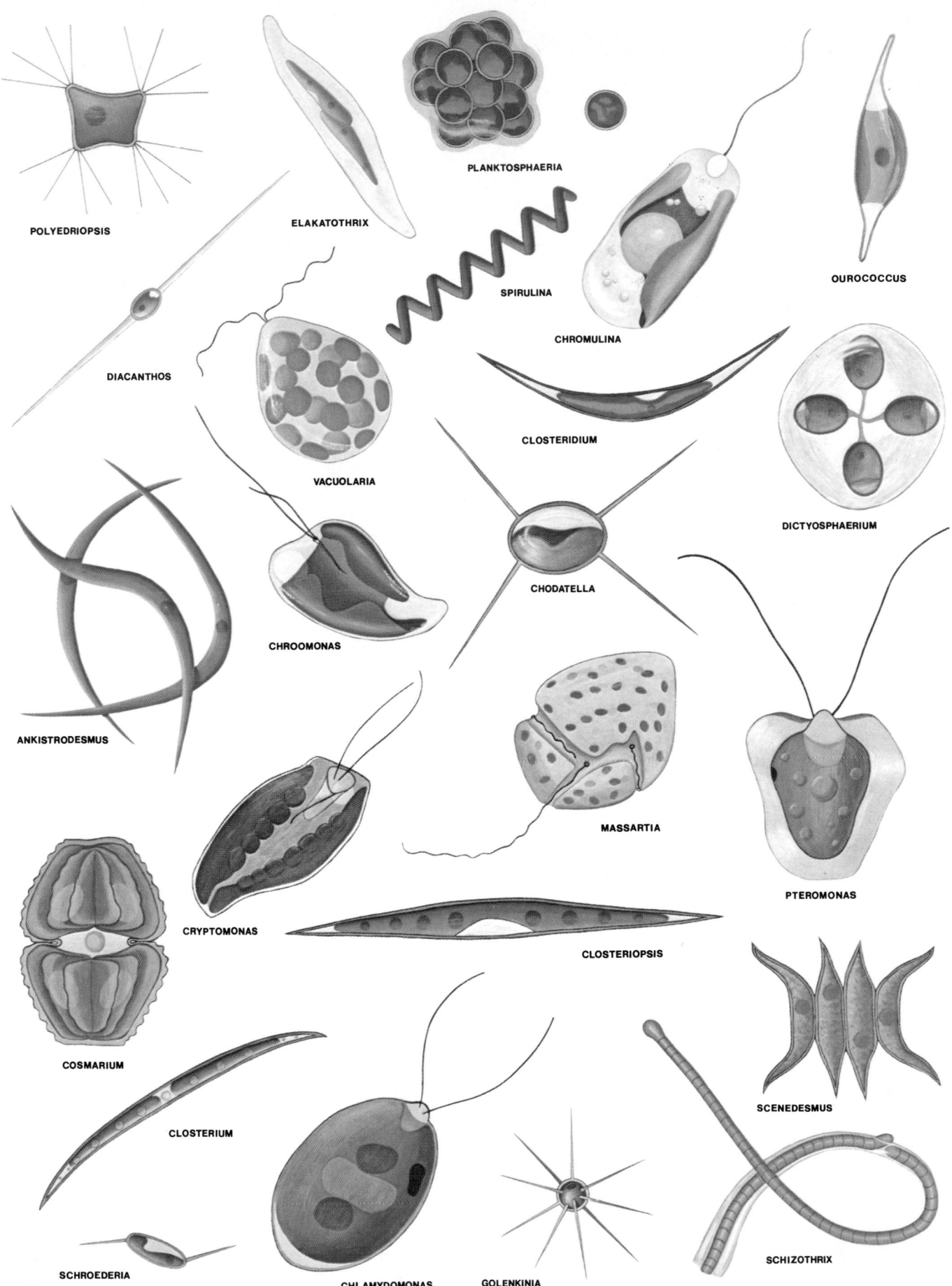

POLYEDRIOPSIS

ELAKATOTHRIX

PLANKTOSPHAERIA

SPIRULINA

CHROMULINA

OUROCOCCUS

DIACANTHOS

VACUOLARIA

CLOSTERIDIUM

DICTYOSPHAERIUM

ANKISTRODESMUS

CHROOMONAS

CHODATELLA

MASSARTIA

PTEROMONAS

CRYPTOMONAS

CLOSTERIOPSIS

COSMARIUM

SCENEDESMUS

CLOSTERIUM

SCHROEDERIA

CHLAMYDOMONAS

GOLENKINIA

SCHIZOTHRIX

Plate 34. Wastewater-treatment-pond algae.

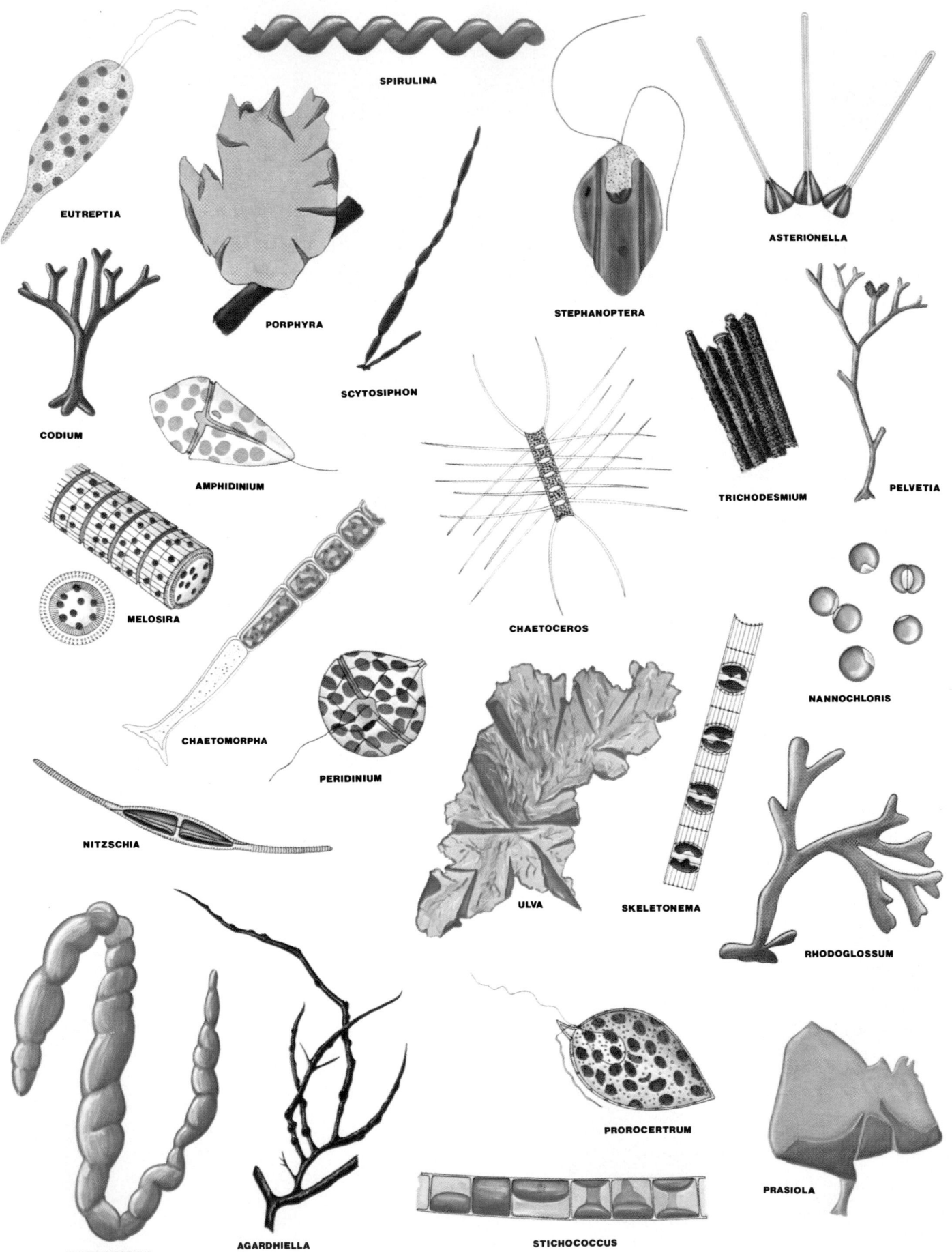

SPIRULINA

EUTREPTIA

PORPHYRA

SCYTOSIPHON

STEPHANOPTERA

ASTERIONELLA

CODIUM

AMPHIDINIUM

PELVETIA

TRICHODESMIUM

MELOSIRA

CHAETOCEROS

NANNOCHLORIS

CHAETOMORPHA

PERIDINIUM

ULVA

SKELETONEMA

RHODOGLOSSUM

NITZSCHIA

ENTEROMORPHA

AGARDHIELLA

PROROCERTRUM

PRASIOLA

STICHOCOCCUS

Plate 35. Estuarine pollution algae.

INDEX

Page numbers followed by "i" and "t" denote illustrations and tables, respectively.

ORDER FORM
(Please print)

Please send me the following items:

_____ copies of **Standard Methods for the Examination of Water and Wastewater, 20th Edition** hardcover
ISBN 0-87553-235-7
- At $155.00 per copy for APHA, AWWA, and WEF members $ _____
- At $200.00 per copy for nonmembers $ _____

_____ copies CD-ROM of **Standard Methods, 20th edition**
- At $235.00 per copy for APHA, AWWA, and WEF members $ _____
- At $285.00 per copy for nonmembers $ _____

_____ copies CD-ROM and book combination
- At $358.00 per copy for APHA, AWWA, and WEF members $ _____
- At $437.00 per copy for nonmembers $ _____

Subtotal $ _____

U.S., add $12.00 per book for shipping and handling.
DC Residents add 5.75% sales tax, MD residents add 5% sales tax.
Non-U.S., add $15.00 per book for shipping and handling.

TOTAL amount $ _____ (including shipping and handling).

❑ Payment enclosed (make check payable to APHA)
❑ VISA   ❑ MasterCard   ❑ American Express

Credit Card Number _____ Expiration Date _____

Cardholder's Signature _____

Name (please print) _____

Member No. _____

Organization _____

Address _____

City/State/ZIP _____

Country _____

Phone ( ) _____

Is this a business address? _____ Yes _____ No

Mail order form and payment to:
American Public Health Association
Publication Sales
P.O. Box 753
Waldorf, Maryland 20604-0753

To order or charge by phone, call 301-893-1894
To fax, call 301-843-0159
On-line order form http://www.apha.org

Standard Methods for the Examination of Water and Wastewater is jointly published by three leading scientific organizations: American Public Health Association, American Water Works Association, and the Water Environment Federation.

THE COPUBLISHERS

To obtain complete publications catalogs, membership benefits, or other information, please contact each organization directly:

American Public Health Association (APHA), founded in 1872, is the largest organization of public health professionals in the world, representing more than 50,000 members from 77 public health occupations. The mission of APHA is to protect and promote personal and environmental health. APHA brings together researchers, practitioners, administrators, teachers, and other health professionals in a unique multidisciplinary environment of professional exchange, study, and action.

> Contact information: APHA
> 1015 15th Street, NW
> Washington, DC 20005
> Phone: (202) 789-5600
> Fax: (202) 789-5661

American Water Works Association (AWWA)

The American Water Works Association (AWWA) is a nonprofit scientific and educational association dedicated to providing safe drinking water to the public. Founded in 1881, AWWA comprises more than 54,000 water professionals worldwide. Through educational and technical information programs, AWWA promotes the improvement of the quality and quantity of the drinking water supply.

> Contact information: AWWA
> 6666 West Quincy Avenue
> Denver, CO 80235
> Phone: (303) 794-7711
> Fax: (303) 794-7310

Water Environment Federation (WEF)

The Water Environment Federation is a not-for-profit technical and educational organization, which was founded in 1928. Its mission is to preserve and enhance the global water environment. Federation members are 41,000 water quality specialists from around the world, including environmental, civil, and chemical engineers, biologists, government officials, treatment plant managers and operators, laboratory technicians, college professors, students, and equipment manufacturers and distributors.

> Contact information: WEF
> 601 Wythe Street
> Alexandria, VA 22314-1994
> Phone: (703) 684-2400
> Fax: (703) 684-2492

Abbreviations

The following symbols and abbreviations are used throughout this book:

Abbreviation	Referent	Abbreviation	Referent
AA	atomic absorption	mol wt	molecular weight
A or amp	ampere(s)	MPN	most probable number
AC	alternating current	MS	mass spectrometer
ACS	American Chemical Society	mV	millivolt(s)
amu	atomic mass units	μA	microampere(s)
APHA	American Public Health Association	μCi	microcurie(s)
ASTM	American Society for Testing and Materials	μg	microgram(s)
AWWA	American Water Works Association	μL	microliter(s)
		μm	micrometer(s)
BOD	biochemical oxygen demand		
		N	normal
°C	degree(s) Celsius	nCi	nanocurie(s)
c	count(s)	ng	nanogram(s)
Ci	curie(s)	NIST	National Institute of Standards and Technology
cm, cm², cm³	centimeter(s), square centimeter(s), cubic centimeter(s)	No.	number
COD	chemical oxygen demand	NTU	nephelometric turbidity unit(s)
conc	concentrated		
cpm	counts per minute	OD	outside diameter
cps	counts per second		
		Pa	pascal
d	day	pCi	picocurie(s)
DC	direct current	pg	picogram(s)
diam	diameter	PTFE	polytetrafluoroethylene
DO	dissolved oxygen	PVC	polyvinyl chloride
DOX	dissolved organic halogen		
dpm	disintegrations per minute	rpm	revolution(s) per minute
		rps	revolution(s) per second
g	gram(s)		
g	gravity, unit acceleration of	SD	standard deviation
GC	gas chromatograph	SDI	sludge density index
GC/MS	gas chromatograph/mass spectrometer	s	second(s)
		sp., spp.	species
h	hour	sp gr	specific gravity
HPLC	high-performance liquid chromatography	ST	standard taper
		SVI	sludge volume index
IC	ion chromatograph		
ICP	inductively coupled plasma	TFE	tetrafluoroethylene
ID	inside diameter	THM	trihalomethane(s)
IU	international unit(s)	TOC	total organic carbon
		TON	threshold odor number
KeV	kiloelectron volt(s)	TOX	total organic halogen
kg	kilogram(s)	Toxicity terms	*see* Section 8010B
kPa	kilopascal		
		U	unit(s)
L	liter(s)	USGS	U.S. Geological Survey
		USP	United States Pharmacopoeia
M	mole or molar	UV	ultraviolet
m, m², m³	meter(s), square meter(s), cubic meter(s)		
MCL	maximum contaminant level	V	volt(s)
MDL	method detection level	v/v	volume ratio
me	milliequivalent(s)		
meV	megaelectron volt(s)	W	watt(s)
mg	milligram(s)	WEF	Water Environment Federation
min	minute(s)	WPCF	*see* WEF
mL	milliliter(s)		
mm, mm², mm³	millimeter(s), square millimeter(s), cubic millimeter(s)		

Abbreviations of periodical titles in reference lists and bibliographies are based on those given in *Biosis. List of Serials with Title Abbreviations*, 1970. Biosciences Information Service of Biological Abstracts, Philadelphia, Pa.